IMPORTANT

D1103560

ources

HERE IS YOUR REGISTRATION CODE TO ACCESS

YOUR PREMIUM McGRAW-HILL ONLINE RESOURCES.

For key premium online resources you need THIS CODE to gain access. Once the code is entered, you will be able to use the Web resources for the length of your course.

If your course is using **WebCT** or **Blackboard**, you'll be able to use this code to access the McGraw-Hill content within your instructor's online course.

Access is provided if you have purchased a new book. If the registration code is missing from this book, the registration screen on our Website, and within your WebCT or Blackboard course, will tell you how to obtain your new code.

Registering for McGraw-Hill Online Resources

TO gain access to your McGraw-Hill web resources simply follow the steps below:

(1) USE YOUR WEB BROWSER TO GO TO: **www.mhhe.com/raven7**

(2) CLICK ON **FIRST TIME USER**.

(3) ENTER THE REGISTRATION CODE* PRINTED ON THE TEAR-OFF BOOKMARK ON THE RIGHT.

(4) AFTER YOU HAVE ENTERED YOUR REGISTRATION CODE, CLICK **REGISTER**.

(5) FOLLOW THE INSTRUCTIONS TO SET-UP YOUR PERSONAL UserID AND PASSWORD.

(6) WRITE YOUR UserID AND PASSWORD DOWN FOR FUTURE REFERENCE.
KEEP IT IN A SAFE PLACE.

TO GAIN ACCESS to the McGraw-Hill content in your instructor's **WebCT** or **Blackboard** course simply log in to the course with the UserID and Password provided by your instructor. Enter the registration code exactly as it appears in the box to the right when prompted by the system. You will only need to use the code the first time you click on McGraw-Hill content.

Thank you, and welcome to your McGraw-Hill online resources!

0-07-291843-8 T/A RAVEN/JOHNSON: BIOLOGY, 7/E

REGISTRATION CODE

7C09-0IIN-0VRD-5D5T-0B7Z

BIOLOGY

Seventh Edition

Peter H. Raven

Director, Missouri Botanical Gardens;
Engelmann Professor of Botany
Washington University

George B. Johnson

Professor Emeritus of Biology
Washington University

Jonathan B. Losos

Professor of Biology
Washington University

Susan R. Singer

Professor of Biology
Carleton College

Illustration Authors
William C. Ober, M.D.
and
Claire W. Garrison, R.N.

 **Higher Education**

Boston   Burr Ridge, IL   Dubuque, IA   Madison, WI   New York   San Francisco   St. Louis
Bangkok   Bogotá   Caracas   Kuala Lumpur   Lisbon   London   Madrid   Mexico City
Milan   Montreal   New Delhi   Santiago   Seoul   Singapore   Sydney   Taipei   Toronto

Higher Education

BIOLOGY, SEVENTH EDITION

Published by McGraw-Hill, a business unit of The McGraw-Hill Companies, Inc., 1221 Avenue of the Americas, New York, NY 10020. Copyright © 2005, 2002, 1999, 1996 by The McGraw-Hill Companies, Inc. All rights reserved. No part of this publication may be reproduced or distributed in any form or by any means, or stored in a database or retrieval system, without the prior written consent of The McGraw-Hill Companies, Inc., including, but not limited to, in any network or other electronic storage or transmission, or broadcast for distance learning.

Some ancillaries, including electronic and print components, may not be available to customers outside the United States.

♻ This book is printed on recycled, acid-free paper containing 10% postconsumer waste.

International   1 2 3 4 5 6 7 8 9 0 VNH/VNH 0 9 8 7 6 5 4 3
Domestic   1 2 3 4 5 6 7 8 9 0 VNH/VNH 0 9 8 7 6 5 4 3

ISBN 0–07–243731–6
ISBN 0–07–111182–4 (ISE)

Publisher: *Martin J. Lange*
Senior sponsoring editor: *Patrick E. Reidy*
Developmental editor: *Anne L. Winch*
Marketing manager: *Tami Petsche*
Lead project manager: *Peggy J. Selle*
Production supervisor: *Kara Kudronowicz*
Senior media project manager: *Jodi K. Banowetz*
Senior media technology producer: *John J. Theobald*
Senior coordinator of freelance design: *Michelle D. Whitaker*
Cover/interior designer: *Christopher Reese*
Senior photo research coordinator: *Lori Hancock*
Photo research: *Meyers Photo-Art*
Supplement producer: *Brenda A. Ernzen*
Compositor: *Carlisle Communications, Ltd.*
Typeface: *10/12 Janson Text Roman*
Printer: *Von Hoffmann Corporation*

Cover images: DNA: © *Doug Struthers/Getty Images*; Pollen: *S. Lowry, Univ. Ulster/Getty Images*; Beetle: © *Davies + Starr/Getty Images*; Leopard: © *Ryan McVay/Getty Images*; Man's Profile: © *Suza Scalora/Getty Images*; Leaf: *Christopher Reese*

The credits section for this book begins on page C-1 and is considered an extension of the copyright page.

Library of Congress Cataloging-in-Publication Data

Biology / Peter H. Raven . . . [et al.]. 7th ed.
   p. cm.
 Rev. ed. of: Biology / Peter H. Raven and George B. Johnson. 6th ed., © 2002.
 ISBN 0–07–243731–6 (hard : alk. paper)
 1. Biology. I. Raven, Peter H. II. Raven, Peter H. Biology.

QH308.2.R38   2005
570—dc22   2003016998
                        CIP

INTERNATIONAL EDITION ISBN 0–07–111182–4
Copyright © 2005. Exclusive rights by The McGraw-Hill Companies, Inc., for manufacture and export. This book cannot be re-exported from the country to which it is sold by McGraw-Hill. The International Edition is not available in North America.

www.mhhe.com

Brief Contents

Contents

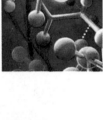
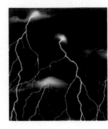
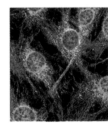
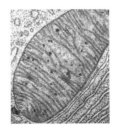

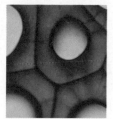

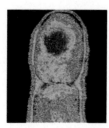

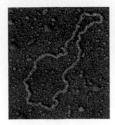

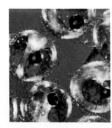

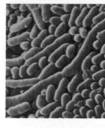
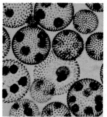

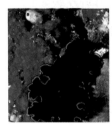

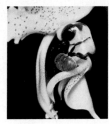
Part VII Animal Form and Function

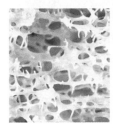

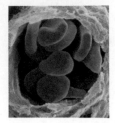

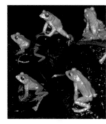

About the Authors

Dr. Peter Raven is director of the Missouri Botanical Garden and Engelmann Professor of Botany at Washington University. A distinguished scientist, Dr. Raven is a member of the National Academy of Sciences, the National Research Council, and is a MacArthur and a Guggenheim fellow. He has received numerous honors and awards for his botanical research and work in tropical conservation, including the National Medal of Science. In addition to coauthoring this text, Raven has authored twenty other books and several hundred scientific articles.

George Johnson is professor emeritus of biology at Washington University in St. Louis, where he has taught genetics and general biology to undergraduates for 30 years. Also professor of genetics at Washington University School of Medicine, he is a student of population genetics and evolution. He has authored more than fifty scientific publications, and several high school and college texts, including *The Living World*, a very successful non-majors college biology text. He has pioneered the development of interactive CD-ROM and web-based investigations for biology teaching.

Jonathan Losos is a professor in the Department of Biology at Washington University and is also chair of the undergraduate Environmental Studies program. An evolutionary biologist, Losos's research has focused on studying patterns of adaptive radiation and evolutionary diversification in lizards. The recipient of several awards including the prestigious Theodosius Dobzhanksy and David Starr Jordan Prizes for outstanding young evolutionary biologists, Losos has published more than eighty scientific articles. He is currently the editor of the *American Naturalist*, a leading journal integrating the fields of evolutionary biology, behavior and ecology.

Susan Singer is professor of biology at Carleton College in Northfield, Minnesota, where she has taught introductory biology, plant biology, plant development, and developmental genetics for 18 years. Her research interests are focused on the development and evolution of flowering plants. Singer has authored numerous scientific publications on plant development, contributed chapters to developmental biology texts, and been actively involved with the education efforts of several professional societies. She serves on the NRC Committee on Undergraduate Science Education.

Preface

We first began work on this text in 1982, over twenty years ago. We set out to write a text that explained biology the way we taught it in the classroom—as the product of evolution. Most texts in 1982 relegated evolution to a few chapters in the diversity section. But evolution pervades biology, and is just as evident in the bacterial character of the mitochondria within your cells, in the biochemical similarities of photosynthesis and glycolysis, in the evolution of genes that control development—everywhere you look in biology, you see Darwin staring back at you. This evolutionary approach has proven popular among our nation's biology faculty, and most texts to greater or lesser degrees now adopt it.

Our text has changed a lot over twenty years, reflecting great changes in biology. The book has become more molecular, as biology has. In particular, a lot more is said about cell biology and development, areas where biology has made enormous strides. But our text remains fundamentally an evolutionary explanation of biology. In this edition, for example, while there is a new chapter on genomes, there is also a new "evo-devo" chapter that examines how genomes and developmental control mechanisms have evolved. This is just one example of our efforts to integrate biological questions and approaches at multiple levels of organization throughout the text, as we strive to guide students toward a connected understanding of biology.

This new seventh edition marks perhaps the greatest change in our text: the addition of two new biologists to the author team, Jonathan Losos, also at Washington University, and Susan Singer of Carleton College. Both made major contributions to the previous edition—Jonathan to the chapter on evolution and ecology, and Susan to the chapters on plant biology—and we are delighted to welcome them as full-fledged authors. In this edition their responsibilities have broadened to include the revolution that is ongoing in our understanding of systematics and evolution at the DNA level, matters that affect many chapters of this book.

Text development today involves an even greater number of people, as instructors from across the country are continually invited to share their knowledge and experience with us through reviews and focus groups. All of the feedback we have received has shaped this edition, resulting in new chapters, reorganization of the table of contents, and expanded coverage in key areas. This edition also incorporated the expertise of three consultants: Randy Di Domenico, University of Colorado—Boulder; Kenneth Mason, Purdue University; and Randall Phillis, University of Massachusetts—Amherst, who provided detailed suggestions for improving the clarity, flow and accuracy of large portions of the text.

How We Have Responded to You

Perhaps more than any other text on the market, this text has continued to evolve as a result of feedback from instructors teaching majors' biology. Overwhelmingly, they have told us that up-to-date content, a clear writing style, quality illustrations and dynamic presentation materials are the most important factors they consider when evaluating textbooks. We have let those values guide our revision of the text, as McGraw-Hill Education worked with those same instructors to create supplements that will help them in the classroom.

Up-to-date Treatment

The core of any majors' biology course is the exploration of cells and genetics, always covered in the first half of any majors' text. This book has been particularly aggressive in keeping its treatment of cell biology and genetics comprehensive and up-to-date. It was the first to present a chapter on cell communication, for example (other books soon followed). We are continuing in that tradition by incorporating such cutting edge topics as the structure of ATP synthetase, small RNAs and RNA editing. This edition also includes a new chapter 17 that explores what we can learn about genomes, covering topics ranging from human health issues and concerns about privacy, agricultural applications, and the potential of genomics in minimizing bioterrorism.

We did not contain this revision to a few select chapters in the first half of the text, however, and it's possible to point to many areas where treatment of recent breakthroughs has been integrated. By concluding our evolution section with a comparative approach to genomes and evolution of development, we were able to connect new breakthroughs in these areas and provide a springboard into the diversity section. Major changes in our understanding of phylogenetic relationships among land plants, protists, and fungi along with other major groups, are reflected in the extensive revision of the diversity chapters. Rapid advances in our knowledge of plant defense responses led to a new chapter on this topic.

Writing Style

Students of biology are responsible for an ever-growing volume of information, and that amount of detail is reflected in today's textbooks, which are increasingly becoming encyclopedic references as opposed to teaching texts.

But students are more likely to succeed with a text that they enjoy reading, that gives them a sense of the wonderment that inspired their own instructors to study biology. For this reason, we have endeavored to strike a balance; an inviting and accessible writing style with the level of authority and rigor expected of a majors' level text.

To further aid the student, every page or two-page spread in this book functions as a semi-independent learning module, organized under its own heading at the top of the left-hand page, with its own summary at the bottom of the right-hand page. This modular presentation makes the conceptual organization of the chapter clear, greatly enhancing student learning.

Illustrations

This book is set apart form others in that its artists, William Ober, M.D. and Claire Garrison, R.N., are part of the author team. Their respective backgrounds as a practicing physician and pediatric registered nurse, and their experience creating art for highly successful anatomy and physiology, zoology and marine biology textbooks, bring an invaluable contribution to the text. The close collaboration between text and illustration authors results in dynamic, accurate figures that aid student understanding and instructor presentation.

- **Combination Figures** These pieces combine a photo or micrograph with a line drawing, to make the connection between conceptual figures and what the student may encounter in lab (Figure 5.10, page 88).
- **Biochemistry Pathway Icons** Found in the discussion of metabolism, these icons help students follow complex processes by highlighting the step currently under discussion (Figure 9.15, page 174).
- **Phylogeny Guideposts** This icon is presented as each group is introduced, to remind students of relationships among diverse organisms (TA 32.1, page 636).
- **Process Boxes** These figures include step-by-step descriptions to walk the student through a compact summary of important concepts (Figure 6.18, page 121)

We have also been fortunate in that this collaboration has allowed us to carefully integrate explanatory text into the figures. The end results are uncluttered, easy to follow illustrations that guide a student through a concept. They also benefit the instructor, as figures without distracting captions can be used for presentation while still allowing instructors to tell their own story.

What Sets this Book Apart

Those who have not used or reviewed previous editions will want to know how this book differs from others.

Evolutionary Focus

The treatment of evolution in this book differs from others in a simple but very important way: Evolution is the organizing principle guiding the teaching of each chapter. Instead of leaping from chemistry directly to cell structure as in other books, this book uses the chemistry of the first chapters to examine the origin of life and the evolution of cells; the cell chapters that follow can then be seen in a broader evolutionary context. Similarly, the treatment of animal anatomy and physiology in other books is largely limited to structure and function—this is the organ and this is how it functions. This book examines each animal body system in terms of how it has evolved. Every section of this book, whether it is genetics or plant biology, presents biology from an evolutionary perspective.

Chemistry in a Biological Context

In talking to students over 30 years of teaching freshman biology, a consistent student complaint has been that the introductory biology course begins with a heavy jolt of chemistry. In other books, only after as many as 100 pages of chemistry do students encounter any biology. This is very off-putting for many students, and gets the course off to a rocky start. This book, by contrast, integrates the chemistry of the first section with biological themes. The treatment of macromolecules in Chapter 3 starts with proteins, which can be easily understood without detailed knowledge of carbohydrates. This arrangement has the distinct advantage of starting the student off with material of obvious relevance to biology.

A Modern Approach

Some of the most obvious differences between this book and others can be seen in the second half of the book, that part devoted to coverage of evolution, diversity, plant biology, anatomy and physiology, and ecology.

Evolution. Our approach to evolutionary biology is unique in two respects. First, we strongly emphasize the role of experiments in studying evolution. Although much of evolutionary biology concerns the study of what happened in the past, that does not mean that experimental approaches are impossible. We emphasize the role that experiments play in studying evolutionary phenomena. More generally, like any detective story, we point out how various approaches must be integrated to fully understand evolutionary diversity.

Second, our book devotes an entire chapter to the evidentiary basis for evolutionary biology. Unlike other aspects of biology, or science in general, the factual basis of evolutionary biology is disputed by some segments of society. Thus, we feel that it is important to clearly present the diversity and depth of evolution that leads almost all biologists to conclude that evolution has occurred. We feel that it is essential for all college biology students—regardless of their own opinions—to understand the scientific basis for this view.

Diversity. Our text has been organized so the diversity section is framed by a discussion of the revolution in taxonomy and phylogenetics (Chapter 25). Complete with vivid examples of dramatic changes in our understanding of relationships among organisms, this chapter can be used alone as an abbreviated approach to diversity or as a foundation for a more comprehensive evolutionary investigation of the diversity of life in the chapters that follow.

The book also differs from other majors' biology textbooks in that its coverage of diversity is more extensive. Consider for example the invertebrates. Other books devote as few as 30 pages to the invertebrates, presenting only the briefest of sketches of what used to be the core of traditional biology courses. This book devotes more than twice as many pages to the invertebrates, followed by a more comprehensive chapter on the vertebrates than is found in other books. Why is this more extensive treatment of diversity important? Even in courses that don't cover diversity in detail, it is important that students be able to uncover for themselves the relations among animal groups.

Plant Biology. The plant biology chapters have undergone extensive revision, and are now organized to lead the student through the plant life cycle. In addition, we have carefully integrated both developmental and genetic perspectives, a fusion not found in other texts. For example, Chapter 36, Vegetative Plant Development, explores root formation in the context of the *monopterous* mutant of *Arabidopsis* that fails to make a root. The shift from its developmental role to its functional role as an auxin receptor begins to move students toward a physiological understanding of plant function.

Anatomy and Physiology. Most books devote nearly the same amount of space to anatomy and physiology, about 250 pages. The differences lie primarily in approach, this book having a more evolutionary focus than others, and in its emphasis on fundamentals.

Ecology and Behavior. We take an integrative view to understanding how the environment functions and how organisms interact with it. This section is broken into different chapters, such as behavioral ecology, population ecology, and community ecology, but the topics are carefully integrated. Moreover, we apply this information extensively in Chapters 56 and 57 (The Biosphere and Conservation Biology) to address the environmental issues facing our planet. We believe it is of the utmost importance that students understand the scientific bases to current problems so they can evaluate efforts to solve them.

Changes to the Seventh Edition

The seventh edition of *Biology* is the result of extensive analysis of the text and evaluation of input from biology instructors who conscientiously reviewed chapters during various stages of this revision. We have utilized the constructive comments provided by these professionals in our continuing efforts to enhance the strengths of the text. Listed first are general changes that have been made to the entire text, which is then followed by specific changes for each part.

End-of-chapter Pedagogy

The end-of-chapter student review has been greatly expanded, offering students a full-page chapter review and three assessment tools: Self Test, multiple choice questions; Test Your Visual Understanding, questions based on a figure from the chapter; and Apply Your Knowledge, critical thinking questions. The assessment doesn't end there, however. These tools are carried over to the web where the student can take an interactive version of the test and receive instructional feedback.

Inquiry Questions

In this edition we have developed Inquiry Questions, which follow the legend in figures presenting graphed data. These questions require the student to think about the information contained in the graph in even greater depth, increasing their understanding of, and facility with, the material.

Answers to both end of chapter questions and Inquiry Questions are found on the Online Learning Center at www.mhhe.com/raven7.

Volumes

We recognize that instructors don't always use the entire text, so we now offer *Biology* in the following volumes:

Volume 1 Chapters 1–20 Chemistry, Cell and Genetics

Volume 2 Chapters 35–51 Plant Biology and Animal Biology

Volume 3 Chapters 21–34, 52–57 Evolution, Diversity and Ecology

Content Changes by Part

Part I The Origin of Living Things

Part I was revised with the intention of creating a more solid foundation of key concepts in biology, which students can then build on in later chapters. Discussions are now clearer and better supported with illustrations.

New Topics and Revised Treatments

Chapter 1 Properties of life, hierarchical organization *Revised;* Additional topics in evolution *New*

Chapter 2 The Nature of Molecules *Entire chapter revised*

Chapter 3 Figures on chaperons and protein denaturing *New;* Protein folding, lipids *Revised*

Chapter 4 Figures on endosymbiosis, Domains/ Kingdoms, phylogenetic tree of life *New* Bacteria and archaebacteria, microfossils *Revised*

Part II Biology of the Cell

Randall Phillis assisted in the revision of Part II by directing the authors to concepts that needed additional detail, and by providing suggestions for improving the accuracy and parsimony of the narrative. Concepts that were covered too briefly in previous editions are now supported with more extensive discussion and new illustrations.

New Topics and Revised Treatments

Chapter 6 Membrane microdomains *New* Osmosis; coupled transport *Revised*

Chapter 7 Signal amplification, expression of cellular identification *Revised*

Chapter 8 Redox reactions, ATP functioning *Revised*

Chapter 9 ATP synthetase *New;* Electron transport; reducing power; chemiosmosis *Revised*

Chapter 10 The Calvin cycle *Revised*

Chapter 11 Cell cycle control *New;* Chromosome structure *Revised*

Part III Genetic and Molecular Biology

With the help of Kenneth Mason, Part III was carefully updated to incorporate the most current research. Chapter 13, Patterns of Inheritance, was rewritten for better organization and clearer presentation. Two new chapters provide expanded discussion in fields where our knowledge has grown exponentially.

Chapter 17, "Genomes" integrates plant and animal genomics, functional genomics and proteomics in a chapter that is inquiry driven rather than a discussion of techniques.

Chapter 20, "Cancer Biology and Cell Technology" explores two areas where recent advances in cell and molecular biology have the potential of revolutionizing medicine. The first is cancer, where research into the molecular events leading to cancer is beginning to suggest effective therapies. The second is cell technology, including cloning, embryonic stem cells, and the exciting and controversial proposal of therapeutic cloning.

New Topics and Revised Treatments

Chapter 12 Meiotic prophase *Revised*

Chapter 13 Patterns of inheritance *Entire chapter revised*

Chapter 14 Eukaryotic DNA replication *Revised*

Chapter 15 Eukaryotic transcription *New*

Chapter 16 The tools of genetic engineering *New*

Chapter 18 Small RNAs, iRNA, RNA editing *New* Transcriptional control in prokaryotes *Revised*

Chapter 19 Vertebrate embryonic axis formation, evolution of homeotic genes *New* Cell movement; cell induction; embryonic determination; pattern formation *Revised*

Part IV Evolution

The Evolution section has been revised to bring more experimental data and analysis into the discussions. Because presentation of the experimental data used to derive conclusions and concepts is key to understanding how the concepts arose from the research, you will see that graphs and charts have become more plentiful in these chapters. The evolution of many groups is reassessed in light of new molecular data.

Chapter 24, "Evolution of Genomes and Developmental Mechanisms" is a new comparative genomics chapter that addresses our emerging understanding of the evolution of development, and helps to provide a conceptual framework for the diversity chapters that follow. The chapter was developed in conjunction with Chapter 17, Genomes, to first provide students with an understanding of what we can learn about genomes, and then having learned about evolution, delve into a deeper discussion of how development has evolved to yield novel phenotypes.

New Topics and Revised Treatments

Chapter 21 Measuring fitness, components of fitness, role of selection in maintaining variation *New* Hardy-Weinberg *Revised*

Chapter 22 Darwin's finches, industrial melanism *New* Evidence from developmental biology for evolution *Revised*

Chapter 23 Plant speciation by chromosomal change, the future of evolution *New*

Part V Diversity of Life on Earth

The fungi chapter has been moved to the phylogenetically appropriate place in the diversity section following plant diversity. Where appropriate, chapters in the diversity unit have been updated to reflect phylogenetic changes. The thoroughly revised and rewritten **Chapter 25, "Systematics and the Phylogenetic Revolution"** addresses the current tension between taxonomy and systematics. The chapter can be used alone, to teach the basic concepts of diversity, or can be used as a starting place for a more in-depth study of this area of biology.

New Topics and Revised Treatments

Chapter 25 Phylogenetics and classification *Entire chapter revised*

Chapter 26 Virus genomes *New;* Viral diseases and HIV *Revised*

Chapter 27 Term "eubacteria" replaced with "bacteria," figures of cell structure and clades *New*

Chapter 28 Phylogenetic approach *New;* Protist disease in South, Central and North America, relationships between algae and land plants *Revised*

Chapter 29 Fossil evidence of ancient angiosperm Archaefructacea, evolution of triploid endosperm *New*
Monophyletic relationships between ferns and horsetails *Revised*

Chapter 30 Fungi *Entire chapter revised*

Chapter 31 Protostomes and deuterostomes *New and expanded*
Classification *Revised*

Chapter 32 Protostome phylogeny, rotifers and cycliophora *Revised*

Chapter 33 Mollusks, annelids, arthropods, and echinoderms combined into one chapter *New*

Chapter 34 Characteristics and phylogeny illustration, primate evolution *New*

Part VI Plant Form and Function

The plant biology chapters have been revised so that the traditional discussion of evolutionary influences on plant form and function are brought into a developmental context. Evolution is still presented as the underlying explanation for the character of vascular tissues, seeds, flowers, and fruits, however the developmental processes that produce these organs are now given more emphasis. Two previously combined topics, transport and nutrition, have been split into separate chapters allowing for more in-depth discussion of both topics.

Chapter 39, "Plant Defense Responses" is a new chapter that provides a thorough discussion of secondary compounds and their roles in both plant defense and human applications. Wound responses and R gene mediated responses are explored in depth with an emphasis on signaling pathways.

New Topics and Revised Treatments

Chapter 35 Updated photographs, discussion of genetic regulation of trichomes *Revised*

Chapter 36 Discussion of signal transduction in germination, comparison of roles of *Hox* genes in plant and animal development *Revised*

Chapter 37 Water relations problems, mRNA transport in phloem *New*

Chapter 38 Newly expanded chapter on plant nutrition. Effects of global change on photosynthesis and balance of plant nutrients, phytoremediation *New*
Nutritional symbioses *Revised and expanded*

Chapter 40 Signal transduction mediated by light including phot1 *New*
Light responses *Revised*

Chapter 41 Plant reproduction *Entire chapter revised and reorganized*

Part VII Animal Form and Function

With the assistance of Randy DiDomenico, many discussions were rewritten for better organization and clarity. Previous chapters on circulation and respiration were combined into one chapter, as were chapters on body organization and locomotion.

New Topics and Revised Treatments

Chapter 42 Combined organization of the animal body and locomotion into one chapter, coordination of organ systems *New*

Chapter 43 Neural, hormonal and accessory organ regulation *New*
Small intestine discussion reorganized to group all functions together *Revised*

Chapter 44 Maximizing rate of gas diffusion *New*
Integration of circulation and respiration chapter *Revised*

Chapter 45 Graded potentials *New;* Membrane and action potentials, synapses and drug addiction *Revised*

Chapter 46 Sensory transduction *Revised*

Chapter 47 The Endocrine System *Entire chapter revised*

Chapter 48 Immunoglobulins, illustrated table *New*
AIDS *Revised*

Chapter 49 Discussion of ammonia, urea and uric acid reorganized, nephron *Revised*

Chapter 50 Sex determination, reptiles and birds *Revised*
Human intercourse *Omitted*

Chapter 51 Combined discussion of chick and mammalian extraembryonic membranes, neurulation *Revised*

Part VIII Ecology and Behavior

The ecology and behavior chapters were moved to follow diversity and physiology, where these chapters are more often taught. There is now an even greater emphasis on experimental data and analysis.

New Topics and Revised Treatments

Chapter 52 Animal cognition *New;* Integration of animal behavior and behavioral ecology chapters *Revised*

Chapter 53 Introduction to ecology, integration of autoecology and population ecology *New*
Population regulation and limitation, human population growth *Revised*

Chapter 54 Introduction, definition of community *New*
Parasitism, succession, disturbance *Revised*

Chapter 55 Geochemical cycles, energy flow, species richness *Revised*

Chapter 56 Differences between aquatic and terrestrial ecosystems *New*
Integration of biosphere and future of the biosphere chapters, global climate change, El Niño *Revised*

Chapter 57 Chapter organization, biodiversity hotspots, amphibian extinctions, invasive species *New*
Economic benefits of biodiversity *Revised*

Overview of Changes to BIOLOGY, Seventh Edition

All Cell & Genetics Chapters Extensively Revised
In addition to discussing important advances, many sections have been reworked for improved clarity.

Chapter 17, "Genomes"
This new chapter describes how researchers sequence entire genomes, and how the information is being used.

Chapter 20, "Cancer Biology and Cell Technology"
This new chapter updates progress in understanding cancer, and introduces many new advances in cloning and stem cell technology.

Evolution & Diversity Sections Extensively Revised
New RNA and genomic information is leading to a reassessment of traditional evolutionary phylogenies.

Chapter 24, "Evolution of Genomes and Developmental Mechanisms"
This new chapter explores the wealth of new information on genome sequences, and introduces the new and exciting field of "evo/devo", the evolution of development.

Treatment of Plant Biology Expanded
A total of seven plant chapters provide extensive plant biology coverage with a molecular development point of view. Chapters have been organized to lead the student through the life cycle of a plant.

Chapter 39, "Plant Defense Responses"
This new chapter captures the excitement of this area of biology, which has seen rapid advances and recent breakthroughs in understanding plant defense responses.

Ecology Chapters Updated and Expanded
Up-to-date examples have been integrated into all chapters; note the use of case histories in Chapter 57.

Physiology Chapters Reworked
Discussions of processes like nervous conduction reworked for increased clarity, and related subjects like circulation and respiration treated together.

End-of-Chapter Assessment
Two pages are now devoted to student review and assessment. A full-page chapter summary is followed by multiple choice, illustration-based and application questions.

Illustrations
Many new illustrations clarify difficult concepts; others illustrate tables to aid understanding. Wherever data are presented in graphs, the figure is accompanied by an Inquiry Question to test the student's understanding.

Teaching and Learning Supplements

McGraw-Hill offers various tools and technology products to support *Biology*. Students can order supplemental study materials by contacting their local bookstore or by calling 800-262-4729. Instructors can obtain teaching aids by calling the Customer Service Department at 800-338-3987, visiting our website at www.mhhe.com/biology, or contacting their local McGraw-Hill sales representative.

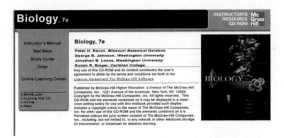

For the Instructor:

Digital Content Manager CD-ROM

This multimedia collection of visual resources allows instructors to utilize artwork from the text in multiple formats to create customized classroom presentations, visually based tests and quizzes, dynamic course website content, or attractive printed support material. The digital assets on this cross-platform CD-ROM include:

Art Library   Color-enhanced, digital files of all illustrations in the book, plus the same art saved in unlabeled and gray scale versions, can be readily incorporated into lecture presentations, exams, or custom-made classroom materials. Upsized labels make the images appropriate for use in large lecture halls.

TextEdit Art Library   Every line art piece is placed into a PowerPoint presentation that allows the user to revise, move, or delete labels as desired for creation of customized presentations or for testing purposes.

Active Art Library   Active Art consists of art files that have been converted to a format that allows the artwork to be edited inside of PowerPoint. Each piece can be broken down to its core elements, grouped or ungrouped, and edited to create customized illustrations.

Animations Library   Full color presentations involving key process figures in the book have been brought to life via animation. These animations offer flexibility for instructors and were designed to be used in lecture. Instructors can pause, rewind, fast forward, and turn audio off/on to create dynamic lecture presentations.

PowerPoint Lecture Outlines   These ready-made presentations combine art and lecture notes for each of the 57 chapters of the book. The presentations can be used as they are, or can be customized to reflect your preferred lecture topics and organization.

PowerPoint Outlines   The art, photos, and tables for each chapter are inserted into blank PowerPoint presentations to which you can add your own notes.

Photo Library   Like the Art Library, digital files of all photographs from the book are available.

Table Library   Every table that appears in the book is provided in electronic form.

Video Library   Contains digitized video clips that can be inserted into a PowerPoint lecture.

Additional Photo Library   Over 700 photos, not found in *Biology*, are available for use in creating lecture presentations.

Instructor's Testing and Resource CD-ROM

The cross-platform CD-ROM contains the Instructor's Manual and Test Item File, both available in both Word and PDF formats. The manual contains chapter synopses, objectives, key terms, outlines, instructional strategies and sources for additional visual resources. The Test Bank offers questions that can be used for homework assignments or the preparation of exams. The computerized test bank utilizes Brownstone Diploma testing software, which allows the user to quickly create customized exams. This user-friendly program allows instructors to search questions by topic, format, or difficulty level; edit existing questions or add new ones; and scramble questions and answer keys for multiple versions of the same test.

Transparencies

A set of 1300 transparency overheads includes every piece of line art and table in the text. The images are printed with better visibility and contrast than ever before, and labels are large and bold for clear projection.

Online Learning Center
www.mhhe.com/raven7

Instructor resources at this site include access to online laboratories, course-specific current articles, real-time news feeds, course updates and research links.

Course Delivery Systems

With help from our partners, WebCT, Blackboard, Top-Class, eCollege, and other course management systems, instructors can take complete control over their course content. These course cartridges also provide online testing and powerful student tracking features. The *Biology* Online Learning Center is available within all of these platforms.

For the Student:

Online Learning Center
www.mhhe.com/raven7

The site includes quizzes for each chapter, interactive activities, and answers to questions from the text. Turn to the inside cover of the text to learn more about the exciting features provided for students through the enhanced *Biology* Online Learning Center.

Student Study Guide

This student resource contains activities and questions to help reinforce chapter concepts. The guide provides students with tips and strategies for mastering the chapter content, concept outlines, concept maps, key terms and sample quizzes.

Acknowledgements

Our goal for *Biology* has always been to present the science in an interesting and engaging way while maintaining a comprehensive and authoritative text. This is a lofty goal considering the mountain of information and research we must go through just to update the text from one edition to the next. This seventh edition would not have been possible without the contributions of many. We are indebted to our colleagues across the country and around the globe that provided numerous suggestions on how to improve on the sixth edition. We wish particularly to thank Kenneth Mason of Purdue University, Randy DiDomenico of the University of Colorado, Boulder, and Randall Phyllis of the University of Massachusetts, Amherst for very detailed advice on how to improve large sections of the text.

As any author knows, a textbook is made not only by an author team aided by their colleagues, but also by a publishing team, a group of people that guide the raw book written by the authors through a yearlong process of reviewing, editing, fine-tuning and production. This edition was particularly fortunate in its book team, led by Patrick Reidy, sponsoring editor; Anne Winch, developmental editor; Tami Petsche, marketing manager; Peggy Selle, project manager; Michelle Whitaker, designer; Megan Jackman and Elizabeth Sievers, off-site editors; Kennie Harris, copy editor, and many more people behind the scenes.

The illustrations are critically important to a biology text, and ours continue to be superbly conceived and rendered by Bill Ober and Claire Garrison.

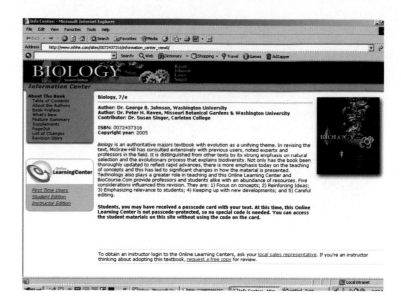

As always, we have had the support of spouses and children, who have seen less of us than they might have liked because of the pressures of getting this revision completed. They have adapted to the many hours this book draws us away from them and, even more than us, look forward to its completion.

As with every edition, acknowledgements would not be complete without thanking the generations of students who have used the many editions of this text. They have taught us as least as much as we have taught them, and, thanks to e-mail, are an increasing part of our lives.

Finally, we need to thank our reviewers. Every text owes a great deal to those instructors across the country who review it. Serving as sensitive antennae for errors and omissions, and as sounding boards for new approaches, reviewers are among the most valuable tools at an author's disposal. Many improvements in this edition are the direct result of their suggestions. Every one of them has our heartfelt thanks.

Reviewers of the Seventh Edition

Heather Addy   *University of Calgary*
Lawrence A. Alice   *Western Kentucky University*
Terry C. Allison   *The University of Texas–Pan American*
Loran C. Anderson   *Florida State University*
Mohammad Ashraf   *City Colleges of Chicago*
Ellen Baker   *Santa Monica College*
R. Neal Band   *Michigan State University*
Dale L. Barnard   *Utah State University*
Diane C. Bassham   *Iowa State University*
Wayne M. Becker   *University of Wisconsin–Madison*
Robert L. Beckmann   *North Carolina State University*
Gerald Bergtrom   *University of Wisconsin–Milwaukee*
Cheryl Briggs   *University of California–Berkeley*
Trey Broadhurst   *Montgomery College*
Arthur L. Buikema, Jr.   *Virginia Tech*
Ann B. Burgess   *University of Wisconsin–Madison*
Carol A. Burkart   *Mountain Empire Community College*
D. Brent Burt   *Stephen F. Austin State University*
David Byres   *Florida Community College at Jacksonville*
Les Chappell   *University of Aberdeen, UK*

Jung Choi   *Georgia Institute of Technology*
Don Cipollini   *Wright State University*
Richard J. Cogdell   *University of Glasgow*
Jerry Cook   *Sam Houston State University*
David T. Corey   *Midlands Technical College*
George Cornwall   *University of Colorado–Boulder/Metropolitan State College of Denver*
Francie Smith Cuffney   *Meredith College*
Paul V. Cupp, Jr.   *Eastern Kentucky University*
James A. Danoff-Burg   *Columbia University*
Sandra L. Davis   *University of Louisiana at Monroe*
Mark D. Decker   *University of Minnesota*
Mary B. Dettman   *Seminole Community College*
John Dickerman   *Northern Illinois University*
Cathy Donald-Whitney   *Collin County Community College*
Thomas W. Dreschel   *Brevard Community College, Kennedy Space Center*
Carolyn S. Dunn   *University of North Carolina at Wilmington*
Roland R. Dute   *Auburn University*
Frederick B. Essig   *University of South Florida*
Bruce E. Felgenhauer   *University of Louisiana at Lafayette*
James Franzen   *University of Pittsburgh*
Andrea Gargas   *University of Wisconsin–Madison*
John V. Gartner, Jr.   *St. Petersburg College*
John R. Geiser   *Western Michigan University*
Florence K. Gleason   *University of Minnesota*
John S. Graham   *Bowling Green State University*
John S. Greenwood   *University of Guelph*
Peggy J. Guthrie   *University of Central Oklahoma*
Adrian Hailey   *University of Bristol*
Dana Brown Haine   *Central Piedmont Community College*
Robert O. Hall   *University of Wyoming*
Robert W. Hamilton   *Loyola University Chicago*
David S. Hibbett   *Clark University*
Leland N. Holland, Jr.   *Pasco-Hernando Community College*
Eva A. Horne   *Kansas State University*
Jeffrey Jack   *University of Louisville*
Lee F. Johnson   *The Ohio State University*
Gregory A. Jones   *Santa Fe Community College*
Walter S. Judd   *University of Florida*
Richard R. Jurin   *University of Northern Colorado*
Thomas C. Kane   *University of Cincinnati*
Ronald Keiper   *Valencia Community College*
John J. Kelly   *Loyola University Chicago*
Cheryl A. Kerfeld   *University of California, Los Angeles*
David J. Kittlesen   *University of Virginia*
William Kroll   *Loyola University Chicago*
Harry D. Kurtz, Jr.   *Clemson University*
Roberta Lammers-Campbell   *Loyola University Chicago*
Peter Lavrentyev   *The University of Akron*
Michael Lawson   *Missouri Southern State College*
Roger M. Lloyd   *Florida Community College at Jacksonville*
David Magrane   *Morehead State University*
Richard Malkin   *University of California–Berkeley*
Terry C. Maxwell   *Angelo State University*
Michael McLeod   *Belmont Abbey College*
Frank J. Messina   *Utah State University*
Sandra Millward   *University of Cincinnati*
Jacalyn S. Newman   *University of Pittsburgh*
Janice Moore   *Colorado State University*
Deborah A. Neher   *University of Toledo*

Erik T. Nilsen   *Virginia Tech*
T. Mark Olsen   *University of Notre Dame*
John C. Osterman   *University of Nebraska–Lincoln*
Daniel M. Pavuk   *Bowling Green State University*
Andrew J. Pease   *Villa Julie College*
Rhoda F. Perozzi   *Virginia Commonwealth University*
Carolyn Peters   *Spoon River College*
Susan Phillips   *Brevard Community College*
Eric R. Pianka   *University of Texas at Austin*
Aleksandar Popadic   *Wayne State University*
Angela R. Porta   *Kean University*
Calvin A. Porter   *Xavier University of Louisiana*
Elena Pravosudova   *Sierra College*
Linda R. Richardson   *Blinn College*
Laurel Roberts   *University of Pittsburgh*
Charles L. Rutherford   *Virginia Tech University*
Erik P. Scully   *Towson University*
Wendy E. Sera   *Seton Hall University*
Alison M. Shakarian   *Salve Regina University*
Neil F. Shay   *University of Notre Dame*
Shree R. Singh   *Alabama State University*
David A. Smith   *Lock Haven University of Pennsylvania*
Willie Smith   *Brevard Community College*
Nancy G. Solomon   *Miami University*
Alan J. Spindler   *Brevard Community College*
Ann Springer   *Hillsborough Community College*
Amy C. Sprinkle   *Jefferson Community College Southwest*
Bruce Stallsmith   *University of Alabama in Huntsville*
John D. Story   *North West Arkansas Community College*
Robert Sullivan   *Marist College*
Marshall D. Sundberg   *Emporia State University*
Pamela S. Thomas   *University of Central Florida*
Patrick A. Thorpe   *Grand Valley State University*
Rani Vajravelu   *University of Central Florida*
Carol M. F. Wake   *South Dakota State University*
Jane Waterman   *University of Central Florida*
Cindy Martinez Wedig   *University of Texas–Pan American*
Olivia Masih White   *University of North Texas*
Lance R. Williams   *The Ohio State University*
Michael Zimmerman   *University of Wisconsin–Oshkosh*

General Biology Symposium

Each year McGraw-Hill holds a General Biology Symposium, which is attended by instructors from across the country. These events are an opportunity for editors from McGraw-Hill to gather information about the needs and challenges of instructors teaching the major's biology course, however it also offers professors a forum for exchanging ideas and experiences with colleagues they might not have otherwise met. The feedback we have received has been invaluable, and has contributed to the success of *Biology* and its supplements.

2003

Marc Ammerlaan   *University of Michigan–Ann Arbor*
Scott Chandler   *University of California, Los Angeles*
Bill Collins   *SUNY at Stony Brook*
Elizabeth Connor   *University of Massachusetts–Amherst*
Steve Connor   *University of South Florida*

Robert Fulginiti   *Xavier University*
Florence Gleason   *University of Minnesota*
Carla Haas   *Penn State University*
David Julian   *University of Florida*
Steve Kelso   *University of Illinois at Chicago*
Bob Locy   *Auburn University*
Kenneth Mason   *Purdue University*
Nancy Solomon   *Miami University–Oxford*
Bill Stein   *SUNY at Binghamton*
Sally Swain   *Middle Tennessee State University*
Linda Waters   *University of Central Florida*

2002

Richard J. Cyr   *Penn State University*
Randy DiDomenico   *University of Colorado–Boulder*
Doug Gaffin   *University of Oklahoma*
Marielle Hoefnagels   *University of Oklahoma*
Jan Jenner   *Science Writer*
Cheryl A. Kerfeld   *University of California, Los Angeles*
Kenneth Mason   *Purdue University*
Michael Meighan   *University of California–Berkeley*
Jane Phillips   *University of Minnesota*
Randall W. Phillis   *University of Massachusetts–Amherst*
Joelle Presson   *University of Maryland*
Leslie Winemiller   *Texas A & M University*
Denise Woodward   *Penn State University*

2001

Mark Ammerlaan   *University of Michigan–Ann Arbor*
Doug Gaffin   *University of Oklahoma*
Jon C. Glase   *Cornell University*
Richard Hallik   *University of Arizona*
Marielle Hoefnagels   *University of Oklahoma*
Fernan Jaramillo   *Carleton College*
Randall Johnson   *University of California, San Diego*
Kenneth Mason   *Purdue University*
Sally Frost-Mason   *Purdue University*
Jorge Moreno   *University of Colorado–Boulder*
Tom Owens   *Cornell University*
Deanna Raineri   *University of Illinois at Urbana-Champaign*
Jon Ruehle   *University of Central Arkansas*
Steven A. Wasserman   *University of California, San Diego*

Reviewers of the Sixth Edition

Michael Adams   *Pasco-Hernando Community College*
Sylvester Allred   *Northern Arizona University*
Lon Alterman   *Clarke College*
Elena Amesbury   *University of Florida*
William Anyonge   *University of California–Los Angeles*
Amir Assadirad   *Delta College*
Gary I. Baird   *Brigham Young University*
Ellen Baker   *Santa Monica College*
Stephen W. Banks   *Louisiana State University–Shreveport*
Ruth Beattie   *University of Kentucky*
Samuel N. Beshers   *University of Illinois*
Christine Konicki Bieszczad   *Saint Joseph College*
John Birdsell   *University of Arizona*
Brenda C. Blackwelder   *Central Piedmont Community College*

Sandra Bobrick   *Community College of Allegheny County
  Allegheny Campus*
Randall Breitwisch   *University of Dayton*
Mark Browning   *Purdue University*
Roger Buckanan   *Arkansas State University*
Theodore Burk   *Creighton University*
John S. Campbell   *Northwest College*
John R. Capeheart   *University of Houston–Downtown*
Michael S. Capp   *Carlow College*
Jeff Carmichael   *University of North Dakota*
George P. Chamuris   *Bloomsburg University*
Susan Cockayne   *Brigham Young University*
William Cohen   *University of Kentucky*
W. Wade Cooper   *Shelton State Community College*
Lisa M. Coussens   *University of California–San Francisco,
  Cancer Research Institute*
Wilson Crone   *Hudson Valley Community College*
Paul V. Cupp Jr.   *Eastern Kentucky University*
Richard Cyr   *The Pennsylvania State University*
Grayson Davis   *Trinity University*
Mark A. DeCrosta   *University of Tampa*
David L. Denlinger   *Ohio State University*
C. Lynn Dorn   *Valencia Community College*
Charles D. Drewes   *Iowa State University*
Sondra Dubowsky   *Allen County Community College*
Peter I. Ekechukwu   *Horry-Georgetown Technical College*
Dennis Emery   *Iowa State University*
Frederick B. Essig   *University of South Florida*
Bruce Evans   *Huntington College*
Deborah Fahey   *Wheaton College*
Linda E. Fisher   *University of Michigan–Dearborn*
Rob Fitch   *Wenatchee Valley College*
Robert Fogel   *University of Michigan*
James Franzen   *University of Pittsburgh–Pittsburgh Campus*
William Friedman   *University of Colorado*
Lawrence Fritz   *Northern Arizona University*
Bernard Frye   *University of Texas at Arlington*
Robert J. Full   *University of California–Berkeley*
Warren Gallin   *University of Alberta*
Darrell Galloway   *The Ohio State University*
Ted Gish   *St. Mary's College*
Donald Glassman   *Des Moines Area Community College*
Jim Glenn   *Red Deer College*
Jim R. Goetze   *Laredo Community College*
Jack M. Goldberg   *University of California–Davis*
Elizabeth Godrick   *Boston University*
Dalton Gossett   *Louisiana State University–Shreveport*
John Griffis   *Joliet Junior College*
Kathryn Gronlund   *New Mexico State University–Carlsbad*
Elizabeth L. Gross   *The Ohio State University*
Patricia A. Grove   *College of Mount St. Vincent*
Randolph Hampton   *University of California–San Diego*
Sehoya E. Harris   *The Pennsylvania State University*
Carla Ann Hass   *The Pennsylvania State University*
Chris Haynes   *Shelton State Community College*
Albert A. Herrera   *University of Southern California*
Pamela Higgins   *Allentown College of St. Francis DeSales*
Richard Hill   *Michigan State University*
Phyllis Hirsch   *East Los Angeles College*
Victoria Hittinger   *Rhode Island College*
Nan Ho   *Las Positas College*

Leland N. Holland, Jr.   *Pasco-Hernando Community College–West Campus*
Elisabeth A. Hooper   *Truman State University*
Terry L. Hufford   *The George Washington University*
Allen Hunt   *Elizabethtown Community College*
Sobrasua E. M. Ibin   *Morris Brown College*
Louis Irwin   *University of Texas at El Paso*
Laurie E. Iten   *Purdue University*
Jeffrey Jack   *College of Arts & Sciences*
James B. Jensen   *Brigham Young University*
Judy Jernstedt   *University of California - Davis*
George P. Johnson   *Arkansas Tech University*
Kenneth V. Kardong   *Washington State University*
Cheryl Kerfeld   *University of California–Los Angeles*
Joanne M. Kilpatrick   *Auburn University at Montgomery*
Peter King   *Francis Marion University*
Edward C. Kisailus   *Canisius College*
Robert M. Kitchin   *University of Wyoming*
Will Kleinelp   *Middlesex County College*
Kenton Ko   *Queen's University*
Ross E. Koning   *Eastern Connecticut State University*
Karen L. Koster   *University of South Dakota*
V.A. Langman   *Louisiana State University–Shreveport*
Simon Lawrance   *Otterbein College*
Jeffrey N. Lee   *Essex County College*
Laura G. Leff   *Kent State University*
Mary E. Lehman   *Longwood College*
Niles Lehman   *University at Albany   SUNY*
Michael Lema   *Midlands Technical College*
Charles Kingsley Levy   *Boston University*
Leslie Lichtenstein   *Massasoit Community College*
Harvey Liftin   *Broward Community College*
Richard Londraville   *University of Akron*
Sonja L. Maki   *Clemson University*
Bradford D. Martin   *La Sierra University*
Barbara Maynard   *Colorado State University*
Deanna McCullough   *University of Houston Downtown*
L. R. McEdward   *University of Florida*
Michael Ray Meighan   *University of California–Berkeley*
John Merrill   *Michigan State University*
Harry A. Meyer   *McNeese State University*
Dennis J. Minchella   *Purdue University*
Jonathan D. Monroe   *James Madison University*
David L. Moore   *Utica College of Syracuse University*
Tony E. Morris   *Fairmont State College*
Roger N. Morrissette   *Framingham State College*
Richard Mortensen   *Albion College*
William H. Nelson   *Morgan State University*
Peter H. Niewiarowski   *University of Akron*
Colleen J. Nolan   *St. Mary's University*
John C. Osterman   *University of Nebraska–Lincoln*
Thomas G. Owens   *Cornell University*
Bruce Parker   *Utah Valley State University*
Dustin Penn   *University of Utah*
Stacia Pieffer-Schneider   *Marquette University*
Carl S. Pike   *Franklin and Marshall College*
Nancy A. Perigo   *Willamette University*
Greg Phillips   *Blinn College–Brenham Campus*
Jon Pigage   *University of Colorado at Colorado Springs*
Barbara Pleasants   *Iowa State University*
John Pleasants   *Iowa State University*

Peggy Pollack   *Northern Arizona University*
Mitch Price   *The Pennsylvania State University*
Margene Ranieri   *Bob Jones University*
Arthur Raske   *Northland Baptist Bible College*
Keith Redetzke   *University of Texas at El Paso*
Peter J. Rizzo   *Texas A&M University*
Ellison Robinson   *Midlands Technical College*
Lyndell P. Robinson   *Lincoln Land Community College*
Angel M. Rodriguez   *Broward Community College*
June R. P. Ross   *Western Washington University*
Patricia Rugaber   *Coastal Georgia Community College*
Connie Rye   *Bevill State Community College*
Nancy K. Sanders   *Truman State University*
Robert B. Sanders   *University of Kansas–Main Campus*
Lisa M. Sardinia   *Pacific University*
Brian W. Schwartz   *Columbus State University*
Bruce S. Serlin   *DePauw University*
Mark A. Sheridan   *North Dakota State University*
Janet Anne Sherman   *Penn College of Technology*
Louis Sherman   *Purdue University*
Jim Shinkle   *Trinity University*
Richard Shippee   *Vincennes University*
Brian Shmaefsky   *Kingwood College*
Michele Shuster   *University of Pittsburgh*
Robert C. Sizemore   *Alcorn State University*
Mark Smith   *Victor Valley College*
Nancy Solomon   *Miami University*
Norm Stacey   *University of Alberta*
Ruth Stutts-Moseley   *Bishop State Community College*
Kathy Sympson   *Florida Keys Community College*
Stan Szarek   *Arizona State University*
Robert H. Tamarin   *University of Massachusetts Lowell*
Michael Tenneson   *Evangel University*
Sharon Thoma   *Edgewood College*
Joanne Kivela Tillotson   *Purchase College   State University of New York*
Maurice Thomas   *Palm Beach Atlantic College*
Thomas Tomasi   *Southwest Missouri State University*
Leslie Towill   *Arizona State University*
Akif Uzman   *University of Houston–Downtown*
Thomas J. Volk   *University of Wisconsin–La Crosse*
Keith D. Waddington   *University of Miami*
D. Alexander Wait   *Southwest Missouri State University*
Timothy S. Wakefield   *Auburn University*
Charles Walcott   *Cornell University*
Eileen Walsh   *Westchester Community College*
Frederick Wasserman   *Boston University*
Steven A. Wasserman   *University of California–San Diego*
Robert F. Weaver   *University of Kansas*
Andrew N. Webber   *Arizona State University*
Harold J. Webster   *Penn State DuBois*
Mark Wheelis   *University of California–Davis*
Lynn D. Wike   *University of South Carolina at Aiken*
William Williams   *Saint Mary's College of Maryland*
Mary L. Wilson   *Gordon College*
Kevin Winterling   *Emory & Henry College*
E. William Wischusen   *Louisiana State University and Agricultural and Mechanical College*
Kenneth Wunch   *Tulane University*
Mark L. Wygoda   *McNeese State University*
Roger Young   *Drury College*

Instructive Art Program

The core of every biology textbook is its art program, and the text and illustration authors of *Biology* have worked together to create a dynamic program of full-color illustrations and photographs that support and further clarify the text explanations. Brilliantly rendered and meticulously reviewed for accuracy and consistency, the carefully conceived illustrations and accompanying photos provide concrete, visual reinforcement of the topics discussed throughout the text.

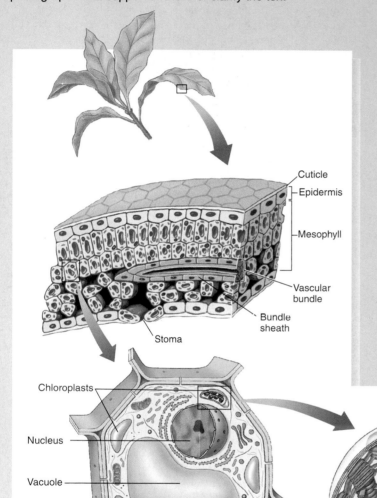

Cuticle
Epidermis
Mesophyll
Vascular bundle
Bundle sheath
Stoma

Chloroplasts
Nucleus
Vacuole
Cell wall

Inner membrane
Outer membrane
Granum
Stoma
Thylakoid

Multi-Level Perspective

Illustrations depicting complex structures or processes combine macroscopic and microscopic views to help you see the relationship between increasingly detailed images.

Light micrographs, as well as scanning and transmission electron micrographs, are used in conjunction with illustrations to present a true picture of what you would encounter in lab. A micron bar is added whenever the magnification is known.

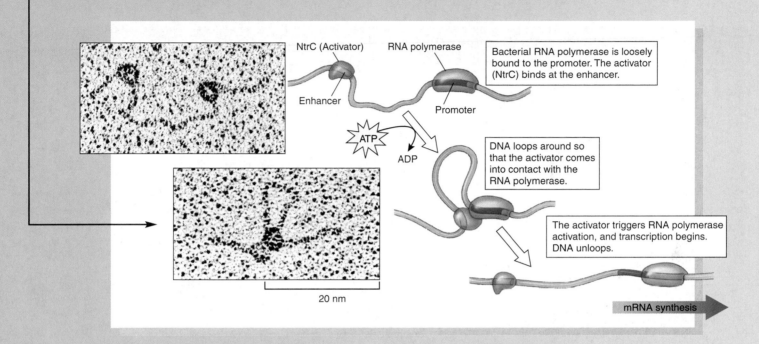

NtrC (Activator)   RNA polymerase

Enhancer   Promoter

ATP   ADP

Bacterial RNA polymerase is loosely bound to the promoter. The activator (NtrC) binds at the enhancer.

DNA loops around so that the activator comes into contact with the RNA polymerase.

The activator triggers RNA polymerase activation, and transcription begins. DNA unloops.

mRNA synthesis

20 nm

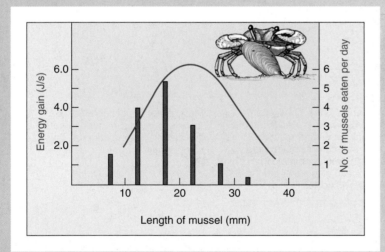

Energy gain (J/s)

No. of mussels eaten per day

Length of mussel (mm)

FIGURE 52.26
Optimal diet.   The shore crab selects a diet of energetically profitable prey. The curve describes the net energy gain (equal to energy gained minus energy expended) derived from feeding on different sizes of mussels. The bar graph shows the numbers of mussels of each size in the diet. Shore crabs tend to feed on those mussels that provide the most energy.
What factors might be responsible for the slight difference in peak prey length relative to the length optimal for maximum energy gain?

Instructive Art Program

Explanatory text boxes describe the action depicted in each step. The discrete, carefully placed boxes guide you through the process, without cluttering the image.

Instructors benefit from this style as well, as the images can be used for presentation without the distraction of extraneous captions.

THE CALVIN CYCLE

1

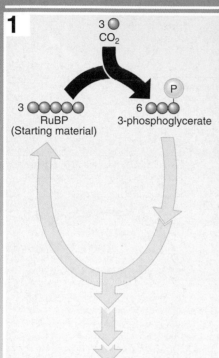

The Calvin cycle begins when a carbon atom from a CO_2 molecule is added to a five-carbon molecule (the starting material). The resulting six-carbon molecule is unstable and immediately splits into three-carbon molecules.

2

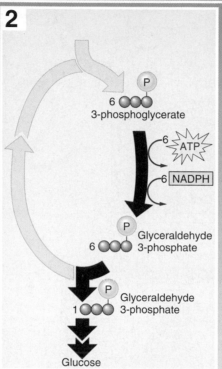

Then, through a series of reactions, energy from ATP and hydrogens from NADPH (the products of the light-dependent reactions) are added to the three-carbon molecules. The now-reduced three-carbon molecules either combine to make glucose or are used to make other molecules.

3

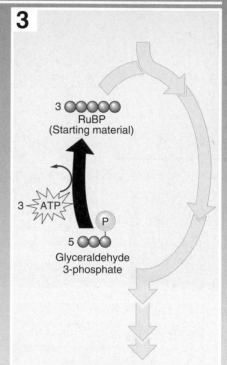

Most of the reduced three-carbon molecules are used to regenerate the five-carbon starting material, thus completing the cycle.

FIGURE 10.17
How the Calvin cycle works.

Process Boxes

Process Boxes break down complex processes into a series of small steps, allowing you to track the key occurrences and learn them as you go.

Phylogeny Guideposts

Phylogeny Guideposts are used in the diversity chapters to help you track relationships among diverse organisms. As each group is introduced, the appropriate branch on the phylogenetic tree is highlighted.

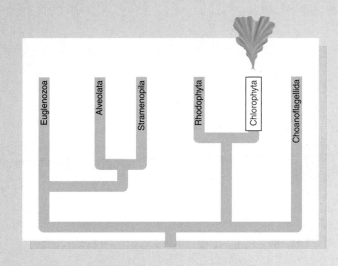

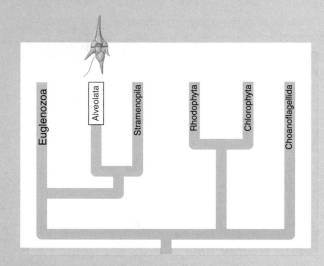

Biochemistry Pathway Icons

These icons are paired with more detailed illustrations to assist you in keeping the big picture in mind when learning complex metabolic processes. The icon highlights which step the main illustration represents, and where that step occurs in the complete process.

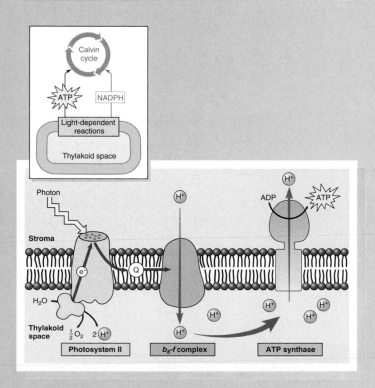

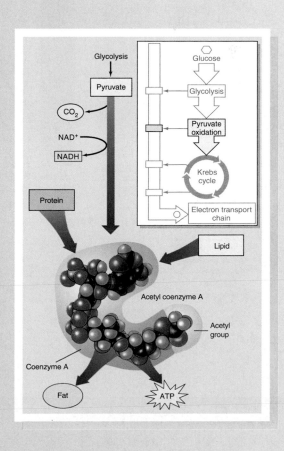

The Learning System

This text is designed to help you learn in a systematic fashion. Simple facts are the building blocks for developing explanations of more complex concepts. The text discussion is presented within a supporting framework of learning aids that help organize studying, reinforce learning, and promote problem-solving skills.

Numbered Headings

The numbered headings employed in the modules form the backbone of the Concept Outline. This consistency makes it easier to identify the key concepts for each chapter, and to then manage the supporting details for each concept.

11

How Cells Divide

Concept Outline

11.1 Prokaryotes divide far more simply than do eukaryotes.

    Cell Division in Prokaryotes. Prokaryotic cells divide by splitting in two.

11.2 The chromosomes of eukaryotes are highly ordered structures.

    Discovery of Chromosomes. All eukaryotic cells contain chromosomes, but different organisms possess differing numbers of chromosomes.

    The Structure of Eukaryotic Chromosomes. Proteins play an important role in packaging DNA in chromosomes.

11.3 Mitosis is a key phase of the cell cycle.

    The Cell Cycle. The cell cycle consists of three growth phases, a nuclear division phase, and a cytoplasmic division stage.

    Interphase: Preparing for Mitosis. In interphase, the cell grows, replicates its DNA, and prepares for cell division.

    Mitosis. In prophase, the chromosomes condense, and microtubules attach sister chromosomes to opposite poles of the cell. In metaphase, the chromosomes align along the center of the cell. In anaphase, the chromosomes separate; in telophase, the spindle dissipates and the nuclear envelope re-forms.

    Cytokinesis. In cytokinesis, the cytoplasm separates into two roughly equal halves.

11.4 The cell cycle is carefully controlled.

    General Strategies of Cell Cycle Control. At three points in the cell cycle, feedback from the cell determines whether the cycle will continue.

    Molecular Mechanisms of Cell Cycle Control. Special proteins regulate the checkpoints of the cell cycle.

    Cancer and the Control of Cell Proliferation. Cancer results from damage to genes encoding proteins that regulate the cell division cycle.

FIGURE 11.1
Cell division in prokaryotes. It's hard to imagine fecal coliform bacteria as being beautiful, but here is *Escherichia coli*, inhabitant of the large intestine and the biotechnology lab, spectacularly caught in the act of fission.

All species of organisms—bacteria, alligators, the weeds in a lawn—grow and reproduce. From the smallest creature to the largest, all species produce offspring like themselves and pass on the hereditary information that makes them what they are. In this chapter, we examine how cells divide and reproduce (figure 11.1). The mechanism of cell reproduction and its biological consequences have changed significantly during the evolution of life on earth. The process is complex in eukaryotes, involving both the replication of chromosomes and their separation into daughter cells. Much of what we are learning about the causes of cancer relates to how cells control this process, and in particular their propensity to divide, a mechanism that in broad outline remains the same in all eukaryotes.

207

11.1 Prokaryotes divide far more simply than do eukaryotes.

Cell Division in Prokaryotes

The end result of cell division in both prokaryotic and eukaryotic cells is two daughter cells, each with the same genetic information as the original cell. The differences between these two basic cell types lead to large differences in how this process occurs. Despite these differences, the essentials of the process are the same: duplication and segregation of genetic information into daughter cells, and division of cellular contents. We will begin by looking at the simpler process, which occurs in prokaryotes: division by **binary fission.**

Most prokaryotes have a genome made up of a single, circular DNA molecule. Despite its apparent simplicity, the DNA molecule of the bacterium *Escherichia coli* is actually on the order of 500 times longer than the cell itself! Thus, this "simple" structure is actually exquisitely packaged to fit into the cell. Although not found in a nucleus, the DNA is in a compacted form called a *nucleoid* that is distinct from the cytoplasm around it.

For many years, it was believed that the *E. coli* DNA molecule was passively segregated by attachment to the membrane and growth of the membrane as the cell elongates. More recently, a more complex picture is emerging that involves both active partitioning of the DNA and formation of a septum that divides the elongated cell in half. Although the details differ, species as different as *E. coli* and *Bacillus subtilis* both exhibit active partitioning of the newly replicated DNA molecules during the division process. This requires both specific sites on the chromosomes and a number of proteins actively involved in the process.

Binary fission begins with the replication of the prokaryotic DNA at a specific site—the origin of replication (see chapter 15)—and proceeds bidirectionally around the circular DNA to a specific site of termination (figure 11.2). Growth of the cell results in elongation, and the newly replicated DNA molecules are actively partitioned to one-fourth and three-quarters of the cell length. This process requires sequences near the origin of replication and results in these sequences being attached to the membrane. The cell itself is partitioned by the growth of new membrane and cell material called a septum (see figure 11.2). This process of septation is complex and under control of the cell as well.

The site of septation is usually the midpoint of the cell and begins with the formation of a ring composed of the molecule FtsZ (figure 11.3). This then results in the accumulation of a number of other proteins, including ones embedded in the membrane. The exact mechanism of septation is not known, but this structure grows inward radially until the cells pinch off into new cells.

FIGURE 11.2
Binary fission. Prior to cell division, the prokaryotic DNA molecule replicates. The replication of the double-stranded, circular DNA molecule *(blue)* that constitutes the genome of a prokaryote begins at a specific site, called the origin of replication. The replication enzymes move out in both directions from that site and make copies *(red)* of each strand in the DNA duplex. The enzymes continue until they meet at another specific site, the terminus of replication. After the DNA is replicated, the cell elongates, and the DNA is partitioned in the cell. Septation then begins, in which new cell membrane material begins to grow and form a septum at approximately the midpoint of the cell. A protein molecule called FtsZ facilitates this process. When the septum is complete, the cell pinches in two, and two daughter cells are formed, each containing a prokaryotic DNA molecule.

208   Part II   Biology of the Cell

Concept Outline

Each chapter begins with an outline that gives you an overview of the content contained within that chapter. Reviewing the concept outline before reading the chapter will help focus your attention on the major concepts you should take away from the chapter.

Modular Format

Each page or two-page spread in *Biology* is organized as an independent module, with its own numbered heading at the top of the left-hand page, and a highlighted summary at the bottom of the right-hand page. This system organizes the information in the chapter within a clear conceptual framework, which in turn helps you learn and retain the material.

Section Summaries

Each module ends with a summary intended to reinforce the key concepts from that section. Reviewing the summary after reading the section will indicate whether you learned the main ideas presented in the module.

Vocabulary Boxes

These boxes are found throughout the text, in chapters that require you to learn many new terms. This saves you time when studying, by placing the definitions you need in one location. It is also a handy study tool, as it reinforces the key terms for that chapter.

FIGURE 11.3
The FtsZ protein. In these dividing *E. coli* bacteria, the FtsZ protein is fluorescent, and its location during binary fission can be seen. The protein assembles into a ring at approximately the midpoint of the cell, where it facilitates septation and cell division. Bacteria in which the *ftsZ* gene is mutated cannot divide.

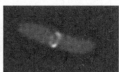

The FtsZ molecule is interesting for a number of reasons. It is highly conserved evolutionarily, having been identified in most prokaryotes, including archaebacteria. It shows some small similarity to eukaryotic tubulin and can form filaments and rings. Recent 3-D crystals show similarity to tubulin as well. It is interesting to speculate that the elaborate spindle found in eukaryotic division may be related to this simple prokaryotic precursor (figure 11.4).

The evolution of eukaryotic cells led to much more complex genomes composed of multiple linear chromosomes housed in a membrane-bounded nucleus. These chromosomes contain even more DNA, and thus pose packaging problems that are solved by DNA being complexed with protein and packaged into functionally distinct chromosomes. This creates more challenges both for the replication of the genome and for its accurate segregation during cell division. The process that evolved to accomplish this segregation of chromosomes is called mitosis.

> Prokaryotes divide by binary fission. Fission begins in the middle of the cell. An active partitioning process ensures that one genome will end up in each daughter cell.

FIGURE 11.4
A comparison of protein assemblies during cell division among different organisms. The prokaryotic protein FtsZ has a structure that is similar to that of the eukaryotic protein tubulin. Tubulin is the protein component of microtubules, which are fibers that play an important role in eukaryotic cell division.

Prokaryotes
No nucleus; single circular chromosome. After DNA is replicated, it is partitioned in the cell. After cell elongation, FtsZ protein assembles into a ring and facilitates septation and cell division.

Some protists
Nucleus present and nuclear envelope remains intact during cell division. Chromosomes linear. Fibers called microtubules, composed of the protein tubulin, pass through tunnels in the nuclear membrane and set up an axis for separation of replicated chromosomes, and cell division.

Other protists
A spindle of microtubules forms between two pairs of centrioles at opposite ends of the cell. The spindle passes through one tunnel in the intact nuclear envelope. Kinetochore microtubules form between kinetochores on the chromosomes and the spindle poles and pull the chromosomes to each pole.

Yeasts
Nuclear envelope remains intact; spindle microtubules form inside the nucleus between spindle pole bodies. A single kinetochore microtubule attaches to each chromosome and pulls each to a pole.

Animals
Spindle microtubules begin to form between centrioles outside of nucleus. As these centrioles move to the poles, the nuclear envelope breaks down, and kinetochore microtubules attach kinetochores of chromosomes to spindle poles. Polar microtubules extend toward the center of the cell and overlap.

(figure labels: Chromosome, Septum (forming), FtsZ protein, Nucleus, Microtubules, Chromosomes, Central spindle of microtubules, Centrioles, Kinetochore microtubules, Kinetochore, Spindle pole body, Spindle microtubules, Fragments of nuclear envelope, Centrioles, Kinetochore and polar microtubules)

A Vocabulary of Cell Division

binary fission Reproduction of a cell by division into two equal or nearly equal parts. Prokaryotes divide by binary fission.

centromere A constricted region of a chromosome about 220 nucleotides in length, composed of highly repeated DNA sequences. During mitosis, the centromere joins the two sister chromatids and is the site to which the kinetochores are attached.

chromatid One of the two copies of a replicated chromosome, joined by a single centromere to the other strand.

chromatin The complex of DNA and proteins of which eukaryotic chromosomes are composed.

chromosome The structure within cells that contains the genes. In eukaryotes, it consists of a single linear DNA molecule associated with proteins. The DNA replicates during S phase, and the replicas separate during M phase.

cytokinesis Division of the cytoplasm of a cell after nuclear division.

euchromatin The portion of a chromosome that is extended except during cell division, and from which RNA is transcribed.

heterochromatin The portion of a chromosome that remains permanently condensed and, therefore, is not transcribed into RNA. Most centromere regions are heterochromatic.

homologues Homologous chromosomes; in diploid cells, one of a pair of chromosomes that carry equivalent genes.

kinetochore A disk of protein bound to the centromere and attached to microtubules during mitosis, linking each chromatid to the spindle apparatus.

microtubule A hollow cylinder, about 25 nanometers in diameter, composed of subunits of the protein tubulin. Microtubules lengthen by the addition of tubulin subunits to their end(s) and shorten by the removal of subunits.

mitosis Nuclear division in which replicated chromosomes separate to form two genetically identical daughter nuclei. When accompanied by cytokinesis, it produces two identical daughter cells.

nucleosome The basic packaging unit of eukaryotic chromosomes, in which the DNA molecule is wound around a cluster of histone proteins. Chromatin is composed of long strings of nucleosomes that resemble beads on a string.

The Learning System

Concept Review

An expanded version of the Concept Outline, the Concept Review details each numbered section head followed by its supporting ideas. Each supporting idea is page referenced to allow you to focus your time on areas where you need additional study.

Concept Review

For interactive testing, visit the Online Learning Center with PowerWeb at www.mhhe.com/Raven7

11.1 Prokaryotes divide far more simply than do eukaryotes.

Cell Division in Prokaryotes

- Most prokaryotes have a genome made up of a single, circular DNA molecule, and replicate via binary fusion. (p. 208)
- Binary fusion begins with DNA replication, which starts at the origin site and proceeds bidirectionally around the circular DNA to a specific site of termination. (p. 208)
- The evolution of eukaryotic cells led to much more complex genomes and, thus, new and different ways to replicate and segregate the genome during cell division. (p. 209)

11.2 The chromosomes of eukaryotes are highly ordered structures.

Discovery of Chromosomes

- Chromosomes were first discovered in 1882 by Walther Fleming. (p. 210)
- The number of chromosomes varies from one species to another. Humans have 23 nearly identical pairs for a total of 46 chromosomes. (p. 210)

The Structure of Eukaryotic Chromosomes

- The DNA is a very long, double-stranded fiber extending unbroken through the entire length of the chromosome. A typical human chromosome contains about 140 million nucleotides. (p. 211)
- Every 200 nucleotides, the DNA duplex is coiled around a core of eight histone proteins, forming a nucleosome. (p. 211)
- The particular array of chromosomes an individual possesses is its karyotype. (p. 212)
- The number of different chromosomes a species contains is known as its haploid (n) number, and is considered one complete set of chromosomes. (p. 212)
- Humans are diploid, with homologues coming from both

Interphase: Preparing for Mitosis

- The cell grows throughout interphase. The G_1 and G_2 phases are periods of protein synthesis and organelle production, while the S phase is when DNA replication occurs. (p. 214)

Mitosis

- Chromatin condensation continues into prophase. The spindle apparatus is assembled, and sister chromatids are linked to opposite poles of the cell by microtubules. The nuclear envelope breaks down. (p. 215)
- During metaphase, chromosomes align in the center of the cell along the metaphase plate. (p. 215)
- Anaphase begins when centromeres divide, freeing the two sister chromatids from each other. Sister chromatids are pulled to opposite poles as the attached microtubules shorten. (pp. 216–217)
- In telophase, the spindle apparatus disassembles, and the nuclear membrane begins to re-form. (p. 217)

Cytokinesis

- Cytokinesis is the phase of the cell cycle when the cell actually divides. Cytokinesis generally involves the cleavage of the cell into roughly equal halves, forming two daughter cells. (p. 218)

11.4 The cell cycle is carefully controlled.

General Strategies of Cell Cycle Control

- A cell uses three main checkpoints to both assess the internal state of the cell and integrate external signals. The G_1/S checkpoint is the primary point at which the cell decides to divide; the G_2/M checkpoint represents a commitment to mitosis; and the spindle checkpoint ensures that all chromosomes are attached to the spindle in preparation for anaphase. (p. 219)

Molecular Mechanisms of Cell Cycle Control

- Two groups of proteins, cyclins and Cdk's, interact and regulate the cell cycle. (p. 220)
- Cells also receive protein signals (growth factors) that affect cell division. (p. 222)

Cancer and the Control of Cell Proliferation

- Cancer is failure of cell division control. (p. 223)
- It is believed that a malfunction in the *p53* gene may allow cells to go through repeated cell division without being stopped at the appropriate checkpoints. (p. 223)
- Proto-oncogenes are normal cellular genes that become oncogenes when mutated. Proto-oncogenes can encode growth factors, protein relay switches, and kinase enzyme. (p. 224)
- Tumor-suppressor genes can also lead to cancer when they are mutated. (p. 224)

Test Your Understanding

For interactive testing, visit the Online Learning Center with PowerWeb at www.mhhe.com/Raven7

Self Test

1. Bacterial cells divide by
   a. mitosis.
   b. replication.
   c. cytokinesis.
   d. binary fission.
2. Most eukaryotic organisms have _____ chromosomes in their cells.
   a. 1–5
   b. 10–50
   c. 100–500
   d. over 1000
3. Replicate copies of each chromosome are called _____ and are joined at the _____.
   a. homologues/centromere
   b. sister chromatids/kinetochore
   c. sister chromatids/centromere
   d. homologues/kinetochore
4. During which phase of the cell cycle is DNA synthesized?
   a. G_1
   b. G_2
   c. S
   d. M
5. Chromosomes are visible under a light microscope
   a. during mitosis.
   b. during interphase.
   c. when they are attached to their sister chromatids.
   d. All of these are correct.
6. During mitosis, the sister chromatids are separated and pulled to opposite poles during which stage?
   a. interphase
   b. metaphase
   c. anaphase
   d. telophase
7. Cytokinesis is
   a. the same process in plant and animal cells.
   b. the separation of cytoplasm and the formation of two cells.
   c. the final stage of mitosis.
   d. the movement of kinetochores.
8. The eukaryotic cell cycle is controlled at several points; which of these statements is *not* true?
   a. Cell growth is assessed at the G_1/S checkpoint.
   b. DNA replication is assessed at the G_2/M checkpoint.
   c. Environmental conditions are assessed at the G_0 checkpoint.
   d. The chromosomes are assessed at the spindle checkpoint.
9. What proteins are used to control cell growth specifically in *multicellular* eukaryotic organisms?
   a. Cdk
   b. MPF
   c. cyclins
   d. growth factors
10. What causes cancer in cells?
    a. damage to genes
    b. chemical damage to cell membranes
    c. UV damage to transport proteins
    d. All of these cause cancer in cells.

Test Your Visual Understanding

a   b   c   d   e

1. Match the mitotic and cell cycle phases with the appropriate figure.
   anaphase
   interphase
   metaphase
   prophase
   telophase

Apply Your Knowledge

1. An ancient plant called horsetail contains 216 chromosomes. How many homologous pairs of chromosomes does it contain? How many chromosomes are present in its cells during metaphase?
2. Colchicine is a poison that binds to tubulin and prevents its assembly into microtubules; cytochalasins are compounds that bind to the ends of actin filaments and prevent their elongation. What effects would these two substances have on cell division in animal cells?
3. If you could construct an artificial chromosome, what elements would you introduce into it, at a minimum, so that it could function normally in mitosis?

Testing Yourself

Each chapter concludes with a set of questions designed to test your knowledge of the content, including multiple choice questions, illustration-based questions, and application questions. Answers to these questions are found on the *Biology* Online Learning Center at: *www.mhhe.com/raven7*. At the site you can take an interactive version of the end-of-chapter quiz that provides you with hints and instructional feedback.

BIOLOGY

1

The Science of Biology

Concept Outline

1.1 Biology is the science of life.

Organization of Living Things. Biology is the science that studies living organisms and how they interact with one another and with their environment. There are several characteristics that define life.

1.2 Scientists form generalizations from observations.

The Nature of Science. Science employs both deductive reasoning and inductive reasoning.

How Science Is Done. Scientists construct hypotheses from systematically collected objective data. They then perform experiments designed to disprove the hypotheses.

1.3 Darwin's theory of evolution illustrates how science works.

Charles Darwin. On a round-the-world voyage, Darwin made observations that eventually led him to formulate the hypothesis of evolution by natural selection.

Darwin's Evidence. The fossil and geographic patterns of life he observed convinced Darwin that a process of evolution had occurred.

Inventing the Hypothesis of Natural Selection. The Malthus idea that populations cannot grow unchecked led Darwin, and another naturalist named Wallace, to propose the hypothesis of natural selection.

Evolution After Darwin: More Evidence. In the century since Darwin, a mass of scientific advances and discoveries have supported his theory of evolution, which is now accepted by practically all practicing biologists.

1.4 Four themes unify biology as a science.

Core Themes Unite Biology. Living things all exhibit cellular organization, a mechanism for heredity (DNA), adaptation to produce unique features as the result of evolution, and the conservation of key features during evolution.

FIGURE 1.1
A replica of the *Beagle*, off the southern coast of South America. The famous English naturalist Charles Darwin set forth on H.M.S. *Beagle* in 1831, at the age of 22.

You are about to embark on a journey—a journey of discovery about the nature of life. Nearly 180 years ago, a young English naturalist named Charles Darwin set sail on a similar journey on board H.M.S. *Beagle;* figure 1.1 shows a replica of the *Beagle*. What Darwin learned on his five-year voyage led directly to his development of the theory of evolution by natural selection, a theory that has become the core of the science of biology. Darwin's voyage seems a fitting place to begin our exploration of biology, the scientific study of living organisms and how they have evolved. Before we begin, however, let's take a moment to think about what biology is and why it's important.

Part I *The Origin of Living Things*

1

1.1 Biology is the science of life.

Organization of Living Things

In its broadest sense, biology is the study of living things—*the science of life.* Living things come in an astounding variety of shapes and forms, and biologists study life in many different ways. They live with gorillas, collect fossils, and listen to whales. They read the messages encoded in the long molecules of heredity and count how many times a hummingbird's wings beat each second.

Properties of Life

What makes something "alive"? Anyone could deduce that a galloping horse is alive and a car is not, but why? We cannot say, "If it moves, it's alive," because a car can move, and gelatin can wiggle in a bowl. They certainly are not alive. What characteristics *do* define life? All living organisms share a family of basic characteristics:

1. **Cellular organization.** All organisms consist of one or more cells. Often too tiny to see, cells carry out the basic activities of living. Each cell is bounded by a membrane that separates it from its surroundings.
2. **Order.** All living things are highly ordered. Your body is composed of many different kinds of cells, each containing many complex molecular structures.
3. **Sensitivity.** All organisms respond to stimuli. Plants grow toward a source of light, and your pupils dilate when you walk into a dark room.
4. **Growth, development, and reproduction.** All organisms are capable of growing and reproducing, and they all possess hereditary molecules that are passed to their offspring, ensuring that the offspring are of the same species.
5. **Energy utilization.** All organisms take in energy and use it to perform many kinds of work. Every muscle in your body is powered with energy you obtain from the food you eat.
6. **Evolutionary adaptation.** All organisms interact with other organisms and the environment in ways that influence survival, and as a consequence, organisms evolve adaptations to their environments.
7. **Homeostasis.** All organisms maintain relatively constant internal conditions, different from their environment, a process called homeostasis.

Hierarchical Organization

The organization of the biological world is hierarchical—that is, each level builds on the level below it.

The Cellular Level. At the cellular level (figure 1.2), atoms, the fundamental elements of matter, are joined together into clusters called **molecules.** Complex biological molecules are assembled into tiny structures called **organelles** within membrane-bounded units we call **cells.**

The cell is the basic unit of life. Many organisms are composed of single cells. Bacteria are single cells, for example. All animals and plants, as well as most fungi and algae, are multicellular—composed of more than one cell.

The Organismal Level. Cells are organized into three levels of organization. The most basic level is that of **tissues,** which are groups of similar cells that act as a functional unit. Tissues, in turn, are grouped into **organs,** which are body structures composed of several different tissues grouped together in a structural and functional unit. Your brain is an organ composed of nerve cells and a variety of connective tissues that form protective coverings and contribute blood. At the third level of organization, organs are grouped into **organ systems.** The nervous system, for example, consists of sensory organs, the brain and spinal cord, and neurons that convey signals to and from them.

The Populational Level. Individual organisms are organized into several hierarchical levels within the living world. The most basic of these is the **population,** which is a group of organisms of the same species living in the same place. All the populations of a particular kind of organism together form a **species,** its members similar in appearance and able to interbreed. At a higher level of biological organization, a **biological community** consists of all the populations of different species living together in one place.

At the highest tier of biological organization, a biological community and the physical habitat within which it lives together constitute an ecological system, or **ecosystem.** For example, the soil and water of a mountain ecosystem interact with the biological community of a mountain meadow in many important ways.

Emergent Properties

At each higher level in the living hierarchy, novel properties emerge. These **emergent properties** result from the way in which components interact, and often cannot be guessed just by looking at the parts themselves. Examining the cells gives little clue of what the animal is like. You have the same array of cell types as a giraffe. It is because the living world exhibits many emergent properties that it is difficult to define "life."

All living things share certain key characteristics including: cellular organization, sensitivity, growth, development and reproduction, adaptation, and homeostasis.

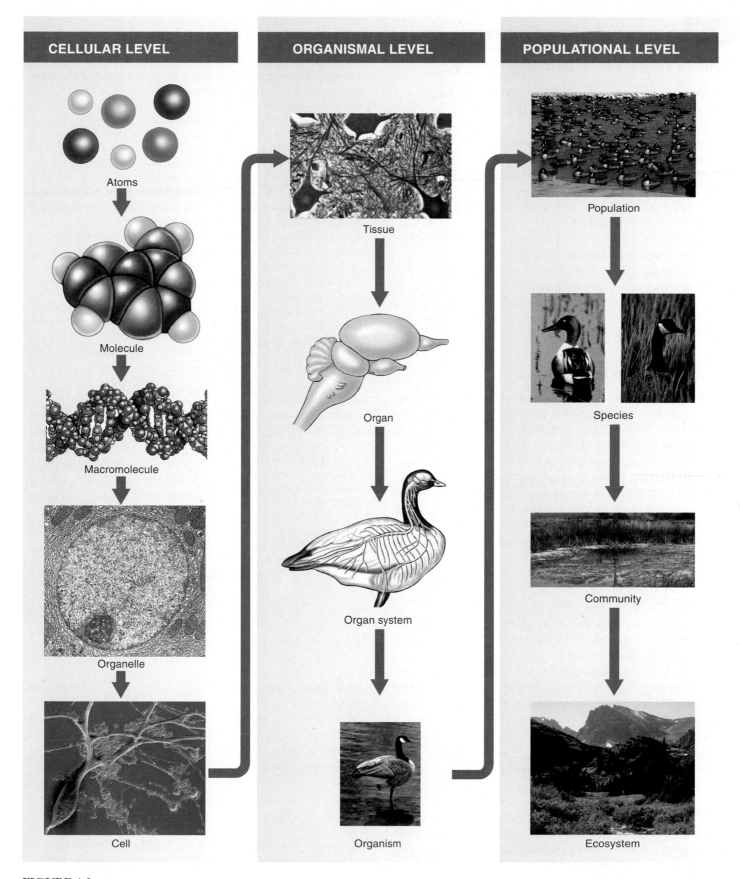

CELLULAR LEVEL

Atoms

Molecule

Macromolecule

Organelle

Cell

ORGANISMAL LEVEL

Tissue

Organ

Organ system

Organism

POPULATIONAL LEVEL

Population

Species

Community

Ecosystem

FIGURE 1.2

Hierarchical organization of living things. Life is highly organized—from small and simple to large and complex, within cells, within multicellular organisms, and among populations of organisms.

1.2 Scientists form generalizations from observations.

The Nature of Science

Biology is a fascinating and important subject because it dramatically affects our daily lives and our futures. Many biologists are working on problems that critically affect our lives, such as the world's rapidly expanding population and diseases like cancer and AIDS. The knowledge these biologists gain will be fundamental to our ability to manage the world's resources in a suitable manner, to prevent or cure diseases, and to improve the quality of our lives and those of our children and grandchildren.

Biology is one of the most successful of the "natural sciences," those devoted to explaining what our world is like. To understand biology, you must first understand the nature of science. Because the basic tool a scientist uses is thought, to understand the nature of science, it is useful to focus for a moment on how scientists think. They reason in two ways: deductively and inductively.

Deductive Reasoning

Deductive reasoning applies general principles to predict specific results. The logic flows from the general to the specific. Over 2200 years ago, the Greek Eratosthenes used Euclidean geometry and deductive reasoning to accurately estimate the circumference of the earth (figure 1.3). This sort of analysis of specific cases using general principles is an example of deductive reasoning. It is the reasoning of mathematics and philosophy and is used to test the validity of general ideas in all branches of knowledge. A biologist uses deductive reasoning to infer the species of a specimen from its characteristics.

Inductive Reasoning

In **inductive reasoning,** the logic flows in the opposite direction, from the specific to the general. Inductive reasoning uses specific observations to construct general scientific principles. If cats possess hair, and dogs possess hair, and every other mammal you observe has hair, then you may infer that perhaps *all* mammals have hair. Inductive reasoning leads to generalizations that can then be tested.

Webster's Dictionary defines science as systematized knowledge derived from observation and experiment carried on to determine the principles underlying what is being studied. In other words, a scientist determines principles from observations, discovering general principles by carefully examining specific cases. Inductive reasoning first became important to science in the 1600s in Europe, when Francis Bacon, Isaac Newton, and others began to use the results of particular experiments to infer general principles about how the world operates. If you release an apple from

FIGURE 1.3
Deductive reasoning: How Eratosthenes estimated the circumference of the earth using deductive reasoning. **1.** On a day when sunlight shone straight down a deep well at Syene in Egypt, Eratosthenes measured the length of the shadow cast by a tall obelisk in the city of Alexandria, about 800 kilometers away. **2.** The shadow's length and the obelisk's height formed two sides of a triangle. Using the recently developed principles of Euclidean geometry, Eratosthenes calculated the angle, a, to be 7° and 12′, exactly $\frac{1}{50}$ of a circle (360°). **3.** If angle $a = \frac{1}{50}$ of a circle, then the distance between the obelisk (in Alexandria) and the well (in Syene) must equal $\frac{1}{50}$ of the circumference of the earth. **4.** Eratosthenes had heard that it was a 50-day camel trip from Alexandria to Syene. Assuming that a camel travels about 18.5 kilometers per day, he estimated the distance between obelisk and well as 925 kilometers (using different units of measure, of course).
5. Eratosthenes thus deduced the circumference of the earth to be 50 × 925 = 46,250 kilometers. Modern measurements put the distance from the well to the obelisk at just over 800 kilometers. Employing a distance of 800 kilometers, Eratosthenes's value would have been 50 × 800 = 40,000 kilometers. The actual circumference is 40,075 kilometers.

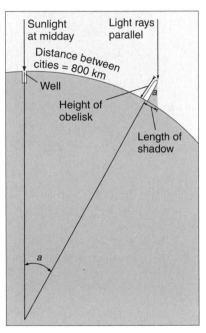

your hand, what happens? The apple falls to the ground. From a host of simple, specific observations like this, Newton inferred a general principle: All objects fall toward the center of the earth. What Newton did was construct a mental model of how the world works, a family of general principles consistent with what he could see and learn. Scientists do the same today. They use specific observations to build general models, and then test the models to see how well they work.

Science is a way of viewing the world that focuses on objective information, putting that information to work to build understanding.

How Science Is Done

How do scientists establish which general principles are true from among the many that might be true? They do this by systematically testing alternative proposals. If these proposals prove inconsistent with experimental observations, they are rejected as untrue. Figure 1.4 illustrates the process. After making careful observations concerning a particular area of science, scientists construct a **hypothesis,** which is a suggested explanation that accounts for those observations. A hypothesis is a proposition that might be true. Those hypotheses that have not yet been disproved are retained. They are useful because they fit the known facts, but they are always subject to future rejection if, in the light of new information, they are found to be incorrect.

Testing Hypotheses

We call the test of a hypothesis an **experiment.** Suppose that a room appears dark to you. To understand why it appears dark, you propose several hypotheses. The first might be, "There is no light in the room because the light switch is turned off." An alternative hypothesis might be, "There is no light in the room because the lightbulb is burned out." And yet another alternative hypothesis might be, "I am going blind." To evaluate these hypotheses, you would conduct an experiment designed to eliminate one or more of the hypotheses. For example, you might test your hypotheses by reversing the position of the light switch. If you do so and the light does not come on, you have disproved the first hypothesis. Something other than the setting of the light switch must be the reason for the darkness. Note that a test such as this does not prove that any of the other hypotheses are true; it merely demonstrates that one of them is not. A successful experiment is one in which one or more of the alternative hypotheses is demonstrated to be inconsistent with the results and is thus rejected.

As you proceed through this text, you will encounter many hypotheses that have withstood the test of experiment. Many will continue to do so; others will be revised as new observations are made by biologists. Biology, like all science, is in a constant state of change, with new ideas appearing and replacing old ones.

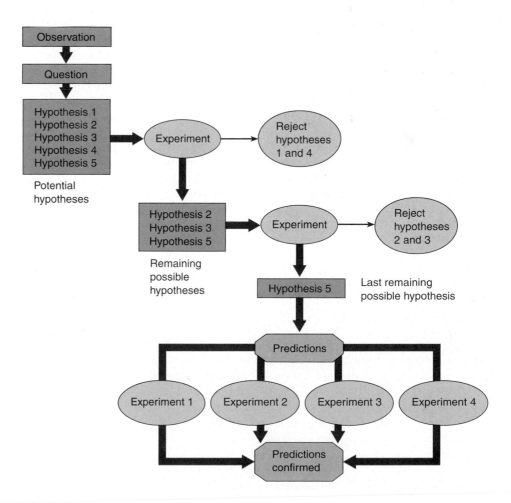

FIGURE 1.4

How science is done. This diagram illustrates how scientific investigations proceed. First, scientists make observations that raise a particular question. They develop a number of potential explanations (hypotheses) to answer the question. Next, they carry out experiments in an attempt to eliminate one or more of these hypotheses. Then, predictions are made based on the remaining hypotheses, and further experiments are carried out to test these predictions. As a result of this process, the least unlikely hypothesis is selected.

Establishing Controls

Often we are interested in learning about processes that are influenced by many factors, or **variables.** To evaluate alternative hypotheses about one variable, all other variables must be kept constant. This is done by carrying out two experiments in parallel: In the first experiment, one variable is altered in a specific way to test a particular hypothesis; in the second experiment, called the **control experiment,** that variable is left unaltered. In all other respects the two experiments are identical, so any difference in the outcomes of the two experiments must result from the influence of the variable that was changed. Much of the challenge of experimental science lies in designing control experiments that isolate a particular variable from other factors that might influence a process.

Using Predictions

A successful scientific hypothesis needs to be not only valid but useful—it needs to tell you something you want to know. A hypothesis is most useful when it makes predictions, because those predictions provide a way to test the validity of the hypothesis. If an experiment produces results inconsistent with the predictions, the hypothesis must be rejected. On the other hand, if the predictions are supported by experimental testing, the hypothesis is supported. The more experimentally supported predictions a hypothesis makes, the more valid the hypothesis is. For example, Einstein's hypothesis of relativity was at first provisionally accepted because no one could devise an experiment that invalidated it. The hypothesis made a clear prediction: that the sun would bend the path of light passing by it. When this prediction was tested in a total eclipse, the light from background stars was indeed bent. Because this result was unknown when the hypothesis was being formulated, it provided strong support for the hypothesis, which was then accepted with more confidence.

Developing Theories

Scientists use the word **theory** in two main ways. A "theory" is a proposed explanation for some natural phenomenon, often based on some general principle. Thus we speak of the principle first proposed by Newton as the "theory of gravity." Such theories often bring together concepts that were previously thought to be unrelated, and offer unified explanations of different phenomena. Newton's theory of gravity provided a single explanation for objects falling to the ground and the orbits of planets around the sun. "Theory" is also used to mean the body of interconnected concepts, supported by scientific reasoning and experimental evidence, that explains the facts in some area of study. Such a theory provides an indispensable framework for organizing a body of knowledge. For example, quantum theory in physics brings together a set of ideas about the nature of the universe, explains experimental facts, and serves as a guide to further questions and experiments.

To a scientist, theories are the solid ground of science, that of which we are most certain. In contrast, to the general public, "theory" implies just the opposite—a *lack* of knowledge, or a guess. Not surprisingly, this difference often results in confusion. In this text, theory will always be used in its scientific sense, in reference to an accepted general principle or body of knowledge.

To suggest, as many critics outside of science do, that evolution is "just a theory" is misleading. The hypothesis that evolution has occurred is an accepted scientific fact; it is supported by overwhelming evidence. Modern evolutionary theory is a complex body of ideas whose importance spreads far beyond explaining evolution; its ramifications permeate all areas of biology, and it provides the conceptual framework that unifies biology as a science.

Research and the Scientific Method

It used to be fashionable to speak of the "scientific method" as consisting of an orderly sequence of logical, either/or steps. Each step would reject one of two mutually incompatible alternatives, as if trial-and-error testing would inevitably lead a researcher through the maze of uncertainty that always impedes scientific progress. If this were indeed so, a computer would make a good scientist. But science is not done this way. As the British philosopher Karl Popper has pointed out, successful scientists without exception design their experiments with a pretty fair idea of how the results are going to come out. They have what Popper calls an "imaginative preconception" of what the truth might be. A hypothesis that a successful scientist tests is not just any hypothesis; rather, it is an educated guess or a hunch, in which the scientist integrates all that he or she knows and allows his or her imagination full play, in an attempt to get a sense of what *might* be true (see Box: How Biologists Do Their Work). It is because insight and imagination play such a large role in scientific progress that some scientists are so much better at science than others, just as Beethoven and Mozart stand out among most other composers.

Some scientists perform what is called *basic research*, which is intended to extend the boundaries of what we know. These individuals typically work at universities, and their research is usually financially supported by their institutions and by external sources, such as the government, industry, and private foundations. Basic research is as diverse as its name implies. Some basic scientists attempt to find out how certain cells take up specific chemicals, while others count the number of dents in tiger teeth. The information generated by basic research contributes to the growing body of scientific knowledge, and it provides the scientific foundation utilized by *applied research*. Scientists who conduct applied research are often employed in some kind of industry. Their work may

How Biologists Do Their Work

The Consent

Late in November, on a single night
Not even near to freezing, the ginkgo trees
That stand along the walk drop all their leaves
In one consent, and neither to rain nor to wind
But as though to time alone: the golden and
green
Leaves litter the lawn today, that yesterday
Had spread aloft their fluttering fans of light.
What signal from the stars? What senses took it
in?
What in those wooden motives so decided
To strike their leaves, to down their leaves,
Rebellion or surrender? And if this
Can happen thus, what race shall be exempt?
What use to learn the lessons taught by time,
If a star at any time may tell us: Now.

Howard Nemerov

What is bothering the poet Howard Nemerov is that life is influenced by forces he cannot control or even identify. It is the job of biologists to solve puzzles such as the one he poses, to identify and try to understand those things that influence life.

Nemerov asks why ginkgo trees (figure 1.A) drop all their leaves at once. To find an answer to questions such as this, biologists and other scientists pose *possible* answers and then try to determine which answers are false. Tests of alternative possibilities are

FIGURE 1.A
A ginkgo tree.

called experiments. To learn why the ginkgo trees drop all their leaves simultaneously, a scientist would first formulate several possible answers, called hypotheses:

Hypothesis 1: Ginkgo trees possess an internal clock that times the release of leaves to match the season. On the day Nemerov describes, this clock sends a "drop" signal (perhaps a chemical) to all the leaves at the same time.

Hypothesis 2: The individual leaves of ginkgo trees are each able to sense day length, and when the days get short enough in the fall, each leaf responds independently by falling.

Hypothesis 3: A strong wind arose the night before Nemerov made his observation, blowing all the leaves off the ginkgo trees.

Next, the scientist attempts to eliminate one or more of the hypotheses by conducting an experiment. In this case, one might cover some of the leaves so that they cannot use light to sense day length. If hypothesis 2 is true, then the covered leaves should not fall when the others do, because they are not receiving the same information. Suppose, however, that despite the covering of some of the leaves, all the leaves still fall together. This result would eliminate hypothesis 2 as a possibility. Either of the other hypotheses, and many others, remain possible.

This simple experiment with ginkgoes points out the essence of scientific progress: Science does not prove that certain explanations are true; rather, it proves that others are not. Hypotheses that are inconsistent with experimental results are rejected, while hypotheses that are not proven false by an experiment are provisionally accepted. However, hypotheses may be rejected in the future when more information becomes available, if they are inconsistent with the new information. Just as you can find the correct path through a maze by trying and eliminating false paths, scientists work to find the correct explanations of natural phenomena by eliminating false possibilities.

involve the manufacture of food additives, the creation of new drugs, or the testing of environmental quality.

After developing a hypothesis and performing a series of experiments, a scientist writes a paper carefully describing the experiment and its results. He or she then submits the paper for publication in a scientific journal, but before it is published, it must be reviewed and accepted by other scientists who are familiar with that particular field of research. This process of careful evaluation, called *peer review*, lies at the heart of modern science, fostering careful work, precise description, and thoughtful analysis. When an important discovery is announced in a paper, other scientists attempt to reproduce the result, providing a check on accuracy and honesty. Nonreproducible results are not taken seriously for long.

The explosive growth in scientific research during the second half of the twentieth century is reflected in the enormous number of scientific journals now in existence. Although some, such as *Science* and *Nature*, are devoted to a wide range of scientific disciplines, most are extremely specialized: *Cell Motility and the Cytoskeleton, Glycoconjugate Journal, Mutation Research*, and *Synapse* are just a few examples.

The scientific process involves rejecting hypotheses that are inconsistent with experimental results or observations. Hypotheses that are consistent with available data are conditionally accepted. The formulation of a hypothesis often involves creative insight.

1.3 Darwin's theory of evolution illustrates how science works.

Charles Darwin

Darwin's theory of evolution explains and describes how organisms on earth have changed over time and acquired a diversity of new forms. This famous theory provides a good example of how a scientist develops a hypothesis and how a scientific theory grows and wins acceptance.

Charles Robert Darwin (1809–1882; figure 1.5) was an English naturalist who, after 30 years of study and observation, wrote one of the most famous and influential books of all time. This book, *On the Origin of Species by Means of Natural Selection, or The Preservation of Favoured Races in the Struggle for Life*, created a sensation when it was published, and the ideas Darwin expressed in it have played a central role in the development of human thought ever since.

In Darwin's time, most people believed that the various kinds of organisms and their individual structures resulted from direct actions of the Creator (and to this day many people still believe this). Species were thought to be specially created and unchangeable, or immutable, over the course of time. In contrast to these views, a number of earlier philosophers had presented the view that living things must have changed during the history of life on earth. Darwin proposed a concept he called natural selection as a coherent, logical explanation for this process, and he brought his ideas to wide public attention.

Darwin's book, as its title indicates, presented a conclusion that differed sharply from conventional wisdom. Although his theory did not directly challenge the existence of a Divine Creator, Darwin argued that the operation of *natural* laws produced change over time, or **evolution.** These views put Darwin at odds with most people of his day, who believed in a literal interpretation of the Bible and thus accepted the idea of a fixed and constant world, largely unchanged since it was created by God.

The story of Darwin and his theory begins in 1831, when he was 22 years old. On the recommendation of

FIGURE 1.5
Charles Darwin.   This newly rediscovered photograph taken in 1881, the year before Darwin died, appears to be the last ever taken of the great biologist.

one of his professors at Cambridge University, he was selected to serve as naturalist on a five-year navigational mapping expedition around the coasts of South America (figure 1.6), aboard H.M.S. *Beagle* (figure 1.7). During this long voyage, Darwin had the chance to study a wide variety of plants and animals on continents and islands

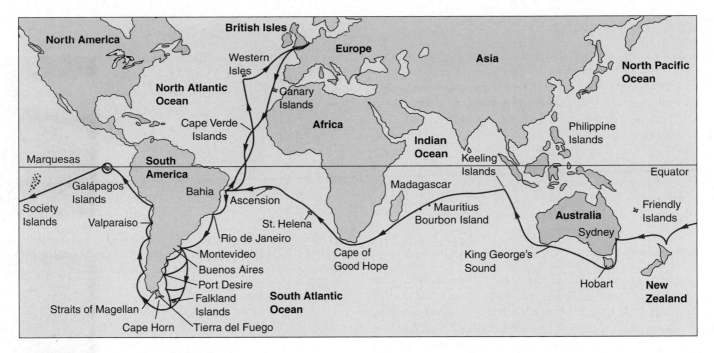

FIGURE 1.6
The five-year voyage of H.M.S. *Beagle*. Most of the time was spent exploring the coasts and coastal islands of South America, such as the Galápagos Islands. Darwin's studies of the animals of the Galápagos Islands played a key role in his eventual development of the concept of evolution by means of natural selection.

FIGURE 1.7
Cross section of the *Beagle*. A 10-gun brig of 242 tons, only 90 feet in length, the *Beagle* had a crew of 74 people! After he first saw the ship, Darwin wrote to his college professor Henslow: "The absolute want of room is an evil that nothing can surmount."

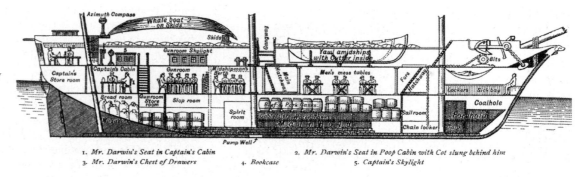

1. *Mr. Darwin's Seat in Captain's Cabin*   2. *Mr. Darwin's Seat in Poop Cabin with Cot slung behind him*
3. *Mr. Darwin's Chest of Drawers*   4. *Bookcase*   5. *Captain's Skylight*

and in distant seas. He was able to explore the biological richness of the tropical forests, examine the extraordinary fossils of huge extinct mammals in Patagonia at the southern tip of South America, and observe the remarkable series of related but distinct forms of life on the Galápagos Islands, off the west coast of South America. Such an opportunity clearly played an important role in the development of his thoughts about the nature of life on earth.

When Darwin returned from the voyage at the age of 27, he began a long period of study and contemplation. During the next 10 years, he published important books on several different subjects, including the formation of oceanic islands from coral reefs and the geology of South

America. He also devoted eight years of study to barnacles, a group of small marine animals with shells that inhabit rocks and pilings, eventually writing a four-volume work on their classification and natural history. In 1842, Darwin and his family moved out of London to a country home at Down, in the county of Kent. In these pleasant surroundings, Darwin lived, studied, and wrote for the next 40 years.

Darwin was the first to propose natural selection as an explanation for the mechanism of evolution that produced the diversity of life on earth. His hypothesis grew from his observations on a five-year voyage around the world.

Darwin's Evidence

One of the obstacles that had blocked the acceptance of any theory of evolution in Darwin's day was the incorrect notion, widely believed at that time, that the earth was only a few thousand years old. Evidence discovered during Darwin's time made this assertion seem less and less likely. The great geologist Charles Lyell (1797–1875), whose *Principles of Geology* (1830) Darwin read eagerly as he sailed on the *Beagle*, outlined for the first time the story of an ancient world of plants and animals in flux. In this world, species were constantly becoming extinct while others were emerging. It was this world that Darwin sought to explain.

What Darwin Saw

When the *Beagle* set sail, Darwin was fully convinced that species were immutable. Indeed, it was not until two or three years after his return that he began to consider seriously the possibility that they could change. Nevertheless, during his five years on the ship, Darwin observed a number of phenomena that were of central importance to him in reaching his ultimate conclusion. For example, in the rich fossil beds of southern South America, he observed fossils of extinct armadillos similar to the armadillos that still lived in the same area (figure 1.8). Why would similar living and fossil organisms be in the same area unless the earlier form had given rise to the other?

Repeatedly, Darwin saw that the characteristics of similar species varied somewhat from place to place. These geographical patterns suggested to him that organismal lineages change gradually as species migrate from one area to another. On the Galápagos Islands, 900 kilometers (540 miles) off the coast of Ecuador, Darwin encountered a variety of different finches on the various islands. The 14 species, although related, differed slightly in appearance, particularly in their beaks (figure 1.9). Darwin felt it most reasonable to assume all these birds had descended from a common ancestor blown by winds from the South Ameri-

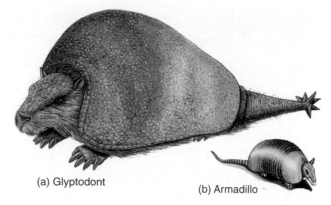

(a) Glyptodont   **(b) Armadillo**

FIGURE 1.8
Fossil evidence of evolution. The now-extinct glyptodont (*a*) was a 2000-kilogram South American armadillo, much larger than the modern armadillo (*b*), which weighs an average of about 4.5 kilograms. (Drawings are not to scale.)

can mainland several million years ago. Eating different foods on different islands, the species had changed during their descent—"descent with modification," or evolution. These finches are discussed in more detail on pages 454 and 483.

In a more general sense, Darwin was struck by the fact that the plants and animals on these relatively young volcanic islands resembled those on the nearby coast of South America. If each one of these plants and animals had been created independently and simply placed on the Galápagos Islands, why didn't they resemble the plants and animals of islands with similar climates, such as those off the coast of Africa, for example? Why did they resemble those of the adjacent South American coast instead?

> The fossils and patterns of life that Darwin observed on the voyage of the *Beagle* eventually convinced him that evolution had taken place.

Large ground finch (seeds)

Cactus ground finch (cactus fruits and flowers)

Vegetarian finch (buds)

Woodpecker finch (insects)

FIGURE 1.9
Four Galápagos finches and what they eat. On the Galápagos Islands, Darwin observed 14 different species of finches differing mainly in their beaks and feeding habits. These four finches eat very different food items, and Darwin surmised that the different shapes of their bills represented evolutionary adaptations that improved their ability to eat the foods available in their specific habitats.

Inventing the Hypothesis of Natural Selection

It is one thing to observe the results of evolution, but quite another to understand how it happens. Darwin's great achievement lies in his formulation of the hypothesis that evolution occurs because of natural selection.

Darwin and Malthus

Of key importance to the development of Darwin's insight was his study of Thomas Malthus's *Essay on the Principle of Population* (1798). In his book, Malthus pointed out that populations of plants and animals (including human beings) tend to increase geometrically, while humans are able to increase their food supply only arithmetically. A *geometric progression* is one in which the elements increase by a constant *factor*; for example, in the progression 2, 6, 18, 54, . . . , each number is three times the preceding one. An *arithmetic progression*, in contrast, is one in which the elements increase by a constant *difference*; in the progression 2, 4, 6, 8, . . . , each number is two greater than the preceding one (figure 1.10).

Because populations increase geometrically, virtually any kind of animal or plant, if it could reproduce unchecked, would cover the entire surface of the world within a surprisingly short time. Instead, populations of species remain fairly constant year after year, because death limits population numbers. Malthus's conclusion provided the key ingredient that was necessary for Darwin to develop the hypothesis that evolution occurs by natural selection.

Sparked by Malthus's ideas, Darwin saw that although every organism has the potential to produce more offspring than can survive, only a limited number actually do survive and produce further offspring. Combining this observation with what he had seen on the voyage of the *Beagle*, as well as with his own experiences in breeding domestic animals, Darwin made an important association (figure 1.11): Those individuals that possess superior physical, behavioral, or other attributes are more likely to survive than those that are not so well endowed. By surviving, they gain the opportunity to pass on their favorable characteristics to their offspring. As the frequency of these characteristics increases in the population, the nature of the population as a whole will gradually change. Darwin called this process selection. The driving force he identified has often been referred to as survival of the fittest.

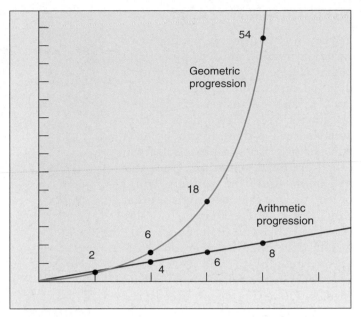

FIGURE 1.10

Geometric and arithmetic progressions. A geometric progression increases by a constant factor (for example, × 2 or × 3 or × 4), while an arithmetic progression increases by a constant difference (for example, units of 1 or 2 or 3). Malthus contended that the human growth curve was geometric, but the human food production curve was only arithmetic.

What is the effect of reducing the constant factor by which the geometric progression increases? Might this effect be achieved with humans? How?

"Can we doubt . . . that individuals having any advantage, however slight, over others, would have the best chance of surviving and procreating their kind? On the other hand, we may feel sure that any variation in the least degree injurious would be rigidly destroyed. This preservation of favorable variations, I call Natural Selection."

FIGURE 1.11

An excerpt from Charles Darwin's *On the Origin of Species*.

Natural Selection

Darwin was thoroughly familiar with variation in domesticated animals and began *On the Origin of Species* with a detailed discussion of pigeon breeding. He knew that breeders selected certain varieties of pigeons and other animals, such as dogs, to produce certain characteristics, a process Darwin called **artificial selection.** Once this had been done, the animals would breed true for the characteristics that had been selected. Darwin had also observed that the differences purposely developed between domesticated races or breeds were often greater than those that separated wild species. Domestic pigeon breeds, for example, show much greater variety than all of the hundreds of wild species of pigeons found throughout the world. Such relationships suggested to Darwin that evolutionary change could occur in nature too. Surely if pigeon breeders could foster such variation by artificial selection, nature could do the same, playing the breeder's role in selecting the next generation—a process Darwin called **natural selection.**

Darwin's theory incorporates the hypothesis of evolution, the process of natural selection, and the mass of new evidence for both evolution and natural selection that he had compiled. Thus, Darwin's theory provides a simple and direct explanation of biological diversity, or why animals are different in different places. Because habitats differ in their requirements and opportunities, the organisms with characteristics favored locally by natural selection will tend to vary in different places.

Darwin Drafts His Argument

Darwin drafted the overall argument for evolution by natural selection in a preliminary manuscript in 1842. After showing the manuscript to a few of his closest scientific friends, however, Darwin put it in a drawer, and for 16 years turned to other research. No one knows for sure why Darwin did not publish his initial manuscript—it is very thorough and outlines his ideas in detail. Some historians have suggested that Darwin was shy of igniting public criticism of his evolutionary ideas because there could have been little doubt in his mind that his hypothesis of evolution by natural selection would spark controversy. Others have proposed that Darwin was simply re-

FIGURE 1.12
Darwin greets his monkey ancestor. In his time, Darwin was often portrayed unsympathetically, as in this drawing from an 1874 publication.

fining his hypothesis all those years, although there is little evidence that he altered his initial manuscript in all that time.

Wallace Has the Same Idea

The stimulus that finally brought Darwin's hypothesis into print was an essay he received in 1858. A young English naturalist named Alfred Russel Wallace (1823–1913) sent the essay to Darwin from Malaysia; it concisely set forth the hypothesis of evolution by means of natural selection, a hypothesis Wallace had developed independently of Darwin. Like Darwin, Wallace had been greatly influenced by Malthus's 1798 essay. Colleagues of Wallace, knowing of Darwin's work, encouraged him to communicate with Darwin. After receiving Wallace's essay, Darwin arranged for a joint presentation of their ideas at a seminar in London. Darwin then completed his own book, expanding the 1842 manuscript he had written so long ago, and submitted it for publication.

Publication of Darwin's Hypothesis

Darwin's book appeared in November 1859 and caused an immediate sensation. Many people were deeply disturbed by the suggestion that human beings were descended from the same ancestor as apes (figure 1.12). Darwin did not actually discuss this idea in his book, but it followed directly from the principles he outlined. In a subsequent book, *The Descent of Man*, Darwin presented the argument directly, building a powerful case that humans and living apes have common ancestors. Although people had long accepted that humans closely resembled apes in many characteristics, the possibility that a direct evolutionary relationship might exist was unacceptable to many. Darwin's arguments for the theory of evolution by natural selection were so compelling, however, that his views were almost completely accepted within the intellectual community of Great Britain after the 1860s.

The fact that populations do not really expand geometrically implies that nature acts to limit population numbers. The traits of organisms that survive to produce more offspring will be more common in future generations—a process Darwin called natural selection.

Evolution After Darwin: More Evidence

More than a century has elapsed since Darwin's death in 1882. During this period, the evidence supporting his theory has grown progressively stronger. Also, many significant advances in our understanding of how evolution works have occurred. Although these advances have not altered the basic structure of Darwin's theory, they have taught us a great deal more about the mechanisms by which evolution occurs. We will briefly explore some of this evidence here; in chapter 22, we will return to the theory of evolution and examine the evidence in more detail.

The Fossil Record

Darwin predicted that the fossil record would yield intermediate links between the great groups of organisms—for example, between fishes and the amphibians thought to have arisen from them, and between reptiles and birds. We now know the fossil record to a degree that was unthinkable in the nineteenth century. Recent discoveries of microscopic fossils have extended the known history of life on earth back to about 2.5 billion years ago. The discovery of other fossils has supported Darwin's predictions and has shed light on how organisms have, over this enormous time span, evolved from the simple to the complex. For vertebrate animals especially, the fossil record is rich and exhibits a graded series of changes in form, with the evolutionary parade visible for all to see.

The Age of the Earth

In Darwin's day, some physicists argued that the earth was only a few thousand years old. This bothered Darwin, because the evolution of all living things from some single original ancestor would have required a great deal more time. Using evidence obtained by studying the rates of radioactive decay, we now know that the physicists of Darwin's time were wrong, very wrong: The earth was formed about 4.5 billion years ago.

The Mechanism of Heredity

Darwin received some of his sharpest criticism in the area of heredity. At that time, no one had any concept of genes or of how heredity works, so it was not possible for Darwin to explain completely how evolution occurs. Theories of heredity in Darwin's day seemed to rule out the possibility of genetic variation in nature, a critical requirement of Darwin's theory. Genetics was established as a science only at the start of the twentieth century, 40 years after the publication of Darwin's *On the Origin of Species*. When scientists began to understand the laws of inheritance (discussed in chapter 13), the heredity problem with Darwin's theory vanished. Genetics accounts in a neat and orderly way for the production of new variations in organisms.

Comparative Anatomy

Comparative studies of animals have provided strong evidence for Darwin's theory. In many different types of vertebrates, for example, the same bones are present, indicating their evolutionary past. Thus, the forelimbs shown in figure 1.13 are all constructed from the same basic array of bones, modified in one way in the wing of a bat, in another way in the fin of a porpoise, and in yet another way in the leg of a horse. The bones are said to be **homologous** in the different vertebrates; that is, they have the same evolutionary origin, but they now differ in structure and function. This contrasts with **analogous** structures, such as the wings of birds and butterflies, which have similar structure and function but different evolutionary origins.

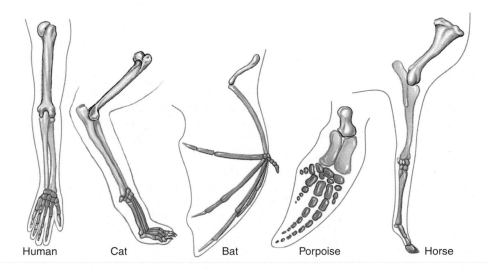

Human   Cat   Bat   Porpoise   Horse

FIGURE 1.13
Homology among vertebrate limbs. The forelimbs of these five vertebrates show the ways in which the relative proportions of the forelimb bones have changed in relation to the particular way of life of each organism.

Molecular Evidence

Evolutionary patterns are also revealed at the molecular level. By comparing the genomes (that is, the sequences of all the genes) of different groups of animals or plants, we can specify the degree of relationship among the groups more precisely than by any other means. A series of evolutionary changes over time should involve a continual accumulation of genetic changes in the DNA. Organisms that are more distantly related should have accumulated a greater number of these evolutionary differences, while two species that are more closely related will share a greater proportion of unchanged DNA. Thus gorillas, which the fossil record indicates diverged from humans between 6 and 8 million years ago, differ from the human genome in 1.6% of its DNA, while chimpanzees, which diverged about 5 million years ago, differ by only 1.2%. This same difference can be seen clearly in the protein hemoglobin (figure 1.14). The macaques, which like humans are primates, have fewer differences from humans in the 146 amino acid hemoglobin beta chain than do more distantly related mammals, such as dogs. Nonmammalian vertebrates, such as birds and frogs, differ even more.

Molecular Clocks. The consistent pattern emerging from a growing mountain of data (discussed in detail in chapter 24) is one of progressive change over time, with more distantly related species showing more differences in their DNA than closely related ones, just as Darwin's theory predicts. For example, the longer the time since two organisms diverged, the greater the number of differences in the nucleotide sequence of the cytochrome *c* gene (figure 1.15). This gene plays a key role in the oxidative metabolism of all terrestrial vertebrates, and appears to have accumulated changes at a constant rate, a phenomenon sometimes referred to as a **molecular clock**. All proteins for which data are available appear to accumulate changes over time, although different proteins evolve at different rates.

Phylogenetic Trees. The sequences of some genes, such as the ones specifying the hemoglobin proteins, have been determined in many organisms, and the entire time course of their evolution can be laid out with confidence by tracing the origins of particular nucleotide changes in the gene sequence. The pattern of descent obtained is called a **phylogenetic tree**. It represents the evolutionary history of the gene, its "family tree." Molecular phylogenetic trees agree well with those derived from the fossil record, which is strong direct evidence of evolution. The pattern of accumulating DNA changes represents, in a real sense, the footprints of evolutionary history.

Since Darwin's time, new discoveries in the fossil record, genetics, anatomy, and molecular biology all strongly support Darwin's theory.

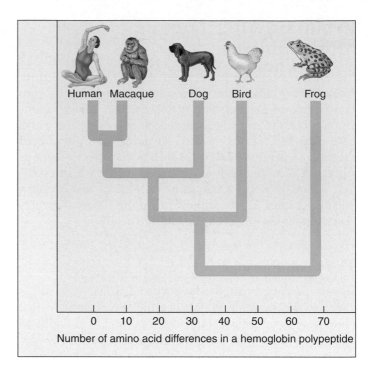

FIGURE 1.14
Molecules reflect evolutionary patterns. Vertebrates that are more distantly related to humans have a greater number of amino acid differences in this vertebrate hemoglobin polypeptide.
Where do you imagine a snake might fall on the graph? Why?

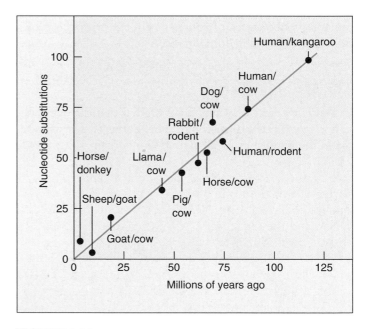

FIGURE 1.15
The molecular clock of cytochrome *c*. When the time since each pair of organisms diverged in the fossil record is plotted against the number of nucleotide differences in the cytochrome *c* gene, the result is a straight line, suggesting this gene is evolving at a constant rate.
Do you think dog differs from cow more than might be expected based on other vertebrates? What might explain this?

1.4 Four themes unify biology as a science.

Core Themes Unite Biology

Organization of Life: The Cell Theory

As was stated at the beginning of this chapter, all organisms are composed of cells, life's basic units (figure 1.16). Cells were discovered by Robert Hooke in England in 1665. Hooke was using one of the first microscopes, one that magnified 30 times. Looking through a thin slice of cork, he observed many tiny chambers, which reminded him of monks' cells in a monastery. Not long after that, the Dutch scientist Anton van Leeuwenhoek used microscopes capable of magnifying 300 times, and discovered an amazing world of single-celled life in a drop of pond water. However, it took almost two centuries before biologists fully understood the significance of cells. In 1839, the German biologists Matthias Schleiden and Theodor Schwann, summarizing a large number of observations by themselves and others, concluded that all living organisms consist of cells. Their conclusion forms the basis of what has come to be known as the **cell theory.** Later, biologists added the idea that all cells come from other cells. The cell theory, one of the basic ideas in biology, is the foundation for understanding the reproduction and growth of all organisms.

Continuity of Life: The Molecular Basis of Inheritance

Even the simplest cell is incredibly complex—more intricate than a computer. The information that specifies what a cell is like—its detailed plan—is encoded in a long, cable-like molecule called **DNA (deoxyribonucleic acid).** Each DNA molecule is formed from two long chains of building blocks, called nucleotides, wound around each other (figure 1.17). The two chains face each other, like two lines of people holding hands. The chains contain information in the same way this sentence does—as a sequence of letters. There are four different nucleotides in DNA, and the sequence in which they occur encodes this information. Specific sequences of several hundred to many thousand nucleotides make up a **gene,** a discrete unit of information. A gene might encode a particular protein or a different kind of unique molecule called RNA, or a gene might act to regulate other genes. The proteins and RNA molecules that are produced determine what the cell will be like.

The continuity of life from one generation to the next—heredity—depends upon the faithful copying of a cell's DNA into daughter cells. The entire set of DNA instructions that specifies a cell is called its **genome.** The sequence of the human genome, 3 billion nucleotides long, was decoded in rough draft form in 2001, a triumph of scientific investigation.

51 µm

FIGURE 1.16
Life in a drop of pond water. All organisms are composed of cells. Some organisms, including these protists, are single-celled, while others, such as plants, worms, and mushrooms, consist of many cells.

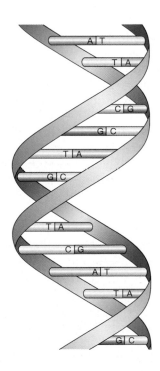

FIGURE 1.17
Genes are made of DNA. Winding around each other like the rails of a spiral staircase, the two strands of DNA make a double helix. Because of their size and shape, the nucleotide represented by the letter A can only pair with the nucleotide represented by the letter T, and likewise, the letter G and the letter C. This means that whatever the sequence is on one strand, the other strand will be its mirror image. From each strand, the other can be easily assembled.

Diversity of Life: Evolutionary Change

The unity of life that we see in the retention of certain key characteristics among many related life-forms contrasts with the incredible diversity of living things that have evolved to fill the varied environments of earth. Biologists divide life's great diversity into three great groups, called domains: Bacteria, Archaea, and Eukarya. The domains Bacteria and Archaea are composed of prokaryotes (single-celled organisms with little internal structure), while domain Eukarya is made up of eukaryotes, organisms composed of a complexly organized cell or multiple complex cells. However, Archaea seem more closely related to Eukarya than to Bacteria. Within the Eukarya are four main groups called kingdoms (figure 1.18). Kingdom Protista consists of all the unicellular eukaryotes except yeasts, as well as the multicellular algae. Because of the great diversity among the protists, many biologists feel kingdom Protista should be split into several kingdoms. Kingdom Plantae consists of organisms that have cell walls of cellulose and obtain energy by photosynthesis. Organisms in the kingdom Fungi have cell walls of chitin and obtain energy by secreting digestive enzymes onto organisms and then absorbing the products they release. Kingdom Animalia contains organisms that lack cell walls and obtain energy by first ingesting other organisms and then digesting them internally.

Unity of Life: Evolutionary Conservation

Biologists believe that all living things have descended from some simple cellular creature that arose about 2.5 billion years ago. Some of the characteristics of that earliest organism have been preserved in all things alive today. The storage of hereditary information in DNA, for example, is common to all living things. Also, all eukaryotes possess a nucleus that contains chromosomes, and flagellae throughout the animal kingdom possess the same 9 + 2 arrangement of microtubules. The retention of these conserved characteristics in a long line of descent usually reflects a fundamental role in the biology of the organism, one not easily changed once adopted. A good example is provided by the homeodomain proteins, proteins that play a critical role in early development in eukaryotes. Conserved characteristics can be seen in approximately 1850 homeodomain proteins, distributed among three different kingdoms of organisms (figure 1.19)! The homeodomain proteins are powerful developmental tools that evolved early, and for which no better alternative has arisen.

Cellular organisms store hereditary information in DNA. Sometimes DNA alterations occur, which when preserved result in evolutionary change. Today's biological diversity is the product of a long evolutionary journey.

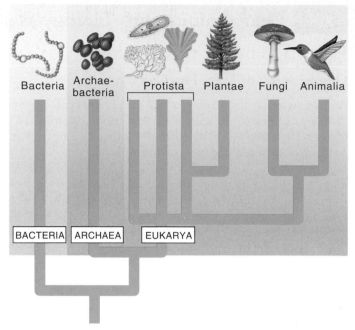

FIGURE 1.18
The diversity of life. Biologists categorize all living things into three overarching groups called domains: Bacteria, Archaea, and Eukarya. Domain Eukarya is composed of four kingdoms: Protista, Plantae, Fungi, and Animalia.

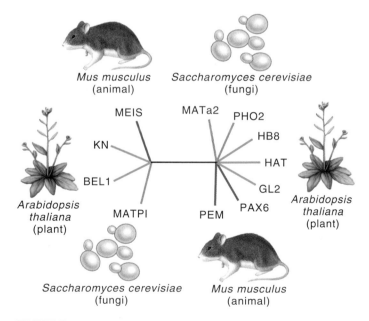

FIGURE 1.19
Tree of homeodomain proteins. Homeodomain proteins are found in fungi (*brown*), plants (*green*), and animals (*blue*). Based on their sequence similarities, these 11 different homeodomain proteins (uppercase letters at the ends of branches) fall into two groups, with representatives from each kingdom in each group. That means, for example, the mouse homeodomain protein PAX6 is more closely related to fungal and flowering plant proteins, such as PHO2 and GL2, than it is to the mouse protein MEIS.

For interactive testing, visit the Online Learning Center with PowerWeb at www.mhhe.com/Raven7

1.1   Biology is the science of life.

Organization of Living Things

- All living organisms share a collection of characteristics: cellular organization, order, sensitivity, growth, development and reproduction, energy utilization, evolutionary adaptation, and homeostasis. (p. 2)
- The biological world is hierarchical from the cellular level to the population level, with emergent properties entering at each higher level. (p. 2)

1.2   Scientists form generalizations from observations.

The Nature of Science

- Deductive reasoning applies general principles to predict specific results, while inductive reasoning uses specific observations to construct general principles. (p. 4)

How Science Is Done

- Scientists make observations and then construct a suggested explanation (hypothesis) to account for those observations. (p. 5)
- A successful experiment rejects one or more alternative hypotheses. (p. 5)
- Often scientists will conduct two experiments in parallel. In one experiment, all the variables are kept constant except one, while in the other experiment, the control experiment, that variable is left unaltered. (p. 6)
- Predictions provide a way to test the validity of a hypothesis. (p. 6)
- A theory is a proposed explanation for some natural phenomenon, but the term is also used to refer to a body of interconnected concepts supported by scientific reasoning and experimental evidence. (p. 6)
- After developing a hypothesis and performing a series of experiments, a paper is written describing the experiment and the results; the paper is then submitted for peer review. (p. 7)

1.3   Darwin's theory of evolution illustrates how science works.

Charles Darwin

- In Darwin's time, most people believed species to be specially created and unchangeable. (p. 8)

- Darwin proposed the concept of natural selection to account for his view that living things must have changed during their history on earth. (p. 8)
- Darwin served as naturalist on a five-year navigational mapping expedition around the coasts of South America. (p. 8)

Darwin's Evidence

- Repeatedly, Darwin found that characteristics of similar species varied from place to place, suggesting that organismal lineages change gradually as species migrate, and that plants and animals on young volcanic islands resembled those on nearby coasts of South America. (p. 10)

Inventing the Hypothesis of Natural Selection

- Darwin was influenced by Thomas Malthus and his idea that plant and animal populations increase geometrically. (p. 11)
- Darwin reasoned that because only a limited number of offspring can survive in each generation, those possessing superior physical or behavioral attributes are more likely to survive. (p. 11)
- Darwin knew of extensive examples of artificial selection in domesticated animals, and he reasoned that such evolutionary change could also occur in nature. (p. 12)

Evolution After Darwin: More Evidence

- The increasingly complete fossil record, the determination that the earth is 4.5 billion years old, the discovery of heredity mechanisms, comparative anatomy studies differentiating homologous and analogous structures, and molecular evidence such as the discovery of molecular clocks all support Darwin's ideas. (pp. 13–14)

1.4   Four themes unify biology as a science.

Core Themes Unite Biology

- The four themes uniting the field of biology are the cell theory, the molecular basis of inheritance, evolutionary change, and evolutionary conservation. (pp. 15–16)

Self Test

1. Which of the following is *not* a property of life?
   a. responding to stimuli
   b. dividing in two
   c. regulating internal conditions
   d. All of these are characteristics that define life.

2. The process of inductive reasoning
   a. involves the observation of specific occurrences to construct a general principle.
   b. involves taking a general principle and applying it to a specific situation.
   c. is not used very often in the study of biology.
   d. is all of the above.

3. A goal of a scientist is to formulate a hypothesis
   a. that will never be proven false.
   b. that is essentially a theory explaining an observation.
   c. that is just a wild guess.
   d. that will be tested by experimentation.

4. An experiment testing a hypothesis will
   a. always support the hypothesis.
   b. always disprove the hypothesis.
   c. include a variable and a control.
   d. not be successful if the hypothesis is rejected.

5. Darwin's proposal that evolution occurs through natural selection caused controversy because
   a. it challenged the existence of a Divine Creator.
   b. it challenged the views of earlier philosophers.
   c. it challenged a literal interpretation of the Bible.
   d. the explanation wasn't based on any observations.

6. Darwin was convinced that evolution had occurred based on his observations that
   a. armadillos on the different islands off the coast of South America had slightly different physical characteristics.
   b. tortoises found on the different Galápagos Islands had different and identifiable shells.
   c. bird fossils showed modifications from modern-day birds found in the same areas.
   d. All of these observations were made by Darwin.

7. A key piece of information for Darwin's hypothesis that evolution occurs through natural selection was
   a. the fossil evidence.
   b. his work breeding pigeons through artificial selection.
   c. Malthus's proposal that death, due to limited food supply, restricts population size.
   d. geographic distribution of similar animals with slight variations in physical characteristics.

8. Which of the following "new" areas of scientific study could Darwin have used to strengthen his hypothesis?
   a. the age of the earth
   b. the mechanism of inheritance.
   c. the expanded fossil record
   d. the geographic distribution of animal species

9. What are homologous anatomical structures?
   a. structures that look different but have similar evolutionary origins
   b. structures that have similar functions but different evolutionary origins
   c. a bat's wing and a butterfly's wing
   d. a bat's wing and a human's leg

10. The "themes" of biology include all of the following except
    a. chemistry.   c. genetics.
    b. cell biology.   d. evolution.

Test Your Visual Understanding

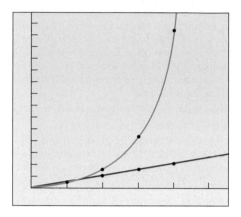

1. Which line on the graph indicates the growth of a population if resources were unlimited? Name three factors that limit population growth.

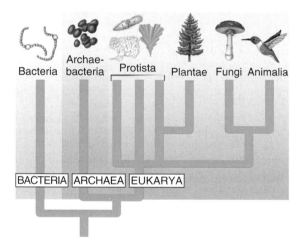

2. By following the general path of evolution, would you say that fungi are more closely related to Archaebacteria or Bacteria? Why?

Apply Your Knowledge

1. When breeding hunting dogs, the parents are selected for their keen sense of smell, which allows them to find the game (i.e., birds) faster. Breeders predict that this trait will be passed on to the dogs' offspring, producing a family of good hunting dogs. This is artificial selection. Explain how this same trait could be selected through natural selection in a population of wild dogs.

2. Assume that four people are stranded on a deserted island and they reproduce once a year. How many people will be on the island after ten years using:
   a. a geometric progression with a factor of 2.
   b. an arithmetic progression with a factor of 2.

2

The Nature of Molecules

Concept Outline

2.1 Atoms are nature's building material.

Atoms. All substances are composed of tiny particles called atoms, each a positively charged nucleus around which negative electrons orbit.

Electrons Determine the Chemical Behavior of Atoms. The closer an electron's orbit is to the nucleus, the lower is its energy level.

2.2 The atoms of living things are among the smallest.

Kinds of Atoms. Of the 92 naturally occurring elements, only 11 occur in organisms in significant amounts.

2.3 Chemical bonds hold molecules together.

Ionic Bonds Form Crystals. Atoms are linked together into molecules, joined by chemical bonds that result from forces such as the attraction of opposite charges or the sharing of electrons.

Covalent Bonds Build Stable Molecules. Chemical bonds formed by the sharing of electrons can be very strong, and require much energy to break.

2.4 Water is the cradle of life.

Chemistry of Water. Water forms weak chemical associations that are responsible for much of the organization of living chemistry.

Water Atoms Act Like Tiny Magnets. Because electrons are shared unequally by the hydrogen and oxygen atoms of water, a partial charge separation occurs. Each water atom acquires a positive and negative pole and is said to be "polar."

Water Clings to Polar Molecules. Because the opposite partial charges of polar molecules attract one another, water tends to cling to itself and to other polar molecules while excluding nonpolar molecules.

Water Ionizes. Because its covalent bonds occasionally break, water contains a low concentration of hydrogen (H^+) and hydroxide (OH^-) ions, the fragments of broken water molecules.

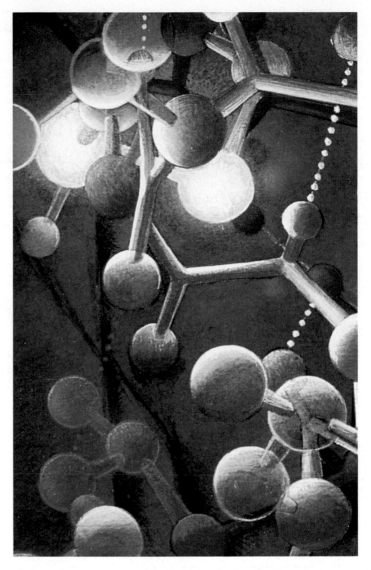

FIGURE 2.1
Cells are made of molecules. Specific, often simple combinations of atoms yield an astonishing diversity of molecules within the cell, each with unique functional characteristics.

About 14 billion years ago, an enormous explosion likely marked the beginning of the universe. With this explosion began a process of star building and planetary formation that eventually led to the formation of earth about 4.5 billion years ago. Starting about 2.5 billion years ago, life began on earth and started to diversify. When viewed from the perspective of 14 billion years, life within our solar system is a recent development, but to understand the origin of life, we need to consider events that took place much earlier. The same processes that led to the evolution of life were responsible for the evolution of molecules (figure 2.1). Thus, our study of life on earth begins with physics and chemistry. As chemical machines ourselves, we must understand chemistry to begin to understand our origins.

2.1 Atoms are nature's building material.

Atoms

Any substance in the universe that has mass and occupies space is defined as **matter.** All matter is composed of extremely small particles called **atoms.** Because of their size, atoms are difficult to study. Not until early in the last century did scientists carry out the first experiments suggesting what an atom is like.

The Structure of Atoms

Objects as small as atoms can be "seen" only indirectly, by using very complex technology such as tunneling microscopy. We now know a great deal about the complexities of atomic structure, but the simple view put forth in 1913 by the Danish physicist Niels Bohr provides a good starting point. Bohr proposed that every atom possesses an orbiting cloud of tiny subatomic particles called **electrons** whizzing around a core like the planets of a miniature solar system. At the center of each atom is a small, very dense nucleus formed of two other kinds of subatomic particles, **protons** and **neutrons** (figure 2.2).

Within the nucleus, the cluster of protons and neutrons is held together by a force that works only over short subatomic distances. Each proton carries a positive (+) charge, and each electron carries a negative (−) charge. Typically, an atom has one electron for each proton. The number of protons an atom has is called its **atomic number.** This number indirectly determines the chemical character of the atom, because it dictates the number of electrons orbiting the nucleus that are available for chemical activity. Neutrons, as their name implies, possess no charge.

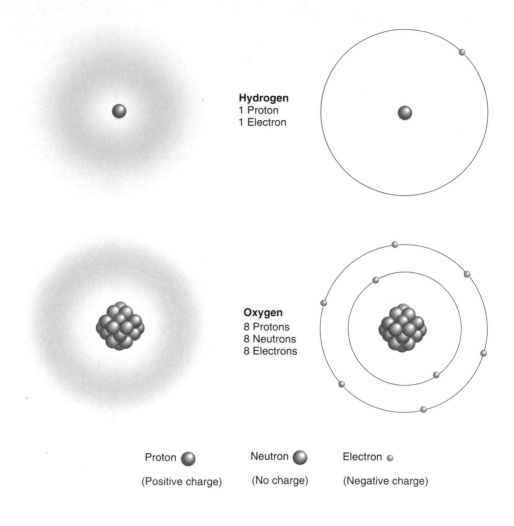

Hydrogen
1 Proton
1 Electron

Oxygen
8 Protons
8 Neutrons
8 Electrons

Proton (Positive charge) Neutron (No charge) Electron (Negative charge)

FIGURE 2.2
Basic structure of atoms.   All atoms have a nucleus consisting of protons and neutrons, except hydrogen, the smallest atom, which has only one proton and no neutrons in its nucleus. Oxygen, for example, has eight protons and eight neutrons in its nucleus. In the simple "Bohr model" of atoms pictured here, electrons spin around the nucleus at a relatively far distance.

Atomic Mass

The terms mass and weight are often used interchangeably, but they have slightly different meanings. *Mass* refers to the amount of a substance, while *weight* refers to the force gravity exerts on a substance. Hence, an object has the same mass whether it is on the earth or the moon, but its weight will be greater on the earth because the earth's gravitational force is greater than the moon's. The **atomic mass** of an atom is equal to the sum of the masses of its protons and neutrons. Atoms that occur naturally on earth contain from 1 to 92 protons and up to 146 neutrons.

The mass of atoms and subatomic particles is measured in units called *daltons.* To give you an idea of just how small these units are, note that it takes 602 million million billion (6.02×10^{23}) daltons to make 1 gram. A proton weighs approximately 1 dalton (actually 1.009 daltons), as does a neutron (1.007 daltons). In contrast, electrons weigh only $\frac{1}{1840}$ of a dalton, so their contribution to the overall mass of an atom is negligible.

FIGURE 2.3
The three most abundant isotopes of carbon. Isotopes of a particular element have different numbers of neutrons.

Carbon-12
6 Protons
6 Neutrons
6 Electrons

Carbon-13
6 Protons
7 Neutrons
6 Electrons

Carbon-14
6 Protons
8 Neutrons
6 Electrons

Isotopes

Atoms with the same atomic number (that is, the same number of protons) have the same chemical properties and are said to belong to the same **element.** Formally speaking, an element is any substance that cannot be broken down to any other substance by ordinary chemical means. However, while all atoms of an element have the same number of protons, they may not all have the same number of neutrons. Atoms of an element that possess different numbers of neutrons are called **isotopes** of that element. Most elements in nature exist as mixtures of different isotopes. Carbon (C), for example, has three isotopes, all containing six protons (figure 2.3). Over 99% of the carbon found in nature exists as an isotope with six neutrons. Because its total mass is 12 daltons (6 from protons plus 6 from neutrons), this isotope is referred to as carbon-12, and symbolized ^{12}C. Most of the rest of the naturally occurring carbon is carbon-13, an isotope with seven neutrons. The rarest carbon isotope is carbon-14, with eight neutrons. Unlike the other two isotopes, carbon-14 is unstable: Its nucleus tends to break up into elements with lower atomic numbers. This nuclear breakup, which emits a significant amount of energy, is called radioactive decay, and isotopes that decay in this fashion are **radioactive isotopes.**

Some radioactive isotopes are more unstable than others and therefore decay more readily. For any given isotope, however, the propensity to decay is constant. The decay time is usually expressed as the **half-life,** the time it takes for one-half of the atoms in a sample to decay. Carbon-14, for example, has a half-life of about 5600 years. A sample of carbon containing 1 gram of carbon-14 today would contain 0.5 gram of carbon-14 after 5600 years, 0.25 gram 11,200 years from now, 0.125 gram 16,800 years from now, and so on. By determining the ratios of the different isotopes of carbon and other elements in biological samples and in rocks, scientists are able to accurately determine when these materials formed.

While there are many useful applications of radioactivity, there are also harmful side effects that must be considered in any planned use of radioactive substances. Radioactive substances emit energetic subatomic particles that have the potential to severely damage living cells, producing mutations in their genes and, at high doses, cell death. Consequently, exposure to radiation is now very carefully controlled and regulated. Scientists who work with radioactivity (basic researchers as well as applied scientists such as X-ray technologists) wear radiation-sensitive badges to monitor the total amount of radioactivity to which they are exposed. Each month the badges are collected and scrutinized. Thus, employees whose work places them in danger of excessive radioactive exposure are equipped with an "early warning system."

Electrons

The positive charges in the nucleus of an atom are counterbalanced by negatively charged electrons located in regions called orbitals at varying distances around the nucleus. Thus, atoms with the same number of protons and electrons are electrically neutral, having no net charge, and are called **neutral atoms.**

Electrons are maintained in their orbits by their attraction to the positively charged nucleus. Sometimes other forces overcome this attraction and an atom loses one or more electrons. In other cases, atoms gain additional electrons. Atoms in which the number of electrons does not equal the number of protons are known as **ions,** and they carry a net electrical charge. An atom that has more protons than electrons has a net positive charge and is called a **cation.** For example, an atom of sodium (Na) that has lost one electron becomes a sodium ion (Na^+), with a charge of $+1$. An atom that has fewer protons than electrons carries a net negative charge and is called an **anion.** A chlorine atom (Cl) that has gained one electron becomes a chloride ion (Cl^-), with a charge of -1.

An atom consists of a nucleus of protons and neutrons surrounded by a cloud of electrons. The number of its electrons largely determines the chemical properties of an atom. Atoms that have the same number of protons but different numbers of neutrons are called isotopes. Isotopes of an atom differ in atomic mass but have similar chemical properties.

Electrons Determine the Chemical Behavior of Atoms

The key to the chemical behavior of an atom lies in the number and arrangement of its electrons in their orbits. It is convenient to visualize individual electrons as following discrete circular orbits around a central nucleus, as in the Bohr model of the atom. However, such a simple picture is not realistic. It is not possible to precisely locate the position of any individual electron at any given time. In fact, a particular electron can be anywhere at a given instant, from close to the nucleus to infinitely far away from it.

Nevertheless, a particular electron is more likely to be located in some positions than in others. The area around a nucleus where an electron is most likely to be found is called the **orbital** of that electron. Some electron orbitals near the nucleus are spherical (*s* orbitals), while others are dumbbell-shaped (*p* orbitals) (figure 2.4). Still other orbitals, more distant from the nucleus, may have different shapes. Regardless of its shape, no orbital may contain more than two electrons.

Almost all of the volume of an atom is empty space, because the electrons are quite far from the nucleus relative to its size. If the nucleus of an atom were the size of a golf ball, the orbit of the nearest electron would be more than a kilometer away. Consequently, the nuclei of two atoms never come close enough in nature to interact with each other. It is for this reason that an atom's electrons, not its protons or neutrons, determine its chemical behavior. This

also explains why the isotopes of an element, all of which have the same arrangement of electrons, behave the same way chemically.

Energy Within the Atom

All atoms possess *energy*, defined as the ability to do work. Because electrons are attracted to the positively charged nucleus, it takes work to keep them in orbit, just as it takes work to hold a grapefruit in your hand against the pull of gravity. The grapefruit is said to possess *potential energy* because of its position; if you were to release it, the grapefruit would fall, and its energy would be reduced. Conversely, if you were to move the grapefruit to the top of a building, you would increase its potential energy. Similarly, electrons have potential energy of position. To oppose the attraction of the nucleus and move the electron to a more distant orbital requires an input of energy and results in an electron with greater potential energy. This is how chlorophyll captures energy from light during photosynthesis (see chapter 10)—the light excites electrons in the chlorophyll. Moving an electron closer to the nucleus has the opposite effect: Energy is released, usually as heat, and the electron ends up with less potential energy (figure 2.5).

A given atom can possess only certain discrete amounts of energy. Like the potential energy of a grapefruit in your hand, the potential energy contributed by the position of an electron in an atom can have only certain values. Every

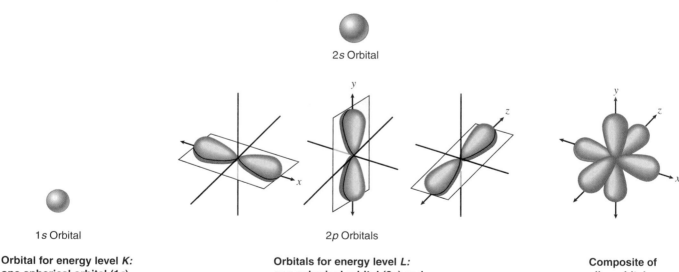

2*s* Orbital

1*s* Orbital

2*p* Orbitals

Orbital for energy level *K*:
one spherical orbital (1*s*)

Orbitals for energy level *L*:
one spherical orbital (2*s*) and
three dumbbell-shaped orbitals (2*p*)

Composite of
all *p* orbitals

FIGURE 2.4
Electron orbitals. The lowest energy level or electron shell—the one nearest the nucleus—is level *K*. It is occupied by a single *s* orbital, referred to as 1*s*. The next highest energy level, *L*, is occupied by four orbitals: one *s* orbital (referred to as the 2*s* orbital) and three *p* orbitals (each referred to as a 2*p* orbital). The four *L*-level orbitals compactly fill the space around the nucleus, like two pyramids set base-to-base. Thus, the lowest energy level, *K*, is populated by two electrons, while the next highest energy level, *L*, is populated by a total of eight electrons.

FIGURE 2.5

Atomic energy levels. When an electron absorbs energy, it moves to higher energy levels, farther from the nucleus. When an electron releases energy, it falls to lower energy levels, closer to the nucleus.

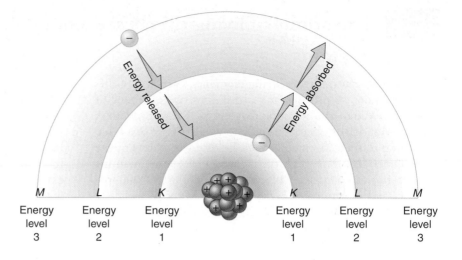

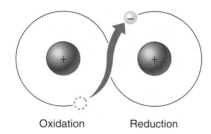

Oxidation   Reduction

FIGURE 2.6

Oxidation and reduction. Oxidation is the loss of an electron; reduction is the gain of an electron. In either reaction, the electron keeps its energy of position.

atom exhibits a ladder of potential energy values, a discrete set of orbits at particular distances from the nucleus.

During some chemical reactions, electrons are transferred from one atom to another. In such reactions, the loss of an electron is called **oxidation,** and the gain of an electron is called **reduction** (figure 2.6). It is important to realize that when an electron is transferred in this way, it keeps its energy of position. In organisms, chemical energy is stored in high-energy electrons that are transferred from one atom to another in reactions involving oxidation and reduction.

Because the amount of energy an electron possesses is related to its distance from the nucleus, electrons that are the same distance from the nucleus have the same energy, even if they occupy different orbitals. Such electrons are said to occupy the same **energy level.** In a schematic diagram of an atom (figure 2.7), the nucleus is represented as a small circle, and the electron energy levels are drawn as concentric rings, with the energy level increasing with distance from the nucleus. Be careful not to confuse energy levels, which are drawn as rings to indicate an electron's *energy,* with orbitals, which have a variety of three-dimensional shapes and indicate an electron's most likely *location.*

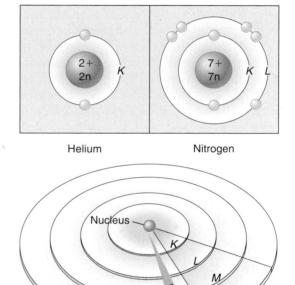

Helium   Nitrogen

FIGURE 2.7

Electron energy levels for helium and nitrogen. *Top:* Gold balls represent electrons. *Bottom:* Each concentric circle represents a different distance from the nucleus and so a different electron energy level.

Electrons are localized about a nucleus in regions called orbitals. No orbital can contain more than two electrons, but many orbitals may be the same distance from the nucleus and, thus, contain electrons of the same energy.

2.2 The atoms of living things are among the smallest.

Kinds of Atoms

There are 92 naturally occurring elements, each with a different number of protons and a different arrangement of electrons. When the nineteenth-century Russian chemist Dmitri Mendeleev arranged the known elements in a table according to their atomic number, he discovered one of the great generalizations in all of science: The elements in his table exhibited a pattern of chemical properties that repeated itself in groups of eight elements. This periodically repeating pattern lent the table its name: the periodic table of elements (figure 2.8).

The Periodic Table

The eight-element periodicity that Mendeleev found is based on the interactions of the electrons in the outer energy levels of the different elements. These electrons are called **valence electrons,** and their interactions are the basis for the differing chemical properties of the elements. For most of the atoms important to life, an outer energy level can contain no more than eight electrons; the chemical behavior of an element reflects how many of the eight positions are filled. Elements possessing all eight electrons in their outer energy level (two for helium) are **inert,** or nonreactive; they include helium (He), neon (Ne), argon (Ar), krypton (Kr), xenon (Xe), and radon (Rn). In sharp contrast, elements with seven electrons (one fewer than the maximum number of eight) in their outer energy level, such as fluorine (F), chlorine (Cl), and bromine (Br), are highly reactive. They tend to gain the extra electron needed to fill the energy level. Elements with only one electron in their outer energy level, such as lithium (Li), sodium (Na), and potassium (K), are also very reactive; they tend to lose the single electron in their outer level.

Mendeleev's periodic table thus leads to a useful generalization, the **octet rule** (Latin *octo*, "eight"), or *rule of eight*: Atoms tend to establish completely full outer energy levels. Most chemical behavior of biological interest can be predicted quite accurately from this simple rule, combined with the tendency of atoms to balance positive and negative charges.

Only 11 elements are found in significant amounts in living organisms.

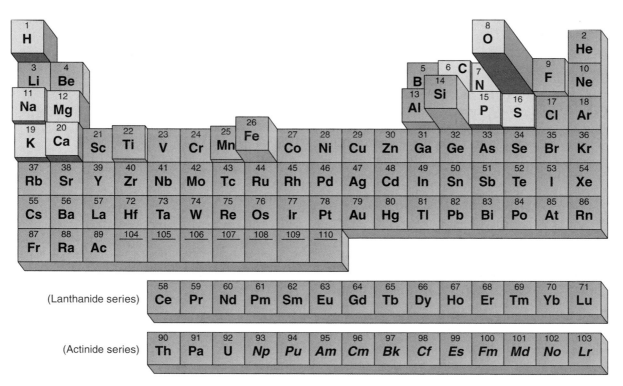

FIGURE 2.8
Periodic table of the elements. In this representation, the frequency of elements that occur in the earth's crust is indicated by the height of the block. Of the 92 naturally occurring elements on earth, only 11 are found in organisms in more than trace amounts (0.01% or higher). These elements, shaded in *blue*, have atomic numbers less than 21 and, thus, have low atomic masses. Four elements—nitrogen, oxygen, carbon, and hydrogen (NOCH for short)—constitute 96.3% of the weight of your body.

2.3 Chemical bonds hold molecules together.

Ionic Bonds Form Crystals

A group of atoms held together by energy in a stable association is called a **molecule.** When a molecule contains atoms of more than one element, it is called a **compound.** The atoms in a molecule are joined by **chemical bonds;** these bonds can result when atoms with opposite charges attract (ionic bonds), when two atoms share one or more pairs of electrons (covalent bonds), or when atoms interact in other ways. We will start by examining **ionic bonds,** which form when atoms with opposite electrical charges (ions) attract.

A Closer Look at Table Salt

Common table salt, the molecule sodium chloride (NaCl), is a lattice of ions in which the atoms are held together by ionic bonds (figure 2.9). Sodium has 11 electrons: 2 in the inner energy level, 8 in the next level, and 1 in the outer (valence) level. The valence electron is unpaired (free) and has a strong tendency to join with another electron. A stable configuration can be achieved if the valence electron is lost to another atom that also has an unpaired electron. The loss of this electron results in the formation of a positively charged sodium ion, Na^+.

The chlorine atom has 17 electrons: 2 in the inner energy level, 8 in the next level, and 7 in the outer level. Hence, one of the orbitals in the outer energy level has an unpaired electron. The addition of another electron to the outer level fills that level and causes a negatively charged chloride ion, Cl^-, to form.

When placed together, metallic sodium and gaseous chlorine react swiftly and explosively, as the sodium atoms donate electrons to chlorine, forming Na^+ and Cl^- ions. Because opposite charges attract, the Na^+ and Cl^- remain associated in an **ionic compound,** NaCl, which is electrically neutral. However, the electrical attractive force holding NaCl together is not directed specifically between particular Na^+ and Cl^- ions, and no discrete sodium chloride molecules form. Instead, the force exists between any one ion and all neighboring ions of the opposite charge, and the ions aggregate in a crystal matrix with a precise geometry. Such aggregations are what we know as salt crystals. If a salt such as NaCl is placed in water, the electrical attraction of the water molecules, for reasons we will point out later in this chapter, disrupts the forces holding the ions in their crystal matrix, causing the salt to dissolve into a roughly equal mixture of free Na^+ and Cl^- ions.

An ionic bond is an attraction between ions of opposite charge in an ionic compound. Such bonds are not formed between particular ions in the compound; rather, they exist between an ion and all of the oppositely charged ions in its immediate vicinity.

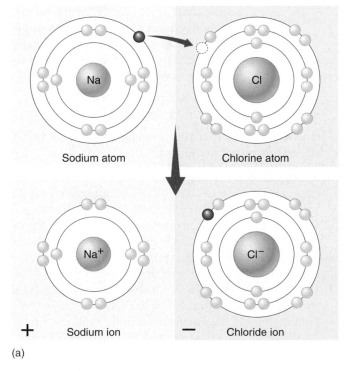

(a)

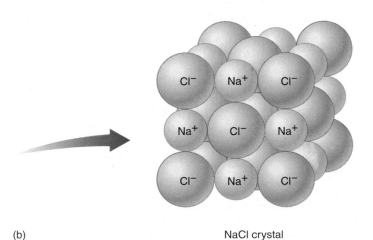

(b)   NaCl crystal

FIGURE 2.9
The formation of ionic bonds by sodium chloride.   (*a*) When a sodium atom donates an electron to a chlorine atom, the sodium atom becomes a positively charged sodium ion, and the chlorine atom becomes a negatively charged chloride ion. (*b*) Sodium chloride (NaCl) forms a highly regular lattice of alternating sodium ions and chloride ions.

Covalent Bonds Build Stable Molecules

Covalent bonds form when two atoms share one or more pairs of valence electrons. Consider hydrogen (H) as an example. Each hydrogen atom has an unpaired electron and an unfilled outer energy level; for these reasons, the hydrogen atom is unstable. However, when two hydrogen atoms are close to each other, each atom's electron is attracted to both nuclei. In effect, the nuclei are able to share their electrons. The result is a diatomic (two-atom) molecule of hydrogen gas (figure 2.10a).

The molecule formed by the two hydrogen atoms is stable for three reasons:

1. **It has no net charge.** The diatomic molecule formed as a result of this sharing of electrons is not charged because it still contains two protons and two electrons.
2. **The octet rule is satisfied.** Each of the two hydrogen atoms can be considered to have two orbiting electrons in its outer energy level. This satisfies the octet rule, because each shared electron orbits both nuclei and is included in the outer energy level of *both* atoms.
3. **It has no free electrons.** The bond between the two atoms also pairs the two free electrons.

Unlike ionic bonds, covalent bonds are formed between two specific atoms, giving rise to true, discrete molecules.

The Strength of Covalent Bonds

The strength of a covalent bond depends on the number of shared electrons. Thus **double bonds,** which satisfy the octet rule by allowing two atoms to share two pairs of electrons, are stronger than **single bonds,** in which only one electron pair is shared. This means more chemical energy is required to break a double bond than a single bond. The strongest covalent bonds are **triple bonds,** such as those that link the two nitrogen atoms of nitrogen gas molecules. Covalent bonds are represented in chemical formulations as lines connecting atomic symbols, where each line between two bonded atoms represents the sharing of one pair of electrons. The **structural formulas** of hydrogen gas and oxygen gas are H—H and O=O, respectively, while their **molecular formulas** are H_2 and O_2.

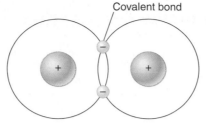

Covalent bond

H_2 (hydrogen gas)

(a)

(b)

FIGURE 2.10
Hydrogen gas. (*a*) Hydrogen gas is a diatomic molecule, meaning that it is composed of two hydrogen atoms, each sharing its electron with the other. (*b*) The flash of fire that consumed the *Hindenburg* occurred when the hydrogen gas used to inflate the dirigible combined explosively with oxygen gas in the air to form water.

Molecules with Several Covalent Bonds

Molecules often consist of more than two atoms. One reason that larger molecules may be formed is that a given atom is able to share electrons with more than one other atom. An atom that requires two, three, or four additional electrons to fill its outer energy level completely may acquire them by sharing its electrons with two or more other atoms.

For example, the carbon atom (C) contains six electrons, four of which are in its outer energy level. To satisfy the octet rule, a carbon atom must gain access to four additional electrons; that is, it must form four covalent bonds. Because four covalent bonds may form in many ways, carbon atoms are found in many different kinds of molecules.

Chemical Reactions

The formation and breaking of chemical bonds, the essence of chemistry, is called a **chemical reaction.** All chemical reactions involve the shifting of atoms from one molecule or ionic compound to another, without any change in the number or identity of the atoms. For convenience, we refer to the original molecules before the reaction starts as **reactants,** and the molecules resulting from the chemical reaction as **products.** For example:

$$\underset{\text{reactants}}{A\text{—}B + C\text{—}D} \rightarrow \underset{\text{products}}{A\text{—}C + B + D}$$

The extent to which chemical reactions occur is influenced by several important factors:

1. **Temperature.** Heating the reactants increases the rate of a reaction (as long as the temperature isn't so high as to destroy the molecules).
2. **Concentration of reactants and products.** Reactions proceed more quickly when more reactants are available. An accumulation of products typically speeds reactions in the reverse direction.
3. **Catalysts.** A catalyst is a substance that increases the rate of a reaction. It doesn't alter the reaction's equilibrium between reactants and products, but it does shorten the time needed to reach equilibrium, often dramatically. In organisms, proteins called enzymes catalyze almost every chemical reaction.

A covalent bond is a stable chemical bond formed when two atoms share one or more pairs of electrons.

(a)

(b)

(c)

FIGURE 2.11

Water takes many forms. (*a*) When water cools below 0° C, it forms beautiful crystals, familiar to us as snow and ice. (*b*) Ice turns to liquid when the temperature is above 0°C; in icy water, this orca must find holes in the covering of ice in order to surface and breathe. (*c*) Liquid water becomes steam when the temperature rises above 100°C, as seen in this hot spring at Yellowstone National Park.

Of all the molecules that are common on earth, only **water** exists as a liquid at the relatively low temperatures that prevail on the earth's surface, three-fourths of which is covered by liquid water (figure 2.11). When life was originating, water provided a medium in which other molecules could move around and interact without being held in place by strong covalent or ionic bonds. Life evolved in water for 2 billion years before spreading to land. And even today, life is inextricably tied to water. About two-thirds of any organism's body is composed of water, and no organism can grow or reproduce in any but a water-rich environment. It is no accident that tropical rain forests are bursting with life, while dry deserts appear almost lifeless except when water becomes temporarily plentiful, such as after a rainstorm.

Chemistry of Water

Water has a simple atomic structure. It consists of an oxygen atom bound to two hydrogen atoms by two single covalent bonds (figure 2.12). The resulting molecule is stable: It satisfies the octet rule, has no unpaired electrons, and carries no net electrical charge.

The single most outstanding chemical property of water is its ability to form weak chemical associations with only 5 to 10% of the strength of covalent bonds. This property, which derives directly from the structure of water, is responsible for much of the organization of living chemistry.

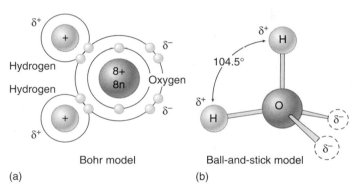
(a) Bohr model   (b) Ball-and-stick model

FIGURE 2.12

Water has a simple molecular structure. (*a*) Each water molecule is composed of one oxygen atom and two hydrogen atoms. The oxygen atom shares one electron with each hydrogen atom. (*b*) The greater electronegativity of the oxygen atom makes the water molecule polar: Water carries two partial negative charges (δ^-) near the oxygen atom and two partial positive charges (δ^+), one on each hydrogen atom.

The chemistry of life is water chemistry. The way life first evolved was determined in large part by the chemical properties of the liquid water in which that evolution occurred.

Water Atoms Act Like Tiny Magnets

In a water molecule, both the oxygen and the hydrogen atoms attract the electrons they share in the covalent bonds; this attraction is called **electronegativity**. However, the oxygen atom is more electronegative than the hydrogen atoms, so it attracts the electrons more strongly than do the hydrogen atoms. As a result, the shared electrons in a water molecule are far more likely to be found near the oxygen nucleus than near the hydrogen nuclei. This stronger attraction for electrons gives the oxygen atom two partial negative δ^- charges, as though the electron cloud were denser near the oxygen atom than around the hydrogen atoms. The Greek letter delta (δ) signifies a partial charge, much weaker than the full unit charge of an ion. Because the water molecule as a whole is electrically neutral, each hydrogen atom carries a partial positive charge (δ^+).

What would you expect the shape of a water molecule to be? Each of water's two covalent bonds has a partial charge at each end, δ^- at the oxygen end and δ^+ at the hydrogen end. The most stable arrangement of these charges is a *tetrahedron*, in which the two negative and two positive charges are approximately equidistant from one another. The oxygen atom lies at the center of the tetrahedron, the hydrogen atoms occupy two of the apexes, and the partial negative charges occupy the other two apexes (figure 2.12b). This results in a bond angle of 104.5° between the two covalent oxygen-hydrogen bonds. (In a regular tetrahedron, the bond angles would be 109.5°; in water, the partial negative charges occupy more space than the hydrogen atoms, and therefore, they compress the oxygen-hydrogen bond angle slightly.)

The water molecule, thus, has distinct "ends," each with a partial charge. A rough analogy might be the two poles of a magnet. (These partial charges are much less than the unit charges of ions, however.) Molecules that exhibit charge separation are called **polar molecules** because of these partially charged poles, and water is one of the most polar molecules known. *The polarity of water underlies its chemistry and the chemistry of life.*

Polar molecules interact with one another, as the δ^- of one molecule is attracted to the δ^+ of another. Many of these interactions involve bridging hydrogen atoms; those that do are called **hydrogen bonds** (figure 2.13). Each hydrogen bond is individually very weak and transient, lasting on average only $\frac{1}{100,000,000,000}$ second (10^{-11} sec). However, the cumulative effects of large numbers of these bonds can be enormous. Water forms an abundance of hydrogen bonds, which are responsible for many of its important physical properties (table 2.1).

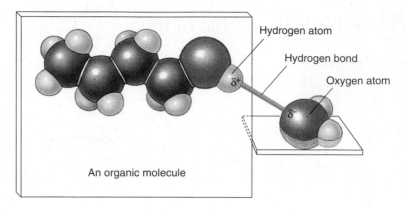

Hydrogen atom
Hydrogen bond
Oxygen atom
δ^+
δ^-
An organic molecule

FIGURE 2.13
Structure of a hydrogen bond. The polar end of this organic molecule is interacting with a water molecule. This interaction bridges a hydrogen atom and is called a hydrogen bond.

The water molecule is very polar, with ends that exhibit partial positive and negative charges. Opposite charges attract, forming weak linkages called hydrogen bonds.

Table 2.1	The Properties of Water	
Property	**Explanation**	**Example of Benefit to Life**
Cohesion	Hydrogen bonds hold water molecules together	Leaves pull water upward from the roots; seeds swell and germinate
High specific heat	Hydrogen bonds absorb heat when they break, and release heat when they form, minimizing temperature changes	Water stabilizes the temperature of organisms and the environment
High heat of vaporization	Many hydrogen bonds must be broken for water to evaporate	Evaporation of water cools body surfaces
Lower density of ice	Water molecules in an ice crystal are spaced relatively far apart because of hydrogen bonding	Because ice is less dense than water, lakes do not freeze solid
Solubility	Polar water molecules are attracted to ions and polar compounds, making them soluble	Many kinds of molecules can move freely in cells, permitting a diverse array of chemical reactions

Water Clings to Polar Molecules

The polarity of water causes it to be attracted to other polar molecules. When the other molecules are also water, the attraction is referred to as **cohesion.** When the other molecules are of a different substance, the attraction is called **adhesion.** Because water is cohesive, it is a liquid, not a gas, at moderate temperatures.

The cohesion of liquid water is also responsible for its **surface tension.** Small insects can walk on water (figure 2.14) because at the air-water interface all of the hydrogen bonds in water face downward, causing the molecules of the water surface to cling together. Water is adhesive to any substance with which it can form hydrogen bonds. That is why substances containing polar molecules get "wet" when they are immersed in water, while those that are composed of nonpolar molecules (such as oils) do not.

The attraction of water to substances such as glass having surface electrical charges is responsible for capillary action: If a glass tube with a narrow diameter is lowered into a beaker of water, water will rise in the tube above the level of the water in the beaker, because the adhesion of water to the glass surface, drawing it upward, is stronger than the force of gravity, drawing it down. The narrower the tube, the greater the electrostatic forces between the water and the glass, and the higher the water rises (figure 2.15).

Water Stores Heat

Water moderates temperature through two properties: its high specific heat and its high heat of vaporization. The temperature of any substance is a measure of how rapidly its individual molecules are moving. Because of the many hydrogen bonds that water molecules form with one another, a large input of thermal energy is required to break these bonds before the individual water molecules can begin moving about more freely and thus have a higher temperature. Therefore, water is said to have a high **specific heat,** which is defined as the amount of heat that must be absorbed or lost by 1 gram of a substance to change its temperature by 1 degree Celsius (°C). Specific heat measures the extent to which a substance resists changing its temperature when it absorbs or loses heat. Because polar substances tend to form hydrogen bonds, and energy is needed to break these bonds, the more polar a substance is, the higher is its specific heat. The specific heat of water (1 calorie/gram/°C) is twice that of most carbon compounds and nine times that of iron. Only ammonia, which is more polar than water and forms very strong hydrogen bonds, has a higher specific heat than water (1.23 calories/gram/°C). Still, only 20% of the hydrogen bonds are broken as water heats from 0° to 100°C.

Because of its high specific heat, water heats up more slowly than almost any other compound and holds its temperature longer when heat is no longer applied. This characteristic enables organisms, which have a high water con-

FIGURE 2.14
Cohesion. Some insects, such as this water strider, literally walk on water. In this photograph you can see how the insect's feet dimple the water as its weight bears down on the surface. Because the surface tension of the water is greater than the force that one foot brings to bear, the strider glides atop the surface of the water rather than sinking.

FIGURE 2.15
Capillary action. Capillary action causes the water within a narrow tube to rise above the surrounding water; the adhesion of the water to the glass surface, which draws water upward, is stronger than the force of gravity, which tends to draw it down. The narrower the tube, the greater the surface area available for adhesion for a given volume of water, and the higher the water rises in the tube.

tent, to maintain a relatively constant internal temperature. The heat generated by the chemical reactions inside cells would destroy the cells if not for the high specific heat of the water within them.

A considerable amount of heat energy (586 calories) is required to change 1 gram of liquid water into a gas. Hence, water also has a high **heat of vaporization.** Because the transition of water from a liquid to a gas requires the input of energy to break its many hydrogen bonds, the evaporation of water from a surface causes cooling of that surface. Many organisms dispose of excess body heat by evaporative cooling; for example, humans and many other vertebrates sweat.

At low temperatures, water molecules are locked into a crystal-like lattice of hydrogen bonds, forming the solid we call ice (figure 2.16). Interestingly, ice is less dense than liquid water because the hydrogen bonds in ice space the water molecules relatively far apart. This unusual feature enables icebergs to float. Were it otherwise, ice would cover nearly all bodies of water, with only the shallow surface melting annually.

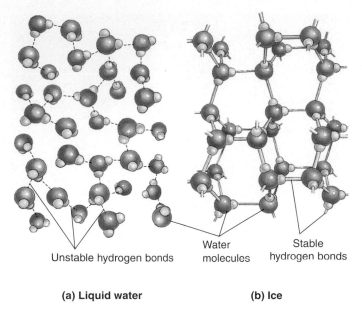

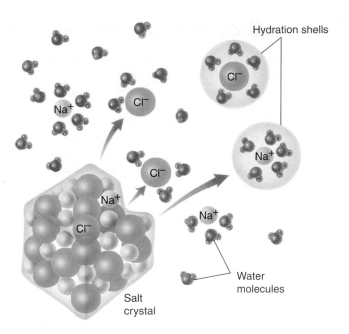

Hydration shells

Cl⁻

Na⁺

Na⁺

Cl⁻

Cl⁻

Na⁺

Na⁺

Water
molecules

Salt
crystal

Unstable hydrogen bonds

Water
molecules

Stable
hydrogen bonds

(a) Liquid water   **(b) Ice**

FIGURE 2.16
The role of hydrogen bonds in an ice crystal. (*a*) In liquid water, hydrogen bonds are not stable and constantly break and re-form. (*b*) When water cools below 0°C, the hydrogen bonds are more stable, and a regular crystalline structure forms in which the four partial charges of one water molecule interact with the opposite charges of other water molecules. Because water forms a particularly open crystal latticework, ice is less dense than liquid water and floats. If it did not, inland bodies of water far from the earth's equator might never fully thaw.

FIGURE 2.17
Why salt dissolves in water. When a crystal of table salt dissolves in water, individual Na⁺ and Cl⁻ ions break away from the salt lattice and become surrounded by water molecules. Water molecules orient around Cl⁻ ions so that their partial positive poles face toward the negative Cl⁻ ion; water molecules surrounding Na⁺ ions orient in the opposite way, with their partial negative poles facing the positive Na⁺ ion. Surrounded by hydration shells, Na⁺ and Cl⁻ ions never reenter the salt lattice.

Water Is a Powerful Solvent

Water is an effective solvent because of its ability to form hydrogen bonds. Water molecules gather closely around any substance that bears an electrical charge, whether that substance carries a full charge (ion) or a charge separation (polar molecule). For example, sucrose (table sugar) is composed of molecules that contain slightly polar hydroxyl (OH) groups. A sugar crystal dissolves rapidly in water because water molecules can form hydrogen bonds with individual hydroxyl groups of the sucrose molecules. Therefore, sucrose is said to be *soluble* in water. Sugar water is a solution of sugar (the solute) dissolved in water (the solvent). Every time a sucrose molecule dissociates or breaks away from the crystal, water molecules surround it in a cloud, forming a **hydration shell** and preventing it from associating with other sucrose molecules. Hydration shells also form around ions such as Na⁺ and Cl⁻ (figure 2.17).

Water Organizes Nonpolar Molecules

Water molecules always tend to form the maximum possible number of hydrogen bonds. When nonpolar molecules such as oils, which do not form hydrogen bonds, are

placed in water, the water molecules act to exclude them. The nonpolar molecules are forced into association with one another, thus minimizing their disruption of the hydrogen bonding of water. In effect, they shrink from contact with water, and for this reason they are referred to as **hydrophobic** (Greek *hydros*, "water," and *phobos*, "fearing"). In contrast, polar molecules, which readily form hydrogen bonds with water, are said to be **hydrophilic** ("water-loving").

The tendency of nonpolar molecules to aggregate in water is known as **hydrophobic exclusion.** By forcing the hydrophobic portions of molecules together, water causes these molecules to assume particular shapes. Different molecular shapes have evolved by alteration of the location and strength of nonpolar regions. As you will see, much of the evolution of life reflects changes in molecular shape that can be induced in just this way.

Because polar water molecules cling to one another, it takes considerable energy to separate them. Water also clings to other polar molecules, causing them to be soluble in a water solution, but water tends to exclude nonpolar molecules.

Water Ionizes

The covalent bonds within a water molecule sometimes break spontaneously. In pure water at 25°C, only 1 out of every 550 million water molecules undergoes this process. When it happens, the proton (hydrogen atom nucleus) dissociates from the molecule. Because the dissociated proton lacks the negatively charged electron it was sharing in the covalent bond with oxygen, its own positive charge is no longer counterbalanced, and it becomes a positively charged **hydrogen ion,** H^+. The rest of the dissociated water molecule, which has retained the shared electron from the covalent bond, is negatively charged and forms a **hydroxide ion,** OH^-. This process of spontaneous ion formation is called **ionization:**

$$H_2O \rightarrow OH^- + H^+$$
$$\text{water} \quad \text{hydroxide ion} \quad \text{hydrogen ion (proton)}$$

At 25°C, a liter of water contains $\frac{1}{10,000,000}$ (or 10^{-7}) mole of H^+ ions. (A **mole** is defined as the weight in grams that corresponds to the summed atomic masses of all of the atoms in a molecule. In the case of H^+, the atomic mass is 1, and a mole of H^+ ions would weigh 1 gram. One mole of any substance always contains 6.02 $\times 10^{23}$ molecules of the substance.) Therefore, the **molar concentration** of hydrogen ions (represented as $[H^+]$) in pure water is 10^{-7} mole/liter. Actually, the hydrogen ion usually associates with another water molecule to form a hydronium (H_3O^+) ion.

pH

A more convenient way to express the hydrogen ion concentration of a solution is to use the **pH scale** (figure 2.18). This scale defines pH as the negative logarithm of the hydrogen ion concentration in the solution:

$$pH = -\log [H^+]$$

Because the logarithm of the hydrogen ion concentration is simply the exponent of the molar concentration of H^+, the pH equals the exponent times -1. Thus, pure water, with an $[H^+]$ of 10^{-7} mole/liter, has a pH of 7. Recall that for every H^+ ion formed when water dissociates, an OH^- ion is also formed, meaning that the dissociation of water produces H^+ and OH^- in equal amounts. Therefore, a pH value of 7 indicates neutrality—a balance between H^+ and OH^-—on the pH scale.

Note that, because the pH scale is *logarithmic*, a difference of 1 on the scale represents a tenfold change in hydrogen ion concentration. This means that a solution with a pH of 4 has 10 times the concentration of H^+ of a solution with a pH of 5.

Acids. Any substance that dissociates in water to increase the concentration of H^+ ions is called an acid. Acidic solutions have pH values below 7. The stronger an acid is, the more H^+ ions it produces and the lower its pH. For example, hydrochloric acid (HCl), which is abundant in your stomach, ionizes completely in water. This means a dilution of 10^{-1} mole per liter of HCl will dissociate to form 10^{-1} mole per liter of H^+ ions, giving the solution a pH of 1. The pH of champagne, which bubbles because of the carbonic acid dissolved in it, is about 2.

Bases. A substance that combines with H^+ ions when dissolved in water is called a base. By combining with H^+ ions, a base lowers the H^+ ion concentration in the solution. Therefore, basic (or alkaline) solutions have pH values above 7. Very strong bases, such as sodium hydroxide (NaOH), have pH values of 12 or more.

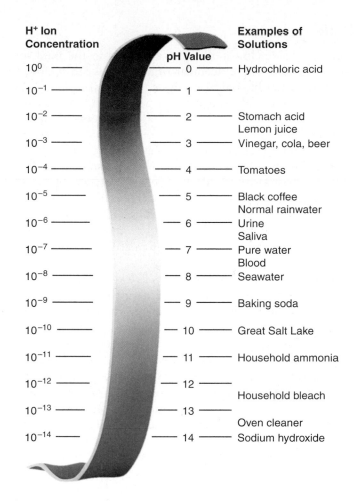

FIGURE 2.18
The pH scale. The pH value of a solution indicates its concentration of hydrogen ions. Solutions with a pH less than 7 are acidic, while those with a pH greater than 7 are basic. The scale is logarithmic, so that a pH change of 1 means a tenfold change in the concentration of hydrogen ions. Thus, lemon juice is 100 times more acidic than tomato juice, and seawater is 10 times more basic than pure water, which has a pH of 7.

Buffers

The pH inside almost all living cells, and in the fluid surrounding cells in multicellular organisms, is fairly close to 7. Most of the biological catalysts (enzymes) in living systems are extremely sensitive to pH; often even a small change in pH will alter their shape, thereby disrupting their activities and rendering them useless. For this reason, it is important that a cell maintain a constant pH level.

Yet the chemical reactions of life constantly produce acids and bases within cells. Furthermore, many animals eat substances that are acidic or basic; cola, for example, is a moderately strong (although dilute) acidic solution. Despite such variations in the concentrations of H^+ and OH^-, the pH of an organism is kept at a relatively constant level by buffers (figure 2.19).

A **buffer** is a substance that acts as a reservoir for hydrogen ions, donating them to the solution when their concentration falls and taking them from the solution when their concentration rises. What sort of substance will act in this way? Within organisms, most buffers consist of pairs of substances, one an acid and the other a base. The key buffer in human blood is an acid-base pair consisting of carbonic acid (acid) and bicarbonate (base). These two substances interact in a pair of reversible reactions. First, carbon dioxide (CO_2) and H_2O join to form carbonic acid (H_2CO_3), which in a second reaction dissociates to yield bicarbonate ion (HCO_3^-) and H^+ (figure 2.20). If some acid or other substance adds H^+ ions to the blood, the HCO_3^- ions act as a base and remove the excess H^+ ions by forming H_2CO_3. Similarly, if a basic substance removes H^+ ions from the blood, H_2CO_3 dissociates, releasing more H^+ ions into the blood. The forward and reverse reactions that interconvert H_2CO_3 and HCO_3^- thus stabilize the blood's pH.

The reaction of carbon dioxide and water to form carbonic acid is important because it permits carbon, essential to life, to enter water from the air. The earth's oceans are rich in carbon because of the reaction of carbon dioxide with water.

In a condition called blood acidosis, human blood, which normally has a pH of about 7.4, drops 0.2 to

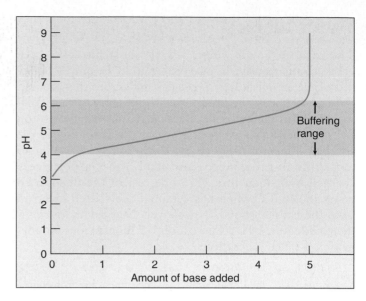

FIGURE 2.19
Buffers minimize changes in pH. Adding a base to a solution neutralizes some of the acid present, and so raises the pH. Thus, as the curve moves to the right, reflecting more and more base, it also rises to higher pH values. A buffer makes the curve rise or fall very slowly over a portion of the pH scale, called the "buffering range" of that buffer.
For this buffer, adding base raises pH more rapidly below pH 4 than above it; what might account for this behavior?

0.4 points on the pH scale. This condition is fatal if not treated immediately. The reverse condition, blood alkalosis, involves an increase in blood pH of a similar magnitude and is just as serious.

The pH of a solution is the negative logarithm of the H^+ ion concentration in the solution. Thus, low pH values indicate high H^+ concentrations (acidic solutions), and high pH values indicate low H^+ concentrations (basic solutions). Even small changes in pH can be harmful to life.

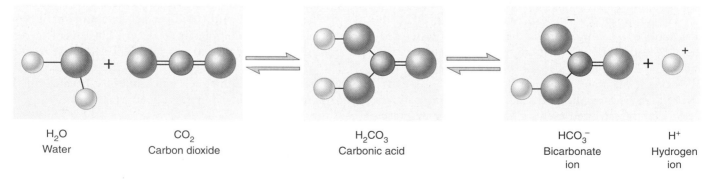

H_2O	CO_2	H_2CO_3	HCO_3^-	H^+
Water	Carbon dioxide	Carbonic acid	Bicarbonate ion	Hydrogen ion

FIGURE 2.20
Buffer formation. Carbon dioxide and water combine chemically to form carbonic acid (H_2CO_3). The acid then dissociates in water, freeing H^+ ions. This reaction makes carbonated beverages acidic, and produced the carbon-rich early oceans where many scientists believe life began.

For interactive testing, visit the Online Learning Center with PowerWeb at www.mhhe.com/Raven7

2.1 Atoms are nature's building material.

Atoms

• All substances are composed of matter, and all matter is composed of atoms. (p. 20)
• Negatively charged electrons circle the nucleus of the atom, while the nucleus is composed of positively charged protons and neutral neutrons. (p. 20)
• An atom's atomic number refers to its number of protons, while its atomic mass refers to the number of protons and neutrons. (p. 20)
• Isotopes are atoms of an element that possess different numbers of neutrons. (p. 21)
• Radioactive isotopes contain nuclei that spontaneously break up into elements with lower atomic numbers. The decay rate is expressed in terms of a half-life. (p. 21)

Electrons Determine the Chemical Behavior of Atoms

• Electrons circle the nucleus of an atom in orbitals. Because the orbitals are so large and are mostly composed of empty space, the nuclei of two atoms never come close enough in nature to interact with each other. (p. 22)
• Because electrons are attracted to the positively charged nucleus, work is necessary so that they stay in their orbits, and thus they have potential energy of position. Moving an electron to a more distant orbital requires energy, while moving an electron closer to the nucleus releases energy. (p. 22)
• Oxidation refers to the loss of an electron; reduction refers to the gain of an electron. (p. 23)

2.2 The atoms of living things are among the smallest.

Kinds of Atoms

• There are 92 naturally occurring elements, arranged in a periodic table based on the interactions of their valence electrons. (p. 24)

2.3 Chemical bonds hold molecules together.

Ionic Bonds Form Crystals

• A molecule is a group of atoms held together by energy in a stable association and joined by chemical bonds. (Compounds are composed of atoms of more than one element.) (p. 25)
• Ionic bonds are formed as attractions between ions of opposite charge, such as those in sodium chloride. (p. 25)

Covalent Bonds Build Stable Molecules

• Covalent bonds form when two atoms share one or more pairs of valence electrons and give rise to true, discrete molecules. (p. 26)

• Covalent bonds are relatively strong, and the strength increases with the number of shared electron pairs. (p. 26)
• A chemical reaction is formed during the formation or breaking of chemical bonds. A reaction may be influenced by several factors, including temperature, concentration of reactants and products, and the presence of catalysts. (p. 26)

2.4 Water is the cradle of life.

Chemistry of Water

• The most outstanding chemical property of water is its ability to form weak chemical associations. (p. 27)

Water Atoms Act Like Tiny Magnets

• A water molecule exhibits electronegativity as both the oxygen and the hydrogen atoms attract the electrons they share in covalent bonds. However, the oxygen atom is more electronegative than the hydrogen atoms. (p. 28)
• Water molecules are polar and exhibit distinct ends with partial charges. (p. 28)
• Hydrogen bonds are formed as opposite charges of bridging hydrogen atoms are attracted, and although each hydrogen bond is relatively weak, the cumulative effect of large numbers of them can be very strong. (p. 28)

Water Clings to Polar Molecules

• Cohesion refers to the attraction of water molecules to other water molecules; adhesion refers to the attraction of other molecules to water molecules. (p. 29)
• The cohesion of water is responsible for its surface tension. (p. 29)
• Water moderates temperatures through its high specific heat and its high heat of vaporization. (p. 29)
• Water is an effective solvent because of its ability to form hydrogen bonds. (p. 30)

Water Ionizes

• A solution's pH is defined as the negative logarithm of its H^+ ion concentration. (p. 31)
• Acids are solutions with high H^+ concentrations (pH < 7), while bases are solutions with low H^+ concentrations (pH > 7). (p. 31)
• Buffers are hydrogen ion reservoirs that either accept or donate H^+ as needed. (p. 32)

Self Test

1. An atom with a neutral charge must contain
   a. the same number of protons and neutrons.
   b. the same number of protons and electrons.
   c. more neutrons because they are neutral.
   d. the same number of neutrons and electrons.
2. Electrons determine the chemical behavior of an atom because
   a. they interact with other atoms.
   b. they determine the charge of the atom (positive, negative, or neutral).
   c. they can be exchanged between atoms.
   d. All of these are correct.
3. The elements within the periodic table are organized by
   a. the number of protons.
   b. the number of neutrons.
   c. the mass of protons and neutrons.
   d. the mass of electrons.
4. Which of the following statements is *not* true?
   a. Molecules held together by ionic bonds are called ionic compounds.
   b. In NaCl, both sodium and chloride have completely filled outer energy levels of 8 electrons.
   c. A sodium atom is able to form an ionic bond with chloride because sodium gives up an electron and chloride gains an electron.
   d. Ionic bonds can form between any two atoms.
5. Oxygen has 6 electrons in its outer energy level; therefore,
   a. it has a completely filled outer energy level.
   b. it can form one double covalent bond or two single covalent bonds.
   c. it does not react with any other atom.
   d. it has a positive charge.
6. The atomic structure of water satisfies the octet rule by
   a. filling the hydrogen atoms' outer energy levels with 8 electrons each.
   b. having electrons shared between the two hydrogen atoms.
   c. having oxygen form covalent bonds with two hydrogen atoms.
   d. having each hydrogen atom give up an electron to the outer energy level of the oxygen atom.
7. The partial charge separation of H_2O results from
   a. the electrons' greater attraction to the oxygen atom.
   b. oxygen's higher electronegativity.
   c. a denser electron cloud near the oxygen atom.
   d. All of these are correct.
8. The attraction of water molecules to other water molecules is
   a. cohesion.
   b. adhesion.
   c. capillary action.
   d. surface tension.
9. What two properties of water help it to moderate changes in temperature?
   a. formation of hydration shells and high specific heat
   b. high heat of vaporization and hydrophobic exclusion
   c. high specific heat and high heat of vaporization
   d. formation of hydration shells and hydrophobic exclusion
10. A substance with a high concentration of hydrogen ions is
    a. called a base.
    b. capable of acting as a buffer.
    c. called an acid.
    d. said to have a high pH.

Test Your Visual Understanding

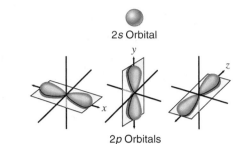

2s Orbital

1s Orbital

2p Orbitals

Orbital for energy level *K*: one spherical orbital (1*s*)

Orbitals for energy level *L*: one spherical orbital (2*s*) and three dumbbell-shaped orbitals (2*p*)

1. This figure shows two energy levels (*K*, which has an *s* orbital labeled 1*s*, and *L*, which has an *s* orbital labeled 2*s* and three *p* orbitals labeled 2*p*). Knowing that the lower level fills with electrons first, followed by the 2*s* orbital and then the *p* orbitals, indicate the number and placement of electrons for the following elements: carbon (C), hydrogen (H), fluorine (F), and neon (Ne).

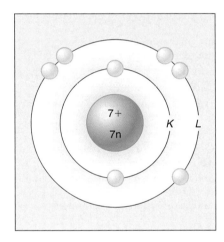

2. This atom has 7 protons and 7 neutrons. What is its atomic number? What is its atomic mass? Can you predict the number of covalent bonds it will form and explain why? What is this element?

Apply Your Knowledge

1. The half-life of radium-226 is 1620 years. If a sample of material contains 16 milligrams of radium-226, how much will it contain in 1620 years? How much will it contain in 3240 years? How long will it take for the sample to contain 1 milligram of radium-226?
2. Scientists find a fossil of a marine animal in the middle of a desert. They use carbon-14 dating to determine how old the fossil is, which would indicate when an ocean covered the desert. There was a 10% (0.10) ratio of ^{14}C compared to ^{12}C in the sample (N_f/N_o) with a ^{14}C half-life ($t_{\frac{1}{2}}$) of 5730 years. Using the equation: $t = [\ln(N_f/N_o)/(-0.693)] \times t_{\frac{1}{2}}$, find the age of the fossil.

3

The Chemical Building Blocks of Life

Concept Outline

3.1 Molecules are the building blocks of life.

Carbon: The Framework of Biological Molecules.
Because individual carbon atoms can form multiple
covalent bonds, biological molecules can be quite complex.

3.2 Proteins perform the chemistry of the cell.

The Many Functions of Proteins. Proteins can be
catalysts, transporters, supporters, and regulators.
Amino Acids: The Building Blocks of Proteins. Proteins
are long chains of various combinations of amino acids.
Protein Structure. A protein's shape is determined by its
amino acid sequence.
How Proteins Fold into Their Functional Shapes.
Chaperone proteins assist in protein folding.
How Proteins Unfold. When conditions such as pH or
temperature fluctuate, proteins may denature or unfold.

3.3 Nucleic acids store and transfer genetic information.

Information Molecules. Nucleic acids store information
in cells. RNA is a single-chain polymer of nucleotides,
while DNA consists of a double helix of nucleotides.

3.4 Lipids make membranes and store energy.

Phospholipids Form Membranes. The aggregation of
phospholipids in water forms biological membranes.
Fats and Other Kinds of Lipids. Organisms utilize a wide
variety of water-insoluble molecules.
Fats as Energy-Storage Molecules. Fats are very efficient
energy-storage molecules because of their high
proportion of C—H bonds.

**3.5 Carbohydrates store energy and provide building
materials.**

Sugars. Sugars are simple carbohydrates, often consisting
of six-carbon rings. The rings can be linked together to
form sugar polymers, or polysaccharides.
Transport and Storage Carbohydrates. Sugars can be
transported as disaccharides and can be stored using the
complex polysaccharides starch and glycogen.
Structural Carbohydrates. Structural carbohydrates, such
as cellulose, are chains of sugars linked in a way that
enzymes cannot easily attack.

FIGURE 3.1
A macromolecule: tRNA. Some of the macromolecules that
are critical to living things play unique roles as molecular tools.
Many of these are proteins that catalyze chemical reactions, but
increasingly we are coming to appreciate the diverse roles of
nucleic acids, such as this transfer RNA, or tRNA, molecule.
tRNA is a type of RNA, or ribonucleic acid, that assists in the
assembly of proteins.

Molecules are extremely small compared to the familiar objects we see around us. Imagine: There are
more molecules in a cup of water than there are stars in the
sky. But some molecules are much larger than water molecules; they consist of thousands of atoms, forming hundreds of molecules that are linked together into long
chains. These enormous assemblies, almost always synthesized by living things, are macromolecules. As we shall see,
there are four types of biological macromolecules, and they
are the basic chemical building blocks from which all organisms are assembled (figure 3.1).

3.1 Molecules are the building blocks of life.

Carbon: The Framework of Biological Molecules

In chapter 2, we discussed how atoms combine to form molecules. In this chapter, we will focus on biological molecules, those chemical compounds that contain carbon. The frameworks of biological molecules consist predominantly of carbon atoms bonded to other carbon atoms or to atoms of oxygen, nitrogen, sulfur, or hydrogen. Because carbon atoms possess four valence electrons and so can form four covalent bonds, molecules containing carbon can form straight chains, branches, or even rings.

Biological molecules consisting only of carbon and hydrogen are called **hydrocarbons.** Covalent bonds between carbon and hydrogen are energy-rich. Propane gas, for example, is a hydrocarbon consisting of a chain of three carbon atoms, with eight hydrogen atoms bound to it:

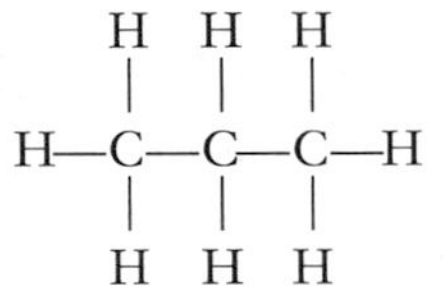

$$
\begin{array}{ccccc}
& H & & H & & H \\
& | & & | & & | \\
H - & C & - & C & - & C - H \\
& | & & | & & | \\
& H & & H & & H \\
\end{array}
$$

Because carbon-hydrogen covalent bonds store considerable energy, hydrocarbons make good fuels. Gasoline, for example, is rich in hydrocarbons.

Functional Groups

Carbon and hydrogen atoms both have very similar electronegativities, so electrons in C—C and C—H bonds are evenly distributed, and there are no significant differences in charge over the molecular surface. For this reason, hydrocarbons are nonpolar. Most biological molecules that are produced by cells, however, also contain other atoms. Because these other atoms often have different electronegativities, molecules containing them exhibit regions of positive or negative charge, and so are polar. These molecules can be thought of as a C—H core to which specific groups of atoms called **functional groups** are attached. For example, a hydrogen atom bonded to an oxygen atom (—OH) is a functional group called a *hydroxyl group.*

Functional groups have definite chemical properties that they retain no matter where they occur. The hydroxyl group, for example, is polar, because its oxygen atom, being very electronegative, draws electrons toward itself (as described in chapter 2). Figure 3.2 illustrates the hydroxyl group and other biologically important functional groups.

Building Biological Macromolecules

Some biological molecules in organisms are small and simple, containing only one or a few functional groups. Others are large, complex assemblies called **macromolecules.** Biological macromolecules are traditionally grouped into four major categories: proteins, nucleic acids, lipids, and carbohydrates (table 3.1). In many cases, these macromolecules are **polymers.** A polymer is a long molecule built by linking together a large number of small, similar chemical subunits, like railroad cars coupled to form a train. For example, complex carbohydrates such as starch are polymers of simple ring-shaped sugars, proteins are polymers of amino acids, and nucleic acids (DNA and RNA) are polymers of nucleotides (see table 3.2).

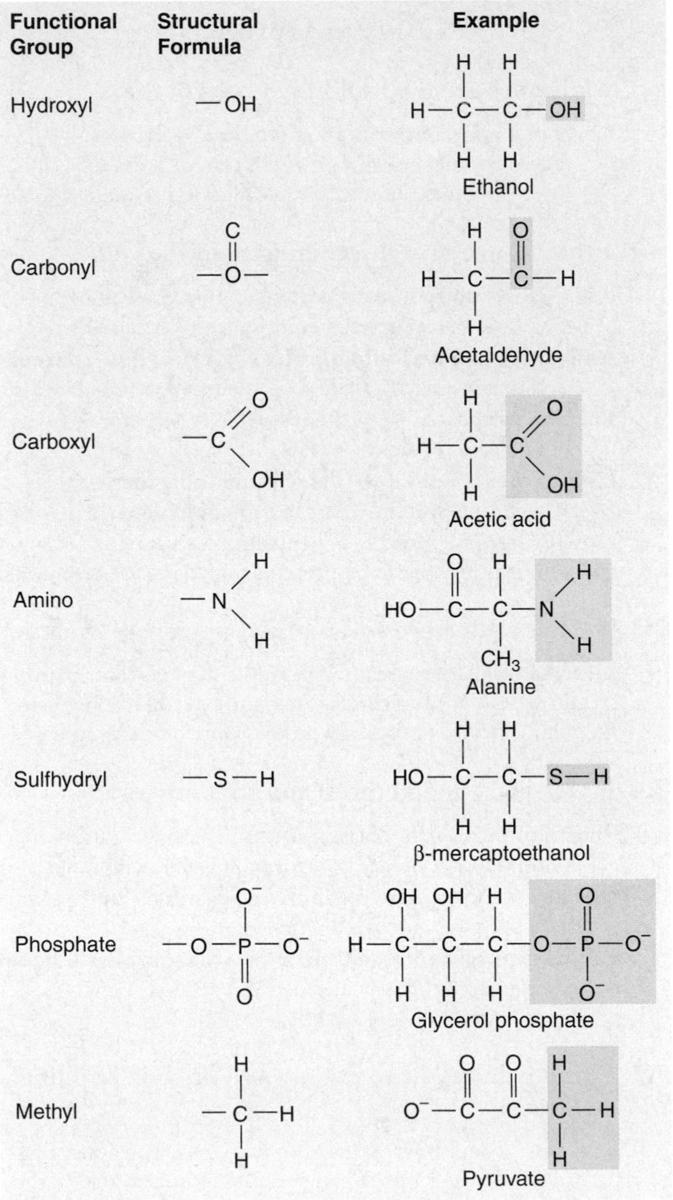

FIGURE 3.2
The primary functional chemical groups.   These groups tend to act as units during chemical reactions and confer specific chemical properties on the molecules that possess them. Amino groups, for example, make a molecule more basic, while carboxyl groups make a molecule more acidic.

Table 3.1 Macromolecules

Macromolecule	Subunit	Function	Example
PROTEINS			
Functional	Amino acids	Catalysis; transport	Hemoglobin
Structural	Amino acids	Support	Hair; silk
NUCLEIC ACIDS			
DNA	Nucleotides	Encodes genes	Chromosomes
RNA	Nucleotides	Needed for gene expression	Messenger RNA
LIPIDS			
Fats	Glycerol and three fatty acids	Energy storage	Butter; corn oil; soap
Phospholipids	Glycerol, two fatty acids, phosphate, and polar R groups	Cell membranes	Lecithin
Prostaglandins	Five-carbon rings with two nonpolar tails	Chemical messengers	Prostaglandin E (PGE)
Steroids	Four fused carbon rings	Membranes; hormones	Cholesterol; estrogen
Terpenes	Long carbon chains	Pigments; structural	Carotene; rubber
CARBOHYDRATES			
Starch, glycogen	Glucose	Energy storage	Potatoes
Cellulose	Glucose	Cell walls	Paper; strings of celery
Chitin	Modified glucose	Structural support	Crab shells

Although the four categories of macromolecules contain different kinds of subunits, they are all assembled in the same fundamental way: To form a covalent bond between two subunit molecules, an —OH group is removed from one subunit and a hydrogen atom (H) is removed from the other (figure 3.3a). This condensation reaction is called **dehydration synthesis,** because the removal of the —OH group and H during the synthesis of a new molecule in effect constitutes the removal of a molecule of water (H_2O). For every subunit that is added to a macromolecule, one water molecule is removed. These and other biochemical reactions require that the reacting substances be held close together and that the correct chemical bonds be stressed and broken. This process of positioning and stressing, termed *catalysis*, is carried out in cells by a special class of proteins known as enzymes.

Cells disassemble macromolecules into their constituent subunits by performing reactions that are essentially the reverse of dehydration—a molecule of water is added instead of removed (figure 3.3b). In this process, which is called **hydrolysis** (Greek *hydro*, "water," + *lyse*, "break"), a hydrogen atom is attached to one subunit and a hydroxyl group to the other, breaking a specific covalent bond in the macromolecule.

Polymers are large molecules consisting of long chains of similar subunits joined by dehydration reactions.

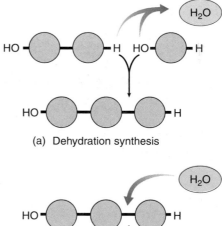

(a) Dehydration synthesis

(b) Hydrolysis

FIGURE 3.3
Making and breaking macromolecules. (*a*) Biological macromolecules are polymers formed by linking subunits together. The covalent bond between the subunits is formed by dehydration synthesis, a process that creates a water molecule for every bond formed. (*b*) Breaking the bond between subunits requires the returning of a water molecule, a process called hydrolysis.

Table 3.2 Polymer Macromolecules

Monomer	Polymer	Cellular structure
Amino acid	Polypeptide	Intermediate filament
Nucleotide	DNA strand	Chromosome
Fatty acid	Fat molecule	Adipose cells with fat droplets
Monosaccharide	Starch	Starch grains in a chloroplast

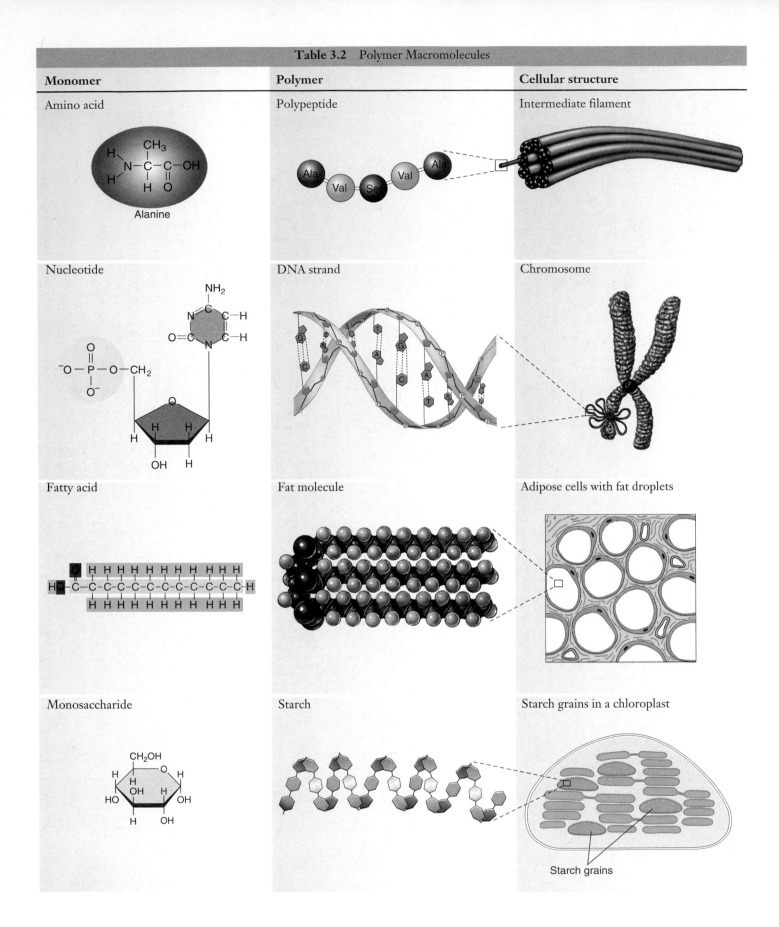

Alanine

Starch grains

3.2 Proteins perform the chemistry of the cell.

The Many Functions of Proteins

We will begin our discussion of biological macromolecules with proteins. The proteins within living organisms are immensely diverse in structure and function. Here we break down their functions into seven categories (table 3.3).

1. **Enzyme catalysis.** We have already encountered one class of proteins, enzymes, which are biological catalysts that facilitate specific chemical reactions. Because of this property, the appearance of enzymes was one of the most important events in the evolution of life. Enzymes are globular proteins, with a three-dimensional shape that fits snugly around the chemicals they work on, facilitating chemical reactions by stressing particular chemical bonds.

2. **Defense.** Other globular proteins use their shapes to "recognize" foreign microbes and cancer cells. These cell surface receptors form the core of the body's hormone and immune systems.

3. **Transport.** A variety of globular proteins transport specific small molecules and ions. The transport protein hemoglobin, for example, transports oxygen in the blood, and myoglobin, a similar protein, transports oxygen in muscle. Iron is transported in blood by the protein transferrin.

Table 3.3 The Many Functions of Proteins

Function	Class of Protein	Examples	Examples of Use
Enzyme catalysis	Enzymes	Hydrolytic enzymes	Cleave polysaccharides
		Proteases	Break down proteins
		Polymerases	Synthesize nucleic acids
		Kinases	Phosphorylate sugars and proteins
Defense	Immunoglobulins	Antibodies	Mark foreign proteins for elimination
	Toxins	Snake venom	Blocks nerve function
	Cell surface antigens	MHC proteins	"Self" recognition
Transport	Circulating transporters	Hemoglobin	Carries O_2 and CO_2 in blood
		Myoglobin	Carries O_2 and CO_2 in muscle
		Cytochromes	Electron transport
	Membrane transporters	Sodium-potassium pump	Excitable membranes
		Proton pump	Chemiosmosis
		Glucose transporter	Transports sugar into cells
Support	Fibers	Collagen	Forms cartilage
		Keratin	Forms hair, nails
		Fibrin	Forms blood clots
Motion	Muscle	Actin	Contraction of muscle fibers
		Myosin	Contraction of muscle fibers
Regulation	Osmotic proteins	Serum albumin	Maintains osmotic concentration of blood
	Gene regulators	*lac* repressor	Regulates transcription
	Hormones	Insulin	Controls blood glucose levels
		Vasopressin	Increases water retention by kidneys
		Oxytocin	Regulates uterine contractions and milk production
Storage	Ion binding	Ferritin	Stores iron, especially in spleen
		Casein	Stores ions in milk
		Calmodulin	Binds calcium ions

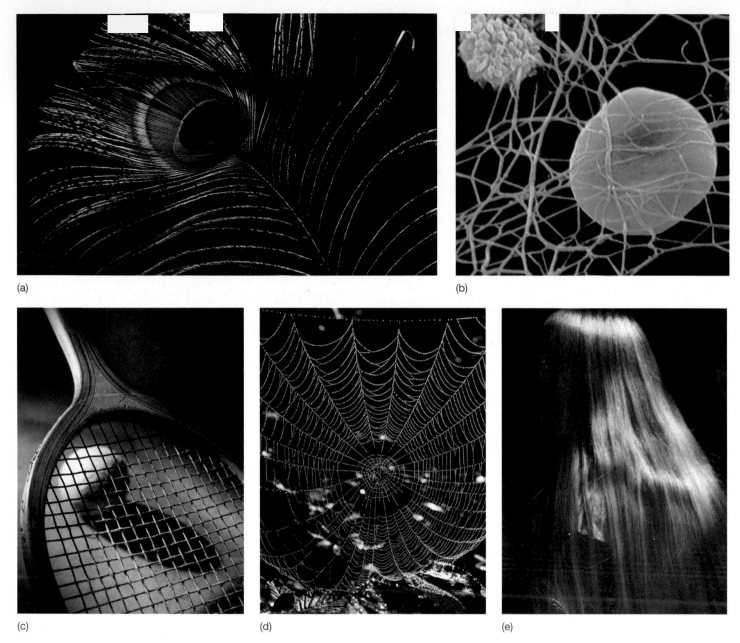

(a)

(b)

(c)

(d)

(e)

FIGURE 3.4
Support. A single functional class of proteins may contain very different members. (*a*) keratin: a peacock feather; (*b*) fibrin: scanning electron micrograph of a blood clot (3000×); (*c*) collagen: strings of a tennis racket from gut tissue; (*d*) silk: a spider's web; (*e*) keratin: human hair.

4. **Support.**   Protein fibers play structural roles (figure 3.4). These fibers include keratin in hair, fibrin in blood clots, and collagen, which forms the matrix of skin, ligaments, tendons, and bones, and is the most abundant protein in a vertebrate body.
5. **Motion.**   Muscles contract through the sliding motion of two kinds of protein filaments: actin and myosin. Contractile proteins also play key roles in the cell's cytoskeleton and in moving materials in cells.
6. **Regulation.**   Small proteins called hormones serve as intercellular messengers in animals. Proteins also play many regulatory roles within the cell, turning on and shutting off genes during development, for example. In addition, proteins also receive information, acting as cell surface receptors.
7. **Storage.**   Calcium and iron are stored by binding as ions to specific storage proteins.

Proteins carry out a diverse array of functions, including enzyme catalysis, defense, transport, support, motion, regulation, and storage.

Amino Acids: The Building Blocks of Proteins

Although proteins are complex and versatile molecules, they are all polymers of only 20 different kinds of amino acids, in a specific order. Many scientists believe amino acids were among the first molecules formed on the early earth. It seems highly likely that the oceans that existed early in the history of the earth contained a wide variety of amino acids.

Amino Acid Structure

An **amino acid** molecule contains an amino group (—NH_2), a carboxyl group (—COOH), and a hydrogen atom, all bonded to a central carbon atom:

$$H_2N—\overset{\displaystyle R}{\underset{\displaystyle H}{C}}—COOH$$

Each amino acid has unique chemical properties determined by the nature of the side group (indicated by R) covalently bonded to the central carbon atom. For example, when the side group is —CH_2OH, the amino acid (serine) is polar, but when the side group is —CH_3, the amino acid (alanine) is nonpolar. The 20 common amino acids are grouped into five chemical classes, based on their side groups:

1. Nonpolar amino acids, such as leucine, often have R groups that contain —CH_2 or —CH_3.
2. Polar uncharged amino acids, such as threonine, have R groups that contain oxygen (or only —H).
3. Charged amino acids, such as glutamic acid, have R groups that contain acids or bases.
4. Aromatic amino acids, such as phenylalanine, have R groups that contain an organic (carbon) ring with alternating single and double bonds.
5. Special-function amino acids have unique individual properties; methionine is often the first amino acid in a chain of amino acids, proline causes kinks in chains, and cysteine links chains together.

Each amino acid affects the shape of a protein differently, depending on the chemical nature of its side group. For example, portions of a protein chain with numerous nonpolar amino acids tend to fold into the interior of the protein by hydrophobic exclusion.

Proteins Are Polymers of Amino Acids

In addition to its R group, each amino acid, when ionized, has a pronated amino (NH_3^+) group at one end and a negative carboxyl (COO^-) group at the other end. The amino and carboxyl groups on a pair of amino acids can undergo a

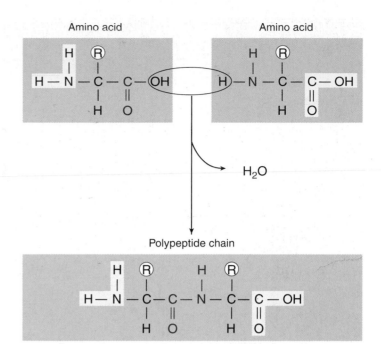

FIGURE 3.5
The peptide bond. A peptide bond forms when the —NH_2 end of one amino acid joins to the —COOH end of another. Because of the partial double-bond nature of peptide bonds, the resulting peptide chain cannot rotate freely around these bonds.

condensation reaction, losing a molecule of water and forming a covalent bond. A covalent bond that links two amino acids is called a **peptide bond** (figure 3.5). The two amino acids linked by such a bond are not free to rotate around the N—C linkage because the peptide bond has a partial double-bond character, unlike the N—C and C—C bonds to the central carbon of the amino acid. The stiffness of the peptide bond is one factor that determines the structural character of the coils and other regular shapes formed by chains of amino acids.

A protein is composed of one or more long chains, or **polypeptides,** composed of amino acids linked by peptide bonds. It was not until the pioneering work of Frederick Sanger in the early 1950s that it became clear that each kind of protein had a specific amino acid sequence. Sanger succeeded in determining the amino acid sequence of insulin and in so doing demonstrated clearly that this protein had a defined sequence, the same for all insulin molecules in the solution. Although many different amino acids occur in nature, only 20 commonly occur in proteins. Figure 3.6 illustrates these 20 "common" amino acids and their side groups.

A protein is a polymer containing a combination of up to 20 different kinds of amino acids. The amino acids fall into five chemical classes, each with different properties. These properties determine the nature of the resulting protein.

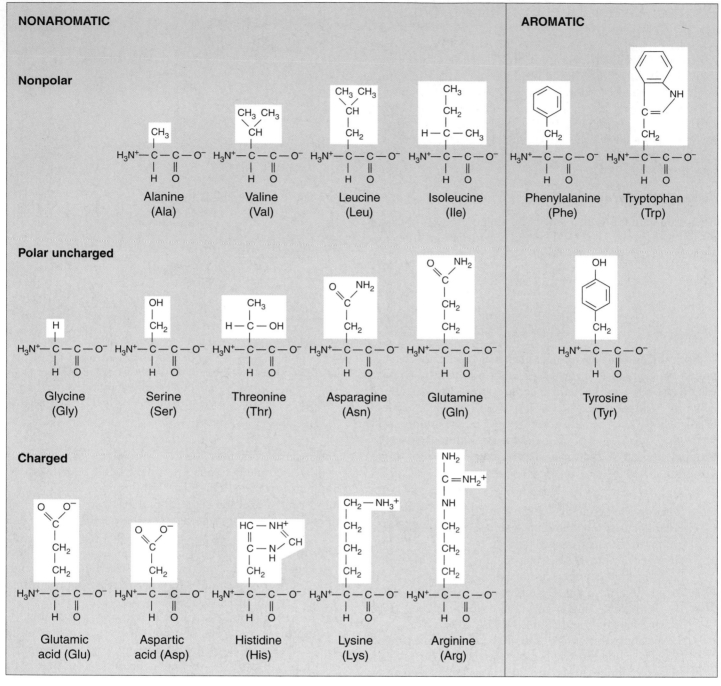

NONAROMATIC

AROMATIC

Nonpolar

Alanine
(Ala)

Valine
(Val)

Leucine
(Leu)

Isoleucine
(Ile)

Phenylalanine
(Phe)

Tryptophan
(Trp)

Polar uncharged

Glycine
(Gly)

Serine
(Ser)

Threonine
(Thr)

Asparagine
(Asn)

Glutamine
(Gln)

Tyrosine
(Tyr)

Charged

Glutamic
acid (Glu)

Aspartic
acid (Asp)

Histidine
(His)

Lysine
(Lys)

Arginine
(Arg)

SPECIAL FUNCTION

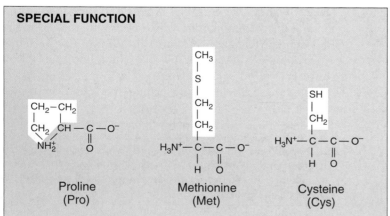

Proline
(Pro)

Methionine
(Met)

Cysteine
(Cys)

FIGURE 3.6

The 20 common amino acids. Each amino acid has the same chemical backbone, but differs in the side, or R, group it possesses. Six of the amino acids are nonpolar because they have —CH_2 or —CH_3 in their R groups, which are not electronegative. Two of the six are bulkier because they contain ring structures, which classifies them also as aromatic. Another six are polar because they have oxygen or just hydrogen in their R groups; these amino acids, which are uncharged, differ from one another in how polar they are. Five other polar amino acids have a terminal acid or base in their R group, and so are capable of ionizing to a charged form. The remaining three special-function amino acids have chemical properties that allow them to help form links between protein chains or kinks in proteins.

Protein Structure

The shape of a protein is very important because it determines the protein's function. Proteins consist of long amino acid chains folded into complex shapes. What do we know about the shape of these proteins? One way to study the shape of something as small as a protein is to look at it with very short wavelength energy—in other words, with X rays. X-ray diffraction is a painstaking procedure that allows the investigator to build up a three-dimensional picture of the position of each atom. The first protein to be analyzed in this way was myoglobin, soon followed by the related protein hemoglobin. As more and more proteins were studied, a general principle became evident: In every protein studied, essentially all the internal amino acids are nonpolar ones—amino acids such as leucine, valine, and phenylalanine. Water's tendency to hydrophobically exclude nonpolar molecules literally shoves the nonpolar portions of the amino acid chain into the protein's interior (figure 3.7). This positions the nonpolar amino acids in close contact with one another, leaving little empty space inside. Polar and charged amino acids are restricted to the surface of the protein except for the few that play key functional roles.

Levels of Protein Structure

The structure of proteins has traditionally been discussed in terms of four levels of structure: *primary, secondary, tertiary,* and *quaternary.* Because of progress in our knowledge of protein structure, two additional levels of structure are increasingly distinguished by molecular biologists: *motifs* and *domains* (figure 3.8), two elements that play important roles in coming chapters.

Primary Structure. The specific amino acid sequence of a protein is its primary structure. This sequence is determined by the nucleotide sequence of the gene that encodes the protein. Because the R groups that distinguish the various amino acids play no role in the peptide backbone of proteins, a protein can consist of any sequence of amino acids. Thus, because any of 20 different amino acids might appear at any position, a protein containing 100 amino acids could form any of 20^{100} different amino acid sequences (that's the same as 10^{130}, or 1 followed by 130 zeros—more than the number of atoms known in the universe). This is an important property of proteins because it permits great diversity.

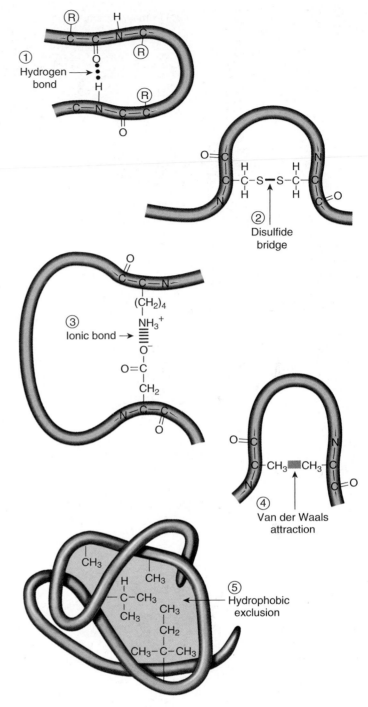

FIGURE 3.7
Interactions that contribute to a protein's shape. Aside from the bonds that link the amino acids in a protein together, several other weaker forces and interactions determine how a protein will fold. (*1*) Hydrogen bonds can form between the different amino acids. (*2*) Fairly strong disulfide bridges can form between two cysteine side chains. (*3*) Ionic bonds can form. (*4*) Van der Waals attractions occur—that is, weak attractions between atoms due to oppositely polarized electron clouds. (*5*) Polar portions of the protein tend to gather on the outside of the protein and interact with water, while the hydrophobic portions of the protein, including nonpolar amino acid chains, are shoved toward the interior of the protein.

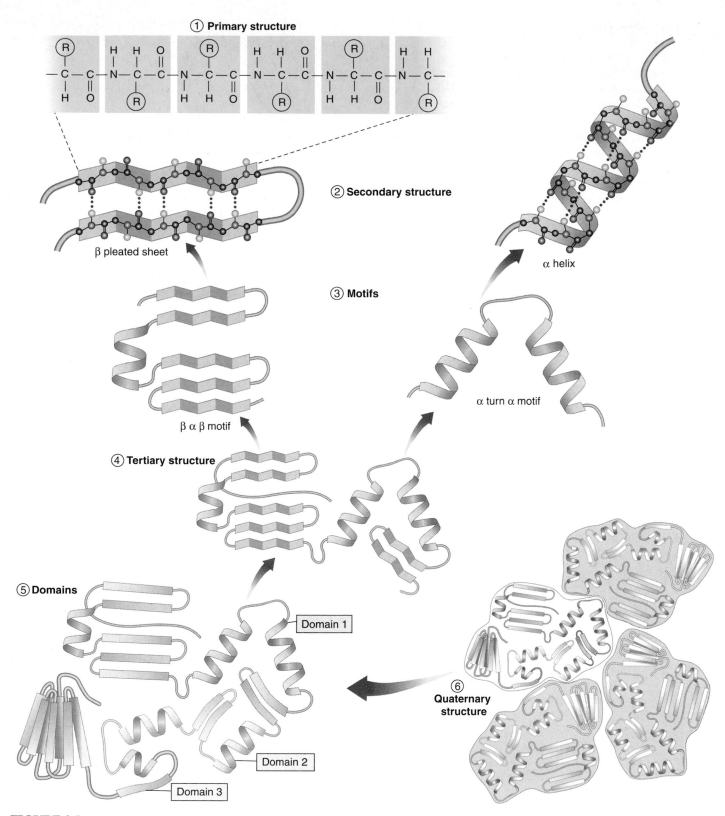

FIGURE 3.8
Six levels of protein structure. The amino acid sequence of a protein is called its (*1*) primary structure. Hydrogen bonds form between nearby amino acids, producing (*2*) foldbacks called beta (β) pleated sheets and coils called alpha (α) helices. These foldbacks and coils constitute the protein's secondary structure. The elements of secondary structure can combine, fold, or crease to form (*3*) motifs. A globular protein folds up on itself further to assume a three-dimensional (*4*) tertiary structure, several of which can be linked together in a protein (*5*) domain. Many proteins aggregate with other polypeptide chains in clusters; this clustering is called the (*6*) quaternary structure of the protein.

Secondary Structure. The amino acid side groups are not the only portions of proteins that form hydrogen bonds. The peptide groups of the main chain also form quite good hydrogen bonds—so good that their interactions with water might be expected to offset the tendency of nonpolar side groups to be forced into the protein interior. Inspection of the protein structures determined by X-ray diffraction reveals why they don't—the polar groups of the main chain form hydrogen bonds with each other! Two patterns of hydrogen bonding occur. In one, hydrogen bonds form along a single chain, linking one amino acid to another farther down the chain. This tends to pull the chain into a coil called an alpha (α) helix. In the other pattern, hydrogen bonds occur across two chains, linking the amino acids in one chain to those in the other. Often, many parallel chains are linked, forming a pleated, sheet-like structure called a beta (β) pleated sheet. The folding of the amino acid chain by hydrogen bonding into these characteristic coils and pleats is called a protein's secondary structure.

Motifs. The elements of secondary structure can combine in proteins in characteristic ways called motifs, or sometimes "supersecondary structure." One very common motif is the $\beta \alpha \beta$ motif, which creates a fold or crease; the so-called "Rossmann fold" at the core of nucleotide binding sites in a wide variety of proteins is a $\beta \alpha \beta \alpha \beta$ motif. A second motif that occurs in many proteins is the β barrel, a β sheet folded around to form a tube. A third type of motif, the α turn α motif, is important because many proteins use it to bind the DNA double helix.

Tertiary Structure. The final folded shape of a globular protein, which positions the various motifs and folds nonpolar side groups into the interior, is called a protein's tertiary structure. A protein is initially driven into its tertiary structure by hydrophobic exclusion from water. Ionic bonds between oppositely charged side groups bring regions into close proximity, and disulfide bonds (covalent links between two cysteine side groups) lock particular regions together. The final folding of a protein is determined by its primary structure—the chemical nature of its side groups (see figure 3.7). Many small proteins can be fully unfolded ("denatured") and will spontaneously refold into their characteristic shape.

The stability of a protein, once it has folded into its 3-D shape, is strongly influenced by how well its interior fits together. When two nonpolar chains in the interior are in very close proximity, they experience a form of molecular attraction called van der Waal's forces. Individually quite weak, these forces can add up to a strong attraction when many of them come into play, like the combined strength of hundreds of hooks and loops on a strip of Velcro. They are effective forces only over short distances, however; no "holes" or cavities exist in the interior of proteins. That is why there are so many different nonpolar amino acids (ala-

nine, valine, leucine, isoleucine). Each has a different-sized R group, allowing very precise fitting of nonpolar chains within the protein interior. Now you can understand why a mutation that converts one nonpolar amino acid within the protein interior (alanine) into another (leucine) very often disrupts the protein's stability; leucine is a lot bigger than alanine and disrupts the precise way the chains fit together within the protein interior. A change in even a single amino acid can have profound effects on protein shape and can result in lost or altered function of the protein.

Domains. Many proteins in your body are encoded within your genes in functional sections called exons (exons will be discussed in detail in chapter 15). Each exon-encoded section of a protein, typically 100 to 200 amino acids long, folds into a structurally independent functional unit called a **domain.** As the polypeptide chain folds, the domains fold into their proper shape, each more-or-less independent of the others. This can be demonstrated experimentally by artificially producing the fragment of a polypeptide that forms the domain in the intact protein, and showing that the fragment folds to form the same structure as it does in the intact protein.

A single polypeptide chain connects the domains of a protein, like a rope tied into several adjacent knots. Often the domains of a protein have quite separate functions—one domain of an enzyme might bind a cofactor, for example, and another the enzyme's substrate.

Quaternary Structure. When two or more polypeptide chains associate to form a functional protein, the individual chains are referred to as subunits of the protein. The subunits need not be the same. Hemoglobin, for example, is a protein composed of two α chain subunits and two β chain subunits. A protein's subunit arrangement is called its quaternary structure. In proteins composed of subunits, the interfaces where the subunits contact one another are often nonpolar, and play a key role in transmitting information between the subunits about individual subunit activities.

A change in the identity of one of these amino acids can have profound effects. Sickle cell hemoglobin is a mutation that alters the identity of a single amino acid at the corner of the β subunit from polar glutamate to nonpolar valine. Putting a nonpolar amino acid on the surface creates a "sticky patch" that causes one hemoglobin molecule to stick to another, forming long, nonfunctional chains and leading to the cell sickling characteristic of the hereditary disorder sickle cell anemia.

Protein structure can be viewed at six levels: (1) the amino acid sequence, or primary structure; (2) coils and sheets, called secondary structure; (3) folds or creases, called motifs; (4) the three-dimensional shape, called tertiary structure; (5) functional units, called domains; and (6) individual polypeptide subunits associated in a quaternary structure.

How Proteins Fold into Their Functional Shapes

How does a protein fold into a specific shape? Nonpolar amino acids play a key role. Until recently, investigators thought that newly made proteins fold spontaneously as hydrophobic interactions with water shove nonpolar amino acids into the protein interior. We now know this view is too simple. Protein chains can fold in so many different ways that trial and error would simply take too long. In addition, as the open chain folds its way toward its final form, nonpolar "sticky" interior portions are exposed during intermediate stages. If these intermediate forms are placed in a test tube in the same protein environment that occurs in a cell, they stick to other unwanted protein partners, forming a gluey mess.

Chaperone Proteins

How do cells avoid this? A vital clue came in studies of unusual mutations that prevented viruses from replicating in bacterial cells—it turned out that the virus proteins could not fold properly! Further study revealed that normal cells contain special proteins called **chaperone proteins,** which help new proteins fold correctly. When the bacterial gene encoding its chaperone protein is disabled by mutation, the bacteria die, clogged with lumps of incorrectly folded proteins. Fully 30% of the bacterial proteins fail to fold to the right shape.

Molecular biologists have now identified more than 17 kinds of proteins that act as molecular chaperones. Many are heat shock proteins, produced in greatly elevated amounts if a cell is exposed to elevated temperature; high temperatures cause proteins to unfold, and heat shock chaperone proteins help the cell's proteins refold (figure 3.9).

There is considerable controversy about how chaperone proteins work. It was first thought that they provided a protected environment within which folding could take place unhindered by other proteins, but it now seems more likely that chaperone proteins rescue proteins that are caught in a wrongly folded state, giving them another chance to fold correctly. When investigators "fed" a deliberately misfolded protein, malate dehydrogenase, to chaperone proteins, the protein was rescued, refolding to its active shape.

Protein Folding and Disease

There are tantalizing suggestions that chaperone protein deficiencies may play a role in certain diseases by failing to facilitate the intricate folding of key proteins. Cystic fibrosis is a hereditary disorder in which a mutation disables a protein that plays a vital part in moving ions across cell membranes. In at least some cases, the vital membrane protein appears to have the correct amino acid sequence, but fails to fold to its final form. It has also been speculated that chaperone deficiency may be a cause of the protein clumping in brain cells that produces the amyloid plaques characteristic of Alzheimer disease.

Chaperone proteins help newly produced proteins fold properly.

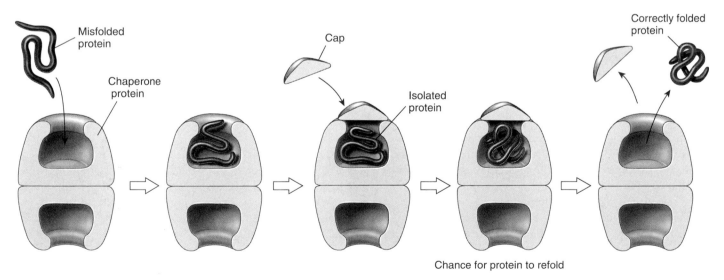

FIGURE 3.9
How one type of chaperone protein works. This barrel-shaped chaperone protein is a heat shock protein, produced in elevated amounts at high temperatures. An incorrectly folded protein enters one chamber of the barrel, and a cap seals the chamber and confines the protein. The isolated protein is now prevented from aggregating with other misfolded proteins, and it has a chance to refold properly. After a short time period, the protein is ejected, folded or unfolded, and the cycle can repeat itself.

How Proteins Unfold

If a protein's environment is altered, the protein may change its shape or even unfold. This process is called **denaturation** (figure 3.10). Proteins can be denatured when the pH, temperature, or ionic concentration of the surrounding solution is changed. When proteins are denatured, they are usually rendered biologically inactive. This is particularly significant in the case of enzymes. Because practically every chemical reaction in a living organism is catalyzed by a specific enzyme, it is vital that a cell's enzymes remain functional. That is the rationale behind traditional methods of salt-curing and pickling: Prior to the ready availability of refrigerators and freezers, the only practical way to keep microorganisms from growing in food was to keep the food in a solution containing a high concentration of salt or vinegar, which denatured the enzymes of microorganisms and kept them from growing on the food.

Most enzymes function within a very narrow range of physical parameters. Blood-borne enzymes that course through a human body at a pH of about 7.4 would rapidly become denatured in the highly acidic environment of the stomach. On the other hand, the protein-degrading enzymes that function at a pH of 2 or less in the stomach would be denatured in the basic pH of the blood. Similarly, organisms that live near oceanic hydrothermal vents have enzymes that work well at the temperature of this extreme environment (over 100°C). They cannot survive in cooler waters, because their enzymes do not function properly at lower temperatures. Any given organism usually has a tolerance range of pH, temperature, and salt concentration. Within that range, its enzymes maintain the proper shape to carry out their biological functions.

When a protein's normal environment is reestablished after denaturation, a small protein may spontaneously refold into its natural shape, driven by the interactions between its nonpolar amino acids and water (figure 3.11). Larger proteins can rarely refold spontaneously because of the complex nature of their final shape. It is important to distinguish denaturation from **dissociation**. The four subunits of hemoglobin may dissociate into four individual molecules (two α-globin and two β-globin) without denaturation of the folded globin proteins, and will readily reassume their four-subunit quaternary structure.

Every globular protein has a narrow range of conditions in which it folds properly; outside that range, proteins tend to unfold.

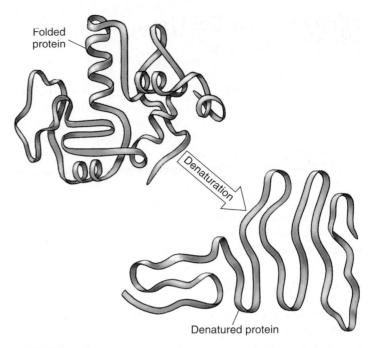

FIGURE 3.10
Protein denaturation. Changes in a protein's environment, such as variations in temperature or pH, can cause a protein to unfold and lose its shape in a process called denaturation. In this denatured state, proteins are biologically inactive.

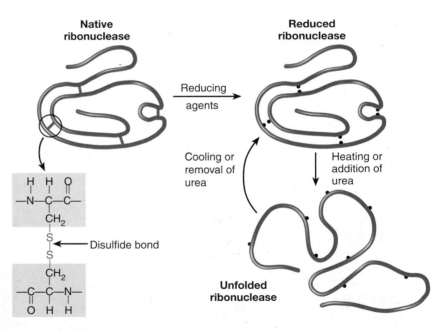

FIGURE 3.11
Primary structure determines tertiary structure. When the protein ribonuclease is treated with reducing agents to break the covalent disulfide bonds that cross-link its chains, and then placed in urea or heated, the protein denatures (unfolds) and loses its enzymatic activity. Upon cooling or removal of urea, it refolds and regains its enzymatic activity. This demonstrates that no information except the amino acid sequence of the protein is required for proper folding: The primary structure of the protein determines its tertiary structure.

3.3 Nucleic acids store and transfer genetic information.

Information Molecules

The biochemical activity of a cell depends on production of a large number of proteins, each with a specific sequence. The ability to produce the correct proteins is passed between generations of organisms, even though the protein molecules themselves are not.

Nucleic acids are the information storage devices of cells, just as disks store the information that computers use, blueprints store the information that builders use, and road maps store the information that tourists use. There are two varieties of nucleic acids: **deoxyribonucleic acid (DNA;** figure 3.12) and **ribonucleic acid (RNA).** The way DNA encodes the information used to assemble proteins is similar to the way the letters on a page encode information (see chapter 14). Unique among macromolecules, nucleic acids are able to serve as templates to produce precise copies of themselves, so that the information that specifies what an organism is can be copied and passed down to its descendants. For this reason, DNA is often referred to as the hereditary material. Cells use the alternative form of nucleic acid, RNA, to read the cell's DNA-encoded information and direct the synthesis of proteins. RNA is similar to DNA in structure and is made as a transcribed copy of portions of the DNA. This transcript passes out into the rest of the cell, where it serves as a blueprint specifying a protein's amino acid sequence. This process will be described in detail in chapter 15.

"Seeing" DNA

DNA molecules cannot be seen with an optical microscope, which is incapable of resolving anything smaller than 1000 atoms across. An electron microscope can image structures as small as a few dozen atoms across, but still cannot resolve the individual atoms of a DNA strand. This limitation was finally overcome in the mid-1980s with the introduction of the scanning-tunneling microscope (figure 3.13).

How do these microscopes work? Imagine you are in a dark room with a chair. To determine the shape of the chair, you could shine a flashlight on it, so that the light bounces off the chair and forms an image on your eye. That's what optical and electron microscopes do; in the latter, the "flashlight" emits a beam of electrons instead of light. You could, however, also reach out and feel the chair's surface with your hand. In effect, you would be putting a probe (your hand) near the chair and measuring how far away the surface is. In a scanning-tunneling microscope, computers advance a probe over the surface of a molecule in steps smaller than the diameter of an atom.

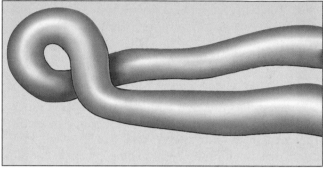

FIGURE 3.12
The first photograph of a DNA molecule. This micrograph, with sketch below, shows a section of DNA magnified a million times. The molecule is so slender that it would take 50,000 of them to equal the diameter of a human hair.

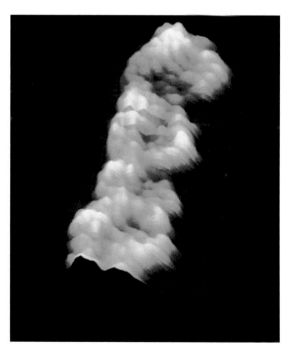

FIGURE 3.13
A scanning-tunneling micrograph of DNA (false color; 2,000,000×). The micrograph shows approximately three turns of the DNA double helix (see figure 3.16).

The Structure of Nucleic Acids

Nucleic acids are long polymers of repeating subunits called **nucleotides.** Each nucleotide consists of three components: a five-carbon sugar (ribose in RNA and deoxyribose in DNA); a phosphate (—PO$_4$) group; and an organic nitrogenous (nitrogen-containing) base (figure 3.14). When a nucleic acid polymer forms, the phosphate group of one nucleotide binds to the hydroxyl group of another, releasing water and forming a phosphodiester bond. A **nucleic acid,** then, is simply a chain of five-carbon sugars linked together by phosphodiester bonds with an organic base protruding from each sugar (figure 3.15).

Two types of organic bases occur in nucleotides. The first type, *purines*, are large, double-ring molecules found in both DNA and RNA; they are adenine (A) and guanine (G). The second type, *pyrimidines*, are smaller, single-ring molecules; they include cytosine (C, in both DNA and RNA), thymine (T, in DNA only), and uracil (U, in RNA only).

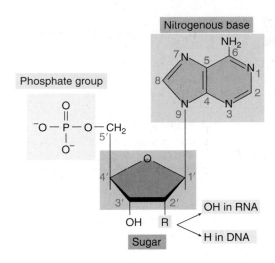

FIGURE 3.14
Structure of a nucleotide. The nucleotide subunits of DNA and RNA are made up of three elements: a five-carbon sugar, an organic nitrogenous base (adenine is shown here), and a phosphate group.

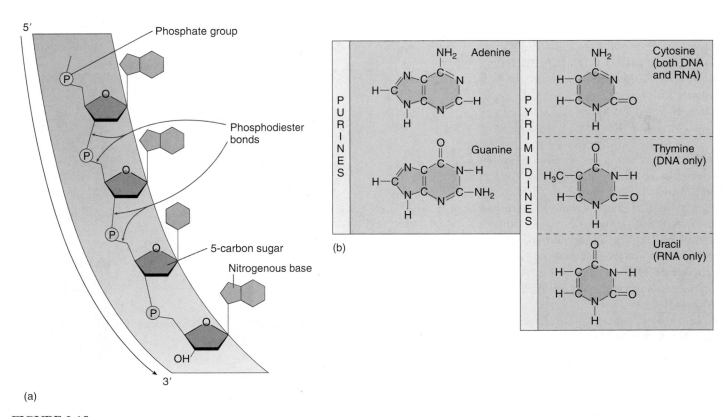

FIGURE 3.15
The structure of a nucleic acid and the organic nitrogen-containing bases. (*a*) In a nucleic acid, nucleotides are linked to one another via phosphodiester bonds, with organic bases protruding from the chain. (*b*) The organic nitrogenous bases can be either purines or pyrimidines. In DNA, thymine replaces the uracil found in RNA.

DNA

Organisms encode the information specifying the amino acid sequences of their proteins as sequences of nucleotides in the DNA. This method of encoding information is very similar to that by which the sequences of letters encode information in a sentence. While a sentence written in English consists of a combination of the 26 different letters of the alphabet in a specific order, the code of a DNA molecule consists of different combinations of the four types of nucleotides in specific sequences, such as CGCTTACG. The information encoded in DNA is used in the everyday metabolism of the organism and is passed on to the organism's descendants.

DNA molecules in organisms exist not as single chains folded into complex shapes, like proteins, but rather as double chains. Two DNA polymers wind around each other like the outside and inside rails of a spiral staircase. Such a winding shape is called a helix, and a helix composed of two chains winding about one another, as in DNA, is called a **double helix.** Each step of DNA's helical staircase is a base-pair, consisting of a base in one chain attracted by hydrogen bonds to a base opposite it on the other chain. These hydrogen bonds hold the two chains together as a duplex (figure 3.16). The base-pairing rules are rigid: Adenine can pair only with thymine (in DNA) or with uracil (in RNA), and cytosine can pair only with guanine. The bases that participate in base-pairing are said to be **complementary** to each other. Additional details of the structure of DNA and how it interacts with RNA in the production of proteins are presented in chapters 14 and 15.

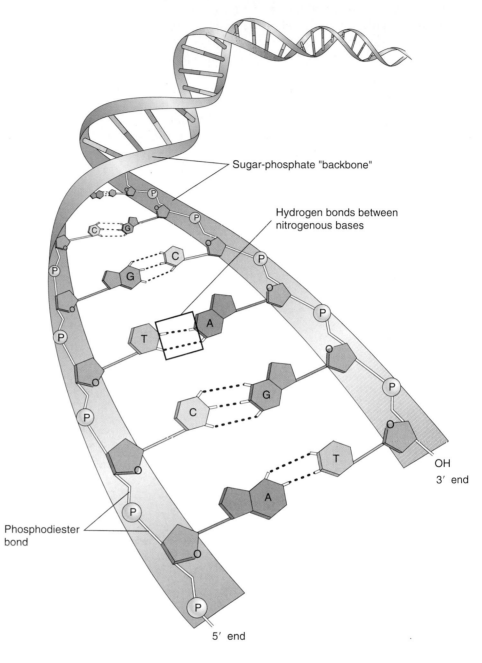

Sugar-phosphate "backbone"

Hydrogen bonds between nitrogenous bases

Phosphodiester bond

OH
3′ end

5′ end

FIGURE 3.16
The structure of DNA. Hydrogen bond formation (dashed lines) between the organic bases, called base-pairing, causes the two chains of a DNA duplex to bind to each other and form a double helix.

RNA

RNA is similar to DNA, but with two major chemical differences. First, RNA molecules contain ribose sugars in which the number 2 carbon is bonded to a hydroxyl group. In DNA, this hydroxyl group is replaced by a hydrogen atom. Second, RNA molecules utilize uracil in place of thymine. Uracil has the same structure as thymine, except that one of its carbons lacks a methyl (—CH₃) group.

Transcribing the DNA message into a chemically different molecule such as RNA allows the cell to tell which is the original information storage molecule and which is the

DNA

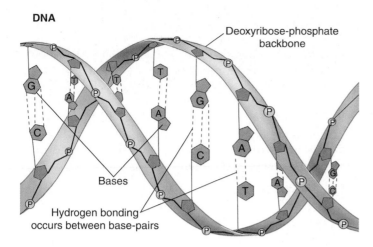

Deoxyribose-phosphate backbone

Bases

Hydrogen bonding occurs between base-pairs

RNA

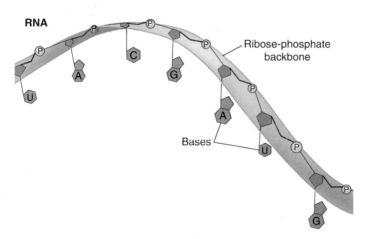

Ribose-phosphate backbone

Bases

FIGURE 3.17
DNA versus RNA. DNA forms a double helix, uses deoxyribose as the sugar in its sugar-phosphate backbone, and utilizes thymine among its nitrogenous bases. RNA, on the other hand, is usually single-stranded, uses ribose as the sugar in its sugar-phosphate backbone, and utilizes uracil in place of thymine.

transcript. DNA molecules are always double-stranded (except for a few single-stranded DNA viruses that will be discussed in chapter 26), while the RNA molecules transcribed from DNA are typically single-stranded (figure 3.17). RNA cannot form double helices as DNA does, because of the steric hindrance of the 2′ —OH group. Using two different molecules, one single-stranded and the other double-stranded, separates the role of DNA in storing hereditary information from the role of RNA in using this information to specify protein structure.

Which Came First, DNA or RNA?

The information necessary for the synthesis of proteins is stored in the cell's double-stranded DNA base sequences. The cell uses this information by first making an RNA

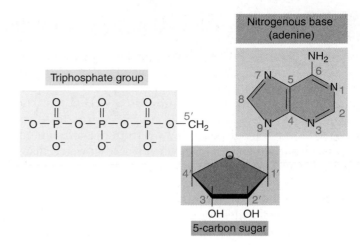

Nitrogenous base (adenine)

Triphosphate group

5-carbon sugar

FIGURE 3.18
ATP. Adenosine triphosphate (ATP) contains adenine, a five-carbon sugar, and three phosphate groups. This molecule serves to transfer energy rather than store genetic information.

transcript of it: RNA nucleotides pair with complementary DNA nucleotides. By storing the information in DNA while using a complementary RNA sequence to actually direct protein synthesis, the cell does not expose the information-encoding DNA chain to the dangers of single-strand cleavage every time the information is used. Therefore, DNA is thought to have evolved from RNA as a means of preserving the genetic information, protecting it from the ongoing wear and tear associated with cellular activity. This genetic system has come down to us from the very beginnings of life.

The cell uses the single-stranded, short-lived RNA transcript to direct the synthesis of a protein with a specific sequence of amino acids. Thus, the information flows from DNA to RNA to protein, a process that has been termed the "central dogma" of molecular biology.

ATP

In addition to serving as subunits of DNA and RNA, nucleotide bases play other critical roles in the life of a cell. For example, adenine is a key component of the molecule *adenosine triphosphate* (ATP; figure 3.18), the energy currency of the cell. It also occurs in the molecules *nicotinamide adenine dinucleotide* (NAD$^+$) and *flavin adenine dinucleotide* (FAD), which carry electrons whose energy is used to make ATP.

A nucleic acid is a long chain of five-carbon sugars with an organic base protruding from each sugar. DNA is a double-stranded helix that stores hereditary information as a specific sequence of nucleotide bases. RNA is a single-stranded molecule that transcribes this information to direct protein synthesis.

3.4 Lipids make membranes and store energy.

Lipids are a loosely defined group of molecules with one main characteristic: They are insoluble in water. The most familiar lipids are fats and oils. Lipids have a very high proportion of nonpolar carbon-hydrogen (C—H) bonds, and so long-chain lipids cannot fold up like a protein to sequester their nonpolar portions away from the surrounding aqueous environment. Instead, when placed in water, many lipid molecules spontaneously cluster together and expose what polar (hydrophilic) groups they have to the surrounding water while sequestering the nonpolar (hydrophobic) parts of the molecules together within the cluster. This spontaneous assembly of lipids is of paramount importance to cells, as it underlies the structure of cellular membranes.

Phospholipids Form Membranes

Complex lipid molecules called **phospholipids** are among the most important molecules of the cell, because they form the core of all biological membranes. An individual phospholipid is a composite molecule, made up of three kinds of subunits:

1. *Glycerol*, a three-carbon alcohol, with each carbon bearing a hydroxyl group. Glycerol forms the backbone of the phospholipid molecule.

2. *Fatty acids*, long chains of —CH$_2$ groups (hydrocarbon chains) ending in a carboxyl (—COOH) group. Two fatty acids are attached to the glycerol backbone in a phospholipid molecule.

3. *Phosphate group*, attached to one end of the glycerol. The charged phosphate group usually has a charged organic molecule linked to it, such as choline, ethanolamine, or the amino acid serine.

The phospholipid molecule can be thought of as having a polar "head" at one end (the phosphate group) and two long, very nonpolar "tails" at the other (figure 3.19). In water, the nonpolar tails of nearby lipid molecules aggregate away from the water, forming spherical *micelles*, with the tails inward. Phospholipids form two layers with tails pointed toward each other—a lipid bilayer (figure 3.20). Lipid bilayers are the basic framework of biological membranes, discussed in detail in chapter 6.

Because the C—H bonds in lipids are very nonpolar, they are not water-soluble, and aggregate together in water. This kind of aggregation by phospholipids forms biological membranes.

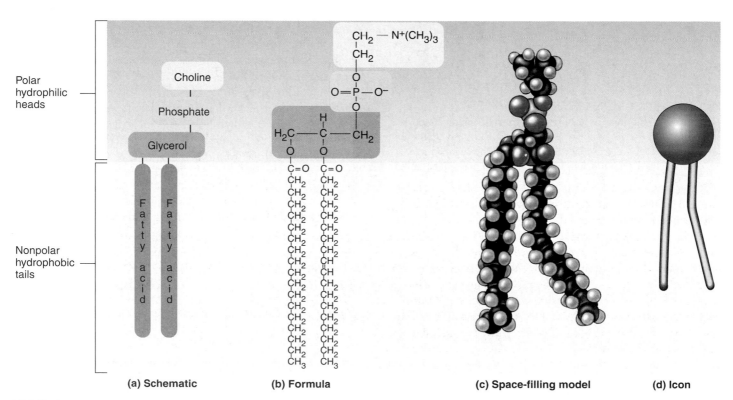

FIGURE 3.19
Phospholipids. The phospholipid phosphatidylcholine is shown as (*a*) a schematic, (*b*) a formula, (*c*) a space-filling model, and (*d*) an icon.

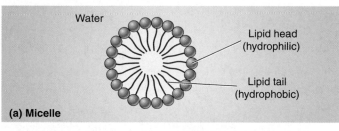

(a) Micelle

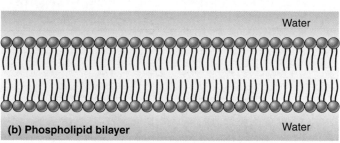

(b) Phospholipid bilayer

FIGURE 3.20
Lipids spontaneously form micelles or lipid bilayers in water.
In an aqueous environment, lipid molecules will orient so that their polar (hydrophilic) heads are in the polar medium, water, and their nonpolar (hydrophobic) tails are held away from the water. (*a*) Droplets called micelles can form, or (*b*) phospholipid molecules can arrange themselves into two layers; in both structures, the hydrophilic heads extend outward and the hydrophobic tails inward.

Fats and Other Kinds of Lipids

Fats are another kind of lipid, but unlike phospholipids, fat molecules do not have a polar end. **Fats** consist of a glycerol molecule to which is attached three fatty acids, one to each carbon of the glycerol backbone. Because it contains three fatty acids, a fat molecule is called a **triglyceride,** or more properly, a **triacylglycerol** (see figure 3.22). The three fatty acids of a triglyceride need not be identical, and often they differ markedly from one another. Organisms store the energy of certain molecules for long periods in the many C—H bonds of fats.

Because triglyceride molecules lack a polar end, they are not soluble in water. Placed in water, they spontaneously clump together, forming fat globules that are very large relative to the size of the individual molecules. Because fats are insoluble, they can be deposited at specific locations within an organism.

Storage fats such as animal fat are one kind of lipid. Oils such as olive oil, corn oil, and coconut oil are also lipids, as are waxes such as beeswax and earwax. The hydrocarbon chains of fatty acids vary in length; the most common are even-numbered chains of 14 to 20 carbons. If all of the internal carbon atoms in the fatty acid chains are bonded to at least two hydrogen atoms, the fatty acid is said to be

(a) Terpene (citronellol)

(b) Steroid (cholesterol)

FIGURE 3.21
Other kinds of lipids. (*a*) Terpenes are found in biological pigments, such as chlorophyll and retinal, and (*b*) steroids play important roles in membranes and in chemical signaling.

saturated, because it contains the maximum possible number of hydrogen atoms (see figure 3.22). If a fatty acid has double bonds between one or more pairs of successive carbon atoms, the fatty acid is said to be **unsaturated.** If a given fatty acid has more than one double bond, it is said to be **polyunsaturated.** Fats made from polyunsaturated fatty acids have low melting points because their fatty acid chains bend at the double bonds, preventing the fat molecules from aligning closely with one another. Consequently, a polyunsaturated fat such as corn oil is usually liquid at room temperature and is called an oil. In contrast, most saturated fats such as those in butter are solid at room temperature.

Organisms contain many other kinds of lipids besides fats (figure 3.21). *Terpenes* are long-chain lipids that are components of many biologically important pigments, such as chlorophyll and the visual pigment retinal. Rubber is also a terpene. *Steroids,* another type of lipid found in membranes, are composed of four carbon rings. Most animal cell membranes contain the steroid cholesterol. Other steroids, such as testosterone and estrogen, function in multicellular organisms as hormones. *Prostaglandins* are a group of about 20 lipids that are modified fatty acids, with two nonpolar "tails" attached to a five-carbon ring. Prostaglandins act as local chemical messengers in many vertebrate tissues.

Cells contain a variety of different lipids, in addition to membrane phospholipids, that play many important roles in cell metabolism.

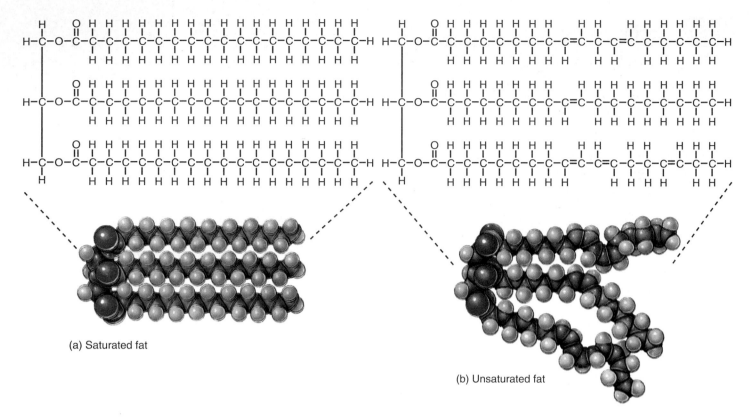

(a) Saturated fat

(b) Unsaturated fat

FIGURE 3.22

Saturated and unsaturated fats. (*a*) A saturated triacylglycerol contains three saturated fatty acids, those with no double bonds and, thus, a maximum number of hydrogen atoms bonded to the carbon chain. Many animal triacylglycerols (fats) are saturated. Because their fatty acid chains can fit closely together, these triacylglycerols form immobile arrays called hard fat. (*b*) An unsaturated triacylglycerol contains three unsaturated fatty acids, those with one or more double bonds and, thus, fewer than the maximum number of hydrogen atoms bonded to the carbon chain. Plant fats are typically unsaturated. The many kinks the double bonds introduce into the fatty acid chains prevent the triacylglycerols from closely aligning and instead produce oils, which are liquid at room temperature.

Fats as Energy-Storage Molecules

Most fats contain over 40 carbon atoms. The ratio of energy-storing C—H bonds to carbon atoms in fats is more than twice that of carbohydrates (see section 3.5), making fats much more efficient molecules for storing chemical energy. On the average, fats yield about 9 kilocalories (kcal) of chemical energy per gram, as compared with somewhat less than 4 kcal per gram for carbohydrates.

All fats produced by animals are saturated (except some fish oils), while most plant fats are unsaturated (figure 3.22). The exceptions are the tropical oils (palm oil and coconut oil), which are saturated despite their fluidity at room temperature. It is possible to convert an oil into a solid fat by adding hydrogen. Peanut butter sold in stores is usually artificially hydrogenated to make the peanut fats solidify, preventing them from separating out as oils while the jar sits on the store shelf. However, artificially hydrogenating unsaturated fats seems to eliminate the health advantage they have over saturated fats, by making both equally rich in C—H bonds. Therefore, it now appears that margarine made from hydrogenated corn oil is no better for your health than butter.

When an organism consumes excess carbohydrate, it is converted into starch, glycogen, or fats and reserved for future use. The reason that many humans gain weight as they grow older is that the amount of energy they need decreases with age, while their intake of food does not. Thus, an increasing proportion of the carbohydrate they ingest is available to be converted into fat.

A diet rich in fats is one of several factors that are thought to contribute to heart disease, particularly atherosclerosis, a condition in which deposits of fatty tissue called plaque adhere to the lining of blood vessels, blocking the flow of blood. Fragments of plaque, breaking off from a deposit, are a major cause of strokes.

Fats are efficient energy-storage molecules because of their high concentration of C—H bonds.

3.5 Carbohydrates store energy and provide building materials.

Sugars

The **carbohydrates** are a loosely defined group of molecules that contain carbon, hydrogen, and oxygen in the molar ratio 1:2:1. Their empirical formula (which lists the atoms in the molecule with subscripts to indicate how many there are of each) is $(CH_2O)_n$, where n is the number of carbon atoms. Because they contain many carbon-hydrogen (C—H) bonds, which release energy when they are oxidized, carbohydrates are well suited for energy storage. Among the most important energy-storage molecules are sugars, which exist in several different forms.

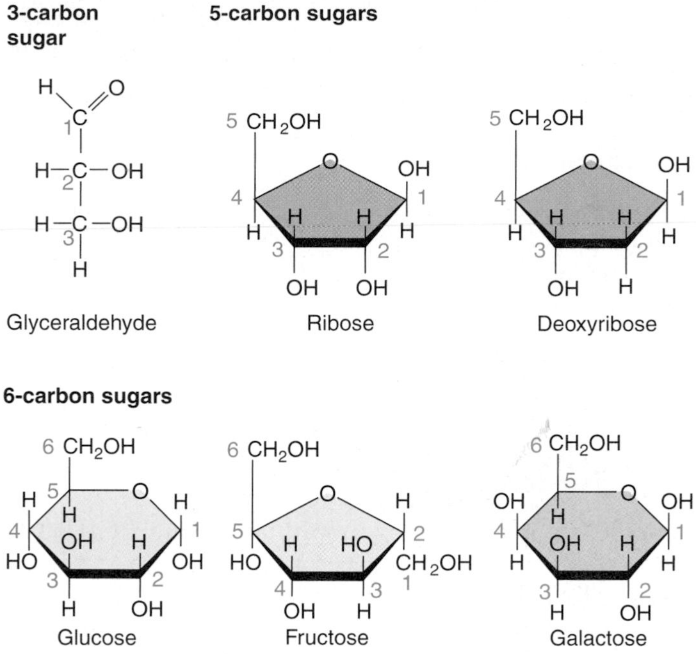

Monosaccharides

The simplest of the carbohydrates are the simple sugars, or **monosaccharides** (Greek *mono*, "single," + Latin *saccharum*, "sugar"). Simple sugars may contain as few as three carbon atoms, but those that play the central role in energy storage have six (figure 3.23). The empirical formula of six-carbon sugars is:

$$C_6H_{12}O_6, \text{ or } (CH_2O)_6$$

Six-carbon sugars can exist in a straight-chain form, but in an aqueous environment they almost always form rings. The most important of these for energy storage is *glucose* (figure 3.24), a six-carbon sugar that has seven energy-storing C—H bonds.

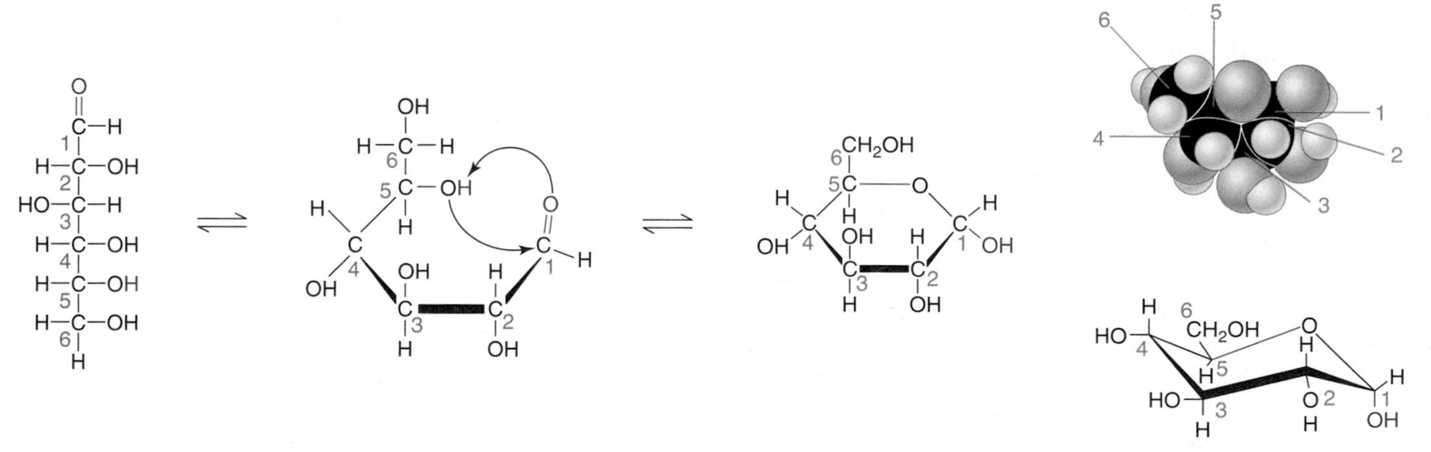

FIGURE 3.23
Monosaccharides. Monosaccharides, or simple sugars, can contain as few as three carbon atoms and are often used as building blocks to form larger molecules. The five-carbon sugars ribose and deoxyribose are components of nucleic acids (see figure 3.17). The six-carbon sugar glucose is a component of large energy-storage molecules. The green numbers refer to the carbon atoms; monosaccharides are conventionally numbered from the more oxidized end.

FIGURE 3.24
Structure of the glucose molecule. Glucose is a linear, six-carbon molecule that forms a ring shape in solution. The structure of the ring can be represented in many ways; the ones shown here are the most common, with the carbons conventionally numbered (in *green*) so that the forms can be compared easily. The bold, darker lines represent portions of the molecule that are projecting out of the page toward you—remember, these are three-dimensional molecules!

Disaccharides

Many familiar sugars, such as sucrose, are "double sugars," two monosaccharides joined by a covalent bond (figure 3.25). Called **disaccharides,** they often play a role in the transport of sugars, as we will discuss shortly.

Polysaccharides

Polysaccharides are macromolecules made up of monosaccharide subunits. Starch is a polysaccharide used by plants to store energy. It consists entirely of glucose molecules, linked one after another in long chains. Cellulose is a polysaccharide that serves as a structural building material in plants. It too consists entirely of glucose molecules linked together into chains, and special enzymes are required to break the links.

Sugar Isomers

Glucose is not the only sugar with the formula $C_6H_{12}O_6$. Other common six-carbon sugars such as fructose and galactose also have this empirical formula (figure 3.26). These sugars are **isomers,** or alternative forms, of glucose or other six-carbon monosaccharides. Even though isomers have the same empirical formula, their atoms are arranged in different ways; that is, their three-dimensional structures are different. These structural differences often account for substantial functional differences between the isomers. Glucose and fructose, for example, are *structural isomers*. In fructose, the double-bonded oxygen is attached to an internal carbon rather than to a terminal one. Your taste buds can tell the difference, as fructose tastes much sweeter than glucose, despite the fact that both sugars have the same chemical composition. This structural difference also has an important chemical consequence: The two sugars form different polymers.

Unlike fructose, galactose has the same bond structure as glucose; the only difference between galactose and glucose is the orientation of one hydroxyl group. Because the hydroxyl group positions are mirror images of each other, galactose and glucose are called *stereoisomers*.

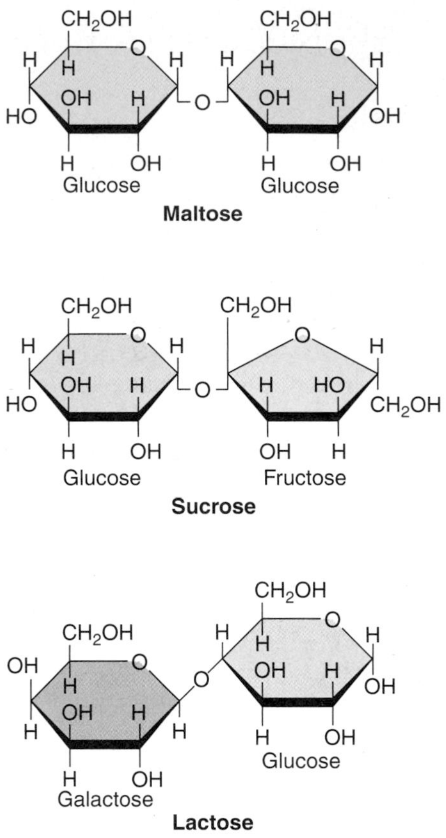

FIGURE 3.25
Disaccharides. Sugars such as maltose, sucrose, and lactose are disaccharides, composed of two monosaccharides linked by a covalent bond.

Sugars are among the most important energy-storage molecules in organisms, containing many energy-storing C—H bonds. The structural differences among sugar isomers can confer substantial functional differences upon the molecules.

FIGURE 3.26
Isomers and stereoisomers. Glucose, fructose, and galactose are isomers with the empirical formula $C_6H_{12}O_6$. A structural isomer of glucose, such as fructose, has identical chemical groups bonded to different carbon atoms, while a stereoisomer of glucose, such as galactose, has identical chemical groups bonded to the same carbon atoms but in different orientations.

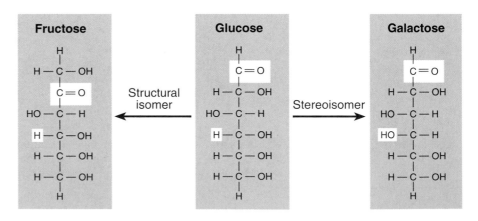

Transport and Storage Carbohydrates

Transport Disaccharides

Most organisms transport sugars within their bodies. In humans, the glucose that circulates in the blood does so as a simple monosaccharide. In plants and many other organisms, however, glucose is converted into a transport form before it is moved from place to place within the organism. In such a form, it is less readily metabolized (used for energy) during transport. Transport forms of sugars are commonly made by linking two monosaccharides together to form a disaccharide (Greek *di*, "two"). Disaccharides serve as effective reservoirs of glucose because the normal glucose-utilizing enzymes of the organism cannot break the bond linking the two monosaccharide subunits. Enzymes that can do so are typically present only in the tissue where the glucose is to be used.

Transport forms differ depending on which monosaccharides link to form the disaccharide. Glucose forms transport disaccharides with itself and many other monosaccharides, including fructose and galactose. When glucose forms a disaccharide with its structural isomer, fructose, the resulting disaccharide is *sucrose*, or table sugar (figure 3.27a). Sucrose is the form in which most plants transport glucose and the sugar that most humans (and other animals) eat. Sugarcane and sugar beets are rich in sucrose.

When glucose is linked to its stereoisomer, galactose, the resulting disaccharide is *lactose*, or milk sugar. Many mammals supply energy to their young in the form of lactose. Adults have greatly reduced levels of lactase, the enzyme required to cleave lactose into its two monosaccharide components, and thus cannot metabolize lactose

efficiently. Most of the energy that is channeled into lactose production is therefore reserved for their offspring.

Storage Polysaccharides

Organisms store the metabolic energy contained in monosaccharides by converting them into disaccharides, such as *maltose* (figure 3.27b), which are then linked together into insoluble forms that are deposited in specific storage areas in their bodies. These insoluble polysaccharides are long polymers of monosaccharides formed by dehydration synthesis. Plant polysaccharides formed from glucose are called **starches.**

The starch with the simplest structure is *amylose*, which is composed of many hundreds of glucose molecules linked together in long, unbranched chains. Each linkage occurs between the number 1 carbon of one glucose molecule and the number 4 carbon of another, so that amylose is, in effect, a longer form of maltose. The long chains of amylose tend to coil up in water, a property that renders amylose insoluble. Potato starch is about 20% amylose.

Most plant starch, including the remaining 80% of potato starch, is a somewhat more complicated variant of amylose called *amylopectin*. Pectins are branched polysaccharides with short, linear amylose branches consisting of 20 to 30 glucose subunits.

The animal version of starch is **glycogen.** Like amylopectin, glycogen is an insoluble polysaccharide containing branched amylose chains. In glycogen, the average chain length is much greater, and there are more branches than in plant starch.

Starches are glucose polymers. Most starches are branched, rendering the polymer insoluble.

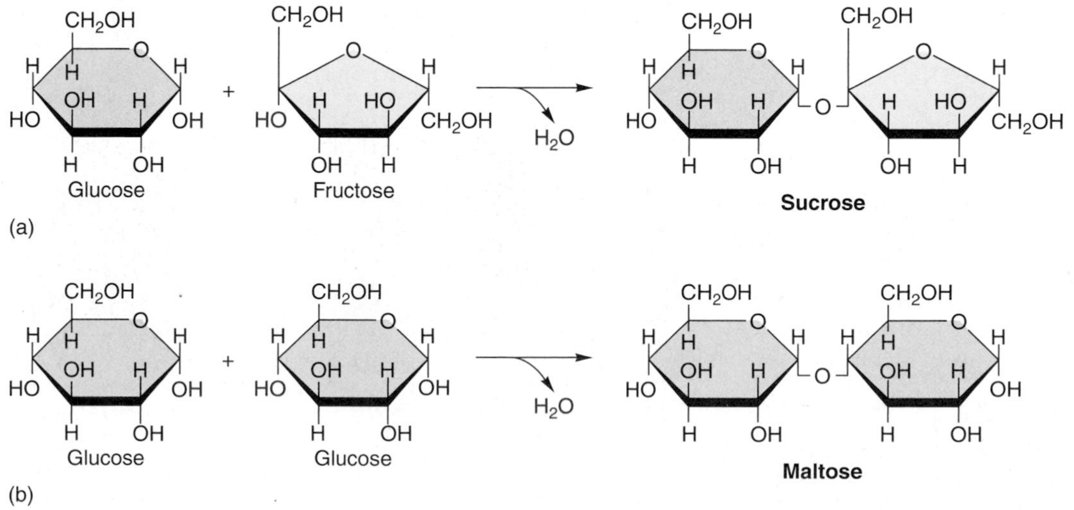

(a)

(b)

FIGURE 3.27
How disaccharides form.
Some disaccharides are used to transport glucose from one part of an organism's body to another; one example is sucrose (*a*), which is found in sugarcane. Other disaccharides, such as maltose in grain (*b*), are used for storage.

Structural Carbohydrates

Cellulose

While some chains of sugars store energy, others serve as structural material for cells. For two glucose molecules to link together, the glucose subunits must be the same form. Glucose can form a ring in two ways, with the hydroxyl group attached to the carbon where the ring closes, being locked into place either below or above the plane of the ring. If below, it is called the **alpha form (α),** and if above, the **beta form (β).** All of the glucose subunits of the starch chain are alpha-glucose (figure 3.28). When a chain of glucose molecules consists of all beta-glucose subunits, a polysaccharide with very different properties results. This structural polysaccharide is *cellulose,* the chief component of plant cell walls (figure 3.28). Cellulose is chemically similar to amylose, with one important difference: the starch-degrading enzymes that occur in most organisms cannot break the bond between two beta-glucose sugars. This is not because the bond is stronger, but rather because its cleavage requires an enzyme most organisms lack. Because cellulose cannot be broken down readily, it works well as a biological structural material and occurs widely in this role in plants. Those few animals able to break down cellulose find it a rich source of energy. Certain vertebrates, such as cows, can digest cellulose by means of bacteria and protists harbored in their digestive tracts that provide the necessary enzymes.

Chitin

The structural building material in arthropods and many fungi is called chitin. *Chitin* is a modified form of cellulose with a nitrogen group added to the glucose units. When cross-linked by proteins, it forms a tough, resistant surface material that serves as the hard exoskeleton of arthropods such as insects and crustaceans (figure 3.29; see chapter 33). Few organisms are able to digest chitin and use it as a major source of nutrition, but most possess a chitinase enzyme, probably to protect against fungi.

Structural carbohydrates are chains of sugars that are not easily digested. They include cellulose in plants and chitin in arthropods and fungi.

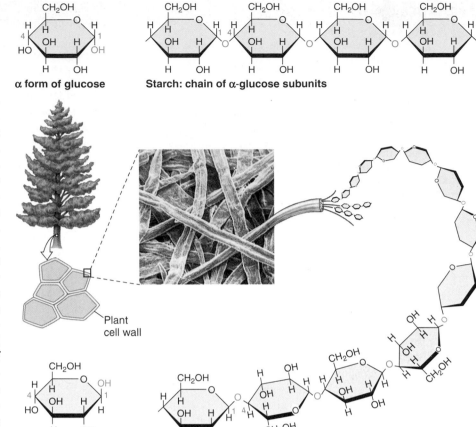

α form of glucose

Starch: chain of α-glucose subunits

Plant cell wall

β form of glucose

Cellulose: chain of β-glucose subunits

FIGURE 3.28
Polymers of glucose. While starch chains consist of alpha-glucose subunits, cellulose chains consist of beta-glucose subunits. Cellulose fibers can be very strong and are quite resistant to metabolic breakdown, which is one reason wood is such a good building material.

FIGURE 3.29
Chitin. Chitin, which might be considered a modified form of cellulose, is the principal structural element in the external skeletons of many invertebrates, such as this lobster.

Concept Review

For interactive testing, visit the Online Learning Center with PowerWeb at www.mhhe.com/Raven7

3.1 Molecules are the building blocks of life.

Carbon: The Framework of Biological Molecules

- Biological molecules consist predominantly of carbon atoms bonded to other carbon atoms or to atoms of oxygen, nitrogen, sulfur, or hydrogen. (p. 36)
- Hydrocarbons are molecules consisting only of carbon and hydrogen; thus, they store considerable energy. (p. 36)
- Functional groups have definite chemical properties that are retained no matter where they occur, and tend to act as units during chemical reactions, conferring specific chemical properties to molecules that possess them. (p. 36)
- Polymers are composed of long chains of similar subunits joined by dehydration synthesis. (pp. 36–37)

3.2 Proteins perform the chemistry of the cell.

The Many Functions of Proteins

- The major functions of proteins within living organisms include enzyme catalysis, defense, transport, support, motion, regulation, and storage. (pp. 39–40)

Amino Acids: The Building Blocks of Proteins

- Proteins are polymers of 20 different kinds of amino acids, arranged in specific orders. (p. 41)
- The 20 common amino acids are grouped into five chemical classes: nonpolar, polar uncharged, charged, aromatic, and special function. (p. 41)
- A peptide bond is a covalent bond that links two amino acids together. (p. 41)

Protein Structure

- Proteins consist of long amino acid chains folded into complex shapes, and the shape of a protein determines its function. (p. 43)
- The specific amino acid sequence of a protein forms its primary structure. (p. 43)
- The folding of the amino acid chain by hydrogen bonding into coils and pleats forms a protein's secondary structure. (p. 45)
- The elements of secondary structure can combine in proteins in characteristic ways (motifs). (p. 45)
- The final folded shape of a globular protein forms a protein's tertiary structure. (p. 45)
- Each exon-encoded section of a protein folds into a structurally independent functional unit (domain). (p. 45)
- Quaternary structure refers to individual polypeptide subunits associated to form functional proteins. (p. 45)

How Proteins Fold into Their Functional Shapes

- Chaperone proteins help new proteins fold correctly. (p. 46)

How Proteins Unfold

- Denaturation occurs when a protein changes shape, or unfolds, and can be caused by a change in environmental pH, temperature, or ionic concentration. (p. 47)
- Denaturation usually renders a protein biologically inactive. (p. 47)

3.3 Nucleic acids store and transfer genetic information.

Information Molecules

- Deoxyribonucleic acid (DNA) is often referred to as heredity material, and encodes the information used to assemble proteins. (p. 48)
- Cells use ribonucleic acid (RNA) to read the DNA-encoded information and direct protein synthesis. (p. 48)
- Nucleic acids are long polymers of repeating subunits (nucleotides) that consist of a five-carbon sugar, a phosphate, and a nitrogen-containing base. (p. 49)
- DNA exists in a double-stranded helix, while RNA is a single-stranded molecule. (pp. 50–51)

3.4 Lipids make membranes and store energy.

Phospholipids Form Membranes

- A phospholipid is a composite molecule made up of glycerol, fatty acids, and a phosphate group. It contains a polar head and two nonpolar tails, forming the core of all biological membranes. (p. 52)

Fats and Other Kinds of Lipids

- A triglyceride is a fat, consisting of a glycerol molecule and three fatty acids. Because triglyceride molecules lack a polar end, they are not soluble in water. (p. 53)
- Other kinds of lipids include terpenes, steroids, and prostaglandins. (p. 53)

Fats as Energy-Storage Molecules

- Fats have a high concentration of C—H bonds, and are thus efficient energy-storage molecules. (p. 54)

3.5 Carbohydrates store energy and provide building materials.

Sugars

- Carbohydrates are a loosely defined group of molecules containing carbon, hydrogen, and oxygen in 1:2:1 ratio. (p. 55)
- Structural differences between sugar isomers confer functional differences between the molecules. (p. 56)

Transport and Storage Carbohydrates

- Starches are glucose polymers found in plants, while glycogen serves as the animal version of starch. (p. 57)

Structural Carbohydrates

- Structural carbohydrates such as cellulose (in plants) and chitin (in arthropods) are chains of sugars that resist digestion because most organisms lack the necessary enzymes. (p. 58)

Test Your Understanding

Self Test

1. Proteins, nucleic acids, lipids, and carbohydrates all have certain characteristics in common. Which of the following is *not* a common characteristic?
   a. They are organic, which means they are all living substances.
   b. They all contain the element carbon.
   c. They contain simpler units that are linked together, making larger molecules.
   d. They all contain functional groups.

2. A peptide bond forms by
   a. a condensation reaction.
   b. dehydration synthesis.
   c. the formation of a covalent bond.
   d. all of these.

3. Protein motifs are considered a type of
   a. primary structure.
   b. secondary structure.
   c. tertiary structure.
   d. quaternary structure.

4. The substitution of one amino acid for another
   a. will change the primary structure of the polypeptide.
   b. can change the secondary structure of the polypeptide.
   c. can change the tertiary structure of the polypeptide.
   d. All of these are correct.

5. Chaperone proteins function by
   a. providing a protective environment in which proteins can fold properly.
   b. degrading proteins that have folded improperly.
   c. rescuing proteins that folded incorrectly and allowing them to refold into the proper configuration.
   d. providing a template for how the proteins should fold.

6. Which of the following lists the purine nucleotides?
   a. adenine and cytosine
   b. guanine and thymine
   c. cytosine and thymine
   d. adenine and guanine

7. The two strands of a DNA molecule are held together through base-pairing. Which of the following best describes the base-pairing in DNA?
   a. Adenine forms two hydrogen bonds with thymine.
   b. Adenine forms two hydrogen bonds with uracil.
   c. Cytosine forms two hydrogen bonds with guanine.
   d. Cytosine forms two hydrogen bonds with thymine.

8. A characteristic common to all lipids is
   a. that they contain long chains of C—H bonds.
   b. that they are insoluble in water.
   c. that they have a glycerol backbone.
   d. All of these are characteristics of all lipids.

9. Carbohydrates have many functions in the cell. Which of the following is an *incorrect* match of the carbohydrate with its function?
   a. sugar transport in plants—disaccharides
   b. energy storage in plants—starches
   c. energy storage in plants—lactose
   d. sugar transport in humans—glucose

10. Which of the following carbohydrates is associated with plants?
    a. glycogen
    b. amylopectin
    c. chitin
    d. levo-glucose

Test Your Visual Understanding

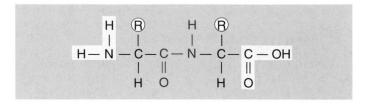

1. The amino acids of a polypeptide affect the shape of the protein. Assume this is a section of a longer polypeptide chain. Predict where each of the following amino acid pairs would be found in the protein—facing toward the outside or folded toward the interior—and explain why:
   a. Both animo acids are valine.
   b. One amino acid is aspartic acid, and the other is serine.
   c. Both amino acids are glycine.
   d. One amino acid is alanine, and the other is isoleucine.

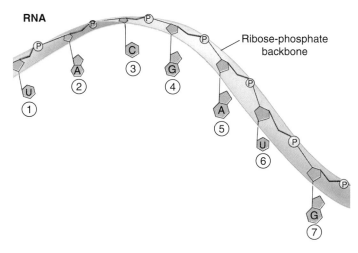

2. Describe the DNA template that produced this molecule of RNA by indicating the DNA bases that are complementary to the numbered RNA bases.

Apply Your Knowledge

1. How many molecules of water are used up in the breakdown of a polypeptide 15 amino acids in length?
2. Why do you suppose the monosaccharide glucose is circulated in the blood of humans rather than a disaccharide, such as sucrose, which is the transport sugar found in plants?

4

The Origin and Early History of Life

Concept Outline

4.1 All living things share key characteristics.

What Is Life? All known organisms share certain general properties, and to a large degree these properties define what we mean by life.

4.2 There is considerable disagreement about the origin of life.

Hypotheses About the Origin of Life. There are both religious and scientific views about the origin of life. This text treats only the latter—the scientifically testable.

Scientists Disagree About Where Life Started. The atmosphere of the early earth was rich in hydrogen, providing a ready supply of energetic electrons with which to build organic molecules.

The Miller-Urey Experiment. Experiments by Miller and Urey and others have attempted to duplicate the conditions of the early earth and produce many of the key molecules of living organisms.

4.3 There are many hypotheses about the origin of cells.

Ideas About the Origin of Cells. The first cells are thought to have arisen spontaneously, but there is little agreement as to the mechanism.

4.4 Cells became progressively more complex as they evolved.

The Earliest Cells. The earliest fossils are of prokaryotes too small to see with the unaided eye.

The First Eukaryotic Cells. Fossils of the first eukaryotic cells do not appear in rocks until 1.5 billion years ago, over 1 billion years after prokaryotes appear. Multicellular life is restricted to the four eukaryotic kingdoms of life.

4.5 Scientists are beginning to take the possibility of extraterrestrial life seriously.

Has Life Evolved Elsewhere? It seems probable that life has evolved on other planets besides our own. The possible presence of life in the warm waters beneath the surface of Europa, a moon of Jupiter, is a source of current speculation.

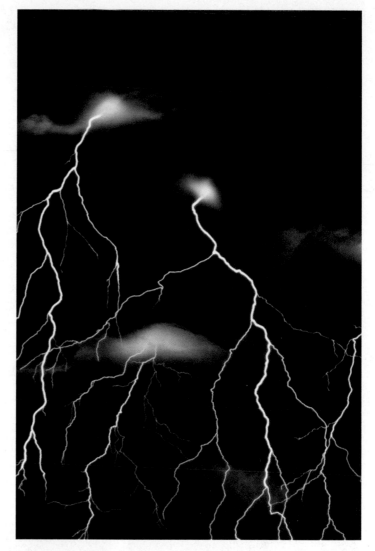

FIGURE 4.1
The origin of life. The fortuitous mix of physical events and chemical elements at the right place and time created the first living cells on earth.

A great many scientists have intriguing ideas that explain how life may have originated on earth, but we know very little for sure. New hypotheses are being proposed constantly, and old ones reevaluated. By the time this text is published, some of the ideas presented here about the origin of life will surely be obsolete. Thus, the contesting ideas are presented in this chapter in an open-ended format, attempting to make clear that there is as yet no single answer to the question of how life originated on earth. Although recent photographs taken by the Hubble Space Telescope have revived controversy about the age of the universe, it seems clear that the earth itself was formed about 4.5 billion years ago. The oldest clear evidence of life—microfossils in ancient rock—are 2.5 billion years old. The origin of life seems to have been sparked by just the right combination of physical events and chemical processes (figure 4.1).

4.1 All living things share key characteristics.

The earth formed as a hot mass of molten rock about 4.5 billion years ago. As the earth cooled, much of the water vapor present in its atmosphere condensed into liquid water that accumulated on the surface in chemically rich oceans. One scenario for the origin of life is that it originated in this dilute, hot, smelly soup of ammonia, formaldehyde, formic acid, cyanide, methane, hydrogen sulfide, and organic hydrocarbons. Whether at the oceans' edge, in hydrothermal deep-sea vents, or elsewhere, the general consensus among researchers is that life arose spontaneously from these early waters. While the way in which this happened remains a puzzle, we cannot escape a certain curiosity about the earliest steps that eventually led to the origin of all living things on earth, including ourselves. How did organisms evolve from the complex molecules that swirled in the early oceans?

What Is Life?

Before we can address the question "What *is* life?", we must first consider what qualifies something as "living." This is not a simple concept because of the loose manner in which the term "alive" is used. Imagine a situation in which two astronauts encounter a large, amorphous blob on the surface of a planet. How would they determine whether it is alive?

Movement.   One of the first things the astronauts might do is observe the blob to see if it moves. Most animals move about (figure 4.2), but movement from one place to another in itself is not diagnostic of life. Most plants and even some animals do not move about, while numerous nonliving objects, such as clouds, do move. The criterion of movement is thus neither *necessary* (possessed by all life) nor *sufficient* (possessed only by life).

Sensitivity.   The astronauts might prod the blob to see if it responds. All living things respond to stimuli (figure 4.3). Plants grow toward light, and animals retreat from fire. Not all stimuli produce responses, however. Imagine kicking a redwood tree or singing to a hibernating bear. This criterion, although superior to the first, is still inadequate to define life.

Death.   The astronauts might attempt to kill the blob. All living things die, while inanimate objects do not. Death is not easily distinguished from disorder, however; a car that breaks down has not died because it was never alive. Death is simply the loss of life, so this is a circular definition at best. Unless one can detect life, death is a meaningless concept, and hence a very inadequate criterion for defining life.

Complexity.   Finally, the astronauts might cut up the blob, to see if it is complexly organized. All living things are complex. Even the simplest bacteria contain

FIGURE 4.2
Movement.   Animals have evolved mechanisms that allow them to move about in their environment. While some animals, like this giraffe, move on land, others move through water or air.

FIGURE 4.3
Sensitivity.   This father lion is responding to a stimulus: He has just been bitten on the rump by his cub. As far as we know, all organisms respond to stimuli, although not always to the same ones or in the same way. Had the cub bitten a tree instead of its father, the response would not have been as dramatic.

a bewildering array of molecules, organized into many complex structures. However, a computer is also complex, but not alive. Complexity is a *necessary* criterion of life, but it is not *sufficient* in itself to identify living things because many complex things are not alive.

To determine whether the blob is alive, the astronauts would have to learn more about it. Probably the best thing they could do would be to examine it more carefully and determine whether it resembles the organisms we are familiar with, and if so, how.

Fundamental Properties of Life

As we discussed in chapter 1, all known organisms share certain general properties. To a large degree, these properties define what we mean by life. The following fundamental properties are shared by all organisms on earth.

Cellular organization.   All organisms consist of one or more *cells*—complex, organized assemblages of molecules enclosed within membranes (figure 4.4).

Sensitivity.   All organisms respond to stimuli—though not always to the same stimuli in the same ways.

Growth.   All living things assimilate energy and use it to maintain order and grow, a process called **metabolism.** Plants, algae, and some bacteria use sunlight to create covalent carbon–carbon bonds from CO_2 and H_2O through photosynthesis. This transfer of the energy in covalent bonds is essential to all life on earth.

Development.   Both unicellular and multicellular organisms undergo systematic, gene-directed changes as they grow and mature.

Reproduction.   All living things reproduce, passing on individuals from one generation to the next.

Regulation.   All organisms have regulatory mechanisms that coordinate internal processes.

Homeostasis.   All living things maintain relatively constant internal conditions, different from their environment.

The Key Role of Heredity

Are these properties adequate to define life? Is a membrane-enclosed entity that grows and reproduces alive? Not necessarily. Soap bubbles and protein microspheres spontaneously form hollow bubbles that enclose a small volume of air or water. These spheres can enclose energy-processing molecules, and they may also grow and subdivide. Despite these features, they are certainly not alive. Therefore, the criteria just listed, although necessary for life, are not sufficient to define life. One ingredient is missing—a mechanism for the preservation of improvement.

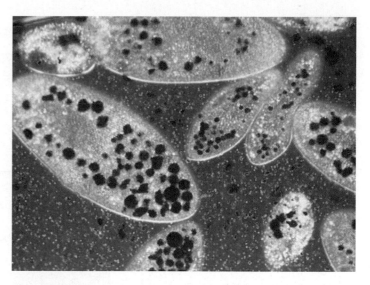

FIGURE 4.4
Cellular organization (150×).   These *Paramecia*, which are complex, single-celled organisms called protists, have just ingested several yeast cells. The yeasts, stained red in this photograph, are enclosed within membrane-bounded sacs called digestive vacuoles. A variety of other organelles are also visible.

Heredity.   All organisms on earth possess a *genetic system* that is based on the replication of a long, complex molecule called DNA. This mechanism allows for adaptation and evolution over time, and is a distinguishing characteristic of living organisms.

To understand the role of heredity in our definition of life, let us return for a moment to protein microspheres. When we examine an individual microsphere, we see it at that precise moment in time but learn nothing of its predecessors. It is likewise impossible to guess what future droplets will be like. The droplets are the passive prisoners of a changing environment, and in this sense they are not alive. The essence of being alive is the ability to encompass change and to reproduce the results of change permanently. Heredity, therefore, provides the basis for the great division between the living and the nonliving. Viruses, while possessing nucleic acids, do not possess a functional genetic system, as they cannot use or reproduce their genes outside of a living cell. A genetic system is the sufficient condition of life. Some changes are preserved because they increase the chances of survival in a hostile world, while others are lost. Not only did life evolve, but evolution is the very essence of life.

All living things on earth are characterized by heredity and possess a handful of other characteristics that serve to define the term life.

4.2 There is considerable disagreement about the origin of life.

Hypotheses About the Origin of Life

The question of how life originated is not easy to answer because it is impossible to go back in time and observe life's beginnings; nor are there any witnesses. Testimony exists in the rocks of the earth, but it is not easily read, and often it is silent on issues crying out for answers. There are, in principle, at least three possibilities: special creation, extraterrestrial origin, and spontaneous origin.

Special Creation

Life-forms may have been put on earth by supernatural or divine forces. The hypothesis that a divine God created life is at the core of most major religions. The oldest hypothesis about life's origins, it is also the most widely accepted. Far more Americans, for example, believe that God created life on earth than believe in the other two hypotheses. Many take an extreme position, accepting the biblical account of life's creation as factually correct. This viewpoint forms the basis for the very unscientific "scientific creationism" viewpoint discussed in chapter 22.

Extraterrestrial Origin

Life may not have originated on earth at all; instead, life may have infected earth from some other planet. This hypothesis, called **panspermia,** proposes that meteors or cosmic dust may have carried significant amounts of complex organic molecules to earth, kicking off the evolution of life. Hundreds of thousands of meteorites and comets are known to have slammed into the early earth, and recent findings suggest that at least some may have carried organic materials. Nor is life on other planets ruled out. For example, the discovery of liquid water under the surface of Jupiter's ice-shrouded moon Europa and suggestions of fossils in rocks from Mars lend some credence to this idea. The hypothesis that an early source of carbonaceous material is extraterrestrial is testable, although it has not yet been proven. Indeed, NASA is plan-

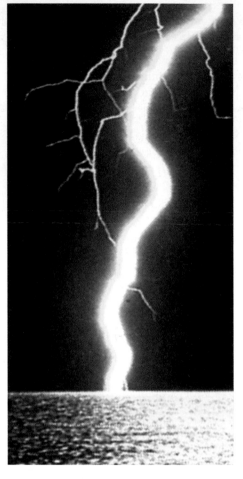

FIGURE 4.5
Lightning. Before life evolved, the simple molecules in the earth's atmosphere combined to form more complex molecules. The energy that drove these chemical reactions may have come from lightning and forms of geothermal energy.

ning to land on Europa, drill through the surface, and send a probe down to see if there is life.

Spontaneous Origin

Most scientists tentatively accept the hypothesis of spontaneous origin—that life evolved from inanimate matter as associations among molecules became more and more complex. In this view, the force leading to life was selection. As changes in molecules increased their stability and caused them to persist longer, these molecules could initiate more and more complex associations, culminating in the evolution of cells.

The Scientific Viewpoint

In this book, we focus on the second and third possibilities, attempting to understand whether natural forces could have led to the origin of life and, if so, how the process might have occurred. This is not to say that the first possibility is definitely incorrect. Any one of the three possibilities might be true. Nor do the second and third possibilities preclude religion—for example, a divine agency might have acted via evolution. However, we are limiting the scope of our inquiry to scientific matters, and only the second and third possibilities permit testable hypotheses to be constructed—that is, explanations that can be tested and potentially disproved.

In our search for understanding, we must look back to early times. Fossils of simple living things have been found in rocks 2.5 billion years old. They tell us that life originated during the early years of the history of our planet. As we attempt to determine how this process took place, we will first focus on how organic molecules may have originated (figure 4.5), and then we will consider how those molecules might have become organized into living cells.

Panspermia and spontaneous origin are the only testable hypotheses of life's origin currently available.

Scientists Disagree About Where Life Started

While most researchers agree that life first appeared as the primitive earth cooled and its rocky crust formed, there is little agreement as to just where this occurred.

Did Life Originate in a Reducing Atmosphere?

The more we learn about the earth's early history, the more likely it seems that earth's first organisms emerged and lived at very high temperatures. Rubble from the forming solar system slammed into the early earth beginning about 4.6 billion years ago, keeping the surface molten hot. As the bombardment slowed down, temperatures dropped. By about 3.8 billion years ago, ocean temperatures are thought to have dropped to a hot 49° to 88°C (120° to 190°F). Between 3.8 and 2.5 billion years ago, life first appeared, promptly after the earth was inhabitable. Thus, as intolerable as early earth's infernal temperatures seem to us today, they gave birth to life.

Very few geochemists agree on the exact composition of the early atmosphere. One popular view is that it contained principally carbon dioxide (CO_2) and nitrogen gas (N_2), along with significant amounts of water vapor (H_2O). It is possible that the early atmosphere also contained hydrogen gas (H_2) and compounds in which hydrogen atoms were bonded to the other light elements (sulfur, nitrogen, and carbon), producing hydrogen sulfide (H_2S), ammonia (NH_3), and methane (CH_4).

We refer to such an atmosphere as a *reducing atmosphere* because of the ample availability of hydrogen atoms and their electrons. In a reducing atmosphere, it would not take as much energy as it would today to form the carbon-rich molecules from which life evolved.

The key to this reducing-atmosphere hypothesis is the assumption that there was very little oxygen around. In an atmosphere with oxygen, amino acids and sugars react spontaneously with the oxygen to form carbon dioxide and water. Therefore, the building blocks of life, the amino acids, would not last long, and the spontaneous formation of complex macromolecules could not occur. Our atmosphere changed once organisms began to carry out photosynthesis, harnessing the energy in sunlight to split water molecules and form complex carbon molecules, giving off gaseous oxygen molecules in the process. The earth's atmosphere is now approximately 21% oxygen.

Critics of the reducing-atmosphere hypothesis point out that no carbonates have been found in rocks dating back to the early earth. This suggests that at that time CO_2 was locked up in the atmosphere, and if that was the case, the prebiotic atmosphere would not have been reducing.

Another problem for the reducing-atmosphere hypothesis is that because a prebiotic reducing atmosphere would have been oxygen free, there would have been no ozone. Without the protective ozone layer, any organic compounds that might have formed would have been broken down quickly by ultraviolet radiation.

Where on Earth Did Life Originate?

Many ideas have been advanced about where on earth life first appeared.

At the ocean's edge. Some scientists have suggested that life arose in bubbles that are constantly forming at the ocean's edge.

Under frozen oceans. One hypothesis proposes that life originated under a frozen ocean, not unlike the one that covers Jupiter's moon Europa today. All evidence suggests, however, that the early earth was quite warm and frozen oceans therefore quite unlikely.

Deep in the earth's crust. Another hypothesis is that life originated deep in the earth's crust. In 1988 Gunter Wachtershauser proposed that life might have formed as a by-product of volcanic activity, with iron and nickel sulfide minerals acting as chemical catalysts to recombine gases spewing from eruptions into the building blocks of life. In later work, he and co-workers were able to use this unusual chemistry to build precursors for amino acids (although they did not actually succeed in making amino acids), and to link amino acids together to form peptides. Critics of this hypothesis point out that the concentrations of chemicals used in their experiments greatly exceed those thought to have been present in nature 3–4 billion years ago.

Within clay. Other researchers have proposed the unusual hypothesis that life is the result of silicate surface chemistry. The surfaces of clays have positive charges to attract organic molecules and exclude water, providing a potential catalytic surface on which life's early chemistry might have occurred. While interesting conceptually, there is little evidence that this sort of process could actually occur.

At deep-sea vents. Becoming more popular is the hypothesis that life originated at deep-sea hydrothermal vents, with the necessary prebiotic molecules being synthesized on metal sulfides in the vents. The positive charge of the sulfides would have acted as a magnet for negatively charged biological molecules. In part, the current popularity of this hypothesis comes from the new science of genomics, which suggests that the ancestors of today's prokaryotes are most closely related to the archaebacteria that live on the deep-sea vents.

No one is sure whether life originated in a reducing atmosphere, under frozen ocean, deep in the earth's crust, within clay, or at deep-sea vents. Perhaps one of these hypotheses will be proven correct or perhaps the correct theory has not yet been proposed.

When life first appeared on earth, the environment was very hot. Most researchers assume that the organic chemicals that were the building blocks of life arose spontaneously at that time. How is a matter of considerable disagreement.

The Miller-Urey Experiment

An early attempt to see what kinds of organic molecules might have been produced on the early earth was carried out in 1953 by Stanley L. Miller and Harold C. Urey. In what has become a classic experiment, they attempted to reproduce the conditions in the earth's primitive oceans under a reducing atmosphere. Even if this assumption proves incorrect—the jury is still out on this—their experiment is critically important because it ushered in the whole new field of prebiotic chemistry.

To carry out their experiment, Miller and Urey (1) assembled a reducing atmosphere rich in hydrogen and excluding gaseous oxygen; (2) placed this atmosphere over liquid water; (3) maintained this mixture at a temperature somewhat below 100°C; and (4) simulated lightning by bombarding it with energy in the form of sparks (figure 4.6).

They found that within a week, 15% of the carbon originally present as methane gas (CH_4) had converted into other simple carbon compounds. Among these compounds were formaldehyde (CH_2O) and hydrogen cyanide (HCN; figure 4.7). These compounds then combined to form simple molecules, such as formic acid ($HCOOH$) and urea (NH_2CONH_2), and more complex molecules containing carbon–carbon bonds, including the amino acids glycine and alanine.

In similar experiments performed later by other scientists, more than 30 different carbon compounds were identified, including the amino acids glycine, alanine, glutamic acid, valine, proline, and aspartic acid. As we saw in chapter 3, amino acids are the basic building blocks of proteins, and proteins are one of the major kinds of molecules of which organisms are composed. Other biologically important molecules were also formed in these experiments. For example, hydrogen cyanide contributed to the production of a complex ring-shaped molecule called adenine—one of the bases found in DNA and RNA. Thus, the key molecules of life could have formed in the atmosphere of the early earth.

The Path of Chemical Evolution

A raging debate among biologists who study the origin of life concerns which organic molecules came first, the nucleic acid RNA or proteins. Scientists are divided into three camps, those that focus on RNA, those that support protein, and those that believe in a combination of the two. All three arguments have their strong points. Like the hypotheses that try to account for where life originated, these competing hypotheses are diverse and speculative.

An RNA World. The "RNA world" group feels that without a hereditary molecule, other molecules could not have formed consistently. This argument earned support when Thomas Cech at the University of Colorado discovered ribozymes, RNA molecules that can behave as enzymes, catalyzing their own assembly. Recent work has

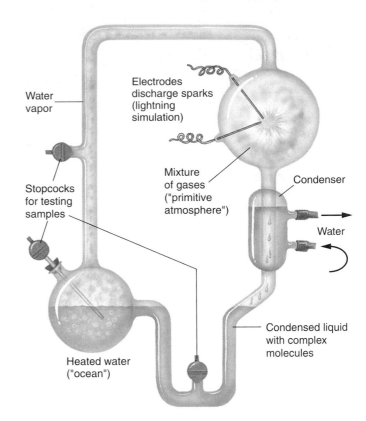

FIGURE 4.6
The Miller-Urey experiment. The apparatus consisted of a closed tube connecting two chambers. The upper chamber contained a mixture of gases thought to resemble the primitive earth's atmosphere. Electrodes discharged sparks through this mixture, simulating lightning. Condensers then cooled the gases, causing water droplets to form, which passed into the second heated chamber, the "ocean." Any complex molecules formed in the atmosphere chamber would be dissolved in these droplets and carried to the ocean chamber, from which samples were withdrawn for analysis.

shown that the RNA contained in ribosomes (discussed in chapter 5) catalyzes the chemical reaction that links amino acids to form proteins. Therefore, the RNA in ribosomes also functions as an enzyme. If RNA has the ability to pass on inherited information and the capacity to act like an enzyme, were proteins really needed?

A Protein World. The "protein-first" group argues that without enzymes (which are proteins), nothing could replicate at all, heritable or not. The protein-first proponents argue that nucleotides, the individual units of nucleic acids such as RNA, are too complex to have formed spontaneously, and certainly too complex to form spontaneously again and again. While there is no doubt that simple proteins are easier to synthesize from abiotic components than nucleotides, both can form in the laboratory under the right conditions. Deciding which came first is a chicken-and-egg paradox. In an effort to shed light on this problem, Julius Rebek and a number of

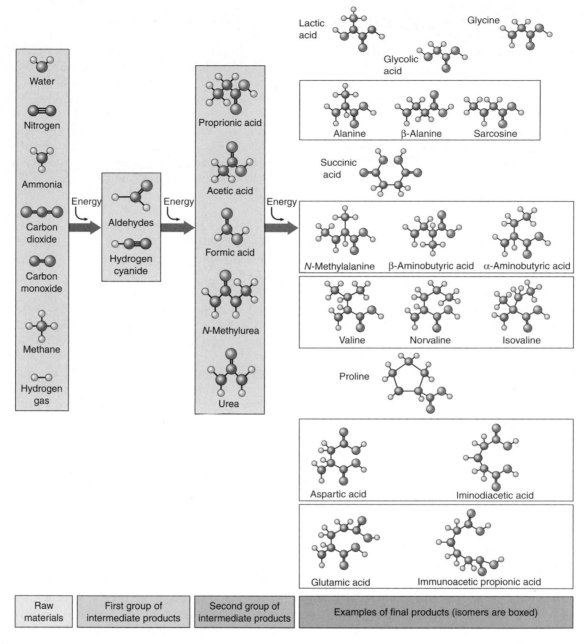

FIGURE 4.7
Results of the Miller-Urey experiment. Seven simple molecules, all gases, were included in the original mixture. Note that oxygen was not among them; instead, the atmosphere is thought to have been rich in hydrogen. At each stage of the experiment, more complex molecules were formed: first aldehydes, then simple acids, then more complex acids. Among the final products, the molecules that are structural isomers of one another are grouped together in boxes. In most cases, only one isomer of a compound is found in living systems today, although many may have been produced in the Miller-Urey experiment.

other chemists have created synthetic nucleotide-like molecules in the laboratory that are able to replicate. Moving even further, Rebek and his colleagues have created synthetic molecules that could replicate and "make mistakes." This simulates mutation, a necessary ingredient for the process of evolution.

A Peptide–Nucleic Acid World. Another important and popular theory about the first organic molecules assumes key roles for both protein and nucleic acids. Because RNA is so complex and unstable, this theory assumes there must have been a pre-RNA world where the protein-nucleic acid (PNA) was the basis for life. PNA is stable and simple enough to have formed spontaneously, and is also a self-replicator.

> **Molecules that are the building blocks of living organisms have been shown to form spontaneously under conditions designed to simulate those of the primitive earth.**

4.3 There are many hypotheses about the origin of cells.

Ideas About the Origin of Cells

The evolution of cells required early organic molecules to assemble into a functional, interdependent unit. Cells, discussed in chapter 5, are essentially little bags of fluid. What the fluid contains depends on the individual cell, but every cell's contents differ from the environment outside the cell. Therefore, an early cell may have floated along in a dilute "primordial soup," but its interior would have had a higher concentration of specific organic molecules.

The Importance of Bubbles

How did these "bags of fluid" evolve from simple organic molecules? As you can imagine, this question is a matter for debate. Scientists favoring an "ocean's edge" scenario for the origin of life have proposed that bubbles may have played a key role in this evolutionary step. A bubble is a hollow, spherical structure. Certain molecules, particularly those with hydrophobic regions, spontaneously form bubbles in water. The structure of the bubble shields the hydrophobic regions of the molecules from contact with water. If you have ever watched the ocean surge upon the shore, you may have noticed the foamy froth created by the agitated water. The edges of the primitive oceans were more than likely very frothy places bombarded by ultraviolet and other ionizing radiation, and exposed to an atmosphere that may have contained methane and other simple organic molecules.

Oparin's Bubble Hypothesis. The first bubble hypothesis is attributed to Alexander Oparin, a Russian chemist with extraordinary insight. In the mid-1930s, Oparin suggested that the present-day atmosphere was incompatible with the creation of life. He proposed that life must have arisen from nonliving matter under a set of very different environmental circumstances some time in the distant history of the earth. His was the theory of **primary abiogenesis** (primary because all living cells are now known to come from previously living cells, except in that first case). At the same time, J. B. S. Haldane, a British geneticist, was independently espousing the same views.

Oparin decided that in order for cells to evolve, they must have had some means of developing chemical complexity, separating their contents from their environment by means of a cell membrane, and concentrating materials within themselves. He termed these early, chemical-concentrating, bubblelike structures **protobionts.**

Oparin's hypothesis was published in English in 1938, and for awhile most scientists ignored them. However, Harold Urey, an astronomer at the University of Chicago, was quite taken with Oparin's ideas. He convinced one of his graduate students, Stanley Miller, to follow Oparin's rationale and see if he could "create" life. The Urey-Miller experiment has proven to be one of the most significant experiments in the history of science. As a result, Oparin's ideas became better known and more widely accepted.

A Host of Bubble Hypotheses. Different versions of "bubble hypotheses" have been championed by numerous scientists since Oparin. The bubbles they propose go by a variety of names; they may be called *microspheres, protocells, protobionts, micelles, liposomes,* or *coacervates,* depending on the composition of the bubbles (lipid or protein) and how they form. In all cases, the bubbles are hollow spheres, and they exhibit a variety of cell-like properties. For example, the lipid bubbles called **coacervates** form an outer boundary with two layers that resembles a biological membrane. They grow by accumulating more subunit lipid molecules from the surrounding medium, and they can form budlike projections and divide by pinching in two, like bacteria. They also can contain amino acids and use them to facilitate various acid-base reactions, including the decomposition of glucose. Although they are not alive, they obviously have many of the characteristics of cells.

A Bubble Scenario. It is not difficult to imagine that a process of chemical evolution involving bubbles or microdrops preceded the origin of life (figure 4.8). The early oceans must have contained untold numbers of these microdrops, billions in a spoonful, each one forming spontaneously, persisting for a while, and then dispersing. Some would, by chance, have contained amino acids with side groups able to catalyze growth-promoting reactions. Those microdrops would have survived longer than ones that lacked those amino acids, because the persistence of both protein microspheres and lipid coacervates is greatly increased when they carry out metabolic reactions such as glucose degradation and when they are actively growing.

Over millions of years, then, the complex bubbles that were better able to incorporate molecules and energy from the lifeless oceans of the early earth would have tended to persist longer than the others. Also favored would have been the microdrops that could use these molecules to expand in size, growing large enough to divide into "daughter"

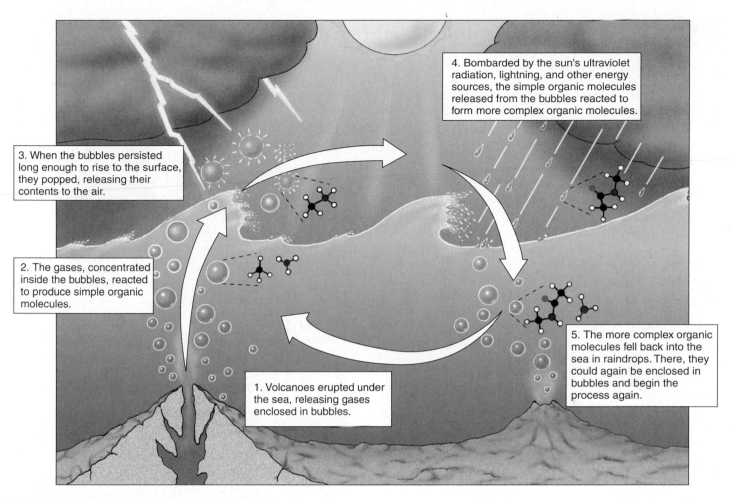

FIGURE 4.8
A current bubble hypothesis. In 1986, geophysicist Louis Lerman proposed that the chemical processes leading to the evolution of life took place within bubbles on the ocean's surface.

The boxes in the figure read:

1. Volcanoes erupted under the sea, releasing gases enclosed in bubbles.

2. The gases, concentrated inside the bubbles, reacted to produce simple organic molecules.

3. When the bubbles persisted long enough to rise to the surface, they popped, releasing their contents to the air.

4. Bombarded by the sun's ultraviolet radiation, lightning, and other energy sources, the simple organic molecules released from the bubbles reacted to form more complex organic molecules.

5. The more complex organic molecules fell back into the sea in raindrops. There, they could again be enclosed in bubbles and begin the process again.

microdrops with features similar to those of their "parent" microdrop. The daughter microdrops have the same favorable combination of characteristics as their parent, and would have grown and divided, too. When a way to facilitate the reliable transfer of new ability from parent to offspring developed, heredity—and life—began.

Current Thinking

Whether the early bubbles that gave rise to cells were lipid or protein remains an unresolved argument. While it is true that lipid microspheres (coacervates) will form readily in water, there appears to be no mechanism for their heritable replication. On the other hand, we *can* imagine a heritable mechanism for protein microspheres. Although protein microspheres do not form readily in water, Sidney Fox and his colleagues at the University of Miami have shown that they can form under dry conditions.

The discovery that RNA can act as an enzyme to assemble new RNA molecules on an RNA template has raised the interesting possibility that neither coacervates nor protein microspheres were the first step in the evolution of life. Perhaps the first components were RNA molecules, and the initial steps on the evolutionary journey led to increasingly complex and stable RNA molecules. Later, stability might have improved further when a lipid (or possibly protein) microsphere surrounded the RNA. At present, those studying this problem have not arrived at a consensus about whether RNA evolved before or after a bubblelike structure that likely preceded cells.

Eventually, DNA took the place of RNA as the replicator in the cell and the storage molecule for genetic information. DNA, because it is a double helix, stores information in a more stable fashion than RNA, which is single-stranded.

Little is known about how the first cells originated. Current hypotheses involve chemical evolution within bubbles, but there is no general agreement about their composition or about how the process occurred.

4.4 Cells became progressively more complex as they evolved.

The Earliest Cells

What do we know about the earliest life-forms? The fossils found in ancient rocks show an obvious progression from simple to complex organisms.

Microfossils

Paleontologists have found uncontestable microfossils in rocks as old as 2.5 billion years (figure 4.9). **Microfossils** are fossilized forms of microscopic life. What do we know about these early microscopic life-forms? Many microfossils are small (1 to 2 micrometers in diameter) and single-celled, lack external appendages, and have little evidence of internal structure. Thus, the organisms of microfossils often resemble present-day bacteria. We call organisms with this simple body plan **prokaryotes,** from the Greek words meaning "before," and "kernel" or "nucleus." The name reflects their lack of a **nucleus,** a spherical organelle characteristic of the more complex cells of **eukaryotes,** to be discussed later in this section.

The record of microfossils older than 2.5 billion years is sparse and controversial. The oldest of them, microscopic squiggles in 3.5-billion-year-old rock from western Australia, proved not to be ancient filamentous cyanobacteria as had been claimed, but lifeless mineral artifacts. The half-dozen other groups of microfossils from this time period are only poorly characterized. Life may have been present earlier than 2.5 billion years ago, but the fossil record provides us with scant evidence.

Judging from the fossil record, eukaryotes did not appear until about 1.5 billion years ago. Therefore, for at least 1 billion years, prokaryotes were the only organisms that existed.

Ancient Prokaryotes: Archaebacteria

Most organisms living today are adapted to the relatively mild conditions of present-day earth. However, if we look in unusual environments, we encounter organisms that are quite remarkable, differing in form and metabolism from other living things. Sheltered from evolutionary alteration in unchanging habitats that resemble earth's early environment, these living relics are the surviving representatives of the first ages of life on earth. In places such as the oxygen-free depths of the Black Sea or the boiling waters of hot springs and deep-sea vents, we can find prokaryotes living at very high temperatures without oxygen.

These unusual prokaryotes are called **archaebacteria,** from the Greek word for "ancient ones." Among the first to be studied in detail have been the **methanogens,** or methane-producing archaebacteria, which are among the most primitive prokaryotes that exist today. These organisms are typically simple in form and able to grow only in an oxygen-free environment; in fact, oxygen poisons them.

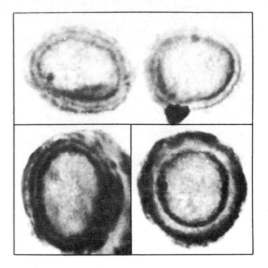

FIGURE 4.9
Cross sections of fossil prokaryotes. These microfossils from the Bitter Springs formation of Australia are of ancient cyanobacteria, far too small to be seen with the unaided eye. In these electron micrographs, the cell walls are clearly evident.

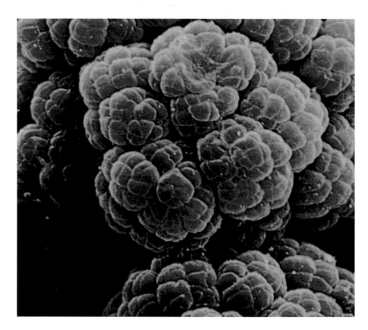

FIGURE 4.10
Methanogens. These archaebacteria (*Methanosarcina barkeri*) are called methanogens because their metabolic activities produce methane gas.

For this reason, they are said to grow "without air," or **anaerobically** (Greek *an*, "without," + *aer*, "air," + *bios*, "life"). Methanogens convert CO_2 and H_2 into methane gas (CH_4) (figure 4.10). Although primitive, they resemble all other prokaryotes in having DNA, a lipid cell membrane, an exterior cell wall, and a metabolism based on an energy-carrying molecule called ATP.

Unusual Cell Structures

When the details of cell wall and membrane structure of the methanogens were examined, they proved to be different from those of all other prokaryotes. Archaebacteria are characterized by a conspicuous lack of a protein cross-linked carbohydrate material called **peptidoglycan** in their cell walls, a key compound in the cell walls of most modern prokaryotes. Archaebacteria also have unusual lipids in their cell membranes that are not found in any other group of organisms as well as major differences in some of their fundamental biochemical processes of metabolism.

Earth's First Organisms?

Other archaebacteria are some of those that live in very salty environments, such as the Dead Sea (**extreme halophiles**—"salt lovers"), or very hot environments, such as hydrothermal volcanic vents under the ocean (**extreme thermophiles**—"heat lovers"). Thermophiles have been found living comfortably in boiling water. Indeed, many kinds of thermophilic archaebacteria thrive at temperatures of 110°C (230°F). Because these thermophiles live at high temperatures similar to those that may have existed in the earth's early oceans, microbiologists speculate that thermophilic archaebacteria may be relics of earth's first organisms.

Just how different are extreme thermophiles from other organisms? A methane-producing type of archaebacteria called *Methanococcus* isolated from deep-sea vents provides a startling picture. These prokaryotes thrive at temperatures of 88°C (185°F) and crushing pressures 245 times greater than at sea level. In 1996 molecular biologists announced that they had succeeded in determining the full nucleotide sequence of *Methanococcus*. This was possible because archaebacterial DNA is relatively small—it has only 1700 genes, coded in a DNA molecule only 1,739,933 nucleotides long (a human cell has 2000 times more). The thermophile nucleotide sequence proved to be astonishingly different from the DNA sequence of any other organism ever studied; fully two-thirds of its genes are unlike any ever known to science before. Clearly, these archaebacteria separated from other life on earth a long time ago. Preliminary comparisons to the gene sequences of other prokaryotes suggest that archaebacteria split from a second larger group of prokaryotes over 2 billion years ago, soon after life began.

Bacteria

The second major group of prokaryotes, the **bacteria,** have very strong cell walls and a simpler gene architecture. Most prokaryotes living today are bacteria. Included in this group are bacteria that have evolved the ability to capture the energy of light and transform it into the en-

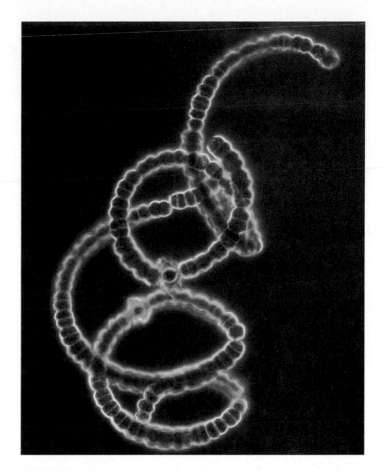

FIGURE 4.11
Living cyanobacteria. Although not multicellular, these bacteria often aggregate into chains such as those seen here.

ergy of chemical bonds within cells. These organisms are *photosynthetic*, as are plants and algae.

One type of photosynthetic bacteria that has been important in the history of life on earth are the cyanobacteria, sometimes called "blue-green algae" (figure 4.11). They have the same kind of chlorophyll pigment that is most abundant in plants and algae, as well as other pigments that are blue or red. Cyanobacteria produce oxygen as a result of their photosynthetic activities, and when they appeared at least 2 billion years ago, they played a decisive role in increasing the concentration of free oxygen in the earth's atmosphere from below 1% to the current level of 21%. As the concentration of oxygen increased, so did the amount of ozone in the upper layers of the atmosphere. The thickening ozone layer afforded protection from most of the ultraviolet radiation from the sun, radiation that is highly destructive to proteins and nucleic acids. Certain cyanobacteria are also responsible for the accumulation of massive limestone deposits.

All prokaryotes now living are members of either archaebacteria or bacteria.

The First Eukaryotic Cells

All fossils more than 1.5 billion years old are generally similar to one another structurally. They are small, simple cells; most measure 0.5 to 2 micrometers in diameter, and none are more than 6 micrometers in diameter. These simple cells eventually evolved into larger, more complex forms—the first eukaryotic cells.

Indirect chemical traces hint that eukaryotes may go as far back as 2.7 billion years, but no fossils as yet support such an early appearance. In rocks about 1.5 billion years old, we begin to see the first microfossils that are noticeably different in appearance from the earlier, simpler forms (figure 4.12). These cells are much larger than those of prokaryotes and have internal membranes and thicker walls. Evidence indicates that cells more than 10 micrometers in diameter rapidly increased in abundance. Some fossilized cells 1.4 billion years old are as much as 60 micrometers in diameter; others, 1.5 billion years old, contain what appear to be small, membrane-bounded structures.

These early fossils mark a major event in the evolution of life: a new kind of organism had appeared. These new cells are called **eukaryotes,** from the Greek words for "true" and "nucleus," because they possess an internal structure called a nucleus. All organisms other than prokaryotes are eukaryotes.

Origin of the Nucleus and ER

Many prokaryotes have infoldings of their outer membranes extending into the cytoplasm and serving as passageways to the surface. The network of internal membranes in eukaryotes called endoplasmic reticulum (ER)

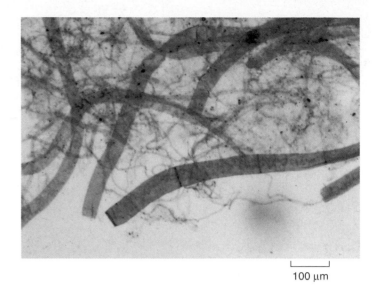

100 μm

FIGURE 4.12
Microfossil of a primitive eukaryote. This multicellular alga is between 900 million and 1 billion years old.

and the nuclear envelope, an extension of the ER network that isolates and protects the nucleus, are thought to have evolved from such infoldings (figure 4.13).

Origin of Mitochondria and Chloroplasts

Bacteria that live within other cells and perform specific functions for their host cells are called *endosymbiotic bacteria.* Their widespread presence in nature led Lynn Margulis in the early 1970s to champion the theory of **endosymbiosis** (Latin, *endo,* "inside," + Greek, *syn,* "together with," + *bios,* "life"), which means living together in close association.

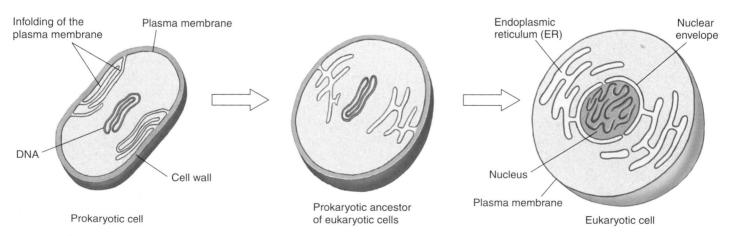

FIGURE 4.13
Origin of the nucleus and endoplasmic reticulum. Many prokaryotes today have infoldings of the plasma membrane (see also figure 27.6). The eukaryotic internal membrane system, called the endoplasmic reticulum (ER), and the nuclear envelope may have evolved from such infoldings of the plasma membrane encasing prokaryotic cells that gave rise to eukaryotic cells.

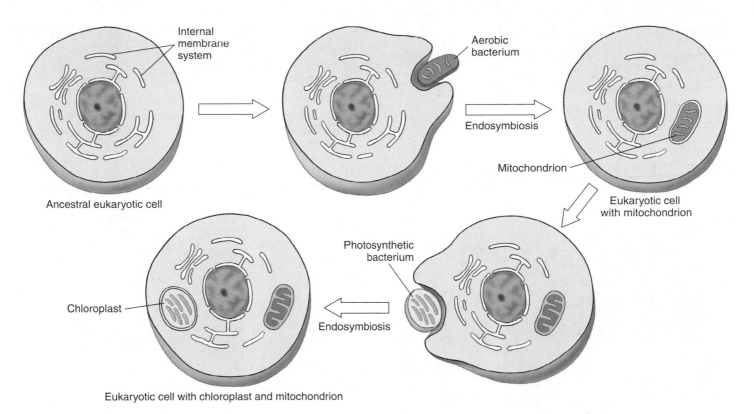

FIGURE 4.14
The theory of endosymbiosis. Scientists propose that ancestral eukaryotic cells, which already had an internal system of membranes, engulfed aerobic bacteria, which then became mitochondria in the eukaryotic cell. Chloroplasts may also have originated this way, with eukaryotic cells engulfing photosynthetic bacteria.

This theory, now widely accepted, suggests that a critical stage in the evolution of eukaryotic cells involved endosymbiotic relationships with prokaryotic organisms. According to this theory, energy-producing bacteria may have come to reside within larger bacteria, eventually evolving into what we now know as mitochondria. Similarly, photosynthetic bacteria may have come to live within other larger bacteria, leading to the evolution of chloroplasts, the photosynthetic organelles of plants and algae (figure 4.14). Bacteria with flagella, long whiplike cellular appendages used for propulsion, may have become symbiotically involved with nonflagellated bacteria to produce larger, motile cells. The fact that we now witness so many symbiotic relationships lends general support to this theory. Even stronger support comes from the observation that present-day organelles such as mitochondria, chloroplasts, and centrioles contain their own DNA, which is remarkably similar to the DNA of bacteria in size and character.

Sexual Reproduction

Most eukaryotic cells also possess the ability to reproduce sexually, something prokaryotes cannot do effectively. **Sexual reproduction** is the process of producing offspring by fertilization, the union of two cells that each have one copy of each chromosome. The great advantage of sexual reproduction is that it allows for frequent genetic recombination, which generates the variation that is the raw material for evolution. Not all eukaryotes reproduce sexually, but most have the capacity to do so. The evolution of meiosis and sexual reproduction (discussed in chapter 12) led to the tremendous explosion of diversity among the eukaryotes.

Multicellularity

Diversity was also promoted by the development of **multicellularity.** Some single eukaryotic cells began living in association with others, in colonies. Eventually, individual members of the colony began to assume different duties, and the colony began to take on the characteristics of a single individual. Multicellularity has arisen many times among the eukaryotes. Practically every organism big enough to be seen with the unaided eye is multicellular, including all animals and plants. The great advantage of multicellularity is that it fosters specialization; some cells devote all of their energies to one task, other cells to another. Few innovations have had as great an impact on the history of life as the specialization made possible by multicellularity.

The Diversity of Life

Confronted with the great diversity of life on earth today, biologists have attempted to categorize similar organisms in order to better understand them, giving rise to the science of taxonomy. In later chapters, we will discuss taxonomy and classification in detail, but for now we can generalize that all living things fall into one of three domains (not to be confused with protein domains described in chapter 3), which include six kingdoms (figure 4.15).

As more is learned about living things, particularly from the newer evidence that DNA studies provide, scientists will continue to reevaluate the relationships among the kingdoms of life (figure 4.16).

For at least the first 1 billion years of life on earth, all organisms were prokaryotes. About 1.5 billion years ago, the first eukaryotes appeared. Biologists place living organisms into six general categories called kingdoms.

Domain Bacteria

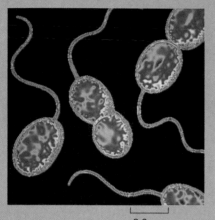

3.8 μm

Kingdom Bacteria. Prokaryotic organisms with a peptidoglycan cell wall, including cyanobacteria, soil bacteria, nitrogen-fixing bacteria, and pathogenic (disease-causing) bacteria.

Domain Archaea

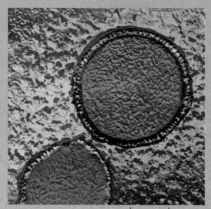

1.7 μm

Kingdom Archaebacteria. Prokaryotes that lack a peptidoglycan cell wall, including the methanogens and extreme halophiles and thermophiles.

Domain Eukarya

Kingdom Protista. Eukaryotic, primarily unicellular (although some algae are multicellular), photosynthetic or heterotrophic organisms, such as amoebas and paramecia.

Kingdom Fungi. Eukaryotic, mostly multicellular (although yeasts are unicellular), heterotrophic, usually nonmotile organisms, with cell walls of chitin, such as mushrooms.

Kingdom Plantae. Eukaryotic, multicellular, nonmotile, usually terrestrial, photosynthetic organisms, such as trees, grasses, and mosses.

Kingdom Animalia. Eukaryotic, multicellular, motile, heterotrophic organisms, such as sponges, spiders, newts, penguins, and humans.

FIGURE 4.15
The three domains of life.

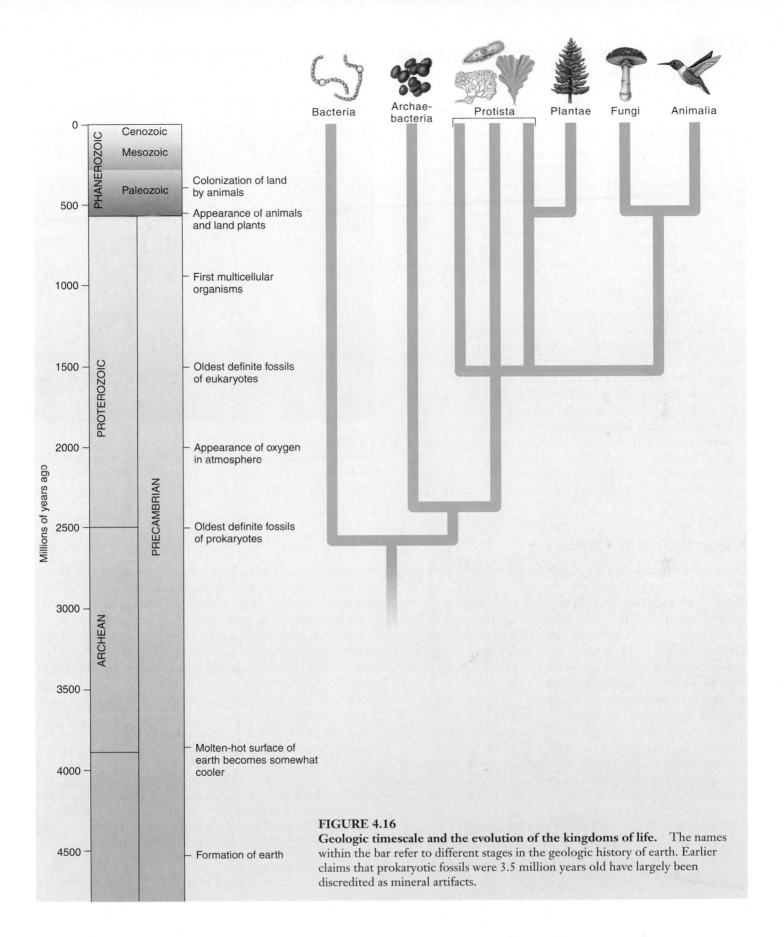

FIGURE 4.16
Geologic timescale and the evolution of the kingdoms of life. The names within the bar refer to different stages in the geologic history of earth. Earlier claims that prokaryotic fossils were 3.5 million years old have largely been discredited as mineral artifacts.

4.5 Scientists are beginning to take the possibility of extraterrestrial life seriously.

Has Life Evolved Elsewhere?

We should not overlook the possibility that life processes might have evolved in different ways on other planets. A functional genetic system, capable of accumulating and replicating changes and thus adapting and evolving, could theoretically evolve from molecules other than carbon, hydrogen, nitrogen, and oxygen in a different environment. Silicon, like carbon, needs four electrons to fill its outer energy level, and ammonia is even more polar than water. Perhaps under radically different temperatures and pressures, these elements might form molecules as diverse and flexible as those carbon has formed on earth.

The universe has 10^{20} (100,000,000,000,000,000,000) stars similar to our sun. We don't know how many of these stars have planets, but it seems increasingly likely that many do. Since 1996, astronomers have been detecting planets orbiting distant stars. At least 10% of stars are thought to have planetary systems. If only 1 in 10,000 of these planets is the right size and at the right distance from its star to duplicate the conditions in which life originated on earth, the "life experiment" will have been repeated 10^{15} times (that is, a million billion times). It does not seem likely that we are alone in the universe. In fact, it seems very possible that life has evolved on other worlds in addition to our own.

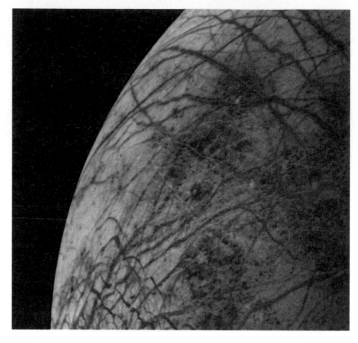

FIGURE 4.17
Is there life elsewhere? Currently the most likely candidate for life elsewhere within the solar system is Europa, one of the many moons of the large planet Jupiter.

Ancient Bacteria on Mars?

A dull gray chunk of rock collected in 1984 in Antarctica ignited an uproar about ancient life on Mars after it was reported to contain evidence of possible life. Analysis of gases trapped within small pockets of the rock indicate it is a meteorite from Mars. It is, in fact, the oldest rock known to science—fully 4.5 billion years old. Back then, when this rock formed on Mars, that cold, arid planet was much warmer, flowed with water, and had a carbon dioxide atmosphere—conditions not too different from those that may have spawned life on earth.

When examined with powerful electron microscopes, carbonate patches within the meteorite exhibit what look like microfossils, some 20 to 100 nanometers in length—100 times smaller than any known bacteria. It is not clear whether they actually are fossils, but the resemblance to bacteria is striking.

Viewed as a whole, the evidence of bacterial life associated with the Mars meteorite is not compelling. Clearly, more painstaking research remains to be done before the discovery can claim a scientific consensus.

Other Planets

There are planets other than ancient Mars with conditions not unlike those on earth. Europa, a large moon of Jupiter, is a promising candidate (figure 4.17). Europa is covered with ice, and photos taken in close orbit in the winter of 1998 reveal seas of liquid water beneath a thin skin of ice. Additional satellite photos taken in 1999 suggest that a few miles under the ice lies a liquid ocean of water larger than earth's, warmed by the push and pull of the gravitational attraction of Jupiter's many large satellite moons. The conditions on Europa now are far less hostile to life than the conditions that existed in the oceans of the primitive earth. In coming decades, satellite missions are scheduled to explore this ocean for life.

Because there are so many stars similar to our sun, life may have evolved many times. Although evidence for life on Mars is not compelling, the seas of Jupiter's moon Europa offer a promising candidate that scientists are eager to investigate.

Concept Review

4.1 **All living things share key characteristics.**

What Is Life?

- All organisms on the earth share several fundamental properties of life. The foremost of these properties is heredity; others include cellular organization, sensitivity, growth, development, reproduction, regulation, and homeostasis. (p. 63)

4.2 **There is considerable disagreement about the origin of life.**

Hypotheses About the Origin of Life

- In principle, the three possible explanations for the origin of life are special creation, extraterrestrial origin, and spontaneous origin. The scientific viewpoint can only address the second and third possibilities because they permit testable hypotheses. (p. 64)

Scientists Disagree About Where Life Started

- Many scientists believe the earth's early atmosphere was a reducing atmosphere of carbon dioxide and nitrogen gas, with very little oxygen. (p. 65)
- Other locations where life may have originated include at the ocean's edge, under frozen seas, deep in the earth's crust, within clay, and at deep-sea vents. (p. 65)

The Miller-Urey Experiment

- Miller and Urey's classic experiment attempted to reproduce a model of earth's possible reducing atmosphere and produce complex organic molecules. (p. 66)
- Subsequent chemical evolution on the earth is argued among many groups, including the RNA-first group, the protein-first group, and a peptide-nucleic acid group. (pp. 66–67)

4.3 **There are many hypotheses about the origin of cells.**

Ideas About the Origin of Cells

- Many different versions of bubble hypotheses have been championed by numerous scientists. (p. 68)
- Current hypotheses encompass chemical evolution within bubbles, but whether early bubbles were lipid or protein remains unsolved. (p. 69)

4.4 **Cells became progressively more complex as they evolved.**

The Earliest Cells

- Documented microfossils have been found that are as old as 2.5 billion years. (p. 70)
- Currently, prokaryotes are either archaebacteria or bacteria. (pp. 70–71)

The First Eukaryotic Cells

- Eukaryotes seemed to appear about 1.5 billion years ago. They seem to be much larger than prokaryotes and contain internal membranes and thicker cell walls. (p. 72)
- Endosymbiotic theory suggests that energy-producing bacteria may have come to reside within larger bacteria, eventually evolving into mitochondria. (pp. 72–73)
- Sexual reproduction allows for frequent genetic recombination and serves as the raw material for evolution. (p. 73)
- Multicellularity has arisen many times among the eukaryotes, and has promoted the development of diversity. (p. 73)

4.5 **Scientists are beginning to take the possibility of extraterrestrial life seriously.**

Has Life Evolved Elsewhere?

- Because at least 10% of stars are thought to have planetary systems, it is theoretically possible that life may have evolved many times. (p. 76)
- Currently, the Jupiter moon Europa is the most promising candidate to hold life because satellite images have revealed liquid oceans under the ice. (p. 76)

Self Test

1. Which of the following is necessary and sufficient for life?
   a. complexity
   b. heredity
   c. growth
   d. all of these
2. Which of the following most accurately describes the origins of life on earth?
   a. Special creation: Life-forms were put on earth by supernatural or divine forces.
   b. Extraterrestrial origin: The earth may have been infected by life from another planet.
   c. Spontaneous origin: Life evolved from inanimate matter to increasing levels of complexity.
   d. None of these are accurate.
3. Which of the following proposals assumes that the earth had a reducing atmosphere low in oxygen?
   a. life evolved deep in the earth's crust
   b. life evolved under frozen oceans
   c. life evolved at the ocean's edge
   d. life evolved at deep-sea vents
4. Scientists have created synthetic nucleotide-like molecules in the laboratory that are able to replicate. This seems to support which hypothesis of chemical evolution?
   a. an RNA-first hypothesis
   b. a protein-first hypothesis
   c. a peptide–nucleic acid hypothesis
   d. All of the hypotheses are supported by these results.
5. A key link needed between bubbles and cells is
   a. that they needed a hereditary molecule.
   b. that they needed to grow and bud off new bubbles.
   c. that they needed to incorporate enzymes.
   d. that they needed to form spontaneously.
6. Prokaryotes were the only life-form on the earth
   a. for about 1.5 billion years.
   b. for about 1 billion years.
   c. for about 2.5 billion years.
   d. for more than 2.5 billion years.
7. The first organisms may have been a type of archaebacteria similar to present-day organisms called
   a. extreme halophiles.
   b. prokaryotes.
   c. extreme thermophiles.
   d. eubacteria.
8. The infolding of outer membranes seen in prokaryotes is believed to have given rise to which of the following?
   a. mitochondrion
   b. chloroplast
   c. endoplasmic reticulum
   d. all of these structures
9. All organisms fall into one of _____ domains, which include _____ kingdoms.
   a. six/three
   b. six/six
   c. three/three
   d. three/six
10. Which of the following statements is false?
    a. Life may have evolved on Mars.
    b. In order for life to evolve anywhere, carbon is required.
    c. A large ocean exists under the icy surface of Europa.
    d. It is possible that life evolved on some other planet.

Test Your Visual Understanding

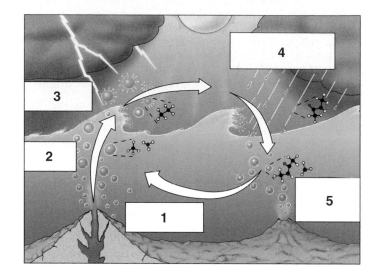

1. Match the following descriptions with the appropriate numbered step in the figure that explains a bubble hypothesis.
   a. Bombarded by the sun's ultraviolet radiation, lightning, and other energy sources, the simple organic molecules released from the bubbles reacted to form more complex organic molecules.
   b. Volcanoes erupted under the sea, releasing gases enclosed in bubbles.
   c. The more complex organic molecules fell back into the sea in raindrops. There, they could be enclosed in bubbles and begin the process again.
   d. When the bubbles persisted long enough to rise to the surface, they popped, releasing their contents to the air.
   e. The gases, concentrated inside the bubbles, reacted to produce simple organic molecules.

Apply Your Knowledge

1. In Fred Hoyle's science fiction novel *The Black Cloud*, the earth is approached by a large interstellar cloud of gas that orients itself around the sun. Scientists soon discover that the cloud is feeding on the sun, absorbing the sun's energy through the excitation of electrons in the outer energy levels of cloud molecules, in a process similar to the photosynthesis that occurs on earth. Different portions of the cloud are isolated from each other by associations of ions created by this excitation. Electron currents pass between these portions, much as they do on the surface of the human brain, endowing the cloud with self-awareness, memory, and the ability to think. Using electrical discharges, the cloud is able to communicate with humans and describe its history. It tells scientists that it originated as a small extrusion from an ancestral cloud, and that since then, it has grown by absorbing molecules and energy from stars such as our sun, on which it has been feeding. Soon the cloud moves off in search of other stars. Is it alive?

5

Cell Structure

Concept Outline

5.1 All organisms are composed of cells.

Characteristics of Cells. A cell is a membrane-bounded unit that contains DNA and cytoplasm. The greater relative surface area of small cells allows more rapid communication between the cell interior and the environment.

Visualizing Cells. Microscopes enable us to see cells and their components, which are very small.

5.2 Eukaryotic cells are more structurally complex than prokaryotic cells.

Prokaryotes: Simple Cells with Complex Physiology. Prokaryotic cells are small and lack membrane-bounded organelles.

Eukaryotes: Cells with Complex Interiors. Eukaryotic cells are compartmentalized by membranes.

5.3 Take a tour of a eukaryotic cell.

The Nucleus: Information Center for the Cell. The nucleus of a eukaryotic cell isolates the cell's DNA.

The Endomembrane System. An extensive system of membranes called the endoplasmic reticulum (ER) subdivides the cell interior, and the Golgi apparatus is a system of membrane channels that collects, modifies, packages, and distributes molecules within the cell.

Ribosomes: Sites of Protein Synthesis. An RNA-protein complex directs the production of proteins.

Organelles That Contain DNA. Some organelles contain their own DNA.

The Cytoskeleton: Interior Framework of the Cell. A network of protein fibers supports the shape of the cell and anchors organelles, as well as facilitating the movement of molecules within the cell. Eukaryotic cell movement utilizes cytoskeletal elements and sometimes flagella and cilia.

5.4 Not all eukaryotic cells are the same.

Vacuoles and Cell Walls. Plant cells have a large central vacuole and strong, multilayered cell walls. Cell walls and vacuoles are also found in some protists and fungi.

The Extracellular Matrix. A complex web of carbohydrate and protein encases animal cells and helps coordinate the activities of tissues.

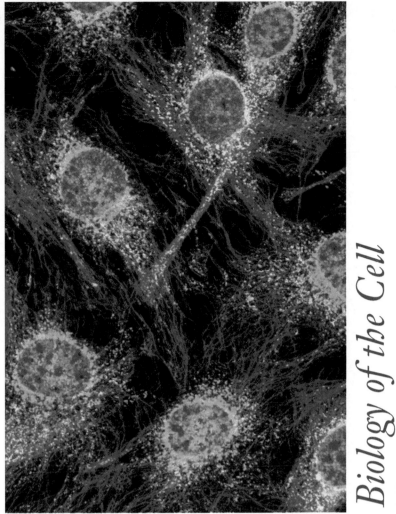

FIGURE 5.1
Fluorescence image of a cell. This amazing image of mammalian fibroblast cells that are found in connective tissue was obtained using a technique called fluorescence microscopy. The use of a variety of fluorescent stains makes it possible to distinguish different components of the cell through a microscope. In this image, the nuclei and DNA are stained blue, and cytoskeletal proteins are stained red.

Part II Biology of the Cell

All organisms are composed of cells. The gossamer wing of a butterfly is a thin sheet of cells, and so is the glistening outer layer of your eyes. The hamburger or tomato you eat is composed of cells, and its contents soon become part of your cells. Some organisms consist of a single cell too small to see with the unaided eye, while others, such as humans, are composed of many cells (figure 5.1). Cells are so much a part of life as we know it that we cannot imagine an organism that is not cellular in nature. In this chapter, we take a close look at the internal structure of cells. In chapters 6–11, we will focus on cells in action—how they communicate with their environment, grow, and reproduce.

5.1 All organisms are composed of cells.

Characteristics of Cells

What is a typical cell like, and what would we find inside it? The general plan of cellular organization varies in the cells of different organisms, but despite these modifications, all cells resemble each other in certain fundamental ways. Before we begin our detailed examination of cell structure, let's first summarize three major features all cells have in common: a nucleoid or nucleus, cytoplasm, and a plasma membrane.

The Central Portion of the Cell Contains the Genetic Material

Every cell contains DNA, the hereditary molecule. In **prokaryotes,** most of the genetic material lies in a single circular molecule of DNA. It typically resides near the center of the cell in an area called the **nucleoid,** but this area is not segregated from the rest of the cell's interior by membranes. By contrast, the DNA of **eukaryotes** is contained in the **nucleus** (figure 5.2), which is surrounded by a double membrane structure called the **nuclear envelope.** In both types of organisms, the DNA contains the genes that code for the proteins synthesized by the cell.

The Cytoplasm Comprises the Rest of the Cell's Interior

A semifluid matrix called the **cytoplasm** fills the interior of the cell, exclusive of the nucleus (nucleoid in prokaryotes) lying within it. The cytoplasm contains the chemical wealth of the cell: the sugars, amino acids, and proteins the cell uses to carry out its everyday activities. In eukaryotic cells, the cytoplasm also contains specialized membrane-bounded compartments called **organelles.**

The Plasma Membrane Surrounds the Cell

The **plasma membrane** encloses a cell and separates its contents from its surroundings. The plasma membrane is a phospholipid bilayer about 5 to 10 nanometers (5 to 10 billionths of a meter) thick, with proteins embedded in it. Viewed in cross section with the electron microscope, such membranes appear as two dark lines separated by a lighter area. This distinctive appearance arises from the tail-to-tail packing of the phospholipid molecules that make up the membrane (figure 5.2). The proteins of a membrane may have large hydrophobic domains, which associate with and become embedded in the phospholipid bilayer.

The proteins of the plasma membrane are in large part responsible for a cell's ability to interact with its environment. *Transport proteins* help molecules and ions move

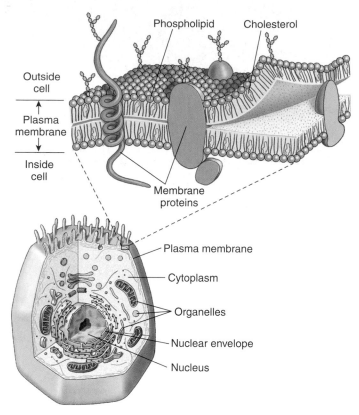

FIGURE 5.2
A generalized eukaryotic cell.   In eukaryotes, a nucleus houses the DNA and is located near the center of the cell. Cytoplasm fills the rest of the cell's interior, which also contains specialized organelles. The plasma membrane encases the cell and consists of a phospholipid bilayer embedded with proteins.

across the plasma membrane, either from the environment to the interior of the cell or vice versa. *Receptor proteins* induce changes within the cell when they come in contact with specific molecules in the environment, such as hormones, or with molecules on the surface of neighboring cells. These molecules can function as *markers* that identify the cell as a particular type. This interaction between cell surface molecules is especially important in multicellular organisms, whose cells must be able to recognize each other as they form tissues.

We'll examine the structure and function of cell membranes more thoroughly in chapter 6.

The Cell Theory

A general characteristic of cells is their microscopic size. While there are a few exceptions—the marine alga *Acetabularia* can be up to 5 centimeters long—a typical eukaryotic cell is 10 to 100 micrometers (10 to

FIGURE 5.3
Surface area-to-volume ratio. As a cell gets larger, its volume increases at a faster rate than its surface area. If the cell radius increases by 10 times, the surface area increases by 100 times, but the volume increases by 1000 times. A cell's surface area must be large enough to meet the needs of its volume.

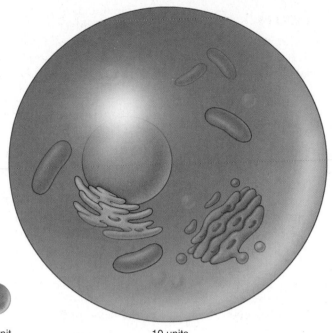

Cell radius (r)	1 unit	10 units
Surface area ($4\pi r^2$)	12.57 units^2	1257 units^2
Volume ($\frac{4}{3}\pi r^3$)	4.189 units^3	4189 units^3

100 millionths of a meter) in diameter; most prokaryotic cells are only 1 to 10 micrometers in diameter.

Because cells are so small, no one observed them until microscopes were invented in the mid-seventeenth century. Robert Hooke first described cells in 1665, when he used a microscope he had built to examine a thin slice of cork, a nonliving tissue found in the bark of certain trees. Hooke observed a honeycomb of tiny, empty (because the cells were dead) compartments. He called the compartments in the cork *cellulae* (Latin, "small rooms"), and the term has come down to us as *cells.* The first living cells were observed a few years later by the Dutch naturalist Antonie van Leeuwenhoek, who called the tiny organisms that he observed "animalcules," meaning little animals. For another century and a half, however, biologists failed to recognize the importance of cells. In 1838, botanist Matthias Schleiden stated that all plants "are aggregates of fully individualized, independent, separate beings, namely the cells themselves." In 1839, Theodor Schwann reported that all animal tissues also consist of individual cells.

These were the earliest statements of the **cell theory,** which in its modern form, includes the following three principles:

1. All organisms are composed of one or more cells, and the life processes of metabolism and heredity occur within these cells.
2. Cells are the smallest living things, the basic units of organization of all organisms.
3. Cells arise only by division of a previously existing cell. Although life likely evolved spontaneously in the environment of the early earth, biologists have

concluded that no additional cells are originating spontaneously at present. Rather, life on earth represents a continuous line of descent from those early cells.

Why Aren't Cells Larger?

Most cells are not large for practical reasons. Proteins and organelles are being synthesized, and materials are continually entering and leaving the cell. All of these processes involve the diffusion of substances at some point, and the larger a cell is, the longer it takes for substances to diffuse from the plasma membrane to the center of the cell. For this reason, an organism made up of many relatively small cells has an advantage over one composed of fewer, larger cells.

The advantage of small cell size is readily visualized in terms of the **surface area-to-volume ratio.** As a cell's size increases, its volume increases much more rapidly than its surface area. For a spherical cell, the increase in surface area is equal to the square of the increase in radius, while the increase in volume is equal to the cube of the increase in radius. Thus, if two cells differ by a factor of 10 in radius, the larger cell will have 10^2, or 100 times, the surface area, but 10^3, or 1000 times, the volume of the smaller cell (figure 5.3). A cell's surface provides its only opportunity for interaction with the environment, since all substances enter and exit a cell via the plasma membrane. This membrane plays a key role in controlling cell function, and because small cells have more surface area per unit of volume than large ones, the control is more effective when cells are relatively small.

Although most cells are small, some cells are quite large and have apparently overcome the surface area-to-volume problem by one or more adaptive mechanisms. For example, some cells, such as muscle cells, have more than one nucleus, allowing genetic information to be spread around a large cell. Some other large cells, such as your own neurons, are long and skinny so that any given point in the cytoplasm is close to the plasma membrane, and thus diffusion between the inside and outside of the cell can still be rapid.

A cell is a membrane-bounded unit that contains the DNA hereditary machinery and cytoplasm. All organisms are cells or aggregates of cells. Multicellular organisms usually consist of many small cells rather than a few large ones because small cells allow more rapid movement of molecules between the center of the cell and the environment.

Visualizing Cells

How many cells are big enough to see with the unaided eye? Other than egg cells, not many (figure 5.4). Most are less than 50 micrometers in diameter, far smaller than the period at the end of this sentence.

The Resolution Problem

How do we study cells if they are too small to see? The key is to understand why we can't see them. The reason we can't see such small objects is the limited resolution of the human eye. **Resolution** is defined as the minimum distance two points can be apart and still be distinguished as two separated points. When two objects are closer together than about 100 micrometers, the light reflected from each strikes the same "detector" cell at the rear of the eye. Only when the objects are farther than 100 micrometers apart will the light from each strike different cells, allowing your eye to resolve them as two objects rather than one.

Microscopes

One way to increase resolution is to increase magnification so that small objects appear larger. Robert Hooke and Antonie van Leeuwenhoek used glass lenses to magnify small cells and cause them to appear larger than the 100-micrometer limit imposed by the human eye. The glass lens adds additional focusing power. Because the glass lens makes the object appear closer, the image on the back of the eye is bigger than it would be without the lens.

Modern light microscopes use two magnifying lenses (and a variety of correcting lenses) to achieve very high magnification and clarity (table 5.1). The first lens focuses the image of the object on the second lens, which magnifies it again and focuses it on the back of the eye. Microscopes that magnify in stages using several lenses are called **compound microscopes.** They can resolve structures that are separated by more than 200 nanometers (nm).

Increasing Resolution

Light microscopes, even compound ones, are not powerful enough to resolve many of the structures within cells. For example, a membrane is only 5 nanometers thick. Why not just add another magnifying stage to the microscope and so increase its resolving power? Because when two objects are closer than a few hundred nanometers, the light beams reflecting from the two images start to overlap. The only way two light beams can get closer together and still be resolved is if their wavelengths are shorter.

One way to avoid overlap is by using a beam of electrons rather than a beam of light. Electrons have a much shorter wavelength, and a microscope employing electron beams has 1000 times the resolving power of a light microscope. **Transmission electron microscopes,** so called because the electrons used to visualize the specimens are transmitted through the material, are capable of resolving objects only 0.2 nanometer apart—just twice the diameter of a hydrogen atom!

A second kind of electron microscope, the **scanning electron microscope,** beams the electrons onto the surface of the specimen. The electrons reflected back from the surface of the specimen, together with other electrons that the specimen itself emits as a result of the bombardment, are amplified and transmitted to a screen, where the image can be viewed and photographed.

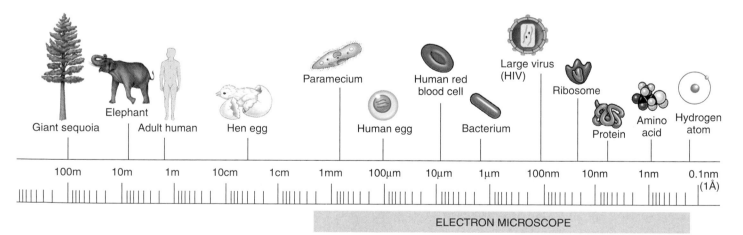

FIGURE 5.4
The size of cells and their contents. Most cells are microscopic in size, although vertebrate eggs are typically large enough to be seen with the unaided eye. Prokaryotic cells are generally 1 to 10 micrometers (μm) across.

Table 5.1 Types of Microscopes

Light Microscopes

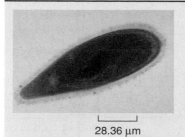

28.36 μm

Bright-field microscope: Light is simply transmitted through a specimen in culture, giving little contrast. Staining specimens improves contrast but requires that cells be fixed (not alive), which can cause distortion or alteration of components.

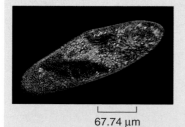

67.74 μm

Dark-field microscope: Light is directed at an angle toward the specimen; a condenser lens transmits only light reflected off the specimen. The field is dark, and the specimen is light against this dark background.

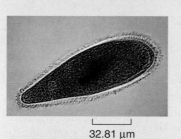

32.81 μm

Phase-contrast microscope: Components of the microscope bring light waves out of phase, which produces differences in contrast and brightness when the light waves recombine.

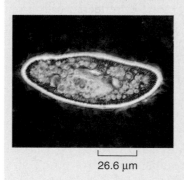

26.6 μm

Differential-interference-contrast microscope: Out-of-phase light waves to produce differences in contrast are combined with two beams of light traveling close together, which create even more contrast, especially at the edges of structures.

Fluorescence microscope: A set of filters transmits only light that is emitted by fluorescently stained molecules or tissues.

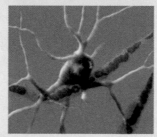

Confocal microscope: Light from a laser is focused to a point and scanned across the specimen in two directions. Clear images of one plane of the specimen are produced, while other planes of the specimen are excluded and do not blur the image. Fluorescent dyes and false coloring enhances the image.

Electron Microscopes

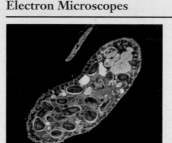

2.56 μm

Transmission electron microscope: A beam of electrons is passed through the specimen. Electrons that pass through are used to form an image. Areas of the specimen that scatter electrons appear dark. False coloring enhances the image.

6.76 μm

Scanning electron microscope: An electron beam is scanned across the surface of the specimen, and electrons are knocked off the surface. Thus, the surface topography of the specimen determines the contrast and the content of the image. False coloring enhances the image.

Scanning electron microscopy yields striking three-dimensional images and has improved our understanding of many biological and physical phenomena (see table 5.1).

Visualizing Cell Structure by Staining Specific Molecules

A powerful tool for analyzing cell structure has been the use of stains that bind to specific molecular targets. Staining has been used in the analysis of tissue samples, or histology, for many years and has been improved dramatically with the use of antibiotics that bind to very specific molecular structures. This process, called *immunocytochemistry*, uses antibodies generated in animals such as rabbits or mice. When these animals are injected with specific pro-

teins, they produce antibodies that specifically bind to the injected protein, which can be purified from their blood. These purified antibodies can then be chemically bonded to enzymes, stains, or fluorescent molecules that glow when exposed to specific wavelengths of light. When cells are washed in a solution containing the antibodies, they bind to cellular structures that contain the target molecule and can be seen with light microscopy. This approach has been used extensively in the analysis of cell structure and function.

Most cells and their components are so small they can only be viewed using microscopes. Different kinds of microscopes and different staining procedures are used, depending on what kind of image is desired.

5.2 Eukaryotic cells are more structurally complex than prokaryotic cells.

Prokaryotes: Simple Cells with Complex Physiology

Prokaryotes are the simplest organisms. Prokaryotic cells are small, consisting of cytoplasm surrounded by a plasma membrane and encased within a rigid cell wall, with no distinct interior compartments (figure 5.5). A prokaryotic cell is like a one-room cabin in which eating, sleeping, and watching TV all occur. Prokaryotes arc very important in the economy of living organisms. They harvest light in photosynthesis, break down dead organisms and recycle their components, cause disease, and are involved in many important industrial processes. As we mentioned in chapter 4, there are two main groups of prokaryotes: archaebacteria and bacteria; prokaryotes are the subject of chapter 27.

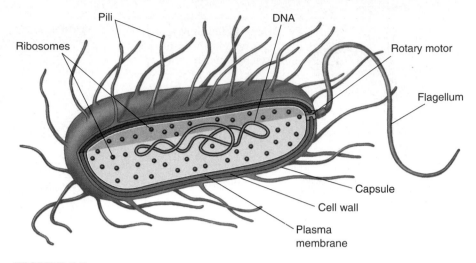

FIGURE 5.5
Structure of a prokaryotic cell.   Generalized cell organization of a prokaryote. Some prokaryotes have hairlike growths on the outside of the cell called pili.

Strong Cell Walls

Most prokaryotic cells are encased by a strong **cell wall.** The cell wall of bacteria is composed of *peptidoglycan*, which consists of a carbohydrate matrix (polymers of sugars) that is cross-linked by short polypeptide units. The cell walls of archaebacteria have varying chemical compositions, but they all lack peptidoglycan. Cell walls protect the cell, maintain its shape, and prevent excessive uptake of water. Plants, fungi, and most protists also have cell walls of a different chemical structure that lack peptidoglycan, which we will discuss in later chapters. With a few exceptions such as the TB- and leprosy-causing bacteria, all bacteria may be classified into two types based on differences in their cell walls detected by the Gram staining procedure. The name refers to the Danish microbiologist Hans Christian Gram, who developed the procedure to detect the presence of certain disease-causing bacteria. **Gram-positive** bacteria have a thick, single-layered cell wall that retains a violet dye from the Gram stain procedure, causing the stained cells to appear purple under a microscope. More complex cell walls have evolved in other groups of bacteria. In them, the wall is multilayered and does not retain the purple dye after Gram staining; such bacteria exhibit the background red dye and are characterized as **gram-negative.**

The susceptibility of bacteria to antibiotics often depends on the structure of their cell walls. The drugs penicillin and vancomycin, for example, interfere with the ability of bacteria to cross-link the peptide units that hold the carbohydrate chains of the wall together. Like removing all the nails from a wooden house, this destroys the integrity of the matrix, which can no longer prevent water from rushing in, swelling the cell to bursting.

Long chains of sugars called polysaccharides cover the cell walls of many bacteria. They enable a bacterium to adhere to teeth, skin, food—or to practically any surface that will support their growth. Many disease-causing bacteria secrete a jellylike protective capsule of polysaccharide around the cell.

Rotating Flagella

Some prokaryotes use a **flagellum** (plural, *flagella*) to move. Flagella are long, threadlike structures protruding from the surface of a cell that are used in locomotion. Prokaryotic flagella are protein fibers that extend out from a bacterial cell. There may be one or more per cell, or none, depending on the species. Bacteria can swim at speeds of up to 20 cell diameters per second by rotating their flagella like screws (figure 5.6). The rotary motor uses the energy stored in a gradient of protons across the plasma membrane to power the movement of the flagellum. Interestingly, the same principle, in which a proton gradient powers the rotation of a molecule, is used in eukaryotic mitochondria and chloroplasts by an enzyme that synthesizes ATP.

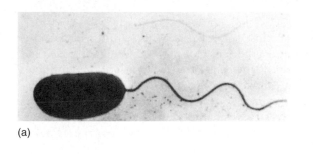

(a)

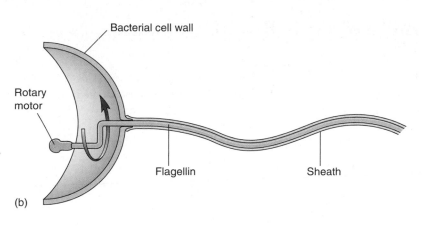

Bacterial cell wall

Rotary motor

Flagellin

Sheath

(b)

(c)

FIGURE 5.6

Prokaryotes swim by rotating their flagella. (*a*) The photograph shows *Vibrio cholerae*, the microbe that causes the serious disease cholera. The unsheathed core visible at the top of the photograph is composed of a single crystal of the protein flagellin. (*b*) In intact flagella, the flagellin core is surrounded by a flexible sheath. (*c*) Imagine that you were standing inside the *Vibrio* cell, turning the flagellum like a turbine. You would create a spiral wave that travels down the flagellum, just as if you were turning a wire within a flexible tube. This movement propels the cell forward when it swims.

Simple Interior Organization

If you were to look at an electron micrograph of a prokaryotic cell, you would be struck by the cell's simple organization. There are few, if any, internal compartments, and while prokaryotic cells contain complex structures like ribosomes, most have no membrane-bounded organelles, the kinds so characteristic of eukaryotic cells. Nor do prokaryotes have a true nucleus. The entire cytoplasm of a prokaryotic cell is one unit with no internal support structure. Consequently, the strength of the cell comes primarily from its rigid cell wall (see figure 5.5).

The plasma membrane of a prokaryotic cell carries out some of the functions organelles perform in eukaryotic cells. For example, before a prokaryotic cell divides, the prokaryotic hereditary material, a simple circle of DNA, replicates. The two DNA molecules that result from the replication attach to the plasma membrane at different points, ensuring that each daughter cell will contain one of the identical units of DNA. Moreover, some photosynthetic bacteria, such as cyanobacteria and *Prochloron* (figure 5.7), have an extensively folded plasma membrane, with the folds extending into the cell's interior. These membrane folds contain the bacterial pigments connected with photosynthesis.

Because a prokaryotic cell contains no membrane-bounded organelles, the DNA, enzymes, and other cytoplasmic constituents have access to all parts of the cell. Reactions are not compartmentalized as they are in eukaryotic cells, and the whole prokaryote operates as a single unit.

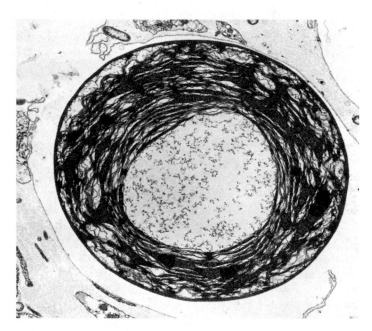

FIGURE 5.7

Electron micrograph of a photosynthetic bacterial cell. Extensive folded photosynthetic membranes are visible in this *Prochloron* cell (14,500×).

Prokaryotes are small cells that lack complex interior organization. They are encased by an exterior wall composed of carbohydrates cross-linked by short polypeptides, and some are propelled by rotating flagella.

Eukaryotes: Cells with Complex Interiors

Eukaryotic cells (figures 5.8 and 5.9) are far more complex than prokaryotic cells. The hallmark of the eukaryotic cell is compartmentalization, achieved by an extensive **endomembrane system** that weaves through the cell interior and by numerous **organelles,** membrane-bounded structures that close off compartments within which multiple biochemical processes can proceed simultaneously and independently. Plant cells often have a large, membrane-bounded sac called a **central vacuole,** which stores proteins, pigments, and waste materials. Both plant and animal cells contain **vesicles,** smaller sacs that store and transport a variety of materials. Inside the nucleus, the DNA is

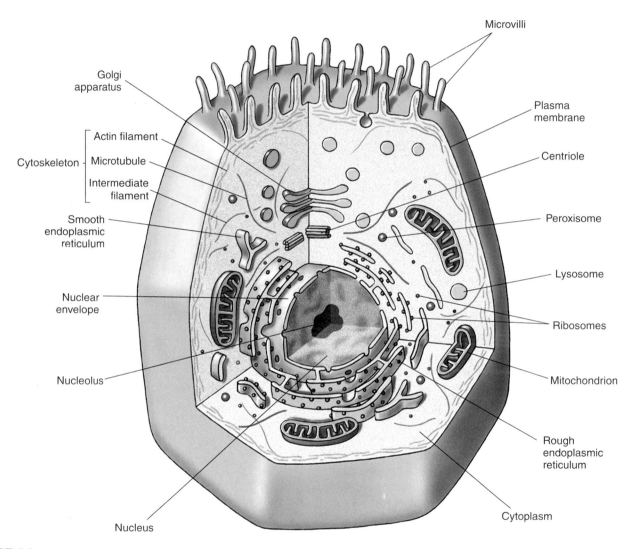

FIGURE 5.8

Structure of an animal cell.   In this generalized diagram of an animal cell, the plasma membrane encases the cell, which contains the cytoskeleton and various cell organelles and interior structures suspended in a semifluid matrix called the cytoplasm. Some kinds of animal cells possess fingerlike projections called microvilli. Other types of eukaryotic cells—for example, many protist cells—may possess flagella, which aid in movement, or cilia, which can have many different functions.

wound tightly around proteins and packaged into compact units called **chromosomes.** All eukaryotic cells are supported by an internal protein scaffold, the **cytoskeleton.** While the cells of animals and some protists lack cell walls, the cells of fungi, plants, and many protists have strong **cell walls** composed of cellulose or chitin fibers embedded in a matrix of other polysaccharides and proteins. In the remainder of this chapter, we will examine the internal components of eukaryotic cells in more detail.

Eukaryotic cells contain membrane-bounded organelles that carry out specialized functions.

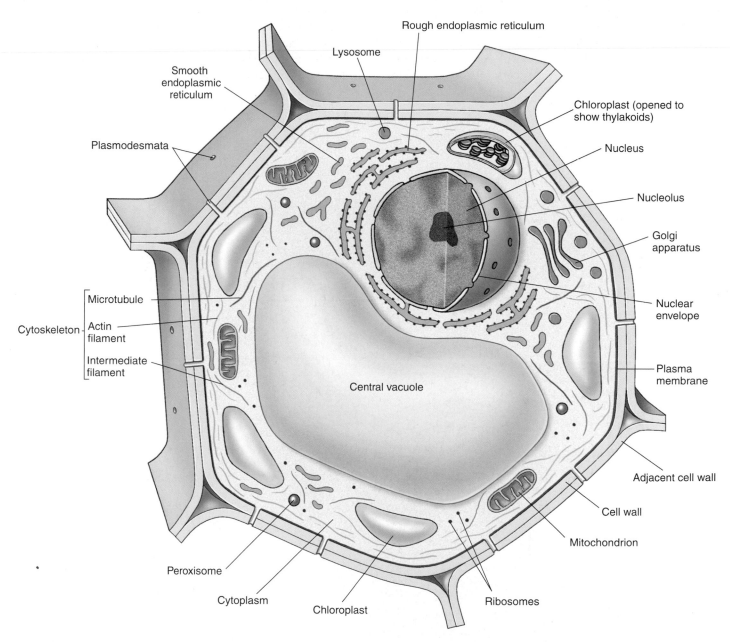

FIGURE 5.9
Structure of a plant cell.   Most mature plant cells contain large central vacuoles, which occupy a major portion of the internal volume of the cell, and organelles called chloroplasts, within which photosynthesis takes place. The cells of plants, fungi, and some protists have cell walls, although the composition of the walls varies among the groups. Plant cells have cytoplasmic connections through openings in the cell wall called plasmodesmata. Flagella occur in sperm of a few plant species, but are otherwise absent in plant and fungal cells. Centrioles are also absent in plant and fungal cells.

5.3 Take a tour of a eukaryotic cell.

The Nucleus: Information Center for the Cell

The largest and most easily seen organelle within a eukaryotic cell is the **nucleus** (Latin, "kernel" or "nut"), first described by the English botanist Robert Brown in 1831. Nuclei are roughly spherical in shape, and in animal cells, they are typically located in the central region of the cell (figure 5.10a). In some cells, a network of fine cytoplasmic filaments seems to cradle the nucleus in this position. The nucleus is the repository of the genetic information that directs the activities of a living eukaryotic cell. Most eukaryotic cells possess a single nucleus, although the cells of fungi and some other groups may have several to many nuclei. Mammalian erythrocytes (red blood cells) lose their nuclei when they mature. Many nuclei exhibit a dark-staining zone called the **nucleolus,** which is a region where intensive synthesis of ribosomal RNA is taking place.

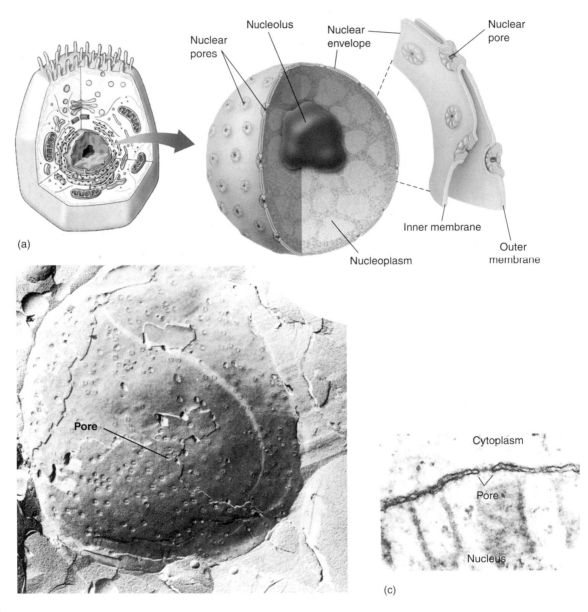

(a)

(b)

(c)

FIGURE 5.10

The nucleus. (*a*) The nucleus is composed of a double membrane, called a nuclear envelope, enclosing a fluid-filled interior containing the chromosomes. In cross section, the individual nuclear pores are seen to extend through the two membrane layers of the envelope; the dark material within the pore is protein, which acts to control access through the pore. (*b*) A freeze-fracture electron micrograph (see figure 6.6) of a cell nucleus, showing nuclear pores (9500×). (*c*) A transmission electron micrograph of the nuclear membrane, showing a nuclear pore.

The Nuclear Envelope: Getting In and Out

The surface of the nucleus is bounded by *two* phospholipid bilayer membranes, which together make up the **nuclear envelope** (see figure 5.10). The outer membrane of the nuclear envelope is continuous with the cytoplasm's interior membrane system, called the endoplasmic reticulum. Scattered over the surface of the nuclear envelope, like craters on the moon, are shallow depressions called **nuclear pores** (figure 5.10*b,c*). These pores form 50 to 80 nanometers apart at locations where the two membrane layers of the nuclear envelope pinch together. Rather than being empty, nuclear pores are filled with proteins that act as gatekeepers, permitting certain molecules to pass into and out of the nucleus. Passage is restricted primarily to two kinds of molecules: (1) proteins moving into the nucleus to be incorporated into nuclear structures or to catalyze nuclear activities; and (2) RNA and protein-RNA complexes formed in the nucleus and exported to the cytoplasm.

The Chromosomes: Packaging the DNA

In both prokaryotes and eukaryotes, DNA contains the hereditary information specifying cell structure and function. However, unlike the circular DNA of prokaryotes, the DNA of eukaryotes is divided into several linear **chromosomes.** Except when a cell is dividing, its chromosomes are extended into threadlike strands, called **chromatin,** of DNA complexed with protein. This open arrangement allows proteins to attach to specific nucleotide sequences along the DNA and regulate gene expression. Without this access, DNA could not direct the day-to-day activities of the cell. During all cellular activities, the chromosomes are associated with packaging proteins called **histones.** When the cell is functioning normally, the DNA is loosely coiled around clusters of histones called **nucleosomes.** These structures resemble beads on a string (figure 5.11). Configuring chromosomes into a more extended form permits enzymes to make RNA copies of DNA. These RNA copies of the information in the DNA direct the synthesis of proteins. When a cell prepares to divide, the DNA coils up around the histones into a much more highly condensed form until the DNA is in a compact mass. Under a light microscope, these fully condensed chromosomes are readily seen in dividing cells as densely staining rods (figure 5.12). After cell division, eukaryotic chromosomes uncoil and can no longer be individually distinguished with a light microscope.

> The nucleus of a eukaryotic cell contains the cell's genetic information and isolates it from the rest of the cell. A distinctive feature of eukaryotes is the organization of their DNA into complex chromosomes.

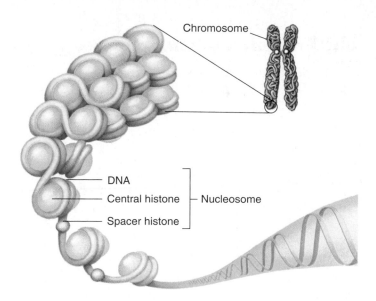

FIGURE 5.11
Nucleosomes. Each nucleosome is a region in which the DNA is wrapped tightly around a cluster of histone proteins.

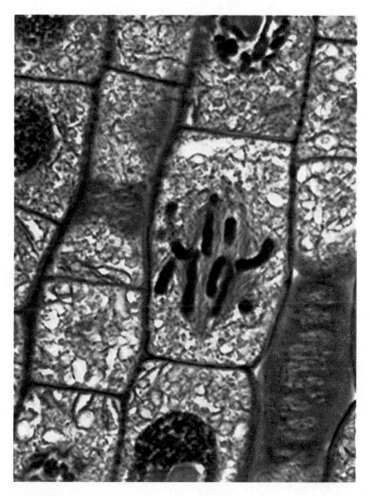

FIGURE 5.12
Eukaryotic chromosomes. These condensed chromosomes within an onion root tip are visible under the light microscope (500×).

The Endomembrane System

The Endoplasmic Reticulum: Compartmentalizing the Cell

The interior of a eukaryotic cell is packed with membranes (table 5.2) so thin that they are invisible under the low resolving power of light microscopes. This **endomembrane system** fills the cell, dividing it into compartments, channeling the passage of molecules through the interior of the cell, and providing surfaces for the synthesis of lipids and some proteins. The presence of these membranes in eukaryotic cells constitutes one of the most fundamental distinctions between eukaryotes and prokaryotes.

The largest of the internal membranes is called the **endoplasmic reticulum (ER)**. *Endoplasmic* means "within the cytoplasm," and *reticulum* is Latin for "a little net." Like the plasma membrane, the ER is composed of a lipid bilayer embedded with proteins. It weaves in sheets through the interior of the cell, creating a series of channels between its folds (figure 5.13). Of the many compartments in eukaryotic cells, the two largest are the inner region of the ER, called the **cisternal space,** and the region exterior to it, the **cytosol.**

Table 5.2		Eukaryotic Cell Structures and Their Functions	
Structure		**Description**	**Function**
Cell wall		Outer layer of cellulose or chitin; or absent	Protection; support
Cytoskeleton		Network of protein filaments	Structural support; cell movement
Flagella (cilia)		Cellular extensions with 9 + 2 arrangement of pairs of microtubules	Motility or moving fluids over surfaces
Plasma membrane		Lipid bilayer with embedded proteins	Regulates what passes into and out of cell; cell-to-cell recognition
Endoplasmic reticulum (ER)		Network of internal membranes	Forms compartments and vesicles; participates in protein and lipid synthesis
Nucleus		Structure (usually spherical) that contains chromosomes and is surrounded by double membrane	Control center of cell; directs protein synthesis and cell reproduction
Golgi apparatus		Stacks of flattened vesicles	Packages proteins for export from cell; forms secretory vesicles
Lysosomes		Vesicles derived from Golgi apparatus that contain hydrolytic digestive enzymes	Digest worn-out organelles and cell debris; play role in cell death
Microbodies		Vesicles that are formed from incorporation of lipids and proteins and that contain oxidative and other enzymes	Isolate particular chemical activities from rest of cell
Mitochondria		Bacteria-like elements with double membrane	"Power plants" of the cell; sites of oxidative metabolism
Chloroplasts		Bacteria-like elements with membranes containing chlorophyll, a photosynthetic pigment	Sites of photosynthesis
Chromosomes		Long threads of DNA that form a complex with protein	Contain hereditary information
Nucleolus		Site of genes for rRNA synthesis	Assembles ribosomes
Ribosomes		Small, complex assemblies of protein and RNA, often bound to endoplasmic reticulum	Sites of protein synthesis

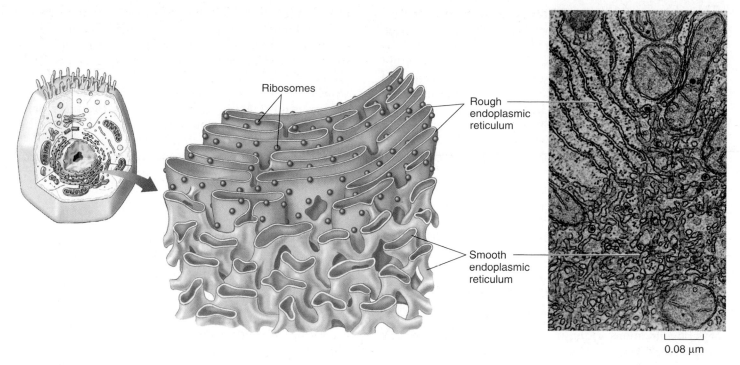

Ribosomes

Rough
endoplasmic
reticulum

Smooth
endoplasmic
reticulum

0.08 µm

FIGURE 5.13
The endoplasmic reticulum.   Rough ER has ribosomes associated with only one side of the endomembrane; the other side is the boundary of a separate compartment within the cell into which the ribosomes extrude newly made proteins destined for secretion. Smooth endoplasmic reticulum has few to no bound ribosomes.

Rough ER: Manufacturing Proteins for Export.   The ER surface regions that are devoted to protein synthesis are heavily studded with **ribosomes,** large molecular aggregates of protein and ribonucleic acid (RNA) that translate RNA copies of genes into protein. (We will examine ribosomes in detail later in this chapter.) Through the electron microscope, these ribosome-rich regions of the ER appear pebbly, like the surface of sandpaper, and they are therefore called **rough ER** (see figure 5.13).

The proteins synthesized on the surface of the rough ER are destined to be exported from the cell, sent to lysosomes or vacuoles, or embedded in the plasma membrane. Proteins to be exported contain special amino acid sequences called **signal sequences.** As a new protein is made by a free ribosome (one not attached to a membrane), the signal sequence of the growing polypeptide attaches to a recognition factor that carries the ribosome and its partially completed protein to a "docking site" on the surface of the ER. As the protein is assembled, it passes through the ER membrane into the interior ER compartment, the cisternal space, from which it is transported by vesicles to the Golgi apparatus (figure 5.14). The protein then travels within vesicles to the inner surface of the plasma membrane, where it is released to the outside of the cell.

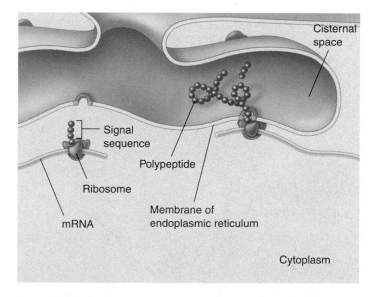

Cisternal
space

Signal
sequence

Polypeptide

Ribosome

Membrane of
endoplasmic reticulum

mRNA

Cytoplasm

FIGURE 5.14
Signal sequences direct proteins to their destinations in the cell.   In this example, a sequence of hydrophobic amino acids (the signal sequence) on a secretory protein attaches them (and the ribosomes making them) to the membrane of the ER. As the protein is synthesized, it passes into the cisternal space of the ER. The signal sequence is clipped off after the leading edge of the protein enters the cisternal space.

Smooth ER: Organizing Internal Activities. Regions of the ER with relatively few bound ribosomes are referred to as **smooth ER.** The membranes of the smooth ER contain many embedded enzymes. Enzymes anchored within the ER, for example, catalyze the synthesis of a variety of carbohydrates and lipids. In cells that carry out extensive lipid synthesis, such as those in the testes, intestine, and brain, smooth ER is particularly abundant. In the liver, the enzymes of the smooth ER are involved in the detoxification of drugs, including amphetamines, morphine, codeine, and phenobarbital.

Some vesicles form at the plasma membrane by budding inward, a process called *endocytosis* (see chapter 6). Some then move into the cytoplasm and fuse with the smooth endoplasmic reticulum. Others form secondary lysosomes or other interior vesicles.

The Golgi Apparatus: Delivery System of the Cell

At various locations within the endomembrane system, flattened stacks of membranes called **Golgi bodies** occur, often interconnected with one another. These structures are named for Camillo Golgi, the nineteenth-century Italian physician who first called attention to them. The number of Golgi bodies a cell contains ranges from 1 or a few in protists to 20 or more in animal cells and several hundred in plant cells. They are especially abundant in glandular cells, which manufacture and secrete substances. Collectively, the Golgi bodies are referred to as the **Golgi apparatus** (figure 5.15).

The Golgi apparatus functions in the collection, packaging, and distribution of molecules synthesized at one place in the cell and utilized at another location in the cell. A Golgi body has a front and a back, with distinctly different membrane compositions at the opposite ends. The front, or receiving end, is called the *cis* face, and is usually located near ER. Materials move to the *cis* face in transport vesicles that bud off the ER. These vesicles fuse with the *cis* face, emptying their contents into the interior, or lumen, of the Golgi apparatus. These ER-synthesized molecules then pass through the channels of the Golgi apparatus until they reach the back, or discharging end, called the *trans* face, where they are discharged in secretory vesicles (figure 5.16).

Proteins and lipids manufactured on the rough and smooth ER membranes are transported into the Golgi apparatus and modified as they pass through it. The most common alteration is the addition or modification of short sugar chains, forming a *glycoprotein* when sugars are complexed to a protein and a *glycolipid* when sugars are bound to a lipid. In many instances, enzymes in the Golgi apparatus modify existing glycoproteins and glycolipids made in the ER by cleaving a sugar from their sugar chain, or modifying one or more of the sugars.

The newly formed or altered glycoproteins and glycolipids collect at the ends of the Golgi bodies in flattened, stacked membrane folds called **cisternae** (Latin, "collecting vessels"). Periodically, the membranes of the cisternae push together, pinching off small, membrane-bounded secretory vesicles containing the glycoprotein and glycolipid molecules. These vesicles then move to other locations in the cell, distributing the newly synthesized molecules to their appropriate destinations.

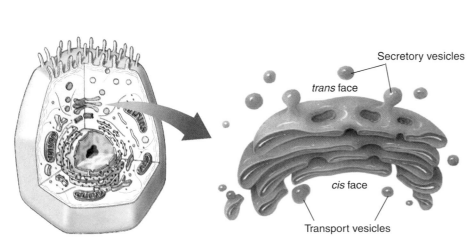

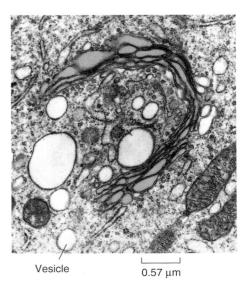

Secretory vesicles

trans face

cis face

Transport vesicles

Vesicle

0.57 μm

FIGURE 5.15
The Golgi apparatus. The Golgi apparatus is a smooth, concave, membranous structure located near the middle of the cell. It receives material for processing in transport vesicles on the *cis* face and sends the material packaged in secretory vesicles off the *trans* face. The substance in a vesicle could be for export out of the cell or for distribution to another region within the same cell.

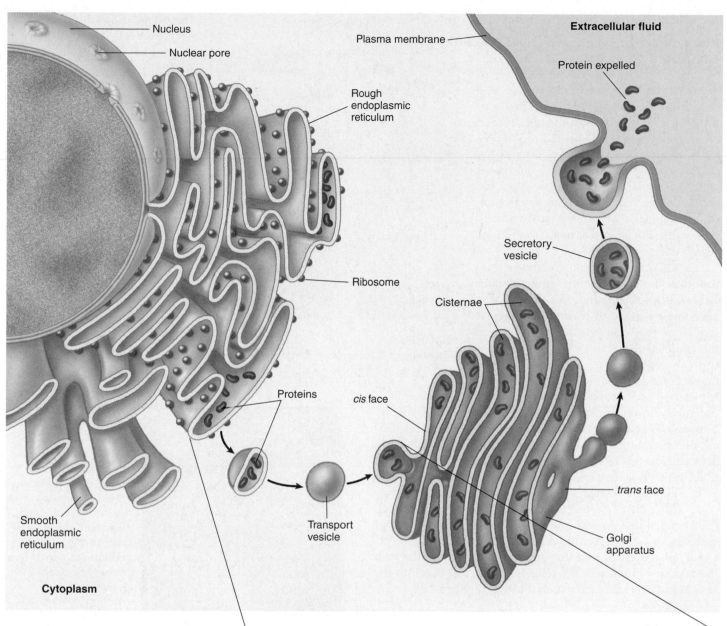

Nucleus

Nuclear pore

Plasma membrane

Extracellular fluid

Protein expelled

Rough endoplasmic reticulum

Ribosome

Secretory vesicle

Cisternae

Proteins

cis face

Transport vesicle

trans face

Golgi apparatus

Smooth endoplasmic reticulum

Cytoplasm

FIGURE 5.16

How proteins are transported within the cell. Proteins are manufactured at the ribosome and then released into the internal compartments of the rough ER. If the newly synthesized proteins are to be used at a distant location in or outside of the cell, they are transported within vesicles that bud off the rough ER and travel to the *cis* face, or receiving end, of the Golgi apparatus. There they are modified and packaged into secretory vesicles. The secretory vesicles then migrate from the *trans* face, or discharging end, of the Golgi apparatus to other locations in the cell, or they fuse with the plasma membrane, releasing their contents to the external cellular environment.

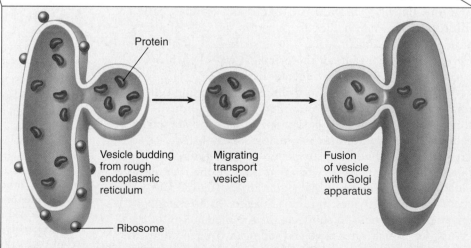

Protein

Vesicle budding from rough endoplasmic reticulum

Migrating transport vesicle

Fusion of vesicle with Golgi apparatus

Ribosome

Vesicles: Enzyme Storehouses

Lysosomes: Intracellular Digestion Centers. Lysosomes, membrane-bounded digestive vesicles, are also components of the endomembrane system that arise from the Golgi apparatus. They contain high levels of degrading enzymes, which catalyze the rapid breakdown of proteins, nucleic acids, lipids, and carbohydrates. Throughout the lives of eukaryotic cells, lysosomal enzymes break down old organelles, recycling their component molecules and making room for newly formed organelles. For example, mitochondria are replaced in some tissues every 10 days.

The digestive enzymes in lysosomes function best in an acidic environment. Lysosomes actively engaged in digestion keep their battery of hydrolytic enzymes (enzymes that catalyze the hydrolysis of molecules) fully active by pumping protons into their interiors and thereby maintaining a low internal pH. Lysosomes that are not functioning actively do not maintain an acidic internal pH and are called *primary lysosomes.* When a primary lysosome fuses with a food vesicle or other organelle, proton pumps in the lysosome are activatcd, its pH falls, and its arsenal of hydrolytic enzymes is activated; it is then called a *secondary lysosome.*

In addition to breaking down organelles and other structures within cells, lysosomes eliminate other cells that the cell has engulfed in a process called *phagocytosis,* a specific type of endocytosis (see chapter 6). When a white blood cell, for example, phagocytizes a passing pathogen, lysosomes fuse with the resulting "food vesicle," releasing their enzymes into the vesicle and degrading the material within (figure 5.17).

Microbodies. Eukaryotic cells contain a variety of enzyme-bearing, membrane-enclosed vesicles called **microbodies.** Microbodies are found in the cells of plants, animals, fungi, and protists. The distribution of enzymes into microbodies is one of the principal ways by which eukaryotic cells organize their metabolism.

While lysosomes bud from the endomembrane system, microbodies grow by incorporating lipids and protein, and then dividing. Plant cells have a special type of microbody called a **glyoxysome,** which contains enzymes that convert fats into carbohydrates. Another type of microbody, a **peroxisome,** contains enzymes that catalyze the removal of electrons and associated hydrogen atoms (figure 5.18). If these oxidative enzymes were not isolated within microbodies, they would tend to short-circuit the metabolism of the cytoplasm, which often involves adding hydrogen atoms to oxygen. The name *peroxisome* refers to the hydrogen perox-

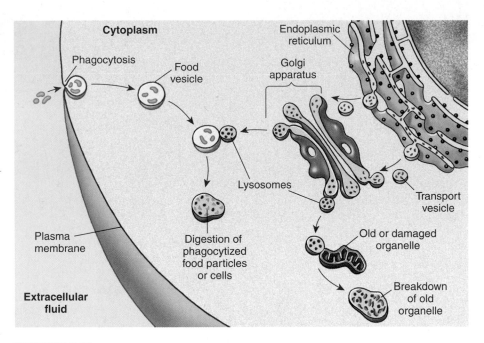

FIGURE 5.17
Lysosomes. Lysosomes contain hydrolytic enzymes that digest particles or cells taken into the cell by phagocytosis and break down old organelles.

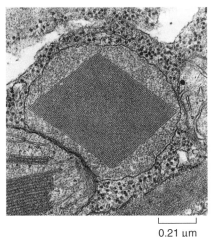

FIGURE 5.18
A peroxisome. Peroxisomes are spherical organelles that may contain a large, diamond-shaped crystal composed of protein. Peroxisomes contain digestive and detoxifying enzymes that produce hydrogen peroxide as a by-product.

0.21 μm

ide produced as a by-product of the activities of the oxidative enzymes in the microbody. Hydrogen peroxide is dangerous to cells because of its violent chemical reactivity. However, peroxisomes also contain the enzyme catalase, which breaks down hydrogen peroxide into harmless water and oxygen.

> The endoplasmic reticulum (ER) is an extensive system of folded membranes that spatially organize the cell's biosynthetic activities. The Golgi apparatus collects, packages, modifies, and distributes molecules. Lysosomes and peroxisomes are vesicles that contain digestive and detoxifying enzymes. The isolation of these enzymes in vesicles protects the rest of the cell from the very reactive chemistry occurring inside.

Ribosomes: Sites of Protein Synthesis

Although the DNA in a cell's nucleus encodes the amino acid sequence of each protein in the cell, the proteins are not assembled there. A simple experiment demonstrates this: If a brief pulse of radioactive amino acid is administered to a cell, the radioactivity shows up associated with newly made protein, not in the nucleus, but in the cytoplasm. When investigators first carried out these experiments, they found that protein synthesis is associated with large RNA-protein complexes outside the nucleus called **ribosomes.**

Ribosomes are made up of several molecules of a special form of RNA called ribosomal RNA, or rRNA, bound within a complex of several dozen different proteins. Ribosomes are among the most complex molecular assemblies found in cells. Each ribosome is composed of two subunits (figure 5.19). The subunits join to form a functional ribosome only when they attach to another kind of RNA, called messenger RNA (mRNA), which is the transcribed copy of the coding information on the DNA in the nucleus. Ribosomes use the information in mRNA to direct the synthesis of a protein.

Proteins that function in the cytoplasm are made by free ribosomes that are not associated with organelle membranes, while proteins bound within membranes or destined for export from the cell are assembled by ribosomes bound to rough ER.

The Nucleolus Manufactures Ribosomal Subunits

When cells are synthesizing a large number of proteins, they must first make a large number of ribosomes. To facilitate this, many hundreds of copies of the genes encoding the ribosomal RNAs are clustered together on the chromosome. By transcribing RNA molecules from this cluster, the cell rapidly generates large numbers of the molecules needed to produce ribosomes.

The clusters of ribosomal RNA genes, the RNAs they produce, and the ribosomal proteins all combine within the nucleus during ribosome production. These areas where ribosomes are being assembled are easily visible within the nucleus as one or more dark-staining regions, called **nucleoli** (singular, *nucleolus;* figure 5.20). Nucleoli can be seen under the light microscope even when the chromosomes are extended, unlike the rest of the chromosomes, which are visible only when condensed.

Ribosomes are the sites of protein synthesis in the cytoplasm.

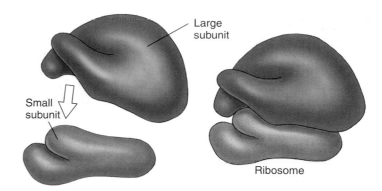

FIGURE 5.19
A ribosome. Ribosomes consist of a large and a small subunit composed of rRNA and protein. The individual subunits are synthesized in the nucleolus and then move through the nuclear pores to the cytoplasm, where they assemble to translate mRNA. Ribosomes serve as sites of protein synthesis.

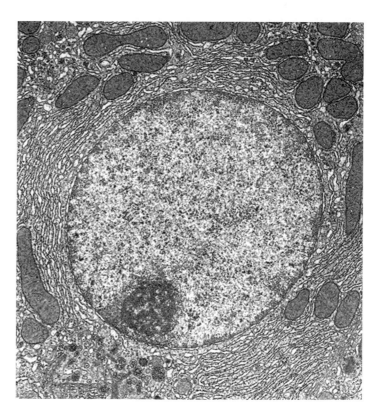

FIGURE 5.20
The nucleolus. This is the interior of a rat liver cell, magnified about 6000 times. A single large nucleus occupies the center of the micrograph. The electron-dense area in the lower left of the nucleus is the nucleolus, the area where the major components of the ribosomes are produced.

6.2 Proteins embedded within cell membranes determine their character.

The Fluid Mosaic Model

A plasma membrane is composed of both lipids and globular proteins. For many years, biologists thought the protein covered the inner and outer surfaces of the phospholipid bilayer like a coat of paint. The widely accepted Davson-Danielli model, proposed in 1935, portrayed the membrane as a sandwich: a phospholipid bilayer between two layers of globular protein. This model, however, was not consistent with what researchers were learning in the 1960s about the structure of membrane proteins. Unlike most proteins found within cells, membrane proteins are not very soluble in water—they possess long stretches of nonpolar hydrophobic amino acids. If such proteins indeed coated the surface of the lipid bilayer, as the Davson-Danielli model suggested, then their nonpolar portions would separate the polar portions of the phospholipids from water, causing the bilayer to dissolve! Because this doesn't happen, there is clearly something wrong with the model.

In 1972, S. Singer and G. J. Nicolson revised the model in a simple but profound way: They proposed that the globular proteins are *inserted* into the lipid bilayer, with their nonpolar segments in contact with the nonpolar interior of the bilayer and their polar portions protruding out from the membrane surface. In this model, called the **fluid mosaic model,** a mosaic of proteins float in the fluid lipid bilayer like boats on a pond (figure 6.5).

Components of the Cell Membrane

A eukaryotic cell contains many membranes. While they are not all identical, they share the same fundamental architecture. Cell membranes are assembled from four components (table 6.1):

1. **Phospholipid bilayer.** Every cell membrane is composed of phospholipids in a bilayer. The other components of the membrane are embedded within the bilayer, which provides a flexible matrix and, at the same time, imposes a barrier to permeability.
2. **Transmembrane proteins.** A major component of every membrane is a collection of proteins that float on or in the lipid bilayer. These proteins provide passageways that allow substances and information to cross the membrane. Many membrane proteins are not fixed in position; they can move about, just as the phospholipid molecules do. Some membranes are crowded with proteins, while in others, the proteins are more sparsely distributed.

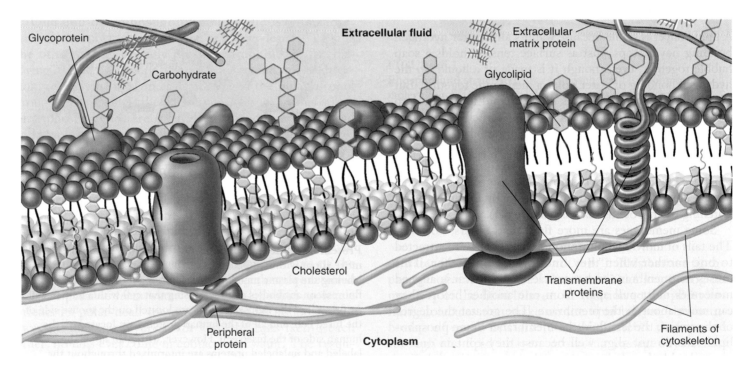

FIGURE 6.5

The fluid mosaic model of cell membranes.   A variety of proteins protrude through the plasma membrane of animal cells, and nonpolar regions of the proteins tether them to the membrane's nonpolar interior. Three important classes of membrane proteins are transport proteins, receptors, and cell surface markers. Carbohydrate chains are often bound to the extracellular portion of these proteins, as well as to the membrane phospholipids. These chains serve as distinctive identification tags, unique to particular cells.

Table 6.1 Components of the Cell Membrane

Component	Composition	Function	How It Works	Example
Phospholipid bilayer	Phospholipid molecules	Provides permeability barrier, matrix for proteins	Excludes water-soluble molecules from nonpolar interior of bilayer	Bilayer of cell is impermeable to water-soluble molecules, such as glucose
Transmembrane proteins	Carriers	Actively and passively transport molecules across membrane	"Escort" molecules through the membrane in a series of conformational changes	Glycophorin carrier for sugar transport; sodium-potassium pump
	Channels	Passively transport molecules across membrane	Create a tunnel that acts as a passage through membrane	Sodium and potassium channels in nerve cells
	Receptors	Transmit information into cell	Signal molecules bind to cell surface portion of the receptor protein; this alters the portion of the receptor protein within the cell, inducing activity	Specific receptors bind peptide hormones and neurotransmitters
Interior protein network	Spectrins	Determine shape of cell	Form supporting scaffold beneath membrane, anchored to both membrane and cytoskeleton	Red blood cell
	Clathrins	Anchor certain proteins to specific sites, especially on the exterior plasma membrane in receptor-mediated endocytosis	Proteins line coated pits and facilitate binding to specific molecules	Localization of low-density lipoprotein receptor within coated pits
Cell surface markers	Glycoproteins	"Self" recognition	Create a protein/carbohydrate chain shape characteristic of individual	Major histocompatibility complex protein recognized by immune system
	Glycolipid	Tissue recognition	Create a lipid/carbohydrate chain shape characteristic of tissue	A, B, O blood group markers

3. **Interior protein network.** Membranes are structurally supported by intracellular proteins that reinforce the membrane's shape. For example, a red blood cell has a characteristic biconcave shape because a scaffold of proteins called spectrin links proteins in the plasma membrane with actin filaments in the cell's cytoskeleton. Membranes use networks of other proteins to control the lateral movements of some key membrane proteins, anchoring them to specific sites.

4. **Cell surface markers.** Membrane sections assemble in the endoplasmic reticulum, transfer to the Golgi apparatus, and then are transported to the plasma membrane. The endoplasmic reticulum adds chains of sugar molecules to membrane proteins and lipids, creating a "sugar coating" called the glycocalyx that extends from the membrane on the outside of the cell. Different cell types exhibit different varieties of these glycoproteins and glycolipids on their surfaces, which act as cell identity markers.

Originally, it was believed the plasma membrane was uniform because of being highly fluid, with lipids and proteins free to diffuse rapidly in the plane of the membrane. However, in the last decade evidence has accumulated suggesting the plasma membrane is not homogeneous, and contains microdomains with distinct lipid and protein composition. One type of microdomain, the *lipid raft*, is heavily enriched in cholesterol and saturated fatty acids, and thus more tightly packed than the surrounding membrane. Lipid rafts appear to be involved in many important biological processes, including signal reception and cell movement. The structural proteins of replicating HIV viruses are targeted to lipid raft regions of the plasma membrane during virus assembly within infected cells.

The fluid mosaic model proposes that membrane proteins are embedded within the lipid bilayer. Membranes are composed of a lipid bilayer within which proteins are anchored. Plasma membranes are supported by a network of fibers and coated on the exterior with cell identity markers.

Examining Cell Membranes

Biologists examine the delicate, filmy structure of a cell membrane using electron microscopes that provide clear magnification to several thousand times. We discussed two types of electron microscopes in chapter 5: the transmission electron microscope (TEM) and the scanning electron microscope (SEM). When examining cell membranes with electron microscopy, specimens must be prepared for viewing.

In one method of preparing a specimen, the tissue of choice is embedded in a hard matrix, usually some sort of epoxy. The epoxy block is then cut with a microtome, a machine with a very sharp blade that makes incredibly thin slices. The knife moves up and down as the specimen advances toward it, causing transparent "epoxy shavings" less than 1 micrometer thick to peel away from the block of tissue. These shavings are placed on a grid, and a beam of electrons is directed through the grid with the TEM. At the high magnification an electron microscope provides, resolution is good enough to reveal the double layers of a membrane.

Freeze-fracturing a specimen is another way to visualize the inside of the membrane (figure 6.6). The tissue is embedded in a medium and quick-frozen with liquid nitrogen. The frozen tissue is then "tapped" with a knife, causing a crack between the phospholipid layers of membranes. Proteins, carbohydrates, pits, pores, channels, or any other structure affiliated with the membrane will pull apart (whole, usually) and stick with one side of the split membrane. A very thin coating of platinum is then evaporated onto the fractured surface, forming a replica or "cast" of the surface. Once the topography of the membrane has been preserved in the cast, the actual tissue is dissolved away, and the cast is examined with electron microscopy, creating a strikingly different view of the membrane.

Visualizing a plasma membrane requires a very powerful electron microscope. Freeze-fracturing reveals the interior of the lipid bilayer.

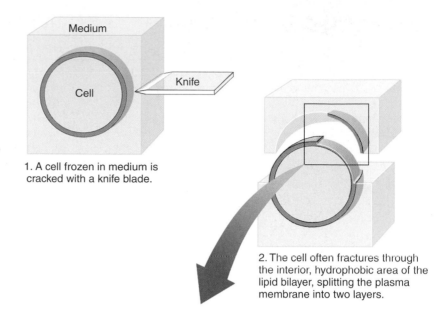

1. A cell frozen in medium is cracked with a knife blade.

2. The cell often fractures through the interior, hydrophobic area of the lipid bilayer, splitting the plasma membrane into two layers.

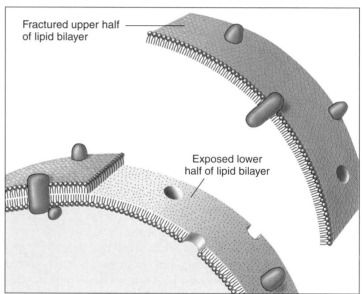

3. The plasma membrane separates such that proteins and other embedded membrane structures remain within one of the layers of the membrane.

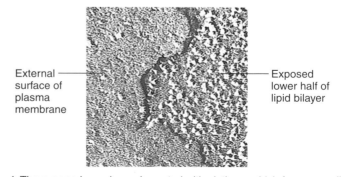

4. The exposed membrane is coated with platinum which forms a replica of the membrane. The underlying membrane is dissolved away, and the replica is then viewed with electron microscopy.

FIGURE 6.6
Viewing a plasma membrane with freeze-fracture microscopy.

Kinds of Membrane Proteins

As we've seen, cell membranes are composed of a complex assembly of proteins enmeshed in a fluid array of phospholipid molecules. This enormously flexible design permits a broad range of interactions with the environment, some directly involving membrane proteins (figure 6.7). Although cells interact with their environment through their plasma membranes in many ways, we will focus on six key classes of membrane protein in this and the following chapter (chapter 7).

1. **Transporters.** Membranes are very selective, allowing only certain substances to enter or leave the cell, either through channels or carriers.
2. **Enzymes.** Cells carry out many chemical reactions on the interior surface of the plasma membrane, using enzymes attached to the membrane.
3. **Cell surface receptors.** Membranes are exquisitely sensitive to chemical messages, detecting them with receptor proteins on their surfaces.

4. **Cell surface identity markers.** Membranes carry cell surface markers that identify them to other cells. Most cell types carry their own ID tags, specific combinations of cell surface proteins characteristic of that cell type.
5. **Cell adhesion proteins.** Cells use specific proteins to glue themselves to one another. Some act by forming temporary interactions, while others form a more permanent bond.
6. **Attachments to the cytoskeleton.** Surface proteins that interact with other cells are often anchored to the cytoskeleton by linking proteins.

The many proteins embedded within a membrane carry out a host of functions, many of which are associated with transport of materials or information across the membrane.

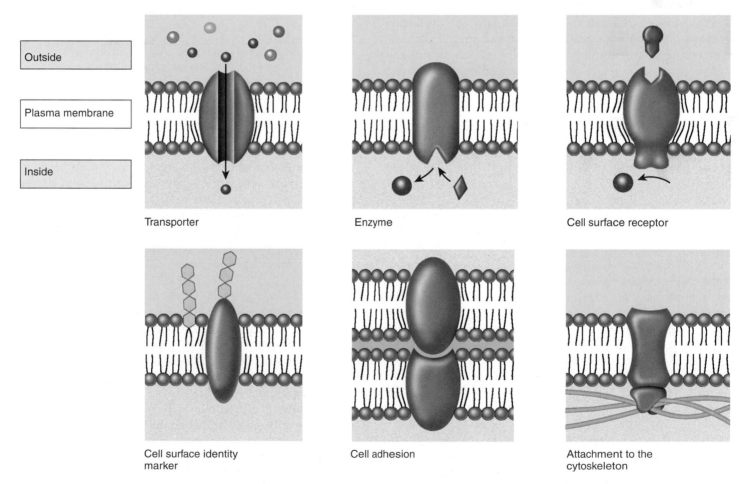

| Outside |
| Plasma membrane |
| Inside |

Transporter Enzyme Cell surface receptor

Cell surface identity marker Cell adhesion Attachment to the cytoskeleton

FIGURE 6.7
Functions of plasma membrane proteins. Membrane proteins act as transporters, enzymes, cell surface receptors, and cell surface markers, as well as aiding in cell-to-cell adhesion and securing the cytoskeleton.

Structure of Membrane Proteins

How can some proteins be anchored into particular positions on the cell membrane while others can move freely?

Anchoring Proteins in the Bilayer

Many membrane proteins are attached to the surface of the membrane by special molecules that associate with phospholipids and thereby anchor the protein to the membrane. Like a ship tied up to a floating dock, these proteins are free to move about on the surface of the membrane tethered to a phospholipid. These anchoring molecules are modified lipids that have (1) nonpolar regions that insert into the internal region of the lipid bilayer, and (2) chemical bonding domains that link directly to proteins.

In contrast, other proteins actually traverse the lipid bilayer. The part of the protein that extends through the lipid bilayer, in contact with the nonpolar interior, consists of nonpolar amino acid helices or β-pleated sheets (figure 6.8). Because water avoids nonpolar amino acids much as it does nonpolar lipid chains, the nonpolar portions of the protein are held within the interior of the lipid bilayer. Although the polar ends of the protein protrude from both sides of the membrane, the protein itself is locked into the membrane by its nonpolar segments. Any movement of the protein out of the membrane, in either direction, brings the nonpolar regions of the protein into contact with water, which "shoves" the protein back into the interior.

Extending Proteins Across the Bilayer

Cell membranes contain a variety of different **transmembrane proteins,** which differ in the way they traverse the lipid bilayer.

Single-Pass Anchors. A single nonpolar segment is adequate to anchor a protein in the membrane. For example, linking proteins of this sort attach the spectrin network of the cytoskeleton to the interior of the plasma membrane (figure 6.9). Many proteins that function as receptors for extracellular signals also have "single-pass" anchors that pass through the membrane only once. The portion of the receptor that extends out from the cell surface binds to specific hormones or other molecules, inducing changes at the other end of the protein in the cell's interior. In this way, information outside the cell is translated into action within the cell. The mechanisms of cell signaling will be addressed in detail in chapter 7.

Multiple-Pass Channels and Carriers. Other proteins have several helical segments that thread their way back and forth through the membrane, forming a channel like the hole in a doughnut. For example, bacteriorhodopsin is one of the key transmembrane proteins that carries out photosynthesis in bacteria. It contains seven nonpolar heli-

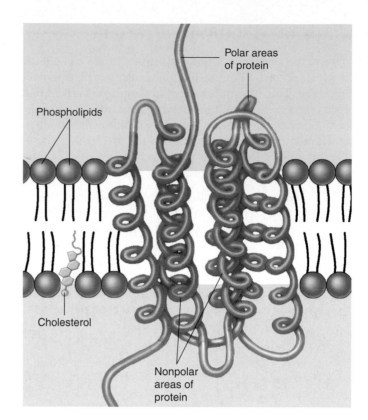

FIGURE 6.8
How nonpolar regions lock proteins into membranes.
A spiral helix of nonpolar amino acids (*red*) extends across the nonpolar lipid interior, while polar (*purple*) portions of the protein protrude out from the bilayer. The protein cannot move in or out because such a movement would drag nonpolar segments of the protein into contact with water.

cal segments that traverse the membrane, forming a channel through which protons pass during the light-driven pumping of protons (figure 6.10). Other transmembrane proteins act as carriers to transport molecules across the membrane using different mechanisms (see section 6.5). All water-soluble molecules or ions that enter or leave the cell are either transported by carriers or pass through channels.

Pores. Some transmembrane proteins have extensive nonpolar regions with secondary configurations of β-pleated sheets instead of α helices. The β sheets form a characteristic motif, folding back and forth in a circle so the sheets come to be arranged like the staves of a barrel. This so-called β barrel, open on both ends, is a common feature of the porin class of proteins that are found within the outer membrane of some bacteria, where they allow molecules to pass through the membrane (figure 6.11).

Transmembrane proteins are anchored into the bilayer by their nonpolar segments. Some proteins pass through the bilayer only once; many channels pass back and forth through the bilayer repeatedly, creating a circular hole in the bilayer.

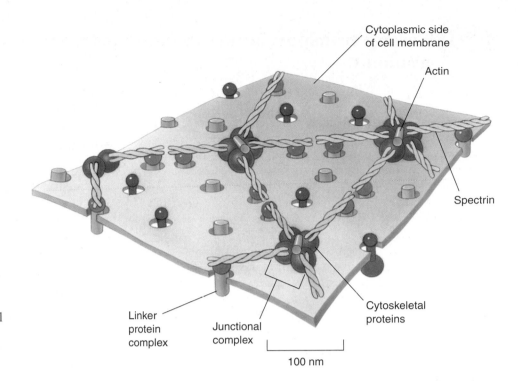

FIGURE 6.9
Linking proteins. Spectrin extends as a mesh anchored to the cytoplasmic side of a red blood cell plasma membrane. The spectrin protein is represented as a twisted dimer, attached to the membrane by special linker proteins. This cytoskeletal protein network confers resiliency to cells, such as red blood cells.

Cytoplasmic side of cell membrane

Actin

Spectrin

Cytoskeletal proteins

Junctional complex

Linker protein complex

100 nm

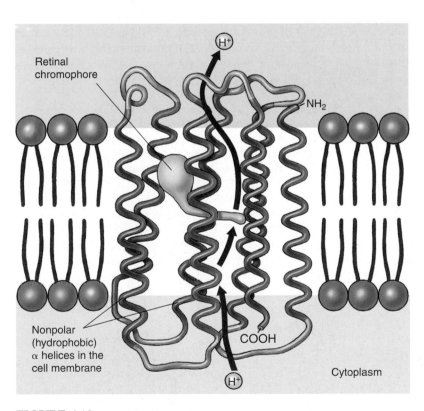

Retinal chromophore

H⁺

NH₂

COOH

Nonpolar (hydrophobic) α helices in the cell membrane

Cytoplasm

H⁺

FIGURE 6.10
A channel protein. This transmembrane protein mediates photosynthesis in the bacterium *Halobacterium halobium*. The protein traverses the membrane seven times with hydrophobic helical strands that are within the hydrophobic center of the lipid bilayer. The helical regions form a channel across the bilayer through which protons are pumped by the retinal chromophore (*green*).

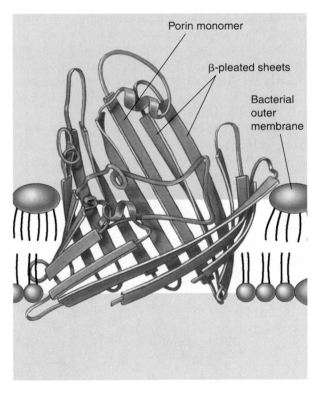

Porin monomer

β-pleated sheets

Bacterial outer membrane

FIGURE 6.11
A pore protein. The bacterial transmembrane protein porin creates large open tunnels called pores in the outer membrane of a bacterium. Sixteen strands of β-pleated sheets run antiparallel to each other, creating a β barrel in the bacterial outer cell membrane. The tunnel allows water and other materials to pass through the membrane.

6.3 Passive transport across membranes moves down the concentration gradient.

Diffusion

Molecules and ions dissolved in water are in constant motion, moving about randomly. This random motion causes a net movement of these substances from regions where their concentration is high to regions where their concentration is lower, a process called **diffusion** (figure 6.12). Net movement driven by diffusion will continue until the concentrations in all regions are the same. You can demonstrate diffusion by filling a jar to the brim with ink, capping it, placing it at the bottom of a bucket of water, and then carefully removing the cap. The ink molecules will slowly diffuse out from the jar until there is a uniform concentration in the bucket and the jar. This uniformity in the concentration of molecules is a type of equilibrium.

Membrane Transport Is Selective

Many molecules that cells require are polar and cannot pass through the nonpolar interior of the phospholipid bilayer. These molecules enter the cell by diffusion through specific channels in the plasma membrane. The inside lining of the channel is polar and thus "friendly" to the polar molecules, facilitating their transport across the membrane. Each type of biomolecule that is transported across the plasma membrane has its own type of transporter (that is, it has its own channel, which fits it like a glove and cannot be used by other molecules). Each channel is said to be selective for that type of molecule, and thus the cell is said to be **selectively permeable,** because only molecules admitted by the channels it possesses can enter it. The plasma membrane of a cell has many types of channels, each selective for a different type of molecule.

Diffusion of Ions Through Channels

Ions are solutes (substances dissolved in water) with an unequal number of protons and electrons. Those with an excess of protons are positively charged and called *cations.* Ions with more electrons than protons are negatively charged and called *anions.* Because they are charged, ions interact well with polar molecules such as water but are repelled by the nonpolar interior of a phospholipid bilayer. Therefore, ions cannot move between the cytoplasm of a cell and the extracellular fluid without the assistance of membrane transport proteins. **Ion channels** possess a hydrated interior that spans the membrane. Ions can diffuse through the channel in either direction without coming into contact with the hydrophobic tails of the phospholipids in the membrane. Two conditions determine the direction of net movement of the ions: their relative concentrations on either side of the membrane, and the voltage across the membrane (a topic we'll explore in chapter 45). Each type of channel is specific for a particular ion, such as calcium (Ca^{++}), sodium (Na^+), potassium (K^+), or chloride (Cl^-), or in some cases, for a few kinds of ions. Ion channels play an essential role in signaling by the nervous system.

Diffusion is the net movement of substances to regions of lower concentration as a result of random motion. It tends to distribute substances uniformly. Membrane transport proteins allow only certain molecules and ions to diffuse through the plasma membrane.

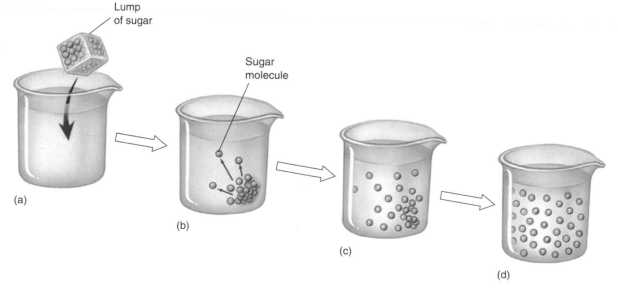

FIGURE 6.12

Diffusion. If a lump of sugar is dropped into a beaker of water (*a*), its molecules dissolve (*b*) and diffuse (*c*). Eventually, diffusion results in an even distribution of sugar molecules throughout the water (*d*).

Facilitated Diffusion

Carriers, another class of membrane proteins, transport ions as well as other solutes, such as sugars and amino acids, across the plasma membrane. Like channels, carriers are specific for a certain type of solute and can transport substances in either direction across the membrane. Unlike channels, however, they facilitate the movement of solutes across the membrane by physically binding to them on one side of the membrane and releasing them on the other. Again, the direction of the solute's net movement simply depends on its *concentration gradient* across the membrane. If the concentration is greater in the cytoplasm than outside the cell, the carrier will transport the molecule from the inside to the outside of the cell. This occurs because the solute is more likely to bind to the carrier on the cytoplasmic side of the membrane. If the concentration is greater in the extracellular fluid, the net movement will be from outside to inside. Thus, the net movement always occurs from areas of high concentration to low, just as it does in simple diffusion, but carriers facilitate the process. For this reason, this mechanism of transport is sometimes called **facilitated diffusion** (figure 6.13).

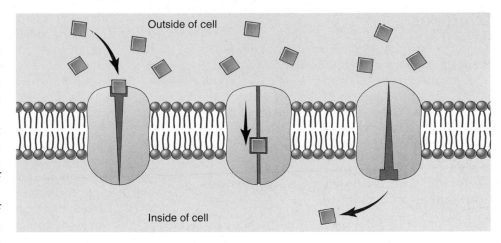

FIGURE 6.13
Facilitated diffusion is a carrier-mediated transport process. Molecules bind to a receptor on the extracellular side of the cell and are conducted through the plasma membrane by a membrane protein.

Facilitated Diffusion in Red Blood Cells

Several examples of facilitated diffusion by carrier proteins can be found in the plasma membranes of vertebrate red blood cells (RBCs). One RBC carrier protein, for example, transports a different molecule in each direction: Cl^- in one direction and bicarbonate ion (HCO_3^-) in the opposite direction. As you will learn in chapter 44, this carrier is important in transporting carbon dioxide in the blood.

A second important facilitated diffusion carrier in RBCs is the glucose transporter. Red blood cells keep their internal concentration of glucose low through a chemical trick: They immediately add a phosphate group to any entering glucose molecule, converting it to a highly charged glucose phosphate that cannot pass back across the membrane. This maintains a steep concentration gradient for unphosphorylated glucose, favoring its entry into the cell. The glucose transporter that carries glucose into the cell does not appear to form a channel in the membrane for the glucose to pass through. Instead, the transmembrane protein appears to bind the glucose and then flip its shape, dragging the glucose through the bilayer and releasing it on the inside of the plasma membrane. Once it releases the glucose, the glucose transporter reverts to its original shape. It is then available to bind the next glucose molecule that approaches the outside of the cell.

Saturation of Transport Through Carriers

A characteristic feature of transport through carriers is that its rate can become saturated. In other words, as the concentration gradient of a substance is progressively increased, its rate of transport will also increase to a certain point and then level off. Further increases in the gradient will produce no additional increase in rate. The explanation for this observation is that the membrane contains a limited number of carriers. When the concentration of the transported substance rises high enough, all of the carriers will be in use, and the capacity of the transport system will be saturated. In contrast, substances that move across the membrane by simple diffusion (diffusion through channels in the bilayer without the assistance of carriers) do not show saturation.

Facilitated diffusion provides the cell with a ready way to prevent the buildup of unwanted molecules within the cell or to take up needed molecules, such as sugars, that may be present outside the cell in high concentrations. Facilitated diffusion has three essential characteristics:

1. **It is specific.** A given carrier transports only certain molecules or ions.
2. **It is passive.** The direction of net movement is determined by the relative concentrations of the transported substance inside and outside the cell. It always moves from high concentration to low concentration.
3. **It saturates.** If all relevant protein carriers are in use, increases in the concentration gradient do not increase the transport rate.

Facilitated diffusion is the transport of molecules and ions across a membrane by specific carriers in the direction of lower concentration of those molecules or ions.

Osmosis

The cytoplasm of a cell contains ions and molecules, such as sugars and amino acids, dissolved in water. The mixture of these substances and water is called an *aqueous solution.* Water, the most common of the molecules in the mixture, is the **solvent,** and the substances dissolved in the water are **solutes.** The ability of water and solutes to diffuse across membranes has important consequences.

Molecules Diffuse down a Concentration Gradient

Both water and solutes diffuse from regions of high concentration to regions of low concentration; that is, they diffuse down their concentration gradients. When two regions are separated by a membrane, what happens depends on whether or not the solutes can pass freely through that membrane. Most solutes, including ions and sugars, are not lipid-soluble and, therefore, are unable to cross the lipid bilayer of the membrane.

Even water molecules, which are very polar, cannot readily cross a lipid bilayer. Water flow is facilitated by **aquaporins,** which are specialized channels for water. A simple experiment demonstrates this. If you place an amphibian egg in hypotonic spring water, where the solute concentration in the cell is higher than the surrounding water, it does not swell. If you then inject aquaporin mRNA into the egg, the channel proteins are expressed, and water will diffuse into the egg, causing it to swell.

Dissolved solutes interact with water molecules, which form hydration shells about the charged solute molecules. When a membrane separates two solutions with different concentrations of solutes, different concentrations of *free* water molecules exist on the two sides of the membrane. The side with higher solute concentration ties up more water molecules in hydration shells. As a consequence, free water molecules move down their concentration gradient, toward the higher solute concentration. This net water movement across a membrane toward a higher solute concentration by diffusion is called **osmosis** (figure 6.14).

The concentration of *all* solutes in a solution determines the **osmotic concentration** of the solution. If two solutions have unequal osmotic concentrations, the solution with the higher concentration is **hyperosmotic** (Greek *hyper,* "more than"), and the solution with the lower concentration is **hypoosmotic** (Greek *hypo,* "less than"). If the osmotic concentrations of two solutions are equal, the solutions are **isosmotic** (Greek *iso,* "the same").

In cells, a plasma membrane separates two aqueous solutions, one inside the cell (the cytoplasm) and one outside (the extracellular fluid). The direction of the net diffusion of water across this membrane is determined by the osmotic concentrations of the solutions on either side. For example, if the cytoplasm of a cell is *hypoosmotic* to the extracellular fluid, water diffuses out of the cell, toward the solution with the higher concentration of solutes (and, therefore, the lower concentration of unbound water molecules). This loss of water from the cytoplasm causes the cell to shrink until the osmotic concentrations of the cytoplasm and the extracellular fluid become equal.

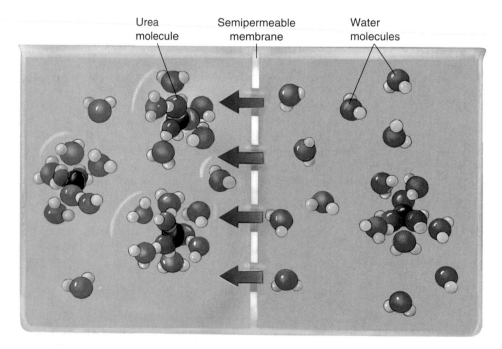

Urea molecule   Semipermeable membrane   Water molecules

FIGURE 6.14
Osmosis.   Charged or polar substances are soluble in water because they form hydrogen bonds with water molecules clustered around them, forming hydration shells. When a polar solute (illustrated here with urea) is added to the solution on one side of a membrane, the water molecules that gather around each urea molecule are no longer free to diffuse across the membrane; in effect, the polar solute has reduced the concentration of free water molecules, creating a gradient. Water moves down the gradient by diffusion from the right to the left.

Osmotic Pressure

What happens if the cell's cytoplasm is *hyper*osmotic to the extracellular fluid? In this situation, water diffuses into the cell from the extracellular fluid, causing the cell to swell. The pressure of the cytoplasm pushing out against the cell membrane, or **hydrostatic pressure,** increases. On the other hand, the **osmotic pressure,** defined as the pressure that must be applied to stop the osmotic movement of water across a membrane, will also be at work (figure 6.15). If the membrane is strong enough, the cell reaches an equilibrium, at which the osmotic pressure, which tends to drive water into the cell, is exactly counterbalanced by the hydrostatic pressure, which tends to drive water back out of the cell. However, a plasma membrane by itself cannot withstand large internal pressures, and an isolated cell under such conditions would burst like an overinflated balloon. Accordingly, it is important for animal cells, which only have plasma membranes, to maintain isosmotic conditions. In contrast, the cells of prokaryotes, fungi, plants, and many protists are surrounded by strong cell walls, which can withstand high internal pressures without bursting.

Maintaining Osmotic Balance

Organisms have developed many solutions to the dilemma posed by being hyperosmotic to their environment.

Extrusion.   Some single-celled eukaryotes, such as the protist *Paramecium*, use organelles called contractile vacuoles to remove water. Each vacuole collects water from various parts of the cytoplasm and transports it to the central part of the vacuole, near the cell surface. The vacuole possesses a small pore that opens to the outside of the cell. By contracting rhythmically, the vacuole pumps out through this pore the water that is continuously seeping into the cell by osmosis.

Isosmotic Solutions.   Some organisms that live in the ocean adjust their internal concentration of solutes to match that of the surrounding seawater. Because they are isosmotic with respect to their environment, no net flow of water occurs into or out of these cells. Many terrestrial animals solve the problem in a similar way, by circulating a fluid through their bodies that bathes cells in an isosmotic solution. The blood in your body, for example, contains a high concentration of the protein albumin, which elevates the solute concentration of the blood to match that of your cells.

Turgor.   Most plant cells are hyperosmotic to their immediate environment, containing a high concentration of solutes in their central vacuoles. The resulting internal hydrostatic pressure, known as **turgor pressure,** presses the plasma membrane firmly against the interior of the cell

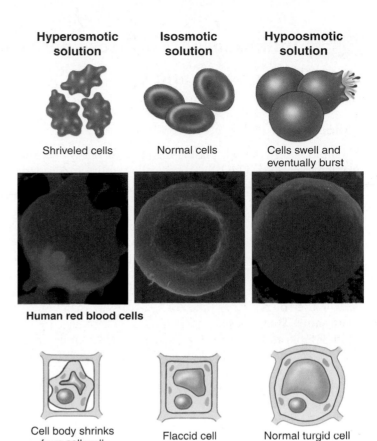

Human red blood cells

Plant cells

FIGURE 6.15
How solutes create osmotic pressure.   In a hyperosmotic solution, water moves out of the cell toward the higher concentration of solutes, causing the cell to shrivel. In an isosmotic solution, the concentration of solutes on either side of the plasma membrane is the same. Osmosis still occurs, but water diffuses into and out of the cell at the same rate, and the cell doesn't change size. In a hypoosmotic solution, the concentration of solutes is higher within the cell than outside, so the net movement of water is into the cell. A cell is an enclosed structure, and so as water enters the cell from a hypoosmotic solution, pressure is applied to the plasma membrane until the cell ruptures. This hydrostatic pressure is counterbalanced by osmotic pressure, the force required to stop the flow of water into the cell. Plant cells have strong cell walls that can apply adequate osmotic pressure to keep the cell from rupturing. This is not the case with animal cells.

wall, making the cell rigid. Most green plants depend on turgor pressure to maintain their shape, and thus they wilt when they lack sufficient water.

Osmosis is the diffusion of water, but not solutes, across a membrane.

6.4 | Bulk transport utilizes endocytosis.

Bulk Passage Into and Out of the Cell

Endocytosis

The lipid nature of their plasma membranes raises a second problem for cells. The substances cells require for growth are for the most part large, polar molecules that cannot cross the hydrophobic barrier a lipid bilayer creates. How do organisms get these substances into their cells? One process is **endocytosis,** in which the plasma membrane envelops food particles. Cells use three major types of endocytosis: phagocytosis, pinocytosis, and receptor-mediated endocytosis (figure 6.16).

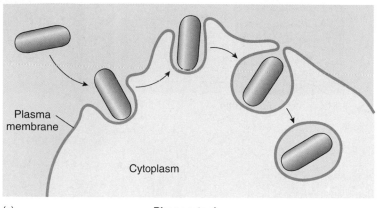

Plasma membrane

Cytoplasm

(a) **Phagocytosis**

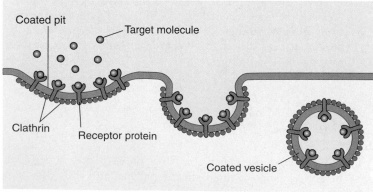

Plasma membrane

Cytoplasm

(b) **Pinocytosis**

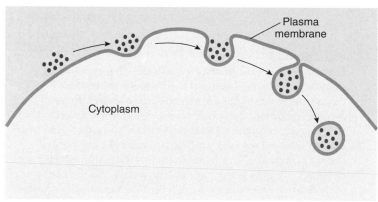

Coated pit

Target molecule

Clathrin

Receptor protein

Coated vesicle

(c) **Receptor-mediated endocytosis**

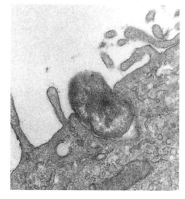

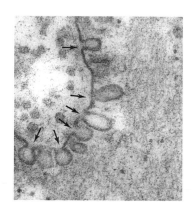

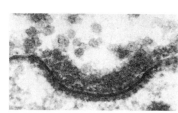

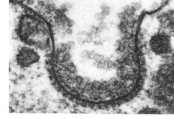

**FIGURE 6.16
Endocytosis.** Both (*a*) phagocytosis and (*b*) pinocytosis are forms of endocytosis. (*c*) In receptor-mediated endocytosis, cells have pits coated with the protein clathrin that initiate endocytosis when target molecules bind to receptor proteins in the plasma membrane. Photo inserts: (*a*) A TEM of phagocytosis of a bacterium, *Rickettsia tsutsugamushi,* by a mouse peritoneal mesothelial cell. The bacterium enters the host cell by phagocytosis and replicates in the cytoplasm. (*b*) A TEM of pinocytosis in a smooth muscle cell. (*c*) A coated pit appears in the plasma membrane of a developing egg cell, covered with a layer of proteins (80,000×). When an appropriate collection of molecules gathers in the coated pit, the pit deepens and will eventually seal off to form a vesicle.

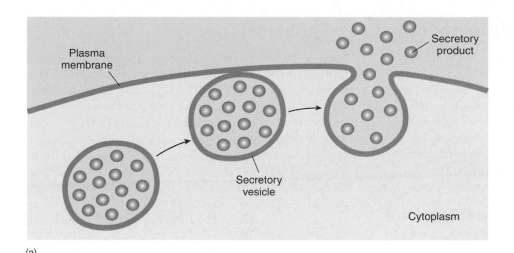
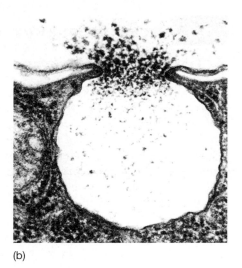

(a)

(b)

FIGURE 6.17
Exocytosis. (*a*) Proteins and other molecules are secreted from cells in small packets called vesicles, whose membranes fuse with the plasma membrane, releasing their contents to the cell surface. (*b*) A transmission electron micrograph showing exocytosis.

Phagocytosis and Pinocytosis. If the material the cell takes in is particulate (made up of discrete particles), such as an organism or some other fragment of organic matter (figure 6.16*a*), the process is called **phagocytosis** (Greek *phagein*, "to eat," + *cytos*, "cell"). If the material the cell takes in is liquid (figure 6.16*b*), it is called **pinocytosis** (Greek *pinein*, "to drink"). Pinocytosis is common among animal cells. Mammalian egg cells, for example, "nurse" from surrounding cells; the nearby cells secrete nutrients that the maturing egg cell takes up by pinocytosis. Virtually all eukaryotic cells constantly carry out these kinds of endocytosis, trapping particles and extracellular fluid in vesicles and ingesting them. Endocytosis rates vary from one cell type to another. They can be surprisingly high; some types of white blood cells ingest 25% of their cell volume each hour!

Receptor-Mediated Endocytosis. Specific molecules are often transported into eukaryotic cells through **receptor-mediated endocytosis.** Molecules to be transported first bind to specific receptors in the plasma membrane. The transport process is specific only to molecules having a shape that fits snugly into the receptor. The plasma membrane of a particular kind of cell contains a characteristic battery of receptor types, each for a different kind of molecule.

The portion of the receptor molecule inside the membrane is trapped in an indented pit coated with the protein clathrin. Each pit acts like a molecular mousetrap, closing over to form an internal vesicle when the right molecule enters the pit (figure 6.16*c*). The trigger that releases the trap is the binding of the properly fitted target molecule to a receptor embedded in the membrane of the pit. When binding occurs, the cell reacts by initiating endocytosis. The process is highly specific and very fast.

One type of molecule that is taken up by receptor-mediated endocytosis is called a low-density lipoprotein (LDL). LDL molecules bring cholesterol into the cell where it can be incorporated into membranes. Cholesterol plays a key role in determining the stiffness of the body's membranes. In the human genetic disease hypercholesterolemia, the receptors lack tails, so they are never caught in the clathrin-coated pits and, thus, never taken up by the cells. The cholesterol stays in the bloodstream of affected individuals, coating their arteries and leading to heart attacks.

It is important to understand that endocytosis in itself does not bring substances directly into the cytoplasm of a cell. The material taken in is still separated from the cytoplasm by the membrane of the vesicle.

Exocytosis

The reverse of endocytosis is **exocytosis,** the discharge of material from vesicles at the cell surface (figure 6.17). In plant cells, exocytosis is an important means of exporting the materials needed to construct the cell wall through the plasma membrane. Among protists, contractile vacuole discharge is a form of exocytosis. In animal cells, exocytosis provides a mechanism for secreting many hormones, neurotransmitters, digestive enzymes, and other substances.

Cells import bulk materials by engulfing them with their plasma membranes in a process called endocytosis; similarly, they extrude or secrete material through exocytosis.

6.5 Active transport across membranes requires energy.

Active Transport

While diffusion, facilitated diffusion, and osmosis are passive transport processes that move materials down their concentration gradients, cells can also move substances across a cell membrane *up* their concentration gradients. This process requires the expenditure of energy, typically from ATP, and is therefore called **active transport.** Like facilitated diffusion, active transport involves highly selective protein carriers within the membrane. These carriers bind to the transported substance, which could be an ion or a simple molecule, such as a sugar, an amino acid, or a nucleotide to be used in the synthesis of DNA.

Active transport is one of the most important functions of any cell. It enables a cell to take up additional molecules of a substance that is already present in its cytoplasm in concentrations higher than in the extracellular fluid. Without active transport, for example, liver cells would be unable to accumulate glucose molecules from the blood plasma, because the glucose concentration is often higher inside the liver cells than it is in the plasma. Active transport also enables a cell to move substances from its cytoplasm to the extracellular fluid despite higher external concentrations.

The Sodium-Potassium Pump

The use of ATP in active transport may be direct or indirect. Let's first consider how ATP is used directly to move ions against their concentration gradient. More than one-third of all of the energy expended by an animal cell that is not actively dividing is used in the active transport of sodium (Na^+) and potassium (K^+) ions. Most animal cells have a low internal concentration of Na^+, relative to their surroundings, and a high internal concentration of K^+. They maintain these concentration differences by actively pumping Na^+ out of the cell and K^+ in. The remarkable protein that transports these two ions across the cell membrane is known as the **sodium-potassium pump** (figure 6.18). The cell obtains the energy it needs to operate the pump from adenosine triphosphate (ATP), a molecule we'll learn more about in chapter 8.

The important characteristic of the sodium-potassium pump is that it is an active transport process, transporting Na^+ and K^+ from areas of low concentration to areas of high concentration. This transport up concentration gradients is the opposite of the passive transport in diffusion; it is achieved only by the constant expenditure of metabolic energy. The sodium-potassium pump works through the following series of conformational changes in the transmembrane protein:

Step 1. Three sodium ions bind to the cytoplasmic side of the protein, causing the protein to change its conformation.

Step 2. In its new conformation, the protein binds a molecule of ATP and cleaves it into adenosine diphosphate and phosphate (ADP + P_i). ADP is released, but the phosphate group remains bound to the protein. The protein is now phosphorylated.

Step 3. The phosphorylation of the protein induces a second conformational change in the protein. This change translocates the three Na^+ across the membrane, so they now face the exterior. In this new conformation, the protein has a low affinity for Na^+, and the three bound Na^+ dissociate from the protein and diffuse into the extracellular fluid.

Step 4. The new conformation has a high affinity for K^+, two of which bind to the extracellular side of the protein as soon as it is free of the Na^+.

Step 5. The binding of the K^+ causes another conformational change in the protein, this time resulting in the dissociation of the bound phosphate group.

Step 6. Freed of the phosphate group, the protein reverts to its original conformation, exposing the two K^+ to the cytoplasm. This conformation has a low affinity for K^+, so the two bound K^+ dissociate from the protein and diffuse into the interior of the cell. The original conformation has a high affinity for Na^+; when these ions bind, they initiate another cycle.

In every cycle, three Na^+ leave the cell and two K^+ enter. The changes in protein conformation that occur during the cycle are rapid, enabling each carrier to transport as many as 300 Na^+ per second. The sodium-potassium pump appears to be ubiquitous in animal cells, although cells vary widely in the number of pump proteins they contain.

Coupled Transport

Some molecules are moved up their concentration gradient by using the energy stored in a gradient of a different molecule. In this process, the energy released as one molecule moves down its concentration gradient is captured and used to move a different molecule against its gradient. As we just learned, the energy stored in ATP molecules can be used to create a gradient of Na^+ and K^+ ions across the membrane. These gradients can then be used to power the transport of other molecules across the membrane.

For example, the transport of glucose across the membrane in animal cells requires energy for two reasons. First, glucose is a large polar molecule that cannot move through the membrane by itself, and second, the concentration of glucose inside the cell is frequently higher than the concentration outside the cell. The glucose transporter uses the Na^+ gradient produced by the

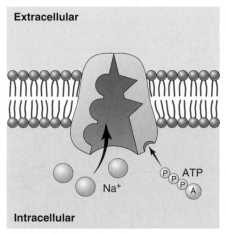

1. Protein in membrane binds intracellular sodium.

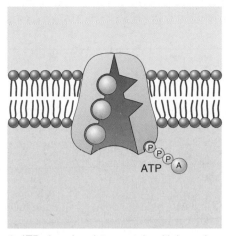

2. ATP phosphorylates protein with bound sodium.

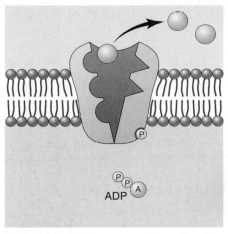

3. Phosphorylation causes conformational change in protein, allowing sodium to leave.

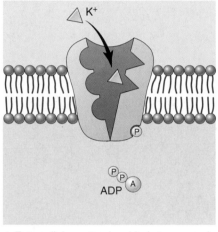

4. Extracellular potassium binds to exposed sites.

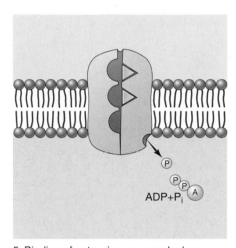

5. Binding of potassium causes dephosphorylation of protein.

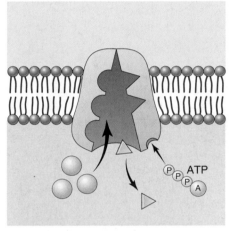

6. Dephosphorylation of protein triggers change back to original conformation, potassium moves into cell, and the cycle repeats.

FIGURE 6.18

The sodium-potassium pump. The protein carrier known as the sodium-potassium pump transports sodium (Na^+) and potassium (K^+) ions across the plasma membrane. For every three Na^+ transported out of the cell, two K^+ are transported into the cell. The sodium-potassium pump is fueled by ATP.

sodium-potassium pump as a source of energy to power the movement of glucose into the cell. In this system, both glucose and Na^+ bind to the transport protein, which allows Na^+ to pass down its concentration gradient, capturing the energy and using it to move glucose into the cell. In this kind of cotransport, both molecules are moving in the same direction across the membrane, and hence it is called *symport* (figure 6.19).

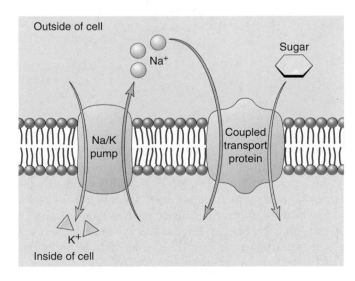

FIGURE 6.19

Coupled transport. A membrane protein transports sodium ions into the cell, down their concentration gradient, at the same time it transports a glucose molecule into the cell. The gradient driving the Na^+ entry is so great that sugar molecules can be brought in against their concentration gradient.

Table 6.2 Mechanisms for Transport Across Cell Membranes

Process	How It Works	Example
PASSIVE PROCESSES		
Diffusion		
Direct	Random molecular motion produces net migration of molecules toward region of lower concentration	Movement of oxygen into cells
Protein channel	Polar molecules pass through a protein channel	Movement of ions in or out of cell
Facilitated diffusion		
Protein carrier	Molecule binds to carrier protein in membrane and is transported across; net movement is toward region of lower concentration	Movement of glucose into cells
Osmosis		
Aquaporins	Diffusion of water across differentially permeable membrane	Movement of water into cells placed in a hypotonic solution
ACTIVE PROCESSES		
Endocytosis		
Membrane vesicle		
Phagocytosis	Particle is engulfed by membrane, which folds around it and forms a vesicle	Ingestion of bacteria by white blood cells
Pinocytosis	Fluid droplets are engulfed by membrane, which forms vesicles around them	"Nursing" of human egg cells
Receptor-mediated endocytosis	Endocytosis triggered by a specific receptor	Cholesterol uptake
Exocytosis		
Membrane vesicle	Vesicles fuse with plasma membrane and eject contents	Secretion of mucus
Active transport		
Protein carrier		
Na^+/K^+ pump	Carrier expends energy to export a substance across a membrane against its concentration gradient	Na^+ and K^+ against their concentration gradients
Coupled transport	Molecules are transported across a membrane against their concentration gradients by the cotransport of sodium ions or protons down their concentration gradients	Coupled uptake of glucose into cells against its concentration gradient

In a related process, called **countertransport,** the inward movement of Na^+ is coupled with the outward movement of another substance, such as Ca^{++} or H^+. As in cotransport, both Na^+ and the other substance bind to the same transport protein, which in this case is called an **antiport,** but they bind on opposite sides of the membrane and are moved in opposite directions. In countertransport, the cell uses the energy released as Na^+ moves down its concentration gradient into the cell to extrude a substance up its concentration gradient.

The mechanisms for transport across cell membranes are summarized in table 6.2.

Active transport moves a solute across a membrane up its concentration gradient, using protein carriers driven by the expenditure of chemical energy. Many molecules are cotransported into cells up their concentration gradients by coupling their movement to that of sodium ions or protons moving down their concentration gradients.

Concept Review

For interactive testing, visit the Online Learning Center with PowerWeb at www.mhhe.com/Raven7

6.1 Biological membranes are fluid layers of lipid.

The Phospholipid Bilayer

- Biological membranes are composed of phospholipid bilayers with hydrophilic, polar heads and hydrophobic, nonpolar, fatty acid tails. (p. 106)
- Phospholipid molecules orient their polar heads toward water and their nonpolar tails away from water. (p. 106)
- The nonpolar interior of a lipid bilayer impedes the passage of any water-soluble substances through the bilayer. (p. 106)

The Lipid Bilayer Is Fluid

- A lipid bilayer is stable because water's affinity for hydrogen bonding never stops, and the hydrogen bonding holds the membrane together in a liquid form. (p. 107)

6.2 Proteins embedded within cell membranes determine their character.

The Fluid Mosaic Model

- In the fluid mosaic model, a mosaic of proteins floats in the fluid lipid bilayer. (p. 108)
- Cell membranes are assembled from four components: phospholipid bilayer, transmembrane proteins, interior protein network, and cell surface markers. (pp. 108–109)

Examining Cell Membranes

- Electron microscopes must be used to visualize the structure of a cell membrane. (p. 110)
- Different procedures can be employed for preparing a specimen for viewing, including embedding tissue in a hard matrix and freeze-fracturing a specimen. (p. 110)

Kinds of Membrane Proteins

- Six key classes of membrane proteins include transporters, enzymes, cell surface receptors, cell surface identity markers, cell adhesion proteins, and attachments to the cytoskeleton. (p. 111)

Structure of Membrane Proteins

- Transmembrane proteins are anchored into the bilayer by nonpolar segments. (p. 112)
- Cell membranes contain a variety of different transmembrane proteins, including single-pass anchors, multiple-pass channels and carriers, and pores. (p. 112)

6.3 Passive transport across membranes moves down the concentration gradient.

Diffusion

- Random motion of molecules causes a net movement of substances from regions of high concentration to regions of lower concentration, and this movement continues until all regions exhibit the same concentration. (p. 114)
- Each transport protein in the plasma membrane is selectively permeable, and thus only allows certain molecules to diffuse through. (p. 114)

Facilitated Diffusion

- Facilitated diffusion occurs as molecules move from an area of higher concentration to an area of lower concentration via specific carriers. (p. 115)
- The essential characteristics of facilitated diffusion are specificity, passivity, and saturation. (p. 115)

Osmosis

- During osmosis, water moves across a membrane toward a solution with a higher solute concentration. (p. 116)
- The direction of net diffusion of water across the membrane is determined by the osmotic concentrations of the solutions on either side. (p. 116)
- Organisms have developed many solutions to being hyperosmotic to their environment, including extrusion, isomotic solutions, and turgor. (p. 117)

6.4 Bulk transport utilizes endocytosis.

Bulk Passage Into and Out of the Cell

- Endocytosis occurs when the plasma membrane envelops food particles and brings them into the cell interior. Three major forms of endocytosis are phagocytosis, pinocytosis, and receptor-mediated endocytosis. (pp. 118–119)
- Exocytosis refers to the discharge of materials from vesicles at the cell surface. (p. 119)

6.5 Active transport across membranes requires energy.

Active Transport

- Active transport is the movement of a solute across a membrane against its concentration gradient, requiring the use of protein carriers with the expenditure of ATP. (p. 120)
- Active transport involves highly selective membrane protein carriers. (p. 120)

Self Test

1. Why is the phospholipid molecule so appropriate as the primary structural component of plasma membranes?
   a. Phospholipids are completely insoluble in water.
   b. Phospholipids form strong chemical bonds between the molecules, forming a stable structure.
   c. Phospholipids form a selectively permeable structure.
   d. Phospholipids form chemical bonds with membrane proteins that keep the proteins within the membrane.

2. Which increases the fluidity of the plasma membrane?
   a. having a large number of membrane proteins
   b. the tight alignment of phospholipids
   c. cholesterol present in the membrane
   d. double bonds between carbon atoms in the fatty acid tails

3. Which best describes the structure of a plasma membrane?
   a. proteins embedded within two layers of phospholipids
   b. phospholipids sandwiched between two layers of proteins
   c. proteins sandwiched between two layers of phospholipids
   d. a layer of proteins on top of a layer of phospholipids

4. What locks all transmembrane proteins in the bilayer?
   a. chemical bonds that form between the phospholipids and the proteins
   b. hydrophobic interactions between nonpolar amino acids of the proteins and the aqueous environments of the cell
   c. attachment to the cytoskeleton
   d. the addition of sugar molecules to the protein surface facing the external environment

5. The movement of sodium ions from an area of higher concentration to an area of lower concentration is called
   a. active transport.
   b. osmosis.
   c. diffusion.
   d. phagocytosis.

6. A cell placed in distilled water will
   a. shrivel up.
   b. swell.
   c. lose water.
   d. result in no net diffusion of water molecules.

7. Sucrose cannot pass through the membrane of a red blood cell (RBC), but water and glucose can. Which solution would cause the RBC to shrink the most?
   a. a hyperosmotic sucrose solution
   b. a hyperosmotic glucose solution
   c. a hypoosmotic sucrose solution
   d. a hypoosmotic glucose solution

8. Which of the following processes requires membrane proteins?
   a. exocytosis
   b. phagocytosis
   c. receptor-mediated endocytosis
   d. pinocytosis

9. Exocytosis involves
   a. the ingestion of large organic molecules or organisms.
   b. the use of ATP.
   c. the uptake of fluids from the environment.
   d. the discharge of materials from cellular vesicles.

10. Molecules that are transported into the cell *up* their concentration gradients do so by
    a. facilitated diffusion.
    b. osmosis.
    c. coupled transport.
    d. none of these.

Test Your Visual Understanding

1. In this figure of a transmembrane protein, what colored area of the protein contains nonpolar amino acids? What colored area contains polar amino acids? What colored area contains amino acids carrying a positive or negative charge?

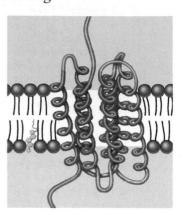

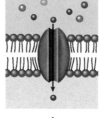 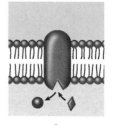 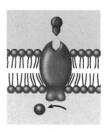

   1   **2**   **3**

2. Match the function of the membrane protein with the appropriate numbered figure:
   a. cell surface receptor
   b. transporter
   c. cell surface identity marker
   d. enzyme

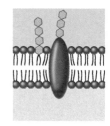

   4

Apply Your Knowledge

1. If during the action of the sodium-potassium pump, 150 molecules of ATP are used, how many sodium ions are transported across the membrane?

2. If a cell's cytoplasm were hyperosmotic to the extracellular fluid, how would the concentration of solutes in the cytoplasm compare with that in the extracellular fluid? Assuming the membrane was permeable only to water, in which direction would water move?

3. Cholera, a disease caused by a bacterial infection of *Vibrio cholerae*, results in severe diarrhea leading to dehydration. A toxin released by the bacterium causes the release of chloride ions (Cl^-) from cells lining the small intestine and inhibits the uptake of sodium ions (Na^+) by these cells. Explain how this disruption of cellular ion concentrations would result in extreme dehydration.

4. Cystic fibrosis is a genetic disease that results in thick, mucous secretions that clog air passages in the lungs. Faulty chloride ion channels keep Cl^- and Na^+ in the cells that line the airways, increasing the intracellular ion concentrations. How does this cause the mucus in the airways to become thick?

7

Cell–Cell Interactions

Concept Outline

7.1 Cells signal one another with chemicals.

Receptor Proteins and Signaling Between Cells.
Receptor proteins embedded in the plasma membrane
change shape when they bind specific signal molecules.
The process of isolating and identifying these receptors
has been aided by new techniques.

Types of Cell Signaling. Cell signaling can occur between
adjacent cells, although chemical signals called
hormones act over long distances.

7.2 Proteins in the cell and on its surface receive signals from other cells.

Intracellular Receptors. Some receptors are located within
the cell cytoplasm. These receptors respond to lipid-
soluble signals, such as steroid hormones.

Cell Surface Receptors. Many cell-to-cell signals are water-
soluble and cannot penetrate membranes. Instead, the
signals are received by transmembrane proteins
protruding from the cell surface.

7.3 Follow the journey of information into the cell.

Initiating the Intracellular Signal. Cell surface receptors
often use "second messengers" to transmit a signal to
the cytoplasm.

Amplifying the Signal: Protein Kinase Cascades.
Surface receptors and second messengers amplify
signals as they travel into the cell, often toward the
cell nucleus.

7.4 Cell surface proteins mediate cell–cell interactions.

The Expression of Cell Identity. Cells possess on their
surfaces a variety of tissue-specific identity markers that
identify both the tissue and the individual.

Intercellular Adhesion. Cells attach themselves to one
another with protein links. Some of the links are very
strong, others more transient. The three categories of
cell junctions are tight junctions, anchoring junctions,
and communicating junctions. Each serves a different
function in the cell.

FIGURE 7.1
Persimmon cells in close contact with one another. These
plant cells and all cells, no matter what their function, interact
with their environment, including the cells around them.

Did you know that each of the 100 trillion cells of your
body shares one key feature with the cells of tigers,
bumblebees, and persimmons (figure 7.1)—a feature that
most prokaryotes and protists lack? Your cells touch and
communicate with one another. Sending and receiving a variety of chemical signals, they coordinate their behavior so that
your body functions as an integrated whole, rather than as a
massive collection of individual cells acting independently.
The ability of cells to communicate with one another is the
hallmark of multicellular organisms. In this chapter, we look
in detail at how the cells of multicellular organisms interact
with one another, first exploring how they signal one another
with chemicals and then examining the ways in which their
cell surfaces interact to organize tissues and body structures.

Cells signal one another with chemicals.

Receptor Proteins and Signaling Between Cells

Communication between cells is common in nature. Cell signaling occurs in all multicellular organisms, providing an indispensable mechanism for cells to influence one another. The cells of multicellular organisms use a variety of molecules as signals, including not only peptides, but also large proteins, individual amino acids, nucleotides, steroids, and other lipids.

Even dissolved gases are used as signals. Nitric oxide (NO) plays a wide variety of roles, including mediating male erections (Viagra functions by stimulating NO release).

Some of these molecules are attached to the surface of the signaling cell; others are secreted through the plasma membrane or released by exocytosis.

Cell Surface Receptors

Any given cell of a multicellular organism is exposed to a constant stream of signals. At any time, hundreds of different chemical signals may be in the environment surrounding the cell. However, each cell responds only to certain signals and ignores the rest, like a person following the conversation of one or two individuals in a noisy, crowded room. How does a cell "choose" which signals to respond to? Located on or within the cell are **receptor proteins,** each with a three-dimensional shape that fits the shape of a specific signal molecule. When a signal molecule approaches a receptor protein of the right shape, the two can bind. This binding induces a change in the receptor protein's shape, ultimately producing a response in the cell. Hence, a given cell responds to the signal molecules that fit the particular set of receptor proteins it possesses, and ignores those for which it lacks receptors.

The Hunt for Receptor Proteins

The characterization of receptor proteins has presented a very difficult technical problem because of their relative scarcity in the cell. Because these proteins may constitute less than 0.01% of the total mass of protein in a cell, purifying them is analogous to searching for a particular grain of sand in a sand dune! To make matters worse, any specific receptor may only be expressed in a small subset of cells in the body. However, two recent techniques—immunochemistry and molecular genetics—have enabled cell biologists to make rapid progress in this area.

Immunochemistry. The first method uses antibodies that recognize specific molecules that bind to receptors.

Using these tools, researchers can selectively purify specific receptor molecules from the complex mix of proteins found in a population of cells. Once purified, the specific chemistry and structure of the receptor molecules can be studied in detail.

Antibodies are special proteins produced by the immune system that, like receptors, bind very specifically to particular target molecules. The body naturally produces antibodies that bind to any foreign molecule entering the body as part of the immune defense. Each antibody-producing cell of the immune system makes its own unique antibody molecule with its own specific ability to bind to the target molecule. Researchers have taken advantage of this by isolating a single immune cell that is producing an antibody that recognizes a particular receptor or other target molecule. By growing a population of cells from that one starting cell (or cloning that cell), a monoclonal antibody can be purified that specifically binds to the target molecule.

Molecular Genetics. A second powerful approach has also allowed great advances in the study of the structure and function of receptors. Although receptors are commonly in low abundance and difficult to purify directly, the cellular defects caused by receptor malfunction can be very pronounced. Researchers have taken advantage of this by intentionally creating mutations in genes that cause significant cellular defects characteristic of receptor malfunction. The genes altered in these mutants are then purified using a variety of techniques, many of which exploit the wealth of information now available from the sequence of entire genomes of organisms (see chapter 17). By characterizing the structure of the genes, much can be learned about the structure and function of the proteins they encode. More importantly, by analyzing the sites within the gene that can be mutated to produce specific defects in the function of the receptor, the relationship between specific protein structures and their cellular functions can be determined.

Remarkably, these techniques have revealed that the enormous number of receptor proteins can be grouped into just a handful of "families" containing many related receptors. Later in this chapter, we will meet some of the members of these receptor families.

Cells in a multicellular organism communicate with others by releasing signal molecules that bind to receptor proteins on the surface of the other cells. Recent advances in protein isolation have yielded a wealth of information about the structure and function of these proteins.

Types of Cell Signaling

Cells can communicate through any of four basic mechanisms, depending primarily on the distance between the signaling and responding cells (figure 7.2). In addition to using these four basic mechanisms, some cells actually send signals to themselves, secreting signals that bind to specific receptors on their own plasma membranes. This process, called *autocrine signaling*, is thought to play an important role in reinforcing developmental changes.

Direct Contact

As we saw in chapter 6, the surface of a eukaryotic cell is richly populated with proteins, carbohydrates, and lipids attached to and extending outward from the plasma membrane. When cells are very close to one another, some of the molecules on the plasma membrane of one cell can be recognized by receptors on the plasma membrane of an adjacent cell. Many of the important interactions between cells in early development occur by means of *direct contact* between cell surfaces (figure 7.2*a*). We'll examine contact-dependent interactions more closely later in this chapter.

Paracrine Signaling

Signal molecules released by cells can diffuse through the extracellular fluid to other cells. If those molecules are taken up by neighboring cells, destroyed by extracellular enzymes, or quickly removed from the extracellular fluid in some other way, their influence is restricted to cells in the immediate vicinity of the releasing cell. Signals with such short-lived, local effects are called *paracrine signals* (figure 7.2*b*). Like direct contact, paracrine signaling plays an important role in early development, coordinating the activities of clusters of neighboring cells.

Endocrine Signaling

If a released signal molecule remains in the extracellular fluid, it may enter the organism's circulatory system and travel widely throughout the body. These longer-lived signal molecules, which may affect cells very distant from the releasing cell, are called **hormones,** and this type of intercellular communication is known as *endocrine signaling* (figure 7.2*c*).

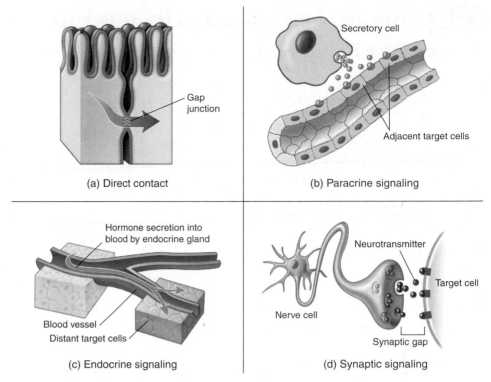

FIGURE 7.2
Four kinds of cell signaling. Cells communicate in several ways. (*a*) Two cells in direct contact with each other may send signals across gap junctions. (*b*) In paracrine signaling, secretions from one cell have an effect only on cells in the immediate area. (*c*) In endocrine signaling, hormones are released into the organism's circulatory system, which carries them to the target cells. (*d*) Chemical synapse signaling involves transmission of signal molecules, called neurotransmitters, from a neuron over a small synaptic gap to the target cell.

Chapter 47 discusses endocrine signaling in detail. Both animals and plants use this signaling mechanism extensively.

Synaptic Signaling

In animals, the cells of the nervous system provide rapid communication with distant cells. Their signal molecules, **neurotransmitters,** do not travel to the distant cells through the circulatory system as hormones do. Rather, the long, fiberlike extensions of nerve cells release neurotransmitters from their tips very close to the target cells (figure 7.2*d*). The association of a neuron and its target cell is called a **chemical synapse,** and this type of intercellular communication is called *synaptic signaling*. While paracrine signals move through the fluid between cells, neurotransmitters cross the synaptic gap and persist only briefly. We will examine synaptic signaling more fully in chapter 45.

A cell can signal others by direct contact, by locally diffusing paracrine signals, by hormones that travel to distant cells, and by synaptic signaling between a nerve cell and other nerve or muscle cells.

7.2 Proteins in the cell and on its surface receive signals from other cells.

Intracellular Receptors

All cell signaling pathways share certain common elements, including a chemical signal that passes from one cell to another and a receptor that receives the signal in or on the target cell. We've looked at the sorts of signals that pass from one cell to another. Now let's consider the nature of the receptors that receive the signals. Table 7.1 summarizes the types of receptors we will discuss in this chapter.

Many cell signals are lipid-soluble or very small molecules that can readily pass across the plasma membrane of the target cell and into the cell, where they interact with a receptor inside. Some bind to protein receptors located in the cytoplasm; others pass across the nuclear membrane as well and bind to receptors within the nucleus. These **intracellular receptors** (figure 7.3) may trigger a variety of responses in the cell, depending on the receptor.

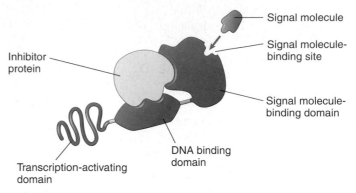

FIGURE 7.3

Basic structure of a gene-regulating intracellular receptor. These receptors are located within the cell and function in the reception of signals such as steroid hormones, vitamin D, and thyroid hormone.

Table 7.1 Cell-Communicating Mechanisms			
Mechanism	**Structure**	**Function**	**Example**
INTRACELLULAR RECEPTORS	No extracellular signal-binding site	Receives signals from lipid-soluble or noncharged, nonpolar small molecules	Receptors for NO, steroid hormone, vitamin D, and thyroid hormone
CELL SURFACE RECEPTORS			
Chemically gated ion channels	Multipass transmembrane protein forming a central pore	Molecular "gates" triggered chemically to open or close	Neurons
Enzymic receptors	Single-pass transmembrane protein	Binds signal extracellularly; catalyzes response intracellularly	Phosphorylation of protein kinases
G-protein-linked receptors	Seven-pass transmembrane protein with cytoplasmic binding site for G protein	Binding of signal to receptor causes GTP to bind a G protein; G protein, with attached GTP, detaches to deliver the signal inside the cell	Peptide hormones, rod cells in the eyes
PHYSICAL CONTACT WITH OTHER CELLS			
Surface markers	Variable; integral proteins or glycolipids in plasma membrane	Identify the cell	MHC complexes, blood groups, antibodies
Tight junctions	Tightly bound, leakproof, fibrous protein seal that surrounds cell	Organizing junction: holds cells together such that materials pass *through* but not *between* the cells	Junctions between epithelial cells in the gut
Desmosomes	Intermediate filaments of cytoskeleton linked to adjoining cells through cadherins	Anchoring junction: binds cells together	Epithelium
Adherens junctions	Transmembrane fibrous proteins	Anchoring junction: connects extracellular matrix to cytoskeleton	Tissues with high mechanical stress, such as the skin
Gap junctions	Six transmembrane connexon proteins creating a pore that connects cells	Communicating junction: allows passage of small molecules from cell to cell in a tissue	Excitable tissue such as heart muscle
Plasmodesmata	Cytoplasmic connections between gaps in adjoining plant cell walls	Communicating junction between plant cells	Plant tissues

Receptors That Act as Gene Regulators

Some intracellular receptors act as regulators of gene expression. Among them are the receptors for steroid hormones, such as cortisol, estrogen, and progesterone, as well as the receptors for a number of other small, lipid-soluble signal molecules, such as vitamin D and thyroid hormone. All of these receptors have similar structures; the genes that code for them may well be the evolutionary descendants of a single ancestral gene. Because of their structural similarities, they are all part of the *intracellular receptor superfamily*.

Each of these receptors has a binding site for DNA. In its inactive state, the receptor typically cannot bind DNA because an inhibitor protein occupies the binding site. When the signal molecule binds to another site on the receptor, the inhibitor is released, and the DNA-binding site is exposed (figure 7.4). The receptor then binds to specific nucleotide sequences on the DNA, which activates (or, in a few instances, suppresses) particular genes, usually located adjacent to the regulatory sites.

The lipid-soluble signal molecules that intracellular receptors recognize tend to persist in the blood far longer than water-soluble signals. Most water-soluble hormones break down within minutes, and neurotransmitters break down within seconds or even milliseconds. On the other hand, a steroid hormone such as cortisol or estrogen persists for hours.

The target cell's response to a lipid-soluble cell signal can vary enormously, depending on the nature of the cell. This is true even when different target cells have the same intracellular receptor, for two reasons: First, the binding site for the receptor on the target DNA differs from one cell type to another, so that different genes are affected when the signal-receptor complex binds to the DNA. Second, most eukaryotic genes have complex controls. We will discuss these controls in detail in chapter 18, but for now it is sufficient to note that several different regulatory proteins are usually involved in reading a eukaryotic gene. Thus, the intracellular receptor interacts with different signals in different tissues. Depending on the cell-specific controls operating in different tissues, the effect the intracellular receptor produces when it binds with DNA will vary.

Receptors That Act as Enzymes

Other intracellular receptors act as enzymes. A very interesting example is the receptor for the signal molecule nitric oxide (NO). The small gas molecule NO diffuses readily out of the cells where it is produced and passes directly into neighboring cells, where it binds to the enzyme guanylyl cyclase. Binding of NO activates the enzyme, enabling it to catalyze the synthesis of cyclic guanosine monophosphate (GMP), an intracellular messenger molecule that produces cell-specific responses such as the relaxation of smooth muscle cells.

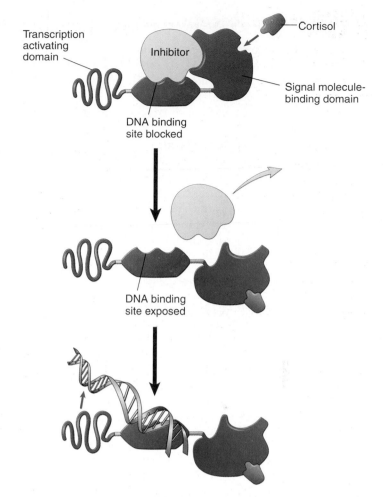

FIGURE 7.4
How intracellular receptors regulate gene transcription. In this model, the binding of the steroid hormone cortisol to a DNA regulatory protein causes it to alter its shape. The inhibitor is released, exposing the DNA-binding site of the regulatory protein. The DNA binds to the site, positioning a specific nucleotide sequence over the transcription-activating domain of the receptor and initiating transcription.

NO has only recently been recognized as a signal molecule in vertebrates. Already, however, a wide variety of roles have been documented. For example, when the brain sends a nerve signal relaxing the smooth muscle cells lining the walls of vertebrate blood vessels, the signal molecule acetylcholine released by the nerve cell near the muscle does not interact with the muscle cell directly. Instead, it causes nearby epithelial cells to produce NO, which then causes the smooth muscle to relax, allowing the vessel to expand and thereby increasing blood flow.

Many target cells possess intracellular receptors, which are activated by substances that pass through the plasma membrane.

Cell Surface Receptors

Most signal molecules are water-soluble, including neurotransmitters, peptide hormones, and the many proteins that multicellular organisms employ as growth factors during development. Water-soluble signals cannot diffuse through plasma membranes. Therefore, to trigger responses in cells, they must bind to receptor proteins on the outer surface of the cell. These **cell surface receptors** (figure 7.5) convert the extracellular signal to an intracellular one, responding to the binding of the signal molecule to the cell's outside by producing a change inside the cell. Most of a cell's receptors are cell surface receptors, and almost all of them belong to one of three receptor superfamilies: chemically gated ion channels, enzymic receptors, and G-protein-linked receptors.

Chemically Gated Ion Channels

Chemically gated ion channels are receptor proteins that ions pass through. The receptor proteins that bind many neurotransmitters have the same basic structure (figure 7.5a). Each is a "multipass" transmembrane protein, meaning that the chain of amino acids threads back and forth across the plasma membrane several times. In the center of the protein is a pore that connects the extracellular fluid with the cytoplasm. The pore is big enough for ions to pass through, so the protein functions as an **ion channel.** The channel is said to be chemically gated because it opens only when a chemical (the neurotransmitter) binds to it. The type of ion (sodium, potassium, calcium, or chloride, for example) that flows across the membrane when a chemically gated ion channel opens depends on the specific shape and charge structure of the channel.

Enzymic Receptors

Many cell surface receptors either act as enzymes or are directly linked to enzymes (figure 7.5b). When a signal molecule binds to the receptor, it activates the enzyme. In almost all cases, these enzymes are **protein kinases,** enzymes that add phosphate groups to proteins. Most enzymic receptors have the same general structure. Each is a single-pass transmembrane protein (the amino acid chain passes through the plasma membrane only once); the portion that binds the signal molecule lies outside the cell, and the portion that carries out the enzyme activity is exposed to the cytoplasm.

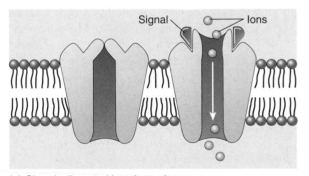

(a) Chemically gated ion channel

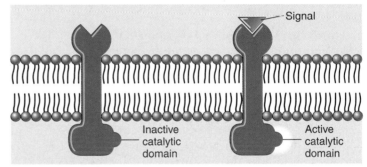

(b) Enzymic receptor

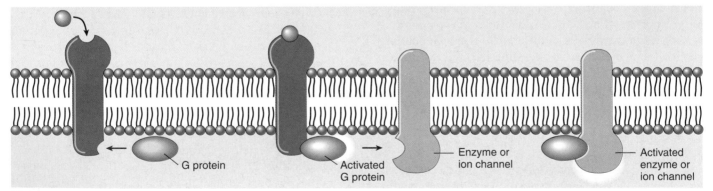

(c) G-protein-linked receptor

FIGURE 7.5

Cell surface receptors. (*a*) Chemically gated ion channels are multipass transmembrane proteins that form a pore in the plasma membrane. This pore is opened or closed by chemical signals. (*b*) Enzymic receptors are single-pass transmembrane proteins that bind the signal on the extracellular surface. A catalytic region on their cytoplasmic portion then initiates enzymatic activity inside the cell. (*c*) G-protein-linked receptors bind to the signal outside the cell and to G proteins inside the cell. The G protein then activates an enzyme or ion channel, mediating the passage of a signal from the cell's surface to its interior.

G-Protein-Linked Receptors

A third class of cell surface receptors acts indirectly on enzymes or ion channels in the plasma membrane with the aid of an assisting protein, called a *guanosine triphosphate* (GTP)-*binding protein*, or **G protein** (figure 7.5c). Receptors in this category use G proteins to mediate passage of the signal from the membrane surface into the cell interior.

How G-Protein-Linked Receptors Work. G proteins are mediators that initiate a diffusible signal in the cytoplasm. They form a transient link between the receptor on the cell surface and the signal pathway within the cytoplasm. Importantly, this signal has a relatively short duration. When a signal arrives, it finds the G protein nestled into the G-protein-linked receptor on the cytoplasmic side of the plasma membrane. Once the signal molecule binds to the receptor, the G-protein-linked receptor changes shape. This change in receptor shape twists the G protein, causing it to bind GTP. The G protein can now diffuse away from the receptor. The "activated" complex of a G protein with attached GTP is then free to initiate a number of events. However, this activation is short-lived, because GTP has a relatively short life span (seconds to minutes) before it is broken down into GDP + P_i. This elegant arrangement allows the G proteins to activate numerous pathways, but only in a transient manner. In order for a pathway to "stay on," there must be a continuous source of incoming extracellular signals. When the rate of external signal drops off, the pathway shuts down.

Scientists have identified more than 100 different G-protein-linked receptors, more than any other kind of cell surface receptor. These receptors mediate signaling, utilizing an incredible range of molecules, including peptide hormones, neurotransmitters, fatty acids, and amino acids. Despite this great variation in specificity, however, all G-protein-linked receptors whose amino acid sequences are known have a similar structure. They are almost certainly closely related in an evolutionary sense, probably arising from a single ancestral sequence. Each of these G-protein-linked receptors is a seven-pass transmembrane protein (figure 7.6); that is, a single polypeptide chain threads back and forth across the lipid bilayer seven times, creating a channel through the membrane.

Evolutionary Origin of G-Protein-Linked Receptors. As research revealed the structure of G-protein-linked receptors, an interesting pattern emerged. The same seven-pass structural motif is seen again and again: in sensory receptors such as the light-activated rhodopsin protein in the vertebrate eye, in the light-activated bacteriorhodopsin proton pump that plays a key role in bacterial photosynthesis, in the receptor that recognizes the yeast mating factor protein, and in many other sensory receptors. Vertebrate

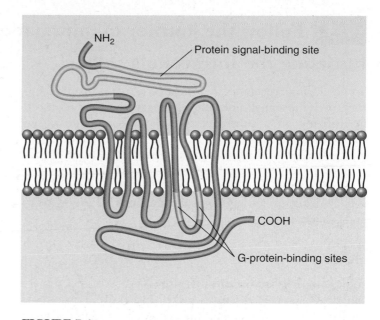

FIGURE 7.6
The G-protein-linked receptor. Each G-protein-linked receptor is a seven-pass transmembrane protein.

rhodopsin is in fact a G-protein-linked receptor and utilizes a G protein. Bacteriorhodopsin is not. The occurrence of the seven-pass structural motif in both, and in so many other G-protein-linked receptors, suggests that this motif is a very ancient one, and that G-protein-linked receptors may have evolved from sensory receptors of single-celled ancestors.

Discovery of G Proteins. Martin Rodbell of the National Institute of Environmental Health Sciences and Alfred Gilman of the University of Texas Southwestern Medical Center received the 1994 Nobel Prize for Medicine or Physiology for their work on G proteins. Rodbell and Gilman's work has proven to have significant ramifications. G proteins are involved in the mechanism employed by over half of all medicines in use today. Studying G proteins will vastly expand our understanding of how these medicines work. Furthermore, the investigation of G proteins should help elucidate how cells communicate in general and how they contribute to the overall physiology of an organism. As Gilman says, G proteins are "involved in everything from sex in yeast to cognition in humans."

Most receptors are located on the surface of the plasma membrane. Chemically gated ion channels open or close when signal molecules bind to the channel, allowing specific ions to diffuse through. Enzymic receptors typically activate intracellular proteins by phosphorylation. G-protein-linked receptors activate an intermediary protein, which then effects the intracellular change.

7.3 Follow the journey of information into the cell.

Initiating the Intracellular Signal

Some enzymic receptors and most G-protein-linked receptors carry the signal molecule's message into the target cell by utilizing other substances to relay the message within the cytoplasm. These other substances, small molecules or ions called **second messengers**, alter the behavior of particular proteins by binding to them and changing their shape. The two most widely used second messengers are cyclic adenosine monophosphate (**cyclic AMP, or cAMP**) and calcium ions.

cAMP

All animal cells studied thus far use cAMP as a second messenger (chapter 47 discusses cAMP in detail). To see how cAMP typically works as a messenger, let's examine what happens when the hormone epinephrine binds to a particular type of G-protein-linked receptor called the β-adrenergic receptor in muscle cells (figure 7.7). When epinephrine binds with this receptor, it activates a G protein, which then stimulates the enzyme **adenylyl cyclase** to produce large amounts of cAMP within the cell (figure 7.8a). The cAMP then binds to and activates the enzyme α-kinase, which adds phosphates to specific proteins in the cell. The effect of this phosphorylation on cell function depends on the identity of the cell and the proteins that are phosphorylated. In muscle cells, for example, the α-kinase phosphorylates and thereby activates enzymes that stimulate the breakdown of glycogen into glucose and inhibit the synthesis of glycogen from glucose. More glucose is then available to the muscle cells.

Calcium

Calcium ions (Ca^{++}) serve widely as second messengers. Ca^{++} levels inside the cytoplasm are normally very low (less than 10^{-7} M), while outside the cell and in the endoplasmic reticulum, Ca^{++} levels are quite high (about 10^{-3} M). Chemically gated calcium channels in the endoplasmic reticulum membrane act as switches; when they open, Ca^{++} rushes into the cytoplasm and triggers proteins sensitive to Ca^{++} to initiate a variety of activities. The efflux of Ca^{++} from the endoplasmic reticulum causes skeletal muscle cells to contract and endocrine cells to secrete hormones.

Gated Ca^{++} channels are opened by a G-protein-linked receptor. In response to signals from other cells,

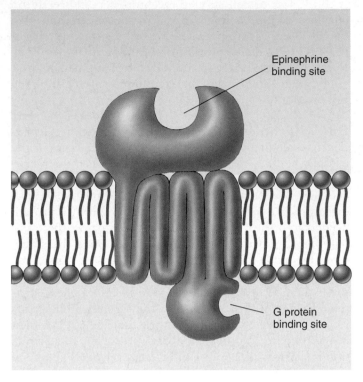

FIGURE 7.7
Structure of the β-adrenergic receptor. The receptor is a G-protein-linked molecule that, when it binds to an extracellular signal molecule, stimulates voluminous production of cAMP inside the cell, which then effects the cellular change.

the receptor activates its G protein, which in turn activates the enzyme *phospholipase C*. This enzyme catalyzes the production of *inositol trisphosphate* (IP₃) from phospholipids in the plasma membrane. The IP₃ molecules diffuse through the cytoplasm to the endoplasmic reticulum and bind to the Ca^{++} channels. This opens the channels and allows Ca^{++} to flow from the endoplasmic reticulum into the cytoplasm (figure 7.8b).

Ca^{++} initiates some cellular responses by binding to *calmodulin*, a 148-amino-acid cytoplasmic protein that contains four binding sites for Ca^{++} (figure 7.9). When four Ca^{++} ions are bound to calmodulin, the calmodulin/Ca^{++} complex binds to other proteins and activates them.

> Cyclic AMP and Ca^{++} often behave as second messengers, relaying messages from receptors to target proteins.

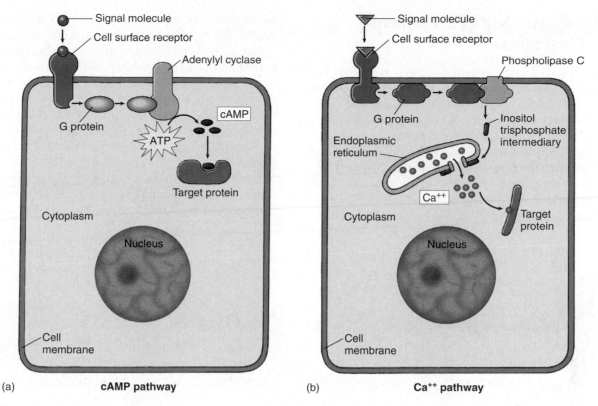

(a)   **cAMP pathway**   (b)   **Ca⁺⁺ pathway**

FIGURE 7.8

How second messengers work.   (*a*) The cyclic AMP (cAMP) pathway. An extracellular receptor binds to a signal molecule and, through a G protein, activates the membrane-bounded enzyme adenylyl cyclase. This enzyme catalyzes the synthesis of cAMP, which binds to the target protein to initiate the cellular change. (*b*) The calcium (Ca^{++}) pathway. An extracellular receptor binds to another signal molecule and, through another G protein, activates the enzyme phospholipase C. This enzyme stimulates the production of inositol trisphosphate, which binds to and opens calcium channels in the membrane of the endoplasmic reticulum. Ca^{++} is released into the cytoplasm, effecting a change in the cell.

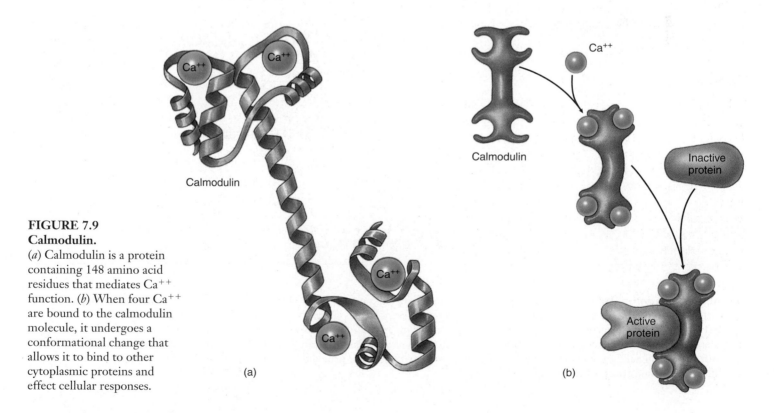

FIGURE 7.9
Calmodulin.

(*a*) Calmodulin is a protein containing 148 amino acid residues that mediates Ca^{++} function. (*b*) When four Ca^{++} are bound to the calmodulin molecule, it undergoes a conformational change that allows it to bind to other cytoplasmic proteins and effect cellular responses.

Amplifying the Signal: Protein Kinase Cascades

Both enzyme-linked and G-protein-linked receptors receive signals at the surface of the cell, but as we've seen, the target cell's response rarely takes place there. In most cases, the signals are relayed to the cytoplasm or the nucleus by second messengers, which influence the activity of one or more enzymes or genes and so alter the behavior of the cell. But most signaling molecules are found in such low concentrations that their diffusion across the cytoplasm would take a great deal of time unless the signal were amplified. Therefore, most enzyme-linked and G-protein-

linked receptors use a chain of other protein messengers to amplify the signal as it is being relayed to the nucleus.

How is the signal amplified? Imagine a relay race where, at the end of each stage, the finishing runner tags five new runners to start the next stage. The number of runners would increase dramatically as the race progresses: 1, then 5, 25, 125, and so on. The same sort of process takes place as a signal is passed from the cell surface to the cytoplasm or nucleus. First, the receptor activates a stage-one protein, almost always by phosphorylating it. The receptor either adds a phosphate group directly, or it activates a G protein that goes on to activate a second protein that does the phosphorylation. Once activated, each of these stage-one

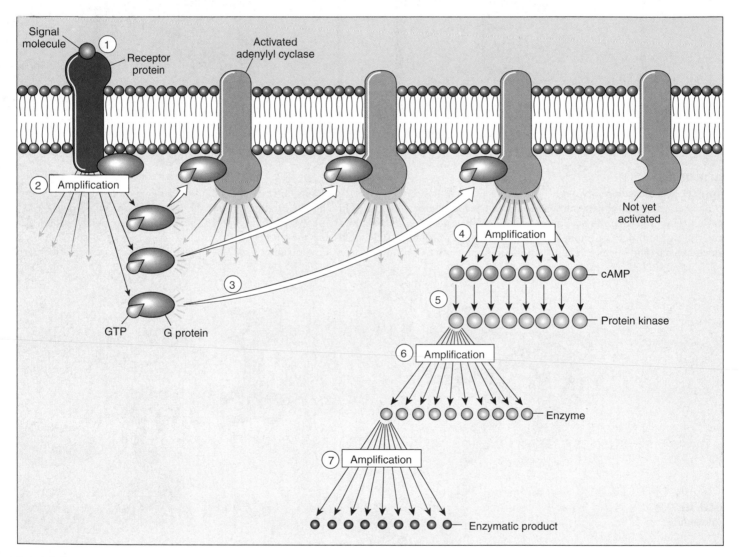

FIGURE 7.10

Signal amplification. Amplification at many stages of the cell-signaling process can ultimately produce a large response by the cell. One cell surface receptor (*1*), for example, may activate many G protein molecules (*2*), each of which activates a molecule of adenylyl cyclase (*3*), yielding an enormous production of cAMP (*4*). Each cAMP molecule in turn activates a protein kinase (*5*), which can phosphorylate and thereby activate several copies of a specific enzyme (*6*). Each of *those* enzymes can then catalyze many chemical reactions (*7*). Starting with 10^{-10} M of signaling molecule, one cell surface receptor can trigger the production of 10^{-6} M of one of the products, an amplification of four orders of magnitude.

proteins in turn activates a large number of stage-two proteins; then each of them activates a large number of stage-three proteins; and so on (figure 7.10). A single cell surface receptor can thus stimulate a cascade of protein kinases to amplify the signal and cause a strong response inside the cell to a weak signal outside.

This kind of chain reaction or signaling cascade is common after receptors are triggered. It can involve several different proteins, but commonly involves G proteins, phospholipase C, IP$_3$, calcium, and sometimes protein kinases. Although each receptor-triggered signal transduction event may employ some of the same intermediate molecules, they ultimately activate particular sets of targets to cause a cellular response that is unique and specific to the particular receptor triggered.

The Vision Amplification Cascade

Let's trace a protein amplification cascade to see exactly how it works. In vision, a single, light-activated rhodopsin (a G-protein-linked receptor) activates hundreds of molecules of the G protein transducin in the first stage of the relay. In the second stage, each transducin causes an enzyme to modify thousands of molecules of a special inside-the-cell messenger called cyclic GMP. (We will discuss cyclic GMP in more detail later.) In about 1 second, a single rhodopsin signal passing through this two-step cascade splits more than 10^5 (100,000) cyclic GMP molecules (figure 7.11). The rod cells of humans are sufficiently sensitive to detect brief flashes of just 5 photons.

The Cell Division Amplification Cascade

The general principle of signal amplification occurs in many cell types and for many receptor types. However, the specific mechanisms of how that amplification is achieved vary considerably. Cell division, for example, is regulated by integrating many signals from outside the cell. One of the receptors involved acts as a protein kinase. The receptor responds to growth-promoting signals by phosphorylating an intracellular protein kinase called ras, which then activates a series of interacting phosphorylation cascades, some with five or more stages. If the ras protein becomes hyperactive for any reason, the cell acts as if it is being constantly stimulated to divide. Ras proteins were first discovered in cancer cells. A mutation of the gene that encodes ras had caused it to become hyperactive, resulting in unrestrained cell proliferation. Almost one-third of human cancers have such a mutation in a *ras* gene.

A small number of surface receptors can generate a vast intracellular response as each stage of the pathway amplifies the next.

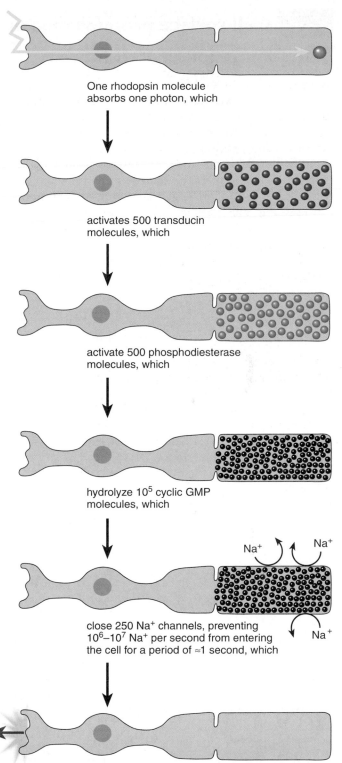

One rhodopsin molecule absorbs one photon, which

activates 500 transducin molecules, which

activate 500 phosphodiesterase molecules, which

hydrolyze 10^5 cyclic GMP molecules, which

close 250 Na$^+$ channels, preventing 10^6–10^7 Na$^+$ per second from entering the cell for a period of ≈1 second, which

hyperpolarizes the rod cell membrane by 1 mV, sending a visual signal to the brain.

FIGURE 7.11
The role of signal amplification in vision. In this vertebrate light-detecting rod cell, *one* single rhodopsin pigment molecule, when excited by a photon, ultimately yields *100,000* split cyclic GMP molecules, which will then effect a change in the membrane of the rod cell that is interpreted by the organism as a visual event.

7.4 Cell surface proteins mediate cell–cell interactions.

The Expression of Cell Identity

With the exception of a few primitive types of organisms, the hallmark of multicellular life is the development of highly specialized groups of cells called **tissues,** such as blood and muscle. Remarkably, each cell within a tissue performs the functions of that tissue and no other, even though all cells of the body are derived from a single fertilized cell and contain the same genetic information—all of the genes found in the genome. However, as an organism develops, the cells acquire their specific identities by carefully controlling the *expression* of those genes, turning on the specific set of genes that encode the particular functions of each cell type. How do cells sense where they are, and how do they "know" which type of tissue they belong to?

Tissue-Specific Identity Markers

One key set of genes function to mark the surfaces of cells to identify them as being of a particular type. When cells make contact, they "read" each others' cell surface markers and react accordingly. When cells that are part of the same tissue type recognize each other, they frequently respond by forming connections between their cell surfaces to better coordinate their functions.

Glycolipids. Most tissue-specific cell surface markers are glycolipids, lipids with carbohydrate heads. The glycolipids on the surface of red blood cells are also responsible for the differences among A, B, and O blood types.

MHC Proteins. One important example of the function of cell surface markers is the recognition of "self" and "nonself" cells by the immune system. This is an important function for multicellular organisms, where the ability to distinguish between the cells of the organism and foreign cells is important.

The immune system of vertebrates uses a particular set of markers to distinguish self from nonself cells, encoded by genes of the *major histocompatibility complex* (MHC). These genes encode a set of cell surface proteins that are unique to the individual and serve as effective identity tags. The MHC proteins and other self-identifying markers are single-pass proteins anchored in the plasma membrane, and many of them are members of a large superfamily of receptors, the immunoglobulins (figure 7.12). Cells of the immune system continually inspect other cells they encounter in the body, triggering the destruction of cells that display foreign or nonself identity markers.

Every cell contains a specific array of marker proteins on its surface. These markers identify each type of cell in a very precise way.

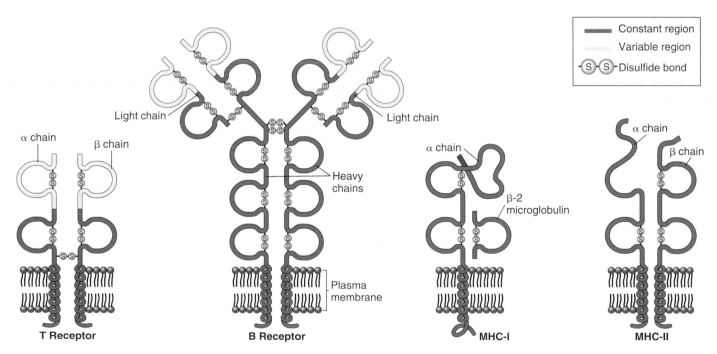

FIGURE 7.12
Structure of the immunoglobulin family of cell surface marker proteins. T and B cell receptors help mediate the immune response in organisms by recognizing and binding to foreign cell markers. MHC antigens label cells as "self," so that the immune system attacks only invading entities, such as bacteria, viruses, and often the cells of transplanted organs that have different MHC antigens.

Intercellular Adhesion

Not all physical contacts between cells in a multicellular organism are fleeting touches. In fact, most cells are in physical contact with other cells at all times, usually as members of organized tissues such as those in a leaf, or in your lungs, heart, or gut. These cells and the mass of other cells clustered around them form long-lasting or permanent connections with each other called **cell junctions** (figure 7.13). The nature of the physical connections between the cells of a tissue in large measure determines what the tissue is like. Indeed, a tissue's proper functioning often depends critically upon how the individual cells are arranged within it. Just as a house cannot maintain its structure without nails and cement, so a tissue cannot maintain its characteristic architecture without the appropriate cell junctions.

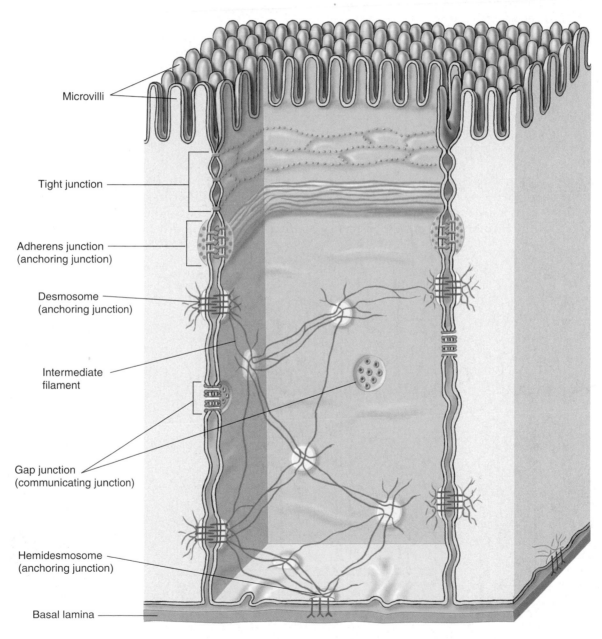

Microvilli

Tight junction

Adherens junction
(anchoring junction)

Desmosome
(anchoring junction)

Intermediate
filament

Gap junction
(communicating junction)

Hemidesmosome
(anchoring junction)

Basal lamina

FIGURE 7.13
An overview of cell junction types. Here, gut epithelial cells illustrate the comparative structures and locations of common cell junctions.

Cell junctions are divided into three categories, based upon their functions: tight junctions, anchoring junctions, and communicating junctions (figure 7.14).

Tight Junctions

Sometimes called occluding junctions, tight junctions connect the plasma membranes of adjacent cells in a sheet, preventing small molecules from leaking between the cells. This allows the sheet of cells to act as a wall within the organ, keeping molecules on one side or the other.

Creating Sheets of Cells. The cells that line an animal's digestive tract are organized in a sheet only one cell thick. One surface of the sheet faces the inside of the tract, and the other faces the extracellular space where blood vessels are located. Tight junctions encircle each cell in the sheet, like a belt cinched around a person's waist. The junctions between neighboring cells are so securely attached that there is no space between them for leakage. Hence, nutrients absorbed from the food in the digestive tract must pass directly through the cells in the sheet to enter the blood because they cannot pass through spaces between cells.

Partitioning the Sheet. The tight junctions between the cells lining the digestive tract also partition the plasma membranes of these cells into separate compartments. Transport proteins in the membrane facing the inside of the tract carry nutrients from that side to the cytoplasm of the cells. Other proteins, located in the membrane on the opposite side of the cells, transport those nutrients from the cytoplasm to the extracellular fluid, where they can enter the blood. For the sheet to absorb nutrients properly, these proteins must remain in the correct locations within the fluid membrane. Tight junctions effectively segregate the proteins on opposite sides of the sheet, preventing them from drifting within the membrane from one side of the sheet to the other. When tight junctions are experimentally disrupted, just this sort of migration occurs.

Anchoring Junctions

Anchoring junctions mechanically attach the cytoskeleton of a cell to the cytoskeletons of other cells or to the extracellular matrix. These junctions are most common in tissues subject to mechanical stress, such as muscle and skin epithelium.

Cadherin and Intermediate Filaments: Desmosomes. Anchoring junctions called **desmosomes** connect the cytoskeletons of adjacent cells (figure 7.15), while *hemidesmosomes* anchor epithelial cells to a basement membrane. Proteins called **cadherins,** most of which are single-pass transmembrane glycoproteins, create the critical link. A variety of attachment proteins link the short cytoplasmic end

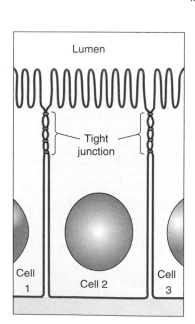

(a) Tight junction

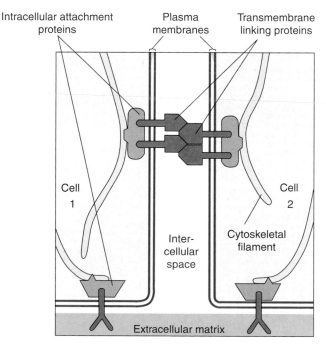

(b) Anchoring junction

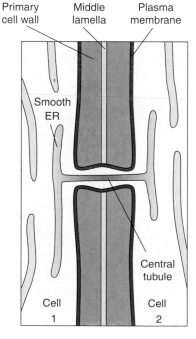

(c) Communicating junction

FIGURE 7.14
The three types of cell junctions. These models represent current thinking on how the structures of the three major types of cell junctions facilitate their function: (*a*) tight junction; (*b*) anchoring junction; (*c*) communicating junction.

of a cadherin to the intermediate filaments in the cytoskeleton. The other end of the cadherin molecule projects outward from the plasma membrane, joining directly with a cadherin protruding from an adjacent cell in a firm handshake, binding the cells together.

Connections between proteins tethered to the intermediate filaments are much more secure than connections between free-floating membrane proteins. As described in chapter 6, proteins are suspended within the membrane by relatively weak interactions between the nonpolar portions of the protein and the membrane lipids. It would not take much force to pull an untethered protein completely out of the membrane.

Cadherin and Actin Filaments. Cadherins can also connect the actin frameworks of cells in cadherin-mediated junctions (figure 7.16). When they do, they form less stable links between cells than when they connect intermediate filaments. Many kinds of actin-linking cadherins occur in different tissues. During vertebrate development, the migration of neurons in the embryo is associated with changes in the type of cadherin expressed on their plasma membranes. This suggests that gene-controlled changes in cadherin expression may provide the migrating cells with a "roadmap" to their destination.

Integrin-Mediated Links. Anchoring junctions called **adherens junctions** connect the actin filaments of one cell with those of neighboring cells or with the extracellular matrix. The linking proteins in these junctions are members of a large superfamily of cell surface receptors called integrins that bind to a protein component of the extracellular matrix. There appear to be many different kinds of integrins (cell biologists have identified 20), each with a slightly different-shaped binding domain. The exact component of the matrix that a given cell binds to depends on which combination of integrins that cell has in its plasma membrane.

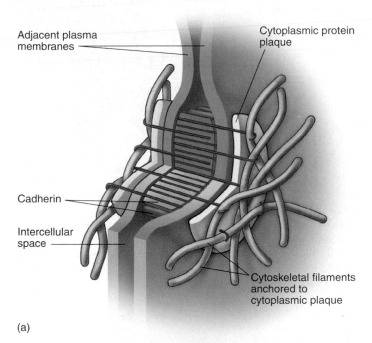

Adjacent plasma membranes

Cytoplasmic protein plaque

Cadherin

Intercellular space

Cytoskeletal filaments anchored to cytoplasmic plaque

(a)

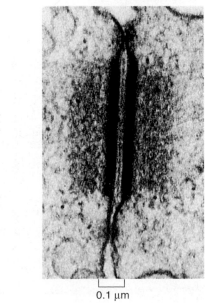

(b)   0.1 μm

FIGURE 7.15
Desmosomes. (*a*) Cadherin proteins create the adhering link between adjoining cells. (*b*) A desmosome anchors adjacent cells to one another.

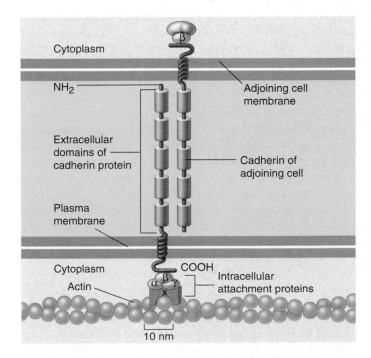

Cytoplasm

NH₂

Extracellular domains of cadherin protein

Plasma membrane

Cytoplasm

Actin

β

Adjoining cell membrane

Cadherin of adjoining cell

COOH

Intracellular attachment proteins

10 nm

FIGURE 7.16
A cadherin-mediated junction. The cadherin molecule is anchored to actin in the cytoskeleton and passes through the membrane to interact with the cadherin of an adjoining cell.

Communicating Junctions

Many cells communicate with adjacent cells through direct connections called **communicating junctions.** In these junctions, a chemical or electrical signal passes directly from one cell to an adjacent one. Communicating junctions establish direct physical connections that link the cytoplasms of two cells together, permitting small molecules or ions to pass from one to the other. In animals, these direct communication channels between cells are called gap junctions, and in plants, plasmodesmata.

Gap Junctions in Animals. Gap junctions are composed of structures called connexons, complexes of six identical transmembrane proteins (figure 7.17). The proteins in a connexon are arranged in a circle to create a channel through the plasma membrane that protrudes several nanometers from the cell surface. A gap junction forms when the connexons of two cells align perfectly, creating an open channel that spans the plasma membranes of both cells. Gap junctions provide passageways large enough to permit small substances, such as simple sugars and amino acids, to pass from the cytoplasm of one cell to that of the next, yet small enough to prevent the passage of larger molecules such as proteins. The connexons hold the plasma membranes of the paired cells about 4 nanometers apart, in marked contrast to the more-or-less direct contact between the lipid bilayers in a tight junction.

Gap junction channels are dynamic structures that can open or close in response to a variety of factors, including Ca^{++} and H^+ ions. This gating serves at least one important function. When a cell is damaged, its plasma membrane often becomes leaky. Ions in high concentrations outside the cell, such as Ca^{++}, flow into the damaged cell and shut its gap junction channels. This isolates the cell and so prevents the damage from spreading to other cells.

Plasmodesmata in Plants. In plants, cell walls separate every cell from all others. Cell–cell junctions occur only at holes or gaps in the walls, where the plasma membranes of adjacent cells can come into contact with each other. Cytoplasmic connections that form across the touching plasma membranes are called **plasmodesmata** (figure 7.18). The majority of living cells within a higher plant are connected with their neighbors by these junctions. Plasmodesmata function much like gap junctions in animal cells, although their structure is more complex. Unlike gap junctions, plasmodesmata are lined with plasma membrane and contain a central tubule that connects the endoplasmic reticula of the two cells.

Cells attach themselves to one another with long-lasting bonds called cell junctions. Tight junctions connect the plasma membranes of adjacent cells in sheets. Anchoring junctions attach to the cells' cytoskeletons. Communicating junctions permit passage of substances between cells.

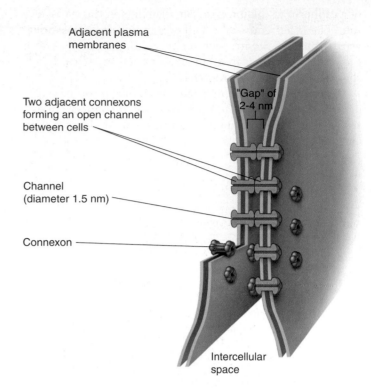

FIGURE 7.17
Gap junctions. Connexons in gap junctions create passageways that connect the cytoplasms of adjoining cells. Gap junctions readily allow the passage of small molecules and ions required for rapid communication (such as in heart tissue), but do not allow the passage of larger molecules such as proteins.

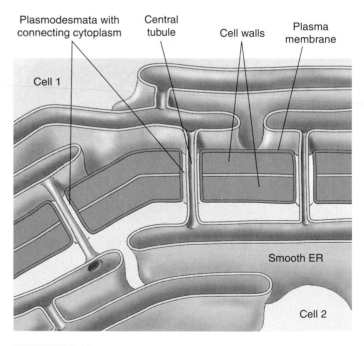

FIGURE 7.18
Plasmodesmata. Plant cells can communicate through specialized openings in their cell walls, called plasmodesmata, where the cytoplasms of adjoining cells are connected (also see figure 7.14c).

For interactive testing, visit the Online Learning Center with PowerWeb at www.mhhe.com/Raven7

7.1 Cells signal one another with chemicals.

Receptor Proteins and Signaling Between Cells

- Receptor proteins are located on or within the cell, and have three-dimensional shapes that fit the shape of specific signal molecules. (p. 126)
- Binding of the signal molecule with the receptor protein induces a change in the protein's shape and produces a cellular response. (p. 126)
- Immunochemistry and molecular genetics are being used to locate and characterize receptor proteins. (p. 126)

Types of Cell Signaling

- Cells can communicate through any of four basic mechanisms: direct contact, paracrine signaling, endocrine signaling, or synaptic signaling. (p. 127)

7.2 Proteins in the cell and on its surface receive signals from other cells.

Intracellular Receptors

- All cell-signaling pathways share certain common elements, including a chemical signal that passes from one cell to another and a receptor that receives the signal in or on the target cell. (p. 128)
- Intracellular receptors may trigger a variety of responses in the cell, dependent on the receptor. (pp. 128–129)

Cell Surface Receptors

- Cell surface receptors convert the extracellular signal to an intercellular one, responding to the binding of the signal molecule to the cell's outside by producing a change inside the cell. (p. 130)
- Many cell surface receptors either act as enzymes or are directly linked to enzymes. (p. 130)
- G-protein-linked receptors activate an intermediary protein, which then effects the intercellular change. (p. 131)

7.3 Follow the journey of information into the cell.

Initiating the Intracellular Signal

- Second messengers, such as cAMP and calcium ions, relay messages from receptors to target proteins. (p. 132)

Amplifying the Signal: Protein Kinase Cascades

- Some surface receptors generate large intracellular responses because each stage of the pathway amplifies the next, causing a cascading effect. (pp. 134–135)

7.4 Cell surface proteins mediate cell–cell interactions.

The Expression of Cell Identity

- As an organism develops, its cells acquire their specific identities by controlling gene expression, turning on the specific set of genes that encode the particular functions of each cell type. (p. 136)
- Every cell contains surface marker proteins that uniquely identify each cell type. (p. 136)

Intercellular Adhesion

- Cells attach to one another using cell junctions. (p. 137)
- Tight junctions connect the plasma membranes of adjacent cells in a sheet. (p. 138)
- Anchoring junctions mechanically attach the cytoskeleton of a cell to the cytoskeletons of other cells or to the extracellular matrix. (p. 138)
- Communication junctions allow communication with adjacent cells through direct connections. (p. 140)

Self Test

1. Which of the following techniques has recently aided the study of receptor proteins?
   a. protein purification
   b. monoclonal antibodies
   c. isolation of cell signal molecules
   d. all of these
2. Which of the following describes autocrine signaling?
   a. Signal molecules released by cells diffuse through the extracellular fluid to other cells.
   b. Signal molecules enter the organism's circulatory system and travel throughout the body.
   c. Signal molecules are released from a cell and bind to receptors on its own plasma membrane.
   d. Signal molecules are released into a narrow space between cells called a synapse.
3. Intracellular receptors usually bind
   a. water-soluble signals.
   b. large molecules that act as signals.
   c. signals on the cell surface.
   d. lipid-soluble signals.
4. Which of the following is *not* a type of cell surface receptor?
   a. chemically gated ion channels
   b. intracellular receptors
   c. enzymic receptors
   d. G-protein-linked receptors
5. Which of the following is *not* a second messenger?
   a. adenylyl cyclase
   b. cyclic adenosine monophosphate
   c. calcium ions
   d. cAMP
6. The amplification of a cellular signal requires all but which of the following?
   a. a second messenger
   b. DNA
   c. a signal molecule
   d. a cascade of protein kinases
7. MHC proteins are
   a. molecules that determine a person's blood type.
   b. large molecules that pass through the membrane many times.
   c. identity markers present on the surface of an individual's cells.
   d. different for each type of tissue in the body.
8. Sheets of cells are formed from which type of cell junctions?
   a. tight junctions
   b. anchoring junctions
   c. communication junctions
   d. none of these
9. Cadherin can be found in which of the following?
   a. tight junctions
   b. anchoring junctions
   c. communication junctions
   d. adherens junctions
10. Plasmodesmata are a type of
    a. gap junction.
    b. anchoring junction.
    c. communicating junction.
    d. tight junction.

Test Your Visual Understanding

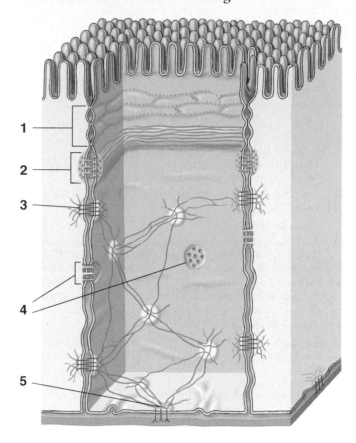

1. Match the following labels with the appropriate structures in the figure, and explain where each type of junction is found.
   adherens junction
   desmosome
   gap junctions
   hemidesmosome
   tight junction

Apply Your Knowledge

1. In paracrine signaling, the signal molecule is destroyed by enzymes in 6 milliseconds. The diffusion rate of the signal through the extracellular fluid is 2 nanometer/1 millisecond. How long will the signal last in the extracellular fluid, and what is the farthest distance a target cell can be from the releasing cell to be affected by the signal?
2. At first glance, the signaling systems that involve cell surface receptors may appear rather complex and indirect, with their use of G proteins, second messengers, and often multiple stages of enzymes. What are the advantages of such seemingly complex response systems?
3. *Shigella flexneri* is one of several species of bacteria that cause shigellosis, or bacillary dysentery. Recent evidence has shown that *S. flexneri* cannot spread between the epithelial cells of the intestines without the expression of cadherin by those cells. Why do you suppose it cannot?

8
Energy and Metabolism

Concept Outline

8.1 The laws of thermodynamics describe how energy changes.

The Flow of Energy in Living Things. Potential energy is present in the electrons of atoms, and so can be transferred from one molecule to another.

The Laws of Thermodynamics. Energy is never lost, but as it is transferred, more and more of it dissipates as heat, a disordered form of energy.

Free Energy. In a chemical reaction, the energy released or supplied is the difference between the bond energies of the reactants and the products, corrected for disorder.

Activation Energy. To start a chemical reaction, an input of energy is required to destabilize existing chemical bonds. Catalysts lower this activation energy, allowing exergonic reactions to proceed faster.

8.2 Enzymes are biological catalysts.

Enzymes. Globular proteins called enzymes catalyze chemical reactions within cells.

How Enzymes Work. Enzymes have sites on their surface shaped to fit their substrates snugly, forcing chemically reactive groups close enough to facilitate a reaction.

Enzymes Take Many Forms. Some enzymes are associated in complex groups; others are not even proteins.

Factors Affecting Enzyme Activity. Each enzyme works most efficiently at its optimal temperature and pH. Metal ions or other substances often help enzymes carry out catalysis.

8.3 ATP is the energy currency of life.

ATP. Cells store and release energy from the phosphate bonds of ATP, the energy currency of the cell.

8.4 Metabolism is the chemical life of a cell.

Biochemical Pathways: The Organizational Units of Metabolism. Biochemical pathways, where the product of one reaction becomes the substrate for the next, are the organizational units of metabolism.

FIGURE 8.1
Lion at lunch.   Energy that this lion extracts from its meal of giraffe will be used to power its roar, fuel its running, and build a bigger lion.

Life can be viewed as a constant flow of energy, channeled by organisms to do the work of living. Each of the significant properties by which we define life—order, growth, reproduction, responsiveness, and internal regulation—requires a constant supply of energy (figure 8.1). Deprived of a source of energy, life stops. Therefore, a comprehensive study of life would be impossible without discussing *bioenergetics*, the analysis of how energy powers the activities of living systems. In this chapter, we focus on energy—what it is and how organisms capture, store, and use it.

The laws of thermodynamics describe how energy changes.

The Flow of Energy in Living Things

Energy is defined as the capacity to do work. It can be considered to exist in two states (figure 8.2). **Kinetic energy** is the energy of motion. Moving objects perform work by causing other matter to move. **Potential energy** is stored energy. Objects that are not actively moving but have the capacity to do so possess potential energy. A boulder perched on a hilltop has potential energy; as it begins to roll downhill, some of its potential energy is converted into kinetic energy. Much of the work that living organisms carry out involves transforming potential energy to kinetic energy.

Energy can take many forms: mechanical energy, heat, sound, electric current, light, or radioactive radiation. Because it can exist in so many forms, there are many ways to measure energy. The most convenient is in terms of heat, because all other forms of energy can be converted into heat. In fact, the study of energy is called **thermodynamics,** meaning "heat changes." The unit of heat most commonly employed in biology is the **kilocalorie** (kcal). One kilocalorie is equal to 1000 calories (cal), and one calorie is the heat required to raise the temperature of one gram of water one degree Celsius (°C). (It is important not to confuse calories with a term related to diets and nutrition, the Calorie with a capital C, which is actually another term for kilocalorie.) Another energy unit, often used in physics, is the **joule;** one joule equals 0.239 cal.

Oxidation-Reduction

Energy flows into the biological world from the sun, which shines a constant beam of light on the earth. It is estimated that the sun provides the earth with more than 13×10^{23} calories per year, or 40 million billion calories per second! Plants, algae, and certain kinds of bacteria capture a fraction of this energy through photosynthesis. In photosynthesis, energy garnered from sunlight is used to combine small molecules (water and carbon dioxide) into more complex molecules (sugars). The energy is stored as potential energy in the covalent bonds between atoms in the sugar molecules. Recall from chapter 2 that an atom consists of a central nucleus surrounded by one or more orbiting electrons, and a covalent bond forms when two atomic nuclei share valence electrons. Breaking such a bond requires energy to pull the nuclei apart. Indeed, the strength of a covalent bond is measured by the amount of energy required to break it. For example, it takes 98.8 kcal to break one mole (6.023×10^{23}) of carbon-hydrogen (C—H) bonds.

During a chemical reaction, the energy stored in chemical bonds may transfer to new bonds. In some of these reactions, electrons actually pass from one atom or molecule to another. When an atom or molecule loses an electron, it is said to be oxidized, and the process by which this occurs is called **oxidation.** The name reflects the fact that in biological systems oxygen, which attracts electrons strongly, is the most common electron accep-

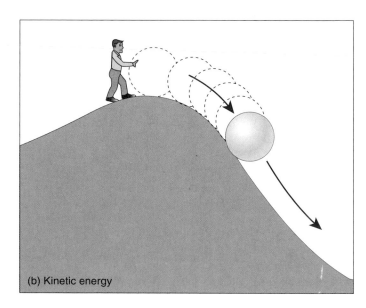

(a) Potential energy

(b) Kinetic energy

FIGURE 8.2

Potential and kinetic energy. (*a*) Objects that have the capacity to move but are not moving have potential energy. The energy required to move the ball up the hill is stored as potential energy. (*b*) Objects that are in motion have kinetic energy. The stored energy is released as kinetic energy as the ball rolls down the hill.

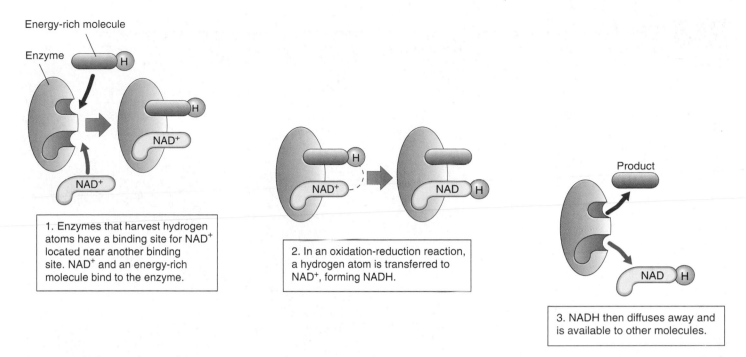

Energy-rich molecule

Enzyme

NAD+

1. Enzymes that harvest hydrogen atoms have a binding site for NAD+ located near another binding site. NAD+ and an energy-rich molecule bind to the enzyme.

NAD+

NAD H

2. In an oxidation-reduction reaction, a hydrogen atom is transferred to NAD+, forming NADH.

Product

NAD H

3. NADH then diffuses away and is available to other molecules.

FIGURE 8.3

Oxidation-reduction reactions often employ cofactors. Cells use a chemical cofactor called NAD+ to carry out oxidation-reduction reactions. Energetic electrons are often paired with a proton as a hydrogen atom. Molecules that gain energetic electrons are said to be reduced, while ones that lose energetic electrons are said to be oxidized. NAD+ oxidizes energy-rich molecules by acquiring their hydrogens (in the figure, this proceeds 1→2→3) and then reduces other molecules by giving the hydrogens to them (in the figure, this proceeds 3→2→1).

tor. Conversely, when an atom or molecule gains an electron, it is said to be reduced, and the process is called **reduction.** Oxidation and reduction always take place together, because every electron that is lost by an atom through oxidation is gained by some other atom through reduction. Therefore, chemical reactions of this sort are called **oxidation-reduction (redox) reactions** (figure 8.3). Energy is transferred from one molecule to another via redox reactions. The reduced form of a molecule thus has a higher level of energy than the oxidized form (figure 8.4). The ability of organisms to store energy in molecules by transferring electrons to them is referred to as *reducing power,* and is a fundamental property of living systems.

Oxidation-reduction reactions play a key role in the flow of energy through biological systems because the electrons that pass from one atom to another carry energy with them. The amount of energy an electron possesses depends on how far it is from the nucleus and how strongly the nucleus attracts it. Generally, electrons that are close to the nucleus and held tightly by it have less energy than electrons that are farther from the nucleus and held loosely by it. Light (and other forms of energy) can add energy to an electron and boost it to a higher energy level. When this electron departs from one atom (oxidation) and moves to another (reduction), the electron's added energy is transferred with it, and the electron orbits the second atom's nucleus at the higher en-

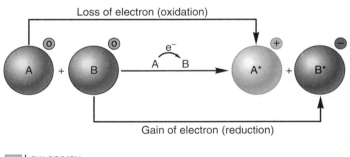

Loss of electron (oxidation)

Gain of electron (reduction)

Low energy
High energy

FIGURE 8.4

Redox reactions. Oxidation is the loss of an electron; reduction is the gain of an electron. In this example, the charges of molecules A and B are shown in small circles to the upper right of each molecule. Molecule A loses energy as it loses an electron, while molecule B gains energy as it gains an electron.

ergy level. The added energy is stored as potential chemical energy that the atom can later release when the electron returns to its original energy level.

> **Energy is the capacity to do work, either actively (kinetic energy) or stored for later use (potential energy). Energy is transferred with electrons. Oxidation is the loss of an electron; reduction is the gain of one.**

The Laws of Thermodynamics

All activities of living organisms—running, thinking, singing, reading these words—involve changes in energy. A set of two universal laws we call the laws of thermodynamics govern all energy changes in the universe, from nuclear reactions to the buzzing of a bee.

The First Law of Thermodynamics

The **First Law of Thermodynamics** concerns the amount of energy in the universe. It states that energy cannot be created or destroyed; it can only change from one form to another (from potential to kinetic, for example). The total amount of energy in the universe remains constant.

The lion eating a giraffe in figure 8.1 is in the process of acquiring energy. Rather than creating new energy or capturing the energy in sunlight, the lion is merely transferring some of the potential energy stored in the giraffe's tissues to its own body (just as the giraffe obtained the potential energy stored in the plants it ate while it was alive). Within any living organism, this chemical potential energy can be shifted to other molecules and stored in different chemical bonds, or it can convert into other forms, such as kinetic energy, light, or electricity. During each conversion, some of the energy dissipates into the environment as **heat,** a measure of the random motion of molecules (and, hence, a measure of one form of kinetic energy). Energy continuously flows through the biological world in one direction, with new energy from the sun constantly entering the system to replace the energy dissipated as heat.

Heat can be harnessed to do work only when there is a heat gradient—that is, a temperature difference between two areas (this is how a steam engine functions). Cells are too small to maintain significant internal temperature differences, so heat energy is incapable of doing the work of cells. Thus, although the total amount of energy in the universe remains constant, the energy available to do work decreases as more of it progressively dissipates as heat.

The Second Law of Thermodynamics

The **Second Law of Thermodynamics** concerns the transformation of potential energy into heat, or random molecular motion. It states that the disorder (more for-

Disorder happens "spontaneously"

Organization requires energy

FIGURE 8.5
Entropy in action. As time elapses, a child's room becomes more disorganized. It takes effort to clean it up.

mally called *entropy*) in the universe is continuously increasing. Put simply, disorder is more likely than order. For example, it is much more likely that a column of bricks will tumble over than that a pile of bricks will arrange themselves spontaneously to form a column. In general, energy transformations proceed spontaneously to convert matter from a more ordered, less stable form to a less ordered, more stable form (figure 8.5).

Entropy

Entropy is a measure of the disorder of a system, so the Second Law of Thermodynamics can also be stated simply as "entropy increases." When the universe formed, it held all the potential energy it will ever have. It has become progressively more disordered ever since, with every energy exchange increasing the amount of entropy.

The First Law of Thermodynamics states that energy cannot be created or destroyed; it can only undergo conversion from one form to another. The Second Law of Thermodynamics states that disorder (entropy) in the universe is increasing. As energy is used, more and more of it is converted to heat, the energy of random molecular motion.

Free Energy

It takes energy to break the chemical bonds that hold the atoms in a molecule together. Heat energy, because it increases atomic motion, makes it easier for the atoms to pull apart. Both chemical bonding and heat have a significant influence on a molecule, the former reducing disorder and the latter increasing it. The net effect, the amount of energy actually available to break and subsequently form other chemical bonds, is called the **free energy** of that molecule. In a more general sense, free energy is defined as the energy available to do work in any system. In a molecule within a cell, where pressure and volume usually do not change, the free energy is denoted by the symbol G (for "Gibbs' free energy," which limits the system being considered to the cell). G is equal to the energy contained in a molecule's chemical bonds (called *enthalpy* and designated **H**) minus the energy unavailable because of disorder (called *entropy* and symbolized as **S**) times the absolute temperature, **T,** in degrees Kelvin (K = °C + 273):

$$G = H - TS$$

Chemical reactions break some bonds in the reactants and form new bonds in the products. Consequently, reactions can produce changes in free energy. When a chemical reaction occurs under conditions of constant temperature, pressure, and volume—as do most biological reactions—the change Δ in free energy (ΔG) is simply:

$$\Delta G = \Delta H - T\Delta S$$

The change in free energy, or ΔG, is a fundamental property of chemical reactions. In some reactions, the ΔG is positive. This means that the products of the reaction contain *more* free energy than the reactants; the bond energy (H) is higher or the disorder (S) in the system is lower. Such reactions do not proceed spontaneously because they require an input of energy. Any reaction that requires an input of energy is said to be **endergonic** ("inward energy").

In other reactions, the ΔG is negative. The products of the reaction contain less free energy than the reactants; either the bond energy is lower or the disorder is higher, or both. Such reactions tend to proceed spontaneously. Any chemical reaction tends to proceed spontaneously if the difference in disorder (TΔS) is *greater* than the difference in bond energies between reactants and products (ΔH). Note that spontaneous does not mean the same thing as instantaneous. A spontaneous reaction may proceed very slowly. These reactions release the excess free energy as heat and are thus said to be **exergonic** ("outward energy"). Figure 8.6 sums up endergonic and exergonic reactions.

Free energy is the energy available to do work. Within cells, the change in free energy (ΔG) is the difference in bond energies between reactants and products (ΔH), minus any change in the degree of disorder of the system (TΔS). Any reaction whose products contain less free energy than the reactants (ΔG is negative) tends to proceed spontaneously.

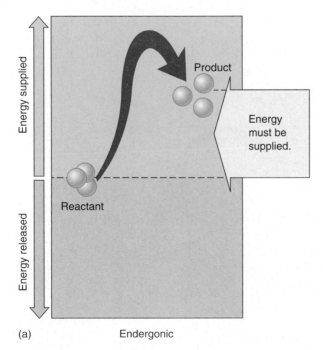

(a) Endergonic

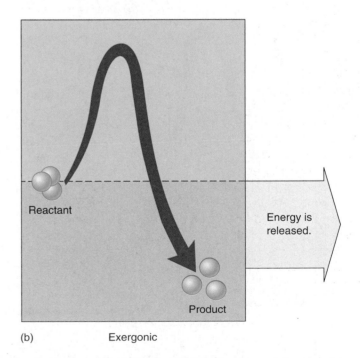

(b) Exergonic

FIGURE 8.6

Energy in chemical reactions. (*a*) In an endergonic reaction, the products of the reaction contain more energy than the reactants, and the extra energy must be supplied for the reaction to proceed. (*b*) In an exergonic reaction, the products contain less energy than the reactants, and the excess energy is released.

Activation Energy

If all chemical reactions that release free energy tend to occur spontaneously, why haven't all such reactions already occurred? One reason is that most reactions require an input of energy to get started. Before it is possible to form new chemical bonds, even bonds that contain less energy, it is first necessary to break the existing bonds, and that takes energy. The extra energy required to destabilize existing chemical bonds and initiate a chemical reaction is called **activation energy** (figure 8.7*a*).

The rate of an exergonic reaction depends on the activation energy required for the reaction to begin. Reactions with larger activation energies tend to proceed more slowly because fewer molecules succeed in overcoming the initial energy hurdle. Activation energies are not constant, however. Stressing particular chemical bonds can make them easier to break. The process of influencing chemical bonds in a way that lowers the activation energy needed to initiate a reaction is called **catalysis,** and substances that accomplish this are known as catalysts (figure 8.7*b*).

Catalysts cannot violate the basic laws of thermodynamics; they cannot, for example, make an endergonic reaction proceed spontaneously. By reducing the activation energy, a catalyst accelerates both the forward and the reverse reactions by exactly the same amount. Hence, it does not alter the proportion of reactant ultimately converted into product.

To grasp this, imagine a bowling ball resting in a shallow depression on the side of a hill. Only a narrow rim of dirt below the ball prevents it from rolling down the hill. Now imagine digging away that rim of dirt. If you remove enough dirt from below the ball, it will start to roll down the hill—but removing dirt from below the ball will *never* cause the ball to roll up the hill. Removing the lip of dirt simply allows the ball to move freely; gravity determines the direction it then travels. Lowering the resistance to the ball's movement will promote the movement dictated by its position on the hill.

Similarly, the direction in which a chemical reaction proceeds is determined solely by the difference in free energy between reactants and products. Like digging away the soil below the bowling ball on the hill, catalysts reduce the energy barrier that is preventing the reaction from proceeding. Catalysts don't favor endergonic reactions any more than digging makes the hypothetical bowling ball roll uphill. Only exergonic reactions can proceed spontaneously, and catalysts cannot change that. What catalysts *can* do is make a reaction proceed much faster.

The rate of a reaction depends on the activation energy necessary to initiate it. Catalysts reduce the activation energy and so increase the rates of reactions, although they do not change the final proportions of reactants and products.

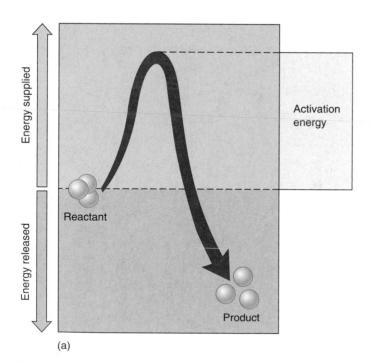

(a)

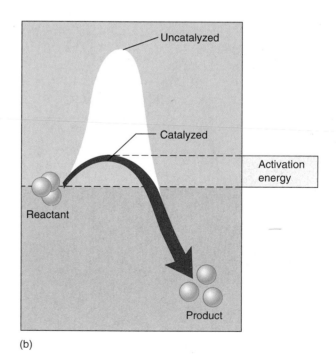

(b)

FIGURE 8.7

Activation energy and catalysis. (*a*) Exergonic reactions do not necessarily proceed rapidly because energy must be supplied to destabilize existing chemical bonds. This extra energy is the activation energy for the reaction. (*b*) Catalysts accelerate particular reactions by lowering the amount of activation energy required to initiate the reaction. Catalysts do not change the free energy of the reactants or products and therefore do not alter the free energy change produced by the reaction.

8.2 Enzymes are biological catalysts.

Enzymes

The chemical reactions within living organisms are regulated by controlling the points at which catalysis takes place. Life itself, therefore, is regulated by catalysts. The agents that carry out most of the catalysis in living organisms are proteins called **enzymes.** (There is increasing evidence that some types of biological catalysis are carried out by RNA molecules.) The unique three-dimensional shape of an enzyme enables it to stabilize a temporary association between **substrates,** the molecules that will undergo the reaction. By bringing two substrates together in the correct orientation, or by stressing particular chemical bonds of a substrate, an enzyme lowers the activation energy required for new bonds to form. The reaction thus proceeds much more quickly than it would without the enzyme. Because the enzyme itself is not changed or consumed in the reaction, only a small amount of an enzyme is needed, and it can be used over and over.

As an example of how an enzyme works, let's consider the reaction of carbon dioxide and water to form carbonic acid. This important enzyme-catalyzed reaction occurs in vertebrate red blood cells:

$$CO_2 + H_2O \rightleftharpoons H_2CO_3$$

carbon water   carbonic
dioxide   acid

This reaction may proceed in either direction, but because it has a large activation energy, the reaction is very slow in the absence of an enzyme: Perhaps 200 molecules of carbonic acid form in an hour in a cell. Reactions that proceed this slowly are of little use to a cell. Cells overcome this problem by employing an enzyme within their cytoplasm called *carbonic anhydrase* (enzyme names usually end in "–ase"). Under the same conditions, but in the presence of carbonic anhydrase, an estimated 600,000 molecules of carbonic acid form every *second!* Thus, the enzyme increases the reaction rate more than one million times. (See Box: Catalysis: A Closer Look at Carbonic Anhydrase.)

Thousands of different kinds of enzymes are known, each catalyzing one or a few specific chemical reactions. By facilitating particular chemical reactions, the enzymes in a cell determine the course of metabolism—the collection of all chemical reactions—in that cell. Different types of cells contain different sets of enzymes, and this difference contributes to structural and functional variations among cell types. For example, the chemical reactions taking place within a red blood cell differ from those that occur within a nerve cell, in part because the cytoplasm and membranes of red blood cells and nerve cells contain different arrays of enzymes.

Cells use proteins called enzymes as catalysts to lower activation energies.

Catalysis: A Closer Look at Carbonic Anhydrase

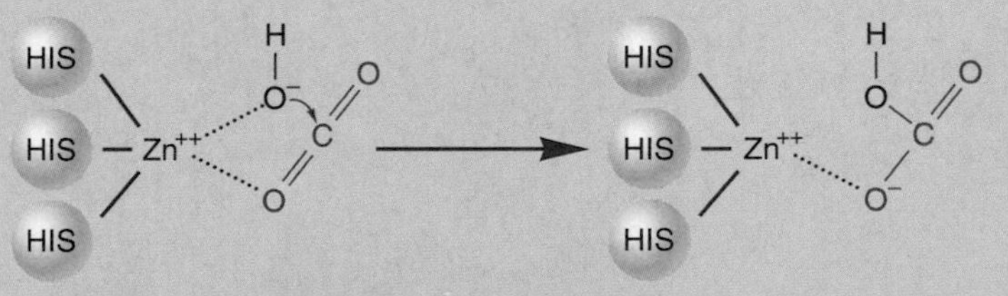

FIGURE 8.A

One of the most rapidly acting enzymes in the human body is carbonic anhydrase, which plays a key role in blood by converting dissolved carbon dioxide (CO_2) into carbonic acid, which dissociates into bicarbonate and hydrogen ions:

$$CO_2 + H_2O \rightarrow H_2CO_3 \rightarrow HCO_3^- + H^+$$

Fully 70% of the CO_2 transported by the blood is transported as bicarbonate ion. This reaction is exergonic and can proceed spontaneously, but its energy of activation is significant, so that little conversion to bicarbonate occurs spontaneously. In the presence of the enzyme carbonic anhydrase, however, the rate of the reaction accelerates by a factor of more than one million!

How does carbonic anhydrase catalyze this reaction so effectively? The active site of the enzyme is a deep cleft traversing the enzyme, as if it had been cut with the blade of an ax. Deep within the cleft, some 1.5 nanometers from the surface, are three histidines, their imidazole (nitrogen ring) groups all pointed at the same place in the center of the cleft. Together, they hold a zinc ion (Zn^{++}) firmly in position. This zinc ion will be the cutting blade of the catalytic process.

Here is how the zinc catalyzes the reaction: Immediately adjacent to the position of the zinc atom in the cleft are a group of amino acids that recognize and bind CO_2. When the CO_2 binds to this site, it interacts with the Zn^{++} in the cleft, orienting in the plane of the cleft. Meanwhile, water bound to the zinc is rapidly converted to hydroxide ion. This hydroxide ion is now precisely positioned to attack the CO_2. When it does so, HCO_3^- is formed—and the enzyme is unchanged (figure 8.A).

Carbonic anhydrase is an effective catalyst because it brings its two substrates into close proximity and optimizes their orientation for reaction. Other enzymes use other mechanisms. Many, for example, use charged amino acids to polarize substrates or electronegative amino acids to stress particular bonds. Whatever the details of the reaction, however, the precise positioning of substrates achieved by the particular shape of the enzyme always plays a key role.

How Enzymes Work

Most enzymes are globular proteins with one or more pockets or clefts on their surface called **active sites** (figure 8.8). Substrates bind to the enzyme at these active sites, forming an **enzyme-substrate complex** (figure 8.9). For catalysis to occur within the complex, a substrate molecule must fit precisely into an active site. When that happens, amino acid side groups of the enzyme end up in close proximity to certain bonds of the substrate. These side groups interact chemically with the substrate, usually stressing or distorting a particular bond and consequently lowering the activation energy needed to break the bond. Once the bonds of the substrates are broken, or new bonds are formed, the substrates have been converted to products. These products then dissociate from the enzyme.

Proteins are not rigid. The binding of a substrate induces the enzyme to adjust its shape slightly, leading to a better *induced fit* between enzyme and substrate. This interaction may also facilitate the binding of other substrates; in such cases, one substrate "activates" the enzyme to receive other substrates.

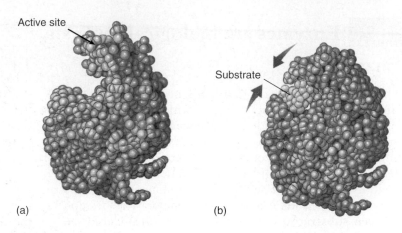

(a)   (b)

FIGURE 8.8
The active site of the enzyme lysozyme. (*a*) The active site, a groove running through lysozyme, fits the shape of the polysaccharide (a chain of sugars) that makes up bacterial cell walls. (*b*) When such a chain of sugars, indicated in yellow, slides into the groove, its entry induces the protein to alter its shape slightly and embrace the substrate more intimately. This induced fit positions a glutamic acid residue in the protein next to the bond between two adjacent sugars, and the glutamic acid "steals" an electron from the bond, causing it to break.

> Enzymes are specific in their choice of substrates. This specificity is due to the active site of the enzyme, which is shaped so that only a certain substrate molecule will fit into it.

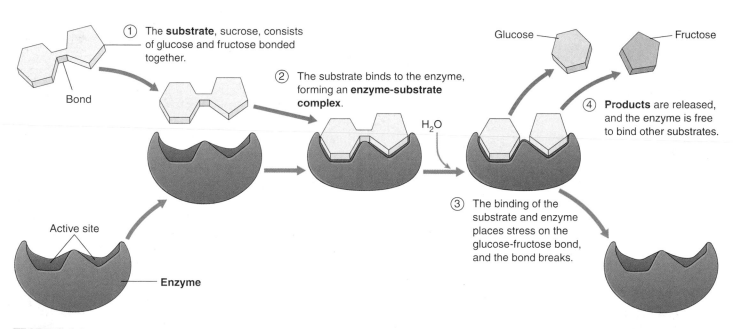

FIGURE 8.9
The catalytic cycle of an enzyme. Enzymes increase the speed with which chemical reactions occur, but they are not altered themselves as they do so. In the reaction illustrated here, the enzyme sucrase is splitting the sugar sucrose (present in most candy) into two simpler sugars: glucose and fructose. (*1*) First, the sucrose substrate binds to the active site of the enzyme, fitting into a depression in the enzyme surface. (*2*) The binding of sucrose to the active site forms an enzyme-substrate complex and induces the sucrase molecule to alter its shape, fitting more tightly around the sucrose. (*3*) Amino acid residues in the active site, now in close proximity to the bond between the glucose and fructose components of sucrose, break the bond. (*4*) The enzyme releases the resulting glucose and fructose fragments, the products of the reaction, and is then ready to bind another molecule of sucrose and run through the catalytic cycle once again. This cycle is often summarized by the equation: $E + S \leftrightarrow [ES] \leftrightarrow E + P$, where E = enzyme, S = substrate, ES = enzyme-substrate complex, and P = products.

Enzymes Take Many Forms

While many enzymes are suspended in the cytoplasm of cells, not attached to any structure, other enzymes function as integral parts of cell membranes and organelles.

Multienzyme Complexes

Often in cells, the several enzymes catalyzing the different steps of a sequence of reactions are associated with one another in noncovalently bonded assemblies called *multienzyme complexes*. The bacterial pyruvate dehydrogenase multienzyme complex seen in figure 8.10 contains enzymes that carry out three sequential reactions in oxidative metabolism. Each complex has multiple copies of each of the three enzymes—60 protein subunits in all. The many subunits work in concert, like a tiny factory.

Multienzyme complexes offer the following significant advantages in catalytic efficiency:

1. The rate of any enzyme reaction is limited by the frequency with which the enzyme collides with its substrate. If a series of sequential reactions occurs within a multienzyme complex, the product of one reaction can be delivered to the next enzyme without releasing it to diffuse away.
2. Because the reacting substrate never leaves the complex during its passage through the series of reactions, the possibility of unwanted side reactions is eliminated.
3. All of the reactions that take place within the multienzyme complex can be controlled as a unit.

In addition to pyruvate dehydrogenase, which controls entry to the Krebs cycle (see chapter 9), several other key processes in the cell are catalyzed by multienzyme complexes. One well-studied system is the fatty acid synthetase complex that catalyzes the synthesis of fatty acids from two-carbon precursors. There are seven different enzymes in this multienzyme complex, and the reaction intermediates remain associated with the complex for the entire series of reactions.

Not All Biological Catalysts Are Proteins

Until a few years ago, most biology textbooks contained statements such as "Proteins called enzymes are the catalysts of biological systems." We can no longer make that statement without qualification. As discussed in chapter 4, Tom Cech and his colleagues at the University of Col-

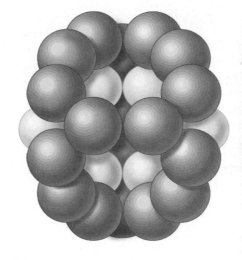

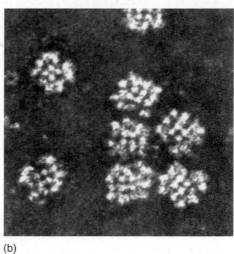

(a)   (b)

FIGURE 8.10
The enzyme pyruvate dehydrogenase. The enzyme (model, *a*) that carries out the oxidation of pyruvate is one of the most complex enzymes known. It has 60 protein subunits, many of which can be seen in (*b*) the electron micrograph (200,000×).

orado reported in 1981 that certain reactions involving RNA molecules appear to be catalyzed in cells by RNA itself, rather than by enzymes. This initial observation has been corroborated by additional examples of RNA catalysis in the last few years. Like enzymes, these RNA catalysts, which are loosely called "ribozymes," greatly accelerate the rate of particular biochemical reactions and show extraordinary specificity with respect to the substrates on which they act.

There appear to be at least two sorts of ribozymes. Some ribozymes have folded structures and catalyze reactions on themselves, a process called *intra*molecular catalysis. Other ribozymes act on other molecules without themselves being changed, a process called *inter*molecular catalysis. Many important cellular reactions involve small RNA molecules, including reactions that remove unnecessary sections from RNA copies of genes, that prepare ribosomes for protein synthesis, and that facilitate the replication of DNA within mitochondria. In all of these cases, the possibility of RNA catalysis is being actively investigated. It seems likely, particularly in the complex process of photosynthesis, that both enzymes and RNA play important catalytic roles.

The ability of RNA, an informational molecule, to act as a catalyst has stirred great excitement among biologists because it appears to provide a potential answer to the question posed in chapter 4: Which came first, the protein or the nucleic acid? It now seems at least possible that RNA may have evolved first and catalyzed the formation of the first proteins.

Not all biological catalysts float freely in the cytoplasm. Some are part of other structures, and others are not even proteins.

Factors Affecting Enzyme Activity

The rate of an enzyme-catalyzed reaction is affected by the concentrations of both the substrate and the enzyme that works on it. In addition, any chemical or physical factor that alters the enzyme's three-dimensional shape—such as temperature, pH, salt concentration, and the binding of specific regulatory molecules—can affect the enzyme's ability to catalyze the reaction.

Temperature

Increasing the temperature of an uncatalyzed reaction will increase its rate because the additional heat represents an increase in random molecular movement. This can add stress to molecular bonds and affect the activation energy of a reaction. The rate of an enzyme-catalyzed reaction also increases with temperature, but only up to a point called the *optimum temperature* (figure 8.11*a*). Below this temperature, the hydrogen bonds and hydrophobic interactions that determine the enzyme's shape are not flexible enough to permit the induced fit that is optimum for catalysis. Above the optimum temperature, these forces are too weak to maintain the enzyme's shape against the increased random movement of the atoms in the enzyme. At these higher temperatures, the enzyme denatures, as we described in chapter 3. Most human enzymes have temperature optima between 35°C and 40°C, a range that includes normal body temperature. Prokaryotes that live in hot springs have more stable enzymes (that is, enzymes held together more strongly), so the temperature optima for those enzymes can be 70°C or higher.

pH

Ionic interactions between oppositely charged amino acid residues, such as glutamic acid (−) and lysine (+), also hold enzymes together. These interactions are sensitive to the hydrogen ion concentration of the fluid the enzyme is dissolved in, because changing that concentration shifts the balance between positively and negatively charged amino acid residues. For this reason, most enzymes have an *optimum pH* that usually ranges from pH 6 to 8. Those enzymes able to function in very acidic environments are proteins that maintain their three-dimensional shape even in the presence of high levels of hydrogen ion. The enzyme pepsin, for example, digests proteins in the stomach at pH 2, a very acidic level (figure 8.11*b*).

Inhibitors and Activators

Enzyme activity is sensitive to the presence of specific substances that bind to the enzyme and cause changes in its shape. Through these substances, a cell is able to regulate which of its enzymes are active and which are inactive at a

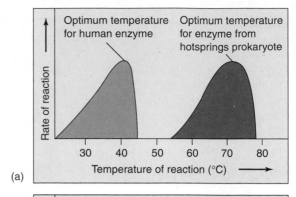

(a)

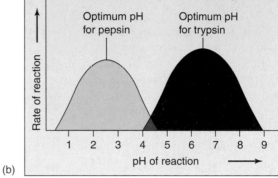

(b)

FIGURE 8.11
Enzyme sensitivity to the environment. The activity of an enzyme is influenced by both (*a*) temperature and (*b*) pH. Most human enzymes, such as the protein-degrading enzyme trypsin, work best at temperatures of about 40°C and within a pH range of 6 to 8.

particular time. This allows the cell to increase its efficiency and to control changes in its characteristics during development. A substance that binds to an enzyme and *decreases* its activity is called an **inhibitor.** Very often, the end product of a biochemical pathway acts as an inhibitor of an early reaction in the pathway, a process called *feedback inhibition* (to be discussed later).

Enzyme inhibition occurs in two ways: **Competitive inhibitors** compete with the substrate for the same active site, displacing a percentage of substrate molecules from the enzymes; **noncompetitive inhibitors** bind to the enzyme in a location other than the active site, changing the shape of the enzyme and making it unable to bind to the substrate (figure 8.12). Most noncompetitive inhibitors bind to a specific portion of the enzyme called an **allosteric site** (Greek *allos*, "other," + *steros*, "form"). These sites serve as chemical on/off switches; the binding of a substance to the site can switch the enzyme between its active and inactive configurations. A substance that binds to an allosteric site and reduces enzyme activity is called an **allosteric inhibitor** (figure 8.12*b*). Alternatively, **activators** bind to allosteric sites and keep the enzymes in their active configurations, thereby *increasing* enzyme activity.

Enzyme Cofactors

Enzyme function is often assisted by additional chemical components known as **cofactors.** For example, the active sites of many enzymes contain metal ions that help draw electrons away from substrate molecules. Zinc is used by some enzymes, such as protein-digesting carboxypeptidase, to draw electrons away from their position in covalent bonds, making the bonds less stable and easier to break. Other elements, such as molybdenum and manganese, are also used as cofactors. Like zinc, these substances are required in the diet in small amounts. When the cofactor is a nonprotein organic molecule, it is called a **coenzyme.** Many vitamins are parts of coenzymes.

In numerous oxidation-reduction reactions that are catalyzed by enzymes, the electrons pass in pairs from the active site of the enzyme to a coenzyme that serves as the electron acceptor. The coenzyme then transfers the electrons to a different enzyme, which releases them (and the energy they bear) to the substrates in another reaction. Often, the electrons combine with protons (H^+) as hydrogen atoms. In this way, coenzymes shuttle energy in the form of hydrogen atoms from one enzyme to another in a cell.

One of the most important coenzymes is the hydrogen acceptor **nicotinamide adenine dinucleotide (NAD^+)** (figure 8.13). The NAD^+ molecule is composed of two nucleotides bound together. As you may recall from chapter 3, a nucleotide is a five-carbon sugar with one or more phosphate groups attached to one end and an organic base attached to the other end. The two nucleotides that make up NAD^+, nicotinamide monophosphate (NMP) and adenine monophosphate (AMP), are joined head-to-head by their phosphate groups. The two nucleotides serve different functions in the NAD^+ molecule: AMP acts as the core, providing a shape recognized by many enzymes; NMP is the active part of the molecule, contributing a site that is readily reduced (that is, easily accepts electrons).

When NAD^+ acquires an electron and a hydrogen atom (actually, two electrons and a proton) from the active site of an enzyme, it is reduced to NADH. The NADH molecule now carries the two energetic electrons and the proton, and can supply them to other molecules and reduce them. The oxidation of energy-containing molecules, which provides energy to cells, involves stripping electrons from those molecules and donating them to NAD^+. As we'll see, much of the energy of NADH is transferred to other molecules.

Enzymes have an optimum temperature and pH, at which the enzyme functions most effectively. Inhibitors decrease enzyme activity, while activators increase it. The activity of enzymes is often facilitated by cofactors, which can be metal ions or other substances. Cofactors that are nonprotein organic molecules are called coenzymes.

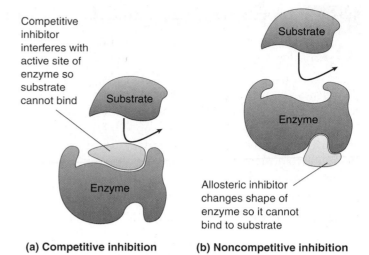

(a) Competitive inhibition   **(b) Noncompetitive inhibition**

FIGURE 8.12
How enzymes can be inhibited. (*a*) In competitive inhibition, the inhibitor interferes with the active site of the enzyme. (*b*) In noncompetitive inhibition, the inhibitor binds to the enzyme at a place away from the active site, effecting a conformational change in the enzyme so that it can no longer bind to its substrate.

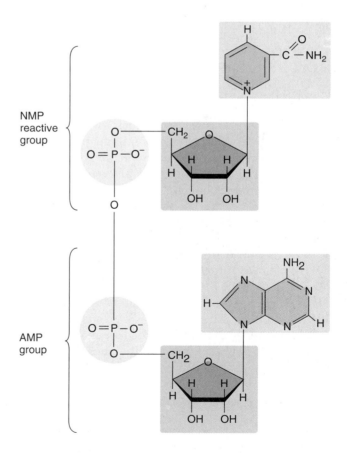

FIGURE 8.13
The chemical structure of nicotinamide adenine dinucleotide (NAD^+). This key coenzyme is composed of two nucleotides, NMP and AMP, bound together.

8.3 ATP is the energy currency of life.

ATP

The chief energy currency all cells use is a molecule called **adenosine triphosphate (ATP).** Cells use their supply of ATP to power almost every energy-requiring process they carry out, from making sugars, to supplying activation energy for chemical reactions, to actively transporting substances across membranes, to moving through their environment and growing.

Structure of the ATP Molecule

Each ATP molecule is a nucleotide composed of three smaller components (figure 8.14). The first component is a five-carbon sugar, ribose, which serves as the backbone to which the other two subunits are attached. The second component is adenine, an organic molecule composed of two carbon-nitrogen rings. Each of the nitrogen atoms in the ring has an unshared pair of electrons and weakly attracts hydrogen ions. Adenine, therefore, acts chemically as a base and is usually referred to as a nitrogenous base (it is one of the four nitrogenous bases found in DNA and RNA). The third component of ATP is a triphosphate group (a chain of three phosphates).

How ATP Stores Energy

The key to how ATP stores energy lies in its triphosphate group. Phosphate groups are highly negatively charged, so they repel one another strongly. Because of this electrostatic repulsion, the covalent bonds joining the phosphates are unstable. The molecule is often referred to as a "coiled spring," the phosphates straining away from one another.

The unstable bonds holding the phosphates together in the ATP molecule have a low activation energy and are easily broken. When they break, they can transfer a considerable amount of energy. In most reactions involving ATP, only the outermost high-energy phosphate bond is hydrolyzed, cleaving off the phosphate group on the end. When this happens, ATP becomes **adenosine diphosphate (ADP),** and energy equal to 7.3 kcal/mole is released under standard conditions. The liberated phosphate group usually attaches temporarily to some intermediate molecule. When that molecule is dephosphorylated, the phosphate group is released as inorganic phosphate (P_i).

How ATP Powers Energy-Requiring Reactions

Cells use ATP to drive endergonic reactions. Such reactions do not proceed spontaneously, because their products possess more free energy than their reactants. However, if the cleavage of ATP's terminal high-energy bond releases more energy than the other reaction consumes, the two re-

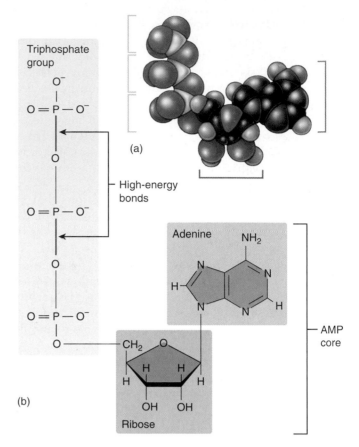

FIGURE 8.14
The ATP molecule. (*a*) The model and (*b*) the structural diagram both show that, like NAD^+, ATP has a core of AMP. In ATP, the reactive group added to the end of the AMP phosphate group is not another nucleotide but rather a chain of two phosphate groups.

actions can be coupled so that the energy released by the hydrolysis of ATP can be used to supply the second endergonic reaction with the energy it needs. Together, these reactions have a net release of energy and are therefore exergonic and will proceed spontaneously. Because almost all endergonic reactions require less energy than is released by the cleavage of ATP, ATP can provide most of the energy a cell needs.

The same feature that makes ATP an effective energy donor—the instability of its phosphate bonds—precludes it from being a good long-term energy storage molecule. Fats and carbohydrates serve that function better. Most cells do not maintain large stockpiles of ATP. Instead, they typically have only a few seconds' supply of ATP at any given time, and they continually produce more from ADP and P_i.

The instability of its phosphate bonds make ATP an excellent energy donor.

8.4 Metabolism is the chemical life of a cell.

Biochemical Pathways: The Organizational Units of Metabolism

Living chemistry, the total of all chemical reactions carried out by an organism, is called **metabolism** (Greek *metabole*, "change"). Those reactions that expend energy to make or transform chemical bonds are called *anabolic* reactions, or **anabolism.** Reactions that harvest energy when chemical bonds are broken are called *catabolic* reactions, or **catabolism.**

Organisms contain thousands of different kinds of enzymes that catalyze a bewildering variety of reactions. Many of these reactions in a cell occur in sequences called **biochemical pathways.** In such pathways, the product of one reaction becomes the substrate for the next (figure 8.15). Biochemical pathways are the organizational units of metabolism, the elements an organism controls to achieve coherent metabolic activity. Most sequential enzyme steps in biochemical pathways take place in specific compartments of the cell; for example, the steps of the krebs cycle (chapter 9), occur inside mitochondria. By determining where many of the enzymes that catalyze these steps are located, we can "map out" a model of metabolic processes in the cell.

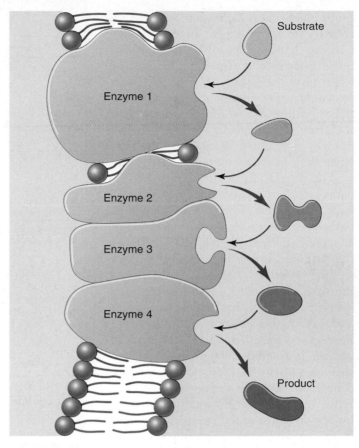

FIGURE 8.15
A biochemical pathway. The original substrate is acted on by enzyme 1, changing the substrate to a new form recognized by enzyme 2. Each enzyme in the pathway acts on the product of the previous stage.

How Biochemical Pathways Evolved

In the earliest cells, the first biochemical processes probably involved energy-rich molecules scavenged from the environment. Most of the molecules necessary for these processes are thought to have existed in the "organic soup" of the early oceans. The first catalyzed reactions were probably simple, one-step reactions that brought these molecules together in various combinations. Eventually, the energy-rich molecules became depleted in the external environment, and only organisms that had evolved some means of making those molecules from other substances in the environment could survive. Thus, a hypothetical reaction,

$$\begin{matrix} F \\ + \to H \\ G \end{matrix}$$

where two energy-rich molecules (F and G) react to produce compound H and release energy, became more complex when the supply of F in the environment ran out. A new reaction was added in which the depleted molecule, F, is made from another molecule, E, which was also present in the environment:

$$\begin{matrix} E \to F \\ + \to H \\ G \end{matrix}$$

When the supply of E in turn became depleted, organisms that were able to make it from some other available precursor, D, survived. When D became depleted, those organisms in turn were replaced by ones able to synthesize D from another molecule, C:

$$\begin{matrix} C \to D \to E \to F \\ + \to H \\ G \end{matrix}$$

This hypothetical biochemical pathway would have evolved slowly through time, with the final reactions in the pathway evolving first and earlier reactions evolving later. Looking at the pathway now, we would say that the organism, starting with compound C, is able to synthesize H by means of a series of steps. This is how the biochemical pathways within organisms are thought to have evolved—not all at once, but one step at a time, backward.

How Biochemical Pathways Are Regulated

For a biochemical pathway to operate efficiently, its activity must be coordinated and regulated by the cell. Not only is it unnecessary to synthesize a compound when plenty is already present, but doing so would waste energy and raw materials that could be put to use elsewhere. It is, therefore, advantageous for a cell to temporarily shut down biochemical pathways when their products are not needed.

The regulation of simple biochemical pathways often depends on an elegant feedback mechanism: The end product of the pathway binds to an allosteric site on the enzyme that catalyzes the first reaction in the pathway. In the hypothetical pathway we just described, the enzyme catalyzing the reaction C → D would possess an allosteric site for H, the end product of the pathway. As the pathway churned out its product and the amount of H in the cell increased, it would become increasingly likely that one of the H molecules would encounter the allosteric site on the C → D enzyme. If the product H functioned as an allosteric inhibitor of the enzyme, its binding to the enzyme would essentially shut down the reaction C → D. Shutting down this reaction, the first reaction in the pathway, effectively shuts down the whole pathway. Hence, as the cell produces increasing quantities of the product H, it automatically inhibits its ability to produce more. This mode of regulation is called **feedback inhibition** (figure 8.16).

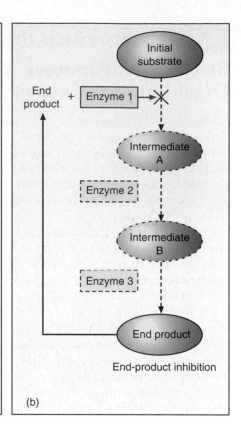

FIGURE 8.16

Feedback inhibition. (*a*) A biochemical pathway with no feedback inhibition. (*b*) A biochemical pathway in which the final end product becomes the allosteric inhibitor for the first enzyme in the pathway. In other words, the formation of the pathway's final end product stops the pathway.

> **A biochemical pathway is an organized series of reactions, often regulated as a unit.**

A Vocabulary of Metabolism

activation energy The energy required to destabilize chemical bonds and to initiate a chemical reaction.

catalysis Acceleration of the rate of a chemical reaction by lowering the activation energy.

coenzyme A nonprotein organic molecule that plays an accessory role in enzyme-catalyzed reactions, often by acting as a donor or acceptor of electrons. NAD^+ is a coenzyme.

endergonic reaction A chemical reaction to which energy from an outside source must be added before the reaction proceeds; the opposite of an exergonic reaction.

entropy A measure of the randomness or disorder of a system. In cells, it is a measure of how much energy has become so dispersed (usually as evenly distributed heat) that it is no longer available to do work.

exergonic reaction An energy-yielding chemical reaction. Exergonic reactions tend to proceed spontaneously, although activation energy is required to initiate them.

free energy Energy available to do work.

kilocalorie 1000 calories. A calorie is the heat required to raise the temperature of 1 gram of water by 1°C.

metabolism The sum of all chemical processes occurring within a living cell or organism.

oxidation The loss of an electron by an atom or molecule. It occurs simultaneously with reduction of some other atom or molecule because an electron that is lost by one is gained by another.

reduction The gain of an electron by an atom or molecule. Oxidation-reduction reactions are an important means of energy transfer within living systems.

substrate A molecule on which an enzyme acts; the initial reactant in an enzyme-catalyzed reaction.

For interactive testing, visit the Online Learning Center with PowerWeb at www.mhhe.com/Raven7

8.1 The laws of thermodynamics describe how energy changes.

The Flow of Energy in Living Things

- Energy is the capacity to do work. Kinetic energy is energy of motion, while potential energy is stored energy. (p. 144)
- Oxidation is the process whereby an atom or molecule loses an electron, and reduction is the process whereby an atom or molecule gains an electron. Oxidation-reduction reactions always take place together. (p. 145)

The Laws of Thermodynamics

- The First Law of Thermodynamics states that energy cannot be created or destroyed; it can only change from one form to another. (p. 146)
- During each energy conversion, some energy dissipates into the environment as heat. (p. 146)
- The Second Law of Thermodynamics states that disorder (entropy) in the universe is continuously increasing. (p. 146)
- Entropy is always increasing because, as more energy is used, more energy is converted to heat. (p. 146)

Free Energy

- The amount of energy available to break and form chemical bonds is referred to as a molecule's free energy. (p. 147)
- Reactions in which the products contain less free energy than the reactants tend to proceed spontaneously. (p. 147)

Activation Energy

- Activation energy is the extra energy required to destabilize existing chemical bonds. (p. 148)
- Catalysts increase the rate of reaction by lowering the activation energy, but do not alter the proportions of the reactants or products. (p. 148)

8.2 Enzymes are biological catalysts.

Enzymes

- Cells use proteins (enzymes) to lower activation energies. Thousands of different enzymes are known, each catalyzing one or a few specific chemical reactions. (p. 149)

How Enzymes Work

- Most enzymes are globular proteins with one or more active sites where specific substrates can bind, forming an enzyme-substrate complex. (p. 150)

Enzymes Take Many Forms

- Multienzyme complexes are noncovalently bonded assemblies of enzymes that catalyze different steps of a reaction sequence, and can offer significant advantages in catalytic efficiency. (p. 151)
- Not all biological catalysts are proteins; for example, certain reactions can be catalyzed by ribosomes. (p. 151)

Factors Affecting Enzyme Activity

- Several factors can affect the activity of an enzyme, including temperature, pH, inhibitors and activators, and enzyme cofactors. (pp. 152–153)

8.3 ATP is the energy currency of life.

ATP

- The chief energy currency all cells use is a molecule called adenosine triphosphate (ATP). (p. 154)
- Because of electrostatic repulsion, the covalent bonds between phosphates are unstable; thus, ATP is an excellent energy donor. (p. 154)

8.4 Metabolism is the chemical life of a cell.

Biochemical Pathways: The Organizational Units of Metabolism

- Metabolism refers to the total of all chemical reactions carried out by an organism. (p. 155)
- Biochemical pathways occur when the product of one reaction becomes the substrate for the next reaction. (p. 155)
- To operate efficiently, a biochemical pathway must be coordinated and regulated by the cell, and regulation often takes place as feedback inhibition. (p. 156)

Self Test

1. An atom gains energy when
   a. an electron is lost from it.
   b. it undergoes oxidation.
   c. it undergoes reduction.
   d. it undergoes an oxidation-reduction reaction.
2. Which of the following is concerned with the amount of energy in the universe?
   a. the First Law of Thermodynamics
   b. the Second Law of Thermodynamics
   c. thermodynamics
   d. entropy
3. In a chemical reaction, if ΔG is negative, it means that
   a. the products contain more free energy than the reactants.
   b. an input of energy is required to break the bonds.
   c. the reaction will proceed spontaneously.
   d. the reaction is endergonic.
4. A catalyst
   a. allows an endergonic reaction to proceed more quickly.
   b. increases the activation energy so that a reaction can proceed more quickly.
   c. lowers the amount of energy needed for a reaction to proceed.
   d. is required for an exergonic reaction to occur.
5. Which of the following statements about enzymes is false?
   a. Enzymes are catalysts within cells.
   b. All the cells of an organism contain the same enzymes.
   c. Enzymes bring substances together so that they undergo a reaction.
   d. Enzymes lower the activation energy of spontaneous reactions in the cell.
6. A multienzyme complex contains
   a. many copies of just one enzyme.
   b. one enzyme and its substrate.
   c. enzymes that catalyze a series of reactions.
   d. side reactions on a substrate.
7. Which of the following has *no* effect on the rate of enzyme-catalyzed reactions?
   a. temperature
   b. pH
   c. concentration of substrate
   d. none of these
8. How is ATP used in the cell to produce cellular energy?
   a. ATP provides energy to drive exergonic reactions.
   b. ATP hydrolysis is coupled to endergonic reactions.
   c. A liberated phosphate group attaches to another molecule, which generates energy.
   d. ATP stores energy by the repulsion of the negatively charged phosphates.
9. Anabolic reactions are reactions that
   a. break chemical bonds.
   b. make chemical bonds.
   c. harvest energy.
   d. occur in a sequence.
10. How is a biochemical pathway regulated?
    a. The product of one reaction becomes the substrate for the next.
    b. The end product replaces the initial substrate in the pathway.
    c. The end product inhibits the first enzyme in the pathway by binding to an allosteric site.
    d. All of these are correct.

Test Your Visual Understanding

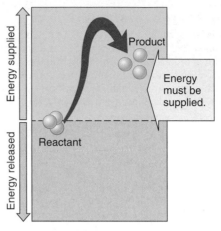

Panel a

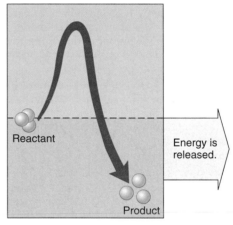

Panel b

1. In which of these two reactions would the change in free energy, ΔG, be positive? Would the product of this reaction have more or less free energy compared to the reactant? Name this type of reaction.
2. In which of these two reactions could a catalyst help speed the reaction? How would the catalyst speed the reaction? Name this type of reaction.

Apply Your Knowledge

1. In a biochemical pathway, three ATP molecules are hydrolyzed. The endergonic reactions in the pathway require a total of 17.3 kcal/mole of energy to drive the reactions of the pathway. What is the overall change in free energy of the biochemical pathway? Is the overall pathway endergonic or exergonic?
2. Oxidation-reduction reactions can involve a wide variety of molecules. Why are those involving hydrogen and oxygen of paramount importance in biological systems?
3. Almost no sunlight penetrates the deep ocean. However, many fish that live there attract prey and potential mates by producing their own light. Where does that light come from? Does its generation require energy?

9

How Cells Harvest Energy

Concept Outline

9.1 Cells harvest the energy in chemical bonds.

Using Chemical Energy to Drive Metabolism. The energy in C—H, C—O, and other chemical bonds can be captured and used to fuel the synthesis of ATP.

9.2 Cellular respiration oxidizes food molecules.

An Overview of Glucose Catabolism. The chemical energy in sugar is harvested by both substrate-level phosphorylation and aerobic respiration.

Stage One: Glycolysis. The 10 reactions of glycolysis capture energy from glucose by reshuffling the bonds.

Stage Two: The Oxidation of Pyruvate. Pyruvate, the product of glycolysis, is oxidized to acetyl-CoA.

Stage Three: The Krebs Cycle. In a series of reactions, electrons are stripped from acetyl-CoA.

Harvesting Energy by Extracting Electrons. A series of oxidation-reduction reactions strip electrons from glucose and use the energy of these electrons to power the synthesis of ATP.

Stage Four: The Electron Transport Chain. The electrons harvested from glucose pass through a chain of membrane proteins that use the energy to pump protons, driving the synthesis of ATP by ATP synthase.

Summarizing the Yield of Aerobic Respiration. The oxidation of glucose by aerobic respiration in eukaryotes produces nearly three dozen ATP molecules.

Regulating Aerobic Respiration. High levels of ATP tend to shut down cellular respiration by inhibiting key reactions.

9.3 Catabolism of proteins and fats can yield considerable energy.

Glucose Is Not the Only Source of Energy. Proteins and fats are dismantled and the products fed into cellular respiration.

9.4 Cells can metabolize food without oxygen.

Fermentation. Fermentation allows continued metabolism in the absence of oxygen by donating the electrons harvested in glycolysis to organic molecules.

9.5 The stages of cellular respiration evolved over time.

The Evolution of Metabolism. The major metabolic processes evolved over a long period, building on what had come before.

FIGURE 9.1
Harvesting chemical energy. Organisms such as these harvest mice depend on the energy stored in the chemical bonds of the food they eat to power their life processes.

Life is driven by energy. All the activities organisms carry out—the swimming of bacteria, the purring of a cat, your thinking about these words—use energy. In this chapter, we discuss the processes all cells use to derive chemical energy from organic molecules and to convert that energy to ATP. Then, in chapter 10, we will examine photosynthesis, which uses light energy rather than chemical energy. The reason we consider the conversion of chemical energy to ATP first is that all organisms, both photosynthesizers and the organisms that feed on them (including the field mice in figure 9.1), are capable of harvesting energy from chemical bonds. However, as you will see, this process and photosynthesis have much in common.

9.1 Cells harvest the energy in chemical bonds.

Using Chemical Energy to Drive Metabolism

Plants, algae, and some bacteria harvest the energy of sunlight through photosynthesis, converting radiant energy into chemical energy. These organisms, along with a few others that use chemical energy in a similar way, are called **autotrophs** ("self-feeders"). All other organisms live on the energy autotrophs produce, using them as food, and are called **heterotrophs** ("fed by others"). At least 95% of the kinds of organisms on earth—all animals and fungi, and most protists and prokaryotes—are heterotrophs.

Where is the chemical energy in food, and how do heterotrophs harvest it to carry out the many tasks of living? Most foods contain a variety of carbohydrates, proteins, and fats, all rich in energy-laden chemical bonds. Carbohydrates and fats, for example, possess many carbon-hydrogen (C—H), as well as carbon-oxygen (C—O) bonds. The job of extracting energy from this complex organic mixture is tackled in stages. First, enzymes break down the large molecules into smaller ones, a process called **digestion.** Then, other enzymes dismantle these fragments a little at a time, harvesting energy from C—H and other chemical bonds at each stage. This process is called **catabolism.**

Cellular Respiration

The energy of a chemical bond is contained in the potential energy of the electrons that make up the bond. Cells harvest this energy by breaking the bonds and shifting the electrons from one molecule to the next. During each transfer, the electrons lose some of their energy. Some of this energy may be captured and used to make ATP or form other chemical bonds; the rest is lost as heat. At the end of this process, the high-energy electrons from the initial chemical bonds have lost much of their energy, and these depleted electrons are transferred to a final electron acceptor. When this acceptor molecule is oxygen, the process is called **aerobic respiration.** When the final electron acceptor is an inorganic molecule other than oxygen, the process is called **anaerobic respiration,** and when it is an organic molecule, the process is called **fermentation.**

Chemically, there is little difference between the catabolism of carbohydrates in a cell and the burning of wood in a fireplace. In both instances, the reactants are carbohydrates and oxygen, and the products are carbon dioxide, water, and energy:

$$C_6H_{12}O_6 + 6\,O_2 \rightarrow 6\,CO_2 + 6\,H_2O + \text{energy (heat or ATP)}$$

The change in free energy in this reaction is -720 kilocalories (-3012 kilojoules) per mole of glucose under the conditions found within a cell. (The traditional value of -686 kilocalories, or -2870 kJ, per mole refers to standard conditions—room temperature, one atmosphere of pressure, etc.) This change in free energy results largely from the breaking of the six C—H bonds in the glucose molecule. The negative sign indicates that the products possess *less* free energy than the reactants. The same amount of energy is released whether glucose is catabolized or burned, but when it is burned, most of the energy is released as heat. This heat cannot be used to perform work in cells. The key to a cell's ability to harvest useful energy through the catabolism of food molecules such as glucose is its conversion of a portion of the energy into a more useful form. Cells do this by using some of the energy to drive the production of ATP, a molecule that can power cellular activities.

The ATP Molecule

Adenosine triphosphate (ATP) is the energy currency of the cell, the molecule that transfers the energy captured during cellular respiration to the many sites that use energy in the cell. How is ATP able to transfer energy so readily? Recall from chapter 8 that ATP is composed of a sugar (ribose) bound to an organic base (adenine) and a chain of three phosphates (a triphosphate group). As shown in figure 9.2, each phosphate group is negatively charged. Because like charges repel each other, the linked phosphate groups push against each other and stress the bond that holds them together. The linked phosphates store the energy of their electrostatic repulsion in the bonds that hold the phosphates together. Transferring a phosphate group to another molecule relaxes the electrostatic spring of ATP, at the same time cocking the spring of the molecule that is phosphorylated. This molecule can then use the energy to undergo some change that requires work.

How Cells Use ATP

Cells use ATP to do most of those activities that require work. One of the most obvious is movement. Tiny fibers within muscle cells pull against one another when muscles contract. Mitochondria move a meter or more along the narrow nerve cells that connect your feet with your spine. Chromosomes are pulled by microtubules during cell division. All of these movements by cells require the expenditure of ATP energy.

A second major way cells use ATP is to drive endergonic reactions. Many of the synthetic activities of the cell are endergonic, because building molecules takes energy. The chemical bonds of the products of these reactions contain more energy, or are more organized, than the reactants. The reaction can't proceed until that extra energy is supplied to the reaction. It is ATP that provides this needed energy.

How ATP Drives Endergonic Reactions

How does ATP drive an endergonic reaction? The enzyme that catalyzes the endergonic reaction has two binding sites on its surface, one for the reactant and another for ATP. The ATP site splits the ATP molecule, liberating over 7 kilocalories (30 kJ) of chemical energy. This energy pushes the reactant at the second site "uphill," driving the endergonic reaction.

When the splitting of ATP molecules drives an energy-requiring reaction in a cell, the two parts of the reaction—ATP-splitting and endergonic—take place in concert. In some cases, the two parts both occur on the surface of the same enzyme; they are physically linked, or "coupled." In other cases, a high-energy phosphate from ATP attaches to the protein catalyzing the endergonic process, activating it. Coupling energy-requiring reactions to the splitting of ATP in this way is one of the key tools cells use to manage energy.

ATP Synthase: A Micro-motor Responsible for the Production of Most ATP

Because of its ability to supply energy for such a broad range of metabolic reactions, ATP is in very high demand in cells and must be synthesized in very large quantities. Although the enzymes that break down fats and sugars help produce a few molecules of ATP directly from the energy supplied by the high-energy bonds of the substrates, the vast majority of ATP produced in the cell is made by the enzyme **ATP synthase,** one of the most important enzymes in all living systems.

ATP synthase catalyzes the synthesis of ATP by using the energy stored in a gradient of protons that is built across membranes, either the plasma membrane of prokaryotes or the inner membranes of mitochondria or chloroplasts. This proton gradient is produced by pumping protons across the membrane using the energy from a series of redox reactions. The electrons driving these reactions are extracted from the high-energy molecules broken down during catabolism, or are energized by light striking chlorophyll in photosynthesis.

ATP synthase uses a fascinating molecular mechanism to perform ATP synthesis (figure 9.3). The enzyme is embedded in the membrane and provides a channel through which protons can move across the membrane down their concentration gradient. As they do so, the energy they release causes components of the enzyme complex to rotate. The mechanical energy of this rotation is then converted to the chemical bond that holds the third phosphate on ATP. Thus, the synthesis of ATP is achieved by a tiny rotary motor whose rotation is driven directly by a gradient of protons. Much of the metabolic machinery a cell uses to break down glucose or other energy-rich molecules is devoted to harvesting the high-energy electrons that supply the power needed to pump protons and create this gradient.

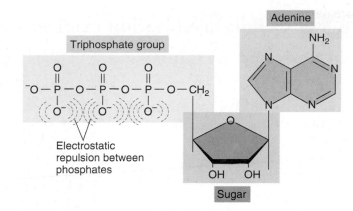

FIGURE 9.2
Structure of the ATP molecule.

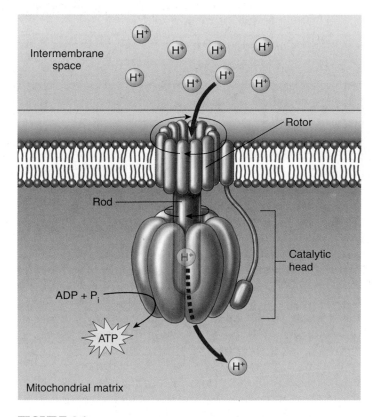

FIGURE 9.3
ATP synthase at work. Protons move across the membrane down their concentration gradient. The energy released causes the rotor and rod structures to rotate. This mechanical energy is converted to chemical energy with the formation of ATP.

The catabolism of glucose into carbon dioxide and water in living organisms releases about 720 kilocalories (3012 kJ) of energy per mole of glucose. This energy is captured in ATP, which stores the energy by linking charged phosphate groups near one another. When the phosphate bonds in ATP are hydrolyzed, energy is released and available to do work.

9.2 Cellular respiration oxidizes food molecules.

An Overview of Glucose Catabolism

Cells are able to make ATP from the catabolism of organic molecules in two different ways.

1. In **substrate-level phosphorylation**, ATP is formed by transferring a phosphate group directly to ADP from a phosphate-bearing intermediate (figure 9.4). During **glycolysis**, discussed below, the chemical bonds of glucose are shifted around in reactions that provide the energy required to form ATP.

2. In **aerobic respiration**, ATP is synthesized by ATP synthase powered by a proton gradient formed from electrons harvested from organic molecules. These electrons, with their energy depleted, are then donated to oxygen gas. Eukaryotes and aerobic prokaryotes produce the vast majority of their ATP this way.

In most organisms, these two processes are combined. To harvest energy to make ATP from the sugar glucose in the presence of oxygen, the cell carries out a complex series of enzyme-catalyzed reactions that occur in four stages: The first stage captures energy by substrate-level phosphorylation through glycolysis; the next three stages carry out aerobic respiration by oxidizing the end product of glycolysis. In this section, we first provide an overview of these stages; then we return to each topic for an in-depth examination.

Glycolysis

Stage One: Glycolysis. The first stage of extracting energy from glucose is a 10-reaction biochemical pathway called glycolysis that produces ATP by substrate-level phosphorylation. The enzymes that catalyze the glycolytic reactions are in the cytoplasm of the cell, not bound to any membrane or organelle. Two ATP molecules are used up early in the pathway, and four ATP molecules are formed by substrate-level phosphorylation. This yields a net of two ATP molecules for each molecule of glucose catabolized. In addition, four electrons are harvested from the chemical bonds of glucose and carried by NADH so they can be used to form ATP by aerobic respiration. Still, the total yield of ATP is small. When the glycolytic process is completed, the two molecules of pyruvate that are formed still contain most of the energy the original glucose molecule held.

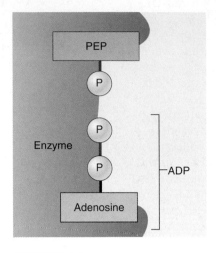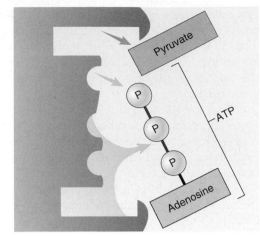

FIGURE 9.4
Substrate-level phosphorylation. Some molecules, such as phosphoenolpyruvate (PEP), possess a high-energy phosphate bond similar to the bonds in ATP. When PEP's phosphate group is transferred enzymatically to ADP, the energy in the bond is conserved, and ATP is created.

Aerobic Respiration

Stage Two: Pyruvate Oxidation. In the second stage, pyruvate, the end product of glycolysis, is converted into carbon dioxide and a two-carbon molecule called acetyl-CoA. For each molecule of pyruvate converted, one molecule of NAD^+ is reduced to NADH, again to carry electrons that can be used to make ATP.

Stage Three: The Krebs Cycle. The third stage introduces this acetyl-CoA into a cycle of nine reactions called the Krebs cycle, named after the British biochemist Sir Hans Krebs, who discovered it. (The Krebs cycle is also called the *citric acid cycle*, for the citric acid, or citrate, formed in its first step, and less commonly, the *tricarboxylic acid cycle*, because citrate has three carboxyl groups.) In the Krebs cycle, two more ATP molecules are extracted by substrate-level phosphorylation, and a large number of electrons are removed by the reduction of NAD^+ to NADH.

Stage Four: The Electron Transport Chain. In the fourth stage, the energetic electrons carried by NADH are transferred to a series of electron carriers that progressively extract the energy the electrons possess and use it to pump protons across a membrane. The resulting proton gradient is used by ATP synthase to produce ATP.

Pyruvate oxidation, the reactions of the Krebs cycle, and ATP production by electron transport chains occur within many forms of prokaryotes and inside the mitochondria of all eukaryotes. Recall from chapter 5 that mitochondria are thought to have evolved from bacteria. Figure 9.5 provides an overview of aerobic respiration.

Anaerobic Respiration

As just noted, in the presence of oxygen, cells can respire aerobically, using oxygen to accept the electrons harvested from food molecules. But even when no oxygen is present to accept the electrons, some organisms can still respire *anaerobically*, using inorganic molecules to accept the electrons. For example, many prokaryotes use sulfur, nitrate, or other inorganic compounds as the electron acceptor in place of oxygen.

Methanogens.
Among the heterotrophs that practice anaerobic respiration are primitive archaebacteria such as the thermophiles and methanogens discussed in chapter 4. Methanogens use carbon dioxide (CO_2) as the electron acceptor, reducing CO_2 to CH_4 (methane) with the hydrogens derived from organic molecules produced by other organisms.

Sulfur Bacteria.
Evidence of a second anaerobic respiratory process among primitive bacteria is seen in a group of rocks about 2.7 billion years old, known as the Woman River iron formation. Organic material in these rocks is enriched for the light isotope of sulfur, ^{32}S, relative to the heavier isotope, ^{34}S. No known geochemical process produces such enrichment, but biological sulfur reduction does, in a process still carried out today by certain primitive prokaryotes. In this sulfate respiration, the prokaryotes derive energy from the reduction of inorganic sulfates (SO_4) to hydrogen sulfide (H_2S). The hydrogen atoms are obtained from organic molecules other organisms produce. These prokaryotes thus do the same thing methanogens do, but they use SO_4 as the oxidizing (that is, electron-accepting) agent in place of CO_2.

The sulfate reducers set the stage for the evolution of photosynthesis, creating an environment rich in H_2S. As discussed in chapter 10, the first form of photosynthesis obtained hydrogens from H_2S using the energy of sunlight.

In aerobic respiration, the cell harvests energy from glucose molecules in a sequence of four major pathways: glycolysis, pyruvate oxidation, the Krebs cycle, and the electron transport chain. Oxygen is the final electron acceptor. Anaerobic respiration donates the harvested electrons to other inorganic compounds.

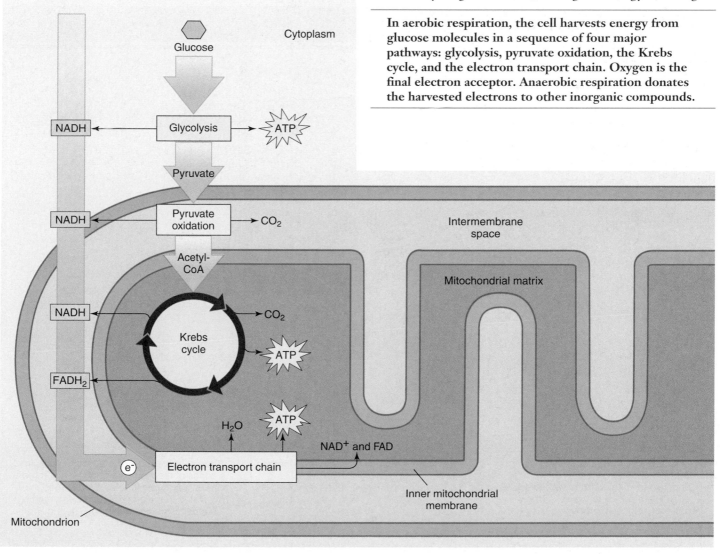

FIGURE 9.5
An overview of aerobic respiration.

Stage One: Glycolysis

The metabolism of primitive organisms focused on glucose. Glucose molecules can be dismantled in many ways, but primitive organisms evolved a glucose-catabolizing process that releases enough free energy to drive the synthesis of ATP in coupled reactions. This process, called glycolysis, occurs in the cytoplasm and involves a sequence of 10 reactions that convert glucose into 2 three-carbon molecules of pyruvate (figure 9.6). For each molecule of glucose that passes through this transformation, the cell nets two ATP molecules by substrate-level phosphorylation.

Priming

The first half of glycolysis consists of five sequential reactions that convert one molecule of glucose into two molecules of the three-carbon compound glyceraldehyde 3-phosphate (G3P). These reactions demand the expenditure of ATP, so they are an energy-requiring process.

Step A: Glucose priming. Three reactions "prime" glucose by changing it into a compound that can be cleaved readily into 2 three-carbon phosphorylated molecules. Two of these reactions require the cleavage of ATP, so this step requires the cell to use 2 ATP molecules.

Step B: Cleavage and rearrangement. In the first of the remaining pair of reactions, the six-carbon product of step A is split into 2 three-carbon molecules. One is G3P, and the other is then converted to G3P by the second reaction (figure 9.7).

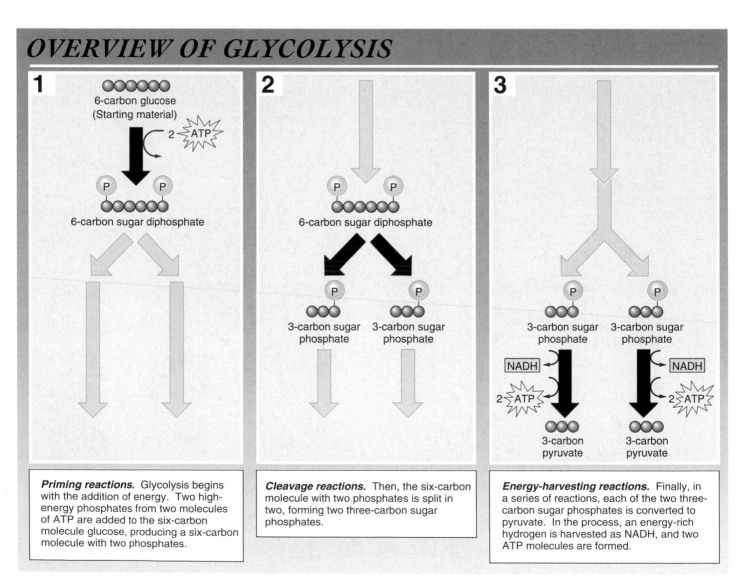

OVERVIEW OF GLYCOLYSIS

1

6-carbon glucose
(Starting material)

2 ATP

6-carbon sugar diphosphate

Priming reactions. Glycolysis begins with the addition of energy. Two high-energy phosphates from two molecules of ATP are added to the six-carbon molecule glucose, producing a six-carbon molecule with two phosphates.

2

6-carbon sugar diphosphate

3-carbon sugar phosphate   3-carbon sugar phosphate

Cleavage reactions. Then, the six-carbon molecule with two phosphates is split in two, forming two three-carbon sugar phosphates.

3

3-carbon sugar phosphate   3-carbon sugar phosphate

NADH   NADH

2 ATP   2 ATP

3-carbon pyruvate   3-carbon pyruvate

Energy-harvesting reactions. Finally, in a series of reactions, each of the two three-carbon sugar phosphates is converted to pyruvate. In the process, an energy-rich hydrogen is harvested as NADH, and two ATP molecules are formed.

FIGURE 9.6
How glycolysis works.

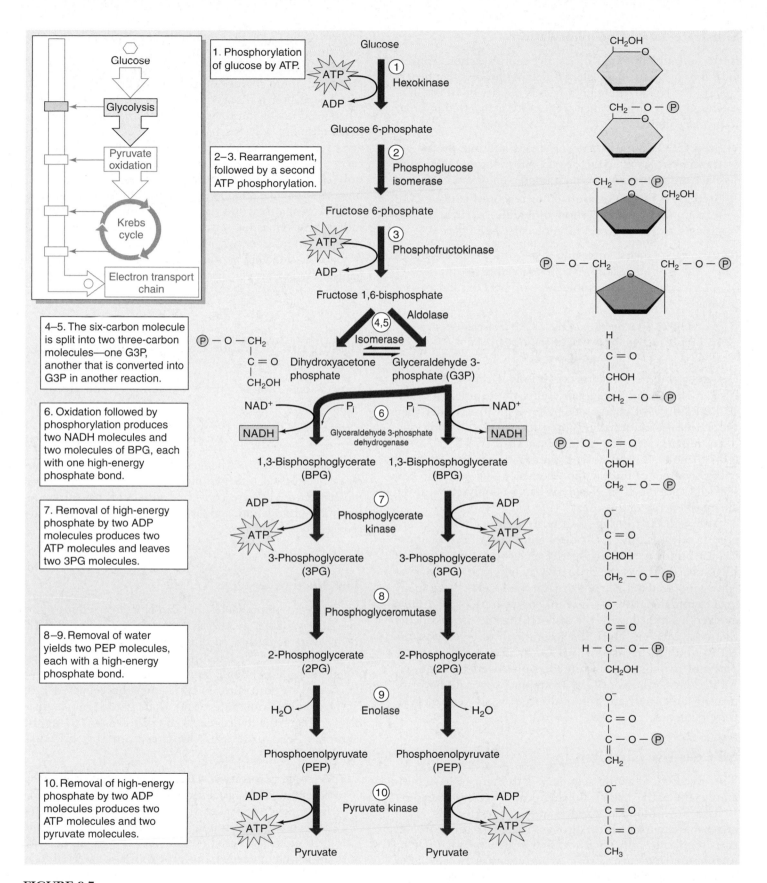

FIGURE 9.7
The glycolytic pathway. The first five reactions convert a molecule of glucose into two molecules of G3P. The second five reactions convert G3P into pyruvate.

Substrate-Level Phosphorylation

In the second half of glycolysis, five more reactions convert G3P into pyruvate in an energy-yielding process that generates ATP. Overall, then, glycolysis is a series of 10 enzyme-catalyzed reactions in which some ATP is invested in order to produce more.

Step C: Oxidation. Two electrons and one proton are transferred from G3P to NAD^+, forming NADH. Both electrons in the new covalent bond come from G3P.

Step D: ATP generation. Four reactions convert G3P into another three-carbon molecule, pyruvate. This process generates two ATP molecules (see figure 9.4).

Because each glucose molecule is split into two G3P molecules, the overall reaction sequence yields two molecules of ATP, as well as two molecules of NADH and two of pyruvate:

$$\begin{array}{l} 4\ \text{ATP} \ (2\ \text{ATP for each of the 2 G3P molecules in step D)} \\ \underline{-\ 2\ \text{ATP} \ (\text{used in the two reactions in step A})} \\ 2\ \text{ATP} \end{array}$$

Under the nonstandard conditions within a cell, each ATP molecule produced represents the capture of about 12 kcal (50 kJ) of energy per mole of glucose, rather than the 7.3 kcal traditionally quoted for standard conditions. This means glycolysis harvests about 24 kcal/mole (100 kJ/mole). This is not a great deal of energy. The total energy content of the chemical bonds of glucose is 686 kcal (2870 kJ) per mole, so glycolysis harvests only 3.5% of the chemical energy of glucose.

Although far from ideal in terms of the amount of energy it releases, glycolysis does generate ATP. For more than a billion years during the anaerobic first stages of life on earth, it was the primary way heterotrophic organisms generated ATP from organic molecules. Like many biochemical pathways, glycolysis is believed to have evolved backward, with the last steps in the process being the most ancient. Thus, the second half of glycolysis, the ATP-yielding breakdown of G3P, may have been the original process early heterotrophs used to generate ATP. The synthesis of G3P from glucose would have appeared later, perhaps when alternative sources of G3P were depleted.

All Cells Use Glycolysis

The glycolytic reaction sequence is thought to have been among the earliest of all biochemical processes to evolve. It uses no molecular oxygen and occurs readily in an anaerobic environment. All of its reactions occur free in the cytoplasm; none is associated with any organelle or membrane structure. Every living creature is capable of carrying out glycolysis. Most present-day organisms, however, can extract considerably more energy from glucose through aerobic respiration.

Why does glycolysis take place even now, since its energy yield in the absence of oxygen is comparatively so paltry? The answer is that evolution is an incremental process: Change occurs by improving on past successes. In catabolic metabolism, glycolysis satisfied the one essential evolutionary criterion—it was an improvement. Cells that could not carry out glycolysis were at a competitive disadvantage, and only cells capable of glycolysis survived. Later improvements in catabolic metabolism built on this success. Glycolysis was not discarded during the course of evolution; rather, it served as the starting point for the further extraction of chemical energy. Metabolism evolved as one layer of reactions added to another. Nearly every present-day organism carries out glycolysis, as a metabolic memory of its evolutionary past.

Closing the Metabolic Circle: The Regeneration of NAD^+

Inspect for a moment the net reaction of the glycolytic sequence:

$$\begin{aligned} \text{glucose} + 2\ \text{ADP} + 2\ P_i + 2\ NAD^+ &\rightarrow \\ 2\ \text{pyruvate} + 2\ \text{ATP} + 2\ \text{NADH} + 2\ H^+ + 2\ H_2O \end{aligned}$$

You can see that three changes occur in glycolysis: (1) Glucose is converted into two molecules of pyruvate; (2) two molecules of ADP are converted into ATP via substrate-level phosphorylation; and (3) two molecules of NAD^+ are reduced to NADH.

The Need to Recycle NADH

As long as food molecules that can be converted into glucose are available, a cell can continually churn out ATP to drive its activities. In doing so, however, it accumulates NADH and depletes the pool of NAD^+ molecules. A cell does not contain a large amount of NAD^+, and for glycolysis to continue, NADH must be recycled into NAD^+. Some molecule other than NAD^+ must ultimately accept the hydrogen atom taken from G3P and be reduced. Two processes can carry out this key task (figure 9.8):

1. **Aerobic respiration.** Oxygen is an excellent electron acceptor. Through a series of electron transfers, the hydrogen atom taken from G3P can be donated to oxygen, forming water. This is what happens in the cells of eukaryotes in the presence of oxygen. Because air is rich in oxygen, this process is also referred to as aerobic metabolism.

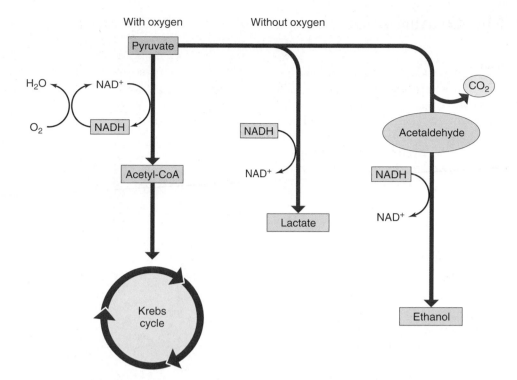

With oxygen   Without oxygen

FIGURE 9.8

What happens to pyruvate, the product of glycolysis?   In the presence of oxygen, pyruvate is oxidized to acetyl-CoA, which enters the Krebs cycle. In the absence of oxygen, pyruvate is instead reduced, accepting the electrons extracted during glycolysis and carried by NADH. When pyruvate is reduced directly, as in muscle cells, the product is lactate. When carbon dioxide is first removed from pyruvate and the product, acetaldehyde, is then reduced, as in yeast cells, the product is ethanol.

2. Fermentation. When oxygen is unavailable, an organic molecule, such as acetaldehyde in wine fermentation, can accept the hydrogen atom instead (figure 9.9). Such fermentation plays an important role in the metabolism of most organisms, even those capable of aerobic respiration.

The fate of the pyruvate that is produced by glycolysis depends upon which of these two processes takes place. The aerobic respiration path starts with the oxidation of pyruvate to a molecule called acetyl-CoA, which is then further oxidized in a series of reactions called the Krebs cycle. The fermentation path, by contrast, involves the reduction of all or part of pyruvate. We will start by examining aerobic respiration (stages two through four), and then look briefly at fermentation.

Glycolysis generates a small amount of ATP by reshuffling the bonds of glucose molecules. In glycolysis, two molecules of NAD⁺ are reduced to NADH. NAD⁺ must be regenerated for glycolysis to continue unabated.

FIGURE 9.9

How wine is made.   The conversion of pyruvate to ethanol takes place naturally in grapes left to ferment on vines, as well as in fermentation vats of crushed grapes. Yeasts carry out the process, but when their conversion increases the ethanol concentration to about 12%, the toxic effects of the alcohol kill the yeast cells. What is left is wine.

Stage Two: The Oxidation of Pyruvate

In the presence of oxygen, the oxidation of glucose that begins in glycolysis continues where glycolysis leaves off—with pyruvate. In eukaryotic organisms, the extraction of additional energy from pyruvate takes place exclusively inside mitochondria. The cell harvests pyruvate's considerable energy in two steps: first, by oxidizing pyruvate to form acetyl-CoA, and then by oxidizing acetyl-CoA in the Krebs cycle.

Producing Acetyl-CoA

Pyruvate is oxidized in a single "decarboxylation" reaction that cleaves off one of pyruvate's three carbons. This carbon then departs as CO_2 (figure 9.10, *top*), a waste molecule that must be "exhaled" from the organism. This reaction produces a two-carbon fragment called an acetyl group, as well as a pair of electrons and their associated hydrogen, which reduce NAD^+ to NADH. The reaction is complex, involving three intermediate stages, and is catalyzed within mitochondria by a *multienzyme complex*. As chapter 8 noted, such a complex organizes a series of enzymatic steps so that the chemical intermediates do not diffuse away or undergo other reactions. Within the complex, component polypeptides pass the substrates from one enzyme to the next, without releasing them. *Pyruvate dehydrogenase*, the complex of enzymes that removes CO_2 from pyruvate, is one of the largest enzymes known; it contains 60 subunits! In the course of the reaction, the acetyl group removed from pyruvate combines with a cofactor called coenzyme A (CoA), forming a compound known as **acetyl-CoA**:

$$\text{pyruvate} + NAD^+ + CoA \rightarrow \text{acetyl-CoA} + NADH + CO_2$$

This reaction produces a molecule of NADH, which is later used to produce ATP. Of far greater significance than the reduction of NAD^+ to NADH, however, is the production of acetyl-CoA (figure 9.10, *bottom*). Acetyl-CoA is important because so many different metabolic processes generate it. Not only does the oxidation of pyruvate, an intermediate in carbohydrate catabolism, produce it, but the metabolic breakdown of proteins, fats, and other lipids also generates acetyl-CoA. Indeed, almost all molecules catabolized for energy are converted into acetyl-CoA. Acetyl-CoA is then channeled into fat synthesis or into ATP production, depending on the organism's energy requirements. Acetyl-CoA is a key point for the many catabolic processes of the eukaryotic cell.

Using Acetyl-CoA

Although the cell forms acetyl-CoA in many ways, only a limited number of processes use acetyl-CoA. Most of it is either directed toward energy storage (lipid synthesis, for example) or oxidized in the Krebs cycle to produce ATP.

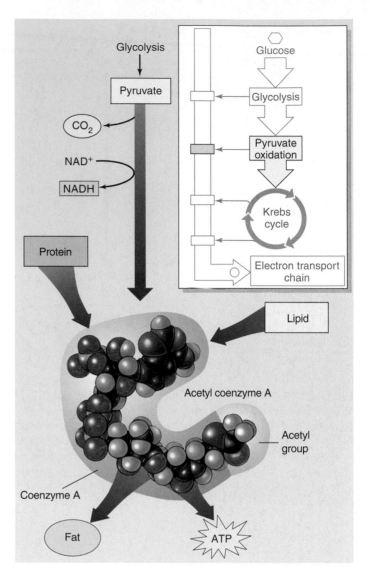

FIGURE 9.10
The oxidation of pyruvate. This complex reaction involves the reduction of NAD^+ to NADH and is thus a significant source of metabolic energy. Its product, acetyl-CoA, is the starting material for the Krebs cycle. Almost all molecules that are catabolized for energy are converted into acetyl-CoA, which is then channeled into fat synthesis or into ATP production.

Which of these two options is taken depends on the level of ATP in the cell. When ATP levels are high, the oxidative pathway is inhibited, and acetyl-CoA is channeled into fatty acid synthesis. This explains why many animals (humans included) develop fat reserves when they consume more food than their bodies require. Alternatively, when ATP levels are low, the oxidative pathway is stimulated, and acetyl-CoA flows into energy-producing oxidative metabolism.

In the second energy-harvesting stage of glucose catabolism, pyruvate is decarboxylated, yielding acetyl-CoA, NADH, and CO_2. This process occurs within the mitochondrion.

Stage Three: The Krebs Cycle

After glycolysis catabolizes glucose to produce pyruvate, and pyruvate is oxidized to form acetyl-CoA, the third stage of extracting energy from glucose begins. In this third stage, acetyl-CoA is oxidized in a series of nine reactions called the **Krebs cycle.** These reactions occur in the matrix of mitochondria. In this cycle, the two-carbon acetyl group of acetyl-CoA combines with a four-carbon molecule called oxaloacetate (figure 9.11). The resulting six-carbon molecule then goes through a sequence of electron-yielding oxidation reactions, during which two CO_2 molecules split off, restoring oxaloacetate. The oxaloacetate is then recycled to bind to another acetyl group. In each turn of the cycle, a new acetyl group replaces the two

CO_2 molecules lost, and more electrons are extracted to drive *proton pumps* that generate ATP.

Overview of the Krebs Cycle

The nine reactions of the Krebs cycle occur in two steps:

Step A: Priming. Three reactions prepare the six-carbon molecule for energy extraction. First, acetyl-CoA joins the cycle, and then chemical groups are rearranged.

Step B: Energy extraction. Four of the six reactions in this step are oxidations in which electrons are removed, and one generates an ATP equivalent directly by substrate-level phosphorylation.

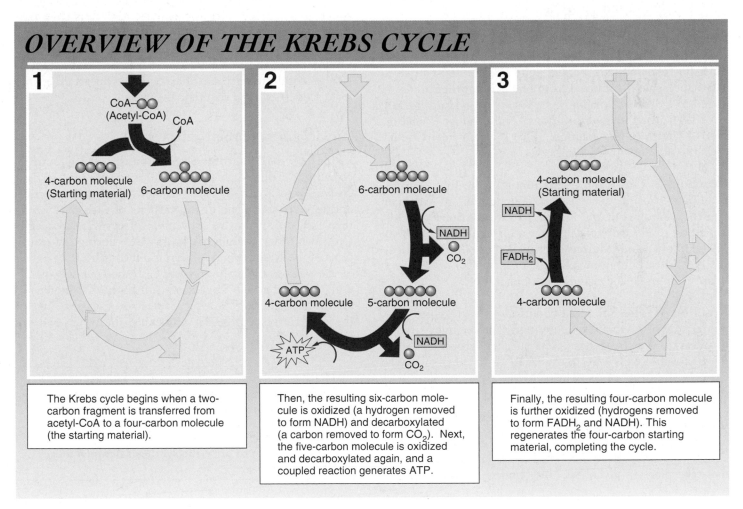

OVERVIEW OF THE KREBS CYCLE

1

CoA—OO
(Acetyl-CoA) CoA

4-carbon molecule
(Starting material)   6-carbon molecule

The Krebs cycle begins when a two-carbon fragment is transferred from acetyl-CoA to a four-carbon molecule (the starting material).

2

6-carbon molecule

NADH
CO_2

4-carbon molecule   5-carbon molecule

ATP

NADH

CO_2

Then, the resulting six-carbon molecule is oxidized (a hydrogen removed to form NADH) and decarboxylated (a carbon removed to form CO_2). Next, the five-carbon molecule is oxidized and decarboxylated again, and a coupled reaction generates ATP.

3

4-carbon molecule
(Starting material)

NADH

$FADH_2$

4-carbon molecule

Finally, the resulting four-carbon molecule is further oxidized (hydrogens removed to form $FADH_2$ and NADH). This regenerates the four-carbon starting material, completing the cycle.

FIGURE 9.11
How the Krebs cycle works.

The Reactions of the Krebs Cycle

The Krebs cycle consists of nine sequential reactions that cells use to extract energetic electrons and drive the synthesis of ATP (figure 9.12). A two-carbon group from acetyl-CoA enters the cycle at the beginning, and two CO_2 molecules and eight electrons are given off during the cycle.

Reaction 1: Condensation. The two-carbon group from acetyl-CoA joins with a four-carbon molecule, oxaloacetate, to form a six-carbon molecule, *citrate*. This condensation reaction is irreversible, committing the two-carbon acetyl group to the Krebs cycle. The reaction is inhibited when the cell's ATP concentration is high and stimulated when it is low. Hence, when the cell possesses ample amounts of ATP, the Krebs cycle shuts down, and acetyl-CoA is channeled into fat synthesis.

Reactions 2 and 3: Isomerization. Before the oxidation reactions can begin, the hydroxyl (—OH) group of citrate must be repositioned. This is done in two steps: First, a water molecule is removed from one carbon; then water is added to a different carbon. As a result, an —H group and an —OH group change positions. The product is an isomer of citrate called isocitrate. This rearrangement facilitates the subsequent reactions.

Reaction 4: The First Oxidation. In the first energy-yielding step of the cycle, isocitrate undergoes an oxidative decarboxylation reaction. First, isocitrate is oxidized, yielding a pair of electrons that reduce a molecule of NAD^+ to NADH. Then the oxidized intermediate is decarboxylated; the central carbon atom splits off to form CO_2, yielding a five-carbon molecule called α-ketoglutarate.

Reaction 5: The Second Oxidation. Next, α-ketoglutarate is decarboxylated by a multienzyme complex similar to pyruvate dehydrogenase. The succinyl group left after the removal of CO_2 joins to coenzyme A, forming succinyl-CoA. In the process, two electrons are extracted, and they reduce another molecule of NAD^+ to NADH.

Reaction 6: Substrate-Level Phosphorylation. The linkage between the four-carbon succinyl group and CoA is a high-energy bond. In a coupled reaction similar to those that take place in glycolysis, this bond is cleaved, and the energy released drives the phosphorylation of guanosine diphosphate (GDP), forming guanosine triphosphate (GTP). GTP is readily converted into ATP, and the four-carbon fragment that remains is called succinate.

Reaction 7: The Third Oxidation. Next, succinate is oxidized to fumarate. The free energy change in this reaction is not large enough to reduce NAD^+. Instead, flavin adenine dinucleotide (FAD) is the electron acceptor. Unlike NAD^+, FAD is not free to diffuse within the mitochondrion; it is an integral part of the inner mitochondrial membrane. Its reduced form, $FADH_2$, can only contribute electrons to the electron transport chain in the membrane.

Reactions 8 and 9: Regeneration of Oxaloacetate. In the final two reactions of the cycle, a water molecule is added to fumarate, forming malate. Malate is then oxidized, yielding a four-carbon molecule of oxaloacetate and two electrons that reduce a molecule of NAD^+ to NADH. Oxaloacetate, the molecule that began the cycle, is now free to combine with another two-carbon acetyl group from acetyl-CoA and reinitiate the cycle.

The Products of the Krebs Cycle

In the process of aerobic respiration, glucose is entirely consumed. The six-carbon glucose molecule is first cleaved into a pair of three-carbon pyruvate molecules during glycolysis. One of the carbons of each pyruvate is then lost as CO_2 in the conversion of pyruvate to acetyl-CoA; two other carbons are lost as CO_2 during the oxidations of the Krebs cycle. All that is left to mark the passing of the glucose molecule into six CO_2 molecules is its energy, some of which is preserved in four ATP molecules and in the reduced state of 12 electron carriers. Ten of these carriers are NADH molecules; the other two are $FADH_2$.

The Krebs cycle generates two ATP molecules per molecule of glucose, the same number generated by glycolysis. More importantly, the Krebs cycle and the oxidation of pyruvate harvest many energized electrons, which can be directed to the electron transport chain to drive the synthesis of much more ATP.

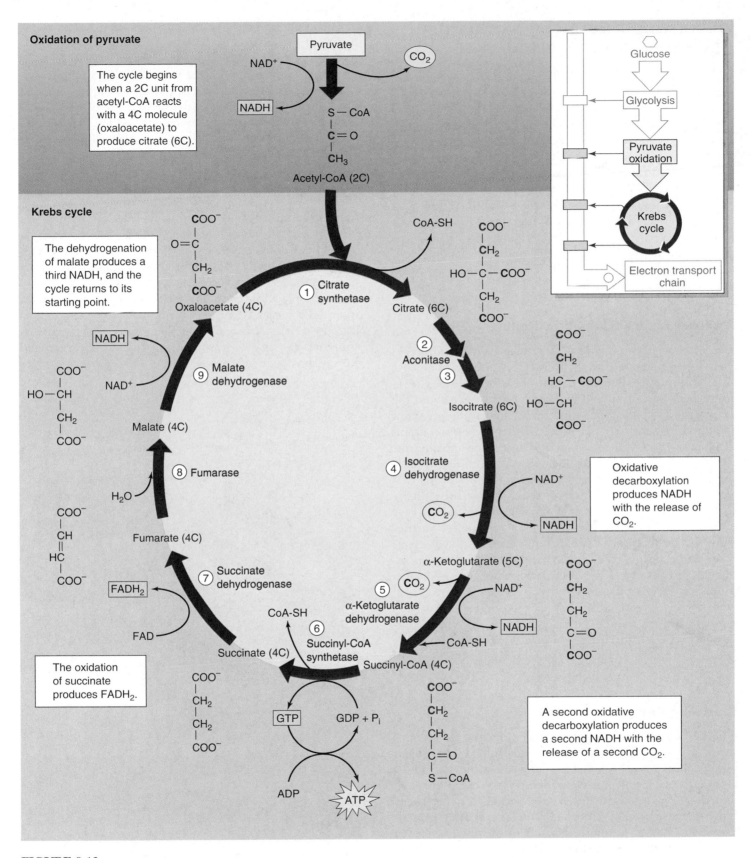

Oxidation of pyruvate

The cycle begins when a 2C unit from acetyl-CoA reacts with a 4C molecule (oxaloacetate) to produce citrate (6C).

Pyruvate

NAD^+

CO_2

NADH

S — CoA

C = O

CH_3

Acetyl-CoA (2C)

Glucose

Glycolysis

Pyruvate oxidation

Krebs cycle

Electron transport chain

Krebs cycle

The dehydrogenation of malate produces a third NADH, and the cycle returns to its starting point.

COO^-
$O = C$
CH_2
COO^-
Oxaloacetate (4C)

① Citrate synthetase

CoA-SH

COO^-
CH_2
$HO - C - COO^-$
CH_2
COO^-

Citrate (6C)

NADH

NAD⁺

⑨ Malate dehydrogenase

COO^-
$HO - CH$
CH_2
COO^-
Malate (4C)

② Aconitase

③

COO^-
CH_2
$HC - COO^-$
$HO - CH$
COO^-

Isocitrate (6C)

⑧ Fumarase

H_2O

④ Isocitrate dehydrogenase

Oxidative decarboxylation produces NADH with the release of CO_2.

NAD^+

CO_2

NADH

COO^-
CH
HC
COO^-
Fumarate (4C)

⑦ Succinate dehydrogenase

$FADH_2$

FAD

⑤ CO_2

α-Ketoglutarate (5C)

COO^-
CH_2
CH_2
$C = O$
COO^-

NAD^+

α-Ketoglutarate dehydrogenase

NADH

The oxidation of succinate produces $FADH_2$.

CoA-SH

⑥ Succinyl-CoA synthetase

CoA-SH

COO^-
CH_2
CH_2
COO^-
Succinate (4C)

Succinyl-CoA (4C)

GTP

GDP + P_i

ADP

ATP

COO^-
CH_2
CH_2
$C = O$
$S - CoA$

A second oxidative decarboxylation produces a second NADH with the release of a second CO_2.

FIGURE 9.12
The Krebs cycle. This series of reactions takes place within the matrix of the mitochondrion. For the complete breakdown of a molecule of glucose, the two molecules of acetyl-CoA produced by glycolysis and pyruvate oxidation each have to make a trip around the Krebs cycle. Follow the different carbons through the cycle, and notice the changes that occur in the carbon skeletons of the molecules as they proceed through the cycle.

Harvesting Energy by Extracting Electrons

To understand how cells direct some of the energy released during glucose catabolism into ATP production, we need to take a closer look at the electrons in the C—H bonds of the glucose molecule. We stated in chapter 8 that when an electron is removed from one atom and donated to another, the electron's potential energy of position is also transferred. In this process, the atom that receives the electron is reduced. We spoke of reduction in an all-or-none fashion, as if it involved the complete transfer of an electron from one atom to another. Often this is just what happens. However, sometimes a reduction simply changes the *degree of sharing* within a covalent bond. Let us now revisit that discussion and consider what happens when the transfer of electrons is incomplete.

A Closer Look at Oxidation-Reduction

The catabolism of glucose is an oxidation-reduction reaction. The covalent electrons in the C—H bonds of glucose are shared approximately equally between the C and H atoms because carbon and hydrogen nuclei have about the same affinity for valence electrons (that is, they exhibit similar *electronegativity*). However, when the carbon atoms of glucose react with oxygen to form carbon dioxide, the electrons in the new covalent bonds take a different position. Instead of being shared equally, the electrons that were associated with the carbon atoms in glucose shift far toward the oxygen atoms in CO_2 because oxygen is very electronegative. Since these electrons are pulled farther from the carbon atoms, the carbon atoms of glucose have been oxidized (loss of electrons) and the oxygen atoms reduced (gain of electrons). Similarly, when the hydrogen atoms of glucose combine with oxygen atoms to form water, the oxygen atoms draw the shared electrons strongly toward them; again, oxygen is reduced and glucose is oxidized. In this reaction, oxygen is an oxidizing (electron-attracting) agent because it oxidizes the atoms of glucose.

Releasing Energy

The key to understanding the oxidation of glucose is to focus on the energy of the shared electrons. In a covalent bond, energy must be added to remove an electron from an atom, just as energy must be used to roll a boulder up a hill. The more electronegative the atom, the steeper the energy hill that must be climbed to pull an electron away from it. However, energy is released when an electron is shifted away from a less electronegative atom and *closer* to a more electronegative atom, just as energy is released when a boulder is allowed to roll down a hill. In the catabolism of glucose, energy is released when glucose is oxidized, as electrons relocate closer to oxygen (figure 9.13).

Glucose is an energy-rich food because it has an abundance of C—H bonds. Viewed in terms of oxidation-reduction, glucose possesses a wealth of electrons held far from their atoms, all with the potential to move closer toward oxygen. In oxidative respiration, energy is released not simply because the hydrogen atoms of the C—H bonds are transferred from glucose to oxygen, but because the positions of the valence electrons shift. This shift releases energy that can be used to make ATP.

Reducing Power

Just as electrons are donated to NAD^+, reducing it to NADH and allowing it to carry the associated energy, NADH can donate its electrons to other molecules, reducing them and being oxidized itself. This ability to supply high-energy electrons is critical to the biosynthesis of many organic molecules, including fats and sugars. In animals, when ATP is plentiful, the reducing power of the accumulated NADH is diverted to supplying fatty acid precursors with high-energy electrons, reducing them to form fats and storing the energy of the electrons.

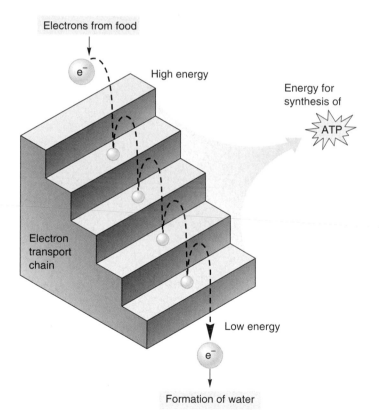

FIGURE 9.13
How electron transport works. This diagram shows how ATP is generated when electrons transfer from one energy level to another. Rather than releasing a single explosive burst of energy, electrons "fall" to lower and lower energy levels in steps, releasing stored energy with each fall as they tumble to the lowest (most electronegative) electron acceptor.

Harvesting the Energy in Stages

It is generally true that the larger the release of energy in any single step, the more of that energy is released as heat (random molecular motion) and the less is available to be channeled into more useful paths. In the combustion of gasoline, the same amount of energy is released whether all of the gasoline in a car's gas tank explodes at once or whether the gasoline burns in a series of very small explosions inside the cylinders. By releasing the energy in gasoline a little at a time, the harvesting efficiency is greater, and more of the energy can be used to push the pistons and move the car.

The same principle applies to the oxidation of glucose inside a cell. If all of the hydrogens were transferred to oxygen in one explosive step, releasing all of the free energy at once, the cell would recover very little of that energy in a useful form. Instead, cells burn their fuel much as a car does, a little at a time. The six hydrogens in the C—H bonds of glucose are stripped off in stages in the series of enzyme-catalyzed reactions collectively referred to as glycolysis and the Krebs cycle. We have had a great deal to say about these reactions already in this chapter. Recall that the hydrogens are removed by transferring them to a coenzyme carrier, NAD^+ (figure 9.14). As discussed in chapter 8, NAD^+ is a very versatile electron acceptor, shuttling energy-bearing electrons throughout the cell. In harvesting the energy of glucose, NAD^+ acts as the primary electron acceptor.

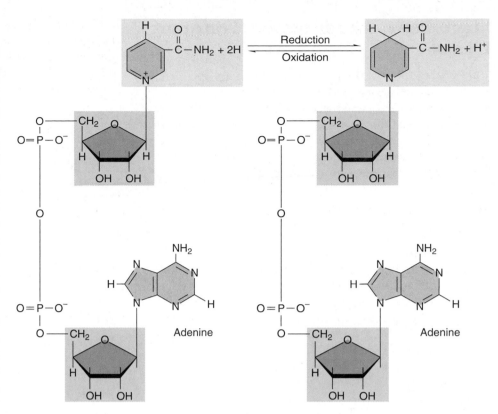

NAD⁺: oxidized form of nicotinamide   **NADH: reduced form of nicotinamide**

FIGURE 9.14
NAD⁺ and NADH. This dinucleotide serves as an "electron shuttle" during cellular respiration. NAD^+ accepts electrons from catabolized macromolecules and is reduced to NADH.

Following the Electrons

As you examine these reactions, try not to become confused by the changes in electrical charge. Always *follow the electrons*. Enzymes extract two hydrogens—that is, two electrons and two protons—from glucose and transfer both electrons and one of the protons to NAD^+. The other proton is released as a hydrogen ion, H^+, into the surrounding solution. This transfer converts NAD^+ into NADH; that is, two negative electrons and one positive proton are added to one positively charged NAD^+ to form NADH, which is electrically neutral.

Energy captured by NADH is not harvested all at once. Instead of being transferred directly to oxygen, the two electrons carried by NADH are passed along the **electron transport chain.** This chain consists of a series of molecules, mostly proteins, embedded within the inner membranes of mitochondria. NADH delivers electrons to the top of the electron transport chain, and oxygen captures them at the bottom. The oxygen then joins with hydrogen ions to form water. At each step in the chain, the electrons move to a slightly more electronegative carrier, and their positions shift slightly. Thus, the electrons move *down* an energy gradient. The entire process releases a total of 53 kcal/mole (222 kJ/mole) under standard conditions. The transfer of electrons along this chain allows the energy to be extracted gradually. Next, we will discuss how this energy is put to work to drive the production of ATP.

> The catabolism of glucose involves a series of oxidation-reduction reactions that release energy by repositioning electrons closer to oxygen atoms. Energy is thus harvested from glucose molecules in gradual steps, using NAD^+ as an electron carrier.

Stage Four: The Electron Transport Chain

The NADH and FADH$_2$ molecules formed during the first three stages of aerobic respiration each contain a pair of electrons that were gained when NAD$^+$ and FAD were reduced. The NADH molecules carry their electrons to the inner mitochondrial membrane, where they transfer the electrons to a series of membrane-associated proteins collectively called the **electron transport chain.**

Moving Electrons Through the Electron Transport Chain

The first of the proteins to receive the electrons is a complex, membrane-embedded enzyme called **NADH dehydrogenase.** A carrier called ubiquinone then passes the electrons to a protein-cytochrome complex called the *bc$_1$ complex.* This complex, along with others in the chain, operates as a proton pump, driving a proton out across the membrane (figure 9.15).

The electron is then carried by another carrier, *cytochrome c*, to the cytochrome oxidase complex. This complex uses four such electrons to reduce a molecule of oxygen. Each oxygen then combines with two hydrogen ions to form water:

$$O_2 + 4\,H^+ + 4\,e^- \rightarrow 2\,H_2O$$

While NADH contributes its electrons to the first protein of the electron transport chain, NADH dehydrogenase, FADH$_2$, which is always attached to the inner mitochondrial membrane, feeds its electrons into the electron transport chain later, to ubiquinone.

It is the availability of a plentiful electron acceptor (often oxygen) that makes oxidative respiration possible. As we'll see in chapter 10, the electron transport chain used in aerobic respiration is similar to, and may well have evolved from, the chain employed in aerobic photosynthesis.

Building an Electrochemical Gradient

In eukaryotes, aerobic metabolism takes place within the mitochondria present in virtually all cells. The internal compartment, or matrix, of a mitochondrion contains the enzymes that carry out the reactions of the Krebs cycle. As

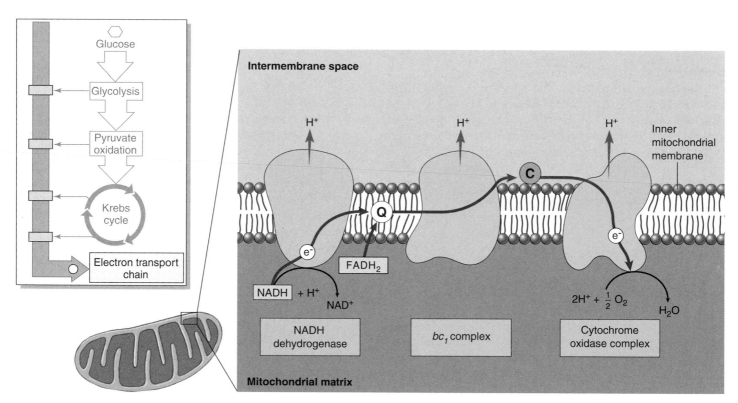

FIGURE 9.15
The electron transport chain.   High-energy electrons harvested from catabolized molecules are transported (*red arrows*) by mobile electron carriers (ubiquinone, marked Q, and cytochrome *c*, marked C) along a chain of membrane proteins. Three proteins use portions of the electrons' energy to pump protons (*blue arrows*) out of the matrix and into the intermembrane space. The electrons are finally donated to oxygen to form water.

the electrons harvested by oxidative respiration are passed along the electron transport chain, the energy they release transports protons out of the matrix and into the outer compartment, sometimes called the intermembrane space. Three transmembrane proteins in the inner mitochondrial membrane (see figure 9.15) actually accomplish the transport. The flow of excited electrons induces a change in the shape of these pump proteins, which causes them to transport protons across the membrane. The electrons contributed by NADH activate all three of these proton pumps, while those contributed by $FADH_2$ activate only two.

Producing ATP: Chemiosmosis

As the proton concentration in the intermembrane space rises above that in the matrix, the matrix becomes slightly negative in charge. This internal negativity attracts the positively charged protons and induces them to reenter the matrix. The higher outer concentration tends to drive protons back in by diffusion; since membranes are relatively impermeable to ions, most of the protons that reenter the matrix pass through ATP synthase, which uses the energy of the gradient to catalyze the synthesis of ATP from ADP and P_i. Thus, the vast majority of the energy originally contained in glucose is ultimately used to generate a gradient of protons in the inner membrane of the mitochondrion. This gradient is then used to drive the synthesis of ATP, which is transported by facilitated diffusion to the many places in the cell where enzymes require energy to drive their reactions. Because the chemical formation of ATP is driven by a diffusion force similar to osmosis, this process is referred to as **chemiosmosis** (figure 9.16).

Thus, the electron transport chain uses electrons harvested in aerobic respiration to pump a large number of protons across the inner mitochondrial membrane. Their subsequent reentry into the mitochondrial matrix through ATP synthase drives the synthesis of ATP by chemiosmosis. Figure 9.17 summarizes the overall process.

The electron transport chain is a series of five membrane-associated proteins. Electrons delivered by NADH and $FADH_2$ are passed from protein to protein along the chain, like a baton in a relay race. These electrons are used to pump protons out of the mitochondrial matrix via the electron transport chain. The return of the protons into the matrix generates ATP.

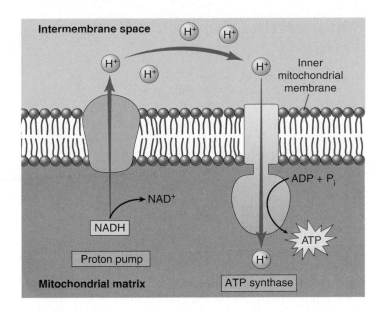

FIGURE 9.16
Chemiosmosis. NADH transports high-energy electrons harvested from the catabolism of macromolecules to "proton pumps" that use the energy to pump protons out of the mitochondrial matrix. As a result, the concentration of protons in the intermembrane space rises, inducing protons to diffuse back into the matrix. Many of the protons pass through ATP synthase, which couples the reentry of protons to the production of ATP.

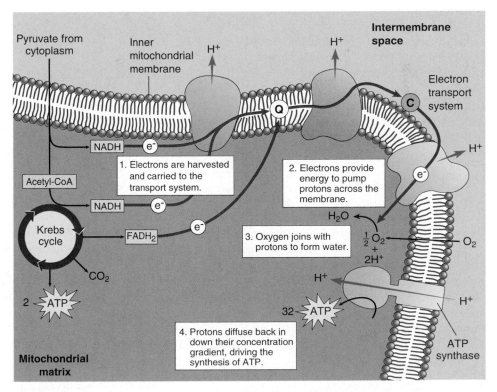

FIGURE 9.17
ATP generation during the Krebs cycle and electron transport chain. This process begins with pyruvate, the product of glycolysis, and ends with the synthesis of ATP.

Summarizing the Yield of Aerobic Respiration

How much metabolic energy does a cell actually gain from the electrons harvested from a molecule of glucose, using the electron transport chain to produce ATP by chemiosmosis?

Theoretical Yield

The chemiosmotic model suggests that one ATP molecule is generated for each proton pump activated by the electron transport chain. As the electrons from NADH activate three pumps and those from $FADH_2$ activate two, we would expect each molecule of NADH and $FADH_2$ to generate three and two ATP molecules, respectively. However, because eukaryotic cells carry out glycolysis in their cytoplasm and the Krebs cycle within their mitochondria, they must transport the two molecules of NADH produced during glycolysis across the mitochondrial membranes, which requires one ATP per molecule of NADH. Thus, the net ATP production is decreased by two. Therefore, the overall ATP production resulting from aerobic respiration *theoretically* should be: 4 (from substrate-level phosphorylation during glycolysis) + 30 (3 from each of 10 molecules of NADH) + 4 (2 from each of 2 molecules of $FADH_2$) − 2 (for transport of glycolytic NADH) = 36 molecules of ATP (figure 9.18).

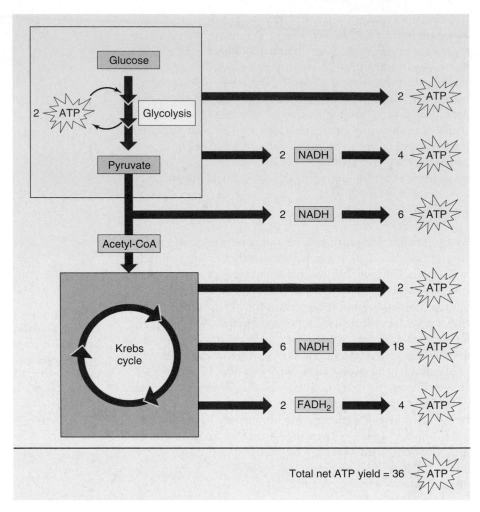

FIGURE 9.18
Theoretical ATP yield.   The theoretical yield of ATP harvested from glucose by aerobic respiration totals 36 molecules.

Actual Yield

The amount of ATP *actually* produced in a eukaryotic cell during aerobic respiration is somewhat lower than 36, for two reasons. First, the inner mitochondrial membrane is somewhat "leaky" to protons, allowing some of them to reenter the matrix without passing through ATP synthase. Second, mitochondria often use the proton gradient generated by chemiosmosis for purposes other than ATP synthesis (such as transporting pyruvate into the matrix). Consequently, the actual measured values of ATP generated by NADH and $FADH_2$ are closer to 2.5 for each NADH and 1.5 for each $FADH_2$. With these corrections, the overall harvest of ATP from a molecule of glucose in a eukaryotic cell is closer to: 4 (from substrate-level phosphorylation) + 25 (2.5 from each of 10 molecules of NADH) + 3 (1.5 from each of 2 molecules of $FADH_2$) − 2 (transport of glycolytic NADH) = 30 molecules of ATP.

The catabolism of glucose by aerobic respiration, in contrast to glycolysis, is quite efficient. Aerobic respiration in a eukaryotic cell harvests about $(7.3 \times 30)/686 = 32\%$ of the energy available in glucose. (By comparison, a typical car converts only about 25% of the energy in gasoline into useful energy.)

The higher efficiency of aerobic respiration was one of the key factors that fostered the evolution of heterotrophs. With this mechanism for producing ATP, it became feasible for nonphotosynthetic organisms to derive metabolic energy exclusively from the oxidative breakdown of other organisms. As long as some organisms captured energy by photosynthesis, others could exist solely by feeding on them.

Aerobic respiration produces approximately 30 molecules of ATP from each molecule of glucose in a eukaryotic cell. This represents about one-third of the energy in the chemical bonds of glucose.

Regulating Aerobic Respiration

When cells possess plentiful amounts of ATP, the key reactions of glycolysis, the Krebs cycle, and fatty acid breakdown are inhibited, slowing ATP production. The regulation of these biochemical pathways by the level of ATP is an example of feedback inhibition. Conversely, when ATP levels in the cell are low, ADP levels are high, and ADP activates enzymes in the pathways of carbohydrate catabolism to stimulate the production of more ATP.

Control of glucose catabolism occurs at two key points in the catabolic pathway (figure 9.19). The control point in glycolysis is the enzyme phosphofructokinase, which catalyzes the conversion of fructose phosphate to fructose bisphosphate. This is the first reaction of glycolysis that is not readily reversible, committing the substrate to the glycolytic sequence. High levels of ADP relative to ATP (implying a need to convert more ADP to ATP) stimulate phosphofructokinase, committing more sugar to the catabolic pathway; so do low levels of citrate (implying the Krebs cycle is not running at full tilt and needs more input). The main control point in the oxidation of pyruvate occurs at the committing step in the Krebs cycle with the enzyme pyruvate decarboxylase. This enzyme is inhibited by high levels of NADH, a key product of the Krebs cycle, implying that no more is needed.

Another control point in the Krebs cycle is the enzyme citrate synthetase, which catalyzes the first reaction, the conversion of oxaloacetate and acetyl-CoA into citrate. High levels of ATP inhibit citrate synthetase (as well as pyruvate decarboxylase and two other Krebs cycle enzymes), shutting down the catabolic pathway.

Relative levels of ADP and ATP regulate the catabolism of glucose at key committing reactions.

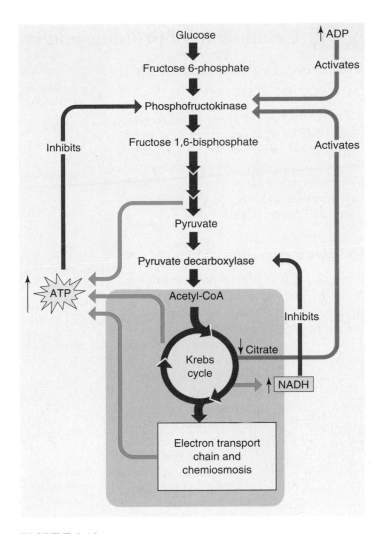

FIGURE 9.19
Control of glucose catabolism. The relative levels of ADP and ATP control the catabolic pathway at two key points: the committing reactions of glycolysis and the Krebs cycle.

A Vocabulary of ATP Generation

aerobic respiration The portion of cellular respiration that requires oxygen as an electron acceptor; it includes pyruvate oxidation, the Krebs cycle, and the electron transport chain.

anaerobic respiration Cellular respiration in which inorganic electron acceptors other than oxygen are used; it includes glycolysis.

cellular respiration The anaerobic oxidation of organic molecules to produce ATP in which the final electron acceptor is organic; it includes aerobic and anaerobic respiration.

chemiosmosis The passage of high-energy electrons along the electron transport chain, which is coupled to the pumping of protons across a membrane and the return of protons to the original side of the membrane through ATP synthase, driving the production of ATP.

fermentation Alternative ATP-producing pathway performed by some cells in the absence of oxygen, in which the final electron acceptor is an organic molecule.

oxidation The loss of an electron. In cellular respiration, high-energy electrons are stripped from food molecules, oxidizing them. Any oxidation must be accompanied by a corresponding reduction.

photosynthesis The chemiosmotic generation of ATP and complex organic molecules powered by the energy derived from light.

substrate-level phosphorylation The generation of ATP by the direct transfer of a phosphate group to ADP from another phosphorylated molecule.

9.3 Catabolism of proteins and fats can yield considerable energy.

Glucose Is Not the Only Source of Energy

Thus far we have discussed the aerobic respiration of glucose, which organisms obtain from the digestion of carbohydrates or from photosynthesis. Organic molecules other than glucose, particularly proteins and fats, are also important sources of energy (figure 9.20).

Cellular Respiration of Protein

Proteins are first broken down into their individual amino acids. The nitrogen-containing side group (the amino group) is then removed from each amino acid in a process called **deamination**. A series of reactions convert the carbon chain that remains into a molecule that takes part in glycolysis or the Krebs cycle. For example, alanine is converted into pyruvate, glutamate into α-ketoglutarate (figure 9.21), and aspartate into oxaloacetate. The reactions of glycolysis and the Krebs cycle then extract the high-energy electrons from these molecules and put them to work making ATP.

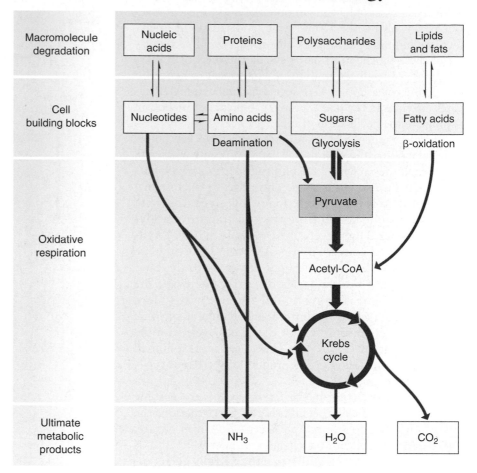

FIGURE 9.20
How cells extract chemical energy. All eukaryotes and many prokaryotes extract energy from organic molecules by oxidizing them. The first stage of this process, breaking down macromolecules into their constituent parts, yields little energy. The second stage, oxidative or aerobic respiration, extracts energy, primarily in the form of high-energy electrons, and produces water and carbon dioxide.

FIGURE 9.21
Deamination. After proteins are broken down into their amino acid constituents, the amino groups are removed from the amino acids to form molecules that participate in glycolysis and the Krebs cycle. For example, the amino acid glutamate becomes α-ketoglutarate, a Krebs cycle molecule, when it loses its amino group.

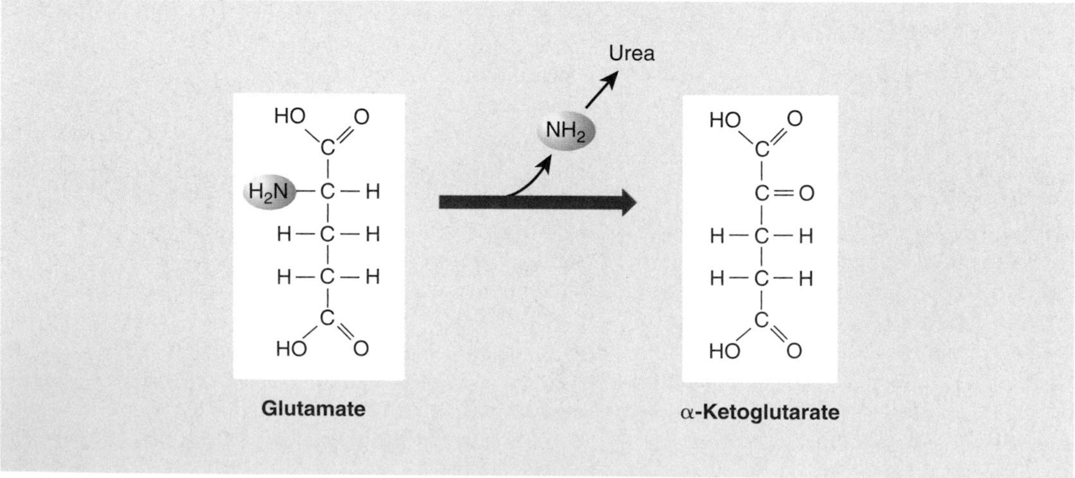

Cellular Respiration of Fat

Fats are broken down into fatty acids plus glycerol. The tails of fatty acids typically have 16 or more —CH_2 links, and the many hydrogen atoms in these long tails provide a rich harvest of energy. Fatty acids are oxidized in the matrix of the mitochondrion. Enzymes there remove the two-carbon acetyl groups from the end of each fatty acid tail until the entire fatty acid is converted into acetyl groups (figure 9.22). Each acetyl group then combines with coenzyme A to form acetyl-CoA. This process is known as **β oxidation.**

How much ATP does the catabolism of fatty acids produce? Let's compare a hypothetical six-carbon fatty acid with the six-carbon glucose molecule, which we've said yields about 30 molecules of ATP in a eukaryotic cell. Two rounds of β oxidation would convert the fatty acid into three molecules of acetyl-CoA. Each round requires one molecule of ATP to prime the process, but it also produces one molecule of NADH and one of $FADH_2$. These molecules together yield four molecules of ATP (assuming 2.5 ATPs per NADH and 1.5 ATPs per $FADH_2$). The oxidation of each acetyl-CoA in the Krebs cycle ultimately produces an additional 10 molecules of ATP. Overall, then, the ATP yield of a six-carbon fatty acid would be approximately: 8 (from two rounds of β oxidation) − 2 (for priming those two rounds) + 30 (from oxidizing the three acetyl-CoAs) = 36 molecules of ATP. Therefore, the respiration of a six-carbon fatty acid yields 20% more ATP than the respiration of glucose. Moreover, a fatty acid of that size would weigh less than two-thirds as much as glucose, so a gram of fatty acid contains more than twice as many kilocalories as a gram of glucose. That is why fat is a storage molecule for excess energy in many types of animals. If excess energy were stored instead as carbohydrate, as it is in plants, animal bodies would be much bulkier.

Proteins, fats, and other molecules are also metabolized for energy. The amino acids of proteins are first deaminated, and fats undergo a process called β oxidation.

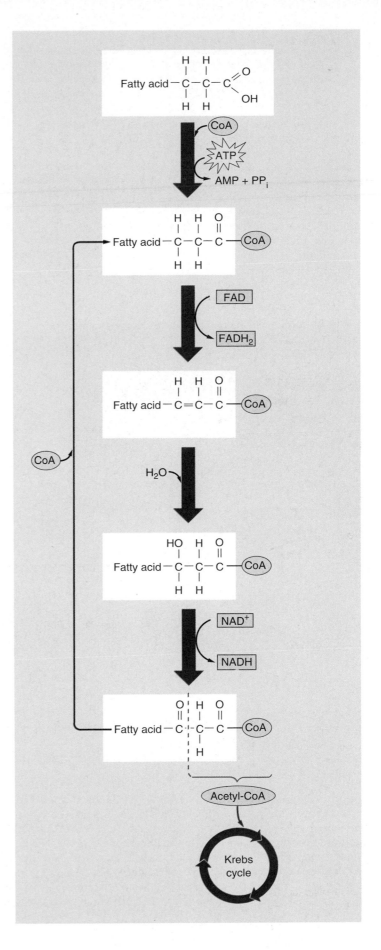

FIGURE 9.22

β oxidation. Through a series of reactions known as β oxidation, the last two carbons in a fatty acid tail combine with coenzyme A to form acetyl-CoA, which enters the Krebs cycle. The fatty acid, now two carbons shorter, enters the pathway again and keeps reentering until all its carbons have been used to form acetyl-CoA molecules. Each round of β oxidation uses one molecule of ATP and generates one molecule each of $FADH_2$ and NADH.

Metabolic Efficiency and the Length of Food Chains

It has been estimated that a heterotroph limited to glycolysis captures only 3.5% of the energy in the autotrophs it consumes as food. Hence, any other heterotrophs that consume the first heterotroph will capture through glycolysis 3.5% of the energy in it, or 0.12% of the energy available in the original autotrophs. A very large base of autotrophs would thus be needed to support a small number of heterotrophs.

When organisms became able to extract energy from organic molecules by oxidative metabolism, this constraint became far less severe, because the efficiency of oxidative respiration is estimated to be about 32%. This increased efficiency results in the transmission of much more energy from one trophic level to another than does glycolysis. (A trophic level is a step in the movement of energy through an ecosystem.) The efficiency of oxidative metabolism has made possible the evolution of food chains in which autotrophs are consumed by heterotrophs, which are consumed by other heterotrophs, and so on. You will read more about food chains in chapter 55.

Even with oxidative metabolism, approximately two-thirds of the available energy is lost at each trophic level, and that limits how long a food chain can be. Most food chains, like the one illustrated in figure 9.A, involve only three or rarely four trophic levels. Too much energy is lost at each transfer to allow chains to be much longer than that. For example, it would be impossible for a large human population to subsist by eating lions captured from the Serengeti Plain of Africa; the amount of grass available there would not support enough zebras and other herbivores to maintain the number of lions needed to feed the human population. Thus, the ecological complexity of our world is fixed in a fundamental way by the chemistry of oxidative respiration.

Stage 1: Photosynthesizers

Stage 2: Herbivores

Stage 3: Carnivore

Stage 4: Scavengers

Stage 5: Refuse utilizers

FIGURE 9.A

A food chain in the savannas, or open grasslands, of East Africa. Stage 1: *Photosynthesizer*. The grass under this yellow fever tree grows actively during the hot, rainy season, capturing the energy of the sun and storing it in molecules of glucose, which are then converted into starch and stored in the grass. Stage 2: *Herbivore*. This mother zebra and her baby consume the grass and transfer some of its stored energy into their own bodies. Stage 3: *Carnivore*. The lion feeds on zebras and other animals, capturing part of their stored energy and storing it in its own body. Stage 4: *Scavenger*. This hyena and the vultures occupy the same level in the food chain as the lion. They are also consuming the body of the dead zebra, which has been abandoned by the lion. Stage 5: *Refuse utilizer*. These butterflies, mostly *Precis octavia*, are feeding on the material left in the hyena's dung after the food the hyena consumed had passed through its digestive tract. At each of these four levels, only about one-third or less of the energy present is used by the recipient.

9.4 Cells can metabolize food without oxygen.

Fermentation

In the absence of oxygen, aerobic metabolism cannot occur, and cells must rely exclusively on glycolysis to produce ATP. Under these conditions, the hydrogen atoms generated by glycolysis are donated to organic molecules in a process called **fermentation** that recycles NAD$^+$, the electron acceptor required for glycolysis to proceed.

Bacteria carry out more than a dozen kinds of fermentations, all using some form of organic molecule to accept the hydrogen atom from NADH and thus recycle NAD$^+$:

$$\text{organic molecule} \rightarrow \text{reduced organic molecule}$$
$$+ \text{ NADH} \qquad\qquad + \text{ NAD}^+$$

Often the reduced organic compound is an organic acid—such as acetic acid, butyric acid, propionic acid, or lactic acid—or an alcohol.

Ethanol Fermentation

Eukaryotic cells are capable of only a few types of fermentation. In one type, which occurs in single-celled fungi called yeast, the molecule that accepts hydrogen from NADH is pyruvate, the end product of glycolysis itself. Yeast enzymes remove a terminal CO_2 group from pyruvate through decarboxylation, producing a two-carbon molecule called acetaldehyde. The CO_2 released causes bread made with yeast to rise, while bread made without yeast (unleavened bread) does not. The acetaldehyde accepts a hydrogen atom from NADH, producing NAD$^+$ and ethanol (ethyl alcohol) (figure 9.23). This particular type of fermentation is of great interest to humans, since it is the source of the ethanol in wine and beer. Ethanol is a by-product of fermentation that is actually toxic to yeast; as it approaches a concentration of about 12%, it begins to kill the yeast. That explains why naturally fermented wine contains only about 12% ethanol.

Lactic Acid Fermentation

Most animal cells regenerate NAD$^+$ without decarboxylation. Muscle cells, for example, use an enzyme called lactate dehydrogenase to transfer a hydrogen atom from NADH back to the pyruvate that is produced by glycolysis. This reaction converts pyruvate into lactic acid and regenerates NAD$^+$ from NADH. It therefore closes the metabolic circle, allowing glycolysis to continue as long as glucose is available. Circulating blood removes excess lactate (the ionized form of lactic acid) from muscles, but when removal cannot keep pace with production, the accumulating lactic acid interferes with muscle function and contributes to muscle fatigue.

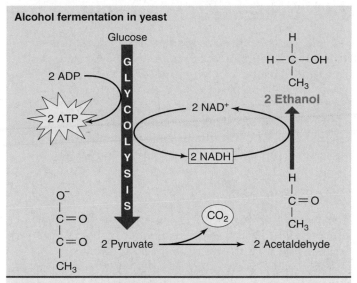

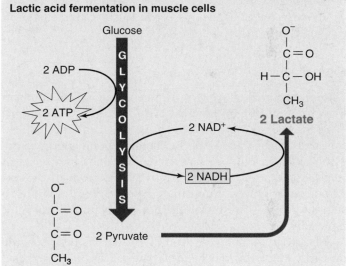

FIGURE 9.23
Fermentation. Yeasts carry out the conversion of pyruvate to ethanol (see also figure 9.9). Muscle cells convert pyruvate into lactate, which is less toxic than ethanol. However, lactate is still toxic enough to produce a painful sensation in muscles during heavy exercise, when oxygen in the muscles is depleted.

In fermentation, which occurs in the absence of oxygen, the electrons that result from the glycolytic breakdown of glucose are donated to an organic molecule, regenerating NAD$^+$ from NADH.

9.5 The stages of cellular respiration evolved over time.

The Evolution of Metabolism

We talk about cellular respiration as a continuous series of stages, but it is important to note that these stages evolved over time. Metabolism has changed a great deal as life on earth has evolved. This has been particularly true of the reactions organisms use to capture energy from the sun to build organic molecules (anabolism) and then break down organic molecules to obtain energy (catabolism). These processes evolved in concert with each other.

Degradation

The most primitive forms of life are thought to have obtained chemical energy by degrading, or breaking down, organic molecules that were abiotically produced.

The first major event in the evolution of metabolism was the origin of the ability to harness chemical bond energy. At an early stage, organisms began to store this energy in the bonds of ATP.

Glycolysis

The second major event in the evolution of metabolism was glycolysis, the initial breakdown of glucose. As proteins evolved diverse catalytic functions, it became possible to capture a larger fraction of the chemical bond energy in organic molecules by breaking chemical bonds in a series of steps. For example, the progressive breakdown of the six-carbon sugar glucose into three-carbon molecules in the 10 steps of glycolysis results in the net production of two ATP molecules.

Glycolysis undoubtedly evolved early in the history of life on earth, since this biochemical pathway has been retained by all living organisms. It is a chemical process that does not appear to have changed for well over 2 billion years.

Anaerobic Photosynthesis

The third major event in the evolution of metabolism was anaerobic photosynthesis. Early in the history of life, some organisms evolved a different way of generating ATP, called photosynthesis. Instead of obtaining energy for ATP synthesis by reshuffling chemical bonds, as in glycolysis, these organisms developed the ability to use light to pump protons out of their cells, and to use the resulting proton gradient to power the production of ATP through chemiosmosis.

Photosynthesis evolved in the absence of oxygen and works well without it. Dissolved H_2S, present in the oceans beneath an atmosphere free of oxygen gas, served as a ready source of hydrogen atoms for building organic molecules. Free sulfur was produced as a by-product of this reaction.

Oxygen-Forming Photosynthesis

The substitution of H_2O for H_2S in photosynthesis was the fourth major event in the history of metabolism. Oxygen-forming photosynthesis employs H_2O rather than H_2S as a source of hydrogen atoms and their associated electrons. Because it garners its hydrogen atoms from reduced oxygen rather than from reduced sulfur, it generates oxygen gas rather than free sulfur.

More than 2 billion years ago, small cells capable of carrying out this oxygen-forming photosynthesis, such as cyanobacteria, became the dominant forms of life on earth. Oxygen gas began to accumulate in the atmosphere. This was the beginning of a great transition that changed conditions on earth permanently. Our atmosphere is now 20.9% oxygen, every molecule of which is derived from an oxygen-forming photosynthetic reaction.

Nitrogen Fixation

Nitrogen fixation was the fifth major step in the evolution of metabolism. Proteins and nucleic acids cannot be synthesized from the products of photosynthesis because both of these biologically critical molecules contain nitrogen. Obtaining nitrogen atoms from N_2 gas, a process called *nitrogen fixation*, requires breaking an $N\equiv N$ triple bond. This important reaction evolved in the hydrogen-rich atmosphere of the early earth, where no oxygen was present. Oxygen acts as a poison to nitrogen fixation, which today occurs only in oxygen-free environments or in oxygen-free compartments within certain prokaryotes.

Aerobic Respiration

Aerobic respiration is the sixth and final event in the history of metabolism. This cellular process harvests energy by stripping energetic electrons from organic molecules. Aerobic respiration employs the same kind of proton pumps as photosynthesis, and is thought to have evolved as a modification of the basic photosynthetic machinery.

Biologists think that the ability to carry out photosynthesis without H_2S first evolved among purple nonsulfur bacteria, which obtain their hydrogens from organic compounds instead. It was perhaps inevitable that among the descendants of these respiring photosynthetic bacteria, some would eventually do without photosynthesis entirely, subsisting only on the energy and hydrogens derived from the breakdown of organic molecules. The mitochondria within all eukaryotic cells are thought to be descendants of these bacteria.

Six major innovations highlight the evolution of metabolism as we know it today.

Concept Review

For interactive testing, visit the Online Learning Center with PowerWeb at www.mhhe.com/Raven7

9.1 Cells harvest the energy in chemical bonds.

Using Chemical Energy to Drive Metabolism

- Autotrophs produce their own chemical energy, while heterotrophs live on the energy autotrophs produce. (p. 160)
- The energy of a chemical bond is contained in the potential energy of the electrons that make up the bond. (p. 160)
- Cells use some of the energy gained by catabolizing food to drive ATP production. (p. 160)
- ATP stores energy by linking charged phosphate groups near one another. (p. 160)
- Cells use ATP to facilitate movement and to drive endergonic reactions. (p. 160)
- The vast majority of ATP produced in the cell is made by ATP synthase. (p. 161)

9.2 Cellular respiration oxidizes food molecules.

An Overview of Glucose Catabolism

- Cells can make ATP from the catabolism of organic molecules two ways: substrate-level phosphorylation and aerobic respiration. (p. 162)
- In many organisms, cells harvest energy from glucose molecules in a sequence of four pathways: glycolysis, pyruvate oxidation, the Krebs cycle, and the electron transport chain. (pp. 162–163)
- Anaerobic respiration donates harvested electrons to inorganic compounds other than oxygen. (p. 163)

Stage One: Glycolysis

- Glycolysis generates ATP by shuffling the bonds in glucose molecules. Two molecules of NAD^+ are reduced to NADH. (p. 164)
- NAD^+ must be regenerated for glycolysis to continue. (pp. 166–167)

Stage Two: The Oxidation of Pyruvate

- Pyruvate is decarboxylated within the mitochondrion, yielding acetyl-CoA, NADH, and CO_2. (p. 168)

Stage Three: The Krebs Cycle

- The Krebs cycle is a series of nine reactions that oxidize acetyl-CoA in the matrix of a mitochondrion. (p. 169)
- The Krebs cycle yields two molecules of ATP per molecule of glucose. (p. 170)

Harvesting Energy by Extracting Electrons

- Glucose catabolism involves a series of oxidation-reduction reactions that release energy by repositioning electrons closer to oxygen atoms. (pp. 172–173)
- Energy is harvested in gradual steps, using NAD^+ as an electron carrier. (p. 172)

Stage Four: The Electron Transport Chain

- The electron transport chain is a series of membrane-associated proteins. Electrons delivered by NADH and $FADH_2$ are passed along the protein chain, and are used to pump protons out of the mitochondrial matrix via the electron transport chain. The return of protons into the matrix through ATP synthase generates ATP. (p. 175)

Summarizing the Yield of Aerobic Respiration

- The theoretical yield of aerobic respiration is 36 molecules of ATP, while the actual yield is around 30 molecules of ATP. (p. 176)
- Aerobic respiration harvests about 32% of the energy available in glucose. (p. 176)

Regulating Aerobic Respiration

- Relative levels of ADP and ATP control the catabolic pathway at the committing reactions of glycolysis and the Krebs cycle. (p. 177)

9.3 Catabolism of proteins and fats can yield considerable energy.

Cellular Respiration of Protein

- Proteins are first broken down into their individual amino acids, and then the amino group is removed from each amino acid by deamination. (p. 178)
- Glycolysis and the Krebs cycle then extract high-energy electrons from the molecules and use them in producing ATP. (p. 178)

Cellular Respiration of Fat

- Fats are metabolized for energy by β oxidation. (p. 179)

9.4 Cells can metabolize food without oxygen.

Fermentation

- Fermentation occurs in the absence of oxygen as electrons from the glycolytic breakdown of glucose are donated to an organic molecule, regenerating NAD^+ from NADH. (p. 181)

9.5 The stages of cellular respiration evolved over time.

The Evolution of Metabolism

- The six major innovations of metabolism are degradation, glycolysis, anaerobic photosynthesis, oxygen-forming photosynthesis, nitrogen fixation, and aerobic respiration. (p. 182)

Self Test

1. In cellular respiration, energy-depleted electrons are donated to an inorganic molecule. In fermentation, what molecule accepts these electrons?
   a. oxygen
   b. an organic molecule
   c. sulfur
   d. an inorganic molecule other than O_2

2. Which of the following is *not* a stage of aerobic respiration?
   a. glycolysis
   b. pyruvate oxidation
   c. the Krebs cycle
   d. the electron transport chain

3. Which steps in glycolysis require the input of energy?
   a. the glucose priming steps
   b. the phosphorylation of glucose
   c. the phosphorylation of fructose 6-phosphate
   d. All of these steps require the input of energy.

4. Pyruvate dehydrogenase is a multienzyme complex that catalyzes a series of reactions. Which of the following is *not* carried out by pyruvate dehydrogenase?
   a. a decarboxylation reaction
   b. the production of ATP
   c. producing an acetyl group from pyruvate
   d. combining the acetyl group with a cofactor

5. How many molecules of CO_2 are produced for each molecule of glucose that passes through glycolysis and the Krebs cycle?
   a. 2
   b. 3
   c. 6
   d. 7

6. The electrons generated from the Krebs cycle are transferred to _____ and then are shuttled to _____.
   a. NAD^+ / oxygen
   b. NAD^+ / electron transport chain
   c. NADH / oxygen
   d. NADH / electron transport chain

7. The electron transport chain pumps protons
   a. out of the mitochondrial matrix.
   b. out of the intermembrane space and into the matrix.
   c. out of the mitochondrion and into the cytoplasm.
   d. out of the cytoplasm and into the mitochondrion.

8. What process of cellular respiration generates the most ATP?
   a. glycolysis
   b. oxidation of pyruvate
   c. Krebs cycle
   d. chemiosmosis

9. Oxidizing which of the following substances yields the most energy?
   a. proteins
   b. glucose
   c. fatty acids
   d. water

10. The final electron acceptor in lactic acid fermentation is
    a. pyruvate.
    b. NAD^+.
    c. lactic acid.
    d. O_2.

Test Your Visual Understanding

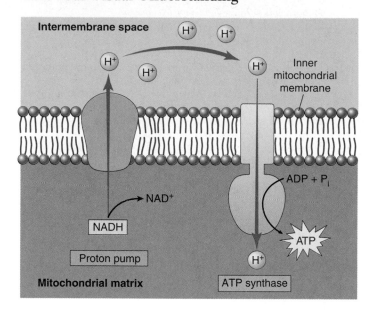

Answer the following questions related to this figure, which shows the process of chemiosmosis.

1. In the figure, proton pumps are transporting hydrogen ions across the membrane. What is the driving force of this pump? What type of membrane transport is the proton pump? Explain.

2. Why is this process called chemiosmosis? What is the force driving the synthesis of ATP? How could the process of ATP synthesis be inhibited or shut down?

Apply Your Knowledge

1. How much energy would be generated in the cells of a person who consumed a diet of pyruvate instead of glucose? Calculate the energy generated on a per molecule basis.

2. As explained in chapter 5, mitochondria are thought to have evolved from bacteria that were engulfed by and lived symbiotically within early eukaryotic cells. Why haven't present-day eukaryotic cells dispensed with mitochondria, placing all of the mitochondrial genes in the nucleus and carrying out all of the metabolic functions of the mitochondria within the cytoplasm?

3. Why do plants typically store their excess energy as carbohydrates rather than fat?

4. If you poke a hole in a mitochondrion, can it still perform oxidative respiration? Can fragments of mitochondria perform oxidative respiration?

10

Photosynthesis

Concept Outline

10.1 What is photosynthesis?

The Chloroplast as a Photosynthetic Machine. The highly organized system of membranes in chloroplasts is essential to the functioning of photosynthesis.

10.2 Learning about photosynthesis: An experimental journey.

The Role of Soil and Water. The added mass of a growing plant comes mostly from photosynthesis. In plants, water supplies the electrons used to reduce carbon dioxide.

Discovery of the Light-Independent Reactions. Photosynthesis is a multistage process. Only the first stages directly require light.

The Role of Light and Reducing Power. The oxygen released during green plant photosynthesis comes from water, and carbon atoms from carbon dioxide are incorporated into organic molecules.

10.3 Pigments capture energy from sunlight.

The Biophysics of Light. The energy in sunlight occurs in "packets" called photons, which are absorbed by pigments.

Chlorophylls and Carotenoids. Photosynthetic pigments absorb light and harvest its energy.

Organizing Pigments into Photosystems. A photosystem uses light energy to eject an energized electron.

How Photosystems Convert Light to Chemical Energy. Some bacteria rely on a single photosystem to produce ATP. Plants use two photosystems in series to generate enough energy to reduce $NADP^+$ and generate ATP.

How the Two Photosystems of Plants Work Together. Photosystems II and I drive the synthesis of the ATP and NADPH needed to form organic molecules.

10.4 Cells use the energy and reducing power captured by the light-dependent reactions to make organic molecules.

The Calvin Cycle. The Calvin cycle uses ATP and NADPH produced in the light-dependent reactions to build organic molecules, a process reversed in mitochondria.

Photorespiration. The enzyme that catalyzes carbon fixation also affects CO_2 release.

FIGURE 10.1
Capturing energy. These sunflower plants, growing vigorously in the August sun, are capturing light energy for conversion into chemical energy through photosynthesis.

The rich diversity of life that covers our earth would be impossible without photosynthesis. Every oxygen atom in the air we breathe was once part of a water molecule, liberated by photosynthesis. All the energy released by the burning of coal, firewood, gasoline, and natural gas, and by our bodies' burning of all the food we eat—directly or indirectly—has been captured from sunlight by photosynthesis. It is vitally important, then, that we understand photosynthesis. Research may enable us to improve crop yields and land use, important goals in an increasingly crowded world. In chapter 9, we described how cells extract chemical energy from food molecules and use that energy to power their activities. In this chapter, we examine photosynthesis, the process by which organisms capture energy from sunlight and use it to build food molecules that are rich in chemical energy (figure 10.1).

10.1 What is photosynthesis?

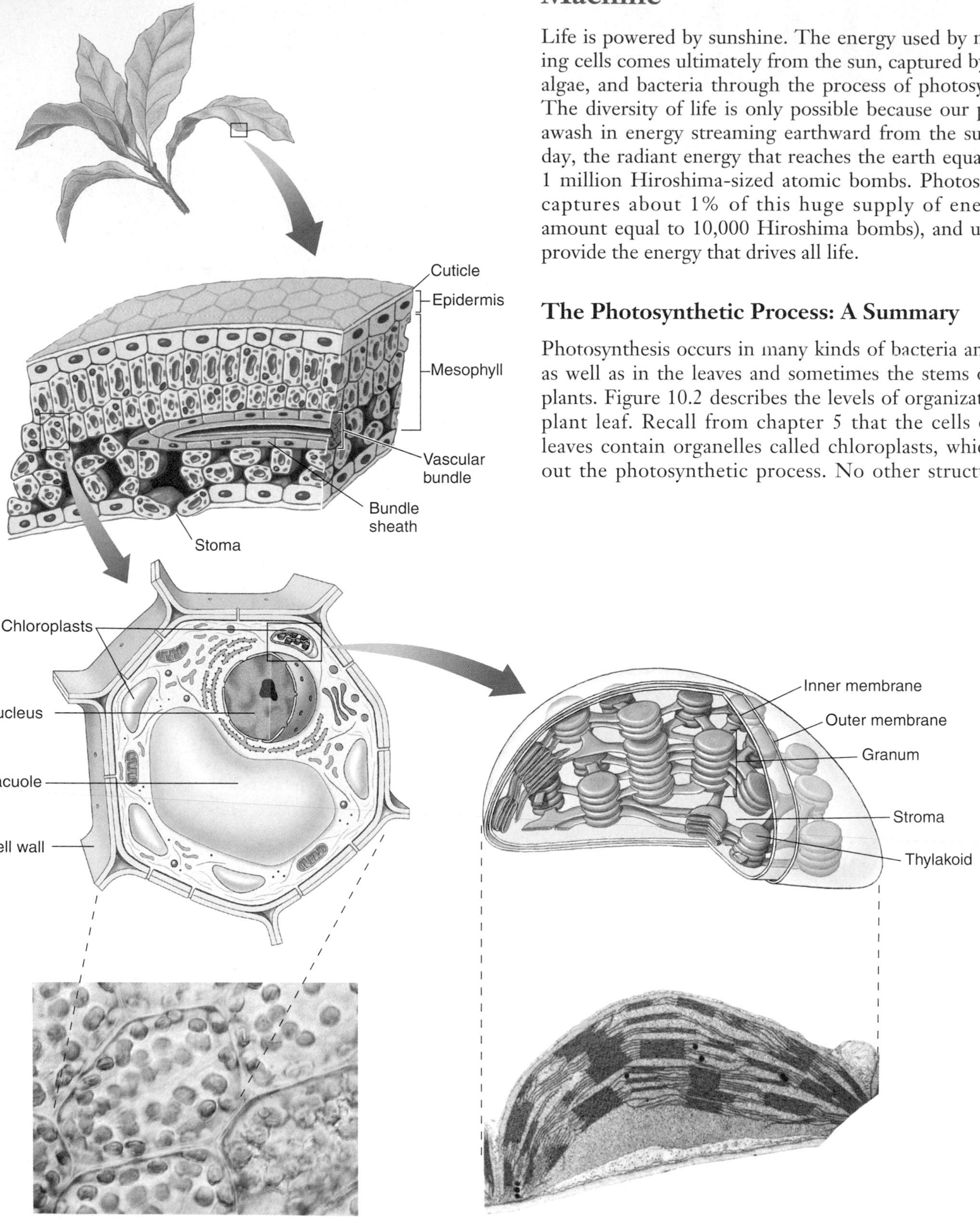

Cuticle
Epidermis

Mesophyll

Vascular
bundle

Bundle
sheath

Stoma

Chloroplasts

Nucleus

Vacuole

Cell wall

Inner membrane

Outer membrane

Granum

Stroma

Thylakoid

The Chloroplast as a Photosynthetic Machine

Life is powered by sunshine. The energy used by most living cells comes ultimately from the sun, captured by plants, algae, and bacteria through the process of photosynthesis. The diversity of life is only possible because our planet is awash in energy streaming earthward from the sun. Each day, the radiant energy that reaches the earth equals about 1 million Hiroshima-sized atomic bombs. Photosynthesis captures about 1% of this huge supply of energy (an amount equal to 10,000 Hiroshima bombs), and uses it to provide the energy that drives all life.

The Photosynthetic Process: A Summary

Photosynthesis occurs in many kinds of bacteria and algae, as well as in the leaves and sometimes the stems of green plants. Figure 10.2 describes the levels of organization in a plant leaf. Recall from chapter 5 that the cells of plant leaves contain organelles called chloroplasts, which carry out the photosynthetic process. No other structure in a

FIGURE 10.2
Journey into a leaf. A plant leaf possesses a thick layer of cells (the mesophyll) rich in chloroplasts. The flattened thylakoids in the chloroplast are stacked into columns called grana. The light-dependent reactions take place on the thylakoid membrane and generate the

plant cell is able to carry out photosynthesis. Photosynthesis takes place in three stages: (1) capturing energy from sunlight; (2) using the energy to make ATP and reducing power in the form of a compound called NADPH; and (3) using the ATP and NADPH to power the synthesis of organic molecules from CO_2 in the air (carbon fixation).

The first two stages can only take place in the presence of light and are commonly called the **light-dependent reactions.** The third stage, the formation of organic molecules from atmospheric CO_2, occurs in a process called the **Calvin cycle.** As long as ATP and NADPH are available, the Calvin cycle can occur either in the presence or in the absence of light, and so the reactions of the Calvin cycle are also called the **light-independent reactions.**

The following simple equation summarizes the overall process of photosynthesis:

$$6\ CO_2 + 12\ H_2O + light \rightarrow C_6H_{12}O_6 + 6\ H_2O + 6\ O_2$$

carbon   water   glucose   water   oxygen
dioxide

Inside the Chloroplast

The internal membranes of chloroplasts, called *thylakoids*, are organized into sacs stacked on one another in columns called **grana** (singular, *granum*). The thylakoid membranes house the photosynthetic pigments for capturing light energy and the machinery to make ATP. Surrounding the thylakoid membrane system is a semiliquid substance called *stroma.* The stroma houses the enzymes needed to assemble organic molecules from CO_2 using energy from ATP and reducing power in NADPH. In the membranes of thylakoids, photosynthetic pigments are clustered together to form a **photosystem.**

Each pigment molecule within the photosystem is capable of capturing *photons*, which are packets of energy. A lattice of proteins holds the pigments in close contact with one another. When light of a proper wavelength strikes a pigment molecule in the photosystem, the resulting excitation passes from one chlorophyll molecule to another. The excited electron is not transferred physically— rather, its *energy* passes from one molecule to another. A crude analogy to this form of energy transfer is the initial "break" in a game of pool. If the cue ball squarely hits the point of the triangular array of 15 pool balls, the two balls at the far corners of the triangle fly off, but none of the central balls move. The energy passes through the central balls to the most distant ones.

Eventually, the energy arrives at a key chlorophyll molecule that is touching a membrane-bound protein. The energy is transferred as an excited electron to that protein, which passes it on to a series of other membrane proteins that put the energy to work making ATP and NADPH and building organic molecules. The photosystem thus acts as a large antenna, gathering the light energy harvested by many individual pigment molecules.

The light-dependent reactions of photosynthesis take place within thylakoid membranes within chloroplasts in leaf cells.

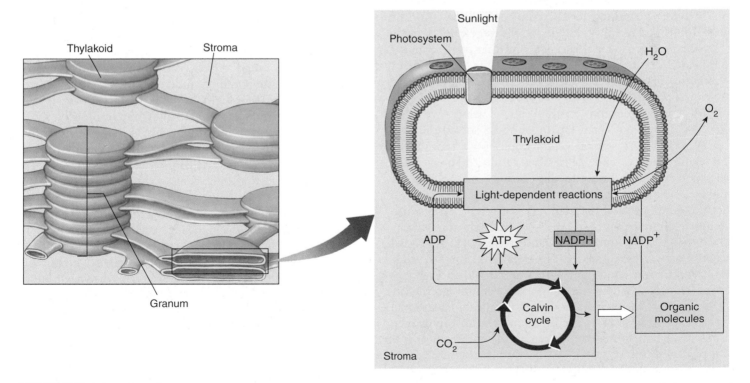

FIGURE 10.2 (continued)
ATP and NADPH that fuel the Calvin cycle. The fluid interior matrix of a chloroplast, the stroma, contains the enzymes that carry out the Calvin cycle.

10.2 Learning about photosynthesis: an experimental journey.

The Role of Soil and Water

The story of how we learned about photosynthesis is one of the most interesting in science and serves as a good introduction to this complex process. It begins over 300 years ago, with a simple but carefully designed experiment by a Belgian doctor, Jan Baptista van Helmont (1577–1644). From the time of the Greeks, plants were thought to obtain their food from the soil, literally sucking it up with their roots; van Helmont thought of a simple way to test this idea. He planted a small willow tree in a pot of soil, after first weighing the tree and the soil. The tree grew in the pot for several years, during which time van Helmont added only water. At the end of five years, the tree was much larger, its weight having increased by 74.4 kilograms. However, the soil in the pot weighed only 57 grams less than it had five years earlier, so obviously all of this added mass could not have come from the soil. With this experiment, van Helmont demonstrated that the substance of the plant was not produced only from the soil. He incorrectly concluded that the water he had been adding mainly accounted for the plant's increased mass.

A hundred years passed before the story became clearer. The key clue was provided by the English scientist Joseph Priestly, in his pioneering studies of the properties of air. On the 17th of August, 1771, Priestly "accidentally hit upon a method of restoring air that had been injured by the burning of candles." He "put a [living] sprig of mint into air in which a wax candle had burnt out and found that, on the 27th of the same month, another candle could be burned in this same air." Somehow, the vegetation seemed to have restored the air! Priestly found that while a mouse could not breathe candle-exhausted air, air "restored" by vegetation was not "at all inconvenient to a mouse." The key clue was that *living vegetation adds something to the air.*

How does vegetation "restore" air? Twenty-five years later, the Dutch physician Jan Ingenhousz solved the puzzle. Working over several years, Ingenhousz reproduced and significantly extended Priestly's results, demonstrating that air was restored only in the presence of sunlight and only by a plant's green leaves, not by its roots. He proposed that the green parts of the plant carry out a process (which we now call photosynthesis) that uses sunlight to split carbon dioxide (CO_2) into carbon and oxygen. He suggested that the oxygen was released as O_2 gas into the air, while the carbon atom combined with water to form carbohydrates. His proposal was a good guess, even though the latter step was subsequently modified. Chemists later found that the proportions of carbon, oxygen, and hydrogen atoms in carbohydrates are indeed about one atom of carbon per molecule of water (as the term *carbohydrate* indicates). A Swiss botanist found in 1804 that water was a necessary reactant. By the end of that century, the overall reaction for photosynthesis could be written as:

$$CO_2 + H_2O + \text{light energy} \rightarrow (CH_2O) + O_2$$

It turns out, however, that there's more to it than that. When researchers began to examine the process in more detail in the twentieth century, the role of light proved to be unexpectedly complex.

> **Van Helmont showed that soil did not add mass to a growing plant. Priestly, Ingenhousz, and others then worked out the basic chemical reaction.**

Discovery of the Light-Independent Reactions

Ingenhousz's early equation for photosynthesis includes one factor we have not discussed: light energy. What role does light play in photosynthesis? At the beginning of the twentieth century, the English plant physiologist F. F. Blackman began to question the role of light in photosynthesis. In 1905, he came to the startling conclusion that photosynthesis is in fact a multistage process, only one of which uses light directly.

Blackman measured the effects of different light intensities, CO_2 concentrations, and temperatures on photosynthesis. As long as light intensity was relatively low, he found photosynthesis could be accelerated by increasing the amount of light, but not by increasing the temperature or CO_2 concentration (figure 10.3). At high light intensities, however, an increase in temperature or CO_2 concentration greatly accelerated photosynthesis. Blackman concluded that photosynthesis consists of an initial set of what he called "light" reactions, that are largely independent of temperature, and a second set of "dark" reactions (also called light-independent reactions), that seemed to be independent of light but limited by CO_2. Do not be confused by Blackman's labels—the so-called "dark" reactions occur in the light (in fact, they require the products of the light-dependent reactions); their name simply indicates that light is not *directly* involved in those reactions.

Blackman found that increased temperature speeds the rate of the dark carbon-reducing reactions, but only up to about 35°C. Higher temperatures cause the rate to fall off rapidly. Because 35°C is the temperature at which many plant enzymes begin to be denatured (the hydrogen bonds that hold an enzyme in its particular catalytic shape begin to be disrupted), Blackman concluded that enzymes must carry out the light-independent reactions.

> **Blackman showed that capturing photosynthetic energy requires sunlight, while building organic molecules does not.**

The Role of Light and Reducing Power

The role of light in the so-called light-dependent and light-independent reactions was worked out in the 1930s by C. B. van Niel, then a graduate student at Stanford University. While studying photosynthesis in bacteria, van Niel discovered that purple sulfur bacteria do not release oxygen during photosynthesis; instead, they convert hydrogen sulfide (H_2S) into globules of pure elemental sulfur that accumulate inside themselves. The process van Niel observed was:

$$CO_2 + 2\ H_2S + \text{light energy} \rightarrow (CH_2O) + H_2O + 2\ S$$

The striking parallel between this equation and Ingenhousz's equation led van Niel to propose that the generalized process of photosynthesis is in fact:

$$CO_2 + 2\ H_2A + \text{light energy} \rightarrow (CH_2O) + H_2O + 2\ A$$

In this equation, the substance H_2A serves as an electron donor. In photosynthesis performed by green plants, H_2A is water, while among purple sulfur bacteria, H_2A is hydrogen sulfide. The product, A, comes from the splitting of H_2A. Therefore, the O_2 produced during green plant photosynthesis results from splitting water, not carbon dioxide.

When isotopes came into common use in biology in the early 1950s, it became possible to test van Niel's revolutionary proposal. Investigators examined photosynthesis in green plants supplied with ^{18}O water; they found that the ^{18}O label ended up in oxygen gas rather than in carbohydrate, just as van Niel had predicted:

$$CO_2 + 2\ H_2^{18}O + \text{light energy} \rightarrow (CH_2O) + H_2O + {}^{18}O_2$$

In algae and green plants, the carbohydrate typically produced by photosynthesis is the sugar glucose, which has six carbons. The complete balanced equation for photosynthesis in these organisms thus becomes:

$$6\ CO_2 + 12\ H_2O + \text{light energy} \rightarrow C_6H_{12}O_6 + 6\ O_2 + 6\ H_2O$$

We now know that the light-dependent reactions of photosynthesis use the energy of light to reduce NADP (an electron carrier molecule) to NADPH and to manufacture ATP. The NADPH and ATP from this stage of photosynthesis are then used in a subsequent stage, the Calvin cycle, to reduce the carbon in carbon dioxide and form a simple sugar whose carbon skeleton can be used to synthesize other organic molecules.

In his pioneering work on the light-dependent reactions, van Niel had further proposed that the reducing power (H^+) generated by the splitting of water was used to convert CO_2 into organic matter in a process he called **carbon fixation**. This reducing power supplies high-energy electrons from NADPH to replace the low-energy electrons in the C—O bonds of carbon dioxide. These high-energy electrons form the C—H bonds of the newly synthesized organic molecules.

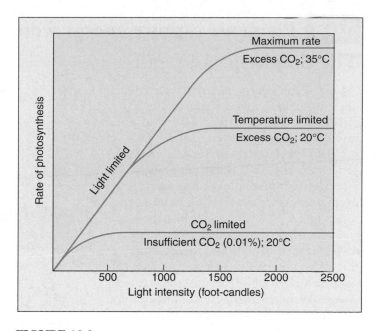

FIGURE 10.3

Discovery of the light-independent reactions. Blackman measured photosynthesis rates under differing light intensities, CO_2 concentrations, and temperatures. As this graph shows, light is the limiting factor at low light intensities, while temperature and CO_2 concentration are the limiting factors at higher light intensities. **Blackman found that increasing light intensity above 2000 foot-candles did not lead to any further increase in the rate of photosynthesis. Can you suggest a hypothesis that would explain why?**

In the 1950s, Robin Hill demonstrated that van Niel was indeed right, and that light energy could be used to generate reducing power. Chloroplasts isolated from leaf cells were able to reduce a dye and release oxygen in response to light. Later experiments showed that the electrons released from water were transferred to $NADP^+$. Arnon and co-workers showed that illuminated chloroplasts deprived of CO_2 accumulate ATP. If CO_2 is then introduced, neither ATP nor NADPH accumulate, and the CO_2 is assimilated into organic molecules. These experiments are important for three reasons: First, they firmly demonstrate that photosynthesis occurs only within chloroplasts. Second, they show that the light-dependent reactions use light energy to reduce $NADP^+$ and to manufacture ATP. Third, they confirm that the ATP and NADPH from this early stage of photosynthesis are then used in the later light-independent reactions to reduce carbon dioxide, forming simple sugars.

Van Niel discovered that photosynthesis splits water molecules, incorporating the carbon atoms of carbon dioxide gas and the hydrogen atoms of water into organic molecules and leaving oxygen gas.
Hill showed that plants can use light energy to generate reducing power. The incorporation of carbon dioxide into organic molecules in the light-independent reactions is called carbon fixation.

10.3 Pigments capture energy from sunlight.

The Biophysics of Light

Where is the energy in light? What is there in sunlight that a plant can use to reduce carbon dioxide? This is the mystery of photosynthesis, the one factor fundamentally different from processes such as cellular respiration. To answer these questions, we will need to consider the physical nature of light. James Clerk Maxwell had theorized that light was an electromagnetic wave– that is, that light moves through the air as oscillating electric and magnetic fields. Proof of this came in a curious experiment carried out in a laboratory in Germany in 1887. A young physicist, Heinrich Hertz, was attempting to verify a highly mathematical theory that predicted the existence of electromagnetic waves. To see whether such waves existed, Hertz designed a clever experiment. On one side of a room, he constructed a powerful spark generator that consisted of two large, shiny metal spheres standing near each other on tall, slender rods. When a very high static electrical charge was built up on one sphere, sparks jumped across to the other sphere.

After constructing this device, Hertz set out to investigate whether the sparking would create invisible electromagnetic waves, so-called radio waves, as predicted by the mathematical theory. On the other side of the room, he placed the world's first radio receiver, a thin metal hoop on an insulating stand. There was a small gap at the bottom of the hoop, so that the hoop did not quite form a complete circle. When Hertz turned on the spark generator across the room, he saw tiny sparks passing across the gap in the hoop! This was the first demonstration of radio waves. But Hertz noted another curious phenomenon. When UV light was shining across the gap on the hoop, the sparks were produced more readily. This unexpected facilitation, called the *photoelectric effect*, puzzled investigators for many years.

The photoelectric effect was finally explained using a concept proposed by Max Planck in 1901. Planck developed an equation that predicted the blackbody radiation curve based upon the assumption that light and other forms of radiation behave as units of energy called *photons*. In 1905, Albert Einstein explained the photoelectric effect utilizing the photon concept. Ultraviolet light has photons of sufficient energy that, when they fell on the loop, electrons were ejected from the metal surface. The photons had transferred their energy to the electrons, literally blasting them from the ends of the hoop and thus facilitating the

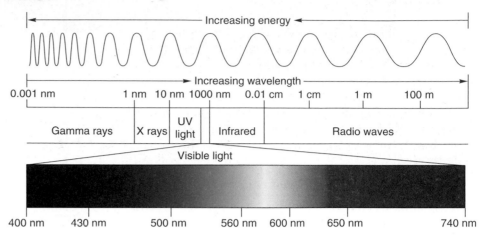

FIGURE 10.4

The electromagnetic spectrum. Light is a form of electromagnetic energy conveniently thought of as a wave. The shorter the wavelength of light, the greater its energy. Visible light represents only a small part of the electromagnetic spectrum between 400 and 740 nanometers.

passage of the electric spark induced by the radio waves. Visible wavelengths of light were unable to remove the electrons because their photons did not have enough energy to free the electrons from the metal surface at the ends of the hoop.

The Energy in Photons

Not all photons possess the same amount of energy. Instead, the energy content of a photon is inversely proportional to the wavelength of the light: Short-wavelength light contains photons of higher energy than long-wavelength light (figure 10.4). X rays, which contain a great deal of energy, have very short wavelengths—much shorter than those of visible light, making them ideal for use in high-resolution microscopes.

Hertz had noted that the strength of the photoelectric effect depends on the wavelength of light; that is, short wavelengths are much more effective than long ones in producing the photoelectric effect. Einstein's theory of the photoelectric effect provides an explanation: Sunlight contains photons of many different energy levels, only some of which our eyes perceive as visible light. The highest energy photons, at the short-wavelength end of the electromagnetic spectrum (see figure 10.4), are gamma rays, with wavelengths of less than 1 nanometer; the lowest energy photons, with wavelengths of up to thousands of meters, are radio waves. Within the visible portion of the spectrum, violet light has the shortest wavelength and the most energetic photons, and red light has the longest wavelength and the least energetic photons.

Ultraviolet Light

The sunlight that reaches the earth's surface contains a significant amount of ultraviolet (UV) light, which, because of its shorter wavelength, possesses considerably more energy than visible light. UV light is thought to have been an important source of energy on the primitive earth when life originated. Today's atmosphere contains ozone (derived from oxygen gas), which absorbs most of the UV photons in sunlight, but a considerable amount of UV light still manages to penetrate the atmosphere. This UV light is a potent force in disrupting the bonds of DNA, causing mutations that can lead to skin cancer. As we will describe in chapter 56, loss of atmospheric ozone due to human activities threatens to cause an enormous jump in the incidence of human skin cancers throughout the world.

Absorption Spectra and Pigments

How does a molecule "capture" the energy of light? A photon can be envisioned as a very fast-moving packet of energy. When it strikes a molecule, its energy is either lost as heat or absorbed by the electrons of the molecule, boosting those electrons into higher energy levels. Whether or not the photon's energy is absorbed depends on how much energy it carries (defined by its wavelength) and also on the chemical nature of the molecule it hits. As we saw in chapter 2, electrons occupy discrete energy levels in their orbits around atomic nuclei. To boost an electron into a different energy level requires just the right amount of energy, just as reaching the next rung on a ladder requires you to raise your foot just the right distance. A specific atom can, therefore, absorb only certain photons of light—namely, those that correspond to the atom's available electron energy levels. As a result, each molecule has a characteristic **absorption spectrum,** the range and efficiency of photons it is capable of absorbing.

Molecules that are good absorbers of light in the visible range are called **pigments.** Organisms have evolved a variety of different pigments, but only two general types are used in green plant photosynthesis: carotenoids and chlorophylls. Chlorophylls absorb photons within narrow energy ranges. Two kinds of chlorophyll in plants, chlorophylls *a* and *b*, preferentially absorb violet-blue and red light (figure 10.5). Neither of these pigments absorbs photons with wavelengths between about 500 and 600 nanometers, and light of these wavelengths is, therefore, reflected by plants.

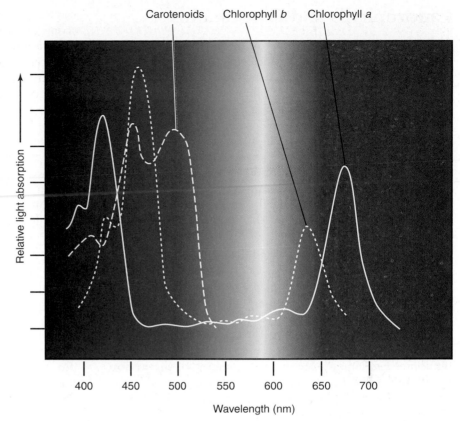

FIGURE 10.5
The absorption spectrum of chlorophyll and carotenoids. The peaks represent wavelengths of sunlight absorbed by the two common forms of photosynthetic pigment, chlorophylls *a* and *b*, and by the carotenoids. Chlorophylls absorb predominantly violet-blue and red light in two narrow bands of the spectrum and reflect green light in the middle of the spectrum. Carotenoids absorb mostly blue and green light and reflect orange and yellow light.

When these photons are subsequently absorbed by the pigment in our eyes, we perceive them as green.

Chlorophyll *a* is the main photosynthetic pigment and the only pigment that can act directly to convert light energy to chemical energy. However, chlorophyll *b*, acting as an *accessory* or secondary light-absorbing pigment, complements and adds to the light absorption of chlorophyll *a*. Chlorophyll *b* has an absorption spectrum shifted toward the green wavelengths. Therefore, chlorophyll *b* can absorb photons that chlorophyll *a* cannot, greatly increasing the proportion of the photons in sunlight that plants can harvest. An important group of accessory pigments, the carotenoids, assist in photosynthesis by capturing energy from light composed of wavelengths that are not efficiently absorbed by either chlorophyll.

In photosynthesis, photons of light are absorbed by pigments; the wavelength of light absorbed depends upon the specific pigment.

Chlorophylls and Carotenoids

Chlorophylls absorb photons by means of an excitation process analogous to the photoelectric effect. These pigments contain a complex ring structure, called a porphyrin ring, with alternating single and double bonds. At the center of the ring is a magnesium atom. Photons absorbed by the pigment molecule excite electrons in the ring, which are then channeled away through the alternating carbon-bond system. Several small side groups attached to the outside of the ring alter the absorption properties of the molecule in different kinds of chlorophyll (figure 10.6). The precise absorption spectrum is also influenced by the local microenvironment created by the association of chlorophyll with specific proteins.

Once Ingenhousz demonstrated that only the green parts of plants can "restore" air, researchers suspected chlorophyll was the primary pigment that plants employ to absorb light in photosynthesis. Experiments conducted in the 1800s clearly verified this suspicion. One such experiment, performed by T. W. Englemann in 1882 (figure 10.7), serves as a particularly elegant example, simple in design and clear in outcome. Englemann set out to characterize the **action spectrum** of photosynthesis—that is, the relative effectiveness of different wavelengths of light in promoting photosynthesis. He carried out the entire experiment utilizing a single slide mounted on a microscope. To obtain different wavelengths of light, he placed a prism under his microscope, splitting the light that illuminated the slide into a spectrum of colors. He then arranged a filament of green algal cells across the spectrum, so that different parts of the filament were illuminated with different wavelengths, and allowed the algae to carry out photosynthesis. To assess how fast photosynthesis was proceeding, Englemann chose to monitor the rate of oxygen production. Lacking a mass spectrometer and other modern instruments, he added aerotactic (oxygen-seeking) bacteria to the slide; he knew they would gather along the filament at locations where oxygen was being produced. He found that the bacteria accumulated in areas illuminated by red and violet light, the two colors most strongly absorbed by chlorophyll.

All plants, algae, and cyanobacteria use chlorophyll *a* as their primary pigments. It is reasonable to ask why these photosynthetic organisms do not use a pigment like retinal (the pigment in our eyes), which has a broad absorption spectrum that covers the range of 500 to 600 nanometers. The most likely hypothesis involves *photoefficiency*. Although retinal absorbs a broad range of wavelengths, it does so with relatively low efficiency. Chlorophyll, in contrast, absorbs in only two narrow bands, but does so with high efficiency. Therefore, plants and most other photosynthetic organisms achieve far higher overall photon capture rates with chlorophyll than with other pigments.

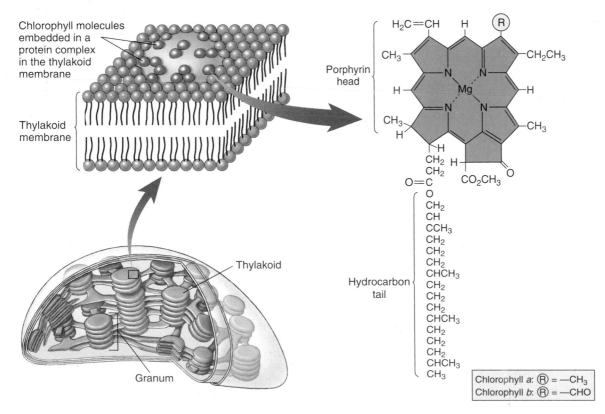

**FIGURE 10.6
Chlorophyll.**
Chlorophyll molecules consist of a porphyrin head and a hydrocarbon tail that anchors the pigment molecule to hydrophobic regions of proteins embedded within the membranes of thylakoids. The only difference between the two chlorophyll molecules is the substitution of a —CHO (aldehyde) group in chlorophyll *b* for a —CH₃ (methyl) group in chlorophyll *a*.

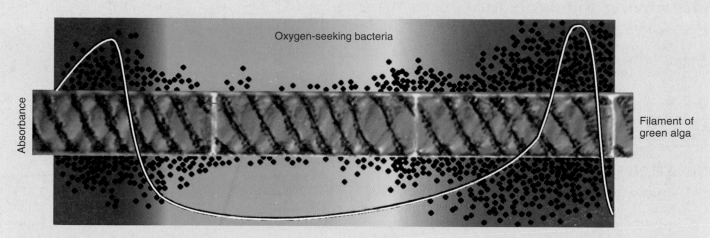

Oxygen-seeking bacteria

Absorbance

Filament of green alga

T.W. Englemann revealed the action spectrum of photosynthesis in the filamentous alga *Spirogyra* in 1882. Englemann used the rate of oxygen production to measure the rate of photosynthesis. As his oxygen indicator, he chose bacteria that are attracted by oxygen. In place of the mirror and diaphragm usually used to illuminate objects under view in his microscope, he substituted a "microspectral apparatus," which, as its name implies, produced a tiny spectrum of colors that it projected upon the slide under the microscope. Then he arranged a filament of algal cells parallel to the spread of the spectrum. The oxygen-seeking bacteria congregated mostly in the areas where the violet and red wavelengths fell upon the algal filament.

FIGURE 10.7
Constructing an action spectrum for photosynthesis. As you can see, the action spectrum for photosynthesis that Englemann revealed in his experiment parallels the absorption spectrum of chlorophyll (see figure 10.5).

Oak leaf in summer

Oak leaf in autumn

FIGURE 10.8
Fall colors are produced by carotenoids and other accessory pigments. During the spring and summer, chlorophyll in leaves masks the presence of carotenoids and other accessory pigments. When cool fall temperatures cause leaves to cease manufacturing chlorophyll, the chlorophyll is no longer present to reflect green light, and the leaves reflect the orange and yellow light that carotenoids and other pigments do not absorb.

Carotenoids consist of carbon rings linked to chains with alternating single and double bonds. They can absorb photons with a wide range of energies, although they are not always highly efficient in transferring this energy. Carotenoids assist in photosynthesis by capturing energy from light composed of wavelengths that are not efficiently absorbed by chlorophylls (figure 10.8; see figure 10.5).

A typical carotenoid is β-carotene, whose two carbon rings are connected by a chain of 18 carbon atoms with alternating single and double bonds. Splitting a molecule of β-carotene into equal halves produces two molecules of vitamin A. Oxidation of vitamin A produces retinal, the pigment used in vertebrate vision. This explains why carrots, which are rich in β-carotene, enhance vision.

A pigment is a molecule that absorbs light. The wavelengths absorbed by a particular pigment depend on the available energy levels to which light-excited electrons can be boosted in the pigment.

Organizing Pigments into Photosystems

The light-dependent reactions of photosynthesis occur in membranes. In photosynthetic bacteria, the plasma membrane itself is the photosynthetic membrane. In plants and algae, by contrast, photosynthesis is carried out by organelles that are the evolutionary descendants of photosynthetic bacteria, chloroplasts—in other words, the photosynthetic membranes exist *within* the chloroplasts. The light-dependent reactions take place in four stages:

1. **Primary photoevent.** A photon of light is captured by a pigment. The result of this primary photoevent is the excitation of an electron within the pigment.
2. **Charge separation.** This excitation energy is transferred to a specialized chlorophyll pigment termed a **reaction center,** which reacts by transferring an energetic electron to an acceptor molecule, thus initiating electron transport.
3. **Electron transport.** The excited electron is shuttled along a series of electron-carrier molecules embedded within the photosynthetic membrane. Several of them react by transporting protons across the membrane, generating a gradient of proton concentration. Its arrival at the pump induces the transport of a proton across the membrane. The electron is then passed to an acceptor.
4. **Chemiosmosis.** The protons that accumulate on one side of the membrane now flow back across the membrane through ATP synthase where chemiosmotic synthesis of ATP takes place, just as it does in aerobic respiration.

Discovery of Photosystems

One way to study how pigments absorb light is to measure the dependence of the output of photosynthesis on the intensity of illumination—that is, how much photosynthesis is produced by how much light. When experiments of this sort are done on plants, they show that the output of photosynthesis increases linearly at low intensities but lessens at higher intensities, finally saturating at high-intensity light. Saturation occurs because all of the light-absorbing capacity of the plant is in use; additional light doesn't increase the output because there is nothing to absorb the added photons.

It is tempting to think that at saturation, all of a plant's pigment molecules are in use. In 1932, plant physiologists Robert Emerson and William Arnold set out to test this hypothesis using an organism in which they could measure both the number of chlorophyll molecules and the output of photosynthesis. In their experiment, they measured the oxygen yield of photosynthesis when a culture of *Chlorella* (a unicellular green alga) was exposed to very brief light flashes lasting only a few microseconds. Assuming the hy-

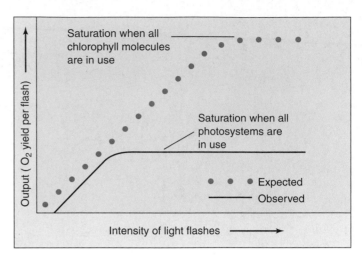

FIGURE 10.9
Emerson and Arnold's experiment.   When photosynthetic saturation is achieved, further increases in intensity cause no increase in output.
Under what experimental conditions would you expect the saturation levels for a given number of chlorophyll molecules to be higher?

pothesis of pigment saturation to be correct, they expected to find that as they increased the intensity of the flashes, the yield per flash would increase, until each chlorophyll molecule absorbed a photon, which would then be used in the light-dependent reactions, producing a molecule of O_2.

Unexpectedly, this is not what happened. Instead, saturation was achieved much earlier, with only one molecule of O_2 per 2500 chlorophyll molecules (figure 10.9)! This led Emerson and Arnold to conclude that light is absorbed not by independent pigment molecules, but rather by clusters of chlorophyll and accessory pigment molecules, which have come to be called *photosystems* (described in more detail below). Light is absorbed by any one of the hundreds of pigment molecules in a photosystem, which transfer their excitation energy to one with a lower energy level than the others. This reaction center of the photosystem acts as an energy sink, trapping the excitation energy. It was the saturation of these reaction centers, not individual molecules, that was observed by Emerson and Arnold.

Architecture of a Photosystem

In chloroplasts and all but the most primitive prokaryotes, light is captured by **photosystems**. Each photosystem is a network of chlorophyll *a* molecules, accessory pigments, and associated proteins held within a protein matrix on the surface of the photosynthetic membrane. Like a magnifying glass focusing light on a precise point, a photosystem channels the excitation energy gathered by any one of its pigment molecules to a specific molecule, the reaction center chlorophyll. This molecule then passes the energy out of the photosystem so it can be put to work driving the synthesis of ATP and molecules.

A photosystem thus consists of two closely linked components: (1) an *antenna complex* of hundreds of pigment molecules that gather photons and feed the captured light energy to the reaction center; and (2) a *reaction center*, consisting of one or more chlorophyll *a* molecules in a matrix of protein, that passes the energy out of the photosystem.

The Antenna Complex.   The antenna complex captures photons from sunlight (figure 10.10). In chloroplasts, the antenna complex is a web of chlorophyll molecules linked together and held tightly on the thylakoid membrane by a matrix of proteins. Varying amounts of carotenoid accessory pigments may also be present. The protein matrix serves as a sort of scaffold, holding individual pigment molecules in orientations that are optimal for energy transfer. The excitation energy resulting from the absorption of a photon passes from one pigment molecule to an adjacent molecule on its way to the reaction center. After the transfer, the excited electron in each molecule returns to the low-energy level it had before the photon was absorbed. Consequently, it is energy, not the excited electrons themselves, that passes from one pigment molecule to the next. The antenna complex funnels the energy from many electrons to the reaction center.

The Reaction Center.   The reaction center is a transmembrane protein-pigment complex. In the reaction center of purple photosynthetic bacteria, which is simpler than in chloroplasts but better understood, a pair of chlorophyll *a* molecules acts as a trap for photon energy, passing an excited electron to an acceptor precisely positioned as its neighbor. Note that here the excited electron itself is transferred, not just the energy as we saw in pigment-pigment transfers. This allows the photon excitation to move away from the chlorophylls and is the key conversion of light to chemical energy.

Figure 10.11 shows the transfer of energy from the reaction center to the primary electron acceptor. By energizing an electron of the reaction center chlorophyll, light creates a strong electron donor where none existed before. The chlorophyll transfers the energized electron to the primary acceptor, a molecule of quinone, reducing the quinone and converting it to a strong electron donor. A weak electron donor then donates a low-energy electron to the chlorophyll, restoring it to its original condition. In plant chloroplasts, water serves as this weak electron

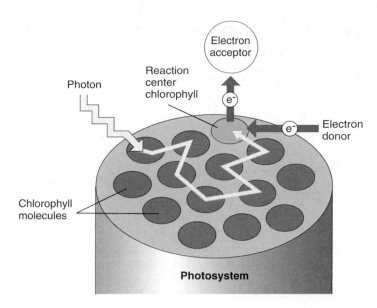

FIGURE 10.10
How the antenna complex works.   When light of the proper wavelength strikes any pigment molecule within a photosystem, the light is absorbed by that pigment molecule. The excitation energy is then transferred from one molecule to another within the cluster of pigment molecules until it encounters the reaction center chlorophyll *a*. When excitation energy reaches the reaction center chlorophyll, electron transfer is initiated.

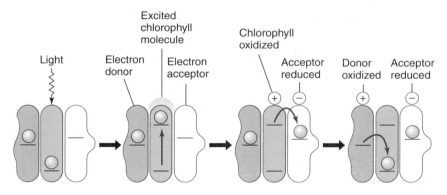

FIGURE 10.11
Converting light to chemical energy.   The reaction center chlorophyll donates a light-energized electron to the primary electron acceptor, reducing it. The oxidized chlorophyll then fills its electron "hole" by oxidizing a donor molecule.

donor. When water is oxidized this way, oxygen is released along with two protons.

> Photosystems contain pigments that capture photon energy from light. The pigments transfer the energy to reaction centers. There, the energy excites electrons, which are channeled away to do chemical work.

How Photosystems Convert Light to Chemical Energy

Bacteria Use a Single Photosystem

Photosynthetic pigment arrays are thought to have evolved more than 2 billion years ago in bacteria similar to the sulfur bacteria alive today. A two-stage process takes place within bacterial photosystems.

1. Electron is joined with a proton to make hydrogen.

In these bacteria, the absorption of a photon of light at a peak absorption of 870 nanometers (near infrared, not visible to the human eye) by the photosystem results in the transmission of an energetic electron along an electron transport chain, eventually combining with a proton to form a hydrogen atom. In the sulfur bacteria, the electron is extracted from hydrogen sulfide, leaving elemental sulfur and protons as by-products. In bacteria that evolved later, as well as in plants and algae, the electron comes from water, producing oxygen and protons as by-products.

2. Electron is recycled to chlorophyll.

The ejection of an electron from the bacterial reaction center leaves it short one electron. Before the photosystem of the sulfur bacteria can function again, an electron must be returned. These bacteria channel the electron back to the pigment through an electron transport system similar to the one described in chapter 9; the electron's passage drives a proton pump that promotes the chemiosmotic synthesis of ATP. One molecule of ATP is produced for every three electrons that follow this path. Viewed overall (figure 10.12), the path of the electron is thus a circle. Chemists therefore call the electron transfer process leading to ATP formation **cyclic photophosphorylation**. Note, however, that the electron that left the P_{870} reaction center was a high-energy electron, boosted by the absorption of a photon of light, while the electron that returns has only as much energy as it had before the photon was absorbed. The difference in the energy of that electron is the photosynthetic payoff, the energy that drives the proton pump.

For more than a billion years, cyclic photophosphorylation was the only form of photosynthetic light-dependent reaction that organisms used. However, its major limitation is that it is geared only toward energy production, not toward biosynthesis. Most photosynthetic organisms incorporate atmospheric carbon dioxide into carbohydrates. Because the carbohydrate molecules are more reduced (have more hydrogen atoms) than carbon dioxide, a source of reducing power (that is, hydrogens) must be provided. Cyclic photophosphorylation does not do this. The hydrogen atoms extracted from H_2S are used as a source of protons, and are not available to join to carbon. Thus, bacteria that are restricted to this process must scavenge hydrogens from other sources, an inefficient undertaking.

Plants Use Two Photosystems

After the sulfur bacteria appeared, other kinds of bacteria evolved an improved version of the photosystem that overcame the limitation of cyclic photophosphorylation in a neat and simple way: A second, more powerful photosystem using another arrangement of chlorophyll a was combined with the original.

In this second photosystem, called **photosystem II**, molecules of chlorophyll a are arranged with a different geometry, so that more shorter-wavelength, higher-energy photons are absorbed than in the ancestral photosystem, which is called **photosystem I**. As in the ancestral photosystem, energy is transmitted from one pigment molecule to another within the antenna complex of these photosystems until it reaches the reaction center, a particular pigment molecule positioned near a strong membrane-bound electron acceptor. In photosystem II, the absorption peak (that is, the wavelength of light most strongly absorbed) of the pigments is approximately 680 nanometers; therefore, the reaction center pigment is called P_{680}. The absorption peak of photosystem I pigments in plants is 700 nanometers, so its reaction center pigment is called P_{700}. Working together, the two photosystems carry out a noncyclic electron transfer.

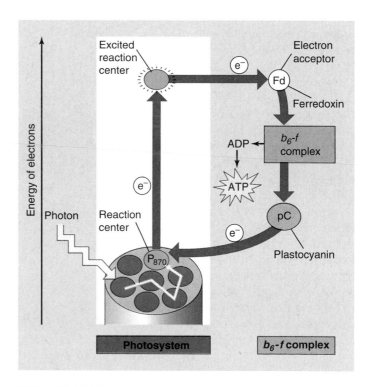

FIGURE 10.12

The path of an electron in sulfur bacteria. When a light-energized electron is ejected from the photosystem reaction center (P_{870}), it passes in a circle, eventually returning to the photosystem from which it was ejected.

When the rate of photosynthesis was measured using two light beams of different wavelengths (one red and the other far-red), the rate was greater than the sum of the rates using individual beams of red and far-red light (figure 10.13). This surprising result, called the **enhancement effect,** can be explained by a mechanism involving two photosystems acting in series (that is, one after the other), one of which absorbs preferentially in the red, the other in the far-red.

The use of two photosystems solves the problem of obtaining reducing power in a simple and direct way, by harnessing the energy of two photosystems. The scheme shown in figure 10.14, called a *Z diagram,* illustrates the two electron-energizing steps, one catalyzed by each photosystem. The electrons originate from water, which holds onto its electrons very tightly (redox potential = +820 mV), and end up in NADPH, which holds its electrons much more loosely (redox potential = −320 mV).

In sulfur bacteria, excited electrons ejected from the reaction center travel a circular path, driving a proton pump and then returning to their original photosystem. Plants employ two photosystems in series, which generates power to reduce $NADP^+$ to NADPH with enough left over to make ATP.

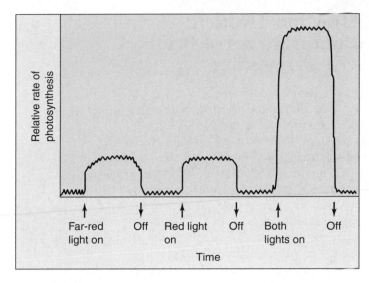

FIGURE 10.13
The enhancement effect. The rate of photosynthesis when red and far-red light are provided together is greater than the sum of the rates when each wavelength is provided individually. This result baffled researchers in the 1950s. Today, it provides the key evidence that photosynthesis is carried out by two photochemical systems with slightly different wavelength optima.
What would you conclude if "both lights on" did not change the relative rate of photosynthesis?

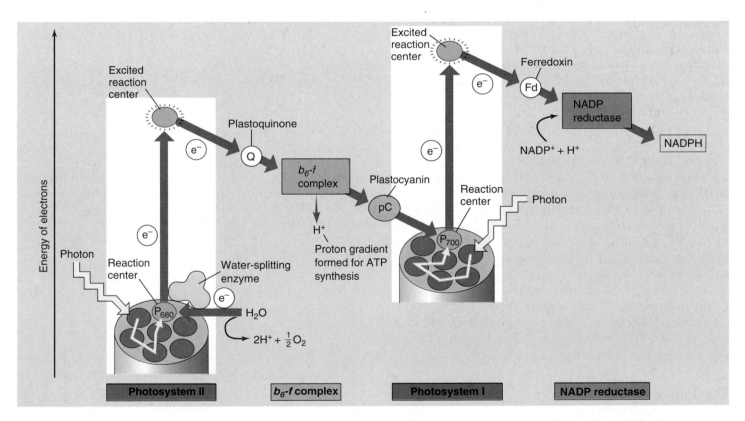

FIGURE 10.14
A Z diagram of photosystems I and II. Two photosystems work sequentially. First, a photon of light ejects a high-energy electron from photosystem II; that electron is used to pump a proton across the membrane, contributing chemiosmotically to the production of a molecule of ATP. The ejected electron then passes along a chain of cytochromes to photosystem I. When photosystem I absorbs a photon of light, it ejects a high-energy electron used to drive the formation of NADPH.

How the Two Photosystems of Plants Work Together

Plants use photosystems II and I in series, first one and then the other, to produce both ATP and NADPH. This two-stage process is called **noncyclic photophosphorylation,** because the path of the electrons is not a circle— the electrons ejected from the photosystems do not return to them, but rather end up in NADPH. The photosystems are replenished instead with electrons obtained by splitting water. Photosystem II acts first. High-energy electrons generated by photosystem II are used to synthesize ATP and then passed to photosystem I to drive the production of NADPH. For every pair of electrons obtained from water, one molecule of NADPH and slightly more than one molecule of ATP are produced.

Photosystem II

The reaction center of photosystem II, called P_{680}, closely resembles the reaction center of purple bacteria. It consists of more than 10 transmembrane protein subunits. The light-harvesting antenna complex consists of some 250 molecules of chlorophyll *a* and accessory pigments bound to several protein chains. In photosystem II, the oxygen atoms of two water molecules bind to a cluster of manganese atoms that are embedded within an enzyme and bound to the reaction center. In a way that is poorly understood, this enzyme splits water, removing electrons one at a time to fill the holes left in the reaction center by departure of light-energized electrons. As soon as four electrons have been removed from the two water molecules, O_2 is released.

The Path to Photosystem I

The primary electron acceptor for the light-energized electrons leaving photosystem II is a quinone molecule, as it was in the bacterial photosystem described earlier. The reduced quinone that results (*plastoquinone,* symbolized Q) is a strong electron donor; it passes the excited

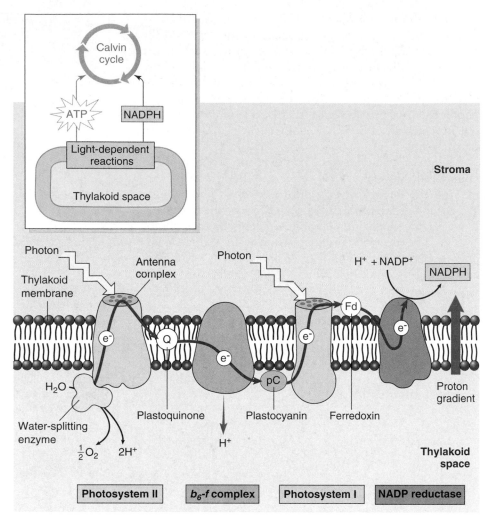

FIGURE 10.15

The photosynthetic electron transport system. When a photon of light strikes a pigment molecule in photosystem II, it excites an electron. This electron is coupled to a proton stripped from water by an enzyme and is passed along a chain of membrane-bound cytochrome electron carriers (*red arrow*). When water is split, oxygen is released from the cell, and the hydrogen ions remain in the thylakoid space. At the proton pump (b_6-f complex), the energy supplied by the photon is used to transport a proton across the membrane into the thylakoid space. The concentration of hydrogen ions within the thylakoid thus increases further. When photosystem I absorbs another photon of light, its pigment passes a second high-energy electron to a reduction complex, which generates NADPH.

electron to a proton pump called the b_6-f *complex* embedded within the thylakoid membrane (figure 10.15). This complex closely resembles the bc_1 complex in the respiratory electron transport chain of mitochondria discussed in chapter 9. Arrival of the energetic electron causes the b_6-f complex to pump a proton into the thylakoid space. A small, copper-containing protein called *plastocyanin* (symbolized pC) then carries the electron to photosystem I.

Making ATP: Chemiosmosis

Each thylakoid is a closed compartment into which protons are pumped from the stroma by the b_6-f complex. The splitting of water also produces added protons that contribute to the gradient. The thylakoid membrane is impermeable to protons, so protons cross back out almost exclusively via the channels provided by *ATP synthases*. These channels protrude like knobs on the external surface of the thylakoid membrane. As protons pass out of the thylakoid through the ATP synthase channel, ADP is phosphorylated to ATP and released into the stroma, the fluid matrix inside the chloroplast (figure 10.16). The stroma contains the enzymes that catalyze the reactions of carbon fixation.

Photosystem I

The reaction center of photosystem I, called P_{700}, is a transmembrane complex consisting of at least 13 protein subunits. Energy is fed to it by an antenna complex consisting of 130 chlorophyll *a* and accessory pigment molecules. Photosystem I accepts an electron from plastocyanin into the hole created by the exit of a light-energized electron. This arriving electron has by no means lost all of its light-excited energy; almost half remains. Thus, the absorption of a photon of light energy by photosystem I boosts the electron leaving the reaction center to a very high energy level. Unlike photosystem II and the bacterial photosystem, photosystem I does not rely on quinones as electron acceptors. Instead, it passes electrons to an iron-sulfur protein called *ferredoxin* (Fd).

Making NADPH

Photosystem I passes electrons to ferredoxin on the stromal side of the membrane (outside the thylakoid). The reduced ferredoxin carries a very-high-potential electron. Two of them, from two molecules of reduced ferredoxin, are then donated to a molecule of $NADP^+$ to form NADPH. The reaction is catalyzed by the membrane-bound enzyme *NADP reductase*. Because the reaction occurs on the stromal side of the membrane and involves the uptake of a proton in forming NADPH, it contributes further to the proton gradient established during photosynthetic electron transport.

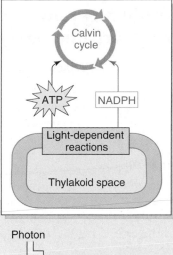

FIGURE 10.16

Chemiosmosis in a chloroplast. The b_6-f complex embedded in the thylakoid membrane pumps protons into the interior of the thylakoid. ATP is produced on the outside surface of the membrane (stroma side), as protons diffuse back out of the thylakoid through ATP synthase channels.

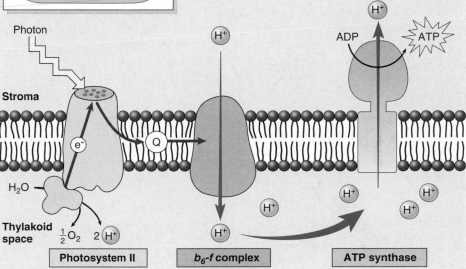

Making More ATP

The passage of an electron from water to NADPH in the noncyclic photophosphorylation described previously generates one molecule of NADPH and slightly more than one molecule of ATP. However, as you will learn later in this chapter, building organic molecules takes more energy than that—it takes one-and-a-half ATP molecules per NADPH molecule to fix carbon. To produce the extra ATP, many plant species are capable of short-circuiting photosystem I, switching photosynthesis into a *cyclic photophosphorylation* mode, so that the light-excited electron leaving photosystem I is used to make ATP instead of NADPH. The energetic electron is simply passed back to the b_6-f complex rather than passing on to $NADP^+$. The b_6-f complex pumps out a proton, adding to the proton gradient driving the chemiosmotic synthesis of ATP. The relative proportions of cyclic and noncyclic photophosphorylation in these plants determine the relative amounts of ATP and NADPH available for building organic molecules.

> The electrons that photosynthesis strips from water molecules provide the energy to form ATP and NADPH. The residual oxygen atoms of the water molecules combine to form oxygen gas.

10.4 Cells use the energy and reducing power captured by the light-dependent reactions to make organic molecules.

The Calvin Cycle

Photosynthesis is a way of making organic molecules from carbon dioxide. These molecules contain many C—H bonds and are highly reduced compared with CO_2. To build them, cells use raw materials provided by light-dependent reactions:

1. **Energy.** ATP (provided by cyclic and noncyclic photophosphorylation) drives the endergonic reactions.
2. **Reducing power.** NADPH (provided by photosystem I) provides a source of hydrogens and the energetic electrons needed to bind them to carbon atoms. Much of the light energy captured in photosynthesis ends up invested in the energy-rich C—H bonds of sugars.

Discovering the Calvin Cycle

Nearly 100 years ago, Blackman concluded that, because of its temperature dependence, photosynthesis might involve enzyme-catalyzed reactions. These reactions form a cycle of enzyme-catalyzed steps similar to the Krebs cycle. This cycle of reactions is called the **Calvin cycle,** after its discoverer, Melvin Calvin of the University of California, Berkeley. As shown by the overview in figure 10.17, the cycle begins when CO_2 binds RuBP to form PGA. Because PGA contains three carbon atoms, this process is also called **C$_3$ photosynthesis.**

Carbon Fixation

The key step in the Calvin cycle—the event that makes the reduction of CO_2 possible—is the attachment of CO_2 to a

THE CALVIN CYCLE

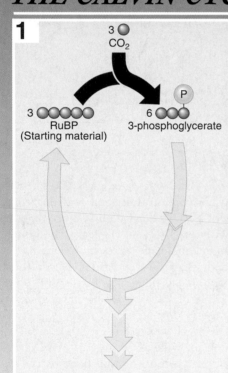

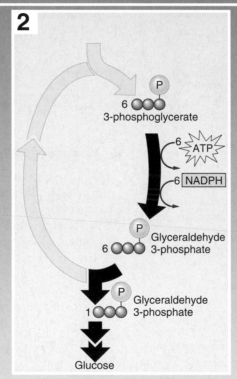

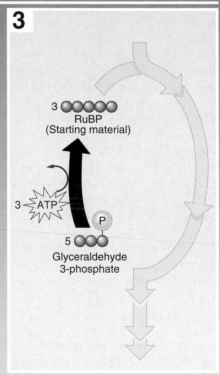

The Calvin cycle begins when a carbon atom from a CO_2 molecule is added to a five-carbon molecule (the starting material). The resulting six-carbon molecule is unstable and immediately splits into three-carbon molecules.

Then, through a series of reactions, energy from ATP and hydrogens from NADPH (the products of the light-dependent reactions) are added to the three-carbon molecules. The now-reduced three-carbon molecules either combine to make glucose or are used to make other molecules.

Most of the reduced three-carbon molecules are used to regenerate the five-carbon starting material, thus completing the cycle.

FIGURE 10.17
How the Calvin cycle works.

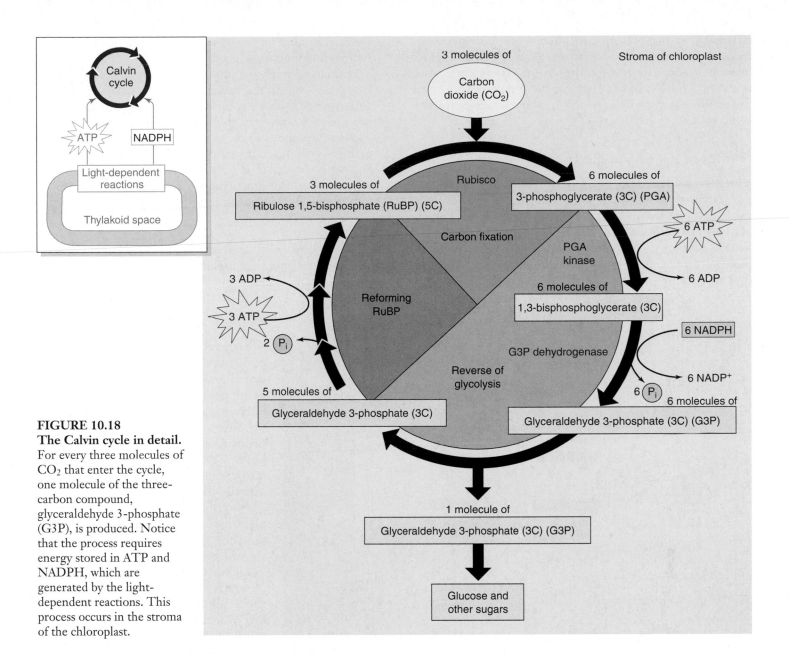

FIGURE 10.18
The Calvin cycle in detail.
For every three molecules of CO_2 that enter the cycle, one molecule of the three-carbon compound, glyceraldehyde 3-phosphate (G3P), is produced. Notice that the process requires energy stored in ATP and NADPH, which are generated by the light-dependent reactions. This process occurs in the stroma of the chloroplast.

very special organic molecule. Photosynthetic cells produce this molecule by reassembling the bonds of two intermediates in glycolysis, fructose 6-phosphate and glyceraldehyde 3-phosphate, to form the energy-rich five-carbon sugar *ribulose 1,5-bisphosphate (RuBP)*.

CO_2 binds to RuBP in the key process called **carbon fixation**, forming two three-carbon molecules of *3-phosphoglycerate (PGA)*. The enzyme that carries out this reaction, *ribulose bisphosphate carboxylase/oxygenase* (usually abbreviated *rubisco*) is a very large, four-subunit enzyme present in the chloroplast stroma.

Reactions of the Calvin Cycle

In a series of reactions, three molecules of CO_2 are fixed by rubisco to produce six molecules of PGA (containing $6 \times 3 = 18$ carbon atoms in all, three from CO_2 and 15 from

RuBP). The 18 carbon atoms then undergo a cycle of reactions that regenerates the three molecules of RuBP used in the initial step (containing $3 \times 5 = 15$ carbon atoms). This leaves one molecule of glyceraldehyde 3-phosphate (three carbon atoms) as the net gain.

The net equation of the Calvin cycle is:

$3\,CO_2 + 9\,ATP + 6\,NADPH + \text{water} \rightarrow$
   glyceraldehyde 3-phosphate $+ 8\,P_i + 9\,ADP + 6\,NADP^+$

With three full turns of the cycle, three molecules of carbon dioxide enter, a molecule of glyceraldehyde 3-phosphate (G3P) is produced, and three molecules of RuBP are regenerated (figure 10.18).

We now know that light is required *indirectly* for different segments of the CO_2 reduction reactions. Five of the Calvin cycle enzymes—including rubisco—are light activated; that is, they become functional or operate more efficiently in the presence of light. Light also promotes transport of three-carbon intermediates across chloroplast membranes that are required for Calvin cycle reactions. And finally, light promotes the influx of Mg^{++} into the chloroplast stroma, which further activates the enzyme rubisco.

Output of the Calvin Cycle

The glyceraldehyde 3-phosphate that is the product of the Calvin cycle is a three-carbon sugar, a key intermediate in glycolysis. Much of it is exported from the chloroplast to the cytoplasm of the cell, where the reversal of several reactions in glycolysis allows it to be converted to fructose 6-phosphate and glucose 1-phosphate, and from that to sucrose, a major transport sugar in plants. (Su-

crose, common table sugar, is a disaccharide made of fructose and glucose.)

In times of intensive photosynthesis, glyceraldehyde 3-phosphate levels in the stroma of the chloroplast rise. As a consequence, some glyceraldehyde 3-phosphate in the chloroplast is converted to glucose 1-phosphate, in a set of reactions analogous to those done in the cytoplasm, by reversing several reactions similar to those of glycolysis. The glucose 1-phosphate is then combined into an insoluble polymer, forming long chains of starch stored as bulky starch grains in chloroplasts.

The Energy Cycle

The energy-capturing metabolisms of the chloroplasts studied in this chapter and the mitochondria studied in chapter 9 are intimately related (figure 10.19). Photosynthesis uses the products of respiration as its starting substrates, and respiration uses the products of photosynthesis as its starting substrates. The Calvin cycle even uses part of the ancient glycolytic pathway, run in reverse, to produce glucose. Also, the principal proteins involved in electron transport in plants are related to those in mitochondria, and in many cases are actually the same.

Photosynthesis is but one aspect of plant biology, although it is an important one. In chapters 35 through 41, we will examine plants in more detail. We have treated photosynthesis here, in a part devoted to cell biology, because photosynthesis arose long before plants did, and all organisms depend directly or indirectly on photosynthesis for the energy that powers their lives.

Plants incorporate carbon dioxide into sugars by means of the Calvin cycle, which is driven by the ATP and NADPH produced in the light-dependent reactions. The Calvin cycle essentially reverses the breakdown of such molecules that occurs in the mitochondrion, forming a cycle in which energy enters from the sun and leaves as heat and work.

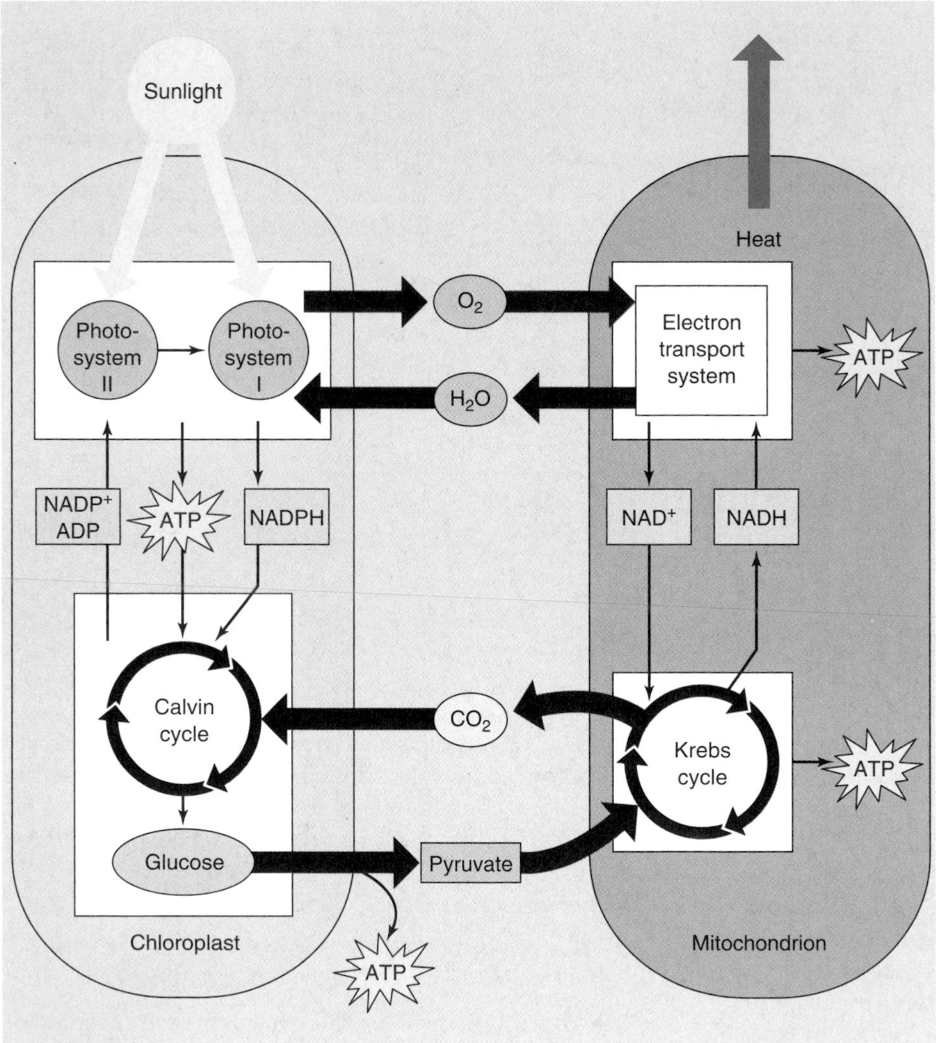

FIGURE 10.19
Chloroplasts and mitochondria: Completing an energy cycle.
Water and oxygen gas cycle between chloroplasts and mitochondria within a plant cell, as do glucose and CO_2. Cells with chloroplasts require an outside source of CO_2 and water and generate glucose and oxygen. Cells without chloroplasts, such as animal cells, require an outside source of glucose and oxygen and generate CO_2 and water.

Photorespiration

Evolution does not necessarily result in optimum solutions. Rather, it favors workable solutions that can be derived from others that already exist. Photosynthesis is no exception. Rubisco, the enzyme that catalyzes the key carbon-fixing reaction of photosynthesis, provides a decidedly suboptimal solution. This enzyme has a second enzyme activity that interferes with the Calvin cycle, *oxidizing* RuBP. In this process, called **photorespiration,** O_2 is incorporated into RuBP, which undergoes additional reactions that actually release CO_2. Hence, photorespiration releases CO_2—essentially undoing the Calvin cycle, which reduces CO_2 to carbohydrate.

The carboxylation and oxidation of RuBP are catalyzed at the same active site on rubisco, and CO_2 and O_2 compete with each other at this site. Under normal conditions at 25°C, the rate of the carboxylation reaction is four times that of the oxidation reaction, meaning that 20% of photosynthetically fixed carbon is lost to photorespiration. This loss rises substantially as temperature increases, because under hot, arid conditions, specialized openings in the leaf called *stomata* (singular, *stoma*) close to conserve water. This closing also cuts off the supply of CO_2 entering the leaf and does not allow O_2 produced by photosynthesis to exit the leaf (figure 10.20). As a result, the low CO_2 and high O_2 conditions within the leaf favor photorespiration.

Plants that fix carbon using only C_3 photosynthesis (the Calvin cycle) are called C_3 plants. In C_3 photosynthesis, RuBP is carboxylated to form a three-carbon compound via the activity of rubisco. Other plants use **C_4 photosynthesis,** in which phosphoenolpyruvate, or PEP, is carboxylated to form a four-carbon compound using the enzyme PEP carboxylase. This enzyme has no oxidation activity, and thus no photorespiration. Furthermore, PEP carboxylase has a much greater affinity for CO_2 than does rubisco. In the C_4 pathway, the four-carbon compound undergoes further modification, only to be decarboxylated. The CO_2 that is released is then captured by rubisco and drawn into the Calvin cycle. Because an organic compound is donating the CO_2, the effective concentration of CO_2 relative to O_2 is increased, and photorespiration is minimized.

The reduction in the yield of fixed carbon as a result of photorespiration is not trivial. C_3 plants lose between 25 and 50% of their photosynthetically fixed carbon in this way. The rate depends largely upon the temperature. In tropical climates, especially those in which the temperature is often above 28°C, the problem is severe, and it has a major impact on tropical agriculture.

The C_4 Pathway

Plants adapted to warmer environments have evolved two principal ways that use the C_4 pathway to deal with the problem of losing large amounts of fixed carbon. In one ap-

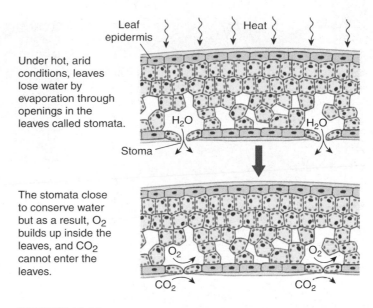

FIGURE 10.20
Conditions favoring photorespiration. In hot, arid environments, stomata close to conserve water, which also prevents CO_2 from entering the leaf and O_2 from exiting the leaf. The high O_2/low CO_2 conditions favor photorespiration.

proach, plants conduct C_4 photosynthesis in the mesophyll cells and the Calvin cycle in the bundle-sheath cells. This creates high local levels of CO_2 to favor the carboxylation reaction of rubisco. These plants, called C_4 plants, include corn, sugarcane, sorghum, and a number of other grasses. In the C_4 pathway, the three-carbon metabolite phosphoenolpyruvate is carboxylated to form the four-carbon molecule oxaloacetate, which is the first product of CO_2 fixation (figure 10.21*a*). In C_4 plants, oxaloacetate is in turn converted into the intermediate malate, which is transported to an adjacent bundle-sheath cell. Inside the bundle-sheath cell, malate is decarboxylated to produce pyruvate, releasing CO_2. Because bundle-sheath cells are impermeable to CO_2, the CO_2 is retained within them in high concentrations. Pyruvate returns to the mesophyll cell, where two of the high-energy bonds in an ATP molecule are split to convert the pyruvate back into phosphoenolpyruvate, thus completing the cycle.

The enzymes that carry out the Calvin cycle in a C_4 plant are located within the bundle-sheath cells, where the increased CO_2 concentration decreases photorespiration. Because each CO_2 molecule is transported into the bundle-sheath cells at a cost of two high-energy ATP bonds, and because six carbons must be fixed to form a molecule of glucose, 12 additional molecules of ATP are required to form a molecule of glucose. In C_4 photosynthesis, the energetic cost of forming glucose is almost twice that of C_3 photosynthesis: 30 molecules of ATP versus 18. Nevertheless, C_4 photosynthesis is advantageous in a hot climate; otherwise, photorespiration would remove more than half of the carbon fixed.

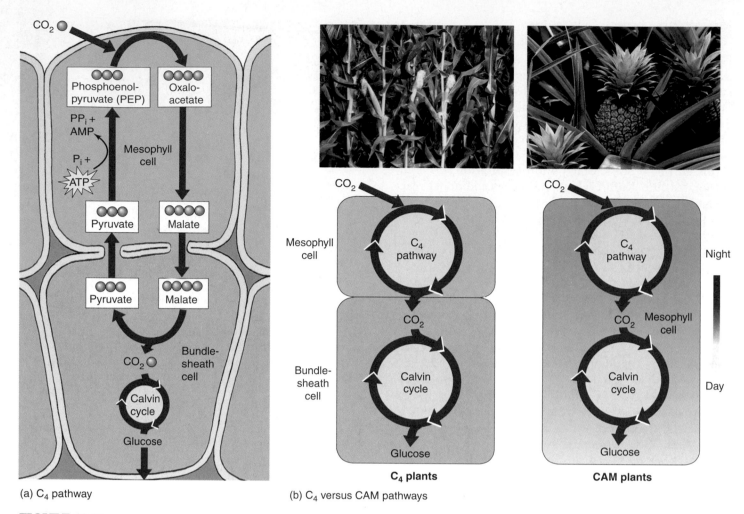

(a) C$_4$ pathway

(b) C$_4$ versus CAM pathways

FIGURE 10.21
Carbon fixation in C$_4$ plants. (*a*) This process is called the C$_4$ pathway because the first molecule formed, oxaloacetate, is a molecule containing four carbons. (*b*) A comparison of C$_4$ and CAM plants. Both C$_4$ and CAM plants utilize the C$_4$ and the C$_3$ pathways. In C$_4$ plants, the pathways are separated spatially: The C$_4$ pathway takes place in the mesophyll cells, and the C$_3$ pathway in the bundle-sheath cells. In CAM plants, the two pathways are separated temporally: The C$_4$ pathway is utilized at night, and the C$_3$ pathway during the day.

The Crassulacean Acid Pathway

A second strategy to decrease photorespiration in hot regions has been adopted by many succulent (water-storing) plants, such as cacti, pineapples, and some members of about two dozen other plant groups. This mode of initial carbon fixation is called **crassulacean acid metabolism (CAM),** after the plant family Crassulaceae (the stonecrops, or hens-and-chicks), in which it was first discovered. In these plants, the stomata, openings in leaves through which CO$_2$ enters and water vapor is lost, open during the night and close during the day. This pattern of stomatal opening and closing is the reverse of that in most plants. CAM plants open stomata at night and initially fix CO$_2$ into organic compounds using the C$_4$ pathway. These organic compounds accumulate throughout the night and are decarboxylated during the day to yield high levels of CO$_2$. In the day, these high levels of CO$_2$ drive the Calvin cycle and minimize photorespiration. Like C$_4$ plants, CAM plants use both C$_4$ and C$_3$ pathways. They differ from C$_4$ plants in that they use the C$_4$ pathway at night and the C$_3$ pathway during the day *within the same cells.* In C$_4$ plants, the two pathways take place in different cells (figure 10.21*b*).

Photorespiration results in decreased yields of photosynthesis. C$_4$ and CAM plants circumvent this problem through modifications of leaf architecture and photosynthetic chemistry that locally increase CO$_2$ concentrations. C$_4$ plants isolate CO$_2$ production spatially, while CAM plants isolate it temporally.

Concept Review

For interactive testing, visit the Online Learning Center with PowerWeb at www.mhhe.com/Raven7

10.1 What is photosynthesis?

The Chloroplast as a Photosynthetic Machine

- The equation for photosynthesis is: carbon dioxide + water + light yields glucose + water + oxygen. (p. 187)
- The light-dependent reactions occur within thylakoid membranes within chloroplasts. (p. 187)

10.2 Learning about photosynthesis: An experimental journey.

The Role of Soil and Water

- Jan Baptista van Helmont demonstrated that soil did not add mass to a growing plant, while Priestly, Ingenhousz, and other chemists worked out the basic formula for photosynthesis: carbon dioxide + water + light yields sugar and oxygen. (p. 188)

Discovery of the Light-Independent Reactions

- In the early 1900s, Blackman showed that capturing photosynthetic energy requires the input of sunlight, but building organic molecules does not. (p. 188)

The Role of Light and Reducing Power

- Van Niel discovered that photosynthesis splits water molecules, incorporating the carbon atoms of carbon dioxide gas and the hydrogen atoms of water into organic molecules and oxygen gas. (p. 189)
- Hill showed that plants can use light energy to generate reducing power. (p. 189)
- Carbon fixation refers to the incorporation of carbon dioxide into organic molecules in the light-independent reactions. (p. 189)

10.3 Pigments capture energy from sunlight.

The Biophysics of Light

- Short-wavelength light contains photons of higher energy than long-wavelength light. (p. 190)
- Sunlight reaching the earth's surface contains a significant amount of ultraviolet light, which possesses considerably more energy than visible light. (p. 191)
- In photosynthesis, photons are absorbed by plant pigments, and specific pigments absorb specific wavelengths. (p. 191)
- Chlorophyll *a* is the main photosynthetic pigment, although chlorophyll *b* and carotenoids also play important roles. (p. 191)

Chlorophylls and Carotenoids

- Chlorophylls absorb photons by means of an excitation process. (p. 192)
- The wavelengths absorbed by a pigment depend on the available energy level to which light-excited electrons can be boosted. (p. 193)

Organizing Pigments into Photosystems

- The light-dependent reactions take place in four stages: primary photoevent, charge separation, electron transport, and chemiosmosis. (p. 194)
- Pigments within photosystems transfer energy to reaction centers where the energy excites electrons that are channeled to perform chemical work. (p. 195)

How Photosystems Convert Light to Chemical Energy

- Plants employ two photosystems in series, which generate power to reduce $NADP^+$ to NADPH. (p. 196)

How the Two Photosystems of Plants Work Together

- Plants use photosystems II and I in series (noncyclic phosphorylation). (p. 198)
- High-energy electrons generated by photosystem II are used to synthesize ATP and then passed to photosystem I to drive the production of NADPH. (p. 199)

10.4 Cells use the energy and reducing power captured by the light-dependent reactions to make organic molecules.

The Calvin Cycle

- The Calvin cycle is also known as C_3 photosynthesis. (p. 200)
- CO_2 binds to RuBP in carbon fixation, forming two three-carbon molecules of PGA. (pp. 200–201)
- Plants incorporate carbon dioxide into sugars in the Calvin cycle, which is driven by the ATP and NADPH produced in the light-dependent reactions. (p. 202)

Photorespiration

- Photorespiration releases CO_2 and results in decreased yields of photosynthesis. (p. 203)
- C_4 photosynthesis circumvents photorespiration by creating high local levels of CO_2 in bundle sheath cells. (p. 203)
- CAM plants isolate CO_2 temporally by opening stomata at night instead of during the day. (p. 204)

Self Test

1. Within chloroplasts, the semiliquid matrix in which the Calvin cycle occurs is called
   a. stroma.
   b. thylakoids.
   c. grana.
   d. photosystem.
2. Visible light occupies what part of the electromagnetic spectrum?
   a. the entire spectrum
   b. the entire upper half (with longer wavelengths)
   c. a small portion in the middle
   d. the entire lower half (with shorter wavelengths)
3. The colors of light that are most effective for photosynthesis are
   a. red, blue, and violet.
   b. green, yellow, and orange.
   c. infrared and ultraviolet.
   d. All colors of light are equally effective.
4. A photosystem consists of
   a. a group of chlorophyll molecules, all of which contribute excited electrons to the synthesis of ATP.
   b. a pair of chlorophyll a molecules.
   c. a group of chlorophyll molecules held together by proteins.
   d. a group of chlorophyll molecules that funnels light energy toward a single chlorophyll b molecule.
5. Which photosystem is believed to have evolved first?
   a. photosystem I
   b. photosystem II
   c. cyclic photophosphorylation
   d. All photosystems evolved at the same time, but in different organisms.
6. Oxygen is produced during photosynthesis when
   a. the carbon is removed from carbon dioxide to make carbohydrates.
   b. hydrogen from water is added to carbon dioxide to make carbohydrates.
   c. water molecules are split to provide electrons for photosystem I.
   d. water molecules are split to provide electrons for photosystem II.
7. During photosynthesis, ATP molecules are generated by
   a. the Calvin cycle.
   b. chemiosmosis.
   c. the electron transport chain.
   d. light striking the chlorophyll molecules.
8. The overall purpose of the Calvin cycle is to
   a. generate molecules of ATP.
   b. generate NADPH.
   c. give off oxygen for animal use.
   d. build organic (carbon) molecules.
9. The final product of the Calvin cycle is
   a. RuBP.
   b. G3P.
   c. glucose.
   d. PGA.
10. C_4 photosynthesis is an adaptation to hot, dry conditions in which
    a. CO_2 is fixed and stored in the leaf.
    b. water is stored in the stem.
    c. oxygen is stored in the root.
    d. light energy is stored in chloroplasts.

Test Your Visual Understanding

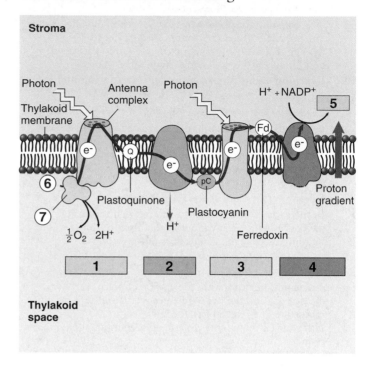

1. Match the following labels with their appropriate location on the figure.
   a. b_6-f complex
   b. H_2O
   c. NADP reductase
   d. NADPH
   e. photosystem I
   f. photosystem II
   g. water-splitting enzyme

Apply Your Knowledge

1. To reduce six molecules of carbon dioxide to glucose via photosynthesis, how many molecules of NADPH and ATP are required?
2. What is the advantage of having many pigment molecules in each photosystem but only one reaction center chlorophyll? In other words, why not couple every pigment molecule directly to an electron acceptor?
3. The two photosystems, P_{680} and P_{700}, of cyanobacteria, algae, and plants yield an oxidant capable of cleaving water. How might the subsequent evolution of cellular respiration have been different if the two photosystems had not evolved, and all photosynthetic organisms were restricted to the cyclic photophosphorylation used by the sulfur bacteria?
4. In theory, a plant kept in total darkness could still manufacture glucose, if it were supplied with which molecules?

11

How Cells Divide

Concept Outline

11.1 Prokaryotes divide far more simply than do eukaryotes.

    Cell Division in Prokaryotes. Prokaryotic cells divide by splitting in two.

11.2 The chromosomes of eukaryotes are highly ordered structures.

    Discovery of Chromosomes. All eukaryotic cells contain chromosomes, but different organisms possess differing numbers of chromosomes.

    The Structure of Eukaryotic Chromosomes. Proteins play an important role in packaging DNA in chromosomes.

11.3 Mitosis is a key phase of the cell cycle.

    The Cell Cycle. The cell cycle consists of three growth phases, a nuclear division phase, and a cytoplasmic division stage.

    Interphase: Preparing for Mitosis. In interphase, the cell grows, replicates its DNA, and prepares for cell division.

    Mitosis. In prophase, the chromosomes condense, and microtubules attach sister chromosomes to opposite poles of the cell. In metaphase, the chromosomes align along the center of the cell. In anaphase, the chromosomes separate; in telophase, the spindle dissipates and the nuclear envelope re-forms.

    Cytokinesis. In cytokinesis, the cytoplasm separates into two roughly equal halves.

11.4 The cell cycle is carefully controlled.

    General Strategies of Cell Cycle Control. At three points in the cell cycle, feedback from the cell determines whether the cycle will continue.

    Molecular Mechanisms of Cell Cycle Control. Special proteins regulate the checkpoints of the cell cycle.

    Cancer and the Control of Cell Proliferation. Cancer results from damage to genes encoding proteins that regulate the cell division cycle.

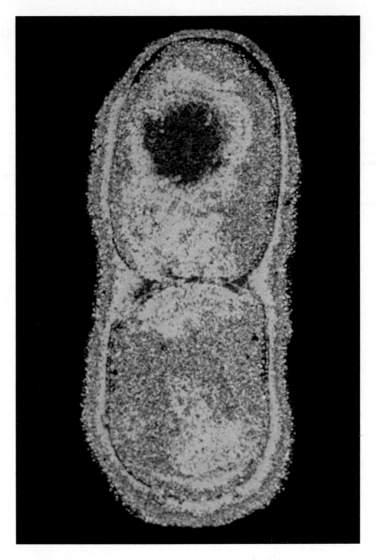

FIGURE 11.1
Cell division in prokaryotes. It's hard to imagine fecal coliform bacteria as being beautiful, but here is *Escherichia coli*, inhabitant of the large intestine and the biotechnology lab, spectacularly caught in the act of fission.

All species of organisms—bacteria, alligators, the weeds in a lawn—grow and reproduce. From the smallest creature to the largest, all species produce offspring like themselves and pass on the hereditary information that makes them what they are. In this chapter, we examine how cells divide and reproduce (figure 11.1). The mechanism of cell reproduction and its biological consequences have changed significantly during the evolution of life on earth. The process is complex in eukaryotes, involving both the replication of chromosomes and their separation into daughter cells. Much of what we are learning about the causes of cancer relates to how cells control this process, and in particular their propensity to divide, a mechanism that in broad outline remains the same in all eukaryotes.

207

11.1 Prokaryotes divide far more simply than do eukaryotes.

Cell Division in Prokaryotes

The end result of cell division in both prokaryotic and eukaryotic cells is two daughter cells, each with the same genetic information as the original cell. The differences between these two basic cell types lead to large differences in how this process occurs. Despite these differences, the essentials of the process are the same: duplication and segregation of genetic information into daughter cells, and division of cellular contents. We will begin by looking at the simpler process, which occurs in prokaryotes: division by **binary fission.**

Most prokaryotes have a genome made up of a single, circular DNA molecule. Despite its apparent simplicity, the DNA molecule of the bacterium *Escherichia coli* is actually on the order of 500 times longer than the cell itself! Thus, this "simple" structure is actually exquisitely packaged to fit into the cell. Although not found in a nucleus, the DNA is in a compacted form called a *nucleoid* that is distinct from the cytoplasm around it.

For many years, it was believed that the *E. coli* DNA molecule was passively segregated by attachment to the membrane and growth of the membrane as the cell elongates. More recently, a more complex picture is emerging that involves both active partitioning of the DNA and formation of a septum that divides the elongated cell in half. Although the details differ, species as different as *E. coli* and *Bacillus subtilis* both exhibit active partitioning of the newly replicated DNA molecules during the division process. This requires both specific sites on the chromosomes and a number of proteins actively involved in the process.

Binary fission begins with the replication of the prokaryotic DNA at a specific site—the origin of replication (see chapter 15)—and proceeds bidirectionally around the circular DNA to a specific site of termination (figure 11.2). Growth of the cell results in elongation, and the newly replicated DNA molecules are actively partitioned to one-fourth and three-quarters of the cell length. This process requires sequences near the origin of replication and results in these sequences being attached to the membrane. The cell itself is partitioned by the growth of new membrane and cell material called a septum (see figure 11.2). This process of septation is complex and under control of the cell as well.

The site of septation is usually the midpoint of the cell and begins with the formation of a ring composed of the molecule FtsZ (figure 11.3). This then results in the accumulation of a number of other proteins, including ones embedded in the membrane. The exact mechanism of septation is not known, but this structure grows inward radially until the cells pinch off into new cells.

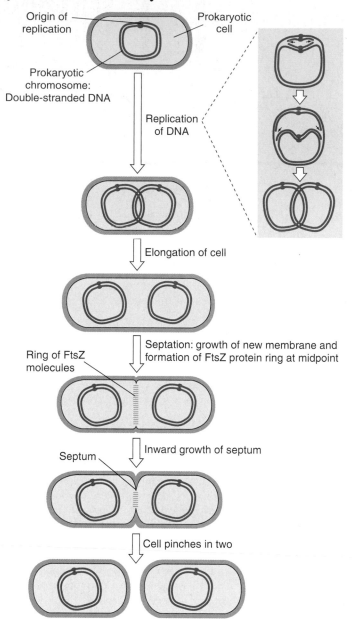

FIGURE 11.2
Binary fission. Prior to cell division, the prokaryotic DNA molecule replicates. The replication of the double-stranded, circular DNA molecule *(blue)* that constitutes the genome of a prokaryote begins at a specific site, called the origin of replication. The replication enzymes move out in both directions from that site and make copies *(red)* of each strand in the DNA duplex. The enzymes continue until they meet at another specific site, the terminus of replication. After the DNA is replicated, the cell elongates, and the DNA is partitioned in the cell. Septation then begins, in which new cell membrane material begins to grow and form a septum at approximately the midpoint of the cell. A protein molecule called FtsZ facilitates this process. When the septum is complete, the cell pinches in two, and two daughter cells are formed, each containing a prokaryotic DNA molecule.

FIGURE 11.3

The FtsZ protein. In these dividing *E. coli* bacteria, the FtsZ protein is fluorescent, and its location during binary fission can be seen. The protein assembles into a ring at approximately the midpoint of the cell, where it facilitates septation and cell division. Bacteria in which the *ftsZ* gene is mutated cannot divide.

The FtsZ molecule is interesting for a number of reasons. It is highly conserved evolutionarily, having been identified in most prokaryotes, including archaebacteria. It shows some small similarity to eukaryotic tubulin and can form filaments and rings. Recent 3-D crystals show similarity to tubulin as well. It is interesting to speculate that the elaborate spindle found in eukaryotic division may be related to this simple prokaryotic precursor (figure 11.4).

The evolution of eukaryotic cells led to much more complex genomes composed of multiple linear chromosomes housed in a membrane-bounded nucleus. These chromosomes contain even more DNA, and thus pose packaging problems that are solved by DNA being complexed with protein and packaged into functionally distinct chromosomes. This creates more challenges both for the replication of the genome and for its accurate segregation during cell division. The process that evolved to accomplish this segregation of chromosomes is called mitosis.

> **Prokaryotes divide by binary fission. Fission begins in the middle of the cell. An active partitioning process ensures that one genome will end up in each daughter cell.**

FIGURE 11.4

A comparison of protein assemblies during cell division among different organisms. The prokaryotic protein FtsZ has a structure that is similar to that of the eukaryotic protein tubulin. Tubulin is the protein component of microtubules, which are fibers that play an important role in eukaryotic cell division.

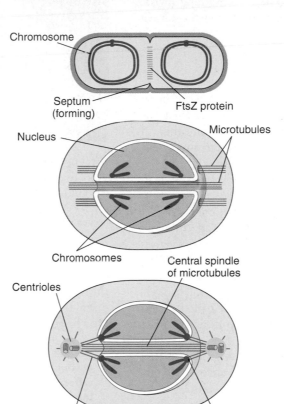

Prokaryotes
No nucleus; single circular chromosome. After DNA is replicated, it is partitioned in the cell. After cell elongation, FtsZ protein assembles into a ring and facilitates septation and cell division.

Some protists
Nucleus present and nuclear envelope remains intact during cell division. Chromosomes linear. Fibers called microtubules, composed of the protein tubulin, pass through tunnels in the nuclear membrane and set up an axis for separation of replicated chromosomes, and cell division.

Other protists
A spindle of microtubules forms between two pairs of centrioles at opposite ends of the cell. The spindle passes through one tunnel in the intact nuclear envelope. Kinetochore microtubules form between kinetochores on the chromosomes and the spindle poles and pull the chromosomes to each pole.

Yeasts
Nuclear envelope remains intact; spindle microtubules form inside the nucleus between spindle pole bodies. A single kinetochore microtubule attaches to each chromosome and pulls each to a pole.

Animals
Spindle microtubules begin to form between centrioles outside of nucleus. As these centrioles move to the poles, the nuclear envelope breaks down, and kinetochore microtubules attach kinetochores of chromosomes to spindle poles. Polar microtubules extend toward the center of the cell and overlap.

11.2 The chromosomes of eukaryotes are highly ordered structures.

Discovery of Chromosomes

Chromosomes were first observed by the German embryologist Walther Fleming in 1882, while he was examining the rapidly dividing cells of salamander larvae. When Fleming looked at the cells through what would now be a rather primitive light microscope, he saw minute threads within their nuclei that appeared to be dividing lengthwise. Fleming called their division **mitosis,** based on the Greek word *mitos,* meaning "thread."

Chromosome Number

Since their initial discovery, chromosomes have been found in the cells of all eukaryotes examined. Their number may vary enormously from one species to another. A few kinds of organisms have only 1 pair of chromosomes, while some ferns have more than 500 pairs (table 11.1). Most eukaryotes have between 10 and 50 chromosomes in their body cells.

Human cells each have 46 chromosomes, consisting of 23 nearly identical pairs (figure 11.5). Each of these 46 chromosomes contains hundreds or thousands of genes that play important roles in determining how a person's body develops and functions. For this reason, possession of all the chromosomes is essential to survival. Humans missing even one chromosome, a condition called *monosomy,* do not survive embryonic development in most cases. Nor does the human embryo develop properly with an extra copy of any one chromosome, a condition called *trisomy.* For all but a few of the smallest chromosomes, trisomy is fatal, and even in those few cases, serious problems result. For example, individuals with an extra copy of

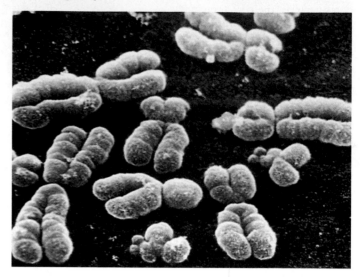

FIGURE 11.5
Human chromosomes. This photograph (950×) shows human chromosomes as they appear immediately before nuclear division. Each DNA molecule has already replicated, forming identical copies held together by a constriction called the centromere.

the very small chromosome 21 develop more slowly than normal and are mentally retarded, a condition called Down syndrome.

> All eukaryotic cells store their hereditary information in chromosomes, but different species utilize very different numbers of chromosomes to store this information.

	Table 11.1 Chromosome Number in Selected Eukaryotes				
Group	**Total Number of Chromosomes**	**Group**	**Total Number of Chromosomes**	**Group**	**Total Number of Chromosomes**
FUNGI		**PLANTS**		**VERTEBRATES**	
Neurospora (haploid)	7	*Haplopappus gracilis*	2	Opossum	22
Saccharomyces (a yeast)	16	Garden pea	14	Frog	26
		Corn	20	Mouse	40
INSECTS		Bread wheat	42	Human	46
Mosquito	6	Sugarcane	80	Chimpanzee	48
Drosophila	8	Horsetail	216	Horse	64
Honeybee	32	Adder's tongue fern	1262	Chicken	78
Silkworm	56			Dog	78

The Structure of Eukaryotic Chromosomes

In the century since chromosomes were discovered, we have learned a great deal about their structure and composition.

Composition of Chromatin

Chromosomes are composed of **chromatin,** a complex of DNA and protein; most chromosomes are about 40% DNA and 60% protein. A significant amount of RNA is also associated with chromosomes because chromosomes are the sites of RNA synthesis. The DNA of a chromosome is one very long, double-stranded fiber that extends unbroken through the entire length of the chromosome. A typical human chromosome contains about 140 million (1.4×10^8) nucleotides in its DNA. The amount of information one chromosome contains would fill about 280 printed books of 1000 pages each, if each nucleotide corresponded to a "word" and each page had about 500 words on it. Furthermore, if the strand of DNA from a single chromosome were laid out in a straight line, it would be about 5 centimeters (2 inches) long. Fitting such a strand into a nucleus is like cramming a string the length of a football field into a baseball—and that's only 1 of 46 chromosomes! In the cell, however, the DNA is coiled, allowing it to fit into a much smaller space than would otherwise be possible.

Chromosome Coiling

How can this long DNA fiber coil so tightly? If we gently disrupt a eukaryotic nucleus and examine the DNA with an electron microscope, we find that it resembles a string of beads (figure 11.6). Every 200 nucleotides, the DNA duplex is coiled around a core of eight histone proteins, forming a complex known as a **nucleosome.** Unlike most proteins, which have an overall negative charge, histones are positively charged, due to an abundance of the basic amino acids arginine and lysine. Thus, they are strongly attracted to the negatively charged phosphate groups of the DNA, and the histone cores act as "magnetic forms" that promote and guide the coiling of the DNA.

Further coiling occurs when the string of nucleosomes wraps up into higher-order coils called solenoids. This 30-nm solenoid forms the basis for interphase chromatin and is the starting point for the further compaction that occurs during mitosis. Chromatin appears to have some organization in the interphase nucleus, but it is not well understood. Further, geneticists have recognized for years that there are domains of chromatin that are not expressed, called **heterochromatin,** and domains of chromatin that are expressed, called **euchromatin.** During mitosis, the chromatin in the solenoid is further arranged around a scaffold of protein that is assembled at this time. The exact nature of this compaction is not known, but it includes radial looping of the solenoid about the protein scaffold.

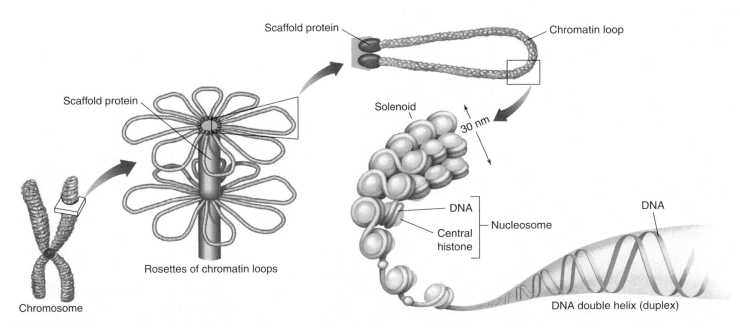

FIGURE 11.6
Levels of eukaryotic chromosomal organization.   Nucleotides assemble into long double strands of DNA molecules. These strands require further packaging to fit into the cell nucleus. The DNA duplex is tightly bound to and wound around proteins called histones. The DNA-wrapped histones are called nucleosomes. The nucleosomes are further coiled into a solenoid. This solenoid is then organized into looped domains. The final organization of the chromosome is not known, but it appears to involve further radial looping into rosettes around a preexisting scaffolding of protein. The arrangement illustrated here is one of many possibilities.

Chromosome Karyotypes

Chromosomes may differ widely among species and sometimes even among individuals of the same species. They vary in size, staining properties, the location of the centromere (a constriction found on all chromosomes), the relative length of the two arms on either side of the centromere, and the positions of constricted regions along the arms. The particular array of chromosomes an individual possesses is called its **karyotype.** The karyotype in figure 11.7 shows marked differences in the chromosomes of humans.

When defining the number of chromosomes in a species, geneticists count the **haploid (*n*)** number of chromosomes. This refers to one complete set of chromosomes necessary to define an organism. For humans and many other species, the normal number of chromosomes in a cell is called the **diploid (2*n*)** number, which is twice the haploid number. This is a reflection of the equal genetic contribution that parents make to human offspring: one complete set of chromosomes from mom and one from dad. We refer to the maternal and paternal chromosomes as being **homologous,** and say that each chromosome has two **homologues.** So you received 23 chromosomes from your mother and 23 chromosomes from your father to produce the diploid number of 46. The timing of the events of meiosis and syngamy can lead to a variety of life cycles with varying amounts of time spent as haploid and diploid (see figure 12.3 and chapters 28, 29, and 30). Humans have a diploid dominant life cycle, meaning that the cells in our bodies that are haploid are gametes found in the gonads.

Chromosomes as seen in a karyotype are only present for a brief period during cell division. Prior to replicating, each chromosome has one **centromere,** a condensed area found on all eukaryotic chromosomes. After replication, each chromosome has two sister **chromatids** that appear to share a common centromere (figure 11.8). At the molecular level, this is probably not true, because centromeric DNA is not late-replicating. However, we can still call this one chromosome because, by convention, we count chromosomes by counting centromeres. Thus, entering division, a human cell has 46 chromosomes composed of 92 chromatids and 46 centromeres.

Eukaryotic genomes are larger and more complex than those of prokaryotes. Eukaryotic DNA is packaged tightly into chromosomes, enabling it to fit inside cells. Haploid cells contain one set of chromosomes, while diploid cells contain two sets.

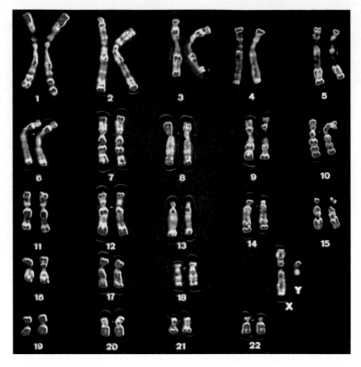

FIGURE 11.7
A human karyotype. The individual chromosomes that make up the 23 pairs differ widely in size and in centromere position. In this preparation, the chromosomes have been specifically stained to indicate further differences in their composition and to distinguish them clearly from one another.

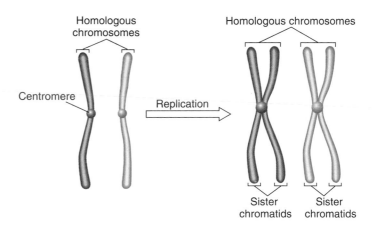

FIGURE 11.8
The difference between homologous chromosomes and sister chromatids. Homologous chromosomes are a pair of the same chromosome—say, chromosome number 16. Sister chromatids are the two replicas of a single chromosome held together by the centromeres after DNA replication.

11.3 Mitosis is a key phase of the cell cycle.

The Cell Cycle

The increased size and more complex organization of eukaryotic genomes over those of prokaryotes required radical changes in the process by which the two replicas of the genome are partitioned into the daughter cells during cell division. This division process is diagrammed as a **cell cycle** consisting of five phases (figure 11.9).

The Five Phases

G_1 is the primary growth phase of the cell. For many organisms, this encompasses the major portion of the cell's life span. **S** is the phase in which the cell synthesizes a replica of the genome. G_2 is the second growth phase, in which preparations are made for genomic separation. During this phase, mitochondria and other organelles replicate, chromosomes condense, and microtubules begin to assemble at a spindle. G_1, S, and G_2 together constitute **interphase,** the portion of the cell cycle between cell divisions.

M (for **mitosis**) is the phase of the cell cycle in which the microtubular apparatus assembles, binds to the chromosomes, and moves the sister chromatids apart. Mitosis is the essential step in the separation of the two daughter genomes. In this section, we will discuss mitosis as it occurs in animals and plants, where the process does not vary much; it is somewhat different among fungi and some protists. Although mitosis is a continuous process, it is traditionally subdivided into four stages: prophase, metaphase, anaphase, and telophase.

C (for **cytokinesis**) is the phase of the cell cycle when the cytoplasm divides, creating two daughter cells. In animal cells, the microtubule spindle helps position a contracting ring of actin that constricts like a drawstring to pinch the cell in two. In cells with a cell wall, such as plant cells, a plate forms between the dividing cells.

Duration of the Cell Cycle

The time it takes to complete a cell cycle varies greatly among organisms. Cells in growing animal embryos can complete their cell cycle in under 20 minutes; the shortest known animal nuclear division cycles occur in fruit fly embryos (8 minutes). Cells such as these simply divide their nuclei as quickly as they can replicate their DNA, without cell growth. Half of the cycle is taken up by S, half by M, and essentially none by G_1 or G_2. Because mature cells require time to grow, most of their cycles are much longer

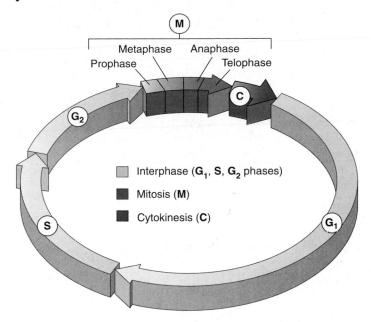

FIGURE 11.9
The cell cycle. This circle represents the 22-hour cell cycle in human cells growing in culture. G_1 represents the primary growth phase of the cell cycle, S the phase during which a replica of the genome is synthesized, and G_2 the second growth phase.

than those of embryonic tissue. Typically, a dividing mammalian cell completes its cell cycle in about 24 hours, but some cells, such as certain cells in the human liver, have cell cycles lasting more than a year. During the cycle, growth occurs throughout the G_1 and G_2 phases (referred to as "gap" phases because they separate S from M), as well as during the S phase. The M phase takes only about an hour, a small fraction of the entire cycle.

Most of the variation in the length of the cell cycle from one organism or tissue to the next occurs in the G_1 phase. Cells often pause in G_1 before DNA replication and enter a resting state called the **G_0 phase;** they may remain in this phase for days to years before resuming cell division. At any given time, most of the cells in an animal's body are in G_0 phase. Some, such as muscle and nerve cells, remain there permanently; others, such as liver cells, can resume G_1 phase in response to factors released during injury.

Most eukaryotic cells repeat a process of growth and division referred to as the cell cycle. The cycle can vary in length from a few minutes to several years.

Interphase: Preparing for Mitosis

The events that occur during interphase—the G_1, S, and G_2 phases—are very important for the successful completion of mitosis. During G_1, cells undergo the major portion of their growth. During the S phase, each chromosome replicates to produce two sister chromatids, which remain attached to each other at the **centromere**. The centromere is a point of constriction on the chromosome, containing a specific DNA sequence to which is bound a disk of protein called a **kinetochore**. This disk functions as an attachment site for fibers called microtubules that assist in cell division (figure 11.10). Each chromosome's centromere is located at a characteristic site.

The cell grows throughout interphase. The G_1 and G_2 segments of interphase are periods of active growth, during which proteins are synthesized and cell organelles produced. The cell's DNA replicates only during the S phase of the cell cycle.

After the chromosomes have replicated in S phase, they remain fully extended and uncoiled. This makes them invisible under the light microscope. In G_2 phase, they begin the long process of **condensation,** coiling ever more tightly. Special *motor proteins* are involved in the rapid final condensation of the chromosomes that occurs early in mitosis. Also during G_2 phase, the cells begin to assemble the machinery they will later use to move the chromosomes to opposite poles of the cell. In animal cells, a pair of microtubule-organizing centers called **centrioles** replicate. All eukaryotic cells undertake an extensive synthesis of *tubulin*, the protein of which microtubules are formed.

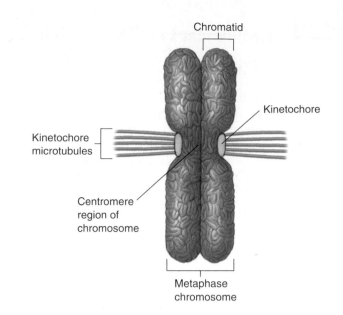

FIGURE 11.10
Kinetochores. In a metaphase chromosome, kinetochore microtubules are anchored to proteins at the centromere.

Interphase is that portion of the cell cycle in which the chromosomes are invisible under the light microscope because they are not yet condensed. It includes the G_1, S, and G_2 phases. In the G_2 phase, the cell mobilizes its resources for cell division.

A Vocabulary of Cell Division

binary fission Reproduction of a cell by division into two equal or nearly equal parts. Prokaryotes divide by binary fission.

centromere A constricted region of a chromosome about 220 nucleotides in length, composed of highly repeated DNA sequences. During mitosis, the centromere joins the two sister chromatids and is the site to which the kinetochores are attached.

chromatid One of the two copies of a replicated chromosome, joined by a single centromere to the other strand.

chromatin The complex of DNA and proteins of which eukaryotic chromosomes are composed.

chromosome The structure within cells that contains the genes. In eukaryotes, it consists of a single linear DNA molecule associated with proteins. The DNA replicates during S phase, and the replicas separate during M phase.

cytokinesis Division of the cytoplasm of a cell after nuclear division.

euchromatin The portion of a chromosome that is extended except during cell division, and from which RNA is transcribed.

heterochromatin The portion of a chromosome that remains permanently condensed and, therefore, is not transcribed into RNA. Most centromere regions are heterochromatic.

homologues Homologous chromosomes; in diploid cells, one of a pair of chromosomes that carry equivalent genes.

kinetochore A disk of protein bound to the centromere and attached to microtubules during mitosis, linking each chromatid to the spindle apparatus.

microtubule A hollow cylinder, about 25 nanometers in diameter, composed of subunits of the protein tubulin. Microtubules lengthen by the addition of tubulin subunits to their end(s) and shorten by the removal of subunits.

mitosis Nuclear division in which replicated chromosomes separate to form two genetically identical daughter nuclei. When accompanied by cytokinesis, it produces two identical daughter cells.

nucleosome The basic packaging unit of eukaryotic chromosomes, in which the DNA molecule is wound around a cluster of histone proteins. Chromatin is composed of long strings of nucleosomes that resemble beads on a string.

Mitosis

Prophase: Formation of the Mitotic Apparatus

When the chromosome condensation initiated in G₂ phase reaches the point at which individual condensed chromosomes first become visible with the light microscope, the first stage of mitosis, **prophase,** has begun. The condensation process continues throughout prophase; consequently, some chromosomes that start prophase as minute threads appear quite bulky before its conclusion. Ribosomal RNA synthesis ceases when the portion of the chromosome bearing the rRNA genes is condensed.

Assembling the Spindle Apparatus.

The assembly of the microtubular apparatus that will later separate the sister chromatids also continues during prophase. In animal cells, the two centriole pairs formed during G₂ phase begin to move apart early in prophase, forming between them an axis of microtubules referred to as spindle fibers. By the time the centrioles reach the opposite poles of the cell, they have established a bridge of microtubules called the spindle apparatus between them. In plant cells, a similar bridge of microtubular fibers forms between opposite poles of the cell, although centrioles are absent in plant cells.

During the formation of the spindle apparatus, the nuclear envelope breaks down, and the endoplasmic reticulum reabsorbs its components. At this point, the microtubular spindle fibers extend completely across the cell, from one pole to the other. Their orientation determines the plane in which the cell will subsequently divide, through the center of the cell at right angles to the spindle apparatus.

In animal cell mitosis, the centrioles extend a radial array of microtubules toward the plasma membrane when they reach the poles of the cell. This arrangement of microtubules is called an **aster.** Although the aster's function is not fully understood, it probably braces the centrioles against the membrane and stiffens the point of microtubular attachment during the retraction of the spindle. Plant cells, which have rigid cell walls, do not form asters.

Linking Sister Chromatids to Opposite Poles.

Each chromosome possesses two kinetochores, one attached to the centromere region of each sister chromatid (see figure 11.10). As prophase continues, a second group of microtubules appears to grow from the poles of the cell toward the centromeres. These microtubules connect the kinetochores on each pair of sister chromatids to the two poles of the spindle. Because microtubules extending from the two poles attach to opposite sides of the centromere, they attach one sister chromatid to one pole and the other sister chromatid to the other pole. This arrangement is absolutely critical to the process of mitosis; any mistakes in microtubule positioning can be disastrous. For example, the attachment of the two sides of a centromere to the same pole leads to a failure of the sister chromatids to separate, so that they end up in the same daughter cell.

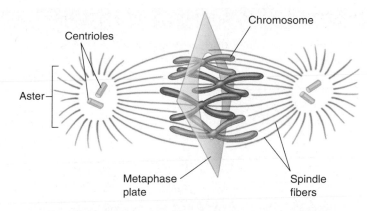

FIGURE 11.11
Metaphase. In metaphase, the chromosomes array themselves in a circle around the spindle midpoint.

Metaphase: Alignment of the Centromeres

In the second stage of mitosis, **metaphase,** the chromosomes align in the center of the cell. When viewed with a light microscope, the chromosomes appear to array themselves in a circle along the inner circumference of the cell, just as the equator girdles the earth (figure 11.11). An imaginary plane perpendicular to the axis of the spindle that passes through this circle is called the *metaphase plate.* The metaphase plate is not an actual structure, but rather an indication of the future axis of cell division. Positioned by the microtubules attached to the kinetochores of their centromeres, all of the chromosomes line up on the metaphase plate. At this point, which marks the end of metaphase, their centromeres are neatly arrayed in a circle, equidistant from the two poles of the cell, with microtubules extending back toward the opposite poles of the cell. Because of its shape, this arrangement is called a spindle.

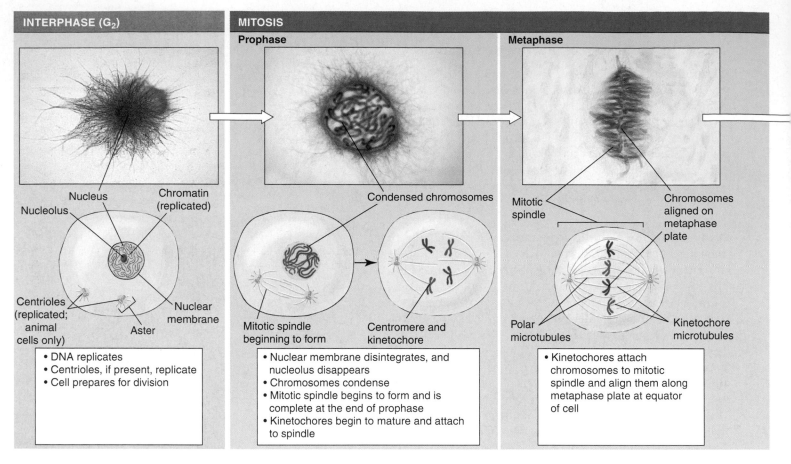

INTERPHASE (G₂)	MITOSIS	
	Prophase	**Metaphase**

Interphase (G₂):
Nucleus
Nucleolus
Chromatin (replicated)
Centrioles (replicated; animal cells only)
Aster
Nuclear membrane

- DNA replicates
- Centrioles, if present, replicate
- Cell prepares for division

Prophase:
Condensed chromosomes
Mitotic spindle beginning to form
Centromere and kinetochore

- Nuclear membrane disintegrates, and nucleolus disappears
- Chromosomes condense
- Mitotic spindle begins to form and is complete at the end of prophase
- Kinetochores begin to mature and attach to spindle

Metaphase:
Mitotic spindle
Chromosomes aligned on metaphase plate
Polar microtubules
Kinetochore microtubules

- Kinetochores attach chromosomes to mitotic spindle and align them along metaphase plate at equator of cell

FIGURE 11.12

Mitosis and cytokinesis in plants (photos) and in animals (drawings). Mitosis (separation of the two genomes) occurs in four stages—prophase, metaphase, anaphase, and telophase—and is followed by cytokinesis (division into two separate cells). In the photos, the chromosomes of the African blood lily, *Haemanthus katharinae*, are stained blue, and microtubules are stained red.

Anaphase and Telophase: Separation of the Chromatids and Reformation of the Nuclei

Of all the stages of mitosis, shown in figure 11.12, **anaphase** is the shortest and the most beautiful to watch. It starts when the centromeres divide. Each centromere splits in two, freeing the two sister chromatids from each other.

To this point in mitosis, sister chromatids have been held together by a complex of proteins called cohesin. In yeasts, cleavage of the cohesin subunit Scc1p by a *separase* proteolytic enzyme is thought to be the event that actually releases the chromatids for migration to daughter cells in anaphase. In vertebrate cells, most of the cohesin dissociates from the chromosomes before the onset of metaphase. However, a small amount remains, locking the two chromatids together. Of particular importance is a subunit dubbed SCC1. Cleavage

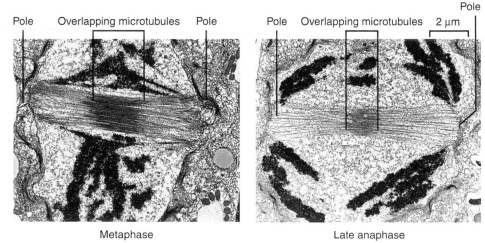

Pole Overlapping microtubules Pole
Pole Overlapping microtubules Pole
2 µm

Metaphase   Late anaphase

FIGURE 11.13

Microtubules slide past each other as the chromosomes separate. In these electron micrographs of dividing diatoms, the overlap of the microtubules lessens markedly during spindle elongation as the cell passes from metaphase to anaphase.

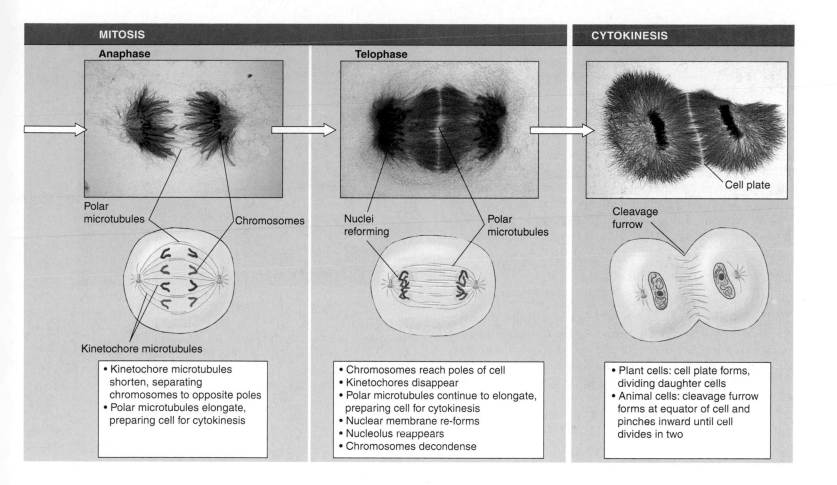

Anaphase

Polar
microtubules

Chromosomes

Kinetochore microtubules

- Kinetochore microtubules
  shorten, separating
  chromosomes to opposite poles
- Polar microtubules elongate,
  preparing cell for cytokinesis

Telophase

Nuclei
reforming

Polar
microtubules

- Chromosomes reach poles of cell
- Kinetochores disappear
- Polar microtubules continue to elongate,
  preparing cell for cytokinesis
- Nuclear membrane re-forms
- Nucleolus reappears
- Chromosomes decondense

Cell plate

Cleavage
furrow

- Plant cells: cell plate forms,
  dividing daughter cells
- Animal cells: cleavage furrow
  forms at equator of cell and
  pinches inward until cell
  divides in two

of the SCC1 cohesin subunit by separase enzymes is essential for centromere division and sister chromatid separation. The centromeres of all the chromosomes divide simultaneously. While the mechanism that achieves this synchrony is not well understood, it likely involves the timing of separase activation by a control element called the anaphase-promoting complex (APC), to be described in more detail later.

Freed from each other, the sister chromatids are pulled rapidly toward the poles to which their kinetochores are attached. In the process, two forms of movement take place simultaneously, each driven by microtubules.

First, *the poles move apart* as microtubular spindle fibers physically anchored to opposite poles slide past each other, away from the center of the cell (figure 11.13). Because another group of microtubules attach the chromosomes to the poles, the chromosomes move apart too. If a flexible membrane surrounds the cell, it becomes visibly elongated.

Second, *the centromeres move toward the poles* as the microtubules that connect them to the poles shorten. This shortening process is not a contraction; the microtubules do not get any thicker. Instead, tubulin subunits are removed from the kinetochore ends of the microtubules by the organizing center. As more subunits are removed, the chromatid-bearing microtubules are progressively disassembled, and the chromatids are pulled ever closer to the poles of the cell.

When the sister chromatids separate in anaphase, the accurate partitioning of the replicated genome—the essential element of mitosis—is complete. In **telophase,** the spindle apparatus disassembles as the microtubules are broken down into tubulin monomers that can be used to construct the cytoskeletons of the daughter cells. A nuclear envelope forms around each set of sister chromatids, which can now be called chromosomes because each has its own centromere. The chromosomes soon begin to uncoil into the more extended form that permits gene expression. One of the early group of genes expressed are the rRNA genes, resulting in the reappearance of the nucleolus.

During prophase, microtubules attach the centromeres joining pairs of sister chromatids to opposite poles of the spindle apparatus. During metaphase, each chromosome is drawn to a ring along the inner circumference of the cell by the microtubules extending from the centromere to the two poles of the spindle apparatus. During anaphase, the poles of the cell are pushed apart by microtubular sliding, and the sister chromatids are drawn to opposite poles by the shortening of the microtubules attached to them. During telophase, the spindle is disassembled, nuclear envelopes are reestablished, and the normal expression of genes present in the chromosomes is reinitiated.

Cytokinesis

Mitosis is complete at the end of telophase. The eukaryotic cell has partitioned its replicated genome into two nuclei positioned at opposite ends of the cell. While mitosis was going on, the cytoplasmic organelles, including mitochondria and chloroplasts (if present), were reassorted to areas that will separate and become the daughter cells. The replication of organelles takes place before cytokinesis, often in the S or G_2 phase. Cell division is still not complete at the end of mitosis, however, because the division of the cell proper has not yet begun. The phase of the cell cycle when the cell actually divides is called **cytokinesis.** It generally involves the cleavage of the cell into roughly equal halves.

Cytokinesis in Animal Cells

In animal cells and the cells of all other eukaryotes that lack cell walls, cytokinesis is achieved by means of a constricting belt of actin filaments. As these filaments slide past one another, the diameter of the belt decreases, pinching the cell and creating a *cleavage furrow* around the cell's circumference (figure 11.14*a*). As constriction proceeds, the furrow deepens until it eventually slices all the way into the center of the cell. At this point, the cell is divided in two (figure 11.14*b*).

Cytokinesis in Plant Cells

Plant cell walls are far too rigid to be squeezed in two by actin filaments. Instead, these cells assemble membrane components in their interior, at right angles to the spindle apparatus. This expanding membrane partition, called a **cell plate,** continues to grow outward until it reaches the interior surface of the plasma membrane and fuses with it, effectively dividing the cell in two (figure 11.15). Cellulose is then laid down on the new membranes, creating two new cell walls. The space between the daughter cells becomes impregnated with pectins and is called a **middle lamella.**

Cytokinesis in Fungi and Protists

In fungi and some groups of protists, the nuclear membrane does not dissolve, and as a result, all the events of mitosis occur entirely *within* the nucleus. Only after mitosis is complete in these organisms does the nucleus divide into two daughter nuclei; then, during cytokinesis, one nucleus goes to each daughter cell. This separate nuclear division phase of the cell cycle does not occur in plants, animals, or most protists.

After cytokinesis in any eukaryotic cell, the two daughter cells contain all the components of a complete cell. While mitosis ensures that both daughter cells contain a full complement of chromosomes, no similar mechanism ensures that organelles such as mitochondria and chloroplasts are

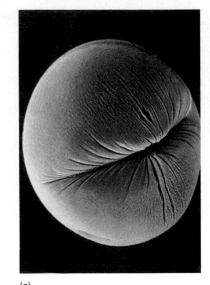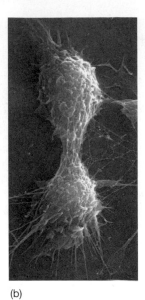

(a)   (b)

FIGURE 11.14
Cytokinesis in animal cells. (*a*) A cleavage furrow forms around a dividing sea urchin egg (30×). (*b*) The completion of cytokinesis in an animal cell. The two daughter cells are still joined by a thin band of cytoplasm occupied largely by microtubules.

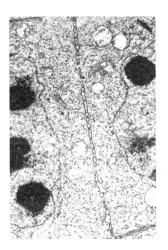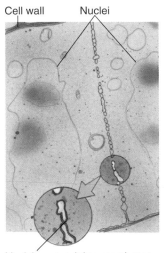

Cell wall   Nuclei

Vesicles containing membrane components fusing to form cell plate

FIGURE 11.15
Cytokinesis in plant cells. In this photograph and companion drawing, a cell plate is forming between daughter nuclei. Once the plate is complete, there will be two cells.

distributed equally between the daughter cells. However, as long as some of each organelle are present in each cell, the organelles can replicate to reach the number appropriate for that cell.

> Cytokinesis is the physical division of the cytoplasm of a eukaryotic cell into two daughter cells.

11.4 The cell cycle is carefully controlled.

General Strategies of Cell Cycle Control

Our knowledge of how the cell cycle is controlled, while still incomplete, has grown enormously in the past 20 years. Our current view integrates two basic concepts: First, there are two irreversible points in the cell cycle—the replication of genetic material and the separation of the sister chromatids. Second, the cell cycle can be put on hold at specific points called *checkpoints*. At any of these checkpoints, the process is assayed for accuracy and can be halted if there are errors. This leads to extremely high fidelity overall for the entire process. The checkpoint organization of the cell cycle also makes it responsive to both the internal state of the cell, including nutritional state and integrity of genetic material, and to signals from the environment, which are integrated at major checkpoints.

Architecture of the Control System

A cell uses three main checkpoints to both assess the internal state of the cell and integrate external signals (figure 11.16): G_1/S, G_2/M, and late metaphase (the spindle checkpoint). Passage through these checkpoints is controlled by kinase enzymes composed of an enzymatic subunit partnered with a protein called cyclin. These enzymes are thus called **cyclin-dependent kinases,** or **Cdk's.** Cyclins are produced and degraded cyclically with mitosis, but it is the activity of Cdk's that drives the cell cycle.

G_1/S Checkpoint

The G_1/S checkpoint is the primary point at which the cell "decides" whether or not to divide. This checkpoint is therefore the primary point at which external signals can influence events of the cycle. It is the phase during which growth factors (discussed later in this chapter) affect the cycle. It is also the phase that links cell division to cell growth and nutrition. In yeast systems, where the majority of the genetic analysis of the cell cycle has been performed, this checkpoint is called "start." In animals, it is called the restriction point (R point). In all systems, once a cell has made this irreversible commitment to replicate its genome, it has committed to divide. The decision made at this checkpoint depends on external signals such as growth factors and internal signals such as nutritional state; it also depends on the genome being intact. Damage to DNA can halt the cycle at this point, as can starvation conditions or lack of growth factors.

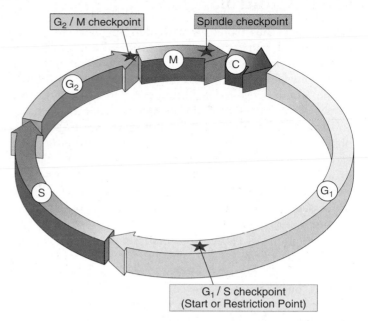

FIGURE 11.16
Control of the cell cycle. Cells use a centralized control system to check whether proper conditions have been achieved before passing three key checkpoints in the cell cycle.

G_2/M Checkpoint

The G_2/M checkpoint has received a large amount of attention due to its complexity and its importance as the stimulus for the events of mitosis. Historically, Cdk's active at this checkpoint were first identified as a class of substance that could be added to frog oocytes arrested in G_2 of meiosis to relieve this G_2 arrest. Because this was a normal step in the maturation of frog eggs, this substance was called "maturation promoting factor," a term that has now evolved into *M-phase-promoting factor* (MPF).

Passage through this checkpoint represents the commitment to mitosis. This checkpoint assesses the success of DNA replication and can stall the cycle if DNA has not been accurately replicated. DNA-damaging agents result in arrest at this checkpoint as well as at the G_1/S checkpoint.

Spindle Checkpoint

The spindle checkpoint ensures that all of the chromosomes are attached to the spindle in preparation for anaphase. The second irreversible step in the cycle is the separation of chromosomes during anaphase, and it is critical that they are properly arrayed at the metaphase plate.

The cell cycle is controlled at three checkpoints.

Molecular Mechanisms of Cell Cycle Control

On the previous page, we provided an overview of cell cycle control strategies; now we examine those processes in detail. The primary molecular mechanism of cell cycle control is phosphorylation, the addition of a phosphate group to the amino acids serine, threonine, and tyrosine in proteins. The enzymes that add phosphates are called kinases, and the enzymes that remove phosphates are called phosphatases. The phosphorylation of a protein can either activate or inactivate it, depending on the protein. Similarly, a protein that is inactivated by phosphorylation will be activated by dephosphorylation. The control of cellular processes by phosphorylation will be a continuing theme throughout this book.

The actions of cell-cycle-specific kinases drive the different stages of the cycle by phosphorylating a wide variety of cellular proteins. All of these protein targets are not known, but the importance of cell cycle kinases is clear. The enzymes that accomplish this phosphorylation are the cyclin-dependent kinases (Cdk's), consisting of an enzymatic subunit partnered with the protein cyclin (figure 11.17). Cyclins are proteins that display characteristic patterns of synthesis and degradation that coincide with phases of the cell cycle. The Cdk's are only active when the kinase is combined with the cyclin. The most important cell cycle kinase was identified in fission yeast and named cdc2. This Cdk can partner with different cyclins at different points in the cell cycle. Thus, the signal to begin comes from this Cdk combined with one cyclin, and the signal for the initiation of mitosis comes from this Cdk combined with a different cyclin.

The important question is: What controls the activity of the Cdk's during the cycle? For many years, a common view was that cyclins drove the cell cycle—that is, the periodic synthesis and destruction of cyclins acted as a clock for the cell cycle. More recently, it has become clear that the cdc2 kinase is controlled by phosphorylation: Phosphorylation at one site activates cdc2, and phosphorylation of cdc2 at another site inactivates it (see figure 11.17). Full activation of the cdc2 kinase requires complexing with the cyclin, and an appropriate pattern of phosphorylation. The exact molecular mechanisms of Cdk control are still an area of active investigation, but the main points are clear.

G₁/S Checkpoint

The G₁/S checkpoint integrates a number of signals, both internal and external (figure 11.18). The internal signals include the nutritional state of the cell and the size of the cell. The external signals include factors that promote cell growth and division (see the discussion of tumor suppressors later in this chapter). In mammalian cells, this is the checkpoint that involves the action of retinoblastoma pro-

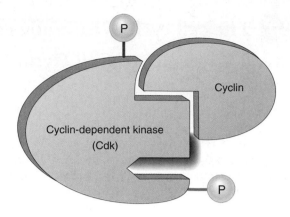

FIGURE 11.17

A complex of two proteins triggers passage through cell cycle checkpoints. Cdk is a protein kinase that activates numerous cell proteins by phosphorylating them. Cyclin is a regulatory protein required to activate Cdk. The activity of Cdk is also controlled by the pattern of phosphorylation: phosphorylation at one site (represented by the red site) inactivates the Cdk, and phosphorylation at another site (represented by the green site) activates the Cdk.

tein (discussed later). In yeasts, the molecular nature of the signal for this checkpoint (called "start") appears to be the accumulation of an S-phase-specific cyclin. This complexes with the cdc2 kinase to form the G₁/S Cdk. This Cdk phosphorylates a number of targets to result in the synthesis of proteins necessary for S phase.

G₂/M Checkpoint

The G₂/M checkpoint integrates a number of signals to trigger the commitment to mitosis. The Cdk that acts at this checkpoint, MPF, has been extensively analyzed in a number of different experimental systems. The control of MPF is sensitive to agents that disrupt or delay replication and to agents that damage DNA. It was once thought that MPF was controlled by the level of the M-phase-specific cyclins, but it has now become clear that this is not the case. While M-phase cyclin is necessary for MPF function, activity is controlled by inhibitory phosphorylation of cdc2. The critical signal in this process is the removal of inhibitor phosphates. This forms a molecular switch based on positive feedback as the active MPF further activates its own activating phosphatase. The checkpoint assesses the balance of kinases that add phosphates with the phosphatase that removes them. Damage to DNA acts through a complex pathway that includes damage sensing and response to tip the balance toward the inhibitory phosphorylation of MPF. In animal cells, the damage-sensing pathway involves the p53 protein, which has been found to be mutated in a variety of human cancers (see discussion of tumor suppressors later in this chapter).

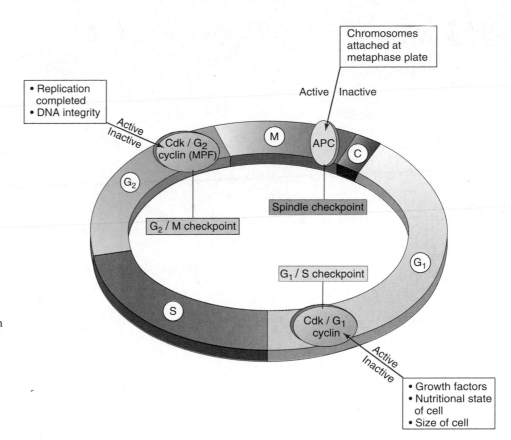

FIGURE 11.18
Checkpoints of the cell cycle. The cell cycle is controlled through three main checkpoints. These integrate internal and external signals to control progress through the cycle. These inputs control the state of two Cdk/cyclin complexes and the anaphase-promoting complex (APC). The arrows represent inputs, which can be complex networks, such as the signal transduction cascade seen in growth factor signaling (see figure 11.19).

Spindle Checkpoint

The spindle checkpoint is before the second irreversible step in the cell cycle, the anaphase separation of sister chromatids. This checkpoint ensures that all chromosomes are present at the metaphase plate and aligned with the centromeres of each sister chromatid oriented to opposite poles. The molecular details of the nature of this checkpoint are not clear, but both the presence of all chromosomes and the tension between opposite poles appear to be important. The signal is transmitted through the anaphase-promoting complex (APC). The sister chromatids at metaphase are still held together by a protein complex called cohesin. The APC acts by removing inhibitors of a protease that destroys the cohesin complex. This simultaneous removal of sister chromatid cohesin results in the separation of sister chromatids during anaphase.

Controlling the Cell Cycle in Multicellular Eukaryotes

Much of the work that led to our current understanding of cell cycle control came out of studies on unicellular fungi: budding yeast and fission yeast. Amazingly, most of these findings also hold true for animal cells, which have much greater constraints on growth due to their or-

ganization into tissues and organs. The major difference between animals and eukaryotes such as fungi and protists is twofold: First, multiple Cdk's control the cycle as opposed to the single Cdk in yeasts, and second, animal cells respond to a greater variety of external signals than yeasts, which primarily respond to signals necessary for mating.

The cells of multicellular eukaryotes are not free to make individual "decisions" about cell division, as yeast cells are. A multicellular body's organization cannot be maintained without severely limiting cell proliferation so that only certain cells divide, and only at appropriate times. The way cells inhibit individual growth of other cells is apparent in mammalian cells growing in tissue culture: A single layer of cells expands over a culture plate until the growing border of cells comes into contact with neighboring cells, and then the cells stop dividing. If a sector of cells is cleared away, neighboring cells rapidly refill that sector and then stop dividing again. How are cells able to sense the density of the cell culture around them? Each growing cell apparently binds minute amounts of positive regulatory signals called **growth factors**, proteins that stimulate cell division (such as MPF). When neighboring cells have used up what little growth factor is present, not enough is left to trigger cell division in any one cell.

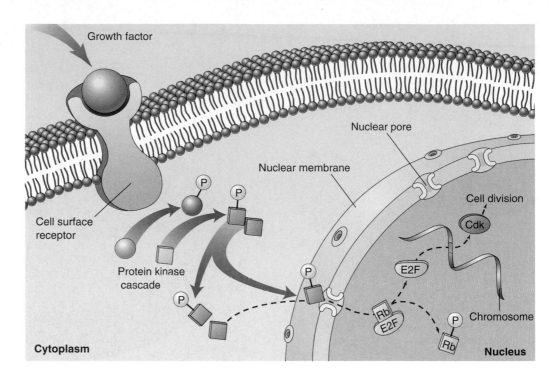

FIGURE 11.19
The cell proliferation-signaling pathway. Binding of a growth factor sets in motion a cascading intracellular signaling pathway (described in chapter 7), which activates nuclear regulatory proteins that trigger cell division. In this example, when the nuclear protein Rb is phosphorylated, another nuclear protein (the transcription factor E2F) is released and is then able to stimulate the production of Cdk proteins.

Growth Factors and the Cell Cycle

As you may recall from chapter 7 (cell–cell interactions), growth factors work by triggering intracellular signaling systems. Fibroblasts, for example, possess numerous receptors on their plasma membranes for one of the first growth factors to be identified, platelet-derived growth factor (PDGF). When PDGF binds to a membrane receptor, it initiates an amplifying chain of internal cell signals that stimulates cell division.

PDGF was discovered when investigators found that fibroblasts would grow and divide in tissue culture only if the growth medium contained blood serum (the liquid that remains after blood clots); blood plasma (blood from which the cells have been removed without clotting) would not work. The researchers hypothesized that platelets in the blood clots were releasing into the serum one or more factors required for fibroblast growth. Eventually, they isolated such a factor and named it PDGF.

Growth factors such as PDGF override cellular controls that otherwise inhibit cell division. When a tissue is injured, a blood clot forms, and the release of PDGF triggers neighboring cells to divide, helping to heal the wound. Only a tiny amount of PDGF (approximately 10^{-10} M) is required to stimulate cell division.

Characteristics of Growth Factors. Over 50 different proteins that function as growth factors have been isolated, and more undoubtedly exist. A specific cell surface receptor "recognizes" each growth factor, its shape fitting that growth factor precisely. When the growth fac-

tor binds with its receptor, the receptor reacts by triggering events within the cell (figure 11.19). The cellular selectivity of a particular growth factor depends upon which target cells bear its unique receptor. Some growth factors, such as PDGF and epidermal growth factor (EGF), affect a broad range of cell types, while others affect only specific types. For example, nerve growth factor (NGF) promotes the growth of certain classes of neurons, and erythropoietin triggers cell division in red blood cell precursors. Most animal cells need a combination of several different growth factors to overcome the various controls that inhibit cell division.

The G_0 Phase. If cells are deprived of appropriate growth factors, they stop at the G_1 checkpoint of the cell cycle. With their growth and division arrested, they remain in the G_0 phase, as we discussed earlier. This nongrowing state is distinct from the interphase stages of the cell cycle, G_1, S, and G_2.

The ability to enter G_0 accounts for the incredible diversity seen in the length of the cell cycle among different tissues. Epithelial cells lining the human gut divide more than twice a day, constantly renewing the lining of the digestive tract. By contrast, liver cells divide only once every year or two, spending most of their time in the G_0 phase. Mature neurons and muscle cells usually never leave G_0.

Two groups of proteins, cyclins and Cdk's, interact to regulate the cell cycle. Cells also receive protein signals called growth factors that affect cell division.

Cancer and the Control of Cell Proliferation

The unrestrained, uncontrolled growth of cells, called cancer, is addressed more fully in chapter 20. However, cancer certainly deserves mention in this chapter because it is essentially a disease of cell division—a failure of cell division *control*. Recent work has identified one of the culprits. Working independently, cancer scientists have repeatedly identified what has proven to be the same gene! Officially dubbed *p53* (researchers italicize the gene symbol to differentiate it from the protein), this gene plays a key role in the G_1 checkpoint of cell division. The gene's product, the p53 protein, monitors the integrity of DNA, checking that it is undamaged. If the p53 protein detects damaged DNA, it halts cell division and stimulates the activity of special enzymes to repair the damage. Once the DNA has been repaired, *p53* allows cell division to continue. In cases where the DNA damage is irreparable, *p53* then directs the cell to kill itself, activating an apoptosis (cell suicide) program (see chapter 19 for a discussion of apoptosis).

By halting division in damaged cells, *p53* prevents the development of many mutated cells, and it is therefore considered a tumor-suppressor gene (even though its activities are not limited to cancer prevention). Scientists have found that *p53* is entirely absent or damaged beyond use in the majority of cancerous cells they have examined. It is precisely because *p53* is nonfunctional that these cancer cells are able to repeatedly undergo cell division without being halted at the G_1 checkpoint (figure 11.20). To test this, scientists administered healthy p53 protein to rapidly dividing cancer cells in a petri dish: The cells soon ceased dividing and died.

Scientists at Johns Hopkins University School of Medicine have further reported that cigarette smoke causes mutations in the *p53* gene. This study, published in 1995, reinforced the strong link between smoking and cancer described in chapter 20.

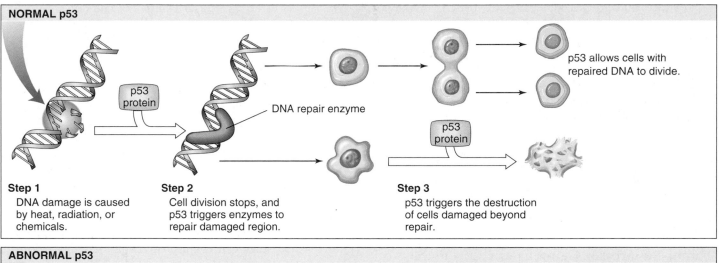

NORMAL p53

p53 protein

DNA repair enzyme

p53 allows cells with repaired DNA to divide.

p53 protein

Step 1
DNA damage is caused by heat, radiation, or chemicals.

Step 2
Cell division stops, and p53 triggers enzymes to repair damaged region.

Step 3
p53 triggers the destruction of cells damaged beyond repair.

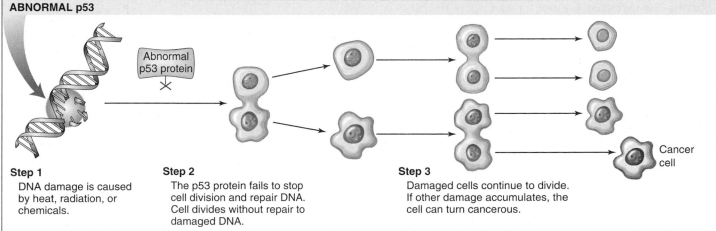

ABNORMAL p53

Abnormal p53 protein

Cancer cell

Step 1
DNA damage is caused by heat, radiation, or chemicals.

Step 2
The p53 protein fails to stop cell division and repair DNA. Cell divides without repair to damaged DNA.

Step 3
Damaged cells continue to divide. If other damage accumulates, the cell can turn cancerous.

FIGURE 11.20
Cell division and p53 protein.   Normal p53 protein monitors DNA, destroying cells that have irreparable damage to their DNA. Abnormal p53 protein fails to stop cell division and repair DNA. As damaged cells proliferate, cancer develops.

Growth Factors and Cancer

The disease we call cancer is actually many different diseases, depending on the tissue affected. The common theme in all cases is the loss of control over the cell cycle.

Proto-oncogenes

Research has identified numerous so-called *oncogenes*, genes that can, when introduced into a cell, cause it to become a cancer cell. This identification then led to the discovery of **proto-oncogenes,** which are the normal cellular genes that become oncogenes when mutated. These proto-oncogenes interact with growth factors in several ways: They can be the receptors for growth factors, and they can be any of the many factors downstream of the signaling by the growth factor receptors. A receptor for a growth factor that is mutated so that it is permanently "on" leaves the cell no longer dependent on the growth factor. This is analogous to a light switch that is stuck on: The light will always be on. PDGF and EGF receptors both fall into the category of proto-oncogenes.

Downstream signaling genes include many that code for proteins acting between the initial growth factor/receptor interaction and the ultimate action of the Cdk's that control the cell cycle. Examples of this category include proteins involved in signal transduction pathways, such as ras, and the transcription factors myc, fos, and jun at the end of such pathways. All of these factors are produced as an early response to growth factor signaling, and themselves stimulate the production of a number of important proteins, including cyclins and Cdk's.

The number of proto-oncogenes identified has grown to more than 50 over the years. This line of research connects our understanding of cancer with our understanding of the molecular mechanisms governing cell cycle control.

Tumor-Suppressor Genes

After the discovery of proto-oncogenes, a second category of genes related to cancer was identified: tumor-suppressor genes. These very interesting genes normally act to inhibit the cell cycle; thus, loss of function of both copies of such genes can lead to cancer by releasing the cell cycle from inhibition. This is in contrast to the proto-oncogenes, in which only one mutant copy is necessary to produce the cancerous phenotype. That is, the proto-oncogenes act in a dominant fashion while tumor suppressors act in a recessive fashion (see chapter 13 for a discussion of dominant and recessive). The first such gene identified was the retinoblastoma susceptibility gene (*Rb*), which predisposes individuals for a rare form of cancer that affects the retina of the eye. Despite the recessive nature of *Rb* at the level of the cell, it is inherited as a dominant in families. This is because inheriting a single mutant copy of *Rb* means that you only have one "good"

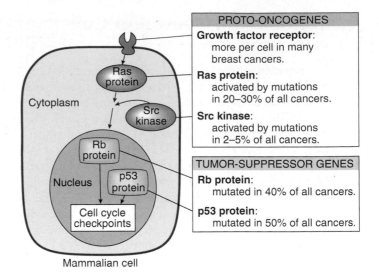

FIGURE 11.21
Key proteins associated with human cancers. Mutations in genes encoding key components of the cell division-signaling pathway are responsible for many cancers. Among them are proto-oncogenes encoding growth factor receptors, protein relay switches such as ras protein, and kinase enzymes such as src, that aid ras function. Mutations that disrupt tumor-suppressor proteins, such as Rb and p53, also foster cancer development.

copy left, and during the hundreds of thousands of divisions that occur to produce the retina, any error that results in loss or damage to the remaining good copy leads to a cancerous cell. A single cancerous cell in the retina then leads to the formation of a retinoblastoma.

The role of the Rb protein in the cell cycle is to integrate signals from growth factors. The Rb protein is called a "pocket protein" because it has binding pockets for other proteins. Its role is therefore to bind important regulatory proteins and prevent them from stimulating the production of the necessary cell cycle proteins discussed previously, such as cyclins or Cdk's. The binding of Rb to other proteins is controlled by phosphorylation such that when it is dephosphorylated, it can bind a variety of regulatory proteins, but loses this binding capacity when phosphorylated. The action of growth factors activates signaling pathways that result in the phosphorylation of Rb protein by a Cdk. This then brings us full circle, because the phosphorylation of Rb releases regulatory proteins that result in the production of S-phase cyclins that are necessary for the action of the Cdk required to pass the G_1/S boundary.

Figure 11.21 summarizes the types of genes that can cause cancer when mutated.

Some proto-oncogenes accelerate the cell cycle by promoting cyclins and Cdk's. Others, tumor-suppressor genes, suppress it by inhibiting their action. Cancer can result if mutations occur in either type of growth factor gene.

For interactive testing, visit the Online Learning Center with PowerWeb at www.mhhe.com/Raven7

11.1 Prokaryotes divide far more simply than do eukaryotes.

Cell Division in Prokaryotes

- Most prokaryotes have a genome made up of a single, circular DNA molecule, and replicate via binary fusion. (p. 208)
- Binary fusion begins with DNA replication, which starts at the origin site and proceeds bidirectionally around the circular DNA to a specific site of termination. (p. 208)
- The evolution of eukaryotic cells led to much more complex genomes and, thus, new and different ways to replicate and segregate the genome during cell division. (p. 209)

11.2 The chromosomes of eukaryotes arc highly ordered structures.

Discovery of Chromosomes

- Chromosomes were first discovered in 1882 by Walther Fleming. (p. 210)
- The number of chromosomes varies from one species to another. Humans have 23 nearly identical pairs for a total of 46 chromosomes. (p. 210)

The Structure of Eukaryotic Chromosomes

- The DNA is a very long, double-stranded fiber extending unbroken through the entire length of the chromosome. A typical human chromosome contains about 140 million nucleotides. (p. 211)
- Every 200 nucleotides, the DNA duplex is coiled around a core of eight histone proteins, forming a nucleosome. (p. 211)
- The particular array of chromosomes an individual possesses is its karyotype. (p. 212)
- The number of different chromosomes a species contains is known as its haploid (*n*) number, and is considered one complete set of chromosomes. (p. 212)
- Humans are diploid, with homologues coming from both the maternal and paternal lineages. (p. 212)

11.3 Mitosis is a key phase of the cell cycle.

The Cell Cycle

- The typical cell cycle consists of five phases: G_1 is the primary growth phase; S is the replication phase; and G_2 is the second growth phase. (G_1, S, and G_2 combined constitute interphase.) M (mitosis) is the phase during which the microtubular apparatus separates sister chromatids, and C (cytokinesis) is the phase during which the cytoplasm divides, creating two daughter cells. (p. 213)
- The duration of the cell cycle varies from species to species, and can range from about 8 minutes to over a year. (p. 213)

Interphase: Preparing for Mitosis

- The cell grows throughout interphase. The G_1 and G_2 phases are periods of protein synthesis and organelle production, while the S phase is when DNA replication occurs. (p. 214)

Mitosis

- Chromatin condensation continues into prophase. The spindle apparatus is assembled, and sister chromatids are linked to opposite poles of the cell by microtubules. The nuclear envelope breaks down. (p. 215)
- During metaphase, chromosomes align in the center of the cell along the metaphase plate. (p. 215)
- Anaphase begins when centromeres divide, freeing the two sister chromatids from each other. Sister chromatids are pulled to opposite poles as the attached microtubules shorten. (pp. 216–217)
- In telophase, the spindle apparatus disassembles, and the nuclear membrane begins to re-form. (p. 217)

Cytokinesis

- Cytokinesis is the phase of the cell cycle when the cell actually divides. Cytokinesis generally involves the cleavage of the cell into roughly equal halves, forming two daughter cells. (p. 218)

11.4 The cell cycle is carefully controlled.

General Strategies of Cell Cycle Control

- A cell uses three main checkpoints to both assess the internal state of the cell and integrate external signals. The G_1/S checkpoint is the primary point at which the cell decides to divide; the G_2/M checkpoint represents a commitment to mitosis; and the spindle checkpoint ensures that all chromosomes are attached to the spindle in preparation for anaphase. (p. 219)

Molecular Mechanisms of Cell Cycle Control

- Two groups of proteins, cyclins and Cdk's, interact and regulate the cell cycle. (p. 220)
- Cells also receive protein signals (growth factors) that affect cell division. (p. 222)

Cancer and the Control of Cell Proliferation

- Cancer is failure of cell division control. (p. 223)
- It is believed that a malfunction in the *p53* gene may allow cells to go through repeated cell division without being stopped at the appropriate checkpoints. (p. 223)
- Proto-oncogenes are normal cellular genes that become oncogenes when mutated. Proto-oncogenes can encode growth factors, protein relay switches, and kinase enzyme. (p. 224)
- Tumor-suppressor genes can also lead to cancer when they are mutated. (p. 224)

For interactive testing, visit the Online Learning Center with PowerWeb at www.mhhe.com/Raven7

Self Test

1. Bacterial cells divide by
   a. mitosis.
   b. replication.
   c. cytokinesis.
   d. binary fission.
2. Most eukaryotic organisms have _____ chromosomes in their cells.
   a. 1–5
   b. 10–50
   c. 100–500
   d. over 1000
3. Replicate copies of each chromosome are called _____ and are joined at the _____.
   a. homologues/centromere
   b. sister chromatids/kinetochore
   c. sister chromatids/centromere
   d. homologues/kinetochore
4. During which phase of the cell cycle is DNA synthesized?
   a. G_1
   b. G_2
   c. S
   d. M
5. Chromosomes are visible under a light microscope
   a. during mitosis.
   b. during interphase.
   c. when they are attached to their sister chromatids.
   d. All of these are correct.
6. During mitosis, the sister chromatids are separated and pulled to opposite poles during which stage?
   a. interphase
   b. metaphase
   c. anaphase
   d. telophase
7. Cytokinesis is
   a. the same process in plant and animal cells.
   b. the separation of cytoplasm and the formation of two cells.
   c. the final stage of mitosis.
   d. the movement of kinetochores.
8. The eukaryotic cell cycle is controlled at several points; which of these statements is *not* true?
   a. Cell growth is assessed at the G_1/S checkpoint.
   b. DNA replication is assessed at the G_2/M checkpoint.
   c. Environmental conditions are assessed at the G_0 checkpoint.
   d. The chromosomes are assessed at the spindle checkpoint.
9. What proteins are used to control cell growth specifically in *multicellular* eukaryotic organisms?
   a. Cdk
   b. MPF
   c. cyclins
   d. growth factors
10. What causes cancer in cells?
    a. damage to genes
    b. chemical damage to cell membranes
    c. UV damage to transport proteins
    d. All of these cause cancer in cells.

Test Your Visual Understanding

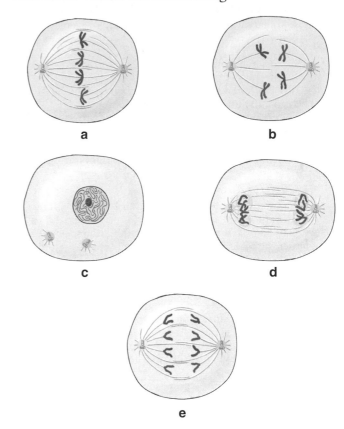

1. Match the mitotic and cell cycle phases with the appropriate figure.
   anaphase
   interphase
   metaphase
   prophase
   telophase

Apply Your Knowledge

1. An ancient plant called horsetail contains 216 chromosomes. How many homologous pairs of chromosomes does it contain? How many chromosomes are present in its cells during metaphase?
2. Colchicine is a poison that binds to tubulin and prevents its assembly into microtubules; cytochalasins are compounds that bind to the ends of actin filaments and prevent their elongation. What effects would these two substances have on cell division in animal cells?
3. If you could construct an artificial chromosome, what elements would you introduce into it, at a minimum, so that it could function normally in mitosis?

12

Sexual Reproduction and Meiosis

Concept Outline

12.1 Meiosis produces haploid cells from diploid cells.

Discovery of Reduction Division. Sexual reproduction does not increase chromosome number because gamete production by meiosis involves a decrease in chromosome number. Individuals produced from sexual reproduction inherit chromosomes from two parents.

12.2 Meiosis has unique features.

Meiosis. Three unique features of meiosis are synapsis, homologous recombination, and reduction division.

12.3 The sequence of events during meiosis involves two nuclear divisions.

Prophase I. Homologous chromosomes pair intimately and undergo crossing over that locks them together.

Metaphase I. Spindle microtubules align the chromosomes in the central plane of the cell.

Completing Meiosis. The second meiotic division is like a mitotic division, but has a very different outcome. Homologues separate at anaphase I, but sister chromatids do not. Sister chromatids separate at anaphase II, producing four haploid cells.

12.4 The evolutionary origin of sex is a puzzle.

Why Sex? Sex may have evolved as a mechanism to repair DNA, or perhaps as a means for contagious elements to spread. Sexual reproduction increases genetic variability by shuffling combinations of genes.

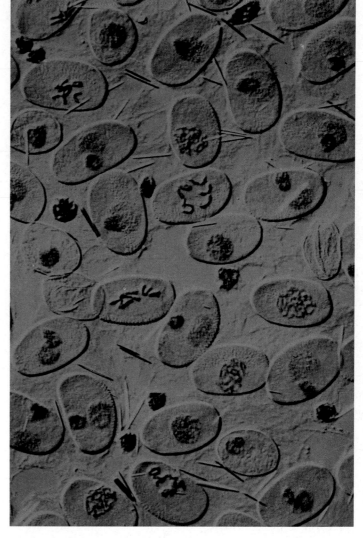

FIGURE 12.1
Plant cells undergoing meiosis. This preparation of pollen cells of a spiderwort, *Tradescantia*, was made by freezing the cells and then fracturing them. It shows several stages of meiosis (600×).

Most animals and plants reproduce sexually. Gametes of opposite sex unite to form a cell that, dividing repeatedly by mitosis, eventually gives rise to an adult body with some 100 trillion cells. The gametes that form the initial cell are the products of a special form of cell division called meiosis (figure 12.1), the subject of this chapter. Meiosis is far more intricate than mitosis, and the details behind it are not as well understood. The basic process, however, is clear. Also clear are the profound consequences of sexual reproduction: It plays a key role in generating the tremendous genetic diversity that is the raw material of evolution.

Part III *Genetic and Molecular Biology*

227

12.1 Meiosis produces haploid cells from diploid cells.

Discovery of Reduction Division

Only a few years after Walther Fleming's discovery of chromosomes in 1882, Belgian cytologist Pierre-Joseph van Beneden was surprised to find different numbers of chromosomes in different types of cells in the roundworm *Ascaris*. Specifically, he observed that the **gametes** (eggs and sperm) each contained two chromosomes, while the **somatic cells** (nonreproductive cells) of embryos and mature individuals each contained four.

Fertilization

From his observations, van Beneden proposed in 1887 that an egg and a sperm, each containing half the complement of chromosomes found in other cells, fuse to produce a single cell called a **zygote.** The zygote, like all of the somatic cells ultimately derived from it, contains two copies of each chromosome. The fusion of gametes to form a new cell is called **fertilization,** or **syngamy.**

Reduction Division

It was clear even to early investigators that gamete formation must involve some mechanism that reduces the number of chromosomes to half the number found in other cells. If it did not, the chromosome number would double with each fertilization, and after only a few generations, the number of chromosomes in each cell would become impossibly large. For example, in just 10 generations, the 46 chromosomes present in human cells would increase to over 47,000 (46×2^{10}).

The number of chromosomes does not explode in this way because of a special reduction division that occurs during gamete formation, producing cells with half the normal number of chromosomes. The subsequent fusion of two of these cells ensures a consistent chromosome number from one generation to the next. This reduction division process, known as **meiosis,** is the subject of this chapter.

The Sexual Life Cycle

Meiosis and fertilization together constitute a cycle of reproduction. Two sets of chromosomes are present in the somatic cells of adult individuals, making them **diploid** cells (Greek *diploos,* "double," + *eidos,* "form"), but only one set is present in the gametes, which are thus **haploid** (Greek *haploos,* "single," + *ploion,* "vessel"). Reproduction that involves this alternation of meiosis and fertilization is called **sexual reproduction.** Its outstanding characteristic is that offspring inherit chromosomes from *two* parents (figure 12.2). You, for example, inherited 23 chromosomes from your mother (maternal homologues), contributed by the egg fertilized at your conception, and 23 from your father (paternal homologues), contributed by the sperm that fertilized that egg.

The life cycles of all sexually reproducing organisms follow a pattern of alternation between diploid and haploid chromosome numbers, but there is some variation in the life cycles. Many types of algae spend the majority of their life cycle in a haploid state, the zygote undergoing meiosis to produce haploid cells that then undergo mitosis (figure 12.3*a*). In most animals, the diploid state dominates; the zygote first undergoes mitosis to produce diploid cells, and later in the life cycle, some of these diploid cells undergo meiosis to produce haploid gametes (figure 12.3*b*). Some plants and some algae alternate between a multicellular haploid phase and a multicellular diploid phase (figure 12.3*c*).

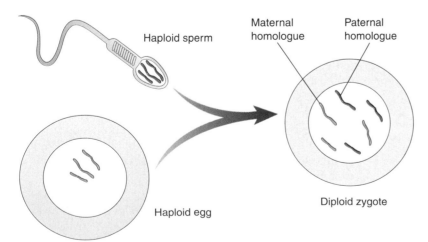

FIGURE 12.2
Diploid cells carry chromosomes from two parents.
A diploid cell contains two versions of each chromosome, a maternal homologue contributed by the haploid egg of the mother, and a paternal homologue contributed by the haploid sperm of the father.

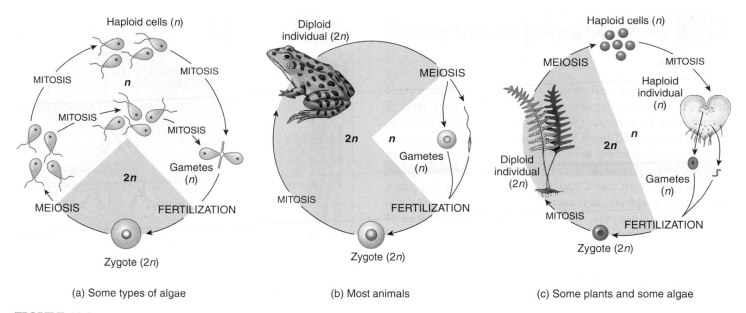

(a) Some types of algae

(b) Most animals

(c) Some plants and some algae

FIGURE 12.3
Three types of sexual life cycles. In sexual reproduction, haploid cells or organisms alternate with diploid cells or organisms.

Germ-Line Cells. In animals, the single diploid zygote undergoes mitosis to give rise to all of the cells in the adult body. The cells not destined to form gametes are called somatic cells, from the Latin word for "body." The cells that will eventually undergo meiosis to produce gametes are set aside from somatic cells early in the course of development. These cells are often referred to as germ-line cells. Both the somatic cells and the gamete-producing germ-line cells are diploid, but while somatic cells undergo mitosis to form genetically identical, diploid daughter cells, gamete-producing germ-line cells undergo meiosis to produce haploid gametes (figure 12.4).

> Meiosis is a process of cell division in which the number of chromosomes in certain cells is halved during gamete formation. In the sexual life cycle, alternation of diploid and haploid generations occurs.

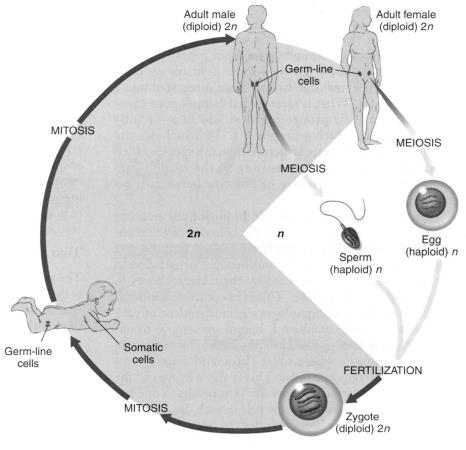

FIGURE 12.4
The sexual life cycle in animals. In animals, the zygote undergoes mitotic divisions and gives rise to all the cells of the adult body. Germ-line cells are set aside early in development and undergo meiosis to form the haploid gametes (eggs or sperm). The rest of the body cells are called somatic cells.

The sequence of events during meiosis involves two nuclear divisions.

Prophase I

In prophase I of meiosis, the DNA coils tighter, and individual chromosomes first become visible under the light microscope as a matrix of fine threads. Because the DNA has already replicated before the onset of meiosis, each of these threads actually consists of two sister chromatids joined at their centromeres. In prophase I, homologous chromosomes become closely associated in synapsis, exchange segments by crossing over, and then separate.

Synapsis

During prophase, the ends of the chromatids attach to the nuclear envelope at specific sites. The sites the homologues attach to are adjacent, so that the members of each homologous pair of chromosomes are brought close together. They then line up side by side, apparently guided by heterochromatin sequences, in the process called synapsis. This association is referred to as *sister chromatid cohesion*.

Crossing Over

Along with the synaptonemal complex that forms during prophase I (figure 12.7), another kind of structure appears that correlates in timing with the recombination process. These are called *recombination nodules* and are thought to contain the enzymatic machinery necessary to break and rejoin homologous chromatids. The details of crossing over are not well understood, but the process involves a complex series of events in which DNA segments are exchanged between nonsister chromatids (figure 12.8). Crossing over between sister chromatids is suppressed during meiosis. Reciprocal crossovers between nonsister chromatids are controlled such that each chromosome arm usually has one or a few crossovers per meiosis, no matter what the size of the chromosome. Human chromosomes typically have two or three.

When crossing over is complete, the synaptonemal complex breaks down, and the homologous chromosomes are released from the nuclear envelope and begin to move away from each other. At this point, there are four chromatids for each type of chromosome (two homologous chromosomes, each of which consists of two sister chromatids). The four chromatids do not separate completely, however, because they are held together in two ways: (1) The two sister chromatids of each homologue, recently created by DNA replication, are held together by their common centromeres; and (2) the paired homologues are held together at the points where crossing over occurred within the synaptonemal complex.

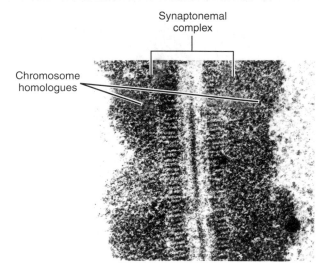

FIGURE 12.7
Structure of the synaptonemal complex. A portion of the synaptonemal complex of the ascomycete *Neotiella rutilans*, a cup fungus.

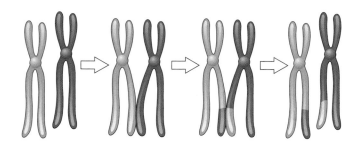

FIGURE 12.8
The results of crossing over. During crossing over, nonsister chromatids may exchange segments.

Chiasma Formation

Evidence of crossing over can often be seen under the light microscope as an X-shaped structure known as a **chiasma** (Greek, "cross"; plural, *chiasmata*). A chiasma is a crossover between nonsister chromatids, stabilized by sister chromatid cohesion. The presence of a chiasma indicates that two chromatids (one from each homologue) have exchanged parts. Like small rings moving down two strands of rope, the chiasmata move to the end of the chromosome arm before metaphase I.

Synapsis takes place early in prophase I of meiosis. Crossing over occurs between the paired DNA strands, creating the chromosomal configurations known as chiasmata. The two homologues are locked together by these exchanges, and they do not disengage readily.

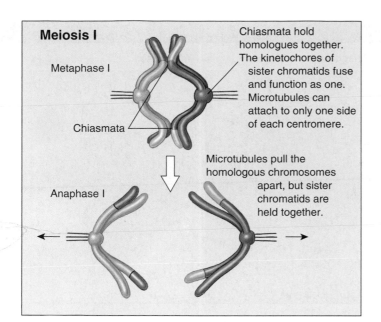
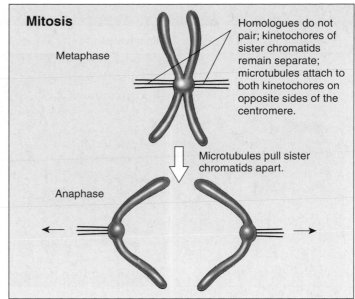

FIGURE 12.9

Chiasmata created by crossing over have a key impact on how chromosomes align in metaphase I. In metaphase I, the chiasmata hold homologous chromosomes together; consequently, the spindle microtubules bind to only one side of each centromere. By the end of meiosis I, mocrotubules shorten, breaking the chiasmata and pulling the homologous chromosomes apart, but sister chromatids are held together by their centromeres. In mitosis, microtubules attach to both sides of each centromere; when the microtubules shorten, the sister chromatids are split and drawn to opposite poles.

Metaphase I

By metaphase I, the second stage of meiosis I, the nuclear envelope has dispersed, and the microtubules form a spindle, just as in mitosis. During prophase I, the chiasmata move down the paired chromosomes from their original points of crossing over, eventually reaching the ends of the chromosomes. At this point, they are called *terminal chiasmata*. Terminal chiasmata hold the homologous chromosomes together in metaphase I, so that only one side of each centromere faces outward from the complex; the other side is turned inward toward the other homologue (figure 12.9). Consequently, spindle microtubules are able to attach to kinetochore proteins only on the outside of each centromere, and the centromeres of the two homologues attach to microtubules originating from opposite poles. This one-sided attachment is in marked contrast to the attachment in mitosis, when kinetochores on *both* sides of a centromere bind to microtubules.

Each joined pair of homologues then lines up on the metaphase plate. The orientation of each pair on the spindle axis is random; either the maternal or the paternal homologue may orient toward a given pole (figure 12.10; see also figure 12.11).

Chiasmata play an important role in aligning the chromosomes on the metaphase plate.

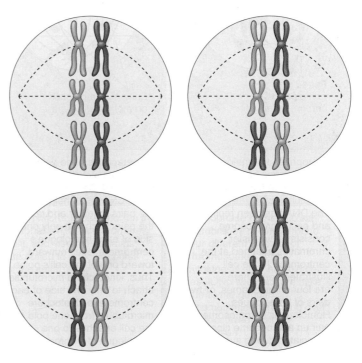

FIGURE 12.10

Random orientation of chromosomes on the metaphase plate. The number of possible chromosome orientations equals 2 raised to the power of the number of chromosome pairs. In this hypothetical cell with three chromosome pairs, eight (2^3) possible orientations exist, four of them illustrated here. Each orientation produces gametes with different combinations of parental chromosomes.

Completing Meiosis

Together, prophase and metaphase take up 90% or more of the duration of meiosis I. After that, meiosis I rapidly concludes. Anaphase I and telophase I proceed quickly, followed—without an intervening period of DNA synthesis—by the second meiotic division.

Anaphase I

In anaphase I, the microtubules of the spindle fibers begin to shorten. As they shorten, they break the chiasmata and pull the centromeres toward the poles, dragging the chromosomes along with them. Anaphase I comes about by the release of sister chromatid cohesion along the chromosome arms but not the centromeres. As a result of this release of cohesion, the homologues are pulled apart. Each centromere moves to one pole, taking both sister chromatids with it. When the spindle fibers have fully contracted, each pole has a complete haploid set of chromosomes consisting of one member of each homologous pair. Because of the random orientation of homologous chromosomes on the metaphase plate, a pole may receive either the maternal or the paternal homologue from each chromosome pair. As a result, the genes on different chromosomes assort independently; that is, meiosis I results in the **independent assortment** of maternal and paternal chromosomes into the gametes.

Telophase I

By the beginning of telophase I, the chromosomes have segregated into two clusters, one at each pole of the cell. Now the nuclear membrane re-forms around each daughter nucleus. Because each chromosome within a daughter nucleus replicated before meiosis I began, each now contains two sister chromatids attached by a common centromere. Importantly, *the sister chromatids are no longer identical* because of the crossing over that occurred in prophase I (figure 12.12). Cytokinesis may or may not occur after telophase I. The second meiotic division, meiosis II, occurs after an interval of variable length.

The Second Meiotic Division

After a typically brief interphase, in which no DNA synthesis occurs, the second meiotic division begins.

Meiosis II resembles a normal mitotic division. Prophase II, metaphase II, anaphase II, and telophase II follow in quick succession (see figure 12.11).

Prophase II. At the two poles of the cell, the clusters of chromosomes enter a brief prophase II, each nuclear envelope breaking down as a new spindle forms.

Metaphase II. In metaphase II, spindle fibers bind to both sides of the centromeres.

Anaphase II. The spindle fibers contract, and centromeric sister chromatid cohesion is released, splitting

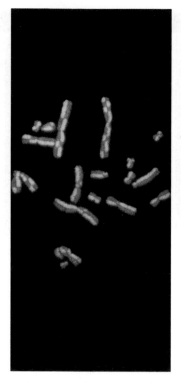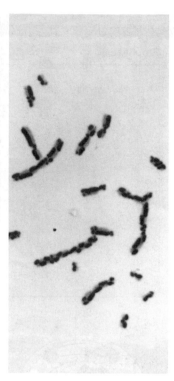

FIGURE 12.12
Crossing over between sister chromatids. Exchange between sister chromatids is usually suppressed during meiosis. Without this suppression, the reciprocal exchange of genetic material occurs between sister chromatids during meiosis I, producing so-called "harlequin" chromosomes, each containing one fluorescent DNA strand.

the centromeres and moving the sister chromatids to opposite poles.

Telophase II. Finally, the nuclear envelope re-forms around the four sets of daughter chromosomes.

The final result of this division is four cells containing haploid sets of chromosomes. No two are alike, because of the crossing over in prophase I. Nuclear envelopes then form around each haploid set of chromosomes. The cells that contain these haploid nuclei may develop directly into gametes, as they do in animals. Alternatively, they may themselves divide mitotically, as they do in plants, fungi, and many protists, eventually producing greater numbers of gametes or, as in some plants and insects, adult individuals with varying numbers of chromosome sets.

During meiosis I, homologous chromosomes move toward opposite poles in anaphase I, and individual chromosomes cluster at the two poles in telophase I. At the end of meiosis II, each of the four haploid cells contains one copy of every chromosome in the set, rather than two. Because of crossing over, no two cells are the same. These haploid cells may develop directly into gametes, as in animals, or they may divide by mitosis, as in plants, fungi, and many protists.

12.4 The evolutionary origin of sex is a puzzle.

Why Sex?

Not all reproduction is sexual. In **asexual reproduction,** an individual inherits all of its chromosomes from a single parent and is, therefore, genetically identical to its parent. Prokaryotic cells reproduce asexually, undergoing binary fission to produce two daughter cells containing the same genetic information (see chapter 11). Most protists reproduce asexually except under conditions of stress; then they switch to sexual reproduction. Among plants, asexual reproduction is common, and many other multicellular organisms are also capable of reproducing asexually. In animals, asexual reproduction often involves the budding off of a localized mass of cells, which grows by mitosis to form a new individual.

Even when meiosis and the production of gametes occur, reproduction may still take place without sex. The development of an adult from an unfertilized egg, called **parthenogenesis,** is a common form of reproduction in arthropods. Among bees, for example, fertilized eggs develop into diploid females, but unfertilized eggs develop into haploid males. Parthenogenesis even occurs among the vertebrates. Some lizards, fishes, and amphibians are capable of reproducing in this way; their unfertilized eggs undergo a mitotic nuclear division without cell cleavage to produce a diploid cell, which then develops into an adult.

Recombination Can Be Destructive

If reproduction can occur without sex, why does sex occur at all? This question has generated considerable discussion, particularly among evolutionary biologists. Sex is of great evolutionary advantage for populations or species, which benefit from the variability generated in meiosis by random orientation of chromosomes and by crossing over. However, evolution occurs because of changes at the level of *individual* survival and reproduction, rather than at the population level, and no obvious advantage accrues to the progeny of an individual that engages in sexual reproduction. In fact, recombination is a destructive as well as a constructive process in evolution. The segregation of chromosomes during meiosis tends to disrupt advantageous combinations of genes more often than it creates new, better-adapted combinations; as a result, some of the diverse progeny produced by sexual reproduction will not be as well adapted as their parents were. In fact, the more complex the adaptation of an individual organism, the less likely that recombination will improve it, and the more likely that recombination will disrupt it. It is, therefore, a puzzle to know what a well-adapted individual gains from participating in sexual reproduction, because *all* of its progeny could maintain its successful gene combinations if that individual simply reproduced asexually.

The Origin and Maintenance of Sex

There is no consensus among evolutionary biologists regarding the evolutionary origin or maintenance of sex. Conflicting hypotheses abound. Alternative hypotheses seem to be correct to varying degrees in different organisms.

The DNA Repair Hypothesis. If recombination is often detrimental to an individual's progeny, then what benefit promoted the evolution of sexual reproduction? Although the answer to this question is unknown, we can gain some insight by examining the protists. Meiotic recombination is often absent among the protists, which typically undergo sexual reproduction only occasionally. Often the fusion of two haploid cells occurs only under stress, creating a diploid zygote.

Why do some protists form a diploid cell in response to stress? Several geneticists have suggested that this occurs because only a diploid cell can effectively repair certain kinds of chromosome damage, particularly double-strand breaks in DNA. Both radiation and chemical events within cells can induce such breaks. As organisms became larger and lived longer, it must have become increasingly important for them to be able to repair such damage. The synaptonemal complex, which in early stages of meiosis precisely aligns pairs of homologous chromosomes, may well have evolved originally as a mechanism for repairing double-strand damage to DNA, using the undamaged homologous chromosome as a template to repair the damaged chromosome. A transient diploid phase would have provided an opportunity for such repair. In yeasts, mutations that inactivate the repair system for double-strand breaks of the chromosomes also prevent crossing over, suggesting a common mechanism for both synapsis and repair processes.

The Contagion Hypothesis. An unusual and interesting alternative hypothesis to explain the origin of sex is that it arose as a secondary consequence of the infection of eukaryotes by mobile genetic elements. Suppose a replicating transposable element were to infect a eukaryotic lineage. If it possessed genes promoting fusion with uninfected cells and synapsis, the transposable element could readily copy itself onto homologous chromosomes. It would rapidly spread by infection through the population, until all members contained it. The bizarre mating type "alleles" found in many fungi are very nicely explained by this hypothesis. Each of several mating types is in fact not an allele but an "idiomorph." Idiomorphs are genes occupying homologous positions on the chromosome but having such dissimilar sequences that they cannot be of homologous origin. These idiomorph genes may simply be the relics of several ancient infections by transposable elements.

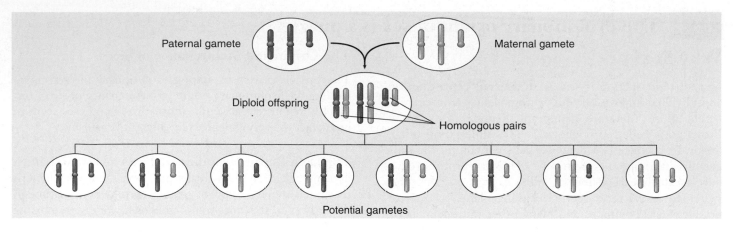

FIGURE 12.13

Independent assortment increases genetic variability. Independent assortment contributes new gene combinations to the next generation because the orientation of chromosomes on the metaphase plate is random. For example, in cells with three chromosome pairs, eight different gametes can result, each with different combinations of parental chromosomes.

The Red Queen Hypothesis. One evolutionary advantage of sex may be that it allows populations to "store" recessive alleles that are currently bad but have promise for reuse at some time in the future. Because populations are constrained by a changing physical and biological environment, selection is constantly acting against such alleles, but sexual species can never get rid of those sheltered in heterozygotes. Thus, the evolution of most sexual species, most of the time, manages to keep pace with ever-changing physical and biological constraints. This "treadmill evolution" is sometimes called the "Red Queen hypothesis," after the Queen of Hearts in Lewis Carroll's *Through the Looking Glass*, who tells Alice, "Now, here, you see, it takes all the running you can do, to keep in the same place."

Muller's Ratchet. The geneticist Herman Muller pointed out in 1965 that asexual populations incorporate a kind of mutational ratchet mechanism: Once harmful mutations arise, asexual populations have no way of eliminating them, and they accumulate over time, like turning a ratchet. Sexual populations, on the other hand, can employ recombination to generate individuals carrying fewer mutations, which selection can then favor. Sex may just be a way to keep the mutational load down.

The Evolutionary Consequences of Sex

While our knowledge of how sexual reproduction evolved is sketchy, it clearly has had an enormous impact on how species evolve, because of its ability to rapidly generate new genetic combinations. Independent assortment (figure 12.13), crossing over, and random fertilization each help generate genetic diversity.

Whatever the forces that led to sexual reproduction, its evolutionary consequences have been profound. No genetic process generates diversity more quickly. And, as you will see in later chapters, genetic diversity is the raw material of evolution—the fuel that drives it and determines its potential directions. In many cases, the pace of evolution appears to increase as the level of genetic diversity increases. For example, programs for selecting larger stature in domesticated animals such as cattle and sheep proceed rapidly at first, but then slow as the existing genetic combinations are exhausted; further progress must then await the generation of new gene combinations. Racehorse breeding provides another graphic example: Thoroughbred racehorses are all descendants of a small initial number of individuals, and selection for speed has accomplished all it can with this limited amount of genetic variability. As proof, the winning times in major races ceased to improve decades ago.

Paradoxically, the evolutionary process is both revolutionary and conservative. It is revolutionary in that the pace of evolutionary change is quickened by genetic recombination, much of which results from sexual reproduction. It is conservative in that evolutionary change is not always favored by selection, which may instead preserve existing combinations of genes. These conservative pressures appear greatest in certain asexually reproducing organisms that do not move around freely and that live in especially demanding habitats. In vertebrates, on the other hand, the evolutionary premium appears to have been on versatility, and sexual reproduction is the predominant mode of reproduction by an overwhelming margin.

The close association between homologous chromosomes that occurs during meiosis may have evolved as a mechanism to repair chromosomal damage, although several alternative mechanisms have also been proposed.

For interactive testing, visit the Online Learning Center with PowerWeb at www.mhhe.com/Raven7

12.1 Meiosis produces haploid cells from diploid cells.

Discovery of Reduction Division

- The fusion of gametes to form a new cell is referred to as fertilization, or syngamy. (p. 228)
- Gamete formation must involve some mechanism to reduce the number of chromosomes to half that found in other cells. (p. 228)
- Sexual reproduction involves an alternation of meiosis and fertilization, and follows a pattern of alternation between diploid and haploid chromosome numbers. (p. 228)

12.2 Meiosis has three unique features.

Meiosis

- Meiosis in diploid organisms consists of two rounds of division, meiosis I and meiosis II, but replication only occurs at the beginning of meiosis I. (p. 230)
- In prophase I of meiosis, homologues pair in synapsis, and crossing over occurs, allowing nonsister chromatids to exchange chromosomal material. (p. 230)

12.3 The sequence of events during meiosis involves two nuclear divisions.

Prophase I

- During prophase, the ends of the chromatids attach to the nuclear envelope at specific sites, and synapsis occurs. (p. 232)
- Crossing over occurs between nonsister chromatids, forming a chiasma. (p. 232)

Metaphase I

- By metaphase I, the nuclear envelope has dispersed, and microtubules form a spindle. (p. 233)
- Terminal chiasmata hold the homologous chromosomes together, and each joined pair of homologues lines up on the metaphase plate. (p. 233)

Completing Meiosis

- In anaphase I, the microtubules of the spindle fibers begin to shorten, pulling the centromeres toward the poles, and thus dragging the chromosomes toward the poles as well. (p. 236)
- Because of the random orientation of homologous chromosomes on the metaphase plate, meiosis I results in independent assortment of maternal and paternal chromosomes into the gametes. (p. 236)
- In telophase I, the chromosomes have segregated into clusters at each pole, and the nuclear membrane re-forms around each new daughter nucleus. (p. 236)
- Meiosis II resembles a normal mitotic division, except that at the end, each of the four haploid cells contains only one set of every chromosome instead of two sets. (p. 236)

12.4 The evolutionary origin of sex is a puzzle.

Why Sex?

- Several hypotheses exist as to the origin and maintenance of sex, including the DNA repair, contagion, Red Queen, and Muller's Ratchet hypotheses. (pp. 237–238)
- Paradoxically, the evolutionary process is both revolutionary and conservative. (p. 238)

Self Test

1. Fertilization results in
   a. a zygote.
   b. a diploid cell.
   c. a cell with a new genetic combination.
   d. All of these are correct.
2. The diploid number of chromosomes in humans is 46. The haploid number is
   a. 138.
   b. 92.
   c. 46.
   d. 23.
3. After chromosome replication and during synapsis,
   a. homologous chromosomes pair along their lengths.
   b. sister chromatids pair at the centromeres.
   c. homologous chromosomes pair at their ends.
   d. sister chromatids pair along their lengths.
4. During which stage of meiosis does crossing over occur?
   a. prophase I
   b. anaphase I
   c. prophase II
   d. telophase II
5. Synapsis is the process whereby
   a. homologous pairs of chromosomes separate and migrate toward a pole.
   b. homologous chromosomes exchange chromosomal material.
   c. homologous chromosomes become closely associated.
   d. the daughter cells contain half of the genetic material of the parent cell.
6. Terminal chiasmata are seen during which phase of meiosis?
   a. anaphase I
   b. prophase I
   c. metaphase I
   d. metaphase II
7. Which of the following occurs during anaphase I?
   a. Chromosomes cluster at the two poles of the cell.
   b. Crossing over occurs.
   c. Chromosomes align down the center of the cell.
   d. One version of each chromosome moves toward a pole.
8. Mitosis results in two _____ cells, while meiosis results in _____ haploid cells.
   a. haploid/four
   b. diploid/two
   c. diploid/four
   d. haploid/two
9. Genetic diversity is greatest in
   a. parthenogenesis.
   b. sexual reproduction.
   c. asexual reproduction.
   d. binary fission.
10. Which of the following is not a hypothesis about the evolution of sex?
    a. It evolved to repair damaged DNA.
    b. It evolved as a way to eliminate individuals.
    c. It evolved as a way to eliminate mutations.
    d. It evolved as a way to "store" recessive alleles that may prove beneficial in the future.

Test Your Visual Understanding

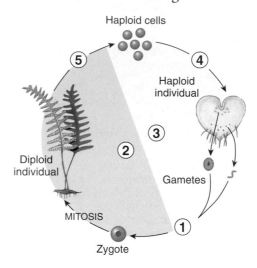

1. Match the following labels with the correct numbers on the figure. (Some labels may be used more than once.)
   a. diploid stage (2n)
   b. meiosis
   c. fertilization
   d. haploid stage (n)
   e. mitosis
2. Human cells spend most of their life cycles in a diploid stage, with only a selected number of cells undergoing meiosis to become haploid. Other organisms spend most of their life cycles in the haploid stage. Look ahead in this text to figures 28.14, 29.13, 29.15, and 30.10. Which of these organisms has a life cycle that is primarily haploid?

Apply Your Knowledge

1. An organism has 56 chromosomes in its diploid stage. Indicate how many chromosomes are present in the following, and explain your reasoning:
   a. somatic cells
   b. metaphase (mitosis)
   c. metaphase I (meiosis)
   d. metaphase II (meiosis)
   e. gametes
2. Humans have 23 pairs of chromosomes. Ignoring the effects of crossing over, what proportion of a woman's eggs contains only chromosomes she received from her mother?
3. Many sexually reproducing lizard species are able to generate local populations that reproduce by parthenogenesis. Would the sex of these local parthenogenetic populations be male, female, or neuter? Explain your reasoning.

13

Patterns of Inheritance

FIGURE 13.1
Human beings are extremely diverse in appearance. The
differences among us are partly inherited and partly the result of
environmental factors we encounter.

Every living creature is a product of the long evolutionary history of life on earth. While all organisms
share this history, only humans wonder about the
processes that led to their origin. We are still far from
understanding everything about our origins, but we have
learned a great deal. Like a partially completed jigsaw
puzzle, the boundaries have fallen into place, and much
of the internal structure is becoming apparent. In this
chapter, we discuss one piece of the puzzle—the enigma
of heredity. Why do groups of people from different
parts of the world often differ in appearance (figure
13.1)? Why do the members of a family tend to resemble
one another more than they resemble members of other
families?

13.1 Mendel solved the mystery of heredity.

Early Ideas About Heredity: The Road to Mendel

As far back as written records go, patterns of resemblance among the members of particular families have been noted and commented on (figure 13.2). Some familial features are unusual, such as the protruding lower lip of the European royal family Hapsburg, evident in pictures and descriptions of family members from the thirteenth century onward. Other characteristics, such as the occurrence of redheaded children within families of redheaded parents, are more common (figure 13.3). Inherited features, the building blocks of evolution, are our concern in this chapter.

Classical Assumption 1: Constancy of Species

Before the twentieth century, two concepts provided the basis for most of the thinking about heredity. The first is that *heredity occurs within species*. For a very long time, people believed it was possible to obtain bizarre composite animals by breeding (crossing) widely different species. The minotaur of Cretan mythology, a creature with the body of a bull and the torso and head of a man, is one example. The giraffe was thought to be another; its scientific name, *Giraffa camelopardalis*, arose from the belief that it had resulted from a cross between a camel and a leopard. From the Middle Ages onward, however, people discovered that such extreme crosses were not possible and that variation and heredity occur mainly within the boundaries of a particular species. In addition, species were thought to have been maintained without significant change from the time of their creation.

Classical Assumption 2: Direct Transmission of Traits

The second early concept related to heredity is that *traits are transmitted directly*. When variation is inherited by offspring from their parents, *what* is transmitted? The ancient Greeks suggested that elements from each of the parents' body parts were transmitted directly to their offspring. Hippocrates called this type of reproductive material *gonos*, meaning "seed." Hence, a characteristic such as a misshapen limb was the result of material that came from the misshapen limb of a parent. Information from each part of the body was supposedly passed along independently of the information from the other parts, and the child was formed after the hereditary material from all parts of the parents' bodies had come together.

This idea was predominant until fairly recently. For example, in 1868, Charles Darwin proposed that all cells

FIGURE 13.2
Heredity and family resemblance. Family resemblances are often strong—a visual manifestation of the mechanism of heredity. This is the Johnson family, the wife and daughters of one of the authors. While each daughter is different, all clearly resemble their mother.

FIGURE 13.3
An inherited trait. Many different traits are inherited in human families. This redhead is exhibiting one of these traits.

and tissues excrete microscopic granules, or "gemmules," that are passed to offspring, guiding the growth of the corresponding part in the developing embryo. Most similar theories of the direct transmission of hereditary material assumed that the male and female contributions *blend* in the offspring. Thus, parents with red and brown hair would produce children with reddish-brown hair, and tall and short parents would produce children of intermediate height.

Koelreuter Demonstrates Hybridization Between Species

Taken together, however, these two classical assumptions lead to a paradox. If no variation enters a species from outside and if the variation within each species blends in every generation, then all members of a species should soon have the same appearance. Obviously, this does not happen. Individuals within most species differ widely from each other, and they differ in characteristics that are transmitted from generation to generation.

How could this paradox be resolved? Actually, the resolution was provided long before Darwin, in the work of the German botanist Josef Koelreuter. In 1760, Koelreuter carried out successful **hybridizations** of plant species by crossing different strains of tobacco and obtaining fertile offspring. The hybrids differed in appearance from both parent strains. When individuals within the hybrid generation were crossed, their offspring were highly variable. Some of these offspring resembled plants of the hybrid generation (their parents), but a few resembled the original strains (their grandparents).

The Classical Assumptions Fail

Koelreuter's work represents the beginning of modern genetics, the first clues pointing to the modern theory of heredity. Koelreuter's experiments provided an important clue about how heredity works: The traits he was studying could be masked in one generation, only to reappear in the next. This pattern contradicts the theory of direct transmission. How could a trait that is transmitted directly disappear and then reappear? Nor were the traits of Koelreuter's plants blended. A contemporary account stated that the traits reappeared in the third generation "fully restored to all their original powers and properties."

It is worth repeating that the offspring in Koelreuter's crosses were not identical to one another. Some resembled

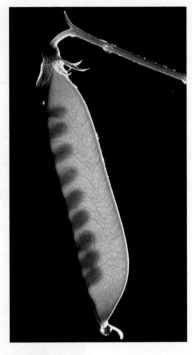

FIGURE 13.4
The garden pea, *Pisum sativum*. Easy to cultivate and able to produce many distinctive varieties, the garden pea was a popular experimental subject in investigations of heredity as long as a century before Gregor Mendel's experiments.

the hybrid generation, while others did not. The alternative forms of the characters Koelreuter was studying were distributed among the offspring. Referring to a heritable feature as a **character,** a modern geneticist would say the alternative forms of each character were **segregating** among the progeny of a mating, meaning that some offspring exhibited one alternative form of a character (for example, hairy leaves), while other offspring from the same mating exhibited a different alternative (smooth leaves). This segregation of alternative forms of a character, or **traits,** provided the clue that led Gregor Mendel to his understanding of the nature of heredity.

Knight Studies Heredity in Peas

Over the next hundred years, other investigators elaborated on Koelreuter's work. Prominent among them were English gentleman farmers trying to improve varieties of agricultural plants. In one such series of experiments, carried out in the 1790s, T. A. Knight crossed two true-breeding varieties (varieties that remain uniform from one generation to the next) of the garden pea, *Pisum sativum* (figure 13.4). One of these varieties had purple flowers, and the other had white flowers. All of the progeny of the cross had purple flowers. Among the offspring of these hybrids, however, were some plants with purple flowers and others, less common, with white flowers. Just as in Koelreuter's earlier studies, a trait from one of the parents had disappeared in one generation only to reappear in the next.

Within these deceptively simple results were the makings of a scientific revolution. Nevertheless, another century passed before the process of gene segregation was fully appreciated. Why did it take so long? One reason is that early workers did not quantify their results. A numerical record of results proved to be crucial to understanding the process. Knight and later experimenters who carried out other crosses with pea plants noted that some traits had a "stronger tendency" to appear than others, but they did not record the numbers of the different classes of progeny. Science was young then, and it was not obvious that the numbers were important.

Early geneticists demonstrated that some forms of an inherited character can: (1) disappear in one generation only to reappear unchanged in future generations; (2) segregate among the offspring of a cross; and (3) be more likely to be represented than their alternatives.

Mendel and the Garden Pea

The first quantitative studies of inheritance were carried out by Gregor Mendel, an Austrian monk (figure 13.5). Born in 1822 to peasant parents, Mendel was educated in a monastery and went on to study science and mathematics at the University of Vienna, where he failed his examinations for a teaching certificate. He returned to the monastery and spent the rest of his life there, eventually becoming abbot. In the garden of the monastery (figure 13.6), Mendel initiated a series of experiments on plant hybridization. The results of these experiments would ultimately change our views of heredity irrevocably.

Why Mendel Chose the Garden Pea

For his experiments, Mendel chose the garden pea, the same plant Knight and many others had studied. The choice was a good one for several reasons. First, many earlier investigators had produced hybrid peas by crossing different varieties. Mendel knew that he could expect to observe segregation of traits among the offspring. Second, a large number of true-breeding varieties of peas were available. Mendel initially examined 32. Then, for further study, he selected lines that differed with respect to seven easily distinguishable traits, such as round versus wrinkled seeds and purple versus white flowers, the latter a character that Knight had studied. Third, pea plants are small and easy to grow, and they have a relatively short generation time. Thus, a researcher can conduct experiments involving numerous plants, grow several generations in a single year, and obtain results relatively quickly.

A fourth advantage of studying peas is that both the male and female sexual organs are enclosed within the pea flower (figure 13.7), as are those of many flowering plants. Furthermore, the gametes produced by the male and female parts of the same flower, unlike those of many flowering plants, can fuse to form viable offspring. Fertilization takes place automatically within an individual flower if it is not disturbed, resulting in offspring that are the progeny from a single individual. Therefore, one can either let individual flowers engage in **self-fertilization,** or remove the flower's male parts before fertilization and introduce pollen from a strain with a different trait, thus performing *cross-pollination* that results in **cross-fertilization.**

FIGURE 13.5
Gregor Johann Mendel. Cultivating his plants in the garden of a monastery in Brunn, Austria (now Brno, Czech Republic), Mendel studied how differences among varieties of peas were inherited when the varieties were crossed. Similar experiments had been done before, but Mendel was the first to quantify the results and appreciate their significance.

FIGURE 13.6
Where Mendel worked. Gregor Mendel carried out his key plant-breeding experiments in this small garden in a monastery.

Mendel's Experimental Design

Mendel was careful to focus on only a few specific differences between the plants he was using and to ignore the countless other differences he must have seen. He also had the insight to realize that the differences he selected to analyze must be comparable. For example, he appreciated that trying to study the inheritance of round seeds versus tall height would be useless.

Mendel usually conducted his experiments in three stages:

1. First, he allowed pea plants of a given variety to produce progeny by self-fertilization for several generations. Mendel thus was able to assure himself that the traits he was studying were indeed **pure-breeding,** transmitted unchanged from generation to generation. Pea plants with white flowers, for example, when crossed with each other, produced only offspring with white flowers, regardless of the number of generations.

2. Mendel then performed crosses between varieties exhibiting alternative forms of characters. For example, he removed the male parts from the flower of a plant that produced white flowers and fertilized it with pollen from a purple-flowered plant. He also carried out the reciprocal cross, using pollen from a white-flowered individual to fertilize a flower on a pea plant that produced purple flowers (figure 13.8).

3. Finally, Mendel permitted the hybrid offspring produced by these crosses to self-fertilize for several generations. By doing so, he allowed the alternative forms of a character to segregate among the progeny. This was the same experimental design that Knight and others had used much earlier. But Mendel went an important step further: He counted the numbers of offspring exhibiting each trait in each succeeding generation. No one had ever done that before. The quantitative results Mendel obtained proved to be of supreme importance in revealing the process of heredity.

Mendel's experiments with the garden pea involved crosses between pure-breeding varieties, followed by a generation or more of self-fertilizing.

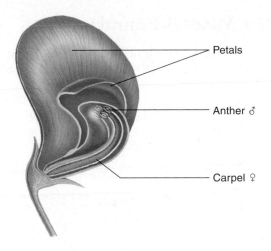

FIGURE 13.7
Structure of the pea flower (longitudinal section). In a pea plant flower, the petals enclose the male anther (containing pollen grains, which give rise to haploid sperm) and the female carpel (containing ovules, which give rise to haploid eggs). This ensures that self-fertilization will take place unless the flower is disturbed.

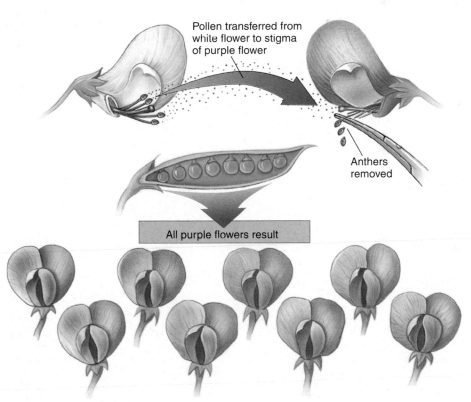

FIGURE 13.8
How Mendel conducted his experiments. Mendel pushed aside the petals of a white flower and collected pollen from the anthers. He then placed that pollen onto the stigma (part of the carpel) of a purple flower whose anthers had been removed, causing cross-fertilization to take place. All the seeds in the pod that resulted from this pollination were hybrids of the white-flowered male parent and the purple-flowered female parent. After planting these seeds, Mendel observed the pea plants they produced. All of the progeny of this cross had purple flowers.

What Mendel Found

The seven characters Mendel studied in his experiments possessed two variants that differed from one another in ways that were easy to recognize and score (table 13.1). We will examine in detail Mendel's crosses with flower color. His experiments with other characters were similar, and they produced similar results.

The F_1 Generation

When Mendel crossed two contrasting varieties of peas, such as white-flowered and purple-flowered plants, the hybrid offspring he obtained did not have flowers of intermediate color, as a hypothesis of blending inheritance would predict. Instead, in every case the flower color of the offspring resembled that of one of their parents. It is customary to refer to these offspring as the **first filial generation,**

Table 13.1 Seven Characters Mendel Studied and His Experimental Results

Character				F_2 Generation	
DOMINANT FORM	×	**RECESSIVE FORM**		**DOMINANT:RECESSIVE**	**RATIO**
Purple flowers	×	White flowers		705:224	3.15:1
Yellow seeds	×	Green seeds		6022:2001	3.01:1
Round seeds	×	Wrinkled seeds		5474:1850	2.96:1
Green pods	×	Yellow pods		428:152	2.82:1
Inflated pods	×	Constricted pods		882:299	2.95:1
Axial flowers	×	Terminal flowers		651:207	3.14:1
Tall plants	×	Dwarf plants		787:277	2.84:1

or F_1 (*filius* is Latin for "son"). Thus, in a cross of white-flowered with purple-flowered plants, the F_1 offspring all had purple flowers, just as Knight and others had reported earlier.

Mendel referred to the form or trait expressed in the F_1 plants as **dominant** and to the alternative form that was not expressed in the F_1 plants as **recessive.** For each of the seven pairs of contrasting traits that Mendel examined, one of the pair proved to be dominant and the other recessive.

The F_2 Generation

After allowing individual F_1 plants to mature and self-fertilize, Mendel collected and planted the seeds from each plant to see what the offspring in the **second filial generation,** or **F_2**, would look like. He found, just as Knight had earlier, that some F_2 plants exhibited white flowers, the recessive trait. After being hidden in the F_1 generation, the recessive form reappeared among some F_2 individuals.

Believing the proportions of the F_2 types would provide some clue about the mechanism of heredity, Mendel counted the numbers of each type among the F_2 progeny (figure 13.9). In the cross between the purple-flowered F_1 plants, he counted a total of 929 F_2 individuals (see table 13.1). Of these, 705 (75.9%) had purple flowers, and 224 (24.1%) had white flowers. Approximately $\frac{1}{4}$ of the F_2 individuals exhibited the recessive form of the character. Mendel obtained the same numerical result with the other six characters he examined: $\frac{3}{4}$ of the F_2 individuals exhibited the dominant trait, and $\frac{1}{4}$ displayed the recessive trait. In other words, the dominant:recessive ratio among the F_2 plants was always close to 3:1. Mendel carried out similar experiments with other traits, such as wrinkled versus round seeds (figure 13.10), and obtained the same result.

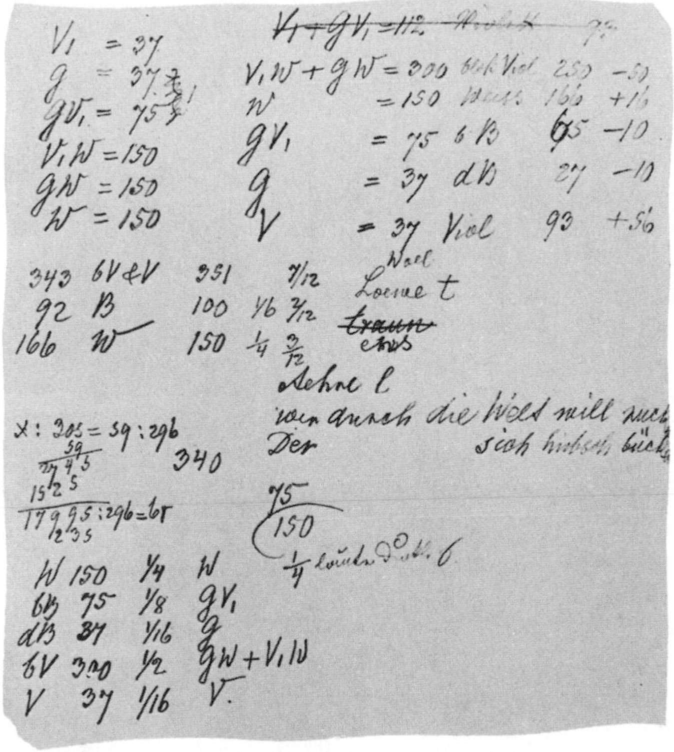
V1 = 37
g = 37
gV1 = 75
V1W = 150
gW = 150
W = 150
V1W + gW = 300
= 150
gV1 = 75
g = 37
V = 37 Viol 93 +56
343 bV&V 351 7/12
92 B 100 1/6 3/12
166 W 150 1/4 3/12
W 150 1/4 W
bB 75 1/8 gV1
dB 37 1/16 g
bV 300 1/2 gW+V1W
V 37 1/16 V.

FIGURE 13.9
A page from Mendel's notebook.

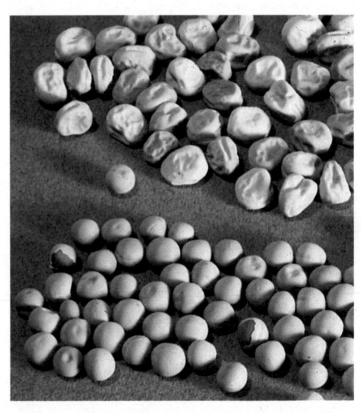

FIGURE 13.10
Seed shape: a Mendelian character. One of the differences Mendel studied involved the shape of pea plant seeds. In some varieties, the seeds were round, while in others, they were wrinkled.

A Disguised 1:2:1 Ratio

Mendel went on to examine how the F_2 plants passed traits to subsequent generations. He found that the recessive $\frac{1}{4}$ were always pure-breeding. For example, in the cross of white-flowered with purple-flowered plants, the white-flowered F_2 individuals reliably produced white-flowered offspring when they were allowed to self-fertilize. By contrast, only $\frac{1}{3}$ of the dominant, purple-flowered F_2 individuals ($\frac{1}{4}$ of all F_2 offspring) proved pure-breeding, while $\frac{2}{3}$ were not. This last class of plants produced dominant and recessive individuals in the third filial (F_3) generation in a 3:1 ratio. This result suggested that, for the entire sample, the 3:1 ratio that Mendel observed in the F_2 generation was really a disguised 1:2:1 ratio: $\frac{1}{4}$ pure-breeding dominant individuals, $\frac{1}{2}$ not-pure-breeding dominant individuals, and $\frac{1}{4}$ pure-breeding recessive individuals (figure 13.11).

Mendel's Model of Heredity

From his experiments, Mendel was able to understand four things about the nature of heredity: (1) The plants he crossed did not produce progeny of intermediate appearance, as a hypothesis of blending inheritance would have predicted. Instead, different plants inherited each alternative intact, as a discrete characteristic that either was or was not visible in a particular generation. (2) For each pair of alternative forms of a character, one alternative was not expressed in the F_1 hybrids, although it reappeared in some F_2 individuals. *The trait that "disappeared" must therefore be latent (present but not expressed) in the F_1 individuals.* (3) The pairs of alternative traits examined segregated among the progeny of a particular cross, some individuals exhibiting one trait and some the other. (4) These alternative traits were expressed in the F_2 generation in the ratio of $\frac{3}{4}$ dominant to $\frac{1}{4}$ recessive. This characteristic 3:1 segregation is often referred to as the **Mendelian ratio.**

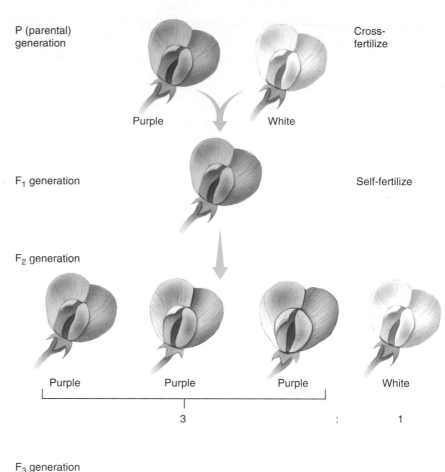

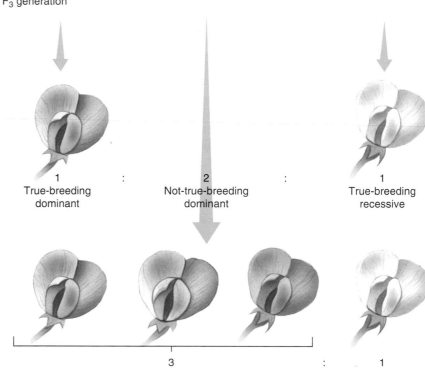

FIGURE 13.11

The F_2 generation is a disguised 1:2:1 ratio. By allowing the F_2 generation to self-fertilize, Mendel found from the offspring (F_3) that the ratio of F_2 plants was one pure-breeding dominant, two not-pure-breeding dominant, and one pure-breeding recessive.

Table 13.2 Some Dominant and Recessive Traits in Humans

Recessive Traits	Phenotypes	Dominant Traits	Phenotypes
Albinism	Lack of melanin pigmentation	Mid-digital hair	Presence of hair on middle segment of fingers
Alkaptonuria	Inability to metabolize homogentisic acid	Brachydactyly	Short fingers
Red-green color blindness	Inability to distinguish red or green wavelengths of light	Huntington disease	Degeneration of nervous system, starting in middle age
Cystic fibrosis	Abnormal gland secretion, leading to liver degeneration and lung failure	Phenylthiocarbamide (PTC) sensitivity	Ability to taste PTC as bitter
Duchenne muscular dystrophy	Wasting away of muscles during childhood	Camptodactyly	Inability to straighten the little finger
Hemophilia	Inability to form blood clots	Hypercholesterolemia (the most common human Mendelian disorder—1 in 500)	Elevated levels of blood cholesterol and risk of heart attack
Sickle cell anemia	Defective hemoglobin that causes red blood cells to curve and stick together	Polydactyly	Extra fingers and toes

To explain these results, Mendel proposed a simple model that has become one of the most famous in the history of science, containing simple assumptions and making clear predictions. The model has five elements:

1. Parents do not transmit physiological traits directly to their offspring. Rather, they transmit discrete information about the traits, what Mendel called "factors." These factors later act in the offspring to produce the trait. In modern terms, we would say that information about the alternative forms of characters that an individual expresses is *encoded* by the factors it receives from its parents.

2. Each individual receives two factors that may code for the same trait or for two alternative traits for a character. We now know that two factors for each character are present in each individual because these factors are carried on chromosomes, and each adult individual is *diploid*. When the individual forms gametes (eggs or sperm), the gametes contain only one of each kind of chromosome (see chapter 12); the gametes are *haploid*. Therefore, only one factor for each character of the adult organism is contained in the gamete. Which of the two factors ends up in a particular gamete is randomly determined.

3. Not all copies of a factor are identical. In modern terms, the alternative forms of a factor, leading to alternative forms of a character, are called **alleles**. When two haploid gametes containing exactly the same allele of a factor fuse during fertilization to form a zygote, the offspring that develops from that zygote is said to be **homozygous;** when the two haploid gametes contain different alleles, the individual offspring is **heterozygous.**

In modern terminology, Mendel's factors are called **genes.** We now know that each gene is composed of a particular DNA nucleotide sequence (see chapter 3). The particular location of a gene on a chromosome is referred to as the gene's **locus** (plural, *loci*).

4. The two alleles, one contributed by the male gamete and one by the female, do not influence each other in any way. In the cells that develop within the new individual, these alleles remain discrete. They neither blend with nor alter each other. (Mendel referred to them as "uncontaminated.") Thus, when the individual matures and produces its own gametes, the alleles for each gene segregate randomly into these gametes.

5. The presence of a particular allele does not ensure that the trait encoded by it will be expressed in an individual carrying that allele. In heterozygous individuals, only one allele (the dominant one) is expressed, while the other (recessive) allele is present but unexpressed. To distinguish between the presence of an allele and its expression, modern geneticists refer to the totality of alleles that an individual contains as the individual's **genotype** and to the physical appearance of that individual as its **phenotype.** The phenotype of an individual is the observable outward manifestation of its genotype, the result of the functioning of the enzymes and proteins encoded by the genes it carries. In other words, the genotype is the blueprint, and the phenotype is the visible outcome.

These five elements, taken together, constitute Mendel's model of the hereditary process. Many traits in humans also exhibit dominant or recessive inheritance, similar to the traits Mendel studied in peas (table 13.2).

When Mendel crossed two contrasting varieties, he found that in the second generation 25% of the offspring were pure-breeding for the dominant trait, 50% were hybrid for the two traits and exhibited the dominant trait, and 25% were pure-breeding for the recessive trait.

How Mendel Interpreted His Results

Does Mendel's model predict the results he actually obtained? To test his model, Mendel first expressed it in terms of a simple set of symbols, and then used the symbols to interpret his results. It is very instructive to do the same. Consider again Mendel's cross of purple-flowered with white-flowered plants. We will assign the symbol P to the dominant allele, associated with the production of purple flowers, and the symbol p to the recessive allele, associated with the production of white flowers. By convention, genetic traits are usually assigned a letter symbol referring to their more common forms, in this case "P" for purple flower color. The dominant allele is written in uppercase, as P; the recessive allele (white flower color) is assigned the same symbol in lower case, p.

In this system, the genotype of an individual that is pure-breeding for the recessive white-flowered trait would be designated pp. In such an individual, both copies of the allele specify the white-flowered phenotype. Similarly, the genotype of a pure-breeding purple-flowered individual would be designated PP, and a heterozygote would be designated Pp (dominant allele first). Using these conventions and denoting a cross between two strains with ×, we can symbolize Mendel's original cross as $pp \times PP$.

The F₁ Generation

Employing these simple symbols, we can now go back and reexamine the crosses Mendel carried out. Because a white-flowered parent (pp) can produce only p gametes, and a pure purple-flowered (homozygous dominant) parent (PP) can produce only P gametes, the union of an egg and a sperm from these parents can produce only heterozygous Pp offspring in the F₁ generation. Because the P allele is dominant, all of these F₁ individuals are expected to have purple flowers. The p allele is present in these heterozygous individuals, but it is not phenotypically expressed. This is the basis for the latency Mendel saw in recessive traits.

The F₂ Generation

When F₁ individuals are allowed to self-fertilize, the P and p alleles segregate randomly during gamete formation. Their subsequent union at fertilization to form F₂ individuals is also random, not being influenced by which alternative alleles the individual gametes carry. What will the F₂ individuals look like? The possibilities may be visualized in a simple diagram called a **Punnett square,** named after its originator, the English geneticist Reginald Crundall Punnett (figure 13.12). Mendel's model, analyzed in terms of a Punnett square, clearly predicts that the F₂

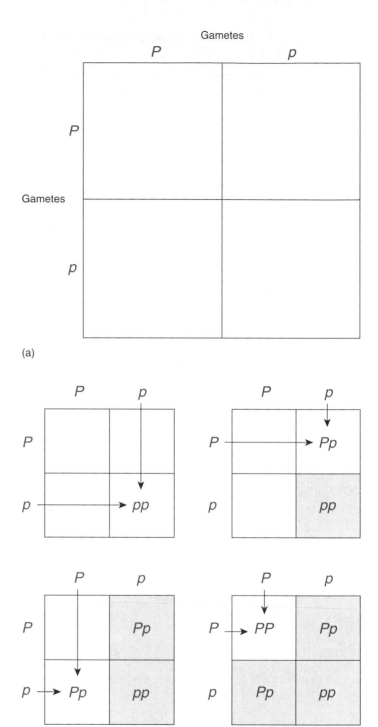

FIGURE 13.12
A Punnett square. (*a*) To make a Punnett square, place the different possible types of female gametes along the top of a square and the different possible types of male gametes along the side. (*b*) Each potential zygote can then be represented as the intersection of a vertical line and a horizontal line.

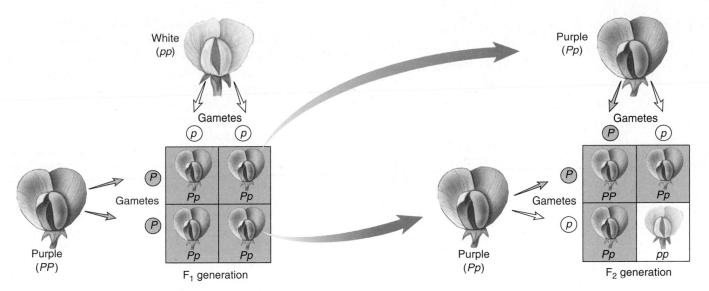

FIGURE 13.13

Mendel's cross of pea plants differing in flower color. All of the offspring of the first cross (the F_1 generation) are Pp heterozygotes with purple flowers. When two heterozygous F_1 individuals are crossed or self-fertilized, three kinds of F_2 offspring are possible: PP homozygotes (purple flowers); Pp heterozygotes (also purple flowers); and pp homozygotes (white flowers). Therefore, in the F_2 generation, the ratio of dominant to recessive phenotypes is 3:1. However, the ratio of genotypes is 1:2:1 (1 PP: 2 Pp: 1 pp).

generation should consist of $\frac{3}{4}$ purple-flowered plants and $\frac{1}{4}$ white-flowered plants, a phenotypic ratio of 3:1 (figure 13.13).

The Laws of Probability Can Predict Mendel's Results

A different way to express Mendel's result is to say that there are three chances in four ($\frac{3}{4}$) that any particular F_2 individual will exhibit the dominant trait, and one chance in four ($\frac{1}{4}$) that an F_2 individual will express the recessive trait. Stating the results in terms of probabilities allows simple predictions to be made about the outcomes of crosses. If both F_1 parents are Pp (heterozygotes), the probability that a particular F_2 individual will be pp (homozygous recessive) is the probability of receiving a p gamete from the male ($\frac{1}{2}$) times the probability of receiving a p gamete from the female ($\frac{1}{2}$), or $\frac{1}{4}$. This is the same operation we perform in the Punnett square illustrated in figure 13.12. The ways probability theory can be used to analyze Mendel's results are discussed in detail on page 253 (see Box: Probability and Allele Distribution).

As you can see in figure 13.13, there are really three kinds of F_2 individuals: $\frac{1}{4}$ are pure-breeding, white-flowered individuals (pp); $\frac{1}{2}$ are heterozygous, purple-flowered individuals (Pp); and $\frac{1}{4}$ are pure-breeding, purple-flowered individuals (PP). The 3:1 phenotypic ratio is really a disguised 1:2:1 genotypic ratio.

Mendel's First Law of Heredity: Segregation

Mendel's model thus accounts in a neat and satisfying way for the segregation ratios he observed. Its central assumption—that alternative alleles of a character segregate from each other in heterozygous individuals and remain distinct—has since been verified in many other organisms. It is commonly referred to as Mendel's first law of heredity, or the **Law of Segregation.** As we saw in chapter 12, the segregational behavior of alternative alleles has a simple physical basis, the random alignment of chromosomes on the metaphase plate during meiosis I, and the subsequent disjunction of the homologues in anaphase I of meiosis. It is a tribute to the intellect of Mendel's analysis that he arrived at the correct scheme with no knowledge of the cellular mechanisms of inheritance; neither chromosomes nor meiosis had yet been described.

The Testcross

To test his hypothesis further, Mendel devised a simple and powerful procedure called the **testcross.** Consider a purple-flowered plant. It is impossible to tell whether such a plant is homozygous or heterozygous simply by looking at its phenotype. To learn its genotype, you must cross it with some other plant. What kind of cross would provide the answer? If you cross it with a homozygous dominant individual, all of the progeny will show the dominant phenotype whether the test plant is homozygous or heterozygous. It is also difficult (but not impossible) to distinguish between the two possible test plant genotypes by crossing with a heterozygous individual. However, if you cross the test plant with a homozygous recessive individual, the two possible test plant genotypes will give totally different results (figure 13.14):

Alternative 1:   Unknown individual homozygous dominant (*PP*)
PP × *pp:* All offspring have purple flowers (*Pp*).

Alternative 2:   Unknown individual heterozygous (*Pp*)
Pp × *pp:* $\frac{1}{2}$ of offspring have white flowers (*pp*), and $\frac{1}{2}$ have purple flowers (*Pp*).

To perform his testcross, Mendel crossed heterozygous F_1 individuals back to the parent homozygous for the recessive trait. He predicted that the dominant and recessive traits would appear in a 1:1 ratio, and that is what he observed. For each pair of alleles he investigated, Mendel observed phenotypic F_2 ratios of 3:1 (see figure 13.13) and testcross ratios very close to 1:1, just as his model had predicted.

Testcrosses can also be used to determine the genotype of an individual when two genes are involved. Mendel carried out many two-gene crosses, some of which we will discuss. He often used testcrosses to verify the genotypes of particular dominant-appearing F_2 individuals. Thus, an F_2 individual showing both dominant traits (*A_ B_*) might have any of the following genotypes: *AABB, AaBB, AABb,* or *AaBb.* By crossing dominant-appearing F_2 individuals with homozygous recessive individuals (that is, *A_ B_* × *aabb*), Mendel was able to determine if either or both of the traits bred pure among the progeny, and so to determine the genotype of the F_2 parent:

AABB	Trait A breeds pure	Trait B breeds pure
AaBB	———	Trait B breeds pure
AABb	Trait A breeds pure	———
AaBb	———	———

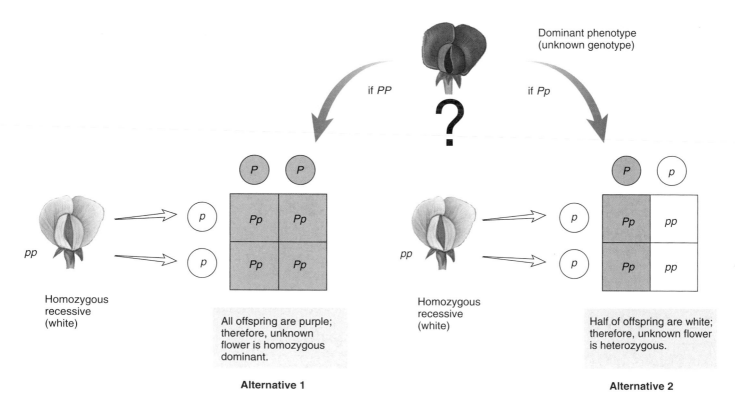

Alternative 1

All offspring are purple; therefore, unknown flower is homozygous dominant.

Alternative 2

Half of offspring are white; therefore, unknown flower is heterozygous.

FIGURE 13.14

A testcross. To determine whether an individual exhibiting a dominant phenotype, such as purple flowers, is homozygous or heterozygous for the dominant allele, Mendel crossed the individual in question with a plant that he knew to be homozygous recessive—in this case, a plant with white flowers.

Probability and Allele Distribution

Many, although not all, alternative alleles produce discretely different phenotypes. Mendel's pea plants were tall or dwarf, had purple or white flowers, and produced round or wrinkled seeds. The eye color of a fruit fly may be red or white, and the skin color of a human may be pigmented or albino. When only two alternative alleles exist for a given character, the distribution of phenotypes among the offspring of a cross is referred to as a **binomial distribution.**

As an example, consider the distribution of sexes in humans. Imagine that a couple has chosen to have three children. How likely is it that two of the children will be boys and one will be a girl? The frequency of any particular possibility is referred to as its *probability* of occurrence. Let p symbolize the probability of having a boy at any given birth and q symbolize the probability of having a girl. Since any birth is equally likely to produce a girl or boy:

$$p = q = \tfrac{1}{2}$$

Table 13.A shows eight possible gender combinations among the three children. The sum of the probabilities of the eight possible combinations must equal one. Thus:

$$p^3 + 3p^2q + 3pq^2 + q^3 = 1$$

The probability that the three children will be two boys and one girl is:

$$3p^2q = 3 \times (\tfrac{1}{2})^2 \times (\tfrac{1}{2}) = \tfrac{3}{8}$$

To test your understanding, try to estimate the probability that two parents heterozygous for the recessive allele producing albinism (*a*) will have one albino child in a family of three. First, set up a Punnett square:

		Mother's Gametes	
		A	**a**
Father's	**A**	AA	Aa
Gametes	**a**	Aa	aa

You can see that one-fourth of the children are expected to be albino (*aa*). Thus, for any given birth, the probability of an albino child is $\tfrac{1}{4}$. This probability can be symbolized by q. The probability of a nonalbino child is $\tfrac{3}{4}$, symbolized by p. Therefore, the probability that there will be one albino child among the three children is:

$$3p^2q = 3 \times (\tfrac{3}{4})^2 \times (\tfrac{1}{4}) = \tfrac{27}{64}, \text{ or } 42\%$$

This means that the chance of having one albino child in the three is 42%.

Table 13.A Binomial Distribution of the Sexes of Children in Human Families

Composition of Family	Order of Birth	Calculation	Probability	
3 boys	bbb	$p \times p \times p$	p^3	
2 boys and 1 girl	bbg	$p \times p \times q$	p^2q	
	bgb	$p \times q \times p$	p^2q	$3p^2q$
	gbb	$q \times p \times p$	p^2q	
1 boy and 2 girls	ggb	$q \times q \times p$	pq^2	
	gbg	$q \times p \times q$	pq^2	$3pq^2$
	bgg	$p \times q \times q$	pq^2	
3 girls	ggg	$q \times q \times q$	q^3	

Mendel's Second Law of Heredity: Independent Assortment

After Mendel had demonstrated that different traits of a given character (alleles of a given gene) segregate independently of each other in crosses, he asked whether different genes also segregate independently. Mendel set out to answer this question in a straightforward way. He first established a series of pure-breeding lines of peas that differed in just two of the seven characters he had studied. He then crossed contrasting pairs of the pure-breeding lines to create heterozygotes. In a cross involving different seed shape alleles (round, *R*, and wrinkled, *r*) and different seed color alleles (yellow, *Y*, and green, *y*), all the F$_1$ individuals were identical, each one heterozygous for both seed shape (*Rr*) and seed color (*Yy*). The F$_1$ individuals of such a cross are **dihybrids,** individuals heterozygous for both genes.

The third step in Mendel's analysis was to allow the dihybrids to self-fertilize. If the alleles affecting seed shape and seed color were segregating independently, then the probability that a particular pair of seed shape alleles would occur together with a particular pair of seed color alleles would be simply the product of the individual probabilities that each pair would occur separately. Thus, the probability that an individual with wrinkled green seeds (*rryy*) would appear in the F$_2$ generation would be equal to the probability of observing an individual with wrinkled seeds ($\tfrac{1}{4}$) times the probability of observing one with green seeds ($\tfrac{1}{4}$), or $\tfrac{1}{16}$.

Because the gene controlling seed shape and the gene controlling seed color are each represented by a pair of alternative alleles in the dihybrid individuals, four types of gametes are expected: *RY, Ry, rY,* and *ry*. Therefore, in the F$_2$ generation, there are 16 possible combinations of alleles, each of them equally probable. Of these, 9 possess at least one dominant allele for each gene (signified *R__Y__*, where the underscore indicates the presence of either allele) and, thus, should have round, yellow seeds. Of the rest, 3 possess at least one dominant *R* allele but are homozygous

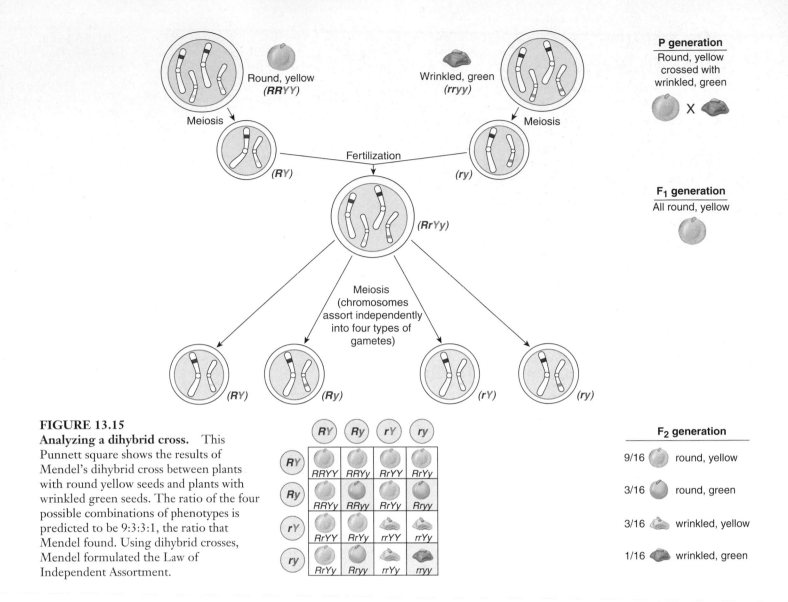

P generation
Round, yellow
crossed with
wrinkled, green

Round, yellow
(**RRYY**)

Wrinkled, green
(**rryy**)

Meiosis

Meiosis

Fertilization

(**RY**)

(**ry**)

(**RrYy**)

F₁ generation
All round, yellow

Meiosis
(chromosomes
assort independently
into four types of
gametes)

(**RY**)

(**Ry**)

(**rY**)

(**ry**)

FIGURE 13.15
Analyzing a dihybrid cross. This Punnett square shows the results of Mendel's dihybrid cross between plants with round yellow seeds and plants with wrinkled green seeds. The ratio of the four possible combinations of phenotypes is predicted to be 9:3:3:1, the ratio that Mendel found. Using dihybrid crosses, Mendel formulated the Law of Independent Assortment.

	RY	Ry	rY	ry
RY	RRYY	RRYy	RrYY	RrYy
Ry	RRYy	RRyy	RrYy	Rryy
rY	RrYY	RrYy	rrYY	rrYy
ry	RrYy	Rryy	rrYy	rryy

F₂ generation

9/16 round, yellow

3/16 round, green

3/16 wrinkled, yellow

1/16 wrinkled, green

recessive for color (R__yy); 3 others possess at least one dominant Y allele but are homozygous recessive for shape (rrY__); and 1 combination among the 16 is homozygous recessive for both genes (rryy) (figure 13.15). The hypothesis that color and shape genes assort independently thus predicts that the F₂ generation will display a 9:3:3:1 phenotypic ratio: nine individuals with round, yellow seeds, three with round, green seeds, three with wrinkled, yellow seeds, and one with wrinkled, green seeds.

What did Mendel actually observe? From a total of 556 seeds from dihybrid plants he had allowed to self-fertilize, he observed: 315 round yellow (R__Y__), 108 round green (R__yy), 101 wrinkled yellow (rrY__), and 32 wrinkled green (rryy). These results are very close to a 9:3:3:1 ratio (which would be 313:104:104:35). Consequently, the two genes appeared to assort completely independently of each other. Note that this independent assortment of different genes in no way alters the independent segregation of individual pairs of alleles. Round versus wrinkled seeds occur in a

ratio of approximately 3:1 (423:133); so do yellow versus green seeds (416:140). Mendel obtained similar results for other pairs of traits.

Mendel's discovery is often referred to as Mendel's second law of heredity, or the **Law of Independent Assortment.** Like segregation, independent assortment can be understood by examining the behavior of chromosomes during meiosis (see figure 13.15). At metaphase I, homologues of each chromosome are oriented toward opposite poles, leading to segregation of alleles in a heterozygote. The independent behavior of each chromosome pair leads to the independent assortment of loci on different chromosomes.

Mendel summed up his discoveries about heredity in two laws. Mendel's first law of heredity states that alternative alleles of a trait segregate independently; his second law of heredity states that genes located on different chromosomes assort independently.

Mendelian Inheritance Is Not Always Easy to Analyze

Although Mendel's results did not receive much notice during his lifetime, three different investigators independently rediscovered his pioneering paper in 1900, 16 years after his death. They came across it while searching the literature in preparation for publishing their own findings, which closely resembled those Mendel had presented more than three decades earlier. In the decades following the rediscovery of Mendel's ideas, many investigators set out to test them. However, scientists attempting to confirm Mendel's theory often had trouble obtaining the same simple ratios he had reported. Often, the expression of the genotype is not straightforward. Most phenotypes reflect the action of many genes, and the phenotype can be affected by alleles that lack complete dominance as well as by the environment.

Continuous Variation

Few phenotypes result from the action of only one gene. Instead, most characters reflect multiple additive contributions to the phenotype by several genes. When multiple genes act jointly to influence a character such as height or weight, the character often shows a range of small differences. Because all of the genes that play a role in determining phenotypes such as height or weight segregate independently of one another, we see a gradation in the degree of difference when many individuals are examined (figure 13.16). We call this gradation **continuous variation,** and we call such traits *quantitative traits.* The greater the number of genes that influence a character, the more continuous the expected distribution of the versions of that character.

How can we describe the variation in a character such as the height of the individuals in figure 13.16? Humans range from quite short to very tall, with average heights more common than either extreme. Often, variations are grouped into categories—in this case, by measuring the heights of the individuals in inches and rounding fractions of an inch to the nearest whole number. Each height, in inches, is a separate phenotypic category. Plotting the numbers in each height category produces a histogram, such as that in figure 13.16. The histogram approximates an idealized bell-shaped curve, and the variation can be characterized by the mean and spread of that curve.

Pleiotropic Effects

Often, an individual allele will have more than one effect on the phenotype. Such an allele is said to be **pleiotropic.** When the pioneering French geneticist Lucien Cuenot studied yellow fur in mice, a dominant trait, he was unable to obtain a pure-breeding yellow strain by crossing individual yellow mice with each other. Individuals homozygous

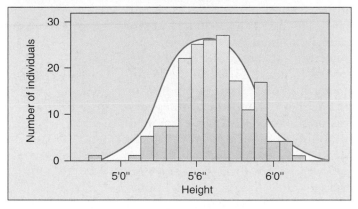

FIGURE 13.16

Height is a continuously varying trait. The photo shows variation in height among students of the 1914 class at the Connecticut Agricultural College. Because many genes contribute to height and tend to segregate independently of one another, the cumulative contribution of different combinations of alleles to height forms a *continuous* distribution of possible heights, in which the extremes are much rarer than the intermediate values.

for the yellow allele died, because the yellow allele was pleiotropic: One effect was yellow coat color, but another was a lethal developmental defect. A pleiotropic allele may be dominant with respect to one phenotypic consequence (yellow fur) and recessive with respect to another (lethal developmental defect). In pleiotropy, one gene affects many traits, in marked contrast to polygeny, where many genes affect one trait. Pleiotropic effects are difficult to predict, because the genes that affect a trait often perform other functions we may know nothing about.

Pleiotropic effects are characteristic of many inherited disorders, including cystic fibrosis and sickle cell anemia, both discussed in section 13.2. In these disorders, multiple symptoms can be traced back to a single gene defect. Cystic fibrosis patients exhibit clogged blood vessels, overly sticky mucus, salty sweat, liver and pancreas failure, and a battery of other symptoms. All are pleiotropic effects of a single defect, a mutation in a gene that encodes a chloride ion transmembrane channel. In sickle cell anemia patients, a defect in the oxygen-carrying hemoglobin molecule causes anemia, heart failure, increased susceptibility to pneumonia, kidney failure, enlargement of the spleen, and many other symptoms. It is usually difficult to deduce the nature of the primary defect from the range of a gene's pleiotropic effects.

Incomplete Dominance

Not all alternative alleles are fully dominant or fully recessive in heterozygotes. Instead, some pairs of alleles produce a heterozygous phenotype that is either intermediate between those of the parents (incomplete dominance) or representative of both parental phenotypes (codominance, discussed in section 13.2). For example, in the cross of red and white flowering Japanese four o'clocks described in figure 13.17, all the F_1 offspring had pink flowers—indicating that neither red nor white flower color was dominant. Does this example of incomplete dominance argue that Mendel was wrong? Not at all. When two of the F_1 pink flowers were crossed, they produced red-, pink-, and white-flowered plants in a 1:2:1 ratio. Heterozygotes are simply intermediate in color.

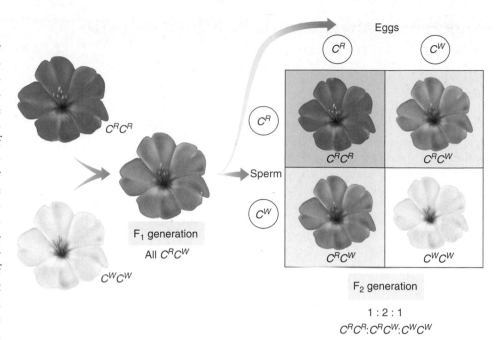

F_1 generation

All C^RC^W

Eggs

Sperm

F_2 generation

1 : 2 : 1

C^RC^R:C^RC^W:C^WC^W

FIGURE 13.17

Incomplete dominance. In a cross between a red-flowered Japanese four o'clock (genotype C^RC^R) and a white-flowered one (C^WC^W), neither allele is dominant. The heterozygous progeny have pink flowers and the genotype C^RC^W. If two of these heterozygotes are crossed, the phenotypes of their progeny occur in a ratio of 1:2:1 (red:pink:white).

Environmental Effects

The degree to which an allele is expressed may depend on the environment. Some alleles are heat-sensitive, for example. Traits influenced by such alleles are more sensitive to temperature or light than are the products of other alleles. The arctic foxes in figure 13.18 make fur pigment only when the weather is warm. Similarly, the *ch* allele in Himalayan rabbits and Siamese cats encodes a heat-sensitive version of tyrosinase, one of the enzymes mediating the production of melanin, a dark pigment. The ch version of the enzyme is inactivated at temperatures above about 33°C. At the surface of the body and head, the temperature is above 33°C and the tyrosinase enzyme is inactive, while it is more active at body extremities such as the tips of the ears and tail, where the temperature is below 33°C. The dark melanin pigment this enzyme produces causes the ears, snout, feet, and tail of Himalayan rabbits and Siamese cats to be black.

(a)

(b)

FIGURE 13.18

Environmental effects on an allele. (*a*) An arctic fox in winter has a coat that is almost white, so it is difficult to see the fox against a snowy background. (*b*) In summer, the same fox's fur darkens to a reddish brown, so that it resembles the color of the surrounding tundra. Heat-sensitive alleles control this color change.

Epistasis

In the tests of Mendel's ideas that followed the rediscovery of his work, scientists had trouble obtaining Mendel's simple ratios, particularly with dihybrid crosses (when individuals heterozygous for two different genes mate). In a dihybrid cross, four different phenotypes are possible among the progeny: Offspring may display the dominant phenotype for both genes, for either one of the genes, or for neither gene. Sometimes, however, it is not possible for an investigator to identify successfully each of the four phenotypic classes, because two or more of the classes look alike. Such situations proved confusing to investigators following Mendel.

One example of such difficulty in identification is seen in the analysis of particular varieties of corn, *Zea mays*. Some commercial varieties exhibit a purple pigment called anthocyanin in their seed coats, while others do not. In 1918, geneticist R. A. Emerson crossed two pure-breeding corn varieties, neither exhibiting anthocyanin pigment. Surprisingly, all of the F_1 plants produced purple seeds.

When two of these pigment-producing F_1 plants were crossed to produce an F_2 generation, 56% were pigment producers and 44% were not. What was happening? Emerson correctly deduced that two genes were involved in producing pigment, and that the second cross had thus been a dihybrid cross. Mendel had predicted 16 equally possible ways gametes could combine. How many of these were in each of the two types Emerson obtained? He multiplied the fraction that were pigment producers (0.56) by 16 to obtain 9, and multiplied the fraction that were not (0.44) by 16 to obtain 7. Thus, Emerson had a **modified ratio** of 9:7 instead of the usual 9:3:3:1 ratio.

Why Was Emerson's Ratio Modified?   When genes act sequentially, as in a biochemical pathway, an allele expressed as a defective enzyme early in the pathway blocks the flow of material through the rest of the pathway. This makes it impossible to judge whether the later steps of the pathway are functioning properly. Such gene interaction, in which one gene can interfere with the expression of another gene, is the basis of the phenomenon called **epistasis**.

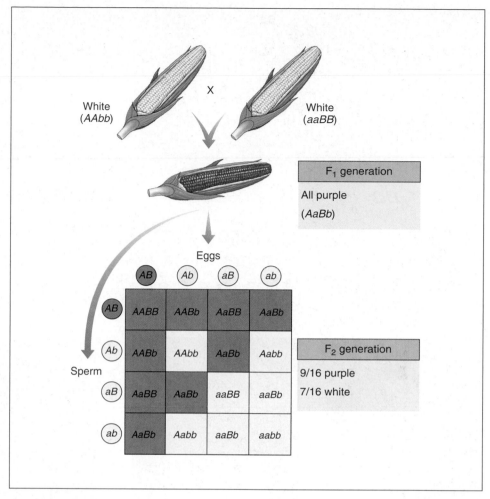

FIGURE 13.19

How epistasis affects grain color.   The purple pigment found in some varieties of corn is the product of a two-step biochemical pathway. Unless both enzymes are active (the plant has a dominant allele for each of the two genes, $A\_B\_$), no pigment is expressed.

The pigment anthocyanin is the product of a two-step biochemical pathway:

$$\text{starting molecule} \xrightarrow{\text{enzyme 1}} \text{intermediate} \xrightarrow{\text{enzyme 2}} \text{anthocyanin}$$
$$\text{(colorless)} \qquad\qquad \text{(colorless)} \qquad\qquad \text{(Purple)}$$

To produce pigment, a plant must possess at least one functional copy of each enzyme gene (figure 13.19). The dominant alleles encode functional enzymes, but the recessive alleles encode nonfunctional enzymes. Of the 16 genotypes predicted by random assortment, 9 contain at least one dominant allele of both genes; they produce purple progeny. The remaining 7 genotypes lack dominant alleles at either or both loci (3 + 3 + 1 = 7) and so are phenotypically the same (nonpigmented), giving the phenotypic ratio of 9:7 that Emerson observed. The inability to see the effect of enzyme 2 when enzyme 1 is nonfunctional is an example of epistasis.

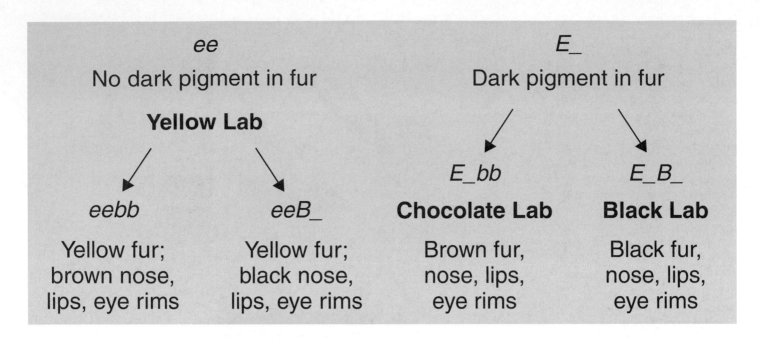

ee No dark pigment in fur **Yellow Lab**		*E_* Dark pigment in fur	
eebb	*eeB_*	*E_bb* **Chocolate Lab**	*E_B_* **Black Lab**
Yellow fur; brown nose, lips, eye rims	Yellow fur; black nose, lips, eye rims	Brown fur, nose, lips, eye rims	Black fur, nose, lips, eye rims

FIGURE 13.20
The effect of epistatic interactions on coat color in dogs. The coat color seen in Labrador retrievers is an example of the interaction of two genes, each with two alleles. The *E* gene determines if the pigment will be deposited in the fur, and the *B* gene determines how dark the pigment will be.

Other Examples of Epistasis

In many animals, coat color is the result of epistatic inter-actions among genes. Coat color in Labrador retrievers, a breed of dog, is due primarily to the interaction of two genes. The *E* gene determines whether dark pigment (eu-melanin) will be deposited in the fur or not. If a dog has the genotype *ee*, no pigment will be deposited in the fur, and it will be yellow. If a dog has the genotype *EE* or *Ee* (*E_*), pigment will be deposited in the fur.

A second gene, the *B* gene, determines how dark the pigment will be. This gene controls the distribution of melanosomes in a hair. Dogs with the genotype *E__bb* will have brown fur and are called chocolate labs. Dogs with the genotype *E__B__* will have black fur. But, even in yellow dogs, the *B* gene does have some effect. Yellow dogs with

the genotype *eebb* will have brown pigment on their nose, lips, and eye rims, while yellow dogs with the genotype *eeB__* will have black pigment in these areas. The interac-tion among these alleles is illustrated in figure 13.20. The genes for coat color in this breed have been found, and a genetic test is available to determine the coat colors in a lit-ter of puppies.

A variety of factors can disguise the Mendelian segregation of alleles. Among them are the continuous variation that results when many genes contribute to a trait, incomplete dominance that produces heterozygotes unlike either parent, environmental influences on the expression of phenotypes, and gene interactions that produce epistasis.

13.2 Human genetics follows Mendelian principles.

Random mutations are constantly occurring in the genes of any population. These gene changes rarely improve the functioning of the proteins encoded by genes, but they are the source of all new alleles. As we have learned more about the genomes of humans and other organisms, a key lesson has been the amount of variation that exists. The database of human mutations, the Human Gene Mutation Database, contains a catalog of 1163 mutant genes that result in clinical manifestations—some 3% of the genome. The vast majority of human mutations have not been studied because they do not produce clinical symptoms. Similar arrays of mutations occur in other organisms.

Most Gene Disorders Are Rare

Although the preceding description may make it seem that humans are largely at risk for genetic diseases of all kinds, the reality is that genetic diseases are actually rare. This is because even though many possible disease alleles may be present in a population, the actual frequency with which any one allele occurs is very low. To examine this, consider *Tay-Sachs disease*, which is an incurable genetic disorder affecting the nervous system. Affected children are normal at birth, first developing symptoms at about 8 months of age when mental deterioration appears. As degeneration progresses, blindness and death follow, usually by age 5. The disease is caused by a single Mendelian recessive allele. The Tay-Sachs allele encodes a mutant form of the enzyme hexosaminidase A that is involved in the breakdown of gangliosides, a class of lipids found in lysosomes in brain cells. The loss of the ability to break down this class of lipids leads to their buildup in lysosomes until the lysosomes burst, releasing their toxic contents into the cell and killing it.

Tay-Sachs is rare in most populations, occurring in only 1 in 300,000 live births in the United States. However, the disease has a high incidence among Ashkenazi Jews, those who are originally from eastern and central Europe; this group includes many of the Jewish people in the United States. In these populations, the incidence rises to as high as 1 in 3500, with an estimated 1 in 28 who actually carry the Tay-Sachs allele. This illustrates two important points about genetic diseases: Their frequencies can vary in different populations with different histories, and even lethal recessive alleles are not entirely removed from a population by natural selection acting against them. The Tay-Sachs allele remains in the population because of a threshold effect: The heterozygote has 50% enzyme activity remaining, compared to the almost complete lack of activity in the homozygous Tay-Sachs individual. This level of enzyme activity is enough to prevent the buildup of gangliosides and subsequent cell death in heterozygotes.

Not All Gene Defects Are Recessive

Most genetic disorders are recessive. A recent cataloging of human disease-causing genes according to biological function sheds some light on this tendency: Enzyme-encoding genes are the largest category by far (over 30% of some 900 genes analyzed). In the case of many enzymes, the amount of residual activity left from a single good allele in a heterozygote is enough to get the job done, as is seen with Tay-Sachs. These mutant alleles are thus recessive.

Despite this tendency, some genetic diseases are due to the action of dominant alleles. *Huntington disease* is a genetic disorder caused by a dominant allele that encodes a variant form of a protein (called huntingtin) found in nerve cells of the brain. In individuals who carry the Huntington allele, these neurons are gradually lost by a mechanism that is thought to be due to deposition of fragments of the altered protein. This leads to neural degeneration, and eventually to death. The search for the Huntington gene took many years of work by many researchers. The approach concentrated on a population in Venezuela with a high incidence of the disease allele and used genetic mapping techniques. It led to a surprise. When the mutant allele was located and analyzed, it proved to be the first case of a new kind of mutation—a set of three bases (CAG) that were duplicated many times over relative to the nonmutant allele. Now known as "triplet expansion," these mutations are discussed in chapter 20.

About 1 in 24,000 individuals in the United States develops Huntington disease. Because it is caused by a dominant allele that is lethal, this raises an obvious question: How can a dominant lethal allele be maintained in a population? In the case of Huntington, it is due to the age of onset of the disease. Individuals carrying the Huntington allele do not show any clinical symptoms until early middle age. Therefore, an individual who is carrying the dominant lethal allele has ample time to reproduce. Given the dominant nature of inheritance, any child of a Huntington sufferer has a 50% probability of receiving the Huntington allele from the affected parent. This can lead to an ethical dilemma for a carrier. In a famous case, Woody Guthrie died of Huntington, and thus his son Arlo, not knowing if he was a carrier, decided not to have children as a young man. We can now test for the presence of the most common Huntington allele, which raises an entirely different dilemma for a person faced with the possibility of carrying the allele: Do you want to know, and if you do, how will it affect your behavior? We will return to the idea of genetic testing later in this chapter.

Most gene defects are rare recessives, although some are dominant.

Human Chromosomes

Each human somatic cell normally has 46 chromosomes, which in meiosis form 23 pairs. By convention, the chromosomes are divided into seven groups (designated A through G), each characterized by a different size, shape, and appearance. The differences among the chromosomes are most clearly visible when the chromosomes are arranged in order in a karyotype (figure 13.34). Techniques that stain individual segments of chromosomes with different-colored dyes make the identification of chromosomes unambiguous. Like a fingerprint, each chromosome always exhibits the same pattern of colored bands.

Human Sex Chromosomes

Of the 23 pairs of human chromosomes, 22 are perfectly matched in both males and females and are called **autosomes**. The remaining pair, the **sex chromosomes**, consist of two similar chromosomes in females and two dissimilar chromosomes in males. In humans, females are designated XX and males XY. One of the sex chromosomes in the male (the Y chromosome) is highly condensed. Because few genes on the Y chromosome are expressed, recessive alleles on a male's single X chromosome have no *active* counterpart on the Y chromosome. Some of the active genes on the Y chromosome are responsible for the features associated with "maleness" in humans. Consequently, any individual with *at least* one Y chromosome is a male.

Sex Chromosomes in Other Organisms

The structure and number of sex chromosomes vary in different organisms (table 13.4). In the fruit fly, *Drosophila*, females are XX and males XY, as in humans and most other vertebrates. However, in birds, the male has two Z chromosomes, and the female has a Z and a W chromosome. Some insects, such as grasshoppers, have no Y chromosome—females are XX and males are characterized as XO (the O indicating the absence of a chromosome).

Sex Determination

In humans, a specific gene on the Y chromosome known as *SRY* plays a key role in the development of male sexual characteristics. This gene is expressed early in development, and acts to masculinize genitalia and secondary sexual organs that would otherwise be female. Lacking a Y chromosome, females fail to undergo these changes.

Among fishes and some species of reptiles, environmental factors can cause changes in the expression of this sex-determining gene and thus in the sex of the adult individual.

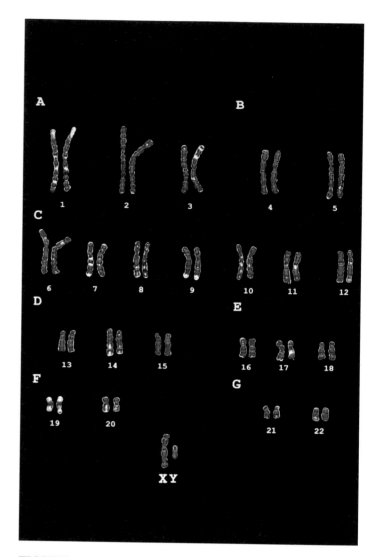

FIGURE 13.34
A human karyotype. This karyotype shows the colored banding patterns, arranged by class, A–G.

Table 13.4 Sex Determination in Some Organisms		Female	Male
Humans, *Drosophila*		XX	XY
Birds		ZW	ZZ
Grasshoppers		XX	XO
Honeybees		Diploid	Haploid

Barr Bodies

Although males have only one copy of the X chromosome and females have two, female cells do not produce twice as much of the proteins encoded by genes on the X chromosome. Instead, one of the X chromosomes in females is inactivated early in embryonic development, shortly after the embryo's sex is determined. Which X chromosome is inactivated varies randomly from cell to cell. If a woman is heterozygous for a sex-linked trait, some of her cells will express one allele and some the other. The inactivated and highly condensed X chromosome is visible as a darkly staining **Barr body** attached to the nuclear membrane.

Why Are X and Y Chromosomes so Different?

Like the other 22 pairs of human chromosomes, the X chromosome is packed with some 1000 genes. Biologists surmise that the reason females possess two copies of the X chromosome and of autosomes is to allow for the repair of the inevitable gene damage that occurs over time due to wear-and-tear, chemical damage, and mistakes in copying.

How can a cell detect and edit out a mutation involving only one or a few nucleotides in one strand of DNA? How does it know which of the two strands is the "correct" version and which the altered one? This neat trick is achieved by every cell having two nearly-identical copies of each chromosome. By comparing the two versions with each other, a cell can identify the "typos" and fix them.

Here is how it works: When a cell detects a chromosomal DNA duplex with a difference between its two DNA strands, that duplex is "repaired" by the rather Draconian expedient of chopping out the entire region, on both strands of the DNA molecule. No effort is made by the cell to determine which strand is correct—both are discarded. The gap that this creates is filled by copying off the sequence present at that region on the other chromosome, a process called **gene conversion.** All this editing happens when the two versions of the chromosome are paired closely together in the early stages of meiosis.

Most biologists think the need for DNA repair is the reason sex evolved. Asexual organisms, which do not undergo meiosis, accumulate mutational damage in a process of irreversible genetic decay that population geneticists call Muller's Ratchet, a progressive loss of genes that can lead to eventual extinction (see chapter 12).

So what are we to make of males? Males, in contrast to females, have only one copy of the X chromosome, not two. The other chromosome of the pair in males, the Y chromosome, is much smaller than the X. Biologists thought until very recently that the Y chromosome had only a few active genes. Because there is no other Y to serve as a pairing partner in meiosis, most of its genes had been thought to have decayed, the victims of Muller's ratchet, leaving the Y chromosome a genetic wasteland with only a very few active genes surviving on it.

We now know this view to have been way too simple. In June of 2003, researchers reported the full gene sequence of the human Y chromosome, and it was nothing like biologists had expected. The human Y chromosome contains not one or two active genes, but 78! One is concerned with male development, most of the others with sperm production and fertility. A few have no obvious role in sex. One of this latter group makes a component of ribosomes (complex tiny engines in the cell which assemble proteins), meaning that every ribosome in a man's body is slightly different from those in a woman's.

Taking all these genes into account, geneticists conclude that men and women differ by 1 to 2 percent of their genomes—which is the same as the difference between a man and a male chimpanzee (or a woman and a female chimpanzee). So we are going to have to reexamine the basis of the differences between the sexes. A lot more of it may be built into the genes than we had supposed.

The Y chromosome is much smaller than the X, and can only pair up with the X at the tips. Thus, there can be no close pairing between X and Y during meiosis, the sort of pairing that allows the proofreading and editing discussed above. We can now see that there is a very good reason evolution has acted to prevent the close pairing of X and Y—those 78 Y chromosome genes. As close pairing allows the exchange of large as well as small segments, any association of X and Y would lead to gene swapping, and the male-determining genes of the Y chromosome would sneak into the X, making everybody male.

One mystery remains. If the Y chromosome cannot pair with the X chromosome, how does it make do without copy-editing to prevent the accumulation of mutations? Why hasn't Muller's ratchet long ago driven males to extinction? The answer is right there for us to see in the Y chromosome sequence, and an elegant answer it is. Most of the 78 active genes on the Y chromosome lie within eight vast palindromes, regions of the DNA sequence that repeat the same sequence twice, running in opposite directions, like the sentence "Madam, I'm Adam," or Napoleon's quip "Able was I ere I saw Elba."

A palindrome has a very neat property: it can bend back on itself, forming a hairpin in which the two strands are aligned with nearly identical DNA sequences. This is the same sort of situation—alignment of nearly identical stretches of chromosomes—which permits the copy-edit of the X chromosome during meiosis. Thus, in the Y chromosome, mutations can be "corrected" by conversion to the undamaged sequence preserved on the other arm of the palindrome. Damage does not accumulate, Muller's ratchet is avoided, and males persist.

In humans, sex is determined by the presence of the Y chromosome, which contains at least 78 transcribed genes.

Human Abnormalities Due to Alterations in Chromosome Number

The failure of homologues or sister chromatids to separate properly during meiosis is called **nondisjunction.** This leads to gametes with the gain or loss of a chromosome, a condition called **aneuploidy.** The frequency of aneuploidy in humans is surprisingly high, being estimated to occur in 5% of conceptions.

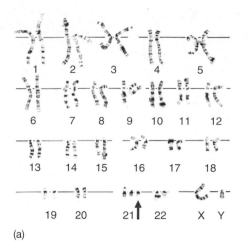

(a)   (b)

FIGURE 13.35
Down syndrome.   (*a*) As shown in this male karyotype, Down syndrome is associated with trisomy of chromosome 21. (*b*) A child with Down syndrome sitting on his father's knee.

Nondisjunction Involving Autosomes

Almost all humans of the same sex have the same karyotype, for the same reason that all automobiles have engines, transmissions, and wheels: Other arrangements don't work well. Humans who have lost even one copy of an autosome are called **monosomics** and do not survive development. In all but a few cases, humans who have gained an extra autosome (called **trisomics**) also do not survive. Data from clinically recognized spontaneous abortions indicate levels of aneuploidy as high as 35%. However, five of the smallest autosomes—those numbered 13, 15, 18, 21, and 22—can be present in humans as three copies and still allow the individual to survive for a time. The presence of an extra chromosome 13, 15, or 18 causes severe developmental defects, and infants with such a genetic makeup die within a few months. In contrast, individuals who have an extra copy of chromosome 21 or, more rarely, chromosome 22, usually survive to adulthood. In these people, the maturation of the skeletal system is delayed, so they generally are short and have poor muscle tone. Their mental development is also affected, and children with trisomy 21 or trisomy 22 are always mentally retarded.

Down Syndrome.   The developmental defect produced by trisomy 21 (figure 13.35*a*) was first described in 1866 by J. Langdon Down; for this reason, it is called **Down syndrome** (formerly "Down's syndrome"). About 1 in every 750 children exhibits Down syndrome, and the frequency is comparable in all racial groups. Similar conditions also occur in chimpanzees and other related primates. In humans, the defect occurs when a particular small portion of chromosome 21 is present in three copies instead of two. In 97% of the human cases examined, all of chromosome 21 is present in three copies. In the other 3%, a small portion of chromosome 21 containing the critical segment has been added to another chromosome by a process called *translocation* (see chapter 20); it exists along with the normal two copies of chromosome 21. This condition is known as *translocation Down syndrome.*

Not much is known about the developmental role of the genes whose extra copies produce Down syndrome, although clues are beginning to emerge from current research. Some researchers suspect that the gene or genes that produce Down syndrome are similar or identical to some of the genes associated with cancer and Alzheimer disease. The reason for this suspicion is that one of the human cancer-causing genes (to be described in chapter 20) and the gene causing Alzheimer disease are located on the segment of chromosome 21 associated with Down syndrome. Moreover, cancer is more common in children with Down syndrome. The incidence of leukemia, for example, is 11 times higher in children with Down syndrome than in unaffected children of the same age.

How does Down syndrome arise? In humans, it comes about almost exclusively as a result of primary nondisjunction of chromosome 21 during egg formation. The cause of these primary nondisjunctions is not known, but their incidence, like that of cancer, increases with maternal age (figure 13.36). In mothers younger than 20 years of age, the risk of giving birth to a child with Down syndrome is about 1 in 1700; in mothers 20 to 30 years old, the risk is only about 1 in 1400. However, in mothers 30 to 35 years old, the risk rises to 1 in 750, and by age 45, the risk is as high as 1 in 16!

Primary nondisjunctions are far more common in women than in men because all of the eggs a woman will ever produce have developed to the point of prophase in meiosis I by the time she is born. By the time she has children, her eggs are as old as she is. In contrast, men produce new sperm daily. Therefore, there is a much greater chance for problems of various kinds, including those that cause primary nondisjunction, to accumulate over time in the gametes of women than in those of men. For this reason, the age of the mother is more critical than that of the father for couples contemplating childbearing.

Nondisjunction Involving the Sex Chromosomes

Individuals who gain or lose a sex chromosome do not generally experience the severe developmental abnormalities caused by similar changes in autosomes. While such individuals have somewhat abnormal features, they often reach maturity.

The X Chromosome. When X chromosomes fail to separate during meiosis, some of the gametes produced possess both X chromosomes, and so are XX gametes; the other gametes that result from such an event have no sex chromosome and are designated "O" (figure 13.37).

If an XX gamete combines with an X gamete, the resulting XXX zygote develops into a female with one functional X chromosome and two Barr bodies. She is sterile but usually normal in other respects. If an XX gamete instead combines with a Y gamete, the effects are more serious. The resulting XXY zygote develops into a sterile male who has many female body characteristics and, in some cases, diminished mental capacity. This condition, called *Klinefelter syndrome*, occurs in about 1 out of every 500 male births.

If an O gamete fuses with a Y gamete, the resulting OY zygote is nonviable and fails to develop further because humans cannot survive when they lack the genes on the X chromosome. If, on the other hand, an O gamete fuses with an X gamete, the XO zygote develops into a sterile female of short stature, with a webbed neck and sex organs that never fully mature during puberty. The mental abilities of an XO individual are in the low-normal range. This condition, called *Turner syndrome*, occurs roughly once in every 5000 female births.

The Y Chromosome. The Y chromosome can also fail to separate in meiosis, leading to the formation of YY gametes. When these gametes combine with X gametes, the XYY zygotes develop into fertile males of normal appearance. The frequency of the XYY genotype (Jacob syndrome) is about 1 per 1000 newborn males, but it is approximately 20 times higher among males in penal and mental institutions. This observation has led to the highly controversial suggestion that XYY males are inherently antisocial, a suggestion supported by some studies but not by others. In any case, most XYY males do not develop patterns of antisocial behavior.

> Gene dosage plays a crucial role in development, so humans do not tolerate the loss or addition of chromosomes well. Autosome loss is always lethal, and an extra autosome is, with few exceptions, lethal too. Additional sex chromosomes have less serious consequences, although they can lead to sterility.

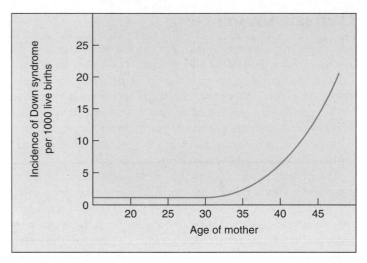

FIGURE 13.36
Correlation between maternal age and the incidence of Down syndrome. As women age, the chances they will bear a child with Down syndrome increase. After a woman reaches 35, the frequency of Down syndrome rises rapidly.
Over a five-year period between ages 20 and 25, the incidence of Down syndrome increases 0.1 per thousand; over a five-year period between ages 35 and 40, the incidence increases to 8.0 per thousand, 80 times as great. The period of time is the same in both instances. What has changed?

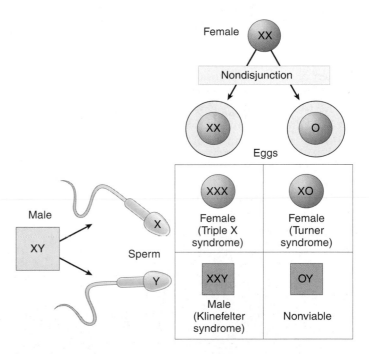

FIGURE 13.37
How nondisjunction can produce abnormalities in the number of sex chromosomes. When nondisjunction occurs in the production of female gametes, the gamete with two X chromosomes (XX) produces Klinefelter males (XXY) and triple-X females (XXX). The gamete with no X chromosome (O) produces Turner females (XO) and nonviable OY males lacking any X chromosome.

Genetic Counseling

Although most genetic disorders cannot yet be cured, we are learning a great deal about them, and progress toward successful therapy is being made in many cases. In the absence of a cure, however, the only recourse is to try to avoid producing children with these conditions. The process of identifying parents at risk for producing children with genetic defects and assessing the genetic state of early embryos is called *genetic counseling.*

If a genetic defect is caused by a recessive allele, how can potential parents determine the likelihood that they carry the allele? One way is through pedigree analysis, often employed as an aid in genetic counseling. By analyzing a person's pedigree, it is sometimes possible to estimate the likelihood that the person is a carrier for certain disorders. For example, if one of your relatives has been afflicted with a recessive genetic disorder such as cystic fibrosis, it is possible that you are a heterozygous carrier of the recessive allele for that disorder. When a couple is expecting a child, and pedigree analysis indicates that both of them have a significant probability of being heterozygous carriers of a recessive allele responsible for a serious genetic disorder, the pregnancy is said to be high-risk. In such cases, there is a significant probability that the child will exhibit the clinical disorder.

Another class of high-risk pregnancy is that in which the mothers are more than 35 years old. As we have seen, the frequency of infants with Down syndrome increases dramatically in the pregnancies of older women (see figure 13.36).

When a pregnancy is diagnosed as high-risk, many women elect to undergo **amniocentesis,** a procedure that permits the prenatal diagnosis of many genetic disorders. In the fourth month of pregnancy, a sterile hypodermic needle is inserted into the expanded uterus of the mother, removing a small sample of the amniotic fluid that bathes the fetus (figure 13.38). Within the fluid are free-floating cells derived from the fetus; once removed, these cells can be grown in cultures in the laboratory. During amniocentesis, the position of the needle and that of the fetus are usually observed by means of *ultrasound.* The sound waves used in ultrasound are not harmful to mother or fetus, and they permit the person withdrawing the amniotic fluid to do so without damaging the fetus. In addition, ultrasound can be used to examine the fetus for signs of major abnormalities.

In recent years, physicians have increasingly turned to a new, less invasive procedure for genetic screening called **chorionic villi sampling.** Using this method, the physician removes cells from the chorion, a membranous part of the placenta that nourishes the fetus. This procedure can be used earlier in pregnancy (by the eighth week) and yields results much more rapidly than does amniocentesis.

To test for certain genetic disorders, genetic counselors look for three characteristics in the cultures of cells obtained from amniocentesis or chorionic villi sampling. First, analysis of the karyotype can reveal aneuploidy (extra or missing chromosomes) and gross chromosomal alterations. Second, in many cases it is possible to test directly for the proper functioning of enzymes involved in genetic disorders. The lack of normal enzymatic activity signals the presence of the disorder. Thus, the lack of the enzyme responsible for breaking down phenylalanine signals PKU (phenylketonuria); the absence of the enzyme responsible for the breakdown of gangliosides indicates Tay-Sachs disease; and so forth.

With the changes in human genetics brought about by the Human Genome Project (see chapter 17), it is possible to design tests for many more diseases. Difficulties still exist in discerning the number and frequency of disease-causing alleles, but these are not insurmountable. At present, there are tests for at least 13 genes with alleles that lead to clinical syndromes. This number is bound to rise and include alleles that do not directly lead to disease states but predispose a person for a particular disease. Chapter 17 treats the many ethical questions that arise.

> **Many gene defects can be detected early in pregnancy, allowing for appropriate planning by the prospective parents.**

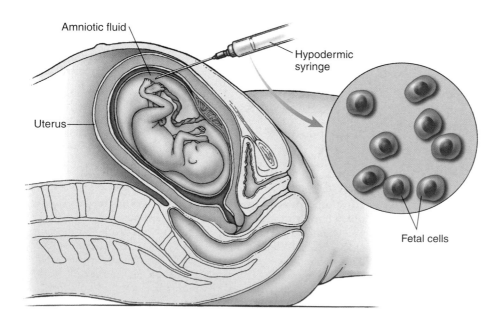

Amniotic fluid

Hypodermic syringe

Uterus

Fetal cells

FIGURE 13.38
Amniocentesis. A needle is inserted into the amniotic cavity, and a sample of amniotic fluid, containing some free cells derived from the fetus, is withdrawn into a syringe. The fetal cells are then grown in culture, and their karyotype and many of their metabolic functions are examined.

For interactive testing, visit the Online Learning Center with PowerWeb at www.mhhe.com/Raven7

13.1 Mendel solved the mystery of heredity.

Early Ideas About Heredity: The Road to Mendel

- Two classical assumptions regarding heredity prior to Mendel were that species were constant and that traits were transmitted directly. (p. 242)
- Early researchers discovered that some inherited characters can disappear and appear again in later generations; some characters can segregate among different offspring; and some forms are more likely to appear than others. (p. 243)

Mendel and the Garden Pea

- Gregor Mendel, an Austrian monk, conducted hybridization studies on garden peas at his monastery. (p. 244)
- Mendel chose garden peas for several reasons: He knew he could produce hybrids by crossing varieties; many varieties were available; and pea plants are easy to grow and have short generation times. In addition, if left alone, the flowers can self-fertilize. (p. 244)
- Mendel conducted his experiment in three stages: (1) He allowed all plants to self-fertilize to produce pure-breeding individuals; (2) he performed crosses between varieties exhibiting alternate character forms; and (3) he allowed hybrid offspring to self-fertilize for several generations. (p. 245)

What Mendel Found

- From his experiments, Mendel derived four conclusions concerning heredity: (1) The plant crosses did not produce progeny of intermediate appearance. (2) For each pair of alternative forms of a character, one alternative was not expressed in the F_1 generation, but did reappear in some F_2 individuals. (3) Pairs of alternative traits segregated among progeny of a particular cross. (4) Alternative traits were expressed in a 3:1 ratio in the F_2 generation, which actually was a 1:2:1 ratio for pure-breeding dominant: non-pure-breeding dominant: pure-breeding recessive. (p. 248)
- Mendel's model of heredity contains five basic elements: (1) Parents do not transmit physiological traits directly to their offspring. (2) Each individual receives two factors that may code for the same trait or two alternative traits. (3) Not all copies of a factor are identical. (4) The alleles from each parent do not influence each other in any way. (5) The presence of a particular allele does not ensure the expression of that trait. (p. 249)

How Mendel Interpreted His Results

- By convention, dominant traits are assigned capital letters and recessive traits lowercase letters, which can easily be put into a Punnett square for analysis. (p. 250)
- Mendel proposed two laws of heredity: the Law of Segregation (alternative alleles segregate independently) and the Law of Independent Assortment (genes on different chromosomes assort independently). (pp. 251–253)

Mendelian Inheritance Is Not Always Easy to Analyze

- Most phenotypes reflect the action of many genes. (p. 255)
- Quantitative traits produce continuous variation. (p. 255)
- Pleiotropic alleles have more than one effect on the phenotype. (p. 255)
- Incomplete dominance occurs when alternative alleles are not fully dominant or recessive, and offspring express both parental phenotypes. (p. 256)

- Environmental effects can affect allele expression to produce varying phenotypes. (p. 256)
- Epistasis can occur when one gene interferes with the expression of another gene. (pp. 257–258)

13.2 Human genetics follows Mendelian principles.

Most Gene Disorders Are Rare

- Most genetic disorders are recessive (e.g., Tay-Sachs), although some are dominant (e.g., Huntington disease). (p. 259)

Multiple Alleles: The ABO Blood Groups

- Some genes, such as those determining ABO blood groups and Rh blood groups, exhibit codominance in which each allele has its own effect. (p. 260)

Patterns of Inheritance Can Be Deduced from Pedigrees

- Family pedigrees can be used to deduce the mode of inheritance of a trait through a family, such as hemophilia in European royal families. (p. 261)

Gene Disorders Can Be Due to Simple Alterations of Proteins

- Sickle cell anemia is caused by a single-nucleotide change in the gene coding for hemoglobin. (p. 262)

Some Defects May Soon Be Curable: Gene Therapy

- Theoretically, some heredity disorders should be able to be cured by inserting a healthy version of a damaged gene into an affected individual. Early attempts have met with varied results. (p. 263)

More Promising Vectors

- New vectors such as AAV may offer greater promise in gene therapy. (p. 264)

13.3 Genes are on chromosomes.

Chromosomes: The Vehicles of Mendelian Inheritance

- Mendelian traits assort independently because the genes determining them are located on chromosomes that assort independently. (p. 266)

Genetic Recombination

- Crossing over creates new genetic combinations. If two different genes are located relatively far apart on a chromosome, crossing over is more likely to occur somewhere between them than if they are located closer together. (p. 267)
- A genetic map can be constructed that measures the distance between genes in terms of the frequency of recombination. (p. 268)

Human Chromosomes

- The twenty-third pair of chromosomes in humans carries the genes that determine sex. The Y chromosome and appears to contain 78 genes. (pp. 270–271)

Human Abnormalities Due to Alterations in Chromosome Number

- Nondisjunction refers to the failure of homologues to separate properly during meiosis and can lead to aneuploidy (incorrect number of chromosomes), which can cause a number of genetic conditions, such as Down syndrome. (p. 272)

Genetic Counseling

- Many genetic abnormalities can be detected early in pregnancy using amniocentesis, chorionic villi sampling, and many newer techniques. (p. 274)

Self Test

1. In order to ensure that he had pure-breeding plants for his experiments, Mendel
   a. cross-fertilized each variety with each other.
   b. let each variety self-fertilize for several generations.
   c. removed the female parts of the plants.
   d. removed the male parts of the plants.
2. When two parents are crossed, the offspring are referred to as the
   a. recessives.
   b. testcross.
   c. F_1 generation.
   d. F_2 generation.
3. A cross between two individuals results in a ratio of 9:3:3:1 for four possible phenotypes. This is an example of a
   a. dihybrid cross.
   b. monohybrid cross.
   c. testcross.
   d. none of these.
4. Human height shows a continuous variation from the very short to the very tall. Height is most likely controlled by
   a. epistatic genes.
   b. environmental factors.
   c. sex-linked genes.
   d. multiple genes.
5. In the human ABO blood grouping, the four basic blood types are type A, type B, type AB, and type O. The blood proteins A and B are
   a. simple dominant and recessive traits.
   b. incomplete dominant traits.
   c. codominant traits.
   d. sex-linked traits.
6. Which of the following describes symptoms of sickle cell anemia?
   a. poor blood circulation due to abnormal hemoglobin molecules
   b. sterility in females
   c. failure of blood to clot
   d. failure of chloride ion transport mechanism
7. What finding finally determined that genes were carried on chromosomes?
   a. heat sensitivity of certain enzymes that determined coat color
   b. sex-linked eye color in fruit flies
   c. the finding of complete dominance
   d. establishing pedigrees
8. A genetic map can be used to determine
   a. the relative position of alleles on chromosomes.
   b. restriction sites on chromosomes.
   c. the frequency of recombination between two genes.
   d. all of these.
9. A Barr body is a(n)
   a. result of primary nondisjunction.
   b. inactivated Y chromosome.
   c. gene that plays a key role in male development.
   d. inactivated X chromosome.
10. Down syndrome in humans is due to
    a. three copies of chromosome 21.
    b. monosomy.
    c. two Y chromosomes.
    d. three X chromosomes.

Test Your Visual Understanding

F$_2$ generation

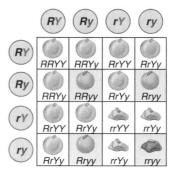

9/16	round, yellow
3/16	round, green
3/16	wrinkled, yellow
1/16	wrinkled, green

1. The figure shows the results of a dihybrid cross in which Mendel was examining the inheritance of two traits: seed shape (R and r) and seed color (Y and y). Consider the results if he had examined three traits. Using these two traits and plant height (T and t), predict:
   a. the genotypes and phenotypes of the parents.
   b. the genotypic and phenotypic ratios of the F_1 generation.
   c. the genotypes of the eggs and sperm.
   d. the phenotypic ratios of the F_2 generation.

Apply Your Knowledge

1. Phenylketonuria (PKU) is a genetic disorder caused by the mutation of an enzyme, phenylalanine hydroxylase, that breaks down phenylalanine in the cell. If identified at birth, dietary restrictions can control the disease. If not identified, buildup of phenylalanine interferes with brain development. The frequency of PKU in the population is 1 in 12,000 births. How many people with PKU would you expect to find in a population of 250,000?
2. How might Mendel's results and the model he formulated have been different if the traits he chose to study were governed by alleles exhibiting incomplete dominance or codominance?

Mendelian Genetics Problems

1. The following illustration describes Mendel's cross of *wrinkled* and *round* seed characters. What is wrong with this diagram? (*Hint:* Do you expect all the seeds in a pod to be the same?)

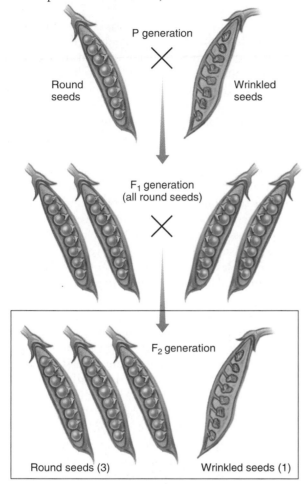

P generation

Round seeds × Wrinkled seeds

F₁ generation (all round seeds)

F₂ generation

Round seeds (3)   Wrinkled seeds (1)

2. The annual plant *Haplopappus gracilis* has two pairs of chromosomes (1 and 2). In this species, the probability that two characters, *a* and *b*, selected at random will be on the same chromosome is equal to the probability that they will both be on chromosome 1 ($\frac{1}{2} \times \frac{1}{2} = \frac{1}{4}$, or 0.25), plus the probability that they will both be on chromosome 2 (also $\frac{1}{2} \times \frac{1}{2} = \frac{1}{4}$, or 0.25), for an overall probability of $\frac{1}{2}$, or 0.5. In general, the probability that two randomly selected characters will be on the same chromosome is equal to $\frac{1}{n}$ where *n* is the number of chromosome pairs. Humans have 23 pairs of chromosomes. What is the probability that any two human characters selected at random will be on the same chromosome?

3. Among Hereford cattle, there is a dominant allele called *polled;* the individuals having this allele lack horns. Suppose you acquire a herd consisting entirely of polled cattle, and you carefully determine that no cow in the herd has horns. Some of the calves born that year, however, grow horns. You remove those calves from the herd and make certain that no horned adult has gotten into your pasture. Despite your efforts, more horned calves are born the next year. What is the reason for the appearance of the horned calves? If your goal is to maintain a herd consisting entirely of polled cattle, what should you do?

4. An inherited trait among humans in Norway causes affected individuals to have very wavy hair, not unlike that of a sheep. The trait, called *woolly*, is very evident when it occurs in families; no child possesses woolly hair unless at least one parent does. Imagine that you are a Norwegian judge, and a woolly-haired man comes before you to sue his normal-haired wife for divorce because their first child has woolly hair but their second child has normal hair. The husband claims this constitutes evidence of his wife's infidelity. Do you accept his claim? Justify your decision.

5. In human beings, Down syndrome, a serious developmental abnormality, results from the presence of three copies of chromosome 21 rather than the usual two copies. If a female exhibiting Down syndrome mates with a normal male, what proportion of her offspring would you expect to be affected?

6. Many animals and plants bear recessive alleles for *albinism*, a condition in which homozygous individuals lack certain pigments. An albino plant, for example, lacks chlorophyll and is white, and an albino human lacks melanin. If two normally pigmented persons heterozygous for the same albinism allele marry, what proportion of their children would you expect to be albino?

7. You inherit a racehorse and decide to put him out to stud. In looking over the studbook, however, you discover that the horse's grandfather exhibited a rare disorder that causes brittle bones. The disorder is hereditary and results from homozygosity for a recessive allele. If your horse is heterozygous for the allele, it will not be possible to use him for stud because the genetic defect may be passed on. How would you determine whether your horse carries this allele?

8. In the fruit fly *Drosophila*, the allele for dumpy wings (*d*) is recessive to the normal long-wing allele (*d⁺*), and the allele for white eye (*w*) is recessive to the normal red-eye allele (*w⁺*). In a cross of $d^+d^+w^+w \times d^+dww$, what proportion of the offspring are expected to be "normal" (long wings, red eyes)? What proportion are expected to have dumpy wings and white eyes?

9. Your instructor presents you with a *Drosophila* with red eyes, as well as a stock of white-eyed flies and another stock of flies homozygous for the red-eye allele. You know that the presence of white eyes in *Drosophila* is caused by homozygosity for a recessive allele. How would you determine whether the single red-eyed fly was heterozygous for the white-eye allele?

10. Some children are born with recessive traits (and, therefore, must be homozygous for the recessive allele specifying the trait), even though neither of the parents exhibits the trait. What can account for this?

11. You collect two individuals of *Drosophila*, one a young male and the other a young, unmated female. Both are normal in appearance, with the red eyes typical of *Drosophila*. You keep the two flies in the same bottle, where they mate. Two weeks later, the offspring they have produced all have red eyes. From among the offspring, you select 100 individuals, some male and some female. You cross each individually with a fly you know to be homozygous for the recessive allele *sepia*, which produces black eyes when homozygous. Examining the results of your 100 crosses, you observe that in about half of the crosses, only red-eyed flies were produced. In the other half, however, the progeny of each cross consists of about 50% red-eyed flies and 50% black-eyed flies. What were the genotypes of your original two flies?

12. Hemophilia is a recessive sex-linked human blood disease that leads to failure of blood to clot normally. One form of hemophilia has been traced to the royal family of England, from which it spread throughout the royal families of Europe. For the purposes of this problem, assume that it originated as a mutation either in Prince Albert or in his wife, Queen Victoria.

  a. Prince Albert did not have hemophilia. If the disease is a sex-linked recessive abnormality, how could it have originated in Prince Albert, a male, who would have been expected to exhibit sex-linked recessive traits?

  b. Alexis, the son of Czar Nicholas II of Russia and Empress Alexandra (a granddaughter of Victoria), had hemophilia, but their daughter Anastasia did not. Anastasia died, a victim of the Russian revolution, before she had any children. Can we assume that Anastasia would have been a carrier of the disease? Would your answer be different if the disease had been present in Nicholas II or in Alexandra?

13. In 1986, *National Geographic* magazine conducted a survey of its readers' abilities to detect odors. About 7% of Caucasians in the United States could not smell the odor of musk. If neither parent could smell musk, none of their children were able to smell it. On the other hand, if the two parents could smell musk, their children generally could smell it too, but a few of the children in those families were unable to smell it. Assuming that a single pair of alleles governs this trait, is the ability to smell musk best explained as an example of dominant or recessive inheritance?

14. A couple with a newborn baby is troubled that the child does not resemble either of them. Suspecting that a mixup occurred at the hospital, they check the blood type of the infant. It is type O. Because the father is type A and the mother type B, they conclude that a mixup must have occurred. Are they correct?

15. Mabel's sister died of cystic fibrosis as a child. Mabel does not have the disease, and neither do her parents. Mabel is pregnant with her first child. If you were a genetic counselor, what would you tell her about the probability that her child will have cystic fibrosis?

16. How many chromosomes would you expect to find in the karyotype of a person with Turner syndrome?

17. A woman is married for the second time. Her first husband has blood type A, and her child by that marriage has type O. Her new husband has type B blood, and when they have a child, its blood type is AB. What is the woman's blood genotype and blood type?

18. Two intensely freckled parents have five children. Three eventually become intensely freckled, and two do not. Assuming this trait is governed by a single pair of alleles, is the expression of intense freckles best explained as an example of dominant or recessive inheritance?

19. Total color blindness is a rare hereditary disorder among humans. Affected individuals can see no colors, only shades of gray. It occurs in individuals homozygous for a recessive allele, and it is not sex-linked. A man whose father is totally color blind intends to marry a woman whose mother is totally color blind. What are the chances they will produce offspring who are totally color blind?

20. A normally pigmented man marries an albino woman. They have three children, one of whom is an albino. What is the genotype of the father?

21. Four babies are born in a hospital, and each has a different blood type: A, B, AB, and O. The parents of these babies have the following pairs of blood groups: A and B, O and O, AB and O, and B and B. Which baby belongs to which parents?

22. A couple both work in an atomic energy plant, and both are exposed daily to low-level background radiation. After several years, they have a child who has Duchenne muscular dystrophy, a recessive genetic defect caused by a mutation on the X chromosome. Neither the parents nor the grandparents have the disease. The couple sues the plant, claiming that the abnormality in their child is the direct result of radiation-induced mutation of their gametes, and that the company should have protected them from this radiation. Before reaching a decision, the judge hearing the case insists on knowing the sex of the child. Which sex would be more likely to result in an award of damages, and why?

14

DNA: The Genetic Material

Concept Outline

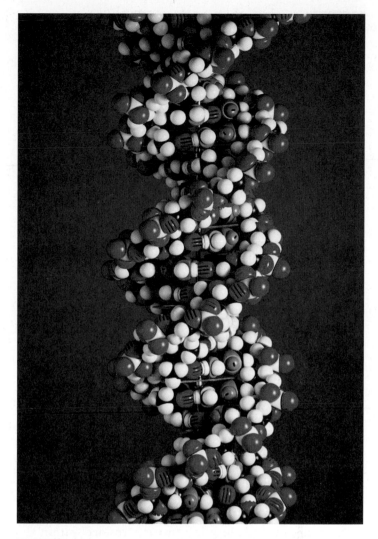

FIGURE 14.1
DNA. The hereditary blueprint in each cell of all living
organisms is a very long, slender molecule called deoxyribonucleic
acid (DNA).

The realization that patterns of heredity can be explained by the segregation of chromosomes in meiosis raised a question that occupied biologists for over 50
years: What is the exact nature of the connection between
hereditary traits and chromosomes? This chapter describes the chain of experiments that have led to our current understanding of DNA and the molecular mechanisms of heredity (figure 14.1). These experiments are
among the most elegant in science. Just as in a good detective story, each discovery has led to new questions. The
intellectual path has not always been a straight one, the
best questions not always obvious. But however erratic
and lurching the course of the experimental journey, our
picture of heredity has become progressively clearer, the
image more sharply defined.

14.1 What is the genetic material?

The Hammerling Experiment: Cells Store Hereditary Information in the Nucleus

Perhaps the most basic question we can ask about hereditary information is where it is stored in the cell. To answer this question, Danish biologist Joachim Hammerling, working at the Max Planck Institute for Marine Biology in Berlin in the 1930s, cut cells into pieces and observed the pieces to see which were able to express hereditary information. For this experiment, Hammerling needed cells large enough to operate on conveniently and differentiated enough to distinguish the pieces. He chose the unicellular green alga *Acetabularia*, which grows up to 5 centimeters, as a model organism for his investigations. Just as Mendel used pea plants and Sturtevant used fruit flies, Hammerling picked a model organism that was suited to the specific experimental question he wanted to answer, assuming that what he learned could then be applied to other organisms.

Individuals of the genus *Acetabularia* have distinct foot, stalk, and cap regions; all are differentiated parts of a single cell. The nucleus is located in the foot. As a preliminary experiment, Hammerling amputated the caps of some cells and the feet of others. He found that when he amputated the cap, a new cap regenerated from the remaining portions of the cell (foot and stalk). When he amputated the foot, however, no new foot regenerated from the cap and stalk. Hammerling, therefore, hypothesized that the hereditary information resided within the foot of *Acetabularia*.

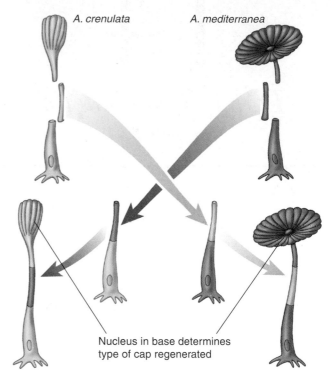

FIGURE 14.2
Hammerling's *Acetabularia* reciprocal graft experiment. Hammerling grafted a stalk of each species of *Acetabularia* onto the foot of the other species. In each case, the cap that eventually developed was dictated by the nucleus-containing foot rather than by the stalk.

Surgery on Single Cells

To test his hypothesis, Hammerling selected individuals from two species of the genus *Acetabularia* in which the caps look very different from one another. *A. mediterranea* has a disk-shaped cap, and *A. crenulata* has a branched, flowerlike cap. Hammerling grafted a stalk from *A. crenulata* to a foot from *A. mediterranea* (figure 14.2). The cap that regenerated looked somewhat like the cap of *A. crenulata*, though not exactly the same.

Hammerling then cut off this regenerated cap and found that a disk-shaped cap exactly like that of *A. mediterranea* formed in the second regeneration and in every regeneration thereafter. This experiment supported Hammerling's hypothesis that the instructions specifying the kind of cap are stored in the foot of the cell, and that these instructions must pass from the foot through the stalk to the cap.

In this experiment, the initial flower-shaped cap was somewhat intermediate in shape, unlike the disk-shaped caps of subsequent generations. Hammerling speculated that this initial cap, which resembled that of *A. crenulata*, was formed from instructions already present in the transplanted stalk when it was excised from the original *A. crenulata* cell. In contrast, all of the caps that subsequently regenerated used new information derived from the foot of the *A. mediterranea* cell the stalk had been grafted onto. In some unknown way, the original instructions that had been present in the stalk were eventually "used up." We now understand that genetic instructions (in the form of messenger RNA, discussed in chapter 15) pass from the nucleus in the foot upward *through the stalk* to the developing cap.

Hereditary information in *Acetabularia* is stored in the foot of the cell, where the nucleus resides.

Transplantation Experiments: Each Cell Contains a Full Set of Genetic Instructions

Because the nucleus is in the foot of *Acetabularia*, Hammerling's experiments suggested that the nucleus is the repository of hereditary information in a cell. A direct test of this hypothesis was carried out in 1952 by the American embryologists Robert Briggs and Thomas King. Using a glass pipette drawn to a fine tip and working with a microscope, Briggs and King removed the nucleus from a frog egg. Without the nucleus, the egg did not develop. However, when they replaced the nucleus with one removed from a more advanced frog embryo cell, the egg developed into an adult frog. Clearly, the nucleus was directing the egg's development (figure 14.3).

Successfully Transplanting Nuclei

Can every nucleus in an organism direct the development of an entire adult individual? The experiment of Briggs and King did not answer this question definitively, because the nuclei they transplanted from frog embryos into eggs often caused the eggs to develop abnormally. Two experiments performed soon afterward gave a clearer answer to the question. In the first, John Gurdon, working with another species of frog at Oxford and Yale, transplanted nuclei from tadpole cells into eggs from which the nuclei had been removed. The experiments were difficult—it was necessary to synchronize the division cycles of donor and host. However, in many experiments, the eggs went on to develop normally, indicating that the nuclei of cells in later stages of development retain the genetic information necessary to direct the development of all other cells in an individual.

Totipotency in Plants

In the second experiment, F. C. Steward at Cornell University in 1958 placed small fragments of fully developed carrot tissue (isolated from a part of the vascular system called the phloem) in a flask containing liquid growth medium. Steward observed that when individual cells broke away from the fragments, they often divided and developed into multicellular roots. When he immobilized the roots by placing them in a solid growth medium, they went on to develop normally into entire, mature plants. Steward's experiment makes it clear that, even in adult tissues, the nuclei of individual plant cells are "totipotent," meaning that each contains a full set of hereditary instructions and can generate an entire adult individual. As you will learn in chapter 19, animal cells, like plant cells, can be totipotent, and a single adult animal cell can generate an entire adult animal.

Hereditary information is stored in the nucleus of eukaryotic cells. Each nucleus in any eukaryotic cell contains a full set of genetic instructions.

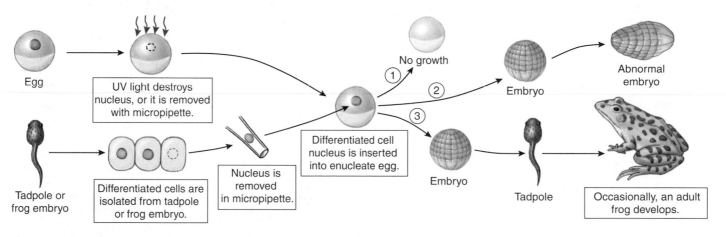

FIGURE 14.3

Nuclear transplant experiments. The nucleus was removed from a frog egg, either by sucking the egg nucleus into a micropipette or, more simply, by destroying it with ultraviolet light. A nucleus obtained from a differentiated cell from a frog embryo or tadpole was then injected into this enucleate egg. The hybrid egg was allowed to develop. One of three results was obtained: (1) No growth occurred, perhaps reflecting damage to the egg cell during the nuclear transplant operation; (2) normal growth and development occurred up to an early embryo stage, but subsequent development was not normal and the embryo did not survive; and (3) normal growth and development occurred, eventually leading to the development of an adult frog. That frog was of the strain that contributed the nucleus and not of the strain that contributed the egg. Only a few experiments gave this third result, but they clearly establish that the nucleus directs frog development.

The Griffith Experiment: Hereditary Information Can Pass Between Organisms

The identification of the nucleus as the repository of hereditary information focused attention on the chromosomes, which were already suspected to be the vehicles of Mendelian inheritance. Specifically, biologists wondered how the **genes,** the units of hereditary information studied by Mendel, were actually arranged in the chromosomes. They knew that chromosomes contained both protein and deoxyribonucleic acid (DNA). Which of these held the genes? Starting in the late 1920s and continuing for about 30 years, a series of investigations addressed this question.

In 1928, the British microbiologist Frederick Griffith made a series of unexpected observations while experimenting with pathogenic (disease-causing) bacteria. When he infected mice with a virulent strain of *Streptococcus pneumoniae* bacteria (then known as *Pneumococcus*), the mice died of blood poisoning. However, when he infected similar mice with a mutant strain of *S. pneumoniae* that lacked the virulent strain's polysaccharide coat, the mice showed no ill effects. The coat was apparently necessary for virulence. The normal pathogenic form of this bacterium is referred to as the S form because it forms smooth colonies on a culture dish. The mutant form, which lacks an enzyme needed to manufacture the polysaccharide coat, is called the R form because it forms rough colonies.

To determine whether the polysaccharide coat itself had a toxic effect, Griffith injected heat-killed bacteria of the virulent S strain into mice; the mice remained perfectly healthy. As a control, he injected mice with a mixture containing heat-killed S bacteria of the virulent strain and live coatless R bacteria, each of which by itself did not harm the mice (figure 14.4). Unexpectedly, the mice developed disease symptoms, and many of them died. The blood of the dead mice was found to contain high levels of live, virulent *Streptococcus* type S bacteria, which had surface proteins characteristic of the live (previously R) strain. Somehow, the information specifying the polysaccharide coat had passed from the dead, virulent S bacteria to the live, coatless R bacteria in the mixture, permanently transforming the coatless R bacteria into the virulent S variety. This transfer of genetic material from one cell to another, called **transformation,** can alter the genetic makeup of the recipient cell.

Hereditary information can pass from dead cells to living ones, transforming them.

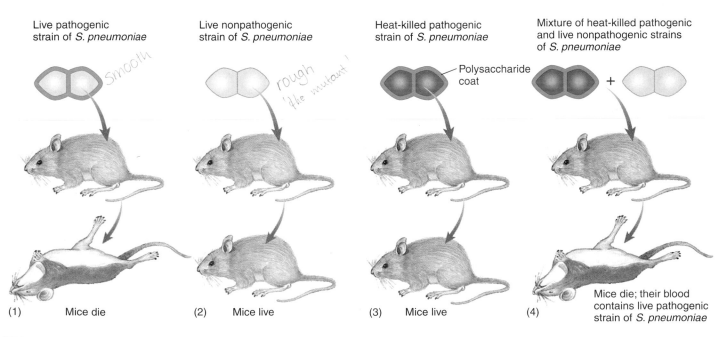

FIGURE 14.4
Griffith's discovery of transformation. (*1*) The pathogenic strain of the bacterium *Streptococcus pneumoniae* kills many of the mice it is injected into. The bacterial cells are covered with a polysaccharide coat, which the bacteria themselves synthesize. (*2*) Interestingly, an injection of live, coatless bacteria produced no ill effects. However, the coat itself is not the agent of disease. (*3*) When Griffith injected mice with dead bacteria that possessed polysaccharide coats, the mice were unharmed. (*4*) But when Griffith injected a mixture of dead bacteria with polysaccharide coats and live bacteria without such coats, many of the mice died, and virulent bacteria with coats were recovered. Griffith concluded that the live cells had been "transformed" by the dead ones; that is, genetic information specifying the polysaccharide coat had passed from the dead cells to the living ones.

The Avery and Hershey–Chase Experiments: The Active Principle Is DNA

The Avery Experiments

The agent responsible for transforming *Streptococcus* went undiscovered until 1944. In a classic series of experiments, Oswald Avery and his co-workers Colin MacLeod and Maclyn McCarty characterized what they referred to as the "transforming principle" from Griffith's experiment. They first prepared the mixture of dead S *Streptococcus* and live R *Streptococcus* that Griffith had used. Then Avery and his colleagues removed as much of the protein as they could from their preparation, eventually achieving 99.98% purity. Despite the removal of nearly all protein, the transforming activity was not reduced. Moreover, the properties of the transforming principle resembled those of DNA in several ways:

1. When the purified principle was analyzed chemically, the array of elements agreed closely with DNA.
2. When spun at high speeds in an ultracentrifuge, the transforming principle migrated to the same level (density) as DNA.
3. Extracting the lipid and protein from the purified transforming principle did not reduce its activity.
4. Protein-digesting enzymes did not affect the principle's activity, nor did RNA-digesting enzymes.
5. The DNA-digesting enzyme DNase destroyed all transforming activity.

The evidence was overwhelming. The researchers concluded that "a nucleic acid of the deoxyribose type is the fundamental unit of the transforming principle of *Pneumococcus* Type III"—in essence, that DNA is the hereditary material for this bacterial species.

The Hershey–Chase Experiment

Avery's results were not widely accepted at first, because many biologists preferred to believe that proteins were the repository of hereditary information. Additional evidence supporting Avery's conclusion was provided in 1952 by Alfred Hershey and Martha Chase, who experimented with **bacteriophages,** viruses that attack bacteria. Viruses, described in more detail in chapter 33, consist of either DNA or RNA (ribonucleic acid) surrounded by a protein coat. When a *lytic* (potentially cell-rupturing) bacteriophage infects a bacterial cell, it first binds to the cell's outer surface and then injects its hereditary information into the cell. There, the hereditary information directs the production of thousands of new viruses within the bacterium. The bacterial cell eventually ruptures, or *lyses*, releasing the newly made viruses.

To identify the hereditary material injected into bacterial cells at the start of an infection, Hershey and Chase used the bacteriophage T2, which contains DNA rather than RNA. They labeled the two parts of the viruses—the DNA and the

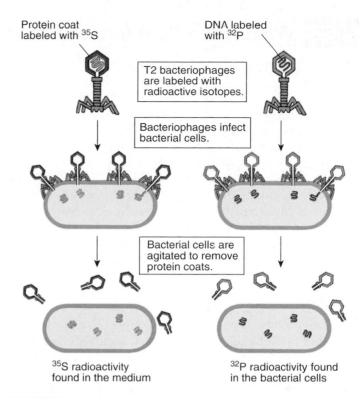

Protein coat labeled with ^{35}S

DNA labeled with ^{32}P

T2 bacteriophages are labeled with radioactive isotopes.

Bacteriophages infect bacterial cells.

Bacterial cells are agitated to remove protein coats.

^{35}S radioactivity found in the medium

^{32}P radioactivity found in the bacterial cells

FIGURE 14.5
The Hershey–Chase experiment. Hershey and Chase found that ^{35}S radioactivity did not enter infected bacterial cells and ^{32}P radioactivity did. They concluded that viral DNA, not protein, was responsible for directing the production of new viruses.

protein coat—with different radioactive isotopes that would serve as tracers. In some experiments, the viruses were grown on a medium containing an isotope of phosphorus, ^{32}P, and the isotope was incorporated into the phosphate groups of newly synthesized DNA molecules. Because phosphorus is not found in protein, this label is specific for DNA. In other experiments, the viruses were grown on a medium containing ^{35}S, an isotope of sulfur, which is incorporated into the amino acids of newly synthesized protein coats. Because sulfur is not found in DNA, this label is specific for protein. The ^{32}P and ^{35}S isotopes are easily distinguished from each other because they emit particles with different energies when they decay.

After the labeled viruses were permitted to infect bacteria, the bacterial cells were agitated violently to remove the protein coats of the infecting viruses from the surfaces of the bacteria. This procedure removed nearly all of the ^{35}S label (and thus nearly all of the viral protein) from the bacteria. However, the ^{32}P label (and thus the viral DNA) had transferred to the interior of the bacteria (figure 14.5) and was found in viruses subsequently released from the infected bacteria. Hence, the hereditary information injected into the bacteria that specified the new generation of viruses was DNA, not protein.

Avery's experiments demonstrate conclusively that DNA is Griffith's transforming material. The hereditary material of bacteriophages is DNA and not protein.

14.2 What is the structure of DNA?

The Chemical Nature of Nucleic Acids

A German chemist, Friedrich Miescher, discovered DNA in 1869, only four years after Mendel's work was published. Miescher extracted a white substance from the nuclei of human cells and fish sperm. The proportion of nitrogen and phosphorus in the substance was different from that in any other known constituent of cells, which convinced Miescher that he had discovered a new biological substance. He called this substance "nuclein," because it seemed to be specifically associated with the nucleus.

Levene's Analysis: DNA Is a Polymer

Because Miescher's nuclein was slightly acidic, it came to be called **nucleic acid.** For 50 years, biologists did little research on the substance because nothing was known of its function in cells. In the 1920s, the basic structure of nucleic acids was determined by the biochemist P. A. Levene, who found that a DNA molecule contains three main components (figure 14.6): (1) a five-carbon sugar; (2) a phosphate (PO_4) group; (3) a nitrogen-containing (nitrogenous) base. The base may be a **purine** (adenine, symbolized as A, or guanine, G) or a **pyrimidine** (thymine, T, or cytosine, C); RNA contains uracil, (U) instead of thymine. From the roughly equal proportions of these components, Levene concluded correctly that DNA and RNA molecules are made of repeating units of the three components. Each unit, consisting of a sugar attached to a phosphate group and a base, is called a **nucleotide.** The identity of the base distinguishes one nucleotide from another.

To identify the various chemical groups in DNA and RNA, it is customary to number the carbon atoms of the base and the sugar and then refer to any chemical group attached to a carbon atom by that number. In the sugar, four of the carbon atoms together with an oxygen atom form a five-membered ring. As illustrated in figure 14.7, the carbon atoms are numbered 1′ to 5′, proceeding clockwise from the oxygen atom; the prime symbol (′) indicates that the number refers to a carbon in a sugar rather than a base. Under this numbering scheme, the phosphate group is attached to the 5′ carbon atom of the sugar, and the base is attached to the 1′ carbon atom. In addition, a free hydroxyl (—OH) group is attached to the 3′ carbon atom.

The 5′ phosphate and 3′ hydroxyl groups allow DNA and RNA to form long chains of nucleotides, because these two groups can react chemically with each other. The reaction between the phosphate group of one nucleotide and the hydroxyl group of another is a dehydration synthesis, eliminating a water molecule and forming a covalent bond

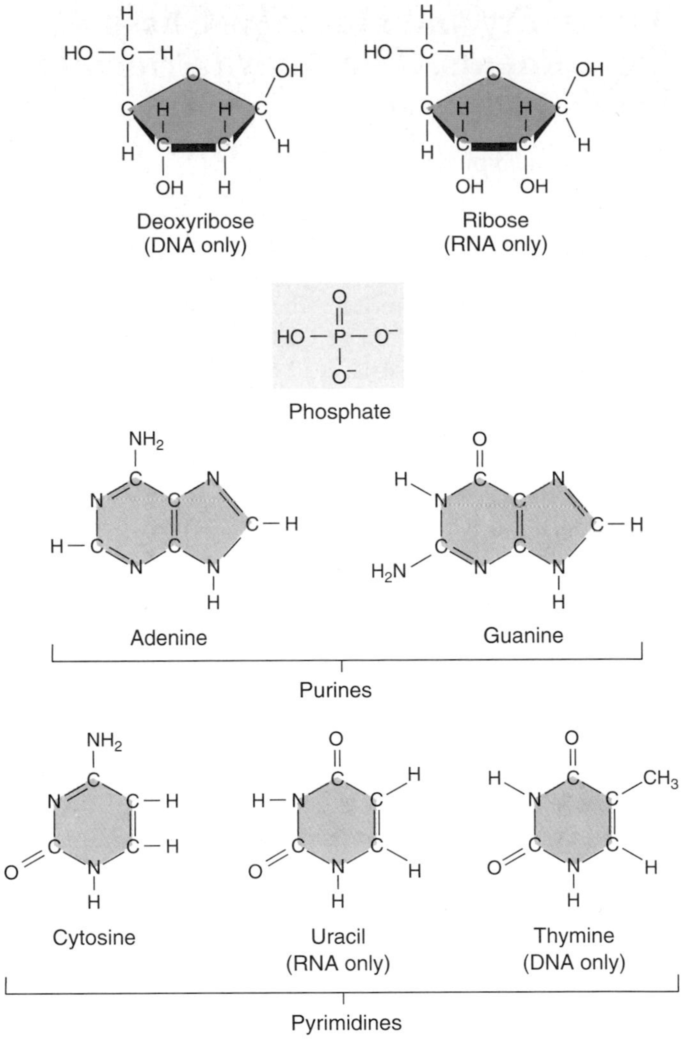

FIGURE 14.6
Nucleotide subunits of DNA and RNA. The nucleotide subunits of DNA and RNA are composed of three elements: (*top*) a five-carbon sugar (deoxyribose in DNA and ribose in RNA); (*middle*) a phosphate group; and (*bottom*) a nitrogenous base (either a purine or a pyrimidine).

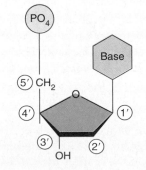

FIGURE 14.7
Numbering the carbon atoms in a nucleotide. The carbon atoms in the sugar of the nucleotide are numbered 1′ to 5′, proceeding clockwise from the oxygen atom. The "prime" symbol (′) indicates that the carbon belongs to the sugar rather than to the base.

Table 14.1 Chargaff's Analysis of DNA Nucleotide Base Compositions				
	Base Composition (mole percent)			
Organism	A	T	G	C
Escherichia coli (K12)	26.0	23.9	24.9	25.2
Mycobacterium tuberculosis	15.1	14.6	34.9	35.4
Yeast	31.3	32.9	18.7	17.1
Herring	27.8	27.5	22.2	22.6
Rat	28.6	28.4	21.4	21.5
Human	30.9	29.4	19.9	19.8

Source: Data from E. Chargaff and J. Davidson (editors), *The Nucleic Acids*, 1955, Academic Press, New York, NY.

that links the two groups. The linkage is called a **phosphodiester bond** because the phosphate group is now linked to the two sugars by means of a pair of ester (P—O—C) bonds (figure 14.8). The two-unit polymer resulting from this reaction still has a free 5′ phosphate group at one end and a free 3′ hydroxyl group at the other, so it can link to other nucleotides. In this way, many thousands of nucleotides can join together in long chains.

Linear strands of DNA or RNA, no matter how long, will almost always have a free 5′ phosphate group at one end and a free 3′ hydroxyl group at the other. Therefore, every DNA and RNA molecule has an intrinsic directionality, and we can refer unambiguously to each end of the molecule. By convention, the sequence of bases is usually expressed in the 5′ to 3′ direction. Thus, the base sequence GTCCAT refers to the sequence:

5′ pGpTpCpCpApT—OH 3′

where the phosphates are indicated by "p." Note that this is not the same molecule as that represented by the reverse sequence:

5′ pTpApCpCpTpG—OH 3′

Levene's early studies indicated that all four types of DNA nucleotides were present in roughly equal amounts. This result, which later proved to be erroneous, led to the mistaken idea that DNA was a simple polymer in which the four nucleotides merely repeated (for instance, GCAT. . . . GCAT. . . . GCAT. . . . GCAT. . . .). If the sequence never varied, it was difficult to see how DNA might contain the hereditary information; this is why Avery's conclusion that DNA is the transforming principle was not readily accepted at first. It seemed more plausible that DNA was simply a structural element of the chromosomes, with proteins playing the central genetic role.

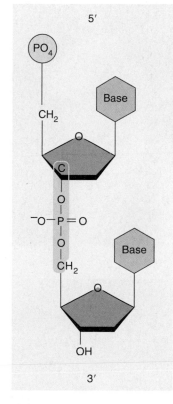

FIGURE 14.8
A phosphodiester bond.

Chargaff's Analysis: DNA Is Not a Simple Repeating Polymer

When Levene's chemical analysis of DNA was repeated using more sensitive techniques that became available after World War II, quite a different result was obtained. The four nucleotides were *not* present in equal proportions in DNA molecules after all. A careful study carried out by Erwin Chargaff showed that the nucleotide composition of DNA molecules varied in complex ways, depending on the source of the DNA (table 14.1). This strongly suggested that DNA was not a simple repeating polymer and might have the information-encoding properties genetic material requires. Despite DNA's complexity, however, Chargaff observed an important underlying regularity in double-stranded DNA: *The amount of adenine present in DNA always equals the amount of thymine, and the amount of guanine always equals the amount of cytosine.* These findings are commonly referred to as **Chargaff's rules:**

1. The proportion of A always equals that of T, and the proportion of G always equals that of C:

   A = T, and G = C

2. It follows that there is always an equal proportion of purines (A and G) and pyrimidines (C and T).

A single strand of DNA or RNA consists of a series of nucleotides joined together in a long chain. In all natural double-stranded DNA molecules, the proportion of A equals that of T, and the proportion of G equals that of C.

The Three-Dimensional Structure of DNA

As it became clear that DNA was the molecule that stored the hereditary information, investigators began to puzzle over how such a seemingly simple molecule could carry out such a complex function.

Franklin: X-ray Diffraction Patterns of DNA

The significance of the regularities pointed out by Chargaff were not immediately obvious, but they became clear when the British chemist Rosalind Franklin (figure 14.9a) carried out an X-ray diffraction analysis of DNA. In X-ray diffraction, a molecule is bombarded with a beam of X rays. When individual rays encounter atoms, their path is bent or diffracted, and the diffraction pattern is recorded on photographic film. The patterns resemble the ripples created by tossing a rock into a smooth lake (figure 14.9b). When carefully analyzed, they yield information about the three-dimensional structure of a molecule.

X-ray diffraction works best on substances that can be prepared as perfectly regular crystalline arrays. However, it was impossible to obtain true crystals of natural DNA at the time Franklin conducted her analysis, so she had to use DNA in the form of fibers. Franklin worked in the laboratory of the British biochemist Maurice Wilkins, who was able to prepare more uniformly oriented DNA fibers than anyone had previously. Using these fibers, Franklin succeeded in obtaining crude diffraction information on natural DNA. The diffraction patterns she obtained suggested that the DNA molecule had the shape of a helix, or corkscrew, with a diameter of about 2 nanometers and a complete helical turn every 3.4 nanometers (figure 14.9c).

Watson and Crick: A Model of the Double Helix

Learning informally of Franklin's results before they were published in 1953, James Watson and Francis Crick, two young investigators at Cambridge University, quickly worked out a likely structure for the DNA molecule (figure 14.10a), which we

(a)

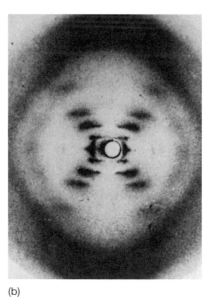

(b)

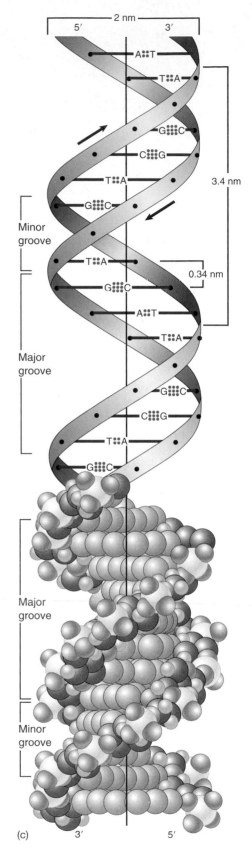

(c)

FIGURE 14.9
Rosalind Franklin's X-ray diffraction patterns. (*a*) Rosalind Franklin. (*b*) This telltale X-ray diffraction photograph of DNA fibers was made in 1953 by Rosalind Franklin in the laboratory of Maurice Wilkins. (*c*) The X-ray diffraction studies of Rosalind Franklin suggested the dimensions of the DNA double helix.

now know was substantially correct. The key to their model was Watson and Crick's understanding that each DNA molecule is actually made up of *two* chains of nucleotides that are intertwined—the double helix.

Backbone.

As described in chapter 3, the two strands of the double helix are made up of long polymers of nucleotides. Each strand is made up of repeating sugar and phosphate units joined by phosphodiester bonds. These two strands are then wrapped about a common axis forming a double helix (see figure 14.9c). The helix is often compared to a spiral staircase, where the two strands of the double helix are the handrails on the staircase.

Complementarity.

What holds the two strands together? Watson and Crick proposed that the bases can form particular sets of hydrogen bonds that lead to specific "base pairs." Thus, adenine (A) can form two hydrogen bonds with thymine (T) to form an A–T base-pair, and guanine (G) can form three hydrogen bonds with cytosine (C) to form a G–C base-pair (figure 14.10b). Note that this configuration also pairs a two-ringed purine with a single-ringed pyrimidine in each case, so that the diameter of each base-pair is the same. We refer to these combinations as *complementary*—for example, a strand that is ATGC would be complementary to a strand that is TACG. This seemingly simple concept has some profound implications. If we know the sequence of one strand, we automatically know the sequence of the other strand. So wherever there is an A in one strand, there must be a T in the other, and wherever there is a G in one strand, there must be a C in the other. This will become critical later when we look at how DNA is replicated and expressed.

The Watson–Crick model explained Chargaff's results: In a double helix, adenine forms two hydrogen bonds with thymine, but it will not form hydrogen bonds properly with cytosine. Similarly, guanine forms three hydrogen bonds with cytosine, but it will not form hydrogen bonds properly with thymine. Because of this base-pairing, adenine and thymine will always occur in the same proportions in any DNA molecule, as will guanine and cytosine.

Antiparallel Configuration.

A single phosphodiester strand has an inherent polarity, meaning that one end terminates in a 3′ OH and the other end terminates in a 5′ PO_4. We thus refer to strands as having either a 5′ to 3′ or 3′ to 5′ polarity. There are obviously two ways we could put together two strands, with the polarity the same in each (parallel) or with the polarity opposite (antiparallel). Native double-stranded DNA always has the antiparallel configuration, with one strand running 5′ to 3′ and the other running 3′ to 5′ (figure 14.10b). This proves to have important implications for DNA replication.

The Watson–Crick DNA Molecule.

To put together the entire structure of a DNA molecule as suggested by Watson and Crick, each DNA molecule is composed of two complementary phosphodiester strands that each form a helix with a common axis. These strands are antiparallel, with the bases extending into the interior of the helix. The bases from opposite strands form base-pairs with each other to join the two complementary strands (figures 14.9c and 14.10b). Although each individual base-pair is of low energy, the sum of many base-pairs has enough energy that the molecule is very stable. To return to our spiral staircase analogy, where the backbone is the handrails, the base-pairs are the stairs.

The DNA molecule is a double helix, the strands held together by base-pairing.

FIGURE 14.10
The DNA double helix.
(*a*) James Watson (*left*) and Francis Crick (*right*) deduced the structure of DNA in 1953 from Chargaff's rules and Franklin's diffraction studies. (*b*) In a DNA duplex molecule, only two base-pairs are possible: Adenine (A) can pair with thymine (T), and guanine (G) can pair with cytosine (C). An A–T base-pair has two hydrogen bonds, while a G–C base-pair has three.

(a)

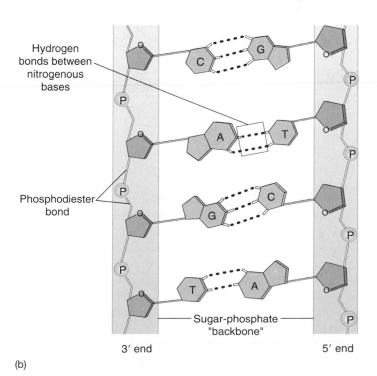
(b)

14.3 How does DNA replicate?

The Meselson–Stahl Experiment: DNA Replication Is Semiconservative

The Watson–Crick model immediately suggested that the basis for copying the genetic information is **complementarity**. One chain of the DNA molecule may have any conceivable base sequence, but this sequence completely determines the sequence of its partner in the duplex. For example, if the sequence of one chain is 5′-ATTGCAT-3′, the sequence of its partner *must* be 3′-TAACGTA-5′. Thus, each chain in the duplex is a complement of the other.

The complementarity of the DNA duplex provides a ready means of accurately duplicating the molecule. If we were to "unzip" the molecule, we would need only to assemble the appropriate complementary nucleotides on the exposed single strands to form two daughter duplexes with the same sequence. This form of DNA replication is called **semiconservative** because, although the sequence of the original duplex is conserved after one round of replication, the duplex itself is not. Instead, each strand of the duplex becomes part of another duplex.

Besides the semiconservative model, two other hypotheses of gene replication were also proposed. The *conservative* model stated that the parental double helix would remain intact and generate DNA copies consisting of entirely new molecules. The *dispersive* model predicted that parental DNA would become dispersed throughout the new copy so that each strand of all the daughter molecules would be a mixture of old and new DNA.

The three hypotheses of DNA replication were evaluated in 1958 by Matthew Meselson and Franklin Stahl of the California Institute of Technology. These two scientists grew bacteria in a medium containing the heavy isotope of nitrogen, ^{15}N, which became incorporated into the bases of the bacterial DNA. After several generations, the DNA of these bacteria was denser than that of bacteria grown in a medium containing the lighter isotope of nitrogen, ^{14}N. Meselson and Stahl then transferred the bacteria from the ^{15}N medium to the ^{14}N medium and collected the DNA at various intervals.

By dissolving the DNA they had collected in a heavy salt called cesium chloride and then spinning the solution at very high speeds in an ultracentrifuge, Meselson and Stahl were able to separate DNA strands of different densities. The enormous centrifugal forces generated by the ultracentrifuge caused the cesium ions to migrate toward the bottom of the centrifuge tube, creating a gradient of cesium concentration, and thus of density. Each DNA strand floats or sinks in the gradient until it reaches the position where its density exactly matches the density of

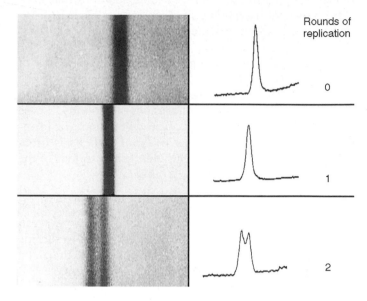

FIGURE 14.11
The key result of the Meselson–Stahl experiment. These bands of DNA, photographed on the left and scanned on the right, are from the density-gradient centrifugation experiment of Meselson and Stahl. At 0 generation, all DNA is heavy; after one replication, all DNA has a hybrid density; after two replications, all DNA is hybrid or light.

the cesium there. Because ^{15}N strands are denser than ^{14}N strands, they migrate farther down the tube to a denser region of the cesium gradient.

The DNA collected immediately after the transfer was all dense. However, after the bacteria completed their first round of DNA replication in the ^{14}N medium, the density of their DNA had decreased to a value intermediate between ^{14}N-DNA and ^{15}N-DNA. After the second round of replication, two density classes of DNA were observed, one intermediate and one equal to that of ^{14}N-DNA (figure 14.11).

Meselson and Stahl interpreted their results as follows: After the first round of replication, each daughter DNA duplex was a hybrid possessing one of the heavy strands of the parent molecule and one light strand; when this hybrid duplex replicated, it contributed one heavy strand to form another hybrid duplex and one light strand to form a light duplex (figure 14.12). Thus, this experiment clearly confirmed the prediction of the Watson–Crick model that DNA replicates in a semiconservative manner.

The basis for the great accuracy of DNA replication is complementarity. A DNA molecule is a duplex, containing two strands that are complementary mirror images of each other, so either one can be used as a template to reconstruct the other.

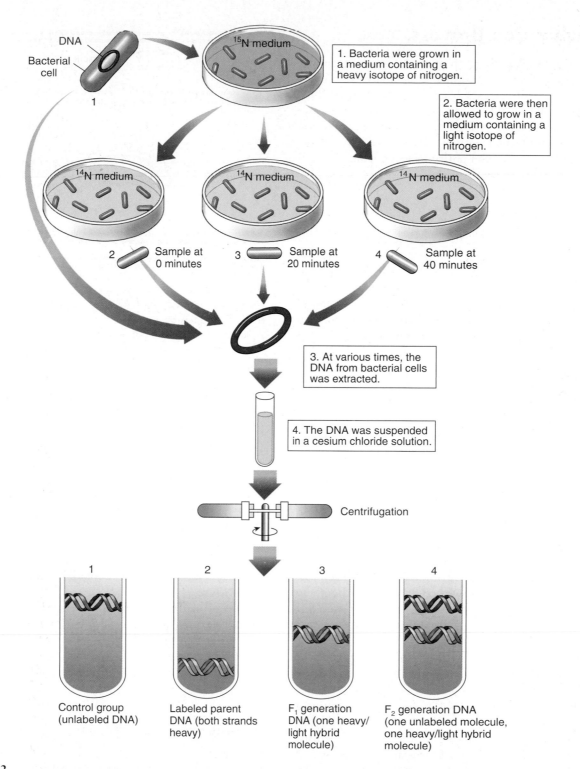

FIGURE 14.12

The Meselson–Stahl experiment: evidence demonstrating semiconservative replication. Bacterial cells were grown for several generations in a medium containing a heavy isotope of nitrogen (^{15}N) and then transferred to a new medium containing the normal lighter isotope (^{14}N). At various times thereafter, samples of the bacteria were collected, and their DNA was dissolved in a solution of cesium chloride, which was spun rapidly in a centrifuge. Because the cesium ion is so massive, it tends to settle toward the bottom of the spinning tube, establishing a gradient of cesium density. DNA molecules sink in the gradient until they reach a place where their density equals that of the cesium; they then "float" at that position. DNA containing ^{15}N is denser than that containing ^{14}N, so it sinks to a lower position in the cesium gradient. After one generation in ^{14}N medium, the bacteria yielded a single band of DNA with a density between that of ^{14}N-DNA and ^{15}N-DNA, indicating that only one strand of each duplex contained ^{15}N. After two generations in ^{14}N medium, two bands were obtained: one of intermediate density (in which one of the strands contained ^{15}N) and one of low density (in which neither strand contained ^{15}N). Meselson and Stahl concluded that replication of the DNA duplex involves building new molecules by separating strands and assembling new partners on each of these templates.

The Replication Process

To be effective, DNA replication must be fast and accurate. The process of DNA replication has been analyzed in great detail in *E. coli* and its viruses over the last 40 years. This has involved work by many researchers using all of the tools available to modern biologists: biochemistry, genetics, microscopy, and molecular biology. These researchers have built a detailed picture of the entire process, which we will present here in an overview.

Origin of Replication

Replication in *E. coli* begins at a specific site, the origin (called *OriC*), and ends at a specific site, the terminus. If you think about coordinating replication with cell division, it makes a lot of sense to have replication begin at a specific site to control the number of initiations and thus "rounds" of replication. The sequence of *OriC* consists of repeated nucleotides that bind a protein involved in initiation of replication, and an AT-rich sequence that can be opened easily during initiation of replication. (Remember, A–T base-pairs have only two hydrogen bonds compared with the three hydrogen bonds in G–C base-pairs.) After initiation, replication proceeds bidirectionally from this unique origin to the unique terminus (figure 14.13). The complete chromosome plus the origin can be thought of as a functional unit called a *replicon*.

Polymerases

The first polymerase, an enzyme that synthesizes nucleic acids, isolated in *E. coli* was called DNA polymerase I (pol I). It was assumed at first that this was the polymerase responsible for the bulk synthesis of DNA during replication. However, a mutant was isolated with no pol I activity that still showed replicative polymerase activity. Two additional polymerases were isolated from this strain of *E. coli* called DNA polymerase II (pol II) and DNA polymerase III (pol III). All three of these enzymes share a number of features, including the ability to synthesize polynucleotide strands. This activity itself has some features that were not expected. First, all known DNA polymerases require a **primer,** a short stretch of DNA or RNA nucleotides hydrogen-bonded to its complementary strand. Thus, DNA polymerases cannot begin synthesis of a new strand of DNA, but need a primer, an existing segment of nucleotides, on which to build. Second, these enzymes can only synthesize in one direction; they extend strands in a 5′ to 3′ direction by copying a template that is 3′ to 5′ (remember complementarity) (figure 14.14). The action of a DNA polymerase is to add new nucleotides to the 3′ OH of a primer. Interestingly, these enzymes also have the ability to remove nucleotides, and so are called "nucleases." Nucleases are classified as either **endonucleases** (cut DNA internally) or **exonucleases** (chew away at an end of DNA).

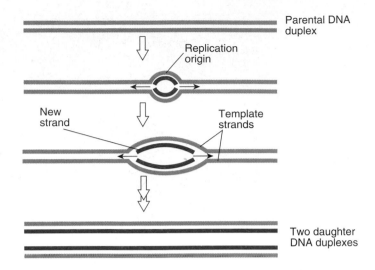

FIGURE 14.13
Origins of replication. At a site called the replication origin, the DNA duplex opens to create two separate strands, each of which can be used as a template for a new strand.

Each of these enzymes has a 3′ to 5′ exonuclease activity that is called a *proofreading function*. Mispaired bases can be removed by this activity, increasing the accuracy of the DNA replication process. The pol I enzyme also has a 5′ to 3′ exonuclease activity that is important in DNA replication, as we shall see later.

For many years, these three polymerases were thought to be the only DNA polymerases in *E. coli*, but recently several new polymerases have been identified. There are now five known polymerases, although all are not active in DNA replication. We will return to the roles of the different polymerases after considering another feature of replication.

Given the directionality of the polymerase enzymes and the antiparallel structure of DNA, it is obvious that replication cannot be the same on each strand. As the DNA double helix opens up, the strands will be antiparallel, so *both* cannot be replicated in a 5′ to 3′ direction as single continuous strands.

Leading and Lagging Strands

One strand, the one oriented in the 3′ to 5′ direction, can have a primer added complementary to its 3′ end, and then be extended 5′ to 3′ as one continuous strand. We call this the **leading strand.** The other strand must be replicated in short 5′ to 3′ stretches, filling in the open helix back to the end. As the helix is opened further, it can be replicated further, but the entire strand can only be replicated in this discontinuous fashion. We call this the **lagging strand.** The short stretches of newly synthesized DNA on the lagging strand have been named **Okazaki fragments** in honor of

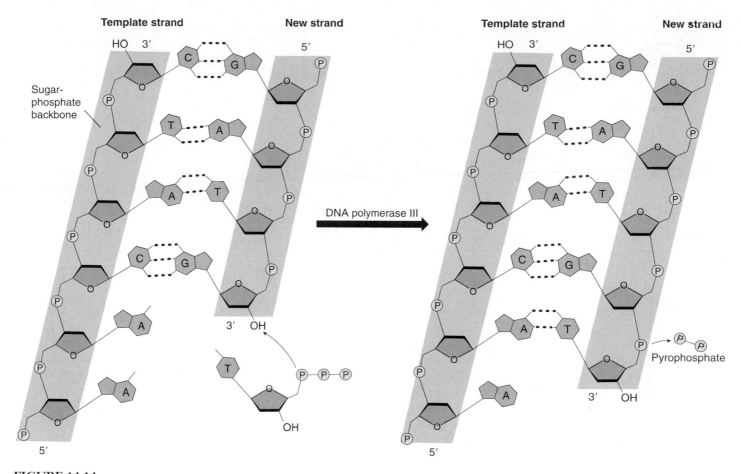

FIGURE 14.14
How nucleotides are added in DNA replication. DNA polymerase III, along with other enzymes, catalyzes the addition of nucleotides to the growing complementary strand of DNA. When a nucleotide is added, two of its phosphates are lost as pyrophosphate.

the man who first was able to detect them experimentally (figure 14.15).

Other Enzymatic Activities

A number of other enzymatic activities are needed to replicate DNA in addition to DNA polymerase. First, an enzyme is required to synthesize a primer to begin replication on the leading strand. There is also a need for continuous priming on the lagging strand. The enzyme responsible for these activities is *DNA primase*, which creates a short RNA primer complementary to a DNA template. It makes sense that the primer be RNA because RNA polymerases (enzymes that synthesize RNA molecules) do not require a primer to begin synthesis, so DNA primase is actually an RNA polymerase. Second, we need an enzyme to open up the helix in front of the polymerase. The enzyme responsible for this is *DNA helicase*, which can employ the energy from the hydrolysis of ATP to unwind DNA strands. Third, we need an enzyme that can remove the torsional strain introduced by opening the double helix. Think of opening the

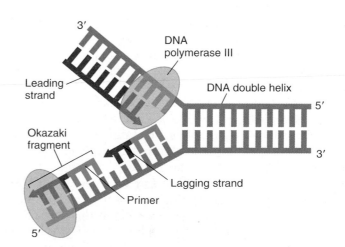

FIGURE 14.15
Synthesis of the leading and lagging strands.
DNA polymerase III synthesizes the leading strand as a continuous strand, but the lagging strand is synthesized in segments, called Okazaki fragments, each beginning with a primer.

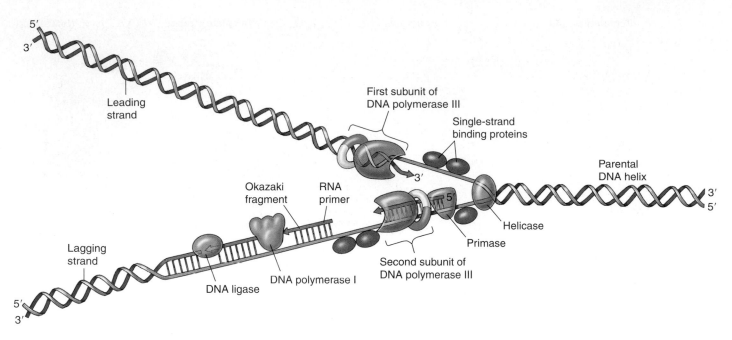

FIGURE 14.16

A DNA replication fork. Helicase enzymes separate the strands of the double helix, and single-stranded binding proteins stabilize the single-stranded regions. Replication occurs by two mechanisms. (1) *Continuous synthesis:* After primase adds a short RNA primer, DNA polymerase III adds nucleotides to the 3′ end of the leading strand. DNA polymerase I then replaces the RNA primer with DNA nucleotides. (2) *Discontinuous synthesis:* Primase adds a short RNA primer (*green*) ahead of the 5′ end of the lagging strand. DNA polymerase III then adds nucleotides to the primer until the gap is filled in. DNA polymerase I replaces the primer with DNA nucleotides, and DNA ligase attaches the short segment of nucleotides to the lagging strand.

double helix as unraveling a rope. As you pull apart the two strands of a rope, torsional strain causes the rope to coil up or form a knot. This strain must be relieved in order to open the DNA helix. The enzyme responsible for this is *DNA gyrase,* a form of topoisomerase, which is a group of enzymes that can alter the topological state of DNA. Fourth, the unwound single strands are not as stable in the aqueous environment of the cell as the double-stranded molecule, because the bases, which are highly hydrophobic, are exposed to the water. To solve this problem, as the DNA is unwound, the exposed single strands are covered by *single-strand binding protein* (ssb). Fifth, we need to remove the primers that are present at the beginning of the leading strand and in each Okazaki fragment on the lagging strand, and we need to join the fragments along the lagging strand. These two jobs are accomplished by DNA pol I and *DNA ligase,* respectively. DNA pol I removes primers with its 3′ to 5′ exonuclease activity and then replaces this with DNA, while DNA ligase creates a phosphodiester bond between adjacent Okazaki fragments. The various enzymes involved in replication are summarized in table 14.2.

The Replication Fork

How can we make sense of all these different activities for what had initially appeared to be a simple process? Instead of considering these many enzymes in isolation, we usually

think of the **replication fork,** the site of opening of the DNA strands where active replication occurs. All of the activities we have just discussed are found and are necessary for the replication fork. Figure 14.16 is meant to show schematically all the proteins in terms of function and how they are now thought to associate with each other. It is also important that all of the proteins involved act in a coordinated manner so that the whole process can proceed smoothly, an idea that will be developed further in the subsequent discussion.

Roles of the Polymerases

The three standard DNA polymerases are not all necessary for DNA replication during cell division. Pol III and pol I both have important roles to play, but pol II is thought to function primarily during DNA repair and not during replication. DNA pol I is made up of a single protein, and although it was the first identified polymerase, it is not the main polymerase at the replication fork. Rather, its role is to remove the repeated DNA primers that are necessary for DNA synthesis. The polymerase found at the replication fork is pol III, a large, multisubunit enzyme. The main catalytic subunit is called the α subunit, which along with two other proteins forms the catalytic core of the enzyme. While the core enzyme can synthesize DNA, it does not remain attached to its template for long stretches, a feature called *processivity.* The addition of the β subunit greatly increases the processivity of the core enzyme. The structure of the β subunit has been worked out, and it is a fascinating structure that has been called a sliding clamp. It literally forms a ring around the template strand, holding the catalytic core to the template (figure 14.17). The complete pol III enzyme is a large multiprotein complex made up of at least seven proteins.

The Replisome

Thus, DNA replication in *E. coli* is accomplished by a large, macromolecular assembly that is really a replication organelle, just as the ribosome is the protein synthetic organelle. We call this assembly the **replisome.** The replisome is a macromolecular protein machine that accomplishes the fast and accurate replication of DNA during cell division. The replisome has two main subcomponents, the *primosome* and a complex of two DNA pol III enzymes, one pol III for each strand. The primosome is composed of primase and helicase and a number of accessory proteins (see figure 14.16). The two pol III complexes include two synthetic core subunits, each with its own β subunit and a number of other proteins that hold the entire

Protein	Role	Size (kd)	Molecules per Cell
Helicase	Unwinds the double helix	300	20
Primase	Synthesizes RNA primers	60	50
Single-strand binding protein	Stabilizes single-stranded regions	74	300
DNA gyrase	Relieves torque	400	250
DNA polymerase III	Synthesizes DNA	≈900	20
DNA polymerase I	Erases primer and fills gaps	103	300
DNA ligase	Joins the ends of DNA segments; DNA repair	74	300

Table 14.2 DNA Replication Enzymes of *E. coli*

FIGURE 14.17

The DNA polymerase III complex. (*a*) The complex contains 10 kinds of protein chains. The protein is a dimer because both strands of the DNA duplex must be replicated simultaneously. The catalytic (α) subunits, the proofreading (ε) subunits, and the "sliding clamp" (β₂) subunits (*yellow and blue*) are labeled. (*b*) The "sliding clamp" units encircle the DNA template and (*c*) move it through the catalytic subunit like a rope drawn through a ring.

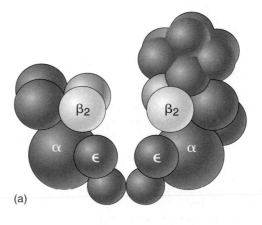

(a)

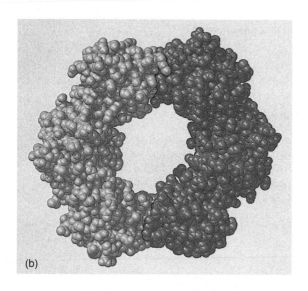

(b)

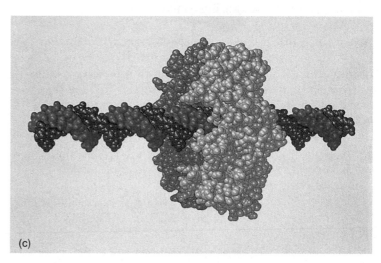

(c)

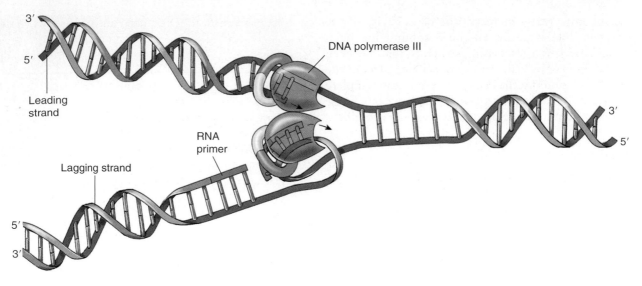

FIGURE 14.18

How DNA polymerase III works. This diagram presents a current view of how DNA polymerase III works. Note that the DNA on the lagging strand is folded to allow the dimeric DNA polymerase III molecule to replicate both strands of the parental DNA duplex simultaneously. This brings the 3′ end of each completed Okazaki fragment close to the start site for the next fragment.

complex together. The pol III form found in the replisome and at the replication fork is made up of over 14 individual proteins (see figure 14.17*a*). Even given the difficulties with lagging-strand synthesis, the two pol III enzymes in the replisome are active on both leading and lagging strands simultaneously. The need for constant priming on the lagging strand explains the need for the primosome complex as part of the entire replisome at the replication fork. How can the two strands be synthesized in the same direction? The model first proposed, still with us in some form, indicates that a loop forms in the lagging strand so that the polymerases can move in the same direction (figure 14.18).

Stages of Replication

Although a continuous process, replication can be logically broken down into stages, much as we earlier broke the continuous process of mitosis into stages.

Initiation. Initiation always occurs at the same site, *OriC*. It is a complex process that involves first the recognition of specific sites within *OriC* by an initiator protein. This initiator protein is then joined by other proteins that are necessary only for initiation to open the helix at the AT-rich region within *OriC*. The primosome is then assembled onto the opened DNA strands, followed by the assembly of the entire replisome complex to complete the replication fork. Because replication is bidirectional from *OriC*, this is happening in both directions such that two replication forks proceed in opposite directions around the chromosome.

Elongation. The majority of the time during replication is spent in elongation, during which the pol III molecules add successive new nucleotides based on complementarity to the template strand. This process is simple on the leading strand, because a single primer is all that is absolutely necessary (although multiple primings may occur). On the lagging strand, the process is considerably more complex. As the helix is opened, a new primer is synthesized, and pol III must release the template it has completed and begin with the "new" template. As the complex proceeds, primers are removed by pol I with its 5′ to 3′ exonuclease activity, and the strands are sealed by DNA ligase. Synthesis proceeds bidirectionally around the circular *E. coli* chromosome, creating two new DNA molecules.

Termination. The exact details of the termination process are not clear, but it is known that a specific termination site is located roughly opposite *OriC* on the circular chromosome. There is also a need to keep the two new molecules from becoming intertwined by the replication process. This requires a topoisomerase enzyme such as DNA gyrase.

DNA replication in *E. coli* involves many different proteins that open and unwind the DNA double helix, stabilize the single strands, synthesize RNA primers, assemble new complementary strands on each exposed parental strand (one of them discontinuously), remove the RNA primer, and join new discontinuous segments on the lagging strand.

Eukaryotic DNA Replication

The biggest difference between eukaryotic and prokaryotic replication is the sheer amount of DNA and how it is packaged (figure 14.19). Eukaryotes usually have multiple chromosomes that are each larger than the *E. coli* chromosome. Thus, the machinery could in principle be the same; however, if there were only a single unique origin for each chromosome, the length of the time necessary for replication would be prohibitive. This problem is solved by the use of multiple origins of replication for each chromosome, resulting in multiple *replicons*, which are sections of DNA replicated from individual origins (figure 14.20). The origins are not as sequence-specific as *OriC*, and seem to depend on chromatin structure as well as on sequence. The number of origins that "fire" can also be adjusted during the course of development so that early, when divisions need to be rapid, more origins are activated.

Although there are some differences in details, such as the names of the enzymes, the replication fork in eukaryotes appears substantially similar to the picture researchers have developed for *E. coli*. The main replication polymerase in eukaryotes is called DNA polymerase α; the sliding clamp subunit also exists in eukaryotes, but it is called PCNA (for Proliferating Cell Nuclear Antigen). This name reflects the fact that PCNA was first identified as an antibody-inducing protein in proliferating cells. The makeup of the primosome in eukaryotes is different in detail from that of the *E. coli* primosome, but its role is the same. Similarly, all the rest of the actors at the replication fork that we have seen in *E. coli* are also present in eukaryotes.

Eukaryotic chromosomes have multiple origins of replication.

FIGURE 14.19
DNA of a single human chromosome. This chromosome has been "exploded," or relieved of most of its packaging proteins. The residual protein scaffolding appears as the dark material in the lower part of the micrograph.

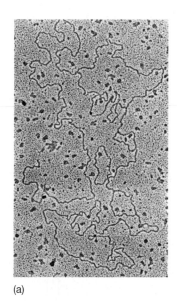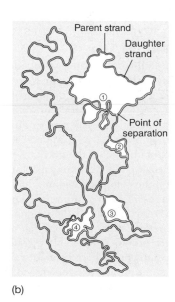

(a) (b)

FIGURE 14.20
Eukaryotic chromosomes possess numerous replication forks spaced along their length. Four replication units (each with two replication forks) are producing daughter strands in (*a*) the electron micrograph. This is indicated in red in (*b*) the corresponding drawing.

14.4 What is a gene?

The One-Gene/One-Polypeptide Hypothesis

As the structure of DNA was being investigated, other biologists continued to puzzle over how the genes described by Mendel were related to DNA. What is the relationship between genotype and phenotype?

Garrod: Inherited Disorders Can Involve Specific Enzymes

In 1902, the British physician Archibald Garrod was working with one of the early Mendelian geneticists, his countryman William Bateson, when he noted that certain diseases among his patients seemed to be more prevalent in particular families. By examining several generations of these families, he found that some of the diseases behaved as if they were the product of simple recessive alleles. Garrod concluded that these disorders were Mendelian traits and that they had resulted from changes in the hereditary information in an ancestor of the affected families.

Garrod investigated several of these disorders in detail. In alkaptonuria, the patients produced urine that contained homogentisic acid (alkapton). This substance oxidized rapidly when exposed to air, turning the urine black. In normal individuals, homogentisic acid is broken down into

simpler substances. With considerable insight, Garrod concluded that patients suffering from alkaptonuria lack the enzyme necessary to catalyze this breakdown. He speculated that many other inherited diseases might also reflect enzyme deficiencies.

Beadle and Tatum: Genes Specify Enzymes

From Garrod's finding, it took but a short leap of intuition to surmise that the information encoded within the DNA of chromosomes acts to specify particular enzymes. This point was not actually established, however, until 1941, when a series of experiments by Stanford University geneticists George Beadle and Edward Tatum provided definitive evidence. Beadle and Tatum deliberately set out to create Mendelian mutations in chromosomes and then studied the effects of these mutations on the organism (figure 14.21).

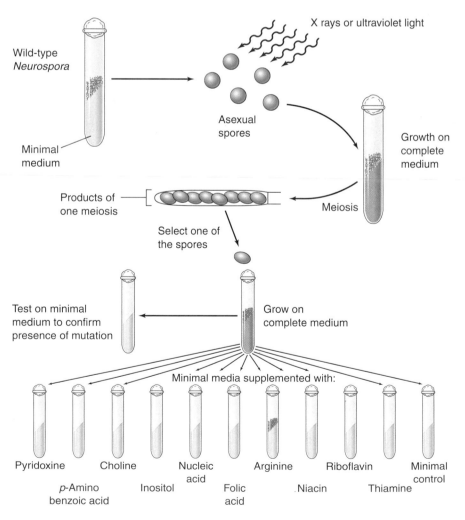

FIGURE 14.21

Beadle and Tatum's procedure for isolating nutritional mutants in *Neurospora*. This fungus grows easily on an artificial medium in test tubes. In this experiment, spores were irradiated to increase the frequency of mutation; they were then placed on a "complete" medium that contained all of the nutrients necessary for growth. Once the fungal colonies were established on the complete medium, individual spores were transferred to a "minimal" medium that lacked various substances the fungus could normally manufacture. Any spore that would not grow on the minimal medium but would grow on the complete medium contained one or more mutations in genes needed to produce the missing nutrients. To determine which gene had mutated, the minimal medium was supplemented with particular substances. The mutation illustrated here produced an arginine mutant, a collection of cells that had lost the ability to manufacture arginine. These cells will not grow on minimal medium but will grow on minimal medium with only arginine added.

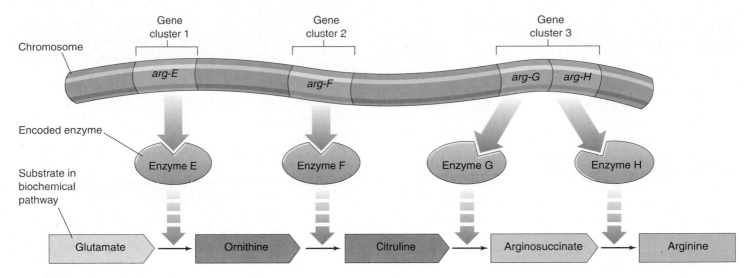

FIGURE 14.22

Evidence for the "one-gene/one-polypeptide" hypothesis. The chromosomal locations of the many arginine mutants isolated by Beadle and Tatum cluster in three places. These clusters correspond to the locations of the genes encoding the enzymes that carry out arginine biosynthesis.

A Defined System. One of the reasons Beadle and Tatum's experiments produced clear-cut results is that the researchers made an excellent choice of experimental organism. They chose the bread mold *Neurospora*, a fungus that can be grown readily in the laboratory on a defined medium (a medium that contains only known substances such as glucose and sodium chloride, rather than some uncharacterized mixture of substances, such as ground-up yeasts). Beadle and Tatum exposed *Neurospora* spores to X rays, expecting that the DNA in some of the spores would experience damage in regions encoding the ability to make compounds needed for normal growth (see figure 14.21).

Isolating Growth-Deficient Mutants. To determine whether any of the progeny of the irradiated spores had mutations causing metabolic deficiencies, Beadle and Tatum placed subcultures of individual fungal cells on a "minimal" medium that contained only sugar, ammonia, salts, a few vitamins, and water. Cells that had lost the ability to make other compounds necessary for growth would not survive on such a medium. Using this approach, Beadle and Tatum succeeded in identifying and isolating many growth-deficient mutants.

Identifying the Deficiencies. Next, the researchers added various chemicals to the minimal medium in an attempt to find one that would enable a given mutant strain to grow. This procedure allowed them to pinpoint the nature of the strain's biochemical deficiency. The addition of arginine, for example, permitted several mutant strains, dubbed *arg* mutants, to grow. When their chromosomal positions were located, the *arg* mutations were found to cluster in three areas.

One-Gene/One-Polypeptide

For each enzyme in the arginine biosynthetic pathway, Beadle and Tatum were able to isolate a mutant strain with a defective form of that enzyme, and the mutation was always located at one of a few specific chromosomal sites. Most importantly, they found there was a different site for each enzyme. Thus, each of the mutants they examined had a defect in a single enzyme, caused by a mutation at a single site on one chromosome. Beadle and Tatum concluded that genes produce their effects by specifying the structure of enzymes and that each gene encodes the structure of one enzyme. They called this relationship the **one-gene/one-enzyme hypothesis** (figure 14.22). Because many enzymes contain multiple protein or polypeptide subunits, each encoded by a separate gene, the relationship is today more commonly referred to as **one-gene/one-polypeptide.** This is a clear statement of the molecular relationship between genotype and phenotype.

Enzymes are responsible for catalyzing the synthesis of all the parts of an organism. They mediate the assembly of nucleic acids, proteins, carbohydrates, and lipids. Therefore, by encoding the structure of enzymes and other proteins, DNA specifies the structure of the organism itself.

As we learn more about genomes and gene expression, this clear relationship may not hold in all cases. In chapter 15, we will see that mRNA can be spliced differently to produce different products from the same gene, and that mRNAs can be edited. The basic principle that remains is: Genes encode proteins, and any single mRNA species encodes a single protein product.

Genetic traits are expressed largely as a result of the activities of enzymes. Organisms store hereditary information by encoding the structures of enzymes and other proteins in their DNA.

How DNA Encodes Protein Structure

What kind of information must a gene encode to specify a protein? For some time, the answer to that question was not clear, because protein structure seemed impossibly complex.

Sanger: Proteins Consist of Defined Sequences of Amino Acids

The picture changed in 1953, the same year Watson and Crick unraveled the structure of DNA. The English biochemist Frederick Sanger, after many years of work, announced the complete sequence of amino acids in the protein insulin, a small protein hormone. This was the first protein for which the amino acid sequence was determined. Sanger's achievement was extremely significant because it demonstrated for the first time that proteins consist of definable sequences of amino acids—that is, for any given form of insulin, every molecule has the same amino acid sequence. Sanger's work soon led to the sequencing of many other proteins, and it became clear that all enzymes and other proteins are strings of amino acids arranged in a certain definite order. Therefore, the information needed to specify a protein such as an enzyme is an ordered list of amino acids.

Ingram: Single Amino Acid Changes in a Protein Can Have Profound Effects

Following Sanger's pioneering work, Vernon Ingram in 1956 discovered the molecular basis of sickle cell anemia, a protein defect inherited as a Mendelian disorder. By analyzing the structures of normal and sickle cell hemoglobin, Ingram, working at Cambridge University, showed that sickle cell anemia is caused by a change from glutamic acid to valine at a single position in the protein (figure 14.23). The alleles of the gene encoding hemoglobin differ only in their specification of this one amino acid in the hemoglobin amino acid chain.

These experiments and other related ones have finally brought us a clear understanding of the unit of heredity. The characteristics of sickle cell anemia and most other hereditary traits are defined by changes in protein structure brought about by an alteration in the sequence of amino acids that make up the protein. This sequence in turn is dictated by the order of nucleotides in a particular region of the chromosome. For example, the critical change leading to sickle cell disease is a mutation that replaces a single thymine with an adenine at the position that codes for glutamic acid, converting the position to valine. The sequence of nucleotides that determines the amino acid sequence of a protein is called a **gene**. Although most genes encode proteins or subunits of proteins, some genes are devoted to producing special forms of RNA, many of which play important roles in protein synthesis themselves.

A half-century of experimentation has made it clear that DNA is the molecule responsible for the inheritance of traits, and that this molecule is divided into functional units called genes.

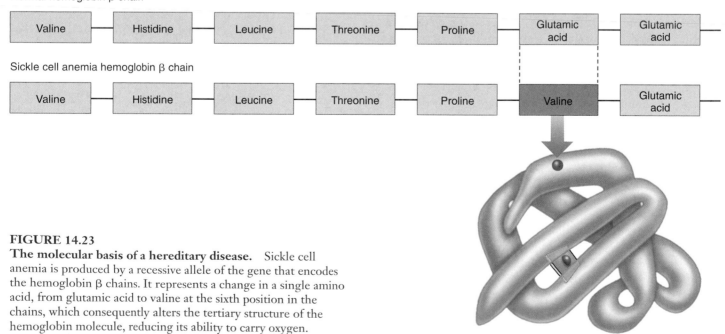

FIGURE 14.23
The molecular basis of a hereditary disease. Sickle cell anemia is produced by a recessive allele of the gene that encodes the hemoglobin β chains. It represents a change in a single amino acid, from glutamic acid to valine at the sixth position in the chains, which consequently alters the tertiary structure of the hemoglobin molecule, reducing its ability to carry oxygen.

Concept Review

For interactive testing, visit the Online Learning Center with PowerWeb at www.mhhe.com/Raven7

14.1 What is the genetic material?

The Hammerling Experiment: Cells Store Hereditary Information in the Nucleus

- Hammerling conducted a series of experiments and discovered that hereditary information in *Acetabularia* resided in the foot, which is also the location of the nucleus. (p. 280)

Transplantation Experiments: Each Cell Contains a Full Set of Genetic Instructions

- Later experiments in the mid-1950s showed that the nucleus of eukaryotic cells includes a full set of genetic information. (p. 281)

The Griffith Experiment: Hereditary Information Can Pass Between Organisms

- Griffith found that transformation occurs when genetic material is transferred from one cell to another, and that live cells can be transformed by dead cells. (p. 282)

The Avery and Hershey–Chase Experiments: The Active Principle Is DNA

- Avery provided conclusive evidence that DNA is the heredity material for the bacterial specimens under investigation. (p. 283)
- Hershey and Chase provided further evidence that heredity material in bacteriophages was found in DNA, not in proteins. (p. 283)

14.2 What is the structure of DNA?

The Chemical Nature of Nucleic Acids

- Both DNA and RNA are formed of nucleotides joined together in series. Each nucleotide is composed of a five-carbon sugar, a phosphate group, and a nitrogen-containing base. (p. 284)
- Chargaff's Rule states that in reference to the nitrogen-containing bases, adenine always pairs with thymine, and guanine always pairs with cytosine. Thus, there are always equal proportions of purines and pyrimidines. (p. 285)

The Three-Dimensional Structure of DNA

- Franklin was able to obtain the first glimpse of DNA using X-ray diffraction in 1953, while Watson and Crick theorized that DNA exists in a double-helical, antiparallel configuration. (pp. 286–287)
- Using a spiral staircase analogy, the handrails of the staircase represent the sugar-phosphate backbone of the DNA double helix, and the steps represent the hydrogen-bonded base pairs. (p. 287)

14.3 How does DNA replicate?

The Meselson–Stahl Experiment: DNA Replication Is Semiconservative

- Meselson and Stahl demonstrated that DNA replication is semiconservative because each strand of the original duplex becomes one of the two strands in each new duplex. (p. 288)

The Replication Process

- Replication of *E. coli* begins at a specific origin, proceeds bidirectionally, and ends at a specific terminus. (p. 290)
- Many enzymes function in DNA replication, including DNA primase, which creates a short RNA primer complementary to a DNA template; DNA helicase, which unwinds the helix in front of DNA polymerase, which then synthesizes new DNA by adding nucleotides to the growing strands; and DNA ligase, which creates phosphodiester bonds between adjacent Okazaki fragments. (pp. 292–293)
- Replication can be divided into three stages: initiation, elongation, and termination. (p. 294)

Eukaryotic DNA Replication

- The major difference between prokaryotic and eukaryotic replication is that eukaryotic chromosomes have multiple replication origins, whereas prokaryotic chromosomes have a single point of origin. (p. 295)

14.4 What is a gene?

The One-Gene/One-Polypeptide Hypothesis

- Beadle and Tatum concluded that genes produce their effects by specifying the structure of enzymes, and that each gene encodes the structure of one enzyme. Today, this is commonly referred to as the one-gene/one-polypeptide relationship. (p. 297)

How DNA Encodes Protein Structure

- Over 50 years of research has yielded clear evidence that DNA is the molecule responsible for the inheritance of traits from one generation to the next, and that DNA is divided into functional subunits, or genes, located on chromosomes. (p. 298)

Self Test

1. Which of the following experiments suggested that the nucleus is the repository for genetic information?
   a. Hammerling's experiment using *Acetabularia*
   b. Griffith's experiment using *S. pneumoniae* and mice
   c. the Hershey and Chase experiment using bacteriophages
   d. Franklin's X-ray diffraction

2. When Hershey and Chase differentially tagged the DNA and proteins of bacteriophages and allowed them to infect bacteria, what did the viruses transfer to the bacteria?
   a. radioactive phosphorus and sulfur
   b. radioactive sulfur
   c. DNA
   d. Both b and c are correct.

3. If one strand of a DNA molecule has the base sequence ATTGCAT, its complementary strand will have the sequence
   a. ATTGCAT
   b. TAACGTA
   c. GCCATGC
   d. CGGTACG

4. DNA is made up of building blocks called
   a. proteins.
   b. bases.
   c. nucleotides.
   d. deoxyribose.

5. X-ray diffraction experiments conducted by _____ led to the determination of the structure of DNA.
   a. Francis Crick
   b. James Watson
   c. Erwin Chargaff
   d. Rosalind Franklin

6. Meselson and Stahl proved that
   a. DNA is the genetic material.
   b. DNA is made from nucleotides.
   c. DNA replicates in a semiconservative manner.
   d. DNA is a double helix held together with base-pairing.

7. DNA polymerase III can only add nucleotides to an existing chain, so _____ is required.
   a. an RNA primer
   b. DNA polymerase I
   c. helicase
   d. a DNA primer

8. Okazaki fragments are
   a. synthesized in the 3′ to 5′ direction.
   b. found on the lagging strand.
   c. found on the leading strand.
   d. assembled as continuous replication.

9. Beadle and Tatum's experiment showed that each enzyme is specified by a single
   a. chromosome.
   b. gene.
   c. nucleotide.
   d. mutation.

10. What was significant about Ingram's work on sickle cell anemia?
    a. The gene for hemoglobin was missing.
    b. Proteins consisted of sequences of amino acids.
    c. A change in one amino acid can affect a protein's structure.
    d. One gene encodes one protein.

Test Your Visual Understanding

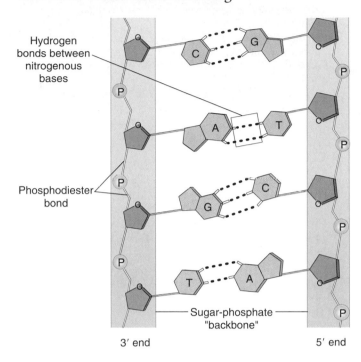

Hydrogen bonds between nitrogenous bases

Phosphodiester bond

Sugar-phosphate "backbone"

3′ end   5′ end

1. Based on the information in this figure, can you explain why pyrimidines don't base-pair with each other, and why purines don't base-pair with each other?

2. Can you also explain why adenine does not base-pair with cytosine, and why thymine does not base-pair with guanine?

Apply Your Knowledge

1. The human genome contains approximately 3 billion (3×10^9) nucleotide base-pairs, and each nucleotide in a strand of DNA takes up about 0.34 nanometers (0.34×10^{-9} meters). How long would the human genome be if it were fully extended?

2. From an extract of human cells growing in tissue culture, you obtain a white, fibrous substance. How would you distinguish whether it was DNA, RNA, or protein?

3. Cells were obtained from a patient with a viral infection. The DNA extracted from these cells consisted of two forms: double-stranded human DNA and single-stranded viral DNA. The base compositions of these two forms of DNA were as follows:

	A	C	G	T
Form 1	22.1%	27.9%	27.9%	22.1%
Form 2	31.3%	31.3%	18.7%	18.7%

Which form was the viral DNA, and which form was the human DNA? Explain your reasoning.

15

Genes and How They Work

Concept Outline

15.1 The Central Dogma traces the flow of gene-encoded information.

Cells Use RNA to Make Protein. The information in genes is expressed in two steps. First it is transcribed into RNA, and then the RNA is translated into protein.

15.2 Genes encode information in three-nucleotide code words.

The Genetic Code. The sequence of amino acids in a protein is encoded in the sequence of nucleotides in DNA such that three nucleotides encode each amino acid.

15.3 Genes are first transcribed, then translated.

Transcription in Prokaryotes. The enzyme RNA polymerase unwinds the DNA helix and synthesizes an RNA copy of one strand.

Transcription in Eukaryotes. Transcription in eukaryotes is more complex than that occurring in prokaryotes, requiring different types of enzymes and posttranscriptional modifications to the RNA.

Translation. mRNA is translated by activating enzymes that select tRNAs to match amino acids. Proteins are synthesized on ribosomes, which provide a framework for the interaction of tRNA and mRNA.

15.4 Eukaryotic gene transcripts are spliced.

The Discovery of Introns. Eukaryotic genes contain extensive material that is not translated.

Differences Between Prokaryotic and Eukaryotic Gene Expression. Gene expression is broadly similar in prokaryotes and eukaryotes, but it differs in some respects.

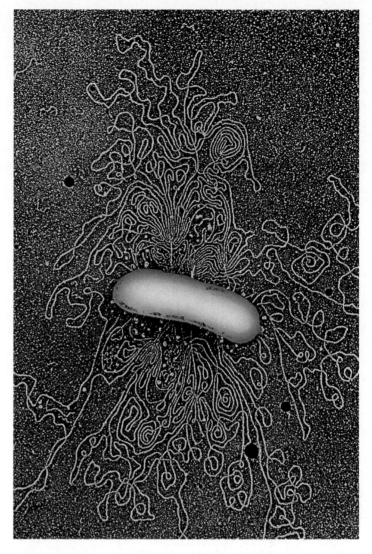

FIGURE 15.1
The unraveled chromosome of an *E. coli* bacterium. This complex tangle of DNA represents the full set of assembly instructions for the living organism *E. coli*.

Every cell in your body contains the hereditary instructions specifying that you will have arms rather than fins, hair rather than feathers, and two eyes rather than one. The color of your eyes, the texture of your fingernails, and all of the other traits you receive from your parents are recorded in the cells of your body. As we have seen, this information is contained in long molecules of DNA (figure 15.1). The essence of heredity is the ability of cells to use the information in their DNA to produce particular proteins, thereby affecting what the cells will be like. In that sense, proteins are the tools of heredity. In this chapter, we will examine how proteins are synthesized from the information in DNA, using both prokaryotes and eukaryotes as models.

15.1 The Central Dogma traces the flow of gene-encoded information.

Cells Use RNA to Make Protein

To find out how a eukaryotic cell uses its DNA to direct the production of particular proteins, you must first ask where in the cell the proteins are made. We can answer this question by placing cells in a medium containing radioactively labeled amino acids for a short time. The cells will take up the labeled amino acids and incorporate them into proteins. If we then look to see where in the cells radioactive proteins first appear, we will find them not in the nucleus, where the DNA is, but rather in the cytoplasm, on large RNA-protein aggregates called **ribosomes** (figure 15.2). These polypeptide-making factories are very complex, composed of two subunits, a small subunit (called 30S in prokaryotes and 40S in eukaryotes) and a large subunit (called 40S in prokaryotes and 60S in eukaryotes). Each subunit is composed of RNA and protein. The small subunit has over 20 proteins and one strand of RNA, and the large subunit has over 30 proteins and two strands of RNA (figure 15.3). Recently, the three-dimensional structure of both prokaryotic subunits and the complete prokaryotic ribosome was solved. These structures support the idea that the RNAs are the main catalytic units and the ribosomal proteins have a more structural role. Protein synthesis involves three different sites on the ribosome surface, called the P, A, and E sites, discussed later in this chapter.

Kinds of RNA

The class of RNA found in ribosomes is called **ribosomal RNA (rRNA).** During polypeptide synthesis, rRNA provides the site where polypeptides are assembled. In addition to rRNA, there are two other major classes of RNA in cells. **Transfer RNA (tRNA)** molecules both transport the amino acids to the ribosome for use in building the polypeptides and position each amino acid at the correct place on the elongating polypeptide chain (figure 15.4). Human cells contain about 45 different kinds of tRNA molecules. **Messenger RNA (mRNA)** molecules are long strands of RNA that are transcribed from DNA and travel to the ribosomes to direct precisely *which* amino acids are assembled into polypeptides.

These RNA molecules, together with ribosomal proteins and certain enzymes, constitute a system that reads the genetic messages encoded by nucleotide sequences in the DNA and produces the polypeptides that those sequences specify. As we will see in chapter 18, biologists have recently discovered that RNA also plays many important roles in controlling how and when genes are used. Having learned to read the many messages of RNA, biologists are gaining a clearer picture of what genes are, how they are able to dictate what a protein will be like, and when the protein will be made.

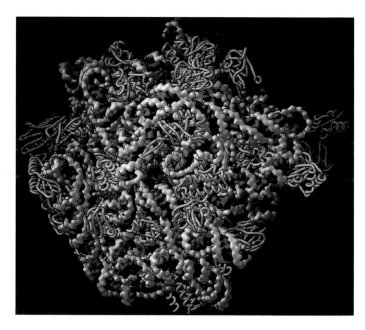

FIGURE 15.3
Ribosomes are very complex machines. The complete atomic structure of a prokaryotic large ribosomal subunit has been determined at 2.4 Å resolution. The RNA of the subunit is shown in gray and the proteins in gold. The subunit's RNA is twisted into irregular shapes that fit together like a three-dimensional jigsaw puzzle. The chemical reactions that form the peptide bond in protein synthesis are carried out deep in the interior by RNA. The ribosome is thus a ribozyme. Proteins are absent from the active site but abundant everywhere on the surface. The proteins stabilize the structure by interacting with adjacent RNA strands.

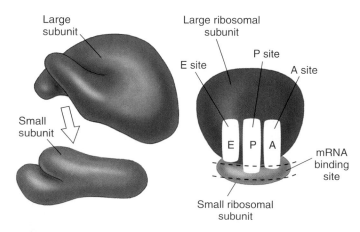

FIGURE 15.2
A ribosome is composed of two subunits. The smaller subunit fits into a depression on the surface of the larger one. The A, P, and E sites on the ribosome, discussed later in this chapter, play key roles in protein synthesis.

The Central Dogma

All organisms, from the simplest bacteria to humans, use the same basic mechanism of reading and expressing genes, a mechanism so fundamental to life as we know it that it is often referred to as the **Central Dogma**: Information passes from the genes (DNA) to an RNA copy of the gene, and the RNA copy directs the sequential assembly of a chain of amino acids (figure 15.5). Stated briefly,

$$DNA \rightarrow RNA \rightarrow protein$$

Transcription: An Overview

The first step of the Central Dogma is the transfer of information from DNA to RNA, which occurs when an mRNA copy of the gene is produced. Because the DNA sequence in the gene is transcribed into an RNA sequence, this stage is called **transcription**. Transcription is initiated when the enzyme **RNA polymerase** binds to a **promoter** binding site located at the beginning of a gene. Starting there, the RNA polymerase moves along the strand into the gene. As it encounters each DNA nucleotide, it adds the corresponding complementary RNA nucleotide to a growing mRNA strand. Thus, guanine (G), cytosine (C), thymine (T), and adenine (A) in the DNA would signal the addition of C, G, A, and uracil (U), respectively, to the mRNA.

When the RNA polymerase arrives at a transcriptional "stop" signal at the opposite end of the gene, it disengages from the DNA and releases the newly assembled RNA chain. This chain is a complementary transcript of the gene from which it was copied.

Translation: An Overview

The second step of the Central Dogma is the transfer of information from RNA to protein, which occurs when the information contained in the mRNA transcript is used to direct the sequence of amino acids during the synthesis of polypeptides by ribosomes. This process is called **translation** because the nucleotide sequence of the mRNA transcript is translated into an amino acid sequence in the polypeptide. Translation begins when an rRNA molecule within the ribosome recognizes and binds to a "start" sequence on the mRNA. The ribosome then moves along the mRNA molecule, three nucleotides at a time. Each group of three nucleotides is a code word that specifies which amino acid will be added to the growing polypeptide chain. The ribosome continues in this fashion until it encounters a translational "stop" signal; then it disengages from the mRNA and releases the completed polypeptide.

The two steps of the Central Dogma, taken together, are a concise summary of the events involved in the expression of an active gene. Biologists refer to this process as **gene expression**.

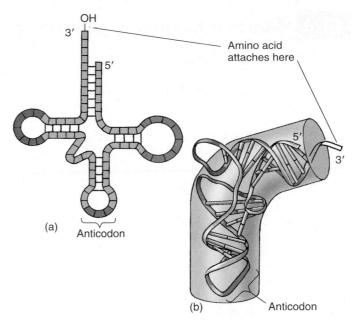

FIGURE 15.4

The structure of tRNA. (*a*) In this two-dimensional schematic, the three loops of tRNA are unfolded. Two of the loops bind to the ribosome during polypeptide synthesis, and the third loop contains the anticodon sequence, which is complementary to a three-base sequence on messenger RNA. Amino acids attach to the free, single-stranded —OH end. (*b*) In this three-dimensional structure, the loops of tRNA are folded.

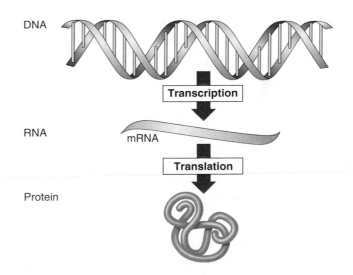

FIGURE 15.5

The Central Dogma of gene expression. DNA is transcribed to make mRNA, which is translated to make a protein.

The information encoded in genes is expressed in two phases: transcription, which produces an mRNA molecule whose sequence is complementary to the DNA sequence of the gene; and translation, which assembles a polypeptide whose amino acid sequence is specified by the nucleotide sequence in the mRNA.

15.2 Genes encode information in three-nucleotide code words.

The Genetic Code

The essential question of gene expression is, "How does the *order* of nucleotides in a DNA molecule encode the information that specifies the order of amino acids in a polypeptide?" The answer came in 1961, through an experiment led by Francis Crick. That experiment was so elegant and the result so critical to understanding the genetic code that we will describe it in detail.

Proving Code Words Have Only Three Letters

Crick and his colleagues reasoned that the genetic code most likely consisted of a series of blocks of information called **codons,** each corresponding to an amino acid in the encoded protein. They further hypothesized that the information within one codon was probably a sequence of three nucleotides specifying a particular amino acid. They arrived at the number three because a two-nucleotide codon would not yield enough combinations to code for the 20 different amino acids that commonly occur in proteins. With four DNA nucleotides (G, C, T, and A), only 4^2, or 16, different pairs of nucleotides could be formed. However, these same nucleotides can be arranged in 4^3, or 64, different combinations of three, more than enough to code for the 20 amino acids.

In theory, the codons in a gene could lie immediately adjacent to each other, forming a continuous sequence of transcribed nucleotides. Alternatively, the sequence could be punctuated with untranscribed nucleotides between the codons, like the spaces that separate the words in this sentence. It was important to determine which method cells employ because these two ways of transcribing DNA imply different translating processes.

To choose between these alternative mechanisms, Crick and his colleagues used a chemical to delete one, two, or three nucleotides from a viral DNA molecule and then asked whether a gene downstream of the deletions was transcribed correctly. When they made a single deletion or two deletions near each other, the **reading frame** of the genetic message shifted, and the downstream gene was transcribed as nonsense. However, when they made three deletions, the correct reading frame was restored, and the sequences downstream were transcribed correctly. They obtained the same results when they made additions to the DNA consisting of one, two, or three nucleotides. As shown in figure 15.6, these results could not have been obtained if the codons were punctuated by untranscribed nucleotides. Thus, Crick and his colleagues concluded that the genetic code is read in increments consisting of three nucleotides (in other words, it is a **triplet code**) and that reading occurs continuously without punctuation between the three-nucleotide units.

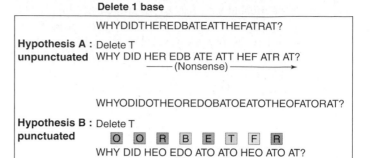

Delete 1 base

WHYDIDTHEREDBATEATTHEFATRAT?

Hypothesis A : Delete T
unpunctuated WHY DID HER EDB ATE ATT HEF ATR AT?
————————(Nonsense)————————→

WHYODIDOTHEOREDOBATOEATOTHEOFATORAT?

Hypothesis B : Delete T
punctuated   O O R B E T F R
WHY DID HEO EDO ATO ATO HEO ATO AT?
————————(Nonsense)————————→

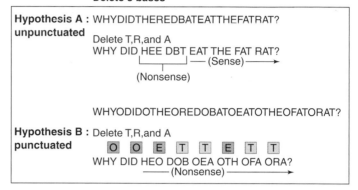

Delete 3 bases

Hypothesis A : WHYDIDTHEREDBATEATTHEFATRAT?
unpunctuated
Delete T,R,and A
WHY DID HEE DBT EAT THE FAT RAT?
———— (Sense)————→
(Nonsense)

WHYODIDOTHEOREDOBATOEATOTHEOFATORAT?

Hypothesis B : Delete T,R,and A
punctuated   O O E T T E T T
WHY DID HEO DOB OEA OTH OFA ORA?
————————(Nonsense)————————→

FIGURE 15.6
Using frameshift alterations of DNA to determine if the genetic code is punctuated.   The hypothetical genetic message presented here is, "Why did the red bat eat the fat rat?" Under hypothesis B, which proposes that the message is punctuated, the three-letter words are separated by nucleotides that are not read (indicated by the letter "O").

Breaking the Genetic Code

Within a year of Crick's experiment, other researchers succeeded in determining the amino acids specified by particular three-nucleotide units. Marshall Nirenberg discovered in 1961 that adding the synthetic mRNA molecule polyU (an RNA molecule consisting of a string of uracil nucleotides) to cell-free systems resulted in the production of the polypeptide polyphenylalanine (a string of phenylalanine amino acids). Therefore, one of the three-nucleotide sequences specifying phenylalanine is UUU. In 1964, Nirenberg and Philip Leder developed a powerful **triplet binding assay** in which a specific triplet was tested to see which radioactive amino acid (complexed to tRNA) it would bind. Some 47 of the 64 possible triplets gave unambiguous results. Har Gobind Khorana decoded the remaining 17 triplets by constructing artificial mRNA molecules of defined sequence and examining what polypeptides they directed. In these ways, all 64 possible three-nucleotide sequences were tested, and the full genetic code was determined (table 15.1).

Table 15.1 The Genetic Code

Table 15.1 The Genetic Code

First Letter	Second Letter								Third Letter
	U		**C**		**A**		**G**		
U	UUU	Phenylalanine	UCU	Serine	UAU	Tyrosine	UGU	Cysteine	U
	UUC		UCC		UAC		UGC		C
	UUA	Leucine	UCA		UAA	Stop	UGA	Stop	A
	UUG		UCG		UAG	Stop	UGG	Tryptophan	G
C	CUU	Leucine	CCU	Proline	CAU	Histidine	CGU	Arginine	U
	CUC		CCC		CAC		CGC		C
	CUA		CCA		CAA	Glutamine	CGA		A
	CUG		CCG		CAG		CGG		G
A	AUU	Isoleucine	ACU	Threonine	AAU	Asparagine	AGU	Serine	U
	AUC		ACC		AAC		AGC		C
	AUA		ACA		AAA	Lysine	AGA	Arginine	A
	AUG	Methionine; Start	ACG		AAG		AGG		G
G	GUU	Valine	GCU	Alanine	GAU	Aspartate	GGU	Glycine	U
	GUC		GCC		GAC		GGC		C
	GUA		GCA		GAA	Glutamate	GGA		A
	GUG		GCG		GAG		GGG		G

A codon consists of three nucleotides read in the sequence shown. For example, ACU codes for threonine. The first letter, A, is in the First Letter column; the second letter, C, is in the Second Letter column; and the third letter, U, is in the Third Letter column. Each of the mRNA codons is recognized by a corresponding anticodon sequence on a tRNA molecule. Some tRNA molecules recognize more than one codon in mRNA, but they always code for the same amino acid. In fact, most amino acids are specified by more than one codon. For example, threonine is specified by four codons, which differ only in the third nucleotide (ACU, ACC, ACA, and ACG).

The Code Is Practically Universal . . .

The genetic code is the same in almost all organisms. For example, the codon AGA specifies the amino acid arginine in bacteria, humans, and all other organisms whose genetic code has been studied. The universality of the genetic code is among the strongest evidence that all living things share a common evolutionary heritage. Because the code is universal, genes transcribed from one organism can be translated in another; the mRNA is fully able to dictate a functionally active protein. Similarly, genes can be transferred from one organism to another and be successfully transcribed and translated in their new host. This universality of gene expression is central to many of the advances of genetic engineering. Many commercial products, such as the insulin used to treat diabetes, are now manufactured by placing human genes into bacteria, which then serve as tiny factories to turn out prodigious quantities of insulin.

. . . But Not Quite

In 1979, investigators began to determine the complete nucleotide sequences of the mitochondrial genomes in humans, cattle, and mice. It came as something of a shock when these investigators learned that the genetic code used by these mammalian mitochondria was not quite the same as the "universal code" that has become so familiar to biologists. In the mitochondrial genomes, what should have been a "stop" codon, UGA, was instead read as the amino acid tryptophan; AUA was read as methionine rather than isoleucine; and AGA and AGG were read as "stop" rather than arginine. Furthermore, minor differences from the universal code have also been found in the genomes of chloroplasts and ciliates (certain types of protists).

Thus, it appears that the genetic code is not quite universal. Some time ago, presumably after they began their endosymbiotic existence, mitochondria and chloroplasts began to read the code differently, particularly the portion of the code associated with "stop" signals.

Within genes that encode proteins, the nucleotide sequence of DNA is read in blocks of three consecutive nucleotides, without punctuation between the blocks. Each block, or codon, codes for one amino acid.

15.3 Genes are first transcribed, then translated.

Transcription in Prokaryotes

The first step in gene expression is the production of an RNA copy of the DNA sequence encoding the gene, a process called **transcription**. To understand the mechanism behind transcription, it is useful to focus first on prokaryotes, where the process is well understood (figure 15.7).

RNA Polymerase

Prokaryotic RNA polymerase is very large and complex, consisting of five subunits: Two α subunits bind regulatory proteins, a β′ subunit binds the DNA template, a β subunit binds RNA nucleoside subunits, and a σ subunit recognizes the promoter and initiates synthesis. Only one of the two strands of DNA, called the **template strand**, is transcribed (that is, copied during transcription). The RNA transcript's sequence is complementary to the template strand. The strand of DNA that is not transcribed is called the **coding strand**. It has the same sequence as the RNA transcript, except T takes the place of U. The coding strand is also known as the *sense (+) strand*, and the template strand as the *antisense (−) strand*.

In prokaryotes, the polymerase adds ribonucleotides to the growing 3′ end of an RNA chain. No primer is needed, and synthesis proceeds in the 5′ to 3′ direction as in DNA replication.

Promoter

Transcription starts at RNA polymerase binding sites called **promoters** on the DNA template strand. A promoter is a short sequence that is not itself transcribed by the polymerase that binds to it. Striking similarities are evident in the sequences of different promoters. For example, two six-base sequences are common to many bacterial promoters: a TTGACA sequence called the **−35 sequence,** located 35 nucleotides upstream of the position where transcription actually starts, and a TATAAT sequence called the **−10 sequence,** located 10 nucleotides upstream of the "start" site.

Bacterial promoters differ widely in efficiency. Strong promoters cause frequent initiations of transcription—as often as every 2 seconds in some bacteria. Weak promoters may transcribe only once every 10 minutes. Most strong promoters have unaltered –35 and –10 sequences, while weak promoters often have substitutions within these sites.

Initiation

The binding of RNA polymerase to the promoter is the first step in gene transcription. In prokaryotes, a subunit of RNA polymerase called σ (sigma) recognizes the –10 sequence in

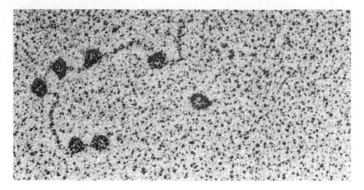

FIGURE 15.7
RNA polymerase.   In this electron micrograph, the dark circles are RNA polymerase molecules bound to several promoter sites on bacterial virus DNA.

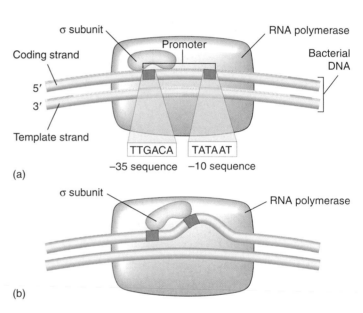

FIGURE 15.8
Initiation of transcription by the σ subunit.   (*a*) RNA polymerase with its σ subunit slides along the DNA until it recognizes the promoter and binds to it. (*b*) Upon binding, the σ subunit begins unwinding the DNA double helix at the −10 sequence, exposing the template strand to be transcribed.

the promoter and binds RNA polymerase there. Importantly, this subunit can detect the –10 sequence without unwinding the DNA double helix.

Once bound to the promoter, the RNA polymerase begins to unwind the DNA helix (figure 15.8). Measurements indicate that bacterial RNA polymerase unwinds a segment approximately 17 base-pairs long, nearly two turns of the DNA double helix. This sets the stage for the assembly of the RNA chain.

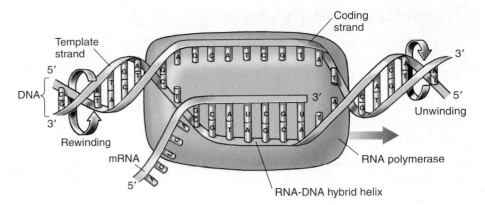

FIGURE 15.9
Model of a transcription bubble. The DNA duplex unwinds as it enters the RNA polymerase complex and rewinds as it leaves. One of the strands of DNA functions as a template, and nucleotide building blocks are assembled into RNA from this template.

Elongation

In prokaryotes, the transcription of the RNA chain usually starts with ATP or GTP. One of these forms the 5′ end of the chain, which grows in the 5′ to 3′ direction as ribonucleotides are added. Unlike DNA synthesis, a primer is not required. The region containing the RNA polymerase, DNA, and growing RNA transcript is called the **transcription bubble** because it contains a locally unwound "bubble" of DNA (figure 15.9). Within the bubble, the first 12 bases of the newly synthesized RNA strand temporarily form a helix with the template DNA strand. Corresponding to not quite one turn of the helix, this stabilizes the positioning of the 3′ end of the RNA so it can interact with an incoming ribonucleotide. The RNA-DNA hybrid helix rotates each time a nucleotide is added so that the 3′ end of the RNA stays at the catalytic site.

The transcription bubble moves down the bacterial DNA at a constant rate, about 50 nucleotides per second, leaving the growing RNA strand protruding from the bubble. After the transcription bubble passes, the now transcribed DNA is rewound as it leaves the bubble.

Unlike DNA polymerase, RNA polymerase has no proofreading capability. Transcription thus produces many more copying errors than replication. These mistakes, however, are not transmitted to progeny. Most genes are transcribed many times, so a few faulty copies are not harmful.

Termination

At the end of a bacterial gene are "stop" sequences that cause the formation of phosphodiester bonds to cease, the RNA-DNA hybrid within the transcription bubble to dissociate, the RNA polymerase to release the DNA, and the DNA within the transcription bubble to rewind. The simplest "stop" signal is a series of G–C base-pairs followed by a series of A–T base-pairs. The RNA transcript of this stop region forms a GC hairpin (figure 15.10),

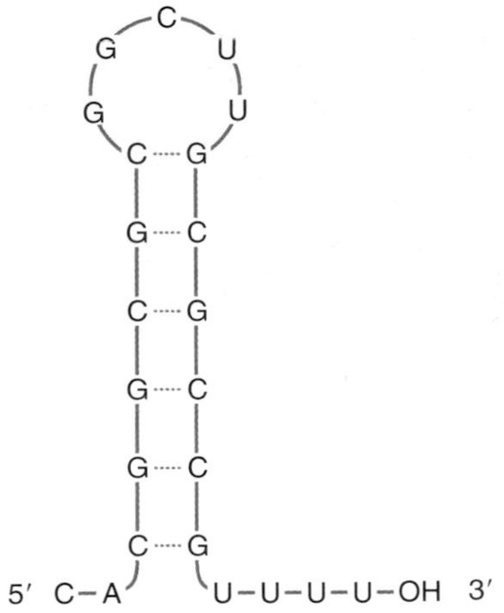

FIGURE 15.10
A GC hairpin. This structure stops gene transcription.

followed by four or more U ribonucleotides. How does this structure terminate transcription? The hairpin causes the RNA polymerase to pause immediately after the polymerase has synthesized it, placing the polymerase directly over the run of four uracils. The pairing of U with DNA's A is the weakest of the four hybrid base-pairs and is not strong enough to hold the hybrid strands together during the long pause. Instead, the RNA strand dissociates from the DNA within the transcription bubble, and transcription stops. A variety of protein factors aid hairpin loops in terminating transcription of particular genes.

Transcription in prokaryotes is carried out by the enzyme RNA polymerase, which unwinds and transcribes the DNA.

Transcription in Eukaryotes

The basic mechanism of transcription by RNA polymerase is the same in eukaryotes as in prokaryotes. However, a number of details about the two processes differ enough that it is necessary to consider them separately. Here we will concentrate only on how eukaryotic systems differ from prokaryotic systems, such as the bacterial system just discussed. All other features may be assumed to be the same.

Multiple RNA Polymerases

Unlike prokaryotes, which have a single RNA polymerase enzyme, eukaryotes have three different RNA polymerases that are distinguished by both structure and function. The enzyme RNA polymerase I is specialized to transcribe rRNA and recognizes promoters that are unique to this enzyme. RNA polymerase II is specialized to transcribe mRNA and some small nuclear RNAs. We will concentrate more on this enzyme later, due to its importance in transcribing all mRNAs in the cell. RNA polymerase III transcribes tRNA and other small RNAs. It too has a specific promoter structure. Together, these three enzymes accomplish all transcription in the nucleus of eukaryotic cells.

Promoters

One way the activities of the three different RNA polymerases are controlled is by the differences in the structure of their promoters. RNA pol I promoters at first puzzled biologists, because comparisons of rRNA genes between species showed no similarities outside the coding region. The current view is that these promoters are specific for each species, and for this reason, cross-species comparisons do not yield similarities.

The RNA pol II promoters are the most complex of the three eukaryotic promoters, probably a reflection of the huge diversity of genes that are transcribed by this polymerase. A "core promoter" contains a so-called **TATA box** that resembles the −10 sequence from bacteria. Other conserved elements are found in many but not all RNA pol II transcribed genes, such as a CAAT box that makes up a second core promoter found in most genes. General transcription factors bind in this region to assemble the initiation complex. Additional upstream elements unique to individual genes are regulatory elements, binding sites for transcription factors that control tissue and developmentally specific expression.

Promoters for RNA pol III also were a source of surprise for biologists examining the control of gene expression in the early days of molecular biology. A common technique for analyzing regulatory regions was to make successive deletions from the 5′ end of genes to see when

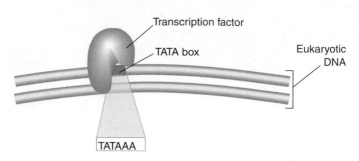

1. A transcription factor recognizes and binds to the TATA box sequence which is part of the core promoter.

2. Other transcription factors are then recruited, and the initiation complex begins to build.

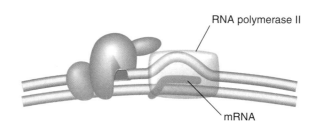

3. Ultimately, RNA polymerase II associates with the transcription factors and the DNA, forming the initiation complex, and transcription begins.

FIGURE 15.11
Eukaryotic initiation complex.   Unlike transcription in prokaryotic cells, where the RNA polymerase recognizes and binds to the promoter, eukaryotic transcription requires the binding of transcription factors to the promoter before RNA polymerase II binds to the DNA. The association of transcription factors and RNA polymerase II at the promoter is called the initiation complex.

these deletions abolished specific transcription. The logic was conditioned from experiences with prokaryotes in which the regulatory regions had been found at the 5′ end of genes. In the case of tRNA genes, the 5′ deletions had no effect on expression! The promoters were found to actually be internal to the gene itself.

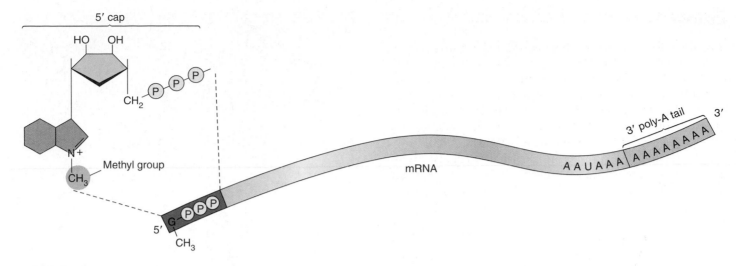

FIGURE 15.12
Posttranscriptional modifications. Eukaryotic mRNA molecules are modified in the nucleus with the addition of a methylated GTP to the 5' end of the transcript, called the 5' cap, and a long chain of adenine residues to the 3' end of the transcript, called the poly-A tail.

Initiation

The initiation of transcription at RNA pol II promoters is analogous to prokaryotic initiation but much more complex. Instead of the polymerase simply recognizing and binding to the promoter, a host of general transcription factors interact at the promoter and RNA pol II to form an *initiation complex* (figure 15.11). We will explore this in detail in chapter 18 when we discuss the control of gene expression.

Posttranscriptional Modifications

There are a number of differences between the transcription of mRNAs in prokaryotes and eukaryotes. Some of these involve the packaging of eukaryotic DNA in the nucleus of the cell, but others were unexpected when first discovered. Between transcription in the nucleus and export of a mature mRNA to the cytoplasm, a number of modifications are performed in the initial transcripts made by RNA pol II.

5' Cap. Eukaryotic transcripts have an unusual structure that is added to the 5' end of mRNAs. The first base in the transcript is usually an adenine (A) or a guanine (G), and this is further modified by the addition of GTP to the 5' PO₄ group, called a **5' cap.** (figure 15.12). This unusual 5' to 5' linkage is found only in this "cap" structure. The G in the GTP is also modified by the addition of a methyl group, so it is often called a methyl-G cap. The cap is added while transcription is still in progress. This cap protects the 5' end of the mRNA from degradation and is also involved in translation initiation.

3' Poly-A Tail. A major difference between transcription in prokaryotes and eukaryotes is that in eukaryotes, the end of the transcript is not the end of the mRNA. The eukaryotic transcript is cleaved downstream of a specific site (AAUAAA) that is not the termination site for trancription. A series of adenine (A) residues, called the **3' poly-A tail,** are added after this cleavage by an enzyme other than RNA polymerase II such that the end of the mRNA is not the end of the transcript and is not created by RNA pol II (figure 15.12). The enzyme that adds the A's is called, appropriately, poly-A polymerase. The poly-A tail appears to play a role in the stability of mRNAs by protecting them from degradation. We will discuss mRNA stability in chapter 18 as well.

Transcription in eukaryotes differs from that in prokaryotes in that there are three RNA polymerase enzymes, an initiation complex forms at the promoter, and the RNAs are modified after transcription.

Translation

In prokaryotes, translation begins when the initial portion of an mRNA molecule binds to an rRNA molecule in a ribosome. The mRNA lies on the ribosome in such a way that only one of its codons is exposed at the polypeptide-making site at any time. A tRNA molecule possessing the complementary three-nucleotide sequence, or anticodon, binds to the exposed codon on the mRNA.

Because this tRNA molecule carries a particular amino acid, that amino acid and no other is added to the polypeptide in that position. As the mRNA molecule moves through the ribosome, successive codons on the mRNA are exposed, and a series of tRNA molecules bind one after another to the exposed codons. Each of these tRNA molecules carries an attached amino acid, which it adds to the end of the growing polypeptide chain (figure 15.13).

There are about 45 different kinds of tRNA molecules. Why are there 45 and not 64 tRNAs (one for each codon)? Because some tRNAs recognize more than one codon due to the fact that the third base-pair of a tRNA anticodon allows some "wobble."

How do particular amino acids become associated with particular tRNA molecules? The key translation step, which pairs the three-nucleotide sequences with appropriate amino acids, is carried out by a remarkable set of enzymes called *activating enzymes*.

Activating Enzymes

Particular tRNA molecules become attached to specific amino acids through the action of activating enzymes called **aminoacyl-tRNA synthetases,** one of which exists for each of the 20 common amino acids (figure 15.14). Therefore, these enzymes must correspond to specific anticodon sequences on a tRNA molecule as well as particular amino acids. Some activating enzymes correspond to only one anticodon and thus to only one tRNA molecule. Others recognize two, three, four, or six different tRNA molecules, each with a different anticodon but coding for the same amino acid (see table 15.1). If we consider the nucleotide sequence of mRNA a coded message, then the 20 activating enzymes are responsible for decoding that message. Or, to put it another way, the attachment of specific amino acids to the appropriate tRNA is the actual translation step.

"Start" and "Stop" Signals

For three of the 64 codons (UAA, UAG, and UGA), there exists no tRNA with a complementary anticodon. These codons, called **nonsense codons,** serve as "stop" signals in the mRNA message, marking the end of a polypeptide. The "start" signal that marks the beginning of a polypeptide within an mRNA message is the codon AUG, which also encodes the amino acid methionine. The ribosome will usually use the first AUG it encounters in the mRNA to signal the start of translation.

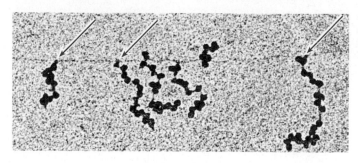

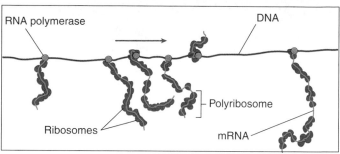

FIGURE 15.13

Translation in action. Bacteria have no nucleus and hence no membrane barrier between the DNA and the cytoplasm. In this electron micrograph of genes being transcribed in the bacterium *Escherichia coli*, you can see every stage of the process. The arrows point to RNA polymerase enzymes. From each mRNA molecule dangling from the DNA, a series of ribosomes is assembling polypeptides. These clumps of ribosomes are sometimes called polyribosomes.

Initiation

In prokaryotes, polypeptide synthesis begins with the formation of an **initiation complex.** First, a tRNA molecule carrying a chemically modified methionine called N-formylmethionine (tRNA^{fMet}) binds to the small ribosomal subunit. Proteins called **initiation factors** position the tRNA^{fMet} on the ribosomal surface at the *P site* (for peptidyl), where peptide bonds will form. Nearby, two other sites will form: the *A site* (for aminoacyl), where successive amino-acid-bearing tRNAs will bind, and the *E site* (for exit), where empty tRNAs will exit the ribosome (figure 15.15). This initiation complex, guided by another initiation factor, then binds to the anticodon AUG on the mRNA. Proper positioning of the mRNA is critical because it determines the reading frame—that is, which groups of three nucleotides will be read as codons. Moreover, the complex must bind to the beginning of the mRNA molecule, so that all of the transcribed gene will be translated. In prokaryotes, the beginning of each mRNA molecule is marked by a *leader sequence* complementary to one of the rRNA molecules on the ribosome. This complementarity ensures that the mRNA is read from the beginning. Prokaryotes often include several genes within a single mRNA transcript (polycistronic mRNA), while each eukaryotic gene is transcribed on a separate mRNA (monocistronic mRNA).

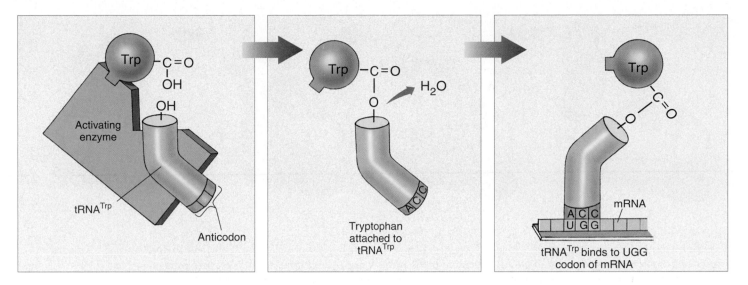

FIGURE 15.14
Activating enzymes "read" the genetic code. Each kind of activating enzyme recognizes and binds to a specific amino acid, such as tryptophan; it also recognizes and binds to the tRNA molecules with anticodons specifying that amino acid, such as ACC for tryptophan. In this way, activating enzymes link the tRNA molecules to specific amino acids.

Initiation in eukaryotes is similar, although it differs in two important ways. First, in eukaryotes, the initiating amino acid is methionine rather than *N*-formylmethionine. Second, the initiation complex is far more complicated than in prokaryotes, containing nine or more protein factors, many consisting of several subunits. Eukaryotic initiation complexes are discussed in detail in chapter 18.

Elongation

After the initiation complex has formed, the large ribosomal subunit binds, exposing the mRNA codon adjacent to the initiating AUG codon, and so positioning it for interaction with another amino-acid-bearing tRNA molecule. When a tRNA molecule with the appropriate anticodon appears, proteins called *elongation factors* assist in binding it to the exposed mRNA codon at the A site. When the second tRNA binds to the ribosome, it places its amino acid directly adjacent to the initial methionine, which is still attached to its tRNA molecule, which in turn is still bound to the ribosome. The two amino acids undergo a chemical reaction, catalyzed by the large ribosomal subunit, which releases the initial methionine from its tRNA and attaches it instead by a peptide bond to the second amino acid.

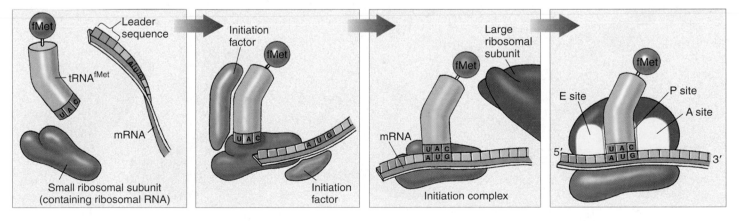

FIGURE 15.15
Formation of the initiation complex. In prokaryotes, proteins called initiation factors play key roles in positioning the small ribosomal subunit and the *N*-formylmethionine tRNA, or tRNA^fMet, molecule at the beginning of the mRNA. When the tRNA^fMet is positioned over the first AUG codon of the mRNA, the large ribosomal subunit binds, forming the P, A, and E sites where successive tRNA molecules bind to the ribosomes, and polypeptide synthesis begins.

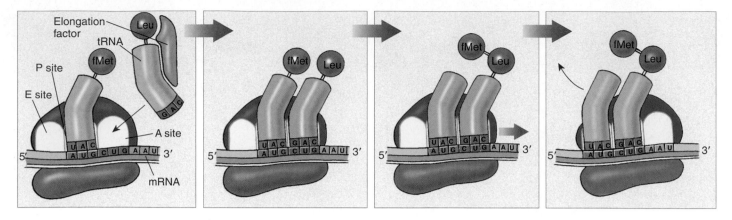

FIGURE 15.16

Translocation. The initiating tRNA^fMet in prokaryotes (tRNA^Met in eukaryotes) occupies the P site, and a tRNA molecule with an anticodon complementary to the exposed mRNA codon binds at the A site. fMet is transferred to the incoming amino acid (Leu in this case), as the ribosome moves three nucleotides to the right along the mRNA. The empty tRNA^fMet moves to the E site to exit the ribosome, the growing polypeptide chain moves to the P site, and the A site is again exposed and ready to bind the next amino-acid-laden tRNA.

Translocation

In a process called **translocation** (figure 15.16), the ribosome now moves (translocates) three more nucleotides along the mRNA molecule in the 5′ to 3′ direction, guided by other elongation factors. This movement relocates the initial tRNA to the E site and ejects it from the ribosome, repositions the growing polypeptide chain (at this point containing two amino acids) to the P site, and exposes the next codon on the mRNA at the A site. When a tRNA molecule recognizing that codon appears, it binds to the codon at the A site, placing its amino acid adjacent to the growing chain. The chain then transfers to the new amino acid, and the entire process is repeated.

Termination

Elongation continues in this fashion until a chain-terminating nonsense codon is exposed (for example, UAA in figure 15.17). Nonsense codons do not bind to tRNA, but they are recognized by **release factors,** proteins that release the newly made polypeptide from the ribosome.

The first step in protein synthesis is the formation of an initiation complex. Each step of the ribosome's progress exposes a codon to which a tRNA molecule with the complementary anticodon binds. The amino acid carried by each tRNA molecule is added to the end of the growing polypeptide chain.

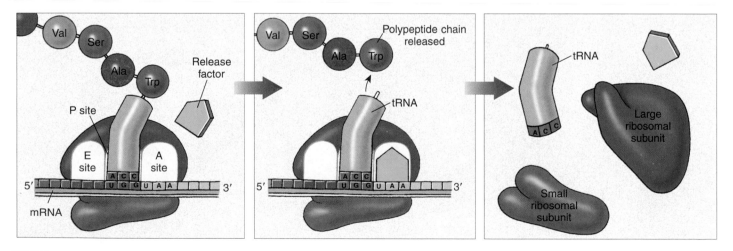

FIGURE 15.17

Termination of protein synthesis. There is no tRNA with an anticodon complementary to any of the three termination signal codons, such as the UAA nonsense codon illustrated here. When a ribosome encounters a termination codon, it therefore stops translocating. A specific release factor facilitates the release of the polypeptide chain by breaking the covalent bond that links the polypeptide to the P-site tRNA.

15.4 Eukaryotic gene transcripts are spliced.

The Discovery of Introns

The first genes isolated were prokaryotic genes found in *E. coli* and its viruses. A clear picture of the nature and some of the control of gene expression had been built up in these systems before the advent of the molecular cloning of eukaryotic genes. It was expected that while details would differ, the outline of gene expression would be similar. The world of biology was in for a shock when the first genes were isolated from eukaryotic organisms. Many of these genes appeared to contain sequences that were not represented in the mRNA. It is hard to exaggerate how unexpected this finding was. A basic tenant of molecular biology that had grown out of work on *E. coli* was that a gene was colinear with its protein product. This means that the sequence of bases in the gene corresponds to the bases in the mRNA, which in turn corresponds to the sequence of amino acids in the protein. In the case of eukaryotic genes, the genes are interrupted by sequences that are not represented in the mRNA and the protein. The term "split genes" was used at the time, but the nomenclature that has stuck describes the unexpected nature of sequences within genes. We call the noncoding DNA that interrupts the sequence of the gene "intervening sequences," or **introns,** and we call the coding sequences **exons** (figure 15.18).

Thus, the basic structure of eukaryotic genes is radically different from that of prokaryotic genes. It is still true that translated eukaryotic mRNA is colinear with its protein product, but the gene may not be. Imagine looking at an interstate highway from a satellite. Scattered randomly along the thread of concrete would be cars, some moving in clusters, others individually; most of the road would be bare. That is what a eukaryotic gene is like—scattered exons embedded within much longer sequences of introns. In humans, only 1 to 1.5% of the genome is devoted to the exons that encode proteins, while 24% is devoted to the noncoding introns within which these exons are embedded.

RNA Splicing

This raises the obvious question of how eukaryotic cells deal with the noncoding introns. The answer is that the product of transcription, called the primary transcript, in addition to the capping and poly-A tail reactions, is cut and put back together to produce the mature mRNA. We call this latter process *RNA splicing*, and it occurs in the nucleus prior to the export of the mRNA to the cytoplasm. The intron/exon junctions are recognized by small nuclear ribonuclearproteins called snRNPs (pronounced "snurps"). These snurps then cluster together forming a

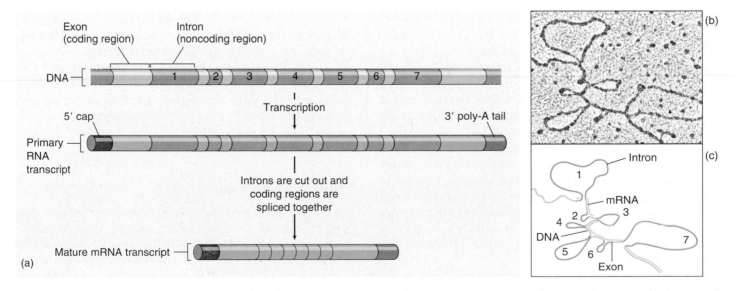

FIGURE 15.18
The eukaryotic ovalbumin gene is fragmented. (*a*) The ovalbumin gene and its primary RNA transcript contain seven segments not present in the mRNA that the ribosomes use to direct protein synthesis. Enzymes cut these segments (introns) out and splice together the remaining segments (exons). (*b*) By hybridizing the processed transcript to the DNA, introns within the DNA sequence can be visualized directly. The seven loops in the electron micrograph are the seven introns represented in (*c*) the schematic drawing.

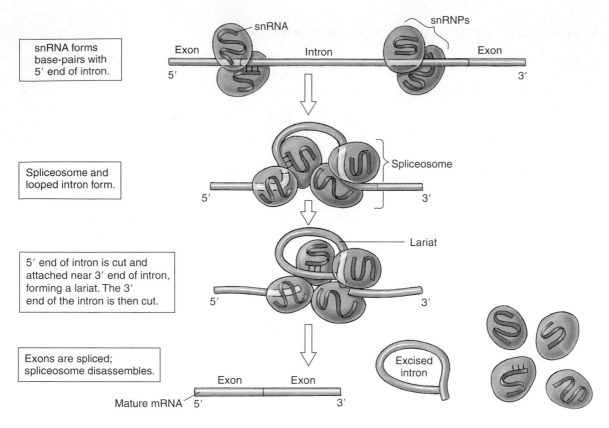

snRNA forms base-pairs with 5′ end of intron.

Spliceosome and looped intron form.

5′ end of intron is cut and attached near 3′ end of intron, forming a lariat. The 3′ end of the intron is then cut.

Exons are spliced; spliceosome disassembles.

snRNA

snRNPs

Exon   Intron   Exon

5′   3′

Spliceosome

5′   3′

Lariat

5′   3′

Excised intron

Exon   Exon

Mature mRNA   5′   3′

FIGURE 15.19

How spliceosomes process RNA.   Particles called snRNPs contain snRNA that interacts with the 5′ end of an intron. Several snRNPs come together and form a spliceosome. As the intron forms a loop, the 5′ end is cut and linked to a site near the 3′ end of the intron. The intron forms a lariat that is excised, and the exons are spliced together. The spliceosome then disassembles and releases the mature mRNA.

larger complex called the **spliceosome,** which is responsible for the splicing, or removal, of the introns. The mechanism of splicing includes cleavage of the 5′ end of the intron such that this 5′ end becomes attached to the 2′ OH of an internal A nucleotide, forming a looped structure called a *lariat* (figure 15.19). The 3′ end of the first exon is then used to displace the 3′ end of the intron, joining the two exons together and releasing the intron.

There are no rules for the numbers of introns per gene or the sizes of introns and exons. Some genes have no introns; others may have 50 introns. The sizes of exons range from a few nucleotides to 7500 nucleotides, and the sizes of introns are equally variable. The presence of introns partly explains why so little of the genome is actually composed of "coding sequences" (see chapter 17 for results from the Human Genome Project).

The "how" of splicing is a lot easier to answer than the "why" of introns. In fact, "why are there introns?" may not even be a valid question. There is no consensus as to whether introns arrived early or late in the evolution of eukaryotic lineages; it may not be an all-or-none issue. Researchers have proposed a number of explanations for how and why introns evolved. One of the more provocative explanations suggests that exons represent functional domains of proteins, and that the intron-exon arrangements found in genes represent the shuffling of these functional units over long periods of evolutionary time. This hypothesis, called *exon shuffling,* has support based on the structure of some genes, but is not supported by the structure of other genes. The degree to which shuffling has been an evolutionary mechanism remains unclear.

Alternative Splicing

One consequence of the splicing process is greater complexity in gene expression in eukaryotes. A single primary transcript can be spliced into different mRNAs by the inclusion of different sets of exons, a process called **alternative splicing.** Although many cases of alternative splicing have been documented, the recent completion of the draft sequence of the human genome, along with other large data sets of expressed sequences, now allow large-scale comparisons between sequences found in mRNAs and the genome. Three different computer-based

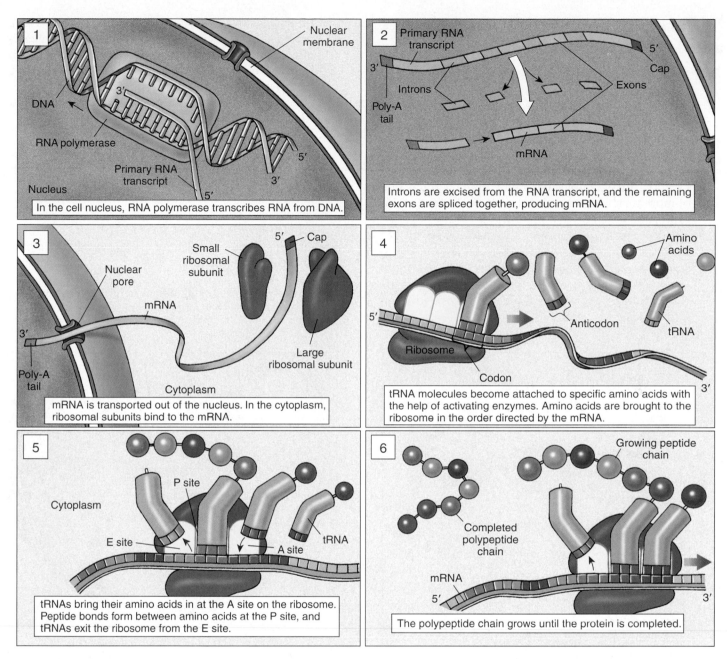

FIGURE 15.20
An overview of gene expression in eukaryotes.

analyses have been performed, producing results that are in rough agreement. These initial genomic assessments indicate a range of 35 to 59% for human genes that exhibit some form of alternative splicing. If we pick the middle ground of around 40%, this still vastly increases the number of potential proteins encoded by the 30,000 genes in the human genome. It is important to point out that these analyses are entirely computer-based and have not been experimentally verified, nor have the functions of the possible spliced products been investigated. However, this does explain how the 30,000 genes

found in the human genome are able to encode the 120,000 different translated mRNAs reported to exist in human cells. The emerging field of proteomics will address the number and functioning of proteins encoded by the human genome. Figure 15.20 summarizes eukaryotic transcription and translation.

Much of the eukaryotic gene is not translated. Noncoding segments called introns, scattered throughout the gene, are removed from the primary transcript before the mRNA is translated.

Differences Between Prokaryotic and Eukaryotic Gene Expression

1. Most eukaryotic genes possess introns. With the exception of a few genes in the Archaebacteria, prokaryotic genes lack introns (figure 15.21).

2. Individual prokaryotic mRNA molecules often contain transcripts of several genes. By placing genes with related functions on the same mRNA, prokaryotes coordinate the regulation of those functions. Eukaryotic mRNA molecules rarely contain transcripts of more than one gene. Regulation of eukaryotic gene expression is achieved in other ways.

3. Because eukaryotes possess a nucleus, their mRNA molecules must be completely formed and must pass across the nuclear membrane before they are translated. Prokaryotes, which lack nuclei, often begin translation of an mRNA molecule before its transcription is completed.

4. In prokaryotes, translation begins at an AUG codon preceded by a special nucleotide sequence. In eukaryotic cells, mRNA molecules are modified at the 5′ leading end after transcription, adding a 5′ cap, a methylated guanosine triphosphate. The cap initiates translation by binding the mRNA, usually at the first AUG, to the small ribosomal subunit.

5. Eukaryotic mRNA molecules are modified before they are translated: Introns are cut out, and the remaining exons are spliced together; a 5′ cap is added; and a 3′ poly-A tail consisting of some 200 adenine (A) nucleotides is added. These modifications can delay the destruction of the mRNA by cellular enzymes.

6. The ribosomes of eukaryotes are a little larger than those of prokaryotes.

Gene expression is broadly similar in prokaryotes and eukaryotes, although it differs in some details.

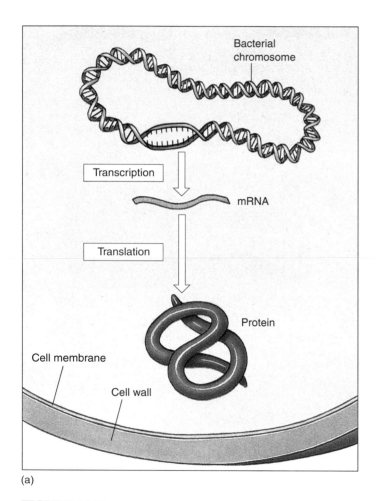

(a)

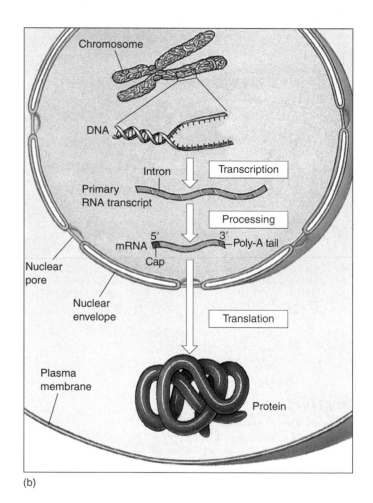

(b)

FIGURE 15.21

Gene information is processed differently in prokaryotes and eukaryotes. (*a*) Prokaryotic genes are transcribed into mRNA, which is translated immediately. Hence, the sequence of DNA nucleotides corresponds exactly to the sequence of amino acids in the encoded polypeptide. (*b*) Eukaryotic genes are typically different, containing long stretches of nucleotides called introns that do not correspond to amino acids within the encoded polypeptide. Introns are removed from the primary RNA transcript of the gene, and a 5′ cap and 3′ poly-A tail are added before the mRNA directs the synthesis of the polypeptide.

15.1 The Central Dogma traces the flow of gene-encoded information.

Cells Use RNA to Make Protein

- Ribosomes are composed of a large and a small subunit. (p. 302)
- Ribosomal RNA (rRNA) functions in polypeptide synthesis; transfer RNA (tRNA) transports amino acids to the ribosome for use in building polypeptides; and messenger RNA (mRNA) is transcribed from DNA and travels to ribosomes to direct polypeptide assembly. (p. 302)
- The Central Dogma is that information passes from DNA to mRNA (transcription), and then mRNA directs the sequential assembly of amino acids into proteins (translation). (p. 303)

15.2 Genes encode information in three-nucleotide code words.

The Genetic Code

- The genetic code is nearly universal. It is read in codons or triplets, increments of three nucleotides on mRNA, with each triplet coding for an amino acid. (pp. 304–305)
- There are 64 possible codons that code for 20 amino acids. (p. 304)

15.3 Genes are first transcribed, then translated.

Transcription in Prokaryotes

- Only the template strand of DNA is transcribed, while the coding strand is not. (p. 306)
- Transcription begins at the promoter site, and the transcription bubble moves along the DNA segment until the termination sequence is reached; the RNA strand then dissociates, and the DNA rewinds. (pp. 306–307)

Transcription in Eukaryotes

- Eukaryotic transcription differs from prokaryotic transcription because there are three different RNA polymerase enzymes, an initiation complex forms at the promoter, and posttranscriptional modification of the RNA occurs. (pp. 308–309)

Translation

- Translation begins when the initial portion of an mRNA molecule binds to an rRNA molecule in a ribosome. (p. 310)
- Each step of the ribosome's progress exposes a codon to which a tRNA molecule with the complementary anticodon binds. The amino acid carried by each tRNA molecule is added to the end of the growing polypeptide chain. (pp. 311–312)

15.4 Eukaryotic gene transcripts are spliced.

The Discovery of Introns

- Eukaryotic genes are interrupted by sequences not represented in the mature mRNA and the protein. Introns are noncoding sequences that interrupt coding sequences, or exons. (p. 313)
- Spliceosomes are responsible for splicing and removing the introns prior to mRNA translation. (p. 314)

Differences Between Prokaryotic and Eukaryotic Gene Expression

- Gene expression is similar in both prokaryotes and eukaryotes, although some differences do exist, such as ribosomal size and the possession of introns. (p. 316)

Self Test

1. The bases of RNA are the same as those of DNA with the exception that RNA contains
   a. cysteine instead of cytosine.
   b. uracil instead of thymine.
   c. cytosine instead of guanine.
   d. uracil instead of adenine.
2. Which one of the following is not a type of RNA?
   a. nRNA (nuclear RNA)
   b. mRNA (messenger RNA)
   c. rRNA (ribosomal RNA)
   d. tRNA (transfer RNA)
3. Each amino acid in a protein is specified by
   a. several genes.
   b. a promoter.
   c. an mRNA molecule.
   d. a codon.
4. The three-nucleotide codon system can be arranged into _____ combinations.
   a. 16
   b. 20
   c. 64
   d. 128
5. The TATA box in eukaryotes is a
   a. core promoter.
   b. −35 sequence.
   c. −10 sequence.
   d. 5′ cap.
6. The site where RNA polymerase attaches to the DNA molecule to start the formation of RNA is called a(n)
   a. promoter.
   b. exon.
   c. intron.
   d. GC hairpin.
7. When mRNA leaves the cell's nucleus, it next becomes associated with
   a. proteins.
   b. a ribosome.
   c. tRNA.
   d. RNA polymerase.
8. If an mRNA codon reads UAC, its complementary anticodon will be
   a. TUC.
   b. ATG.
   c. AUG.
   d. CAG.
9. The nucleotide sequences on DNA that actually have information encoding a sequence of amino acids are
   a. introns.
   b. exons.
   c. UAA.
   d. UGA.
10. Which of the following statements is correct about prokaryotic gene expression?
    a. Prokaryotic mRNAs must have introns spliced out.
    b. Prokaryotic mRNAs are often translated before transcription is complete.
    c. Prokaryotic mRNAs contain the transcript of only one gene.
    d. All of these statements are correct.

Test Your Visual Understanding

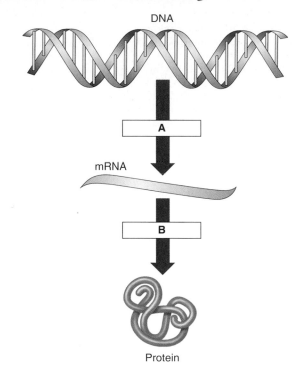

1. Match the correct labels from the following list with the lettered boxes on the figure (not all labels will be used).
   replication
   RNA processing
   transcription
   translation
   a. What is this figure illustrating?
   b. Where does process A occur in the eukaryotic cell?
   c. Where does process B occur in the eukaryotic cell?

Apply Your Knowledge

1. Assume you had four copies per cell of mature mRNA for insulin, a protein made of two polypeptide chains with a total of 51 amino acids. Assume one ribosome can attach onto an mRNA every 20 nucleotides. One amino acid can be translated in 60 milliseconds. How many copies of insulin could be made in three minutes?
2. The nucleotide sequence of a hypothetical eukaryotic gene is:
   TACATACTAGTTAC G TCGCCCGGAAATATC

   If a mutation in this gene were to change the fifteenth nucleotide (blue boxed nucleotide) from guanine to thymine, what effect might it have on the expression of this gene?

16

Gene Technology

Concept Outline

16.1 Molecular biologists can manipulate DNA to clone genes.

The DNA Manipulator's Toolbox. Enzymes that cleave DNA at specific sites allow DNA segments from different sources to be spliced together.

Host/Vector Systems. Fragments produced by cleaving DNA with restriction endonucleases can be spliced into plasmids, phages, or other vectors, which can be used to insert the DNA into host cells.

Using Vectors to Transfer Genes. Plasmids are used to transfer smaller fragments of DNA into host cells. Other vectors are used to transfer larger segments of DNA into hosts.

DNA Libraries. A DNA library is a collection of DNA fragments that comprise an organism's genome.

16.2 Genetic engineering involves easily understood procedures.

The Four Stages of a Genetic Engineering Experiment. Gene engineers cut DNA into fragments, which they then splice into vectors that carry the fragments into cells. The cells are then screened for the gene of interest.

Working with Gene Clones. Gene technology is used in a variety of procedures involving DNA manipulation.

16.3 Biotechnology is producing a scientific revolution.

Medical Applications. Many drugs and vaccines are now produced using gene technology, and gene technology is also being used in the treatment of genetic diseases.

Agricultural Applications. Genetic engineers have developed crops resistant to pesticides and pests, as well as commercially superior animals.

Risk and Regulation. Genetic engineering raises important questions about danger and privacy.

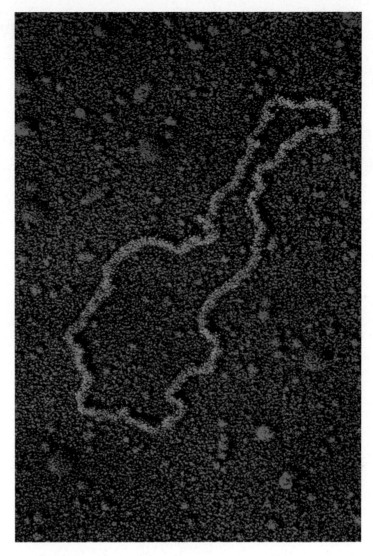

FIGURE 16.1
A famous plasmid. The circular molecule in this electron micrograph is pSC101, the first plasmid used successfully to clone a vertebrate gene. Its name comes from the fact that it was the 101st plasmid isolated by Stanley Cohen.

Over the past decades, the development of new and powerful techniques for studying and manipulating DNA has revolutionized genetics. The cloning plasmid shown in figure 16.1 is one example of the tools that are now available. These techniques have allowed biologists to intervene directly in the genetic fate of organisms for the first time. In this chapter, we explore these technologies and consider how they apply to specific problems of great practical importance. Few areas of biology will have as great an impact on our future lives.

16.1 Molecular biologists can manipulate DNA to clone genes.

The ability to directly isolate and manipulate genetic material was one of the most profound changes in the field of biology in the late twentieth century. Now, in the twenty-first century, we are already looking at entire genomes. So how did we get from cloning individual genes to determining the sequence of the human genome? We will begin to answer this question by describing the technology necessary to clone genes and how this technology is being applied.

The DNA Manipulator's Toolbox

Development of the technology behind cloning and manipulating DNA sequences began in 1975 with the construction of the first recombinant DNA molecules. From these first tentative steps, a technology evolved that has altered all of biology. To understand this technology, we need to first consider the tools required to work directly with DNA. From this, we will step through a simple genetic engineering experiment, and then look at how these techniques are being applied in medicine and agriculture.

One of the most basic requirements for manipulating DNA is a set of enzymes. The toolbox of the molecular biologist is filled with such enzymes. This toolbox has become extremely sophisticated over the years as more and more enzymes that alter DNA have been isolated. Notice that all of these enzymes already exist in nature, so that all of our manipulations are at some level merely imitations of what cells do naturally.

Restriction Endonucleases

The enzymes that catalyzed the revolution in molecular biology were those able to cleave DNA at specific sites: **restriction endonucleases.** Nucleases are enzymes that degrade DNA, and many were known prior to the isolation of the first restriction enzyme. But restriction endonucleases are different because they are able to fragment DNA at specific sites. This kind of activity, long sought by molecular biologists, eventually came out of basic research on why bacterial viruses can infect some cells and not others. The molecular basis for this "host restriction" proved to be enzymes that could cleave DNA at specific sequences. The host cells are able to avoid being cleaved by modifying their own DNA at the same sites by methylation. Since the initial discovery of these restriction endonucleases, hundreds more have been isolated that recognize and cleave different **restriction sites.**

The ability to cut DNA at specific sites is significant in two ways: First, it allows a form of physical mapping that was previously not possible, and second, it allows the creation of recombinant molecules. Physical maps can be constructed based on the positioning of cleavage sites for restriction enzymes. Such restriction maps provide crucial data for identifying and working with DNA molecules. The ability to construct recombinant molecules is even more important, because many steps in the process of cloning and manipulating DNA require the ability to put molecules from different sources together.

How do restriction enzymes allow the creation of recombinant molecules? The answer lies in the nature of the enzymes themselves. There are two types of restriction enzymes: type I and type II. Type I enzymes make simple cuts across both DNA strands, and are not often used in cloning and manipulating DNA. It is type II enzymes that allow creation of recombinant molecules. Type II enzymes recognize a specific DNA sequence, ranging from 4 bases to 12 bases, and cleave the DNA at a particular base within this sequence. Type II sites show what is called *dyad symmetry*, or twofold rotational symmetry. This can be seen in the sequence itself: It reads the same from 5′ to 3′ in one direction as it does on the other strand in the opposite direction 5′ to 3′. Given this kind of sequence, cutting the DNA at the same base on either strand can lead to staggered cuts that will produce "sticky ends." This short, unpaired sequence will be the same for any DNA that is cut by the enzyme. Thus, these sticky ends allow DNAs from different sources to be easily joined together (figure 16.2).

Ligase

The two ends of a DNA molecule cut by a type II restriction enzyme have complementary sequences as we have seen, and so can pair to form a duplex. But to form a stable DNA molecule from the two fragments, we still need an enzyme to join the molecules. The enzyme **DNA ligase** catalyzes the formation of a phosphodiester bond between adjacent phosphate and hydroxyl groups of DNA nucleotides. For a proper link to form, there can be no gaps in either strand, just a "nick" in each strand. A nick is a break in one strand with no gap. The action of ligase is to seal nicks in one or both strands. This is the same enzyme that joins Okazaki fragments on the lagging strand during DNA replication (see chapter 14). In the toolbox of the molecular biologist, the action of ligase is necessary to create stable recombinant molecules from the fragments that restriction enzymes make possible.

The tools of the molecular biologist are enzymes that cut DNA into fragments and other enzymes that join the DNA fragments into complete molecules.

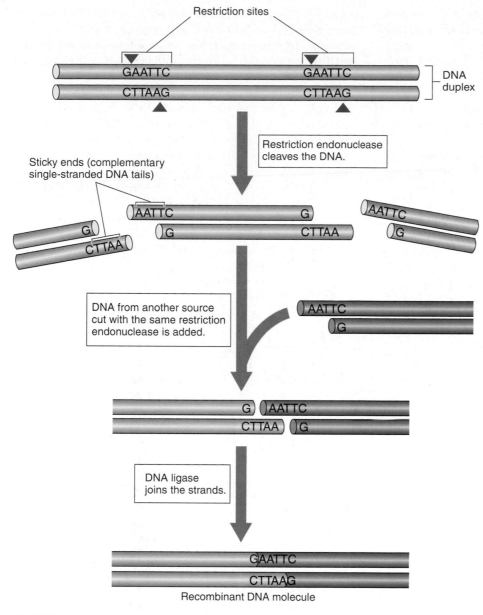

Restriction sites

GAATTC   GAATTC
CTTAAG   CTTAAG   } DNA duplex

Restriction endonuclease cleaves the DNA.

Sticky ends (complementary single-stranded DNA tails)

AATTC   G
G   CTTAA

AATTC
G

DNA from another source cut with the same restriction endonuclease is added.

AATTC
G

G   AATTC
CTTAA   G

DNA ligase joins the strands.

GAATTC
CTTAAG

Recombinant DNA molecule

FIGURE 16.2
Many restriction endonucleases produce DNA fragments with "sticky ends." The restriction endonuclease *Eco*RI always cleaves the sequence GAATTC between G and A. Because the same sequence occurs on both strands, both are cut. However, the two sequences run in opposite directions on the two strands. As a result, single-stranded tails are produced that are complementary to each other, or "sticky."

Host/Vector Systems

Although we can synthesize short sequences of DNA in vitro, to clone large unknown sequences, it is important to be able to propagate recombinant DNA molecules in a cell. Thus, along with a toolbox of enzymes, molecular biologists use cells as factories to produce large amounts of recombinant DNA. The ability to propagate DNA in a host cell requires a **vector** (something to carry the recombinant DNA molecule) that can enter and replicate in the host.

Such host/vector systems are crucial to molecular biology.

The most flexible and common host used for routine cloning is the bacterium *E. coli*. This is not the only host, of course. We now routinely clone eukaryotic DNA, using mammalian tissue culture cells, yeast cells, and insect cells as host systems. Each kind of host/vector system allows particular uses of the cloned DNA.

The two most commonly used vectors are plasmids and phages. *Plasmids* are small, extrachromosomal DNAs that are dispensable to the cell. *Phages* are viruses that infect bacterial cells. The key to a vector is that it is not required by the cell, but it can be directly selected for by including a marker such as antibiotic resistance.

Plasmids

Plasmid vectors (small, circular chromosomes) are typically used to clone relatively small pieces of DNA, up to a maximum of about 10 kilobases (kb). A plasmid vector must have (1) an origin of replication to allow it to be replicated in *E. coli* independently of the chromosome, and (2) a selectable marker, usually antibiotic resistance. The selectable marker allows the presence of the plasmid to be easily identified through selection; cells that contain the marker will live when plated on antibiotic-containing growth media, and cells that lack the plasmid will not live (they are killed by the antibiotic). A fragment of DNA is inserted by the techniques described in figure 16.2 in a region of the plasmid called the **multiple cloning site (MCS).** This region contains a number of unique restriction sites such that when the plasmid is cut with any of these enzymes, it will produce a linear plasmid. The DNA of interest can then be ligated into this site, after which plasmid DNA is introduced into cells.

Next, the presence of inserted DNA must be confirmed. This is typically done by screening for inactivation of another gene. For example, in one method, the presence of inserted DNA results in inactivation of the *LacZ'* gene. This event can be detected by plating cells on medium containing X-gal, a substrate for the enzyme encoded by *LacZ'*. Cells with an active version of the *LacZ'* gene will produce the enzyme and will be blue on this substrate, while cells that have an inactive *LacZ'* will lack the enzyme and will be white (figure 16.3*a*). Note that this is a "screen," not a "selection," because all cells survive. The use of selection or screening depends on the situation; the two approaches differ in their relative power. Selection allows the identification of extremely rare events because only the desired phenotype survives. Genetic screening preserves all the cells in an experiment, but is limited by the ability of the investigator to distinguish between two phenotypes—that is, all cells have to be checked, as opposed to looking for a survivor.

Phages

Phage vectors are larger than plasmid vectors, and can take inserts up to 40 kb. Most phage vectors are based on the well-studied virus phage lambda. Lambda-based vectors are used today primarily for cDNA libraries (that is, collections of DNA fragments complementary to the DNA being investigated). Lambda's use as a vector, while useful for cloning large fragments, has two requirements not shared with plasmid vectors: First, lambda requires live cells to replicate, so it is necessary to infect cells to recover lambda. Second, the lambda genome is linear, so instead of "opening up" a circle, the middle part of the lambda genome is removed and replaced with the inserted DNA. This means that after the inserted DNA is ligated to the two lambda "arms," it must be packaged into a phage head in vitro, and then used to infect *E. coli*. The two "arms" without any inserted DNA are not packaged into phage heads. This is a kind of selection for recombinant phages, because the arms alone will not be propagated (figure 16.3*b*).

Producing large quantities of recombinant DNA is accomplished by using vectors such as plasmids and phages. A gene of interest can be inserted into vector DNA, which infects and is replicated in a cell.

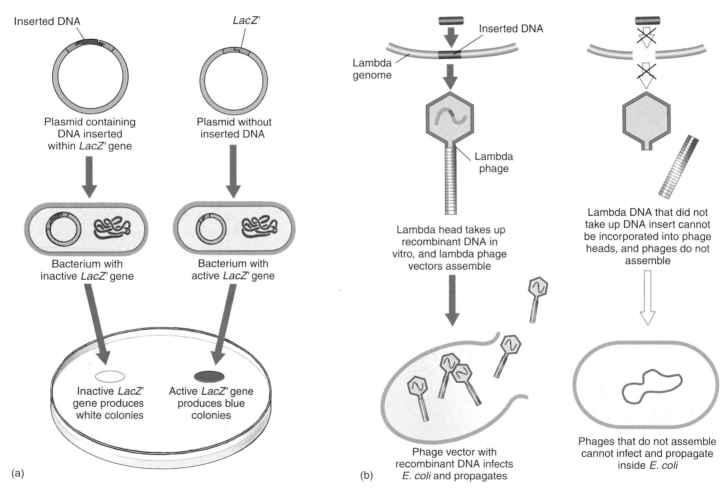

(a)

(b)

FIGURE 16.3
Using plasmid and phage vectors. (*a*) Host bacteria can be screened for the presence of recombinant plasmids by observing the presence of a phenotype in the host, such as the lack of production of blue color when the bacteria are plated in a medium containing X-gal. (*b*) Phage vectors can be selected for the presence of recombinant DNA by the ability of the phage to assemble in vitro, infect a host, and propagate inside its host.

Using Vectors to Transfer Genes

A chimera is a mythical creature with the head of a lion, the body of a goat, and the tail of a serpent. Although no such creatures ever existed in nature, biologists have made chimeras of a more modest kind through genetic engineering.

Constructing pSC101

One of the first genetically engineered **chimeras** was manufactured from a bacterial plasmid called a resistance transfer factor by American geneticists Stanley Cohen and Herbert Boyer in 1973. Cohen and Boyer used a restriction endonuclease called *Eco*RI, which is obtained from *Escherichia coli*, to cut the plasmid into fragments. One fragment, 9000 nucleotides in length, contained both the origin of replication necessary for replicating the plasmid and a gene that conferred resistance to the antibiotic tetracycline (tet^R). Because both ends of this fragment were cut by the same restriction endonuclease, they could be ligated to form a circle, a smaller plasmid Cohen dubbed pSC101.

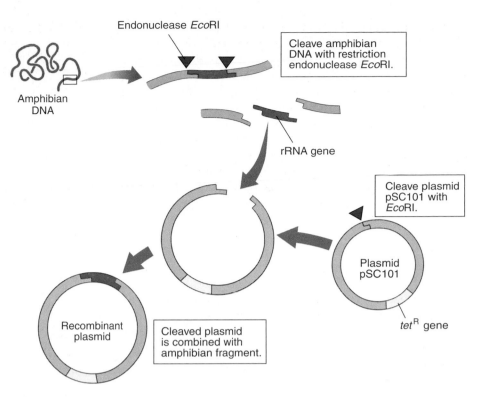

FIGURE 16.4
One of the first genetic engineering experiments. This diagram illustrates how Cohen and Boyer inserted an amphibian gene encoding rRNA into pSC101. The plasmid contains a single site cleaved by the restriction endonuclease *Eco*RI; it also contains tet^R, a gene that confers resistance to the antibiotic tetracycline. The rRNA-encoding gene was inserted into pSC101 by cleaving the amphibian DNA and the plasmid with *Eco*RI and allowing the complementary sequences to pair.

Using pSC101 to Make Recombinant DNA

Cohen and Boyer also used *Eco*RI to cleave DNA that coded for rRNA that they had isolated from an adult amphibian, the African clawed frog, *Xenopus laevis*. They then mixed the fragments of *Xenopus* DNA with pSC101 plasmids that had been "reopened" by *Eco*RI, and allowed bacterial cells to take up DNA from the mixture (figure 16.4). Some of the bacterial cells immediately became resistant to tetracycline, indicating that they had incorporated the pSC101 plasmid with its antibiotic-resistance gene. Furthermore, some of these pSC101-containing bacteria also began to produce frog ribosomal RNA! Cohen and Boyer concluded that the frog rRNA gene must have been inserted into the pSC101 plasmids in those bacteria. In other words, the two ends of the pSC101 plasmid, produced by cleavage with *Eco*RI, had joined to the two ends of a frog DNA fragment that contained the rRNA gene, also cleaved with *Eco*RI.

The pSC101 plasmid containing the frog rRNA gene is a true chimera, an entirely new genome that never existed in nature and never would have evolved by natural means. It is a form of **recombinant DNA**—that is, DNA created in the laboratory by joining together pieces of different genomes to form a novel combination.

Other Vectors

The introduction of foreign DNA fragments into host cells has become common in molecular genetics. As mentioned previously, the genome that carries the foreign DNA into the host cell is called a vector. Plasmids, with names such as pUC18, can be induced to make hundreds of copies of themselves and thus of the foreign genes they contain. Much larger pieces of DNA can be introduced using a YAC (yeast artificial chromosome) as a vector instead of a plasmid. Not all vectors have bacterial targets. Animal viruses are serving as vectors to carry genes into monkey and human cells, and animal genes have even been introduced into plant cells.

One of the first recombinant genomes produced by genetic engineering was a bacterial plasmid into which an amphibian ribosomal RNA gene was inserted. Viruses or artificial chromosomes can also be used as vectors to insert foreign DNA into host cells and create recombinant genomes.

DNA Libraries

In order to place a particular gene or DNA sequence into a vector, you must first have a DNA source that contains that sequence. Typically, the source is a collection of DNA fragments representing all of the DNA from an organism. We call this a **DNA library.** A DNA library is a collection of DNAs from a specific source in a form that can be propagated in a host. In the case of a plasmid, this would be a collection of cells all harboring plasmids with different inserted DNAs such that the whole genome is represented. In the case of a phage, this would be a collection of phages, each carrying a different insert such that the entire genome is represented (figure 16.5).

Genomic Libraries

The simplest kind of DNA library that can be made is a **genomic library,** a representation of the entire genome in a vector. This entire genome is randomly fragmented, each fragment having been inserted into a vector and introduced into the host. Creating a genomic library is harder than it sounds, because it is difficult to truly randomly fragment DNA. The best methods use hydrodynamic shear forces, created by passing the DNA through a syringe. The experimenter then adds small synthetic linker sequences to the ends of the random fragments. These linkers contain a restriction site that can be used to connect the inserts to the vector. Genomic libraries are usually made in phage lambda due to the larger insert sizes. A genomic library would be a set of phage with random inserts that cover the entire genome.

cDNA Libraries

In addition to genomic libraries, investigators often wish to limit an experiment to all *expressed* genes. This is a much smaller amount of DNA than the entire genome. Such a gene-expression library is made possible by the use of another enzyme: **reverse transcriptase.** Reverse transcriptase was isolated from a class of viruses called retroviruses. The life cycle of a retrovirus requires making a DNA copy from the RNA genome of the virus. We can take advantage of the activity of the retrovirus enzyme to make DNA copies from mRNA. Such DNA copies of mRNA are called cDNA (figure 16.6). A cDNA library is made by first isolating mRNA from genes being expressed, and then using mRNA to make cDNA with the reverse transcriptase enzyme. The cDNA is then used to make a library, usually in phage lambda. These cDNA libraries are extremely useful and are often made for the genes expressed in specific tissues or cells.

> **DNA libraries are produced using plasmids or phages as vectors. The DNA library consists of all host cells that have taken up the recombinant DNA, representing all the DNA from the source DNA.**

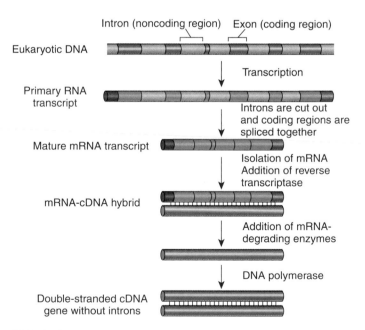

FIGURE 16.6
The formation of cDNA. A mature mRNA transcript is isolated from the cytoplasm of a cell. The enzyme reverse transcriptase is then used to make a DNA strand complementary to the processed mRNA. That newly made strand of DNA is the template for the enzyme DNA polymerase, which assembles a complementary DNA strand along it, producing cDNA, a double-stranded DNA version of the intron-free mRNA.

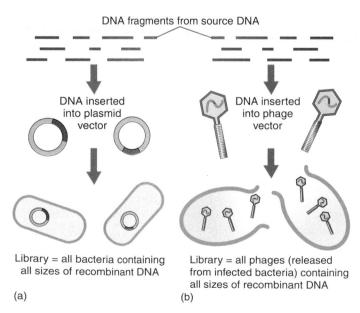

FIGURE 16.5
Creating DNA libraries. DNA libraries can be produced using (*a*) plasmid vectors or (*b*) phage vectors.

The Four Stages of a Genetic Engineering Experiment

Like the experiment of Cohen and Boyer, most genetic engineering experiments consist of four stages: DNA cleavage, production of recombinant DNA, cloning, and screening.

Stage 1: DNA Cleavage

A restriction endonuclease is used to cleave the source DNA into fragments. Because the endonuclease's recognition sequence is likely to occur many times within the source DNA, cleavage will produce a large number of different fragments. A different set of fragments will be obtained by employing endonucleases that recognize different sequences. The fragments can be separated from one another according to their size by gel electrophoresis, a procedure illustrated in figure 16.7.

Stage 2: Production of Recombinant DNA

The fragments of DNA are inserted into plasmids or viral vectors, which have been cleaved with the same restriction endonuclease as the source DNA.

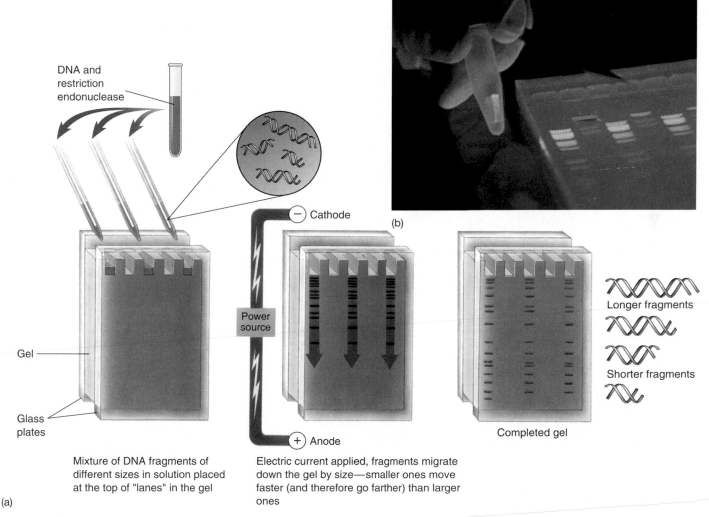

FIGURE 16.7

Gel electrophoresis. (*a*) After restriction endonucleases have cleaved the DNA, the fragments are loaded on a gel, and an electric current is applied. The DNA fragments migrate through the gel, with bigger ones moving more slowly. The fragments can be visualized easily because the migrating bands fluoresce in UV light when stained with ethidium bromide. (*b*) In the photograph, one band of DNA has been excised from the gel for further analysis and can be seen glowing in the tube the technician holds.

Stage 3: Cloning

The plasmids or viruses serve as vectors that can introduce the DNA fragments into cells—usually, but not always, bacteria (figure 16.8). As each cell reproduces, it forms a clone of cells that all contain the fragment-bearing vector. Each clone is maintained separately, and all of them together constitute a clone library of the original source DNA as described in section 16.1.

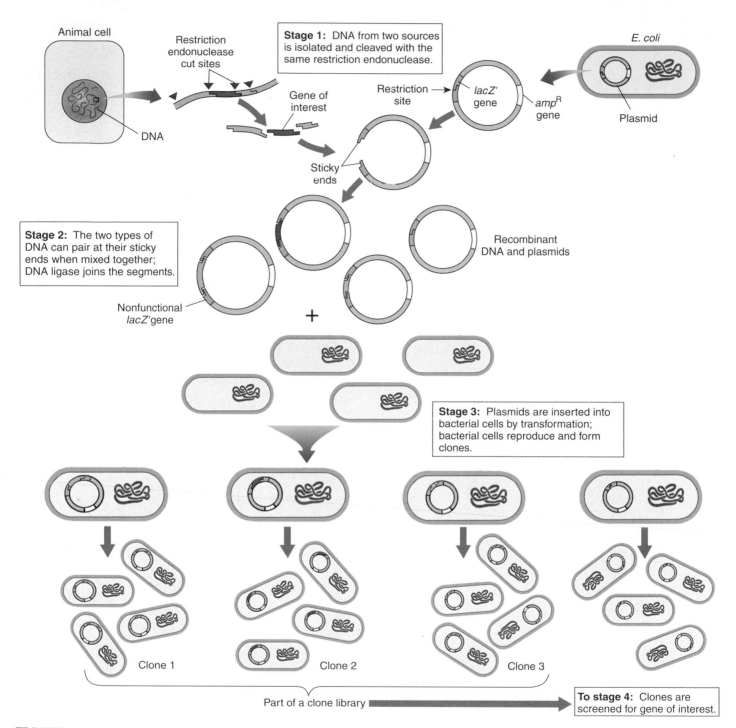

FIGURE 16.8
Stages in a genetic engineering experiment. In stage 1, DNA containing the gene of interest (in this case, from an animal cell) and DNA from a plasmid are cleaved with the same restriction endonuclease. The genes *amp*^R and *lacZ′* are contained within the plasmid and used for screening a clone (stage 4). In stage 2, the two cleaved sources of DNA are mixed together and pair at their sticky ends. In stage 3, the recombinant DNA is inserted into a bacterial cell, which reproduces and forms clones. In stage 4, the bacterial clones will be screened for the gene of interest.

Stage 4: Screening

The clones containing a specific DNA fragment of interest, often a fragment that includes a particular gene, are identified from the clone library. Let's examine this stage in more detail, because it is generally the most challenging in any genetic engineering experiment.

4–I: Preliminary Screening of Clones. Investigators initially try to eliminate from the library any clones that do not contain vectors, as well as clones whose vectors do not contain fragments of the source DNA. The first category of clones can be eliminated by employing a vector with a gene that confers resistance to a specific antibiotic, such as tetracycline, penicillin, or ampicillin. In figure 16.9*a*, the gene *amp*^R is incorporated into the plasmid and confers resistance to the antibiotic ampicillin. When the clones are exposed to a medium containing that antibiotic, only clones that contain the vector will be resistant to the antibiotic and able to grow. This is an example of genetic selection for plasmids containing cells. It is always desirable to design experiments able to provide at least one round of selection.

One way to eliminate clones with vectors lacking an inserted DNA fragment is to use a vector that, in addition to containing antibiotic-resistance genes, contains the *lacZ'* gene, which is required to produce β-galactosidase, an enzyme that enables the cells to metabolize the sugar X-gal. Metabolism of X-gal results in the formation of a blue reaction product, so any cells whose vectors contain a functional version of this gene will turn blue in the presence of X-gal (figure 16.9*b*). However, if we use a restriction endonuclease whose recognition sequence lies within the *lacZ'* gene, the gene will be interrupted when recombinants are formed, and the cell will be unable to metabolize X-gal. Therefore, cells with vectors that contain a fragment of source DNA should remain colorless in the presence of X-gal. This is an example of a genetic screen. Because all cells, both with and without an insert, survive, it is not selection. We often screen based on presence or absence of a particular phenotype (blue color on X-gal in this case).

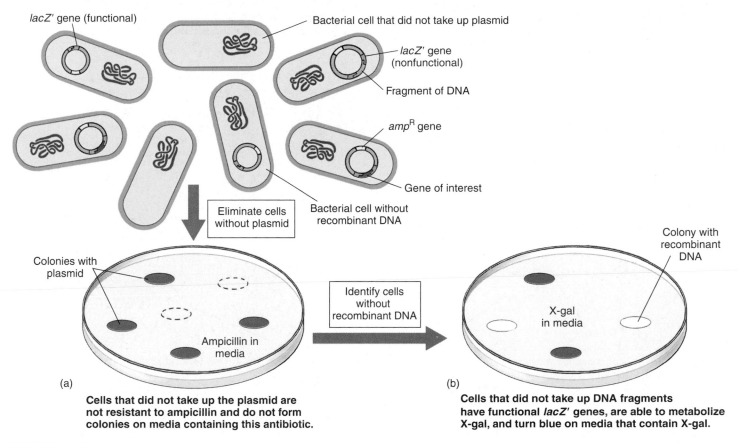

(a) Cells that did not take up the plasmid are not resistant to ampicillin and do not form colonies on media containing this antibiotic.

(b) Cells that did not take up DNA fragments have functional *lacZ'* genes, are able to metabolize X-gal, and turn blue on media that contain X-gal.

FIGURE 16.9

Stage 4-I: Identifying recombinant clones. Bacteria are transformed with recombinant plasmids that contain the gene for ampicillin resistance (*amp*^R). This allows selection on medium containing ampicillin. The plasmid also contains the *lacZ'* gene that encodes β-galactosidase, an enzyme involved in the metabolism of the sugar lactose. An artificial substrate called X-gal can be cleaved by the enzyme to produce a blue color. (*a*) Selection for *amp*^R: Only those bacteria that have been transformed and contain a plasmid will be able to grow on ampicillin. (*b*) Screening for β-galactosidase activity: The plasmid has been constructed such that inserted DNA will interrupt the *lacZ'* gene, making the β-galactosidase enzyme nonfunctional. Thus, when plated on medium containing X-gal, bacteria harboring plasmids with no insert will be blue, while cells that harbor plasmids that have inserts will not (they will be white).

Any cells that are able to grow in a medium containing the antibiotic but don't turn blue in the medium with X-gal must have incorporated a vector with a fragment of source DNA. Identifying cells that have a *specific* fragment of the source DNA is the next step in screening clones.

4–II: Finding the Gene of Interest. A clone library may contain anywhere from a few dozen to many thousand individual fragments of source DNA. Many of those fragments will be identical, so to assemble a complete library of the entire source genome, several hundred thousand clones could be required. A complete *Drosophila* (fruit fly) library, for example, contains more than 40,000 different clones; a complete human library consisting of fragments 20 kilobases long would require close to a million clones. To search such an immense library for a clone that contains a fragment corresponding to a particular gene requires ingenuity, but many different approaches have been successful.

The most general procedure for screening clone libraries to find a particular gene is **hybridization** (figure 16.10). In this method, the cloned genes form base-pairs with complementary sequences on another nucleic acid. The complementary nucleic acid is called a **probe** because it is used to probe for the presence of the gene of interest.

At least part of the nucleotide sequence of the gene of interest must be known to be able to construct the probe.

In this method of screening, bacterial colonies containing an inserted gene are grown on agar. Some cells are transferred to a filter pressed onto the colonies, forming a replica of the plate. The filter is then treated with a solution that denatures the bacterial DNA and contains a radioactively labeled probe. The probe hybridizes with complementary single-stranded sequences on the bacterial DNA.

When the filter is laid over photographic film, areas that contain radioactivity will expose the film (autoradiography). Only colonies that contain the gene of interest hybridize with the radioactive probe and emit radioactivity onto the film. The pattern on the film is then compared to the original master plate, and the gene-containing colonies may be identified.

Genetic engineering generally involves four stages: cleaving the source DNA; making recombinants; cloning copies of the recombinants; and screening the cloned copies for the desired gene. Screening can be achieved by making the desired clones resistant to certain antibiotics and giving them other properties that make them readily identifiable.

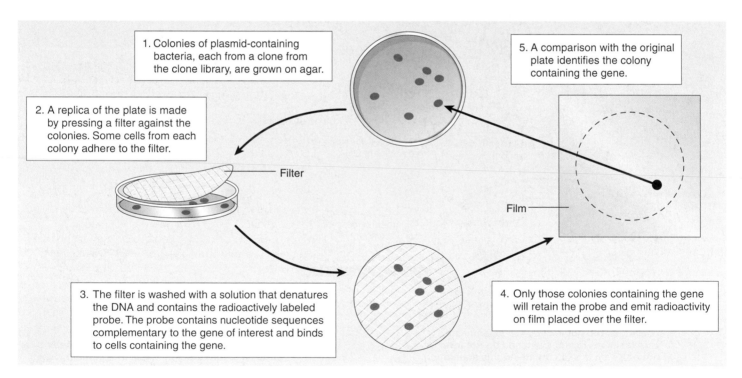

FIGURE 16.10
Stage 4-II: Using hybridization to identify the gene of interest. (*1*) Each of the colonies on these bacterial culture plates represents millions of clones descended from a single cell. To test whether a certain gene is present in any particular clone, it is necessary to identify colonies whose cells contain DNA that hybridizes with a probe containing DNA sequences complementary to the gene. (*2*) Pressing a filter against the master plate causes some cells from each colony to adhere to the filter. (*3*) The filter is then washed with a solution that denatures the DNA and contains the radioactively labeled probe. (*4*) Only those colonies that contain DNA that hybridizes with the probe, and thus contain the gene of interest, will expose film in autoradiography. (*5*) The film is then compared to the master plate to identify the gene-containing colony.

Working with Gene Clones

Once a gene has been successfully cloned, a variety of procedures are available to characterize it.

Getting Enough DNA to Work With: The Polymerase Chain Reaction

Once a particular gene is identified within the library of DNA fragments, the final requirement is to make multiple copies of it. One way to do this is to insert the identified fragment into a bacterium; after repeated cell divisions, millions of cells will contain copies of the fragment. A far more direct approach, however, is to use DNA polymerase to copy the gene sequence of interest through the **polymerase chain reaction** (**PCR**; figure 16.11). Kary Mullis developed PCR in 1983 while he was a staff chemist at the Cetus Corporation; in 1993, his discovery won him the Nobel Prize in chemistry. PCR can amplify specific sequences or add sequences (such as endonuclease recognition sequences) as primers to cloned DNA. There are three steps in PCR:

Step 1: Denaturation. First, an excess of primer (typically, a synthetic sequence of 20 to 30 nucleotides) is mixed with the DNA fragment to be amplified. This mixture of primer and fragment is heated to about 98°C. At this temperature, the double-stranded DNA fragment dissociates into single strands.

Step 2: Annealing of Primers. Next, the solution is allowed to cool to about 60°C. As it cools, the single strands of DNA reassociate into double strands. However, because of the large excess of primer, each strand of the fragment base-pairs with a complementary primer flanking the region to be amplified, leaving the rest of the fragment single-stranded.

Step 3: Primer Extension. Now a very heat-stable type of DNA polymerase, called Taq polymerase (after the thermophilic bacterium *Thermus aquaticus*, from which Taq is extracted) is added, along with a supply of all four nucleotides. Using the primer, the polymerase copies the rest of the fragment as if it were replicating DNA. When it is done, the primer has been lengthened into a complementary copy of the entire single-stranded fragment. Because *both* DNA strands are replicated, there are now two copies of the original fragment.

Steps 1 to 3 are now repeated, and the two copies become four. It is not necessary to add any more polymerase, because the heating step does not harm this particular enzyme. Each heating and cooling cycle, which can be as short as 1 or 2 minutes, doubles the number of DNA molecules. After 20 cycles, a single fragment produces more than one million (2^{20}) copies! In a few hours, 100 billion copies of the fragment can be manufactured.

PCR, now fully automated, has revolutionized many aspects of science and medicine because it allows the investigation of minute samples of DNA. In criminal investiga-

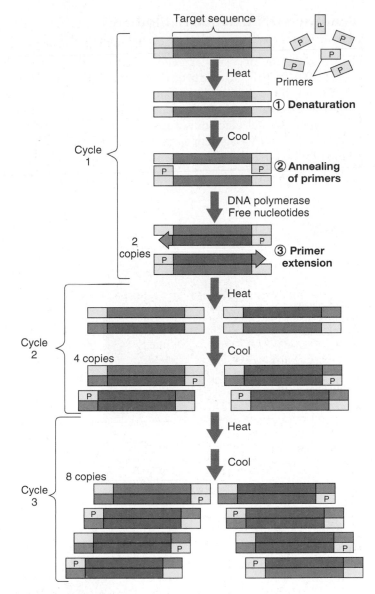

FIGURE 16.11

The polymerase chain reaction. (*1*) *Denaturation:* A solution containing primers and the DNA fragment to be amplified is heated so that the DNA dissociates into single strands. (*2*) *Annealing of primers:* The solution is cooled, and the primers bind to complementary sequences on the DNA flanking the region to be amplified. (*3*) *Primer extension:* DNA polymerase then copies the remainder of each strand, beginning at the primer. Steps 1–3 are then repeated with the replicated strands. This process is repeated many times, each time doubling the number of copies, until enough copies of the DNA fragment exist for analysis.

tions, *DNA fingerprints* are prepared from the cells in a tiny speck of dried blood or at the base of a single human hair. In pediatric medicine, physicians can detect genetic defects in very early embryos by collecting a few sloughed-off cells and amplifying their DNA. PCR could also be used to examine the DNA of historical figures, such as Abraham Lincoln, or that of now-extinct species, as long as even a minuscule amount of their DNA remains intact.

Identifying DNA: Southern Blotting

Once a gene has been cloned, it may be used as a probe to identify the same or a similar gene in another sample (figure 16.12). In this procedure, called a **Southern blot**, DNA from the sample is cleaved into restriction fragments with a restriction endonuclease, and the fragments are spread apart by gel electrophoresis. The double-stranded helix of each DNA fragment is then denatured into single strands by making the pH of the gel basic, and the gel is "blotted" with a sheet of nitrocellulose, transferring some of the DNA strands to the sheet. Next, a probe consisting of purified, single-stranded DNA corresponding to a specific gene (or mRNA transcribed from that gene) is poured over the sheet. Any fragment that has a nucleotide sequence complementary to the probe's sequence will hybridize (base-pair) with the probe (figure 16.13). If the probe has been labeled with ^{32}P, it will be radioactive, and the sheet will show a band of radioactivity where the probe hybridized with the complementary fragment.

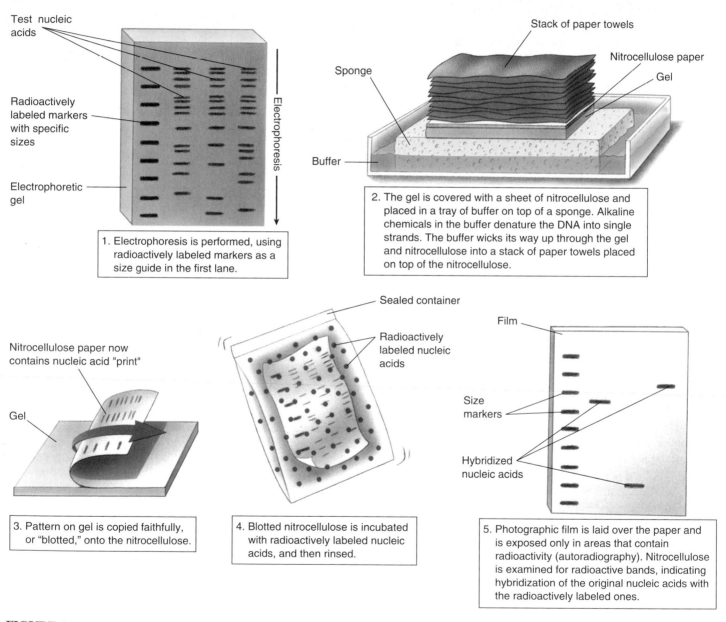

FIGURE 16.12

The Southern blot procedure. E. M. Southern developed this procedure in 1975 to enable DNA fragments of interest to be visualized in a complex sample containing many other fragments of similar size. The DNA is separated on a gel, and then transferred ("blotted") onto a solid support medium such as nitrocellulose paper or a nylon membrane. Next, it is incubated with a radioactive single-stranded copy of the gene of interest, which hybridizes to the blot at the location(s) where there is a fragment with a complementary sequence. The positions of radioactive bands on the blot identify the fragments of interest.

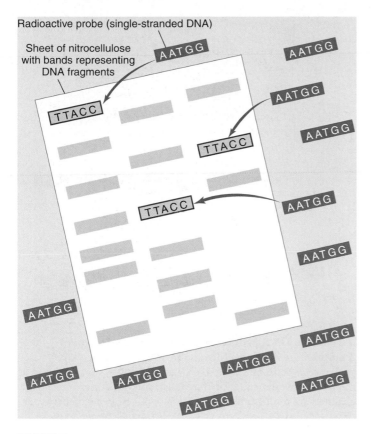

Radioactive probe (single-stranded DNA)

Sheet of nitrocellulose with bands representing DNA fragments

FIGURE 16.13
The hybridization step in Southern blot procedure. A probe (usually several hundred nucleotides in length) of single-stranded DNA (or an mRNA complementary to the gene of interest) is poured over the nitrocellulose sheet containing the DNA fragments. All DNA fragments that contain nucleotide sequences complementary to the probe will bind the probe.

Distinguishing Differences in DNA: RFLP Analysis

Often a researcher wishes not to find a specific gene, but rather to identify a particular *individual* using a specific gene as a marker. One powerful way to do this is by analyzing **restriction fragment length polymorphisms,** or **RFLPs** (figure 16.14). Point mutations, sequence repetitions, and transposons (see chapter 20) that occur within or between the restriction endonuclease recognition sites will alter the length of the DNA fragments (restriction fragments) the restriction endonucleases produce. DNA from different individuals rarely has exactly the same array of restriction sites and distances between sites, so the population is said to be polymorphic (having many forms) for their restriction fragment patterns. By cutting a DNA sample with a particular restriction endonuclease, separating the fragments according to length on an electrophoretic gel, and then using a radioactive probe to identify the fragments on the gel, we can obtain a pattern of bands often unique for each region of DNA analyzed. These are the **DNA fingerprints** mentioned previously as being used in forensic analysis during criminal investigations. RFLPs are also useful as markers to identify particular groups of people at risk for certain genetic disorders. This technique requires a large amount of DNA relative to PCR-based techniques, but is very reliable. The use of PCR has facilitated comparisons with smaller samples.

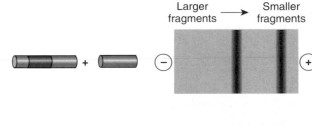

Restriction endonuclease cutting sites

Larger fragments → Smaller fragments

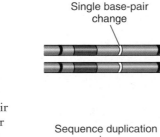

Single base-pair change

FIGURE 16.14
Restriction fragment length polymorphism (RFLP) analysis.
(*a*) Three samples of DNA differ in their restriction sites due to a single base-pair substitution in one case and a sequence duplication in another case. (*b*) When the samples are cut with a restriction endonuclease, different numbers and sizes of fragments are produced. (*c*) Gel electrophoresis separates the fragments, and different banding patterns result.

Sequence duplication

(a) Three different DNA duplexes

(b) Cut DNA

(c) Gel electrophoresis of restriction fragments

DNA Fingerprinting

As stated previously, two individuals rarely produce identical RFLP analyses; therefore, these DNA fingerprints can be used in criminal investigations. Figure 16.15 shows the DNA fingerprints a prosecuting attorney presented in a rape trial in 1987. They consist of autoradiographs, parallel bars on X-ray film resembling the line patterns of the universal price code found on groceries. Each bar represents the position of a DNA restriction endonuclease fragment produced by techniques similar to those described in figures 16.7 and 16.14. The lane with many bars represents a standardized control. Two different probes were used to identify the restriction fragments. A vaginal swab had been taken from the victim within hours of her attack; from it, semen was collected and its DNA analyzed for restriction endonuclease patterns.

Compare the restriction endonuclease patterns of the semen to that of the suspect, Andrews. You can see that the suspect's two patterns match that of the rapist (and are not at all like those of the victim). Clearly, the semen collected from the rape victim and the blood sample from the suspect came from the same person. The suspect was Tommie Lee Andrews, and on November 6, 1987, the jury returned a verdict of guilty. Andrews became the first person in the United States to be convicted of a crime based on DNA evidence.

Since the Andrews verdict, DNA fingerprinting has been admitted as evidence in more than 2000 court cases (figure 16.16). While some probes highlight profiles shared by many people, others are quite rare. Using several probes, identity can be clearly established or ruled out.

Just as fingerprinting revolutionized forensic evidence in the early 1900s, so DNA fingerprinting is revolutionizing it today. A hair, a minute speck of blood, or a drop of semen can all serve as sources of DNA to damn or clear a suspect. As the man who analyzed Andrews's DNA says, "It's like leaving your name, address, and social security number at the scene of the crime. It's that precise." Of course, laboratory analyses of DNA samples must be carried out properly—sloppy procedures could lead to a wrongful conviction. After widely publicized instances of questionable lab procedures, national standards are being developed.

Techniques such as Southern blotting and PCR enable investigators to identify specific genes and produce them in large quantities, while RFLP analysis and DNA fingerprinting identify individuals and unknown gene sequences.

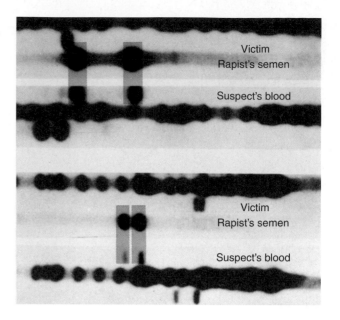

FIGURE 16.15

Two of the DNA profiles that led to the conviction of Tommie Lee Andrews for rape in 1987. The two DNA probes seen here were used to characterize DNA isolated from the victim, the semen left by the rapist, and the suspect. The dark channels are multiband controls. There is a clear match between the suspect's DNA and the DNA of the rapist's semen in these two profiles.

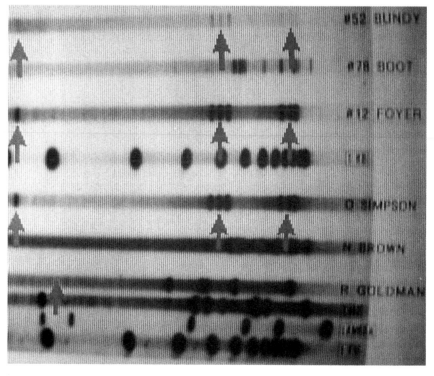

FIGURE 16.16

DNA profiles of O. J. Simpson and blood samples from the murder scene of his former wife. These patterns were used as evidence in the highly publicized and controversial 1995 murder trial.

16.3 Biotechnology is producing a scientific revolution.

Medical Applications

Pharmaceuticals

The first and perhaps most obvious commercial application of genetic engineering was the introduction of genes that encode clinically important proteins into bacteria. Because bacterial cells can be grown cheaply in bulk (fermented in giant vats, as is done with the yeasts that make beer), bacteria that incorporate recombinant genes can synthesize large amounts of the proteins those genes specify. This method has been used to produce several forms of human insulin and interferon, as well as other commercially valuable proteins, such as human growth hormone (figure 16.17) and erythropoietin, which stimulates red blood cell production.

Among the medically important proteins now manufactured by these approaches are **atrial peptides,** small proteins that may provide a new way to treat high blood pressure and kidney failure. Another is **tissue plasminogen activator,** a human protein synthesized in minute amounts that causes blood clots to dissolve and may be effective in preventing and treating heart attacks and strokes.

A problem with this general approach has been the difficulty of separating the desired protein from the others the bacteria make. The purification of proteins from such complex mixtures is both time-consuming and expensive, but it is still easier than isolating the proteins from the tissues of animals (for example, insulin from hog pancreases), which is how such proteins were formerly obtained. Recently, however, researchers have succeeded in producing RNA transcripts of cloned genes; they can then use the transcripts to produce only those proteins in a test tube containing the transcribed RNA, ribosomes, cofactors, amino acids, tRNA, and ATP.

Gene Therapy

In 1990, researchers first attempted to combat genetic defects by the transfer of human genes. When a hereditary disorder is the result of a single defective gene, an obvious way to cure the disorder is to add a working copy of the gene. This approach is being used in an attempt to combat cystic fibrosis (see chapter 13), and it offers potential for treating muscular dystrophy and a variety of other disorders (table 16.1). One of the first successful attempts was the transfer of a gene encoding the enzyme adenosine deaminase into the bone marrow of two girls suffering from a rare blood disorder caused by the lack of this enzyme. However, while many clinical trials are under way, no others have yet proven successful. This extremely promising approach will require a lot of additional effort. A more detailed account of gene therapy is presented in chapter 13.

FIGURE 16.17
Genetically engineered human growth hormone. These two mice are genetically identical, but the large one has one extra gene: the gene encoding human growth hormone. The gene was added to the mouse's genome by genetic engineers and is now a stable part of the mouse's genetic endowment.

Table 16.1 Diseases Being Treated in Clinical Trials of Gene Therapy
Disease
Cancer (melanoma, renal cell, ovarian, neuroblastoma, brain, head and neck, lung, liver, breast, colon, prostate, mesothelioma, leukemia, lymphoma, multiple myeloma)
SCID (severe combined immunodeficiency)
Cystic fibrosis
Gaucher disease
Familial hypercholesterolemia
Hemophilia
Purine nucleoside phosphorylase deficiency
Alpha-1 antitrypsin deficiency
Fanconi anemia
Hunter syndrome
Chronic granulomatous disease
Rheumatoid arthritis
Peripheral vascular disease
AIDS

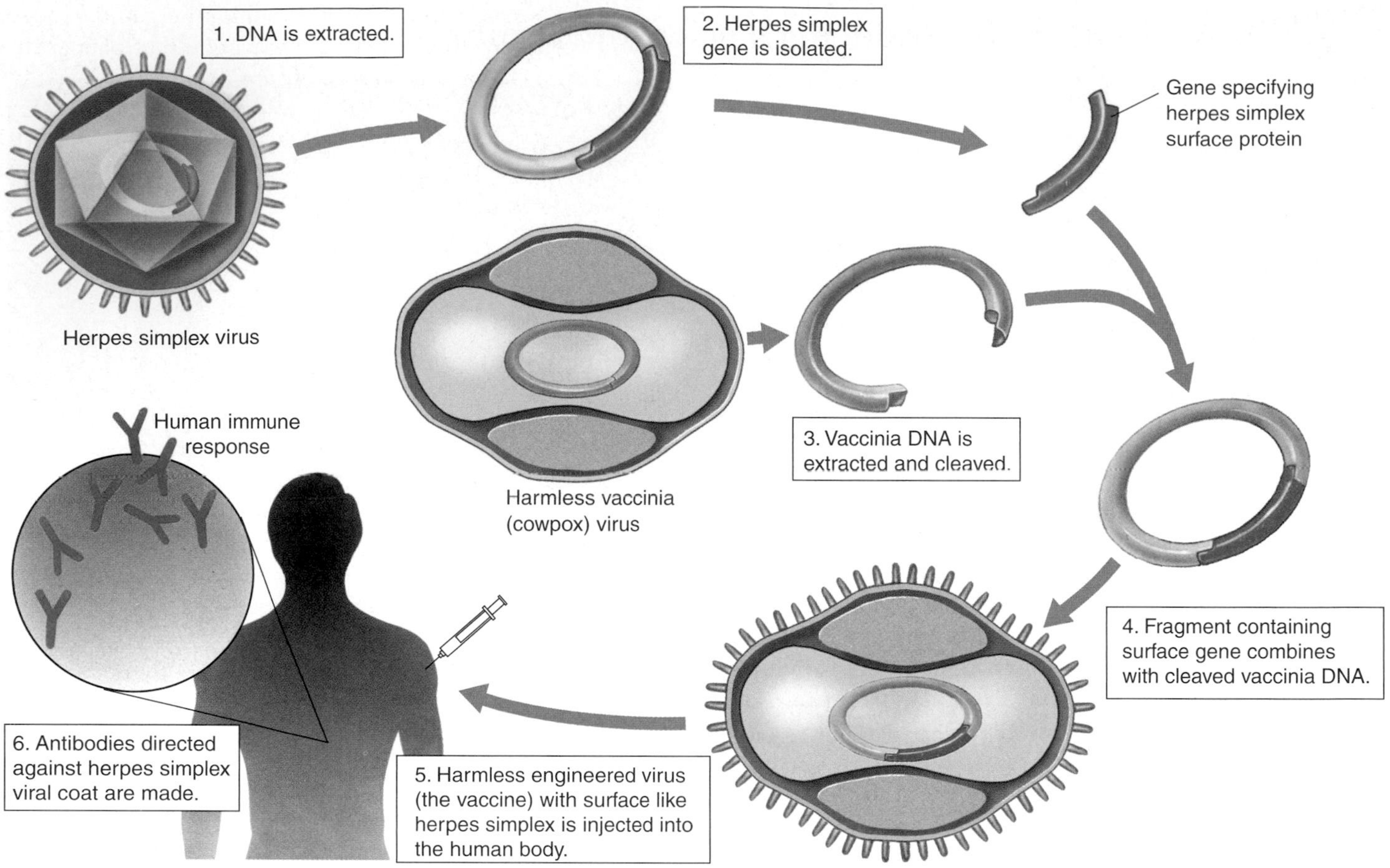
1. DNA is extracted.
2. Herpes simplex gene is isolated.
Gene specifying herpes simplex surface protein
Herpes simplex virus
Human immune response
Harmless vaccinia (cowpox) virus
3. Vaccinia DNA is extracted and cleaved.
4. Fragment containing surface gene combines with cleaved vaccinia DNA.
6. Antibodies directed against herpes simplex viral coat are made.
5. Harmless engineered virus (the vaccine) with surface like herpes simplex is injected into the human body.

FIGURE 16.18
Strategy for constructing a subunit vaccine for herpes simplex.

Piggyback Vaccines

Another area of potential significance involves the use of genetic engineering to produce **subunit vaccines** against viruses such as those that cause herpes and hepatitis. Genes encoding part of the protein-polysaccharide coat of the herpes simplex virus or hepatitis B virus are spliced into a fragment of the vaccinia (cowpox) virus genome (figure 16.18). The vaccinia virus, which British physician Edward Jenner used almost 200 years ago in his pioneering vaccinations against smallpox, is now used as a vector to carry the herpes or hepatitis viral coat gene into cultured mammalian cells. These cells produce many copies of the recombinant virus, which has the outside coat of a herpes or hepatitis virus. When this recombinant virus is injected into a mouse or rabbit, the immune system of the infected animal produces antibodies directed against the coat of the recombinant virus. It therefore develops an immunity to herpes or hepatitis virus. Vaccines produced in this way are harmless because the vaccinia virus is benign, and only a small fragment of the DNA from the disease-causing virus is introduced via the recombinant virus.

The great attraction of this approach is that it does not depend upon the nature of the viral disease. In the future, similar recombinant viruses may be injected into humans to confer resistance to a wide variety of viral diseases.

In 1995, the first clinical trials began to test a novel new kind of **DNA vaccine,** one that depends not on antibodies but rather on the second arm of the body's immune defense, the so-called cellular immune response, in which blood cells known as killer T cells attack infected cells. The infected cells are attacked and destroyed when they stick fragments of foreign proteins onto their outer surfaces, which the T cells then detect. (The discovery by Peter Doherty and Rolf Zinkernagel that infected cells do this led to their receiving the Nobel Prize in Physiology or Medicine in 1996.) The first DNA vaccines spliced an influenza virus gene encoding an internal nucleoprotein into a plasmid, which was then injected into mice. The mice developed strong cellular immune responses to influenza. Although new and controversial, the approach offers great promise.

Genetic engineering has produced commercially valuable proteins and gene therapies, and will possibly give rise to new and powerful vaccines as well.

Agricultural Applications

Another major area of genetic engineering is manipulation of the genes of key crop plants. In plants, the primary experimental difficulty has been identifying a suitable vector for introducing recombinant DNA. Plant cells do not possess the many plasmids that bacteria have, so the choice of potential vectors is limited. The most successful results thus far have been obtained with the **Ti** (tumor-inducing) **plasmid** of the plant bacterium *Agrobacterium tumefaciens*, which infects broadleaf plants such as tomato, tobacco, and soybean. Part of the Ti plasmid integrates into the plant DNA, and researchers have succeeded in attaching other genes to this portion of the plasmid (figure 16.19). The characteristics of a number of plants have been altered using this technique, which should be valuable in improving crops and forests. Among the features scientists would like to affect are resistance to disease, frost, and other forms of stress; nutritional balance and protein content; and herbicide resistance. Unfortunately, *Agrobacterium* generally does not infect cereals such as corn, rice, and wheat, but alternative methods can be used to introduce new genes into them.

A recent advance in genetically manipulated fruit is Calgene's "Flavr Savr" tomato, which has been approved for sale by the United States Department of Agriculture (USDA).

The tomato has been engineered to inhibit genes that cause cells to produce ethylene. In tomatoes and other plants, ethylene acts as a hormone to speed fruit ripening. In Flavr Savr tomatoes, inhibition of ethylene production delays ripening. The result is a tomato that can stay on the vine longer and resists overripening and rotting during transport to market. A variety used for tomato paste has been engineered for anti-sense expression of the gene encoding polygalactaronidase to increase yield and reduce spoilage.

Nitrogen Fixation

A long-range goal of agricultural genetic engineering is to introduce the genes that allow soybeans and other legume plants to "fix" nitrogen into key crop plants. These so-called *nif* **genes** are found in certain symbiotic root-colonizing bacteria. Living in the root nodules of legumes, these bacteria break the powerful triple bond of atmospheric nitrogen gas, converting N_2 into NH_3 (ammonia). The plants then use the ammonia to make amino acids and other nitrogen-containing molecules. Other plants lack these bacteria and cannot fix nitrogen, so they must obtain their nitrogen from the soil. Farmland where these crops are grown soon becomes depleted of nitrogen, unless nitrogenous fertilizers are applied. Worldwide, farmers applied over 60 million metric tons of such

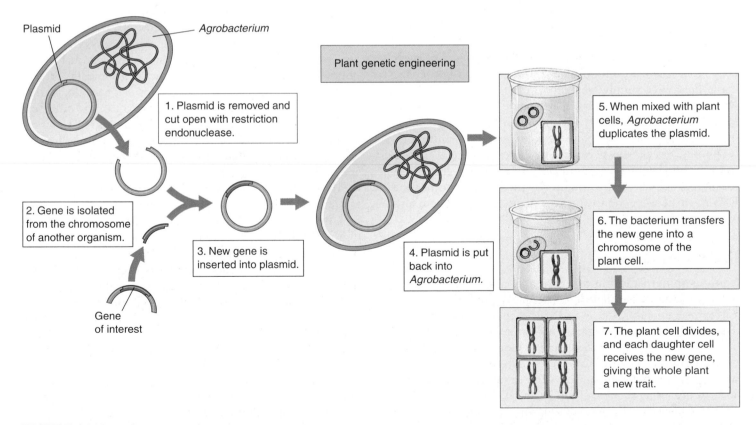

FIGURE 16.19
The Ti plasmid. This *Agrobacterium tumefaciens* plasmid is used in plant genetic engineering.

fertilizers in 1987, an expensive undertaking. Farming costs would be much lower if major crops such as wheat and corn could be engineered to carry out biological nitrogen fixation. However, introducing the nitrogen-fixing genes from bacteria into plants has proved difficult because these genes do not seem to function properly in eukaryotic cells, due to problems protecting nitrogenase from O_2. Researchers are actively experimenting with other species of nitrogen-fixing bacteria whose genes might function better in plant cells.

Herbicide Resistance

Recently, broadleaf plants have been genetically engineered to be resistant to **glyphosate,** the active ingredient in Roundup, a powerful, biodegradable herbicide that kills most actively growing plants (figure 16.20). Glyphosate works by inhibiting an enzyme called EPSP synthetase, which plants require to produce aromatic amino acids. Humans do not make aromatic amino acids; they get them from their diet, so they are unaffected by glyphosate. To make glyphosate-resistant plants, agricultural scientists used a Ti plasmid to insert extra copies of the EPSP synthetase genes into plants. These engineered plants produce 20 times the normal level of EPSP synthetase, enabling them to synthesize proteins and grow despite glyphosate's suppression of the enzyme. In later experiments, a bacterial form of the EPSP synthetase gene that differs from the plant form by a single nucleotide was introduced into plants via Ti plasmids; the bacterial enzyme in these plants is not inhibited by glyphosate.

These advances are of great interest to farmers because a crop resistant to Roundup would never have to be weeded if the field were simply treated with the herbicide. Because Roundup is a broad-spectrum herbicide, farmers would no longer need to employ a variety of different herbicides, most of which kill only a few kinds of weeds. Furthermore, glyphosate breaks down readily in the environment, unlike many other herbicides commonly used in agriculture. A plasmid is actively being sought for the introduction of the EPSP synthetase gene into cereal plants, making them also glyphosate-resistant.

Insect Resistance

Many commercially important plants are attacked by insects, and the traditional defense against such attacks is to apply insecticides. Over 40% of the chemical insecticides used today are targeted against boll weevils, bollworms, and other insects that eat cotton plants. Genetic engineers are now attempting to produce plants that are resistant to insect pests and thus remove the need to use many externally applied insecticides.

The approach is to insert into crop plants genes encoding proteins that are harmful to the insects that feed on the plants but harmless to other organisms. One such insecticidal

FIGURE 16.20
Genetically engineered herbicide resistance. All four of these petunia plants were exposed to equal doses of the herbicide Roundup. The two on top were genetically engineered to be resistant to glyphosate, the active ingredient in Roundup, while the two on the bottom were not.

protein has been identified in *Bacillus thuringiensis,* a soil bacterium. When the tomato hornworm caterpillar ingests this protein, enzymes in the caterpillar's stomach convert it into an insect-specific toxin, causing paralysis and death. Because these enzymes are not found in other animals, the protein is harmless to them. Using the Ti plasmid, scientists have transferred the gene encoding this protein into tomato and tobacco plants. They have found that these **transgenic** plants are indeed protected from attack by the insects that would normally feed on them. In 1995, the Environmental Protection Agency (EPA) approved altered forms of potato, cotton, and corn. The genetically altered potato can kill the Colorado potato beetle, a common pest. The altered cotton is resistant to cotton bollworm, budworm, and pink bollworm. The corn has been altered to resist the European corn borer and other mothlike insects.

Monsanto scientists screening natural compounds extracted from plant and soil samples have recently isolated a new insect-killing compound from a fungus, the enzyme cholesterol oxidase. Apparently, the enzyme disrupts membranes in the insect gut. The fungus gene, called the Bollgard gene after its discoverer, has been successfully inserted into a variety of crops. It kills a wide range of insects, including the cotton boll weevil and the Colorado potato beetle, both serious agricultural pests. Field tests began in 1996.

Some insect pests attack plant roots, and *B. thuringiensis* is being employed to counter that threat as well. This bacterium does not normally colonize plant roots, so biologists have introduced the *B. thuringiensis* insecticidal protein gene into root-colonizing bacteria, especially strains of *Pseudomonas.* Field testing of this promising procedure has been approved by the EPA.

The Real Promise of Plant Genetic Engineering

In the last decade, the cultivation of genetically modified crops of corn, cotton, and soybeans has become commonplace in the United States. In 1999, over half of the 72 million acres planted with soybeans in the United States were planted with seeds genetically modified to be herbicide resistant, with the result that less tillage has been needed and soil erosion has been greatly lessened. These improved crop varieties have mainly benefited farmers, enabling them to cultivate their crops more cheaply and efficiently. The food the public gets is the same; it just costs less to get it to the table.

Like the first act of a play, these developments have served mainly to set the stage for the real action, which is only now beginning to happen. The promise of plant genetic engineering is to produce genetically modified plants with desirable traits that directly benefit the consumer.

One recent advance, nutritionally improved rice, gives us a hint of what is to come. In developing countries, large numbers of people live on simple diets that are poor sources of vitamins and minerals (what botanists called "micronutrients"). Worldwide, the two major micronutrient deficiencies are iron, affecting 1.4 billion women (24% of the world population), and vitamin A, affecting 40 million children (7% of the world population). The deficiencies are especially severe in developing countries where the major staple food is rice. In recent research, Swiss bioengineer Ingo Potrykus and his team at the Institute of Plant Sciences, Zurich, have gone a long way toward solving this problem. Supported by the Rockefeller Foundation and with results to be made free to developing countries, their development of "golden rice" is a model of what plant genetic engineering can achieve.

To solve the problem of dietary iron deficiency among rice eaters, Potrykus first asked why rice is such a poor source of dietary iron. The problem, and the answer, proved to have three parts:

1. **Too little iron.** The proteins of rice endosperm contain unusually low amounts of iron. To solve this problem, a ferritin gene was transferred into rice from beans (figure 16.21). Ferritin is a protein with an extraordinarily high iron content, so its addition greatly increased the iron content of the rice.
2. **Inhibition of iron absorption by the intestine.** Rice contains an unusually high concentration of a chemical called phytate, which inhibits iron reabsorption in the intestine, meaning that it stops your body from taking up the iron in the rice. To solve this problem, a gene encoding an enzyme that destroys phytate was transferred into rice from a fungus.
3. **Too little sulfur for efficient iron absorption.** Sulfur is required for iron uptake, and rice has very little of it. To solve this problem, a gene encoding a particularly sulfur-rich metallothionin protein was transferred into rice from wild rice.

To solve the problem of vitamin A deficiency, the same approach was taken. First, the problem was identified. It turns out that rice only goes partway toward making β-carotene (provitamin A); there are no enzymes in rice to catalyze the last four steps. Thus, genes encoding these four enzymes were added to rice from a familiar flower, the daffodil.

Potrykus's development of transgenic rice to combat dietary deficiencies involved no subtle tricks, just straightforward bioengineering and the will to get the job done. The transgenic "golden" rice he has developed will directly improve the lives of millions of people. His work is

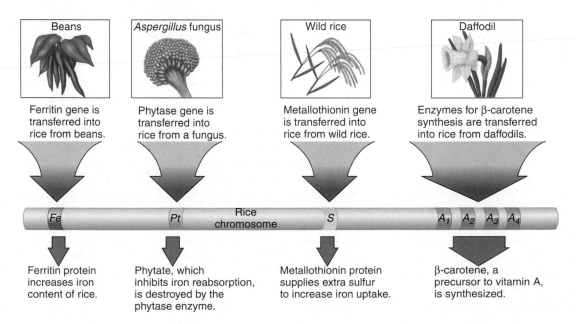

FIGURE 16.21
Transgenic rice.
Developed by Swiss bioengineer Ingo Potrykus, transgenic rice offers the promise of improving the diets of people in rice-consuming developing countries, where iron and vitamin A deficiencies are serious problems.

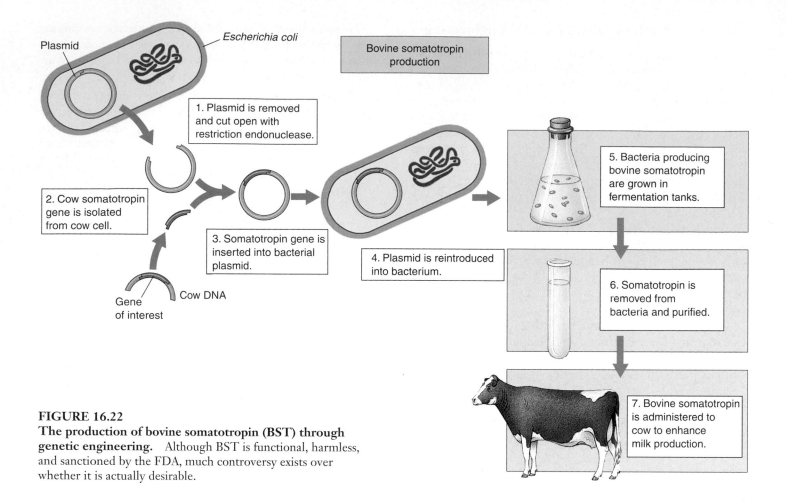

FIGURE 16.22
The production of bovine somatotropin (BST) through genetic engineering. Although BST is functional, harmless, and sanctioned by the FDA, much controversy exists over whether it is actually desirable.

representative of the very real promise of genetic engineering to help meet the challenges of the new millennium.

The list of gene modifications that directly aid consumers will only grow. In Holland, Dutch bioengineers have announced that they are genetically engineering plants to act as vaccine-producing factories! To petunias they have added a gene for a vaccine against dog parvovirus, hiding the gene within the petunia genes that direct nectar production. The drug is produced in the nectar, collected by bees, and extracted from the honey. It is hard to believe this isn't science fiction. Clearly, the real promise of plant genetic engineering lies ahead, and not very far.

Farm Animals

The gene encoding the growth hormone somatotropin was one of the first to be cloned successfully. In 1994, Monsanto received federal approval to make its recombinant bovine somatotropin (BST) commercially available, and dairy farmers worldwide began to add the hormone as a supplement to their cows' diets, increasing the animals' milk production (figure 16.22). Genetically engineered somatotropin is also being tested to see if it increases the

muscle weight of cattle and pigs, and as a treatment for human disorders, such as dwarfism, in which the pituitary gland fails to make adequate levels of somatotropin. BST ingested in milk or meat has no effect on humans, because it is a protein and is digested in the stomach. Nevertheless, BST has met with some public resistance, due primarily to generalized fears about gene technology. Some people mistrust milk produced through genetic engineering, even though the milk itself is identical to other milk. Problems concerning public perception are not uncommon as gene technology makes an even greater impact on our lives.

Transgenic animals engineered to have specific desirable genes are becoming increasingly available to breeders. Now, instead of selectively breeding for several generations to produce a racehorse or a stud bull with desirable qualities, the process can be shortened by simply engineering such an animal right at the start.

Gene technology is revolutionizing agriculture, increasing yields and resistance to pests, improving nutritional value, and producing animals with desirable traits.

Risk and Regulation

The advantages afforded by genetic engineering are revolutionizing our lives. But what are the disadvantages—the potential costs and dangers of genetic engineering? Many people, including influential activists and members of the scientific community, have expressed concern that genetic engineers are "playing God" by tampering with genetic material. For instance, what would happen if we fragmented the DNA of a cancer cell, and then incorporated the fragments at random into vectors that were propagated within bacterial cells? Might there not be a danger that some of the resulting bacteria would transmit an infective form of cancer? Could genetically engineered products administered to plants or animals turn out to be dangerous for consumers after several generations? What kind of unforeseen impact on the ecosystem might "improved" crops have? Is it ethical to create "genetically superior" organisms, including humans?

How Do We Measure the Potential Risks of Genetically Modified Crops?

While the promise of genetic engineering is very much in evidence, this same technology has been the cause of outright war between researchers and protesters. In June 1999, British protesters attacked an experimental plot of genetically modified (GM) sugar beets; the following August, they destroyed a test field of GM canola (used for cooking oil and animal feed). The contrast could not be more marked between American acceptance of genetically modified crops on the one hand, and European distrust of genetically modified foods on the other. The intense feelings generated by this dispute point to the need to understand how we measure the risks associated with the genetic engineering of plants.

Two sets of risks need to be considered. The first stems from eating genetically modified foods, and the other concerns potential ecological effects.

Is Eating Genetically Modified Food Dangerous?

Protesters worry that genetically modified food may have been rendered somehow dangerous. To sort this out, it is useful to bear in mind that bioengineers modify crops in two quite different ways. One class of gene modification makes the crop easier to grow; a second class of modification is intended to improve the food itself.

The introduction of Roundup-resistant soybeans to Europe is an example of the first class of modification. This modification has been very popular with farmers in the United States, who planted half their crop with these soybeans in 1999. They like GM soybeans because the beans can be raised without intense cultivation (weeds are killed with Roundup herbicide instead), which both saves money and lessens soil erosion. But is the soybean that results nutritionally different? No. The gene that confers Roundup resistance in soybeans does so by protecting the plant's ability to manufacture so-called "aromatic" amino acids. In unprotected weeds, by contrast, Roundup blocks this manufacturing process, killing the weed. Because humans don't make any aromatic amino acids anyway (we get them in our diets), Roundup doesn't hurt us. The GM soybean we eat is nutritionally the same as an "organic" one, just cheaper to produce.

In the second class of modification, where a gene is added to improve the nutritional character of some food, the food will be nutritionally different. In each of these instances, it is necessary to examine the possibility that consumers may prove allergic to the product of the introduced gene. In one instance, for example, addition of a methionine-enhancing gene from the Brazil nut into soybeans (which are deficient in this amino acid) was discontinued when six of eight individuals allergic to Brazil nuts produced antibodies to the GM soybeans, suggesting the possibility of a reverse reaction. Instead, methionine levels in GM crops are being increased with genes from sunflowers. Screening for allergy problems is now routine.

On both scores, then, the risk of bioengineering to the food supply seems very slight. GM foods to date seem completely safe.

Are GM Crops Harmful to the Environment?

What are we to make of the much-publicized report that Monarch butterflies might be killed by eating pollen blowing out of fields planted with GM corn? First, it should come as no surprise. The GM corn (so-called Bt corn) was engineered to contain an insect-killing toxin (harmless to people) in order to combat corn borer pests. Of course it will kill any butterflies or other insects in the immediate vicinity of the field. However, focus on the fact that the GM cornfields do not need to be sprayed with pesticide to control the corn borer. An estimated $9 billion in damage is caused annually by the application of pesticides in the United States, and billions of insects and other animals, including an estimated 67 million birds, are killed each year. This pesticide-induced murder of wildlife is far more damaging ecologically than any possible effects of GM crops on butterflies.

Will pests become resistant to the GM toxin? Not nearly as fast as they now become resistant to the far higher levels of chemical pesticide we spray on crops.

How about the possibility that introduced genes will pass from GM crops to their wild or weedy relatives? This sort of gene flow happens naturally all the time, and so this is a legitimate question. But if genes for resistance to Roundup herbicide spread from cultivated sugar beets to wild populations of sugar beets in Europe, why would that be a problem? Besides, there is almost never a potential relative around to receive the modified gene from the GM crop; no wild relatives of soybeans exist in Europe, for example. Thus, no genes can escape from GM soybeans in Europe, any more than genes can flow from you to other kinds of animals.

Calvin and Hobbes

by Bill Watterson

On either score, then, the risk of bioengineering to the environment seems very slight. Indeed, in some cases, it lessens the serious environmental damage produced by cultivation and agricultural pesticides.

Should We Label Genetically Modified Foods?

While there seems little tangible risk in the genetic modification of crops, it is important to assure the public that these risks are being carefully assessed. Few issues manage to raise the temperature of discussions about plant genetic engineering more than the labeling of genetically modified (GM) crops. Agricultural producers have argued that there are no demonstrable risks, so a GM label can only have the function of scaring off wary consumers. Consumer advocates respond that consumers have every right to make that decision and to have access to the information necessary to make it.

In considering this matter, it is important to separate two quite different issues, the *need* for a label and the *right* of the public to have one. Every serious scientific investigation of the risks of GM foods has concluded that they are safe—indeed, in the case of soybeans and many other crops modified to improve cultivation, the foods themselves are not altered in any detectable way, and no nutritional test could distinguish them from "organic" varieties. So there seems to be little if any health need for a GM label for genetically engineered foods.

The right of people to know what they are eating is a very different issue. Europeans fear genetic manipulation because it is unfamiliar. People there don't trust their regulatory agencies as we do in the United States, because their agencies have a poor track record of protecting them. When they look at genetically modified foods, they are haunted by past experiences of regulatory ineptitude. In England, people remember British regulators' failure to protect consumers from meat infected with mad cow disease.

It does no good whatsoever to tell a fearful European that there is no evidence to warrant fear, no trace of data supporting danger from GM crops. A European consumer will simply respond that the harm is not yet evident—that we don't know enough to see the danger lurking around the corner. "Slow down," the European consumers say. "Give research a chance to look around all the corners. Let's be sure." No one can argue against caution, but it is difficult to imagine what else researchers can look into, because safety has already been explored very thoroughly. The fear remains, though, for the simple reason that no amount of information can remove it. Like a child scared of a monster under the bed, looking under the bed again doesn't help—the monster still might be there next time. And that means we are going to have to have GM labels, for people have every right to be informed about something they fear.

What should these labels be like? A label that only says "GM FOOD" simply acts as a brand. Like a POISON label, it shouts a warning to the public of lurking danger. Why not instead have a GM label that provides information to the consumer—that tells what regulators know about that product? Here are some examples:

For Bt corn: The production of this food was made more efficient by the addition of genes that made plants resistant to pests so that fewer pesticides were required to grow the crop.

For Roundup-ready soybeans: Genes have been added to this crop to render it resistant to herbicides. This reduces soil erosion by lessening the need for weed-removing cultivation.

For high β-carotene rice: Genes have been added to this food to enhance its β-carotene content and thus combat vitamin A deficiency.

GM food labels that in each instance actually tell consumers what has been done to the gene-modified crop would go a long way toward hastening public acceptance of gene technology in the kitchen.

Genetic engineering affords great opportunities for progress in medicine and food production, although many people are concerned about possible risks. On balance, the risks appear slight, and the potential benefits substantial.

Concept Review

For interactive testing, visit the Online Learning Center with PowerWeb at www.mhhe.com/Raven7

16.1 Molecular biologists can manipulate DNA to clone genes.

The DNA Manipulator's Toolbox

- Restriction endonucleases are able to cleave DNA at specific restriction sites. (p. 320)
- DNA ligase joins DNA fragments by catalyzing the formation of phosphodiester bonds between DNA nucleotides. (p. 320)

Host/Vector Systems

- The ability to propagate DNA in a host cell requires a vector that can enter and replicate in the host. The two most commonly used vectors are plasmids and phages. (p. 321, p. 324)
- Using these vectors, the gene or genes of interest are introduced into the target organisms, where they infect cells and replicate the altered vector DNA. (p. 322)

Using Vectors to Transfer Genes

- Viruses and artificial chromosomes can also be used as vectors to insert foreign DNA into host cells, creating recombinant DNA and recombinant genomes. (p. 323)

DNA Libraries

- A DNA library is a collection of DNA fragments from a specific source in a form that can be propagated in a host. (p. 324)
- A genomic library is a representation of the entire genome in a vector. (p. 324)

16.2 Genetic engineering involves easily understood procedures.

The Four Stages of a Genetic Engineering Experiment

- The four stages of a genetic engineering experiment are (1) cleaving the source DNA, (2) producing recombinant DNA, (3) cloning copies of the recombinants, and (4) screening the cloned copies for the desired genes. (pp. 325–328)

Working with Gene Clones

- One approach to making enough DNA to work with is to use DNA polymerase to copy the gene of interest through the polymerase chain reaction (PCR). (p. 329)
- The three steps in PCR are denaturation, primer annealing, and primer extension. (p. 329)
- After 20 cycles, a single fragment produces more than one million copies of the target sequence. (p. 329)
- In the Southern Blotting process, DNA is cleaved into restriction fragments, and the fragments are then separated by gel electrophoresis. Fragments that contain the gene of interest can then be identified when a radioactively labeled probe is washed over paper that contains a copy of the gel's pattern (p. 330)
- Restriction fragment length polymorphisms (RFLP) analysis and DNA fingerprinting are both used to identify individuals and specific unknown gene sequences. (pp. 331–332)

16.3 Biotechnology is producing a scientific revolution.

Medical Applications

- Pharmaceuticals, gene therapy, and piggyback vaccines are all medical applications that have been developed from biotechnology advances. (pp. 333–334)

Agricultural Applications

- Increased nitrogen fixation, herbicide resistance, insect resistance, and dietary deficiencies are all agricultural problems that are currently being addressed via gene technology and transgenic species. (pp. 335–338)

Risk and Regulation

- Genetic engineering has the potential to create great medical and agricultural advances, although many people are concerned with possible risks of genetic engineering, especially those associated with transgenic species. (p. 340)
- In short, the potential benefits of genetic engineering appear to outweigh the potential risks. (p. 340)

Self Test

1. Cutting certain genes out of molecules of DNA requires the use of special
   a. degrading nucleases.
   b. restriction endonucleases.
   c. eukaryotic enzymes.
   d. viral enzymes.
2. Which of the following cannot be used as a vector?
   a. phage
   b. plasmid
   c. bacterium
   d. All can be used as vectors.
3. In Cohen and Boyer's recombinant DNA experiments, restriction endonucleases were used to
   a. isolate fragments of cloned bacterial plasmids.
   b. isolate fragments of frog DNA that contained an rRNA gene.
   c. cleave the bacterial plasmid.
   d. All of these are correct.
4. A DNA library is
   a. a general collection of all genes sequenced thus far.
   b. a collection of DNA fragments that make up the entire genome of a particular organism.
   c. a DNA fragment inserted into a vector.
   d. all DNA fragments identified with a probe.
5. A probe is used in which stage of genetic engineering?
   a. cleaving DNA
   b. recombining DNA
   c. cloning
   d. screening
6. The enzyme used in the polymerase chain reaction is
   a. restriction endonuclease.
   b. reverse transcriptase.
   c. DNA polymerase.
   d. RNA polymerase.
7. A method used to distinguish DNA of one individual from another is
   a. polymerase chain reaction.
   b. cDNA.
   c. reverse transcriptase.
   d. restriction fragment length polymorphism.
8. Inserting a gene encoding a pathogenic microbe's surface protein into a harmless virus produces a:
   a. piggyback vaccine
   b. virulent virus
   c. active disease-causing pathogen
   d. pharmaceutical human protein
9. Although the Ti plasmid has revolutionized plant genetic engineering, one limitation of its use is that it
   a. cannot infect broadleaf plants.
   b. cannot be used on fruit-bearing plants.
   c. cannot transmit prokaryotic genes.
   d. does not infect cereal plants such as corn and rice.
10. Which of the following is *not* an application of genetic engineering in plants?
    a. nitrogen fixation
    b. DNA vaccines
    c. resistance to glyphosate
    d. production of insecticidal proteins in plants

Test Your Visual Understanding

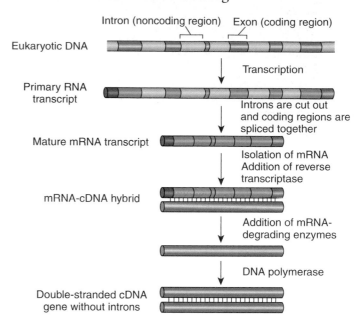

1. What process is illustrated in this figure? Where in the cell would you look to find the primary RNA transcript, and where would you look to find the mature mRNA transcript?
2. When might researchers use this process—that is, what would they be trying to accomplish?

Apply Your Knowledge

1. The human genome has about 3 billion base-pairs. Assume that you want to clone the entire human genome using various vectors, but there is a limit to the size of a DNA fragment that can be inserted in a vector. Following is a list of vectors along with their size limit of DNA fragment. Calculate how many vectors of each type would be needed to generate a library of the human genome.
   a. bacterial plasmid—18 kilo base-pairs
   b. phages—25 kilo base-pairs
   c. YACs—250 kilo base-pairs
2. A major focus of genetic engineering has been on attempting to produce large quantities of scarce human proteins by placing the appropriate genes into bacteria and thus turning the bacteria into protein production machines. Human insulin and many other proteins are produced this way. However, this approach does not work for producing human hemoglobin. Even if the proper clone is identified, the fragments containing the hemoglobin genes are successfully incorporated into bacterial plasmids, and the bacteria are infected with the plasmids, no hemoglobin is produced by the bacteria. Why doesn't this experiment work?

17

Genomes

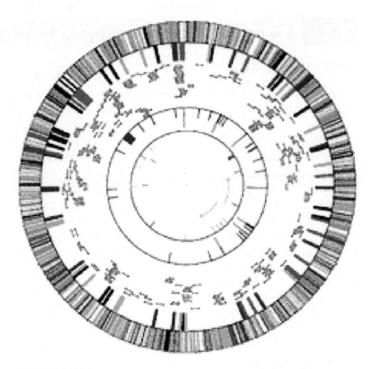

Concept Outline

17.1 Genomes can be mapped both genetically and physically.

Genome Maps. The first genetic maps revealed the positional relationship among genes on chromosomes, but did not link them to specific DNA sequences. In physical maps, physical pieces of DNA are mapped to specific regions on chromosomes.

17.2 Genome sequencing produces the ultimate physical map.

Sequencing. Automated sequencers and computer programs can be used to determine the entire sequence of a genome using either a clone-by-clone or shotgun method. Sequencers can only sequence relatively short segments of the genome at a time. Assembling these segments into a whole genome is like putting a jigsaw puzzle together.

17.3 Being more complex does not necessarily require more genes.

Human Genome Project. The draft sequence for the human genome was released in February 2001. Perhaps the greatest surprise was how few genes comprise our genome.

Genome Geography. In addition to functional genes, genomes contain a variety of repetitive DNA patterns that may have originally come from other organisms.

Comparative Genomics. Synteny, conservation of large DNA sequences among related organisms, makes it possible to extend information about sequenced genomes to other genomes.

17.4 Genomics is opening a new window on life.

Functional Genomics. Using DNA microarrays and inserting genes into organisms allow biologists to study the function of newly identified genes.

Proteomics. The genome produces the proteome, all the proteins in an organism. Proteomics is the study of these proteins and how they function.

Using Genomic Information. Intra- and interspecific genome comparisons can yield insight into evolution, crop improvement, and genetic disease.

FIGURE 17.1
Haemophilus influenzae **genome.** The 1.8 million base-pairs of *H. influenzae* were sequenced in 1995. This bacterium causes respiratory infections and meningitis in children. It was the first free-living organism to be sequenced; its genome was found to be 10 times larger than any of the viral genomes that had previously been sequenced. Just five years later, the human genome was sequenced. Each color represents genes with similar functions.

Reprinted with permission from Fleishmann, et al., "Whole-Genome Random Sequencing LOOK UP," Science, 269: 496–512. Copyright © 1995 American Association for the Advancement of Science.

Maps help you find specific locations. Individual countries can be located on a map of the world, and if there is enough detail, a city or perhaps even a famous landmark can be pinpointed. Genome maps work in a similar way, but are linear. Your genome is all of the DNA in the nucleus of one of your cells. Within this DNA world are chromosomes containing individual genes with specific base-pair sequences. For almost a century, geneticists have identified and located genes one by one. Creating a more comprehensive and detailed map required a breakthrough in DNA sequencing technology. The genomes of almost 100 organisms and many more viruses have been sequenced (figure 17.1). Now, more landmarks are needed to navigate genome maps and understand how the genome is used to build a functioning organism.

17.1 Genomes can be mapped both genetically and physically.

Genome Maps

News headlines celebrate the mapping of many genomes, including the human and rice genomes. Mapping genes is not new, but mapping an entire genome is a recent and astounding accomplishment (see figure 17.1). To understand how genome maps are made, we must first consider the different possible types of genome maps. Each type provides different information. **Genetic maps** are linkage maps (see chapter 13). They show the relative location of genes on a chromosome as determined by recombination frequencies. **Physical maps** are diagrams showing the relative positions of **landmarks** within specific DNA sequences. Landmarks include specific sequences of DNA and sites where restriction enzymes cut DNA. These landmarks help you find your way through the genome and identify parts of the genome that interest you.

Genetic Maps

The first genetic (linkage) map was made in 1911 when Sturtevant mapped five genes in *Drosophila* (see chapter 13, figure 13.32). Distances on a genetic map are measured in centimorgans (cM) in honor of the geneticist Morgan. One cM represents a 0.01% recombination frequency. Today, 13,744 genes have been mapped on the *Drosophila* genome. Linkage mapping can be done without knowing the DNA sequence of a gene. Computer programs make it possible to create a linkage map for 1000 genes at a time. There are a few limitations to genetic maps. Distances between genes determined by recombination frequencies between traits do not directly correspond with physical distance on a chromosome. The conformation of DNA between genes varies and can affect the frequency of recombination. Also, not all genes have obvious phenotypes that can be followed in segregating crosses.

Physical Maps

Distances between landmarks on a physical map are measured in base-pairs (1000 base-pairs equal 1 kilobase, kb). It is not necessary to know the DNA sequence of a segment of DNA in order to create a physical map or to know if that DNA codes for a specific gene. The first physical maps were created by cutting genomic DNA with different restriction enzymes (figure 17.2) and then putting the pieces back together, based on size and overlap, into a contiguous segment of the genome called a **contig**. Each restriction enzyme recognizes a very specific sequence of DNA that it cleaves, as described in chapter 16. The very first restriction enzymes to be isolated came from *Haemophilus*, which was also the first free-living genome to be sequenced (see figure 17.1).

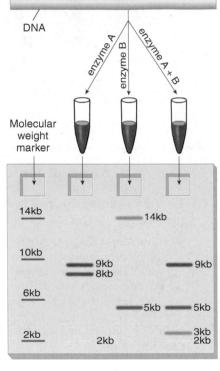

1. Multiple copies of a segment of DNA are cut with restriction enzymes.

2. The fragments produced by enzyme A only, by enzyme B only, and by enzymes A and B simultaneously are run out side-by-side on a gel, which separates them according to size, smaller fragments running faster.

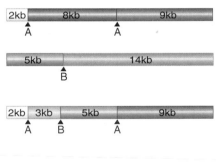

3. The fragments are arranged so that the smaller ones produced by the simultaneous cut can be grouped to generate the larger ones produced by the individual enzymes.

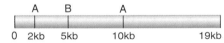

4. A physical map is constructed.

FIGURE 17.2
Restriction enzymes can be used to create a physical map. Comparing the sizes of DNA fragments produced by two restriction enzymes individually with the sizes of those produced by the two enzymes acting jointly allows you to deduce the locations where the enzymes cleave the DNA.

Homologous regions of DNA can vary slightly in base-pair composition among individuals in a population. When a homologous stretch of DNA is cut with restriction enzymes in different individuals, fragments of different lengths may be produced. These fragments are called restriction fragment length polymorphisms (RFLPs, see figure 16.14). When a RFLP segregates with a genetic trait, it becomes possible to link the genetic map (represented by the heritable trait) and the physical map (repre-

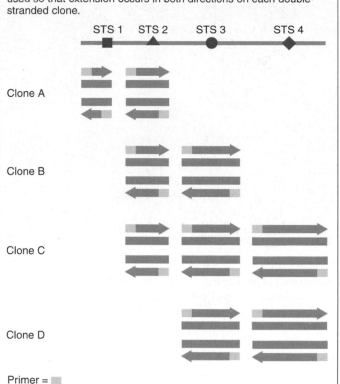

Step 1: Amplify parts of a cloned DNA segment with PCR, using primers that recognize particular STSs. Two primers per STS are used so that extension occurs in both directions on each double-stranded clone.

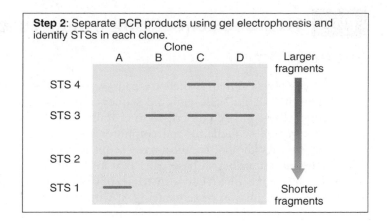

Step 2: Separate PCR products using gel electrophoresis and identify STSs in each clone.

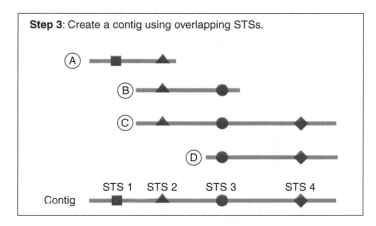

Step 3: Create a contig using overlapping STSs.

FIGURE 17.3

Creating a physical map with sequence-tagged sites. The presence of landmarks called sequence-tagged sites, or STSs, in the human genome made it possible to begin creating a physical map large enough in scale to provide a foundation for sequencing the entire genome. (*1*) Primers (shown as arrows) that recognize unique STSs are added to a cloned segment of DNA, followed by DNA replication via PCR. (*2*) The PCR products from each reaction are separated based on size on a DNA gel, and the STSs contained in each clone are identified. (*3*) The cloned DNA segments are then aligned based on overlapping STSs to create a contig.

sented by the specific RFLP). RFLPs are one of many types of physical landmarks in genomes.

Constructing a physical map for large genomes requires the efforts of many research labs. Suppose that two groups are each working on a segment of DNA that has been cloncd (inserted into the genome of a virus or bacterium that can replicate), but has not been sequenced. How can the two groups tell if their clones contain overlapping DNA sequences? First, it would be helpful if everyone used the same type of landmark. Second, it would be helpful if these landmarks or tags could be completely described in a shared database so that researchers did not have to mail DNA clones back and forth. The solution for groups working on cloned DNA from the same species is to use **sequenced-tagged sites (STSs).** An STS is a 100 to 500 base-pair sequence of a clone. The clone itself is much larger than the length of the tag, and it would take a much longer time to sequence this complete DNA segment. The clone needs to be from single-copy DNA so that the STS is unique within the genome. PCR (polymerase chain reaction, see figure 16.11) primers can be designed based on the STSs. Once you know the sequence of part of the land-

mark, you can use polymerase chain reactions (PCR) to determine if a cloned piece of DNA contains this STS. Primers are designed for PCR amplification of each known STS. Using these primers, sequences containing this STS can be amplified. The fragments produced by the extension of primers are separated using gel electrophoresis and stained with ethidium bromide to identify fragments based on size. The fragments can be pieced together to make a physical map by identifying the STSs in each fragment and overlapping them. Because of the high density of STSs in the human genome and the relative ease of identifying an STS in a DNA clone, it became possible to develop physical maps on the large scale of the 3.2-gigabase genome in the mid-1990s (figure 17.3). STSs make a scaffold for genome sequences.

Genetic maps provide information on the relative distances between genes. Distances are measured in centimorgans. Physical maps locate landmarks in the genome. Distance is measured in base-pairs. Linked physical and genetic maps provide a scaffold for whole genome sequencing.

17.2 Genome sequencing produces the ultimate physical map.

Sequencing

The ultimate physical map is the base-pair sequence of the entire genome. Large-scale genome sequencing uses the basic sequencing approach described in chapter 16, but relies on automated sequencing and computer analysis (figure 17.4). In a few hours, an automated sequencer can sequence the same number of base-pairs that a person could manually sequence in a year—up to 50,000 base-pairs. Without the automation of sequencing, it would not have been possible to sequence large, eukaryotic genomes like that of humans.

Automated Sequencers

While it would be ideal to isolate DNA from an organism, add it to a sequencer, and then come back in a week or two to pick up a computer-generated printout of the genome sequence for that organism, life is not quite that simple. Sequencers provide accurate sequences for DNA segments up to 500 base-pairs (bp) long. Even then, errors are possible. So, five to ten copies of a genome are sequenced to reduce errors.

DNA is prepared for sequencing by making many copies of it with fluorescently labeled adenines (A), cytosines (C), thymines (T), and guanines (G), mixed in with unlabeled nucleotides (you might want to review DNA replication in chapter 14). The fluorescent label for each nucleotide is distinct in color. These labeled nucleotides also lack an OH group in the 3′ position. Without the 3′ OH group, no more nucleotides can be added. As the DNA is replicated, the labeled nucleotides are incorporated randomly, resulting in sequences of DNA of different lengths. This labeled DNA is then added to one of 96 wells of a gel that has been

FIGURE 17.4
Automated sequencing. This sequence facility simultaneously runs multiple automated sequencers, each processing 96 samples at a time.

poured between two vertical glass plates spaced less than 0.5 millimeters apart. Each well is filled with different batches of labeled DNA. (Newer machines use 96 capillary tubes the diameter of a human hair.) The DNA then separates according to size, with the shorter sequences passing a laser beam detector first. The detector sends color information to a computer that translates the data into a sequence. If the smallest band terminates with a labeled A, it passes the detector first and A is recorded as the first base. If a labeled G is found in the largest band, that is recorded as the last base. Work through figure 17.5 to be sure you understand how a sequence is determined.

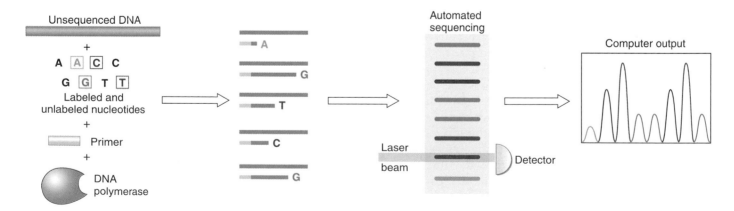

FIGURE 17.5
Sequencing DNA. DNA is replicated, and fluorescent nucleotides that halt replication are randomly inserted into different sequences. An automated sequencer determines the base-pair sequence.
Why is it important to know whether or not your primer hybridizes to the template strand in the replication step?

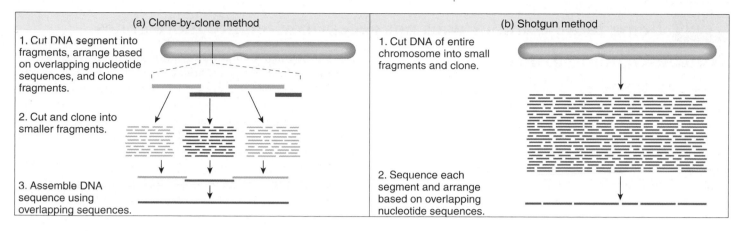

(a) Clone-by-clone method	(b) Shotgun method
1. Cut DNA segment into fragments, arrange based on overlapping nucleotide sequences, and clone fragments.	1. Cut DNA of entire chromosome into small fragments and clone.
2. Cut and clone into smaller fragments.	
3. Assemble DNA sequence using overlapping sequences.	2. Sequence each segment and arrange based on overlapping nucleotide sequences.

FIGURE 17.6

Comparison of sequencing methods. (*a*) The clone-by-clone method involves many rounds of cloning and arranging the fragments of a segment of DNA before assembling the final DNA sequence. (*b*) The shotgun method involves fragmenting and cloning the entire chromosome and then using a computer to assemble the final DNA sequence based on overlapping nucleotide sequences.

Artificial Chromosomes

The development of a new type of vector, called artificial chromosomes, has allowed scientists to clone larger pieces of DNA. The first generation of these new vectors were **yeast artificial chromosomes,** called **YACs**. These were constructed by using a yeast origin of replication and a yeast centromere sequence, with foreign DNA added to this construct. The origin of replication allows the artificial chromosome to replicate independently of the rest of the genome, and the centromere sequences make them mitotically stable. YACs were useful for cloning larger pieces of DNA but had many drawbacks, including a tendency to rearrange, or to lose portions of DNA by deletion. Despite the difficulties, the YACs were used early on to construct physical maps by restriction enzyme digestion of the YAC DNA.

The artificial chromosomes most commonly used now, particularly for large-scale sequencing, are made in *E. coli.* These are **bacterial artificial chromosomes,** called **BACs**. They are a logical extension of the use of bacterial plasmids to include much larger pieces of DNA that behave like a second bacterial chromosome. These BAC vectors accept DNA inserts between 100 and 200kb.

Sequencing by Whole Genomes

Clone-by-Clone Sequencing. The cloning of large inserts in BACs facilitates the analysis of entire genomes. The strategy most commonly pursued is to construct a physical map first, and then use it to place the site of BAC clones for later sequencing. The alignment of large portions of a chromosome requires identifying regions that overlap between clones. This can be done either by constructing restriction maps of each BAC clone or by identifying STSs between clones. If two BAC clones have the same STS, this means they must overlap.

Once a number of BAC clones have been aligned, this forms a contig, or contiguous stretch of DNA. The individual BAC clones can then be sequenced 500 base-pairs at a time to produce the sequence of the entire contig (figure 17.6*a*). This strategy of physical mapping followed by sequencing is called clone-by-clone sequencing.

Shotgun Sequencing. The idea of shotgun sequencing is to simply sequence all cloned fragments and use a computer to put together the overlaps. This approach is much less labor-intensive than the clone-by-clone method, but it requires much greater computer power to assemble the final sequence and very efficient algorithms to find overlaps. Unlike the clone-by-clone approach, shotgun sequencing does not tie the sequence to any other information about the genome. Many investigators have used both clone-by-clone and shotgun sequencing techniques, and hybrid approaches that use both are becoming the norm. This has the strength of tying the sequence to a physical map while greatly reducing the labor involved. The two methods are shown graphically in figure 17.6.

Assembler programs compare multiple copies of sequenced regions in order to *assemble* a **consensus sequence.** Consensus sequences reduce error. Although computer assemblers are incredibly powerful, final human analysis is required after both clone-by-clone and shotgun cloning to determine when a genome sequence is sufficiently accurate to be useful to researchers.

Genomes are cut into small pieces of DNA that are sequenced in automated sequencers. Shotgun cloning starts with small pieces of sequenced DNA and utilizes extensive computational analysis to determine the whole sequence from overlapping pieces. The clone-by-clone strategy starts with a physical map with many landmarks. Pieces of DNA from known sites on the map are sequenced.

17.3 Being more complex does not necessarily require more genes.

Human Genome Project

One of the defining characteristics of genome science (**genomics**) is the scale of the projects. In an era when large numbers are commonplace (for example, computers with 120-gigabyte hard drives), it is easy to underestimate the enormity of the task of mapping and sequencing the 3.2-gigabase (3,200,000,000-nucleotide) human genome. Conceptually, the approach is not all that different from the manual sequencing of the 5375-nucleotide genome of the X174 virus in 1977. As you saw in section 17.2, the real challenge is being able to rapidly sequence and assemble sequences for very large genomes. The limiting factor for sequencing projects has been genome size. Of the 100 or so sequenced genomes of free-living organisms, most are bacterial. Automated sequencing, new computer algorithms, STS maps, and BACs that can hold large DNA inserts have all made it possible to sequence our gigantic genome and, more recently, the even larger genome of rice. Perhaps the most startling finding in the analysis of the human genome is that the complexity of an organism is not necessarily reflected in the number of genes it contains. Humans have about twice as many genes as *Drosophila* and fewer genes than rice (figure 17.7).

The vast scale of genomics has ushered in a new way of doing biological research, involving large teams. While a single individual can clone a small genome manually, a huge genome like ours requires the collaborative efforts of hundreds of researchers. One example of such teamwork is the Human Genome Project, the results of which have illustrated key genomic concepts and offered tantalizing possibilities for understanding the genetic basis of disease and unlocking the mysteries of evolution.

The Human Genome Project originated in 1990 when a group of American scientists formed the International Human Genome Sequencing Consortium. The goal of this publicly funded effort was to use a clone-by-clone approach to sequence the human genome. Both genetic and physical maps were enhanced and published in the 1990s, and used as scaffolding to sequence each chromosome. Then, in May 1998, Craig Venter, who had sequenced *Haemophilus influenzae*, announced that he had formed a private company to sequence the human genome. He proposed to shotgun-clone the 3.2-gigabase genome in only two years. The consortium rose to the challenge, and the race to sequence the human genome began. The upshot was a tie of sorts. On June 26, 2000, the two groups jointly announced success, and each published its findings simultaneously in 2001. The consortium's draft alone included 248 names on its partial list of authors.

The draft sequence of the human genome is more of a beginning than an ending. Gaps in the sequence are still being filled, and the map is still being finished. More significantly, research on the whole genome can move ahead. Now that the ultimate physical map is in place and being integrated with the genetic map, diseases such as diabetes that result from defects in more than one gene can be addressed. Comparisons with other genomes are already changing our understanding of genome evolution (see chapter 24).

Sequencing of large genomes, including our own via the Human Genome Project, is changing a long-held belief that organisms with complex body plans and behaviors have more genes than simpler organisms.

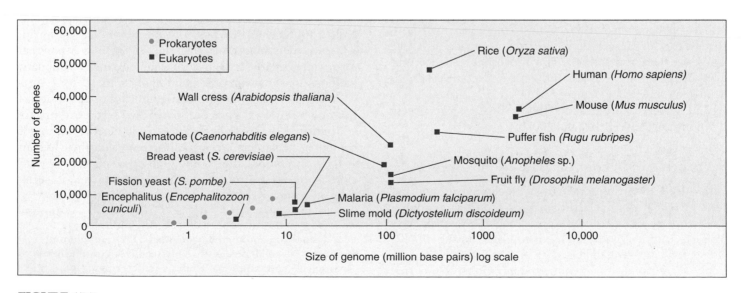

FIGURE 17.7
Size and complexity of genomes.   In general, eukaryotic genomes are larger and have more genes than prokaryotic genomes, although the size of the organism is not the determining factor. The mouse genome is nearly as large as the human genome, and the rice genome contains more genes that the human genome.

Genome Geography

Finding Genes in the Genome

Once you have sequenced a genome, the next step is to determine which regions of the genome contain which genes and what those genes do. There is a lot of information to be mined from the sequence data. Using markers from physical maps and information from genetic maps, it is possible to find the sequence of the small percentage of genes that are identified by mutations with an observable effect. Information in the nucleotide sequence itself can also be used in the search for genes. A gene begins with a "start" codon such as ATG and contains no "stop" codons (UAA, UGA, or UAG) for a distance long enough to encode a protein. This coding region is referred to as an **open reading frame (ORF).** While these sequences are likely to be genes, they may or may not actually be translated into a functional protein. Potential gene sequences need to be tested experimentally to determine whether they have a function. It is also possible to search genome databases for sequences that have homology to known genes in other species. Using computer programs to search for genes, to compare genomes, and to assemble genomes are only a few of the new genomics approaches falling under the heading of **bioinformatics.**

Another way to find genes that are actually expressed in an organism is to use a type of STS called an **expressed sequence tag (EST).** ESTs are made by isolating mRNA, using reverse transcriptase to make cDNA copies of the mRNA, and then sequencing one or both ends of each cDNA. Just like STSs, PCR primers can be designed to make copies of genomic DNA or cDNA containing the EST. Unlike some STSs that serve only as physical landmarks on the genome, ESTs can identify functional genes that are transcribed to produce RNA. ESTs have been used to identify 87,000 cDNAs in different human tissues. About 80% of these cDNAs were previously unknown. This gene discovery approach allows the researcher to directly relate the new information to the map of the genome.

How is it possible that 87,000 cDNAs representing 87,000 different mRNAs could be identified in human tissue when there only appear to be 30,000 genes in the human genome? This was a puzzle for genome researchers who had predicted that the human genome would contain close to 100,000 genes. It turns out that the answer lies in the modularity of eukaryotic genes, which consist of exons interspersed with introns. Following transcription, the introns are removed, and exons are spliced together. In some cells, some of the splice sites are skipped, and one or more exons is removed along with the introns. This process, called **alternative splicing** (figure 17.8), yields different proteins that can have different functions. Thus, the added complexity of proteins in the human genome comes not from additional genes, but from new ways to put existing gene parts together.

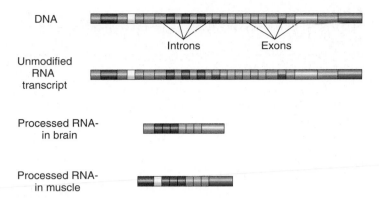

FIGURE 17.8
Alternative splicing can result in the transcription of different mRNAs from the same coding sequence. In some cells, exons can be excised along with neighboring introns, resulting in different proteins. Alternative splicing explains why 30,000 human genes can code for three to four times as many proteins.

Gene Organization in Eukaryotic Genomes

Four different classes of protein-encoding genes are found in the eukaryotic genomes, differing largely in gene copy number.

Single-copy genes. Many genes exist as single copies on a particular chromosome. Most mutations in these genes result in recessive Mendelian inheritance. Silent copies, inactivated by mutation, are called *pseudogenes.*

Segmental duplications. Sometimes whole blocks of genes are copied from one chromosome to another, resulting in *segmental duplication.* Blocks of similar genes in the same order are found throughout the human genome. Chromosome 19 seems to have been the biggest borrower, sharing blocks of genes with 16 other chromosomes.

Multigene families. As we have learned more about eukaryotic genomes, it has become apparent that many genes exist as parts of *multigene families*, groups of related but distinctly different genes that often occur together in clusters. These genes appear to have arisen from a single ancestral gene that duplicated during an uneven meiotic crossover in which genes were added to one chromosome and subtracted from the other.

Tandem clusters. Identical copies of genes can also be found in *tandem clusters.* These genes are all transcribed simultaneously, which increases the amount of mRNA available for protein production. For example, the genes encoding rRNA are typically present in clusters of several hundred copies.

Bioinformatic approaches can identify gene sequences in a sequenced genome. Genes are found as single copies as well as in arrays of closely related or exact copies in eukaryotic genomes.

Table 17.1 Classes of DNA Sequences Found in the Human Genome

Class	Frequency	Description
Protein-encoding genes	1%	Translated portions of the 30,000 genes scattered about the chromosomes
Introns	24%	Noncoding DNA that comprises the great majority of each human gene
Segmental duplications	5%	Regions of the genome that have been duplicated
Pseudogenes (inactive genes)	2%	Sequence that has characteristics of a gene, but is not a functional gene
Structural DNA	20%	Constitutive heterochromatin, localized near centromeres and telomeres
Simple sequence repeats	3%	Stuttering repeats of a few nucleotides such as CGG, repeated thousands of times
Transposable elements	45%	21%: Long interspersed elements (LINEs), which are active transposons
		13%: Short interspersed elements (SINEs), which are active transposons
		8%: Retrotransposons, which contain long terminal repeats (LTRs) at each end
		3%: DNA transposon fossils

Noncoding DNA in Eukaryotic Genomes

Sequencing of several eukaryotic genomes has now been completed, and one of the most notable characteristics is the amount of noncoding DNA they possess. The recent sequencing of the human genome reveals a particularly startling picture. Each of your cells has about 6 feet of DNA stuffed into it, but of that, less than 1 inch is devoted to genes! Nearly 99% of the DNA in your cells has little or nothing to do with the instructions that make you *you*.

True genes are scattered about the human genome in clumps among the much larger amount of noncoding DNA, like isolated hamlets in a desert. There are six major sorts of noncoding human DNA (table 17.1).

Noncoding DNA within genes. As discussed in chapter 15, a human gene is not simply a stretch of DNA, like the letters of a word. Instead, a human gene is made up of numerous fragments of protein-encoding information (exons) embedded within a much larger matrix of noncoding DNA (introns). Together, introns make up about 24% of the human genome, and exons, less than 1.5%.

Structural DNA. Some regions of the chromosomes remain highly condensed, tightly coiled, and untranscribed throughout the cell cycle. Called *constitutive heterochromatin*, these portions tend to be localized around the centromere or located near the ends of the chromosome, at the telomeres.

Simple sequence repeats. Scattered about chromosomes are *simple sequence repeats* (SSRs). An SSR is a one- to three-nucleotide sequence such as CA or CGG, repeated like a broken record thousands and thousands of times. SSRs can arise from DNA replication errors. SSRs make up about 3% of the human genome.

Segmental duplications. Blocks of genomic sequences composed of from 10,000 to 300,000 base-pairs have duplicated and moved either within a chromosome or to a nonhomologous chromosome.

Pseudogenes. These are inactive genes that may have lost function because of mutation.

Transposable elements. Fully 45% of the human genome consists of mobile bits of DNA called transposable elements. Discovered by Barbara McClintock in 1950 (she received the Nobel Prize for her discovery in 1983), transposable elements are bits of DNA that are able to jump from one location on a chromosome to another.

How do transposable elements accomplish this remarkable feat? In some cases, the transposon is duplicated, and the duplicated DNA moves to a new place in the genome, so the number of copies of the transposon increase. Other types of transposons are excised without duplication and insert themselves elsewhere in the genome.

Human chromosomes contain four sorts of transposable elements. Fully 21% of the genome consists of *long interspersed elements* (LINEs). An ancient and very successful element, LINEs are about 6000 base-pairs long, and contain all the equipment needed for transposition. LINEs encode a reverse transcriptase enzyme that can make a cDNA copy of the transcribed LINE RNA. The result is a double-stranded segment that can reinsert into the genome rather than undergo translation into a protein.

Short interspersed elements (SINEs) are similar to LINEs, but cannot transpose without using the transposition machinery of LINEs. Nested within the genome's LINEs are over half a million copies of a SINE element called ALu, 10% of the human genome. Like a flea on a dog, ALu moves with the LINE it resides within. Just as a flea sometimes jumps to a different dog, so ALu sometimes uses the enzymes of its LINE to move to a new chromosome location. Often ALu jumps right into genes, causing harmful mutations.

Two other sorts of transposable elements are also found in the human genome: 8% of the human genome is devoted to retrotransposons called *long terminal repeats* (LTRs). Although the transposition mechanism is a bit different from that of LINEs, LTRs also use reverse transcriptase

A Vocabulary of Genomics

alternative splicing Exons are combined in different ways so that one gene can produce multiple proteins.

annotate To determine what different regions of a sequenced genome, especially coding regions, do.

assembly Arranging sequenced pieces of DNA in the correct order in the genome.

BAC Bacterial artificial chromosome. Genomic DNA (about 150,000 bp) is inserted into bacterial DNA where many copies can be made.

clone-by-clone sequencing Hierarchical sequencing strategy that starts with physical and genetic maps and breaks DNA sequences from known locations into smaller pieces for sequencing and assembly.

contig Sequence of DNA assembled by identifying overlaps among smaller DNA segments. The term contig comes from "contiguous."

draft sequence An initial genome sequence that has some gaps but is useful to researchers.

EST Expressed sequence tag. Partially sequenced cDNAs used to identify genes in a genomic sequence.

finished sequence A genome sequence with minimal gaps, sequences in the right order, and no more than 1 error in 10,000 nucleotides.

genetic map Map of the relative position of genes based on linkage analysis—that is, frequency of recombination during meiosis.

genome All the genetic information (DNA) in a cell. In eukaryotes, the genome is equivalent to all of the chromosomes in a haploid cell. These cells also have a mitochondrial genome and, in the case of plants, a chloroplast genome.

PCR Polymerase chain reaction. Method for making many copies of a piece of DNA very rapidly.

proteome All the proteins that can be produced by a genome.

repetitive DNA Multiple copies of the same sequence of DNA. The length of the sequence varies. Large eukaryotic genomes consist mostly of repetitive DNA.

restriction enzyme An enzyme that recognizes a specific DNA sequence and cuts the DNA there.

RFLP Restriction fragment length polymorphism. Variation in a given DNA sequence in a population can result in fragments of different lengths when the DNA is cut with a restriction enzyme. RFLPs make good landmarks for physical maps.

shotgun sequencing All the DNA in a genome is broken into small pieces and directly sequenced. Computer programs assemble the sequences in order.

SNP Single nucleotide polymorphism. A single base-pair variation that can be used to differentiate between individuals of the same species.

STS Sequenced-tagged site. A unique sequence of nucleotides in the genome that is used for mapping. ESTs and SNPs are types of STSs.

synteny Blocks of DNA with genes in the same order that are conserved between species. Synteny makes it possible to use genome-mapped organisms to study related organisms.

---

to ensure that copies are double-stranded and can reintegrate into the genome. Some 3% is devoted to dead transposons, elements that have lost the signals for replication and can no longer jump.

Variation in the Human Genome

One fact that is becoming clear from analysis of the human genome is the huge amount of genetic variation that exists in our species. This information has practical use as we identify a specific kind of variation: **single nucleotide polymorphisms (SNPs)** in the human genome. SNPs are sites where individuals differ by only a single nucleotide. To be classified as a polymorphism, a SNP must be present in at least 1% of the population. At present, the International SNP Map Working Group has identified 50,000 SNPs in coding regions of the genome and an additional 1.4 million in noncoding DNA. It is estimated that this represents about 10% of the variation available.

These SNPs are being used to look for associations between genes. We expect that the genetic recombination occurring during meiosis randomizes all but the most tightly linked genes. We call the tendency for genes to not be randomized **linkage disequilibrium.** This kind of association can be used to map genes. The preliminary analysis of SNPs shows that many are in linkage disequilibrium. This unexpected result has led to the idea of genomic haplotypes, or regions of chromosomes that are not being exchanged by recombination. If these haplotypes stand up to further analysis, they could greatly aid in mapping the genetic basis of disease. The Human Genome Project is now working on a haplotype map of the genome.

Gene sequences in eukaryotes vary greatly in copy number, some occurring many thousands of times and others only once. Only about 1% of the human genome is devoted to protein-encoding genes. Much of the rest is composed of transposable elements.

Comparative Genomics

With the large number of sequenced genomes, it is now possible to make comparisons at both the gene and genome level (see figure 17.7). One of the striking lessons learned from the sequence of the human genome is how very like other organisms humans are. More than half of the genes of *Drosophila* have human counterparts. Among mammals, the differences are even fewer. Humans have only 300 genes that have no counterpart in the mouse genome.

The flood of information from different genomes has given rise to a new field: **genomics.** At this point, we have the complete sequences for close to 100 bacterial genomes. Among eukaryotes, we have the full genome sequences of both types of yeast used in genetics, *S. cerevisiae* and *S. pombe*, as well as the protist *Plasmodium*, the invertebrate animals *Drosophila* and *C. elegans*, and the vertebrates puffer fish, mouse, and human. In the plant kingdom, the genomes for *Arabidopsis* and rice have been completed. It is important to remember that most of these genomes are **draft sequences** that include many gaps in regions of highly repetitive DNA.

One active field of genomics is to determine the minimal genome that can support a cell. Studies have used a variety of approaches, including the analysis of parasitic bacteria with greatly reduced genomes, coupled with the analysis of cognate genes in known genomes. We have not yet been able to estimate the number of genes that is the minimal number necessary to support life, but we are closing in on that goal.

Another related ongoing project is to learn the true "language" of proteins. As more and more genomic sequences are available to analyze, it is becoming clear that only a limited number of basic kinds of proteins are encoded by genomes and that a limited number of domains have been used over and over by evolution.

The use of comparative genomics to ask evolutionary questions is also a field of great promise. For this reason, the next round of animal genomes being sequenced includes the chimp, our closest living relative. Comparison of human and chimp genomes will provide a clearer picture of our relationship. The comparison of the many prokaryotic genomes already indicates a greater degree of lateral gene transfer than was previously suspected.

Synteny

Similarities and differences between highly conserved genes in different species can be investigated on a gene-by-gene basis. Genome science allows for a much larger-scale approach to comparing genomes by taking advantage of **synteny.** Synteny refers to the conserved arrangements of segments of DNA in related genomes. Physical mapping can be used to look for conserved segments of DNA in genomes that have not been sequenced. Comparisons with the sequenced syntenous segment in another species can be very helpful. To illustrate this, consider rice, already sequenced, and its grain relatives maize, barley, and wheat, none of which have been fully sequenced. Even though these plants diverged more than 50 million years ago, the chromosomes of rice, corn, wheat, and other grass crops show extensive conserved arrangements of segments (synteny) (figure 17.9). In a genomic sense, "rice is wheat." Interestingly, the rice genome has more genes than the human genome. However, rice still has a much smaller genome than its grain relatives that represent a major food source for humans. By understanding the rice genome at the level of its DNA sequence, it should be much easier to identify and isolate genes from grains with larger genomes. DNA sequence analysis of cereal grains will be important for identifying genes associated with disease resistance, crop yield, nutritional quality, and growth capacity.

Organelle Genomes

Mitochondria and chloroplasts are bacterial relatives living in eukaryotes as a result of endosymbiosis. Their genomes have been sequenced in some species and are most like prokaryotic genomes. The chloroplast genome, having about 100 genes, is minute compared to the rice genome, with 32,000 to 55,000 genes.

The chloroplast is a plant organelle that functions in photosynthesis. It can independently replicate in the plant cell because it has its own genome. The DNA in the chloroplasts of all land plants have about the same number of genes, and they are present in about the same order. In contrast to the evolution of the DNA in the plant cell nucleus, chloroplast DNA has evolved at a more conservative pace, and therefore shows a more interpretable evolutionary pattern when scientists study DNA sequence similarities. Chloroplast DNA is also not subject to modification caused by transposable elements and mutations due to recombination.

Over time, there appears to have been some genetic exchange between the nuclear and chloroplast genomes. For example, the key enzyme (RUBISCO) in the Calvin cycle of photosynthesis consists of large and small subunits. The small subunit is encoded in the nuclear genome. The protein it encodes has a targeting sequence that allows it to enter the chloroplast and combine with large subunits, which are coded for and produced by the chloroplast. The evolutionary history of the localization of these genes is a puzzle. Comparative genomics and their evolutionary implications are explored in detail in chapter 24, after we have established the fundamentals of evolutionary theory.

Comparisons of whole genome maps reveal a surprising commonality of genes among organisms. Only a small percentage of the DNA of eukaryotic genomes codes for functional proteins.

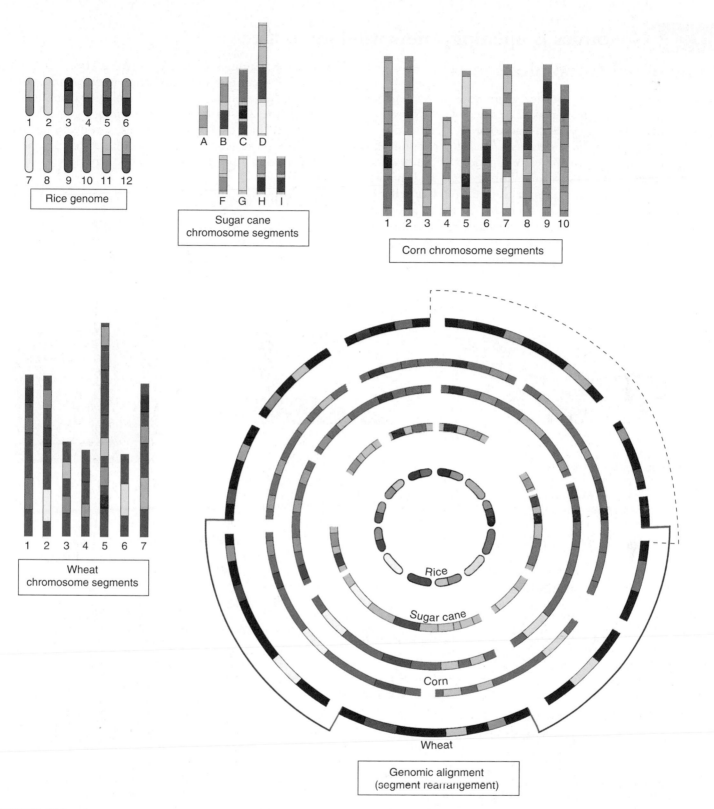

FIGURE 17.9

Grain genomes are rearrangements of similar chromosome segments. Shades of the same color represent pieces of DNA that are conserved among the different species but have been rearranged. By splitting the individual chromosomes of major grass species into segments and rearranging the segments, researchers have found that the genome components of rice, sugarcane, corn, and wheat are highly conserved. This implies that the order of the segments in the ancestral grass genome has been rearranged by recombination as the grasses have evolved.

17.4 Genomics is opening a new window on life.

Functional Genomics

Sequencing the human and rice genomes represents major technological accomplishment. The new field of bioinformatics takes advantage of high-end computer technology to analyze the growing gene databases, look for relationships among genomes, and hypothesize functions of genes based on sequence. Genomics is now shifting gears and moving back to hypothesis-driven science, to **functional genomics,** the study of the function of genes and their products. Bioinformatics approaches are being combined with newly developed technologies for analyzing genes and their functions. Like sequencing whole genomes, finding

how these genomes work requires the efforts of a large team. For example, an international community of researchers have come together with a plan to assign function to all of the 20,000 to 25,000 *Arabidopsis* genes by 2010 (Project 2010). One of the first steps is to determine when and where these genes are expressed. Each step beyond that will require additional enabling technology.

DNA Microarrays

How can DNA sequences be made available to researchers, other than as databases of electronic information? DNA microarrays (figure 17.10) are a way to make DNA se-

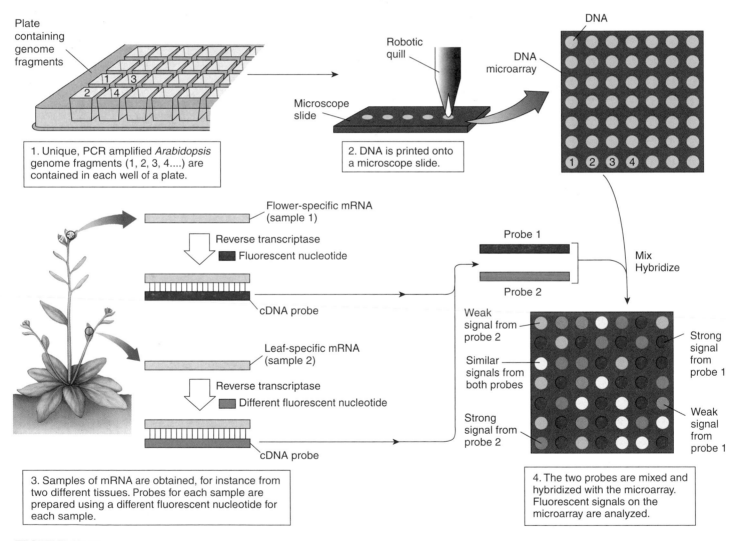

1. Unique, PCR amplified *Arabidopsis* genome fragments (1, 2, 3, 4....) are contained in each well of a plate.

2. DNA is printed onto a microscope slide.

3. Samples of mRNA are obtained, for instance from two different tissues. Probes for each sample are prepared using a different fluorescent nucleotide for each sample.

4. The two probes are mixed and hybridized with the microarray. Fluorescent signals on the microarray are analyzed.

FIGURE 17.10
Microarrays. Microarrays are created by robotically placing DNA onto a microscope slide. The microarray can then be probed with RNA from tissues of interest to identify expressed DNA. The microarray with hybridized probes is analyzed and often displayed as a false-color image. If a gene is frequently expressed in one of the samples, the fluorescent signal will be strong (*red or green*) where the gene is located on the microarray. If a gene is rarely expressed in one of the samples, the signal will be weak (*pink or light green*). A yellow color indicates genes that are expressed at similar levels in each sample.

FIGURE 17.11

Growth of a transgenic plant. DNA containing a gene for herbicide resistance was transferred into wheat (*Triticum aestivum*). The DNA also contains the *GUS* gene, which is used as a tag or label. The *GUS* gene produces an enzyme that catalyzes the conversion of a staining solution from clear to blue. (*a*) Embryonic tissue just prior to insertion of foreign DNA. (*b*) Following DNA transfer, callus cells containing the foreign DNA are indicated by color from the *GUS* gene (*blue spots*). (*c*) Shoot formation in the transgenic plants growing on a selective medium. Here, the gene for herbicide resistance in the transgenic plants allows growth on the selective medium containing the herbicide. (*d*) Comparison of growth on the selection medium for transgenic plants bearing the herbicide resistance gene (*left*) and a nontransgenic plant (*right*).

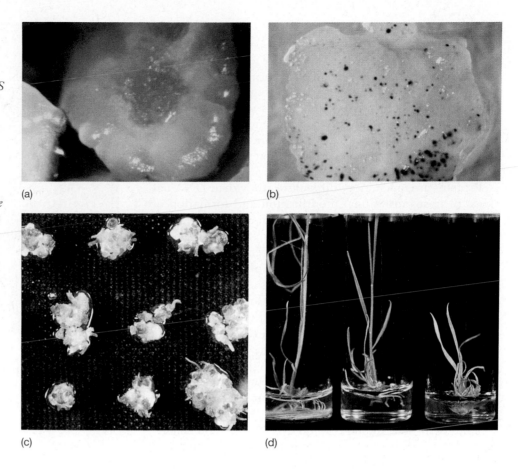

(a)　(b)　(c)　(d)

quences available to many researchers for the purpose of investigating gene function. To prepare a particular microarray, fragments of DNA are deposited on a microscope slide by a robot at indexed locations. Silicon chips instead of slides can also be arrayed. Researchers are currently using a chip with 24,000 *Arabidopsis* genes on it to identify genes that are expressed developmentally in certain tissues or in response to environmental factors. RNA from these tissues can be isolated and used as a probe for these microarrays. Only those sequences that are expressed in the tissues will be present and will hybridize to the microarray (see figure 17.10).

Microarrays are also being used to discern genetic differences among individuals of the same species. As we mentioned earlier, every human has a unique SNP profile that can be screened on arrays or screened using other emerging technology. A huge SNP database is being constructed for the human genome. About 1.42 million SNPs are distributed throughout our genome. On average, that is about one SNP in every 1200 base-pairs. SNPs can alter gene function or serve as markers embedded within repetitive DNA. About 85% of all human exons are less than 5 kilobases from a SNP. In addition to being fairly ubiquitous, SNPs have a very low rate of mutation. Medical applications of SNPs are discussed in chapter 20. SNPs are also valuable in following human migration and evolution.

Transgenics

How do you determine whether two genes that are from different species but have similar sequences have the same function? How can you be sure that a gene identified by an annotation program actually functions as a gene in the organism? One way to address these questions is to create transgenic organisms. The technology for creating transgenic organisms was discussed in chapter 16 and is illustrated for plants in figure 17.11. To test whether an *Arabidopsis* gene is homologous to a rice gene, it can be inserted into cells of rice, which are then regenerated into a rice plant. Different markers can be incorporated into the gene so that its protein product can be visualized or isolated in the transgenic plant. In some cases, the transgene (inserted foreign gene) may affect a visible phenotype. Of course, transgenics are but one of many ways to address questions about gene function.

Genomics has provided us with millions of new genes to investigate. Finding the function of these previously unknown genes depends on technologies, including microarrays, that can screen large numbers of genes very rapidly.

Proteomics

To fully understand how genes work, we need to characterize the proteins they produce. This information is essential in understanding cell biology, physiology, development, and evolution. How are similar genes used in different plants to create biochemically and morphologically distinct organisms? In many ways, we continue to ask the same questions that even Mendel asked, but at a much different level of organization.

Proteins are much more difficult to study than DNA because of posttranslational modification and formation of complexes of proteins. As we have seen, a single gene can code for multiple proteins using alternative splicing (see figure 17.8). While all the DNA in a genome can be isolated from a single cell, only a portion of the *proteome* (all the proteins coded for by the genome) is expressed in a single cell or tissue. There is an additional level of complexity to consider in **proteomics** (the study of the proteome). Just because a gene is being transcribed into RNA in a cell at a specific time, this does not mean that the transcript is necessarily being translated into a protein right then. Thus, an intermediate step in characterizing the proteome is studying the *transcriptome*, all the RNA present in a cell or tissue at a specific time.

The use of new methods to quickly identify and characterize large numbers of proteins is the distinguishing feature between traditional protein biochemistry and proteomics. As with genomics, the challenge is one of scale. Ideally, a researcher would like to be able to examine a nucleotide sequence and know what sort of functional protein the gene specifies. Databases of protein structures in different organisms can be searched to predict the structure and function of genes known only by sequence, as identified in genome projects. We are getting a clearer picture of how gene sequence relates to protein shape and function. Having a greater number of DNA sequences available allows for more extensive comparisons and identification of common structural patterns as groups of proteins continue to emerge.

Fortunately, while there may be as many as a million different proteins, most are just variations on a handful of themes. The same shared structural motifs—barrels, helices, molecular zippers—are found in the proteins of plants, insects, and humans (figure 17.12; also see chapter 3 for more information on protein motifs). The maximum number of distinct motifs has been estimated to be fewer than 5000. About 1000 of these motifs have already been cataloged. Both publicly and privately financed efforts are now under way to detail the shapes of all the common motifs.

Protein arrays, comparable to arrays of DNA, are being used to analyze large numbers of proteins simultaneously. Making a protein array starts with isolating the transcriptome of a cell or tissue. Then cDNAs are constructed and reproduced by cloning them into bacteria or viruses. Tran-

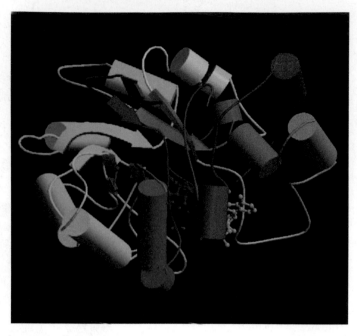

FIGURE 17.12
Computer-generated model of an enzyme. Searchable databases contain known protein structures, including human aldose reductase shown here.

scription and translation occur in the prokaryotic host, and micromolar quantities of protein are isolated and purified. These are then spotted onto glass slides. Protein arrays can be probed in at least three different ways. They can be screened with antibodies to specific proteins. Antibodies are labeled so that they can be detected, and the patterns on the protein array can be determined by computer analysis. An array of proteins can also be screened with another protein to detect binding or other protein interactions. Thousands of interactions can be tested simultaneously. For example, calmodulin (which mediates Ca^{++} function; see figure 7.9) was labeled and used to probe a yeast proteome array with 5800 proteins. The screen revealed 39 proteins that bound calmodulin. Of those 39, 33 were previously unknown! A third type of screen uses small molecules to assess whether or not they will bind to any of the proteins on the array. This approach shows promise for discovering new drugs that will inhibit proteins involved in disease.

Proteomics is a large-scale effort to identify and study the proteins coded for by the genome. Web-based databases allow researchers to compare new DNA sequences with known sequences and to predict the structure of the protein encoded in the DNA. New proteins are being identified by screening arrays of proteins.

Using Genomic Information

Swimming in a Sea of Genes

The genomics revolution has yielded millions of new genes to be investigated. The potential of genomics to improve human health through medical diagnostics and to improve nutrition through agriculture is enormous. Mutations in a single gene can explain some, but not most, heritable diseases. With entire genomes to search, the probability of unraveling human, animal, and plant diseases is greatly improved. While proteomics will lead to new pharmaceuticals, the immediate impact of genomics is being seen in diagnostics. Both improved technology and gene discovery are enhancing diagnosis of genetic abnormalities. Diagnostics are also being used to identify individuals. For example, SNPs were among the forensic diagnostic tools used to identify remains of victims of the terrorist attack on the World Trade Center in New York City.

The September 11 attacks were followed by an increased awareness and concern about biological weapons. When cases of anthrax began appearing in the fall of 2001, genome sequencing made it possible to explore possible sources of the deadly bacteria and determine whether or not they had been genetically engineered to increase their lethality. Substantial effort turned toward the use of genomic tools to distinguish between naturally occurring infections and intentional outbreaks of disease. The Centers for Disease Control and Prevention (CDC) prioritized bacteria and viruses that are likely targets for bioterrorism (table 17.2). Prokaryotes, with their small genomes, can be rapidly sequenced and readily altered to make them even more virulent. The focus on pathogens extends beyond bioterrorism. The spread of emerging viruses, including Ebola and Severe Acute Respiratory Syndrome (SARS), is increasingly problematic as global interactions among humans become more frequent.

Human evolutionary history and patterns of migration are subject to lively debate among scientists. The fossil record, mitochondrial DNA sequences, and now SNP comparisons are beginning to clarify a rather fuzzy picture. The worldwide pattern of SNPs to date indicates that Eurasian populations differ from African populations. However, the Eurasian SNP patterns in noncoding regions of DNA appear to be a subset of the African population. This is consistent with other work supporting an African origin for humans. The SNP data also raise an important caution about medical diagnostics based on SNP profiles; that is, the geographic origins of an individual must be considered.

Table 17.2 High-Priority Pathogens for Genomic Research

Pathogen	Disease	Genome[*]
Variola major	Smallpox	Complete
Bacillus anthracis	Anthrax	Complete
Yersinia pestis	Plague	In progress
Clostridium botulinum	Botulism	In progress
Francisella tularensis	Tularemia	Complete
Filoviruses	Ebola and Marburg hemorrhagic fever	Both are complete
Arenaviruses	Lassa fever and Argentine hemorrhagic fever	Both are complete

[*]There are multiple strains of these viruses and bacteria. "Complete" indicates that at least one has been sequenced. For example, the Florida strain of anthrax was the first to be sequenced.

FIGURE 17.13
Rice field.   Most of the rice grown globally is directly consumed by humans and is the dietary mainstay of 2 billion people.

Feeding the World

Globally, nutrition is the greatest impediment to human health. Much of the excitement about the rice genome project is based on its potential for improving the yield and nutritional quality of rice and other cereals worldwide. The development of golden rice discussed in chapter 16 is a promising example of improved nutrition through genetic approaches. About one-third of the world population obtains half its calories from rice (figure 17.13). In some regions, individuals consume up to 1.5 kilograms of rice daily. More than 500 million tons of rice are produced each year, but this may not be enough food in the future. The current world population of 6 billion is expected to peak at 9 billion by 2070 and then decline to around 8.4 billion by 2100. Producing enough food to feed 9 billion people would be a major challenge. Food distribution can be problematic given the uneven distribution of good farmland throughout the world. One solution is to breed crops that are better adapted to regional environments with less than ideal soil and climate.

Due in large part to scientific advances in crop breeding and farming techniques, world food production has more than doubled, while world cropland has increased by only 9%. The world now farms an area the size of South America, but without the scientific advances of the past 40 years, the entire western hemisphere would need to be farmed to produce enough food for the world. Unfortunately, water usage for crops has tripled in that time period, and quality farmland is being lost to soil erosion. Many believe that conventional crop breeding programs may have reached their limit. The question is, How best to feed billions of additional people without destroying much of the planet in the process? Scientists are also concerned about the effects of global climate change on agriculture worldwide. Increasing the yield and quality of crops grown for both human and livestock consumption, especially on more marginal farmland, will depend on many factors. But genetic engineering built on the findings of genomics projects can contribute significantly to the solution.

Most crops grown in the United States produce less than half of their genetic potential because of environmental stresses (salt, water, and temperature), herbivores, and pathogens (figure 17.14). Identifying genes that can provide stress and pest protection is the focus of many current genomics research projects. Most likely, multiple genes will be involved. Having access to entire genomic sequences will enhance the probability of identifying critical genes.

Who Owns a Genome?

Genome science is also a source of ethical challenges and dilemmas. One example is the issue of gene patents. Actually, it is the use of a gene, not the gene itself, that is patentable. For a gene-related patent, the product and its function must be known. The public genome consortia, supported by federal funding, have been driven by the belief that the sequence of genomes should be freely available to all and should not be patented. Private companies patent gene functions, but often make sequence data available with certain restrictions. The physical sciences have negotiated the landscape of public and for-profit research for decades, but this is relatively new territory for biologists.

Another ethical issue involves privacy. How sequence data are used is the focus of thoughtful and ongoing discussions. The Universal Declaration on the Human Genome and Human Rights states, "The human genome underlies the fundamental unity of all members of the human family, as well as the recognition of their inherent dignity and diversity. In a symbolic sense, it is the heritage of humanity." While we talk about "the" human genome, each of us has subtly different genomes that can be used to identify us. We can already screen for genetic disorders such as cystic fibrosis and Huntington disease, but genomics will greatly increase the number of screenable traits. Behavioral genomics is an area that is also rich with possibilities and

FIGURE 17.14

Corn crop productivity well below its genetic potential due to drought stress. Corn production can be limited by water deficiencies due to the drought that occurs during the growing season in dry climates. Global climate change may increase drought stress in areas where corn is the major crop.
The corn genome has not been sequenced. How could you use information from the rice genome sequence to try to improve drought tolerance in corn?

dilemmas. Very few behavioral traits can be accounted for by single genes. Two genes have been associated with fragile-X mental retardation, and three with early-onset Alzheimer disease. Comparisons of multiple genomes will likely lead to the identification of multiple genes controlling a range of behaviors. Will this change the way we view acceptable behavior?

What if employers or insurance companies gain access to your personal SNP profile? Could you be discriminated against because you have a genetic tendency toward chemical addiction or heart disease? On a more positive note, the U.S. Armed Forces require DNA samples from members for possible casualty identification, and DNA-based identification brought peace of mind to some families of the World Trade Center victims.

In Iceland, the parliament has voted to have a private company create a database from pooled medical, genetic, and genealogical information about all Icelanders, a particularly fascinating population from a genetic perspective. Because minimal migration or immigration has occurred there over the last 800 years, the information that can be mined from the Icelandic database is phenomenal. Ultimately, the value of that information has to be weighed against any possible discrimination or stigmatization of individuals or groups.

Genomics offers tremendous benefits for human health, but also raises new questions about privacy and other issues.

Concept Review

For interactive testing, visit the Online Learning Center with PowerWeb at www.mhhe.com/Raven7

17.1 Genomes can be mapped both genetically and physically.

Genome Maps

- Genetic maps show the relative location of genes on a chromosome as determined by recombination frequencies. Genetic maps measure distance in centimorgans. (p. 344)
- Physical maps show relative positions of landmarks within specific DNA sequences and measure distance in base-pairs. (p. 344)

17.2 Genome sequencing produces the ultimate physical map.

Sequencing

- Large-scale genome sequencing relies on an automated sequencer and computer analysis, but to reduce errors, genomes are cut into short segments for replication. (p. 346)
- Clone-by-clone sequencing starts by constructing a physical map with many landmarks and then sequencing known sites. (p. 347)
- Shotgun cloning sequences cloned fragments, and then uses computational analysis to determine the whole sequence from the overlapping sequences. (p. 347)

17.3 Being more complex does not necessarily require more genes.

Human Genome Project

- The Human Genome Project originated in 1990, and success was announced in 2001. (p. 348)
- Eukaryotic genomes appear to be larger than prokaryotic genomes, but the size of the organism does not appear to play a determining role. (p. 348)

Genome Geography

- Bioinformatic approaches can be used to identify gene sequences. (p. 349)
- Alternative splicing yields different proteins that can have different functions. (p. 349)
- Four different classes of protein-encoding genes found in eukaryotic genomes are (1) single-copy genes, (2) segmental duplications, (3) multigene families, and (4) tandem clusters. (p. 349)

- Only about 1% of the human genome is devoted to protein-encoding genes. (p. 350)
- Six major sorts of noncoding human DNA are (1) noncoding DNA within genes, (2) structural DNA, (3) simple sequence repeats, (4) segmental duplications, (5) pseudogenes, and (6) transposable elements. (p. 350)

Comparative Genomics

- Synteny refers to the conserved arrangements of segments of DNA in related genomes. (p. 352)
- Comparisons with the sequenced syntenous segment in another species can help in determining evolutionary relationships. (p. 352)

17.4 Genomics is opening a new window on life.

Functional Genomics

- Genomics is shifting toward functional genomics, the study of the function of genes and their products. (p. 354)
- DNA microarrays can screen large numbers of genes very rapidly. (p. 355)

Proteomics

- Proteins are much more difficult to study than DNA because of the posttranslational modification and formation of protein complexes. (p. 356)
- Proteonomics is an effort to identify and study proteins coded for by the genome, and is distinguished from traditional protein biochemistry by the use of new methods to quickly identify and characterize large numbers of proteins. (p. 356)

Using Genomic Information

- The potential of genomics to improve human health through medical diagnostics and to improve nutrition through agriculture is enormous. (p. 357)
- Genomic science is also a source of ethical challenges and dilemmas, such as patient rights and personal privacy issues. (p. 358)

Self Test

1. Researchers from many labs collaborated to determine the sequence of the human genome. How did labs avoid sequencing the same fragments multiple times?
   a. Each lab could isolate DNA from one particular chromosome to divide the sequencing projects.
   b. Using restriction fragment length polymorphisms, labs could ensure that were not sequencing the same fragments.
   c. Using short sequences from their respective clones, sequenced-tagged sites (STSs), researchers could check to make sure their fragments were not already being sequenced by another group.
   d. By comparing sequences from collaborating labs, researchers could ensure that they were not sequencing the same fragments.

2. Some of your friends are trying to make sense of their genome. To help them out, you draw an analogy between our chromosomes and the interstate highway system. In this analogy, every interstate represents a single chromosome. How could you describe the relationship between chromosomes and genes to your friends using this analogy?
   a. Every mile marker would represent a gene.
   b. Every town would represent a gene.
   c. Every state would represent a gene.
   d. Genes would be defined by the twists and turns on the highway.

3. Imagine that you broke your mother's favorite vase and had to reconstruct it from the shattered pieces. To do this, you would have to look for pieces with similar ends to join and then progressively glue every piece together. What sequencing strategy does this most closely represent?
   a. shotgun sequencing
   b. contig sequencing
   c. clone-by-clone sequencing
   d. manual sequencing

4. Knowing the sequence of an entire genome
   a. completes our understanding of every gene's function in the organism.
   b. allows us to predict the genetic cause of every disease in the organism.
   c. provides a template for constructing an artificial life-form.
   d. provides the raw data that can then be used to identify specific genes.

5. If you were to look at the sequence of an entire chromosome, how could you identify which segments might contain a gene?
   a. You could identify large protein-coding regions (open reading frames).
   b. You could look for a match with an expressed sequence tag (EST).
   c. You could look for consensus regulatory sequences that could initiate transcription.
   d. All of these strategies could be used to identify possible genes.
   e. It is impossible to predict genes from sequence data alone.

6. You have been hired to characterize the genome of a novel organism, *Undergraduatus genomicus*. After fully sequencing the 10^6 base-pairs in the genome, you predict that this organism has approximately 10,000 genes. You have a collaborator on this project, however, who has identified 20,000 different expressed sequence tags from this organism. How can you resolve this conflict?
   a. You suggest that your collaborator is an idiot who counted every gene twice!
   b. You suggest that your collaborator may have identified multiple isoforms of the same gene that could arise by alternative splicing.
   c. You fear that you may have underestimated the number of genes, because you forgot that the organism is diploid and you did not count both copies of every gene in your total.
   d. You only identified genes with open reading frames, but most genes do not encode proteins, so your number will be low.

7. In addition to coding sequences, our genome contains
   a. noncoding DNA within genes (i.e., introns).
   b. structural DNA involved in telomeres and centromeres.
   c. simple repetitive DNA.
   d. DNA from transposable elements that have jumped around in the genome.
   e. All of these are present in genomic DNA.

8. Natural variation in the length of tandem repeat sequences (VNTRs) found in the genome can be used to identify individual people by their DNA fingerprint. Why is this possible?
   a. The statement is not true; such variability prevents this from being a useful identification tool.
   b. The changes in repeat length change the DNA synthesis pattern, so the cell cycles have different lengths, making the cells of different people different sizes.
   c. The changes in repeat length occur very infrequently, so there is only one pattern that everybody shares.
   d. The changes in repeat length occur very frequently, so everybody has a unique pattern of different lengths when several repeats are examined.

Test Your Visual Understanding

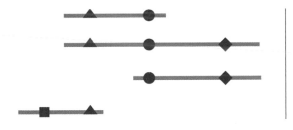

1. From the information given in the above diagram, construct a contig map of the region presented.

Apply Your Knowledge

1. Every cell in your body contains the same genomic DNA, yet the proteome of different tissues is unique. How can you explain this?

2. Chromosomes are much like interstate highways. Develop this analogy by assigning a chromosomal counterpart to the following:
   a. the beginning and end of a particular highway
   b. towns along the highway
   c. stretches of highway that pass through wilderness

3. If you are given a sample of DNA from an unknown organism, how could you determine the origin of the DNA sample?

18

Control of Gene Expression

Concept Outline

18.1 Gene expression is controlled by regulating transcription.

An Overview of Transcriptional Control. In prokaryotes, transcription is regulated by controlling access of RNA polymerase to the promoter in a flexible and reversible way; eukaryotes, by contrast, regulate many of their genes by turning them on and off in a more permanent fashion.

18.2 Regulatory proteins read DNA without unwinding it.

How to Read a Helix Without Unwinding It. Regulatory proteins slide special segments called DNA-binding motifs along the major groove of the DNA helix, reading the sides of the bases.

Four Important DNA-Binding Motifs. DNA-binding proteins contain structural motifs, such as the helix-turn-helix motif, which fit into the major groove of the DNA helix.

18.3 Prokaryotes regulate genes by controlling transcription initiation.

Prokaryotic Gene Regulation. Repressor proteins inhibit RNA polymerase's access to the promoter, while activators facilitate its binding.

18.4 Transcriptional control in eukaryotes operates at a distance.

Transcriptional Control in Eukaryotes. Eukaryotic genes use a complex collection of transcription factors and enhancers to aid the polymerase in transcription.

The Effect of Chromatin Structure on Gene Expression. The tight packaging of eukaryotic DNA into nucleosomes does not interfere with gene expression.

Posttranscriptional Control in Eukaryotes. Gene expression can be controlled at a variety of levels after transcription. Both proteins and small RNAs play major roles in regulating the stability and expression of gene transcripts.

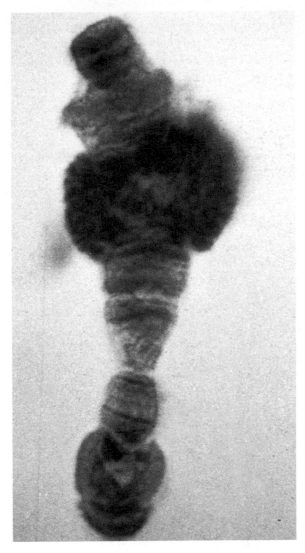

FIGURE 18.1
Chromosome puffs. In this chromosome of the fly *Drosophila melanogaster*, individual active genes can be visualized as "puffs" on the chromosomes. The RNA being transcribed from the DNA template has been radioactively labeled, and the dark specks indicate its position on the chromosome.

In an orchestra, all of the instruments do not play at the same time; if they did, all that would be produced is noise. Instead, a musical score determines which instruments in the orchestra play when. Similarly, all of the genes in an organism are not expressed at the same time, each gene producing the protein it encodes full tilt. Instead, different genes are expressed at different times, with a genetic score written in regulatory regions of the DNA determining which genes are active when (figure 18.1).

18.1 Gene expression is controlled by regulating transcription.

An Overview of Transcriptional Control

Control of gene expression is essential to all organisms. In prokaryotes, it allows the cell to take advantage of changing environmental conditions. In multicellular eukaryotes, it is critical for directing development and maintaining homeostasis.

Regulating Promoter Access

One way to control gene expression is to regulate the initiation of transcription. In order for a gene to be transcribed, RNA polymerase must have access to the DNA helix and must be capable of binding to the gene's **promoter,** a specific sequence of nucleotides at one end of the gene that tells the polymerase where to begin transcribing. How is the initiation of transcription regulated? Protein-binding nucleotide sequences on the DNA regulate the initiation of transcription by modulating the ability of RNA polymerase to bind to the promoter. These protein-binding sites are usually only 10 to 15 nucleotides in length (even a large regulatory protein has a "footprint," or binding area, of only about 20 nucleotides). Hundreds of these regulatory sequences have been characterized, and each provides a binding site for a specific protein able to recognize the sequence. Binding the protein to the regulatory sequence either *blocks* transcription by getting in the way of RNA polymerase, or *stimulates* transcription by facilitating the binding of RNA polymerase to the promoter.

Transcriptional Control in Prokaryotes

Control of gene expression is accomplished very differently in prokaryotes than in the cells of eukaryotes. Prokaryotic cells have been shaped by evolution to grow and divide as rapidly as possible, enabling them to exploit transient resources. Proteins in prokaryotes turn over rapidly. This allows them to respond quickly to changes in their external environment by changing patterns of gene expression. In prokaryotes, the primary function of gene control is to adjust the cell's activities to its immediate environment. Changes in gene expression alter which enzymes are present in the cell in response to the quantity and type of available nutrients and the amount of oxygen present. Almost all of these changes are fully reversible, allowing the cell to adjust its enzyme levels up or down as the environment changes.

Transcriptional Control in Eukaryotes

The cells of multicellular organisms, on the other hand, have been shaped by evolution to be protected from transient changes in their immediate environment. Most of them experience fairly constant conditions. Indeed, **homeostasis—** the maintenance of a constant internal environment—is considered by many to be the hallmark of multicellular organisms. Although cells in such organisms still respond to signals in their immediate environment (such as growth factors and hormones) by altering gene expression, in doing so they participate in regulating the body as a whole. In multicellular organisms with relatively constant internal environments, the primary function of gene control in a cell is not to respond to that cell's immediate environment, but rather to participate in regulating the body as a whole.

Some of these changes in gene expression compensate for changes in the physiological condition of the body. Others mediate the decisions that *produce* the body, ensuring that the right genes are expressed in the right cells at the right time during development. The growth and development of multicellular organisms entail a long series of biochemical reactions, each catalyzed by a specific enzyme. Once a particular developmental change has occurred, these enzymes cease to be active, lest they disrupt the events that must follow. To produce these enzymes, genes are transcribed in a carefully prescribed order, each for a specified period of time, following a fixed genetic program that may even lead to programmed cell death. The one-time expression of the genes that guide such a program is fundamentally different from the reversible metabolic adjustments prokaryotic cells make to the environment. In all multicellular organisms, changes in gene expression within particular cells serve the needs of the whole organism, rather than the survival of individual cells.

Posttranscriptional Control

Gene expression can be regulated at many levels. By far the most common form of regulation in both prokaryotes and eukaryotes is **transcriptional control**—that is, control of the transcription of particular genes by RNA polymerase. Other less common forms of control occur after transcription, influencing the mRNA that is produced from the genes or the activity of the proteins encoded by the mRNA. These controls, collectively referred to as **posttranscriptional controls,** will be discussed later.

Gene expression is controlled at the transcriptional and posttranscriptional levels. Transcriptional control, the more common type, is effected by the binding of proteins to regulatory sequences within the DNA.

18.2 Regulatory proteins read DNA without unwinding it.

How to Read a Helix Without Unwinding It

The ability of certain proteins to bind to *specific* DNA regulatory sequences provides the basic tool of gene regulation, the key ability that makes transcriptional control possible. To understand how cells control gene expression, it is first necessary to gain a clear picture of this molecular recognition process.

Looking into the Major Groove

Molecular biologists used to think that the DNA helix had to unwind before proteins could distinguish one DNA sequence from another; only in this way, they reasoned, could regulatory proteins gain access to the hydrogen bonds between base-pairs. We now know it is unnecessary for the helix to unwind, because proteins can bind to its outside surface, where the edges of the base-pairs are exposed. Careful inspection of a DNA molecule reveals two helical grooves winding around the molecule, one deeper than the other. Within the deeper groove, called the **major groove**, the nucleotides' hydrophobic methyl groups, hydrogen atoms, and hydrogen bond donors and acceptors protrude. The pattern created by these chemical groups is unique for each of the four possible base-pair arrangements, providing a ready way for a protein nestled in the groove to read the sequence of bases (figure 18.2).

DNA-Binding Motifs

Protein-DNA recognition is an area of active research; so far, the structures of over 30 regulatory proteins have been analyzed. Although each protein is unique in its fine details, the part of the protein that actually binds to the DNA is much less variable. Almost all of these proteins employ one of a small set of **structural**, or **DNA-binding**, **motifs**, particular bends of the protein chain that permit it to interlock with the major groove of the DNA helix.

Regulatory proteins identify specific sequences on the DNA double helix, without unwinding it, by inserting DNA-binding motifs into the major groove of the double helix where the edges of the bases protrude.

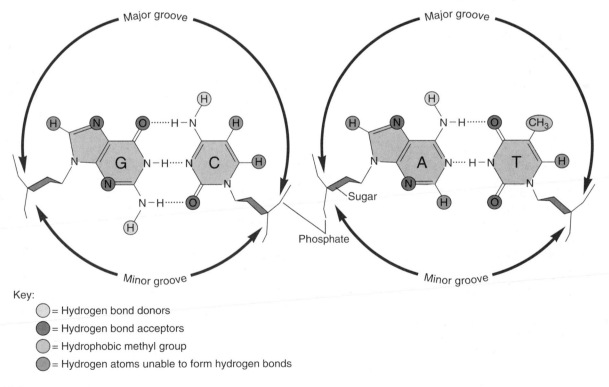

Key:
- = Hydrogen bond donors
- = Hydrogen bond acceptors
- = Hydrophobic methyl group
- = Hydrogen atoms unable to form hydrogen bonds

FIGURE 18.2
Reading the major groove of DNA. Looking down into the major groove of a DNA helix, we can see the edges of the bases protruding into the groove. Each of the four possible base-pair arrangements (two are shown here) extends a unique set of chemical groups into the groove, indicated in this diagram by differently colored balls. A regulatory protein can identify the base-pair arrangement by this characteristic signature.

Four Important DNA-Binding Motifs

The Helix-Turn-Helix Motif

The most common DNA-binding motif is the **helix-turn-helix,** constructed from two α-helical segments of the protein linked by a short, nonhelical segment, the "turn" (figure 18.3). As the first DNA-binding motif recognized, the helix-turn-helix motif has since been identified in hundreds of DNA-binding proteins.

A close look at the structure of a helix-turn-helix motif reveals how proteins containing such motifs are able to interact with the major groove of DNA. Interactions between the helical segments of the motif hold them at roughly right angles to each other. When this motif is pressed against DNA, one of the helical segments (called the recognition helix) fits snugly in the major groove of the DNA molecule, while the other butts up against the outside of the DNA molecule, helping to ensure the proper positioning of the recognition helix. Most DNA regulatory sequences recognized by helix-turn-helix motifs occur in symmetrical pairs. Such sequences are bound by proteins containing two helix-turn-helix motifs separated by 3.4 nanometers (nm), the distance required for one turn of the DNA helix (figure 18.4). Having *two* protein-DNA-binding sites doubles the zone of contact between protein and DNA, and so greatly strengthens the bond that forms between them.

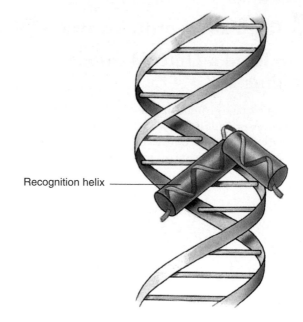

FIGURE 18.3
The helix-turn-helix motif. The recognition helix, one helical region of the motif, actually fits into the major groove of DNA. There it contacts the edges of base-pairs, enabling it to recognize specific sequences of DNA bases.

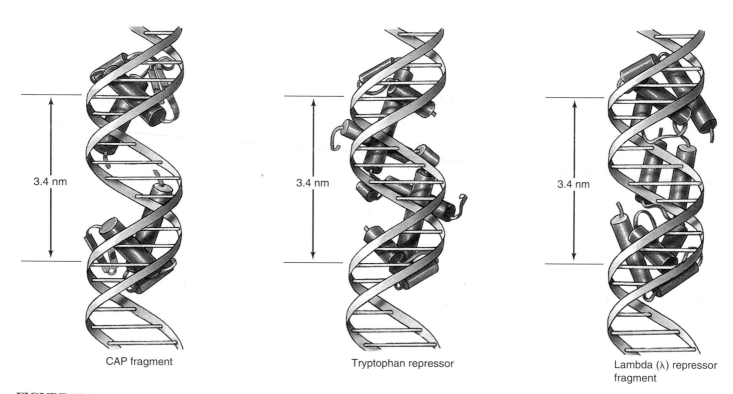

CAP fragment

Tryptophan repressor

Lambda (λ) repressor fragment

FIGURE 18.4
How the helix-turn-helix binding motif works. The three regulatory proteins illustrated here (*purple*) all bind to DNA using a pair of helix-turn-helix binding motifs. In each case, the two copies of the motif (*red*) are separated by 3.4 nanometers, precisely the spacing of one turn of the DNA helix. This allows the regulatory proteins to slip into two adjacent portions of the major groove in DNA, providing a strong attachment.

The Homeodomain Motif

A special class of helix-turn-helix motifs plays a critical role in development in a wide variety of eukaryotic organisms, including humans. These motifs were discovered when researchers began to characterize a set of homeotic mutations in *Drosophila* (mutations that alter how the parts of the body are assembled). They found that the mutant genes encoded regulatory proteins whose normal function was to initiate key stages of development by binding to developmental switch-point genes. More than 50 of these regulatory proteins have been analyzed, and they all contain a nearly identical sequence of 60 amino acids, the **homeodomain** (figure 18.5*b*). The center of the homeodomain is occupied by a helix-turn-helix motif that binds to the DNA. Surrounding this motif within the homeodomain is a region that always presents the motif to the DNA in the same way.

The Zinc Finger Motif

A different kind of DNA-binding motif uses one or more zinc atoms to coordinate its binding to DNA. Called **zinc fingers** (figure 18.5*c*), these motifs exist in several forms. In one form, a zinc atom links an α-helical segment to a β sheet segment so that the helical segment fits into the major groove of DNA. This sort of motif often occurs in clusters, the β sheets spacing the helical segments so that each helix contacts the major groove. The more zinc fingers in the cluster, the stronger the protein binds to the DNA. In other forms of the zinc finger motif, the β sheet's place is taken by another helical segment.

The Leucine Zipper Motif

In yet another DNA-binding motif, two different protein subunits cooperate to create a single DNA-binding site. This motif is created where a region on one of the subunits containing several hydrophobic amino acids (usually leucines) interacts with a similar region on the other subunit. This interaction holds the two subunits together at those regions, while the rest of the subunits are separated. Called a **leucine zipper,** this structure has the shape of a Y, with the two arms of the Y being helical regions that fit into the major groove of DNA (figure 18.5*d*). Because the two subunits can contribute quite different helical regions to the motif, leucine zippers allow for great flexibility in controlling gene expression.

Regulatory proteins bind to the edges of base-pairs exposed in the major groove of DNA. Most contain structural motifs such as the helix-turn-helix, homeodomain, zinc finger, or leucine zipper.

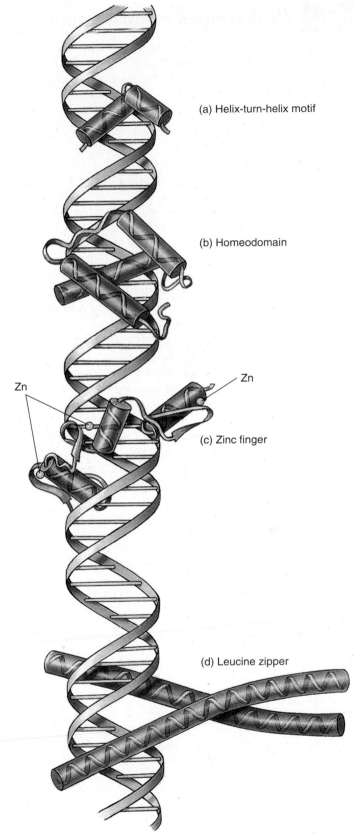

(a) Helix-turn-helix motif

(b) Homeodomain

Zn   Zn

(c) Zinc finger

(d) Leucine zipper

FIGURE 18.5
Major DNA-binding motifs.

18.3 Prokaryotes regulate genes by controlling transcription initiation.

Most organisms regulate gene expression by controlling the initiation of transcription. This is accomplished by the binding of proteins to regulatory sequences on the DNA. Prokaryotes and eukaryotes share some common themes, but they have some profound differences as well. We will focus first on prokaryotic systems and then look at how eukaryotic systems differ from this.

Prokaryotic Gene Regulation

Changes in the environments that bacteria and other prokaryotes encounter often result in changes in gene expression. How prokaryotes respond to changes depends in each instance on the nature of the proteins encoded by relevant genes. In general, genes encoding proteins involved in anabolic pathways (building up molecules) respond oppositely from genes encoding proteins involved in catabolic pathways (breaking down molecules). If a bacterium encounters a molecule in the environment that can be broken down for energy, such as the sugar lactose, then the bacterium will start to make the proteins necessary to utilize lactose. When lactose is not present, however, there is no need to make these proteins. Thus, we say that the synthesis of the proteins is *induced* by the presence of lactose. If a molecule available in the environment is the end product of a biosynthetic pathway, such as the amino acid tryptophan, then a bacterium in that environment will not synthesize the proteins necessary to make tryptophan. If tryptophan ceases to be available, then the bacterium will begin to make the proteins necessary for its synthesis. When bacteria *don't* make the proteins necessary for biosynthesis, we call this *repression*. In the case of both induction and repression, the bacterium is adjusting to produce the proteins that are optimal for its immediate environment.

Induction and repression are observed in a variety of prokaryotic gene systems. What is the nature of the genetic control circuits responsible for these adjustments? Knowing that control is probably at the level of initiation of transcription does not tell us the nature of the control. It might be either positive or negative. A regulatory molecule can increase the rate of initiation (positive) or decrease the rate of initiation (negative). On the surface, repression may appear to be negative and induction positive, but in fact both involve negative control.

For either mechanism to work, the molecule in the environment, such as lactose or tryptophan, must produce the proper effect on the gene being regulated. In the case of induction, it is necessary for the presence of lactose to prevent a negative regulator from binding to its regulatory sequence. In the case of repression, by contrast, the presence of tryptophan must make a negative regulator bind its regulatory sequence. These responses are opposite because the needs of the cell are opposite for anabolic versus catabolic pathways.

Operons

Prokaryotic genes are often organized into operons. **Operons** are multiple genes that are part of a single gene expression unit. The genes of an operon are all part of the same mRNA and thus are controlled by the same promoter. In bacteria, genes that are involved in the same metabolic pathway are often organized in this fashion. For example, the proteins necessary for the utilization of lactose are encoded by the *lac* operon (figure 18.6), and the proteins necessary for the synthesis of tryptophan are encoded by the *trp* operon. Both of these operons are regulated by systems of negative regulation.

FIGURE 18.6

The *lac* region of the *Escherichia coli* chromosome. The *lac* operon consists of a promoter, an operator, and three genes that code for proteins required for the metabolism of lactose. In addition, there is a binding site for the catabolite activator protein (CAP), which affects whether or not RNA polymerase will bind to the promoter. Gene *I* codes for a repressor protein, which will bind to the operator and block transcription of the *lac* genes. The genes *Z*, *Y*, and *A* encode the two enzymes and the permease involved in the metabolism of lactose.

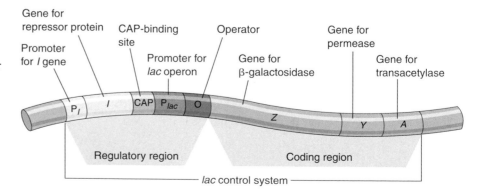

FIGURE 18.7

How the *trp* operon is controlled.
The tryptophan repressor cannot bind the operator (which is located *within* the promoter) unless tryptophan first binds to the repressor. Therefore, in the absence of tryptophan, the promoter is free to function, and RNA polymerase transcribes the operon. In the presence of tryptophan, the tryptophan-repressor complex binds tightly to the operator, preventing RNA polymerase from initiating transcription.

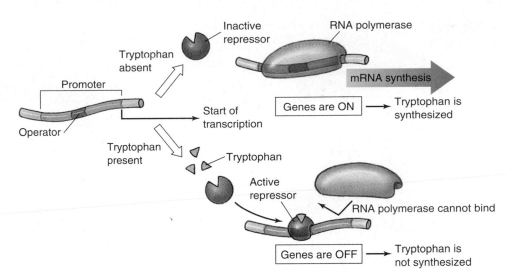

FIGURE 18.8

How the tryptophan repressor works. The binding of tryptophan to the repressor increases the distance between the two recognition helices in the repressor, allowing the repressor to fit snugly into two adjacent portions of the major groove in DNA.

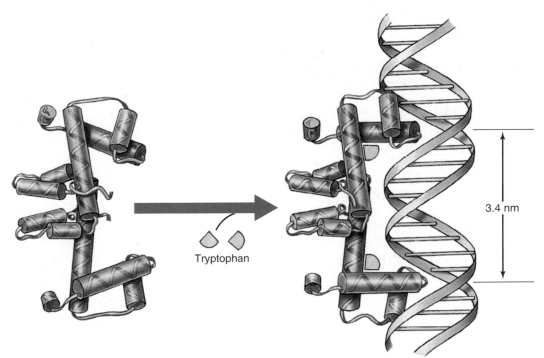

Repressors Are Off Switches

Repressors are proteins that bind to regulatory sites on DNA and prevent or decrease the initiation of transcription. Induction and repression are both mediated by a repressor molecule. These repressors do not act alone; each responds to specific effector molecules. Effector binding can alter the conformation of the repressor to either enhance or abolish its binding to DNA. These repressor proteins are allosteric proteins with an active site that binds DNA and a regulatory site that binds effectors.

trp Operon

In the *trp* operon, the operon is repressed in the presence of tryptophan and unrepressed in the absence of tryptophan. When tryptophan is present in the medium sur-

rounding the bacterium, the cell shuts off transcription of the *trp* genes by means of a tryptophan repressor, a helix-turn-helix regulatory protein that binds to the operator site located within the *trp* promoter (figure 18.7). Binding of the repressor to the operator prevents RNA polymerase from binding to the promoter. The key to the functioning of this control mechanism is that the tryptophan repressor cannot bind to DNA unless it has first bound to two molecules of tryptophan. The binding of tryptophan to the repressor alters the orientation of a pair of helix-turn-helix motifs in the repressor, causing their recognition helices to fit into adjacent major grooves of the DNA (figure 18.8).

Thus, the bacterial cell's synthesis of tryptophan depends upon the absence of tryptophan in the environment. When the environment lacks tryptophan, there is nothing to

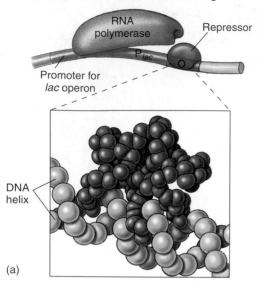

RNA polymerase cannot transcribe *lac* genes

(a)

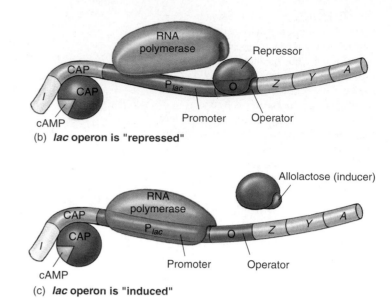

(b) **lac operon is "repressed"**

(c) **lac operon is "induced"**

FIGURE 18.9

How the *lac* repressor works. (*a*) The *lac* repressor. Because the repressor fills the major groove of the DNA helix, RNA polymerase cannot fully attach to the promoter, and transcription is blocked. (*b*) The *lac* operon is shut down (repressed) when the repressor protein is bound to the operator site. Because promoter and operator sites overlap, RNA polymerase and the repressor cannot functionally bind at the same time, any more than two people can sit in the same chair at once. (*c*) The *lac* operon is transcribed (induced) when CAP is bound and when allolactose binding to the repressor changes its shape so that it can no longer sit on the operator site and block RNA polymerase activity.

activate the repressor, so the repressor cannot prevent RNA polymerase from binding to the *trp* promoter. The *trp* genes are transcribed, and the cell proceeds to manufacture tryptophan from other molecules. On the other hand, when tryptophan is present in the environment, it binds to the repressor, which is then able to bind to the *trp* promoter. This blocks transcription of the *trp* genes, and the cell's synthesis of tryptophan halts.

lac Operon

The enzymes encoded by the *lac* operon are necessary only when lactose is available. The *lac* repressor binds DNA in the absence of lactose, but not in the presence of lactose. The DNA binding of the repressor is mediated by a metabolite of lactose, allolactose. In the absence of lactose, the operon is repressed, while in the presence of lactose, the synthesis of the proteins encoded by the operon is induced. As the level of lactose falls, the allolactose will no longer bind repressor, allowing the repressor to bind DNA again (figure 18.9).

Activators Are "On" Switches

In addition to repressors, there are proteins called **activators** that can bind DNA to stimulate the initiation of transcription. These allosteric proteins are the logical and physical opposites to repressors. All of the logic just ex-

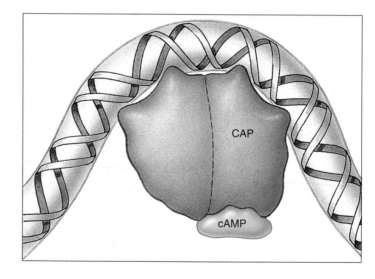

FIGURE 18.10

How CAP works. Binding of the catabolite activator protein (CAP) to DNA causes the DNA to bend around it. This increases the activity of RNA polymerase.

plained about repressors is reversed for activators: Effector molecules enhance activator binding to catabolic operons, increasing transcription.

Activators play a key role in the phenomenon of glucose repression, the preferential use of glucose in the presence of other sugars. Despite the name, glucose repression is

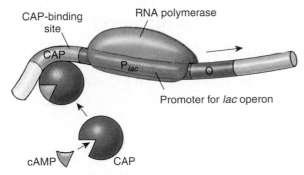

(a) **Glucose low, promoter activated**

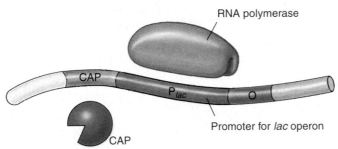

(b) **Glucose high, promoter not activated**

FIGURE 18.11
How the CAP site works. The CAP molecule can attach to the CAP-binding site only when the molecule is bound to cAMP. (*a*) When glucose levels are low, cAMP is abundant and binds to CAP. The cAMP-CAP complex binds to the CAP site, bends in the DNA, and gives RNA polymerase access to the promoter. (*b*) When glucose levels are high, cAMP is scarce, and CAP is unable to activate the promoter.

Glucose	Lactose	
		CAP-binding site — RNA-polymerase binding site (promoter) — Operator — *lacZ* gene
+	+	Operon OFF because CAP is not bound
+	−	Repressor. Operon OFF both because *lac* repressor is bound and CAP is not
−	−	RNA polymerase. CAP. Operon OFF because *lac* repressor is bound
−	+	RNA polymerase. CAP. mRNA synthesis. Operon ON because CAP is bound and *lac* repressor is not

FIGURE 18.12
Two regulatory proteins control the *lac* operon. Together, the *lac* repressor and CAP provide a very sensitive response to the cell's need to utilize lactose-metabolizing enzymes.

mediated by an activator protein, the **catabolite activator protein (CAP),** which stimulates the transcription of all operons encoding proteins necessary for the catabolism of sugars. The CAP protein is another allosteric protein whose binding is controlled by an effector—in this case, cAMP (figure 18.10). Amounts of cAMP in the cell are inversely related to the level of glucose. Thus, when glucose levels are high, cAMP levels are low, and vice versa (figure 18.11). Since CAP must be bound to cAMP to bind to DNA, this prevents activation of other catabolic operons (glucose repression).

Combinations of Switches

From what we have seen, the *lac* operon is actually under both positive and negative control (figure 18.12). The negative control is due to the *lac* repressor protein, and the positive control is due to CAP. This combination of positive and negative control gives the cell fine control over the levels of the proteins needed to utilize lactose in a variety of environments. In the absence of lactose, the

proteins are not made at all because the repressor will prevent initiation of transcription. In the presence of lactose, the repressor will no longer bind DNA, allowing transcription to initiate. However, if glucose is also present, then the level of cAMP will be low and CAP will be unable to bind the *lac* promoter. Transcription will occur but at a low level. In the absence of glucose, cAMP levels will be high and CAP will bind the *lac* promoter, resulting in maximal transcription.

Bacteria and other prokaryotes regulate gene expression transcriptionally through the use of repressor and activator "switches." The transcription of some clusters of genes is regulated by both repressors and activators.

Transcriptional Control in Eukaryotes

The control of transcription in eukaryotes is much more complex than in prokaryotes. The basic concepts of protein-DNA interactions are still valid, but the nature and number of interacting proteins is much greater due to some obvious differences in eukaryotes. First, eukaryotes have their DNA organized into chromatin, complicating protein-DNA interactions considerably. Second, eukaryotic transcription occurs in the nucleus, and translation occurs in the cytoplasm; this is in stark contrast to prokaryotes, where these processes are spatially and temporally coupled. As a consequence, the eukaryotic transcription apparatus is considerably more complex than the prokaryotic RNA polymerase, and the amount of DNA involved in regulating eukaryotic genes is much greater. The need for a fine degree of flexible control is important for eukaryotes with complex developmental programs and multiple tissue types. We will see, however, that general themes emerge from this complexity.

Eukaryotic Transcription Factors

Eukaryotic transcription requires a variety of proteins, or factors, that fall into two categories: basal transcription factors and specific transcription factors. The basal factors are necessary for the assembly of a transcription apparatus and recruitment of RNA pol II (see chapter 15) to a promoter. The specific factors increase the level of transcription in certain cell types or in response to specific signals.

Basal Transcription Factors. Transcription of RNA pol II templates (that is, genes that encode protein products) requires more than just RNA pol II to initiate transcription. A host of **basal transcription factors** are also necessary to establish productive initiation. These factors are required for transcription to occur, but do not increase the rate above this basal rate. They are given letter designations that follow the designator TFII, for transcription factor RNA pol II. The most important of these factors, TFIID, contains the TATA-binding protein that recognizes the TATA box sequence in the promoter (figure 18.13). Binding of TFIID is followed by binding of TFIIE, TFIIF, TFIIA, TFIIB, and TFIIH and a host of accessory factors called *transcription-associated factors*, TAFs. The *initiation complex* that results (figure 18.14) is clearly much more complex than a bacterial RNA polymerase, a single pol holoenzyme. And there is yet another level of complexity. The initiation complex, while capable of initiating synthesis at a basal level, will not achieve transcription at a high level without the participation of other specific factors.

Specific Transcription Factors. **Specific transcription factors** act in a tissue- or time-dependent manner to stimulate higher levels of transcription than the basal level. The number and diversity of these factors are overwhelming. Some sense can be made of this proliferation of factors by concentrating on the DNA-binding motif, as opposed to the specific factors. A key common theme that emerges from the study of these factors is that specific transcription factors, called **activators,** have a domain organization. Each factor consists of a DNA-binding domain

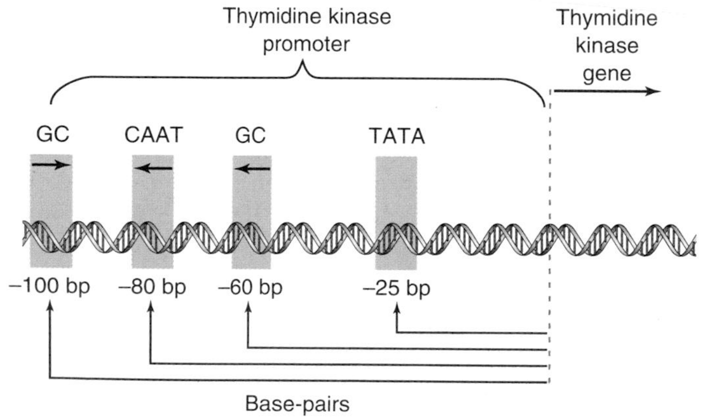

FIGURE 18.13

A eukaryotic promoter. This promoter for the gene encoding the enzyme thymidine kinase contains the TATA box that a transcription factor binds to, as well as three other DNA sequences that direct the binding of other elements of the transcription apparatus.

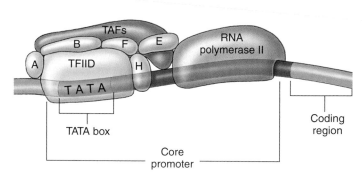

FIGURE 18.14

Formation of a eukaryotic initiation complex. The basal transcription factor, TFIID, binds to the TATA box and is joined by the other basal factors, TFIIE, TFIIF, TFIIA, TFIIB, and TFIIH. This complex is added to by a number of transcription-associated factors (TAFs) that together recruit the RNA pol II molecule to the core promoter.

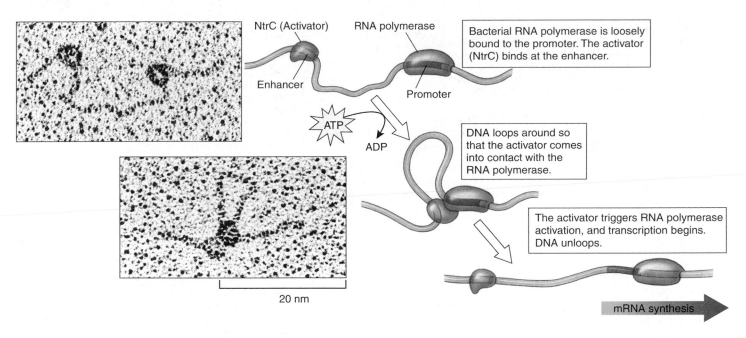

NtrC (Activator) RNA polymerase

Enhancer

Promoter

Bacterial RNA polymerase is loosely bound to the promoter. The activator (NtrC) binds at the enhancer.

ATP

ADP

DNA loops around so that the activator comes into contact with the RNA polymerase.

The activator triggers RNA polymerase activation, and transcription begins. DNA unloops.

mRNA synthesis

20 nm

FIGURE 18.15

An enhancer in action. When the bacterial activator NtrC binds to an enhancer, it causes the DNA to loop over to a distant site where RNA polymerase is bound, activating transcription. While such enhancers are rare in prokaryotes, they are common in eukaryotes.

and a separate activating domain that interacts with the transcription apparatus. These domains are essentially independent in the protein, such that they can be "swapped" between different factors and retain their function. This has been demonstrated by domain-swapping experiments in which the activation domain of one factor is spliced to the DNA-binding domain of another factor. The resulting hybrid factor shows the DNA-binding specificity expected of the DNA-binding domain, while activating transcription to the level expected of the activating domain.

Enhancers. **Enhancers** were originally defined as DNA sequences necessary for high levels of transcription that can act in a position- and orientation-independent manner. At first, this seems counterintuitive, especially since we had been conditioned by prokaryotic systems to expect control regions to be immediately upstream of the coding region of genes. It turns out that enhancers are the binding site of the specific transcription factors. The ability of enhancers to act over large distances was at first puzzling to molecular biologists. We now think this is accomplished by DNA bending to form a loop, positioning the enhancer close to the promoter (figure 18.15). Distance separating two sites on the chromosome does not have to translate to great physical distance, because the flexibility of DNA allows bending and looping to bring the enhancer and the activator into contact with the transcription factors (figure 18.16).

Coactivators and Mediators. There are other factors that specifically mediate transcription factor action. These *coactivators* and *mediators* are also necessary for activation of transcription by the transcription factor. They act by bind-

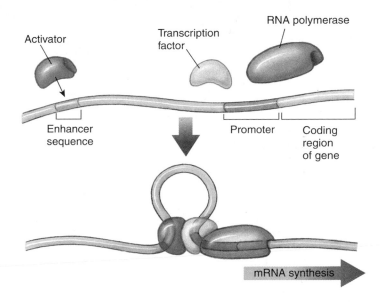

Activator

Transcription factor

RNA polymerase

Enhancer sequence

Promoter

Coding region of gene

mRNA synthesis

FIGURE 18.16

How enhancers work. The enhancer site is located far away from the gene being regulated. Binding of an activator (*red*) to the enhancer allows the activator to interact with the transcription factors (*green*) associated with RNA polymerase, activating transcription.

ing the transcription factor and then binding to another part of the transcription apparatus. These mediators are essential to the function of some transcription factors, but not all transcription factors require them. There is a much smaller number of coactivators than transcription factors because the same coactivator can be used with multiple transcription factors.

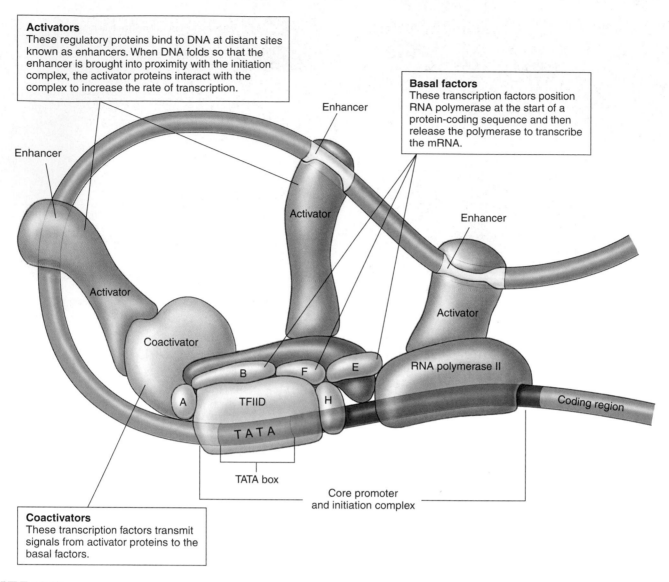

Activators
These regulatory proteins bind to DNA at distant sites known as enhancers. When DNA folds so that the enhancer is brought into proximity with the initiation complex, the activator proteins interact with the complex to increase the rate of transcription.

Basal factors
These transcription factors position RNA polymerase at the start of a protein-coding sequence and then release the polymerase to transcribe the mRNA.

Enhancer

Enhancer

Enhancer

Activator

Activator

Activator

Activator

Coactivator

B F E

A TFIID H

RNA polymerase II

Coding region

T A T A

TATA box

Core promoter
and initiation complex

Coactivators
These transcription factors transmit signals from activator proteins to the basal factors.

FIGURE 18.17

Interactions of various factors within the transcription complex. All specific transcription factors bind to enhancer sequences that may be distant from the promoter. These proteins can then interact with the initiation complex by DNA looping to bring the factors into proximity with the initiation complex. As detailed in the text, some transcription factors, called activators, can directly interact with the RNA polymerase II or the initiation complex, while others require additional coactivators.

Tying It All Together. How can we make sense of this extremely complicated situation? It is important to realize that although a few general principles apply to a broad range of situations, nearly every eukaryotic gene or group of coordinately regulated genes represents a unique case. Virtually all genes that are transcribed by RNA pol II need the same suite of basal factors to assemble an initiation complex, but the assembly of this complex and its ultimate level of transcription depend in each instance on the other specific factors involved that make up the *transcription complex* (figure 18.17). The makeup of eukaryotic promoters is thus either very simple, if you only consider that which is

needed for the initiation complex, or very complex, if you consider all factors that may bind and affect transcription. This kind of combinatorial gene regulation leads to great flexibility in the control of gene expression because it provides great flexibility in responding to the many signals affecting transcription that a cell may receive, allowing integration of these signals.

Transcription factors and enhancers confer great flexibility on the control of gene expression in eukaryotes.

The Effect of Chromatin Structure on Gene Expression

As if the conditions described so far were not complex enough, eukaryotes have the additional hurdle of possessing DNA that is packaged into chromatin. The packaging of DNA first into nucleosomes (figure 18.18) and then into higher-order chromatin structures is now thought to be directly related to the control of gene expression. Chromatin structure at its lowest level is the organization of DNA and histone proteins into nucleosomes. These nucleosomes may block binding of transcription factors and RNA pol II at the promoter. The higher-order organization of chromatin, which is not completely understood, appears to depend on the state of the histones in nucleosomes. Histones can be modified to result in a greater condensation of chromatin-making promoters, even less accessible for protein-DNA interactions. There is also a chromatin remodeling complex that can make DNA more accessible.

DNA Methylation

Chemical **methylation** of the DNA was once thought to play a major role in gene regulation in vertebrate cells. The addition of a methyl group to cytosine creates 5-methylcytosine but has no effect on base-pairing with guanine (figure 18.19), just as the addition of a methyl group to uracil produces thymine without affecting base-pairing with adenine. Many inactive mammalian genes are methylated, and it was tempting to conclude that methylation caused the inactivation. However, methylation is now viewed as having a less direct role, blocking accidental transcription of "turned-off" genes. Vertebrate cells apparently possess a protein that binds to clusters of 5-methylcytosine, preventing transcriptional activators from gaining access to the DNA. DNA methylation in vertebrates thus ensures that once a gene is turned off, it stays off.

Chromatin Structure and Transcriptional Activators

As in prokaryotes, not all gene regulation involves repression of transcription. In at least some instances, the transcription of specific genes is activated. The activation of transcription by the activating domains of transcription factors has not been well characterized. Some activators seem to act by interacting directly with the initiation complex or with coactivators that interact with the initiation complex. Other cases are not so clear. The emerging consensus is that coactivators act by modifying the structure of chromatin by adding acetyl groups to amino acids, making DNA accessible to transcription factors. Thus, transcribed DNA is correlated with the presence of nucleosomes having histones that have been acetylated. Recently, some coactivators have been shown to be histone acetylases. In these cases, it appears that transcription is increased by removing higher-

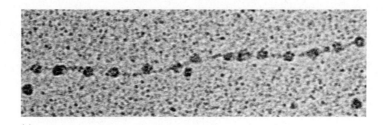

(a)

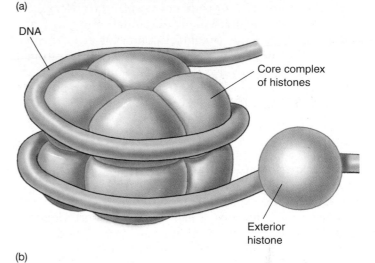

(b)

FIGURE 18.18
Nucleosomes. (*a*) In the electron micrograph, the individual nucleosomes have diameters of about 10 nanometers. (*b*) In the diagram of a nucleosome, the DNA double helix is wound around a core complex of eight histones; one additional histone binds to the outside of the nucleosome, exterior to the DNA.

Cytosine → Methylation → 5-methylcytosine

FIGURE 18.19
DNA methylation. Cytosine is methylated, creating 5-methylcytosine. Because the methyl group is positioned to the side, it does not interfere with the hydrogen bonds of a G–C base-pair.

order chromatin structure that would prevent transcription. Some corepressors have been shown to be histone deacetylases as well. This fits well with the view that histone modification affects chromatin structure, and that coactivators and corepressors then modulate this effect.

Transcriptional control of gene expression occurs in eukaryotes despite the tight packaging of DNA into nucleosomes.

Posttranscriptional Control in Eukaryotes

FIGURE 18.20
Small RNAs form double-stranded loops. These three RNA molecules fold back to form hairpin loops because the sequences of the left and right halves are complementary, and form base-pairs.

Thus far we have discussed gene regulation entirely in terms of transcription initiation—that is, when and how often RNA polymerase starts "reading" a particular gene. Most gene regulation appears to occur at this point. However, there are many other points after transcription where gene expression could be regulated in principle, and all of them serve as control points for at least some eukaryotic genes. In general, these posttranscriptional control processes involve the recognition of specific sequences on the primary RNA transcript by regulatory proteins and small RNA molecules.

Small RNAs

Recent experiments indicate that a class of RNA molecules loosely called *small RNAs* may play a major role in regulating gene expression by interacting directly with primary gene transcripts. Small RNAs are short segments of RNA ranging in length from 21 to 28 nucleotides. Researchers focusing on far larger messenger RNA (mRNA), transfer RNA (tRNA), and ribosomal RNA (rRNA) had not noticed these far smaller bits, tossing them out during experiments. The first hints of the existence of small RNAs emerged in 1993, when researchers reported in the nematode *Caenorhabditis elegans* the presence of tiny RNA molecules that don't encode any protein. These small RNAs appeared to regulate the activity of specific *C. elegans* genes.

Soon researchers found evidence of similar small RNAs in a wide range of other organisms. In the plant *Arabidopsis thaliana*, small RNAs seemed to be involved in the regulation of genes critical to early development, while in yeasts they were identified as the agents that silence genes in tightly packed regions of the genome. In the ciliated protozoan *Tetrahymena thermophila*, the loss of major blocks of DNA during development seems guided by small RNA molecules.

RNA Interference. What is going on here? How do small fragments of RNA act to regulate gene expression? The first clue emerged in 1998, when researchers injected small stretches of double-stranded RNA into *C. elegans*. Double-stranded RNA forms when a single strand folds back in a hairpin loop; this occurs because the two ends of the strand have a complementary nucleotide sequence, so that base-pairing holds the strands together much as it does the strands of a DNA duplex (figure 18.20). The result? The double-stranded RNA strongly inhibited the expression of the genes from which the double-stranded RNA had been generated. This kind of gene silencing, since seen in *Drosophila* and other organisms, is called **RNA interference.**

How Small RNAs Regulate Gene Expression. In 2001, researchers identified an enzyme, dubbed *dicer*, that appears to generate the small RNAs in the cell. Dicer chops double-stranded RNA molecules into little pieces. Two types of small RNA result: **microRNAs (miRNAs)** and **small interfering RNAs (siRNAs).**

miRNAs appear to act by binding directly to mRNAs and preventing their translation into protein. Researchers have identified over 100 different miRNAs, and are still trying to sort out how each functions and which miRNAs occur in which species.

siRNAs appear to be the main agents of RNA interference, acting to degrade particular messenger RNAs after they have been transcribed but before they can be translated by the ribosomes. The exact way they achieve this degradation of selected gene transcripts is not yet known. Current data suggest that dicer delivers siRNAs to an enzyme complex called RISC, which searches out and degrades any mRNA molecules with a complementary sequence (figure 18.21).

RNA interference appears to play a major role in *epigenetic change*, a change in gene expression that is passed from one generation to another but is not caused by changes in the DNA sequence of genes. Epigenetic regulation is in many cases the result of alterations in DNA packaging. As we saw in chapter 11, the DNA of eukaryotes is packaged in a highly compact form that enables it to fit into the cell nucleus. DNA is wrapped tightly around histone proteins to form nucleosomes (see figure 18.18), and then the strand of nucleosomes is twisted into higher-order filaments. By altering how the strands are twisted, siRNAs can alter which genes are accessible for gene expression. Just how siRNAs achieve this change in chromatin shape is not known.

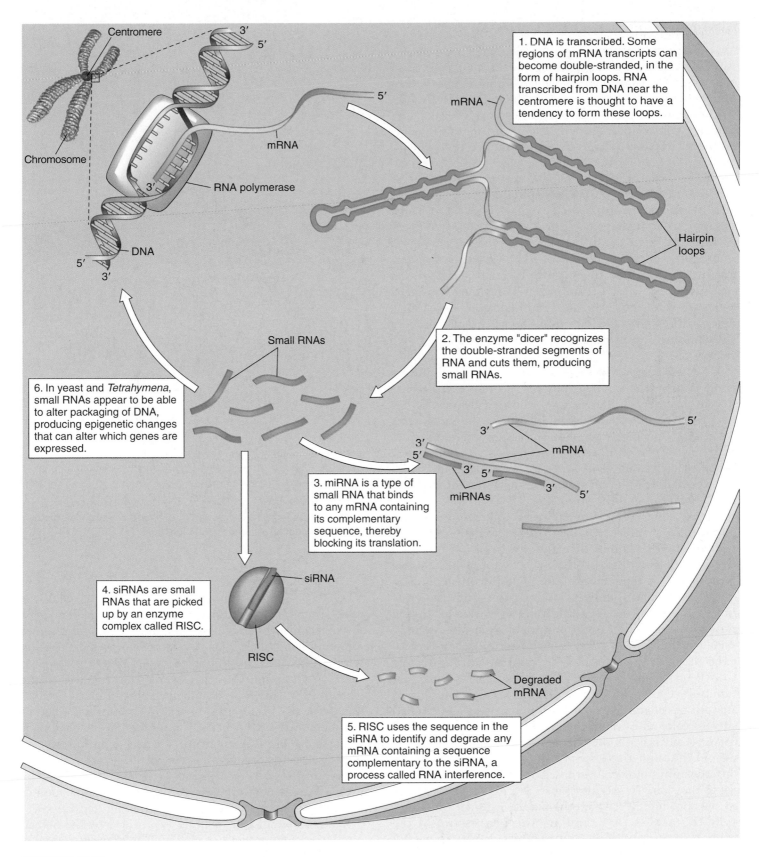

Centromere

Chromosome

5'
3'

mRNA

RNA polymerase

3'

5'
DNA
3'

1. DNA is transcribed. Some regions of mRNA transcripts can become double-stranded, in the form of hairpin loops. RNA transcribed from DNA near the centromere is thought to have a tendency to form these loops.

mRNA

5'

Hairpin loops

Small RNAs

6. In yeast and *Tetrahymena*, small RNAs appear to be able to alter packaging of DNA, producing epigenetic changes that can alter which genes are expressed.

2. The enzyme "dicer" recognizes the double-stranded segments of RNA and cuts them, producing small RNAs.

3'
5'

mRNA

5'

3'
5'
3'
5'

miRNAs

3. miRNA is a type of small RNA that binds to any mRNA containing its complementary sequence, thereby blocking its translation.

siRNA

4. siRNAs are small RNAs that are picked up by an enzyme complex called RISC.

RISC

Degraded mRNA

5. RISC uses the sequence in the siRNA to identify and degrade any mRNA containing a sequence complementary to the siRNA, a process called RNA interference.

FIGURE 18.21
How small RNAs may act to regulate gene expression. Small RNAs are produced when double-stranded hairpin loops of an RNA transcript are cut. Although the details are not well understood, two types of small RNA, miRNA and siRNA, are thought to block gene expression within the nucleus at the level of the mRNA gene transcript, a process called RNA interference. As the above diagram suggests, small RNAs are also thought to influence chromatin packaging in some organisms.

Alternative Splicing of the Primary Transcript

As we learned in chapter 15, most eukaryotic genes have a patchwork structure, being composed of numerous short coding sequences (exons) embedded within long stretches of noncoding sequences (introns). The initial mRNA molecule copied from a gene by RNA polymerase, the **primary transcript,** is a faithful copy of the entire gene, including introns as well as exons. Before the primary transcript is translated, the introns, which comprise on average 90% of the transcript, are removed in a process called *RNA processing*, or **RNA splicing.** Particles called *small nuclear ribonucleoproteins*, or *snRNPs* (more informally, **snurps**), are thought to play a role in RNA splicing. These particles reside in the nucleus of a cell and are composed of proteins and a special type of RNA called *small nuclear RNA*, or *snRNA*. One kind of snRNP contains snRNA that can bind to the 5′ end of an intron by forming base-pairs with complementary sequences on the intron. When multiple snRNPs combine to form a larger complex called a **spliceosome,** the intron loops out and is excised (see chapter 15).

RNA splicing provides a potential point at which the expression of a gene can be controlled, because exons can be spliced together in different ways, allowing a variety of different polypeptides to be assembled from the same gene. Alternative splicing is common in insects and vertebrates, with two or three different proteins produced from one gene. In many cases, gene expression is regulated by changing which splicing event occurs during different stages of development or in different tissues.

An excellent example of alternative splicing in action is found in two different human organs, the thyroid gland and the hypothalamus. The thyroid gland (see chapter 47) is responsible for producing hormones that control processes such as metabolic rate. The hypothalamus, located in the brain, collects information from the body (for example, salt balance) and releases hormones that in turn regulate the release of hormones from other glands, such as the pituitary gland (see chapter 47). The two organs produce two distinct hormones, calcitonin and CGRP (calcitonin gene-related peptide) as part of their function. Calcitonin is responsible for controlling the amount of calcium we take up from our food and the balance of calcium in tissues such as bone and teeth. CGRP is involved in a number of neural and endocrine functions. Although these two hormones are used for very different physiological purposes, the hormones are made using the same transcript (figure 18.22). The appearance of one product versus another is determined by tissue-specific factors that regulate the processing of the primary transcript. This ability offers another powerful way to control the expression of gene products, ranging from proteins with subtle differences to totally unrelated proteins.

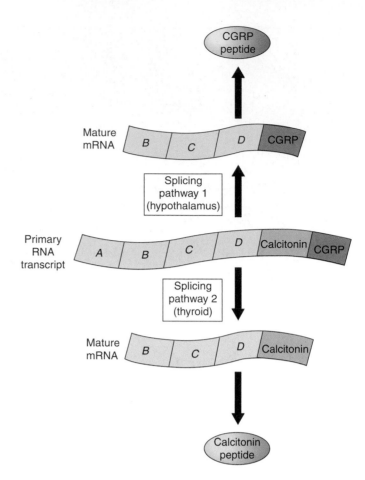

FIGURE 18.22
Alternative splicing products.   The same transcript made from one gene can be spliced differently to give rise to two very distinct protein products, calcitonin and CGRP.

RNA Editing

The generation of multiple transcripts from the same gene by alternative splicing allows for the production of more proteins than "genes." An even more unexpected alteration of the message is the editing of mature transcripts to produce an altered mRNA that is not truly encoded in the genome. First discovered as the insertion of uracil residues into some RNA transcripts in protozoa, this was thought to be an anomaly. RNA editing of a different sort has since been found in mammalian species, including humans. In this case, the editing involves the chemical modification of a base to change its base-pairing properties. This occurs by either deamination of cytosine to uracil, or deamination of adenine to inosine (inosine pairs as G during translation).

An example of RNA editing is provided by the human protein apolipoprotein B, which is involved in the transport of cholesterol and triglycerides. The gene that encodes this protein (*apoB*) is large and complex, consisting of 39 introns scattered across almost 50 kilobases of DNA.

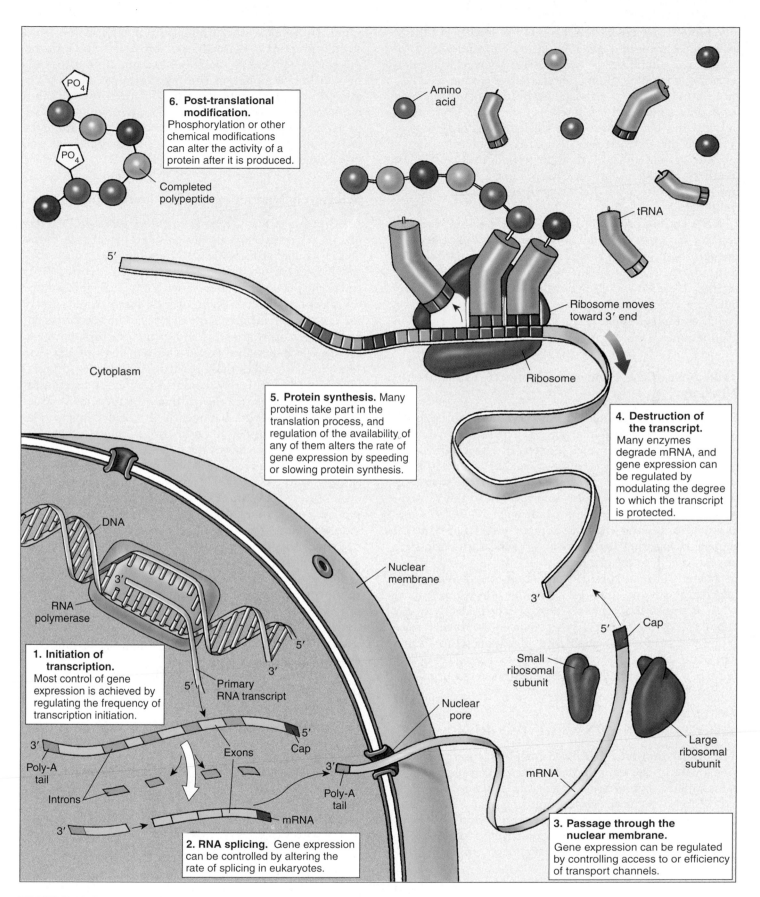

6. Post-translational modification. Phosphorylation or other chemical modifications can alter the activity of a protein after it is produced.

Completed polypeptide

PO₄

Amino acid

tRNA

5′

Ribosome moves toward 3′ end

Ribosome

Cytoplasm

5. Protein synthesis. Many proteins take part in the translation process, and regulation of the availability of any of them alters the rate of gene expression by speeding or slowing protein synthesis.

4. Destruction of the transcript. Many enzymes degrade mRNA, and gene expression can be regulated by modulating the degree to which the transcript is protected.

DNA

RNA polymerase

3′

Nuclear membrane

3′

5′

Cap

1. Initiation of transcription. Most control of gene expression is achieved by regulating the frequency of transcription initiation.

5′

Primary RNA transcript

3′

Small ribosomal subunit

mRNA

5′

3′

Exons

Cap

Large ribosomal subunit

3′

Poly-A tail

Introns

3′

Poly-A tail

Nuclear pore

mRNA

3′

mRNA

2. RNA splicing. Gene expression can be controlled by altering the rate of splicing in eukaryotes.

3. Passage through the nuclear membrane. Gene expression can be regulated by controlling access to or efficiency of transport channels.

FIGURE 18.23
Six levels where gene expression can be controlled in eukaryotes.

The protein exists in two isoforms: a full-length APOB100 form, and a truncated APOB48 form. The truncated form is due to an alteration of the mRNA that changes a codon encoding glutamine to one that is a "stop" codon. Further, this editing occurs in a tissue-specific manner; the edited form only appears in the intestine, while the liver makes only the full-length form. The full-length APOB100 form is part of the LDL particle that carries cholesterol. High levels of serum LDL are thought to be a major predictor of atherosclerosis in humans. It does not appear that editing has any effect on the levels of the intestine-specific transcript.

RNA editing has also been observed in some brain receptors for opiates in humans. One of these receptors, the serotonin (5-HT) receptor, is edited at multiple sites to produce a total of 12 different isoforms of the protein. It is not clear how widespread these forms of editing are, but they are further evidence that the information encoded within genes is not the end of the story for protein production.

Transport of the Processed Transcript Out of the Nucleus

Processed mRNA transcripts exit the nucleus through the nuclear pores described in chapter 5. The passage of a transcript across the nuclear membrane is an active process that requires the transcript to be recognized by receptors lining the interior of the pores. Specific portions of the transcript, such as the poly-A tail, appear to play a role in this recognition. The transcript cannot move through a pore as long as any of the splicing enzymes remain associated with the transcript, ensuring that partially processed transcripts are not exported into the cytoplasm.

There is little hard evidence that gene expression is regulated at this point, although it could be. On average, about 10% of transcribed genes are exon sequences, but only about 5% of the total mRNA produced as primary transcript ever reaches the cytoplasm. This suggests that about half of the exon primary transcripts never leave the nucleus, but it is not clear whether the disappearance of this mRNA is selective.

Selecting Which mRNAs Are Translated

The translation of a processed mRNA transcript by the ribosomes in the cytoplasm involves a complex of proteins called *translation factors*. In at least some cases, gene expression is regulated by modification of one or more of these factors. In other instances, **translation repressor proteins** shut down translation by binding to the beginning of the transcript, so that it cannot attach to the ribo-

some. In humans, the production of ferritin (an iron-storing protein) is normally shut off by a translation repressor protein called aconitase. Aconitase binds to a 30-nucleotide sequence at the beginning of the ferritin mRNA, forming a stable loop to which ribosomes cannot bind. When the cell encounters iron, the binding of iron to aconitase causes the aconitase to dissociate from the ferritin mRNA, freeing the mRNA to be translated and increasing ferritin production 100-fold.

Selectively Degrading mRNA Transcripts

Another aspect that affects gene expression is the stability of mRNA transcripts in the cell cytoplasm (see figure 18.21). Unlike prokaryotic mRNA transcripts, which typically have a half-life of about 3 minutes, eukaryotic mRNA transcripts are very stable. For example, β-globin gene transcripts have a half-life of over 10 hours, an eternity in the fast-moving metabolic life of a cell. The transcripts encoding regulatory proteins and growth factors, however, are usually much less stable, with half-lives of less than 1 hour. What makes these particular transcripts so unstable? In many cases, they contain specific sequences near their 3′ ends that make them attractive targets for enzymes that degrade mRNA. A sequence of A and U nucleotides near the 3′ poly-A tail of a transcript promotes removal of the tail, which destabilizes the mRNA. Histone transcripts, for example, have a half-life of about 1 hour in cells that are actively synthesizing DNA; at other times during the cell cycle, the poly-A tail is lost, and the transcripts are degraded within minutes. Other mRNA transcripts contain sequences near their 3′ ends that are recognition sites for endonucleases, which causes these transcripts to be digested quickly. The short half-lives of the mRNA transcripts of many regulatory genes are critical to the function of those genes because they enable the levels of regulatory proteins in the cell to be altered rapidly. There are multiple pathways of decay, including deadenylation-dependent decay. *Cis*-acting elements can either stabilize or destabilize a transcript.

A review of various methods of posttranscriptional control of gene expression is provided in figure 18.23.

Posttranscriptional control of gene expression is exercised by both proteins and small RNAs. Proteins interacting with small nuclear RNA carry out alternative splicing of RNA transcripts. Chemical alterations of specific bases lead to RNA transcript editing. After the primary transcript is modified, further control of gene expression occurs by translation repression and selective degradation of mRNA transcripts.

Concept Review

18.1 Gene expression is controlled by regulating transcription.

An Overview of Transcriptional Control

- One way to control gene expression is to regulate the initiation of transcription by regulating promoter access. (p. 362)
- Gene expression is controlled at the transcriptional and posttranscriptional levels, with transcriptional control being the most common form of control (p. 362)

18.2 Regulatory proteins read DNA without unwinding it.

How to Read a Helix Without Unwinding It

- Regulatory proteins identify DNA sequences without unwinding the helix by inserting DNA-binding motifs into the major groove where the edges of the base-pairs are exposed. (p. 363)

Four Important DNA-Binding Motifs

- The four most important DNA-binding motifs are the helix-turn-helix motif, the homeodomain motif, the zinc finger motif, and the leucine zipper motif. (pp. 364–365)

18.3 Prokaryotes regulate genes by controlling transcription initiation.

Prokaryotic Gene Regulation

- How prokaryotes respond to changes depends on the nature of the proteins encoded by relevant genes. (p. 366)
- Prokaryotic genes are often organized into operons, multiple genes that are part of a single gene expression unit. (p. 366)
- Repressors are proteins that bind to regulatory sites on DNA and prevent or decrease initiation of transcription. (p. 367)
- Activators can bind to DNA to stimulate the initiation of transcription. (p. 368)

18.4 Transcriptional control in eukaryotes operates at a distance.

Transcriptional Control in Eukaryotes

- Eukaryotic transcription factors fall into two categories: basal transcriptional factors and specific transcriptional factors. Along with enhancers, coactivators, and mediators, these factors provide flexibility in controlling eukaryotic gene expression. (pp. 370–372)

The Effect of Chromatin Structure on Gene Expression

- Despite the additional complexity of DNA packaging in eukaryotes, transcriptional control of gene expression can still occur in eukaryotes. (p. 373)

Posttranscriptional Control in Eukaryotes

- Posttranscriptional control of gene expression is exercised by proteins and small RNAs. (p. 374)
- Small RNAs such as miRNAs and siRNAs are thought to regulate gene expression through RNA interference and, in some cases, through the alteration of DNA packaging. (pp. 374–375)
- Proteins interact with small nuclear RNA and carry out alternative splicing of RNA transcripts. (p. 376)
- RNA transcript editing can be caused by chemical alterations of specific bases. (pp. 376–378)
- Following modification of the primary transcript, further gene expression control occurs by translation repression and selective degradation of mRNA transcripts. (p. 378)

Self Test

1. Prokaryotes and eukaryotes use several methods to regulate gene expression, but the most common method is
   a. translational control.
   b. transcriptional control.
   c. posttranscriptional control.
   d. control of mRNA passage from the nucleus.

2. The two protein subunits of the leucine zipper are held together
   a. in the shape of a Y.
   b. by the interaction of leucine amino acids.
   c. by hydrophobic interactions.
   d. All of these are correct.

3. The helix-turn-helix motif contains two helical segments, and in order for the motif to bind DNA, the _____ fits into the major groove of the DNA.
   a. homeodomain
   b. recognition helix
   c. zinc finger
   d. leucine zipper

4. A(n) _____ is a piece of DNA with a group of genes that are transcribed together as a unit.
   a. promoter
   b. repressor
   c. operator
   d. operon

5. What effect would the addition of lactose have on a repressed *lac* operon?
   a. The operator site on the operon would move.
   b. It would reinforce the repression of that gene.
   c. The *lac* operon would be transcribed.
   d. It would have no effect whatsoever.

6. A type of DNA sequence that is located far from a gene but can promote its expression is a(n)
   a. promoter.
   b. activator.
   c. enhancer.
   d. TATA box.

7. Which of the following is *not* found in a eukaryotic transcription complex?
   a. activator
   b. RNA
   c. enhancer
   d. TATA-binding protein

8. DNA methylation of genes
   a. inhibits transcription by blocking the base-pairing between methylated cytosine and guanine.
   b. inhibits transcription by blocking the base-pairing between uracil and adenine.
   c. prevents transcription by blocking the TATA sequence.
   d. makes sure genes that are turned off remain turned off.

9. Which of the following are *not* matched correctly?
   a. RNA splicing—occurs in the nucleus
   b. snRNP—splicing out exons from the transcript
   c. poly-A tail—increased transcript stability
   d. All are matched correctly.

10. Which of the following is *not* a method of posttranscriptional control in eukaryotic cells?
    a. processing the transcript
    b. selecting the mRNA molecules that are translated
    c. digesting the DNA immediately after translation
    d. selectively degrading the mRNA transcripts

Test Your Visual Understanding

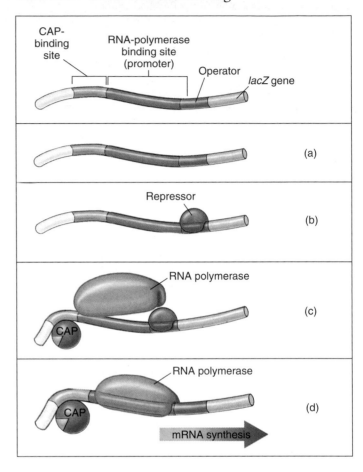

1. Match the following descriptions with the appropriate lettered panels in the figure, and explain what would be needed to activate the operon if it is not activated.
   i. Operon is OFF because *lac* repressor is bound.
   ii. Operon is OFF because CAP is not bound.
   iii. Operon is ON because CAP is bound and *lac* repressor is not.
   iv. Operon is OFF both because *lac* repressor is bound and because CAP is not.

Apply Your Knowledge

1. A method of posttranscriptional control is the selective degradation of mRNA transcripts. A transcript encoding a growth factor contains the terminal sequence AAGCUUGAAU and has a half-life of 40 minutes. Another transcript that encodes an immunoglobin has a terminal sequence of GGAUCGCCAGG and has a half-life of about 2 hours. The half-life is related to the rate of degradation by the equation: $t_{1/2} = 0.693/K$, where $t_{1/2}$ is the half-life, and K is the rate of degradation. Compare the degradation rates of the two transcripts.

2. All human beings have a rich growth of *E. coli* bacteria in their large intestine. Will the *lac* operon in the bacteria present in a lactose-intolerant individual who is careful never to consume anything containing lactose (milk sugar) be activated or repressed? Explain.

19

Cellular Mechanisms of Development

Concept Outline

19.1 Development is a regulated process.

Overview of Development. Studies of cellular mechanisms have focused on mice, fruit flies, flowering plants, and nematodes.

Vertebrate Development. Vertebrates develop in a highly orchestrated fashion.

Insect Development. Insect development is highly specialized, with many key events occurring in a fused mass of cells.

Plant Development. Unlike animal development, which is buffered from the environment, plant development is sensitive to environmental influences.

Nematode Development. The developmental history of every somatic cell of this animal is known.

19.2 Multicellular organisms employ the same basic mechanisms of development.

Cell Movement and Induction. Animal cells move by extending protein cables that they use to pull themselves past surrounding cells. Transcription within cells is influenced by signal molecules from other cells.

Determination. Cells become reversibly committed to particular developmental paths.

Pattern Formation. Diffusion of chemical inducers governs pattern formation in fly embryos.

Homeotic Genes. Master genes determine the form body segments will take.

Programmed Cell Death. Some genes, when activated, kill their cells.

19.3 Aging can be considered a developmental process.

Theories of Aging. While there are many ideas about why cells age, no single theory of aging is widely accepted.

FIGURE 19.1
A collection of future fish undergo embryonic development. Inside a transparent fish egg, a single cell becomes millions of cells that form eyes, fins, gills, and other body parts.

In chapter 18, we explored gene expression from the perspective of an individual cell, examining the diverse mechanisms that a cell may employ to control the transcription of particular genes. Now we will broaden our perspective and look at the unique challenge posed by the development of a cell into a multicellular organism (figure 19.1). In the course of this developmental journey, a pattern of decisions about transcription are made that cause particular lines of cells to proceed along different paths, spinning an incredibly complex web of cause and effect. Yet, for all its complexity, this developmental program works with impressive precision. In this chapter, we explore the mechanisms multicellular organisms use to control their development and achieve this precision.

19.1 Development is a regulated process.

Overview of Development

Organisms in all three multicellular kingdoms—fungi, plants, and animals—realize cell specialization by orchestrating gene expression. That is, different cells express different genes at different times. To understand development, we need to focus on how cells determine which genes to activate, and when.

Among the fungi, the specialized cells are largely limited to reproductive cells. In basidiomycetes and ascomycetes (the so-called higher fungi), certain cells produce pheromones that influence other cells, but the basic design of all fungi is quite simple. For most of its life, a fungus has a two-dimensional body, consisting of long filaments of cells that are only imperfectly separated from each other. Fungal maturation is primarily a process of growth rather than specialization.

Development is far more complex in plants, in which the adult individuals contain a variety of specialized cells organized into tissues and organs. A hallmark of plant development is flexibility; as a plant develops, the precise array of tissues it achieves is greatly influenced by its environment.

In animals, development is complex and rigidly controlled, producing a bewildering array of specialized cell types through mechanisms that are much less sensitive to the environment. The subject of intensive study, animal development has in the last decades become relatively well understood.

Here we will focus on four developmental systems that researchers have studied intensively: (1) an animal with a very complexly arranged body, a vertebrate; (2) a less complex animal with an intricate developmental cycle, an insect; (3) a flowering plant; and (4) a very simple animal, a nematode (figure 19.2).

To begin our investigation, we will first examine the overall process of development in these four quite different organisms so that we can sort through differences in the gross process to uncover basic similarities in underlying mechanisms. We will start by describing the overall process in vertebrates, because it is the best understood among the animals. Then we will examine the very different developmental process carried out by insects, in which genetics has allowed us to gain detailed knowledge of many aspects. Finally, we will look at development in a third very different organism, a flowering plant, and then we will review development in a nematode worm.

Almost all multicellular organisms undergo development. The process has been well studied in plants and in animals, especially in vertebrates, insects, and nematodes.

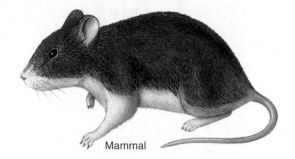
Mammal

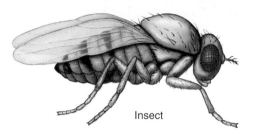
Insect

Flowering plant

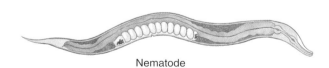
Nematode

FIGURE 19.2
Four developmental systems. Researchers studying the cellular mechanisms of development have focused on these four organisms.

Vertebrate Development

Vertebrate development is a dynamic process in which cells divide rapidly and move over each other as they first establish the basic geometry of the body (figure 19.3). At different sites, particular cells then proceed to form the body's organs, and then the body grows to a size and shape that will allow it to survive after birth. The entire process, described more fully in chapter 51, is traditionally divided into phases. As in mitosis, however, the boundaries between phases are somewhat artificial, and the phases, in fact, grade into one another.

Cleavage

Vertebrates begin development as a single fertilized egg, the zygote. Within an hour after fertilization, the zygote begins to divide rapidly into a larger and larger number of smaller and smaller cells called **blastomeres,** until a solid ball of cells is produced (figure 19.4). This initial period of cell division, termed **cleavage,** is not accompanied by any increase in the overall size of the embryo; rather, the contents of the zygote are simply partitioned into the daughter cells. The two ends of the zygote are traditionally referred to as the **animal pole** and the **vegetal pole.** In general, the blastomeres of the animal pole go on to form the external tissues of the body, while those of the vegetal pole form the internal tissues. The initial top-bottom (dorsal-ventral) orientation of the embryo is determined at fertilization by the location where the sperm nucleus enters the egg, a point that corresponds roughly to the future belly. After about 12 divisions, the burst of cleavage divisions slows, and transcription of key genes begins within the embryo cells.

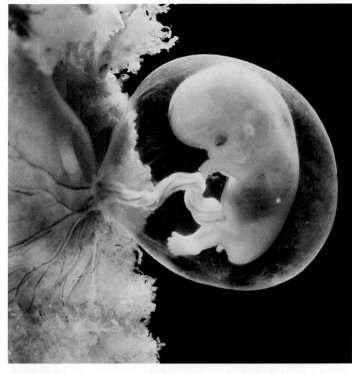

FIGURE 19.3
The miracle of development. This nine-week-old human fetus started out as a single cell: a fertilized egg, or zygote. The zygote's daughter cells have been repeatedly dividing and specializing to produce the distinguishable features of a fetus.

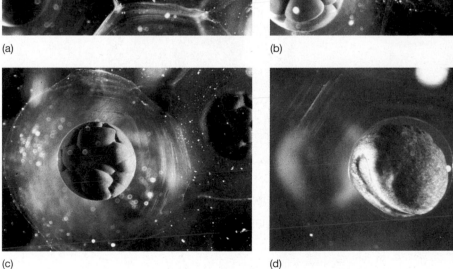

(a)

(b)

FIGURE 19.4
Cleavage divisions producing a frog embryo. (*a*) The initial divisions are, in this case, on the side of the embryo facing you, producing (*b*) a cluster of cells on this side of the embryo, which soon expands to become (*c*) a compact mass of cells. (*d*) This mass eventually invaginates into the interior of the embryo, forming a gastrula, and then a neurula.

(c)

(d)

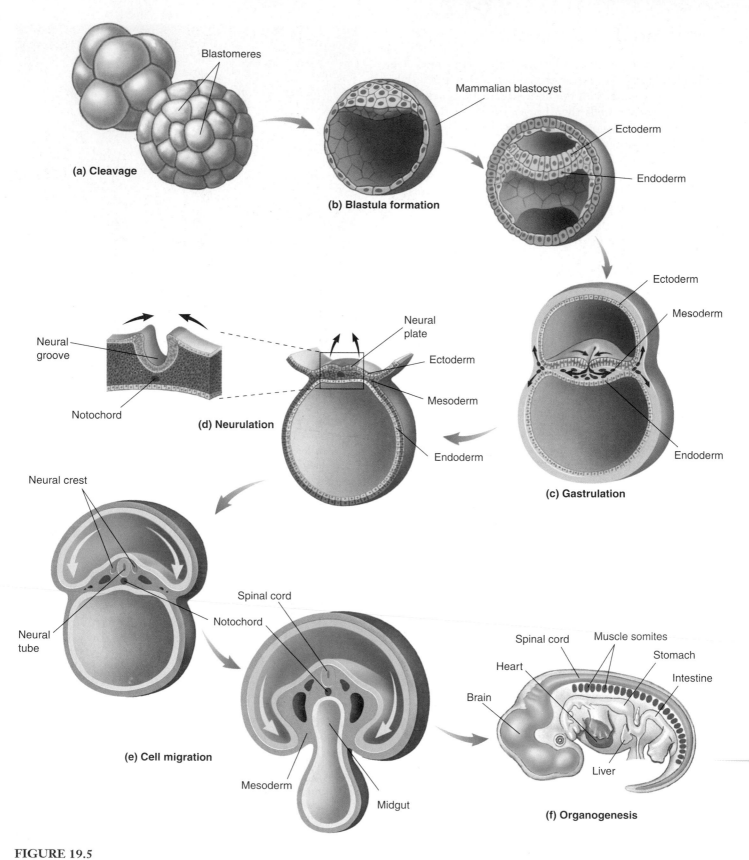

FIGURE 19.5

The path of vertebrate development. The major events in the development of *Mus musculus*, the house mouse, are (*a*) cleavage, (*b*) formation of blastula, (*c*) gastrulation, (*d*) neurulation, (*e*) cell migration, (*f*) organogenesis, and (*g*) growth.

Formation of the Blastula

The outermost blastomeres (figure 19.5*a*) in the ball of cells produced during cleavage are joined to one another by tight junctions, which, as you may recall from chapter 7, are belts of protein that encircle a cell and weld it firmly to its neighbors. These tight junctions create a seal that isolates the interior of the cell mass from the surrounding medium. At about the 16-cell stage, the cells in the interior of the mass begin to pump Na$^+$ from their cytoplasm into the spaces between cells. The resulting osmotic gradient causes water to be drawn into the center of the cell mass, enlarging the intercellular spaces. Eventually, the spaces coalesce to form a single large cavity within the cell mass. The resulting hollow ball of cells is called a **blastula,** or **blastocyst** in mammals (figure 19.5*b*).

Gastrulation

Some cells of the blastula then push inward, forming a **gastrula** that is invaginated. Cells move by using extensions called lamellipodia to crawl over neighboring cells, which respond by forming lamellipodia of their own. Soon a sheet of cells contracts on itself and shoves inward, starting the invagination. Called **gastrulation** (figure 19.5*c*), this process creates the main axis of the vertebrate body, converting the blastula into a bilaterally symmetrical embryo with a central gut. From this point on, the embryo has three **germ layers** whose organization foreshadows the future organization of the adult body. The cells that invaginate and form the tube of the primitive gut are endoderm; they give rise to the stomach, lungs, liver, and most of the other internal organs. The cells that remain on the exterior are ectoderm, and their derivatives include the skin on the outside of the body and the nervous system. The cells that break away from the invaginating cells and invade the space between the gut and the exterior wall are mesoderm; they eventually form the notochord, bones, blood vessels, connective tissues, and muscles.

Neurulation

Soon after gastrulation is complete, a broad zone of ectoderm begins to thicken on the dorsal surface of the embryo, an event triggered by the presence of the notochord beneath it. The thickening is produced by the elongation of certain ectodermal cells. Those cells then assume a wedge shape by contracting bundles of actin filaments at one end. This change in shape causes the neural tissue to roll up into a tube, which eventually pinches off from the rest of the ectoderm and gives rise to the brain and spinal cord. This tube is called the **neural tube,** and the process by which it forms is termed **neurulation** (figure 19.5*d*).

Cell Migration

During the next stage of vertebrate development, a variety of cells migrate to form distant tissues, following specific paths through the embryo to particular locations (figure 19.5*e*). These migrating cells include those of the **neural crest,** which pinch off from the neural tube and form a number of structures, including some of the body's sense organs; cells that migrate from central blocks of muscle tissue called **somites** and form the skeletal muscles of the body; and the precursors of blood cells and gametes. When a migrating cell reaches its destination, receptor proteins on its surface interact with proteins on the surfaces of cells in the destination tissue, triggering changes in the cytoskeleton of the migrating cell that cause it to cease moving.

Organogenesis and Growth

At the end of this wave of cell migration and colonization, the basic vertebrate body plan has been established, although the embryo is only a few millimeters long and has only about 10^5 cells. Over the course of subsequent development, tissues will develop into organs (figure 19.5*f*), and the embryo will grow to be a hundred times larger, with a million times as many cells (figure 19.5*g*).

Vertebrates develop in a highly orchestrated fashion. The zygote divides rapidly, forming a hollow ball of cells that then pushes inward, forming the main axis of an embryo that goes on to develop tissues and, after a process of cell migration, organs.

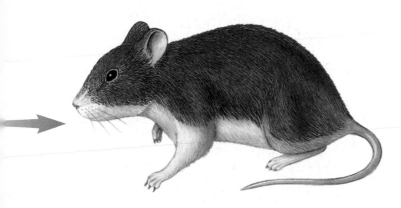

(g) Growth

Insect Development

Like all animals, insects develop through an orchestrated series of cell changes, but the path of development is quite different from that of a vertebrate. Many insects produce two different kinds of bodies during their development, the first a tubular eating machine called a **larva,** and the second a flying sex machine with legs and wings. The passage from one body form to the other, called **metamorphosis,** involves a radical shift in development. Here we describe development in the fruit fly *Drosophila* (figure 19.6), which is the subject of much genetic research.

Maternal Genes

The development of an insect like *Drosophila* begins before fertilization, with the construction of the egg. Specialized nurse cells that help the egg grow move some of their own mRNA into the end of the egg nearest them (figure 19.7*a*). As a result, mRNAs produced by maternal genes are positioned in particular locations in the egg, so that after repeated divisions subdivide the fertilized egg, different daughter cells will contain different maternal products. Thus, the action of maternal (rather than zygotic) genes determines the initial course of development.

Syncytial Blastoderm

After fertilization, 12 rounds of nuclear division without cytokinesis produce about 6000 nuclei, all within a single cytoplasm. All of the nuclei within this **syncytial blastoderm** (figure 19.7*b*) can freely communicate with one another, but nuclei located in different sectors of the egg experience different maternal products. The nuclei then space themselves evenly along the surface of the blastoderm, and membranes grow between them. Folding of the embryo and primary tissue development soon follow, in a process fundamentally similar to that seen in vertebrate development. The tubular body that results within a day of fertilization is a larva.

Larval Instars

The larva begins to feed immediately, and as it does so, it grows. However, its chitinous exoskeleton cannot stretch much, and within a day it sheds the exoskeleton. Before the

FIGURE 19.6
The fruit fly, *Drosophila melanogaster*. A dorsal view of *Drosophila*, one of the most intensively studied animals in development.

new exoskeleton has had a chance to harden, the larva expands in size. A total of three larval stages, or **instars,** are produced over a period of four days (figure 19.7*c*).

Imaginal Discs

During embryonic growth, about a dozen groups of cells called **imaginal discs** are set aside in the body of the larva (figure 19.7*d*). Imaginal discs play no role in the life of the larva, but are committed to form key parts of the adult fly's body.

Metamorphosis

After the last larval stage, a hard outer shell forms, and the larva is transformed into a **pupa** (figure 19.7*e*). Within the pupa, the larval cells break down and release their nutrients, which are used in the growth and development of the various imaginal discs (eye discs, wing discs, leg discs, and so on). The imaginal discs then associate with one another, assembling themselves into the body of the adult fly (figure 19.7*f*). The metamorphosis of a *Drosophila* larva into a pupa and then into an adult fly takes about four days, after which the pupal shell splits and the fly emerges.

Drosophila development proceeds through two discrete phases, the first a larval phase that gathers food, and the second an adult phase that is capable of flight and reproduction.

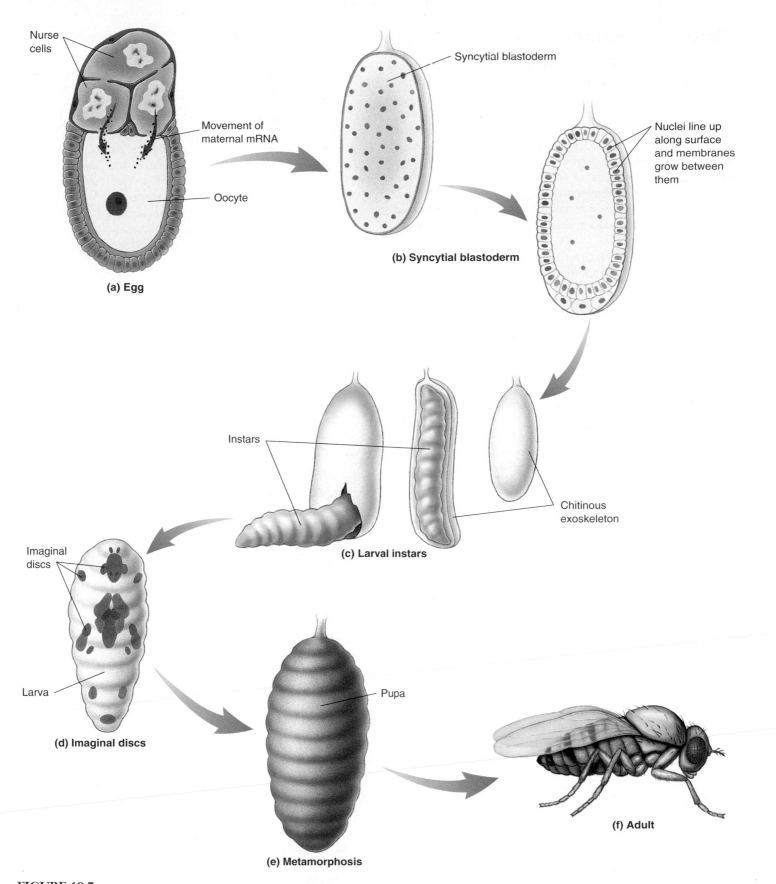

FIGURE 19.7
The path of insect development. The major events in the development of *Drosophila melanogaster* are (*a*) egg, (*b*) syncytial blastoderm, (*c*) larval instars, (*d*) imaginal discs, (*e*) metamorphosis, and (*f*) adult.

Plant Development

At the most basic level, the developmental paths of plants and animals share many key elements. However, the mechanisms used to achieve body form are quite different. While animal cells follow an orchestrated series of movements during development, plant cells are encased within stiff cellulose walls, and therefore cannot move. Each cell in a plant is fixed into position when it is created. Instead of using cell migration, plants develop by building their bodies outward, creating new parts from special groups of self-renewing cells called **meristems.** As meristem cells continually divide, they produce cells that can differentiate into the tissues of the plant.

Another major difference between animals and plants is that most animals are mobile and can move away from unfavorable circumstances, while plants are anchored in position and must simply endure whatever environment they experience. Plants compensate for this restriction by relaxing the rules of development to accommodate local circumstances. Instead of creating a body in which every part is specified to have a fixed size and location, a plant assembles its body from a few types of modules, such as leaves, roots, branch nodes, and flowers. Each module has a rigidly controlled structure and organization, but how the modules are utilized is quite flexible. As a plant develops, it simply adds more modules, with the environment having a major influence on the type, number, size, and location of what is added. In this way, the plant is able to adjust the path of its development to local circumstances.

Early Cell Division

The first division of the fertilized egg in a flowering plant is off-center, so that one of the daughter cells is small, with dense cytoplasm (figure 19.8a). That cell, the future embryo, begins to divide repeatedly, forming a ball of cells. The other daughter cell also divides repeatedly, forming an elongated structure called a **suspensor,** which links the embryo to the nutrient tissue of the seed. The suspensor also provides a route for nutrients to reach the developing embryo. Just as the animal embryo acquires its initial axis as a cell mass formed during cleavage divisions, so the plant embryo forms its root-shoot axis at this time. Cells near the suspensor are destined to form a root, while those at the other end of the axis ultimately become a shoot.

Tissue Formation

Three basic tissues differentiate while the plant embryo is still a ball of cells (figure 19.8b), analogous to the formation of the three germ layers in animal embryos, although in plants, no cell movements are involved. The outermost cells in a plant embryo become **epidermal cells.** The bulk of the embryonic interior consists of **ground tissue** cells that eventually function in food and water storage. Finally, cells at the core of the embryo are destined to form the future **vascular tissue.**

Seed Formation

Soon after the three basic tissues form, a flowering plant embryo develops one or two seed leaves called **cotyledons.** At this point, development is arrested, and the embryo is either surrounded by nutritive tissue or has amassed stored food in its cotyledons (figure 19.8c). The resulting package, known as a *seed*, is resistant to drought and other unfavorable conditions; in its dormant state, it is a vehicle for dispersing the embryo to distant sites and allows a plant embryo to survive in environments that might kill a mature plant.

Germination

A seed germinates in response to changes in its environment brought about by water, temperature, or other factors. The embryo within the seed resumes development and grows rapidly, its roots extending downward and its leaf-bearing shoots extending upward (figure 19.8d).

Meristematic Development

Plant development exhibits its great flexibility during the assembly of the modules that make up a plant body. Apical meristems at the root and shoot tips generate the large numbers of cells needed to form leaves, flowers, and all other components of the mature plant (figure 19.8e). At the same time, meristems ensheathing the stems and roots produce the wood and other tissues that allow growth in circumference. A variety of hormones produced by plant tissues influence meristem activity and, thus, the development of the plant body. Plant hormones (see chapter 41) are the tools that allow plant development to adjust to the environment.

Morphogenesis

The form of a plant body is largely determined by controlled changes in cell shape as cells expand osmotically after they form (see figure 19.8e). Plant growth-regulating hormones and other factors influence the orientation of bundles of microtubules on the interior of the plasma membrane. These microtubules seem to guide cellulose deposition as the cell wall forms around the outside of a new cell. The orientation of the cellulose fibers, in turn, determines how the cell will elongate as it increases in volume, and so determines the cell's final shape.

In a developing plant, leaves, flowers, and branches are added to the growing body in ways that are strongly influenced by the environment.

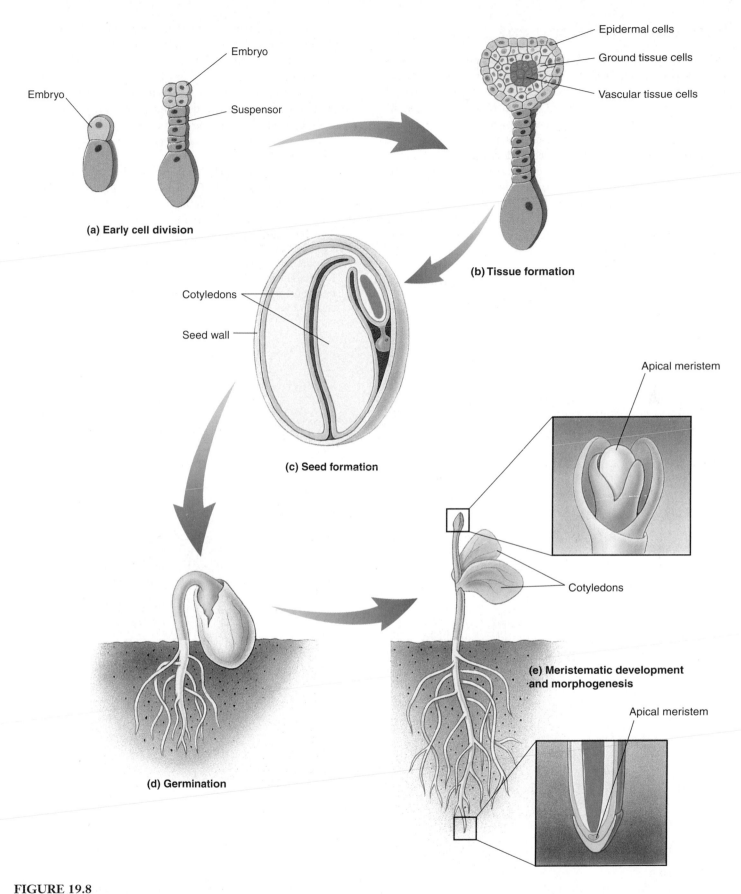

(a) Early cell division

Embryo

Embryo

Suspensor

Epidermal cells

Ground tissue cells

Vascular tissue cells

(b) Tissue formation

Cotyledons

Seed wall

(c) Seed formation

Apical meristem

Cotyledons

(e) Meristematic development and morphogenesis

Apical meristem

(d) Germination

FIGURE 19.8
The path of plant development. The developmental stages of *Arabidopsis thaliana* are (*a*) early cell division, (*b*) tissue formation, (*c*) seed formation, (*d*) germination, and (*e*) meristematic development and morphogenesis.

Nematode Development

One of the most completely described models of development is the tiny nematode *Caenorhabditis elegans*. Only about 1 millimeter long, it consists of 959 somatic cells and has about the same amount of DNA as *Drosophila*. The entire genome has been mapped as a series of overlapping fragments, and a serious effort is under way to determine the complete DNA sequence of the genome.

Because *C. elegans* is transparent, individual cells can be followed as they divide. By observing them, researchers have learned how each of the cells that make up the adult worm is derived from the fertilized egg. As shown on the lineage map in figure 19.9*a*, the egg divides into two, and then its daughter cells continue to divide. Each horizontal line on the map represents one round of cell division. The length of each vertical line represents the time between cell divisions, and the end of each vertical line represents one fully differentiated cell. In figure 19.9*b*, the major organs of

the worm are color-coded to match the colors of the corresponding groups of cells on the lineage map.

Some of these differentiated cells, such as some of the cells that generate the worm's external cuticle, are "born" after only 8 rounds of cell division; other cuticle cells require as many as 14 rounds. The cells that make up the worm's pharynx, or feeding organ, are born after 9 to 11 rounds of division, while cells in the gonads require up to 17 divisions.

Exactly 302 nerve cells are destined for the worm's nervous system. Exactly 131 cells are programmed to die, mostly within minutes of their birth. The fate of each cell is the same in every *C. elegans* individual, except for the cells that will become eggs and sperm.

The nematode develops 959 somatic cells from a single fertilized egg in a carefully orchestrated series of cell divisions, which have been carefully mapped by researchers.

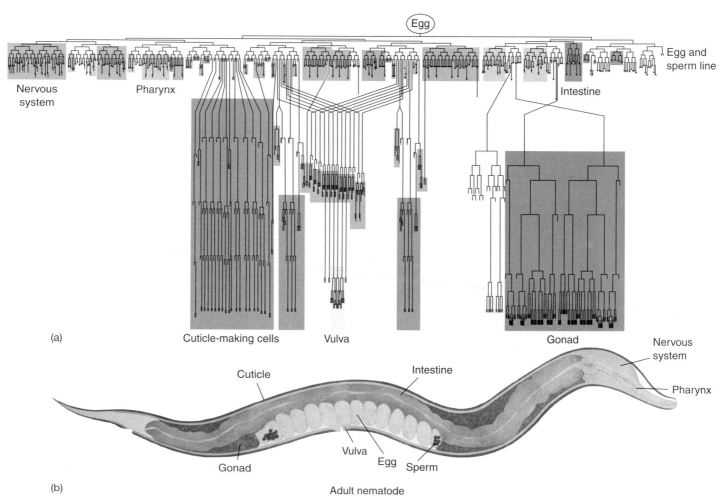

FIGURE 19.9
Studying development in the nematode. Development in *C. elegans* has been mapped out such that the fate of each cell from the single egg cell has been determined. (*a*) The lineage map shows the number of cell divisions from the egg, and the color coding links their placement in (*b*) the adult organism.

19.2 Multicellular organisms employ the same basic mechanisms of development.

Despite the many differences in the four developmental paths we have just discussed, it is becoming increasingly clear that most multicellular organisms develop according to molecular mechanisms that are fundamentally very similar. This observation suggests that these mechanisms evolved very early in the history of multicellular life. Here, we focus on six mechanisms that seem to be of particular importance in the development of a wide variety of organisms: cell movement, induction, determination, pattern formation, homeotic genes, and programmed cell death. We will consider them in roughly the order in which they first become important during development.

Cell Movement and Induction

Cell Movement

The migration of cells is important during many stages of development, from the early events of gastrulation to the formation of the nervous system during morphogenesis. The movement of cells involves both adhesion and loss of adhesion. Adhesion is necessary for cells to get "traction," but cells that are initially attached must lose this adhesion to be able to leave a site. Cell movement also involves both cell-to-cell interactions and cell-to-substrate interactions. Cell-to-cell interactions are often mediated through cadherins (introduced in chapter 7), while cell-to-substrate interactions often involve integrin-to-extracellular-matrix (ECM) interactions. Integrin was initially identified as the receptor for the ECM molecule fibronectin.

Cadherins have proved to be a large gene family, with over 80 members identified in humans. In the genomes of *Drosophila*, *C. elegans*, and humans, the cadherins can be sorted into four subfamilies that exist in all three genomes. These subfamilies can be further subdivided by both their function and species distribution. The cadherin family is not found in yeast or *Dictyostelium*, implying that they are a metazoan innovation. These proteins are all transmembrane proteins that share a common motif, the *cadherin domain*, a 110-amino-acid domain in the extracellular portion of the protein that mediates Ca^{++}-dependent binding between like cadherins.

Experiments in which cells are allowed to sort in vitro illustrate the function of cadherins. Cells with the same cadherins will adhere specifically while not adhering to other cells with different cadherins. If cell populations with different cadherins are disaggregated and then allowed to reaggregate, they will sort into two populations of cells based on the nature of the cadherins on their surface. This experiment can be extended by using cells with different amounts of the same cadherins: The cells will sort such that the cells with the most cadherins will be in the interior of the cell mass while the cells on the outside will have fewer cadherins.

An example of the action of cadherins can be seen in the development of the vertebrate nervous system. The neural-specific cadherin N-cadherin is expressed in the neuro-ectoderm during neurulation. The expression of N-cadherin is not essential for this process, but is involved in the sorting of future neural cells versus ectoderm. During later events in neurogenesis, a number of neural-specific cadherins are expressed in different subsets of neural tissue. These cadherins appear to be involved in the sorting of different subsets of the developing nervous system. This pattern of cadherins with spatially constrained expression is a common theme in development. It is not clear whether the pattern is causal or a correlation. The experiments being attempted to test this causality involve removing gene function. However, because "knocking out" a gene can cause multiple defects, the experimental results are as yet difficult to interpret.

In some tissues, such as connective tissue, much of the volume of the tissue is taken up by the spaces *between* cells. These spaces are not vacant, however. Rather, they are filled with a network of molecules secreted by surrounding cells, principally a matrix of long polysaccharide chains covalently linked to proteins (proteoglycans), within which are embedded strands of fibrous protein (collagen, elastin, and fibronectin). Migrating cells traverse this matrix by binding to it with cell surface proteins called integrins, which were also described in chapter 7. Integrins are attached to actin filaments of the cytoskeleton and protrude out from the cell surface in pairs, like two hands. The "hands" grasp a specific component of the matrix such as collagen or fibronectin, thus linking the cytoskeleton to the fibers of the matrix. In addition to providing an anchor, this binding can initiate changes within the cell, alter the growth of the cytoskeleton, and change the way the cell secretes materials into the matrix.

Thus, cell migration is largely a matter of changing patterns of cell adhesion. As a migrating cell travels, it continually extends projections that probe the nature of its environment. Tugged this way and that by different tentative attachments, the cell literally feels its way toward its ultimate target site.

Induction

In *Drosophila*, the initial cells created by cleavage divisions contain different developmental signals (called **determinants**) from the egg, setting individual cells off on different developmental paths. This pattern of development is called **mosaic development.** In mammals, by contrast, all of the blastomeres receive equivalent sets of determinants; body form is determined by cell–cell interactions, a pattern called **regulative development.**

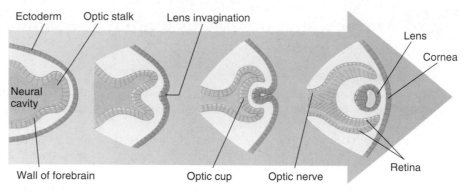

We can demonstrate the importance of cell–cell interactions in development by separating the cells of an early blastula and allowing them to develop independently. Under these conditions, animal pole blastomeres develop features of ectoderm, and vegetal pole blastomeres develop features of endoderm, but none of the cells ever develop features characteristic of mesoderm. However, if animal pole and vegetal pole cells are placed next to each other, some of the animal pole cells will develop as mesoderm. The interaction between the two cell types triggers a switch in the developmental path of the cells! When a cell switches from one path to another as a result of interaction with an adjacent cell, **induction** has taken place (figure 19.10).

How do cells induce developmental changes in neighboring cells? Apparently, the inducing cells secrete proteins that act as intercellular signals. Signal molecules, which we discussed in detail in chapter 7, are capable of producing abrupt changes in the patterns of gene transcription.

In some cases, particular groups of cells called **organizers** produce diffusible signal molecules that convey positional information to other cells. Organizers can have a profound influence on the development of surrounding tissues (see chapter 51). Working as signal beacons, they inform surrounding cells of their distance from the organizer. The closer a particular cell is to an organizer, the higher the concentration of the signal molecule, or **morphogen,** it experiences (figure 19.11). Although only a few morphogens have been isolated, they are thought to be part of a widespread mechanism for determining relative position during development.

A single morphogen can have different effects, depending upon how far away from the organizer the affected cell is located. Thus, low levels of the morphogen activin will cause cells of the animal pole of an early *Xenopus* embryo to develop into epidermis, while slightly higher levels will induce the cells to develop into muscles, and levels a little higher than that will induce them to form notochord (figure 19.12).

Cells migrate by extending probes to neighboring cells, which they use to pull themselves along. Interactions between cells strongly influence the developmental paths they take. Signal molecules from an inducing cell alter patterns of transcription in the cells that come in contact with it.

FIGURE 19.10
Development of the vertebrate eye by induction. The eye develops as an extension of the forebrain called the optic stalk that grows out until it contacts the ectoderm. This contact induces the formation of a lens from the ectoderm.

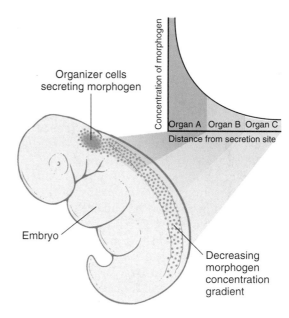

FIGURE 19.11
An organizer creates a morphogen gradient. As a morphogen diffuses from the organizer site, it becomes less concentrated. Different concentrations of the morphogen stimulate the development of different organs.

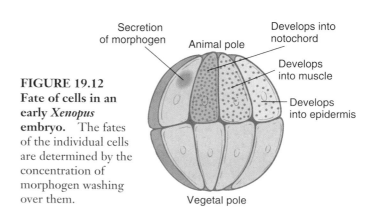

FIGURE 19.12
Fate of cells in an early *Xenopus* embryo. The fates of the individual cells are determined by the concentration of morphogen washing over them.

Determination

The vertebrate egg is symmetrical in its contents as well as its shape, so that all of the cells of an early blastoderm are equivalent up to the eight-cell stage. The cells are said to be **totipotent,** meaning that they are potentially capable of expressing all of the genes of their genome. If they are separated from one another, any one of them can produce a completely normal individual. Indeed, just this sort of procedure has been used to produce sets of four or eight identical offspring in the commercial breeding of particularly valuable lines of cattle. The reverse process works too; if cells from two different eight-cell-stage embryos are combined, a single normal individual results. Such an individual is called a **chimera,** because it contains cells from different genetic lines (figure 19.13).

Vertebrate cells start to become different after the eight-cell stage as a result of cell–cell interactions like those we just discussed. At this point, the pathway that will influence the future developmental fate of the cells is determined. The commitment of a particular cell to a specialized developmental path is called **determination.** A cell in the prospective brain region of an amphibian embryo at the early gastrula stage has not yet been determined; if transplanted elsewhere in the embryo, it will develop like its new neighbors (see chapter 51). By the late gastrula stage, however, determination has taken place, and the cell will develop as neural tissue no matter where it is transplanted. Determination must be carefully distinguished from **differentiation,** which is the cell specialization that occurs at the end of the developmental path. Cells may become *determined* to give rise to particular tissues long before they actually *differentiate* into those tissues. For example, the cells of a *Drosophila* eye imaginal disc are fully determined to produce an eye, but they remain totally undifferentiated during most of the course of larval development.

The Mechanism of Determination

What is the molecular mechanism of determination? The gene regulatory proteins discussed in detail in chapter 16 are the tools cells use to initiate developmental changes. When genes encoding these proteins are activated, one of their effects is to reinforce their own activation. This makes the developmental switch deterministic, initiating a chain of events that leads down a particular developmental pathway. Cells in which a set of regulatory genes have been activated may not actually undergo differentiation until some time later, when other factors interact with the regulatory protein and cause it to activate still other genes. Nevertheless, once the initial "switch" is thrown, the cell is fully committed to its future developmental path.

Often, before a cell becomes fully committed to a particular developmental path, it first becomes partially commit-

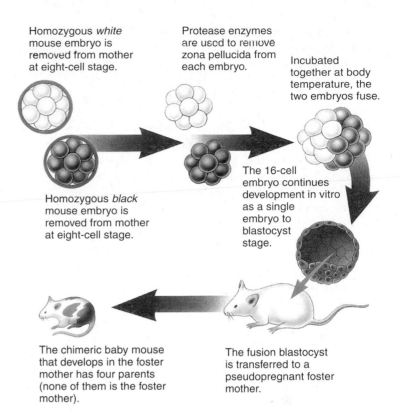

FIGURE 19.13
Constructing a chimeric mouse. Cells from two eight-cell individuals fuse to form a single individual.

ted, acquiring **positional labels** that reflect its location in the embryo. These labels can have a great influence on how the pattern of the body subsequently develops. In a chicken embryo, if tissue at the base of the leg bud (which would normally give rise to the thigh) is transplanted to the tip of the identical-looking wing bud (which would normally give rise to the wing tip), that tissue will develop into a toe rather than a thigh! The tissue has already been determined as leg but is not yet committed to being a particular part of the leg. Therefore, it can be influenced by the positional signaling at the tip of the wing bud to form a tip (in this case, a tip of leg).

Is Determination Irreversible?

Until very recently, biologists thought determination was irreversible. Experiments carried out in the 1950s and 1960s by John Gurdon and others made what seemed a convincing case: Using very fine pipettes (hollow glass tubes) to suck the nucleus out of a frog or toad egg, these researchers replaced the egg nucleus with a nucleus sucked out of a body cell taken from another individual (see figure 14.3). If the transplanted nucleus was obtained from an advanced embryo, the egg went on to develop into a tadpole, but most died before becoming an adult.

Nuclear transplant experiments were attempted without success by many investigators, until finally, in 1984, Steen Willadsen, a Danish embryologist working in Texas, succeeded in cloning a sheep using the nucleus from a cell of an early embryo. The key to his success was in picking a cell very early in development. This exciting result was soon replicated by others in a host of other organisms, including pigs and monkeys.

Only early embryo cells seemed to work, however. Researchers became convinced, after many attempts to transfer older nuclei, that animal cells become irreversibly committed after the first few cell divisions of the developing embryo.

We now know this conclusion to have been unwarranted. The key advance unraveling this puzzle was made in Scotland by geneticists Keith Campbell and Ian Wilmut, who reasoned that perhaps the egg and the donated nucleus needed to be at the same stage in the cell cycle. They removed mammary cells from the udder of a six-year-old sheep. The cells were grown in tissue culture; then, in preparation for cloning, the researchers substantially reduced for five days the concentration of serum nutrients on which the sheep mammary cells were subsisting. Starving the cells caused them to pause at the beginning of the cell cycle. In parallel preparation, eggs obtained from a ewe were enucleated (figure 19.14).

Mammary cells and egg cells were surgically combined in January of 1996. A little over five months later, on July 5, 1996, one sheep gave birth to a lamb named Dolly, the first clone generated from a fully differentiated animal cell. Dolly established beyond all dispute that determination is reversible—that with the right techniques, the fate of a fully differentiated cell *can* be altered.

Since Dolly, clones have been created for the following organisms: mouse, cow, pig, goat, cat, and rabbit. All of these used some form of adult cell. The efficiency in all cases is quite low, on the order of 3 to 5% viable adults from transferred nuclei. As discussed in chapter 20, we have much to learn about the events necessary to reprogram an adult nucleus to act as an embryonic nucleus.

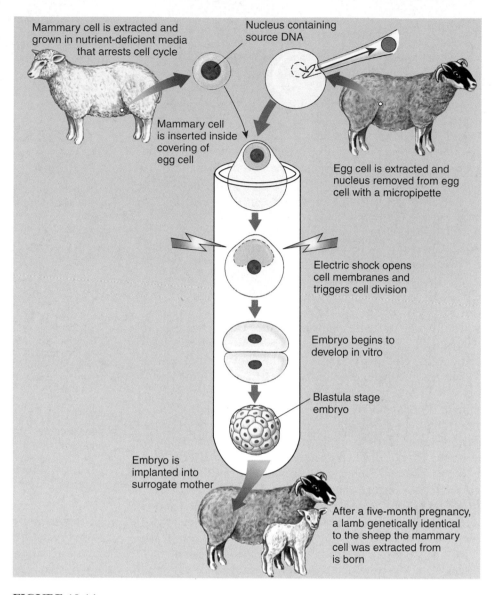

FIGURE 19.14
Proof that determination is reversible.   This experiment by Campbell and Wilmut, the first successful cloning of an adult animal, shows that a differentiated adult cell can be used to drive all of development.

The commitment of particular cells to certain developmental fates is fully reversible.

Pattern Formation

All animals seem to use positional information to determine the basic pattern of body compartments and, thus, the overall architecture of the adult body. How is positional information encoded in labels and read by cells? To answer this question, let us consider how positional labels are used in pattern formation in *Drosophila*. The Nobel Prize in Physiology or Medicine was awarded in 1995 for the unraveling of this puzzle (figure 19.15).

Pattern formation is an unfolding process during development. In the later stages, it may involve morphogenesis of organs, but during the earliest events of development, it lays down the basic body plan, the establishment of the

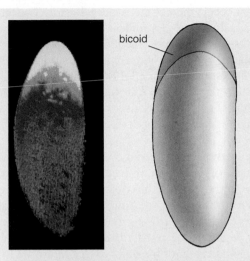

bicoid

Establishing the polarity of the embryo: Fertilization of the egg triggers the production of bicoid protein from maternal RNA in the egg. The bicoid protein diffuses through the egg, forming a gradient. This gradient determines the polarity of the embryo, with the head and thorax developing in the zone of high concentration (*yellow* through *red* in photo).

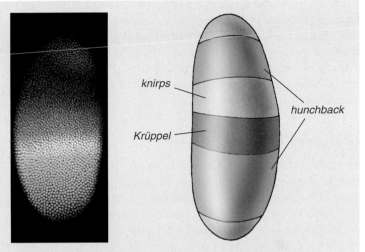

knirps

Krüppel

hunchback

Setting the stage for segmentation: About 2½ hours after fertilization, bicoid protein turns on a series of brief signals from so-called gap genes. The gap proteins act to divide the embryo into large blocks. In this photo, fluorescent dyes in antibodies that bind to the gap proteins Krüppel *(red)* and hunchback *(green)* make the blocks visible; the region of overlap is yellow.

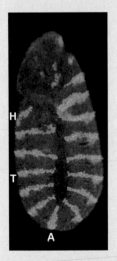

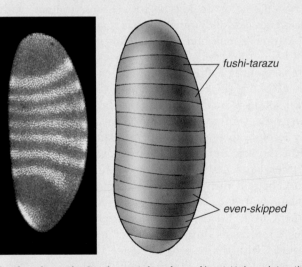

fushi-tarazu

even-skipped

Laying down the fundamental regions: About ½ hour later, the gap genes switch on a so-called "pair-rule" gene called *hairy*. Hairy produces a series of boundaries within each block, dividing the embryo into seven fundamental regions.

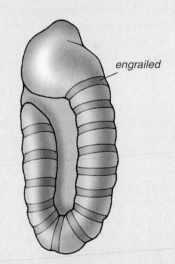

engrailed

H

T

A

Forming the segments: The final stage of segmentation occurs when a "segment-polarity" gene called *engrailed* divides each of the seven regions into halves, producing 14 narrow compartments. Each compartment corresponds to one segment of the future body. There are three head segments (H, top left), three thoracic segments (T, lower left), and eight abdominal segments (A, from bottom left to upper right).

FIGURE 19.15
Body organization in an early *Drosophila* embryo. In these images by 1995 Nobel laureate Christiane Nüsslein-Volhard and Sean Carroll, we watch a *Drosophila* egg pass through the early stages of development, in which the basic segmentation pattern of the embryo is established. The drawings help illustrate what is occurring in the photos.

anterior/posterior (A/P) axis and the dorsal/ventral (D/V) axis. Thus, pattern formation can be considered the process of taking a radially symmetrical cell and imposing two perpendicular axes to define the basic body plan, which in this way becomes bilaterally symmetrical. The differences in the development of protostomes and deuterostomes initially led investigators to the idea that pattern formation must be fundamentally different processes in these divergent species. Even among vertebrates, there appeared to be significant differences between the amphibian *Xenopus*, which has an egg cell with significant polarity, and mammals, which exhibit highly regulative development and egg cells with no apparent polarity. Recently, some interesting parallels have been observed, although the mechanisms in different lineages remain distinctive.

Drosophila Pattern Formation

Forming the Axis. The invertebrate *Drosophila* is the best understood in terms of the events of early patterning. As will be described later, a hierarchy of gene expression that begins with maternally expressed genes controls the development of *Drosophila*. Two signaling pathways control the establishment of A/P and D/V polarity. The A/P axis is formed based on a gradient of the **bicoid** protein. This protein gradient is established by an interesting mechanism: Follicle cells provide maternally produced *bicoid* mRNA that becomes anchored at the anterior pole. Translation of this spatially localized message then produces an anterior-to-posterior gradient of the protein. The posterior determining protein **oskar** is also localized based on prior localization of the *oskar* mRNA. The localization of both of these mRNAs is achieved by attachment to microtubules, as shown by the action of agents that disrupt microtubules. The motor protein dynein appears to be necessary for the localization of bicoid, while oskar localization appears to be dependent on the motor protein kinesin. The action of the bicoid protein is then to control the expression of so-called gap genes that are expressed in the zygote (discussed later).

The dorsal/ventral axis in *Drosophila* is controlled by a different mechanism that does not involve localized cytoplasmic determinants like bicoid. Instead, an accumulation of message on one side of the nucleus activates a transcription factor that results in directional translation and the accumulation of protein product on that surface of the ooctye. The *gurken* mRNA accumulates on one surface of the nucleus such that, when translated, it accumulates on the membrane on one side of the cell. This will be the future dorsal side of the embryo; however, it is not directly specified by gurken—rather, *gurken* signals cause overlying follicle cells to adopt a dorsal fate. These cells then produce a signal that results in the selected transport of the transcription factor *dorsal* into the nucleus of ventral cells. This transcription factor then acts to produce a ventral to dorsal protein gradient similar to the signaling by bicoid.

While utilizing profoundly different mechanisms, the unifying factor controlling the establishment of both A/P and D/V polarity is that *gurken* and *bicoid* are both maternally expressed genes so that the polarity of the future embryo in both instances is laid down in the oocyte using information coming from the maternal genome. Both cases also involve an interaction between the future oocyte and the follicle cells. While the preceding is in some respects an oversimplification of the events that occur, the outline is clear: Polarity is established by an interaction between follicle cells and the oocyte, an interaction that results in the creation of morphogen gradients in the oocyte based on maternal information. These gradients then drive the expression of the zygotic genes that will actually pattern the embryo. This reliance on a hierarchy of regulatory genes, and on response to the interactions of cells, are unifying themes for all of development.

Producing the Body Plan. Bicoid protein exerts this profound effect on the organization of the embryo by activating genes that encode the first mRNAs to be transcribed after fertilization. Within the first two hours, before cellularization of the syncytial blastoderm, a group of six genes called the **gap genes** begins to be transcribed. These genes map out the coarsest subdivision of the embryo (see figure 19.15). One of them is a gene called *hunchback* (because an embryo without *hunchback* lacks a thorax and so, takes on a hunched shape). Although *hunchback* mRNA is distributed throughout the embryo, its translation is controlled by the protein product of another maternal mRNA called *nanos* (named after the Greek word for "dwarf," because mutants without *nanos* genes lack abdominal segments and, hence, are small). The **nanos** protein binds to *hunchback* mRNA, preventing it from being translated. The only place in the embryo where there is too little nanos protein to block translation of *hunchback* mRNA is the far anterior end. Consequently, hunchback protein is made primarily at the anterior end of the embryo. As it diffuses back toward the posterior end, it sets up a second morphogen gradient responsible for establishing the thoracic and abdominal segments.

Other gap genes act in more posterior regions of the embryo. They, in turn, activate 11 or more **pair-rule genes.** (When mutated, each of these genes alters every other body segment.) One of the pair-rule genes, named *hairy*, produces seven bands of protein, which look like stripes when visualized with fluorescent markers. These bands establish boundaries that divide the embryo into seven zones. Finally, a group of 16 or more **segment polarity genes** subdivide these zones. The *engrailed* gene, for example, divides each of the seven zones established by *hairy* into anterior and posterior compartments. The 14 compartments that result correspond to the three head segments, three thoracic segments, and eight abdominal segments of the embryo.

Thus, within three hours after fertilization, a highly orchestrated cascade of segmentation gene activity produces the fly embryo's basic body plan. The activation of these and other developmentally important genes (figure 19.16)

FIGURE 19.16

A gene controlling organ formation in *Drosophila*. Called *tinman*, this gene is responsible for the formation of gut musculature and the heart. The dye shows expression of *tinman* in (*a*) five-hour and (*b*) seventeen-hour *Drosophila* embryos. (*c*) The gut musculature then appears along the edges of normal embryos, but (*d*) is not present in embryos in which the gene has been mutated. (*e*) The heart tissue develops along the center of normal embryos but (*f*) is missing in *tinman* mutant embryos.

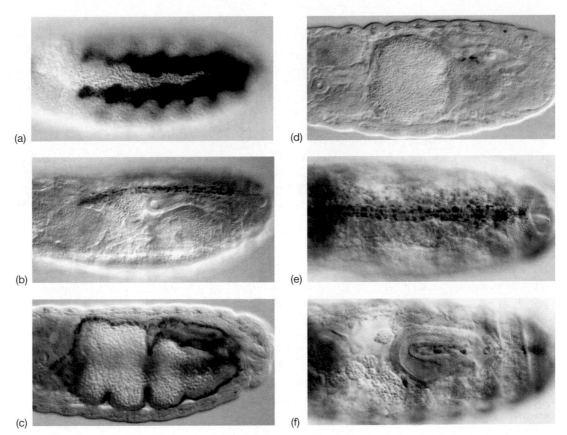

(a) (b) (c) (d) (e) (f)

depends upon the free diffusion of morphogens that is possible within a syncytial blastoderm. In mammalian embryos with cell partitions, other mechanisms must operate.

Vertebrate Axis Formation

Development of at least one vertebrate, the frog *Xenopus*, now appears to have more in common with *Drosophila* than anticipated. The *Xenopus* oocyte has intrinsic polarity, an animal pole and a vegetal pole; these are not directly related to the future A/P or D/V axes but do imply the existence of cytoplasmic determinants. The D/V axis has always been the most interesting to embryologists because it contains the famous Spemann organizer. In experiments done in the 1920s, Spemann and Mangold showed that a bisected embryo that contained the dorsal half developed normally. This was refined to show that the dorsal lip of the blastopore alone could be transplanted to produce a second embryonic axis. The molecular nature of this organizer has been of interest ever since its discovery. The earliest events that lead to the determination of the D/V axis are related to the fertilization event. The side of the embryo opposite the sperm entry point will become the future dorsal side. The egg reorganizes based on a signal from the point of entry of the sperm. The egg rotates, and microtubules emanating from the point of sperm entry result in the movement of dorsal determinants from the vegetal pole to the side of the egg opposite the site of sperm entry. This

ultimately results in the formation of the organizer on the dorsal surface. The signaling events following this involve a transforming growth factor pathway, and the so-called *wnt* pathway. These pathways are also involved in *Drosophila* patterning downstream of the dorsal transcription factor.

The mammalian embryo has long been known to lack obvious cytoplasmic determinants. Mammals are an example of regulative development, in which development unfolds based on interactions between cells. Yet, even in this case, new evidence indicates that patterning may arise earlier than previously thought. The mouse embryo can be manipulated in a number of ways that do not perturb development: disruption of blastomeres, removal of polar cytoplasm, even removal of blastomeres up to the 8-cell stage. However, recent evidence indicates that the site of sperm entry is "remembered" by the developing embryo and influences later events. The site of sperm entry is later translated into the proximo-distal axis by signaling from the extraembryonic cells of the embryo during early development. All of this indicates that the differences in development we see represent evolution in different lineages, but that a limited number of actual mechanisms may have been used in many different ways. This will be amplified later in this chapter when we discuss homeobox-containing genes.

In *Drosophila*, diffusion of chemical inducers produces the embryo's basic body plan, a cascade of genes dividing it into 14 compartments.

Homeotic Genes

Role of Homeotic Genes

With the basic body plan laid down by the mechanisms described earlier, the next step is to give identity to the segments of the embryo. The outline of *Drosophila* pattern formation produces a segmented embryo with A/P and D/V polarity. The genes that then give rise to segment identity are the homeobox-containing genes. Mutations in homeotic genes lead to the appearance of perfectly normal body parts in unusual places. For example, mutations in *bithorax* (figure 19.17) cause a fly to grow an extra pair of wings, as if it had a double thoracic segment, and mutations in *Antennapedia* cause legs to grow out of the head in place of antennae! In the early 1950s, geneticist Edward Lewis discovered that several homeotic genes, including *bithorax*, map together on the third chromosome of *Drosophila* in a tight cluster called the **bithorax complex.** Mutations in these genes all affect body parts of the thoracic and abdominal segments, and Lewis concluded that the genes of the bithorax complex control the development of body parts in the rear half of the thorax and all of the abdomen. Most interestingly, the order of the genes in the bithorax complex mirrors the order of the body parts they control, as if the genes are activated serially! Genes at the beginning of the cluster switch on development of the thorax; those in the middle control the anterior part of the abdomen; and those at the end affect the tip of the abdomen. A second cluster of homeotic genes, the **Antennapedia complex,** was discovered in 1980 by Thomas Kaufmann. The Antennapedia complex governs the anterior end of the fly, and the order of genes in it also corresponds to the order of segments they control (figure 19.18).

FIGURE 19.17
Mutations in homeotic genes.   Three separate mutations in the *bithorax* gene caused this fruit fly to develop an extra thoracic segment, with accompanying wings. Compare this photograph with that of the normal fruit fly in figure 19.6.

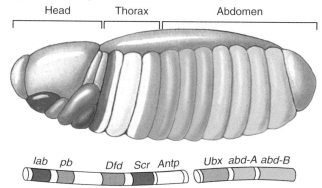

Drosophila embryo

Head   Thorax   Abdomen

lab   pb   Dfd   Scr Antp   Ubx abd-A abd-B

Drosophila HOM genes

FIGURE 19.18
***Drosophila* homeotic genes.**   Called the homeotic gene complex, or HOM complex, the genes are grouped into two clusters, the Antennapedia complex (anterior) and the bithorax complex (posterior).

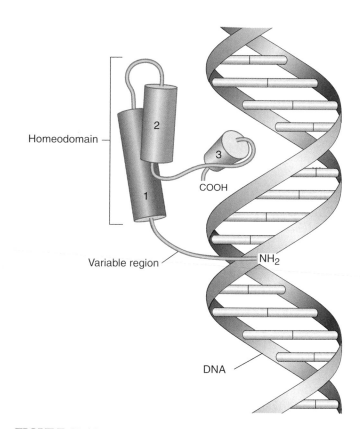

FIGURE 19.19
Homeodomain protein.   This protein plays an important regulatory role when it binds to DNA and regulates expression of specific genes. The variable region of the protein determines the specific activity of the protein. The homeodomain, a 60-amino-acid sequence common to all proteins of this type, is coded for by the homeobox region of genes and is composed of three α-helices. One of the helices (number 3) recognizes and binds to a specific DNA sequence in target genes.

The Homeobox

Drosophila homeotic genes typically contain the **homeobox,** a sequence of 180 nucleotides that codes for a 60-amino-acid, DNA-binding peptide domain called the homeodomain (figure 19.19). As we saw in chapter 18, proteins that contain the homeodomain function as transcription factors, ensuring that developmentally related genes are transcribed at the appropriate time. Segmentation genes such as *bicoid* and *engrailed* also contain the homeobox sequence. Clearly, the homeobox distinguishes those portions of the genome devoted to pattern formation.

Evolution of Homeobox-Containing Genes

A large amount of research has been devoted to analyzing homeobox-containing genes, or **Hox genes,** in many different organisms. These investigations have led to a fairly coherent view of homeotic gene evolution. It is now clear that the *Drosophila* bithorax and Antennapedia complexes represent two parts of a single cluster of genes. There are four copies of *Hox* gene clusters in vertebrate species. As in *Drosophila*, the spatial domains of *Hox* gene expression correlate with the order of the genes on the chromosome (figure 19.20). The existence of four *Hox* clusters in vertebrates is viewed by many as evidence that two duplication events of the entire genome must have occurred in the vertebrate lineage. This raises the issue of when the original cluster arose. To answer this question, researchers have turned to more and more primitive organisms, first to the primitive chordate *Amphioxus* (now called *Branchiostoma*). The finding of one cluster of *Hox* genes in *Amphioxus* implies that indeed there have been two duplications of the vertebrate lineage, at least of the *Hox* cluster in the vertebrates. Given the single cluster in arthropods, this implies that the common ancestor to all bilaterians had a single *Hox* cluster as well. The next logical step is to look at the next available outgroup: cnidarians and ctenophores. Thus far, *Hox* genes have been found in a number of cnidarian species, but their organization is not yet clear. The analysis is confused by the presence of genes that contain homeoboxes but are not in a *Hox* cluster. Such genes exist in vertebrate genomes and *Drosophila* as well, but their relationship to the ancestral *Hox* cluster is not known.

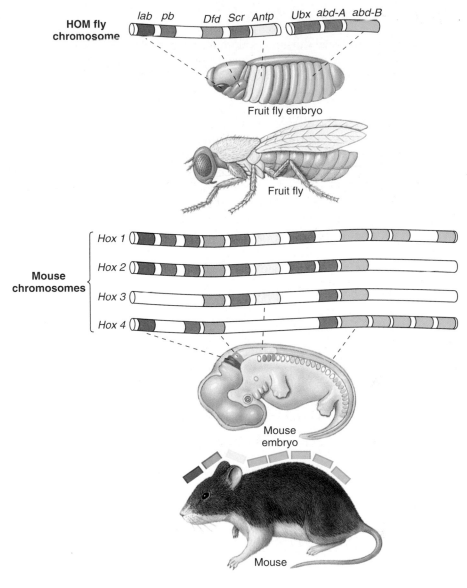

FIGURE 19.20
A comparison of homeotic gene clusters in the fruit fly *Drosophila melanogaster* and the mouse *Mus musculus*. Similar genes, the *Drosophila* HOM genes and the mouse *Hox* genes, control the development of front and back parts of the body. These genes are located on a single chromosome in the fly and on four separate chromosomes in mammals. On this illustration, the genes are color-coded to match the parts of the body in which they are expressed.

Homeotic genes encode transcription factors that activate blocks of genes specifying particular body parts.

Programmed Cell Death

Not every cell produced during development is destined to survive. For example, the cells between your fingers and toes die; if they did not, you would have paddles rather than digits. Vertebrate embryos produce a very large number of neurons, ensuring that there are enough neurons to make all of the necessary synaptic connections, but over half of these neurons never make connections and die in an orderly way as the nervous system develops. Unlike accidental cell deaths due to injury, these cell deaths are planned—and indeed required—for proper development. Cells that die due to injury typically swell and burst, releasing their contents into the extracellular fluid. This form of cell death is called **necrosis.** In contrast, cells programmed to die shrivel and shrink in a process called **apoptosis** (Greek, "shedding of leaves in autumn"), and their remains are taken up by surrounding cells.

Gene Control of Apoptosis

This sort of developmentally regulated cell suicide occurs when a "death program" is activated. All animal cells appear to possess such programs. In the nematode worm, for example, the same 131 cells always die during development in a predictable and reproducible pattern of apoptosis. Three genes govern this process. Two (*ced-3* and *ced-4*) constitute the death program itself; if either is mutant, those 131 cells do not die, and go on instead to form nervous and other tissue. The third gene (*ced-9*) represses the death program encoded by the other two (figure 19.21*a*). The same sorts of apoptosis programs occur in human cells: The *bax* gene encodes the cell death program, and another, an oncogene called *bcl-2*, represses it (figure 19.21*b*). The mechanism of apoptosis appears to have been highly conserved during the course of animal evolution. The protein made by *bcl-2* is 25% identical in amino acid sequence to that made by *ced-9*. If a copy of the human *bcl-2* gene is transferred into a nematode with a defective *ced-9* gene, *bcl-2* suppresses the cell death program of *ced-3* and *ced-4!*

How does *bax* kill a cell? The bax protein seems to induce apoptosis by binding to the permeability pores of the cell's mitochondria, increasing the mitochondrion's permeability and, in doing so, triggering cell death. How does *bcl-2* prevent cell death? One suggestion is that it prevents damage from *free radicals*, highly reactive fragments of atoms that can damage cells severely. Proteins or other molecules that destroy free radicals are called **antioxidants.** Antioxidants are almost as effective as *bcl-2* in blocking apoptosis.

Animal development involves programmed cell death (apoptosis), in which particular genes, when activated, kill their cells.

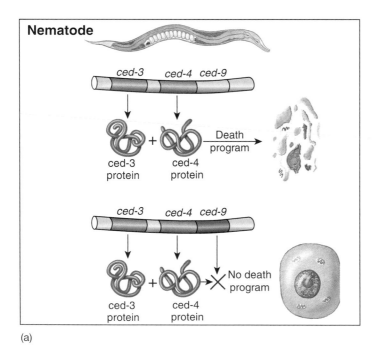

(a)

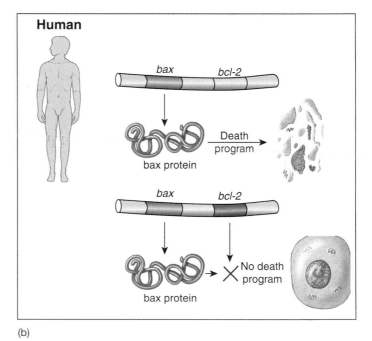

(b)

FIGURE 19.21

Programmed cell death.   Apoptosis, or programmed cell death, is necessary for the normal development of all animals. (*a*) In the developing nematode, for example, two genes, *ced-3* and *ced-4*, code for proteins that cause the programmed cell death of 131 specific cells. In the other cells of the developing nematode, the product of a third gene, *ced-9*, represses the death program encoded by *ced-3* and *ced-4*. (*b*) In developing humans, the product of a gene called *bax* causes a cell death program in some cells and is blocked by the *bcl-2* gene in other cells.

19.3 Aging can be considered a developmental process.

Theories of Aging

All humans age, and eventually die. As you can see in figure 19.22, the "safest" age is around puberty: 10- to 15-year-olds have the lowest risk of dying. The death rate begins to increase rapidly after puberty as an exponential function of increasing age. Plotted on a log scale as in figure 19.22 (a so-called Gompertz plot), the mortality rate increases as a straight line from about 15 to 90 years, doubling about every eight years (the "Gompertz number"). By the time we reach 100, age has taken such a toll that the risk of dying reaches 50% per year.

A wide variety of theories have been advanced to explain why humans and other animals age. No single theory has gained general acceptance, but the following five are being intensively investigated.

Accumulated Mutation Hypothesis

The oldest general theory of aging is that cells accumulate mutations as they age, leading eventually to lethal damage. Careful studies have shown that somatic mutations do indeed accumulate during aging. As cells age, for example, they tend to accumulate the modified base 8-hydroxyguanine, in which an OH group is added to the base guanine. There is little direct evidence, however, that these mutations *cause* aging. No acceleration in aging occurred among survivors of Hiroshima and Nagasaki despite their enormous added mutation load, arguing against any general relationship between mutation and aging.

Telomere Depletion Hypothesis

In a seminal experiment carried out in 1961, Leonard Hayflick demonstrated that fibroblast cells growing in tissue culture will divide only a certain number of times (figure 19.23). After about 50 population doublings, cell division stops—the cell cycle is blocked just before DNA replication. If a cell sample is taken after 20 doublings and frozen, when thawed it resumes growth for 30 more doublings, and then stops.

An explanation of the "Hayflick limit" was suggested in 1986 when Howard Cooke first glimpsed an extra length of DNA at the ends of chromosomes. These **telomeres**, repeats of the sequence TTAGGG, were found to be substantially shorter in older somatic tissue, and Cooke speculated that a 100-base-pair portion of the telomere cap was lost by a chromosome during each cycle of DNA replication. Eventually, after some 50 replication cycles, the protective telomeric cap would be used up, and the cell line would then enter senescence, no longer able to proliferate. Cancer cells appear to avoid telomeric shortening.

Research reported in 1998 has confirmed Cooke's hypothesis, providing direct evidence for a causal relationship between telomeric shortening and cell senescence. Using genetic engineering, researchers transferred into human primary cell cultures a gene that leads to expression of telomerase, an enzyme that builds TTAGGG telomeric caps.

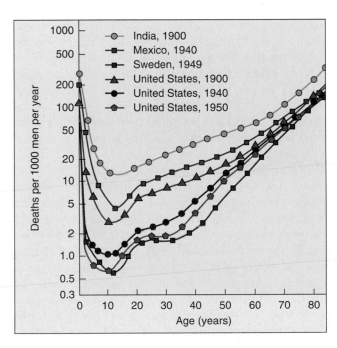

FIGURE 19.22
Gompertz curves. While human populations may differ 25-fold in their mortality rates before puberty, the slopes of their Gompertz curves are about the same in later years.
How do you explain this difference?

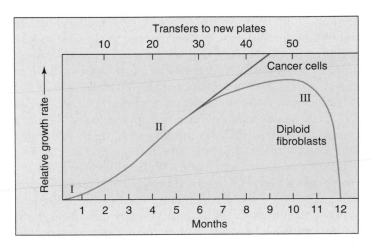

FIGURE 19.23
Hayflick's experiment. Fibroblast cells stop growing after about 50 doublings. Growth is rapid in phases I and II, but slows in phase III, as the culture becomes senescent, until the final doubling. Cancer cells, by contrast, do not "age."
How do cancer cells overcome this 50-division "Hayflick limit?"

The result was unequivocal. New telomeric caps were added to the chromosomes of the cells, and the cells with the artificially elongated telomeres did not senesce at the Hayflick limit, continuing to divide in a healthy and vigorous manner for more than 20 additional generations.

Wear-and-Tear Hypothesis

Numerous theories of aging focus in one way or another on the general idea that cells wear out over time, accumulating damage until they are no longer able to function. Loosely dubbed the "wear-and-tear" hypothesis, this idea implies that there is no inherent designed-in limit to aging, just a statistical one—that is, disruption, wear, and damage over time erode a cell's ability to function properly.

Considerable evidence indicates that aging cells do accumulate damage. Some of the most interesting evidence concerns free radicals, fragments of molecules or atoms that contain an unpaired electron. Free radicals are very reactive chemically and can be quite destructive in a cell. Free radicals are produced as natural by-products of oxidative metabolism, but most are mopped up by special enzymes that function to sweep the cell interior free of their destructive effects.

One of the most damaging free radical reactions that occurs in cells causes glucose to become linked to proteins, a nonenzymatic process called glycation. Two of the most commonly glycated proteins are collagen and elastin, key components of the connective tissues in our joints. Glycated proteins can cross-link to one another, reducing the flexibility of connective tissues in the joints and producing many of the other characteristic symptoms of aging.

Gene Clock Hypothesis

There is very little doubt that at least some aspects of aging are under the direct control of genes. Just as genes regulate the body's development, so they appear to regulate its rate of aging. Some genes appear to promote longevity. For example, people over 100 years old are five times as likely to carry a mutation in mitochondrial DNA called C150T.

Other genes produce premature aging. For example, in 1996, a gene was identified that is responsible for Werner's syndrome, which produces premature aging in the young and affects some 10 people per million worldwide. The syndrome is named after Otto Werner, who in 1904 reported on a German family affected by premature aging and believed a genetic component was at work. Werner's syndrome appears in adolescence, usually producing death before age 50 of heart attack or one of a variety of rare connective tissue cancers. Located on the short arm of chromosome 8, the gene seems to affect a helicase enzyme involved in the repair of DNA. The gene, which codes for a 1432-amino-acid protein, has been fully sequenced, and four mutant alleles identified. Helicase enzymes are needed to unwind the DNA double helix whenever DNA has to be replicated, repaired, or transcribed. The high incidence of certain cancers among Werner's syndrome patients leads investigators to speculate that the mutant helicase may fail to activate critical tumor-suppressor genes. The potential role of helicases in aging is the subject of heated research.

Research on aging in other animals strongly supports the hypothesis that genes regulate the rate of aging. A *Drosophila melanogaster* gene mutation called *Indy* ("I'm not dead yet") doubles the fruit fly life span from the usual 37 days to an average of 70 days. When researchers isolated the DNA of the *Indy* gene and compared its DNA sequence with the Human Genome Project sequences, they found that the *Indy* gene is 50% similar to a human gene called *dicarboxylate cotransporter*. In humans, dicarboxylate cotransporter proteins move preliminary products of food metabolism (dicarboxylic acids of the Krebs cycle) across membranes to where the food's processing takes place. In mutant *Indy* flies, poor dicarboxylic acid pumping means that less metabolic energy can be gleaned from the fly's food. In essence, the *Indy* mutation is the genetic equivalent of caloric restriction. Starving is known to prolong life in the nematode *Caenorhabditis elegans*, but *Indy*'s caloric restriction does not involve the unpleasantness of starving. The *Indy* mutation in effect puts flies on a severe diet, while the flies eat as much as normal and lead a normal vigorous life—for far longer.

Current Theories of Aging

An intriguing connection has been found recently that may shed light on the calorie restriction observations. In both *C. elegans* and *Drosophila*, there appears to be a connection between aging and signaling by insulin-like receptors. In both of these organisms, mutants that reduce signaling through insulin-like receptors lead to an increased life span. The relationship between calorie restriction experiments and the insulin-deficient mutants is that insulin is involved in a complex pathway that controls caloric intake. The effect in *Drosophila* is limited to females, implying that trade-offs between fecundity and life span may be related to this pathway.

Work in both of these systems has failed to find genetic evidence for a clocklike mechanism similar to what has been found for circadian rhythms. A number of genes have been identified that either extend or shorten life span, but these seem to be like the insulin case, in which there is no direct counting mechanism. The shortening of chromosomes due to lack of telomerase activity is a form of counting, but only for replications and not for a senescence program. Current thinking is that trade-offs between body size and life span, and between fecundity and life span (at least in females), may be at the heart of the genetics of aging.

Among the theories advanced to explain aging, many involve the progressive accumulation of damage to DNA. When genes affecting aging have been isolated, they affect DNA repair processes.

Concept Review

For interactive testing, visit the Online Learning Center with PowerWeb at www.mhhe.com/Raven7

19.1 Development is a regulated process.

Overview of Development

- Organisms in all three multicellular kingdoms utilize gene expression (i.e., different cells express different genes at different times) to achieve cell specialization. (p. 382)

Vertebrate Development

- At different sites in the vertebrate embryo, particular cells proceed to form the body's tissues and organs, and then the body grows to a size and shape that will allow survival after birth. (p. 383)
- Vertebrate zygotes divide rapidly to form a blastula. Subsequent stages in vertebrate development include gastrulation, neurulation, cell migration, organogenesis, and growth. (pp. 383–385)

Insect Development

- Many insects go through metamorphosis from a larva to a fully functioning, reproductive adult. (p. 386)
- The development of *Drosophila melanogaster* proceeds from an egg to a syncytial blastoderm, to larval instars, which have imaginal discs, and finally through metamorphosis to an adult. (pp. 386–387)

Plant Development

- Plant cells are encased in cellulose walls. Instead of using cell migration, plants develop by building their bodies outward, creating new parts from meristems. (p. 388)
- Plants cope with environmental change by adjusting the type, number, size, and location of its structures to accommodate local circumstances. (p. 388)
- Three basic tissue types of plants are epidermal cells, ground tissue, and vascular tissue. (p. 388)
- Seeds germinate in response to environmental changes due to water, temperature, and other factors. (p. 388)

Nematode Development

- Nematodes develop a known number of somatic cells from a single fertilized egg in a complex series of cell divisions. (p. 390)

19.2 Multicellular organisms employ the same basic mechanisms of development.

Cell Movement and Induction

- Cell movement is largely a matter of changing patterns of cell adhesion. As a migrating cell travels, it continuously extends projections that probe the environment, and thus the cell feels its way toward a target. (p. 391)
- In some cases, particular groups of cells called organizers produce diffusible signal molecules that convey positional information to other cells. (p. 392)
- Induction occurs when a cell switches from one developmental path to another as a result of interactions with an adjacent cell. (p. 392)

Determination

- Totipotent cells are potentially capable of expressing all the genes of their genome. (p. 393)
- Determination is the commitment of a particular cell to a specialized developmental path. (p. 393)
- Recent experiments have shown that determination is reversible. (p. 394)

Pattern Formation

- All animals appear to use positional information to determine the basic pattern of body compartments and the overall body architecture. (p. 395)

Homeotic Genes

- Homeotic genes encode transcription factors that activate blocks of genes specifying particular body parts. (p. 399)

Programmed Cell Death

- Necrosis is cell death due to injury-caused swelling and bursting, while apoptosis refers to programmed cell death. (p. 400)
- The mechanism of apoptosis appears to have been highly conserved during the course of animal evolution. (p. 400)

19.3 Aging can be considered a developmental process.

Theories of Aging

- Several theories of aging that are being intensively investigated include the accumulated mutation hypothesis, the telomere depletion hypothesis, the wear-and-tear hypothesis, the gene clock hypothesis, and trade-offs between body size and life span, and between fecundity and life span. (p. 402)

Self Test

1. Which of the following series of events represents the path of vertebrate development?
   a. formation of blastula, cleavage, neurulation, cell migration, gastrulation, organogenesis, growth
   b. formation of blastula, cleavage, gastrulation, neurulation, cell migration, organogenesis, growth
   c. cleavage, formation of blastula, gastrulation, neurulation, cell migration, organogenesis, growth
   d. cleavage, gastrulation, formation of blastula, neurulation, cell migration, organogenesis, growth

2. Which of the following statements about *Drosophila* development is false?
   a. *Drosophila* go through four larval instar stages before undergoing metamorphosis.
   b. During the syncytial blastoderm stage, nuclei line up along the surface of the egg.
   c. Imaginal discs are groups of cells set aside that will give rise to key parts of the adult fly.
   d. Maternal, rather than zygotic, genes govern early *Drosophila* development.

3. If a plant embryo failed to form enough ground tissue, what function(s) would likely be directly affected in the corresponding mature plant?
   a. seed formation
   b. meristem development
   c. cotyledon formation
   d. food and water storage

4. *C. elegans* is a powerful developmental model because
   a. these nematodes are very small, so it is easy to maintain a large population in a laboratory.
   b. the fate of every cell has been mapped.
   c. the fates of cells that will become eggs and sperm are predetermined.
   d. these nematodes have the same amount of DNA as *Drosophila*.

5. Which of the following best describes a morphogen?
   a. a cell that secretes diffusible signaling molecules that play a role in specifying cell fate
   b. a diffusible signaling molecule that plays a role in specifying cell fate
   c. a protein that helps mediate direct cell–cell interaction
   d. a protein that enables cells to become totipotent

6. What would happen as a result of a transplantation experiment in a chick embryo in which cells determined to become a forelimb were replaced by cells determined to become a hindlimb?
   a. A hindlimb would form in the region where the forelimb should be.
   b. A forelimb would form in the region where the hindlimb should be.
   c. Nothing; the forelimb would form normally.
   d. Neither a forelimb nor a hindlimb would form because the cells were already determined.

7. Which group of genes, identified by Nusslein-Volhard and Caroll, is responsible for the final stages of segmentation in *Drosophila* embryos?
   a. morphogen gradient genes
   b. gap genes
   c. segment-polarity genes
   d. pair-rule genes

8. Suppose that during a mutagenesis screen to isolate mutations in *Drosophila*, you come across a fly with legs growing out of its head. What gene cluster is likely affected?
   a. *Bicoid*
   b. *Hunchback*
   c. *Bithorax*
   d. *Antennapedia*

9. What would be the likely result of a mutation of the *bcl-2* gene on the level of apoptosis?
   a. no change
   b. a decrease in apoptosis
   c. an increase in apoptosis
   d. First, it would increase, but later it would decrease.

10. The gene clock hypothesis is best described by which of the following explanations?
    a. Mutations accumulate partially through the addition of an −OH group to the base guanine.
    b. Specific genes exist to promote longevity.
    c. Free radicals can cause genetic mutations, particularly when we are sleeping.
    d. Calorie restriction leads to an increased life span.

Test Your Visual Understanding

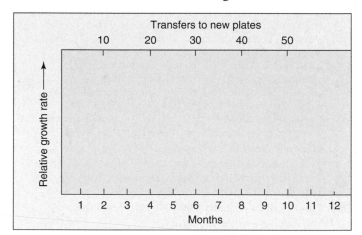

1. Hayflick's experiment revealed that noncancerous cells have a definitive life span, whereas cancerous cells do not have the same restrictions. Draw curves that represent the growth patterns of each cell type.

Apply Your Knowledge

1. You have generated a cell line that expresses an altered form of cadherin. This mutant cadherin has the 110-amino-acid extracellular domain that is required for interaction with other cadherins, but lacks a transmembrane domain. If you were to mix this cell population with other cells expressing a wild-type, or normal, form of cadherin, would you expect these two cell populations to aggregate with each other? Why or why not? Would the mutant cells be able to aggregate with other mutant cells?

20

Cancer Biology and Cell Technology

Concept Outline

20.1 Recombination alters gene location.

Gene Transfer. Plasmids can move between bacterial cells and carry bacterial genes.

Reciprocal Recombination. Reciprocal recombination can alter genes in several ways.

20.2 Mutations are changes in the genetic message.

Kinds of Mutation. Some mutations alter genes themselves; others alter the positions of genes.

DNA Repair. Cells repair damage to DNA effectively in a variety of ways.

20.3 Most cancer results from mutation of growth-regulating genes

What Is Cancer? Cancer is a growth disorder of cells.

Causes of Cancer. Many cancers are caused by chemicals that alter DNA.

Cancer and the Cell Cycle. Cancer results from mutation of genes regulating cell proliferation.

Smoking and Cancer. Smoking causes lung cancer.

Curing Cancer. New approaches offer promise of a cure.

20.4 Reproductive cloning of animals, once thought impossible, isn't.

The Challenge of Cloning. As recently as 1997, scientists thought it was impossible to clone an animal from an adult cell. They were wrong.

20.5 Therapeutic cloning is a promising but controversial possibility.

Stem Cells. Embryonic stem cells are capable of forming any tissue of the body. Tissue-specific adult stem cells may provide an alternative to embryonic stem cells.

Therapeutic Cloning. Somatic nuclear transfer of human embryonic stem cells seems to offer significant therapeutic promise.

Grappling with the Ethics of Stem Cell Research. Few issues in modern science raise as many difficult ethical issues as stem cell research.

FIGURE 20.1
Embryonic stem cells growing in tissue culture. Embryonic stem cells derived from early human embryos will grow indefinitely in tissue culture. When transplanted, they can sometimes be induced to form new cells of the adult tissue into which they have been placed. This suggests exciting therapeutic uses.

The last decade has seen remarkable advances in our understanding of the molecular events leading to cancer. In essence, cancer results from damage to grow-controlling genes. This damage can result from positional changes induced by recombination or, more commonly, from chemical changes induced by mutation. After reviewing these two processes briefly, this chapter explores in depth what we have learned about cancer. It then turns to another area in which landmark progress is being made—cell technology. While animal cloning, gene therapy, and stem cell research (figure 20.1) might have been treated within three different chapters, this chapter pulls them all together so that the broad sweep of what is occurring is most apparent. Advances in cell technology hold the promise of literally revolutionizing biology and medicine.

20.1 Recombination alters gene location.

Gene Transfer

Two very different sorts of recombination processes alter genes: gene transfer and reciprocal recombination. We will consider gene transfer first.

Genes are not fixed in their locations on chromosomes or on the circular DNA molecules of prokaryotes; they can move around. Some genes move because they are part of small, circular, extrachromosomal DNA segments called **plasmids.** Plasmids enter and leave the main genome at specific places where a nucleotide sequence matches one present on the plasmid. Plasmids occur primarily in prokaryotes, in which the main genomic DNA can interact readily with other DNA fragments. About 5% of the DNA that occurs in a bacterium is plasmid DNA. Some plasmids are very small, containing only one or a few genes, while others are quite complex and contain many genes. Other genes move within **transposons,** which jump from one genomic position to another at random in both prokaryotes and eukaryotes.

Gene transfer by plasmid movement was discovered by Joshua Lederberg and Edward Tatum in 1947. Three years later, transposons were discovered by Barbara McClintock. However, her work implied that the position of genes in a genome need not be constant. Researchers accustomed to viewing genes as fixed entities, like beads on a string, did not readily accept the idea of transposons. Therefore, while Lederberg and Tatum were awarded a Nobel Prize for their discovery in 1958, McClintock did not receive a Nobel Prize for hers until 1983.

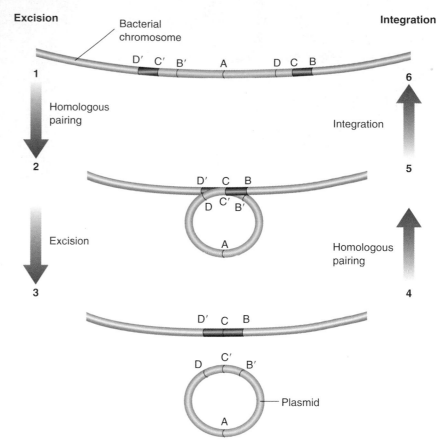

FIGURE 20.2

Excision and integration of a plasmid. Because the ends of the two sequences in the bacterial genome are the same (D′, C′, B′, and D, C, B), it is possible for the two ends to pair. Steps 1–3 show the sequence of events if the strands exchange during the pairing. The result is excision of the loop and a free circle of DNA—a plasmid. Steps 4–6 show the sequence when a plasmid integrates itself into a bacterial genome.

Plasmid Creation

To understand how plasmids arise, consider a hypothetical stretch of bacterial DNA that contains two copies of the same nucleotide sequence. It is possible for the two copies to base-pair with each other and create a transient "loop," or double duplex. All cells have recombination enzymes that can cause such double duplexes to undergo a **reciprocal exchange,** in which they exchange strands. As a result of the exchange, the loop is freed from the rest of the DNA molecule and becomes a plasmid (figure 20.2, steps 1–3). Any genes between the duplicated sequences (such as gene A in figure 20.2) are transferred to the plasmid.

Once a plasmid has been created by reciprocal exchange, DNA polymerase will replicate it if it contains a replication origin, often without the controls that restrict the main genome to one replication per cell division.

Integration

A plasmid created by recombination can reenter the main genome the same way it left. Sometimes the region of the plasmid DNA that was involved in the original exchange, called the **recognition site,** aligns with a matching sequence on the main genome. If a recombination event occurs anywhere in the region of alignment, the plasmid will integrate into the genome (figure 20.2, steps 4–6). Integration can occur wherever any shared sequences exist, so plasmids may be integrated into the main genome at positions other than the one from which they arose. If a plasmid is integrated at a new position, it transfers its genes to that new position.

Gene Transfer by Conjugation

One of the startling discoveries Lederberg and Tatum made was that plasmids can pass from one bacterium to another. The plasmid they studied was part of the genome of *Escherichia coli*. It was given the name F, for fertility factor, because only cells having that plasmid integrated into their DNA could act as plasmid donors. These cells are called Hfr cells (for "high-frequency recombination"). The F plasmid contains a DNA replication origin and several genes that promote its transfer to other cells. These genes encode protein subunits that assemble on the surface of the bacterial cell, forming a hollow tube called a **pilus.**

When the pilus of one cell (F$^+$) contacts the surface of another cell that lacks a pilus, and therefore does not contain an F plasmid (F$^-$), the pilus draws the two cells close together so that DNA can be exchanged (figure 20.3). First, the F plasmid binds to a site on the interior of the F$^+$ cell just beneath the pilus now called a *conjugation bridge*. Then, by a process called **rolling-circle replication,** the F plasmid begins to copy its DNA at the binding point. As it is replicated, the single-stranded copy of the plasmid passes into the other cell. There, a complementary strand is added, creating a new, stable F plasmid (figure 20.4). In this way, genes are passed from one bacterium to another. This transfer of genes between bacteria is called **conjugation.**

In an Hfr cell, with the F plasmid integrated into the main bacterial genome rather than free in the cytoplasm, the F plasmid can still organize the transfer of genes. In this case, the integrated F region binds beneath the pilus and initiates the *replication of the bacterial genome,* transferring the newly replicated portion to the recipient cell. Transfer proceeds as if the bacterial genome were simply a part of the F plasmid. By studying this phenomenon, researchers have been able to locate the positions of different genes in bacterial genomes (figure 20.5).

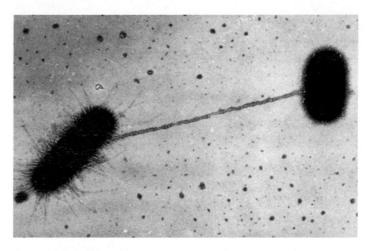

FIGURE 20.3
Contact by a pilus. The pilus of an F$^+$ cell connects to an F$^-$ cell and draws the two cells close together so that DNA transfer can occur.

Gene Transfer by Transposition

Transposons (figure 20.6) are small segments of DNA capable of moving from one location to another in the genome. After spending many generations in one position, a transposon may abruptly move to a new position in the genome, carrying various genes along with it. Transposons encode an enzyme called **transposase,** which inserts the transposon into the genome, a process known as **transposition** (figure 20.7). Because this enzyme usually does not recognize any particular sequence on the genome, transposons appear to move to random destinations.

The movement of any given transposon is relatively rare: It may occur perhaps once in 100,000 cell generations. Although low, this rate is still about 10 times as frequent as the rate at which random mutational changes

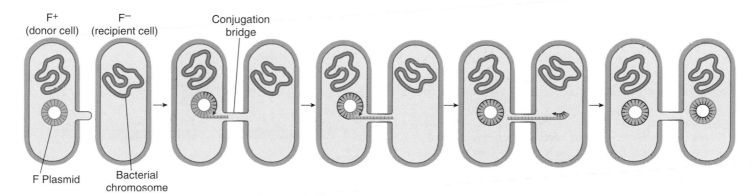

FIGURE 20.4
Gene transfer between bacteria. Donor cells (F$^+$) contain an F plasmid that recipient cells (F$^-$) lack. The F plasmid replicates itself and transfers the copy across a conjugation bridge. The remaining strand of the plasmid serves as a template to build a replacement. When the single strand enters the recipient cell, it serves as a template to assemble a double-stranded plasmid. When the process is complete, both cells contain a complete copy of the plasmid.

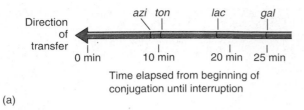

(a)

FIGURE 20.5
A conjugation map of the *E. coli* chromosome. Scientists have been able to break the *Escherichia coli* conjugation bridges by agitating the cell suspension rapidly in a blender. By agitating at different intervals after the start of conjugation, investigators can locate the positions of various genes along the bacterial genome. (*a*) The closer the genes are to the origin of replication, the sooner one has to turn on the blender to block their transfer. (*b*) Map of the *E. coli* genome developed using this method.

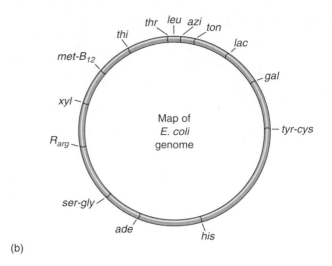

(b)

occur. Furthermore, most cells contain many transposons. Hence, over long periods of time, transposition can have an enormous evolutionary impact.

One way this impact can be felt is through mutation. The insertion of a transposon into a gene often destroys the gene's function, resulting in what is termed **insertional inactivation.** This phenomenon is thought to cause a significant number of the spontaneous mutations observed in nature, and is thus a potential cause of cancer.

Transposition can also facilitate **gene mobilization,** the bringing together in one place of genes that are usually located at different positions in the genome. In bacteria, for example, a number of genes encode enzymes that make the bacteria resistant to antibiotics such as penicillin, and many of these genes are located on plasmids. The simultaneous exposure of bacteria to multiple antibiotics, a common medical practice some years ago, favors the persistence of plasmids that have managed to acquire several resistance genes. Transposition can rapidly generate such composite plasmids, called **resistance transfer factors (RTFs),** by moving antibiotic-resistance genes from several plasmids to one. Bacteria possessing RTFs are thus able to survive treatment with a wide variety of antibiotics. RTFs are thought to be responsible for much of the recent difficulty in treating hospital-engendered *Staphylococcus aureus* infections and the new drug-resistant strains of tuberculosis.

> Plasmids transfer copies of bacterial genes (and even entire genomes) from one bacterium to another. Transposition is the one-way transfer of genes to a randomly selected location in the genome. The genes move because they are associated with mobile genetic elements called transposons.

FIGURE 20.6
Transposon. Transposons form characteristic stem-and-loop structures called "lollipops" because their two ends have the same nucleotide sequence as inverted repeats. These ends pair together to form the stem of the lollipop.

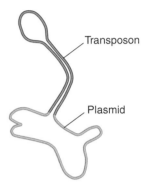

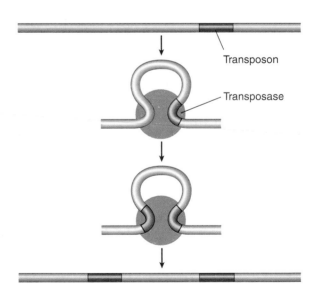

FIGURE 20.7
Transposition. Transposase does not recognize any particular DNA sequence; rather, it selects one at random, moving the transposon to a random location. Some transposons leave a copy of themselves behind when they move.

Reciprocal Recombination

In the second major mechanism for producing genetic recombination, reciprocal recombination among eukaryotes, two homologous chromosomes exchange all or part of themselves during the process of meiosis.

Crossing Over

As we saw in chapter 12, crossing over occurs in the first prophase of meiosis, when two homologous chromosomes line up side by side within the synaptonemal complex. At this point, the homologues exchange DNA strands at one or more locations. This exchange of strands can produce chromosomes with new combinations of alleles.

Imagine, for example, that a giraffe has genes encoding neck length and leg length at two different loci on one of its chromosomes. Imagine further that a recessive mutation occurs at the neck length locus, leading after several rounds of independent assortment to some individuals that are homozygous for a variant "long-neck" allele. Similarly, a recessive mutation at the leg length locus leads to homozygous "long-leg" individuals.

It is very unlikely that these two mutations would arise at the same time in the same individual because the probability of two independent events occurring together is the product of their individual probabilities. If the spontaneous occurrence of both mutations in a single individual were the only way to produce a giraffe with both a long neck and long legs, it would be extremely unlikely that such an individual would ever occur. Because of recombination, however, a crossover in the interval between the two genes could in one meiosis produce a chromosome bearing both variant alleles. This ability to reshuffle gene combinations rapidly is what makes recombination so important to the production of natural variation.

Unequal Crossing Over

Reciprocal recombination can occur in any region along two homologous chromosomes with sequences similar enough to permit close pairing. Mistakes in pairing occasionally happen when several copies of a sequence exist in different locations on a chromosome. In such cases, one copy of a sequence may line up with one of the duplicate copies instead of with its homologous copy. Such misalignment causes slipped mispairing, which, as we will discuss later, can lead to small deletions and frameshift mutations. If a crossover occurs in the pairing region, it will result in unequal crossing over because the two homologues will exchange segments of unequal length.

In unequal crossing over, one chromosome gains extra copies of the multicopy sequences, while the other chromosome loses them (figure 20.8). This process can generate a chromosome with hundreds of copies of a particular gene, lined up side by side in tandem array.

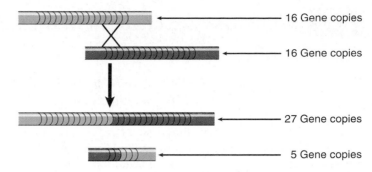

FIGURE 20.8
Unequal crossing over. When a repeated sequence pairs out of register, a crossover within the region will produce one chromosome with fewer gene copies and one with more. Much of the gene duplication that has occurred in eukaryotic evolution may well be the result of unequal crossing over.

Because the genomes of most eukaryotes possess multiple copies of transposons scattered throughout the chromosomes, unequal crossing over between copies of transposons located in different positions has had a profound influence on gene organization in eukaryotes. As we shall see later, most of the genes of eukaryotes appear to have been duplicated one or more times during their evolution.

Gene Conversion

Because the two homologues that pair within a synaptonemal complex are not identical, some nucleotides in one homologue are not complementary to their counterpart in the other homologue. These occasional nonmatching pairs of nucleotides are called **mismatch pairs.**

As you might expect, the cell's error-correcting machinery is able to detect mismatch pairs. If a mismatch is detected during meiosis, the enzymes that "proofread" new DNA strands during DNA replication correct it. The mismatched nucleotide in one of the homologues is excised and replaced with a nucleotide complementary to the one in the other homologue. Its base-pairing partner in the first homologue is then replaced, producing two chromosomes with the same sequence. This error correction causes one of the mismatched sequences to convert into the other, a process called **gene conversion.**

Unequal crossing over occurs between chromosomal regions that are similar in nucleotide sequence but not homologous. Gene conversion is the alteration of one homologue by the cell's error-detection and repair system to make it resemble the other homologue.

20.2 Mutations are changes in the genetic message.

Kinds of Mutation

Mutations are changes in the hereditary message of an organism. Mutations can have significant effects on the individual when they occur in somatic tissue, but are only inherited if they occur in germ-line tissue. Because mutations can occur randomly anywhere in a cell's DNA, mutations are often detrimental, just as making a random change in a computer program or a musical score usually worsens performance. The consequences of a detrimental mutation may be minor or catastrophic, depending on the function of the altered gene.

The effect of a mutation depends critically on the identity of the cell in which the mutation occurs. During the embryonic development of all multicellular organisms, there comes a point when cells destined to form gametes (germ-line cells) are segregated from those that will form the other cells of the body (somatic cells). Only when a mutation occurs within a germ-line cell is it passed to subsequent generations as part of the hereditary endowment of the gametes derived from that cell.

Mutations in germ-line tissue are of enormous biological importance because they provide the raw material from which natural selection produces evolutionary change. In a somatic cell, mutation may have drastic effects on the individual organism in which it occurs, because it is passed on to all the cells that are descended from the original mutant cell. Thus, if a mutant lung cell divides, all cells derived from it will carry the mutation. Somatic mutations of lung cells are, as we shall see, the principal cause of lung cancer in humans.

One category of mutational changes affects the message itself, producing alterations in the sequence of DNA nucleotides. A second category of mutation arises as the result of changes in gene location. Table 20.1 summarizes the sources and types of mutations.

Mutations Altering the Sequence of DNA

Mutational alterations of DNA arise in at least five ways:

1. **Base substitution.** Spontaneous pairing errors or polymerase mistakes may occur during DNA replication. These changes typically involve only one or a few base-pairs in the coding sequence, a so-called point mutation.
2. **Chemical modification.** A base may be chemically altered by a mutagenic chemical, another class of point mutation.
3. **DNA breaks.** Ionizing radiation can cause double-strand breaks in DNA, often resulting in the deletion (loss) of short segments.
4. **Slipped mispairing.** Deletions may also result from so-called slipped mispairing. When a sequence present in more than one copy on a chromosome pairs out of

register—like a shirt buttoned wrong—the loop this mistake produces is sometimes excised by the cell's repair enzymes, producing a short deletion. Many of these deletions start or end in the middle of a codon, thereby shifting the reading frame by one or two bases. These so-called *frameshift mutations* cause the gene to be read in the wrong 3-base groupings, distorting the genetic message.
5. **Triplet expansion.** When a 3-base sequence is repeated several times in tandem within a gene, the number of repeats may become expanded. Thus, a CAG repeat, when expanded, produces a polyglutamine region within the resulting mutant protein. More than 15 human disease states are associated with triplet expansion within a gene.

Table 20.1 Types of Mutation

Mutation	Example Result
NO MUTATION	Normal B protein is produced by the *B* gene.
SEQUENCE CHANGES	
Base substitution — Substitution of one or a few bases due to pairing error or chemical modification	B protein is inactive because changed amino acid disrupts function.
Triplet expansion X 200 — Additional copies of a repeated 3-base sequence	B protein is inactive because inserted material disrupts proper shape.
Deletion — Loss of one or a few bases due to ionizing radiation or slipped mispairing	B protein is inactive because portion of protein is missing.
CHANGES IN GENE POSITION	
Chromosomal rearrangement	*B* gene is inactivated or regulated differently in its new location on chromosome.
Insertional inactivation — Addition of a transposon within a gene	B protein is inactive because inserted material disrupts gene translation or protein function.

Mutations Arising from Changes in Gene Position

Chromosome location is an important factor in determining whether genes are transcribed. Some genes cannot be transcribed if they are adjacent to a tightly coiled region of the chromosome, even though the same gene can be transcribed normally in any other location. Transcription of many chromosomal regions appears to be regulated in this manner; the binding of specific proteins regulates the degree of coiling in local regions of the chromosome, determining the accessibility RNA polymerase has to genes located within those regions.

Chromosomal Rearrangements. Chromosomes undergo several different kinds of gross physical alterations that have significant effects on the locations of their genes. There are two main kinds of alteration: translocations, in which a segment of one chromosome becomes part of another chromosome, and inversions, in which the orientation of a portion of a chromosome is reversed. Translocations often have significant effects on gene expression. Inversions, on the other hand, usually do not alter gene expression, but lead to serious problems in meiosis (figure 20.9): None of the gametes that contain chromatids produced following such a crossover event will have a complete set of genes.

Other chromosomal alterations change the number of gene copies an individual possesses. These include **duplication,** in which entire segments of the chromosome become duplicated; **deletions,** like the small deletions discussed earlier but involving much larger amounts of genetic material; **aneuploidy,** in which whole chromosomes are lost or gained; and **polyploidy,** in which entire sets of chromosomes are added. Even in diploid organisms, most deletions are harmful because they halve the number of gene copies within a diploid genome and thus seriously affect the level of transcription. Duplications cause gene imbalance and are also usually harmful.

Insertional Inactivation. As discussed in section 20.1, transposons are capable of moving from one location to another in the genome, using an enzyme to cut and paste themselves into new genetic neighborhoods. Transposons select their new locations at random, and are as likely to enter one segment of a chromosome as another. Inevitably, some transposons end up inserted into genes, and this almost always inactivates the gene. The encoded protein now has a large meaningless chunk within it, disrupting its structure. This form of mutation, called *insertional inactivation*, is common in nature. Other mutations result when insertion of a transposon at a new site affects the regulation of neighboring genes because it contains a strong promoter that turns on genes that would otherwise be silent.

As you might expect, a variety of human gene disorders are the result of transposition. The human transposon called *Alu*, for example, is responsible for an X-linked hemophilia, inserting into clotting factor IX and placing a premature "stop" codon there. It also causes inherited high levels of cholesterol (hypercholesterolemia) when *Alu* elements insert into the gene encoding the low-density lipoprotein (LDL) receptor. In one very interesting case, a *Drosophila* transposon called *Mariner* proves responsible for a rare human neurological disorder called Charcot-Marie-Tooth disease, in which the muscles and nerves of the legs and feet gradually wither away. The *Mariner* transposon is inserted into a key gene called *CMT* on chromosome 17, creating a weak site where the chromosome can break. No one knows how the *Drosophila* transposon got into the human genome.

Mutations result both from changes in the nucleotide sequence of genes and from the movement of genes to new locations on the chromosomes.

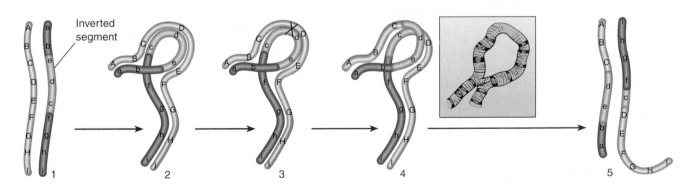

FIGURE 20.9

The consequence of inversion. (*1*) When a segment of a chromosome is inverted, (*2*) it can pair in meiosis only by forming an internal loop. (*3*) Any crossing over that occurs within the inverted segment during meiosis will result in nonviable gametes; some genes are lost from each chromosome, while others are duplicated (*4* and *5*). For clarity, only two strands are shown, although crossing over occurs in the four-strand stage. The pairing that occurs between inverted segments is sometimes visible under the microscope as a characteristic loop (*inset*).

DNA Repair

Given the possible number of errors that can occur during replication and the number of possible agents that can damage DNA, it is not surprising that systems have evolved to repair damaged DNA. These systems operate in a number of ways to safeguard the cell's hereditary information.

Mismatch Repair

First discovered in the bacterium *E. coli*, mismatch repair has now been found to be common in the cells of most organisms. These systems are designed to correct errors occurring during DNA replication. Even with their proofreading activities, the error rate of DNA polymerase is relatively high. Mismatch repair systems were first identified in strains of *E. coli* with unusually high mutation rates. A small number of genes proved to be responsible for the high mutation rates. These genes, labeled *Mut* for mutations, led to the identification of the activities involved in mismatch repair.

The **mismatch repair system** uses a clever strategy to identify sites of mutation. Specific enzymes in the cell recognize the sequence GATC on cellular DNA and add a methyl (CH_3) group to the adenine. Many mismatch repair mutants proved unable to methylate this adenine, and are dubbed *dam* mutations ("deficient in adenine methylation"). Methylation of GATC sequences allows the identification of the parental strand after DNA replication—it is the GATC-methylated strand. A little later, a maintenance methylase acts to methylate the adenines on the GATC sequences of the new daughter strand. However, in the short interval before it does so, the cell's mismatch repair enzymes act: (1) They locate replication errors as sites where the base mismatch between the two strands distorts the helix; (2) they identify the parental strand (the methylated one); and (3) they correct the opposite strand so that it is complementary to the methylated one.

Specific Repair Systems

Cells contain a number of repair systems that are specific for particular lesions in DNA. One such **UVR photorepair system** corrects damage caused by UV light. The energy in ultraviolet light is absorbed by thymine, causing the formation of cyclobutane links between adjacent thymines; because the resulting thymine dimers cannot be replicated, their creation leads to mutations. The damage is repaired by an enzyme that is able to absorb a photon of visible light and use the energy from this to cleave the cyclobutane bond.

Another specific repair system repairs any instances where uracil (an RNA nucleotide) is present in DNA by mistake, instead of thymine. An enzyme, uracil-N-glycosylase, cleaves the uracil side group from the nucleotide base without disrupting the DNA chain. This leaves an **apyrmidinic (AP) site** with a nucleotide that has no attached base. A specific nuclease then recognizes this AP site and removes the truncated nucleotide. The resulting one-base gap is then restored by the repair polymerase. Other chemically damaged bases are also repaired by similar two-step processes.

Excision Repair

DNA replication errors and chemical damage to DNA can also be repaired by **excision repair,** a nonspecific system that can repair a wide variety of lesions to DNA. In broad outline, excision repair involves two steps: (1) The damaged region of DNA is recognized and removed, (2) the repair polymerase fills in the resulting gap with newly synthesized DNA. In *E. coli*, this excision repair process is controlled by the UVR photorepair system. The genes that encode the relevant proteins were identified based on their unusual UV sensitivity, with the genes named *uvr* A, B, C, and D.

Because this nonspecific system responds only to distortions of the helix caused by damaged bases, excision repair can handle a wide variety of DNA damage and is generally "error-free"—that is, the DNA is restored to the wild-type condition after the damaged bases are removed.

Post-Replication Repair

Post-replication repair is the only repair system able to handle double-strand breaks in DNA. Repairing this sort of damage requires a second complete DNA duplex molecule with the same sequence as the damaged DNA. Because prokaryotes have only one copy of their DNA except during DNA replication, this form of repair is only possible during replication in a prokaryotic cell. In eukaryotes, this process is called **recombinational repair,** and occurs during prophase of meiosis I, when the two chromosomes are paired in close register. Many of the enzymes responsible for recombinational repair are also active in the recombination events of meiosis.

Cells are able to repair many potential mutations by recognizing the pairing mismatch that a base substitution creates or by recognizing damaged or inappropriate bases.

20.3 Most cancer results from mutation of growth-regulating genes.

What Is Cancer?

Cancer is a growth disorder of cells. It starts when an apparently normal cell begins to grow in an uncontrolled and invasive way (figure 20.10). The result is a cluster of cells, called a **tumor,** that constantly expands in size. Cells that leave the tumor and spread throughout the body, forming new tumors at distant sites, are called **metastases** (figure 20.11). Cancer is perhaps the most pernicious disease. Of the children born in 1999, one-third will contract cancer at some time during their lives; one-fourth of the male children and one-third of the female children will someday die of cancer. Most of us have had family or friends affected by the disease. In 2002, an estimated 550,000 Americans died of cancer.

Not surprisingly, researchers are expending a great deal of effort to learn the cause of this disease. Scientists have made a great deal of progress in the last 20 years using molecular biological techniques, and the rough outlines of understanding are now emerging. We now know that cancer is a gene disorder of somatic tissue, in which damaged genes fail to properly control cell proliferation. The cell division cycle is regulated by a sophisticated group of proteins described in chapter 11. Cancer results from the mutation of the genes encoding these proteins.

Cancer can be caused by chemicals that mutate DNA or, in some instances, by viruses that circumvent the cell's normal proliferation controls. Whatever the immediate cause, however, all cancers are characterized by unrestrained growth and division. Cell division never stops in a cancerous line of cells. Cancer cells are virtually immortal—until the body in which they reside dies.

Cancer is unrestrained cell proliferation caused by damage to genes regulating the cell division cycle.

FIGURE 20.10
Lung cancer cells. These cells (530×) are from a tumor located in the alveolus (air sac) of a lung.

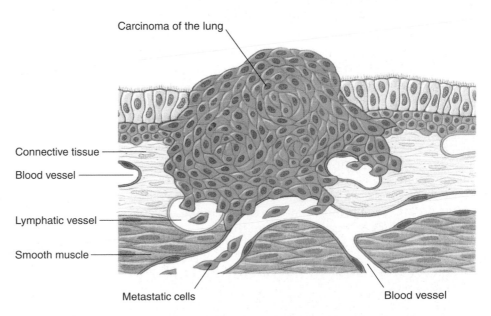

Carcinoma of the lung

Connective tissue

Blood vessel

Lymphatic vessel

Smooth muscle

Metastatic cells

Blood vessel

FIGURE 20.11
Portrait of a cancer. This ball of cells is a carcinoma (cancer tumor) developing from epithelial cells that line the interior surface of a human lung. As the mass of cells grows, it invades surrounding tissues, eventually penetrating lymphatic and blood vessels, both plentiful within the lung. These vessels carry metastatic cancer cells throughout the body, where they lodge and grow, forming new masses of cancerous tissue.

Causes of Cancer

Cancer can occur in almost any tissue, so a bewildering number of different cancers are possible. Tumors arising from cells in connective tissue, bone, or muscle are known as **sarcomas**, while those that originate in epithelial tissue such as skin are called **carcinomas.** In the United States, three of the deadliest human cancers are lung cancer, cancer of the colon and rectum, and breast cancer (table 20.2). Lung cancer, responsible for the most cancer deaths, is largely preventable; most cases result from smoking cigarettes. Colorectal cancers appear to be fostered by the high-meat diets so favored in the United States. The cause of breast cancer is still a mystery, although in 1994 and 1995 researchers isolated two genes responsible for hereditary susceptibility to breast cancer, *BRCA1* and *BRCA2* (breast cancer genes #1 and #2 located on human chromosomes 17 and 13); their discovery offers hope that researchers will soon be able to unravel the fundamental mechanism leading to hereditary breast cancer, which makes up about one-third of all breast cancers.

The association of particular chemicals with cancer, especially those that are potent mutagens, led researchers early on to suspect that cancer might be caused, at least in part, by chemicals, the so-called **chemical carcinogenesis theory.** Agents thought to cause cancer are called **carcinogens.** A simple and effective way to test whether a chemical is mutagenic is the Ames test (figure 20.12), named for its developer, Bruce Ames. The test uses a strain of *Salmonella* bacteria that has a defective histidine-synthesizing gene. Because these bacteria cannot make histidine, they cannot grow on media without it. Only a back-mutation that restores the ability to manufacture histidine will permit growth. Thus, the number of colonies of these bacteria that grow on histidine-free medium is a measure of the frequency of back-mutation. A majority of chemicals that cause back-mutations in this test are carcinogenic, and vice versa. To increase the sensitivity of the test, the strains of bacteria are altered to disable their DNA repair machinery. The search for the cause of cancer has focused in part on chemical carcinogens and other environmental factors, including ionizing radiation such as X rays.

Table 20.2 Incidence of Cancer in the United States in 2002*			
Type of Cancer	**New Cases**	**Deaths**	**% of Cancer Deaths**
Lung	169,400	154,900	28
Colon and rectum	148,300	56,600	10
Leukemia/lymphoma	91,700	47,500	9
Breast	205,000	40,000	7
Prostate	189,000	30,200	5
Pancreas	30,300	29,700	5
Ovary	23,300	13,900	3
Liver	16,600	14,100	3
Nervous system/eye	19,200	13,300	2
Stomach	13,100	12,600	2
Bladder	56,500	12,600	2
Kidney	31,800	11,600	2
Cervix/uterus	52,300	10,700	2
Oral cavity	28,900	7,400	1
Malignant melanoma	53,600	7,400	1
Sarcoma (connective tissue)	10,400	5,800	1
All other cancers	145,500	87,200	16

*In the United States in 2002, there were an estimated 1,284,900 reported cases of new cancers and 555,500 cancer deaths, indicating that roughly half the people who develop cancer die from it.

Source: Data from the American Cancer Society, Inc., 2002.

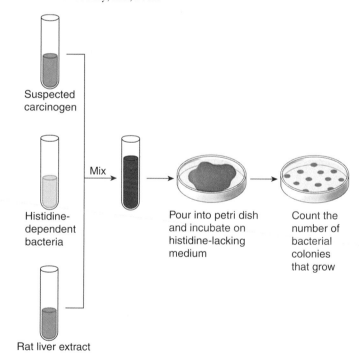

FIGURE 20.12
The Ames test. This test uses a strain of *Salmonella* bacteria that requires histidine in the growth medium due to a mutated gene. If a suspected carcinogen is mutagenic, it can reverse this mutation. Rat liver extract is added because it contains enzymes that can convert carcinogens into mutagens. The mutagenicity of the carcinogen can be quantified by counting the number of bacterial colonies that grow on a medium lacking histidine.

Some Tumors Are Caused by Chemicals

The chemical carcinogenesis theory was first advanced over 200 years ago in 1761 by Dr. John Hill, an English physician, who noted unusual tumors of the nose in heavy snuff users and suggested that tobacco had produced these cancers. In 1775, a London surgeon, Sir Percivall Pott, made a similar observation, noting that men who had been chimney sweeps exhibited frequent cancer of the scrotum, and suggesting that soot and tars might be responsible. British sweeps washed themselves infrequently and always seemed covered with soot. Chimney sweeps on the continent, who washed daily, had much less of this scrotal cancer. These and many other observations led to the hypothesis that cancer results from the action of chemicals on the body.

Demonstrating That Chemicals Can Cause Cancer

Over a century passed before this theory was directly tested. In 1915, the Japanese doctor Katsusaburo Yamagiwa applied extracts of coal tar to the skin of 137 rabbits every two or three days for three months. Then he waited to see what would happen. After a year, cancers appeared at the site of application in seven of the rabbits. Yamagiwa had induced cancer with the coal tar, the first direct demonstration of chemical carcinogenesis. In the decades that followed, this approach demonstrated that many chemicals were capable of causing cancer. Importantly, most of them were potent mutagens.

Because these were lab studies, many people did not accept that the results applied to real people. Do tars in fact induce cancer in humans? In 1949, the American physician Ernst Winder and the British epidemiologist Richard Doll independently reported that lung cancer showed a strong link to cigarette smoking, which introduces tars into the lungs. Winder interviewed 684 lung cancer patients and 600 normal controls, asking whether each had ever smoked. Cancer rates were 40 times higher in heavy smokers than in nonsmokers. Doll's study was even more convincing. He interviewed a large number of British physicians, noting which ones smoked, and then waited to see which would develop lung cancer. Many did. Overwhelmingly, those who did were smokers. From these studies, it seemed likely as long as 50 years ago that tars and other chemicals in cigarette smoke induce cancer in the lungs of persistent smokers. While this suggestion was (and is) resisted by the tobacco industry, the evidence that has accumulated since these pioneering studies makes a clear case, and there is no longer any real doubt. Chemicals in cigarette smoke cause cancer.

Carcinogens Are Common

In ongoing investigations over the last 50 years, many hundreds of synthetic chemicals have been shown capable of causing cancer in laboratory animals. Among them are

Table 20.3 Chemical Carcinogens in the Workplace

Chemical	Cancer	Workers at Risk for Exposure
COMMON EXPOSURE		
Benzene	Myelogenous leukemia	Painters; dye users; furniture finishers
Diesel exhaust	Lung	Railroad and bus-garage workers; truckers; miners
Mineral oils	Skin	Metal machinists
Pesticides	Lung	Sprayers
Cigarette tar	Lung	Smokers
UNCOMMON EXPOSURE		
Asbestos	Mesothelioma, lung	Brake-lining, insulation workers
Synthetic mineral fibers	Lung	Wall and pipe insulation and duct wrapping users
Hair dyes	Bladder	Hairdressers and barbers
Paint	Lung	Painters
Polychlorinated biphenyls	Liver, skin	Users of hydraulic fluids and lubricants, inks, adhesives, insecticides
Soot	Skin	Chimney sweeps; bricklayers; firefighters; heating-unit service workers
RARE EXPOSURE		
Arsenic	Lung, skin	Insecticide/herbicide sprayers; tanners; oil refiners
Formaldehyde	Nose	Wood product, paper, textile, and metal product workers.

trichloroethylene, asbestos, benzene, vinyl chloride, arsenic, arylamide, and a host of complex petroleum products with chemical structures resembling chicken wire. People in the workplace encounter chemicals daily (table 20.3).

In addition to identifying potentially dangerous substances, what have the studies of potential carcinogens told us about the nature of cancer? What do these cancer-causing chemicals have in common? *They are all mutagens, each capable of inducing changes in DNA.*

Cancers occur in all tissues. Chemicals that produce mutations in DNA are often potent carcinogens. Tars in cigarette smoke, for example, are the direct cause of most lung cancers.

Cancer and the Cell Cycle

In chapter 11, we learned about the cell cycle and its control. Given the intricate interplay of regulatory elements involved in the cell cycle, it is not surprising that mutations that affect this regulatory system can lead to uncontrolled cell growth: cancer. Two general categories of genes are affected by such mutations: proto-oncogenes and tumor-suppressor genes. Before examining current attempts to counteract the effects of cancer-causing mutations to somatic DNA, we will briefly consider how mutations in cell cycle control genes can lead to the malignant transformation that is the basis of many forms of cancer in humans (table 20.4).

Oncogenes

Genes that when introduced into a normal cell cause it to become a cancer cell are called **oncogenes** (Greek *onco*, "tumor"). Oncogenes were originally discovered by the process of **transfection,** in which DNA isolated from one cell is introduced into another cell. In transfection experiments, DNA isolated from tumor-derived cells led to the transformation of normal cells into cancer cells. These same genes, in mutated form, were soon isolated from a variety of tumors. The normal, nonmutated forms of these genes came to be called **proto-oncogenes,** or genes that can be mutated to produce a tumor-forming oncogene. Subsequent study has shown that these genes all encode proteins that are necessary for the cell to interpret external signals for growth (figure 20.13). As discussed in chapter 11, these signals are necessary for a cell to leave G_0 and to pass the G_1 checkpoint. Mutations in proto-oncogenes accelerate the cell cycle by amplifying these signals.

If a growth factor receptor is mutated such that it is always "on" regardless of binding to the growth factor, the gene encoding the receptor will act as an oncogene. Similarly, all of the genes encoding the proteins necessary to carry this signal from the surface of the cell to the nucleus can be mutated to greatly amplify the signal, effectively converting the signal pathway genes to oncogenes.

As oncogenes produce their effect in the absence of the normal "divide" signal, they will be inherited as dominant alleles. That is, they will exert their cancer-causing effect even in cells heterozygous for the mutant oncogene.

Chromosomal abnormalities can lead to the activation of proto-oncogenes. A rearrangement that moves a strong enhancer near a proto-oncogene can lead either to overexpression or to expression in a tissue where the proto-oncogene normally is not expressed. This was first observed in chronic myelogenous leukemia where an abnormal chromosome, called the Philadelphia chromosome, is produced when chromosomes 9 and 22 exchange genetic information. The translocation of a portion of the chromosome creates a chimeric gene that includes the proto-oncogene *c-ABL*, causing *c-ABL* to be expressed and producing leukemia.

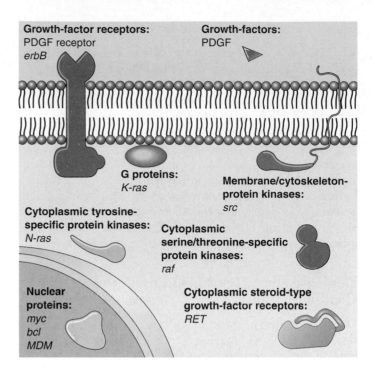

FIGURE 20.13
The main classes of oncogenes.   Before they are altered by mutation to their cancer-causing condition, oncogenes are called proto-oncogenes (that is, genes able to become oncogenes). Illustrated here are the principal classes of proto-oncogenes, with some typical representatives indicated.

Among the most widely studied oncogenes are *myc* and *ras.* Expression of *myc* stimulates the production of cyclins and cyclin-dependent protein kinases (Cdk's), key elements in regulating the checkpoints of cell division.

The *ras* gene product is involved in the cellular response to a variety of growth factors, including EGF, an intercellular signal that normally initiates cell proliferation. When EGF binds to a specific receptor protein on the plasma membrane of epithelial cells, the portion of the receptor that protrudes into the cytoplasm stimulates the Ras protein to bind to GTP. The Ras protein/GTP complex in turn recruits and activates a protein called Raf to the inner surface of the plasma membrane, which in turn activates cytoplasmic kinases and so triggers an intracellular signaling system (see chapter 7). The final step is the activation of transcription factors that trigger cell proliferation. Cancer-causing mutations in *ras* greatly reduce the amount of EGF necessary to initiate cell proliferation.

Tumor-Suppressor Genes

If the first class of cancer-inducing mutations "steps on the accelerator" of cell division, the second class of cancer-inducing mutations "removes the brakes." Cell division is normally turned off in healthy cells by proteins that pre-

Table 20.4 Some Genes Implicated in Human Cancers

Gene	Product	Cancer
ONCOGENES		
Genes Encoding Growth Factors or Their Receptors		
erb-B	Receptor for epidermal growth factor	Glioblastoma (a brain cancer); breast cancer
erb-B2	A growth factor receptor (gene also called *neu*)	Breast cancer; ovarian cancer; salivary gland cancer
PDGF	Platelet-derived growth factor	Glioma (a brain cancer)
RET	A growth factor receptor	Thyroid cancer
Genes Encoding Cytoplasmic Relays in Intracellular Signaling Pathways		
K-ras	Protein kinase	Lung cancer; colon cancer; ovarian cancer; pancreatic cancer
N-ras	Protein kinase	Leukemias
Genes Encoding Transcription Factors That Activate Transcription of Growth-Promoting Genes		
c-myc	Transcription factor	Lung cancer; breast cancer; stomach cancer; leukemias
L-myc	Transcription factor	Lung cancer
N-myc	Transcription factor	Neuroblastoma (a nerve cell cancer)
Genes Encoding Other Kinds of Proteins		
bcl-2	Protein that blocks cell suicide	Follicular B cell lymphoma
bcl-1	Cyclin D1, which stimulates the cell cycle clock (gene also called *PRAD1*)	Breast cancer; head and neck cancers
MDM2	Protein antagonist of p53 tumor-suppressor protein	Wide variety of sarcomas (connective tissue cancers)
TUMOR-SUPPRESSOR GENES		
Genes Encoding Cytoplasmic Proteins		
APC	Step in a signaling pathway	Colon cancer; stomach cancer
DPC4	A relay in signaling pathway that inhibits cell division	Pancreatic cancer
NF-1	Inhibitor of Ras, a protein that stimulates cell division	Neurofibroma; myeloid leukemia
NF-2	Inhibitor of Ras	Meningioma (brain cancer); schwannoma (cancer of cells supporting peripheral nerves)
Genes Encoding Nuclear Proteins		
MTS1	p16 protein, which slows the cell cycle clock	A wide range of cancers
p53	p53 protein, which halts cell division at the G_1 checkpoint	A wide range of cancers
Rb	Rb protein, which acts as a master brake of the cell cycle	Retinoblastoma; breast cancer; bone cancer; bladder cancer
Genes Encoding Proteins of Unknown Cellular Locations		
BRCA1	?	Breast cancer; ovarian cancer
BRCA2	?	Breast cancer
VHL	?	Renal cell cancer

vent cyclins from binding to Cdk's. The genes that encode these proteins are called **tumor-suppressor genes.** Their mutant alleles are genetically recessive.

One of the first tumor-suppressor genes to be discovered was the *Rb* gene discussed in chapter 11. The role of *Rb* is to bind to the transcription factor E2F, preventing its activity. The E2F transcription factor is necessary for the expression of a number of cell-cycle-specific genes. In the normal course of the cell cycle, the action of cyclin-dependent kinase acts to phosphorylate Rb. Because phosphorylated Rb cannot bind E2F, this releases E2F from its inhibition and allows the cell cycle to progress.

A loss of normal Rb function leads to the loss of control over the cell cycle (figure 20.14). However, both normally functioning copies of *Rb* must be lost before control is removed. Thus, the cancer-inducing mutant alleles of *Rb* are genetically recessive.

Another well-characterized tumor-suppressor gene is *p53*, a gene found to be mutated in a large proportion of human cancers. The role of *p53* is to integrate signals that sense DNA damage during G_1 and G_2 (figure 20.15). When significant DNA damage is detected, *p53* induces apoptosis to remove the damaged cell. Loss of *p53* leads to accumulation of mutations that would have otherwise been removed, and thus failure to prevent transformation to cancerous growth. One of the reasons repeated smoking leads inexorably to lung cancer is that it induces *p53* mutations. Indeed, almost half of all cancers involve mutations of the *p53* gene.

Epigenetic Events and Cancer

Recent research on the relationship between chromatin structure and the regulation of gene expression also shed light on the process of malignant transformation. The role of methylation and histone deacetylation discussed in chapter 18 has a bearing on cancer as well. The majority of methylation of mammalian DNA is of cytosine at sites

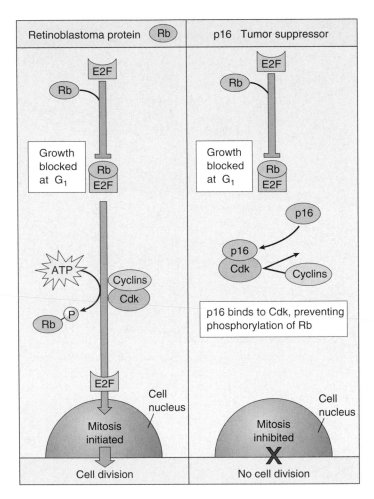

FIGURE 20.14
How the tumor-suppressor genes *Rb* and *p16* interact to block cell division. The retinoblastoma protein (Rb) binds to the transcription factor (E2F) that activates genes in the nucleus, preventing this factor from initiating mitosis. The G_1 checkpoint is passed when Cdk interacts with cyclins to phosphorylate Rb, releasing E2F. The p16 tumor-suppressor protein reinforces Rb's inhibitory action by binding to Cdk so that Cdk is not available to phosphorylate Rb.

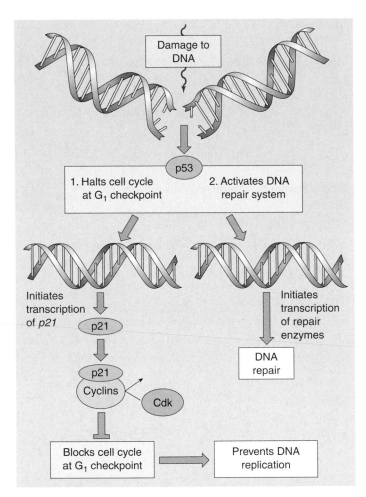

FIGURE 20.15
The role of tumor-suppressor *p53* in regulating the cell cycle. The p53 protein works at the G_1 checkpoint to check for DNA damage. If the DNA is damaged, p53 activates the DNA repair system and stops the cell cycle at the G_1 checkpoint (before DNA replication). This allows time for the damage to be repaired. p53 stops the cell cycle by inducing the transcription of *p21*. The p21 protein then binds to cyclins and prevents them from complexing with Cdk.

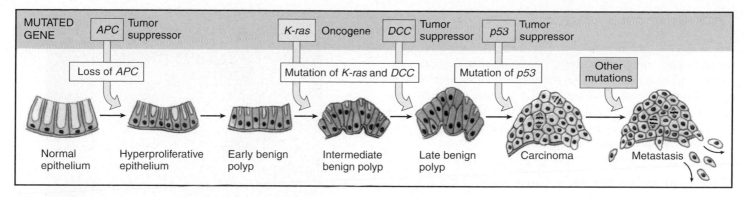

FIGURE 20.16

The progression of mutations that commonly lead to colorectal cancer. The fatal metastasis is the last of six serial changes that the epithelial cells lining the rectum undergo. One of these changes is brought about by mutation of a proto-oncogene, and three of them involve mutations that inactivate tumor-suppressor genes.

where cytosine is followed by guanine in the DNA sequence (called CpG repeat sites). Most of the CpG repeats found in humans are methylated. Those that are not, so-called "CpG islands," correlate with genes that are expressed. Recently, hypermethylation of CpG islands has been observed in a number of human cancers. Although it is not yet entirely clear that hypermethylation is a direct cause of cancer, rather than a result of it, the finding that key tumor-suppressor genes may exhibit this hypermethylation strengthens the case for this epigenetic process as a contributing cause. And even if the hypermethylation is not the cancer-initiating event, aberrant methylation patterns in cancer cells may shut off additional genes, accelerating the process. This provides another potential mechanism of inactivation of tumor-suppressor genes to add to our growing list: point mutation, structural alterations of chromosomes, and now epigenetic control of gene expression.

Cancer Is a Multistep Process

Cells control proliferation at several checkpoints, and all of these controls must be inactivated for cancer to be initiated. Therefore, the induction of most cancers involves the mutation of multiple genes; four to six is a typical number (figure 20.16). In many of the tissue culture cell lines used to study cancer, most of the controls are already inactivated, so that mutations in only one or a few genes transform the line into cancerous growth. The need to inactivate several regulatory genes almost certainly explains why most cancers occur in people over 40 years of age (figure 20.17); in older persons, individual cells have had more time to accumulate multiple mutations. It is now clear that mutations, including those in potentially cancer-causing genes, do accumulate over time. Using the polymerase chain reaction (PCR), researchers in 1994 searched for a certain cancer-associated gene mutation in the blood cells of 63 cancer-free people. They found that the mutation occurred 13 times more often in people over 60 years old than in people under 20.

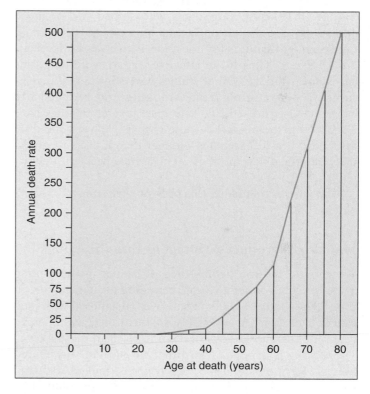

FIGURE 20.17

The annual death rate from cancer climbs with age. The rate of cancer deaths increases steeply after age 40 and even more steeply after age 60, suggesting that several independent mutations must accumulate to give rise to cancer.

Cancer is a disease in which the controls that normally restrict cell proliferation do not operate. In some cases, cancerous growth is initiated by the inappropriate activation of proteins that regulate the cell cycle; in other cases, it is initiated by the inactivation of proteins that normally suppress cell division.

Smoking and Cancer

How can we prevent cancer? The most obvious strategy is to minimize mutational insult. Anything that decreases exposure to mutagens can decrease the incidence of cancer because exposure has the potential to mutate a normal gene into an oncogene. It is no accident that the most reliable tests for the carcinogenicity of a substance are those that measure the substance's mutagenicity.

The Association between Smoking and Cancer

About one-third of all cases of cancer in the United States are directly attributable to cigarette smoking. The association between smoking and cancer is particularly striking for lung cancer (figure 20.18). Studies of male smokers show a highly positive correlation between the number of cigarettes smoked per day and the incidence of lung cancer (figure 20.19). For individuals who smoke two or more packs a day, the risk of contracting lung cancer is at least 40 times greater than it is for nonsmokers, whose risk level approaches zero. Clearly, an effective way to avoid lung cancer is not to smoke. Other studies have shown a clear relationship between cigarette smoking and reduced life expectancy (figure 20.20). Life insurance companies have calculated that smoking a single cigarette lowers one's life expectancy by 10.7 minutes—longer than it takes to smoke the cigarette! Every pack of 20 cigarettes bears an unwritten label:

"The price of smoking this pack of cigarettes is 3½ hours of your life."

Smoking Introduces Mutagens into the Lungs

Over half a million people died of cancer in the United States in 2002; about 28% of them died of lung cancer. In the 1980s, about 140,000 persons were diagnosed with lung cancer each year. Around 90% of them died within three years after diagnosis; 96% of them were cigarette smokers.

Smoking is a popular pastime. In 2001 in the United States, 23.5% of adults and 31% of teens smoked, and U.S. smokers consumed 420 billion cigarettes in 2000. The smoke emitted from these cigarettes contains some 3000 chemical components, including vinyl chloride, benzo[a]pyrenes, and nitroso-nor-nicotine, all potent mutagens. Smoking places these mutagens into direct contact with the tissues of the lungs.

Mutagens in the Lungs Cause Cancer

Introducing powerful mutagens into the lungs causes considerable damage to the genes of the epithelial cells that line the lungs and are directly exposed to the chemicals. Among the genes that are mutated as a result are some whose normal function is to regulate cell proliferation. When these genes are damaged, lung cancer results.

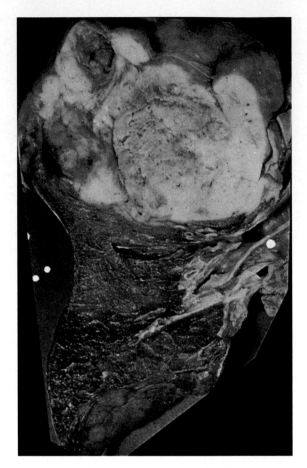

FIGURE 20.18
Photo of a cancerous human lung. The bottom half of the lung is normal, while a cancerous tumor has completely taken over the top half. The cancer cells will eventually break through into the lymph and blood vessels and spread through the body.

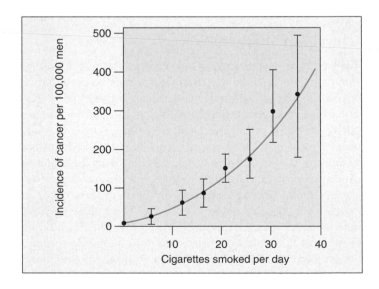

FIGURE 20.19
Smoking causes cancer. The annual incidence of lung cancer per 100,000 men clearly increases with the number of cigarettes smoked per day.

This process has been clearly demonstrated for benzo[*a*]pyrene (BP), one of the potent mutagens released into cigarette smoke from tars in the tobacco. The epithelial cells of the lung absorb BP from tobacco smoke and chemically alter it to a derivative form. This derivative form, benzo[*a*]pyrene-diolepoxide (BPDE), binds directly to the tumor-suppressor gene *p53* and mutates it to an inactive form. The protein encoded by *p53* oversees the G_1 cell cycle checkpoint described in chapter 11 and is one of the body's key mechanisms for preventing uncontrolled cell proliferation. The destruction of *p53* in lung epithelial cells greatly hastens the onset of lung cancer; *p53* is mutated to an inactive form in over 70% of lung cancers. When examined, the *p53* mutations in cancer cells almost all occur at one of three "hotspots." The key evidence linking smoking and cancer is that when the mutations of *p53* caused by BPDE from cigarettes are examined, they occur at the same three specific "hotspots!"

The Incidence of Cancer Reflects Smoking

Cigarette manufacturers argue that the causal connection between smoking and cancer has not been proved, and that somehow the relationship is coincidental. Look carefully at the data presented in figure 20.21, and see if you agree. The upper graph, compiled from data on American men, shows the incidence of smoking from 1900 to 1990 and the incidence of lung cancer over the same period. Note that as late as 1920, lung cancer was a rare disease. About 20 years after the incidence of smoking began to increase among men, lung cancer also started to become more common.

Now look at the lower graph, which presents data on American women. Because of social mores, significant numbers of American women did not smoke until after World War II, when many social conventions changed. As late as 1963, when lung cancer among males was near current levels, this disease was still rare in women. In the United States that year, only 6588 women died of lung cancer. But as more women smoked, more developed lung cancer, again with a lag of about 20 years. American women today have achieved equality with men in the numbers of cigarettes they smoke, and their lung cancer death rates are today approaching those for men. In 2002, more than 65,000 women died of lung cancer in the United States. The current annual rate of deaths from lung cancer in male and female smokers is 180 per 100,000, or about 2 out of every 1000 smokers *each year*.

The easiest way to avoid cancer is to avoid exposure to mutagens. The single greatest contribution one can make to a longer life is not to smoke.

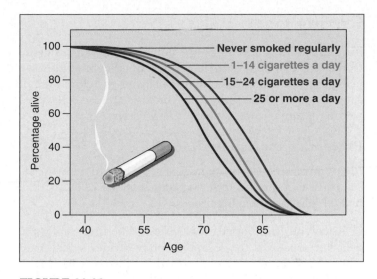

FIGURE 20.20
Tobacco reduces life expectancy. The world's longest-running survey of smoking, begun in 1951 in Britain, revealed that by 1994 the death rate for smokers had climbed to three times the rate for nonsmokers among men 35 to 69 years of age. *Source:* Data from *New Scientist*, October 15, 1994.

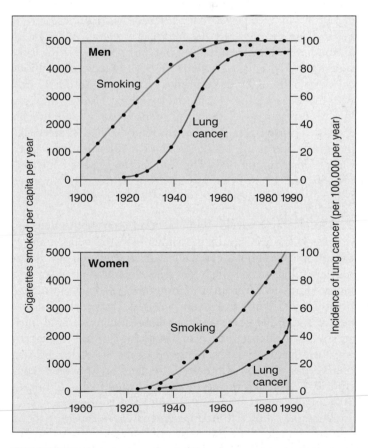

FIGURE 20.21
The incidence of lung cancer in men and women. What do these graphs indicate about the connection between smoking and lung cancer?

Curing Cancer

Potential cancer therapies are being developed on many fronts; eight of these targets are described here and shown in figure 20.22. Some therapies act to prevent the start of cancer within cells. Others act outside cancer cells, preventing tumors from growing and spreading.

Preventing the Start of Cancer

Many promising cancer therapies act within potential cancer cells, focusing on different stages of the cell's "Shall I divide?" decision-making process.

1. Receiving the Signal to Divide. The first step in the decision process is the reception of a "divide" signal, usually a small protein called a growth factor released from a neighboring cell. The growth factor is received by a protein receptor on the cell surface. Mutations that increase the number of receptors on the cell surface amplify the division signal and so lead to cancer. For example, over 20% of breast cancer tumors have been proven to overproduce a protein called HER2 associated with the receptor for epidermal growth factor.

Therapies directed at this stage of the decision process utilize the human immune system to attack cancer cells. Special protein molecules called *monoclonal antibodies*, created by genetic engineering, are the therapeutic agents. Monoclonal antibodies are designed to seek out and stick to HER2. Like waving a red flag, the presence of the monoclonal antibody calls down attack by the immune system on the HER2 cell. Because breast cancer cells overproduce HER2, they are killed preferentially. The drug manufacturer Genentech's recently approved monoclonal antibody, called "herceptin," has given promising results in clinical tests. In other tests, the monoclonal antibody C225, directed against epidermal growth factor receptors, has succeeded in curing advanced colon cancer. Clinical trials of C225 have begun.

2. The Relay Switch. The second step in the decision process is the passage of the signal into the cell's interior, the cytoplasm. This is carried out in normal cells by a protein called Ras that acts as a relay switch. When growth factor binds to a receptor such as EGF, the adjacent Ras protein acts like it has been "goosed," contorting into a new shape. This new shape is chemically active, and initiates a chain of reactions that passes the "divide" signal inward toward the nucleus. Mutated forms of the Ras protein behave like a relay switch stuck in the "on" position, continually instructing the cell to divide when it should not. In 30% of all cancers, a mutant form of Ras is present.

Therapies directed at this stage of the decision process take advantage of the fact that normal Ras proteins are inactive when made. Only after it has been modified by the special enzyme *farnesyl transferase* does Ras protein become able to function as a relay switch. In tests on animals, farnesyl transferase inhibitors induce the regression of tumors and prevent the formation of new ones.

3. Amplifying the Signal. The third step in the decision process is the amplification of the signal within the cytoplasm. Just as a TV signal needs to be amplified in order to be received at a distance, so a "divide" signal must be amplified if it is to reach the nucleus at the interior of the cell, a very long journey at a molecular scale. Cells use an ingenious trick to amplify the signal. Ras, when "on," activates an enzyme, a protein kinase. This protein kinase activates other protein kinases that in their turn activate still others. The trick is that once a protein kinase enzyme is activated, it goes to work like a demon, activating hoards of others every second! And each and every one it activates behaves the same way, activating still more, in a cascade of ever-widening effect. At each stage of the relay, the signal is amplified 1000-fold. Mutations stimulating any of the protein kinases can dangerously increase the already amplified signal and lead to cancer. For example, 5% of all cancers have a mutant hyperactive form of the protein kinase Src.

Therapies directed at this stage of the decision process employ so-called "antisense RNA" directed specifically against Src or other cancer-inducing kinase mutations. The idea is that the *src* gene uses a complementary copy of itself to manufacture the Src protein (the "sense" RNA or messenger RNA), and a mirror image complementary copy of the sense RNA ("antisense" RNA) will stick to it, gumming it up so it can't be used to make Src protein. The approach appears promising. In tissue culture, antisense RNAs inhibit the growth of cancer cells, and some also appear to block the growth of human tumors implanted in laboratory animals. Human clinical trials are under way.

4. Releasing the Brake. The fourth step in the decision process is the removal of the "brake" the cell uses to restrain cell division. In healthy cells, this brake, a tumor-suppressor protein called Rb, blocks the activity of a transcription factor protein called E2F. When free, E2F enables the cell to copy its DNA. Normal cell division is triggered to begin when Rb is inhibited, unleashing E2F. Mutations that destroy Rb release E2F from its control completely, leading to ceaseless cell division. A defective form of Rb is found in 40% of all cancers.

Therapies directed at this stage of the decision process are only now being attempted. They focus on drugs able to inhibit E2F, which should halt the growth of tumors arising from inactive Rb. Experiments in mice in which the E2F genes have been destroyed provide a model system to study such drugs, which are being actively investigated.

5. Checking That Everything Is Ready. The final step in the decision process is the mechanism used by the cell to ensure that its DNA is undamaged and ready to divide. This job is carried out in healthy cells by the tumor-suppressor protein p53, which inspects the integrity of the DNA. When it detects damaged or foreign DNA, p53 stops cell division and activates the cell's DNA repair systems. If the damage doesn't get repaired in a reasonable time, p53 pulls the plug, triggering events that kill the cell. In this way, mutations such as those that cause cancer are either repaired, or the cells containing them eliminated. If p53 is itself destroyed by mutation, future damage accumulates unrepaired. Among this damage are mutations that lead to cancer. Approximately 50% of all cancers have a disabled p53. Fully 70 to 80% of lung cancers have a mutant inactive p53; the chemical benzo[*a*]pyrene in cigarette smoke is a potent mutagen of p53.

A promising new therapy using adenovirus (responsible for mild colds) is being targeted at cancers with a mutant p53. To grow in a host cell, adenovirus must use the product of its gene, *E1B*, to block the host cell's p53, thereby enabling replication of the adenovirus DNA. This means that while mutant adenovirus without *E1B* cannot grow in healthy cells, the mutants should be able to grow in, and destroy, cancer cells with defective p53. When human colon and lung cancer cells are introduced into mice lacking an immune system and allowed to produce substantial tumors, 60% of the tumors simply disappear when treated with E1B-deficient adenovirus, and do not reappear later. Initial clinical trials are very encouraging.

6. Stepping on the Gas.   Cell division starts with replication of the DNA. In healthy cells, another tumor suppressor "keeps the gas tank nearly empty" for the DNA replication process by inhibiting production of an enzyme called telomerase. Without this enzyme, a cell's chromosomes lose material from their tips, called telomeres. Every time a chromosome is copied, more tip material is lost. After some 30 divisions, so much is lost that copying is no longer possible. Cells in the tissues of an adult human have typically undergone 25 or more divisions. Cancer can't get very far with only the five remaining cell divisions, so inhibiting telomerase is a very effective natural brake on the cancer process. It is thought that almost all cancers involve a mutation that destroys the telomerase inhibitor, releasing this brake and making cancer possible. It should be possible to block cancer by reapplying this inhibition. Cancer therapies that inhibit telomerase are just beginning clinical trials.

Preventing the Spread of Cancer

7. Tumor Growth.   Once a cell begins cancerous growth, it forms an expanding tumor. As the tumor grows ever larger, it requires an increasing supply of food and nutrients, obtained from the body's blood. To facilitate this necessary grocery shopping, tumors leak substances into the surrounding tissues that encourage angiogenesis, the formation of small blood vessels. Chemicals that inhibit this process are called angiogenesis inhibitors. In mice, two such angiogenesis inhibitors, angiostatin and endostatin, caused tumors to regress to microscopic size. This very exciting result has proven controversial, but initial human trials seem promising.

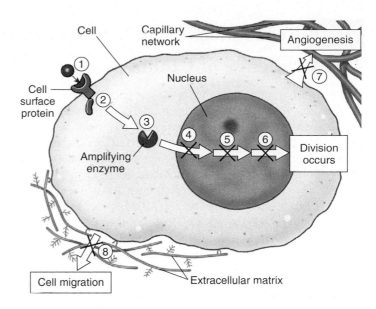

FIGURE 20.22
Potential targets for cancer therapies. New molecular therapies target eight different stages in the cancer process. (*1*) On the cell surface, a growth factor signals the cell to divide. (*2*) Just inside the cell, a protein relay switch passes on the divide signal. (*3*) In the cytoplasm, enzymes amplify the signal. In the nucleus, (*4*) a "brake" preventing DNA replication is released, (*5*) proteins check that the replicated DNA is not damaged, and (*6*) other proteins rebuild chromosome tips so DNA can replicate. (*7*) The new tumor promotes angiogenesis, the formation of growth-promoting blood vessels. (*8*) Some cancer cells break away from the extracellular matrix and invade other parts of the body.

8. Metastasis.   If cancerous tumors simply continued to grow where they form, many could be surgically removed, and far fewer would prove fatal. Unfortunately, many cancerous tumors eventually metastasize; that is, individual cancer cells break their moorings to the extracellular matrix and spread to other locations in the body where they initiate formation of secondary tumors. This process involves (1) metal-requiring protease enzymes that cleave the cell-matrix linkage, (2) components of the extracellular matrix such as fibronectin that also promote the migration of several noncancerous cell types, and (3) RhoC, a GTP-hydrolyzing enzyme that promotes cell migration by providing needed GTP. All of these components are necessary for metastasis to occur, and all offer promising targets for future anticancer therapy.

Therapies such as those described here are only part of a wave of potential treatments under development and clinical trial. The clinical trials will take years to complete, but in the coming decade we can expect cancer to become a curable disease.

> **Our understanding of how mutations produce cancer has progressed to the point where promising potential therapies can be tested.**

20.4 Reproductive cloning of animals, once thought impossible, isn't.

The Challenge of Cloning

The promise of genetic engineering has had little impact on the one area of medicine where it might be expected to have had the greatest impact—reproduction. This is now changing, led by surprising advances in cloning commercial animals. The difficulty in using genetic engineering to improve livestock is in getting enough animals. Breeding genetically improved individuals produces offspring only slowly. Ideally, one would like to "Xerox" many exact genetic copies of the desirable strain, but adult animals can't be cloned—or can they? At one time, it was commonly accepted that they couldn't be, but we now know that conclusion was unwarranted.

Wilmut's Lamb

Although earlier attempts to clone animals using adult cells generally failed, the key advance was made in Scotland by Keith Campbell, a geneticist studying the cell cycle of agricultural animals. Campbell reasoned, "Maybe the egg and the donated nucleus need to be at the same stage in the cell cycle." This proved to be a key insight. In 1994, researcher Neil First, and in 1995 Campbell himself working with reproductive biologist Ian Wilmut, succeeded in cloning farm animals from advanced embryos by first starving the cells, so that they paused at the beginning of the cell cycle at the G_1 checkpoint. Two starved cells are thus synchronized at the same point in the cell cycle.

Wilmut then attempted the key breakthrough, the experiment that had eluded researchers since Spemann proposed it 59 years before: He set out to transfer the nucleus from an adult differentiated cell into an enucleated egg, and to allow the resulting embryo to grow and develop in a surrogate mother, hopefully producing a healthy animal.

Wilmut removed mammary cells from the udder of a six-year-old sheep (figure 20.23). (With tongue in cheek, the clone was later named "Dolly" after the country singer Dolly Parton.) The cells were grown in tissue culture; some were frozen so that in the future it would be possible with genetic fingerprinting to prove that a clone was indeed genetically identical to the six-year-old ewe.

In preparation for cloning, Wilmut's team reduced for five days the concentration of serum on which the sheep mammary cells were subsisting. In parallel preparation, eggs obtained from another ewe were enucleated, the nucleus of each egg carefully removed with a micropipette.

Mammary cells and egg cells were then surgically combined in January of 1996, a mammary cell being inserted inside the covering around each egg cell. Wilmut then applied a brief electrical shock. A neat trick, this causes the plasma membranes surrounding the two cells to become leaky, so that the contents of the mammary cell pass into the egg cell. The shock also jump-starts the cell cycle, initiating cell division.

After six days, in 30 of 277 tries, the dividing embryo reached the hollow-ball blastula stage, and 29 of these were transplanted into surrogate mother sheep. Approximately five months later, on July 5, 1997, one sheep gave birth to a lamb. This lamb, "Dolly," was the first successful clone generated from a differentiated animal cell.

Wilmut's successful cloning of fully differentiated sheep cells is a milestone event in gene technology. Even though his procedure proved inefficient (only one of 277 trials succeeded), it established the point beyond all doubt that adult animal cells *can* be cloned. In subsequent years, researchers

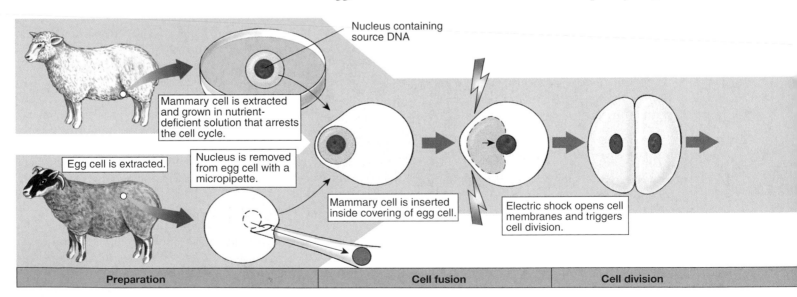

Nucleus containing source DNA

Mammary cell is extracted and grown in nutrient-deficient solution that arrests the cell cycle.

Egg cell is extracted.

Nucleus is removed from egg cell with a micropipette.

Mammary cell is inserted inside covering of egg cell.

Electric shock opens cell membranes and triggers cell division.

| Preparation | Cell fusion | Cell division |

FIGURE 20.23

Wilmut's animal cloning experiment. Wilmut combined a nucleus from a mammary cell with an enucleated egg cell to successfully clone a sheep, named Dolly, that grew to be a normal adult and bore healthy offspring before dying of lung disease in 2003.

have succeeded in greatly improving the efficiency of cloning. Unfortunately, as we will see, problems often develop in the cloned animals.

Problems with Reproductive Cloning

Since Dolly's birth in 1997, scientists have successfully cloned sheep, mice, cattle, goats, and pigs. However, only a small percentage of the transplanted embryos survive to term; most die late in pregnancy. Those that survive to be born usually die soon thereafter. Many become oversized, a condition known as *large offspring syndrome.*

The few cloned offspring that reach childhood face an uncertain future, as their development into adults tends to go unexpectedly haywire.

The Importance of Genomic Imprinting

What is going wrong? It turns out that as human eggs and sperm mature, their DNA is conditioned by the parent female or male, a process called reprogramming. Chemical changes are made to the DNA that alter when particular genes are expressed without changing DNA sequences.

In the years since Dolly, scientists have learned a lot about reprogramming. It appears to occur by a process called *genomic imprinting.* While the details are complex, the basic mechanism of genomic imprinting is simple.

Like a book, a gene can have no impact unless it is read. Genomic imprinting works by blocking the cell's ability to read certain genes. A gene is locked in the "off" position by chemically altering some of the cytosine DNA units. Because this involves adding a $-CH_3$ group (a methyl group), the process is called *methylation.* After a gene has been methylated, the polymerase protein that is supposed to "read" the gene can no longer recognize it. The gene has been shut off.

Genomic imprinting can also lock genes in the "on" position, permanently activating them. This process also uses methylation; in this case, however, it is not the gene that is blocked. Rather, a DNA sequence that normally would have prevented the gene from being read is blocked.

Why Cloning Fails

Normal human development depends on precise genomic imprinting. This chemical reprogramming of the DNA, which occurs in adult reproductive tissue, takes months for sperm and years for eggs.

During cloning, by contrast, the reprogramming of the donor DNA must occur within a few minutes. After the donor nucleus is added to an egg whose nucleus has been removed, the reconstituted egg begins to divide within minutes, starting the process of making a new individual.

Cloning fails because there is simply not enough time in these few minutes to get the reprogramming job done properly. For example, Lorraine Young of the Roslin Institute in Scotland (Dolly's birthplace) reported in 2001 that in Large Offspring Syndrome sheep, many genes have failed to become properly methylated.

Human cloning will not be practical until scientists figure out how to reprogram a donor nucleus, as occurs in the DNA of sperm or eggs in our bodies. This reprogramming may be as simple as finding a way to postpone the onset of cell division after adding a donor nucleus to the enucleated egg, or may prove to be a much more complex process.

Recent experiments have demonstrated the possibility of cloning differentiated mammalian tissue. Reproductive cloning of animals from adult tissue usually fails for lack of proper gene conditioning.

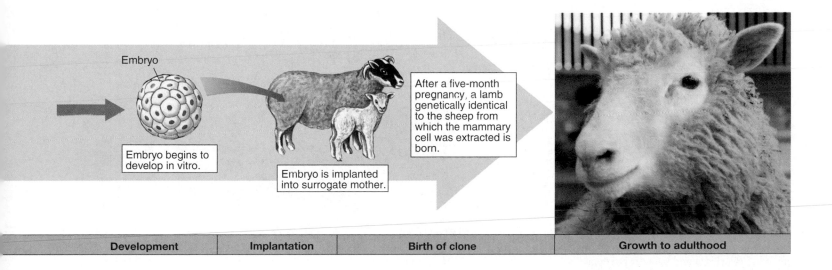

Embryo

Embryo begins to develop in vitro.

Embryo is implanted into surrogate mother.

After a five-month pregnancy, a lamb genetically identical to the sheep from which the mammary cell was extracted is born.

| Development | Implantation | Birth of clone | Growth to adulthood |

20.5 Therapeutic cloning is a promising but controversial possibility.

Stem Cells

Embryonic Stem Cells

In 1981, mouse stem cells were first discovered to be *pluripotent*—to have the ability to form any body tissue, and even an adult animal. This finding launched the era of stem cell research. After many years of failed attempts, human embryonic stem cells were isolated by James Thomson of the University of Wisconsin in 1998.

What is an embryonic stem cell? At the dawn of a human life, a sperm fertilizes an egg to create a single cell destined to become a child. As development commences, that cell begins to divide, producing after five or six days a small ball of a few hundred cells called a blastocyst. Described in chapter 51, a blastocyst consists of a protective outer layer destined to form the placenta, enclosing an inner cell mass of **embryonic stem cells** (figure 20.24). Each embryonic stem cell is capable by itself of developing into a healthy individual. In cattle breeding, for example, these cells are frequently separated by the breeder and used to produce multiple clones of valuable offspring.

Because they can develop into any tissue, these embryonic stem cells offer the exciting possibility of restoring damaged tissues, such as muscle or nerve tissue (figure 20.25). Experiments have already been tried successfully in mice. Heart muscle cells grown from mouse embryonic stem cells have been successfully integrated with the heart tissue of a living mouse. This suggests that the damaged heart muscle of heart attack victims might be repairable with stem cells. In other experiments with mice, damaged spinal neurons have been partially repaired, suggesting a

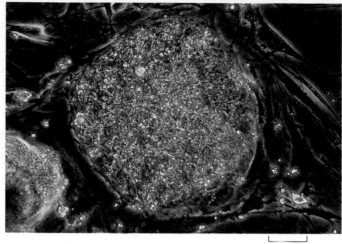

0.60 μm

FIGURE 20.24
Human embryonic stem cells (20×). Stem cells removed from a six-day blastocyst can be established in culture and then main-tained indefinitely. This mass is a colony of undifferentiated human embryonic stem cells surrounded by fibroblasts (elongated cells) that serve as a "feeder layer."

path to treating spinal injuries. Dopamine-producing neurons of mouse brains whose progressive loss is responsible for Parkinson disease have been successfully replaced with embryonic stem cells, as have the islet cells of the pancreas whose loss leads to juvenile diabetes.

These experiments in mice suggest that embryonic stem cell therapy may hold great promise in treating a wide variety of human illnesses involving damaged or lost tissues. This research, however, is quite controversial be-

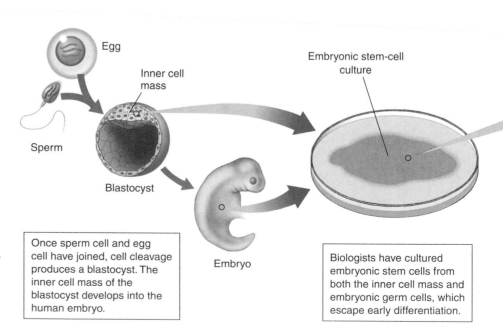

FIGURE 20.25 Using embryonic stem cells to restore damaged tissue.
Embryonic stem cells can develop into any body tissue. Methods are being developed for growing the tissue and using it to repair damaged tissue in adults, such as the brain cells of multiple sclerosis patients, heart muscle, and spinal nerves.

Once sperm cell and egg cell have joined, cell cleavage produces a blastocyst. The inner cell mass of the blastocyst develops into the human embryo.

Biologists have cultured embryonic stem cells from both the inner cell mass and embryonic germ cells, which escape early differentiation.

cause embryonic stem cells are typically isolated from embryos discarded by reproductive clinics, where in vitro fertilization procedures typically produce many more human embryos than can be successfully implanted in the prospective mother's womb. This raises serious ethical issues among those citizens who believe life starts at fertilization.

A second serious problem is associated with using embryonic stem cells to replace defective or lost tissues. All of the successful experiments just described were carried out in mice whose immune systems had been disabled. Had these mice possessed fully functional immune systems, they almost certainly would have rejected the implanted stem cells as foreign. For such stem cell therapy to work in humans, this problem needs to be addressed and solved.

Tissue-Specific Adult Stem Cells

New experimental results hint at a way around the ethical maze presented by the use of stem cells derived from embryos. Go back for a moment to our explanation of how a vertebrate develops. What happens to the embryonic stem cells? They start to take different developmental paths. Some become destined to form nerve tissue and, after this decision is taken, can never produce any other kind of cell. They are then called nerve stem cells. Others become specialized to produce blood, and still others muscle. Each major tissue is represented by its own kind of **tissue-specific stem cell.** Now here's the key point: As development proceeds, these tissue-specific stem cells persist—even in adults. So why not use these adult cells, rather than embryonic stem cells?

Transplanted Tissue-Specific Stem Cells Cure Multiple Sclerosis in Mice

In pathfinding 1999 laboratory experiments by Dr. Evan Snyder of Harvard Medical School, tissue-specific stem cells were able to restore lost brain tissue. He and his co-workers injected neural stem cells (immediate descendants of embryonic stem cells able to become any kind of neural cell) into the brains of newborn mice with a disease resembling multiple sclerosis (MS). These mice lacked the cells that maintain the layers of myelin insulation around signal-conducting nerves. The injected stem cells migrated all over the brain and were able to convert themselves into the missing type of cell. The new cells then proceeded to repair the ravages of the disease by replacing the lost insulation of signal-conducting nerve cells. Many of the treated mice fully recovered. In mice at least, tissue-specific stem cells offer a treatment for MS.

The approach seems very straightforward and should apply to humans. Indeed, blood stem cells are already routinely used in humans to replenish the bone marrow of cancer patients after marrow-destroying therapy. The problem with extending the approach to other kinds of tissue-specific stem cells is that it has not always been possible to find the kind of tissue-specific stem cell needed.

Human embryonic stem cells offer the possibility of replacing damaged or lost human tissues, although the procedures are controversial. Transplanted tissue-specific stem cells may allow us to replace damaged or lost tissue, offering cures for many disorders that cannot now be treated, while avoiding the ethical problems posed by the use of embryonic stem cells.

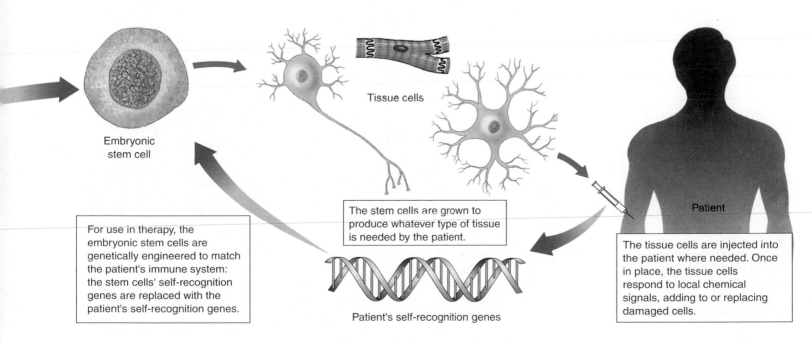

Embryonic stem cell

Tissue cells

Patient

For use in therapy, the embryonic stem cells are genetically engineered to match the patient's immune system: the stem cells' self-recognition genes are replaced with the patient's self-recognition genes.

The stem cells are grown to produce whatever type of tissue is needed by the patient.

The tissue cells are injected into the patient where needed. Once in place, the tissue cells respond to local chemical signals, adding to or replacing damaged cells.

Patient's self-recognition genes

Therapeutic Cloning

By surgically transplanting embryonic stem cells, scientists have performed the remarkable feat of repairing disabled body tissues in mice. The basic strategy for repairing damaged tissues is to surgically transfer embryonic stem cells to the damaged area, where the stem cells can form healthy replacement cells. Stem cells transferred into mouse heart muscle develop into heart muscle cells, replacing cells dead from heart attack. Stem cells transferred into a mouse brain form neurons, offering hope that we eventually will learn to use embryonic stem cells to repair spinal injury. Embryonic stem cells of mice have been induced to become insulin-secreting pancreas cells. The new cells produce only about 2% as much insulin as normal cells do, so there is still plenty to learn, but the take-home message is clear: Transplanted embryonic stem cells offer a path toward curing type-1 diabetes.

While exciting, these advances in stem cell research were all experiments carried out in strains of mice without functioning immune systems. This prevents the mice from rejecting transplanted stem cells as "foreign." A human with a normal immune system might well refuse to accept transplanted stem cells simply because they are from another individual.

Early in 2001, a research team at Rockefeller University reported a way around this potentially serious problem. Their solution? First, they isolate skin cells; then, using the same procedure that created Dolly, they create an embryo from those cells. After removing the nucleus from the skin cell, they insert it into an egg whose nucleus has already been removed. The egg with its skin cell nucleus is allowed to form a 120-cell embryo. The embryo is then destroyed, and its cells are used as embryonic stem cells for transfer to injured tissue.

Using this procedure, which they called **therapeutic cloning,** the researchers succeeded in making cells from the tail of a mouse convert into the dopamine-producing cells of the brain that are lost in Parkinson disease (figure 20.26).

Therapeutic cloning successfully addresses the key problem that must be solved before stem cells can be used to repair human tissues damaged by heart attack, nerve injury, diabetes, or Parkinson disease—the problem of immune acceptance. Since stem cells are cloned from the body's own tissues in therapeutic cloning, they pass the immune system's "self" identity check, and the body readily accepts them.

Two key problems remain unsolved, however. First, a means must be found to achieve proper genomic imprinting. Second, investigators must learn how to preserve the egg's mitotic machinery. When the first partially successful cloning of a human embryo from adult skin cells was reported in November 2001, the cloned cells did not survive long enough to make stem cells. In fact, no primate has ever been successfully cloned. From the very first step, the cells don't divide properly. Why? Chromosomes within the egg carry proteins that act as molecular motors during spindle formation. In humans and other primates, these proteins are so tightly bound to the chromosomes that cloning's first step of DNA removal pulls them out too, dooming hope of normal blastocyst formation.

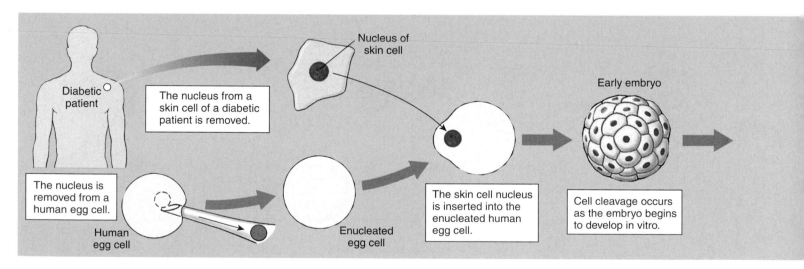

FIGURE 20.26

How human embryos might be used for therapeutic cloning.   Therapeutic cloning differs from reproductive DNA cloning in that after the initial similar stages, the embryo is destroyed and its embryonic stem cells are extracted, grown in culture, and added to a tissue of the individual who provided the DNA. In reproductive cloning, by contrast, the embryo is preserved to be implanted and grown to term in a surrogate mother. This latter procedure was used in cloning Dolly the sheep. Human cells were first cloned in November 2001 in a failed attempt to obtain stem cells for therapeutic cloning procedures such as those outlined in this figure.

The difficult ethical issues raised by human therapeutic cloning could be largely avoided if the stem cells did not have to be harvested from an embryo. Imagine, for example, that it were possible to find pluripotent stem cells somewhere in the body of an *adult* human. In 2001, researchers claimed that they had found just such cells in the bone marrow of mice. They transplanted single stem cells from mouse bone marrow into the marrow of individuals whose marrow had been destroyed. After 11 months, the initial stem cell had given rise to descendant cells that had migrated throughout the body, forming new bone, blood, lung, esophagus, stomach, intestine, liver, and skin cells (figure 20.27). The bone marrow stem cells appear to have the properties of the long-sought pluripotent adult stem cells. Many labs are trying to repeat this preliminary result.

Therapeutic cloning involves initiating blastocyst development from a patient's tissue using nuclear transplant procedures, and then using these embryonic stem cells to replace the patient's damaged or lost tissue.

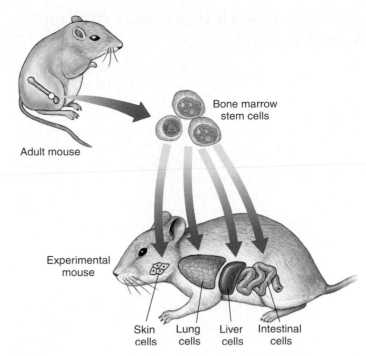

FIGURE 20.27
Pluripotent stem cells. In May 2001, a single cell from the bone marrow of a mouse was claimed to have added functional cells to the lungs, liver, intestine, and skin of an experimental mouse.

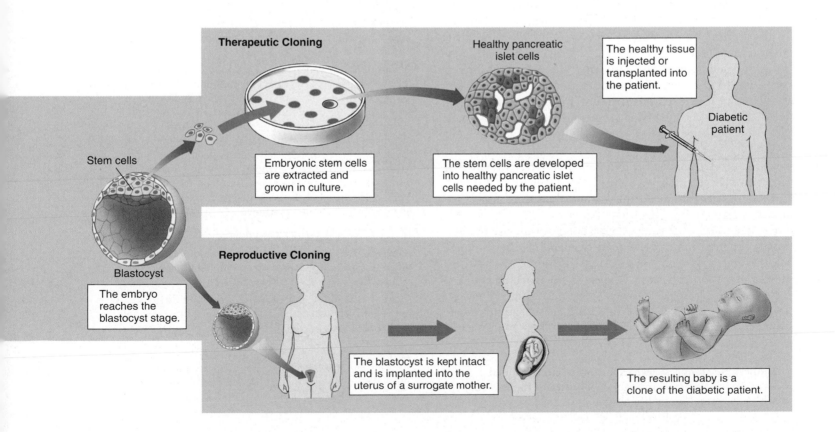

Grappling with the Ethics of Stem Cell Research

Human embryonic stem cells retain the potential to become any tissue in the body and thus have enormous promise for treating a wide range of diseases. Human embryonic stem cells are very difficult to isolate and establish in culture, but a few dozen lines have been successfully obtained from the inner cell mass of six-day blastocysts. It is important to isolate the cells at this early stage, before development begins the process of restricting what sorts of tissues the stem cells can become. The blastocysts are obtained from reproductive clinics, which routinely produce excess embryos in the process of helping infertile couples have children by in vitro fertilization.

However, obtaining embryonic stem cells destroys the early embryo in the process; for this reason, stem cell research raises profound ethical issues. The timeless question of when human life begins cannot be avoided when human embryos are being deliberately destroyed. What is the moral standing of a six-day human embryo? Resolving the tension between scientific knowledge and moral sensibilities brings into play religious, philosophical, and cultural issues. Table 20.5 illustrates the range of issues being discussed.

It will come as no surprise that government, which funds much of modern biomedical research, has become embroiled in the controversy. In Britain, reproductive cloning is banned, but stem cell research and therapeutic cloning to obtain clinically useful stem cells are both permitted. Because the research is funded by the government, there is careful ethical supervision of all research by a variety of governmental oversight committees. Britain's Human Fertilization and Embryology Authority (HFEA), for example, is a panel of scientists and ethicists accountable to Parliament, which oversees government-funded stem cell research. Similar arrangements are being established in Japan and France. Germany, by contrast, discourages all stem cell research.

In the United States, the situation is ambivalent. American stem cell research is chiefly carried out in private research labs using no government funds and thus subject to no ethical oversight. This leaves American scientists pretty much free to do what they want, as long as they use private money. Federal funds were made available in the summer of 2001 for research on the small number of existing human embryonic stem cell lines. In what has become a very political contest between those favoring increased stem cell research and those opposing all such research, it seems certain that federal government policies with regard to stem cell research will fluctuate for some time to come.

Embryonic stem cell research is quite controversial because it involves many ethical issues, not the least of which is when human life begins.

Table 20.5 The Ethics of Stem Cell Research

1. DESTRUCTION OF HUMAN EMBRYOS

Opponents: A human life begins at the moment egg and sperm are united, so destroying an embryo in order to harvest embryonic stem cells is simply murder, and morally wrong. Benefits to others, however great, cannot justify the destruction of a human life.

Proponents: While human embryos should be treated with respect, the potential for saving lives that embryonic stem cell research offers is also a strong moral imperative. The blastocysts being used to obtain stem cells were created to help infertile couples conceive, and would have been destroyed in any event. Besides, it is not clear that an individual life begins at fertilization. An early embryo can split, leading to the birth of identical twins, so it can be argued that individuality begins some days after fertilization.

2. POSSIBILITY OF FUTURE ABUSE

Opponents: Permitting embryonic stem cell research may open the door to further ethically objectionable research. Certainly the development of embryonic stem cell therapies will lead to a cry for therapeutic cloning, so that the therapies can actually be employed in clinical situations. This creation of embryos specifically for the production of embryonic stem cells is morally wrong. In addition, therapeutic cloning to obtain clinical embryonic stem cells is unnatural because it involves producing a viable human embryo without fertilization. If therapeutic use of the results of stem cell research is unacceptable, there is little use in carrying out the research in the first place. It only delays and excalibrates the morally difficult choice posed by therapeutic cloning. Even more disturbing, it opens the door to reproductive cloning— the production of human babies from cloned embryos. This moral nightmare will always be a threat if an absolute line is not drawn preventing all cloning.

Proponents: In practice, only a few hundred cell lines are likely to be required to carry out embryonic stem cell research. The derivation of human cell lines, which is difficult and expensive to do, would be limited to these few lines. Continuous destruction of embryos would be neither desirable nor likely. Therapeutic cloning presents a separate and more complex ethical issue. Because fertilization is not involved, the blastocyst might better be thought of as an "activated egg" rather than an embryo. Such a distinction has biological merit and avoids the ethical issues posed by human reproductive cloning, which should be banned.

3. ALTERNATIVE SOURCES OF STEM CELLS

Opponents: Why not use stem cells derived from adult tissues? These stem cells raise no difficult ethical issues, and can lead to the same medical benefits.

Proponents: Adult stem cells simply cannot do the job. By the time embryonic stem cells have developed into adult stem cells, they have lost much of their developmental versatility, and so lack the range of medical capabilities necessary for regenerative medicine. Also, adult stem cells are not very prolific, and have proven to be difficult to use in therapeutic procedures on experimental laboratory animals.

20.1 Recombination alters gene location.

Gene Transfer

- Some genes move because they are part of plasmids that transfer copies of bacterial genes from one bacterium to another, while others move within transposons that jump from one genomic position to another at random. (pp. 406–408)

Reciprocal Recombination

- Unequal crossing over occurs when homologues exchange segments of unequal length. One chromosome gains extra copies of the multicopy sequences, while the other chromosome loses them. (p. 409)
- Gene conversion occurs as error correction systems in a cell convert one of the mismatched sequences to resemble the other. (p. 409)

20.2 Mutations are changes in the genetic message.

Kinds of Mutation

- Mutations are changes in the hereditary message of an organism, and germ-line mutations provide raw material for evolutionary change. (p. 410)
- DNA mutational alterations arise in at least five ways: base substitution, chemical modification, DNA breaks, slipped mispairing, and triplet expansion. (p. 410)
- Some chromosomal alterations, such as duplications, deletions, aneuploidy, and polyploidy, can change the number of gene copies an individual possesses. (p. 411)

DNA Repair

- Cells can repair many potential genetic mutations by recognizing pairing mismatches created by a base substitution or by recognizing inappropriate bases. (p. 412)

20.3 Most cancer results from mutation of growth-regulating genes.

What Is Cancer?

- Cancer occurs when a cell begins to grow in an uncontrolled and invasive way. It can result in a cluster of cells (tumor) that might metastasize and form new tumors at distant sites. (p. 413).
- Cancer is a gene disorder of somatic tissue, in which damaged genes fail to properly control cell proliferation. (p. 413)

Causes of Cancer

- Agents thought to cause cancer are called carcinogens; for example, some chemicals are carcinogens. (p. 414)
- Chemicals in cigarette smoke, such as tar, have clearly been demonstrated to cause cancer. (p. 415)

Cancer and the Cell Cycle

- Oncogenes are genes that, when introduced, cause a normal cell to become cancerous. (p. 416)
- Proto-oncogenes are normal forms of genes that can be mutated to produce a tumor-forming oncogene. (p. 416)
- Chromosomal abnormalities can lead to the activation of proto-oncogenes. (p. 416)

- Cell division is normally turned off in healthy cells by tumor-suppressor genes. (pp. 416–417)
- Cells control proliferation at several checkpoints, and all of these controls must be inactivated for cancer to be initiated. Therefore, the induction of cancer typically involves the mutation of four to six genes. (p. 419)

Smoking and Cancer

- About one-third of all cancer cases in the United States are directly attributable to cigarette smoking. (p. 420)
- Introducing mutagens into the lungs causes damage to the genes of the epithelial cells that line the lungs and are directly exposed to the chemicals. (p. 420)
- The destruction of *p53* in lung epithelial cells hastens the onset of lung cancer; *p53* is mutated to an inactive form in over 70% of lung cancers. (p. 421)

Curing Cancer

- Cancer therapies are being developed on many fronts, which can be consolidated into two broad categories: preventing the start of cancer and preventing the spread of cancer. (pp. 422–423)

20.4 Reproductive cloning of animals, once thought impossible, isn't.

The Challenge of Cloning

- Cloning differentiated mammalian tissue is possible, but most attempts have failed due to improper gene reprogramming. (pp. 424–425)

20.5 Therapeutic cloning is a promising but controversial possibility.

Stem Cells

- Each embryonic stem cell is capable, by itself, of developing into a healthy individual. (p. 426)
- Because embryonic stem cells can develop into any tissue, they offer the possibility of replacing damaged or lost tissues. (p. 426)
- Stem cell research is very controversial because embryonic stem cells are typically isolated from embryos discarded by reproductive clinics. (p. 426)
- Tissue-specific stem cells may allow the replacement of damaged or lost tissues while avoiding the ethical considerations involved in using embryonic tissues. (p. 427)

Therapeutic Cloning

- Therapeutic cloning involves initiating blastocyst development from a patient's tissue, and then using these embryonic stem cells to replace the patient's damaged or lost tissue. (pp. 428–429)

Grappling with the Ethics of Stem Cell Research

- One of the most controversial issues involving embryonic stem cell research is deciding when human life begins. (p. 430)

Self Test

1. Tumor-suppressor genes includes *p53* and *Rb*. How would a "gain-of-function" mutation likely affect the cell?
   a. The cell would divide constantly because of the loss of cell cycle repression.
   b. The cell would divide much less frequently because of the extra cell cycle repression.
   c. The cell would divide normally because these genes have no effect on cell cycle control.
   d. The cell would commit suicide by apoptosis.

2. In lab, you are studying cell cycle control in the fission yeast *S. pombe*. A student finds a new mutant that she wants to call "giant" because the cells are much larger than normal (suggesting that it is not dividing normally). What type of mutation do you think the student has isolated?
   a. a loss-of-function mutation in a tumor-suppressor gene
   b. a loss-of-function mutation in a cellular proto-oncogene
   c. a gain-of-function mutation in a tumor-suppressor gene
   d. a gain-of-function mutation in a cellular proto-oncogene
   e. Both *a* and *d* are possible.
   f. Both *b* and *c* are possible.

3. Which of the following would be an effective approach to a new cancer therapy?
   a. finding a way to stabilize *p53* specifically in tumor cells
   b. preventing nucleotide synthesis in tumor cells
   c. inactivating the HER2 receptor on tumor cells
   d. inhibiting growth of new blood vessels with endostatin
   e. All of these would help fight cancer.

4. How would the cell cycle be affected if you removed the phosphorylation sites in the Rb protein?
   a. The cell cycle would not be affected because pRb is not phosphorylated normally.
   b. The cell cycle would be blocked in G_1.
   c. The cell cycle would be blocked in G_2.
   d. The cell cycle would be shorter.

5. Embryonic stem (ES) cells are an attractive source of material for therapeutic cloning because
   a. they can be induced to assume any cell fate.
   b. ES cells are not targets for the host immune response, so tissue rejection is not an issue.
   c. there are no other sources of stem cells to use for therapeutic cloning, so ES cells are the only solution.
   d. ES cells will not work as a source of tissue for cloning.

6. How would growing cells in the presence of methyl-adenosine affect the mismatch repair system?
   a. The repair system would only repair half of the errors introduced by DNA polymerase.
   b. There would be no repair of mismatched DNA.
   c. Mismatch repair would be normal, but excision repair would fail.
   d. Methyladenosine would prevent DNA replication, so there would be no need for mismatch repair.

7. Too much time in a tanning booth probably causes DNA damage to epithelial cells. The most likely effect would be
   a. depurination.
   b. pyrimidine dimers.
   c. deamination.
   d. single-stranded nicks in the phosphodiester backbone.

8. Using a "car and driver" analogy, which of the following accurately describes the role of tumor-suppressor genes and proto-oncogenes in normal cells?
   a. Tumor-suppressor genes are the gas pedal, while proto-oncogenes are the brakes.
   b. Tumor-suppressor genes are the brakes, while proto-oncogenes are the gas.
   c. Both tumor-suppressor genes and proto-oncogenes are like the gas, but tumor-suppressors are like turbo and proto-oncogenes are like a regular carburetor.
   d. Tumor-suppressor genes are like the steering wheel, and proto-oncogenes are like the turn signals.

9. During the early years of cancer research, there were two schools of thought regarding the causes of cancer: 1) that cancer was caused entirely by environmental factors, and 2) that cancer was caused by genetic factors. Which was correct?
   a. #1 because we have identified many potential carcinogens
   b. #2 because we know of many proto-oncogenes
   c. #2 because we know of many tumor suppressor genes
   d. Both were correct; most chemical carcinogens function by altering genes.

10. If you found a specific chromosomal deletion in the genome from a tumor, what could be the cause of this specific cancer?
    a. The deletion likely affected a tumor-suppressor gene, leading to a loss of function in the tumor cells.
    b. The deletion likely affected a proto-oncogene, leading to a loss of function in the tumor cells.
    c. The deletion likely affected a tumor-suppressor gene, leading to a gain of function in the tumor cells.
    d. The deletion likely affected a proto-oncogene, leading to a gain of function in the tumor cells.

Test Your Visual Understanding

1. If you were to observe two bacterial cells as shown here, what would you suggest is happening?

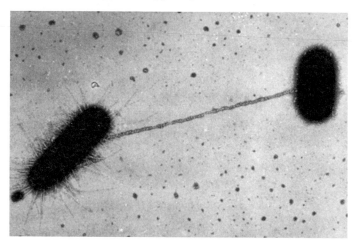

Apply Your Knowledge

1. The data in table 20.3 show the incidence of specific cancers following exposure to environmental carcinogens. Discuss how this type of chemical exposure leads to such a high proportion of skin and lung tumors. (*Hint:* Think about the points of contact with environmental chemicals.)

2. Pretend that you are preparing for a debate about the use of embryonic stem cells for therapeutic cloning, and list three pros and three cons of this technology.

21

Genes Within Populations

FIGURE 21.1
Genetic variation. Natural populations contain abundant
genetic variation for many traits, such as flower color.

Part IV *Evolution*

No other human being is exactly like you (unless you
have an identical twin). Often the particular characteristics of an individual have an important bearing on its
survival, on its chances to reproduce, and on the success of
its offspring. Evolution is driven by such consequences.
Genetic variation that influences these characteristics provides the raw material for natural selection, and natural
populations contain a wealth of such variation. In plants
(figure 21.1), insects, and vertebrates, many genes exhibit
some level of variation. In this chapter, we explore genetic
variation in natural populations and consider the evolutionary forces that cause allele frequencies in natural populations to change. These deceptively simple matters lie at the
core of evolutionary biology, which is the topic of this
chapter and chapters 22–24.

21.1 Genes vary in natural populations.

Genetic Variation Is the Raw Material of Evolution

Evolution: Descent with Modification

The word *evolution* is widely used in the natural and social sciences. It refers to how an entity—be it a social system, a gas, or a planet—changes through time. Although development of the modern concept of evolution in biology can be traced to Darwin's *On the Origin of Species*, the first five editions of this book never actually used the term! Rather, Darwin used the phrase "descent with modification." Although many more complicated definitions have been proposed, Darwin's phrase probably best captures the essence of biological evolution: Through time, species accumulate differences; as a result, when new species are formed, the descendant species differ from their ancestors.

Natural Selection: An Important Mechanism of Evolutionary Change

Darwin was not the first to propose a theory of evolution. Rather, he followed a long line of earlier philosophers and naturalists who deduced that the many kinds of organisms around us were produced by a process of evolution. Unlike his predecessors, however, Darwin proposed **natural selection** as the mechanism of evolution. Natural selection produces evolutionary change when some individuals in a population possess certain inherited characteristics and produce more surviving offspring than individuals lacking these characteristics. As a result, the population gradually comes to include more and more individuals with the advantageous characteristics. In this way, the population evolves and becomes better adapted to its local circumstances.

Natural selection was by no means the only evolutionary mechanism proposed. A rival theory, championed by the prominent biologist Jean-Baptiste Lamarck, was that evolution occurred by the **inheritance of acquired characteristics.** According to Lamarck, individuals passed on to their offspring body and behavior changes acquired during their lives. Thus, Lamarck proposed that ancestral giraffes with short necks tended to stretch their necks to feed on tree leaves, and this extension of the neck was passed on to subsequent generations, leading to the long-necked giraffe (figure 21.2*a*). In Darwin's theory, by contrast, the variation is not created by experience, but is the result of preexisting genetic differences among individuals (figure 21.2*b*).

Although the efficacy of natural selection is now widely accepted, it is not the only process that can lead to changes in the genetic makeup of populations. Allele frequencies can also change when mutations occur repeatedly from one

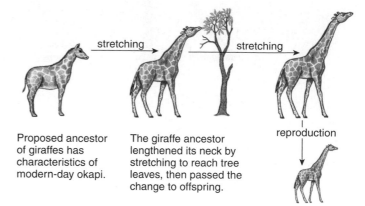

Proposed ancestor of giraffes has characteristics of modern-day okapi.

The giraffe ancestor lengthened its neck by stretching to reach tree leaves, then passed the change to offspring.

(a) Lamarck's theory: variation is acquired.

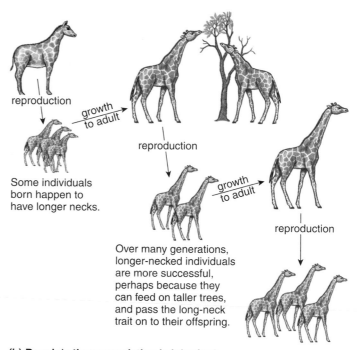

Some individuals born happen to have longer necks.

Over many generations, longer-necked individuals are more successful, perhaps because they can feed on taller trees, and pass the long-neck trait on to their offspring.

(b) Darwin's theory: variation is inherited.

FIGURE 21.2
Two ideas of how giraffes might have evolved long necks.

allele to another and when migrants bring alleles into a population. In addition, when populations are small, the frequencies of alleles can change randomly as the result of chance events. The relationship between natural selection and other processes varies. Sometimes, natural selection overwhelms the effects of other processes, but as we shall see later in this chapter, this is not always the case.

Darwin proposed that natural selection on variants within populations leads to evolutionary change.

Genetic Variation in Nature

Evolution within a species can result from any process that causes a change in the genetic composition of a population, and thus we cannot talk about evolution without also considering **population genetics,** the study of the properties of genes in populations. It is best to start by looking at the genetic variation present among individuals within a species. This is the raw material available for the selective process.

Measuring Levels of Genetic Variation

As we saw in chapter 13, a natural population can contain a great deal of genetic variation. This is true not only of humans, but of all organisms. How much variation usually occurs? Biologists have looked at many different genes in an effort to answer this question:

1. **Genes that influence blood groups.** Chemical analysis has revealed the existence of more than 30 blood group genes in humans, in addition to the ABO locus. At least one-third of these genes are routinely found in several alternative allelic forms in human populations. In addition to these, more than 45 variable genes encode other proteins in human blood cells and plasma, which are not considered blood groups. Thus, many genetically variable genes are present in this one system alone.
2. **Genes that influence enzymes.** Alternative alleles of genes specifying particular enzymes are easy to distinguish by measuring how fast the alternative proteins migrate in an electrical field (a process called **electrophoresis**). A great deal of variation exists at enzyme-specifying loci. About 5% of the enzyme loci of a typical human are heterozygous: If you picked an individual at random, and in turn selected one of the enzyme-encoding genes of that individual at random, the chances are 1 in 20 (5%) that the gene you selected would be heterozygous in that individual.

Considering the entire genome, it is fair to say that almost all humans are different from one another. This is also true of other organisms, except for those that reproduce asexually. In nature, genetic variation is the rule.

Enzyme Polymorphism

Many loci in a given population have more than one allele at frequencies significantly greater than would occur due to mutation alone. Researchers refer to a locus with more variation than can be explained by mutation as **polymorphic** (Greek *poly*, "many," + *morphe*, "forms") (figure 21.3). The extent of such variation within natural populations was not even suspected a few decades ago, until modern techniques such as protein electrophoresis made it possible to examine enzymes and other proteins directly. We now know that most populations of insects and plants are polymorphic (that

FIGURE 21.3
Polymorphic variation. These Australian snails, all of the species *Bankivia fasciata*, exhibit considerable variation in pattern and color. Individual differences are inherited and passed on to offspring.

is, have more than one allele occurring at a frequency greater than 5%) at more than half of their enzyme-encoding loci, although vertebrates are somewhat less polymorphic. **Heterozygosity** (the probability that a randomly selected gene will be heterozygous for a randomly selected individual) is about 15% in *Drosophila* and other invertebrates, between 5% and 8% in vertebrates, and around 8% in outcrossing plants. These high levels of genetic variability provide ample supplies of raw material for evolution.

DNA Sequence Polymorphism

The advent of gene technology has made it possible to assess genetic variation even more directly by sequencing the DNA itself. In a pioneering study in 1989, Martin Kreitman sequenced ADH genes isolated from 11 individuals of the fruit fly *Drosophila melanogaster*. He found 43 variable sites, only one of which had been detected by protein electrophoresis! Since then, numerous other studies of variation at the DNA level have confirmed these findings: Abundant variation exists in both the coding regions of genes and in their nontranslated introns—considerably more variation than we can detect by examining enzymes with electrophoresis.

Natural populations contain considerable amounts of genetic variation—more than can be accounted for by mutation alone.

21.2 Why do allele frequencies change in populations?

Genetic variation within natural populations was a puzzle to Darwin and his contemporaries. The way in which meiosis produces genetic segregation among the progeny of a hybrid had not yet been discovered. Selection, scientists then thought, should always favor an optimal form, and so tend to eliminate variation. Moreover, the theory of **blending inheritance**—in which offspring were expected to be phenotypically intermediate relative to their parents—was widely accepted. If blending inheritance were correct, then the effect of any new genetic variant would quickly be diluted to the point of disappearance in subsequent generations.

The Hardy–Weinberg Principle

Following the rediscovery of Mendel's research, two people in 1908 independently solved the puzzle of why genetic variation persists—G. H. Hardy, an English mathematician, and W. Weinberg, a German physician. They pointed out that the original proportions of the genotypes in a population will remain constant from generation to generation, as long as the following assumptions are met:

1. The population size is very large.
2. Random mating is occurring.
3. No mutation takes place.
4. No genes are input from other sources (no immigration takes place).
5. No selection occurs.

Because their proportions do not change, the genotypes are said to be in **Hardy–Weinberg equilibrium.**

In algebraic terms, the Hardy–Weinberg principle is written as an equation. Consider a population of 100 cats in which 84 are black and 16 are white. The frequencies of the two phenotypes would be 0.84 (or 84%) black and 0.16 (or 16%) white. Based on these phenotypic frequencies, can we deduce the underlying frequency of genotypes? If we assume that the white cats are homozygous recessive for an allele we designate b, and the black cats are therefore either homozygous dominant BB or heterozygous Bb, we can calculate the **allele frequencies** of the two alleles in the population from the proportion of black and white individuals, assuming the population is in Hardy–Weinberg equilibrium. Let the letter p designate the frequency of the B allele and the letter q the frequency of the alternative allele. Because there are only two alleles, p plus q must always equal 1.

The Hardy–Weinberg equation can now be expressed in the form of what is known as a binomial expansion:

$$(p + q)^2 = \underset{\substack{\text{(Individuals} \\ \text{homozygous} \\ \text{for allele } B; \\ \text{color black)}}}{p^2} + \underset{\substack{\text{(Individuals} \\ \text{heterozygous} \\ \text{with alleles } B + b; \\ \text{color black)}}}{2pq} + \underset{\substack{\text{(Individuals} \\ \text{homozygous} \\ \text{for allele } b; \\ \text{color white)}}}{q^2}$$

If $q^2 = 0.16$ (the frequency of white cats), then $q = 0.4$. Therefore, p, the frequency of allele B, would be 0.6 (1.0 − 0.4 = 0.6). We can now easily calculate the **genotype frequencies:** There are $p^2 = (0.6)^2 = 0.36$, or 36 homozygous dominant BB individuals in a population of 100 cats. The heterozygous individuals have the Bb genotype, and would have a frequency of $2pq$, or $(2 \times 0.6 \times 0.4) = 0.48$, or 48 heterozygous Bb individuals.

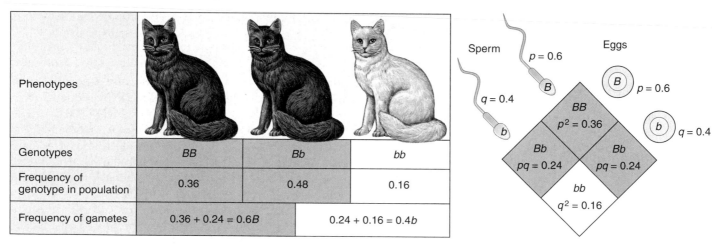

FIGURE 21.4
The Hardy–Weinberg equilibrium.   In the absence of factors that alter them, the frequencies of gametes, genotypes, and phenotypes remain constant generation after generation.
If all white cats died, what proportion of the kittens in the next generation would be white?

Using the Hardy–Weinberg Equation

The Hardy–Weinberg equation is a simple extension of the Punnett square described in chapter 13, with two alleles assigned frequencies, p and q. Figure 21.4 allows you to trace genetic reassortment during sexual reproduction and see how it affects the frequencies of the B and b alleles during the next generation. In constructing this diagram, we have assumed that the union of sperm and egg in these cats is random, so that all combinations of b and B alleles occur. For this reason, the alleles are mixed randomly and represented in the next generation in proportion to their original representation. Each individual egg or sperm in each generation has a 0.6 chance of receiving a B allele ($p = 0.6$) and a 0.4 chance of receiving a b allele ($q = 0.4$).

In the next generation, therefore, the chance of combining two B alleles is p^2, or 0.36 (that is, 0.6×0.6), and approximately 36% of the individuals in the population will continue to have the BB genotype. The frequency of bb individuals is q^2 (0.4×0.4) and so will continue to be about 16%, and the frequency of Bb individuals will be $2pq$ ($2 \times 0.6 \times 0.4$), or on average, 48%. Phenotypically, if the population size remains at 100 cats, we will still see approximately 84 black individuals (with either BB or Bb genotypes) and 16 white individuals (with the bb genotype) in the population. Allele, genotype, and phenotype frequencies have remained unchanged from one generation to the next, despite the reshuffling of genes that occurs during mei_____sive-
ness_____ed in
an _____ange
thro_____

[handwritten note:]
heterozygous
= having two different
alleles of the same
gene

_____iberg
equ_____es of
the_____b was
0.10_____d ab-
sen_____s are
that_____vidu-
als_____(Be-
cau_____gous
offs_____gous
offs_____ss of
hon_____

[handwritten note:]
homozygous
= having two identical
alleles of the same
gene

_____iberg
equ_____s are
met_____hich
populations are not at equilibrium are the most interesting, because they indicate that one or more evolutionary processes are at work.

Why Do Allele Frequencies Change?

According to the Hardy–Weinberg principle, both the allele and genotype frequencies in a large, random-mating population will remain constant from generation to gen-

Table 21.1	Agents of Evolutionary Change
Factor	**Description**
Mutation	The ultimate source of variation. Individual mutations occur so rarely that mutation alone usually does not change allele frequency much.
Gene flow	A very potent agent of change. Populations exchange members or gametes.
Nonrandom mating	Inbreeding is the most common form. It does not alter allele frequency but changes the proportion of heterozygotes.
Genetic drift	Statistical accidents. The random fluctuation in allele frequencies increases as population size decreases.
Selection	The only agent that produces *adaptive* evolutionary changes.

eration if no mutation, no gene flow, and no selection occur. The stipulations tacked onto the end of this statement are important. In fact, they are the key to the importance of the Hardy–Weinberg principle, because individual allele frequencies often change in natural populations, with some alleles becoming more common and others decreasing in frequency. The Hardy–Weinberg principle establishes a convenient baseline against which to measure such changes. By looking at how various factors alter the proportions of homozygotes and heterozygotes, we can identify the forces affecting particular situations we observe.

Many factors can alter allele frequencies. Only five, however, alter the proportions of homozygotes and heterozygotes enough to produce significant deviations from the proportions predicted by the Hardy–Weinberg principle: mutation, gene flow (including both immigration into and emigration out of a given population), nonrandom mating, genetic drift (random change in allele frequencies, which is more likely in small populations), and selection (table 21.1). Of these, only selection produces adaptive evolutionary change because only in selection does the result depend on the nature of the environment. The other factors operate relatively independently of the environment, so the changes they produce are not shaped by environmental demands.

The Hardy–Weinberg principle states that in a large population mating at random and in the absence of other forces that would change the proportions of the different alleles at a given locus, the process of sexual reproduction (meiosis and fertilization) alone will not change these proportions.

Five Agents of Evolutionary Change

1. Mutation

Mutation from one allele to another can obviously change the proportions of particular alleles in a population. Mutation rates are generally so low that they have little effect on the Hardy–Weinberg proportions of common alleles. A typical gene mutates about once per 100,000 cell divisions. Because this rate is so low, other evolutionary processes are usually more important in determining how allele frequencies change. Nonetheless, mutation is the ultimate source of genetic variation and thus makes evolution possible (figure 21.5a). It is important to remember, however, that the likelihood of a particular mutation occurring is not affected by natural selection; that is, mutations do not occur more frequently in situations in which they would be favored by natural selection.

2. Gene Flow

Gene flow is the movement of alleles from one population to another. It can be a powerful agent of change because members of two different populations may exchange genetic material. Sometimes gene flow is obvious, as when an animal moves from one place to another. If the characteristics of the newly arrived animal differ from those of the animals already there, and if the newcomer is adapted well enough to the new area to survive and mate successfully, the genetic composition of the receiving population may be altered. Other important kinds of gene flow are not as obvious. These subtler movements include the drifting of gametes or immature stages of plants or marine animals from one place to another (figure 21.5b). Pollen, the male gamete of flowering plants, is often carried great distances by insects and other animals that visit flowers. Seeds may also blow in the wind or be carried by animals to new populations far from their place of origin. In addition, gene flow may also result from the mating of individuals belonging to adjacent populations.

Consider two populations initially different in allele frequencies: In population 1, $p = 0.2$ and $q = 0.8$; in population 2, $p = 0.8$ and $q = 0.2$. Gene flow will tend to bring the rarer allele into each population. Thus, allele frequencies will change from generation to generation, and the populations will not be in Hardy–Weinberg equilibrium. Only when allele frequencies reach 0.5 for both alleles in both populations will equilibrium be attained. This example also indicates that gene flow tends to homogenize allele frequencies among populations.

3. Nonrandom Mating

Individuals with certain genotypes sometimes mate with one another more commonly than would be expected on a random basis, a phenomenon known as nonrandom mating (fig-

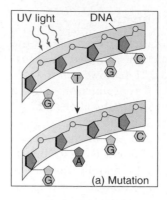

(a) Mutation

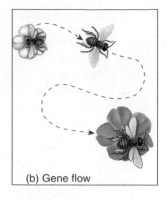

(b) Gene flow

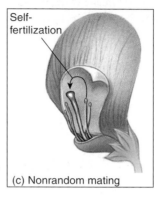

(c) Nonrandom mating

(d) Genetic drift

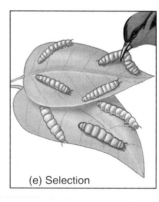

(e) Selection

FIGURE 21.5
Five agents of evolutionary change. (*a*) Mutation, (*b*) gene flow, (*c*) nonrandom mating, (*d*) genetic drift by founder effect, and (*e*) selection.

ure 21.5c). **Assortative mating,** in which phenotypically similar individuals mate, is a type of nonrandom mating that causes the frequencies of particular genotypes to differ greatly from those predicted by the Hardy–Weinberg principle. Assortative mating does not change the frequency of the alleles, but rather increases the proportion of homozygous individuals because phenotypically similar individuals are likely to be genetically similar and thus produce offspring with two copies of the same allele. This is why populations of self-fertilizing plants consist primarily of homozygous individuals. By contrast, **disassortative mating,** in which phenotypically different individuals mate, produces an excess of heterozygotes.

4. Genetic Drift

In small populations, frequencies of particular alleles may change drastically by chance alone. Such changes in allele frequencies occur randomly, as if the frequencies were drifting, and are thus known as **genetic drift** (figure 21.5*d*). For this reason, a population must be large to be in Hardy–Weinberg equilibrium. If the gametes of only a few individuals form the next generation, the alleles they carry may by chance not be representative of the parent population from which they were drawn, as illustrated in figure 21.6, where a small number of individuals are removed from a bottle containing many. By chance, most of the individuals removed are blue, so the new population has a much higher population of blue individuals than the parent generation had.

A set of small populations that are isolated from one another may come to differ strongly as a result of genetic drift even if the forces of natural selection do not differ between the populations. Indeed, because of genetic drift, harmful alleles may increase in frequency in small populations, despite selective disadvantage, and favorable alleles may be lost even though they are selectively advantageous. It is interesting to realize that humans have lived in small groups for much of the course of their evolution; consequently, genetic drift may have been a particularly important factor in the evolution of our species.

Even large populations may feel the effect of genetic drift. Large populations may have been much smaller in the past, and genetic drift may have greatly altered allele frequencies at that time. Imagine a population containing only two alleles of a gene, *B* and *b*, in equal frequency (that is, $p = q = 0.5$). In a large Hardy–Weinberg population, the genotype frequencies are expected to be 0.25 *BB*, 0.50 *Bb*, and 0.25 *bb*. If only a small sample produces the next generation, large deviations in these genotype frequencies can occur by chance. Imagine, for example, that four individuals form the next generation, and that by chance they are two *Bb* heterozygotes and two *BB* homozygotes—that is, the allele frequencies in the next generation are $p = 0.75$ and $q = 0.25$! If you were to replicate this experiment 100 times, each time randomly drawing four individuals from the parental population, one of the two alleles would be missing entirely from about 8 of the 100 populations. This leads to an important conclusion: Genetic drift results in the loss of alleles in isolated populations. Although genetic drift occurs in any population, it is particularly likely in situations with founder effects and bottlenecks.

Founder Effects. Sometimes one or a few individuals disperse and become the founders of a new, isolated population at some distance from their place of origin. These pioneers are not likely to have all the alleles present in the source population. Thus, some alleles may be lost from the new population, and others may change drastically in frequency. In some cases, previously rare alleles in the source population may be a significant fraction of the new population's genetic endowment. This phenomenon is called the **founder effect.** Founder effects are not rare in nature. Many self-pollinating plants start new populations from a single seed.

Founder effects have been particularly important in the evolution of organisms on distant oceanic islands, such as

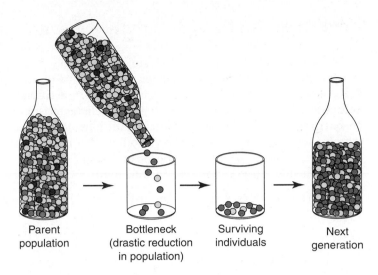

Parent population → Bottleneck (drastic reduction in population) → Surviving individuals → Next generation

FIGURE 21.6
Genetic drift: A bottleneck effect. The parent population contains roughly equal numbers of blue and yellow individuals and a small number of red individuals. By chance, the few remaining individuals that contribute to the next generation are mostly blue. The bottleneck occurs because so few individuals form the next generation, as might happen after an epidemic or a catastrophic storm.

the Hawaiian Islands and the Galápagos Islands visited by Darwin. Most of the organisms in such areas probably derive from one or a few initial "founders." In a similar way, isolated human populations founded by relatively few individuals are often dominated by genetic features characteristic of their particular founders. Amish populations in the United States, for example, have unusually high frequencies of a number of conditions, such as polydactylism (the presence of a sixth finger).

The Bottleneck Effect. Even if organisms do not move from place to place, occasionally their populations may be drastically reduced in size. This may result from flooding, drought, epidemic disease, and other natural forces, or from progressive changes in the environment. The few surviving individuals may constitute a random genetic sample of the original population (unless some individuals survive specifically because of their genetic makeup). The resultant alterations and loss of genetic variability have been termed the bottleneck effect.

Some living species appear to be severely depleted genetically and have probably suffered from a bottleneck effect in the past. For example, the northern elephant seal, which breeds on the western coast of North America and nearby islands, was nearly hunted to extinction in the nineteenth century and was reduced to a single population containing perhaps no more than 20 individuals on the island of Guadalupe off the coast of Baja, California. As a result of this bottleneck, even though the seal populations have rebounded and now number in the tens of thousands, this species has lost almost all of its genetic variation.

5. Selection

As Darwin pointed out, some individuals leave behind more progeny than others, and the rate at which they do so is affected by phenotype and behavior. We describe the results of this process as **selection** (figure 21.5e). In **artificial selection,** a breeder selects for the desired characteristics. In **natural selection,** environmental conditions determine which individuals in a population produce the most offspring. For natural selection to occur and result in evolutionary change, three conditions must be met:

a. **Variation must exist among individuals in a population.** Natural selection works by favoring individuals with some traits over individuals with alternative traits. If no variation exists, natural selection cannot operate.

b. **Variation among individuals results in differences in the number of offspring surviving in the next generation.** This is the essence of natural selection. Because of their phenotype or behavior, some individuals are more successful than others in producing offspring. Although many traits are phenotypically variable, varying individuals do not always differ in survival and reproductive success.

c. **Variation must be genetically inherited.** For natural selection to result in evolutionary change, the selected differences must have a genetic basis. However, not all variation has a genetic basis—even genetically identical individuals may be phenotypically quite distinctive if they grow up in different environments. Such environmental effects are common in nature. In many turtles, for example, individuals that hatch from eggs laid in moist soil are heavier, with longer and wider shells, than individuals from nests in drier areas. When phenotypically different individuals do not differ genetically, then differences in the number of their offspring will not alter the genetic composition of the population in the next generation, and thus, no evolutionary change will have occurred.

It is important to remember that natural selection and evolution are not the same—the two concepts often are incorrectly equated. Natural selection is a process, whereas evolution is the historical record of change through time. Evolution is an outcome, not a process. Natural selection (the process) can lead to evolution (the outcome), but natural selection is only one of several processes that can pro-duce evolutionary change. Moreover, natural selection can occur without producing evolutionary change; only if variation is genetically based will natural selection lead to evolution.

Selection to Avoid Predators. Many of the most dramatic documented instances of adaptation involve genetic changes that decrease the probability of capture by a predator. The caterpillar larvae of the common sulphur butterfly *Colias eurytheme* usually exhibit a dull, kelly green color, providing excellent camouflage against the alfalfa plants on which they feed. An alternative bright blue color morph is kept at very low frequency because this color renders the larvae highly visible on the food plant, making it easier for bird predators to see them (see figure 21.5e). In a similar fashion, the way the shell markings in the land snail *Cepaea nemoralis* match its background habitat reflects the same pattern of avoiding predation by camouflage.

One of the most dramatic examples of background matching involves ancient lava flows in the middle of deserts in the American Southwest. In these areas, the black rock formations produced when the lava cooled contrast starkly with the surrounding bright glare of the desert sand. Populations of many species of animals occurring on these rocks—including lizards, rodents, and a variety of insects—are dark in color, whereas sand-dwelling populations in surrounding areas are much lighter (figure 21.7). Predation is the likely cause selecting for these differences in color. Laboratory studies have confirmed that predatory birds are adept at picking out individuals occurring on backgrounds to which they are not adapted.

FIGURE 21.7
Pocket mice from the Tularosa Basin of New Mexico whose color matches their background. Black lava formations are surrounded by desert, and selection favors coat color in pocket mice that matches their surroundings.

Selection to Match Climatic Conditions. Many studies of selection have focused on genes encoding enzymes, because in such cases the investigator can directly assess the consequences to the organism of changes in the frequency of alternative enzyme alleles. Often investigators find that enzyme allele frequencies vary latitudinally, with one allele more common in northern populations but progressively less common at more southern locations. A superb example is seen in studies of a fish, the mummichog *(Fundulus heteroclitus)*, which ranges along the eastern coast of North America. In this fish, allele frequencies vary geographically for the gene that produces the enzyme lactate dehydrogenase, which catalyzes the conversion of pyruvate to lactate (figure 21.8). Biochemical studies show that the enzymes formed by these alleles function differently at different temperatures, thus explaining their geographic distributions. For example, the form of the enzyme that is more frequent in the north is a better catalyst at low temperatures than the enzyme from the south. Moreover, studies indicate that at low temperatures, individuals with the northern allele swim faster, and presumably survive better, than individuals with the alternative allele.

Selection for Pesticide Resistance. A particularly clear example of selection in natural populations is provided by studies of pesticide resistance in insects. The widespread use of insecticides has led to the rapid evolution of resistance in more than 500 pest species. For example, in the housefly, the resistance allele at the *pen* gene decreases the uptake of insecticide, whereas alleles at the *kdr* and *dld-r* genes decrease the number of target sites, thus decreasing the binding ability of the insecticide (figure 21.9). Other alleles enhance the ability of the insects' enzymes to identify and detoxify insecticide molecules.

Single genes are also responsible for resistance in other organisms. The pigweed, *Amaranthus hybridus*, is one of about 28 agricultural weeds that have evolved resistance to the herbicide triazine. Triazine inhibits photosynthesis by binding to a protein in the chloroplast membrane, causing the plant to die. Single amino acid substitutions in the gene encoding the protein diminish the ability of triazine to decrease the plant's photosynthetic capabilities. Similarly, Norway rats are normally susceptible to the pesticide warfarin, which diminishes the clotting ability of the rat's blood and leads to fatal hemorrhaging. However, a resistance allele at a single gene alters a metabolic pathway and renders warfarin ineffective.

Five factors can bring about a deviation from the proportions of homozygotes and heterozygotes predicted by the Hardy–Weinberg principle. Only selection regularly produces adaptive evolutionary change, but the genetic constitution of populations, and thus the course of evolution, can also be affected by mutation, gene flow, nonrandom mating, and genetic drift.

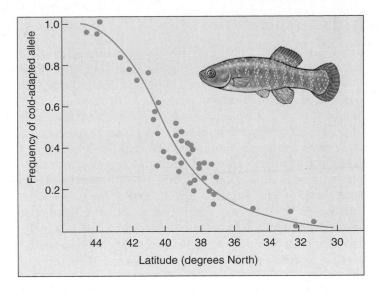

FIGURE 21.8
Selection to match climatic conditions. The frequency of the cold-adapted allele for lactate dehydrogenase in a type of fish (the mummichog, *Fundulus heteroclitus*) decreases at lower latitudes, which are warmer.
Why does the allele frequency change from north to south?

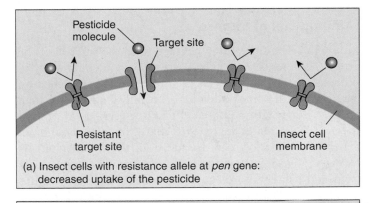

(a) Insect cells with resistance allele at *pen* gene: decreased uptake of the pesticide

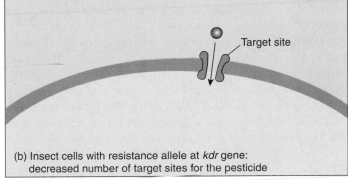

(b) Insect cells with resistance allele at *kdr* gene: decreased number of target sites for the pesticide

FIGURE 21.9
Selection for pesticide resistance. Resistance alleles at genes such as *pen* and *kdr* allow insects to be more resistant to pesticides. Insects that possess these resistance alleles have become more common through selection.

Measuring Fitness

Selection occurs when individuals with one phenotype leave more surviving offspring in the next generation than individuals with an alternative phenotype. Evolutionary biologists quantify reproductive success as **fitness,** the number of surviving offspring left in the next generation. It is important to realize that fitness is a relative concept; the most fit phenotype is simply the one that produces, on average, the greatest number of offspring. Suppose, for example, that in a population of toads, two phenotypes exist: green and brown. Suppose, further, that green toads leave, on average, 4.0 offspring in the next generation, but brown toads leave only 2.5. By custom, the most fit phenotype is assigned a fitness value of 1.0, and other phenotypes are expressed as relative proportions. In this case, the fitness of the brown phenotype would be 2.5/4.0 = 0.625. A difference in fitness of 0.375 is quite large; natural selection in this case strongly favors the green phenotype. If differences in color have a genetic basis, then we would expect evolutionary change to occur; the frequency of green toads should be substantially greater in the next generation. Further, if the fitness of two phenotypes remained unchanged, we would expect alleles for the brown phenotype to disappear from the population.

Components of Fitness

Although selection is often characterized as "survival of the fittest," differences in survival are only one component of fitness. Even if no differences in survival occur, selection may operate if some individuals are more successful than others in attracting mates. In many territorial animal species, large males mate with many females, and small males rarely get to mate. In addition, the number of offspring produced per mating is also important. Large female frogs and fish lay more eggs than smaller females and thus may leave more offspring in the next generation. Fitness is therefore a combination of survival, mating success, and number of offspring per mating.

Selection favors phenotypes with the greatest fitness, but predicting fitness from a single component can be tricky because traits favored for one component of fitness may be at a disadvantage for others. For example, in water striders, larger females lay more eggs per day (figure 21.10). Thus, natural selection at this stage favors large size. However, larger females also die younger and thus have fewer opportunities to reproduce; consequently, smaller females have a survival advantage. Overall, the two opposing directions of selection cancel each other out, and intermediate-sized females, on average, leave the most offspring in the next generation.

> An organism's reproductive success is affected by how long it survives, how often it mates, and how many offspring it produces per mating.

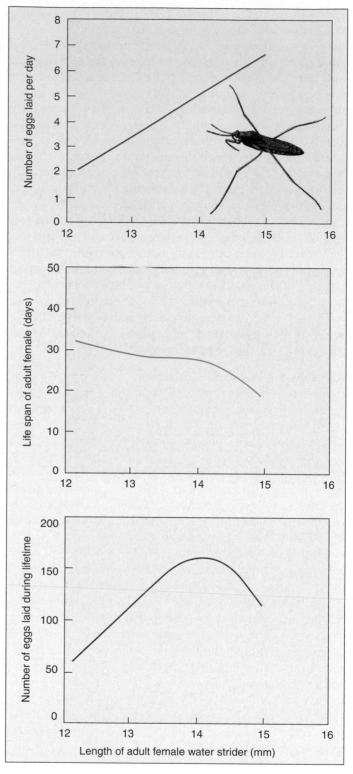

FIGURE 21.10

Body size and egg-laying in water striders. Larger female water striders lay more eggs per day, but also survive for a shorter period of time. As a result, intermediate-sized females produce the most offspring over the course of their entire lives and thus have the highest fitness.

What evolutionary change in body size might you expect? If the number of eggs laid per day was not affected by body size, would your prediction change?

Interactions Among Evolutionary Forces

Levels of variation retained in a population may be determined by the relative strength of different evolutionary processes. In theory, for example, if allele B mutates to allele b at a high enough rate, allele b could be maintained in the population even if natural selection strongly favored allele B. In nature, however, mutation rates are rarely high enough to counter the effects of natural selection.

The effect of natural selection also may be countered by genetic drift. Both processes may act to remove variation from a population. However, whereas selection is a deterministic process that operates to increase the representation of alleles that enhance survival and reproductive success, drift is a random process. Thus, in some cases, drift may lead to a decrease in the frequency of an allele that is favored by selection. In some extreme cases, drift may even lead to the loss of a favored allele from a population. Remember, however, that the magnitude of drift is negatively related to population size; consequently, natural selection is expected to overwhelm drift except when populations are very small.

Gene Flow Versus Natural Selection

Gene flow can be either a constructive or a constraining force. On one hand, gene flow can increase the adaptedness of a species by spreading a beneficial mutation that arises in one population to other populations within a species. On the other hand, gene flow can impede adaptation within a population by continually importing inferior alleles from other populations. Consider two populations of a species that live in different environments. In this situation, natural selection might favor different alleles—B and b—in the different populations. In the absence of gene flow and other evolutionary processes, the frequency of B would be expected to reach 100% in one population and 0% in the other. However, if gene flow were occurring between the two populations, then the less favored allele would continually be reintroduced into each population. As a result, the frequency of the two alleles in each population would reflect a balance between the rate at which gene flow brings the inferior allele into a population and the rate at which natural selection removes it.

A classic example of gene flow opposing natural selection occurs on abandoned mine sites in Great Britain. Although mining activities ceased hundreds of years ago, the

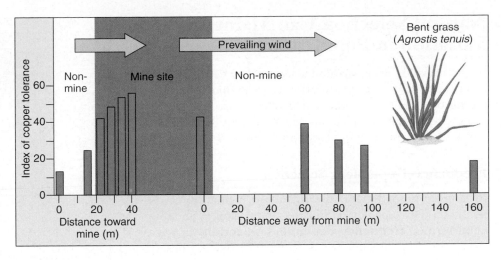

FIGURE 21.11

Degree of copper tolerance in grass plants on and near ancient mine sites.
Individuals with tolerant alleles have decreased growth rates on unpolluted soil. Thus, we would expect copper tolerance to be 100% on mine sites and 0% on non-mine sites. However, prevailing winds blow pollen containing nontolerant alleles onto the mine site and tolerant alleles beyond the site's borders.
Would you expect the frequency of copper tolerance to be affected by distance from the mine site? How would your answer change depending on whether you were upwind or downwind from the mine site?

concentration of metal ions in the soil is still much greater than in surrounding areas. Heavy metal concentrations are generally toxic to plants, but alleles at certain genes confer resistance. The ability to tolerate heavy metals comes at a price, however; individuals with the resistance allele exhibit lower growth rates on nonpolluted soil. Consequently, we would expect the resistance allele to occur with a frequency of 100% on mine sites and 0% elsewhere. Heavy metal tolerance has been studied particularly intensively in the slender bent grass *Agrostis tenuis*, in which researchers have found that the resistance allele occurs at intermediate levels in many areas (figure 21.11). The explanation relates to the reproductive system of this grass in which pollen, the male gamete (that is, the floral equivalent of sperm), is dispersed by the wind. As a result, pollen—and the alleles it carries—can be blown for great distances, leading to levels of gene flow between mine sites and unpolluted areas high enough to counteract the effects of natural selection.

In general, the extent to which gene flow can hinder the effects of natural selection should depend on the relative strengths of the two processes. In species in which gene flow is generally strong, such as birds and wind-pollinated plants, the frequency of the less favored allele may be relatively high, whereas in more sedentary species that exhibit low levels of gene flow, such as salamanders, the favored allele should occur at a frequency near 100%.

> Allele frequencies sometimes reflect a balance between opposing processes, such as gene flow and natural selection. In such cases, observed frequencies will depend on the relative strength of the processes.

Natural Selection Can Maintain Variation in Populations

In the previous pages, natural selection has been discussed as a process that removes variation from a population by favoring one allele over others at a gene locus. However, in some circumstances, selection can do exactly the opposite and actually maintain population variation.

Frequency-Dependent Selection

In some circumstances, the fitness of a phenotype depends on its frequency within the population, a phenomenon termed **frequency-dependent selection.** In negative frequency-dependent selection, rare phenotypes are favored by selection. Assuming a genetic basis for phenotypic variation, such selection will have the effect of making rare alleles more common, thus maintaining variation.

Negative frequency-dependent selection can occur for many reasons. For example, it is well-known that animals or people searching for something form a "search image." That is, they become particularly adept at picking out certain objects. Consequently, predators may form a search image for common prey phenotypes. Rare forms may thus be preyed upon less frequently. An example is fish predation on an insect, the water boatman, which occurs in three different colors. Experiments indicate that each of the color types is preyed upon disproportionately when it is the most common one (figure 21.12a). Another cause of negative frequency dependence is resource competition. If genotypes differ in their resource requirements, as occurs in many plants, then the rarer genotype will have fewer competitors. If the different resource types are equally abundant, the rarer genotype will be at an advantage relative to the more common genotype.

Positive frequency-dependent selection has the opposite effect; by favoring common forms, it tends to eliminate variation from a population. For example, predators don't always select common individuals. In some cases, "oddballs" stand out from the rest and attract attention (figure 21.12b).

Oscillating Selection

In some cases, selection favors one phenotype at one time and another phenotype at another time, a phenomenon called **oscillating selection.** If selection repeatedly oscillates in this fashion, the effect will be to maintain genetic variation in the population. One example, discussed in chapter 22, concerns the medium ground finch in the Galápagos Islands. In times of drought, the supply of small, soft seeds is depleted, but there are still enough large seeds around. Consequently, birds with big bills are favored. However, when wet conditions return, the ensuing abundance of small seeds favors birds with smaller bills.

Oscillating selection and frequency-dependent selection are similar because in both cases, the form of selection

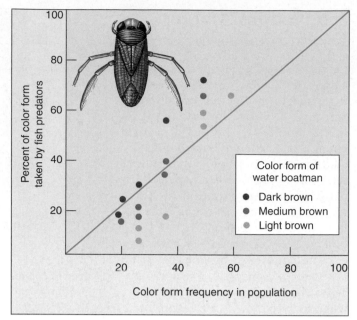

(a)

(b)

FIGURE 21.12
Frequency-dependent selection. (*a*) Predators often form search images for the most common prey. In one experiment, fish were placed in an enclosure with water boatmen (aquatic insects) of three different colors. When each color was common, the fish disproportionately captured boatmen of that color. By contrast, when each color form was uncommon, it was rarely taken (negative frequency-dependent selection). (*b*) However, in some cases, rare individuals stand out from the rest and draw the attention of predators; thus, in these cases, common phenotypes have the advantage (positive frequency-dependent selection).

changes through time. However, it is important to recognize that they are not the same: In oscillating selection, at any one time, the fitness of a phenotype does not depend on its frequency; rather, environmental changes lead to the oscillation in selection. In contrast, in frequency-dependent selection, it is the change in frequencies themselves that leads to the changes in fitness of the different phenotypes.

Heterozygote Advantage

If heterozygotes are favored over homozygotes, then natural selection actually tends to maintain variation in the population. Such **heterozygote advantage** will favor individuals with copies of both alleles, and thus will work to maintain both alleles in the population. Some evolutionary biologists believe that heterozygote advantage is pervasive and can explain the high levels of polymorphism observed in natural populations. Others, however, believe that it is relatively rare.

The best-documented example of heterozygote advantage is sickle cell anemia, a hereditary disease affecting hemoglobin in humans. Individuals with sickle cell anemia exhibit symptoms of severe anemia and abnormal red blood cells that are irregular in shape, with a great number of long, sickle-shaped cells (figure 21.13a). Chapter 13 discusses why the sickle cell mutation (S) causes red blood cells to sickle.

The average incidence of the S allele in central African populations is about 0.12, far higher than that found among African Americans. From the Hardy–Weinberg principle, you can calculate that 1 in 5 central African individuals is heterozygous at the S allele, and 1 in 100 is homozygous and develops the fatal form of the disorder. People who are homozygous for the sickle cell allele almost never reproduce because they usually die before they reach reproductive age. Why is the S allele not eliminated from the central African population by selection, rather than being maintained at such high levels? People who are heterozygous for the sickle cell allele (and thus do not suffer from sickle cell anemia) are much less susceptible to malaria—one of the leading causes of illness and death in central Africa, especially among young children. The reason is that when the parasite that causes malaria, *Plasmodium falciparum*, enters a red blood cell, it causes extremely low oxygen tension in the cell, which leads to cell sickling even in heterozygotes (but not in individuals that do not have the sickle cell allele). Such cells are quickly filtered out of the bloodstream by the spleen, thus eliminating the parasite. (The spleen's filtering effect is what leads to anemia in homozygotes as large numbers of red blood cells are removed; in the case of malaria, only cells with the *Plasmodium* parasite sickle).

Consequently, even though most homozygous recessive individuals die before they have children, the sickle cell allele is maintained at high levels in these populations (it is selected for) because it is associated with resistance to malaria in heterozygotes and also, for reasons not yet fully understood, with increased fertility in female heterozygotes.

For people living in areas where malaria is common, having the sickle cell allele in the heterozygous condition has adaptive value (figure 21.13b). Among African Americans, however, many of whose ancestors have lived for some 15 generations in a country where malaria has been relatively rare and is now essentially absent, the environment does not place a premium on resistance to malaria. Consequently, no adaptive value counterbalances the ill effects of the disease; in this nonmalarial environment, selection is acting to eliminate the S allele. Only 1 in 375 African Americans develops sickle cell anemia, far less than in central Africa.

Selection can maintain variation within populations in a number of ways.

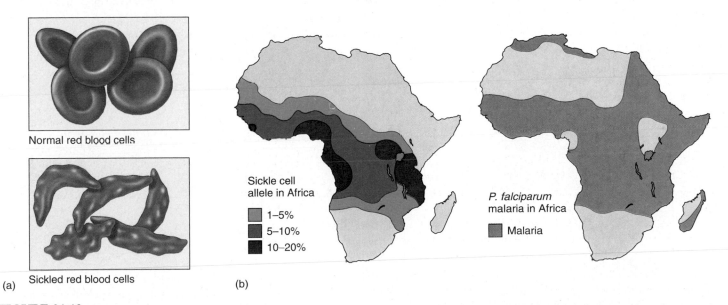

(a) Normal red blood cells

Sickled red blood cells

Sickle cell allele in Africa

■ 1–5%
■ 5–10%
■ 10–20%

P. falciparum malaria in Africa

■ Malaria

(b)

FIGURE 21.13
Frequency of sickle cell allele and distribution of *Plasmodium falciparum* malaria. (*a*) The red blood cells of people homozygous for the sickle cell allele collapse into sickled shapes when the oxygen level in the blood is low. (*b*) The distribution of the sickle cell allele in Africa coincides closely with that of *P. falciparum* malaria.

21.3 Selection can act on traits affected by many genes.

Forms of Selection

In nature, many traits—perhaps most—are affected by more than one gene. The interactions between genes are typically complex, as you saw in chapter 13. For example, alleles of many different genes play a role in determining human height (see figure 13.16). In such cases, selection operates on all the genes, influencing most strongly those that make the greatest contribution to the phenotype. How selection changes the population depends on which genotypes are favored.

Disruptive Selection

In some situations, selection acts to eliminate intermediate types, a phenomenon called **disruptive selection** (figure 21.14a). A clear example is the different beak sizes of the African fire-bellied seedcracker finch *Pyronestes ostrinus* (figure 21.15). Populations of these birds contain individuals with large and small beaks, but very few individuals with intermediate-sized beaks. As their name implies, these birds feed on seeds, and the available seeds fall into two size categories: large and small. Only large-beaked birds can open the tough shells of large seeds, whereas birds with the smaller beaks are more adept at handling small seeds. Birds with intermediate-sized beaks are at a disadvantage with both seed types: unable to open large seeds and too clumsy to efficiently process small seeds. Consequently, selection acts to eliminate the intermediate phenotypes, in effect partitioning (or "disrupting") the population into two phenotypically distinct groups.

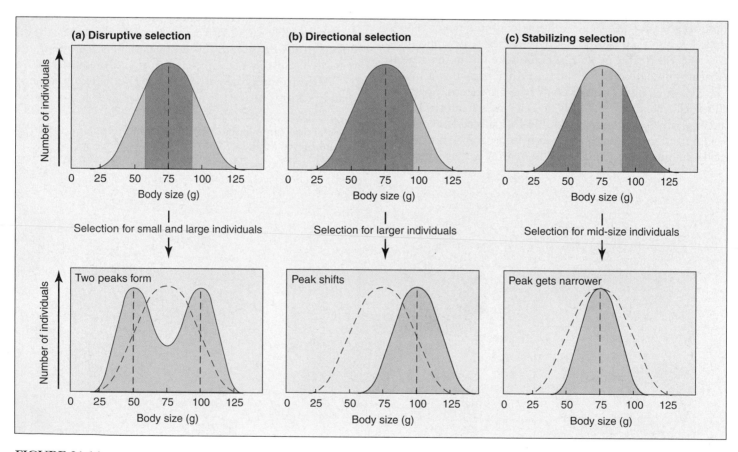

FIGURE 21.14

Three kinds of selection. The top panels show the populations before selection has occurred, with the forms that will be selected against shaded red and the forms that will be favored shaded blue. The bottom panels indicate what the populations will look like after selection has occurred, assuming that variation in the traits is genetically based. (*a*) In disruptive selection, individuals in the middle of the range of phenotypes of a certain trait are selected against (*red*), and the extreme forms of the trait are favored (*blue*). (*b*) In directional selection, individuals concentrated toward one extreme of the array of phenotypes are favored. (*c*) In stabilizing selection, individuals with midrange phenotypes are favored, with selection acting against both ends of the range of phenotypes.

FIGURE 21.15
Disruptive selection for large and small beaks. Differences in beak size in the fire-bellied seedcracker finch of west Africa are the result of disruptive selection.

Directional Selection

When selection acts to eliminate one extreme from an array of phenotypes, the genes promoting this extreme become less frequent in the population. This form of selection is called **directional selection** (figure 21.14*b*). Thus, in the *Drosophila* population illustrated in figure 21.16, the elimination of flies that move toward light causes the population to contain fewer individuals with alleles promoting such behavior. If you were to pick an individual at random from the new fly population, there is a smaller chance that it would spontaneously move toward light than if you had selected a fly from the old population. Artificial selection has changed the population in the direction of lower light attraction.

Stabilizing Selection

When selection acts to eliminate both extremes from an array of phenotypes, the result is to increase the frequency of the already common intermediate type. This form of selection is called **stabilizing selection** (figure 21.14*c*). In effect, selection is operating to prevent change away from this middle range of values. Selection does not change the most common phenotype of the population, but rather makes it even more common by eliminating extremes. Many examples are known. In humans, infants with intermediate weight at birth have the highest survival rate (figure 21.17). In ducks and chickens, eggs of intermediate weight have the highest hatching success.

> Selection on traits affected by many genes can favor both extremes of the trait, or intermediate values, or only one extreme.

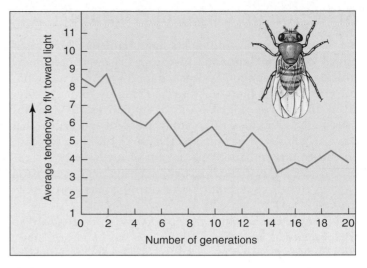

FIGURE 21.16
Directional selection for negative phototropism in *Drosophila*. Flies that moved toward light were discarded, and only flies that moved away from light were used as parents for the next generation. This procedure was repeated for 20 generations, producing substantial evolutionary change.
What would happen if after 20 generations, experimenters started keeping flies that moved *toward* the light and discarded the others?

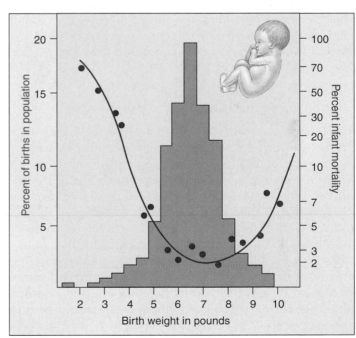

FIGURE 21.17
Stabilizing selection for birth weight in human beings. The death rate among babies (*red curve; right y-axis*) is lowest at an intermediate birth weight; both smaller and larger babies have a greater tendency to die than those around the most frequent weight (*blue area; left y-axis*) of between 7 and 8 pounds.
As improved medical technology leads to decreased infant mortality rates, how would you expect the distribution of birthrates in the population to change?

Selection on Color in Guppies

To study evolution, biologists have traditionally investigated what has happened in the past, sometimes many millions of years ago. To learn about dinosaurs, a paleontologist looks at dinosaur fossils. To study human evolution, an anthropologist looks at human fossils and, increasingly, examines the "family tree" of mutations that have accumulated in human DNA over millions of years. In this traditional approach, evolutionary biology is similar to astronomy and history, relying on observation rather than experiment to examine ideas about past events.

Nonetheless, evolutionary biology is not entirely an observational science. Darwin was right about many things, but one area in which he was mistaken concerns the pace at which evolution occurs. Darwin thought that evolution occurred at a very slow, almost imperceptible pace. However, in recent years many case studies have demonstrated that in some circumstances evolutionary change can occur rapidly. Consequently, it is possible to establish experimental studies to test evolutionary hypotheses. Although laboratory studies on fruit flies and other organisms have been common for more than 50 years, it has only been recently that scientists have started conducting experimental studies of evolution in nature. One excellent example of how observations of the natural world can be combined with rigorous experiments in the lab and in the field concerns research on the guppy, *Poecilia reticulata*.

Guppies Live in Different Environments

The guppy is a popular aquarium fish because of its bright coloration and prolific reproduction. In nature, guppies are found in small streams in northeastern South America and in many mountain streams on the nearby island of Trinidad. One interesting feature of several of the streams is that they have waterfalls. Amazingly, guppies and some other fish are capable of colonizing portions of the stream above the waterfall. The killifish, *Rivulus hartii*, is a particularly good colonizer; apparently on rainy nights, it will wriggle out of the stream and move through the damp leaf litter. Guppies are not so proficient, but they are good at swimming upstream. During flood seasons, rivers sometimes overflow their banks, creating secondary channels that move through the forest. On these occasions, guppies may be able to move upstream and invade the pools above waterfalls. By contrast, some species are not capable of such dispersal and thus are only found in streams below the first waterfall. One species whose distribution is restricted by waterfalls is the pike cichlid, *Crenicichla alta*, a voracious predator that feeds on other fish, including guppies.

Because of these barriers to dispersal, guppies can be found in two very different environments. In pools just

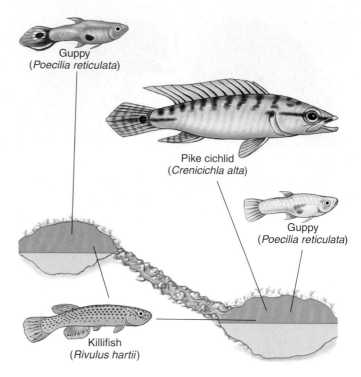

FIGURE 21.18
The evolution of protective coloration in guppies.   In pools below waterfalls where predation is high, male guppies (*Poecilia reticulata*) are drab colored. In the absence of the highly predatory pike cichlid (*Crenicichla alta*), male guppies in pools above waterfalls are much more colorful and attractive to females. The killifish (*Rivulus hartii*) is also a predator, but it only rarely eats guppies. The evolution of these differences in guppies can be experimentally tested.

below the waterfalls, predation by the pike cichlid is a substantial risk, and rates of survival are relatively low. By contrast, in similar pools just above the waterfall, the only predator present is the killifish, which rarely preys on guppies. Guppy populations above and below waterfalls exhibit many differences. In the high-predation pools, guppies exhibit drab coloration. Moreover, they tend to reproduce at a younger age and attain relatively smaller adult sizes. By contrast, male fish above the waterfall display gaudy colors that they use to court females (figure 21.18). Adults mature later and grow to larger sizes.

These differences suggest the function of natural selection. In the low-predation environment, males display gaudy colors and spots that they use to court females. Moreover, larger males are most successful at holding territories and mating with females, and larger females lay more eggs. Thus, in the absence of predators, larger and more colorful fish may have produced more offspring, leading to the evolution of those traits. In pools below the waterfall, however, natural selection would favor different traits. Colorful males are likely to attract the attention of

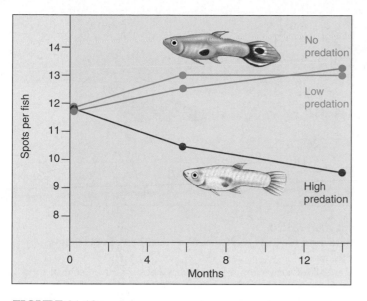

FIGURE 21.19

Evolutionary change in spot number.   Guppies raised in low-predation or no-predation environments in laboratory greenhouses had a greater number of spots, whereas selection in more dangerous environments, such as the pools with the highly predatory pike cichlid, led to less conspicuous fish. The same results are seen in field experiments conducted in pools above and below waterfalls (*photo*).
How do these results depend on the manner by which the guppy predators locate their prey?

the pike cichlid, and high predation rates mean that most fish live short lives; thus, individuals that are more drab and shunt energy into early reproduction, rather than growth to a larger size, are likely to be favored by natural selection.

The Experiments

Although the differences between guppies living above and below the waterfalls suggest that they represent evolutionary responses to differences in the strength of predation, alternative explanations are possible. Perhaps, for example, only very large fish are capable of crawling past the waterfall to colonize pools. If this were the case, then a founder effect would occur in which the new population was established solely by individuals with genes for large size.

Laboratory Experiment.   The only way to rule out such alternative possibilities is to conduct a controlled experiment. John Endler, now at the University of California, Santa Barbara, conducted the first experiments in large pools in laboratory greenhouses. At the start of the experiment, a group of 2000 guppies was divided equally among 10 large pools. Six months later, pike cichlids were added to four of the pools and killifish to another four, with the remaining two pools left to serve as "no-predation" controls. Fourteen months later (which corresponds to 10 guppy generations), the scientists compared the populations. The guppies in the killifish and control pools were indistinguishable—brightly colored and large. In contrast, the guppies in the pike cichlid pools were smaller and drab in coloration. These results established

that predation can lead to rapid evolutionary change, but do these laboratory experiments reflect what occurs in nature?

Field Experiment.   To find out, Endler and colleagues—including David Reznick, now at the University of California, Riverside—located two streams that had guppies in pools below a waterfall, but not above it. As in other Trinidadian streams, the pike cichlid was present in the lower pools, but only the killifish was found above the waterfalls. The scientists then transplanted guppies to the upper pools and returned at several-year intervals to monitor the populations. Despite originating from populations in which predation levels were high, the transplanted populations rapidly evolved the traits characteristic of low-predation guppies: They matured late, attained greater size, and had brighter colors. Control populations in the lower pools, by contrast, continued to be drab and to mature early and at a smaller size (figure 21.19). Laboratory studies confirmed that the variations between the populations were the result of genetic differences. These results demonstrate that substantial evolutionary change can occur in less than 12 years. More generally, these studies indicate how scientists can formulate hypotheses about how evolution occurs and then test these hypotheses in natural conditions. The results give strong support to the theory of evolution by natural selection.

Evolutionary biology is a historical science. Nonetheless, in some circumstances, experiments can be conducted in nature to test hypotheses about how evolution occurs. Such studies often reveal that natural selection can lead to rapid evolutionary change.

Limits to What Selection Can Accomplish

Genes Have Multiple Effects

Although selection is the most powerful of the principal agents of genetic change, there are limits to what it can accomplish. These limits arise for a variety of reasons. For example, alleles often affect multiple aspects of the phenotype (the phenomenon of pleiotropy; see chapter 13). These multiple effects tend to set limits on how much a phenotype can be altered. For example, selecting for large clutch size in chickens eventually leads to eggs with thinner shells that break more easily. For this reason, we could never produce chickens that lay eggs twice as large as the best layers do now, gigantic cattle that yield twice as much meat as our leading strains, or corn with an ear at the base of every leaf, instead of just at the base of a few leaves.

Evolution Requires Genetic Variation

Over 80% of the gene pool of the thoroughbred horses racing today goes back to 31 known ancestors from the late eighteenth century. Despite intense directional selection on thoroughbreds, their performance times have not improved for more than 50 years (figure 21.20). Years of intense selection presumably have removed variation from the population at a rate greater than it could be replenished by mutation, such that now no genetic variation remains and evolutionary change is not possible.

In some cases, phenotypic variation for a trait may never have had a genetic basis. The compound eyes of insects are made up of hundreds of visual units, termed ommatidia (see chapter 33). In some individuals, the left eye contains more ommatidia than the right eye. In other individuals, the right eye contains more than the left. However, despite intense selection in the laboratory, scientists have never been able to produce a line of fruit flies that consistently have more ommatidia in the left eye than in the right. The reason is that separate genes do not exist for the left and right eyes. Rather, the same genes affect both eyes, and differences in the number of ommatidia result from differences that occur as the eyes are formed in the development process (figure 21.21). Thus, despite the existence of phenotypic variation, no genetic variation is available for selection to favor.

Gene Interactions Affect Fitness of Alleles

As discussed in chapter 13, **epistasis** is the phenomenon in which an allele for one gene may have different effects, depending on the alleles present at other genes. Because of epistasis, the selective advantage of an allele at one gene may vary among genotypes. If a population is polymorphic for a second gene, then selection on the first gene may be constrained because different alleles are favored in different individuals of the same population.

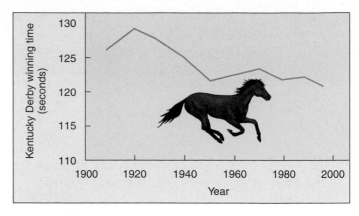

FIGURE 21.20
Selection for increased speed in racehorses is no longer effective.
Kentucky Derby winning speeds have not improved significantly since 1950.
What might explain the lack of change in winning speeds?

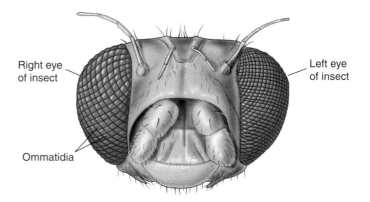

FIGURE 21.21
Phenotypic variation in insect ommatidia. In some individuals, the number of ommatidia in the left eye is greater than the number in the right eye.

Dan Hartl, now of Harvard University, and colleagues grew bacteria with different alleles of the enzyme 6-PGD on gluconate, the enzyme's substrate. Under normal conditions, Hartl found that bacteria with the different alleles all grew at the same rate; natural selection thus did not favor one allele over another. However, a second gene controls an alternative biochemical pathway for the metabolism of gluconate. An uncommon allele at this gene disables this second pathway so that only 6-PGD mediates the utilization of gluconate. When Hartl studied bacteria with this uncommon allele, he found that bacteria with several 6-PGD alleles were markedly superior to others. Thus, epistatic interactions exist between the two genes, and the outcome of natural selection on one gene depends on which alleles are present at the second gene.

The ability of selection to produce evolutionary change is hindered by a variety of factors.

21.1 Genes vary in natural populations.

Genetic Variation Is the Raw Material of Evolution

- Evolution refers to change in a species over time as genetic differences accumulate. (p. 434)
- Darwin proposed that natural selection was the mechanism of evolution, while Jean-Baptiste Lamarck believed individuals acquired characteristics during their lifetime, and then passed those onto their offspring. (p. 434)

Genetic Variation in Nature

- Evolution can result from any process causing change in a population's genetic variation. (p. 435)
- DNA analysis has found abundant evidence of polymorphic genes (loci with more than one allele). (p. 435)

21.2 Why do allele frequencies change in populations?

The Hardy–Weinberg Principle

- Original genotypic proportions in a population will remain constant as long as the population is large, mates at random, experiences no mutation or immigration, and is not affected by selection. (p. 436)

Five Agents of Evolutionary Change

- Mutation, gene flow, nonrandom mating, genetic drift (including founder effects and bottleneck effects), and selection (either natural or artificial) are the major factors driving evolutionary change. (pp. 438–441)

Measuring Fitness

- Fitness is measured as the number of living offspring in the next generation. Maximum fitness is defined as 1.0. (p. 442)

Interactions Among Evolutionary Forces

- Population variation may be determined by counterbalancing forces such as gene flow and natural selection. The extent of interaction depends on the relative strength of each force. (p. 443)

Natural Selection Can Maintain Variation in Populations

- Negative frequency-dependent selection will favor rare phenotypes, while positive frequency-dependent selection will favor common phenotypes. (p. 444)
- Oscillating selection will favor different phenotypes at different times. (p. 444)
- Variation can be maintained if heterozygotes are favored over homozygotes, as in the case of sickle cell anemia and malaria. (p. 445)

21.3 Selection can act on traits affected by many genes.

Forms of Selection

- Disruptive selection acts to eliminate intermediate phenotypes. (p. 446)
- Directional selection acts to eliminate one phenotypic extreme. (p. 447)
- Stabilizing selection acts to eliminate both phenotypic extremes. (p. 447)

Selection on Color in Guppies

- Guppies live in different natural environments. Populations under predation pressure tend to exhibit drab colors and to be smaller than guppies not under predation pressure. (p. 448)
- Laboratory and field experiments have shown that predation can lead to rapid evolutionary change and that the differences between populations are genetically based. (p. 449)

Limits to What Selection Can Accomplish

- Multiple effects of the same gene can limit the level to which a phenotype can be altered. (p. 450)
- Selection on phenotypic variation cannot lead to evolutionary change without genetic variation. (p. 450)
- Due to epistasis, the effect of an allele at one gene may depend on what alleles are present at other genes. (p. 450)

Self Test

1. Which of the following is *not* an assumption of the Hardy–Weinberg equilibrium?
   a. Mating occurs preferentially.
   b. The size of the population is large.
   c. There is no migration.
   d. There are no mutations.
2. In a population of red (dominant) and white flowers, the frequency of red flowers is 91%. What is the frequency of the red allele?
   a. 9%
   b. 30%
   c. 91%
   d. 70%
3. Which of the following best describes gene flow?
   a. random mating
   b. migration
   c. genetic drift
   d. selection
4. Which of the following conditions is *not* needed for natural selection to occur in a population?
   a. Individuals must be able to move between populations.
   b. Variation must be genetically inherited.
   c. Certain variations allow an individual to produce more offspring that survive in the next generation.
   d. There must be variations in the phenotypes of individuals.
5. Which of the following is the ultimate source of genetic variation in a population?
   a. gene flow
   b. assortative mating
   c. mutation
   d. selection
6. Natural selection can be countered by which of the following?
   a. genetic drift
   b. gene flow
   c. mutation
   d. All of these can counter natural selection in some way.
7. The maintenance of the sickle cell allele in human populations in central Africa is an example of
   a. gene flow.
   b. heterozygote advantage.
   c. genetic drift.
   d. nonrandom mating.
8. What would happen in the U.S. if malaria once again became a widespread disease?
   a. Over time, the sickle cell allele would become more prevalent in the population.
   b. Many individuals in the population would die of malaria.
   c. Individuals who were heterozygous for the sickle cell allele would be less susceptible to malaria.
   d. All of these events would occur.
9. _____ operates to eliminate intermediate phenotypes.
   a. Directional selection
   b. Disruptive selection
   c. Stabilizing selection
   d. Random chance
10. When transplanted to streams above waterfalls, guppy populations evolved more spots because
    a. predators are not present.
    b. they consumed different sources of food.
    c. spots make guppies harder to see against their background.
    d. all of the above

Test Your Visual Understanding

(a) Gene flow

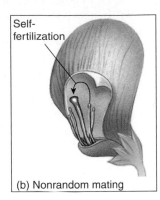

Self-fertilization

(b) Nonrandom mating

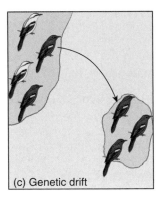

(c) Genetic drift

(d) Selection

1. Match the following descriptions with the correct panels in the figure.
   a. Phenotypically similar individuals mate.
   b. Individuals migrate.
   c. Space exploration expands with the settlement of a population of 100 individuals on Mars.
   d. A nuclear power plant dumps boiling water into a reservoir, killing all bacteria except those that contain a heat shock gene.

Applying Your Knowledge

1. Consider a human population that is similar to the ideal Hardy–Weinberg population in that it is very large and generally random-mating. Although mutations occur, they alone do not lead to great changes in allele frequencies. However, migration occurs at relatively high levels—perhaps 1% per year. The following data describe relative numbers of individuals bearing the two alleles of the MN blood group:

	MM	MN	NN	Total
Individuals	1787	3037	1305	6129

Do these data suggest that some factor is disrupting the Hardy–Weinberg proportions of the three genotypes? What are the allele frequencies of M and N?

22

The Evidence for Evolution

Concept Outline

22.1 Evidence indicates that natural selection can produce evolutionary change.

The Beaks of Darwin's Finches. Natural selection favors stouter bills in dry years, when large tough-to-crush seeds are the only food available to finches.

Peppered Moths and Industrial Melanism. Natural selection favors dark-colored moths in areas of heavy pollution, while light-colored moths survive better in unpolluted areas.

Artificial Selection. Artificial selection practiced in laboratory studies, agriculture, and domestication demonstrate that selection can produce substantial evolutionary change.

22.2 Fossil evidence indicates that evolution has occurred.

The Fossil Record. When fossils are arranged in the order of their age, a continual series of change is seen, with new changes added at each stage.

The Evolution of Horses. The record of horse evolution is particularly well-documented and instructive.

22.3 Evidence for evolution can be found in other fields of biology.

The Anatomical Record. When the anatomical features of living animals are examined, evidence of shared ancestry is often apparent.

The Molecular Record. When gene or protein sequences from organisms are arranged, species thought to be closely related based on fossil evidence are seen to be more similar than species thought to be distantly related.

Convergent Evolution and the Biogeographical Record. Natural selection favors similar forms under similar circumstances, and the geographic distribution of species reflects evolutionary divergence.

22.4 The theory of evolution has proven controversial.

Darwin's Critics. Critics have raised seven objections to Darwin's theory of evolution by natural selection.

FIGURE 22.1
Fossil remains of trilobites. Trilobites were a diverse class of arthropods that lived 245–540 million years ago. The rich fossil record of these primarily bottom-dwelling marine organisms clearly documents the origins of new species and reveals how species evolve through time.

As we discussed in chapter 1, when Darwin proposed his revolutionary theory of evolution by natural selection, little actual evidence existed to bolster his case. Instead, Darwin relied on observations of the natural world, logic, and results obtained by breeders working with domestic animals. Since his day, however, the evidence for Darwin's theory has become overwhelming. The case is built upon two pillars: first, evidence that natural selection can produce evolutionary change, and second, evidence from the fossil record that evolution has occurred (figure 22.1). In addition, information from many different areas of biology—fields as different as anatomy, molecular biology, and biogeography—is only interpretable scientifically as the outcome of evolution.

22.1 Evidence indicates that natural selection can produce evolutionary change.

As we saw in chapter 21, a variety of processes produce evolutionary change. Nonetheless, in agreement with Darwin, most evolutionary biologists believe that natural selection is the primary process responsible for major evolutionary changes. Although we cannot travel back through time, a variety of modern-day evidence confirms the power of natural selection as an agent of evolutionary change. These data come from both the field and the laboratory and from both natural and human-altered situations.

The Beaks of Darwin's Finches

Darwin's finches are a classic example of evolution by natural selection. When he visited the Galápagos Islands off the coast of Ecuador in 1835, Darwin collected 31 specimens of finches from three islands. Darwin, not an expert on birds, had trouble identifying the specimens, believing by examining their bills that his collection contained wrens, "gross-beaks," and blackbirds.

The Importance of the Beak

Upon Darwin's return to England, ornithologist John Gould recognized that Darwin's collection was in fact a closely related group of distinct species, all similar to one another except for their bills. In all, 14 species are now recognized. The ground finches with the larger bills in figure 22.2 feed on seeds that they crush in their beaks, whereas the ones with narrower bills eat insects. Other species include fruit and bud eaters, insect eaters, and species that feed on cactus fruits and the insects they attract; some populations of the sharp-beaked ground finch even include "vampires" that creep up on seabirds and use their sharp beaks to drink their blood. Perhaps most remarkable are the tool users, woodpecker finches that pick up a twig, cactus spine, or leaf stalk, trim it into shape with their bills, and then poke it into dead branches to pry out grubs.

The correspondence between the beaks of the 14 finch species and their food source immediately suggested to Darwin that natural selection had shaped them. In *The*

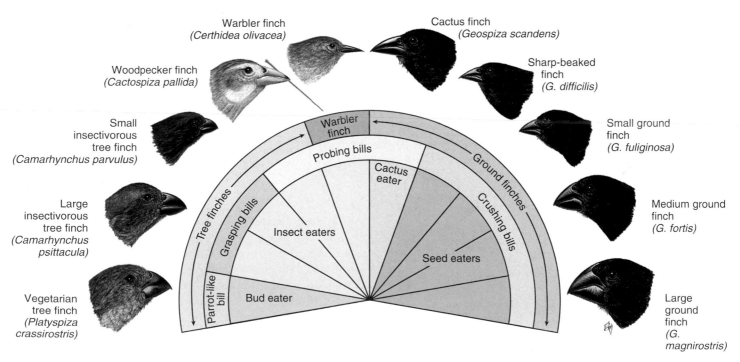

FIGURE 22.2
Darwin's finches. Ten species of Darwin's finches occur on Santa Cruz, one of the Galápagos Islands. These species show differences in bills and feeding habits. This condition presumably arose when the finches evolved new species in habitats lacking small birds. The bills of several species resemble those of distinct families of birds on the mainland. For example, the woodpecker finch uses cactus spines to probe in crevices of bark and rotten wood for food. All of these species are derived from a single common ancestor.

Voyage of the Beagle, Darwin wrote, "Seeing this gradation and diversity of structure in one small, intimately related group of birds, one might really fancy that from an original paucity of birds in this archipelago, one species has been taken and modified for different ends."

A Closer Look

Darwin's observations suggest that differences among species in beak size and shape have evolved as the species adapted to use different food resources, but can this hypothesis be tested? As we discussed in chapter 21, the theory of evolution by natural selection requires that three conditions be met: (1) Variation must exist in the population; (2) this variation must lead to differences among individuals in survival and reproductive success; and (3) variation among individuals must be genetically transmitted to the next generation.

The key to successfully testing Darwin's proposal proved to be patience. Starting in 1973, Peter and Rosemary Grant of Princeton University and generations of their students have studied the medium ground finch, *Geospiza fortis*, on a tiny island in the center of the Galápagos called Daphne Major. These finches feed preferentially on small, tender seeds, produced in abundance by plants in wet years. The birds resort to larger, drier seeds, which are harder to crush, only when small seeds become depleted during long periods of dry weather, when plants produce few seeds.

The Grants quantified beak shape among the medium ground finches of Daphne Major by carefully measuring beak depth (width of beak, from top to bottom, at its base) on individual birds. Measuring many birds every year, they were able to assemble for the first time a detailed portrait of evolution in action. The Grants found that not only did a great deal of variation in beak depth exist among members of the population, but the average beak depth changed from one year to the next in a predictable fashion. During droughts, plants produced few seeds, and all available small seeds were quickly eaten, leaving large seeds as the major remaining source of food. As a result, birds with large beaks survived better, because they were better able to break open these large seeds. Consequently, the average beak depth of birds in the population increased the next year, only to decrease again when wet seasons returned (figure 22.3a).

Could these changes in beak dimension reflect the action of natural selection? An alternative possibility might be that the changes in beak depth do not reflect changes in gene frequencies, but rather are simply a response to diet; for example, perhaps crushing large seeds causes a growing bird to develop a larger beak. To rule out this possibility, the Grants measured the relation of parent beak size to offspring beak size, examining many broods over several years. The depth of the beak was passed down faithfully from one generation to the next, regardless of environ-

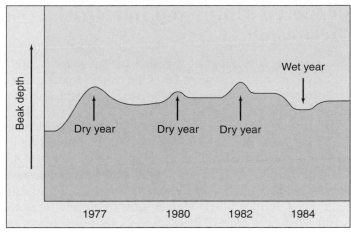
(a)

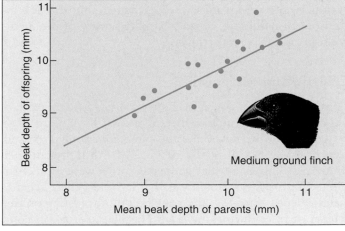
(b)

FIGURE 22.3
Evidence that natural selection alters beak size in *Geospiza fortis*. (*a*) In dry years, when only large, tough seeds are available, the mean beak size increases. In wet years, when many small seeds are available, smaller beaks become more common. (*b*) Beak depth is inherited from parents to offspring.
Suppose a bird with a large bill mates with a bird with a small bill. Would the bills of the pair's offspring tend to be larger or smaller than the bills of offspring from a pair of birds with medium-sized bills?

mental conditions (figure 22.3b), suggesting that the differences among individuals in beak size reflect genetic differences, and thus, that the year-to-year changes in average beak depth represent evolutionary change resulting from natural selection.

Among Darwin's finches, natural selection adjusts the shape of the beak in response to the nature of the available food supply. Such adjustments can be seen to occur even today.

Peppered Moths and Industrial Melanism

When the environment changes, natural selection often may favor new traits in a species. One classic example concerns the peppered moth, *Biston betularia*. Adults come in a range of shades, from light gray with black speckling (hence the name "peppered" moth) to jet black (melanic). Extensive genetic analysis has shown that the moth's body color is a genetic trait, reflecting different alleles of a single gene. Black individuals have a dominant allele, one that was present but very rare in populations before 1850. From that time on, dark individuals increased in frequency in moth populations near industrialized centers until they made up almost 100% of these populations. Biologists soon noticed that in industrialized regions where the dark moths were common, the tree trunks were darkened almost black by the soot of pollution, which also killed many of the light-colored lichens on tree trunks.

Selection for Melanism

Why did dark moths gain a survival advantage around 1850? An amateur moth collector named J. W. Tutt proposed in 1896 what became the most commonly accepted hypothesis explaining the decline of the light-colored moths. He suggested that peppered forms were more visible to predators on sooty trees that have lost their lichens. Consequently, birds ate the peppered moths resting on the trunks of trees during the day. The black forms, in contrast, were at an advantage because they were camouflaged (figure 22.4). Although Tutt initially had no evidence, British ecologist Bernard Kettlewell tested the hypothesis in the 1950s by releasing equal numbers of dark and light individuals into two sets of woods: one near heavily polluted Birmingham, and the other in unpolluted Dorset. Kettlewell then set up traps in the woods to see how many of both kinds of moths survived. To evaluate his results, he had marked the released moths with a dot of paint on the underside of their wings, where birds could not see it.

In the polluted area near Birmingham, Kettlewell trapped only 19% of the light moths, but 40% of the dark ones. This indicated that dark moths had a far better chance of surviving in these polluted woods, where the tree trunks were dark. In the relatively unpolluted Dorset woods, Kettlewell recovered 12.5% of the light moths but only 6% of the dark ones. This indicated that where the tree trunks were still light-colored, light moths had a much better chance of survival. Kettlewell later solidified his argument by placing moths on trees and filming birds looking for food. Sometimes the birds actually passed right over a moth that was the same color as its background. Kettlewell's finding—that birds more frequently detect moths whose color does not match their background—has subsequently been confirmed in eight separate field studies, with a variety of experimental designs and corrections for defi-

FIGURE 22.4
Tutt's hypothesis explaining industrial melanism. These photographs show dead specimens of the peppered moth (*Biston betularia*) placed on trees. Tutt proposed that the dark melanic variant of the moth is more visible to predators on unpolluted trees (*top*), while the light "peppered" moth is more visible to predators on bark blackened by industrial pollution (*bottom*).

ciencies in Kettlewell's initial experimental design. These results, combined with the recapture studies, provide strong evidence for the action of natural selection and implicate birds as the agent of selection in the case of the peppered moth.

Industrial Melanism

The term **industrial melanism** refers to the process by which darker individuals come to predominate over lighter individuals. In industrialized areas throughout Eurasia and North America, dozens of other species of moths have changed in the same way as the peppered moth.

Could it be that melanism confers an inherent physiological advantage over lighter-colored forms? While consistent with the worldwide success of industrial melanism, this possibility can be eliminated because of the subsequent reversal of melanism. That is, natural selection is now acting *against* the melanic allele.

Selection Against Melanism

In the second half of the twentieth century, with the widespread implementation of pollution controls, the trend toward melanism began reversing for many species of moths throughout the northern continents.

In England, the air pollution promoting industrial melanism began to reverse following enactment of the Clean Air Act in 1956. Beginning in 1959, the *Biston* population at Caldy Common outside Liverpool has been sampled each year. The frequency of the melanic (dark) form has dropped from a high of 93% in 1959 to a low of 15% in 1995 (figure 22.5). The drop correlates well with a significant drop in air pollution, particularly with a lowering of the levels of sulfur dioxide and suspended particulates, both of which act to darken trees. The drop is consistent with a 15% selective disadvantage acting against moths with the dominant melanic allele.

Interestingly, the same reversal of melanism occurred in the United States. Of 576 peppered moths collected at a field station near Detroit from 1959 to 1961, 515 were melanic, a frequency of 89%. The American Clean Air Act, passed in 1963, led to significant reductions in air pollution. Resampled in 1994, the Detroit field station peppered moth population had only 15% melanic moths (figure 22.5)! The moths in Liverpool and Detroit, both part of the same natural experiment, exhibit strong evidence for natural selection.

Reconsidering the Agent of Natural Selection

While the evidence for natural selection in the case of the peppered moth is strong, Tutt's hypothesis about the agent of selection is currently being reevaluated. Researchers have noted that the recent selection against melanism does not appear to correlate with changes in tree lichens. At Caldy Common, the light form of the peppered moth began to increase in frequency long before lichens began to reappear on the trees. At the Detroit field station, the lichens never changed significantly as the dark moths first became dominant and then declined over the last 30 years. In fact, investigators have not been able to find peppered moths on Detroit trees at all, whether covered with lichens or not. Wherever the moths rest during the day, it does not appear to be on tree bark. Some evidence suggests they rest on leaves on the treetops, but no one is sure. Could poisoning by pollution rather than predation by birds be the agent of natural selection on the moths? Perhaps—but to date, only predation by birds is backed by experimental evidence.

FIGURE 22.5

Selection against melanism. The circles indicate the frequency of melanic *Biston betularia* moths at Caldy Common in England, sampled continuously from 1959 to 1995. Diamonds indicate frequencies of melanic *B. betularia* in Michigan from 1959 to 1962 and from 1994 to 1995.

What can you conclude from the fact that the frequency of melanic moths decreased to the same degree in the two locations?

Researchers supporting bird predation as the agent of natural selection point out that a bird's ability to detect moths may depend less on the presence or absence of lichens, and more on other ways in which the environment is darkened by industrial pollution. Pollution tends to cover all objects in the environment with a fine layer of particulate dust, which tends to decrease how much light surfaces reflect. In addition, pollution has a particularly severe effect on birch trees, which are light in color. Both effects would tend to make the environment darker, and thus would favor darker moths by protecting them from predation by birds.

Despite this uncertainty over the agent of selection, the overall pattern is clear. Kettlewell's experiments establish indisputably that selection favors dark moths in polluted habitats and light moths in pristine areas. The increase and subsequent decrease in the frequency of melanic moths, correlated with levels of pollution independently on two continents, demonstrates clearly that this selection drives evolutionary change. The current reconsideration of the agent of natural selection illustrates well the way in which scientific progress is achieved: Hypotheses, such as Tutt's, are put forth and then tested. If rejected, new hypotheses are formulated, and the process begins anew.

Natural selection has favored the dark form of the peppered moth in areas subject to severe air pollution, perhaps because on darkened trees they are less easily seen by moth-eating birds. Selection has in turn favored the light form as pollution has abated.

Artificial Selection

Humans have imposed selection upon plants and animals since the dawn of civilization. Just as in natural selection, artificial selection operates by favoring individuals with certain phenotypic traits, allowing them to reproduce and pass their genes on to the next generation. Assuming that phenotypic differences are genetically determined, such directional selection should lead to evolutionary change, and indeed, it has. Artificial selection, imposed in laboratory experiments, agriculture, and the domestication process, has produced substantial change in almost every case in which it has been applied. This success is strong proof that selection is an effective evolutionary process.

Laboratory Experiments

With the rise of genetics as a field of science in the 1920s and 1930s, researchers began conducting experiments to test the hypothesis that selection can produce evolutionary change. A favorite subject was the laboratory fruit fly, *Drosophila melanogaster*. Geneticists have imposed selection on just about every conceivable aspect of the fruit fly—including body size, eye color, growth rate, life span, and exploratory behavior—with a consistent result: Selection for a trait leads to strong and predictable evolutionary response.

In one classic experiment, scientists selected for fruit flies with many bristles (stiff, hairlike structures) on their abdomen. At the start of the experiment, the average number of bristles was 9.5. Each generation, scientists picked out the 20% of the population with the greatest number of bristles and allowed them to reproduce, thus establishing the next generation. After 86 generations of such selection, the average number of bristles had quadrupled, to nearly 40! In another experiment, fruit flies were selected for either the most or the fewest numbers of bristles. Within 35 generations, the populations did not overlap at all in range of variation (figure 22.6).

Similar experiments have been conducted on a wide variety of other laboratory organisms. For example, by selecting for rats that were resistant to tooth decay, scientists were able to increase in less than 20 generations the average time for onset of decay from barely over 100 days to greater than 500 days.

Agriculture

Similar methods have been practiced in agriculture for many centuries. Familiar livestock, such as cattle and pigs, and crops, such as corn and strawberries, are greatly different from their wild ancestors (figure 22.7). These differences have resulted from generations of selection for desirable traits, such as milk production and corn ear size.

An experiment with corn demonstrates the ability of artificial selection to rapidly produce major change in crop plants. In 1896, agricultural scientists began selecting on

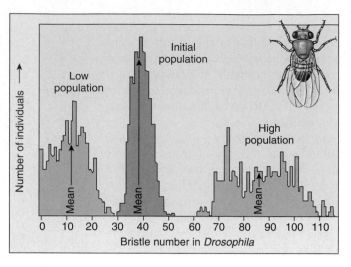

FIGURE 22.6
Artificial selection in the laboratory. In this experiment, one population of *Drosophila* was selected for low numbers of bristles and the other for high numbers. Note that not only did the means of the populations change greatly in 35 generations, but also all individuals in both experimental populations lie outside the range of the initial population.
What would happen if, within a population, both small and large individuals were allowed to breed, but middle-sized ones were not?

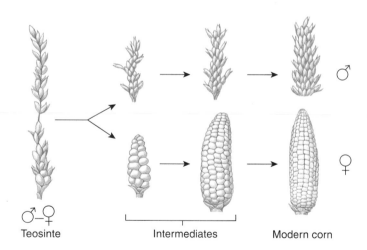

FIGURE 22.7
Corn looks very different from its ancestor. The tassels and seeds of a wild grass, such as teosinte, evolved into the male tassels and female ears of modern corn, which are on different parts of the plant, in contrast to the ancestral condition in teosinte.

the oil content of corn kernels, which initially was 4.5%. As in the fruit fly experiments, the top 20% of all individuals were allowed to reproduce. By 1986, at which time 90 generations had passed, average oil content had increased approximately 450% in the corn kernels.

Domestication

Artificial selection has also been responsible for the great variety of breeds of cats, dogs (figure 22.8), pigeons, and other domestic animals. In some cases, breeds have been developed for particular purposes. Greyhound dogs, for example, were bred by selecting for maximal running abilities, resulting in an animal with long legs, a long tail (for use as a rudder), an arched back (to increase stride length), and great muscle mass. By contrast, the odd proportions of the ungainly dachshund resulted from selection for dogs that could enter narrow holes in pursuit of badgers. In other cases, breeds have been developed primarily for their appearance, such as the many colorful and ornamented breeds of pigeons or cats.

Domestication also has led to unintentional selection for some traits. In recent years, as part of an attempt to domesticate the silver fox, Russian scientists have chosen the most docile animals in each generation and allowed them to reproduce. Within 40 years, most foxes were exceptionally docile, not only allowing themselves to be petted, but also whimpering to get attention and sniffing and licking their caretakers (figure 22.9). In many respects, they had become no different than domestic dogs! However, it was not only their behavior that changed. These foxes also began to exhibit different color patterns, such as the floppy ears, curled tails, and shorter legs and tails seen in some dog breeds. Presumably, the genes responsible for docile behavior either affect these traits as well or are closely linked to the gene for these other traits (the phenomena of pleiotropy and linkage discussed in chapter 13). As selection has favored docile animals, it has also led to the evolution of these other traits.

Can Selection Produce Major Evolutionary Changes?

Given that we can observe the results of selection operating over relatively short periods of time, most scientists believe that natural selection is the process responsible for the evolutionary changes documented in the fossil record. Some critics of evolution accept that selection can lead to changes within a species, but contend that such changes are relatively minor in scope and not equivalent to the substantial changes documented in the fossil record. In other words, it is one thing to change the number of bristles on a fruit fly or the size of an ear of corn, and quite another to produce an entirely new species.

This argument does not fully appreciate the extent of change produced by artificial selection. Consider, for example, the existing breeds of dogs, all of which have been produced since wolves were first domesticated, perhaps 10,000 years ago. If the various dog breeds did not exist and a paleontologist found fossils of animals similar to dachshunds, greyhounds, mastiffs, and chihuahuas, there is no question that they would be considered different species. Indeed, the

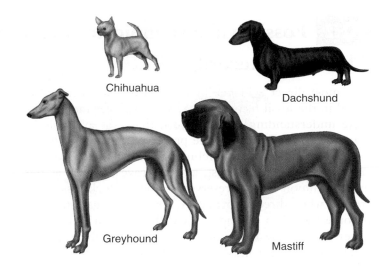

FIGURE 22.8
Breeds of dogs. The differences between these dogs are greater than the differences displayed in any wild species of canids.

FIGURE 22.9
Domesticated foxes. After 40 years of selectively breeding the tamest individuals, artificial selection has produced silver foxes that are not only as friendly as domestic dogs, but also exhibit many physical traits seen in dog breeds.

differences in size and shape exhibited by these breeds are greater than those between members of different genera in the family Canidae—such as coyotes, jackals, foxes, and wolves—which have been evolving separately for 5–10 million years. Consequently, the claim that artificial selection produces only minor changes is clearly incorrect. If selection operating over a period of only 10,000 years can produce such substantial differences, it should be powerful enough, over the course of many millions of years, to produce the diversity of life we see around us today.

> Artificial selection often leads to rapid and substantial results over short periods of time, thus demonstrating the power of selection to produce major evolutionary change.

The Evolution of Horses

One of the best-studied cases in the fossil record concerns the evolution of horses. Modern-day members of the Equidae include horses, zebras, donkeys, and asses, all of which are large, long-legged, fast-running animals adapted to living on open grasslands. These species, all classified in the genus *Equus*, are the last living descendants of a long lineage that has produced 34 genera since its origin in the Eocene period, approximately 55 million years ago. Examination of these fossils has provided a particularly well-documented case of how evolution has proceeded through adaptation to changing environments.

The First Horse

The earliest known members of the horse family, species in the genus *Hyracotherium*, didn't look much like modern-day horses at all. Small, with short legs and broad feet (figure 22.14), these species occurred in wooded habitats, where they probably browsed on leaves and herbs and escaped predators by dodging through openings in the forest vegetation. The evolutionary path from these diminutive creatures to the workhorses of today has involved changes in a variety of traits, including size, toe reduction, and tooth size and shape.

Size. The first species of horses were the size of dogs or smaller. By contrast, modern equids can weigh more than a half ton. Examination of the fossil record reveals that horses changed little in size for their first 30 million years, but since then, a number of different lineages have exhibited rapid and substantial increases. However, trends toward decreased size were also exhibited among some branches of the equid evolutionary tree (figure 22.15).

Toe Reduction. The feet of modern horses have a single toe enclosed in a tough, bony hoof. By contrast, *Hyracotherium* had four toes on its front feet and three on its hind feet. Rather than hooves, these toes were encased in fleshy pads like those of dogs and cats. Examination of the fossils clearly shows the transition through time: increase in length of the central toe, development of the bony hoof, and reduction and loss of the other toes (figure 22.16). As with body size, these trends occurred concurrently on several different branches of the horse evolutionary tree. At the same time, horses were evolving changes in the length and skeletal structure of their limbs, leading to animals capable of running long distances at high speeds.

Tooth Size and Shape. The teeth of *Hyracotherium* were small and relatively simple in shape. Through time, horse teeth have increased greatly in length and have developed a complex pattern of ridges on their molars and premolars (figure 22.16). The effect of these changes is to produce teeth better capable of chewing tough and gritty vegeta-

FIGURE 22.14
Hyracotherium sandrae, **one of the earliest horses.** This species was the size of a housecat.

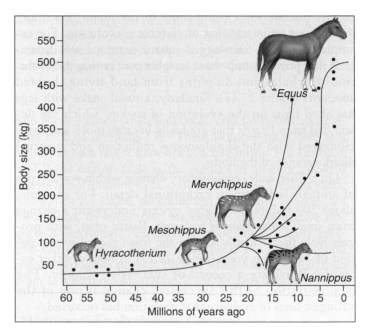

FIGURE 22.15
Evolutionary change in body size of horses. Lines indicate evolutionary relationships and reveal that although most of the change involved increases in size, some decreases also occurred. **Why might the evolutionary line leading to *Nannippus* have experienced an evolutionary decrease in body size?**

tion, such as grass, which tends to wear teeth down. Accompanying these changes have been alterations in the shape of the skull that strengthened the skull to withstand the stresses imposed by continual chewing. As with body size, evolutionary change has not been constant through time. Rather, much of the change in tooth shape has occurred within the past 20 million years.

All of these changes may be understood as adaptations to changing global climates. In particular, during the late Miocene and early Oligocene epochs (approximately 20 to 25 million years ago), grasslands became widespread in

North America, where much of horse evolution occurred. As horses adapted to these habitats, high-speed locomotion probably became more important to escape predators. By contrast, the greater flexibility provided by multiple toes and shorter limbs, which was advantageous for ducking through complex forest vegetation, was no longer beneficial. At the same time, horses were eating grasses and other vegetation that contained more grit and other hard substances, thus favoring teeth and skulls better suited for withstanding such materials.

Evolutionary Trends

For many years, horse evolution was held up as an example of constant evolutionary change through time. Some even saw in the record of horse evolution evidence for a progressive, guiding force, consistently pushing evolution in a single direction. We now know that such views are misguided; evolutionary change over millions of years is rarely so simple.

Rather, the fossils demonstrate that, although overall trends have been evident in a variety of characteristics, evolutionary change has been far from constant and uniform through time. Instead, rates of evolution have varied widely, with long periods of little observable change and some periods of great change. Moreover, when changes happen, they often occur simultaneously in different lineages of the horse evolutionary tree. Finally, even when a trend exists, exceptions, such as the evolutionary decrease in body size exhibited by some lineages, are not uncommon. These patterns, evident in our knowledge of horse evolution, are usually discovered for any group of plants and animals for which we have an extensive fossil record, as we shall see when we discuss human evolution in chapter 34.

Horse Diversity

One reason that horse evolution was originally conceived of as linear through time may be that modern horse diversity is relatively limited. Thus, it is easy to mentally picture a straight line from *Hyracotherium* to modern day *Equus*. However, today's limited horse diversity—only one surviving genus—is unusual. Indeed, at the peak of horse diversity in the Miocene epoch, 13 genera of horses could be found in North America alone. These species differed in body size and in a wide variety of other characteristics. Presumably, they lived in different habitats and exhibited different dietary preferences. Had this diversity existed to modern times, early evolutionary biologists presumably would have had a different outlook on horse evolution.

The extensive fossil record for horses provides a detailed view of the evolutionary diversification of this group, from small forest dwellers to large and fast modern grassland species.

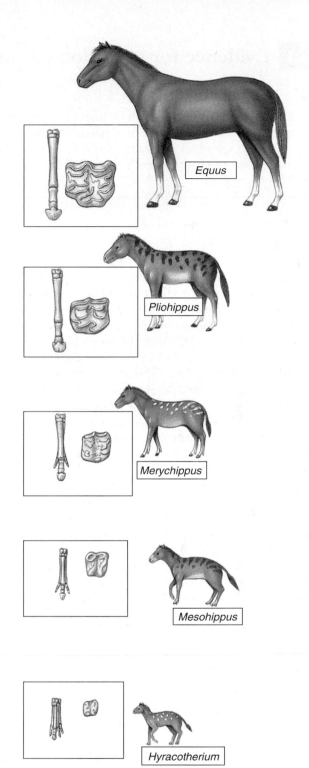

FIGURE 22.16
Evolutionary trends in horses through time. Five representative species illustrate that body size increased over time in horses. The inset boxes show the decrease in number of toes, producing the single hoof of modern horses, and the increased size and complexity of the molar teeth (viewed from the top). As figure 22.15 demonstrates, not all evolutionary changes conformed to these trends.

22.3 Evidence for evolution can be found in other fields of biology.

The Anatomical Record

Much of the power of the theory of evolution is its ability to provide a sensible framework for understanding the diversity of life. Many observations from throughout biology simply cannot be understood in any meaningful way except as a result of evolution.

Homology

As vertebrates evolved, the same bones were sometimes put to different uses. Yet the bones are still seen, their presence betraying their evolutionary past. For example, the forelimbs of vertebrates are all **homologous structures**—structures with different appearances and functions that all derived from the same body part in a common ancestor. You can see in figure 22.17 how the bones of the forelimb have been modified in different ways for different mammals. Why should these very different structures be composed of the same bones? If evolution had not occurred, this would indeed be a riddle. But when we consider that all of these animals are descended from a common ancestor, it is easy to understand that natural selection has modified the same initial starting blocks to serve very different purposes.

Development

Some of the strongest anatomical evidence supporting evolution comes from comparisons of how organisms develop. Embryos of different types of vertebrates, for example, often are similar early on, but become more different as they develop. Early in their development, human and fish embryos both possess pharyngeal pouches, which in humans develop into various glands and ducts and in fish turn into gill slits. At a later stage, every human embryo has a long bony tail, the vestige of which we carry to adulthood as the coccyx at the end of our spine. Human fetuses even possess a fine fur (called *lanugo*) during the fifth month of development. These relict developmental forms suggest strongly that our development has evolved, with new instructions modifying ancestral developmental patterns.

Imperfect Structures

Because natural selection can only work on the variation present in a population, it should not be surprising that some organisms do not appear perfectly adapted to their environments. For example, most animals with long necks have many neck vertebrae for enhanced flexibility: Geese have up to 25, and plesiosaurs, the long-necked reptiles that patrolled the seas during the age of dinosaurs, had as many as 76. By contrast, almost all mammals have only 7 neck vertebrae, and so does the giraffe! In the absence of variation in vertebrae number, selection led to evolutionary increase in vertebrae size to produce the long neck of the giraffe.

An excellent example of an imperfect design is the eye of vertebrate animals, in which the photoreceptors face backward, toward the wall of the eye (figure 22.18a). As a result, the nerve fibers extend not backward, toward the brain, but forward into the eye chamber, where they slightly obstruct light. Moreover, these fibers bundle together to form the optic nerve, which exits through a hole at the back of the eye, creating a blind spot. By contrast, the eye of mollusks—such as squid and octopuses—are more optimally designed: The photoreceptors face for-

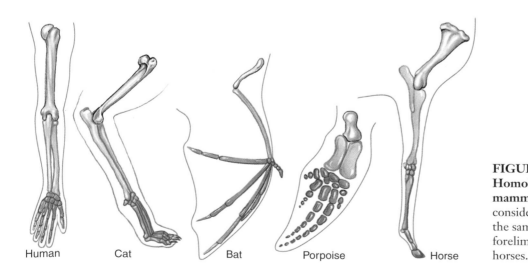

Human   Cat   Bat   Porpoise   Horse

FIGURE 22.17
Homology of the bones of the forelimb of mammals.   Although these structures show considerable differences in form and function, the same basic bones are present in the forelimbs of humans, cats, bats, porpoises, and horses.

ward, and the nerve fibers exit at the back, neither obstructing light nor creating a blind spot (figure 22.18*b*). Such examples illustrate that natural selection is like a tinkerer, working with whatever material is available to craft a workable solution, rather than like an engineer, who can design and build the best possible structure for a given task. Workable, but imperfect structures such as the vertebrate eye are an expected outcome of evolution by natural selection.

Vestigial Structures

Many organisms possess **vestigial structures** that have no apparent function but resemble structures their presumed ancestors had. Humans, for example, possess a complete set of muscles for wiggling their ears, just like many other mammals do. Boa constrictors have hip bones and rudimentary hind legs. Manatees (a type of aquatic mammal often referred to as "sea cows") have fingernails on their fins (which evolved from legs). Blind cave fish, which never see the light of day, have small, nonfunctional eyes. Figure 22.19 illustrates the skeleton of a baleen whale, which contains pelvic bones, as other mammal skeletons do, even though such bones serve no known function in the whale. The human vermiform appendix is apparently vestigial; it represents the degenerate terminal part of the cecum, the blind pouch or sac in which the large intestine begins. In other mammals, such as mice, the cecum is the largest part of the large intestine and functions in storage—usually of bulk cellulose in herbivores. Although some functions have been suggested, it is difficult to assign any current function to the vermiform appendix. In many respects, it is a dangerous organ: Quite often, it becomes infected, leading to an inflammation called appendicitis. Without surgical removal, the appendix may burst, allowing the contents of the gut to come in contact with the lining of the body cavity, a potentially fatal event.

It is difficult to understand vestigial structures such as these as anything other than evolutionary relicts, holdovers from the past. However, the existence of vestigial structures argues strongly for the common ancestry of the members of the groups that share them, regardless of how different those groups have subsequently become.

Comparisons of the anatomy of different living animals often reveal evidence of shared ancestry. In some instances, the same organ has evolved to carry out different functions; in others, an organ loses its function altogether.

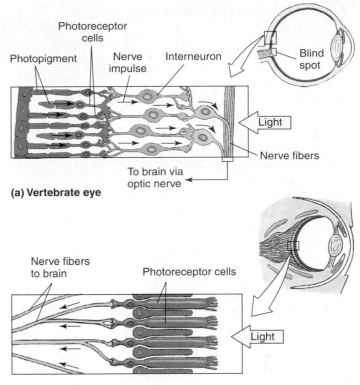

(a) Vertebrate eye

(b) Mollusk eye

FIGURE 22.18
The eyes of vertebrates and mollusks. (*a*) Photoreceptors of vertebrates point backward, whereas (*b*) those of mollusks face forward. As a result, vertebrate nerve fibers pass in front of the photoreceptor; where they bundle together and exit the eye, a blind spot is created. Mollusks' eyes have neither of these problems.

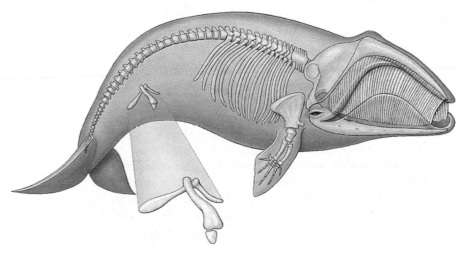

FIGURE 22.19
Vestigial structures. The skeleton of a whale reveals the presence of pelvic bones. These bones resemble those of other mammals, but are only weakly developed in the whale and have no apparent function.

The Molecular Record

Traces of our evolutionary past are also evident at the molecular level. The fact that organisms have evolved successively from relatively simple ancestors implies that a record of evolutionary change is present in the cells of each of us—in our DNA. When an ancestral species gives rise to two or more descendants, those descendants initially exhibit fairly high overall similarity in their DNA. However, as the descendants evolve independently, they accumulate more and more differences in their DNA. Consequently, organisms that are more distantly related would be expected to accumulate a greater number of evolutionary differences, whereas two species that are more closely related should share a greater portion of their DNA.

To examine this hypothesis, we need an estimate of evolutionary relationships that has been developed from data other than DNA. (It would be a circular argument to use DNA to estimate relationships and then conclude that closely related species are more similar in their DNA than are distantly related species.) Such a hypothesis of evolutionary relationships is provided by the fossil record, which indicates when particular types of organisms evolved. In addition, by comparing the anatomical structures of fossils and of modern species, we can infer how closely species are related to each other.

When degree of genetic similarity is compared with our ideas of evolutionary relationships based on fossils, a close match is evident. For example, when the human hemoglobin polypeptide is compared to the corresponding molecule in other species, closely related species are found to be more similar. Chimpanzees, gorillas, orangutans, and macaques, vertebrates thought to be more closely related to humans, have fewer differences from humans in the 146-amino-acid hemoglobin β chain than do more distantly related mammals, such as dogs. Nonmammalian vertebrates differ even more, and nonvertebrate hemoglobins are the most different of all (figure 22.20). Similar patterns are also evident when the DNA itself is compared. For example, chimps and humans, which are thought to have descended from a common ancestor that lived approximately 6 million years ago, exhibit few differences in their DNA.

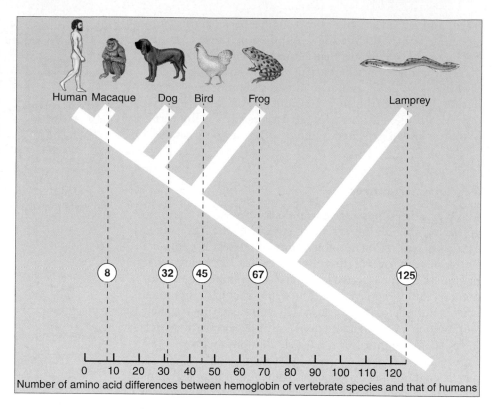

FIGURE 22.20
Molecules reflect evolutionary divergence. The greater the evolutionary distance from humans (as revealed by the white evolutionary tree which is based on the fossil record), the greater is the number of amino acid differences in the vertebrate hemoglobin polypeptide.

Why should closely related species be similar in DNA? Because DNA is the genetic code that produces the structure of living organisms, one might expect species that are similar in overall appearance and structure, such as humans and chimpanzees, to be more similar in DNA than are more dissimilar species, such as humans and frogs. This expectation would hold true even if evolution had not occurred. However, as we saw in chapter 17, large portions of the genome are composed of noncoding stretches of DNA (sometimes called "junk DNA") that have no known function and appear to serve no purpose. If evolution had not occurred, there would be no reason to expect similar-appearing species to also be similar in their junk DNA. However, comparisons of such stretches of DNA provide the same results as for other parts of the genome: More closely related species are more similar, an observation that only makes sense if evolution has occurred.

> Comparison of the DNA of different species provides strong evidence for evolution. Species deduced from the fossil record to be closely related are more similar in their DNA than are species thought to be more distantly related.

Convergent Evolution and the Biogeographical Record

Biogeography, the study of the geographic distribution of species, reveals that different geographical areas sometimes exhibit groups of plants and animals of strikingly similar appearance, even though the organisms may be only distantly related. It is difficult to explain so many similarities as the result of coincidence. Instead, natural selection appears to have favored parallel evolutionary adaptations in similar environments. Because selection in these instances has tended to favor changes that made the two groups more alike, their phenotypes have converged. This form of evolutionary change is referred to as **convergent evolution**.

The Marsupial-Placental Convergence

In the best-known case of convergent evolution, two major groups of mammals, marsupials and placentals, have evolved in a very similar way, even though the two lineages have been living independently on separate continents. Australia separated from the other continents more than 70 million years ago, after marsupials had evolved but before placental mammals had appeared. As a result, the only placental mammals in Australia are bats and a few colonizing rodents, and Australia is dominated by marsupials, members of a group in which the young are born in a very immature condition and held in a pouch until they are ready to emerge into the outside world. Thus, even though placental mammals are the dominant mammalian group throughout most of the world, marsupials retained supremacy in Australia.

What are the Australian marsupials like? To an astonishing degree, they resemble the placental mammals living today on the other continents (figure 22.21). The similarity between some individual members of these two sets of mammals argues strongly that they are the result of convergent evolution, similar forms having evolved in different, isolated areas because of similar selective pressures in similar environments.

Island Evolution

The geographical distribution of species provides evidence for evolution in other ways. Darwin was the first to present evidence that animals and plants living on oceanic islands resemble most closely the forms on the nearest continent—a relationship that only makes sense as reflecting common ancestry. For example, Galápagos tortoises and finches are more similar to South American tortoises and finches than to those of more distant continents or islands, even though the environment in the Galápagos is quite different from nearby parts of South America. This relationship strongly suggests that the island forms evolved from individuals that came from the adjacent mainland at some time in the past. In the absence of evolution, there seems to be no logical explanation for why individual kinds of island plants and animals would be clearly related to others on the nearest mainland, rather than to species occupying similar habitats on more distant landmasses.

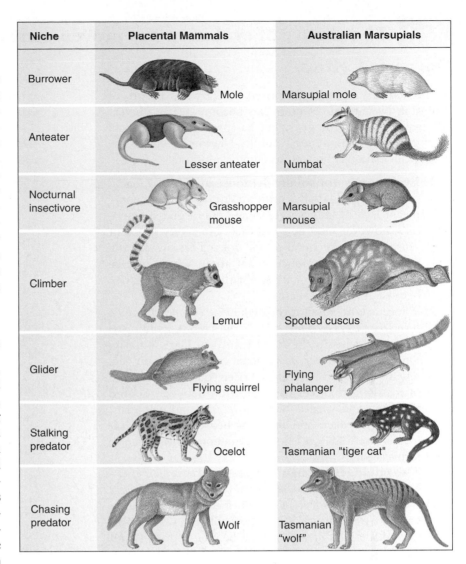

FIGURE 22.21
Convergent evolution. Marsupials in Australia resemble placental mammals occupying similar ecological niches elsewhere in the rest of the world. They evolved in isolation after Australia separated from other continents.

Convergence is the evolution of similar forms in different lineages when exposed to the same selective pressures. The biogeographical distribution of species often reflects the outcome of evolutionary diversification.

The theory of evolution has proven controversial.

Darwin's Critics

In the century since he proposed it, Darwin's theory of evolution by natural selection has become nearly universally accepted by biologists, but has proven controversial among some of the general public. Darwin's critics raise seven principal objections to teaching evolution:

1. **Evolution is not solidly demonstrated.** *"Evolution is just a theory,"* Darwin's critics point out, as if "theory" meant lack of knowledge, some kind of guess. Scientists, however, use the word theory in a very different sense than the general public does. Theories are the solid ground of science—that about which we are most certain. Few of us doubt the theory of gravity because it is "just a theory."

2. **There are no fossil intermediates.** *"No one ever saw a fin on the way to becoming a leg,"* critics claim, pointing to the many gaps in the fossil record in Darwin's day. Since then, however, many fossil intermediates in vertebrate evolution have indeed been found. A clear line of fossils now traces the transition between hoofed mammals and whales, between reptiles and mammals, between dinosaurs and birds, and between apes and humans. The fossil evidence of evolution between major forms is compelling.

3. **The intelligent design argument.** *"The organs of living creatures are too complex for a random process to have produced—the existence of a clock is evidence of the existence of a clockmaker."* The intermediates in the evolution of the mammalian ear can be seen in fossils, and many intermediate "eyes" are known in various invertebrates. These intermediate forms arose because they have value—being able to detect light a little is better than not being able to detect it at all. Complex structures such as eyes evolved as a progression of slight improvements.

4. **Evolution violates the Second Law of Thermodynamics.** *"A jumble of soda cans doesn't by itself jump neatly into a stack—things become more disorganized due to random events, not more organized."* Biologists point out that this argument ignores what the second law really says: Disorder increases in a closed system, which the earth most certainly is not. Energy continually enters the biosphere from the sun, fueling life and all the processes that organize it.

5. **Proteins are too improbable.** *"Hemoglobin has 141 amino acids. The probability that the first one would be leucine is 1/20, and that all 141 would be the ones they are by chance is $(1/20)^{141}$, an impossibly rare event."* This argument illustrates a lack of understanding of probability and statistics—you cannot use probability to argue backwards. The probability that a student in a classroom has a particular birthday is 1/365; arguing this way, the probability that everyone in a class of 50 would have the birthdays they do is $(1/365)^{50}$, and yet there the class sits.

6. **Natural selection does not imply evolution.** *"No scientist has come up with an experiment in which fish evolve into frogs and leap away from predators."* Is microevolution (evolution within a species) the mechanism that has produced macroevolution (evolution among species)? Most biologists who have studied the problem think so. The differences between breeds produced by artificial selection—such as chihuahuas, mastiffs, and greyhounds—are more distinctive than the differences between some wild species, and laboratory selection experiments sometimes create forms that cannot interbreed and thus would in nature be considered different species. Thus, production of radically different forms has indeed been observed, repeatedly. To object that evolution still does not explain really major differences, such as those between fish and amphibians, simply takes us back to point 2. These changes take millions of years, and are seen clearly in the fossil record.

7. **The irreducible complexity argument.** *The intricate molecular machinery of the cell cannot be explained by evolution from simpler stages. For example, because each part of a complex cellular process such as blood clotting is essential to the overall process, how can natural selection fashion any one part?* What's wrong with this argument is that each part of a complex molecular machine evolves as part of the system. Natural selection can act on a complex system because at every stage of its evolution, the system functions. Parts that improve function are added, and because of later changes, become essential. The mammalian blood clotting system, for example, has evolved from much simpler systems. The core clotting system evolved at the dawn of the vertebrates more than 500 million years ago, and is found today in primitive fish such as lampreys. One hundred million years later, as vertebrates evolved, proteins were added to the clotting system, making it sensitive to substances released from damaged tissues. Fifty million years later, a third component was added, triggering clotting by contact with the jagged surfaces produced by injury. At each stage, as the clotting system evolved to become more complex, its overall performance came to depend on the added elements. Thus, blood clotting has become "irreducibly complex" as the result of Darwinian evolution.

Darwin's theory of evolution has proven controversial among some of the general public, although the commonly raised objections are without scientific merit.

For interactive testing, visit the Online Learning Center with PowerWeb at www.mhhe.com/Raven7

22.1 Evidence indicates that natural selection can produce evolutionary change.

The Beaks of Darwin's Finches

- Darwin collected 31 finch specimens from three Galápagos Islands. (p. 454)
- Darwin found correspondence between 14 finch species and their food sources (p. 454)
- Observations suggest that beak differences evolved as adaptations to different food sources. (p. 455)

Peppered Moths and Industrial Melanism

- Adults range in color from light gray to black (dominant). (p. 456)
- Black moths were rare before 1850, but common after that. (p. 456)
- The Industrial Revolution caused trees to become sooty, making light-colored moths more visible to predators. Thus, darker moths became more prevalent than lighter moths. (p. 456)
- The second half of the twentieth century saw widespread pollution control, and subsequently a reversed trend away from melanism. (p. 457)
- No matter the agent or the selective force, selection favored dark moths in polluted habitats and light moths in pristine areas. (p. 457)

Artificial Selection

- Artificial selection has produced substantial change in almost every case to which it has been applied. (p. 458)
- Selection for a trait in a lab animal such as the fruit fly leads to strong and predictable evolutionary response. (p. 458)
- Differences in agricultural crops and the domestication of animals have resulted from generations of selection for desirable traits. (pp. 458–459)
- By extrapolation, selection, over the course of many millions of years, likely has the power to produce the current diversity of life. (p. 459)

22.2 Fossil evidence indicates that evolution has occurred.

The Fossil Record

- Rock fossils are created when an organism becomes buried in sediments, and hard tissue mineralizes. (p. 460)
- Fossils can be dated in a relative manner by looking at their relative position in rock strata, and dated in an absolute manner using radioactive isotopes and the ratio of derivative isotopes. (p. 460).
- At its largest scale, the fossil record documents the course of life through time. (p. 460)

- Demonstration of successive change is one of the strongest lines of evolutionary evidence. (p. 461)

The Evolution of Horses

- The evolution of horses is one of the best-studied cases in the fossil record. (p. 462)
- The earliest horses were small, with short legs and broad feet. (p. 462)
- Most changes to horses have been explained as adaptations to changing global climates. (p. 463)
- Modern horse diversity is relatively limited. (p. 463)

22.3 Evidence for evolution can be found in other fields of biology.

The Anatomical Record

- Homologous structures have different appearances and functions but are derived from the same body part in a common ancestor. (p. 464)
- Relict developmental forms in a wide variety of vertebrate embryos suggest that evolution has modified ancestral developmental patterns. (p. 464)
- Vestigial structures have no apparent function, but resemble structures of presumed ancestors. (p. 465)

The Molecular Record

- Recent descendants from a common ancestor initially exhibit relatively high DNA similarity. (p. 466)
- The more distantly related two organisms are, the larger the number of DNA differences, in both coding and noncoding regions, that evolve. (p. 466)

Convergent Evolution and the Biogeographical Record

- Different areas can exhibit groups of organisms that resemble one another. (p. 467)
- Natural selection appears to have favored parallel evolutionary adaptations within similar environments.
- Island forms are often closely related to species on the nearest mainland. (p. 467)

22.4 The theory of evolution has proven controversial.

Darwin's Critics

- Darwin's critics raise several objections to teaching evolution: evolution is not solidly demonstrated; there is a lack of fossil intermediates; the intelligent design argument; evolution violates the Second Law of Thermodynamics; proteins are too improbable; natural selection does not imply evolution; and the irreducible complexity argument. None of these objections has scientific merit. (p. 468)

Self Test

1. Which of the following best describes the correlation between beak size and the amount of rain on Daphne Major?
   a. Birds with small beaks are favored in dry years.
   b. All birds are favored equally in wet years.
   c. Birds with large beaks are favored during wet years.
   d. Birds with large beaks are favored during dry years.

2. In peppered moths, the black coloration is selected when soot covers tree bark; this phenomenon is called
   a. artificial selection.
   b. convergent evolution.
   c. industrial melanism.
   d. none of these.

3. Evolutionary change through artificial selection has been demonstrated in all but which of the following?
   a. Galápagos finches
   b. *Drosophila*
   c. corn
   d. dog breeding

4. Darwin's examinations of fossils relied on _____ dating to determine the evolution of species.
   a. absolute
   b. carbon
   c. relative
   d. radioactive isotope

5. The missing links between whales and their hoofed ancestors include
   a. *Pakicetus.*
   b. *Archaeopteryx.*
   c. *Equus.*
   d. all of these.

6. Evolution has occurred in the horse as seen by
   a. a reduction in body size.
   b. an increase in complexity of the ridges on teeth.
   c. an increase in the number of toes.
   d. all of these.

7. Over time, the same bones in different vertebrates were put to different uses. This falls under the category of
   a. missing links.
   b. vestigial structures.
   c. analogous structures.
   d. homologous structures.

8. After examining the evidence related to the evolution of hemoglobin, you might conclude that
   a. bird hemoglobin evolved prior to lamprey hemoglobin.
   b. frogs are more closely related to lampreys than to birds.
   c. evolutionary changes occur at the molecular level.
   d. only DNA can be examined for establishing evolutionary differences.

9. An example of convergent evolution is
   a. Australian marsupials and placental mammals.
   b. the flippers in fish, penguins, and dolphins.
   c. the wings in birds, bats, and insects.
   d. all of these.

10. The shape of the beaks of Darwin's finches, industrial melanism, and the changes in horse teeth are all examples of
    a. artificial selection.
    b. natural selection.
    c. convergent evolution.
    d. homologous structures.

Test Your Visual Understanding

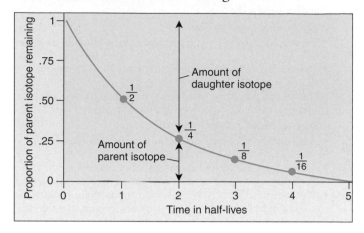

1. The graph illustrates how a radioactive isotope decays over time. Some isotopes decay more quickly than others do (they have shorter half-lives), but all isotope decay follows this same scale—that is, half of the isotope atoms decay with each half-life. For each of the isotopes in the following list, calculate how long it will take each parent sample to decay to 12.5% of the original amount. Also, graph three of the isotopes, plotting the proportion of parent isotope remaining to the number of half-lives, and compare these three graphs with the figure. Are they similar, or are they different?

Isotope	Half-life
a. beryllium-11	13.81 seconds
b. oxygen-15	2 minutes
c. sodium-24	15 hours
d. phosphorus-32	14.3 days
e. carbon-14	5,730 years
f. plutonium-239	24,110 years

Apply Your Knowledge

1. In a laboratory experiment, researchers selected for an increase and a decrease in protein content of corn seeds. The initial population contained an average of 9.5% protein by weight. As with the artificial selection experiments described in this chapter, corn seeds with the top 20% protein content were crossed and corn seeds with the lowest 20% protein contents were crossed. After 50 generations, the high-protein offspring averaged 19.2% protein, and the low-protein offspring averaged 5.4% protein.
   a. What percentage of change was recorded for the high-protein and low-protein populations?
   b. Which trait, the high- or the low-protein level, was modified more because of selection? Can you explain why one trait was modified more?

2. Why is it incorrect to think of evolution as progressive (i.e., proceeding from lowest or simplest to highest or most complex)?

23

The Origin of Species

Concept Outline

23.1 Species are the basic units of evolution.

    The Nature of Species. Species are groups of actually or potentially interbreeding natural populations that are reproductively isolated from other such groups and that maintain connectedness over geographic distances.

23.2 Species maintain their genetic distinctiveness through barriers to reproduction.

    Prezygotic Isolating Mechanisms. Some breeding barriers prevent the formation of zygotes.

    Postzygotic Isolating Mechanisms. Other breeding barriers prevent the proper development or reproduction of the zygote after it forms.

    Problems with the Biological Species Concept. Hybridization and other difficulties have prompted some scientists to propose alternative species concepts.

23.3 We have learned a great deal about how species form.

    Reproductive Isolation May Evolve as a By-Product of Evolutionary Change. Speciation can occur in the absence of natural selection, but reproductive isolation generally occurs more quickly when populations are adapting to different environments.

    The Geography of Speciation. Speciation occurs most readily when populations are geographically isolated.

23.4 Clusters of species reflect rapid evolution.

    Hawaiian *Drosophila*. More than one-quarter of the world's fruit fly species are found on the Hawaiian Islands.

    Darwin's Finches. Thirteen species of finches, all descendants of one ancestral finch, occupy diverse niches.

    Lake Victoria Cichlid Fishes. Isolation has led to extensive species formation among these small fishes.

    New Zealand Alpine Buttercups. Repeated glaciations have fostered waves of species formation.

    The Pace of Evolution. Evolutionary change can be slow and gradual or rapid and discontinuous.

    Speciation and Extinction Through Time. The number of species has increased through time.

    The Future of Evolution. Evolution continues in human-altered environments.

FIGURE 23.1
A Galápagos marine iguana basks in the sun on its isolated island. How does geographic isolation contribute to the formation of new species?

Although Darwin titled his book *On the Origin of Species*, he never actually discussed what he referred to as that "mystery of mysteries"—how one species gives rise to another. Rather, his argument concerned evolution by natural selection; that is, how one species evolves through time to adapt to its changing environment. Although of fundamental importance to evolutionary biology, the process of adaptation does not explain how one species becomes another (figure 23.1). Much less can it explain how one species can give rise to many descendant species, a process we call **speciation**. As we shall see, adaptation may be involved in the speciation process, but it need not be.

23.1 Species are the basic units of evolution.

The Nature of Species

Before we can discuss how one species gives rise to another, we need to understand exactly what a species is. Even though the definition of a species is of fundamental importance to evolutionary biology, this issue has still not been completely settled and is currently the subject of considerable research and debate. However, any concept of a species must account for two phenomena: the distinctiveness of species that occur together at a single locality, and the connection that exists among populations of the same species that are geographically separated.

The Distinctiveness of Sympatric Species

Put out a birdfeeder on your balcony or back porch and you will attract a wide variety of different types of birds (especially if you put out a variety of different kinds of foods). In the midwestern United States, for example, you might routinely see cardinals, blue jays, downy woodpeckers, house finches—even hummingbirds in the summer. Although it might take a few days of careful observation, you would soon be able to readily distinguish the many different species. The reason is that species that occur to-

gether (termed **sympatric**) are distinctive entities that are phenotypically different, utilize different parts of the habitat, and behave separately. This observation is generally true not only for birds, but also for most other types of organisms.

Occasionally, two species occur together that appear to be nearly identical. In most cases, however, our inability to distinguish the two reflects our own reliance on vision as our primary sense. When the mating calls or chemicals exuded by such species are examined, they usually reveal great differences. In other words, even though we have trouble separating them, the animals themselves have no such difficulties!

Geographic Variation Within Species

Within the units classified as species, populations that occur in different areas may be more or less distinct from one another. Such groups of distinctive individuals may be classified taxonomically as **subspecies,** or **varieties.** (The vague term "race" has a similar connotation, but is no longer commonly used.) In areas where these populations approach one another, individuals often exhibit combinations of features characteristic of both populations (fig-

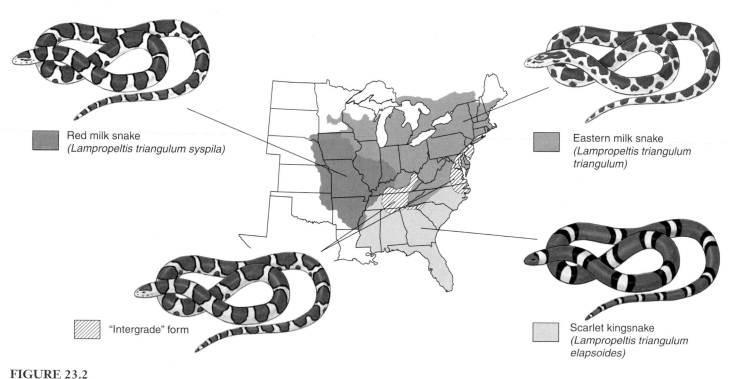

Red milk snake
(*Lampropeltis triangulum syspila*)

"Intergrade" form

Eastern milk snake
(*Lampropeltis triangulum triangulum*)

Scarlet kingsnake
(*Lampropeltis triangulum elapsoides*)

FIGURE 23.2
Geographic variation in the milk snake, *Lampropeltis triangulum.* Although subspecies appear phenotypically quite distinctive from each other, they are connected by populations that are phenotypically intermediate.

ure 23.2). In other words, even though geographically distant populations may appear distinct, they are usually connected by intervening populations that are intermediate in their characteristics.

The Biological Species Concept

What can account both for the distinctiveness of sympatric species and for the connectedness of geographic populations of the same species? One obvious possibility is that each species exchanges genetic material only with other members of its species. If sympatric species commonly exchanged genes, we might expect such species to rapidly lose their distinctions as the **gene pools** (that is, all of the alleles present in a species) of the different species became homogenized. Thus, the inability of sympatric species to exchange genes may allow them to remain distinct. Conversely, the ability of geographically distant populations of the same species to share genes through the process of gene flow may keep these populations integrated as members of the same species. Based on these ideas, the evolutionary biologist Ernst Mayr coined the **biological species concept,** which defines species as ". . . groups of actually or potentially interbreeding natural populations which are reproductively isolated from other such groups."

In other words, the biological species concept says that a species is composed of populations whose members mate with each other and produce fertile offspring—or would do so if they came into contact. Conversely, populations whose members do not mate with each other or who cannot produce fertile offspring are said to be **reproductively isolated** and, thus, members of different species.

What causes reproductive isolation? If organisms cannot interbreed or cannot produce fertile offspring, they clearly belong to different species. However, some populations that are considered separate species can interbreed and produce fertile offspring, but they ordinarily do not do so under natural conditions. They are still considered reproductively isolated in that genes from one species generally will not be able to enter the gene pool of the other species. Table 23.1 summarizes the steps at which barriers to successful reproduction may occur. Such barriers are termed **reproductive isolating mechanisms** because they prevent genetic exchange between species. We will discuss examples of these, beginning with those that prevent the formation of zygotes, which are called prezygotic isolating mechanisms. Postzygotic isolating mechanisms prevent the proper functioning of zygotes after they form.

Species are groups of organisms that are distinct from other, co-occurring species and that are interconnected geographically. The ability to exchange genes appears to be a hallmark of such species.

Table 23.1 Reproductive Isolating Mechanisms

Mechanism		Description
PREZYGOTIC ISOLATING MECHANISMS		
Geographic isolation		Species occur in different areas, which are often separated by a physical barrier such as a river or mountain range.
Ecological isolation		Species occur in the same area, but they occupy different habitats and rarely encounter each other.
Temporal isolation		Species reproduce in different seasons or at different times of the day.
Behavioral isolation		Species differ in their mating rituals.
Mechanical isolation	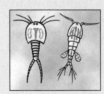	Structural differences between species prevent mating.
Prevention of gamete fusion	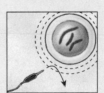	Gametes of one species function poorly with the gametes of another species or within the reproductive tract of another species.
POSTZYGOTIC ISOLATING MECHANISMS		
Hybrid inviability or infertility		Hybrid embryos do not develop properly, hybrid adults do not survive in nature, or hybrid adults are sterile or have reduced fertility.

Species maintain their genetic distinctiveness through barriers to reproduction.

How do species keep their separate identities? Reproductive isolating mechanisms fall into two categories: **prezygotic isolating mechanisms,** which prevent the formation of zygotes, and **postzygotic isolating mechanisms,** which prevent the proper functioning of zygotes after they form. In the following discussion, we examine various isolating mechanisms in these two categories and offer examples that illustrate how the isolating mechanisms operate to help species retain their identities.

Prezygotic Isolating Mechanisms

Ecological Isolation

Even if two species occur in the same area, they may utilize different portions of the environment and thus not hybridize because they do not encounter each other. For example, in India, the ranges of lions and tigers overlapped until about 150 years ago. Even so, there were no records of natural hybrids. Lions stayed mainly in the open grassland and hunted in groups called prides; tigers tended to be solitary creatures of the forest (figure 23.3). Because of their ecological and behavioral differences, lions and tigers rarely came into direct contact with each other, even though their ranges overlapped over thousands of square kilometers.

In another example, the ranges of two toads, *Bufo woodhousei* and *B. americanus*, overlap in some areas. Although these two species can produce viable hybrids, they usually do not interbreed because they utilize different portions of the habitat for breeding. Whereas *B. woodhousei* prefers to breed in streams, *B. americanus* breeds in rainwater puddles. Similarly, the ranges of two species of dragonflies overlap in Florida. However, the dragonfly *Progomphus obscurus* lives near rivers and streams, while *P. alachuenis* lives near lakes.

Similar situations occur among plants. Two species of oaks occur widely in California: the valley oak, *Quercus lobata*, and the scrub oak, *Q. dumosa*. The valley oak, a graceful deciduous tree that can be as tall as 35 meters, occurs in the fertile soils of open grassland on gentle slopes and valley floors. In contrast, the scrub oak is an evergreen shrub, usually only 1 to 3 meters tall, which often forms the kind of dense scrub known as chaparral. The scrub oak is found on steep slopes in less fertile soils. Hybrids between these different oaks do occur and are fully fertile, but they are rare. The sharply distinct habitats of their parents limit their occurrence together, and there is no intermediate habitat where the hybrids might flourish.

FIGURE 23.3

Lions and tigers are ecologically isolated. The ranges of lions and tigers used to overlap in India. However, lions and tigers do not hybridize in the wild because they utilize different portions of the habitat. Lions live in open grassland, whereas tigers are solitary animals that live in the forest. Hybrids, such as this tiglon, have been successfully produced in captivity, but hybridization does not occur in the wild.

Behavioral Isolation

Chapter 52 describes the often elaborate courtship and mating rituals of some groups of animals. Related species of organisms such as birds often differ in their courtship rituals, which tends to keep these species distinct in nature even if they inhabit the same places (figure 23.4). For example, mallard and pintail ducks are perhaps the two most common freshwater ducks in North America. In captivity, they produce completely fertile offspring, but in nature they nest side-by-side and only rarely hybridize.

Sympatric species avoid mating with members of the wrong species in a variety of ways; every mode of communication imaginable appears to be used by some species. Differences in visual signals, as we have just discussed, are common. However, other types of animals rely more on other sensory modes for communication. Many species, such as frogs, birds and a variety of insects, use vocalizations to attract mates. Predictably, sympatric species of these animals produce different calls (figure 23.5).

Other species rely on the detection of chemical signals, called **pheromones.** The use of pheromones in moths has been particularly well studied. When they are ready to mate, female moths emit a pheromone that males can detect at great distances. Sympatric species differ in the pheromone they produce: Either they use different chemical compounds, or if using the same compounds, they differ in the proportions used. Laboratory studies indicate that males are remarkably adept at distinguishing the pheromones of their own species from those of other species or even from synthetic compounds similar, but not identical, to that of their own species.

Some species even use electroreception. African electric fish have specialized organs in their tails that produce electrical discharges and electroreceptors on their skins that detect electrical charges. These discharges are used to communicate in social interactions; field experiments indicate that males can distinguish between signals produced by their own and other species, probably on the basis of differences in the timing of the electrical pulses.

Other Prezygotic Isolating Mechanisms

Temporal Isolation.   *Lactuca graminifolia* and *L. canadensis*, two species of wild lettuce, grow together along roadsides throughout the southeastern United States. Hybrids between these two species are easily made experimentally and are completely fertile. But such hybrids are rare in nature because *L. graminifolia* flowers in early spring and *L. canadensis* flowers in summer. When their blooming periods overlap, as happens occasionally, the two species do form hybrids, which may become locally abundant.

Many species of closely related amphibians have different breeding seasons that prevent hybridization between the species. For example, five species of frogs of the genus *Rana* occur together in most of the eastern United States, but hybrids are rare because the peak breeding time is different for each of them.

Mechanical Isolation.   Structural differences prevent mating between some related species of animals. Aside from such obvious features as size, the structure of the male and female copulatory organs may be incompatible. In many insect and other arthropod groups, the sexual organs, particularly those of the male, are so diverse that they are used as a primary basis for distinguishing species.

Similarly, flowers of related species of plants often differ significantly in their proportions and structures. Some of these differences limit the transfer of pollen from one plant species to another. For example, bees may carry the pollen of one species on a certain place on their bodies; if this area does not come into contact with the receptive structures of the flowers of another plant species, the pollen is not transferred.

Prevention of Gamete Fusion.   In animals that shed their gametes directly into water, eggs and sperm derived from different species may not attract one another. Many land animals may not hybridize successfully because the sperm of one species function so poorly within the reproductive tract of another that fertilization never takes place. In plants, the growth of pollen tubes may be impeded in

FIGURE 23.4
Differences in courtship rituals can isolate related bird species.   These Galápagos blue-footed boobies select their mates only after an elaborate courtship display. This male is lifting his feet in a ritualized high-step that shows off his bright blue feet. The display behavior of the two other species of boobies that occur in the Galápagos is much different.

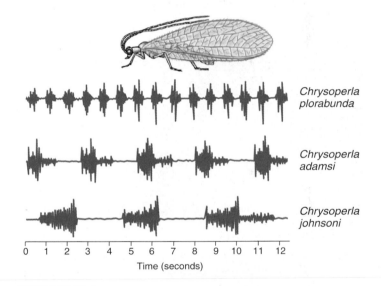

FIGURE 23.5
Differences in courtship song of sympatric species of lacewings.   Lacewings are small insects that rely on auditory signals produced by vibrating their abdomens together to attract mates. As these sound recordings indicate, the sounds produced by sympatric species differ greatly. Females, which detect the calls as they are transmitted through solid surfaces such as branches, are able to distinguish calls of different species and only respond to individuals producing their own species' call.

hybrids between different species. In both plants and animals, such isolating mechanisms prevent the union of gametes, even following successful mating.

Prezygotic isolating mechanisms lead to reproductive isolation by preventing the formation of hybrid zygotes.

Postzygotic Isolating Mechanisms

All of the factors we have discussed so far tend to prevent hybridization. If hybrid matings do occur and zygotes are produced, many factors may still prevent those zygotes from developing into normally functioning, fertile individuals. As we saw in chapter 19, development in any species is a complex process. In hybrids, the genetic complements of two species may be so different that they cannot function together normally in embryonic development. For example, hybridization between sheep and goats usually produces embryos that die in the earliest developmental stages.

Leopard frogs (*Rana pipiens* complex) of the eastern United States are a group of similar species, assumed for a long time to constitute a single species (figure 23.6). However, careful examination revealed that although the frogs appear similar, successful mating between them is rare because of problems that occur as the fertilized eggs develop. Many of the hybrid combinations cannot be produced even in the laboratory.

Examples of this kind, in which similar species have been recognized only as a result of hybridization experi-ments, are common in plants. Sometimes the hybrid embryos can be removed at an early stage and grown in an artificial medium. When these hybrids are supplied with extra nutrients or other supplements that compensate for their weakness or inviability, they may complete their development normally.

Even if hybrids survive the embryo stage, however, they may not develop normally. If the hybrids are weaker than their parents, they will almost certainly be eliminated in nature. Even if they are vigorous and strong, as in the case of the mule, a hybrid between a female horse and a male donkey, they may still be sterile and thus incapable of contributing to succeeding generations. Hybrids may be sterile because the development of sex organs is abnormal, because the chromosomes derived from the respective parents cannot pair properly, or due to a variety of other causes.

Postzygotic isolating mechanisms are those in which hybrid zygotes fail to develop or develop abnormally, or in which hybrids cannot become established in nature.

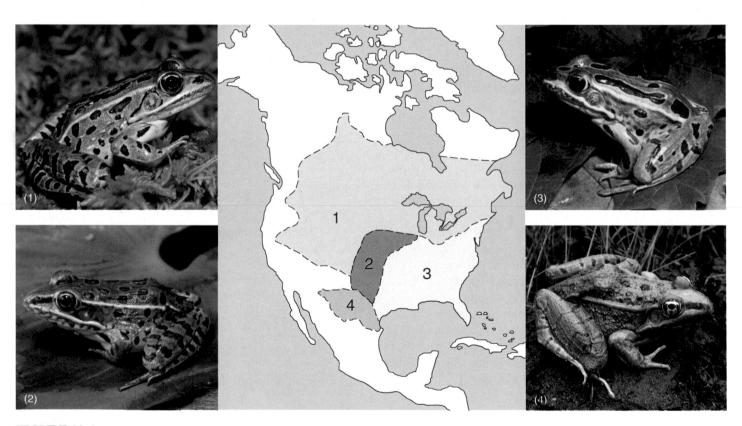

FIGURE 23.6
Postzygotic isolation in leopard frogs. Numbers indicate the following species in the geographical ranges shown: (*1*) *Rana pipiens;* (*2*) *Rana blairi;* (*3*) *Rana sphenocephala;* (*4*) *Rana berlandieri.* These four species resemble one another closely in their external features. Their status as separate species was first suspected when hybrids between some pairs of these species were found to produce defective embryos in the laboratory. Subsequent research revealed that the mating calls of the four species differ substantially, indicating that the species have both pre- and postzygotic isolating mechanisms.

Problems with the Biological Species Concept

The biological species concept has proven to be an effective way of understanding the existence of species in nature. Nonetheless, it has a number of problems that have led some scientists to propose alternative species concepts.

One criticism concerns the extent to which all species truly are reproductively isolated. By definition, under the biological species concept, species should not interbreed and produce fertile offspring. Nonetheless, in recent years biologists have detected much greater amounts of hybridization than previously realized between populations that seem to coexist as distinct biological entities. Botanists have always been aware that species can often experience substantial amounts of hybridization. For example, more than 50% of California plant species included in one study were not well defined by genetic isolation. Such coexistence without genetic isolation can be long-lasting: Fossil data show that balsam poplars and cottonwoods have been phenotypically distinct for 12 million years but have routinely produced hybrids throughout this time. Consequently, many botanists have long felt that the biological species concept only applies to animals.

What is becoming increasingly evident, however, is that hybridization is not all that uncommon in animals, either. In recent years, many cases of substantial hybridization between animal species have been documented. One recent survey indicated that almost 10% of the world's 9500 bird species are known to have hybridized in nature. Galápagos finches provide a particularly well-studied example. Three species on the island of Daphne Major—the medium ground finch, the cactus finch, and the small ground finch—are clearly distinct morphologically, and occupy different ecological niches. Studies over the past 20 years by Peter and Rosemary Grant found that, on average, 2% of the medium ground finches and 1% of the cactus ground finches mated with other species every year. Furthermore, hybrid offspring appeared to be at no disadvantage in terms of survival or subsequent reproduction. This is not a trivial amount of genetic exchange, and one might expect to see the species coalesce into one genetically variable population, but the species are maintaining their distinctiveness.

This is not to say that hybridization is rampant throughout the animal world. Most bird species do not hybridize, and even fewer probably experience significant amounts of hybridization. Still, it is common enough to cast doubt about whether reproductive isolation is the only force maintaining the integrity of species.

Natural Selection and the Ecological Species Concept

An alternative hypothesis proposes that the distinctions among species are maintained by natural selection. The idea is that each species has adapted to its own specific part of the environment. Stabilizing selection then maintains the species' adaptations; hybridization has little effect because alleles introduced into the gene pool from other species are quickly eliminated by natural selection.

We have already seen in chapter 21 that the interaction between gene flow and natural selection can have many outcomes. In some cases, strong selection can overwhelm any effects of gene flow, but in other situations, gene flow can prevent populations from eliminating less successful alleles from a population. As a general explanation, then, natural selection is not likely to have any fewer exceptions than the biological species concept, although it may prove more successful for certain types of organisms or habitats.

Other Problems with the Biological Species Concept

The biological species concept has been criticized for other reasons as well. For example, it can be difficult to apply the concept to populations that do not occur together in nature. Because individuals of these populations do not encounter each other, it is not possible to observe whether they would interbreed naturally. Although experiments can determine whether fertile hybrids can be produced, this information is not enough because many species that coexist without interbreeding in nature will readily hybridize in the artificial settings of the laboratory or zoo. Consequently, evaluating whether such populations constitute different species is ultimately a judgment call. In addition, the concept is more limited than its name would imply. Many organisms are asexual and reproduce without mating; reproductive isolation has no meaning for such organisms.

For these reasons, a variety of other ideas have been put forward to establish criteria for defining species. Many of these are specific to a particular type of organism, and none has universal applicability. In truth, there may be no single explanation for what maintains the identity of species. Given the incredible variation evident in plants, animals, and microorganisms in all aspects of their biology, it is perhaps not surprising that different processes are operating in different organisms. In addition, some scientists have turned from emphasizing the processes that maintain species distinctions to examining the evolutionary history of populations. These genealogical species concepts are currently a topic of great debate. The study of species concepts is thus an area of active research that demonstrates the dynamic nature of the field of evolutionary biology.

The surprisingly high incidence of hybridization in plants and animals is causing some researchers to seek alternative species concepts. Because of the diversity of living organisms, no single definition of what constitutes a species may be universally applicable.

23.3 We have learned a great deal about how species form.

One of the oldest questions in the field of evolution is: How does one ancestral species become divided into two descendant species? If species are defined by the existence of reproductive isolation, then the process of speciation equates with the evolution of reproductive isolating mechanisms. How do reproductive isolating mechanisms evolve?

Reproductive Isolation May Evolve as a By-Product of Evolutionary Change

Most reproductive isolating mechanisms initially arise for some reason other than to provide reproductive isolation. For example, a population that colonizes a new habitat may evolve adaptations for living in that habitat. As a result, individuals from that population might never encounter individuals from the ancestral population. Even if they do meet, the population in the new habitat may have evolved new phenotypes or behaviors so that members of the two populations no longer recognize each other as potential mates. For this reason, some biologists believe that the term "isolating mechanisms" is misguided, because it implies that the traits evolved specifically for the purpose of genetically isolating a species, which in most cases is probably incorrect.

Selection May Reinforce Isolating Mechanisms

The formation of species is a continuous process, one that we can understand because of the existence of intermediate stages at all levels of differentiation. If populations that are partly differentiated come into contact with one another, they may still be able to interbreed freely, and the differences between them may disappear over the course of time as genetic exchange homogenizes the populations. Conversely, if the populations are reproductively isolated, then no genetic exchange will occur, and the two populations will be different species.

However, there is an intermediate situation in which reproductive isolation has partially evolved, but is not complete. As a result, hybridization will occur at least occasionally. If the hybrids are partly sterile, or not as well adapted to the existing habitats as their parents, they will be at a disadvantage. As a result, selection would favor any alleles in the parental populations that prevented hybridization because individuals that avoided hybridizing would produce more successful offspring and thus pass more of their genes on to subsequent generations. The result would be the continual improvement of prezygotic isolating mechanisms until the two populations were completely reproductively isolated. This process is

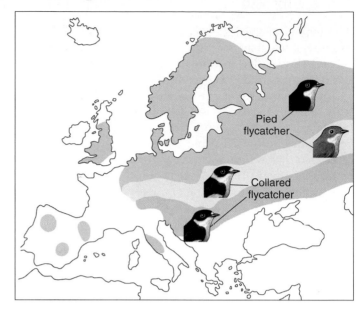

FIGURE 23.7
Reinforcement in European flycatchers. The pied flycatcher and the collared flycatcher appear very similar when they occur alone. However, in places where the two species occur sympatrically (indicated by the tan color), they have evolved differences in color, which allow individuals to choose mates from their own species and thus avoid hybridizing.

termed **reinforcement** because initially incomplete isolating mechanisms are reinforced by natural selection until they are completely effective.

An example of reinforcement is provided by pied and collared flycatchers. Throughout much of eastern and central Europe, these two species are geographically separated (allopatric) and are very similar in color (figure 23.7). However, in the Czech Republic and Slovakia, the two species occur together and occasionally hybridize, producing offspring that usually have very low fertility. At those sites, the species have evolved to look very different, and birds prefer to mate with individuals with their own species' color, in contrast to birds from allopatric populations, which prefer the allopatric color pattern. As a consequence of the color differences in sympatry, the rate of hybridization is extremely low. These results indicate that when populations of the two species came into contact, natural selection led to differences in color patterns, resulting in the evolution of pre-mating isolation.

Reinforcement is by no means inevitable, however. When incompletely isolated populations come together, gene flow immediately begins to occur between the species. Although hybrids may be inferior, they are not completely inviable or infertile (if they were, the species would already be completely reproductively isolated); hence, when these hybrids reproduce with members of either population, they will serve as a conduit of genetic exchange from one population to the other. As a result, the two populations will tend to lose their genetic distinctiveness. Thus, a race ensues: Can reproductive isolation be perfected before gene flow destroys the differences between the populations? Experts disagree on the likely outcome, but many believe that reinforcement is the much less common outcome.

The Role of Natural Selection in Speciation

What role does natural selection play in the speciation process? Certainly, the process of reinforcement is driven by natural selection favoring the perfection of reproductive isolation. But, as we have seen, reinforcement may not be common. Is natural selection necessarily involved in the initial evolution of isolating mechanisms?

Random Changes May Cause Reproductive Isolation

As we discussed in chapter 21, populations may diverge for purely random reasons. Genetic drift in small populations, founder effects, and population bottlenecks all may lead to changes in traits that cause reproductive isolation. For example, in the Hawaiian Islands, closely related species of *Drosophila* often differ greatly in their courtship behavior. Colonization of new islands by these fruit flies probably involves a founder effect, in which one or a few fruit flies—perhaps only a single pregnant female—is blown by strong winds to a new island. Changes in courtship behavior between ancestor and descendant populations may be the result of such founder events. Given long enough periods of time, any two isolated populations will diverge due to genetic drift. (Remember that even large populations experience drift, but at a lower rate than small populations do.) In some cases, this random divergence may affect traits responsible for reproductive isolation, and speciation will have occurred.

Adaptation and Speciation

Nonetheless, adaptation and speciation are probably related in many cases. As species adapt to different circumstances, they will accumulate many differences that may lead to reproductive isolation. For example, if one population of flies adapts to wet conditions and another to dry conditions, the populations will evolve a variety of differences in physiological and sensory traits; these differences may promote ecological and behavioral isolation,

FIGURE 23.8
Dewlaps of different species of Caribbean *Anolis* lizards. Males use their dewlaps in both territorial and courtship displays. Coexisting species almost always differ in their dewlaps, which are used in species recognition. Some dewlaps are easier to see in open habitats, whereas others are more visible in shaded environments.

and may cause any hybrids they produce to be poorly adapted to either habitat.

Selection might also act directly on mating behavior. Male *Anolis* lizards, for example, court females by extending a colorful flap of skin, called a "dewlap," located under their throat (figure 23.8). The ability of one lizard to see the dewlap of another lizard depends not only on the color of the dewlap, but on the environment in which the lizards occur. Thus, a light-colored dewlap is most effective in reflecting light in a dim forest, whereas dark colors are more apparent in the bright glare of open habitats. As a result, when these lizards occupy new habitats, natural selection will favor evolutionary change in dewlap color because males whose dewlaps cannot be seen will attract few mates. However, the lizards also distinguish members of their own species from those of other species by the color of the dewlap. Hence, adaptive change in mating behavior could have the incidental consequence of causing speciation.

Laboratory scientists have conducted experiments on fruit flies and other organisms in which they isolate populations in different laboratory chambers and measure how much reproductive isolation evolves. These experiments indicate that genetic drift by itself can lead to some degree of reproductive isolation, but in general, reproductive isolation evolves more rapidly when the populations are forced to adapt to different laboratory environments (such as temperature or food type).

Reproductive isolating mechanisms can evolve either through random changes or as an incidental by-product of adaptive evolution. Under some circumstances, however, natural selection can directly select for traits that increase the reproductive isolation of a species.

The Geography of Speciation

Speciation is a two-part process. First, initially identical populations must diverge, and second, reproductive isolation must evolve to maintain these differences. The difficulty with this process, as we have seen, is that the homogenizing effect of gene flow between populations will constantly be acting to erase any differences that may arise, either by genetic drift or natural selection. However, gene flow only occurs between populations that are in contact, and populations can become isolated for a variety of reasons (figure 23.9). Consequently, evolutionary biologists have long recognized that speciation is much more likely in geographically isolated, or allopatric, populations.

Allopatric Speciation

Ernst Mayr was the first biologist to demonstrate that geographically separated, or **allopatric,** populations appear much more likely to have evolved substantial differences leading to speciation. Marshalling data from a wide variety of organisms and localities, Mayr made a strong case for allopatric speciation as the primary means of speciation. For example, the Papuan kingfisher, *Tanysiptera hydrocharis,* varies little throughout its wide range in New Guinea despite the great variation in the island's topography and climate. By contrast, isolated populations on nearby islands are strikingly different from each other and from the mainland population (figure 23.10).

Many other examples indicate that speciation can occur in allopatry. Given that one would expect isolated populations to diverge over time by either drift or selection, this result is not surprising. Rather, the question becomes: Is geographic isolation *required* for speciation to occur?

Sympatric Speciation

For decades, biologists have debated whether one species could split into two at a single locality, without the two new species ever having been geographically separated. Scientists have suggested that such sympatric speciation could occur either instantaneously or over the course of multiple generations. Although most of the hypotheses suggested so far are highly controversial, one type of instantaneous sympatric speciation is known to occur commonly, as the result of polyploidy.

Instantaneous Speciation Through Polyploidy. Instantaneous sympatric speciation occurs when an individual is born that is reproductively isolated from all other members of its species. In most cases, a mutation that would cause an individual to be so different from others of its species would have many adverse pleiotropic side-effects, and the individual would not survive. One exception, however, occurs through the process of **polyploidy,** commonly seen in plants. A polyploid individual has more than two

FIGURE 23.9
Populations can become geographically isolated for a variety of reasons. (*a*) Colonization of distant areas by one or a few individuals can establish populations in a distant place. (*b*) Barriers to movement can split an ancestral population into two isolated populations. (*c*) Extinction of intermediate populations can leave the remaining populations isolated from each other.

sets of chromosomes. Polyploids can arise in two ways. In **autopolyploidy,** all of the chromosomes may arise from a single species. This might happen, for example, due to an error in meiosis that causes an individual to have four sets of chromosomes. Such individuals, termed tetraploids, could fertilize themselves or mate with other tetraploids, but could not mate and produce fertile offspring with normal diploids. The reason is that the offspring from such a mating would be triploid (having three sets of chromosomes) and would be sterile due to problems with chromosome pairing during meiosis.

A more common type of polyploid speciation is **allopolyploidy,** which occurs sometimes when two species hybridize. The resulting offspring, having one copy of the chromosomes of each species, is usually infertile because the chromosomes do not pair correctly in meiosis. However, such individuals are often otherwise healthy, can reproduce asexually, and can even become fertile through a variety of events. For example, if the chromosomes of such an individual were to spontaneously

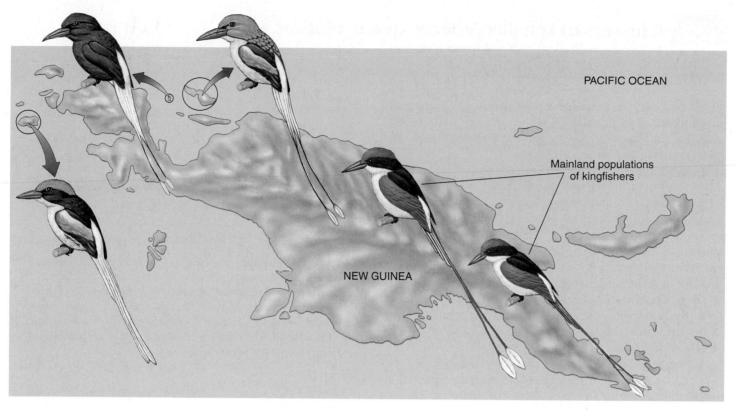

PACIFIC OCEAN

Mainland populations
of kingfishers

NEW GUINEA

FIGURE 23.10
Phenotypic differentiation in the Papuan kingfisher in New Guinea. Isolated island populations (*left*) are quite distinctive, showing variation in tail feather structure and length, plumage coloration, and bill size, whereas kingfishers on the mainland (*right*) show little variation.

double, as just described, the resulting tetraploid would have two copies of each set of chromosomes. Consequently, pairing would no longer be a problem in meiosis: Each set of chromosomes could pair with itself. As a result, such tetraploids would be able to intermate, and a new species would have been created.

It is estimated that about half of the approximately 260,000 species of plants have a polyploid episode in their history, including many of great commercial importance, such as bread wheat, cotton, tobacco, sugarcane, bananas, and potatoes. Speciation by polyploidy is also known to occur in a variety of animals, including insects, fish, and salamanders, although much more rarely than in plants.

Sympatric Speciation by Disruptive Selection. Some investigators believe that sympatric speciation can occur over the course of multiple generations through the process of disruptive selection. As we saw in chapter 21, disruptive selection can cause a population to contain individuals exhibiting two different phenotypes. One might think that if selection were strong enough, these two phenotypes would evolve over a number of generations into different species. However, before the two phenotypes could become different species, they would have to evolve

reproductive isolating mechanisms. Because the two phenotypes would initially not be reproductively isolated at all, genetic exchange between individuals of the two phenotypes would tend to prevent genetic divergence in mating preferences or other isolating mechanisms. As a result, the two phenotypes would be retained as polymorphisms within a single population. For this reason, most biologists consider sympatric speciation of this type to be a rare event.

Nonetheless, in recent years, a number of cases have appeared that are difficult to interpret in any way other than as sympatric speciation. For example, Lake Barombi Mbo in Cameroon is an extremely small and ecologically homogeneous crater lake, with no opportunity for within-lake isolation. Nonetheless, 11 species of closely related cichlid fish occur in the lake; all of the species are more closely related evolutionarily to each other than to any species outside of the crater. The most reasonable explanation is that an ancestral species colonized the crater and subsequently speciated in sympatry multiple times.

Speciation occurs much more readily in the absence of gene flow among populations. However, speciation can occur in sympatry by means of polyploidy and perhaps by disruptive selection.

23.4 Clusters of species reflect rapid evolution.

One of the most visible manifestations of evolution is the existence of groups of closely related species that have recently evolved from a common ancestor by adapting to different habitats. Such **adaptive radiations** are particularly common on oceanic islands, where the original colonist probably encountered an environment with few species and many available resources.

Adaptive radiation requires both speciation and adaptation to different habitats. A classic model postulates that a species colonizes multiple islands in an archipelago. Speciation subsequently occurs allopatrically, and then the newly arisen species colonize other islands, producing multiple species per island (figure 23.11). Adaptation to new habitats can either occur during the allopatric phase as the species respond to different environments on the different islands, or after two species become sympatric. In the latter case, this adaptation may be driven by the need to minimize competition for available resources with other species, a process termed **character displacement.**

An alternative possibility is that adaptive radiation occurs through repeated instances of sympatric speciation, producing a suite of species adapted to different habitats. As we just discussed, such scenarios are hotly debated.

Hawaiian *Drosophila*

More than one-third of the world's species in the fly genus *Drosophila* occur on the Hawaiian Islands. New species of *Drosophila* are still being discovered in Hawaii, although the rapid destruction of the native vegetation is making the search more difficult. Aside from their sheer number, Hawaiian *Drosophila* species are unusual because of their incredible diversity of morphology and behavioral traits (figure 23.12). Evidently, when their ancestors first reached these islands, they encountered many "empty" habitats that other kinds of insects and other animals occupied elsewhere. As a result, the species have adapted to all manners of fruit fly life, and include predators, parasites, and herbivores, as well as species specialized for eating the detritus in leaf litter and the nectar of flowers. The larvae of various species live in rotting stems, fruits, bark, leaves, or roots, or feed on sap. No comparable diversity of *Drosophila* species is found anywhere else in the world.

A second, closely related genus of flies, *Scaptomyza*, also forms a species cluster in Hawaii, where it is represented by as many as 300 species. The genera *Scaptomyza* and *Drosophila* are so closely related that scientists have suggested that all of the estimated 800 species of these two genera that occur in Hawaii may have derived from a single common ancestor.

The great diversity of Hawaiian species is a result of the geological history of these islands. New islands have continually arisen from the sea in the region of the

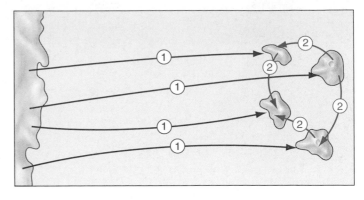

FIGURE 23.11
Classic model adaptive radiation on island archipelagoes.
(*1*) An ancestral species colonizes islands in an archipelago. Subsequently, the populations speciate in allopatry. (*2*) Then some of these new species colonize other islands, leading to local communities of two or more species. Ecological specialization can either occur when species are in allopatry (*1*) or as the result of ecological interactions between species after (*2*).

(a)   (b)

FIGURE 23.12
Hawaiian *Drosophila*. The hundreds of species that have evolved on the Hawaiian Islands are extremely variable in appearance, although genetically almost identical. (*a*) *Drosophila heteroneura.* (*b*) *Drosophila digressa.*

Hawaiian Islands. As they have done so, they appear to have been invaded successively by the various *Drosophila* groups present on the older islands. New species thus have evolved as new islands have been colonized. In addition, the Hawaiian Islands are among the most volcanically active islands in the world. Periodic lava flows often have created patches of habitat within an island, termed *kipukas*, in a sea of barren rock. *Drosophila* populations isolated in these kipukas often speciate. In these ways, rampant speciation combined with ecological opportunity has led to an unparalleled diversity of insect life.

> The adaptive radiation of about 800 species of the flies *Drosophila* and *Scaptomyza* on the Hawaiian Islands, probably from a single common ancestor, is one of the most remarkable examples of intensive species formation found anywhere on earth.

Darwin's Finches

We have already mentioned the diversity of Darwin's finches on the Galápagos Islands in chapter 22. Presumably, the ancestor of Darwin's finches reached these islands before other land birds, and many of the types of habitats where birds occur on the mainland were unoccupied. As the new arrivals moved into these vacant ecological niches and adopted new lifestyles, they were subjected to diverse sets of selective pressures. Under these circumstances, and aided by the geographic isolation afforded by the many islands of the Galápagos archipelago, the ancestral finches rapidly split into a series of diverse populations, some of which evolved into separate species. These species now occupy many different kinds of habitats on the Galápagos Islands, which are comparable to the habitats several distinct groups of birds occupy on the mainland. As illustrated in figure 23.13, the 14 species comprise four groups:

1. **Ground finches.** There are six species of *Geospiza* ground finches. Most of the ground finches feed on seeds. The size of their bills is related to the size of the seeds they eat. Some of the ground finches feed primarily on cactus flowers and fruits, and have a longer, larger, and more pointed bill than the others.
2. **Tree finches.** There are five species of insect-eating tree finches. Four species have bills suitable for feeding on insects. The woodpecker finch has a chisel-like beak. This unusual bird carries around a twig or a cactus spine, which it uses to probe for insects in deep crevices.
3. **Warbler finches.** These unusual birds play the same ecological role in the Galápagos woods that warblers play on the mainland, searching continually over the leaves and branches for insects. They have slender, warbler-like beaks.
4. **Vegetarian finch.** The very heavy bill of this bud-eating bird is used to wrench buds from branches.

Recently, scientists have examined the DNA of Darwin's finches to study their evolutionary history. These studies suggest that the deepest branches in the finch evolutionary tree lead to warbler finches, which implies that warbler finches were among the first types to evolve after colonization of the islands.

Darwin's finches, all derived from one similar mainland species, have radiated widely on the Galápagos Islands in the absence of competition from other types of birds.

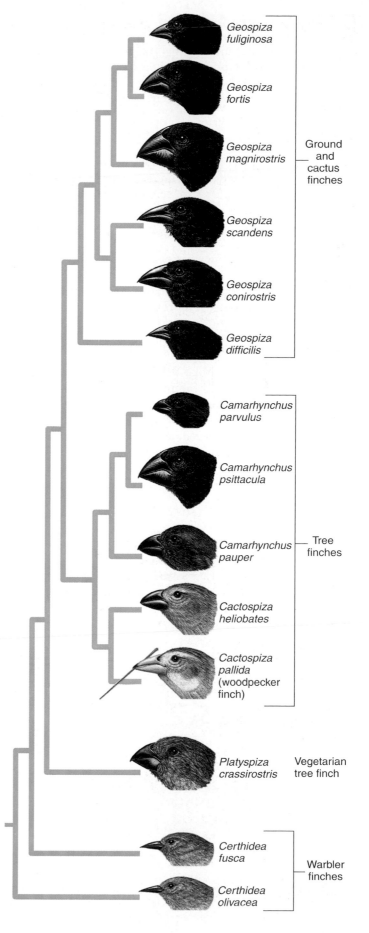

FIGURE 23.13
An evolutionary tree of Darwin's finches. Their position at the base of the evolutionary tree of Darwin's finches suggests that warbler finches were among the first ecological types to evolve in the Galápagos.

Lake Victoria Cichlid Fishes

Lake Victoria is an immense, shallow, freshwater sea about the size of Switzerland in the heart of equatorial East Africa. Until recently, the lake was home to an incredibly diverse collection of over 300 species of cichlid fishes.

Recent Radiation

The cluster of cichlid species appears to have evolved recently and quite rapidly. By sequencing the cytochrome *b* gene in many of the lake's fish, scientists have been able to estimate that the first cichlids entered Lake Victoria only 200,000 years ago, colonizing from the Nile. Dramatic changes in water level encouraged species formation. As the lake rose, it flooded new areas and opened up new habitats. Many of the species may have originated after the lake dried down 14,000 years ago, isolating local populations in small lakes until the water level rose again.

Cichlid Diversity

Cichlids are small, perchlike fishes ranging from 5 to 25 centimeters in length, and the males come in endless varieties of colors. The ecological and morphological diversity of these fish is remarkable, particularly given the short span over which they have evolved. We can gain some sense of the vast range of types by looking at how different species eat. There are mud biters, algae scrapers, leaf chewers, snail crushers, zooplankton eaters, insect eaters, prawn eaters, and fish eaters. Snail shellers pounce on slow-crawling snails and spear their soft parts with long, curved teeth before the snail can retreat into its shell. Scale scrapers rasp slices of scales off other fish. There are even cichlid species that are "pedophages," eating the young of other cichlids.

Cichlid fish have a remarkable trait that may have been instrumental in their evolutionary radiation: A second set of functioning jaws occurs in the throats of cichlid fish (figure 23.14)! The ability of these jaws to manipulate and process food has freed the oral jaws to evolve for other purposes, and the result has been the incredible diversity of ecological roles filled by these fish.

Abrupt Extinction

Recently, much of the cichlid radiation has disappeared. In the 1950s, the Nile perch, a commercial fish with a voracious appetite, was introduced on the Ugandan shore of Lake Victoria. Since then, it has spread through the lake, eating its way through the cichlids. By 1990, many of the open-water cichlid species had become extinct, as well as others living in rocky shallow regions. Over 70% of all the named Lake Victoria cichlid species had disappeared, as well as untold numbers of species that had yet to be described.

Very rapid speciation occurred among cichlid fishes isolated in Lake Victoria, but widespread extinction followed with the introduction of a predator into the lake.

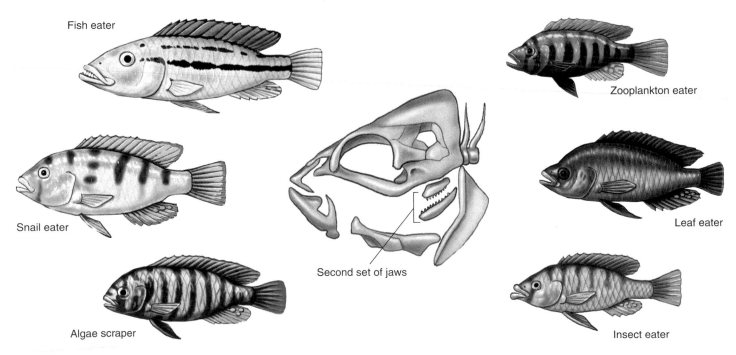

Fish eater

Zooplankton eater

Snail eater

Leaf eater

Algae scraper

Second set of jaws

Insect eater

FIGURE 23.14
Cichlid fishes of Lake Victoria. These fishes have evolved adaptations to use a variety of different habitats. The second set of jaws located in the throat of these fish has provided evolutionary flexibility, allowing oral jaws to be modified in many ways.

New Zealand Alpine Buttercups

Adaptive radiations such as those we have described in Galápagos finches, Hawaiian *Drosophila*, and cichlid fishes seem to be favored by *periodic isolation*. For example, finches and *Drosophila* invade new islands, local species evolve, and they in turn reinvade the home island, in a cycle of expanding diversity. Similarly, cichlids become isolated by falling water levels, evolving separate species in isolated populations that later are merged when the lake's water level rises again.

A clear example of the role periodic isolation plays in species formation can be seen in the alpine buttercups (genus *Ranunculus*) that grow among the glaciers of New Zealand (figure 23.15). More species of alpine buttercup grow on the two islands of New Zealand than in all of North and South America combined. Detailed studies by the Canadian taxonomist Fulton Fisher revealed that the evolutionary mechanism responsible for inducing this diversity is recurrent isolation associated with the recession of glaciers. The 14 species of alpine *Ranunculus* occupy five distinctive habitats within glacial areas: *snowfields* (rocky crevices among outcrops in permanent snowfields at 2130 to 2740 meters elevation); *snowline fringe* (rocks at lower margin of snowfields between 1220 and 2130 m); *stony debris* (slopes of exposed loose rocks at 610 to 1830 m); *sheltered situations* (shaded by rock or shrubs at 305 to 1830 m); and *boggy habitats* (sheltered slopes and hollows, poorly drained tussocks at elevations between 760 and 1525 m).

Ranunculus speciation and diversification have been promoted by repeated cycles of glacial advance and retreat. As the glaciers retreat, populations become isolated on mountain peaks, permitting speciation (figure 23.16). In the next advance, these new species can expand throughout the mountain range, coming into contact with their close relatives. In this way, one initial species could give rise to many

FIGURE 23.15
A New Zealand alpine buttercup. Fourteen species of alpine *Ranunculus* grow among the glaciers and mountains of New Zealand, including this *R. lyallii*, the giant buttercup.

descendants. Moreover, on isolated mountaintops during glacial retreats, species have convergently evolved to occupy similar habitats; these distantly related but ecologically similar species have then been brought back into contact in subsequent glacial advances.

Recurrent isolation promotes species formation.

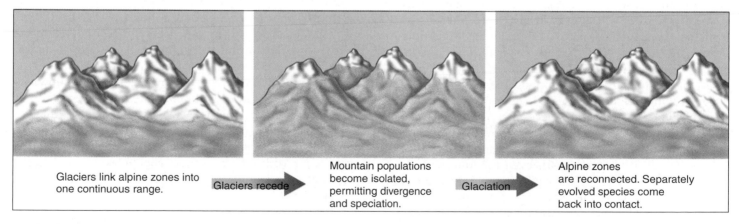

Glaciers link alpine zones into one continuous range.

Glaciers recede

Mountain populations become isolated, permitting divergence and speciation.

Glaciation

Alpine zones are reconnected. Separately evolved species come back into contact.

FIGURE 23.16
Periodic glaciation encouraged species formation among alpine buttercups in New Zealand. The formation of extensive glaciers during the Pleistocene epoch linked the alpine zones (*white*) of many mountains together. When the glaciers receded, these alpine zones were isolated from one another, only to become reconnected with the advent of the next glacial period. During periods of isolation, populations of alpine buttercups diverged in the isolated habitats.

The Pace of Evolution

We have discussed the manner in which speciation may occur, but we haven't yet considered the relationship between speciation and the evolutionary change that occurs within a species. For more than a century after the publication of *On the Origin of Species*, the standard view was that evolutionary change occurred extremely slowly. Such change would be nearly imperceptible from generation to generation, but would accumulate such that, over the course of thousands and millions of years, major changes could occur. This view is termed **gradualism** (figure 23.17*a*).

Evolution in Spurts?

Gradualism was challenged in 1972 by paleontologists Niles Eldredge of the American Museum of Natural History in New York and Stephen Jay Gould of Harvard University, who argued that species experience long periods of little or no evolutionary change (termed **stasis**), punctuated by bursts of evolutionary change occurring over geologically short time intervals. They called this phenomenon **punctuated equilibrium** (figure 23.17*b*), and argued that these periods of rapid change occurred only during the speciation process.

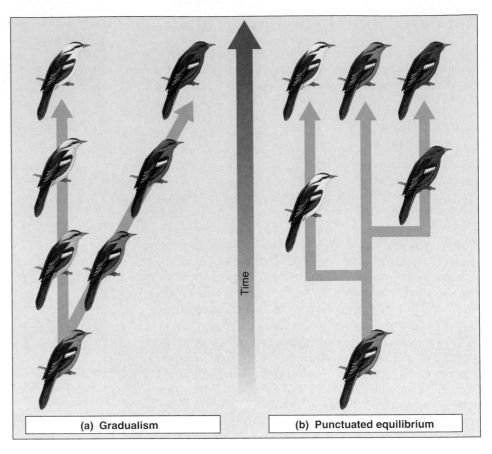

(a) Gradualism

(b) Punctuated equilibrium

FIGURE 23.17
Two views of the pace of macroevolution. (*a*) Gradualism suggests that evolutionary change occurs slowly through time and is not linked to speciation, whereas (*b*) punctuated equilibrium surmises that phenotypic change occurs in bursts associated with speciation, separated by long periods of little or no change.

Initial criticism of the punctuated equilibrium hypothesis focused on whether rapid change could occur over short periods of time. As we have seen in the last two chapters, however, when natural selection is strong, rapid and substantial evolutionary change can occur. A more difficult question involves the long periods of lack of change: Why would species exist for thousands, even millions, of years without changing? Although a number of possible reasons have been suggested, most researchers now believe that a combination of stabilizing and oscillating selection is responsible for stasis. If the environment does not change over long periods of time, or if environmental changes oscillate back-and-forth, then selection may favor stasis, even for long periods of time. One factor that may enhance this stasis is the ability of species to shift their ranges; for example, during the ice ages, when the global climate cooled, the geographic ranges of many species shifted southward, so that the species continued to experience similar environmental conditions.

Eldredge and Gould's proposal prompted a great deal of research. Some well-documented groups, such as African mammals, clearly have evolved gradually, not in spurts. Other groups, such as marine bryozoa, seem to show the irregular pattern of evolutionary change predicted by the punctuated equilibrium model. It appears, in fact, that gradualism and punctuated equilibrium are two ends of a continuum. Although some groups appear to have evolved solely in a gradual manner and others only in a punctuated mode, many other groups show evidence of both gradual and punctuated episodes at different times in their evolutionary history. However, the idea that speciation is necessarily linked to phenotypic change has not been supported: It is now clear that speciation can occur without phenotypic change, and that phenotypic change can occur within species in the absence of speciation.

> Evolutionary change can be slow and gradual. Or it may be rapid and discontinuous, separated by long periods of stasis that may result from a combination of stabilizing and oscillating selection.

Speciation and Extinction Through Time

Biological diversity has increased vastly since the Cambrian period. However, the trend has been far from consistent (figure 23.18). After a rapid rise, diversity reached a plateau for about 200 million years, but since then has risen steadily. Because changes in the number of species reflect the rate of origin of new species relative to the rate at which existing species disappear, this long-term trend reveals that speciation has, in general, outpaced extinction.

Nonetheless, speciation has not always outpaced extinction. In particular, interspersed in the long-term increase in species diversity have been a number of sharp declines, termed **mass extinctions.** Five major mass extinctions have been identified, the most severe one occurring at the end of the Permian period, approximately 250 million years ago. At that time, more than half of all families and as many as 96% of all species may have perished.

The most famous and well-studied extinction, though not as drastic, occurred at the end of the Cretaceous period (65 million years ago), at which time the dinosaurs and a variety of other organisms went extinct. Recent findings have supported the hypothesis that this extinction event was triggered when a large asteroid slammed into the earth, perhaps causing global forest fires and obscuring the sun for months by throwing particles into the air. This mass extinction did have one positive effect, though: Once the dinosaurs disappeared, mammals, which previously had been small and inconspicuous, diversified explosively, ultimately producing a wide variety of organisms, including elephants, tigers, whales, and humans.

Although species diversity rebounds after mass extinctions, the recovery is not rapid. Examination of the fossil record indicates that rates of speciation do not immediately increase after an extinction pulse, but rather take about 10 million years to reach their maximum. As a result, species diversity may require 10 million years, or even much longer, to attain its previous level.

A Sixth Extinction

The number of species in the world in recent times is greater than it has ever been. Unfortunately, that number is decreasing at an alarming rate due to human activities (see chapter 56). Some estimate that as many as one-fourth of all species will become extinct in the near future, a rate of extinction not seen on earth since the Cretaceous mass extinction. Moreover, the rebound in species diversity may be even slower than in previous mass extinction events because, instead of the ecologically depauperate, energy-rich environment that has occurred in the aftermath of previous events, a large proportion of the world's resources will be already taken up by human activities.

> The number of species has increased through time, although not at a constant rate. Several major extinction events have substantially, though briefly, reduced the number of species. Diversity rebounds, but the recovery is not rapid, and the organisms making up that diversity are not the same as those that existed before the extinction event.

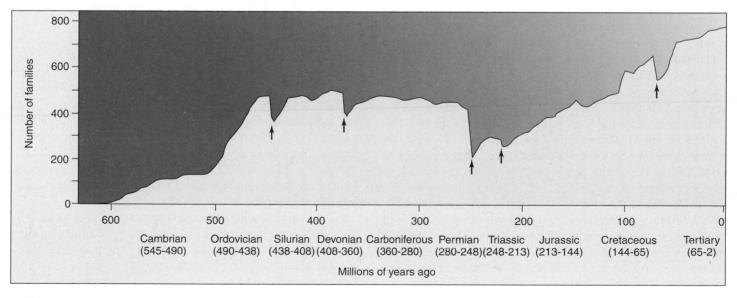

FIGURE 23.18

Diversity through time. The taxonomic diversity of families of marine animals has increased since the Cambrian period, although occasional dips have occurred. The fossil record is most complete for marine organisms because they are more readily fossilized than terrestrial species. Families are shown, rather than species, because many species are known from only one specimen, thus introducing error into estimates of the time of extinction. Arrows indicate the five major mass extinction events.

The Future of Evolution

In this chapter and chapters 21 and 22, we have discussed the results of evolution through time. What does the future hold? As mentioned on the previous page and discussed in greater detail in chapters 56 and 57, global biodiversity seems headed for a major extinction event from which recovery will be slow. Does this mean the end of evolution? We can use what we know about evolutionary processes to predict how evolution will occur in the future, both for diversity in general and for the human species in particular.

The Future Operation of Evolutionary Processes

Human impacts on the environment will affect the evolutionary process in many ways. Most obviously, by changing the environment, humans are changing the patterns of natural selection. In many cases, these changes will be so drastic that populations will be unable to adapt. But for those species that can survive, natural selection will act on genetic variation to produce evolutionary change. Global climate change, in particular, will be a major challenge, leading either to evolutionary change or extinction for many species.

Other factors will also lead to evolutionary change. Decreased population sizes will increase the likelihood of genetic drift, and geographic isolation of formerly connected populations will remove the homogenizing effect of gene flow, allowing these populations to evolve differences as adaptations to local environments. Chemicals and radiation released into the environment could even increase the mutation rate.

Consequently, for those species able to survive, evolutionary processes will continue and in some cases will even be accelerated. But what about species diversity? Extinction rates are increasing vastly, but it is also possible that speciation rates may increase, at least in some circumstances. The reason is that many formerly widespread species now exist only as geographically isolated populations (figure 23.19). Moreover, humans have introduced species to localities in which they formerly didn't occur, thus creating isolated populations. Given the importance of allopatry in the speciation process, these actions are likely to increase speciation rates for some species. This is not to say that geographic fragmentation is a good thing: Many small populations will go extinct before they can speciate, and this increased rate of speciation is unlikely to make up for the vastly increased rate of extinction, at least not for a very long time.

Human Evolutionary Future

Many science fiction writers have speculated on where evolution will take the human species, but consideration of the evolutionary process suggests that these ideas are fanciful.

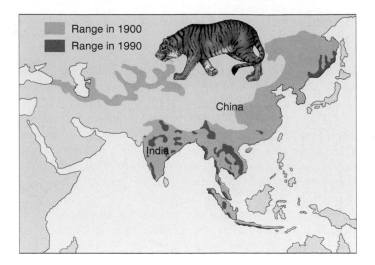

FIGURE 23.19
Tigers now exist in geographically isolated populations. Human activities, such as hunting and habitat destruction, have greatly reduced populations of the tiger, and have fragmented the species' range into many small and isolated populations.

In recent times, the movement of people around the globe has begun to erase human differences, a clear example of the homogenizing effect of gene flow. Moreover, to an ever-increasing extent, people with different ethnic origins are intermating, further diminishing differentiation among human populations. Because of the huge size of the human population, genetic drift is unlikely to be very important. Assuming that mutations don't increase too much, that leaves only natural selection as an engine of evolutionary change in humans.

The question then becomes: Do the conditions necessary for evolution by natural selection exist in humans? That is, are there phenotypic traits that both affect the number of surviving offspring and are genetically inherited from parent to offspring? Certainly, in one way, human populations will evolve because many genetic diseases that used to be fatal, and would thus remove alleles from the population, can now be treated successfully. As a result, we can expect the frequency of such alleles to increase in future generations. Other than such obvious examples, though, we leave it to the reader to ponder whether and in what cases the conditions for evolution by natural selection are likely to be met in future human populations. Of course, the advent of the genomics revolution (see chapter 17) adds another dimension to this discussion. Will future technological advances allow us to alter the human gene pool directly? And, if so, will that be a good idea?

For those species able to avoid extinction, human-caused changes should lead to evolutionary adaptation and, in some cases, to speciation.

For interactive testing, visit the Online Learning Center with PowerWeb at www.mhhe.com/Raven7

23.1 Species are the basic units of evolution.

The Nature of Species

- Any species concept must be able to account for species that occur together and for populations of the same species that are geographically separated. (p. 472)
- Although sympatric species occur together, they are usually phenotypically different and utilize different parts of the shared habitat. (p. 472)
- Ernst Mayr developed the biological species concept which defines species as: "groups of actually or potentially inter-breeding natural populations which are reproductively isolated from other such groups." (p. 473)
- Reproductive isolating mechanisms prevent genetic exchange between species. (p. 473)

23.2 Species maintain their genetic distinctiveness through barriers to reproduction.

Prezygotic Isolating Mechanisms

- Prezygotic isolating mechanisms prevent the formation of zygotes. (p. 474)
- Two species may not hybridize because they occupy different portions of an area and do not encounter each other. (p. 474)
- Related species often maintain distinctiveness due to behavioral differences such as courtship rituals. (p. 474)
- Sympatric species may utilize visual and sensory communication, pheromone reception, and electroreception to avoid mating with the wrong species. (pp. 474–475)
- Temporal isolation, mechanical isolation, and prevention of gamete fusion are other types of prezygotic isolating mechanisms. (p. 475)

Postzygotic Isolating Mechanisms

- Postzygotic isolating mechanisms prevent proper functioning of zygotes after they form. (p. 476)
- Hybrids that are weaker than their parents will almost certainly be eliminated in nature. (p. 476)
- Some hybrids are healthy, but infertile. (p. 476)

Problems with the Biological Species Concept

- Hybridization is more common than previously believed. (p. 477)
- Reproductive isolation may not be the only force maintaining species integrity. (p. 477)
- An alternative explanation is that species distinctions are maintained by natural selection. (p. 477)
- Due to extreme diversity, no single species definition may be adequate. (p. 477)

23.3 We have learned a great deal about how species form.

Reproductive Isolation May Evolve as a By-Product of Evolutionary Change

- Partial reproductive isolation might lead to reinforcement. Individuals in parental populations that avoid hybridization would be more successful at transmitting their genes to future generations. This would lead to continual improvement of prezygotic isolating mechanisms until the populations are completely reproductively isolated. (p. 478)
- Partial isolation might also lead to the loss of genetic distinctiveness due to gene flow. (pp. 478–479)
- Genetic drift may also cause reproductive isolation, but such isolation usually evolves more rapidly in the presence of selective pressures. (p. 479)

The Geography of Speciation

- In order for speciation to occur, similar populations must diverge, and then reproductive isolation must evolve. (p. 480)
- Allopatric populations are much more likely to diverge into separate species. (p. 480)
- Polyploidy, either by autopolyploidy or allopolyploidy, leads to immediate sympatric speciation in many plants and some animals. (pp. 480–481)
- Disruptive selection may lead to sympatric speciation, but this is controversial. (p. 481)

23.4 Clusters of species reflect rapid evolution.

- Adaptive radiation occurs when groups of closely related species evolve from a common ancestor by adapting to different habitats. (p. 482)

Hawaiian Drosophila

- When the ancestors of *Drosophila* reached the Hawaiian Islands, they encountered many empty habitats. Rampant speciation and ecological opportunity led to intense insect diversity. (p. 482)

Darwin's Finches

- Darwin's finches are derived from a single mainland species. The 14 species found on the islands comprise four groups: ground finches, tree finches, warbler finches, and vegetarian finches. (p. 483)

Lake Victoria Cichlid Fishes

- Many of the varied species originated after the lake dried down and isolated local populations. (p. 484)
- Ecological and morphological diversity was quite high. (p. 484)
- Widespread extinction occurred after the predatory Nile perch was introduced into the lake. (p. 484)

New Zealand Alpine Buttercups

- *Ranunculus* speciation and diversification have been fostered by repeated cycles of glacial advance and retreat. (p. 485)

The Pace of Evolution

- Darwin advocated slow evolutionary change (gradualism). (p. 486)
- Niles Eldredge and Stephen Jay Gould argued that species went through long periods of little change (stasis) and then experienced rapid bursts of evolutionary change over short periods of time during speciation events. (p. 486)

Speciation and Extinction Through Time

- Over the long term, speciation has outpaced extinction. (p. 487)
- Five major mass extinctions have interspersed the long-term increase, and have been followed by large-scale, although slow, periods of diversification. (p. 487)
- Human activities may be leading to a sixth major extinction. (p. 487)

The Future of Evolution

- Humans are changing the patterns of natural selection, and some of these changes will be so drastic that many species will not be able to adapt quickly enough to survive. (p. 488)
- Decreased population size and increased geographic isolation could also remove the homogenizing effect of gene flow and lead to moderate to high levels of speciation in species that are able to avoid extinction. (p. 488)

Self Test

1. Prezygotic isolating mechanisms include all of the following except
   a. hybrid sterility.
   b. courtship rituals.
   c. habitat separation.
   d. seasonal reproduction.
2. Which of the following is an example of mechanical isolation?
   a. Two species of birds live in the same habitat; one mates in spring and the other in summer.
   b. Two species of frogs have different mating calls.
   c. The flower structure of one species prevents the transfer of pollen from another species.
   d. One species of lizards inhabits the trees, and another species inhabits the ground cover.
3. _____ isolating mechanisms include improper development of hybrids and failure of hybrids to become established in nature.
   a. Prezygotic
   b. Postzygotic
   c. Temporal
   d. Mechanical
4. Reproductive isolation and the evolution of species could occur through which of the following?
   a. founder effect
   b. reinforcement
   c. adaptation
   d. all of these
5. Speciation occurs most frequently in populations that are
   a. sympatric.
   b. undergoing disruptive selection.
   c. allopatric.
   d. not geographically separated.
6. The large number of Hawaiian *Drosophila* species has likely resulted from
   a. adaptive radiation.
   b. a single common ancestor.
   c. geographic isolation.
   d. all of these.
7. The finch species of the Galápagos Islands are grouped according to their food sources. Which of the following is *not* a finch food source?
   a. seeds
   b. carrion
   c. insects
   d. tree buds
8. Cichlid diversity can be attributed to
   a. adaptive radiation.
   b. new habitats and geographic isolation.
   c. a second set of jaws in the throat of the fish.
   d. All of these factors contributed to cichlid diversity.
9. The hypothesis that evolution occurs in spurts, with great amounts of evolutionary change followed by periods of stasis, is
   a. punctuated equilibrium.
   b. allopatric speciation.
   c. gradualism.
   d. Hardy–Weinberg equilibrium.
10. Biological diversity through time has
   a. gradually increased.
   b. been constant.
   c. increased overall despite periodic drops.
   d. both increased and decreased with no overall change.

Test Your Visual Understanding

1. In all of the examples in the figure, one interbreeding population has been divided, which results in two or more geographically isolated populations. Over time, these isolated populations can undergo little or no evolutionary change, can undergo speciation, or can become extinct. Explain under what conditions each scenario can occur:
   a. Population undergoes little or no evolutionary change.
   b. Population undergoes speciation.
   c. Population becomes extinct.

Applying Your Knowledge

1. Adaptive radiation results when an ancestral species gives rise to many descendants which are adapted to many different parts of the environment. How would scenarios for adaptive radiation differ if speciation occurred allopatrically versus sympatrically?
2. Polyploid animals are far less common than polyploid plants. Why do you think this might be so? (*Hint:* Refer to the discussion of nondisjunction in chapter 13).

24

Evolution of Genomes and Developmental Mechanisms

Concept Outline

24.1 Evolutionary history is written in genomes.

Comparative Genomics.   Comparisons among genomes are addressing the differences among genomes and what those differences mean in terms of genome evolution.

Origins of Genomic Differences.   In addition to single genes mutating, whole genomes, large chunks of DNA, and individual genes have all duplicated over evolutionary time.

24.2 Developmental mechanisms are evolving.

Evolution of Development.   The fields of evolution and development are intersecting to address common questions in light of growing information about the genetic and genomic bases of both evolution and development. Comparative genomics can identify genes with similar sequences, but actual experiments are needed to test the function of similar genes in different organisms.

Diversity of Eyes in the Natural World.   Eyes appear to have evolved independently multiple times based on morphological traits. On the other hand, the *Pax6* gene has been found to initiate eye development in organisms as far-ranging as ribbon worms, fruit flies, and mice.

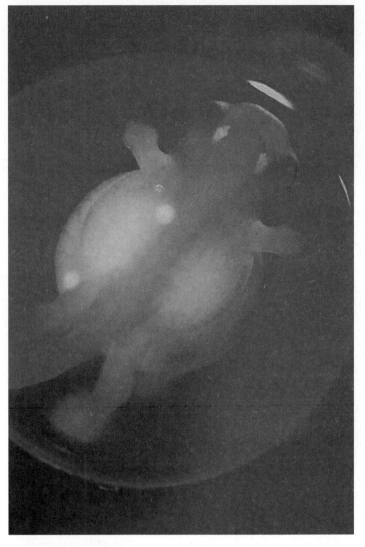

FIGURE 24.1
Frogs without polliwogs.   The Puerto Rican tree frog, *Eleutheradactylus coqui*, develops on land in a huge egg. There is no tadpole stage, and an adult frog emerges from the egg. How the tree frog lost its tadpole is a fascinating evolutionary question.

How have complex traits evolved? Many clues are hidden in genomes. As more genomes are sequenced, the field of comparative genomics is yielding some surprising answers and many, many questions. Genetic and genomic tools make it possible to investigate how development has evolved, yielding novel phenotypes built from highly conserved gene families. How is it that sister species of frogs can have completely different patterns of development (figure 24.1)? One frog goes from fertilized egg to adult frog with no intermediate tadpole stage. The sister species has an extra developmental stage neatly slipped in between early development and the formation of limbs—the tadpole stage. Genomic findings are accentuating the biological paradox that many genes are highly conserved, and the tremendous diversity of life shares, at some point, a common ancestral genome.

24.1 Evolutionary history is written in genomes.

Comparative Genomics

A key challenge of modern evolutionary biology is to find a way to link the evolution of DNA sequences, which we are now able to study in great detail, with the evolution of the complex morphological characters used to construct a traditional phylogeny. Many different genes contribute to complex characters, such as feathers, and making the connection between a specific change in one of the genes and a modification in a morphological character is particularly difficult. Comparing genomes (entire DNA sequences) of different species provides a powerful new tool to explore these relationships. Genomes are more than instruction books for building and maintaining an organism; they contain vast amounts of information on the history of life. As you saw in chapter 17, the growing number of fully sequenced genomes in all kingdoms is leading to a revolution in comparative evolutionary biology (table 24.1). It is now possible to explore the genetic differences between species in a very direct way, examining one-by-one the footprints on the evolutionary path between different species.

Over 100 different prokaryotic genomes have been completed, and at least 18 different eukaryotic genomes have

Table 24.1 Milestones for Comparative Eukaryotic Genomics

Organism		Estimated Genome Size (Mb)	Estimated Number of Genes	Year Sequenced
VERTEBRATES				
Homo sapiens (human)		3,200	30,000	2001
Mus musculus (mouse)		2,500	30,000	2002
Fugu rubripes (pufferfish)		365	33,609	2002
INVERTEBRATES				
Drosophila melanogaster (fruit fly)		137	13,600	2000

Table 24.1   Milestones for Comparative Eukaryotic Genomics *Continued*

Organism		Estimated Genome Size (Mb)	Estimated Number of Genes	Year Sequenced
Anopheles gambiae (mosquito)		278	46,000–56,000	2002
FUNGI *Schizosaccharomyces pombe* (fission yeast)		13.8	4,824	2002
Saccharomyces cerevisiae (brewer's yeast)		12.7	5,805	1997
PLANTS *Arabidopsis thaliana* (wall cress)		125	25,498	2000
Oryza sativa (rice)		430	32,000–55,000	2002
PROTISTS *Plasmodium falciparum* (malaria parasite)		23	5,300	2002

218 μm

been or are being sequenced. As the draft (preliminary) sequences of these and other genomes become available in the next few years, our view of the evolution of life on earth should become far clearer and our knowledge of phylogenetic relationships far more certain. The few genomic sequences that are already completed give exciting clues as to what is to come.

The Tiger Pufferfish

The draft sequence of the tiger pufferfish (*Fugu rubripes*) was completed in 2002, only the second vertebrate genome to be fully sequenced. For the first time, we were able to compare the genomes of two vertebrates. Some human and pufferfish genes have been conserved during 450 million years of evolution, while other genes are unique to each species. About 25% of human genes have no counterparts in *Fugu*. Also, extensive rearrangements have occurred during the 450 million years since mammals and teleost fish diverged, indicating a considerable scrambling of gene order. Finally, the human genome has more repetitive DNA, which counts for less than one-sixth of the *Fugu* sequence.

The pufferfish genome, 365 million base-pairs (Mb) in length, has only one-ninth of the DNA of humans, although both vertebrate species have approximately the same number of genes. Why do humans have so much extra DNA? Much of it appears to be in the form of introns that are substantially bigger than those in pufferfish. The *Fugu* genome has a handful of "giant" genes containing long introns; studying them should provide insight into the evolutionary forces that have driven the change in genome size during vertebrate evolution.

Sequences that are conserved between humans and pufferfish provide valuable clues for understanding the genetic basis of many human diseases. Amino acids critical to protein function tend to be preserved over the course of evolution, and changes at such sites within genes are more likely to cause disease. It is difficult to distinguish functionally conserved sites in the protein sequences of humans when comparing human proteins with those of other mammals because not enough time has elapsed for sufficient changes to accumulate at nonconserved sites. Because the pufferfish genome is only distantly related to humans, conserved sequences are far more easily distinguished.

The Mouse

Later in 2002, a draft sequence of the mouse (*Mus musculus*) genome was completed by an international consortium of investigators, allowing for the first time a comparison of two mammalian genomes. The human genome has about 400 million more nucleotides than that of the mouse. A comparison of the two genomes reveals that both have about 30,000 genes, and they share the bulk of them; in fact, the human genome shares 99% of its genes with mice. Humans and mice diverged about 75 million years ago, too little time for many evolutionary differences to accumulate. There are only 300 genes unique to either organism, comprising about 1% of the genome. Most of the nearly 150 genes unique to mice are linked with either the sense of smell, which is highly developed in rodents, or with reproduction. Comparing these similar genomes is already yielding information on the function of genes. It is much easier to design experiments to identify gene function in an experimental system like the mouse than in humans. Once the mouse sequence was known, the function of 1000 previously unidentified human genes was understood.

The genomes of humans and mice are so similar that one wonders why mice and humans are so different. The best explanation for why a mouse develops into a mouse and not a human is that the genes are expressed at different times and possibly in different tissues. This may be the case for the cystic fibrosis gene, which has been identified in both species. Defects in the human cystic fibrosis gene cause especially devastating effects in the lungs, but mice with cystic fibrosis do not have lung symptoms.

Comparison of the mouse and human genomes reveals that since mice and humans last shared a common ancestor about 75 million years ago (MYA), mouse DNA has mutated about twice as fast as human DNA. This is a fascinating observation in search of an explanation. The difference in generation time between mice and humans could account for some of this distinct mutation rate, because mice would have had more opportunities to mix and match genomic components during meiosis.

Perhaps the most unexpected finding in comparing the mouse and human genomes lies in the similarities between the "junk" DNA, mostly retrotransposons (refer to chapter 17), in the two species. This DNA does not code for proteins. A survey of the location of retrotransposon DNA in both species shows that it has independently ended up in comparable regions of the genome. It's beginning to look like this "junk" DNA may have more of a function than was previously assumed. The possibility that it is rich in regulatory RNA sequences, such as those described in chapter 18, is being actively investigated. In one such study, researchers collected almost all of the RNA transcripts made by mouse cells taken from every tissue. While most of the transcripts code for mouse proteins, as many as 4280 could not be matched to any known mouse protein. This suggests that a large part of the transcribed genome consists of nonprotein-encoding genes—that is, transcripts that function as RNA. Perhaps this can explain why a single retrotransposon can cause heritable differences in coat color in mice.

A draft of the rat genome has just been completed, and even more exciting news about the evolution of mammalian genomes may come forth. One of the most exciting aspects of comparing rat and mice genomes is the potential to capitalize on the extensive research on rat physiology, especially heart disease, and the long history of genetics in mice. Linking genes to disease just became easier.

Variation in the organization of genomes is as intriguing as gene sequence differences. Over long segments of chromosomes, the linear order of mouse and human genes is the same—the common ancestral sequence has been preserved in both species. This **conservation of synteny** (see chapter 17) was anticipated from earlier gene mapping studies, and provides strong evidence that evolution actively shapes the organization of the mammalian genome.

The Chimpanzee

The chimpanzee genome project is still under way. Humans and chimps diverged from a common ancestor only about 5 million years ago. This is too little time for much genetic differentiation to evolve between the two species, but enough for significant morphological and behavioral differences to have evolved. Preliminary sequence compar-

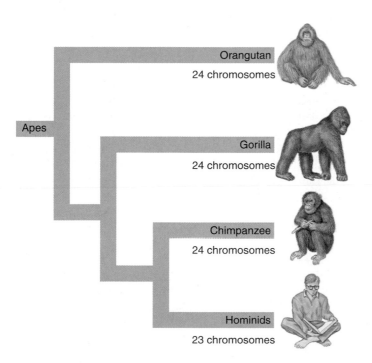

FIGURE 24.2
Living great apes. All living great apes, with the exception of humans, have a haploid chromosome number of 24. Humans have not lost a chromosome; rather, two smaller chromosomes fused to make a single chromosome.

isons indicate that chimp DNA is 98.7% identical to human DNA. If just the gene sequences encoding proteins are considered, the similarity increases to 99.2%. How could two species differ so much in body and behavior, and yet have almost equivalent sets of genes?

One potential answer to this question is based on the observation that chimp and human genomes show very different patterns of gene transcription activity, at least in brain cells. Investigators used microarrays (see Fig. 17.10) containing up to 18,000 human genes. Fluid extracted from living brain cells that contained expressed RNA transcripts was washed over the array of genes. Each gene lights up if a transcript of that gene is present in the fluid. The more copies of the gene, the more intense the signal. Because the chimp genome is so similar to that of humans, the microarray could detect the activity of chimp genes reasonably well. While the same genes were transcribed in chimp and human brain cells, the levels of transcription varied widely. It would seem that much of the difference between humans and chimps lies in which genes are transcribed, and when.

Humans have one fewer chromosome than chimpanzees, gorillas, and orangutans (figure 24.2). It's not that we have lost a chromosome. Rather, at some point in time, two mid-sized ape chromosomes fused to make what is now human chromosome 2, the second largest chromosome in our genome.

The fusion leading to human chromosome 2 is an example of the sort of genome reorganization that has occurred in many species. Rearrangements like this can provide evolutionary clues, but are not always definitive proof of how closely related two species are. Consider the organization of known **orthologs** (that is, genes with the same ancestral sequence) shared by humans, chickens, and mice. One study estimated that 72 chromosome arrangements had occurred since the chicken and human last shared a common ancestor. This is substantially less than the estimated 128 rearrangements between chicken and mouse or 171 between mouse and human. Does this mean that chickens and humans are more closely related than mice and humans or mice and chickens? No, what these data actually show is that chromosome rearrangements have occurred at a much lower frequency in humans and chickens than in mice. Chromosomal rearrangements in mice seem to have occurred at twice the rate seen in humans.

This finding of marked differences in the rate of chromosomal rearrangement among vertebrates raises new questions about genome evolution that are currently being explored. Identifying genomes that have undergone relatively slow chromosome change is most helpful in reconstructing the hypothetical genomes of ancestral vertebrates. If regions of chromosomes have changed little in distantly related vertebrates over the last 300 million years, it is reasonable to hypothesize that the common ancestor had genomic similarities.

Insects: *Drosophila* and *Anopheles*

The insects are the most species-rich and morphologically diverse animal group on earth. Two insect genomes have been sequenced. *Drosophila melanogaster* has been a laboratory model for genetic studies for much of the last century, and is arguably the best-understood gene system in biology. *Anopheles gambiae*, the malaria mosquito, along with *Plasmodium falciparum*, the protistan parasite it transmits, have together caused an enormous impact on the world's health, resulting in 1.7 to 2.5 million deaths each year. The genomes of both *Anopheles* and *Plasmodium* were sequenced in 2002.

The fruit fly *Drosophila* and the mosquito *Anopheles* are separated by approximately 250 million years of evolution, and appear to have evolved more rapidly over that interval than vertebrates. The extent of similarity between these two insects is comparable to that between humans and pufferfish, which diverged 450 million years ago. The organization of genes on the chromosomes has undergone significant shuffling between the two insect species. Interestingly, *Drosophila* exhibits less noncoding DNA than *Anopheles*, although the evolutionary force driving this reduction in noncoding regions is not clear.

The Protist *Plasmodium*

The parasitic protist *Plasmodium falciparum*, which causes malaria, has a relatively small genome of 24.6 million base-pairs that proved very difficult to sequence. It has an unusually high proportion of adenine and thymine, making it hard to distinguish one portion of the genome from the next. The project took five years to complete. *P. falciparum* appears to have about 5300 genes, with those of related function clustered together, suggesting that they might share the same regulatory DNA.

P. falciparum is a particularly crafty organism that hides from our immune system inside red blood cells, regularly changing the proteins it presents on the surface of the red blood cell. This has made it particularly difficult to develop a vaccine or other treatment for malaria. Now, a link to chloroplast-like structures in *P. falciparum* has raised other possibilities for treatment. An odd subcellular component called the apicoplast, found only in *Plasmodium* and its relatives, appears to be derived from a chloroplast appropriated from algae consumed by the parasite's ancestor. Analysis of the *Plasmodium* genome reveals that about 12% of all the parasite's proteins, encoded by the nuclear genome, head for the apicoplast. These proteins act there to produce fatty acids. The apicoplast is the only place the parasite makes the fatty acids it needs to survive, suggesting that drugs targeted at this biochemical pathway might be very effective against malaria. Another possibility is to look at chloroplast-specific herbicides, which might kill *Plasmodium* by targeting the chloroplast-derived apicoplast.

Flowering Plants: Rice and *Arabidopsis*

Few plant genomes have been sequenced, the first being *Arabidopsis thaliana*, the wall cress, a tiny member of the mustard family often used as a model organism for studying plant molecular genetics and development. Its genome sequence, largely completed in 2000, revealed 25,948 genes, about as many as humans have.

Arabidopsis is mainly an experimental model, with no commercial significance. However, the second plant genome for which a draft sequence has been prepared—rice—is of enormous economic significance. *Oryza sativa* belongs to the grass family, which includes maize (corn), wheat, barley, sorghum, and sugarcane. Together, these crops provide most of the world's food and animal feed. Unlike most grasses, rice has a relatively small genome of 430 million base-pairs (the maize genome is 2500 Mb, and barley's is an enormous 4900 Mb). Two different subspecies of rice were sequenced, yielding similar results. The proportion of the rice nuclear genome devoted to repetitive DNA, for example, was 42% in one variety, and 45% in the other. Retrotransposons, the most numerous large repeats, account for more than 15% of the rice genome in each study.

The rice genome has proven to contain a surprisingly large number of genes. Both rice draft sequences place the gene number for *Oryza* higher than for any other genome yet sequenced. One study suggests 53,000 to 63,000 genes; the other, using more conservative criteria, indicates 33,000 to 50,000 genes. These numbers will become more precise as genome annotation continues.

More than 80% of the genes found in rice are also found in *Arabidopsis*. Among the other 20% must be the genes responsible for the many physiological and morphological differences between rice (a monocot) and *Arabidopsis* (a dicot), two very different kinds of flowering plants. About one-third of the genes in *Arabidopsis* and rice appear to be in some sense "plant" genes—that is, genes not found in any animal or fungal genome sequenced so far. These include the many thousands of genes involved in photosynthesis and photosynthetic anatomy. Among the other plant genes are many that are very similar to those found in animal and fungi genomes, particularly genes involved in basic intermediary metabolism, in genome replication and repair, and in protein synthesis.

Both rice and *Arabidopsis* have higher *copy numbers* for gene families (multiple slightly divergent copies of a gene) than are seen in animals or fungi, spread among different chromosomes amid clusters of other duplicated genes. This suggests that these plants have undergone numerous episodes of polyploidy and/or segmental duplication during the 150 to 200 million years since rice and *Arabidopsis* diverged from a common ancestor.

Genome sequences are being determined for a wide variety of organisms, leading to the new field of comparative genomics.

Origins of Genomic Differences

Comparison of even the limited number of genomes discussed in this chapter reveals that genomes evolve dynamically. The genome changes that provide the raw material for this evolution arise in at least six ways:

1. mutation of a single gene;
2. when regions of DNA duplicate;
3. when large chunks of chromosomes rearrange;
4. when individual chromosomes duplicate;
5. when whole genomes duplicate or combine with the genome of another species to create polyploids (cells with three or more complete copies of a genome);
6. when DNA from other species becomes integrated into genomes.

We will explore genome changes beginning with large-scale changes, including duplication of whole genomes.

Polyploidization

We will start our exploration of genome evolution with gene duplication. Duplication of entire genomes by polyploidization obviously represents a huge change in genomes. There is growing evidence that duplication of fewer genes in organisms has also been a major contributor to morphological diversity. In both cases, the duplicated genes have the opportunity to acquire new functions because the duplication provides a functioning backup gene.

Humans have nine times as much DNA as pufferfish because of longer introns, but why do tulips contain over 170 times as much DNA as the small weed *Arabidopsis thaliana*? Large-scale evolutionary change, including duplication of entire genomes, certainly provides at least part of the explanation (figure 24.3). Because meiosis requires an even number of chromosome sets, species with ploidy levels that are multiples of two can reproduce sexually. However, meiosis would be a disaster in a $3n$ organism such as the banana, since three sets of chromosomes can't be evenly divided between two cells. (Sketch out what would happen in meiosis in a $3n$ banana cell, referring back to chapter 12 if necessary.) Bananas have to rely on asexual means of propagation.

Polyploidy ($3n$ or greater) can result from either genome duplication in one species or from hybridization of two different species. Hybridization may be followed by genome duplication, providing an even number of chromosome sets and so allowing meiosis to occur. For example, bread wheat arose from two hybridization and whole-genome duplication events (figure 24.4). Such major changes in the genome have often resulted in new species.

Whole-genome duplication, however, is insufficient to explain the size of some genomes. Wheat and rice are very closely related and have similar gene content. Yet, the wheat genome is 40 times larger than the rice genome.

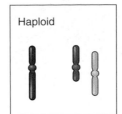  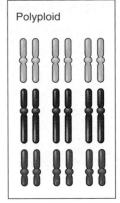

FIGURE 24.3
Chromosome numbers possible in plant genomes. *Haploid:* a set of chromosomes without their pairs; for example, the chromosome number present in a gamete. *Diploid:* a single set of chromosome pairs. *Polyploid:* multiple sets of chromosome pairs; for example, bananas have a triple set of chromosomes and are therefore polyploid.

This difference in genome size cannot be explained solely by the fact that bread wheat is a hexaploid ($6n$) and rice is a diploid ($2n$). The fact that the wheat genome contains lots of repetitive DNA has increased its DNA content, but not necessarily its gene content. Now that the rice genome is fully sequenced, attention has shifted to sequencing the other cereal grains, especially wheat. Comparisons between the rice and wheat genomes should provide clues about the genome of their common ancestor.

Segmental Duplication

One of the greatest sources of novel traits in genomes is duplication of segments of DNA. When a gene duplicates, the two most likely fates of the duplicate gene are: (1) losing function through subsequent mutation, and (2) gaining a novel function through subsequent mutation. In fact, most duplicate genes lose function, and the average half-life of duplicates is about 4 million years. This is a huge amount of time on a human scale, but very short in terms of evolutionary change. How then can researchers claim that gene duplication is a major evolutionary force for gene innovation (genes gaining new function)? One piece of evidence can be found by asking where in the genome gene duplication is most likely to occur. In humans, the highest rates of duplication have occurred in the three most gene-rich chromosomes of the genome. The seven chromosomes with the fewest genes also had the least amount of duplication. Remember, having fewer genes does not mean that there is less total DNA. Even more compelling, certain types of human genes were more likely to be duplicated: growth and development genes, immune system genes, and cell-surface receptors. About 5% of the human genome

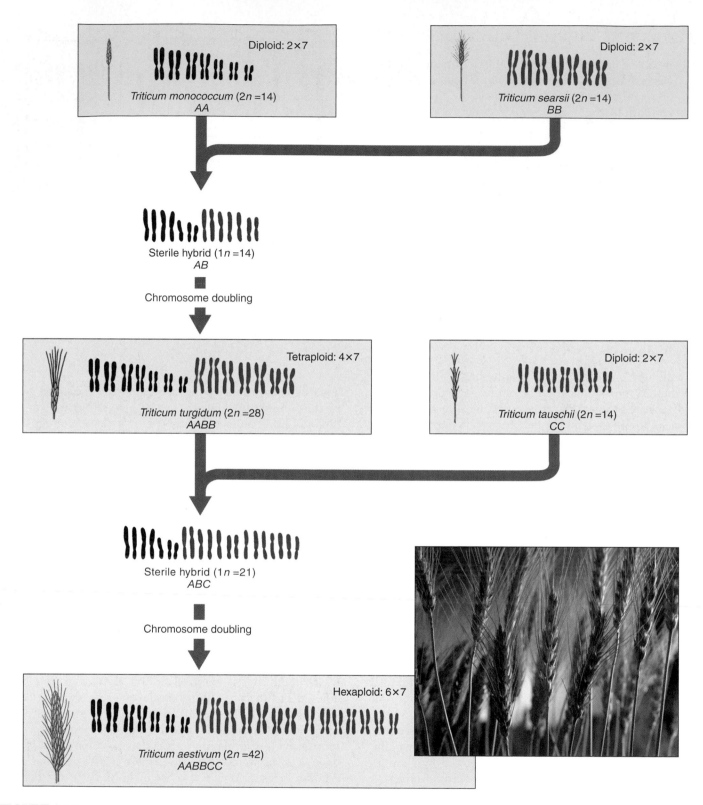

FIGURE 24.4

Evolutionary history of wheat. Domestic wheat arose in southwestern Asia in the hilly country of what is now Iraq. This region contains a rich assembly of grasses of the genus *Triticum*. Domestic wheat (*T. aestivum*) is a polyploid species of *Triticum* that arose through two so-called "allopolyploid" events. (*1*) Two different diploid species, symbolized here as *AA* and *BB*, hybridized to form an *AB* polyploid; the species were so different that *A* and *B* chromosomes could not pair in meiosis, so the *AB* polyploid was sterile. However, in some plants the chromosome number spontaneously doubled due to a failure of chromosomes to separate in meiosis, producing a fertile tetraploid species, *AABB*. This wheat is used in the production of pasta. (*2*) In a similar fashion, the tetraploid species *AABB* hybridized with another diploid species, *CC*, to produce, after another doubling event, the hexaploid *T. aestivum*, *AABBCC*. This bread wheat is commonly used throughout the world.

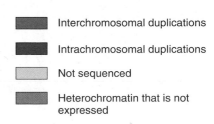

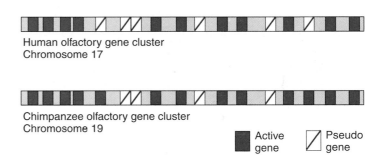

Interchromosomal duplications

Intrachromosomal duplications

Not sequenced

Heterochromatin that is not expressed

FIGURE 24.5
Segmental duplication on the human Y chromosome. Each red region has 98% sequence similarity with a sequence on a different human chromosome. Each blue region has 98% sequence similarity with a sequence elsewhere on the Y chromosome. The Y chromosome is one continuous piece of DNA shown in segments here for illustrative purposes only.

Human olfactory gene cluster
Chromosome 17

Chimpanzee olfactory gene cluster
Chromosome 19

■ Active gene

◫ Pseudo gene

FIGURE 24.6
Gene inactivation. While almost all mouse olfactory receptor genes are functional, an evolutionary loss of olfactory receptors has occurred in primates, which rely less on their sense of smell. Comparisons of the olfactory receptor genes in humans and chimpanzees reveal that humans have more pseudogenes (inactive genes) than chimpanzees have.

consists of segmental duplications (figure 24.5). As more species are compared, gene duplication rates appear to vary. *Drosophila* has about 31 new duplicates per genome per million years, which is equivalent to 0.0023 duplications per gene per million years. The rate is about 10 times faster for the nematode *Caenorhabditis elegans*. Two genes within an organism that arose from the duplication of one gene in an ancestor are called **paralogs.** Orthologs, as defined earlier, reflect the conservation of a single gene from a common ancestor.

Gene Inactivation

The loss of gene function is another important way genomes evolve. Consider the olfactory receptor (OR) genes that are responsible for our sense of smell. These genes code for receptors that bind odorants, initiating a cascade of signaling events that eventually lead to our perception of scents. Gene inactivation seems to have been the most frequent explanation for our reduced sense of smell relative to that of the great apes and other mammals. Primate genomes have over 1000 copies of OR genes (figure 24.6). An estimated 70% of human OR genes are inactive **pseudogenes** (sequences of DNA that are very similar to functional genes, but do not produce a functional product). Half the chimpanzee and gorilla OR genes function effectively, while half are pseudogenes. Over 95% of New World monkey OR genes and probably all mouse OR genes are working quite well. What can we conclude from this high rate of inactivation of human OR genes? Most likely, humans came to rely on other senses, reducing the selection pressure against loss of OR gene function by random mutation.

Lateral Gene Transfer

Evolutionary biologists build phylogenies on the assumption that genes are passed from generation to generation, a process called **vertical gene transfer.** Hitchhiking genes from other species, referred to as **lateral gene transfer,** lead to phylogenetic complexity. Lateral gene transfer was most prevalent very early in the history of life, when the boundaries between individual cells and species seem to have been less firm than they are now. Earlier in the history of life, gene swapping between species was rampant.

The extensive gene swapping among early organisms is causing many researchers to reexamine the base of the tree of life. Early phylogenies based on ribosomal RNA (rRNA) sequences indicate that an early prokaryote gave rise to two major domains, the Bacteria and the Archaea. From one of these lineages, the domain Eukarya emerged, its organelles produced by engulfing specialized prokaryotes (figure 24.7).

This somewhat straightforward rRNA phylogeny is being revised as more microbial genomes are sequenced; by 2002, 87 microbial genomes had been sequenced. Gene swapping appears to have occurred frequently early in the history of life. Exchanges between organisms were

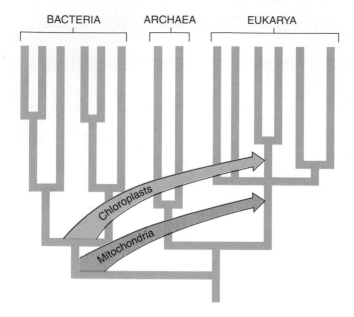

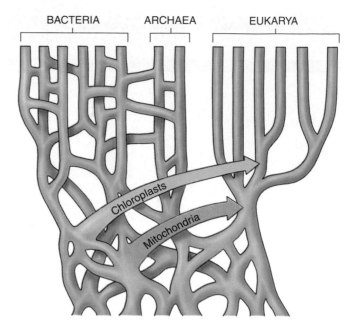

FIGURE 24.7
Phylogeny based on a universal common ancestor. The three domains share a common ancestor, and the tree of life is firmly rooted.

FIGURE 24.8
Lateral gene transfer. Early in the history of life, organisms may have freely exchanged genes. To a lesser extent, this transfer continues today. The tree of life may be more like a web or a net.

so substantial then that the base and the deepest branches of the tree of life are being reevaluated. Phylogenies built with rRNA sequences suggest that the domain Archaea is more closely related to the Eukarya than to the Bacteria. However, as more microbial genomes are sequenced, bacterial and archaeal genes are showing up in the same organism! The most likely conclusion is that organisms swapped genes, possibly even absorbing DNA obtained from a food source. Perhaps the base of the tree of life is better viewed as a web than a branch (figure 24.8).

Let's move a bit closer to home and look at the human genome, which is riddled with foreign DNA, often in the form of transposons (jumping genes, discussed in chapter 20). The many transposons of the human genome provide a paleontological record over several hundred million years. Comparisons among versions of a transposon that has duplicated many times allow researchers to construct a "family tree" to identify the ancestral form of the transposon. The percent sequence divergence of duplicates allows the researcher to estimate the time when that particular transposon originally invaded the human genome. In humans, most of the DNA hitchhiking seems to have occurred mil-

lions of years ago. Our genome carries many ancient transposons, making it quite different from other genomes that have been studied, such as those of *Drosophila*, *C. elegans*, and *Arabidopsis*. One explanation for the observed low level of transposons in *Drosophila* is that fruit flies somehow eliminate unnecessary DNA from their genome 75 times faster than humans do. Our genome has simply hung on to hitchhiking DNA more often. While the human genome has had minimal transposon activity in the past 50 million years, mice by contrast are continuing to acquire new transposable elements. This may explain in part the more rapid change in chromosome organization in mice than in humans that we discussed earlier in this chapter.

Whole-genome duplication, segmental duplication, and loss of gene function have all contributed to the evolution of genomes. Lateral gene transfer has led to an unexpected mixing of genes among organisms. This creates phylogenetic dilemmas at the base of the tree of life and leaves humans wondering whether their genome is really their own or a set of DNA shared with others.

24.2 Developmental mechanisms are evolving.

Evolution of Development

The era of comparative genomics is making it possible to bring distinct fields of biology together to address fundamental questions about evolution and development. Closely related sea urchins have been discovered that have very distinctive developmental patterns (figure 24.9). For example, the direct-developing urchin never makes a pluteus larva—it just jumps ahead to its adult form. Sequencing the sea urchin genome, coupled with new genomics techniques such as microarrays (discussed in chapter 17), should help us unravel the changes in patterns of gene expression that most likely account for these dramatic differences in the development of complex organisms that end up with nearly identical adult bodies. These and other questions about the relationship between development and evolution are leading to exciting experiments and findings. Next, we will investigate ways that new traits have arisen by adapting a gene for a new function or by having different genes converge on a similar function.

Same Gene, New Function

If all but 300 of humans' 30,000 genes are shared with mice, why are mice and humans so different? Part of the answer is that genes with similar sequences in two different species may work in slightly or even dramatically different ways. For example, the evolution of vertebrates can partially be explained by the co-option of an existing gene for a new function. Ascidians are basal chordates that have a notochord but no vertebrae. The *Brachyury* gene of ascidians codes for a transcription factor and is expressed in the developing notochord (figure 24.10). *Brachyury* is not a novel gene that appeared in animals as vertebrates evolved. It is found in invertebrates as well, where it has a different function. Most likely, an ancestral *Brachyury* gene was co-opted for a new role in notochord development.

How does a gene gain a different function? One scenario is that it is a regulatory gene that turns on a different set of genes in different organisms. A particularly intriguing example of this is eye development. Phylogenetic analysis shows that the eyes of vertebrates and insects are analogous, not homologous, structures. Molecular genetic analysis, however, reveals that the highly conserved *Pax6* gene launches eye development in vertebrates and insects alike. In this case, the compound eye of a fruit fly and the simple eye of a mouse develop from cells that express *Pax6*. Jumping from gene to phenotype leaves out an important step—development. Development is the black box where linear sequences of DNA are translated into three-dimensional form. To fully understand evolution, we must consider development.

The use of genomics to investigate how development has evolved is a newly emerging field that combines genetics, genomics, evolution, and development to question how the diversity of life has arisen. There are some two dozen conserved gene families that regulate development in animals. Some of these gene families have ancestral genes that predate the origins of animals. These genes serve as a toolkit to build an animal.

The paradox that puzzles evolutionary developmental biologists is how this same toolkit can be used to build an insect, a bird, a bat, a whale, or a human. This paradox is illustrated with the *Brachyury* and *Pax6* genes just described. One explanation is that these genes turn on different genes or combinations of genes in different animals. For example, all tetrapods

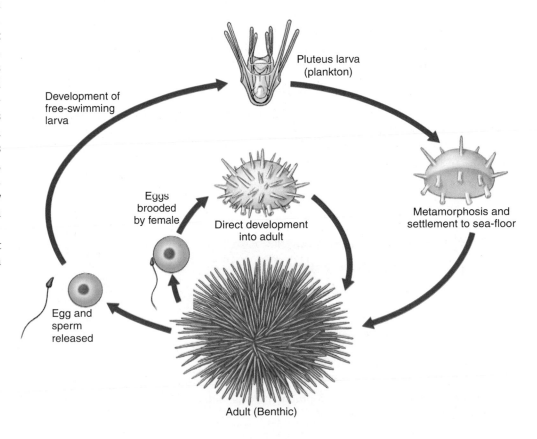

FIGURE 24.9
Direct and indirect sea urchin development. Phylogenetic analysis shows that indirect development was the ancestral state. Direct-developing sea urchins have lost an intermediate stage of development.

Pluteus larva (plankton)

Development of free-swimming larva

Eggs brooded by female

Direct development into adult

Metamorphosis and settlement to sea-floor

Egg and sperm released

Adult (Benthic)

have four limbs—two hindlimbs and two forelimbs. The forelimb in a bird is actually the wing. Our forelimb is the arm. Clearly, these are two very different structures, but they have a common evolutionary origin. That is, they are homologous structures.

At the genetic level, humans and birds both express the *Tbx5* gene in developing limb buds. *Tbx5* is a member of a gene family with a specific *domain*, a conserved sequence of base-pairs. *Tbx5* encodes a protein domain called the T-box, which is a transcription factor. So, *Tbx5*-encoded protein turns on a gene or genes that are needed to make a limb. What seems to have changed as birds and humans evolved are the genes that are transcribed because of the Tbx5 protein (figure 24.11). In the ancestral tetrapod, perhaps Tbx5 protein bound to only one gene and triggered transcription. In humans and birds, a few genes are expressed in response to Tbx5 protein, but they are different genes.

The story of the evolution of limb development is far more complex than *Tbx5*, which gets limb development started. The genes that *Tbx5* regulate in turn may affect the expression of other genes. Protein

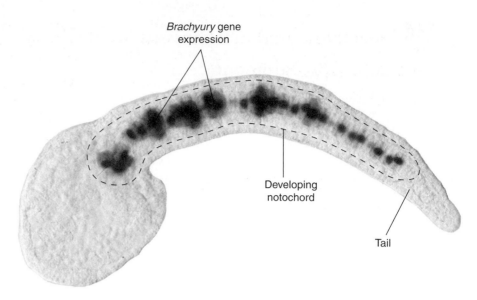

FIGURE 24.10
Co-opting a gene for a new function. *Brachyury* is a gene found in invertebrates that has been used for notochord development in this ascidian, a basal chordate. By attaching the *Brachyury* promoter to a gene with a protein product that stains blue, it is possible to see that *Brachyury* gene expression in ascidians is associated with the development of the notochord, a novel function compared to its function in organisms lacking a notochord.

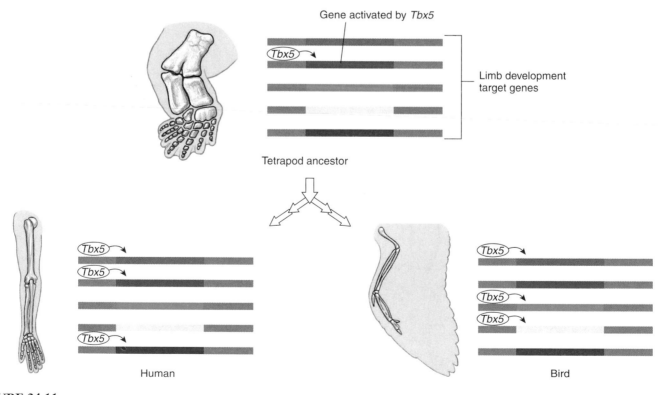

FIGURE 24.11
***Tbx5* regulates wing and arm development.** Wings and arms are very different, but the development of each depends on *Tbx5*. Why the difference? *Tbx5* turns on different genes in birds and humans.

products serve as enzymes in biochemical pathways and provide the building blocks for other protein structures. Mining genome sequences for different organisms will be essential in identifying all the genes involved. Also, development occurs in four dimensions—three-dimensional space over time. Changing the timing of gene expression, as well as the genes that are expressed, can result in dramatic changes in shape.

Different Genes, Convergent Function

Insect wings, especially those of moths and butterflies, have beautiful patterns that can protect them from predation and allow them to thermoregulate (figure 24.12). The origins of these patterns are best explained by the recruitment of existing regulatory programs for new functions. In butterflies, bristles that were linked to neurons and played a sensory role have become scales on wings that produce amazing colors. Structures that have their origins in bristles are inverted, and the daughter cell that gave rise to neuronal connections dies in the development of scales. Later in development, pigment production is triggered. Not all insects have co-opted the same sets of genes for new functions, but all the evolutionary pathways have converged around these novel, highly patterned wings.

Functional Genomics

Sequence comparisons among organisms are essential for both phylogenetic and comparative developmental studies. Careful analysis is needed to distinguish paralogs from orthologs.

Rapidly evolving research on bioinformatics, which utilizes computer programming to analyze DNA and protein data, leads to hypotheses that can be tested experimentally. You have already seen how this can work with highly conserved genes such as *Pax6* and *Tbx5*. However, a single base mutation can change an active gene into an inactive or pseudogene. This means that, although function can be inferred from sequence data, experiments are necessary to demonstrate the actual function of the gene. This type of work is called functional genomics and is explained in chapter 17.

Tools for functional analysis exist in model systems, but need to be developed in other organisms on the tree of life if we are going to piece together evolutionary history. Model systems such as yeast, *Arabidopsis*, the nematode worm, the fruit fly, and the mouse have been selected because they are easy to manipulate in some ways. These models all have good genetic systems in which mutant genes can be studied by making crosses. They have fairly short life cycles (imagine trying to study genetics in redwood trees or elephants!). They can be maintained and will reproduce in the laboratory. Also, it is possible to visualize gene expression within parts of the organism using labeled markers and to create transgenic organisms that contain and may express foreign genes.

Many factors have contributed to the evolution of development, including the acquisition of new functions by the same genes. Functional genomic analysis is necessary to determine the actual function of similar genes in different species.

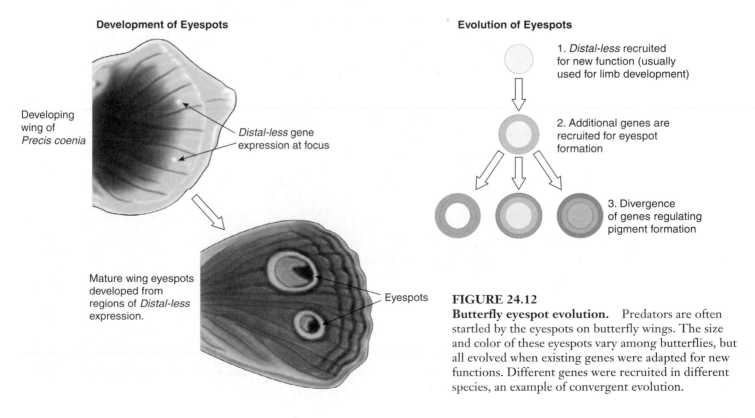

Development of Eyespots

Developing wing of *Precis coenia*

Distal-less gene expression at focus

Mature wing eyespots developed from regions of *Distal-less* expression.

Eyespots

Evolution of Eyespots

1. *Distal-less* recruited for new function (usually used for limb development)

2. Additional genes are recruited for eyespot formation

3. Divergence of genes regulating pigment formation

FIGURE 24.12
Butterfly eyespot evolution. Predators are often startled by the eyespots on butterfly wings. The size and color of these eyespots vary among butterflies, but all evolved when existing genes were adapted for new functions. Different genes were recruited in different species, an example of convergent evolution.

Diversity of Eyes in the Natural World

The eye is one of the most complex organs, and biologists have studied it for centuries. Indeed, explaining how such a complicated structure could evolve was one of the great challenges facing Darwin: If all parts of a structure such as an eye are required for proper functioning, how could natural selection build such a structure? Darwin's response was that even intermediate structures—which provide, for example, the ability to distinguish light from dark—would be advantageous compared with the ancestral state of no visual capability whatsoever, and thus would be favored by natural selection. In this way, by incremental improvements in function, natural selection could build a complicated structure.

Comparative anatomists long have noted that the structures of the eyes of different types of animals are quite different. Consider, for example, the difference in the eyes of a vertebrate, an insect, a mollusk (octopus), and a planarian (figure 24.13). The eyes of these organisms are extremely different in many ways, ranging from compound eyes, to simple eyes, to mere eyespots. Consequently, these eyes are examples of convergent evolution and are analogous, rather than homologous (that is, they are constructed from different ancestral structures). For this reason, evolutionary biologists traditionally viewed the eyes of different organisms as independently evolved, perhaps as many as 20 times! Moreover, this view holds that the most recent common ancestor of these forms was a primitive animal with no ability to detect light.

Unexpected Findings from Molecular Developmental Biology

In the early 1990s, biologists studied the development of the eye in both vertebrates and insects. In each case, a gene was discovered that codes for a transcription factor important in lens formation; the mouse gene was given the name *Pax6*, whereas the fly gene was called *eyeless* (because a mutation in that gene led to a lack of production of the transcription factor and thus the absence of eye development). When these genes were sequenced, it became apparent that they were highly similar; in essence, the homologous gene was responsible for triggering lens formation in both insects and vertebrates.

A stunning demonstration of the homology of these genes was conducted by the Swiss biologist Walter Gehring, who inserted the mouse version of the *Pax6* into the genome of a fruit fly, creating a transgenic fly. In the fly, the *Pax6* gene was turned on by regulatory factors in the leg of the fly. *Pax6* was expressed, and an eye formed on the leg of the fly (figure 24.14)!

These results were truly shocking to the evolutionary biology community. Insects and vertebrates diverged from a common ancestor more than 500 million years ago. Moreover, given the differences in structure of the vertebrate and insect eye, the standard assumption was that the eyes were independently evolved, and thus that the development of those eyes would be controlled by completely different genes. That eye development was affected by the same gene, and that the genes were so similar that the vertebrate gene seemed to function normally in the insect genome, was completely unexpected.

The *Pax6* story extends to eyeless fish found in caves (figure 24.15). Fish that live in dark caves need to rely on senses other than sight. In cavefish, *Pax6* gene expression is greatly reduced. Eyes start to develop, but then degenerate.

How ancient is the *Pax6* gene? Has it always been responsible for initiating eye development, or did it have an

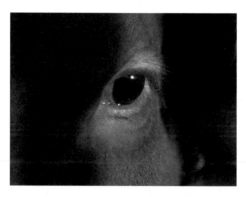

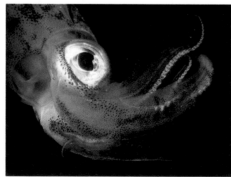

FIGURE 24.13
A diversity of eyes. Morphological and anatomical comparisons of eyes are consistent with the hypothesis of independent, convergent evolution of eyes in diverse species such as flies and humans.

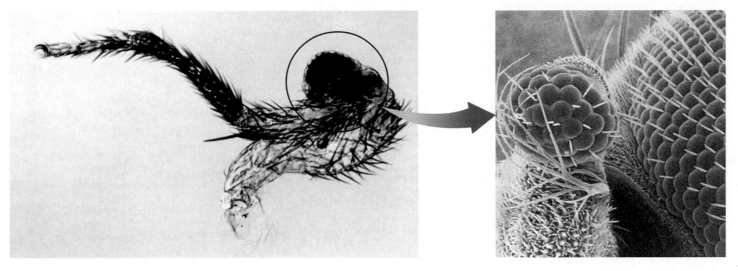

FIGURE 24.14
Mouse *Pax6* makes an eye on the leg of a fly.   *Pax6* and *eyeless* are functional homologs. The *Pax6* master regulator gene can initiate compound eye development in a fruit fly or simple eye development in a mouse.

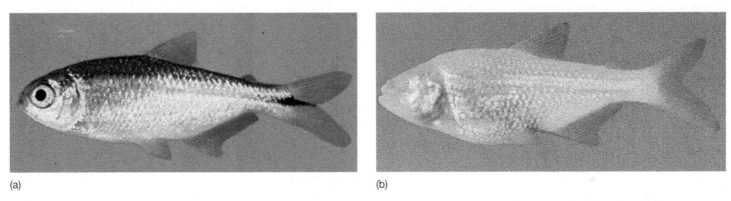

(a)   (b)

FIGURE 24.15
Cavefish have lost their sight.   Mexican tetras (*Astyanax mexicanus*) have (*a*) surface-dwelling members and (*b*) cave-dwelling members of the same species. The cavefish have very tiny eyes, partly because of reduced expression of *Pax6*.

ancestral function? Recent discoveries have yielded further surprises. Even the very simple ribbon worm, *Lineus sanguineus*, relies on *Pax6* for development of its eyespots. A *Pax6* homolog has been cloned and shown to be expressed at the sites where eyespots develop. This simple marine invertebrate evolved later than the flatworm planaria. Just like planaria, ribbon worms can regenerate their head region if it is removed. In an elegant experiment, the head of the ribbon worm was removed, and the regeneration of eyespots was followed while the expression of the *Pax6* homolog was observed using in situ hybridization. To observe *Pax6* gene expression, an antisense RNA sequence of the *Pax6* was made and labeled with a color marker. When the regenerating ribbon worms were exposed to the antisense *Pax6* probe, the antisense RNA paired with expressed *Pax6* RNA transcripts and could be seen as colored spots under the microscope (figure 24.16).

Similar experiments were tried with planaria species, but the conclusion was quite different from in ribbon worms. If a planaria is cut in half lengthwise, it can regenerate its missing half, including the second eyespot, but no *Pax6* gene expression is associated with regenerating the eyespots. Planaria have *Pax6*-related genes, but inactivating those genes does not stop eye regeneration (figure 24.17). These *Pax6*-related genes are, however, expressed in the central nervous system. Perhaps some clues as to the origin of *Pax6*'s role in eye development will be uncovered as comparisons between ribbon worm and planaria regeneration continue.

Evolution of the Eye Reconsidered

How can these findings be explained? One possibility is that eyes in different types of animals are truly independently

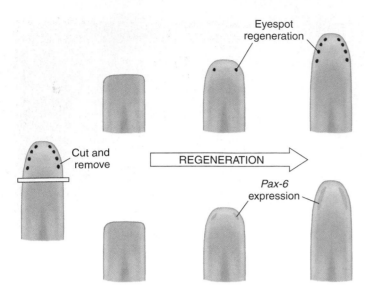

FIGURE 24.16
Pax6 **needed for ribbon worm eyespot regeneration.** *Pax6* is expressed at the same time and place as eyespots on regenerating ribbon worms. The presence of *Pax6* transcripts was visualized through in situ hybridization with a colored antisense probe.

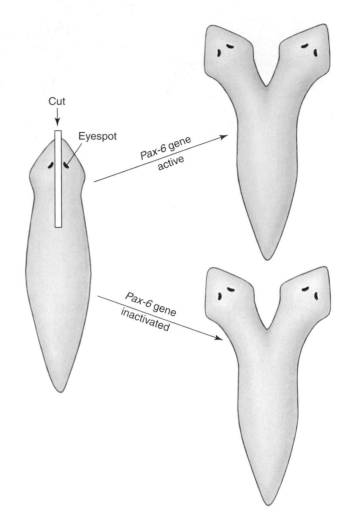

FIGURE 24.17
Pax6 **not required for planaria eyespot regeneration.** Planaria can regenerate their heads and eyespots when cut in half longitudinally. Unlike ribbon worms, *Pax6* does not appear to play a role in planaria eyespot regeneration. When both *Pax6*-related genes in planaria were prevented from producing a protein product, eyespots still formed.

evolved as originally believed. If this is the case, why is *Pax6* so similar and able to play a similar role in the development of the eye in so many different groups? Proponents of this viewpoint point out that *Pax6* is involved not only in development of the eye, but also in development of the entire forehead region of many organisms. Consequently, it is possible that if *Pax6* had a regulatory role in the forehead of early animals, perhaps it has been independently co-opted time and time again to serve a role in eye development. This role would be consistent with the data on planarians (figure 24.17).

Many other biologists find this interpretation unlikely. The consistent use of *Pax6* in eye development in so many organisms, the fact that it functions in the same role in each case, and the great similarity in DNA sequence and even functional replaceability suggest to many that *Pax6* only acquired its evolutionary role in eye development a single time, in the common ancestor of all extant organisms that use *Pax6* in eye development.

Given the great dissimilarity among eyes of different groups, how can this be? One hypothesis is that the common ancestor of these groups was not completely blind, as traditionally assumed. Rather, that organism may have had some sort of rudimentary visual system—maybe no more than a pigmented photoreceptor cell, maybe a slightly more elaborate organ that could distinguish light from dark. Whatever the exact phenotype, the important point is that some sort of basic visual system existed that used *Pax6* in its development. Subsequently, the descendants of this ancestor diversified independently, evolving the sophisticated and complex image-forming eyes exhibited by different animal groups today.

Most evolutionary and developmental biologists today support some form of this hypothesis. Nonetheless, it is important to keep in mind that there is no independent evidence that the common ancestor of most of today's animal groups, a primitive form that lived probably more than 500 million years ago, had any ability to detect light. The reason for this belief comes, not from the fossil record, but from a synthesis of phylogenetic and molecular developmental data.

Understanding the evolution of eyes illustrates the power of multidisciplinary approaches to elucidate the evolutionary history of the world's biological diversity.

24.1 Evolutionary history is written in genomes.

Comparative Genomics

- An important challenge of modern evolutionary biology is to link the evolution of DNA sequences with the evolution of complex morphological characters used to construct phylogenies. (p. 492)
- Over 100 prokaryotic genomes and at least 18 eukaryotic genomes have either been completed or are currently being sequenced. (pp. 492–494)
- The pufferfish genome has 365 million base-pairs—basically the same number of genes as the human genome, but only one-ninth of the DNA. (p. 494)
- Humans share 99% of their genes with mice, and thus only diverged about 75 MYA. (p. 494)
- Differences in development are explained by genes being expressed at different times and/or in different tissues. (p. 494)
- Humans and chimps diverged from a common ancestor about 5 MYA, and their sequences are 98.7% identical. (p. 495)
- If only small chromosomal differences exist between distantly related vertebrates over the past 300 million years, a reasonable hypothesis is that the common ancestor had a similar genome. (p. 495)
- The fruit fly (*Drosophila*) and the mosquito (*Anopheles*) are separated by about 250 million years of evolution, and appear to have evolved more rapidly during that time than vertebrates did. (p. 496)

Origins of Genomic Differences

- Genomic changes can be made by at least six factors: the mutation of a single gene, duplicated regions of DNA, rearrangements of large chunks of DNA, chromosome duplication, polyploidy, and interspecies gene integration. (p. 497)
- Growing evidence indicates that duplication of just a few genes has been a major contributor to morphological diversity. (p. 497)
- Whole-genome duplication is insufficient to explain the size of some genomes. (p. 497)

- One of the greatest sources of novel genomic traits is the duplication of DNA segments. The fate of the duplicate gene is most likely either the gaining of a novel function or the loss of function through subsequent mutations. (p. 497)
- About 5% of the human genome consists of segmental duplications. (pp. 497–499)
- Genomes also evolve due to the loss of gene function. (p. 499)
- Vertical gene transfer refers to genes passing from generation to generation, while lateral gene transfer refers to genes moving from one species to another. (p. 499)
- Gene swapping appears to have occurred frequently early in the history of life. (pp. 499–500)
- Most human lateral gene transfer appears to have occurred millions of years ago. (p. 500)

24.2 Developmental mechanisms are evolving.

Evolution of Development

- Genes with similar sequences in two different species may work in different ways. (p. 501)
- Regulatory genes may turn on different sets of genes in different organisms. (p. 501)
- About two dozen conserved gene families regulate animal development. (p. 501)
- Changing the timing of gene expression, and/or the genes expressed, can result in dramatic changes in form. (p. 503)
- New functions can arise for existing structures by the recruitment of existing regulatory programs. (p. 503)
- Although gene function can be inferred from sequence data, actual gene function must be exhibited through experimentation. (p. 503)

Diversity of Eyes in the Natural World

- Natural selection can build very complicated structures via incremental improvements in function. (p. 504)
- Eye development in many different animal groups is an example of convergent evolution and represents analogous structures, but a common gene appears to initiate eye development in many of these animals. (p. 504)

Self Test

1. Humans and pufferfish have a similar number of genes, yet the human genome is approximately nine times larger than the pufferfish genome. In what form is much of this extra DNA?
   a. introns
   b. exons
   c. retrotransposons
   d. RNA
2. Genome comparisons have suggested that mouse DNA has mutated about twice as fast as human DNA. What is a possible explanation for this discrepancy?
   a. Mice are much smaller than humans.
   b. Mice live in much less sanitary conditions than humans and are therefore exposed to a wider range of mutation-causing substances.
   c. Mice have a smaller genome size.
   d. Mice have a much shorter generation time.
3. How many pairs of chromosomes do chimpanzees carry?
   a. 23
   b. 46
   c. 24
   d. 48
4. Why was the genome of the protist *P. falciparum* difficult to sequence?
   a. This protist has a large genome.
   b. This organism hides inside red blood cells, making it difficult to obtain enough DNA for the sequencing project.
   c. The apicoplast structures in this protist inhibit sequencing reactions.
   d. The genome contains a high proportion of adenine and thymine.
5. All of the following are believed to contribute to genomic diversity among various species, *except*
   a. gene duplication.
   b. gene transcription.
   c. lateral gene transfer.
   d. chromosomal rearrangements.
6. What is the fate of *most* duplicated genes?
   a. gene inactivation
   b. gain of a novel function through subsequent mutation
   c. They are transferred to a new organism using lateral gene transfer.
   d. They become orthologs.
7. Which of the following best describes pseudogenes?
   a. two functional genes within an organism that arose from the duplication of one gene
   b. genes that share the same ancestral sequence, but are found in different organisms
   c. sequences of DNA that are very similar to functional genes, but do not produce a functional product
   d. sequences of DNA that are very similar to inactive genes, but do produce a functional product
8. The *Tbx5* gene is known to play a role in which process?
   a. notochord development
   b. limb formation
   c. eye formation
   d. sexual reproduction
9. Which of the following organisms is not considered a model genetic system?
   a. mice
   b. fruit flies
   c. humans
   d. yeast
10. Which of the following statements about *Pax6* is false?
    a. *Pax6* has a similar function in mice and flies.
    b. *Pax6* is involved in eyespot formation in ribbon worms.
    c. *Pax6* is required for eye formation in *Drosophila*.
    d. *Pax6* is required for eyespot formation in planaria.

Test Your Visual Understanding

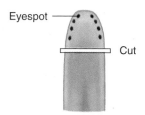

1. *Pax6* is known to play a role in the formation of eyespots during regeneration of the ribbon worm. Beginning with the removal of the head in the above diagram, outline the regeneration process and eyespot formation in a normal, wild-type ribbon worm and one that lacks *Pax6* expression. How are they different?

Apply Your Knowledge

1. How might knowledge of the *Oryza sativa* genome help combat world hunger?
2. Can a human embryo that exhibits polyploidy survive until birth? What is the difference between polyploidy and trisomy?
3. In a paper by Halder et al., 1995, the authors use the inducible GAL4-UAS expression system to artificially overexpress the *eyeless* (*Pax6*) gene in tissues where it is not normally expressed. Why does this system use the yeast transcriptional activator, GAL4, which is not normally present in *Drosophila*?

25

Systematics and the Phylogenetic Revolution

Concept Outline

25.1 Biologists name organisms in a systematic way.

The Classification of Organisms. Biologists group organisms based on shared characteristics.

25.2 Scientists construct phylogenies to understand the evolutionary relationships among species.

Systematics. Evolutionary relationships can be reconstructed.

25.3 Phylogenetics is the basis of all comparative biology.

Analogy Versus Homology. Two species may possess similar traits either because both inherited them from a common ancestor or because they have independently evolved them.

25.4 All living organisms are grouped into one of a few major categories.

The Kingdoms of Life. Organisms are grouped into three great groups called domains, and within domains into kingdoms.

Domain Archaea (Archaebacteria). The Archae are primitive prokaryotes that may live in extreme environments.

Domain Bacteria (Bacteria). Bacteria are more numerous than any other organism.

Domain Eukarya (Eukaryotes). There are four kingdoms of eukaryotes, three of them entirely or predominantly multicellular.

Viruses: A Special Case. Viruses are not organisms, and thus do not belong in any kingdom.

25.5 Molecular data are revolutionizing taxonomy.

The Impact of Molecular Cladistics. DNA sequence comparisons among organisms have led to new phylogenies.

Making Sense of the Protists. The protists are paraphyletic.

Origin of Land Plants. Molecular phylogenetics has identified the closest living relatives of land plants.

Sorting Out the Animals. Relationships among animal groups are being redefined.

FIGURE 25.1
Biological diversity. For more than 2000 years, living things have been categorized based on common characteristics. Amid the diverse life-forms in the Great Barrier Reef in Australia, different corals share a similar anatomy, pattern of development, mode of nutrition, level of organization, and biochemical composition.

Part V Diversity of Life on Earth

All organisms share many biological characteristics. They are composed of one or more cells, carry out metabolism and transfer energy with ATP, and encode hereditary information in DNA. Yet, there is also a tremendous diversity of life, ranging from bacteria and amoebas to blue whales and sequoia trees (figure 25.1). For generations, biologists have tried to group organisms based on shared characteristics. The most meaningful groupings are based on the study of evolutionary relationships among organisms. New methods for constructing evolutionary trees and a sea of molecular sequence data are leading to new evolutionary hypotheses to explain life's diversification.

25.1 Biologists name organisms in a systematic way.

The Classification of Organisms

Organisms were first classified more than 2000 years ago by the Greek philosopher Aristotle, who categorized living things as either plants or animals. The Greeks and Romans expanded this simple system and grouped animals and plants into basic units such as cats, horses, and oaks. Eventually, these units began to be called **genera** (singular, *genus*), the Latin word for "groups." Starting in the Middle Ages, these names began to be systematically written down in Latin, the language used by scholars at that time. Thus, cats were assigned to the genus *Felis*, horses to *Equus*, and oaks to *Quercus*.

Species Names

Until the mid-1700s, whenever biologists wanted to refer to a particular kind of organism, which they called a **species,** they added a series of descriptive terms to the name of the genus; this was a *polynomial*, or "many names" system. A much simpler system of naming organisms stems from the work of the Swedish biologist Carolus Linnaeus (1707–1778). In the 1750s, Linnaeus used the polynomial names *Apis pubescens, thorace subgriseo, abdomine fusco, pedibus posticis glabris utrinque margine ciliates* to denote the European honeybee. But as a kind of shorthand, he also included a two-part name for each species. For example, the honeybee became *Apis mellifera*. These two-part names, or *binomials* (*bi*, "two") have become our standard way of designating species.

Taxonomy (Greek, "arranging rules") is the science of classifying living things, and a group of organisms at a particular level in a classification system is called a **taxon** (plural, *taxa*). By agreement among taxonomists throughout the world, no two organisms can have the same name, and all names are in Latin. The scientific name of an organism is the same anywhere in the world and avoids the confusion caused by

(a)

(b)

FIGURE 25.2

Common names make poor labels. In North America, the common names "corn" (*a*) and "bear" (*b*) bring clear images to our minds (*photos on left*), but the images are very different for someone living in Europe or Australia (*photos on right*).

common names (figure 25.2). Also by agreement, the first word of the binomial name is the genus to which the organism belongs. This word is always capitalized. The second word refers to the particular species and is not capitalized. The two words together are called the species name (or scientific name) and are written in italics or distinctive

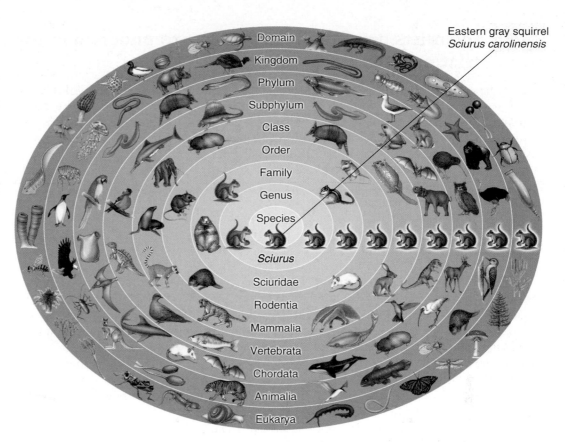

FIGURE 25.3
The hierarchical system used in classifying an organism. The organism is first recognized as a eukaryote (domain Eukarya). Within this domain, it is an animal (kingdom Animalia). Among the different phyla of animals, it is a vertebrate (phylum Chordata, subphylum Vertebrata). The organism's fur characterizes it as a mammal (class Mammalia). Within this class, it is distinguished by its gnawing teeth (order Rodentia). Next, because it has four front toes and five back toes, it is a squirrel (family Sciuridae). Within this family, it is a tree squirrel (genus *Sciurus*), with gray fur and white-tipped hairs on the tail (species *Sciurus carolinensis*, the eastern gray squirrel).

Labels on figure (outer to inner): Domain, Kingdom, Phylum, Subphylum, Class, Order, Family, Genus, Species, *Sciurus*, Sciuridae, Rodentia, Mammalia, Vertebrata, Chordata, Animalia, Eukarya

Eastern gray squirrel
Sciurus carolinensis

print—for example, *Homo sapiens*. Once a genus has been used in the body of a text, it is often abbreviated in later uses. For example, the dinosaur *Tyrannosaurus rex* becomes *T. rex*.

The Taxonomic Hierarchy

In the decades following Linnaeus, taxonomists began to group organisms into larger, more inclusive categories. Genera with similar properties were grouped into a cluster called a **family,** and similar families were placed into the same **order** (figure 25.3). Orders with common properties were placed into the same **class,** and classes with similar characteristics into the same **phylum** (plural, *phyla*). Finally, the phyla were assigned to one of several great groups, the **kingdoms.** Biologists currently recognize six kingdoms: two kinds of prokaryotes (Archaebacteria and Bacteria), a largely unicellular group of eukaryotes (Protista), and three multicellular groups (Fungi, Plantae, and Animalia).

In addition, an eighth level of classification, called a *domain*, is sometimes used. Biologists recognize three domains, which will be discussed in section 25.4. The names of the taxonomic units higher than the genus level are capitalized but *usually* not printed distinctively, italicized, or underlined.

The categories at the different levels may include many, a few, or only one taxon. For example, there is only one liv-ing genus of the family Hominidae, but several living genera of Fagaceae. To someone familiar with classification or having access to the appropriate reference books, each taxon implies both a set of characteristics and a group of organisms belonging to the taxon. For example, a honeybee has the species (level 1) name *Apis mellifera*. Its genus (level 2), *Apis*, is a member of the family Apidae (level 3). All members of this family are bees—some solitary, others living in hives as *A. mellifera* does. Knowledge of its order (level 4), Hymenoptera, tells you that *A. mellifera* is likely able to sting and may live in colonies. Its class (level 5), Insecta, indicates that *A. mellifera* has three major body segments, with wings and three pairs of legs attached to the middle segment. Its phylum (level 6), Arthropoda, tells us that the honeybee has a hard cuticle of chitin and jointed appendages. Its kingdom (level 7), Animalia, indicates that *A. mellifera* is a multicellular heterotroph whose cells lack cell walls.

By convention, the first part of a binomial species name identifies the genus to which the species belongs, and the second part distinguishes that particular species from other species in the genus. Genera are grouped into families, families into orders, orders into classes, and classes into phyla. Phyla are the basic units within kingdoms; such a system is hierarchical.

25.2 Scientists construct phylogenies to understand the evolutionary relationships among species.

One of the great challenges of modern science is to understand the history of life on earth, from the earliest single-celled organisms to the complex organisms we see around us today. Unfortunately, we cannot go back in time to retrace the course of evolution. If the fossil record were perfect, we could trace the evolutionary history of species and examine how each species arose and proliferated. However, as we discussed in chapter 22, the fossil record is far from complete. Although it answers many questions about life's diversification, it leaves many others unsettled.

Consequently, scientists must rely on other types of evidence to establish the best hypothesis of evolutionary relationships. It is important to bear in mind that the outcomes of such studies *are* hypotheses. As such, they require further testing with new data; if disproven, they lead to the formulation of new, better-established hypotheses, and the cycle begins anew.

Systematics

The reconstruction and study of evolutionary relationships is called **systematics.** By looking at the similarities and differences between species, systematics can construct an evolutionary tree, or **phylogeny,** which represents a hypothesis about which species are closely related and in what order related species evolved.

Because species evolve from other species, we would expect descendant species to be fairly similar to their ancestors. More generally, we would expect that the greater the time since two species shared a common ancestor, the more different they would be. Early systematists relied on this reasoning and constructed phylogenies based on overall similarity. If, in fact, species diverged at a constant rate, then the amount of divergence between two species would be a function of how long they had been diverging, and thus phylogenies based on degree of similarity would be accurate.

However, as chapter 21 revealed, evolution can occur very rapidly at some times and very slowly at others. Moreover, evolution is not unidirectional—sometimes species evolve in one direction and then back the other way (the phenomenon of oscillating selection). Species invading new habitats are likely to experience new selective pressures and may change greatly; those staying in the same habitats as their ancestors may change only a little. For this reason, similarity is not necessarily a good predictor of how long it has been since two species shared a common ancestor.

A second fundamental problem exists as well: Evolution is not always divergent. In chapter 22, we discussed **convergent evolution,** in which two species independently evolve to become more similar. Often, this is because species evolving the same adaptation use similar habitats. As a result, two species that are not closely related may end up being more similar to each other than they are to close relatives. Similarly, some species may reverse evolutionary course and evolve to be similar to distant ancestral species, rather than to more closely related species.

Cladistics

For these reasons, most systematists no longer construct their phylogenetic hypotheses solely on the basis of similarity. Rather, they distinguish similarity that is inherited from the common ancestor of an entire group, and is called **ancestral,** from similarity that arose within the group and thus is termed **derived.** In this method, termed **cladistics,** only shared derived characters are considered informative in determining evolutionary relationships.

To employ this method, systematists first gather data on a number of characters for all the species in the analysis. Characters can be any aspect of the phenotype, including morphology, physiology, behavior, and DNA. As chapters 17 and 24 showed, the revolution in genomics will soon provide a vast body of data that may revolutionize our ability to identify and study character variation across species.

To be useful, species must vary in *character states.* For example, consider the character "teeth" in amniote vertebrates (birds, reptiles, and mammals). This character has two states: presence in most mammals and reptiles and absence in birds and a few other groups (e.g., turtles). Once the data are assembled, the first step is to *polarize* the characters—that is, to determine whether particular character states are ancestral or derived. To polarize the character "teeth," we must determine which state was exhibited by the most recent common ancestor of this group.

Usually, we do not have a fossil that we are confident is the most recent ancestor. As a result, the method of **outgroup comparison** is used to assign character polarity. To use this method, a species or group of species that is closely related to, but not a member of, the group under study is designated as the **outgroup.** Character states exhibited by the outgroup are assumed to be ancestral, and other states are considered derived. Polarity assignments are most effective when several different outgroups are used. In the preceding example, teeth are generally present in the closest relatives of amniotes—amphibians and fish. Consequently, the presence of teeth in mammals and reptiles is considered ancestral, and their absence in birds is considered derived.

Once all characters have been polarized, systematists use this information to construct a **cladogram,** which depicts a hypothesis of evolutionary relationships. Species that share derived characters belong to a **clade.** A derived character shared by clade members is called a **synapomorphy** of that clade. Clades are thus evolutionary units and refer to all the descendants of a particular common ancestor.

Figure 25.4 illustrates that a simple cladogram is a nested set of clades, each characterized by its own synapomorphies. For example, amniotes are a clade for which the evolution of an amniotic membrane is a synapomorphy. Within that clade, mammals are a clade, with hair as a synapomorphy, and so on.

Several points need to be kept in mind when examining a cladogram. First, although the cladogram does not contain ancestral species, each node (the point where two branches of the tree come together) represents a hypothetical ancestral species. Thus, to tell how closely related two species are, we simply look to see how recently they shared a common ancestor. For example, humans and gorillas are relatively closely related because they share a recent common ancestor. By contrast, humans and sharks are not closely related because their most recent common ancestor occurs deep in the cladogram. A second consequence of this way of portraying relationships is that species that occur next to each other at the top of the cladogram are not necessarily closely related. Lizards and salamanders are right next to each other in figure 25.4, yet lizards share a more recent ancestor with—and thus are more closely related to—humans than they are to salamanders. Finally, each ancestral node has two branches, and it is arbitrary which one is put on the left and which on the right. For example, the cladogram in figure 25.4 could just as correctly have placed the branch containing the shark on the far right and the branch leading to the salamander, lizard, tiger, gorilla, and human on the left.

Ancestral states are also called **plesiomorphies,** and shared ancestral states are called **symplesiomorphies.** In contrast to synapomorphies, symplesiomorphies are not informative about phylogenetic relationships. Consider, for

Traits: Organism	Jaws	Lungs	Amniotic membrane	Hair	No tail	Bipedal
Lamprey	0	0	0	0	0	0
Shark	1	0	0	0	0	0
Salamander	1	1	0	0	0	0
Lizard	1	1	1	0	0	0
Tiger	1	1	1	1	0	0
Gorilla	1	1	1	1	1	0
Human	1	1	1	1	1	1

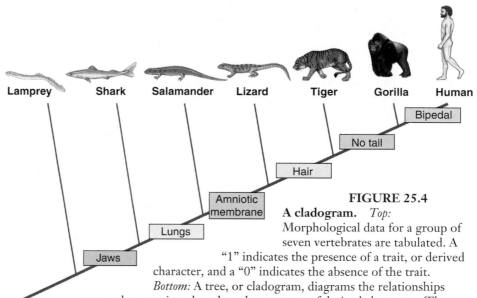

FIGURE 25.4
A cladogram. *Top:* Morphological data for a group of seven vertebrates are tabulated. A "1" indicates the presence of a trait, or derived character, and a "0" indicates the absence of the trait. *Bottom:* A tree, or cladogram, diagrams the relationships among the organisms based on the presence of derived characters. The derived characters between the cladogram branch points are shared by all organisms above the branch points and are not present in any below it. The outgroup (in this case, the lamprey) does not possess any of the derived characters.

example, the character state "presence of a tail," which is exhibited by lampreys, sharks, salamanders, lizards, and tigers. Does this mean that tigers are more closely related—and shared a more recent common ancestor with—lizards and sharks than with apes and humans, their fellow mammals? The answer, of course, is no: Because symplesiomorphies only reflect character states inherited from a distant ancestor, they do not imply that species with that state are closely related.

In real examples, however, phylogenetic studies are rarely so simple. The reason is that in some cases, shared derived characters have evolved independently and thus are false signals of close evolutionary relationship. In addition, derived characters may sometimes be lost as species within a clade re-evolve the ancestral state. Such characteristics are referred to as **homoplasies.** For example, frogs do not have a tail. Thus, absence of a tail is a synapomorphy that unites not only gorillas and humans, but also frogs. However, frogs have neither an amniotic membrane nor hair, both of which are synapomorphies for clades that contain gorillas and humans. In cases such as this, when there are conflicts among the characters, systematists rely on the **principle of parsimony,** which favors the hypothesis that requires the fewest assumptions. As a result, the phylogeny that requires the fewest evolutionary events is considered the best hypothesis of phylogenetic relationships. Thus, in the example just stated, grouping frogs with salamanders is favored because it requires only one instance of homoplasy (the multiple origins of taillessness), whereas a phylogeny in which frogs were most closely related to humans and gorillas would require two homoplastic evolutionary events (the loss of both amniotic membranes and hair in frogs).

If characters evolve from one state to another at a slow rate compared to the frequency of speciation events, the principle of parsimony works well in reconstructing evolutionary relationships because its underlying assumption—that shared derived similarity is indicative of recent common ancestry—is usually correct. In fact, this assumption seems likely to be correct for many types of data.

However, in recent years systematists have realized that some characters evolve so rapidly that the principle of parsimony may be misleading. Of particular interest is the rate at which some parts of the DNA evolve. As discussed in chapter 17, some stretches of DNA do not appear to have any function. As a result, mutations that occur in these parts of the DNA are not removed by natural selection, and thus the rate of evolution can be quite high. Moreover, because only four character states are possible for any nucleotide base (A, C, G, or T), the probability that two species may independently evolve the same derived character state is quite high. If, in fact, such homoplasy dominates the character data set, then the assumptions of the principle of parsimony are violated, and phylogenies inferred using this method are unlikely to be accurate.

For this reason, systematists in recent years have been exploring other methods based on statistical approaches, such as maximum likelihood, to infer phylogenies. These methods start with an assumption of the rate at which characters evolve and then fit the data to these models to derive the phylogeny that best accords with these assumptions. One advantage of these methods is that different assumptions of rate of evolution can be used for different characters. Thus, if some DNA characters are more constrained to evolve slowly than are other parts of the DNA, the methods can employ different models of evolution for the different character states.

In general, cladograms such as the one in figure 25.4 only indicate the order of evolutionary branching events; they do not contain information about the timing of these events. In some cases, however, branching events can be timed, either by reference to fossils, or by making assumptions about the rate at which characters change. One widely used, but controversial, method is the **molecular clock,** which states that the rate of evolution of a molecule is constant through time, such that divergence in DNA in this model can be used to calibrate when branching events occur. Although the molecular clock appears to hold in some cases, in many others the data indicate that rates of evolution are not constant enough to be used in timing evolutionary events. Systematists are still debating how often, and under what circumstances, the molecular clock holds.

Systematics and Classification

While systematics is the reconstruction and study of evolutionary relationships, *classification* refers to how we place species and higher groups in the taxonomic hierarchy. To understand why the two are not always congruent, we need to consider how species may be grouped based on their phylogenetic relationships. A **monophyletic** group includes the most recent common ancestor of the group and all of its descendants. By definition, a clade is a monophyletic group. A **paraphyletic** group includes the most recent common ancestor of the group, but not all its descendants, and a **polyphyletic** group does not include the most recent common ancestor of all members of the group (figure 25.5).

Taxonomic hierarchies are based on shared traits and ideally should reflect evolutionary relationships. Traditional taxonomic groups, however, do not always fit well with new understandings of phylogenetic relationships. Historically, birds are placed in the class Aves, whereas other dinosaurs are in the Reptilia. But recent phylogenetic advances make clear that birds evolved from dinosaurs. The last common ancestor of all birds and a dinosaur was a meat-eating dinosaur. There is no way to arrange the evolutionary tree to support separate monophyletic groups for reptiles (including dinosaurs and crocodiles, as well as lizards, snakes, and turtles) and for birds. Yet, the terms are so familiar that suddenly referring to birds as a type of dinosaur, and thus a type of reptile, is confusing to some. Biologists are currently grappling with the extent to which our classification system should be changed to accommodate new understandings of phylogenetic relationships.

> Cladistics is the most commonly used method to reconstruct evolutionary relationships. In cladistics, shared derived characters are used to construct a cladogram, which indicates phylogenetic relationships among species.

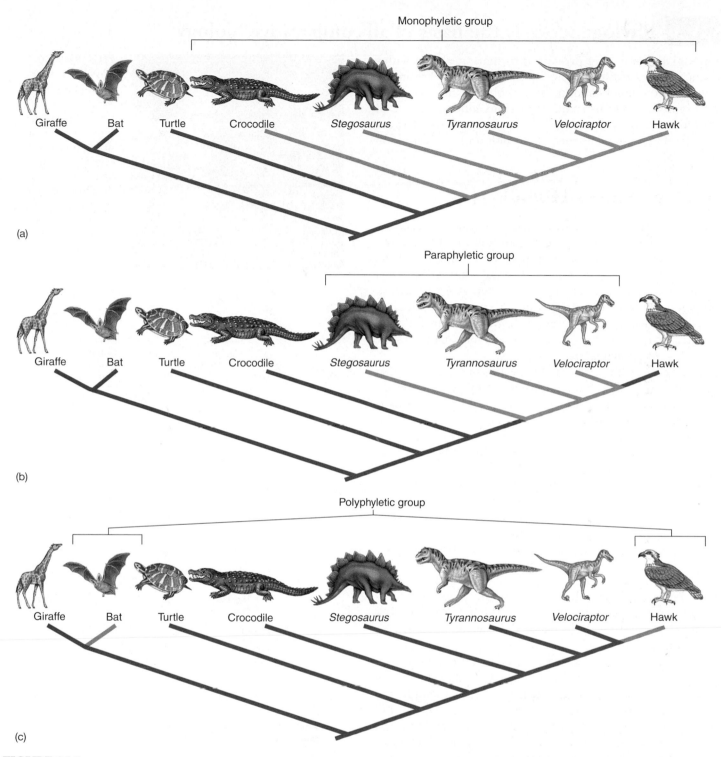

FIGURE 25.5

Monophyletic, paraphyletic, and polyphyletic groups. (*a*) A monophyletic group consists of the most recent common ancestor and all of its descendants. For example, the name "Archosaurs" is given to the monophyletic group that includes a crocodile, *Stegosaurus*, *Tyrannosaurus*, *Velociraptor*, and a hawk. (*b*) A paraphyletic group consists of the most recent common ancestor and some of its descendants. For example, some, but not all, taxonomists traditionally give the name "dinosaurs" to the paraphyletic group that includes *Stegosaurus*, *Tyrannosaurus*, and *Velociraptor*. This group is paraphyletic because one descendant of the most recent ancestor of these species, the bird, is not included in the group. Other taxonomists include birds within the Dinosauria because *Tyrannosaurus* and *Velociraptor* are more closely related to birds than to other dinosaurs. (*c*) A polyphyletic group does not contain the most recent common ancestor of the group, and taxonomists do not assign taxa to polyphyletic groups. For example, bats and birds could be classified in the same group because they have similar shapes, anatomical features, and habitats. However, their similarities reflect convergent evolution, not common ancestry.

25.3 Phylogenetics is the basis of all comparative biology.

Phylogenies not only provide information about evolutionary relationships among species, but also are indispensable for understanding how evolution has occurred. By examining the distribution of traits among species in the context of the phylogenetic relationships of these species, much can be learned about how and why evolution proceeds.

Analogy Versus Homology

In chapter 22, we pointed out that homologous structures are those that are derived from the same body part in a common ancestor. Thus, the flipper of a dolphin and the leg of a horse are homologous because they are derived from the same bone in an ancestral vertebrate. By contrast, the wings of birds and dragonflies are analogous, because they are derived from different ancestral structures. Phylogenetic analysis can help determine whether structures are homologous or analogous.

For example, recent fossil discoveries have revealed that many species of dinosaurs exhibited parental care. They incubated eggs laid in nests and took care of growing baby dinosaurs, many of which could not have fended for themselves; in fact, some recent fossils show dinosaurs sitting on a nest in exactly the same posture used by birds today (figure 25.6)! Initially, these discoveries were treated as remarkable and unexpected—dinosaurs apparently had independently evolved behaviors similar to those of modern-day organisms. However, examination of the phylogenetic position of dinosaurs indicates that they are most closely related to two living groups of animals—crocodiles and birds—both of which exhibit parental care (see figure 25.5). Thus, it would appear likely that the similar parental care exhibited by crocodiles, dinosaurs, and birds did not evolve convergently; rather, the behaviors are homologous, inherited by each of these groups from their common ancestor.

In other cases, by contrast, phylogenetic analysis can indicate that similar traits have evolved independently in different clades and thus are analogous. For example, the fossil record reveals that extremely elongated canines (saber teeth) occurred in a number of different groups of extinct carnivorous mammals. Examination of saber-toothed-ness in a phylogenetic context reveals that it most likely evolved independently at least three times (figure 25.7). Although how these teeth were actually used is still debated, all saber-toothed carnivores had body proportions similar to those of cats, which suggests that these different types of carnivores all evolved a similar predatory lifestyle.

FIGURE 25.6
Fossil dinosaur incubating its eggs. This remarkable fossil of *Oviraptor* shows the dinosaur sitting on its nest of eggs just as chickens do today. Not only is the dinosaur squatting on the nest, but its forelimbs are outstretched, perhaps to shade the eggs.

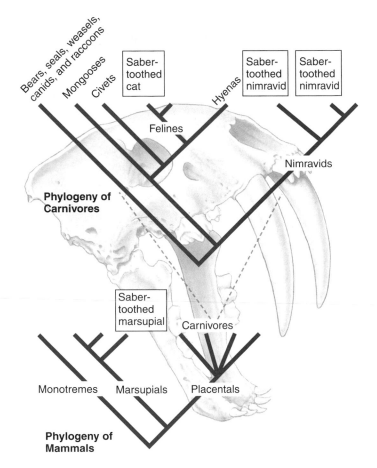

FIGURE 25.7
Distribution of saber-toothed mammals. Saber teeth have evolved at least three times in mammals—once within marsupials, once in felines, and at least once in a now-extinct group of catlike carnivores called nimravids. It is possible that the condition evolved twice in nimravids, but another, less likely, possibility is that saber teeth evolved only once and were subsequently lost in other taxa. (Not all of the branches within marsupials and placentals are shown).

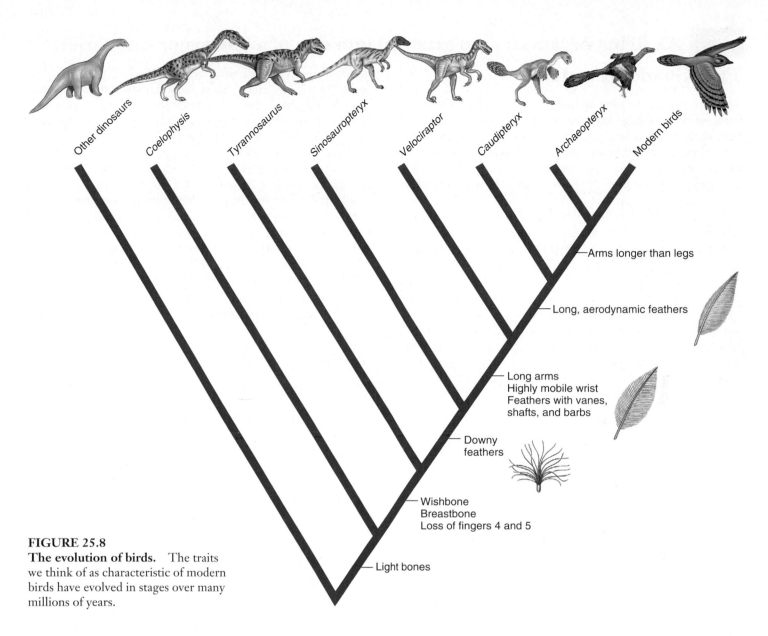

FIGURE 25.8
The evolution of birds. The traits we think of as characteristic of modern birds have evolved in stages over many millions of years.

Labels on cladogram (top to bottom): Arms longer than legs; Long, aerodynamic feathers; Long arms / Highly mobile wrist / Feathers with vanes, shafts, and barbs; Downy feathers; Wishbone / Breastbone / Loss of fingers 4 and 5; Light bones

Taxa: Other dinosaurs, Coelophysis, Tyrannosaurus, Sinosauropteryx, Velociraptor, Caudipteryx, Archaeopteryx, Modern birds

Evolution of Complex Characters

Most complex characters do not evolve, fully formed, in one step. Rather, they are often built up, step-by-step, in a series of evolutionary transitions. Phylogenetic analysis can help discover these evolutionary sequences. Modern-day birds—with their wings, feathers, light bones, and breastbone—appear to be exquisitely adapted flying machines. Fossil discoveries in recent years now allow us to reconstruct the evolution of these features. When the fossils are arranged phylogenetically, it becomes clear that the features characterizing living birds did not appear simultaneously. Figure 25.8 shows how the features important to flight evolved sequentially, probably over a long period of time, in the ancestors of modern birds.

One important finding often revealed by studies of the evolution of complex characters is that the initial stages of a character evolved as an adaptation to some environmental selective pressure other than that for which the character is currently adapted. Examination of figure 25.8 reveals that the first feathery structures evolved deep in theropod phylogeny, in animals whose forearms clearly were not modified for flight. Thus, these structures must have evolved for some other purpose, perhaps to serve as insulation. Through time, these structures were modified, to the extent that modern feathers are well designed to maximize aerodynamic performance.

Examination of the characters on a cladogram can provide great insight on how they evolved, including whether they have evolved multiple times and how complex characters were constructed evolutionarily.

25.4 All living organisms are grouped into one of a few major categories.

The Kingdoms of Life

The earliest classification systems recognized only two kingdoms of living things: animals and plants. But as biologists discovered microorganisms and learned more about other organisms, they added kingdoms in recognition of certain fundamental differences discovered among organisms. Most biologists now use a six-kingdom system first proposed by Carl Woese of the University of Illinois (figure 25.9*a*).

In this system, four kingdoms consist of eukaryotic organisms. The two most familiar kingdoms, **Animalia** and **Plantae,** contain only organisms that are multicellular during most of their life cycle. The kingdom **Fungi** contains multicellular forms and single-celled yeasts, which are thought to have multicellular ancestors. Fundamental differences divide these three kingdoms. Plants are mainly stationary, but some have motile sperm; most fungi lack motile cells; animals are mainly motile. Animals ingest their food, plants manufacture it, and fungi digest it by means of secreted extracellular enzymes. Each of these kingdoms probably evolved from a different single-celled ancestor.

The large number of eukaryotes that do not fit in any of the three eukaryotic kingdoms are arbitrarily grouped into a single kingdom called **Protista** (see chapter 28). Most protists are unicellular or, in the case of some algae, have a unicellular phase in their life cycle. This kingdom reflects the current controversy between taxonomic and phylogenetic approaches. Growing knowledge about evolutionary relationships challenges classification systems, even at as high a level of organization as the kingdom. Protists are polyphyletic, meaning that the group does not contain the most recent ancestor of all members of the group. Another way to look at this is that there are several protist lineages with distinct evolutionary origins. These groups are not directly related to each other. Some green algae, the streptophytes (formerly the charophyceans), are more closely related to members of the kingdom Plantae than to most other protist lineages. Most systematists concur that the kingdom Plantae should be extended to include the green algae. To accomodate this new phylogentic understanding, Kingdom Plantae has been renamed the green plant kingdom, or the Virdiplantae.

The remaining two kingdoms, **Archaebacteria** and **Bacteria,** consist of prokaryotic organisms, which are vastly different from all other living things (see chapter 27). Archaebacteria are a diverse group that includes the methanogens and extreme thermophiles, and its members differ from the other prokaryotes, Bacteria. Archaebacteria are not necessarily more ancient than bacteria, as their name erroneously implies. Initially thought to inhabit only extreme environments, archaebacteria have since been found in more diverse habitats.

Domains

As biologists have learned more about the archaebacteria, it has become increasingly clear that this group is very different from all other organisms. When the full genomic DNA sequences of an archaebacterium and a bacterium were first compared in 1996, the differences proved striking. Archaebacteria are as different from bacteria as bacteria are from eukaryotes. Recognizing this, biologists are increasingly adopting a classification of living organisms that recognizes three **domains,** a taxonomic level higher than kingdom (figure 25.9*b*). Archaebacteria are in one domain, bacteria in a second, and eukaryotes in the third.

Living organisms are grouped into three general categories called domains. One of the domains, the eukaryotes, is subdivided into four kingdoms: protists, fungi, plants, and animals.

(a) A six-kingdom system

| Bacteria | Archaebacteria | Protista | Fungi | Plantae (virdiplantae) | Animalia |

(b) A three-domain system

| Bacteria | Archaea | | | Eukarya | |

FIGURE 25.9
Different approaches to classifying living organisms. (*a*) Bacteria and Archaebacteria are so distinct that they have been assigned to separate kingdoms. (*b*) Even at the domain level, these two groups of prokaryotes are distinct.

Domain Archaea (Archaebacteria)

The archaebacteria (Greek *archaio*, "ancient") seem to have diverged very early from the bacteria and are more closely related to eukaryotes than to bacteria (figure 25.10). This conclusion comes largely from comparisons of genes that encode ribosomal RNAs. The last several years have seen an explosion of DNA sequence information from microorganisms, information that paints a more complex picture. It had been thought that by sequencing numerous microbes we could eventually come up with an accurate picture of the phylogeny of the earliest organisms on earth. The new whole-genome DNA sequence data described in chapter 24 tells us that it will not be that simple. Comparing whole-genome sequences leads evolutionary biologists to a variety of phylogenetic trees, some of which contradict each other. It appears that, during their early evolution, microorganisms have swapped genetic information (lateral gene transfer), making constructing phylogenetic trees very difficult.

As an example of the problem, we can look at *Thermotoga*, a thermophile found on Volcano Island off the coast of Italy. The sequence of one of its RNAs places it squarely within the bacteria near an ancient microbe called *Aquifex*. Recent DNA sequencing, however, fails to support any consistent relationship between the two microbes. There is disagreement as to the effect of lateral gene transfer on the ability of evolutionary biologists to provide accurate phylogenies from molecular data. For now, we will provisionally accept the tree presented in figure 25.10. Over the next few years, we can expect to see considerable change in accepted viewpoints as more and more data are brought to bear.

Although they are a diverse group, all archaebacteria share certain key characteristics (see table 25.1). Their cell walls lack peptidoglycan (an important component of the cell walls of bacteria); the lipids in the cell membranes of archaebacteria have a different structure from those in all other organisms; and archaebacteria have distinctive ribosomal RNA sequences. Some of their genes possess introns, unlike those of bacteria.

The archaebacteria are grouped into three general categories—methanogens, extremophiles, and nonextreme archaebacteria—based primarily on the environments in which they live or their specialized metabolic pathways.

Methanogens obtain their energy by using hydrogen gas (H_2) to reduce carbon dioxide (CO_2) to methane gas (CH_4). They are strict anaerobes, poisoned by even traces of oxygen. They live in swamps, marshes, and the intestines of mammals. Methanogens release about 2 billion tons of methane gas into the atmosphere each year.

Extremophiles are able to grow under conditions that seem extreme to us. There are several types of extremophiles:

Thermophiles ("heat lovers") live in very hot places, typically in temperatures ranging from 60° to 80°C. Many thermophiles are autotrophs, and their metabolisms are

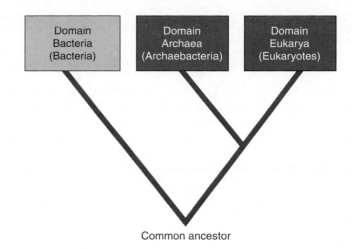

FIGURE 25.10
An evolutionary relationship among the three domains. Members of the domain Bacteria are thought to have diverged early from the evolutionary line that gave rise to the archaebacteria and eukaryotes.

based on sulfur. Some thermophilic archaebacteria form the basis of food webs around deep-sea thermal vents, where they must withstand extreme temperatures and pressures. Other types, such as *Sulfolobus*, inhabit the hot sulfur springs of Yellowstone National Park at 70° to 75°C. The recently described *Pyrolobus fumarii* holds the current record for heat stability, with a 106°C temperature optimum and 113°C maximum. It is so heat-tolerant that it is not killed by a one-hour treatment in an autoclave (121°C)!

Halophiles ("salt lovers") live in very salty places, including the Great Salt Lake in Utah, Mono Lake in California, and the Dead Sea in Israel. Whereas the salinity of seawater is around 3%, these archaebacteria thrive in, and indeed require, water with a salinity of 15 to 20%.

pH-tolerant archaebacteria grow in highly acidic (pH = 0.7) and very basic (pH = 11) environments.

Pressure-tolerant archaebacteria, isolated from ocean depths, require at least 300 atmospheres of pressure to survive; they tolerate up to 800 atmospheres!

Nonextreme archaebacteria grow in the same environments bacteria do. As the genomes of archaebacteria have become better known, microbiologists have been able to identify **signature sequences** of DNA present only in archaebacteria. The newly discovered microbe *Nanoarchaeum equitens* was identified as an archaebacterium based on a signature sequence. This odd Icelandic microbe may have the smallest known genome, only 500 base-pairs.

Archaebacteria are poorly understood prokaryotes that inhabit diverse environments, some of them extreme.

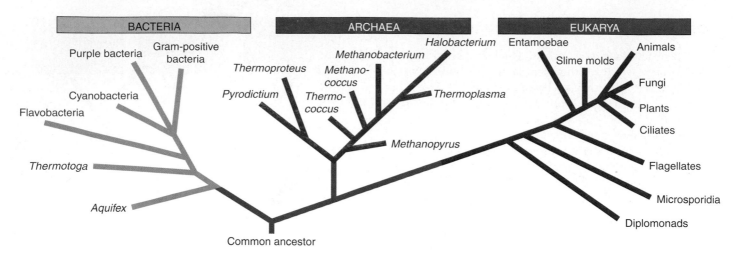

FIGURE 25.11

A tree of life. This phylogeny, prepared from rRNA analyses, shows the evolutionary relationships among the three domains. The base of the tree was determined by examining genes that are duplicated in all three domains, the duplication presumably having occurred in the common ancestor. Archaebacteria and eukaryotes diverged later than bacteria, and are more closely related to each other than either is to bacteria. Bases of trees constructed with other traits are often less clear because of lateral gene transfer (see chapter 24).

Domain Bacteria (Bacteria)

The bacteria are the most abundant organisms on earth. There are more living bacteria in your mouth than there are mammals living on earth. Although too tiny to see with the unaided eye, bacteria play critical roles throughout the biosphere. They extract from the air all the nitrogen used by organisms, and play key roles in cycling carbon and sulfur. Much of the world's photosynthesis is carried out by bacteria. However, certain groups of bacteria are also responsible for many forms of disease. Understanding their metabolism and genetics is a critical part of modern medicine.

There are many different kinds of bacteria, and the evolutionary links between them are not well understood. While taxonomists disagree about the details of bacterial classification, most recognize 12 to 15 major groups of bacteria. Comparisons of the nucleotide sequences of ribosomal RNA (rRNA) molecules are beginning to reveal how these groups are related to one another and to the other two domains. One view of our current understanding of the "tree of life" is presented in figure 25.11. The oldest divergences represent the deepest-rooted branches in the tree. The archaebacteria and eukaryotes are more closely related to each other than to bacteria and are on a separate evolutionary branch of the tree, even though archaebacteria and bacteria are both prokaryotes.

Bacteria are as different from archaebacteria as they are from eukaryotes.

Table 25.1 Features of the Domains of Life

| Feature | Domain | | |
	Archaea	Bacteria	Eukarya
Amino acid that initiates protein synthesis	Methionine	Formyl-methionine	Methionine
Introns	Present in some genes	Absent	Present
Membrane-bounded organelles	Absent	Absent	Present
Membrane lipid structure	Branched	Unbranched	Unbranched
Nuclear envelope	Absent	Absent	Present
Number of different RNA polymerases	Several	One	Several
Peptidoglycan in cell wall	Absent	Present	Absent
Response to the antibiotics streptomycin and chloramphenicol	Growth not inhibited	Growth inhibited	Growth not inhibited

Domain Eukarya (Eukaryotes)

For at least 1 billion years, prokaryotes ruled the earth. No other organisms existed to eat them or compete with them, and their tiny cells formed the world's oldest fossils. Members of the third great domain of life, the eukaryotes, appear in the fossil record much later, only about 2.5 billion years ago. However, despite the metabolic similarity of eukaryotic cells to prokaryotic cells, their structure and function enabled cells to be larger, and eventually, allowed multicellular life to evolve.

Four Kingdoms of Eukaryotes

The first eukaryotes were unicellular organisms. A wide variety of unicellular eukaryotes exist today, grouped together in the kingdom Protista (along with some multicellular descendants) on the basis that they do not fit into any of the other three kingdoms of eukaryotes. Protists vary from the relatively simple, single-celled amoeba to multicellular organisms such as kelp, which can be 20 meters long.

Fungi, plants, and animals are largely multicellular kingdoms, each a distinct evolutionary line from a single-celled ancestor that would be classified in the kingdom Protista. Because of the size and ecological dominance of plants, animals, and fungi, and because they are predominantly multicellular, we recognize them as kingdoms distinct from Protista, even though the amount of diversity among the protists is much greater than that within or between the fungi, plants, and animals.

Endosymbiosis and the Origin of Eukaryotes

The hallmark of eukaryotes is complex cellular organization, highlighted by an extensive endomembrane system that subdivides the eukaryotic cell into functional compartments. Not all cellular compartments, however, are derived from the endomembrane system. With few exceptions, modern eukaryotic cells possess energy-producing organelles called mitochondria, and some eukaryotic cells possess chloroplasts, which are energy-harvesting organelles. Mitochondria and chloroplasts are both believed to have entered early eukaryotic cells by a process called **endosymbiosis.** We discussed the theory of the endosymbiotic origin of mitochondria and chloroplasts in chapters 4 and 5; their origin is also diagrammed in figure 25.12. Both organelles contain their own ribosomes, which are more similar to bacterial ribosomes than to eukaryotic cytoplasmic ribosomes. They manufacture their own inner membranes. They divide independently of the cell and contain chromosomes similar to those in bacteria. Comparison of the nucleotide sequence of this DNA with that of a variety of organisms indicates clearly that mitochondria are the descendants of purple nonsulfur bacteria that were incorporated into eukaryotic cells early in the history of the group, while chloroplasts are derived from cyanobacteria.

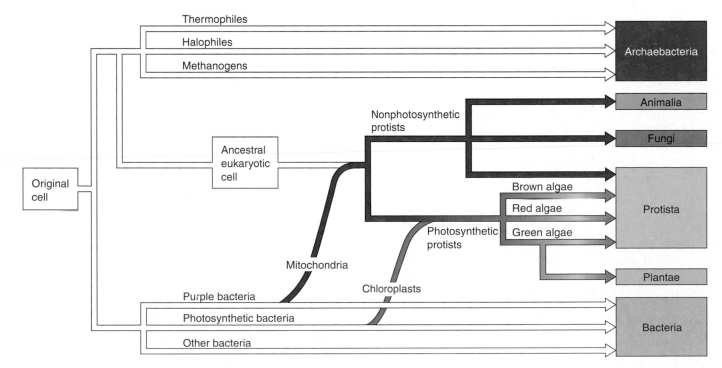

FIGURE 25.12
Diagram of the evolutionary relationships among the six kingdoms of organisms. The colored lines indicate symbiotic events.

Key Characteristics of Eukaryotes

Although eukaryotic organisms are extraordinarily diverse, they share three characteristics that distinguish them from prokaryotes: compartmentalization, multicellularity in many, but not all, eukaryotes, and sexuality.

Compartmentalization. Discrete compartments provide evolutionary opportunities for increased specialization within the cell, as we see with chloroplasts and mitochondria. The evolution of a nuclear membrane also accounts for increased complexity in eukaryotes. For example, a discrete nucleus led to new ways to control gene expression. In prokaryotes, both transcription and translation occur at the same time, and often a protein is being translated from a strand of RNA that is still being transcribed from its template DNA (see figure 15.13). In eukaryotes, RNA transcripts are processed and transported across the nuclear membrane into the cytosol, where ribosomes and other translational machinery are present to initiate the translation of the RNA message into protein. The physical separation of transcription and translation in eukaryotes adds additional levels of control to the process of gene expression.

Multicellularity. The unicellular body plan has been tremendously successful, with unicellular prokaryotes and eukaryotes constituting about half of the biomass on earth. Yet, a single cell has limits. The evolution of multicellularity allowed organisms to deal with their environments in novel ways. Distinct types of cells, tissues, and organs can be differentiated within the complex bodies of multicellular organisms. With such a functional division within its body, a multicellular organism can do many things, including protect itself, resist drought efficiently, regulate its internal conditions, move about, seek mates and prey, and carry out other activities on a scale and with a complexity that would be impossible for its unicellular ancestors. With all these advantages, it is not surprising that multicellularity has arisen independently so many times.

True multicellularity, in which the activities of individual cells are coordinated and the cells themselves are in contact, occurs only in eukaryotes and is one of their major characteristics. The cell walls of bacteria occasionally adhere to one another, and bacterial cells may also be held together within a common sheath. Some bacteria form filaments, sheets, or three-dimensional aggregates, but the individual cells remain independent of each other, reproducing and carrying on their metabolic functions without coordinating with the other cells. Such bacteria are considered colonial, but none are truly multicellular. Many protists also form similar colonial aggregates of many cells with little differentiation or integration.

Other protists—the red, brown, and green algae, for example—have independently attained multicellularity.

One lineage of multicellular green algae was the ancestor of the plants (see chapters 28 and 29), and most taxonomists now place its members in the green plant kingdom, (called the Virdiplantae). Historically, the plant kingdom included only multicellular land plants, a group that arose from a single ancestor in terrestrial habitats and that has a unique set of characteristics. Aquatic plants are recent derivatives of land plants.

The multiple origins of multicellularity are also seen in the fungi and animals that arose from unicellular protist ancestors with different characteristics. As we will see in subsequent chapters, the groups that seem to have given rise to each of these kingdoms are still in existence.

Sexuality. Another major characteristic of eukaryotic organisms as a group is sexuality. Although some interchange of genetic material occurs in bacteria (see chapter 27), it is certainly not a regular, predictable mechanism in the same sense that sex is in eukaryotes. The sexual cycle characteristic of eukaryotes alternates between **syngamy,** the union of male and female gametes producing a cell with two sets of chromosomes, and **meiosis,** cell division producing daughter cells with one set of chromosomes. This cycle differs sharply from any exchange of genetic material occurring in prokaryotes. The evolution of meiosis correlates with a huge increase in the number of species on earth.

As we have seen, in diploid cells, one set of chromosomes comes from the male parent and one from the female parent. These chromosomes segregate during meiosis. Because crossing over frequently occurs during meiosis (see chapter 12), no two products of a single meiotic event are ever identical. As a result, the offspring of sexual, eukaryotic organisms vary widely, thus providing the raw material for evolution. Sexual reproduction, with its regular alternation between syngamy and meiosis, produces genetic variation. Sexual organisms can adapt to the demands of their environments because they produce a variety of progeny.

In many of the unicellular phyla of protists, sexual reproduction occurs only occasionally. The first eukaryotes were probably haploid. Diploids seem to have arisen on a number of separate occasions by the fusion of haploid cells, which then eventually divided by meiosis.

The characteristics of the six kingdoms are outlined in table 25.2; note that the archaebacteria and bacteria are grouped in the same column.

Eukaryotic cells are highly compartmentalized, and they acquired mitochondria and chloroplasts by endosymbiosis. The complex differentiation we associate with many life-forms depends on multicellularity and sexuality, which must have been highly advantageous to have evolved independently so often.

Table 25.2 Characteristics of the Six Kingdoms

	Archaebacteria and Bacteria	Protista	Plantae	Fungi	Animalia
Cell Type	Prokaryotic	Eukaryotic	Eukaryotic	Eukaryotic	Eukaryotic
Nuclear Envelope	Absent	Present	Present	Present	Present
Transcription and Translation	Occur in same compartment	Occur in different compartments	Occur in different compartments	Occur in different compartments	Occur in different compartments
Histone Proteins Associated with DNA	Absent	Present	Present	Present	Present
Cytoskeleton	Absent	Present	Present	Present	Present
Mitochondria	Absent	Present (or absent)	Present	Present	Present
Chloroplasts	None (photosynthetic membranes in some types)	Present (some forms)	Present	Absent	Absent
Cell Wall	Noncellulose (polysaccharide plus amino acids)	Present in some forms, various types	Cellulose and other polysaccharides	Chitin and other noncellulose polysaccharides	Absent
Means of Genetic Recombination, If Present	Conjugation, transduction, transformation	Fertilization and meiosis	Fertilization and meiosis	Fertilization and meiosis	Fertilization and meiosis
Mode of Nutrition	Autotrophic (chemosynthetic, photosynthetic) or heterotrophic	Photosynthetic or heterotrophic, or combination of both	Photosynthetic, chlorophylls *a* and *b*	Absorption	Ingestion
Motility	Bacterial flagella, gliding or nonmotile	9 + 2 cilia and flagella; amoeboid, contractile fibrils	None in most forms; 9 + 2 cilia and flagella in gametes of some forms	Both motile and nonmotile	9 + 2 cilia and flagella, contractile fibrils
Multicellularity	Absent	Absent in most forms	Present in all forms	Present in most forms	Present in all forms
Nervous System	None	Primitive mechanisms for conducting stimuli in some forms	A few have primitive mechanisms for conducting stimuli	None	Present (except sponges), often complex

Viruses: A Special Case

Viruses possess only a portion of the properties of organisms. **Viruses** are literally "parasitic" chemicals, segments of DNA or RNA wrapped in a protein coat. They cannot reproduce on their own, and for this reason they are not considered alive by biologists. They can, however, reproduce within cells, often with disastrous results to the host organism. Earlier theories that viruses represent a kind of halfway point between life and nonlife have largely been abandoned. Instead, viruses are now viewed as detached fragments of the genomes of organisms due to the high degree of similarity found among some viral and eukaryotic genes.

Viruses thus present a special classification problem. Because they are not organisms, we cannot logically place them in any of the kingdoms. Viruses are really just complicated associations of molecules, bits of nucleic acids usually surrounded by a protein coat. But, despite their simplicity, viruses are able to invade cells and direct the genetic machinery of these cells to manufacture more of the molecules that make up the virus. Viruses can infect organisms at all taxonomic levels.

Viruses vary greatly in appearance and size. The smallest are only about 17 nanometers in diameter, and the largest are up to 1000 nanometers (1 micrometer) in their greatest dimension (figure 25.13). The largest viruses are barely visible with a light microscope, but viral morphology is best revealed using the electron microscope. Viruses are so small that they are comparable to molecules in size; a hydrogen atom is about 0.1 nanometer in diameter, and a large protein molecule is several hundred nanometers in its greatest dimension.

Biologists first began to suspect the existence of viruses near the end of the nineteenth century. European scientists attempting to isolate the infectious agent responsible for hoof-and-mouth disease in cattle concluded that it was smaller than a bacterium. Investigating the agent further, the scientists found that it could not multiply in solution—it could only reproduce itself within living host cells that it infected. The infecting agents were called viruses.

The true nature of viruses was discovered in 1933, when the biologist Wendell Stanley prepared an extract of a plant virus called tobacco mosaic virus (TMV) and attempted to purify it. To his great surprise, the purified TMV preparation precipitated (that is, separated from solution) in the form of crystals. This was surprising because precipitation is something that only chemicals do—the TMV virus was acting like a chemical off the shelf rather than like an organism. Stanley concluded that TMV is best regarded as just that—chemical matter rather than a living organism.

Within a few years, scientists disassembled the TMV virus and found that Stanley was right. TMV was not cellular but chemical. Each particle of TMV virus is in fact a mixture of two chemicals: RNA and protein. The TMV virus has the structure of a twinkie, a tube made of an RNA core surrounded by a coat of protein. Later workers were able to separate the RNA from the protein and purify and store each chemical. Then, when they reassembled the two components, the reconstructed TMV particles were fully able to infect healthy tobacco plants, so clearly the particles were the virus itself, not merely chemicals derived from it. Further experiments carried out on other viruses yielded similar results.

> **Viruses are chemical assemblies that can infect cells and replicate within them. They are not organisms, and are not classified in the kingdoms of life.**

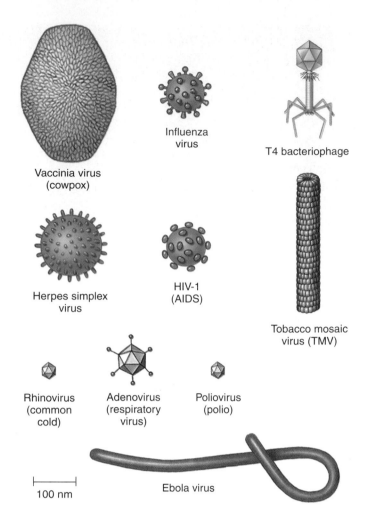

FIGURE 25.13
Viral diversity. Viruses exhibit extensive diversity in shape and size. At the scale these sample viruses are shown, a human hair would be nearly 8 meters thick.

25.5 Molecular data are revolutionizing taxonomy.

The Impact of Molecular Cladistics

In reading this chapter and chapter 24, you may have sensed some tension between traditional classification systems and systems based on evolutionary relationships at the molecular as well as the cellular and organismal levels. Traditional classification systems group organisms based on morphological similarities that seem the most important, but there is no clear basis for determining which traits matter most. As a result of this taxonomic tradition, some species have widely used scientific names that do not reflect accurate evolutionary relationships. In chapter 24, you saw how molecular data are being added to character sets and used in phylogenetic analyses aimed at identifying common ancestry. As more is learned about organisms, taxonomic groupings are changing rapidly.

Systematic phylogeneticists use cladistics to build taxonomic hierarchies. Each clade has a single common ancestor. A taxonomic hierarchy arises as systematists nest cladograms together into a phylogeny. It is important to remember that, even with exponentially increasing molecular sequence data, these phylogenies are hypotheses. Cladistic approaches emphasize the order of evolutionary events, but do not provide information about how highly diverged two groups are or are not. Lively debate about classification continues and is fueled not only by DNA sequences, but by efforts to integrate fossil data into phylogenies. In this chapter, we highlight a few of the major changes in classification. As you proceed with the subsequent chapters, you will gain an appreciation for both the diversity of living organisms and the challenges in understanding their evolutionary relationships.

Traditional classification systems are based on similar traits, but do not take into account evolutionary relationships. Systematic phylogenetics is based on evolutionary relationships established using cladistics. Traits can be morphological or molecular.

Making Sense of the Protists

The weakest part of the six-kingdom classification system shown in figure 25.9 is the kingdom Protista. Eukaryotes diverged rapidly in a world that was shifting from anaerobic to aerobic conditions. We may never be able to completely sort out the relationships among different lineages during this major evolutionary transition. However, molecular systematics clearly shows that the protists are paraphyletic (figure 25.14). While we continue to use the term protist as a catchall for any eukaryote that is not a plant, fungus, or animal, this grouping is not based on evolutionary relationships.

What to do with the 200,000 protists? The six main branchings of protists, shown at the base of figure 25.14, represent a current working hypothesis, although at least 60 protists do not seem to fit in any of the six groups. Choanoflagellates are most closely related to sponges, and indeed to all animals. The green algae can be split into two monophyletic groups, one of which gave rise to land plants (the evolution of land plants is explored in more detail on the next page). Many systematists are calling for the new kingdom called Virdiplantae, or the green plant kingdom, includes all the green algae (not the red or brown algae) and the land plants. Thus the definition of a plant has been expanded beyond those species that made it onto land. Although the kingdom Protista is in ruins, our understanding of the evolutionary relationships among these early eukaryotes is growing exponentially.

Molecular systematics and cladistics have led to a new understanding of the relationships among organisms formerly classified as members of kingdom Protista.

FIGURE 25.14
The fall of kingdom Protista.
Systematists have shown that protists as a group are not monophyletic. Note how some lineages are actually more closely related to plants or animals than they are to other protists.

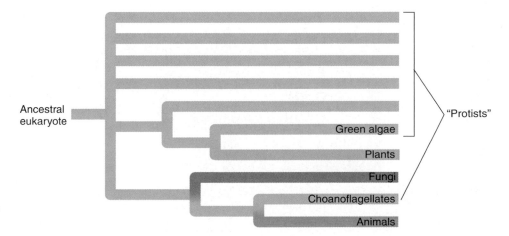

Ancestral eukaryote / Green algae / Plants / Fungi / Choanoflagellates / Animals / "Protists"

Origin of Land Plants

The origin of land plants from a green algal ancestor has long been recognized as a major evolutionary event. The phylogenetic relationships among the algae and the first land plants have been fuzzy and subject to long debate. Cell biology, biochemistry, and molecular systematics have provided surprising answers. The green algae consist of two monophyletic groups, the Chlorophyta and the Streptophyta. The land plants are actually members of the Streptophyta, rather than a separate kingdom. Within the Linnaean hierarchy, this means that new phylogenetic information demoted the land plants from a kingdom to a branch within the algal group Streptophyta. The Streptophyta, along with the sister green algal clade Chlorophyta, are now considered by most to make up the kingdom Virdiplantae, mentioned previously (figure 25.15).

In the past decade, new hypotheses about relationships within the Streptophyta have emerged, with the current phylogeny composed of seven clades as shown in figure 25.15. What was the earliest streptophyte? Conflicting answers have been obtained with different phylogenetic analyses, but growing evidence supports the hypothesis that the scaly, unicellular flagellate *Mesostigma* (order Mesostigmatales) represents the earliest streptophyte branch.

Which of the Streptophyta clades contains the closest living relative of land plants? The two contenders have been the Charales, with about 300 species, and the Coleochaetales, with about 30 species. Both lineages are freshwater algae, but the Charales are huge compared to the microscopic Coleochaetales. At the moment, the Charales appear to be the sister clade to land plants, with the Coleochaetales the next closest relatives. Charales fossils dating back 420 million years indicate that the common ancestor of land plants was a relatively complex freshwater alga.

Molecular phylogenetics has changed the way algae and plants are classified. The challenges to existing taxonomies have been sweeping—from kingdom-level changes to a reexamination of the relationships among orders.

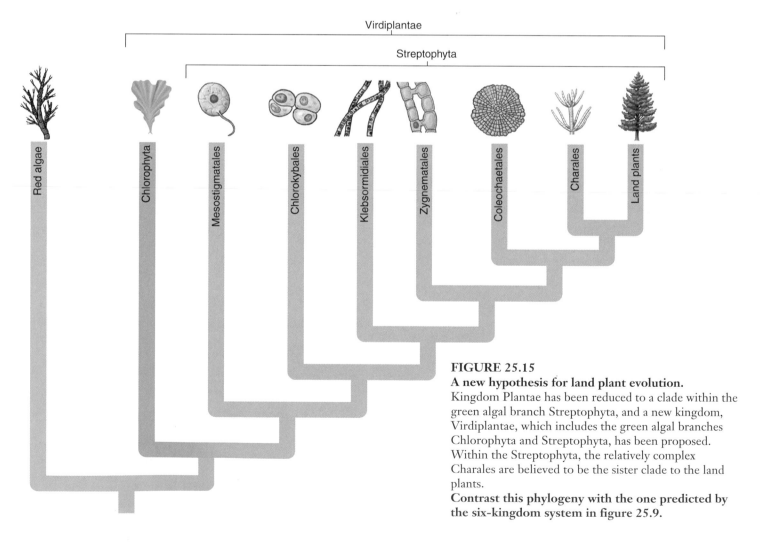

FIGURE 25.15
A new hypothesis for land plant evolution.
Kingdom Plantae has been reduced to a clade within the green algal branch Streptophyta, and a new kingdom, Virdiplantae, which includes the green algal branches Chlorophyta and Streptophyta, has been proposed. Within the Streptophyta, the relatively complex Charales are believed to be the sister clade to the land plants.
Contrast this phylogeny with the one predicted by the six-kingdom system in figure 25.9.

Sorting Out the Animals

Molecular systematics is leading to a revision of evolutionary history in all kingdoms, including the animals. Some phylogenies are changing, and others, including mammalian phylogenies, are actually being written for the first time. Next, we will explore three examples: the relationship between annelids and arthropods, relationships within the arthropods, and the discovery of phylogenetic relationships among mammals.

The Problem Posed by Segmentation

Morphological traits such as segmentation have been used to group arthropods and annelids together, but comparisons of rRNA sequences are raising questions about how closely the segmented worms (annelids) are related to arthropods, which include insects, spiders, and crustaceans (e.g., lobsters). As rRNA sequences are obtained, it is becoming increasingly clear that annelids and arthropods are more distantly related than we previously believed.

Chordates, including humans, also have segmented body plans that evolved independently of segmentation in annelids and arthropods. The vertebrae in your backbone reflect segmentation. However, developmental differences have long been used to separate chordates from the annelids and arthropods. For example, annelids and arthropods are called *protostomes* because the mouth forms before the anus, while chordates are grouped with the *deuterostomes*, in which the mouth develops second. Until molecular data became available, segmentation was thought to have evolved twice, once in the protostomes and once in the deuterostomes.

With the addition of molecular traits, annelids and arthropods fell into two distinct protostome branches (figure 25.16): lophotrochozoans and ecdysozoans. These two branches have been evolving independently since ancient times. Lophotrochozoans include flatworms, mollusks, and annelids. Two ecdysozoan phyla have been particularly successful, roundworms (nematodes) and arthropods.

In the new protostome phylogeny, annelids and arthropods do not constitute a monophyletic group. This means that segmentation arose twice, not once, in the protostomes and then once again in the chordates (figure 25.16). How did this happen? The most likely explanation is that members of the same family of genes were co-opted at least three times to create segmented body plans. Segmentation is regulated by the *Hox* gene family that contains a homeodomain region (review chapter 19). The *Hox* ancestral genes predate the ecdysozoans and lophotrochozoans. The ancestor of the lophotrochozoans, ecdysozoans, and deuterostomes most likely already had seven *Hox* genes. Some of these genes appear to have evolved a role in segmentation.

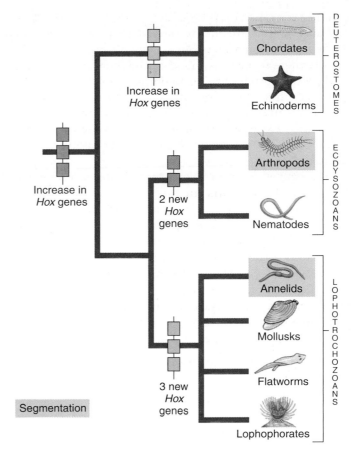

FIGURE 25.16
Multiple origins of segmentation. New phylogenies based on ribosomal RNA show that segmentation in arthropods and annelids arose independently. Segmentation in both appears to be regulated by some of the *Hox* genes.

Is an Insect a Flying Crustacean?

Arthropods are the most diverse of all the animal phyla, with more species than all other animal phyla combined, most of them insects. Insects have traditionally been set apart from the crustaceans, grouped instead with the myriapods (centipedes and millipedes) in a taxon called Tracheata. This phylogeny, still widely employed, dates back to benchmark work by Robert Snodgrass in the 1930s. He pointed out that insects, centipedes and millipedes are united by several seemingly powerful attributes, including uniramous (single-branched) legs. All crustacean appendages, by contrast, are basically biramous, or "two-branched" (figure 25.17), although some of these appendages have become single-branched by reduction in the course of their evolution.

Taxonomists have traditionally assumed a character such as two-branched appendages to be a fundamental one, conserved over the course of evolution, and thus suitable for making taxonomic distinctions. The patterning of appendages among arthropods is orchestrated by *Hox* genes. A single one of these *Hox* genes, called *Distal-less*, has been shown to initiate development of unbranched limbs in insects and branched limbs in crustaceans. The same *Distal-less* gene is found in many animal phyla, including the vertebrates. *Distal-less* appears to be necessary to initiate limb development, and it turns on genes that are more directly involved in the development of the limb itself. Evolutionary changes in the genes that *Distal-less* acts upon most likely account for differences in limb morphology.

In recent years, a mass of accumulating morphological and molecular data has led many taxonomists to suggest new arthropod phylogenies. The most revolutionary of these, championed by Richard Brusca of Columbia University, considers crustaceans to be the basic arthropod group, and insects a close sister group. Molecular phylogenies all place insects as a sister group to crustaceans, not myriapods. In conflict with 150 years of morphology-based thinking, these conclusions are certain to engender lively discussion.

Sorting Out the Mammalian Family Tree

In the preceding arthropod examples, evolutionary history has been rewritten. In mammals, however, parts of the phylogeny are just emerging, based on molecular data. Among the vertebrate classes, mammals are unique because they have mammary glands to feed their young. Mammals fall into three major groups. Monotremes, such as the platypus, are the only egg-laying mammals. Marsupials, such as kangaroos, are born quite early in development and continue development while nursing in a pouch. The majority of mammals—over 90%—are eutherians, or placental mammals. There are at least 18 extant (living) orders of eutherians, which tend to be divided into four major groups. The first major split occurred between the African clade (figure 25.18) and the other placental mammals when South America and Africa separated about 103 million years ago. Aardvarks and elephants are part of this African lineage, a clade we did not even know existed a decade ago. In South America, anteaters and armadillos soon appeared. Then two other branches arose—one including whales, horses, and carnivores, and the other, primates and rodents. Sorting out the relationships within these branches is an ongoing challenge.

Where does an aquatic mammal such as the whale fit? Whale fossils from 55 million years ago have hindlimbs. Whales were initially thought to be relatives of horses, camels, deer, antelope, pigs, and hippopotamuses based on morphological information from fossils and extant animals. DNA sequence data, however, revealed a particularly close relationship between whales and hippopotamuses. With this new phylogenetic information, the possibility arises that some adaptations to aquatic environments in both species had a common origin. Understanding evolutionary

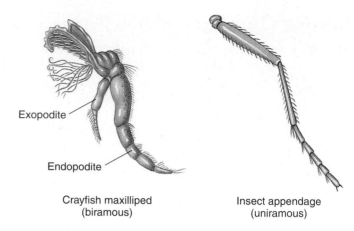

Exopodite

Endopodite

Crayfish maxilliped
(biramous)

Insect appendage
(uniramous)

FIGURE 25.17
Branched and single appendages. A biramous leg in a crustacean (crayfish) and a uniramous leg in an insect.

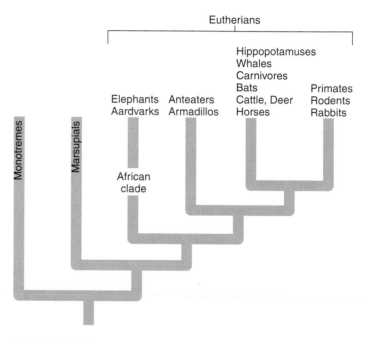

FIGURE 25.18
Major groups of mammals.

relationships among organisms does more than provide biologists with a sense of order and a logical way to name organisms. A phylogenetically based taxonomy allows researchers to ask important questions about physiology, behavior, and development using information already known about a related species.

Molecular systematics and cladistic approaches are providing new information about the evolutionary relationships among animals, including members of our own class, the mammals.

25.1 Biologists name organisms in a systematic way.

The Classification of Organisms

- Organisms were first classified by Aristotle over 2000 years ago. (p. 510)
- Similar individuals were eventually classified into units called genera (groups). (p. 510)
- Many descriptive terms were used in naming until the mid-1700s, when Linnaeus began using a binomial system. (p. 510)
- Taxonomy is the science of classifying living organisms; a group of similar organisms is put into a shared taxon. (p. 510)
- Taxonomic classifications are composed of (in ascending order): species, genus, family, order, class, phylum, kingdom, and domain. Each level may include anywhere from one to several taxa. (p. 511)

25.2 Scientists construct phylogenies to understand the evolutionary relationships among species.

Systematics

- The field of systematics constructs and studies evolutionary relationships (phylogenies). (p. 512).
- Descendant species should be relatively similar to their ancestors, and the amount of time since separation should be positively correlated with the degree of genetic divergence. But evolution is not constant in its rate, not unidirectional, and not always divergent. (p. 512)
- Cladistics distinguishes between ancestral and derived similarity, and only considers shared derived characters informative. (p. 512)
- Cladograms are constructed to show hypothesized evolutionary relationships, with species sharing derived characteristics put into the same clade. (p. 513)
- Systematists use the principle of parsimony and favor hypotheses with the fewest assumptions and the fewest evolutionary events. (p. 514)
- A monophyletic group contains the most recent common ancestor and all its descendants; a paraphyletic group does not include all the descendants; and a polyphyletic group does not include the most recent ancestor or all the members of the group. (p. 514)

25.3 Phylogenetics is the basis of all comparative biology.

Analogy Versus Homology

- Homologous structures are derived from the same body part in a common ancestor, while analogous structures are derived from different ancestral structures. (p. 516)
- Most complex characters are built up in a stepwise fashion, often with a series of evolutionary transitions. (p. 517)

25.4 All living organisms are grouped into one of a few major categories.

The Kingdoms of Life

- Most biologists now use a six-kingdom system. The kingdoms Animalia, Plantae (Virdiplantae), Fungi, and Protista contain eukaryotic organisms, while Archaebacteria and Bacteria contain prokaryotic organisms. (p. 518)

- Three different domains, a classification level higher than kingdom, are now recognized as well: Archaebacteria, Bacteria, and Eukarya. (p. 518)

Domain Archaea (Archaebacteria)

- Archaebacteria appear to have diverged from bacteria early, and are more closely related to eukaryotes than to bacteria. (p. 519)
- The archaebacteria are grouped into three general categories based on their environment: methanogens, extremophiles (e.g., thermophiles and halophiles), and nonextreme archaebacteria. (p. 519)

Domain Bacteria (Bacteria)

- Bacteria are the most abundant organisms on earth, and they carry out much of the earth's photosynthesis. (p. 520)
- Most taxonomists recognize 12–15 major groups of bacteria. (p. 520)

Domain Eukarya (Eukaryotes)

- Eukaryotes appear in the fossil record only about 2.5 billion years ago. (p. 521)
- Complex cellular organization with an extensive system of intracellular membranes is the most distinctive characteristic of eukaryotes. (p. 521)
- Compartmentalization, multicellularity (in many eukaryotes), and sexuality are all key characteristics of eukaryotes. (p. 522)

Viruses: A Special Case

- Viruses are segments of DNA or RNA wrapped in a protein coat. They can infect cells and use cellular machinery to replicate but are not considered living organisms. (p. 524)
- Viruses defy taxonomic efforts. (p. 524)

25.5 Molecular data are revolutionizing taxonomy.

The Impact of Molecular Cladistics

- Within traditional classification systems, no clear basis exists for determining the most important traits. (p. 525)
- Systematic phylogenies use cladistics to build taxonomic hierarchies. (p. 525)

Making Sense of the Protists

- Molecular systematics demonstrates that protists are paraphyletic. (p. 525)
- "Protist" is a catchall word encompassing 200,000 species. (p. 525)

Origin of Land Plants

- Although the origin of land plants from a green algal ancestor has long been recognized, new evolutionary relationships have been hypothesized. (p. 526)

Sorting Out the Animals

- New RNA investigations are raising questions about how closely annelids are related to arthropods. (p. 527)
- Arthropods are the most diverse of all the animal phyla. (p. 527)
- New DNA evidence is also bringing into question the traditional evolutionary paths of some mammals. (p. 528)
- Over 90% of mammals are placental mammals. (p. 528)

Test Your Understanding

For interactive testing, visit the Online Learning Center with PowerWeb at www.mhhe.com/Raven7

Self Test

1. There are over 300 different species of fiddler crabs. In all of these species, the adult males possess an enlarged front claw. Based on this feature, biologists group these different species together in the same genus, *Uca*. Therefore, the large claw is
   a. an outgroup.
   b. an analogous trait.
   c. a homoplasy.
   d. a synapomorphy.

2. The domestic dog, *Canis familiaris*, is related to wild dog species such as the gray wolf, *Canis lupus*. To which most exclusive category do the wolf and dog both belong?
   a. order
   b. family
   c. genus
   d. species

3. The simplest animals, the sponges (Parazoa), lack true tissues. All other animals (Eumetazoa) have distinct tissues. The presence of true tissues is a
   a. useful character for distinguishing among the eumetazoans.
   b. shared ancestral state.
   c. shared derived state.
   d. plesiomorphy.

4. Originally, organisms were classified according to a two-domain system that consisted of the bacteria and the eukaryotes. Recently, a three-domain system of classification has been proposed in which domain Bacteria was divided into domain Archaea and domain Bacteria. This is because
   a. there are so many different species of prokaryotes.
   b. of the large genetic differences between the archaebacteria and the bacteria.
   c. the archaebacteria have a nucleus like that of eukaryotes.
   d. the archaebacteria live in inhospitable environments, so they should be put in a different domain.

5. Which of the following is the most abundant group of organisms on earth?
   a. archaebacteria
   b. bacteria
   c. protists
   d. fungi

6. Biologists do not consider viruses to be alive. This is because viruses
   a. are unable to reproduce on their own.
   b. do not contain genetic material.
   c. do not contain proteins.
   d. are smaller than living organisms.

7. As more is learned about the molecular and genetic aspects of organisms, taxonomic groupings are changing. Which group of eukaryotes is the focus of the most controversy related to their taxonomic groupings?
   a. bacteria
   b. protists
   c. animals
   d. fungi

8. According to new molecular and biochemical information, should the plants still be recognized as a distinct kingdom?
   a. Yes, the green algae are now considered to be in the kingdom Plantae.
   b. Yes, the green algae and the plants are now recognized to be unrelated.
   c. No, the plants and the green algae are now classified together in the kingdom Virdiplantae.
   d. No, the plants are now classified together within the Chlorophyta.

9. The relationship between the annelids and the arthropods is now thought to be
   a. monophyletic.
   b. polyphyletic.
   c. paraphyletic
   d. unrelated.

Test Your Visual Understanding

1. Using the image from figure 25.8, fill in the following table to indicate the presence or absence of the derived character.

	Light Bones	Breastbone	Downy Feathers	Feathers with vanes, shafts, and barbs	Aerodynamic feathers	Arms longer than legs
Other Dinosaurs						
Coelophysis						
Tyrannosaurus						
Sinosauropteryx						
Velociraptor						
Caudipteryx						
Archaeopteryx						
Modern Birds						

Apply Your Knowledge

1. Most biologists consider viruses to be nonliving because they are unable to replicate outside of a host cell. How might you argue against this position? What characteristics do viruses have them make them lifelike?

2. Phylogenetic relationships are established based on assumptions about evolutionary relatedness. It is assumed that the more similar two organisms are, the more closely they are related. What problems does the process of convergent evolution cause for systematists?

3. Ethnobotanical research involves the exploration of plants that may have uses, particularly medical, for humans. Of what value is knowledge of systematics to an ethnobotanist?

26

Viruses

Concept Outline

26.1 Viruses are strands of nucleic acid encased within a protein coat.

The Nature of Viruses. Viruses occur in all organisms. Able to reproduce only within living cells, viruses are not themselves alive.

26.2 Bacterial viruses exhibit two sorts of reproductive cycles.

Bacteriophages. Some bacterial viruses, called bacteriophages, rupture the cells they infect, while others integrate themselves into the bacterial chromosome to become a stable part of the bacterial genome.

Cell Transformation and Phage Conversion. Integrated bacteriophages sometimes modify the host bacterium they infect.

26.3 HIV is a complex animal virus.

AIDS. The animal virus HIV infects certain key cells of the immune system, destroying the body's ability to defend itself from cancer and disease. The HIV infection cycle is typically a lytic cycle, in which the HIV RNA first directs the production of a corresponding DNA, and this DNA then directs the production of progeny virus particles.

The Future of HIV Treatment. Combination therapies and chemokines offer promising avenues of AIDS therapy.

26.4 Nonliving infectious agents are responsible for many human diseases.

Disease Viruses. Some of the most serious viral diseases have only recently infected human populations, as the result of transfer from other hosts.

Prions and Viroids. In some instances, proteins and "naked" RNA molecules can also transmit diseases.

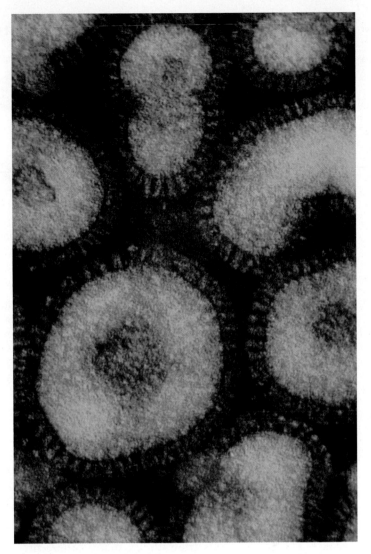

FIGURE 26.1
Influenza viruses (30,000×). A virus has been referred to as "a piece of bad news wrapped up in a protein." How can something as "simple" as a virus have such a profound effect on living organisms?

We start our exploration of the diversity of life with viruses. Viruses are genetic elements enclosed in protein and are not considered organisms, because they cannot reproduce independently. However, due to their disease-producing potential, viruses are important biological entities. The virus particles you see in figure 26.1 produce the important disease influenza. Other viruses cause such diseases as AIDS, polio, and smallpox, and some can lead to cancer. Many scientists have attempted to unravel the nature of viral genes and to discover how they work. For more than four decades, viral studies have been thoroughly intertwined with those of genetics and molecular biology. In the future, viruses are expected to be one of the principal tools used to experimentally carry genes from one organism to another. Already, viruses are being employed in the treatment of human genetic diseases.

26.1 Viruses are strands of nucleic acid encased within a protein coat.

The Nature of Viruses

Viral Structure

All viruses have the same basic structure: a core of nucleic acid surrounded by protein. Individual viruses contain only a single type of nucleic acid, either DNA or RNA. The DNA or RNA genome may be linear or circular, and single-stranded or double-stranded. Viruses are frequently classified by the nature of their genomes.

Nearly all viruses form a protein sheath, or **capsid**, around their nucleic acid core (figure 26.2). The capsid is composed of one to a few different protein molecules repeated many times. In some viruses, specialized enzymes are stored within the capsid. Many animal viruses form an **envelope** around the capsid that is rich in proteins, lipids, and glycoprotein molecules. While some of the material of the envelope is derived from the host cell's membrane, the envelope does contain proteins derived from viral genes as well.

Viruses occur in virtually every kind of organism that has been investigated for their presence. However, each type of virus can replicate in only a very limited number of cell types. The suitable cells for a particular virus are collectively referred to as its **host range.** Some viruses wreak havoc on the cells they infect; many others produce no disease or other outward sign of their infection. Still other viruses remain dormant for years until a specific signal triggers their expression. A given organism often has more than one kind of virus. This suggests that there may be many more kinds of viruses than there are kinds of organisms—perhaps millions of them. Only a few thousand viruses have been described at this point.

Viral Replication

An infecting virus can be thought of as a set of instructions, not unlike a computer program. A computer's operation is directed by the instructions in its operating program, just as a cell is directed by DNA-encoded instructions. A new program introduced into the computer can cause the computer to cease what it is doing and devote all of its energies to another activity, such as making copies of the introduced program. The new program is not itself a computer and cannot make copies of itself when it is outside the computer. Like a virus, the introduced program is simply a set of instructions.

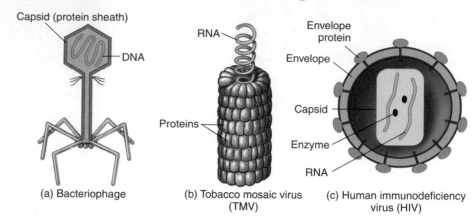

FIGURE 26.2

The structure of a bacterial, plant, and animal virus. (*a*) Bacterial viruses, called bacteriophages, often have a complex structure. (*b*) TMV infects plants and consists of 2130 identical protein molecules (*purple*) that form a cylindrical coat around the single strand of RNA (*green*). The RNA backbone determines the shape of the virus and is protected by the identical protein molecules packed tightly around it. (*c*) In the human immunodeficiency virus (HIV), the two-stranded RNA core is held within a capsid that is encased by a protein envelope.

Viruses can reproduce only when they enter cells and utilize the cellular machinery of their hosts. Viruses code their genes on a single type of nucleic acid, either DNA or RNA, but viruses lack ribosomes and the enzymes necessary for protein synthesis. Viruses are able to reproduce because their genes are translated into proteins by the cell's genetic machinery. These proteins lead to the production of more viruses.

Viral Shape

Most viruses have an overall structure that is either **helical** or **isometric.** Helical viruses, such as the tobacco mosaic virus in figure 26.2*b*, have a rodlike or threadlike appearance. Isometric viruses have a roughly spherical shape whose geometry is revealed only under the highest magnification.

The only structural pattern found so far among isometric viruses is the **icosahedron,** a structure with 20 equilateral triangular facets. Most viruses are icosahedral in basic structure. The icosahedron is the basic design of the geodesic dome and is the most efficient symmetrical arrangement that linear subunits can take to form a shell with maximum internal capacity.

Viral Genome Structure

Viral genomes vary greatly (table 26.1). Some viruses, including those that cause flu, measles, and AIDS, possess RNA genomes. Other viruses have DNA genomes, such as the viruses causing smallpox and herpes. The nucleic acid

Table 26.1 Important Human Viral Diseases

Disease	Pathogen		Genome	Vector/Epidemiology
Chicken pox	Varicella zoster		Double-stranded DNA	Spread through contact with infected individuals. No cure. Rarely fatal. Vaccine approved in U.S. in early 1995.
Hepatitis B (viral)	Hepadnavirus		Double-stranded DNA	Highly infectious through contact with infected body fluids. Approximately 1% of U.S. population infected. Vaccine available. No cure. Can be fatal.
Herpes	Herpes simplex virus		Double-stranded DNA	Fever blisters; spread primarily through contact with infected saliva. Very prevalent worldwide. No cure. Exhibits latency— the disease can be dormant for several years.
Mononucleosis	Epstein-Barr virus		Double-stranded DNA	Spread through contact with infected saliva. May last several weeks; common in young adults. No cure. Rarely fatal.
Smallpox	Variola virus		Double-stranded DNA	Historically a major killer; the last recorded case of smallpox was in 1977. A worldwide vaccination campaign wiped out the disease completely.
AIDS	HIV		(+) Single-stranded RNA (two segments)	Destroys immune defenses, resulting in death by infection or cancer. Over 42 million cases worldwide by 2002.
Polio	Enterovirus		(+) Single-stranded RNA	Acute viral infection of the CNS that can lead to paralysis and is often fatal. Prior to the development of Salk's vaccine in 1954, 60,000 people a year contracted the disease in the U.S. alone.
Yellow fever	Flavivirus		(+) Single-stranded RNA	Spread from individual to individual by mosquito bites; a notable cause of death during the construction of the Panama Canal. If untreated, this disease has a peak mortality rate of 60%.
Ebola	Filoviruses		(−) Single-stranded RNA	Acute hemorrhagic fever; virus attacks connective tissue, leading to massive hemorrhaging and death. Peak mortality is 50–90% if untreated. Outbreaks confined to local regions of central Africa.
Influenza	Influenza viruses		(−) Single-stranded RNA	Historically a major killer (22 million died in 18 months in 1918–19); wild Asian ducks, chickens, and pigs are major reservoirs. The ducks are not affected by the flu virus, which shuffles its antigen genes while multiplying within them, leading to new flu strains.
Measles	Paramyxoviruses		(−) Single-stranded RNA	Extremely contagious through contact with infected individuals. Vaccine available. Usually contracted in childhood, when it is not serious; more dangerous to adults.
SARS	Coronavirus		(−) Single-stranded RNA	Acute respiratory infection; an emerging disease, can be fatal, especially in the elderly.
Pneumonia	Influenza virus		(−) Single-stranded RNA	Acute infection of the lungs; often fatal without treatment.
Rabies	Rhabdovirus		(−) Single-stranded RNA	An acute viral encephalomyelitis transmitted by the bite of an infected animal. Fatal if untreated.

in some viruses is single-stranded, while in others, it is double-stranded. In single-stranded RNA viruses, the genome can contain the same base sequence as the mRNA used to produce the viral proteins, in which case the RNA strand can serve as the mRNA, and the virus is called a positive-strand virus. Or, the genome can contain bases complementary to the viral mRNA, in which case the virus is called a negative-strand virus.

Viruses occur in all organisms and can only reproduce within living cells.

26.2 Bacterial viruses exhibit two sorts of reproductive cycles.

Bacteriophages

Bacteriophages are viruses that infect bacteria. They are diverse both structurally and functionally, and are united solely by their occurrence in bacterial hosts. Many of these bacteriophages, called *phages* for short, are large and complex, with relatively large amounts of DNA and proteins. Some of them have been named as members of a "T" series (T1, T2, and so forth); others have been given different kinds of names. To illustrate the diversity of these viruses, T3 and T7 phages are icosahedral and have short tails. In contrast, the so-called T-even phages (T2, T4, and T6) have an icosahedral head, a capsid that consists primarily of three proteins, a connecting neck with a collar and long "whiskers," a long tail, and a complex base plate (figure 26.3).

The Lytic Cycle

During the process of bacterial infection by T4 phage, at least one of the tail fibers of the phage—they are normally held near the phage head by the "whiskers"—contacts the lipoproteins of the host bacterial cell wall. The other tail fibers set the phage perpendicular to the surface of the bacterium and bring the base plate into contact with the cell surface. The tail contracts, and the tail tube passes through an opening that appears in the base plate, piercing the bacterial cell wall. The contents of the head, mostly DNA, are then injected into the host cytoplasm.

When a virus kills the infected host cell in which it is replicating, the reproductive cycle is referred to as a **lytic** cycle (figure 26.4, *top*). The T-series bacteriophages are all **virulent viruses,** multiplying within infected cells and eventually lysing (rupturing) them. However, they vary considerably as to when they become virulent within their host cells.

The Lysogenic Cycle

Many bacteriophages do not immediately kill the cells they infect, instead integrating their nucleic acid into the genome of the infected host cell. While residing there, the virus's DNA is called a **prophage.** Among the bacteriophages that do this is the lambda (λ) phage of *Escherichia coli*. We know as much about this bacteriophage as we do about virtually any other biological particle; the complete sequence of its 48,502 bases has been determined. At least 23 proteins are associated with the development and maturation of lambda phage, and many other enzymes are involved in integrating these viruses into the host genome.

The integration of a virus into a cellular genome is called **lysogeny.** At a later time, the prophage may exit the genome and initiate virus replication. This sort of reproductive cycle, involving a period of genome integration, is called a **lysogenic cycle** (figure 26.4, *bottom*). Viruses that become stably integrated within the genome of their host cells are called **lysogenic viruses,** or **temperate viruses.**

> Bacteriophages are a diverse group of viruses that attack bacteria. Some kill their host in a lytic cycle; others integrate into the host's genome, initiating a lysogenic cycle.

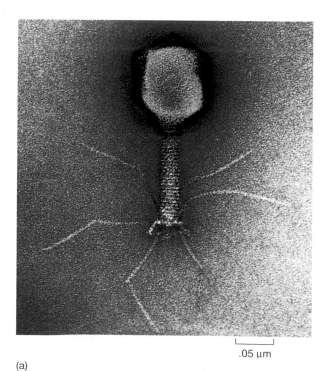

.05 μm

(a)

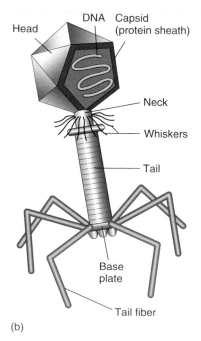

DNA Capsid
(protein sheath)

Head

Neck

Whiskers

Tail

Base
plate

Tail fiber

(b)

FIGURE 26.3
A bacterial virus. Bacteriophages exhibit a complex structure. (*a*) Electron micrograph and (*b*) diagram of the structure of a T4 bacteriophage (some facets removed to reveal the interior).

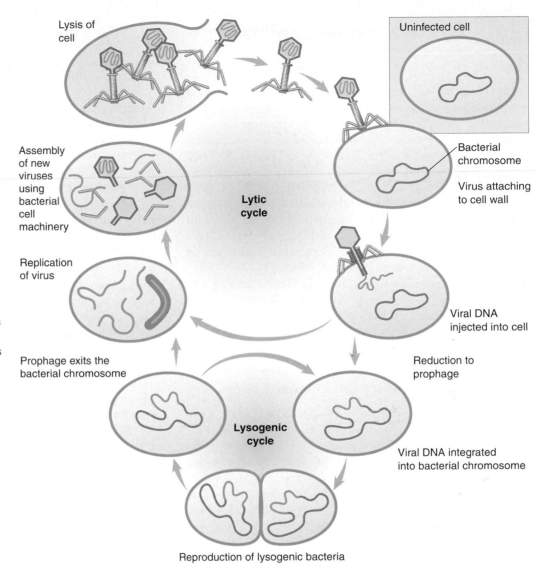

Lysis of cell

Uninfected cell

Assembly of new viruses using bacterial cell machinery

Bacterial chromosome

Lytic cycle

Virus attaching to cell wall

Replication of virus

Viral DNA injected into cell

Prophage exits the bacterial chromosome

Reduction to prophage

Lysogenic cycle

Viral DNA integrated into bacterial chromosome

Reproduction of lysogenic bacteria

FIGURE 26.4
Lytic and lysogenic cycles of a bacteriophage. In the lytic cycle, the bacteriophage exists as viral DNA free in the bacterial host cell's cytoplasm; the viral DNA directs the production of new viral particles by the host cell until the virus kills the cell by lysis. In the lysogenic cycle, the bacteriophage DNA is integrated into the large, circular DNA molecule of the host bacterium and is reproduced along with the host DNA as the bacterium replicates. It may continue to replicate and produce lysogenic bacteria or enter the lytic cycle and kill the cell. Bacteriophages are much smaller relative to their hosts than illustrated in this diagram.

Cell Transformation and Phage Conversion

During the integrated portion of a lysogenic reproductive cycle, viral genes are often expressed. The RNA polymerase of the host cell reads the viral genes just as if they were host genes. Sometimes, expression of these genes has an important effect on the host cell, altering it in novel ways. The genetic alteration of a cell's genome by the introduction of foreign DNA is called **transformation.** When the foreign DNA is contributed by a bacterial virus, the alteration is called **phage conversion.**

Phage Conversion of the Cholera-Causing Bacterium

An important example of the sort of phage conversion directed by viral genes is provided by the bacterium responsible for an often-fatal human disease. The disease-causing bacterium *Vibrio cholerae* usually exists in a harmless form, but a second disease causing, virulent form also occurs. In this latter form, the bacterium causes the deadly disease cholera, but how the bacteria changed from harmless to deadly was not known until recently. Research now shows that a bacteriophage that infects *V. cholerae* introduces into the host bacterial cell a gene that codes for the cholera toxin. This gene becomes incorporated into the bacterial chromosome, where it is translated along with the other host genes, thereby converting the benign bacterium to a disease-causing agent. The transfer occurs through bacterial pili (see chapter 27); in further experiments, mutant bacteria that did not have pili were resistant to infection by the bacteriophage. This discovery has important implications in efforts to develop vaccines against cholera, which have been unsuccessful up to this point.

Bacteriophages convert *Vibrio cholerae* bacteria from a harmless form into disease-causing agents.

26.3 HIV is a complex animal virus.

AIDS

A diverse array of viruses occur among animals. A good way to gain a general idea of what they are like is to look at one animal virus in detail. Here we will examine the virus responsible for a comparatively new and fatal viral disease, acquired immunodeficiency syndrome (AIDS). AIDS was first reported in the United States in 1981. It was not long before the infectious agent, a **retrovirus** (an RNA virus that transcribes its genes into DNA) called human immunodeficiency virus (HIV), was identified by laboratories in France. Study of HIV revealed it to be closely related to a chimpanzee virus, suggesting a recent host expansion to humans from chimpanzees in central Africa.

Infected humans have little resistance to infection, and nearly all of them eventually die of diseases that noninfected individuals easily ward off. Few who contract AIDS survive more than a few years untreated. The AIDS epidemic is discussed further in chapter 48.

How HIV Compromises the Immune System

In normal individuals, an army of specialized cells (white blood cells) patrols the bloodstream, attacking and destroying any invading bacteria or viruses. In AIDS patients, this army of defenders is vanquished. One special kind of white blood cell, called a CD4$^+$ cell (discussed further in chapter 48), is required to rouse the defending cells to action. In AIDS patients, the virus homes in on CD4$^+$ cells, infecting and killing them until none are left. Without these crucial immune system cells, the body cannot mount a defense against invading bacteria or viruses. AIDS patients die of infections that a healthy person could fight off.

Clinical symptoms typically do not begin to develop until after a long latency period, generally 8 to 10 years after the initial infection with HIV. Although carriers of HIV have no clinical symptoms during this long interval, they are apparently fully infectious, which makes the spread of HIV very difficult to control. The reason HIV remains hidden for so long seems to be that its infection cycle continues throughout the 8- to 10-year latent period without doing serious harm to the infected person. Eventually, however, a random mutational event in the virus allows it to quickly overcome the immune defense, starting AIDS.

The HIV Infection Cycle

The HIV virus infects and eliminates key cells of the immune system, destroying the body's ability to defend itself from cancer and infection. The way HIV infects humans (figure 26.5) provides a good example of how animal viruses replicate. Most other viral infections follow a similar course, although the details of entry and replication differ in individual cases.

Attachment. When HIV is introduced into the human bloodstream, the virus particle circulates throughout the entire body but will only infect CD4$^+$ cells. Most other animal viruses are similarly narrow in their requirements; hepatitis goes only to the liver, and rabies to the brain.

How does a virus such as HIV recognize a specific kind of target cell? Recall from chapter 7 that every kind of cell in the human body has a specific array of cell-surface glycoprotein markers that serve to identify them to other, similar cells. Each HIV particle possesses a glycoprotein (called gp120) on its surface that precisely fits a cell-surface marker protein called CD4 on the surfaces of immune system cells called macrophages and T cells. Macrophages are infected first.

Entry into Macrophages. After docking onto the CD4 receptor of a macrophage, HIV requires a second macrophage receptor, called CCR5, to pull itself across the cell membrane. After gp120 binds to CD4, it goes through a conformational change that allows it to bind to CCR5. The current model suggests that after the conformational change, the second receptor passes the gp120-CD4 complex through the cell membrane, triggering passage of the contents of the HIV virus into the cell by endocytosis, with the cell membrane folding inward to form a deep cavity around the virus.

Replication. Once inside the macrophage, the HIV particle sheds its protective coat. This leaves virus RNA floating in the cytoplasm, along with a virus enzyme that was also within the virus shell. This enzyme, called **reverse transcriptase,** synthesizes a double strand of DNA complementary to the virus RNA, often making mistakes and so introducing new mutations. This double-stranded DNA may integrate itself into the host cell's DNA; it directs the host cell machinery to produce many copies of the virus. HIV does not rupture and kill the macrophage cells it infects. Instead, the new viruses are released from the cell by *budding*, a process much like exocytosis. HIV synthesizes large numbers of viruses in this way, challenging the immune system over a period of years.

Entry into T Cells. During this time, HIV is constantly replicating and mutating. Eventually, by chance, HIV alters the gene for gp120 in a way that causes the gp120 protein to change its second-receptor allegiance. This new form of gp120 protein prefers to bind instead to a different second receptor, CXCR4, a receptor that occurs on the surface of T lymphocyte CD4$^+$ cells. Soon the body's T lymphocytes become infected with HIV. This has deadly consequences, as new viruses exit the cell by rupturing the cell, thus

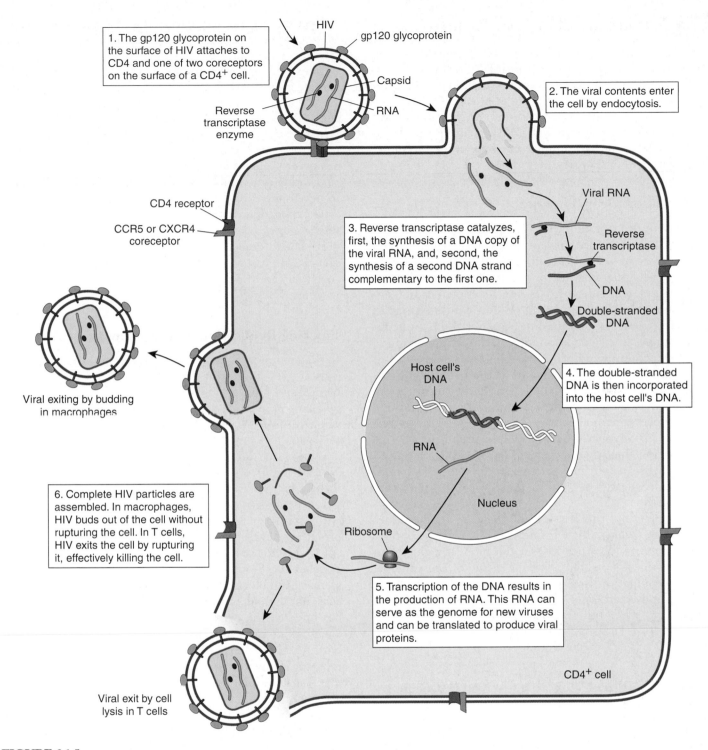

1. The gp120 glycoprotein on the surface of HIV attaches to CD4 and one of two coreceptors on the surface of a CD4+ cell.

HIV

gp120 glycoprotein

Capsid

Reverse transcriptase enzyme

RNA

2. The viral contents enter the cell by endocytosis.

CD4 receptor

CCR5 or CXCR4 coreceptor

Viral RNA

3. Reverse transcriptase catalyzes, first, the synthesis of a DNA copy of the viral RNA, and, second, the synthesis of a second DNA strand complementary to the first one.

Reverse transcriptase

DNA

Double-stranded DNA

Host cell's DNA

4. The double-stranded DNA is then incorporated into the host cell's DNA.

Viral exiting by budding in macrophages

RNA

Nucleus

6. Complete HIV particles are assembled. In macrophages, HIV buds out of the cell without rupturing the cell. In T cells, HIV exits the cell by rupturing it, effectively killing the cell.

Ribosome

5. Transcription of the DNA results in the production of RNA. This RNA can serve as the genome for new viruses and can be translated to produce viral proteins.

CD4+ cell

Viral exit by cell lysis in T cells

FIGURE 26.5

The HIV infection cycle.   The cycle begins and ends with free HIV particles present in the bloodstream of its human host. These free viruses infect white blood cells that have CD4 receptors, cells called CD4+ cells.

killing the T cell. The shift to the CXCR4 second receptor is followed swiftly by a steep drop in the number of T cells. This destruction of the body's T cells blocks the immune response and leads directly to the onset of AIDS, with cancers and opportunistic infections free to invade the defenseless body.

HIV, the virus that causes AIDS, is an RNA virus that replicates inside human cells by first making a DNA copy of itself. It is only able to gain entrance to those cells possessing a particular cell-surface marker recognized by a glycoprotein on its own surface.

The Future of HIV Treatment

New discoveries about how HIV works continue to fuel research on devising ways to counter HIV. For example, scientists are testing drugs and vaccines that act on HIV receptors, researching the possibility of blocking CCR5, and looking for defects in the structures of HIV receptors in individuals who are infected with HIV but have not developed AIDS. Figure 26.6 summarizes some of the recent developments and discoveries.

Combination Drug Therapy

A variety of drugs inhibit HIV in the test tube. These include AZT and similar nucleoside analogs (which inhibit viral nucleic acid replication) and protease inhibitors (which inhibit the cleavage of the large polyproteins encoded by *gag, pol,* and *env* genes into functional capsid, enzyme, and envelope segments). When combinations of these drugs were administered to people with HIV in controlled studies, their condition improved. A combination of a protease inhibitor and two AZT-type drugs entirely eliminated the HIV virus from many of the patients' bloodstreams. Importantly, all of these patients began to receive the drug therapy within three months of contracting the virus, before their bodies had an opportunity to develop tolerance to any one of them. Widespread use of this **combination therapy** has cut the U.S. AIDS death rate by three-fourths since its introduction in the mid-1990s, from 49,000 AIDS deaths in 1995 to 36,000 in 1996, and about 9000 in 2001.

Unfortunately, this sort of combination therapy does not appear to actually eliminate HIV from the body. While the virus disappears from the bloodstream, traces of it can still be detected in lymph tissue of the patients. When combination therapy is discontinued, virus levels in the bloodstream once again rise. Because of demanding therapy schedules and many side effects, long-term combination therapy does not seem a promising approach.

Vaccine Therapy: Using a Defective HIV Gene to Combat AIDS

Recently, five people in Australia who are HIV-positive but have not developed AIDS in 14 years were found to have all received a blood transfusion from the same HIV-positive person, who also has not developed AIDS. This led scientists to believe that the strain of virus transmitted to these people has some sort of genetic defect that prevents it from effectively disabling the human immune system. In subsequent research, a defect was found in one of the nine genes present in this strain of HIV. This gene is called *nef,* named for "negative factor," and the defective version of *nef* in the HIV strain that infected the six Australians seems to be missing some pieces. Viruses with the defective gene may have reduced reproductive capability, allowing the immune system to keep the virus in check.

This finding has exciting implications for developing a vaccine against AIDS. Before this, scientists have been unsuccessful in trying to produce a harmless strain of AIDS that can elicit an effective immune response. The Australian strain with the defective *nef* gene has the potential to be used in a vaccine that would arm the immune system against this and other strains of HIV.

Another potential application of this discovery is its use in developing drugs that inhibit HIV proteins that speed virus replication. It seems that the protein produced from the *nef* gene is one of these critical HIV proteins, because viruses with defective forms of *nef* do not reproduce, as seen in the cases of the six Australians. Research is currently under way to develop a drug that targets the *nef* protein.

Blocking Replication: Chemokines and CAF

In the laboratory, chemicals called **chemokines** appear to inhibit HIV infection by binding to and blocking the CCR5 and CXCR4 coreceptors. As you might expect, people long infected with the HIV virus who have not developed AIDS prove to have high levels of chemokines in their blood.

The search for HIV-inhibiting chemokines is intense. Not all results are promising. Researchers report that in their tests, the levels of chemokines were not different between patients in which the disease was not progressing and those in which it was rapidly progressing. More promising, levels of another factor called CAF (CD8+ cell antiviral factor) *are* different between these two groups. Researchers have not yet succeeded in isolating CAF, which seems not to block receptors that HIV uses to gain entry to cells, but instead, to prevent replication of the virus once it has infected the cells. Research continues on the use of chemokines in treatments for HIV infection, either increasing the amount of chemokines or disabling the CCR5 receptor. However, promising research on CAF suggests that it may be an even better target for treatment and prevention of AIDS.

One problem with using chemokines as drugs is that they are also involved in the inflammatory response of the immune system. Chemokines attract white blood cells to areas of infection. Chemokines work beautifully in small amounts and in local areas, but chemokines in mass numbers can cause an inflammatory response that is worse than the original infection. Injections of chemokines may hinder the immune system's ability to respond to local chemokines, or may even trigger an out-of-control inflammatory response. Thus, scientists caution that injection of chemokines could make patients *more* susceptible to infections, and they continue to research other methods of using chemokines to treat AIDS.

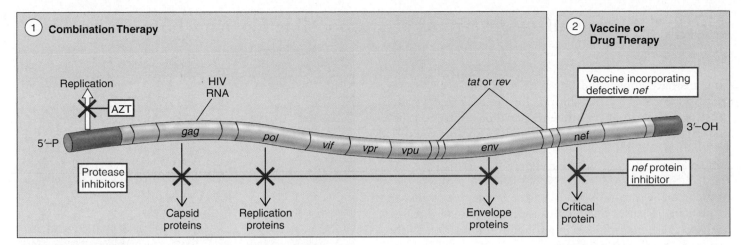

1 Combination Therapy

Replication

AZT

HIV RNA

5'–P

gag

pol

vif

vpr

vpu

tat or *rev*

env

nef

3'–OH

Protease inhibitors

Capsid proteins

Replication proteins

Envelope proteins

2 Vaccine or Drug Therapy

Vaccine incorporating defective *nef*

nef protein inhibitor

Critical protein

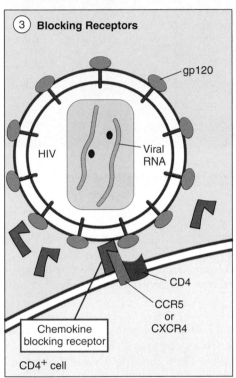

3 Blocking Receptors

gp120

HIV

Viral RNA

CD4

CCR5 or CXCR4

Chemokine blocking receptor

CD4+ cell

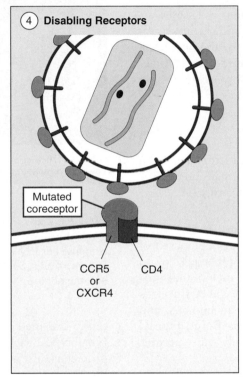

4 Disabling Receptors

Mutated coreceptor

CCR5 or CXCR4

CD4

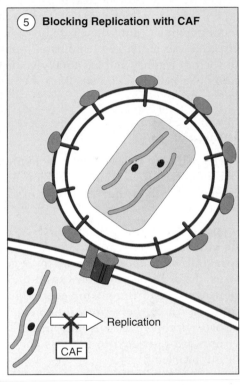

5 Blocking Replication with CAF

Replication

CAF

FIGURE 26.6

Potential new treatments for HIV. Research is currently under way in the following five areas: (*1*) Combination therapy involves using two drugs, AZT to block replication of the virus and protease inhibitors to block the production of critical viral proteins. (*2*) Using a defective form of the viral gene *nef*, scientists may be able to construct an HIV vaccine. Also, drug therapy that inhibits *nef*'s protein product is being tested. (*3*) Other research focuses on the use of chemokine chemicals to block receptors (CXCR4 and CCR5), thereby disabling the mechanism HIV uses to enter CD4+ cells. (*4*) Producing mutations that will disable receptors may also be possible. (*5*) Finally, CAF, an antiviral factor that acts inside the CD4+ cell, may be able to block replication of HIV.

Blocking or Disabling Receptors

A 32-base-pair deletion in the gene that codes for the CCR5 receptor appears to block HIV infection. Individuals at high risk of HIV infection who are homozygous for this mutation do not seem to develop AIDS. In one study of 1955 people, scientists found no individuals who were infected and homozygous for the mutated allele. The allele seems to be more common in Caucasian populations (10 to 11%) than in African-American populations (2%), and absent in African and Asian populations. Treatment for AIDS involving disruption of CCR5 looks promising, because research indicates that people live perfectly well without CCR5. Attempts to block or disable CCR5 are being sought in numerous laboratories.

A cure for AIDS is not yet in hand, but many new approaches look promising.

26.4 Nonliving infectious agents are responsible for many human diseases.

Disease Viruses

Humans have known and feared diseases caused by viruses for thousands of years. Among the diseases that viruses cause (see table 26.1) are influenza, smallpox, infectious hepatitis, yellow fever, polio, AIDS, and SARS (figure 26.7). In addition, viruses have been implicated in some cancers and leukemias. In view of their effects, it is easy to see why the late Sir Peter Medawar, Nobel laureate in physiology or medicine, wrote, "A virus is a piece of bad news wrapped in protein." Viruses not only cause many human diseases, but also cause major losses in agriculture, forestry, and the productivity of natural ecosystems.

Polio and smallpox were very serious human diseases, affecting millions, until they were brought under control. An effective vaccine largely eliminated polio in the latter half of the last century and has totally eradicated natural infections of smallpox. The smallpox virus now exists only in laboratories.

Influenza

Perhaps the most lethal virus in human history has been the influenza virus. Some 22 million Americans and Europeans died of flu within 18 months in 1918 and 1919.

Types.
Flu viruses are animal RNA viruses. An individual flu virus resembles a rod studded with spikes composed of two kinds of protein. There are three general "types" of flu virus, distinguished by their capsid (inner membrane) protein, which is different for each type: Type A flu virus causes most of the serious flu epidemics in humans, and also occurs in mammals and birds. Type B and type C viruses are restricted to humans and rarely cause serious health problems.

Subtypes.
Different strains of flu virus, called subtypes, differ in their protein spikes. One of these proteins, hemagglutinin (H), aids the virus in gaining access to the cell interior. The other, neuraminidase (N), helps the daughter virus break free of the host cell once virus replication has been completed. Parts of the H molecule contain "hotspots" that display an unusual tendency to change as a result of mutation of the viral RNA during imprecise replication. Point mutations cause changes in these spike proteins in 1 of 100,000 viruses during the course of each generation. These highly variable segments of the H molecule are targets against which the body's antibodies are directed. The high variability of these targets improves the reproductive capacity of the virus and hinders our ability to make perfect vaccines. Because of accumulating changes in the H and N molecules, different flu vaccines are required to protect against different subtypes. Type A flu viruses are cur-

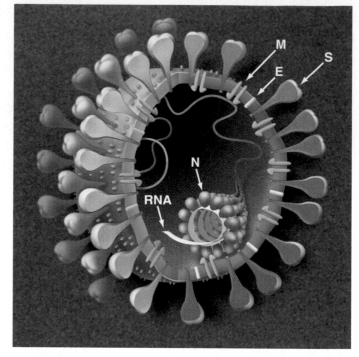

SARS genome

R_A R_B S E M N

FIGURE 26.7
SARS coronavirus. The 29,751-nucleotide SARS genome is composed of RNA and contains six principle genes: R_A and R_B—replicases; S—spike proteins; E—envelope glycoproteins; M—membrane glycoprotein; N—nucleocapsid protein.

rently classified into 13 distinct H subtypes and 9 distinct N subtypes, each of which requires a different vaccine to protect against infection. Thus, the type A virus that caused the Hong Kong flu epidemic of 1968 has type 3 H molecules and type 2 N molecules, and is called A(H3N2).

Importance of Recombination.
The greatest problem in combating flu viruses arises not through mutation, but through recombination. Viral genes are readily reassorted by genetic recombination, sometimes putting together novel combinations of H and N spikes unrecognizable by human antibodies specific for the old configuration. Viral recombination of this kind seems to have been responsible for the three major flu pandemics (that is, worldwide epidemics) that occurred in the twentieth century, by producing drastic shifts in H-N combinations. The "killer flu" of 1918, A(H1N1), killed 40 million people worldwide. The Asian flu of 1957, A(H2N2), killed over 100,000 Americans. The Hong Kong flu of 1968, A(H3N2), infected 50 million people in the United States alone, of which 70,000 died.

It is no accident that new strains of flu usually originate in the Far East. The most common hosts of influenza virus are ducks, chickens, and pigs, which in Asia often live in close proximity to each other and to humans. Pigs are subject to infection by both bird and human strains of the virus, and individual animals are often simultaneously infected with multiple strains. This creates conditions favoring genetic recombination between strains, producing new combinations of H and N subtypes. The Hong Kong flu, for example, arose from recombination between A(H3N8) from ducks and A(H2N2) from humans. The new strain of influenza, in this case A(H3N2), then passed back to humans, creating an epidemic because the human population has never experienced that H-N combination before.

Emerging Viruses

Sometimes viruses that originate in one organism pass to another, thus expanding their host range. Often, this expansion is deadly to the new host. HIV, for example, arose in chimpanzees and relatively recently passed to humans. Influenza is fundamentally a bird virus. Viruses that originate in one organism and then pass to another and cause disease are called **emerging viruses.** They represent a considerable threat in an age when airplane travel potentially allows infected individuals to move about the world quickly, spreading an infection.

An emerging virus caused a sudden outbreak of a hemorrhagic-type infection in the southwestern United States in 1993. This highly fatal disease was soon attributed to a species of **hantavirus,** a single-stranded RNA virus associated with rodents. This species was eventually traced to deer mice. The deer mouse hantavirus is transmitted to humans through fecal contamination in areas of human habitation. Controlling the deer mouse population limited the disease.

Sometimes the origin of an emerging virus is unknown, making an outbreak more difficult to control. Among the most lethal of emerging viruses are a collection of filamentous viruses arising in central Africa that cause severe hemorrhagic fever. With lethality rates in excess of 50%, these so-called filoviruses are among the most lethal infectious diseases known. One, **Ebola virus** (figure 26.8), has exhibited lethality rates in excess of 90% in isolated outbreaks in central Africa. The outbreak of Ebola virus in the summer of 1995 in Zaire killed 245 people out of 316 infected—a mortality rate of 78%. The latest outbreak occurred in Gabon, West Africa, in February 1996. The natural host of Ebola is unknown.

A recently emerged species of coronavirus was responsible for a worldwide outbreak in 2003 of **severe acute respiratory syndrome (SARS),** a respiratory infection with pneumonia-like symptoms that is fatal in over 8% of cases. When the 29,751-nucleotide RNA genome of the SARS coronavirus was sequenced, it proved to be a completely new form of coronavirus, not closely related to any

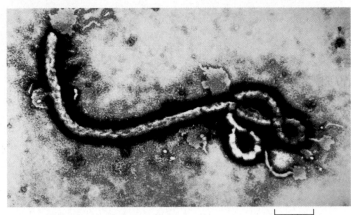

0.3 µm

FIGURE 26.8
The Ebola virus. This virus, with a fatality rate that can exceed 90%, appears sporadically in West Africa. Health professionals are scrambling to identify the natural host of the virus, which is unknown, so they can devise strategies to combat transmission of the disease.

of the three previously described forms. Virologists suspect that the SARS coronavirus most likely came from civets (a weasel-like mammal) and other wild animals that live in China and are eaten as delicacies. If the SARS virus indeed exists in natural populations, it will be difficult to prevent future outbreaks without an effective vaccine. Genome sequences have been analyzed from SARS patients at various stages of the outbreak, and these analyses indicate that the virus's mutation rate is low, in marked contrast to another RNA virus, HIV. The stable genome of the SARS virus should make development of a SARS vaccine practical.

Viruses and Cancer

Through epidemiological studies and research, scientists have established a link between some viral infections and the subsequent development of cancer. Examples include the association between chronic hepatitis B infections and the development of liver cancer and the development of cervical carcinoma following infections with certain strains of papillomaviruses. It has been suggested that viruses contribute to about 15% of all human cancer cases worldwide. Viruses are capable of altering the growth properties of human cells they infect by triggering the expression of cancer-causing genes (see chapter 20). Virus-induced cancer is not simply a matter of infection; it involves complex interactions with cellular genes and requires a series of events in order to develop.

Some of the most lethal diseases of humans are caused by viruses, particularly viruses that have transferred to humans from some other host, like those causing influenza (a bird virus), Ebola (animal host unknown), and SARS (small mammal virus).

Prions and Viroids

For decades, scientists have been fascinated by a peculiar group of fatal brain diseases. These diseases have an unusual property: It is years and often decades after infection before the disease is detected in infected individuals. The brains of infected individuals develop numerous small cavities as neurons die, producing a marked spongy appearance. Called **transmissible spongiform encephalopathies (TSEs),** these diseases include scrapie in sheep, mad cow disease in cattle, and kuru and Creutzfeldt-Jakob disease in humans.

TSEs can be transmitted experimentally by injecting infected brain tissue into a recipient animal's brain. TSEs can also spread via tissue transplants and, apparently, food. The disease kuru was common in the Fore people of Papua, New Guinea, because they practiced ritual cannibalism, literally eating the brains of infected individuals. Mad cow disease spread widely among the cattle herds of England in the 1990s because cows were fed bonemeal prepared from cattle carcasses to increase the protein content of their diet. Like the Fore, the British cattle were literally eating the tissue of cattle that had died of the disease.

A Heretical Suggestion

In the 1960s, British researchers T. Alper and J. Griffith noted that infectious TSE preparations remained infectious even after exposure to radiation that would destroy DNA or RNA. They suggested that the infectious agent was a protein. Perhaps, they speculated, the protein usually preferred one folding pattern, but could sometimes misfold, and then catalyze other proteins to do the same, the misfolding spreading like a chain reaction. This heretical suggestion was not accepted by the scientific community, because it violates a key tenet of molecular biology: Only DNA or RNA act as hereditary material, transmitting information from one generation to the next.

Prusiner's Prions

In the early 1970s, physician Stanley Prusiner, moved by the death of a patient from Creutzfeldt-Jakob disease, began to study TSEs. Prusiner became fascinated with Alper and Griffith's hypothesis. Try as he might, Prusiner could find no evidence of nucleic acids or viruses in the infectious TSE preparations. He concluded, as Alper and Griffith had, that the infectious agent was a *protein*, which in a 1982 paper he named a **prion,** for "proteinaceous infectious particle."

Prusiner went on to isolate a distinctive prion protein, and for two decades continued to amass evidence that prions play a key role in triggering TSEs. The scientific community resisted Prusiner's renegade conclusions, but eventually experiments done in Prusiner's and other laboratories began to convince many. For example, when Prusiner in-

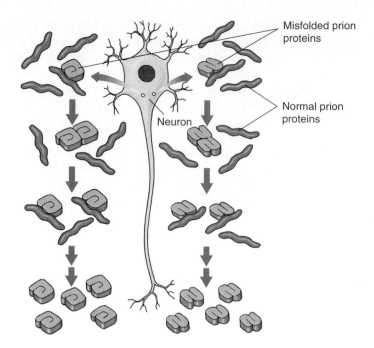

FIGURE 26.9
How prions arise. Misfolded prions seem to cause normal prion protein to misfold simply by contacting them. When prions misfolded in different ways (*blue*) contact normal prion protein (*purple*), the normal prion protein misfolds in the same way.

jected prions of a different abnormal conformation into several different hosts, these hosts developed prions with the same abnormal conformations as the parent prions (figure 26.9). In another important experiment, Charles Weissmann showed that mice genetically engineered to lack Prusiner's prion protein are immune to TSE infection. However, if brain tissue with the prion protein is grafted into the mice, the grafted tissue—but not the rest of the brain—can then be infected with TSE. In 1997, Prusiner was awarded the Nobel Prize in physiology or medicine for his work on prions.

Viroids

Viroids are tiny, naked molecules of RNA, only a few hundred nucleotides long, that are important infectious disease agents in plants. A recent viroid outbreak killed over ten million coconut palms in the Philippines. It is not clear how viroids cause disease. One clue is that viroid nucleotide sequences resemble the sequences of introns within ribosomal RNA genes. These sequences are capable of catalyzing excision from DNA—perhaps the viroids are catalyzing the destruction of chromosomal integrity.

Prions are infectious proteins that some scientists believe are responsible for serious brain diseases. In plants, naked RNA molecules called viroids can also transmit disease.

For interactive testing, visit the Online Learning Center with PowerWeb at www.mhhe.com/Raven7

26.1 Viruses are strands of nucleic acid encased within a protein coat.

The Nature of Viruses

- All viruses have the same basic structure: a nucleic acid core surrounded by a protein. (p. 532)
- Nearly all viruses form a capsid. (p. 532)
- Each virus can only replicate in a limited number of cell types, and the suitable cells are referred to as the host range. (p. 532)
- Viruses must enter cells and utilize cellular machinery to reproduce. (p. 532)
- The overall structure of a virus is either helical or isometric. (p. 532)

26.2 Bacterial viruses exhibit two sorts of reproductive cycles.

Bacteriophages

- Bacteriophages are viruses that infect bacteria. (p. 534)
- The lytic cycle refers to the reproductive cycle when a virus kills an infected host. Virulent viruses multiply within infected cells and eventually lyse the cells. (p. 534)
- Most bacteriophages do not immediately kill their hosts; these are referred to as prophages during lysogeny, the integration of a virus into a cellular genome. (p. 534)

Cell Transformation and Phage Conversion

- Transformation refers to the alteration of a cell's genome by the introduction of foreign DNA. (p. 535)

26.3 HIV is a complex animal virus.

AIDS

- In AIDS patients, the human immunodeficiency virus infects and kills CD4$^+$ cells, eventually causing the patients to die of an infection that would normally be fought off by a healthy immune system. (p. 536)
- HIV can only enter cells that possess a cell-surface marker recognized by a glycoprotein on the HIV particle surface. (p. 536)
- The HIV infection cycle includes the following steps: attachment, entry into cells, replication, integration into the host cell's DNA, production of new viruses, and finally exit from the cells. (p. 536)

The Future of HIV Treatment

- The future of HIV treatment includes using combination drug therapy such as AZT analogs and protease inhibitors, using a defective HIV gene, using chemokines and CAF, and blocking or disabling receptors. (pp. 538–539)

26.4 Nonliving infectious agents are responsible for many human diseases.

Disease Viruses

- Viruses are known to cause many human diseases, including influenza, smallpox, infectious hepatitis, yellow fever, polio, AIDS, and SARS. (p. 540)
- Viruses have also been implicated in some cancers and leukemias. (p. 540)
- Emerging viruses originate in one organism and then pass to another. (p. 541)

Prions and Viroids

- Brain diseases known as transmissible spongiform encephalopathies are known to be caused by a protein called a prion (proteinaceous infectious particle). (p. 542)
- Viroids, naked RNA molecules, can transmit diseases in plants. (p. 542)

Self Test

1. Viruses consist of a _____ core surrounded by a protein coat.
   a. RNA
   b. DNA
   c. chromosome
   d. nucleic acid

2. Phages infect bacterial cells by
   a. poking holes in the cell and injecting their DNA.
   b. destroying the bacterial cell wall.
   c. receptor-mediated endocytosis.
   d. exocytosis.

3. The result of the viral lytic cycle is
   a. the release of new viruses.
   b. the incorporation of the viral genome into the host genome.
   c. the conversion of the virus into a prophage.
   d. Both b and c are correct.

4. The alteration of a cell's genome by the incorporation of foreign DNA is called
   a. genetic conversion.
   b. mutation.
   c. transformation.
   d. reverse transcription.

5. Which of the following is a viral glycoprotein that plays a role in the infection of human cells by HIV?
   a. gp120
   b. CD4
   c. CCR5
   d. Both b and c are correct.

6. The HIV enzyme, _____, produces a DNA copy of the viral genome once it is inside the host cell.
   a. DNA polymerase
   b. reverse transcriptase
   c. RNA polymerase
   d. helicase

7. The drug AZT and its analogs function by
   a. inhibiting the replication of viral nucleic acid.
   b. blocking the production of envelope proteins.
   c. blocking the production of capsid proteins.
   d. blocking the binding of the virus to human cell receptors.

8. What is the function of the $CD8^+$ cell antiviral factor (CAF) in human white blood cells?
   a. disables the CD4 receptor
   b. blocks replication of the HIV virus
   c. interferes with the production of viral proteins
   d. blocks the CCR5 or CXCR4 receptors

9. New strains of flu viruses are most likely to arise in the Far East because
   a. of overpopulation in these areas.
   b. adequate sanitation is lacking.
   c. common hosts (ducks, chickens, pigs) live in close proximity to humans.
   d. of environmental mutagens that cause changes in viral RNA.

10. Which of the following infectious agents does *not* contain protein?
    a. viruses
    b. viroids
    c. prions
    d. none of these

Test Your Visual Understanding

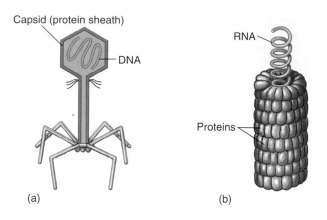

(a)

(b)

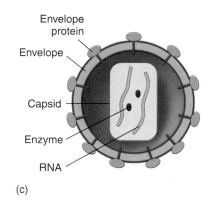

(c)

1. Describe the overall shape of each of the viruses in the figure, either as helical, isometric, or icosahedral.
2. List which of these viruses attacks:
   a. animals
   b. plants
   c. bacteria

Apply Your Knowledge

1. AIDS was first identified in 1981 in the United States, and in 2002, an estimated 495,000 cases of AIDS and/or HIV had been reported. The total U.S. population at the end of 2002 was approximately 290 million. What percent of the population was living with AIDS and/or HIV?

2. In what ways might the early, self-replicating particles that gave rise to the first organisms have resembled, or differed from, viruses?

3. If a viral disease such as influenza could kill 22 million people over 18 months in 1918–1919, why do you suppose the killing ever stopped? What prevented the flu virus from continuing its lethal assault on humanity?

27

Prokaryotes

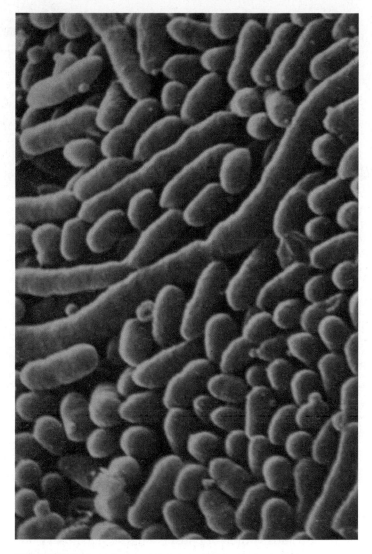

FIGURE 27.1
A colony of prokaryotes. With their enormous adaptability and metabolic versatility, prokaryotes are found in every habitat on earth, carrying out many of the vital processes of ecosystems, including photosynthesis, nitrogen fixation, and decomposition.

Concept Outline

27.1 Prokaryotes are the smallest and most numerous organisms.

The Prevalence of Prokaryotes. The simplest and most abundant of organisms, prokaryotes are also thought to be the most ancient. Prokaryotes lack the high degree of internal compartmentalization characteristic of eukaryotes.

27.2 Prokaryotes exhibit considerable diversity in both structure and metabolism.

Prokaryotic Diversity. There are at least 17 major groups of prokaryotes, although many more remain to be discovered.

27.3 Prokaryotes are more complex than commonly supposed.

The Prokaryotic Cell Surface. Prokaryotes have a cell wall and may contain structures such as pili and flagella.
The Cell Interior. While prokaryotes lack extensive internal compartments, they may have complex internal membranes.
Prokaryotic Variation. Mutation and recombination generate enormous variation within bacterial populations.
Prokaryotic Metabolism. Prokaryotes obtain carbon atoms and energy from a wide array of sources, including other organisms.

27.4 Prokaryotes are responsible for many diseases, but they also make important contributions to ecosystems.

Human Bacterial Diseases. Many serious human diseases are caused by bacteria, and some of these diseases are responsible for millions of deaths each year.
Benefits of Prokaryotes. Prokaryotes have had a profound impact on the world's ecology, and play a major role in modern medicine and agriculture.

The simplest organisms living on earth today are prokaryotes, and biologists think they closely resemble the first organisms to evolve on earth. Too small to see with the unaided eye, prokaryotes are the most abundant of all organisms (figure 27.1) and are the only ones characterized by prokaryotic cellular organization. Life on earth could not exist without prokaryotes because they make possible many of the essential functions of ecosystems, including the capture of nitrogen from the atmosphere, decomposition of organic matter, and in many aquatic communities, photosynthesis. Indeed, prokaryotic photosynthesis is thought to have been the source for much of the oxygen in the earth's atmosphere. Prokaryotic research continues to provide extraordinary insights into genetics, ecology, and disease. Thus, an understanding of prokaryotes is essential.

27.1 Prokaryotes are the smallest and most numerous organisms.

The Prevalence of Prokaryotes

About 5000 different kinds of prokaryotes are currently recognized, but there are doubtless many thousands more awaiting proper identification. Everyplace microbiologists look, new species are being discovered, in some cases altering the way we think about prokaryotes. In the 1970s and 1980s, a new type of prokaryote was analyzed that eventually led to the division of prokaryotes into two groups, the archaebacteria (or Archaea) and the bacteria.

Archaebacteria and bacteria are the oldest, structurally simplest, and most abundant forms of life on earth. They are also the only organisms with prokaryotic cellular organization. Represented in the oldest rocks from which fossils have been obtained, 2.5 billion years old, prokaryotes were abundant for over a billion years before eukaryotes appeared in the world. Early photosynthetic bacteria (cyanobacteria) altered the earth's atmosphere by producing oxygen, which led to extreme bacterial and eukaryotic diversity. Bacteria play a vital role both in productivity and in cycling the substances essential to all other life-forms. Bacteria are the only organisms capable of fixing atmospheric nitrogen.

Prokaryotes are ubiquitous on earth, and live everywhere eukaryotes do. Many of the other, more extreme environments in which prokaryotes are found would be lethal to any other form of life. Prokaryotes live in hot springs that would cook other organisms, in hypersaline environments that would dehydrate other cells, and in atmospheres rich in such toxic gases as methane or hydrogen sulfide that would kill most other organisms. These harsh environments may be similar to the conditions present on the early earth when life first began. It is likely that prokaryotes evolved to dwell in these harsh conditions early on and have retained the ability to exploit these areas as the rest of the atmosphere has changed.

Prokaryotic Form

Prokaryotes are mostly simple in form and exhibit one of three basic structures: **bacillus** (plural, *bacilli*), straight and rod-shaped; **coccus** (plural, *cocci*), spherical-shaped; and **spirillum** (plural, *spirilla*), long and helical-shaped, also called spirochetes. Spirilla generally do not form associations with other cells; they swim singly through their environments. A complex structure within their cell wall allows them to spin their corkscrew-shaped bodies, propelling them along. Some rod-shaped and spherical bacteria form colonies, adhering end-to-end after they have divided, forming chains. Some bacterial colonies change into stalked structures, grow long, branched filaments, or form erect structures that release **spores,** single-celled bodies that grow into new bacterial individuals. Some filamentous

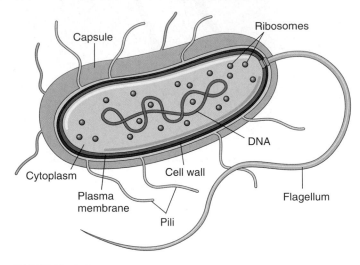

FIGURE 27.2
The structure of a prokaryotic cell. Prokaryotic cells, such as this bacterium, are small and lack interior organization. The plasma membrane is encased within a rigid cell wall, and DNA is not contained in a membrane-bounded nucleus. In addition to a flagellum, prokaryotes may also possess outgrowths called pili that aid in attachment to surfaces or other cells.

bacteria are capable of a gliding motion, often combined with rotation around a longitudinal axis. Biologists have not yet determined the mechanism by which they move.

Prokaryotes Versus Eukaryotes

Prokaryotes differ from eukaryotes in numerous important features (table 27.1). These differences represent some of the most fundamental distinctions that separate any groups of organisms.

1. **Unicellularity.** All prokaryotes are fundamentally single-celled (figure 27.2). In some types, individual cells adhere to each other within a matrix and form filaments; however, the cells retain their individuality. Cyanobacteria, in particular, are likely to form such associations, but their cytoplasm is not directly interconnected, as often is the case in multicellular eukaryotes. The activities of a bacterial colony are less integrated and coordinated than those of multicellular eukaryotes. A primitive form of colonial organization occurs in gliding bacteria, which move together and form spore-bearing structures. Such coordinated multicellular forms are rare among prokaryotes.
2. **Cell size.** As new species of prokaryotes are discovered, we are finding that the size of prokaryotic cells varies tremendously, by as much as five orders of

magnitude. Most prokaryotic cells are only 1 micrometer or less in diameter. Most eukaryotic cells are well over 10 times that size.

3. **Chromosomes.** Eukaryotic cells have a membrane-bounded nucleus containing chromosomes made up of both nucleic acids and proteins. Prokaryotes do not have membrane-bounded nuclei, nor do they have chromosomes of the kind present in eukaryotes, in which DNA forms a structural complex with proteins. Instead, their naked circular DNA is localized in a zone of the cytoplasm called the **nucleoid.**

4. **Cell division and genetic recombination.** Cell division in eukaryotes takes place by mitosis and involves spindles made up of microtubules. Cell division in prokaryotes takes place mainly by binary fission (see chapter 11). True sexual reproduction occurs only in eukaryotes and involves syngamy and meiosis, with an alternation of diploid and haploid forms. Despite their asexual mode of reproduction, prokaryotes do have mechanisms that lead to the transfer of genetic material. These mechanisms are far less regular than those of eukaryotes and do not involve the equal participation of the individuals between which the genetic material is transferred (see chapter 20).

5. **Internal compartmentalization.** In eukaryotes, the enzymes for cellular respiration are packaged in mitochondria. In prokaryotes, the corresponding enzymes are not packaged separately but are bound to the cell membranes (see chapters 5 and 9). The cytoplasm of prokaryotes, unlike that of eukaryotes, contains no internal compartments or cytoskeleton and no organelles except ribosomes.

6. **Flagella.** Prokaryotic flagella are simple in structure, composed of a single fiber of the protein flagellin (see chapter 5). Eukaryotic flagella and cilia are complex and have a 9 + 2 structure of microtubules (see figure 5.27). Bacterial flagella also function differently, spinning like propellers, while eukaryotic flagella have a whiplike motion.

7. **Metabolic diversity.** Only one kind of photosynthesis occurs in eukaryotes, and it involves the release of oxygen. Photosynthetic bacteria have several different patterns of anaerobic and aerobic photosynthesis, involving the formation of end products such as sulfur, sulfate, and oxygen (see chapter 10). Prokaryotic cells can also be chemoautotrophic, using the energy stored in chemical bonds of inorganic molecules to synthesize carbohydrates; eukaryotes are not capable of this metabolic process.

Prokaryotes are the oldest and most abundant organisms on earth. Prokaryotes differ from eukaryotes in a wide variety of characteristics, a degree of difference as great as that separating any other groups of organisms.

Table 27.1 Prokaryotes Compared to Eukaryotes

Feature	Example
Unicellularity. All prokaryotes are basically single-celled. Even though some bacteria may adhere together or form filaments, their cytoplasm is not directly interconnected, and their activities are not integrated and coordinated as is the case in multicellular eukaryotes.	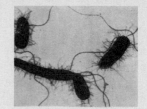 Unicellular bacteria
Cell Size. Most bacterial cells are only about 1 micrometer in diameter, while most eukaryotic cells are over 10 times that size.	Bacterial cell Eukaryotic cell
Chromosomes. Prokaryotic DNA exists as a single circle in the cytoplasm, while in eukaryotes, proteins are complexed with the DNA into multiple chromosomes.	Bacterial genome Eukaryotic chromosomes
Cell Division. Prokaryotic cells divide by binary fission (see chapter 11). The cells simply pinch in two. In eukaryotes, microtubules pull chromosomes to opposite poles during the cell division process, called mitosis.	Binary fission in bacteria — Mitosis in eukaryotes
Internal Compartmentalization. Unlike eukaryotic cells, bacterial cells contain no internal compartments, no internal membrane system, and no cell nucleus.	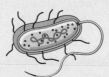 Bacterial cell
Flagella. Prokaryotic flagella are simple, composed of a single fiber of protein that spins like a propeller. Flagella in eukaryotes are complex structures that whip back and forth rather than rotating.	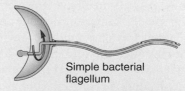 Simple bacterial flagellum
Metabolic Diversity. Prokaryotes possess many metabolic abilities that eukaryotes do not; some prokaryotes can perform several different kinds of anaerobic and aerobic photosynthesis, obtain their energy from oxidizing inorganic compounds, or fix atmospheric nitrogen.	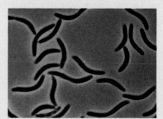 Chemoautotrophs

27.2 Prokaryotes exhibit considerable diversity in both structure and metabolism.

Prokaryotic Diversity

Prokaryotes are not easily classified according to their forms, and only recently has enough been learned about their biochemical and metabolic characteristics to develop a satisfactory overall classification comparable to that used for other organisms. Early systems for classifying prokaryotes relied on differential stains such as the Gram stain. Key characteristics used in classifying prokaryotes were:

1. photosynthetic or nonphotosynthetic;
2. motile or nonmotile;
3. unicellular or colony-forming or filamentous;
4. formation of spores or division by transverse binary fission.

With the development of genetic and molecular approaches, prokaryotic classifications can at last reflect true evolutionary relatedness. Molecular approaches include: (1) the analysis of the amino acid sequences of key proteins; (2) the analysis of nucleic acid base sequences by establishing the percent of guanine (G) and cytosine (C); (3) nucleic acid hybridization, which is essentially the mixing of single-stranded DNA from two species and determining the amount of base-pairing (closely related species will have more bases pairing); (4) gene and RNA sequencing, especially looking at ribosomal RNA; and (5) whole-genome sequencing.

Based on these sorts of molecular data, several groupings of prokaryotes have been proposed. The most widely accepted is that presented in *Bergey's Manual*, second edition, 2001 (figure 27.3).

Kinds of Prokaryotes

Although they lack the structural complexity of eukaryotes, prokaryotes have diverse internal chemistries, varying metabolisms, and unique functions. Prokaryotes have adapted to many kinds of environments, including some you might consider harsh. They have successfully invaded very salty waters, very acidic or alkaline environments, and very hot or cold areas. They are found in hot springs where the temperatures exceed 78°C (172°F) and have been recovered living beneath 435 meters of ice in Antarctica!

Much of what we know about bacteria has been learned from studies in the laboratory, which has placed limits on our knowledge. Field studies suggest that bacteria able to be cultured in the lab represent but a small fraction of the kinds of prokaryotes that occur in soil, most of which cannot be cultured with existing techniques. We clearly have only scraped the surface of prokaryotic diversity.

Prokaryotes split into two lines early in the history of life. Archaebacteria and bacteria are as different in structure and metabolism from each other as either is from the eukaryotes. The domain Archaea consists of the archaebacteria ("ancient bacteria"—although they are actually not as ancient as the other prokaryotic domain). It was once thought that survivors of this group were confined to extreme environments that may resemble habitats on the early earth. However, the use of genetic screening has revealed that these "ancient" prokaryotes live in nonextreme environments as well. The other, more ancient domain, the Bacteria, includes nearly all of the named species of prokaryotes.

Comparing Archaebacteria and Bacteria

Archaebacteria and bacteria are similar in that both have a prokaryotic cellular structure, but they vary considerably at the biochemical and molecular levels. There are four key areas in which they differ:

1. **Plasma membranes.** All prokaryotes have plasma membranes with a lipid-bilayer architecture (as described in chapter 6). The plasma membranes of bacteria and archaebacteria, however, are made of very different kinds of lipids.
2. **Cell wall.** Both kinds of prokaryotes typically have cell walls covering the plasma membrane that strengthen the cell. The cell walls of bacteria are constructed of carbohydrate-protein complexes called peptidoglycan, which link together to create a strong mesh that gives the bacterial cell wall great strength. The cell walls of archaebacteria lack peptidoglycan.
3. **Gene translation machinery.** Bacteria possess ribosomal proteins and an RNA polymerase that are distinctly different from those of eukaryotes. However, the ribosomal proteins and RNA of archaebacteria are very similar to those of eukaryotes.
4. **Gene architecture.** The genes of bacteria are not interrupted by introns, while at least some of the genes of archaebacteria do possess introns.

While superficially similar, archaebacteria and bacteria differ from one another in a wide variety of characteristics.

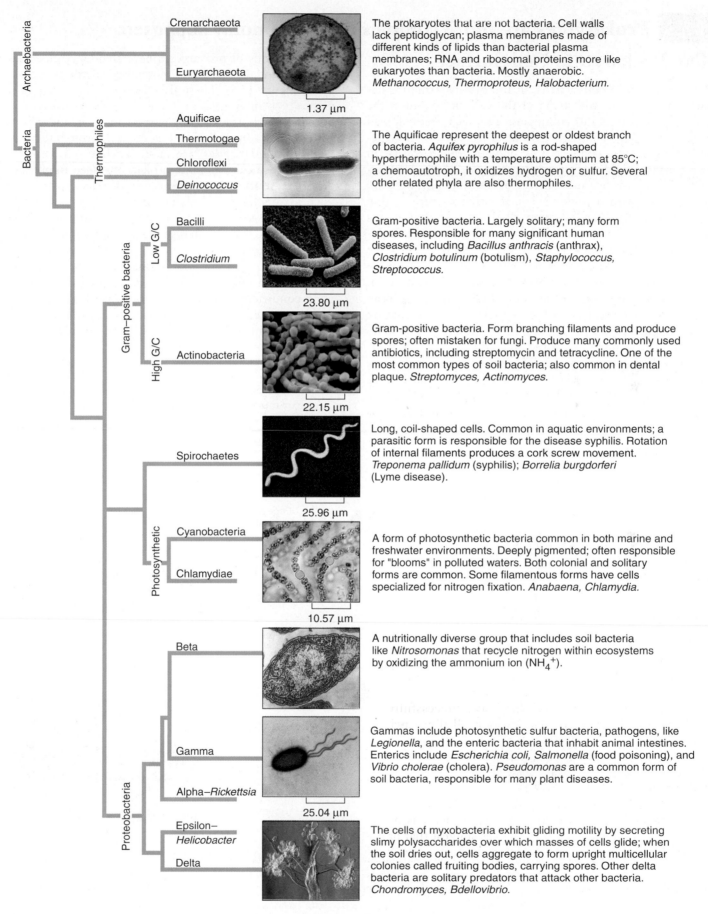

Archaebacteria

Crenarchaeota

Euryarchaeota

The prokaryotes that are not bacteria. Cell walls lack peptidoglycan; plasma membranes made of different kinds of lipids than bacterial plasma membranes; RNA and ribosomal proteins more like eukaryotes than bacteria. Mostly anaerobic. *Methanococcus, Thermoproteus, Halobacterium.*

1.37 μm

Bacteria

Thermophiles

Aquificae

Thermotogae

Chloroflexi

Deinococcus

The Aquificae represent the deepest or oldest branch of bacteria. *Aquifex pyrophilus* is a rod-shaped hyperthermophile with a temperature optimum at 85°C; a chemoautotroph, it oxidizes hydrogen or sulfur. Several other related phyla are also thermophiles.

Gram-positive bacteria

Low G/C

Bacilli

Clostridium

Gram-positive bacteria. Largely solitary; many form spores. Responsible for many significant human diseases, including *Bacillus anthracis* (anthrax), *Clostridium botulinum* (botulism), *Staphylococcus, Streptococcus.*

23.80 μm

High G/C

Actinobacteria

Gram-positive bacteria. Form branching filaments and produce spores; often mistaken for fungi. Produce many commonly used antibiotics, including streptomycin and tetracycline. One of the most common types of soil bacteria; also common in dental plaque. *Streptomyces, Actinomyces.*

22.15 μm

Spirochaetes

Long, coil-shaped cells. Common in aquatic environments; a parasitic form is responsible for the disease syphilis. Rotation of internal filaments produces a cork screw movement. *Treponema pallidum* (syphilis); *Borrelia burgdorferi* (Lyme disease).

25.96 μm

Photosynthetic

Cyanobacteria

Chlamydiae

A form of photosynthetic bacteria common in both marine and freshwater environments. Deeply pigmented; often responsible for "blooms" in polluted waters. Both colonial and solitary forms are common. Some filamentous forms have cells specialized for nitrogen fixation. *Anabaena, Chlamydia.*

10.57 μm

Beta

A nutritionally diverse group that includes soil bacteria like *Nitrosomonas* that recycle nitrogen within ecosystems by oxidizing the ammonium ion (NH_4^+).

Gamma

Alpha–*Rickettsia*

Gammas include photosynthetic sulfur bacteria, pathogens, like *Legionella*, and the enteric bacteria that inhabit animal intestines. Enterics include *Escherichia coli, Salmonella* (food poisoning), and *Vibrio cholerae* (cholera). *Pseudomonas* are a common form of soil bacteria, responsible for many plant diseases.

25.04 μm

Proteobacteria

Epsilon–*Helicobacter*

Delta

The cells of myxobacteria exhibit gliding motility by secreting slimy polysaccharides over which masses of cells glide; when the soil dries out, cells aggregate to form upright multicellular colonies called fruiting bodies, carrying spores. Other delta bacteria are solitary predators that attack other bacteria. *Chondromyces, Bdellovibrio.*

FIGURE 27.3

The major clades of prokaryotes. The classification adopted here is that of *Bergey's Manual of Systematic Bacteriology*, second edition, 2001.

27.3 Prokaryotes are more complex than commonly supposed.

The Prokaryotic Cell Surface

The prokaryotic **cell wall** is an important structure because it maintains the shape of the cell and protects the cell from swelling and rupturing. The bacterial cell wall usually consists of **peptidoglycan,** a network of polysaccharide molecules connected by polypeptide crosslinks. In some bacteria, the peptidoglycan forms a thick, complex network around the outer surface of the cell. This network is interlaced with peptide chains. In other bacteria, a thin layer of peptidoglycan is sandwiched between two plasma membranes. The outer membrane contains large molecules of lipopolysaccharide, lipids with polysaccharide chains attached. These two major types of bacteria can be identified using a staining process called a **Gram stain.** **Gram-positive** bacteria have the thicker peptidoglycan wall and stain a purple color, while the more common **gram-negative** bacteria contain less peptidoglycan and do not retain the purple-colored dye (figure 27.4). Gram-negative bacteria stain red. The outer membrane layer makes gram-negative bacteria resistant to many antibiotics that interfere with cell wall synthesis in gram-positive bacteria. In some kinds of bacteria, an additional gelatinous layer, the capsule, surrounds the cell wall.

Many kinds of prokaryotes have slender, rigid, helical **flagella** (singular, *flagellum*) composed of the protein flagellin (figure 27.5). These flagella range from 3 to 12 micrometers in length and are very thin—only 10 to 20 nanometers thick. They are anchored in the cell wall and spin like a propeller, pulling the cell through the water.

Pili (singular, *pilus*) are other hairlike structures that occur on the cells of some prokaryotes (see figure 27.2). They are shorter than prokaryotic flagella (up to several micrometers long) and about 7.5 to 10 nanometers thick. Pili help the prokaryotic cells attach to appropriate substrates and exchange genetic information (see chapter 20).

Some prokaryotes form thick-walled **endospores** around their genome and a small portion of the surrounding cytoplasm when they are exposed to nutrient-poor conditions. These endospores are highly resistant to environmental stress, especially heat, and can germinate to form new individuals after decades or even centuries.

Prokaryotes are encased within a cell wall composed of one or more layers. They also may contain external structures such as flagella and pili.

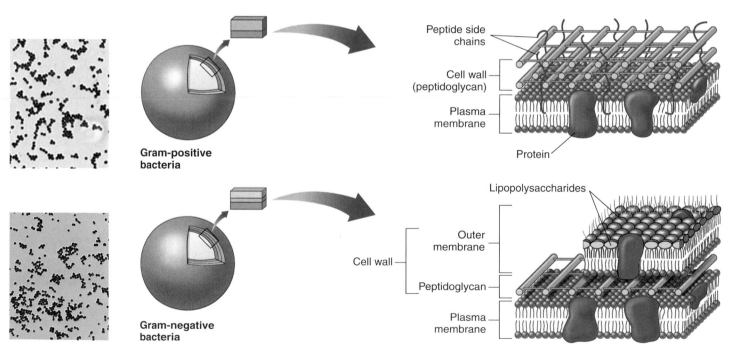

FIGURE 27.4
The Gram stain. The peptidoglycan layer encasing gram-positive bacteria traps crystal violet dye, so the bacteria appear purple in a Gram-stained smear (named after Hans Christian Gram, who developed the technique). Because gram-negative bacteria have much less peptidoglycan (located between the plasma membrane and an outer membrane), they do not retain the crystal violet dye and so exhibit the red background stain (usually a safranin dye).

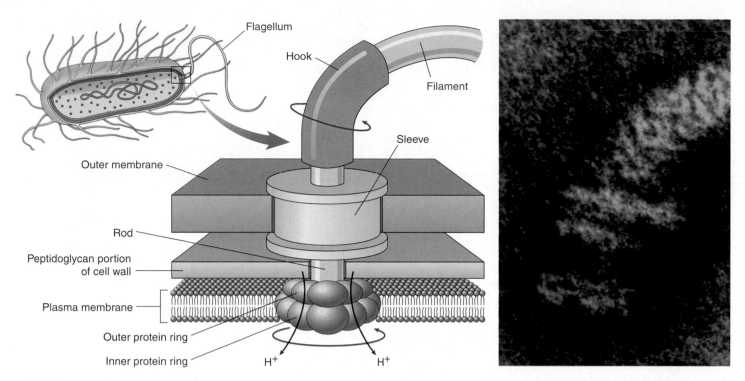

FIGURE 27.5
The flagellar motor of a gram-negative bacterium. A protein filament, composed of the protein flagellin, is attached to a protein rod that passes through a sleeve in the outer membrane and through a hole in the peptidoglycan layer to rings of protein anchored in the cell wall and plasma membrane, like rings of ballbearings. The rod rotates when the inner protein ring attached to the rod turns with respect to the outer ring fixed to the cell wall. The inner ring is an H^+ ion channel, a proton pump that uses the passage of protons into the cell to power the movement of the inner ring past the outer one.

The Cell Interior

The most fundamental characteristic of prokaryotic cells is their simple interior organization. Prokaryotic cells lack the extensive functional compartmentalization seen within eukaryotic cells, but they do have the following structures:

Internal membranes. Many prokaryotes possess invaginated regions of the plasma membrane that function in respiration or photosynthesis (figure 27.6).

Nucleoid region. Prokaryotes lack nuclei and do not possess the complex chromosomes characteristic of eukaryotes. Instead, their genes are encoded within a single double-stranded ring of DNA that is crammed into one region of the cell known as the **nucleoid region.** Many prokaryotic cells also possess small, independently replicating circles of DNA called plasmids. Plasmids contain only a few genes, which are usually not essential for the cell's survival. They are best thought of as an excised portion of the prokaryotic genome.

Ribosomes. Prokaryotic ribosomes are smaller than those of eukaryotes and differ in protein and RNA content. Antibiotics such as tetracycline and chloramphenicol can tell the difference—they bind to prokaryotic ribosomes and block protein synthesis, but do not bind to eukaryotic ribosomes.

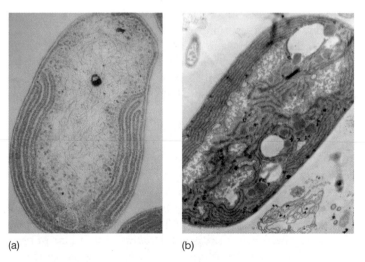

(a)   (b)

FIGURE 27.6
Prokaryotic cells often have complex internal membranes.
(*a*) This aerobic bacterium exhibits extensive respiratory membranes within its cytoplasm not unlike those seen in mitochondria. (*b*) This cyanobacterium has thylakoid-like membranes that provide a site for photosynthesis.

The interior of a prokaryotic cell may possess functional membranes and a nucleoid region.

Prokaryotic Variation

Prokaryotes reproduce rapidly, allowing genetic variations to spread quickly through a population. Two processes create variation among prokaryotes; mutation and genetic recombination.

Mutation

Mutations can arise spontaneously in bacteria as errors in DNA replication occur. Certain factors tend to increase the likelihood of errors occurring, such as radiation, ultraviolet light, and various chemicals. A typical bacterium such as *Escherichia coli* contains about 5000 genes. It is highly probable that one mutation will occur by chance in one out of every million copies of a gene. With 5000 genes in a bacterium, the laws of probability predict that 1 out of every 200 bacteria will have a mutation (figure 27.7). A spoonful of soil typically contains over a billion bacteria and therefore should contain something on the order of 5 million mutant individuals!

With adequate food and nutrients, a population of *E. coli* can double in under 20 minutes. Because bacteria multiply so rapidly, mutations can spread rapidly in a population and can change the characteristics of that population.

The ability of prokaryotes to change rapidly in response to new challenges often has adverse effects on humans. Recently, a number of strains of the bacterium *Staphylococcus aureus* associated with serious infections in hospitalized patients have appeared, some of them with alarming frequency. Unfortunately, these strains have acquired resistance to penicillin and a wide variety of other antibiotics, so that infections caused by them are very difficult to treat. *Staphylococcus* infections provide an excellent example of the way in which mutation and intensive selection can bring about rapid change in bacterial populations. Such changes have serious medical implications when, as in the case of *Staphylococcus*, strains of bacteria emerge that are resistant to a variety of antibiotics.

Recently, concern has arisen over the prevalence of antibacterial soaps in the marketplace. They are marketed as a means of protecting your family from harmful bacteria; however, it is likely that their routine use will favor bacteria that have mutations, making them immune to the antibiotics. Ultimately, extensive use of antibacterial soaps could have an adverse effect on our ability to treat common bacterial infections.

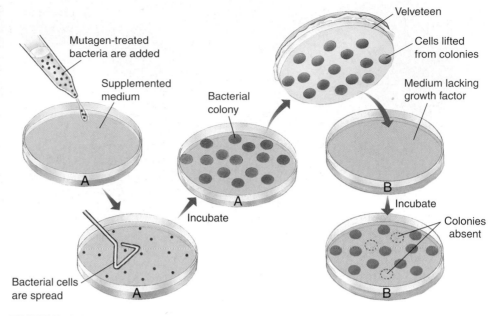

FIGURE 27.7

A mutant hunt in bacteria. Mutations in bacteria can be detected by a technique called replica plating, which allows the genetic characteristics of the colonies to be investigated without destroying them. The bacterial colonies, growing on a semisolid agar medium, are transferred from A to B using a sterile velveteen disc pressed on the plate. Plate A has a medium that includes special growth factors, while plate B has a medium that lacks some of these growth factors. Bacteria that are not mutated can produce their own growth factors and do not require them to be added to the medium. The colonies absent in B were unable to grow on the deficient medium and were thus mutant colonies; they were already present but undetected in A.

Genetic Recombination

Another source of genetic variation in populations of prokaryotes is recombination, discussed in detail in chapter 20. Recombination occurs in prokaryotes when genes are transferred from one cell to another by viruses, or through conjugation. The rapid transfer of newly produced, antibiotic-resistant genes by plasmids has been an important factor in the appearance of the resistant strains of *S. aureus* discussed earlier. An even more important example in terms of human health involves the Enterobacteriaceae, the family of bacteria to which the common intestinal bacterium *Escherichia coli* belongs. In this family are many important pathogenic bacteria, including the organisms that cause dysentery, typhoid, and other major diseases. At times, some of the genetic material from these pathogenic species is exchanged with or transferred to *E. coli* by plasmids. Because of its abundance in the human digestive tract, *E. coli* poses a special threat if it acquires harmful traits.

The short generation time of prokaryotes allows much mutation and recombination to occur, leading to genetic diversity.

Prokaryotic Metabolism

Prokaryotes have evolved many mechanisms to acquire the energy and nutrients they need for growth and reproduction. Many are **autotrophs**, organisms that obtain their carbon from inorganic CO_2. Autotrophs that obtain their energy from sunlight are called *photoautotrophs*, while those that harvest energy from inorganic chemicals are called *chemoautotrophs*. Other prokaryotes are **heterotrophs**, organisms that obtain at least some of their carbon from organic molecules such as glucose. Heterotrophs that obtain their energy from sunlight are called *photoheterotrophs*, while those that harvest energy from organic molecules are called *chemoheterotrophs*.

Photoautotrophs.   Many bacteria carry out photosynthesis, using the energy of sunlight to build organic molecules from carbon dioxide. The cyanobacteria use chlorophyll *a* as the key light-capturing pigment and H_2O as an electron donor, releasing oxygen gas as a by-product. Other bacteria use bacteriochlorophyll as their pigment and H_2S as an electron donor, leaving elemental sulfur as the by-product.

Chemoautotrophs.   Some prokaryotes obtain their energy by oxidizing inorganic substances. Nitrifiers, for example, oxidize ammonia or nitrite to obtain energy, producing the nitrate that is taken up by plants. This process is called nitrification and is essential in terrestrial ecosystems because plants can only absorb nitrogen in the form of nitrate. Other prokaryotes oxidize sulfur, hydrogen gas, and other inorganic molecules. On the dark ocean floor at depths of 2500 meters, entire ecosystems subsist on prokaryotes that oxidize hydrogen sulfide as it escapes from thermal vents.

Photoheterotrophs.   The so-called purple nonsulfur bacteria use light as their source of energy but obtain carbon from organic molecules, such as carbohydrates or alcohols that have been produced by other organisms.

Chemoheterotrophs.   Most prokaryotes obtain both carbon atoms and energy from organic molecules. These include decomposers and most pathogens.

How Heterotrophs Infect Host Organisms

In the 1980s, researchers studying the disease-causing species of *Yersinia*, a group of gram-negative bacteria, found that they produced and secreted large amounts of proteins. Most proteins secreted by gram-negative bacteria have special signal sequences that allow them to pass through the bacterium's double membrane. The proteins secreted by *Yersinia* lacked a key signal sequence that two known secretion mechanisms require for transport across the double membrane of gram-negative bacteria. The proteins must therefore have been secreted by a third type of system, which researchers called the *type III system*.

As more bacterial species are studied, the genes coding for the type III system are turning up in other gram-negative animal pathogens, and even in more distantly related plant pathogens. The genes seem more closely related to one another than do the bacteria. Furthermore, the genes are similar to those that code for bacterial flagella.

The role of these proteins is still under investigation, but it seems that some of the proteins are used to transfer other virulence proteins into nearby eukaryotic cells. Given the similarity of the type III genes to the genes that code for flagella, some scientists hypothesize that the transfer proteins may form a flagellum-like structure that shoots virulence proteins into the host cells. Once in the eukaryotic cells, the virulence proteins may determine the host's response to the pathogen. In *Yersinia*, proteins secreted by the type III system are injected into macrophages; they disrupt signals that tell the macrophages to engulf bacteria. *Salmonella* and *Shigella* use their type III proteins to enter the cytoplasm of eukaryotic cells and thus are protected from the immune system of their host. The proteins secreted by *E. coli* alter the cytoskeleton of nearby intestinal eukaryotic cells, resulting in a bulge onto which the bacterial cells can tightly bind.

Currently, researchers are looking for a way to disarm the bacteria using knowledge of their internal machinery, possibly by causing the bacteria to release the virulence proteins before they are near eukaryotic cells. Others are studying the eukaryotic target proteins and the process by which they are affected.

Bacteria as Plant Pathogens

Many costly diseases of plants are associated with particular heterotrophic bacteria. Almost every kind of plant is susceptible to one or more kinds of bacterial disease, including blights, soft rots, and wilts. The symptoms of these plant diseases vary, but they are commonly manifested as spots of various sizes on the stems, leaves, flowers, or fruits. Fire blight, which destroys pear and apple trees and related plants, is a well-known example of bacterial disease. Most bacteria that cause plant diseases are members of the group of rod-shaped bacteria known as pseudomonads.

While prokaryotes obtain carbon and energy in many ways, most are chemoheterotrophs. Some heterotrophs have evolved sophisticated ways to infect their hosts.

27.4 Prokaryotes are responsible for many diseases, but they also make important contributions to ecosystems.

Human Bacterial Diseases

Bacteria cause many diseases in humans, including cholera, leprosy, tetanus, bacterial pneumonia, whooping cough, diphtheria, and Lyme disease (table 27.2). Members of the genus *Streptococcus* are associated with scarlet fever, rheumatic fever, pneumonia, and other infections. Tuberculosis (TB), another bacterial disease, is still a leading cause of death in humans. TB and some of the other bacterial diseases are mostly spread through the air in water vapor, while such diseases as typhoid fever, paratyphoid fever, and bacillary dysentery are dispersed in food or water. Typhus is spread among rodents and humans by insect vectors.

Tuberculosis

Tuberculosis has been one of the great killer diseases for thousands of years. TB afflicts the respiratory system and is easily transmitted from person to person through the air. Currently, about one-third of all people worldwide are infected with *Mycobacterium tuberculosis*, the tuberculosis bacterium (figure 27.8). Eight million new cases crop up each year, with about 3 million people dying from the disease annually; the World Health Organization predicts 4 million deaths a year by 2005. In fact, in 1997, TB was the leading cause of death from a *single infectious agent* worldwide. Since the mid-1980s, the United States has been experiencing a dramatic resurgence of tuberculosis. The causes of this resurgence include social factors such as poverty, crowding, homelessness, and incarceration (the same factors that have always promoted the spread of TB). The increasing prevalence of HIV infections is also a significant contributing factor. People with AIDS are much more likely to develop TB than people with healthy immune systems.

In addition to the increased numbers of cases—more than 15,900 nationally as of 2001—alarming outbreaks of multidrug-resistant strains of tuberculosis have occurred. These strains are resistant to the best available anti-TB medications. Thus, multidrug-resistant TB is particularly concerning because it requires much more time to treat, is more expensive to treat, and may prove fatal.

The basic principles of TB treatment and control are to make sure all patients complete a full course of medication so that all of the bacteria causing the infection are killed and drug-resistant strains do not develop. Great efforts are being made to ensure that high-risk individuals who are infected but not yet sick receive preventative therapy, which is 90% effective in reducing the likelihood of developing active TB.

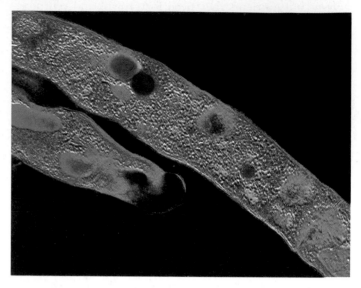

FIGURE 27.8
Mycobacterium tuberculosis. This color-enhanced image shows the rod-shaped bacterium responsible for tuberculosis in humans.

Dental Caries

One human disease we do not usually consider bacterial in origin arises in the film on our teeth. This film, or plaque, consists largely of bacterial cells surrounded by a polysaccharide matrix. Most of the bacteria in plaque are filaments of rod-shaped cells classified as various species of *Actinomyces*, which extend out perpendicular to the surface of the tooth. Many other bacterial species are also present in plaque. Tooth decay, or **dental caries,** is caused by the bacteria present in the plaque, which persists especially in places that are difficult to reach with a toothbrush. Diets that are high in sugars are especially harmful to teeth because certain bacteria (especially *Streptococcus sanguis* and *S. mutans*) ferment the sugars to lactic acid, a substance that reduces the pH of the mouth, causing the local loss of calcium from the teeth. Eating sugary snacks frequently or sucking on candy over a period of time keeps the pH level of the mouth low, resulting in the steady degeneration of the tooth enamel. As the calcium is removed from the tooth, the remaining soft matrix of the tooth becomes vulnerable to attack by bacteria, which begin to break down its proteins, and tooth decay progresses rapidly. Fluoride makes the teeth more resistant to decay because it retards the loss of calcium. It was first realized that bacteria cause tooth decay when germ-free animals were raised. Their teeth do not decay even if they are fed sugary diets.

Table 27.2 Important Human Bacterial Diseases

Disease	Pathogen	Vector/Reservoir	Epidemiology
Anthrax	*Bacillus anthracis*	Animals, including processed skins	Bacterial infection that can be transmitted through contact or ingestion. Rare except in sporadic outbreaks. May be fatal.
Botulism	*Clostridium botulinum*	Improperly prepared food	Contracted through ingestion or contact with wound. Produces acute toxic poison; can be fatal.
Chlamydia	*Chlamydia trachomatis*	Humans, STD	Urogenital infections with possible spread to eyes and respiratory tract. Occurs worldwide; increasingly common over past 20 years.
Cholera	*Vibrio cholerae*	Human feces, plankton	Causes severe diarrhea that can lead to death by dehydration; 50% peak mortality if the disease goes untreated. A major killer in times of crowding and poor sanitation; over 100,000 died in Rwanda in 1994 during a cholera outbreak.
Dental caries	*Streptococcus*	Humans	A dense collection of this bacteria on the surface of teeth leads to secretion of acids that destroy minerals in tooth enamel; sugar alone will not cause caries.
Diphtheria	*Corynebacterium diphtheriae*	Humans	Acute inflammation and lesions of mucous membranes. Spread through contact with infected individual. Vaccine available.
Gonorrhea	*Neisseria gonorrhoeae*	Humans only	STD, on the increase worldwide. Usually not fatal.
Hansen disease (leprosy)	*Mycobacterium leprae*	Humans, feral armadillos	Chronic infection of the skin; worldwide incidence about 10–12 million, especially in southeast Asia. Spread through contact with infected individuals.
Lyme disease	*Borrelia burgdorferi*	Ticks, deer, small rodents	Spread through bite of infected tick. Lesion followed by malaise, fever, fatigue, pain, stiff neck, and headache.
Peptic ulcers	*Helicobacter pylori*	Humans	Originally thought to be caused by stress or diet, most peptic ulcers now appear to be caused by this bacterium; good news for ulcer sufferers because it can be treated with antibiotics.
Plague	*Yersinia pestis*	Fleas of wild rodents: rats and squirrels	Killed ¼ of the population of Europe in the 14th century; endemic in wild rodent populations of the western U.S. today.
Pneumonia	*Streptococcus, Mycoplasma, Chlamydia*	Humans	Acute infection of the lungs; often fatal without treatment.
Tuberculosis	*Mycobacterium tuberculosis*	Humans	An acute bacterial infection of the lungs, lymph, and meninges. Its incidence is on the rise, complicated by the development of new strains of the bacterium that are resistant to antibiotics.
Typhoid fever	*Salmonella typhi*	Humans	A systemic bacterial disease of worldwide incidence. Less than 500 cases a year are reported in the U.S. The disease is spread through contaminated water or foods (such as improperly washed fruits and vegetables). Vaccines are available for travelers.
Typhus	*Rickettsia typhi*	Lice, rat fleas, humans	Historically a major killer in times of crowding and poor sanitation; transmitted from human to human through the bite of infected lice and fleas. Typhus has a peak untreated mortality rate of 70%.

Sexually Transmitted Diseases

A number of bacteria cause sexually transmitted diseases (STDs). Three that are particularly important are gonorrhea, syphilis, and chlamydia (figure 27.9).

Gonorrhea. **Gonorrhea** is one of the most prevalent communicable diseases in North America. Caused by the bacterium *Neisseria gonorrhoeae*, gonorrhea can be transmitted through sexual intercourse or any other sexual contacts in which body fluids are exchanged, such as oral or anal intercourse. Gonorrhea can infect the throat, urethra, cervix, or rectum and can spread to the eyes and internal organs, causing conjunctivitis (a severe infection of the eyes) and arthritic meningitis (an infection of the joints). Left untreated in women, gonorrhea can cause pelvic inflammatory disease (PID), a condition in which the fallopian tubes become scarred and blocked. PID can eventually lead to sterility. The incidence of gonorrhea has been on the decline in the United States, but it remains a serious threat worldwide.

Syphilis. **Syphilis,** a very destructive STD, was once prevalent but is now less common due to the advent of blood-screening procedures and antibiotics. Syphilis is caused by a spirochete bacterium, *Treponema pallidum*, that is transmitted during sexual intercourse or through direct contact with an open syphilis sore. The bacterium can also be transmitted from a mother to her fetus, often causing damage to the heart, eyes, and nervous system of the baby.

Once inside the body, the disease progresses in four distinct stages. The first, or primary stage, is characterized by the appearance of a small, painless, often unnoticed sore called a chancre. The chancre resembles a blister and occurs at the location where the bacterium entered the body about three weeks following exposure. This stage of the disease is highly infectious, and an infected person may unwittingly transmit the disease to others.

The second stage of syphilis is marked by a rash, a sore throat, and sores in the mouth. The bacteria can be transmitted at this stage through kissing or contact with an open sore. The third stage of syphilis is symptomless. This stage may last for several years, and at this point, the person is no longer infectious, but the bacteria are still present in the body, attacking the internal organs. The fourth and final stage of syphilis is the most debilitating, as the damage done by the bacteria in the third stage becomes evident. Sufferers at this stage of syphilis experience heart disease, mental deficiency, and nerve damage, which may include loss of motor functions or blindness.

Chlamydia. Sometimes called the "silent STD," **chlamydia** is caused by an unusual bacterium, *Chlamydia trachomatis*, that has both bacterial and viral characteristics. Like a bacterium, it is susceptible to antibiotics, and like a virus, it depends on its host to replicate its genetic material, meaning that it is an obligate internal parasite.

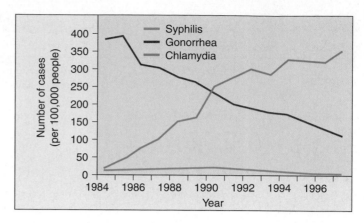

FIGURE 27.9
Trends in sexually transmitted diseases in the United States. How is it possible for the incidence of one STD (chlamydia) to rise as another (gonorrhea) falls?

The bacterium is transmitted through vaginal, anal, or oral intercourse with an infected person.

Chlamydia is called the "silent STD" because women usually experience no symptoms until after the infection has become established. In part because of this symptomless nature, the incidence of chlamydia has skyrocketed, increasing by more than sevenfold nationally since 1984. The effects of an established chlamydia infection on the female body are extremely serious. Chlamydia can cause pelvic inflammatory disease (PID), which can lead to sterility.

It has recently been established that infection of the reproductive tract by *Chlamydia* can cause heart disease. *Chlamydia* produce a peptide similar to one produced by cardiac muscle. As the body's immune system tries to fight off the infection, it recognizes this peptide. The similarity between the bacterial and cardiac peptides confuses the immune system, and T cells attack cardiac muscle fibers, inadvertently causing inflammation of the heart and other problems.

Within the last few years, two types of tests for chlamydia have been developed that look for the presence of the bacteria in the discharge from men and women. The treatment for chlamydia is antibiotics, usually tetracycline (penicillin is not effective against chlamydia). Any woman who experiences the symptoms associated with this STD should be tested for the presence of the chlamydia bacterium; otherwise, her fertility may be at risk.

This discussion of STDs may give the impression that sexual activity is fraught with danger, and in a way, it is. It is folly not to take precautions to avoid STDs. The best way to do this is to know one's sexual partners well enough to discuss the possible presence of an STD. Condom use can also prevent transmission of most of the diseases. Responsibility for protection lies with each individual.

Bacterial diseases have a major impact worldwide. Sexually transmitted diseases (STDs) are becoming increasingly widespread among Americans as sexual activity increases.

Benefits of Prokaryotes

Prokaryotes, the first organisms to evolve on earth, were largely responsible for creating the properties of the atmosphere and the soil over billions of years. Today, they affect our lives in many important ways.

Prokaryotes and the Environment

Life on earth depends critically on the cycling of chemical elements between organisms and the physical environments in which they live—that is, between the biological and physical elements of ecosystems. Prokaryotes and fungi play many key roles in this chemical cycling, a process discussed in detail in chapter 55.

Decomposition. The carbon, nitrogen, phosphorus, sulfur, and other atoms of which your body is built all have come from the physical environment, and when you die and decay, they all return to it. The prokaryotes and fungi that carry out the decomposition portion of chemical cycles, releasing a dead body's atoms to the environment, are called decomposers.

Fixation. Other prokaryotes play important roles in fixation, the other half of chemical cycles, helping to return elements from the nonliving environment to organisms. The role of photosynthetic prokaryotes in recycling carbon is obvious. The organic compounds that plants, algae, and photosynthetic prokaryotes produce from CO_2 pass up through food chains to form the bodies of all the ecosystem's heterotrophs—all the animals and fungi and nonphotosynthetic protists. Cyanobacteria are thought to have added oxygen to the earth's atmosphere as a by-product of their photosynthesis.

Less obvious, but no less critical to life on earth, is the role of prokaryotes in recycling nitrogen. The nitrogen in the earth's atmosphere is in the form of N_2 gas. A triple covalent bond links the two nitrogen atoms and is not easy to break. Among the earth's organisms, only prokaryotes are able to accomplish this feat, reducing N_2 to ammonia (NH_3), which is used to build amino acids and other nitrogen-containing biological molecules. When the organisms that contain these molecules die, other prokaryotes, called *denitrifiers*, return the nitrogen to the atmosphere, completing the cycle.

To fix atmospheric nitrogen, prokaryotes employ an enzyme complex called nitrogenase, encoded by a set of genes called *nif* ("nitrogen fixation") genes. The *nif* gene complex is found in a wide range of free-living prokaryotes. Under anaerobic conditions, the most important free-living nitrogen fixers are members of the gram-positive bacterial genus *Clostridium*. In aquatic environments, nitrogen fixation is carried out largely by cyanobacteria such as *Anabaena*. Because the nitrogen fixation process is strictly anaerobic and extremely sensitive to oxygen, cyanobacteria possess het-

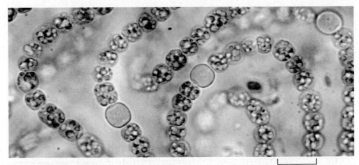

22.70 µm

FIGURE 27.10
Cyanobacteria carry out nitrogen fixation. Cells of this cyanobacterium, *Anabaena*, adhere in chains, and nitrogen is fixed in the enlarged, specialized cells, which are called heterocysts.

erocysts, specialized cells impermeable to oxygen (figure 27.10). In soil, nitrogen fixation occurs in the roots of plants that harbor symbiotic colonies of nitrogen-fixing bacteria. These associations include *Rhizobium* (a genus of proteobacteria; see figure 27.3) with legumes, *Frankia* (an actinomycete) with many woody shrubs, and *Anabaena* with water ferns.

Symbiotic Prokaryotes

Many prokaryotes live in symbiotic association with eukaryotes. Symbiosis (Greek, "living together") refers to the ecological relationship between different species that live in direct contact with each other. The symbiotic association of nitrogen-fixing bacteria with plant roots is an example of *mutualism*, a form of symbiosis in which both parties benefit. The bacteria supply the plant with useful nitrogen, and the plant supplies the bacteria with sugars and other organic nutrients.

Many bacteria live symbiotically within the digestive tracts of animals. Cows and other grazing mammals are unable to digest cellulose (much of the bulk of the grass they eat) because they lack the required cellulase enzyme. However, living in a special section of the cow's gut called the rumen are colonies of cellulase-producing bacteria that make the cellulase enzyme, allowing the cow to digest its food (see chapter 43 for a fuller account). Similarly, humans maintain large colonies of bacteria in the large intestine that produce vitamins—particularly B_{12} and K—which the human body cannot make itself.

Many bacteria inhabit the outer surfaces of animals and plants without doing damage. These are examples of *commensalism*, in which one organism (the bacterium) receives benefits while the animal or plant is neither benefited nor harmed. As we have seen in this chapter, some pathogenic bacteria do harm their hosts. *Parasitism* is a form of symbiosis in which one member (in this case, the bacterium) is benefited, and the other (the infected animal or plant) is harmed.

Bacteria and Genetic Engineering

Applying genetic engineering methods to produce improved strains of bacteria for commercial use holds enormous promise for the future. For example, bacteria are under intense investigation as nonpolluting insect control agents. *Bacillus thuringiensis* attacks insects in nature, and improved, highly specific strains of *B. thuringiensis* have greatly increased its usefulness as a biological control agent. Genetically modified bacteria have also been extraordinarily useful in producing insulin and other therapeutic proteins.

The use of organisms to remove pollutants from water, air, and soil is called *bioremediation*. In sewage treatment plants, the solid matter from raw sewage is allowed to settle, and then gradually mixed with a growing culture of bacteria and archaebacteria. Genetically modified bacteria—in this case, pseudomonads (proteobacteria)—are being employed successfully to remove petroleum products from beaches after oil tanker spills (figure 27.11).

Bacteria are now widely used as "biofactories" in the commercial production of a variety of enzymes, vitamins, and antibiotics, an approach discussed in detail in chapter 16. Most of the insulin used to treat diabetes is produced this way. Immense cultures of bacteria, often genetically modified to enhance performance, are used to produce commercial acetone, and of course to ferment sugars into alcohol in the production of beer.

Bacteria and Bioweapons

With the success of antibiotics in treating bacterial killer diseases such as typhus and cholera, many of us have been lulled into thinking that the battle against infectious bacteria has been won. However, to defeat an infectious disease, you must control its transmission, and unfortunately the new century has seen the introduction of a more deadly way for disease to spread—by the deliberate action of people using disease as a weapon. For several decades, the United States and Russia carried out extensive bioweapons programs, and while the U.S. program was discontinued in 1969, the Russian bioweapons program continued for another two decades. In 2001, bioterrorists struck at the United States with anthrax spores developed in the American bioweapons program, adding them to letters sent through the mail. While few died, the attack points out the dangerous potential of biological weapons. Anthrax and the smallpox virus pose the greatest immediate threats, although gene-modified pathogens offer an even greater future danger. Bioweapons are discussed at length in the enhancement chapter *Infectious Disease and Bioterrorism*, which you can read at www.mhhe.com/raven7.

Prokaryotes make many important contributions to the world ecosystem, playing key roles as cyclers of carbon and nitrogen, as symbionts, as bioremediation agents, and as aids in the commercial production of medicinal drugs and other products.

FIGURE 27.11
Using bacteria to clean up oil spills.   Bacteria can often be used to remove environmental pollutants, such as petroleum hydrocarbons and chlorinated compounds. In rocky areas contaminated by the *Exxon Valdez* oil spill (*left*), oil-degrading bacteria produced dramatic results (*right*).

27.1 Prokaryotes are the smallest and most numerous organisms.

The Prevalence of Prokaryotes

- Archaebacteria and bacteria are the oldest, structurally simplest, and most abundant forms of life on earth. (p. 546)
- About 5000 different kinds of prokaryotes are currently recognized. (p. 546)
- In the 1980s, a new discovery led to the division of prokaryotes into two groups: archaebacteria and bacteria. (p. 546)
- Prokaryotes are ubiquitous on earth and exist in some of the most hostile environments on the planet. (p. 546)
- Bacteria exhibit one of three basic structures: bacillus (straight and rod-shaped), coccus (spherical-shaped), and spirillum (long and helical-shaped). (p. 546)
- Prokaryotes differ from eukaryotes in many important features, including unicellularity, cell size, chromosomes, cell division and genetic recombination, internal compartmentalization, flagella, and metabolic diversity. (pp. 546–547)

27.2 Prokaryotes exhibit considerable diversity in both structure and metabolism.

Prokaryotic Diversity

- Key characteristics used in classifying bacteria are photosynthetic or nonphotosynthetic, motile or nonmotile, unicellular or colony-forming or filamentous, and formation of spores or division by transverse binary fission. (p. 548)
- Molecular approaches to classification include analysis of amino acid sequences of key proteins, analysis of nucleic acid base sequences, nucleic acid hybridization, ribosomal RNA sequencing, and whole-genome sequencing. (p. 548)
- Archaebacteria and bacteria differ in four key areas: plasma membrane, cell wall, gene translation machinery, and gene architecture. (p. 548)

27.3 Prokaryotes are more complex than commonly supposed.

The Prokaryotic Cell Surface

- A prokaryotic cell wall usually consists of peptidoglycan, forming either thick or thin walls; prokaryotes can be classified and identified by Gram staining as either gram-positive or gram-negative bacteria. (p. 550)
- Many prokaryotes have slender, rigid flagella for propulsion, and some prokaryotes have pili to help in attachment and genetic exchange. (p. 550)

The Cell Interior

- Prokaryotic cells lack the extensive functional compartmentalization found in eukaryotic cells, but do possess internal membranes, a nucleoid region, and ribosomes. (p. 551)

Prokaryotic Variation

- Bacteria have a very short generation time, thus mutation and genetic recombination play an important role in producing and maintaining genetic diversity. (p. 552)

Prokaryotic Metabolism

- Autotrophs can be broken down into photoautotrophs and chemoautotrophs; heterotrophs can be broken down into photoheterotrophs and chemoheterotrophs. Chemoheterotrophs appear to be the most common. (p. 553)
- Almost every kind of plant is susceptible to one or more bacterial diseases. (p. 553)

27.4 Prokaryotes are responsible for many diseases, but they also make important contributions to ecosystems.

Human Bacterial Diseases

- Tuberculosis and dental caries are just two of the many human diseases caused by bacteria. A few others are cholera, leprosy, tetanus, bacterial pneumonia, whooping cough, diptheria, and Lyme disease. (p. 554)
- A number of bacteria also cause sexually transmitted diseases such as gonorrhea, syphilis, and chlamydia. (p. 556)

Benefits of Prokaryotes

- Prokaryotes affect everyday life in many ways, including decomposition of dead organisms, chemical fixation, symbiotic relationships (such as that between nitrogen-fixing bacteria and plant roots), genetic engineering improvements in agricultural crops, commercial production of pharmaceuticals, and bioremediation. (pp. 557–558)
- Some disease-causing bacteria can even be used as forms of bioterrorism. (p. 558)

Self Test

1. Once they evolved, _____ forever changed the atmosphere on earth.
   a. bacteria
   b. archaebacteria
   c. cyanobacteria
   d. eukaryotes
2. Which of the following statements is *not* true of prokaryotic cells?
   a. Prokaryotic cells are multicellular.
   b. Prokaryotic cells do not have a nucleus.
   c. Prokaryotic cells have circular DNA.
   d. Prokaryotic cells have flagella.
3. When bacterial cell walls are covered with an outer membrane of lipopolysaccharide, they are
   a. gram-positive.
   b. gram-negative.
   c. encapsulated.
   d. endospores.
4. The prokaryotic genome is contained in the
   a. plasmid.
   b. endospore.
   c. pilus.
   d. nucleoid region.
5. Archaebacteria and bacteria differ in all of the following ways except
   a. the structure of the cell wall.
   b. their presence in nonextreme environments.
   c. the structure of the plasma membrane.
   d. the kinds of ribosomal proteins they possess.
6. Genetic recombination has led to antibiotic resistance through the transfer of
   a. pili.
   b. endospores.
   c. plasmids.
   d. bacterial chromosomes.
7. Prokaryotic organisms that obtain their energy by oxidizing inorganic substances are called
   a. chemoautotrophs.
   b. photoautotrophs.
   c. chemoheterotrophs.
   d. photoheterotrophs.
8. What disease is experiencing new outbreaks because of antibiotic resistance?
   a. smallpox
   b. cholera
   c. tuberculosis
   d. diphtheria
9. The disease sometimes referred to as the "silent STD" because it is usually asymptomatic in women early on is
   a. chlamydia.
   b. gonorrhea.
   c. syphilis.
   d. pelvic inflammatory disease.
10. Which of the following can be attributed to bacteria?
    a. decomposition of dead organic matter
    b. increasing oxygen levels in the atmosphere
    c. production of antibiotics
    d. all of these

Test Your Visual Understanding

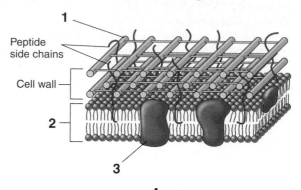

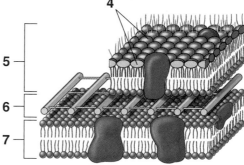

1. This figure shows two kinds of bacterial cell walls. Match the following labels with the appropriate numbered structures.
   lipopolysaccharides
   outer membrane
   peptidoglycan
   plasma membrane
   protein
2. What type of bacteria would have the cell wall structure shown in the upper figure? What type of bacteria would have the cell wall structure shown in the lower figure?

Apply Your Knowledge

1. Bacterial populations grow through binary fission, which means the bacteria divide in half such that each bacterium gives rise to two bacteria. Assume that a bacterium is placed in culture and undergoes binary fission every 30 minutes. The addition of a competing bacterium reduces the population size by 25%. How many bacteria from the original culture will be present after 24 hours?
2. Justify the assertion that life on earth could not exist without prokaryotes.
3. What are the functions of antibiotics in the prokaryotes that produce them?
4. What are endospores? Why would they be an advantage? Would a bacterium with the ability to form endospores have a greater or lesser chance of extinction? Why?

28

Protists

Concept Outline

28.1 Eukaryotes probably arose by endosymbiosis.

Endosymbiosis. Mitochondria and chloroplasts are thought to have arisen by engulfing aerobic and photosynthetic bacteria.

28.2 The kingdom Protista is by far the most diverse of any eukaryotic kingdom.

The Challenge of Classifying the Protists. Groups of protists are evolutionarily unrelated, which presents many challenges for taxonomists trying to classify them.

General Biology of the Protists. The kingdom Protista contains members exhibiting a wide range of methods of locomotion, nutrition, and reproduction.

28.3 Protists can be categorized into six groups.

Euglenozoa. Euglenoids and kinetoplastids were among the first free-living eukaryotes to contain mitochondria.

Alveolata. Characterized by a space under the plasma membrane, the dinoflagellates, apicomplexes and ciliates, have varied modes of locomotion.

Stramenopila and Rhodophyta. These brown and red algae have origins that are distinct from those of the green plants.

Chlorophyta. Chlorophyta is one of two green plant lineages. The other lineage (Streptophyta) gave rise to the land plants.

Choanoflagellida and Protists That Are Difficult to Categorize. Choanoflagellates are most like the common ancestor of all animals. Around 60 other lineages of protists have been identified, but there is not yet enough information to determine their relationships to other protist lineages.

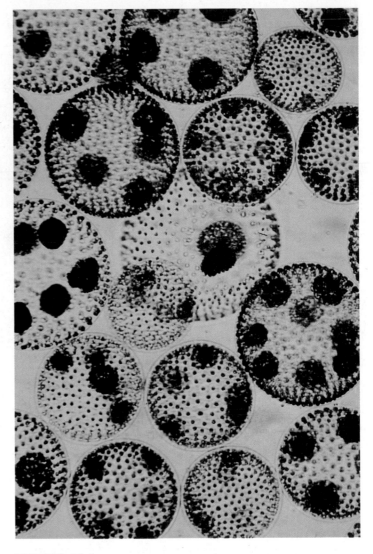

FIGURE 28.1
Volvox, **a colonial protist.** The protists are a large, diverse group of primarily single-celled organisms, a group from which the other three eukaryotic kingdoms each evolved.

For more than half of the long history of life on earth, all life was microscopic in size. The biggest organisms that existed for over 2 billion years were single-celled bacteria fewer than 6 micrometers thick. These prokaryotes lacked internal membranes, except for invaginations of surface membranes in photosynthetic bacteria. The first evidence of a different kind of organism is found in tiny fossils in rock 2.5 billion years old. These fossil cells are much larger than bacteria (up to ten times larger) and contain internal membranes and what appear to be small, membrane-bounded structures. The complexity and diversity of form among these single cells is astonishing. The step from relatively simple to quite complex cells marks one of the most important events in the evolution of life, the appearance of a new kind of organism, the eukaryote (figure 28.1). Eukaryotes that are clearly not animals, plants, or fungi have been lumped together and called protists.

28.1 Eukaryotes probably arose by endosymbiosis.

Endosymbiosis

What was the first eukaryote like? We cannot be sure, but a good model is *Pelomyxa palustris*, a single-celled, nonphotosynthetic organism that appears to represent an early stage in the evolution of eukaryotic cells (figure 28.2). The cells of *Pelomyxa* are much larger than bacterial cells and contain a complex system of internal membranes. Although they resemble some of the largest early fossil eukaryotes, these cells are unlike those of any other eukaryote because *Pelomyxa* lacks mitochondria. No clear alternative to nuclear mitosis (chromosome replication) has been described for *Pelomyxa*. Its nuclei divide by pinching apart into two daughter nuclei, around which new membranes form. Although *Pelomyxa* cells lack mitochondria, two kinds of bacteria living within them may play the same role that mitochondria do in most other eukaryotes.

Biologists know very little about the origin of *Pelomyxa*, except that in many of its fundamental characteristics it resembles the archaebacteria far more than the bacteria. Because of this general resemblance, it is widely assumed that the first eukaryotic cells were nonphotosynthetic descendants of archaebacteria.

What about the wide gap between *Pelomyxa* and all other eukaryotes? Where did mitochondria come from? Most biologists agree with the theory of **endosymbiosis,** which proposes that mitochondria originated as symbiotic, aerobic (oxygen-requiring) bacteria (figure 28.3). Symbiosis (Greek *syn*, "together with," + *bios*, "life") means living together in close association. Recall from chapter 5 that mitochondria are sausage-shaped organelles 1 to 3 micrometers long, about the same size as most bacteria. Mitochondria are bounded by two membranes. Aerobic bacteria are thought to have become mitochondria when they were engulfed by ancestral eukaryotic cells, much like *Pelomyxa*, early in the history of eukaryotes.

The most similar bacteria to mitochondria today are the nonsulfur purple bacteria, which are able to carry out oxidative metabolism (described in chapter 9). In mitochondria, the outer membrane is smooth and is thought to be derived from the endoplasmic reticulum of the host cell, which, like *Pelomyxa*, may have already contained a complex system of internal membranes. The inner membrane of a mitochon-

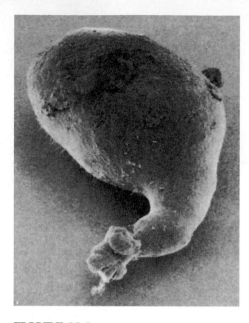

FIGURE 28.2
Pelomyxa palustris. This unique, amoeba-like protist lacks mitochondria. No alternative to nuclear mitosis in *Pelomyxa* has been described. *Pelomyxa* may represent a very early stage in the evolution of eukaryotic cells.

drion is folded into numerous layers, resembling the folded membranes of nonsulfur purple bacteria; embedded within this membrane are the proteins that carry out oxidative metabolism. The engulfed bacteria became the interior portion of the mitochondria we see today. Host cells were unable to carry out the Krebs cycle or other metabolic reactions necessary for living in an atmosphere that contained increasing amounts of oxygen before they had acquired these bacteria.

Pelomyxa is not the only protist to lack mitochondria. These anaerobic protists are not necessarily closely related. *Giardia*, another protist without mitochondria, can pass from human to human via contaminated water and cause diarrhea. Mitochondrial genes are found in their nuclei, leading to the conclusion that *Giardia* evolved from aerobes. Thus, *Giardia* is unlikely to represent an early protist.

During the billion and a half years in which mitochondria have existed as endosymbionts within eukaryotic cells, most of their genes have been transferred to the chromosomes of the host cells—but not all. Each mitochondrion still has its own genome, a circular, closed molecule of DNA similar to that found in bacteria, on which are located genes encoding the essential proteins of oxidative metabolism. These genes are transcribed within the mitochondrion, using mitochondrial ribosomes that are smaller than those of eukaryotic cells, very much like bacterial ribosomes in size and structure. Mitochondria divide by simple fission, just as bacteria do, and they also replicate and sort their DNA much as bacteria do. However, nuclear genes direct the process, and mitochondria cannot be grown outside of the eukaryotic cell, in cell-free culture.

The mechanisms of mitosis and cytokinesis, now so common among eukaryotes, did not evolve all at once. Traces of very different, and possibly intermediate, mechanisms survive today in some of the eukaryotes. In fungi and some groups of protists, for example, the nuclear membrane does not dissolve, and mitosis is confined to the nucleus. When mitosis is complete in these organisms, the nucleus divides into two daughter nuclei, and only then does the rest of the cell divide. In most protists, or in plants or animals, the nuclear membrane breaks down during mitosis. We do not know if mitosis without nuclear membrane dissolution represents an intermediate step on the

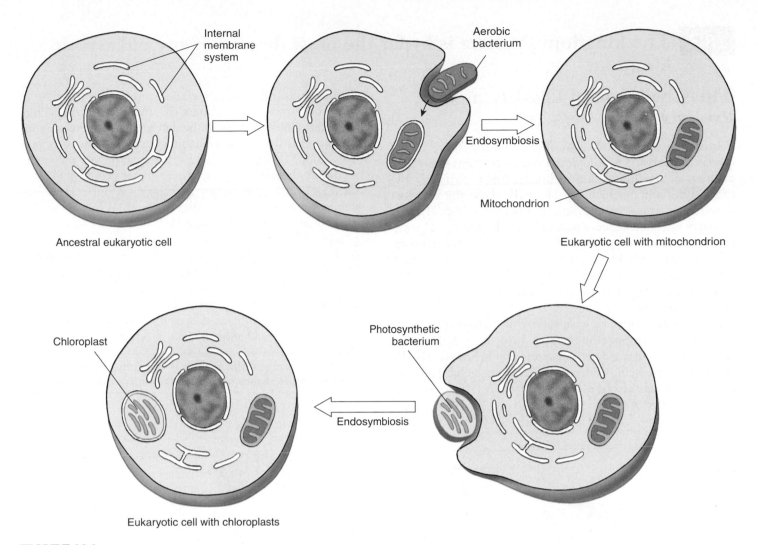

FIGURE 28.3
The theory of endosymbiosis.   Scientists propose that ancestral eukaryotic cells, which already had an internal system of membranes, engulfed aerobic bacteria, which then became mitochondria in the eukaryotic cell. Chloroplasts may also have originated this way, with eukaryotic cells engulfing photosynthetic bacteria.

evolutionary journey to the form of mitosis that is characteristic of most eukaryotes today, or if it is simply a different way of solving the same problem. There are no fossils in which we can see the interiors of dividing cells well enough to be able to trace the history of mitosis.

Endosymbiosis Is Not Rare

Many eukaryotic cells contain other endosymbiotic bacteria in addition to mitochondria. Plants and algae contain chloroplasts, bacteria-like organelles that were apparently derived from symbiotic photosynthetic bacteria. Chloroplasts have a complex system of inner membranes and a circle of DNA. Centrioles, organelles associated with the assembly of microtubules, resemble in many respects spirochete bacteria, and they contain bacteria-like DNA involved in the production of their structural proteins. While all mitochondria are thought to have arisen from a single symbiotic event, it is difficult to be sure with chloroplasts. Three biochemically distinct classes of chloroplasts exist, each resembling a different bacterial ancestor. Red algae possess pigments similar to those of cyanobacteria; plants and green algae more closely resemble the photosynthetic bacterium *Prochloron*; and brown algae and other photosynthetic protists resemble a third group of bacteria. This diversity of chloroplasts has led to the widely held belief that eukaryotic cells acquired chloroplasts by endosymbiosis at least three different times. Recent comparisons of chloroplast DNA sequences, however, suggest a single origin of chloroplasts, followed by very different post-endosymbiotic histories. For example, in each of the three main lines, different genes became relocated to the nucleus, lost, or modified.

The theory of endosymbiosis proposes that mitochondria and chloroplasts originated as symbiotic bacteria.

28.2 The kingdom Protista is by far the most diverse of any eukaryotic kingdom.

The Challenge of Classifying the Protists

Protists are the most diverse of the four kingdoms in the domain Eukarya. The kingdom Protista contains many unicellular, colonial, and multicellular groups. Probably the most important statement we can make about the kingdom Protista is that it is paraphyletic and not a kingdom at all; as a matter of convenience, single-celled eukaryotic organisms have typically been grouped together and called protists. This lumps 200,000 different and only distantly related forms together. The "single-kingdom" classification of the Protista is artificial and not representative of any evolutionary relationships.

New applications of a wide variety of molecular methods are providing important insights into the relationships among the protists. Many questions about how to classify the protists are being addressed with these techniques. Are protists best considered as several different kingdoms, each of equal rank with animals, plants, and fungi? Are some of the protists actually members of other kingdoms? While these questions continue to be debated, ever-increasing information is becoming available concerning which organisms among the protists are most likely to be monophyletic.

In this chapter, we group the 15 major protist phyla into six major monophyletic groups, based on our current understanding of phylogeny (figure 28.4). While these lineages may change, this approach allows us to examine groups with many shared traits. Keep in mind that about 60 of the protist lineages cannot yet be placed on the tree of life with any confidence! Protists exemplify the challenges and excitement of the revolutionary changes in taxonomy and phylogeny we explored in chapter 25. Understanding the evolution of protists is key to understanding the origins of plants, fungi, and animals.

The taxonomy of the protists is in a state of flux as new information shapes our understanding of this kingdom.

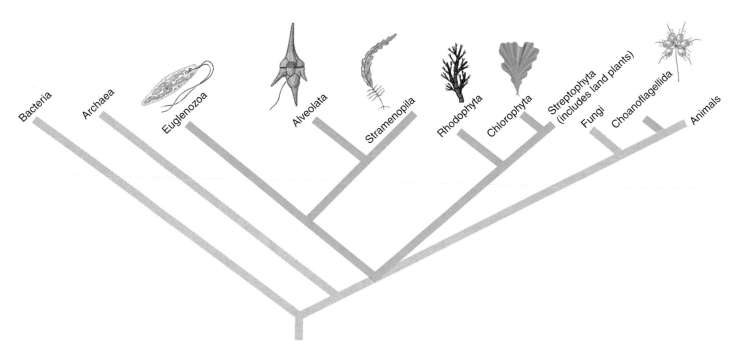

FIGURE 28.4
The challenge of protistan classification. Our understanding of the evolutionary relationships among protists is currently in flux. The most recent data support six major, monophyletic groups within the protists. Consider this a working model, not fact. The green algae (Chlorophyta) are not truly monophyletic in that another branch, Streptophyta, gave rise to the land plants. Protist lineages are shaded in blue.

General Biology of the Protists

Protists are united on the basis of a single negative characteristic: They are eukaryotes that are not fungi, plants, or animals. In all other respects, they are highly variable, with no uniting features. Many are unicellular (figure 28.5), but numerous colonial and multicellular groups exist. Most are microscopic, but some are as large as trees. They represent all symmetries, and exhibit all types of nutrition.

The Cell Surface

Protists possess a varied array of cell surfaces. Some protists, such as amoebas, are surrounded only by their plasma membrane. All other protists have a plasma membrane with extracellular material (ECM) deposited on the outside of the membrane. Some ECM forms strong cell walls. Diatoms and forams secrete glassy shells of silica.

Locomotor Organelles

Movement in protists is also accomplished by diverse mechanisms. Protists move chiefly by either flagellar rotation or pseudopodial movement. Many protists wave one or more flagella to propel themselves through the water, while others use banks of short, flagella-like structures called cilia to create water currents for their feeding or propulsion. Pseudopods (Greek *pseudo-*, "false," + *podos*, "foot") are the chief means of locomotion among amoebas (see chapter 5), whose pseudopods are large, blunt extensions of the cell body called lobopodia. Other related protists extend thin, branching protrusions called filopodia. Still other protists extend long, thin pseudopods called axopodia supported by axial rods of microtubules. Axopodia can be extended or retracted. Because the tips can adhere to adjacent surfaces, the cell can move by a rolling motion, shortening the axopodia in front and extending those in the rear.

Cyst Formation

Many protists with delicate surfaces are successful in quite harsh habitats. How do they manage to survive so well? They form cysts, dormant forms of a cell with resistant outer coverings in which cell metabolism is more or less completely shut down. Not all cysts are so sturdy. Vertebrate parasitic amoebas, for example, form cysts that are quite re-

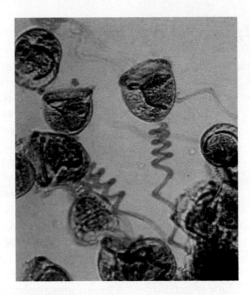

FIGURE 28.5
A unicellular protist. Kingdom Protista is a catch-all kingdom for many different groups of unicellular organisms, such as this *Vorticella* (a ciliate), which is heterotrophic, feeds on bacteria, and has a retractable stalk.

sistant to gastric acidity, but will not tolerate desiccation or high temperature.

Nutrition

Protists employ every form of nutritional acquisition except the chemoautotrophic form, which has so far been observed only in prokaryotes. Some protists are photosynthetic and are called **phototrophs.** Others are heterotrophs that obtain energy from organic molecules synthesized by other organisms. Among the heterotrophic protists, those that ingest visible particles of food are called **phagotrophs.** Phagotrophs ingest food particles into intracellular vesicles called food vacuoles or phagosomes. Lysosomes fuse with the food vacuoles, introducing enzymes that digest the food particles within. Digested molecules are absorbed across the vacuolar membrane. Protists that ingest food in soluble form are called **osmotrophs.**

Reproduction

Protists typically reproduce asexually, although some protists have an obligate sexual reproductive phase and others undergo sexual reproduction at times of stress, including food shortages. Asexual reproduction involves mitosis, but the process is often somewhat different from the mitosis that occurs in multicellular animals. For example, the nuclear membrane often persists throughout mitosis, with the microtubular spindle forming within it. One category of asexual reproduction is fission. The most common type of fission is **binary fission,** in which a cell simply splits into nearly equal halves. When the progeny cell is considerably smaller than its parent and then grows to adult size, the fission is called **budding.** In multiple fission, or **schizogony,** common among some protists, fission is preceded by several nuclear divisions, so that fission produces several individuals almost simultaneously.

Sexual reproduction also takes place in many forms among the protists. In ciliates and some flagellates, meiosis results in gamete formation (gametic meiosis). Thus, these protists are diploid for most of their life cycle, as are metazoans. In spore-producing protists, meiosis occurs directly *after* fertilization, and all the individuals that are produced are haploid until the next zygote is formed. In algae, both haploid and diploid cells undergo mitosis, producing an alternation of generations similar to that seen in plants.

Protists exhibit a wide range of forms, locomotion, nutrition, and reproduction.

28.3 Protists can be categorized into six groups.

Six lineages of protists have been identified—Euglenozoa, Alveolata, Stramenopila, Rhodophyta, Chlorophyta, and Choanoflagellida. However, many other protists are not close relatives of any of these groups.

Euglenozoa

Euglenoids

Euglenoids diverged early and were among the earliest free-living eukaryotes to possess mitochondria. Euglenoids clearly illustrate the impossibility of distinguishing "plants" from "animals" among the protists. About one-third of the approximately 40 genera of euglenoids have chloroplasts and are fully autotrophic; the others lack chloroplasts, ingest their food, and are heterotrophic.

Some euglenoids with chloroplasts may become heterotrophic in the dark; the chloroplasts become small and nonfunctional. If they are put back in the light, they may become green within a few hours. Photosynthetic euglenoids may sometimes feed on dissolved or particulate food.

Individual euglenoids range from 10 to 500 micrometers long and are highly variable in form. Interlocking proteinaceous strips arranged in a helical pattern form a flexible structure called the pellicle, which lies within the plasma membrane of the euglenoids. Because its pellicle is flexible, a euglenoid is able to change its shape.

Reproduction in this phylum occurs by mitotic cell division. The nuclear envelope remains intact throughout the process of mitosis. No sexual reproduction is known to occur in this group.

In *Euglena* (figure 28.6), the genus for which the phylum is named, two flagella are attached at the base of a flask-shaped opening called the reservoir, which is located at the anterior end of the cell. One of the flagella is long and has a row of very fine, short, hairlike projections along one side. A second, shorter flagellum is located within the reservoir but does not emerge from it. Contractile vacuoles collect excess water from all parts of the organism and empty it into the reservoir, which apparently helps regulate the osmotic pressure within the organism. The stigma, an organ that also occurs in the green algae (phylum Chlorophyta), is light-sensitive and helps these photosynthetic organisms move toward light.

Cells of *Euglena* contain numerous small chloroplasts. These chloroplasts, like those of the green algae and plants, contain chlorophylls *a* and *b*, together with carotenoids. Although the chloroplasts of euglenoids differ somewhat in structure from those of green algae, they probably had a common origin. It seems likely that euglenoid chloroplasts ultimately evolved from a symbiotic relationship through ingestion of green algae. Recent phylogenetic evidence indicates that *Euglena* had multiple origins within the Euglenoids, and the concept of a single *Euglena* genus is now being debated.

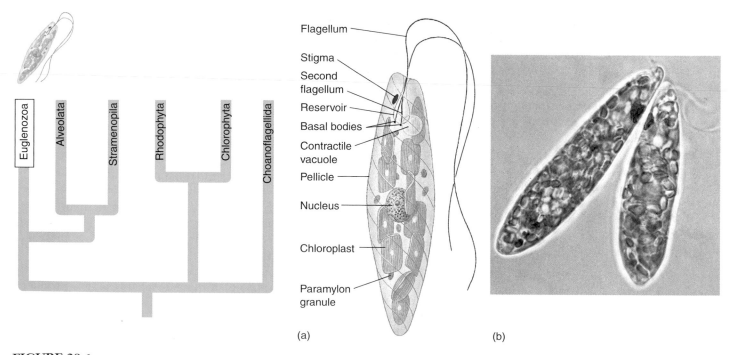

(a)

(b)

FIGURE 28.6

Euglenoids. (*a*) Diagram of *Euglena*. Paramylon granules are areas where food reserves are stored. (*b*) Micrograph of individuals of the genus *Euglena*.

Kinetoplastids

A second major group within the Euglenozoa is the kinetoplastids. The name kinetoplastid refers to a unique, single mitochondrion in each cell. The mitochondria have two types of DNA—mini-circles and maxi-circles. (Remember that prokaryotes have circular DNA, and mitochondria had prokaryotic origins.) This mitochondrial DNA is responsible for very rapid glycolysis and also for an unusual kind of editing of the DNA by guide RNAs encoded in the mini-circles.

Parasitism has evolved multiple times within the kinetoplastids. Trypanosomes are kinetoplastids that cause many serious human diseases, the most familiar being trypanosomiasis, also known as African sleeping sickness, which causes extreme lethargy and fatigue (figure 28.7). Other diseases caused by trypanosomes include East Coast fever, leishmaniasis, and Chagas disease, all of great importance in tropical areas where they afflict millions of people each year. Leishmaniasis, which is transmitted by sand flies, causes skin sores and in some cases can affect internal organs, leading to death. About 1.5 million new cases are reported each year. The rise in leishmaniasis in South America correlates with the move of infected individuals from rural to urban environments, where there is a greater chance of spreading the parasite. Chagas disease is caused by *Trypanosoma cruzi*. At least 90 million people, from the southern United States to Argentina, are at risk of contracting *T. cruzi* from small wild mammals that carry the parasite and can spread it to other mammals and humans through skin contact with urine and feces. Blood transfusions have also increased the spread of the infection. Chagas disease can lead to severe cardiac and digestive problems in humans and domestic animals, but appears to be tolerated in the wild mammals.

These diseases make it impossible to raise domestic cattle for meat or milk in a large portion of Africa and are resulting in high medical expenses and loss of time from the workforce in South America. Control is especially difficult because of the unique attributes of these organisms. For example, tsetse fly–transmitted trypanosomes have evolved an elaborate genetic mechanism for repeatedly changing the antigenic nature of their protective glycoprotein coat, thus dodging the antibodies their hosts produce against them (see chapter 48). Only a single one out of some 1000 to 2000 variable antigen genes is expressed at a time. Rearrangements of these genes during the asexual cycle of the organism allow for the expression of a seemingly endless variety of different antigen genes that maintain infectivity by the trypanosomes.

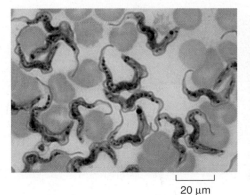

20 μm

(a)

(b)

FIGURE 28.7
A kinetoplastid. (*a*) *Trypanosoma* among red blood cells. The nuclei (dark-staining bodies), anterior flagella, and undulating, changeable shape of the trypanosomes are visible in this photograph (500×). (*b*) The tsetse fly, shown here sucking blood from a human arm, can carry trypanosomes.

When the trypanosomes are ingested by a tsetse fly, they embark on a complicated cycle of development and multiplication, first in the fly's gut and later in its salivary glands. It is their position in the salivary glands that allows them to move into their vertebrate host. Recombination has been observed between different strains of trypanosomes introduced into a single fly, suggesting that mating, syngamy, and meiosis occur, even though they have not been observed directly. Although most trypanosome reproduction is asexual, this sexual cycle, reported for the first time in 1986, affords still further possibilities for recombination in these organisms.

In the guts of the flies that spread them, trypanosomes are noninfective. When they are ready to transfer to the skin or bloodstream of their host, trypanosomes migrate to the salivary glands and acquire the thick coat of glycoprotein antigens that protect them from the host's antibodies. When they are taken up by a fly, the trypanosomes again shed their coats. The production of vaccines against such a system is complex, but tests are under way. Releasing sterilized flies to impede the reproduction of populations is another technique attempted to control the fly population. Traps made of dark cloth and scented like cows, but poisoned with insecticides, have likewise proved effective. Research is proceeding rapidly because the presence of tsetse flies with their associated trypanosomes blocks the use of some 11 million square kilometers of potential grazing land in Africa.

Euglenozoa include free-living and parasitic protists that move with flagella. *Euglena* have chloroplasts obtained via endosymbiosis, and trypanosomes have unusual mitochondria that use RNA editing.

Alveolata

Members of the Alveolata include the dinoflagellates, apicomplexes, and ciliates, all of which have a common lineage despite their diverse modes of locomotion. One common trait is the presence of either a space or alveoli (hence the name alveolata) below their plasma membranes.

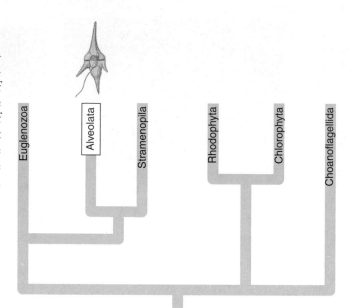

Dinoflagellates

Most dinoflagellates are photosynthetic unicells with two flagella. Dinoflagellates live in both marine and freshwater environments. Some dinoflagellates are luminous and contribute to the twinkling or flashing effects we sometimes see in the sea at night, especially in the tropics.

The flagella, protective coats, and biochemistry of dinoflagellates are distinctive, and the dinoflagellates do not appear to be directly related to any other phylum. Plates made of a cellulose-like material, often encrusted with silica, encase the dinoflagellate cells (figure 28.8). Grooves at the junctures of these plates usually house the flagella, one encircling the cell like a belt, and the other perpendicular to it. By beating in their grooves, these flagella cause the dinoflagellate to spin as it moves. Most have chlorophylls *a* and *c*, in addition to

carotenoids, so that in the biochemistry of their chloroplasts, they resemble the diatoms and the brown algae, possibly acquiring such chloroplasts by forming endosymbiotic relationships with members of those groups.

The poisonous and destructive "red tides" that occur frequently in coastal areas are often associated with great population explosions, or "blooms," of dinoflagellates, whose pigments color the water. Red tides have a profound, detrimental effect on the fishing industry in the United States. Some 20 species of dinoflagellates produce powerful toxins that inhibit the diaphragm and cause respiratory failure in many vertebrates. When the toxic dinoflagellates are abundant, many fishes, birds, and marine mammals may die.

Although sexual reproduction occurs under starvation conditions, dinoflagellates reproduce primarily by asexual cell division. Asexual cell division relies on a unique form of mitosis in which the permanently condensed chromosomes divide longitudinally within a permanent nuclear envelope. After the numerous chromosomes duplicate, the nucleus divides into two daughter nuclei. Also, the dinoflagellate chromosome is unique among eukaryotes in that the DNA is not generally complexed with histone proteins. In all other eukaryotes, the chromosomal DNA is complexed with histones to form nucleosomes, structures that represent the first order of DNA packaging in the nucleus. How dinoflagellates maintain distinct chromosomes with a small amount of histones remains a mystery.

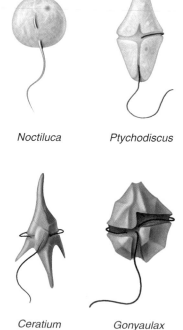

FIGURE 28.8
Some dinoflagellates:
Noctiluca, Ptychodiscus, Ceratium,* and *Gonyaulax.
Noctiluca, which lacks the heavy cellulose armor characteristic of most dinoflagellates, is one of the bioluminescent organisms that cause the waves to sparkle in warm seas. In the other three genera, the shorter, encircling flagellum is seen in its groove, with the longer one projecting away from the body of the dinoflagellate. (Not drawn to scale.)

Noctiluca   *Ptychodiscus*

Ceratium   *Gonyaulax*

Apicomplexes

Apicomplexes are spore-forming parasites of animals. They are called apicomplexes (short for apical complex) because of a unique arrangement of fibrils, microtubules, vacuoles, and other cell organelles at one end of the cell. The best-known apicomplex is the malarial parasite *Plasmodium*. *Plasmodium* glides inside the red blood cells of its host with amoeboid-like contractility. Like other apicomplexes, *Plasmodium* has a complex life cycle involving sexual and asexual phases and alternation between different hosts, in this case mosquitoes and humans (figure 28.9). Even though *Plasmodium* has mitochondria, it is a *microaerophil*, growing best in a low-O_2, high-CO_2 environment.

Efforts to eradicate malaria have focused on (1) eliminating the mosquito vectors; (2) developing drugs to poison the parasites that have entered the human body; and (3) developing vaccines. From the 1940s to the 1960s, wide-scale applications of DDT killed mosquitoes in the United States, Italy, Greece, and certain areas of Latin America. For a time, the worldwide elimination of malaria appeared possible. But this hope was soon crushed by the development of DDT-resistant mosquitoes in many regions. Further, there are serious environmental concerns about the use of DDT. In addition to the problems with resistant strains of mosquitoes, strains of *Plasmodium* have appeared that are resistant to the drugs historically used to kill them, including quinine.

An experimental vaccine containing a surface protein of one malaria-causing parasite, *P. falciparum*, seems to induce the immune system to defend against future infections. In tests, six out of seven vaccinated people did not get malaria after being fed upon by mosquitoes that carried *P. falciparum*. Many are hopeful that this new vaccine may be able to fight malaria.

Gregarines are another group of apicomplexes that use their distinctive apical complex to attach themselves in the intestinal epithelium of arthropods, annelids, and mollusks. Most of the gregarine body, aside from the apical complex, is in the intestinal cavity, and nutrients appear to be obtained through the apicomplex attachment to the cell. One of the larger gregarines is frequently hosted by earthworms.

Ciliates

As the name indicates, most ciliates feature large numbers of cilia. These heterotrophic, unicellular protists are 10 to 3000 micrometers long. Their cilia are usually arranged either in longitudinal rows or in spirals around the cell. Cilia are anchored to microtubules beneath the plasma membrane, and they beat in a coordinated fashion. In some groups, the cilia have specialized functions, becoming fused into sheets, spikes, and rods that may then function as mouths, paddles, teeth, or feet. The ciliates have a tough but flexible outer covering called the *pellicle* that enables them to squeeze through or move around obstacles.

All known ciliates have two different types of nuclei within their cells—small micronuclei and larger macronuclei (figure 28.10). Macronuclei divide by mitosis and are essential for the physiological function of the well-known ciliate *Paramecium*. The micronucleus of some *Tetrahymena pyriformis*, a common laboratory species, was experimentally removed in the 1930s, and their descendants continue to reproduce asexually to this day! *Paramecium*, however, is not immortal. The cells divide asexually for about 700 generations and then die if sexual reproduction has not occurred. The micronucleus in ciliates is needed only for sexual reproduction.

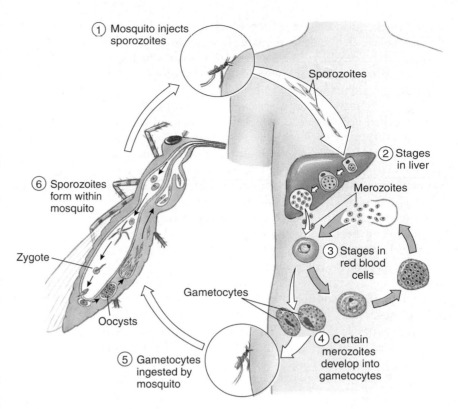

FIGURE 28.9
The life cycle of *Plasmodium*. *Plasmodium*, the apicomplex that causes malaria, has a complex life cycle that alternates between mosquitoes and mammals.

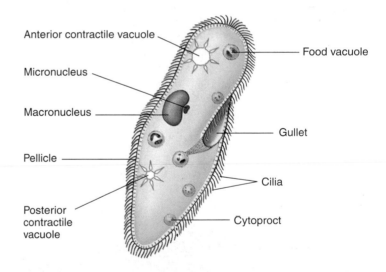

FIGURE 28.10
Paramecium. The main features of this familiar ciliate include cilia, two nuclei, and numerous specialized organelles.

Ciliates form vacuoles for ingesting food and regulating water balance. Food first enters the gullet, which in *Paramecium* is lined with cilia fused into a membrane (figure 28.10). From the gullet, the food passes into food vacuoles, where enzymes and hydrochloric acid aid in its digestion. Afterward, the vacuole empties its waste contents through a special pore in the pellicle called the **cytoproct,** which is essentially an exocytotic vesicle that appears periodically when solid particles are ready to be expelled. The contractile vacuoles, which regulate water balance, periodically expand and contract as they empty their contents to the outside of the organism.

Like most ciliates, *Paramecium* undergoes a sexual process called *conjugation,* in which two individual cells remain attached to each other for up to several hours (figure 28.11). Paramecia have multiple mating types. Only cells of two different genetically determined mating types can conjugate. Meiosis in the micronuclei of each individual produces several haploid micronuclei, and the two partners exchange a pair of these micronuclei through a cytoplasmic bridge between the two partners.

In each conjugating individual, the new micronucleus fuses with one of the micronuclei already present in that individual, resulting in the production of a new diploid micronucleus. After conjugation, the macronucleus in each cell disintegrates, while the new diploid micronucleus undergoes mitosis, thus giving rise to two new identical diploid micronuclei in each individual. One of these micronuclei becomes the precursor of the future micronuclei of that cell, while the other micronucleus undergoes multiple rounds of DNA replication, becoming the new macronucleus. This complete segregation of the genetic material is unique to the ciliates and makes them ideal organisms for the study of certain aspects of genetics.

Paramecium strains that kill other, sensitive strains of *Paramecium* long puzzled researchers. Initially, killer strains were believed to have genes coding for a substance toxic to sensitive strains. The true source of the toxin turned out to be an endosymbiotic bacterium in the "killer" strains. If this bacterium is engulfed by a "nonkiller" strain, the toxin is released, and the sensitive *Paramecium* dies.

Alveolata comprise what is believed to be a monophyletic group of organisms with varied forms of locomotion and reproduction.

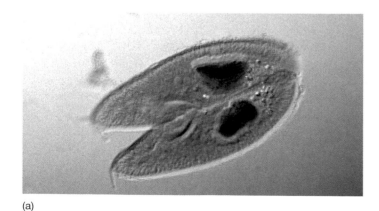

(a)

FIGURE 28.11
Life cycle of *Paramecium*. (*a,b*) In sexual reproduction, two mature cells fuse in a process called conjugation (100×).

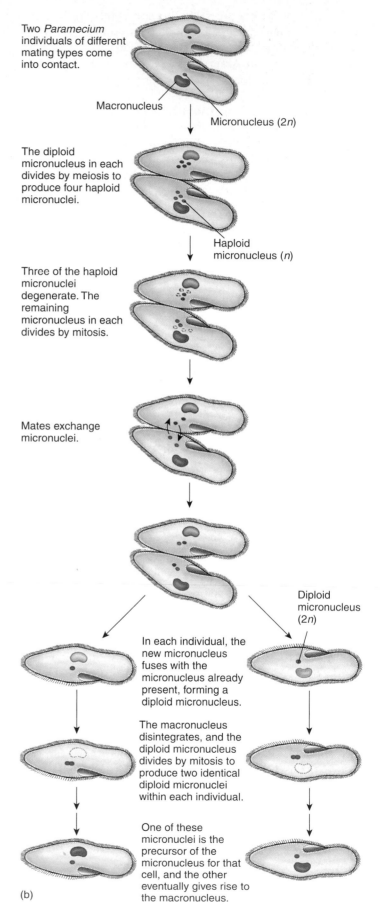

Two *Paramecium* individuals of different mating types come into contact.

Macronucleus

Micronucleus (2*n*)

The diploid micronucleus in each divides by meiosis to produce four haploid micronuclei.

Haploid micronucleus (*n*)

Three of the haploid micronuclei degenerate. The remaining micronucleus in each divides by mitosis.

Mates exchange micronuclei.

Diploid micronucleus (2*n*)

In each individual, the new micronucleus fuses with the micronucleus already present, forming a diploid micronucleus.

The macronucleus disintegrates, and the diploid micronucleus divides by mitosis to produce two identical diploid micronuclei within each individual.

One of these micronuclei is the precursor of the micronucleus for that cell, and the other eventually gives rise to the macronucleus.

(b)

Stramenopila and Rhodophyta

Stramenopila

Stramenopiles include **brown algae, diatoms,** and the **oomycetes** (water molds). Brown algae are the most conspicuous seaweeds in many northern regions (figure 28.12). The life cycle of the brown alga is marked by an alternation of generations between a sporophyte (diploid) and a gametophyte (haploid). Some sporophyte cells go through meiosis and produce spores. These spores germinate and undergo mitosis to produce the large individuals we recognize, such as the kelps. The gametophytes are often much smaller, filamentous individuals, perhaps a few centimeters in width.

Diatoms, members of the phylum Chrysophyta, are photosynthetic, unicellular organisms with unique double shells made of opaline silica, which are often strikingly and characteristically marked (figure 28.13). The shells of diatoms are like small boxes with lids, one half of the shell fitting inside the other. Their chloroplasts, containing chlorophylls *a* and *c*, as well as carotenoids, resemble those of the brown algae and dinoflagellates. Diatoms produce a unique carbohydrate called chrysolaminarin. Some diatoms move by using two long grooves, called raphes, which are lined with vibrating fibrils. The exact mechanism is still being unraveled and may involve the ejection of mucopolysaccharide streams from the raphe that propel the diatom. Pencil-shaped diatoms can slide back and forth on each other, creating an ever-changing shape.

All oomycetes are either parasites or saprobes (organisms that live by feeding on dead organic matter). They are distinguished from other protists by the structure of

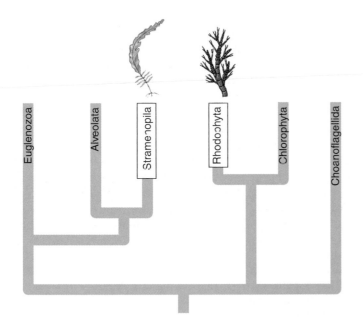

FIGURE 28.12
Brown algae. The massive "groves" of giant kelp that occur in relatively shallow water along the coasts throughout the world provide food and shelter for many different kinds of organisms.

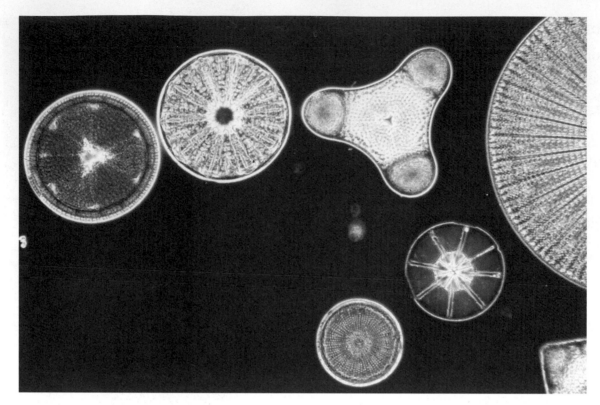

FIGURE 28.13
Diatoms.   These different radially symmetrical diatoms have unique silica double shells.

their motile spores, or zoospores, which bear two unequal flagella, one pointed forward and the other backward. Zoospores are produced asexually in a sporangium. Sexual reproduction involves the formation of male and female reproductive organs that produce gametes. Most oomycetes are found in water, but their terrestrial relatives are plant pathogens. *Phytophthora infestans*, which causes late blight of potatoes, was responsible for the Irish Potato famine of 1845 and 1847. During the famine, about 400,000 people starved to death or died of diseases complicated by starvation. Millions of Irish people emigrated to the United States and elsewhere as a result of this disaster.

Rhodophyta

Rhodophyta, the red algae, range from microscopic organisms to those rivaling the brown algae in size. Sushi rolls are wrapped in nori, a red alga. Red algal polysaccharides are used commercially to thicken ice cream and cosmetics. This lineage lacks flagella and centrioles, and has the accessory photosynthetic pigments phycoerythrin, phycocyanin, and allophycocyanin, which are arranged within structures called phycobilisomes. They reproduce using alternation of generations.

The origin of the over 7000 species of Rhodophyta has been a source of controversy. Evidence supporting both very early eukaryotic origins and a common ancestry with green algae has been considered. Molecular comparisons of the chloroplasts in red and green algae support a single endosymbiotic origin for both. This would have been an ancient event whereby a cyanobacterium was engulfed by a host cell. Chloroplast genome comparisons tell us about the cyanobacteria symbiont, but nuclear DNA contains information about host cell origins. Comparisons of the nuclear DNA coding for the large subunit of RNA polymerase II from two red algae, a green alga, and another protist support the conclusion that the Rhodophyta emerged before the evolutionary lineage that led to plants, animals, and fungi. How can we reconcile the data from plastid and nuclear DNA? The host cells and the cyanobacterial symbionts probably did not follow congruent evolutionary pathways. The host cell that gave rise to red algae may have been distinct from the one that gave rise to plants. One possibility is that different host cells engulfed the same bacterial symbiont. What other evolutionary scenarios could explain these results? Tentatively, we will treat Rhodophyta and Chlorophyta (the green algae) as sister clades based on the substantial amount of chloroplast data.

Red and brown algae are marine or freshwater organisms. The oomycetes lack photosynthetic capabilities.

Chlorophyta

Green algae are of special interest, because of their unusual diversity and because the ancestors of the plant kingdom were clearly multicellular green algae. There are two distinct lineages of green algae—the chlorophytes discussed here, and another lineage (Streptophyta) that gave rise to the land plants (see chapter 29). The chlorophytes have an extensive fossil record dating back 900 million years. Modern chlorophytes closely resemble land plants, especially in their chloroplasts, which are biochemically similar to those of the plants. They contain chlorophylls *a* and *b*, as well as carotenoids.

Chlamydomonas probably represents a primitive state for green algae (figure 28.14). Individuals are microscopic (usually less than 25 micrometers long), green, and rounded, and they have two flagella at the anterior end. They move rapidly in water by beating their flagella in opposite directions. Each individual has an eyespot, which contains about 100,000 molecules of rhodopsin, the same pigment employed in vertebrate eyes. Light received by this eyespot is used by the alga to help direct its swimming. Most individuals of *Chlamydomonas* are haploid. *Chlamydomonas* reproduces asexually as well as sexually (figure 28.14).

Several lines of evolutionary specialization have been derived from organisms such as *Chlamydomonas*. The first is the evolution of nonmotile, unicellular green algae. *Chlamydomonas* is capable of retracting its flagella and settling down as an immobile unicellular organism if the pond in which it lives dries out. Some common algae found in soil and bark, such as *Chlorella*, are essentially like *Chlamydomonas* in this trait, but do not have the ability to form flagella. *Chlorella* is widespread in both fresh and salt water as well as in soil, and is only known to reproduce asexually.

Another major line of specialization from cells like those of *Chlamydomonas* concerns the formation of motile, colonial organisms. In these genera of green algae, the *Chlamydomonas*-like cells retain some of their individuality. The most elaborate of these organisms is *Volvox* (see figure 28.1), a hollow sphere made up of a single layer of 500 to 60,000 individual cells, each cell having two flagella. Only a small number of the cells are reproductive. Some reproductive cells may divide asexually, bulge inward, and give rise to new colonies that initially remain within the parent colony. Others produce gametes.

Nonmotile, unicellular algae and multicellular, flagellated colonies have been derived from green algae such as *Chlamydomonas*—a biflagellated, unicellular organism. Chlorophytes did not give rise to land plants.

FIGURE 28.14

Life cycle of *Chlamydomonas* (Chlorophyta). Individual cells of this microscopic, biflagellated alga, which are haploid, divide asexually, producing identical copies of themselves. At times, such haploid cells act as gametes—fusing, as shown in the lower right-hand side of the diagram, to produce a zygote. The zygote develops a thick, resistant wall, becoming a zygospore; this is the only diploid cell in the entire life cycle. Within this diploid zygospore, meiosis takes place, ultimately resulting in the release of four haploid individuals. Because of the segregation during meiosis, two of these individuals are called the (+) strain and the other two, the (−) strain. Only + and − individuals are capable of mating with each other when syngamy does take place, although both may divide asexually to reproduce themselves.

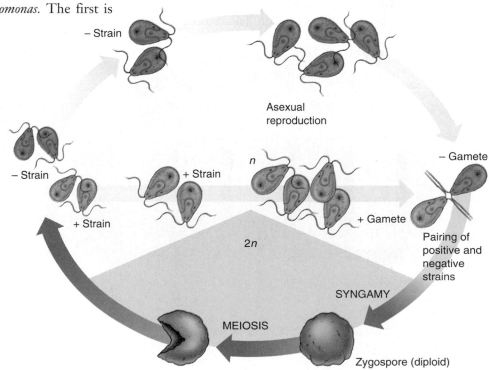

– Strain

Asexual reproduction

– Gamete

n

– Strain

+ Strain

+ Gamete

+ Strain

Pairing of positive and negative strains

2*n*

SYNGAMY

MEIOSIS

Zygospore (diploid)

Choanoflagellida and Protists That Are Difficult to Categorize

Choanoflagellida

Choanoflagellates are most like the common ancestor of the sponges and, indeed, all animals. Choanoflagellates have a single emergent flagellum surrounded by a funnel-shaped, contractile collar composed of closely placed filaments, a structure that is exactly matched in the sponges. Colonial forms resemble freshwater sponges. These protists feed on bacteria strained out of the water by their collar. The close relationship of choanoflagellates to animals was further demonstrated by the strong homology between a surface receptor (a tyrosine kinase receptor) found in choanoflagellates and sponges. This surface receptor initiates a signaling pathway involving phosphorylation (see chapter 7).

Amoebas

So far, we have organized the protists based on their closest relatives. Some lineages vary tremendously if you consider just a single trait. For example, the stramenopiles include autotrophic, marine algae and terrestrial plant pathogens. As seen in chapter 25, it is also possible for unrelated organisms to acquire similar traits. That is the case with amoebas, which have similar cell morphology, but are not monophyletic (figure 28.15).

Amoebas move from place to place by means of their pseudopods. Pseudopods are flowing projections of cytoplasm that extend and pull the amoeba forward or engulf food particles, a process called cytoplasmic streaming. An amoeba puts a pseudopod forward and then flows into it. Microfilaments of actin and myosin similar to those found in muscles are associated with these movements. The pseudopods can form at any point on the cell body so that it can move in any direction. The pseudopods of amoeboid cells give them truly amorphous bodies. One group, however, has more distinct structures. Members of the phylum Actinopoda, often called radiolarians, secrete glassy exoskeletons made of silica. These skeletons give the unicellular organisms a distinct shape, exhibiting either bilateral or radial symmetry. The shells of different species form many elaborate and beautiful shapes, with pseudopods extruding outward along spiky projections of the skeleton (figure 28.16). Microtubules support these cytoplasmic projections.

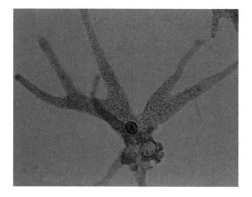

FIGURE 28.15
Amoeba proteus. The projections are pseudopods; an amoeba moves by flowing into them.

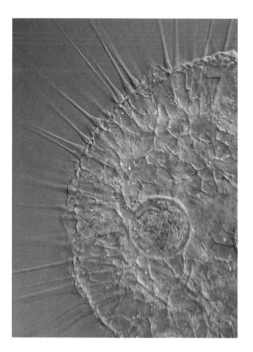

FIGURE 28.16
Actinosphaerium with needlelike pseudopods.

Foraminifera

Members of the phylum Foraminifera are heterotrophic marine protists. They range in diameter from about 20 micrometers to several centimeters. They resemble tiny snails and can form 3-meter-deep layers in marine sediments. Characteristic of the group are pore-studded shells (called *tests*) composed of organic materials usually reinforced with grains of inorganic matter. These grains may be calcium carbonate, sand, or even plates from the shells of echinoderms or spicules (minute needles of calcium carbonate) from sponge skeletons. Depending on the building materials they use, foraminifera—often informally called "forams"—may have shells of very different appearance. Some of them are brilliantly colored red, salmon, or yellow-brown.

Most foraminifera live in sand or are attached to other organisms, but two families consist of free-floating planktonic organisms. Their tests may be single-chambered, but are more often multichambered, and they sometimes have a spiral shape resembling that of a tiny snail. Thin cytoplasmic projections called *podia* emerge through openings in the tests (figure 28.17). Podia are used for swimming, gathering materials for the tests, and feeding. Forams eat a wide variety of small organisms.

The life cycles of foraminifera are extremely complex, involving alternation between haploid and diploid generations. Forams have contributed massive accumulations of their tests to the fossil record for more than 200 million years. Because of the excellent preservation of their tests and the striking differences among them, forams are very important as geological markers. The pattern of occurrence of different forams is often used as a guide in searching for oil-bearing strata. Limestones all over the world, including the famous White Cliffs of Dover in southern England, are often rich in forams (figure 28.18).

Slime Molds

Slime molds originated at least three distinct times, and the three lineages are very distantly related. We will explore two lineages—the plasmodial slime molds, which are huge, single-celled, multinucleate, oozing masses, and the cellular slime molds, in which single cells combine and differentiate, creating an early model of multicellularity.

Plasmodial slime molds stream along as a **plasmodium,** a nonwalled, multinucleate mass of cytoplasm that resembles a moving mass of slime (figure 28.19). This is called the *feeding phase,* and the plasmodia may be orange, yellow, or another color. Plasmodia show a back-and-forth streaming of cytoplasm that is very conspicuous, especially under a microscope. They are able to pass through the mesh in cloth or simply to flow around or through other obstacles. As they move, they engulf and digest bacteria, yeasts, and other small particles of organic matter. A multinucleated *Plasmodium* cell undergoes mitosis synchronously, with the

FIGURE 28.17
A representative of the foraminifera. Podia, thin cytoplasmic projections, extend through pores in the calcareous test, or shell, of this living foram (90×).

FIGURE 28.18
White Cliffs of Dover. The limestone that forms these cliffs is composed almost entirely of fossil shells of protists, including foraminifera.

FIGURE 28.19
A plasmodial protist. This multinucleate plasmodium moves about in search of the bacteria and other organic particles that it ingests.

nuclear envelope breaking down, but only at late anaphase or telophase. Centrioles are absent in cellular slime molds.

When either food or moisture is in short supply, the plasmodium migrates relatively rapidly to a new area. Here it stops moving and either forms a mass in which spores differentiate or divides into a large number of small mounds, each of which produces a single, mature **sporangium,** the structure in which spores are produced. These sporangia are often beautiful and extremely complex in form (figure 28.20). The spores are highly resistant to unfavorable environmental influences and may last for years if kept dry.

The cellular slime molds have become an important group for the study of cell differentiation because of their

FIGURE 28.20
Sporangia of a plasmodial slime mold. These *Arcyria* sporangia are found in the phylum Myxomycota.

relatively simple developmental systems (figure 28.21). The individual organisms behave as separate amoebas, moving through the soil and ingesting bacteria. When food becomes scarce, the individuals aggregate to form a moving "slug." Cyclic adenosine monophosphate (cAMP) is sent out in pulses by some of the cells, and other cells move in the direction of the cAMP to form the slug. In the cellular slime mold *Dictyostelium discoideum*, this slug goes through morphogenesis to make stalk and spore cells. The spores then go on to form a new amoeba if they land in a moist habitat.

> **Choanoflagellates are the most like the common ancestor of all animals. The evolutionary origins of some protists, including amoebas and slime molds, are less well understood, and these organisms may have arisen independently more than once.**

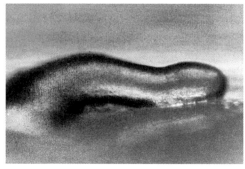

FIGURE 28.21
Development in *Dictyostelium discoideum*, a cellular slime mold. (*a*) First, a spore germinates, forming an amoeba. The amoebas feed and reproduce until the food runs out. (*b*) The amoebas aggregate and move toward a fixed center. (*c*) Next, they form a multicellular "slug," 2 to 3 millimeters long, that migrates toward light. (*d*) The slug stops moving and begins to differentiate into (*e*) a spore-forming body called a sorocarp. (*f*) Within the heads of the sorocarps, amoebas become encysted as spores.

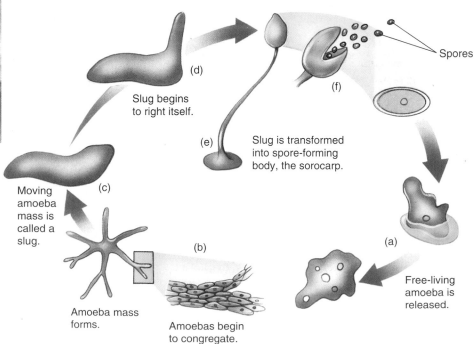

Spores

(d) Slug begins to right itself.

(e) Slug is transformed into spore-forming body, the sorocarp.

(c) Moving amoeba mass is called a slug.

Amoeba mass forms.

(b) Amoebas begin to congregate.

(a) Free-living amoeba is released.

28.1 Eukaryotes probably arose by endosymbiosis.

Endosymbiosis

- Single-celled, nonphotosynthetic *Pelomyxa palustris* appears to represent an early stage in eukaryotic evolution. (p. 562)
- *Pelomyxa* lacks mitochondria, but contains two kinds of bacteria that may play a similar role. (p. 562)
- Most biologists believe mitochondria originated as symbiotic, aerobic bacteria; this theory is called endosymbiosis. (p. 562)
- Aerobic bacteria are believed to have evolved into mitochondria when they were engulfed by ancestral eukaryotic cells. (p. 562)
- Although most mitochondrial genes have been transferred to the host cell's chromosomes, each mitochondrion still has its own circular, closed molecule of DNA. (p. 562)
- Many eukaryotic cells contain other endosymbiotic bacteria in addition to mitochondria. (p. 563)
- Three biochemically distinct classes of chloroplasts exist, with each resembling a different bacterial ancestor. (p. 563)
- Recent investigations of chloroplast DNA suggest a single origin of chloroplasts. (p. 563)

28.2 The kingdom Protista is by far the most diverse of any eukaryotic kingdom.

The Challenge of Classifying the Protists

- Protists are the most diverse of the four kingdoms in the domain Eukarya. (p. 564)
- The classification of Protista is artificial and not representative of any evolutionary relationships. (p. 564)

General Biology of the Protists

- Protists are united on the basis that they are eukaryotes that are not fungi, plants, or animals. (p. 565)
- Protists exhibit a varied array of cell surfaces and diverse mechanisms for locomotion. (p. 565)
- Many protists form cysts with resistant coverings; they lack cell metabolism. (p. 565)
- Protists are known to be phototrophs, phagotrophs, and osmotrophs. They typically reproduce asexually through binary fission. (p. 565)

28.3 Protists can be categorized into six groups.

Euglenozoa

- Euglenoids were among the earliest free-living eukaryotes to possess mitochondria. (p. 566)
- It seems likely that chloroplasts ultimately evolved from a symbiotic relationship through ingestion of green algae. (p. 566)
- Kinetoplastids have a unique, single mitochondrion in each cell. (p. 567)

- Trypanosomes cause multiple serious diseases in humans, including trypanosomiasis (African sleeping sickness), East Coast fever, and Chagas disease. (p. 567)
- Rearrangement of genes during the asexual cycle allows for the expression of a large number of varieties of different antigen genes. (p. 567)

Alveolata

- Most dinoflagellates are photosynthetic unicells possessing two flagella. (p. 568)
- Protists cannot be grouped together based solely on the presence of flagella. (p. 568)
- At least 20 dinoflagellate species are known to produce powerful toxins that provoke respiratory failure in vertebrates. (p. 568)
- Apicomplexes are spore-forming animal parasites, such as the malarial parasite *Plasmodium*, that have a unique arrangement of organelles at one end of the cell. (pp. 568–569)
- Ciliates are extremely complex organisms with cilia usually arranged in longitudinal rows or in spirals around the cell. (pp. 569–570)
- All known ciliates have both a micro- and a macronucleus. (pp. 569–570)

Stramenopila and Rhodophyta

- Stramenopiles include brown algae, diatoms, and oomycetes. (pp. 571–572)
- Rhodophyta is made up of over 7000 species of red algae. (p. 572)
- Evidence exists supporting both early eukaryotic origins of Stramenopila and Rhodophyta and a common ancestry with green algae. (p. 572)
- Tentatively, Rhodophyta and Chlorophyta will be treated as sister clades. (p. 572)

Chlorophyta

- Chlorophytes make up one distinct lineage of green algae, while the other lineage, the streptophytes, gave rise to land plants. (p. 573)
- *Chlamydomonas* most likely represents a primitive state for green algae. (p. 573)
- Nonmotile, unicellular green algae and motile, colonial organisms are two lines of evolutionary specialization derived from early green algae. (p. 573)

Choanoflagellida and Protists That Are Difficult to Categorize

- Choanoflagellates are most similar to the sponges and to all animals. (p. 574)
- Amoebas and slime molds are very difficult to categorize because amoebas have similar cell morphology but are not monophyletic, and because slime molds originated at least three distinct times and the three lineages are only distantly related. (pp. 574–576)

Self Test

1. The mitochondria of eukaryotic cells most likely arose as a result of endosymbiosis between a eukaryotic cell and a
   a. blue-green alga.
   b. nonsulfur purple bacterium.
   c. red alga.
   d. cyanobacterium.

2. The protists are a paraphyletic group that have traditionally been grouped together because
   a. they are all genetically similar to each other.
   b. they all have very similar morphological characters.
   c. they are not fungi, animals, or plants.
   d. they all have similar nutritional modes and live in similar environments.

3. Many protists are able to resist harsh environmental conditions by
   a. forming a cyst and slowing metabolism during times of stress.
   b. utilizing a wide variety of nutritional modes.
   c. reproducing asexually through the process of budding.
   d. moving away from a harsh environment by using pseudopodia.

4. The light-sensing organ in *Euglena* is a
   a. flagellum.
   b. contractile vacuole.
   c. pellicle.
   d. stigma.

5. The parasitic kinetoplast that causes leishmaniasis must spend part of its life cycle in a nonhuman host. What organism(s) serve(s) as the vector for this life cycle?
   a. small mammals
   b. a sand fly
   c. a mosquito
   d. a tse-tse fly

6. You are examining cells from an unknown organism under the microscope. You note that the cells have a membrane-bounded nucleus and that there are small cavities in the membranes along the internal cell surface. Based upon this information alone, you conclude that these most likely are _____ cells.
   a. bacterial
   b. diatom
   c. dinoflagellate
   d. kinetoplastid

7. The parasitic protist that causes malaria, *Plasmodium*, must spend part of its life cycle in a nonhuman host. What organism(s) serve(s) as the vector for this life cycle?
   a. small mammals
   b. a sand fly
   c. a mosquito
   d. a tse-tse fly

8. *Phytopthora infestans* is the oomycete that causes
   a. Chagas disease.
   b. potato blight.
   c. red tides.
   d. African sleeping sickness.

9. A biologist discovers an alga that is marine, multicellular, lacks flagella and centrioles, and contains phycobilisomes. It probably belongs to which group?
   a. Rhodophyta
   b. brown algae
   c. Chlorophyta
   d. Foraminifera

10. The green algae gave rise to which modern group of organisms?
    a. photosynthetic bacteria
    b. plants
    c. photosynthetic euglenoids
    d. animals

Test Your Visual Understanding

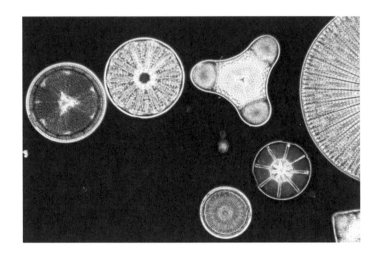

1. This image shows several different types of protists that are characterized by a rigid double wall of silica. To what group do these organisms belong?

Apply Your Knowledge

1. Protists typically reproduce asexually. However, some protists undergo sexual reproduction during times of environmental stress. What advantage does sexual reproduction give a protist during these times?

2. Scientists are working to develop a vaccine against African sleeping sickness. Use what you know about the trypanosome responsible for this disease to determine why it is so difficult to develop a vaccine.

29

Overview of Plant Diversity

Concept Outline

29.1 Plants have multicellular haploid and diploid stages in their life cycles.

The Evolutionary Origins of Plants. Plants evolved from freshwater green algae and eventually developed cuticles, stomata, conducting systems, and reproductive strategies that adapt them well for life on land.

Plant Life Cycles. Plants have haplodiplontic life cycles. Diploid sporophytes produce haploid spores by meiosis. Spores develop into haploid gametophytes by mitosis and produce haploid gametes.

29.2 Nonvascular plants are relatively unspecialized, but successful in many terrestrial environments.

Mosses, Liverworts, and Hornworts. The most conspicuous part of a nonvascular plant is the green photosynthetic gametophyte, which supports the smaller sporophyte nutritionally.

29.3 Seedless vascular plants have well-developed conducting tissues in their sporophytes.

Features of Vascular Plants. In vascular plants, specialized tissue called xylem conducts water and dissolved minerals within the plant, and tissue called phloem conducts sucrose and hormones within the plant.

Seedless Vascular Plants. Seedless vascular plants have a much more conspicuous sporophyte than nonvascular plants do, and many have well-developed conducting systems in the stems, roots, and leaves.

29.4 Seeds protect and aid in the dispersal of plant embryos.

Seed Plants. In seed plants, the sporophyte is dominant. Male and female gametophytes develop within the sporophyte and depend on it for food. Seeds allow embryos to germinate when conditions are favorable.

Gymnosperms. In gymnosperms, the female gametophyte (ovule) is not completely enclosed by sporophyte tissue at the time of pollination.

Angiosperms. In angiosperms, the ovule is completely enclosed by sporophyte tissue at the time of pollination. Angiosperms, by far the most successful plant group, produce flowers.

FIGURE 29.1
An arctic tundra. Tundra is one of the harshest environments on earth, and yet many diverse plants have made it their home. These ecosystems are fragile and particularly susceptible to global change.

Plant evolution is the story of the conquest of land by green algal ancestors. For about 500 million years, algae were confined to a watery domain, limited by the need for water to reproduce, provide structural support, prevent water loss, and provide some protection from the sun's ultraviolet irradiation. Numerous evolutionary solutions to these challenges have resulted in over 300,000 species of plants dominating all terrestrial communities today, from forests to alpine tundra (figure 29.1), and from agricultural fields to deserts. Most plants are photosynthetic, converting light energy into chemical-bond energy and providing oxygen for all aerobic organisms. We rely on plants for food, clothing, wood for shelter and fuel, chemicals, and many medicines. This chapter explores the evolutionary history and strategies that have allowed plants to inhabit most terrestrial environments over millions of years.

29.1 Plants have multicellular haploid and diploid stages in their life cycles.

The Evolutionary Origins of Plants

What is a plant? We will use the term plant to refer to a group of organisms that share a freshwater algal ancestor and have evolved over a 470-million-year period. This chapter will explore land plants. The defining characteristic of land plants is the protection of their embryos, essential for survival in a terrestrial environment. It is surprising that just a single species of green algae gave rise to the entire terrestrial plant lineage, from mosses through the flowering plants (angiosperms). Exactly what this ancestral alga was is still a mystery, but close relatives, the Charales, exist in freshwater lakes today (see chapter 25 for more information on the evolution of land plants from Charales species). DNA sequence data are consistent with the claim that a single "Eve" gave rise to all plants. The shared evolutionary history with green algae has led biologists to rename kingdom Plantae as kingdom Virdiplantae to include the green algae. Fungi are not a part of this scheme. They are more closely related to metazoan animals (see chapter 31).

Land plants, though diverse, have certain characteristics in common. For example, all of them afford some protection to their embryos, and all have multicellular haploid and diploid phases. Over time, the trend has been toward more embryo protection and a smaller haploid stage in the life cycle. In addition, land plants can be compared based on the presence or absence of conducting systems, which facilitate the transport of water and nutrients. **Nonvascular plants** lack vascular tissue, while **vascular plants** have water-conducting xylem and food-conducting phloem strands of tissues in their stems, roots, and leaves. For purposes of discussion in this chapter, we subdivide the plants into the four major categories shown in figure 29.2, based on common characteristics and the order in which the types of plants are thought to have evolved not all of the groups are clades:

- The *nonvascular land plants*, which are not monophyletic, include three phyla: mosses, liverworts, and hornworts.
- Recent evidence supports two distinct monophyletic lineages of *seedless vascular plants:* (1) club mosses, and (2) ferns, whisk ferns, and horsetails. Ferns and horsetails are the closest relatives to the seed plants (the gymnosperms and angiosperms).
- *Gymnosperms* have seeds that protect their embryos. Included in this group are the conifers and cycads.
- *Angiosperms* arose about 150 million years ago with further innovations. Their distinguishing adaptations are flowers, which may attract pollinators, and fruits surrounding the seeds, protecting the embryos and aiding in seed dispersal.

FIGURE 29.2
Four major groups of land plants. In this chapter, we discuss four major groups of plants. There are actually five distinct lineages; seedless vascular plants have two different origins but are grouped here for simplicity. The ancestral green algae are discussed in chapters 25 and 28.

Adaptations to Land

Unlike their freshwater ancestors, most land plants have only limited amounts of water available. As an adaptation to living on land, most plants are protected from **desiccation**—the tendency of organisms to lose water to the air—by a waxy **cuticle** that is secreted onto their exposed surfaces. The cuticle is relatively impermeable, preventing water loss. However, this solution limits the gas exchange essential for respiration and photosynthesis. Gas diffusion into and out of a plant occurs through tiny mouth-shaped openings called **stomata** (singular, *stoma*).

Two additional adaptations allowed larger land plants to flourish. The evolution of leaves which may have occured multiple times resulted in increased photosynthetic surface area. The shift to a dominant diploid generation, accompanied by the structural support of vascular tissue, allowed plants to take advantage of the vertical dimension of the terrestrial environment, making the evolution of trees possible.

Plants evolved from freshwater green algae and eventually developed reproductive strategies, conducting systems, stomata, and cuticles that adapt them well for life on land.

Plant Life Cycles

All plants undergo mitosis after both gamete fusion and meiosis. The result is a multicellular haploid and a multicellular diploid individual, unlike the human life cycle, in which gamete fusion directly follows meiosis. Humans have a **diplontic** life cycle, meaning that only the diploid stage is multicellular, but the plant life cycle is **haplodiplontic**, having multicellular haploid and diploid stages. The basic haplodiplontic cycle is summarized in figure 29.3. Brown, red, and green algae are also haplodiplontic (see chapter 28). While humans produce gametes via meiosis, land plants actually produce gametes by mitosis in a multicellular, haploid individual. The diploid generation, or **sporophyte**, alternates with the haploid generation, or **gametophyte**. Sporophyte means "spore plant," and gametophyte means "gamete plant." These terms indicate the kinds of reproductive cells the respective generations produce.

The diploid sporophyte produces haploid spores (not gametes) by meiosis. Meiosis takes place in structures called **sporangia**, where diploid **spore mother cells (sporocytes)** undergo meiosis, each producing four haploid **spores.** Spores divide by mitosis, producing a multicellular, haploid gametophyte. Spores are the first cells of the gametophyte generation.

In turn, the haploid gametophyte is produced by mitosis and is the source of gametes. When the gametes fuse, the zygote they form is diploid and is the first cell of the next sporophyte generation. The zygote grows into a diploid sporophyte that produces sporangia in which meiosis ultimately occurs.

While all plants are haplodiplontic, the haploid generation consumes a much larger portion of the life cycle in mosses than in gymnosperms and angiosperms. In mosses, liverworts, and ferns, the gametophyte is photosynthetic and free-living; in other plants, it is usually nutritionally dependent on the sporophyte. When you look at moss, what you see is largely gametophyte tissue; the sporophytes are usually smaller, brownish or yellowish structures attached to the tissues of the gametophyte. In all vascular plants, the gametophytes are much smaller than the sporophytes. In seed plants, the gametophytes are nutritionally dependent on the sporophytes and are enclosed within their tissues. When you look at a gymnosperm or angiosperm, what you see, with rare exceptions, is a sporophyte.

While the sporophyte generation can get very large, the size of the gametophyte is limited in all plants. What we identify as a moss plant is a gametophyte, and it produces gametes at its tips. The egg is stationary, and sperm lands near the egg in a droplet of water. If the moss were the height of a sequoia, not only would it need vascular tissue for conduction and support, but the sperm would have to swim up the tree! In contrast, the fern gametophyte develops on the forest floor where gametes can meet. Fern trees are especially abundant in Australia; the haploid spores fall to the ground and develop into gametophytes.

Having completed our overview of plant life cycles, we will consider the major plant groups. As we do so, we will see a reduction of the gametophyte from group to group, a loss of multicellular **gametangia** (structures in which gametes are produced), and increasing specialization for life on land, including the remarkable structural adaptations of the flowering plants, the dominant plants today. Similar trends must have characterized the evolution of seed plants over the hundreds of millions of years since a freshwater alga first moved onto land.

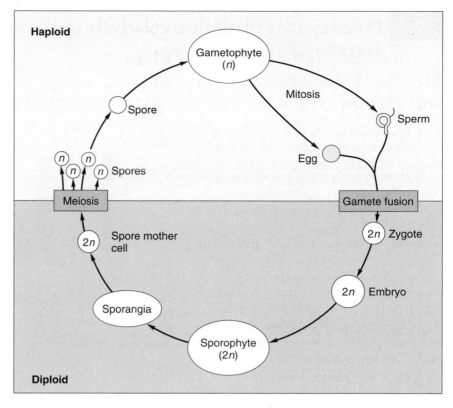

FIGURE 29.3
A generalized plant life cycle. Note that both haploid and diploid individuals can be multicellular. Also, spores are produced by meiosis, while gametes are produced by mitosis.

Plants have haplodiplontic life cycles. Diploid sporophytes produce haploid spores by meiosis. Spores develop into haploid gametophytes by mitosis and produce haploid gametes.

29.2 Nonvascular plants are relatively unspecialized, but successful in many terrestrial environments.

Mosses, Liverworts, and Hornworts

The approximately 24,700 species of **bryophytes** are simple but highly adapted to a diversity of terrestrial environments (even deserts!). Scientists now agree that bryophytes consist of three quite distinct phyla of relatively unspecialized plants—mosses, liverworts, and hornworts. Their gametophytes are photosynthetic. Sporophytes are attached to the gametophytes and depend on them nutritionally to varying degrees. Like ferns and certain other vascular plants, bryophytes require water (e.g., rainwater) to reproduce sexually. It is not surprising that they are especially common in moist places, both in the tropics and temperate regions.

Most bryophytes are small; few exceed 7 centimeters in height. The gametophytes are more conspicuous than the sporophytes. Some of the sporophytes are completely enclosed within gametophyte tissue; others are not and usually turn brownish or straw-colored at maturity.

Mosses (Bryophyta)

The gametophytes of mosses typically consist of small, leaflike structures (not true leaves, which contain vascular tissue) arranged spirally or alternately around a stemlike axis (figure 29.4); the axis is anchored to its substrate by means of **rhizoids.** Each rhizoid consists of several cells that absorb water, but not nearly the volume of water that is absorbed by a vascular plant root. Moss "leaves" have little in common with leaves of vascular plants, except for the superficial appearance of the green, flattened blade and slightly thickened midrib that runs lengthwise down the middle. Only one cell layer thick (except at the midrib), they lack vascular strands and stomata, and all the cells are haploid.

Water may rise up a strand of specialized cells in the center of a moss gametophyte axis. Some mosses also have specialized food-conducting cells surrounding those that conduct water.

Multicellular gametangia are formed at the tips of the leafy gametophytes (figure 29.5). Female gametangia (**archegonia**) may develop either on the same gametophyte as the male gametangia (**antheridia**) or on separate plants. A single egg is produced in the swollen lower part of an archegonium, while numerous sperm are produced in an antheridium. When sperm are released from an antheridium, they swim with the aid of flagella through a film of dew or rainwater to the archegonia. One sperm (which is haploid) unites with an egg (also haploid), forming a diploid zygote. The zygote divides by mitosis and develops into the sporophyte, a slender, basal stalk with a swollen capsule, the *sporangium*, at its tip. As the sporophyte develops, its base is embedded in gametophyte tissue, its nutritional source. The sporangium is often cylindrical or club-shaped. Spore mother cells within the sporangium undergo meiosis, each producing four haploid spores. In many mosses at maturity, the top of the sporangium pops off, and the spores are released. A spore that lands in a suitable damp location may germinate and grow into a threadlike structure, which branches to form rhizoids and "buds" that grow upright. Each bud develops into a new gametophyte plant consisting of a leafy axis.

In the Arctic and the Antarctic, mosses are the most abundant plants, boasting not only the largest number of individuals in these harsh regions. The greatest diversity of moss species, however, is found in the tropics. Many mosses are able to withstand prolonged periods of drought, although mosses are not common in deserts. Most are remarkably sensitive to air pollution and are rarely found in abundance in or near cities or other areas with high levels of air pollution. Some mosses, such as the peat mosses (*Sphagnum*), can absorb up to 25 times their weight in water and are valuable commercially as a soil conditioner or as a fuel when dry.

FIGURE 29.4
A hair-cup moss, *Polytrichum* (phylum Bryophyta). The leaflike structures belong to the gametophyte. Each of the yellowish-brown stalks with a capsule, or sporangium, at its summit is a sporophyte.

Liverworts (Hepaticophyta)

The Old English word wyrt means "plant" or "herb." Some common liverworts have flattened gametophytes with lobes resembling those of liver—hence the name "liverwort." Although the lobed liverworts are the best-known representatives of this phylum, they constitute only about 20% of the species (figure 29.6). The other 80% are leafy and superficially resemble mosses. The gametophytes are prostrate instead of erect, and the rhizoids are one-celled.

Some liverworts have air chambers containing upright, branching rows of photosynthetic cells, each chamber having a pore at the top to facilitate gas exchange. Unlike stomata, the pores are fixed open and cannot close.

Sexual reproduction in liverworts is similar to that in mosses. Lobed liverworts may form gametangia in umbrella-like structures. Asexual reproduction occurs when lens-shaped pieces of tissue that are released from the gametophyte grow to form new gametophytes.

Hornworts (Anthocerotophyta)

The origin of hornworts is a puzzle. They are most likely among the earliest land plants, yet the earliest hornwort fossil spores date from the Cretaceous period, 65 to 145 million years ago, when angiosperms were emerging.

The small hornwort sporophytes resemble tiny green broom handles rising from filmy gametophytes usually less than 2 centimeters in diameter (figure 29.7). The sporophyte base is embedded in gametophyte tissue, from which it derives some of its nutrition. However, the sporophyte has stomata, is photosynthetic, and provides much of the energy needed for growth and reproduction. Hornwort cells usually have a single chloroplast.

The three major phyla of nonvascular plants are all relatively unspecialized, but well suited for diverse terrestrial environments.

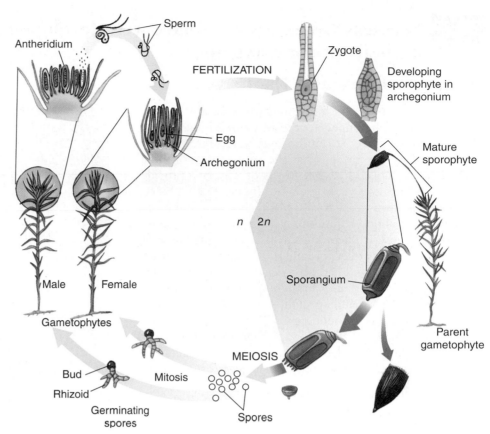

FIGURE 29.5
Life cycle of a typical moss. The majority of the life cycle of a moss is in the haploid state. The leafy gametophyte is photosynthetic, while the smaller sporophyte is not, and is nutritionally dependent on the gametophyte. Water is required to carry sperm to the egg.

FIGURE 29.6
A common liverwort, *Marchantia* (phylum Hepaticophyta). The sporophytes are formed by fertilization within the tissues of the umbrella-shaped structures that arise from the surface of the flat, green, creeping gametophyte.

FIGURE 29.7
Hornworts (phylum Anthocerotophyta). Hornwort sporophytes are seen in this photo. Unlike the sporophytes of other bryophytes, most hornwort sporophytes are photosynthetic.

29.3 Seedless vascular plants have well-developed conducting tissues in their sporophytes.

Features of Vascular Plants

The first vascular plants for which we have a relatively complete record belonged to the phylum Rhyniophyta. They flourished some 410 million years ago but are now extinct. We are not certain what the very earliest of these vascular plants looked like, but fossils of *Cooksonia* provide some insight into their characteristics (figure 29.8). *Cooksonia*, the first known vascular land plant, appeared in the late Silurian period about 420 million years ago. It was successful partly because it encountered little competition as it spread out over vast tracts of land. The plants were only a few centimeters tall and had no roots or leaves. They consisted of little more than a branching axis, the branches forking evenly and expanding slightly toward the tips. They were **homosporous** (producing only one type of spore). Sporangia formed at branch tips. Other ancient vascular plants that followed evolved more complex arrangements of sporangia. Leaves began to appear as protuberances from stems.

Cooksonia and the other early plants that followed it became successful colonizers of the land by developing efficient water- and food-conducting systems known as **vascular tissues** (Latin *vasculum*, "vessel" or "duct"). These tissues consist of strands of specialized cylindrical or elongated cells that form a network throughout a plant, extending from near the tips of the roots, through the stems, and into true leaves. One type of vascular tissue, *xylem*, conducts water and dissolved minerals upward from the roots; another type of tissue, *phloem*, conducts sucrose and hormonal signals throughout the plant. It is important to note that vascular tissue develops in the sporophyte, but (with few exceptions) not in the gametophyte. (See the discussion of vascular tissue structure in chapter 35.) The presence of a cuticle and stomata are also characteristic of vascular plants.

Three clades of vascular plants exist today: (1) lycophytes (club mosses), (2) pterophytes (ferns and their relatives), and (3) seed plants. Advances in molecular systematics have changed the way we view the evolutionary history of vascular plants. Whisk ferns and horsetails were long believed to be distinct phyla that were transitional between bryophytes and vascular plants. Phylogenetic evidence now shows they are the closest living relatives to ferns.

The seven living phyla of vascular plants (table 29.1) dominate terrestrial habitats everywhere, except for the highest mountains and the tundra. The haplodiplontic life cycle persists, but the gametophyte has been reduced during the evolution of some phyla. A similar reduction in multicellular gametangia has occurred.

Sporangia

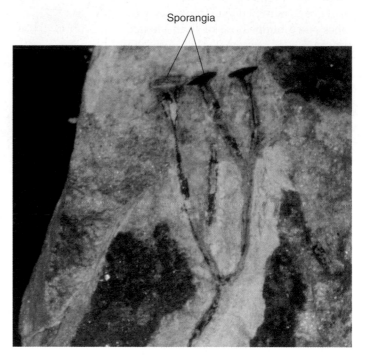

FIGURE 29.8
***Cooksonia*, the first known vascular land plant.** This fossil represents a plant that lived some 410 million years ago. *Cooksonia* belongs to phylum Rhyniophyta, consisting entirely of extinct plants. Its upright, branched stems, which were no more than a few centimeters tall, terminated in sporangia, as seen here. It probably lived in moist environments such as mudflats, had a resistant cuticle, and produced spores typical of vascular plants.

Accompanying this reduction in size and complexity of the gametophytes has been the appearance of the seed. Seeds are highly resistant structures well suited to protect a plant embryo from drought and to some extent from predators. In addition, almost all seeds contain a supply of food for the young plant. Seeds occur only in **heterosporous** plants (plants that produce two types of spores). Heterospory is believed to have arisen multiple times in the plants. Fruits in the flowering plants add a layer of protection to seeds and attract animals that assist in seed dispersal, expanding the potential range of the species. Flowers, which evolved among the angiosperms, attract pollinators. Flowers allow plants to secure the benefits of wide outcrossing in promoting genetic diversity.

Most vascular plants have well-developed conducting tissues, specialized stems, leaves, roots, cuticles, and stomata. Many have seeds, which protect embryos until conditions are suitable for further development.

Table 29.1 The Seven Phyla of Extant Vascular Plants

Phylum	Examples		Key Characteristics	Approximate Number of Living Species
SEED PLANTS				
Anthophyta	Flowering plants (angiosperms)		Heterosporous. Sperm not motile; conducted to egg by a pollen tube. Seeds enclosed within a fruit. Leaves greatly varied in size and form. Herbs, vines, shrubs, trees. About 14,000 genera.	250,000
Coniferophyta	Conifers (including pines, spruces, firs, yews, redwoods, and others)		Heterosporous seed plants. Sperm not motile; conducted to egg by a pollen tube. Leaves mostly needlelike or scalelike. Trees, shrubs. About 50 genera.	601
Cycadophyta	Cycads		Heterosporous. Sperm flagellated and motile but confined within a pollen tube that grows to the vicinity of the egg. Palmlike plants with pinnate leaves. Secondary growth slow compared to that of the conifers. 10 genera. Seeds in cones.	206
Gnetophyta	Gnetophytes		Heterosporous. Sperm not motile; conducted to egg by a pollen tube. The only gymnosperms with vessels. Trees, shrubs, vines. Three very diverse genera (*Ephedra*, *Gnetum*, *Welwitschia*).	65
Ginkgophyta	*Ginkgo*		Heterosporous. Sperm flagellated and motile but conducted to the vicinity of the egg by a pollen tube. Deciduous tree with fan-shaped leaves that have evenly forking veins. Seeds resemble a small plum with fleshy, ill-scented outer covering. One genus.	1
SEEDLESS VASCULAR PLANTS				
Pterophyta	Ferns		Primarily homosporous (a few heterosporous). Sperm motile. External water necessary for fertilization. Leaves are megaphylls that uncoil as they mature. Sporophytes and virtually all gametophytes are photosynthetic. About 365 genera.	11,000
	Horsetails		Homosporous. Sperm motile. External water necessary for fertilization. Stems ribbed, jointed, either photosynthetic or nonphotosynthetic. Leaves scalelike, in whorls; nonphotosynthetic at maturity. One genus.	15
	Whisk ferns		Homosporous. Sperm motile. External water necessary for fertilization. No differentiation between root and shoot. No leaves; one of the two genera has scalelike extensions and the other leaflike appendages.	6
Lycophyta	Club mosses		Homosporous or heterosporous. Sperm motile. External water necessary for fertilization. Leaves are microphylls. About 12–13 genera.	1,150

Seedless Vascular Plants

The earliest vascular plants lacked seeds. Members of four phyla of living vascular plants lack seeds, as do at least three other phyla known only from fossils. As we explore the adaptations of the vascular plants, we focus on both reproductive strategies and the advantages of increasingly complex transport systems.

Club Mosses (Lycophyta)

The club mosses are relics of an ancient past when vascular plants first evolved (figure 29.9). They are the sister group to all vascular plants. Several genera of club mosses, some of them treelike, became extinct about 270 million years ago. Today, club mosses are worldwide in distribution but are most abundant in the tropics and moist temperate regions. Members of the 12–13 genera and about 1150 living species of club mosses superficially resemble true mosses, but once their internal structure and reproductive processes became known, it was clear that these vascular plants are quite unrelated to mosses. Modern club mosses are either homosporous or heterosporous. The sporophytes have leafy stems that are seldom more than 30 centimeters long.

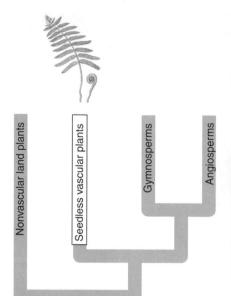

FIGURE 29.9
A club moss. *Lycopodium clavatum*, a club moss, is representative of the vascular plant phylum most closely related to the nonvascular bryophytes.

Whisk Ferns, Horsetails, and Ferns (Pterophyta)

Whisk ferns and horsetails are close relatives of ferns. They all form antheridia and archegonia. Free water is required for the process of fertilization, during which the sperm, which have flagella, swim to and unite with the eggs. In contrast, most seed plants have nonflagellated sperm.

Whisk Ferns. Whisk ferns, which occur in the tropics and subtropics, consist merely of evenly forking green stems without roots (figure 29.10). The two or three species of the genus *Psilotum* do, however, have tiny, green, spirally arranged flaps of tissue lacking veins and stomata. Another genus, *Tmespiteris*, has more leaflike appendages. Given the simple structure of whisk ferns, it was particularly surprising to learn that they are monophyletic with ferns. The gametophytes of whisk ferns are essentially colorless and are less than 2 millimeters in diameter, but they can be up to 18 millimeters long. They form parasitic associations with fungi, which furnish their nutrients. Some develop elements of vascular tissue and have the distinction of being the only gametophytes known to do so.

FIGURE 29.10
A whisk fern. Whisk ferns have no roots or leaves.

Horsetails. The 15 living species of horsetails are all homosporous. They constitute a single genus, *Equisetum*. Fossil forms of *Equisetum* extend back 300 million years to an era when some of their relatives were treelike. Today, they are widely scattered around the world, mostly in damp places. Some that grow among the coastal redwoods of California may reach a height of 3 meters, but most are less than a meter tall (figure 29.11).

Horsetail sporophytes consist of ribbed, jointed, photosynthetic stems that arise from branching underground rhizomes with roots at their nodes. A whorl of nonphotosynthetic, scalelike leaves emerges at each node. The stems, which are hollow in the center, have silica deposits in the epidermal cells of the ribs, and the interior parts of the stems have two sets of vertical, tubular canals. The larger outer canals, which alternate with the ribs, contain air, while the smaller inner canals opposite the ribs contain water. Horsetails are also called scouring rush because pioneers used them to scrub pans.

Ferns. Ferns are the most abundant group of seedless vascular plants, with about 11,000 living species. Recent research indicates that they may be the closest relatives to the seed plants. The fossil record indicates that ferns originated during the Devonian period about 350 million years ago and became abundant and varied in form during the next 50 million years. Their apparent ancestors had no broad leaves and were established on land as much as 375 million years ago. Today, ferns flourish in a wide range of habitats throughout the world; however, about 75% of the species occur in the tropics.

The conspicuous sporophytes may be less than a centimeter in diameter—as seen in small aquatic ferns such as *Azolla*—or more than 24 meters tall, with leaves up to 5 meters or longer in the tree ferns (figure 29.12). The sporophytes and the smaller gametophytes, which rarely reach 6 millimeters in diameter, are both photosynthetic. The fern life cycle differs from that of a moss primarily in the much greater development, independence, and dominance of the fern's sporophyte. The fern sporophyte is structurally more complex than the moss sporophyte; the fern sporophyte has vascular tissue and well-differentiated roots, stems, and leaves. The gametophyte, however, lacks vascular tissue.

Fern sporophytes typically have a horizontal underground stem called a *rhizome*, with roots emerging from the sides. The leaves, referred to as *fronds*, usually develop at the tip of the rhizome as tightly rolled-up coils ("fiddleheads") that unroll and expand. Many fronds are highly dissected and feathery, making the ferns that produce them prized as ornamentals. Some ferns, such as *Marsilea*, have fronds that resemble a four-leaf clover, but *Marsilea* fronds still begin as coiled fiddleheads. Other ferns produce a mixture of photosynthetic fronds and nonphotosynthetic reproductive fronds that tend to be brownish in color.

Most ferns are homosporous, producing distinctive *sporangia*, usually in clusters called *sori*, typically on the underside of the fronds. Sori are often protected during their development by a transparent, umbrella-like covering. (At first glance, one might mistake the sori for an infection on the plant.) Diploid *spore mother cells* in each sporangium undergo meiosis, producing haploid spores. At maturity, the spores are catapulted from the sporangium by a snapping action, and those that land in suitable damp locations may germinate, producing gametophytes that are often heart-shaped, are only one cell layer thick (except in the center), and have rhizoids that anchor them to their substrate. These rhizoids are not true roots because they lack vascular tissue, but they do aid in transporting water and nutrients from the soil. Flask-shaped

FIGURE 29.11 A horsetail, *Equisetum telmateia*. This species forms two kinds of erect stems; one is green and photosynthetic, and the other, which terminates in a spore-producing "cone," is mostly light brown.

FIGURE 29.12 A tree fern (phylum Pterophyta) in the forests of Malaysia. The ferns are by far the largest group of seedless vascular plants.

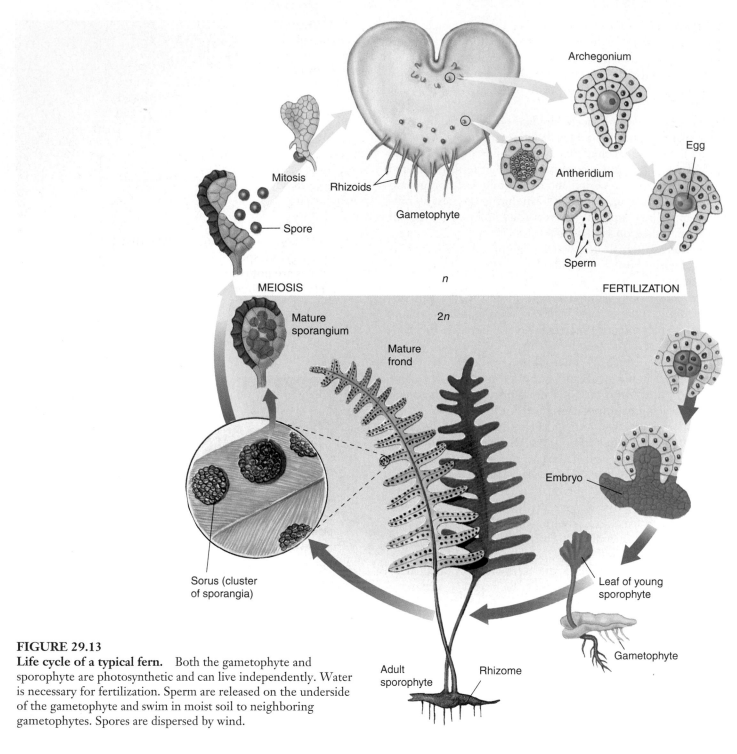

FIGURE 29.13

Life cycle of a typical fern. Both the gametophyte and sporophyte are photosynthetic and can live independently. Water is necessary for fertilization. Sperm are released on the underside of the gametophyte and swim in moist soil to neighboring gametophytes. Spores are dispersed by wind.

archegonia and globular *antheridia* are produced on either the same or a different gametophyte.

The sperm formed in the antheridia have flagella, with which they swim toward the archegonia when water is present, often in response to a chemical signal secreted by the archegonia. One sperm unites with the single egg toward the base of an archegonium, forming a *zygote*. The zygote then develops into a new sporophyte, completing the life cycle (figure 29.13). Multicellular gametangia still develop. As discussed earlier, the shift to a dominant sporo-

phyte generation allows ferns to achieve significant height without interfering with sperm swimming to the egg. The multicellular archegonia provide some protection for the developing embryo.

> The two seedless vascular plant phyla have a large and conspicuous sporophyte, with vascular tissue. Many have well-differentiated roots, stems, and leaves. The shift to a dominant sporophyte led to the evolution of trees.

29.4 Seeds protect and aid in the dispersal of plant embryos.

Seed Plants

Seed plants first appeared about 425 million years ago. Their ancestors appear to have been spore-bearing plants known as progymnosperms. Progymnosperms shared several features with modern gymnosperms, including secondary xylem and phloem (which allows for an increase in girth later in development). Some progymnosperms had leaves. Their reproduction was very simple, and it is not certain which particular group of progymnosperms gave rise to seed plants.

From an evolutionary and ecological perspective, the seed represents an important advance. The embryo is protected by an extra layer of sporophyte tissue, creating the ovule. During development, this tissue hardens to produce the seed coat. In addition to protecting the embryo from drought, the seed can be easily dispersed. Perhaps even more significantly, the presence of seeds introduces into the life cycle a dormant phase that allows the embryo to survive until environmental conditions are favorable for further growth.

Seed plants produce two kinds of gametophytes—male and female, each of which consists of just a few cells. Pollen grains, multicellular male gametophytes, are conveyed to the egg in the female gametophyte by wind or a pollinator. In some seed plants, the sperm move toward the egg through a growing pollen tube. This eliminates the need for external water. In contrast to the seedless plants, the whole male gametophyte, rather than just the sperm, moves to the female gametophyte. A female gametophyte develops within an ovule. In flowering plants (angiosperms), the ovules are completely enclosed within diploid sporophyte tissue (ovaries that develop into the fruit). In gymnosperms (mostly cone-bearing seed plants), the ovules are not completely enclosed by sporophyte tissue at the time of pollination.

A common ancestor that had seeds gave rise to the gymnosperms and the angiosperms. Seeds can allow for a pause in the life cycle until environmental conditions are more optimal.

A Vocabulary of Plant Terms

androecium   The stamens of a flower.

anther   The pollen-producing portion of a stamen. This is a sporophyte structure where microspores produced by meiosis develop into male gametophytes.

antheridium   The sperm-producing structure found in the gametophytes of seedless plants and certain fungi.

archegonium   The multicellular egg-producing structure in the gametophytes of seedless plants and gymnosperms.

carpel   A leaflike organ in angiosperms that encloses one or more ovules; a unit of a gynoecium.

double fertilization   The process in which one sperm fuses with the egg, forming a zygote, and the other sperm fuses with the two polar nuclei, forming the primary endosperm nucleus in angiosperms.

embryo sac   The female gametophyte in flowering plants.

endosperm   The usually triploid (although it can have a much higher ploidy level) food supply of some angiosperm seeds.

filament   The stalklike structure that supports the anther of a stamen.

gametophyte   The multicellular, haploid phase of a plant life cycle, in which gametes are produced by mitosis.

gynoecium   The carpel(s) of a flower.

heterosporous   Refers to a plant that produces two types of spores: microspores and megaspores.

homosporous   Refers to a plant that produces only one type of spore.

integument   The outer layer(s) of an ovule; integuments become the seed coat of a seed.

micropyle   The opening in the ovule integument through which the pollen tube grows.

nucellus   The tissue of an ovule in which an embryo sac develops.

ovary   The basal, swollen portion of a carpel; it contains the ovules and develops into the fruit.

ovule   A seed plant structure within an ovary; it contains a female gametophyte surrounded by the nucellus and one or two integuments. At maturity, an ovule becomes a seed.

pollen grain   Male gametophyte in seed plants; binucleate or trinucleate seed plant structure produced from a microspore in a microsporangium.

pollination   The transfer of a pollen grain from an anther to a stigma in angiosperms, or to the vicinity of the ovule in gymnosperms.

primary endosperm nucleus   The triploid nucleus resulting from the fusion of a single sperm with the two polar nuclei.

seed   A reproductive structure that develops from an ovule in seed plants. It consists of an embryo and a food supply surrounded by a seed coat.

seed coat   The protective layer of a seed; it develops from the integument(s).

spore   A haploid reproductive cell, produced when a diploid spore mother cell undergoes meiosis; it gives rise by mitosis to a gametophyte.

sporophyte   The multicellular, diploid phase of a plant life cycle; it is the generation that ultimately produces spores.

stamen   A unit of an androecium; it consists of a pollen-bearing anther and usually a stalklike filament.

stigma   The uppermost pollen-receptive portion of a gynoecium.

Gymnosperms

There are four groups of living gymnosperms (conifers, cycads, gnetophytes, and *Ginkgo*), all of which lack the flowers and fruits of angiosperms. In all of them, the **ovule,** which becomes a seed, rests exposed on a scale (modified shoot or leaf) and is not completely enclosed by sporophyte tissues at the time of pollination. The name gymnosperm combines the Greek root *gymnos*, or "naked," with *sperma*, or "seed." In other words, gymnosperms are naked-seeded plants. However, although the ovules are naked at the time of pollination, the seeds of gymnosperms are sometimes enclosed by other sporophyte tissues by the time they are mature.

Details of reproduction vary somewhat in gymnosperms, and their forms vary greatly. For example, cycads and *Ginkgo* have motile sperm, even though the sperm are carried within a pollen tube, while conifers and gnetophytes have sperm with no flagella. The female cones range from tiny, woody structures weighing less than 25 grams and having a diameter of a few millimeters, to massive structures weighing more than 45 kilograms and growing to lengths of more than a meter.

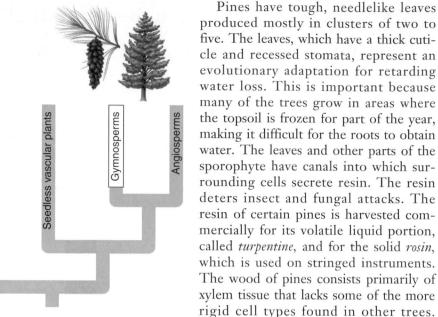

Conifers (Coniferophyta)

The most familiar gymnosperms are conifers (phylum Coniferophyta), which include pines (figure 29.14), spruces, firs, cedars, hemlocks, yews, larches, cypresses, and others. The coastal redwood (*Sequoia sempervirens*), a conifer native to northwestern California and southwestern Oregon, is the tallest living vascular plant; it may attain nearly 100 meters (300 feet) in height. Another conifer, the bristlecone pine (*Pinus longaeva*) of the White Mountains of California, is the oldest living tree; one specimen is 4900 years of age. Conifers are found in the colder temperate and sometimes drier regions of the world. They are sources of timber, paper, resin, taxol (used to treat cancer), and other economically important products.

Pines.   More than 100 species of pines exist today, all native to the northern hemisphere, although the range of one species does extend a little south of the equator. Pines and spruces are members of the vast coniferous forests that lie between the arctic tundra and the temperate deciduous forests and prairies to their south. During the past century, pines have been extensively planted in the southern hemisphere.

Pines have tough, needlelike leaves produced mostly in clusters of two to five. The leaves, which have a thick cuticle and recessed stomata, represent an evolutionary adaptation for retarding water loss. This is important because many of the trees grow in areas where the topsoil is frozen for part of the year, making it difficult for the roots to obtain water. The leaves and other parts of the sporophyte have canals into which surrounding cells secrete resin. The resin deters insect and fungal attacks. The resin of certain pines is harvested commercially for its volatile liquid portion, called *turpentine*, and for the solid *rosin*, which is used on stringed instruments. The wood of pines consists primarily of xylem tissue that lacks some of the more rigid cell types found in other trees. Thus, it is considered a "soft" rather than a "hard" wood. The thick bark of pines represents another adaptation for surviving fires and subzero temperatures. Some cones actually depend on fire to open them, releasing seed to reforest burnt areas.

FIGURE 29.14
Conifers.   Slash pines, *Pinus palustris*, in Florida are representative of the Coniferophyta, the largest phylum of gymnosperms.

As mentioned earlier, all seed plants are heterosporous, so the spores give rise to two types of gametophytes (figure 29.15). The male gametophytes (pollen grains) of pines develop from microspores, which are produced in male cones that develop in clusters of 30 to 70, typically at the tips of the lower branches; there may be hundreds of such clusters on any single tree. The male cones generally are 1 to 4 centimeters long and consist of small, papery scales arranged in a spiral or in whorls. A pair of microsporangia form as sacs within each scale. Numerous microspore mother cells in the microsporangia undergo meiosis, each becoming four microspores. The microspores develop into four-celled pollen grains with a pair of air sacs that give them added buoyancy when released into the air. A single cluster of male pine cones may produce more than 1 million pollen grains.

Female cones typically are produced on the upper branches of the same tree that produces male cones. Female cones are larger than male cones, and their scales become woody. Two ovules develop toward the base of each scale. Each ovule contains a megasporangium called the **nucellus.** The nucellus itself is completely surrounded by a thick layer of cells called the integument that has a small opening (the **micropyle**) toward one end. One of the layers of the integument later becomes the seed coat. A single megaspore mother cell within each megasporangium undergoes meiosis, becoming a row of four megaspores. Three of the megaspores break down, but the remaining one, over the better part of a year, slowly develops into a female gametophyte. The female gametophyte at maturity may consist of thousands of cells, with two to six archegonia formed at the micropylar end. Each archegonium contains an egg so large it can be seen without a microscope.

Female cones usually take two or more seasons to mature. At first they may be reddish or purplish in color, but they soon turn green, and during the first spring, the scales spread apart. While the scales are open, pollen grains carried by the wind drift down between them, some catching in sticky fluid oozing out of the micropyle. The pollen grains within the sticky fluid are slowly drawn down through the micropyle to the top of the nucellus, and the scales close shortly thereafter. The archegonia and the rest of the female gameto-

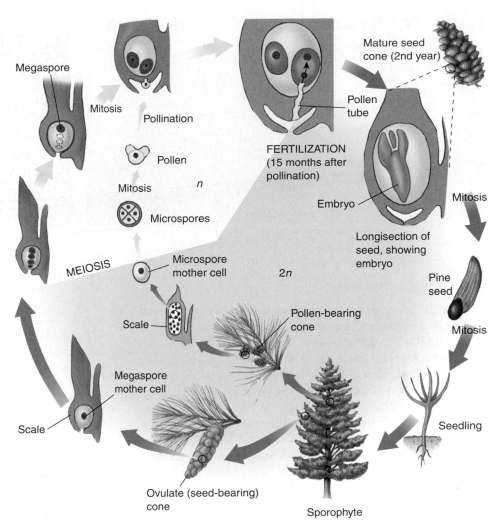

FIGURE 29.15
Life cycle of a typical pine. The male and female gametophytes are dramatically reduced in size in these plants. Wind generally disperses the male gametophyte (pollen), which produces sperm. Pollen tube growth delivers the sperm to the egg on the female cone. Additional protection for the embryo is provided by the integument, which develops into the seed coat.

phyte are not mature until about a year later. While the female gametophyte is developing, a pollen tube emerges from a pollen grain at the bottom of the micropyle and slowly digests its way through the nucellus to the archegonia. While the pollen tube is growing, one of the pollen grain's four cells, the generative cell, divides by mitosis, with one of the resulting two cells dividing once more. These last two cells function as sperm. The germinated pollen grain with its two sperm is the mature male gametophyte.

About 15 months after pollination, the pollen tube reaches an archegonium, and discharges its contents into it. One sperm unites with the egg, forming a zygote. The other sperm and cells of the pollen grain degenerate. The zygote develops into an embryo within a seed. After dispersal and germination of the seed, the young sporophyte of the next generation develops into a tree.

(a) (b) (c)

FIGURE 29.16
Three phyla of gymnosperms. (*a*) An African cycad, *Encephalartos transvenosus*. (*b*) *Welwitschia mirabilis* represents one of the three genera of gnetophytes. (*c*) Maidenhair tree, *Ginkgo biloba*, the only living representative of the phylum Ginkgophyta.

Cycads (Cycadophyta)

Cycads are slow-growing gymnosperms of tropical and subtropical regions. The sporophytes of most of the 100 known species resemble palm trees (figure 29.16*a*) with trunks that can attain heights of 15 meters or more. Unlike palm trees, which are flowering plants, cycads produce cones and have a life cycle similar to that of pines. The female cones, which develop upright among the leaf bases, are huge in some species and can weigh up to 45 kilograms. The sperm of cycads, although formed within a pollen tube, are released within the ovule to swim to an archegonium. These sperm are the largest sperm cells among all living organisms. Several species of cycads are facing extinction in the wild and soon may exist only in botanical gardens.

Gnetophytes (Gnetophyta)

There are three genera and about 65 living species of Gnetophyta. They are the only gymnosperms with vessels in their xylem, a particularly efficient conducting cell type that is a common feature in angiosperms. The members of the three genera differ greatly from one another in form. One of the most bizarre of all plants is *Welwitschia*, which occurs in the Namib and Mossamedes deserts of southwestern Africa (figure 29.16*b*). The stem is shaped like a large, shallow cup that tapers into a taproot below the surface. It has two strap-shaped, leathery leaves that grow continuously from their base, splitting as they flap in the wind. The reproductive structures of *Welwitschia* are conelike, appear toward the bases of the leaves around the rims of the stems, and are produced on separate male and female plants.

More than half of the gnetophyte species are in the genus *Ephedra*, which is common in arid regions of the western United States and Mexico. Species are found on every continent except Australia. The plants are shrubby, with stems that superficially resemble those of horsetails,

being jointed and having tiny, scalelike leaves at each node. Male and female reproductive structures may be produced on the same or different plants. The drug ephedrine, widely used in the treatment of respiratory problems, was in the past extracted from Chinese species of *Ephedra*, but it has now been largely replaced with synthetic preparations. Ephedrine found in herbal remedies for weight loss has been linked to strokes and heart attacks.

The best-known species of *Gnetum* is a tropical tree, but most species are vinelike. All species have broad leaves similar to those of angiosperms. One *Gnetum* species is cultivated in Java for its tender shoots, which are cooked as a vegetable.

Ginkgo (Ginkgophyta)

The fossil record indicates that members of the Ginkgophyta were once widely distributed, particularly in the northern hemisphere; today, only one living species, the maidenhair tree (*Ginkgo biloba*), remains. The tree, which sheds its leaves in the fall, was first encountered by Europeans in cultivation in Japan and China; it apparently no longer exists in the wild. The common name comes from the resemblance of its fan-shaped leaves to the leaflets of maidenhair ferns (figure 29.16*c*). Like the sperm of cycads, those of *Ginkgo* have flagella. The ginkgo is dioecious—that is, the male and female reproductive structures are produced on separate trees. The fleshy outer coverings of the seeds of female ginkgo plants exude the foul smell of rancid butter, caused by the presence of butyric and isobutyric acids. As a result, male plants vegetatively propagated from shoots are preferred for cultivation. Because of its beauty and resistance to air pollution, *Ginkgo* is commonly planted along city streets.

Gymnosperms are mostly cone-bearing seed plants. In gymnosperms, the ovules are not completely enclosed by sporophyte tissue at pollination.

Angiosperms

The 250,000 known species of flowering plants are called angiosperms because their ovules, unlike those of gymnosperms, are enclosed within diploid tissues at the time of pollination. The name *angiosperm* derives from the Greek words *angeion*, "vessel," and *sperma*, "seed." The "vessel" in this instance refers to the carpel, which is a modified leaf that encapsulates seeds. The carpel develops into the fruit, a unique angiosperm feature. While some gymnosperms, including yew, have fleshlike tissue around their seeds, it is of a different origin and not a true fruit.

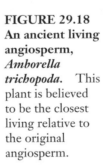

FIGURE 29.17
Fossil of basal angiosperm.
Archaefructus fossil with multiseeded carpels (fruits) and stamens. This is the oldest known angiosperm in the fossil record.

Angiosperm Origins

The origins of the angiosperms puzzled even Darwin (his "abominable mystery"). Recent fossil and molecular sequence data have provided exciting clues about basal angiosperms. In the remote Liaoning province of China, a complete angiosperm fossil that is at least 125 million years old has been found (figure 29.17). The fossil may represents a new, basal, and extinct angiosperm family, Archaefructaceae, with two species, *Archaefructus liaoningensis* and *A. sinensis*. *Archaefructus* was an herbaceous, aquatic plant. This family is proposed to be the sister clade to all other angiosperms but there is a lively debate about the validity of this claim. How do we know that *Archaefructus* is an angiosperm? The fossils have both male and female reproductive structures. They lack the sepals and petals that evolved in later angiosperms and attract pollinators. The fossils were so well preserved that fossil pollen could be examined using scanning electron microscopy. Although *Archaefructus* is ancient, it is unlikely to be the very first angiosperm. Still, the incredibly well-preserved fossils provide valuable detail on angiosperms in the Upper Jurassic/Lower Cretaceous period, when dinosaurs roamed the earth.

Consensus has also been growing on the most basal living angiosperm—*Amborella trichopoda* (figure 29.18). *Amborella*, with small, cream-colored flowers, is even more primitive than magnolias or water lilies. This small shrub, found only on the island of New Caledonia in the South Pacific, is the last remaining species of the earliest extant lineage of the angiosperms, arising about 135 million years ago. While *Amborella* is not the original angiosperm, it is sufficiently close that studying its reproductive biology may help us understand the early radiation of the angiosperms. The angiosperm phylogeny reflects an evolutionary hypothesis that is driving new research on angiosperm origins (figure 29.19).

FIGURE 29.18
An ancient living angiosperm, *Amborella trichopoda*. This plant is believed to be the closest living relative to the original angiosperm.

Monocots and Eudicots

Biologists have long grouped angiosperms into two classes, dicots and monocots. Although phylogenetic evidence indicates that much evolutionary diversity is masked by such grouping (see figure 29.19). More meaningful comparisons are made among eudicots and the lineage that gave rise to monocots and magnolias.

Eudicots (about 175,000 species) are the more primitive of the two classes, with monocots apparently having derived from early dicots. Included in the dicots are the great majority of familiar angiosperms—almost all kinds of trees and shrubs, snapdragons, mints, peas, sunflowers, and other

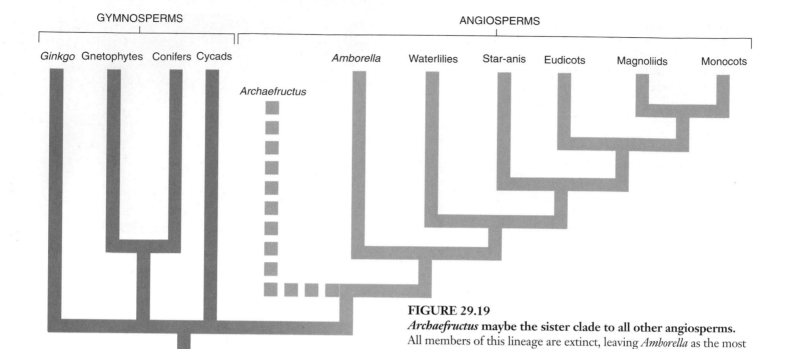

GYMNOSPERMS

ANGIOSPERMS

Ginkgo Gnetophytes Conifers Cycads

Amborella Waterlilies Star-anis Eudicots Magnoliids Monocots

Archaefructus

FIGURE 29.19
***Archaefructus* maybe the sister clade to all other angiosperms.**
All members of this lineage are extinct, leaving *Amborella* as the most basal, living angiosperm. Gymnosperms are colored dark green.

plants. Monocots (about 65,000 species) include the lilies, grasses, cattails, palms, agaves, yuccas, pondweeds, orchids, and irises.

Monocots and eudicots differ from each other in a number of features, some of which are listed in figure 29.20. Monocots and eudicots differ fundamentally in other ways as well. For example, about one-sixth of all eudicot species are **annuals** (plants that complete their entire growth cycle within a year); there are, however, very few annual monocots. Underground swollen storage organs, such as bulbs, occur much more frequently in monocots than they do in eudicots. There are many species of woody dicots (mostly trees or shrubs), but no monocots have true wood; however, a few monocots, such as *agave* and *aloe*, produce extra bundles of conducting tissues that give them a woody texture. Endosperm, which is usually present in mature monocot seeds, is largely absent in mature eudicot seeds. Other specific differences will be presented in chapter 35.

The Structure of Flowers

Flowers are considered to be modified stems bearing modified leaves. Regardless of their size and shape, they all share certain features (figure 29.21). Each flower originates as a **primordium** that develops into a bud at the end of a stalk called a **pedicel.**

MONOCOTS

1. Seed with one cotyledon ("seed leaf").
2. Leaves with parallel veins.
3. Vascular cambium rarely occur.
4. Flower parts mostly in threes or multiples of three.

EUDICOTS

1. Seed with two cotyledons ("seed leaves").
2. Leaves with a network of veins.
3. Lateral meristems (cambia) present.
4. Flower parts mostly in fours or fives or multiples of four or five.

FIGURE 29.20
Comparison of monocots and eudicots.

The pedicel expands slightly at the tip to form the **receptacle,** to which the remaining flower parts are attached. The other flower parts typically are attached in circles called *whorls.* The outermost whorl is composed of **sepals.** Most flowers have three to five sepals, which are green and somewhat leaflike; they often function in protecting the immature flower and in some species may drop off as the flower opens. The next whorl consists of **petals** that are often colored, attracting pollinators such as insects and birds. The petals, which commonly number three to five, may be separate, fused together, or missing altogether in *wind-pollinated flowers.*

The third whorl consists of **stamens,** collectively called the **androecium** (Greek *andros,* "male," + *oikos,* "house"). Each stamen consists of a pollen-bearing **anther** and a stalk called a **filament,** which may be missing in some flowers. At the center of the flower is the **gynoecium** (Greek *gynos,* "female," + *oikos,* "house"), consisting of one or more **carpels.** The first carpel is believed to have been formed from a leaflike structure with ovules along its margins. The edges of the blade then rolled inward and fused together, forming a carpel. Primitive flowers can have several to many separate carpels, but in most flowers, two to several carpels are fused together. Such fusion can be seen in an orange sliced in half; each segment represents one carpel.

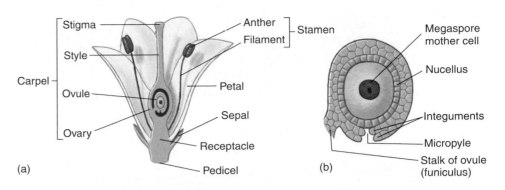

FIGURE 29.21
Diagram of an angiosperm flower.
(*a*) The main structures of the flower are labeled. (*b*) Details of an ovule. The ovary as it matures will become a fruit; as the ovule's outer layers (integuments) mature, they will become a seed coat.

A carpel has three major regions (figure 29.21*a*). The **ovary** is the swollen base, which contains from one to hundreds of **ovules**; the ovary later develops into a **fruit**. The tip of the carpel is called a **stigma**. Most stigmas are sticky or feathery, causing pollen grains that land on them to adhere. Typically, a neck or stalk called a **style** connects the stigma and the ovary; in some flowers, the style may be very short or even missing. Many flowers have nectar-secreting glands called *nectaries*, often located toward the base of the ovary. Nectar is a fluid containing sugars, amino acids, and other molecules that attracts insects, birds, and other animals to flowers.

The Angiosperm Life Cycle

While a flower bud is developing, a single megaspore mother cell in the ovule undergoes meiosis, producing four megaspores (figure 29.22). In most flowering plants, three of the megaspores soon disappear while the nucleus of the remaining one divides mitotically, and the cell slowly expands until it becomes many times its original size. While this expansion is occurring, each of the daughter nuclei divides twice, resulting in eight haploid nuclei arranged in two groups of four. At the same time, two layers of the ovule, the **integuments**, differentiate and become the *seed coat* of a seed. The integuments, as they develop, leave a small gap or pore at one end called the *micropyle* (see figure 29.21*b*). One nucleus from each group of four migrates toward the center, where they function as **polar nuclei.** Polar nuclei may fuse together, forming a single diploid nucleus, or they may form a single cell with two haploid nuclei. Cell walls also form around the remaining nuclei. In the group closest to the micropyle, one cell functions as the **egg;** the other two nuclei are called **synergids.** At the other end, the three cells are now called **antipodals;** they have no apparent function and eventually break down and disappear. The large sac with eight nuclei in seven cells is called an **embryo sac;** it constitutes the female gametophyte. Although it is completely dependent on the sporophyte for nutrition, it is a multicellular, haploid individual.

While the female gametophyte is developing, a similar but less complex process takes place in the anthers. Most anthers have patches of tissue (usually four) that eventually become chambers lined with nutritive cells. The tissue in each patch is composed of many diploid microspore mother cells that undergo meiosis more or less simultaneously, each producing four microspores. The four microspores at first remain together as a quartet or tetrad, and the nucleus of each microspore divides once; in most species, the microspores of each quartet then separate. At the same time, a two-layered wall develops around each microspore. As the anther matures, the wall between adjacent pairs of chambers breaks down, leaving two larger sacs. At this point, the binucleate microspores have become **pollen grains.** The outer pollen grain wall layer often becomes beautifully sculptured, and it contains chemicals that may react with others in a stigma to signal whether or not development of the male gametophyte should proceed to completion. The pollen grain has areas called *apertures*, through which a pollen tube may later emerge.

Pollination is simply the mechanical transfer of pollen from its source (an anther) to a receptive area (the stigma of a flowering plant). Most pollination takes place between flowers of different plants and is brought about by insects, wind, water, gravity, bats, and other animals. In as many as one-quarter of all angiosperms, however, a pollen grain may be deposited directly on the stigma of its own flower, and self-pollination occurs. Pollination may or may not be followed by *fertilization*, depending on the genetic compatibility of the pollen grain and the flower on whose stigma it has landed. (In some species, complex, genetically controlled mechanisms prevent self-fertilization to enhance genetic diversity in the progeny.) If the stigma is receptive, the pollen grain's dense cytoplasm absorbs substances from the stigma and bulges through an aperture. The bulge develops into a *pollen tube* that responds to chemical and mechanical stimuli that guide it to the embryo sac. It follows a diffusion gradient of the chemicals and grows down through the style and into the micropyle. The pollen tube usually takes several hours to two days to reach the micropyle, but in a few instances, it may take up to a year. One of the pollen grain's two cells, the *generative cell*, lags behind. Its nucleus divides in the pollen grain or in the pollen tube, producing two sperm cells. Unlike sperm in mosses, ferns, and some gymnosperms, the sperm of flowering plants have no flagella. At this point, the pollen grain with its tube and sperm has become a mature male gametophyte.

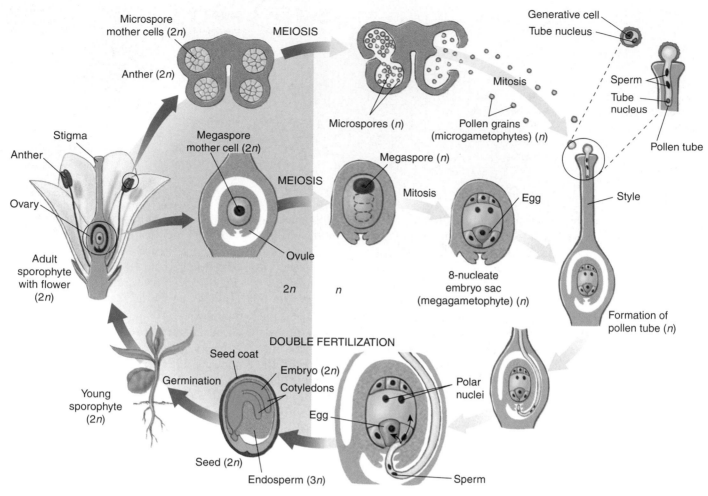

FIGURE 29.22
Life cycle of a typical angiosperm. As in pines, external water is no longer required for fertilization. In most species of angiosperms, animals carry pollen to the carpel. The outer wall of the carpel forms the fruit, which often entices animals to disperse the seed.

As the pollen tube enters the embryo sac, it destroys a synergid in the process and then discharges its contents. Both sperm are functional, and an event called **double fertilization** follows. One sperm unites with the egg and forms a zygote, which develops into an embryo sporophyte plant. The other sperm and the two polar nuclei unite, forming a triploid primary endosperm nucleus. The primary endosperm nucleus begins dividing rapidly and repeatedly, becoming triploid **endosperm tissue** that may soon consist of thousands of cells. Endosperm tissue can become an extensive part of the seed in grasses such as corn (see figure 36.7).

In most flowering plants, endosperm provides nutrients for the embryo that develops from the zygote; in many species, such as peas and beans, it disappears completely by the time the seed is mature. Following double fertilization, the integuments harden and become the seed coat of a seed. The haploid cells remaining in the embryo sac (antipodals, synergid, tube nucleus) degenerate.

Until recently, the nutritional, triploid endosperm was believed to be the ancestral state in angiosperms. A recent analysis of extant, basal angiosperms revealed that diploid endosperms were also common. The female gametophyte in these species has four, not eight nuclei. At the moment, it is unclear whether diploid or triploid endosperms are the most primitive.

Angiosperms are characterized by ovules that at pollination are enclosed within an ovary at the base of a carpel, a structure unique to the phylum; a fruit develops from the ovary. Evolutionary innovations—including flowers to attract pollinators, fruits to protect and aid in embryo dispersal, and double fertilization providing additional nutrients for the embryo—all have contributed to the widespread success of this phylum.

For interactive testing, visit the Online Learning Center with PowerWeb at www.mhhe.com/Raven7

29.1 Plants have multicellular haploid and diploid stages in their life cycles.

The Evolutionary Origins of Plants

- The term plant will be used to refer to a group of organisms sharing a freshwater ancestor that have evolved over a 470-million-year period. (p. 580)
- The defining characteristic of plants is the protection of their embryos. (p. 580)
- All plants have multicellular haploid and diploid phases, with the trend over time toward increasing embryo protection and a smaller haploid stage. (p. 580)
- Most plants are protected from desiccation by a waxy cuticle, and have stomata through which gases diffuse into and out of the plant. (p. 580)

Plant Life Cycles

- Plants have haplodiplontic life cycles. (p. 581)
- The diploid sporophyte produces haploid spores by meiosis, and spores divide by mitosis, producing a multicellular, haploid gametophyte. (p. 581)

29.2 Nonvascular plants are relatively unspecialized, but successful in many terrestrial environments.

Mosses, Liverworts, and Hornworts

- Nonvascular plants are divided into three major phyla— Bryophyta, Hepaticophyta, and Anthocerophyta— with all members being relatively unspecialized, but collectively able to inhabit diverse environments. (pp. 582–583)

29.3 Seedless vascular plants have well-developed conducting tissues in their sporophytes.

Features of Vascular Plants

- The seven living phyla of vascular plants dominate almost all terrestrial habitats. (p. 584)
- In modern vascular plants, the gametophytes have been reduced in size and complexity; individuals exhibit highly specialized conductive tissues. (p. 584)

Seedless Vascular Plants

- The two seedless vascular plant phyla, Lycophyta (club mosses) and Pterophyta (whisk ferns, horsetails, and ferns), have a large and conspicuous sporophyte. Many also have well-differentiated roots, stems, and leaves. (pp. 586–588)

29.4 Seeds protect and aid in the dispersal of plant embryos.

Seed Plants

- Seed plants first appeared about 425 MYA. (p. 589)
- The seed provides an extra layer of sporophyte tissue and a dormant phase that allows a pause in the life cycle to await more favorable environmental conditions. (p. 589)
- Seed plants produce male and female gametophytes. (p. 589)

Gymnosperms

- Gymnosperms represent cone-bearing, naked-seeded plants and are composed of four living groups (conifers, cycads, gnetophytes, and ginkgo). (p. 590)
- Conifers, the most familiar gymnosperms, include pines, spruces, firs, cedars, hemlocks, yews, larches, and cypresses. (p. 590)
- During pollination, the ovules are not completely enclosed by protective sporophyte tissue. (pp. 590–592)

Angiosperms

- During pollination, ovules are enclosed within diploid tissue (an ovary). (p. 593)
- The approximately 250,000 known species of flowering plants have long been grouped into two classes—monocots and eudicots. These groups are not monophyletic. (pp. 593–594)
- Monocots and eudicots differ according to the number of cotyledons, leaf venation, presence of lateral meristems, and number of flower parts. (p. 594)
- Flowers, which may attract pollinators, are modified stems bearing modified leaves, with flower parts attached in whorls. (p. 594)
- Fruits protect embryos and seeds, and aid in seed dispersal. (pp. 593–596)
- In most flowering plants, double fertilization provides nutrients for the developing embryo. (p. 596)

Self Test

1. All plants exhibit alternation of generations. This means their life cycle
   a. includes both haploid and diploid gametes.
   b. shows only asexual reproduction.
   c. has both a multicellular haploid stage and a multicellular diploid stage.
   d. does not include meiosis.

2. The plant life cycle has both a sporophyte and a gametophyte generation. In the sporophyte stage,
   a. gametes are produced.
   b. meiosis occurs.
   c. only mitosis takes place.
   d. gametophytes form.

3. In plants,
   a. gametes are produced directly after meiosis.
   b. gametes are produced directly after mitosis.
   c. no gametes are motile.
   d. seeds are always produced.

4. Which of the following is true of the bryophytes?
   a. It is the only group that shows an alternation of generations.
   b. Bryophytes exhibit extensive vascular tissue.
   c. The sporophyte (multicellular diploid) is the conspicuous stage.
   d. The gametophyte (multicellular haploid) is the conspicuous stage.

5. A plant's vascular tissue is composed of xylem and phloem. The xylem generally transports _____, whereas the phloem transports _____.
   a. water/sugar
   b. sugar/water
   c. water/water
   d. sugar/sugar

6. In the conifers, the
   a. gametophyte is prominent, and the sporophyte is dependent upon the gametophyte.
   b. sporophyte is prominent, with the sporophyte and gametophyte living independently.
   c. sporophyte is prominent, and the gametophyte is dependent upon the sporophyte.
   d. gametophyte is prominent, and the sporophyte stage has disappeared.

7. Which of the following characters is seen in the gymnosperms, but is not seen in other seeded vascular plants?
   a. alternation of generations
   b. exposed seeds
   c. sporophyte stage
   d. pollen

8. All flowering plants (angiosperms)
   a. produce exposed seeds.
   b. are nonvascular.
   c. have flagellated sperm.
   d. have fruit.

9. In the angiosperms, the
   a. gametophyte is prominent, and the sporophyte is dependent upon the gametophyte.
   b. sporophyte is prominent, with the sporophyte and gametophyte living independently.
   c. sporophyte is prominent, and the gametophyte is dependent upon the sporophyte.
   d. gametophyte is prominent, and the sporophyte stage has disappeared.

10. Which of the following is *not* characteristic of a monocot?
    a. leaves with parallel veins
    b. flower parts usually in threes or multiples of three
    c. lateral meristems occurring rarely
    d. seed with two cotyledons

Test Your Visual Understanding

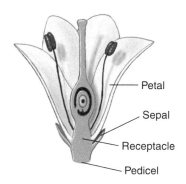

1. Label this diagram with the following terms: stigma, style, ovule, ovary, carpel, anther, filament, stamen.

Apply Your Knowledge

1. A major factor in life on land is coping with ultraviolet radiation from the sun. Early in the evolution of land plants, there was a change from having a prominent haploid gametophyte to a prominent diploid sporophyte. Explain why this change may have occurred in reference to ultraviolet radiation.

2. Plants provide many vital products for humans. Make a list of all the plant-based products that you use in a single day.

3. Many flowering plants have adapted in response to a particular animal pollinator, and they rely on that animal for proper pollination to occur. What will happen to the plant population if the animal pollinator goes extinct?

30

Fungi

Concept Outline

30.1 The fungi share several key characteristics.

Distinctive Fungal Features. All fungi are heterotrophs by absorption. Fungal cell walls include chitin.

The Body of a Fungus. Some fungi have filamentous bodies (hyphae), and others have yeast morphology.

How Fungi Reproduce. Fungi reproduce both asexually and sexually. In sexual reproduction, hyphae attracted by chemical signaling fuse.

How Fungi Obtain Nutrients. Fungi obtain nutrients by external digestion of dead or living organisms.

Metabolic Pathways. Fungal metabolic pathways provide resources for humans.

Ecology of Fungi. Fungi are among the most important decomposers in terrestrial ecosystems.

30.2 There are four major groups of fungi.

Phylogenetic Relationships. Molecular data are revising the evolutionary history of the fungi.

Chytridiomycota. Chytrids are motile fungi.

Zygomycota. In zygomycetes, the fusion of hyphae leads to the formation of a zygote inside a zygosporangium.

Basidiomycota. Basidiomycetes include mushrooms and single-celled fungi. There is only a single diploid cell in the life cycle. Meiosis immediately follows karyogamy and produces haploid spores outside a basidium.

Ascomycota. Ascomycetes range from unicells to filamentous fungi. Hyphal fusion leads to karyogamy within the ascus. The diploid cell immediately undergoes meiosis to produce spores.

30.3 Fungi participate in many symbioses.

Lichens. A lichen is a mutualistic association between a fungus and an alga or cyanobacterium.

Mycorrhizae. Mycorrhizae are mutualistic associations between fungi and the roots of plants.

Endophytes. Most likely, all plants have fungi living within them that afford protection from herbivores.

Mutualistic Animal Symbioses. Symbiotic relationships with animals range from partnerships in the guts of grazing animals to being farmed by ants.

Fungal Parasites and Pathogens. Agriculture and human health are frequently threatened by fungal parasites.

FIGURE 30.1
Spores ejected from a pore on the surface of a puffball fungus. The fungi constitute a unique kingdom of heterotrophic organisms. Along with bacteria, they are important decomposers and disease-causing organisms. Some puffballs are almost 1 meter in diameter and may contain 7 trillion spores—enough to circle the earth's equator!

The fungi, an often-overlooked group of unicellular and multicellular organisms, have a profound impact on ecology and human health. Fungi are found everywhere—from the tropics to the tundra and in both terrestrial and aquatic environments (figure 30.1). Fungi made it possible for plants to colonize land by associating with rootless stems and aiding in the uptake of nutrients and water. Mushrooms and toadstools are fungi, multicellular creatures that grow so rapidly that they seem to appear overnight on our lawns. A single *Armillaria* fungus can cover 15 hectares underground and weigh 100 tons. Some yeasts are used to make bread and beer. Other fungi cause disease in plants and animals. These fungal killers are particularly problematic because fungi are animals' closest relatives. Drugs that can kill fungi often have toxic effects on animals, including humans.

599

30.1 The fungi share several key characteristics.

Distinctive Fungal Features

Mycologists, scientists who study fungi, believe there may be as many as 1.5 million fungal species. Currently, taxonomists classify the fungi into four main groups: the chytrids, the zygomycetes, the basidiomycetes, and the ascomycetes (figure 30.2). Recent phylogenetic analysis of DNA sequences and protein sequences indicates that fungi are more closely related to animals than to plants. The last common ancestor was a single cell, and unique solutions to multicellularity evolved in fungi and in animals. While the fungi are incredibly diverse, they share some characteristics:

1. **Fungi arc hctcrotrophs.** Fungi obtain their food by secreting digestive enzymes into substrates, including fallen logs and the skin of frogs. Then they absorb the organic molecules released by the enzymes. Thus, fungi actually live in their food.
2. **Fungi have several cell types.** Multicellular fungi are primarily filamentous in their growth form (that is, their bodies consist of long, slender filaments called hyphae), even though these hyphae may be packed together to form complex structures such as a mushroom. Hyphal cells have a range of morphologies. Some unicellular fungi have a flagellum.
3. **Some fungi have a dikaryon stage.** Many sexually reproducing fungi undergo a stage in which two haploid cells coexist in a single cell (dikaryon) for some period of time before fusing to make a diploid nucleus.
4. **Fungi have cell walls that include chitin.** The cell walls of fungi are built of polysaccharides (chains of sugars) and chitin, the same tough material a crab shell is made of.
5. **Fungi undergo nuclear mitosis.** Mitosis in fungi is different from that in plants and animals in one key respect: The nuclear envelope does not break down and re-form. Instead, mitosis takes place *within* the nucleus. A spindle apparatus forms there, dragging chromosomes to opposite poles of the *nucleus* (not the cell, as in most other eukaryotes). This type of mitosis is found in some protists as well (see chapter 28).

Fungi absorb their food after digesting it with secreted enzymes. Besides this mode of nutrition, the fungi are characterized by combined dikaryon stages, nuclear mitosis, and other traits.

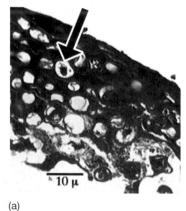   

(a)   (b)   (c)   (d)

FIGURE 30.2

Representatives of the four major groups of fungi. (*a*) The parasitic chytrid *Batrachochytrium dendrobatidis* is causing a worldwide decline in amphibians. This stained skin sample from an infected frog shows sporangia of *B. dendrobatidis* (*arrow*). (*b*) *Pilobolus*, a zygomycete, grows on animal dung. Stalks about 10 millimeters long contain dark, spore-bearing sacs. (*c*) *Amanita muscaria*, the fly agaric, is a toxic basidiomycete. (*d*) The cup fungus *Cookeina tricholoma* is an ascomycete from the rain forest of Costa Rica. In the cup fungi, the spore-producing structures line the cup; in basidiomycetes that form mushrooms such as *Amanita*, they line the gills beneath the cap of the mushroom. All visible structures of fungi, including the ones shown here, arise from an extensive network of filamentous hyphae that penetrates and is interwoven with the substrate on which they grow.

The Body of a Fungus

Fungi exist either as single-celled yeasts or in multicellular form with several different cell types. We will focus on cells called **hyphae** (singular, *hypha*), which are slender filaments, barely visible to the naked eye. Some hyphae are continuous or branching tubes filled with cytoplasm and multiple nuclei. Other hyphae are typically made up of long chains of cells joined end-to-end and divided by cross-walls called **septa** (singular, *septum*). The septa rarely form a complete barrier, except when they separate the reproductive cells. Even fungi with septa can be considered one long cell. Cytoplasm characteristically flows or streams freely throughout the hyphae, passing through major pores in the septa (figure 30.3). Because of this streaming, proteins synthesized throughout the hyphae may be carried to their actively growing tips. As a result, fungal hyphae may grow very rapidly when food and water are abundant and the temperature is optimum.

A mass of connected hyphae is called a **mycelium** (plural, *mycelia*). This word and the term *mycology* are both derived from the Greek word for "fungus," *mykes*. The mycelium of a fungus (figure 30.4) constitutes a system that may, in the aggregate, be many meters long. Growth is in two dimensions, not three. This mycelium grows into the soil, wood, or other material in which the fungus is growing. Digestion begins, and all parts of such a fungus are metabolically active.

In two of the four major groups of fungi, reproductive structures formed of interwoven hyphae, such as mushrooms, puffballs, and morels, are produced at certain stages of the life cycle. These structures expand rapidly because of rapid inflation of the hyphae. For this reason, mushrooms that are ready to reproduce can appear suddenly on your lawn.

The cell walls of fungi are formed of polysaccharides and chitin, not cellulose as are those of plants and many groups of protists. Chitin is the same material that makes up the major portion of the hard shells, or exoskeletons, of arthropods, a group of animals that includes insects and crustaceans (see chapter 33). Chitin is one of the shared traits that has led scientists to believe that fungi and animals are more closely related than fungi and plants.

Mitosis in multicellular fungi differs from that in most other organisms. Because of the linked nature of the cells, the cell itself is not the relevant unit of reproduction; instead, the nucleus is. The nuclear envelope does not break down and re-form; instead, the spindle apparatus is formed *within* it. Centrioles are absent in all fungi; instead, fungi regulate the formation of microtubules during mitosis with small, relatively amorphous structures called *spindle plaques*. This unique combination of features strongly suggests that fungi originated from some unknown group of single-celled eukaryotes with these characteristics.

Multicellular fungi exist primarily in the form of filamentous hyphae, either with or without septa. These and other unique features support placing fungi in a separate kingdom.

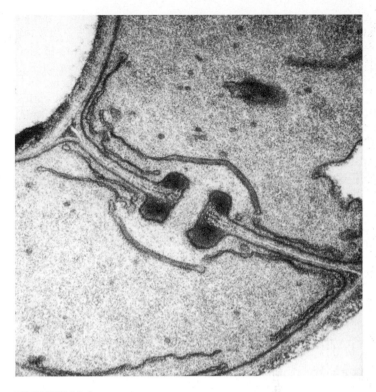

FIGURE 30.3
A septum. This transmission electron micrograph of a section through a hypha of the basidiomycete *Inonotus tomentosus* shows a pore through which the cytoplasm streams (45,000×).

FIGURE 30.4
Fungal mycelium. This mycelium, composed of hyphae, is growing through leaves on the forest floor in Maryland.

How Fungi Reproduce

Fungi are different from most animals and plants in that each cell (or hypha) can house one, two, or more nuclei. If each compartment of a hypha has only one nucleus, it is called **monokaryotic**; a hypha with two nuclei is called **dikaryotic.** In a dikaryotic cell, the two haploid nuclei exist independently. Dikaryotic hyphae have some of the genetic properties of diploids, because both genomes are transcribed.

Sometimes, many nuclei intermingle in the common cytoplasm of a fungal mycelium, which can lack distinct cells. If a dikaryotic or multinucleate hypha has nuclei that are derived from two genetically distinct individuals, the hypha is called **heterokaryotic.** Hyphae whose nuclei are genetically similar to one another are called **homokaryotic.**

These distinctions are important in understanding the life cycles of the individual groups, which can be quite complicated. Many fungi are capable of producing both sexual and asexual spores. When a fungus reproduces sexually, two haploid hyphae of compatible mating types come together and fuse. In animals, plants, and some fungi, the fusion of two haploid cells immediately results in a diploid cell ($2n$). However, in other fungi (basidiomycetes and ascomycetes), an intervening dikaryotic stage ($1n + 1n$) occurs before the parental nuclei fuse and form a diploid nucleus. In ascomycetes, this dikaryotic stage is brief, occuring in only a few cells of the sexual reproductive structure. In basidiomycetes, however, it can last for most of the life of the fungus, including both the feeding and sexual spore-producing structures.

The cytoplasm in fungal hyphae normally flows through perforated septa, or moves freely in their absence. Reproductive structures are an important exception to this general pattern. When reproductive structures form, they are cut off by complete septa that lack perforations or have perforations that soon become blocked.

Spores are the most common means of reproduction among fungi. They may form as a result of either asexual or sexual processes, and they are often dispersed by the wind. When spores land in a suitable place, they germinate, giving rise to a new fungal mycelium. Because the

47.85 µm

FIGURE 30.5
Fungal spores. Scanning electron micrograph of fungal spores that infect rose plants.

spores are very small, between 2 and 75 microns in diameter (figure 30.5), they can remain suspended in the air for long periods of time. Unfortunately, many of the fungi that cause diseases in plants and animals are spread rapidly by such means. The spores of other fungi are routinely dispersed by insects or other small animals. Only one group of fungi, the chytrids, retain the ancestral flagella and have motile zoospores.

It had long been believed that the worldwide presence of fungal species could be accounted for, on an evolutionary timescale, by the almost limitless, long-distance dispersal of fungal spores. However, recent biogeographic studies have examined the phylogenetic relationships among fungi in distant parts of the world and disproved this long-held assumption.

Fungi reproduce sexually after two hyphae of opposite mating types fuse. Asexual reproduction by spores is a common means of reproduction.

FIGURE 30.6
A carnivorous fungus. The oyster mushroom *Pleurotus ostreatus* not only decomposes wood but also immobilizes nematodes, which the fungus uses as a source of nitrogen.

How Fungi Obtain Nutrients

All fungi obtain their food by secreting digestive enzymes into their surroundings and then absorbing back into the fungus the organic molecules produced by this *external digestion.* The fungal body plan reflects this approach. Unicellular fungi have the greatest surface area-to-volume ratio of any fungus, maximizing the surface area for absorption. Extensive networks of hyphae provide an enormous surface area for absorption in a fungal mycelium. Many fungi are able to break down the cellulose in wood, cleaving the linkages between glucose subunits and then absorbing the glucose molecules as food. Fungi also digest lignin, which strengthens plant cell walls. Fungi's specialized metabolic pathways allow them to obtain nutrients from dead trees and from an extraordinary range of organic compounds, including tiny roundworms called nematodes.

The mycelium of the edible oyster mushroom *Pleurotus ostreatus* (figure 30.6) excretes a substance that paralyzes nematodes (see chapter 32) that feed on the fungus. When the worms become sluggish and inactive, the fungal hyphae envelop and penetrate their bodies. Then the fungus secretes digestive juices and absorbs the nematode's nutritious contents, just like it would from a plant source. The fungus usually grows within living trees or on old stumps, obtaining the bulk of its glucose through the enzymatic digestion of cellulose and lignin from plant cell walls, so that the nematodes it consumes apparently serve mainly as a source of nitrogen—a substance almost always in short supply in biological systems. Other fungi are even more active predators than *Pleurotus,* snaring, trapping, or firing projectiles into nematodes, rotifers, and other small animals on which they prey.

> Fungi secrete digestive enzymes onto organic matter and then absorb the products of the digestion.

Metabolic Pathways

Humans use the metabolic pathways of fungi for commercial purposes. Anaerobic fermentation (see chapter 9) in some fungi gives wine and certain cheeses (Roquefort and Brie) their delicate flavors. Other fungi make possible the manufacture of soy sauce, miso, and other fermented foods. Vast industries depend on the biochemical manufacture of organic substances such as citric acid by fungi in culture. The ability of **yeasts** (single-celled as opposed to filamentous fungi) to ferment carbohydrates, breaking down glucose to ethanol and carbon dioxide, is fundamental in the production of bread, beer, and wine. Many different strains of yeast have been domesticated and selected for these processes. Wild yeasts—those that occur naturally in the areas where wine is made—were historically important in wine-making, but domesticated yeasts are normally used now. The most important yeast in all these processes is *Saccharomyces cerevisiae,* a yeast that has been used by humans throughout recorded history. Yeasts are now used on a large scale to produce protein for the enrichment of animal food. Some shamanic cultures have used hallucinogenic compounds produced by fungi. Many antibiotics, including the first used on a wide scale (penicillin from the asexual "mold" *Penicillium*) are derived from fungi. Fungi are also a source of steroids for medicinal use.

If water is present, some fungi can break down almost any carbon-containing compound—even jet fuel! For this reason, there is much interest in using fungi for **bioremediation,** cleaning up soils or waters that are environmentally contaminated, using organisms able to degrade the toxin. Some fungi can even remove inorganic substances from soil. For example, high levels of selenium in the soil are accumulated in local plants and can be toxic to grazing animals. At least three species of fungi have been isolated that combine selenium, accumulated at the San Luis National Wildlife Refuge in California's San Joaquin Valley, with harmless volatile chemicals—thus removing excess selenium from the soil. Bioremediation solutions will be more complex than just adding specific fungi to a waste side. For example, soils may need to have additional nutrients added to sustain restoration efforts.

> Fungal metabolic pathways are used to produce food and pharmaceuticals. The fungi's amazing ability to break down almost any carbon-containing product makes members of this kingdom excellent candidates for bioremediation.

Ecology of Fungi

Fungi, together with bacteria, are the principal decomposers in the biosphere. They break down organic materials and return the substances locked in those molecules to circulation in the ecosystem. Fungi are virtually the only organisms capable of breaking down lignin, one of the major constituents of wood. By breaking down such substances, fungi release critical building blocks, such as carbon, nitrogen, and phosphorus, from the bodies of living or dead organisms and make them available to other organisms.

The interactions, or **symbioses,** between fungi and living organisms fall into a broad range of categories. In some cases, the symbiosis is obligate (essential for survival), and in other cases it is facultative (the fungus can survive without the host). Within a group of closely related fungi, several different types of symbiosis can be found. *Pathogens* and *parasites* gain resources from their host, but have a negative effect on the host that can even lead to death. The difference between pathogens and parasites is that pathogens cause disease while parasites do not. *Commensal* relationships benefit one partner and do not harm the other. Fungi that are in a *mutualistic* relationship benefit both themselves and their hosts. The varied forms of fungal symbiosis are discussed in section 30.3. The range of these interactions, along with decomposition of nonliving organic materials, contributes to the diversity and importance of fungi in ecosystems.

Fungal species cause disease in plants (figure 30.7), and they are responsible for billions of dollars in agricultural losses every year. Not only are fungi among the most harmful pests of living plants, but they also spoil food products that have been harvested and stored. In addition, fungi often secrete substances into the foods they are attacking that make these foods unpalatable, carcinogenic, or poisonous.

Human and animal diseases can also be fungal in origin. Ringworm and athlete's foot can be treated with topical antifungal ointments. *Pneumocystis jiroveci* (formerly *P. carinii*) invades the lungs, disrupting breathing, and can spread to other organs. In immune-suppressed AIDS patients, this can lead to death. Another fungus is threatening amphibian populations globally (see section 30.3).

Many mutualistic associations between fungi and autotrophic organisms are ecologically important. **Lichens** are mutualistic symbiotic associations between fungi and either green algae or cyanobacteria. They are prominent nearly everywhere in the world, especially in unusually harsh habitats such as bare rock. **Mycorrhizae,** specialized mutualistic symbiotic associations between the roots of plants and fungi, are characteristic of about 90% of all plants. In each of them, the photosynthetic organisms fix atmospheric carbon dioxide and thus make organic material available to the fungi. The metabolic activities of the fungi, in turn, enhance the overall ability of the symbiotic association to exist in a particular habitat. In the case of mycorrhizae, the fungal partner expedites the plant's absorption of essential nutrients such as phosphorus. These associations will be discussed further in section 30.3.

Fungi are key decomposers and symbionts within almost all terrestrial ecosystems and play many other important ecological roles.

FIGURE 30.7
World's largest organism? *Armillaria*, a pathogenic fungus shown here afflicting three discrete regions of coniferous forest in Montana, grows out from a central focus as a single, circular clone. The large patch at the bottom of the picture is almost 8 hectares. The largest clone measured so far has been 15 hectares—pretty impressive for a single individual!

30.2 There are four major groups of fungi.

Phylogenetic Relationships

Our understanding of fungal phylogeny is going through rapid and exciting changes, aided by increasing molecular sequence data (figure 30.8 and table 30.1). We now believe that fungi are more closely related to animals than to plants. The relationships among fungi are less clear. To illustrate the diversity within this kingdom, we will look at four groups of fungi—Chytridiomycota, Zygomycota, Basidiomycota, and Ascomycota. The Ascomycota and Basidiomycota are monophyletic, but the other two groups are not. Two other monophyletic groups are tentative in terms of phylogenetic placement. The Glomales evolved as a clade from one of the groups within the Zygomycota, and a provisional phylum, Glomeromycota, has been proposed to accommodate these organisms. Glomales are ecologically the most important fungi within the Zygomycota because they are the fungal partners in arbuscular mycorrhizal symbioses. The Microsporidia, also monophyletic, are obligate animal parasites that may belong to the kingdom Fungi. Several other groups that historically were considered fungi, such as the slime molds and water molds (see chapter 29), now are classed as protists, not fungi. As additional molecular evidence is acquired, analyzed, and integrated with other traits, a new fungal phylogeny is likely to appear.

Fungal phylogenies are changing rapidly. These close relatives of animals are provisionally placed in four groups—Chytridiomycota, Zygomycota, Basidiomycota, and Ascomycota.

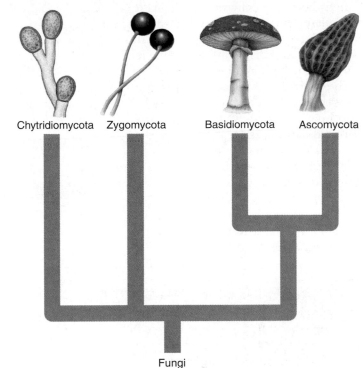

FIGURE 30.8
The four major groups of fungi. Chytridiomycota and Zygomycota are not monophyletic, but Ascomycota and Basidiomycota are. Two additional monophyletic branches have been proposed, but are still somewhat tentative and not shown here.

Table 30.1 Fungi			
Group	**Typical Examples**	**Key Characteristics**	**Approximate Number of Living Species**
Chytridiomycota	*Allomyces*	Aquatic, flagellated fungi that produce haploid gametes in sexual reproduction or diploid zoospores in asexual reproduction	1,000
Zygomycota	*Rhizopus, Pilobolus*	Multinucleate hyphae lack septa, except for reproductive structures; fusion of hyphae leads directly to formation of a zygote in zygosporangium, in which meiosis occurs just before it germinates; asexual reproduction is most common	1,050
Basidiomycota	Mushrooms, toadstools, rusts	In sexual reproduction, basidiospores are borne on club-shaped structures called basidia; asexual reproduction occurs occasionally	22,000
Ascomycota	Truffles, morels	In sexual reproduction, ascospores are formed inside a sac called an ascus; asexual reproduction is also common	45,000

Chytridiomycota

Chytridiomycota (chytrids) are aquatic, flagellated fungi that are most closely related to ancestral fungi. Motile zoospores are a distinguishing character of this fungal group (figure 30.9). Since only chytrids have flagella, this trait must have been lost among ancestors of modern groups. Also, since most chytrids are aquatic, it is likely that fungi first appeared in the water, as did plants and animals. The presence of chitin in the cell walls of chytrids is a unifying feature of all fungi.

Fossils and molecular data indicate that animals and fungi last shared a common ancestor at least 460 million

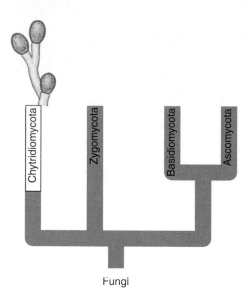

years ago (MYA), but inconsistencies remain. The oldest fossil resembles extant members of the genus *Glomus* that arose within one of the Zygomycota lineages. One DNA analysis placed the animal/fungal divergence at 1500 million years ago (MYA). This latter estimate is not generally accepted, but many researchers believe that the last common ancestor existed around close to 670 million years ago (MYA) based on an analysis of multiple genes.

Regardless of the time of origin, chytrids play a major role in ecosystems. Details of their symbiotic relationships are considered in section 30.3.

Chytrids are the closest living relatives of the first fungal ancestors.

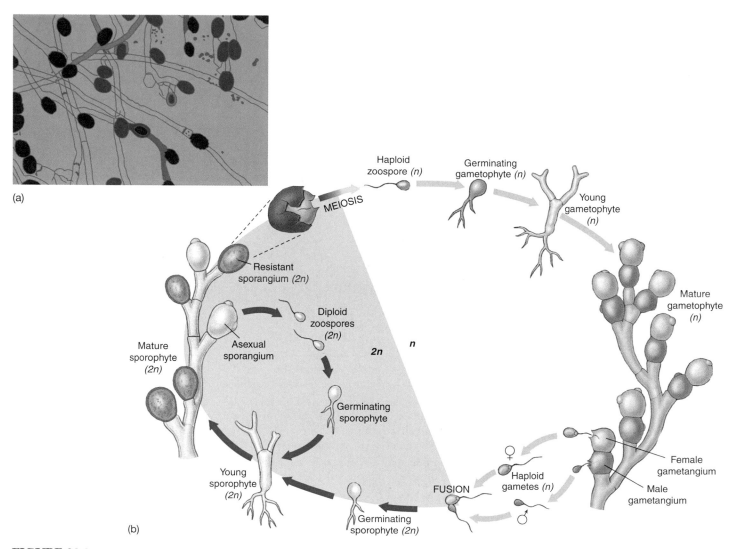

FIGURE 30.9

***Allomyces*, a chytrid that grows in the soil.** (*a*) The spherical sporangia can produce either diploid zoospores via mitosis or haploid zoospores via meiosis. (*b*) Life cycle of *Allomyces*, which has both haploid and diploid multicellular stages (alternation of generations).

Zygomycota

Zygomycetes include only about 1050 named species, but they are incredibly diverse. Among them are some of the more common bread molds (figure 30.10), as well as a variety of others found on decaying organic material, including strawberries and other fruits. A few human pathogens are in this group. So too are the Glomales, which perhaps made the evolution of terrestrial plants possible through arbuscular mycorrhizal symbiosis with roots, enhancing uptake of minerals and water from the soil. Today, they may help maintain plant biodiversity.

The zygomycetes lack septa in their hyphae except when they form sporangia or gametangia. The group is named after a characteristic feature of the sexual phase of the life cycle, a structure called a **zygosporangium** (figure 30.10*b*). Sexual reproduction occurs by the fusion of gametangia, which contain numerous nuclei. The gametangia are cut off from the hyphae by complete septa. These gametangia may be formed on hyphae of different mating types or on a single hypha. Once the haploid nuclei have fused, forming diploid zygote nuclei, the area where the fusion has taken place develops into a zygosporangium. The zygospo-

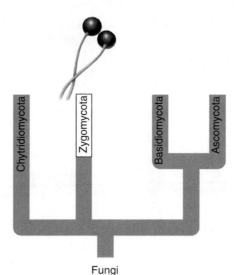

rangium, which may contain one or more diploid nuclei, acquires a thick coat that helps the fungus survive conditions not favorable for growth. Meiosis, followed by mitosis, occurs during the germination of the zygosporangium, which releases haploid spores. Haploid hyphae grow when these haploid spores germinate. Except for the zygote nuclei, all nuclei of the zygomycetes are haploid.

Asexual reproduction occurs much more frequently than sexual reproduction in the zygomycetes. During asexual reproduction, hyphae on the surface of the fungus' food produce clumps of erect stalks, called **sporangiophores.** The tips of the sporangiophores form **sporangia,** which are separated by septa. Thin-walled haploid spores are produced within the sporangia. These spores are thus shed above the substrate, in a position where they may be picked up by the wind and dispersed to a new food source.

> **Many zygomycetes form characteristic resting structures, called zygosporangia, which contain one or more diploid nuclei. The hyphae of zygomycetes are multinucleate, with septa only where gametangia or sporangia are separated.**

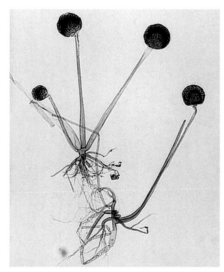

(a)

FIGURE 30.10
***Rhizopus*, a zygomycete that grows on simple sugars.** This fungus is often found on moist bread or fruit. (*a*) The dark, spherical, spore-producing sporangia are on hyphae about 1 centimeter tall. The rootlike hyphae anchor the sporangia. (*b*) Life cycle of *Rhizopus*. The Zygomycota group is named for the zygosporangia characteristic of *Rhizopus*.

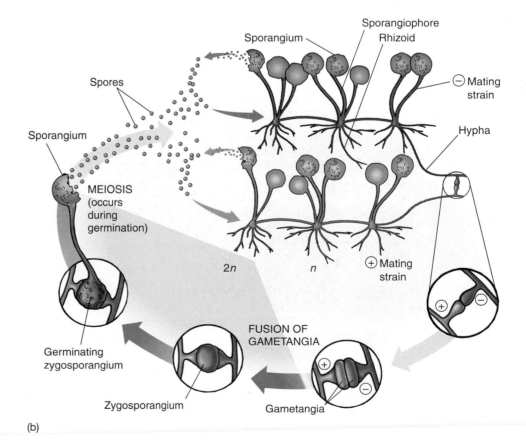

(b)

Basidiomycota

The basidiomycetes include some of the most familiar fungi. Among the basidiomycetes are not only the mushrooms, toadstools, puffballs, jelly fungi, and shelf fungi, but also many important plant pathogens, including rusts and smuts (figure 30.11). Rust infections resemble rusting metals, while smut infections appear black and powdery due to the spores. Many mushrooms are used as food, but others are hallucinogenic or deadly poisonous.

Basidiomycetes are named for their characteristic sexual reproductive structure, the club-shaped **basidium** (plural, *basidia*). Karyogamy occurs within the basidium, giving rise to the only diploid cell of the life cycle (figure 30.11*b*). Meiosis occurs immediately after karyogamy. In the basidiomycetes, the four haploid products of meiosis are incorporated into **basidiospores.** In most members of this phylum, the basidiospores are borne at the end of the basidia on slender projections. Thus the structure of a basidium differs from that of an ascus, although functionally the two are identical. Recall that the ascospores of the ascomycetes are borne internally in asci instead of externally as in basidiospores.

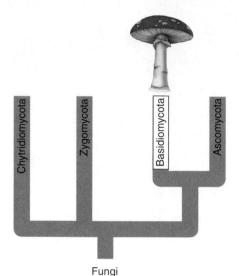

Chytridiomycota

Zygomycota

Basidiomycota

Ascomycota

Fungi

The life cycle of a basidiomycete continues with the production of monokaryotic hyphae after spore germination. These hyphae lack septa early in development. Eventually, septa form between the nuclei of the monokaryotic hyphae. A basidiomycete mycelium made up of monokaryotic hyphae is called a *primary mycelium*. Different mating types of monokaryotic hyphae may fuse, forming a dikaryotic mycelium, or *secondary mycelium*. Such a mycelium is heterokaryotic, with two nuclei representing the two different mating types, between each pair of septa. The maintenance of two genomes in the heterokaryon allows for more genetic plasticity than in a diploid cell with one nucleus. One genome may compensate for mutations in the other. The **basidiocarps,** or mushrooms, are formed entirely of secondary (dikaryotic) mycelium. Gills on the undersurface of the cap of a mushroom form vast numbers of minute spores. It has been estimated that a mushroom with a cap measuring 7.5 centimeters in diameter produces as many as 40 million spores per hour!

Most basidiomycete hyphae are dikaryotic. After karyogamy, meiosis occurs within basidia.

(a)

FIGURE 30.11
Basidiomycetes. (*a*) The death cap mushroom, *Amanita phalloides.* When eaten, these mushrooms are usually fatal. (*b*) Life cycle of a basidiomycete. The basidium is the reproductive structure.

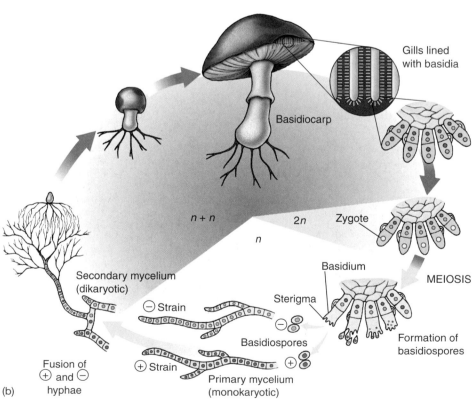

(b)

Ascomycota

The ascomycetes (phylum Ascomycota) contain about 75% of the known fungi. Among the ascomycetes are such familiar and economically important fungi as bread yeasts, common molds, morels (figure 30.12*a*), and truffles. Also included in this phylum are many serious plant pathogens, including the chestnut blight, *Cryphonectria parasitica*, and Dutch elm disease, *Ophiostoma ulmi*. Penicillin-producing ascomycetes are in the genus *Penicillium*.

The ascomycetes are named for their characteristic reproductive structure, the microscopic, saclike **ascus** (plural, *asci*). Karyogamy, resulting in the only diploid nucleus of the ascomycete life cycle (figure 30.12*c*), occurs within the ascus. Asci are differentiated within a structure made up of densely interwoven hyphae, corresponding to the visible portions of a morel or cup fungus, called the **ascocarp**. Meiosis immediately follows karyogamy, forming four haploid daughter nuclei. These usually divide again by mitosis,

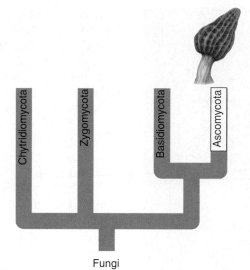

producing eight haploid nuclei that become walled **ascospores**. In many ascomycetes, the ascus becomes highly turgid at maturity and ultimately bursts, often at a preformed area. When this occurs, the ascospores may be thrown as far as 30 centimeters, an amazing distance considering that most ascospores are only about 10 micrometers long. This would be equivalent to throwing a baseball (diameter 7.5 centimeters) 1.25 kilometers—about 10 times the length of a home run!

Asexual reproduction is very common in the ascomycetes. It takes place by means of **conidia** (singular, *conidium*), spores cut off by septa at the ends of modified hyphae called **conidiophores**. Conidia allow for the rapid colonization of a new food source. Many conidia are multinucleate. The hyphae of ascomycetes are divided by septa, but the septa are perforated, and the cytoplasm flows along the length of each hypha. The septa that cut off the asci and conidia are initially perforated, but later become blocked.

(a)

(b)

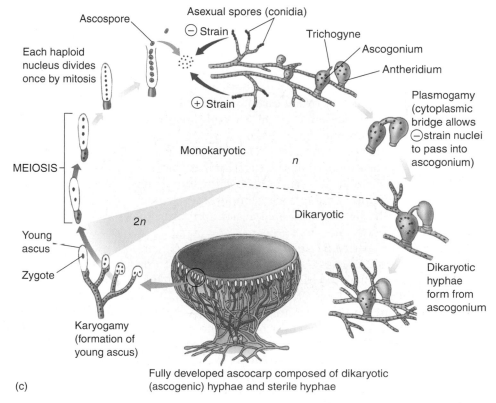

(c)

FIGURE 30.12
Ascomycetes. (*a*) This morel, *Morchella esculenta*, is a delicious edible ascomycete that appears in early spring. (*b*) A cup fungus. (*c*) Life cycle of an ascomycete. Haploid ascospores form within the ascus.

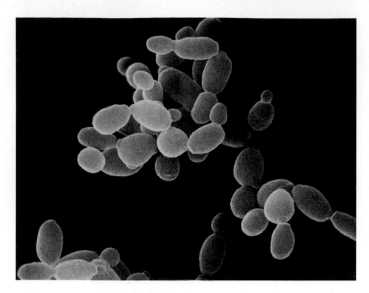

FIGURE 30.13
Budding in *Saccharomyces*.   As shown in this scanning electron micrograph (19,000×), the cells tend to hang together in chains, a feature that calls to mind the derivation of single-celled yeasts from multicellular ancestors.

Yeast morphology is found among some ascomycetes. Most of their reproduction is asexual and takes place by cell fission or budding, when a smaller cell forms from a larger one (figure 30.13). Sometimes two yeast cells fuse, forming one cell containing two nuclei. This cell may then function as an ascus, with karyogamy followed immediately by meiosis. The resulting ascospores function directly as new yeast cells.

The ability of yeasts to ferment carbohydrates, breaking down glucose to produce ethanol and carbon dioxide, is fundamental in the production of bread, beer, and wine. Many different strains of yeast have been domesticated and selected for these processes using the sugars in rice, barley, wheat, and corn. Wild yeasts—ones that occur naturally in the areas where wine is made—were important in wine making historically, but domesticated yeasts are normally used now.

Wild yeast, often *Candida milleri*, is still important in making sourdough bread. Unlike most breads that are made with pure cultures of yeast, sourdough uses an active culture of wild yeast and bacteria that generate acid. This culture is maintained and small amounts—starter culture—are used for each batch of bread. The combination of yeast and acid-producing bacteria is needed for fermentation and gives sourdough bread its unique flavor.

Whether bread is made with wild or cultured yeast, fermentation occurs as the dough rises, releasing carbon dioxide and ethanol. The carbon dioxide is captured in the dough and causes it to rise.

The most important yeast in baking, brewing, and wine making is *Saccharomyces cerevisiae*. This yeast has been used by humans throughout recorded history. Yeast is used as a nutritional supplement because it contains high levels of B vitamins and about fifty percent of yeast is protein. Other yeasts are important pathogens and cause disease.

Over the past few decades, yeasts have become increasingly important in genetic research. They were the first eukaryotes to be manipulated extensively by the techniques of genetic engineering, and they still play the leading role as models for research in eukaryotic cells. In 1996, the genome sequence of *S. cerevisiae*, the first eukaryote to be sequenced entirely, was completed. It contains 6000 genes. With their rapid generation time and a rapidly increasing pool of genetic and biochemical information, the yeasts in general and *S. cerevisiae* in particular are becoming the eukaryotic cells of choice for many types of experiments in molecular and cellular biology. *S. cerevisiae* has become, in this respect, comparable to *Escherichia coli* among the bacteria, and is continuing to provide significant insights into the functioning of eukaryotic systems.

The fungal genome initiative is now under way to provide sequence information on other fungi. Initially fifteen fungi were selected for whole genome sequencing. These fungi were selected based on impact on human health, including plant pathogens that threaten our food supply. The ascomycete *Coccidiodes posadasii* was included because it is endemic in soil in the southwestern portion of the United States, can cause a fatal infection, and has been considered a possible bioterrorism threat. In the U.S. the annual infection rate is about 100,000 individuals, although only a small percentage of infected individuals die. A second important criterion in selecting which fungi to sequence was the potential to provide information on fungal evolution. Key evolutionary branch points were selected and additional genomes are now being considered. Extensive comparisons with the yeast genome are already yielding intriguing information. This new information will complement and expand our understanding of the diverse fungus kingdom.

Ascomycetes undergo karyogamy within a characteristic saclike structure, the ascus. Meiosis follows, resulting in the production of ascospores. Yeasts within this group generally reproduce asexually by budding.

30.3 Fungi participate in many symbioses.

Lichens

Lichens (figure 30.14) are symbiotic associations between a fungus and a photosynthetic partner. The term symbiosis was coined to describe this relationship. While many lichens are excellent examples of mutualism, the kind of symbiotic association that benefits both partners, some fungi are parasitic on their photosynthetic host. Ascomycetes are the fungal partners in all but about 20 of the approximately 15,000 species of lichens estimated to exist. Most of the visible body of a lichen consists of its fungus, but between the filaments of that fungus are cyanobacteria, green algae, or sometimes both (figure 30.15). Specialized fungal hyphae penetrate or envelop the photosynthetic cell walls within them and transfer nutrients directly to the fungal partner. Note that while fungi penetrate the cell wall, they do not penetrate the plasma membrane. Biochemical signals sent out by the fungus apparently direct its cyanobacterial or green algal component to produce metabolic substances that it does not produce when growing independently of the fungus. The fungi in lichens are unable to grow normally without their photosynthetic partners, and the fungi protect their partners from strong light and desiccation. When fungal components of lichens have been experimentally isolated from their photosynthetic partner, they survive, but grow very slowly.

The durable construction of the fungus combined with the photosynthetic properties of its partner have enabled lichens to invade the harshest habitats—the tops of mountains, the farthest northern and southern latitudes, and dry, bare rock faces in the desert. In harsh, exposed areas, lichens are often the first colonists, breaking down the rocks and setting the stage for the invasion of other organisms.

Lichens are often strikingly colored because of the presence of pigments that probably play a role in protecting the photosynthetic partner from the destructive action of the sun's rays. These same pigments may be extracted from the lichens and used as natural dyes. The traditional method of manufacturing Scotland's famous Harris tweed used fungal dyes.

Lichens vary in sensitivity to pollutants in the atmosphere, and some species are used as bioindicators of air quality. Their sensitivity results from their ability to absorb substances dissolved in rain and dew. Lichens are generally absent in and around cities because of automo-

(a)

(b)

FIGURE 30.14
Lichens are found in a variety of habitats. (*a*) A fruticose lichen, *Cladina evansii*, growing on the ground in Florida. (*b*) A foliose ("leafy") lichen, *Parmotrema gardneri*, growing on the bark of a tree in a mountain forest in Panama.

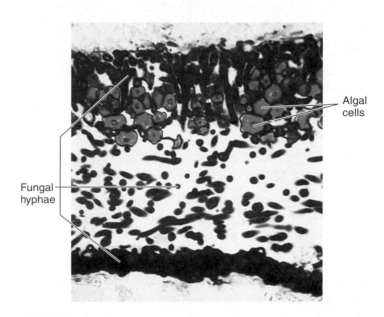
Algal cells

Fungal hyphae

FIGURE 30.15
Stained section of a lichen. This section (250×) shows fungal hyphae (*purple*) more densely packed into a protective layer on the top and, especially, the bottom layer of the lichen. The blue cells near the upper surface of the lichen are those of a green alga. These cells supply carbohydrate to the fungus.

bile traffic and industrial activity, but some are adapted to these conditions. As pollution decreases, lichen populations tend to increase.

Lichens are symbiotic associations between a fungus—an ascomycete in all but a very few instances—and a photosynthetic partner, which may be a green alga or a cyanobacterium or both.

Mycorrhizae

The roots of about 90% of all known species of vascular plants normally are involved in mutualistic symbiotic relationships with certain kinds of fungi. It has been estimated that these fungi probably amount to 15% of the total weight of the world's plant roots. Associations of this kind are termed **mycorrhizae** (Greek *mykes*, "fungus," + *rhiza*, "roots"). The fungi in mycorrhizal associations function as extensions of the root system. The fungal hyphae dramatically increase the amount of soil contact and total surface area for absorption. When mycorrhizae are present, they aid in the direct transfer of phosphorus, zinc, copper, and other nutrients from the soil into the roots. The plant, on the other hand, supplies organic carbon to the fungus, so the system is an example of mutualism.

There are two principal types of mycorrhizae (figure 30.16). In **arbuscular mycorrhizae,** the fungal hyphae penetrate the outer cells of the plant root, forming coils, swellings, and minute branches, and also extend out into the surrounding soil. In **ectomycorrhizae,** the hyphae surround but do not penetrate the cell walls of the roots. In both kinds of mycorrhizae, the mycelium extends far out into the soil. A single root may associate with many fungal species, dividing the root at a millimeter-by-millimeter level.

Arbuscular Mycorrhizae

Arbuscular mycorrhizae are by far the more common of the two types, involving roughly 70% of all plant species. The fungal component in them are Glomales, a monophyletic group that arose within one of the zygomycete lineages. The Glomales are associated with more than 200,000 species of plants. Unlike mushrooms, none of the Glomales produce aboveground fruiting structures, and as a result, it is difficult to arrive at an accurate count of the number of extant species. Arbuscular mycorrhizal fungi are being studied intensively because they are potentially capable of increasing crop yields with lower phosphate and energy inputs.

The earliest fossil plants often show arbuscular mycorrhizal roots. Such associations may have played an important role in allowing plants to colonize land. The soils available at such times would have been sterile and completely lacking in organic matter. Plants that form mycorrhizal associations are particularly successful in infertile soils; considering the fossil evidence, the suggestion that mycorrhizal associations found in the earliest plants helped them succeed on such soils seems reasonable. In addition, the closest living relatives of early vascular plants surviving today continue to depend strongly on mycorrhizae.

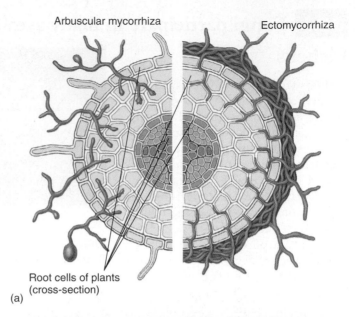

Arbuscular mycorrhiza   Ectomycorrhiza

Root cells of plants (cross-section)

(a)

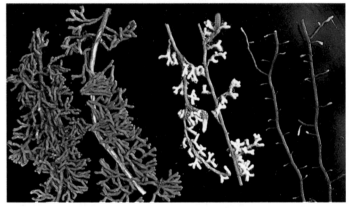

(b)

FIGURE 30.16
Arbuscular mycorrhizae and ectomycorrhizae. (*a*) In arbuscular mycorrhizae, fungal hyphae penetrate the root cells of plants. In ectomycorrhizae, fungal hyphae do not penetrate root cells, but grow around and extend between the cells.
(*b*) Ectomycorrhizae on the roots of pines. From left to right are yellow-brown mycorrhizae formed by *Pisolithus*, white mycorrhizae formed by *Rhizopogon*, and pine roots not associated with a fungus.

Some nonphotosynthetic plants cheat their fungal partner. The symbiosis is one-way because the plant has no photosynthetic resources to offer. Instead of a two-partner symbiosis, a tripartite symbiosis is established. The fungal mycelium extends between a photosynthetic plant and a nonphotosynthetic, parasitic plant. This third member of the symbiosis is called an *epiparasite*. Not only does it obtain phosphate from the fungus, but it uses the fungus to channel carbohydrates from the photosynthetic plant to itself. Epiparasitism also occurs in ectomycorrhizal symbiosis. Symbiotic relationships shift from mutualism to parasitism or from parasitism to mutualism over evolutionary time among even closely related fungi.

Ectomycorrhizae

Ectomycorrhizae (see figure 30.16*b*) involve far fewer kinds of plants than do arbuscular mycorrhizae—perhaps a few thousand. Most ectomycorrhizal hosts are forest trees, such as pines, oaks, birches, willows, eucalyptus, dipterocarps, and caesalpinioid legumes. The fungal components in most ectomycorrhizae are basidiomycetes, but some are ascomycetes. Most ectomycorrhizal fungi are not restricted to a single species of plant, and most ectomycorrhizal plants form associations with many ectomycorrhizal fungi. Different combinations have different effects on the physiological characteristics of the plant and its ability to survive under different environmental conditions. At least 5000 species of fungi are involved in ectomycorrhizal relationships.

Orchids are absolutely dependent on symbiosis with ectomycorrhizal fungi, especially in the early stages of development. In this relationship, the orchid has become parasitic upon the fungus.

Mycorrhizae are symbiotic associations between plants and fungi.

Endophytes

Endophytic fungi live inside plants, actually in the intercellular spaces. Found throughout the plant kingdom, many of these relationships may be examples of parasitism or commensalism. However, there is growing evidence that some of these fungi protect their hosts from herbivores by producing chemical toxins or deterrents. Most often, the fungus synthesizes alkaloids that protect the plant. As you will see in chapter 39, plants also synthesize a wide range of alkaloids, many of which serve to defend the plant.

One way to assess whether or not an endophyte is enhancing the health of its host plant is to grow plots of plants with and without the same endophyte. An experiment with Italian ryegrass, *Lolium multiflorum*, demonstrated that it is more resistant to aphid parasitism when an endophytic fungus, *Neotyphodium*, is present.

Endophytes can protect their hosts from parasites by producing chemical toxins.

Mutualistic Animal Symbioses

A range of mutualistic fungal–animal symbioses has been identified. For example, ruminant animals host fungi in their gut. Grassy diets have a high content of cellulose and lignin that cannot be digested by the grazing animal. Fungal enzymes, however, release nutrients that would otherwise be unavailable to the animal. The fungus gains a nutrient-rich environment in exchange.

To further explore fungal–animal mutualisms, we will focus on a tripartite symbiosis involving ants, plants, and fungi. Leaf-cutter ants are the dominant herbivore in the New World tropics. These ants, members of the tribe Attini, have an obligate symbiosis with specific fungi that they have domesticated and maintain in an underground garden. The ants provide fungi with leaves to eat and protection from pathogens and other predators (figure 30.17). The fungi are the ants' food source. Depending on the species of ant, the ant nest can be as small as a golf ball or as large as 50 centimeters in diameter and many feet deep. Some nests are inhabited by millions of leaf-cutter ants that maintain fungal gardens. These social insects have a caste system, and different ants have specific roles. Traveling on trails as long as 200 meters, leaf-cutter ants search for foliage for their fungi. A colony of ants can defoliate an entire tree in a day. This ant farmer–fungi symbiosis evolved multiple times and may have occurred as early as 50 million years ago.

Fungi have coevolved with animals in mutualistic relationships.

FIGURE 30.17
Ant–fungal symbiosis. Ants farming their fungal garden.

Fungal Parasites and Pathogens

Animal Parasites and Pathogens

Fungi can create devastating human diseases that are often difficult to treat because of their close phylogenetic relationship with animals. Yeast ascomycetes are important pathogens that cause diseases such as thrush; *Candida*, for example, causes common oral or vaginal infections. Mold allergies are common, and mold-infested "sick" buildings pose concerns for inhabitants. Individuals with suppressed immune systems and people undergoing steroid treatments for inflammation disorders are particularly at risk for fungal disease.

An example of a parasitic fungal–animal symbiosis is chytridiomycosis, first identified in 1998 as an emerging infectious disease of amphibians. Amphibian populations have been declining worldwide for over three decades. The decline correlates with the presence of the chytrid *Batrachochytrium dendrobatidis* (see figure 30.2*a*), identified after extensive studies of frog carcasses. Sick and dead frogs were more likely than healthy frogs to have flasklike structures encased in their skin (figure 30.18). The connection with *B. dendrobatidis* has been supported by DNA sequence data, by isolating and culturing the chytrid, and by infecting healthy frogs with the organism and replicating disease symptoms. Bathing frogs in antifungal drugs can halt the disease or even eliminate the chytrids. How the disease emerged simultaneously on different continents is a yet unsolved mystery. Both environmental change and carriers are being considered.

Plant Parasites and Pathogens

Pathogenic fungal–plant symbioses are numerous. Crop losses are extensive throughout the world. Fungal pathogens of plants can also harm the animals that consume the plants. *Fusarium* species growing on spoiled food produce highly toxic substances, including vomitoxin, which has been implicated in brain damage in the southwestern United States.

Aflatoxins, which are among the most carcinogenic compounds known, are produced by some *Aspergillus flavus* strains growing on corn, peanuts, and cotton seed (figure 30.19).

FIGURE 30.18
Frog killed by chytridiomycosis. Lesions formed by the chytrid can be seen on the abdomen of this frog.

Aflatoxins can also damage the kidneys and the nervous system. Most developed countries have legal limits on the concentration of aflatoxin permitted in different foods. More recently, aflatoxins have been considered as possible bioterrorism agents. Hot, humid summers are especially conducive to the growth of this fungus. Monoculture has enhanced its spread, while crop rotation with resistant crops can help control the spread of *Aspergillus*.

Fungi can severely harm or kill both plants and animals.

(a)

(b)

4.50 μm

FIGURE 30.19
Aspergillus flavus **infects maize and can produce aflatoxins that are harmful to animals.** (*a*) Maize (corn) infected with the fungus. (*b*) A photomicrograph of *Aspergillus flavus* conidia.

30.1 The fungi share several key characteristics.

Distinctive Fungal Features

- Recent analysis indicates that fungi are more closely related to animals than to plants. (p. 600)
- Fungi are very diverse, but share several characteristics: Fungi are heterotrophs; they have several cell types although they are primarily filamentous; some have a dikaryon stage; they have cell walls that include chitin; and they exhibit nuclear mitosis. (p. 600)

The Body of a Fungus

- Many fungi have slender filaments called hyphae that are continuous, composed of branching tubes, or made up of long chains of cells joined end-to-end and divided by septa. (p. 601)
- A mycelium is a mass of connected hyphae that grows into the material in which the fungus is feeding. (p. 601)
- Cell walls are formed of polysaccharides and chitin. (p. 601)

How Fungi Reproduce

- Fungi exhibit both sexual and asexual reproduction. (p. 602)
- Sexual reproduction occurs when the haploid hyphae of two individuals fuse. (p. 602)
- Most fungi use spores to reproduce. (p. 602)

How Fungi Obtain Nutrients

- All fungi secrete digestive enzymes into their surroundings and absorb the organic molecules produced via external digestion. (p. 603)

Metabolic Pathways

- Humans use the metabolic pathways of fungi for commercial products such as cheese, soy sauce, bread, beer, and wine. (p. 603)
- Fungi are also used in bioremediation to rid water or soils of environmental contamination. (p. 603)

Ecology of Fungi

- Fungi are the principal decomposers in the biosphere and are nearly the only organisms capable of breaking down lignin. (p. 604)
- Symbioses fall into many categories, including obligate and facultative. (p. 604)
- Fungal species are responsible for billions of dollars of agricultural losses annually, and cause many human and animal diseases, including ringworm and athlete's foot. (p. 604)
- Lichens are mutualistic associations between fungi and algae or cyanobacteria, while mycorrhizae are specialized mutualistic associations between plant roots and fungi. (p. 604)

30.2 There are four major groups of fungi.

Phylogenetic Relationships

- Although the phylogeny of fungi changes rapidly, currently they are divided into four groups (Chytridiomycota, Zygomycota, Basidiomycota and Ascomycota. (p. 605)

Chytridiomycota

- The chytrids are aquatic, flagellated fungi that are the closest living relatives to the first fungi. (p. 606)

Zygomycota

- The zygomycetes are a very diverse group that includes many of the common bread molds. (p. 607)
- Zygomycotes are named after a characteristic feature of their sexual phase, the zygosporangium. (p. 607)
- In zygomycetes, asexual reproduction occurs much more frequently than sexual reproduction. (p. 607)

Basidiomycota

- Basidiomycetes include mushrooms, toadstools, puffballs, and many plant pathogens, such as smuts and rusts. (p. 608)
- Basidiomycetes are named for their characteristic sexual reproductive structure, the basidium. (p. 608)

Ascomycota

- The ascomycetes contain about 75% of known fungi, including bread yeasts, common molds, morels and truffles, and many plant pathogens, such as chestnut blight. (p. 609)
- Ascomycetes are named for their characteristic reproductive structure, the ascus. (p. 609)
- In ascomycetes, asexual reproduction is very common and takes place by conidia produced at the ends of conidiophores. (p. 609)

30.3 Fungi participate in many symbioses.

Lichens

- Lichens are symbiotic relationships between a fungus and a photosynthetic partner that colonize even the harshest habitats on the planet. (p. 611)
- Lichens vary in their sensitivity to atmospheric pollutants, and some are used as air quality indicators. (p. 611)

Mycorrhizae

- Mycorrhizae are mutualistic relationships between plant roots and certain fungi. (p. 612)
- The fungi in mycorrhizal associations increase the amount of soil contact and the total absorption area. (p. 612)
- Two principal types of mycorrhizae are arbuscular and ectomycorrhizae. (p. 612)

Endophytes

- Endophytic fungi live in the intercellular spaces inside living plants. (p. 613)

Mutualistic Animal Symbioses

- A range of fungal–animal symbioses has been identified. These include ruminants and the fungi in their gut, and leaf-cutter ants and specific fungal species. (p. 613)

Fungal Parasites and Pathogens

- *Candida* can cause common oral or vaginal infections. (p. 614)
- A chytrid can cause chytridiomycosis in amphibians.
- Crop losses due to parasitic or pathogenic fungi are extensive throughout the world. (p. 614)
- Aflatoxins can cause kidney or nervous system damage. (p. 614)

Self Test

1. How many species of fungi are thought to exist?
   a. 1500
   b. 150,000
   c. 1.5 million
   d. 1.5 billion
2. Which of the following is *not* a characteristic of the fungi?
   a. They are all absorptive heterotrophs.
   b. They have cell walls made of chitin.
   c. Mitosis takes place within the nuclear membrane.
   d. They are all motile.
3. A mycelium is
   a. a specialized reproductive structure of a fungus.
   b. a mass of connected fungal hyphae.
   c. a mutualistic relationship between a fungus and a plant.
   d. a partition between the cells of fungal hyphae.
4. Which of the following statements best describes fungi?
   a. All are eukaryotic, multicellular autotrophs.
   b. All are eukaryotic heterotrophs that feed by absorption.
   c. All are prokaryotic, multicellular autotrophs.
   d. All are eukaryotic heterotrophs that feed by ingestion.
5. A wildlife pathologist is examining some skin tissue from a dead frog. She notes the presence of a fungus. She cultures some of the fungal cells and notices that some of the cells are flagellated. She concludes that the frog has a fungal disease caused by
   a. an ascomycete.
   b. a zygomycete.
   c. a basidiomycete.
   d. a chytrid.
6. You are walking in the woods and see a fungus that is unfamiliar to you. You remove a reproductive structure, and take it home to examine further. When you look at it under the microscope, you find a zygosporangium. Based on this information alone, this fungus is a(n)
   a. zygomycete.
   b. chytrid.
   c. basidiomycete.
   d. ascomycete.
7. A basidium is typically observed in the common
   a. bread mold.
   b. gilled mushroom.
   c. lichen.
   d. chytrid.
8. An ascomycete can be distinguished from other fungi
   a. because ascomycetes are mainly diploid.
   b. because ascomycetes lack a dikaryotic phase.
   c. by the presence of eight sexual spores in an ascus.
   d. by the presence of gills on the mycelium.
9. A lichen can be described as a mutualistic symbiosis between an ascomycete and a(n)
   a. chytrid.
   b. archaebacterium.
   c. green alga.
   d. angiosperm root.

Test Your Visual Understanding

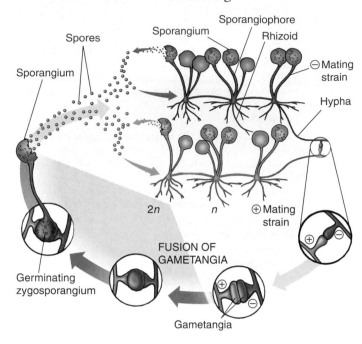

1. This is a diagram of the life cycle of a typical bread mold.
   a. Is this sexual or asexual reproduction?
   b. On this diagram, indicate where in the life cycle meiosis takes place.
   c. Label the reproductive structures of this fungus.
   d. To what group of fungi does this organism belong?

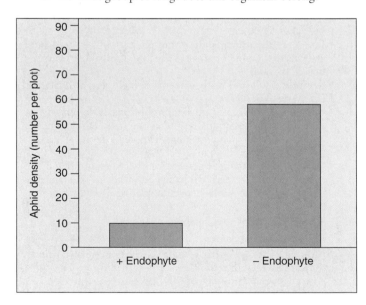

2. Two equal-sized plots of Italian ryegrass were grown, one with and one without endophytes. Each plot was exposed to aphids in the experiment shown above.
   a. What conclusion would you draw from the experimental results?
   b. What advantages and disadvantages does the ryegrass have in an endophytic relationship?

31

Overview of Animal Diversity

Concept Outline

31.1 Animals are multicellular heterotrophs without cell walls.

    Some General Features of Animals. Animals lack cell walls and move more rapidly and in more complex ways than other organisms.

31.2 Animals are a very diverse kingdom.

    The Traditional Classification of Animals. The 35 animal phyla have traditionally been arrayed on the animal family tree according to certain key features of the animal body plan.

31.3 The animal body plan has undergone many changes.

    Five Key Transitions in Body Plan. Over the course of animal evolution, five key transitions in body organization have occurred.

31.4 The way we classify animals is being reevaluated.

    A New Look at the Metazoan Family Tree. Traditional and molecular approaches lead to somewhat different views of the animal family tree.

    "Evo-Devo" and the Roots of the Animal Family Tree. The 35 traditional animal phyla all evolved before the Cambrian period.

FIGURE 31.1
Biologists used to think that sponges were marine plants. Perhaps the earliest animals to evolve, sponges have motile larvae that may travel thousands of miles, but as adults, sponges are sessile and remain attached to one spot on the ocean floor.

We will now explore the great diversity of animals, the result of a long evolutionary history. Animals, constituting millions of species, are among the most abundant living things. Found in every conceivable habitat, they bewilder us with their diversity. More than a million species have been described, and several million more are thought to await discovery. Despite their great diversity, all animals have much in common. For example, locomotion is a distinctive characteristic of animals, although not all animals move about. Early naturalists thought that sponges (figure 31.1) and corals were plants because the adults seemed rooted to one place and did not move about. Biologists have long debated the evolutionary relationships among animal groups. In recent years, molecular tools are providing new insights and leading biologists to reconsider the animal tree of life.

31.1 Animals are multicellular heterotrophs without cell walls.

Some General Features of Animals

Animals are the eaters, or consumers, of the earth. Animals are a very diverse group—no one criterion fits all—but several characteristics are of major importance (table 31.1): (1) Animals are heterotrophs and must ingest plants, algae, or other animals for nourishment. (2) All animals are multicellular, and unlike plants and protists, animal cells lack cell walls. (3) Animals are able to move from place to place.

(4) Animals are very diverse in form and habitat. (5) Most animals reproduce sexually. (6) Animals have a characteristic pattern of embryonic development and possess unique tissues.

Animals are complex multicellular organisms typically characterized by high mobility and heterotrophy. Most animals also possess internal tissues and reproduce sexually.

Table 31.1 General Features of Animals

Heterotrophs. Unlike autotrophic plants and algae, animals cannot construct organic molecules from inorganic chemicals. All animals are heterotrophs—that is, they obtain energy and organic molecules by ingesting other organisms. Some animals (herbivores) consume autotrophs; other animals (carnivores) consume heterotrophs; and still others (detritivores) consume decomposing organisms.

Multicellular. All animals are multicellular, often with complex bodies like that of this brittlestar *(right)*. The unicellular heterotrophic organisms called protozoa, which were at one time regarded as simple animals, are now considered members of the large and diverse kingdom Protista, discussed in chapter 28.

No Cell Walls. Animal cells are distinct among those of multicellular organisms because they lack rigid cell walls and are usually quite flexible, as are these cancer cells. The many cells of animal bodies are held together by extracellular lattices of structural proteins such as collagen. Other proteins form a collection of unique intercellular junctions between animal cells.

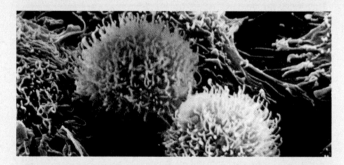

Active Movement. The ability of animals to move more rapidly and in more complex ways than members of other kingdoms is perhaps their most striking characteristic, one that is directly related to the flexibility of their cells and the evolution of nerve and muscle tissues. A remarkable form of movement unique to animals is flying, an ability that is well developed among vertebrates and insects such as this butterfly. The only terrestrial vertebrate group never to have evolved flight is the amphibians.

Table 31.1 General Features of Animals *Continued.*

Diverse in Form. Almost all animals (99%) are **invertebrates,** which, like this millipede, lack a backbone. Of the estimated 10 million living animal species, only 42,500 have a backbone and are referred to as **vertebrates.** Animals are very diverse in form, ranging in size from organisms too small to see with the unaided eye to enormous whales and giant squids.

Diverse in Habitat. The animal kingdom includes about 35 phyla, most of which, like these jellyfish (phylum Cnidaria), occur in the sea. Far fewer phyla occur in fresh water, and fewer still occur on land. Members of three successful marine phyla, Arthropoda (insects), Mollusca (snails), and Chordata (vertebrates), also dominate animal life on land.

Sexual Reproduction. Most animals reproduce sexually, as these tortoises are doing. Animal eggs, which are nonmotile, are much larger than the small, usually flagellated sperm. In animals, cells formed in meiosis function directly as gametes. The haploid cells do not divide by mitosis first, as they do in plants and fungi, but rather fuse directly with each other to form the zygote. Consequently, with a few exceptions, there is no counterpart among animals to the alternation of haploid (gametophyte) and diploid (sporophyte) generations characteristic of plants.

Embryonic Development. Most animals have a similar pattern of embryonic development. The zygote first undergoes a series of mitotic divisions, called *cleavage,* and like this dividing frog egg, becomes a solid ball of cells, the **morula,** and then a hollow ball of cells, the **blastula.** In most animals, the blastula folds inward at one point to form a hollow sac with an opening at one end called the **blastopore.** An embryo at this stage is called a **gastrula.** The subsequent growth and movement of the cells of the gastrula differ widely from one phylum of animals to another.

Unique Tissues. The cells of all animals except sponges are organized into structural and functional units called **tissues,** collections of cells that have joined together and are specialized to perform a specific function. Animals are unique in having two tissues associated with movement: (1) muscle tissue, which powers animal movement, and (2) nervous tissue, which conducts signals among cells. Neuromuscular junctions, where nerves connect with muscle tissue, are shown here.

31.2 Animals are a very diverse kingdom.

The Traditional Classification of Animals

The multicellular animals, or metazoans, are traditionally divided into 35 distinct phyla. The diversity of animals can be clearly seen in table 31.2, which describes key characteristics of 20 of the more widely studied animal phyla.

How can we make sense of this great diversity? Taxonomists, attempting to sort out which phyla are related, have traditionally created phylogenies (family trees) by comparing anatomical features and aspects of embryological development. Little other data was available, and a broad consensus emerged over the last century concerning the main branches of the animal family tree.

The First Branch: Tissues

Taxonomists traditionally divide the kingdom Animalia into two main branches: **Parazoa** ("beside animals")—animals that for the most part lack a definite symmetry and possess neither tissues nor organs, mostly composed of the sponges, phylum Porifera; and **Eumetazoa** ("true animals")—animals that have a definite shape and symmetry and, in most cases, tissues organized into organs and organ systems.

The Second Branch: Symmetry

The eumetazoan branch of the animal family tree itself has two principal branches, differing in the nature of the embryonic layers that form during development and go on to differentiate into the tissues of the adult animal: Eumetazoans of the subgroup **Radiata** (having radial symmetry) have two layers, an outer *ectoderm* and an inner *endoderm*, and thus are called diploblastic. All other eumetazoans, the **Bilateria** (having bilateral symmetry), are triploblastic and produce a third layer, the *mesoderm*, between the ectoderm and endoderm (figure 31.2).

Further Branches

Further branches of the animal family tree were assigned by taxonomists by comparing traits that seemed profoundly important to the evolutionary history of phyla, key features of the body plan shared by all animals belonging to that branch. Thus, the bilateral animals were split into groups with a body cavity and those without; animals with a body cavity were split into those with a coelom (body cavity enclosed by mesoderm) and those without; animals with a

(a) Parazoa: no tissues (b) Eumetazoan: tissues

(c) Radiata: radial symmetry (d) Bilateria: bilateral symmetry

FIGURE 31.2
Broad groupings of the kingdom Animalia. (*a, b*) Animals are divided into Parazoa, animals that lack tissues and a definite symmetry, and Eumetazoa, animals that have symmetry and organized tissues. (*c, d*) Eumetazoans are further divided into Radiata, animals with radial symmetry, and Bilateria, animals with bilateral symmetry.

coelom were split into those whose coelom derived from the digestive tube and those that did not; and so on.

Because of the either-or nature of the categories set up by traditional taxonomists, this approach has produced a family tree with a lot of paired branches (figure 31.3). The arbitrary nature of the divisions has always been obvious to biologists, but until recently most biologists felt this phylogeny faithfully represented the general nature of the evolutionary history of metazoans. Although new molecular techniques have now suggested other ways of classifying animals (see section 31.4), the traditional phylogeny is still employed by most taxonomists and will provide the framework for discussion in this book.

Animals are traditionally classified into some 35 phyla. The evolutionary relationships among the animal phyla have been inferred until recent years by assuming relatedness of phyla that share certain fundamental morphological characters, which are assumed to have arisen only once.

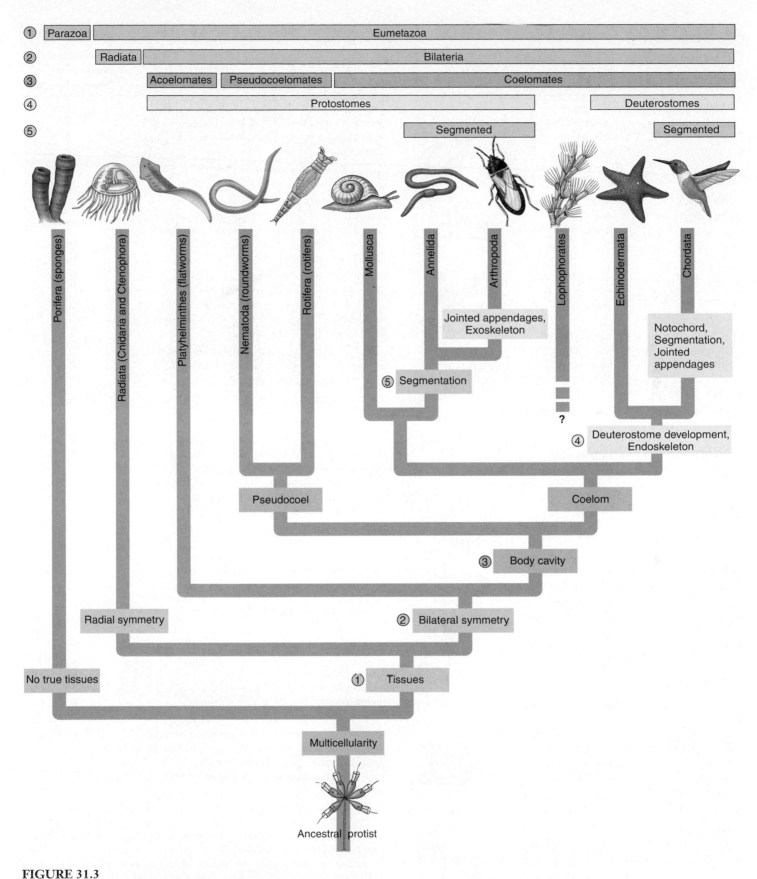

FIGURE 31.3

A possible phylogeny of the major groups of the kingdom Animalia. Transitions in the animal body plan are identified along the branches; five key advances are the evolution of tissues, bilateral symmetry, a body cavity, deuterostome development, and segmentation.

Table 31.2   The Major Animal Phyla

Phylum	Typical Examples		Key Characteristics	Approximate Number of Named Species
Arthropoda (arthropods)	Beetles, other insects, crabs, spiders		Most successful of all animal phyla; chitinous exoskeleton covering segmented bodies with paired, jointed appendages; many insect groups have wings.	1,000,000
Mollusca (mollusks)	Snails, oysters, octopuses, nudibranchs		Soft-bodied coelomates whose bodies are divided into three parts: head-foot, visceral mass, and mantle; many have shells; almost all possess a unique rasping tongue, called a radula; 35,000 species are terrestrial.	110,000
Chordata (chordates)	Mammals, fish, reptiles, birds, amphibians		Segmented coelomates with a notochord; possess a dorsal nerve cord, pharyngeal slits, and a tail at some stage of life; in vertebrates, the notochord is replaced during development by the spinal column; 20,000 species are terrestrial.	56,000
Platyhelminthes (flatworms)	Planaria, tapeworms, liver flukes		Solid, unsegmented, bilaterally symmetrical worms; no body cavity; digestive cavity, if present, has only one opening.	20,000
Nematoda (roundworms)	*Ascaris*, pinworms, hookworms, *Filaria*		Pseudocoelomate, unsegmented, bilaterally symmetrical worms; tubular digestive tract passing from mouth to anus; tiny; without cilia; live in great numbers in soil and aquatic sediments; some are important animal parasites.	20,000
Annelida (segmented worms)	Earthworms, polychaetes, tube worms, leeches		Coelomate, serially segmented, bilaterally symmetrical worms; complete digestive tract; most have bristles called setae on each segment that anchor them during crawling.	12,000
Cnidaria (cnidarians)	Jellyfish, hydra, corals, sea anemones		Soft, gelatinous, radially symmetrical bodies whose digestive cavity has a single opening; possess tentacles armed with stinging cells called cnidocytes that shoot sharp harpoons called nematocysts; almost entirely marine.	10,000
Echinodermata (echinoderms)	Sea stars, sea urchins, sand dollars, sea cucumbers		Deuterostomes with radially symmetrical adult bodies; endoskeleton of calcium plates; five-part body plan and unique water-vascular system with tube feet; able to regenerate lost body parts; marine.	6,000
Porifera (sponges)	Barrel sponges, boring sponges, basket sponges, vase sponges		Asymmetrical bodies without distinct tissues or organs; saclike body consists of two layers breached by many pores; internal cavity lined with food-filtering cells called choanocytes; most are marine (150 species live in fresh water).	5,150

Table 31.2 The Major Animal Phyla *Continued.*

Phylum	Typical Examples		Key Characteristics	Approximate Number of Named Species
Bryozoa (moss animals)	*Bowerbankia, Plumatella,* sea mats, sea moss		Microscopic, aquatic deuterostomes that form branching colonies; possess circular or U-shaped row of ciliated tentacles for feeding called a lophophore that usually protrudes through pores in a hard exoskeleton; also called Ectoprocta because the anus, or proct, is external to the lophophore; marine or freshwater.	4,000
Rotifera (wheel animals)	Rotifers		Small, aquatic pseudocoelomates with a crown of cilia around the mouth resembling a wheel; almost all live in fresh water.	2,000
Five Phyla of Minor Worms	Velvet worms, acorn worms, arrow worms, giant tube worms		**Chaetognatha** (arrow worms): coelomate deuterostomes; bilaterally symmetrical; large eyes (some) and powerful jaws.	980
			Hemichordata (acorn worms): marine worms with dorsal *and* ventral nerve cords.	
			Onychophora (velvet worms): protostomes with a chitinous exoskeleton; evolutionary relics.	
			Pogonophora (tube worms): sessile deep-sea worms with long tentacles; live within chitinous tubes attached to the ocean floor.	
			Nemertea (ribbon worms): acoelomate, bilaterally symmetrical marine worms with long, extendable proboscis.	
Brachiopoda (lamp shells)	*Lingula*		Like bryozoans, possess a lophophore, but within two clamlike shells; more than 30,000 species known as fossils.	300
Ctenophora (sea walnuts)	Comb jellies, sea walnuts		Gelatinous, almost transparent, often bioluminescent marine animals; eight bands of cilia; largest animals that use cilia for locomotion; complete digestive tract with anal pore.	100
Phoronida (phoronids)	*Phoronis*		Lophophorate tube worms; often live in dense populations; unique U-shaped gut, instead of the straight digestive tube of other tube worms.	12
Loricifera (loriciferans)	*Nanaloricus mysticus*		Tiny, bilaterally symmetrical, marine pseudocoelomates that live in spaces between grains of sand; mouthparts include a unique flexible tube. A recent addition to the traditional 35 animal phyla, loriciferans were discovered in 1983.	6

31.3 The animal body plan has undergone many changes.

Five Key Transitions in Body Plan

1. Evolution of Tissues

The simplest animals, the Parazoa, lack both defined tissues and organs. Characterized by the sponges, these animals exist as aggregates of cells with minimal intercellular coordination. All other animals, the Eumetazoa, have distinct tissues with highly specialized cells. The evolution of tissues is the first key transition in the animal body plan.

2. Evolution of Bilateral Symmetry

Sponges also lack any definite symmetry, growing asymmetrically as irregular masses. Virtually all other animals have a definite shape and symmetry that can be defined along an imaginary axis drawn through the animal's body. Animals with symmetry belong to either the Radiata, animals with radial symmetry, or the Bilateria, animals with bilateral symmetry.

Radial Symmetry. Symmetrical bodies first evolved in marine animals belonging to two phyla: Cnidaria (jellyfish, sea anemones, and corals) and Ctenophora (comb jellies). The bodies of members of these two phyla, the Radiata, exhibit **radial symmetry,** a body design in which the parts of the body are arranged around a central axis in such a way that any plane passing through the central axis divides the organism into halves that are approximate mirror images (figure 31.4*a*).

Bilateral Symmetry. The bodies of all other animals, the Bilateria, are marked by a fundamental **bilateral symmetry,** a body design in which the body has a right and a left half that are mirror images of each other (figure 31.4*b*). A bilaterally symmetrical body plan has a top and a bottom, better known respectively as the *dorsal* and *ventral* portions of the body. It also has a front, or *anterior* end, and a back, or *posterior* end. In some higher animals, such as echinoderms (starfish), the adults are radially symmetrical, but even in them the larvae are bilaterally symmetrical.

Bilateral symmetry constitutes the second major evolutionary advance in the animal body plan. This unique form of organization allows parts of the body to evolve in different ways, permitting different organs to be located in different parts of the body. Also, bilaterally symmetrical animals move from place to place more efficiently than radially symmetrical ones, which, in general, lead a sessile or passively floating existence. Due to their increased mobility, bilaterally symmetrical animals are efficient in seeking food, locating mates, and avoiding predators.

The bilaterally symmetrical eumetazoans produce three germ layers: an outer **ectoderm,** an inner **endoderm,** and a

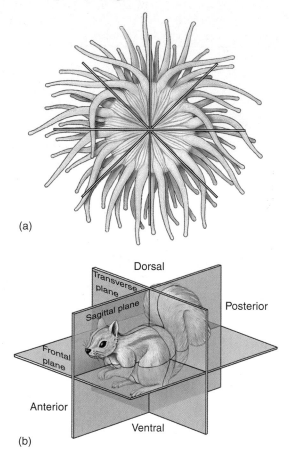

FIGURE 31.4
A comparison of radial and bilateral symmetry. (*a*) Radially symmetrical animals, such as this sea anemone, can be bisected into equal halves in any two-dimensional plane. (*b*) Bilaterally symmetrical animals, such as this squirrel, can only be bisected into equal halves in one plane (the sagittal plane).

third layer, the **mesoderm,** between the ectoderm and endoderm. In general, the outer coverings of the body and the nervous system develop from the ectoderm; the digestive organs and intestines develop from the endoderm; and the skeleton and muscles develop from the mesoderm. The radially symmetrical animals have only two layers, the endoderm and the ectoderm, and sponges lack any germ layers.

Much of the nervous system in bilaterally symmetrical animals is in the form of major longitudinal nerve cords. In a very early evolutionary advance, nerve cells became grouped in the anterior end of the body. These nerve cells probably first functioned mainly to transmit impulses from the anterior sense organs to the rest of the nervous system. This trend ultimately led to the evolution of a definite head and brain area, a process called **cephalization,** as well as to the increasing dominance and specialization of these organs in the more advanced animal phyla.

3. Evolution of a Body Cavity

A third key transition in the animal body plan was the evolution of the body cavity. The evolution of efficient organ systems within the animal body was not possible until a body cavity evolved for supporting organs, distributing materials, and fostering complex developmental interactions.

The presence of a body cavity enables the digestive tract to be larger and longer. This longer passage allows for storage of undigested food, longer exposure to enzymes for more complete digestion, and even storage and final processing of food remnants. With such an arrangement, an animal can eat a great deal when it is safe to do so and then hide during the digestive process, thus limiting the animal's exposure to predators. The tube within the body cavity architecture is also more flexible, thus allowing the animal greater freedom to move.

An internal body cavity also provides space within which the gonads (ovaries and testes) can expand, allowing large numbers of eggs and sperm to accumulate. Such storage capacity makes possible the diverse modifications of breeding strategy that characterize the more advanced phyla of animals. Furthermore, large numbers of gametes can be stored and released when the conditions are as favorable as possible for the survival of the young animals.

Kinds of Body Cavities.   Three basic kinds of body plans evolved in the Bilateria (figure 31.5). **Acoelomates** have no body cavity. **Pseudocoelomates** have a body cavity called the **pseudocoel** located between the mesoderm and endoderm. In animals having the third type of body plan, called **coelomates,** a fluid-filled body cavity develops not between the endoderm and mesoderm, but rather entirely within the mesoderm. Such a body cavity is called a **coelom.** In coelomates, the gut is suspended, along with other organ systems of the animal, within the coelom; the coelom, in turn, is surrounded by a layer of epithelial cells entirely derived from the mesoderm (figure 31.5). A major advantage of the coelomate body plan is that it allows contact between mesoderm and endoderm, so that tissues can interact during development. For example, contact between mesoderm and endoderm permits localized portions of the digestive tract to develop into complex, highly specialized regions such as the stomach. In pseudocoelomates, mesoderm and endoderm are separated by the body cavity, limiting developmental interactions between these tissues, which ultimately limits tissue specialization and development.

The development of the coelom poses a problem—circulation—which is solved in pseudocoelomates by churning the fluid within the body cavity. In coelomates, the gut is again surrounded by tissue that presents a barrier to diffusion, just as it was in the solid bodies of acoelomates. This problem is solved among coelomates by the development of a **circulatory system,** a network of vessels that carry fluids to parts of the body. The circulating fluid, or blood, carries nutrients and oxygen to the tissues and re-

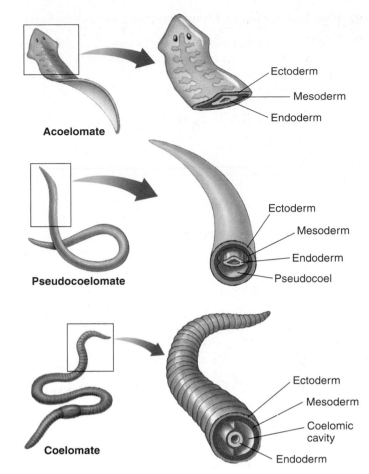

FIGURE 31.5
Three body plans for bilaterally symmetrical animals.
Acoelomates, such as flatworms, have no body cavity between the digestive tract (endoderm) and the outer body layer (ectoderm). Pseudocoelomates have a body cavity, the pseudocoel, between the endoderm and the mesoderm. Coelomates have a body cavity, the coelom, that develops entirely within the mesoderm, and so is lined on both sides by mesoderm tissue.

moves wastes and carbon dioxide. Blood is usually pushed through the circulatory system by contraction of one or more muscular hearts. In an **open circulatory system,** the blood passes from vessels into sinuses, mixes with body fluid, and then reenters the vessels later in another location. In a **closed circulatory system,** the blood is physically separated from other body fluids and can be separately controlled. Also, blood moves through a closed circulatory system faster and more efficiently than it does through an open system.

The evolutionary relationship among coelomates, pseudocoelomates, and acoelomates is not clear. Acoelomates, for example, could have given rise to coelomates, but scientists also cannot rule out the possibility that acoelomates were derived from coelomates. The different phyla of pseudocoelomates form two groups that do not appear to be closely related.

4. The Evolution of Deuterostome Development

Bilaterally symmetrical animals exhibit a pattern of embryonic development that begins with mitotic cell divisions of the egg that lead to the formation of a hollow ball of cells, called the **blastula.** The blastula indents to form a two-layer-thick ball with a **blastopore** opening to the outside and a primitive gut cavity called the **archenteron.** Bilaterians can be divided into two groups based on differences in the basic pattern of development. One group is called the **protostomes** (Greek *protos,* "first," + *stoma,* "mouth") and includes the flatworms, nematodes, mollusks, annelids, and arthropods. Two outwardly dissimilar groups, the echinoderms and the chordates, together with a few other smaller related phyla, comprise the second group, the **deuterostomes** (Greek, *deuteros,* "second," + *stoma,* "mouth"). Protostomes and deuterostomes differ in several aspects of embryo growth.

Cleavage. The progressive division of cells during embryonic growth is called *cleavage.* The cleavage pattern relative to the embryo's polar axis determines how the cells will array. In nearly all protostomes, each new cell buds off at an angle oblique to the polar axis. As a result, a new cell nestles into the space between the older ones in a closely packed array. This pattern is called **spiral cleavage** because a line drawn through a sequence of dividing cells spirals outward from the polar axis (figure 31.6 *top*). In deuterostomes, the cells divide parallel to and at right angles to the polar axis. As a result, the pairs of cells from each division are positioned directly above and below one another; this process gives rise to a loosely packed array of cells. This pattern is called **radial cleavage** because a line drawn through a sequence of dividing cells describes a radius outward from the polar axis (figure 31.6 *bottom*).

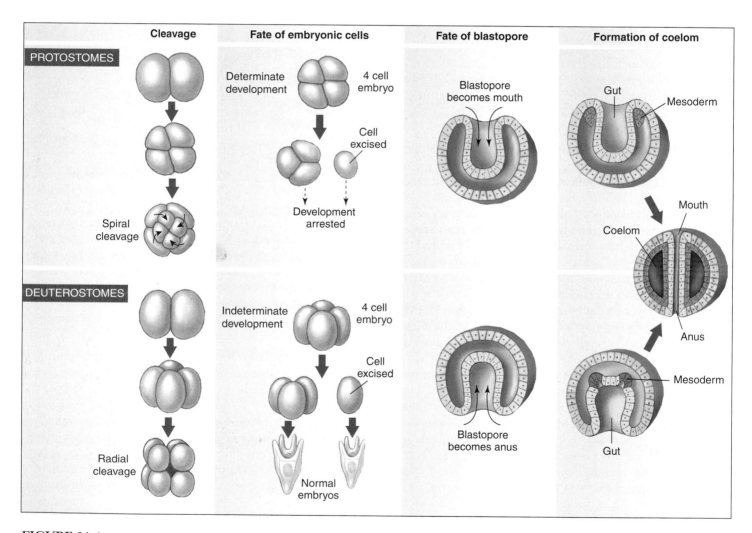

FIGURE 31.6

Embryonic development in protostomes and deuterostomes. In protostomes, embryonic cells cleave in a spiral pattern and exhibit determinate development; the blastopore becomes the animal's mouth, and the coelom originates from a mesodermal split. In deuterostomes, embryonic cells cleave radially and exhibit indeterminate development; the blastopore becomes the animal's anus, and the coelom originates from an evagination, or outpouching, of the archenteron.

Fate of Embryonic Cells. Protostomes exhibit **determinate development.** In this type of development, each embryonic cell has a predetermined fate in terms of what kind of tissue it will form in the adult. Before cleavage begins, the chemicals that act as developmental signals are localized in different parts of the egg. Consequently, the cell divisions that occur after fertilization separate different signals into different daughter cells. This process specifies the fate of even the very earliest embryonic cells. Deuterostomes, on the other hand, display **indeterminate development.** The first few cell divisions of the egg produce identical daughter cells. Any one of these cells, if separated from the others, can develop into a complete organism. This is possible because the chemicals that signal the embryonic cells to develop differently are not localized until later in the animal's development.

Fate of Blastopore. In protostomes, the mouth (stoma) of the animal develops from or near the blastopore. If such an animal has a distinct anus or anal pore, it develops later in another region of the embryo. This same pattern of development, in a general sense, is seen in all noncoelomate animals. In deuterostomes, the blastopore gives rise to the organism's anus, and the mouth develops from a second pore that arises in the blastula later in development.

Formation of the Coelom. In all coelomates, the coelom originates from mesoderm. In protostomes, this occurs simply and directly: The cells simply move away from one another as the coelomic cavity expands within the mesoderm. However, in deuterostomes, whole groups of cells usually move around to form new tissue associations. The coelom is normally produced by an evagination of the archenteron, the central tube within the gastrula, also called the primitive gut. This tube, lined with endoderm, opens to the outside via the blastopore and eventually becomes the gut cavity.

Deuterostomes evolved from protostomes more than 630 million years ago, and the consistency of deuterostome development, along with its distinctiveness from that of the protostomes, suggests that it evolved once, in a common ancestor to all of the phyla that exhibit it.

5. The Evolution of Segmentation

The fifth key transition in the animal body plan involved the subdivision of the body into segments. Just as it is efficient for workers to construct a tunnel from a series of identical prefabricated parts, so segmented animals are "assembled" from a succession of identical segments. During the animal's early development, these segments become most obvious in the mesoderm but later are reflected in the ectoderm and endoderm as well. Two advantages result from early embryonic segmentation:

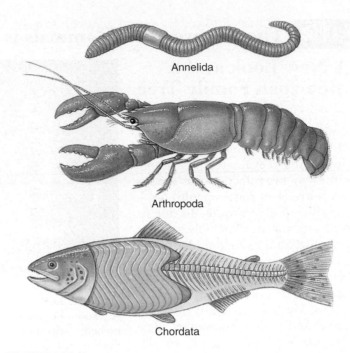

FIGURE 31.7
Segmentation. Annelids, arthropods, and chordates exhibit segmentation.

1. In annelids and other highly segmented animals, each segment may go on to develop a more or less complete set of adult organ systems. Damage to any one segment need not be fatal to the individual, since the other segments duplicate that segment's functions.
2. Locomotion is far more effective when individual segments can move independently because the animal as a whole has more flexibility of movement. Because the separations isolate each segment into an individual skeletal unit, each is able to contract or expand autonomously in response to changes in hydrostatic pressure. Therefore, a long body can move in ways that are often quite complex.

Segmentation, also referred to as *metamerism*, underlies the organization of all advanced animal body plans. In some adult arthropods, the segments are fused, but segmentation is usually apparent in their embryological development. In vertebrates, the backbone and muscular areas are segmented, although segmentation is often disguised in the adult form. True segmentation is found in only three phyla: the annelids, the arthropods, and the chordates (figure 31.7), although this trend is evident in many phyla.

Five key transitions in body design are responsible for most of the differences we see among the major animal phyla: the evolution of (1) tissues, (2) bilateral symmetry, (3) a body cavity, (4) deuterostome development, and (5) segmentation.

31.4 The way we classify animals is being reevaluated.

A New Look at the Metazoan Family Tree

The traditional animal phylogeny, while accepted by a broad consensus of biologists for almost a century, is now being reevaluated. Its simple either-or organization has always presented certain problems—puzzling minor groups do not fit well into the standard scheme. For example, the myzostomids (figure 31.8), an enigmatic and anatomically bizarre group of marine animals, are parasites or symbionts of echinoderms. Myzostomid fossils are found associated with echinoderms since the Ordovician period, so the myzostomid-echinoderm relationship is a very ancient one. Their long history of obligate association has led to the loss or simplification of many myzostomid body elements, leaving them, for example, with no body cavity (they are acoelomates) and only incomplete segmentation.

This character loss has led to considerable disagreement among taxonomists. However, while taxonomists have disagreed about the details, all have generally allied myzostomids in some fashion with the annelids—sometimes within the polychaetes and sometimes as a separate phylum closely allied to the annelids.

Recently, this view has been challenged. New taxonomical comparisons using molecular data have come to very different conclusions. Researchers have examined two components of the protein synthesis machinery—the small ribosomal subunit rRNA gene and an elongation factor gene (the one called 1 alpha). The phylogeny they obtain does not place the myzostomids with the annelids. Indeed, they find that the myzostomids have no close links to the annelids at all. Instead, surprisingly, they are more closely allied with the flatworms!

This result hints strongly that the key morphological characters that biologists have traditionally used to construct animal phylogenies—segmentation, coeloms, jointed appendages, and the like—are not the conservative characters we had supposed. Among the myzostomids, these features appear to have been gained and lost again during the course of their evolution. If this unconservative evolutionary pattern should prove general, our view of the evolution of the animal body plan, and how the various animal phyla relate to one another, will soon be in need of major revision.

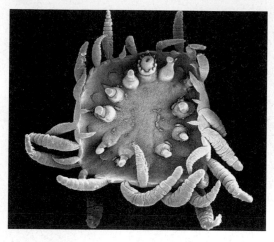

FIGURE 31.8
A taxonomic puzzle. *Myzostoma mortenensi* has no body cavity and incomplete segmentation. Such animals present a classification challenge, causing taxonomists to reconsider traditional animal phylogenies and the characters they are based on.

The last decade has seen a wealth of new molecular RNA and DNA sequence data on the various animal groups. The new field of **molecular systematics** uses unique sequences within certain genes to identify clusters of related groups. Expressed in the terminology of cladistic taxonomy, the sharing of derived sequence characters unique to a group and its ancestors defines clusters of monophyletic taxa that make up clades. The animal phylogenetic tree viewed in these terms is a hierarchy of clades nested within larger clades.

Using these sorts of molecular data, a variety of molecular phylogenies have been produced in the last decade. While differing from one another in many important respects, the new molecular phylogenies have the same deep branch structure as the traditional animal family tree. However, most agree on one revolutionary difference from the traditional phylogeny used in this text and presented in figure 31.3: The protostomes are broken into two distinct clades. Different molecular taxonomists distribute lophophorates, pseudocoelomates, and acoelomates among the two clades in different ways.

At present, molecular phylogenetic analysis of the animal kingdom is in its infancy. Phylogenies developed from different molecules sometimes suggest quite different evolutionary relationships. However, the childhood of this approach is likely to be short. Over the next few years, a mountain of additional molecular data can be anticipated. As more data are brought to bear, the confusion can be expected to lessen.

A year-2003 consensus molecular phylogeny, developed from DNA, ribosomal RNA, and protein studies, is presented in figure 31.9. In it, the traditional protostome group is broken up into Ecdysozoa and Lophotrochozoa. This new view of the metazoan Tree of Life is only a rough outline; in the future, more data should allow us to more confidently resolve relationships within groupings. Still, it is already clear that major groups are related in very different ways in molecular phylogenies than in the more traditional morphological one.

The use of molecular data to construct phylogenies is likely to significantly alter our understanding of relationships among the animal phyla.

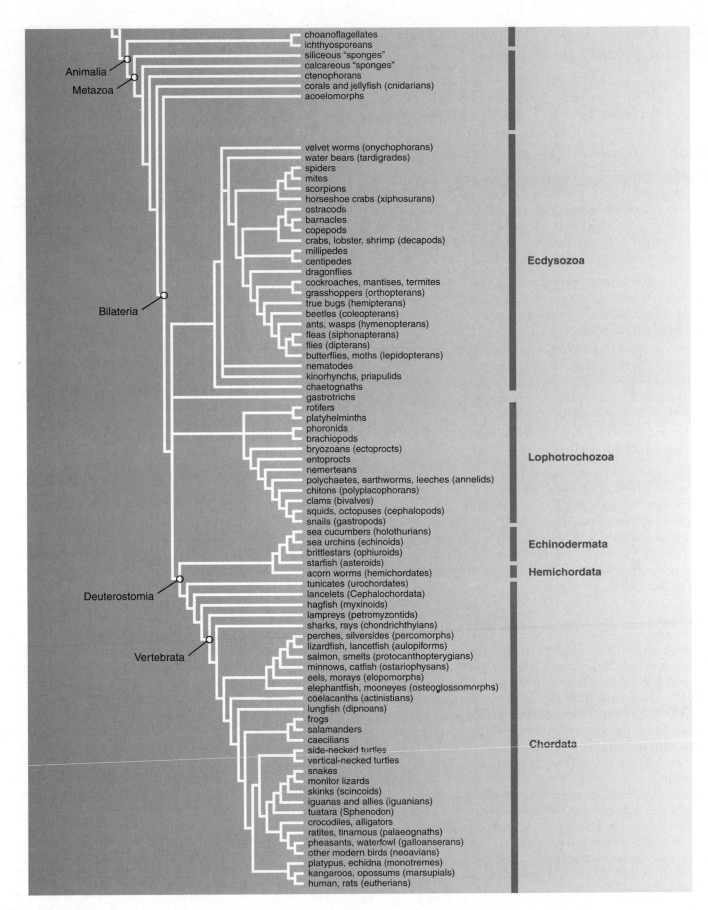

FIGURE 31.9

Proposed revision of the animal tree of life. This molecular phylogeny, based on comparisons of DNA, ribosomal RNA, and protein sequence differences, was assembled in 2003 by the journal *Science* from discussions with numerous taxonomists. It reflects an emerging consensus among molecular investigators that traditional phylogenies based on shared morphological and embryological characters do not reflect true ancestral relationships.

"Evo-Devo" and the Roots of the Animal Family Tree

Some of the most exciting contributions of molecular systematics are being made to our understanding of the base of the animal family tree.

Origin of Metazoans

Most taxonomists agree that the animal kingdom is monophyletic—that is, that parazoans and eumetazoans have a common ancestor. This ancestor was presumably a protist, but from which line of protists did animals evolve? There are three prominent hypotheses for the origin of metazoans from single-celled protists:

The **multinucleate hypothesis** suggests that metazoans arose from a multinuclear protist similar to today's ciliates. The cells later became compartmentalized into the multicellular condition.

The **colonial flagellate hypothesis,** first proposed by Haeckel in 1874, states that metazoans descended from colonial protists, hollow spherical colonies of flagellated cells. Some of the cells of sponges are strikingly like those of choanoflagellate protists.

The **polyphyletic origin hypothesis** proposes that sponges evolved independently from eumetazoans.

Molecular systematics based on ribosomal RNA sequences settles this argument clearly in favor of the colonial flagellate hypothesis. The molecular evidence excludes the multinucleate ciliate hypothesis because metazoans are closer to eukaryotic algae than to ciliates. The polyphyletic origin hypothesis is excluded because metazoans represent a monophyletic assembly (the sister group of metazoans appears to be fungi).

Early Diversification of the Animal Family Tree

A second contentious issue in animal phylogeny is being addressed successfully with molecular systematics. Study of the fossil record reveals that the great diversity of animals evolved quite rapidly in geological terms around the beginning of the Cambrian period. Nearly all the major animal body plans can be seen in Cambrian rocks dating from 543 to 525 million years ago.

In rock from the earlier Ediacaran period as old as 565 million years, fossil cnidarians are found, along with what appear to be fossil mollusks and the burrows of worms. This implies that the earliest branches of the animal family tree arose before the Cambrian period.

In the half-billion years since the early Cambrian, no significant new innovations in animal body plan have occurred. Biologists have long debated what caused this **Cambrian explosion** of animal diversity (figure 31.10).

FIGURE 31.10
Diversity of animals that evolved during the Cambrian explosion. In addition to the ancestors of many present-day groups, such as insects and vertebrates, a variety of bizarre creatures evolved that left no descendants, such as: (1) *Amiskwia*, (2) *Odontogriphus*, (3) *Eldonia*, (4) *Halichondrites*, (5) *Anomalocaris canadensis*, (6) *Pikaia*, (7) *Canadia*, (8) *Marrella splendens*, (9) *Opabinia*, (10) *Ottoia*, (11) *Wiwaxia*, (12) *Yohoia*, (13) *Xianguangia*, (14) *Aysheaia*, (15) *Sidneyia*, (16) *Dinomischus*, and (17) *Hallucigenia*. The natural history of these species is open to speculation.

Many have argued that the emergence of new body plans was the consequence of the emergence of predatory lifestyles, which encouraged an arms race between defenses, such as armor, and innovations that improved mobility and hunting success. Others have attributed the rapid diversification in body plans to geological factors, such as the buildup of dissolved oxygen and minerals in the oceans. A third possibility arises from molecular studies being carried out by biologists in the new field of **"evo-devo,"** a synthesis of evolutionary biology and developmental biology. Much of the variation in animal body plan is associated with changes in the location or time of expression of *Hox* genes within developing animal embryos (see chapters 19 and 24). Perhaps the Cambrian explosion reflects the evolution of the *Hox* developmental gene complex, which provides a tool that can produce rapid changes in body plan.

The animal kingdom is monophyletic and arose from a colonial flagellated protist. The great diversity in the animal body plan arose quickly, possibly as the result of the evolution of *Hox* genes.

Concept Review

For interactive testing, visit the Online Learning Center with PowerWeb at www.mhhe.com/Raven7

31.1 Animals are multicellular heterotrophs without cell walls.

Some General Features of Animals

- Animals are multicellular heterotrophs that are diverse in form and habitat, are mobile, reproduce sexually, and have characteristic embryonic development. (p. 618)

31.2 Animals are a very diverse kingdom.

The Traditional Classification of Animals

- Taxonomists have traditionally created phylogenies by comparing anatomical features and embryological development. (p. 620)
- Kingdom Animalia is traditionally divided into the Parazoa, which lack a definite symmetry and organized tissues, and the Eumetazoa, which have a definite symmetry and organized tissues. (p. 620)
- The eumetazoan branch is divided into Radiata and Bilateria. (p. 620)
- Bilateral animals further split into groups with and without a body cavity. (p. 620)

31.3 The animal body plan has undergone many changes.

Five Key Transitions in Body Plan

- The evolution of tissues involved cell specialization. (p. 624)
- The evolution of bilateral symmetry allowed organization of body parts, including cephalization, and increased motility. Bilaterally symmetrical animals produce three germ layers—ectoderm, endoderm, and mesoderm. (p. 624)
- The evolution of a body cavity allowed the subsequent development of efficient organ systems, such as an open or a closed circulatory system. Three basic types of body plans are (1) the acoelomates with no body cavity, (2) the pseudocoelomates with a cavity between the mesoderm and endoderm, and (3) the coelomates with a fluid-filled body cavity entirely within the mesoderm. (p. 625)
- The evolution of deuterostome development involved changes in cleavage patterns, determination, the fate of the blastopore, and the formation of the coelom. (pp. 626–627)
- The evolution of segmentation allowed the duplication of body organs and more effective locomotion. (p. 627)

31.4 The way we classify animals is being reevaluated.

A New Look at the Metazoan Family Tree

- Traditional characters used to construct phylogenies are not as conservative as once supposed. (p. 628)
- Molecular systematics uses unique sequences within certain genes to identify shared characteristics that define the monophyletic taxa making up clades. (p. 628)

"Evo-Devo" and the Roots of the Animal Family Tree

- The multinucleate hypothesis, the colonial flagellate hypothesis, and the polyphyletic origin hypothesis all try to account for the origin of metazoans from single-celled protists. (p. 630)
- A large diversity of animal body plans occurred around the Cambrian period, with no new innovations since. (p. 630)
- The emergence of body plans has been hypothesized to be caused by the emergence of predatory lifestyles, geological factors, and changes in the location or time of expression of *Hox* genes within developing animal embryos. (p. 630)

Self Test

1. Which feature is *not* characteristic of the animals?
   a. multicellular heterotrophs
   b. sexual reproduction
   c. embryonic development
   d. cell walls
2. Most of the phyla of the animal kingdom are found
   a. on land.
   b. in the ocean.
   c. burrowing underground.
   d. in freshwater habitats
3. The subkingdom of animals that lack symmetry and have no true tissues or organs is the
   a. Eumetazoa.
   b. Radiata.
   c. Parazoa.
   d. Bilateria.
4. Except for the Radiata, the eumetazoans
   a. lack true tissues and organs.
   b. are monoblastic.
   c. are diploblastic.
   d. are triploblastic.
5. Cnidarians and ctenophores differ from other eumetazoans by having
   a. radial symmetry.
   b. bilateral symmetry.
   c. major organ systems.
   d. three tissue layers.
6. The advantage of bilateral symmetry is that it allowed for the evolution of
   a. appendages.
   b. cephalization.
   c. reproductive structures.
   d. parasites.
7. Members of which group are *not* deuterostomes?
   a. chordates
   b. echinoderms
   c. arthropods
   d. none of these; all are deuterostomes.
8. The evolution of an internal body cavity offered an advantage in animal body design in all areas except
   a. circulation.
   b. digestion.
   c. freedom of movement.
   d. gamete storage.
9. The segments of annelids are
   a. apparent in the embryo but not in the adult.
   b. specialized for different functions.
   c. present in the mesoderm (muscles) but not in the ectoderm.
   d. repetitive—each able to develop a complete set of adult organs.
10. Which of the following hypotheses about the origin of metazoans is supported by ribosomal RNA analysis?
    a. the multinucleate hypothesis
    b. the colonial flagellate hypothesis
    c. the polyphyletic origin hypothesis
    d. None of these; rRNA analysis doesn't support any specific hypothesis.

Test Your Visual Understanding

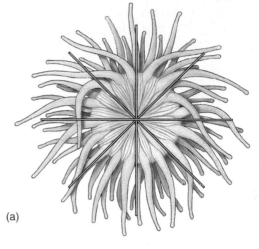

(a)

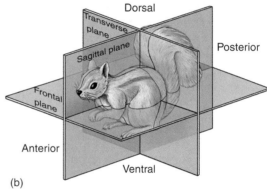

(b)

1. Which of these drawings (*a* or *b*) depicts a radially symmetrical animal? Which depicts a bilaterally symmetrical animal? Although a radially symmetrical animal can be bisected into equal halves in any two-dimensional plane, can you describe a plane of orientation of a radially symmetrical animal that produces dissimilar halves? (Refer to the planes of bisection in *b*.)

Apply Your Knowledge

1. In what ways is an earthworm more complex than a flatworm?
2. Why is it believed that echinoderms and chordates, which are so dissimilar, are members of the same evolutionary line?

32

Noncoelomate Invertebrates

Concept Outline

32.1 The classification of invertebrates is currently being reevaluated.

An Uproar Over Invertebrate Phylogeny. The bilaterally symmetrical invertebrates have traditionally been separated according to the nature of their body cavity. However, comparisons of ribosomal RNA are suggesting a very different invertebrate phylogeny.

32.2 The simplest animals are not bilaterally symmetrical.

Parazoa. Sponges are the most primitive animals, without either tissues or, for the most part, symmetry.

Radiata. Cnidarians and ctenophores have distinct tissues and radial symmetry.

32.3 Acoelomates are solid worms that lack a body cavity.

The Bilaterian Acoelomates. Flatworms are the simplest bilaterally symmetrical animals; they are solid worms that lack a body cavity, but possess true organs.

32.4 Pseudocoelomates have a simple body cavity.

The Pseudocoelomates. Nematodes and rotifers possess a simple body cavity.

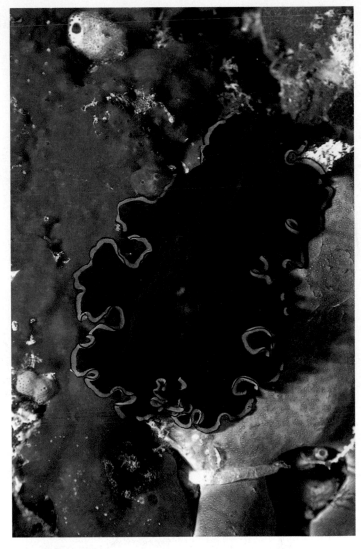

FIGURE 32.1
A noncoelomate: a marine flatworm. Some of the earliest invertebrates to evolve, marine flatworms possess internal organs but lack a mesoderm-encased body cavity called a coelom.

We will start our exploration of the great diversity of animals with the simplest members of the animal kingdom—sponges, jellyfish, and simple worms. These animals lack a body cavity called a coelom, and are thus called noncoelomates (figure 32.1). The major organization of the animal body first evolved in these animals, a basic body plan upon which all the rest of animal evolution has depended. In chapter 33, we consider the invertebrate animals that have a coelom, and in chapter 34 the vertebrates. Despite their great diversity, you will see that all animals have much in common.

32.1 The classification of invertebrates is currently being reevaluated.

An Uproar Over Invertebrate Phylogeny

There is little disagreement among biologists about the classification of animals. The 35 traditional animal phyla are firmly established in the biological lexicon—for example, an annelid worm would be classified in the phylum Annelida by any competent taxonomist. However, there is great disagreement about how the 35 animal phyla are related to one another. Depending on what aspects of the phyla are compared, different biologists draw quite different family trees.

Two Approaches

For many years, biologists have based their reconstructions of the animal family tree on key aspects of body architecture, lumping together those phyla that share fundamental aspects of body plan. For the better part of a century, biologists agreed on the principal aspects of this tree, basing the phylogeny largely on anatomical and embryological comparisons

However, as we learned in chapter 31, a variety of new, quite different, animal family trees have been proposed in the last decade by researchers employing molecular rather than anatomical comparisons, focusing particularly on differences in ribosomal RNA sequences.

Like the traditional tree based on body plan, these new trees based on rRNA comparisons place the sponges (phylum Porifera), the only animals without tissues, in the category Parazoa, and place all animals with tissues in the category Eumetazoa. Among the eumetazoans, both approaches place the radially symmetrical cnidarians and ctenophores in the category Radiata. All other eumetazoans are assigned to one of two groups of bilaterally symmetrical animals that differ in their embryological development—protostomes or deuterostomes.

It is in how they construct the protostome branch of the animal family tree that the new rRNA phylogenies differ radically from the traditional one.

The Traditional Protostome Phylogeny

As noted in chapter 31, the traditional animal family tree divides the bilaterally symmetrical animals into three great branches, based on the nature of the body cavity: (1) acoelomates (such as phylum Platyhelminthes), which have no body cavity; (2) pseudocoelomates (such as phylum Nematoda), which have a pseudocoel body cavity separating mesoderm from endoderm; and (3) coelomates (such as phyla Arthropoda, Annelida, and Mollusca), which have a coelom body cavity encased within mesoderm (figure 32.2). Coelomates can be either protostomes or deuterostomes.

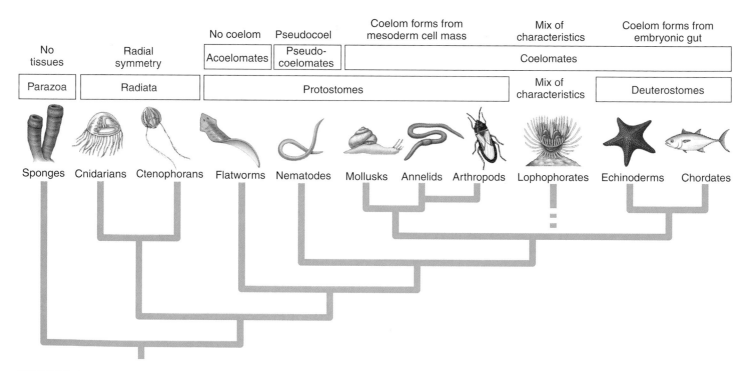

FIGURE 32.2
The traditional protostome phylogeny. Biologists have traditionally separated the bilaterally symmetrical animals into three groups that differ with respect to their body cavity: acoelomates, pseudocoelomates, and coelomates.

The Novel rRNA Protostome Phylogenies

Phylogenies based on rRNA sequences suggest a very different lineage of the protostome phyla. Two major clades are recognized as having evolved independently since ancient times: the lophotrochozoans and the ecdysozoans (figure 32.3).

Lophotrochozoans. Lophotrochozoan animals grow the same way you do, by adding additional mass to an existing body. Most live in water, and propel themselves through it using cilia. Many, although not all, lophotrochozoans have a type of free-living larva known as a **trochophore**.

Four major kinds of protostomes are assigned to the Lophotrochozoa:

The **lophophorate phyla,** which include bryozoans, phoronids, and brachiopods, are all coelomates with a horseshoe-shaped crown of ciliated tentacles called a **lophophore** around their mouths, used for feeding.

Flatworms, phylum Platyhelminthes, are acoelomates. They have a simple body plan with no enclosed body cavity and no organs for transporting oxygen to the body's interior.

Mollusks, phylum Mollusca, are unsegmented coelomates. They display a wide variety of body forms; included in this group are octopuses, snails, and clams.

Annelids, phylum Annelida, are segmented coelomate worms. This group includes both marine polychaetes and terrestrial earthworms.

Ecdysozoans. Ecdysozoans are the molting animals. They increase in size by molting their external skeletons, an ability that seems to have evolved only once. Molting animals have been successful in nearly all environments, and all move about by means other than ciliary action. Ecdysozoans all share a common set of homeobox genes, regulatory genes that govern the pattern of embryonic development.

Of the numerous phyla of protostomes assigned to the Ecdysozoa, two have been particularly successful:

Roundworms, phylum Nematoda, are pseudocoelomate worms that shed their hard cuticle four times as they grow to adult size. They are one of the most abundant and widely distributed of all animal phyla.

Arthropods, phylum Arthropoda, are coelomate animals with segmented external skeletons and jointed appendages. The most successful of all animal phyla, arthropods include insects, spiders, and crustaceans.

As more complete genomic comparisons become available, our picture of the evolutionary lineage of the protostomes will undoubtedly become clearer. For now, until a consensus emerges, this text will continue to adopt the traditional arrangement of the phyla.

The protostomes are traditionally grouped according to the nature of their body cavity, but recent rRNA evidence suggests they might be better grouped according to whether or not they molt.

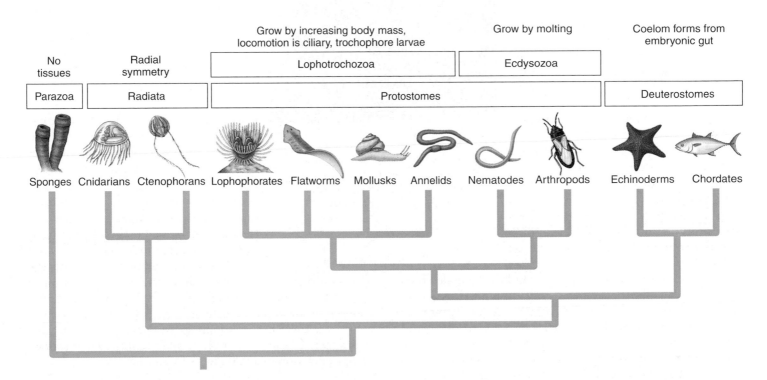

FIGURE 32.3

An rRNA protostome phylogeny. New phylogenies based on differences in ribosomal RNA genes suggest that protostomes might better be grouped according to whether they grow by molting (ecdysozoans) or by adding mass to an existing body (lophotrochozoans).

32.2 The simplest animals are not bilaterally symmetrical.

Parazoa

The sponges are parazoans, animals that lack tissues and organs and a definite symmetry. However, sponges, like all animals, have true, complex *multicellularity*. The body of a sponge contains several distinctly different types of cells whose activities are loosely coordinated with one another. As we will see, the coordination between cell types in the eumetazoans increases and becomes quite complex.

The Sponges

There are perhaps 5000 species of marine sponges, phylum Porifera, and about 150 species that live in fresh water. In the sea, sponges are abundant at all depths. Although some sponges are tiny (no more than a few millimeters across), others, such as the loggerhead sponges, may reach 2 meters or more in diameter. A few small sponges are radially symmetrical, but most members of this phylum completely lack symmetry. Many sponges are colonial. Some have a low and encrusting form, while others may be erect and lobed, sometimes in complex patterns (figure 32.4). Although larval sponges are free-swimming, adults are **sessile,** or anchored in place to submerged objects.

Although sponges, like all animals, are composed of multiple cell types (figure 32.5), there is relatively little coordination among sponge cells. A sponge seems to be little more than a mass of cells embedded in a gelatinous matrix, but these cells are specialized for different functions and recognize one another with a high degree of fidelity.

The basic structure of a sponge can best be understood by examining the form of a young individual. A small, anatomically simple sponge first attaches to a substrate and then grows into a vaselike shape. The walls of the "vase" have three functional layers. First, facing into the internal cavity are specialized flagellated cells called **choanocytes,** or collar cells. These cells line either the entire body interior or, in many large and more complex sponges, specialized chambers. Second, the bodies of sponges are bounded

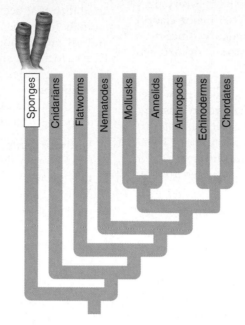

FIGURE 32.4
Aplysina longissima. This beautiful, bright orange and red elongated sponge is found on deep regions of coral reefs. The osculum is an opening at the top.

by an outer epithelial layer consisting of flattened cells somewhat like those that make up the epithelia, or outer layers, of animals in other phyla. Some portions of this layer contract when touched or exposed to appropriate chemical stimuli, and this contraction may cause some of the pores to close. Third, between these two layers, sponges consist mainly of a gelatinous, protein-rich matrix called the **mesohyl,** within which various types of amoeboid cells occur. In addition, many kinds of sponges have minute needles of calcium carbonate or silica known as **spicules,** or fibers of a tough protein called **spongin,** or both, within this matrix. Spicules and spongin strengthen the bodies of the sponges in which they occur. A spongin skeleton is the model for the bathtub sponge, once the skeleton of a real animal, but now usually a cellulose or plastic mimic.

Sponges feed in a unique way. The beating of flagella that line the inside of the sponge draws water in through numerous small pores; the name of the phylum, Porifera, refers to this system of pores. Plankton and other small organisms are filtered from the water, which flows through passageways and eventually is forced out through an **osculum,** a specialized, larger pore.

Choanocytes. Each choanocyte closely resembles a protist with a single flagellum (figure 32.5), a similarity that reflects its evolutionary derivation. The beating of the flagella of the many choanocytes that line the body interior draws water in through the pores and through the sponge, thus bringing in food and oxygen and expelling wastes. Each choanocyte flagellum beats independently, and the pressure they create collectively in the cavity forces water out of the osculum. In some sponges, the inner wall of the body interior is highly convoluted, increasing the surface area and, therefore, the number of flagella that can drive the water. In such a sponge, 1 cubic centimeter of sponge can propel more than 20 liters of water per day.

Reproduction in Sponges. Some sponges will re-form themselves once they have passed through a silk mesh.

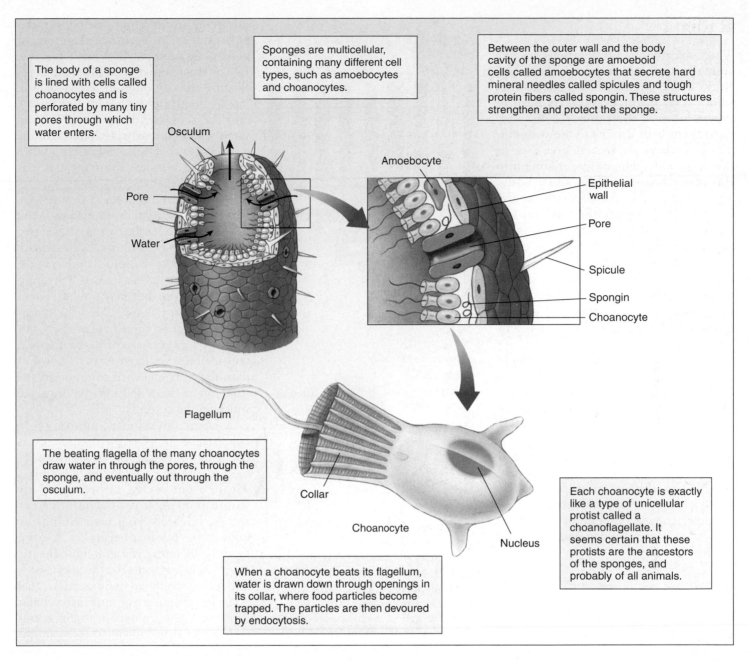

The body of a sponge is lined with cells called choanocytes and is perforated by many tiny pores through which water enters.

Sponges are multicellular, containing many different cell types, such as amoebocytes and choanocytes.

Between the outer wall and the body cavity of the sponge are amoeboid cells called amoebocytes that secrete hard mineral needles called spicules and tough protein fibers called spongin. These structures strengthen and protect the sponge.

Osculum

Pore

Water

Amoebocyte

Epithelial wall

Pore

Spicule

Spongin

Choanocyte

Flagellum

The beating flagella of the many choanocytes draw water in through the pores, through the sponge, and eventually out through the osculum.

Collar

Choanocyte

Nucleus

Each choanocyte is exactly like a type of unicellular protist called a choanoflagellate. It seems certain that these protists are the ancestors of the sponges, and probably of all animals.

When a choanocyte beats its flagellum, water is drawn down through openings in its collar, where food particles become trapped. The particles are then devoured by endocytosis.

FIGURE 32.5

Phylum Porifera: Sponges. Sponges are composed of several distinctly different cell types, whose activities are coordinated with each other. The sponge body is not symmetrical and has no organized tissues.

Thus, as you might suspect, sponges frequently reproduce by simply breaking into fragments. If a sponge breaks up, the resulting fragments usually are able to reconstitute whole new individuals. Sexual reproduction is also exhibited by sponges, with some mature individuals producing eggs and sperm. Larval sponges may undergo their initial stages of development within the parent. They have numerous external, flagellated cells and are free-swimming. After a short planktonic stage, they settle down on a suitable substrate, where they begin their transformation into adults.

Sponges probably represent the most primitive animals, possessing multicellularity but neither tissue-level development nor body symmetry. Their cellular organization hints at the evolutionary ties between the unicellular protists and the multicellular animals. Sponges are unique in the animal kingdom in possessing choanocytes, special flagellated cells whose beating drives water through the body interior.

Radiata

The subkingdom Eumetazoa contains animals that evolved the first key transition in the animal body plan: distinct *tissues*. Two distinct cell layers form in the embryos of these animals: an outer ectoderm and an inner endoderm. These embryonic tissues give rise to the basic body plan, differentiating into the many tissues of the adult body. Typically, the outer covering of the body (called the epidermis) and the nervous system develop from the ectoderm, and the layer of digestive tissue (called the **gastrodermis**) develops from the endoderm. A layer of gelatinous material, called the **mesoglea**, lies between the epidermis and gastrodermis and contains the muscles in most eumetazoans.

Eumetazoans also evolved true body symmetry and are divided into two major groups. The Radiata includes two phyla of radially symmetrical organisms: Phylum Cnidaria (pronounced ni-DAH-ree-ah), or the cnidarians, is composed of hydroids, jellyfish, sea anemones, and corals. Phylum Ctenophora (pronounced tea-NO-fo-rah) is composed of the comb jellies, also called the ctenophores. All other eumetazoans are in the Bilateria and exhibit a fundamental bilateral symmetry.

The Cnidarians

Cnidarians are nearly all marine, although a few live in fresh water. These fascinating and simply constructed animals are basically gelatinous in composition. They differ markedly from the sponges in organization; their bodies are made up of distinct tissues, although they have not evolved true organs. These animals are carnivores. For the most part, they do not actively move from place to place, but rather capture their prey (which includes fishes, crustaceans, and many other kinds of animals) with the tentacles that ring their mouths.

Cnidarians may have two basic body forms, polyps and medusae (figure 32.6). Polyps are cylindrical and are usually found attached to a firm substrate. They may be solitary or colonial. In a polyp, the mouth faces away from the substrate on which the animal is growing, and therefore often faces upward. Many polyps build up a chitinous or

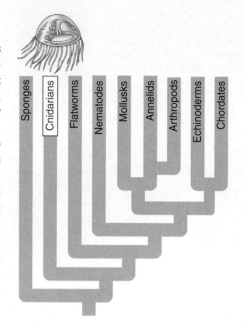

Sponges • Cnidarians • Flatworms • Nematodes • Mollusks • Annelids • Arthropods • Echinoderms • Chordates

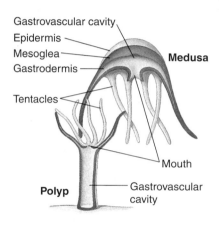

Gastrovascular cavity
Epidermis
Mesoglea
Gastrodermis
Tentacles
Medusa
Mouth
Polyp
Gastrovascular cavity

FIGURE 32.6
Two body forms of cnidarians, the medusa and the polyp. These two phases alternate in the life cycles of many cnidarians. But a number—including the corals and sea anemones—exist only as polyps, and some exist only as medusae. Both forms have two fundamental layers of cells, separated by a jellylike layer called the mesoglea.

calcareous (made up of calcium carbonate) external or internal skeleton, or both. Only a few polyps are free-floating. In contrast, most medusae are free-floating and are often umbrella-shaped. Their mouths usually point downward, and the tentacles hang down around them. Medusae, particularly those of the class Scyphozoa, are commonly known as jellyfish because their mesoglea is thick and jellylike.

Many cnidarians occur only as polyps, while others exist only as medusae; still others alternate between these two phases during their life cycles. Both phases consist of diploid individuals. Polyps may reproduce asexually by budding; if they do, they may produce either new polyps or medusae. Medusae reproduce sexually. In most cnidarians, fertilized eggs give rise to free-swimming, multicellular, ciliated larvae known as **planulae.** Planulae are common in the plankton at times and may be dispersed widely in the currents.

A major evolutionary innovation in cnidarians, compared with sponges, is the internal extracellular digestion of food (figure 32.7). Digestion takes place within a gut cavity, rather than only within individual cells. Digestive enzymes, released from cells lining the walls of the cavity, partially break down food. Cells lining the gut subsequently engulf food fragments by phagocytosis.

The extracellular fragmentation that precedes phagocytosis and intracellular digestion allows cnidarians to digest animals larger than individual cells, an important improvement over the strictly intracellular digestion that occurs in sponges.

Nets of nerve cells coordinate contraction of cnidarian muscles, apparently with little central control. Cnidarians have no blood vessels, respiratory system, or excretory organs.

On their tentacles and sometimes on their body surface, cnidarians bear specialized cells called **cnidocytes.** The name of the phylum, Cnidaria, refers to these cells, which are highly distinctive and occur in no other group of organisms. Within each cnidocyte is a **nematocyst,** a small but powerful "harpoon." Each nematocyst features a coiled, threadlike tube. Lining the inner wall of the tube is a series of barbed spines. Cnidarians use the threadlike tube to spear their prey; then they draw the harpooned

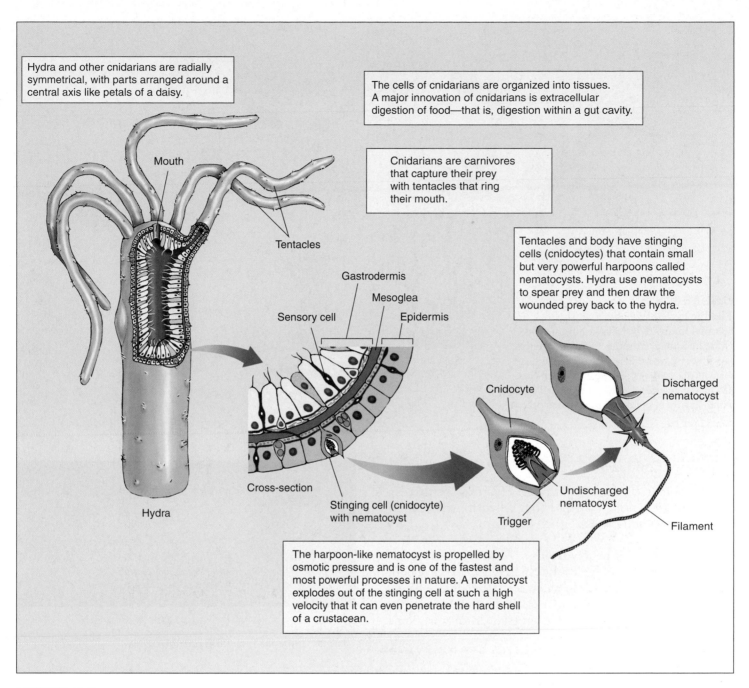

Hydra and other cnidarians are radially symmetrical, with parts arranged around a central axis like petals of a daisy.

The cells of cnidarians are organized into tissues. A major innovation of cnidarians is extracellular digestion of food—that is, digestion within a gut cavity.

Cnidarians are carnivores that capture their prey with tentacles that ring their mouth.

Tentacles and body have stinging cells (cnidocytes) that contain small but very powerful harpoons called nematocysts. Hydra use nematocysts to spear prey and then draw the wounded prey back to the hydra.

Mouth

Tentacles

Gastrodermis

Mesoglea

Sensory cell

Epidermis

Cnidocyte

Discharged nematocyst

Cross-section

Hydra

Stinging cell (cnidocyte) with nematocyst

Undischarged nematocyst

Trigger

Filament

The harpoon-like nematocyst is propelled by osmotic pressure and is one of the fastest and most powerful processes in nature. A nematocyst explodes out of the stinging cell at such a high velocity that it can even penetrate the hard shell of a crustacean.

FIGURE 32.7

Phylum Cnidaria: cnidarians. The cells of a cnidarian such as this *Hydra* are organized into specialized tissues. The interior gut cavity is specialized for extracellular digestion—that is, digestion within a gut cavity rather than within individual cells. Cnidarians are radially symmetrical, with parts arranged around a central axis like the petals of a daisy.

prey back with the tentacle containing the cnidocyte. Nematocysts may also serve a defensive purpose. To propel the harpoon, the cnidocyte uses water pressure. Before firing, the cnidocyte builds up a very high internal osmotic pressure. This is done by using active transport to build a high concentration of ions inside, while keeping the cnidocyte's cell wall impermeable to water. Within the undischarged nematocyst, osmotic pressure reaches about 140 atmospheres.

When a flagellum-like trigger on the cnidocyte is stimulated to discharge, its walls become permeable to water, which rushes inside and violently pushes out the barbed filament. Nematocyst discharge is one of the fastest cellular processes in nature. The nematocyst is pushed outward so explosively that the barb can penetrate even the hard shell of a crab. A toxic protein often produces a stinging sensation, causing some cnidarians to be called "stinging nettles."

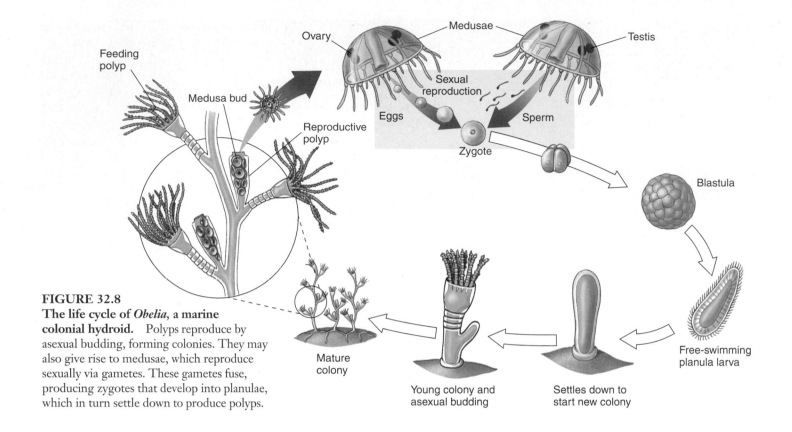

FIGURE 32.8
The life cycle of *Obelia*, a marine colonial hydroid. Polyps reproduce by asexual budding, forming colonies. They may also give rise to medusae, which reproduce sexually via gametes. These gametes fuse, producing zygotes that develop into planulae, which in turn settle down to produce polyps.

Classes of Cnidarians

There are four classes of cnidarians: Hydrozoa (hydroids), Scyphozoa (jellyfish), Cubozoa (box jellyfish), and Anthozoa (anemones and corals).

Class Hydrozoa: The Hydroids. Most of the approximately 2700 species of hydroids (class Hydrozoa) have both polyp and medusa stages in their life cycle (figure 32.8). Most of these animals are marine and colonial, such as *Obelia* and the very unusual Portuguese man-of-war. Some of the marine hydroids are bioluminescent.

A well-known hydroid is the abundant freshwater genus *Hydra*, which is exceptional in having no medusa stage and existing as a solitary polyp. Each polyp sits on a basal disk, which the hydra can use to glide around, aided by mucous secretions. It can also move by somersaulting—bending over and attaching itself to the substrate by its tentacles, and then looping over to a new location. If the polyp detaches itself from the substrate, it can float to the surface.

Class Scyphozoa: The Jellyfish. The approximately 200 species of jellyfish (class Scyphozoa) are transparent or translucent marine organisms, some of a striking orange, blue, or pink color (figure 32.9). These animals spend most of their time floating near the surface of the sea. In all of them, the medusa stage is dominant—much larger and more complex than the polyp stage. The medusae are bell-shaped, with hanging tentacles around

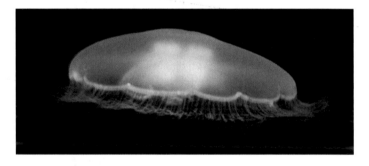

FIGURE 32.9
Class Scyphozoa. Jellyfish, *Aurelia aurita.*

their margins. The polyp stage is small, inconspicuous, and simple in structure.

The outer layer, or epithelium, of a jellyfish contains a number of specialized epitheliomuscular cells, each of which can contract individually. Together, the cells form a muscular ring around the margin of the bell that pulses rhythmically and propels the animal through the water. Jellyfish have separate male and female individuals. After fertilization, planulae form, which then attach and develop into polyps. The polyps can reproduce asexually as well as budding off medusae. In some jellyfish that live in the open ocean, the polyp stage is suppressed, and planulae develop directly into medusae.

Class Cubozoa: The Box Jellyfish.
Until recently, the cubozoans were considered an order of Scyphozoa. As their name implies, they are box-shaped medusae; the polyp stage is inconspicuous and in many cases not known. Most are only a few centimeters in height, although some are 25 centimeters tall. A tentacle or group of tentacles is found at each corner of the box (figure 32.10). Box jellies are strong swimmers and voracious predators of fish. The stings of some species can be fatal to humans.

Class Anthozoa: The Sea Anemones and Corals. By far the largest class of cnidarians is Anthozoa, the "flower animals" (Greek *anthos,* "flower"). The approximately 6200 species of this group are solitary or colonial marine animals. They include stonelike corals, soft-bodied sea anemones, and other groups known by such fanciful names as sea pens, sea pansies, sea fans, and sea whips (figure 32.11). All of these names reflect a plantlike body topped by a tuft or crown of hollow tentacles. Like other cnidarians, anthozoans use these tentacles in feeding. Nearly all members of this class that live in shallow waters harbor symbiotic algae, which supplement the nutrition of their hosts through photosynthesis. Fertilized eggs of anthozoans usually develop into planulae that settle and develop into polyps; no medusae are formed.

Sea anemones are a large group of soft-bodied anthozoans that live in coastal waters all over the world and are especially abundant in the tropics. When touched, most sea anemones retract their tentacles into their bodies and fold up. Sea anemones are highly muscular and relatively complex organisms, with greatly divided internal cavities. These animals range from a few millimeters to about 10 centimeters in diameter and are perhaps twice that high.

Corals are another major group of anthozoans. Many of them secrete tough outer skeletons, or exoskeletons, of calcium carbonate and are thus stony in texture. Others, includ-

FIGURE 32.10
Class Cubozoa. Box jellyfish, *Chironex fleckeri.*

FIGURE 32.11
Class Anthozoa. The sessile, soft-bodied sea anemone.

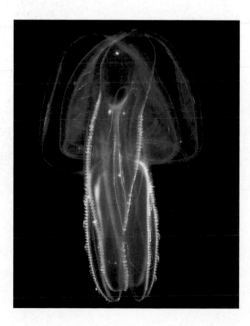

FIGURE 32.12
A comb jelly (phylum Ctenophora). Note the comblike plates along the ridges of the base.

ing the gorgonians, or soft corals, do not secrete exoskeletons. Some of the hard corals help form coral reefs, which are shallow-water limestone ridges that occur in warm seas. Although the waters where coral reefs develop are often nutrient-poor, the coral animals are able to grow actively because of the abundant algae found within them.

The Ctenophores (Comb Jellies)

The members of the small phylum Ctenophora range from spherical to ribbonlike and are known as comb jellies, or sea walnuts. Traditionally, the roughly 90 marine species of ctenophores were considered closely related to the cnidarians. However, ctenophores are structurally more complex than cnidarians. They have anal pores, so that water and other substances pass completely through the animal. Comb jellies, abundant in the open ocean, are transparent and usually only a few centimeters long. The members of one group have two long, retractable tentacles that they use to capture their prey.

Ctenophores propel themselves through water with eight comblike plates of fused cilia that beat in a coordinated fashion (figure 32.12). They are the largest animals that use cilia for locomotion. Many ctenophores are bioluminescent, giving off bright flashes of light particularly evident in the open ocean at night.

Cnidarians and ctenophores have tissues and radial symmetry. Cnidarians have a specialized kind of cell called a cnidocyte. Ctenophores propel themselves through the water by means of eight comblike plates of fused cilia.

32.3 Acoelomates are solid worms that lack a body cavity.

The Bilaterian Acoelomates

The Bilateria are characterized by the second key transition in the animal body plan, *bilateral symmetry*, which allowed animals to achieve high levels of specialization within parts of their bodies. The simplest bilaterians are the acoelomates; they lack any internal cavity other than the digestive tract. As discussed earlier, all bilaterians have three embryonic germ layers during development: ectoderm, endoderm, and mesoderm. We will focus our discussion of the acoelomates on the largest phylum of the group, the flatworms.

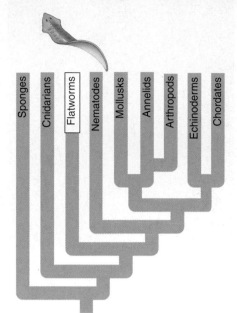

Phylum Platyhelminthes: The Flatworms

Phylum Platyhelminthes consists of some 20,000 species. These ribbon-shaped, soft-bodied animals are flattened dorsoventrally, from top to bottom. Flatworms are among the simplest of bilaterally symmetrical animals, but they do have a definite head at the anterior end and they do possess organs. Their bodies are solid; the only internal space consists of the digestive cavity (figure 32.13).

Flatworms range in length from a millimeter or less to many meters, as in some tapeworms. Most species of flatworms are parasitic, occurring within the bodies of many other kinds of animals (figure 32.14). Other flatworms are free-living, occurring in a wide variety of marine and freshwater habitats, as well as in moist places on land. Free-living flatworms are carnivores and scavengers; they eat various small animals and bits of organic debris. They move from place to place by means of ciliated epithelial cells, which are particularly concentrated on their ventral surfaces.

Those flatworms that have a digestive cavity have an incomplete gut, one with only one opening. As a result, they cannot feed, digest, and eliminate undigested particles of food simultaneously, and thus, flatworms cannot feed continuously, as more advanced animals can. Muscular contractions in the upper end of the gut cause a strong sucking force, allowing flatworms to ingest their food and tear it into small bits. The gut is branched and extends throughout the body, functioning in both digestion and transport of food. Cells that line the gut engulf most of the food particles by phagocytosis and digest them; but, as in the cnidarians, some of these particles are partly digested extracellularly. Tapeworms, which are parasitic flatworms, lack digestive systems. They absorb their food directly through their body walls.

Unlike cnidarians, flatworms have an excretory system, which consists of a network of fine tubules (little tubes) that

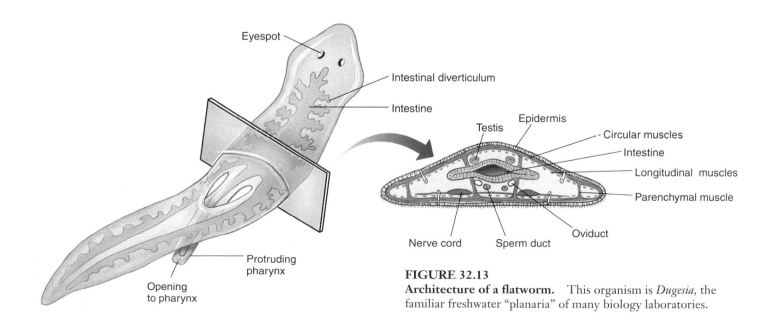

FIGURE 32.13
Architecture of a flatworm. This organism is *Dugesia*, the familiar freshwater "planaria" of many biology laboratories.

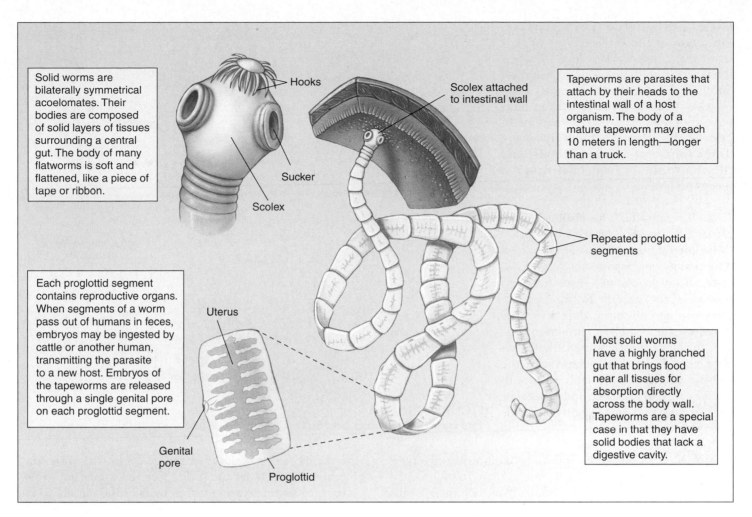

Solid worms are bilaterally symmetrical acoelomates. Their bodies are composed of solid layers of tissues surrounding a central gut. The body of many flatworms is soft and flattened, like a piece of tape or ribbon.

Hooks

Sucker

Scolex

Scolex attached to intestinal wall

Tapeworms are parasites that attach by their heads to the intestinal wall of a host organism. The body of a mature tapeworm may reach 10 meters in length—longer than a truck.

Repeated proglottid segments

Each proglottid segment contains reproductive organs. When segments of a worm pass out of humans in feces, embryos may be ingested by cattle or another human, transmitting the parasite to a new host. Embryos of the tapeworms are released through a single genital pore on each proglottid segment.

Uterus

Genital pore

Proglottid

Most solid worms have a highly branched gut that brings food near all tissues for absorption directly across the body wall. Tapeworms are a special case in that they have solid bodies that lack a digestive cavity.

FIGURE 32.14

Phylum Platyhelminthes: Flatworms.   Acoelomate flatworms such as this beef tapeworm, *Taenia saginata*, are bilaterally symmetrical solid worms. In addition, all bilaterians have three embryonic layers and a distinct head.

runs throughout the body. Cilia line the hollow centers of bulblike **flame cells** located on the side branches of the tubules. Cilia in the flame cells move water and excretory substances into the tubules and then to exit pores located between the epidermal cells. Flame cells were named because of the flickering movements of the tuft of cilia within them. They primarily regulate the water balance of the organism. The excretory function of flame cells appears to be secondary. A large proportion of the metabolic wastes excreted by flatworms diffuses directly into the gut and is eliminated through the mouth.

Like sponges, cnidarians, and ctenophores, flatworms lack circulatory systems for the transport of oxygen and food molecules. Consequently, all flatworm cells must be within diffusion distance of oxygen and food. Flatworms have thin bodies and highly branched digestive cavities that make such a relationship possible.

The nervous system of flatworms is very simple. Like cnidarians, some primitive flatworms have only a nerve net.

However, most members of this phylum have longitudinal nerve cords that constitute a simple central nervous system.

Free-living members of this phylum have eyespots on their heads. These are inverted, pigmented cups containing light-sensitive cells connected to the nervous system. These eyespots enable the worms to distinguish light from dark; worms move away from strong light.

The reproductive systems of flatworms are complex. Most flatworms are **hermaphroditic,** with each individual containing both male and female sexual structures. In many of them, fertilization is internal. When they mate, each partner deposits sperm in the copulatory sac of the other. The sperm travel along special tubes to reach the eggs. In most free-living flatworms, fertilized eggs are laid in cocoons strung in ribbons and hatch into miniature adults. Some parasitic flatworms undergo a complex succession of distinct larval forms. Flatworms are also capable of asexual regeneration. In some genera, when a single individual is divided into two or more parts, each part can regenerate an entirely new flatworm.

Class Turbellaria: Turbellarians.
Only one of the three classes of flatworms, the turbellarians (class Turbellaria), are free-living. One of the most familiar is the freshwater genus *Dugesia*, the common planaria used in biology laboratory exercises. Other members of this class are widespread and often abundant in lakes, ponds, and the sea. Some also occur in moist places on land.

Class Trematoda: The Flukes.
Two classes of parasitic flatworms live within the bodies of other animals: flukes (class Trematoda) and tapeworms (class Cestoda). Both groups of worms have epithelial layers resistant to the digestive enzymes and immune defenses produced by their hosts—an important feature in their parasitic way of life. However, they lack certain features of the free-living flatworms, such as cilia in the adult stage, eyespots, and other sensory organs that lack adaptive significance for an organism that lives within the body of another animal.

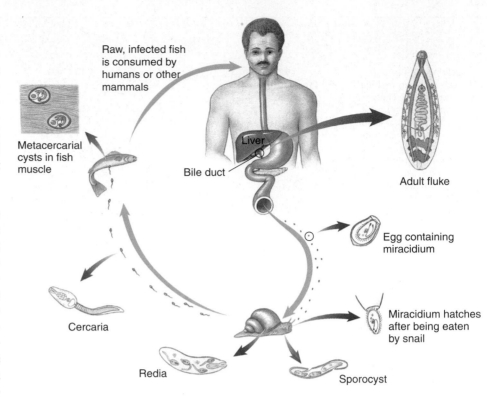

FIGURE 32.15
Life cycle of the human liver fluke, *Clonorchis sinensis*.

Flukes take in food through their mouth, just like their free-living relatives. There are more than 10,000 named species of flukes, ranging in length from less than 1 millimeter to more than 8 centimeters. Flukes attach themselves within the bodies of their hosts by means of suckers, anchors, or hooks. Some have a life cycle that involves only one host, usually a fish. Most have life cycles involving two or more hosts. Their larvae almost always occur in snails, and there may be other intermediate hosts. The final host of these flukes is almost always a vertebrate.

To human beings, one of the most important flatworms is the human liver fluke, *Clonorchis sinensis*. It lives in the bile passages of the liver of humans, cats, dogs, and pigs. It is especially common in Asia. The worms are 1 to 2 centimeters long and have a complex life cycle. Although they are hermaphroditic, cross-fertilization usually occurs between different individuals. Eggs, each containing a complete, ciliated first-stage larva, or **miracidium,** are passed in the feces (figure 32.15). If they reach water, they may be ingested by a snail. Within the snail, an egg transforms into a *sporocyst*, a baglike structure with embryonic germ cells. Within the sporocysts are produced **rediae,** which are elongated, nonciliated larvae. These larvae continue growing within the snail, giving rise to several individuals of the tadpole-like next larval stage, **cercariae.**

Cercariae escape into the water, where they swim about freely. If they encounter a fish of the family Cyprinidae—

the family that includes carp and goldfish—they bore into the muscles or under the scales, lose their tails, and transform into **metacercariae** within cysts in the muscle tissue. If a human being or other mammal eats raw infected fish, the cysts dissolve in the intestine, and the young flukes migrate to the bile duct, where they mature. An individual fluke may live for 15 to 30 years in the liver. In humans, a heavy infestation of liver flukes may cause cirrhosis of the liver and death.

Other very important flukes are the blood flukes of genus *Schistosoma*. They afflict about 1 in 20 of the world's population, more than 200 million people throughout tropical Asia, Africa, Latin America, and the Middle East. Three species of *Schistosoma* cause the disease called schistosomiasis, or bilharzia. Some 800,000 people die each year from this disease.

Recently, a great deal of effort has gone into controlling schistosomiasis. The worms protect themselves from the body's immune system in part by coating themselves with a variety of the host's own antigens that effectively render the worm immunologically invisible (see chapter 48). Despite this difficulty, the search is on for a vaccine that would cause the host to develop antibodies to one of the antigens of the young worms before they protect themselves with host antigens. This vaccine would protect humans from infection. The disease can be cured with drugs after infection.

Class Cestoda: The Tapeworms. Class Cestoda is the third class of flatworms; like flukes, they live as parasites within the bodies of other animals. In contrast to flukes, tapeworms simply hang on to the inner walls of their hosts by means of specialized terminal attachment organs and absorb food through their skins. Tapeworms lack digestive cavities as well as digestive enzymes. They are extremely specialized in relation to their parasitic way of life. Most species of tapeworms occur in the intestines of vertebrates, about a dozen of them regularly in humans.

The long, flat bodies of tapeworms are divided into three zones: the **scolex,** or attachment organ; the unsegmented **neck;** and a series of repetitive segments, the **proglottids** (see figure 32.14). The scolex usually bears several suckers and may also have hooks. Each proglottid is a complete hermaphroditic unit, containing both male and female reproductive organs. Proglottids are formed continuously in an actively growing zone at the base of the neck, with maturing ones moving farther back as new ones are formed in front of them. Ultimately, the proglottids near the end of the body form mature eggs. As these eggs are fertilized, the zygotes in the very last segments begin to differentiate, and these segments fill with embryos, break off, and leave their host with the host's feces. Embryos, each surrounded by a shell, emerge from the proglottid through a pore or the ruptured body wall. They are deposited on leaves, in water, or in other places where they may be picked up by another animal.

The beef tapeworm *Taenia saginata* occurs as a juvenile in the intermuscular tissue of cattle but as an adult in the intestines of human beings. A mature adult beef tapeworm may reach a length of 10 meters or more. These worms attach themselves to the intestinal wall of their host by a scolex with four suckers. Segments shed from the end of the worm pass from the human in the feces and may crawl onto vegetation. These segments ultimately rupture and scatter the embryos. Embryos may remain viable for up to five months. If ingested by cattle, they burrow through the wall of the intestine and ultimately reach muscle tissues through the blood or lymph vessels. About 1% of the cattle in the United States are infected, and some 20% of the beef consumed is not federally inspected. Thus, when humans eat infected beef that is cooked "rare," infection by these tapeworms is likely. As a result, the beef tapeworm is a frequent parasite of humans.

Phylum Nemertea: The Ribbon Worms

The phylogenetic relationship of phylum Nemertea (figure 32.16) to other free-living flatworms is unclear. Nemerteans are often called ribbon worms, or proboscis worms. These aquatic worms have the body plan of a flatworm, but also possess a fluid-filled sac that may be a primitive coelom. This sac serves as a hydraulic power source for their proboscis, a long muscular tube that can be thrust out quickly from a sheath to capture prey. Shaped like a thread or a ribbon, ribbon worms are mostly marine and consist of about 900 species. Ribbon worms are large, often 10 to 20 centimeters, and sometimes many meters, in length. They are the simplest animals that possess a **complete digestive system,** one that has two separate openings, a mouth and an anus. Ribbon worms also exhibit a circulatory system in which blood flows in vessels. Many important evolutionary trends that become fully developed in more advanced animals make their first appearance in the Nemertea.

> The acoelomates, typified by flatworms, are the most primitive bilaterally symmetrical animals and the simplest animals in which organs occur.

FIGURE 32.16
A ribbon worm, *Lineus* (phylum Nemertea). This is the simplest animal with a complete digestive system.

The Pseudocoelomates

All bilaterians except solid worms possess an internal body cavity, the third key transition in the animal body plan. Seven phyla are characterized by their possession of a pseudocoel (see figure 31.5). Their evolutionary relationships remain unclear, with the possibility that the pseudocoelomate condition arose independently many times. The pseudocoel serves as a hydrostatic skeleton—one that gains its rigidity from being filled with fluid under pressure. The animals' muscles can work against this "skeleton," thus making the movement of the pseudocoelomates far more efficient than that of the acoelomates. Pseudocoelomates lack a defined circulatory system; this role is performed by fluids that move within the pseudocoel. In this section, we focus on three significant pseudocoelomate phyla.

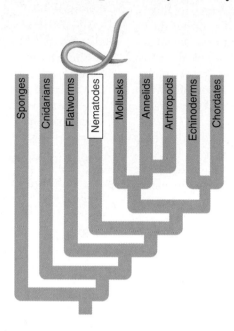

Phylum Nematoda: The Roundworms

Nematodes, eelworms, and other roundworms constitute a large phylum, Nematoda, with some 20,000 recognized species. Scientists estimate that the actual number might approach 100 times that many. Members of this phylum are found everywhere. Nematodes are abundant and diverse in marine and freshwater habitats, and many members of this phylum are parasites of animals (figure 32.17) and plants. Many nematodes are microscopic and live in soil. A spadeful of fertile soil may contain, on the average, a million nematodes.

Nematodes are bilaterally symmetrical, unsegmented worms. They are covered by a flexible, thick cuticle, which is molted four times as they grow. Their muscles constitute a layer beneath the epidermis and extend along the length of the worm, rather than encircling its body. These longitudinal muscles pull against both the cuticle and the pseudocoel, which forms a hydrostatic skeleton. When nematodes move, their bodies whip about from side to side.

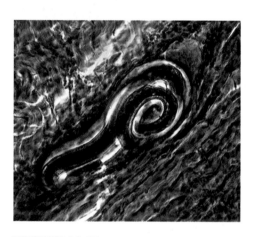

FIGURE 32.17
***Trichinella* nematode encysted in pork.** The serious disease trichinosis can result from eating undercooked pork or bear meat containing such cysts.

Nematodes gather nutrients and exchange oxygen through their cuticle, but they also possess a well-developed digestive system. Near the mouth of a nematode, at its anterior end, are usually 16 raised, hairlike, sensory organs. The mouth is often equipped with piercing organs called **stylets.** Food passes through the mouth as a result of the sucking action produced by the rhythmic contraction of a muscular chamber called the **pharynx** at the worm's anterior end. After passing through a short corridor into the pharynx, food continues through the other portions of the digestive tract, where it is broken down and then digested. The wall of the intestine is only one cell layer thick. Some of the water with which the food has been mixed is reabsorbed near the end of the digestive tract, and material that has not been digested is eliminated through the anus (figure 32.18).

Nematodes completely lack flagella or cilia, even on sperm cells. Reproduction in nematodes is sexual, with the sexes usually separate. Their development is simple, and the adults consist of very few cells. For this reason, nematodes have become extremely important subjects for genetic and developmental studies (see chapter 19). The 1-millimeter-long *Caenorhabditis elegans* matures in only three days, its body is transparent, and it has only 959 cells. It is the only animal whose complete developmental cellular anatomy is known.

The diets of nematodes vary greatly. Many are active hunters, preying on protists and other small animals. Many species of nematodes are parasites, living within the bodies of larger animals. Almost every species of plant and animal that has been studied has been found to have at least one parasitic species of nematode living in it. The largest known nematode, reaching a length of 9 meters, is a parasite in the placenta of female sperm whales. About 50 species of nematodes, including several that are rather common in the United States, regularly parasitize human beings. For example, hookworms, mostly of the genus *Necator*, can be common in southern states. By sucking blood through the intestinal wall, they can produce anemia if untreated.

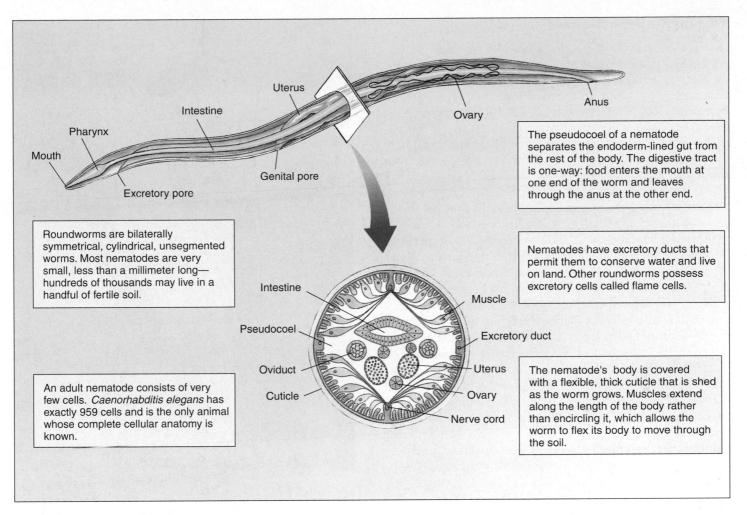

Mouth
Pharynx
Intestine
Uterus
Ovary
Anus
Excretory pore
Genital pore

The pseudocoel of a nematode separates the endoderm-lined gut from the rest of the body. The digestive tract is one-way: food enters the mouth at one end of the worm and leaves through the anus at the other end.

Roundworms are bilaterally symmetrical, cylindrical, unsegmented worms. Most nematodes are very small, less than a millimeter long—hundreds of thousands may live in a handful of fertile soil.

Nematodes have excretory ducts that permit them to conserve water and live on land. Other roundworms possess excretory cells called flame cells.

Intestine
Pseudocoel
Oviduct
Cuticle
Muscle
Excretory duct
Uterus
Ovary
Nerve cord

An adult nematode consists of very few cells. *Caenorhabditis elegans* has exactly 959 cells and is the only animal whose complete cellular anatomy is known.

The nematode's body is covered with a flexible, thick cuticle that is shed as the worm grows. Muscles extend along the length of the body rather than encircling it, which allows the worm to flex its body to move through the soil.

FIGURE 32.18
Phylum Nematoda: Roundworms.   Roundworms such as this nematode possess a body cavity between the gut and the body wall called the pseudocoel. It allows nutrients to circulate throughout the body and prevents organs from being deformed by muscle movements.

Nematode-Caused Diseases

The most serious and common nematode-caused disease in temperate regions is trichinosis, caused by worms of the genus *Trichinella*. These worms live in the small intestine of pigs, where fertilized female worms burrow into the intestinal wall. Once it has penetrated these tissues, each female produces about 1500 live young. The young enter the lymph channels and travel to muscle tissue throughout the body, where they mature and form highly resistant, calcified cysts. Infection in human beings or other animals arises from eating undercooked or raw pork in which the cysts of *Trichinella* are present. If the worms are abundant, a fatal infection can result, but such infections are rare; in the United States, only about 20 deaths have been attributed to trichinosis during the past decade.

Pinworms, *Enterobius*, are abundant throughout the United States, where it is estimated they infect about 30% of all children and about 16% of adults. Adult pinworms live in the human large intestine. Fortunately, the symptoms they cause are not severe, and the worms can easily be controlled by drugs.

The intestinal roundworm, *Ascaris*, infects approximately one of six people worldwide but is rare in areas with modern plumbing. Like pinworms, these worms live in the intestine. Their fertilized eggs are spread in feces, and can remain viable for years in the soil. Adult females, which are up to 30 centimeters long, contain up to 30 million eggs, and can lay up to 20,000 of them each day!

Other nematode-caused diseases are extremely serious in the tropics. *Filaria* infects at least 250 million people worldwide. Up to 10 centimeters long, they live in the lymphatic system, which they may seriously obstruct, causing severe inflammation and swelling. Extreme filariasis produces the condition known as elephantiasis.

Phylum Rotifera: Rotifers

Rotifers (phylum Rotifera) are bilaterally symmetrical, unsegmented pseudocoelomates. Although they are pseudocoelomates, rotifers are very unlike nematodes. Several features suggest their ancestors may have resembled flatworms. Rotifers are very small—at 50–500 micrometers long, they are smaller than some ciliate protists. But rotifers have complex bodies with three cell layers and highly developed internal organs. A complete gut passes from mouth to anus. An extensive pseudocoel acts as a hydroskeleton to provide body rigidity.

Rotifers are basically aquatic animals that propel themselves through the water by rapidly beating cilia, like a boat with oars. There are about 1800 named species of this phylum. While a few rotifers live in soil or in the capillary water in cushions of mosses, most occur in fresh water, and they are common everywhere. Most live no longer than 1 or 2 weeks, but some species can survive in a desiccated inactive state on the leaves of plants; when rain falls, the rotifers regain activity and actively feed in the film of water that temporarily covers the leaf. Very few rotifers are marine.

Rotifers have a well-developed food-processing apparatus. A conspicuous organ on the tip of the head called the corona gathers food (figure 32.19). It is composed of a circle of cilia that sweeps food into the rotifer's mouth and is also used for locomotion. Rotifers are often called "wheel animals" because the cilia, when beating together, resemble the movement of spokes radiating from a wheel.

A Relatively New Phylum: Cycliophora

In December 1995, two Danish biologists reported the discovery of a strange new kind of creature, smaller than a period on a printed page. The tiny organism had a striking circular mouth surrounded by a ring of fine, hairlike cilia, and its life cycle is so unusual that they assigned it to an entirely new phylum, Cycliophora (Greek, "carrying a small wheel") (figure 32.20). There are 35 traditional animal phyla. A 36th, the loricifera, was discovered in 1983. Cycliophora is the 37th. Cycliophorans live on the mouthparts of lobsters. When the lobster to which it is attached starts to molt, the tiny symbiont begins a bizarre form of sexual reproduction. Dwarf males emerge, composed of nothing but brains and reproductive organs. Each dwarf male seeks out another female symbiont on the molting lobster and fertilizes its eggs, generating free-swimming individuals that can seek out another lobster and renew the life cycle.

All pseudocoelomates have fluid-filled body cavities that separate endoderm from mesoderm. Nematodes are bilaterally symmetrical, unsegmented worms. Other pseudocoelomates have very different body plans. Rotifers, while tiny, are very complex, as are cycliophorans.

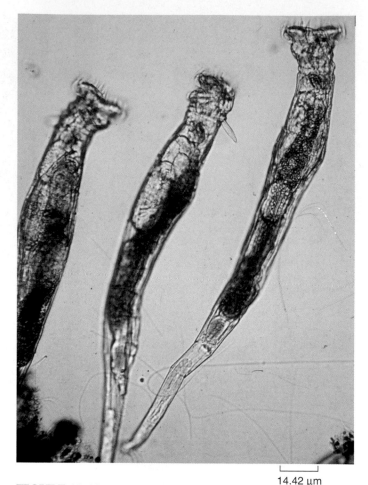

FIGURE 32.19
A rotifer. Microscopic in size, rotifers are smaller than some ciliate protists, and yet have complex internal organs.

14.42 µm

FIGURE 32.20
Cycliophorans. Tinier than the period at the end of this sentence, these pseudocoelomates live on the mouthparts of lobsters. Many individuals of the species *Symbion pandora* are shown attached to a cilia-covered mouthpart of a lobster (50×).

For interactive testing, visit the Online Learning Center with PowerWeb at www.mhhe.com/Raven7

32.1 The classification of invertebrates is currently being reevaluated.

An Uproar Over Invertebrate Phylogeny

- Great disagreement exists as to how the 35 animal phyla are related. (p. 634)
- Animal family trees have historically been based on lumping key aspects of body architecture, and more recently by focusing on differences in ribosomal RNA sequences. (p. 634)
- In the traditional animal family tree, bilaterally symmetrical animals are divided into acoelomates, pseudocoelomates, and coelomates. Phylogenies based on rRNA sequences recognize two clades: lophotrochozoans and ecdysozoans. (pp. 634–635)

32.2 The simplest animals are not bilaterally symmetrical.

Parazoa

- Sponges are multicellular animals that lack tissue-level development and body symmetry, and uniquely contain choanocytes with flagella, which line the internal cavity of the sponge. The beating of these flagella draws water in and through the sponge, thus bringing in food and oxygen and expelling wastes. (p. 636)

Radiata

- Distinct tissues are seen in the eumetazoans. Different cell layers in the embryo form different tissues and give rise to the basic body plan. (p. 638)
- Cnidarians are gelatinous organisms made of distinct tissues but lacking true organs. They have two basic body forms— polyps and medusae—and contain specialized cells called cnidocytes on their tentacles. (p. 638)
- There are four basic classes of cnidarians: Hydrozoa, Scyphozoa, Cubozoa, and Anthozoa. (pp. 640–641)
- Ctenophores propel themselves through the water with fused cilia. (p. 641)

32.3 Acoelomates are solid worms that lack a body cavity.

The Bilaterian Acoelomates

- Bilateria are characterized by bilateral symmetry, allowing high levels of specialization. (p. 642)
- Phylum Platyhelminthes, the flatworms, contains class Turbellaria, (Turbellarians), class Trematoda (flukes), and class Cestoda (tapeworms). (pp. 642–645)
- Phylum Nemertea contains the ribbon worms. (p. 645)

32.4 Pseudocoelomates have a simple body cavity.

The Pseudocoelomates

- In the pseudocoelomates, circulation occurs in a pseudocoel, not in a defined circulatory system. (p. 646)
- Phylum Nematoda contains roundworms, which are bilaterally symmetrical and unsegmented. (p. 646)
- Many diseases are caused by nematodes, such as *Trichinella, Enterobius, Ascaris,* and *Filaria.* (p. 647)
- Phylum Rotifera is composed of aquatic animals that propel themselves through the water by beating cilia. (p. 648)
- Phylum Cycliophora is composed of tiny organisms that are so unusual they were assigned a new phylum. (p. 648)

Self Test

1. The traditional animal family tree places more significance on body plan (the presence of a coelom), while the rRNA tree places more significance on
   a. embryological development patterns.
   b. whether or not the animal molts.
   c. symmetry.
   d. all of these.
2. A key evolutionary development seen for the first time in the sponges is
   a. a complete digestive system.
   b. tissues.
   c. body symmetry.
   d. multicellularity.
3. All of the following are found in sponges except
   a. spicules.
   b. choanocytes.
   c. a digestive tract.
   d. sexual and/or asexual reproduction.
4. The first animal group to show extracellular digestion was the
   a. sponges.
   b. cnidarians.
   c. flatworms.
   d. roundworms.
5. Cnidarians project a nematocyst to capture their prey by
   a. building up a high internal osmotic pressure.
   b. ejecting it with a jet of water.
   c. using a springlike apparatus.
   d. muscle contractions that "throw" the nematocyst.
6. Which of the following is an example of an organism with the medusa body form?
   a. a hydra
   b. a coral
   c. an anemone
   d. a jellyfish
7. Key evolutionary advances of the flatworms are bilateral symmetry and
   a. a coelom.
   b. internal organs.
   c. a one-way digestive tract.
   d. a body cavity.
8. For excretion, flatworms use
   a. a miracidium.
   b. osmosis.
   c. flame cells.
   d. proglottids.
9. The type of body cavity seen in the roundworms is called a(n)
   a. coelom.
   b. acoelom.
   c. pseudocoel.
   d. gastrovascular cavity.
10. The type of pseudocoelomates found in soil, freshwater and marine environments, and as parasites, are
    a. nematodes.
    b. *Trichinella*.
    c. rotifers.
    d. Cycliophora.

Test Your Visual Understanding

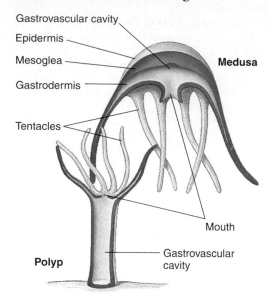

1. What phylum is represented by the two organisms pictured in this figure, and what are some general characteristics of this phylum?
2. What type of organism spends most of its life cycle in the medusa body form? What type of organism spends most of its life cycle in the polyp body form?

Apply Your Knowledge

1. A sponge is a filter feeder, which means it eats by filtering food from the water that passes through its pores and out its osculum. Assume that a sponge filters 1.8 milliliters of water per second. How much water is filtered in an hour? In a day?
2. Coral reefs are often found in nutrient-poor waters. What type of symbiotic relationship helps these coral animals grow actively? What are the advantages for each member in the relationship? What would happen to the coral reefs if the water they are growing in became cloudy with pollution?
3. Parasites, especially those that require two or more hosts to complete their life cycles, often produce very large numbers of offspring. What advantage would this present to them?

33

Coelomate Invertebrates

Concept Outline

33.1 Mollusks were among the first coelomates.

Mollusk Body Plan.   The mollusk body plan is characterized by three distinct sections.

Major Classes of Mollusks.   The three major classes of mollusks are gastropods, bivalves, and cephalopods.

33.2 Annelids were the first segmented animals.

Segmented Bodies.   Annelids are segmented coelomate worms, most of which live in the sea.

Major Classes of Annelids.   The three major classes of annelids are the polychaetes (marine worms), the oligochaetes (earthworms and related freshwater worms), and the hirudines (leeches).

33.3 Lophophorates appear to be a transitional group.

Lophophorates.   The three phyla of lophophorates share a unique ciliated feeding structure.

33.4 Arthropods are the most diverse of all animal groups.

Arthropod Body Plan.   Arthropods have segmented bodies and open circulatory systems.

A Major Group of Arthropods: Crustaceans.   Crustaceans are a large, diverse, primarily marine subphylum of arthropods.

Major Classes of Arthropods: Arachnids.   Spiders have eight legs.

Major Classes of Arthropods: Centipedes and Millipedes.   Centipedes and millipedes have 30+ legs.

Major Classes of Arthropods: Insects.   Insects, the most diverse class of animals, are the only arthropods that can fly.

33.5 Echinoderms are radially symmetrical as adults.

Deuterostome Development and an Endoskeleton.   All echinoderms are marine deuterostomes with an endoskeleton.

Echinoderm Body Plan.   Echinoderms are radially symmetrical as adults.

Major Classes of Echinoderms.   All six classes of echinoderms exhibit a five-part body plan.

FIGURE 33.1
Coelomate invertebrates have evolved a very diverse array of body plans.   This insect (*Polistes*, the common paper wasp) has a segmented body, jointed appendages, and an exoskeleton.

Although acoelomates and pseudocoelomates have proven very successful, a third way of organizing the animal body has also evolved, one that occurs in the bulk of the animal kingdom. We will begin our discussion of the coelomate invertebrate animals with mollusks, which include such animals as clams, snails, slugs, and octopuses. Annelids, such as earthworms, leeches, and seaworms, are also coelomates, but in addition, were the earliest group of animals to evolve segmented bodies. Arthropods, such as the paper wasp in figure 33.1, evolved jointed appendages, and have become the most successful of all animal groups. Echinoderms are exclusively marine animals that exhibit deuterostome development and endoskeletons, two evolutionary innovations that they share with the chordates, which are the subject of chapter 34.

33.1 Mollusks were among the first coelomates.

Mollusk Body Plan

The evolution of the *coelom* was a significant advance in the structure of the animal body. Coelomates have a new body design that repositions the fluid and allows complex tissues and organs to develop. This new body plan also made it possible for animals to evolve a wide variety of different body architectures and to grow to much larger sizes than acoelomate animals. Among the earliest groups of coelomates were the mollusks and the annelids.

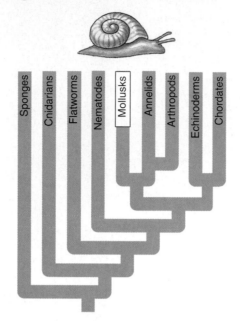

Mollusks

Mollusks (phylum Mollusca) are an extremely diverse animal phylum, second only to the arthropods, with over 110,000 described species. Mollusks exhibit a variety of body forms and live in many different environments. Mollusks include snails, slugs, clams, scallops, oysters, cuttlefish, octopuses, and many other familiar animals (figure 33.2). The durable shells of some mollusks are often beautiful and elegant; they have long been favorite objects for professional scientists and amateurs alike to collect, preserve, and study. Chitons and nudibranchs are less familiar marine mollusks. Mollusks are characterized by a coelom, and while there is extraordinary diversity in this phylum, many of the basic components of the mollusk body plan can be seen in figure 33.3.

Mollusks evolved in the oceans, and most groups have remained there. Marine mollusks are widespread and often abundant. Some groups of mollusks that have invaded freshwater and terrestrial habitats include the snails and slugs that live in your garden. Terrestrial mollusks are often abundant in places that are at least seasonally moist. Some of these places, such as the crevices of desert rocks, may appear very dry, but even these habitats have at least a temporary supply of water at certain times.

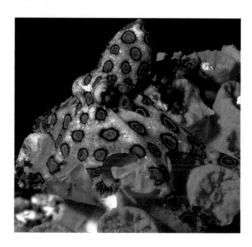

FIGURE 33.2
A mollusk. The blue-ringed octopus is one of the few mollusks dangerous to humans. Strikingly beautiful, it is equipped with a sharp beak and poison glands—divers give it a wide berth!

As a group, mollusks are an important source of food for humans. Oysters, clams, scallops, mussels, octopuses, and squids are among the culinary delicacies that belong to this large phylum. Mollusks are also economically significant in many other ways. For example, pearls are produced in oysters, and the material called mother-of-pearl, often used in jewelry and other decorative objects, is produced in the shells of a number of different mollusks, but most notably in the snail called abalone. Mollusks are not wholly beneficial to humans, however. Bivalve mollusks called shipworms burrow through wood submerged in the sea, damaging boats, docks, and pilings. The zebra mussel has recently invaded North American ecosystems via the ballast water of cargo ships from Europe, wreaking havoc in many aquatic ecosystems. Slugs and terrestrial snails often cause extensive damage to garden flowers, vegetables, and crops. Other mollusks serve as hosts to the intermediate stages of many serious parasites, including several nematodes and flatworms, discussed in chapter 32.

Mollusks range in size from almost microscopic to huge. Although most measure a few centimeters in their largest dimension, the giant squid, which is occasionally cast ashore but has rarely been observed in its natural environment, may grow up to 21 meters long! Weighing up to 250 kilograms, the giant squid is the largest invertebrate and one of the heaviest. Millions of giant squid probably inhabit the deep regions of the ocean, even though they are seldom caught. Another large mollusk is the bivalve *Tridacna maxima*, the giant clam, which may be as long as 1.5 meters and may weigh as much as 270 kilograms (figure 33.4).

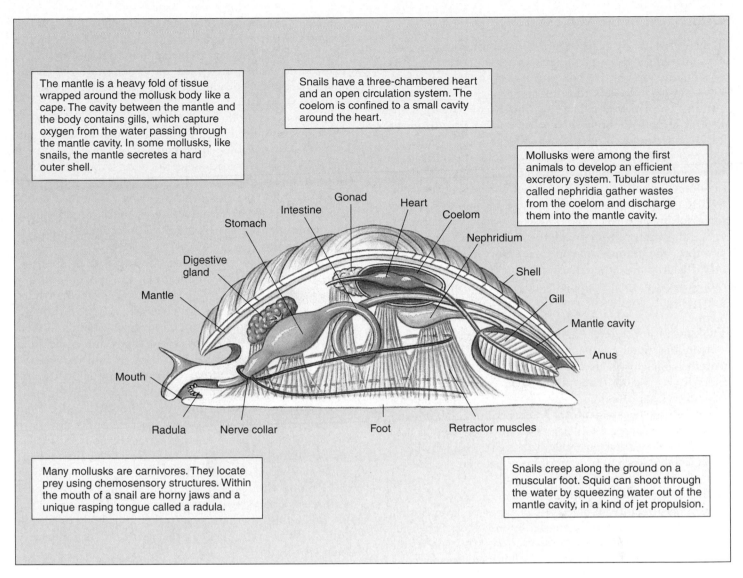

The mantle is a heavy fold of tissue wrapped around the mollusk body like a cape. The cavity between the mantle and the body contains gills, which capture oxygen from the water passing through the mantle cavity. In some mollusks, like snails, the mantle secretes a hard outer shell.

Snails have a three-chambered heart and an open circulation system. The coelom is confined to a small cavity around the heart.

Mollusks were among the first animals to develop an efficient excretory system. Tubular structures called nephridia gather wastes from the coelom and discharge them into the mantle cavity.

Many mollusks are carnivores. They locate prey using chemosensory structures. Within the mouth of a snail are horny jaws and a unique rasping tongue called a radula.

Snails creep along the ground on a muscular foot. Squid can shoot through the water by squeezing water out of the mantle cavity, in a kind of jet propulsion.

Labels: Stomach, Intestine, Gonad, Heart, Coelom, Digestive gland, Nephridium, Mantle, Shell, Gill, Mantle cavity, Anus, Mouth, Radula, Nerve collar, Foot, Retractor muscles

FIGURE 33.3
Phylum Mollusca: mollusks. As shown in this generalized molluscan body plan, the body cavity of a mollusk is a coelom, which is completely enclosed within the mesoderm. This allows physical contact between the mesoderm and the endoderm, permitting interactions that lead to development of highly specialized organs such as a stomach.

FIGURE 33.4
Giant clam. Second only to the arthropods in number of described species, members of the phylum Mollusca occupy almost every habitat on earth. This giant clam, *Tridacna maxima*, has a green color caused by the presence of symbiotic dinoflagellates (zooxanthellae). Through photosynthesis, the dinoflagellates probably contribute most of the food supply of the clam, although it remains a filter feeder like most bivalves. Some individual giant clams may be nearly 1.5 meters long and weigh up to 270 kilograms.

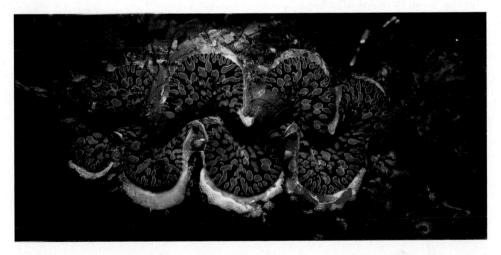

Architecture of the Mollusk Body

In their basic body plan (figure 33.5), mollusks have distinct bilateral symmetry. Their digestive, excretory, and reproductive organs are concentrated in a **visceral mass,** and a muscular **foot** is their primary mechanism of locomotion. They may also have a differentiated **head** at the anterior end of the body. Folds (often two) arise from the dorsal body wall and enclose a cavity between themselves and the visceral mass; these folds constitute the **mantle.** In some mollusks, the mantle cavity acts as a lung; in others, it contains gills. **Gills** are specialized portions of the mantle that usually consist of a system of filamentous projections rich in blood vessels. These projections greatly increase the surface area available for gas exchange and, therefore, the animal's overall respiratory potential. Mollusk gills are very efficient, and many gilled mollusks extract 50% or more of the dissolved oxygen from the water that passes through the mantle cavity. Finally, in most members of this phylum, the outer surface of the mantle also secretes a protective shell.

A mollusk shell consists of a horny outer layer, rich in protein, which protects the two underlying calcium-rich layers from erosion. The middle layer consists of densely packed crystals of calcium carbonate. The inner layer is pearly and increases in thickness throughout the animal's life. When it reaches a sufficient thickness, this layer is used as mother-of-pearl. Pearls themselves are formed when a foreign object, such as a grain of sand, becomes lodged between the mantle and the inner shell layer of **bivalve mollusks** (two-shelled), including clams and oysters. The mantle coats the foreign object with layer upon layer of shell material to reduce irritation caused by the object. The shell of mollusks serves primarily for protection. Many species can withdraw for protection into their shell if they have one.

In aquatic mollusks, a continuous stream of water passes into and out of the mantle cavity, drawn by the cilia on the gills. This water brings in oxygen and, in the case of the bivalves, food; it also carries out waste materials. When the gametes are being produced, they are frequently carried out in the same stream.

The foot of a mollusk is muscular and may be adapted for locomotion, attachment, food capture (in squids and octopuses), or various combinations of these functions. Some mollusks secrete mucus, forming a path that they glide along on their foot. In cephalopods—squids and octopuses–the foot is divided into arms, also called tentacles. In some *pelagic* forms, mollusks that are perpetually free-swimming, the foot is modified into winglike projections or thin fins.

One of the most characteristic features of all the mollusks except the bivalves is the **radula,** a rasping, tonguelike organ used for feeding. The radula consists primarily of dozens to thousands of microscopic, chitinous teeth arranged in rows (figure 33.6). Gastropods (snails and their relatives) use their radula to scrape algae and other food materials off their substrates and then to convey this food to the digestive tract. Other gastropods are active predators, some using a modified radula to drill through the shells of prey and extract the food. The small holes often seen in oyster shells are produced by gastropods that have bored holes to kill the oyster and extract its body for food.

The circulatory system of all mollusks except cephalopods consists of a heart and an open system in which blood circulates freely. The mollusk heart usually has three chambers, two that collect aerated blood from the gills, and a third that pumps it to the other body tissues. In mollusks, the coelom takes the form of a small cavity around the heart.

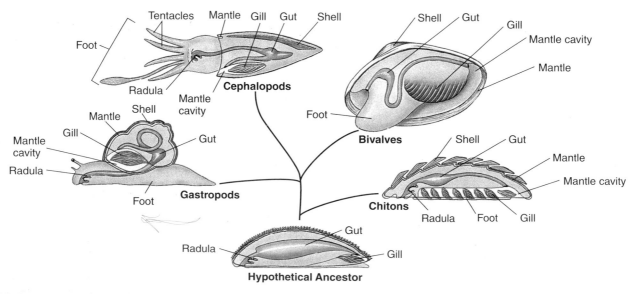

FIGURE 33.5
Body plans among the mollusks.

Nitrogenous wastes are removed from the mollusk by one or two tubular structures called **nephridia.** A typical nephridium has an open funnel, the **nephrostome,** which is lined with cilia. A coiled tubule runs from the nephrostome into a bladder, which in turn connects to an excretory pore. Wastes are gathered by the nephridia from the coelom and discharged into the mantle cavity. The wastes are then expelled from the mantle cavity by the continuous pumping of the gills. Sugars, salts, water, and other materials are reabsorbed by the walls of the nephridia and returned to the animal's body as needed to achieve an appropriate osmotic balance.

In animals with a closed circulatory system, such as annelids, cephalopod mollusks, and vertebrates, the coiled tubule of a nephridium is surrounded by a network of capillaries. Wastes are extracted from the circulatory system through these capillaries, transferred into the nephridium, and then subsequently discharged. Salts, water, and other associated materials may also be reabsorbed from the tubule of the nephridium back into the capillaries. For this reason, the excretory systems of these coelomates are much more efficient than the flame cells of the acoelomates, which pick up substances only from the body fluids. Mollusks were one of the earliest evolutionary lines to develop an efficient excretory system. Other than chordates, coelomates with closed circulation have similar excretory systems.

Reproduction in Mollusks

Most mollusks have distinct male and female individuals, although a few bivalves and many gastropods are hermaphroditic. Even in hermaphroditic mollusks, cross-fertilization is most common. Remarkably, some sea slugs and oysters are able to change from one sex to the other several times during a single season.

Most aquatic mollusks engage in external fertilization. The males and females release their gametes into the water, where they mix and fertilization occurs. However, gastropods more often have internal fertilization, with the male inserting sperm directly into the female's body. Internal fertilization is one of the key adaptations that allowed gastropods to colonize the land.

Many marine mollusks have free-swimming larvae called **trochophores** (figure 33.7a), which closely resemble the larval stage of many marine annelids. Trochophores swim by means of a row of cilia that encircles the middle of their body. In most marine snails and in bivalves, a second free-swimming stage, the veliger, follows the trochophore stage. This **veliger** stage has the beginnings of a foot, shell, and mantle (figure 33.7b). Trochophores and veligers drift widely in the ocean currents, dispersing mollusks to new areas.

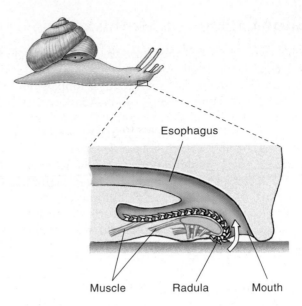

FIGURE 33.6
Structure of the radula in a snail. The radula consists of chitin and is covered with rows of teeth that extend backward. As the animal feeds, its mouth opens, and the radula brings food in by scraping backward.

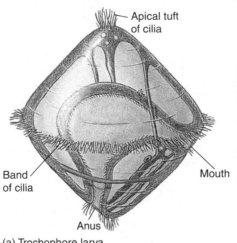

(a) Trochophore larva

(b) Veliger stage

FIGURE 33.7
Stages in the molluscan life cycle. (a) The trochophore larva of a mollusk. Similar larvae, as you will see, are characteristic of some annelid worms as well as a few other phyla. (b) The veliger stage of a mollusk.

Mollusks were among the earliest animals to evolve an efficient excretory system. The mantle of mollusks not only secretes their protective shell, but also forms a cavity that is essential to respiration.

Major Classes of Mollusks

Of the seven classes of mollusks, we will examine four as representatives of the phylum: (1) Polyplacophora—chitons; (2) Gastropoda—snails, slugs, limpets, and their relatives; (3) Bivalvia—clams, oysters, scallops, and their relatives; and (4) Cephalopoda—squids, octopuses, cuttlefishes, and nautilus. By studying living mollusks and the fossil record, some scientists have deduced that the ancestral mollusk was probably a dorsoventrally flattened, unsegmented, wormlike animal that glided on its ventral surface. This animal may also have had a chitinous cuticle and overlapping calcareous scales. Other scientists believe that mollusks arose from segmented ancestors and became unsegmented secondarily.

Class Polyplacophora: The Chitons

Chitons are marine mollusks that have oval bodies with eight overlapping calcareous plates. Underneath the plates, the body is not segmented. Chitons creep along using a broad, flat foot surrounded by a groove or mantle cavity in which the gills are arranged. Most chitons are grazing herbivores that live in shallow marine habitats, but some live at depths of more than 7000 meters.

Class Gastropoda: The Snails and Slugs

The class Gastropoda contains about 40,000 described species of snails, slugs, and similar animals. This class is primarily a marine group, but it also contains many freshwater and terrestrial mollusks (figure 33.8). Most gastropods have a shell, but some, such as slugs and nudibranchs, have lost their shells through the course of evolution. Gastropods generally creep along on a foot, which may be modified for swimming.

The heads of most gastropods have a pair of tentacles with eyes at the ends. These tentacles have been lost in some of the more advanced forms of the class. Within the mouth cavity of many members of this class are horny jaws and a radula.

During embryological development, gastropods undergo torsion. **Torsion** is the process by which the mantle cavity and anus are moved from a posterior location to the front of the body, where the mouth is located. Torsion is brought about by a disproportionate growth of the lateral muscles; that is, one side of the larva grows much more rapidly than the other. A 120° rotation of the visceral mass brings the mantle cavity above the head and twists many internal structures. In some groups of gastropods, varying degrees of detorsion have taken place. The **coiling**, or spiral winding, of the shell is a separate process. This process has led to the loss of the right gill and right nephridium in most gastropods. Thus, the visceral mass of gastropods has become bilaterally asymmetrical during the course of evolution.

FIGURE 33.8
A gastropod mollusk. The terrestrial snail *Allogona townsendiana*.

Gastropods display extremely varied feeding habits. Some are predatory, others scrape algae off rocks (or aquarium glass), and others are scavengers. Many are herbivores, and some terrestrial ones are serious garden and agricultural pests. The radula of oyster drills is used to bore holes in the shells of other mollusks, through which the contents of the prey can be removed. In cone shells, the radula has been modified into a kind of poisonous harpoon, which is shot with great speed into the prey.

Sea slugs, or nudibranchs, are active predators; a few species of nudibranchs have the extraordinary ability to extract the nematocysts from the cnidarian polyps they eat, transfer them through their digestive tract to the surface of their gills intact, and use them for their own protection. Nudibranchs get their name from their gills, which instead of being enclosed within the mantle cavity are exposed along the dorsal surface (*nudi*, "naked," + *branch*, "gill").

In terrestrial gastropods, the empty mantle cavity, which was occupied by gills in their aquatic ancestors, is extremely rich in blood vessels and serves as a lung, in effect. This structure evolved in animals living in environments with plentiful oxygen; it absorbs oxygen from the air much more effectively than a gill could, but is not as effective under water.

Class Bivalvia: The Bivalves

Members of the class Bivalvia include the clams, scallops, mussels, and oysters. Bivalves have two lateral (left and right) shells (valves) hinged together dorsally (figure 33.9). A ligament hinges the shells together and causes them to gape open. Pulling against this ligament are one or two large adductor muscles that can draw the shells together.

The mantle secretes the shells and ligament and envelops the internal organs within the pair of shells. The mantle is frequently drawn out to form two siphons, one for an incoming and one for an outgoing stream of water. The siphons often function as snorkels to allow bivalves to filter water through their body while remaining almost completely buried in sediments. A complex folded gill lies on each side of the visceral mass. These gills consist of pairs of filaments that contain many blood vessels. Rhythmic beating of cilia on the gills creates a pattern of water circulation. Most bivalves are sessile filter feeders. They extract small organisms from the water that passes through their mantle cavity.

Bivalves do not have distinct heads or radulas, differing from gastropods in this respect (see figure 33.5). However, most have a wedge-shaped foot that may be adapted, in different species, for creeping, burrowing, cleansing the animal, or anchoring it in its burrow. Some species of clams can dig into sand or mud very rapidly by means of muscular contractions of their foot.

Bivalves disperse from place to place largely as larvae. While most adults are adapted to a burrowing way of life, some genera of scallops can move swiftly through the water by using their large adductor muscles to clap their shells together. These muscles are what we usually eat as "scallops." The edge of a scallop's body is lined with tentacle-like projections tipped with complex eyes.

There are about 10,000 species of bivalves. Most species are marine, although many also live in fresh water. Over 500 species of pearly freshwater mussels, or naiads, occur in the rivers and lakes of North America.

Class Cephalopoda: The Octopuses, Squids, and Nautiluses

The more than 600 species of the class Cephalopoda—octopuses, squids, and nautiluses—are the most intelligent of the invertebrates. They are active marine predators that swim, often swiftly, and compete successfully with fish. The foot has evolved into a series of tentacles equipped with suction cups, adhesive structures, or hooks that seize prey efficiently. Squids have 10 tentacles; octopuses, as indicated by their name, have eight (figure 33.10); and the nautilus has about 80 to 90. After snaring prey with its tentacles, a cephalopod bites the prey with its strong, beaklike paired jaws and then pulls it into its mouth by the tonguelike action of the radula.

Cephalopods have highly developed nervous systems, and their brains are unique among mollusks. Their eyes are very elaborate, and have a structure much like that of vertebrate eyes, although they evolved separately (see chapter 46). Many cephalopods exhibit complex patterns of behavior and a high level of intelligence; octopuses can be easily trained to distinguish among classes of objects. Most members of this class have closed circulatory systems and are the only mollusks that do.

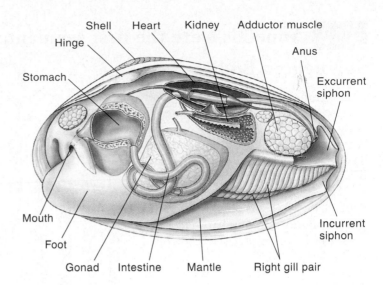

FIGURE 33.9
Diagram of a clam.   Internal organs and the foot are shown. The left shell and mantle are removed. Bivalves such as this clam circulate water through their gills and filter out food particles.

FIGURE 33.10
A cephalopod.   Pacific giant octopus (*Octopus defleini*).

Although they evolved from shelled ancestors, living cephalopods, except for the few species of nautilus, lack an external shell. Like other mollusks, cephalopods take water into the mantle cavity and expel it through a siphon. Cephalopods have modified this system into a means of jet propulsion. When threatened, they eject water violently and shoot themselves through the water.

Most octopuses and squids are capable of changing color to suit their background or display messages to one another. They accomplish this feat using their *chromatophores*, pouches of pigments embedded in the epithelium.

Gastropods typically live in a hard shell. Bivalves have hinged shells but do not have a distinct head area. Cephalopods possess well-developed brains and are the most intelligent invertebrates.

33.2 Annelids were the first segmented animals.

Segmented Bodies

A key transition in the animal body plan was *segmentation*, the building of a body from a series of similar segments. The first segmented animals to evolve were most likely **annelid worms**, phylum Annelida (figure 33.11). One advantage of having a body built from repeated units (segments) is that the development and function of these units can be more precisely controlled, at the level of individual segments or groups of segments. For example, different segments may possess different combinations of organs or perform different functions relating to reproduction, feeding, locomotion, respiration, or excretion.

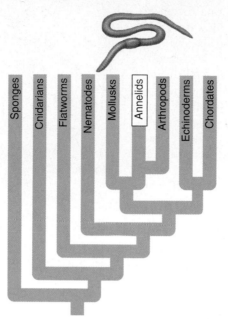

Sponges | Cnidarians | Flatworms | Nematodes | Mollusks | Annelids | Arthropods | Echinoderms | Chordates

FIGURE 33.11
A polychaete annelid. *Nereis virens* is a wide-ranging, predatory, marine polychaete worm equipped with feathery parapodia for movement and respiration, as well as jaws for hunting. You may have purchased *Nereis* as fishing bait!

Annelids

Two-thirds of all annelids live in the sea (about 8000 species), and most of the rest, some 3100 species, are earthworms. Annelids are characterized by three principal features:

1. **Repeated segments.** The body of an annelid worm is composed of a series of ringlike segments running the length of the body, looking like a stack of doughnuts or a roll of coins (figure 33.12). Internally, the segments are divided from one another by partitions called **septa,** just as bulkheads separate the segments of a submarine. In each of the cylindrical segments, the excretory and locomotor organs are repeated. The fluid within the coelom of each segment creates a hydrostatic (liquid-supported) skeleton that gives the segment rigidity, like an inflated balloon. Muscles within each segment push against the fluid in the coelom. Because each segment is separate, each can expand or contract independently. This lets the worm move in complex ways.
2. **Specialized segments.** The anterior (front) segments of annelids have become modified to contain specialized sensory organs. Some are sensitive to light, and elaborate eyes with lenses and retinas have evolved in some annelids. A well-developed cerebral ganglion, or brain, is contained in one anterior segment.
3. **Connections.** Although partitions separate the segments, materials and information do pass between segments. Annelids have a closed circulatory system that carries blood from one segment to another. A ventral nerve cord connects the nerve centers or ganglia in each segment with one another and with the brain. These neural connections are critical features that allow the worm to function and behave as a unified and coordinated organism.

Body Plan of the Annelids

The basic annelid body plan is a tube within a tube, with the internal digestive tract—a tube running from mouth to anus—suspended within the coelom. The tube that makes up the digestive tract has several portions—the pharynx, esophagus, crop, gizzard, and intestine—that are specialized for different functions.

Annelids make use of their hydrostatic skeleton for locomotion. To move, annelids contract circular muscles running around each segment. Doing so squeezes the segment, causing the coelomic fluid to squirt outward, like a tube of toothpaste. Because the fluid is trapped in the segment by the septa, instead of escaping like toothpaste, the fluid causes the segment to elongate and get much thinner. By then contracting longitudinal muscles that run along the length of the worm, the segment is returned to its original shape. In most annelid groups, each segment typically possesses **setae,** bristles of chitin that help anchor the worms during locomotion. By extending the setae in some segments so that they anchor in the substrate and retracting them in other segments, the worm can squirt its body, section by section, in either direction.

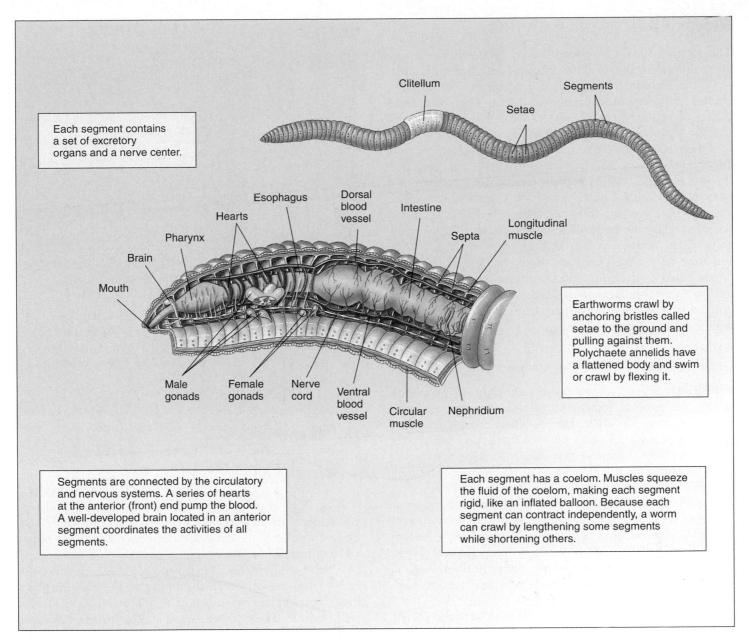

Each segment contains a set of excretory organs and a nerve center.

Clitellum

Setae

Segments

Esophagus

Dorsal blood vessel

Intestine

Hearts

Pharynx

Septa

Longitudinal muscle

Brain

Mouth

Earthworms crawl by anchoring bristles called setae to the ground and pulling against them. Polychaete annelids have a flattened body and swim or crawl by flexing it.

Male gonads

Female gonads

Nerve cord

Ventral blood vessel

Circular muscle

Nephridium

Segments are connected by the circulatory and nervous systems. A series of hearts at the anterior (front) end pump the blood. A well-developed brain located in an anterior segment coordinates the activities of all segments.

Each segment has a coelom. Muscles squeeze the fluid of the coelom, making each segment rigid, like an inflated balloon. Because each segment can contract independently, a worm can crawl by lengthening some segments while shortening others.

FIGURE 33.12
Phylum Annelida: annelids. Marine polychaetes and earthworms (phylum Annelida) were most likely the first organisms to evolve a body plan based on partly repeated body segments. Segments are separated internally from each other by septa.

Unlike the arthropods and most mollusks, most annelids have a closed circulatory system. Annelids exchange oxygen and carbon dioxide with the environment through their body surfaces; most lack gills or lungs. However, much of their oxygen supply reaches the different parts of their bodies through their blood vessels. Some of these vessels at the anterior end of the worm body are enlarged and heavily muscular, serving as hearts that pump the blood. Earthworms have five pulsating blood vessels on each side that serve as hearts, helping to pump blood from the main dorsal vessel, which is their major pumping structure, to the main ventral vessel.

The excretory system of annelids consists of ciliated, funnel-shaped nephridia generally similar to those of mollusks. These nephridia—each segment has a pair—collect waste products and transport them out of the body through the coelom by way of specialized excretory tubes.

Annelids are a diverse group of coelomate animals characterized by serial segmentation. Each segment in the annelid body has its own excretory and locomotor elements, and is connected to other segments by a common circulatory and neural system.

Major Classes of Annelids

The roughly 12,000 described species of annelids occur in many different habitats. They range in length from as little as 0.5 millimeter to more than 3 meters in some polychaetes and giant Australian earthworms. There are three classes of annelids: (1) Polychaeta, free-living, almost entirely marine bristleworms, comprising some 8000 species; (2) Oligochaeta, terrestrial earthworms and related marine and freshwater worms, with some 3100 species; and (3) Hirudinea, leeches, mainly freshwater predators or bloodsuckers, with about 500 species. The annelids are believed to have evolved in the sea, with polychaetes the most primitive class. Oligochaetes seem to have evolved from polychaetes, perhaps by way of brackish water to estuaries and then to streams. Leeches share with oligochaetes an organ called a **clitellum,** which secretes a cocoon specialized to receive the eggs. It is generally agreed that leeches evolved from oligochaetes, specializing in their bloodsucking lifestyle as external parasites.

Class Polychaeta: The Polychaetes

Polychaetes (class Polychaeta) include clamworms, plume worms, scaleworms, lugworms, twin-fan worms, sea mice, peacock worms, and many others. These worms are often surprisingly beautiful, with unusual forms and sometimes iridescent colors (figure 33.13). Polychaetes are often a crucial part of marine food chains, because they are extremely abundant in certain habitats.

Some polychaetes live in tubes or permanent burrows of hardened mud, sand, mucuslike secretions, or calcium carbonate. From the tubes in which they live, these sedentary polychaetes project a set of feathery tentacles that sweep the water for food, making them primarily filter feeders. Other polychaetes are active swimmers, crawlers, or burrowers. Many are active predators.

Polychaetes have a well-developed head with specialized sense organs; they differ from other annelids in this respect. Their bodies are often highly organized into distinct regions formed by groups of segments related in function and structure. Their sense organs include eyes, which range from simple eyespots to quite large and conspicuous stalked eyes.

Another distinctive characteristic of polychaetes is the paired, fleshy, paddlelike flaps, called **parapodia,** on most of their segments. These parapodia, which bear bristlelike setae, are used in swimming, burrowing, or crawling. They also play an important role in gas exchange because they greatly increase the surface area of the body. Some polychaetes that live in burrows or tubes may have parapodia featuring hooks to help anchor the worm. Slow crawling is carried out by means of the parapodia. Rapid crawling and swimming are accomplished by undulating the body. In addition, the polychaete epidermis often includes ciliated cells that aid in respiration and food procurement.

FIGURE 33.13
A polychaete. The shiny bristleworm, *Oenone fulgida.*

The sexes of polychaetes are usually separate, and fertilization is often external, occurring in the water and away from both parents. Unlike other annelids, polychaetes usually lack permanent **gonads,** the sex organs that produce gametes. They produce their gametes directly from germ cells in the lining of the coelom or in their septa. Fertilization results in the production of ciliated, mobile trochophore larvae similar to the larvae of mollusks. The trochophores develop for long periods in the plankton before beginning to add segments and thus changing to a juvenile form that more closely resembles the adult form.

Class Oligochaeta: The Earthworms

The body of an earthworm (class Oligochaeta) consists of 100 to 175 similar segments, with a mouth on the first and an anus on the last. Earthworms seem to eat their way through the soil because they suck in organic and other material by expanding their strong pharynx. Everything they ingest passes through their long, straight digestive tract. One region of this tract, the gizzard, grinds up the organic material with the help of soil particles.

The material that passes through an earthworm is deposited outside the opening of its burrow in the form of castings that consist of irregular mounds. In this way, earthworms aerate and enrich the soil. A worm can eat its own weight in soil every day.

In view of the underground lifestyle that earthworms have evolved, it is not surprising that they have no eyes. However, earthworms do have numerous light-, chemo-, and touch-sensitive cells, mostly concentrated in segments near each end of the body—those regions most likely to encounter light or other stimuli. Earthworms have fewer setae than polychaetes and no parapodia or head region.

Earthworms are hermaphroditic, another way in which they differ from most polychaetes. When they mate (figure 33.14), their anterior ends point in opposite directions, and their ventral surfaces touch. The *clitellum* is a thickened band on an earthworm's body; the mucus it secretes holds the worms together during copulation. Sperm cells are released from pores in specialized segments of one partner into the sperm receptacles of the other, the process going in both directions simultaneously.

Two or three days after the worms separate, the clitellum of each worm secretes a mucous cocoon, surrounded by a protective layer of chitin. As this sheath passes over the female pores of the body—a process that takes place as the worm moves—it receives eggs. As it subsequently passes along the body, it incorporates the sperm that were deposited during copulation. Fertilization of the eggs takes place within the cocoon. When the cocoon finally passes over the end of the worm, its ends pinch together. Within the cocoon, the fertilized eggs develop directly into young worms similar to adults.

Class Hirudinea: The Leeches

Leeches (class Hirudinea) occur mostly in fresh water, although a few are marine and some tropical leeches occupy terrestrial habitats. Most leeches are 2 to 6 centimeters long, but one tropical species reaches up to 30 centimeters. Leeches are usually flattened dorsoventrally, like flatworms. They are hermaphroditic, and develop a clitellum during the breeding season; cross-fertilization is obligatory because they are unable to self-fertilize.

A leech's coelom is reduced and continuous throughout the body, not divided into individual segments as in the polychaetes and oligochaetes. Leeches have evolved suckers at one or both ends of the body. Those that have suckers at both ends move by attaching first one and then the other end to the substrate, looping along. Many species are also capable of swimming. Except for one species, leeches have no setae.

Some leeches have evolved the ability to suck blood from animals. Many freshwater leeches live as external parasites. They remain on their hosts for long periods and suck their blood from time to time.

The best-known leech is the medicinal leech, *Hirudo medicinalis* (figure 33.15). Individuals of *Hirudo* are 10 to 12 centimeters long and have bladelike, chitinous jaws

FIGURE 33.14
Earthworms mating. The anterior ends are pointing in opposite directions.

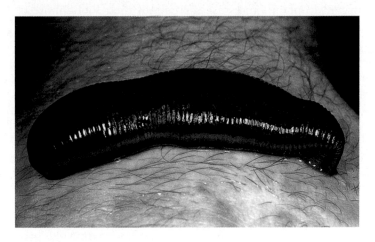

FIGURE 33.15
A leech. *Hirudo medicinalis*, the medicinal leech, is seen here feeding on a human arm. Leeches uses chitinous, bladelike jaws to make an incision to access blood and secrete an anticoagulant to keep the blood from clotting. Both the anticoagulant and the leech itself have made important contributions to modern medicine.

that rasp through the skin of the victim. The leech secretes an anticoagulant into the wound to prevent the blood from clotting as it flows out, and its powerful sucking muscles pump the blood out quickly once the hole has been opened. Leeches were used in medicine for hundreds of years to suck blood out of patients whose diseases were mistakenly believed to be caused by an excess of blood. Today, European pharmaceutical companies still raise and sell leeches, but they are used to remove excess blood after certain surgeries. Following the surgery, blood may accumulate because veins may function improperly and fail to circulate the blood. The accumulating blood "turns off" the arterial supply of fresh blood, and the tissue often dies. When leeches remove the excess blood, new capillaries form in about a week, and the tissues remain healthy.

Segmented annelids evolved in the sea. Earthworms are their descendants, as are parasitic leeches.

33.3 Lophophorates appear to be a transitional group.

Lophophorates

Three phyla of marine animals—Phoronida, Ectoprocta, and Brachiopoda—are characterized by a **lophophore**, a circular or U-shaped ridge around the mouth bearing one or two rows of ciliated, hollow tentacles. Because of this unusual feature, they are thought to be related to one another. The lophophore presumably arose in a common ancestor. The coelomic cavity of lophophorates extends into the lophophore and its tentacles. The lophophore functions as a surface for gas exchange and as a food-collection organ. Lophophorates use the cilia of their lophophore to capture the organic detritus and plankton on which they feed. Lophophorates are attached to their substrate or move slowly.

Lophophorates share some features with mollusks, annelids, and arthropods (all protostomes), and share others with deuterostomes. Cleavage in lophophorates is mostly radial, as in deuterostomes. The formation of the coelom varies; some lophophorates resemble protostomes in this respect, others deuterostomes. In the Phoronida, the mouth forms from the blastopore, while in the other two phyla, it forms from the end of the embryo opposite the blastopore. Molecular evidence shows that the ribosomes of all lophophorates are decidedly protostome-like, lending strength to the argument for placing them within the protostome phyla. Despite the differences among the three phyla, the unique structure of the lophophore seems to indicate that the members share a common ancestor. Their relationships continue to present a fascinating puzzle.

Phylum Phoronida: The Phoronids

Phoronids (phylum Phoronida) superficially resemble the common polychaete tube worms seen on dock pilings, but they have many important differences. Each phoronid secretes a chitinous tube and lives out its life within it (figure 33.16). They also extend tentacles to feed and quickly withdraw them when disturbed, but the resemblance to the tube worm ends there. Instead of a straight, tube-within-a-tube body plan, phoronids have a U-shaped gut. Only about 10 phoronid species are known, ranging in length from a few millimeters to 30 centimeters. Some species lie buried in sand; others are attached to rocks, either singly or in groups. Phoronids develop as protostomes, with radial cleavage and the anus developing secondarily.

Phylum Ectoprocta: The Bryozoans

Ectoprocts (phylum Ectoprocta) look like tiny, short versions of phoronids (figure 33.17). They are small—usually less than 0.5 millimeter long—and live in colonies that look

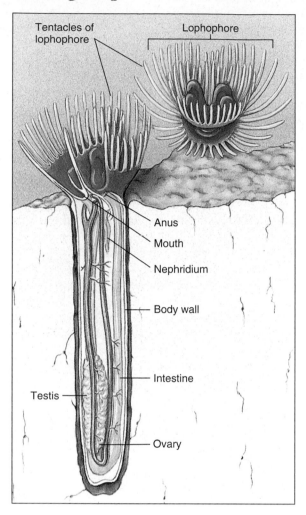

FIGURE 33.16
Phoronids (phylum Phoronida). A phoronid, such as *Phoronis*, lives in a chitinous tube that the animal secretes to form the outer wall of its body. The lophophore consists of two parallel, horseshoe-shaped ridges of tentacles and can be withdrawn into the tube when the animal is disturbed.

like patches of moss on the surfaces of rocks, seaweed, or other submerged objects (in fact, their common name, bryozoans, translates from Greek as "moss-animals"). The name Ectoprocta refers to the location of the anus (proct), which is external to the lophophore. The 4000 species include both marine and freshwater forms—the only nonmarine lophophorates. Individual ectoprocts secrete a tiny chitinous chamber called a **zoecium** that attaches to rocks and to other members of the colony. Individuals communicate chemically through pores between these chambers. Ectoprocts develop as deuterostomes, with the mouth developing secondarily; cleavage is radial.

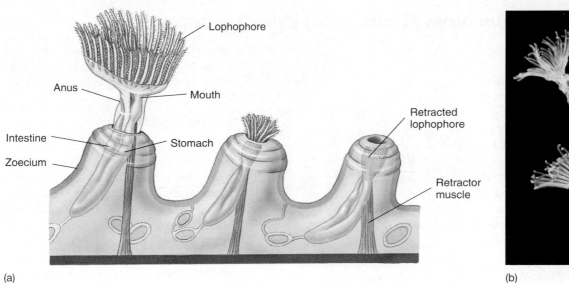

(a)

(b)

FIGURE 33.17

Ectoprocts (phylum Ectoprocta). (*a*) This drawing depicts a small portion of a colony of the freshwater ectoproct *Plumatella*, which grows on the underside of rocks. The individual at the left has a fully extended lophophore, the structure characteristic of the three lophophorate phyla. The tiny individuals of *Plumatella* disappear into their shells when disturbed. (*b*) *Plumatella repens*, a freshwater bryozoan.

Phylum Brachiopoda: The Brachiopods

Brachiopods, or lamp shells, superficially resemble clams, with two calcified shells (figure 33.18). Many species attach to rocks or sand by a stalk that protrudes through an opening in one shell. The lophophore lies within the shell and functions when the brachiopod's shells are opened slightly. Although a little more than 300 species of brachiopods (phylum Brachiopoda) exist today, more than 30,000 species of this phylum are known as fossils. Because brachiopods were common in the earth's oceans for millions of years and because their shells fossilize readily, they are often used as index fossils to define a particular time period or sediment type. Brachiopods develop as deuterostomes and show radial cleavage.

> The three phyla of lophophorates probably share a common ancestor, and they show a mixture of protostome and deuterostome characteristics.

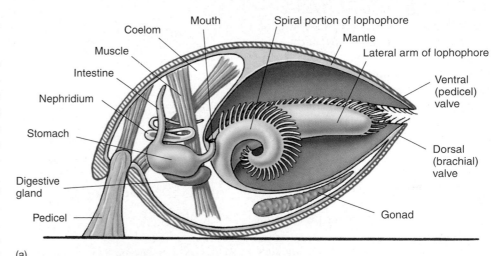

(a)

(b)

FIGURE 33.18

Brachiopods (phylum Brachiopoda). (*a*) The lophophore lies within two calcified shells, or valves. (*b*) The brachiopod *Terebratolina septentrionalis* is shown here slightly opened so that the lophophore is visible.

33.4 Arthropods are the most diverse of all animal groups.

Arthropod Body Plan

With the evolution of the first annelids, many of the major innovations of animal structure had already appeared: the division of tissues into three primary types (endoderm, mesoderm, and ectoderm), bilateral symmetry, a coelom, and segmentation. With arthropods, two more innovations arose—the development of *jointed appendages* and an *exoskeleton*. Jointed appendages and an exoskeleton have allowed arthropods (phylum Arthropoda) to become the most diverse phylum.

The Arthropods

Arthropods, especially the largest class—insects, are by far the most successful of all animals. Well over 1,000,000 species—about two-thirds of all the named species on earth—are members of this phylum (figure 33.19). One scientist recently estimated, based on the number and diversity of insects in tropical forests, that as many as 30 million species may comprise this one class alone. About 200 million insects are alive at any one time for each human! Insects and other arthropods (figure 33.20) abound in every habitat on the planet, but they especially dominate the land, along with flowering plants and vertebrates.

The majority of arthropod species consist of small animals, mostly about a millimeter in length. Members of the phylum range in adult size from about 80 micrometers long (some parasitic mites) to 3.6 meters across (a gigantic crab found in the sea off Japan).

Arthropods, especially insects, are of enormous economic importance and affect all aspects of human life. They compete with humans for food of every kind, play a key role in the pollination of certain crops, and cause billions of dollars of damage to crops, before and after harvest. They are by far the most important herbivores in all terrestrial ecosystems and are a valuable food source as well. Virtually every kind of plant is eaten by one or more species of insect. Diseases spread by insects cause enor-

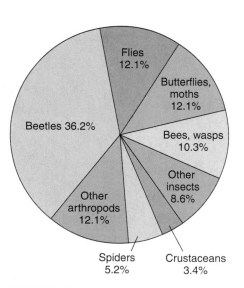

FIGURE 33.19
Arthropods are a successful group.
About two-thirds of all named species are arthropods. About 80% of all arthropods are insects, and about half of the named species of insects are beetles.

mous financial damage each year and strike every kind of domesticated animal and plant, as well as human beings.

Jointed Appendages

The name "arthropod" comes from two Greek words, *arthros*, "jointed," and *podes*, "feet." All arthropods have jointed appendages. The numbers of these appendages are reduced in the more advanced members of the phylum. Individual appendages may be modified into antennae, mouthparts of various kinds, or legs. Some appendages, such as the wings of certain insects, are not homologous to the other appendages; insect wings evolved separately.

To gain some idea of the importance of jointed appendages, imagine yourself without them—no hips, knees, ankles, shoulders, elbows, wrists, or knuckles. Without jointed appendages, you could not walk or grasp any object. Arthropods use jointed appendages such as legs for walking, antennae to sense their environment, and mouthparts for feeding.

An Exterior Skeleton

The arthropod body plan has a second major innovation: a rigid external skeleton, or **exoskeleton,** made of chitin and protein. In any animal, the skeleton functions to provide places for muscle attachment. In arthropods, the muscles attach to the interior surface of the hard exoskeleton, which also protects the animal from predators and impedes water loss. Chitin is chemically similar to cellulose, the dominant structural component of plants, and shares similar properties of toughness and flexibility. Together, the chitin and protein provide an external covering that is very strong but also capable of flexing in response to the contraction of muscles attached to it. In most crustaceans, the exoskeleton is made even tougher, although less flexible, with deposits of calcium salts. However, there is a limitation. The exoskeleton must be much thicker to bear the pull of the muscles in large insects than in small ones.

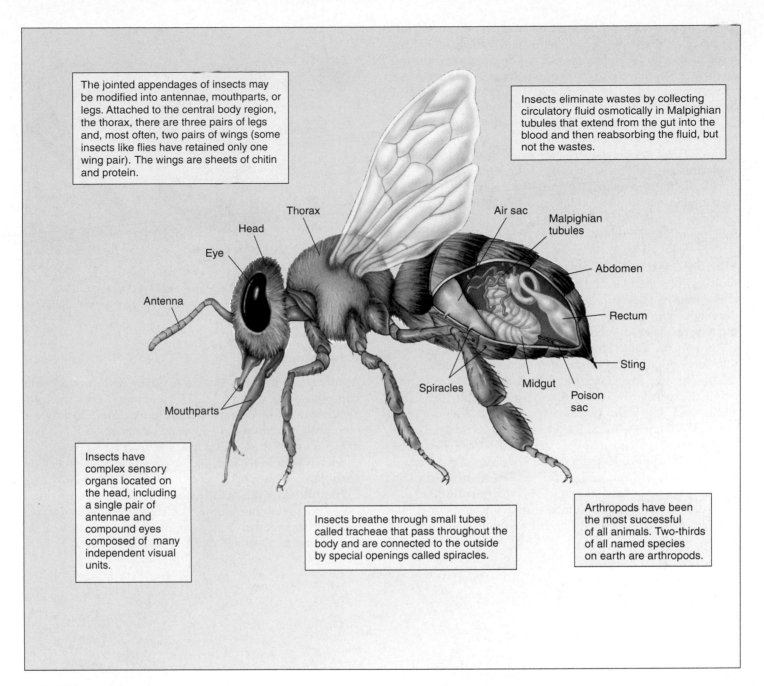

The jointed appendages of insects may be modified into antennae, mouthparts, or legs. Attached to the central body region, the thorax, there are three pairs of legs and, most often, two pairs of wings (some insects like flies have retained only one wing pair). The wings are sheets of chitin and protein.

Insects eliminate wastes by collecting circulatory fluid osmotically in Malpighian tubules that extend from the gut into the blood and then reabsorbing the fluid, but not the wastes.

Thorax

Head

Air sac

Malpighian tubules

Eye

Abdomen

Antenna

Rectum

Sting

Mouthparts

Spiracles

Midgut

Poison sac

Insects have complex sensory organs located on the head, including a single pair of antennae and compound eyes composed of many independent visual units.

Insects breathe through small tubes called tracheae that pass throughout the body and are connected to the outside by special openings called spiracles.

Arthropods have been the most successful of all animals. Two-thirds of all named species on earth are arthropods.

FIGURE 33.20

Phylum Arthropoda: arthropods. Insects and other arthropods (phylum Arthropoda) have a coelom, segmented bodies, and jointed appendages. The three body regions of an insect (head, thorax, and abdomen) are each actually composed of a number of segments that fuse during development. All arthropods have a strong exoskeleton made of chitin. One class, the insects, has evolved wings that permit them to fly rapidly through the air.

That is why you don't see beetles as big as birds, or crabs the size of a cow—the exoskeleton would be so thick the animal couldn't move its great weight. Because this size limitation is inherent in the body design of arthropods, there are practically no large arthropods—few are larger than your thumb.

Arthropod bodies are segmented like those of annelids, a phylum to which at least some arthropods are clearly related. Members of some classes of arthropods have many body segments. In others, the segments have become fused together into functional groups, or **tagmata** (singular, *tagma*), such as the head and thorax of an insect. This fusing process, known as *tagmatization*, is of central importance in the evolution of arthropods. In most arthropods, the original segments can be distinguished during larval development. All arthropods have a distinct head, sometimes fused with the thorax to form a tagma called the **cephalothorax.**

The bodies of all arthropods are covered by an exoskeleton, or cuticle, that contains chitin. This tough outer covering, against which the muscles work, is secreted by the epidermis and fused with it. The exoskeleton remains fairly flexible at specific points, allowing the exoskeleton to bend and appendages to move. The exoskeleton protects arthropods from water loss and helps to protect them from predators, parasites, and injury.

Molting. Arthropods periodically undergo **ecdysis,** or molting, the shedding of the outer cuticular layer. When they outgrow their exoskeleton, they form a new one underneath. This process is controlled by hormones. When the new exoskeleton is complete, it becomes separated from the old one by fluid. This fluid dissolves the chitin and protein (and calcium carbonate, if present) from the old exoskeleton. The fluid increases in volume until, finally, the original exoskeleton cracks open, usually along the back, and is shed. The arthropod emerges, clothed in a new, pale, and still somewhat soft exoskeleton. The arthropod then "puffs itself up," ultimately expanding to full size. The blood circulation to all parts of the body aids in this expansion, and many insects and spiders take in air to assist them as well. The expanded exoskeleton subsequently hardens. While the exoskeleton is soft, the animal is especially vulnerable. At this stage, arthropods often hide under stones, leaves, or branches.

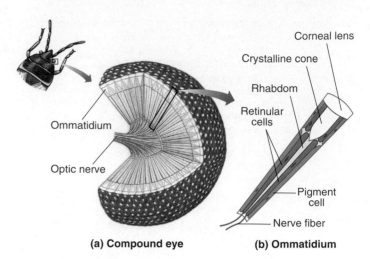

(a) Compound eye **(b) Ommatidium**

FIGURE 33.21
The compound eye. The compound eyes in insects are complex structures composed of many independent visual units called ommatidia.

Compound Eye

Another important structure in many arthropods is the **compound eye** (figure 33.21). Compound eyes are composed of many independent visual units, often thousands of them, called **ommatidia** (singular, *ommatidium*). Each ommatidium is covered with a lens and linked to a complex of eight retinular cells and a light-sensitive central core, or **rhabdom.**

Simple eyes, or **ocelli** (singular, *ocellum*) with single lenses are found in the other arthropod groups, and sometimes occur together with compound eyes, as is often the case in insects. Ocelli function in distinguishing light from

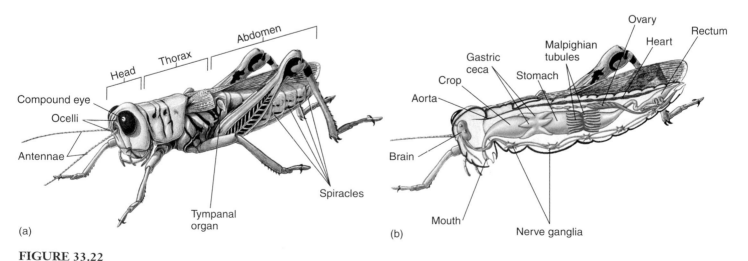

FIGURE 33.22
A grasshopper (order Orthoptera). This grasshopper illustrates the major structural features of the insects, the most numerous group of arthropods. (*a*) External anatomy. (*b*) Internal anatomy.

darkness. The ocelli of some flying insects, namely locusts and dragonflies, function as horizon detectors and help the insect visually stabilize its course in flight.

Circulatory System

In the course of arthropod evolution, the coelom has become greatly reduced, consisting only of cavities that house the reproductive organs and some glands. Arthropods completely lack cilia, both on the external surfaces of the body and on the internal organs. Like annelids, arthropods have a tubular gut that extends from mouth to anus.

The circulatory system of arthropods is open; their blood flows through cavities between the internal organs and not through closed vessels. The principal component of an insect's circulatory system is a longitudinal vessel called the heart. This vessel runs near the dorsal surface of the thorax and abdomen (figure 33.22b). When it contracts, blood flows into the head region of the insect.

When an insect's heart relaxes, blood returns to it through a series of valves. These valves are located in the posterior region of the heart and allow the blood to flow inward only. Thus, blood from the head and other anterior portions of the insect gradually flows through the spaces between the tissues toward the posterior end and then back through the one-way valves into the heart.

Nervous System

The central feature of the arthropod nervous system is a double chain of segmented ganglia running along the animal's ventral surface. At the anterior end of the animal are three fused pairs of dorsal ganglia, which constitute the brain. However, much of the control of an arthropod's activities is relegated to ventral ganglia. Therefore, the animal can carry out many functions, including eating, moving, and copulating, even if the brain has been removed. The brain of arthropods seems to be a control point, or inhibitor, for various actions, rather than a stimulator, as it is in vertebrates.

Respiratory System

Insects depend on their respiratory rather than their circulatory system to carry oxygen to their tissues. All parts of the body need to be near a respiratory passage to obtain oxygen. Along with the thickness of their chitin exoskeletons, this is another feature of arthropod design that places severe limitations on their size compared to that of vertebrates.

Unlike most animals, the arthropods have no single major respiratory organ. The respiratory system of most terrestrial arthropods consists of small, branched, cuticle-lined air ducts called **tracheae** (figure 33.23). These tra-

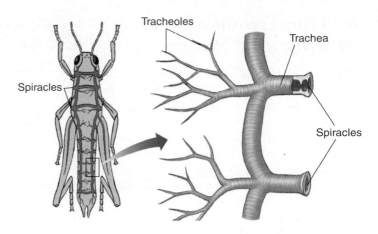

FIGURE 33.23
Tracheae and tracheoles. Tracheae and tracheoles are connected to the exterior by specialized openings called spiracles and carry oxygen to all parts of a terrestrial insect's body.

cheae, which ultimately branch into very small **tracheoles,** are a series of tubes that transmit oxygen throughout the body. Tracheoles are in direct contact with individual cells, and oxygen diffuses directly across the plasma membranes. Air passes into the tracheae by way of specialized openings in the exoskeleton called **spiracles,** which, in most insects, can be opened and closed by valves. The ability to prevent water loss by closing the spiracles was a key adaptation that facilitated the invasion of the land by arthropods.

Excretory System

Although various kinds of excretory systems are found in different groups of arthropods, we will focus here on the unique excretory system consisting of **Malpighian tubules,** which evolved in terrestrial insects. Malpighian tubules are slender projections from the digestive tract that are attached at the junction of the midgut and hindgut (see figure 49.11). Fluid passes through the walls of the Malpighian tubules to and from the blood in which the tubules are bathed. As this fluid passes through the tubules toward the hindgut, nitrogenous wastes are precipitated as concentrated uric acid or guanine. These substances are then emptied into the hindgut and eliminated. Most of the water and salts in the fluid are reabsorbed by the hindgut and rectum and returned to the arthropod's body. Malpighian tubules are an efficient mechanism for water conservation and were another key adaptation facilitating invasion of the land by arthropods.

Arthropods are segmented animals with jointed appendages. All arthropods have a rigid chitin and protein exoskeleton, and an open circulatory system. Insects have many adaptations for terrestrial living.

A Major Group of Arthropods: Crustaceans

The crustaceans (subphylum Crustacea) are a large group of primarily aquatic organisms, consisting of some 35,000 species of crabs, shrimps, lobsters, crayfish, barnacles, water fleas, pillbugs, and related groups (table 33.1). Most crustaceans have two pairs of antennae, three types of chewing appendages, and various numbers of pairs of legs. All crustacean appendages, with the possible exception of the first pair of antennae, are basically biramous ("two-branched"). In some crustaceans, appendages appear to have only a single branch; in those cases, one of the branches has been lost during the course of evolutionary specialization. The **nauplius** larva stage through which all crustaceans pass (figure 33.24) provides evidence that all members of this diverse group are descended from a common ancestor. The nauplius hatches with three pairs of appendages and metamorphoses through several stages before reaching maturity. In many groups, this nauplius stage is passed in the egg, and development of the hatchling to the adult form is direct.

Crustaceans differ from insects but resemble centipedes and millipedes in that they have appendages on their abdomen as well as on their thorax. They are the only arthropods with two pairs of antennae. Their **mandibles** (biting jaws) likely originated from a pair of limbs that took on a chewing function during the course of evolution, a process that apparently occurred independently in the common ancestor of the terrestrial mandibulates. Many crustaceans have compound eyes. In addition, they have delicate tactile hairs that project from the cuticle all over the body. Larger crustaceans have feathery gills near the bases of their legs. In smaller members of this class, gas exchange takes place directly through the thinner areas of the cuticle or the entire body. Most crustaceans have separate sexes. Many different kinds of specialized copulation occur among the crustaceans, and the members of some orders carry their eggs with them, either singly or in egg pouches, until they hatch.

Decapod Crustaceans

Large, primarily marine crustaceans such as shrimps, lobsters, and crabs, along with their freshwater relatives, the crayfish, are collectively called *decapod crustaceans* (figure 33.25). The term *decapod* means "ten-footed." In these animals, the exoskeleton is usually reinforced with calcium carbonate. Most of their body segments are fused into a cephalothorax covered by a dorsal shield, or carapace,

Table 33.1	Major Groups Within the Traditional Classification of the Phylum Arthropoda	
Group	**Characteristics**	**Members**
Crustaceans	Mouthparts are mandibles (biting jaws); appendages are biramous ("two-branched")	Lobsters, crabs, shrimp, isopods, barnacles
Arachnids	Mouthparts are chelicerae (pincers or fangs)	Spiders, mites, ticks, scorpions, daddy longlegs
Centipedes and millipedes	Mouthparts are mandibles; bodies consist of a head and numerous body segments bearing paired uniramous ("single-branched") appendages	Centipedes, millipedes
Insects	Mouthparts are mandibles; appendages are uniramous	Beetles, bees, flies, fleas, true bugs, grasshoppers, butterflies, termites

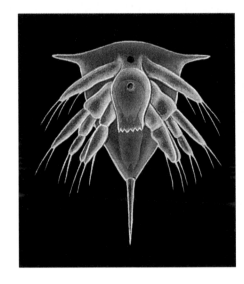

FIGURE 33.24 The nauplius larva. Although crustaceans are diverse, they have fundamentally similar larvae. The nauplius larva of a crustacean is an important unifying feature found in all members of this group.

which arises from the head. The crushing pincers common in many decapod crustaceans are used in obtaining food—for example, by crushing mollusk shells.

In lobsters and crayfish, appendages called **swimmerets** occur in lines along the ventral surface of the abdomen and are used in reproduction and swimming. In addition, flattened appendages known as **uropods** form a kind of compound "paddle" at the end of the abdomen. These animals may also have a **telson,** or tail spine. By snapping its abdomen, the animal propels itself through the water rapidly and forcefully. Crabs differ from lobsters and crayfish in proportion; their carapace is much larger and broader, and the abdomen is tucked under it.

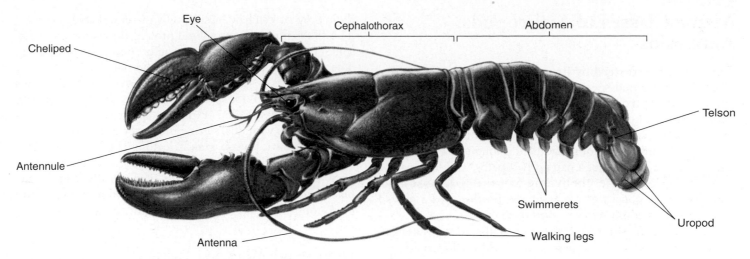

FIGURE 33.25
Decapod crustacean. A lobster, *Homarus americanus*, with its principal features labeled.

Terrestrial and Freshwater Crustaceans

Although most crustaceans are marine, many occur in fresh water, and a few have become terrestrial. These include pillbugs and sowbugs, the terrestrial members of a large order of crustaceans known as the isopods (order Isopoda). About half of the estimated 4500 species of this order are terrestrial and live primarily in places that are moist, at least seasonally. Sand fleas or beach fleas (order Amphipoda) are other familiar crustaceans, many of which are semiterrestrial (intertidal) species.

Along with the larvae of larger species, minute crustaceans are abundant in plankton. Especially significant are the tiny copepods (order Copepoda; figure 33.26), which are among the most abundant multicellular organisms on earth.

Sessile Crustaceans

Barnacles (order Cirripedia; figure 33.27) are a group of crustaceans that are sessile as adults. Barnacles have free-swimming larvae, which ultimately attach their heads to a piling, rock, or other submerged object and then stir food into their mouth with their feathery legs. Calcareous plates protect the barnacle's body, and these plates are usually attached directly and solidly to the substrate. Although most crustaceans have separate sexes, barnacles are hermaphroditic, but they generally cross-fertilize.

> Crustaceans include marine, freshwater, and terrestrial forms. All possess a nauplius larval stage and branched appendages.

FIGURE 33.27
Gooseneck barnacles, *Lepas anatifera*, feeding. These are stalked barnacles; many others lack a stalk.

FIGURE 33.26
Freshwater crustacean. A copepod with attached eggs, a member of an abundant group of marine and freshwater crustaceans (order Copepoda), most of which are a few millimeters long. Copepods are important components of plankton.

Major Classes of Arthropods: Arachnids

The largely terrestrial arthropod class Arachnida, with some 57,000 named species, occupies a distinct evolutionary line of arthropods in which the most anterior appendages, mouthparts called **chelicerae** (singular, *chelicera*), often function as fangs or pincers. Arachnids include spiders, ticks, mites, scorpions, and daddy longlegs. Arachnids have a pair of chelicerae, a pair of pedipalps, and four pairs of walking legs. The chelicerae are the foremost appendages; they consist of a stout basal portion and a movable fang often connected to a poison gland.

The next pair of appendages, **pedipalps,** resemble legs but have one less segment and are not used for locomotion. In male spiders, the pedipalps arc specialized copulatory organs. In scorpions, they are large pincers.

Most arachnids are carnivorous. The main exception is mites, which are largely herbivorous. Most arachnids can ingest only preliquified food, which they often digest externally by secreting enzymes into their prey. They can then suck up the digested material with their muscular, pumping pharynx. Arachnids are primarily, but not exclusively, terrestrial. Some 4000 known species of mites and one species of spider live in fresh water, and a few mites live in the sea.

Many spiders have a unique respiratory system that involves **book lungs,** a series of leaflike plates within a chamber. Air is drawn into and expelled out of this chamber by muscular contraction. Book lungs may exist alongside tracheae, or they may function instead of tracheae.

Order Araneae: The Spiders

There are about 35,000 named species of spiders (order Araneae). These animals play a major role in virtually all terrestrial ecosystems. They are particularly important as predators of insects and other small animals. Spiders hunt their prey or catch it in silk webs of remarkable diversity. The silk is formed from a fluid protein that is forced out of spinnerets on the posterior portion of the spider's abdomen. The webs and habits of spiders are often distinctive.

Many kinds of spiders, including the familiar wolf spiders and tarantulas, do not spin webs but instead hunt their prey actively. Others, called trap-door spiders, construct silk-lined burrows with lids, seizing their prey as it passes by.

Spiders have poison glands leading through their chelicerae, which are pointed and used to bite and paralyze prey. The bites of some members of this order, such as the black widow and brown recluse spiders (figure 33.28), are poisonous to humans and other large mammals.

Order Acari: Mites and Ticks

The order Acari, the mites and ticks, is the largest in terms of number of species and the most diverse of the arachnids. Although only about 30,000 species of mites and ticks have been named, scientists that study the group estimate that a million or more members of this order may exist.

Most mites are small, less than 1 millimeter long, but adults of different species range from 100 nanometers to 2 centimeters. In most mites, the cephalothorax and abdomen are fused into an unsegmented, ovoid body. Respiration occurs either by means of tracheae or directly through the exoskeleton. Many mites pass through several distinct stages during their life cycle. In most, an inactive eight-legged prelarva gives rise to an active six-legged larva, which in turn produces a succession of three eight-legged stages and, finally, the adult males and females.

Mites and ticks are diverse in structure and habitat. They are found in virtually every terrestrial, freshwater, and marine habitat known, and feed on fungi, plants, and animals. They act as predators and as internal and external parasites of both invertebrates and vertebrates.

Ticks are blood-eating parasites that occur on the surface of their host. They are larger than most other mites and cause discomfort by sucking the blood of humans and other animals. Ticks can carry many diseases, including some caused by viruses, bacteria, and protozoa. The spotted fevers (for example, Rocky Mountain spotted fever) are caused by bacteria carried by ticks. Lyme disease is apparently caused by spirochaetes transmitted by ticks. Red-water fever, or Texas fever, is an important tick-borne protozoan disease of cattle, horses, sheep, and dogs.

FIGURE 33.28
Two common poisonous spiders. (*a*) The black widow spider, *Latrodectus mactans.* (*b*) The brown recluse spider, *Loxosceles reclusa.* Both species are common throughout temperate and subtropical North America, but bites are rare in humans.

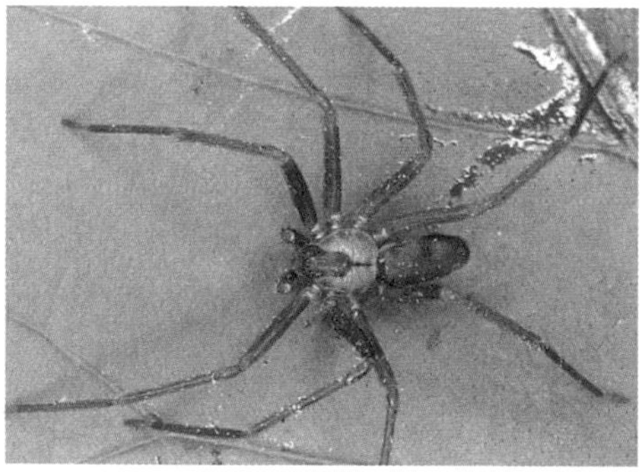

Scorpions, spiders, and mites are all arachnids, and possess chelicerates.

Major Classes of Arthropods: Centipedes and Millipedes

Millipedes, centipedes, and insects all respire by means of tracheae and excrete their waste products through Malpighian tubules. These groups were certainly derived from annelids, probably those similar to the oligochaetes, which they resemble in their embryology.

The centipedes (class Chilopoda) and millipedes (class Diplopoda) both have bodies that consist of a head region followed by numerous segments, all more or less similar and nearly all bearing paired appendages. Although the name *centipede* would imply an animal with 100 legs and the name *millipede* one with 1000, adult centipedes usually have 30 or more legs, while adult millipedes have 60 or. Centipedes have one pair of legs on each body segment (figure 33.29), while millipedes have two (figure 33.30). Each segment of a millipede is a tagma that originated during the group's evolution when two ancestral segments fused. This explains why millipedes have twice as many legs per segment as centipedes.

In both centipedes and millipedes, fertilization is internal and takes place by direct transfer of sperm. The sexes are separate, and all species lay eggs. Young millipedes usually hatch with three pairs of legs; they experience a number of growth stages, adding segments and legs as they mature, but do not change in general appearance.

Centipedes, of which some 2500 species are known, are all carnivorous and feed mainly on insects. The appendages of the first trunk segment are modified into a pair of poison fangs. The poison is often quite toxic to human beings, and many centipede bites are extremely painful, sometimes even dangerous.

In contrast, most millipedes are herbivores, feeding mainly on decaying vegetation. A few millipedes are carnivorous. Many millipedes can roll their bodies into a flat coil or sphere because the dorsal area of each of their body segments is much longer than the ventral one. More than 10,000 species of millipedes have been named, but this is estimated to be no more than one-sixth of the actual number of species that exists. In each segment of their body,

FIGURE 33.29
A centipede. Centipedes, such as this member of the genus *Scolopendra*, are active predators.

FIGURE 33.30
A millipede. Millipedes, such as this *Sigmoria* individual, are herbivores.

most millipedes have a pair of complex glands that produce a bad-smelling fluid. This fluid is exuded for defensive purposes out through openings along the sides of the body. The chemistry of the external secretions of different millipedes has become a subject of considerable interest because of the chemical diversity of the compounds involved and their effectiveness in protecting millipedes from attack. Some species produce cyanide gas from segments near their head end. Millipedes live primarily in damp, protected places, such as under leaf litter, in rotting logs, under bark or stones, or in the soil.

Centipedes are segmented hunters with one pair of legs on each segment. Millipedes are segmented herbivores with two pairs of legs on each segment.

Major Classes of Arthropods: Insects

The insects, class Insecta, are by far the largest group of organisms on earth, whether measured in terms of numbers of species or numbers of individuals. Insects live in every conceivable habitat on land and in fresh water, and a few have even invaded the sea. More than half of all the named animal species are insects, and the actual proportion is doubtless much higher because millions of additional forms await detection, classification, and naming. Approximately 90,000 described species occur in the United States and Canada, and the actual number of species in this area probably approaches 125,000. A hectare of lowland tropical forest is estimated to be inhabited by as many as 41,000 species of insects, and many suburban gardens may have 1500 or more species. It has been estimated that approximately a billion billion (10^{18}) individual insects are alive at any one time. A glimpse at the enormous diversity of insects is presented in figure 33.31 and also later in table 33.2.

(a)

(b)

(c)

(d)

(e)

(f)

FIGURE 33.31

Insect diversity. (*a*) Luna moth, *Actias luna*. Luna moths and their relatives are among the most spectacular insects (order Lepidoptera). (*b*) Soldier fly, *Ptecticus trivittatus* (order Diptera). (*c*) Boll weevil, *Anthonomus grandis*. Weevils are one of the largest groups of beetles (order Coleoptera). (*d*) A thorn-shaped treehopper, *Umbonia crassicornis* (order Homoptera). (*e*) Grasshopper (order Orthoptera). (*f*) Termite, *Macrotermes bellicosus* (order Isoptera). The large, sausage-shaped individual is a queen, specialized for laying eggs; most of the smaller individuals around the queen are nonreproductive workers, but the larger individual at the lower left is a reproductive male.

Order	Typical Examples		Key Characteristics	Approximate Number of Named Species
Coleoptera	Beetles		The most diverse animal order; two pairs of wings; front pair of wings is a hard cover that partially protects the transparent rear pair of flying wings; heavily armored exoskeleton; biting and chewing mouthparts; complete metamorphosis	350,000
Diptera	Flies		Some that bite people and other mammals are considered pests; front flying wings are transparent; hindwings are reduced to knobby balancing organs; sucking, piercing, and lapping mouthparts; complete metamorphosis	120,000
Lepidoptera	Butterflies, moths		Often collected for their beauty; two pairs of broad, scaly, flying wings, often brightly colored; hairy body; tubelike, sucking mouthparts; complete metamorphosis	120,000
Hymenoptera	Bees, wasps, ants		Often social, known to many by their sting; two pairs of transparent flying wings; mobile head and well-developed eyes; often possess stingers; chewing and sucking mouthparts; complete metamorphosis	100,000
Hemiptera and **Homoptera**	True bugs, bedbugs, leafhoppers, aphids, cicades		Often live on blood; some are plant-eaters; two pairs of wings, or wingless; piercing, sucking mouthparts; simple metamorphosis	60,000
Orthoptera	Grasshoppers, crickets, roaches		Known for their jumping; two pairs of wings or wingless; among the largest insects; biting and chewing mouthparts in adults; simple metamorphosis.	20,000
Odonata	Dragonflies		Among the most primitive of the insect order; two pairs of transparent flying wings; large, long, and slender body; chewing mouthparts; simple metamorphosis	5,000
Isoptera	Termites		One of the few types of animals able to eat wood; two pairs of wings, but some stages are wingless; social insects; there are several body types with division of labor; chewing mouthparts; simple metamorphosis	2,000
Siphonaptera	Fleas		Small, known for their irritating bites; wingless; small, flattened body with jumping legs; piercing and sucking mouthparts; complete metamorphosis	1,200

Table 33.2 Major Orders of Insects

External Features

Insects are primarily a terrestrial group, and most, if not all, of the aquatic insects probably had terrestrial ancestors. Most insects are relatively small, ranging in size from 0.1 millimeter to about 30 centimeters in length or wingspan. Insects have three body sections, the head, thorax, and abdomen; three pairs of legs, all attached to the thorax; and one pair of antennae. In addition, they may have one or two pairs of wings. Insect mouthparts all have the same basic structure but are modified in different groups in relation to their feeding habits (figure 33.32). Most insects have compound eyes, and many have ocelli as well. The insect thorax consists of three segments, each with a pair of legs. Legs are completely absent in the larvae of certain groups—for example, in most flies (order Diptera) and mosquitoes (figure 33.33). If two pairs of wings are present, they attach to the middle and posterior segments of the thorax. If only one pair of wings is present, it usually attaches to the middle segment. The thorax is almost entirely filled with muscles that operate the legs and wings.

The wings of insects arise as saclike outgrowths of the body wall. In adult insects, the wings are solid except for the veins. Insect wings are not homologous to the other appendages. Basically, insects have two pairs of wings, but in some groups, such as flies, the second set has been reduced to a pair of balancing knobs called halteres during the course of evolution. Most insects can fold their wings over their abdomen when they are at rest; but a few, such as the dragonflies and damselflies (order Odonata), keep their wings erect or outstretched at all times.

Insect forewings may be tough and hard, as in beetles. If they are, they form a cover for the hindwings and usually open during flight. The tough forewings also serve a protective function in the order Orthoptera, which includes grasshoppers and crickets. The wings of insects are made of sheets of chitin and protein; their strengthening veins are tubules of chitin and protein. Moths and butterflies have wings covered with detachable scales that provide most of their bright colors (figure 33.34). In some wingless insects, such as the springtails or silverfish, wings never evolved. Other wingless groups, such as fleas and lice, are derived from ancestral groups of insects that had wings.

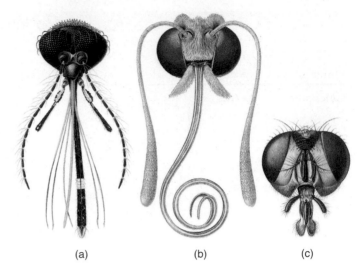

(a)   **(b)**   **(c)**

FIGURE 33.32
Modified mouthparts in three kinds of insects. Mouthparts are modified for (*a*) piercing in the mosquito, *Culex;* (*b*) sucking nectar from flowers in the alfalfa butterfly, *Colias;* and (*c*) sopping up liquids in the housefly, *Musca domestica.*

FIGURE 33.33
Larvae of a mosquito, *Culex pipiens.*
The aquatic larvae of mosquitoes are quite active. They breathe through tubes from the surface of the water, as shown here. Covering the water with a thin film of oil causes them to drown.

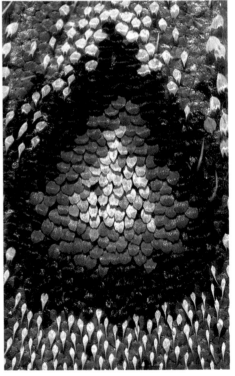

FIGURE 33.34
Scales on the wing of *Parnassius imperator*, a butterfly from China.
Scales of this sort account for most of the colored patterns on the wings of butterflies and moths.

Internal Organization

The internal features of insects resemble those of the other arthropods in many ways. The digestive tract is a tube, usually somewhat coiled. It is often about the same length as the body. However, in the orders Hemiptera and Homoptera, which contain the leafhoppers, cicadas, and related groups, and in many flies (order Diptera), the digestive tube may be greatly coiled and several times longer than the body. Such long digestive tracts are generally found in insects that have sucking mouthparts and feed on juices rather than on protein-rich solid foods because they offer a greater opportunity to absorb fluids and their dissolved nutrients. The digestive enzymes of the insect also are more dilute and thus less effective in a highly liquid medium than in a more solid one. Longer digestive tracts give these enzymes more time to work while food is passing through.

The anterior and posterior regions of an insect's digestive tract are lined with cuticle. Digestion takes place primarily in the stomach, or midgut, and excretion takes place through Malpighian tubules. Digestive enzymes are mainly secreted from the cells that line the midgut, although some are contributed by the salivary glands near the mouth.

The tracheae of insects extend throughout the body and permeate its different tissues. In many winged insects, the tracheae are dilated in various parts of the body, forming air sacs. These air sacs are surrounded by muscles and form a kind of bellows system to force air deep into the tracheal system. The spiracles, a maximum of 10 on each side of the insect, are paired and located on or between the segments along the sides of the thorax and abdomen. In most insects, the spiracles can be opened by muscular action. Closing the spiracles at times may be important in retarding water loss. In some parasitic and aquatic groups of insects, the spiracles are permanently closed. In these groups, the tracheae run just below the surface of the insect, and gas exchange takes place by diffusion.

The **fat body** is a group of cells located in the insect body cavity. This structure may be quite large in relation to the size of the insect, and it serves as a food-storage reservoir, also having some of the functions of a vertebrate liver. It is often more prominent in immature insects than in adults, and it may be completely depleted when metamorphosis is finished. Insects that do not feed as adults retain their fat bodies and live on the food stored in them throughout their adult lives (which may be very short).

Sense Receptors

In addition to their eyes, insects have several characteristic kinds of sense receptors. These include **sensory hairs,** which are usually widely distributed over their bodies. The sensory hairs are linked to nerve cells and are sensitive to mechanical and chemical stimulation. They are particularly abundant on the antennae and legs—the parts of the insect most likely to come into contact with other objects.

Sound, which is of vital importance to insects, is detected by tympanal organs in groups such as grasshoppers and crickets, cicadas, and some moths. These organs are paired structures composed of a thin membrane, the **tympanum,** associated with the tracheal air sacs. In many other groups of insects, sound waves are detected by sensory hairs. Male mosquitoes use thousands of sensory hairs on their antennae to detect the sounds made by the vibrating wings of female mosquitoes.

Sound detection in insects is important both for protection from enemies and for communication. Many insects communicate by making sounds, most of which are quite soft, very high-pitched, or both, and thus inaudible to humans. Only a few groups of insects, especially grasshoppers, crickets, and cicadas, make sounds that people can hear. Male crickets and longhorned grasshoppers produce sounds by rubbing their two front wings together. Short-horned grasshoppers do so by rubbing their hind legs over specialized areas on their wings. Male cicadas vibrate the membranes of air sacs located on the lower side of the most anterior abdominal segment.

In addition to using sound, nearly all insects communicate by means of chemicals or mixtures of chemicals known as **pheromones.** These compounds, extremely diverse in their chemical structure, are sent forth into the environment, where they are active in very small amounts and convey a variety of messages to other individuals. These messages not only convey the attraction and recognition of members of the same species for mating, but they also mark trails for members of the same species, as in the ants.

Insect Life Histories

During the course of their development, many insects undergo **metamorphosis.** In *simple metamorphosis*, seen in grasshoppers, immature stages are quite similar to adults, but gradually get bigger and more developed over a series of molts. In *complete metamorphosis*, immature stages called **larvae** are often wormlike and active in feeding. A resting stage, during which the insect is called a **pupa** or **chrysalis,** immediately precedes the final molt into the adult form.

All insects possess three body segments (tagmata): the head, the thorax, and the abdomen. The three pairs of legs are attached to the thorax. Most insects have compound eyes, and many have one or two pairs of wings. Insects possess sophisticated means of sensing their environment, including sensory hairs, tympanal organs, and chemoreceptors. Many insects undergo simple or complete metamorphosis.

33.5 Echinoderms are radially symmetrical as adults.

Deuterostome Development and an Endoskeleton

The mollusks, annelids, and arthropods discussed so far are protostomes. However, the echinoderms described in this section, are characterized by *deuterostome development*, a key evolutionary advance. The *endoskeleton* makes its first appearance in the echinoderms also.

The Echinoderms

Deuterostomate marine animals called **echinoderms** appeared nearly 600 million years ago. Echinoderms (phylum Echinodermata) are an ancient group of marine animals consisting of about 6000 living species (figure 33.35) and are also well represented in the fossil record. The term *echinoderm* means "spiny skin" and refers to an **endoskeleton** composed of hard, calcium-rich plates just beneath the delicate skin (figure 33.36). When they first form, the plates are enclosed in living tissue and so are truly an endoskeleton, although in adults they frequently fuse, forming a hard shell. Another innovation in echinoderms is the development of a hydraulic system to aid in movement or feeding. Called a **water-vascular system,** this fluid-filled system is composed of a central ring canal from which five radial canals extend out into the body and arms.

Many of the most familiar animals seen along the seashore, sea stars (starfish), brittle stars, sea urchins, sand dollars, and sea cucumbers, are echinoderms. All are radially symmetrical as adults. While some other kinds of animals are radially symmetrical, none have the complex organ systems of adult echinoderms. Echinoderms are well represented not only in the shallow waters of the sea but also in its abyssal depths. In the oceanic trenches, which are the deepest regions of the oceans, sea cucumbers account for more than 90% of the biomass! All of them are bottom-dwellers except for a few swimming sea cucumbers. The adults range from a few millimeters to more than a meter in diameter (for one species of sea star) or in length (for a species of sea cucumber).

There is an excellent fossil record of the echinoderms, extending back into the Cambrian period. However, despite this wealth of information, the origin of echinoderms remains unclear. They are thought to have evolved from bilaterally symmetrical ancestors because echinoderm larvae are bilateral. The radial symmetry that is the hallmark of echinoderms develops later, in the adult body. Many biologists believe that early echinoderms were sessile and evolved radiality as an adaptation to the sessile existence. Bilaterality is of adaptive value to an animal that travels through its environment, while radiality is of value to an animal whose environment meets it on all sides. Echinoderms attached to the sea bottom by a central stalk were once common, but only about 80 such species survive today.

Echinoderms are a unique, exclusively marine group of organisms in which deuterostome development and an endoskeleton are seen for the first time.

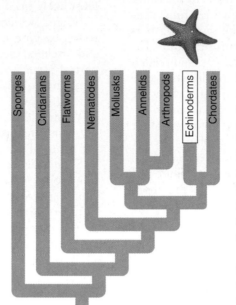

Sponges | Cnidarians | Flatworms | Nematodes | Mollusks | Annelids | Arthropods | Echinoderms | Chordates

(a)   (b)   (c)

FIGURE 33.35
Diversity in echinoderms. (*a*) Sea star, *Oreaster occidentalis* (class Asteroidea), in the Gulf of California, Mexico. (*b*) Warty sea cucumber, *Parastichopus parvimensis* (class Holothuroidea), Philippines. (*c*) Sea urchin (class Echinoidea).

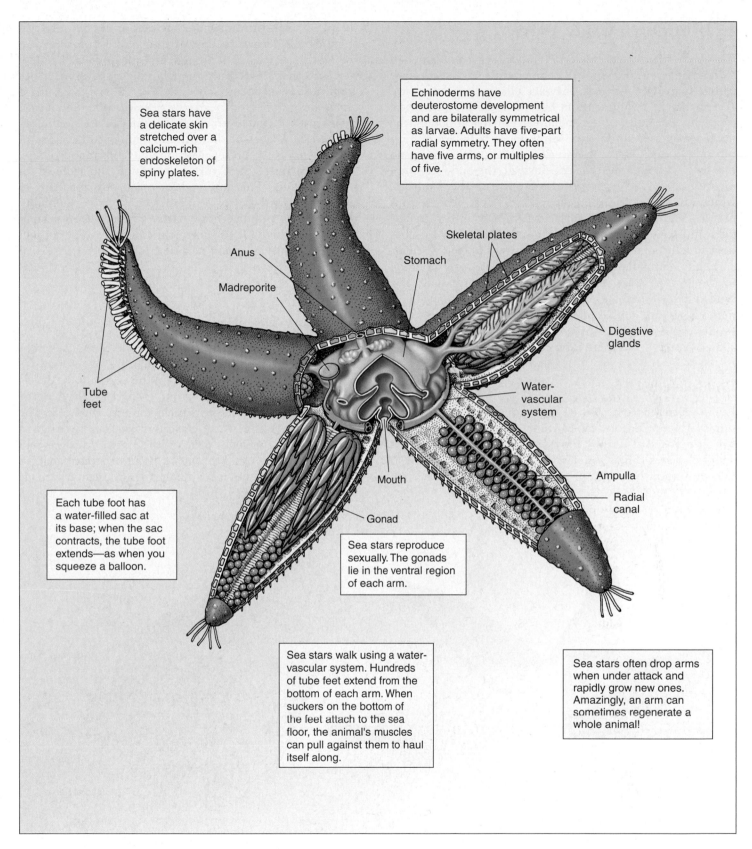

Sea stars have a delicate skin stretched over a calcium-rich endoskeleton of spiny plates.

Echinoderms have deuterostome development and are bilaterally symmetrical as larvae. Adults have five-part radial symmetry. They often have five arms, or multiples of five.

Anus

Madreporite

Skeletal plates

Stomach

Digestive glands

Tube feet

Water-vascular system

Each tube foot has a water-filled sac at its base; when the sac contracts, the tube foot extends—as when you squeeze a balloon.

Mouth

Ampulla

Radial canal

Gonad

Sea stars reproduce sexually. The gonads lie in the ventral region of each arm.

Sea stars walk using a water-vascular system. Hundreds of tube feet extend from the bottom of each arm. When suckers on the bottom of the feet attach to the sea floor, the animal's muscles can pull against them to haul itself along.

Sea stars often drop arms when under attack and rapidly grow new ones. Amazingly, an arm can sometimes regenerate a whole animal!

FIGURE 33.36

Phylum Echinodermata: echinoderms.   Echinoderms, such as sea stars (phylum Echinodermata), are coelomates with a deuterostomate pattern of development. A delicate skin stretches over an endoskeleton made of calcium-rich plates, often fused into a continuous, tough, spiny layer.

Echinoderm Body Plan

The body plan of echinoderms undergoes a fundamental shift during development. All echinoderms have **secondary radial symmetry**—that is, they are bilaterally symmetrical during larval development but become radially symmetrical as adults. Because of their radially symmetrical bodies, the usual terms used to describe an animal's body are not applicable: Dorsal, ventral, anterior, and posterior have no meaning without a head or tail. Instead, the body structure of echinoderms is discussed in reference to their mouths, which are located on the oral surface. Most echinoderms crawl along on their oral surfaces, although in sea cucumbers and some other echinoderms, the animal's axis lies horizontally, and they crawl with the oral surface in front.

Echinoderms have a five-part body plan corresponding to the arms of a sea star or the design on the "shell" of a sand dollar. These animals have no head or brain. Their nervous systems consist of a central **nerve ring** from which branches arise. The animals are capable of complex response patterns, but there is no centralization of function.

Endoskeleton

Echinoderms have a delicate epidermis, containing thousands of neurosensory cells, stretched over an endoskeleton composed of either movable or fixed calcium-rich (calcite) plates called ossicles. The animals grow more or less continuously, but their growth slows with age. When the plates first form, they are enclosed in living tissue. In some echinoderms, such as asteroids and holothuroids, the ossicles are widely scattered, and the body wall is flexible. In others, especially the echinoids, the ossicles become fused and form a rigid shell. In many cases, these plates bear spines. In nearly all species of echinoderms, the entire skeleton, even the long spines of sea urchins, is covered by a layer of skin. Another important feature of this phylum is the presence of mutable collagenous tissue, which can range in texture from tough and rubbery to weak and fluid. This amazing tissue accounts for many of the special attributes of echinoderms, such as the ability to rapidly autotomize body parts. The plates in certain portions of the body of some echinoderms are perforated by pores. Through these pores extend **tube feet,** part of the water-vascular system that is a unique feature of this phylum.

The Water-Vascular System

The water-vascular system of an echinoderm radiates from a ring canal that encircles the animal's esophagus. Five **radial canals,** their positions determined early in the development of the embryo, extend into each of the five parts of the body and determine its basic symmetry (figure 33.37). Water enters the water-vascular system through a **madreporite,** a sievelike plate on the animal's surface, and flows to the ring canal through a tube, or stone canal, so named because of the surrounding rings of calcium carbon-

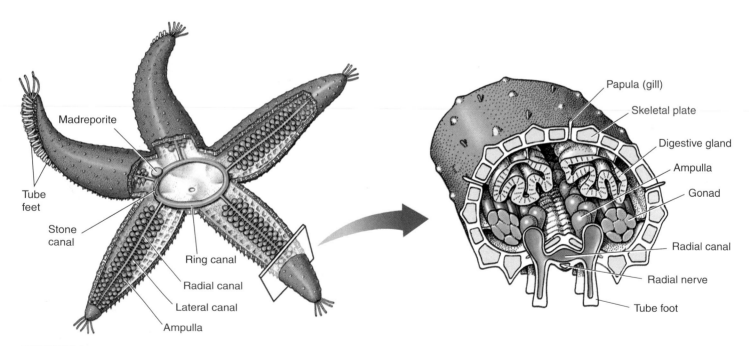

FIGURE 33.37
The water-vascular system of an echinoderm. Radial canals allow water to flow into the tube feet. As the ampulla in each tube foot contracts, the tube extends and can attach to the substrate. Subsequently, muscles in the tube feet contract, bending the tube foot and pulling the animal forward.

FIGURE 33.38
Tube feet. The nonsuckered tube feet of the sea star, *Ludia magnifica*, are extended.

ate. The five radial canals in turn extend out through short side branches into the hollow tube feet (figure 33.38). In some echinoderms, each tube foot has a sucker at its end; in others, suckers are absent. At the base of each tube foot is a muscular sac, the **ampulla,** which contains fluid. When the ampulla contracts, the fluid is prevented from entering the radial canal by a one-way valve and is forced into the tube foot, thus extending it. When extended, the foot can attach itself to the substrate. Longitudinal muscles in the tube foot wall then contract, causing the tube feet to bend. Relaxation of the muscles in the ampulla allows the fluid to flow back into the ampulla, which moves the foot. By the concerted action of a very large number of small, individually weak tube feet, the animal can move across the seafloor.

Sea cucumbers usually have five rows of tube feet on the body surface that are used in locomotion. They also have modified tube feet around their mouth cavity that are used in feeding. In sea lilies, the tube feet arise from the branches of the arms, which extend from the margins of an upward-directed cup. With these tube feet, the animals take food from the surrounding water. In brittle stars (see figure 33.41), the tube feet are pointed and specialized for feeding.

Body Cavity

In echinoderms, the coelom, which is proportionately large, connects with a complicated system of tubes and helps provide circulation and respiration. In many echinoderms, respiration and waste removal occur across the skin through small, fingerlike extensions of the coelom called papulae (see figure 33.37). They are covered with a thin layer of skin and protrude through the body wall to function as gills.

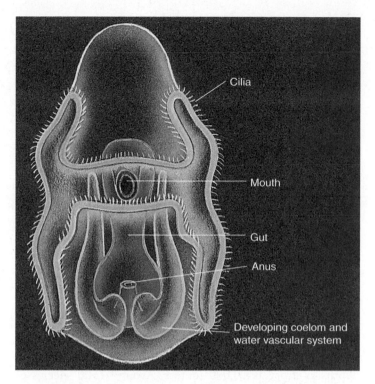

FIGURE 33.39
The free-swimming larva of an echinoderm. The bands of cilia by which the larva moves are prominent in this drawing. Such bilaterally symmetrical larvae suggest that the ancestors of the echinoderms were not radially symmetrical, like the living members of the phylum.

Reproduction

Many echinoderms are able to regenerate lost parts, and some, especially sea stars and brittle stars, drop various parts when under attack. In a few echinoderms, asexual reproduction takes place by splitting, and the broken parts of sea stars can sometimes regenerate whole animals. Some of the smaller brittle stars, especially tropical species, regularly reproduce by breaking into two equal parts; each half then regenerates a whole animal.

Despite the ability of many echinoderms to break into parts and regenerate new animals from them, most reproduction in the phylum is sexual and external. The sexes in most echinoderms are separate, although there are few external differences. Fertilized eggs of echinoderms usually develop into free-swimming, bilaterally symmetrical larvae (figure 33.39), which differ from the trochophore larvae of mollusks and annelids. These larvae form a part of the plankton until they metamorphose through a series of stages into the more sedentary adults.

Echinoderms are characterized by a secondary radial symmetry and a five-part body plan. They have characteristic calcium-rich plates called ossicles and a unique water-vascular system that includes hollow tube feet.

Major Classes of Echinoderms

There are more than 20 extinct classes of echinoderms and an additional 6 with living members: (1) Crinoidea, sea lilies and feather stars; (2) Asteroidea, sea stars, or starfish; (3) Ophiuroidea, brittle stars; (4) Echinoidea, sea urchins and sand dollars; (5) Holothuroidea, sea cucumbers; and (6) Concentricycloidea, sea daisies. All the classes of echinoderms exhibit a basic five-part body plan, although the basic symmetry of the phylum is not obvious in some. Even sea cucumbers, which are shaped like their plant namesake, have five radial grooves running along their soft, sluglike bodies. We describe three of the living classes here.

Class Asteroidea: The Sea Stars

Sea stars, or starfish (class Asteroidea; figure 33.40), are perhaps the most familiar echinoderms. Among the most important predators in many marine ecosystems, they range in size from a centimeter to a meter across. They are abundant in the intertidal zone, but they also occur at depths as great as 10,000 meters. Around 1500 species of sea stars occur throughout the world.

The body of a sea star is composed of a central disc that merges gradually with the arms. Although most sea stars have five arms, members of some families have many more, typically in multiples of five. The body is somewhat flattened, flexible, and covered with a pigmented epidermis.

Class Ophiuroidea: The Brittle Stars

Brittle stars (class Ophiuroidea; figure 33.41) are the largest class of echinoderms in numbers of species (about 2000), and they are probably the most abundant also. Secretive, they avoid light and are more active at night.

Brittle stars have slender, branched arms. The most mobile of echinoderms, brittle stars move by pulling themselves along, "rowing" over the substrate by moving their arms, often in pairs or groups, from side to side. Some brittle stars use their arms to swim, a very unusual habit among echinoderms.

Like sea stars, brittle stars have five arms. Closely related to the sea stars, on closer inspection they are surprisingly different. They have no pedicellariae, as sea stars have, and the groove running down the length of each arm is closed over and covered with ossicles. Their tube feet lack ampullae, have no suckers, and are used for feeding, not locomotion.

Class Echinoidea: The Sea Urchins and Sand Dollars

The members of the class Echinoidea, sand dollars and sea urchins, lack distinct arms but have the same five-part body plan as all other echinoderms. Five rows of tube feet pro-

FIGURE 33.40
Class Asteroidea.   This class includes the familiar starfish, or sea stars.

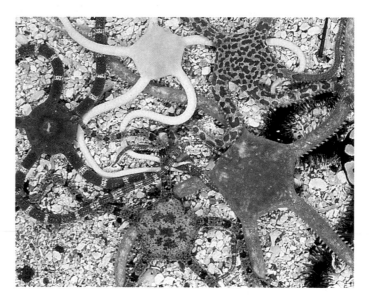

FIGURE 33.41
Class Ophiuroidea.   Brittle stars crawl actively across their marine substrates.

trude through the plates of the calcareous skeleton. About 950 living species constitute the class Echinoidea. Because of their calcareous plates, sea urchins and sand dollars are well preserved in the fossil record, with more than 5000 additional species described.

The six classes of echinoderms have five-part body symmetry, and are all radially symmetrical as adults.

For interactive testing, visit the Online Learning Center with PowerWeb at www.mhhe.com/Raven7

33.1 Mollusks were among the first coelomates.

Mollusk Body Plan

- The evolution of a coelom was a significant advance in animal body structure because it repositions fluid and allows complex tissues and organs to develop. (p. 652)
- Mollusks are extremely diverse; they exhibit a wide range of body forms and inhabit a large variety of habitats. (p. 652)
- The molluscan body plan is bilaterally symmetrical, with an efficient excretory system and a muscular foot for locomotion. (p. 654)
- Most mollusks have distinct sexes, and most aquatic mollusks engage in external fertilization. (p. 655)

Major Classes of Mollusks

- Major classes of mollusks include the class Polyplacophora (chitons), class Gastropoda (snails and slugs), class Bivalvia (bivalves), and class Cephalopoda (octopuses, squids, and nautiluses). (pp. 656–657)

33.2 Annelids were the first segmented animals.

Segmented Bodies

- Segmentation was a key transition in animal body plans, because similar segments can be individually controlled for different functions. (p. 658)
- Three principal features of annelids are repeated segments, specialized segments, and connections between segments. (p. 658)
- The annelid body plan is a tube within a tube. (p. 658)
- Locomotion is carried out by a squeezing motion as the hydrostatic skeleton forces fluid from one segment to the next. (p. 658)

Major Classes of Annelids

- The roughly 12,000 recognized annelid species are divided into three classes: Polychaeta (polychaetes), Oligochaeta (earthworms), and Hirudinea (leeches). (pp. 660–661)

33.3 Lophophorates appear to be a transitional group.

Lophophorates

- The marine phyla Phoronida (phoronids), Ectoprocta (bryozoans), and Brachiopoda (brachiopods) are characterized by a ridge around the mouth bearing ciliated, hollow tentacles (a lophophore) that functions in gas exchange and food collection. (pp. 662–663)

33.4 Arthropods are the most diverse of all animal groups.

Arthropod Body Plan

- Over two-thirds of all named species on the earth are arthropods. (p. 664)
- Jointed appendages used as legs, antennae, and mouthparts, and an exoskeleton of chitin and protein used for protection and water loss prevention, have allowed arthropods to become the most diverse phylum on the planet. (p. 664)
- Arthropods are segmented, with some segments becoming fused into tagmata (functional groups). (pp. 665–666)
- The exoskeleton is secreted by, and fused with, the epidermis, and provides a hard surface for muscle attachment. (p. 666)

- All arthropods have an open circulatory system; some have adaptations such as compound eyes, a respiratory system composed of tracheae, and an excretory system composed of Malpighian tubules. (pp. 666–667)

A Major Group of Arthropods: Crustaceans

- Most crustaceans (35,000 species) have two pairs of antennae, three types of chewing appendages, and various pairs of legs. (p. 668)
- Crustaceans are found in marine, freshwater, and terrestrial habitats. (pp. 668–669)

Major Classes of Arthropods: Arachnids

- Arachnids (57,000 species) have a pair of chelicerae, a pair of pedipalps, and four pairs of walking legs. (p. 670)
- Two major orders are Araneae (spiders) and Acari (mites and ticks). (p. 670)

Major Classes of Arthropods: Centipedes and Millipedes

- Centipedes (class Chilopoda) and millipedes (class Diplopoda) are made of a head region followed by numerous similar segments. Centipedes have one pair of legs per segment, and millipedes have two pairs per segment. (p. 671)

Major Classes of Arthropods: Insects

- Class Insecta is the largest group of organisms on the planet, living in nearly every possible habitat. (p. 672)
- Most are relatively small, and contain three body sections: head, thorax, and abdomen, with three pairs of legs attached to the thorax, and one pair of antennae. (p. 674)
- Most insects have compound eyes. (p. 674)
- Sensory hairs, tympanal organs, and chemoreceptors all act as sense receptors in insects. (p. 675)
- Many insects undergo either simple or complex metamorphosis. (p. 675)

33.5 Echinoderms are radially symmetrical as adults.

Deuterostome Development and an Endoskeleton

- Echinoderms are marine animals with hard calcium plates forming a true endoskeleton in young individuals. (p. 676)

Echinoderm Body Plan

- All echinoderms are bilaterally symmetrical during larval development, and become radially symmetrical as adults. (p. 678)
- Echinoderms have a five-part body plan with a central, branched nerve ring and an endoskeleton composed of calcium-rich plates (ossicles). (p. 678)
- Many echinoderms can regenerate lost parts, but in most of them, reproduction is sexual and external. (p. 679)

Major Classes of Echinoderms

- There are six living classes of Echinoderms: Crinoidea (sea lilies and feather stars), Asteroidea (sea stars or starfish), Ophiuroidea (brittle stars), Echinoidea (sea urchins and sand dollars), Holothuroidea (sea cucumbers), and Concentricycloidea (sea daisies). (p. 680)

Self Test

1. Of the mollusks, snails are in the class of
   a. gastropods.
   b. bivalves.
   c. cephalopods.
   d. chitons.
2. A mantle is
   a. present only in bivalves.
   b. a structure that acts as a lung or contains gills.
   c. a rasping, tonguelike organ in mollusks.
   d. necessary for mollusks to be motile.
3. Segmentation was first apparent in the
   a. flatworms.
   b. annelids.
   c. mollusks.
   d. arthropods.
4. Which of the following is not present in polychaetes?
   a. a coelom
   b. parapodia
   c. permanent gonads
   d. setae
5. The lophophore, the structure characteristic of lophophorates,
   a. functions in gas exchange.
   b. functions in feeding.
   c. can be withdrawn when the animal is disturbed.
   d. All of these are correct.
6. The phylum that shows the greatest diversity, or the greatest number of species, is
   a. Arthropoda.
   b. Brachiopoda.
   c. Echinodermata.
   d. Mollusca.
7. Air for respiration enters the insect body through the
   a. tracheae.
   b. spiracles.
   c. tracheoles.
   d. Malpighian tubules.
8. Arthropods shed their old exoskeleton as they grow in a process known as
   a. tagmatization.
   b. metamorphosis.
   c. chrysalis.
   d. ecdysis.
9. Which animal group has radial symmetry, a water-vascular system, moves with tube feet, and has an endoskeleton?
   a. arachnids
   b. crustaceans
   c. echinoderms
   d. cnidarians
10. The echinoderms that lack distinct arms are the
    a. brittle stars.
    b. sea urchins.
    c. sea stars.
    d. Asteroidea.

Test Your Visual Understanding

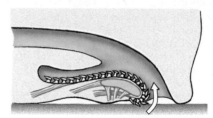

1. The radula shown in the figure above is a feeding structure found in individuals of the phylum Mollusca but is lacking in one group of mollusks. What group of mollusks does not have a radula? How do individuals in this group eat?

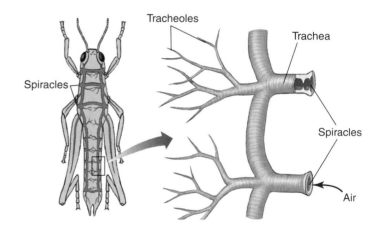

2. Horticultural oils are sometimes used as insecticides to eliminate insect pests from foliage by coating the insect with oil. Referring to the figure above, can you explain how this method of insecticidal control works?

Apply Your Knowledge

1. Freshwater bivalves are an important ecological resource because they filter freshwater systems. A population of freshwater bivalves located in a meter-squared area of substratum filters 10 m³ of water a day. How many liters of water are filtered per day (1 m³ equals 1000 liters)? How many liters of water are filtered per hour? How many liters of water would be filtered by a population filling a 5 m² area of substratum?
2. Scientists believe the ancestral mollusk had a very limited shell, consisting mainly of calcareous plates. The shell became more developed in some groups but was lost in others. What is the evolutionary advantage of having a shell? Of not having one?
3. Although arthropods are very successful in aquatic environments, what are the key adaptations that facilitated the invasion of the land by arthropods?
4. Why is it believed that echinoderms and chordates, which are so dissimilar, are members of the same evolutionary line?

34

Vertebrates

Concept Outline

34.1 Attaching muscles to an internal framework greatly improves movement.

    The Chordates. Chordates have an internal flexible rod, the first stage in the evolution of a truly internal skeleton.

34.2 Nonvertebrate chordates have a notochord but no backbone.

    The Nonvertebrate Chordates. Lancelets are thought to resemble the ancestors of vertebrates.

34.3 The evolution of vertebrates involved invasions of sea, land, and air.

    Characteristics of Vertebrates. Vertebrates have a true, usually bony endoskeleton, with a backbone encasing the spinal column, and a skull-encased brain.

    Fishes. Over half of all vertebrate species are fishes, which include the group from which all other vertebrates evolved.

    Amphibians. The key innovation that made life on land possible for vertebrates was the pulmonary vein.

    Reptiles. Reptiles were the first vertebrates to completely master the challenge of living on dry land.

    The Rise and Fall of Dominant Reptile Groups. Now-extinct forms of reptiles dominated life on land for 250 million years. Four orders survive today.

    Birds. Birds are much like reptiles, but with feathers.

    History of the Birds. Birds are thought to have evolved from dinosaurs with adaptations of feathers and flight.

    Mammals. Mammals are the only vertebrates that possess hair and mammary glands.

    The Orders of Mammals. Mammals only became common when dinosaurs disappeared.

34.4 Evolution among the primates has focused on brain size and locomotion.

    Primates. Only primates have binocular vision and grasping hands.

    Australopithecines. The evolution of bipedalism was the first step on the hominid evolutionary path.

    The Genus *Homo*. The genus of which humans are a member evolved in Africa.

FIGURE 34.1
A large vertebrate. Today, mammals, such as this elephant, dominate vertebrate life on land, but for over 200 million years, mammals were a minor group in a world dominated by reptiles.

Members of the phylum Chordata exhibit great improvements in the endoskeleton over what is seen in echinoderms. As we saw in chapter 33, the endoskeleton of echinoderms is functionally similar to the exoskeleton of arthropods; it is a hard shell that encases the body, with muscles attached to its inner surface. Chordates employ a very different kind of endoskeleton, one that is truly internal. Members of the phylum Chordata are characterized by a flexible rod that develops along the back of the embryo. Muscles attached to this rod allowed early chordates to swing their backs from side to side, swimming through the water. This key evolutionary advance, attaching muscles to an internal element, started chordates along an evolutionary path that led to the vertebrates—and, for the first time, to truly large animals (figure 34.1).

34.1 Attaching muscles to an internal framework greatly improves movement.

The Chordates

Chordates (phylum Chordata) are deuterostome coelomates whose nearest relatives in the animal kingdom are the echinoderms, the only other deuterostomes. However, unlike echinoderms, chordates are characterized by a *notochord, jointed appendages,* and *segmentation.* There are some 56,000 species of chordates, a phylum that includes birds, reptiles, amphibians, fishes, and mammals.

Four features characterize the chordates and have played an important role in the evolution of the phylum (figure 34.2):

1. A single, hollow **nerve cord** runs just beneath the dorsal surface of the animal. In vertebrates, the dorsal nerve cord differentiates into the brain and spinal cord.

2. A flexible rod, the **notochord,** forms on the dorsal side of the primitive gut in the early embryo and is present at some developmental stage in all chordates. The notochord is located just below the nerve cord. The notochord may persist throughout the life cycle of some chordates or be displaced during embryonic development, as in most vertebrates, by the vertebral column that forms around the nerve cord.

3. **Pharyngeal slits** connect the **pharynx,** a muscular tube that links the mouth cavity and the esophagus, with the outside. In terrestrial vertebrates, the slits do not actually connect to the outside and are better termed **pharyngeal pouches.** Pharyngeal pouches are present in the embryos of all vertebrates. They become slits, open to the outside in animals with gills, but disappear in those lacking gills. The presence of these structures in all vertebrate embryos provides evidence of their aquatic ancestry.

4. Chordates have a **postanal tail** that extends beyond the anus, at least during their embryonic development. Nearly all other animals have a terminal anus.

All chordates have all four of these characteristics at some time in their lives. For example, humans have pharyngeal pouches, a dorsal nerve cord, a postanal tail, and a notochord as embryos. As adults, the nerve cord remains, while the notochord is replaced by the vertebral column, and all but one pair of pharyngeal pouches are lost. This remaining pair forms the Eustachian tubes that connect the throat to the middle ear. The postanal tail regresses forming the tail bone.

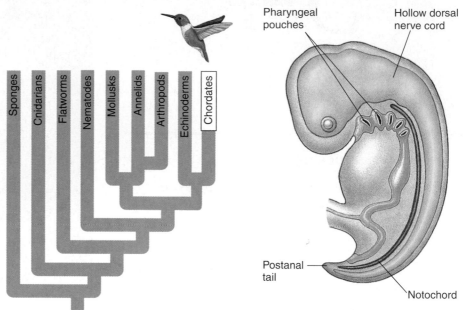

FIGURE 34.2
The four principal features of the chordates, as shown in a generalized embryo.

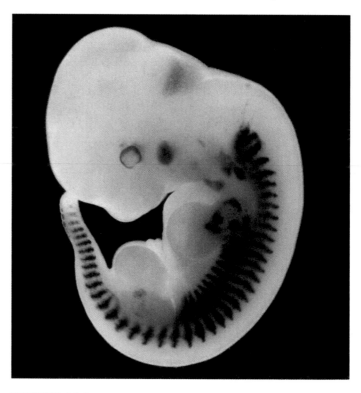

FIGURE 34.3
A mouse embryo. At 11.5 days of development, the mesoderm is already divided into segments called somites (stained dark in this photo), reflecting the fundamentally segmented nature of all chordates.

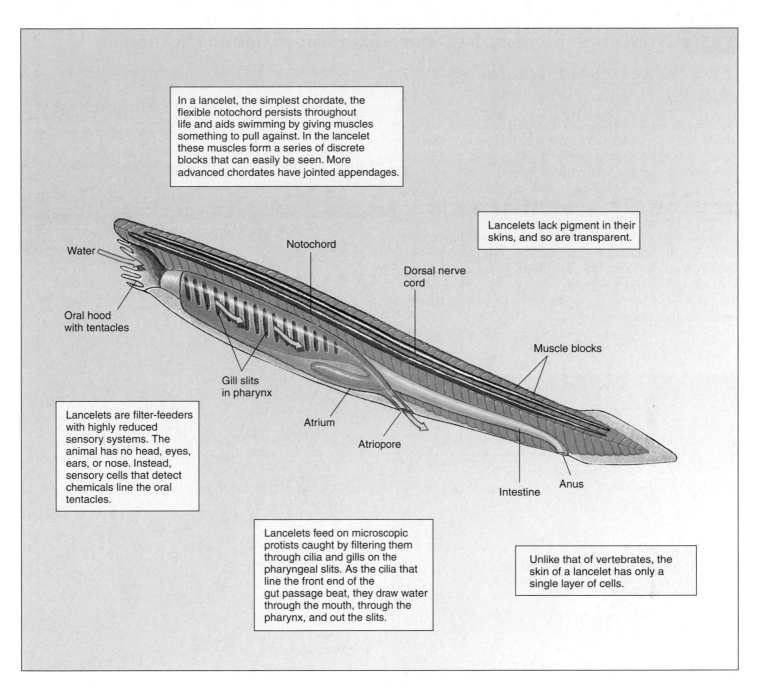

In a lancelet, the simplest chordate, the flexible notochord persists throughout life and aids swimming by giving muscles something to pull against. In the lancelet these muscles form a series of discrete blocks that can easily be seen. More advanced chordates have jointed appendages.

Lancelets lack pigment in their skins, and so are transparent.

Water

Notochord

Dorsal nerve cord

Oral hood with tentacles

Muscle blocks

Gill slits in pharynx

Lancelets are filter-feeders with highly reduced sensory systems. The animal has no head, eyes, ears, or nose. Instead, sensory cells that detect chemicals line the oral tentacles.

Atrium

Atriopore

Intestine

Anus

Lancelets feed on microscopic protists caught by filtering them through cilia and gills on the pharyngeal slits. As the cilia that line the front end of the gut passage beat, they draw water through the mouth, through the pharynx, and out the slits.

Unlike that of vertebrates, the skin of a lancelet has only a single layer of cells.

FIGURE 34.4

Phylum Chordata: chordates. Vertebrates, tunicates, and lancelets are chordates (phylum Chordata), coelomate animals with a flexible rod, the notochord, that provides resistance to muscle contraction and permits rapid lateral body movements. Chordates also possess pharyngeal pouches and slits (reflecting their aquatic ancestry and present habitat in some) and a hollow dorsal nerve cord. In vertebrates, the notochord is replaced during embryonic development by the vertebral column.

A number of other characteristics also distinguish the chordates fundamentally from other animals. Chordates' muscles are arranged in segmented blocks that affect the basic organization of the chordate body and can often be clearly seen in embryos of this phylum (figure 34.3). Most chordates have an internal skeleton against which the muscles work. Either this internal skeleton or the notochord (figure 34.4) makes possible the extraordinary powers of locomotion that characterize the members of this group.

> **Chordates are characterized by a hollow dorsal nerve cord, a notochord, pharyngeal pouches, and a postanal tail at some point in their development. The flexible notochord anchors internal muscles and allows rapid, versatile movement.**

34.2 Nonvertebrate chordates have a notochord but no backbone.

The Nonvertebrate Chordates

Tunicates

The tunicates (subphylum Urochordata) are a group of about 1250 species of marine animals. Most of them are sessile as adults, with only the larvae having a notochord and nerve cord. As adults, they exhibit neither a major body cavity nor visible signs of segmentation (figure 34.5*a,b*). Most species occur in shallow waters, but some are found at great depths. In some tunicates, adults are colonial, living in masses on the ocean floor. The pharynx is lined with numerous cilia, and the animals obtain their food by ciliary action. The cilia beat, drawing a stream of water into the pharynx, where microscopic food particles are trapped in a mucous sheet secreted from a structure called an *endostyle*.

The tadpolelike larvae of tunicates plainly exhibit all of the basic characteristics of chordates and mark the tunicates as having the most primitive combination of features found in any chordate (figure 34.5*c*). The larvae do not feed and have a poorly developed gut. They remain free-swimming for only a few days before settling to the bottom and attaching themselves to a suitable substrate by means of a sucker.

Tunicates change so much as they mature and adjust developmentally to a sessile, filter-feeding existence that it would be difficult to discern their evolutionary relationships by examining an adult. Many adult tunicates secrete a **tunic,** a tough sac composed mainly of cellulose, a substance frequently found in the cell walls of plants and algae but rarely found in animals. The tunic surrounds the animal and gives the subphylum its name. Colonial tuni-

(a)

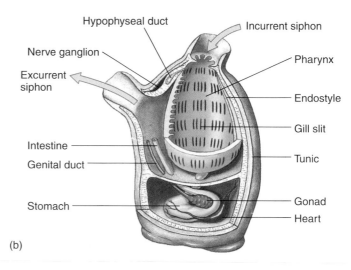

(b)

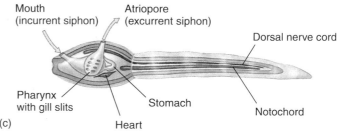

(c)

FIGURE 34.5

Tunicates (phylum Chordata, subphylum Urochordata).
(*a*) The sea peach, *Halocynthia auranthium.* (*b*) Diagram of the structure of an adult tunicate. (*c*) Diagram of the structure of a larval tunicate, showing the characteristic tadpole-like form. Larval tunicates resemble the postulated common ancestor of the chordates.

cates may have a common sac and a common opening to the outside. A group of Urochordates, the Larvacea, retains the tail and notochord into adulthood. One theory of vertebrate origins involves a larval form, perhaps that of a tunicate, which acquires the ability to reproduce.

Lancelets

Lancelets are scaleless, fishlike marine chordates a few centimeters long that occur widely in shallow water throughout the oceans of the world. Lancelets (subphylum Cephalochordata) were given their English name because they resemble a lancet—a small, two-edged surgical knife. There are about 23 species of this subphylum. Most of them belong to the genus *Branchiostoma*, formerly called *Amphioxus*, a name still used widely. In lancelets, the notochord runs the entire length of the dorsal nerve cord and persists throughout the animal's life.

Lancelets spend most of their time partly buried in sandy or muddy substrates, with only their anterior ends protruding (figure 34.6). They can swim, although they rarely do so. Their muscles can easily be seen as a series of discrete blocks. Lancelets have many more pharyngeal gill slits than fishes, which they resemble in overall shape. They lack pigment in their skin, which has only a single layer of cells, unlike the multilayered skin of vertebrates. The lancelet body is pointed at both ends. There is no distinguishable head or sensory structures other than pigmented light receptors.

Lancelets feed on microscopic plankton, using a current created by beating cilia that line the oral hood, pharynx, and gill slits (figure 34.7). The gill slits provide an exit for the water and are an adaptation for filter feeding. The oral hood projects beyond the mouth and bears sensory tentacles, which also ring the mouth. Males and females are separate, but no obvious external differences exist between them.

Biologists are not sure whether lancelets are primitive or are actually degenerate fishes whose structural features have been reduced and simplified during the course of evolution. The fact that lancelets feed by means of cilia and have a single-layered skin, coupled with distinctive features of their excretory systems, suggests that this is an ancient group of chordates. The recent discovery of fossil forms

FIGURE 34.6
Lancelets. Two lancelets, *Branchiostoma lanceolatum* (phylum Chordata, subphylum Cephalochordata), partly buried in shell gravel, with their anterior ends protruding. The muscle segments are clearly visible; the square objects along the side of the body are gonads.

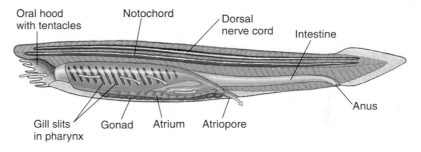

FIGURE 34.7
The structure of a lancelet. This diagram shows the path through which the lancelet's cilia pull water.

similar to living lancelets in rocks 550 million years old—well before the appearance of any fishes—also argues for the antiquity of this group. Recent studies by molecular systematists further support the hypothesis that lancelets are vertebrates' closest ancestors.

Nonvertebrate chordates, including tunicates and lancelets, have notochords but no vertebrae. They are the closest relatives of vertebrates.

34.3 The evolution of vertebrates involved invasions of sea, land, and air.

Characteristics of Vertebrates

Vertebrates (subphylum Vertebrata) are chordates with a spinal column. The name *vertebrate* comes from the individual bony or cartilaginous segments called vertebrae that make up the spine. Vertebrates differ from the tunicates and lancelets in two important respects:

1. **Vertebral column.** In all vertebrates except the earliest fishes, the notochord is replaced during the course of embryonic development by a vertebral column (figure 34.8). The column is a series of bony or cartilaginous vertebrae that enclose and protect the dorsal nerve cord like a sleeve.
2. **Head.** Vertebrates have a distinct and well-differentiated head; the brain is fully encased within a protective box, the skull or cranium, made of bone or cartilage. For this reason, the vertebrates are sometimes called the **craniate chordates.**

In addition to these two key characteristics, vertebrates differ from other chordates in other important respects (figure 34.9):

1. **Neural crest.** A unique group of embryonic cells called the neural crest contributes to the development of many vertebrate structures. These cells develop on the crest of the neural tube as it forms by invagination and pinching together of the neural plate (see chapter 51 for a detailed account). Neural crest cells then migrate to various locations in the developing embryo, where they participate in the development of a variety of structures.
2. **Internal organs.** Internal organs characteristic of vertebrates include a liver, kidneys, and endocrine glands. The ductless endocrine glands secrete hormones that help regulate many of the body's functions. All vertebrates have a heart and a closed circulatory system. In both their circulatory and their excretory functions, vertebrates differ markedly from other animals.
3. **Endoskeleton.** The endoskeleton of most vertebrates is made of cartilage or bone. Cartilage and bone are specialized tissue containing fibers of the protein collagen compacted together. Bone also contains crystals of a calcium phosphate salt. Bone forms in two stages. First, collagen is laid down in a matrix of fibers along stress lines to provide flexibility, and then calcium minerals infiltrate the fibers, providing rigidity. The great advantage of bone over chitin as a structural material is that bone is strong without being brittle. The vertebrate endoskeleton makes possible the great size and extraordinary powers of movement that characterize this group.

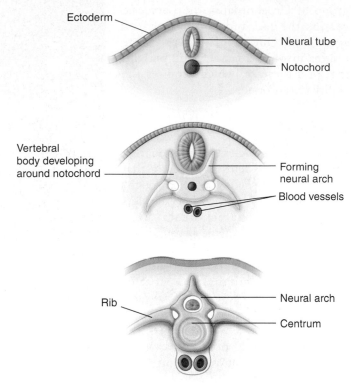

FIGURE 34.8
Embryonic development of a vertebra. During the evolution of animal development, the flexible notochord is surrounded and eventually replaced by a cartilaginous or bony covering, the centrum. The neural tube is protected by an arch above the centrum, and the vertebra may also have a hemal arch, which protects major blood vessels below the centrum. The vertebral column functions as a strong, flexible rod that the muscles pull against when the animal swims or moves.

Overview of the Evolution of Vertebrates

The first vertebrates evolved in the oceans about 470 million years ago. They were jawless fishes with a single caudal fin. Many of them looked like a flat hot dog, with a hole at one end and a fin at the other. The appearance of a hinged jaw was a major advancement, opening up new food options, and jawed fishes became the dominant creatures in the sea. Their descendants, the amphibians, invaded the land. Salamander-like amphibians and other, much larger, now-extinct amphibians were the first vertebrates to live successfully on land. Amphibians, in turn, gave rise to the first reptiles about 300 million years ago. Within 50 million years, reptiles, better suited than amphibians to living out of water, replaced them as the dominant land vertebrates.

With the success of reptiles, vertebrates truly came to dominate the surface of the earth. Many kinds of reptiles evolved, ranging in size from smaller than a chicken to

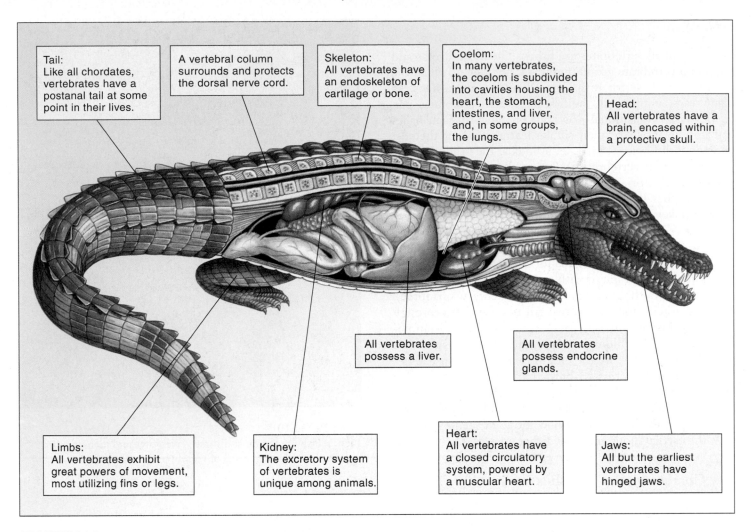

Tail:
Like all chordates, vertebrates have a postanal tail at some point in their lives.

A vertebral column surrounds and protects the dorsal nerve cord.

Skeleton:
All vertebrates have an endoskeleton of cartilage or bone.

Coelom:
In many vertebrates, the coelom is subdivided into cavities housing the heart, the stomach, intestines, and liver, and, in some groups, the lungs.

Head:
All vertebrates have a brain, encased within a protective skull.

All vertebrates possess a liver.

All vertebrates possess endocrine glands.

Limbs:
All vertebrates exhibit great powers of movement, most utilizing fins or legs.

Kidney:
The excretory system of vertebrates is unique among animals.

Heart:
All vertebrates have a closed circulatory system, powered by a muscular heart.

Jaws:
All but the earliest vertebrates have hinged jaws.

FIGURE 34.9
Major characteristics of vertebrates. Adult vertebrates are characterized by an internal skeleton of cartilage or bone, including a vertebral column and a skull. Several other internal and external features are characteristic of vertebrates.

bigger than a truck. Some flew, and others swam. Among them evolved reptiles that gave rise to the two remaining great lines of terrestrial vertebrates, birds (descendants of the dinosaurs) and mammals. Dinosaurs and mammals appear at about the same time in the fossil record, 220 million years ago. For over 150 million years, dinosaurs dominated the face of the earth. Over all these centuries (think of it—over *a million centuries!*), the largest mammal was no bigger than a cat. Then, about 65 million years ago, the dinosaurs abruptly disappeared, for reasons that are still hotly debated. In their absence, mammals and birds quickly took their place, becoming in turn abundant and diverse.

The history of vertebrates has been a series of evolutionary advances that have allowed vertebrates to first invade the sea and then the land. In this chapter, we examine the key evolutionary advances that permitted vertebrates to invade the land successfully. As you will see, this invasion was a staggering evolutionary achievement, involving fundamental changes in many body systems.

Vertebrates are a diverse group, containing members adapted to life in aquatic habitats, on land, and in the air. There are nine principal classes of living vertebrates. Five of the classes are fishes that live in the water, and four are land-dwelling **tetrapods,** animals with four limbs. (The name tetrapod comes from two Greek words meaning "four-footed.") The extant classes of fishes are (1) the class Myxini, the hagfish, and (2) the class Cephalaspidomorphi, the lampreys, which together comprise the superclass Agnatha (the jawless fishes); (3) the class Chondrichthyes, the cartilaginous fishes, sharks, skates, and rays; and (4) class Actinopterygii and (5) class Sarcopterygii, comprising the bony fishes that are dominant today. The four classes of tetrapods are Amphibia, the amphibians; Reptilia, the reptiles; Aves, the birds; and Mammalia, the mammals.

Vertebrates, the principal chordate group, are characterized by a vertebral column and a distinct head.

Fishes

Over half of all vertebrates are fishes. The most diverse and successful vertebrate group (figure 34.10), they provided the evolutionary base for invasion of land by amphibians. In many ways, amphibians, the first terrestrial vertebrates, can be viewed as transitional—fish out of water. In fact, fishes and amphibians share many similar features, among the host of obvious differences. First, let us look at the fishes.

The story of vertebrate evolution started in the ancient seas of the Cambrian period (545 to 490 million years ago), when the first backboned animals appeared. Figure 34.11 shows the key vertebrate characteristics that evolved subsequently. Wriggling through the water, jawless and toothless, the first fishes sucked up small food particles from the ocean floor like miniature vacuum cleaners. Most were less than a foot long, respired with gills, and had no paired fins or vertebrae (although some had rudimentary vertebrae); they had a head and a primitive tail to push them through the water. For 50 million years, during the Ordovician period (490 to 438 million years ago), these simple fishes were the only vertebrates. By the end of this period, fish had developed primitive fins to help them swim and massive shields of bone for protection. Jawed fishes first appeared during the Silurian period (438 to 408 million years ago), and along with them came a new mode of feeding. Later, both the cartilaginous and bony fishes appeared.

FIGURE 34.10

Fish. Fish are the most diverse vertebrates and include more species than all other kinds of vertebrates combined.

Key Characteristics of Fishes

From whale sharks 18 meters long to tiny cichlids no larger than your fingernail, fishes vary considerably in size, shape, color, and appearance. Some live in freezing arctic seas, others in warm, freshwater lakes, and still others spend a lot of time out of water entirely. However varied, all fishes have important characteristics in common :

1. **Vertebral column.** Fish have an internal skeleton with a bony or cartilaginous spine surrounding the dorsal nerve cord, and a bony or cartilaginous skull encasing the brain. An exception is the agnathans, composed of the hagfish and lampreys. In hagfish, a cartilaginous skeleton and skull are present, but vertebrae are not; the notochord persists and provides support. In lampreys, a cartilaginous skeleton and notochord are present, but rudimentary cartilaginous vertebrae also surround the notochord in places.

2. **Jaws and paired appendages.** Fishes other than agnathans all have jaws and paired appendages, features that are seen in all tetrapods too (figure 34.11). Jaws allowed the predation of larger and active prey. Bony fishes and cartilaginous fishes have two pairs of fins—a pair of pectoral fins at the shoulder and a pair of pelvic fins at the hip. Paired appendages greatly improved maneuverability and are the precursors of arms and legs in tetrapods. In the lobe-finned fish, these pairs of fins became jointed.

3. **Gills.** Fishes are water-dwelling creatures and must extract oxygen dissolved in the water around them. They do this by directing a flow of water through their mouths and across their gills. The gills are composed of fine filaments of tissue that are rich in blood vessels. They are located at the back of the pharynx and supported by arches of cartilage. Blood moves through the gills in the opposite direction to the flow of water in order to maximize the efficiency of oxygen absorption (see chapter 44).

4. **Single-loop blood circulation.** Blood is pumped from the heart to the gills. From the gills, the oxygenated blood passes to the rest of the body, and then returns to the heart. The heart is a muscular tube-pump made of four chambers that contract in sequence.

5. **Nutritional deficiencies.** Fishes are unable to synthesize the aromatic amino acids and must consume them in their diet. This inability has been inherited by all their vertebrate descendants.

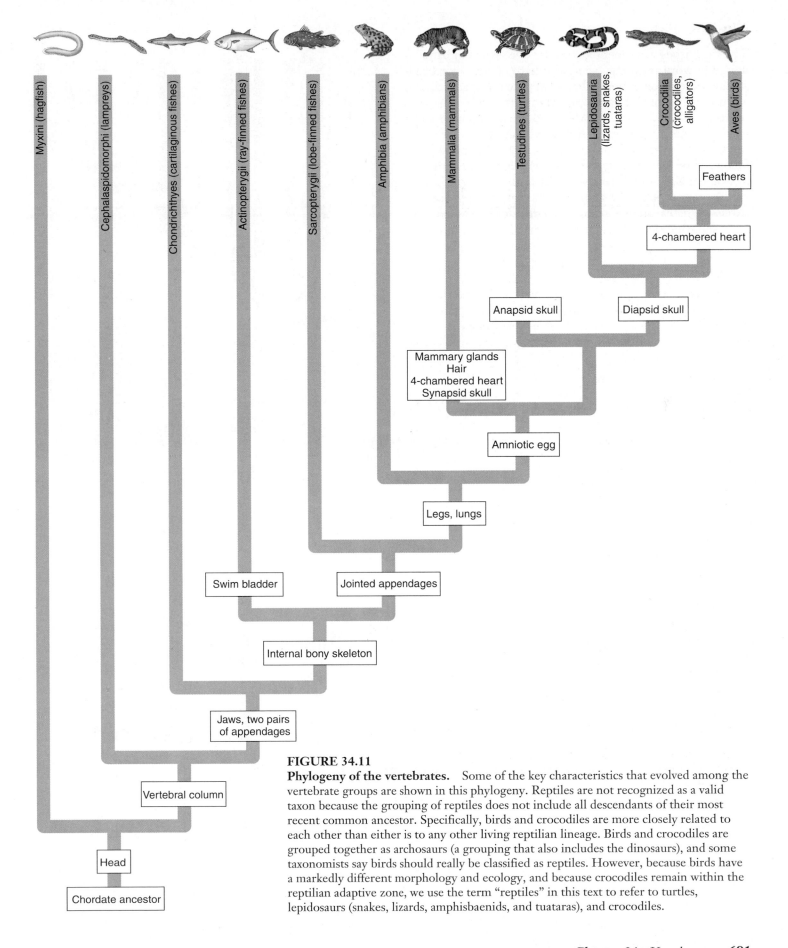

FIGURE 34.11

Phylogeny of the vertebrates. Some of the key characteristics that evolved among the vertebrate groups are shown in this phylogeny. Reptiles are not recognized as a valid taxon because the grouping of reptiles does not include all descendants of their most recent common ancestor. Specifically, birds and crocodiles are more closely related to each other than either is to any other living reptilian lineage. Birds and crocodiles are grouped together as archosaurs (a grouping that also includes the dinosaurs), and some taxonomists say birds should really be classified as reptiles. However, because birds have a markedly different morphology and ecology, and because crocodiles remain within the reptilian adaptive zone, we use the term "reptiles" in this text to refer to turtles, lepidosaurs (snakes, lizards, amphisbaenids, and tuataras), and crocodiles.

The First Fishes

The first fishes were members of the five orders of ostraco-derms (a word meaning "shell-skinned"). Only their head-shields were made of bone; their elaborate internal skeletons were constructed of cartilage. Many ostracoderms were bottom-dwellers, with a jawless mouth underneath a flat head, and eyes on the upper surface. Ostracoderms thrived in the Ordovician period (figure 34.12) and in the period that followed, the Silurian period (438–408 million years ago), only to become almost completely extinct at the close of the Devonian period (408–360 million years ago). One group, the jawless Agnatha, survive today as hagfish (class Myxini; table 34.1) and parasitic lampreys (class Cephalaspidomorphi).

Evolution of the Jaw

A fundamentally important evolutionary advance that oc-curred in the late Silurian period, 410 million years ago, was the development of jaws. Jaws evolved from the most anterior of a series of arch-supports made of cartilage that

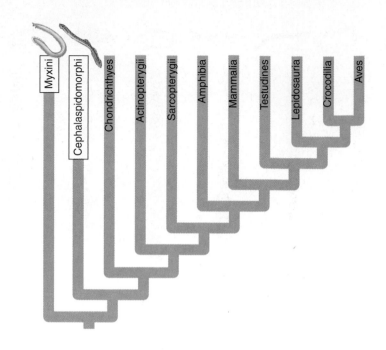

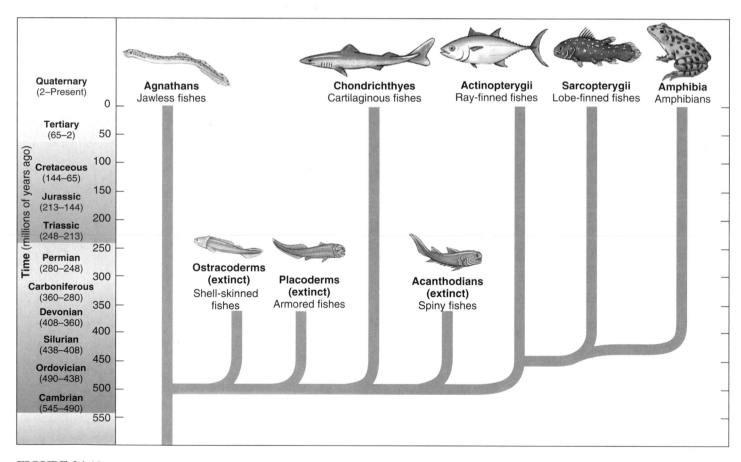

FIGURE 34.12
Evolution of the fishes. The evolutionary relationships among the different groups of fishes as well as between fishes and amphibians are shown. The spiny and armored fishes that dominated the early seas are now extinct.

Table 34.1 Major Classes of Fishes

Class	Typical Examples		Key Characteristics	Approximate Number of Living Species
Actinopterygii	Ray-finned fishes		Most diverse group of vertebrates; swim bladders and bony skeletons; paired fins supported by bony rays	30,000
Sarcopterygii	Lobe-finned fishes		Largely extinct group of bony fishes; ancestral to amphibians; paired lobed fins	7
Chondrichthyes	Sharks, skates, rays		Streamlined hunters; cartilaginous skeletons; no swim bladders; internal fertilization	750
Myxini	Hagfishes		Jawless fishes with no paired appendages; scavengers; mostly blind, but having a well-developed sense of smell	30
Cephalaspidomorphi	Lampreys		Largely extinct group of jawless fishes with no paired appendages; parasitic and nonparasitic types; all breed in fresh water	35
Placodermi	Armored fishes		Jawed fishes with heavily armored heads; many were quite large	Extinct
Acanthodii	Spiny fishes		Fishes with jaws; all now extinct; paired fins supported by sharp spines	Extinct

were used to reinforce the tissue between gill slits, holding the slits open (figure 34.13). This transformation was not as radical as it might at first appear. Each gill arch was formed by a series of several cartilages (later to become bones) arranged somewhat in the shape of a V turned on its side, with the point directed outward. Imagine the fusion of the front pair of arches at top and bottom, with hinges at the points, and you have the primitive vertebrate jaw. The top half of the jaw is not attached to the skull directly except at the rear. Teeth developed on the jaws from modified scales on the skin that lined the mouth.

Armored fishes called placoderms and spiny fishes called acanthodians both had jaws. Spiny fishes were very common during the early Devonian period, largely replacing ostracoderms, but became extinct themselves at the close of the Permian. Like ostracoderms, they had internal skeletons made of cartilage, but their scales contained small plates of bone, foreshadowing the much larger role bone would play in the future of vertebrates. Spiny fishes were predators and far better swimmers than ostracoderms, with as many as seven fins to aid their swimming. All of these fins were reinforced with strong spines, giving these fishes their name. No spiny fishes survive today.

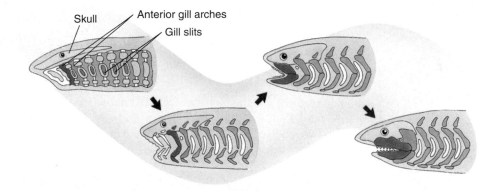

FIGURE 34.13
Evolution of the jaw. Jaws evolved from the anterior gill arches of ancient, jawless fishes.

By the mid-Devonian period, the heavily armored placoderms became common. A very diverse and successful group, seven orders of placoderms dominated the seas of the late Devonian, only to become extinct at the end of that period. The front of the placoderm body was more heavily armored than the rear. The placoderm jaw was much improved over the primitive jaw of spiny fishes, with the upper jaw fused to the skull and the skull hinged on the shoulder. Many of the placoderms grew to enormous sizes, some over 30 feet long, with two-foot skulls that had an enormous bite.

GEOLOGIC TIME SCALE	ERAS	PERIOD	EPOCH	MILLIONS OF YEARS AGO	BIOLOGICAL EVENTS
Cenozoic	Cenozoic	Quaternary	Recent	(.01-0)	Human civilization
Mesozoic			Pleistocene	(1.8-.01)	Ice ages; modern humans (genus *Homo*) appear
		Tertiary	Pliocene	(5-1.8)	Hominids appear; large carnivores
Paleozoic			Miocene	(23-5)	Apes arise; climate cooler; plains and grasslands
			Oligocene	(36-23)	Monkeys appear; mild climate
			Eocene	(57-36)	Radiation of flowering plants; most modern mammalian orders represented
			Paleocene	(65-57)	Radiation of mammals, birds, and insects; tropical conditions
				MASS EXTINCTION	
	Mesozoic	Cretaceous		(144-65)	Flowering plants appear; climax of dinosaurs followed by extinction
		Jurassic		(213-144)	First birds; dinosaurs radiate; lush forests
				MASS EXTINCTION	
Precambrian		Triassic		(248-213)	Conifers dominant; dinosaurs arise; first mammals
				MASS EXTINCTION (96% OF SPECIES DISAPPEAR)	
		Permian		(280-248)	Radiation of reptiles; mammal-like reptiles arise
		Carboniferous		(360-280)	Forest of ferns and conifers; amphibians arise; insects radiate
				MASS EXTINCTION	
	Paleozoic	Devonian		(408-360)	Sharks and bony fishes appear; lobe-finned fishes evolve
		Silurian		(438-408)	Jawless fishes diversify; first jawed fishes; colonization of land by plants and arthropods
				MASS EXTINCTION	
		Ordovician		(490-438)	First vertebrates (jawless fishes); first plants
		Cambrian		(545-490)	Cambrian explosion: nearly all the major animal body plans arise; earliest chordates
	Precambrian			600	Soft-bodied invertebrates and algae diversify
				900	Multicellular organisms appear
				1500	Oldest definite fossils of eukaryotes
				2500	Oldest definite fossils of prokaryotes
				4500	Formation of the Earth

FIGURE 34.14

The geological time scale. Vertebrates appear soon after the Cambrian period, and have played a major role in life's diversity ever since.

The Rise of Active Swimmers

At the end of the Devonian period, essentially all of these pioneer vertebrates disappeared, replaced by sharks and bony fishes in one of several mass extinctions that occurred during earth's geological and biological history (figure 34.14). Sharks and bony fishes first evolved in the early Devonian, 400 million years ago. In these fishes, the jaw was improved even further, with the first gill arch behind the jaws being transformed into a supporting strut or prop, joining the rear of the lower jaw to the rear of the skull. This allowed the mouth to open very wide, into almost a full circle. In a great white shark, this wide-open mouth can be a very efficient weapon.

The major factor responsible for the replacement of primitive fishes by sharks and bony fishes was that they had a superior design for swimming. The typical shark or bony fish is streamlined. The head of the fish acts as a wedge to cleave through the water, and the body tapers back to the tail, allowing the fish to slip through the water with a minimum amount of turbulence.

In addition, sharks and bony fishes have an array of mobile fins that greatly aid swimming. First, a propulsion fin, a large and efficient tail (caudal) fin, helps drive the fish through the water when it is swept side-to-side, pushing against the water and thrusting the fish forward. Second, stabilizing fins, one (or sometimes two) dorsal fins on the back, act as a stabilizer to prevent rolling as the fish swims through the water, while another ventral fin acts as a keel to prevent side-slip. Third, paired fins at shoulder and hip ("a fin at each corner") consist of a front (pectoral) pair and a rear (pelvic) pair. These fins act like the elevator flaps of an airplane to assist the fish in going up or down through the water, as rudders to help it turn sharply left or right, and as brakes to help it stop quickly.

Sharks Become Top Predators

In the period following the Devonian, the Carboniferous period (360 to 280 million years ago), sharks became the dominant predator in the sea. Sharks (class Chondrichthyes) have a skeleton made of cartilage, like primitive fishes, but it is "calcified," strengthened by granules of calcium carbonate deposited in the outer layers of cartilage. The result is a very light and strong skeleton. Streamlined, with paired fins and a light, flexible skeleton, sharks are superior swimmers (figure 34.15). Their pectoral fins are particularly large, jutting out stiffly like airplane wings—and that is how they function, adding lift to compensate for the downward thrust of the tail fin. Sharks are very aggressive predators, and some early sharks reached enormous size.

Sharks were among the first vertebrates to develop teeth. These teeth evolved from rough scales on the skin and are not set into the jaw as human teeth are, but rather sit atop it. The teeth are not firmly anchored and are easily lost. In a shark's mouth, the teeth are arrayed in up to 20 rows; the teeth in front do the biting and cutting, while be-

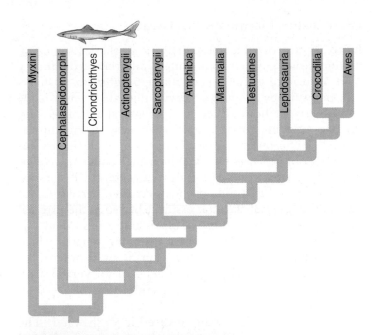

FIGURE 34.15 Chondrichthyes. Members of the class Chondrichthyes, such as this blue shark, are mainly predators or scavengers.

hind them other teeth grow and wait their turn. When a tooth breaks or is worn down, a replacement from the next row moves forward. One shark may eventually use more than 20,000 teeth. This programmed loss of teeth offers a great advantage: The teeth in use are always new and sharp. A shark's skin is covered with tiny, teethlike scales, giving it a rough "sandpaper" texture. Like the teeth, these scales are constantly replaced throughout the shark's life.

Reproduction among the class Chondrichthyes is the most advanced of any fishes. Shark eggs are fertilized internally. During mating, the male grasps the female with modified fins called claspers. Sperm run from the male into the female through grooves in the claspers. Although a few species lay fertilized eggs, the eggs of most species develop within the female's body, and the pups are born alive.

Many of the early evolutionary lines of sharks died out during the great extinction at the end of the Permian period (280 to 248 million years ago). The survivors thrived and underwent a burst of diversification during the Mesozoic era, when most of the modern groups of sharks appeared. Skates and rays (flattened sharks that are bottom-dwellers) evolved at this time, some 200 million years after the sharks first appeared. Sharks competed successfully with the marine reptiles of that time and are still the dominant predators of the sea. Today, there are 275 species of sharks, more kinds than existed in the Carboniferous period.

Bony Fishes Dominate the Water

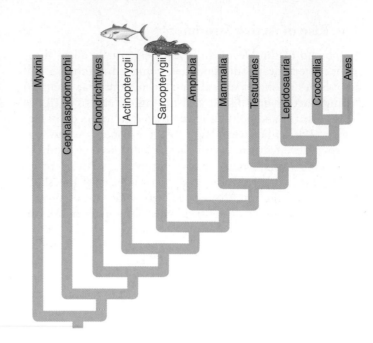

Bony fishes evolved at the same time as sharks, some 400 million years ago, but took quite a different evolutionary road. Instead of gaining speed through lightness, as sharks did, bony fishes adopted a heavy internal skeleton made completely of bone. Such an internal skeleton is very strong, providing a base against which very strong muscles can pull. The process of *ossification* (the evolutionary replacement of cartilage by bone) happened suddenly in evolutionary terms, completing a process started by sharks, who lay down a thin film of bone over their cartilage. Not only is the internal skeleton ossified, but so is the external skeleton, the outer covering of plates and scales. Many scientists believe bony fishes evolved from spiny sharks, which also had bony plates set in their skin. Bony fishes are the most successful of all fishes, indeed of all vertebrates. There are several dozen orders containing more than 30,000 living species.

Unlike sharks, bony fishes evolved in fresh water. The most ancient fossils of bony fishes are found in freshwater lake beds from the middle Devonian period. These first bony fishes were small and possessed paired air sacs connected to the back of the throat. These sacs could be inflated with air to buoy the fish up or deflated to sink it down.

Most bony fishes have highly mobile fins, very thin scales, and completely symmetrical tails (which keep the fish on a straight course as it swims through the water). This is a very successful design for a fish. Two great groups arose from these pioneers: the lobe-finned fishes (class Sarcopterygii), ancestors of the first tetrapods, and the ray-finned fishes (class Actinopterygii), which include the vast majority of today's fishes (figure 34.16).

The characteristic feature of all ray-finned fishes is an internal skeleton of parallel bony rays that support and stiffen each fin. There are no muscles within the fins; they are moved by muscles within the body. In ray-finned fishes, the primitive air sacs are transformed into an air pouch, which provides a remarkable degree of control over buoyancy.

Important Adaptations of Bony Fishes

The remarkable success of the bony fishes has resulted from a series of significant adaptations that have enabled them to dominate life in the water. These include the swim bladder, lateral line system, and gill cover.

Swim Bladder. Although bones are heavier than cartilaginous skeletons, bony fishes are still buoyant because they possess a swim bladder, a gas-filled sac that allows them to regulate their buoyant density and so remain suspended at any depth in the water effortlessly (figure 34.17). Sharks, by contrast, must move through the water or sink, because their bodies are denser than water. In primitive bony fishes, the swim bladder is a dorsal outpocketing of the pharynx behind the throat, and these

species fill the swim bladder by simply gulping air at the surface of the water. In most of today's bony fishes, the swim bladder is an independent organ that is filled and drained of gases, mostly nitrogen and oxygen, internally. How do bony fishes manage this remarkable trick? It turns out that the gases are harvested from their blood by a unique gland that discharges the gases into the bladder when more buoyancy is required. To reduce buoyancy, gas is released by a muscular valve from the bladder back into the blood supply. A variety of physiological factors control the exchange of gases between the blood stream and the swim bladder.

Lateral Line System. Although precursors are found in sharks, bony fishes possess a fully developed lateral line system. The lateral line system consists of a series of sensory organs that project into a canal beneath the surface of the skin. The canal runs the length of the fish's body and is open to the exterior through a series of sunken pits. Movement of water past the fish forces water through the canal. The sensory organs consist of clusters of cells with

FIGURE 34.16 Ray-finned fishes. The ray-finned fishes (class Actinopterygii) are extremely diverse. This Korean angelfish in Fiji is one of the many striking fishes that live around coral reefs in tropical seas.

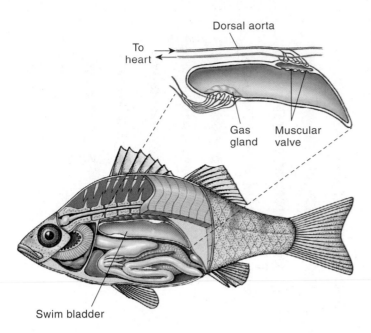

FIGURE 34.17
Diagram of a swim bladder. The bony fishes use this structure, which evolved as a dorsal outpocketing of the pharynx, to control their buoyancy in water. The swim bladder can be filled with or drained of gas to allow the fish to control buoyancy. Gases are taken from the blood, and the gas gland secretes the gases into the swim bladder; gas is released from the bladder by a muscular valve.

FIGURE 34.18
The living coelacanth, *Latimeria chalumnae*, a lobe-finned fish (class Sarcopterygii). Discovered in the western Indian Ocean in 1938, this coelacanth represents a group of fishes thought to have been extinct for about 70 million years. Scientists who studied living individuals in their natural habitat at depths of 100 to 200 meters observed them drifting in the current and hunting other fishes at night. Some individuals are nearly 3 meters long; they have a slender, fat-filled swim bladder. *Latimeria* is a strange animal, and its discovery was a complete surprise.

hairlike projections called cilia, embedded in a gelatinous cap. The hairs are deflected by the slightest movement of water over them. The pits are oriented so that some are stimulated no matter what direction the water moves (see chapter 46). Nerve impulses from these sensory organs permit the fish to assess its rate of movement through water, sensing the movement as pressure waves against its lateral line. This is how a trout orients itself with its head upstream.

The lateral line system also enables a fish to detect motionless objects at a distance by the movement of water reflected off the object. In a very real sense, this is the fish equivalent of hearing. The basic mechanism of cilia deflection by pressure waves is very similar to what happens in human ears (see chapter 46).

Gill Cover. Most bony fishes have a hard plate called the operculum that covers the gills on each side of the head. Flexing the operculum permits bony fishes to pump water over their gills. The gills are suspended in the pharyngeal slits that form a passageway between the pharynx and the outside of the fish's body. When the operculum is closed, it seals off the exit. When the mouth is open, closing the operculum increases the volume of the mouth cavity, so that water is drawn into the mouth. When the mouth is closed, opening the operculum decreases the volume of the mouth

cavity, forcing water past the gills to the outside. Using this very efficient bellows, bony fishes can pass water over the gills while stationary in the water. That is what a goldfish is doing when it seems to be gulping in a fish tank.

The Path to Land

Lobe-finned fishes (class Sarcopterygii) evolved 390 million years ago, shortly after the first bony fishes appeared. Only seven species survive today, a single species of coelacanth (figure 34.18) and six species of lungfish. Lobe-finned fishes have paired fins that consist of a long fleshy muscular lobe (hence their name), supported by a central core of bones that form fully articulated joints with one another. There are bony rays only at the tips of each lobed fin. Muscles within each lobe can move the fin rays independently of one another, a feat no ray-finned fish could match. Although rare today, lobe-finned fishes played an important part in the evolutionary story of vertebrates. Amphibians almost certainly evolved from the lobe-finned fishes.

Fishes were the first vertebrates. Fishes are characterized by gills and a simple, single-loop circulatory system. Cartilaginous fishes, such as sharks, are fast swimmers, while the very successful bony fishes have unique characteristics such as swim bladders and lateral line systems.

Amphibians

Frogs, salamanders, and caecilians, the damp-skinned vertebrates, are direct descendants of fishes. They are the sole survivors of a very successful group, the amphibians, the first vertebrates to walk on land. Most present-day amphibians are small and live largely unnoticed by humans. Amphibians are among the most numerous of terrestrial animals; there are more species of amphibians than of mammals. Throughout the world, amphibians play key roles in terrestrial food chains.

Characteristics of Living Amphibians

Biologists have classified living species of amphibians into three orders (table 34.2): 4200 species of frogs and toads in 22 families make up the order Anura ("without a tail"); 500 species of salamanders and newts in 9 families make up the order Urodela, or Caudata ("visible tail"); and 150 species (6 families) of wormlike, nearly blind organisms called caecilians that live in the tropics make up the order Apoda, or Gymnophiona ("without legs"). These amphibians have several key characteristics in common:

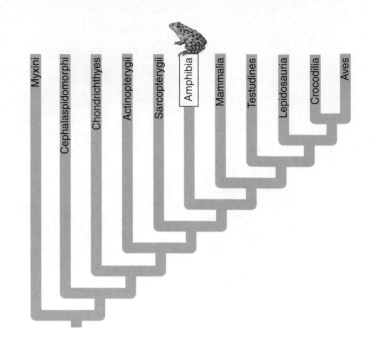

1. **Legs.** Frogs and salamanders have four legs and can move about on land quite well. Legs were one of the key adaptations to life on land. Caecilians have lost their legs during the course of adapting to a burrowing existence.
2. **Cutaneous respiration.** Frogs, salamanders, and caecilians all supplement the use of lungs by respiring directly across their skin, which is kept moist and provides an extensive surface area. This mode of respiration is only efficient for a high surface-to-volume ratio in an animal.
3. **Lungs.** Most amphibians possess a pair of lungs, although the internal surfaces are poorly developed, with much less surface area than reptilian or mammalian lungs have. Amphibians still breathe by lowering the floor of the mouth to suck in air, and then raising it back to force the air down into the lungs (see chapter 44).

4. **Pulmonary veins.** After blood is pumped through the lungs, two large veins called pulmonary veins return the aerated blood to the heart for repumping. This allows the aerated blood to be pumped to the tissues at a much higher pressure than when it leaves the lungs.
5. **Partially divided heart.** The initial chamber of the fish heart is absent in amphibians, and the second and last chambers are separated by a dividing wall that helps prevent aerated blood from the lungs from mixing with nonaerated blood being returned to the heart from the rest of the body. This separates the blood circulation into two separate paths, pulmonary and systemic. The separation is imperfect; the third chamber has no dividing wall (see Chapter 44).

Several other specialized characteristics are shared by all present-day amphibians. In all three orders, there is a zone of weakness between the base and the crown of the teeth. They also have a peculiar type of sensory rod cell in the retina of the eye called a "green rod." The exact function of this rod is unknown.

	Table 34.2	Orders of Amphibians		
Order	**Typical Examples**		**Key Characteristics**	**Approximate Number of Living Species**
Anura	Frogs, toads		Compact, tailless body; large head fused to the trunk; rear limbs specialized for jumping	4200
Urodela (Caudata)	Salamanders, newts		Slender body; long tail and limbs set out at right angles to the body	500
Apoda (Gymnophiona)	Caecilians		Tropical group with a snakelike body; no limbs; little or no tail	150

Origin of Amphibians

The word *amphibia* (Greek, "both lives") nicely describes the essential quality of modern-day amphibians, reflecting their ability to live in two worlds—the aquatic world of their fish ancestors and the terrestrial world they first invaded. Here, we review the checkered history of this group, almost all of whose members have been extinct for the last 200 million years. Then we will examine in more detail what the few kinds of surviving amphibians are like.

Paleontologists (scientists who study fossils) agree that amphibians must have evolved from the lobe-finned fishes, although for some years these scientists have disagreed about whether the direct ancestors were coelacanths, lungfish, or the extinct rhipidistian fishes. Good arguments can be made for each. Many details of amphibian internal anatomy resemble those of the coelacanth. Lungfish and rhipidistians have openings in the tops of their mouths similar to the internal nostrils of amphibians. In addition, lungfish have paired lungs, like those of amphibians. Recent DNA analysis indicates lungfish are in fact far more closely related to amphibians than are coelacanths. Most paleontologists consider that amphibians evolved from rhipidistian fishes, rather than lungfish, because the patterns of bones in the early amphibian skull and limbs bear a remarkable resemblance to those seen in the rhipidistians. They also share a particular tooth structure.

The successful invasion of land by vertebrates involved a number of major adaptations:

1. Legs were necessary to support the body's weight as well as to allow movement from place to place (figure 34.19).
2. Lungs were necessary to extract oxygen from the air. Even though far more oxygen is available to gills in air than in water, the delicate structure of fish gills requires the buoyancy of water to support them, and they will not function in air.
3. The heart had to be redesigned to make full use of new respiratory systems and to deliver the greater amounts of oxygen required by walking muscles.
4. Reproduction had to be carried out in water until methods evolved to prevent eggs from drying out.
5. Most importantly, a system had to be developed to prevent the body itself from drying out.

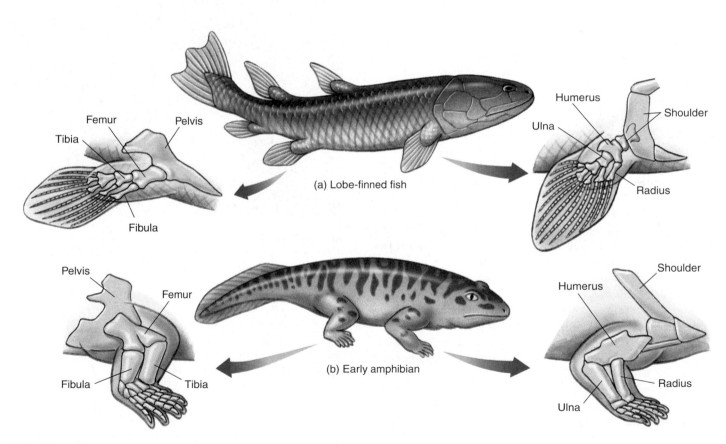

FIGURE 34.19

A comparison between the limbs of a lobe-finned fish and those of a primitive amphibian. (*a*) A lobe-finned fish. Some of these animals could probably move onto land. (*b*) A primitive amphibian. As illustrated by their skeletal structure, the legs of such an animal could clearly function on land much better than the fins of the lobe-finned fish.

The First Amphibian

Amphibians solved these problems only partially, but their solutions worked well enough that amphibians have survived for 350 million years. Evolution does not insist on perfect solutions, only workable ones.

Ichthyostega, the earliest amphibian fossil (figure 34.20), was found in a 370-million-year-old rock in Greenland. At that time, Greenland was part of the North American continent and lay near the equator. All amphibian fossils from the next 100 million years are found in North America. Only when Asia and the southern continents merged with North America to form the supercontinent Pangaea did amphibians spread throughout the world.

Ichthyostega was a strongly built animal, with four sturdy legs well supported by hip and shoulder bones. Unlike the bone structure of fish, the shoulder bones were no longer attached to the skull, and the hipbones were braced against the backbone, so the limbs could support the animal's weight. To strengthen the backbone further, long, broad ribs that overlap each other formed a solid cage for the lungs and heart. The rib cage was so solid that it probably couldn't expand and contract for breathing. Instead, *Ichthyostega* obtained oxygen somewhat as a frog does, by lowering the floor of the mouth to draw air in, and then raising it to push air down the windpipe into the lungs.

The Rise and Fall of Amphibians

Amphibians first became common during the Carboniferous period (360 to 280 million years ago). Fourteen families of amphibians are known from the early Carboniferous, nearly all of them aquatic or semi-aquatic, like *Ichthyostega*. By the late Carboniferous, much of North America was covered by low-lying tropical swamplands, and 34 families of amphibians thrived in this wet terrestrial environment, sharing it with pelycosaurs and other early reptiles. In the early Permian period that followed (280 to 248 million years ago), a remarkable change occurred among amphibians—they began to leave the marshes for dry uplands. Many of these terrestrial amphibians had bony plates and armor covering their bodies and grew to be very large, some as big as a pony (figure 34.21). Both their large size and the complete covering of their bodies indicate that these amphibians did not use the skin respiratory system of present-day amphibians, but rather had an impermeable leathery skin to prevent water loss. By the mid-Permian period, there were 40 families of amphibians. Only 25% of them were still semi-aquatic like *Ichthyostega*; 60% of the amphibians were fully terrestrial, and 15% were semiterrestrial. This was the peak of amphibian success, sometimes called the Age of Amphibians.

By the end of the Permian period, a reptile called a therapsid had become common, ousting the amphibians from their newly acquired niche on land. Following the mass extinction event at the end of the Permian, therap-

FIGURE 34.20
Amphibians were the first vertebrates to walk on land. Reconstruction of *Ichthyostega*, one of the first amphibians with efficient limbs for crawling on land, had an improved olfactory sense associated with a lengthened snout and a relatively advanced ear structure for picking up airborne sounds. Despite these features, *Ichthyostega*, which lived about 350 million years ago, was still quite fishlike in overall appearance and represents a very early amphibian.

FIGURE 34.21
A terrestrial amphibian of the Permian period. *Cacops*, a large, extinct amphibian, had extensive body armor.

sids were the dominant land vertebrate, and most amphibians were aquatic. This trend continued in the following Triassic period (248 to 213 million years ago), which saw the virtual extinction of amphibians from land. By the end of the Triassic, there were only 15 families of amphibians (including the first frog), and almost without exception, they were aquatic. Some of these grew to great size; one was 3 meters long. Only two groups of amphibians are known from the subsequent Jurassic period (213 to 144 million years ago), the anurans (frogs and toads) and the urodeles (salamanders and newts). The Age of Amphibians was over.

Amphibians Today

All of today's amphibians descended from the two families of amphibians that survived the Age of the Dinosaurs. During the Tertiary period (65 to 2 million years ago), these moist-skinned amphibians accomplished a highly successful invasion of wet habitats all over the world, and today there are over 4800 species of amphibians in 37 different families, comprising the orders Anura, Urodela, and Apoda.

Anura. Frogs and toads, amphibians without tails, live in a variety of environments, from deserts and mountains to ponds and puddles (figure 34.22*a*). Frogs have smooth, moist skin, a broad body, and long hind legs that make them excellent jumpers. Most frogs live in or near water, although some tropical species live in trees. Unlike frogs, toads have a dry, bumpy skin and short legs, and are well adapted to dry environments. All adult anurans are carnivores, eating a wide variety of invertebrates.

Most frogs and toads return to water to reproduce, laying their eggs directly in water. Their eggs lack watertight external membranes and would dry out quickly if out of the water. Eggs are fertilized externally and hatch into swimming larval forms called tadpoles. Tadpoles live in the water, where they generally feed on minute algae. After considerable growth, the body of the tadpole gradually changes into that of an adult frog. This process of abrupt change in body form is called **metamorphosis.**

Urodela (Caudata). Salamanders have elongated bodies, long tails, and smooth, moist skin (figure 34.22*b*). They typically range in length from a few inches to a foot, although giant Asiatic salamanders of the genus *Andrias* are as much as 1.5 meters long and weigh up to 33 kilograms. Most salamanders live in moist places, such as under stones or logs, or among the leaves of tropical plants. Some salamanders live entirely in water.

Salamanders lay their eggs in water or in moist places. Fertilization is usually external, although a few species practice a type of internal fertilization in which the female picks up sperm packets deposited by the male. Unlike anurans, the young that hatch from salamander eggs do not undergo profound metamorphosis, but are born looking like small adults and are carnivorous.

Apoda (Gymnophiona). Caecilians, members of the order Apoda (Gymnophiona), are a highly specialized group of tropical burrowing amphibians (figure 34.22*c*). These legless, wormlike creatures average about 30 centimeters long, but can be up to 1.3 meters long. They have very small eyes and are often blind. They resemble worms but have jaws with teeth. They eat worms and other soil invertebrates. The caecilian male deposits sperm directly into the female, and the female usually bears live young. Mud eels, small amphibians with tiny forelimbs and no hindlimbs that live in the eastern United States, are not apodans, but highly specialized urodelians.

(a)

(b)

(c)

FIGURE 34.22
Class Amphibia. (*a*) Red-eyed tree frog, *Agalychnis callidryas* (order Anura). (*b*) An adult barred tiger salamander, *Ambystoma tigrinum* (order Urodela). (*c*) A caecilian, *Caecilia tentaculata* (order Apoda).

Amphibians ventured onto land some 370 million years ago. They are characterized by moist skin, legs (secondarily lost in some species), lungs (usually), and a more complex and divided circulatory system. They are still tied to water for reproduction.

Reptiles

If we think of amphibians as a first draft of a manuscript about survival on land, then reptiles are the finished book. For each of the five key challenges of living on land, reptiles improved on the innovations first seen in amphibians. Legs were arranged to support the body's weight more effectively, allowing reptile bodies to be bigger and to run. Lungs and heart were altered to make them more efficient. The skin was covered with dry plates or scales to minimize water loss, and eggs were encased in watertight covers. Reptiles were the first truly *terrestrial* vertebrates.

Over 7000 species of reptiles (class Reptilia) now live on earth (table 34.3). They are a highly successful group in today's world; there are three reptile species for every two mammal species. While it is traditional to think of reptiles as more primitive than mammals, the great majority of reptiles living today evolved from lines that appeared after therapsids did (the line that leads directly to mammals).

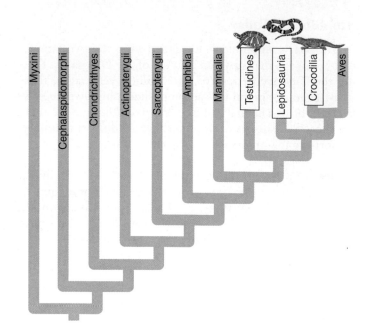

Key Characteristics of Reptiles

All living reptiles share certain fundamental characteristics, features they retain from the time when they replaced amphibians as the dominant terrestrial vertebrates. Among the most important are:

1. **Amniotic egg.** Amphibians never succeeded in becoming fully terrestrial because amphibian eggs must be laid in water to avoid drying out. Most reptiles lay watertight eggs that contain a food source (the yolk) and a series of four membranes—the yolk sac, the amnion, the allantois, and the chorion (figure 34.23). Each membrane plays a role in making the egg an independent life-support system. The outermost membrane of the egg is the **chorion,** which lies just beneath the porous shell. It allows respiratory gases to pass through, but retains water within the egg. The **amnion** encases the developing embryo within a fluid-filled cavity. The **yolk sac** provides food from the yolk for the embryo via blood vessels connecting to the embryo's gut. The **allantois** surrounds a cavity into which waste products from the embryo are excreted. All modern reptiles (as well as birds and mammals) show exactly this same pattern of membranes within the egg. These three classes are called amniotes.

2. **Dry skin.** Living amphibians have moist skin and must remain in moist places to avoid drying out. Reptiles have dry, watertight skin. A layer of scales or armor covers their bodies, preventing water loss. These scales develop as surface cells fill with keratin, the same protein that forms claws, fingernails, hair, and bird feathers.

3. **Thoracic breathing.** Amphibians breathe by squeezing their throat to pump air into their lungs;

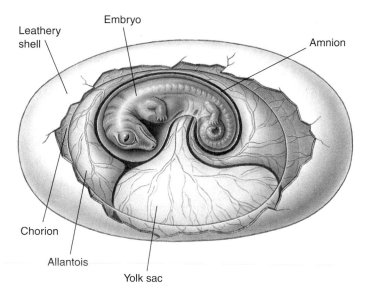

FIGURE 34.23
The watertight egg. The amniotic egg is perhaps the most important feature that allows reptiles to live in a wide variety of terrestrial habitats.

this limits their breathing capacity to the volume of their mouth. Reptiles developed pulmonary breathing, expanding and contracting the rib cage to suck air into the lungs and then force it out. The capacity of this system is limited only by the volume of the lungs.

Reptiles were the first vertebrates to completely master the challenge of living on dry land.

Table 34.3   Major Orders of Reptiles

Order	Typical Examples		Key Characteristics	Approximate Number of Living Species
Squamata, suborder Sauria	Lizards		Lizards; limbs set at right angles to body; anus is in transverse (sideways) slit; most are terrestrial	3800
Squamata, suborder Serpentes	Snakes		Snakes; no legs; move by slithering; scaly skin is shed periodically; most are terrestrial	3000
Chelonia	Turtles, tortoises, sea turtles		Ancient armored reptiles with shell of bony plates to which vertebrae and ribs are fused; sharp, horny beak without teeth	250
Crocodylia	Crocodiles, alligators, gavials, caimans		Advanced reptiles with four-chambered heart and socketed teeth; anus is a longitudinal (lengthwise) slit; closest living relatives to birds	25
Rhynchocephalia	Tuataras		Sole survivors of a once successful group that largely disappeared before dinosaurs; fused, wedgelike, socketless teeth; primitive third eye under skin of forehead	2
Ornithischia	Stegosaur		Dinosaurs with two pelvic bones facing backward, like a bird's pelvis; herbivores, with turtle-like upper beak; legs under body	Extinct
Saurischia	Tyrannosaur		Dinosaurs with one pelvic bone facing forward, the other back, like a lizard's pelvis; both plant- and flesh-eaters; legs under body	Extinct
Pterosauria	Pterosaur		Flying reptiles; wings were made of skin stretched between fourth fingers and body; wingspans of early forms typically 60 centimeters, later forms nearly 8 meters	Extinct
Plesiosaura	Plesiosaur		Barrel-shaped marine reptiles with sharp teeth and large, paddle-shaped fins; some had snakelike necks twice as long as their bodies	Extinct
Ichthyosauria	Ichthyosaur		Streamlined marine reptiles with many body similarities to sharks and modern fishes	Extinct

The Rise and Fall of Dominant Reptile Groups

During the 250 million years that reptiles were the dominant large terrestrial vertebrates, four major forms of reptiles took turns as the dominant type: pelycosaurs, therapsids, thecodonts, and dinosaurs.

Pelycosaurs: Becoming a Better Predator

Early reptiles such as *pelycosaurs* were better adapted to life on dry land than amphibians because they evolved watertight eggs. They had powerful jaws because of an innovation in skull design and muscle arrangement. Pelycosaurs were **synapsids,** meaning that their skulls had a pair of temporal holes behind the openings for the eyes. An important feature of reptile classification is the presence and number of openings behind the eyes (see figure 34.28). Their jaw muscles were anchored to these holes, which allowed them to bite more powerfully. An individual pelycosaur weighed about 200 kilograms. With long, sharp, "steak knife" teeth, pelycosaurs were the first land vertebrates to kill beasts their own size (figure 34.24). Dominant for 50 million years, pelycosaurs once made up 70% of all land vertebrates. They died out about 250 million years ago, replaced by their direct descendants—the therapsids.

Therapsids: Speeding Up Metabolism

Therapsids (figure 34.25) ate ten times more frequently than their pelycosaur ancestors. Evidence indicates that they may have been endotherms, able to regulate their own body temperature. The extra food consumption would have been necessary to produce body heat. This would have permitted therapsids to be far more active than other vertebrates of that time, when winters were cold and long. For 20 million years, therapsids (also called "mammal-like reptiles") were the dominant land vertebrate, until they were largely replaced 230 million years ago by a cold-blooded, or ectothermic, reptile line—the thecodonts. Therapsids became extinct 170 million years ago, but not before giving rise to their descendants—the mammals.

Thecodonts: Wasting Less Energy

Thecodonts were **diapsids,** their skulls having two pairs of temporal holes, and like amphibians and early reptiles, they were ectotherms (figure 34.26). Thecodonts largely replaced therapsids when the world's climate warmed 230 million years ago. In the warm climate, the therapsid's endothermy no longer offered a competitive advantage, and ectothermic thecodonts needed only a tenth as much food. Thecodonts were the first land vertebrates to be bipedal—to stand and walk on two feet. They were dominant through the Triassic period and survived for 15 million years, until they were replaced by their direct descendants—the dinosaurs.

FIGURE 34.24
A pelycosaur. *Dimetrodon*, a carnivorous pelycosaur, had a dorsal sail that is thought to have been used to dissipate body heat or gain it by basking.

FIGURE 34.25
A therapsid. This small, weasel-like cynodont therapsid, *Megazostrodon*, may have had fur. Living in the late Triassic period, this therapsid is so similar to modern mammals that some paleontologists consider it the first mammal.

FIGURE 34.26
A thecodont. *Euparkeria*, a thecodont, had rows of bony plates along the sides of the backbone, as seen in modern crocodiles and alligators.

Dinosaurs: Learning to Run Upright

Dinosaurs evolved from thecodonts about 220 million years ago. Unlike the thecodonts, their legs were positioned directly underneath their bodies, a significant improvement in body design (figure 34.27). This design placed the weight of the body directly over the legs, which allowed dinosaurs to run with great speed and agility. A dinosaur fossil can be distinguished from a thecodont fossil by the presence of a hole in the side of the hip socket. Because the dinosaur leg is positioned underneath the socket, the force is directed upward, not inward, thus requiring no bone on the side of the socket. Dinosaurs went on to become the most successful of all land vertebrates, dominating for 150 million years. All dinosaurs became extinct rather abruptly 65 million years ago, apparently as a result of an asteroid's impact.

Figures 34.28 and 34.29 summarize the evolutionary relationships among the extinct and living reptiles.

FIGURE 34.27
Mounted skeleton of *Afrovenator*. This bipedal carnivore was about 30 feet long and lived in Africa about 130 million years ago.

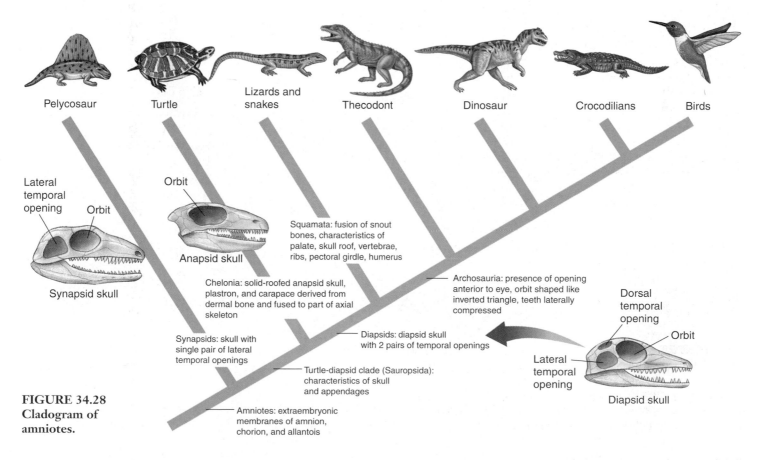

FIGURE 34.28
Cladogram of amniotes.

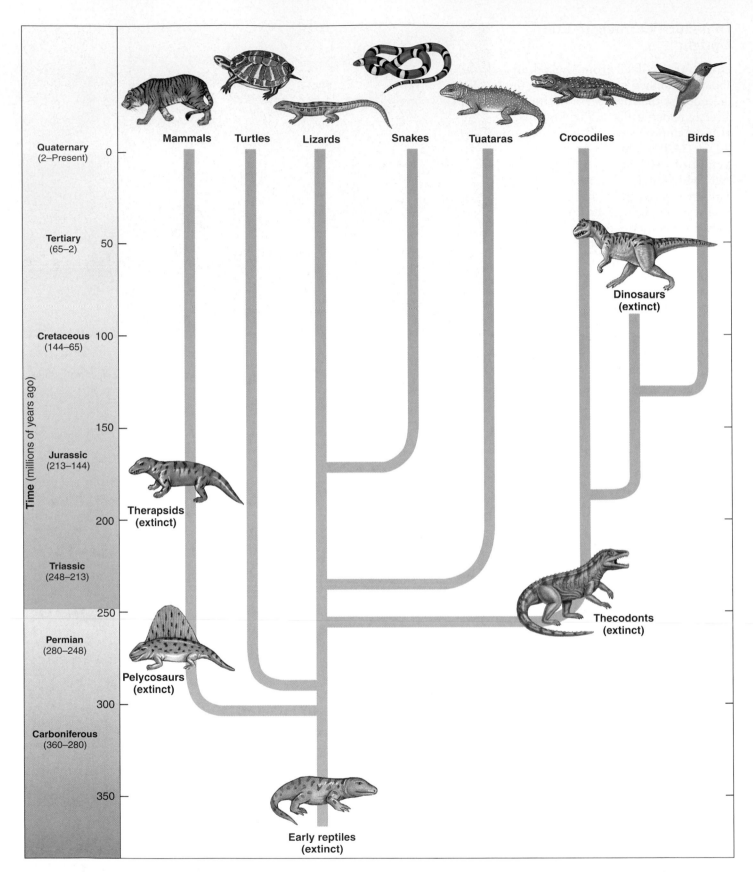

FIGURE 34.29

Evolutionary relationships among the reptiles.   There are four orders of living reptiles: turtles, lizards and snakes, tuataras, and crocodiles. This phylogenetic tree shows how these four orders are related to one another and to dinosaurs, birds, and mammals.

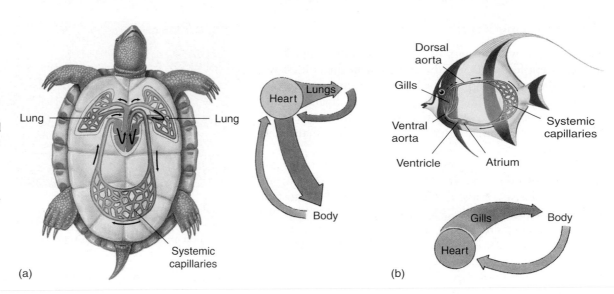

FIGURE 34.30
A comparison of reptile and fish circulation. (*a*) In reptiles such as this turtle, blood is repumped after leaving the lungs, and circulation to the rest of the body remains vigorous. (*b*) The blood in fishes flows from the gills directly to the rest of the body, resulting in slower circulation.

(a)

(b)

Lung

Lung

Systemic capillaries

Lungs

Heart

Body

Dorsal aorta

Gills

Ventral aorta

Ventricle

Atrium

Systemic capillaries

Gills

Body

Heart

Today's Reptiles

Most of the major reptile orders are now extinct. Of the 16 orders of reptiles that have existed, only 4 survive:

1. **Turtles.** The most ancient surviving lineage of reptiles is that of turtles. Turtles have anapsid skulls much like those of the first reptiles. Turtles have changed little in the past 200 million years.

2. **Lizards and snakes.** Most reptiles living today belong to the second lineage to evolve, the lizards and snakes. Lizards and snakes are descended from an ancient lineage of lizardlike reptiles that branched off the main line of reptile evolution in the late Permian period, 250 million years ago, before the thecodonts appeared (see figure 34.29). Throughout the Mesozoic era, during the dominance of the dinosaurs, these reptiles survived as minor elements of the landscape, much as mammals did. Like mammals, lizards and snakes became diverse and common only after the dinosaurs disappeared.

3. **Tuataras.** The third lineage of surviving reptiles to evolve were the Rhynchocephalonts, small diapsid reptiles that appeared shortly before the dinosaurs. They lived throughout the time of the dinosaurs and were common in the Jurassic period. They began to decline in the Cretaceous period, apparently unable to compete with lizards, and were already rare by the time dinosaurs disappeared. Today, only two species of the order Rhynchocephalia survive, both tuataras living on small islands near New Zealand.

4. **Crocodiles.** The fourth lineage of living reptile, crocodiles, appeared on the evolutionary scene much later than other living reptiles. Crocodiles are descended from the same line of thecodonts that gave rise to the dinosaurs, and they resemble dinosaurs in many ways. They have changed very little in over 200 million years. Crocodiles, pterosaurs, thecodonts, and dinosaurs together make up a group called archosaurs ("ruling reptiles").

Other Important Characteristics

As you might imagine from the structure of the amniotic egg, reptiles and other amniotes do not practice external fertilization as most amphibians do. There would be no way for a sperm to penetrate the membrane barriers protecting the egg. Instead, the male places sperm inside the female, where the sperm fertilize the egg before the membranes are formed. This is called internal fertilization.

The circulatory system of reptiles is improved over that of fish and amphibians, providing oxygen to the body more efficiently (figure 34.30). The improvement is achieved by extending the septum within the heart from the atrium partway across the ventricle. This septum creates a partial wall that tends to lessen mixing of oxygen-poor blood with oxygen-rich blood within the ventricle. In crocodiles, the septum completely divides the ventricle, creating a four-chambered heart, just as it does in birds and mammals (and probably in dinosaurs).

All living reptiles are **ectothermic,** obtaining their heat from external sources. In contrast, **endothermic** animals are able to generate their heat internally. In addition, **homeothermic** animals have a constant body temperature, and **poikilothermic** animals have a body temperature that fluctuates with ambient temperature. Reptiles are largely ectothermic poikilotherms; their body temperature is largely determined by their surroundings, and therefore it fluctuates. Reptiles also regulate their temperature through behavior. They may bask in the sun to warm up or seek shade to prevent overheating. The thecodont ancestors of crocodiles were ectothermic, as crocodiles are today. The later dinosaurs from which birds evolved were endothermic. Crocodiles and birds differ in this one important respect. Ectothermy is a principal reason crocodiles have been grouped among the reptiles.

Kinds of Living Reptiles

The four surviving orders of reptiles contain about 7000 species. Reptiles occur worldwide except in the coldest regions, where it is impossible for ectotherms to survive. Reptiles are among the most numerous and diverse of terrestrial vertebrates. The four living orders of the class Reptilia are Chelonia, Rhynchocephalia, Squamata, and Crocodilia.

Order Chelonia: Turtles and Tortoises. The order Chelonia consists of about 250 species of turtles (most of which are aquatic; figure 34.31) and tortoises (which are terrestrial). They differ from all other reptiles because their bodies are encased within a protective shell. Many of them can pull their head and legs into the shell as well, for total protection from predators. Turtles and tortoises lack teeth but have sharp beaks.

Today's turtles and tortoises have changed very little since the first turtles appeared 200 million years ago. Turtles are anapsid, meaning that they lack the temporal openings in the skull characteristic of other living reptiles, which are diapsid. This evolutionary stability of turtles may reflect the continuous benefit of their basic design—a body covered with a shell. In some species, the shell is made of hard plates; in other species, it is a covering of tough, leathery skin. In either case, the shell consists of two basic parts. The carapace is the dorsal covering, while the plastron is the ventral portion. In a fundamental commitment to this shell architecture, the vertebrae and ribs of most turtle and tortoise species are fused to the inside of the carapace. All of the support for muscle attachment comes from the shell.

While most tortoises have a dome-shaped shell into which they can retract their head and limbs, water-dwelling turtles have a streamlined, disc-shaped shell that permits rapid turning in water. Freshwater turtles have webbed toes; in marine turtles, the forelimbs have evolved into flippers. Although marine turtles spend their lives at sea, they must return to land to lay their eggs. Many species migrate long distances to do this. Atlantic green turtles migrate from their feeding grounds off the coast of Brazil to Ascension Island in the middle of the South Atlantic—a distance of more than 2000 kilometers—to lay their eggs on the same beaches where they themselves hatched.

Order Rhynchocephalia: Tuataras. Today, the order Rhynchocephalia contains only two species of tuataras, large, lizardlike animals about half a meter long. The only place in the world where these endangered species are found is a cluster of small islands off the coast of New Zealand. The native Maoris of New Zealand named the tuatara for the conspicuous spiny crest running down its back.

An unusual feature of the tuatara (and some lizards) is the inconspicuous "third eye" on the top of its head, called a parietal eye. Concealed under a thin layer of scales, the

FIGURE 34.31
Red-bellied turtles, *Pseudemys rubriventris*.   This turtle is common in the northeastern United States.

eye has a lens and a retina and is connected by nerves to the brain. Why have an eye, if it is covered up? The parietal eye may function to alert the tuatara when it has been exposed to too much sun, protecting it against overheating. Unlike most reptiles, tuataras are most active at low temperatures. They burrow during the day and feed at night on insects, worms, and other small animals.

Order Squamata: Lizards and Snakes. The order Squamata (figure 34.32) consists of three suborders: Sauria, some 3800 species of lizards; Amphisbaenia, about 135 species of worm lizards; and Serpentes, about 3000 species of snakes. The distinguishing characteristics of this order are the presence of paired copulatory organs in the male and a lower jaw that is not joined directly to the skull. A movable hinge with five joints (a human jaw has only one) allows great flexibility in the movements of the jaw. In addition, the loss of the lower arch of bone below the lower opening in the skull of lizards makes room for large muscles to operate their jaws. Most lizards and snakes are carnivores, preying on insects and small animals, and these improvements in jaw design have made a major contribution to their evolutionary success.

The chief difference between lizards and snakes is that most lizards have limbs, and snakes do not. Snakes also lack movable eyelids and external ears. Lizards are a more ancient group than modern snakes, which evolved only 20 million years ago. Common lizards include iguanas, chameleons, geckos, and anoles. Most are small, measuring less than a foot in length. The largest lizards belong to the monitor family. The largest of all monitors is the Komodo dragon of Indonesia, which reaches 3 meters in length and weighs up to 100 kilograms. Snakes also vary in length from only a few inches to nearly 10 meters .

Lizards and snakes rely on agility and speed to catch prey and elude predators. Only two species of lizard are venomous, the Gila monster of the southwestern United States and the beaded lizard of western Mexico. Similarly, most species of snakes are nonvenomous. Of the 13 families of snakes, only 4 are venomous: the elapids

(a)

FIGURE 34.33
Crocodile. Most crocodiles resemble birds and mammals in having four-chambered hearts; all other living reptiles have three-chambered hearts. Like birds, crocodiles are more closely related to dinosaurs than to any of the other living reptiles.

(b)

FIGURE 34.32
Representatives from the order Squamata. (*a*) An Australian skink, *Sphenomorophus*. Some burrowing lizards lack legs, and the snakes evolved from one line of legless lizards. (*b*) A smooth green snake, *Liochlorophis vernalis*.

(cobras, kraits, and coral snakes); the sea snakes; the vipers (adders, bushmasters, rattlesnakes, water moccasins, and copperheads); and some colubrids (African boomslang and twig snake).

Many lizards, including skinks and geckos, have the ability to lose their tails and then regenerate a new one. This apparently allows these lizards to escape from predators.

Order Crocodilia: Crocodiles and Alligators. The order Crocodilia is composed of 25 species of large, primarily aquatic, primitive-looking reptiles (figure 34.33). In addition to crocodiles and alligators, the order includes two less familiar animals: the caimans and the gavials. Crocodilians have remained relatively unchanged since they first evolved.

Crocodiles are largely nocturnal animals that live in or near water in tropical or subtropical regions of Africa, Asia, and South America. The American crocodile is found in southern Florida, Cuba, and throughout tropical Central America. Nile crocodiles and estuarine crocodiles

can grow to enormous size and are responsible for many human fatalities each year. There are only two species of alligators: one living in the southern United States and the other a rare endangered species living in China. Caimans, which resemble alligators, are native to Central America. Gavials are a group of fish-eating crocodilians with long, slender snouts that live only in India and Burma.

All crocodilians are carnivores. They generally hunt by stealth, waiting in ambush for prey, and then attacking ferociously. Their bodies are well adapted for this form of hunting: Their eyes are on top of their heads, and their nostrils are on top of their snouts, so they can see and breathe while lying quietly submerged in water. They have enormous mouths, studded with sharp teeth, and very strong necks. A valve in the back of the mouth prevents water from entering the air passage when a crocodilian feeds underwater.

In many ways, crocodiles resemble birds far more than they do other living reptiles. For example, alone among living reptiles, crocodiles care for their young (a trait they share with at least some dinosaurs) and have a four-chambered heart, as birds do. Why are crocodiles more similar to birds than to other living reptiles? Most biologists now believe that birds are in fact the direct descendants of dinosaurs. Both crocodiles and birds are more closely related to dinosaurs, and to each other, than to lizards and snakes.

Many major reptile groups that dominated life on land for 250 million years are now extinct. The four living orders of reptiles include the turtles, lizards and snakes, tuataras, and crocodiles.

Birds

Only four groups of animals have evolved the ability to fly—insects, pterosaurs, birds, and bats. Pterosaurs, flying reptiles, evolved from gliding reptiles and flew for 130 million years before becoming extinct with the dinosaurs. The ways these very different animals meet the challenges of flight are startlingly similar. Like water running downhill through similar gullies, evolution tends to seek out similar adaptations, but major differences occur as well. The success of birds lies in the development of a structure unique in the animal world—the feather. Developed from reptilian scales, feathers are the ideal adaptation for flight, serving as lightweight airfoils that are easily replaced if damaged (unlike the vulnerable skin wings of pterosaurs and bats). Today, birds (class Aves) are the most successful and diverse of all terrestrial vertebrates, with 28 orders containing a total of 166 families and about 8600 species (table 34.4).

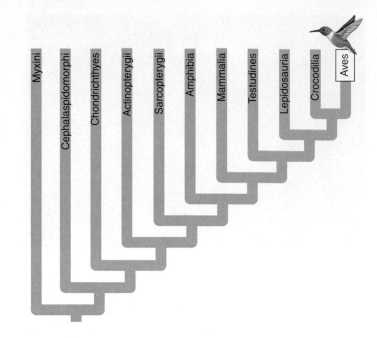

Key Characteristics of Birds

Modern birds lack teeth and have only vestigial tails, but they still retain many reptilian characteristics. For instance, birds lay amniotic eggs, although the shells of bird eggs are hard rather than leathery. Also, reptilian scales are present on the feet and lower legs of birds. What makes birds unique? Two primary characteristics distinguish them from living reptiles:

1. **Feathers.**   Feathers are modified reptilian scales that serve two functions: providing lift for flight and conserving heat. The structure of feathers combines maximum flexibility and strength with minimum weight (figure 34.34). Feathers develop from tiny pits in the skin called follicles. In a typical flight feather, a shaft emerges from the follicle, and pairs of vanes develop from its opposite sides. At maturity, each vane has many branches called barbs. The barbs, in turn, have many projections called barbules that are equipped with microscopic hooks. These hooks link the barbs to one another, giving the feather a continuous surface and a sturdy but flexible shape. Like scales, feathers can be replaced. Among living animals, feathers are unique to birds. Recent fossil finds suggest that some dinosaurs may have had feathers.

2. **Flight skeleton.**   The bones of birds are thin and hollow. Many of the bones are fused, making the bird skeleton more rigid than a reptilian skeleton. The fused sections of backbone and of the shoulder and hip girdles form a sturdy frame that anchors muscles during flight. The power for active flight comes from large breast muscles that can make up 30% of a bird's total body weight. They stretch down from the wing and attach to the breastbone,

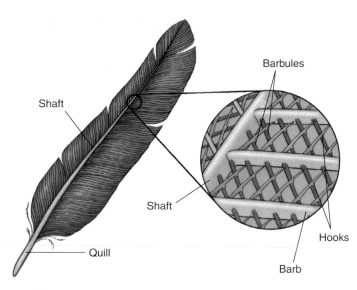

FIGURE 34.34
A feather.   The enlargement shows how the secondary branches and barbs of the vanes are linked together by microscopic barbules.

which is greatly enlarged and bears a prominent keel for muscle attachment. They also attach to the fused collarbones that form the so-called "wishbone." No other living vertebrates have a fused collarbone or a keeled breastbone.

Birds are the most diverse of all terrestrial vertebrates. They are closely related to reptiles, but unlike reptiles or any other animals, birds have feathers.

Order	Typical Examples		Key Characteristics	Approximate Number of Living Species
Passeriformes	Crows, mockingbirds, robins, sparrows, starlings, warblers		*Songbirds* Well-developed vocal organs; perching feet; dependent young	5276 (largest of all bird orders; contains over 60% of all species)
Apodiformes	Hummingbirds, swifts		*Fast fliers* Short legs; small bodies; rapid wing beat	428
Piciformes	Honeyguides, toucans, woodpeckers		*Woodpeckers or toucans* Grasping feet; chisel-like, sharp bills can break down wood	383
Psittaciformes	Cockatoos, parrots		*Parrots* Large, powerful bills for crushing seeds; well-developed vocal organs	340
Charadriiformes	Auks, gulls, plovers, sandpipers, terns		*Shorebirds* Long, stiltlike legs; slender, probing bills	331
Columbiformes	Doves, pigeons		*Pigeons* Perching feet; rounded, stout bodies	303
Falconiformes	Eagles, falcons, hawks, vultures		*Birds of prey* Carnivorous; keen vision; sharp, pointed beaks for tearing flesh; active during the day	288
Galliformes	Chickens, grouse, pheasants, quail		*Gamebirds* Often limited flying ability; rounded bodies	268
Gruiformes	Bitterns, coots, cranes, rails		*Marsh birds* Long, stiltlike legs; diverse body shapes; marsh-dwellers	209
Anseriformes	Ducks, geese, swans		*Waterfowl* Webbed toes; broad bill with filtering ridges	150
Strigiformes	Barn owls, screech owls		*Owls* Nocturnal birds of prey; strong beaks; powerful feet	146
Ciconiiformes	Herons, ibises, storks		*Waders* Long-legged; large bodies	114
Procellariiformes	Albatrosses, petrels		*Seabirds* Tube-shaped bills; capable of flying for long periods of time	104
Sphenisciformes	Emperor penguins, crested penguins		*Penguins* Marine; modified wings for swimming; flightless; found only in southern hemisphere; thick coats of insulating feathers	18
Dinornithiformes	Kiwis		*Kiwis* Flightless; small; primitive; confined to New Zealand	2
Struthioniformes	Ostriches		*Ostriches* Powerful running legs; flightless; only two toes; very large	1

Table 34.4 Major Orders of Birds

History of the Birds

A 150-million-year-old fossil of the first known bird, *Archaeopteryx* (figure 34.35)—pronounced "archie-op-terichs"—was found in 1862 in a limestone quarry in Bavaria, the impression of its feathers stamped clearly into the rocks.

Birds Are Descended from Dinosaurs

The skeleton of *Archaeopteryx* shares many features with small theropod dinosaurs. About the size of a crow, its skull has teeth, and very few of its bones are fused to one another, a feature that some paleontologists consider dinosaurian rather than avian. Its bones are thought to have been solid, not hollow like a bird's. Also, it has a long, reptilian tail and no enlarged breastbone such as modern birds use to anchor flight muscles. Finally, it has the forelimbs of a dinosaur. Because of its many dinosaur features, several *Archaeopteryx* fossils were originally classified as the coelurosaur *Compsognathus*, a small theropod dinosaur of similar size—until feathers were discovered on the fossils. What makes *Archaeopteryx* distinctly avian is the presence of feathers on its wings and tail. It also has other birdlike features, notably a wishbone. Dinosaurs lack a wishbone, although thecodonts had them.

The remarkable similarity of *Archaeopteryx* to *Compsognathus* has led almost all paleontologists to conclude that *Archaeopteryx* is the direct descendant of dinosaurs—indeed, that today's birds are "feathered dinosaurs." Some even speak flippantly of "carving the dinosaur" at Thanksgiving dinner. The recent discovery of feathered dinosaurs in China lends strong support to this inference. The dinosaur

FIGURE 34.35
Archaeopteryx. An artist's reconstruction of *Archaeopteryx*, an early bird about the size of a crow. Closely related to its ancestors among the bipedal dinosaurs, *Archaeopteryx* lived in the forests of central Europe 150 million years ago. The true feather colors of *Archaeopteryx* are not known.

Caudipteryx, for example, is clearly intermediate between *Archaeopteryx* and dinosaurs, having large feathers on its tail and arms but also many features of *Velociraptor* dinosaurs (figure 34.36). Because the arms of *Caudipteryx* were too short to use as wings, feathers probably didn't evolve for flight, but instead served as insulation, much as fur does for animals. Flight is an ability certain kinds of dinosaurs achieved as they evolved longer arms. We call these dinosaurs birds.

Despite their close affinity to dinosaurs, biologists continue to classify birds as Aves, a separate class, because of their unique adaptations and great diversity, including three key evolutionary novelties: feathers, hollow bones, and physiological mechanisms such as superefficient lungs

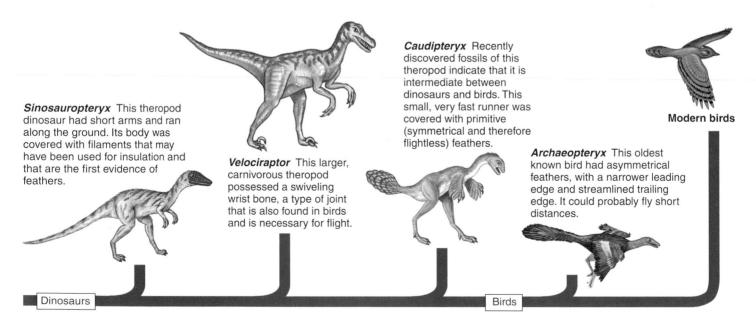

Sinosauropteryx This theropod dinosaur had short arms and ran along the ground. Its body was covered with filaments that may have been used for insulation and that are the first evidence of feathers.

Velociraptor This larger, carnivorous theropod possessed a swiveling wrist bone, a type of joint that is also found in birds and is necessary for flight.

Caudipteryx Recently discovered fossils of this theropod indicate that it is intermediate between dinosaurs and birds. This small, very fast runner was covered with primitive (symmetrical and therefore flightless) feathers.

Archaeopteryx This oldest known bird had asymmetrical feathers, with a narrower leading edge and streamlined trailing edge. It could probably fly short distances.

Modern birds

Dinosaurs

Birds

FIGURE 34.36
The evolutionary path to the birds. Almost all paleontologists now accept the theory that birds are the direct descendants of theropod dinosaurs.

that permit sustained, powered flight. This practical judgment should not conceal the basic agreement among almost all biologists that birds are the direct descendants of theropod dinosaurs, as closely related to coelurosaurs as are other theropods (see figure 34.36).

By the early Cretaceous period, only a few million years after *Archaeopteryx* lived, a diverse array of birds had evolved, with many of the features of modern birds. Fossils in Mongolia, Spain, and China discovered within the last few years reveal a diverse collection of toothed birds with the hollow bones and breastbones necessary for sustained flight. Other fossils reveal highly specialized, flightless diving birds. The diverse birds of the Cretaceous shared the skies with pterosaurs for 70 million years.

Because the impression of feathers is rarely fossilized and modern birds have hollow, delicate bones, the fossil record of birds is incomplete. Relationships among the 166 families of modern birds are mostly inferred from studies of the degree of DNA similarity among living birds. These studies suggest that the most ancient living birds are the flightless birds, such as the ostrich. Ducks, geese, and other waterfowl evolved next, in the early Cretaceous, followed by a diverse group of woodpeckers, parrots, swifts, and owls. The largest of the bird orders, Passeriformes, or songbirds (60% of all species of birds today), evolved in the mid-Cretaceous. The more specialized orders of birds, such as shorebirds, birds of prey, flamingos, and penguins, did not appear until the late Cretaceous. All but a few of the modern orders of toothless birds are thought to have arisen before the disappearance of the pterosaurs and dinosaurs at the end of the Cretaceous period, 65 million years ago.

Birds Today

You can tell a great deal about the habits and food of a bird by examining its beak and feet. For instance, carnivorous birds such as owls have curved talons for seizing prey and sharp beaks for tearing apart their meal. The beaks of ducks are flat for shoveling through mud, while the beaks of finches are short, thick seed-crushers. There are 28 orders of birds, the largest consisting of over 5000 species (figure 34.37).

Many adaptations enabled birds to cope with the heavy energy demands of flight:

1. **Efficient respiration.** Flight muscles consume an enormous amount of oxygen during active flight. The

FIGURE 34.37
Class Aves. This summer tanager, *Piranga rubra*, is a member of the largest order of birds, the Passeriformes, with over 5000 species.

reptilian lung has a limited internal surface area, not nearly enough to absorb all the oxygen needed. Mammalian lungs have a greater surface area, but as we will see in chapter 44, bird lungs satisfy this challenge with a radical redesign. When a bird inhales, the air goes past the lungs to a series of air sacs located near and within the hollow bones of the back; from there, the air travels to the lungs and then to a set of anterior air sacs before being exhaled. Because air always passes through the lungs in the same direction, and blood flows past the lung at right angles to the airflow, gas exchange is highly efficient.

2. **Efficient circulation.** The revved-up metabolism needed to power active flight also requires very efficient blood circulation, so that the oxygen captured by the lungs can be delivered to the flight muscles quickly. In the heart of most living reptiles, oxygen-rich blood coming from the lungs mixes with oxygen-poor blood returning from the body because the wall dividing the ventricle into two chambers is not complete. In birds, the wall dividing the ventricle is complete, and the two blood circulations do not mix, so flight muscles receive fully oxygenated blood.

In comparison with reptiles and most other vertebrates, birds have a rapid heartbeat. A hummingbird's heart beats about 600 times a minute, and an active chickadee's heart beats 1000 times a minute. In contrast, the heart of the large, flightless ostrich averages 70 beats per minute—the same rate as the human heart.

3. **Endothermy.** Birds, like mammals, are endothermic. Many paleontologists believe the dinosaurs that birds evolved from were endothermic as well. Birds maintain body temperatures significantly higher than those of most mammals, ranging from 40° to 42°C (human body temperature is 37°C). Feathers provide excellent insulation, helping to conserve body heat. The high temperatures maintained by endothermy permit metabolism in the bird's flight muscles to proceed at a rapid pace, to provide the ATP necessary to drive rapid muscle contraction.

The class Aves probably debuted 150 million years ago with *Archaeopteryx*. Modern birds are characterized by feathers, scales, a thin, hollow skeleton, auxiliary air sacs, and a four-chambered heart. Birds lay amniotic eggs and are endothermic.

Mammals

There are about 4500 living species of mammals (class Mammalia), the smallest number of species in any of the five classes of vertebrates. Most large, land-dwelling vertebrates are mammals (figure 34.38), and they tend to dominate terrestrial communities, as did the dinosaurs they replaced. When you look out over an African plain, you see the big mammals—the lions, zebras, gazelles, and antelope. Your eye does not as readily pick out the many birds, lizards, and frogs that live in the grassland community with them. But the typical mammal is not all that large. Of the 4500 species of mammals, 3200 are rodents, bats, shrews, or moles.

Key Mammalian Characteristics

Mammals are distinguished from all other classes of vertebrates by two fundamental characteristics—hair and mammary glands—and are marked by several other notable features:

1. **Hair.** All mammals have hair. Even apparently naked whales and dolphins grow sensitive bristles on their snouts. The evolution of fur and the ability to regulate body temperature enabled mammals to invade colder climates that ectothermic reptiles could not inhabit, and the insulation fur provided may have ensured the survival of mammals when the dinosaurs perished. Mammals are endothermic animals, and typically maintain body temperatures higher than the temperature of their surroundings. The dense undercoat of many mammals reduces the amount of body heat that escapes.

   Another function of hair is camouflage. The coloration and pattern of a mammal's coat usually matches its background. A little brown mouse is practically invisible against the brown leaf litter of a forest floor, while the orange and black stripes of a Bengal tiger disappear against the orange-brown color of the tall grass in which it hunts. Hairs also function as sensory structures. The whiskers of cats and dogs are stiff hairs that are very sensitive to touch. Mammals that are active at night or live underground often rely on their whiskers to locate prey or to avoid colliding with objects. Hair can also serve as a defense weapon. Porcupines and hedgehogs protect themselves with long, sharp, stiff hairs called quills.

   Unlike feathers, which evolved from modified reptilian scales, mammalian hair is a completely different form of skin structure. An individual mammalian hair is a long, protein-rich filament that extends like a stiff thread from a bulblike foundation beneath the skin known as a hair follicle. The filament is composed mainly of dead cells filled with the fibrous protein keratin.

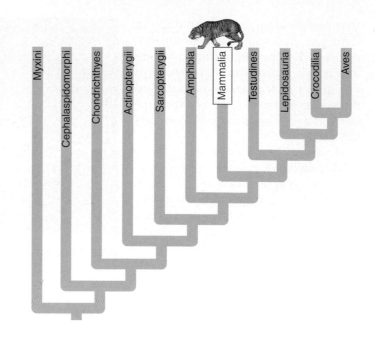

Myxini | Cephalaspidomorphi | Chondrichthyes | Actinopterygii | Sarcopterygii | Amphibia | Mammalia | Testudines | Lepidosauria | Crocodilia | Aves

FIGURE 34.38
Mammals. African elephants, *Loxodonta africana* (order Proboscidea), at a water hole.

2. **Mammary glands.** All female mammals possess mammary glands that secrete milk. Newborn mammals, born without teeth, suckle this milk. Even baby whales are nursed by their mother's milk. Milk is a fluid rich in fat, sugar, and protein. A liter of human milk contains 11 grams of protein, 49 grams of fat, 70 grams of carbohydrate (chiefly the sugar lactose), and 2 grams of minerals critical to early growth, such as calcium. About 95% of the volume is water, critical to avoid dehydration. Milk is a very high calorie food (human milk has 750 kcal per liter), important because of the high energy needs of a rapidly growing

newborn mammal. About 50% of the energy in the milk comes from fat.

3. **Endothermy.**   As stated previously, mammals are endothermic, a crucial adaptation that has allowed them to be active at any time of the day or night and to colonize severe environments, from deserts to ice fields. Many characteristics, including hair that provides insulation, played important roles in making endothermy possible. Also, the more efficient blood circulation provided by the four-chambered heart and the more efficient respiration provided by the *diaphragm* (a special sheet of muscles below the rib cage that aids breathing) make possible the higher metabolic rate upon which endothermy depends.

4. **Placenta.**   In most mammal species, females carry their young in a uterus during development, nourishing them through a placenta, and give birth to live young. The placenta is a specialized organ within the uterus of the pregnant mother that brings the bloodstream of the fetus into close contact with the bloodstream of the mother (figure 34.39). Food, water, and oxygen can pass across from mother to child, and wastes can pass over to the mother's blood and be carried away.

5. **Teeth.**   Reptiles have homodont dentition, meaning that their teeth are all the same. However, mammals have heterodont dentition, with different types of teeth that are highly specialized to match particular eating habits (figure 34.40). It is usually possible to determine a mammal's diet simply by examining its teeth. Compare the skull of a dog (a carnivore) and a deer (an herbivore). The dog's long canine teeth are well suited for biting and holding prey, and some of its premolar and molar teeth are triangular and sharp for ripping off chunks of flesh. In contrast, canine teeth are absent in deer; instead, the deer clips off mouthfuls of plants with flat, chisel-like incisors on its lower jaw. The deer's molars are large and covered with ridges to effectively grind and break up tough plant tissues. Rodents, such as beavers, are gnawers and have long incisors for chewing through branches or stems. These incisors are ever-growing; that is, the ends wear down, but new incisor growth maintains the length.

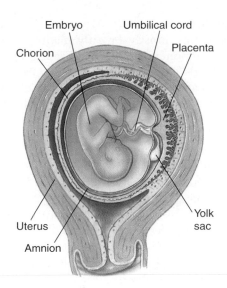

FIGURE 34.39
The placenta.   The placenta is characteristic of the largest group of mammals, the placental mammals. It evolved from membranes in the amniotic egg. The umbilical cord evolved from the allantois. The chorion, or outermost part of the amniotic egg, forms most of the placenta itself. The placenta serves as the provisional lungs, intestine, and kidneys of the embryo, without ever mixing maternal and fetal blood.

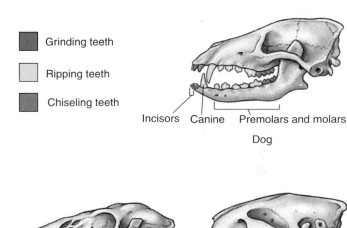

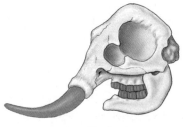

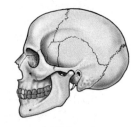

FIGURE 34.40
Mammals have different types of specialized teeth.   While reptiles have all the same kind of teeth, mammals have different types of teeth specialized for different feeding habits. Carnivores, such as dogs, have canine teeth that are able to rip food; some of the premolars and molars in dogs are also ripping teeth. Herbivores, such as deer, have incisors to chisel off vegetation and molars designed to grind up the plant material. In the beaver, the chiseling incisors dominate. In the elephant, the incisors have become specialized weapons, and molars grind up vegetation. Humans are omnivores; we have ripping, chiseling, and grinding teeth.

6. Digestion of plants. Most mammals are herbivores, eating mostly or only plants. Cellulose, the major component of plant cell walls, forms the bulk of a plant's body and is a major source of food for mammalian herbivores. The cellulose molecule has the structure of a pearl necklace, with each pearl a glucose sugar molecule. Mammals do not have enzymes that can break the links between the pearls to release the glucose elements for use as food. Herbivorous mammals rely on a mutualistic partnership with bacteria that have the necessary cellulose-splitting enzymes.

Mammals such as cows, buffalo, antelopes, goats, deer, and giraffes have huge, four-chambered stomachs that function as fermentation vats. The first chamber is the largest and holds a dense population of cellulose-digesting bacteria. Chewed plant material passes into this chamber, where the bacteria attack the cellulose. The material is then digested further in the rest of the stomach.

Rodents, horses, rabbits, and elephants are herbivores that employ mutualistic bacteria to digest cellulose in a different way. They have relatively small stomachs, and instead digest plant material in their large intestine, like a termite. The bacteria that actually carry out the digestion of the cellulose live in a pouch called the cecum that branches from the end of the small intestine.

Even with these complex adaptations for digesting cellulose, a mouthful of plant is less nutritious than a mouthful of flesh. Herbivores must consume large amounts of plant material to gain sufficient nutrition. An elephant eats 135 to 150 kilograms of food (300 to 400 pounds) each day.

7. Hooves and horns. Keratin, the protein of hair, is also the structural building material in claws, fingernails, and hooves. Hooves are specialized keratin pads on the toes of horses, cows, sheep, antelopes, and other running mammals. The pads are hard and horny, protecting the toe and cushioning it from impact.

The horns of cattle, sheep, and antelope are composed of a core of bone surrounded by a sheath of keratin. The bony core is attached to the skull, and the horn is not shed. The horn you see is the outer sheath, made of hairlike fibers of keratin compacted into a very hard structure. Deer antlers are made not of keratin but of bone. Male deer grow and shed a set of antlers each year. While growing during the summer, antlers are covered by a thin layer of skin known as velvet.

8. Flight. Bats are the only mammals capable of powered flight (figure 34.41). Like the wings of birds, bat wings are modified forelimbs. The bat wing is a leathery membrane of skin and muscle stretched over the bones of four fingers. The edges

FIGURE 34.41
Greater horseshoe bat, *Rhinolophus ferrumequinum*. The bat is the only mammal capable of true flight.

of the membrane attach to the side of the body and to the hind leg. When resting, most bats prefer to hang upside down by their toe claws. After rodents, bats are the second largest order of mammals. They have been a particularly successful group because many species have been able to utilize a food resource that most birds do not have access to—night-flying insects.

How do bats navigate in the dark? Late in the eighteenth century, the Italian biologist Lazzaro Spallanzani showed that a blinded bat could fly without crashing into things and still capture insects. Clearly another sense other than vision was being used by bats to navigate in the dark. When Spallanzani plugged the ears of a bat, it was unable to navigate and collided with objects. Spallanzani concluded that bats "hear" their way through the night world.

We now know that bats have evolved a sonar system that functions much like the sonar devices used by ships and submarines to locate underwater objects. As a bat flies, it emits a very rapid series of extremely high-pitched "clicking" sounds well above the range of human hearing. The high-frequency pulses are emitted either through the mouth or, in some cases, through the nose. The soundwaves bounce off obstacles or flying insects, and the bat hears the echo. Through sophisticated processing of this echo within its brain, a bat can determine not only the direction of an object but also the distance to the object (see chapter 46).

Mammals first appeared 220 million years ago, evolving to their present position of dominance in modern terrestrial ecosystems. Mammals are the only vertebrates that possess hair and mammary glands.

The Orders of Mammals

Mammals have been around since the time of the dinosaurs, although they were never common until the dinosaurs disappeared. We have learned a lot about the evolutionary history of mammals from their fossils. The first mammals arose from therapsids in the mid-Triassic period about 220 million years ago, just as the first dinosaurs evolved from thecodonts. Tiny, shrewlike creatures that lived in trees eating insects, the earliest mammals were only a minor element in a land that quickly came to be dominated by dinosaurs. Fossils reveal that these early mammals had large eye sockets, evidence that they may have been active at night. Early mammals had a single lower jawbone. Therapsid fossils show a change from the reptile lower jaw, having several bones, to a jaw closer to the mammalian type. Two of the bones forming the therapsid jaw joint retreated into the middle ear of mammals, linking with a bone already there and producing a three-bone structure that amplifies sound better than the reptilian ear.

Table 34.5 Some Groups of Extinct Mammals

Group	Description
Cave bears	Numerous in the ice ages; this enormous vegetarian bear slept through the winter in large groups.
Irish elk	Neither Irish nor an elk (it is a kind of deer), *Megaloceros* was the largest deer that ever lived, with horns spanning 12 feet. Seen in French cave paintings, they became extinct about 2500 years ago.
Mammoths	Although only two species of elephants survive today, the elephant family was far more diverse during the late Tertiary. Many were cold-adapted mammoths with fur.
Giant ground sloths	*Megatherium* was a giant, 20-foot ground sloth that weighed three tons and was as large as a modern elephant.
Sabertooth cats	The jaws of these large, lionlike cats opened an incredible 120° to allow the animal to drive its huge upper pair of saber teeth into prey.

Early Divergence in Mammals

For 155 million years, while the dinosaurs flourished, mammals were a minor group of small insectivores and herbivores. Only five orders of mammals arose in that time, and their fossils are scarce, indicating that mammals were not abundant. However, the two groups to which present-day mammals belong did appear. The most primitive mammals, direct descendants of therapsids, were members of the subclass Prototheria. Most prototherians were small and resembled modern shrews. All prototherians laid eggs, as did their therapsid ancestors. The only prototherians surviving today are the monotremes—the duckbilled platypus and the echidnas, or spiny anteaters. The other major mammalian group is the subclass Theria. All of the mammals you are familiar with, including humans, are therians. Therians are viviparous (that is, their young are born alive). The two major living therian groups are marsupials, or pouched mammals, and placental mammals. Kangaroos, opossums, and koalas are marsupials. Dogs, cats, humans, horses, and most other mammals are placentals.

The Age of Mammals

At the end of the Cretaceous period 65 million years ago, the dinosaurs and numerous other land and marine animals became extinct, but mammals survived, possibly because of the insulation their fur provided. In the Tertiary period (lasting from 65 million years to 2 million years ago), mammals rapidly diversified, taking over many of the ecological roles once dominated by dinosaurs (table 34.5). Mammals reached their maximum diversity late in the Tertiary period, about 15 million years ago. At that time, tropical conditions existed over much of the world. During the last 15 million years, world climates have deteriorated, and the area covered by tropical habitats has decreased, causing a decline in the total number of mammalian species. Seventeen of them (containing 94% of the species) are placental. The other two are the primitive monotremes and the marsupials.

Today's Mammals

Monotremes: Egg-laying Mammals. The duck-billed platypus and two species of echidna, or spiny anteater, are the only living monotremes (figure 34.42a). Among living mammals, only monotremes lay shelled eggs. The structure of their shoulder and pelvis is more similar to that of the early reptiles than to any other living mammal. Also like reptiles, monotremes have a cloaca, a single opening through which feces, urine, and reproductive products leave the body. Monotremes are more closely related to early mammals than is any other living mammal.

In addition to many reptilian features, monotremes have both of the defining mammalian features: fur and functioning mammary glands. Young monotremes drink their mother's milk after they hatch from eggs. Females lack well-developed nipples, so the babies cannot suckle. Instead, the milk oozes onto the mother's fur, and the babies lap it off with their tongues.

The platypus, found only in Australia, lives much of its life in the water and is a good swimmer. It uses its bill much as a duck does, rooting in the mud for worms and other soft-bodied animals. Echidnas of Australia and New Guinea have very strong, sharp claws, which they use for burrowing and digging. The echidna probes with its long, beaklike snout for insects, especially ants and termites.

Marsupials: Pouched Mammals. The major difference between marsupials (figure 34.42b) and other mammals is their pattern of embryonic development. In marsupials, a fertilized egg is surrounded by chorion and amniotic membranes, but no shell forms around the egg as it does in monotremes. During most of its early development, the marsupial embryo is nourished by an abundant yolk within the egg. Shortly before birth, a short-lived placenta forms from the chorion membrane. Soon after, sometimes within eight days of fertilization, the embryonic marsupial is born. It emerges tiny and hairless, and crawls into the marsupial pouch, where it latches onto a nipple and continues its development.

Marsupials evolved shortly before placental mammals, about 100 million years ago. Today, most species of marsupials live in Australia and South America, areas that have been historically isolated. Marsupials in Australia and New Guinea have diversified to fill ecological positions occupied by placental mammals elsewhere in the world. For example, kangaroos are the Australian grazers, playing the role antelope, horses, and buffalo perform elsewhere. The placental mammals in Australia and New Guinea today arrived relatively recently and include some introduced by humans. The only marsupial found in North America is the Virginia opossum.

Placental Mammals. As stated earlier in this section, mammals that produce a true placenta that nourishes the

(a) (b)

(c)

FIGURE 34.42

Today's mammals. (*a*) This echidna, *Tachyglossus aculeatus*, is a monotreme. (*b*) Marsupials include kangaroos, such as this adult with young in its pouch. (*c*) This female African lion, *Panthera leo* (order Carnivora), is a placental mammal.

embryo throughout its entire development are called placental mammals (figure 34.42c). Most species of mammals living today, including humans, are in this group. Of the 19 orders of living mammals, 17 are placental mammals. Table 34.6 shows some of these orders, in addition to the marsupials. They are a very diverse group, ranging in size from 1.5-gram pygmy shrews to 100,000 kilogram whales.

Early in the course of embryonic development, the placenta forms. Both fetal and maternal blood vessels are abundant in the placenta, and substances can be exchanged efficiently between the bloodstreams of mother and offspring (see figure 34.39). The fetal placenta is formed from the membranes of the chorion and allantois. The maternal side of the placenta is part of the wall of the uterus, the organ in which the young develop. In placental mammals, unlike marsupials, the young undergo a considerable period of development before they are born.

Mammals were not a major group until the dinosaurs disappeared. Mammalian specializations include the placenta, a tooth design suited to diet, and specialized sensory systems.

Table 34.6 Major Orders of Mammals

Order	Typical Examples		Key Characteristics	Approximate Number of Living Species
Rodentia	Beavers, mice, porcupines, rats		*Small plant-eaters* Chisel-like incisor teeth	1814
Chiroptera	Bats		*Flying mammals* Primarily fruit- or insect-eaters; elongated fingers; thin wing membrane; nocturnal; navigate by sonar	986
Insectivora	Moles, shrews		*Small, burrowing mammals* Insect-eaters; the most primitive placental mammals; spend most of their time underground	390
Marsupialia	Kangaroos, koalas		*Pouched mammals* Young develop in abdominal pouch	280
Carnivora	Bears, cats, raccoons, weasels, dogs		*Carnivorous predators* Teeth adapted for shearing flesh; no native families in Australia	240
Primates	Apes, humans, lemurs, monkeys		*Tree-dwellers* Large brain size; binocular vision; opposable thumb; group that evolved from a line that branched off early from other mammals	233
Artiodactyla	Cattle, deer, giraffes, pigs		*Hoofed mammals* With two or four toes; mostly herbivores	211
Cetacea	Dolphins, porpoises, whales		*Fully marine mammals* Streamlined bodies; front limbs modified into flippers; no hindlimbs; blowholes on top of head; no hair except on muzzle	79
Lagomorpha	Rabbits, hares, pika		*Rodent-like jumpers* Four upper incisors (rather than the two seen in rodents); hind legs often longer than forelegs, an adaptation for jumping	69
Suborder Pinnipedia	Sea lions, seals, walruses		*Marine carnivores* Feed mainly on fish; limbs modified for swimming	34
Edentata	Anteaters, armadillos, sloths		*Toothless insect-eaters* Many are toothless, but some have degenerate, peglike teeth	30
Perissodactyla	Horses, rhinoceroses, zebras		*Hoofed mammals with one or three toes* Herbivorous teeth adapted for chewing	17
Proboscidea	Elephants		*Long-trunked herbivores* Two upper incisors elongated as tusks; largest living land animal	2

34.4 Evolution among the primates has focused on brain size and locomotion.

Primates

Primates are mammals with two distinct features that allowed them to succeed in the arboreal, insect-eating environment:

1. **Grasping fingers and toes.** Unlike the clawed feet of tree shrews and squirrels, primates have grasping hands and feet that let them grip limbs, hang from branches, seize food, and in some primates, use tools. The first digit in many primates is opposable and at least some, if not all, of the digits have nails.
2. **Binocular vision.** Unlike the eyes of shrews and squirrels, which sit on each side of the head so that the two fields of vision do not overlap, the eyes of primates are shifted forward to the front of the face. This produces overlapping binocular vision that lets the brain judge distance precisely—important to an animal moving through the trees.

FIGURE 34.43
A prosimian. This tarsier, a prosimian native to tropical Asia, shows the characteristic features of primates: grasping fingers and toes and binocular vision.

Other mammals have binocular vision, but only primates have both binocular vision and grasping hands, making them particularly well adapted to their environment.

The Evolution of Prosimians

About 40 million years ago, the earliest primates split into two groups: the prosimians and the anthropoids. The **prosimians** ("before monkeys") looked something like a cross between a squirrel and a cat and were common in North America, Europe, Asia, and Africa. Only a few prosimians survive today—lemurs, lorises, and tarsiers (figure 34.43). In addition to grasping digits and binocular vision, prosimians have large eyes with increased visual acuity. Most prosimians are nocturnal, feeding on fruits, leaves, and flowers, and many lemurs have long tails for balancing.

Anthropoids

Anthropoids include monkeys, apes, and humans. Anthropoids are almost all diurnal—that is, active during the day—feeding mainly on fruits and leaves. Evolution favored many changes in eye design, including color vision, that were adaptations to daytime foraging. An expanded brain governs the improved senses, with the braincase forming a larger portion of the head. Anthropoids, like the relatively few diurnal prosimians, live in groups with complex social interactions, and they tend to care for their young for prolonged periods, allowing for a long childhood of learning and brain development. About 30 million years ago, some anthropoids migrated to South America. Their descendants, known as the New World monkeys, are easy to identify: All are arboreal; they have flat, spreading noses; and many of them grasp objects with long, prehensile tails. Anthropoids that remained in Africa gave rise to two lineages: the Old World monkeys and the hominoids (apes and humans). Old World monkeys include ground-dwelling as well as arboreal species. None of them have prehensile tails, their nostrils are close together, their noses point downward, and some have toughened pads of skin for prolonged sitting.

The **hominoids** include the apes and the **hominids** (humans and their direct ancestors). The living apes consist of the gibbon (genus *Hylobates*), orangutan (*Pongo*), gorilla (*Gorilla*), and chimpanzee (*Pan*). Apes have larger brains than monkeys, and they lack tails. With the exception of the gibbon, which is small, all living apes are larger than any monkey. Apes exhibit the most adaptable behavior of any mammal except human beings. Once widespread in Africa and Asia, apes are rare today, living in relatively small areas. No apes ever occurred in North or South America.

Studies of ape DNA have explained a great deal about how the living apes evolved. The Asian apes evolved first. The line of apes leading to gibbons diverged from other apes about 15 million years ago, while orangutans split off about 10 million years ago (figure 34.44). Neither group is closely related to humans.

The African apes evolved more recently, between 6 and 10 million years ago. These apes are the closest living relatives to humans; some taxonomists have even advocated placing humans and the African apes in the same zoological family, the Hominidae. Fossils of the earliest hominids (humans and their direct ancestors), described later in this section, suggest that the common ancestor of the hominids was more like a chimpanzee than a gorilla. Based on

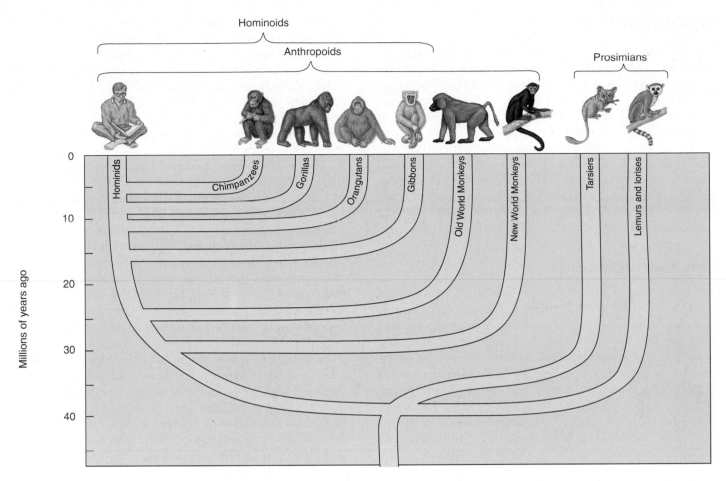

FIGURE 34.44
A primate evolutionary tree. The most ancient of the primates are the prosimians, while the hominids were the most recent to evolve.

genetic differences, scientists estimate that gorillas diverged from the line leading to chimpanzees and humans some 8 million years ago.

Soon after the gorilla lineage diverged, the common ancestor of all hominids split off from the chimpanzee line to begin the evolutionary journey leading to humans. Because this split was so recent, the genes of humans and chimpanzees have not had time to evolve many genetic differences. For example, a human hemoglobin molecule differs from its chimpanzee counterpart in only a single amino acid. In general, humans and chimpanzees exhibit a level of genetic similarity normally found between closely related sibling species of the same genus!

Comparing Apes to Hominids

The common ancestor of apes and hominids is thought to have been an arboreal climber. Much of the subsequent evolution of the hominoids reflected different approaches to locomotion. Hominids became **bipedal,** walking upright, while the apes evolved knuckle-walking, supporting their weight on the back sides of their fingers. (Monkeys, by contrast, use the palms of their hands.)

Humans depart from apes in several areas of anatomy related to bipedal locomotion. Because humans walk on two legs, their vertebral column is more curved than an ape's, and the human spinal cord exits from the bottom rather than the back of the skull. The human pelvis has become broader and more bowl-shaped, with the bones curving forward to center the weight of the body over the legs. The hip, knee, and foot (in which the human big toe no longer splays sideways) have all changed proportions. Being bipedal, humans carry much of the body's weight on the lower limbs, which comprise 32 to 38% of the body's weight and are longer than the upper limbs; human upper limbs do not bear the body's weight and make up only 7 to 9% of human body weight. African apes walk on all fours, with the upper and lower limbs both bearing the body's weight; in gorillas, the longer upper limbs account for 14 to 16% of body weight, the somewhat shorter lower limbs for about 18%.

Primates include the prosimians and anthropoids. The apes, monkeys, and hominids make up the anthropoids, and chimpanzees seem the most closely related to humans.

Australopithecines

Five to 10 million years ago, the world's climate began to get cooler, and the great forests of Africa were largely replaced with savannas and open woodland. In response to these changes, a new kind of hominoid was evolving, one that was bipedal. These new hominoids are classified as hominids—that is, of the human line.

The major groups of hominids include three to seven species of the genus *Homo* (depending how you count them), seven species of the older, smaller-brained genus *Australopithecus*, and several even older lineages. In every case where the fossils allow a determination to be made, the hominids are bipedal, the hallmark of hominid evolution.

Early Australopithecines

Our knowledge of australopithecines is based on hundreds of fossils, all found in South and East Africa (except for one specimen from Chad). It is probable, however, that australopithecines lived over a much broader area of Africa because only in the southern and eastern parts of the continent are sediments of the proper age exposed to fossil hunters. The evolution of hominids seems to have begun with an initial radiation of numerous species. The seven species identified so far provide ample evidence that australopithecines were a diverse group, and additional species will undoubtedly be described by future investigators.

These early hominids weighed about 18 kilograms and were about 1 meter tall. Their dentition was distinctly hominid, but their brains were no larger than those of apes, generally 500 cubic centimeters (cc) or less. *Homo* brains, by comparison, are usually larger than 600 cc; modern *H. sapiens* brains average 1350 cc.

The structure of australopithecine fossils clearly indicates that they walked upright. Evidence of bipedalism includes a set of some 69 hominid footprints found at Laetoli, East Africa. Two individuals, one larger than the other, walked upright side-by-side for 27 meters, their footprints preserved in a layer of 3.7-million-year-old volcanic ash. Importantly, the big toe is not splayed out to the side as in a monkey or ape, indicating that these footprints were clearly made by hominids.

The evolution of bipedalism marks the beginning of hominids. Bipedalism seems to have evolved as australopithecines left dense forests for grasslands and open woodland (figure 34.45). Whether larger brains or bipedalism evolved first was a matter of debate for some time. One school of thought proposed that hominid brains enlarged first, and then hominids became bipedal. Supporters of this view speculated that human intelligence was necessary to make the decision to walk upright and move out of the forests onto the grassland. Another school of thought saw bipedalism as a precursor to larger brains, arguing that bipedalism freed the forelimbs to manufacture and use tools, leading to the evolution of bigger brains. Recently, a treasure trove of fossils unearthed in Africa has settled the debate. These fossils demonstrate that bipedalism extended back 4 million years; knee joint, pelvis, and leg bones all exhibit the hallmarks of an upright stance. Substantial brain expansion, on the other hand, did not appear until roughly 2 million years ago. In hominid evolution, upright walking clearly preceded large brains.

The reason bipedalism evolved in hominids remains a matter of controversy. No tools appeared until 2.5 million years ago, so tool-making seems an unlikely cause. Alternative ideas suggest that walking upright is faster and uses less energy than walking on four legs; that an upright posture

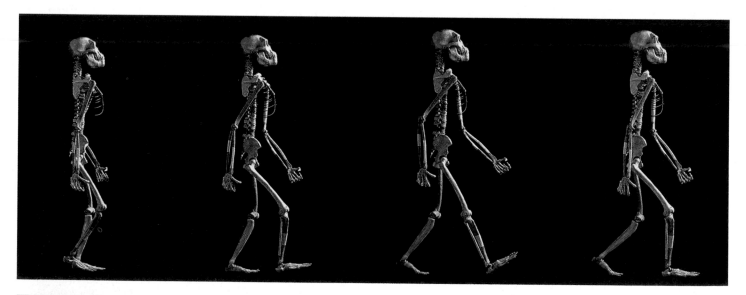

FIGURE 34.45

A reconstruction of an early hominid walking upright. These articulated plaster skeletons, made by Owen Lovejoy and his students at Kent State University, depict an early hominid (*Australopithecus afarensis*) walking upright.

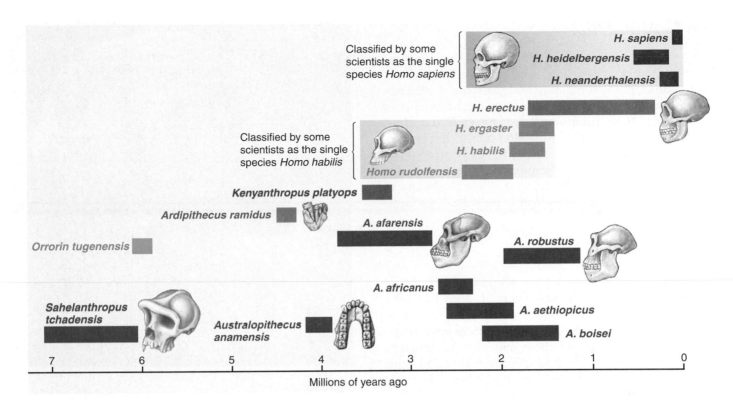

FIGURE 34.46
A hominid evolutionary tree. In this tree, the most widely accepted, the horizontal bars show the dates of first and last appearances of proposed species. Six species of *Australopithecus* and seven of *Homo* are included, as well as four other newly described early hominid genera. The oldest hominid, *Sahelanthropus tchadensis*, was discovered in Chad, Central Africa, in 2002. Some researchers question whether this 6–7 million-year-old fossil is a hominid, suggesting that it may instead be a common ancestor of chimpanzees and hominids.

permits hominids to pick fruit from trees and see over tall grass; that being upright reduces the body surface exposed to the sun's rays; that an upright stance aided the wading of aquatic hominids; and that bipedalism frees the forelimbs of males to carry food back to females, encouraging pair-bonding. All of these suggestions have their proponents, and none are universally accepted. The origin of bipedalism, the key event in the evolution of hominids, remains a mystery.

Early Hominids

In recent years, anthropologists have found a remarkable series of early hominid fossils extending as far back as 6–7 million years. Often displaying a mixture of primitive and modern traits, these fossils have thrown the study of early hominids into turmoil. While the inclusion of these fossils among the hominids seems warranted, only a few specimens of these early genera have been discovered, and they do not provide enough information to determine with any degree of certainty their relationships to australopithecines and humans. The search for additional early hominid fossils continues.

Differing Views of the Hominid Family Tree

Investigators take two different philosophical approaches to characterizing the diverse group of African hominid fossils. One group (the "lumpers") focuses on common elements in different fossils, and tends to lump together fossils that share key characters, attributing differences among the fossils to diversity within the group. Other investigators (the "splitters") are more inclined to assign fossils that exhibit differences to different species. For example, in the hominid evolutionary tree presented in figure 34.46, lumpers recognize three species of *Homo*, while splitters recognize no fewer than seven! At this point, it is not possible to decide which view is correct; more fossils are needed to determine how much the differences between fossils represent within-species variation and how much they characterize between-species differences.

The evolution of bipedalism—walking upright—marks the beginning of hominid evolution, although no one is quite sure why bipedalism evolved. The root of the hominid evolutionary tree is only imperfectly known.

The Genus *Homo*

The first humans evolved from australopithecine ancestors about 2 million years ago. The exact ancestor has not been clearly defined, but is commonly thought to be *A. afarensis*. Only within the last 30 years have a significant number of fossils of early *Homo* been uncovered. An explosion of interest has fueled intensive field exploration in the last few years, and new finds are announced regularly; every year, our picture of the base of the human evolutionary tree grows clearer. The following account will undoubtedly be outdated by future discoveries, but it provides a good example of science at work.

The First Human: *Homo habilis*

In the early 1960s, stone tools were found scattered among hominid bones close to the site where *A. boisei* had been unearthed. Although the fossils were badly crushed, painstaking reconstruction of the many pieces suggested a skull with a brain volume of about 680 cubic centimeters, larger than the australopithecine range of 400 to 550 cubic centimeters. Because of its association with tools, this early human was called *Homo habilis*, meaning "handy man." Partial skeletons discovered in 1986 indicate that *H. habilis* was small in stature, with arms longer than its legs and a skeleton much like that of *Australopithecus*. Because of its general similarity to australopithecines, many researchers at first questioned whether this fossil was human.

How Diverse Was Early *Homo?*

Because so few fossils of early *Homo* have been found, lively debate has ensued concerning whether they should all be lumped into *H. habilis* or split into three species: *H. rudolfensis*, *H. habilis*, and *H. ergaster* (Greek *ergaster*, "workman"). If the three species designations are accepted, as increasing numbers of researchers are doing, it would appear that *Homo* underwent an adaptive radiation, with *H. rudolfensis* the most ancient species, followed by *H. habilis* and then *H. ergaster*. Because of its modern skeleton, *H. ergaster* (figure 34.47) is thought to be the most likely ancestor to later species of *Homo*.

Out of Africa: *Homo erectus*

Our picture of what early *Homo* was like lacks detail, because it is based on only a few specimens. We have much more information about the species that replaced it, *Homo erectus*.

Homo erectus was a lot larger than *Homo habilis*—about 1.5 meters tall. It had a large brain, about 1000 cubic centimeters, and walked erect. Its skull had prominent brow ridges and, like modern humans, a rounded jaw. Most interesting of all, the shape of the skull interior suggests that *H. erectus* was able to talk.

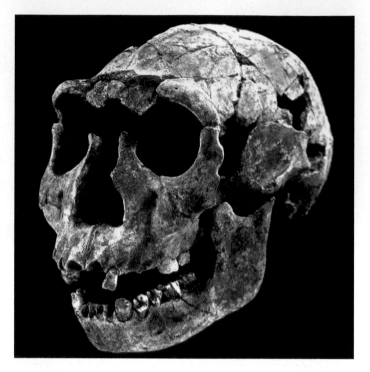

FIGURE 34.47
Early *Homo*. This skull of a boy, who apparently died in early adolescence, is 1.6 million years old and has been assigned to the species *Homo ergaster*. He was about 1.5 meters in height and weighed approximately 47 kilograms.

Where did *H. erectus* come from? It should be no surprise to you that it came out of Africa. In 1976, a complete *H. erectus* skull was discovered in East Africa. It was 1.5 million years old, a million years older than earlier Asian finds. Far more successful than *H. habilis*, *H. erectus* quickly became widespread and abundant in Africa, and within 1 million years had migrated into Asia and Europe. A social species, *H. erectus* lived in tribes of 20 to 50 people, often dwelling in caves. They successfully hunted large animals, butchered them using flint and bone tools, and cooked them over fires—a site in China contains the remains of horses, bears, elephants, and rhinoceroses.

Homo erectus survived for over a million years, longer than any other species of human. These very adaptable humans only disappeared in Africa about 500,000 years ago, as modern humans were emerging. Interestingly, they survived even longer in Asia, until 250,000 years ago.

The evolutionary journey entered its final phase when modern humans first appeared in Africa about 600,000 years ago. Investigators who focus on human diversity denote three species of modern humans: *Homo heidelbergensis*, *H. neanderthalensis*, and *H. sapiens*. Other investigators lump the three species into one, *H. sapiens* ("wise man"). The oldest modern human, *Homo heidelbergensis*, is known from a 600,000-year-old fossil from Ethiopia. Although it coexisted with *H. erectus* in Africa, *H. heidelbergensis* has more

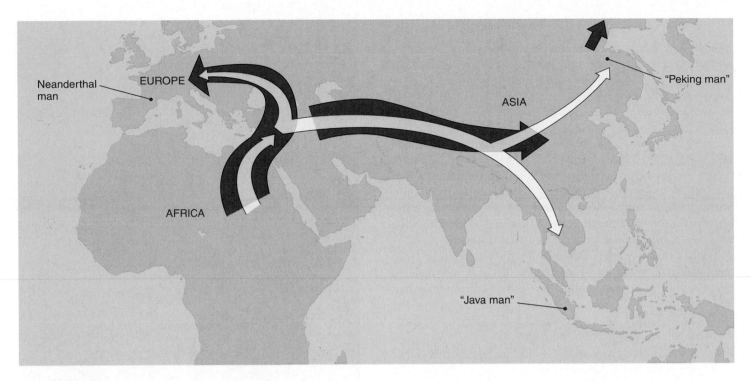

FIGURE 34.48
Out of Africa—many times. A still-controversial theory suggests that *Homo* spread from Africa to Europe and Asia repeatedly. First, *Homo erectus* (*white* arrow) spread as far as Java and China. Later, *H. erectus* was followed and replaced by *Homo neanderthalensis*, a pattern repeated again still later by *Homo sapiens* (*red* arrow).

advanced anatomical features, including a bony keel running along the midline of the skull, a thick ridge over the eye sockets, and a large brain. Also, its forehead and nasal bones are very like those of *H. sapiens*.

As *H. erectus* was becoming rarer, about 130,000 years ago, a new species of human arrived in Europe from Africa. *Homo neanderthalensis* likely branched off the ancestral line leading to modern humans as long as 500,000 years ago. Compared to modern humans, Neanderthals were short, stocky, and powerfully built; their skulls were massive, with protruding faces, heavy, bony ridges over the brows, and larger braincases.

Out of Africa—Again?

The oldest known fossil of *Homo sapiens*, our own species, is from Ethiopia and is about 130,000 years old. Other fossils from Israel appear to be between 100,000 and 120,000 years old. Outside of Africa and the Middle East, no clearly dated *H. sapiens* fossils older than roughly 40,000 years of age have been found. Proponents of a viewpoint known as the Recently-Out-of-Africa Hypothesis interpret this as evidence that *H. sapiens* evolved in Africa and then migrated to Europe and Asia. An opposing view, the Multiregional Hypothesis, argues that the human races independently evolved from *H. erectus* in different parts of the world.

Recently, scientists studying human mitochondrial DNA have added fuel to the fire of this controversy. Because DNA accumulates mutations over time, the oldest populations should show the greatest genetic diversity. Researchers sequencing the entire mDNA from 53 individuals of differing ethnic backgrounds found that modern humans shared a common ancestor 170,000 years ago, confirming that *H. sapiens* originated in Africa at about this time. The data also reveal a distinct branch on the human family tree 52,000 years ago, separating Africans from non-Africans. This is consistent with the hypothesis that humans originated in Africa, from there spreading to all parts of the world and retracing the path taken by *H. erectus* half a million years before (figure 34.48).

Another way to examine the human family tree is to look at genes on the Y chromosome, which does not undergo recombination. Y chromosomes pass down unchanged in males from one generation to the next. Any new changes that arise during evolution are easy to track on the family tree, because they too are passed down unchanged. The researchers in a large study looked at the pattern of gene variation among more than a thousand European males. While they identified many different patterns of variation, fully 80% of European males shared a single pattern, suggesting that modern Europeans have a common ancestor. The data indicate the pattern arose some 40,000 to 50,000 years ago. In other words, *H. sapiens* came to Europe recently.

By both these sets of evidence, the Multiregional Hypothesis is wrong. Our family tree has a single stem.

Cro-Magnons Replace the Neanderthals

The Neanderthals (classified by many paleontologists as a separate species, *Homo neanderthalensis*) were named after the Neander Valley of Germany where their fossils were first discovered in 1856. Rare at first outside of Africa, they became progressively more abundant in Europe and Asia, and by 70,000 years ago had become common. The Neanderthals made diverse tools, including scrapers, spearheads, and hand axes. They lived in huts or caves. Neanderthals took care of their injured and sick and commonly buried their dead, often placing food, weapons, and even flowers with the bodies. Such attention to the dead strongly suggests that they believed in a life after death. This is the first evidence of the symbolic thinking characteristic of modern humans.

Fossils of *H. neanderthalensis* abruptly disappear from the fossil record about 34,000 years ago and are replaced by fossils of *H. sapiens* called the Cro-Magnons (named after the valley in France where their fossils were first discovered). We can only speculate why this sudden replacement occurred, but it was complete all over Europe in a short period. A variety of evidence indicates that Cro-Magnons came from Africa—fossils of essentially modern aspect but as much as 100,000 years old have been found there. Cro-Magnons seem to have replaced the Neanderthals completely in the Middle East by 40,000 years ago, and then spread across Europe, coexisting with the Neanderthals for several thousand years. Recent analyses of Neanderthal DNA reveal it to be quite distinct from Cro-Magnon DNA, indicating the two species did not interbreed. Neanderthals are our cousins, not our ancestors. The Cro-Magnons that replaced the Neanderthals had a complex social organization and are thought to have had full language capabilities. Elaborate and often beautiful cave paintings made by Cro-Magnons can be seen throughout Europe (figure 34.49).

Humans of modern appearance eventually spread across Siberia to North America, where they arrived at least 13,000 years ago, after the ice had begun to retreat and a land bridge still connected Siberia and Alaska. By 10,000 years ago, about 5 million people inhabited the entire world (compared with more than 6 billion today).

Our Own Species: *Homo sapiens*

H. sapiens is the only surviving species of the genus *Homo*, and indeed the only surviving hominid. Some of the best fossils of *Homo sapiens* are 20 well-preserved skeletons with skulls found in a cave near Nazareth in Israel. Modern dating techniques estimate these humans to be between 90,000 and 100,000 years old. The skulls are modern in appearance and size, with high, short braincases, vertical foreheads with only slight brow ridges, and a cranial capacity of roughly 1550 cubic centimeters.

We humans are animals and the product of evolution. Our evolution has been marked by a progressive increase

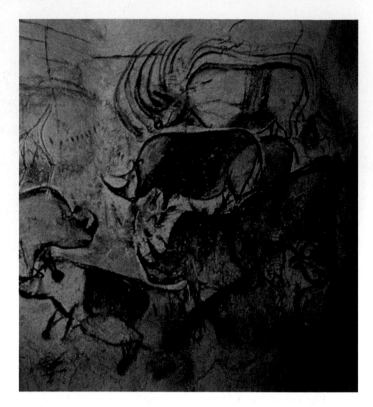

FIGURE 34.49
Cro-Magnon art. Rhinoceroses are among the animals depicted in this remarkable cave painting found in 1995 near Vallon-Pont d'Arc, France.

in brain size, distinguishing us from other animals in several ways. First, humans are able to make and use tools effectively—a capability that, more than any other factor, has been responsible for our dominant position in the animal kingdom. Second, although not the only animal capable of conceptual thought, humans have refined and extended this ability until it has become the hallmark of our species. Finally, we use symbolic language and can, with words, shape concepts out of experience and transmit that accumulated experience from one generation to another. Thus, we have undergone what no other animal ever has: extensive cultural evolution. Through culture, we have found ways to change and mold our environment, rather than changing evolutionarily in response to the demands of the environment. We control our biological future in a way never before possible—an exciting potential and a frightening responsibility.

Several species of *Homo* evolved in Africa, and some migrated from there to Europe and Asia. *Homo sapiens*, our species, seems to have evolved in Africa and then, like *H. erectus* before it, migrated to Europe and Asia. Our species, *Homo sapiens*, is proficient at conceptual thought and tool use, and is the only animal that uses symbolic language.

34.1 Attaching muscles to an internal framework greatly improves movement.

The Chordates

- Four features characterize the chordates: (1) single, hollow nerve cord; (2) a flexible notochord present at some developmental stage; (3) pharyngeal pouches connecting the pharynx and the esophagus; (4) a postanal tail at least during embryonic development. (p. 684)

34.2 Nonvertebrate chordates have a notochord but no backbone.

The Nonvertebrate Chordates

- Tunicates possess a notochord and a nerve cord as larvae, but exhibit no major body cavity or segmentation as adults. (p. 686)
- Lancelets are fishlike marine chordates with a permanent notochord running the entire length of the dorsal nerve cord. (p. 687)

34.3 The evolution of vertebrates involved invasions of sea, land, and air.

Characteristics of Vertebrates

- Vertebrates differ from tunicates and lancelets in that they have a vertebral column instead of a notochord and a distinct, well-differentiated head. (p. 688)
- The history of the vertebrates includes a series of evolutionary advances that allowed them to invade the sea and then the land. (p. 689)

Fishes

- Fish were the first vertebrates and are the most diverse and successful vertebrate group. (p. 690)
- Key characteristics of fish include a vertebral column, jaws and paired appendages, gills, single-loop circulation, and nutritional deficiencies. (p. 690)
- The first fish had heads made of bone and internal skeletons made of cartilage. Jaws later evolved from cartilage arch supports. (pp. 692–693)
- Sharks eventually became dominant sea predators, partially due to a skeleton composed of calcified cartilage. Sharks were also among the first vertebrates to develop teeth. (p. 695)
- Bony fish evolved at the same time as sharks, but adopted a heavy internal skeleton made of bone. Such ossification provided a strong base for muscle attachment and evolved in fresh water. (p. 696)
- Bony fishes also evolved important adaptations such as a swim bladder for buoyancy, a lateral line sensory system, and a gill cover (operculum) to permit water to be pumped over the gills. (pp. 696–697)

Amphibians

- Key characteristics of living amphibians include legs, cutaneous respiration, lungs, pulmonary veins, and a partially divided heart. (p. 698)
- Paleontologists believe amphibians must have evolved from lobe-finned fishes. Amphibians today include frogs and toads, salamanders, and caecilians. (pp. 699-701)

Reptiles

- Key characteristics of reptiles include the amniotic egg, dry skin, and thoracic breathing. (p. 702)

The Rise and Fall of Dominant Reptile Groups

- Four major forms of reptiles took their respective turns as the dominant large terrestrial vertebrates: pelycosaurs, therapsids, theocodonts, and dinosaurs. (pp. 704–705)
- Other important characteristics of reptiles are that they practice internal fertilization, have an improved circulatory system, and are ectothermic. (p. 707)
- Only four reptilian orders survive today: turtles, lizards and snakes, tuataras, and crocodiles. (pp. 707–709)

Birds

- Modern birds retain many reptilian characteristics, but lack teeth and have vestigial tails. They are distinguished from living reptiles by feathers and the presence of a thin, hollow flight skeleton. (p. 710)

History of the Birds

- *Archaeopteryx* was probably the first bird, arising 150 MYA and likely a direct descendant of dinosaurs. (pp. 712–713)
- Aves continues to be listed as a separate class due to the evolutionary novelties of feathers and hollow bones and to physiological mechanisms such as efficient lungs capable of sustaining powered flight. (p. 713)

Mammals

- Key mammalian characteristics include hair, mammary glands, a placenta, heterodont dentition, the ability to digest plant material, keratinized hooves and horns, and flight capability (in bats). (pp. 714–716)

The Orders of Mammals

- Mammals were not common until dinosaurs disappeared. Modern mammals fall into one of three categories: monotremes, egg-laying mammals; marsupials, pouched mammals; and placentals. (pp. 717–718)

34.4 Evolution among the primates has focused on brain size and locomotion.

Primates

- Grasping fingers and toes and binocular vision are two features that allowed primates to flourish. (p. 720)
- Modern prosimians include lemurs, lorises, and tarsiers, while anthropoids include monkeys, apes, and humans. (pp. 720–721)

Australopithecines

- Bipedalism marked the beginning of hominid evolution, although the reason for such evolution remains controversial. (p. 722)
- Currently, two philosophical approaches, lumping and splitting, are used to characterize African hominid fossils. (p. 723)

The Genus Homo

- The first humans (*Homo habilis*) evolved from australopithecine ancestors about 2 MYA. (p. 724)
- *Homo erectus* replaced *H. habilis*, and is believed to have come out of Africa. (pp. 724–725)
- *Homo sapiens* is both the only surviving species of the genus *Homo* and the only surviving hominid. (p. 726)
- Humans are the only animals that can effectively make tools, that have refined and extended the ability to use conceptual thought, and that can use symbolic language and shape concepts and experiences with words. (p. 726)

Self Test

1. Which of the following is a characteristic of chordates but is *not* found in other animals?
   a. a notochord
   b. jointed appendages
   c. an exoskeleton
   d. all of these

2. In which animal(s) does the notochord persist in the adult?
   a. tunicates
   b. lampreys
   c. lancelets
   d. all of these

3. The very first vertebrates were
   a. cartilaginous fish.
   b. fishes with jaws.
   c. amphibians.
   d. jawless fish.

4. Which of the following is *not* a characteristic of fishes?
   a. gills
   b. lungs
   c. single-loop blood circulation
   d. nutritional deficiencies

5. What adaptation of bony fish allows them to detect and orient themselves in the upstream direction?
   a. the swim bladder
   b. lobed fins
   c. the operculum
   d. the lateral line system

6. In order for amphibians to be successful on land, they had to develop which of the following?
   a. a more efficient swim bladder
   b. cutaneous respiration and lungs
   c. more efficient gills
   d. shelled eggs

7. Amniotic eggs evolved as a means to
   a. protect the embryo while the parent sits on the egg.
   b. protect the embryo from predators.
   c. allow the parent to gather food, rather than sitting on the nest.
   d. prevent the embryo from drying out.

8. A group of early reptiles that may have been warm-blooded was the
   a. pelycosaurs.
   b. therapsids.
   c. thecodonts.
   d. all of these.

9. *Archaeopteryx* is believed to be the transition fossil between dinosaurs and birds because, like a bird, *Archaeopteryx* had feathers and
   a. a tail similar to that of modern birds.
   b. scales.
   c. a toothless, elongated mouth like a beak.
   d. a fused collarbone, indicating flying ability.

10. Mammals that have live births but incubate newborns in a pouch through the completion of development are
    a. monotremes.
    b. marsupials.
    c. duck-billed platypuses.
    d. placental mammals.

Test Your Visual Understanding

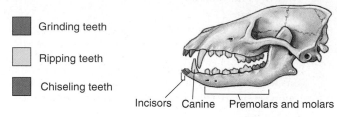

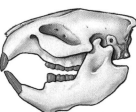

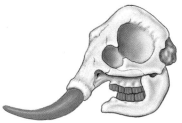

Grinding teeth
Ripping teeth
Chiseling teeth

Incisors Canine Premolars and molars
Dog

Deer

Beaver

Elephant

Human

1. Based on this figure, predict whether each type of consumer is a carnivore, herbivore, or omnivore.
   a. dog   c. beaver   e. human
   b. deer   d. elephant

Apply Your Knowledge

1. Homeothermic animals use 98% of cellular energy in metabolism and "store" 2% for growth. Poikilotherms have lower metabolisms and so are able to store 44% of cellular energy for growth. For homeotherms, 77.5% of chemical energy is converted into cellular energy with an efficiency of 77.5%. By comparison, poikilotherm efficiency is 41.9%. How much food must be consumed by each type of animal to gain one gram of weight?

2. What characteristics allowed vertebrates to attain great sizes?

3. What limits the ability of amphibians to occupy the full range of terrestrial habitats and allows other terrestrial vertebrates to occupy them successfully?

4. List some of the advantages that the early birds, in which flight was not nearly as efficient as it is in most of their modern descendants, might have had as a result of the presence of feathers.

35

Plant Form

Concept Outline

35.1 Meristems elaborate the plant body plan after germination.

    Meristems. Growth occurs in the continually dividing cells that function like the stem cells in animals.

    Organization of the Plant Body. The plant body is a series of reiterative units stacked above and below the ground.

    Primary and Secondary Growth. Different meristems allow plants to grow in both height and circumference.

35.2 Plants have three basic tissues, each composed of several cell types.

    Dermal Tissue. This tissue forms the "skin" of the plant body, protecting it and preventing water loss.

    Ground Tissue. Much of a young plant is ground tissue, which supports the plant body and stores food and water.

    Vascular Tissue. Special conducting tissues move water and sugars through the plant body.

35.3 Root cells differentiate as they become distanced from the dividing root apical meristem.

    Root Structure. Roots have a durable cap, behind which primary growth occurs.

    Modified Roots. Roots can have specialized functions.

35.4 Stems are the backbone of the shoot, transporting nutrients and supporting the aerial plant organs.

    Stem Structure. The stem supports the leaves and is anchored by the roots. Vascular tissues are organized within the stem in different ways.

    Modified Stems. Specialized stems are adapted for storage and for asexual propagation.

35.5 Leaves are adapted to support basic plant functions.

    Leaf External Structure. Leaves have flattened blades.

    Leaf Internal Structure. Leaves contain cells that carry out photosynthesis, gas exchange, and evaporation.

    Modified Leaves. In some plants, leaf development has been modified and meets a unique need.

FIGURE 35.1
All vascular plants share certain characteristics. Vascular plants such as this tree require an elaborate system of mechanical support and fluid transport to grow this large. Smaller plants have similar (though simpler) structures. Much of this support system is actually underground in the form of extensive branching root systems.

Although the similarities among a cactus, an orchid plant, and a tree might not be obvious at first sight, most plants have a basic unity of structure (figure 35.1). This unity is reflected in how the plants are constructed; in how they grow, manufacture, and transport their food; and in how their development is regulated. This chapter addresses the question of how a vascular plant is "built." We will focus on the diversity of cell, tissue, and organ types that compose the adult plant body. The roots and shoots that give the adult plant its distinct above- and below-ground architecture are the final product of a basic body plan first established during embryogenesis, a process we will explore in detail in chapter 36.

35.1 Meristems elaborate the plant body plan after germination.

Meristems

The plant body that develops after germination depends on the activities of meristematic tissues. Meristematic tissues are clumps of small cells with dense cytoplasm and proportionately large nuclei that act as stem cells do in animals. That is, one cell divides to give rise to two cells. One remains meristematic, while the other is free to differentiate and contribute to the plant body. In this way, the population of meristem cells is continually renewed. Molecular genetic evidence supports the hypothesis that stem cells and meristem cells may also share some common pathways of gene expression.

Elongation of both root and shoot takes place as a result of repeated cell divisions and subsequent elongation of the cells produced by the **apical meristems.** In some vascular plants, including shrubs and most trees, **lateral meristems** produce an increase in girth.

Apical Meristems

Apical meristems are located at the tips of stems (figure 35.2) and at the tips of roots (figure 35.3), just behind the root cap. The plant tissues that result from primary growth are called **primary tissues.** During periods of growth, the cells of apical meristems divide and continually add more cells to the tips of a seedling's body. Thus, the seedling lengthens. Primary growth in plants is brought about by the apical meristems. The elongation of the root and stem forms what is known as the **primary plant body,** which is made up of primary tissues. The primary plant body comprises the young, soft shoots and roots of a tree or shrub, or the entire plant body in some plants.

Both root and shoot apical meristems are composed of delicate cells that need protection. The root apical meristem is protected from the time it emerges by the root cap. Root cap cells are produced by the root meristem and are sloughed off and replaced as the root moves through the soil. A variety of adaptive mechanisms protect the shoot apical meristem during germination (figure 35.4). The epicotyl or hypocotyl ("stemlike" tissue above or below the cotyledons) may bend as the seedling emerges to minimize the force on the shoot tip. Monocots (a late-evolving group of angiosperms; see chapter 29) often have a coleoptile (sheath of tissue) that forms a protective tube around the emerging shoot. Later in development, the leaf primordia cover the shoot apical meristem, which is particularly susceptible to desiccation.

The apical meristem gives rise to three types of embryonic tissue systems, called **primary meristems.** Cell division continues in these partly differentiated tissues as they develop into the primary tissues of the plant body. The three

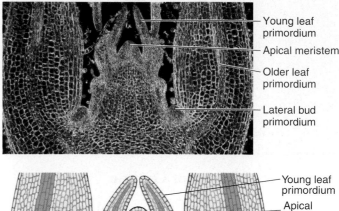

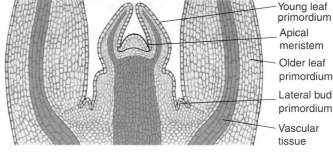

FIGURE 35.2

An apical shoot meristem. *Top:* This longitudinal section through a shoot apex in *Coleus* shows the tip of a stem. Between the young leaf primordia is the apical meristem. *Bottom:* A diagram delineates the areas that make up the tip of a stem.

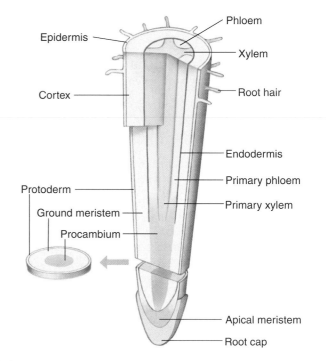

FIGURE 35.3

An apical root meristem. This diagram of meristems in the root shows their relation to the root tip.

(a)   (b)

FIGURE 35.4
Developing seedling. Apical meristems are protected early in development. (*a*) In corn, a monocot, a sleeve of tissue called the coleoptile sheaths the shoot tip until it has made it to daylight. (*b*) In this soybean, a eudicot, a bent epicotyl (stem above the cotyledons), rather than the shoot tip, pushes through the soil before straightening.

primary meristems are the **protoderm,** which forms the epidermis; the **procambium,** which produces primary vascular tissues (primary xylem and primary phloem); and the **ground meristem,** which differentiates further into ground tissue. In some plants, such as horsetails and corn, **intercalary meristems** arise in stem internodes, adding to the internode lengths. If you walk through a cornfield (when the corn is about knee high) on a quiet summer night, you may hear a soft popping sound. This is caused by the rapid growth of intercalary meristems. The amount of stem elongation that occurs in a very short time is quite surprising.

Lateral Meristems

Many herbaceous plants (that is, plants with fleshy, not woody stems) exhibit only primary growth, but others also exhibit **secondary growth.** Most trees, shrubs, and some herbs have active lateral meristems, which are peripheral cylinders of meristematic tissue within the stems and roots (figure 35.5). Although secondary growth increases girth in many nonwoody plants, its effects are most dramatic in woody plants, which have two lateral meristems. Within the bark of a woody stem is the **cork cambium,** a lateral meristem that produces the cork cells of the outer bark. Just beneath the bark is the **vascular cambium,** a lateral meristem that produces secondary vascular tissue. The vascular cambium forms between the xylem and phloem in vascular bundles, adding secondary vascular tissue on opposite sides of the vascular cambium. *Secondary xylem* is the main component of wood. *Secondary phloem* is very close to the outer surface of a woody stem. Removing the bark of a tree damages the phloem and may eventually kill the tree. Tissues formed from lateral meristems, which comprise

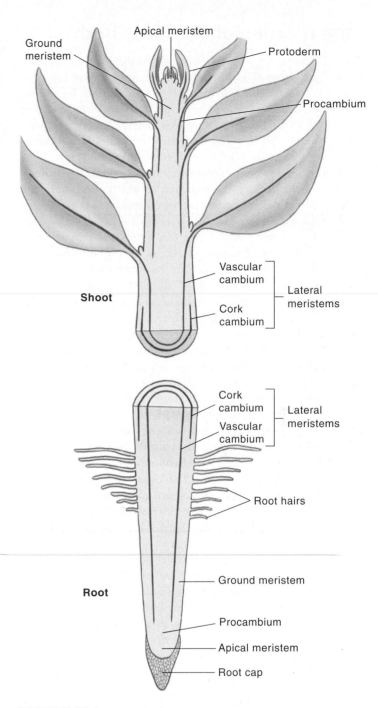

FIGURE 35.5
Apical and lateral meristems. Apical meristems produce primary growth, the elongation of the root and stem. In some plants, the lateral meristems produce an increase in the girth of a plant. This type of growth is secondary because the lateral meristems were not directly produced by apical meristems.

most of the trunk, branches, and older roots of trees and shrubs, are known as **secondary tissues** and are collectively called the secondary plant body.

Meristems are actively dividing, embryonic tissues responsible for both primary and secondary growth.

Organization of the Plant Body

Coordination of primary and secondary meristematic growth produces the body of the adult sporophyte (*2n* generation). Plant bodies do not necessarily have a fixed size. Parts such as leaves, roots, branches, and flowers all vary in size and number from plant to plant—even within a species. The development of the form and structure of plant parts may be relatively rigidly controlled, but some aspects of leaf, stem, and root development are quite flexible. As a plant grows, the number, location, size, and even structure of leaves and roots are often influenced by the environment.

A vascular plant consists of a *root system* and a *shoot system* (figure 35.6). The root system anchors the plant and penetrates the soil, from which it absorbs water and ions crucial to the plant's nutrition. The shoot system consists of the *stems* and their *leaves*. The stem serves as a framework for positioning the leaves, the principal sites of photosynthesis. The arrangement, size, and other features of the leaves are critically important in the plant's production of food. Flowers, other reproductive organs, and ultimately, fruits and seeds are also formed on the shoot (see chapter 41). The iterative unit of the vegetative shoot consists of the internode, node, leaf, and axillary bud, but not reproductive structures. Axillary buds are apical meristems derived from the primary apical meristem that allow the plant to branch or replace the main shoot if it is munched by an herbivore. A vegetative axillary bud has the capacity to reiterate the development of the primary shoot. When the plant has shifted to the reproductive phase of development (see chapter 41), these axillaries may produce flowers or floral shoots.

Three basic types of tissues exist in plants: *dermal tissue*, *ground tissue*, and *vascular tissue*. Each of the three basic tissues has its own distinctive, functionally related cell types, some of which will be discussed later in this chapter. In plants limited to primary growth, the **dermal tissue** system is composed of the **epidermis.** This tissue is one cell layer thick in most plants, and forms the outer protective covering of the plant. In young, exposed parts of the plant, the epidermis is covered with a fatty **cutin** layer constituting the **cuticle;** in plants such as the desert succulents, a layer of wax may be added to the cuticle. In plants with secondary growth, the bark forms the outer protective layer and is considered a part of the dermal tissue system.

Ground tissue consists primarily of thin-walled **parenchyma** cells that are initially (but briefly) more or less spherical. However, the cells, which have living protoplasts, push up against each other shortly after they are produced and assume other shapes, often ending up with 11 to 17 sides. Parenchyma cells may live for many years; they function in storage, photosynthesis, and secretion.

Vascular tissue includes two kinds of conducting tissues: (1) **xylem,** which conducts water and dissolved miner-

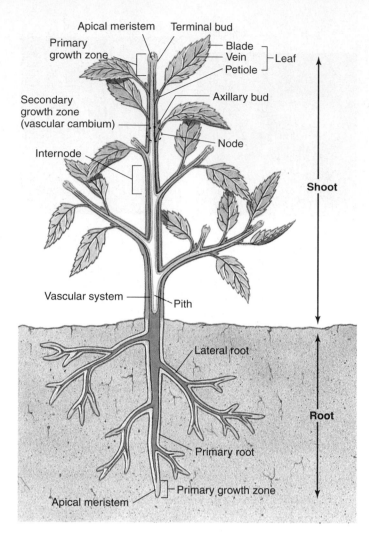

FIGURE 35.6
Diagram of a plant body. Branching in both the root and shoot systems increases the number of apical meristems. A significant increase in stem/root circumference and the formation of bark can only occur if secondary growth is initiated by the vascular and cork cambiums (secondary meristems). The lime green areas are zones of active elongation; secondary growth occurs in the lavender areas.

als, and (2) **phloem,** which conducts the carbohydrates—mainly sucrose—used by plants for food. The phloem also transports hormones, amino acids, and other substances that are necessary for plant growth. Xylem and phloem differ in structure as well as in function.

Root and shoot meristems give rise to a plant body with an extensive underground, branching root system and an aboveground shoot system with iterative units of node, leaf joined at the node, internode, and axillary bud.

Primary and Secondary Growth

Primary growth plays an important role in establishing the basic body plan of the organism. Some plants have secondary growth, allowing for an increase in diameter. Here we will look at how the meristems give rise to highly differentiated tissues that sustain the growing plant body. In the earliest vascular plants, many of which are extinct, the vascular tissues produced by primary meristems played the same conducting roles as they do in contemporary vascular plants. However, there was no differentiation of the plant body into stems, leaves, and roots. The presence of these three kinds of organs is a property of most modern plants and reflects increasing specialization in relation to the demands of a terrestrial existence.

With the evolution of secondary growth, vascular plants could develop thick trunks and become treelike (figure 35.7). This evolutionary advance in the sporophyte generation made possible the development of forests and the domination of the land by plants. Judging from the fossil record, secondary growth evolved independently in several groups of vascular plants by the middle of the Devonian period 380 million years ago.

Two types of conducting systems were present in the earliest plants, and these systems have become characteristic of vascular plants as a group. *Sieve-tube members* conduct carbohydrates away from areas where they are manufactured or stored. *Vessel members* and *tracheids* are thick-walled cells that transport water and dissolved minerals up from the roots. Both kinds of cells are elongated and occur in linked strands, forming tubes. Sieve-tube members are characteristic of phloem tissue; vessel members and tracheids are characteristic of xylem tissue. In primary tissues, which result from primary growth, these two types of tissue are typically associated with each other in the same vascular strands. In secondary growth, the phloem is found on the periphery, while a very thick xylem core develops more centrally. You will see that the roots and shoots of many vascular plants have different patterns of vascular tissue and secondary growth. Keep in mind that water and nutrients travel between the most distant tip of a redwood root and the tip of the shoot. For the system to work, these tissues connect, which they do in the transition zone between the root and the shoot. In section 35.2, we will consider the three tissue systems that are present in all plant organs, whether the plant has secondary growth or not.

> Plants grow as a result of the division of meristematic tissue. Primary growth results from cell division at the apical meristem at the tip of the plant, making the shoot longer. Secondary growth results from cell division at the lateral meristem in a cylinder encasing the shoot, and increases the shoot's girth.

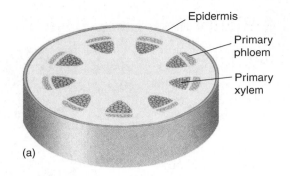

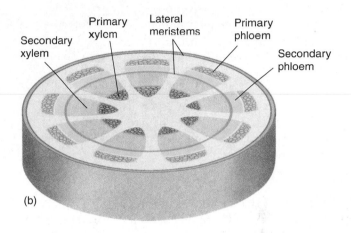

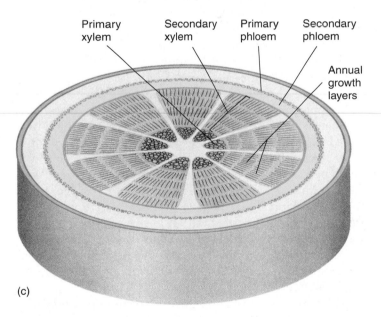

FIGURE 35.7
Secondary growth. (*a*) Before secondary growth begins in eudicot stems, primary tissues continue to elongate as the apical meristems produce primary growth. (*b*) As secondary growth begins, the lateral meristems produce secondary tissues, and the stem's girth increases. (*c*) In this four-year-old stem, the secondary tissues continue to widen, and the trunk has become thick and woody. Note that the lateral meristems form cylinders that run axially (up and down) in the roots and shoots that have them.

Plants have three basic tissues, each composed of several cell types.

Dermal Tissue

Epidermal cells, which originate from the protoderm, cover all parts of the primary plant body. A number of types of specialized cells occur in the epidermis, including **guard cells, trichomes,** and **root hairs.**

Guard cells are paired, sausage-shaped cells flanking a **stoma** (plural, *stomata*), a mouth-shaped epidermal opening. Guard cells, unlike other epidermal cells, contain chloroplasts. Stomata occur in the epidermis of leaves (figure 35.8) and sometimes on other parts of the plant, such as stems or fruits. The passage of oxygen and carbon dioxide, as well as the diffusion of water in vapor form, takes place almost exclusively through the stomata. There are from 1000 to more than 1 million stomata per square centimeter of leaf surface. In many plants, stomata are more numerous on the lower epidermis of the leaf than on the upper epidermis, a factor that minimizes water loss. Some plants have stomata only on the lower epidermis, and a few, such as water lilies, have them only on the upper epidermis.

Guard cell formation is the result of an asymmetrical cell division. The patterning of these asymmetrical divisions that results in stomatal distribution has intrigued developmental biologists. Research on mutants that get "confused" about where to position stomata is providing information on the timing of stomatal initiation and the kind of intercellular communication that triggers guard cell formation. For example, the *too many mouths* mutation that occurs in *Arabidopsis* may be caused by the failure of developing stomata to suppress stomatal formation in neighboring cells (figure 35.9).

Trichomes are hairlike outgrowths of the epidermis (figure 35.10). They occur frequently on stems, leaves, and reproductive organs. A "fuzzy" or "woolly" leaf is covered with trichomes that can be seen clearly with a microscope under low magnification. Trichomes keep leaf surfaces cool and reduce evaporation. Trichomes can vary greatly in form; some consist of a single cell, while others are multicellular. Some are glandular, often secreting sticky or toxic substances to deter herbivory. Genes, including *GLABOROUS3*, that regulate trichome development have been identified (figure 35.11).

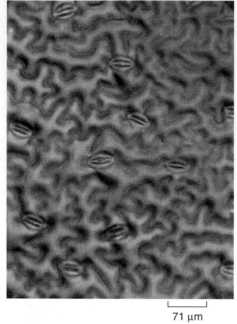

71 μm

71 μm

(a)   (b)

FIGURE 35.8

Epidermis of a eudicot and monocot leaf.   Stomata are evenly distributed over the epidermis of monocots and eudicots, but the patterning is quite different. (*a*) A pea (eudicot) leaf with a random arrangement of stomata. (*b*) A maize (corn, a monocot) leaf with stomata evenly spaced in rows. These photos also show the variety of cell shapes in plants. Some plant cells are boxlike, as seen in maize (*b*), while others are irregularly shaped, as seen in peas (*a*).

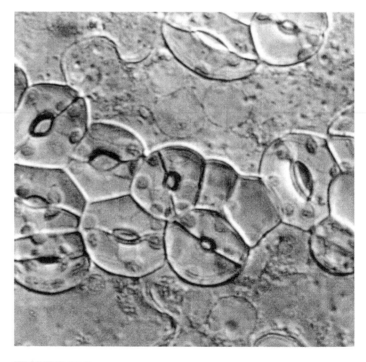

FIGURE 35.9

The *too many mouths* stomatal mutant.   This *Arabidopsis* mutant plant lacks an essential signal for spacing guard cells.

Root hairs, which are tubular extensions of individual epidermal cells, occur in a zone just behind the tips of young, growing roots (see figure 35.3). Because a root hair is simply an extension and not a separate cell, there is no cross-wall isolating it from the epidermal cell. Root hairs keep the root in intimate contact with the surrounding soil particles and greatly increase the root's surface area and the efficiency of absorption. As the root grows, the extent of the root hair zone remains roughly constant as root hairs at the older end slough off while new ones are produced at the other end. Most of the absorption of water and minerals occurs through root hairs, especially in herbaceous plants. Root hairs should not be confused with lateral roots, which are multicellular and originate deep within the root.

In the case of secondary growth, the cork cambium (discussed in section 35.4) produces the bark of a tree trunk or root. This replaces the epidermis, which gets stretched and broken with the radial expansion of the axis.

Some epidermal cells are specialized for protection, others for absorption. The spacing of these specialized cells within the epidermis maximizes their function and is an intriguing developmental puzzle.

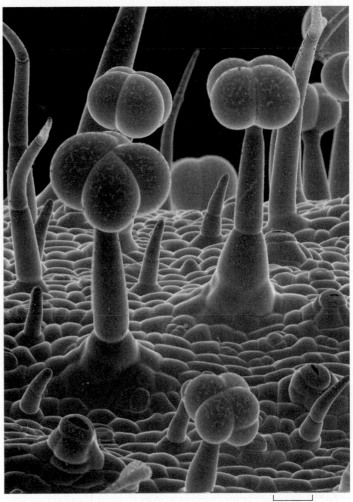

34.62 μm

FIGURE 35.10
Trichomes. The trichomes with tan, bulbous tips on this tomato plant are glandular trichomes. These trichomes secrete oils that deter insects.

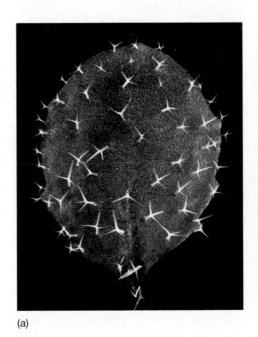

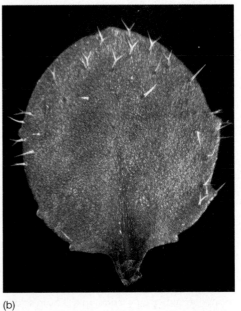

(a)

(b)

FIGURE 35.11
Trichome mutation. Mutants have revealed genes involved in regulating the spacing and development of trichomes in *Arabidopsis*. (*a*) Wild type. (*b*) *glaborous3* (*gl3*), which fails to initiate trichome development.

Ground Tissue

Parenchyma

Parenchyma cells, which have large vacuoles, thin walls, and an average of 14 sides at maturity, are the most common type of plant cell. They are the most abundant cells of primary tissues and may also occur, to a much lesser extent, in secondary tissues (figure 35.12*a*). Most parenchyma cells have only primary walls, which are walls laid down while the cells are still maturing. Parenchyma are less specialized than other plant cells, although there are many variations that do have special functions, such as nectar and resin secretion or storage of latex, proteins, and metabolic wastes.

Parenchyma cells, which have functional nuclei and are capable of dividing, commonly also store food and water, and usually remain alive after they mature; in some plants (for example, cacti), they may live to be over 100 years old. The majority of cells in fruits such as apples are parenchyma. Some parenchyma contain chloroplasts, especially in leaves and in the outer parts of herbaceous stems. Such photosynthetic parenchyma tissue is called *chlorenchyma.*

Collenchyma

Like parenchyma cells, **collenchyma cells** have living protoplasts and may live for many years. The cells, which are usually a little longer than wide, have walls that vary in thickness (figure 35.12*b*). Collenchyma cells, which are relatively flexible, provide support for plant organs, allowing them to bend without breaking. They often form strands or continuous cylinders beneath the epidermis of stems or leaf petioles (stalks) and along the veins in leaves. Strands of collenchyma provide much of the support for stems in which secondary growth has not taken place. The parts of celery that we eat (petioles, or leaf stalks) have "strings" that consist mainly of collenchyma and vascular bundles (conducting tissues).

Sclerenchyma

Sclerenchyma cells have tough, thick walls; they usually lack living protoplasts when they are mature. Their secondary cell walls are often impregnated with **lignin,** a highly branched polymer that makes cell walls more rigid. Cell walls containing lignin are said to be *lignified.* Lignin is common in the walls of plant cells that have a supporting or mechanical function. Some kinds of cells have lignin deposited in primary as well as secondary cell walls.

There are two types of sclerenchyma: fibers and sclereids. *Fibers* are long, slender cells that are usually grouped together in strands. Linen, for example, is woven from strands of sclerenchyma fibers that occur in the phloem of flax. *Sclereids* are variable in shape but often branched. They may occur singly or in groups; they are not elongated, but may have various forms, including that of a star. The gritty texture of a pear is caused by groups of sclereids that occur throughout the soft flesh of the fruit (figure 35.12*c*). Both of these tough, thick-walled cell types serve to strengthen the tissues in which they occur.

Parenchyma cells are the most common type of plant cells and have various functions. Collenchyma cells provide much of the support in young stems and leaves. Sclerenchyma cells strengthen plant tissues and may be nonliving at maturity.

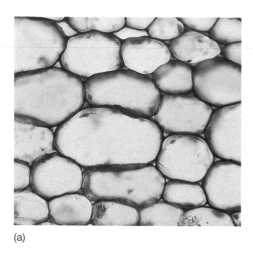
(a)

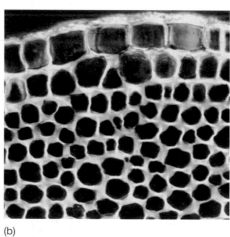
(b)

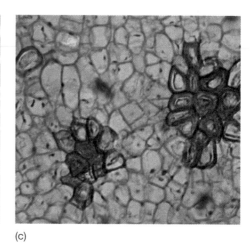
(c)

FIGURE 35.12
The three types of ground tissue. (*a*) Parenchyma cells. Only primary cell walls are seen in this cross section of parenchyma cells from grass. (*b*) Collenchyma cells. Thickened side walls are seen in this cross section of collenchyma cells from a young branch of elderberry (*Sambucus*). In other kinds of collenchyma cells, the thickened areas may occur at the corners of the cells or in other kinds of strips. (*c*) Sclereids. Clusters of sclereids ("stone cells"), stained red in this preparation, in the pulp of a pear. The surrounding thin-walled cells, stained green, are parenchyma. These sclereid clusters give pears their gritty texture.

Vascular Tissue

Xylem

Xylem, the principal water-conducting tissue of plants, usually contains a combination of *vessels*, which are continuous tubes formed from dead, hollow, cylindrical cells (*vessel members*) arranged end-to-end, and *tracheids*, which are dead cells that taper at the ends and overlap one another (figure 35.13). In some plants (but not angiosperms), tracheids are the only water-conducting cells present; water passes in an unbroken stream through the xylem from the roots up through the shoot and into the leaves. When the water reaches the leaves, much of it passes into a film of water on the outside of the parenchyma cells, and then it diffuses in the form of water vapor into the intercellular spaces and out of the leaves into the surrounding air, mainly through the stomata. This diffusion of water vapor from a plant is known as **transpiration.** In addition to conducting water, dissolved minerals, and inorganic ions such as nitrates and phosphates throughout the plant, xylem supplies mechanical support for the plant body.

Primary xylem is derived from the procambium, which comes from the apical meristem. *Secondary xylem* is formed by the vascular cambium, a lateral meristem that develops later. Wood consists of accumulated secondary xylem.

Vessel members are found almost exclusively in angiosperms. In primitive angiosperms, vessel members tend to resemble fibers and are relatively long. In more advanced angiosperms, vessel members tend to be shorter and wider, resembling microscopic, squat coffee cans with both ends removed. Both vessel members and tracheids have thick, lignified secondary walls and no living protoplasts at maturity. Lignin is produced by the cell and secreted to strengthen the cellulose cell walls before the protoplast dies, leaving only the cell wall. When the continuous stream of water in a plant flows through tracheids, it passes through *pits*, which are small, mostly rounded-to-elliptical areas where no secondary wall material has been deposited. The pits of adjacent cells occur opposite one another. In contrast, vessel members, which are joined end-to-end, may be almost completely open or may have bars or strips of wall material across the open ends.

Vessels appear to conduct water more efficiently than do the overlapping strands of tracheids. We know this partly because vessel members have evolved from tracheids independently in several groups of plants, suggesting that they are favored by natural selection. It is also probable that some types of fibers have evolved from tracheids, becoming specialized for strengthening rather than conducting. Some ancient flowering plants have only tracheids, but virtually all modern angiosperms have vessels. Plants having a mutation that prevents the differentiation of vessels, but does not affect tracheids, wilt soon after germination and are unable to transport water efficiently.

In addition to conducting cells, xylem typically includes fibers and parenchyma cells (ground tissue cells). The parenchyma cells, which are usually produced in horizontal rows called *rays* by special *ray initials* of the vascular cambium, function in lateral conduction and food storage. An initial is another term for a meristematic cell. It divides to produce another initial and a cell that differentiates. In cross sections of woody stems and roots, the rays can be seen radiating out from the center of the xylem like the spokes of a wheel. Fibers are abundant in some kinds of wood, such as oak (*Quercus*), and the wood is correspondingly dense and heavy. The arrangements of these and other kinds of cells in the xylem make it possible to identify most plant genera and many species from their wood alone. These fibers are a major component in modern paper. Earlier paper was made from fibers in phloem.

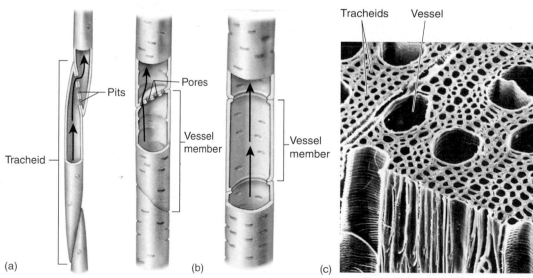

FIGURE 35.13
Comparison between tracheids and vessel members. (*a*) In tracheids, the water passes from cell to cell by means of pits. (*b*) In vessel members, water moves by way of perforation plates or between bars of wall material. In gymnosperm wood, tracheids both conduct water and provide support; in most kinds of angiosperms, vessels are present in addition to tracheids, or present exclusively. These two types of cells conduct water, and fibers provide additional support. (*c*) Scanning electron micrograph of the wood of red maple, *Acer rubrum* (350×).

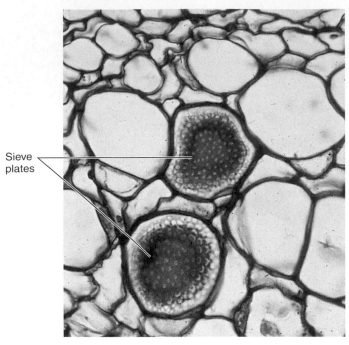

(a)

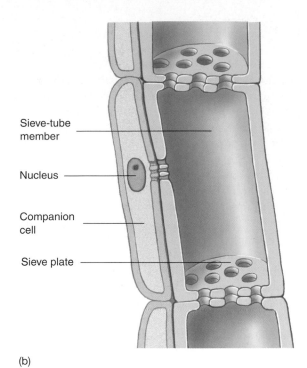

Sieve-tube
member

Nucleus

Companion
cell

Sieve plate

(b)

FIGURE 35.14
A sieve-tube member. (*a*) Looking down into sieve plates in squash phloem reveals the perforations through which sucrose and hormones move. (*b*) Sieve-tube member cells are stacked, with sieve plates forming the connection. The narrow cell with the nucleus at the left of the sieve-tube member is a companion cell. This cell nourishes the sieve-tube members, which have plasma membranes, but no nuclei.

Phloem

Phloem, which is located toward the outer part of roots and stems, is the principal food-conducting tissue in vascular plants. If a plant is *girdled* (by removing a substantial strip of bark down to the vascular cambium), the plant eventually dies from starvation of the roots.

Food conduction in phloem is carried out through two kinds of elongated cells: *sieve cells* and *sieve-tube members*. Seedless vascular plants and gymnosperms have only sieve cells; most angiosperms have sieve-tube members. Both types of cells have clusters of pores known as *sieve areas*. Sieve areas are more abundant on the overlapping ends of the cells and connect the protoplasts of adjoining sieve cells and sieve-tube members. Both of these types of cells are living, but most sieve cells and all sieve-tube members lack a nucleus at maturity.

In sieve-tube members, some sieve areas have larger pores and are called *sieve plates* (figure 35.14). Sieve-tube members occur end-to-end, forming longitudinal series called *sieve tubes*. Sieve cells are less specialized than sieve-tube members, and the pores in all of their sieve areas are

roughly of the same diameter. In an evolutionary sense, sieve-tube members are more advanced, more specialized, and presumably, more efficient than sieve cells.

Each sieve-tube member is associated with an adjacent specialized parenchyma cell known as a *companion cell*. Companion cells apparently carry out some of the metabolic functions needed to maintain the associated sieve-tube member. In angiosperms, a common initial cell divides asymmetrically to produce a sieve-tube member cell and its companion cell. Companion cells have all the components of normal parenchyma cells, including nuclei, and numerous **plasmodesmata** (cytoplasmic connections between adjacent cells) connect their cytoplasm with that of the associated sieve-tube members. Sieve cells in nonflowering plants have albuminous cells that function as companion cells. Fibers and parenchyma cells are often abundant in phloem.

Xylem conducts water and dissolved minerals from the roots to the shoots and the leaves. Phloem carries organic materials from one part of the plant to another.

35.3 Root cells differentiate as they become distanced from the dividing root apical meristem.

Root Structure

The three tissue systems are found in the three kinds of vegetative organs in plants: roots, stems, and leaves. Roots have a simpler pattern of organization and development than stems, and we will consider them first. Four regions are commonly recognized in developing roots: the **root cap,** the **zone of cell division,** the **zone of elongation,** and the **zone of maturation** (figure 35.15). In the last three zones mentioned, the boundaries are not clearly defined. When apical initials divide, daughter cells that end up on the tip end of the root become root cap cells. Cells that divide in the opposite direction pass through the three other zones before they finish differentiating. As you consider the different zones, visualize the tip of the root moving away from the soil surface by growth. This will counter the static image of a root that diagrams and photos convey.

The Root Cap

The root cap has no equivalent in stems. It is composed of two types of cells: the inner *columella cells* (they look like columns), and the outer, lateral *root cap cells,* which are continuously replenished by the root apical meristem. In some plants with larger roots, the root cap is quite obvious. Its main function is to protect the delicate tissues behind it as growth extends the root through mostly abrasive soil particles. Golgi bodies in the outer root cap cells secrete and release a slimy substance that passes through the cell walls to the outside. The root cap cells, which have an average life of less than a week, are constantly being replaced from the inside, forming a mucilaginous lubricant that eases the root through the soil. The slimy mass also provides a medium for the growth of beneficial nitrogen-fixing bacteria in the roots of plants such as legumes.

A new root cap is produced when an existing one is artificially or accidentally removed from a root. The root cap also functions in the *perception of gravity.* The columella cells are highly specialized, with the endoplasmic reticulum in the periphery and the nucleus located at either the middle or the top of the cell. There are no large vacuoles. Columella cells contain amyloplasts (plastids with starch grains) that collect on the sides of cells facing the pull of gravity. When a potted plant is placed on its side, the amyloplasts drift or tumble down to the side nearest the source of gravity, and the root bends in that direction. Lasers have been used to ablate (kill) individual columella cells in *Arabidopsis.* It turns out that only two columella cells are sufficient for gravity sensing! The precise nature of the gravitational response is not known, but some evidence indicates that calcium ions in the amyloplasts influence the distribution of growth hormones (auxin in this case) in the cells. Multiple signaling mechanisms may exist, because bending has been observed in the absence of auxin. A current hypothesis is that an electrical signal moves from the columella cell to cells in the elongation zone (the region closest to the zone of cell division).

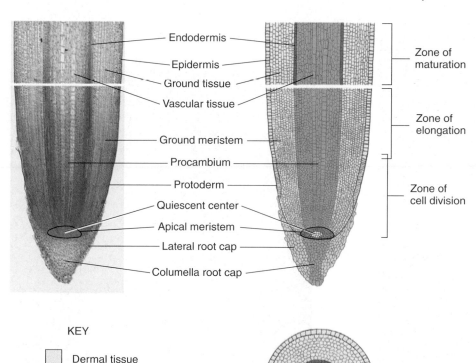

Endodermis
Epidermis
Ground tissue
Vascular tissue

Ground meristem
Procambium
Protoderm
Quiescent center
Apical meristem
Lateral root cap
Columella root cap

Zone of maturation
Zone of elongation
Zone of cell division

KEY

Dermal tissue
Ground tissue
Vascular tissue

Root in cross-section

FIGURE 35.15
Root structure. A root tip in corn, *Zea mays.*

The Zone of Cell Division

The apical meristem is shaped like an inverted, concave dome of cells and is located in the center of the root tip in the area protected by the root cap. Most of the activity in this *zone of cell division* takes place toward the edges of the dome, where the cells divide every 12 to 36 hours, often rhythmically, reaching a peak of division once or twice a day. Most of the cells are essentially cuboidal, with small vacuoles and proportionately large, centrally located nuclei. These rapidly dividing cells are daughter cells of the apical meristem. The *quiescent center* is a group of cells in the center of the root apical meristem that divide very infrequently. This makes sense if you think about a solid ball expanding—the outer surface would have to increase far more rapidly than the very center.

The apical meristem daughter cells soon subdivide into the three primary tissues previously discussed: *protoderm*, *procambium*, and *ground meristem*. Genes have been identified in the relatively simple root of *Arabidopsis* that regulate the patterning of these tissue systems. The patterning of these cells begins in this zone, but not until the cells reach the zone of maturation is the anatomical and morphological expression of this patterning fully revealed. For example, the *WERE-WOLF* gene is required for the patterning of the two root epidermal cell types, those with and without root hairs (figure 35.16*a*). The *SCARECROW* gene is necessary in ground cell differentiation (figure 35.16*b*), causing an asymmetric cell division that gives rise to two cylinders of cells from one. The outer cell layer becomes ground tissue and serves a storage function. The inner cell layer forms the endodermis, which regulates the intercellular flow of water and solutes into the vascular core of the root. Cells in this region develop according to their position. If that position changes because of a mistake in cell division or the ablation of another cell, the cell develops according to its new position.

The Zone of Elongation

In the *zone of elongation*, roots lengthen because the cells produced by the primary meristems become several times longer than wide, and their width also increases slightly. The small vacuoles present merge and grow until they occupy 90% or more of the volume of each cell. No further increase in cell size occurs above the zone of elongation, and the mature parts of the root, except for increasing in girth, remain stationary for the life of the plant.

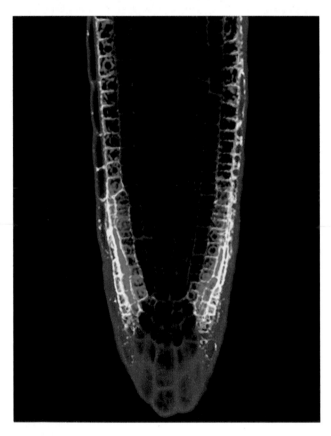

(a)

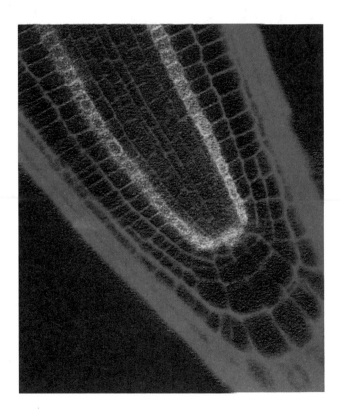

(b)

FIGURE 35.16
Tissue-specific gene expression.   (*a*) Epidermis–specific gene expression. The promoter of the *WEREWOLF* gene of *Arabidopsis* was attached to a green fluorescent protein and used to make a transgenic plant. The green fluorescence shows the epidermal cells where the gene is expressed. The red was used to visually indicate cell boundaries. (*b*) Ground tissue–specific gene expression. The *SCARECROW* gene is needed for an asymmetric cell division, allowing for the formation of side-by-side ground tissue and endodermal cells. These form two layers in wild type plants, but mutants such as the one shown here only have one cell layer (*green*) because the asymmetric cell division does not occur.

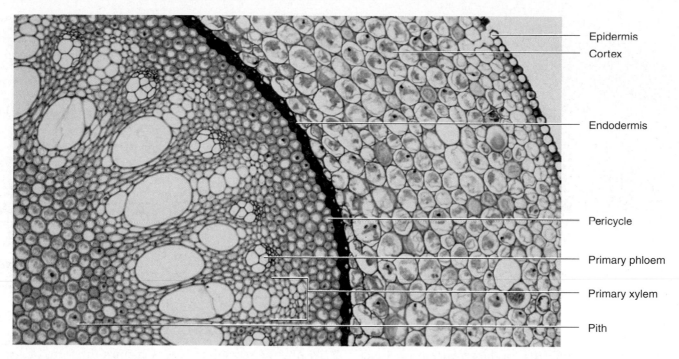

FIGURE 35.17
Cross section of the zone of maturation of a young monocot root. Greenbrier (*Smilax*) is a monocot (100×).

Labels on figure: Epidermis, Cortex, Endodermis, Pericycle, Primary phloem, Primary xylem, Pith

The Zone of Maturation

The cells that have elongated in the zone of elongation become differentiated into specific cell types in the *zone of maturation* (figure 35.17). The cells of the root surface cylinder mature into *epidermal cells*, which have a very thin cuticle. Many of the epidermal cells each develop a **root hair;** the protuberance is not separated by a cross-wall from the main part of the cell, and the nucleus may move into it. Root hairs, which can number over 35,000 per square centimeter of root surface and many billions per plant, greatly increase the surface area and therefore the absorptive capacity of the root. The root hairs usually are alive and functional for only a few days before they are sloughed off at the older part of the zone of maturation, while new ones are being produced toward the zone of elongation. Symbiotic bacteria that fix atmospheric nitrogen into a form usable by legumes enter the plant via root hairs and "instruct" the plant to create a nodule around it.

Parenchyma cells are produced by the ground meristem immediately to the interior of the epidermis. This tissue, called the **cortex,** may be many cell layers wide and functions in food storage. The inner boundary of the cortex differentiates into a single-layered cylinder of **endodermis** (figure 35.17), whose primary walls are impregnated with *suberin*, a fatty substance that is impervious to water. The suberin is produced in bands, called **Casparian strips**, that surround each adjacent endodermal cell wall perpendicular to the root's surface (figure 35.18). This blocks transport

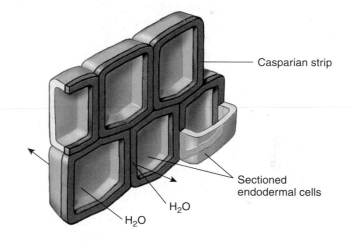

FIGURE 35.18
Casparian strip. The Casparian strip is a water-proofing band that forces water and minerals to pass through the plasma membranes, rather than through the air spaces in the cell walls.

Labels on figure: Casparian strip, Sectioned endodermal cells, H₂O

between cells. The two surfaces that are parallel to the root surface are the only way into the core of the root, and the plasma membranes control what passes through.

All the tissues interior to the endodermis are collectively referred to as the **stele.** Immediately adjacent and interior to the endodermis is a cylinder of parenchyma cells known as the **pericycle.** Pericycle cells can divide, even after they mature. They can give rise to *lateral* (branch) *roots* or, in eudicots, to part of the *vascular cambium.*

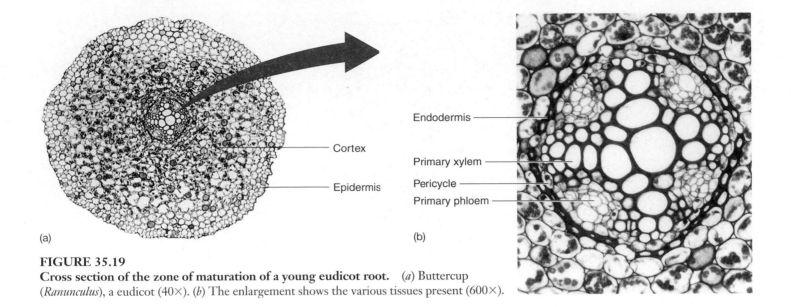

(a)

(b)

Endodermis

Primary xylem

Pericycle

Primary phloem

Cortex

Epidermis

FIGURE 35.19

Cross section of the zone of maturation of a young eudicot root. (*a*) Buttercup (*Ranunculus*), a eudicot (40×). (*b*) The enlargement shows the various tissues present (600×).

The water-conducting cells of the *primary xylem* are differentiated as a solid core in the center of young eudicot roots. In a cross section of a eudicot root, the central core of primary xylem often is somewhat star-shaped, having from two to several radiating arms that point toward the pericycle (figure 35.19). In monocot (and a few eudicot) roots, the primary xylem is in discrete *vascular bundles* arranged in a ring, which surrounds parenchyma cells, called **pith,** at the very center of the root. **Primary phloem,** composed of cells involved in food conduction, is differentiated in discrete groups of cells adjacent to the xylem in both eudicot and monocot roots.

In eudicots and other plants with **secondary growth,** part of the pericycle and the parenchyma cells between the phloem patches and the xylem become the root vascular cambium, which starts producing secondary xylem to the inside and secondary phloem to the outside. Even-

tually, the secondary tissues acquire the form of concentric cylinders. The primary phloem, cortex, and epidermis become crushed and are sloughed off as more secondary tissues are added. In the pericycle of woody plants, the cork cambium contributes to the bark, which will be discussed in section 35.4 (see figure 35.26). In the case of secondary growth in eudicot roots, everything outside the stele is lost and replaced with bark. Figure 35.20 summarizes the process of differentiation that occurs in plant tissue.

Root apical meristems produce a root cap at the tip and root tissue on the opposite side. Cells mature as the root cap and meristem grow away from them. Transport systems, external barriers, and a branching root system develop from the primary root as it matures.

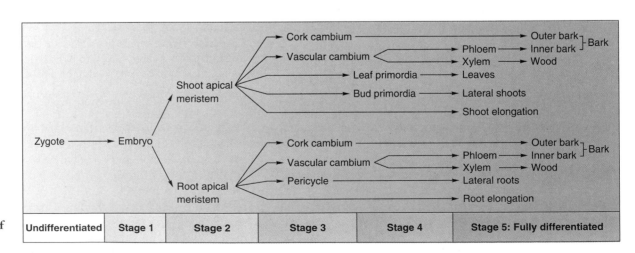

**FIGURE 35.20
Stages in the differentiation of plant tissues.**

Modified Roots

Most plants produce either a *taproot system*, characterized by a single large root with smaller branch roots, or a *fibrous root system*, composed of many smaller roots of similar diameter. Some plants, however, have intriguing root modifications with specific functions in addition to those of anchorage and absorption.

Not all roots are produced by roots. Any root that arises along a stem or in some place other than the root of the plant is called an *adventitious root*. For example, climbing plants such as ivy produce roots from their stems; these can anchor the stems to tree trunks or to a brick wall. Adventitious root formation in ivy depends on the developmental stage of the shoot. When the shoot enters the adult phase of development, it is no longer capable of initiating these roots. Below we investigate functions of modified roots

Prop roots. Some monocots, such as corn, produce thick adventitious roots from the lower parts of the stem. These so-called prop roots grow down to the ground and brace the plants against wind.

(a)

(b)

(c)

FIGURE 35.21
Three types of modified roots. (*a*) Pneumatophores (*foreground*) are spongy outgrowths from the roots below. (*b*) A water storage root weighing over 25 kilograms (60 pounds). (*c*) Buttress roots of a tropical fig tree.

Aerial roots. Some plants, such as epiphytic orchids (orchids that are attached to tree branches and grow unconnected to the ground without being parasitic in any way) have roots that extend into the air. Some aerial roots have an epidermis that is several cell layers thick, an adaptation to reduce water loss. These aerial roots may also be green and photosynthetic, as in the vanilla orchid.

Pneumatophores. Some plants that grow in swamps and other wet places may produce spongy outgrowths called *pneumatophores* from their underwater roots (figure 35.21*a*). The pneumatophores commonly extend several centimeters above water, facilitating oxygen uptake in the roots beneath.

Contractile roots. The roots from the bulbs of lilies and from several other plants, such as dandelions, contract by spiraling to pull the plant a little deeper into the soil each year until they reach an area of relatively stable temperature. The roots may contract to one-third their original length as they spiral like a corkscrew due to cellular thickening and constricting.

Parasitic roots. The stems of certain plants that lack chlorophyll, such as dodder (*Cuscuta*), produce peglike roots called *haustoria* that penetrate the host plants around which they are twined. The haustoria establish contact with the conducting tissues of the host and effectively parasitize their host.

Food storage roots. The xylem of branch roots of sweet potatoes and similar plants produce at intervals many extra parenchyma cells that store large quantities of carbohydrates. Carrots, beets, parsnips, radishes, and turnips have combinations of stem and root that also function in food storage. Cross sections of these roots reveal multiple rings of secondary growth.

Water storage roots. Some members of the pumpkin family (Cucurbitaceae), especially those that grow in arid regions, may produce water storage roots weighing 50 or more kilograms (figure 35.21*b*).

Buttress roots. Certain species of fig and other tropical trees produce huge buttress roots toward the base of the trunk, which provide considerable stability (figure 35.21*c*).

Some plants have modified roots that carry out photosynthesis, gather oxygen, parasitize other plants, store food or water, or support the stem.

35.4 Stems are the backbone of the shoot, transporting nutrients and supporting the aerial plant organs.

Stem Structure

External Form

The shoot apical meristem initiates stem tissue and intermittently produces bulges (**primordia**) that will develop into leaves, other shoots, or even flowers (figure 35.22). The stem is an axis from which organs grow. Leaves may be arranged in a spiral around the stem, or they may be in pairs opposite one another; they also may occur in *whorls* (circles) of three or more. The spiral arrangement is the most common, and for reasons still not understood, sequential leaves tend to be placed 137.5° apart. This angle relates to the golden mean, a mathematical ratio found in nature (the angle of coiling in some shells), in classical architecture (the Parthenon wall dimensions), and even in modern art (Mondrian). This pattern of leaf arrangement, called **phyllotaxy,** may optimize the exposure of leaves to the sun.

The region or area of leaf attachment (no structure is involved) is called a **node;** the area of stem between two nodes is called an **internode.** A leaf usually has a flattened *blade* and sometimes a *petiole* (stalk). When the petiole is missing, the leaf is then said to be *sessile*. Note that the word sessile as applied to plants has a different meaning than when applied to animals; in plants, it means "attached." The angle between a petiole (or blade) and the stem is called an axil. An **axillary bud** is produced in each axil. This bud is a product of the primary shoot apical meristem, which, with its associated leaf primordia, is called a **terminal bud.** Axillary buds frequently develop into branches or may form meristems that will develop into flowers. (Refer back to figure 35.6 to review these terms.)

Neither monocots nor herbaceous eudicot stems produce a cork cambium. The stems are usually green and photosynthetic, with at least the outer cells of the cortex containing chloroplasts. Herbaceous stems commonly have stomata, and may have various types of trichomes (hairs).

Woody stems can persist over a number of years and develop distinctive markings in addition to the original organs that form. Terminal buds usually extend the length of the shoot during the growing season. Some buds, such as those of geraniums, are unprotected, but most buds of woody plants have protective winter *bud scales* that drop off, leaving tiny *bud scale scars* as the buds expand. Some twigs have tiny scars of a different origin. A pair of butterfly-like appendages called *stipules* (part of the leaf) develop at the base of some leaves. The stipules can fall off and leave *stipule scars*. When the leaves of deciduous trees drop in the fall, they leave *leaf scars* with tiny *bundle scars*, marking where vascular connections were. The shapes, sizes, and other features of leaf scars can be distinctive enough to identify the plants in winter (figure 35.23).

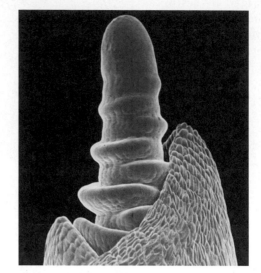

FIGURE 35.22
A shoot apex. Scanning electron micrograph of the apical meristem of wheat (*Triticum*) (200×).

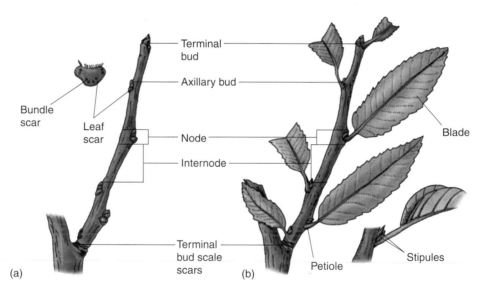

FIGURE 35.23
A woody twig. (*a*) In winter. (*b*) In summer.

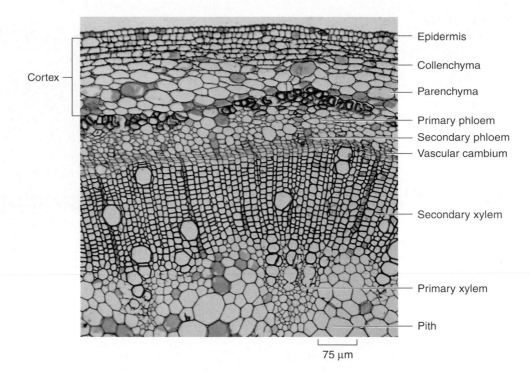

Cortex

Epidermis

Collenchyma

Parenchyma

Primary phloem
Secondary phloem
Vascular cambium

Secondary xylem

Primary xylem

Pith

75 μm

FIGURE 35.24
Early stage in vascular cambium differentiation in the castor bean, *Ricirus.* The outer part of the cortex consists of collenchyma, and the inner part of parenchyma.

Internal Form

As in roots, an *apical meristem* at the tip of each stem produces *primary meristems* that contribute to the stem's increase in length. Three primary meristems develop from the apical meristem. The *protoderm* gives rise to the *epidermis.* The *ground meristem* produces parenchyma cells. Parenchyma cells in the center of the stem constitute the *pith;* parenchyma cells away from the center constitute the *cortex.* The *procambium* produces cylinders of *primary xylem* and *primary phloem*, which are surrounded by ground tissue.

A strand of xylem and phloem called a *trace* branches off from the main cylinder of xylem and phloem and enters the developing leaf, flower, or shoot. Trace formation leaves a space in the main cylinder of conducting tissues called a **gap.** In eudicots, a **vascular cambium** develops between the primary xylem and primary phloem (figure 35.24). In many ways, this is a connect-the-dots game whereby the vascular cambium connects the ring of primary vascular bundles. In monocots, these bundles are scattered throughout the ground tissue (figure 35.25), and there is no logical way to connect them that would allow a uniform increase in girth. Lacking a vascular cambium, monocots do not have secondary growth.

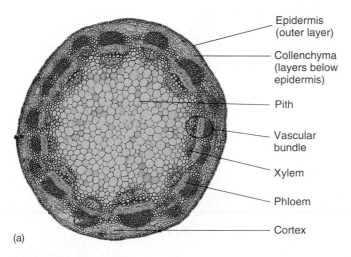

Epidermis (outer layer)

Collenchyma (layers below epidermis)

Pith

Vascular bundle

Xylem

Phloem

Cortex

(a)

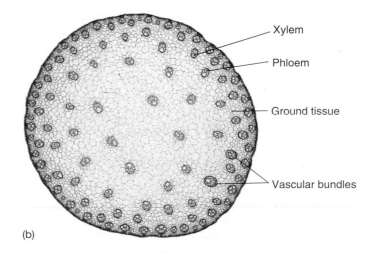

Xylem

Phloem

Ground tissue

Vascular bundles

(b)

FIGURE 35.25
Stems. Transverse sections of a young stem in (*a*) a eudicot, the common sunflower (*Helianthus annuus*), in which the vascular bundles are arranged around the outside of the stem (10×); and (*b*) a monocot, corn (*Zea mays*), with the scattered vascular bundles characteristic of the class (5×).

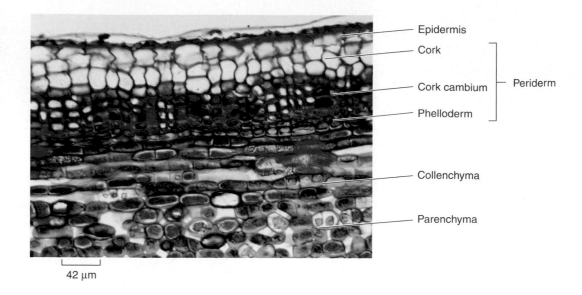

FIGURE 35.26
Section of periderm. An early stage in the development of periderm in cottonwood, *Populus* sp.

Epidermis

Cork

Cork cambium } Periderm

Phelloderm

Collenchyma

Parenchyma

42 µm

The cells of the vascular cambium divide indefinitely, producing **secondary tissues** (mainly secondary xylem and secondary phloem). The production of xylem is extensive in trees, and the resulting built-up tissue is called wood. Rings in the stump of a tree reveal annual patterns of growth; cell size varies, depending on growth conditions. In woody eudicots and gymnosperms, a second cambium, the *cork cambium*, arises in the outer cortex (occasionally in the epidermis or phloem); it produces boxlike *cork cells* to the outside and also may produce parenchyma-like *phelloderm* cells to the inside. The cork cambium, cork, and phelloderm are collectively referred to as the *periderm* (figure 35.26). Cork tissues, whose cells become impregnated with suberin shortly after they are formed and then die, constitute the *outer bark*. The cork tissue, whose suberin is impervious to moisture, cuts off water and food to the epidermis, which dies and sloughs off. In young stems, gas exchange between stem tissues and the air takes place through stomata, but as the cork cambium produces cork, it also produces patches of unsuberized cells beneath the stomata. These unsuberized cells, which permit gas exchange to continue, are called *lenticels* (figure 35.27).

> The stem results from the dynamic growth of the shoot apical meristem, which initiates stem tissue and organs, including leaves. Shoot apical meristems initiate new apical meristems at the junction of leaf and stem. These meristems can form buds that reiterate the growth pattern of the terminal bud, or they can make flowers directly.

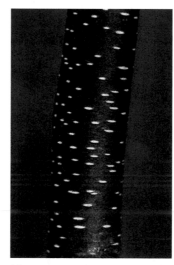

(a)

FIGURE 35.27
Lenticels. (*a*) Lenticels, the numerous small, pale, raised areas shown here on cherry tree bark (*Prunus cerasifera*), allow gas exchange between the external atmosphere and the living tissues immediately beneath the bark of woody plants. Highly variable in form in different species, lenticels are an aid in identifying deciduous trees and shrubs in winter. (*b*) Transverse section through a lenticel (extruding area) in a stem of elderberry, *Sambucus canadensis* (30×).

(b)

Modified Stems

Although most stems grow erect, some have modifications that serve special purposes, including natural *vegetative propagation*. In fact, the widespread artificial vegetative propagation of plants, both commercial and private, frequently involves cutting modified stems into segments, which are then planted, producing new plants. As you become acquainted with the following modified stems, keep in mind that stems have *leaves* at *nodes*, with *internodes* between the nodes, and *buds* in the *axils* of the leaves, while roots have no leaves, nodes, or axillary buds.

Bulbs. Onions, lilies, and tulips have swollen underground stems that are really large buds with adventitious roots at the base (figure 35.28a). Most of a *bulb* consists of fleshy leaves attached to a small, knoblike stem. In onions, the fleshy leaves are surrounded by papery, scalelike leaf bases of the long, green, aboveground leaves.

Corms. Crocuses, gladioluses, and other popular garden plants produce *corms* that superficially resemble bulbs. Cutting a corm in half, however, reveals no fleshy leaves. Instead, almost all of a corm consists of stem, with a few papery, brown nonfunctional leaves on the outside, and adventitious roots below.

Rhizomes. Perennial grasses, ferns, irises, and many other plants produce *rhizomes*, which typically are horizontal stems that grow underground, often close to the surface (figure 35.28b). Each node has an inconspicuous scalelike leaf with an axillary bud; much larger photosynthetic leaves may be produced at the rhizome tip. Adventitious roots are produced throughout the length of the rhizome, mainly on the lower surface.

Runners and stolons. Strawberry plants produce horizontal stems with long internodes, which, unlike rhizomes, usually grow along the surface of the ground. Several *runners* may radiate out from a single plant (figure 35.28c). Some botanists use the term *stolon* synonymously with runner; others reserve the term stolon for a stem with long internodes that grows underground, as seen in Irish (white) potato plants. An Irish potato itself, however, is another type of modified stem—a tuber.

Tubers. In Irish potato plants, carbohydrates may accumulate at the tips of stolons, which swell, becoming *tubers*; the stolons die after the tubers mature (figure 35.28d). The "eyes" of a potato are axillary buds formed in the axils of scalelike leaves. The scalelike leaves, which are present when the potato is starting to form, soon drop off; the tiny ridge adjacent to each "eye" of a mature potato is a leaf scar.

Tendrils. Many climbing plants, such as grapes and Boston ivy, produce modified stems known as *tendrils*, which twine around supports and aid in climbing (figure 35.28e). Some tendrils, such as those of peas and pumpkins, are actually modified leaves or leaflets.

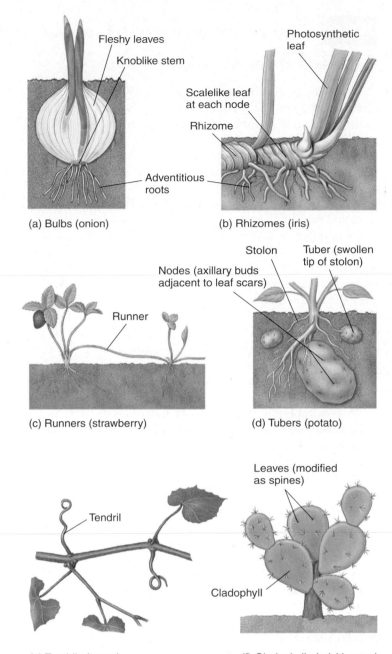

(a) Bulbs (onion)

(b) Rhizomes (iris)

Fleshy leaves
Knoblike stem
Photosynthetic leaf
Scalelike leaf at each node
Rhizome
Adventitious roots

(c) Runners (strawberry)

Runner

(d) Tubers (potato)

Stolon
Tuber (swollen tip of stolon)
Nodes (axillary buds adjacent to leaf scars)

(e) Tendrils (grape)

Tendril

(f) Cladophylls (prickly pear)

Leaves (modified as spines)
Cladophyll

FIGURE 35.28
Types of modified stems.

Cladophylls. Cacti and several other plants produce flattened, photosynthetic stems called *cladophylls* that resemble leaves (figure 35.28f). In cacti, the real leaves are modified as spines.

Some plants possess modified stems that serve special purposes, including food storage, support, or vegetative propagation.

Leaf External Structure

Leaves, which are initiated as *primordia* by the apical meristems (see figure 35.2), are vital to life as we know it. They are the principal sites of photosynthesis on land. Leaves expand primarily by cell enlargement and some cell division. Like arms and legs in humans, they are determinate structures, which means their growth stops at maturity. Because leaves are crucial to a plant, features such as their arrangement, form, size, and internal structure are highly significant and can differ greatly. Different patterns have adaptive value in different environments.

Leaves are really an extension of the shoot apical meristem and stem development. When they first emerge as primordia, they are not committed to being leaves. Experiments in which very young leaf primordia in fern and in coleus are isolated and grown in culture demonstrate this. If the primordia are young enough, they will form an entire shoot rather than a leaf. So, positioning the primordia and beginning the initial cell divisions occur before those cells are committed to the leaf developmental pathway.

Leaves fall into two different morphological groups, which may reflect differences in evolutionary origin. A *microphyll* is a leaf with one vein that does not leave a gap when it branches from the vascular cylinder of the stem; microphylls are mostly small and are associated primarily with the phylum Lycophyta (see chapter 29). Most plants have leaves called *megaphylls*, which have several to many veins; a megaphyll's conducting tissue leaves a gap in the stem's vascular cylinder as it branches from it.

Most eudicot leaves have a flattened *blade* and a slender stalk, the *petiole*. The flattening of the leaf blade reflects a shift from radial symmetry to dorsal-ventral (top-bottom) symmetry. We're just beginning to understand how this shift occurs by analyzing mutants such as *phantastica*, which prevents this transition (figure 35.29). In addition, a pair of *stipules* may be present at the base of the petiole. The stipules, which may be leaflike or modified as *spines* (as in the black locust, *Robinia pseudo-acacia*) or *glands* (as in the cherry trees, *Prunus cerasifera*), vary considerably in size from microscopic to almost half the size of the leaf blade. Development of stipules appears to be independent of development of the rest of the leaf.

Grasses and other monocot leaves usually lack a petiole and tend to sheathe the stem toward the base. *Veins* (a term used for the vascular bundles in leaves), consisting of both xylem and phloem, are distributed throughout the leaf blades. The main veins are parallel in most monocot leaves; the veins of eudicots, on the other hand, form an often intricate network (figure 35.30).

FIGURE 35.29
The *phantastica* mutant in snapdragon. Snapdragon leaves are usually flattened, with a top and bottom side (*left*). In the *phantastica* mutant (*right*), the leaf never completely flattens nor differentiates distinct top and bottom sides.

(a)   (b)

FIGURE 35.30
Eudicot and monocot leaves. (*a*) The leaves of eudicots, such as this African violet relative from Sri Lanka, have netted, or reticulate, veins. (*b*) Those of monocots, such as this cabbage palmetto, have parallel veins. The eudicot leaf has been cleared with chemicals and stained with a red dye to make the veins show more clearly.

(b)

(c)

(a)

FIGURE 35.31

Simple versus compound leaves. (*a*) A simple leaf, its margin deeply lobed, from the tulip tree (*Liriodendron tulipifera*). (*b*) A pinnately compound leaf, from a mountain ash (*Sorbus* sp.). A compound leaf is associated with a single lateral bud, located where the petiole is attached to the stem. (*c*) Palmately compound leaves of a Virginia creeper (*Parthenocissus quinquefolia*).

Leaf blades come in a variety of forms, from oval to deeply lobed to having separate leaflets. In **simple leaves** (figure 35.31*a*), such as those of lilacs or birch trees, the blades are undivided, but simple leaves may have teeth, indentations, or lobes of various sizes, as in the leaves of maples and oaks. In **compound leaves,** such as those of ashes, box elders, and walnuts, the blade is divided into *leaflets.* The relationship between the development of compound and simple leaves is an open question. Two explanations are being debated: (1) A compound leaf is a highly lobed simple leaf, or (2) a compound leaf utilizes a shoot development program. To address this question, researchers are using single mutations that are known to convert compound leaves to simple leaves.

If the leaflets are arranged in pairs along a common axis (the axis is called a *rachis*—the equivalent of the main central vein, or *midrib*, in simple leaves), the leaf is *pinnately compound* (figure 35.31*b*). If, however, the leaflets radiate out from a common point at the blade end of the petiole, the leaf is *palmately compound* (figure 35.31*c*). Palmately compound leaves occur in buckeyes (*Aesculus* spp.) and Virginia creeper (*Parthenocissus quinquefolia*). The leaf blades themselves may have similar arrangements of their veins, and are said to be pinnately or palmately veined.

Leaves, regardless of whether they are simple or compound, may be *alternately* arranged (alternate leaves usually spiral around a shoot, one leaf per node), or they may be in *opposite* pairs (two leaves per node). Less often, three or more leaves may be in a *whorl*, a circle of leaves at the same level at a node (figure 35.32).

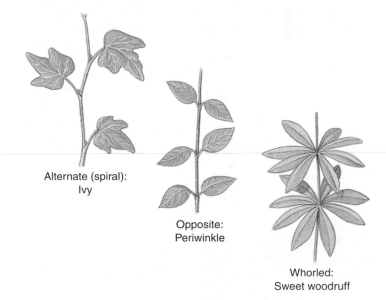

Alternate (spiral):
Ivy

Opposite:
Periwinkle

Whorled:
Sweet woodruff

FIGURE 35.32

Types of leaf arrangements. The three common types of leaf arrangements are alternate, opposite, and whorled.

Leaves are the principal sites of photosynthesis. Their blades may be arranged in a variety of ways. In simple leaves, the blades are undivided, while in compound leaves the leaf is composed of two or more leaflets.

Leaf Internal Structure

The entire surface of a leaf is covered by a transparent epidermis; most of these epidermal cells have no chloroplasts. The epidermis itself has a waxy *cuticle* of variable thickness, and may have different types of glands and trichomes (hairs) present. The lower epidermis (and occasionally the upper epidermis) of most leaves contains numerous slitlike or mouth-shaped *stomata* (figure 35.33). Stomata, as discussed in section 35.2, are flanked by *guard cells* and function in gas exchange and regulation of water movement through the plant.

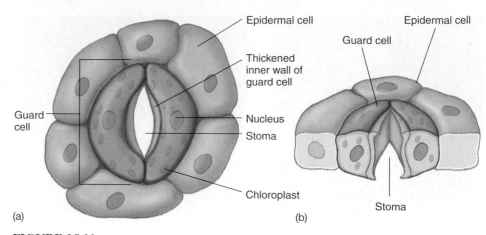

FIGURE 35.33
A stoma. (*a*) Surface view. (*b*) View in cross section.

The tissue between the upper and lower epidermis is called **mesophyll.** Mesophyll is interspersed with veins (vascular bundles) of various sizes. Most eudicot leaves have two distinct types of mesophyll. Closest to the upper epidermis are one to several (usually two) rows of tightly packed, barrel-shaped to cylindrical *chlorenchyma* cells (parenchyma with chloroplasts) that constitute the **palisade mesophyll** (figure 35.34). Some plants, including species of *Eucalyptus*, have leaves that hang down, rather than extending horizontally. They have palisade parenchyma on both sides of the leaf, and there is, in effect, no upper side. Nearly all leaves have loosely arranged **spongy mesophyll** cells between the palisade mesophyll and the lower epidermis, with many air spaces throughout the tissue. The interconnected intercellular spaces, along with the stomata, function in gas exchange and the passage of water vapor from the leaves. The mesophyll of monocot leaves is not differentiated into palisade and spongy layers, and there is often little distinction between the upper and lower epidermis. This anatomical difference often correlates with a modified photosynthetic pathway that maximizes the amount of CO_2 relative to O_2 to reduce energy loss through photorespiration (see chapter 10). The anatomy of a leaf directly relates to its juggling act of balancing water loss, gas exchange, and transport of photosynthetic products to the rest of the plant.

Stomata in leaf epidermis allow gas and vapor exchange between the atmosphere and chloroplast-rich mesophyll cells. Vascular tissue permits transport between the leaf and the rest of the plant.

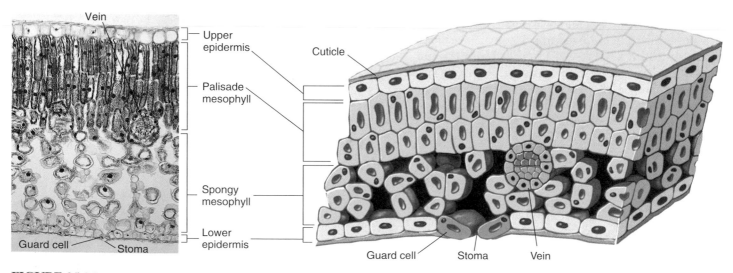

FIGURE 35.34
A leaf in cross section. Transection of a leaf showing the arrangement of palisade and spongy mesophyll, a vascular bundle or vein, and the epidermis with paired guard cells flanking the stoma.

Modified Leaves

As plants colonized a wide variety of environments, from deserts to lakes to tropical rain forests, plant organ modifications arose that would adapt the plants to their specific habitats. Leaves, in particular, have evolved some remarkable adaptations. A brief discussion of a few of these modifications follows:

Floral leaves (bracts). Poinsettias and dogwoods have relatively inconspicuous, small, greenish-yellow flowers. However, both plants produce large modified leaves called **bracts** (mostly colored red in poinsettias and white or pink in dogwoods). These bracts surround the true flowers and perform the same function as showy petals (figure 35.35). It should be noted, however, that bracts can also be quite small and inconspicuous in some plants.

Spines. The leaves of many cacti, barberries, and other plants are modified as *spines* (see figure 35.28*f*). In cacti, having less leaf surface reduces water loss and also may deter predators. Spines should not be confused with *thorns*, such as those on the honey locust (*Gleditsia triacanthos*), which are modified stems, or with the *prickles* on raspberries and rosebushes, which are simply outgrowths from the epidermis or the cortex just beneath it.

Reproductive leaves. Several plants, notably *Kalanchoë*, produce tiny but complete plantlets along their margins. Each plantlet, when separated from the leaf, is capable of growing independently into a full-sized plant. The walking fern (*Asplenium rhizophyllum*) produces new plantlets at the tips of its fronds. Although many species can regenerate a whole plant from isolated leaf tissue, this in vivo regeneration is unique among just a few species.

Window leaves. Several genera of plants growing in arid regions produce succulent, cone-shaped leaves with transparent tips. The leaves often become mostly buried in sand blown by the wind, but the transparent tips, which have a thick epidermis and cuticle, admit light to the hollow interiors. This allows photosynthesis to take place beneath the surface of the ground.

Shade leaves. Leaves produced where they receive significant amounts of shade tend to be larger in surface area, but thinner and with less mesophyll than leaves on the same tree receiving more direct light. This plasticity in development is remarkable, because both types of leaves on the plant have exactly the same genes. Environmental signals can have a major effect on development.

Insectivorous leaves. Almost 200 species of flowering plants are known to have leaves that trap insects; some plants digest the insects' soft parts. Plants with insectivorous leaves often grow in acid swamps deficient in needed elements or containing elements in forms not readily available to the plants; this inhibits the plants' ca-

FIGURE 35.35
Modified leaves. In this dogwood "flower," the white-colored bracts (modified leaves) surround the several true flowers in the center, which have no petals.

pacities to maintain metabolic processes sufficient to meet their growth requirements. Their needs are, however, met by the supplementary absorption of nutrients from the animal kingdom.

Pitcher plants (for example, *Sarracenia*, *Darlingtonia*, or *Nepenthes*) have cone-shaped leaves in which rainwater can accumulate. The insides of the leaves are very smooth, but stiff, downward-pointing hairs line the rim. An insect falling into such a leaf finds it very difficult to escape and eventually drowns. The leaf absorbs the nutrients released when bacteria, and in most species the plant's own digestive enzymes, decompose the insect bodies. Other plants, such as sundews (*Drosera*), have glands that secrete sticky mucilage that traps insects, which are then digested by enzymes. The Venus flytrap (*Dionaea muscipula*) produces leaves that look hinged at the midrib. When tiny trigger hairs on the leaf blade are stimulated by a moving insect, the two halves of the leaf snap shut, and digestive enzymes break down the soft parts of the trapped insect into nutrients that can be absorbed through the leaf surface. Nitrogen is the most common nutrient needed. Curiously, the Venus flytrap cannot survive in a nitrogen-rich environment, perhaps as a result of a trade-off during the intricate evolutionary process that resulted in its ability to capture and digest insects.

The leaves of plants exhibit a variety of adaptations, including spines, vegetative reproduction, and even carnivorous leaves.

35.1 Meristems elaborate the plant body plan after germination.

Meristems

- Meristematic tissues are masses of small cells containing dense cytoplasm and proportionately large nuclei. (p. 730)
- Apical meristems are located at the tips of roots and shoots, and produce primary tissue by dividing and continually adding more cells to the tips of the seedling's body. (p. 730)
- Root apical meristem is protected by a root cap. (p. 730)
- Apical meristem gives rise to three primary meristems: protoderm, procambium, and ground meristem. (pp. 730–731)
- Most woody plants possess active lateral meristems that produce secondary growth and increase a plant's girth with secondary tissues. (p. 731)

Organization of the Plant Body

- A vascular plant consists of a root system for anchoring the plant and a shoot system composed of stems and leaves that serve as the main photosynthetic sites and hold the reproductive organs, such as flowers, fruits, and seeds. (p. 732)
- Three basic tissue types are ground tissue, dermal tissue, and vascular tissue. (p. 732)

Primary and Secondary Growth

- Primary growth plays an important role in establishing the organism's basic body plan. (p. 733)
- Evolution of secondary growth allowed the development of thick trunks and wide, treelike organisms. (p. 733)

35.2 Plants have three basic tissues, each composed of several cell types.

Dermal Tissue

- Epidermal cells arise from the protoderm and cover all primary plant parts. (p. 734)
- Guard cells, trichomes, and root hairs are all specialized epidermal cells that function in absorption or protection. (pp. 734–735)

Ground Tissue

- Parenchyma cells are the most abundant type of plant cells, are relatively unspecialized, and perform a wide variety of functions. (p. 736)
- Collenchyma cells are relatively flexible and provide support for plant organs. (p. 736)
- Sclerenchyma cells have tough, thick walls and are divided into long, slender fibers and variable-sized, branched sclereids. (p. 736)

Vascular Tissue

- Xylem is the main water-conducting tissue. Composed of vessels and tracheids, xylem carries water to the shoots and leaves, and helps drive transpiration. (p. 737)
- Primary xylem is derived from procambium, and secondary xylem is formed by vascular cambium. (p. 737)
- Phloem is the principal food-conducting tissue in vascular plants. (p. 738)
- Food conduction is carried out through sieve cells and sieve-tube members. (p. 738)

35.3 Root cells differentiate as they become distanced from the dividing root apical meristem.

Root Structure

- The four major regions in the root are the root cap, which protects delicate tissues as the roots elongate, and help perceive gravity; the zone of cell division, where the cells multiply; the zone of elongation, where roots lengthen as the cells produced by the primary meristem become several times longer than wide; and the zone of maturation, where cells differentiate into specific cell types. (pp. 739–741)

Modified Roots

- Most plants produce either a taproot system or a fibrous root system, although other root types, such as aerial roots, pneumatophores, contractile roots, parasitic roots, food storage roots, water storage roots, and buttress roots, can also exist. (p. 743)

35.4 Stems are the backbone of the shoot, transporting nutrients and supporting the aerial plant organs.

Stem Structure

- Leaves may be arranged alternately, oppositely, or whorled on the stem. (p. 744)
- Stems are divided into nodes and internodes, while leaves are divided into a blade and often a petiole and stipules. (p. 744)
- Axillary meristems, which are produced in the angle between a petiole and the stem, frequently develop into branches or into flowers. (p. 744)
- In primary meristems, the protoderm gives rise to the epidermis, the ground meristem produces parenchyma cells, and procambium produces primary xylem and primary phloem. (p. 745)

Modified Stems

- Bulbs, corms, rhizomes, runners and stolons, tubers, tendrils, and cladophylls are all modified stems that perform special functions, such as food storage, plant support, or vegetative propagation. (p. 747)

35.5 Leaves are adapted to support basic plant functions.

Leaf External Structure

- Leaves are initiated as primordia by the apical meristems and are the principal sites of photosynthesis on land. (p. 748)
- Leaves are either simple, with undivided blades, or compound, with the blade divided into leaflets. (p. 749)

Leaf Internal Structure

- Leaves are divided into a waxy cuticle, an upper epidermis, palisade mesophyll, spongy mesophyll, and lower epidermis. (p. 750)

Modified Leaves

- Leaves have evolved a variety of adaptations, including floral leaves, spines, reproductive leaves, window leaves, shade leaves, and insectivorous leaves. (p. 751)

Self Test

1. Fifteen years ago, your parents hung a swing from the lower branch of a large tree growing in your yard. When you sit in it today, you realize it is exactly the same height off the ground as it was when you first sat in it 15 years ago. The reason the swing has not grown taller as the tree has grown is that
   a. the tree trunk is showing secondary growth.
   b. the tree trunk is part of the primary growth system of the plant, but elongation is no longer occurring in that part of the tree.
   c. trees lack apical meristems and so do not get taller.
   d. you are hallucinating, because it is impossible for the swing not to have gotten taller as the tree grew.

2. Cloning animals is a relatively new phenomenon, but cloning plants has been done for a long time. Which of the following plant cell types would be the *least* successful to clone a plant from?
   a. a mature xylem vessel element
   b. a mature stomatal guard cell
   c. a quiescent center cell
   d. All of these would work for cloning a plant.
   e. None of these would work for cloning a plant.

3. If you were to relocate the pericycle of a plant root to the epidermal layer, how would it affect root growth?
   a. Secondary growth in the mature region of the root would not occur.
   b. The root apical meristem would produce vascular tissue in place of dermal tissue.
   c. Nothing would change, because the pericycle is normally located near the epidermal layer of the root.
   d. Lateral roots would grow from the outer region of the root and fail to connect with the vascular tissue.

4. In a variation on the old "guess your weight" game, you are playing "guess how big this structure will get" at the yearly carnival. There are a number of bizarre plant structures to choose from, but having read this textbook, you are confident of certain victory. Which of the following plant structures would you choose so that you could accurately predict the final size?
   a. an oak shoot
   b. a lotus flower
   c. a bamboo root
   d. the root of a tomato plant

5. When you peel your Irish potatoes for dinner, you are removing the majority of their:
   a. dermal tissue.
   b. vascular tissue.
   c. ground tissue.
   d. Only a and b are removed with the peel.
   e. All of these are removed with the peel.

6. You can determine the age of an oak tree by counting the annual rings of _____ formed by the _____.
   a. primary xylem/apical meristem
   b. secondary phloem/vascular cambium
   c. dermal tissue/cork cambium
   d. secondary xylem/vascular cambium

7. Which of the following does not arise from meristematic activity in a plant?
   a. secondary xylem
   b. border cells
   c. tendrils
   d. corms
   e. All of these arise from the activity of plant meristems.

8. Mosses are thought to resemble the primitive plants that first inhabited the land. Interestingly, these plants lack a vascular system. Therefore they should lack
   a. mesophyll cells.
   b. shoots.
   c. phloem.
   d. collenchyma.

9. Plant organs form by
   a. cell division in gamete tissue.
   b. cell division in meristematic tissue.
   c. cell migration into the appropriate position in the tissue.
   d. rearranging the genetic material in the precursor cells so that the organ-specific genes are activated.

Test Your Visual Understanding

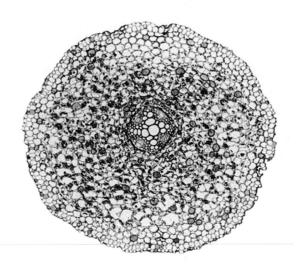

1. From this cross section of a plant organ, predict the identity of this tissue.

Apply Your Knowledge

1. Plant organs undergo many modifications to deal with environmental challenges. Define your favorite modified root, shoot, and leaf, and make a case for why it is the best example of a modified plant organ.

2. You design a grand scheme to construct a "super plant" that can increase photosynthetic productivity well above normal levels. To do this, you pack the leaves with palisade parenchyma (mesophyll) cells. Would this increase the photosynthetic productivity of a leaf? Why or why not? (*Hint:* Think about gas exchange in a leaf, and remember the role of spongy mesophyll.)

3. You have identified a mutant maize plant that cannot differentiate vessel cells. How would this affect the functioning of the plant? Devise an environment that would maximize the growth of this mutant.

36

Vegetative Plant Development

Concept Outline

36.1 Plant embryo development establishes a basic body plan.

Establishing the Root-Shoot Axis. Asymmetric cell division starts patterning the embryo.

Establishing Three Tissue Systems. Three tissue systems (dermal, ground, and vascular tissues) are established without any cell movement. While the embryo is still a round ball, the root-shoot axis is established. The shape of the plant is determined by planes of cell division and direction of cell elongation. Nutrients are used during embryogenesis, but proteins, lipids, and carbohydrates are also set aside to support the plant during germination before it becomes photosynthetic.

36.2 The seed protects the dormant embryo from water loss.

How Seeds Form. Seeds allow plants to survive unfavorable conditions and invade new habitats.

36.3 Fruit formation enhances the dispersal of seeds.

How Fruits Form. Seed-containing fruits are carried far by animals, wind, and water, allowing angiosperms to colonize large areas.

36.4 Germination initiates post-seed development.

Mechanisms of Germination. External signals, including water, light, abrasion, and temperature, can trigger germination. Rupturing the seed coat and acquiring adequate oxygen are essential. Organic reserves stored in the endosperm or cotyledon are made available to the embryo during germination.

FIGURE 36.1
Emerging from the soil is just the beginning. Plant development continues through the entire life cycle.

How does a fertilized egg develop into a complex adult plant body? Because plant cells cannot move, the timing and directionality of each cell division must be carefully orchestrated. Cells need information about their location relative to other cells in order for cell specialization to be coordinated. The developing embryo is quite fragile, and numerous protective structures have evolved since plants first colonized land.

Only a portion of the plant has actually formed when it first emerges from the soil (figure 36.1). New plant organs develop throughout the plant's life.

36.1 Plant embryo development establishes a basic body plan.

Establishing the Root-Shoot Axis

In plants, three-dimensional shape and form arise by regulating the amount and pattern of cell division. The very first cell division results in two different cell types, and early in embryonic development, most cells can give rise to a wide range of cell and organ types, including leaves. As development proceeds, the cells with multiple potentials are mainly restricted to regions called meristems. Many meristems have been established by the time embryogenesis ends and the seed becomes dormant. Apical meristems continue adding cells to the growing root and shoot tips after germination. Apical meristem cells of corn, for example, divide every 12 hours, producing half a million cells per day in an actively growing corn plant. Lateral meristems can cause an increase in the girth of some plants, while intercalary meristems in the stems of grasses allow for elongation.

Apical meristems establish the root-shoot axis in embryogenesis. Also, three basic tissue systems are established: dermal, ground, and vascular tissue. These tissues contain various cell types that can be highly differentiated for specific functions. These tissue systems are organized radially around the root-shoot axis.

While the embryo is developing, two other critical events are occurring. The first event is the establishment of a food supply that will support the embryo during germination while it gains photosynthetic capacity. In angiosperms, double fertilization produces endosperm for nutrition; in gymnosperms, the megagametophyte is the food source. The second critical event is the differentiation of ovule tissue (from the parental sporophyte) to form a hard, protective covering around the embryo. The seed then enters a dormant phase, signaling the end of embryogenesis. Environmental signals, such as water, temperature, and light, can break dormancy and trigger a cascade of internal events resulting in germination.

Early Cell Division and Patterning

The first division of the zygote (fertilized egg) in a flowering plant is asymmetric and generates cells with two different fates (figure 36.2). One daughter cell is small, with dense cytoplasm. That cell, which will become the embryo, begins to divide repeatedly in different planes, forming a ball of cells. The other, larger daughter cell divides repeatedly, forming an elongated structure called a *suspensor*, which links the embryo to the nutrient tissue of the seed. The suspensor also

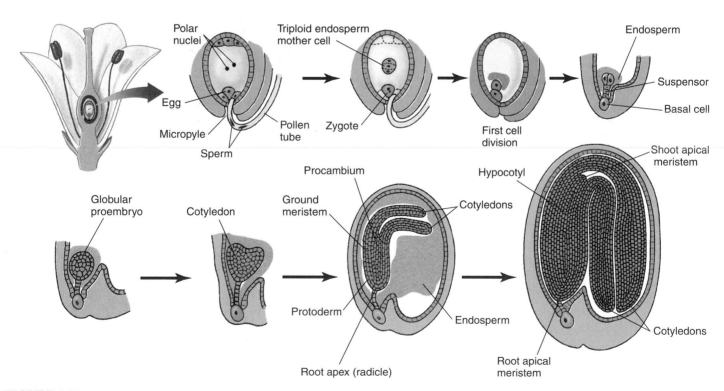

FIGURE 36.2
Stages of development in an angiosperm embryo. The very first cell division is asymmetric. Differentiation begins almost immediately after fertilization.

FIGURE 36.3

Asymmetric cell division in a *Fucus* zygote.
An unequal distribution of material in the zygote leads to a bulge where the first cell division will occur. This division results in a smaller cell that will go on to divide and produce the rhizoid that anchors the alga; the larger cell divides to form the thallus, or main algal body. The point of sperm entry determines where the smaller rhizoid cell will form, but light and gravity can modify this to ensure that the rhizoid will point downward where it can anchor this brown alga. Calcium-mediated currents set up an internal gradient of charged molecules, which leads to a weakening of the cell wall where the rhizoid will form. The fate of the two resulting cells is held "in memory" by cell wall components.

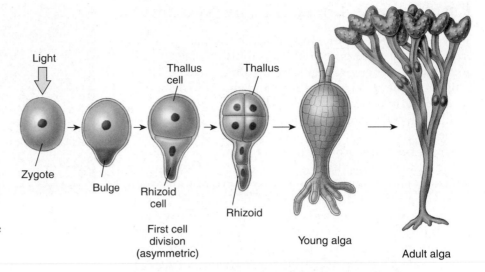

provides a route for nutrients to reach the developing embryo. The root-shoot axis also forms at this time. Cells near the suspensor are destined to form a root, while those at the other end of the axis ultimately become a shoot.

Investigating mechanisms for establishing asymmetry in plant embryo development is difficult because the zygote is embedded within the gametophyte, which is surrounded by sporophyte tissue (ovule and carpel tissue). One approach has been to use the brown alga *Fucus* as a model system. Although there is a huge evolutionary difference between brown algae and the angiosperms, early embryogenesis reveals similarities that may have ancient origins. In *Fucus*, the egg is released prior to fertilization, so no extra tissues surround the zygote. A bulge that develops on one side of the zygote establishes the vertical axis. Cell division occurs, and the original bulge becomes the smaller of the two cells. The smaller cell develops into a rhizoid that anchors the alga, and the larger cell develops into the main body, or thallus, of the sporophyte. This axis is first established by the point of sperm entry, but it can be changed by environmental signals, especially light and gravity, which ensure that the rhizoid is down and the thallus is up. Internal gradients are established that specify where the rhizoid will form in response to environmental signals (figure 36.3). The ability to "remember" where the rhizoid will form depends on the presence of the cell wall. In experiments, enzymatic removal of the cell wall in *Fucus* cells that were already specified to form either rhizoids or plant body resulted in cells that could give rise to both. Cell walls contain a wide variety of carbohydrates and proteins attached to the wall's structural fibers. Attempting to pin down the identities of these suspected developmental signals is an area of active research.

Genetic approaches make it possible to explore asymmetric development in angiosperms. Studies of mutants reveal what is going wrong in development, which often makes it possible to infer normal developmental mechanisms. For example, the *suspensor* mutant in *Arabidopsis* un-

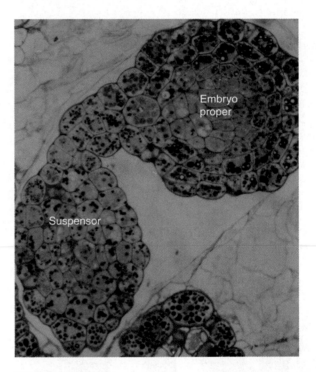

FIGURE 36.4

The embryo suppresses development of the suspensor as a second embryo. This *suspensor* mutant of *Arabidopsis* has a defect in embryo development. Aborted embryo development is followed by embryo-like development of the suspensor.

dergoes aberrant development in the embryo followed by embryo-like development of the suspensor (figure 36.4). Analysis of this mutant led to the conclusion that the embryo normally prevents the suspensor from developing into a second embryo.

Early in embryogenesis, the root-shoot axis is established.

Establishing Three Tissue Systems

Three basic tissues differentiate while the plant embryo is still a ball of cells, called the globular stage (figure 36.5), but no cell movements are involved. The *protoderm* consists of the outermost cells in a plant embryo and will become **dermal tissue.** These cells almost always divide with their cell plate perpendicular to the surface, thus perpetuating a single outer layer of cells. Dermal tissue produces cells that protect the plant from desiccation, including the stomata that open and close to facilitate gas exchange and minimize water loss. A ground meristem gives rise to the bulk of the embryonic interior, consisting of **ground tissue** cells that eventually function in food and water storage. Finally, *procambium* at the core of the embryo is destined to form the future **vascular tissue,** which is responsible for water and nutrient transport.

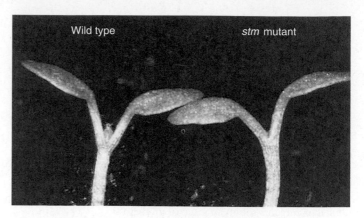

FIGURE 36.6
***SHOOTMERISTEMLESS* is needed for shoot formation.** Shoot-specific genes specify formation of the shoot apical meristem. The *stm* mutant of *Arabidopsis* has a normal root meristem but fails to produce a shoot meristem between its two cotyledons.

Root and Shoot Formation

The root-shoot axis is established during the globular stage of development. The shoot apical meristem later gives rise to leaves and eventually to reproductive structures. While both the shoot and root meristems are apical meristems, their formation is controlled independently. This conclusion is supported by mutant analysis in *Arabidopsis*, in which the *shootmeristemless* (*stm*) mutant fails to produce a viable shoot, but does produce a root (figure 36.6). The *STM* gene has a homeodomain region and shares a common evolutionary origin with the *Hox* genes that are important in establishing animal body plans (see chapters 19 and 25). Unlike animals, however, *Hox*-like genes have a more lim-

ited role in regulating plant body plans. Other gene families, encoding different transcription factors, have been identified as playing key roles in patterning in plants.

For example, *MONOPTEROUS* (*MP*) is necessary for root formation, but not shoot formation, in *Arabidopsis*. This root-specific gene is not a *Hox* relative. Rather, *MP* codes for a transcription factor activated by the hormone auxin. The MP protein is called an auxin response factor (ARF) because it binds auxin. Once activated, MP binds to the promoter of another gene, leading to transcription.

The hormone auxin is important in root-shoot axis formation. Auxin is one of seven classes of hormones that reg-

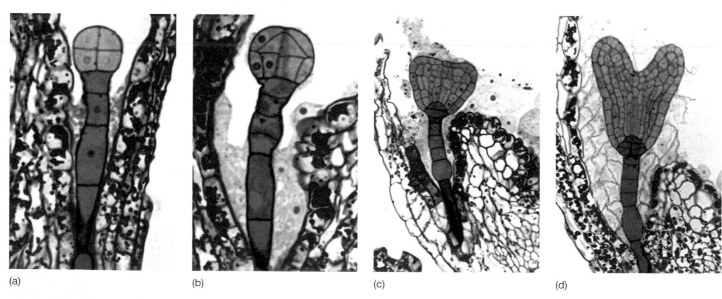

(a)   (b)   (c)   (d)

FIGURE 36.5
Early developmental stages of *Arabidopsis thaliana*. (*a*) Early cell division has produced the embryo and suspensor. (*b*) Globular stage. (*c,d*) Heart-shaped stage.

ulate plant development and function; we will explore each of these classes of hormones in chapter 40.

As you study the development of roots and shoots after germination, you will notice that many of the same patterns of tissue differentiation seen in the embryo are reiterated in the apical meristems. However, cell fates are generally more limited after embryogenesis, when embryo genes are not expressed. For example, the *LEAFY COTYLEDON* gene in *Arabidopsis* is active in early and late embryo development and may be responsible for maintaining an embryonic environment. It is possible to turn this gene on later in development using recombinant DNA techniques (see chapter 16). In that case, embryos can form on leaves!

Morphogenesis

The globular stage gives rise to a heart-shaped embryo with two bulges in one group of angiosperms (the dicots, such as *A. thaliana* in figure 36.5c,d) and a ball with a bulge on a single side in another group (the monocots). The bulges are cotyledons ("first leaves") and are produced by the embryonic cells, not the shoot apical meristem that begins forming during the globular stage. This process, called *morphogenesis* (generation of form), results from changes in planes and rates of cell division. Plant cells cannot move. The form of a plant body is largely determined by the plane in which its cells divide. It is also controlled by changes in cell shape as cells expand osmotically after they form. Both microtubules and actin play a role in establishing the position of the cell plate, which determines the direction of division. Plant hormones and other factors influence the orientation of bundles of microtubules on the interior of the plasma membrane. These microtubules also guide cellulose deposition as the cell wall forms around the outside of a new cell. This process determines the cell's final shape. For example, think of the cell as a cube where four of the six sides are reinforced more heavily with cellulose; the cell will expand and grow in the direction of the two sides having less reinforcement. Much is being learned about morphogenesis at the cell biological level from mutants that divide, but cannot control their plane of cell division or the direction of cell expansion.

Food Storage

Throughout embryogenesis, starch, lipids, and proteins are produced. The seed storage proteins are so abundant that the genes coding for them were the first cloning targets for plant molecular biologists. Providing nutritional resources is part of the evolutionary trend toward enhancing embryo survival (see chapter 38). The sporophyte transfers nutrients via the suspensor in angiosperms. (In gymnosperms, the suspensor serves only to push the embryo closer to the gametophytic nutrient source.) This happens concurrently with the development of the endosperm (present only in angiosperms, although double fertilization has been observed in the gymnosperm *Ephedra*) which may be extensive or minimal. Endosperm in coconut is the "milk," a liquid. In corn, the endosperm is solid, and in popcorn, it expands with heat to form the edible part of popped corn. In peas and beans, the endosperm is used up during embryo development, and nutrients are stored in thick, fleshy cotyledons (figure 36.7). The photosynthetic machinery is built in response to light. So, it is critical that seeds have stored nutrients to aid in germination until the growing sporophyte can photosynthesize. A seed buried too deeply will use up all its reserves in cellular respiration before reaching the surface and sunlight.

After the root-shoot axis is established, a radial, three-tissue system and a stored food supply are formed through controlled cell division and expansion.

Corn

Bean

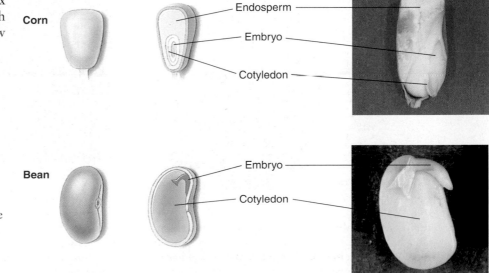

FIGURE 36.7
Endosperm in maize and bean. The maize kernel has endosperm that is still present at maturity, while the endosperm in the bean has disappeared. The bean embryo's cotyledons take over food storage functions.

36.2 The seed protects the dormant embryo from water loss.

How Seeds Form

Early in the development of an angiosperm embryo, a profoundly significant event occurs: The embryo stops developing. In many plants, development of the embryo is arrested soon after the meristems and cotyledons differentiate. The integuments—the outer cell layers of the ovule—develop into a relatively impermeable *seed coat*, which encloses the seed with its dormant embryo and stored food (figure 36.8). The *seed* is a vehicle for dispersing the embryo to distant sites and allows a plant embryo to survive in environments that might kill a mature plant.

Adaptive Importance of Seeds

Seeds are important adaptively in at least four ways:

1. Seeds maintain dormancy under unfavorable conditions and postpone development until better conditions arise. If conditions are marginal, a plant can "afford" to have some seeds germinate, because others will remain dormant.
2. Seeds afford maximum protection to the young plant at its most vulnerable stage of development.
3. Seeds contain stored food that permits a young plant to develop before photosynthetic activity begins.
4. Perhaps most important, seeds are adapted for dispersal, facilitating the migration of plant genotypes into new habitats.

Once a seed coat forms, most of the embryo's metabolic activities cease. A mature seed contains only about 5 to 20% water. Under these conditions, the seed and the young plant within it are very stable; its arrested growth is primarily due to the progressive and severe desiccation of the embryo and the associated reduction in metabolic activity. Germination cannot take place until water and oxygen reach the embryo; meanwhile, the seed coat may crack by abrasion or alternate freezing and thawing. Seeds of some plants have been known to remain viable for hundreds and, in rare instances, thousands of years.

Specific adaptations often help ensure that seeds will germinate only under appropriate conditions. Sometimes, seeds lie within tough cones that do not open until they are exposed to the heat of a fire (figure 36.9). This causes the seed to germinate in an open, fire-cleared habitat where nutrients are relatively abundant, having been released from plants burned in the fire. Seeds of other plants germinate only when inhibitory chemicals leach from their seed coats, thus guaranteeing their germination when sufficient water is available. Still other seeds germinate only after they pass through the intestines of birds or mammals or are regurgitated by them, which both weakens the seed coats and ensures dispersal. Sometimes seeds of plants thought to be extinct in a particular area may germinate under unique or improved environmental circumstances, and the plants may then become reestablished.

> Seed dormancy is an important evolutionary factor in plants, ensuring their survival in unfavorable conditions and allowing them to germinate when the chances of survival for the young plants are the greatest.

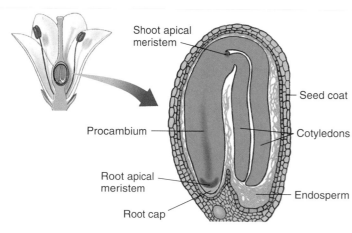

FIGURE 36.8
A mature angiosperm embryo. Note that the two cotyledons have grown in a bent shape to accommodate the tight confines of the seed. In some embryos, the shoot apical meristem will have already initiated a few leaf primordia as well.

Shoot apical meristem

Seed coat

Procambium

Cotyledons

Root apical meristem

Endosperm

Root cap

(a)   (b)

FIGURE 36.9
Fire induces seed germination in some pines. Fire can destroy adult jack pines, but stimulate growth of the next generation. (*a*) The cones of a jack pine are tightly sealed and cannot release the seeds protected by the scales. (*b*) High temperatures lead to the release of the seeds.

36.3 Fruit formation enhances the dispersal of seeds.

How Fruits Form

Survival of angiosperm embryos depends on *fruit* development as well as seed development. Fruits that contain seeds are defined simply as mature ovaries (carpels). During seed formation, the flower ovary begins to develop into fruit (see figure 29.22). Fruits form in many ways and exhibit a wide array of adaptations for dispersal. The differences among some of the fruit types are shown in figure 36.10. Three layers of ovary wall can have distinct fates, which account for the diversity of fruit types from fleshy to dry and hard. An array of mechanisms allow for the release of the seed(s) from the fruits. Developmentally, fruits are fascinating organs

Follicles
Split along one carpel edge only; milkweed, larkspur.

Legumes
Split along two carpel edges with seeds attached to carpel edges; peas, beans.

Samaras
Not split and with a wing formed from the outer tissues; maples, elms, ashes.

Drupes
Single seed enclosed in a hard pit; peaches, plums, cherries.

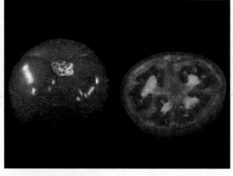

True berries
More than one seed and a thin skin; blueberries, tomatoes, grapes, peppers.

Hesperidia
More than one seed and a leathery skin; oranges, lemons, limes.

Aggregate fruits
Derived from many ovaries of a single flower; strawberries, blackberries.

FIGURE 36.10
Examples of some kinds of fruits. Distinguishing features of each of these fruit types are listed below each photo. Follicles, legumes, and samaras are examples of dry fruits. Drupes, true berries, and hesperidia are simple fleshy fruits; they develop from a flower with a single carpel. Aggregate and multiple fruits are compound fleshy fruits; they develop from flowers with more than one ovary or from more than one flower.

Multiple fruits
Develop from a cluster of flowers; mulberries, pineapples.

that contain three genotypes in one package. The fruit and seed coat are from the prior sporophyte generation. Remnants of the gametophyte generation that produced the egg are found in the seed, and the embryo represents the next sporophyte generation.

The Dispersal of Fruits

Aside from the many ways fruits can form, they also exhibit a wide array of specialized dispersal methods. Fruits with fleshy coverings, often shiny black or bright blue or red, normally are dispersed by birds or other vertebrates (figure 36.11*a*). Like red flowers, red fruits signal an abundant food supply. By feeding on these fruits, birds and other animals may carry seeds from place to place and thus transfer plants from one suitable habitat to another.

Fruits with hooked spines, such as those of burgrass (figure 36.11*b*), are typical of several genera of plants that occur in the northern deciduous forests. Such fruits are often disseminated by mammals, including humans. Squirrels and similar mammals disperse and bury fruits such as acorns and other nuts. Other fruits, such as those of maples, elms, and ashes, have wings that aid in their distribution by the wind. The dandelion provides another familiar example of a fruit type that is dispersed by wind (figure 36.12), and the dispersal of seeds from plants such as milkweeds, willows, and cottonwoods is similar. Orchids have minute, dustlike seeds, which are likewise blown away by the wind.

Coconuts and other plants that characteristically occur on or near beaches are regularly spread throughout a region by water (figure 36.13). This sort of dispersal is especially important in the colonization of distant island groups, such as the Hawaiian Islands. It has been calculated that the seeds of about 175 angiosperms, nearly one-third from North America, must have reached Hawaii to have evolved into the roughly 970 species found there today. Some of these seeds blew through the air, others were transported on the feathers or in the guts of birds, and still others drifted across the Pacific. Although the distances are rarely as great as the distance between Hawaii and the mainland, dispersal is just as important for mainland plant species that have discontinuous habitats, such as mountaintops, marshes, or north-facing cliffs.

Fruits, which are characteristic of angiosperms, are extremely diverse. The evolution of specialized structures allows fruits (and, hence, seeds) to be dispersed by animals, wind, and water.

(a)   (b)

FIGURE 36.11
Animal-dispersed fruits. (*a*) The bright red berries of this honeysuckle, *Lonicera hispidula*, are highly attractive to birds. After eating the fruits, birds may carry the seeds they contain for great distances either internally or, because of their sticky pulp, stuck to their feet or other body parts. (*b*) You will know if you have ever stepped on the fruits of *Cenchrus incertus;* their spines adhere readily to any passing animal.

FIGURE 36.12
Wind-dispersed fruits. False dandelion, *Pyrrhopappus carolinianus.* The "parachutes" widely disperse the fruits of both false and true dandelions in the wind, much to the gardener's despair.

FIGURE 36.13
A water-dispersed fruit. This fruit of the coconut palm, *Cocos nucifera*, is sprouting on a sandy beach. Coconuts, one of the most useful fruits for humans in the tropics, have become established on even the most distant islands by drifting there on the waves.

36.4 Germination initiates post-seed development.

When conditions are satisfactory, the embryo emerges from its previously desiccated state, utilizes food reserves, and resumes growth. Although **germination** is a process characterized by several stages, it is often defined as the emergence of the radicle (first root) through the seed coat. As the sporophyte pushes through the seed coat, it orients with the environment so that the root grows down and the shoot grows up. New growth comes from delicate meristems that are protected from environmental rigors. The shoot becomes photosynthetic, and the postembryonic phase of growth and development is under way. Figure 36.14 shows the process of germination and subsequent development of the plant body in dicots and monocots.

Mechanisms of Germination

Germination is the first step in the development of the plant outside its seed coat. Germination begins when a seed absorbs water and its metabolism resumes. The amount of water a seed can absorb is phenomenal and creates a force strong enough to break the seed coat. At this point, it is important that oxygen be available to the developing embryo because plants, like animals, require oxygen for cellular respiration. Few plants produce seeds that germinate successfully under water, although some, such as rice, have evolved a tolerance to anaerobic conditions.

Even though a dormant seed may have imbibed a full supply of water and may be respiring, synthesizing proteins and RNA, and apparently carrying on normal metabolism, it may fail to germinate without an additional signal from the environment. This signal may be light of the correct wavelengths and intensity, a series of cold days, or simply the passage of time at temperatures appropriate for germination. The seeds of many plants will not germinate unless they have been **stratified**—held for periods of time at low temperatures. This phenomenon prevents the seeds of plants that grow in cold areas from germinating until they have passed the winter, thus protecting their tender seedlings from harsh, cold conditions.

Germination can occur over a wide temperature range (5° to 30°C), although certain species and specific habitats may have relatively narrow optimum ranges. Some seeds will not germinate even under the best conditions. In some species, a significant fraction of a season's seeds remain dormant, providing a gene pool of great evolutionary significance to the future plant population.

The Utilization of Reserves

Germination occurs when all internal and external requirements are met. Germination and early seedling growth require the utilization of metabolic reserves stored in the starch grains of **amyloplasts** (colorless plastids that store starch) and protein bodies. Fats and oils also are important food reserves in some kinds of seeds. These food sources can readily be digested during germination, producing glycerol and fatty acids, which yield energy through cellular respiration; they can also be converted to glucose. Depending on the kind of plant, any of these reserves may be stored in the embryo or in the endosperm.

In the kernels of cereal grains, the single cotyledon is modified into a relatively massive structure called the **scutellum** (figure 36.14b), from the Latin word meaning "shield." The abundant food stored in the scutellum is used up before the endosperm during germination. Later, while the seedling is becoming established, the scutellum serves as a nutrient conduit from the endosperm to the rest of the embryo. This is one of the best examples of how hormones modulate development in plants (figure 36.15). The embryo produces gibberellic acid, a hormone, that signals the outer layer of the endosperm called the **aleurone** to produce α-amylase. This enzyme is responsible for breaking down the endosperm's starch into sugars that are passed by the scutellum to the embryo. Abscisic acid, another plant hormone, which is important in establishing dormancy, can inhibit starch breakdown. Abscisic acid levels may be reduced when a seed absorbs water.

The emergence of the embryonic root and shoot from the seed during germination varies widely from species to species. In most plants, the root emerges before the shoot appears and anchors the young seedling in the soil (see figure 36.14). In plants such as peas and corn, the cotyledons may be held below ground; in other plants, such as beans, radishes, and onions, the cotyledons are held above ground. The cotyledons may or may not become green and contribute to the nutrition of the seedling as it becomes established. The period from the germination of the seed to the establishment of the young plant is very critical for the plant's survival; the seedling is unusually susceptible to disease and drought during this period.

During germination and early seedling establishment, the utilization of food reserves stored in the embryo or the endosperm is mediated by gibberellic acid and abscisic acid.

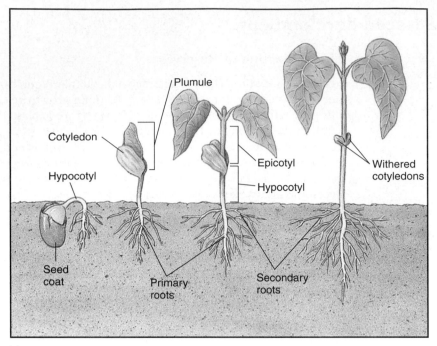

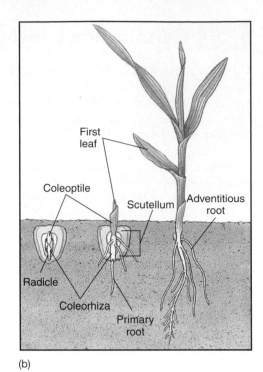

(a)

(b)

FIGURE 36.14
Shoot development. The stages shown are for (*a*) a eudicot, the common bean (*Phaseolus vulgaris*), and (*b*) a monocot, maize (*Zea mays*).

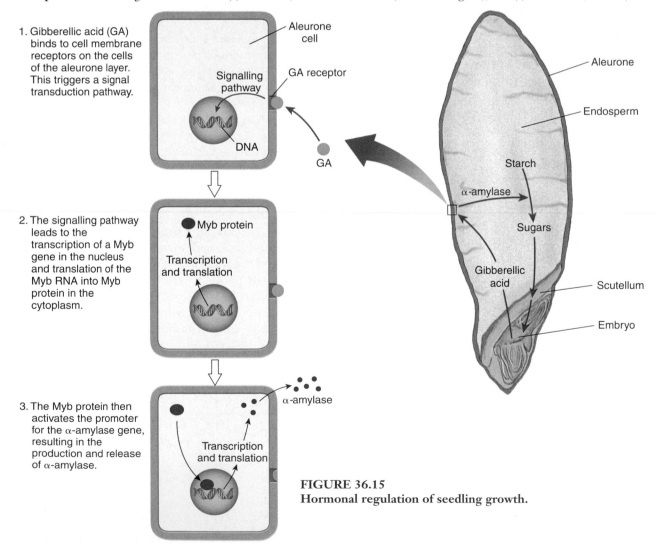

1. Gibberellic acid (GA) binds to cell membrane receptors on the cells of the aleurone layer. This triggers a signal transduction pathway.

2. The signalling pathway leads to the transcription of a Myb gene in the nucleus and translation of the Myb RNA into Myb protein in the cytoplasm.

3. The Myb protein then activates the promoter for the α-amylase gene, resulting in the production and release of α-amylase.

FIGURE 36.15
Hormonal regulation of seedling growth.

For interactive testing, visit the Online Learning Center with PowerWeb at www.mhhe.com/Raven7

36.1 Plant embryo development establishes a basic body plan.

Establishing the Root-Shoot Axis

- As plant development proceeds, cells with multiple potentials are mainly restricted to meristem regions. (p. 756)
- During embryogenesis, the root-shoot axis develops, cell differentiation occurs, and three basic tissue systems are established. Two other critical events also occur: establishment of an embryological food supply, and differentiation of ovule tissue to form a hard, protective embryo covering. (p. 756)

Establishing Three Tissue Systems

- Three basic tissues differentiate while the plant embryo is in the globular stage. (p. 758)
- The root and shoot both arise from apical meristems, but their formation is independently controlled. (p. 758)
- Morphogenesis results from changes in planes and rates of cell division. (p. 759)
- Seed storage molecules such as starch, lipids, and proteins were the first cloning targets because they are so abundant throughout embryogenesis. (p. 759)

36.2 The seed protects the dormant embryo from water loss.

How Seeds Form

- Early in development, the angiosperm embryo stops developing, and the integuments develop into a relatively impermeable seed coat. (p. 760)
- Seeds are adaptively important because they maintain dormancy under unfavorable conditions, protect young plants during vulnerable stages, store food for young plants, and facilitate dispersal. (p. 760)
- Other specific adaptations help ensure that seeds will only germinate under specific conditions. Such adaptations include tough cones and the presence of inhibitory chemicals in the seed coats. (p. 760)

36.3 Fruit formation enhances the dispersal of seeds.

How Fruits Form

- Fruits are mature ovaries (carpels). (p. 761)
- Fruits contain the genotypes of the prior sporophyte generation (fruit and seed coat), the prior gametophyte generation (parts of the seed), and the next sporophyte generation (embryo). (pp. 761–762)
- Fruits can exhibit a wide variety of specialized dispersal methods, including hooked spines, wings, and light feathery appendages. These forms allow dispersal by such methods as animals, wind, and water. (p. 762)

36.4 Germination initiates post-seed development.

Mechanisms of Germination

- Germination begins when a seed absorbs water and resumes metabolism. (p. 763)
- A dormant seed that has imbibed water and is carrying on normal metabolism may still not germinate without an additional environmental signal such as low temperature. (p. 763)
- Gibberellic acid and abscisic acid mediate the utilization of food reserves in the embryo or endosperm. (p. 763)

Self Test

1. If you could use a micro-laser to destroy the larger cell in a two-cell plant embryo, how would it likely affect embryonic development?
   a. The embryo would develop normally except it would not become anchored in the seed wall.
   b. The embryo would develop normally except it would have multiple cotyledons.
   c. The embryo would fail to develop, but a fully functional suspensor would form.
   d. The embryo would immediately be aborted, and the seed would not form.

2. Loss-of-function mutations in the *suspensor* gene in *Arabidopsis* lead to the development of two embryos in a seed. After analyzing the expression of this gene in early embryos, you find high levels of suspensor mRNA in the developing suspensor cells. What is the likely function of the suspensor protein?
   a. Suspensor protein likely stimulates development of the embryonic tissue.
   b. Suspensor protein likely stimulates development of the suspensor tissue.
   c. Suspensor protein likely inhibits embryonic development in the suspensor.
   d. Suspensor protein likely inhibits suspensor development in the embryo.

3. How would plant development change if the functions of *SHOOTMERISTEMLESS (STM)* and *MONOPTEROUS (MP)* were reversed?
   a. The embryo-suspensor axis would be reversed.
   b. The embryo-suspensor axis would be duplicated.
   c. The root-shoot axis would be reversed.
   d. The root-shoot axis would be duplicated.

4. The most obvious difference between plant embryonic development and animal embryonic development is that
   a. plants develop from unfertilized eggs, while animals develop from fertilized eggs.
   b. plant morphogenesis is entirely growth dependent, while animal morphogenesis involves movement of cells within the embryo.
   c. plant embryos have an available source of nutrients, while animal embryos must begin feeding to obtain nutrients.

5. Which of the following is *not* evident from looking at a plant embryo?
   a. You can tell if the plant is a monocot or dicot.
   b. You can tell where the shoot will form.
   c. You can tell where the root will form.
   d. You can tell when the seed will germinate.

6. Both seeds and fruits are well adapted to
   a. provide nutrition to animals.
   b. act as a dispersal mechanism for plants.
   c. allow plant embryos to remain dormant for long periods of time.
   d. all of the above

7. The longest period of time that a seed can remain dormant is
   a. days.
   b. weeks.
   c. months.
   d. years.

8. Fruits are complex organs that are specialized for dispersal of seeds. Which of the following plant tissues does *not* contribute to mature fruit?
   a. sporophytic tissue from the previous generation
   b. gametophytic tissue from the previous generation
   c. sporophytic tissue from the next generation
   d. gametophytic tissue from the next generation

9. If you wanted to ensure that a seed failed to germinate, which of the following strategies would be most effective?
   a. prevent imbibition
   b. prevent desiccation
   c. prevent fertilization
   d. prevent dispersal

10. How would a loss-of-function mutation in the α-amylase gene affect seed germination?
    a. The seed could not imbibe water.
    b. The embryo would starve.
    c. The seed coat would not rupture.
    d. The seed would germinate prematurely.

Test Your Visual Understanding

1. Label this diagram, and predict the future axes of the plant.

Apply Your Knowledge

1. You are writing a science fiction screenplay in which the best of animal and plant lifestyles are to be used in making a "super-species." Discuss the aspects of plant development you would include in this new species. (*Hint:* Think about how cool it would be to get dormant animal embryos that used all kinds of ways to move around.)

2. The oldest known seeds to successfully germinate were found in the Yukon Territory in the Canadian Arctic in the 1950s. Radiocarbon dating found these seeds to be approximately 10,000 years old. Discuss the mechanisms that seeds use to remain dormant for long periods of time. (*Hint:* Do a Web search for "oldest seeds," and you will find many accounts. Think of the most important components of dormancy, and discuss them.)

3. How might the reproductive success of angiosperms have changed if seeds had developed without fruit? (*Hint:* Think about the way fruits help disperse plant seeds.)

37

Transport in Plants

Concept Outline

37.1 Water and minerals move upward through the xylem.

Overview of Water and Mineral Movement Through Plants. The bulk movement of water and dissolved minerals is the result of movement between cells, across plasma membranes, and through tubes of xylem.

Water and Mineral Absorption. Water and minerals enter the plant through the roots.

Water and Mineral Movement. A combination of the properties of water, structure of xylem, and transpiration of water through the leaves results in the passive movement of water to incredible heights. Water leaves the plant through openings in the leaves called stomata. Too much water is harmful to a plant, although many plants have adaptations that make them tolerant of flooding.

37.2 Dissolved sugars and hormones are transported in the phloem.

Phloem Transport Is Bidirectional. Sucrose and hormones can move from shoot to root or from root to shoot in the phloem. Phloem transport requires energy to load and unload sieve tubes.

FIGURE 37.1
How does water get to the top of this tree? We would expect gravity to make such a tall column of water too heavy to be maintained by capillary action. What pulls the water up?

L and plants face two major challenges: maintaining water balance and providing sufficient structural support for upright growth (figure 37.1). The vascular system transports water, minerals, and carbohydrates over great distances. Whereas secondary growth of vascular tissue allows trees to achieve great heights, water balance keeps herbaceous plants upright. Think of a plant cell as a water balloon pressing against the insides of a box. If the balloon springs a leak, the support is gone, and the box can collapse.

Overview of Water and Mineral Movement Through Plants

Local Changes Result in the Long-Distance, Upward Movement of Water

Did you ever wonder how water gets from the roots to the top of a tree 10 stories high (see figure 37.1)? Water moves through the spaces between the protoplasts of cells, through plasmodesmata (connections between cells), through plasma membranes, and through the interconnected, water-conducting xylem elements extending throughout a plant. Water first enters the roots and then moves to the xylem. After that water rises through the xylem because of a combination of factors, and some of that water exits through the stomata in the leaves (figure 37.2).

The greatest distances traveled by water molecules and dissolved minerals are in the xylem. For example, once water enters the xylem of a redwood, it can move upward 100 meters. Some "pushing" from the pressure of water entering the roots is involved. However, most of the force is "pulling" caused by water evaporating (**transpiration**) through stomata. This works because water molecules stick to each other (cohesion) and to the walls of the tracheid or xylem vessel (adhesion). The result is an unusually stable column of liquid reaching great heights.

While this chapter focuses mainly on the mechanics of water transport through xylem, the movement of water at the cellular level plays a significant role in bulk water trans-port in the plant as well, although over much shorter distances. You know that the Casparian strip in the root forces water to move through cells (see figure 35.18). In parenchyma cells, most water moves across membranes rather than in the intercellular spaces. For a long time, it was believed that water moved across plasma membranes only by osmosis through the lipid bilayer. We now know that osmosis is enhanced by water channels called **aquaporins** (figure 37.3). These transport channels occur in both plants and animals; in plants, they exist in vacuole and plasma membranes. At least 30 different genes code for aquaporin-like proteins in *Arabidopsis*. Aquaporins speed up osmosis, but do not change the direction of water movement. They are important in maintaining water balance within a cell and in moving water into the xylem.

Water Potential

Plant biologists often discuss the forces that act on water within a plant in terms of *potentials*. Potentials are a way of representing free energy and predicting which way water will move. The energetics principles in chapter 8 apply to water potential. The key is to remember that water will move from a solution with higher water potential to a solution with lower water potential.

To predict the direction of water movement between two different solutions, you must first determine the water potential of each solution. There are two components to water potential: (1) physical forces, such as a plant cell wall

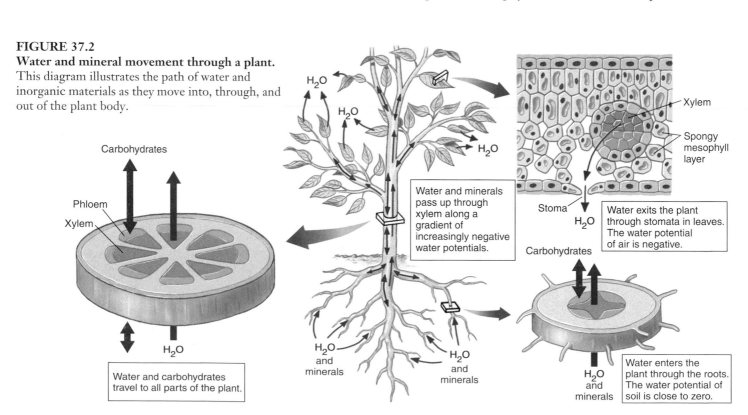

FIGURE 37.2
Water and mineral movement through a plant.
This diagram illustrates the path of water and inorganic materials as they move into, through, and out of the plant body.

Carbohydrates

Phloem

Xylem

Water and carbohydrates travel to all parts of the plant.

H_2O

H_2O

H_2O

H_2O

Water and minerals pass up through xylem along a gradient of increasingly negative water potentials.

H_2O and minerals

H_2O and minerals

Xylem

Spongy mesophyll layer

Stoma

H_2O

Carbohydrates

H_2O and minerals

Water exits the plant through stomata in leaves. The water potential of air is negative.

Water enters the plant through the roots. The water potential of soil is close to zero.

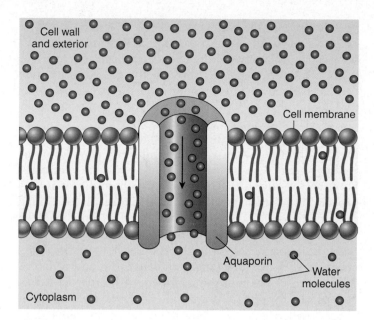

FIGURE 37.3

Aquaporins. Aquaporins are water-selective pores in the plasma membrane that increase the rate of osmosis. They do not alter the direction of water movement, however.

or gravity, and (2) the concentration of solute in each solution. We will start with physical forces. The *turgor pressure*, which is a physical pressure that results as water enters the cell vacuoles, is referred to as **pressure potential** (Ψ_p; figure 37.4*a*). As turgor pressure increases, Ψ_p increases.

There is also a potential caused by an uneven distribution of a solute on either side of a membrane, which will result in osmosis (diffusion of water to the side with the greater concentration of solute). Applying pressure (on the side that has the greater concentration of solute) prevents osmosis. The smallest amount of pressure needed to stop osmosis is the **solute** (or **osmotic**) **potential** (Ψ_s) of the solution (figure 37.4*b*).

Water will enter a cell osmotically until it is stopped by the pressure potential caused by the cell wall. The **water potential** (Ψ) of a plant cell is the sum of its pressure potential and solute potential; it represents the total potential energy of the water in a plant ($\Psi = \Psi_p + \Psi_s$). Pure water without applied pressure has a water potential of zero. Water will move to the cell with the more negative water potential (figure 37.4*c*).

Water potential regulates movement of water through a whole plant, as well as across cell membranes. Water in a plant moves along a gradient between the water potential in the soil (which can be close to zero) to successively more negative water potentials in the roots, stems, leaves, and atmosphere. On the plant surface, water loss through transpiration creates a negative pressure. It depends on its osmotic absorption by the roots and the negative pressures created by water loss from the leaves and other plant surfaces (see figure 37.2). The negative pressure generated by transpiration is largely responsible for the upward movement of water in xylem.

(a) Pressure potential Ψ_p

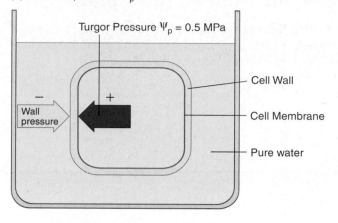

(b) Solute potential Ψ_s

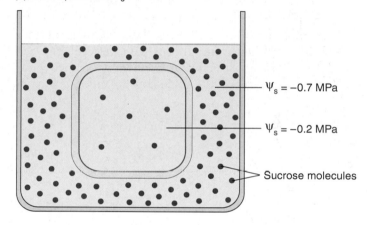

(c) Water potential

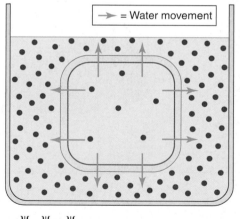

$$\Psi = \Psi_s + \Psi_p$$
$$\Psi_{cell} = -0.2 \text{ MPa} + 0.5 \text{ MPa} = 0.3 \text{ MPa}$$
$$\Psi_{solution} = -0.2 \text{ MPa (solution has no pressure potential)}$$

FIGURE 37.4
Determining water potential.
(*a*) Cell walls exert pressure in the opposite direction of cell turgor pressure. (*b*) Using the given solute potentials, predict the direction of water movement based just on solute potential. (*c*) Water potential is the sum of Ψ_s and Ψ_p. Water will move out of the cell because the Ψ of the solution is more negative.

Aquaporins enhance osmosis, which ultimately affects bulk water transport. Transpiration literally pulls water up the stem from the roots, which have the greater water potential. This works because of the strong cohesive forces between water molecules and adhesion to walls of tracheids and vessels.

Water and Mineral Absorption

Most of the water absorbed by the plant comes in through root hairs, which collectively have an enormous surface area. Root hairs are almost always turgid because their solute potential is greater than that of the surrounding soil due to mineral ions being actively pumped into the cells. Because the mineral ion concentration in the soil water is usually much lower than it is in the plant, an expenditure of energy (supplied by ATP) is required in order for such ions to accumulate in root cells. The plasma membranes of root hair cells contain a variety of protein transport channels, through which *proton pumps* transport specific ions against even large concentration gradients. Once in the roots, the ions, which are plant nutrients, are transported via the xylem throughout the plant.

The ions may follow the cell walls and the spaces between them, but more often they go directly through the plasma membranes and the protoplasm of adjacent cells (figure 37.5). When mineral ions pass between the cell walls, they do so nonselectively. Eventually, on their journey inward, they reach the endodermis, and any further passage through the cell walls is blocked by the Casparian strips. Water and ions must pass through the plasma membranes and protoplasts of the endodermal cells to reach the xylem. However, transport through the cells of the endodermis is selective. The endodermis, with its unique structure, along with the cortex and epidermis, controls which ions reach the xylem.

Transpiration from the leaves (figure 37.6), which creates a pull on the water columns, indirectly plays a role in helping water, with its dissolved ions, enter the root cells. However, at night, when the relative humidity may approach 100%, transpiration may not take place. Under these circumstances, the negative pressure component of water potential becomes small or nonexistent. However, active transport of ions into the roots still continues. This results in an increasingly high ion concentration within the cells, which causes more water to enter the root hair cells by osmosis. In terms of water potential, we say that active transport increases the solute potential of the roots. The result is movement of water into the plant and up the xylem columns despite the absence of transpiration. This phenomenon is called **root pressure.**

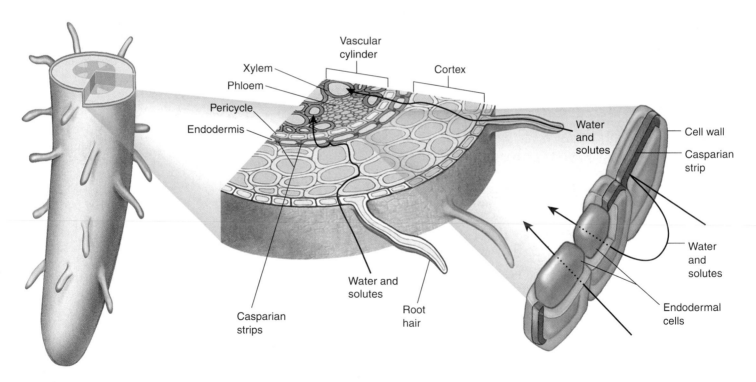

FIGURE 37.5
The pathways of mineral transport in roots. Minerals are absorbed at the surface of the root, mainly by the root hairs. In passing through the cortex, they must either follow the cell walls and the spaces between them or go directly through the plasma membranes and the protoplasts of the cells, passing from one cell to the next by way of the plasmodesmata. When they reach the endodermis, however, their further passage through the cell walls is blocked by the Casparian strips, and they must pass through the membrane and protoplast of an endodermal cell before they can reach the xylem.

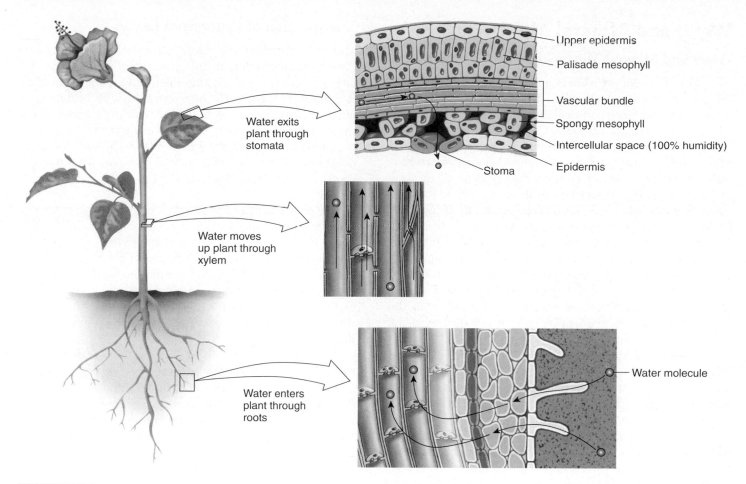

Water exits
plant through
stomata

Upper epidermis

Palisade mesophyll

Vascular bundle

Spongy mesophyll

Intercellular space (100% humidity)

Epidermis

Stoma

Water moves
up plant through
xylem

Water enters
plant through
roots

Water molecule

FIGURE 37.6
Transpiration. Water evaporating from the leaves through the stomata causes additional water to move upward in the xylem and also to enter the plant through the roots.

Under certain circumstances, root pressure is so strong that water will ooze out of a cut plant stem for hours or even days. When root pressure is very high, it can force water up to the leaves, where it may be lost in a liquid form through a process known as **guttation** (figure 37.7). Guttation does not take place through the stomata, but instead occurs through special groups of cells located near the ends of small veins that function only in this process. Root pressure is never sufficient to push water up great distances.

> **Water enters the plant by osmosis. Transport of minerals (ions) across the endodermis is selective. Root pressure, which often occurs at night, is caused by the continued, active accumulation of ions in the roots at times when transpiration from the leaves is very low or absent.**

FIGURE 37.7
Guttation. In herbaceous plants, water passes through specialized groups of cells at the edges of the leaves; it is visible here as small droplets around the edge of the leaf in this strawberry plant (*Fragaria ananassa*).

Water and Mineral Movement

Water and Mineral Movement Through the Xylem

It is clear that root pressure is insufficient to push water to the top of a tall tree, although it can help. So, what does work? Otto Renner proposed the solution in Germany in 1911. Passage of air across leaf surfaces results in loss of water by evaporation, creating a pull at the open, upper end of the "tube." Evaporation from the leaves produces a tension on the entire water column that extends all the way down to the roots. Water has an inherent tensile strength that arises from the *cohesion* of its molecules, their tendency to form hydrogen bonds with one another. The tensile strength of a column of water varies inversely with the diameter of the column; that is, the smaller the diameter of the column, the greater the tensile strength. Because plant vessels are very narrow in diameter, the cohesive forces in them are strong. The water molecules also *adhere* to the sides of the tracheid or xylem vessels, further stabilizing the long column of water.

The water column would fail if air bubbles were inserted (visualize a tower of blocks, and then pull one out in the middle). Anatomical adaptations decrease the probability of this. Individual tracheids and vessel members are connected by one or more *pits* (cavities) in their walls. Air bubbles are generally larger than these openings, so cannot pass through them. Furthermore, the cohesive force of water is so great that the bubbles are forced into rigid spheres that have no plasticity and therefore cannot squeeze through the openings. Freezing or deformation of cells can cause small bubbles of air to form within xylem cells. Any bubbles that do form are limited to the xylem elements where they originate, and water may continue to rise in parallel columns. This is more likely to occur with seasonal temperature changes. As a result, most of the active xylem in woody plants occurs peripherally, toward the vascular cambium.

Most of the minerals the plant needs enter the root through active transport. Ultimately, they are removed from the roots and relocated through the xylem to other metabolically active parts of the plant. Phosphorus, potassium, nitrogen, and sometimes iron may be abundant in the xylem during certain seasons. In many plants, such a pattern of ionic concentration helps conserve these essential nutrients, which may move from mature deciduous parts such as leaves and twigs to areas of active growth. Keep in mind that minerals that are relocated via the xylem must move with the generally upward flow through the xylem. Not all minerals can reenter the xylem conduit. Calcium, an essential nutrient, cannot be transported elsewhere once it has been deposited in a particular plant part.

Transpiration of Water from Leaves

More than 90% of the water taken in by the roots of a plant is ultimately lost to the atmosphere through transpiration from the leaves. Water moves into the pockets of air in the leaf from the moist surfaces of the walls of the mesophyll cells. As discussed in chapter 35, these intercellular spaces are in contact with the air outside the leaf by way of the stomata. Water that evaporates from the surfaces of the mesophyll cells leaves the stomata as vapor. This water is continuously replenished from the tips of the veinlets in the leaves.

Water is essential for plant metabolism, but it is continuously being lost to the atmosphere through the stomata. Photosynthesis requires a supply of CO_2 entering the stomata from the atmosphere. This results in two somewhat conflicting requirements: the need to minimize the loss of water to the atmosphere and the need to admit carbon dioxide. Structural features such as stomata and the cuticle have evolved in response to one or both of these requirements.

The rate of transpiration depends on weather conditions, including humidity and the time of day. As stated earlier in this section, after the sun sets, transpiration from the leaves decreases. The sun is the ultimate source of potential energy for water movement. The water potential that is responsible for water movement is largely the product of negative pressure generated by transpiration, which is driven by the warming effects of sunlight.

Regulation of Transpiration Rate. On a short-term basis, closing the stomata can control water loss. This occurs in many plants when they are subjected to water stress. However, the stomata must be open at least part of the time so that CO_2 can enter. As CO_2 enters the intercellular spaces, it dissolves in water before entering the plant's cells. The gas dissolves mainly in water on the walls of the intercellular spaces below the stomata. The continuous stream of water that reaches the leaves from the roots keeps these walls moist.

Stomata open and close because of changes in the turgor pressure of their guard cells. The sausage-shaped guard cells stand out from other epidermal cells not only because of their shape, but also because they are the only epidermal cells containing chloroplasts. Their distinctive wall construction, which is thicker on the inside and thinner elsewhere, results in a bulging out and bowing when they become turgid. You can make a model of this for yourself by taking two elongated balloons, tying the closed ends together, and inflating both balloons slightly. When you hold the two open ends together, there should be very little space between the two balloons. Now place

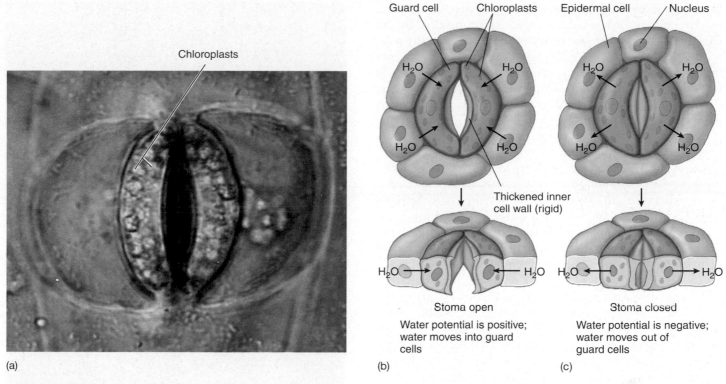

Chloroplasts

Guard cell   Chloroplasts   Epidermal cell   Nucleus

H₂O   H₂O   H₂O   H₂O

H₂O   H₂O   H₂O   H₂O

Thickened inner
cell wall (rigid)

H₂O →   ← H₂O   H₂O →   ← H₂O

Stoma open   Stoma closed
Water potential is positive;   Water potential is negative;
water moves into guard   water moves out of
cells   guard cells

(a)   (b)   (c)

FIGURE 37.8

How a stoma opens and closes.   (*a*) Open stoma (*b*) When potassium ions from surrounding cells are pumped into guard cells, the guard cell turgor pressure increases as water enters by osmosis. The increased turgor pressure causes the guard cells to bulge, with the thick walls on the inner side of each guard cell bowing outward, thereby opening the stoma. (*c*) When the potassium ions passively leave the guard cells, the guard cells lose water and turgor, and the stoma closes.

duct tape on the inside edge of both balloons and inflate each one a bit more. Hold the open ends together. You should now be holding a doughnut-shaped pair of "guard cells" with a "stoma" in the middle. Real guard cells rely on the influx and efflux of water, rather than air, to open and shut.

Turgor in guard cells results from the active uptake of potassium (K^+) ions. An increase in K^+ concentration creates a water potential in the guard cells that causes water to enter osmotically. As a result, these cells accumulate water and become turgid, opening the stomata (figure 37.8*a*). When K^+ passively leaves the guard cells, water also leaves. The guard cells lose turgor, and the stomata close (figure 37.8*b*). In some species, Cl^- accompanies the K^+ in and out of the guard cells, thus maintaining electrical neutrality.

When a whole plant wilts because insufficient water is available, the guard cells may also lose turgor, and as a result, the stomata may close. The guard cells of many plant species regularly become turgid in the morning, when photosynthesis occurs, and lose turgor in the evening, regardless of the availability of water. When they are turgid, the stomata open, and CO_2 enters freely; when they are flaccid, CO_2 is largely excluded, but water loss is also retarded.

Experimental evidence is consistent with several pathways regulating stomatal opening and closing. K^+ transport through a specific channel, against a concentration gradient, is promoted by light. Blue light triggers proton (H^+) transport. The resulting proton gradient drives the opening of K^+ channels. When guard cells lose turgor, sucrose export may be another important factor.

Abscisic acid, a plant hormone discussed in chapter 40, plays a primary role in allowing K^+ to pass rapidly out of guard cells, causing the stomata to close in response to drought. This hormone is released from chloroplasts and produced in leaves. It binds to specific receptor sites in the plasma membranes of guard cells. Plants likely control the duration of stomatal opening through the integration of several stimuli. In chapter 40, we will explore the interactions between the environment and the plant in more detail.

Other Factors Regulating Transpiration.   Factors such as CO_2 concentration, light, and temperature can also affect stomatal opening. When CO_2 concentrations are high, the guard cells of many plant species lose turgor, and their stomata close. Additional CO_2 is not needed at such times, and water is conserved when the guard cells are closed. The stomata also close if the temperature exceeds 30° to 34°C, when transpiration would increase substantially. In the dark, stomata open at low concentrations of CO_2. In chapter 10, we mentioned CAM photosynthesis, which occurs in succulent plants such as cacti. In this process, CO_2 is taken in at night and fixed during the day. CAM photosynthesis conserves water in dry environments where succulent plants grow.

Many mechanisms for regulating the rate of water loss have evolved in plants. One involves dormancy during dry times of the year; another involves loss of leaves. Deciduous plants are common in areas that periodically experience severe drought. Plants are often deciduous in regions where winters are severe, when water is locked up in ice and snow and thus unavailable to them. In a general sense, annual plants conserve water when conditions are unfavorable, simply by going into dormancy as seeds.

Thick, hard leaves, often with relatively few stomata—and frequently with stomata only on the lower side of the leaf—lose water far more slowly than large, pliable leaves with abundant stomata. Temperatures are significantly reduced in leaves covered with masses of woolly-looking trichomes. These trichomes also increase humidity at the leaf surface. Plants in arid or semiarid habitats often have their stomata in crypts or pits in the leaf surface. Within these depressions, the water vapor content of the air may be high, reducing the rate of water loss.

Plant Responses to Flooding

Plants can also receive too much water, and ultimately "drown." Flooding rapidly depletes available oxygen in the soil and interferes with the transport of minerals and carbohydrates in the roots. Abnormal growth often results. Hormone levels change in flooded plants; ethylene (the only hormone that is a gas) increases, while gibberellins and cytokinins usually decrease. Hormonal changes contribute to the abnormal growth patterns.

Oxygen deprivation is among the most significant problems. Standing water has much less oxygen than moving water. Generally, standing-water flooding is more harmful to a plant (riptides excluded). Flooding that occurs when a plant is dormant is much less harmful than flooding when it is growing actively.

Physical changes that occur in the roots as a result of oxygen deprivation may halt the flow of water through the plant. Paradoxically, even though the roots of a plant may be standing in water, its leaves may be drying out. One adaptive solution is that the stomata of flooded plants often close to maintain leaf turgor.

Adapting to Life in Fresh Water. The algal ancestors of plants adapted to a freshwater environment from a saltwater environment before the "move" onto land. This

FIGURE 37.9
Adaptation to flooded conditions. The "knees" of the bald cypress (*Taxodium*) form whenever it grows in wet conditions, increasing its ability to take in oxygen.

involved a major change in controlling salt balance. Since that time, many have "moved" back into fresh water and grow in places that are often or always flooded naturally; they have adapted to these conditions during the course of their evolution (figure 37.9). One of the most frequent adaptations among such plants is the formation of **aerenchyma,** loose parenchymal tissue with large air spaces in it (figure 37.10). Aerenchyma is very prominent in water lilies and many other aquatic plants. Oxygen may be transported from the parts of the plant above water to those below by way of passages in the aerenchyma. This supply of oxygen allows oxidative respiration to take place even in the submerged portions of the plant.

Some plants normally form aerenchyma, whereas others, subject to periodic flooding, can form it when necessary. In corn, ethylene, which becomes abundant under the anaerobic conditions of flooding, induces aerenchyma formation. Plants also respond to flooded conditions by forming larger lenticels (which facilitate gas exchange) and additional adventitious roots.

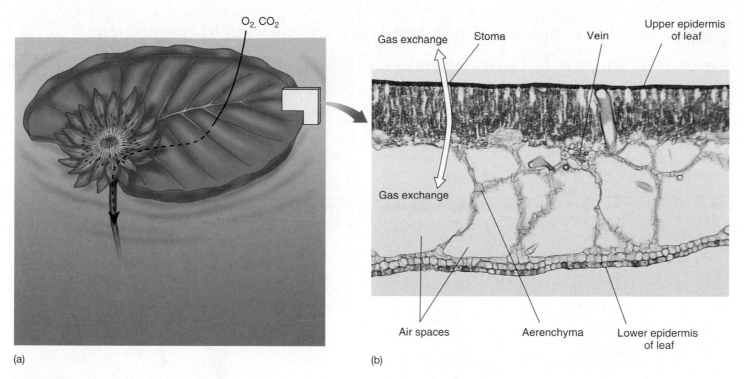

(a)

(b)

FIGURE 37.10
Aerenchyma. This tissue facilitates gas exchange in aquatic plants. (*a*) Water lilies float on the surface of ponds, collecting oxygen and then transporting it to submerged portions of the plant. (*b*) Large air spaces in the leaves of the water lily add buoyancy. The specialized parenchyma tissue that forms these open spaces is called aerenchyma. Gas exchange occurs through stomata found only on the upper surface of the leaf.

Adapting to Life in Salt Water. Plants such as mangroves that grow in areas normally flooded with salt water must not only provide a supply of oxygen to their submerged parts, but also control their salt balance. The salt must be excluded, actively secreted, or diluted as it enters. The arching silt roots of mangroves are connected to long, spongy, air-filled roots that emerge above the mud. These roots, called pneumatophores (see chapter 35), have large lenticels on their above-water portions through which oxygen enters; it is then transported to the submerged roots (figure 37.11). In addition, the succulent leaves of mangroves contain large quantities of water, which dilute the salt that reaches them. Many plants that grow in such conditions also either secrete large quantities of salt or block salt uptake at the root level.

> Transpiration from leaves pulls water and minerals up the xylem. This works because of the physical properties of water and the narrow diameters of the conducting tubes. Stomata open when their guard cells become turgid. Opening and closing of stomata is osmotically regulated. Biochemical, anatomical, and morphological adaptations have evolved to reduce water loss through transpiration. Plants are harmed by excess water. However, plants can survive flooded conditions, and even thrive in them, if they can deliver oxygen to their submerged parts.

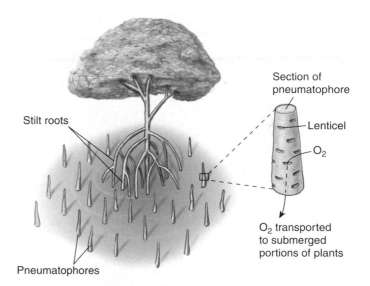

FIGURE 37.11
How mangroves get oxygen to their submerged parts.
Mangrove plants grow in areas that are commonly flooded, and much of each plant is usually submerged. However, modified roots called pneumatophores supply the submerged portions of the plant with oxygen because these roots emerge above the water and have large lenticels. Oxygen can enter the roots through the lenticels, pass into the abundant aerenchyma, and move to the rest of the plant.

37.2 Dissolved sugars and hormones are transported in the phloem.

Phloem Transport Is Bidirectional

Most carbohydrates manufactured in leaves and other green parts are distributed through the phloem to the rest of the plant. This process, known as **translocation,** provides suitable carbohydrate building blocks for the roots and other actively growing regions of the plant. Carbohydrates concentrated in storage organs such as tubers, often in the form of starch, are also converted into transportable molecules, such as sucrose, and moved through the phloem. The pathway that sugars and other substances travel within the plant has been demonstrated precisely by using radioactive tracers, despite the fact that living phloem is delicate and the process of transport within it is easily disturbed. Radioactive carbon dioxide ($^{14}CO_2$) is incorporated into glucose as a result of photosynthesis. The glucose is used to make sucrose, which is transported in the phloem. Such studies have shown that sucrose moves both up and down in the phloem.

Aphids, a group of insects that extract plant sap for food, have been valuable tools in understanding translocation. Aphids thrust their *stylets* (piercing mouthparts) into phloem cells of leaves and stems to obtain the abundant sugars there. When a feeding aphid is removed by cutting its stylet, the liquid from the phloem continues to flow through the detached mouthpart and is thus available in pure form for analysis (figure 37.12). The liquid in the phloem contains 10 to 25% dry matter, almost all of which is sucrose. Using aphids to obtain the critical samples and radioactive tracers to mark them, it has been demonstrated that substances in phloem can move remarkably fast, up to 50 to 100 centimeters per hour.

It is important to note that phloem also transports plant hormones. As we will explore in chapter 40, environmental signals can result in the rapid translocation of hormones in the plant. Recent evidence also indicates that mRNA can move through the phloem, providing a previously unknown mechanism for long-distance communication among cells.

Energy Requirements for Phloem Transport

The most widely accepted model of how carbohydrates in solution move through the phloem has been called the **mass-flow hypothesis,** pressure flow hypothesis, or bulk flow hypothesis. Experimental evidence supports much of this model. Dissolved carbohydrates flow from a **source** and are released at a **sink,** where they are utilized. Carbohydrate sources include photosynthetic tissues, such as the mesophyll of leaves. Food-storage tissues, such as the cortex of roots, can be either sources or sinks. Sinks also occur at the growing tips of roots and stems and in developing fruits.

(a)

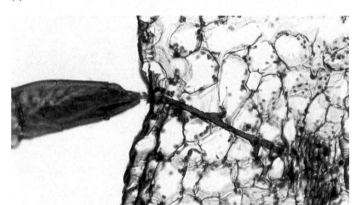

(b)

FIGURE 37.12
Feeding on phloem. (*a*) Aphids, including this individual of *Macrosiphon rosae* shown on the edge of a rose leaf, feed on the food-rich contents of the phloem, which they extract through (*b*) their piercing mouthparts, called stylets. When an aphid is separated from its stylet and the cut stylet is left in the plant, the phloem fluid oozes out of it and can then be collected and analyzed.

In a process known as *phloem loading*, carbohydrates (mostly sucrose) enter the sieve tubes in the smallest veinlets at the source. This is an energy-requiring step because active transport is needed. Companion cells and parenchyma cells adjacent to the sieve tubes provide the ATP energy to drive this transport. Then, because of the

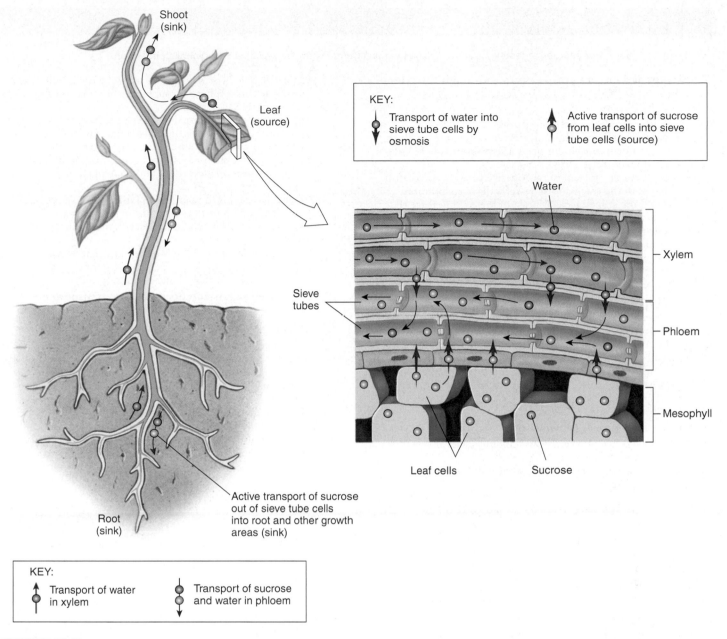

FIGURE 37.13

Diagram of mass flow. In this diagram, *red* dots represent sucrose molecules, and *blue* dots symbolize water molecules. After moving from the mesophyll cells of a leaf or another part of the plant into the conducting cells of the phloem, the sucrose molecules are transported to other parts of the plant by mass flow and unloaded where they are required.

difference between the water potentials in the sieve tubes and in the nearby xylem cells, water flows into the sieve tubes by osmosis. Turgor pressure in the sieve tubes increases. The increased turgor pressure drives the fluid throughout the plant's system of sieve tubes. At the sink, carbohydrates are actively removed. Water moves from the sieve tubes by osmosis, and the turgor pressure there drops, causing a mass flow from the more positive pressure at the source to the more negative pressure at the sink (figure 37.13). Most of the water at the sink then diffuses back into the xylem, where it may either be recirculated or lost through transpiration.

Transport of sucrose and other carbohydrates through sieve tubes does not require energy. The loading and unloading of these substances from the sieve tubes does.

37.1 Water and minerals move upward through the xylem.

Overview of Water and Mineral Movement Through Plants

- Water enters the roots and moves through the xylem because of multiple factors. (p. 768)
- Aquaporins enhance the transport of water at the cellular level, and such movement plays a significant role in bulk water transport. (p. 768)
- Because water molecules are cohesive to each other, and adhesive to the tracheids of vessel walls, transpiration through the stomata exerts most of the pulling force that moves water up the plant. (p. 768)
- Water potential is the combination of a plant's pressure potential and its solute potential. (p. 769)
- Water will move from a solution with higher water potential to a solution with lower water potential. (p. 769)

Water and Mineral Absorption

- Most of a plant's absorbed water enters by root hairs, which are usually turgid because their solute potential is greater than that of their surroundings. (p. 770)
- Plasma membranes of root hair cells contain protein transport channels through which protein pumps transport specific ions against large concentration gradients. (p. 770)
- Root pressure causes the movement of water into the plant and up the xylem despite the absence of transpiration. (pp. 770–771)

Water and Mineral Movement

- The tensile strength of a water column varies inversely with the column diameter. (p. 772)
- The formation of air bubbles in the column would cause it to fail. (p. 772)

- More than 90% of water taken in by the roots is ultimately lost to the atmosphere through transpiration from the leaves. (p. 772)
- Stomata open and close due to changes in guard cell turgor pressure resulting from the active uptake of potassium, which is regulated by abscisic acid. (p. 773)
- Other factors, such as CO_2 concentration, light, and temperature, can also affect and control stomatal opening. (pp. 773–774)
- Plants have evolved many mechanisms to conserve water loss. These include dormancy during dry periods, deciduousness, thick hard leaves, and the presence of trichomes. (p. 774)
- Oxygen deprivation is among the most significant problems facing plants in flooded areas because standing water has much less oxygen than moving water. (p. 774)

37.2 Dissolved sugars and hormones are transported in the phloem.

Phloem Transport Is Bidirectional

- Translocation distributes carbohydrates manufactured in the leaves to other parts of the plant. (p. 776)
- The mass-flow hypothesis suggests that dissolved carbohydrates flow from a source and are released at a sink. (p. 776)
- Carbohydrates enter sieve tubes by osmosis, and the resultant increased turgor pressure drives the water throughout the sieve-tube system. At the sink, carbohydrates are actively removed. (pp. 776–777)

38

Plant Nutrition

FIGURE 38.1
Soil: A rich source of nutrients. Plants rely on extensive root systems to obtain essential nutrients. When there is little topsoil, roots may be found close to the surface.

Concept Outline

38.1 Plants require a variety of nutrients in addition to the direct products of photosynthesis.

Plant Nutrients. Plants require a few macronutrients in large amounts and several micronutrients in trace amounts.

Soil. Plant growth is significantly influenced by the nature of the soil.

38.2 Global change could alter the balance among photosynthesis, respiration, and use of nutrients acquired through the soil.

Global Change. Atmospheric CO_2 concentrations have increased about 31% since 1750 and could potentially alter the relative amounts of nutrients produced in a plant.

38.3 Some plants have novel strategies for obtaining nutrients.

Nitrogen Fixation. Some plants entice bacteria to produce organic nitrogen for them. These bacteria may be free-living or form a symbiotic relationship with a host plant.

Nutritional Adaptations. Venus flytraps and other carnivorous plants lure and capture insects and then digest them to obtain nutrients. Parasitic plants feed off other plants. About 90% of all vascular plants rely on fungal associations to gather essential nutrients, especially phosphorus.

38.4 Plants can remove harmful chemicals from the soil.

Phytoremediation. Heavy metals and toxic organic compounds can be taken up by plant roots and be either isolated or broken down into a nontoxic form in the plant.

Vast energy inputs are required for the ongoing construction of a plant. In this chapter, we ask, what inputs, besides energy from the sun, does a plant need to survive? Plants, like animals, need various nutrients to remain alive and healthy. The lack of an important nutrient may slow a plant's growth or make the plant more susceptible to disease or even death. Plants acquire these nutrients mainly through photosynthesis and from the soil (figure 38.1). Nitrogen is particularly problematic because plants cannot directly convert atmospheric nitrogen into amino acids. A few plants are able to capture animals and secrete digestive juices to make nitrogen available for absorption. In addition to contributing nutrients, the soil hosts bacteria and fungi that aid plants in obtaining nutrients in a usable form.

38.1 Plants require a variety of nutrients in addition to the direct products of photosynthesis.

Plant Nutrients

The major source of plant nutrition is the fixation of atmospheric carbon dioxide (CO_2) into simple sugar using the energy of the sun. CO_2 enters through the stomata (openings on the surface of the leaf). Oxygen (O_2) is a product of photosynthesis and an atmospheric component that also moves through the stomata. Oxygen is used in cellular respiration to release energy from the chemical bonds in the sugar in order to support growth and maintenance in the plant. However, CO_2 and light energy are not sufficient for the synthesis of all the molecules a plant needs. Plants require a number of inorganic nutrients. Some of these are macronutrients, which the plants need in relatively large amounts, and others are micronutrients, which are required in trace amounts (table 38.1). There are nine macronutrients: carbon, oxygen, and hydrogen—the three elements found in all organic compounds—as well as nitrogen (essential for amino acids), potassium, calcium, magnesium (the center of the chlorophyll molecule), phosphorus, and sulfur. Each of these nutrients approaches or, as in the case of carbon, may greatly exceed 1% of the dry weight of a healthy plant. The seven micronutrient elements—chlorine, iron, manganese, zinc, boron, copper, and molybdenum—constitute from less than one to several hundred parts per million in most plants. A deficiency of any one can have severe effects on plant growth (figure 38.2). The macronutrients were generally discovered in the last century, but the micronutrients have been detected much more recently as technology developed to identify and work with such small quantities.

Nutritional requirements are assessed in hydroponic cultures; the plant roots are suspended in aerated water containing nutrients. The solutions contain all the necessary nutrients in the right proportions but with certain known or suspected nutrients left out. The plants are then allowed to grow and are studied for the presence of abnormal symptoms that might indicate a need for the missing element (figure 38.3). However, the water or vessels used often contain enough micronutrients to allow the plants to grow normally, even though these substances were not added deliberately to the solutions. To give an idea of how small the needed quantities of micronutrients may be, the standard dose of molybdenum added to seriously deficient soils in Australia amounts to about 34 grams (about one handful) per hectare (a square

	Table 38.1	Essential Nutrients in Plants	
Element	Principal Form in Which Element Is Absorbed	Approximate Percent of Dry Weight	Examples of Important Functions
MACRONUTRIENTS			
Carbon	(CO_2)	44	Major component of organic molecules
Oxygen	(O_2, H_2O)	44	Major component of organic molecules
Hydrogen	(H_2O)	6	Major component of organic molecules
Nitrogen	(NO_3^-, NH_4^+)	1–4	Component of amino acids, proteins, nucleotides, nucleic acids, chlorophyll, coenzymes, enzymes
Potassium	(K^+)	0.5–6	Protein synthesis, operation of stomata
Calcium	(Ca^{++})	0.2–3.5	Component of cell walls, maintenance of membrane structure and permeability; activates some enzymes
Magnesium	(Mg^{++})	0.1–0.8	Component of chlorophyll molecule; activates many enzymes
Phosphorus	($H_2PO_4^-$, $HPO_4^=$)	0.1–0.8	Component of ADP and ATP, nucleic acids, phospholipids, several coenzymes
Sulfur	($SO_4^=$)	0.05–1	Components of some amino acids and proteins, coenzyme A
MICRONUTRIENTS (CONCENTRATIONS IN PPM)			
Chlorine	(Cl^-)	100–10,000	Osmosis and ionic balance
Iron	(Fe^{++}, Fe^{+++})	25–300	Chlorophyll synthesis, cytochromes, nitrogenase
Manganese	(Mn^{++})	15–800	Activator of certain enzymes
Zinc	(Zn^{++})	15–100	Activator of many enzymes; active in formation of chlorophyll
Boron	(BO_3^- or $B_4O_7^=$)	5–75	Possibly involved in carbohydrate transport, nucleic acid synthesis
Copper	(Cu^{++})	4–30	Activator or component of certain enzymes
Molybdenum	($MoO_4^=$)	0.1–5	Nitrogen fixation, nitrate reduction

FIGURE 38.2
Mineral deficiencies in plants. (*a*) Leaves of a healthy Marglobe tomato (*Lycopersicon esculentum*) plant. (*b*) Chlorine-deficient plants with necrotic leaves (leaves with patches of dead tissue). (*c*) Copper-deficient plant with blue-green, curled leaves. (*d*) Zinc-deficient plant with small, necrotic leaves. The agricultural implications of deficiencies such as these are obvious; a trained observer can determine the nutrient deficiencies affecting a plant simply by inspecting it.

(a)

(b)

(c)

(d)

100 meters on a side—about 2.5 acres), once every 10 years! Most plants grow satisfactorily in hydroponic culture, and the method, although expensive, is occasionally practical for commercial purposes. Analytical chemistry has made it much easier to test plant material for levels of different molecules.

Increasing the nutrients in crop species, especially in developing countries, could have tremendous human health benefits. For example, food fortification is an active area of research focused on ways to increase uptake of minerals by plants and transport them in roots and shoots where they are stored for later human consumption. Phosphate uptake can be increased if it is more soluble in the soil. Plants have been genetically modified to secrete citrate, an organic acid that solubilizes phosphate. As an added benefit, the citrate binds to aluminum, which can be toxic to plants and animals, and limits the uptake of aluminum into plants. For other nutrients, such as iron, manganese, and zinc, plasma membrane transport is a limiting factor. Genes coding for these plasma membrane transporters have been cloned in other species and are being incorporated into crop plants. Eventually, breakfast cereals may be fortified with additional nutrients while the grains are growing in the field as opposed to when they are processed in the factory.

> The plant macronutrients carbon, oxygen, and hydrogen constitute about 94% of a plant's dry weight; the other macronutrients—nitrogen, potassium, calcium, magnesium, phosphorus, and sulfur—each approach or exceed 1% of a plant's dry weight.

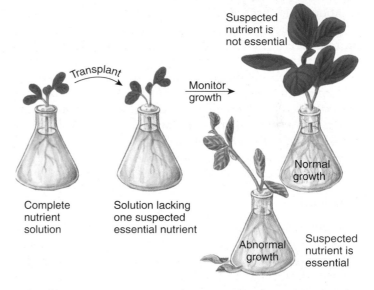

FIGURE 38.3
Identifying nutritional requirements of plants. A seedling is first grown in a complete nutrient solution. The seedling is then transplanted to a solution that lacks one suspected essential nutrient. The growth of the seedling is studied for the presence of abnormal symptoms, such as discolored leaves or stunted growth. If the seedling's growth is normal, the nutrient that was left out may not be essential; if the seedling's growth is abnormal, the lacking nutrient is essential for growth.

Soil

Composition of Soil

Plant growth is affected by soil composition. Soil is the highly weathered outer layer of the earth's crust. It is composed of a mixture of ingredients, which may include sand, rocks of various sizes, clay, silt, humus, and various other forms of mineral and organic matter. Pore spaces containing water and air occur between the particles. The mineral fraction of soils varies according to the composition of the rocks. The earth's crust includes about 92 naturally occurring elements; figure 2.8 in chapter 2 shows the periodic table of elements and the elements' frequency in the earth's crust. Most elements are combined as inorganic compounds called **minerals;** most rocks consist of several different minerals. The soil is also full of microorganisms that break down and recycle organic debris. For example, about 5 metric tons of carbon is tied up in the organisms present in the soil under a hectare of wheat land in England, an amount that approximately equals the weight of 100 sheep!

Most roots are found in **topsoil** (figure 38.4), which is a mixture of mineral particles of varying size (most less than 2 millimeters in diameter), living organisms, and **humus.** Humus consists of partly decayed organic material. When topsoil is lost because of erosion or poor landscaping, both the water-holding capacity and the nutrient relationships of the soil are adversely affected.

About half of the total soil volume is occupied by spaces or pores, which may be filled with air or water, depending on moisture conditions. Some of the soil water is unavailable to plants. Due to gravity, some of the water that reaches a given soil drains through it immediately. Another fraction of the water is held in small soil pores, which are generally less than about 50 micrometers in diameter. This water is readily available to plants. When it is depleted through evaporation or root uptake, the plant will wilt and eventually die unless more water is added to the soil.

Cultivation

In natural communities, nutrients are recycled and made available to organisms on a continuous basis. When these communities are replaced by cultivated crops, the situation changes drastically: The soil is much more exposed to erosion and the loss of nutrients. For this reason, cultivated crops and garden plants usually must be supplied with additional mineral nutrients.

One solution to the problem of nutrient loss is **crop rotation.** For example, a farmer might grow corn in a field one year and soybeans the next year. These crops remove different amounts of various macro- and micronutrients from the soil. Soybean plants even add nitrogen compounds to the soil, released by nitrogen-fixing bacteria growing in nodules on their roots. Sometimes farmers

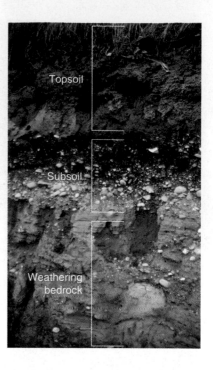

FIGURE 38.4
Most roots occur in the topsoil. The uppermost layer in soil is called topsoil, and it contains organic matter, such as roots, small animals, humus, and mineral particles of various sizes. Subsoil lies underneath the topsoil and contains larger mineral particles and relatively little organic matter. Beneath the subsoil are layers of bedrock, the raw material from which soil is formed over time and through weathering.

allow a field to lie fallow—that is, they do not grow a crop in the field for a year or two. This allows natural processes to rebuild the field's store of nutrients.

Another farming practice that helps maintain soil fertility involves plowing-under plant material left in fields. You can do the same thing in a lawn or garden by leaving grass clippings and dead leaves. Decomposers in the soil do the rest, turning the plant material into humus.

Fertilizers are also used to replace nutrients lost in cultivated fields. The most important mineral nutrients that need to be added to soils are nitrogen (N), phosphorus (P), and potassium (K). All of these elements are needed in large quantities (see table 38.1) and are the most likely to become deficient in the soil. Both synthetic and organic fertilizers are often added in large quantities and can be significant sources of pollution in certain situations (see chapter 56). Organic fertilizers were widely used long before chemical fertilizers became available. Substances such as manure or the remains of dead animals have traditionally been applied to crops, and plowed-under plant material can also increase the soil's fertility. There is no basis for believing that organic fertilizers supply any element to plants that inorganic fertilizers cannot provide. However, organic fertilizers build up the humus content of the soil, which often enhances its water- and nutrient-retaining properties. For this reason, nutrient availability to plants at different times of the year may be improved, under certain circumstances, with organic fertilizers.

Soils contain organic matter and various minerals and nutrients. Farming practices such as crop rotation, plowing crops under, and fertilization are often necessary to maintain soil fertility.

38.2 Global change could alter the balance among photosynthesis, respiration, and use of nutrients acquired through the soil.

Global Change

Why all the fuss about global change? The Intergovernmental Panel on Climate Change has concluded that CO_2 is probably at its highest concentration in the atmosphere in at least 20 million years. Since 1750, atmospheric CO_2 has increased 31%, which correlates with increases in many human activities, including the burning of fossil fuel. The long-term effects of elevated CO_2 are complex and not fully understood. Chapter 56 explores the causal link between elevated CO_2 and global warming. Here, we will consider how increased CO_2 may alter nutrient balance within plants.

Elevated CO^2 and Photosynthesis

First we will investigate the relationship between photosynthesis and the relative concentration of atmospheric CO_2.

You might find it helpful to review chapter 10, including section 10.4, which discusses the importance of the ratio of atmospheric CO_2 and O_2 in determining whether or not energetically wasteful photorespiration will occur. The two questions to be addressed in this section are (1) Does elevated CO_2 increase the rate of photosynthesis, and (2) will elevated CO_2 change the ratio of carbohydrates and proteins in plants?

The Calvin cycle of photosynthesis fixes atmospheric CO_2 into sugar, a major plant nutrient. The first step of the Calvin cycle stars the most abundant protein on earth, ribulose 1,5-bisphosphate carboxylase/oxygenase (RUBISCO). The active site of this enzyme can bind either CO_2 or O_2 and catalyze the addition of either molecule to a five-carbon molecule, ribulose 1,5-bisphosphate, beginning the Calvin cycle (figure 38.5). Whereas CO_2 is used to produce a three-carbon sugar that can in turn be used to synthesize glucose and sucrose, O_2 is used in photorespiration, which results in neither nutrient nor energy production. As the relative amount of CO_2 increases, the Calvin cycle becomes more efficient. Thus, it is reasonable to hypothesize that the global increase in CO_2 should lead to increased photosynthesis and increased plant growth. Assuming that nutrient availability in the soil remains the same, the more rapidly growing plants should have lower levels of nitrogen-containing compounds, such as proteins, and also lower levels of minerals obtained from the soil. How can such hypotheses be tested?

Growing plants in an environment where CO_2 levels can be precisely controlled is the optimal way to determine how CO_2 concentrations affect plant nutrition. Experiments with potted plants in growth chambers are one approach, but far more information can be obtained in natural areas enriched with CO_2. For example, Duke Experimental Forest has rings of towers that blow CO_2 toward the center of the ring

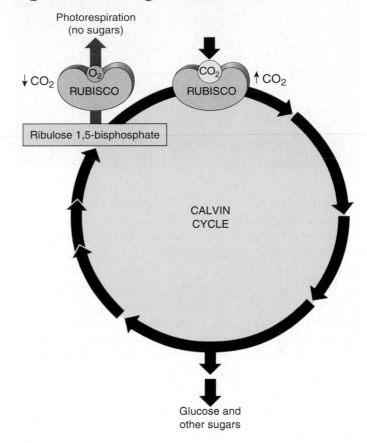

FIGURE 38.5
Photorespiration. CO_2 and O_2 compete for the same site on the enzyme that catalyzes the first step in the Calvin cycle. If CO_2 binds, a three-carbon sugar is produced that can make glucose and sucrose. If O_2 binds, photorespiration occurs, and energy is used to break down a five-carbon molecule without yielding any useful product. As the ratio of CO_2 to O_2 increases, the Calvin cycle can produce more sugar.

(figure 38.6). These rings are 30 meters in diameter and allow studies to be conducted at the ecosystem level. Such facilities allow for long-term studies of the effects of altered atmospheric conditions on ecosystems.

Such studies have complex results. Potatoes grown in a European facility had a 40% higher photosynthetic rate when the concentration of CO_2 was approximately doubled. Potted plants often show an initial increase, followed by a decrease in photosynthesis over time that is associated with lower levels of RUBISCO production. Different species of plants in a Florida oak-shrub system showed different responses to elevated CO_2 levels, while over three years in the Duke Experimental Forest, plants achieved more biomass in the CO_2 enclosures than outside the enclosures if the soil contained sufficient nitrogen availability to support growth.

(a)

(b)

FIGURE 38.6

Experimentally elevating CO$_2$. Enclosures in the Duke Experimental Forest provide ecosystem-level comparisons of plants grown in ambient and elevated CO$_2$ environments. (*a*) Each ring enclosure is 30 meters in diameter. (*b*) Towers surrounding the rings blow CO$_2$ into the enclosure under closely monitored conditions.

As you saw earlier in this chapter, nitrogen availability limits plant growth. As CO$_2$ levels increase, less nitrogen and other macronutrients may be found in leaves. In that event, herbivores would need to eat more biomass to obtain adequate nutrients. This is of significant concern in agriculture and could affect human health. Insect infestations could be more devastating if each herbivore consumed more biomass. Protein deficiencies resulting from decreased nitrogen on crops. Again, it is difficult to generalize among all plants. A study of poplar trees showed no change in the relative amounts of nitrogen in leaves. Here, an increase in root surface area appears to have enhanced nitrogen uptake.

Elevated Temperature and Respiration

As much as half of all the carbohydrates produced from photosynthesis each day can be used in plant respiration that same day. The amount of carbohydrate available for respiration may be affected by atmospheric CO$_2$ and photosynthesis, as we just discussed. Furthermore, the anticipated rise in temperature over the next century may affect the rate of respiration in other ways. Altered respiration rates can affect overall nutrient balance and plant growth. For these reasons, we will consider how elevated temperatures might affect respiration and, as a result, nutrient balance in plants.

The potential changes in respiration and photosynthesis are complex and difficult to predict because of the dependent nature of the two processes. Increased photosynthesis makes more carbohydrate available for respiration, and increased respiration would lead to an increase in atmospheric CO$_2$. Biologists have known for a long time that respiration rates are temperature sensitive in a broad range of

plant species. The current question is how plant respiration is affected by both short-term and long-term temperature changes.

Why does respiration rate change with temperature? One important factor is the effect of temperature on enzyme activity in respiration. As you saw in chapter 3, enzyme activity changes with temperature. This is particularly important at very low temperatures and also very high temperatures that lead to protein denaturation (enzymes are protein). The amount of available carbohydrate can also affect the rate of respiration. For some plants this is the most critical factor in the case of moderate temperature increases.

Many responses of respiration rate to temperature change may be short-term, rather than long term. There is growing evidence that respiration rate acclimates to a temperature increase over time, especially in leaves and roots that develop after the temperature shift. Over a long period of time at an elevated temperature, a plant could end up respiring at the same rate it had previously respired at a lower temperature. Not all plants will necessarily respond to temperature change in the same way or over the same time period. Consider a very fast growing versus a very slow growing plant. The fast growing plant might acclimate more rapidly because it will be producing new leaves and new roots more quickly than the slower growing plant. These new organs will acclimate to the temperature change more fully than already developed organs. If global temperature increases are as rapid as predicted in the decades ahead, fast growing plants could have a selective advantage over plants that grow at a more measured pace.

Elevated CO$_2$ levels and temperature have the potential to alter plant nutrition, which will affect natural ecosystems and agriculture.

38.3 Some plants have novel strategies for obtaining nutrients.

Nitrogen Fixation

Plants need ammonia (NH_3) to build amino acids, but most of the nitrogen in the atmosphere is in the form of N_2. Plants lack the biochemical pathways (including the enzyme nitrogenase) necessary to convert gaseous nitrogen to ammonia, but some bacteria have this capacity. Some of these bacteria live in close association with the roots of plants. Others end up being housed in plant tissues created especially for them called nodules (figure 38.7).

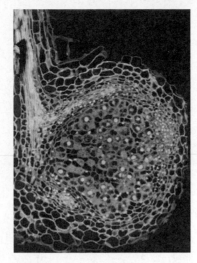

FIGURE 38.7
Nitrogen-fixing nodule. A root hair of alfalfa is invaded by *Rhizobium*, a bacterium (*yellow*) that fixes nitrogen.

Legumes and a few other plants can form root nodules. Hosting these bacteria costs the plant energy, but is well worth it when the soil lacks nitrogen. To conserve energy, legume root hairs will not respond to bacterial signals when nitrogen levels are high.

Just how do legumes and nitrogen-fixing *Rhizobium* bacteria get together (figure 38.8)? Extensive signaling between the bacteria and the legume not only lets each organism know the other is present, but also checks whether the bacterium is the correct species for the specific legume. These highly evolved symbiotic relationships depend on exact species matches. Soybean and garden peas are both legumes, but each has its own species of symbiotic *Rhizobium*.

Nitrogen can be obtained from bacteria living in close association with plant roots.

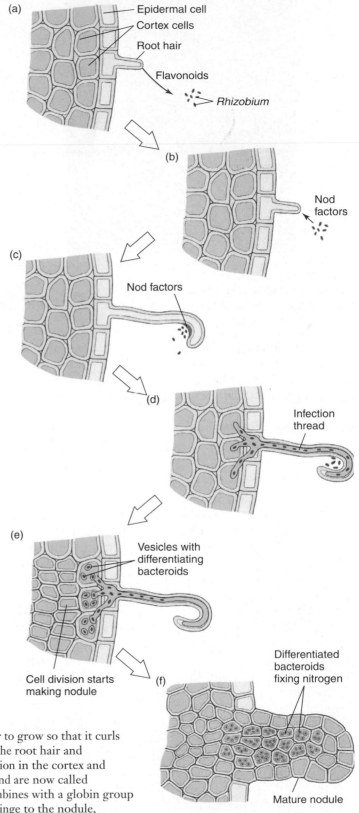

FIGURE 38.8
Nodule formation in *Rhizobium*. (a) Pea roots produce flavonoids (a group of molecules used for plant defense and for making reddish pigment, among numerous other functions). The flavonoids are transported into the rhizobial cells. (b) Flavonoids signal rhizobia to produce sugar-containing compounds called Nod (nodulation) factors. (c) Nod factors bind to the surface of root hairs and signal the root hair to grow so that it curls around the rhizobia. (d) Rhizobia make infection threads that grow in the root hair and move into the cortex of the root. The rhizobia take control of cell division in the cortex and pericycle cells of the root (see chapter 35). (e) Rhizobia change shape and are now called bacteroids. Bacteroid produce an oxygen-binding heme group that combines with a globin group from the pea to make leghemoglobin. Leghemoglobin gives a pinkish tinge to the nodule, and its function is much like that of hemoglobin, bringing oxygen to the rapidly respiring bacteroids. (f) Bacteroids produce nitrogenase and begin fixing atmospheric nitrogen for the plant's use. In return, the plant provides organic compounds.

FIGURE 38.9
Asian pitcher plant, *Nepenthes*.   Insects enter this carnivorous plant and are trapped and digested. Complex communities of invertebrate animals and protists inhabit the pitchers.

Nutritional Adaptations

Carnivorous Plants

Some plants are able to obtain nitrogen directly from other organisms, just as animals do. These carnivorous plants often grow in acidic soils, such as bogs, that lack organic nitrogen. By capturing and digesting small animals directly, such plants obtain adequate nitrogen supplies and thus are able to grow in these seemingly unfavorable environments. Carnivorous plants have modified leaves adapted to lure and trap insects and other small animals. The plants digest their prey with enzymes secreted from various types of glands.

Pitcher plants attract insects by the bright, flowerlike colors within their pitcher-shaped leaves and perhaps also by sugar-rich secretions (figure 38.9). Once inside the pitcher, insects slide down into the cavity of the leaf, which is filled with water and digestive enzymes. This passive mechanism provides pitcher plants with a steady supply of nitrogen.

The Venus flytrap (*Dionaea muscipula*), which grows in the bogs of coastal North and South Carolina, has three sensitive hairs on each side of each leaf that, when touched, trigger the two halves of the leaf to snap together (figure 38.10). Once the Venus flytrap enfolds a prey item within a leaf, enzymes secreted from the leaf surfaces digest the prey. These flytraps use a growth mechanism to close and open. As a result, they can only open and close a limited number of times.

FIGURE 38.10
Venus flytrap, *Dionaea muscipula*.   If this fly touches two of the trichomes (hairs) on this modified leaf in a short time span, the trap will close. The plant will secrete digestive enzymes that release nitrogen compounds from the fly, which will then be absorbed by the flytrap.

In the sundews, glandular trichomes secrete both sticky mucilage, which traps small animals, and digestive enzymes; they do not close rapidly. Venus flytraps and sundews share a common ancestor that lacked the snap-trap mechanism characteristic of flytraps. However, flytraps have an even closer relative, *Aldrovanda vesicular*, the aquatic waterwheel. The waterwheel is a rootless plant that uses trigger hairs and a snap-trap mechanism like that of the Venus flytrap to capture and digest small animals (figure 38.11). Molecular phylogenetic studies indicate that Venus flytraps are sister species with sundews, forming a sister clade. It appears that the snap-trap mechanism evolved only once in descendants of a sundew ancestor.

FIGURE 38.11
Aquatic waterwheel, *Aldrovanda vesicular.* This close relative of the Venus flytrap snaps shut to capture and digest small aquatic animals. This aquatic plant's ancestor was a land dweller.

FIGURE 38.12
Indian pipe, *Hypopitys uniflora.* This plant is called a saprophyte because it lacks chlorophyll and depends completely on decaying organic matter for all its nutrients. Indian pipes are frequently found in the forests of the northeastern United States.

The waterwheel's ancestor must have been a terrestrial plant that made its way back into the water.

Bladderworts (*Utricularia*) are aquatic, but appear to have different origins than the waterwheel, as well as a different mechanism for trapping small animals for lunch. Small animals are swept into their bladderlike leaves by the rapid action of a springlike trapdoor; then the leaves digest these animals.

Mycorrhizae

Nitrogen is not the only nutrient that is difficult for plants to obtain without assistance. While symbiotic relationships with nitrogen-fixing bacteria are rare, symbiotic associations with mycorrhizal fungi are found in about 90% of the vascular plants. These fungi have been described in detail in chapter 30. In terms of plant nutrition, it is important to recognize the significant role these organisms play in enhancing phosphorus transfer to the plant. The uptake of some of the micronutrients is also enhanced. Functionally, the mycorrhizae extend the surface area of nutrient uptake substantially.

Parasitic Plants

Parasitic plants come in photosynthetic and nonphotosynthetic varieties. In total, at least 3000 types of plants are known to tap into the nutrient resources of other plants. Adaptations include structures that tap into the vascular tissue of the host plant so that nutrients can be siphoned into the parasite. One example is dodder, which looks like brown twine wrapped around its host. Dodder lacks chlorophyll and relies totally on its host for all its nutritional needs. Indian pipe also lacks chlorophyll and hooks into host trees through the fungal hyphae of the host's mycorrhizae (figure 38.12).

Carnivorous plants obtain nutrients, especially nitrogen, directly by capturing and digesting insects and other organisms, while parasitic plants prey upon other plants. Mycorrhizal fungi help plants obtain phosphorus and other nutrients from the soil.

38.4 Plants can remove harmful chemicals from the soil.

Phytoremediation
Trichloroethylene

Trichloroethylene (TCE) is a solvent that has been widely used as spot remover in the dry-cleaning industry, for degreasing engine turbines, as an ingredient in paints and cosmetics, and even as an anesthetic in human and veterinary medicine. Unfortunately, TCE is also a confirmed carcinogen, and chronic exposure can damage the liver. In 1980, the Environmental Protection Agency (EPA) established a Superfund to clean up contamination in the United States. Forty percent of all sites funded by the Superfund include TCE contamination. How do you clean up 1900 hectares of soil in a Marine Corps Air Station in Orange County, California, that contain TCE once used to clean fighter jets? Landfills can isolate, but not eliminate, this volatile substance. Burning eliminates it from the site, but may release harmful substances into the atmosphere. A promising approach is to use plants to remove TCE from the soil by exploiting their mechanisms for nutrient uptake. This process of removing contamination from soil or water using plants is called **phytoremediation** (figure 38.13).

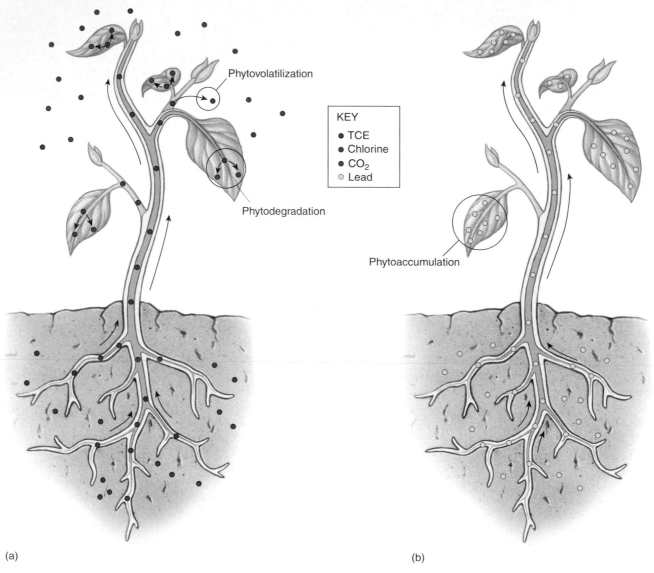

KEY
- TCE
- Chlorine
- CO$_2$
- Lead

FIGURE 38.13
Phytoremediation. Plants can use the same mechanisms to remove both nutrients and toxins from the soil. (*a*) TCE can be taken up by plants and degraded into CO$_2$ and chlorine before being released into the atmosphere. This process is called *phytodegradation*. Some of the TCE moves so rapidly through the xylem that it is not degraded before it is released through the stomata as a gas in a process called *phytovolatilization*. (*b*) Other toxins, including heavy metals such as lead, can be taken up by plants, but not degraded. Such *phytoaccumulation* is particularly effective in removing toxins if they are stored in the shoot, where they can more easily be harvested.

Phytoremediation can work in a number of ways. Plants may secrete a substance from their roots that breaks down the contaminant. More often, the harmful chemical enters the roots and preferably is transported to the shoot system, which is easier to remove from the site. Some substances are simply stored by the plant; later, the plant material is harvested, dried, and removed to a storage site. An even more successful strategy is for the plant to break down the contaminant into nontoxic by-products. Poplar trees (genus *Populus*) may provide just such a solution for TCE-contaminated sites (figure 38.14). Poplars naturally take up TCE from the soil and metabolize it into CO_2 and chlorine. Other plant species can break down TCE, but poplars have the advantage of size and rapid transpiration. A five-year-old poplar can move between 100 and 200 liters of water from its roots out through its leaves in a day. A plant that transpires less would not be able to remove as much TCE in a day.

Although removing TCE with poplar trees sounds like the perfect solution, this method has some limitations. Not all the TCE is metabolized. Given the rapid rate of transpiration in poplar, some of the TCE, which is volatile, enters the atmosphere via the leaves. Once in the air, TCE has a half-life of nine hours (half of it will break down into smaller molecules every nine hours). Clearly, more risk assessment is needed before poplars are planted on every TCE-containing Superfund site. The TCE that remains in the plant is metabolized quickly, and it is possible that the wood could be used after remediation is complete. It has been suggested that any remaining TCE would be eliminated if the wood were processed to make paper. As with any phytoremediation plan, it is both important and challenging to estimate how much of a contaminant can be removed from a site by plants. Genetically modified poplars have been shown to metabolize about four times as much TCE as nonmodified poplars, so perhaps greater metabolic rates can be obtained. How do you weigh the efficiency of the genetically modified poplar against possible risks, if any?

Trinitrotoluene

In addition to volatile chemicals such as TCE, phytoremediation also holds promise for other environmental contaminants, including the explosive TNT (trinitrotoluene) and heavy metals. TNT is a solid, yellow material that was used widely in grenades and bombs until 1980. Contamination is found around factories that made TNT. In some places, there is enough TNT in the soil to detonate, and thus incineration is not a viable option for removing TNT from most sites. Another issue is that TNT can seep into the groundwater; this is a matter of concern because TNT is carcinogenic and associated with liver disease. TNT tends to stay near the top of the soil and to wash away quickly. Bean (*Phaseolus vulgaris*), poplar, and the aquatic parrot feather (*Myriophyllum spicatum*) can uptake and degrade low levels of TNT, but at higher concentrations, TNT is toxic to these plants.

Heavy Metals

Heavy metals, including arsenic, cadmium, and lead, persist in soils and are toxic to animals in even small quantities. Many plants are also susceptible to heavy metal toxicity, but species near mines have evolved strategies to partition certain heavy metals from the rest of the plant (see figure 38.13b). Estimated costs for phytoremediation are 50 to 80% lower than cleanup strategies that involve digging and dumping the contaminated soil elsewhere. For effective phytoextraction of heavy metals from the soil, the metal must be transported to the leaves of the plant, which can be harvested and removed from the site. Four hundred species of plants have been identified that have the ability to hyperaccumulate toxic metals from the soil. For example, *Brassica juncea* (a relative of broccoli and mustard plants) is especially effective at hyperaccumulating lead in the shoots of the plant. Unfortunately, *B. juncea* is a small, slow-growing plant, and eventually it becomes saturated with lead.

How would lead or cadmium travel from the soil into the leaves of a plant? There are some hints that metal transporters exist that load the metal in the soil into the xylem. Citrate, mentioned earlier, can increase the rate of metal transport in xylem. The metals are sequestered inside vacuoles in the leaves. Trichomes, which are modified leaf epidermal cells, can sequester both lead and cadmium. These hyperaccumulating plants are not a panacea for metal-contaminated soil because of concern that animals might move into the site and graze on lead- or cadmium-enriched plants. Harvesting and consolidating dried plant material is not a simple matter, but still phytoremediation is a promising technique.

FIGURE 38.14
Phytoremediation for TCE. The U.S. Air Force is testing phytoremediation technology to clean up TCE at a former Air Force base in Forth Worth, Texas.

(a)

(b)

FIGURE 38.15

Aznalcóllar mine spill. (*a*) When the dike of a holding lagoon for mine waste broke, 5 million cubic meters of black sludge containing heavy metals was released into a national park and (*b*) the Guadiamar River. (*c*) Large amounts of sludge were removed mechanically, and phytoremediation appears to be a promising solution for the remaining heavy metals.

(c)

Phytoremediation may be a solution to the pollution resulting from a 1998 accident at the Aznalcóllar mine in Spain. A dam that contained the sludge from the mining operation broke, releasing 5 million cubic meters of sludge, composed of arsenic, cadmium, lead, and zinc, over 4300 hectares of land (figure 38.15). Much of the sludge was physically removed and dumped into an open mine pit. Phytoremediation solutions are being sought for the remaining contaminated soil. Since the original spill, three plant species with the potential to hyperaccumulate some of the metals have begun growing in the area. These plants are fairly large and can accumulate a substantial amount of metal. They offer the advantage of being native species, thus reducing the dangers of introducing a nonnative, potentially invasive species to clean up the spill.

> **Phytoremediation has the potential to be a safe and cost-effective means of removing toxic metals and organic compounds from water and soil. Plants that can efficiently degrade the toxin and have a large biomass are particularly effective phytoremediators.**

38.1 Plants require a variety of nutrients in addition to the direct products of photosynthesis.

Plant Nutrients

- The major source of plant nutrition is fixation of atmospheric carbon dioxide into simple sugar, using energy from the sun. (p. 782)
- Plants require nutrients in both large quantities (macronutrients) and small quantities (micronutrients). (p. 782)
- There are nine macronutrients: carbon, oxygen, hydrogen, nitrogen, potassium, calcium, phosphorus, magnesium, and sulfur. (p. 782)
- There are seven micronutrients: chlorine, iron, manganese, zinc, boron, copper, and molybdenum. (p. 782)

Soil

- Soil is composed of a mixture of sand, rocks, clay, silt, humus, minerals, organic matter, and living organisms. (p. 784)
- About half of the total soil volume is occupied by spaces or pores, which may be filled with either air or water. (p. 784)
- Natural communities recycle nutrients, but cultivated communities are much more exposed to erosion, and thus often must be supplied with mineral nutrients. (p. 784)
- Other solutions to nutrient loss are crop rotation, planting legumes with nitrogen-fixing bacteria, allowing fields to periodically lie fallow, and plowing-under plant material. (p. 784)
- Nitrogen (N), phosphorus (P), and potassium (K) are the most important minerals that usually need to be added to soils. (p. 784)
- Organic fertilizers build up humus, but do not supply any elements to plants that inorganic fertilizers cannot supply. (p. 784)

38.2 Global change could alter the balance among photosynthesis, respiration, and use of nutrients acquired through the soil.

Global Change

- Carbon dioxide may be at its highest concentration in the atmosphere in 20 million years. (p. 785)

- As carbon dioxide levels increase, less nitrogen and other macronutrients may be found in plant leaves, thus causing herbivores to eat more biomass to obtain the same level of nutrients. (p. 786)

38.3 Some plants have novel strategies for obtaining nutrients.

Nitrogen Fixation

- Plants need ammonia (NH_3) to build amino acids, but most of the nitrogen in the atmosphere exists as N_2, and plants lack the biochemical pathways necessary to convert gaseous nitrogen to ammonia. (p. 787)
- Some bacteria live in close association with plant roots and can fix atmospheric nitrogen. (p. 787)

Nutritional Adaptations

- Some plants can also obtain nitrogen directly from other organisms by adopting a carnivorous feeding strategy. (p. 788)
- Symbiotic associations with mycorrhizal fungi are found in about 90% of vascular plants. (p. 789)
- Parasitic plants exist in photosynthetic and nonphotosynthetic varieties. (p. 789)

38.4 Plants can remove harmful chemicals from the soil.

Phytoremediation

- Phytoremediation is the process of removing contamination from soil or water using chemicals produced by plants. Examples are trichloroethylene (TCE), trinitrotoluene (TNT), and several heavy metals, including arsenic, cadmium, and lead. (pp. 790–791)
- The estimated costs of phytoremediation are 50–80% lower than the costs of more conventional cleanup strategies. (p. 791)

Self Test

1. Phosphate uptake by roots involves
   a. a PO_4 channel in the endoderm of the root.
   b. soil bacteria to make the phosphate bioaccessible.
   c. symbiotic fungal hyphae to increase the effective surface area of the root.
   d. root nodules fixing phosphate from the air.

2. In an attempt to create a nitrogen-fixing bioreactor, you set out to culture *Rhizobium* from the root nodules of pea plants. After months of work, you are still having trouble getting the *Rhizobium* to grow, let alone fix nitrogen. What might explain this failure?
   a. *Rhizobium* only grows in association with plant root hairs.
   b. *Rhizobium* from root nodules lacks a cell wall, so it will not grow in culture.
   c. You accidentally put a fungicide in your growth medium that kills all of the *Rhizobium*.
   d. *Rhizobium* from root nodules lacks chromosomal DNA, so it will not grow in culture.

3. You are performing an experiment to determine the nutrient requirements for a newly discovered plant and find that for some reason your plants die if you leave boron out of the growth medium but do fine with as low as 5 parts per million in solution. This suggests that boron is
   a. an essential macronutrient.
   b. a nonessential micronutrient.
   c. an essential micronutrient.
   d. a nonessential macronutrient.

4. You are setting up some planters to grow flowers in your apartment and discover that you bought sand rather than potting soil. You decide to go ahead and plant the flowers in the sand to save yourself a trip back to the store. Much to your dismay, all of your plants die! How can you explain this failure?
   a. Sand does not pack tightly enough around the roots to hold the plants upright, so they fall over and die.
   b. Sand does not hold enough water to sustain plant metabolism, so all of the plants die from desiccation.
   c. Sand does not contain enough nutrient material to sustain plant growth, so the plants die from starvation.
   d. Sand is too abrasive and damages the roots as they grow through it, so the plants die.

5. If you discover a purely white plant growing on the forest floor, you can safely assume that it is a
   a. mutant plant that uses a colorless pigment to carry out photosynthesis.
   b. parasitic plant that acquires its macronutrients from other plants.
   c. plant that only carries out the dark reactions of photosynthesis and therefore doesn't need green pigments.
   d. plant that does not require macronutrients and therefore does not need photopigments.

6. Growing peas that are defective in flavonoid production would
   a. prevent bioaccumulation of potassium in the roots.
   b. prevent bioaccumulation of nitrogen in the roots.
   c. prevent bioaccumulation of phosphorus in the roots.
   d. prevent the production of flavor molecules in the seeds so that they would be unpalatable.

7. Feeding your Venus flytrap a common brand of all-purpose plant fertilizer would likely cause
   a. it to die from nitrogen overload.
   b. its traps to become large enough to capture small mammals.
   c. no change, because these plants can only use nitrogen from insects.
   d. its traps to fall off, because it would not need to acquire nitrogen from insects.

8. If you were asked how to clean up a trichloroethylene (TCE) spill without having to resort to burning or other chemical methods, how would you do it?
   a. Plant poplar trees to phytoremediate the soil.
   b. Plant bean plants to replace the TCE with fixed nitrogen.
   c. Plant *Brassica* plants to phytoaccumulate the TCE.
   d. Plant Indian pipe because it is not adversely affected by TCE in the soil.

Test Your Visual Understanding

1. If you found a grove of poplar trees like that shown in this image, what might you conclude about the area?

Apply Your Knowledge

1. Match each of the following nutrients with its appropriate function in plants, and discuss whether it is a macro- or micronutrient.
   a. carbon   ___ cell wall formation
   b. nitrogen   ___ nucleic acid formation
   c. phosphorus   ___ amino acid production
   d. iron   ___ nitrogen fixation
   e. molybdenum   ___ chlorophyll production

2. If you were to eat one ton (1000 kg) of potatoes, approximately how much of each of the following minerals would you eat? (*Hint:* Assume that the proportion of these nutrients in potatoes is the same as the average for other plants.)
   a. copper
   b. zinc
   c. potassium
   d. iron

3. Using what you know about phytoremediation, design a strategy to speculate for gold that would not require any digging or disruption of the soil.

39

Plant Defense Responses

Concept Outline

39.1 Morphological and physiological features protect plants from invasion.

Plants Under Attack. The dermal tissue system affords plants some protection against viruses, bacteria, fungi, animals, and other plants.

39.2 Some plant defenses act by poisoning the invader.

Chemical Toxins. Plants have a storehouse of secondary metabolites, chemicals that can drug or kill invaders.

Secondary Metabolites and Humans. Some secondary metabolites have pharmaceutical value for humans.

39.3 Some plants have coevolved with bodyguards.

Animals That Protect Plants. Using morphological and chemical signaling, plants have coevolved with animals that protect them from other pests.

39.4 Systemic responses also protect plants from invaders.

Wound Responses. Generalized responses to wounding result in the production of proteinase inhibitors that interfere with the digestion of an herbivore.

Pathogen-Specific Responses. Viruses, bacteria, fungi, and insects can have genes coding for molecules that bind to specific receptor molecules in a plant to trigger a systemic response to the pathogen.

FIGURE 39.1
Under attack. This fungus (*Botrytis ciberea*) is exploiting the rich source of sugar in these grapes.

Plants are constantly under attack by viruses, bacteria, fungi, animals, and even other plants (figure 39.1). An amazing array of defense mechanisms has evolved to block or temper an invasion. Many plant–pest relationships undergo coevolution, with the plant winning at some times and the pest winning with a new offensive adaptation at other times. The first line of plant defense is thick cell walls covered with a strong cuticle. Bark, thorns, and even trichomes can deter a hungry insect. When that first line of defense fails, a chemical arsenal of toxins is waiting. Many of these molecules have no effect on the plant, but are modified by microbes in the intestine of an herbivore into a poisonous compound. Maintaining a toxin arsenal is energy-intensive; thus, an alternative means of defense uses induced responses to protect and prevent future attacks.

39.1 Morphological and physiological features protect plants from invasion.

Plants Under Attack

There are no tornado shelters for trees. Storms and changing environmental conditions provide life-threatening challenges for plants. Structurally, trees can withstand high winds and the weight of ice and snow, but there are limits. Winds can knock down a tree, and herbivores can clip the shoot off a small plant. Axillary buds give many plants a second chance as they grow out and replace the lost shoot (figure 39.2).

While abiotic factors such as weather present genuine threats to a plant, even greater daily threats exist in the form of viruses, bacteria, fungi, animals, and other plants. These enemies can tap into the nutrient resources of plants or use their DNA-replicating mechanisms to self-replicate. Some invaders kill the plant cells immediately, leading to necrosis (brown, dead tissue). Some insects tap into the phloem of a plant seeking carbohydrates, but leave behind a hitchhiking virus or bacterium. The threat of these invaders is reduced when their own natural predators are around. One of the greatest problems with invasive species such as the alfalfa plant bug (figure 39.3) is its lack of natural predators.

The first defense all plants have is the dermal tissue system. Layers of lipid material protect exposed plant surfaces from water loss and attack. Aboveground plant parts are covered with cutin. Many long-chain fatty acids are linked together to make this macro molecule. Suberin, another type of linked fatty acid chains is found in cell walls of subterranian plant organs. Epidermal cells throughout the plant secrete wax which is a mixture of hydrophobic lipids. Silica, trichomes, bark, and even thorns also protect the nutrient-rich plant interior.

Unfortunately, these defenses can be penetrated in many ways. Mechanical wounds leave an open passageway through which microbial organisms can penetrate. Parasitic nematodes use their sharp mouthparts to get through the plant cell wall. They either trigger the plant cells to divide, forming a tumorous growth, or they attach to a single plant cell, which enlarges and transfers carbohydrates from the plant to the hungry nematode (figure 39.4).

Fungi strategically find the weak spot in the dermal system, the stomatal openings, to enter the plant. Some fungi that have coevolved with a monocot having evenly spaced stomata appear to measure distance to locate stomatal openings. Figure 39.5 shows the phases of fungal invasion, which can include the following:

1. Windblown spores land on leaves. A germ tube emerges from the spore. Host recognition is necessary for the infection to proceed.
2. The spore germinates and forms an adhesion pad, allowing it to stick to the leaf.

FIGURE 39.2
Saved from a hungry deer. These trees have an odd shape because deer chewed back the tender shoot tips during a long, cold winter. Fortunately, some axillary buds grew out and produced new leaves in the spring.

FIGURE 39.3
Alfalfa plant bug. This invasive species is an agricultural nightmare because it arrived without any natural predators.

FIGURE 39.4
Nematodes attack the roots of crop plants. (*a*) A nematode breaks through the epidermis of the root. (*b*) Root-knot nematodes form tumors on roots.

3. Hyphae grow through cell walls and press against the cell membrane.
4. Hyphae differentiate into specialized structures called haustoria. They expand, surrounded by cell membrane, and nutrient transfer begins.

Mutualistic and parasitic relationships are often just opposite sides of the evolutionary coin. In chapters 30 and 38, we saw how mychorrhizal fungi use a mechanism similar to the one just described to the mutual benefit of both the plant and the fungus.

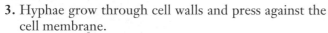

The dermal tissue system is the first line of defense, but invaders have evolved numerous strategies to overcome this barrier.

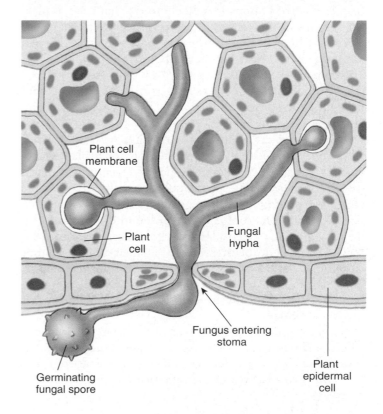

FIGURE 39.5
Fungi sneak in through stomata. Fungal hyphae penetrate cell walls, but not plasma membranes. The close contact between the fungal hyphae and the plant plasma membrane allows for ready transfer of plant nutrients to the fungus.

39.2 Some plant defenses act by poisoning the invader.

Chemical Toxins

Plants can poison their enemies in many ways. Over 3000 species of plants produce cyanide-containing compounds called cyanogenic glycosides that break down into cyanide when ingested. Cyanide stops electron transport, blocking cellular respiration. Cassava, a major food staple for many Africans, is filled with *cyanogenic glycosides* (specifically, manihotoxins) in the outer layers of the edible root. Unless these outer layers are scrubbed off, the cumulative effect of eating primarily cassava can be deadly.

Many plants are filled with toxins that kill herbivores or, at the very least, make them quite ill. How did the biosynthetic pathways that produce these toxins evolve? Growing evidence indicates that the metabolic pathways needed to sustain life in plants have taken some evolutionary side trips, leading to the production of a stockpile of chemicals known as secondary metabolites. Many of these secondary metabolites affect herbivores as well as humans (table 39.1). *Alkaloids*, including caffeine, nicotine, cocaine, and morphine, can affect multiple cellular processes; if a plant cannot kill its enemies, it can overstimulate them with caffeine or sedate them with morphine. The tobacco hornworm, *Manduca sexta*, can level a field of tobacco (figure 39.6); however, wild species of tobacco appear to have elevated levels of nicotine that are lethal to tobacco hornworms. *Tannins* bind to proteins and inactivate them; for example, some act by blocking enzymes from digesting proteins. An insect that gets sick from a dose of tannins is unlikely to try that type of plant for lunch another time. Small doses, equivalent to the amount a human gets from eating apples or blackberries, are unlikely to cause any major digestive difficulties. This example does, however, point out that humans avoid many of the cumulative toxic effects of secondary metabolites by eating a varied diet. *Oils*, including peppermint and sage, repel insects.

Why don't the toxins kill the plant? There are two ways this is prevented. One strategy is for a plant to sequester a toxin in a membrane-bound structure so that it can do no harm. The second solution is for the plant to produce a compound that is not toxic unless it is metabolized, often by microorganisms, in the intestine of an animal. Cyanogenic glycosides are a good example of the latter solution. The plant produces a sugar-bound cyanide compound that does not affect electron transport chains. Once an animal ingests cyanogenic glycoside, the compound is enzymatically broken down, releasing toxic hydrogen cyanide.

Some chemical toxins protect plants from other plants. **Allelopathy** occurs when a chemical signal secreted by the roots of one plant blocks the germination of nearby seeds or inhibits the growth of a neighboring plant. This strat-

FIGURE 39.6
Herbivores can kill plants. Tobacco hornworms consume huge amounts of tobacco leaf tissue.

FIGURE 39.7
Black walnuts are allelopaths. Alfalfa seedlings die when their roots come in contact with the roots of a black walnut tree.

egy minimizes shading, maximizes the ability of a plant to use radiant sunlight for photosynthesis, and minimizes competition for nutrients in the soil. Black walnut trees (*Juglans nigra*) are a good example. Very little vegetation will grow under a black walnut tree because of allelopathy (figure 39.7).

Coevolution has led to defenses against some plant toxins. A tropical butterfly, *Helioconius sara*, can sequester the cyanogenic glycosides it ingests from its sole food source, the passion vine. Even more intriguing is a biochemical pathway that allows the butterfly to safely break down cyanogenic glycosides and use the released nitrogen.

Modified metabolic pathways lead to the production of toxins that protect plants from herbivores.

Table 39.1 Secondary Metabolites

Compound	Source	Structure	Effect on humans
Morphine (alkaloid)	Opium poppy *Papaver somniferum*		Narcotic pain killer
Quinine (alkaloid)	Quinine bark *Cinchona officinalis*		Antimalarial drug
Taxol (terpenoid)	Pacific Yew *Taxus brevifolia*		Anticancer drug
Genistein (phytoestrogen)	Soybean *Glycine max*		Estrogen mimic
Manihotoxin (cyanogenic glycoside)	Cassava *Manihot esculenta*		Metabolized to release lethal cyanide

Secondary Metabolites and Humans

Not only have humans been inadvertently poisoned by plants, but throughout much of human history, they have also been intentionally poisoned by other humans using plant products. Socrates died after he drank a hemlock extract containing an alkaloid that paralyzes motor nerve endings.

Major research efforts on plant secondary metabolites are in progress because of their potential benefits, as well as dangers, to human health (see table 39.1). You are likely to encounter discussion about soy products even in the popular press. Soy, like many plants, contains **phytoestrogens.** Phytoestrogens are very similar to the human hormone estrogen, and some of them can bind to the estrogen receptor. (For more information on estrogens, refer to chapter 47.) In soy, genistein is one of the major phytoestrogens. Comparative studies between Asian populations that consume large amounts of soy and populations with lower dietary intake of soy products are raising intriguing questions and some conflicting results. For example, the lower rate of prostate cancer in Asian males might be accounted for by the down-regulation of androgen and estrogen receptors by a phytoestrogen. Soy is being marketed as a means for minimizing menopausal symptoms caused by declining estrogen levels in older women. In humans, dietary phytoestrogens cross the placenta and are found in the amniotic fluid during the second trimester of pregnancy. Questions have been raised about the effect of phytoestrogens on developing fetuses and even on babies who consume soy-based formula because of allergies to cow's milk formula. Because hormonal signaling is so complex, much more research is needed to fully understand how phytoestrogens affect human physiology and development.

Taxol, a secondary metabolite found in the Pacific yew, is effective in fighting cancer, especially breast cancer. The discovery of taxol's pharmaceutical value raised an environmental challenge. The very existence of the Pacific yew was being threatened as the shrubs were destroyed so that taxol could be extracted. Fortunately, it became possible to synthesize taxol synthetically. Taxol is not an isolated case of drug discovery in plants. The hidden pharmaceutical value of many plants may lead to increased conservation efforts to protect plants that have the potential to make contributions toward human health. Although the plant pharmaceutical industry is growing, it is certainly not a new field.

The Incas of Peru were treating malaria with a drink (tonic water) made from the bark of *Cinchona* trees by the 1600s. Malaria is caused by four types of human malaria parasites, *Plasmodium vivax*, *P. malariae*, *P. ovale* and *P. falciparum*, which are carried by female *Anopholes* mosquitoes. *Plasmodium falciparum* is the most lethal. Symptoms include severe fevers and vomiting. The parasite feeds on red blood cells and death can result from anemia or blocking of blood flow to the brain. By 1820, the active ingredient in the bark of *Cinchona* trees, quinine (see table 39.1), had been identified. In the nineteenth century, British soldiers in India used quinine-containing tonic water to fight malaria. They masked the bitter taste of quinine with gin, creating the first gin and tonic drinks. In 1944 Robert Woodward and William Doering synthesized quinine. Now several other synthetic drugs are available to treat malaria. Exactly how quinine and synthetic versions of this drug family work has puzzled researchers for a long time. Quinine can affect DNA replication. Also, when *P. falciparum* breaks down hemoglobin from red blood cells in its digestive vacuole, an intermediary toxic form of heme is released. Quinine may interfere with the subsequent polymerization of these hemes, leading to a build up of toxic hemes that poison the parasite.

Unfortunately, even today malaria is a major threat to human health with over a million deaths per year. Ninety percent of these deaths are in sub-Saharan Africa. An estimated 300,000,000 individuals are infected. *P. falciparum* strains have acquired resistance to synthetic drugs and quinine is once again the drug of choice in some cases. The recent sequencing of both the *Anopholes* and *P. falciparum* genomes (see chapter 24) may lead to new strategies to combat this disease, but currently insecticide-treated mosquito netting to prevent infection and quinine to treat infections are among the best options.

Herbal remedies have been used for centuries in most cultures. A resurgence of interest in plant-based remedies is resulting in a growing and unregulated industry. Although herbal remedies have great promise, keep in mind that each plant contains many secondary metabolites, many of which have evolved to cause harm to herbivores.

Secondary metabolites have tremendous pharmaceutical potential.

39.3 Some plants have coevolved with bodyguards.

Animals That Protect Plants

Acacia Trees and Ants.   Several species of ants provide small armies to protect some species of *Acacia* from other herbivores. These ants may inhabit a home in an enlarged thorn of the tree. These stinging ants attack other insects (figure 39.8) and sometimes small mammals and epiphytic plants (see chapter 35 for more information on epiphytic plants). Some of the *Acacia* species provide their ants with sugar in nectaries away from the flowers and even with lipid food bodies at the tips of leaves.

The only problem with ants chasing away insects is that acacia trees depend on bees to pollinate their flowers. What keeps the ants from swarming and stinging a bee that stops by to pollinate? Evidence indicates that when a flower opens on an acacia tree, it produces some type of chemical ant deterrent that does not deter the bees. This chemical has not yet been identified.

Parasitoid Wasps, Caterpillars, and Leaves.   Caterpillars fill up on leaf tissue before they metamorphose into a moth or a butterfly. In some cases, proteinase inhibitors in leaves are sufficient to deter very hungry caterpillars. However, another strategy is being used quite successfully. As the caterpillar chews away, a wound response in the plant leads to the release of a volatile compound. This compound wafts through the air, and if a female parasitoid wasp happens to be in the neighborhood, it is immediately attracted to the source. Parasitoid wasps are so named because they are parasitic on caterpil-

lars. The wasp lays her fertilized eggs in the caterpillar that is filling up on the leaf of the plant. These eggs hatch, and the emerging larvae eat the caterpillar, not the leaf (figure 39.9).

Complex coevolution between plants and animals has resulted in associations that protect the plant from other animals.

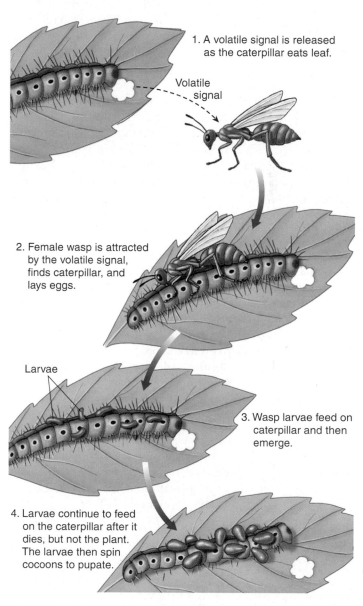

1. A volatile signal is released as the caterpillar eats leaf.

Volatile signal

2. Female wasp is attracted by the volatile signal, finds caterpillar, and lays eggs.

Larvae

3. Wasp larvae feed on caterpillar and then emerge.

4. Larvae continue to feed on the caterpillar after it dies, but not the plant. The larvae then spin cocoons to pupate.

FIGURE 39.9
Parasitoid wasps provide protection from herbivory.   (*1*) A caterpillar eating a leaf triggers a wound response, releasing a volatile compound. (*2*) A wasp is attracted by this compound and lays eggs in the caterpillar. (*3,4*) Wasp larvae consume the caterpillar, but the leaf is unharmed.

FIGURE 39.8
Ants attacking a katydid to protect "their" *Acacia*.   Through coevolution, ants are sheltered by acacia trees and attack otherwise harmful herbivores.

39.4 Systemic responses also protect plants from invaders.

Wound Responses

So far, we have focused mainly on static plant responses to invaders. Other than the induced signal that attracts parasitoid wasps to a munching caterpillar, most of the chemical toxins are maintained at steady-state levels, and morphological structures that help defend plants are part of the normal developmental program. There is an energetic downside to these strategies. If a chemical is needed only when the plant is under siege, resources could be conserved if the response was inducible—that is, if a defense response was launched only when an invader had been recognized. In this section, we explore such defense mechanisms.

Wound responses occur when a leaf is chewed or injured. One outcome is the rapid production of proteinase inhibitors. These chemical toxins aren't in the stockpile of defenses, but are produced in response to wounding. Proteinase inhibitors bind to digestive enzymes in the gut of the inhibitor. The proteinase inhibitors are produced throughout the plant, not just at the wound site. How are cells in distant parts of the plant signaled to produce proteinase inhibitors? In tomato plants, the following sequence of events is responsible for this systemic response (figure 39.10):

1. Wounded leaves produce an 18-amino-acid peptide called *systemin* from a larger precursor protein.
2. Systemin moves through the apoplast (the space between cell walls, not through individual cells) of the wounded tissue and into the nearby phloem. This small peptide-signaling molecule then moves throughout the plant in the phloem.
3. Cells with a systemin receptor bind the systemin, which leads to the production of jasmonic acid.
4. Jasmonic acid signals gene expression, which leads to the production of proteinase inhibitor.

Although we know the most about the signaling pathway involving jasmonic acid, other molecules are involved in wound response as well. For example, salicylic acid, which is the active ingredient in aspirin, and cell wall fragments also appear to be important.

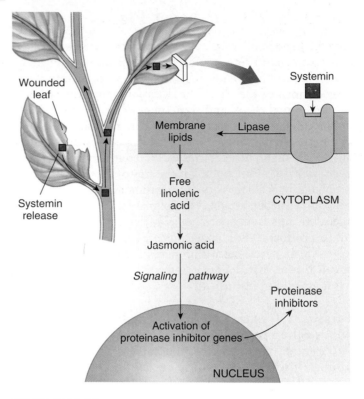

FIGURE 39.10
Wound response in tomato. Wounding a tomato leaf leads to the production of jasmonic acid in other parts of the plants. Jasmonic acid initiates a signaling pathway that turns on genes needed to synthesize proteinase inhibitor.

Mechanical damage also elicits wound responses, which presents a challenge in designing experiments with plants that involve cutting or otherwise mechanically damaging the tissue. It is important to run control experiments to be sure you are answering your question and not observing a wound response.

> Wounding triggers the release of a small peptide signaling molecule called systemin that travels throughout the plant to initiate the production of proteinase inhibitors.

Pathogen-Specific Responses

Recognizing the Invader

Although wound responses are independent of the specific enemy, other responses are triggered by a specific pathogen having a specific allele of a gene. Half a century ago, the geneticist Flor proposed that there is a plant resistance gene (*R*) whose product interacts with that of a pathogen avirulence gene (*avr*). This is called the gene-for-gene hypothesis (figure 39.11), and several pairs of *avr* and *R* genes have been cloned in different species pathogenized by microbes, fungi, and even insects in one case. This research has been motivated partially by the agronomic benefit of identifying genes that can be added to protect crop plants from invaders. Much is now known about the signal transduction pathways that follow the recognition of the pathogen by the *R* gene. These pathways lead to the triggering of the **hypersensitive response (HR),** which leads to rapid cell death around the source of the invasion and also to a longer-term resistance (figure 39.12). A gene-for-gene response does not always occur, but plants still have defense responses to pathogens and also to mechanical wounding. Some of the response pathways may be similar. Also, fragments of cell wall carbohydrates may serve as recognition and signaling molecules.

While our focus so far has been on invaders outside the plant kingdom, more is being learned about how parasitic plants invade other plants. Specific molecules are released from the root hairs of the host that the parasitic plant recognizes and responds to with invasive action. Less is known about the host response, and so far the different defense genes that are activated appear to be ineffective.

Responding to the Invader

When a plant is attacked and a gene-for-gene recognition occurs, the HR leads to very rapid cell death around the site of attack. This seals off the wounded tissue to prevent the pathogen or pest from moving into the rest of the plant. Hydrogen peroxide and nitric oxide are produced and may signal a cascade of biochemical events resulting in the localized death of host cells. These chemicals may also have negative effects on the pathogen, although protective mechanisms have coevolved in pathogens. Other antimicrobial agents produced include the phytoalexins, which are chemical defense agents. A variety of pathogenesis-related genes (*PR* genes) are expressed, and their proteins can function as either antimicrobial agents or signals for other events that protect the plant.

In the case of virulent invaders (no *R* gene recognition), changes in local cell walls at least partially block the pathogen or pest from moving further into the plant. In this case, an HR does not occur, and the local plant cells are not suicidal.

Preparing for Future Attacks

In addition to the HR or other local responses, plants are capable of a systemic response to a pathogen or pest attack.

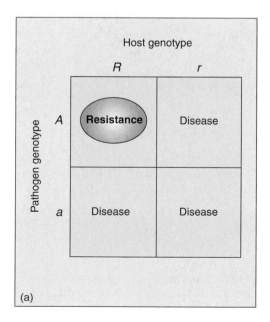

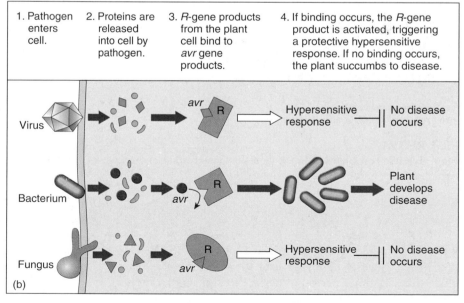

FIGURE 39.11
Gene-for-gene hypothesis.   Flor proposed that pathogens have an avirulence (*avr*) gene that recognizes the product of a plant resistance gene (*R*). If the virus, bacterium, fungus, or insect has an *avr* gene product that matches the *R* gene product, a defense response will occur.

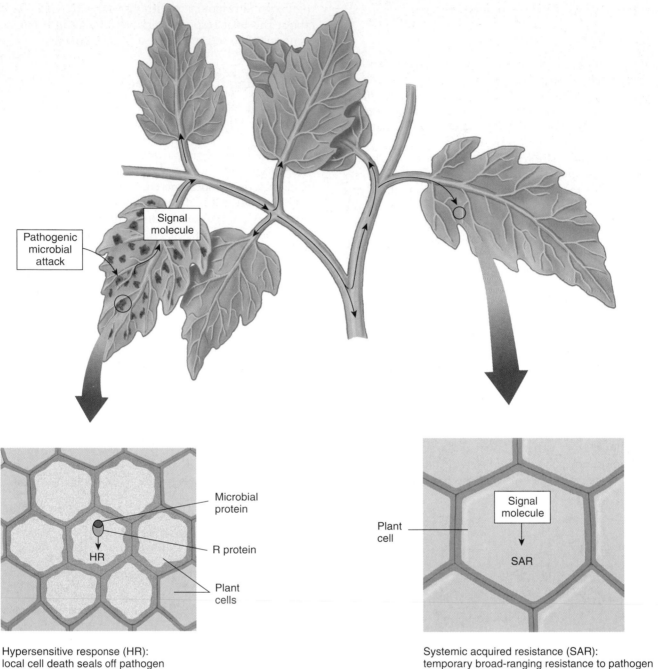

Hypersensitive response (HR):
local cell death seals off pathogen

Systemic acquired resistance (SAR):
temporary broad-ranging resistance to pathogen

FIGURE 39.12

Plant defense responses.   In the gene-for-gene response, a cascade of events is triggered, leading to local cell death (HR) and to longer-term resistance in the rest of the plant (SAR).

This is called a **systemic acquired resistance (SAR)** (figure 39.12). Several pathways lead to broad-ranging resistance that lasts for a period of days. The signals that induce SAR include salicylic acid and jasmonic acid. SAR allows the plant to respond more quickly if it is attacked again. However, this is not the same as the human immune response, in which antibodies (proteins) that recognize specific antigens (foreign proteins) persist in the body. SAR is neither as specific nor as long-lasting.

> **Plants defend themselves from invasion in ways reminiscent of the animal immune system. When an invader is recognized, localized cell death seals off the infected area.**

For interactive testing, visit the Online Learning Center with PowerWeb at www.mhhe.com/Raven7

39.1 Morphological and physiological features protect plants from invasion.

Plants Under Attack

- Because plants are constantly under attack by viruses, bacteria, fungi, animals, and other plants, a wide array of mechanisms has evolved to block such invasions. (p. 796)
- One of the greatest problems with invasive species is the lack of natural predators. (p. 796)
- The first plant defense is the dermal tissue system. (pp. 796–797)
- Fungal invasion can include several phases, such as spores landing on leaves, spore germination, hyphae growing through cell walls, and hyphae differentiating into specialized haustoria. (pp. 796–797)

39.2 Some plant defenses act by poisoning the invader.

Chemical Toxins

- Many plants produce cyanogenic glycosides that break down into cyanide when ingested, which stops electron transport and blocks cellular respiration. (p. 798)
- Alkaloids, including caffeine, nicotine, cocaine, and morphine, can affect a wide variety of cellular processes. (p. 798)
- Tannins bind to, and inactivate, protein. (p. 798)
- Two strategies explain why toxins do not kill the plants. One is for a plant to sequester a toxin in a membrane-bound structure, and the second is to produce a compound that is not toxic unless it is metabolized in the intestine of an animal. (p. 798)
- Allelopathy occurs when a chemical signal secreted by the roots of one plant blocks the germination of nearby seeds or inhibits the growth of a neighboring plant. (p. 798)

Secondary Metabolites and Humans

- Many plant secondary metabolites, including phytoestrogens and taxol, have potential benefits to human health. (p. 800)

39.3 Some plants have coevolved with bodyguards.

Animals That Protect Plants

- Several species of ants provide small armies to protect some acacia trees from herbivory, and receive sugar from nectaries in reward. (p. 801)
- A wound response in some plants releases volatile compounds to attract female parasitoid wasps, which then attack herbivorous caterpillars. (p. 801)

39.4 Systemic responses also protect plants from invaders.

Wound Responses

- Wound responses, which occur when a leaf is chewed or injured, release substances such as proteinase inhibitors. (p. 802)

Pathogen-Specific Responses

- When a plant is attacked and gene-for-gene recognition occurs, the hypersensitive response leads to rapid cell death around the site of attack, which seals off the wounded tissue to prevent the pathogen or pest from moving further into the plant. (p. 803)
- Systemic acquired resistance (SAR) can also be induced by signals such as salicylic acid and jasmonic acid that allow the plant to respond more quickly if it is attacked again. (pp. 803–804)

Self Test

1. If you were a plant pathogen, what would be the first obstacle to invading a host plant that you would have to overcome?
   a. chemical toxins on the surface of a plant
   b. physical barriers on the exterior of a plant
   c. animal guardians of the host plant
   d. immune proteins in the plant tissue

2. Some plants are recognized by fungal pathogens on the basis of their stomatal pores. Which of the following would provide these plants with immunity from fungal infection?
   a. removing all of the stomata from the plant
   b. changing the spacing of stomatal pores in these plants
   c. reinforcing the cell wall in the guard cells of stomatal pores
   d. increasing the number of trichomes on the surfaces of these plants

3. Eating unscrubbed cassava root would likely
   a. lead to indigestion because the skin of the cassava plant is very difficult to digest.
   b. make you sick because the soil on the surface may contain harmful microbes.
   c. make you sick because the skin contains cyanogenic glycosides that would produce cyanide in the digestive tract.
   d. harm your teeth because of the small stones that would be on the surface of the root.

4. You decide to plant a garden in which a beautiful black walnut tree is the centerpiece. You are quite disappointed, however, when none of the seeds you plant around the tree grow. What might explain this observation?
   a. the tree filters out too much light, so the seeds fail to germinate.
   b. the roots of the tree deplete all of the nutrients from the soil, so the new seedlings starve.
   c. the tree produces chemical toxins that prevent seed germination.
   d. the roots deplete all of the water from the soil and thereby prevent seed germination.

5. Vegetarians whose diet consists largely of soy are less likely to develop prostate cancer because
   a. soy contains the anticancer drug taxol.
   b. eating meat increases the probability of developing prostate cancer, so eliminating it from your diet reduces the chances of developing the disease.
   c. soy protein prevents accumulation of the prostate-specific antigen (PSA) associated with prostate cancer.
   d. soy contains a phytoestrogen that may down-regulate estrogen and androgen receptors in males who consume high soy diets.

6. The drink gin and tonic was created by British soldiers in India as a means of
   a. dealing with boredom during the long hours between battles.
   b. increasing revenue by selling the drink in the officer's club.
   c. fighting malaria with the quinine in tonic water.
   d. fighting scurvy with the juniper extract in gin.

7. Tomato plants are not good to eat because
   a. when their tissue is damaged, it generates a foul odor that makes people sick.
   b. they do not contain any useful nutrients for animals.
   c. they contain chemical toxins that will make animals sick.
   d. when their tissue is damaged, they produce proteinase inhibitors that will prevent digestive enzyme function in animals.

8. A plant lacking *R* genes would likely
   a. be unable to carry out photosynthesis.
   b. be susceptible to infection by pathogens.
   c. be susceptible to predation by herbivores.
   d. paralyze animals who ingested it.

9. Both plant and animal immune systems can
   a. develop memory of past pathogens to more effectively deal with subsequent infections.
   b. establish physical barriers to infection.
   c. initiate expression of proteins to help fight the infection.
   d. kill their own cells to prevent spread of the infection.
   e. All of these are true.

Test Your Visual Understanding

1. Label structures *a–e* with the plant from which each comes and the effect each has on humans.

(a)

(b)

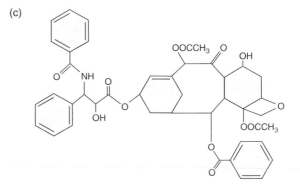

(c)

(d)

(e)

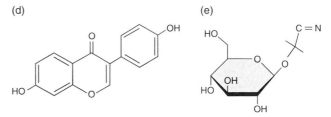

Apply Your Knowledge

1. Herbal medicine has long relied on plant extracts to cure human disease. Find three examples of herbal remedies that have been exploited by the pharmaceutical industry in modern drug production.

2. Diagram the events surrounding infection of a plant by a fungal pathogen. Include strategies used both by the pathogen and by the host plant.

3. If you wanted to find new phytopharmaceuticals that could fight cancer, how would you go about identifying new compounds?

40

Sensory Systems in Plants

Concept Outline

40.1 Plants respond to light.

Photomorphogenesis. Light can initiate developmental pathways, including seed germination and flowering. Uniform red light can trigger photomorphogenesis.

Phototropisms. Phototropisms are directional growth movements affected by unidirectional light, often of specific wavelengths.

40.2 Plants respond to gravity.

Gravitropism. Growth responses to gravity depend on signaling processes that occur after the gravitational field is perceived.

40.3 Plants respond to touch.

Thigmotropism and Thigmonasty. Touch receptors can signal changes in other molecules, leading to the expression of specific genes. Tropic responses depend on the direction of the touch stimulus, while nastic responses do not.

Turgor Movement. Changes in the water pressure within plant cells result in quick and reversible plant movements.

40.4 Water and temperature elicit plant responses.

Dormancy. Dormancy is a response to water, temperature, and light.

Surviving Temperature Extremes. Heat shock proteins protect cells from high temperature damage. Cold temperatures trigger physiological responses that enhance survival.

40.5 The hormones that guide growth are keyed to the environment.

Plant Hormones. Hormones are grouped into seven classes: auxins, cytokinins, gibberellins, brassinosteroids, oligosaccharins, ethylene, and abscisic acid.

FIGURE 40.1
Plant growth is affected by environmental cues. The branches of this fallen tree are growing straight up in response to gravity and light.

All organisms sense and interact with their environments. This is particularly true of plants. Plant survival and growth are critically influenced by abiotic factors, including water, wind, and light. The effect of the local environment on plant growth also accounts for much of the variation in adult form within a species (figure 40.1). In this chapter, we explore how a plant senses such factors and transduces these signals to elicit an optimal physiological, growth, or developmental response. Although responses can be observed on a macroscopic scale, the mechanism of response occurs at the level of the cell. Signals are perceived when they interact with a receptor molecule, causing a shape change and altering the receptor's ability to interact with signaling molecules. Hormones play an important role in the internal signaling that brings about environmental responses and are keyed in many ways to the environment.

40.1 Plants respond to light.

Photomorphogenesis

Several environmental factors, including light, can initiate seed germination, flowering, and other critical developmental events in the life of a plant. **Photomorphogenesis** is nondirectional, light-triggered development. It can result in complex changes in form, including flowering, which is discussed in chapter 41. Red photoreceptors often signal photomorphogenesis, although they can also signal phototropisms. Unlike photomorphogenesis, **phototropisms** (Greek *trope*, "turn") are directional. Both photomorphogenesis and phototropisms compensate for the plant's inability to walk away from unfavorable environmental conditions.

Red Light Receptors

Specific wavelengths of light can change the conformation of light-receptor molecules that initiate biological change. The pigment-containing protein **phytochrome** exists in two interconvertible forms, P_r and P_{fr}. In the first form, phytochrome absorbs red light; in the second, it absorbs far-red light. When P_r absorbs a photon of red light (660 nm), it is instantly converted into P_{fr}, and when P_{fr} absorbs a photon of far-red light (730 nm), it is instantly converted to P_r. P_{fr} is biologically active, and P_r is biologically inactive. In other words, when P_{fr} is present, a given biological reaction that is affected by phytochrome will occur. When most of the P_{fr} has been replaced by P_r, the reaction will not occur (figure 40.2). Although we refer to phytochrome as a single molecule here, several different phytochromes have been identified that appear to have specific biological functions.

The existence of phytochrome was demonstrated in 1959 by Harry A. Borthwick and his collaborators at the U.S. Department of Agriculture. Researchers have since shown that the molecule consists of two parts: a smaller part that is sensitive to light and a larger portion that is a protein. The protein component initiates a signal transduction pathway leading to a particular tropism. Phytochrome is present in all groups of plants and in a few genera of green algae, but not in other protists, bacteria, or fungi Phytochrome systems probably evolved among the green algae and were present in the common ancestor of the plants.

Red Light Responses

Phytochrome is involved in many plant growth responses. For example, seed germination is inhibited by far-red light and stimulated by red light in many plants. Because chlorophyll absorbs red light strongly but does not absorb far-red light, light passing through the green leaves of canopy trees above a seed inhibits seed germination. Consequently, seeds on the ground under deciduous plants, which lose their leaves in winter, are more apt to germinate in the spring after the leaves have decomposed and the seedlings are exposed to direct sunlight. This greatly improves the chances that the seedlings will become established.

Elongation of the shoot in an *etiolated* seedling (one that is pale and slender from having been kept in the dark) is caused by a lack of red light. Such plants become normal when exposed to red light. Etiolation is an energy conservation strategy to help plants growing in the dark reach the light before they die. They don't green up until light becomes available, and they divert energy to internode elongation. The de-etiolated (*det2*) *Arabidopsis* mutant has a poor etiolation response. It does not have elongated internodes, and it greens up a bit in the dark. Apparently, *det2* mutants are defective in an enzyme necessary for brassinosteroid biosynthesis, leading researchers to suspect that brassinosteroids play a role in plant responses to light through phytochrome.

Red and far-red light also signal plant spacing. The closer together plants are, the more likely they are to grow tall and try to outcompete others for the sunshine. Plants somehow measure the amount of far-red light bounced back to them from neighboring plants. If their perception is distorted by putting a light-blocking collar around the stem, the elongation response no longer occurs.

Red light changes the shape of phytochrome and can trigger photomorphogenesis.

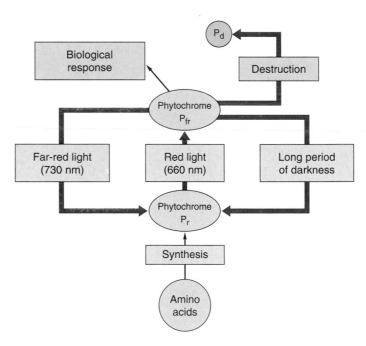

FIGURE 40.2

How phytochrome works. Phytochrome is synthesized in the P_r form from amino acids. When exposed to red light, P_r changes to P_{fr}, the active form that elicits a response in plants. P_{fr} is converted to P_r when exposed to far-red light, and it also converts to P_r or is destroyed in darkness. The destruction product is designated P_d.

Phototropisms

Phototropisms contribute to the variety of branching patterns we see within a species as shoots grow toward light. Tropisms are particularly intriguing because they challenge us to connect environmental signals with cellular perception of the signal, transduction into biochemical pathways, and ultimately an altered growth response. Phototropic responses include the bending of growing stems and other plant parts toward sources of light with blue wavelengths (460 nm range) (figure 40.3). In general, stems are positively phototropic, growing toward a light source, while most roots do not respond to light, or in exceptional cases, exhibit only a weak negative phototropic response. The phototropic reactions of stems are clearly of adaptive value, giving plants greater exposure to available light. They are also important in determining the development of plant organs and, therefore, the appearance of the plant. Individual leaves may display phototropic responses. The position of leaves is important to the photosynthetic efficiency of the plant. A plant hormone called auxin (discussed in section 40.5) is probably involved in most, if not all, of the phototropic growth responses of plants.

The recent identification of blue-light receptors in plants is leading to exciting discoveries of how the light signal can ultimately be connected with a phototropic response. Currently, only the early steps in this signal transduction are understood. We will consider the blue-light receptor phototropin 1 (phot1). Phototropin was identified through the characterization of a nonphototropic mutant. The phot1 protein has two light-sensing regions, and they change conformation in response to blue light. This activates another region of the protein that is a kinase. Both phot1 and a similar receptor, phot2, are receptor kinases unique to plants. A portion of phot1 is a kinase that actually autophosphorylates (figure 40.4). That is, it adds a phosphate group to itself. Refer to chapter 7 for information on the role of phosphorylation in signal transduction. It will be intriguing to watch the story of the phot1 signal transduction pathway unfold, leading to an explanation of why plants grow toward the light.

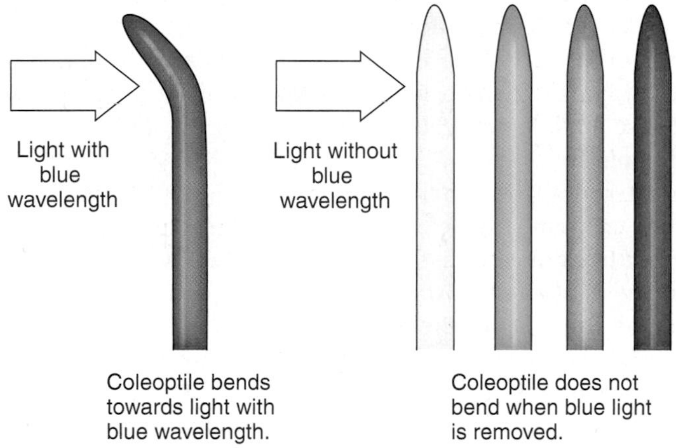

FIGURE 40.3
Phototropism. Oat coleoptiles growing toward light with blue wavelengths. Colors indicate the color of light shining on coleoptiles. Arrows indicate the direction of light.

Although plants cannot move away or toward optimal conditions, they can grow. Phototropisms are growth responses toward a unidirectional, often blue, light source.

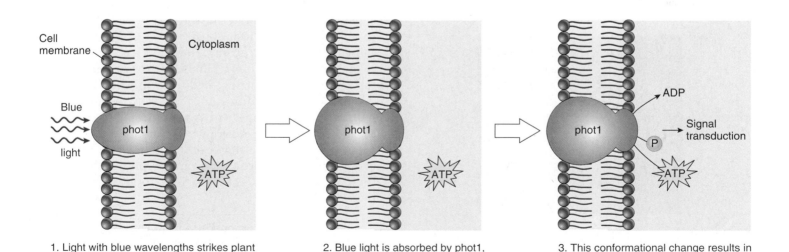

1. Light with blue wavelengths strikes plant cell membrane with phototropin 1 (phot1).

2. Blue light is absorbed by phot1, causing a change in conformation.

3. This conformational change results in autophosphorylation, triggering a signal transduction.

FIGURE 40.4
Blue-light receptor. Blue light activates the light-sensing region of phot1, which in turn stimulates the kinase region of phot1 to autophosphorylate. This is just the first step in a signal transduction pathway that leads to phototropic growth.

40.2 Plants respond to gravity.

Gravitropism

When a potted plant is tipped over, the shoot bends and grows upward. The same thing happens when a storm pushes plants over in a field. These are examples of **gravitropism,** the response of a plant to the gravitational field of the earth (figure 40.5; see figure 40.1). Because plants also grow in response to light, separating out phototropic effects is important in the study of gravitropism.

Gravitropic responses are present at germination when the root grows down and the shoot grows up. Why does a shoot have a *negative gravitropic response* (growth away from gravity), while a root has a positive gravitropic response? The opportunity to experiment on the space shuttle in a gravity-free environment has accelerated research in this area. Auxins play a primary role in gravitropic responses, but they may not be the only way gravitational information is sent through the plant. When John Glenn made his second trip into space, he was accompanied by an experiment designed to test the role of gravity and electrical signaling in root bending. Analysis of gravitropic mutants is also adding to our understanding of gravitropism. Four steps lead to a gravitropic response:

1. Gravity is perceived by the cell.
2. Signals form in the cell that perceives gravity.
3. The signal is transduced intra- and intercellularly.
4. Differential cell elongation occurs between cells in the "up" and "down" sides of the root or shoot.

The first steps in perceiving gravity are being debated. *Amyloplasts*, plastids that contain starch, sink toward the gravitational field and may be involved in sensing gravity. Amyloplasts may interact with the cytoskeleton, but the net effect is that auxin becomes more concentrated on the lower side of the stem axis than on the upper side. The increased auxin concentration on the lower side in stems causes the cells in that area to grow more than the cells on the upper side. The result is a bending upward of the stem against the force of gravity—in other words, a negative gravitropic response. Such differences in hormone concentration have not been as well documented in roots. Nevertheless, the upper sides of roots oriented horizontally grow more rapidly than the lower sides, causing the root ultimately to grow downward; this phenomenon is known as *positive gravitropism.* In shoots, the gravity-sensing cells are in the endodermis. Mutants such as *scarecrow* and *short root* in *Arabidopsis*, which do not undergo normal endodermal development, fail to have a normal gravitropic response. These endodermal cells are the sites of the amyloplasts in the stems.

In roots, the gravity-sensing cells are located in the root cap, and the cells that actually undergo asymmetric growth are in the distal elongation zone, which is closest to the root cap. How the information is transferred over this distance is an intriguing question. Auxin may be involved, but when auxin transport is suppressed, a gravitropic response still occurs in the distal elongation zone. Some type of electrical signaling involving membrane polarization has been hypothesized, and this was tested aboard the space shuttle. So far, the verdict is still not in on the exact mechanism.

It may surprise you to learn that in tropical rain forests, the roots of some plants may grow up the stems of neighboring plants, instead of exhibiting the normal positive gravitropic responses typical of other roots. It appears that the rainwater dissolves nutrients, both while passing through the lush upper canopy of the forest, and subsequently while trickling down the tree trunks. Such water functions as a more reliable source of nutrients for the roots than the nutrient-poor soils in which the plants are anchored. Explaining this in terms of current hypotheses is a challenge. It has been proposed that roots are more sensitive to auxin than are shoots and that auxin may actually inhibit growth on the lower side of a root, resulting in a positive gravitropic response. Perhaps in these tropical plants, the sensitivity to auxin in roots is reduced.

FIGURE 40.5
Plant response to gravity.   This plant (*Zebrina pendula*) was placed horizontally and allowed to grow for seven days. Note the negative gravitational response of the shoot.

Gravitropism, the response of a plant to gravity, generally causes shoots to grow up (negative gravitropism) and roots to grow down (positive gravitropism).

40.3 Plants respond to touch.

Thigmotropism and Thigmonasty

Thigmotropism and thigmonasty are names derived from the Greek root *thigma*, meaning "touch." A **thigmotropism** is a directional growth response of a plant or plant part to contact with an object, animal, plant, or even the wind (figure 40.6). **Thigmonastic** responses are very similar to thigmotropism, except that the direction of the growth response is the same regardless of the direction of the signal. When a tendril makes contact with an object, specialized epidermal cells, whose action is not clearly understood, perceive the contact and promote uneven growth, causing the tendril to curl around the object, sometimes as rapidly as within 3 to 10 minutes. Both auxin and ethylene appear to be involved in tendril movements, and they can induce coiling in the absence of any contact stimulus. In other plants, such as clematis, bindweed, and dodder, leaf petioles or unmodified stems twine around other stems or solid objects. Curiously, the tendrils of some species coil toward the site of the stimulus (thigmotropic growth), while tendrils of other species may always coil clockwise, regardless of the side of the tendril that makes contact with an object.

Arabidopsis is proving valuable as a model system to explore plant responses to touch. A gene has been identified that is expressed in 100-fold higher levels 10 to 30 minutes after touch. Given the value of a molecular genetics approach in dissecting the pathways leading from an environmental signal to a growth response, this gene provides a promising first step in understanding how plants respond to touch.

Perhaps the most dramatic touch response is the snapping of a Venus flytrap (*Dionaea muscipula*) when trigger hairs on the modified leaves are stimulated by touch. As discussed in chapter 38, the modified leaves of the flytrap close in response to a touch stimulus, trapping insects or other potential sources of protein. A flytrap can shut in a mere 0.3 second, and the mechanism of this response has intrigued generations of researchers, including Darwin. If two of the three hairs on a leaf are touched in rapid succession, or if a single hair is touched twice, an electrical signal (action potential) moves through the leaf, leading to an enlargement of the outer cells. The enlarged outer cells of the flytrap cause the trap to close. What is particularly amazing about this response is that the outer cells actually grow. The cell walls soften in response to the electrical signal, and the high pressure (turgor) of the water inside the cells pushes against the softened walls to enlarge the cell. This growth mechanism is distinct from other turgor movements (discussed next) because the water is already within the cell, not transferred into it in response to the electrical signal.

If digestible prey is caught, the trap will open about 24 hours later through the growth of inner cells of the flytrap. This growth response can only be triggered about four times before the leaf dies, presumably because so much energy is required and the individual flytrap (modified leaf) runs out of energy.

Other Tropisms

Thigmotropisms, gravitropisms, and phototropisms are among the best-understood tropisms, but others have been recognized. They include *electrotropism* (response to electricity); *chemotropism* (response to chemicals); *traumotropism* (response to wounding); *thermotropism* (response to temperature); *aerotropism* (response to oxygen); *skototropism* (response to dark); and *geomagnetotropism* (response to magnetic fields). Roots often follow a diffusion gradient of water coming from a cracked pipe and enter the crack. Some investigators call such growth movement *hydrotropism*, but others question whether responses to water and several other "stimuli" are true tropisms.

FIGURE 40.6
Thigmotropism. The thigmotropic response of these twining stems causes them to coil around the object with which they have come in contact.

Turgor Movement

Unlike tropisms, some touch-induced plant movements are not based on growth responses, but instead result from reversible changes in the turgor pressure of specific cells. **Turgor** is pressure within a living cell resulting from diffusion of water into it, as discussed in chapter 37. If water leaves turgid cells (cells having turgor pressure), the cells may collapse, causing plant movement; conversely, water entering a limp cell may also cause movement as the cell once more becomes turgid.

Many other plants, including those of the legume family (Fabaceae), exhibit leaf movements in response to touch or other stimuli. After exposure to a stimulus, the changes in leaf orientation are mostly associated with rapid turgor pressure changes in **pulvini** (singular, *pulvinus*), multicellular swellings located at the base of each leaf or leaflet. When leaves with pulvini, such as those of the sensitive plant (*Mimosa pudica*), are stimulated by wind, heat, touch, or in some instances, intense light, an electrical signal is generated. The electrical signal is translated into a chemical signal, with potassium ions followed by water migrating from the cells in one half of a pulvinus to the intercellular spaces in the other half. The loss of turgor in half of the pulvinus causes the leaf to "fold." The movements of the leaves and leaflets of a sensitive plant are especially rapid;

the folding occurs within a second or two after the leaves are touched (figure 40.7). Over a span of about 15 to 30 minutes after the leaves and leaflets have folded, water usually diffuses back into the same cells from which it left, and the leaf returns to its original position.

Not all turgor movements are stimulated by touch. Some are actually triggered by light. The leaves of some plants may track the sun, with their blades oriented at right angles to it; how their orientation is directed is, however, poorly understood. Such leaves can move quite rapidly (as much as 15° an hour). This movement maximizes photosynthesis and is analogous to solar panels designed to track the sun.

Some of the most familiar reversible changes due to turgor pressure are seen in leaves and flowers that "open" during the day and "close" at night. For example, the flowers of four o'clocks open at 4 P.M., and evening primrose petals open at night. The blades of plant leaves that exhibit such a daily shift in position may not actually fold; instead, their orientation may change as a result of turgor movements. Bean leaves are horizontal during the day when their pulvini are turgid, but become more or less vertical at night as the pulvini lose turgor (figure 40.8). These sleep movements reduce water loss from transpiration during the night, but maximize photosynthetic surface area during the day. In these cases, the movement is closely tied to an internal rhythm.

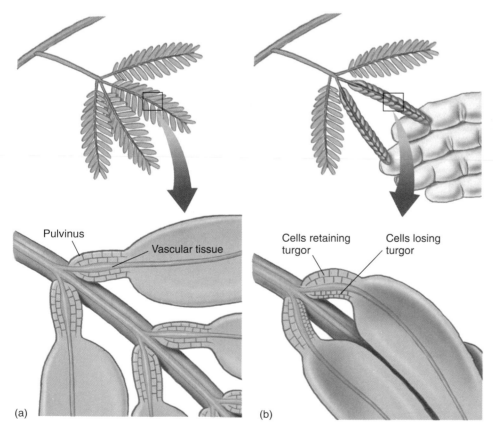

(c)

FIGURE 40.7

Sensitive plant (*Mimosa pudica*).
(*a*) The blades of *Mimosa* leaves are divided into numerous leaflets; at the base of each leaflet is a swollen structure called a pulvinus. (*b*) Changes in turgor cause leaflets to fold in response to a stimulus. (*c*) When leaves are touched (center two leaves), they fold due to loss of turgor.

Circadian Clocks

How do leaves know when to "sleep"? They have endogenous circadian clocks that set a rhythm with a period of about 24 hours (actually, it is closer to 22 or 23 hours). Although shorter and much longer naturally occurring rhythms also exist, circadian rhythms are particularly common and widespread because of the day-night cycle on earth. Jean de Mairan, a French astronomer, first identified circadian rhythms in 1729. He studied the sensitive plant, which in addition to having a touch response, closes its leaflets and leaves at night as does the bean plant just described. When de Mairan put the plants in total darkness, they continued "sleeping" and "waking" just as they had when exposed to night and day. This is one of four characteristics of a circadian rhythm—it must continue to run in the absence of external inputs. In addition, a circadian rhythm must be about 24 hours in duration and can be reset or entrained. (Perhaps you've experienced entrainment when traveling to a different time zone, in the form of jet-lag recovery.) The fourth characteristic is that the clock can compensate for differences in temperature. This is quite unique considering what we know about biochemical reactions because most rates of reactions vary significantly based on temperature. Circadian clocks exist in many organisms and appear to have evolved independently multiple times. The mechanism behind the clocks is not fully understood, but is being actively investigated at the molecular level.

Thigmotropic and thigmonastic movements are growth responses of plants to contact. Turgor movements of plants are reversible and involve changes in the turgor pressure of specific cells. Circadian clocks are endogenous timekeepers that keep plant movements and other responses synchronized with the environment.

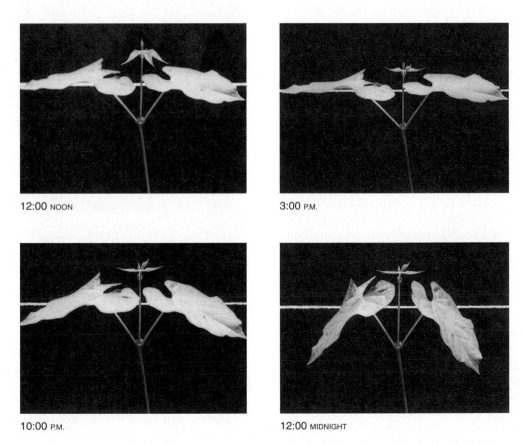

12:00 NOON

3:00 P.M.

10:00 P.M.

12:00 MIDNIGHT

FIGURE 40.8
Sleep movements in bean leaves. In the bean plant, leaf blades are oriented horizontally during the day and vertically at night. The string behind the leaves indicates the horizontal plane.

40.4 Water and temperature elicit plant responses.

Dormancy

Sometimes, modifying the direction of growth is not enough to protect a plant from harsh conditions. The ability to cease growth and go into a dormant stage provides a survival advantage. The extreme example is seed dormancy, but there are intermediate approaches to waiting out the bad times as well. Environmental signals both initiate and end dormant phases in the life of a plant. Temperature, water, and light are examples of such signals.

In temperate regions, we generally associate dormancy with winter, when freezing temperatures and the accompanying unavailability of water make it impossible for plants to grow. During this season, buds of deciduous trees and shrubs remain dormant, and apical meristems remain well protected inside enfolding scales. Perennial herbs spend the winter underground as stout stems or roots packed with stored food. Many other kinds of plants, including most annuals, pass the winter as seeds.

In some seasonally dry climates, seed dormancy occurs primarily during the dry season, often the summer. Rainfalls trigger germination when conditions for survival are more favorable. Annual plants occur frequently in areas of seasonal drought. Seeds are ideal for allowing annual plants to bypass the dry season, when there is insufficient water for growth. When it rains, they can germinate, and the plants can grow rapidly, having adapted to the relatively short periods when water is available. Chapter 36 covered some of the mechanisms involved in breaking seed dormancy and allowing germination under favorable circumstances. These include leaching the chemicals that inhibit germination or mechanically cracking the seed coats, a procedure particularly suitable for promoting growth in seasonally dry areas. Whenever rains occur, they leach out the chemicals from the seed coats, and the hard coats of other seeds may be cracked when they are being washed along temporarily flooded arroyos (figure 40.9).

Seeds may remain dormant for a surprisingly long time. Many legumes (plants of the pea and bean family, Fabaceae) have tough seeds that are virtually impermeable to water and oxygen. These seeds often last decades and even longer without special care; they will eventually germinate when their seed coats have been cracked and water is available. Seeds that are thousands of years old have been successfully germinated!

Favorable temperatures, day length, and amounts of water can release buds, underground stems and roots, and seeds from a dormant state. Requirements vary among species. For example, some weed seeds germinate in cooler parts of the year and are inhibited from germinating by warmer temperatures. Day-length differences near the equator and in more temperate regions have dramatic ef-

(a)

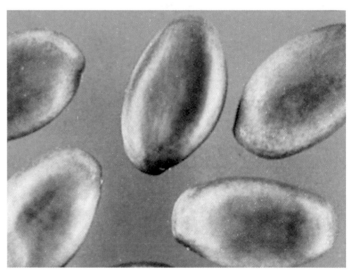
(b)

FIGURE 40.9
Dormancy, a common adaptation to periodic drought.
(*a*) This desert Palo Verde tree (*Cercidium floridum*) has (*b*) tough seeds that germinate only after they are cracked.

fects on dormancy. For example, tree dormancy is common in temperate climates when the days are short but is unusual in tropical trees growing near the equator.

Mature plants may become dormant in dry or cold seasons that are unfavorable for growth. Dormant plants usually lose their leaves, and drought-resistant winter buds are produced. Long unfavorable periods may be bypassed through the production of seeds, which themselves can remain dormant for long periods.

Surviving Temperature Extremes

Sometimes temperatures change rapidly, and dormancy is not possible. How do plants survive temperature extremes? Knowing the lipid composition of a plant's membranes can help predict whether the plant will be sensitive or resistant to chilling. The more unsaturated the membrane lipids are, the more resistant the plant is to chilling. This is because desaturase enzyme levels convert the double bonds in the unsaturated lipids to single bonds. This process lowers the temperature at which the membrane becomes rigid and cannot function properly. Saturated lipids solidify at a higher temperature. *Arabidopsis* plants genetically modified to contain a higher percentage of saturated fatty acids have proved to be more sensitive to chilling.

Even highly unsaturated membranes are not enough to protect plants from freezing temperatures. At freezing, ice crystals form and the cells die from dehydration. Not enough liquid water is available for metabolism. Some plants have the ability to undergo deep **supercooling** and survive temperatures as low as $-40°$ C. Supercooling occurs when ice crystal formation is limited, and the crystals are stored in extracellular spaces where they cannot damage cell organelles. Further, the cells must be able to withstand gradual dehydration.

Acquiring tolerance to chilling or freezing as the temperature drops requires the expression of new genes, including antifreeze proteins that prevent ice crystals from forming. Ice crystals can also form (nucleate) around bacteria naturally found on the leaf surface. Some bacteria have been genetically engineered so that they do not nucleate ice crystals. Spraying leaves with these modified bacteria can provide frost tolerance in some crops.

High temperatures can be equally harmful because proteins denature and lose their function. If temperatures suddenly rise $5°$ to $10°C$, **heat shock proteins (HSPs)** are produced. These proteins can stabilize other proteins so that they don't unfold or misfold at higher temperatures. In some cases, HSPs induced by temperature increases can also protect plants from another stress, including chilling.

Plants can survive otherwise lethal temperatures if they are gradually exposed to increasing temperature. These plants have acquired *thermotolerance*. More is being learned about temperature acclimation by isolating mutants that fail to acquire thermotolerance, including the aptly named *hot* mutants in *Arabidopsis* (figure 41.10). *HOT* codes for a HSP characterization of other *HOT* genes indicates that thermotolerance requires more than the synthesis of HSPs. Some of the other *HOT* genes stabilize membranes and are necessary for protein activity.

Chilling and freezing acclimation rely on high levels of unsaturated fatty acids, supercooling, and synthesis of antifreeze proteins. HSPs stabilize proteins at high temperatures.

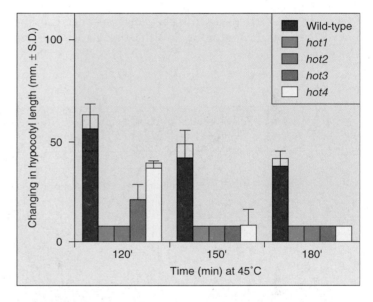

FIGURE 40.10
Temperature acclimation in *Arabidopsis hot* mutants.
Seedlings that grow optimally at 22°C for 90 minutes, followed by a 120 minute recovery at 22°C. Pre-adapted plants were then stressed at 45°C for 120, 150, or 180 minutes before the change in hypocotyl length was recorded.

40.5 The hormones that guide growth are keyed to the environment.

Plant Hormones

Although the initial responses of plants to environmental signals may rely primarily on electrical signaling, longer-term responses that alter morphology rely on complex physiological networks. Many internal signaling pathways involve plant hormones, which are the focus of this section. Hormones are involved in responses to the environment, as well as in internally regulated development (see chapter 37).

Hormones are chemical substances produced in small, often minute quantities in one part of an organism and then transported to another part, where they bring about physiological or developmental responses. The activity of hormones results from their capacity to stimulate certain physiological processes and to inhibit others (figure 40.11). How they act in a particular instance is influenced both by the hormone and the tissue that receives the message.

In animals, hormones are usually produced at definite sites, most commonly in organs. In plants, hormones are not produced in specialized tissues but, instead, in tissues that also carry out other, usually more obvious functions. There are seven major kinds of plant hormones: auxin, cytokinins, gibberellins, brassinosteroids, oligosaccharins, ethylene, and abscisic acid (table 40.1). Current research is focused on the biosynthesis of hormones and on characterizing the hormone receptors that trigger signal transduction pathways. Much of the molecular basis of hormone function remains enigmatic.

Because hormones are involved in so many aspects of plant function and development, we have chosen to integrate examples of hormone activity with specific aspects of plant biology throughout the text. In this section, our goal is to give a brief overview of these hormones.

The seven major kinds of plant hormones are auxin, cytokinins, gibberellins, brassinosteroids, oligosaccharins, ethylene, and abscisic acid.

(a)

(b)

FIGURE 40.11
Plant hormones have profound effects. Plant hormones, often acting together, influence many aspects of plant growth and development, including (*a*) leaf abscission and (*b*) the formation of mature fruit.

Table 40.1 Functions of the Major Plant Hormones

Hormone		Major Functions	Where Produced or Found in Plant
Auxins		Promotion of stem elongation and growth; formation of adventitious roots; inhibition of leaf abscission; promotion of cell division (with cytokinins); inducement of ethylene production; promotion of lateral bud dormancy	Apical meristems; other immature parts of plants
Cytokinins		Stimulation of cell division, but only in the presence of auxin; promotion of chloroplast development; delay of leaf aging; promotion of bud formation	Root apical meristems; immature fruits
Gibberellins		Promotion of stem elongation; stimulation of enzyme production in germinating seeds	Roots and shoot tips; young leaves; seeds
Brassinosteroids		Overlapping functions with auxins and gibberellins	Pollen, immature seeds, shoots, leaves
Oligosaccharins		Pathogen defense, possibly reproductive development	Cell walls
Ethylene		Control of leaf, flower, and fruit abscission; promotion of fruit ripening	Roots, shoot apical meristems; leaf nodes; aging flowers; ripening fruits
Abscisic acid		Inhibition of bud growth; control of stomatal closure; some control of seed dormancy; inhibition of effects of other hormones	Leaves, fruits, root caps, seeds

Auxins

More than a century ago, an organic substance known as **auxin** became the first plant hormone to be discovered. Auxin increases the plasticity of plant cell walls and is involved in elongation of stems. Cells can enlarge in response to changes in turgor pressure. Cell walls must be fairly plastic for this expansion to occur. Auxin plays a role in softening cell walls. The discovery of auxin and its role in plant growth is an elegant example of thoughtful experimental design and is recounted here for that reason. Recent efforts have uncovered an auxin receptor. Transport mechanisms are also being unraveled. As with all the classes of hormones, we are just beginning to understand, at a cellular and molecular level, how hormones regulate growth and development.

Discovery of Auxin. In his later years, the great evolutionist Charles Darwin became increasingly devoted to the study of plants. In 1881, he and his son Francis published a book called *The Power of Movement of Plants*. In this book, the Darwins reported their systematic experiments on the response of growing plants to light—responses that came to be known as phototropisms (discussed in section 40.1). They used germinating oat (*Avena sativa*) and canary grass (*Phalaris canariensis*) seedlings in their experiments and made many observations in this field.

Charles and Francis Darwin knew that if light came primarily from one direction, the seedlings would bend strongly toward it. If they covered the tip of the shoot with a thin glass tube, the shoot would bend as if it were not covered. However, if they used a metal foil cap to exclude light from the plant tip, the shoot would not bend (figure 40.12). They also found that using an opaque collar to exclude light from the stem below the tip did not keep the area above the collar from bending.

In explaining these unexpected findings, the Darwins hypothesized that when the shoots were illuminated from one side, they bent toward the light in response to an "influence" that was transmitted downward from its source at the tip of the shoot. For some 30 years, the Darwins' perceptive experiments remained the sole source of information about this interesting phenomenon. Then the Danish plant physiologist Peter Boysen-Jensen and the Hungarian plant physiologist Arpad Paal independently demonstrated that the substance causing the shoots to bend was a chemical. They showed that if the tip of a germinating grass seedling was cut off and then replaced, with a small block of agar separating it from the rest of the seedling, the seedling would grow as if there had been no change. Something evidently was passing from the tip of the seedling through the agar into the region where the bending occurred. On the basis of these observations under conditions of either uniform illumination or darkness, Paal suggested that an unknown substance continually moves down from the tips of grass seedlings and promotes growth on all sides.

(a)

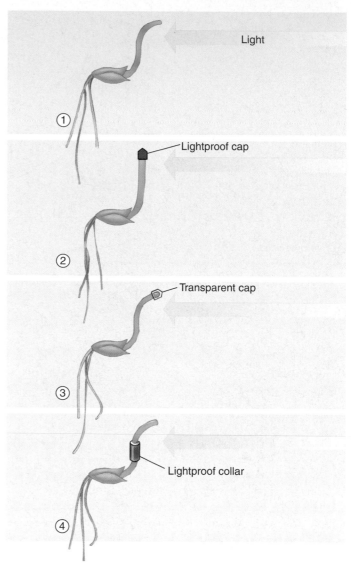

(b)

FIGURE 40.12

The Darwins' experiment. (*a*) Young grass seedlings normally bend toward the light. (*b*) The bending (*1*) did not occur when the tip of a seedling was covered with a lightproof cap (*2*), but did occur when it was covered with a transparent one (*3*). When a collar was placed below the tip (*4*), the characteristic light response took place. From these experiments, the Darwins concluded that, in response to light, an "influence" that caused bending was transmitted from the tip of the seedling to the area below, where bending normally occurs.

Such a light pattern would not, of course, cause the shoot to bend.

Then, in 1926, the Dutch plant physiologist Frits Went carried Paal's experiments an important step further. Went cut off the tips of oat seedlings that had been illuminated normally and set these tips on agar. He then took oat seedlings that had been grown in the dark and cut off their tips in a similar way. Finally, Went cut tiny blocks from the agar on which the tips of the light-grown seedlings had been placed and placed them off-center on the tops of the decapitated dark-grown seedlings (figure 40.13). Even though these seedlings had not been exposed to the light themselves, they bent away from the side on which the agar blocks were placed.

As an experimental control, Went put blocks of pure agar on the decapitated stem tips and noted either no effect or a slight bending toward the side where the agar blocks were placed. Finally, Went cut sections out of the lower portions of the light-grown seedlings. He placed these sections on the tips of decapitated, dark-green oat seedlings and again observed no effect.

As a result of his experiments, Went was able to show that the substance that had diffused into the agar from the tips of light-grown oat seedlings could make seedlings bend when they otherwise would have remained straight. He also showed that this chemical messenger caused the cells on the side of the seedling into which it flowed to grow more than those on the opposite side (figure 40.14). In other words, the chemical enhanced rather than retarded cell elongation. He named the substance that he had discovered auxin (Greek *auxein*, "to increase").

Went's experiments provided a basis for understanding the responses that the Darwins had obtained some 45 years earlier. The oat seedlings bent toward the light because of differences in the auxin concentrations on the two sides of the shoot. The side of the shoot that was in the shade had more auxin, and its cells therefore elongated more than those on the lighted side, bending the plant toward the light.

The Effects of Auxins. Auxin acts to adapt the plant to its environment in a highly advantageous way. It promotes growth and elongation and facilitates the plant's response to its environment. Environmental signals directly influence the distribution of auxin in the plant. How does the environment—specifically, light—exert this influence? Theoretically, it might destroy the auxin, decrease the cells' sensitivity to auxin, or cause the auxin molecules to migrate

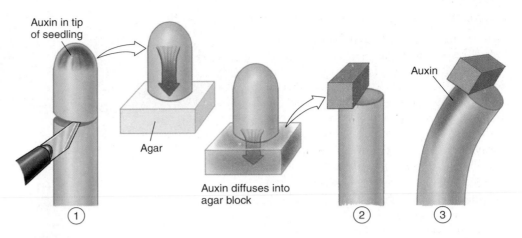

FIGURE 40.13
Frits Went's experiment. (*1*) Went removed the tips of oat seedlings and put them on agar, an inert, gelatinous substance. (*2*) Blocks of agar were then placed off-center on the ends of other oat seedlings from which the tips had been removed. (*3*) The seedlings bent away from the side on which the agar block was placed. Went concluded that a substance he named *auxin* promoted the elongation of the cells and that it accumulated on the side of an oat seedling away from the light.

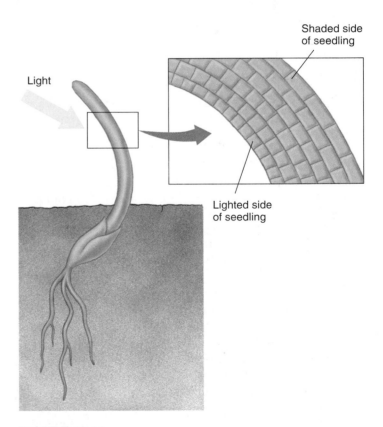

FIGURE 40.14
Auxin causes cells on the dark side to elongate. Plant cells that are in the shade have more auxin and grow faster than cells on the lighted side, causing the plant to bend toward light. Further experiments showed exactly why there is more auxin on the shaded side of a plant.

away from the light into the shaded portion of the shoot. This last possibility has proved to be the case.

In a simple but effective experiment, Winslow Briggs inserted a thin sheet of transparent mica vertically between the half of the shoot oriented toward the light and the half of the shoot oriented away from it (figure 40.15). He found that light from one side does not cause a shoot with such a barrier to bend. When Briggs examined the illuminated plant, he found equal auxin levels on both the light and dark sides of the barrier. He concluded that a normal plant's response to light from one direction involves auxin migrating from the light side to the dark side, and that the mica barrier prevented a response by blocking the migration of auxin.

The effects of auxin are numerous and varied. Auxin promotes the activity of the vascular cambium and the vascular tissues. Also, auxins are present in pollen in large quantities and play a key role in the development of fruits. Synthetic auxins are used commercially for the same purpose. Fruits will normally not develop if fertilization has not occurred and seeds are not present, but frequently they will if auxins are applied. Pollination may trigger auxin release in some species, leading to fruit development even before fertilization.

How Auxin Works. In spite of this long history of research on auxin, its molecular basis of action has been an enigma. The chemical structure of the most common auxin, indoleacetic acid (IAA), resembles that of the amino acid tryptophan, from which it is probably synthesized by plants (figure 40.16). Unlike animal hormones, a specific signal is not sent to specific cells, eliciting a predictable response. There are most likely multiple auxin perception sites. Auxin is also unique among the plant hormones in that it is transported toward the base of the plant. Two families of genes have been identified in *Arabidopsis* that are involved in auxin transport. For example, one protein is involved in the top-to-bottom transport of auxin, while two other proteins function in the root tip to regulate the growth response to gravity. We are still a long way from linking the measurable and observable effects of auxin to events that transpire after it travels to a site and binds to a receptor.

One of the downstream effects of auxin is an increase in the plasticity of the plant cell wall, but this only works on young cell walls lacking extensive secondary cell wall formation. A more plastic wall can stretch more while its

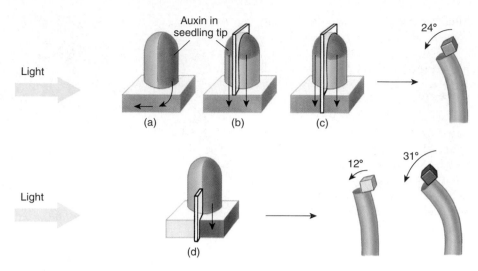

FIGURE 40.15

Phototropism and auxin: The Winslow Briggs experiments. The basic design of these experiments was to place the tip of an oat seedling on an agar block, apply light from one side, and observe the degree of curvature produced when the agar blocks were later placed on the decapitated seedlings. However, Briggs inserted a barrier in various places and noted how this affected the location of auxin. A comparison of (*a*) and (*b*) with similar experiments performed in the dark showed that auxin production does not depend on light; all of the experiments produced approximately 24° of curvature. If a barrier was inserted in the agar block (*d*), light caused the displacement of the auxin away from the light. Finally, experiment (*c*) showed that displacement had occurred, rather than different rates of auxin production on the dark and light sides, because when displacement was prevented with a barrier, both sides of the agar block produced about 24° of curvature.

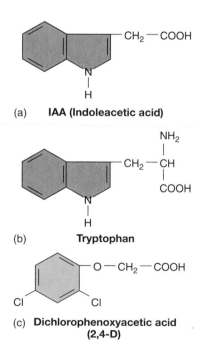

FIGURE 40.16

Auxins. (*a*) Indoleacetic acid (IAA), the principal naturally occurring auxin. (*b*) Tryptophan, the amino acid from which plants probably synthesize IAA. (*c*) Dichlorophenoxyacetic acid (2,4-D), a synthetic auxin, is a widely used herbicide.

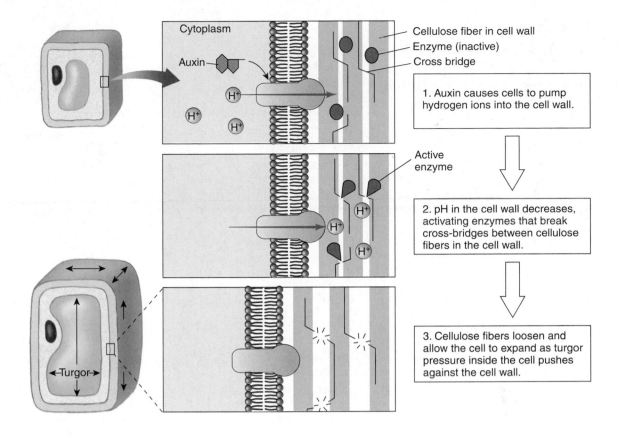

FIGURE 40.17
Acid growth hypothesis. Auxin stimulates the release of hydrogen ions from the target cells, which alters the pH of the cell wall. This optimizes the activity of enzymes that break bonds in the cell wall, allowing them to expand.

Cytoplasm

Auxin

Cellulose fiber in cell wall
Enzyme (inactive)
Cross bridge

1. Auxin causes cells to pump hydrogen ions into the cell wall.

Active enzyme

2. pH in the cell wall decreases, activating enzymes that break cross-bridges between cellulose fibers in the cell wall.

Turgor

3. Cellulose fibers loosen and allow the cell to expand as turgor pressure inside the cell pushes against the cell wall.

protoplast is swelling during active cell growth. The **acid growth hypothesis** provides a model linking auxin to cell wall expansion (figure 40.17). Auxin causes responsive cells to actively transport hydrogen ions from the cytoplasm into the cell wall space. This decreases the pH, which activates enzymes that can break the bonds between cell wall fibers. Remember that different enzymes operate optimally at different pHs. This hypothesis has been experimentally supported in several ways. Buffers that prevent cell wall acidification block cell expansion. Other compounds that release hydrogen ions from the cell can also cause cell expansion. The movement of hydrogen ions has been observed in response to auxin treatment. This mechanism results in a rapid response to an environmental signal. The snapping of the Venus flytrap described in section 40.3 is postulated to involve an acid growth response that allows cells to grow in just 0.3 seconds and close the trap.

Synthetic Auxins. Synthetic auxins, such as NAA (naphthalene acetic acid) and IBA (indolebutyric acid), have many uses in agriculture and horticulture. One of their most important uses is based on their prevention of abscission, the process that causes a leaf or other organ to fall from a plant. Synthetic auxins are used to prevent fruit drop in apples before they are ripe and to hold berries on holly that is being prepared for shipping. Synthetic auxins

are also used to promote flowering and fruiting in pineapples and to induce the formation of roots in cuttings.

Synthetic auxins are routinely used to control weeds. When used as herbicides, they are applied in higher concentrations than IAA would normally occur in plants. One of the most important synthetic auxin herbicides is 2,4-dichlorophenoxyacetic acid, usually known as 2,4-D (see figure 40.16c). It kills weeds in grass lawns by selectively eliminating broad-leaved dicots. The stems of the dicot weeds cease all axial growth.

The herbicide 2,4,5-trichlorophenoxyacetic acid, better known as 2,4,5-T, is closely related to 2,4-D. 2,4,5-T was widely used as a broad-spectrum herbicide to kill weeds and the seedlings of woody plants. It became notorious during the Vietnam War as a component of a jungle defoliant known as Agent Orange and was banned in 1979 for most uses in the United States. When 2,4,5-T is manufactured, it is unavoidably contaminated with minute amounts of dioxin. Dioxin, in doses as low as a few parts per billion, has produced liver and lung diseases, leukemia, miscarriages, birth defects, and even death in laboratory animals.

Auxin is synthesized in the apical meristems of shoots. It causes young stems to bend toward light when it migrates toward the darker side, where it makes young cell walls more plastic and thereby promotes cell elongation. Synthetic auxins have uses in agriculture and horticulture.

Cytokinins

Cytokinins comprise another group of naturally occurring growth hormones in plants. Studies by Gottlieb Haberlandt of Austria around 1913 demonstrated the existence of an unknown chemical in various tissues of vascular plants that, in cut potato tubers, would cause parenchyma cells to become meristematic, and would induce the differentiation of a cork cambium. Coconut milk promoted the differentiation of organs in masses of plant tissue growing in culture. Later, it was discovered that cytokinins in coconut milk promoted the organ differentiation. Subsequent studies have focused on the role cytokinins play in the differentiation of tissues from callus.

A cytokinin is a plant hormone that, in combination with auxin, stimulates cell division and differentiation in plants. Most cytokinins are produced in the root apical meristems and transported throughout the plant. Developing fruits are also important sites of cytokinin synthesis. In mosses, cytokinins cause the formation of vegetative buds on the gametophyte. In all plants, cytokinins, working with other hormones, seem to regulate growth patterns.

Cytokinins are purines that appear to be derivatives of, or have molecule side chains similar to, those of adenine (figure 40.18). Other chemically diverse molecules, not known to occur naturally, have effects similar to those of cytokinins. Cytokinins promote the growth of lateral buds into branches (figure 40.19).

Conversely, cytokinins inhibit the formation of lateral roots, while auxins promote their formation. As a consequence of these relationships, the balance between cytokinins and auxin, along with many other factors, determines the form of a plant. In addition, the application of

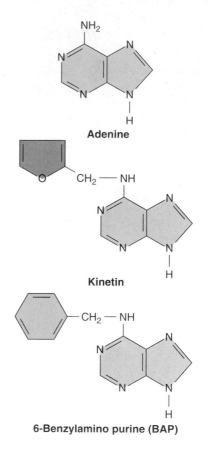

FIGURE 40.18
Some cytokinins. Two commonly used synthetic cytokinins: kinetin and 6-benzylamino purine. Note their resemblance to the purine base adenine.

FIGURE 40.19
Cytokinins stimulate lateral bud growth. (*a*) When the apical meristem of a plant is intact, auxin from the apical bud will inhibit the growth of lateral buds. (*b*) When the apical bud is removed, cytokinins are able to induce the growth of lateral buds into branches. (*c*) When the apical bud is removed and auxin is added to the cut surface, lateral bud outgrowth is suppressed.

(a)

(b)

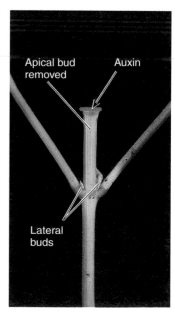

(c)

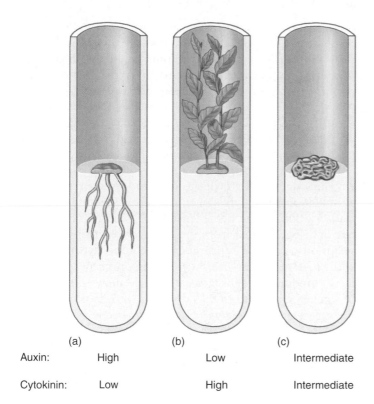

(a)   (b)   (c)

| Auxin: | High | Low | Intermediate |
| Cytokinin: | Low | High | Intermediate |

FIGURE 40.20
Relative amounts of cytokinins and auxin affect organ regeneration in culture. In tobacco, (*a*) high auxin-to-cytokinin ratios favor root development; (*b*) high cytokinin-to-auxin ratios favor shoot development; and (*c*) intermediate concentrations result in the formation of undifferentiated cells. These developmental responses to cytokinin/auxin ratios in culture are species-specific.

FIGURE 40.21
Crown gall tumor. Sometimes cytokinins can be used against the plant by a pathogen. In this case, *Agrobacterium tumefaciens* (a bacterium) has incorporated a piece of its DNA into the plant genome. This DNA contains genes coding for enzymes necessary for cytokinin and auxin biosynthesis. The increased levels of these hormones in the plant cause massive cell division and the formation of a tumor.

cytokinins to leaves detached from a plant retards their yellowing. Therefore, they function as anti-aging hormones.

The action of cytokinins, like that of other hormones, has been studied in terms of its effects on the growth and differentiation of masses of tissue growing in defined media. Plant tissue can form shoots, roots, or an undifferentiated mass of tissues, depending on the relative amounts of auxin and cytokinin (figure 40.20). In the early cell-growth experiments, coconut "milk" was an essential factor. Eventually, researchers discovered that coconut milk is not only rich in amino acids and other reduced nitrogen compounds required for growth, but it also contains cytokinins. Cytokinins seem to be essential for mitosis and cell division. They apparently promote the synthesis or activation of proteins that are specifically required for mitosis.

Cytokinins have also been used against plants by pathogens. The bacterium *Agrobacterium*, for example, introduces genes into the plant genome that increase the rate of cytokinin, as well as auxin, production. This causes massive cell division and the formation of a tumor called crown gall (figure 40.21). How these hormone biosynthesis genes ended up in a bacterium is an intriguing evolutionary question. Coevolution does not always work to the plant's advantage.

> Cytokinins are plant hormones that, in combination with auxin, stimulate cell division and, along with a number of other factors, determine the course of differentiation. In contrast to auxins, cytokinins are purines that are related to or derived from adenine.

Gibberellins

Gibberellins are named after the fungus *Gibberella fujikuroi*, which causes rice plants, on which it is parasitic, to grow abnormally tall. The Japanese plant pathologist Eiichi Kurosawa investigated bakanae ("foolish seedling") disease in the 1920s. He grew *Gibberella* in culture and obtained a substance that, when applied to rice plants, produced bakanae. This substance was isolated and its structural formula identified by Japanese chemists in 1939. British chemists reconfirmed the formula in 1954. Although such chemicals were first thought to be only a curiosity, they have since turned out to belong to a large class of more than 100 naturally occurring plant hormones called gibberellins. All are acidic and are usually abbreviated GA (for gibberellic acid), with a different subscript (GA_1, GA_2, and so forth) to distinguish each one. While gibberellins function endogenously as hormones, they also function as pheromones in ferns. In ferns, gibberellin-like compounds released from one gametophyte can trigger the development of male reproductive structures on a neighboring gametophyte.

Gibberellins, which are synthesized in the apical portions of stems and roots, have important effects on stem elongation. The elongation effect is enhanced if auxin is also present. The application of gibberellins to certain dwarf mutants is known to restore normal growth and development in many plants (figure 40.22). Some dwarf mutants produce insufficient amounts of gibberellin and respond to GA applications; others lack the ability to perceive gibberellin. The large number of gibberellins are all part of a complex biosynthetic pathway that has been unraveled using gibberellin-deficient mutants in maize (corn). While many of these gibberellins are intermediate forms in the production of GA_1, recent work shows that different forms may have specific biological roles.

In chapter 37, we noted the role of gibberellins in stimulating the production of α-amylase and other hydrolytic enzymes needed for utilization of food resources during germination and establishment of cereal seedlings. How is transcription of the genes encoding these enzymes regulated? GA is a signal from the embryo that turns on transcription of one or more genes encoding hydrolytic enzymes in the aleurone layer. GA somehow enhances DNA-binding proteins, which in turn allow transcription of a gene. Synthesis of DNA does not seem to occur during the early stages of seed germination but becomes important when the radicle has grown through the seed coats.

Gibberellins also affect a number of other aspects of plant growth and development. In some cases, GAs hasten seed germination, apparently by substituting for the effects of cold or light requirements. Gibberellins are used commercially to space grape flowers by extending internode length so that the fruits have more room to grow (figure 40.23).

Gibberellins are an important class of plant hormones that are produced in the apical regions of shoots and roots. They play the major role in controlling stem elongation for most plants, acting in concert with auxin and other hormones.

FIGURE 40.22
Effects of gibberellins. This rapid-cycling member of the mustard family *(Brassica rapa)* will "bolt" and flower because of increased gibberellin levels. Mutants such as the rosette mutant *(left)* are defective in producing gibberellins. They can be rescued by applying gibberellins. Other mutants have been identified that are defective in perceiving gibberellins, and they will not respond to gibberellin applications.

FIGURE 40.23
Applications of gibberellins increase the space between grapes. Larger grapes *(right)* develop because there is more room between individual grapes.

Brassinosteroids and Oligosaccharins

Brassinosteroids. Although we've known about **brassinosteroids** for 30 years, it is only recently that they have claimed their place as a class of plant hormones. They were first discovered in *Brassica* pollen, hence the name. Their historical absence in discussions of hormones may be partially due to their functional overlap with other plant hormones, especially auxins and gibberellins. Additive effects among these three classes have been reported. The application of molecular genetics to the study of brassinosteroids has led to tremendous advances in our understanding of how they are made and, to some extent, how they function in signal transduction pathways. What is particularly intriguing about brassinosteroids is their similarity to animal steroid hormones (figure 40.24). One of the genes coding for an enzyme in the brassinosteroid biosynthetic pathway has significant similarity to an enzyme used in the synthesis of testosterone and related steroids. Brassinosteroids have been identified in algae and appear to be ubiquitous among the plants. It is plausible that their evolutionary origin predated the plant–animal split.

Brassinosteroids have a broad spectrum of physiological effects—elongation, cell division, bending of stems, vascular tissue development, delayed senescence, membrane polarization, and reproductive development. Environmental signals can trigger brassinosteroid actions. Mutants have been identified that alter the response to brassinosteroid, but signal transduction pathways remain to be uncovered. From an evolutionary perspective, it will be quite interesting to see how these pathways compare with animal steroid signal transduction pathways.

Oligosaccharins. In addition to cellulose, plant cell walls are composed of numerous complex carbohydrates called oligosaccharides. Some evidence indicates that these cell wall components (when degraded by pathogens) function as signaling molecules as well as structural wall components. Oligosaccharides that are proposed to have a hormonelike function are called **oligosaccharins.** Oligosaccharins can be released from the cell wall by enzymes secreted by pathogens. These carbohydrates are believed to signal defense responses, such as the hypersensitive response discussed in chapter 39. Another oligosaccharin has been shown to inhibit auxin-stimulated elongation of pea stems. These molecules are active at concentrations one to two orders of magnitude less than those of the traditional plant hormones. You have seen how auxin and cytokinin ratios can affect organogenesis in culture. Oligosaccharins also affect the phenotype of regenerated tobacco tissue, inhibiting root formation and stimulating flower production in tissues that are competent to regenerate flowers. How the culture results translate to in vivo systems is an open question. The

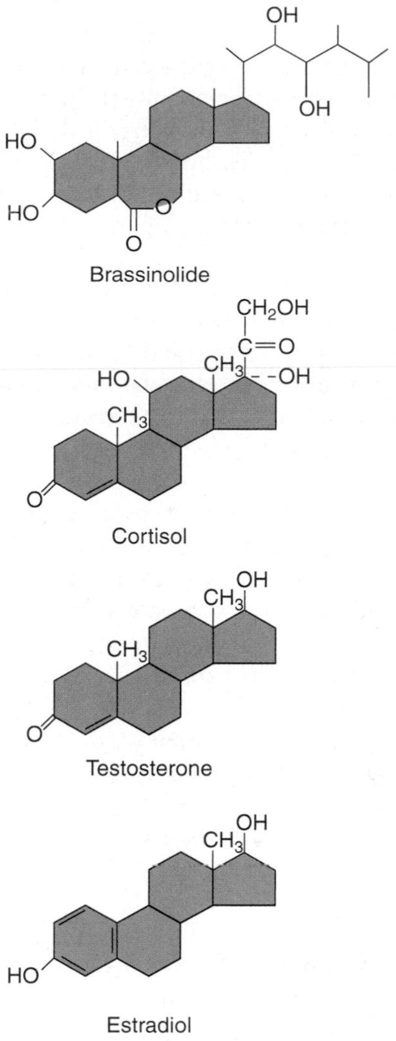

FIGURE 40.24
Brassinosteroids. Brassinolide and other brassinosteroids have structural similarities to animal steroid hormones. Cortisol, testosterone, and estradiol are animal steroid hormones.

structural biochemistry of oligosaccharins makes them particularly challenging molecules to study. Current research focuses on ways oligosaccharins interface with cells and initiate signal transduction pathways.

Brassinosteroids are structurally similar to animal steroid hormones. They have many effects on plant growth and development that parallel those of auxins and gibberellins. Oligosaccharins are complex carbohydrates that are released from cell walls and appear to regulate both pathogen responses and growth and development in some plants.

Ethylene

Long before its role as a plant hormone was appreciated, the simple, gaseous hydrocarbon **ethylene** ($H_2C—CH_2$) was known to defoliate plants when it leaked from gaslights in streetlamps. Ethylene is, however, a natural product of plant metabolism that, in minute amounts, interacts with other plant hormones. When auxin is transported down from the apical meristem of the stem, it stimulates the production of ethylene in the tissues around the lateral buds and thus retards their growth. Ethylene also suppresses stem and root elongation, probably in a similar way. An ethylene receptor has been identified and characterized. It appears to have evolved early in the evolution of photosynthetic organisms, sharing features with environmental-sensing proteins identified in bacteria.

Ethylene plays a major role in fruit development. At first, auxin, which is produced in significant amounts in pollinated flowers and developing fruits, stimulates ethylene production; this, in turn, hastens fruit ripening. Complex carbohydrates are broken down into simple sugars, chlorophylls are broken down, cell walls become soft, and the volatile compounds associated with flavor and scent in ripe fruits are produced.

One of the first observations that led to the recognition of ethylene as a plant hormone was the premature ripening in bananas produced by gases coming from oranges. Such relationships have led to major commercial uses of ethylene. For example, tomatoes are often picked green and artificially ripened later by the application of ethylene. Ethylene is widely used to speed the ripening of lemons and oranges as well. Carbon dioxide has the opposite effect of arresting ripening; fruits are often shipped in an atmosphere of carbon dioxide. Also, a biotechnology solution has been developed in which one of the genes necessary for ethylene biosynthesis has been cloned, and its antisense copy inserted into the tomato genome (figure 40.25). The antisense copy of the gene is a nucleotide sequence that is complementary to the sense copy of the gene. In this transgenic plant, both the sense and antisense sequences for the ethylene biosynthesis gene are transcribed. The sense and antisense mRNA sequences then pair with each other. This blocks translation, which requires single-stranded RNA; ethylene is not synthesized, and the transgenic tomatoes do not ripen. Sturdy green tomatoes can be shipped without ripening and rotting. Exposing these tomatoes to ethylene later allows them to ripen.

Studies have shown that ethylene plays an important ecological role. Ethylene production increases rapidly when a plant is exposed to ozone and other toxic chemicals, temperature extremes, drought, attack by pathogens or herbivores, and other stresses. The increased production of ethylene that occurs can accelerate the loss of leaves or fruits that have been damaged by these stresses. Some of the damage associated with exposure to ozone is due to the ethylene produced by the plants. The production of ethylene by plants attacked by herbivores or infected with pathogens may be a signal to activate the defense mechanisms of the plants. This may include the production of molecules toxic to the pests.

Ethylene, a simple gaseous hydrocarbon, is a naturally occurring plant hormone. Among its numerous effects is the stimulation of ripening in fruit. Ethylene production is also elevated in response to environmental stress.

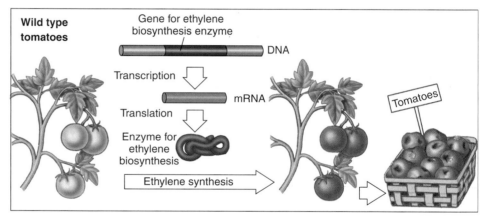

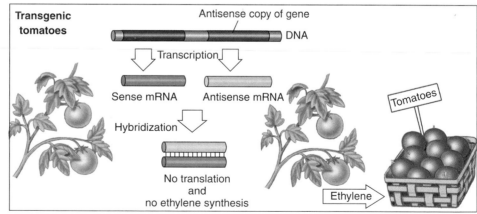

FIGURE 40.25
Genetic regulation of fruit ripening. An antisense copy of the gene for ethylene biosynthesis prevents the formation of ethylene and subsequent ripening of transgenic fruit. The antisense strand is complementary to the sequence for the ethylene biosynthesis gene. After transcription, the antisense mRNA pairs with the sense mRNA, and the double-stranded mRNA cannot be translated into a functional protein. Ethylene is not produced, and the fruit does not ripen. The fruit is sturdier for shipping in its unripened form and can be ripened later with exposure to ethylene. Thus, while wild-type tomatoes may already be rotten and damaged by the time they reach stores, transgenic tomatoes stay fresh longer.

Abscisic Acid

Abscisic acid appears to be synthesized mainly in mature green leaves, fruits, and root caps. The hormone earned its name because applications of it appear to stimulate fruit abscission in cotton, but there is little evidence that it plays an important role in this process. Ethylene is actually the chemical that promotes senescence and abscission.

Abscisic acid probably induces the formation of winter buds—dormant buds that remain through the winter. The conversion of leaf primordia into bud scales follows (figure 40.26*a*). Like ethylene, abscisic acid may also suppress growth of dormant lateral buds. It appears that abscisic acid, by suppressing growth and elongation of buds, can counteract some of the effects of gibberellins (which stimulate growth and elongation of buds); it also promotes senescence by counteracting auxin (which tends to retard senescence). Abscisic acid plays a role in seed dormancy and is antagonistic to gibberellins during germination. Abscisic acid levels in seeds rise during embryogenesis. As maize embryos develop in the kernels on the cob, abscisic acid is necessary to induce dormancy and prevent precocious germination, called **vivipary** (figure 40.26*b*). It is also important in controlling the opening and closing of stomata (figure 40.26*c*).

Abscisic acid occurs in all groups of plants and apparently has been functioning as a growth-regulating substance since early in the evolution of the plant kingdom. Relatively little is known about the exact nature of its physiological and biochemical effects. These effects are very rapid—often taking place within a minute or two—and therefore, they must be at least partly independent of gene expression. Some longer-term effects of abscisic acid involve the regulation of gene expression. This is currently an active area of research. Having all the genes sequenced in a model plant, *Arabidopsis*, makes it easier to identify which genes are transcribed in response to abscisic acid. Abscisic acid levels become greatly elevated when the plant is subject to stress, especially drought. Like other plant hormones, abscisic acid will probably prove to have valuable commercial applications when its mode of action is better understood. It is a particularly strong candidate for understanding desiccation tolerance.

Abscisic acid, produced chiefly in mature green leaves and in fruits, suppresses the growth of buds and promotes leaf senescence. It also plays an important role in controlling the opening and closing of stomata. Abscisic acid may be critical in ensuring survival under environmental stress, especially water stresses.

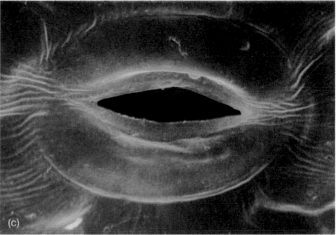

FIGURE 40.26

Effects of abscisic acid. (*a*) Abscisic acid plays a role in the formation of these winter buds of an American basswood. These buds will remain dormant for the winter, and bud scales—modified leaves—will protect the buds from desiccation. (*b*) In addition to bud dormancy, abscisic acid is necessary for dormancy in seeds. This viviparous mutant in maize is deficient in abscisic acid, and the embryos begin germinating on the developing cob. (*c*) Abscisic acid also affects the closing of stomata by influencing the movement of potassium ions out of guard cells.

40.1 Plants respond to light.

Photomorphogenesis

- Photomorphogenesis describes nondirectional, light-triggered development. (p. 808)
- Photomorphogenesis is often triggered by red photoreceptors. (p. 808)
- Phytochrome is a light receptor involved in many plant growth responses, such as seed germination, shoot elongation, and plant spacing. It exists in two forms, red and far-red. (p. 808)

Phototropisms

- Phototropism describes directional growth toward sources of blue-wavelength light. (p. 809)
- In general, stems are positively phototropic while roots are nonresponsive or weakly negatively phototropic. (p. 809)

40.2 Plants respond to gravity.

Gravitropism

- Gravitropism refers to the response of a plant to the earth's gravitational field. (p. 810)
- Gravitropism occurs at germination when shoots grow upward and roots grow downward. (p. 810)
- Four steps lead to a gravitropic response: Gravity is perceived by the cell; signals form in the cell; a signal is transduced intra- and intercellularly; and differential cell elongation occurs. (p. 810)

40.3 Plants respond to touch.

Thigmotropism and Thigmonasty

- Thigmotropism is directional growth due to contact with another object. (p. 811)
- Thigmonastic responses are growth due to contact with another object, but the direction of the growth response is independent of the direction of the contact. (p. 811)
- Coiling tendrils can be examples of either thigmotropism or thigmonasty. (p. 811)
- Other tropisms have been recognized, including electrotropism (electricity), chemotropism (chemicals), traumotropism (wounding), thermotropism (temperature), aerotropism (oxygen), skototropism (darkness), geomagnetotropism (magnetic fields), and hydrotropism (water). (p. 811)

Turgor Movement

- Some plant movements are due to turgor pressure changes in specific cells that may be induced by several factors, including touch and light. (p. 812)
- Circadian clocks are endogenous clocks that keep plant movements and systems synchronized on a 24-hour cycle. (p. 813)

40.4 Water and temperature elicit plant responses.

Dormancy

- The ability to enter a dormant phase provides a survival advantage, especially under harsh conditions. (p. 814)
- Environmental signals trigger both the start and the end of dormancy. (p. 814)

Surviving Temperature Extremes

- Some plants have the ability to undergo deep supercooling where ice crystals are stored in extracellular spaces. (p. 815)
- Heat shock proteins can be produced to stabilize proteins under high temperatures. (p. 815)

40.5 The hormones that guide growth are keyed to the environment.

Plant Hormones

- Hormones are chemical substances produced in small quantities in one part of an organism and then transported to another part, where they bring about a response. (p. 816)
- In animals, hormones are usually produced in specialized tissues, but this is not the case in plants. (p. 816)
- Auxin, the first plant hormone to be discovered, increases the plasticity of plant cell walls and is involved in stem elongation. Auxin is synthesized in the apical meristems of shoots and helps cause stems to bend toward light. (p. 818)
- Cytokinins stimulate plant cell division and differentiation in combination with auxin, promote growth of lateral buds into branches, and inhibit the formation of lateral roots. (p. 822)
- Gibberellins, which are named after the fungus *Gibberella fujikuroi*, function endogenously as hormones. They are synthesized in apical portions of stems and roots and have effects on stem elongation. (p. 824)
- Brassinosteroids, which are very similar to animal steroid hormones, exhibit a broad spectrum of effects that can be triggered by environmental signals. (p. 825)
- Oligosaccharins function as signaling molecules and structural wall components when degraded by pathogens. They signal defense responses and inhibit auxin-stimulated stem elongation. (p. 825)
- Ethylene plays a major role in stimulating fruit development. (p. 826)
- Abscisic acid, which is mainly synthesized in mature green leaves and fruits, suppresses growth of dormant lateral buds and plays a role in promoting senescence, seed dormancy, and stomatal control. Levels of abscisic acid increase in plants under stress, and accordingly may play an important role in desiccation tolerance. (p. 827)

Self Test

1. Which of the following seeds would likely germinate even if it were located on the floor of a densely leaved forest?
   a. a seed lacking chlorophyll *a*
   b. a seed lacking phytochrome P_r
   c. a seed lacking phytochrome P_{fr}
   d. a seed lacking phototropin

2. Which of the following statements provides a true example of both photomorphogenesis and phototropism
   a. phototropism is growth toward blue light, and photomorphogenesis is growth toward red light.
   b. phototropism is growth toward blue light, and photomorphogenesis is germination triggered by near-red light.
   c. phototropism is growth toward red light, and photomorphogenesis is germination triggered by blue light.
   d. phototropism is movement toward blue light, that does not involve growth; photomorphogenesis is movement toward red light that does involve growth.

3. If you were to plant a de-etiolated (*det2*) mutant *Arabidopsis* seed and keep it in a dark box, what would you expect to happen?
   a. The seed would germinate normally, but the plant would not become tall and spindly while it sought a light source.
   b. The seed would fail to germinate because it would not have light.
   c. The seed would germinate, and the plant would become tall and spindly while it sought a light source.
   d. The seed would germinate, and the plant would immediately die because it could not make sugar in the dark.

4. Growing plants in zero gravity on the space shuttle prevents
   a. phototropism because it is dark in space.
   b. photomorphogenesis because only near red-light filters into space.
   c. gravitropism because there is little gravity in space.
   d. aerotropism because there is no oxygen in space.

5. When a Venus flytrap closes, it is an example of
   a. growth.
   b. cell migration.
   c. muscle contraction.
   d. a nervous twitch

6. We often have the misconception that plants are unable to move in their environment. Many plants, however, display daily movements to maximize their capacity to absorb light energy (e.g., bean leaves). These daily changes in shape are caused by
   a. changes in turgor in specific cells.
   b. growth of specific cells.
   c. muscle contraction in the leaves.
   d. temperature changes in the environment.

7. When Charles and Francis Darwin investigated phototropisms in plants, they discovered that
   a. auxin was responsible for light-dependent growth.
   b. light was detected at the tip of a plant.
   c. light was detected along the shoot of a plant.
   d. only red light stimulated phototropism.

8. The chemical defoliant Agent Orange was banned because
   a. the synthetic auxin used in the preparation, 2,4,5-trichlorophenoxyacetic acid (2,4,5-T), caused premature growth in humans.
   b. the synthetic auxin used in the preparation, 2,4,5-trichlorophenoxyacetic acid (2,4,5-T), caused plants to grow more vigorously than the environment could sustain.
   c. a contaminant in the preparation of 2,4,5-trichlorophenoxyacetic acid (2,4,5-T) caused disease and birth defects in humans.
   d. a contaminant in the preparation of 2,4,5-trichlorophenoxyacetic acid (2,4,5-T) killed beneficial plants.

9. You have come up with a brilliant idea to stretch your grocery budget by buying green fruit in bulk and then storing it in a bag that you have blown up like a balloon. As you need fruit, you will take it out of the bag, and it will miraculously ripen. How would this work?
   a. The bag would block light from reaching the fruit, so it would not ripen.
   b. The bag would keep the fruit cool, so it would not ripen.
   c. The high CO_2 levels in the bag would prevent ripening.
   d. The high O_2 levels in the bag would prevent ripening.

10. If you were to accidentally plant a mutant strain of barley that could not synthesize the plant hormone abscisic acid (ABA), what would you expect to happen?
    a. The shoots would elongate too much and fall over because they could not support themselves.
    b. The shoots would not elongate normally, and you would get short plants.
    c. The seeds would germinate prematurely.
    d. The leaves would fall off the plant.

Test Your Visual Understanding

1. Images *a–c* show tobacco cells cultured under different auxin:cytokinin ratios. Label each image with the appropriate hormone ratio.

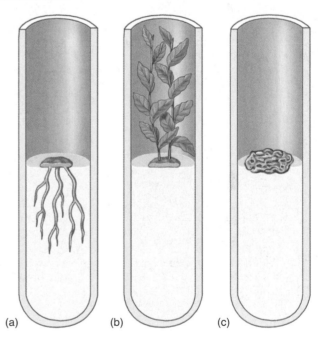

(a)　　　　(b)　　　　(c)

Auxin:

Cytokinin:

Apply Your Knowledge

1. Discuss the similarities and differences between thigmotropism and turgor movement.
2. Compare the mechanisms that animals and plants use to survive harsh environments by thinking of an equivalent animal response for each of the following:
   a. dormancy
   b. thigmomorphogenesis
   c. abscission
   d. phototropism

Chapter 40 Sensory Systems in Plants **829**

41

Plant Reproduction

Concept Outline

41.1 The environment influences reproduction.

Plants Undergo Metamorphosis. The transition of a shoot meristem from vegetative to adult is called phase change.

Pathways Leading to Flower Production. Photoperiod is regulated in complex ways.

Identity Genes and the Formation of Floral Meristems and Floral Organs. Floral meristem identity genes activate floral organ identity genes.

41.2 Flowers are highly evolved for reproduction.

Evolution of the Flower. A complete flower has four whorls, containing protective sepals, petals, male stamens, and female carpels.

Formation of Angiosperm Gametes. The male gametophytes are the pollen grains, and the female gametophyte is the embryo sac.

Pollination. Evolutionary modifications of flowers have enhanced effective pollination.

Self-Pollination. Self-pollination is favored in stable environments, but outcrossing enhances genetic variability.

Fertilization. Angiosperms use two sperm cells, one to fertilize the egg and the other to produce a nutrient tissue called endosperm.

41.3 Many plants can clone themselves by asexual reproduction.

Asexual Reproduction. Some plants do without sexual reproduction, instead cloning new individuals from parts of themselves.

Plant Tissue Culture. Plants can be cloned from isolated cells in the laboratory.

41.4 How long do plants and plant organs live?

The Life Span of Plants. Clonal plants can live indefinitely. Parts of plants senesce and die. Some plants (annuals and biennials) reproduce sexually only once and die. Perennials continue to grow and flower year after year.

FIGURE 41.1
Reproductive success in flowering plants. Unique reproductive systems and strategies have coevolved between plants and animals, accounting for almost 250,000 species of flowering plants inhabiting all but the harshest environments on earth.

The remarkable evolutionary success of flowering plants can be linked to their reproductive strategies. The evolution and development of flowers has been of paramount importance (figure 41.1). In this chapter, we explore reproductive strategies in the angiosperms and how their unique features—flowers and fruits—have contributed to their success. This is, in part, a story of coevolution between plants and animals that ensures greater genetic diversity by dispersing plant gametes widely. However, in a stable environment, there are advantages to maintaining the status quo genetically. Asexual reproduction is a strategy to clonally propagate individuals. An unusual twist to sexual reproduction in some flowering plants is that senescence and death of the parent plant follow.

41.1 The environment influences reproduction.

Plants Undergo Metamorphosis

Overview of Flowering Initiation

Carefully regulated processes determine when and where flowers will form. Plants must often gain competence to respond to internal or external signals regulating flowering. Once plants are competent to reproduce, a combination of factors—including light, temperature, and both promotive and inhibitory internal signals—determines when a flower is produced (figure 41.2). These signals turn on genes that specify where the floral organs—sepals, petals, stamens, and carpels—will form. Once cells have instructions to become a specific floral organ, yet another developmental cascade leads to the three-dimensional construction of flower parts.

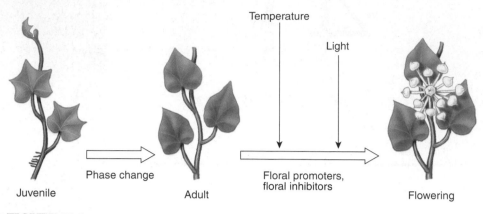

FIGURE 41.2
Factors involved in initiating flowering. This model depicts the environmentally cued and internally processed events that result in a shoot meristem initiating flowers.

Phase Change

Plants go through developmental changes leading to reproductive maturity just as many animals do. This shift from juvenile to adult development is seen in the metamorphosis of a tadpole to an adult frog or a caterpillar to a butterfly that can then reproduce. Plants undergo a similar metamorphosis that leads to the production of a flower. Unlike the juvenile frog, which loses its tail, plants just keep adding structures to existing structures with their meristems. At germination, most plants are incapable of producing a flower, even if all the environmental cues are optimal. Internal developmental changes allow plants to obtain **competence** to respond to external and/or internal signals that trigger flower formation. This transition is referred to as **phase change**. Phase change can be morphologically obvious or very subtle. Take a look at an oak tree in the winter. The lower leaves will still be clinging to the branches, while the upper ones will be gone (figure 41.3a). Those lower branches were initiated by a juvenile meristem. The fact that they did not respond to environmental cues and drop their leaves indicates that they are young branches and have not made a phase change. Ivy also has distinct juvenile and adult phases of growth (figure 41.3b). Stem tissue produced by a juvenile meristem initiates adventitious roots that can cling to walls.

(a)

(b)

FIGURE 41.3
Phase change. (a) The lower branches of this oak tree represent the juvenile phase of development; they cling to their leaves in the winter. The lower leaves are not able to form an abscission layer and break off the tree in the fall. Such visible changes are marks of phase change, but the real test is whether or not the plant is able to flower. (b) Juvenile ivy (left) makes adventitious roots and has an alternating leaf arrangement. Adult ivy (right) cannot make adventitious roots and has leaves with a different morphology arranged on an upright stem in a spiral.

If you look at very old brick buildings covered with ivy, you will notice that the uppermost branches are falling off because they have transitioned to the adult phase of growth and have lost the ability to produce adventitious roots. It is important to remember that even though a plant has reached the adult stage of development, it may or may not produce reproductive structures. Other factors may be necessary to trigger flowering.

Generally it is easier to get a plant to revert from an adult to juvenile state than to induce phase change experimentally. Applications of gibberellins and severe pruning can cause reversion. There is evidence in peas and *Arabidopsis* for genetically controlled repression of flowering. The *embryonic flower* mutant of *Arabidopsis* flowers almost immediately (figure 41.4), which is consistent with the hypothesis that the wild-type allele suppresses flowering. It is possible that flowering is the default state and that mechanisms have evolved to delay flowering. This delay allows the plant to store more energy to be allocated for reproduction.

The best example of inducing the juvenile-to-adult transition comes from overexpressing a gene necessary for flowering that is found in many species. This gene, *LEAFY*, was cloned in *Arabidopsis*, and its promoter was replaced with a viral promoter that results in constant, high levels of *LEAFY* transcription. This gene construct was then introduced into cultured aspen cells that were used to regenerate plants. When *LEAFY* is overexpressed in aspen, flowering occurs in weeks instead of years (figure 41.5). Phase change requires both sufficient signal and the ability to perceive the signal. Some plants acquire competence in the shoot to perceive a signal of a certain intensity. Others acquire competence to produce sufficient promotive signal(s) and/or to decrease inhibitory signal(s).

Plants become reproductively competent through changes in signaling and perception. The transition to the adult stage of development, during which reproduction is possible, is called phase change. Plants in the adult phase of development may or may not produce reproductive structures (flowers), depending on environmental cues.

FIGURE 41.4
EMBRYONIC FLOWER gene prevents early flowering.
The *embryonic flower* mutant flowers upon germination. This is an aberrant flower with malformed carpels and other defective floral structures just above the root system.

(a)

(b)

FIGURE 41.5
Overexpression of a flowering gene can accelerate phase change. (*a*) Normally, an aspen tree grows for several years before producing flowers (see boxed insert).
(*b*) Overexpression of the *Arabidopsis* flowering gene, *LEAFY*, causes rapid flowering in a transgenic aspen (see boxed insert).

Pathways Leading to Flower Production

Three genetically regulated pathways to flowering have been identified: (1) the light-dependent pathway, (2) the temperature-dependent pathway, and (3) the autonomous pathway. Plants can rely primarily on one pathway, but all three pathways can be present. The environment can promote or repress flowering, and in some cases, it can be relatively neutral. For example, light can be a signal that long summer days have arrived in a temperate climate and that conditions are favorable for reproduction. In other cases, plants depend on light to accumulate sufficient amounts of sucrose to fuel reproduction, but flower independently of day length. Temperature can also be used as a signal. Gibberellins have been linked to the pathway called *vernalization*, the requirement for a period of chilling of seeds or shoots for flowering. Clearly, reproductive success would be unlikely in the middle of a blizzard. Assuming that regulation of reproduction first arose in more constant tropical environments, many of the day-length and temperature controls would have evolved as plants colonized more temperate climates. The complexity of the flowering pathways has been dissected physiologically. Now, analysis of flowering mutants is providing insight into the molecular mechanisms of the flowering pathways. The redundancy of pathways to flowering ensures that there will be another generation. We will examine each of the three flowering pathways in turn.

Light-Dependent Pathway

Flowering requires much energy accumulated via photosynthesis. Thus, all plants require light for flowering, but this is distinct from the **photoperiodic,** or light-dependent, flowering pathway. Aspects of growth and development in most plants are keyed to changes in the proportion of light to dark in the daily 24-hour cycle (day length). This provides a mechanism for organisms to respond to seasonal changes in the relative length of day and night. Day length changes with the seasons; the farther a region is from the equator, the greater the variation in day length. The flowering responses of plants to day length fall into several basic categories. When the daylight becomes shorter than a critical length, flowering is initiated in **short-day plants** (figure 41.6). When the daylight becomes longer than a critical length, flowering is initiated in **long-day plants.**

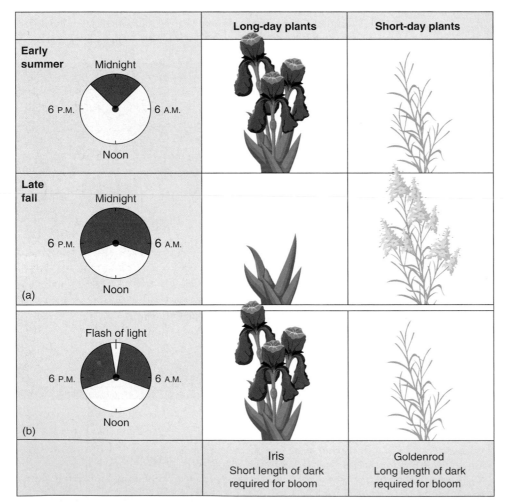

	Long-day plants	Short-day plants
Early summer Midnight 6 P.M. 6 A.M. Noon		
Late fall Midnight 6 P.M. 6 A.M. Noon (a)		
Flash of light 6 P.M. 6 A.M. Noon (b)		
	Iris Short length of dark required for bloom	Goldenrod Long length of dark required for bloom

FIGURE 41.6
How flowering responds to day length. (*a*) This iris is a long-day plant that is stimulated by short nights to flower in the spring. The goldenrod is a short-day plant that, throughout its natural distribution in the northern hemisphere, is stimulated by long nights to flower in the fall. (*b*) If the long night of winter is artificially interrupted by a flash of light, the goldenrod will not flower, and the iris will. Although the terms refer to the length of day, in each case, it is the duration of uninterrupted darkness that determines when flowering will occur.

Other plants, such as snapdragons, roses, and many native to the tropics (for example, tomatoes), flower when mature regardless of day length, as long as they have received enough light for normal growth. These are referred to as **day-neutral plants.** Several grasses (for example, Indian grass, *Sorghastrum nutans*), as well as ivy, have two critical photoperiods; they will not flower if the days are too long, and they also will not flower if the days are too short. In some species, there is a sharp distinction between long and short days. In others, flowering occurs more rapidly or slowly depending on the length of day. These plants, which rely on other flowering pathways as well, are called facultative long- or short-day plants. The garden pea is an example of a facultative long-day plant. In both long- and short-day plants, it is actually the length of darkness (night), not the length of day, that is physiologically significant. Using light as a cue permits plants to flower when abiotic environmental conditions are optimal, pollinators are available, and competition for resources with other plants may be less. For example, the spring ephemerals flower in the woods before the canopy leafs out and blocks the sunlight necessary for photosynthesis.

At middle latitudes, most long-day plants flower in the spring and early summer; examples of such plants include clover, irises, lettuce, spinach, and hollyhocks. Short-day plants usually flower in late summer and fall; these include chrysanthemums, goldenrods, poinsettias, soybeans, and many weeds. Commercial plant growers use these responses to day length to bring plants into flower at specific times. For example, photoperiod is manipulated in greenhouses so that poinsettias flower just in time for the winter holidays (figure 41.7). The geographic distribution of certain plants may be determined by their flowering responses to day length.

Photoperiod is perceived by several different forms of phytochrome and also by a blue-light-sensitive molecule (cryptochrome). The conformational change in a light-receptor molecule triggers a cascade of events that leads to the production of a flower. There is a link between light and the circadian rhythm regulated by an internal clock that facilitates or inhibits flowering. At a molecular level, the gaps between light signaling and production of flowers are rapidly filling in, and the control mechanisms have been found to be quite complex.

As an example of how day length affects a specific flowering gene, we will examine *Arabidopsis,* a facultative long-day plant that flowers in response to both far-red and blue light. Red light inhibits flowering. The gene *CONSTANS* (*CO*) is expressed under long days but not short days. The loss of *CO* product does not change the time when a plant flowers under short days, but delays flowering under long days. What happens is that the gene is positively regulated by cryptochrome that perceives blue light under long days. Active cryptochrome may release *CO* from inhibition by phytochrome B exposed to red light. Simply put, flowering is promoted by repressing a gene that represses flowering!

FIGURE 41.7
Flowering time can be altered. Manipulation of photoperiod in greenhouses ensures that short-day poinsettias flower in time for the winter holidays. Even after flowering is induced, many developmental events must occur in order to produce species-specific flowers.

CO is a transcription factor that turns on other genes, which results in the expression of *LEAFY*. As discussed in connection with phase change earlier in this section, *LEAFY* is one of the key genes that "tells" a meristem to switch over to flowering. We will see that other pathways also converge on this important gene.

The Flowering Hormone: Does It Exist? The "Holy Grail" in plant biology has been a flowering hormone, quested unsuccessfully for more than 50 years. A considerable amount of evidence demonstrates the existence of substances that promote flowering and substances that inhibit it. Grafting experiments have shown that these substances can move from leaves to shoots. The complexity of their interactions, as well as the fact that multiple chemical messengers are evidently involved, has made this scientifically and commercially interesting search very difficult, and to this day, the existence of a flowering hormone remains strictly hypothetical. We do know that *LEAFY* can be expressed in the vegetative as well as the reproductive portions of plants. Clearly, information about day length gathered by leaves is transmitted to shoot apices. Given that there are multiple pathways to flowering, several signals may be facilitating communication between leaves and shoots. We also know that roots can be a source of floral inhibitors affecting shoot development.

Temperature-Dependent Pathway

Cold temperatures can accelerate or permit flowering in many species. As with light, this ensures that plants flower at more optimal times. Some plants require a period of chilling before flowering, called **vernalization.** This phenomenon was discovered by the Russian scientist Lysenko while trying to solve the problem of winter wheat rotting in the fields. Because winter wheat would not flower without a period of chilling, Lysenko chilled the seeds and then successfully planted them in the spring. Although this was a scientifically significant discovery, Lysenko erroneously concluded that he had converted one species, winter wheat, to another, spring wheat, by simply altering the environment. Unfortunately, a great many problems, including mistreatment of Russian geneticists, resulted, and genetics and Darwinian evolution were suspect in Russia for half a century.

Vernalization is necessary for some seeds or plants in later stages of development. Analysis of mutants in *Arabidopsis* and pea indicate that vernalization is a separate flowering pathway that may be linked to the hormone gibberellin. In this pathway, repression may also lead to flowering. High levels of one of the gene products in the pathway may block the promotion of flowering by gibberellins. When plants are chilled, less of this gene product, and gibberellin activity may increase. It is known that gibberellins enhance the expression of *LEAFY*. One proposal is that both vernalization and autonomous pathways (discussed next) share a common intersection affecting gibberellin promotion of flowering. Weigel has shown that gibberellin actually binds the promoter of the *LEAFY* gene, so its effect on flowering is direct. The connection between gibberellin levels and temperature also needs to be understood.

Autonomous Pathway

The autonomous pathway to flowering is independent of external cues except for basic nutrition. Presumably, this was the first pathway to evolve. Day-neutral plants often depend primarily on the autonomous pathway, which allows plants to "count" and "remember." For example, a field of day-neutral tobacco will produce a uniform number of nodes before flowering. If the shoots of these plants are removed at different positions, axillary buds will grow out and produce the same number of nodes as the removed portion of the shoot (figure 41.8). At a certain point in development, the shoots become committed or **determined** to flower (figure 41.9) The upper axillary buds of flowering tobacco will remember their position when rooted or grafted. The terminal shoot tip becomes florally determined about four nodes before it initiates a flower. In some other species, this commitment is less stable and/or occurs later.

How do shoots know where they are and at some point "remember" that information? It is clear that inhibitory

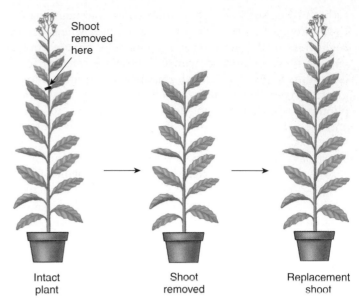

Shoot removed here

Intact plant

Shoot removed

Replacement shoot

(a) Upper axillary bud released from apical dominance

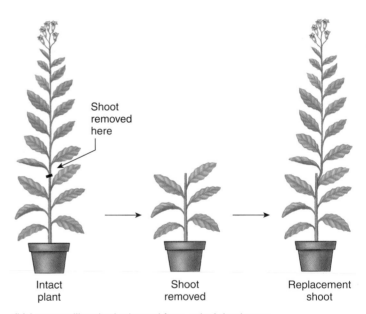

Shoot removed here

Intact plant

Shoot removed

Replacement shoot

(b) Lower axillary bud released from apical dominance

FIGURE 41.8
Plants can "count." When axillary buds of flowering, day-neutral tobacco plants are released from apical dominance by removing the main shoot, they replace the number of nodes that were initiated by the main shoot.
(After McDaniel 1996.)

signals are sent from the roots. If bottomless pots are continuously placed over a growing tobacco plant and filled with soil, flowering is delayed by the formation of adventitious roots (figure 41.10). Control experiments with leaf removal show that it is the addition of roots, not the loss of leaves, that delays flowering. A balance between floral promoting and inhibiting signals may

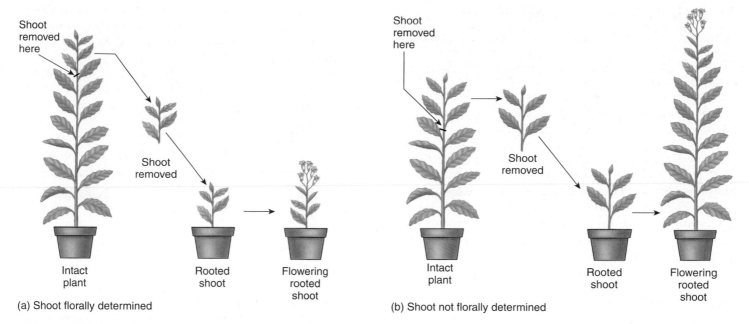

Shoot removed here

Shoot removed

Intact plant

Rooted shoot

Flowering rooted shoot

(a) Shoot florally determined

Shoot removed here

Shoot removed

Intact plant

Rooted shoot

Flowering rooted shoot

(b) Shoot not florally determined

FIGURE 41.9
Plants can "remember." At a certain point in the flowering process, shoots become committed to making a flower. This is called floral determination. (*a*) Florally determined shoots "remember" their position when rooted in a pot. That is, they produce the same number of nodes that they would have if they had grown out on the plant, and then they flower. (*b*) Shoots that are not yet florally determined cannot remember how many nodes they have left, so they start counting again. That is, they develop like a seedling and then flower. (After McDaniel 1996.)

regulate when flowering occurs in the autonomous pathway and the other pathways as well.

Determination for flowering is tested at the organ or whole-plant level by changing the environment and ascertaining whether or not the fate has changed. How does floral determination correlate with molecular changes? In *Arabidopsis*, floral determination correlates with the increase of *LEAFY* gene expression and has occurred by the time a second flowering gene, *APETALA1*, is expressed. Because all three flowering pathways appear to converge with increased levels of *LEAFY*, this determination event should occur in species with a variety of balances among the pathways.

> **In the light-dependent pathway, plants use light-receptor molecules to measure the length of night. This information is then used to signal pathways that promote or inhibit flowering. Light receptors in the leaves trigger events that result in changes in the shoot meristem. The temperature-dependent pathway includes vernalization, the requirement for a period of chilling before a plant can flower. The autonomous pathway leads to flowering independent of environmental cues. Plants integrate information about position in regulating flowering, and both promoters and inhibitors of flowering are important.**

Control plants: no treatment

Experimental plants: pot-on-pot treatment

Control plants: Lower leaves were continually removed

FIGURE 41.10
Roots can inhibit flowering. Adventitious roots formed as bottomless pots were continuously placed over growing tobacco plants, delaying flowering. The delay in flowering is caused by the roots, not by the loss of the leaves. This was shown by removing leaves on control plants at the same time and in the same position as leaves on experimental plants that became buried as pots were added.

Identity Genes and the Formation of Floral Meristems and Floral Organs

Arabidopsis and snapdragon are valuable model systems for identifying flowering genes and understanding their interactions. The three flowering pathways discussed earlier in this section lead to an adult meristem becoming a floral meristem by either activating or repressing the inhibition of floral meristem identity genes (figure 41.11). Two of the key floral meristem identity genes are *LEAFY* and *APETALA1*. These genes establish the meristem as a flower meristem. They then turn on floral organ identity genes. The floral organ identity genes define four concentric whorls, moving inward in the floral meristem, as sepal, petal, stamen, and carpel. Meyerowitz and Coen proposed a model, called the ABC model, to explain how three classes of floral organ identity genes could specify four distinct organ types (figure 41.12). The ABC model proposes that three classes of organ identity genes (*A, B,* and *C*) specify the floral organs in the four floral whorls. By studying mutants, the researchers have determined the following:

1. Class *A* genes alone specify the sepals.
2. Class *A* and class *B* genes together specify the petals.
3. Class *B* and class *C* genes together specify the stamens.
4. Class *C* genes alone specify the carpels.

The beauty of the ABC model is that it is entirely testable by making different combinations of floral organ identity

mutants. Each class of genes is expressed in two whorls, yielding four different combinations of the gene products. When any one class is missing, aberrant floral organs occur in predictable positions.

It is important to recognize that this is actually only the beginning of the making of a flower. These organ identity genes are transcription factors that turn on many more genes that will actually give rise to the three-dimensional flower. Other genes "paint" the petals—that is, complex biochemical pathways lead to the accumulation of anthocyanin pigments in vacuoles. These pigments can be orange, red, or purple, and the actual color is influenced by pH as well.

The Formation of Gametes

The ovule within the carpel has origins more ancient than the angiosperms. Floral parts are modified leaves, and within the ovule is the female gametophyte. This next generation develops from placental tissue in the ovary. A megaspore mother cell develops and meiotically gives rise to the embryo sac. Usually, two layers of integument tissue form around this embryo sac and will become the seed coat. Genes responsible for initiating the integuments have been identified. Some genes also affect leaf structure.

The complex and elegant process that gives rise to the reproductive structure called the flower is often compared with metamorphosis in animals. It is indeed a metamorphosis, but the subtle shift from mitosis to meiosis in the megaspore mother cell leading to the development of a haploid, gamete-producing gametophyte is perhaps even more critical. The same can be said for pollen formation in the anther of the stamen. As we will see in section 41.2, the flower houses the haploid generations that will produce gametes. The flower also functions to increase the probability that male and female gametes from different (or sometimes the same) plants will unite.

Floral structures form as a result of floral meristem identity genes turning on floral organ identity genes that specify where sepals, petals, stamens, and carpels will form. This is followed by organ development, which involves many complex pathways that account for floral diversity among species.

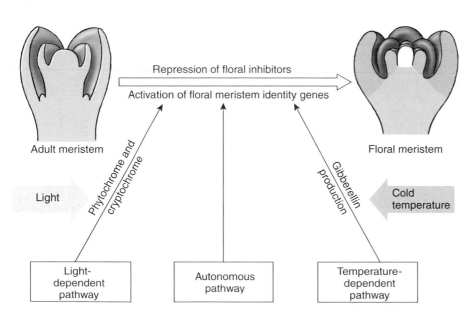

FIGURE 41.11
Model for flowering. The light-dependent, temperature-dependent, and autonomous flowering pathways promote the formation of floral meristems from adult meristems by repressing floral inhibitors and activating floral meristem identity genes.

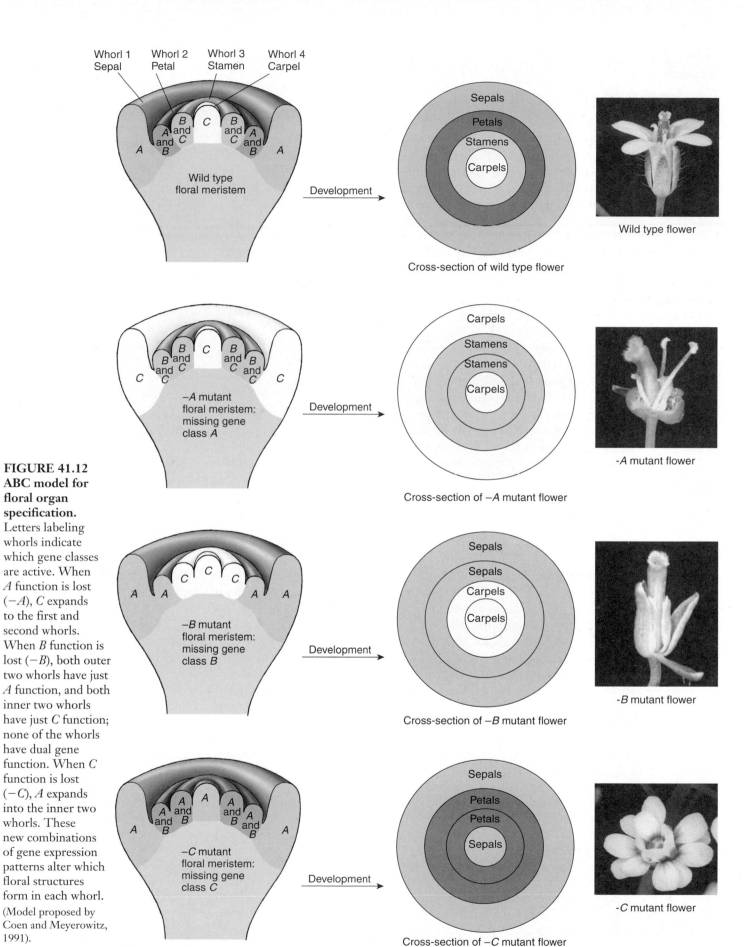

Whorl 1 Sepal   Whorl 2 Petal   Whorl 3 Stamen   Whorl 4 Carpel

Wild type floral meristem

Development →

Cross-section of wild type flower

Wild type flower

−A mutant floral meristem: missing gene class A

Development →

Cross-section of −A mutant flower

-A mutant flower

−B mutant floral meristem: missing gene class B

Development →

Cross-section of −B mutant flower

-B mutant flower

−C mutant floral meristem: missing gene class C

Development →

Cross-section of −C mutant flower

-C mutant flower

FIGURE 41.12 ABC model for floral organ specification. Letters labeling whorls indicate which gene classes are active. When *A* function is lost (−*A*), *C* expands to the first and second whorls. When *B* function is lost (−*B*), both outer two whorls have just *A* function, and both inner two whorls have just *C* function; none of the whorls have dual gene function. When *C* function is lost (−*C*), *A* expands into the inner two whorls. These new combinations of gene expression patterns alter which floral structures form in each whorl. (Model proposed by Coen and Meyerowitz, 1991).

41.2 Flowers are highly evolved for reproduction.

Evolution of the Flower

Pollination in angiosperms does not involve direct contact between the pollen grain and the ovule. Pollen matures within the anthers and is transported, often by insects, birds, or other animals, to the stigma of another flower. When pollen reaches the stigma, it germinates, and a pollen tube grows down, carrying the sperm nuclei to the embryo sac. After double fertilization takes place, development of the embryo and endosperm begins. The seed matures within the ripening fruit; the germination of the seed initiates another life cycle.

Successful pollination in many angiosperms depends on the regular attraction of **pollinators** such as insects, birds, and other animals, so that pollen is transferred between plants of the same species. When animals disperse pollen, they perform the same functions for flowering plants that they do for themselves when they actively search out mates. The relationship between plant and pollinator can be quite intricate. Mutations in either partner can block reproduction. If a plant flowers at the "wrong" time, the pollinator may not be available. If the morphology of the flower or pollinator is altered, there may be physical barriers to pollination. Clearly, floral morphology has coevolved with pollinators, and the result is much more complex and diverse than the initiation of four distinct whorls of organs described in section 41.1.

Characteristics of Floral Evolution

The evolution of the angiosperms is a focus of chapter 29. Here we need to keep in mind that the diversity of angiosperms is partly due to the evolution of a great variety of floral phenotypes that may enhance the effectiveness of pollination. All floral organs are thought to have evolved from leaves. In some early angiosperms, these organs maintain the spiral phyllotaxy often found in leaves. The trend has been toward four distinct whorls. A *complete flower* has four whorls of parts (calyx, corolla, androecium, and gynoecium), while an *incomplete flower* lacks one or more of the whorls (figure 41.13).

In both complete and incomplete flowers, the **calyx** usually constitutes the outermost whorl; it consists of flattened appendages, called **sepals**, which protect the flower in the bud. The petals collectively make up the **corolla** and may be fused. Many petals function to attract pollinators. While these two outer whorls of floral organs are sterile, they can enhance reproductive success.

Androecium (Greek *andros*, "man," + *oikos*, "house") is a collective term for all the **stamens** (male structures) of a flower. Stamens are specialized structures that bear the angiosperm microsporangia. Similar structures bear the microsporangia in the pollen cones of gymnosperms. Most

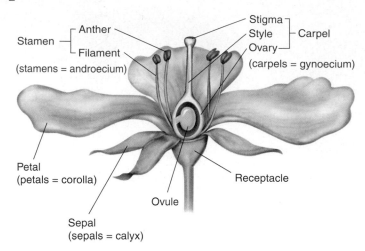

FIGURE 41.13
Structure of an angiosperm flower. This is a complete flower.

living angiosperms have stamens whose **filaments** ("stalks") are slender and often threadlike, and whose four microsporangia are evident at the apex in a swollen portion, the **anther.** Some of the more primitive angiosperms have stamens that are flattened and leaflike, with the sporangia produced from the upper or lower surface.

The **gynoecium** (Greek *gyne*, "woman," + *oikos*, "house") is a collective term for all the female parts of a flower. In most flowers, the gynoecium, which is unique to angiosperms, consists of a single **carpel** or two or more fused carpels. Single or fused carpels are often referred to as simple or compound pistils, respectively. Most flowers with which we are familiar—for example, those of tomatoes and oranges—have a compound pistil. Other, less specialized flowers—for example, buttercups and stonecups—may have several to many separate, simple pistils, each formed from a single carpel. **Ovules** (which develop into seeds) are produced in the pistil's swollen lower portion, the **ovary,** which usually narrows at the top into a slender, necklike **style** with a pollen-receptive **stigma** at its apex. Sometimes the stigma is divided, with the number of stigma branches indicating how many carpels are in the particular pistil. Carpels are essentially inrolled floral leaves with ovules along the margins. It is possible that the first carpels were leaf blades that folded longitudinally; the margins, which had hairs, did not actually fuse until the fruit developed, but the hairs interlocked and were receptive to pollen. In the course of evolution, evidence indicates that the hairs became localized into a stigma, a style was formed, and the fusing of the carpel margins ultimately resulted in a pistil. In many modern flowering plants, the carpels have become highly modified and are not visually distinguishable from one another unless the pistil is cut open.

Trends of Floral Specialization

Two major evolutionary trends led to the wide diversity of modern flowering plants: (1) Separate floral parts have grouped together, or fused, and (2) floral parts have been lost or reduced (figure 41.14). In the more advanced angiosperms, the number of parts in each whorl has often been reduced from many to few. The spiral patterns of attachment of all floral parts in primitive angiosperms have, in the course of evolution, given way to a single whorl at each level. The central axis of many flowers has shortened, and the whorls are close to one another. In some evolutionary lines, the members of one or more whorls have fused with one another, sometimes joining into a tube. In other kinds of flowering plants, different whorls may be fused together. Whole whorls may even be lost from the flower, which may lack sepals, petals, stamens, carpels, or various combinations of these structures. Modifications often relate to pollination mechanisms, and in plants such as the grasses, wind has replaced animals for pollen dispersal.

Although much floral diversity is the result of natural selection related to pollination, it is important to recognize the impact of breeding (artificial selection) on flower morphology. Humans have selected for practical or aesthetic traits that may have little adaptive value to species in the wild. For example, maize (corn) has been selected to satisfy the human palate. Human intervention ensures the reproductive success of each generation; however, in a natural setting, modern corn would not have the same protection from herbivores as its ancestors, and the fruit dispersal mechanism would be quite different (see figure 22.7). Floral shops sell heavily bred species with modified petals, often due to polyploidy, that enhance their economic value but not their ability to attract pollinators. In making inferences about symbioses between flowers and pollinators, be sure to look at native plants that have not been genetically altered by human intervention.

Trends in Floral Symmetry

Other trends in floral evolution have affected the symmetry of the flower. Primitive flowers such as those of buttercups are *radially symmetrical;* that is, one could draw a line anywhere through the center and have two roughly equal halves. Flowers of many advanced groups are *bilaterally symmetrical;* that is, they are divisible into two equal parts along only a single plane. Examples of such flowers are snapdragons, mints, and orchids (figure 41.15). Bilaterally symmetrical flowers are also common among violets and peas. In these groups, they are often associated with advanced and highly precise pollination systems. Bilateral symmetry has arisen independently many times. In snapdragons, the *CYCLOIDIA* gene regulates floral symmetry, and in its absence flowers are more radial (figure 41.16). Here the experimental alteration of a single gene is sufficient to cause a dramatic change in morphology. Whether

FIGURE 41.14 Trends in floral specialization. Wild geranium, *Geranium maculatum.* The petals are reduced to five each, the stamens to ten.

FIGURE 41.15 Bilateral symmetry in an orchid. While primitive flowers are usually radially symmetrical, flowers of many advanced groups, such as the orchid family (Orchidaceae), are bilaterally symmetrical.

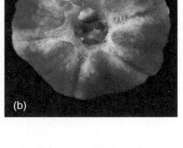

FIGURE 41.16
Genetic regulation of asymmetry in flowers. (*a*) Snapdragon flowers normally have bilateral symmetry. (*b*) The *CYCLOIDIA* gene regulates floral symmetry, and *cycloidia* mutant snapdragons have radially symmetrical flowers.

the same gene or functionally similar genes arose naturally in parallel in other species is an open question.

Some of the first angiosperms likely had numerous free, spirally arranged flower parts. Modification of floral parts appears to be closely tied to pollination mechanisms. More recently, horticulturists have bred plants for aesthetic reasons, resulting in an even greater diversity of flowers.

Formation of Angiosperm Gametes

Reproductive success depends on uniting the gametes (egg and sperm) found in the embryo sacs and pollen grains of flowers. As mentioned in chapter 29, plant sexual life cycles are characterized by an alternation of generations, in which a diploid sporophyte generation gives rise to a haploid gametophyte generation. In angiosperms, the gametophyte generation is very small and is completely enclosed within the tissues of the parent sporophyte. The male gametophytes, or microgametophytes, are **pollen grains.** The female gametophyte, or megagametophyte, is the **embryo sac.** Pollen grains and the embryo sac both are produced in separate, specialized structures of the angiosperm flower.

Like animals, angiosperms have separate structures for producing male and female gametes (figure 41.17), but the reproductive organs of angiosperms are different from those of animals in two ways. First, in angiosperms, both male and female structures usually occur together in the same individual flower. Second, angiosperm reproductive structures are not permanent parts of the adult individual. Angiosperm flowers and reproductive organs develop seasonally, at times of the year most favorable for pollination. In some cases, reproductive structures are produced only once, and the parent plant dies. It is significant that the germ line for angiosperms is not set aside early in development, but forms quite late, as detailed in section 41.1.

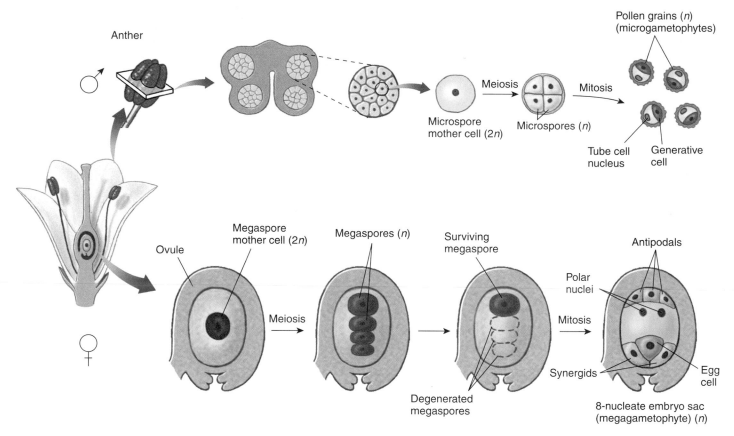

FIGURE 41.17

Formation of pollen grains and the embryo sac. Diploid (*2n*) microspore mother cells are housed in the anther and divide by meiosis to form four haploid (*n*) microspores. Each microspore develops by mitosis into a pollen grain. The generative cell within the pollen grain will later divide to form two sperm cells. Within the ovule, one diploid megaspore mother cell divides by meiosis to produce four haploid megaspores. Usually only one of the megaspores survives, and the other three degenerate. The surviving megaspore divides by mitosis to produce an embryo sac with eight nuclei.

FIGURE 41.18

Pollen grains. (*a*) In the Easter lily, *Lilium candidum*, the pollen tube emerges from the pollen grain through the groove or furrow that occurs on one side of the grain. (*b*) In a plant of the sunflower family, *Hyoseris longiloba*, three pores are hidden among the ornamentation of the pollen grain. The pollen tube may grow out through any one of them.

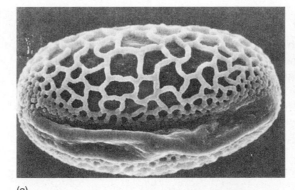
(a)

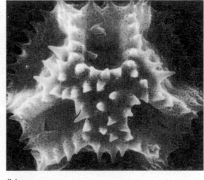
(b)

Pollen Formation

Pollen grains form in the two pollen sacs located in the anther. Each pollen sac contains specialized chambers in which the *microspore mother cells* are enclosed and protected. The microspore mother cells undergo meiosis to form four haploid microspores. Subsequently, mitotic divisions form four pollen grains. Inside each pollen grain is a generative cell; this cell will later divide to produce two sperm cells.

Pollen grain shapes are specialized for specific flower species. As discussed in more detail later in this section, fertilization requires that the pollen grain grow a tube that penetrates the style until it encounters the ovary. Most pollen grains have a furrow from which this pollen tube emerges; some grains have three furrows (figure 41.18).

Embryo Sac Formation

Eggs develop in the ovules of the angiosperm flower. Within each ovule is a megaspore mother cell. Each megaspore mother cell undergoes meiosis to produce four haploid megaspores. In most plants, however, only one of these megaspores survives; the rest are absorbed by the ovule. The lone remaining megaspore undergoes repeated mitotic divisions to produce eight haploid nuclei that are enclosed within a seven-celled embryo sac. Within the embryo sac, the eight nuclei are arranged in precise positions. One nucleus is located near the opening of the embryo sac in the egg cell. Two are located in a single cell in the middle of the embryo sac and are called polar nuclei; two nuclei are contained in cells called synergids that flank the egg cell; and the other three nuclei reside in cells called the antipodals, located at the end of the sac, opposite the egg cell (figure 41.19). The first step in uniting the sperm in the pollen grain with the egg and polar nuclei is to get pollen germinating on the stigma of the carpel and growing toward the embryo sac.

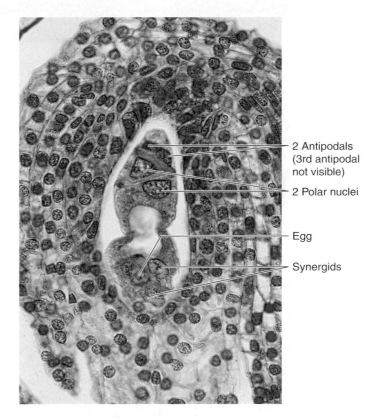

2 Antipodals (3rd antipodal not visible)

2 Polar nuclei

Egg

Synergids

FIGURE 41.19

A mature embryo sac of a lily. Eight nuclei are produced by mitotic divisions of the haploid megaspore. One is in the egg, two are polar nuclei, two occur in synergid cells, and three are in antipodal cells.

In angiosperms, both male and female structures often occur together in the same individual flower. These reproductive structures are not a permanent part of the adult individual, and the germ line is not set aside early in development.

Pollination

Pollination is the process by which pollen is placed on the stigma. Pollen may be carried to the flower by wind or by animals, or it may originate within the individual flower itself. When pollen from a flower's anther pollinates the same flower's stigma, the process is called *self-pollination*.

Pollination in Early Seed Plants

Early seed plants were pollinated passively, by the action of the wind. As in present-day conifers, great quantities of pollen were shed and blown about, occasionally reaching the vicinity of the ovules of the same species. Individual plants of any given species must grow relatively close to one another for such a system to operate efficiently. Otherwise, the chance that any pollen will arrive at the appropriate destination is very small. The vast majority of windblown pollen travels less than 100 meters. This short distance is significant compared with the long distances pollen is routinely carried by certain insects, birds, and other animals.

Pollination by Animals

The spreading of pollen from plant to plant by pollinators visiting flowers of specific angiosperm species has played an important role in the evolutionary success of the group. It now seems clear that the earliest angiosperms, and perhaps their ancestors also, were insect-pollinated, and the coevolution of insects and plants has been important for both groups for over 100 million years. Such interactions have also been important in bringing about increased floral specialization. As flowers become increasingly specialized, so do their relationships with particular groups of insects and other animals.

Bees. Among insect-pollinated angiosperms, the most numerous groups are those pollinated by bees (figure 41.20). Like most insects, bees initially locate sources of food by odor, and then orient themselves on the flower or group of flowers by its shape, color, and texture. Flowers that bees characteristically visit are often blue or yellow. Many have stripes or lines of dots that indicate the location of the nectaries, which often occur within the throats of specialized flowers. Some bees collect nectar, which is used as a source of food for adult bees and occasionally for larvae. Most of the approximately 20,000 species of bees visit flowers to obtain pollen. Pollen is used to provide food in cells where bee larvae complete their development.

Only a few hundred species of bees are social or semi-social in their nesting habits. These bees live in colonies, as

FIGURE 41.20
Pollination by a bumblebee. As this bumblebee, *Bombus* sp., squeezes into the bilaterally symmetrical, advanced flower of a member of the mint family, the stigma contacts its back and picks up any pollen that the bee may have acquired during a visit to a previous flower.

do the familiar honeybee, *Apis mellifera*, and the bumblebee, *Bombus* sp. Such bees produce several generations a year and must shift their attention to different kinds of flowers as the season progresses. To maintain large colonies, they also must use more than one kind of flower as a food source at any given time.

Except for these social and semi-social bees and about 1000 species that are parasitic in the nests of other bees, the great majority of bees—at least 18,000 species—are solitary. Solitary bees in temperate regions characteristically produce only a single generation in the course of a year. Often, they are active as adults for as little as a few weeks a year.

Solitary bees often use the flowers of a given group of plants almost exclusively as sources of their larval food. The highly constant relationships of such bees with those flowers may lead to modifications, over time, in both the flowers and the bees. For example, the time of day when the flowers open may correlate with the time when the bees appear; the mouthparts of the bees may become elongated in relation to tubular flowers; or the bees' pollen-collecting apparatuses may be adapted to the anthers of the plants that they normally visit. When such relationships are established, they provide both an efficient mechanism of pollination for the flowers and a constant source of food for the bees that "specialize" on them.

Insects Other Than Bees. Among flower-visiting insects other than bees, a few groups are especially prominent. Flowers such as phlox, which are visited regularly by butterflies, often have flat "landing platforms" on which butterflies perch. They also tend to have long, slender floral tubes filled with nectar that is accessible to the long, coiled proboscis characteristic of Lepidoptera, the order of insects that includes butterflies and moths. Flowers such as jimsonweed, evening primrose, and others visited regularly by moths are often white, yellow, or some other pale color; they also tend to be heavily scented, making the flowers easy to locate at night.

Birds. Several interesting groups of plants are regularly visited and pollinated by birds, especially the hummingbirds of North and South America and the sunbirds of Africa (figure 41.21). Such plants must produce large amounts of nectar because birds will not continue to visit flowers of that plant if they do not find enough food to maintain themselves. Flowers producing large amounts of nectar have no advantage in being visited by insects because an insect could obtain its energy requirements at a single flower and would not cross-pollinate the flower. How are these different selective forces balanced in flowers that are "specialized" for hummingbirds and sunbirds?

Ultraviolet light is highly visible to insects. *Carotenoids*, yellow or orange pigments frequently found in plants, are responsible for the colors of many flowers, including sunflowers and mustard. Carotenoids reflect both in the yellow range and in the ultraviolet range, the mixture resulting in a distinctive color called "bee's purple." Such yellow flowers may also be marked in distinctive ways normally invisible to us, but highly visible to bees and other insects (figure 41.22). These markings can be in the form of a bull's-eye or a landing strip.

Red does not stand out as a distinct color to most insects, but it is a very conspicuous color to birds. To most insects, the red upper leaves of poinsettias look just like the other leaves of the plant. Consequently, even though the flowers produce abundant supplies of nectar and attract hummingbirds, insects tend to bypass them. Thus, the red color both signals to birds the presence of abundant nectar and makes that nectar as inconspicuous as possible to insects. Red is also seen again in fruits that are dispersed by birds.

Other Animals. Other animals, including bats and small rodents, may aid in pollination. The signals here are also species-specific. These animals assist in dispersing the seeds and fruits that result from pollination. Monkeys are attracted to orange and yellow, and thus can be effective in dispersing those fruits.

Wind-Pollinated Angiosperms

Many angiosperms, representing a number of different groups, are wind-pollinated—a characteristic of early seed plants. Among them are such familiar plants as oaks, birches, cottonwoods, grasses, sedges, and nettles. The flowers of these plants are small, greenish, and odorless; their corollas are reduced or absent (see figures 41.23 and 41.24). Such flowers often are grouped together in fairly large numbers and may hang down in tassels that wave about in the wind and shed pollen freely. Many wind-pollinated plants have stamen- and carpel-containing flowers separated among individuals or on a single individual. If the pollen-producing and ovule-bearing flowers are separated, it is certain that pollen released to the wind will reach a flower other than the one that shed it, a strategy

FIGURE 41.21
Hummingbirds and flowers. A long-tailed hermit hummingbird extracts nectar from the flowers of *Heliconia imbricata* in the forests of Costa Rica. Note the pollen on the bird's beak. Hummingbirds of this group obtain nectar primarily from long, curved flowers that more or less match the length and shape of their beaks.

(a)   (b)

FIGURE 41.22
How a bee sees a flower. (*a*) The yellow flower of *Ludwigia peruviana* (Onagraceae) photographed in normal light and (*b*) with a filter that selectively transmits ultraviolet light. The outer sections of the petals reflect both yellow and ultraviolet, a mixture of colors called "bee's purple"; the inner portions of the petals reflect yellow only and therefore appear dark in the photograph that emphasizes ultraviolet reflection. To a bee, this flower appears as if it has a conspicuous central bull's-eye.

that greatly promotes outcrossing. Some wind-pollinated plants, especially trees and shrubs, flower in the spring, before the development of their leaves can interfere with the wind-borne pollen. Wind-pollinated species do not depend on the presence of a pollinator for species survival.

> Bees are the most frequent and characteristic pollinators of flowers. Insects often are attracted by the odors of flowers. Bird-pollinated flowers are characteristically odorless and red, with the nectar not readily accessible to insects.

Self-Pollination

All of the modes of pollination that we have considered thus far tend to lead to outcrossing, which is as highly advantageous for plants as it is for eukaryotic organisms generally. Nevertheless, self-pollination also occurs among angiosperms, particularly in temperate regions. Most of the self-pollinating plants have small, relatively inconspicuous flowers that shed pollen directly onto the stigma, sometimes even before the bud opens. You might logically ask why many self-pollinated plant species have survived if outcrossing is as important genetically for plants as it is for animals. There are two basic reasons for the frequent occurrence of self-pollinated angiosperms:

1. Self-pollination obviously is ecologically advantageous under certain circumstances because self-pollinators do not need to be visited by animals to produce seed. As a result, self-pollinated plants expend less energy in producing pollinator attractants and can grow in areas where the kinds of insects or other animals that might visit them are absent or very scarce—as in the Arctic or at high elevations.

2. In genetic terms, self-pollination produces progenies that are more uniform than those that result from outcrossing. Remember that because meiosis is involved, recombination still takes place, and the offspring will not be identical to the parent. However, such progenies may contain high proportions of individuals well-adapted to particular habitats. Self-pollination in normally outcrossing species tends to produce large numbers of ill-adapted individuals because it brings together deleterious recessive alleles, but some of these combinations may be highly advantageous in particular habitats. In such habitats, it may be advantageous for the plant to continue self-pollinating indefinitely. This is the main reason many self-pollinating plant species are weeds—not only have humans made weed habitats uniform, but they have also spread the weeds all over the world.

Factors That Promote Outcrossing

Outcrossing, as we have stressed, is critically important for the adaptation and evolution of all eukaryotic organisms. Often, flowers contain both stamens and pistils, which increases the likelihood of self-pollination. One strategy to promote outcrossing is to separate stamens and pistils.

In various species of flowering plants—for example, willows and some mulberries—staminate and pistillate flowers may occur on separate plants. Such plants, which produce only ovules or only pollen, are called **dioecious** (Greek, "two houses"). Obviously, they cannot self-pollinate and must rely exclusively on outcrossing. In other kinds of plants, such as oaks, birches, corn (maize), and pumpkins, separate male and female flowers may both be produced on the same

FIGURE 41.23 Staminate and pistillate flowers of a birch, *Betula* sp. Birches are monoecious; their staminate flowers hang down in long, yellowish tassels, while their pistillate flowers mature into clusters of small, brownish, conelike structures.

FIGURE 41.24 Wind-pollinated flowers. The large yellow anthers, dangling on very slender filaments, are hanging out, about to shed their pollen to the wind. Later, these flowers will become pistillate, with long, feathery stigmas—well suited for trapping windblown pollen—sticking far out of them. Many grasses, like this one, are therefore dichogamous.

plant. Such plants are called **monoecious,** meaning "one house" (figure 41.23). In monoecious plants, the separation of pistillate and staminate flowers, which may mature at different times, greatly enhances the probability of outcrossing.

Even if, as usually is the case, functional stamens and pistils are both present in each flower of a particular plant species, these organs may reach maturity at different times. Plants in which this occurs are called **dichogamous.** If the stamens mature first, shedding their pollen before the stigmas are receptive, the flower is effectively staminate at that time. Once the stamens have finished shedding pollen, the stigma or stigmas may become receptive, and the flower may become essentially pistillate (figures 41.24 and 41.25). This has the same effect as if the flower completely lacked either functional stamens or functional pistils; its outcrossing rate is thereby significantly increased.

Many flowers are constructed such that the stamens and stigmas do not come in contact with each other. With such an arrangement, there is a natural tendency for the pollen to be transferred to the stigma of another flower rather than to the stigma of its own flower, thereby promoting outcrossing.

Even when a flower's stamens and stigma mature at the same time, genetic **self-incompatibility,** which is widespread in flowering plants, increases outcrossing. Self-incompatibility results when the pollen and stigma recognize each other as being genetically related, and pollen tube growth is blocked (figure 41.26). Self-incompatibility is controlled by the *S* (self-incompatibility) locus. There are many alleles at the *S* locus that regulate recognition responses between the pollen and stigma. Researchers have identified two types of self-incompatibility. *Gametophytic self-incompatibility* depends on the haploid *S* locus of the pollen and the diploid *S* locus of the stigma. If either of the *S* alleles in the stigma matches the pollen *S* allele, pollen tube growth stops before it reaches the embryo sac. Petunias have gametophytic self-incompatibility. In *sporophytic self-incompatibility*, as occurs in broccoli, both *S* alleles of the pollen parent are important; if the alleles in the stigma match either of the pollen parent *S* alleles, the haploid pollen will not germinate.

Much is being learned about the molecular and biochemical basis of recognition mechanisms and the signal transduction pathways that block the successful growth of the pollen tube. Pollen-recognition mechanisms may have originated in a common ancestor of the gymnosperms. Fossils with pollen tubes from the Carboniferous period are consistent with the hypothesis that they had highly evolved pollen-recognition systems. Systems that recognized foreign pollen may have predated self-recognition systems.

(a) (b)

FIGURE 41.25

Dichogamy, as illustrated by the flowers of fireweed, *Epilobium angustifolium*. More than 200 years ago (in the 1790s) fireweed, which is outcrossing, was one of the first plant species to have its process of pollination described. First, the anthers shed pollen, and then the style elongates above the stamens while the four lobes of the stigma curl back and become receptive. Consequently, the flowers are functionally staminate at first, becoming pistillate about two days later. The flowers open progressively up the stem, so that the lowest are visited first. Working up the stem, the bees encounter pollen-shedding, staminate-phase flowers and become covered with pollen, which they then carry to the lower, functionally pistillate flowers of another plant. Shown here are flowers in (a) the staminate phase and (b) the pistillate phase.

> Self-pollinated angiosperms are frequent in areas where strong selective pressure produces large numbers of genetically uniform individuals adapted to specific, relatively uniform habitats. Outcrossing in plants may be promoted through dioecism, monoecism, self-incompatibility, or the physical separation or different maturation times of the stamens and pistils. Outcrossing promotes genetic diversity.

FIGURE 41.26

Self-pollination can be genetically controlled so self-pollen is blocked. (*a*) Gametophytic self-incompatibility is determined by the haploid pollen genotype. (*b*) Sporophytic self-incompatibility recognizes the genotype of the diploid pollen parent, not just the haploid pollen genotype. In both cases, the recognition is based on the *S* locus, which has many different alleles. The subscript numbers indicate the *S* allele genotype. In gametophytic self-incompatibility, the block comes after pollen tube germination. In sporophytic self-incompatibility, the pollen tube fails to germinate.

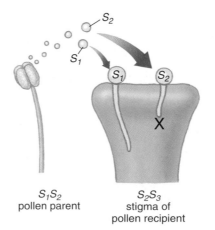
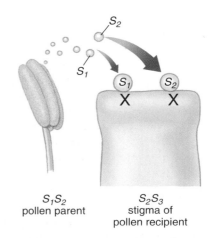

S_1S_2
pollen parent

S_2S_3
stigma of pollen recipient

S_1S_2
pollen parent

S_2S_3
stigma of pollen recipient

(a) Gametophytic self-incompatibility

(b) Sporophytic self-incompatibility

Fertilization

Fertilization in angiosperms is a complex, somewhat unusual process in which two sperm cells are utilized in a unique process called **double fertilization.** Double fertilization results in two key developments: (1) the fertilization of the egg, and (2) the formation of a nutrient substance called endosperm that nourishes the embryo. Once a pollen grain has been spread by wind, by animals, or through self-pollination, it adheres to the sticky, sugary substance that covers the stigma and begins to grow a **pollen tube** that pierces the style (figure 41.27). The pollen tube, nourished by the sugary substance, grows until it reaches the ovule in the ovary. Meanwhile, the generative cell within the pollen grain tube cell divides to form two sperm cells.

The pollen tube eventually reaches the embryo sac in the ovule. At the entry to the embryo sac, one of the nuclei flanking the egg cell degenerates, and the pollen tube enters that cell. The tip of the pollen tube bursts and releases the two sperm cells. One of the sperm cells fertilizes the egg cell, forming a zygote. The other sperm cell fuses with the two polar nuclei located at the center of the embryo sac, forming the triploid (3*n*) primary endosperm nucleus. The primary endosperm nucleus eventually develops into the endosperm.

Once fertilization is complete, the embryo develops as its cells divide numerous times. Meanwhile, protective tissues enclose the embryo, resulting in the formation of the seed. The seed, in turn, is enclosed in another structure, called the fruit. These typical angiosperm structures evolved in response to the need for seeds to be dispersed over long distances to ensure genetic variability.

In double fertilization, angiosperms utilize two sperm cells. One fertilizes the egg, while the other helps form a substance called endosperm that nourishes the embryo.

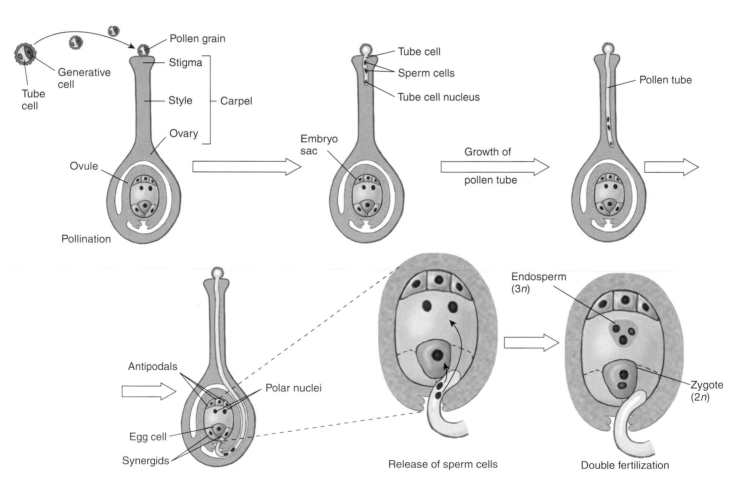

FIGURE 41.27
The formation of the pollen tube and double fertilization. When pollen lands on the stigma of a flower, the pollen tube cell grows toward the embryo sac, forming a pollen tube. While the pollen tube is growing, the generative cell divides to form two sperm cells. When the pollen tube reaches the embryo sac, it enters one of the synergids and releases the sperm cells. In a process called double fertilization, one sperm cell nucleus fuses with the egg cell to form the diploid (2*n*) zygote, and the other sperm cell nucleus fuses with the two polar nuclei to form the triploid (3*n*) endosperm nucleus.

41.3 Many plants can clone themselves by asexual reproduction.

Asexual Reproduction

While self-pollination reduces genetic variability, asexual reproduction results in genetically identical individuals because only mitotic cell divisions occur. In the absence of meiosis, individuals that are highly adapted to a relatively unchanging environment persist for the same reasons that self-pollination is favored. Should conditions change dramatically, there will be less variation in the population for natural selection to act upon, and the species may be less likely to survive. Asexual reproduction is also used in agriculture and horticulture to propagate a particularly desirable plant whose traits would be altered by sexual reproduction or even by self-pollination. Most roses and potatoes, for example, are vegetatively (asexually) propagated.

FIGURE 41.28
Vegetative reproduction. Small plants arise from notches along the leaves of the house plant *Kalanchoë daigremontiana.*

Vegetative Reproduction

In a very common form of asexual reproduction called *vegetative reproduction*, new plant individuals are simply cloned from parts of adults (figure 41.28). The forms of vegetative reproduction in plants are many and varied.

Runners. Some plants reproduce by means of *runners*—long, slender stems that grow along the surface of the soil. In the cultivated strawberry, for example, leaves, flowers, and roots are produced at every other node on the runner. Just beyond each second node, the tip of the runner turns up and becomes thickened. This thickened portion first produces adventitious roots and then a new shoot that continues the runner.

Rhizomes. Underground horizontal stems, or *rhizomes*, are also important reproductive structures, particularly in grasses and sedges. Rhizomes invade areas near the parent plant, and each node can give rise to a new flowering shoot. The noxious character of many weeds results from this type of growth pattern, and many garden plants, such as irises, are propagated almost entirely from rhizomes. Corms, bulbs, and tubers are also stems specialized for storage and reproduction. White potatoes are propagated artificially from tuber segments, each with one or more "eyes." The eyes, or "seed pieces," of potato give rise to the new plant.

Suckers. The roots of some plants—for example, cherry, apple, raspberry, and blackberry—produce *suckers*, or sprouts, which give rise to new plants. Commercial varieties of banana do not produce seeds and are propagated by suckers that develop from buds on underground stems. When the root of a dandelion is broken, as it may be if one attempts to pull it from the ground, each root fragment may give rise to a new plant.

Adventitious Plantlets. In a few plant species, even the leaves are reproductive. One example is the houseplant *Kalanchoë daigremontiana*, familiar to many people as the "maternity plant," or "mother of thousands." The common names of this plant are based on the fact that numerous plantlets arise from meristematic tissue located in notches along the leaves. The maternity plant is ordinarily propagated by means of these small plants, which, when they mature, drop to the soil and take root.

Apomixis

In certain plants, including some citruses, certain grasses (such as Kentucky bluegrass), and dandelions, the embryos in the seeds may be produced asexually from the parent plant. This kind of asexual reproduction is known as *apomixis*. The seeds produced in this way give rise to individuals that are genetically identical to their parents. Thus, although these plants reproduce asexually by cloning diploid cells in the ovule, they also gain the advantage of seed dispersal, an adaptation usually associated with sexual reproduction. Embryos can also form via mitosis when plant tissues are cultured. In general, vegetative reproduction, apomixis, and other forms of asexual reproduction promote the exact reproduction of individuals that are particularly well suited to a certain environment or habitat. Asexual reproduction among plants is far more common in harsh or marginal environments, where there is little leeway for variation. For example, a greater proportion of asexual plants occur in the Arctic than in temperate regions.

Plants that reproduce asexually clone new individuals from portions of the root, stem, leaves, or ovules of adult individuals. The asexually produced progeny are genetically identical to the parent individual.

Plant Tissue Culture

Whole plants can be cloned by regenerating plant cells or tissues on nutrient medium with growth hormones. This is another form of asexual reproduction. Cultured leaf, stem, and root tissues can undergo organogenesis in culture and form roots and shoots. In some cases, individual cells can also give rise to whole plants in culture. In other cases, the cell wall is removed with enzymes, leaving behind the *protoplast*, a plant cell enclosed only by a plasma membrane.

Plant protoplasts are more easily transformed with foreign DNA using approaches such as electroporation. In addition, protoplasts isolated from different plants can be forced to fuse together to form a hybrid. If they are regenerated into whole plants, these hybrids formed from protoplast fusion can represent genetic combinations that would never occur in nature. Hence, protoplast fusion can provide an additional means of genetic engineering, allowing beneficial traits from one plant to be incorporated into another plant despite broad differences between the species. When either single or fused protoplasts are transferred to a culture growth medium, cell wall regeneration takes place. This is followed by cell division to form a callus (figure 41.29). Once a callus is formed, whole plants can be produced in culture.

It is often possible to regenerate an entire plant from one or a few cells.

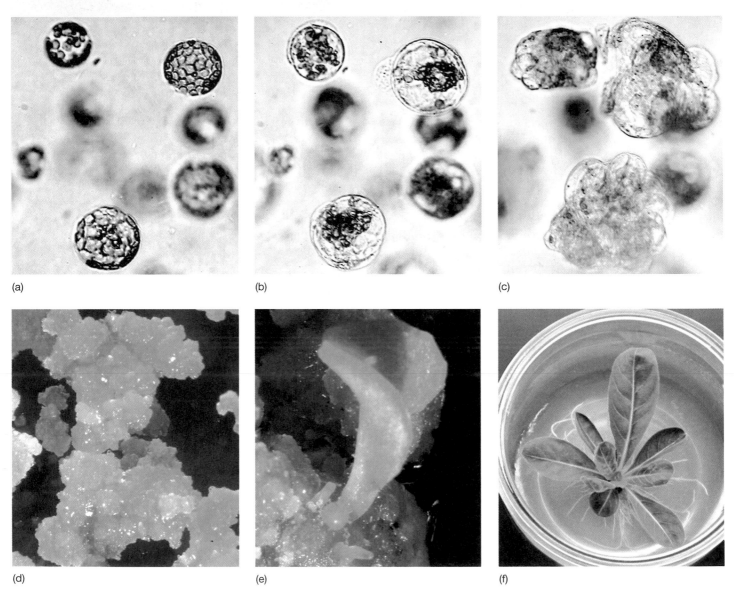

(a)   (b)   (c)

(d)   (e)   (f)

FIGURE 41.29

Protoplast regeneration.   Different stages in the recovery of intact plants from single plant protoplasts of evening primrose. (*a*) Individual plant protoplasts. (*b*) Regeneration of the cell wall and the beginning of cell division. (*c,d*) Aggregates of plant cells resulting from cell division can form a callus. (*e*) Production of somatic cell embryos from the callus. (*f*) Recovery of a plantlet from the somatic cell embryo in culture. The plant can later be rooted in soil.

41.4 How long do plants and plant organs live?

The Life Span of Plants

Once established, plants live for highly variable periods of time, depending on the species. Life span may or may not correlate with reproductive strategy. Woody plants, which have extensive secondary growth, nearly always live longer than herbaceous plants, which have limited or no secondary growth. Bristlecone pine, for example, can live upward of 4000 years. Some herbaceous plants send new stems above the ground every year, producing them from woody underground structures. Others germinate and grow, flowering just once before they die. Shorter-lived plants rarely become very woody because there is not enough time for secondary tissues to accumulate. Depending on the length of their life cycles, herbaceous plants may be annual, biennial, or perennial, while woody plants are generally perennial (figure 41.30). Determining life span is even more complicated for clonally reproducing organisms. Aspen trees form huge clones from asexual reproduction of their roots. Collectively, an aspen clone may form the largest "organism" on earth. Other asexually reproducing plants may cover less territory but live for thousands of years. Creosote bushes in the Mojave Desert have been identified that are up to 12,000 years old!

(a)

(b)

FIGURE 41.30
Annual and perennial plants. Plants live for very different lengths of time. (*a*) Desert annuals complete their entire life span in a few weeks. (*b*) Some trees, such as the giant redwood (*Sequoiadendron giganteum*), which occurs in scattered groves along the western slopes of the Sierra Nevada in California, live 2000 years or more.

Annual Plants

Annual plants grow, flower, and form fruits and seeds within one growing season; they then die when the process is complete. Many crop plants are annuals, including corn, wheat, and soybeans. Annuals generally grow rapidly under favorable conditions and in proportion to the availability of water or nutrients. The lateral meristems of some annuals, such as sunflowers or giant ragweed, do produce poorly developed secondary tissues, but most annuals are entirely herbaceous. Annuals typically die after flowering once, the developing flowers or embryos using hormonal signaling to reallocate nutrients so the parent plant literally starves to death. This can be demonstrated by comparing a population of bean plants in which the beans are continually picked with a population in which the beans are left on the plant. The frequently picked population will continue to grow and yield beans much longer than the untouched population. The process that leads to the death of a plant is called *senescence*.

Biennial Plants

Biennial plants, which are much less common than annuals, have life cycles that take two years to complete. During the first year, biennials store photosynthate in underground storage organs. During the second year of growth, flowering stems are produced using energy stored in the underground parts of the plant. Certain crop plants, including carrots, cabbage, and beets, are biennials, but these plants generally are harvested for food during their first season, before they flower. They are grown for their leaves or roots, not for their fruits or seeds. Wild biennials include evening primroses, Queen Anne's lace, and mullein. Many plants that are considered biennials actually do not flower until they are three or more years of age, but all biennial plants flower only once before they die.

Perennial Plants

Perennial plants continue to grow year after year and may be herbaceous, as are many woodland, wetland, and prairie wildflowers, or woody, as are trees and shrubs. The majority of vascular plant species are perennials. Herbaceous perennials rarely experience any secondary growth in their stems; the stems die each year after a period of relatively rapid growth and food accumulation. Food is often stored in the plants' roots or underground stems, which can become quite large in comparison with their less substantial aboveground counterparts.

Trees and shrubs generally flower repeatedly, but there are exceptions. Bamboo lives for many seasons as a nonreproducing plant, but senesces and dies after flowering. The same is true for at least one tropical tree, which achieves great heights before flowering and senescing. Considering the tremendous amount of energy that goes into the growth of a tree, this particular reproductive strategy is quite curious.

Trees and shrubs are either *deciduous*, with all the leaves falling at one particular time of year and the plants remaining bare for a period, or *evergreen*, with the leaves dropping throughout the year and the plants never appearing completely bare. In northern temperate regions, conifers are the most familiar evergreens, but in tropical and subtropical regions, most angiosperms are evergreen, except where there is severe seasonal drought. In these areas, many angiosperms are deciduous, losing their leaves during the drought and thus conserving water.

Organ Abscission

Senescence is an important developmental process that leads to the death of an organ, a shoot, or the whole plant. Annual and biennial plants undergo whole-plant senescence, but individual organs on any plant can also senesce and be shed. The process by which leaves or petals are shed is called **abscission.**

One advantage to organ senescence is that nutrient sinks can be dispensed with. For example, shaded leaves that are no longer photosynthetically productive can be shed. Petals, which are modified leaves, may senesce once pollination occurs. Orchid flowers remain fresh for long periods of time, even in a florist shop. However, once pollination occurs, a hormonal change is triggered that leads to petal senescence. This makes sense in terms of allocation of energy resources because the petals are no longer necessary to attract a pollinator. On a larger scale, deciduous plants in temperate areas produce new leaves in the spring and then lose them in the fall. In the tropics, however, the production and subsequent loss of leaves in some species is correlated with wet and dry seasons. Evergreen plants, such as most conifers, usually have a complete change of leaves every two to seven years, periodically losing some but not all of their leaves.

Abscission involves changes that take place in an *abscission zone* at the base of the petiole (figure 41.31). Young leaves produce hormones (especially cytokinins) that inhibit the development of specialized layers of cells in the abscission zone. Hormonal changes take place as the leaf ages, however, and two layers of cells become differentiated. (Despite the name, abscisic acid is not involved in this process.) A *protective layer*, which may be several cells wide, develops on the stem side of the petiole base. These cells become impregnated with *suberin*, a fatty substance that is

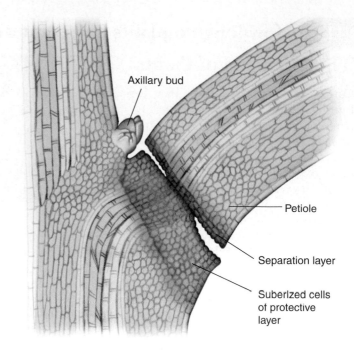

FIGURE 41.31

Leaf abscission. Hormonal changes in the leaf's abscission zone cause abscission. Two layers of cells in the abscission zone differentiate into a protective layer and a separation layer. As pectins in the separation layer break down, wind and rain can easily separate the leaf from the stem.

impervious to moisture. A *separation layer* develops on the side of the leaf blade; the cells of the separation layer sometimes divide, swell, and become gelatinous. When temperatures drop, the duration and intensity of light diminishes as the days grow shorter, or other environmental changes occur, enzymes break down the pectins in the middle lamellae of the separation cells. Wind and rain can then easily separate the leaf from the stem. Left behind is a sealed leaf scar that is protected from bacteria and other disease organisms.

As the abscission zone develops, the green chlorophyll pigments present in the leaf break down, revealing the yellows and oranges of other pigments, such as carotenoids, that previously had been masked by the intense green colors. At the same time, water-soluble red or blue pigments called *anthocyanins* and *betacyanins* may also accumulate in the vacuoles of the leaf cells—all contributing to an array of fall colors in leaves (see figure 40.10*a*).

Annual plants complete their whole growth cycle within a single year. Biennial plants flower only once, normally after two seasons of growth. Perennials flower repeatedly and live for many years. Abscission occurs when a plant sheds its organs.

41.1 The environment influences reproduction.

Plants Undergo Metamorphosis

- Plants go through developmental changes leading to maturity by adding on structures to existing structures with their meristems. (p. 832)
- Phase changes allow plants to obtain competence to respond to external or internal signals triggering flower formation. (p. 832)
- Fully mature plants may still require specific environmental cues in order to flower. (p. 833)

Pathways Leading to Flower Production

- Three genetically regulated flowering pathways have been identified: a light-dependent pathway, a temperature-dependent pathway, and the autonomous pathway. (p. 834)
- Flowering responses to day length can fall into several basic categories, including short-day plants, long-day plants, and day-neutral plants. (pp. 834–835)
- Vernalization, a chilling period, may be necessary for certain seeds or developing plants. (p. 836)
- In some plants, flowering is independent of environmental cues, and at some point in development, shoots become determined to flower. (p. 836)

Identity Genes and the Formation of Floral Meristems and Floral Organs

- The ABC model explains how three classes of floral organ identity genes could specify four distinct organ types. (p. 838)

41.2 Flowers are highly evolved for reproduction.

Evolution of the Flower

- Successful pollination in many angiosperms depends on regular attraction of pollinators, thus floral morphology has coevolved with pollinators. (p. 840)
- The trend in floral evolution has been toward four distinct whorls; a complete flower contains all four whorls, and an incomplete flower lacks one or more whorls. (p. 840)
- The wide diversity of modern flowering plants has been driven by two trends: grouping or fusing of separate floral parts, and loss or reduction of floral parts. (p. 841)
- Artificial selection has had a large impact on flower morphology. (p. 841)
- Primitive flowers are often radially symmetrical, while advanced flowers tend to be bilaterally symmetrical. (p. 841)

Formation of Angiosperm Gametes

- Reproductive success depends on the union of male and female gametes. (p. 842)
- Angiosperm reproductive organs differ from those of animals in that angiosperm male and female structures usually occur in the same individual flower, and angiosperm reproductive structures are not permanent parts of adult individuals. (p. 842)

Pollination

- Pollination occurs when pollen is placed on the stigma. (p. 844)
- Early seed plants were passively pollinated by the wind, but animal pollinators have also played an important role in angiosperm evolution. (p. 844)
- Animal pollinators include bees and other insects, birds, bats, and small rodents, as well as many others. (pp. 844–845)

Self-Pollination

- Self-pollination of angiosperms occurs frequently due to the ecological advantage of independence from a pollinator, and self-pollination produces uniform progeny that may be advantageous in some habitats. (p. 846)
- Outcrossing promotes genetic diversity and may be promoted by several strategies, including physical or temporal separation of stamens and pistils. (pp. 846–847)

Fertilization

- Double fertilization in angiosperms uses two sperm cells; one fertilizes an egg and the other helps form endosperm to nourish the embryo. (p. 848)

41.3 Many plants can clone themselves by asexual reproduction.

Asexual Reproduction

- Asexual reproduction is used to produce individuals genetically identical to the parent. (p. 849)
- Vegetative reproduction can be accomplished via runners, rhizomes, suckers, and adventitious plantlets. (p. 849)

Plant Tissue Culture

- Whole plants can be cloned by regenerating plant cells or tissues on nutrient medium. (p. 850)

41.4 How long do plants and plant organs live?

The Life Span of Plants

- Life span may or may not correlate with reproductive strategy. (p. 851)
- Annual plants grow, flower, form fruits and seeds, and then die, all within one growing season, while biennial plants take two years to complete their life cycle, and perennial plants continue to grow year after year. (p. 851)
- Abscission is the process by which leaves or petals are shed, and is advantageous in that a plant can shed its nutrient sinks. (p. 852)

Self Test

1. Flowers must
   a. be insect pollinated.
   b. show indeterminate growth.
   c. be part of the angiosperm life cycle.
   d. be self-incompatible.

2. In Iowa, a company called Team Corn works to ensure that fields of seed corn outcross so that hybrid vigor can be maintained. They do this by removing the staminate (i.e., pollen-producing) flowers from the corn plants. In an attempt to put Team Corn out of business, you would like to develop genetically engineered corn plants that
   a. contain Z-genes to prevent germination of pollen on the stigma surface.
   b. contain S-genes to stop pollen tube growth during self-fertilization.
   c. express B-type homeotic genes throughout developing flowers.

3. Monoecious plants such as corn have separate staminate and carpelate flowers. Knowing what you do about the molecular mechanisms of floral development, which of the following might explain the development of single-sex flowers?
   a. Expression of B-type genes in presumptive carpels will generate staminate flowers.
   b. Expression of A-type genes in presumptive stamens will generate carpelate flowers.
   c. Restricting B-type gene expression to presumptive petals will generate carpelate flowers.
   d. All of these are correct.

4. You have been asked to collect plant sperm for a new plant breeding program involving in vitro fertilization. Which of the following tissues would be a good source of sperm?
   a. anthers
   b. ovaries
   c. pollen
   d. spores

5. Street trees (those lining the roads) are generally male. This reduces the cleanup of the streets because the trees only produce pollen, not fruits. Therefore, the kinds of trees that are normally chosen as street trees are
   a. dioecious.
   b. monoecious.
   c. gymnosperms.
   d. trees producing perfect flowers.
   e. You cannot have a tree that is only male.

6. If you wanted to create a super tobacco plant to increase the number of leaves/acre on a tobacco farm, which of the following strategies would potentially work?
   a. Supress more root growth in the plants.
   b. Decrease expression of the *LEAFY* gene in the shoot apical meristem.
   c. Harvest lower leaves as the plant is growing so that flowering is delayed.
   d. Remove flowers so that the plant will produce more vegetative internodes than normal.

7. One of the most notable differences between gamete formation in animals and gamete formation in plants is that
   a. plants produce gametes in somatic tissue, while animals produce gametes in germ tissue.
   b. plants produce gametes by mitosis, while animals produce gametes by meiosis.

c. plants produce only one of each gamete, while animals produce many gametes.
   d. plants produce gametes that are diploid, while animals produce gametes that are haploid.

8. If you were to discover a flower that was small, white, and heavily scented, its most likely pollinator would be
   a. bees.
   b. birds.
   c. humans.
   d. moths.

9. Under which of the following conditions would pollen from an S_2S_5 plant successfully pollinate an S_1S_5 flower?
   a. If you used pollen from a carpelate flower, to fertilize a staminate flower, it would be successful.
   b. If the plants used gametophytic self-incompatibility, half of the pollen would be successful.
   c. If the plants used sporophytic self-incompatibility, half of the pollen would be successful.
   d. Pollen from an S_2S_5 plant can never pollinate an S_1S_5 flower.

10. You have been commissioned by kindergartners across the country to develop a dandelion-breeding program to improve the quality of dandelion bouquets to give to mom. Initially, you are very excited about this project until you remember that
   a. dandelions reproduce asexually using runners.
   b. dandelion flowers set seed without fertilization.
   c. if dandelion flowers are not yellow, they will fail to attract the correct pollinators.
   d. dandelions are really gymnosperms.

Test Your Visual Understanding

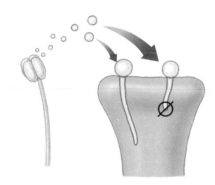

1. Describe what is taking place in this figure, and predict the genotypes of the staminate and carpelate tissue. (*Hint:* Think about the two major genetic strategies for self-incompatibility [gametophytic and sporophytic], and decide which one fits the image.)

Apply Your Knowledge

1. We often have the impression that plants lack the ability to move around in the environment. This, however, is far from the truth. Discuss the variety of ways that plants successfully move throughout the environment. (*Hint:* Think of the different mobile life stages of various plants [i.e., spores, gametophytes, seeds] and how they can be transported.)

Part VII

42

The Animal Body and How It Moves

Concept Outline

42.1 The bodies of vertebrates are organized into functional systems.

    Organization of the Body. Cells are organized into tissues, and tissues are organized into organs.

42.2 Epithelial tissue forms membranes and glands.

    Characteristics of Epithelial Tissue. Epithelial membranes cover all body surfaces.

42.3 Connective tissues contain abundant extracellular material.

    Connective Tissue Proper. Connective tissues come in many forms, but all have abundant extracellular material.

    Special Connective Tissues. Special connective tissues include cartilage, bone, and blood, each with their own unique form of extracellular material.

42.4 Muscle tissue provides for movement, and nerve tissue provides for control.

    Muscle Tissue. Muscle tissue contains the filaments actin and myosin, which enable the muscles to contract.

    Nerve Tissue. Nerve cells, or neurons, have specialized regions that produce and conduct electrical impulses.

42.5 Coordinated efforts of organ systems are necessary for locomotion.

    Organ Systems Involved in Animal Movement. Movement in animals is produced by the coordinated efforts of the nervous, muscular, and skeletal systems

    The Skeletal System Supports Movement. The three types of skeletal systems are hydrostatic skeletons, exoskeletons, and endoskeletons.

    Muscle Contractions Move Bones at Joints. Muscles attach to bones, and when they contract, they move the bones at the joints.

42.6 Muscle contraction powers animal locomotion.

    Muscle Contraction. Thick and thin myofilaments slide past one another to cause muscle shortening.

    Modes of Animal Locomotion. Locomotion requires both propulsion and control mechanisms.

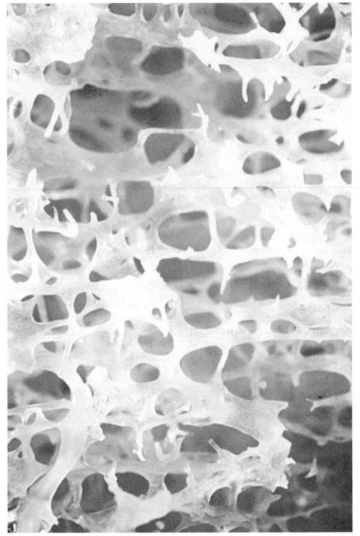

FIGURE 42.1
Bone. Like most of the tissues in the vertebrate body, bone is a dynamic structure, constantly renewing itself.

When most people think of animals, they think of their pet dogs and cats, and of the animals they've seen in a zoo, on a farm, in an aquarium, or out in the wild. When they think about the diversity of animals, they may think of the differences between the predatory lions and tigers and the herbivorous deer and antelope, or between a ferocious-looking shark and a playful dolphin. Despite the differences among these animals, they are all vertebrates. All vertebrates share the same basic body plan, with the same sorts of tissues and organs operating in much the same way. In this chapter, we begin a detailed consideration of the biology of vertebrates and of the fascinating structure and function of their bodies. Further, we will explore how the body functions in a complex activity, locomotion, in which several organ systems—the muscular, skeletal (figure 42.1), and nervous systems—cooperate to propel the body through its environment.

Organization of the Body

The bodies of all vertebrates have the same general architecture. The body plan is basically a tube suspended within a tube. Starting from the inside, it is composed of the digestive tract, a long tube that travels from one end of the body to the other (mouth to anus). This tube is suspended within an internal body cavity, the *coelom*. In fishes, amphibians, and most reptiles, the coelom is subdivided into two cavities, one housing the heart and the other the liver, stomach, and intestines. In mammals and some reptiles, a sheet of muscle, the *diaphragm*, separates the **peritoneal cavity**, which contains the stomach, intestines, and liver, from the **thoracic cavity;** the thoracic cavity is further subdivided into the *pericardial cavity*, which contains the heart, and *pleural cavities*, which contain the lungs (figure 42.2). All vertebrate bodies are supported by an internal **skeleton** made of jointed bones or cartilage blocks that grow as the body grows. A bony *skull* surrounds the brain, and a column of bones, the *vertebrae*, surrounds the dorsal nerve cord, or *spinal cord*.

There are four levels of organization in the vertebrate body: (1) cells, (2) tissues, (3) organs, and (4) organ systems. Like those of all animals, the bodies of vertebrates are composed of different cell types. Depending on the taxon, between 50 and several hundred different kinds of cells contribute to the adult vertebrate body.

Tissues

Groups of cells similar in structure and function are organized into **tissues**. Early in development, the cells of the growing embryo differentiate (specialize) into three fundamental embryonic tissues, called *germ layers*. From the innermost to the outermost layers, these are the **endoderm**, **mesoderm**, and **ectoderm**. These germ layers, in turn, differentiate into the scores of different cell types and tissues that are characteristic of the vertebrate body. In adult vertebrates, there are four principal kinds of tissues, or *primary tissues:* epithelial, connective, muscle, and nervous (figure 42.3), each discussed in separate sections of this chapter.

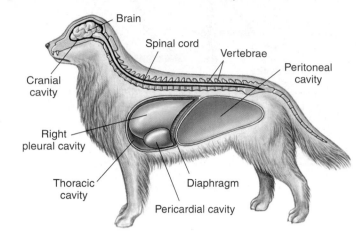

FIGURE 42.2
Architecture of the vertebrate body. All vertebrates have a dorsal central nervous system. In mammals and some reptiles, a muscular diaphragm divides the coelom into the thoracic cavity and the peritoneal cavity.

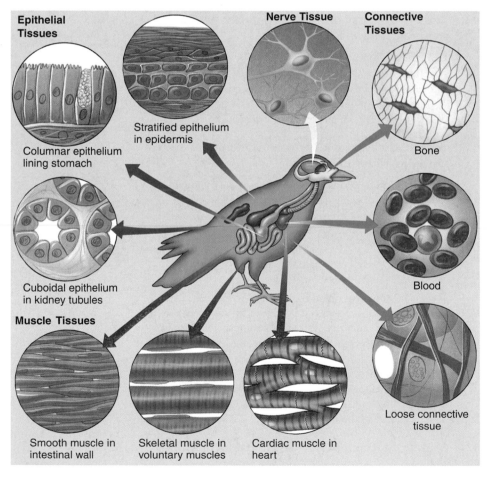

FIGURE 42.3
Vertebrate tissue types. Epithelial tissues are indicated by blue arrows, connective tissues by green arrows, muscle tissues by red arrows, and nervous tissue by a yellow arrow.

Organs and Organ Systems

Organs are body structures composed of several different tissues that form a structural and functional unit (figure 42.4). One example is the heart, which contains cardiac muscle, connective tissue, and epithelial tissue and is laced with nerve tissue that helps regulate the heartbeat. An **organ system** is a group of organs that cooperate to perform the major activities of the body. For example, the digestive system is composed of the digestive tract, liver, gallbladder, and pancreas. These organs cooperate in the digestion of food and the absorption of digestion products into the body. The vertebrate body contains 11 principal organ systems (table 42.1 and figure 42.5).

The bodies of humans and other mammals contain a cavity divided by the diaphragm into thoracic and peritoneal cavities. The body's cells are organized into tissues, which are, in turn, organized into organs and organ systems.

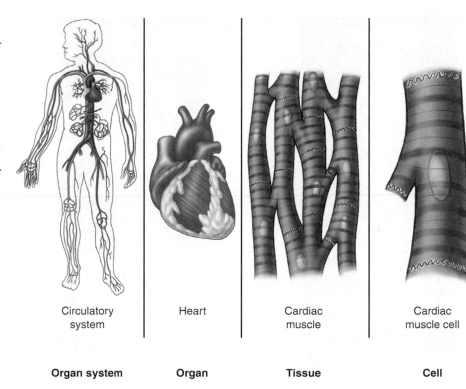

| Circulatory system | Heart | Cardiac muscle | Cardiac muscle cell |
| Organ system | Organ | Tissue | Cell |

FIGURE 42.4
Levels of organization within the body. Similar cell types operate together and form tissues. Tissues functioning together form organs. Several organs working together to carry out a function for the body are called an organ system. An example of an organ system is the circulatory system, which consists of the heart and blood vessels.

Table 42.1 The Major Vertebrate Organ Systems

System	Functions	Components	Detailed Treatment
Circulatory	Transports cells, respiratory gases, and chemical compounds throughout the body	Heart, blood vessels, lymph, and lymph structures	Chapter 44
Digestive	Captures soluble nutrients from ingested food	Mouth, esophagus, stomach, intestines, liver, and pancreas	Chapter 43
Endocrine	Coordinates and integrates the activities of the body	Pituitary, adrenal, thyroid, and other ductless glands	Chapter 47
Integumentary	Covers and protects the body	Skin, hair, nails, scales, feathers, and sweat glands	Chapter 48
Lymphatic/ Immune	Vessels transport extracellular fluid and fat to circulatory system; lymph nodes and lymphatic organs provide defenses against microbial infection and cancer	Lymphatic vessels, lymph nodes, thymus, tonsils, and spleen	Chapter 48
Muscular	Produces body movement	Skeletal muscle, cardiac muscle, and smooth muscle	Chapter 42
Nervous	Receives stimuli, integrates information, and directs the body	Nerves, sense organs, brain, and spinal cord	Chapters 45, 46
Reproductive	Carries out reproduction	Testes, ovaries, and associated reproductive structures	Chapter 50
Respiratory	Captures oxygen and exchanges gases	Lungs, trachea, gills, and other air passageways	Chapter 44
Skeletal	Protects the body and provides support for locomotion and movement	Bones, cartilage, and ligaments	Chapter 42
Urinary/ Osmoregulatory	Removes metabolic wastes from the bloodstream	Kidney, bladder, and associated ducts	Chapter 49

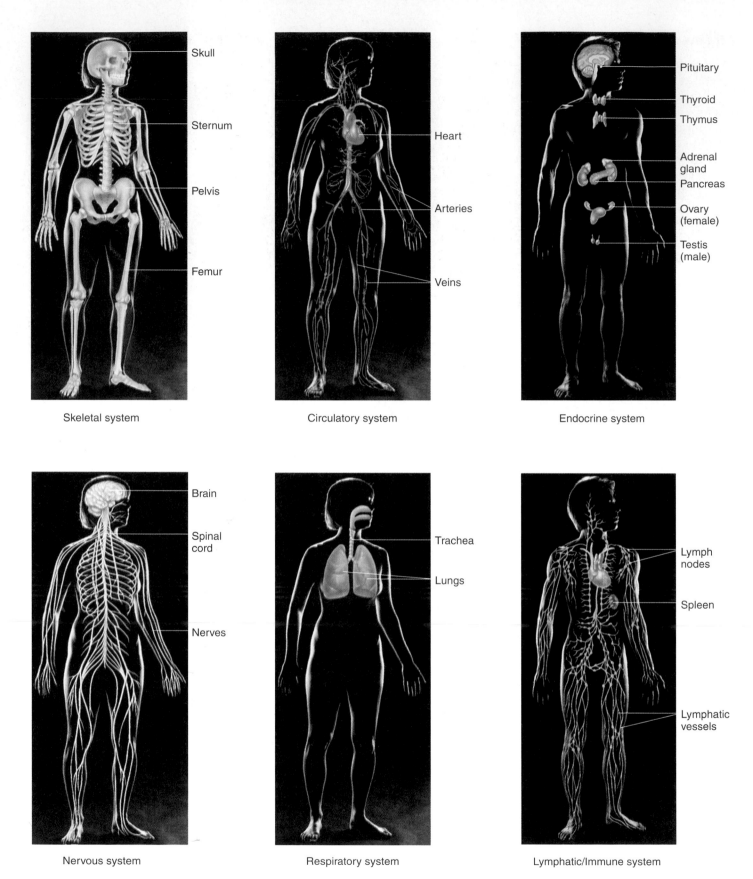

Skull

Sternum

Pelvis

Femur

Skeletal system

Heart

Arteries

Veins

Circulatory system

Pituitary

Thyroid

Thymus

Adrenal gland

Pancreas

Ovary (female)

Testis (male)

Endocrine system

Brain

Spinal cord

Nerves

Nervous system

Trachea

Lungs

Respiratory system

Lymph nodes

Spleen

Lymphatic vessels

Lymphatic/Immune system

FIGURE 42.5
Vertebrate organ systems. The 11 principal organ systems of the human body are shown, including both male and female reproductive systems.

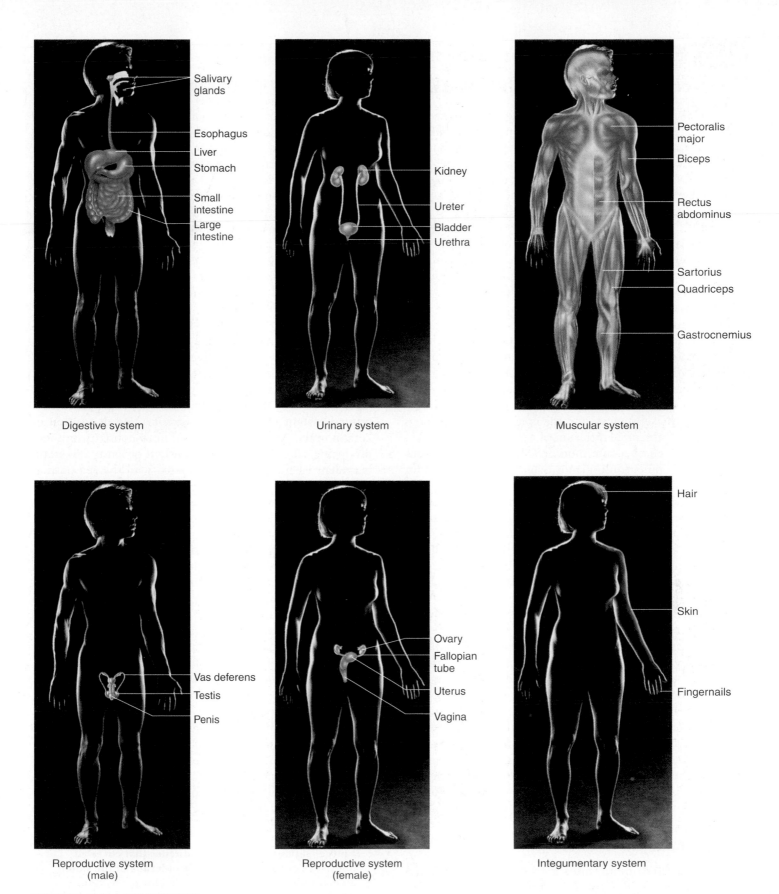

Digestive system

Salivary glands

Esophagus
Liver
Stomach

Small intestine

Large intestine

Urinary system

Kidney

Ureter

Bladder
Urethra

Muscular system

Pectoralis major

Biceps

Rectus abdominus

Sartorius
Quadriceps

Gastrocnemius

Reproductive system (male)

Vas deferens

Testis

Penis

Reproductive system (female)

Ovary
Fallopian tube

Uterus

Vagina

Integumentary system

Hair

Skin

Fingernails

FIGURE 42.5—CONTINUED.

42.2 Epithelial tissue forms membranes and glands.

Characteristics of Epithelial Tissue

An epithelial membrane, or **epithelium,** covers every surface of the vertebrate body. Epithelial membranes are derived from all three germ layers. The epidermis, derived from ectoderm, constitutes the outer portion of the skin. The inner surface of the digestive tract is lined by an epithelium derived from endoderm, and the inner surfaces of the body cavities are lined with an epithelium derived from mesoderm.

Because all body surfaces are covered by epithelial membranes, a substance must pass through an epithelium in order to enter or leave the body. Epithelial membranes thus provide a barrier that can impede the passage of some substances while facilitating the passage of others. For land-dwelling vertebrates, the relative impermeability of the surface epithelium (the epidermis) to water offers essential protection from dehydration and from airborne pathogens (disease-causing organisms). On the other hand, the epithelial lining of the digestive tract must allow selective entry of the products of digestion while providing a barrier to toxic substances, and the epithelium of the lungs must allow for the rapid diffusion of gases.

Some epithelia become modified in the course of embryonic development into glands, which are specialized for secretion. A characteristic of all epithelia is that the cells are tightly bound together, with very little space between them. As a consequence, blood vessels cannot be interposed between adjacent epithelial cells. Therefore, nutrients and oxygen must diffuse to the epithelial cells from blood vessels supplying underlying connective tissues. This places a limit on the thickness of epithelial membranes; most are only one or a few cell layers thick.

Epithelium possesses remarkable regenerative powers, constantly replacing its cells throughout the life of the animal. For example, the liver, a gland formed from epithelial tissue, can readily regenerate after substantial portions of it have been surgically removed. The epidermis is renewed every two weeks, and the epithelium inside the stomach is replaced every two to three days.

There are two general classes of epithelial membranes: simple and stratified. These classes are further subdivided into squamous, cuboidal, and columnar, based upon the shape of the cells (table 42.2). Squamous cells are flat, cuboidal cells are about as wide as they are tall, and columnar cells are taller than they are wide.

Types of Epithelial Tissues

Simple epithelial membranes are one cell layer thick. A *simple, squamous epithelium* is composed of squamous epithelial cells that have an irregular, flattened shape with tapered edges. Examples of such membranes are those that line the lungs and blood capillaries, where the thin, delicate nature of these membranes permits the rapid movement of molecules (such as the diffusion of gases). A *simple cuboidal epithelium* lines the small ducts of some glands, and a *simple columnar epithelium* is found in the airways of the respiratory tract and in the gastrointestinal tract, among other locations. Interspersed among the columnar epithelial cells of mucous membranes are numerous *goblet cells*, which are specialized to secrete mucus. The columnar epithelial cells of the respiratory airways contain cilia on their apical surface (the surface facing the lumen, or cavity), which move mucus toward the throat. In the small intestine, the apical surface of the columnar epithelial cells forms fingerlike projections called *microvilli*, which increase the surface area for the absorption of food.

Stratified epithelial membranes are several cell layers thick and are named according to the features of their uppermost layers. For example, the epidermis is a *stratified squamous epithelium*. In terrestrial vertebrates, it is further characterized as a *keratinized epithelium*, because its upper layer consists of dead squamous cells and is filled with a water-resistant protein called *keratin*. The deposition of keratin in the skin can be increased in response to abrasion, producing calluses. The water-resistant property of keratin is evident when the skin is compared with the red portion of the lips, which can easily become dried and chapped because it is covered by a nonkeratinized, stratified squamous epithelium.

The glands of vertebrates are derived from invaginated epithelium. In **exocrine glands,** the connection between the gland and the epithelial membrane is maintained as a duct. The duct channels the product of the gland to the surface of the epithelial membrane and thus to the external environment (or to an interior compartment that opens to the exterior, such as the digestive tract). Examples of exocrine glands include sweat and sebaceous (oil) glands, which secrete to the external surface of the skin, and accessory digestive glands such as the salivary glands, liver, and pancreas, which secrete to the surface of the epithelium lining the digestive tract.

Endocrine glands are ductless glands; their connections with the epithelium from which they were derived are lost during development. Therefore, their secretions, called *hormones*, are not channeled onto an epithelial membrane. Instead, hormones enter blood capillaries and circulate through the body. Endocrine glands are discussed in more detail in chapter 47.

Epithelial tissues include membranes that cover all body surfaces and glands. The epidermis of the skin is an epithelial membrane specialized for protection, whereas membranes that cover the surfaces of hollow organs are often specialized for transport and secretion.

Table 42.2   Epithelial Tissue

Simple Epithelium

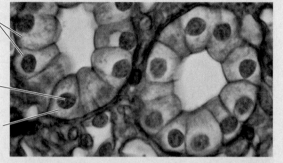

Simple squamous epithelial cell

Nucleus

SQUAMOUS

Typical Location
Lining of lungs, capillary walls, and blood vessels

Function
Cells very thin; provides thin layer across which diffusion can readily occur

Characteristic Cell Types
Epithelial cells

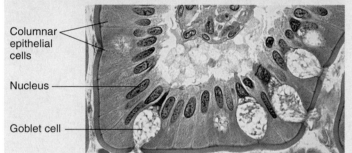

Cuboidal epithelial cells

Nucleus

Cytoplasm

CUBOIDAL

Typical Location
Lining of some glands and kidney tubules; covering of ovaries

Function
Cells rich in specific transport channels; functions in secretion and absorption

Characteristic Cell Types
Gland cells

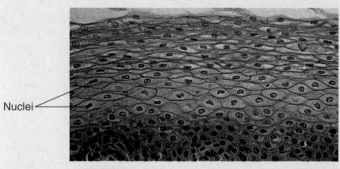

Columnar epithelial cells

Nucleus

Goblet cell

COLUMNAR

Typical Location
Surface lining of stomach, intestines, and parts of respiratory tract

Function
Thicker cell layer; provides protection and functions in secretion and absorption

Characteristic Cell Types
Epithelial cells

Stratified Epithelium

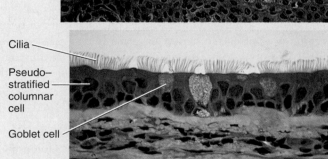

Nuclei

SQUAMOUS

Typical Location
Outer layer of skin; lining of mouth

Function
Tough layer of cells; provides protection

Characteristic Cell Types
Epithelial cells

Cilia

Pseudo–stratified columnar cell

Goblet cell

PSEUDOSTRATIFIED COLUMNAR

Typical Location
Lining of parts of the respiratory tract

Function
Secretes mucus; dense with cilia that aid in movement of mucus; provides protection

Characteristic Cell Types
Gland cells; ciliated epithelial cells

42.3 Connective tissues contain abundant extracellular material.

Connective Tissue Proper

Connective tissues are derived from embryonic mesoderm and occur in many different forms (table 42.3). These various forms are divided into two major classes: **connective tissue proper,** which is further divided into loose and dense connective tissues, and **special connective tissues,** which include cartilage, bone, and blood. At first glance, it may seem odd that such diverse tissues are placed in the same category. Yet all connective tissues do share a common structural feature: They all have abundant extracellular material because their cells are spaced widely apart. This extracellular material is generically known as the **matrix** of the tissue. In bone, the extracellular matrix contains crystals that make the bones hard; in blood, the extracellular matrix is plasma, the fluid portion of the blood.

Loose connective tissue consists of cells scattered within an amorphous mass of proteins that form a **ground substance.** This gelatinous material is strengthened by a loose scattering of protein fibers such as *collagen* (figure 42.6), *elastin*, which makes the tissue elastic, and *reticulin*, which supports the tissue by forming a collagenous meshwork. The flavored gelatin we eat for dessert consists primarily of extracellular material extracted from the loose connective tissues of animals. During the development of both loose and dense connective tissues, the cells that produce and secrete the extracellular matrix are known as *fibroblasts*. Fibroblasts transform into fibrocytes, which maintain the matrix in mature connective tissue proper.

Loose connective tissue contains other cells as well, including *mast cells*, which produce histamine (a blood vessel dilator) and heparin (an anticoagulant), and *macrophages*, the immune system's first defense against invading organisms, as is described in detail in chapter 48.

Adipose cells, or fat cells, important for nutrient storage, can also be found in loose connective tissue. In certain areas of the body, including under the skin, in bone marrow, and around the kidneys as well as in women's hips and breasts, these cells can develop in large groups, forming adipose tissue (figure 42.7). Each adipose cell contains a droplet of fat (triglycerides) within a storage vesicle. When that fat is needed for energy, the adipose cell hydrolyzes its stored triglyceride and secretes fatty acids into the blood for oxidation by the cells of the muscles, liver, and other organs. The number of adipose cells in an adult is generally fixed because they are incapable of dividing. When a person gains weight, the cells become larger, and when weight is lost, the cells shrink.

Dense connective tissue contains tightly packed collagen fibers, making it stronger than loose connective tissue. It consists of two types: regular and irregular. The collagen fibers of **dense regular connective tissue** are lined up in parallel, like the strands of a rope. This is the structure of *tendons*, which bind muscle to bone, and *ligaments*, which

FIGURE 42.6
Collagen fibers.   Each fiber is composed of many individual collagen strands and can be very strong under tension.

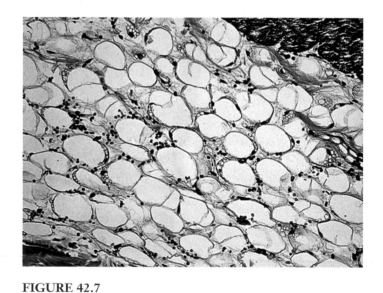

FIGURE 42.7
Adipose tissue.   Fat is stored in globules of adipose tissue, a type of loose connective tissue. As a person gains or loses weight, the size of the fat globules increases or decreases. A person cannot decrease the number of fat cells by losing weight.

bind bone to bone. In contrast, the collagen fibers of **dense irregular connective tissue** have many different orientations. This type of connective tissue produces the tough coverings that package organs, such as the *capsules* of the kidneys and adrenal glands. It also covers muscle as *epimysium*, nerves as *perineurium*, and bones as *periosteum*.

Connective tissues are characterized by abundant extracellular materials forming a matrix between loosely organized cells. Connective tissue proper may be classified as either loose or dense.

Table 42.3 Connective Tissue

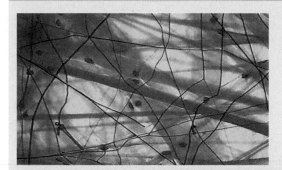

LOOSE CONNECTIVE TISSUE

Typical Location
Beneath skin; between organs

Function
Provides support, insulation, food storage, and nourishment for epithelium

Characteristic Cell Types
Fibroblasts, macrophages, mast cells, fat cells

DENSE CONNECTIVE TISSUE

Typical Location
Tendons; sheath around muscles; kidney; liver; dermis of skin

Function
Provides flexible, strong connections

Characteristic Cell Types
Fibroblasts

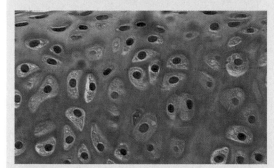

CARTILAGE

Typical Location
Spinal discs; knees and other joints; ear; nose; tracheal rings

Function
Provides flexible support, shock absorption, and reduction of friction on load-bearing surfaces

Characteristic Cell Types
Chondrocytes

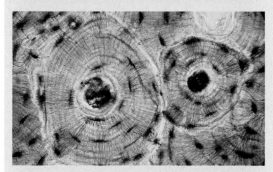

BONE

Typical Location
Most of skeleton

Function
Protects internal organs; provides rigid support for muscle attachment

Characteristic Cell Types
Osteocytes

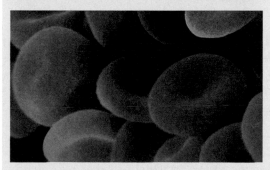

BLOOD

Typical Location
Circulatory system

Function
Functions as highway of immune system and primary means of communication between organs

Characteristic Cell Types
Erythrocytes, leukocytes

Special Connective Tissues

The special connective tissues—cartilage, bone, and blood—each have unique cells and extracellular matrices that allow them to perform their specialized functions.

Cartilage

Cartilage (figure 42.8) is a specialized connective tissue in which the ground substance is formed from a characteristic type of glycoprotein, called chondroitin, and collagen fibers laid down along lines of stress in long, parallel arrays. The result is a firm and flexible tissue that does not stretch, is far tougher than loose or dense connective tissue, and has great tensile strength. Cartilage makes up the entire skeletal system of the modern agnathans and cartilaginous fishes (see chapter 34), replacing the bony skeletons that were characteristic of the ancestors of these vertebrate groups. In most adult vertebrates, however, cartilage is restricted to the articular (joint) surfaces of bones that form freely movable joints and to other specific locations. In humans, for example, the tip of the nose, the pinna (outer ear flap), the intervertebral discs of the backbone, the larynx (voice box), and a few other structures are composed of cartilage.

Chondrocytes, the cells of the cartilage, live within spaces called *lacunae* within the cartilage ground substance. These cells remain alive, even though there are no blood vessels within the cartilage matrix, because they receive oxygen and nutrients by diffusion through the cartilage ground substance from surrounding blood vessels. This diffusion can only occur because the cartilage matrix is well hydrated and not calcified, as is bone.

Bone

In the course of fetal development, the bones of vertebrate fins, arms, and legs, among others, are first "modeled" in cartilage. The cartilage matrix then calcifies at particular locations, so that the chondrocytes are no longer able to obtain oxygen and nutrients by diffusion through the matrix. The dying and degenerating cartilage is then replaced by living bone. Bone cells, or *osteocytes*, can remain alive even though the extracellular matrix becomes hardened

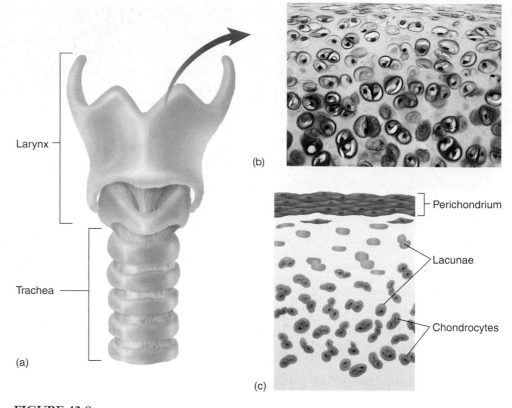

FIGURE 42.8
Cartilage is a strong, flexible tissue found in the larynx (voice box), the trachea, and several other structures in the human body. The larynx (*a*) is seen under the light microscope in (*b*), where the cartilage cells, or chondrocytes, are visible within cavities, or lacunae, in the matrix (extracellular material) of the cartilage. This is diagrammed in (*c*).

with crystals of calcium phosphate. This is because blood vessels travel through central canals into the bone. Osteocytes extend cytoplasmic processes toward neighboring osteocytes through tiny canals, or *canaliculi* (figure 42.9). Osteocytes communicate with the blood vessels in the central canal through this cytoplasmic network.

The Structure of Bone

In bone, an organic extracellular matrix containing collagen fibers is impregnated with small, needle-shaped crystals of calcium phosphate in the form of hydroxyapatite crystals. Hydroxyapatite is rigid but brittle, giving bone great strength. Collagen, on the other hand, is flexible but weak. As a result, bone is both strong and flexible. The collagen acts to spread the stress over many crystals, making bone more resistant to fracture than hydroxyapatite is by itself.

Bone is a dynamic, living tissue that is constantly reconstructed throughout the life of an individual. New bone is formed by *osteoblasts*, which secrete the collagen-containing organic matrix in which calcium phosphate is later deposited. After the calcium phosphate is deposited, the cells, now known as osteocytes, are encased within spaces called *lacunae* in the calcified matrix. Yet another type of bone cells, called osteoclasts, act to dissolve bone and thereby aid in the remodeling of bone in response to physical stress.

Bone is constructed in thin, concentric layers, or *lamellae*, which are laid down around narrow channels called *Haversian canals* that run parallel to the length of the bone. Haversian canals contain nerve fibers and blood vessels, which keep the osteocytes alive even though they are entombed in a calcified matrix. The concentric lamellae of bone, with their entrapped osteocytes, that surround a Haversian canal form the basic unit of bone structure, called a Haversian system.

Bone formation occurs in two ways. In flat bones, such as those of the skull, osteoblasts located in a web of dense irregular connective tissue produce bone within that tissue instead of forming first as a cartilage model. In long bones, the bone is first "modeled" in cartilage. Calcification then occurs, and bone is formed as the cartilage degenerates. At the end of this process, cartilage remains only at the articular (joint) surfaces of the bones and at the growth plates located in the necks of the long bones. A child grows taller as the cartilage thickens in the growth plates and then is partly replaced with bone. A person stops growing (usually by the late teenage years) when the entire cartilage growth plate becomes replaced by bone. At this point, only the articular cartilage at the ends of the bone remains.

The ends and interiors of long bones are composed of an open lattice of bone called *spongy bone*. The spaces within contain marrow, where most blood cells are formed (figure 42.9). Surrounding the spongy bone tissue are concentric layers of *compact bone*, where the bone is much denser. Compact bone tissue gives bone the strength to withstand mechanical stress.

The calcified matrix of bones serves as a dynamic reservoir for the alternate storage and release of calcium and phosphate ions, depending on the needs of body fluids and tissues. The bones of the skeletal system support and protect the body, and serve as levers for the forces produced by contraction of skeletal muscles. The role of bone in locomotion is discussed in section 42.5.

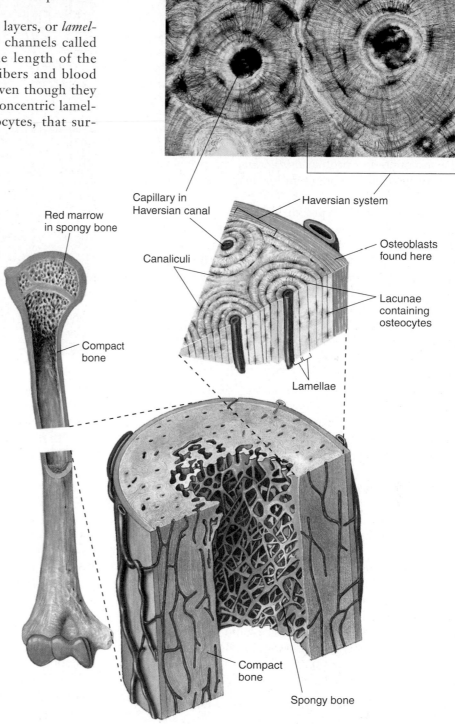

FIGURE 42.9

The structure of bone. Bone cells, or osteocytes, reside in lacunae (cavities) in the bone matrix. Although the bone matrix is calcified, the osteocytes remain alive because they can be nourished by blood vessels in the central Haversian canals. Nourishment is carried between osteocytes through a network of cytoplasmic processes extending through tiny canals, called canaliculi. Some parts of bone are dense and compact, giving the bone strength. Other parts are spongy, with a more open lattice; it is there that most blood cells are formed.

Blood

Blood is classified as a connective tissue because it contains abundant extracellular material, the fluid plasma. The cells of blood are erythrocytes, or red blood cells, and leukocytes, or white blood cells (figure 42.10). Blood also contains platelets, or *thrombocytes*, which are fragments of a type of bone marrow cell.

Erythrocytes are the most abundant blood cells; there are about 5 billion in every milliliter of blood. During their maturation in mammals, they lose their nucleus, mitochondria, and endoplasmic reticulum. As a result, mammalian erythrocytes are relatively inactive metabolically. Each erythrocyte contains about 300 million molecules of the iron-containing protein *hemoglobin*, the principal carrier of oxygen in vertebrates and in many other groups of animals.

There are several types of leukocytes, but together they are only one-thousandth as numerous as erythrocytes. Unlike erythrocytes, leukocytes have nuclei and mitochondria but lack the red pigment hemoglobin. These cells are therefore hard to see under a microscope without special staining. The names *neutrophils*, *eosinophils*, and *basophils* distinguish three types of leukocytes on the basis of their staining properties; other leukocytes include *lymphocytes* and *monocytes*. These different types of leukocytes play critical roles in immunity, as described in chapter 48.

The blood plasma is the "commons" of the body; it (or a derivative of it) travels to and from every cell in the body. As the plasma circulates, it carries nourishment, waste products, heat, and regulatory molecules. Practically every substance used by cells, including sugars, lipids, and amino acids, is delivered by the plasma to the body cells. Waste products from the cells are carried by the plasma to the kidneys, liver, and lungs or gills for disposal, and regulatory molecules (hormones) secreted by endocrine gland cells are carried by the plasma to regulate the activities of most organs of the body. The plasma also contains sodium, calcium, and other inorganic ions that all cells need, as well as numerous proteins. Plasma proteins include *fibrinogen*, produced by the liver, which helps blood to clot; *albumin*, also produced by the liver, which helps balance or stabilize blood osmolarity and serves as an important blood buffer; and *antibodies*, produced by lymphocytes and needed for immunity.

Similarities in Types of Connective Tissues

While the descriptions of the types of connective tissue suggest numerous different functions for these tissues, they have some similarities. As stated earlier, all connective tissues are formed from embryonic mesoderm, and they all contain abundant extracellular material called matrix; however, the extracellular matrix material is dif-

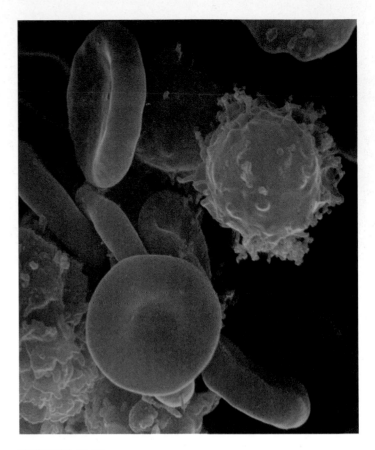

FIGURE 42.10
White and red blood cells. White blood cells, or leukocytes, are roughly spherical and have irregular surfaces with numerous extending pili (500×). Red blood cells, or erythrocytes, are flattened spheres, typically with a depressed center, forming biconcave discs.

ferent in different types of connective tissue. Specialized cells for each tissue type are embedded within the extracellular matrix.

When referring to cells of connective tissue, two suffixes are often used, *-blast* and *-cyte*. The suffix "blast" refers to an embryonic cell type, and "cyte" refers to the mature or functional cell type. Therefore, a fibroblast gives rise to a fibrocyte, a chondroblast gives rise to a chondrocyte, an osteoblast gives rise to an osteocyte, and so on. These similarities in structure and nomenclature help in clarifying the various types and functions of connective tissue.

Special connective tissues each have a unique extracellular matrix between cells. The matrix of cartilage is composed of organic material, whereas that of bone is impregnated with calcium phosphate crystals. The matrix of blood is fluid, the plasma.

42.4 Muscle tissue provides for movement, and nerve tissue provides for control.

Muscle Tissue

Muscle cells are the motors of the vertebrate body. The characteristic that makes them unique is the relative abundance and organization of actin and myosin filaments within them. Although these filaments form a fine network in all eukaryotic cells, where they contribute to cellular movements, they are far more abundant and organized in muscle cells, which are specialized for contraction. Vertebrates possess three kinds of muscle: smooth, skeletal, and cardiac (table 42.4). Skeletal and cardiac muscles are also known as **striated muscles** because their cells appear to have transverse stripes when viewed in longitudinal section under the microscope. The contraction of each skeletal muscle is under voluntary control, whereas the contraction of cardiac and smooth muscles is generally involuntary. Muscles are described in more detail later in this chapter.

Smooth Muscle

Smooth muscle was the earliest form of muscle to evolve, and it is found throughout the animal kingdom. In vertebrates, smooth muscle occurs in the organs of the internal environment, or *viscera*, and is sometimes known as visceral muscle. Smooth muscle tissue is organized into

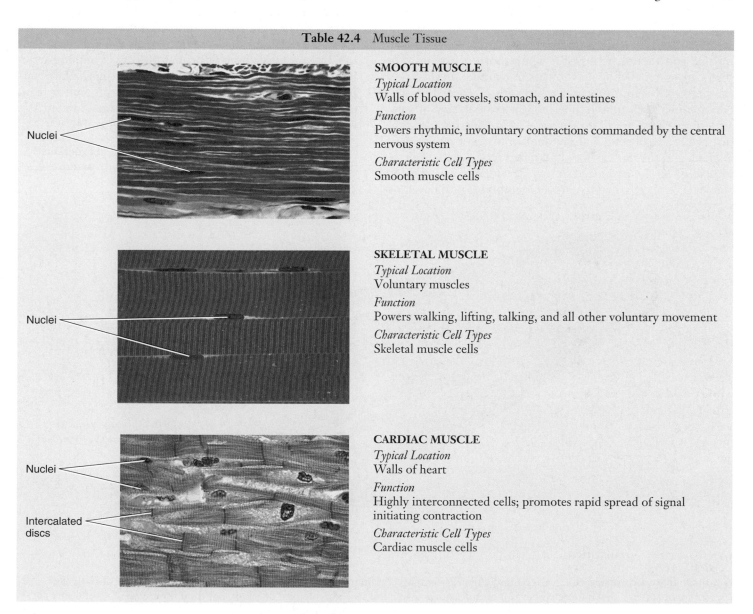

Table 42.4 Muscle Tissue

Nuclei

SMOOTH MUSCLE

Typical Location
Walls of blood vessels, stomach, and intestines

Function
Powers rhythmic, involuntary contractions commanded by the central nervous system

Characteristic Cell Types
Smooth muscle cells

Nuclei

SKELETAL MUSCLE

Typical Location
Voluntary muscles

Function
Powers walking, lifting, talking, and all other voluntary movement

Characteristic Cell Types
Skeletal muscle cells

Nuclei

Intercalated discs

CARDIAC MUSCLE

Typical Location
Walls of heart

Function
Highly interconnected cells; promotes rapid spread of signal initiating contraction

Characteristic Cell Types
Cardiac muscle cells

sheets of long, spindle-shaped cells, each cell containing a single nucleus. In some tissues, the cells contract only when they are stimulated by a nerve, and then all of the cells in the sheet contract as a unit. In vertebrates, muscles of this type line the walls of many blood vessels and make up the iris of the eye. In other smooth muscle tissues, such as those in the wall of the gut, the muscle cells themselves may spontaneously initiate electric impulses and contract, leading to a slow, steady contraction of the tissue. Nerves regulate, rather than cause, this activity.

Skeletal Muscle

Skeletal muscles are usually attached by tendons to bones, so that, when the muscles contract, they cause the bones to move at their joints. A skeletal muscle is made up of numerous, very long muscle cells, called *muscle fibers*, which lie parallel to each other within the muscle and insert into the tendons on the ends of the muscle. Each skeletal muscle fiber is stimulated to contract by a nerve fiber; therefore, a stronger muscle contraction will result when more of the muscle fibers are stimulated by nerve fibers to contract. In this way, the nervous system can vary the strength of skeletal muscle contraction. Each muscle fiber contracts by means of substructures called *myofibrils* (figure 42.11) containing highly ordered arrays of actin and myosin myofilaments that, when aligned, give the muscle fiber its striated appearance. Skeletal muscle fibers are produced during development by the fusion of several cells, end to end. This embryological develop-

ment explains why a mature muscle fiber contains many nuclei. The structure and function of skeletal muscle is explained in more detail later in this chapter.

Cardiac Muscle

The hearts of vertebrates are composed of striated muscle cells arranged very differently from the fibers of skeletal muscle. Instead of having very long, multinucleate cells running the length of the muscle, cardiac muscle is composed of smaller, interconnected cells, each with a single nucleus. The interconnections between adjacent cells appear under the microscope as dark lines called *intercalated discs*. In reality, these lines are regions where adjacent cells are linked by *gap junctions*. As we noted in chapter 7, gap junctions have openings that permit the movement of small substances and electric charges from one cell to another. These interconnections enable the cardiac muscle cells to form a single functioning unit known as a myocardium. Certain specialized cardiac muscle cells can generate electrical impulses spontaneously, but the rate of impulse activity is usually regulated by the nervous system. These impulses spread across the gap junctions from cell to cell, causing all of the cells in the myocardium to contract. We will describe this process more fully in chapter 44.

Smooth muscles provide a variety of visceral functions. Skeletal muscles enable the vertebrate body to move. Cardiac muscle powers the heartbeat.

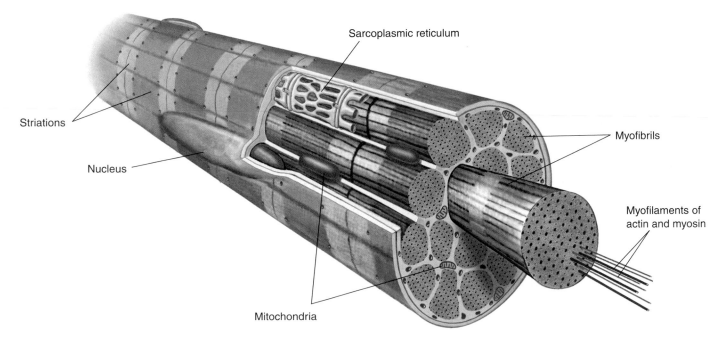

FIGURE 42.11
A muscle fiber, or muscle cell. Each muscle fiber is composed of numerous myofibrils, which in turn, are composed of actin and myosin myofilaments. Each muscle fiber is multinucleate as a result of its embryological development from the fusion of several smaller cells. Muscle cells have a modified endoplasmic reticulum called the sarcoplasmic reticulum.

Nerve Tissue

The fourth major class of vertebrate tissue is nerve tissue (table 42.5). Its cells include neurons and their supporting cells, called neuroglia. Neurons are specialized to produce and conduct electrochemical events, or "impulses." Most neurons consist of three parts: a cell body, dendrites, and an axon (figure 42.12). The cell body of a neuron contains the nucleus. Dendrites are thin, highly branched extensions that receive incoming stimulation and conduct electrical events to the cell body. As a result of this stimulation and the electrical events produced in the cell body, outgoing impulses may be produced at the origin of the axon. The axon is a single extension of cytoplasm that conducts impulses away from the cell body. Some axons can be quite long. For example, the cell bodies of neurons that control the muscles in your feet lie in the spinal cord, and their axons may extend over a meter to your feet.

Neuroglia do not conduct electrical impulses but instead support and insulate neurons and eliminate foreign materials in and around neurons. In many neurons, neuroglia cells associate with the axons and form an insulating covering, a *myelin sheath*, produced by successive wrapping of the membrane around the axon. Adjacent neuroglia cells are separated by gaps known as *nodes of Ranvier*, which serve as sites for accelerating an impulse (see chapter 45).

The nervous system is divided into the central nervous system (CNS), which includes the brain and spinal cord, and the peripheral nervous system (PNS), which includes *nerves* and *ganglia*. Nerves consist of axons in the PNS that are bundled together in much the same way as wires are bundled together in a cable. Ganglia are collections of neuron cell bodies.

> There are different types of neurons, but all are specialized to receive, produce, and conduct electrical signals. Neuroglia do not conduct electrical impulses but have various functions, including insulating axons to accelerate an electrical impulse. Both neurons and neuroglia are present in the CNS and the PNS.

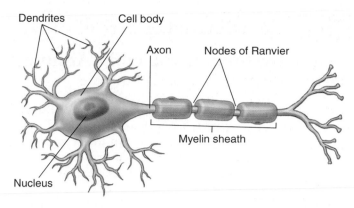

FIGURE 42.12
A neuron has a very long projection called an axon. A nerve impulse is received by the dendrites and then passed to the cell body and out through the axon. In some neurons specialized for rapid signal conduction, the axon is encased in a myelin sheath that is interrupted at intervals. At its far end, the axon may branch to terminate on more than one cell.

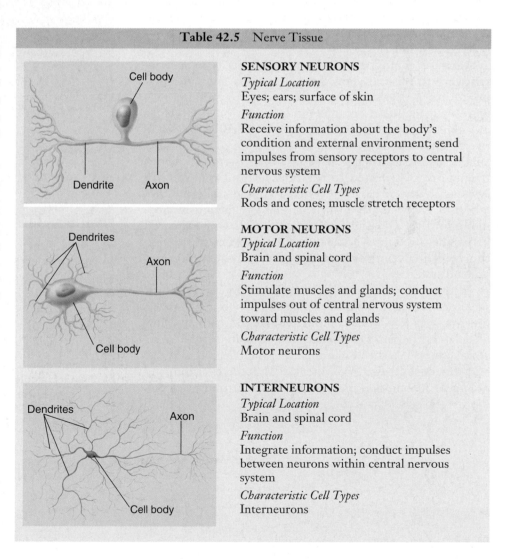

Table 42.5 Nerve Tissue	
	SENSORY NEURONS
	Typical Location
	Eyes; ears; surface of skin
	Function
	Receive information about the body's condition and external environment; send impulses from sensory receptors to central nervous system
	Characteristic Cell Types
	Rods and cones; muscle stretch receptors
	MOTOR NEURONS
	Typical Location
	Brain and spinal cord
	Function
	Stimulate muscles and glands; conduct impulses out of central nervous system toward muscles and glands
	Characteristic Cell Types
	Motor neurons
	INTERNEURONS
	Typical Location
	Brain and spinal cord
	Function
	Integrate information; conduct impulses between neurons within central nervous system
	Characteristic Cell Types
	Interneurons

42.5 Coordinated efforts of organ systems are necessary for locomotion.

Organ Systems Involved in Animal Movement

At the beginning of this chapter, we mentioned the organization of cells into tissues, tissues into organs, and organs cooperating as organ systems to perform major body functions. In the next level of organization, organ systems cooperate in performing complex activities, such as locomotion or body movement.

Plants and fungi move only by growing, or as the passive passengers of wind and water. Of the three multicellular kingdoms, only animals explore their environment in an active way, through locomotion. When a mosquito flies, its wings are moved rapidly through the air by quickly contracting flight muscles. When an earthworm burrows through the soil, its movement is driven by strong muscles pushing its body past the surrounding dirt. The rattlesnake in figure 42.13 slithers across the sand by a rhythmic contraction of the muscles sheathing its body. Humans walk by contracting muscles in their legs. Although our focus is on vertebrates, it is important to realize that essentially all animals employ muscles.

Animal locomotion provides a model (an example body function) to illustrate how the cooperation and coordination of many organ systems are necessary to produce, regulate, and maintain complex body activities. Energy for movement is derived from nutrients acquired by the digestive system, and oxygen needed to metabolize these nutrients is acquired through the respiratory system. These nutrients are distributed to the muscles and other organ systems by the circulatory system. Muscle and bone metabolism is constantly maintained and regulated by hormones of the endocrine system, and so on. Our focus will be on the direct production of animal movement through the coordinated efforts of the nervous, muscular, and skeletal systems.

In the following pages, we specifically examine how skeletons provide support and attachment sites for muscles, how muscles move bones, how muscles contract, and how muscle contraction is activated and regulated by the nervous system. We then examine muscle activity when at rest and during exercise. We end the chapter by exploring modes of locomotion used by animals that move through the water, on land, and across the sky.

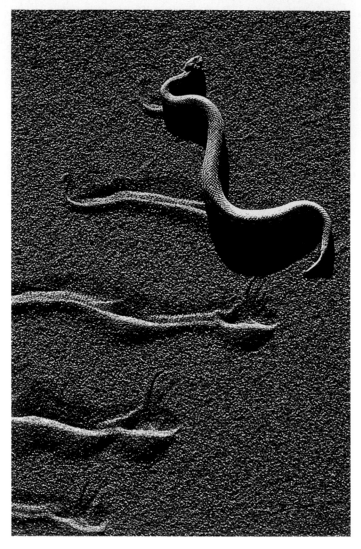

FIGURE 42.13
On the move. The movements made by this sidewinder rattlesnake are the result of strong muscle contractions acting on the bones of the skeleton. Without muscles and some type of skeletal system, complex locomotion as shown here would not be possible.

Locomotion is a complex activity characteristic of animals. A complex activity such as locomotion requires the cooperation and coordination of many organ systems.

The Skeletal System Supports Movement

Types of Skeletons

Animal locomotion is accomplished through the force of muscles acting on a rigid skeletal system. There are three types of skeletal systems in the animal kingdom: hydrostatic skeletons, exoskeletons, and endoskeletons.

Hydrostatic skeletons are primarily found in soft-bodied invertebrates, such as earthworms and jellyfish. In this case, a fluid-filled cavity is encircled by muscle fibers. As the muscles contract, the fluid in the cavity moves and changes the shape of the cavity. In an earthworm, for example, a wave of contraction of circular muscles begins anteriorly and compresses each segment of the body, so that the fluid pressure pushes it forward. Contractions of longitudinal muscles then pull the rear of the body forward (figure 42.14).

Exoskeletons surround the body as a rigid, hard case. Arthropods, such as crustaceans and insects, have exoskeletons made of the polysaccharide *chitin* (figure 42.15a). An exoskeleton offers great protection to internal organs, resists bending, and provides attachment sites for muscles. However, in order to grow, the animal must periodically molt. During molting, the animal is particularly vulnerable to predation because its old exoskeleton has been shed. Having an exoskeleton also limits the size of the animal. An animal with an exoskeleton cannot get too large because its exoskeleton would have to become thicker and heavier in order to prevent collapse as the animal grew larger. If an insect were the size of a human being, its exoskeleton would have to be so thick and heavy that it would be unable to move.

Endoskeletons, found in vertebrates and echinoderms, are rigid internal skeletons to which muscles are attached. Vertebrates have a flexible exterior that accommodates the movements of their skeleton. The endoskeleton of vertebrates is composed of cartilage or bone. Unlike chitin, bone is a cellular, living tissue capable of growth, self-repair, and remodeling in response to physical stresses. A vertebrate endoskeleton (figure 42.15b) is divided into an axial and an appendicular skeleton. The axial skeleton's bones form the axis of the body and support and protect the organs of the head, neck, and chest. The appendicular skeleton's bones include the bones of the limbs and the pectoral and pelvic girdles that attach them to the axial skeleton.

There are three types of animal skeletons: hydrostatic skeleton, exoskeleton, and endoskeleton. The endoskeletons in vertebrates are composed of bone or cartilage and are organized into axial and appendicular portions.

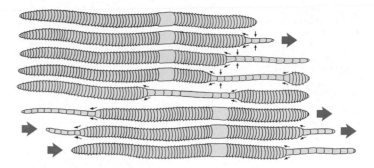

FIGURE 42.14

Locomotion in earthworms.   The hydrostatic skeleton of the earthworm uses muscles to move fluid within the segmented body cavity, changing the shape of the animal. When an earthworm's circular muscles contract, the internal fluid presses on the longitudinal muscles, which then stretch to elongate segments of the earthworm. A wave of contractions down the body of the earthworm produces forward movement.

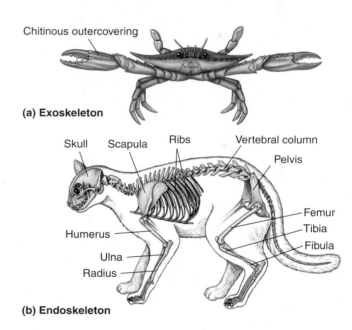

(a) **Exoskeleton**

Skull   Scapula   Ribs   Vertebral column
Pelvis
Humerus
Femur
Tibia
Fibula
Ulna
Radius

(b) **Endoskeleton**

FIGURE 42.15

Exoskeleton and endoskeleton.   (*a*) The hard, tough outer covering of an arthropod, such as this crab, is its exoskeleton. (*b*) Vertebrates, such as this cat, have endoskeletons. The axial skeleton is shown in the peach color, and the appendicular skeleton in the yellow color. Some of the major bones are labeled.

Muscle Contractions Move Bones at Joints

Types of Joints

The skeletal movements of the body are produced by contraction and shortening of muscles. Skeletal muscles are generally attached by tendons to bones, so when the muscles shorten, the attached bones move. These movements of the skeleton occur at joints, or articulations, where one bone meets another. There are three main classes of joints:

1. **Immovable joints** include the *sutures* that join the bones of the skull (figure 42.16*a*).
2. **Slightly movable joints** include those in which the bones are bridged by cartilage. The vertebral bones of the spine are separated by pads of cartilage called *intervertebral discs* (figure 42.16*b*). These *cartilaginous joints* allow some movement, primarily flexibility, while acting as efficient shock absorbers.
3. **Freely movable joints** include many types of joints and are also called synovial joints, because the articulating ends of the bones are located within a *synovial capsule* filled with a lubricating fluid. The ends of the bones are capped with cartilage, and the synovial capsule is strengthened by ligaments that hold the articulating bones in place.

Synovial joints allow the bones to move in directions dictated by the structure of the joint. For example, a joint in the finger allows only a hingelike movement, while the joint between the thigh bone (femur) and pelvis has a ball-and-socket structure that permits a variety of different movements (figure 42.16*c*).

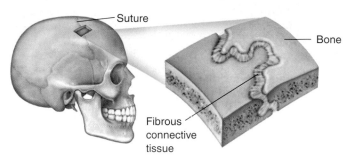

(a) Immovable joint

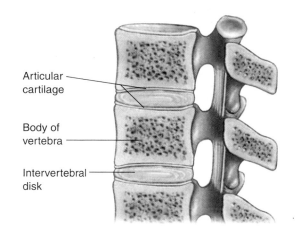

(b) Slightly movable joints

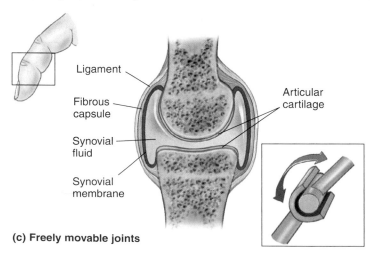

(c) Freely movable joints

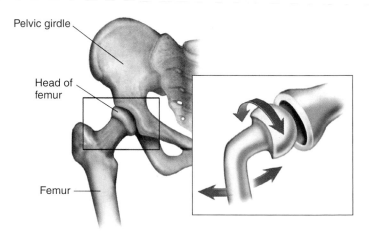

FIGURE 42.16
Three types of joints.   (*a*) Immovable joints include the sutures of the skull; (*b*) slightly movable joints include the cartilaginous joints between the vertebrae; and (*c*) freely movable joints are the synovial joints, such as a finger joint and or a hip joint.

Actions of Skeletal Muscles

Skeletal muscles produce movement of the skeleton when they contract. Usually, the two ends of a skeletal muscle are attached to different bones (although in some cases, one or both ends may be connected to some other kind of structure, such as skin). The attachments to bone are made by means of dense connective tissue straps called *tendons.* Tendons have elastic properties that allow "give-and-take" during muscle contraction. One attachment of the muscle, the **origin,** remains relatively stationary during a contraction. The other end of the muscle, the **insertion,** is attached to the bone that moves when the muscle contracts. For example, contraction of the biceps muscle in the upper arm causes the forearm (the insertion of the muscle) to move toward the shoulder (the origin of the muscle).

Muscles that cause the same action at a joint are **synergists.** For example, the various muscles of the quadriceps group in humans (front thigh muscles) are synergists; they all act to extend the knee joint. Muscles that produce opposing actions are **antagonists.** For example, muscles that flex a joint are antagonistic to muscles that extend that joint (figure 42.17*a*). In humans, when the hamstring muscles contract, they cause flexion of the knee joint (figure 42.17*b*). Therefore, the quadriceps and hamstrings are antagonists to each other. In general, the muscles that antagonize a given movement are relaxed when that movement is performed. Thus, when the hamstrings flex the knee joint, the quadriceps muscles relax.

Isotonic and Isometric Contractions. In order for muscle fibers to shorten when they contract, they must generate a force that is greater than the opposing forces that act to prevent movement of the muscle's insertion. When you lift a weight by contracting muscles in your biceps, for example, the force produced by the muscle is greater than the force of gravity on the object you are lifting. In this case, the muscle and all of its fibers shorten in length. This type of contraction is referred to as **isotonic contraction,** because the force of contraction remains relatively constant throughout the shortening process (*iso,* "same"; *tonic,* "strength").

There are periods during muscle contraction when the tension is not great enough to move the bone or object. During this time, the muscle does not change in length, and so this is called **isometric** (literally, "same length") **contraction.** Isometric contractions occur as a phase of normal muscle contraction. They typically occur preceding isotonic contractions, before enough tension is produced for movement, to hold an object stationary or to provide tautness and stability to the body.

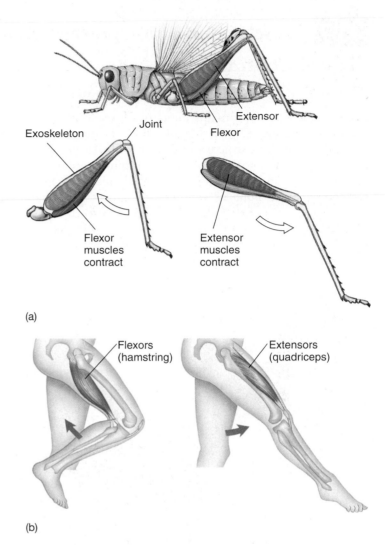

(a)

(b)

FIGURE 42.17
Flexor and extensor muscles of the leg. (*a*) Antagonistic muscles control the movement of an animal with an exoskeleton, such as the jumping of a grasshopper. When the smaller flexor tibia muscle contracts, it pulls the lower leg in toward the upper leg. Contraction of the extensor tibia muscles straightens out the leg and sends the insect into the air. (*b*) Similarly, antagonistic muscles can act on an endoskeleton. In humans, the hamstrings, a group of three muscles, produce flexion of the knee joint, whereas the quadriceps, a group of four muscles, produce extension.

Joints confer flexibility to a rigid skeleton, allowing a range of motion determined by the type of joint. Muscles, positioned across joints, contract and cause the movement of the body. Synergistic muscles have the same action, whereas antagonistic muscles have opposite actions. Both muscle groups are involved in locomotion. Isotonic contractions involve the shortening of muscle, while isometric contractions do not alter the length of the muscle.

42.6 Muscle contraction powers animal locomotion.

Muscle Contraction

The Sliding Filament Mechanism of Contraction

Each skeletal muscle contains numerous **muscle fibers,** as described in section 42.4. Each muscle fiber encloses a bundle of 4 to 20 elongated structures called **myofibrils.** Each myofibril, in turn, is composed of thick and thin **myofilaments** (figure 42.18). The muscle fiber appears striated under a microscope because along its length its myofibrils have alternating dark and light bands. The banding pattern results from the organization of the myofilaments within the myofibril. The thick myofilaments are stacked together to pro-

duce the dark bands, called *A bands*; the thin filaments alone are found in the light bands, or *I bands*.

Each I band in a myofibril is divided in half by a disc of protein, called a *Z line* because of its appearance in electron micrographs. The thin filaments are anchored to these discs of proteins that form the Z lines. If you look at an electron micrograph of a myofibril (figure 42.19), you see that the structure of the myofibril repeats from Z line to Z line. This repeating structure, called a **sarcomere,** is the smallest subunit of muscle contraction.

The thin filaments stick partway into, and overlap with, thick filaments on each side of an A band, but in a

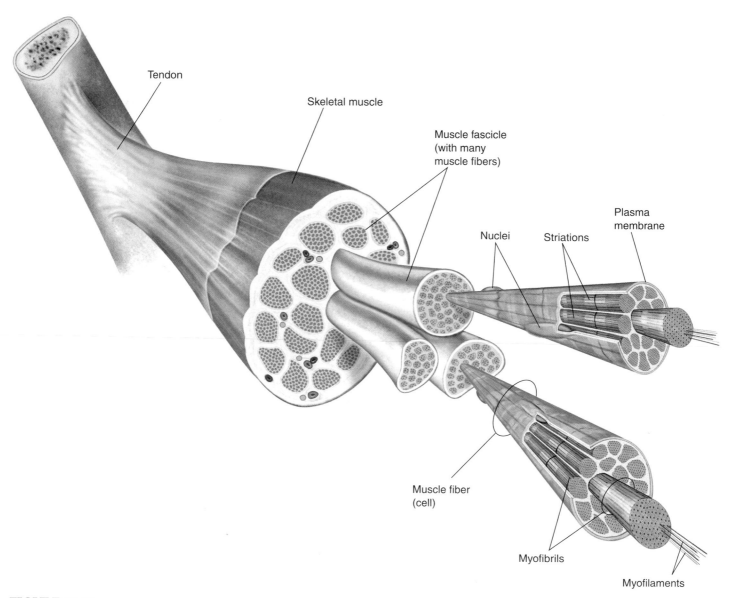

FIGURE 42.18
The organization of skeletal muscle. Each muscle is composed of many fascicles, which are bundles of muscle cells, or fibers. Each fiber is composed of many myofibrils, which are each, in turn, composed of myofilaments.

resting muscle, do not project all the way to the center of the A band. As a result, the center of an A band (called an *H band*) is lighter than each side, with its interdigitating thick and thin filaments. This appearance of the sarcomeres changes when the muscle contracts.

A muscle contracts and shortens because its myofibrils contract and shorten. When this occurs, the myofilaments do *not* shorten; instead, the thin filaments slide deeper into the A bands (figure 42.20). This makes the H bands narrower until, at maximal shortening, they disappear entirely. It also makes the I bands narrower, because the dark A bands are brought closer together. This is the **sliding filament mechanism** of contraction.

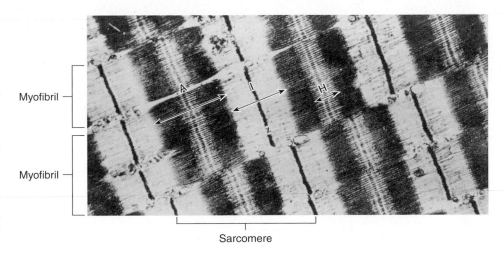

FIGURE 42.19

An electron micrograph of skeletal myofibrils. The Z lines that serve as the borders of the sarcomeres are clearly seen within each myofibril. The thick filaments comprise the A bands; the thin filaments are within the I bands and stick partway into the A bands, overlapping with the thick filaments. There is no overlap of thick and thin filaments at the central region of an A band, which is therefore lighter in appearance. This is the H band.

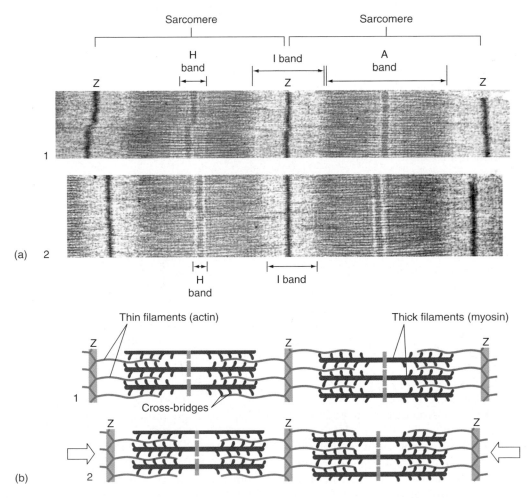

FIGURE 42.20

The sliding filament mechanism of contraction. (*a*) An electron micrograph and (*b*) a diagram show that as the thin filaments slide deeper into the centers of the sarcomeres, the Z lines are brought closer together. (*1*) Relaxed muscle; (*2*) partially contracted muscle.

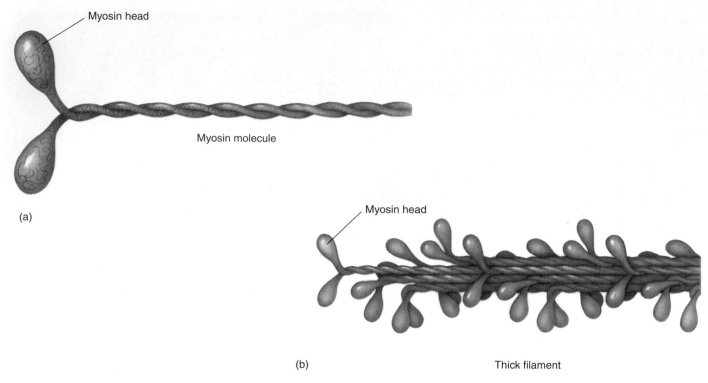

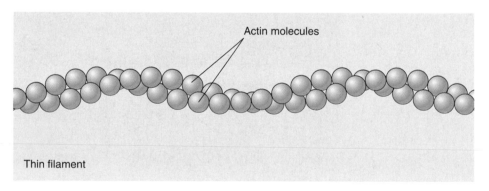

(a)

(b)   Thick filament

FIGURE 42.21
Thick filaments are composed of myosin. (*a*) Each myosin molecule consists of two polypeptide chains wrapped around each other; at the end of each chain is a globular region referred to as the "head." (*b*) Thick filaments consist of myosin molecules combined into bundles from which the heads protrude at regular intervals.

Electron micrographs reveal **cross-bridges** that extend from the thick to the thin filaments, suggesting a mechanism that might cause the filaments to slide. To understand how this is accomplished, we have to examine the thick and thin filaments at a molecular level. Biochemical studies show that each thick filament is composed of many **myosin** proteins packed together, and every myosin molecule has a "head" region that protrudes from the thick filaments (figure 42.21). These myosin heads form the cross-bridges seen in electron micrographs. Biochemical studies also show that each thin filament consists primarily of many globular **actin** proteins twisted into a double helix (figure 42.22). Therefore, if we were able to see a sarcomere at a molecular level, it would have the structure depicted in figure 42.23*a*.

Before the myosin heads bind to the actin of the thin filaments, they act as ATPase enzymes, splitting ATP into ADP and P$_i$. This activates the heads, "cocking" them so that they can bind to actin and form cross-bridges. Once a myosin head binds to actin, it undergoes a conformational

FIGURE 42.22
Thin filaments are composed of globular actin proteins. Two rows of actin proteins are twisted together in a helix to produce the thin filaments. Other proteins, tropomyosin and troponin, associate with the strands of actin and are involved in muscle contraction. These other proteins are discussed later in the chapter.

(shape) change, pulling the thin filament toward the center of the sarcomere (figure 42.23*b*) in a *power stroke*. At the end of the power stroke, the myosin head binds to a new molecule of ATP. This allows the head to detach from actin and continue the **cross-bridge cycle** (figure 42.24), which repeats as long as the muscle is stimulated to contract.

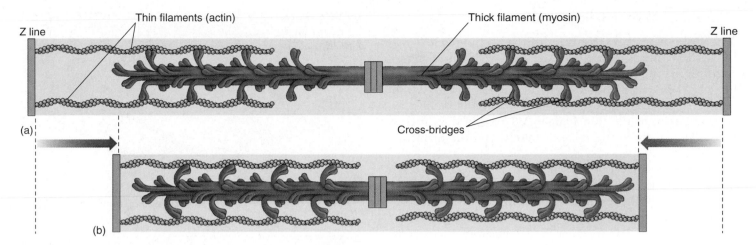

Thin filaments (actin)

Thick filament (myosin)

Z line

Z line

(a)

Cross-bridges

(b)

FIGURE 42.23

The interaction of thick and thin filaments in striated muscle sarcomeres. (*a*) The heads on the two ends of the thick filaments are oriented in opposite directions, so that the cross-bridges pull the thin filaments and the Z lines on each side of the sarcomere toward the center. (*b*) This sliding of the filaments produces muscle contraction.

FIGURE 42.24

The cross-bridge cycle in muscle contraction. (*a*) With ADP and P_i attached to the myosin head, (*b*) the head is in a conformation that can bind to actin and form a cross-bridge. (*c*) Binding causes the myosin head to assume a more bent conformation, moving the thin filament along the thick filament (to the left in this diagram) and releasing ADP and P_i. (*d*) Binding of ATP to the head detaches the cross-bridge; cleavage of ATP into ADP and P_i puts the head into its original conformation, allowing the cycle to begin again.

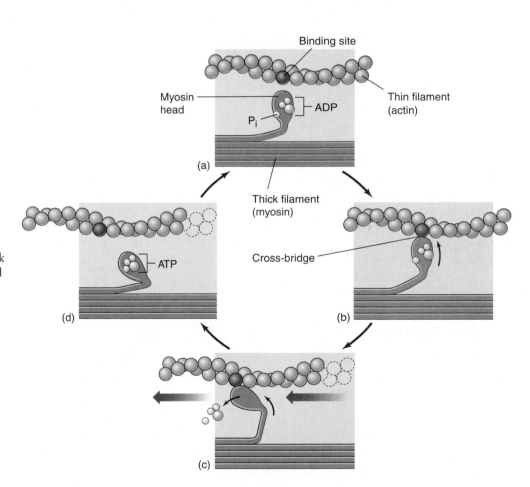

Binding site

Myosin head

ADP

Thin filament (actin)

P_i

(a)

Thick filament (myosin)

Cross-bridge

ATP

(d)

(b)

(c)

In death, the cell can no longer produce ATP, and therefore the cross-bridges cannot be broken—this causes the muscle stiffness of death called *rigor mortis*. A living cell, however, always has enough ATP to allow the myosin heads to detach from actin. How, then, is the cross-bridge cycle arrested so that the muscle can relax? The regulation of muscle contraction and relaxation requires additional factors that we will discuss next.

Control of Muscle Contraction

The Role of Ca++ in Contraction. When a muscle is relaxed, its myosin heads are "cocked" and ready, through the splitting of ATP, but are unable to bind to actin. This is because the attachment sites for the myosin heads on the actin are physically blocked by another protein, known as **tropomyosin,** in the thin filaments. Cross-bridges therefore cannot form in the relaxed muscle, and the filaments cannot slide.

In order to contract a muscle, the tropomyosin must be moved out of the way so that the myosin heads can bind to the uncovered actin attachment sites. This requires the action of **troponin,** a regulatory protein complex that holds tropomyosin and actin together. The regulatory interactions between troponin and tropomyosin are controlled by the calcium ion (Ca++) concentration of the muscle cell cytoplasm.

When the Ca++ concentration of the muscle cell cytoplasm is low, tropomyosin inhibits cross-bridge formation, and the muscle is relaxed (figure 42.25a). When the Ca++ concentration is raised, Ca++ binds to troponin. This causes the troponin-tropomyosin complex to be shifted away from the attachment sites for the myosin heads on the actin. Cross-bridges can thus form, undergo power strokes, and produce muscle contraction (figure 42.25b).

Where does the Ca++ come from? Muscle fibers store Ca++ in a modified endoplasmic reticulum called a sarcoplasmic reticulum, or SR (figure 42.26). When a muscle fiber is stimulated to contract, an electrical impulse travels down into the muscle fiber through invaginations of the cell membrane called the **transverse tubules (T tubules).** This triggers the release of Ca++ from the SR. Ca++ then diffuses into the myofibrils, where it binds to troponin and causes contraction. The contraction of muscles is regulated by nerve activity, and so nerves must influence the distribution of Ca++ in the muscle fiber.

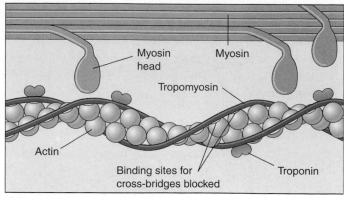

(a)

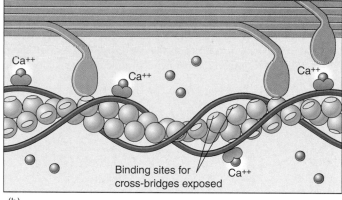

(b)

FIGURE 42.25
How calcium controls striated muscle contraction. (*a*) When the muscle is at rest, a long filament of the protein tropomyosin blocks the myosin-binding sites on the actin molecule. Because myosin is unable to form cross-bridges with actin at these sites, muscle contraction cannot occur. (*b*) When Ca++ binds to another protein, troponin, the Ca++-troponin complex displaces tropomyosin and exposes the myosin-binding sites on actin, permitting cross-bridges to form and contraction to occur.

FIGURE 42.26
The relationship between the myofibrils, transverse tubules, and sarcoplasmic reticulum. Impulses travel down the axon of a motor neuron that synapses with a muscle fiber. The impulses are conducted along the fiber membrane and down the transverse tubules and stimulate the release of Ca++ from the sarcoplasmic reticulum into the cytoplasm. Ca++ diffuses toward the myofibrils and causes contraction.

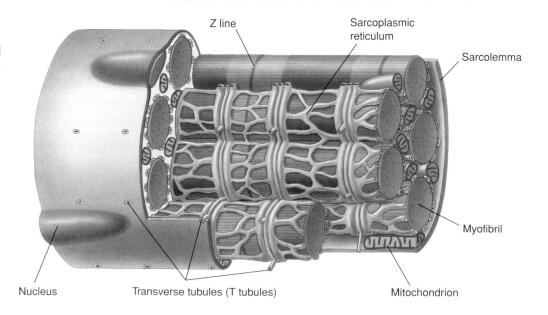

Nerves Stimulate Contraction. Muscles are stimulated to contract by motor neurons. The particular motor neurons that stimulate skeletal muscles, as opposed to cardiac and smooth muscles, are called *somatic motor neurons*. The axon (see figure 42.12) of a somatic motor neuron extends from the neuron cell body and branches to make functional connections, or *synapses*, with a number of muscle fibers. (Synapses are discussed in more detail in chapter 45). One axon can stimulate many muscle fibers, and in some animals a muscle fiber may be innervated by more than one motor neuron. However, in humans, each muscle fiber only has a single synapse with a branch of one axon.

When a somatic motor neuron delivers electrochemical impulses, it stimulates contraction of the muscle fibers it innervates (makes synapses with) through the following events:

1. The motor neuron, at its synapse with the muscle fibers, releases a chemical known as a *neurotransmitter*. The specific neurotransmitter released by somatic motor neurons is **acetylcholine (ACh).** ACh acts on the muscle fiber membrane to stimulate the muscle fiber to produce its own electrochemical impulses.
2. The impulses spread along the membrane of the muscle fiber and are carried into the muscle fibers through the T tubules.
3. The T tubules conduct the impulses toward the sarcoplasmic reticulum, which then releases Ca^{++}. As described earlier, the Ca^{++} binds to troponin, which exposes the cross-bridge binding sites on the actin myofilaments, stimulating muscle contraction.

When impulses from the nerve cease, the nerve stops releasing ACh. This stops the production of impulses in the muscle fiber. When the T tubules no longer produce impulses, Ca^{++} is pumped back into the SR by active transport. Troponin is no longer bound to Ca^{++}, so tropomyosin returns to its inhibitory position, allowing the muscle to relax.

The involvement of Ca^{++} in muscle contraction is called **excitation-contraction coupling** because it is the release of Ca^{++} that links the excitation of the muscle fiber by the motor neuron to the contraction of the muscle.

Motor Units and Recruitment. A single muscle fiber can produce variable tension depending on the frequency of stimulation. The response of an entire muscle depends upon the number of individual fibers involved and their degree of tension. The set of muscle fibers innervated by all axonal branches of a given motor neuron plus the motor neuron itself is defined as a **motor unit** (figure 42.27). Every time the motor neuron produces impulses, all muscle fibers in that motor unit contract together. The division of the muscle into motor units allows the muscle's strength of contraction to be finely graded, a requirement for coordinated movements of the skeleton. Muscles that require a

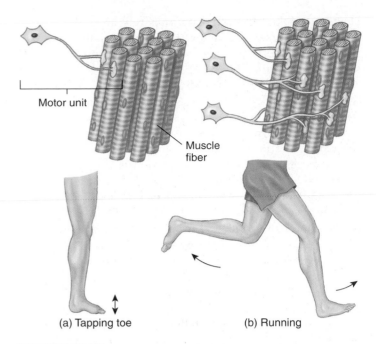

FIGURE 42.27
The number and size of motor units. (*a*) Precise muscle contractions require smaller motor units. (*b*) Large muscle movements require larger motor units. The more motor units activated, the stronger the contraction.

finer degree of control have smaller motor units (fewer muscle fibers per neuron) than muscles that require less precise control but must exert more force. For example, there are only a few muscle fibers per motor neuron in the muscles that move the eyes, while there are several hundred to more than a thousand per motor neuron in the large muscles of the legs.

Most muscles contain motor units in a variety of sizes, which can be selectively activated by the nervous system. The weakest contractions of a muscle involve the activation of a few small motor units. If a slightly stronger contraction is necessary, additional small motor units are also activated. The initial increments to the total force generated by the muscle are therefore relatively small. As ever greater forces are required, more and larger motor units are brought into action, and the force increments become larger. The nervous system's use of increased numbers and sizes of motor units to produce a stronger contraction is termed **recruitment.**

Types of Muscle Fibers

Muscle Fiber Twitches. An isolated skeletal muscle can be studied by stimulating it artificially with electric shocks. If a muscle is stimulated with a single electric shock, it will quickly contract and relax in a response called a **twitch.** Increasing the stimulus voltage increases the strength of the twitch up to a maximum. If a second electric shock is delivered immediately after the first, it will produce a second twitch that may partially "ride piggyback" on the first. This cumulative response is called **summation** (figure 42.28).

If the stimulator is set to deliver an increasing frequency of electric shocks automatically, the relaxation time between successive twitches will get shorter and shorter as the strength of contraction increases. Finally, at a particular frequency of stimulation, no visible relaxation occurs between successive twitches. Contraction is smooth and sustained, as it is during normal muscle contraction in the body. This smooth, sustained contraction is called **tetanus.** (The term *tetanus* should not be confused with the disease of the same name, which is accompanied by a painful state of muscle contracture, or *tetany*.)

Skeletal muscle fibers can be divided on the basis of their contraction speed into **slow-twitch fibers,** or type I fibers, and **fast-twitch fibers,** or type II fibers. The muscles that move the eyes, for example, have a high proportion of fast-twitch fibers and reach maximum tension in about 7.3 milliseconds; the soleus muscle in the leg, by contrast, has a high proportion of slow-twitch fibers and requires about 100 milliseconds to reach maximum tension (figure 42.29).

Muscles such as the soleus must be able to sustain a contraction for a long period of time without fatigue. The resistance to fatigue demonstrated by these muscles is aided by other characteristics of slow-twitch (type I) fibers that endow them with a high capacity for aerobic respiration. Slow-twitch fibers have a rich capillary supply, numerous mitochondria and aerobic respiratory enzymes, and a high concentration of **myoglobin** pigment. Myoglobin is a red pigment, similar to the hemoglobin in red blood cells, but its higher affinity for oxygen improves the delivery of oxygen to the slow-twitch fibers. Because of their high myoglobin content, slow-twitch fibers are also called *red fibers.*

The thicker, fast-twitch (type II) fibers have fewer capillaries and mitochondria than slow-twitch fibers and not as much myoglobin; hence, these fibers are also called *white fibers.* Fast-twitch fibers are adapted to respire anaerobically by using a large store of glycogen and high concentrations of glycolytic enzymes. Fast-twitch fibers are adapted for the rapid generation of power and can grow thicker and stronger in response to weight training. The "dark meat" and "white meat" found in chicken and turkey consists of muscles with primarily red and white fibers, respectively.

In addition to the type I and type II fibers, human muscles also have an intermediate form of fibers that are fast-twitch but also have a high oxidative capacity, and so are more resistant to fatigue. Endurance training increases the proportion of these fibers in muscles.

Muscle Metabolism During Rest and Exercise. Skeletal muscles at rest obtain most of their energy from the aerobic respiration of fatty acids. During exercise, muscle glycogen and blood glucose are also used as energy sources. The energy obtained by cellular respiration is

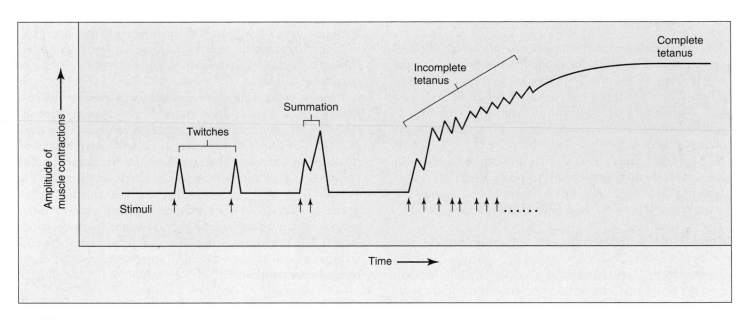

FIGURE 42.28
Summation. Muscle twitches summate to produce a sustained, tetanic contraction. This pattern is produced when the muscle is stimulated electrically or naturally by neurons. Tetanus, a smooth, sustained contraction, is the normal type of muscle contraction in the body.
What determines the maximum amplitude of a summated muscle contraction?

used to make ATP, which is needed for (1) the movement of the cross-bridges during muscle contraction and (2) the pumping of Ca^{++} back into the sarcoplasmic reticulum during muscle relaxation. ATP can be obtained by skeletal muscles quickly by combining ADP with phosphate derived from creatine phosphate. This compound was produced previously in the resting muscle by combining creatine with phosphate derived from the ATP generated in cellular respiration.

Skeletal muscles respire anaerobically for the first 45 to 90 seconds of moderate-to-heavy exercise, because the cardiopulmonary system requires this amount of time to sufficiently increase the oxygen supply to the exercising muscles. If exercise is moderate, aerobic respiration contributes the major portion of the skeletal muscle energy requirements following the first 2 minutes of exercise.

Whether exercise is light, moderate, or intense for a given person depends upon that person's maximal capacity for aerobic exercise. The maximum rate of oxygen consumption in the body (by aerobic respiration) is called the maximal oxygen uptake, or the aerobic capacity. The intensity of exercise can also be defined by the lactate threshold. This is the percentage of the maximal oxygen uptake at which a significant rise in blood lactate levels occurs as a result of anaerobic respiration. For example, in an average healthy person, a significant amount of blood lactate appears when exercise is performed at about 50 to 70% of the maximal oxygen uptake.

Muscle Fatigue and Physical Training.   Muscle fatigue refers to the use-dependent decrease in the ability of a muscle to generate force. The reasons for fatigue are not entirely understood. In most cases, however, muscle fatigue is correlated with the production of lactic acid by the exercising muscles. Lactic acid is produced by the anaerobic respiration of glucose, and glucose is obtained from muscle glycogen and from the blood. Lactate production and muscle fatigue are therefore also related to the depletion of muscle glycogen.

Because the depletion of muscle glycogen and the buildup of lactic acid place a limit on exercise, any adaptation that spares muscle glycogen and/or efficiently removes lactic acid will improve physical endurance. Trained athletes have an increased proportion of energy derived from the aerobic respiration of fatty acids, resulting in a slower

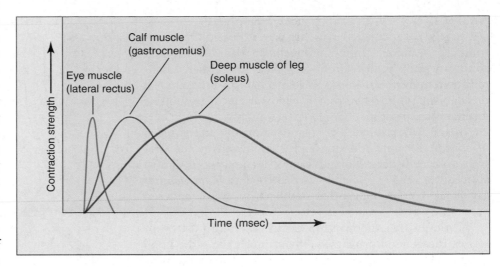

FIGURE 42.29
Skeletal muscles have different proportions of fast-twitch and slow-twitch fibers. The muscles that move the eye contain mostly fast-twitch fibers, whereas the deep muscle of the leg (the soleus) contains mostly slow-twitch fibers. The calf muscle (gastrocnemius) is intermediate in its composition.
How would you determine if the calf muscle contains a mix of fast-twitch and slow-twitch fibers, or instead is composed of an intermediate form of fiber?

depletion of their muscle glycogen reserve. The greater the level of physical training, the higher the proportion of energy derived from the aerobic respiration of fatty acids. Athletes also have greater muscle vascularization. This increased blood flow facilitates both the delivery of oxygen and the removal of lactic acid from active muscle. Because the aerobic capacity of endurance-trained athletes is higher than that of untrained people, athletes can perform more exercise before lactic acid production and glycogen depletion cause muscle fatigue.

Endurance training does not increase muscle size. Muscle enlargement is produced only by frequent periods of high-intensity exercise in which muscles work against high resistance, as in weight lifting. As a result of resistance training, type II (fast-twitch) muscle fibers become thicker due to the increased size and number of their myofibrils. Weight training, therefore, causes skeletal muscles to grow by **hypertrophy** (increased cell size) rather than by cell division and an increased number of cells.

ATP provides the energy for the cross-bridge cycle that drives muscle contraction. In order for a muscle to contract, Ca^{++} must be released from the sarcoplasmic reticulum, where it is stored. Muscle contraction is stimulated by neurons. Recruitment of varying sizes and numbers of motor units is used to produce different types of muscle contraction.

Modes of Animal Locomotion

Animals are unique among multicellular organisms in their ability to actively move from one place to another. Locomotion requires both a propulsive mechanism and a control mechanism. Animals employ a wide variety of propulsive mechanisms, most involving contracting muscles to generate the necessary force. The quantity, quality, and position of contractions are initiated and coordinated by the nervous system. In large animals, active locomotion is almost always produced by appendages that oscillate—*appendicular locomotion*—or by bodies that undulate, pulse, or undergo peristaltic waves—*axial locomotion.*

While animal locomotion occurs in many different forms, the general principles remain much the same in all groups. The physical constraints to movement—gravity and frictional drag—are the same in every environment, differing only in degree. You can conveniently divide the environments through which animals move into three types, each involving its own forms of locomotion: water, land, and air.

Locomotion in Water

Many aquatic and marine invertebrates move along the bottom of a body of water using the same form of locomotion employed by terrestrial animals moving over the land surface. Flatworms employ ciliary activity to brush themselves along, roundworms a peristaltic slither, leeches a contract-anchor-extend creeping. Crabs walk using limbs to pull themselves along; mollusks use a muscular foot, while starfish use unique tube feet to do the same thing.

Moving directly through the water (swimming) presents quite a different challenge. Water's buoyancy reduces the influence of gravity. The primary force retarding forward movement is frictional drag, so body shape is important in reducing the friction and turbulence produced by swimming through the water.

Some marine invertebrates move about using hydraulic propulsion. For example, scallops clap their shells together forcefully, while squids and octopuses squirt water like a marine jet.

All aquatic and marine vertebrates swim. Swimming involves using the body or its appendages to push against the water. An eel swims by sinuous undulations of its whole body (figure 42.30*a*). The undulating body waves of eel-like swimming are created by waves of muscle contraction alternating between the left and right axial musculature. As each body segment in turn pushes against the water, the moving wave forces the eel forward.

Fish, reptiles, and aquatic amphibians swim in a way similar to eels, but only undulate the posterior (back) portion of the body (figure 42.30*b*) and sometimes only the caudal (rear) fin. This allows considerable specialization of the front end of the body, while sacrificing little propulsive force.

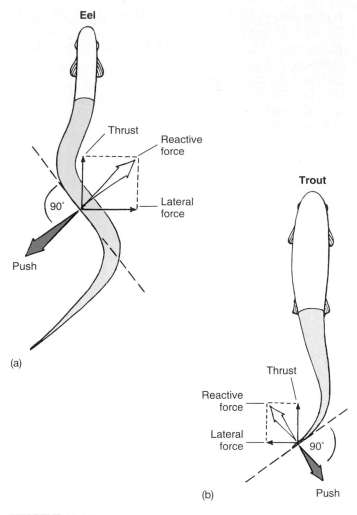

(a)

(b)

FIGURE 42.30
Movements of swimming fishes. (*a*) An eel pushes against the water with its whole body, while (*b*) a trout pushes only with its posterior half.

Whales also swim using undulating body waves, but unlike any of the fishes, the waves pass from top to bottom and not from side to side. The body musculature of eels and fish is highly segmental; that is, a muscle segment alternates with each vertebra. This arrangement permits the smooth passage of undulatory waves along the body. Whales are unable to produce lateral undulations because mammals do not have this arrangement.

Many terrestrial tetrapod vertebrates swim, usually with appendicular locomotion. Most birds that swim, such as ducks and geese, propel themselves through the water by pushing against it with their hind legs, which typically have webbed feet. Frogs, turtles, and most marine mammals also swim with their hind legs and have webbed feet. Tetrapod vertebrates that swim with their forelegs usually have these limbs modified as flippers, and pull themselves through the water; examples include sea turtles, penguins, and fur seals. A few principally terrestrial tetrapod vertebrates, such as polar bears and platypuses, swim with walking forelimbs not modified for swimming.

FIGURE 42.31

Animals that hop or leap use their rear legs to propel themselves through the air. The powerful leg muscles of this frog allow it to explode from a crouched position to a takeoff in about 100 milliseconds.

Locomotion on Land

The three great groups of terrestrial animals—mollusks, arthropods, and vertebrates—each move over land in different ways.

Mollusk locomotion is far less efficient than that of the other groups. Snails, slugs, and other terrestrial mollusks secrete a path of mucus that they glide along, pushing with a muscular foot.

Only vertebrates and arthropods (insects, spiders, and crustaceans) have developed a means of rapid surface locomotion. In both groups, the body is raised above the ground and moved forward by pushing against the ground with a series of jointed appendages, the legs.

Because legs must provide support as well as propulsion, it is important that the sequence of their movements not shove the body's center of gravity outside the legs' zone of support. If they do, the animal loses its balance and falls. The necessity to maintain stability determines the sequence of leg movements, which are similar in vertebrates and arthropods.

The apparent differences in the walking gaits of these two groups reflect the differences in leg number. Vertebrates are tetrapods (four limbs), while all arthropods have six or more limbs. Although having many legs increases stability during locomotion, it also appears to reduce the maximum speed that can be attained.

The basic walking pattern of all tetrapod vertebrates is left hind leg (LH), left foreleg (LF), right hind leg (RH), right foreleg (RF), and then the same sequence again and again. Unlike insects, vertebrates can begin to walk with any of the four legs, and not just the posterior pair. Both arthropods and vertebrates achieve faster gaits by overlapping the leg movements of the left and right sides. For example, a horse can convert a walk to a trot by moving diagonally opposite legs simultaneously.

The highest running speeds of tetrapod vertebrates, such as the gallop of a horse, are obtained with asymmetric gaits. When galloping, a horse is never supported by more than two legs, and occasionally is supported by none. This reduces friction against the ground to an absolute minimum, increasing speed. With their larger number of legs, arthropods cannot have these speedy asymmetric gaits, because the movements of the legs would interfere with each other.

Not all animals walk or run on land. Many insects, such as grasshoppers, leap using strong rear legs to propel themselves through the air. Vertebrates such as kangaroos, rabbits, and frogs are also effective leapers (figure 42.31).

Many invertebrates use peristaltic motion to slide over the surface. Among vertebrates, this form of locomotion is exhibited by snakes and caecilians (legless amphibians). Most snakes employ serpentine locomotion, in which the body is thrown into a series of sinuous curves (see figure 42.13). The movements superficially resemble those of eel-like swimming, but the similarity is more analogous than homologous. Propulsion occurs not by a wave of contraction undulating the body, but by a simultaneous lateral thrust in all segments of the body in contact with the ground. To go forward, it is necessary that the strongest muscular thrust push against the ground opposite the direction of movement. Because of this, thrust tends to occur at the anterior (outside) end of the inward-curving side of the loop of the snake's body.

Self Test

1. Given what you know about the hierarchy of chemical organization, answer the following question. If Los Angeles was a molecule, Orange County was a cell, and California was a tissue, what would the United States be?
   a. a sister cell
   b. an organ system
   c. an organ
   d. an organism

2. Which organ system is primarily responsible for coordinating, regulating, and integrating the various activities of the body?
   a. nervous
   b. endocrine
   c. muscular
   d. respiratory

3. Which of the following statements best describes endocrine glands?
   a. Endocrine glands are ductless glands that secrete hormones.
   b. Endocrine glands are ductless glands that secrete sweat, saliva, and digestive enzymes.
   c. Endocrine glands secrete hormones through a duct.
   d. Endocrine glands secrete sweat, saliva, and digestive enzymes through a duct.

4. Which of the following is *not* considered a connective tissue?
   a. blood
   b. muscle
   c. adipose tissue
   d. cartilage

5. Which of the following statements about nerve tissue is false?
   a. Neurons transmit sensory information to the brain.
   b. Both neurons and neuroglia are present in the CNS and PNS.
   c. Neurons conduct electrical impulses.
   d. All types of cells in nerve tissue conduct electrical impulses.

6. Exoskeletons provide excellent protection to internal organs. However, animals that utilize exoskeletons are usually relatively small. Why?
   a. These animals are only able to produce a limited amount of chitin.
   b. Exoskeletons are not living tissue, and therefore they cannot grow.
   c. A large exoskeleton would be too heavy to move.
   d. During molting, these animals are especially vulnerable to predators and therefore do not usually live long enough to grow bigger.

7. Which of the following statements best describes the sliding filament mechanism of muscle contraction?
   a. Actin and myosin filaments do not shorten, but rather, slide past each other.
   b. Actin and myosin filaments shorten and slide past each other.
   c. As they slide past each other, actin filaments shorten, while myosin filaments do not shorten.
   d. As they slide past each other, myosin filaments shorten, while actin filaments do not shorten.

8. What is the role of Ca^{++} in muscle contraction?
   a. It binds to tropomyosin, enabling troponin to move and reveal binding sites for cross-bridges.
   b. It binds to troponin, enabling tropomyosin to move and reveal binding sites for cross-bridges.
   c. It binds to tropomyosin, enabling troponin to release ATP.
   d. It binds to troponin, enabling tropomyosin to release ATP.

9. Motor neurons stimulate muscle contraction via the release of
   a. Ca^{++}.
   b. ATP.
   c. acetylcholine.
   d. hormones.

10. Which of the following statements about muscle metabolism is false?
    a. Skeletal muscles at rest obtain most of their energy from muscle glycogen and blood glucose.
    b. ATP can be quickly obtained by combining ADP with phosphate derived from creatine phosphate.
    c. Exercise intensity is related to the maximum rate of oxygen consumption.
    d. ATP is required for the pumping of the Ca^{++} back into the sarcoplasmic reticulum.

Test Your Visual Understanding

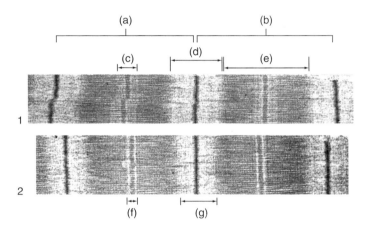

1. Use the following terms to label the structures in the figure.
   sarcomere
   A band
   H band
   I band
   Z line

Apply Your Knowledge

1. You are planning to travel to a hot and dry desert region where little water is available. If you were able to either (a) increase the amount of keratin in your stratified squamous epithelium, or (b) increase the number of your exocrine sweat glands, which would be a better survival strategy?

2. You have just been hired as a personal trainer. One of your clients asks you if sit-ups are considered isometric or isotonic exercises. What is your response?

3. The nerve gas sarin inhibits the enzyme acetylcholinesterase, which is normally present in the neuromuscular junction and required to break down acetylcholine into inactive fragments. Based on this information, what do you think are the likely effects of this nerve gas on muscle function?

43

Fueling Body Activities: Digestion

Concept Outline

43.1 Animals employ a digestive system to prepare food for assimilation by cells.

Types of Digestive Systems. Some animals have a gastrovascular cavity, but others have a digestive tract.
Vertebrate Digestive Systems. Different regions of the gastrointestinal tract are adapted for different functions.

43.2 Food is ingested, swallowed, and transported to the stomach.

Mouth and Teeth. Carnivores, herbivores, and omnivores display differences in tooth structure.
Esophagus and Stomach. The esophagus delivers food to the stomach, which mixes it with gastric juice to form chyme. Chyme is delivered into the small intestine.

43.3 The small and large intestines have very different functions.

Small Intestine. The small intestine is the major region of digestion and absorption.
Large Intestine. The large intestine forms, stores, and expels feces. It also plays a role in absorption.
Variations in Vertebrate Digestive Systems. Digestive systems are adapted to particular diets.

43.4 Neural, hormonal, and accessory organ regulation controls digestion.

Neural and Hormonal Regulation of the Digestive Tract. Both neural and hormonal signals orchestrate digestive function.
Accessory Organ Function in Nutrient Processing and Distribution. Molecules absorbed by the small intestine are transported directly to the liver for processing. Insulin and glucagon regulate blood glucose concentrations.

43.5 All animals require food energy and essential nutrients.

Food Energy and Energy Expenditure. The intake of food energy must balance the energy expended by the body in order to maintain a stable weight.
Essential Nutrients. Food must contain vitamins, minerals, and specific amino acids and fatty acids for health.

FIGURE 43.1
Animals are heterotrophs. All animals must consume plant material or other animals in order to live. The nuts in this chipmunk's cheeks will be consumed and converted to body tissue, energy, and refuse.

Plants and other photosynthetic organisms can produce the organic molecules they need from inorganic components. Therefore, they are autotrophs, or self-sustaining. Animals are heterotrophs: They must consume organic molecules present in other organisms (figure 43.1). The molecules heterotrophs eat must be digested into smaller molecules in order to be absorbed into the animal's body. Once these products of digestion enter the body, the animal can use them for energy in cellular respiration or for the construction of the larger molecules that make up its tissues. The process of animal digestion is the focus of this chapter.

43.1 Animals employ a digestive system to prepare food for assimilation by cells.

Types of Digestive Systems

Heterotrophs are divided into three groups on the basis of their food sources. Animals that eat plants exclusively are classified as **herbivores;** common examples include cows, horses, rabbits, and sparrows. Animals that are meat-eaters, such as cats, eagles, trout, and frogs, are **carnivores.** Animals that eat both plants and other animals are **omnivores.** Humans are omnivores, as are pigs, bears, and crows.

Single-celled organisms (as well as sponges) digest their food intracellularly. Other animals digest their food extracellularly, within a digestive cavity. In this case, the digestive enzymes are released into a cavity that is continuous with the animal's external environment. In cnidarians and in flatworms such as planarians, the digestive cavity has only one opening that serves as both mouth and anus. There is no specialization within this type of digestive system, called a *gastrovascular cavity,* because every cell is exposed to all stages of food digestion (figure 43.2).

Specialization occurs when the digestive tract, or alimentary canal, has a separate mouth and anus, so that transport of food is one-way. The most primitive digestive tract is seen in nematodes (phylum Nematoda), where it is simply a tubular *gut* lined by an epithelial membrane. Earthworms (phylum Annelida) have a digestive tract specialized in different regions for the ingestion, storage, fragmentation, digestion, and absorption of food. All higher animal groups, including all vertebrates, show similar specializations (figure 43.3).

The ingested food may be stored in a specialized region of the digestive tract or may first be subjected to physical fragmentation. This fragmentation may occur through the chewing action of teeth (in the mouth of many vertebrates) or the grinding action of pebbles (in the gizzard of earthworms and birds). Chemical digestion then occurs, breaking down the larger food molecules of polysaccharides and disaccharides, fats, and proteins into their smallest subunits. Chemical digestion involves hydrolysis reactions that liberate the subunit molecules—primarily monosaccharides, amino acids, and fatty acids—from the food. These products of chemical digestion pass through the epithelial lining of the gut into the blood, in a process known as *absorption.* Any molecules in the food that are not absorbed cannot be used by the animal. These waste products are excreted, or defecated, from the anus.

Most animals digest their food extracellularly. The digestive tract, with a one-way transport of food and specialization of regions for different functions, allows food to be ingested, physically fragmented, chemically digested, and absorbed.

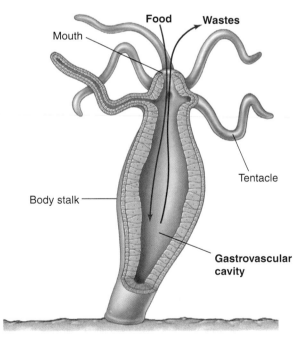

FIGURE 43.2
The gastrovascular cavity of *Hydra,* a cnidarian. In gastrovascular cavities, one common opening serves as both the mouth and the anus. There are no specialized regions, and extracellular digestion occurs throughout the cavity.

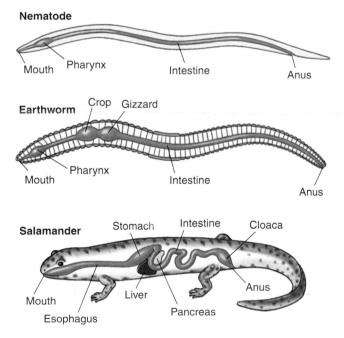

FIGURE 43.3
The one-way digestive tract of nematodes, earthworms, and vertebrates. One-way movement through the digestive tract allows different regions of the digestive system to become specialized for different functions.

Vertebrate Digestive Systems

In humans and other vertebrates, the digestive system consists of a tubular gastrointestinal tract and accessory digestive organs (figure 43.4). The initial components of the gastrointestinal tract are the mouth and the pharynx, which is the common passage of the oral and nasal cavities. The pharynx leads to the esophagus, a muscular tube that delivers food to the stomach, where some preliminary digestion occurs. From the stomach, food passes to the small intestine, where a battery of digestive enzymes continues the digestive process. The products of digestion, together with minerals and most imbibed water, are absorbed across the wall of the small intestine into the bloodstream. The small intestine empties what remains into the large intestine, where some of the remaining water and minerals are absorbed. In most vertebrates other than mammals, the waste products emerge from the large intestine into a cavity called the cloaca (see figure 43.3), which also receives the products of the urinary and reproductive systems. In mammals, the urogenital products are separated from the fecal material in the large intestine; the fecal material enters the rectum and is expelled through the anus.

In general, carnivores have shorter intestines for their size than do herbivores. A short intestine is adequate for a carnivore, but herbivores ingest a large amount of plant cellulose, which resists digestion. These animals have a long, convoluted small intestine. In addition, mammals called *ruminants* (such as cows) that consume grass and other vegetation have stomachs with multiple chambers, where bacteria aid in the digestion of cellulose. Other herbivores, including rabbits and horses, digest cellulose (with the aid of bacteria) in a pouch called the **cecum** located at the beginning of the large intestine.

The accessory digestive organs include the liver, which produces *bile* (a green solution that emulsifies fat), the gallbladder, which stores and concentrates the bile, and the pancreas. The pancreas produces *pancreatic juice*, which contains digestive enzymes and bicarbonate buffer. Both bile and pancreatic juice are secreted into the first region of the small intestine where they aid digestion.

The tubular gastrointestinal tract of a vertebrate has a characteristic layered structure (figure 43.5). The innermost layer is the mucosa, an epithelium that lines the interior of the tract (the lumen). The next major tissue layer, made of connective tissue, is called the submucosa. Just outside the submucosa is the muscularis, which consists of a double layer of smooth muscles. The muscles in the inner layer have a circular orientation, and those in the outer layer are arranged longitudinally. Another connective tissue layer, the serosa, covers the external surface of the tract. Nerves, intertwined in regions called *plexuses*, are located in the submucosa and help regulate the gastrointestinal activities.

The vertebrate digestive system consists of a tubular gastrointestinal tract, which is modified in different animals and composed of a series of tissue layers.

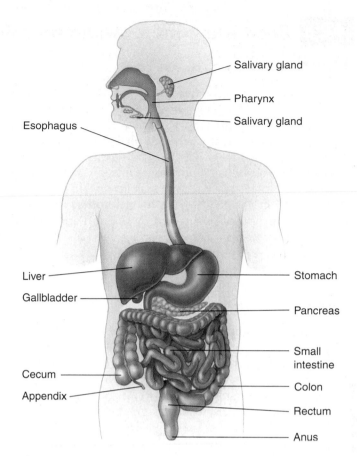

FIGURE 43.4
The human digestive system. The human digestive system consists of the esophagus, stomach, small intestine, large intestine, rectum, and anus, aided by accessory organs.

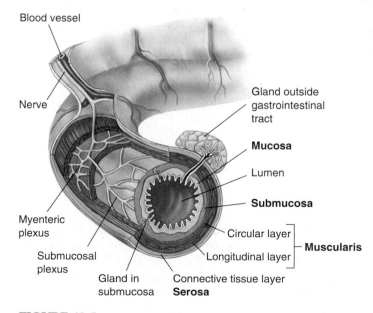

FIGURE 43.5
The layers of the gastrointestinal tract. The mucosa contains a lining epithelium; the submucosa is composed of connective tissue (as is the serosa); and the muscularis consists of smooth muscles. Glands secrete substances via ducts into specific regions of the tract.

Mouth and Teeth

Specializations of the digestive systems in different kinds of vertebrates reflect differences in the way these animals live. Fishes have a large pharynx with gill slits, while air-breathing vertebrates have a greatly reduced pharynx. Many vertebrates have teeth (figure 43.6), and chewing (*mastication*) breaks up food into small particles and mixes it with fluid secretions. Birds, which lack teeth, break up food in their two-chambered stomachs (figure 43.7). In one of these chambers, the gizzard, small pebbles ingested by the bird are churned together with the food by muscular action. This churning grinds up the seeds and other hard plant material into smaller chunks that can be digested more easily.

Vertebrate Teeth

Carnivorous mammals have pointed teeth that lack flat grinding surfaces. Such teeth are adapted for cutting and shearing. Carnivores often tear off pieces of their prey but have little need to chew them, because digestive enzymes can act directly on animal cells. (Recall how a cat or dog gulps down its food.) By contrast, grass-eating herbivores, such as cows and horses, must pulverize the cellulose cell walls of plant tissue before digesting it. These animals have large, flat teeth with complex ridges well-suited to grinding.

Human teeth are specialized for eating both plant and animal food. Viewed simply, humans are carnivores in the front of the mouth and herbivores in the back (figure 43.8). The four front teeth in the upper and lower jaws are sharp, chisel-shaped incisors used for biting. On each side of the incisors are sharp, pointed teeth called cuspids (sometimes referred to as "canine" teeth), which are used for tearing food. Behind the canines are two premolars and three molars, all with flattened, ridged surfaces for grinding and crushing food. Children have only 20 teeth, but these deciduous teeth are lost during childhood and are replaced by 32 adult teeth.

The Mouth

Inside the mouth, the tongue mixes food with a mucous solution, saliva. In humans, three pairs of salivary glands secrete saliva into the mouth through ducts in the mouth's mucosal lining. Saliva moistens and lubricates the food so that it is easier to swallow and does not abrade the tissue it passes on its way through the esophagus. Saliva also contains the hydrolytic enzyme salivary amylase, which initiates the breakdown of the polysaccharide starch into the disaccharide maltose. This digestion is usually minimal in humans, however, because most people don't chew their food very long.

The secretions of the salivary glands are controlled by the nervous system, which in humans maintains a constant

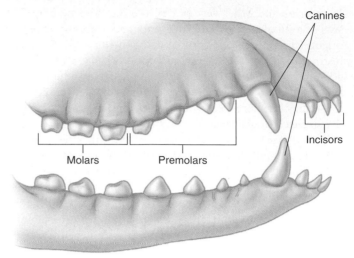

FIGURE 43.6
Diagram of generalized vertebrate dentition. Different vertebrates have specific variations from this generalized pattern, depending on whether the vertebrate is an herbivore, carnivore, or omnivore.

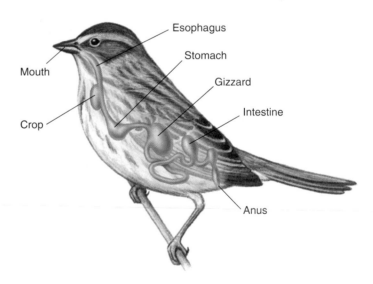

FIGURE 43.7
The digestive tract of birds. Birds lack teeth but have a muscular chamber called the gizzard that works to break down food. Birds swallow gritty objects or pebbles that lodge in the gizzard and pulverize food before it passes into the intestine. Food is stored in the crop.

flow of about half a milliliter per minute when the mouth is empty of food. This continuous secretion keeps the mouth moist. The presence of food in the mouth triggers an increased rate of secretion, as taste-sensitive neurons in the mouth send impulses to the brain, which responds by stimulating the salivary glands. The most potent stimuli are acidic solutions; lemon juice, for example, can increase the rate of salivation eightfold. The sight, sound, or smell of food can stimulate salivation markedly in dogs, but in humans, these stimuli are much less effective than thinking or talking about food.

When food is ready to be swallowed, the tongue moves it to the back of the mouth. In mammals, the process of swallowing begins when the soft palate elevates, pushing against the back wall of the pharynx (figure 43.9). Elevation of the soft palate seals off the nasal cavity and prevents food from entering it. Pressure against the pharynx triggers an automatic, involuntary response called a reflex. In this reflex, pressure on the pharynx stimulates neurons within its walls, which send impulses to the swallowing center in the brain. In response, electrical impulses in motor neurons stimulate muscles to contract and raise the *larynx* (voice box). This pushes the *glottis*, the opening from the larynx into the trachea (windpipe), against a flap of tissue called the *epiglottis*. These actions keep food out of the respiratory tract, directing it instead into the esophagus.

In many vertebrates, ingested food is fragmented through the tearing or grinding action of specialized teeth. In birds, this is accomplished through the grinding action of pebbles in the gizzard. Food mixed with saliva is swallowed and enters the esophagus.

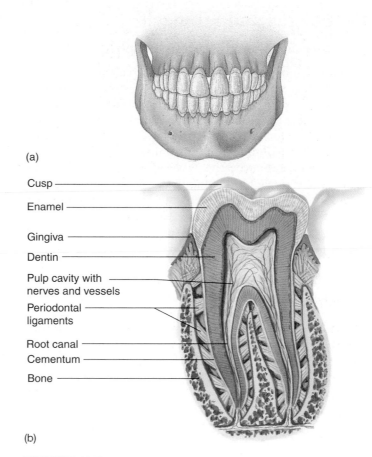

(a)

Cusp

Enamel

Gingiva

Dentin

Pulp cavity with nerves and vessels

Periodontal ligaments

Root canal

Cementum

Bone

(b)

FIGURE 43.8
Human teeth. (*a*) The front six teeth on the upper and lower jaws are cuspids and incisors. The remaining teeth, running along the sides of the mouth, are grinders called premolars and molars. Hence, humans have carnivore-like teeth in the front of their mouths and herbivore-like teeth in the back. (*b*) Each tooth is alive, with a central pulp containing nerves and blood vessels. The actual chewing surface is a hard enamel layered over the softer dentin, which forms the body of the tooth.

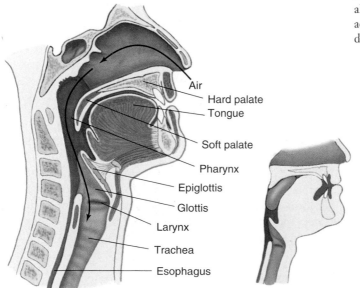

Air

Hard palate

Tongue

Soft palate

Pharynx

Epiglottis

Glottis

Larynx

Trachea

Esophagus

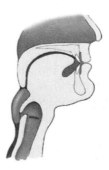

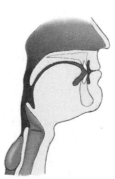

FIGURE 43.9
The human pharynx, palate, and larynx. Food that enters the pharynx is prevented from entering the nasal cavity by elevation of the soft palate, and is prevented from entering the larynx and trachea (the airways of the respiratory system) by elevation of the larynx against the epiglottis.

Esophagus and Stomach

Structure and Function of the Esophagus

Swallowed food enters a muscular tube called the esophagus, which connects the pharynx to the stomach. In adult humans, the esophagus is about 25 centimeters long; the upper third is enveloped in skeletal muscle, for voluntary control of swallowing, while the lower two-thirds is surrounded by involuntary smooth muscle. The swallowing center stimulates successive waves of contraction in these muscles that move food along the esophagus to the stomach. These rhythmic waves of muscular contraction are called peristalsis (figure 43.10); they enable humans and other vertebrates to swallow even if they are upside down.

In many vertebrates, the movement of food from the esophagus into the stomach is controlled by a ring of circular smooth muscle, or a *sphincter*, that opens in response to the pressure exerted by the food. Contraction of this sphincter prevents food in the stomach from moving back into the esophagus. Rodents and horses have a true sphincter at this site and thus cannot regurgitate, while humans lack a true sphincter. Normally, the esophagus is closed off except during swallowing.

Structure and Function of the Stomach

The stomach (figure 43.11) is a saclike portion of the digestive tract. Its inner surface is highly convoluted, enabling it to fold up when empty and open out like an expanding balloon as it fills with food. Thus, while the human stomach has a volume of only about 50 milliliters when empty, it may expand to contain 2 to 4 liters of food when full. Carnivores that engage in sporadic gorging as an important survival strategy possess stomachs that are able to distend much more than that.

Secretory Systems

The stomach contains an extra layer of smooth muscle for churning food and mixing it with *gastric juice*, an acidic secretion of the tubular gastric glands of the mucosa (figure 43.11). These exocrine glands contain two kinds of secretory cells: parietal cells, which secrete hydrochloric acid (HCl), and chief cells, which secrete pepsinogen, a weak protease (protein-digesting enzyme) that requires a very low pH to be active. This low pH is provided by the HCl. Activated pepsinogen molecules then cleave one another at specific sites, producing a much more active protease, pepsin. This process of secreting a relatively inactive enzyme that is then converted into a more active enzyme outside the cell prevents the chief cells from digesting themselves. It should be noted that in adult humans only proteins are partially digested in the stomach—no significant digestion of carbohydrates or fats occurs.

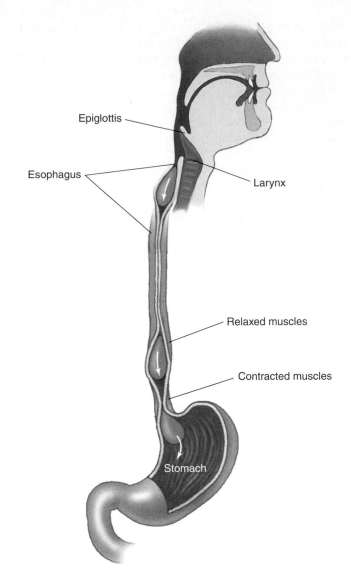

FIGURE 43.10
The esophagus and peristalsis. After food has entered the esophagus, rhythmic waves of muscular contraction, called peristalsis, move the food down to the stomach.

In addition to producing HCl, the parietal cells of the stomach also secrete intrinsic factor, a polypeptide needed for the intestinal absorption of vitamin B_{12}. Because this vitamin is required for the production of red blood cells, persons who lack sufficient intrinsic factor develop a type of anemia (low red blood cell count) called *pernicious anemia*.

Action of Acid

The human stomach produces about 2 liters of HCl and other gastric secretions every day, creating a very acidic solution inside the stomach. The concentration of HCl in this solution is about 10 millimolar, corresponding to a pH

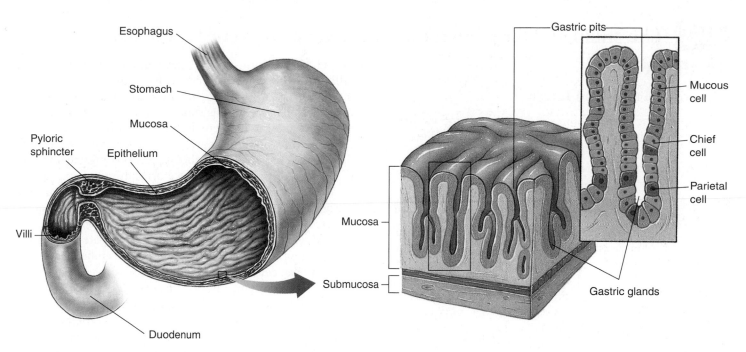

FIGURE 43.11

The stomach and duodenum. Food enters the stomach from the esophagus. A band of smooth muscle called the pyloric sphincter controls the entrance to the duodenum, the upper part of the small intestine. The epithelial walls of the stomach are dotted with gastric pits, which contain gastric glands that secrete hydrochloric acid and the enzyme pepsinogen. The gastric glands consist of mucous cells, chief cells that secrete pepsinogen, and parietal cells that secrete HCl. Gastric pits are the openings of the gastric glands.

of 2. Thus, gastric juice is about 250,000 times more acidic than blood, whose normal pH is 7.4. The low pH in the stomach helps denature food proteins, making them easier to digest, and keeps pepsin maximally active. Active pepsin hydrolyzes food proteins into shorter chains of polypeptides that are not fully digested until the mixture enters the small intestine. The mixture of partially digested food and gastric juice is called chyme.

The acidic solution within the stomach also kills most of the bacteria that are ingested with the food. The few bacteria that survive the stomach and enter the intestine intact are able to grow and multiply there, particularly in the large intestine. In fact, vertebrates harbor thriving colonies of bacteria within their intestines, and bacteria are a major component of feces. As we will discuss later, bacteria that live within the digestive tract of cows and other ruminants play a key role in the ability of these mammals to digest cellulose.

Ulcers

Overproduction of gastric acid can occasionally eat a hole through the wall of the stomach. Such *gastric ulcers* are rare, however, because epithelial cells in the mucosa of the stomach are protected somewhat by a layer of alkaline mucus, and because those cells are rapidly replaced by cell division if they become damaged (gastric epithelial cells are replaced every 2 to 3 days). Over 90% of gastrointestinal ul-

cers are *duodenal ulcers*. These may be produced when excessive amounts of acidic chyme are delivered into the duodenum, so that the acid cannot be properly neutralized through the action of alkaline pancreatic juice (described later). Susceptibility to ulcers is increased when the mucosal barriers to self-digestion are weakened by an infection of the bacterium *Helicobacter pylori*. Indeed, modern antibiotic treatments of this infection can reduce symptoms and often even cure the ulcer.

Leaving the Stomach

Chyme leaves the stomach through the *pyloric sphincter* (see figure 43.11) to enter the small intestine. This is where all terminal digestion of carbohydrates, lipids, and proteins occurs, and where the products of digestion—amino acids, glucose, and so on—are absorbed into the blood. Only some of the water in chyme and a few substances, such as aspirin and alcohol, are absorbed through the wall of the stomach.

Peristaltic waves of contraction propel food along the esophagus to the stomach. Gastric juice contains strong hydrochloric acid and the protein-digesting enzyme pepsin, which begins the digestion of proteins into shorter polypeptides. The acidic chyme is then transferred through the pyloric sphincter to the small intestine.

43.3 The small and large intestines have very different functions.

Small Intestine

Digestion in the Small Intestine

The capacity of the small intestine is limited, and its digestive processes take time. Consequently, efficient digestion requires that only relatively small amounts of chyme be introduced from the stomach into the small intestine at any one time. Coordination between gastric and intestinal activities is regulated by neural and hormonal signals, which we will describe in section 43.4.

The small intestine is approximately 4.5 meters long in a living person, but 6 meters long at autopsy when the muscles relax. The first 25 centimeters is the **duodenum;** the remainder of the small intestine is divided into the **jejunum** and the **ileum.** The duodenum receives acidic chyme from the stomach, digestive enzymes and bicarbonate from the pancreas, and bile from the liver and gallbladder. The pancreatic juice enzymes digest larger food molecules into smaller fragments. This occurs primarily in the duodenum and jejunum.

The epithelial wall of the small intestine is covered with tiny, fingerlike projections called **villi** (singular, *villus;* fig-

ure 43.12). In turn, each of the epithelial cells lining the villi is covered on its apical surface (the side facing the lumen) by many foldings of the plasma membrane that form cytoplasmic extensions called *microvilli*. These are quite tiny and can be seen clearly only with an electron microscope. In a light micrograph, the microvilli resemble the bristles of a brush, and for that reason the epithelial wall of the small intestine is also called a brush border.

The villi and microvilli greatly increase the surface area of the small intestine; in humans, this surface area is 300 square meters! It is over this vast surface that the products of digestion are absorbed. The microvilli also participate in digestion because a number of digestive enzymes are embedded within the epithelial cells' plasma membranes, with their active sites exposed to the chyme. These brush border enzymes include those that hydrolyze the disaccharides lactose and sucrose, among others. Many adult humans lose the ability to produce the brush border enzyme *lactase* and therefore cannot digest lactose (milk sugar), a rather common condition called *lactose intolerance*. The brush border enzymes complete the digestive process that started with the action of the pancreatic enzymes released into the duodenum.

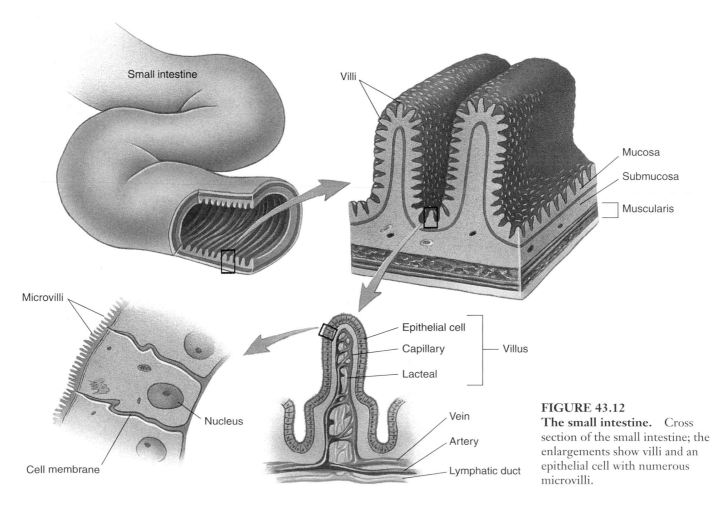

FIGURE 43.12
The small intestine. Cross section of the small intestine; the enlargements show villi and an epithelial cell with numerous microvilli.

Accessory Organs to Small Intestine Function

Secretions of the Pancreas. The pancreas (figure 43.13), a large gland situated near the junction of the stomach and the small intestine, is one of the accessory organs that contribute secretions to the digestive tract. Pancreatic fluid is secreted into the duodenum through the *pancreatic duct*; thus, the pancreas functions as an exocrine organ. This fluid contains a host of enzymes, including trypsin and chymotrypsin, which digest proteins; pancreatic amylase, which digests starch; and lipase, which digests fat. These enzymes are released into the duodenum primarily as inactive enzymes, called zymogens, and are then activated by the brush border enzymes of the intestine. Pancreatic enzymes digest proteins into smaller polypeptides, polysaccharides into shorter chains of sugars, and fat into free fatty acids and other products. The digestion of these molecules is then completed by the brush border enzymes.

Pancreatic fluid also contains bicarbonate, which neutralizes the HCl from the stomach and gives the chyme in the duodenum a slightly alkaline pH. The digestive enzymes and bicarbonate are produced by clusters of secretory cells known as *acini*.

In addition to its exocrine role in digestion, the pancreas also functions as an endocrine gland, secreting several hormones into the blood that control the blood levels of glucose and other nutrients. These hormones are produced in the **islets of Langerhans,** clusters of endocrine cells scattered throughout the pancreas. The two most important pancreatic hormones, insulin and glucagon, are discussed in section 43.4.

Liver and Gallbladder. The liver is the largest internal organ of the body (see figure 43.4). In an adult human, the liver weighs about 1.5 kilograms and is the size of a football. The main exocrine secretion of the liver is bile, a fluid mixture consisting of *bile pigments* and *bile salts* that is delivered into the duodenum during the digestion of a meal. The bile pigments do not participate in digestion; they are waste products resulting from the liver's destruction of old red blood cells and ultimately are eliminated with the feces. If the excretion of bile pigments by the liver is blocked, the pigments can accumulate in the blood and cause a yellow staining of the tissues known as *jaundice*.

In contrast, the bile salts play a very important role in the digestion of fats. Because fats are insoluble in water, they enter the intestine as drops within the watery chyme. The bile salts, which are partly lipid-soluble and partly water-soluble, work like detergents, dispersing the large drops of fat into a fine suspension of smaller droplets. This emulsification process produces a greater surface area of fat upon which the lipase enzymes can act, and thus allows the digestion of fat to proceed more rapidly.

After it is produced in the liver, bile is stored and concentrated in the gallbladder. The arrival of fatty food in the duodenum triggers a neural and endocrine reflex (discussed later) that stimulates the gallbladder to contract, causing bile to be transported through the common bile duct and injected into the duodenum. If the bile duct is blocked by a *gallstone* (formed from a hardened precipitate of cholesterol), contraction of the gallbladder will cause pain generally felt under the right scapula (shoulder blade).

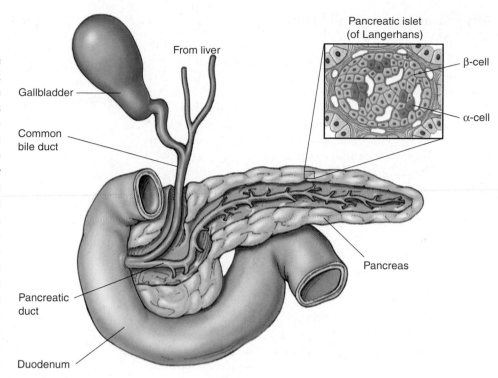

FIGURE 43.13
The pancreas. The pancreatic duct and bile duct empty into the duodenum. The pancreas secretes pancreatic juice into the pancreatic duct. The pancreatic islets of Langerhans secrete hormones into the blood; α cells secrete glucagon, and β cells secrete insulin. The liver secretes bile, which consists of bile pigments (waste products from the liver) and bile salts. Bile salts play a role in the digestion of fats. Bile is stored in the gall bladder until it is needed in the duodenum upon the arrival of fatty food.

Absorption in the Small Intestine

The amino acids and monosaccharides resulting from the digestion of proteins and carbohydrates, respectively, are transported across the brush border into the epithelial cells that line the intestine (figure 43.14*a*). They then move to the other side of the epithelial cells, and from there are transported across the membrane and into the blood capillaries within the villi. The blood carries these products of digestion from the intestine to the liver via the hepatic portal vein. The term *portal* here refers to a special arrangement of vessels, seen only in a couple of instances, where one organ (the liver, in this case) is located "downstream" from another organ (the intestine). As a result, the second organ receives blood-borne molecules from the first. Because of the hepatic portal vein, the liver is the first organ to receive most of the products of digestion. This arrangement is important for the functions of the liver, as is described in section 43.4.

The products of fat digestion are absorbed by a different mechanism (figure 43.14*b*). Fats (triglycerides) are hydrolyzed into fatty acids and monoglycerides, which are absorbed into the intestinal epithelial cells and reassembled into triglycerides. The triglycerides then combine with proteins to form small particles called chylomicrons. Instead of entering the hepatic portal circulation, the chylomicrons are absorbed into lymphatic capillaries (see chapter 44), which empty their contents into the blood in veins near the neck. Chylomicrons can make the blood plasma appear cloudy if a sample of blood is drawn after a fatty meal.

The amount of fluid passing through the small intestine in a day is startlingly large: approximately 9 liters. However, almost all of this fluid is absorbed into the body rather than eliminated in the feces. About 8.5 liters are absorbed in the small intestine and an additional 350 milliliters in the large intestine. Only about 50 grams of solid and 100 milliliters of liquid leave the body as feces. The normal fluid absorption efficiency of the human digestive tract thus approaches 99%, which is very high indeed.

Digestion occurs primarily in the duodenum with the aid of pancreatic and liver secretions. The small intestine provides a large surface area for absorption. Glucose and amino acids from food are absorbed through the small intestine and enter the blood via the hepatic portal vein, going to the liver. Fat from food enters the lymphatic system.

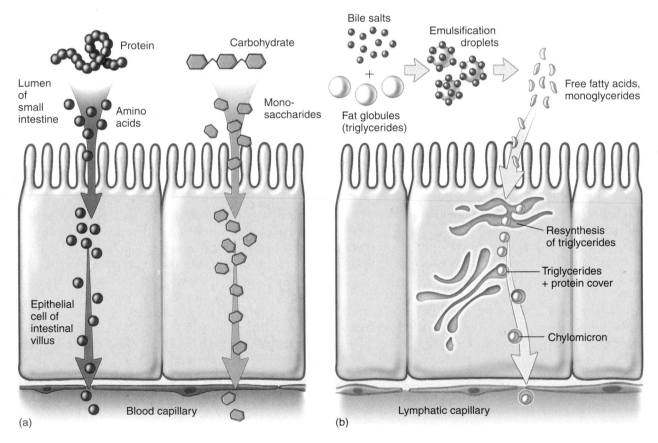

FIGURE 43.14

Absorption of the products of digestion. (*a*) Monosaccharides and amino acids are transported into blood capillaries. (*b*) Fatty acids and monoglycerides within the intestinal lumen are absorbed and converted within the intestinal epithelial cells into triglycerides. These are then coated with proteins to form tiny structures called chylomicrons, which enter lymphatic capillaries.

Large Intestine

The large intestine, or colon, is much shorter than the small intestine, occupying approximately the last meter of the digestive tract; it is called "large" because of its larger diameter, not its length. The small intestine empties directly into the large intestine at a junction where two vestigial structures, the cecum and the appendix, remain (figure 43.15). No digestion takes place within the large intestine, and only about 4% of the absorption of fluids by the intestine occurs there. The large intestine is not as convoluted as the small intestine, and its inner surface has no villi. Consequently, the large intestine has less than one-thirtieth the absorptive surface area of the small intestine. Although some water, sodium, vitamin K, and other products of bacterial metabolism are absorbed across its wall, the primary function of the large intestine is to concentrate waste material. Within it, undigested material, primarily bacterial fragments and cellulose, is compacted and stored. Many bacteria live and reproduce within the large intestine, and the excess bacteria are incorporated into the refuse material, called *feces*. Bacterial fermentation produces gas within the colon at a rate of about 500 milliliters per day. This rate increases greatly after the consumption of beans or other vegetable matter because the passage of undigested plant material (fiber) into the large intestine provides substrates for fermentation.

The human colon has evolved to process food with a relatively high fiber content. Diets that are low in fiber, which are common in the United States, result in a slower passage of food through the colon. Low dietary fiber content is thought to be associated with the level of colon cancer in the United States, which is among the highest in the world.

Compacted feces, driven by peristaltic contractions of the large intestine, pass from the large intestine into a short tube called the rectum. From the rectum, the feces exit the body through the anus. Two sphincters control passage through the anus. The first is composed of smooth muscle and opens involuntarily in response to pressure inside the rectum. The second, composed of striated muscle, can be controlled voluntarily by the brain, thus permitting a conscious decision to delay defecation.

In all vertebrates except most mammals, the reproductive and urinary tracts empty together with the digestive tract into a common cavity, the cloaca. In some reptiles and birds, additional water from either the feces or urine may be absorbed in the cloaca before the products are expelled from the body.

The large intestine concentrates wastes for excretion by absorbing water. Some ions and vitamin K are also absorbed by the large intestine.

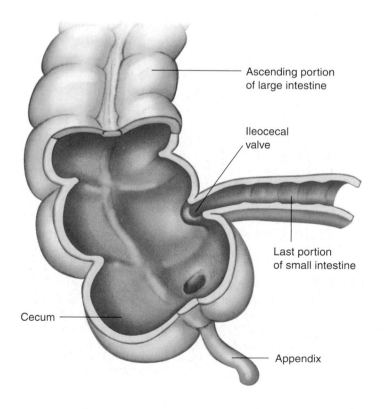

Ascending portion of large intestine

Ileocecal valve

Last portion of small intestine

Cecum

Appendix

FIGURE 43.15
The junction of the small and large intestines in humans. The large intestine, or colon, starts with the cecum, which is relatively small in humans compared with that in other mammals. A vestigial structure called the appendix extends from the cecum.

Variations in Vertebrate Digestive Systems

Most animals lack the enzymes necessary to digest cellulose, the carbohydrate that functions as the chief structural component of plants. The digestive tracts of some animals, however, contain bacteria and protists that convert cellulose into substances the host can digest. Although digestion by gastrointestinal microorganisms plays a relatively small role in human nutrition, it is an essential element in the nutrition of many other kinds of animals, including insects such as termites and cockroaches, and a few groups of herbivorous mammals. The relationships between these microorganisms and their animal hosts are mutually beneficial and provide an excellent example of symbiosis (see chapter 54).

Cows, deer, and other herbivores called ruminants have large, divided stomachs (figure 43.16). The first portion consists of the rumen and a smaller chamber, the reticulum; the second portion consists of two additional chambers: the omasum and abomasum. The rumen, which may hold up to 50 gallons, serves as a fermentation vat in which bacteria and protists convert cellulose and other molecules into a variety of simpler compounds. The location of the rumen at the front of the four chambers is important because it allows the animal to regurgitate and rechew the contents of the rumen, an activity called *rumination*, or "chewing the cud." The cud is then swallowed and enters the reticulum, from which it passes to the omasum and then to the abomasum, where it is finally mixed with gastric juice. Hence, only the abomasum is equivalent to the human stomach in its function. This process leads to far more efficient digestion of cellulose in ruminants than in mammals that lack a rumen, such as horses.

In some animals, such as rodents, horses, deer, and lagomorphs (rabbits and hares), the digestion of cellulose by microorganisms takes place in the cecum, which is greatly enlarged (figure 43.17). Because the cecum is located beyond the stomach, regurgitation of its contents is impossible. However, rodents and lagomorphs have evolved another way to digest cellulose that achieves a degree of efficiency similar to that of ruminant digestion. They do this by eating their feces, thus passing the food through their digestive tract a second time. The second passage makes it possible for the animal to absorb the nutrients produced by the microorganisms in its cecum. Animals that engage in this practice of **coprophagy** (Greek *copros*, "excrement," + *phagein*, "eat") cannot remain healthy if they are prevented from eating their feces. Animals whose diets don't include cellulose, such as insectivores or carnivores, don't have a cecum, or if they do, it is greatly reduced.

Cellulose is not the only plant product that vertebrates can use as a food source because of the digestive activities

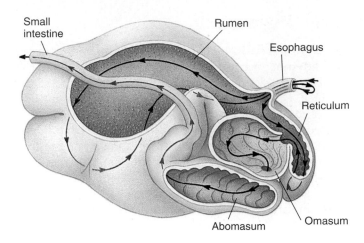

FIGURE 43.16
Four-chambered stomach of a ruminant. The grass and other plants that a ruminant, such as a cow, eats enter the rumen, where they are partially digested. Before moving into a second chamber, the reticulum, the food may be regurgitated and rechewed. The food is then transferred to the rear two chambers: the omasum and abomasum. Only the abomasum is equivalent to the human stomach in its function of secreting gastric juice.

of intestinal microorganisms. Wax, a substance indigestible by most terrestrial animals, is digested by symbiotic bacteria living in the gut of honey guides, African birds that eat the wax in bee nests. In the marine food chain, wax is a major constituent of copepods (crustaceans in plankton), and many marine fish and birds appear able to digest wax with the aid of symbiotic microorganisms.

Another example of the way intestinal microorganisms function in the metabolism of their animal hosts is provided by the synthesis of vitamin K. All mammals rely on intestinal bacteria to synthesize this vitamin, which is necessary for the clotting of blood. Birds, which lack these bacteria, must consume the required quantities of vitamin K in their food. In humans, prolonged treatment with antibiotics greatly reduces the populations of bacteria in the intestine; under such circumstances, it may be necessary to provide supplementary vitamin K.

Much of the food value of plants is tied up in cellulose, and the digestive tract of many animals harbors colonies of cellulose-digesting microorganisms. Intestinal microorganisms also produce molecules such as vitamin K that are important to the well-being of their vertebrate hosts.

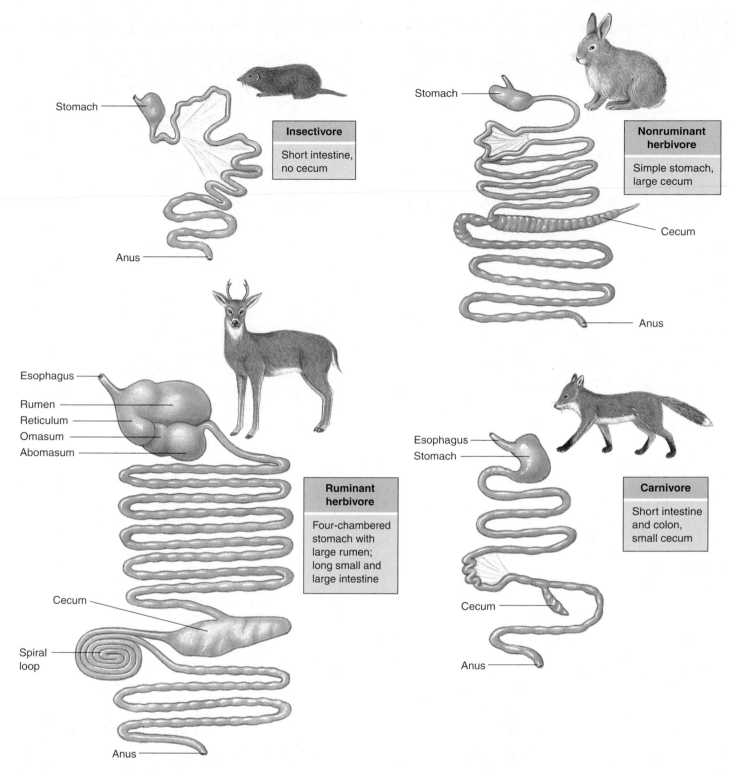

FIGURE 43.17
The digestive systems of different mammals reflect their diets.   Herbivores, such as rabbits and deer, require long digestive tracts with specialized compartments for the breakdown of plant matter. Protein diets are more easily digested; thus, insectivorous and carnivorous mammals, such as voles and foxes, have short digestive tracts with few specialized pouches.

43.4 Neural, hormonal, and accessory organ regulation controls digestion.

Neural and Hormonal Regulation of the Digestive Tract

The activities of the gastrointestinal tract are coordinated by the nervous system and the endocrine system. The nervous system, for example, stimulates salivary and gastric secretions in response to the sight, smell, and consumption of food. When food arrives in the stomach, proteins in the food stimulate the secretion of a stomach hormone called gastrin, which in turn stimulates the secretion of pepsinogen and HCl from the gastric glands (figure 43.18). The secreted HCl then lowers the pH of the gastric juice, which acts to inhibit further secretion of gastrin. Because inhibition of gastrin secretion will reduce the amount of HCl released into the gastric juice, a negative feedback loop is completed. In this way, the secretion of gastric acid is kept under tight control.

The passage of chyme from the stomach into the duodenum of the small intestine inhibits the contractions of the stomach, so that no additional chyme can enter the duodenum until the previous amount can be processed. This stomach or gastric inhibition is mediated by a neural reflex and by duodenal hormones secreted into the blood. These hormones are collectively known as the **enterogastrones** (*entero*, "intestine" + *gastro*, "stomach"). The major enterogas-

trones include cholecystokinin (CCK), secretin, and gastric inhibitory peptide (GIP). Chyme with high fat content is the strongest stimulus for CCK and GIP secretions, whereas increasing chyme acidity primarily influences the release of secretin. All three of these enterogastrones inhibit gastric motility and gastric juice secretions and can result in fatty meals remaining in the stomach longer than nonfatty meals.

In addition to gastric inhibition, CCK and secretin have other important regulatory functions in digestion. CCK also stimulates increased pancreatic secretions of digestive enzymes and gallbladder contractions. Gallbladder contractions inject more bile into the duodenum, which enhances the emulsification and efficient digestion of fats. The other major function of secretin is to stimulate the pancreas to release more bicarbonate, which neutralizes the acidity of the chyme. Secretin has the distinction of being the first hormone ever discovered. Table 43.1 summarizes the actions of the digestive hormones and enzymes.

Neural and hormonal reflexes regulate the activity of the digestive system. The stomach's secretions are regulated by food and by gastrin. Enterogastrones secreted by the duodenum inhibit gastric functions and promote digestion in the duodenum.

Table 43.1	Hormones and Enzymes of Digestion

HORMONES

Hormone	Class	Source	Stimulus	Action	Note
Gastrin	Polypeptide	Pyloric portion of stomach	Entry of food into stomach	Stimulates secretion of HCl and pepsinogen by stomach	Acts on same organ that secretes it
Cholecystokinin (CCK)	Polypeptide	Duodenum	Fatty chyme in duodenum	Stimulates gallbladder contraction and secretion of digestive enzymes by pancreas	Structurally similar to gastrin
Gastric inhibitory peptide (GIP)	Polypeptide	Duodenum	Fatty chyme in duodenum	Inhibits stomach emptying	Also stimulates insulin secretion
Secretin	Polypeptide	Duodenum	Acidic chyme in duodenum	Stimulates secretion of bicarbonate by pancreas	The first hormone to be discovered (1902)

ENZYMES

Location	Enzymes	Substances	Digestion Products
Salivary glands	Amylase	Starch, glycogen	Disaccharides
Stomach	Pepsin	Proteins	Short peptides
Small intestine (brush border)	Peptidases	Short peptides	Amino acids
	Nucleases	DNA, RNA	Sugars, nucleic acid bases
	Lactase, maltase, sucrase	Disaccharides	Monosaccharides
Pancreas	Lipase	Triglycerides	Fatty acids, glycerol
	Trypsin, chymotrypsin	Proteins	Peptides
	DNase	DNA	Nucleotides
	RNase	RNA	Nucleotides

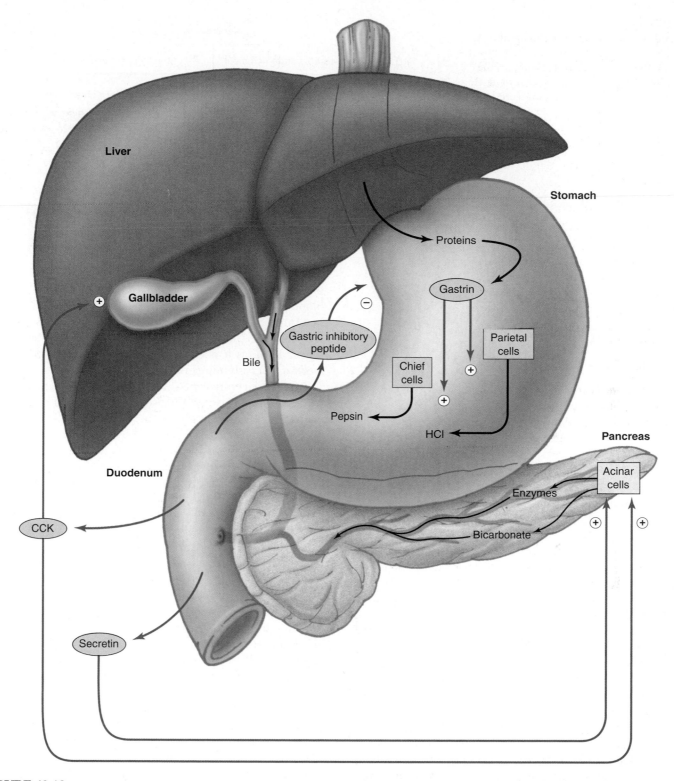

FIGURE 43.18

Hormonal control of the gastrointestinal tract. Gastrin, secreted by the mucosa of the stomach, stimulates the secretion of pepsinogen (which is converted into pepsin) and HCl. The duodenum secretes three hormones: cholecystokinin (CCK), which stimulates contraction of the gallbladder and secretion of pancreatic enzymes; secretin, which stimulates secretion of pancreatic bicarbonate; and gastric inhibitory peptide, which inhibits stomach emptying.

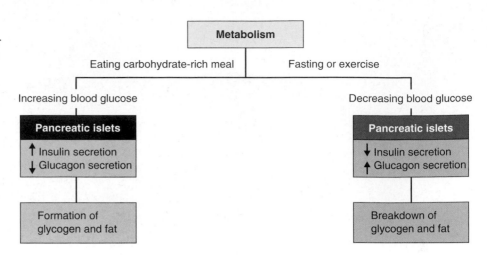

FIGURE 43.19

The actions of insulin and glucagon. After a meal, an increased secretion of insulin by the β cells of the pancreatic islets promotes the deposition of glycogen and fat. During fasting or exercising, increased glucagon secretion by the α cells of the pancreatic islets and decreased insulin secretion promote the breakdown (through hydrolysis reactions) of glycogen and fat.

Accessory Organ Function in Nutrient Processing and Distribution

Regulatory Functions of the Liver

Because the hepatic portal vein carries blood from the stomach and intestine directly to the liver, the liver is in a position to chemically modify the substances absorbed in the gastrointestinal tract before they reach the rest of the body. For example, ingested alcohol and other drugs are taken into liver cells and metabolized; this is why the liver is often damaged as a result of alcohol and drug abuse. The liver also removes toxins, pesticides, carcinogens, and other poisons, converting them into less toxic forms. An important example of this is the liver's conversion of the toxic ammonia produced by intestinal bacteria into urea, a compound that can be contained safely and carried by the blood at higher concentrations.

Similarly, the liver regulates the levels of many compounds produced within the body. Steroid hormones, for instance, are converted into less active and more water-soluble forms by the liver. These molecules are then included in the bile and eliminated from the body in the feces, or carried by the blood to the kidneys and excreted in the urine.

The liver also produces most of the proteins found in blood plasma. The total concentration of plasma proteins is significant because it must be kept within normal limits in order to maintain osmotic balance between blood and interstitial (tissue) fluid. If the concentration of plasma proteins drops too low, as can happen as a result of liver disease such as cirrhosis, fluid accumulates in the tissues, a condition called *edema*.

Regulation of Blood Glucose Concentration

The neurons in the brain obtain their energy primarily from the aerobic respiration of glucose obtained from the blood plasma. It is therefore extremely important that the blood glucose concentration not fall too low, as might happen during fasting or prolonged exercise. It is also important that the blood glucose concentration not stay at too high a level, as it does in people with uncorrected *diabetes mellitus*, because this can lead to tissue damage.

After a carbohydrate-rich meal, the liver and skeletal muscles remove excess glucose from the blood and store it as the polysaccharide glycogen. This process is stimulated by the hormone insulin, secreted by the β *(beta) cells* in the islets of Langerhans of the pancreas (figure 43.19). When blood glucose levels decrease, as they do between meals, during periods of fasting, and during exercise, the liver secretes glucose into the blood. This glucose is obtained in part from the breakdown of liver glycogen to glucose-6-phosphate, a process called glycogenolysis. The phosphate group is then removed, and free glucose is secreted into the blood. Skeletal muscles lack the enzyme needed to remove the phosphate group, and so, even though they have glycogen stores, they cannot secrete glucose into the blood. The breakdown of liver glycogen is stimulated by another hormone, glucagon, which is secreted by the α *(alpha) cells* of the islets of Langerhans in the pancreas (figure 43.19).

If fasting or exercise continues, the liver begins to convert other molecules, such as amino acids and lactic acid, into glucose. This process is called **gluconeogenesis** ("new formation of glucose"). The amino acids used for gluconeogenesis are obtained from muscle protein, which explains the severe muscle wasting that occurs during prolonged fasting.

Pancreatic hormones and the liver regulate blood glucose concentrations. Insulin stimulates the formation of glycogen and fat in the liver. Glucagon stimulates the breakdown of glycogen in the liver, which releases glucose into the blood.

43.5 All animals require food energy and essential nutrients.

Food Energy and Energy Expenditure

The ingestion of food serves two primary functions: It provides a source of energy, and it provides raw materials the animal is unable to manufacture for itself. Even an animal that is completely at rest requires energy to support its metabolism; the minimum rate of energy consumption under defined resting conditions is called the *basal metabolic rate* (BMR). The BMR is relatively constant for a given individual, depending primarily on the person's age, sex, and body size.

Exercise raises the metabolic rate above the basal levels, so the amount of energy the body consumes per day is determined not only by the BMR but also by the level of physical activity. If food energy taken in is greater than the energy consumed per day, the excess energy will be stored in glycogen and fat. Because glycogen reserves are limited, however, continued ingestion of excess food energy results primarily in the accumulation of fat. The intake of food energy is measured in kilocalories (1 kilocalorie = 1000 calories; nutritionists use Calorie with a capital C instead of kilocalorie). The measurement of kilocalories in food is determined by the amount of heat generated when the food is "burned," either literally, when the caloric content of food is measured using a calorimeter, or in the body, when the food is digested and later oxidized during cellular respiration. Caloric intake can be altered by the choice of diet, and the amount of energy expended in exercise can be changed by the choice of lifestyle. The daily energy expenditures (metabolic rates) of people vary between 1300 and 5000 kilocalories per day, depending on the person's BMR and level of physical activity. If the food kilocalories ingested exceed the metabolic rate for a sustained period, the person will accumulate an amount of fat that is deleterious to health, a condition called **obesity.** In the United States, about 30% of middle-aged women and 15% of middle-aged men are classified as obese, which means they weigh at least 20% more than the average weight for their height.

Regulation of Food Intake

Scientists have for years suspected that adipose tissue secretes a hormonal *satiety factor* (a circulating chemical that decreases appetite), because genetically obese mice lose weight when their circulatory systems are surgically joined with those of normal mice. Apparently, some weight-loss hormone is passing into the obese mice! The satiety factor secreted by adipose tissue has recently been identified. It is the product of a gene first observed in a strain of mice known as *ob/ob* (*ob* stands for "obese"; the double symbols indicate that the mice are homozygous for this gene—they inherit it from both parents). The *ob* gene has been cloned

FIGURE 43.20
Effects of the hormone leptin. These two mice are identical twins, both members of a mutant strain of obese mice. The mouse on the right has been injected with the hormone leptin. It lost 30% of its body weight in just two weeks, with no apparent side effects.

in mice and more recently in humans, and has been found to be expressed (that is, to produce mRNA) only in fat cells. The protein product of this gene, the presumed satiety factor, is called leptin. The *ob* mice produce a mutated and ineffective form of leptin, and this defect causes their obesity. When injected with normal leptin, they stop eating and lose weight (figure 43.20).

More recent studies in humans show that the activity of the *ob* gene and the blood concentrations of leptin are actually higher in obese than in lean people, and that the leptin produced by obese people appears to be normal. It has therefore been suggested that most cases of human obesity may result from a reduced sensitivity to the actions of leptin in the brain rather than from reduced leptin production by adipose cells. Aggressive research is ongoing, as might be expected from the possible medical and commercial applications of these findings.

In the United States, serious eating disorders have become much more common since the mid-1970s. The most common of these disorders are *anorexia nervosa*, a condition in which the afflicted individuals literally starve themselves, and *bulimia*, in which individuals gorge themselves and then vomit, so that their weight stays constant. Ninety to 95% of those suffering from these disorders are female; researchers estimate that 2 to 5% of the adolescent girls and young women in the United States have eating disorders.

> The amount of caloric energy expended by the body depends on the basal metabolic rate and the additional calories consumed by exercise. Obesity results if the ingested food energy exceeds the energy expenditure by the body over a prolonged period.

Essential Nutrients

Over the course of their evolution, many animals have lost the ability to synthesize specific substances that nevertheless continue to play critical roles in their metabolism. Substances that an animal cannot manufacture for itself, but that are necessary for its health, must be obtained in the diet and are referred to as **essential nutrients.**

Included among the essential nutrients are *vitamins*, certain organic substances required in trace amounts. For example, humans, apes, monkeys, and guinea pigs have lost the ability to synthesize ascorbic acid (vitamin C). If vitamin C is not supplied in sufficient quantities in their diets, they will develop scurvy, a potentially fatal disease caused by degeneration of connective tissues. Humans require at least 13 different vitamins (table 43.2).

Some essential nutrients are required in more than trace amounts. Many vertebrates, for example, are unable to synthesize 1 or more of the 20 amino acids used in making proteins. These *essential amino acids* must be obtained from proteins in the food they eat. There are nine essential amino acids for humans. People who are vegetarians must choose their foods so that the essential amino acids in one food complement those in another.

In addition, all vertebrates have lost the ability to synthesize certain unsaturated fatty acids and therefore must obtain them in food. On the other hand, some essential nutrients that vertebrates can synthesize cannot be manufactured by the members of other animal groups. For example, vertebrates can synthesize cholesterol, a key component of steroid hormones, but some carnivorous insects cannot.

Food also supplies **essential minerals** such as calcium, phosphorus, and other inorganic substances, including a wide variety of *trace elements* such as zinc and molybdenum, which are required in very small amounts. Animals obtain trace elements either directly from plants or from animals that have eaten plants.

The body requires vitamins and minerals obtained in food. Also, food must provide particular essential amino acids and fatty acids that the body cannot manufacture by itself.

Table 43.2 Major Vitamins Required by Humans

Vitamin	Function	Source	Deficiency Symptoms
Vitamin A (retinol)	Used in making visual pigments, maintaining epithelial tissues	Green vegetables, milk products, liver	Night blindness, flaky skin
B-complex vitamins			
B_1	Coenzyme in CO_2 removal during cellular respiration	Meat, grains, legumes	Beriberi, weakening of heart, edema
B_2 (riboflavin)	Part of coenzymes FAD and FMN, which play metabolic roles	Many different kinds of foods	Inflammation and breakdown of skin, eye irritation
B_3 (niacin)	Part of coenzymes NAD^+ and $NADP^+$	Liver, lean meats, grains	Pellagra, inflammation of nerves, mental disorders
B_5 (pantothenic acid)	Part of coenzyme-A, a key connection between carbohydrate and fat metabolism	Many different kinds of foods	Rare: fatigue, loss of coordination
B_6 (pyridoxine)	Coenzyme in many phases of amino acid metabolism	Cereals, vegetables, meats	Anemia, convulsions, irritability
B_{12} (cyanocobalamin)	Coenzyme in the production of nucleic acids	Red meats, dairy products	Pernicious anemia
Biotin	Coenzyme in fat synthesis and amino acid metabolism	Meat, vegetables	Rare: depression, nausea
Folic acid	Coenzyme in amino acid and nucleic acid metabolism	Green vegetables	Anemia, diarrhea
Vitamin C	Important in forming collagen, cement of bone, teeth, connective tissue of blood vessels; may help maintain resistance to infection	Fruit, green leafy vegetables	Scurvy, breakdown of skin, blood vessels
Vitamin D (calciferol)	Increases absorption of calcium and promotes bone formation	Dairy products, cod liver oil	Rickets, bone deformities
Vitamin E (tocopherol)	Protects fatty acids and cell membranes from oxidation	Margarine, seeds, green leafy vegetables	Rare
Vitamin K	Essential to blood clotting	Green leafy vegetables	Severe bleeding

43.1 Animals employ a digestive system to prepare food for assimilation by cells.

Types of Digestive Systems

- Heterotrophs are divided into three groups based on their primary food source: Herbivores eat plants, carnivores eat meat, and omnivores eat both plants and animals. (p. 888)
- Most organisms digest their food extracellularly within a digestive cavity. (p. 888)
- In digestion, fragmentation of food particles is followed by chemical digestion of food molecules, which involves hydrolysis to separate subunit molecules so they can be absorbed into the bloodstream. (p. 888)

Vertebrate Digestive Systems

- Vertebrate digestive systems consist of a tubular gastrointestinal tract and accessory digestive organs composed of different tissue layers. (p. 889)

43.2 Food is ingested, swallowed, and transported to the stomach.

Mouth and Teeth

- Carnivorous mammals have pointed teeth lacking a flat grinding surface that are adapted for cutting. Herbivores have large, flat teeth with complex ridges for pulverizing plant cells composed of cellulose. (p. 890)
- Birds, which lack teeth, break up food using ingested pebbles inside of a gizzard. (p. 890)
- The tongue mixes food with a mucous solution, and saliva moistens and lubricates food, making it easier to swallow. (pp. 890–891)

Esophagus and Stomach

- The esophagus connects the pharynx to the stomach. (p. 892)
- Successive waves of contraction move food along the esophagus to the stomach, and contraction of a sphincter restricts food from moving backward into the esophagus from the stomach. (p. 892)
- The stomach churns food and mixes it with gastric juices to assist in chemical digestion. Hydrochloric acid creates an acidic environment, and the enzyme pepsin breaks up proteins into polypeptides. (pp. 892–893)
- Dissolved food (chyme) is transferred through the pyloric sphincter into the small intestine. (p. 893)

43.3 The small and large intestines have very different functions.

Small Intestine

- The capacity of the small intestine is limited; thus, for maximal efficiency, only small amounts of chyme must enter at each time interval. (p. 894)
- The small intestine is the site of nutrient absorption and is divided into the duodenum, the jejunum, and the ileum. (p. 894)
- The epithelial wall is covered with villi and microvilli to increase surface area, and consequently increase absorptive capacity. (p. 894)
- The pancreas, liver, and gallbladder are all accessory organs to the small intestine and provide secretions to aid in digestion. (p. 895)
- Amino acids and monosaccharides enter the blood via the hepatic portal vein. The products of fat digestion are first absorbed into the lymphatic capillaries. (p. 896)

Large Intestine

- The large intestine (colon) is much shorter than the small intestine and engages in no digestion and very little absorption. (p. 897)
- The large intestine's primary function is to remove water and concentrate waste for excretion. (p. 897)

Variations in Vertebrate Digestive Systems

- While most animals lack the enzymes necessary to digest cellulose, the digestive tracts of some animals contain bacteria and protists that convert cellulose into digestible substances. (p. 898)
- Ruminants possess multichambered stomachs and regurgitate and rechew rumen contents. (p. 898)
- Horses, deer, and lagomorphs possess a large cecum, which contains microorganisms that digest cellulose. (p. 898)
- Rodents and lagomorphs also engage in coprophagy. (p. 898)

43.4 Neural, hormonal, and accessory organ regulation controls digestion.

Neural and Hormonal Regulation of the Digestive Tract

- The activities of the gastrointestinal tract are coordinated by the nervous system and the endocrine system (p. 900)
- Protein in food stimulates secretion of gastrin, which in turn stimulates secretion of pepsinogen and HCl. (p. 900)
- Enterogastrones are secreted by the duodenum and promote digestion in the duodenum while inhibiting gastric functions. (p. 900)

Accessory Organ Function in Nutrient Processing and Distribution

- The liver chemically modifies absorbed substances in the gastrointestinal tract and converts toxins and poisons into less toxic forms. (p. 902)
- Insulin stimulates the liver to remove excess glucose from the blood and store it as the polysaccharide glycogen. Glucagon stimulates the breakdown of glycogen, which releases glucose into the blood. (p. 902)

43.5 All animals require food energy and essential nutrients.

Food Energy and Energy Expenditure

- The ingestion of food provides a source of energy and the raw materials the animal is unable to manufacture for itself. (p. 903)
- Basal metabolic rate (BMR) is the minimum rate of energy consumption under defined resting conditions. (p. 903)
- The amount of energy consumed by a body per day is determined by the BMR and the level of physical activity. (p. 903)
- Obesity may occur if the amount of food ingested per day exceeds the daily energy expenditure for a long period of time. (p. 903)

Essential Nutrients

- Essential nutrients are substances that an animal cannot manufacture for itself, but that are necessary for the maintenance of the organism's health. (p. 904)
- Many vertebrates cannot synthesize one or more of the 20 essential amino acids, and thus must obtain them from the food they eat. In addition, all vertebrates have lost the ability to synthesize certain unsaturated fatty acids. (p. 904)

Self Test

1. All of the following have one-way digestive systems except
   a. a nematode.
   b. a planarian.
   c. an earthworm.
   d. a human.
2. Intestines of herbivores are _____ compared to carnivores
   a. longer
   b. shorter
   c. about the same size
   d. less convoluted
3. When a mammal swallows, the food is prevented from going up into the nasal cavity by the
   a. esophagus.
   b. tongue.
   c. soft palate.
   d. epiglottis.
4. The first site of protein digestion in the digestive system is
   a. in the mouth.
   b. in the esophagus.
   c. in the stomach.
   d. in the small intestine.
5. How is the digestion of fats different from that of proteins and carbohydrates?
   a. Fat digestion occurs in the small intestine, and the digestion of proteins and carbohydrates occurs in the stomach.
   b. Fats are absorbed into the cells as fatty acids and monoglycerides but are then modified for absorption into the blood; amino acids and glucose are not modified further.
   c. Fats enter the hepatic portal circulation, but proteins and carbohydrates enter the lymphatic system.
   d. Fats are absorbed in the large intestine, and proteins and carbohydrates are absorbed in the small intestine.
6. The primary function of the large intestine is
   a. the breakdown and absorption of fats.
   b. the absorption of vitamin K.
   c. the absorption of water.
   d. the concentration of solid wastes.
7. The _____ secretes digestive enzymes and bicarbonate solution into the small intestine to aid digestion.
   a. pancreas
   b. liver
   c. gallbladder
   d. All of these are correct.
8. Which of the following represents the action of insulin?
   a. increases blood glucose levels by hydrolysis of glycogen
   b. increases blood glucose levels by stimulating glucagon production
   c. decreases blood glucose levels by forming glycogen
   d. increases blood glucose levels by promoting cellular uptake of glucose
9. Gastrin functions by
   a. enhancing the secretion of HCl in the stomach.
   b. enhancing the secretion of pepsinogen in the stomach.
   c. a negative feedback loop.
   d. all of these.
10. Essential organic substances that are used in only tiny amounts by the body are called
    a. trace elements.
    b. vitamins.
    c. hormones.
    d. minerals.

Test Your Visual Understanding

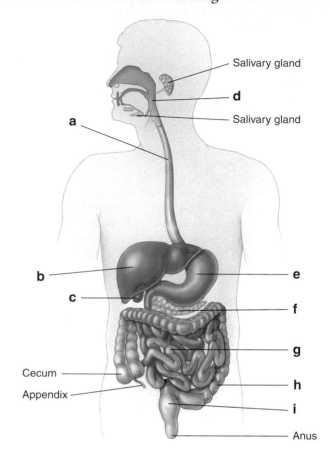

1. Match the following descriptions with the appropriate lettered structures:
   i. site of sodium and vitamin K absorption
   ii. contracts in response to CCK
   iii. connects the mouth with the stomach
   iv. serves as both an endocrine and an exocrine gland
   v. releases brush border enzymes
   vi. storage area at the end of the large intestine
   vii. site of initial carbohydrate digestion
   viii. secretes bile pigments and bile salts
   ix. site of initial protein digestion

Apply Your Knowledge

1. The average length of the small intestine, when fully extended after death, is 6 meters, with a diameter of approximately 2.5 centimeters. Because of folds, villi, and microvilli, the surface area is 2,000,000 square centimeters. What percentage of increase is attributed to the folding, villi, and microvilli in the small intestine?
2. Many birds possess crops, although few mammals do. Suggest a reason for this difference between birds and mammals.

44

Circulation and Respiration

Concept Outline

44.1 Circulatory systems are the transportation highways of the animal body.

Open and Closed Circulatory Systems. All vertebrates have a closed circulation, while many invertebrate animals have open circulatory systems.

Blood. Blood consists of fluid plasma in which circulating blood cells are suspended.

Characteristics of Blood Vessels. Blood leaves the heart in arteries and returns in veins; in between, the blood passes through capillaries.

44.2 The circulatory and respiratory systems evolved together in vertebrates.

Increasing Body Size Requires Special Circulatory and Respiratory Adaptations. As increased body size and physiological abilities evolved in vertebrate animals, so did their need for more efficient gas exchange.

44.3 The cardiac cycle drives the cardiovascular system.

The Cardiac Cycle. The right and left sides of the heart rest and receive blood at the same time; then they pump the blood into arteries at the same time.

Blood Flow and Blood Pressure. Blood flow and blood pressure depend on the diameter of the arterial vessels and the amount of blood pumped by the heart.

44.4 Respiration has evolved to maximize the rate of gas diffusion.

How Animals Maximize the Efficiency of Respiration. The diffusion rate increases when surface area or concentration gradient increases. Aquatic vertebrates use gills for respiration, while terrestrial vertebrates use lungs.

44.5 Mammalian breathing is a dynamic process.

Structures and Mechanisms of Breathing. The rib cage and lung volumes are expanded during inspiration by the contraction of the diaphragm and other muscles.

Hemoglobin and Gas Transport. Hemoglobin loads with oxygen in the lungs and unloads its oxygen in the tissue capillaries. Carbon dioxide is converted into carbonic acid in erythrocytes and is transported as bicarbonate.

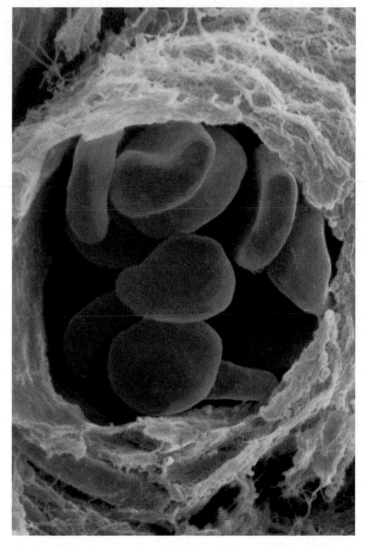

FIGURE 44.1
Red blood cells. This ruptured blood vessel, seen in a scanning electron micrograph, is full of red blood cells, which move through vessels, transporting oxygen from one place to another in the body.

Every cell in the animal body must acquire the energy it needs for living from other molecules in food. Like residents of a city whose food is imported from farms in the countryside, cells in the body need trucks to carry the food, highways for the trucks to travel on, and a way to cook the food when it arrives. In animals, the circulatory system provides blood and blood vessels (the trucks and highways) (figure 44.1), and the respiratory system provides the oxygen necessary for metabolism (fuel to cook the food). These two systems will be discussed together in this chapter, because their functions are so intimately intertwined.

44.1 Circulatory systems are the transportation highways of the animal body.

Open and Closed Circulatory Systems

Among the unicellular protists, oxygen and nutrients are obtained directly from the external aqueous environment by simple diffusion. The body wall is only two cell layers thick in cnidarians, such as *Hydra*, and flatworms, such as planarians. Each cell layer is in direct contact with either the external environment or the gastrovascular cavity (figure 44.2*a*). The gastrovascular cavity of *Hydra* (see chapter 43) extends from the body cavity into the tentacles, and that of a planarian branches extensively to supply every cell with oxygen and nutrients. Larger animals, however, have tissues that are several cell layers thick, so that many cells are too far away from the body surface or digestive cavity to exchange materials directly with the environment. Instead, oxygen and nutrients are transported from the environment and digestive cavity to the body cells by an internal fluid within a *circulatory system.*

The two main types of circulatory systems are *open* and *closed*. In an **open circulatory system,** such as that found in mollusks and arthropods (figure 44.2*b*), there is no distinction between the circulating fluid (blood) and the extracellular fluid of the body tissues (interstitial fluid or lymph). This fluid is thus called **hemolymph.** In insects, the heart is a muscular tube that pumps hemolymph through a network of channels and cavities in the body. The fluid then drains back into the central cavity.

In a **closed circulatory system,** the circulating fluid, or blood, is always enclosed within blood vessels that transport it away from and back to a pump, the **heart** (figure 44.2*c*). Some invertebrates, like annelids (see chapter 33), and all vertebrates have a closed circulatory system. In annelids such as an earthworm, a dorsal vessel contracts rhythmically to function as a pump. Blood is pumped through five small connecting arteries, which also function as pumps, to a ventral vessel, which transports the blood posteriorly until it eventually reenters the dorsal vessel. Smaller vessels branch from each artery to supply the tissues of the earthworm with oxygen and nutrients and to remove waste products.

The Functions of Vertebrate Circulatory Systems

The vertebrate circulatory system is more elaborate than the invertebrate circulatory system. It functions in transporting oxygen and nutrients to tissues by the cardiovascular system. Blood vessels form a tubular network that permits blood to flow from the heart to all the cells of the body and then back to the heart. *Arteries* carry blood away from the heart, whereas *veins* return blood to the heart. Blood passes from the arterial to the venous system through *capillaries*, which are the thinnest and most numerous of the blood vessels.

As blood plasma passes through capillaries, the pressure of the blood forces some of this fluid out of the capillary walls. Fluid derived from plasma that passes out of capillary walls into the surrounding tissues is called **interstitial fluid.** Some of this fluid returns directly to capillaries, and some enters into **lymph vessels,** located in the connective tissues around the blood vessels. This fluid, now called *lymph*, is returned to the venous blood at specific sites. The lymphatic system is considered a part of the circulatory system and is discussed later in this chapter.

The vertebrate circulatory system has three principal functions: transportation, regulation, and protection.

1. **Transportation.** All of the substances essential for cellular metabolism are transported by the circulatory system. These substances can be categorized as follows:

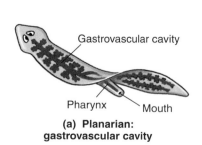

(a) Planarian:
gastrovascular cavity

Gastrovascular cavity

Pharynx   Mouth

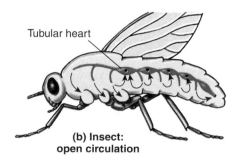

Tubular heart

(b) Insect:
open circulation

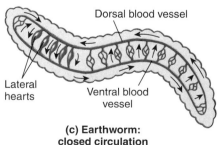

Dorsal blood vessel

Lateral hearts

Ventral blood vessel

(c) Earthworm:
closed circulation

FIGURE 44.2
Circulatory systems of the animal kingdom. (*a*) The gastrovascular cavity of a planarian serves as both a digestive and a circulatory system, delivering nutrients directly to the tissue cells by diffusion from the digestive cavity. (*b*) In the open circulation of an insect, hemolymph is pumped from a tubular heart into cavities in the insect's body; the hemolymph then returns to the blood vessels so that it can be recirculated. (*c*) In the closed circulation of the earthworm, blood pumped from the hearts remains within a system of vessels that returns it to the hearts. All vertebrates also have closed circulatory systems.

a. Respiratory. Red blood cells, or *erythrocytes*, transport oxygen to the tissue cells. In the capillaries of the lungs or gills, oxygen attaches to hemoglobin molecules within the erythrocytes and is transported to the cells for aerobic respiration. Carbon dioxide, a by-product of cellular respiration, is carried by the blood to the lungs or gills for elimination.

b. Nutritive. The digestive system is responsible for the breakdown of food into nutrient molecules that can be absorbed through the intestinal wall and into the blood vessels of the circulatory system. The blood then carries these absorbed products of digestion through the liver and to the cells of the body.

c. Excretory. Metabolic wastes, excessive water and ions, and other molecules in the fluid portion of blood are filtered to the capillaries of the kidneys and excreted in urine.

2. Regulation. The cardiovascular system transports regulatory hormones and participates in temperature regulation.

a. Hormone transport. The blood carries hormones from the endocrine glands, where they are secreted, to the distant target organs they regulate.

b. Temperature regulation. In warm-blooded vertebrates, or **endotherms,** a constant body temperature is maintained regardless of the ambient temperature. This is accomplished in part by blood vessels located just under the epidermis. When the ambient temperature is cold, the superficial vessels constrict to divert the warm blood to deeper vessels. When the ambient temperature is warm, the superficial vessels dilate so that the warmth of the blood can be lost by radiation (figure 44.3).

Some vertebrates also retain heat in a cold environment by using **countercurrent heat exchange**. In this process, a vessel carrying warm blood from deep within the body passes next to a vessel returning cooler blood from the surface of the body (figure 44.4). The warm blood going out heats the cooler blood returning from the body surface, so that this blood is warmer when it reaches the interior of the body.

3. Protection. The circulatory system protects against injury and foreign microbes or toxins introduced into the body.

a. Blood clotting. The clotting mechanism protects against blood loss when vessels are damaged. This clotting mechanism involves both proteins from the blood plasma and cell fragments called platelets (discussed next).

b. Immune defense. The blood contains white blood cells, or leukocytes, that provide immunity against many disease-causing agents. Some white blood cells are phagocytic, some produce antibodies, and some act by other mechanisms to protect the body.

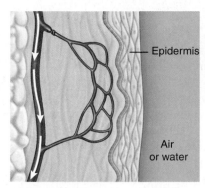
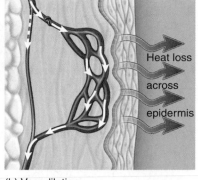

(a) Vasoconstriction   (b) Vasodilation

FIGURE 44.3
Regulation of heat loss. The amount of heat lost at the body's surface can be regulated by controlling the flow of blood to the surface. (*a*) Constriction of surface blood vessels limits flow and heat loss; (*b*) dilation of these vessels increases flow and heat loss.

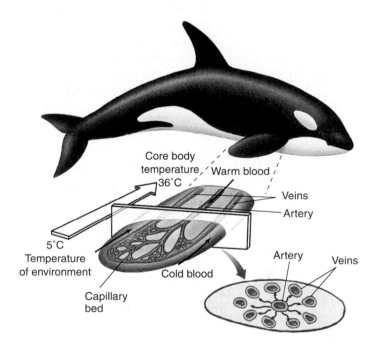

FIGURE 44.4
Countercurrent heat exchange. Many marine animals, such as this killer whale, limit heat loss in cold water using countercurrent heat exchange. The warm blood pumped from within the body in arteries loses heat to the cooler blood returning from the skin in veins. This warms the venous blood so that the core body temperature can remain constant in cold water, and it cools the arterial blood so that less heat is lost when the arterial blood reaches the tip of the extremity.

Circulatory systems may be open or closed. All vertebrates have a closed circulatory system, in which blood circulates away from the heart in arteries and back to the heart in veins. The circulatory system serves a variety of functions, including transportation, regulation, and protection.

Blood

Blood is composed of a fluid **plasma** and several different kinds of cells that circulate within that fluid (figure 44.5). Blood platelets, although included in figure 44.5, are not complete cells; rather, they are fragments of cells that are produced in the bone marrow. Blood plasma is the matrix in which blood cells and platelets are suspended. Interstitial (extracellular) fluids originate from the fluid present in plasma.

Plasma contains the following solutes:

1. **Metabolites, wastes, and hormones.** Dissolved within the plasma are all of the metabolites used by cells, including glucose, amino acids, and vitamins. Also dissolved in the plasma are hormones that regulate cellular activities, wastes such as nitrogen compounds, and CO_2 produced by metabolizing cells. Most CO_2 is carried in the blood as bicarbonate because free carbon dioxide would decrease blood pH.

2. **Ions.** Like the water of the seas in which life arose, blood plasma is a dilute salt solution. The predominant plasma ions are sodium, chloride, and bicarbonate ions. In addition, plasma contains trace amounts of other ions such as calcium, magnesium, copper, potassium, and zinc. The composition of the plasma, therefore, is similar to that of seawater, but plasma has a lower total ion concentration than that of present-day seawater.

3. **Proteins.** The liver produces most of the plasma proteins, including **albumin,** which comprises most of the plasma protein; the alpha (α) and beta (β) **globulins,** which serve as carriers of lipids and steroid hormones; and *fibrinogen*, which is required for blood clotting. Following an injury to a blood vessel, platelets release clotting factors (proteins) into the blood. In the presence of these clotting factors, fibrinogen is converted into insoluble threads of *fibrin*. Fibrin then aggregates to form the clot. Blood plasma with the fibrinogen removed is called **serum.**

The Blood Cells

Red blood cells function in oxygen transport, white blood cells in immunological defenses, and platelets in blood clotting (figure 44.5).

Erythrocytes and Oxygen Transport. Each cubic millimeter of blood contains about 5 million **red blood cells,** or **erythrocytes.** The fraction of the total blood volume that is occupied by erythrocytes is called the blood's *hematocrit;* in humans, it is typically around 45%.

Blood cell	Life span in blood	Function
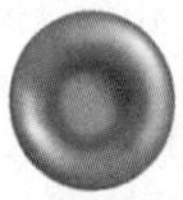 Erythrocyte	120 days	O_2 and CO_2 transport
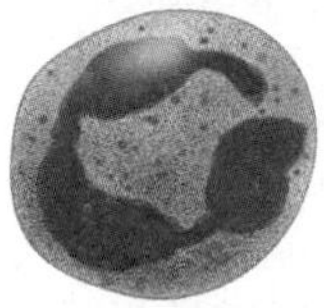 Neutrophil	7 hours	Immune defenses
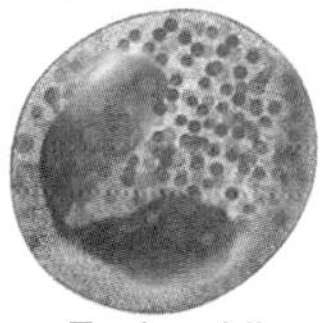 Eosinophil	Unknown	Defense against parasites
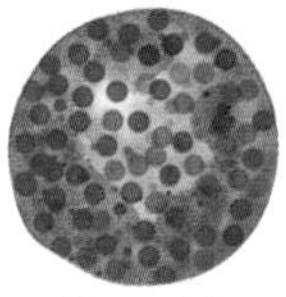 Basophil	Unknown	Inflammatory response
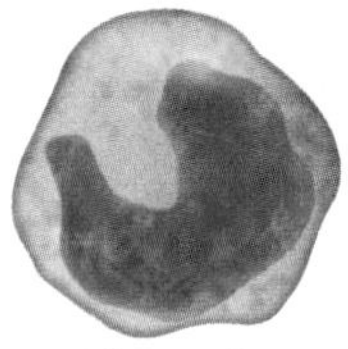 Monocyte	3 days	Immune surveillance (precursor of tissue macrophage)
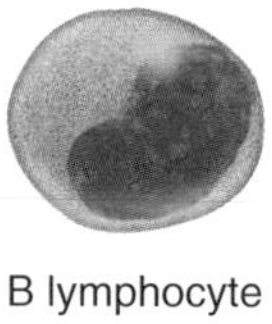 B lymphocyte	Unknown	Antibody production (precursor of plasma cells)
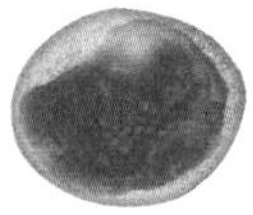 T lymphocyte	Unknown	Cellular immune response
 Platelets	7- 8 days	Blood clotting

FIGURE 44.5
Types of blood cells. Erythrocytes (*top*) are red blood cells, platelets (*bottom*) are fragments of a bone marrow cell, and all the other cells are different types of leukocytes, or white blood cells.

Each erythrocyte resembles a doughnut-shaped disc with a central depression that does not go all the way through. As we've already seen, the erythrocytes of vertebrates contain hemoglobin, a pigment that binds and transports oxygen. In vertebrates, hemoglobin is found only in erythrocytes. In invertebrates, the oxygen-binding pigment (not always hemoglobin) is also present in plasma.

Erythrocytes develop from unspecialized cells, called *stem cells*. When plasma oxygen levels decrease, the kidney converts a plasma protein into the hormone *erythropoietin*. Erythropoietin then stimulates the production of erythrocytes in bone marrow through a process called *erythropoiesis*. In mammals, maturing erythrocytes lose their nuclei. This is different from the mature erythrocytes of all other vertebrates, which remain nucleated. As mammalian erythrocytes age, they are removed from the blood by phagocytic cells of the spleen, bone marrow, and liver. Balancing this loss, new erythrocytes are constantly formed in the bone marrow.

Leukocytes Defend the Body. Less than 1% of the cells in human blood are **leukocytes,** or **white blood cells;** there are only 1 or 2 leukocytes for every 1000 erythrocytes. Leukocytes are larger than erythrocytes and have nuclei. Furthermore, leukocytes are not confined to the blood as erythrocytes are, but can migrate out of capillaries into the interstitial (tissue) fluid.

There are several kinds of leukocytes, each of which plays a specific role in defending the body against invading microorganisms and other foreign substances, as described in chapter 48. **Granular leukocytes** include **neutrophils, eosinophils,** and **basophils,** which are named according to the staining properties of granules in their cytoplasm. **Nongranular leukocytes** include **monocytes** and **lymphocytes.** Neutrophils are the most numerous of the leukocytes, followed in order by lymphocytes, monocytes, eosinophils, and basophils.

Platelets Help Blood to Clot. *Megakaryocytes* are large cells present in bone marrow. Pieces of cytoplasm are pinched off the megakaryocytes and become **platelets.** Platelets enter the blood, where they play an important role in blood clotting. When a blood vessel is broken, smooth muscle in the vessel walls contracts, causing the vessel to constrict. Platelets then accumulate at the injured site and form a plug by sticking to each other and to the surrounding tissues. This plug is reinforced by threads of the protein **fibrin** (figure 44.6), which contract to form a tighter mass. The tightening plug of platelets, fibrin, and often trapped erythrocytes constitutes a blood clot.

Plasma, the liquid portion of the blood, contains different types of proteins, ions, metabolites, wastes, and hormones. Erythrocytes contain hemoglobin and serve in oxygen transport. Leukocytes have specialized functions that protect the body from invading pathogens, and platelets participate in blood clotting.

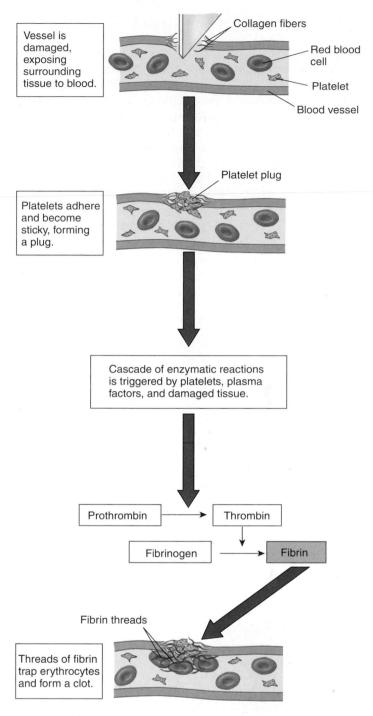

FIGURE 44.6
Blood clotting. Fibrin is formed from a soluble protein, fibrinogen, in the plasma. This reaction is catalyzed by the enzyme thrombin, which is formed from an inactive enzyme called prothrombin. The activation of thrombin is the last step in a cascade of enzymatic reactions that produces a blood clot when a blood vessel is damaged.

Characteristics of Blood Vessels

Blood leaves the heart through vessels known as **arteries.** These continually branch, forming a hollow "tree" that enters each of the organs of the body. The finest, microscopically sized branches of the arterial trees are the **arterioles.** Blood from the arterioles enters the **capillaries** (Latin *capillus*, "a hair"), an elaborate latticework of very narrow, thin-walled tubes. After traversing the capillaries, the blood is collected into **venules;** the venules lead to larger vessels called **veins,** which carry blood back to the heart.

Arteries, arterioles, veins, and venules all have the same basic structure (figure 44.7). The innermost layer is an epithelial sheet called the *endothelium*. Covering the endothelium is a thin layer of elastic fibers, a smooth muscle layer, and a connective tissue layer. The walls of these vessels are thus too thick to permit any exchange of materials between the blood and the tissues outside the vessels. The walls of capillaries, however, are made up of only the endothelium, so molecules and ions can leave the blood plasma by diffusion, by filtration through pores between the cells of the capillary walls, and by transport through the endothelial cells. Therefore, as blood passes through capillaries, gases and metabolites are exchanged with the interstitial fluids and cells of the body.

Arteries and Arterioles

Arteries function in transporting blood away from the heart. The larger arteries contain more elastic fibers in their walls than other blood vessels, allowing them to recoil each time they receive a volume of blood pumped by the heart. Smaller arteries and arterioles are less elastic, but their disproportionately thick smooth muscle layer enables them to resist bursting.

The vast tree of arteries presents a frictional resistance to blood flow. The narrower the vessel, the greater the frictional resistance to flow. In fact, a vessel that is half the diameter of another has 16 times the frictional resistance! This is because the resistance to blood flow is inversely proportional to the radius of the vessel. Therefore, within the arterial tree, it is the small arteries and arterioles that provide the greatest resistance to blood flow. Contraction of the smooth muscle layer of the arterioles results in **vasoconstriction,** which greatly increases resistance and decreases flow. Relaxation of the smooth muscle layer results in **vasodilation,** decreasing resistance and increasing blood flow to an organ (see figure 44.3).

In addition, blood flow through some organs is regulated by rings of smooth muscle around arterioles near the region where they empty into capillaries. These **precapillary sphincters** (figure 44.8) can close off specific capillary beds completely. For example, the closure of precapillary sphincters in the skin contributes to the vasoconstriction that limits heat loss in cold environments.

Exchange in the Capillaries

Each time the heart contracts, it must produce sufficient pressure to pump blood against the resistance of the arterial tree and into the capillaries. The vast number and extensive branching of the capillaries ensure that *every cell in the body is within 100 micrometers (μm) of a capillary*. On the average, capillaries are about 1 millimeter long and 8 micrometers (μm) in diameter, only slightly larger than a red blood cell (5 to 7 μm in diameter). Despite the close fit, red blood cells are flexible enough to squeeze through capillaries without difficulty.

Although each capillary is very narrow, there are so many of them that the capillaries have the greatest *total*

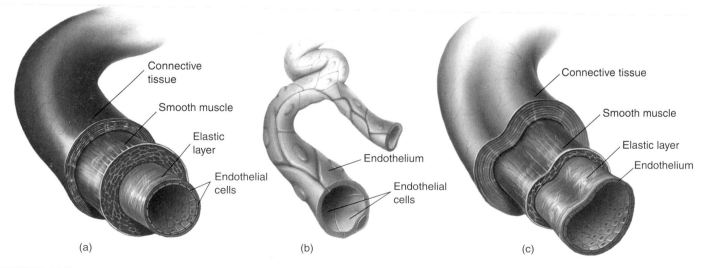

FIGURE 44.7
The structure of blood vessels. (*a*) Arteries and (*c*) veins have the same tissue layers. (*b*) Capillaries are composed of only a single layer of endothelial cells. (Not to scale.)

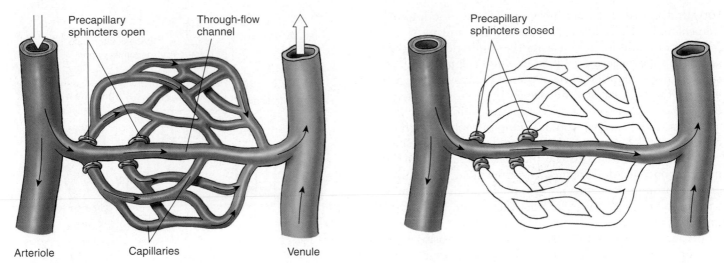

Precapillary sphincters open

Through-flow channel

Precapillary sphincters closed

Arteriole   Capillaries   Venule

(a) Blood flows through capillary network

(b) Blood flow in capillary network is limited

FIGURE 44.8
The capillary network connects arteries with veins. (*a*) Most of the exchange between the blood and the extracellular fluid occurs while the blood is in the capillaries. Entrance to the capillaries is controlled by bands of muscle called precapillary sphincters. (*b*) When a sphincter contracts, it narrows the entrance to the capillary. By contracting these sphincters, the body can limit the amount of blood in the capillary network of a particular tissue, and thus control the rate of exchange in that tissue.

cross-sectional area of any other type of vessel. Consequently, this allows more time for blood to exchange materials with the surrounding extracellular fluid. By the time the blood reaches the end of a capillary, it has released some of its oxygen and nutrients and picked up carbon dioxide and other waste products. Consequently, blood loses most of its pressure and velocity in passing through the vast capillary networks, and so is under very low pressure when it enters the veins.

Venules and Veins

Blood flows from the venules to ever-larger veins, and ultimately back to the heart. Venules and veins have the same tissue layers as arteries, but they have a thinner layer of smooth muscle. Less muscle is needed because the pressure in the veins is only about one-tenth that in the arteries. Most of the blood in the cardiovascular system is contained within veins, which can expand when needed to hold additional amounts of blood. You can see the expanded veins in your feet when you stand for a long time.

When the blood pressure in the veins is so low, how does the blood return to the heart from the feet and legs? The venous pressure alone is not sufficient, but several sources provide help. Most significantly, skeletal muscles surrounding the veins can contract to move blood by squeezing the veins, a mechanism called the **venous pump.** Blood moves in one direction through the veins back to the heart with the help of **venous valves** (figure 44.9). When a person's veins expand too much with blood, the venous valves may no longer work, and the blood may pool in the veins. Veins in this condition are known as varicose veins.

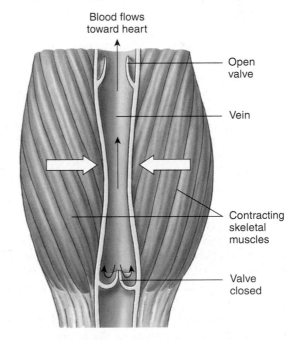

Blood flows toward heart

Open valve

Vein

Contracting skeletal muscles

Valve closed

FIGURE 44.9
One-way flow of blood through veins. Venous valves ensure that blood moves through the veins in only one direction, back to the heart.

The Lymphatic System

The cardiovascular system is considered a closed system because all its vessels are connected with one another—none are simply open-ended. However, some water and solutes in the blood plasma do filter through the walls of the capillaries to form the interstitial (tissue) fluid. This filtration is driven by the pressure of the blood, and it helps supply the tissue cells with oxygen and nutrients. Most of the fluid is filtered from the capillaries near their arteriolar ends, where the blood pressure is higher, and returned to the capillaries near their venular ends (figure 44.10). This return of fluid occurs by osmosis, which is driven by a higher solute concentration within the capillaries. Most of the plasma proteins cannot escape through the capillary pores because of their large size, and so the concentration of proteins in the plasma is greater than the protein concentration in the interstitial fluid. The difference in protein concentration produces an osmotic pressure, called the *oncotic pressure*, that causes osmosis of water into the capillaries.

Because interstitial fluid is the product of plasma fluids flowing from the blood into the tissues as a result of the blood pressure, high capillary blood pressure could cause too much interstitial fluid to accumulate. A common example of this occurs in pregnant women, when the fetus compresses veins and thereby increases the capillary blood pressure in the mother's lower limbs. The increased interstitial fluid can cause swelling of the tissues, or *edema*, of the feet. Edema may also result if the plasma protein concentration (and thus the oncotic pressure) is too low. Fluids will not return to the capillaries but will remain as interstitial fluid. This may be caused either by liver disease, because the liver produces most of the plasma proteins, or by protein malnutrition (*kwashiorkor*).

Even under normal conditions, the amount of fluid filtered out of the capillaries is greater than the amount that returns to the capillaries by osmosis. However, the remainder does eventually return to the cardiovascular system by way of an *open* circulatory system called the **lymphatic system.** The lymphatic system consists of lymphatic capillaries, lymphatic vessels, lymph nodes, and lymphatic organs, including the spleen and thymus. Excess fluid in the tissues drains into blind-ended lymph capillaries with highly permeable walls. This fluid, now called **lymph,** passes into progressively larger lymphatic vessels, which resemble veins and have one-way valves (figure 44.11). The lymph eventually enters two major lymphatic vessels, which drain into veins on each side of the neck.

Movement of lymph in mammals is accomplished by skeletal muscles squeezing against the lymphatic vessels, a mechanism similar to the venous pump that moves blood through veins. In some cases, the lymphatic vessels also contract rhythmically. In many fishes, all amphibians and reptiles, bird embryos, and some adult birds, movement of lymph is propelled by **lymph hearts.**

As the lymph moves through lymph nodes and lymphatic organs, it is modified by phagocytic cells that line the channels of those organs. In addition, the lymph nodes and lymphatic organs contain *germinal centers* where the ac-

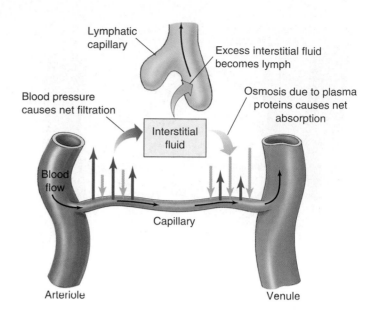

FIGURE 44.10
Plasma fluid, minus proteins, is filtered out of capillaries. This forms interstitial fluid, which bathes the tissues. Much of the interstitial fluid is returned to the capillaries by the osmotic pressure generated by the higher protein concentration in plasma. The excess interstitial fluid is drained into open-ended lymphatic capillaries, which ultimately return the fluid to the cardiovascular system.

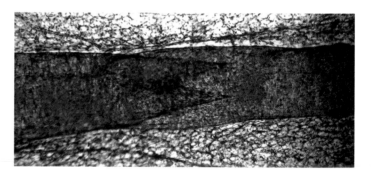

FIGURE 44.11
A lymphatic vessel valve (25×). Valves allow lymph to flow in one direction (from left to right in this figure) but not in the reverse direction.

tivation and proliferation of lymphocytes, a type of white blood cell critically important in immunity, occurs.

Blood is pumped from the heart into the arterial system, which branches into fine arterioles. This blood is delivered into the thinnest and most numerous of vessels, the capillaries, where exchanges with the tissues occur. Blood returns to the heart through veins. Lymphatic vessels return excess interstitial fluid, called lymph, back to the cardiovascular system.

44.2 The circulatory and respiratory systems evolved together in vertebrates.

Increasing Body Size Requires Special Circulatory and Respiratory Adaptations

In chapter 42, we described one of the greatest evolutionary achievements of animals—locomotion. For vertebrates, the combination of large body size and locomotion was possible because of the "coevolution" of the circulatory and respiratory systems. As the body size and physiological abilities of animals increased, so did the need for more efficient mechanisms to both deliver nutrients and oxygen and remove wastes and carbon dioxide from the growing mass of tissues. Animals have evolved a remarkable set of adaptations inextricably linking circulation and respiration. In this context, this chapter examines the evolution of the heart, the evolution of separate pulmonary and systemic circulations, and the principles of gas exchange.

The Fish Heart

Chordates that were ancestral to the vertebrates are thought to have had simple tubular hearts, similar to those now seen in lancelets (see chapter 34). The heart was little more than a specialized zone of the ventral artery, more heavily muscled than the rest of the arteries, which contracted in simple peristaltic waves. A pumping action results because the uncontracted portions of the vessel have a larger diameter than the contracted portion, and thus present less resistance to blood flow.

The development of gills by fishes required a more efficient pump, and in fishes we see the evolution of a true chamber-pump heart. The fish heart is, in essence, a tube with four chambers arrayed one after the other (figure 44.12a). The first two chambers—the **sinus venosus** and **atrium**—are collection chambers, while the second two, the **ventricle** and **conus arteriosus,** are pumping chambers.

As might be expected, the sequence of the heartbeat in fishes is a peristaltic sequence, starting at the rear and moving to the front, similar to the beat of the early chordate heart. The sinus venosus is the first chamber to contract, followed by the atrium, the ventricle, and finally the conus arteriosus. Despite shifts in the relative positions of the chambers in the vertebrates that evolved later, this heartbeat sequence is maintained in all vertebrates. In fish, the electrical impulse that produces the contraction is initiated in the sinus venosus; in other vertebrates, the electrical impulse is initiated by a structure homologous to the sinus venosus.

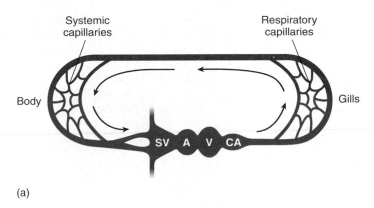

(a)

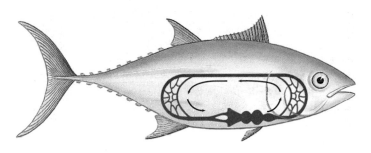

(b)

FIGURE 44.12
The heart and circulation of a fish. (a) Diagram of a fish heart, showing the chambers in series with each other. (SV = sinus venosus; A = atrium; V = ventricles; CA = conus arteriosus.) (b) Diagram of fish circulation, showing that blood is pumped by the ventricle through the gills and then to the body. Blood rich in oxygen (oxygenated) is shown in *red*; blood low in oxygen (deoxygenated) is shown in *blue*.

The fish heart is remarkably well suited to the gill respiratory apparatus and represents one of the major evolutionary innovations in the vertebrates. Perhaps its greatest advantage is that the blood that passes through the gills is fully oxygenated when it moves into the tissues. After blood leaves the conus arteriosus, it moves through the gills, where it becomes oxygenated; from the gills, it flows through a network of arteries to the rest of the body; then it returns to the heart through the veins (figure 44.12b). This arrangement has one great limitation, however. In passing through the capillaries in the gills, the blood loses much of the pressure developed by the contraction of the heart, so the circulation from the gills through the rest of the body is sluggish. This feature limits the rate of oxygen delivery to the rest of the body.

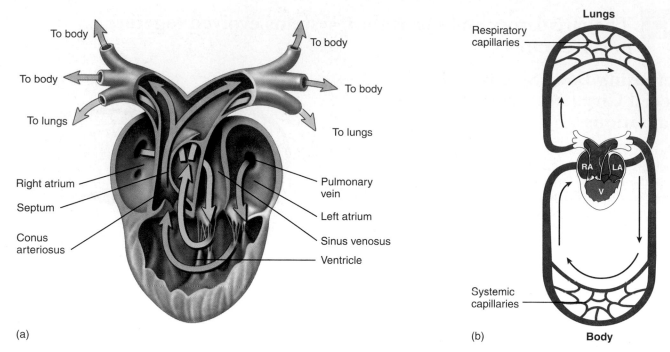

FIGURE 44.13
The heart and circulation of an amphibian. (*a*) The frog heart has two atria but only one ventricle, which pumps blood both to the lungs and to the body. (*b*) Despite the potential for mixing, the oxygenated and deoxygenated bloods (*red* and *blue*, respectively) mix very little as they are pumped to the body and lungs. The ventricle is shown in *purple*. (RA = right atrium; LA = left atrium; V = ventricle.)

Amphibian and Reptile Circulation

The advent of lungs in amphibians involved a major change in the pattern of circulation. After blood is pumped by the heart through the *pulmonary arteries* to the lungs, it does not go directly to the tissues of the body but is instead returned via the *pulmonary veins* to the heart. This results in two circulations: one between heart and lungs, called the **pulmonary circulation,** and one between the heart and the rest of the body, called the **systemic circulation.**

If no changes had occurred in the structure of the heart, the oxygenated blood from the lungs would be mixed in the heart with the deoxygenated blood returning from the rest of the body. Consequently, the heart would pump a mixture of oxygenated and deoxygenated blood rather than fully oxygenated blood. The amphibian heart has two structural features that help reduce this mixing (figure 44.13). First, the atrium is divided into two chambers: The right atrium receives deoxygenated blood from the systemic circulation, and the left atrium receives oxygenated blood from the lungs. These two stores of blood, therefore, do not mix in the atria, and little mixing occurs when the contents of each atrium enter the single, common ventricle, due to internal channels created by recesses in the ventricular wall. The conus arteriosus is partially separated by a dividing wall, which directs deoxygenated blood into the pulmonary arteries to the lungs and oxygenated blood into the *aorta*, the major artery of the systemic circulation, to the body.

Because there is only one ventricle in an amphibian heart, the separation of the pulmonary and systemic circulations is incomplete. Amphibians in water, however, can obtain additional oxygen by diffusion through their skin. This process, called **cutaneous respiration,** helps supplement the oxygenation of the blood in these vertebrates.

Among reptiles, additional modifications have reduced the mixing of blood in the heart still further. In addition to having two separate atria, reptiles have a septum that partially subdivides the ventricle. This results in an even greater separation of oxygenated and deoxygenated blood within the heart. The separation is complete in one order of reptiles, the crocodiles, which have two separate ventricles divided by a complete septum. Crocodiles therefore have completely divided pulmonary and systemic circulations. Another change in the circulation of reptiles is that the conus arteriosus has become incorporated into the trunks of the large arteries leaving the heart.

Mammalian and Bird Hearts

Mammals, birds, and crocodiles have a four-chambered heart with two separate atria and two separate ventricles (figure 44.14). The right atrium receives deoxygenated blood from the body and delivers it to the right ventricle, which pumps the blood to the lungs. The left atrium receives oxygenated blood from the lungs and delivers it to the left ventricle, which pumps the oxygenated blood to the

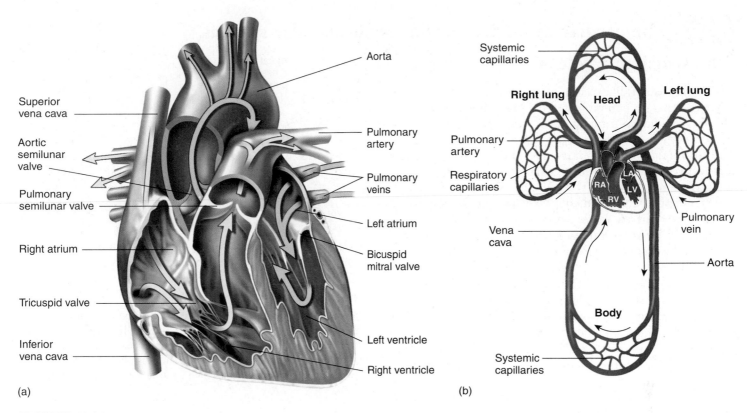

FIGURE 44.14

The heart and circulation of mammals and birds.   (*a*) The path of blood through the four-chambered heart. (*b*) The right side of the heart receives deoxygenated blood and pumps it to the lungs; the left side of the heart receives oxygenated blood and pumps it to the body. In this way, the pulmonary and systemic circulations are kept completely separate. (RA = right atrium; LA = left atrium; RV = right ventricle; LV = left ventricle.)

rest of the body. This completely double circulation is powered by a two-cycle pump. Both atria fill with blood and simultaneously contract, emptying their blood into the ventricles. Both ventricles contract at the same time, pushing blood simultaneously into the pulmonary and systemic circulations. The increased efficiency of the double circulatory system in mammals and birds is thought to have been important in the evolution of endothermy (warm-bloodedness), because a more efficient circulation is necessary to support the high metabolic rate required.

Because the overall circulatory system is closed, the same volume of blood must move through the pulmonary circulation as through the much larger systemic circulation with each heartbeat. Therefore, the right and left ventricles must pump the same amount of blood each time they contract. If the output of one ventricle did not match that of the other, fluid would accumulate and pressure would increase in one of the circuits. The result would be increased filtration out of the capillaries and edema (as occurs in congestive heart failure, for example). Although the volume of blood pumped by the two ventricles is the same, the pressure they generate is not. The left ventricle, which pumps blood through the higher-resistance systemic pathway, is more muscular and generates more pressure than does the right ventricle.

Throughout the evolutionary history of the vertebrate heart, the sinus venosus has served as a pacemaker, the site where the impulses that initiate the heartbeat originate. Although it constitutes a major chamber in the fish heart, it is reduced in size in amphibians and further reduced in reptiles. In mammals and birds, the sinus venosus is no longer evident as a separate chamber, but its disappearance is not really complete. Some of its tissue remains in the wall of the right atrium, near the point where the systemic veins empty into the atrium. This tissue, which is called the *sinoatrial (SA) node*, is still the site where each heartbeat originates.

The fish heart is a modified tube, consisting of a series of four chambers, which pumps with wavelike contractions. Amphibians and reptiles have two circulations, pulmonary and systemic, that deliver blood to the lungs and the rest of the body, respectively. The oxygenated blood from the lungs is kept relatively separate from the deoxygenated blood from the rest of the body by incomplete divisions within the heart. In bird and mammal hearts, the division within the heart is complete; there is no mixing of oxygenated and deoxygenated blood.

44.3 The cardiac cycle drives the cardiovascular system.

The Cardiac Cycle

The human heart, like that of all mammals and birds, is really two separate pumping systems operating within a single organ. The right pump sends blood to the lungs, and the left pump sends blood to the rest of the body.

The heart has two pairs of valves. One pair, the **atrioventricular (AV) valves,** guards the opening between the atria and ventricles. The AV valve on the right side is the **tricuspid valve,** and the AV valve on the left is the **bicuspid,** or **mitral, valve.** Another pair of valves, together called the **semilunar valves,** guard the exits from the ventricles to the arterial system; the **pulmonary valve** is located at the exit of the right ventricle, and the **aortic valve** is located at the exit of the left ventricle. These valves open and close as the heart goes through its **cardiac cycle** of rest (*diastole*) and contraction (*systole*). The closing of these valves produces the "lub-dub" sounds heard with a stethoscope.

Blood returns to the resting heart through veins that empty into the right and left atria. As the atria fill and the pressure in them rises, the AV valves open to admit the blood into the ventricles. The ventricles become about 80% filled during this time. Contraction of the atria wrings out the final 20% of the 80 milliliters of blood the ventricles will receive, on average, in a resting person. These events occur while the ventricles are relaxing, a period called ventricular **diastole.**

After a slight delay, the ventricles contract, a period called ventricular **systole.** Contraction of each ventricle increases the pressure within each chamber, causing the AV valves to forcefully close (the "lub" sound), thereby preventing blood from backing up into the atria. Immediately after the AV valves close, the pressure in the ventricles forces the semilunar valves open so that blood can be pushed out into the arterial system. As the ventricles relax, closing of the semilunar valves prevents backflow (the "dub" sound).

The right and left **pulmonary arteries** deliver oxygen-depleted blood from the right ventricle to the right and left lungs. As previously mentioned, these return blood to the left atrium of the heart via the **pulmonary veins.** The **aorta** and all its branches are systemic arteries, carrying oxygen-rich blood from the left ventricle to all parts of the body. The **coronary arteries** are the first branches off the aorta; these supply the heart muscle itself. Other systemic arteries branch from the aorta as it makes an arch above the heart, and as it descends and traverses the thoracic and abdominal cavities. These branches provide all body organs with oxygenated blood. The blood from the body organs, now lower in oxygen, returns to the heart in the systemic veins. These eventually empty into two major veins: the **superior vena cava,** which drains the upper body, and the **inferior vena cava,** which drains the lower body. These veins empty into the right atrium and thereby complete the systemic circulation.

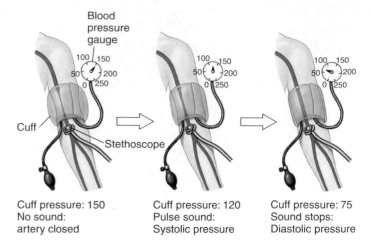

Blood pressure gauge

Cuff

Stethoscope

Cuff pressure: 150
No sound:
artery closed

Cuff pressure: 120
Pulse sound:
Systolic pressure

Cuff pressure: 75
Sound stops:
Diastolic pressure

FIGURE 44.15
Measurement of blood pressure. The blood pressure cuff is tightened to stop the blood flow through the brachial artery. As the cuff is loosened, the systolic pressure is recorded as the pressure at which a pulse is heard. The diastolic pressure is recorded as the pressure at which a sound is no longer heard.

Measuring Arterial Blood Pressure

As the ventricles contract, great pressure is generated in the arteries throughout the body. You can tell this by feeling your pulse, either on the inside of your wrist below the thumb or on the sides of your neck below your ear and jawbone. The contraction of the ventricles has to be strong enough to force blood through capillary beds, but not so strong as to cause damage to smaller arteries and arterioles. Doctors measure your blood pressure to determine how hard your heart is working.

The measuring device used, called a sphygmomanometer, measures the blood pressure in the brachial artery found on the inside part of the arm, at the elbow (figure 44.15). A cuff wrapped around the upper part of the arm is tightened enough to stop the flow of blood to the lower part of the arm. As the cuff is loosened, blood begins pulsating through the artery and can be detected using a stethoscope. Two measurements are recorded: the systolic and the diastolic pressures. The **systolic pressure** is the peak pressure during ventricular systole (contraction of the ventricle). The **diastolic pressure** is the minimum pressure between heartbeats (repolarization of the ventricles). The blood pressure is written as a ratio of systolic over diastolic pressure, and for a healthy person in his or her twenties, a typical blood pressure is 120/75 (measured in millimeters of mercury or mm Hg). A condition called *hypertension* (high blood pressure) occurs when the ventricles experience very strong contractions, and either the systolic pressure is greater than 150 mm Hg or the diastolic pressure is greater than 90 mm Hg.

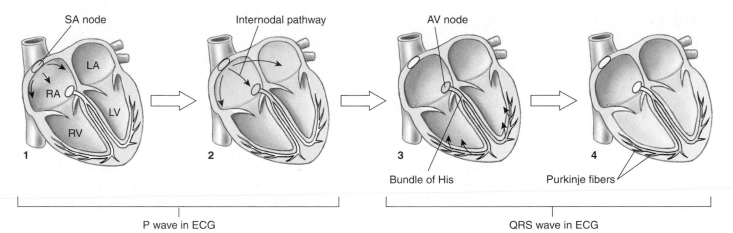

SA node   Internodal pathway   AV node

LA

RA

LV

RV

1   2   3   4

Bundle of His   Purkinje fibers

P wave in ECG   QRS wave in ECG

FIGURE 44.16

The path of electrical excitation in the heart. A wave of depolarization begins at the sinoatrial (SA) node and spreads through the muscle fibers of the atria, causing them to contract (forming the P wave on the ECG). The depolarization also is conducted through the internodal pathway, from the SA node to the atrioventricular (AV) node, from which it passes to the ventricles along the septum by the bundle of His. Large-diameter Purkinje fibers carry the depolarization into the right and left ventricular muscles (forming the QRS wave on the ECG). The T wave on the ECG corresponds to the repolarization of the ventricles.

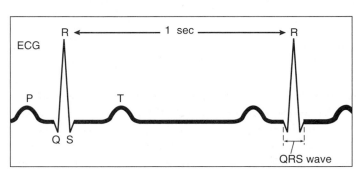

ECG

R   1 sec   R

P   T

Q S

QRS wave

Electrical Excitation and Contraction of the Heart

As in other types of muscle, contraction of heart muscle is stimulated by membrane **depolarization,** a reversal of the electrical polarity that normally exists across the plasma membrane (see chapter 45). In skeletal muscles, only signals from the nervous system can initiate depolarization. However, the heart contains specialized autogenic (self-depolarizing) cells, which can initiate depolarization without neural activation. One such group of autogenic cells, derived from the sinus venosus, is the sinoatrial (SA) node (figure 44.16), located in the wall of the right atrium. The SA node acts as a pacemaker for the rest of the heart by producing spontaneous depolarizations at a faster rate than other autogenic cardiac cells. Each depolarization initiated by this pacemaker follows two tracts of specialized conduction fibers, the interatrial and internodal pathways. The interatrial pathway transmits the depolarizing signal from the SA node to the cardiac muscle fibers of the left atrium, while the internodal pathway activates the muscle of the right atrium and conducts the signals to the atrioventricular (AV) node. Once activated, depolarizations spread quickly from one muscle fiber to another in a wave that envelops the right and left atria nearly simultaneously. The rapid spread of depolarization is made possible because special conducting fibers are present and because the cardiac muscle cells are coupled by gap junction.

A sheet of connective tissue separating the atria from the ventricles blocks the spread of excitation through muscle fibers from one chamber to the other. The AV node provides the only pathway for conduction of the depolarization from the atria to the ventricles. The fibers of the AV node slow down the conduction of the depolarizing signals, delaying the contraction of the ventricle by about 0.1 second. This delay permits the atria to finish contracting and emptying their blood into the ventricles before the ventricles contract.

From the AV node, the wave of depolarization is conducted rapidly over both ventricles by a network of fibers called the **atrioventricular bundle,** or **bundle of His.** These fibers relay the depolarization to **Purkinje fibers,** which directly stimulate the myocardial cells of the left and right ventricles, causing their almost simultaneous contraction.

The spread of electrical activity through the heart creates currents that can be recorded from the surface of the body with electrodes placed on the limbs and chest. The recording, called an **electrocardiogram** (**ECG** or **EKG**), shows how the cells of the heart depolarize and repolarize during the cardiac cycle (figure 44.16). Depolarization causes contraction of the heart, while repolarization causes relaxation. The first peak in the recording, P, is produced by the depolarization of the atria, and thus is associated with atrial systole. The second, larger peak, QRS, is produced by ventricular depolarization; during this time, the ventricles contract (ventricular systole) and eject blood into the arteries. The last peak, T, is produced by ventricular repolarization; at this time, the ventricles begin diastole.

> **The cardiac cycle consists of systole and diastole; the ventricles contract at systole and relax at diastole. The SA node in the right atrium initiates waves of depolarization that stimulate first the atria and then the ventricles to contract.**

Blood Flow and Blood Pressure

Cardiac Output

Cardiac output is the volume of blood pumped by each ventricle per minute. Because humans (like all vertebrates) have a closed circulation, the cardiac output is the same as the volume of blood that traverses the systemic or pulmonary circulations per minute. It is calculated by multiplying the heart rate by the *stroke volume*, which is the volume of blood ejected by each ventricle per beat. For example, if the heart rate is 72 beats per minute and the stroke volume is 70 milliliters, the cardiac output is 5 liters per minute, which is about average in a resting person.

Cardiac output increases during exercise because of an increase in heart rate and stroke volume. When exercise begins, the heart rate increases up to about 100 beats per minute. As exercise becomes more intense, skeletal muscles squeeze on veins more vigorously, returning blood to the heart more rapidly. In addition, the ventricles contract more strongly, so they empty more completely with each beat.

During exercise, the cardiac output increases to a maximum of about 25 liters per minute in an average young adult. Although the cardiac output has increased five times, not all organs receive five times the blood flow; some receive more, others less. This is because the arterioles in some organs, such as in the digestive system, constrict, while the arterioles in the exercising muscles and heart dilate. As previously mentioned, the resistance to flow decreases as the radius of the vessel increases. As a consequence, vasodilation greatly increases blood flow to specific tissues, and vasoconstriction greatly decreases it.

Blood Pressure and the Baroreceptor Reflex

The arterial blood pressure depends on two factors: how much blood the ventricles pump (the cardiac output) and how great a resistance to flow the blood encounters in the entire arterial system. An increased blood pressure, therefore, could be produced by an increase in either heart rate or blood volume (because both increase the cardiac output) or by vasoconstriction, which increases the resistance to blood flow. Conversely, blood pressure will fall if the heart rate slows or if the blood volume is reduced—for example, by dehydration or excessive bleeding (hemorrhage).

Changes in the arterial blood pressure are detected by **baroreceptors** located in the arch of the aorta and in the carotid arteries. These receptors activate sensory neurons that relay information to *cardiovascular control centers* in the medulla oblongata, a region of the brain stem. When the baroreceptors detect a fall in blood pressure, they stimulate neurons that go to blood vessels in the skin and viscera, causing arterioles in these organs to constrict and raise the blood pressure. This baroreceptor reflex therefore completes a negative feedback loop that acts to correct the fall in blood pressure and restore homeostasis.

Blood Volume Reflexes

Blood pressure depends in part on the total blood volume. A decrease in blood volume, therefore, will decrease blood pressure, if all else remains equal. Blood volume regulation involves the effects of four hormones: (1) antidiuretic hormone; (2) aldosterone; (3) atrial natriuretic hormone; and (4) nitric oxide.

Antidiuretic Hormone. *Antidiuretic hormone* (ADH), also called *vasopressin*, is secreted by the posterior pituitary gland in response to an increase in the osmotic concentration of the blood plasma. Dehydration, for example, causes the blood volume to decrease while the remaining plasma becomes more concentrated. This stimulates *osmoreceptors* in the hypothalamus of the brain, a region located immediately above the pituitary. The osmoreceptors promote thirst and stimulate ADH secretion from the posterior pituitary gland. ADH, in turn, stimulates the kidneys to retain more water in the blood, excreting less in the urine (urine is derived from blood plasma—see chapter 49). A dehydrated person thus drinks more and urinates less, helping to raise the blood volume and restore homeostasis.

Aldosterone. If a person's blood volume is lowered (by dehydration, for example), the flow of blood through the organs will be reduced if no compensation occurs. Whenever the kidneys experience a decreased blood flow, a group of kidney cells initiate the release of an enzyme known as renin into the blood. Renin activates a blood protein, angiotensin, that stimulates vasoconstriction throughout the body while it also stimulates the adrenal cortex (the outer region of the adrenal glands) to secrete the hormone *aldosterone*. This important steroid hormone is necessary for life; it acts on the kidneys to promote the retention of Na^+ and water in the blood. This not only increases blood volume, but helps maintain blood osmolarity as well. An animal that lacks aldosterone will die if untreated, because so much of the blood volume is lost in urine that the blood pressure falls too low to sustain life.

Atrial Natriuretic Hormone. When the body needs to eliminate excessive Na^+, less aldosterone is secreted by the adrenals, so that less Na^+ is retained by the kidneys. In recent years, scientists have learned that Na^+ excretion in the urine is promoted by another hormone. Surprisingly, this hormone is secreted by the right atrium of the heart—the heart is an endocrine gland! The right atrium secretes *atrial natriuretic hormone* in response to stretching of the atrium by an increased blood volume. The action of atrial natriuretic hormone completes a negative feedback loop, because it promotes the elimination of Na^+ and water, which will lower the blood volume and pressure.

Nitric Oxide. *Nitric oxide* (NO) is a gas that acts as a hormone in vertebrates, regulating blood pressure and blood flow. As described in chapter 7, nitric oxide gas is a

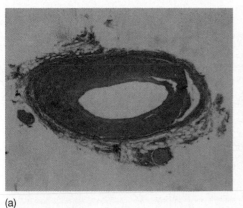 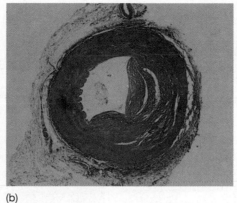 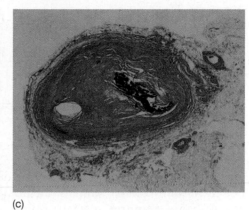

(a)   (b)   (c)

FIGURE 44.17

Atherosclerosis.   (*a*) The coronary artery shows only minor blockage. (*b*) The artery exhibits severe atherosclerosis—much of the passage is blocked by buildup on the interior walls of the artery. (*c*) The coronary artery is essentially completely blocked.

paracrine hormone, produced by one cell, that penetrates through membranes and alters the activities of other neighboring cells. In 1998, the Nobel Prize for medicine was awarded for the discovery of this signal transmission activity. How does NO regulate blood pressure? Nitric oxide gas produced by the surface endothelial cells of blood vessels passes inward through the cell layers of the vessel, causing the smooth muscles that encase it to relax and the blood vessel to dilate (become wider). For over a century, heart patients have been prescribed nitroglycerin to relieve chest pain, but only now has it become clear that nitroglycerin acts by releasing nitric oxide gas.

Cardiovascular Diseases

Cardiovascular diseases are the leading cause of death in the United States; more than 42 million people have some form of cardiovascular disease. Heart attacks are the main cause of cardiovascular deaths in the United States, accounting for about one-fifth of all deaths. Heart attacks result from an insufficient supply of blood reaching one or more parts of the heart muscle, which causes myocardial cells in those parts to die. Heart attacks may be caused by a blood clot forming somewhere in the coronary arteries (the arteries that supply the heart muscle with blood) and blocking the passage of blood through those vessels. They may also result if an artery is blocked by atherosclerosis. Recovery from a heart attack is possible if the portion of the heart that was damaged is small enough that the other blood vessels in the heart can enlarge their capacity and resupply the damaged tissues. **Angina pectoris,** which literally means "chest pain," occurs for reasons similar to those that cause heart attacks, but it is not as severe. The pain may occur in the heart and often also in the left arm and shoulder. Angina pectoris is a warning sign that the blood supply to the heart is inadequate but still sufficient to avoid myocardial cell death.

Strokes are caused by an interference with the blood supply to the brain. They may occur when a blood vessel bursts in the brain, or when blood flow in a cerebral artery is blocked by a thrombus (blood clot) or by atherosclerosis. The effects of a stroke depend on the severity of the damage and where in the brain the stroke occurs.

Atherosclerosis is an accumulation within the arteries of fatty materials, abnormal amounts of smooth muscle, deposits of cholesterol or fibrin, or various kinds of cellular debris. These accumulations cause blood flow to be reduced (figure 44.17). The lumen (interior) of the artery may be further reduced in size by a clot that forms as a result of the atherosclerosis. In the severest cases, the artery may be blocked completely. Atherosclerosis is promoted by genetic factors, smoking, hypertension (high blood pressure), and high blood cholesterol levels. Diets low in cholesterol and saturated fats (from which cholesterol can be made) can help lower the level of blood cholesterol, and therapy for hypertension can reduce that risk factor. Stopping smoking, however, is the single most effective action a smoker can take to reduce the risk of atherosclerosis.

Arteriosclerosis, or hardening of the arteries, occurs when calcium is deposited in arterial walls. It tends to occur when atherosclerosis is severe. Not only do such arteries have restricted blood flow, but they also lack the ability to expand as normal arteries do to accommodate the volume of blood pumped out by the heart. This inflexibility forces the heart to work harder.

Cardiac output depends on the rate of the heartbeat and how much blood is ejected per beat. Blood flow is regulated by the degree of constriction of the arteries, which affects the resistance to flow. Blood pressure is influenced by blood volume; the volume of water retained in the vascular system is regulated by hormones that act on the kidneys and blood vessels.

44.4 Respiration has evolved to maximize the rate of gas diffusion.

How Animals Maximize the Efficiency of Respiration

Connecting Internal Cellular Respiration with External Body System Respiration

One of the major physiological challenges facing all multicellular animals is the acquisition and distribution of oxygen and the disposal of carbon dioxide (figure 44.18). For most multicellular animals, including the vertebrates, gas exchange requires special respiratory organs, which provide intimate contact between the gases in the external environment and the circulatory system. As we have previously discussed (see section 44.2), one of the primary functions of the circulatory system is acquisition, distribution, and removal of gases to and from tissues throughout the body. This is necessary because animals extract energy from food molecules using the biochemical process called cellular respiration, during which oxygen is needed by cells to serve as the final electron acceptor of the electron transport chain and carbon dioxide is the product of the oxidation of organic molecules during pyruvate oxidation and the Krebs cycle. While the term cellular respiration pertains to the use of oxygen and the production of carbon dioxide at the cellular level, the general term *respiration* describes the uptake of oxygen from the environment and the disposal of carbon dioxide into the environment at the body system level. Cellular respiration is sometimes referred to as internal respiration, and gas exchange at the body system level is sometimes called external respiration. Communication between internal respiration and external respiration is provided by the circulatory system.

Respiration at the body system level involves a host of processes not found at the cellular level, ranging from the mechanics of breathing to the exchange of oxygen and carbon dioxide in respiratory organs. Invertebrates display a wide variety of respiratory organs, including the epithelium, tracheae (systems of tubes), and gills. Some vertebrates, such as fish and larval amphibians, also use gills, while other amphibians use their skin or epithelia either as a supplemental or primary external respiratory organ. Many adult amphibians, reptiles, birds, and mammals have lungs to perform external respiration. In both aquatic and terrestrial animals, oxygen diffuses into, and carbon dioxide diffuses out of, the blood in these highly vascularized respiratory organs. This blood is then transported throughout the animal body, servicing all the body tissues other than the gills or lungs. In the body tissues, the direction of gas diffusion is the reverse of that observed in the respiratory organs. The mechanics, design, and evolution of respiratory systems, along with the principles of gas diffusion between the blood and tissues, are the subjects of the following pages.

FIGURE 44.18
Elephant seals are respiratory champions. Diving to depths greater than those reached by all other marine animals, including sperm whales and sea turtles, elephant seals can hold their breath for over two hours, descend and ascend rapidly in the water, and endure repeated dives without suffering any apparent respiratory distress.

The Principles of Gas Exchange in Animals

Respiration involves the diffusion of gases across plasma membranes. Because plasma membranes must be surrounded by water to be stable, the external environment in gas exchange is always aqueous. This is true even in terrestrial animals; in these cases, oxygen from air dissolves in a thin layer of fluid that covers the respiratory surfaces, such as the alveoli in lungs.

In vertebrates, the gases diffuse into the aqueous layer covering the epithelial cells that line the respiratory organs. The diffusion process is passive, driven only by the differ-

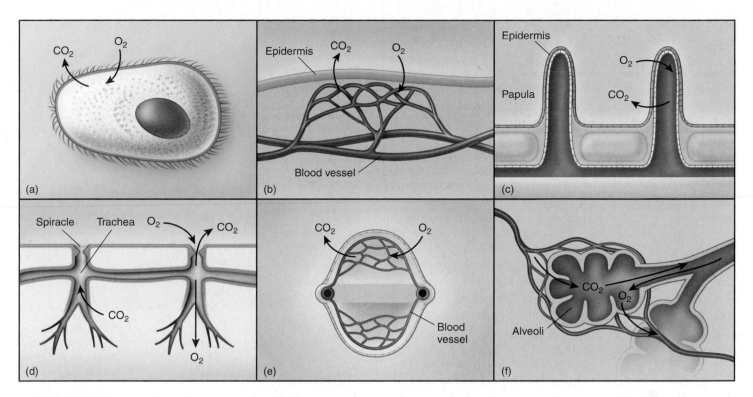

FIGURE 44.19

Gas exchange may take place in a variety of ways. (*a*) Gases diffuse directly into single-celled organisms. (*b*) Amphibians and many other animals respire across their skin. (*c*) Echinoderms have protruding papulae, which provide an increased respiratory surface area. (*d*) Insects respire through an extensive tracheal system. (*e*) The gills of fishes provide a very large respiratory surface area and countercurrent exchange. (*f*) The alveoli in mammalian lungs provide a large respiratory surface area but do not permit countercurrent exchange.

ence in O_2 and CO_2 concentrations on the two sides of the membranes. In general, the rate of diffusion between two regions is governed by a relationship known as **Fick's Law of Diffusion:**

$$R = D \times A \frac{\Delta p}{d}$$

In this equation,

R = the rate of diffusion—the amount of oxygen or carbon dioxide diffusing per unit of time;

D = the diffusion constant;

A = the area over which diffusion takes place;

Δp = the difference in concentration (for gases, the difference in their partial pressures) between the interior of the organism and the external environment;

d = the distance across which diffusion takes place.

Major changes in the mechanism of respiration that have occurred during the evolution of animals (figure 44.19) have tended to optimize the rate of diffusion, R. By inspecting Fick's Law, you can see that natural selection can optimize R by favoring changes that (1) increase the surface area, A; (2) decrease the distance, d; or (3) increase the concentration difference, as indicated by Δp. The evolution of respiratory systems has involved changes in all of these factors.

How Animals Maximize the Rate of Diffusion

The levels of oxygen required by oxidative metabolism cannot be obtained by diffusion alone over distances greater than about 0.5 millimeter. This restriction severely limits the size of organisms that obtain their oxygen entirely by diffusion directly from the environment. Protists are small enough that such diffusion can be adequate (see figure 44.19*a*), but most multicellular animals are much too large.

Most of the more primitive phyla of invertebrates lack special respiratory organs, but they have developed means of improving the movement of water over respiratory structures. In a number of different ways, many of which involve beating cilia, these organisms create a *water current* that continuously replaces the water over the respiratory surfaces. Because of this continuous replenishment with water containing fresh oxygen, *the external oxygen concentration does not decrease along the diffusion pathway*. Although each oxygen molecule that passes into the organism has been removed from the surrounding water, new water continuously replaces the oxygen-depleted water. This increases the rate of diffusion by maximizing the concentration difference—the Δp of the Fick equation.

All of the more advanced invertebrates (mollusks, arthropods, echinoderms), as well as vertebrates, possess respiratory organs that increase the surface area available for diffusion and bring the external environment (either water or air) close to the internal fluid, which is usually circulated throughout the body. The respiratory organs thus increase the rate of diffusion by maximizing surface area and decreasing the distance the diffusing gases must travel (the A and d factors, respectively, in the Fick equation).

Atmospheric Pressure and Partial Pressures. Dry air contains 78.09% nitrogen, 20.95% oxygen, 0.93% argon and other inert gases, and 0.03% carbon dioxide. Convection currents cause air to maintain a constant composition to altitudes of at least 100 kilometers, although the *amount* (number of molecules) of air that is present decreases with altitude (figure 44.20).

Imagine a column of air extending from the ground to the limits of the atmosphere. All of the gas molecules in this column experience the force of gravity, so they have weight and can exert pressure. If this column were on top of one end of a U-shaped tube of mercury at sea level, it would exert enough pressure to raise the other end of the tube 760 millimeters under a set of specified, standard conditions (figure 44.20). An apparatus that measures air pressure is called a barometer, and 760 mm Hg (millimeters of mercury) is the barometric pressure of the air at sea level. A pressure of 760 mm Hg is also defined as **one atmosphere** of pressure.

Each type of gas contributes to the total atmospheric pressure according to its fraction of the total molecules present. That fraction contributed by a gas is called its **partial pressure** and is indicated by P_{N_2}, P_{O_2}, P_{CO_2}, and so on. The total pressure is the sum of the partial pressures of all gases present. For dry air, the partial pressures are calculated simply by multiplying the fractional composition of each gas in the air by the atmospheric pressure. Thus, at sea level, the partial pressures of N_2, O_2, and CO_2 are:

$$P_{N_2} = 760 \times 79.02\% = 600.6 \text{ mm Hg}$$
$$P_{O_2} = 760 \times 20.95\% = 159.2 \text{ mm Hg}$$
$$P_{CO_2} = 760 \times 0.03\% = 0.2 \text{ mm Hg}$$

Humans do not survive long at altitudes above 6000 meters. Although the air at these altitudes still contains 20.95% oxygen, the atmospheric pressure is only about 380 mm Hg, so its P_{O_2} is only 80 mm Hg (380 × 20.95%), only half the amount of oxygen available at sea level.

Gills as Respiratory Structures in Aquatic Vertebrates

Aquatic respiratory organs increase the diffusion surface area by extensions of tissue, called *gills*, that project out into the water. Gills can be simple, as in the papulae of echinoderms (see figure 44.19c), or complex, as in the highly convoluted gills of fish (see figure 44.19e). The great increase in diffusion surface area provided by gills enables aquatic

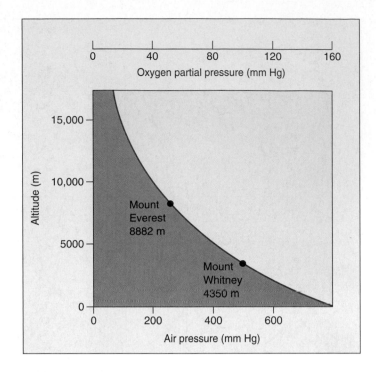

FIGURE 44.20
The relationship between air pressure and altitude above sea level. At the high altitudes characteristic of mountaintops, air pressure is much less than at sea level. At the top of Mount Everest, the world's highest mountain, the air pressure is only one-third that at sea level.
What is the difference in the percentage of oxygen between Mount Everest and Mount Whitney?

organisms to extract far more oxygen from water than would be possible from their body surface alone.

External gills (gills that are not enclosed within body structures) provide a greatly increased surface area for gas exchange. Examples of vertebrates with external gills are the larvae of many fish and amphibians, as well as developmentally arrested (*neotenic*) amphibian larvae that remain permanently aquatic, such as the axolotl. One of the disadvantages of external gills is that they must constantly be moved because the oxygen in stagnant water would quickly become depleted as it diffused from the water into the blood passing through the gills. The highly branched gills, however, offer significant resistance to movement, making this form of respiration ineffective except in smaller animals. Another disadvantage is that external gills are easily damaged. The thin epithelium required for gas exchange does not provide the normal protective external layer of skin.

Other types of aquatic animals evolved specialized *branchial chambers*, which provide a means of pumping water past stationary gills. Mollusks, for example, have an internal *mantle cavity* that opens to the outside and contains the gills. Contraction of the muscular walls of the mantle cavity draws water in and then expels it. In crustaceans, the branchial chamber lies between the bulk of the body and

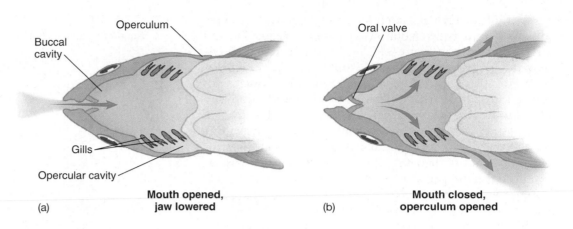

FIGURE 44.21
How most bony fishes respire. The gills are suspended between the buccal (mouth) cavity and the opercular cavity. Respiration occurs in two stages. (*a*) The oral valve in the mouth is opened and the jaw is depressed, drawing water into the buccal cavity while the opercular cavity is closed. (*b*) The oral valve is closed and the operculum is opened, drawing water through the gills to the outside.

(a) Mouth opened, jaw lowered

(b) Mouth closed, operculum opened

the hard exoskeleton of the animal. This chamber contains gills and opens to the surface beneath a limb. Movement of the limb draws water through the branchial chamber, thus creating currents over the gills.

The Gills of Bony Fishes. The gills of bony fishes are located between the *buccal* (mouth) *cavity* and the *opercular cavities* (figure 44.21). The buccal cavity can be opened and closed by opening and closing the mouth, and the opercular cavity can be opened and closed by movements of the **operculum,** or gill cover. The two sets of cavities function as pumps that expand alternately to move water into the mouth, through the gills, and out of the fish through the open operculum. Water is brought into the buccal cavity by lowering the jaw and floor of the mouth, and then moves through the gills toward the opening of the operculum, which lies posterior to the opercular cavity. The lower pressure in the opercular cavity causes water to move in the correct direction across the gills, and tissues that act as valves ensure that the movement is one-way.

Some fishes that swim continuously, such as tuna, have practically immobile opercula. These fishes swim with their mouths partly open, constantly forcing water over the gills in a form of **ram ventilation.** Most bony fishes, however, have flexible gill covers that permit a pumping action. For example, the remora, a fish that rides "piggyback" on sharks, uses ram ventilation while the shark is swimming, but uses the pumping action of its opercula when the shark stops swimming.

There are four **gill arches** on each side of the fish head. Each gill arch is composed of two rows of *gill filaments*, and each gill filament contains thin membranous plates, or *lamellae*, that project out into the flow of water (figure 44.22). Water flows past the lamellae in one direction only. Within each lamella, blood flows in a direction that is *opposite* the direction of water movement. This arrangement is called **countercurrent flow,** and it acts to maximize the oxygenation of the blood by increasing the concentration gradient of oxygen along the pathway for diffusion, increasing Δp in Fick's Law of Diffusion.

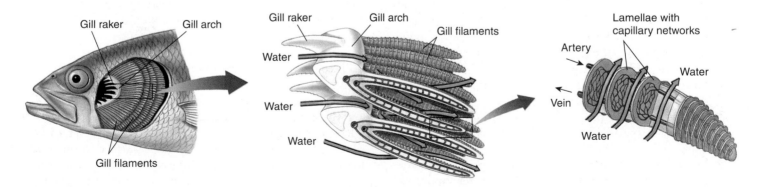

FIGURE 44.22
Structure of a fish gill. Water passes from the gill arch over the filaments (from left to right in the diagram). Water always passes the lamellae in a direction opposite to the direction of blood flow through the lamellae. The success of the gill's operation critically depends on this countercurrent flow of water and blood.

The advantages of a countercurrent flow system were discussed earlier in relation to temperature regulation and are again shown in figure 44.23*a*. Blood low in oxygen enters the back of the lamella, where it comes in close proximity to water that has already had most of its oxygen removed as it flowed through the lamella in the opposite direction. However, the water still has a higher oxygen concentration than the blood at this point, so oxygen diffuses from the water to the blood. As the blood flows toward the front of the lamella, it runs next to water that has a still higher oxygen content, so oxygen continuously diffuses from the water to the blood. Thus, countercurrent flow ensures that a concentration gradient remains between blood and water throughout the flow. This permits oxygen to continue to diffuse all along the lamellae, so that the blood leaving the gills has nearly as high an oxygen concentration as the water entering the gills.

This concept is easier to understand if we look at what would happen if blood and water flowed in the same direction—that is, had a *concurrent flow*. The difference in oxygen concentration would be very high at the front of each lamella, where oxygen-depleted blood would meet oxygen-rich water entering the gill (figure 44.23*b*). The concentration difference would fall rapidly, however, as the water lost oxygen to the blood. Net diffusion of oxygen would cease when the oxygen concentration of the blood matched that of the water. At this point, much less oxygen would have been transferred to the blood than is the case with countercurrent flow. The flow of blood and water in a fish gill is in fact countercurrent, and because of the countercurrent exchange of gases, fish gills are the most efficient of all respiratory organs.

Respiration in Air-Breathing Animals

Despite the high efficiency of gills as respiratory organs in aquatic environments, gills were replaced in terrestrial animals for two principal reasons:

1. **Air is less buoyant than water.** The fine membranous lamellae of gills lack inherent structural strength and rely on water for their support. A fish out of water, although awash in oxygen (water contains only 5 to 10 mL O_2/L, compared with air with 210 mL O_2/L), soon suffocates because its gills collapse into a mass of tissue. This collapse greatly reduces the diffusion surface area of the gills. Unlike gills, internal air passages can remain open, because the body itself provides the necessary structural support.
2. **Water diffuses into air through evaporation.** Atmospheric air is rarely saturated with water vapor, except immediately after a rainstorm. Consequently, terrestrial organisms that are surrounded by air constantly lose water to the atmosphere. Gills would provide an enormous surface area for water loss.

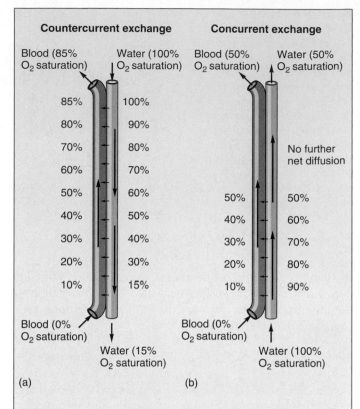

When blood and water flow in *opposite* directions (*a*), the initial oxygen concentration difference between water and blood is not large, but is sufficient for oxygen to diffuse from water to blood. As more oxygen diffuses into the blood, raising the blood's oxygen concentration, the blood encounters water with ever higher oxygen concentrations. At every point, the oxygen concentration is higher in the water, so that diffusion continues. In this example, blood attains an oxygen concentration of 85%. When blood and water flow in the *same* direction (*b*), oxygen can diffuse from the water into the blood rapidly at first, but the diffusion rate slows as more oxygen diffuses from the water into the blood, until finally the concentrations of oxygen in water and blood are equal. In this example, blood's oxygen concentration cannot exceed 50%.

FIGURE 44.23
Countercurrent exchange. This process allows for the most efficient blood oxygenation known in nature.

Two main types of respiratory organs are used by terrestrial animals, and both sacrifice respiratory efficiency to some extent in exchange for reduced evaporation. The first are the **tracheae** of insects (see chapter 33 and figure 44.19*d*). Tracheae comprise an extensive network of air-filled tubular passages connecting the surface of an insect to all portions of its body. Oxygen diffuses from these passages directly into cells, without the intervention of a circulatory system. Piping air directly from the external environment to the cells works very well in insects because their small bodies give them a high surface area-to-volume ratio. Insects prevent excessive water loss by closing the external

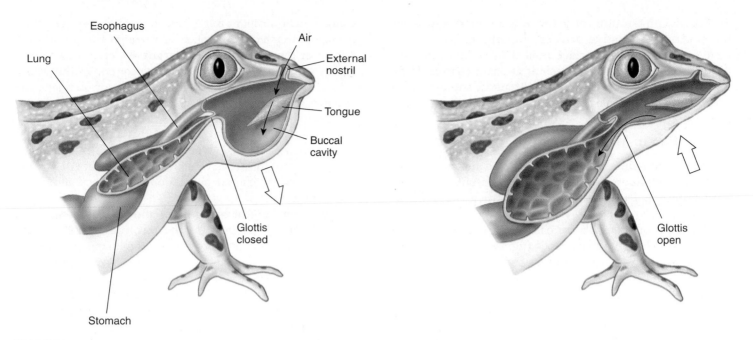

FIGURE 44.24
Amphibian lungs. Each lung of this frog is an outpouching of the gut and is filled with air by the creation of a positive pressure in the buccal cavity. The amphibian lung lacks the structures present in the lungs of other terrestrial vertebrates that provide an enormous surface area for gas exchange, and so are not as efficient as the lungs of other vertebrates.

openings, called spiracles, of the tracheae whenever their internal CO_2 levels fall below a certain point.

The other main type of terrestrial respiratory organ is the **lung.** A lung minimizes evaporation by moving air through a branched tubular passage; the air becomes saturated with water vapor before reaching the portion of the lung where a thin, wet membrane permits gas exchange. The lungs of all terrestrial vertebrates except birds use a **uniform pool** of air that is in contact with the gas exchange surface. Unlike the one-way flow of water that is so effective in the respiratory function of gills, air moves in and out by way of the same airway passages, a two-way flow system. This means that the inflowing and outflowing air follows the same tubular airways. Let us now examine the structure and function of lungs in the four classes of terrestrial vertebrates.

Respiration in Amphibians and Reptiles

Amphibians. The lungs of amphibians are formed as saclike outpouchings of the gut (figure 44.24). Although the internal surface area of these sacs is increased by folds, much less surface area is available for gas exchange in amphibian lungs than in the lungs of other terrestrial vertebrates. Each amphibian lung is connected to the rear of the oral cavity, or pharynx, and the opening to each lung is controlled by a valve, the glottis.

Amphibians do not breathe the same way other terrestrial vertebrates do. Amphibians force air into their lungs by creating a greater-than-atmospheric pressure (positive

pressure) in the air outside their lungs. They do this by filling their buccal cavity with air, closing their mouth and nostrils, and then elevating the floor of their oral cavity. This pushes air into their lungs in the same way that a pressurized tank of air is used to fill balloons. This is called positive pressure breathing; in humans, it would be analogous to forcing air into a victim's lungs by performing mouth-to-mouth resuscitation.

All other terrestrial vertebrates breathe by expanding their lungs and thereby creating a lower-than-atmospheric pressure (a negative pressure) within the lungs. This is called negative pressure breathing and is analogous to taking air into an accordion by pulling the accordion out to a greater volume. In reptiles, birds, and mammals, this is accomplished by expanding the thoracic (chest) cavity through muscular contractions, as will be described in section 44.5.

The oxygenation of amphibian blood by the lungs is supplemented by cutaneous respiration—the exchange of gases across the skin, which is wet and well vascularized in amphibians. Cutaneous respiration is actually more significant than pulmonary (lung) ventilation in frogs during winter, when their metabolisms are slow. Lung function becomes more important during the summer as the frog's metabolism increases. Although not common, some terrestrial amphibians, such as plethodontid salamanders, rely on cutaneous respiration exclusively.

Reptiles. Terrestrial reptiles have dry, tough, scaly skins that not only prevent desiccation, but also prohibit cutaneous respiration. Reptiles expand their rib cages by muscular contraction, and thereby take air into their lungs through negative pressure breathing. Their lungs have somewhat more surface area than the lungs of amphibians and so are more efficient at gas exchange. Cutaneous respiration, however, can occur in some reptiles, such as marine sea snakes.

Respiration in Mammals

Endothermic animals, such as birds and mammals, have higher metabolic rates to produce and sustain their body temperature and thus require a more efficient respiratory system.

The lungs of mammals are packed with millions of *alveoli*, tiny sacs clustered like grapes (figure 44.25). This provides each lung with an enormous surface area for gas ex-

change. Air is brought to the alveoli through a system of air passages. Inhaled air is taken in through the mouth and nose past the pharynx to the **larynx** (voice box), where it passes through an opening in the vocal cords, the *glottis*, into a tube supported by C-shaped rings of cartilage, the **trachea** (windpipe). The trachea bifurcates into right and left **bronchi** (singular, *bronchus*), which enter each lung and further subdivide into **bronchioles** that deliver the air into blind-ended sacs called **alveoli.** The alveoli are surrounded by an extremely extensive capillary network. All gas exchange between the air and blood takes place across the walls of the alveoli.

The branching of bronchioles and the vast number of alveoli combine to increase the respiratory surface area (*A* in Fick's Law) far above that of amphibians or reptiles. In humans, there are about 300 million alveoli in each of the two lungs, and the total surface area available for diffusion can be as much as 80 square meters, or about 42 times the surface area of the body. Respiration in mammals will be considered in more detail in section 44.5.

Respiration in Birds

The avian respiratory system has a unique structure that affords birds the most efficient respiration of all terrestrial vertebrates. Unlike the blind-ended alveoli in the lungs of

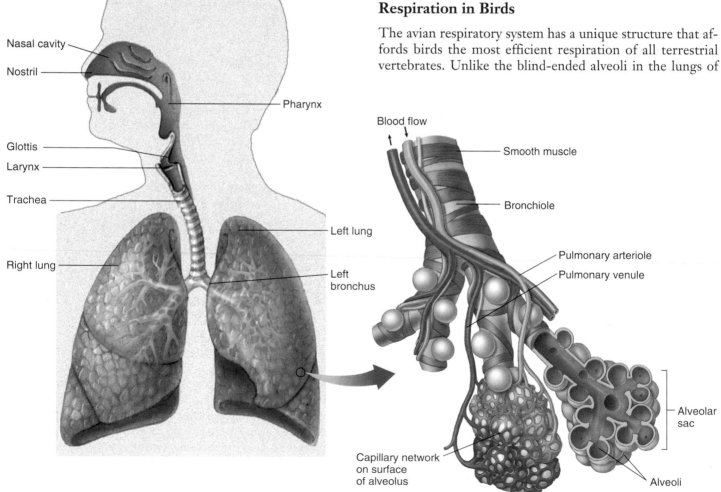

FIGURE 44.25
The human respiratory system and the structure of the mammalian lung. The lungs of mammals have an enormous surface area because of the millions of alveoli that cluster at the ends of the bronchioles. This provides for efficient gas exchange with the blood.

mammals, the bird lung channels air through tiny air vessels called parabronchi, where gas exchange occurs. Air flows through the parabronchi in one direction only; this is similar to the unidirectional flow of water through a fish gill, but markedly different from the two-way flow of air through the airways of other terrestrial vertebrates. In other terrestrial vertebrates, the inhaled fresh air is mixed with "old" oxygen-depleted air that was not exhaled from the previous breathing cycle. In birds, only fresh air enters the parabronchi of the lung, and the old air exits the lung by a different route.

The unidirectional flow of air through the parabronchi of an avian lung is achieved through the action of air sacs, which are unique to birds (figure 44.26a). There are two groups of air sacs, anterior and posterior. When they are expanded during *inspiration*, they take in air, and when they are compressed during *expiration*, they push air into and through the lungs.

If we follow the path of air through the avian respiratory system, we see that respiration occurs in two cycles (figure 44.26b). Each cycle has an inspiration and expiration phase—but the air inhaled in one cycle is not exhaled until the second cycle. Upon inspiration, both the anterior and posterior air sacs expand and take in air. The inhaled air, however, only enters the posterior air sacs; the anterior air sacs fill with air from the lungs. Upon expiration, the air forced out of the anterior air sacs is exhaled, but the air forced out of the posterior air sacs enters the lungs. This process is repeated in the second cycle, so that air flows through the lungs in one direction and is exhaled at the end of the second cycle.

The unidirectional flow of air also permits a second respiratory efficiency: The flow of blood through the avian lung runs at a 90° angle to the air flow. This **cross-current flow** is not as efficient as the 180° countercurrent flow in fish gills, but it has the capacity to extract more oxygen from the air than a mammalian lung can. Because of the unidirectional air flow in the parabronchi and cross-current blood flow, a sparrow can be active at an altitude of 6000 meters, while a mouse, which has a similar body mass and metabolic rate, cannot respire successfully at that elevation.

The exchange of oxygen and carbon dioxide between an organism and its environment is maximized by increasing the concentration gradient and the surface area and by decreasing the distance that the diffusing gases must travel. In the gills of bony fishes, blood flows in an opposite direction to the flow of water. This countercurrent flow maximizes gas exchange, making the fish's gill an efficient respiratory organ. Amphibians, reptiles, and all other terrestrial vertebrates take air into saclike lungs that provide a huge surface area for gas exchange. The avian respiratory system is the most efficient among terrestrial vertebrates because it has unidirectional air flow and cross-current blood flow through the lungs.

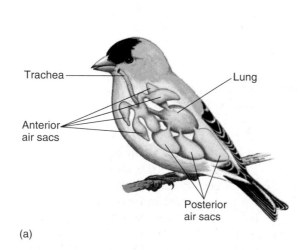

(a)

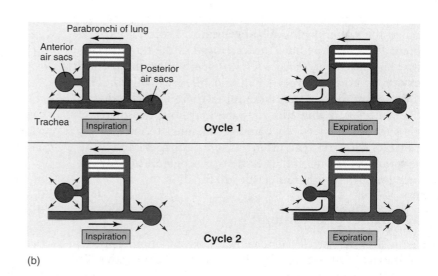

(b)

FIGURE 44.26

How a bird breathes.   (*a*) Birds have a system of air sacs, divided into an anterior group and posterior group, that extend between the internal organs and into the bones. (*b*) Breathing occurs in two cycles. *Cycle 1:* Inhaled air (shown in *red*) is drawn from the trachea into the posterior air sacs and then is exhaled into the lungs. *Cycle 2:* Air is drawn from the lungs into the anterior air sacs and then is exhaled through the trachea. Passage of air through the lungs is always in the same direction, from posterior to anterior (right to left in this diagram).

44.5 Mammalian breathing is a dynamic process.

Structures and Mechanisms of Breathing

In mammals, inspired air travels through the oral or nasal cavities, the pharynx, the trachea, bronchi, and bronchioles to reach the alveoli, where gas exchange occurs. Each alveolus is composed of an epithelium only one cell thick, and is surrounded by blood capillaries with walls that are also only one cell layer thick. There are about 30 billion capillaries in both lungs, or about 100 capillaries per alveolus. Thus, an alveolus can be visualized as a microscopic air bubble whose entire surface is bathed by blood. Because the alveolar air and the capillary blood are separated by only two cell layers, the distance between the air and blood is only 0.5 to 1.5 micrometers, allowing for the rapid exchange of gases by diffusion by decreasing d in Fick's Law.

The blood leaving the lungs, as a result of this gas exchange, normally contains a partial oxygen pressure (P_{O_2}) of about 100 millimeters of mercury. As previously discussed, the P_{O_2} is a measure of the concentration of dissolved oxygen—you can think of it as indicating the plasma oxygen. Because the P_{O_2} of the blood leaving the lungs is close to the P_{O_2} of the air in the alveoli (about 105 mm Hg), the lungs do a very effective, but not perfect, job of oxygenating the blood. After gas exchange in the systemic capillaries, the blood that returns to the right side of the heart is depleted in oxygen, with a P_{O_2} of about 40 millimeters of mercury. These changes in the P_{O_2} of the blood, as well as the changes in plasma carbon dioxide (indicated as the P_{CO_2}), are shown in figure 44.27.

The outside of each lung is covered by a thin membrane called the **visceral pleural membrane**. A second, **parietal pleural membrane** lines the inner wall of the thoracic cavity. The space between these two membranes, the **pleural cavity**, is normally very small and filled with fluid. This fluid causes the two membranes to adhere to each other in the same way a thin film of water can hold two plates of glass together, effectively coupling the lungs to the thoracic cavity. The pleural membranes package each lung separately—if one collapses due to a perforation of the membranes, the other lung can still function.

Mechanics of Breathing

As in all other terrestrial vertebrates except amphibians, air is drawn into the lungs by the creation of a negative, or subatmospheric, pressure. In accordance with *Boyle's Law*, when the volume of a given quantity of gas increases, its pressure decreases. This occurs because the volume of the thorax is increased during inspiration (inhalation), and the lungs likewise expand because of the adherence of the vis-

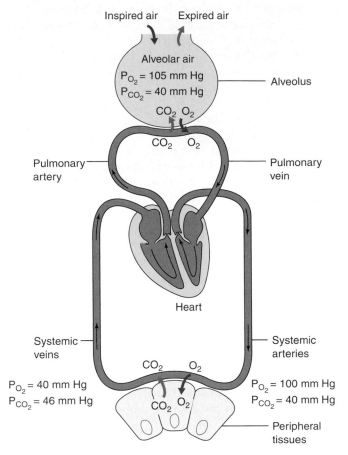

FIGURE 44.27
Gas exchange in the blood capillaries of the lungs and systemic circulation. As a result of gas exchange in the lungs, the systemic arteries carry oxygenated blood with a relatively low carbon dioxide concentration. After the oxygen is unloaded to the tissues, the blood in the systemic veins has a lowered oxygen content and an increased carbon dioxide concentration.

ceral and parietal pleural membranes. When the pressure within the lungs is lower than the atmospheric pressure, air enters the lungs.

The thoracic volume is increased through contraction of two sets of muscles: the *external intercostal muscles* and the *diaphragm*. The diaphragm is the more important of the two during quiet or resting inhalation. During inspiration, contraction of the external intercostal muscles between the ribs raises the ribs and expands the rib cage. Contraction of the diaphragm, a convex sheet of striated muscle separating the thoracic cavity from the abdominal cavity, causes the diaphragm to lower and assume a more flattened shape. This expands the volume of the thorax and lungs while it increases the pressure on the abdomen (causing the belly to protrude). You can force a deeper in-

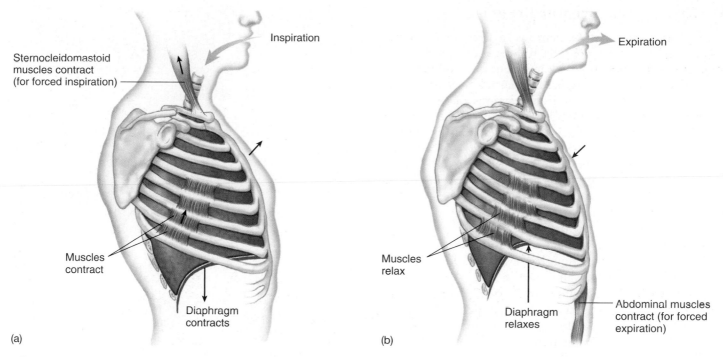

FIGURE 44.28

How a human breathes. (*a*) Inspiration. The diaphragm contracts and the walls of the chest cavity expand, increasing the volume of the chest cavity and lungs. As a result of the larger volume, air is drawn into the lungs. (*b*) Expiration. The diaphragm and chest walls return to their normal positions as a result of elastic recoil, reducing the volume of the chest cavity and forcing air out of the lungs through the trachea. Note that inspiration can be forced by contracting accessory respiratory muscles (such as the sternocleidomastoid), and expiration can be forced by contracting abdominal muscles.

spiration both by contracting the intercostals further and by contracting other muscles that insert on the sternum or rib cage and expand the thoracic cavity and lungs to a greater extent (figure 44.28*a*).

The thorax and lungs have a degree of *elasticity*—that is, they tend to resist distension, and they recoil when the distending force subsides. Expansion of the thorax and lungs during inspiration places these structures under elastic tension. The relaxation of the external intercostal muscles and diaphragm produces unforced expiration, because it relieves that elastic tension and allows the thorax and lungs to recoil. You can force a greater expiration by contracting your abdominal muscles and thereby pressing the abdominal organs up against the diaphragm (figure 44.28*b*).

Breathing Measurements

A variety of terms are used to describe the volume changes of the lung during breathing. At rest, each breath moves a **tidal volume** of about 500 milliliters of air into and out of the lungs. About 150 milliliters of the tidal volume are contained in the tubular passages (trachea, bronchi, and bronchioles), where no gas exchange occurs. The air in this *anatomical dead space* mixes with fresh air during inspiration. This is one of the reasons

respiration in mammals is not as efficient as in birds, where air flow through the lungs is one-way.

The maximum amount of air that can be expired after a forceful, maximum inspiration is called the **vital capacity.** This measurement, which averages 4.6 liters in young men and 3.1 liters in young women, can be clinically important, because an abnormally low vital capacity may indicate damage to the alveoli in various pulmonary disorders. For example, in **emphysema,** a potentially fatal condition usually caused by cigarette smoking, vital capacity is reduced as the alveoli are progressively destroyed.

A person normally breathes at a rate and depth that properly oxygenate the blood and remove carbon dioxide, keeping the blood P_{O_2} and P_{CO_2} within a normal range. If breathing is insufficient to maintain normal blood gas measurements (a rise in the blood P_{CO_2} is the best indicator), the person is **hypoventilating.** If breathing is excessive for a particular metabolic rate, so that the blood P_{CO_2} is abnormally lowered, the person is said to be **hyperventilating.** Perhaps surprisingly, the increased breathing that occurs during moderate exercise is not necessarily hyperventilation, because the faster breathing is matched to the faster metabolic rate, and blood gas measurements remain normal. Next, we will describe how breathing is regulated to keep pace with metabolism.

Mechanisms That Regulate Breathing

Each breath is initiated by neurons in a *respiratory control center* located in the medulla oblongata, a part of the brain stem (see chapter 45). These neurons send impulses to the diaphragm and external intercostal muscles, stimulating them to contract, and contractions of these muscles expand the chest cavity, causing inspiration. When these neurons stop producing impulses, the inspiratory muscles relax and expiration occurs. Although the muscles of breathing are skeletal muscles, they are usually controlled automatically. This control can be voluntarily overridden, however, as in hypoventilation (breath holding) or hyperventilation.

A proper rate and depth of breathing is required to maintain the blood oxygen and carbon dioxide levels in the normal range. Thus, although the automatic breathing cycle is driven by neurons in the brain stem, these neurons must be responsive to changes in blood P_{O_2} and P_{CO_2} in order to maintain homeostasis. You can demonstrate this mechanism by simply holding your breath. Your blood carbon dioxide level immediately rises, and your blood oxygen level falls. After a short time, the urge to breathe induced by the changes in blood gases becomes overpowering. This is due primarily to the rise in blood carbon dioxide, as indicated by a rise in P_{CO_2}, rather than to the fall in oxygen levels.

A rise in P_{CO_2} causes an increased production of carbonic acid (H_2CO_3), which is formed from carbon dioxide and water and acts to lower the blood pH (carbonic acid dissociates into HCO_3^- and H^+, thereby increasing blood H^+ concentration). A fall in blood pH stimulates neurons in the **aortic** and **carotid bodies,** which are sensory structures known as *peripheral chemoreceptors* in the aorta and the carotid artery. These receptors send impulses to the respiratory control center in the medulla oblongata, which then stimulates increased breathing. The brain also contains chemoreceptors, but they cannot be stimulated by blood H^+ because the blood is unable to enter the brain. After a brief delay, however, the increased blood P_{CO_2} also causes a decrease in the pH of the cerebrospinal fluid (CSF) bathing the brain. This occurs because CO_2 does cross the blood brain barrier and combines with water in the CSF to form carbonic acid. Hydrogen ions (H^+) liberated by this acid stimulate the **central chemoreceptors** in the brain (figure 44.29).

The peripheral chemoreceptors are responsible for the immediate stimulation of breathing when the blood P_{CO_2} rises, but this immediate stimulation only accounts for about 30% of increased ventilation. The central chemoreceptors are responsible for the sustained increase in ventilation if P_{CO_2} remains elevated. The increased respiratory rate then acts to eliminate the extra CO_2, bringing the blood pH back to normal (figure 44.30).

A person cannot voluntarily hyperventilate for too long. The decrease in plasma P_{CO_2} and increase in pH of plasma

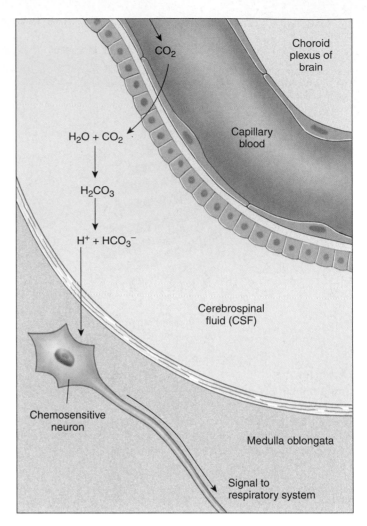

FIGURE 44.29
The effect of blood CO_2 on cerebrospinal fluid (CSF).
Changes in the pH of the CSF are detected by chemosensitive neurons in the brain that help regulate breathing.

and CSF caused by hyperventilation extinguish the reflex drive to breathe. They also lead to constriction of cerebral blood vessels, causing dizziness. People can hold their breath longer if they hyperventilate first, because it takes longer for the CO_2 levels to build back up, not because hyperventilation increases the P_{O_2} of the blood. Actually, in people with normal lungs, P_{O_2} becomes a significant stimulus for breathing only at high altitudes, where the P_{O_2} is low. Low P_{O_2} can also stimulate breathing in patients with emphysema, where the lungs are so damaged that blood CO_2 can never be adequately eliminated.

Humans inspire by contracting muscles that insert on the rib cage and by contracting the diaphragm. Expiration is produced primarily by muscle relaxation and elastic recoil. Breathing serves to keep the blood gases and pH in the normal range and is under the reflex control of peripheral and central chemoreceptors.

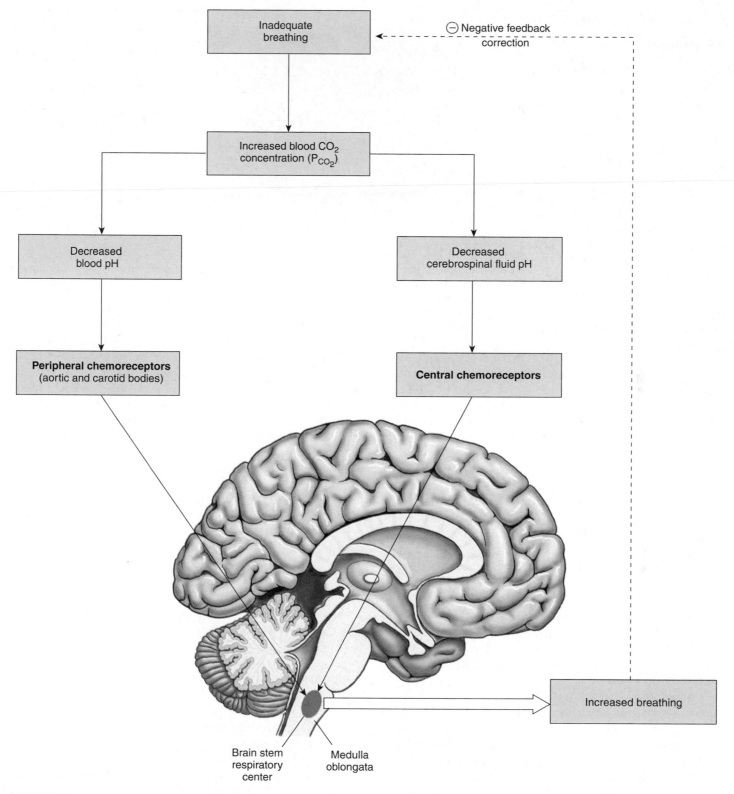

FIGURE 44.30

The regulation of breathing by chemoreceptors. Peripheral and central chemoreceptors sense a fall in the pH of blood and cerebrospinal fluid, respectively, when the blood carbon dioxide levels rise as a result of inadequate breathing. In response, they stimulate the respiratory control center in the medulla oblongata, which directs an increase in breathing. As a result, the blood carbon dioxide concentration is returned to normal, completing the negative feedback loop.

Hemoglobin and Gas Transport

When oxygen diffuses from the alveoli into the blood, its journey is just beginning. The circulatory system delivers oxygen to tissues for respiration and carries away carbon dioxide. The transport of oxygen and carbon dioxide by the blood is itself an interesting and physiologically important process.

The amount of oxygen that can be dissolved in the blood plasma depends directly on the P_{O_2} of the air in the alveoli, as we explained earlier. When the lungs are functioning normally, the blood plasma leaving the lungs has almost as much dissolved oxygen as is theoretically possible, given the P_{O_2} of the air. Because of oxygen's low solubility in water, however, blood plasma can contain a maximum of only about 3 milliliters of O_2 per liter. Yet, whole blood carries almost 200 milliliters of O_2 per liter! Most of the oxygen is bound to molecules of hemoglobin inside red blood cells.

Hemoglobin is a protein composed of four polypeptide chains and four organic compounds called *heme groups*. At the center of each heme group is an atom of iron, which can bind to a molecule of oxygen (figure 44.31). Thus, each hemoglobin molecule can carry up to four molecules of oxygen. Hemoglobin loads up with oxygen in the lungs, forming **oxyhemoglobin.** This molecule has a bright red, tomato juice color. As blood passes through capillaries in the rest of the body, some of the oxyhemoglobin releases oxygen, becoming **deoxyhemoglobin.** Deoxyhemoglobin has a dark red color (the color of blood that is collected from the veins of blood donors), but it imparts a bluish tinge to tissues. Because of these color changes, vessels that carry oxygenated blood are always shown in artwork with a red color, and vessels that carry oxygen-depleted blood are indicated with a blue color.

Hemoglobin is an ancient protein that is not only the oxygen-carrying molecule in all vertebrates, but is also used as an oxygen carrier by many invertebrates, including annelids, mollusks, echinoderms, flatworms, and even some protists. Many other invertebrates, however, employ different oxygen carriers, such as *hemocyanin*. In hemocyanin, the oxygen-binding atom is copper instead of iron. Hemocyanin is not found in blood cells, but is instead dissolved in the circulating fluid (hemolymph) of invertebrates.

Oxygen Transport

The P_{O_2} of the air within alveoli at sea level is approximately 105 millimeters of mercury (mm Hg), which is less

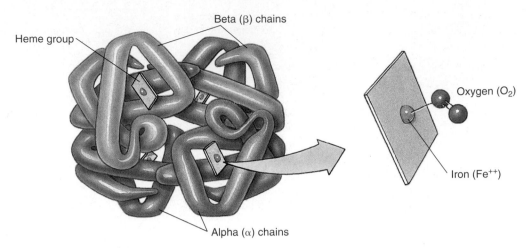

Heme group · Beta (β) chains · Oxygen (O_2) · Iron (Fe^{++}) · Alpha (α) chains

FIGURE 44.31
The structure of the hemoglobin protein. Hemoglobin consists of four polypeptide chains— two alpha (α) chains and two beta (β) chains. Each chain is associated with a heme group, and each heme group has a central iron atom, which can bind to a molecule of O_2.

than the P_{O_2} of the atmosphere (approximately 159 mm Hg) because of the mixing of freshly inspired air with "old" air in the anatomical dead space of the respiratory system. The P_{O_2} of the blood leaving the alveoli is slightly less than this, about 100 mm Hg, because the blood plasma is not completely saturated with oxygen due to slight inefficiencies in lung function. At a blood P_{O_2} of 100 mm Hg, approximately 97% of the hemoglobin within red blood cells is in the form of oxyhemoglobin—indicated as a percent oxyhemoglobin saturation of 97%.

As the blood travels through the systemic blood capillaries, oxygen leaves the blood and diffuses into the tissues. Consequently, the blood that leaves the tissue in the veins has a P_{O_2} that is decreased (in a resting person) to about 40 mm Hg. At this lower P_{O_2}, the percent saturation of hemoglobin is only 75%. A graphic representation of these changes is called an oxyhemoglobin dissociation curve (figure 44.32). In a person at rest, therefore, 22% (97% minus 75%) of the oxyhemoglobin has released its oxygen to the tissues. Put another way, roughly one-fifth of the oxygen is unloaded in the tissues, leaving four-fifths of the oxygen in the blood as a reserve.

This large reserve of oxygen serves an important function. It enables the blood to supply the body's oxygen needs during exercise as well as at rest. During exercise, the muscles' accelerated metabolism uses more oxygen from the capillary blood and thus decreases the venous blood P_{O_2}. For example, the P_{O_2} of the venous blood could drop to 20 mm Hg; in this case, the percent saturation of hemoglobin will be only 35% (figure 44.32). Because arterial blood still contains 97% oxyhemoglobin (ventilation increases proportionately with exercise), the amount of oxygen unloaded is now 62% (97% minus 35%), instead of the 22% at rest. In addition to this function, the oxygen reserve also ensures that the blood contains enough oxygen to maintain life for four to five minutes if breathing is interrupted or if the heart stops pumping.

Oxygen transport in the blood is affected by other conditions. The CO_2 produced by metabolizing tissues as a product of aerobic respiration combines with H_2O to form carbonic acid (H_2CO_3). H_2CO_3 dissociates into bicarbonate and H^+, thereby lowering blood pH. This reaction occurs primarily inside red blood cells, where the lowered pH reduces hemoglobin's affinity for oxygen and thus causes it to release oxygen more readily. The effect of pH on hemoglobin's affinity for oxygen, known as the *Bohr effect*, is the result of H^+ and CO_2 binding to hemoglobin and is shown graphically by a shift of the oxyhemoglobin dissociation curve to the right (figure 44.33*a*). Increasing temperature has a similar effect on hemoglobin's affinity for oxygen (figure 44.33*b*). Because skeletal muscles produce carbon dioxide more rapidly during exercise and active muscles produce heat, the blood unloads a higher percentage of the oxygen it carries during exercise.

Carbon Dioxide Transport

The systemic capillaries deliver oxygen to the tissues and remove carbon dioxide. About 8% of the CO_2 in blood is simply dissolved in plasma; another 20% is bound to hemoglobin. (Because CO_2 binds to the protein portion of hemoglobin and not to the heme irons, it does not compete with oxygen; however, it does cause hemoglobin's shape to change, lowering its affinity for oxygen.) The remaining 72%

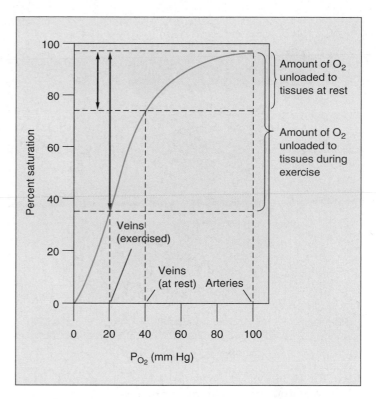

FIGURE 44.32
The oxyhemoglobin dissociation curve. Hemoglobin combines with O_2 in the lungs, and this oxygenated blood is carried by arteries to the body cells. After oxygen is removed from the blood to support cellular respiration, the blood entering the veins contains less oxygen.
How would you determine how much oxygen was unloaded to the tissues?

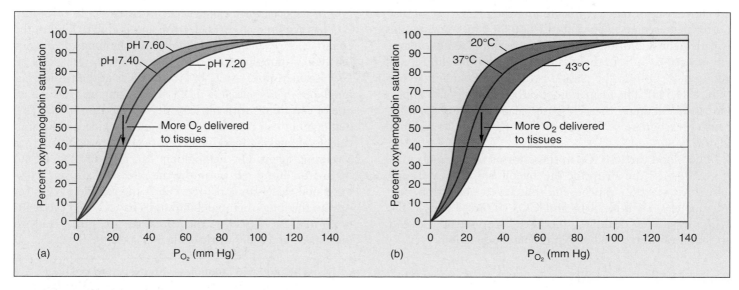

FIGURE 44.33
The effect of pH and temperature on the oxyhemoglobin dissociation curve. (*a*) Lower blood pH and (*b*) higher blood temperatures shift the oxyhemoglobin dissociation curve to the right, facilitating oxygen unloading. In this example, this can be seen as a lowering of the oxyhemoglobin percent saturation from 60% to 40%, indicating that the difference of 20% more oxygen is unloaded to the tissues.
What effect does high blood pressure have on oxygen unloading to the tissues during exercise?

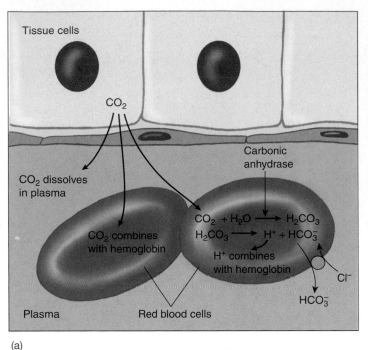

(a)

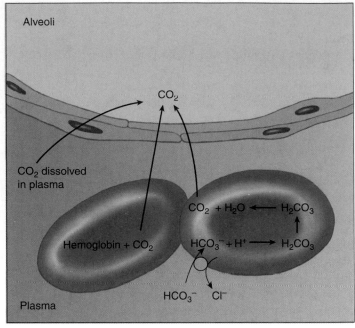

(b)

FIGURE 44.34
The transport of carbon dioxide by the blood. (*a*) Passage into bloodstream. CO_2 is transported in three ways: dissolved in plasma, bound to the protein portion of hemoglobin, and as carbonic acid and bicarbonate, which form in the red blood cells. (*b*) Removal from bloodstream. When the blood passes through the pulmonary capillaries, these reactions are reversed so that CO_2 gas is formed, which is exhaled.

of the CO_2 diffuses into the red blood cells, where the enzyme carbonic anhydrase catalyzes the combination of CO_2 with water to form carbonic acid (H_2CO_3). Carbonic acid dissociates into bicarbonate (HCO_3^-) and hydrogen (H^+) ions. The H^+ binds to deoxyhemoglobin, and the bicarbonate moves out of the erythrocyte into the plasma via a transporter that exchanges one chloride ion for a bicarbonate (this is called the "chloride shift"). This reaction removes large amounts of CO_2 from the plasma, facilitating the diffusion of additional CO_2 into the plasma from the surrounding tissues (figure 44.34*a*). The formation of carbonic acid is also important in maintaining the acid-base balance of the blood, because bicarbonate serves as the major buffer of the blood plasma.

The blood carries CO_2 in these forms to the lungs. The lower P_{CO_2} of the air inside the alveoli causes the carbonic anhydrase reaction to proceed in the reverse direction, converting H_2CO_3 into H_2O and CO_2 (figure 44.34*b*). The CO_2 diffuses out of the red blood cells and into the alveoli, so that it can leave the body in the next exhalation.

Nitric Oxide Transport

Hemoglobin also has the ability to hold and release nitric oxide gas (NO). Although a noxious gas in the atmosphere, nitric oxide has an important physiological role in the body and acts on many kinds of cells to change their shapes and functions. For example, in blood vessels, the presence of

NO causes the blood vessels to expand because it relaxes the surrounding muscle cells (see chapter 7). Thus, blood flow and blood pressure are regulated by the amount of NO released into the bloodstream.

A current hypothesis proposes that hemoglobin carries NO in a special form called super nitric oxide. In this form, NO has acquired an extra electron and is able to bind to the amino acid cysteine in hemoglobin. In the lungs, hemoglobin that is dumping CO_2 and picking up O_2 also picks up NO as super nitric oxide. In blood vessels at the tissues, hemoglobin that is releasing its O_2 and picking up CO_2 can do one of two things with nitric oxide. To increase blood flow, hemoglobin can release the super nitric oxide as NO into the blood, making blood vessels expand because NO acts as a relaxing agent. Or, hemoglobin can trap any excess NO on its iron atoms left vacant by the release of oxygen, causing blood vessels to constrict. When the red blood cells return to the lungs, hemoglobin dumps its CO_2 and the regular form of NO bound to the iron atoms. It is then ready to pick up O_2 and super nitric oxide and continue the cycle.

Deoxyhemoglobin combines with oxygen in the lungs to form oxyhemoglobin, which dissociates in the tissue capillaries to release its oxygen. Carbon dioxide is transported in the blood in three ways: dissolved in the plasma, bound to hemoglobin, and the majority as bicarbonate in the plasma following an enzyme-catalyzed reaction in the red blood cells.

Concept Review

For interactive testing, visit the Online Learning Center with PowerWeb at www.mhhe.com/Raven7

44.1 Circulatory systems are the transportation highways of the animal body.

Open and Closed Circulatory Systems

- In an open circulatory system, there is no distinction between the circulating fluid and the extracellular fluid of body tissues. In a closed circulatory system, the circulating fluid is always enclosed within blood vessels that transport it away from, and back to, a pump. (p. 908)
- The three principal functions of the vertebrate circulatory system are transportation, regulation, and protection. (pp. 908–909)

Blood

- The fluid portion of blood is plasma, and it contains metabolites, wastes, and hormones, as well as ions and proteins. (p. 910)
- Red blood cells (erythrocytes) function in oxygen transport, white blood cells (leukocytes) function in immunological defenses, and platelets help in blood clotting. (pp. 910–911)

Characteristics of Blood Vessels

- Blood leaves the heart through arteries and returns to the heart in veins. (p. 912)
- As blood passes through capillaries, gases and metabolites are exchanged with interstitial fluids and body cells. (p. 912)
- The lymphatic system returns interstitial fluid (lymph) to the vascular system in an open system. The lymphatic system is composed of lymphatic capillaries, lymphatic vessels, lymph nodes, and lymphatic organs such as the spleen and thymus. (pp. 913–914)

44.2 The circulatory and respiratory systems evolved together in vertebrates.

Increasing Body Size Requires Special Circulatory and Respiratory Adaptations

- As animal body size and physiological complexity increase, so does the need for more efficient mechanisms to deliver nutrients and oxygen to the tissues and to remove wastes and carbon dioxide from the tissues. (p. 915)
- The development of gills by fishes required a more efficient pump, and thus a true chamber-pump heart evolved. (p. 915)
- The advent of lungs in reptiles and amphibians results in two circulations: pulmonary circulation between the heart and the lungs, and systemic circulation between the heart and the rest of the body. (p. 916)
- Amphibians can obtain additional oxygen through cutaneous respiration. (p. 916)
- Mammals, birds, and crocodiles have a four-chambered heart with two separate atria and two separate ventricles, and there is no mixing of oxygenated and deoxygenated blood. (pp. 916–917)

44.3 The cardiac cycle drives the cardiovascular system.

The Cardiac Cycle

- The cardiac cycle is composed of two periods: systole (ventricles contract) and diastole (ventricles relax). (p. 918)
- Systolic pressure is the peak pressure during ventricular systole, while diastolic pressure is the minimum pressure between heartbeats. (p. 918)

- The sinoatrial node acts as a pacemaker for the rest of the heart by producing spontaneous depolarizations (reversal of electrical polarity) at a faster rate than other cardiac cells. (p. 919)
- An electrocardiogram records how the cells of the heart depolarize and repolarize during the cardiac cycle. (p. 919)

Blood Flow and Blood Pressure

- Cardiac output refers to the volume of blood pumped by each ventricle per minute, and is calculated by multiplying the heart rate by the stroke volume. (p. 920)
- Blood flow is regulated by artery constriction, and blood pressure is influenced by blood volume. (p. 920)
- Blood volume regulation involves the effects of four hormones: antidiuretic hormone, aldosterone, atrial natriuretic hormone, and nitric oxide. (pp. 920–921)
- Cardiovascular disease, the leading cause of death in the United States, includes such conditions as angina pectoris, strokes, atherosclerosis, and arteriosclerosis. (p. 921)

44.4 Respiration has evolved to maximize the rate of gas diffusion.

How Animals Maximize the Efficiency of Respiration

- Increasing the concentration gradient and surface area and decreasing the distance diffusing gases must travel maximize the exchange of oxygen and carbon dioxide between an organism and its environment. (pp. 923–924)
- The countercurrent flow of blood and water in bony fish maximizes gas exchange. (pp. 925–926)
- Gills were replaced in terrestrial animals because air is less buoyant than water, and water diffuses into the air through evaporation. (p. 926)
- Terrestrial vertebrates take air into saclike lungs that provide a large surface area for gas exchange. (p. 927)
- The respiration system in birds is highly efficient because it has unidirectional air flow and cross-current blood flow through the lungs. (pp. 928–929)

44.5 Mammalian breathing is a dynamic process.

Structures and Mechanisms of Breathing

- When the pressure within the lungs is lower than the atmospheric pressure, air enters the lungs. (p. 930)
- The maximum amount of air that can be expired after a maximum inspiration is the vital capacity. (p. 931)
- Humans breathe in (inspire) by contracting intercostal muscles between the ribs and contracting the diaphragm. They breathe out (expire) by relaxing the intercostal muscles and the diaphragm. (p. 932)
- A proper rate and depth of breathing is required to maintain the blood oxygen and carbon dioxide levels in normal ranges. (p. 932)

Hemoglobin and Gas Transport

- Hemoglobin takes up oxygen in the lungs, forming oxyhemoglobin, which releases oxygen as blood passes through capillaries in the rest of the body, forming deoxyhemoglobin. (p. 934)
- Carbon dioxide is transported in the blood via three methods: dissolved in plasma, bound to hemoglobin, and diffused into red blood cells where carbonic acid is formed (pp. 935–936)

Self Test

1. Which of the following statements is false?
   a. Only arteries carry oxygenated blood.
   b. Both arteries and veins have a layer of smooth muscle.
   c. Both arteries and veins branch out into capillary beds.
   d. Sphincters regulate blood flow through capillaries.

2. The lymphatic system is like the circulatory system in that they both
   a. have nodes that filter out pathogens.
   b. have a network of arteries.
   c. have capillaries.
   d. are closed systems.

3. A molecule of CO_2 that is generated in the cardiac muscle of the left ventricle would *not* pass through which of the following structures before leaving the body?
   a. right atrium
   b. left atrium
   c. right ventricle
   d. left ventricle

4. In vertebrate hearts, atria contract from the top, while ventricles contract from the bottom. How is this accomplished?
   a. The depolarization from the sinoatrial node proceeds across the atria from the top, while the depolarization from the atrioventricular node is carried to the bottom of the ventricles before it emanates over the ventricular tissue.
   b. The depolarization from the sinoatrial node is initiated from motor neurons coming down from our brain, while the depolarization from the atrioventricular node is initiated from motor neurons coming up from our spinal cord.
   c. Gravity carries the depolarization from the sinoatrial node down from the top of the heart, while contraction of the diaphragm forces the depolarization from the atrioventricular node to move from the bottom up.
   d. This statement is false; both contract from the bottom.

5. Throughout evolution, natural selection has favored changes that optimize respiration mechanisms in animals. According to Fick's Law, which of the following changes would optimize the rate of diffusion, R?
   a. decrease the surface area, A, over which diffusion takes place
   b. increase the distance, d, over which diffusion takes place
   c. increase the concentration difference, Δp, between the interior of the organism and the external environment
   d. increase the diffusion constant, D

6. Which type of circulatory system regulation do some vertebrates use to maintain body temperature in a cold environment?
   a. concurrent exchange
   b. countercurrent exchange
   c. vasodilation
   d. nitric oxide production

7. What is the stroke volume of a woman with a heart rate of 66 beats/minute and a cardiac output of 4.7 liters/minute?
   a. 310 milliliters/beat
   b. 14 milliliters/beat
   c. 70 milliliters/beat
   d. 71 milliliters/beat

8. If the total pressure of gas dissolved in blood plasma is 120 mm Hg, what is the partial pressure of nitrogen in the plasma? (Recall that air contains approximately 78% nitrogen.)
   a. 94 mm Hg
   b. 42 mm Hg
   c. 22 mm Hg
   d. 78 mm Hg

9. When you take a deep breath, your stomach moves out because
   a. swallowing air increases the volume of the thoracic cavity.
   b. your stomach shouldn't move out when you take a deep breath because you want the volume of your chest cavity to increase, not your abdominal cavity.
   c. contracting your abdominal muscles pushes your stomach out, generating negative pressure in your lungs.
   d. when your diaphragm contracts, it moves down, pressing your abdominal cavity out.

10. If you hold your breath for a long time, body CO_2 levels are likely to _____, and the pH of body fluids is likely to _____.
    a. increase; increase
    b. decrease; increase
    c. increase; decrease
    d. decrease; decrease

Test Your Visual Understanding

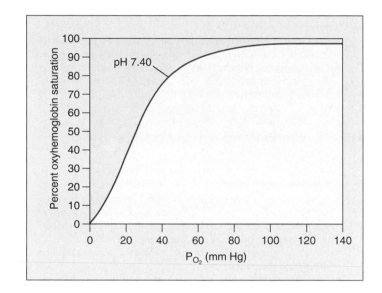

1. The Bohr effect describes the effect of pH on hemoglobin's affinity for oxygen. This figure shows a standard oxyhemoglobin dissociation curve. Draw a curve that represents the effect of a low pH according to the Bohr effect.

Apply Your Knowledge

1. Tissue plasminogen is an enzyme that, when activated, breaks down fibrin. How might the drug tissue plasminogen activator, t-PA, help reduce the damage caused by strokes and heart attacks?

2. Dr. Hearthealth is so impressed with your knowledge of cardiovascular physiology that he asks you to help with a particularly difficult case. He tells you that his patient has a prolonged PQ wave in an electrocardiogram (ECG), which you know means that the delay between atrial contraction and ventricular contraction is too long. What heart structure is likely causing this problem? Why?

45

The Nervous System

Concept Outline

45.1 The nervous system consists of neurons and supporting cells.

    Neuron Organization. Neurons and neuroglia are organized into the central nervous system and the peripheral nervous system.

45.2 Nerve impulses are produced on the axon membrane.

    The Resting Membrane Potential. The inside of the membrane is electrically negative in comparison with the outside.

    Graded Potentials Generate Action Potentials. Sensory and chemical stimuli regulate gated channels that produce depolarizations and hyperpolarizations in the plasma membrane.

45.3 Neurons form junctions called synapses with other cells.

    Structure of Synapses. Neurotransmitters diffuse to the postsynaptic cell and bind to receptor proteins.

    Neurotransmitters and Their Functions. Some neurotransmitters cause a depolarization in the postsynaptic membrane; others produce inhibition by hyperpolarization.

45.4 The central nervous system consists of the brain and spinal cord.

    The Evolution of the Vertebrate Brain. Vertebrate brains include a forebrain, midbrain, and hindbrain.

    The Human Forebrain. The cerebral cortex contains areas specialized for different functions.

    The Spinal Cord. Reflex responses and messages to and from the brain are coordinated by the spinal cord.

45.5 The peripheral nervous system consists of sensory and motor neurons.

    Components of the Peripheral Nervous System. A spinal nerve contains sensory and motor neurons.

    The Autonomic Nervous System. Sympathetic motor neurons arouse the body for fight or flight; parasympathetic motor neurons have antagonistic actions.

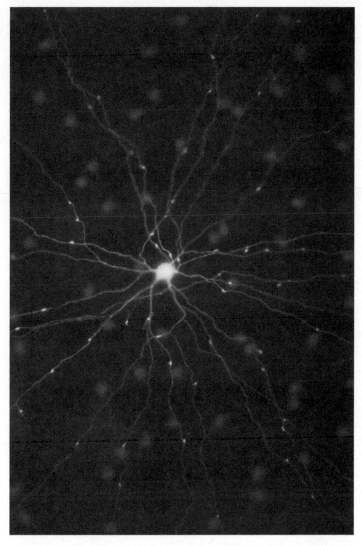

FIGURE 45.1
A neuron in the retina of the eye. This neuron has been injected with a fluorescent dye, making its cell body and long dendrites readily apparent (500×).

All animals except sponges use a network of nerve cells to gather information about the body's condition and the external environment, to process and integrate that information, and to issue commands to the body's muscles and glands. Just as telephone cables run from every compartment of a submarine to the conning tower, where the captain controls the ship, so bundles of nerve cells called neurons (figure 45.1) connect every part of an animal's body to its command and control center, the brain and spinal cord. The animal body is run just like a submarine, with status information about what is happening in organs and outside the body flowing into the command center, which analyzes the data and issues commands to glands and muscles.

45.1 The nervous system consists of neurons and supporting cells.

Neuron Organization

An animal must be able to respond to environmental stimuli. A fly escapes a swat; the antennae of a crayfish detect food, and the crayfish moves toward it. To do this, a fly or a crayfish must have *sensory receptors* that can detect the stimulus and *motor effectors* that can respond to it. In most invertebrate phyla and in all vertebrate classes, sensory receptors and motor effectors are linked by way of the nervous system.

As describe in chapter 42, the nervous system consists of neurons and supporting cells. The three types of neurons are shown in figure 45.2. **Sensory neurons** (or afferent neurons) carry impulses from sensory receptors to the **central nervous system (CNS),** which is composed of the brain and spinal cord. **Motor neurons** (or efferent neurons) carry impulses from the CNS to effectors—muscles and glands. A third type of neuron is present in the nervous systems of most invertebrates and all vertebrates: **interneurons** (or association neurons). These neurons are located in the brain and spinal cord of vertebrates, where they help provide more complex reflexes and higher associative functions, including learning and memory.

Together, sensory and motor neurons constitute the **peripheral nervous system (PNS)** in vertebrates. Motor neurons that stimulate skeletal muscles to contract are **somatic motor neurons,** and those that regulate the activity of the smooth muscles, cardiac muscle, and glands are **autonomic motor neurons.** The autonomic motor neurons are further broken down into the **sympathetic** and **parasympathetic** divisions, which act to counterbalance each other. Figure 45.3 illustrates the relationships among the different parts of the vertebrate nervous system.

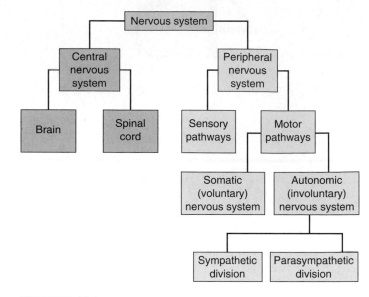

FIGURE 45.3
The divisions of the vertebrate nervous system. The major divisions are the central and peripheral nervous systems.

Despite their varied appearances, most neurons have the same functional architecture (figure 45.4). The **cell body** is an enlarged region containing the nucleus. Extending from the cell body are one or more cytoplasmic extensions called **dendrites.** Motor and association neurons possess a profusion of highly branched dendrites, enabling those cells to receive information from many different sources simultaneously. Some neurons have ex-

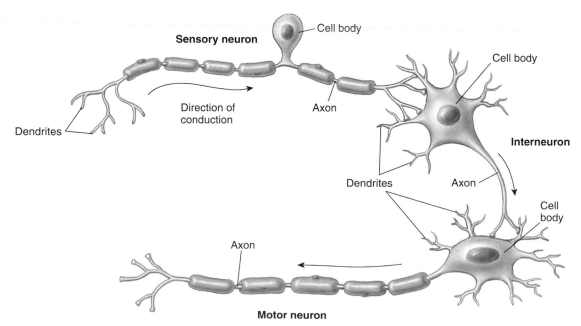

FIGURE 45.2
Three types of neurons.
Sensory neurons carry information about the environment to the brain and spinal cord. *Interneurons,* found in the brain and spinal cord, providing links between sensory and motor neurons. *Motor neurons* carry impulses or "commands" to muscles and glands (effectors).

tensions from the dendrites called *dendritic spines* that increase the surface area available to receive stimuli. The surface of the cell body integrates the information arriving at its dendrites. If the resulting membrane excitation is sufficient, it triggers impulses that are conducted away from the cell body along an **axon.** Each neuron has a single axon leaving its cell body, although an axon may also branch to stimulate a number of cells. An axon can be quite long: The axons controlling the muscles in your feet are more than a meter long, and the axons that extend from the skull to the pelvis in a giraffe are about 3 meters long!

Neurons are supported both structurally and functionally by **supporting cells,** which are called **neuroglia.** These cells are ten times smaller and ten times more numerous than neurons and serve a variety of functions, including supplying the neurons with nutrients, removing wastes from neurons, guiding axon migration, and providing immune functions. Two of the most important kinds of neuroglia in vertebrates are **Schwann cells** and **oligodendrocytes,** which produce **myelin sheaths** that surround the axons of many neurons. Schwann cells produce myelin in the PNS, while oligodendrocytes produce myelin in the CNS. During development, these cells wrap themselves around each axon several times to form the myelin sheath, an insulating covering consisting of multiple layers of compacted membrane (figure 45.5). Axons that have myelin sheaths are said to be myelinated, and those that don't are unmyelinated. In the CNS, myelinated axons form the **white matter,** and the unmyelinated dendrites and cell bodies form the **gray matter.** In the PNS, both myelinated and unmyelinated axons are bundled together, much like wires in a cable, to form **nerves.**

The myelin sheath is interrupted at intervals of 1 to 2 micrometers by small gaps known as **nodes of Ranvier** (see figure 45.4). The role of the myelin sheath in impulse conduction will be discussed later in this chapter.

Neurons and neuroglia make up the central and peripheral nervous systems in vertebrates. Interneurons, sensory, and motor neurons play different roles in the nervous system, and the neuroglia aid their function, in part by producing myelin sheaths.

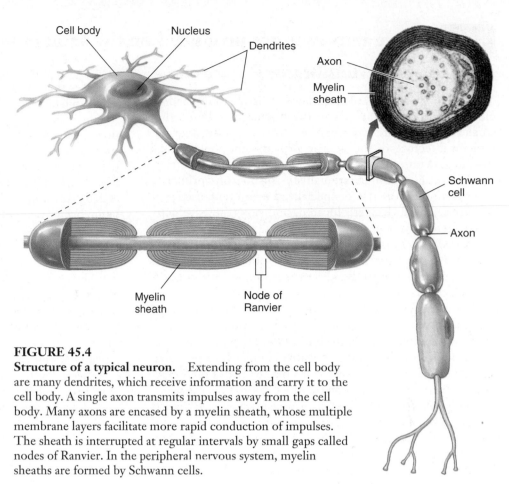

FIGURE 45.4
Structure of a typical neuron. Extending from the cell body are many dendrites, which receive information and carry it to the cell body. A single axon transmits impulses away from the cell body. Many axons are encased by a myelin sheath, whose multiple membrane layers facilitate more rapid conduction of impulses. The sheath is interrupted at regular intervals by small gaps called nodes of Ranvier. In the peripheral nervous system, myelin sheaths are formed by Schwann cells.

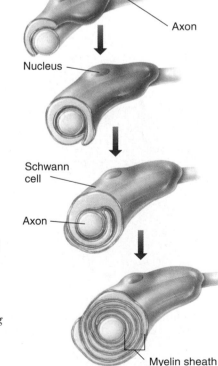

FIGURE 45.5
The formation of the myelin sheath around a peripheral axon. The myelin sheath is formed by successive wrappings of Schwann cell membranes, leaving most of the Schwann cell cytoplasm outside the myelin.

45.2 | Nerve impulses are produced on the axon membrane.

The Resting Membrane Potential

Neuron function depends on transient permeability and/or electrical property changes in the plasma membrane that spread or propagate from one part of the cell to another. The architecture of the neuron provides the mechanisms for the generation and spread of these membrane electrical potentials. The unique mechanisms of neurons primarily depend on the presence of specialized membrane transport proteins, where they are located, and how they are activated. Before we can understand how electrical signals are generated and transmitted between cells within the nervous system, we must first examine some of the basic electrical properties common to the plasma membranes of most animal cells.

The battery in a car or a flashlight separates electrical charges between its two poles. There is said to be a *potential difference*, or voltage, between the poles, with one pole being positive and the other negative. Similarly, a potential difference exists across every cell's plasma membrane. The side of the membrane exposed to the cytoplasm is the negative pole, and the side exposed to the extracellular fluid is the positive pole. The potential difference measured between these two poles is called the membrane potential. When a neuron is not being stimulated, it maintains a *resting membrane potential*. A cell is very small, and so its membrane potential is very small. The electrical potential of a car battery is typically 12 volts, but the resting membrane potential of many vertebrate neurons ranges from -40 to -90 millivolts (mV) or 0.04 to 0.09 volts. For the examples and figures in this chapter, we will use an average resting

membrane potential value of -70 mV. The negative sign indicates that the inside of the cell is negative with respect to the outside.

The inside of the cell is more negatively charged in relation to the outside because of three factors: (1) Molecules such as proteins, carbohydrates, and nucleic acids that carry net negative charges are more abundant inside the cell. Because these molecules are too large to diffuse out, they are called *fixed anions*. (2) The sodium-potassium (Na^+/K^+) pump brings two potassium ions (K^+) into the cell for every three sodium ions (Na^+) it pumps out (figure 45.6). This helps establish and maintain concentration gradients resulting in high K^+ and low Na^+ concentrations inside the cell and high Na^+ and low K^+ concentrations outside the cell. (3) Ion leak channels in the cell membrane are more numerous for K^+ than for Na^+. Ion **leak channels** are membrane proteins, which form pores through the membrane allowing the constant flow of specific ions (such as K^+ or Na^+) in and out of the cell by facilitated diffusion. Because there are more leak channels for K^+, it is easier for this ion to diffuse out of the cell than for Na^+ to diffuse into the cell. Other ions also can cross the membrane either through their own specific leak channels or otherwise, but their contribution to the resting membrane potential is minimal compared to that of K^+ and Na^+ in many neurons.

The average resting membrane potential is -70 mV because of an unequal distribution of electrical charges across the membrane. But why -70 mV rather than -50 mV or -10 mV? To understand this, we need to remember that two major forces act on the ions involved in establishing the resting membrane potential: (1) Due to an electrical

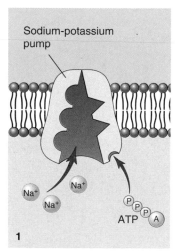 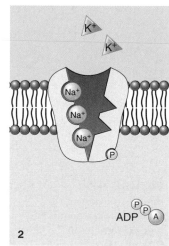 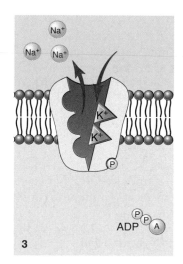 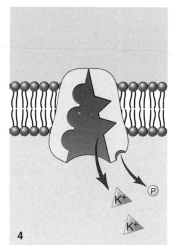

FIGURE 45.6

The sodium-potassium pump. This pump transports three Na^+ to the outside of the cell and simultaneously transports two K^+ to the inside of the cell. This is an active transport carrier requiring the (phosphorylating) energy of ATP.

Table 45.1	The Ionic Composition of Cytoplasm and Extracellular Fluid (ECF)			
Ion	Concentration in Cytoplasm (mM)	Concentration in ECF (mM)	Ratio	Equilibrium Potential (mV)
Na^+	15	150	10:1	+60
K^+	150	5	1:30	−90
Cl^-	7	110	15:1	−70

gradient, ions are attracted to ions or molecules of opposite charge. (2) Due to concentration gradients, each ion type, like all solutes, moves from an area of its highest concentration to an area of its lower concentration.

The positively charged ions, called *cations*, outside the cell are attracted to the negatively charged, fixed anions inside the cell. The two major cations involved in plasma membrane potentials are Na^+ and K^+, although others contribute. The electrical gradient across the membrane favors the flow of cations into the cell. The concentration gradient established by the sodium-potassium pump favors K^+ flow out of the cell and Na^+ flow into the cell. Thus, even though the plasma membrane is more permeable to K^+ than to other cations, little net movement of K^+ occurs because its electrical and chemical gradients are in opposition. Both the electrical and chemical gradients strongly favor the flow of Na^+ into the cell. But the net movement of Na^+ into the cell is minimal because of the lack of Na^+ leak channels, and the little that does diffuse in is pumped out by the Na^+/K^+ pump. Therefore, due to the combined activities of the fixed anions, the sodium-potassium pump, and the membrane leak channels, the electrochemical equilibrium or resting membrane potential for an average neuron can remain stable at −70 mV.

The point where the electrical and chemical forces for a particular ion balance each other is called the **equilibrium potential** (table 45.1). In order to calculate the equilibrium potential for a particular ion, you must assume the plasma membrane being studied is only permeable to that one ion. If K^+ were the only cation involved, the resting membrane potential of the cell would equal the K^+ equilibrium potential, the voltage at which the influx of K^+ equals the efflux of K^+, which would be about −90 mV for the average neuron. At −80 mV, K^+ would diffuse out of the cell, and at −100 mV, K^+ would diffuse into the cell. However, the membrane is also slightly permeable to other ions, especially Na^+, and its equilibrium potential is +60 mV. It is the effects of Na^+ leaking into the cell that make the resting membrane potential less negative. With a membrane potential less negative than −90 mV, K^+ diffuses out of the cell, and the combined effects bring the equilibrium potential for the resting cell to −70 mV. The

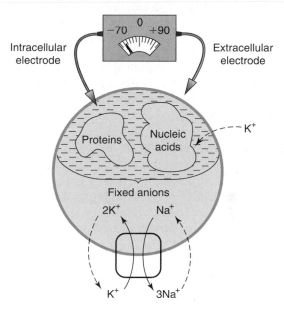

FIGURE 45.7
Establishment of the resting membrane potential. The fixed anions (primarily proteins and nucleic acids) attract cations from the extracellular fluid. If the membrane were only permeable to K^+, an equilibrium would be established, and the membrane potential would be −90 mV. A true resting membrane potential is about −70 mV, because the membrane does allow a low rate of Na^+ diffusion into the cell. This is not in balance with the outward diffusion of K^+, so the cell equilibrium is maintained by the action of the sodium-potassium pumps.

resting membrane potential of a neuron can be seen using a voltmeter and a pair of electrodes, one outside and one inside the cell (figure 45.7).

The resting plasma membrane maintains a potential difference as a result of the uneven distribution of charges, where the inside of the membrane is negatively charged in comparison with the outside (−70 mV). The magnitude, measured in millivolts, of this potential difference primarily reflects the difference in K^+ concentration.

Graded Potentials Generate Action Potentials

Graded Potentials

The uniqueness of neurons in comparison to other cells is not the production and maintenance of the resting membrane potential, but rather the sudden temporary disruptions to the resting membrane potential that occur in response to stimuli. Graded potentials are caused by the activation of specialized membrane transport proteins called **gated ion channels.** Whereas ion leak channels are always open, gated ion channels are closed in the normal resting cell. Gated ion channels in sensory receptors can open in response to sensory stimuli, such as light and heat, and trigger a type of graded potential called a receptor potential. In most neurons, gated ion channels in dendrites respond when chemicals bind to them (figure 45.8). They are referred to as **chemical-** or **ligand-gated channels.** Ligands are chemical groups that attach to larger molecules to regulate or contribute to their function. When they temporarily bind to membrane receptor proteins or channels, they cause the shape of the protein to change, thus opening the ion channel. Membrane ligand-gated channels can be activated or opened by such chemicals as hormones and neurotransmitters, resulting in plasma membrane permeability changes.

The permeability changes can be measured as depolarizations or hyperpolarizations of the membrane potential. If the plasma membrane is depolarized slightly, an oscilloscope will show a small upward deflection of the line that soon decays back to the resting membrane potential. A depolarization makes the membrane potential less negative (more positive), while a hyperpolarization makes the membrane potential more negative. A permeability change that caused the membrane potential to go from -70 mV to -65 mV would be a depolarization, and an increase in negativity to -75 mV or -80 mV would be a hyperpolarization. These small changes in membrane potential are called *graded potentials* because their amplitudes depend on the strength of the stimulus or the amount of ligand available to bind with their receptors. Depolarizing or hyperpolarizing potentials can add together to amplify or reduce their effects, just as two waves add to make one bigger one when they meet in synchrony or cancel each other out when a trough meets with a crest. The ability of graded potentials to combine is called **summation** (figure 45.9). Depolarizations bring a cell closer to threshold potential, and hyperpolarizations move a membrane further from threshold potential. *Threshold* is the level of depolarization needed to produce an action potential.

Generation of Action Potentials

Once a particular level of depolarization is reached (about -55 mV in some mammalian axons), a nerve impulse, or **action potential,** is produced in the region where the axon

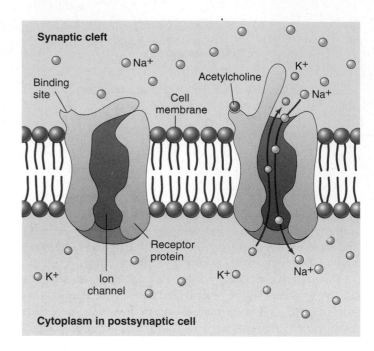

FIGURE 45.8
The binding of acetylcholine (ACh) to its receptor opens ion channels. The chemical or ligand-regulated gates to these channels open when the neurotransmitter ACh binds to the receptor.

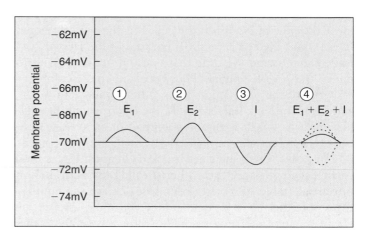

FIGURE 45.9
Graded potentials. Graded potentials are produced by the regulation of chemical or ligand-gated channels. (*1*) A weak excitatory stimulus, E_1, elicits a smaller depolarization than (*2*) a stronger stimulus, E_2. (*3*) An inhibitory stimulus, I, produces a hyperpolarization. (*4*) Because graded potentials can summate, if all three stimuli occur very close together, the resulting polarity change will be the algebraic sum of the three changes individually.

arises from the cell body. A depolarization that reaches or exceeds the threshold opens both the Na^+ and K^+ *voltage-gated ion channels* (figure 45.10). Ion voltage-gated channels are another special type of membrane channel found in muscle cells and neurons composed of a protein or proteins that have *gates*, portions of the channel protein that open or close the channel's pore at certain levels of membrane potential, usually threshold. In other words, in the axons of neurons and in muscle fibers, the voltage gates are closed or open depending on the membrane potential. When neurons or muscle cells are stimulated, Na^+ and K^+ voltage-gated channels open, and the membrane becomes more permeable to Na^+ and K^+, but the Na^+ channels open first. Sodium rushes into the cell, down its concentration gradient. This sudden influx of positive charges reduces the negativity on the inside of the cell and causes the cell to *depolarize* (move toward a polarity above that of the resting potential). After a slight delay, potassium voltage-gated channels also open, and K^+ flows out of the cell down its concentration gradient. Similarly, the membrane becomes more permeable to Cl^-, and Cl^- flows into the cell. But the effects of Cl^- on the membrane potential are far less than those of K^+. The inside of the cell again becomes more negative and causes the cell to *hyperpolarize* (move its polarity below that of the resting potential).

The rapid diffusion of Na^+ into the cell shifts the membrane potential toward the equilibrium potential for Na^+ (+60 mV). Recall that the positive sign indicates that the membrane reverses polarity as Na^+ rushes in. When the action potential is recorded on an oscilloscope, this part of the action potential appears as the *rising phase* of a spike (figure 45.11). The membrane potential never quite reaches +60 mV because the Na^+ channels close too soon,

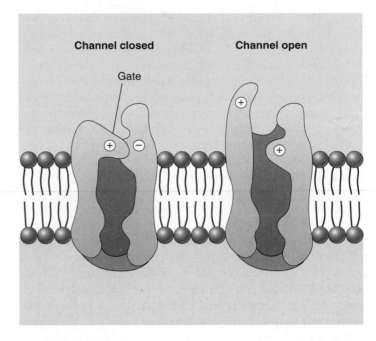

FIGURE 45.10
Voltage-gated ion channels. In neurons and muscle cells, specialized channels for Na^+ and K^+ have gates that are closed at the resting membrane potential but open when a threshold level of depolarization is attained.

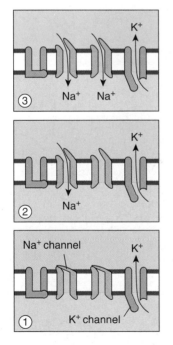
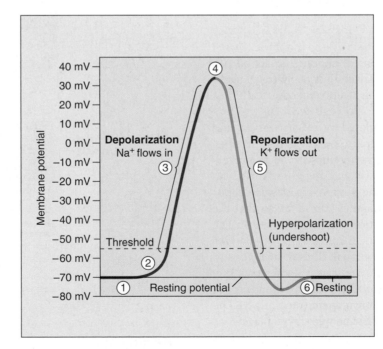
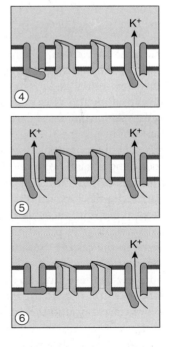

FIGURE 45.11
The action potential. (*1*) At resting membrane potential, voltage-gated ion channels are closed, but there is some leakage of K^+. (*2*) In response to a stimulus, the cell begins to depolarize, and once the threshold level is reached, an action potential is produced. (*3*) Rapid depolarization occurs (the rising portion of the spike) because voltage-gated sodium channels open, allowing Na^+ to diffuse into the axon. (*4*) At the top of the spike, Na^+ channels close, and voltage-gated potassium channels that were previously closed begin to open. (*5*) With the K^+ channels open, repolarization occurs because of the diffusion of K^+ out of the axon. (*6*) An undershoot occurs before the membrane returns to its original resting potential.

and at about the same time, the K$^+$ channels that were previously closed begin to open. The action potential thus peaks at about +30 mV. Opening the K$^+$ channels allows K$^+$ to diffuse out of the cell, repolarizing the plasma membrane. On an oscilloscope, this repolarization of the membrane appears as the *falling phase* of the action potential. In many cases, the repolarization carries the membrane potential to a value slightly more negative than the resting potential for a brief period because K$^+$ channels remain open, resulting in an *undershoot*. The entire sequence of events in an action potential is over in a few milliseconds.

Action potentials have two distinguishing characteristics. First, they follow an all-or-none law: Each threshold depolarization produces either a full action potential, because the voltage-gated Na$^+$ channels open completely at threshold, or none at all. Second, action potentials are always separate events; they cannot add together or interfere with one another as graded potentials can because the membrane enters a brief refractory period after it generates an action potential during which time voltage-gated Na$^+$ channels cannot reopen.

The production of an action potential results entirely from the passive diffusion of ions. However, at the end of each action potential, the cytoplasm has a little more Na$^+$ and a little less K$^+$ than it did at rest. The constant activity of the sodium-potassium pumps compensates for these changes. Thus, although active transport is not required to produce action potentials, it is needed to maintain the ion gradients.

Propagation of Action Potentials

Although we often speak of axons as conducting action potentials (impulses), action potentials do not really travel along an axon—they are events that are reproduced at different points along the axon membrane. This can occur for two reasons: Action potentials are stimulated by depolarization, and an action potential can serve as a depolarization stimulus. Each action potential, during its rising phase, reflects a reversal in membrane polarity (from −70 mV to +30 mV) as Na$^+$ diffuses rapidly into the axon. The positive charges can depolarize the next region of membrane to threshold, so that the next region produces its own action potential (figure 45.12). Meanwhile, the previous region of membrane repolarizes back to the resting membrane potential. This is analogous to people in a stadium performing the "wave": Individuals stay in place as they stand up (depolarize), raise their hands (peak of the action potential), and sit down again (repolarize). The wave travels around the stadium, the people stay in place.

Saltatory Conduction

Action potentials are conducted without decrement (without decreasing in amplitude); thus, the last action potential at the end of the axon is just as large as the first action potential. The velocity of conduction is greater if the diameter of the axon is large or if the axon is myelinated (table 45.2). Myeli-

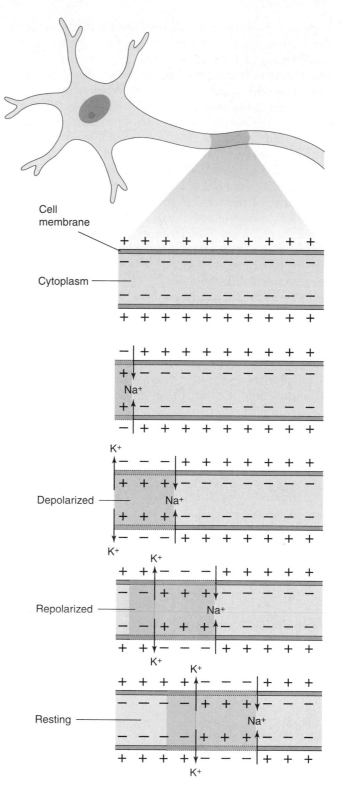

FIGURE 45.12

Propagation of an action potential in an unmyelinated axon. When one region produces an action potential and undergoes a reversal of polarity, it serves as a depolarization stimulus for the next region of the axon. In this way, action potentials are regenerated along each small region of the unmyelinated axon membrane.

nated axons conduct impulses more rapidly than nonmyelinated axons because the action potentials in myelinated axons are only produced at the nodes of Ranvier. One action potential still serves as the depolarization stimulus for the next, but the depolarization at one node spreads quickly to the next opening its voltage-gated channels. The impulses therefore seem to jump from node to node (figure 45.13) in a process called **saltatory conduction** (Latin *saltare*, "to jump").

To see how saltatory conduction speeds nervous transmission, return for a moment to the "wave" analogy used on the previous page to describe propagation of an action potential. The "wave" moves across the seats of a crowded stadium as fans standing up in one section trigger the next section to stand up in turn. Because the "wave" will skip sections of empty bleachers, it actually progresses around the stadium even faster with more empty sections. The wave doesn't have to wait for the missing people to stand, simply "jumping" the gaps—just as saltatory conduction jumps the nonconduction "gaps" of myelin between exposed nodes.

Graded potentials, caused by the opening of ligand-gated channels, can depolarize or hyperpolarize the membrane. Depolarizing graded potentials can summate, resulting in the rapid inward diffusion of Na^+ followed by the outward diffusion of K^+ through voltage-gated channels. This produces a rapid change in the membrane potential called an action potential. Action potentials are all-or-none events and cannot summate. Action potentials are regenerated along an axon as one action potential serves as the depolarization stimulus for the next action potential.

Table 45.2 Conduction Velocities of Some Axons

	Axon Diameter (μm)	Myelin	Conduction Velocity (m/s)
Squid giant axon	500	No	25
Large motor axon to human leg muscle	20	Yes	120
Axon from human skin pressure receptor	10	Yes	50
Axon from human skin temperature receptor	5	Yes	20
Motor axon to human internal organ	1	No	2

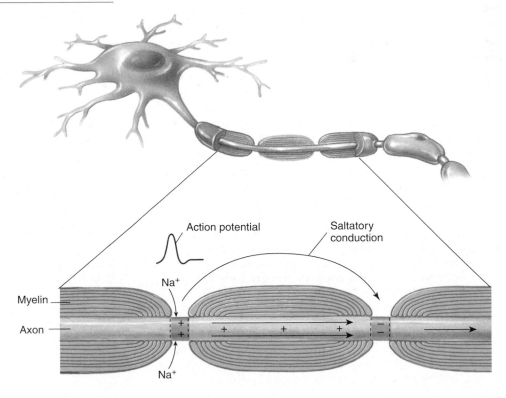

FIGURE 45.13

Saltatory conduction in a myelinated axon. Action potentials are only produced at the nodes of Ranvier in a myelinated axon. One node depolarizes the next node so that the action potentials can skip between nodes. As a result, saltatory ("jumping") conduction in a myelinated axon is more rapid than conduction in an unmyelinated axon.

Action potential

Saltatory conduction

Na$^+$

Myelin

Axon

Na$^+$

45.3 Neurons form junctions called synapses with other cells.

Structure of Synapses

An action potential passing down an axon eventually reaches the end of the axon and all of its branches. These branches may form junctions with the dendrites of other neurons, with muscle cells, or with gland cells. Such intercellular junctions are called **synapses.** The neuron whose axon transmits action potentials to the synapse is the *presynaptic cell*, while the cell receiving the signal on the other side of the synapse is the *postsynaptic cell.* Two basic types of synapses have been observed in the nervous systems of animals: electrical and chemical. Electrical synapses involve direct cytoplasmic connections formed by gap junctions between the pre- and postsynaptic neurons (see figure 7.17). Membrane potential changes, including action potentials, pass directly and rapidly from one cell to the other through the gap junctions. Electrical synapses are characteristic of invertebrate nervous systems, but somewhat rare in vertebrates.

The vast majority of vertebrate synapses are of the chemical type. Although the presynaptic and postsynaptic cells may appear to touch when the synapse is seen under a light microscope, examination with an electron microscope reveals that most synapses have a **synaptic cleft,** a narrow space that separates these two cells (figure 45.14). The end of the presynaptic axon is swollen and contains numerous **synaptic vesicles,** which are each packed with chemicals called **neurotransmitters.** When action potentials arrive at the end of the axon, they stimulate the opening of voltage-gated Ca^{++} channels, causing a rapid inward diffusion of Ca^{++}. This serves as the stimulus for the fusion of the synaptic vesicles' membrane with the plasma membrane of the axon, so that the contents of the vesicles can be released by exocytosis (figure 45.15). The higher the frequency of action potentials in the presynaptic axon, the more vesicles will release their contents of neurotransmitters. The neurotransmitters diffuse rapidly to the other side of the cleft and bind to chemical- or ligand-gated **receptor proteins** in the membrane of the postsynaptic cell. There are different types of neurotransmitters, and they act in different ways. We will next consider the action of a few of the important neurotransmitter chemicals.

Electrical synapses are characteristic of invertebrates, and chemical synapses predominate in vertebrates. Electrical synapses transmit signals through gap junctions. In chemical synapses, the presynaptic axon is separated from the postsynaptic cell by a narrow synaptic cleft. Neurotransmitters diffuse across it to transmit a nerve impulse.

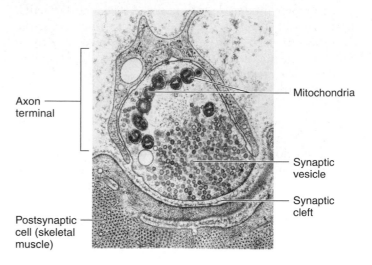

FIGURE 45.14
A synaptic cleft. An electron micrograph showing a neuromuscular synapse.

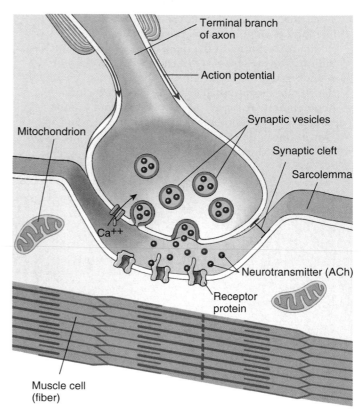

FIGURE 45.15
The release of neurotransmitter. Action potentials arriving at the end of an axon trigger the uptake of Ca^{++}, which causes synaptic vesicles to fuse with the plasma membrane and release their neurotransmitters (acetylcholine [ACh] in this case), which diffuse across the synaptic gap and bind to chemical or ligand-gated receptors in the postsynaptic membrane.

Neurotransmitters and Their Functions

Acetylcholine was the first neurotransmitter chemical to be discovered and is widely used in the nervous system. However, many other neurotransmitter chemicals have been shown to play important roles, and ongoing research continues to produce new information about neurotransmitter function.

Acetylcholine

Acetylcholine (ACh) is the neurotransmitter that crosses the synapse between a motor neuron and a muscle fiber. This synapse is called a *neuromuscular junction* (figure 45.16). Acetylcholine binds to its receptor proteins in the postsynaptic membrane and thereby causes ion channels within these proteins to open (see figure 45.8). Recall from section 45.2 that the gates to these ion channels are said to be ligand-gated because they open in response to the binding of a chemical, ACh in this case, rather than in response to depolarization. The opening of the ligand-gated channels permits Na^+ to diffuse into the postsynaptic cell and K^+ to diffuse out. Although both ions move at the same time, the inward diffusion of Na^+ occurs at a faster rate and has the predominant effect because of its more favorable electrochemical gradient. As a result, that site on the postsynaptic membrane produces a depolarization (figure 45.17a) called an **excitatory postsynaptic potential (EPSP).** The EPSP, if large enough, can open the voltage-gated channels for Na^+ and K^+ that are responsible for action potentials. Because the postsynaptic cell we are discussing is a skeletal muscle cell, the action potentials it produces stimulate muscle contraction through the mechanisms discussed in chapter 42.

If ACh stimulates muscle contraction, we must be able to eliminate ACh from the synaptic cleft in order to relax our muscles. This illustrates a general principle: Molecules such as neurotransmitters and certain hormones must be quickly eliminated after secretion if they are to be

FIGURE 45.16
Neuromuscular junctions. A light micrograph shows axons branching to make contact with several individual muscle fibers.

FIGURE 45.17
Different neurotransmitters can have different effects. (*a*) An excitatory neurotransmitter promotes a depolarization, or excitatory postsynaptic potential (EPSP). (*b*) An inhibitory neurotransmitter promotes a hyperpolarization, or inhibitory postsynaptic potential (IPSP).

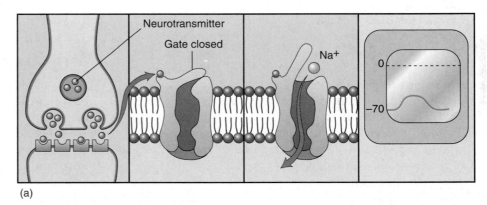

(a)

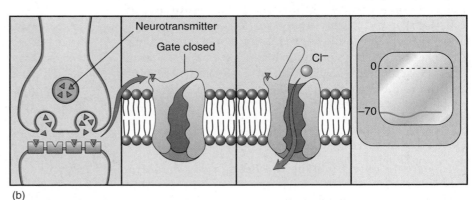

(b)

effective regulators. In the case of ACh, the elimination is achieved by an enzyme in the postsynaptic membrane called **acetylcholinesterase (AChE).** This enzyme is one of the fastest known, cleaving ACh into inactive fragments. Nerve gas and the agricultural insecticide parathion are potent inhibitors of AChE; in humans, they can produce severe spastic paralysis and even death if the respiratory muscles become paralyzed.

Although ACh acts as a neurotransmitter between motor neurons and skeletal muscle cells, many neurons also use ACh as a neurotransmitter at their synapses with other neurons; in these cases, the postsynaptic membrane is generally on the dendrites or cell body of the second neuron. The EPSPs produced must then travel through the dendrites and cell body to the initial segment of the axon, where the first voltage-regulated channels needed for action potentials are located. This is where the first action potentials will be produced, providing that the EPSP depolarization is above the threshold needed to trigger action potentials.

Glutamate, Glycine, and GABA

Glutamate is the major excitatory neurotransmitter in the vertebrate CNS, producing EPSPs and stimulating action potentials in those postsynaptic neurons brought to threshold. Although normal amounts produce physiological stimulation, excessive stimulation by glutamate has been shown to cause neurodegeneration, as in Huntington disease.

Glycine and **GABA** (an acronym for gamma-aminobutyric acid) are inhibitory neurotransmitters. If you remember that action potentials are triggered by a threshold level of depolarization, you will understand that hyperpolarizations of the membrane would cause inhibition by making threshold potential more difficult to reach. These neurotransmitters cause the opening of ligand-gated channels for Cl^-, which has a concentration gradient favoring its diffusion into the neuron. Because Cl^- is negatively charged, it makes the inside of the membrane even more negative than it is at rest—for example, from -70 mV to -85 mV (see figure 45.17b). This hyperpolarization is called an **inhibitory postsynaptic potential (IPSP),** and is very important for neural control of body movements and other brain functions. Interestingly, the drug diazepam (Valium) causes its sedative and other effects by enhancing the binding of GABA to its receptors and thereby increasing the effectiveness of GABA at the synapse.

Biogenic Amines

The *biogenic amines* include the hormone epinephrine (adrenaline), together with the neurotransmitters dopamine, norepinephrine, and serotonin. Epinephrine, norepinephrine, and dopamine are derived from the amino acid tyrosine and are included in the subcategory of *catecholamines*. Serotonin is a biogenic amine derived from a different amino acid, tryptophan.

Dopamine is a very important neurotransmitter used in some areas of the brain controlling body movements and other functions. Degeneration of particular dopamine-releasing neurons produces the resting muscle tremors of Parkinson disease, and people with this condition are treated with L-dopa (an acronym for dihydroxyphenylalanine), a precursor from which dopamine can be produced. Additionally, studies suggest that excessive activity of dopamine-releasing neurons in other areas of the brain is associated with schizophrenia. As a result, patients with schizophrenia are sometimes helped by drugs that block the production of dopamine.

Norepinephrine is used by neurons in the brain and also by particular autonomic neurons, where its action as a neurotransmitter complements the action of the hormone epinephrine, secreted by the adrenal gland. The autonomic nervous system will be discussed in section 45.5.

Serotonin is a neurotransmitter involved in the regulation of sleep and also implicated in various emotional states. Insufficient activity of neurons that release serotonin may be one cause of clinical depression; this is suggested by the fact that antidepressant drugs, particularly fluoxetine (Prozac) and related compounds, specifically block the elimination of serotonin from the synaptic cleft. The drug lysergic acid diethylamide (LSD) specifically blocks serotonin receptors in a region of the brain stem known as the raphe nuclei.

Other Neurotransmitters

Axons also release various polypeptides, called **neuropeptides,** at synapses. These neuropeptides may have a typical neurotransmitter function, or they may have more subtle, long-term action on the postsynaptic neurons. In the latter case, they are often referred to as **neuromodulators.** A given axon generally releases only one kind of neurotransmitter, but many can release both a neurotransmitter and a neuromodulator.

One important neuropeptide is **substance P,** which is released at synapses in the CNS by sensory neurons activated by painful stimuli. The perception of pain, however, can vary depending on circumstances; an injured football player may not feel the full extent of his trauma, for example, until he's taken out of the game. The intensity with which pain is perceived partly depends on the effects of neuropeptides called **enkephalins** and **endorphins.** Enkephalins are released by axons descending from the brain into the spinal cord and inhibit the passage of pain information back up to the brain. Endorphins are released by neurons in the brain stem and also block the perception of pain. Opium and its derivatives,

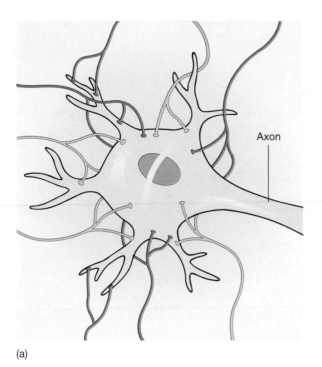

(a)

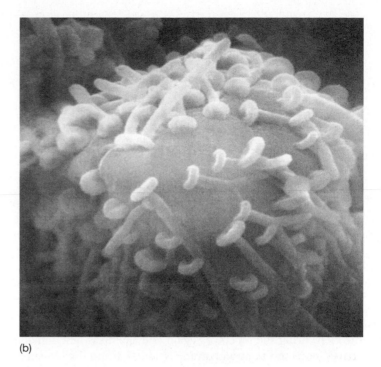

(b)

FIGURE 45.18

Integration of EPSPs and IPSPs takes place on the neuronal cell body. (*a*) The synapses made by some axons are excitatory (*blue*); the synapses made by other axons are inhibitory (*red*). The summed influence of all of these inputs determines whether the axonal membrane of the postsynaptic cell will be sufficiently depolarized to produce an action potential. (*b*) Micrograph of a neuronal cell body with numerous synapses (15,000×).

morphine and heroin, have an analgesic (pain-reducing) effect because they are similar enough in chemical structure to bind to the receptors normally utilized by enkephalins and endorphins. For this reason, the enkephalins and the endorphins are referred to as *endogenous opiates*.

Nitric oxide (NO) is the first gas known to act as a regulatory molecule in the body. Because NO is a gas, it diffuses through membranes, so it cannot be stored in vesicles. It is produced as needed from the amino acid arginine. Nitric oxide's actions are very different from those of the more familiar nitrous oxide (N_2O), or laughing gas, sometimes used by dentists. Nitric oxide diffuses out of the presynaptic axon and into neighboring cells by simply passing through the lipid portions of the plasma membranes. In the PNS, nitric oxide is released by some neurons that innervate the gastrointestinal tract, penis, respiratory passages, and cerebral blood vessels. These are autonomic neurons that cause smooth muscle relaxation in their target organs. This can produce, for example, the engorgement of the spongy tissue of the penis with blood, causing erection. The drug sildenafil (Viagra) increases the release of nitric oxide in the penis, prolonging erection. Nitric oxide is also released as a neurotransmitter in the brain, and has been implicated in the processes of learning and memory.

Synaptic Integration

The activity of a postsynaptic neuron in the brain and spinal cord of vertebrates is influenced by different types of input from a number of presynaptic neurons. For example, a single motor neuron in the spinal cord can have in excess of as many as 50,000 synapses from presynaptic axons! Each postsynaptic neuron may receive both excitatory and inhibitory synapses. The EPSPs (depolarizations) and IPSPs (hyperpolarizations) from these synapses interact with each other when they reach the cell body of the neuron. Small EPSPs add together to bring the membrane potential closer to the threshold, while IPSPs subtract from the depolarizing effect of the EPSPs, deterring the membrane potential from reaching threshold (figure 45.18). This process is called **synaptic integration.**

Neurotransmitters and Drug Addiction

When certain cells of your nervous system are exposed to a constant stimulus that produces a chemically mediated signal for a prolonged period, they may lose their ability to respond to that stimulus, a process called *habituation*. You are familiar with this loss of sensitivity—when you sit in a chair, how long are you aware of the chair? Some nerve cells are particularly prone to this loss of sensitivity. If receptor proteins within synapses are exposed to high levels of neurotransmitter molecules for prolonged periods, that nerve cell often responds by decreasing the number of receptor proteins into the membrane. This feedback is a normal function in all neurons, one of several mechanisms that have evolved to make the cell more efficient; in this case, it adjusts the number of "tools" (receptor proteins) in the membrane "workshop" to suit the workload.

Cocaine. The drug cocaine causes abnormally large amounts of neurotransmitter to remain in the synapses for long periods of time. Cocaine affects nerve cells in the brain's pleasure pathways (the so-called limbic system). These cells transmit pleasure messages using the neurotransmitter dopamine. Using radioactively labeled cocaine molecules, investigators found that cocaine binds tightly to the transporter proteins on presynaptic membranes. These proteins normally remove the neurotransmitter dopamine after it has acted. Like a game of musical chairs in which all the chairs become occupied, eventually no unoccupied transporter proteins are available to the dopamine molecules, so the dopamine stays in the cleft, firing the receptors again and again. As new signals arrive, more and more dopamine is added, firing the pleasure pathway more and more often (figure 45.19).

When receptor proteins on limbic system nerve cells are exposed to high levels of dopamine neurotransmitter molecules for prolonged periods of time, the nerve cells "turn down the volume" of the signal by lowering the number of receptor proteins on their surfaces. They respond to the greater number of neurotransmitter molecules by simply reducing the number of targets available for these molecules to hit (figure 45.20). The cocaine user is now addicted. With so few receptors, the user needs the drug to maintain even normal levels of limbic activity.

Is Nicotine an Addictive Drug? Investigators attempting to explore the habit-forming nature of nicotine used what had been learned about cocaine to carry out what seemed a reasonable experiment. They introduced radioactively labeled nicotine into the brain and looked to see what sort of transporter protein it attached itself to. To their great surprise, the nicotine ignored proteins on the presynaptic membrane and instead bound directly to a specific receptor on the postsynaptic cell! This was totally unexpected: Since nicotine does not normally occur in the brain, why should it have a receptor there?

Intensive research followed, and researchers soon learned that the "nicotine receptors" were a class of receptors that normally served to bind the neurotransmitter acetylcholine. Nicotine has evolved in tobacco plants as a secondary compound because it affects the CNS of herbivorous insects. It was just an accident of nature that nicotine was also able to bind to some human ACh receptors. Other types of ACh receptors don't respond to nicotine. What, then, is the normal function of the nicotine-binding ACh

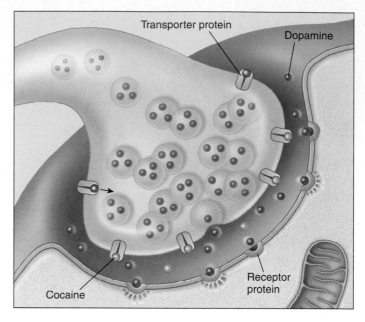

FIGURE 45.19
How cocaine alters events at the synapse. When cocaine binds to the dopamine transporters, the neurotransmitter survives longer in the synapse and continues to stimulate the postsynaptic cell. Cocaine thus acts to intensify pleasurable sensations.

receptors? The target of considerable research, these receptors turned out to be one of the brain's most important tools. The brain uses them to coordinate the activities of many other kinds of receptors, acting to "fine-tune" the sensitivity of a wide variety of behaviors.

When neurobiologists compare the nerve cells in the brains of smokers to those of nonsmokers, they find changes in both the number of nicotine receptors and the levels of RNA used to make the receptors. They have found that the brain adjusts to prolonged exposure to nicotine by "turning down the volume" in two ways: (1) by making fewer receptor proteins to which nicotine can bind; and (2) by altering the pattern of *activation* of the nicotine receptors (that is, their sensitivity to neurotransmitter).

This second adjustment is responsible for the profound effect of smoking on the brain's activities. By overriding the normal system used by the brain to coordinate its many activities, nicotine alters the pattern of release into the synaptic clefts of many neurotransmitters, including acetylcholine, dopamine, serotonin, and many others. As a result, changes in level of activity occur in a wide variety of nerve pathways within the brain.

Addiction occurs when chronic exposure to nicotine induces the nervous system to adapt physiologically. The brain compensates for the many changes nicotine induces by making other changes. Adjustments are made to the numbers and sensitivities of many kinds of receptors within the brain, restoring an appropriate balance of activity.

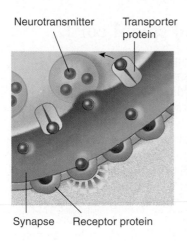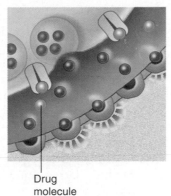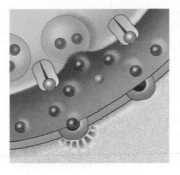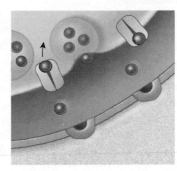

Neurotransmitter   Transporter
                    protein

Synapse   Receptor protein   Drug
                                    molecule

1. Neurotransmitter is reabsorbed at a normal synapse.

2. Drug molecules prevent reabsorption and cause overstimulation of the postsynaptic membrane.

3. The number of receptors decreases.

4. The synapse is less sensitive when the drug is removed.

FIGURE 45.20
Drug addiction. (*1*) In a normal synapse, the neurotransmitter binds to a transporter molecule and is rapidly reabsorbed after it has acted. (*2*) When a drug molecule binds to the transporters, reabsorption of the neurotransmitter is blocked, and the postsynaptic cell is overstimulated by the increased amount of neurotransmitter left in the synapse. (*3*) The central nervous system adjusts to the increased firing by producing fewer receptors in the postsynaptic membrane. The result is addiction. (*4*) When the drug is removed, normal absorption of the neurotransmitter resumes, and the decreased number of receptors creates a less-sensitive nerve pathway. Physiologically, the only way a person can then maintain normal functioning is to continue to take the drug. Only if the drug is removed permanently will the nervous system eventually adjust again and restore the original amount of receptors.

Now, what happens if you stop smoking? Everything is out of whack! The newly coordinated system *requires* nicotine to achieve an appropriate balance of nerve pathway activities. This is *addiction* in any sensible use of the term. The body's physiological response is profound and unavoidable. Using "willpower" cannot prevent addiction to nicotine, any more than willpower can stop a bullet when playing Russian roulette with a loaded gun. If you smoke cigarettes for a prolonged period, you will become addicted.

What do you do if you are addicted to smoking cigarettes and you want to stop? When use of an addictive drug such as nicotine is stopped, the level of signaling will change to levels far from normal. If the drug is not reintroduced, the altered level of signaling will eventually induce the nerve cells to once again make compensatory changes that restore an appropriate balance of activities within the brain. Over time, receptor numbers, their sensitivity, and patterns of release of neurotransmitters all revert to normal, once again producing normal levels of signaling along the pathways. There is no way to avoid the down side of addiction. The pleasure pathways will not function at normal levels until the number of receptors on the affected nerve cells have time to readjust.

Many people attempt to quit smoking by using patches containing nicotine; the idea is that providing gradually lesser doses of nicotine allows the smoker to be weaned of his or her craving for cigarettes. The patches do reduce the craving for cigarettes—as long as you keep using the patches! Actually, using such patches simply substitutes one (admittedly less dangerous) nicotine source for another. If you are going to quit smoking, there is no way to avoid the necessity of eliminating the drug to which you are addicted. As hard as it is to hear the bad news, there is no easy way out. The only way to quit is to quit.

Acetylcholine stimulates the opening of ligand-gated ion channels, causing a depolarization called an excitatory postsynaptic potential (EPSP). Glycine and GABA are inhibitory neurotransmitters that produce hyperpolarization of the postsynaptic membrane. There are also many other neurotransmitters, including the biogenic amines: dopamine, norepinephrine, and serotonin. The effects of many synapses and different neurotransmitters are integrated through summation of depolarizations and hyperpolarizations.

The Evolution of the Vertebrate Brain

Sponges are the only major phylum of multicellular animals that lack nerves. The simplest nervous systems occur among cnidarians (figure 45.21), in which all neurons are similar and linked to one another in a web, or *nerve net*. There is no associative activity, no control of complex actions, and little coordination. The simplest animals with associative activity in the nervous system are the free-living flatworms, phylum Platyhelminthes. Running down the bodies of these flatworms are two nerve cords, from which peripheral nerves extend outward to the muscles of the body. The two nerve cords converge at the front end of the body, forming an enlarged mass of nervous tissue that also contains interneurons with synapses connecting neurons to one another. This primitive "brain" is a rudimentary central nervous system and permits a far more complex control of muscular responses than is possible in cnidarians.

All of the subsequent evolutionary changes in nervous systems can be viewed as a series of elaborations on the characteristics already present in flatworms. For example, earthworms exhibit a central nervous system that is connected to all other parts of the body by peripheral nerves. And, in arthropods, the central coordination of complex responses is increasingly localized in the front end of the nerve cord. As this region evolved, it came to contain a progressively larger number of interneurons, and to develop tracts, which are highways within the brain that connect associative elements.

Casts of the interior braincases of fossil agnathans, fishes that swam 500 million years ago, have revealed much about the early evolutionary stages of the vertebrate brain. Although small, these brains already had the three divisions

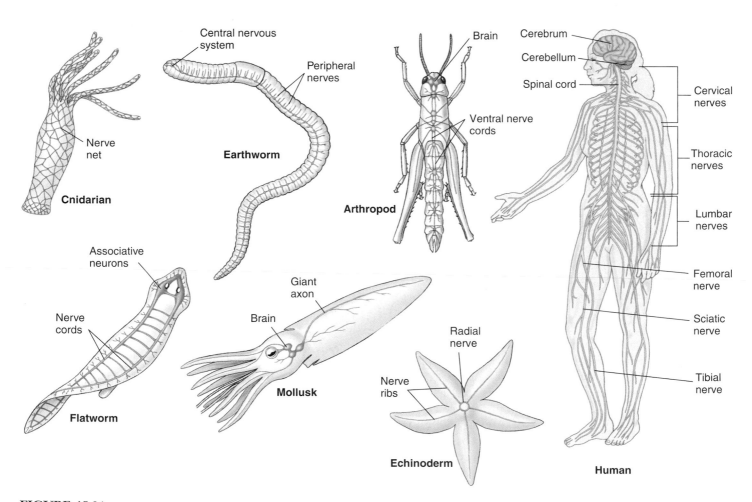

FIGURE 45.21
Evolution of the nervous system. Animals exhibit a progressive elaboration of organized nerve cords and the centralization of complex responses in the front end of the nerve cord. This evolutionary process is referred to as cephalization.

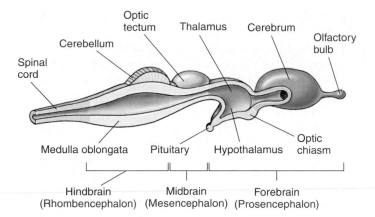

Optic
tectum   Thalamus   Cerebrum
Cerebellum   Olfactory
                                       bulb
Spinal
cord

Medulla oblongata   Pituitary   Hypothalamus   Optic
                                                   chiasm

Hindbrain   Midbrain   Forebrain
(Rhombencephalon)   (Mesencephalon) (Prosencephalon)

FIGURE 45.22
The basic organization of the vertebrate brain can be seen in the brains of primitive fishes. The brain is divided into three regions that are found in differing proportions in all vertebrates: the hindbrain, which is the largest portion of the brain in fishes; the midbrain, which in fishes is devoted primarily to processing visual information; and the forebrain, which is concerned mainly with olfaction (the sense of smell) in fishes. In terrestrial vertebrates, the forebrain plays a far more dominant role in neural processing than it does in fishes.

that characterize the brains of all contemporary vertebrates: (1) the hindbrain, or rhombencephalon; (2) the midbrain, or mesencephalon; and (3) the forebrain, or prosencephalon (figure 45.22 and table 45.3).

The hindbrain was the major component of these early brains, as it still is in fishes today. Composed of the *cerebellum*, *pons*, and *medulla oblongata*, the hindbrain may be considered an extension of the spinal cord devoted primarily to coordinating motor reflexes. Tracts containing large numbers of axons run like cables up and down the spinal cord to the hindbrain. The hindbrain, in turn, integrates the many sensory signals coming from the muscles and coordinates the pattern of motor responses.

Much of this coordination is carried on within a small extension of the hindbrain called the cerebellum ("little cerebrum"). In more advanced vertebrates, the cerebellum plays an increasingly important role as a coordinating center for movement and is correspondingly larger than it is in the fishes. In all vertebrates, the cerebellum processes data on the current position and movement of each limb, the state of relaxation or contraction of the muscles involved, and the general position of the body and its relation to the outside world. These data are gathered and synthesized in the cerebellum, and the resulting commands are issued to efferent pathways.

In fishes, the remainder of the brain is devoted to the reception and processing of sensory information. The midbrain is composed primarily of the **optic lobes,** which receive and process visual information, while the forebrain

Table 45.3	Subdivisions of the Central Nervous System
Major Subdivision	**Function**
SPINAL CORD	Spinal reflexes; relays sensory and motor information
BRAIN	
Hindbrain (Rhombencephalon)	
Medulla oblongata	Sensory nuclei; reticular activating system; autonomic functions
Pons	Reticular activating system; autonomic functions
Cerebellum	Coordination of movements; balance
Midbrain (Mesencephalon)	Reflexes involving eyes and ears
Forebrain (Prosencephalon)	
Diencephalon	
Thalamus	Relay station for ascending sensory and descending motor tracts; autonomic functions
Hypothalamus	Autonomic functions; neuroendocrine control
Telencephalon (cerebrum)	
Basal ganglia	Motor control
Corpus callosum	Connects and relays information between the two hemispheres
Hippocampus (limbic system)	Memory; emotion
Cerebral cortex	Higher cognitive functions; integrates and interprets sensory information; organizes motor output

is devoted to the processing of *olfactory* (smell) information. The brains of fishes continue growing throughout their lives. This continued growth is in marked contrast to the brains of other classes of vertebrates, which generally complete their development by infancy. The human brain continues to develop through early childhood, but no new neurons are produced once development has ceased, except in the tiny hippocampus, which controls which experiences are filed away into long-term memory and which are forgotten.

The Dominant Forebrain

Starting with the amphibians and continuing more prominently in the reptiles, processing of sensory information is increasingly centered in the forebrain. This pattern was the dominant evolutionary trend in the further development of the vertebrate brain (figure 45.23).

The forebrain in reptiles, amphibians, birds, and mammals is composed of two elements that have distinct functions. The *diencephalon* (Greek *dia*, "between") consists of the thalamus and hypothalamus. The **thalamus** is an integrating and relay center between incoming sensory information and the cerebrum. The **hypothalamus** participates in basic drives and emotions and controls the secretions of the pituitary gland. The *telencephalon*, or "end brain" (Greek *telos*, "end"), is located at the front of the forebrain and is devoted largely to associative activity. In mammals, the telencephalon is called the **cerebrum.**

The Expansion of the Cerebrum

In examining the relationship between brain mass and body mass among the vertebrates, you can see a remarkable difference between fishes and reptiles, on the one hand, and birds and mammals, on the other. Mammals have brains that are particularly large relative to their body mass. This is especially true of porpoises and humans; the human brain weighs about 1.4 kilograms. The increase in brain size in the mammals largely reflects the great enlargement of the cerebrum, the dominant part of the mammalian brain. The cerebrum is the center for correlation, association, and learning in the mammalian brain. It receives sensory data from the thalamus and issues motor commands to the spinal cord via descending tracts of axons.

In vertebrates, the central nervous system is composed of the brain and the spinal cord (see table 45.3). These two structures are responsible for most of the information processing within the nervous system and consist primarily of interneurons and neuroglia. Ascending tracts carry sensory information to the brain. Descending tracts carry impulses from the brain to the motor neurons and interneurons in the spinal cord that control the muscles of the body.

The vertebrate brain consists of three primary regions: the forebrain, midbrain, and hindbrain. The hindbrain was the principal component of the brain of early vertebrates; it was devoted to the control of motor activity. In vertebrates more advanced than fishes, the processing of information is increasingly centered in the forebrain.

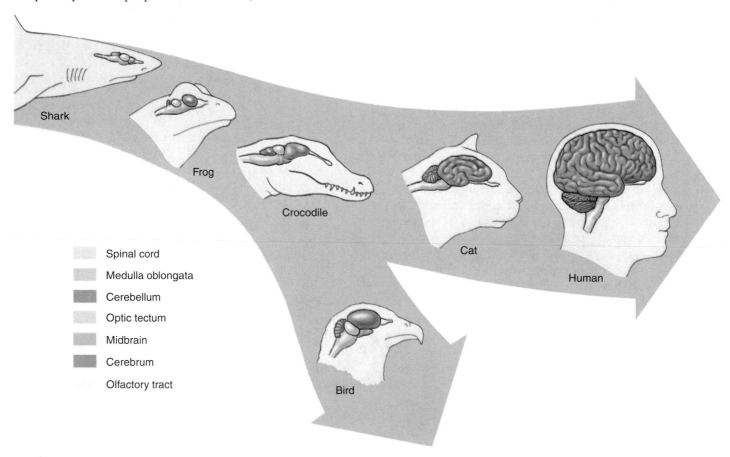

Shark

Frog

Crocodile

Cat

Human

Bird

Spinal cord
Medulla oblongata
Cerebellum
Optic tectum
Midbrain
Cerebrum
Olfactory tract

FIGURE 45.23
Evolution of the vertebrate brain. The relative sizes of different brain regions have changed as vertebrates have evolved. In sharks and other fishes, the hindbrain is predominant, and the rest of the brain serves primarily to process sensory information. In amphibians and reptiles, the forebrain is far larger, and it contains a larger cerebrum devoted to associative activity. In birds, which evolved from reptiles, the cerebrum is even more pronounced. In mammals, the cerebrum covers the optic tectum and is the largest portion of the brain. The dominance of the cerebrum is greatest in humans, in whom it envelops much of the rest of the brain.

The Human Forebrain

The human cerebrum is so large that it appears to envelop the rest of the brain. It is split into right and left cerebral hemispheres, which are connected by a tract called the *corpus callosum* (figure 45.24). The hemispheres are further divided into the *frontal, parietal, temporal,* and *occipital lobes.*

Each hemisphere primarily receives sensory input from the opposite, or contralateral, side of the body and exerts motor control primarily over that side. Therefore, a touch on the right hand, for example, is relayed primarily to the left hemisphere, which may then initiate movement of that hand in response to the touch. Damage to one hemisphere due to a stroke often results in a loss of sensation and paralysis on the contralateral side of the body.

Cerebral Cortex

Much of the neural activity of the cerebrum occurs within a layer of gray matter only a few millimeters thick on its outer surface. This layer, called the **cerebral cortex,** is densely packed with nerve cells. In humans, it contains over 10 billion nerve cells, amounting to roughly 10% of all the neurons in the brain. The surface of the cerebral cortex is highly convoluted; this is particularly true in the human brain, where the convolutions increase the surface area of the cortex threefold.

The activities of the cerebral cortex fall into one of three general categories: motor, sensory, and associative. Each of its regions is associated with a specific function (figure 45.25). The primary motor cortex lies along the *gyrus* (convolution) on the posterior border of the frontal lobe, just in front of the central *sulcus* (crease). Each point on its surface is associated with the movement of a different part of the body (figure 45.26, *right*). Just behind the central sulcus, on the anterior edge of the parietal lobe, lies the primary somatosensory cortex. Each point in this area receives input from sensory neurons serving cutaneous and muscle senses in a particular part of the body (figure 45.26, *left*). Large areas of the primary motor cortex and primary somatosensory cortex are devoted to the fingers, lips, and tongue because of the need for manual dexterity and speech. The auditory cortex lies within the temporal lobe, and different regions of this cortex deal with different sound frequencies. The visual cortex lies on the occipital lobe, with different sites processing information from different positions on the retina, equivalent to particular points in the visual fields of the eyes.

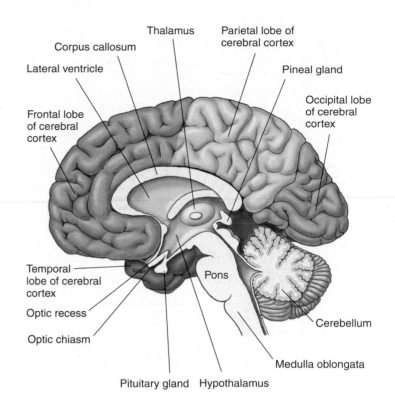

FIGURE 45.24
A section through the human brain. In this sagittal section showing one cerebral hemisphere, the corpus callosum, a fiber tract connecting the two cerebral hemispheres, can be clearly seen.

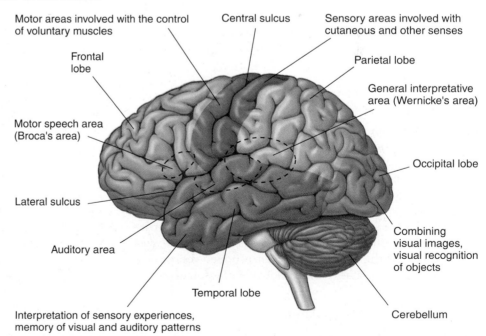

FIGURE 45.25
The cerebrum. This diagram shows the lobes of the cerebrum and indicates some of the known regions of specialization.

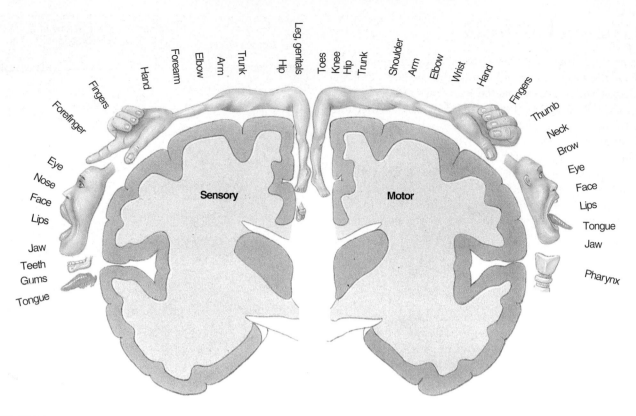

FIGURE 45.26

The primary somatosensory cortex (*left*) and the primary motor cortex (*right*).   Each of these regions of the cerebral cortex is associated with a different region of the body, as indicated in this stylized map. The areas of the body are drawn in relative proportion to the amount of cortex dedicated to their sensation or control. For example, the hands have large areas of sensory and motor control, while the pharynx has a considerable area of motor control but little area devoted to the sensations of the pharynx.

The portion of the cerebral cortex that is not occupied by these motor and sensory cortices is referred to as the *association cortex*. The site of higher mental activities, the association cortex reaches its greatest extent in primates, especially humans, where it makes up 95% of the surface of the cerebral cortex.

Basal Ganglia

Buried deep within the white matter of the cerebrum are several collections of cell bodies and dendrites that produce islands of gray matter. These aggregates of neuron cell bodies, which are collectively termed the *basal ganglia*, receive sensory information from ascending tracts and motor commands from the cerebral cortex and cerebellum. Outputs from the basal ganglia are sent down the spinal cord, where they participate in the control of body movements. Damage to specific regions of the basal ganglia can produce the resting tremor of muscles that is characteristic of Parkinson disease.

Thalamus and Hypothalamus

The thalamus is a primary site of sensory integration in the brain. Visual, auditory, and somatosensory information is sent to the thalamus, where the sensory tracts synapse with association neurons. The sensory information is then relayed via the thalamus to the occipital, temporal, and parietal lobes of the cerebral cortex, respectively. The transfer of each of these types of sensory information is handled by specific aggregations of neuron cell bodies within the thalamus.

The hypothalamus integrates the visceral activities. It helps regulate body temperature, hunger and satiety, thirst, and—along with the limbic system—various emotional states. The hypothalamus also controls the pituitary gland, which in turn regulates many of the other endocrine glands of the body. By means of its interconnections with the cerebral cortex and with control centers in the brain stem (a term used to refer collectively to the midbrain, pons, and medulla oblongata), the hypothalamus helps coordinate the neural and hormonal responses to many internal stimuli and emotions.

The *hippocampus* and *amygdala*, along with the hypothalamus, are the major components of the **limbic system.** This is an evolutionarily ancient group of linked structures deep within the cerebrum that are responsible for emotional responses. The hippocampus is also believed to be important in the formation and recall of memories, a topic we will discuss later.

Language and Other Functions

Arousal and Sleep. The brain stem contains a diffuse collection of neurons referred to as the *reticular formation*. One part of this formation, the reticular activating system, controls consciousness and alertness. All of the sensory pathways feed into this system, which monitors the information coming into the brain and identifies important stimuli. When the reticular activating system has been stimulated to arousal, it increases the level of activity in many parts of the brain. Neural pathways from the reticular formation to the cortex and other brain regions are depressed by anesthetics and barbiturates.

The reticular activating system controls both sleep and the waking state. It is easier to sleep in a dark room than in a lighted one because there are fewer visual stimuli to stimulate the reticular activating system. In addition, activity in this system is reduced by serotonin, a neurotransmitter discussed in section 45.3. Serotonin causes the level of brain activity to fall, bringing on sleep.

Sleep is not the loss of consciousness. Rather, it is an active process whose multiple states can be revealed by recording the electrical activity of the brain in an electroencephalogram (EEG). In a relaxed but awake individual whose eyes are shut, the EEG consists primarily of large, slow waves that occur at a frequency of 8 to 13 hertz (cycles per second). These waves are referred to as *alpha waves*. In an alert subject whose eyes are open, the EEG waves are more rapid (*beta waves* are seen at frequencies of 13 to 30 hertz) and is more desynchronized because multiple sensory inputs are being received, processed, and translated into motor activities.

Theta waves (4 to 7 hertz) and *delta waves* (0.5 to 4 hertz) are seen in various stages of sleep. The first change seen in the EEG with the onset of drowsiness is a slowing and reduction in the overall amplitude of the waves. This slow-wave sleep has several stages but is generally characterized by decreases in arousability, skeletal muscle tone, heart rate, blood pressure, and respiratory rate. During REM sleep (named for the rapid eye movements that occur during this stage), the EEG resembles that of a relaxed, awake individual, and the heart rate, blood pressure, and respiratory rate are all increased. Paradoxically, individuals in REM sleep are difficult to arouse and are more likely to

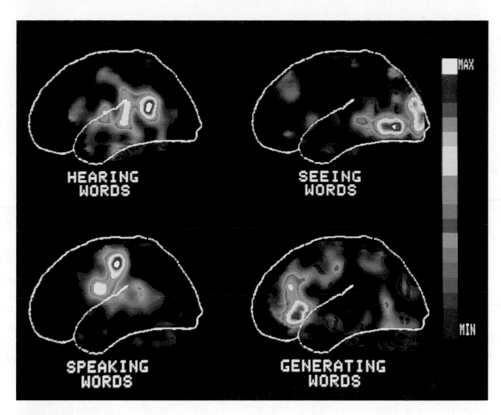

FIGURE 45.27
Different brain regions control various language activities. This illustration shows how the brain reacts in human subjects asked to listen to a spoken word, to read that same word silently, to repeat the word out loud, and then to speak a word related to the first. Regions of white, red, and yellow show the greatest activity. Compare this with figure 45.25 to see how regions of the brain are mapped.

awaken spontaneously. Dreaming occurs during REM sleep, and the rapid eye movements resemble the tracking movements made by the eyes when awake, suggesting that dreamers "watch" their dreams.

Language and Spatial Recognition. Although the two cerebral hemispheres seem structurally similar, they are responsible for different activities. The most thoroughly investigated example of this lateralization of function is language. The left hemisphere is the "dominant" hemisphere for language in 90% of right-handed people and nearly two-thirds of left-handed people. (By *dominant*, we mean it is the hemisphere in which most neural processing related to language is performed.) Different brain regions control language in the dominant hemisphere (figure 45.27). Wernicke's area (see figure 45.25), located in the parietal lobe between the primary auditory and visual areas, is important for language comprehension and the formulation of thoughts into speech. Broca's area, found near the part of the motor cortex controlling the face, is responsible for the generation of motor output needed for language communication. Damage to these brain areas can cause language disorders known as *aphasias*. For example, if Wernicke's area is damaged, the person's speech is rapid and fluid but lacks meaning; words are tossed together as in a "word salad."

While the dominant hemisphere for language is adept at sequential reasoning, like that needed to formulate a sentence, the nondominant hemisphere (the right hemisphere in most people) is adept at spatial reasoning, the type of reasoning needed to assemble a puzzle or draw a picture. It is also the hemisphere primarily involved in musical ability—a person with damage to Broca's speech area in the left hemisphere may not be able to speak but may retain the ability to sing! Damage to the nondominant hemisphere may lead to an inability to appreciate spatial relationships and may impair musical activities such as singing. Even more specifically, damage to the inferior temporal cortex in that hemisphere eliminates the capacity to recall faces. Reading, writing, and oral comprehension remain normal, and patients with this disability can still recognize acquaintances by their voices. The nondominant hemisphere is also important for the consolidation of memories of nonverbal experiences.

Memory and Learning.   One of the great mysteries of the brain is the basis of memory and learning. There is no one part of the brain in which all aspects of a memory appear to reside. Specific cortical sites cannot be identified for particular memories because relatively extensive cortical damage does not selectively remove memories. Although memory is impaired if portions of the brain, particularly the temporal lobes, are removed, it is not lost entirely. Many memories persist in spite of the damage, and the ability to access them gradually recovers with time. Therefore, investigators who have tried to probe the physical mechanisms underlying memory often have felt that they were grasping at a shadow. Although we still do not have a complete understanding of these mechanisms, we have learned a good deal about the basic processes in which memories are formed.

There appear to be fundamental differences between short-term and long-term memory. Short-term memory is transient, lasting only a few moments. Such memories can readily be erased by the application of an electrical shock, leaving previously stored long-term memories intact. This result suggests that short-term memories are stored electrically in the form of a transient neural excitation. Long-term memory, in contrast, appears to involve structural changes in certain neural connections within the brain. Two parts of the temporal lobes, the hippocampus and the amygdala, are involved in both short-term memory and its consolidation into long-term memory. Damage to these

structures impairs the ability to process recent events into long-term memories.

Synapses that are used intensively for a short period of time display more effective synaptic transmission upon subsequent use. This phenomenon is called long-term potentiation (LTP). During LTP, the presynaptic neuron may release increased amounts of neurotransmitter with each action potential, and the postsynaptic neuron may become increasingly sensitive to the neurotransmitter. It is believed that these changes in synaptic transmission may be responsible for some aspects of memory storage.

Mechanism of Alzheimer Disease Still a Mystery

In the past, little was known about *Alzheimer disease*, a condition in which the memory and thought processes of the brain become dysfunctional. Drug companies are eager to develop new products for the treatment of Alzheimer disease, but they have little concrete evidence to go on. Scientists disagree about the biological nature of the disease and its cause. Two hypotheses have been proposed: One suggests that nerve cells in the brain are killed from the outside in, and the other that the cells are killed from the inside out.

In the first hypothesis, external proteins called β-amyloid peptides kill nerve cells. A mistake in protein processing produces an abnormal form of the peptide, which then forms aggregates, or plaques. The plaques begin to fill in the brain and then damage and kill nerve cells. However, these amyloid plaques have been found in autopsies of people who did not have Alzheimer disease.

The second hypothesis maintains that the nerve cells are killed by an abnormal form of an internal protein. This protein, called tau (τ), normally functions to maintain protein transport microtubules. Abnormal forms of τ assemble into helical segments that form tangles, which interfere with the normal functioning of the nerve cells. Researchers continue to study whether tangles and plaques are causes or effects of Alzheimer disease.

Progress has been made in identifying genes that increase the likelihood of developing Alzheimer disease and genes that, when mutated, can cause the disorder. Most Alzheimer patients do not have these mutated genes, but for those that do, the symptoms of Alzheimer disease are expressed much earlier in life.

The cerebrum is composed of two cerebral hemispheres. Each hemisphere consists of the gray matter of the cerebral cortex overlying white matter and islands of gray matter (nuclei) called the basal ganglia. These areas are involved in the integration of sensory information, control of body movements, and such associative functions as learning and memory.

The Spinal Cord

The spinal cord is a cable of neurons extending from the brain down through the backbone (figure 45.28). It is enclosed and protected by the vertebral column and layers of membranes called *meninges*, which also cover the brain. Inside the spinal cord are two zones. The inner zone, called gray matter, primarily consists of the cell bodies of interneurons, motor neurons, and neuroglia. The outer zone, called white matter, contains cables of sensory axons in the dorsal columns and motor axons in the ventral columns. These nerve tracts may also contain the dendrites of other nerve cells. Messages from the body and the brain run up and down the spinal cord, the body's "information highway."

In addition to relaying messages, the spinal cord also functions in reflexes, the sudden, involuntary movement of muscles. A reflex produces a rapid motor response to a stimulus because the sensory neuron passes its information to a motor neuron in the spinal cord, without higher-level processing. One of the most frequently used reflexes in your body is blinking, a reflex that protects your eyes. If an object such as an insect or a cloud of dust approaches your eye, the eyelid blinks before you realize what has happened. The reflex occurs before the cerebrum is aware the eye is in danger.

FIGURE 45.28
A view down the human spinal cord. Pairs of spinal nerves can be seen extending from the spinal cord. Along these nerves, as well as the cranial nerves that arise from the brain, the central nervous system communicates with the rest of the body.

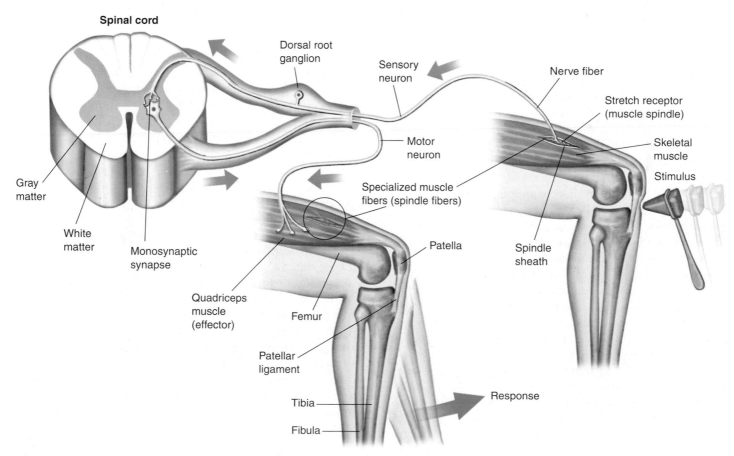

FIGURE 45.29
The knee-jerk reflex. This is the simplest reflex, involving only sensory and motor neurons.

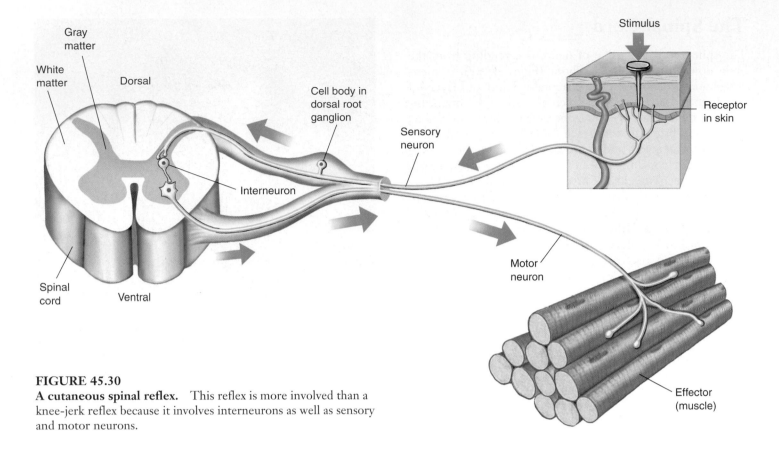

FIGURE 45.30
A cutaneous spinal reflex. This reflex is more involved than a knee-jerk reflex because it involves interneurons as well as sensory and motor neurons.

Because they pass information along only a few neurons, reflexes are very fast. Many reflexes never reach the brain. The nerve impulse travels only as far as the spinal cord and then comes right back as a motor response. A few reflexes, such as the knee-jerk reflex (figure 45.29), are monosynaptic reflex arcs. In these, the sensory nerve cell makes synaptic contact directly with a motor neuron in the spinal cord whose axon travels directly back to the muscle. The knee-jerk reflex is also an example of a *muscle stretch reflex*. When the muscle is briefly stretched by tapping the patellar ligament with a rubber mallet, the *muscle spindle apparatus* is also stretched. The spindle apparatus is embedded within the muscle, and, like the muscle fibers outside the spindle, is stretched along with the muscle. Stretching of the spindle activates sensory neurons that synapse directly with somatic motor neurons within the spinal cord. As a result, the somatic motor neurons conduct action potentials to the skeletal muscle fibers and stimulate the muscle to contract. This reflex is the simplest in the vertebrate body because only one synapse is crossed in the reflex arc.

Most reflexes in vertebrates, however, involve a single connecting interneuron between the sensory and the motor neuron (figure 45.30). The withdrawal of a hand from a hot stove or the blinking of an eye in response to a puff of air involves a relay of information from a sensory neuron through one or more interneurons to a motor neuron. The motor neuron then stimulates the appropriate muscle to contract.

Spinal Cord Regeneration

In the past, scientists have tried to repair severed spinal cords by installing nerves from another part of the body to bridge the gap and act as guides for the spinal cord to regenerate. But most of these experiments have failed because, although axons may regenerate through the implanted nerves, they cannot penetrate the spinal cord tissue once they leave the implant. Also, there is a factor that inhibits nerve growth in the spinal cord. However, after discovering that fibroblast growth factor stimulates nerve growth, neurobiologists working with rats tried gluing on the nerves, from the implant to the spinal cord, with fibrin that had been mixed with the fibroblast growth factor. Three months later, rats with the nerve bridges began to show movement in their lower bodies. In further analyses of the experimental animals, dye tests indicated that the spinal cord nerves had regrown from both sides of the gap. Many scientists are encouraged by the potential to use a similar treatment in human medicine. However, most spinal cord injuries in humans do not involve a completely severed spinal cord; often, nerves are crushed, which results in different tissue damage. Also, while the rats with nerve bridges did regain some locomotory ability, tests indicated that they were barely able to walk or stand.

The spinal cord relays messages to and from the brain and processes some sensory information directly.

45.5 The peripheral nervous system consists of sensory and motor neurons.

Components of the Peripheral Nervous System

The peripheral nervous system consists of nerves and ganglia. Nerves are cablelike collections of axons (figure 45.31), usually containing both sensory and motor neurons. Ganglia are aggregations of neuron cell bodies located outside the central nervous system.

At its origin, a spinal nerve separates into sensory and motor components. The axons of sensory neurons enter the dorsal surface of the spinal cord and form the **dorsal root** of the spinal nerve, whereas motor axons leave from the ventral surface of the spinal cord and form the **ventral root** of the spinal nerve. The cell bodies of sensory neurons are grouped together outside each level of the spinal cord in the **dorsal root ganglia.** The cell bodies of somatic motor neurons, on the other hand, are located within the spinal cord and so are not located in ganglia.

Somatic motor neurons stimulate skeletal muscles to contract, and autonomic motor neurons innervate involuntary effectors—smooth muscles, cardiac muscle, and glands. A comparison of the somatic and autonomic nervous systems is provided in table 45.4, and we will discuss each system in turn. Somatic motor neurons stimulate the skeletal muscles of the body to contract in response to conscious commands and as part of reflexes that do not require conscious control. Conscious control of skeletal muscles is achieved by activation of tracts of axons that descend from the cerebrum to the appropriate level of the spinal cord. Some of these descending axons stimulate spinal cord motor neurons directly, while others activate interneurons that in turn stimulate the spinal motor neurons. When a particular muscle is stimulated to contract, however, its antagonist must be inhibited. In order to flex the arm, for example, the flexor muscles must be stimulated while the antagonistic extensor muscle is inhibited (see figure 42.17). Descending motor axons produce this necessary inhibition by causing hyperpolarizations (IPSPs) of the spinal motor neurons that innervate the antagonistic muscles.

A spinal nerve contains sensory neurons that enter the dorsal root and motor neurons that enter the ventral root of the nerve. Somatic motor neurons innervate skeletal muscles and stimulate the muscles to contract.

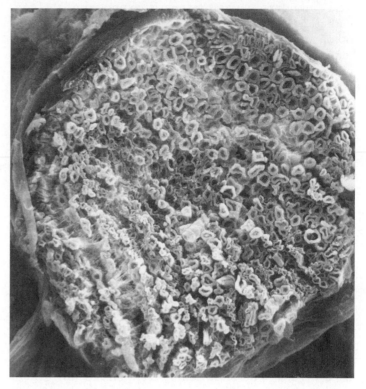

FIGURE 45.31
Nerves in the peripheral nervous system. Photomicrograph (1600×) showing a cross section of a bullfrog nerve. The nerve is a bundle of axons bound together by connective tissue. Many myelinated axons are visible, each looking somewhat like a doughnut.

Table 45.4	Comparison of the Somatic and Autonomic Nervous Systems	
Characteristic	**Somatic**	**Autonomic**
Effectors	Skeletal muscle	Cardiac muscle
		Smooth muscle
		Gastrointestinal tract
		Blood vessels
		Airways
		Exocrine glands
Effect on motor nerves	Excitation	Excitation or inhibition
Innervation of effector cells	Always single	Typically dual
Number of sequential neurons in path to effector	One	Two
Neurotransmitter	Acetylcholine	Acetylcholine, norepinephrine

The Autonomic Nervous System

The autonomic nervous system is composed of the sympathetic and parasympathetic divisions and the medulla oblongata of the hindbrain, which coordinates this system. Although they differ, the sympathetic and parasympathetic divisions share several features. In both, the efferent motor pathway involves two neurons: The first has its cell body in the CNS and sends an axon to an autonomic ganglion, while the second has its cell body in the autonomic ganglion and sends its axon to synapse with a smooth muscle, cardiac muscle, or gland cell (figure 45.32). The first neuron is called a *preganglionic neuron*, and it always releases ACh at its synapse. The second neuron is a *postganglionic neuron*; those in the parasympathetic division release ACh, while those in the sympathetic division release norepinephrine.

In the sympathetic division, the preganglionic neurons originate in the thoracic and lumbar regions of the spinal cord (figure 45.33, *left*). Most of the axons from these neurons synapse in two parallel chains of ganglia immediately outside the spinal cord. These structures are usually called the *sympathetic chain* of ganglia. The sympathetic chain contains the cell bodies of postganglionic neurons, and it is the axons from these neurons that innervate the different visceral organs. There are some exceptions to this general pattern, however. Most importantly, the axons of some preganglionic sympathetic neurons pass through the sym-

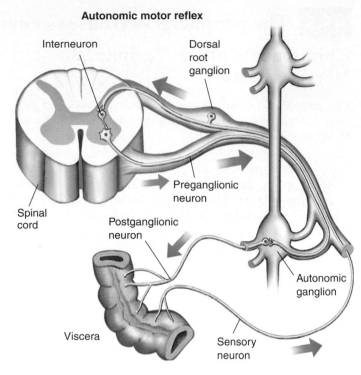

Autonomic motor reflex

FIGURE 45.32

An autonomic neural path. There are two motor neurons in the efferent pathway. The first, or preganglionic neuron, exits the CNS and synapses at an autonomic ganglion. The second, or postganglionic neuron, exits the ganglion and regulates the visceral effectors (smooth muscle, cardiac muscle, or glands).

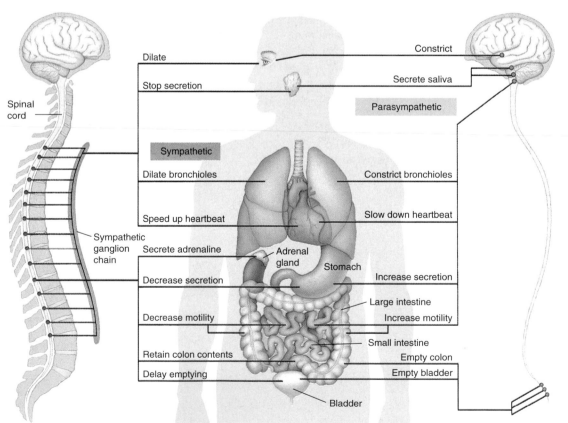

FIGURE 45.33

The sympathetic and parasympathetic divisions of the autonomic nervous system. The preganglionic neurons of the sympathetic division exit the thoracic and lumbar regions of the spinal cord, while those of the parasympathetic division exit the brain and sacral region of the spinal cord. The ganglia of the sympathetic division are located near the spinal cord, while those of the parasympathetic division are located near the organs they innervate. Most of the internal organs are innervated by both divisions.

Table 45.5 Autonomic Innervation of Target Tissues

Target Tissue	Sympathetic Stimulation	Parasympathetic Stimulation
Pupil of eye	Dilation	Constriction
Glands		
Salivary	Vasoconstriction; slight secretion	Vasodilation; copious secretion
Gastric	Inhibition of secretion	Stimulation of gastric activity
Liver	Stimulation of glucose secretion	Inhibition of glucose secretion
Sweat	Sweating	None
Gastrointestinal tract		
Sphincters	Increased tone	Decreased tone
Wall	Decreased tone	Increased motility
Gallbladder	Relaxation	Contraction
Urinary bladder		
Muscle	Relaxation	Contraction
Sphincter	Contraction	Relaxation
Heart muscle	Increased rate and strength	Decreased rate
Lungs	Dilation of bronchioles	Constriction of bronchioles
Blood vessels		
In muscles	Dilation	None
In skin	Constriction	None
In viscera	Constriction	Dilation

pathetic chain without synapsing and, instead, terminate within the adrenal gland. The adrenal gland consists of an outer part, or cortex, and an inner part, or medulla. The adrenal medulla receives sympathetic nerve innervation and secretes the hormone epinephrine (adrenaline) in response.

When the sympathetic division becomes activated, epinephrine is released into the blood as a hormonal secretion, and norepinephrine is released at the synapses of the postganglionic neurons. Epinephrine and norepinephrine act to prepare the body for fight or flight (figure 45.34). Physiological effects on the body include faster and stronger heartbeat, increased blood glucose concentration, blood flow diverted to the muscles and heart, and dilated bronchioles (table 45.5).

These responses are antagonized by the parasympathetic division. Preganglionic parasympathetic neurons originate in the brain and sacral regions of the spinal cord (see figure 45.33, *right*). Because of this origin, there cannot be a chain of parasympathetic ganglia analogous to the sympathetic chain. Instead, the preganglionic axons, many of which travel in the vagus (the tenth cranial

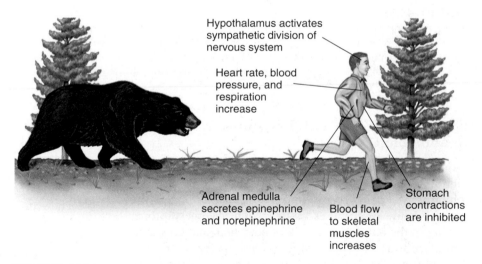

Hypothalamus activates sympathetic division of nervous system

Heart rate, blood pressure, and respiration increase

Adrenal medulla secretes epinephrine and norepinephrine

Blood flow to skeletal muscles increases

Stomach contractions are inhibited

FIGURE 45.34
The sympathetic division of the nervous system in action. To prepare the body for fight or flight, the sympathetic division is activated and causes changes in many organs, glands, and body processes.

nerve, terminate in ganglia located near or even within the internal organs. The postganglionic neurons then regulate the internal organs by releasing ACh at their synapses. Parasympathetic nerve effects include a slowing of the heart, increased secretions and activities of digestive organs, and so on.

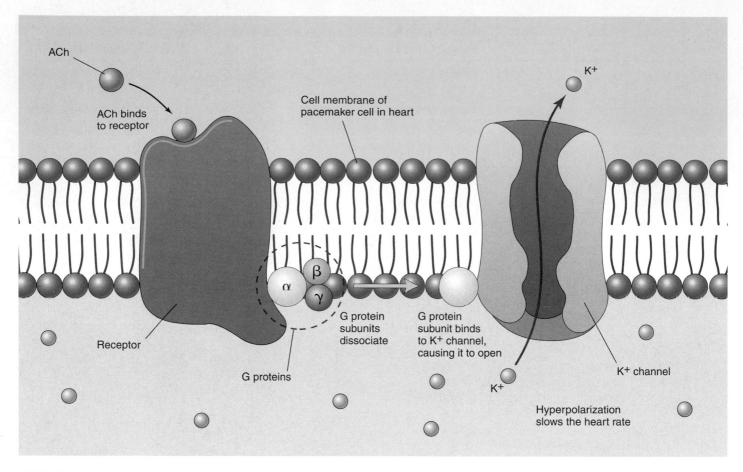

FIGURE 45.35

The parasympathetic effects of ACh require the action of G proteins.   The binding of ACh to its receptor causes dissociation of a G protein complex, releasing some components of this complex to move within the membrane and bind to other proteins that form ion channels. Shown here are the effects of ACh on the heart, where the G protein components cause the opening of potassium channels. This leads to outward diffusion of potassium and hyperpolarization, slowing the heart rate.

G Proteins Mediate Cell Responses to Autonomic Nerves

You might wonder how ACh can slow the heart rate—an inhibitory effect—when it has excitatory effects elsewhere. This inhibitory effect in the pacemaker cells of the heart is produced because ACh causes the potassium channels to open, leading to the outward diffusion of potassium and thus to hyperpolarization. This and other parasympathetic effects of ACh are produced indirectly, using a group of membrane proteins called **G proteins** (so called because they are regulated by guanosine diphosphate and guanosine triphosphate [GDP and GTP]). Because the ion channels are located some distance away from the receptor proteins for ACh, the G proteins are needed to serve as connecting links between them.

There are three G protein subunits, designated α, β, and γ, bound together and attached to the receptor protein for ACh. When ACh, released by parasympathetic endings, binds to its receptor, the G protein subunits dissociate (figure 45.35). Specific G protein components move within the membrane to the potassium channel and cause it to open,

producing hyperpolarization and a slowing of the heart. In other organs, the G proteins have different effects that lead to excitation. In this way, for example, the parasympathetic nerves that innervate the stomach can cause increased gastric secretions and contractions.

The sympathetic nerve effects also involve the action of G proteins. Stimulation by norepinephrine from sympathetic nerve endings and epinephrine from the adrenal medulla requires G proteins to activate the target cells. We will describe this in more detail, together with hormone action, in chapter 47.

The sympathetic division of the autonomic system, together with the adrenal medulla, activates the body for fight-or-flight responses, whereas the parasympathetic division generally promotes relaxation and digestion. When both systems innervate the same effector, they generally have antagonistic effects. The actions of parasympathetic nerves are produced by ACh, whereas the actions of sympathetic nerves are produced by norepinephrine.

45.1 The nervous system consists of neurons and supporting cells.

Neuron Organization

- Sensory neurons carry impulses into the central nervous system (CNS), and motor neurons carry impulses away from the CNS. (p. 940)
- Sensory and motor neurons together form the peripheral nervous system (PNS). (p. 940)
- Neurons are supported by neuroglia. Two of the most important neuroglia in vertebrates are Schwann cells and oligodendrocytes, which produce myelin sheaths. (p. 941)

45.2 Nerve impulses are produced on the axon membrane.

The Resting Membrane Potential

- When a neuron is not being stimulated, it maintains a resting membrane potential. (p. 942)
- The inside of a cell is more negatively charged in relation to the outside of the cell, thus cations outside the cell are attracted to anions inside the cell. (p. 942)
- Due to combined activities of fixed anions, the sodium-potassium pump, and the membrane leak channels, the resting membrane potential for an average neuron remains stable at -70 mV. (pp. 942–943)

Graded Potentials Generate Action Potentials

- Graded potentials are caused by the activation of gated ion channels, and the ability of graded potentials to combine is referred to as summation, which results in the inward diffusion of Na^+ and the outward diffusion of K^+. (p. 944)
- Threshold refers to the level of depolarization required to produce an action potential. (p. 944)
- Action potentials are all-or-none events, and are always separate events. (pp. 944–947)
- Action potentials are propagated as events are reproduced at different points along the membrane. (p. 947)

45.3 Neurons form junctions called synapses with other cells.

Structure of Synapses

- Electrical synapses are characteristic of invertebrate nervous systems, but somewhat rare in vertebrates, while the vast majority of vertebrate synapses are chemical. (p. 948)
- Most synapses have a synaptic cleft separating the two cells. (p. 948)
- Neurotransmitters diffuse rapidly across the cleft and bind to receptor proteins in order to transmit a nerve impulse. (p. 948)

Neurotransmitters and Their Functions

- Acetylcholine binds to receptor proteins in the postsynaptic membrane and causes ion channels to open, producing an excitatory postsynaptic potential (EPSP). (p. 949)
- Glutamate is the major excitatory neurotransmitter in the vertebrate CNS, producing EPSPs, and glycine and GABA are inhibitory neurotransmitters that lead to an inhibitory postsynaptic potential (IPSP). (p. 950)
- Biogenic amines are a class of neurotransmitters that include dopamine, norepinephrine, and serotonin. (p. 950)
- If receptor proteins within synapses are exposed to high levels of neurotransmitter molecules for prolonged periods, the nerve cell often responds by decreasing the number of receptor proteins in the postsynaptic membrane. (pp. 951–953)

45.4 The central nervous system consists of the brain and spinal cord.

The Evolution of the Vertebrate Brain

- Cnidarians exhibit the simplest nervous systems, and in flatworms, there is some associative activity. All subsequent evolutionary changes in nervous systems can be viewed as a series of elaborations on characteristics already present in flatworms. (p. 954)
- The vertebrate brain is composed of three major regions: forebrain, midbrain, and hindbrain. (pp. 954–955)
- The hindbrain was the major component in early brains, while starting with amphibian evolution, processing of sensory information is increasingly centered in the forebrain. (pp. 955–956)

The Human Forebrain

- The cerebrum is divided into right and left cerebral hemispheres, which are further divided into frontal, parietal, temporal, and occipital lobes. (p. 957)
- Much of neural activity occurs within the cerebral cortex. The activities of the cerebral cortex fall within the general categories of motor, sensory, and associative. (p. 957)
- The thalamus is a primary site of sensory integration, receiving visual, auditory, and somatosensory information. (p. 958)
- The hypothalamus integrates visceral activities and controls the pituitary gland. (p. 958)
- The left hemisphere is usually the dominant hemisphere for language. (p. 959)
- No one part of the brain appears to control memory. Short-term memory appears to be transient, while long-term memory appears to involve structural changes in neural connections. (p. 960)

The Spinal Cord

- The spinal cord is a cable of neurons extending from the brain down the backbone. It functions in message relaying and reflexes. (pp. 961–962)

45.5 The peripheral nervous system consists of sensory and motor neurons.

Components of the Peripheral Nervous System

- The peripheral nervous system consists of nerves (cablelike collections of axons) and ganglia (aggregations of neuron cell bodies located outside the central nervous system). (p. 963)
- Somatic motor neurons stimulate skeletal muscles to contract, and autonomic motor neurons innervate smooth muscle, cardiac muscle, and glands. (p. 963)

The Autonomic Nervous System

- The autonomic nervous system is composed of the sympathetic and parasympathetic divisions and the medulla oblongata of the hindbrain. (p. 964)
- When the sympathetic division becomes activated, epinephrine and norepinephrine prepare the body for a fight-or-flight response. Parasympathetic nerve actions produced by ACh slow the heart and increase secretions of the digestive organs, thus causing an antagonistic reaction to the fight-or-flight response. (pp. 964–966)

Self Test

1. _____ nerves carry impulses away from the central nervous system.
   a. Sensory
   b. Motor
   c. Afferent
   d. Association
2. Which of the following best describes the electrical state of a neuron at rest?
   a. The inside of a neuron is more negatively charged than the outside.
   b. The outside of a neuron is more negatively charged than the inside.
   c. The inside and the outside of a neuron have the same electrical charge.
   d. K^+ ions leak into a neuron at rest.
3. A nerve impulse is initiated when
   a. physical disruption of the cell membrane causes some of its contents, including ions, to leak out.
   b. the Schwann cells move into their new positions.
   c. voltage-gated channels close.
   d. a reversal in the polarized state of the cell causes it to reach threshold.
4. An action potential travels from one neuron to the next across the synapse using
   a. calcium ions.
   b. Schwann cells.
   c. neurotransmitters.
   d. All of these are involved.
5. Synapses are excitatory or inhibitory based on
   a. integration.
   b. summation.
   c. autonomic control.
   d. saltatory conduction.
6. Modern-day fish are like early vertebrates in that the dominant part of their brains is the
   a. forebrain.
   b. midbrain.
   c. hindbrain.
   d. optic lobes.
7. Much of the sensory, motor, and associative neural activity of the cerebrum occurs on the surface, a layer called the
   a. corpus callosum.
   b. cerebral cortex.
   c. limbic system.
   d. basal ganglia.
8. Which is not involved in the knee-jerk reflex?
   a. stretching of the muscle
   b. motor neuron
   c. muscle spindle
   d. an interneuron
9. The _____ cannot be controlled by conscious thought.
   a. motor neurons
   b. somatic nervous system
   c. autonomic nervous system
   d. skeletal muscles
10. A fight-or-flight response in the body is controlled by the
    a. sympathetic division of the nervous system.
    b. parasympathetic division of the nervous system.
    c. release of ACh from postganglionic neurons.
    d. somatic nervous system.

Test Your Visual Understanding

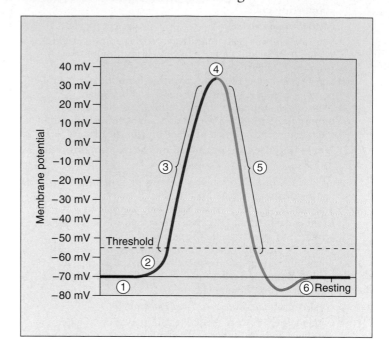

1. Match the following descriptions with the appropriate numbered steps on the figure.
   a. Na^+ channels open, and Na^+ flows into the cell, causing rapid depolarization.
   b. K^+ channels open, and K^+ flows out of the cell, causing rapid repolarization.
   c. Na^+ channels are closed, and some K^+ leaks out to maintain a resting potential.
   d. K^+ continues to flow out of the cell, producing a hyperpolarization of the membrane.
   e. Na^+ channels close, and K^+ channels begin to open.
   f. A stimulus causes a depolarization of the membrane.

Apply Your Knowledge

1. The equilibrium potentials for particular ions are determined by their concentrations inside and outside of the cell and are calculated using the Nernst equation: $V = 58\text{ mV} \times \log_{10} C_o/C_i$ where C_o is the concentration of the ion outside the cell and C_i is the concentration of the ion inside the cell. Using this equation, calculate the equilibrium potential across the plasma membrane assuming that it is due solely to (a) K^+ and (b) Na^+. The concentrations for K^+ are 150 mM inside and 5 mM outside. The concentrations for Na^+ are 15 mM inside and 150 mM outside.
2. Tetraethylammonium (TEA) is a drug that blocks voltage-gated K^+ channels. What effect would TEA have on the action potentials produced by a neuron? If TEA could be applied selectively to a presynaptic neuron that releases an excitatory neurotransmitter, how would it alter the synaptic effect of that neurotransmitter on the postsynaptic cell?

46

Sensory Systems

Concept Outline

46.1 Animals employ a wide variety of sensory receptors.

Categories of Sensory Receptors and Their Actions. Sensory receptors can be classified according to the type of stimuli to which they can respond.

46.2 Mechanical and chemical receptors sense the body's condition.

Detecting Temperature and Pressure. Receptors within the skin respond to touch, pressure, pain, heat, and cold.

Sensing Muscle Contraction and Blood Pressure. A muscle spindle responds to stretching of the muscle; receptors in arteries monitor changes in blood pressure.

Sensing Taste, Smell, and Body Position. Receptors that respond to chemicals produce sensations of taste and smell. Hair cells send nerve impulses when they are bent.

46.3 Auditory receptors detect pressure waves in the air.

The Ears and Hearing. Sound causes vibrations in the ear that bend hair cell processes, initiating a nerve impulse. Some mammals use echolocation, a type of sonar, to locate objects in the dark using sound waves.

46.4 Optic receptors detect light over a broad range of wavelengths.

Evolution of the Eye. True image-forming eyes evolved independently in several phyla.

Vertebrate Photoreceptors. Light causes a pigment molecule in a rod or cone cell to dissociate; this "bleaching" reaction activates the photoreceptor.

Visual Processing in the Vertebrate Retina. Action potentials travel from the retina of the eyes to the brain for visual perception.

46.5 Some vertebrates use heat, electricity, or magnetism for orientation.

Diversity of Sensory Experiences. Special receptors can detect heat, electrical currents, and magnetic fields.

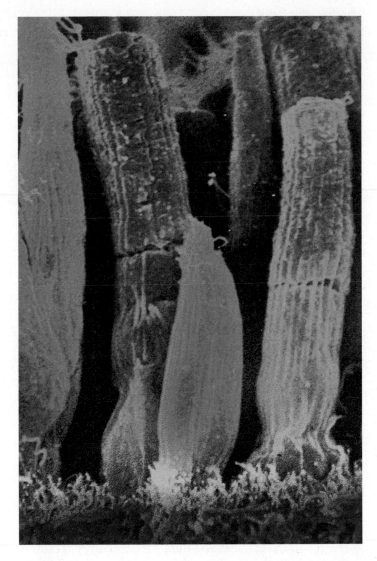

FIGURE 46.1
Photoreceptors in the vertebrate eye. Rods, the broad, tubular cells, allow black-and-white vision, while cones, the short, tapered cells, are responsible for color vision. Not all vertebrates have both types of receptors.

All input from sensory neurons to the central nervous system arrives in the same form, as action potentials propagated by afferent (inward-conducting) sensory neurons. Different sensory neurons project to different brain regions, and so are associated with different sensory modalities (figure 46.1). The intensity of the sensation depends on the frequency of action potentials conducted by the sensory neuron. A sunset, a symphony, and searing pain are distinguished by the brain only in terms of the identity of the sensory neuron carrying the action potentials and the frequency of these impulses. Thus, if the auditory nerve is artificially stimulated, the brain perceives the stimulation as sound. But if the optic nerve is artificially stimulated in exactly the same manner and degree, the brain perceives a flash of light.

46.1 Animals employ a wide variety of sensory receptors.

Categories of Sensory Receptors and Their Actions

Sensory information is conveyed to the central nervous system (CNS) and perceived in a four-step process (figure 46.2):

1. *Stimulation.* A physical stimulus impinges on a sensory neuron or an associated, but separate, sensory receptor.
2. *Transduction.* The stimulus energy is transformed into graded potentials in the dendrites of the sensory neuron.
3. *Transmission.* The axon of the sensory neuron conducts action potentials along an afferent pathway to the CNS.
4. *Interpretation.* The brain creates a sensory perception from the electrochemical events produced by afferent stimulation. We actually see (as well as hear, touch, taste, and smell) with our brains, not with our sense organs.

Sensory receptors differ with respect to the nature of the environmental stimulus that best activates their sensory dendrites. Broadly speaking, we can recognize three classes of environmental stimuli: (1) mechanical forces, which stimulate **mechanoreceptors;** (2) chemicals, which stimulate **chemoreceptors;** and (3) electromagnetic and thermal energy, which stimulate a variety of receptors, including the **photoreceptors** of the eyes (table 46.1).

The simplest sensory receptors are *free nerve endings* that respond to bending or stretching of the sensory neuron membrane, to changes in temperature, or to chemicals such as oxygen in the extracellular fluid. Other sensory receptors are more complex, involving the association of the sensory neurons with specialized epithelial cells.

Sensing the External and Internal Environments

Exteroceptors are receptors that sense stimuli that arise in the external environment. Almost all of a vertebrate's exterior senses evolved in water before vertebrates invaded the land. Consequently, many senses of terrestrial vertebrates emphasize stimuli that travel well in water, using receptors that have been retained in the transition from the sea to the land. Mammalian hearing, for example, converts an airborne stimulus into a waterborne one, using receptors similar to those that originally evolved in the water. A few vertebrate sensory systems that function well in the water, such as the electrical organs of fish, cannot function in the air and are not found among terrestrial vertebrates. On the other hand, some land-dwellers have sensory systems, such as infrared receptors, that could not function in the sea.

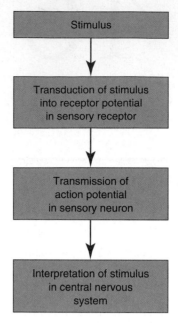

FIGURE 46.2
The path of sensory information. Sensory stimuli are transduced into receptor potentials, which can trigger sensory neuron action potentials that are conducted to the brain for interpretation.

Table 46.1	Classes of Environmental Stimuli	
Mechanical Forces	**Chemicals**	**Electromagnetic and Thermal Energy**
Pressure	Taste	Light
Gravity	Smell	Heat
Inertia	Humidity	Electricity
Sound		Magnetism
Touch		
Vibration		

Sensory systems can provide several levels of information about the external environment. Some sensory systems provide only enough information to determine that an object is present; they call the animal's attention to the object but give little or no indication of where it is located. Other sensory systems provide information about the location of an object, permitting the animal to move toward it. Still other sensory systems enable the brain to construct a three-dimensional image of an object and its surroundings.

Table 46.2 Sensory Transduction Among the Vertebrates

Stimulus	Receptor	Location	Structure	Transduction Process
INTEROCEPTION				
Temperature	Heat receptors and cold receptors	Skin, hypothalamus	Free nerve ending	Temperature change opens/closes ion channels in membrane
Touch	Meissner's corpuscles, Hair follicle receptors, Merkel cells, Ruffinicorpuscles	Skin epithelium	Nerve ending within elastic capsule	Rapid or extended change in pressure deforms membrane
Vibration	Pacinian corpuscles	Deep within skin	Nerve ending within elastic capsule	Severe change in pressure deforms membrane
Pain	Nociceptors	Throughout body	Free nerve ending	Chemicals or changes in pressure or temperature open/close ion channels in membrane
Muscle stretch	Stretch receptors	Within muscles	Spiral nerve endings wrapped around muscle spindle	Stretch of spindle deforms membrane
Blood pressure	Baroreceptors	Arterial branches	Nerve endings over thin part of arterial wall	Stretch of arterial wall deforms membrane
EXTEROCEPTION				
Gravity	Statocysts	Outer chambers of inner ear	Otoliths and hair cells	Otoliths deform hair cells
Motion	Cupula	Semicircular canals of inner ear	Collection of hair cells	Fluid movement deforms hair cells
	Lateral line organ	Within grooves on body surface of fish	Collection of hair cells	Fluid movement deforms hair cells
Taste	Taste bud cells	Mouth; skin of fish	Chemoreceptors: epithelial cells with microvilli	Chemicals bind to membrane receptors
Smell	Olfactory neurons	Nasal passages	Chemoreceptors: ciliated neurons	Chemicals bind to membrane receptors
Hearing	Organ of Corti	Cochlea of inner ear	Hair cells between basilar and tectorial membranes	Sound waves in fluid move hairs, deform membranes
Vision	Rod and cone cells	Retina of eye	Array of photosensitive pigments	Light initiates process that closes ion channels
Heat	Pit organ	Face of snake	Temperature receptors in two chambers	Receptors compare temperatures of surface and interior chambers
Electricity	Ampullae of Lorenzini	Within skin of fishes	Closed vesicles with asymmetrical ion channel distribution	Electrical field alters ion distribution on membranes
Magnetism	Unknown	Unknown	Unknown	Deflection at magnetic field initiates nerve impulses?

Interoceptors sense stimuli that arise from within the body. These internal receptors detect stimuli related to muscle length and tension, limb position, pain, blood chemistry, blood volume and pressure, and body temperature. Many of these receptors are simpler than those that monitor the external environment and are believed to bear a closer resemblance to primitive sensory receptors. In the rest of this chapter, we will consider the different types of interoceptors and exteroceptors according to the kind of stimulus each is specialized to detect (table 46.2).

Sensory Transduction

Sensory cells respond to stimuli because they possess *stimulus-gated ion channels* in their membranes. The sensory stimulus causes these ion channels to open or close, depending on the sensory system involved. In doing so, a sensory stimulus produces a change in the membrane potential of the receptor cell. In most cases, the sensory stimulus produces a depolarization of the receptor cell, analogous to the excitatory postsynaptic potential (EPSP, described in chapter 45) produced in a postsynaptic cell in response to neurotransmitter. A depolarization that occurs in a sensory receptor upon stimulation is referred to as a **receptor potential** (figure 46.3*a*).

Like an EPSP, a receptor potential is graded: the larger the sensory stimulus, the greater the degree of depolarization. Receptor potentials also decrease in size (*decrement*) with distance from their source. This prevents small, irrelevant stimuli from reaching the cell body of the sensory neuron. If the receptor potential or summation of receptor potentials is great enough to generate a threshold level of depolarization, it will stimulate the production of action potentials that are conducted by a sensory axon into the CNS (figure 46.3*b*). The greater the sensory stimulus, the greater the depolarization of the receptor potential and the higher the frequency of action potentials. There is generally a logarithmic relationship between stimulus intensity and action potential frequency—for example, a sensory stimulus that is ten times greater than another stimulus will produce action potentials at twice the frequency of the other stimulus. This allows the brain to interpret the strength of a sensory stimulus based on the frequency of incoming signals.

Sensory receptors transduce stimuli in the internal or external environment into graded depolarizations, which stimulate the production of action potentials. Sensory receptors may be classified on the basis of the type of stimulus energy to which they respond.

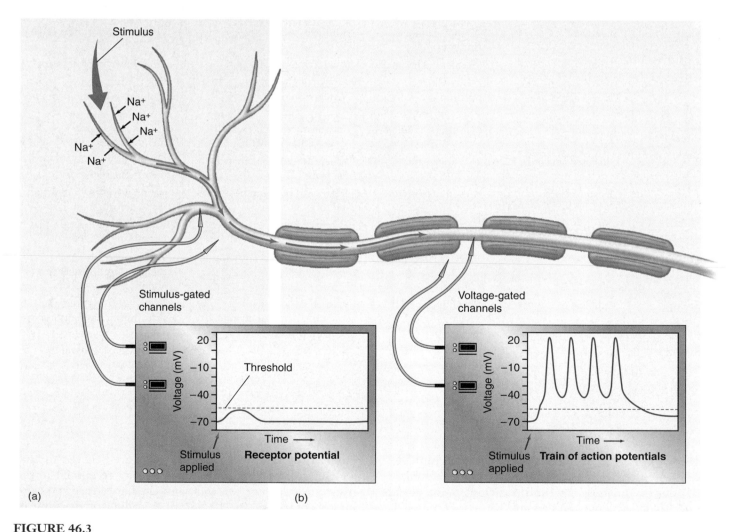

FIGURE 46.3
Events in sensory transduction. (*a*) Depolarization of a free nerve ending leads to a receptor potential that spreads by local current flow to the axon. (*b*) Action potentials are produced in the axon in response to a sufficiently large receptor potential.

46.2 Mechanical and chemical receptors sense the body's condition.

Detecting Temperature and Pressure

While the receptors of the skin, called the **cutaneous receptors,** are classified as interoceptors, they in fact respond to stimuli at the border between the external and internal environments. These receptors serve as good examples of the specialization of receptor structure and function, responding to heat, cold, pain, touch, and pressure.

The skin contains two populations of **thermoreceptors,** which are naked dendritic endings of sensory neurons that are sensitive to changes in temperature. *Cold receptors* are stimulated by a fall in temperature and inhibited by warming, while *warm receptors* are stimulated by a rise in temperature and inhibited by cooling. Cold receptors are located immediately below the epidermis, while warm receptors are located slightly deeper, in the dermis. Thermoreceptors are also found within the hypothalamus of the brain, where they monitor the temperature of the circulating blood and thus provide the CNS with information on the body's internal (core) temperature.

A stimulus that causes or is about to cause tissue damage is perceived as pain. The receptors that transmit impulses that are perceived by the brain as pain are called **nociceptors.** Although there are specific nociceptors, many hyperstimulated sensory receptors can also produce the perception of pain in the brain. Most nociceptors consist of free nerve endings located throughout the body, especially near surfaces where damage is most likely to occur. Different nociceptors may respond to extremes in temperature, very intense mechanical stimulation, or specific chemicals in the extracellular fluid, including some that are released by injured cells. The thresholds of these sensory cells vary; some nociceptors are sensitive only to actual tissue damage, while others respond before damage has occurred.

Several types of mechanoreceptors are present in the skin, some in the dermis and others in the underlying subcutaneous tissue (figure 46.4). **Mechanoreceptors** contain sensory cells with ion channels that are sensitive to a mechanical force applied to the membrane. These channels open in response to mechanical distortion of the membrane, initiating a depolarization (receptor potential) that causes the sensory neuron to generate action potentials. Morphologically specialized receptors that respond to fine touch are most concentrated on areas such as the fingertips and face. They are used to localize cutaneous stimuli very precisely and can be either phasic (intermittently activated) or tonic (continuously activated). The phasic receptors include *hair follicle receptors* and *Meissner's corpuscles*, which are present on body surfaces that do not contain hair, such as the fingers, palms, and nipples. The tonic receptors consist of *Ruffini corpuscles* in the dermis and *touch dome endings (Merkel cells)* located near the surface of the skin. These receptors monitor the duration of a touch and the extent to which it is applied.

Deep below the skin in the subcutaneous tissue lie phasic, pressure-sensitive receptors called **Pacinian corpuscles.** Each of these receptors consists of the end of an afferent axon, surrounded by a capsule of alternating layers of connective tissue cells and extracellular fluid. When sustained pressure is applied to the corpuscle, the elastic capsule absorbs much of the pressure and the axon ceases to produce impulses. Pacinian corpuscles thus monitor only the onset and removal of pressure, as may occur repeatedly when something that vibrates is placed against the skin.

Different cutaneous receptors respond to touch, pressure, heat, cold, and pain. Some of these receptors are naked dendrites of sensory neurons, while others have supporting structures that modify the activities of their sensory dendrites.

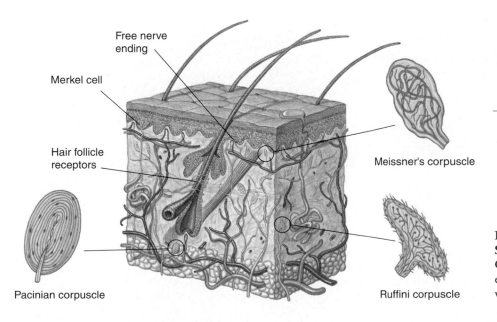

Free nerve ending

Merkel cell

Hair follicle receptors

Pacinian corpuscle

Meissner's corpuscle

Ruffini corpuscle

FIGURE 46.4
Sensory receptors in human skin.
Cutaneous receptors may be free nerve endings or sensory dendrites in association with other supporting structures.

Sensing Muscle Contraction and Blood Pressure

Muscle Length and Tension

Buried within the skeletal muscles of all vertebrates except the bony fishes are muscle spindles, sensory stretch receptors that lie in parallel with the rest of the fibers in the muscle (figure 46.5). Each spindle consists of several thin muscle fibers wrapped together and innervated by a sensory neuron, which becomes activated when the muscle, and therefore the spindle, is stretched. Muscle spindles, together with other receptors in tendons and joints, are known as **proprioceptors,** sensory receptors that provide information about the relative position or movement of the animal's body parts. The sensory neurons conduct action potentials into the spinal cord, where they synapse with somatic motor neurons that innervate the muscle. This pathway constitutes the muscle stretch reflex, including the knee-jerk reflex, discussed in chapter 45.

When a muscle contracts, it exerts tension on the tendons attached to it. The Golgi tendon organs, another type of proprioceptor, monitor this tension; if it becomes too high, they elicit a reflex that inhibits the motor neurons innervating the muscle. This reflex helps ensure that muscles do not contract so strongly that they damage the tendons to which they are attached.

Blood Pressure

Blood pressure is monitored at two main sites in the body. One is the *carotid sinus*, an enlargement of the left and right internal carotid arteries that supply blood to the brain. The other is the *aortic arch*, the portion of the aorta very close to its emergence from the heart. The walls of the blood vessels at both sites contain a highly branched network of afferent neurons called **baroreceptors,** which detect tension or stretch in the walls. When the blood pressure decreases, the frequency of impulses produced by the baroreceptors decreases. The CNS responds to this reduced input by stimulating the sympathetic division of the autonomic nervous system, causing an increase in heart rate and vasoconstriction. Both effects help raise the blood pressure, thus maintaining homeostasis. A rise in blood pressure increases baroreceptor impulses, which conversely reduces sympathetic activity and stimulates the parasympathetic division, slowing the heart and lowering the blood pressure.

Mechanical distortion of the plasma membrane of mechanoreceptors produces nerve impulses that serve to monitor muscle length from skeletal muscle spindles and to monitor blood pressure from baroreceptors within arteries.

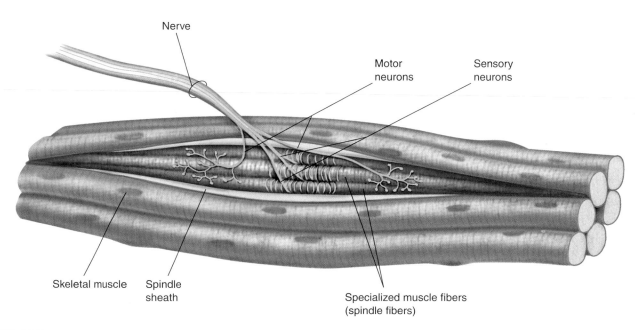

Nerve

Motor neurons

Sensory neurons

Skeletal muscle

Spindle sheath

Specialized muscle fibers (spindle fibers)

FIGURE 46.5

How a muscle spindle works.   A muscle spindle is a stretch receptor embedded within skeletal muscle. Stretching of the muscle elongates the spindle fibers and stimulates the sensory dendritic endings wrapped around them. This causes the sensory neurons to send impulses to the CNS, where they synapse with interneurons and, in some cases, motor neurons.

Sensing Taste, Smell, and Body Position

Some sensory cells, called **chemoreceptors,** contain membrane proteins that can bind to particular chemicals or ligands in the extracellular fluid. In response to this chemical interaction, the membrane of the sensory neuron becomes depolarized, leading to the production of action potentials. Chemoreceptors are used in the senses of taste and smell and are also important in monitoring the chemical composition of the blood and cerebrospinal fluid.

Taste

Taste buds—collections of chemosensitive epithelial cells associated with afferent neurons—mediate the sense of taste in vertebrates. In a fish, the taste buds are scattered over the surface of the body. These are the most sensitive vertebrate chemoreceptors known. They are particularly sensitive to amino acids; a catfish, for example, can distinguish between two different amino acids at a concentration of less than 100 parts per billion (1 g in 10,000 L of water)! The ability to taste the surrounding water is very important to bottom-feeding fish, enabling them to sense the presence of food in an often murky environment.

The taste buds of all terrestrial vertebrates are located in the epithelium of the tongue and oral cavity, within raised areas called *papillae* (figure 46.6). Taste buds are onion-shaped structures of between 50 and 100 taste cells; each cell has fingerlike projections called microvilli that poke up through an opening at the top of the taste bud, called the taste pore (figure 46.6c). Chemicals from food dissolve in saliva and contact the taste cells through the taste pore.

Within a taste bud, the chemicals that produce salty and sour tastes act directly through ion channels, and those responsible for sweet and bitter tastes bind to surface receptor proteins that trigger G proteins to initiate changes in the cell interior that ultimately open and close ion channels. These changes in ion channels initiate signals in the sensory neurons extending from the taste buds to the brain. There they interact with other sensory neurons carrying information related to smell.

Like vertebrates, many arthropods also have taste chemoreceptors. For example, flies, because of their mode of searching for food, have taste receptors in sensory hairs located on their feet. The sensory hairs contain different chemoreceptors that are able to detect sugars, salts, and a wide variety of other tastes by the integration of stimuli from these chemoreceptors (figure 46.7). If they step on potential food, the proboscis (the tubular feeding apparatus) extends to feed.

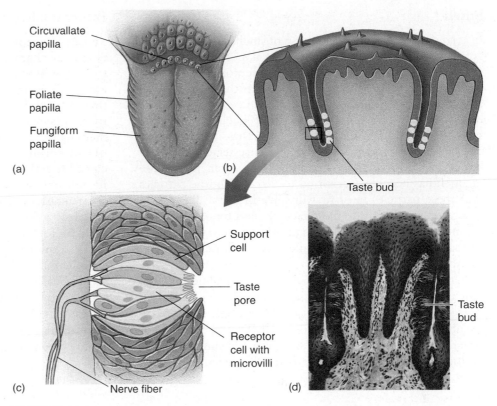

FIGURE 46.6
Taste. (*a*) Human tongues have projections called papillae that bear taste buds. Different sorts of taste buds are located on different regions of the tongue. (*b*) Groups of taste buds are embedded within a papilla. (*c*) Individual taste buds are bulb-shaped collections of chemosensitive receptors that open out into the mouth through a pore. (*d*) Photomicrograph of taste buds in papillae.

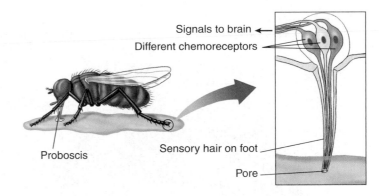

FIGURE 46.7
Many insects taste with their feet. In the blowfly shown here, chemoreceptors extend into the sensory hairs on the foot. Different chemoreceptors detect different types of food molecules. When the fly steps in a food substance, it can taste the different food molecules and extend its proboscis for feeding.

Smell

In terrestrial vertebrates, the sense of smell, or olfaction, involves chemoreceptors located in the upper portion of the nasal passages (figure 46.8). These receptors, whose dendrites end in tassels of cilia, project into the nasal mucosa and their axons project directly into the cerebral cortex. A terrestrial vertebrate uses its sense of smell in much the same way that a fish uses its sense of taste—to sample the chemical environment around it. Because terrestrial vertebrates are surrounded by air rather than water, their sense of smell has become specialized to detect airborne particles (but these particles must first dissolve in extracellular fluid before they can activate the olfactory receptors). The sense of smell can be extremely acute in many mammals, so much so that a single odorant molecule may be all that is needed to excite a given receptor.

Although humans can detect only four modalities of taste, they can discern thousands of different smells. New research suggests that as many as a thousand different genes may code for different receptor proteins for smell. The particular set of olfactory neurons that respond to a given odor might serve as a "fingerprint" the brain can use to identify the odor.

Internal Chemoreceptors

Sensory receptors within the body detect a variety of chemical characteristics of the blood or fluids derived from the blood, including cerebrospinal fluid. Included among these receptors are the *peripheral chemoreceptors* of the aortic and carotid bodies, which are sensitive primarily to plasma pH, and the *central chemoreceptors* in the medulla oblongata of the brain, which are sensitive to the pH of cerebrospinal fluid. These receptors were discussed together with the regulation of breathing in chapter 44. When the breathing rate is too low, the concentration of plasma CO_2 increases, producing more carbonic acid and causing a fall in the blood pH. The carbon dioxide can also enter the cerebrospinal fluid and cause a lowering of the pH, thereby stimulating the central chemoreceptors. This chemoreceptor stimulation indirectly affects the respiratory control center of the brain stem, which increases the breathing rate. The aortic bodies can also respond to a lowering of blood oxygen concentrations, but this effect is normally not significant unless a person goes to a high altitude.

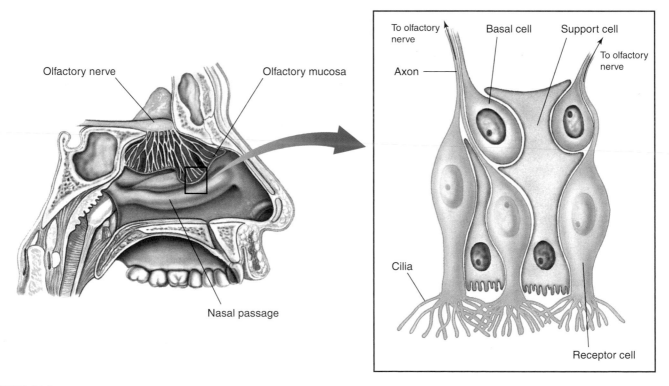

FIGURE 46.8

Smell.   Humans detect smells by means of olfactory neurons (receptor cells) located in the lining of the nasal passages. The axons of these neurons transmit impulses directly to the brain via the olfactory nerve. Basal cells regenerate new olfactory neurons to replace dead or damaged cells. Olfactory neurons typically live about one month.

The Lateral Line System

The lateral line system provides fish with a sense of "distant touch," enabling them to sense objects that reflect pressure waves and low-frequency vibrations. This enables a fish to detect prey, for example, and to swim in synchrony with the rest of its school. It also enables a blind cave fish to sense its environment by monitoring changes in the patterns of water flow past the lateral line receptors. The lateral line system is found in amphibian larvae, but is lost at metamorphosis and is not present in any terrestrial vertebrate. The sense provided by the lateral line system supplements the fish's sense of hearing, which is performed by a different sensory structure. The structures and mechanisms involved in hearing are described in section 46.3.

The lateral line system consists of sensory structures within a longitudinal canal in the fish's skin that extends along each side of the body and within several canals in the head (figure 46.9a). The sensory structures are known as hair cells because they have hairlike processes at their surface that project into a gelatinous membrane called a *cupula* (Latin, "little cup"). The hair cells are innervated by sensory neurons that transmit impulses to the brain.

Hair cells have several hairlike processes of approximately the same length, called *stereocilia*, and one longer process called a *kinocilium* (figure 46.9b). Vibrations carried through the fish's environment produce movements of the cupula, which cause the hairs to bend. When the stereocilia bend in the direction of the kinocilium, the associated sensory neurons are stimulated and generate a receptor potential. As a result, the frequency of action potentials produced by the sensory neuron is increased. On the other hand, if the stereocilia are bent in the opposite direction, the activity of the sensory neuron is inhibited.

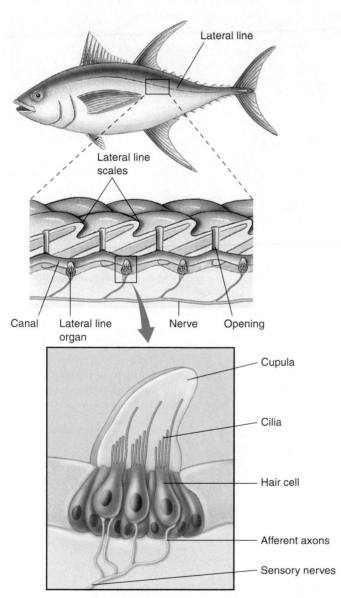

(a)

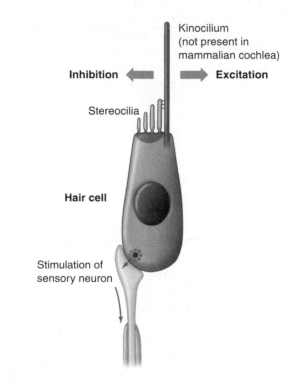

(b)

FIGURE 46.9

The lateral line system. (*a*) This system consists of canals running the length of the fish's body beneath the surface of the skin. Within these canals are sensory structures containing hair cells with cilia that project into a gelatinous cupula. Pressure waves traveling through the water in the canals deflect the cilia and depolarize the sensory neurons associated with the hair cells. (*b*) Hair cells are mechanoreceptors with hairlike cilia that project into a gelatinous membrane. The hair cells of the lateral line system (and the membranous labyrinth of the vertebrate inner ear) have a number of smaller cilia called stereocilia and one larger kinocilium. When the cilia bend in the direction of the kinocilium, the hair cell releases a chemical transmitter that depolarizes the associated sensory neuron. Bending of the cilia in the opposite direction has an inhibitory effect.

Gravity and Angular Acceleration

Most invertebrates can orient themselves with respect to gravity due to a sensory structure called a *statocyst*. Statocysts generally consist of ciliated hair cells with the cilia embedded in a gelatinous membrane containing crystals of calcium carbonate. These "stones," or *statoliths*, increase the mass of the gelatinous membrane so that it can bend the cilia when the animal's position changes. If the animal tilts to the right, for example, the statolith membrane will bend the cilia on the right side and activate associated sensory neurons.

A similar structure is found in the membranous labyrinth of the inner ear of vertebrates. The labyrinth is a system of fluid-filled membranous chambers and tubes that constitute the organs of equilibrium and hearing in vertebrates. This membranous labyrinth is surrounded by bone and perilymph, which is similar in ionic content to interstitial fluid. Inside, the chambers and tubes are filled with endolymph fluid, which is similar in ionic content to intracellular fluid. Though intricate, the entire structure is very small; in a human, it is about the size of a pea.

The receptors for gravity in vertebrates consist of two chambers of the membranous labyrinth called the utricle and saccule (figure 46.10). Within these structures are hair cells with stereocilia and a kinocilium, similar to those in the lateral line system of fish. The hairlike processes are embedded within a gelatinous membrane containing calcium carbonate crystals; this is known as an *otolith membrane*, because of its location in the inner ear (*oto* is derived from the Greek word for ear). Because the otolith organ is oriented differently in the utricle and saccule, the utricle is more sensitive to horizontal acceleration (as in a moving car) and the saccule to vertical acceleration (as in an elevator). In both cases, the

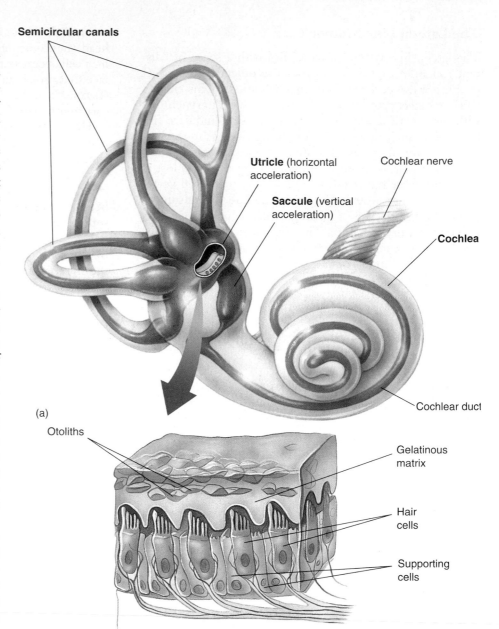

(a)

(b)

FIGURE 46.10

The structure of the utricle and saccule. (*a*) The relative positions of the utricle and saccule within the membranous labyrinth of the human inner ear. (*b*) Enlargement of a section of the utricle or saccule showing the otoliths embedded in the gelatinous matrix that covers the hair cells.

acceleration causes the stereocilia to bend, and consequently produces action potentials in an associated sensory neuron.

The membranous labyrinth of the utricle and saccule is continuous with three semicircular canals, oriented in different planes so that angular acceleration in any direction can be detected (figure 46.11). At the ends of the canals are swollen chambers called *ampullae*, into which protrude the cilia of another group of hair cells. The tips of the cilia are embedded within a sail-like wedge of gelatinous material called a *cupula* (similar to the cupula of the fish lateral line system) that protrudes into the endolymph fluid of each semicircular canal.

When the head rotates, the fluid inside the semicircular canals pushes against the cupula and causes the cilia to bend. This bending either depolarizes or hyperpolarizes the hair cells, depending on the direction in which the cilia are bent. This is similar to the way the lateral line system works in a fish: If the stereocilia are bent in the direction of the kinocilium, a depolarization (receptor potential) is produced, which stimulates the production of action potentials in associated sensory neurons.

The saccule, utricle, and semicircular canals are collectively referred to as the *vestibular apparatus*. While the saccule and utricle provide a sense of linear acceleration, the semicircular canals provide a sense of angular acceleration. The brain uses information that comes from the vestibular apparatus about the body's position in space to maintain balance and equilibrium.

Receptors that sense chemicals originating outside the body are responsible for the senses of odor, smell, and taste. Internal chemoreceptors help to monitor chemicals produced within the body and are needed for the regulation of breathing. Hair cells in the lateral line organ of fishes detect water movements, and hair cells in the vestibular apparatus of terrestrial vertebrates provide a sense of acceleration and balance.

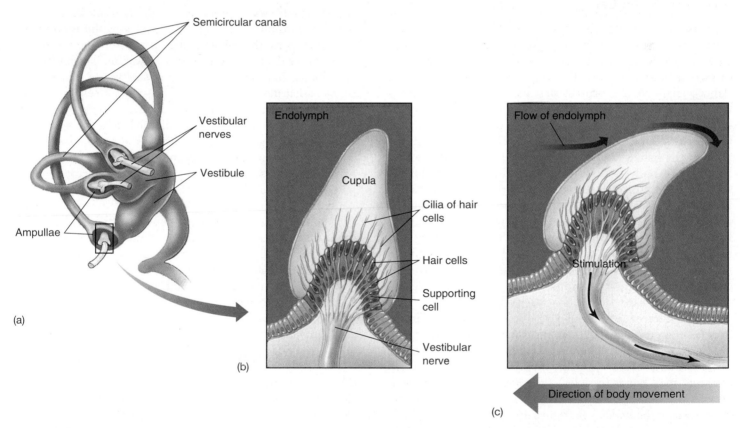

(a)

(b)

(c)

FIGURE 46.11
The structure of the semicircular canals. (*a*) The position of the semicircular canals in relation to the rest of the inner ear. (*b*) Enlargement of a section of one ampulla, showing how hair cell cilia insert into the cupula. (*c*) Angular acceleration in the plane of the semicircular canal causes bending of the cupula, thereby stimulating the hair cells.

46.3 Auditory receptors detect pressure waves in the air.

The Ears and Hearing

As discussed in section 46.2, fish detect vibrational pressure waves in water by means of their lateral line system, and terrestrial vertebrates detect similar vibrational pressure waves in air by means of similar hair cell mechanoreceptors in the inner ear. Hearing actually works better in water than in air because water transmits pressure waves more efficiently. Despite this limitation, hearing is widely used by terrestrial vertebrates to monitor their environments, communicate with other members of the same species, and detect possible sources of danger. Auditory stimuli travel farther and more quickly than chemical ones, and auditory receptors provide better directional information than do chemoreceptors. Auditory stimuli alone, however, provide little information about distance.

Structure of the Ear

Fish use their lateral line system to detect water movements and vibrations emanating from relatively nearby objects, and their hearing system to detect vibrations that originate from a greater distance. The hearing system of fish consists of the otolith organs in the membranous labyrinth (utricle and saccule) previously described, together with a very small outpouching of the membranous labyrinth called the lagena. Sound waves travel through the body of the fish as easily as through the surrounding water, because the body is composed primarily of water. Therefore, an object of different density is needed in order for the sound to be detected. This function is served by the otolith (composed of calcium carbonate crystals) in many fish. In catfish, minnows, and suckers, however, this function is served by an air-filled swim bladder that vibrates with the sound. A chain of small bones, Weberian ossicles, then transmits the vibrations to the saccule in some of these fish.

In the ears of terrestrial vertebrates, vibrations in air may be channeled through an ear canal to the eardrum, or tympanic membrane. These structures are part of the *outer ear*. Vibrations of the tympanic membrane cause movement of three small bones (ossicles)—the *malleus* (hammer), *incus* (anvil), and *stapes* (stirrup)—that are located in a bony cavity known as the *middle ear* (figure 46.12 *a,b*). These middle ear ossicles are analogous to the Weberian ossicles in fish. The middle ear is connected to the throat by the *Eustachian tube*, which equalizes the air pressure between the middle ear and the external environment. The "ear popping" you may have experienced when flying in an airplane or driving on a mountain is caused by pressure equalization between the two sides of the eardrum.

The stapes vibrates against a flexible membrane, the *oval window*, which leads into the *inner ear*. Because the oval window is smaller in diameter than the tympanic membrane, vibrations against it produce more force per unit area, transmitted into the inner ear. The inner ear consists of the cochlea (Latin, "snail"), a bony structure containing part of the membranous labyrinth called the cochlear duct. The cochlear duct is located in the center of the cochlea; the area above the cochlear duct is the *vestibular canal*, and the area below is the *tympanic canal* (figure 46.12*c*). All three chambers are filled with fluid. The oval window opens to the upper vestibular canal, so that when the stapes causes it to vibrate, it produces pressure waves of fluid. These pressure waves travel down to the tympanic canal, pushing another flexible membrane, the *round window*, that transmits the pressure back into the middle ear cavity.

Transduction in the Cochlea

As the pressure waves produced by vibrations of the oval window are transmitted through the cochlea to the round window, they cause the cochlear duct to vibrate. The bottom of the cochlear duct, called the *basilar membrane*, is quite flexible and vibrates in response to these pressure waves. The surface of the basilar membrane contains sensory hair cells, similar to those of the vestibular apparatus and lateral line system but lacking a kinocilium. The cilia from the hair cells project into an overhanging gelatinous membrane, the *tectorial membrane*. This sensory apparatus, consisting of the basilar membrane, hair cells with associated sensory neurons, and tectorial membrane, is known as the **organ of Corti** (46.12*d*).

As the basilar membrane vibrates, the cilia of the hair cells bend in response to the movement of the basilar membrane relative to the tectorial membrane. As in the lateral line organs and the vestibular apparatus, the bending of these cilia depolarizes the hair cells. The hair cells, in turn, stimulate the production of action potentials in sensory neurons that project to the brain, where they are interpreted as sound.

Frequency Localization in the Cochlea

The basilar membrane of the cochlea consists of elastic fibers of varying length and stiffness, like the strings of a musical instrument, embedded in a gelatinous material. At the base of the cochlea (near the oval window), the fibers of the basilar membrane are short and stiff. At the far end of the cochlea (the apex), the fibers are 5 times longer and 100 times more flexible. Therefore, the resonant frequency of the basilar membrane is higher at the base than at the apex; the base responds to higher pitches, the apex to lower.

When a wave of sound energy enters the cochlea from the oval window, it initiates an up-and-down motion that travels

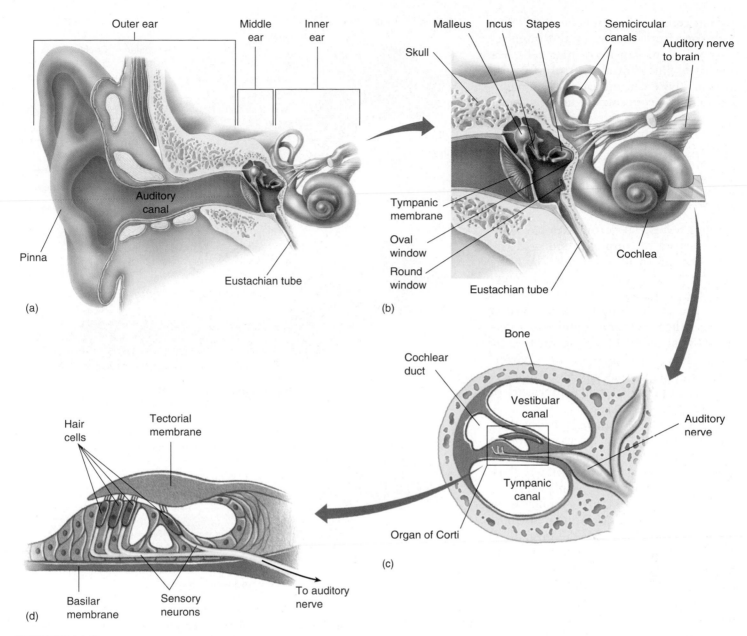

FIGURE 46.12
Structure of the human ear. (*a*) Sound waves passing through the ear canal produce vibrations of the tympanic membrane, which cause movement of the (*b*) middle ear ossicles (the malleus, incus, and stapes) against an inner membrane called the oval window. Vibration of the oval window sets up pressure waves that (*c* and *d*) travel through the fluid in the vestibular and tympanic canals of the cochlea.

the length of the basilar membrane. However, this wave imparts most of its energy to that part of the basilar membrane with a resonant frequency near the frequency of the sound wave, resulting in a maximum deflection of the basilar membrane at that point (figure 46.13). As a result, the hair cell depolarization is greatest in that region, and the afferent axons from that region are stimulated to produce action potentials more than those from other regions. When these action potentials arrive in the brain, they are interpreted as representing a sound of that particular frequency, or pitch.

The flexibility of the basilar membrane limits the frequency range of human hearing to between approximately 20 and 20,000 cycles per second (hertz) in children. Our ability to hear high-pitched sounds decays progressively throughout middle age. Other vertebrates can detect sounds at frequencies lower than 20 hertz and much higher than 20,000 hertz. Dogs, for example, can detect sounds at 40,000 hertz, enabling them to hear high-pitched dog whistles that seem silent to a human listener.

Hair cells are also innervated by efferent axons from the brain, and impulses in those axons can make hair cells less sensitive. This central control of receptor sensitivity can increase an individual's ability to concentrate on a particular auditory signal (for example, a single voice) in the midst of background noise, which is effectively "tuned out" by the efferent axons.

Sonar

Because terrestrial vertebrates have two ears located on opposite sides of the head, the information provided by hearing can be used by the CNS to determine the *direction* of a sound source with some precision. Sound sources vary in strength, however, and sounds are attenuated (weakened) to varying degrees by the presence of objects in the environment. For these reasons, auditory sensors do not provide a reliable measure of *distance*.

A few groups of mammals that live and obtain their food in dark environments have circumvented the limitations of darkness. A bat flying in a completely dark room easily avoids objects placed in its path—even a wire less than a millimeter in diameter. Shrews use a similar form of "lightless vision" beneath the ground, as do whales and dolphins beneath the sea. All of these mammals perceive distance by means of sonar. They emit sounds and then determine the time it takes these sounds to reach an object and return to the animal. This process is called *echolocation*. A bat, for example, produces clicks that last 2 to 3 milliseconds and are repeated several hundred times per second. The three-dimensional imaging achieved with such an auditory sonar system is quite sophisticated.

Being able to "see in the dark" has opened a new ecological niche to bats, one largely closed to birds because birds must rely on vision. There are no truly nocturnal birds; even owls rely on vision to hunt, and do not fly on dark nights. Because bats are able to be active and efficient in total darkness, they are one of the most numerous and widespread of all orders of mammals.

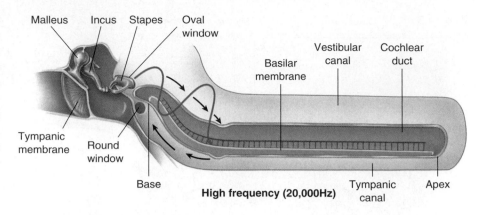

High frequency (20,000Hz)

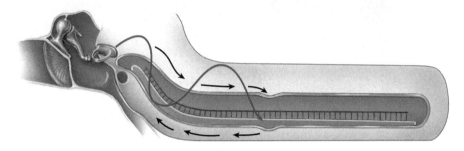

Medium frequency (2000Hz)

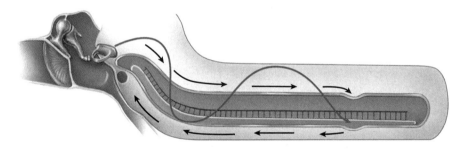

Low frequency (500Hz)

FIGURE 46.13

Frequency localization in the cochlea.   The cochlea is shown unwound, so that the length of the basilar membrane can be seen. The fibers within the basilar membrane vibrate in response to different frequencies of sound, related to the pitch of the sound. Thus, regions of the basilar membrane show maximum vibrations in response to different sound frequencies. Notice that high-frequency (pitch) sounds vibrate the basilar membrane more toward the base (*upper panel*), while low frequencies cause vibrations more toward the apex (*lower panel*).

The middle ear ossicles vibrate in response to sound waves, creating fluid vibrations within the inner ear. This causes the hair cells to bend, transducing the sound into action potentials. Some mammals emit sounds and then determine the time it takes for the sound to return, using the method of sonar to locate themselves and other objects in a totally dark environment by the characteristics of the echo. Bats are the most adept at this echolocation.

46.4 Optic receptors detect light over a broad range of wavelengths.

Evolution of the Eye

Vision begins with the capture of light energy by photoreceptors. Because light travels in a straight line and arrives virtually instantaneously, visual information can be used to determine both the direction and the distance of an object. No other stimulus provides as much detailed information.

Many invertebrates have simple visual systems with photoreceptors clustered in an eyespot. Simple eyespots can be made sensitive to the direction of a light source by the addition of a pigment layer that shades one side of the eye. Flatworms have a screening pigmented layer on the inner and back sides of both eyespots, allowing stimulation of the photoreceptor cells only by light from the front of the animal (figure 46.14). The flatworm will turn and swim in the direction in which the photoreceptor cells are the least stimulated. Although an eyespot can perceive the direction of light, it cannot be used to construct a visual image. The members of four phyla—annelids, mollusks, arthropods, and chordates—have evolved well-developed, image-forming eyes. True image-forming eyes in these phyla, though strikingly similar in structure, are believed to have evolved independently (figure 46.15). Interestingly, the photoreceptors in all of them use the same light-capturing molecule, suggesting that not many alternative molecules are able to play this role.

Structure of the Vertebrate Eye

The eye of a human is typical of the vertebrate eye (figure 46.16). The "white of the eye" is the sclera, formed of tough connective tissue. Light enters the eye through a transparent cornea, which begins to focus the light. This occurs because light is refracted (bent) when it travels into

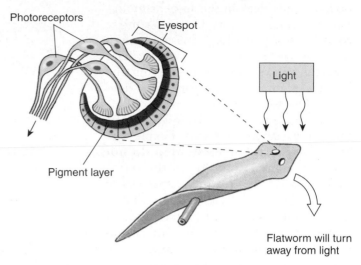

FIGURE 46.14
Simple eyespots in the flatworm. Eyespots can detect the direction of light because a pigmented layer on one side of the eyespot screens out light coming from the back of the animal. Light is thus detected more readily coming from the front of the animal; flatworms respond by turning away from the light.

a medium of different density. The colored portion of the eye is the iris; contraction of the iris muscles in bright light decreases the size of its opening, the pupil. Light passes through the pupil to the lens, a transparent structure that completes the focusing of the light onto the retina at the back of the eye. The lens is attached by the *suspensory ligament* to the ciliary muscles.

The shape of the lens is influenced by the amount of tension in the suspensory ligament, which surrounds the

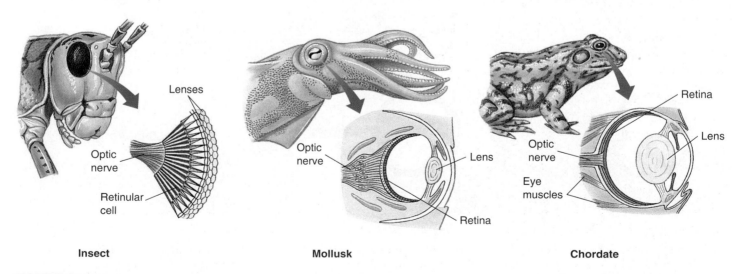

FIGURE 46.15
Eyes in three phyla of animals. Although they are superficially similar, these eyes differ greatly in structure and are not homologous. Each has evolved separately and, despite the apparent structural complexity, has done so from simpler structures.

lens and attaches it to the circular ciliary muscle. When the ciliary muscle contracts, it puts slack in the suspensory ligament, and the lens becomes more rounded and powerful. This is required for close vision; in far vision, the ciliary muscles relax, moving away from the lens and tightening the suspensory ligament. The lens thus becomes more flattened and less powerful, keeping the image focused on the retina. People who are nearsighted or farsighted do not properly focus the image on the retina (figure 46.17). Interestingly, the lens of an amphibian or a fish does not change shape; these animals instead focus images by moving their lens in and out, just as you would do to focus a camera.

> **Annelids, mollusks, arthropods, and chordates have independently evolved image-forming eyes. The vertebrate eye admits light through a pupil and then focuses this light by means of an adjustable lens onto the retina at the back of the eye.**

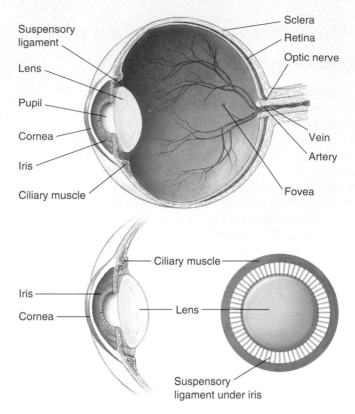

FIGURE 46.16
Structure of the human eye. The transparent cornea and lens focus light onto the retina at the back of the eye, which contains the photoreceptors (rods and cones). The center of each eye's visual field is focused on the fovea. Focusing is accomplished by contraction and relaxation of the ciliary muscle, which adjusts the curvature of the lens.

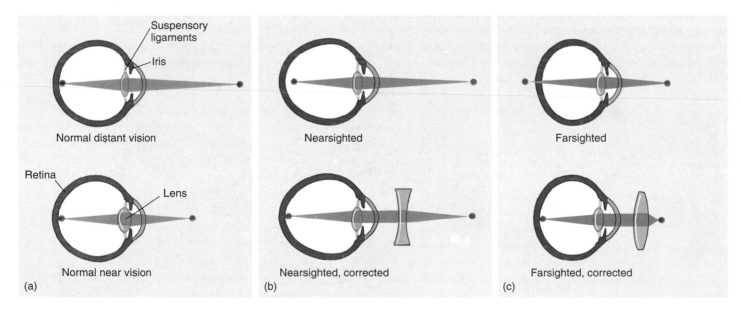

FIGURE 46.17
Focusing the human eye. (*a*) In people with normal vision, the image remains focused on the retina in both near and far vision because of changes produced in the curvature of the lens. When a person with normal vision stands 20 feet or more from an object, the lens is in its least convex form, and the image is focused on the retina. (*b*) In nearsighted people, the image comes to a focus in front of the retina, and the image thus appears blurred. (*c*) In farsighted people, the focus of the image would be behind the retina because the distance from the lens to the retina is too short. Corrective lenses adjust the angle of the light as it enters the eye, focusing it on the retina.

Vertebrate Photoreceptors

The vertebrate retina contains two kinds of photoreceptors, called rods and cones (figure 46.18). Rods are responsible for black-and-white vision when the illumination is dim, while cones are responsible for high visual acuity (sharpness) and color vision. Humans have about 100 million rods and 3 million cones in each retina. Most of the cones are located in the central region of the retina known as the fovea, where the eye forms its sharpest image. Rods are almost completely absent from the fovea.

Rods and cones have the same basic cellular structure. An inner segment rich in mitochondria contains numerous vesicles filled with neurotransmitter molecules. It is connected by a narrow stalk to the outer segment, which is packed with hundreds of flattened discs stacked on top of one another. The light-capturing molecules, or photopigments, are located on the membranes of these discs.

In rods, the photopigment is called rhodopsin. It consists of the protein opsin bound to a molecule of *cis*-retinal, which is produced from vitamin A (figure 46.19). Vitamin A is derived from carotene, a photosynthetic pigment in plants. The photopigments of cones, called photopsins, are structurally very similar to rhodopsin. Humans have three kinds of cones, each of which possesses a photopsin consisting of *cis*-retinal bound to a protein with a slightly different amino acid sequence. These differences shift the *absorption maximum*, the region of the electromagnetic spectrum that is best absorbed by the pigment (figure 46.20). The absorption maximum of the *cis*-retinal in rhodopsin is 500 nanometers (nm); in contrast, the absorption maxima of the three kinds of cone photopsins are 420 nm (blue-absorbing), 530 nm (green-absorbing), and 560 nm (red-absorbing). These differences in the light-absorbing properties of the photopsins are responsible for the different color sensitivities of the three kinds of cones, which are often referred to as simply blue, green, and red cones.

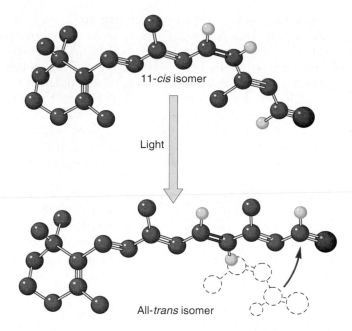

FIGURE 46.19
Absorption of light.   When light is absorbed by a photopigment, the 11-*cis* isomer of retinal, the light-capturing portion of the pigment, undergoes a change in shape: The linear end of the molecule (at the right in this diagram) rotates about a double bond (indicated here in *red*). The resulting isomer is referred to as all-*trans* retinal. This change in retinal's shape initiates a chain of events that leads to hyperpolarization of the photoreceptor.

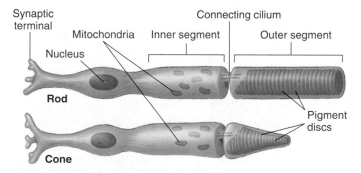

FIGURE 46.18
Rods and cones.   The pigment-containing outer segment in each of these cells is separated from the rest of the cell by a partition through which there is only a narrow passage, the connective cilium.

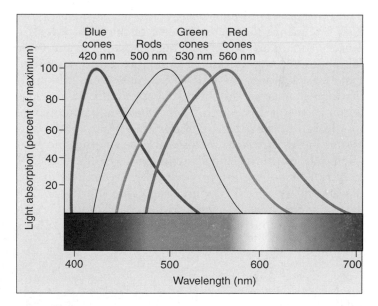

FIGURE 46.20
Color vision.   The absorption maximum of *cis*-retinal in the rhodopsin of rods is 500 nanometers (nm). However, the "blue cones" have their maximum light absorption at 420 nm; the "green cones" at 530 nm; and the "red cones" at 560 nm. The brain perceives all other colors from the combined activities of these three cones' systems.

Most vertebrates, particularly those that are diurnal (active during the day), have color vision, as do many insects. Indeed, honeybees can see light in the near-ultraviolet range, which is invisible to the human eye. Color vision requires the presence of more than one photopigment in different receptor cells, but not all animals with color vision have the three-cone system characteristic of humans and other primates. Fish, turtles, and birds, for example, have four or five kinds of cones; the "extra" cones enable these animals to see near-ultraviolet light. Many mammals (such as squirrels), on the other hand, have only two types of cones.

The retina is made up of three layers of cells (figure 46.21): The layer closest to the external surface of the eyeball consists of the rods and cones, the next layer contains bipolar cells, and the layer closest to the cavity of the eye is composed of ganglion cells. Thus, light must first pass through the ganglion cells and bipolar cells in order to reach the photoreceptors! The rods and cones synapse with the bipolar cells, and the bipolar cells synapse with the ganglion cells, which transmit impulses to the brain via the optic nerve. The flow of sensory information in the retina is therefore opposite to the path of light through the retina. It should also be noted that the retina contains two additional types of neurons, horizontal cells and amacrine cells. Stimulation of horizontal cells by photoreceptors at the center of a spot of light on the retina can inhibit the response of photoreceptors peripheral to the center. This lateral inhibition enhances contrast and sharpens the image.

Sensory Transduction in Photoreceptors

The transduction of light energy into nerve impulses follows a sequence that is the inverse of the usual way that sensory stimuli are detected. This is because, in the dark, the photoreceptors release an inhibitory neurotransmitter that hyperpolarizes the bipolar neurons. Thus inhibited, the bipolar neurons do not release excitatory neurotransmitter to the ganglion cells. Light *inhibits* the photoreceptors from releasing their inhibitory neurotransmitter, and by this means, *stimulates* the bipolar cells and thus the ganglion cells, which transmit action potentials to the brain.

A rod or cone contains many Na^+ channels in the plasma membrane of its outer segment, and in the dark, many of these channels are open. As a consequence, Na^+ continuously diffuses into the outer segment and across the narrow stalk to the inner segment. This flow of Na^+ that occurs in the absence of light is called the *dark current*, and it causes the membrane of a photoreceptor to be somewhat depolarized in the dark. In the light, the Na^+ channels in the outer segment rapidly close, reducing the dark current and causing the photoreceptor to hyperpolarize.

Researchers have discovered that cyclic guanosine monophosphate (cGMP) is required to keep the Na^+ channels open, and that the channels will close if the cGMP is converted into GMP. How does light cause this conversion and consequent closing of the Na^+ channels? When a photopigment absorbs light, *cis*-retinal isomerizes and dissociates from opsin in what is known as the bleaching reaction. As a result of this dissociation, the opsin protein changes shape. Each opsin is associated with over 100 regulatory *G proteins* (see chapters 7 and 45). When the opsin changes shape, the G proteins dissociate, releasing subunits that activate hundreds of molecules of the enzyme *phosphodiesterase*. This enzyme converts cGMP to GMP, thus closing the Na^+ channels at a rate of about 1000 per second and inhibiting the dark current. The absorption of a single photon of light can block the entry of more than a million sodium ions, without changing potassium permeability, thereby causing the photoreceptor to hyperpolarize and release less inhibitory neurotransmitter. Freed from inhibition, the bipolar cells activate ganglion cells, which transmit action potentials to the brain. Ganglion cells are the only retinal neurons capable of generating an action potential.

Photoreceptor rods and cones contain the photopigment *cis*-retinal, which dissociates in response to light and indirectly activates bipolar neurons and then ganglion cells.

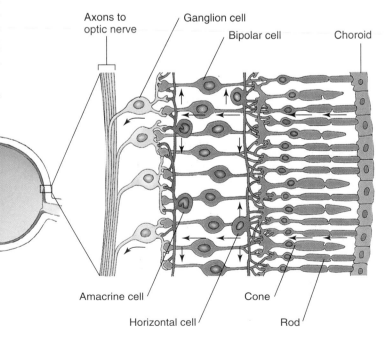

FIGURE 46.21
Structure of the retina. Note that the rods and cones are at the rear of the retina, not the front. Light passes through four other types of cells in the retina before it reaches the rods and cones. Once the photoreceptors are activated, they stimulate bipolar cells, which in turn stimulate ganglion cells. The direction of receptor potentials in the retina is thus opposite to the direction of light.

Visual Processing in the Vertebrate Retina

Action potentials propagated along the axons of ganglion cells are relayed through structures called the lateral geniculate nuclei of the thalamus and projected to the occipital lobe of the cerebral cortex (figure 46.22). There the brain interprets this information as light in a specific region of the eye's receptive field. The pattern of activity among the ganglion cells across the retina encodes a point-to-point map of the receptive field, allowing the retina and brain to image objects in visual space. In addition, the frequency of impulses in each ganglion cell provides information about the light intensity at each point, while the relative activity of ganglion cells connected (through bipolar cells) with the three types of cones provides color information.

The relationship between receptors, bipolar cells, and ganglion cells varies in different parts of the retina. In the fovea, each cone makes a one-to-one connection with a bipolar cell, and each bipolar cell synapses, in turn, with one ganglion cell. This point-to-point relationship is responsible for the high acuity of foveal vision. Outside the fovea, many rods can converge on a single bipolar cell, and many bipolar cells can converge on a single ganglion cell. This convergence permits the summation of neural activity, making the area of the retina outside the fovea more sensitive to dim light than the fovea, but at the expense of acuity and color vision. This is why dim objects, such as faint stars at night, are best seen when you don't look directly at them. It has been said that we use the periphery of the eye as a detector and the fovea as an inspector.

Color blindness is due to an inherited lack of one or more types of cones. People with normal color vision are *trichromats;* those with only two types of cones are *dichromats.* For example, people with color blindness may lack red cones (have *protanopia*), and have difficulty distinguishing red from green. Men are far more likely to be color blind than women, because the trait for color blindness is carried on the X chromosome; men have only one X chromosome per cell, whereas women have two X chromosomes and so can carry the trait in a recessive state.

Binocular Vision

Primates (including humans) and most predators have two eyes, one located on each side of the face. When both eyes are trained on the same object, the image that each sees is slightly different because each eye views the object from a different angle. This slight displacement of the images (an effect called *parallax*) permits **binocular vision,** the ability to perceive three-dimensional images and to sense depth. Having their eyes facing forward maximizes the field of overlap in which this stereoscopic vision occurs.

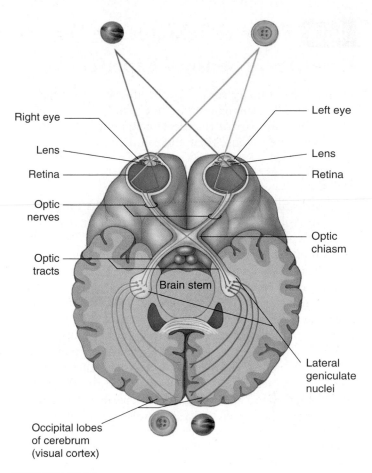

FIGURE 46.22
The pathway of visual information. Action potentials in the optic nerves are relayed from the retina to the lateral geniculate nuclei, and from there to the visual cortex of the occipital lobes. Notice that the medial fibers of the optic nerves, arising from the inner portion of the retinas, cross to the other side at the optic chiasm, so that each hemisphere of the cerebrum receives input from both eyes.

In contrast, prey animals generally have eyes located to the sides of the head, preventing binocular vision but enlarging the overall receptive field. It seems that natural selection has favored the detection of potential predators over depth perception in many prey species. The eyes of the American woodcock, for example, are located at exactly opposite sides of its skull so that it has a 360° field of view without turning its head! Most birds have laterally placed eyes and, as an adaptation, have two foveas in each retina. One fovea provides sharp frontal vision, like the single fovea in the retina of mammals, and the other fovea provides sharper lateral vision.

The axons of ganglion cells transmit action potentials to the thalamus, which in turn relays visual information to the occipital lobe of the brain. The fovea provides high visual acuity, whereas the retina outside the fovea provides high sensitivity to dim light. Binocular vision with overlapping visual fields provides depth perception.

46.5 Some vertebrates use heat, electricity, or magnetism for orientation.

Diversity of Sensory Experiences

Vision is the primary sense used by all vertebrates that live in a light-filled environment, but visible light is by no means the only part of the electromagnetic spectrum that vertebrates use to sense their environment.

Heat

Electromagnetic radiation with wavelengths longer than those of visible light is too low in energy to be detected by photoreceptors. Radiation from this *infrared* ("below red") portion of the spectrum is what we normally think of as radiant heat. Heat is an extremely poor environmental stimulus in water because water has a high thermal capacity and readily absorbs heat. Air, in contrast, has a low thermal capacity, so heat in air is a potentially useful stimulus. However, the only vertebrates known to have the ability to sense infrared radiation are the snakes known as pit vipers.

The pit vipers possess a pair of heat-detecting pit organs located on either side of the head between the eye and the nostril (figure 46.23). The pit organs permit a blindfolded rattlesnake to accurately strike at a warm, dead rat. Each pit organ is composed of two chambers separated by a membrane. The infrared radiation falls on the membrane and warms it. Thermal receptors on the membrane are stimulated. The nature of the pit organ's thermal receptor is not known; it probably consists of temperature-sensitive neurons innervating the two chambers. The two pit organs appear to provide stereoscopic information, in much the same way that two eyes do. Indeed, in snakes the information transmitted from the pit organs is processed in the brain by the homologous structure to the visual center in other vertebrates.

Electricity

While air does not readily conduct an electrical current, water is a good conductor. All aquatic animals generate electrical currents from contractions of their muscles. A number of different groups of fishes can detect these electrical currents. The *electrical fish* even have the ability to produce electrical discharges from specialized electrical organs. Electrical fish use these weak discharges to locate their prey and mates and to construct a three-dimensional image of their environment even in murky water.

The elasmobranchs (sharks, rays, and skates) have electroreceptors called the ampullae of Lorenzini. The receptor cells are located in sacs that open through jelly-filled canals to pores on the body surface. The jelly is a very good conductor, so a negative charge in the opening of the canal can depolarize the receptor at the base, causing the release of neurotransmitter and increased activity of sensory neurons. This allows sharks, for example, to detect the electri-

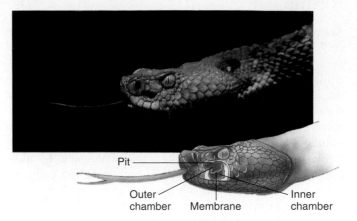

FIGURE 46.23
"Seeing" heat. The depression between the nostril and the eye of this rattlesnake opens into the pit organ. In the cutaway portion of the diagram, you can see that the organ is composed of two chambers separated by a membrane. Snakes known as pit vipers have this unique ability to sense infrared radiation (heat).

cal fields generated by the muscle contractions of their prey. Although the ampullae of Lorenzini were lost in the evolution of teleost fish (most of the bony fish), electroreception reappeared in some groups of teleost fish that use sensory structures analogous to the ampullae of Lorenzini. Electroreceptors evolved yet another time, independently, in the duck-billed platypus, an egg-laying mammal. The receptors in its bill can detect the electrical currents created by the contracting muscles of shrimp and fish, enabling the mammal to detect its prey at night and in muddy water.

Magnetism

Eels, sharks, bees, and many birds appear to navigate along the magnetic field lines of the earth. Even some bacteria use such forces to orient themselves. Birds kept in blind cages, with no visual cues to guide them, will peck and attempt to move in the direction in which they would normally migrate at the appropriate time of the year. They will not do so, however, if the cage is shielded from magnetic fields by steel. Indeed, if the magnetic field of a blind cage is deflected 120° clockwise by an artificial magnet, a bird that normally orients to the north will orient toward the east-southeast. There has been much speculation about the nature of the magnetic receptors in these vertebrates, but the mechanism is still very poorly understood.

Pit vipers can locate warm prey by infrared radiation (heat), and many aquatic vertebrates can locate prey and ascertain the contours of their environment by means of electroreceptors. Magnetic receptors may aid in bird migration.

For interactive testing, visit the Online Learning Center with PowerWeb at www.mhhe.com/Raven7

46.1 Animals employ a wide variety of sensory receptors.

Categories of Sensory Receptors and Their Actions

- Sensory information is conveyed and perceived in a four-step process: stimulation, transduction, transmission, and interpretation. (p. 970)
- Exteroceptors sense stimuli that arise in the external environment, and interoceptors sense stimuli that arise from within the body. (pp. 970–971)
- The greater the sensory stimulus, the greater the depolarization of the receptor potential and the higher the frequency of action potentials. (p. 972)

46.2 Mechanical and chemical receptors sense the body's condition.

Detecting Temperature and Pressure

- Cutaneous receptors respond to stimuli at the border between the external and internal environments. (p. 973)
- Thermoreceptors are naked dendritic endings sensitive to changes in temperature. (p. 973)
- Nociceptors are free nerve endings that perceive tissue damage as pain. (p. 973)
- Mechanoreceptors contain sensory cells with ion channels sensitive to a mechanical force. (p. 973)
- Pacinian corpuscles are phasic, pressure-sensitive receptors in the subcutaneous tissue. (p. 973)

Sensing Muscle Contraction and Blood Pressure

- Proprioceptors are sensory receptors that provide information about the relative position or movements of body parts. (p. 974)
- Distortion of mechanoreceptors produces nerve impulses that monitor muscle length (p. 974)
- Baroreceptors monitor blood pressure by detecting tension in blood vessel walls. (p. 974)

Sensing Taste, Smell, and Body Position

- Chemoreceptors contain membrane proteins that can bind to particular chemicals or ligands in the extracellular fluid. (p. 975)
- The taste buds of terrestrial vertebrates are located within the epithelium of the tongue within papillae, while the sense of smell involves chemoreceptors in the upper portions of the nasal passage. (p. 975)
- Internal chemoreceptors monitor blood and blood fluids. (p. 976)
- The lateral line system in fish consists of sensory structures within a longitudinal canal filled with hair cells that detect water movements. (p. 977)
- Hair cells in the inner ear of terrestrial vertebrates detect gravity, and hair cells in the vestibular apparatus sense angular acceleration and balance. (pp. 977–979)

46.3 Auditory receptors detect pressure waves in the air.

The Ears and Hearing

- Auditory stimuli travel farther and more quickly than chemical stimuli, and auditory receptors provide better directional information than chemoreceptors. (p. 980)
- Vibrations of the tympanic membrane cause movement of ossicles in the middle ear, creating pressure waves in the fluid-filled chambers. (p. 980)
- Cilia from hair cells bend in response to movement, depolarizing hair cells and stimulating the production of action potentials in sensory neurons projecting to the brain where they are interpreted as sound. (pp. 980–982)
- Many mammals use echolocation as a means of perceiving distance. (p. 982)

46.4 Optic receptors detect light over a broad range of wavelengths.

Evolution of the Eye

- Four phyla (annelids, mollusks, arthropods, and chordates) have evolved image-forming eyes. (p. 983)
- The eyes of vertebrates admit light through the pupil and focus the light with an adjustable lens onto the retina. (pp. 983–984)

Vertebrate Photoreceptors

- The vertebrate retina contains two types of photoreceptors: rods, which are responsible for black and white vision, and cones, which are responsible for sharpness and color vision. (p. 985)
- Light inhibits photoreceptors from releasing an inhibitory neurotransmitter, thus stimulating bipolar cells and ganglion cells that transmit action potentials to the brain. (p. 986)

Visual Processing in the Vertebrate Retina

- Action potentials propagated along ganglion cells are relayed through the thalamus and projected to the occipital lobe of the brain. (p. 987)
- The frequency of impulses in each ganglion cell provides information about light intensity, while the relative activity of ganglion cells connected with the three types of cones provides color information. (p. 987)
- Color blindness is due to an inherited lack of at least one type of cone. (p. 987)
- Binocular vision occurs when a displacement of images (parallax) allows the perception of three-dimensional images and depth. (p. 987)

46.5 Some vertebrates use heat, electricity, or magnetism for orientation.

Diversity of Sensory Experiences

- Pit vipers sense heat with a pair of pit organs located on either side of the head. (p. 988)
- All aquatic animals generate electrical currents in their muscles, and some groups can detect electrical currents. (p. 988)
- Many animals, including eels, sharks, bees, and many birds, use the earth's magnetic fields for navigation. (p. 988)

Self Test

1. All sensory receptors are able to initiate nerve impulses by opening or closing
   a. voltage-gated ion channels.
   b. exteroceptors.
   c. interoceptors.
   d. stimulus-gated ion channels.
2. What type of stimulus triggers a response in nociceptors?
   a. pressure
   b. pain
   c. heat
   d. touch
3. What is the function of baroreceptors?
   a. They detect changes in blood pressure.
   b. They detect muscle contractions and the movement of limbs.
   c. They are exteroceptors.
   d. They detect changes in blood chemistry.
4. Cilia from sensory cells that detect a body in motion are located within the
   a. lateral line system.
   b. cochlea.
   c. semicircular canals.
   d. pit organ.
5. Two senses that detect changes in chemical concentrations are
   a. touch and pressure.
   b. sight and smell.
   c. hearing and balance.
   d. taste and smell.
6. A person with defective otolith sensory receptors
   a. is deaf.
   b. has a difficult time maintaining balance.
   c. cannot detect external temperature changes.
   d. has a faulty sense of smell.
7. The ear detects sound by the movement of
   a. the basilar membrane.
   b. the tectorial membrane.
   c. the cupula that surrounds the hair cells.
   d. fluid in the semicircular canals.
8. Which of the following statements is incorrect?
   a. Vertebrates focus the eye by changing the shape of the lens.
   b. All vertebrate and invertebrate eyes use the same light-capturing molecule.
   c. The vertebrate eye adjusts the amount of light entering the eye by contracting the ciliary muscles.
   d. Fish focus the eye by moving the lens closer to or farther from the retina.
9. Most sensory receptors function by producing depolarizing potentials. Which of the following function by hyperpolarization rather than depolarization?
   a. proprioceptors
   b. nociceptors
   c. olfactory receptors
   d. rods and cones
10. Which of the following stimuli is not detected by fish?
    a. infrared radiation
    b. sound
    c. electricity
    d. magnetism

Test Your Visual Understanding

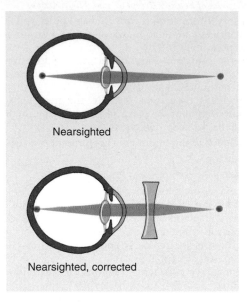

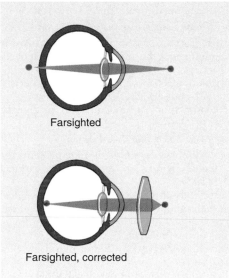

1. Corrective lenses are used to refocus the image correctly on the retina of the eye. Explain how each corrective lens in the figure modifies the refraction of the light from the image and why each is able to correct the respective eye problem.

Apply Your Knowledge

1. The human tongue contains about 10,000 taste buds, with each taste bud housing about 50 taste cells. Each taste cell has a life span of about 10 days. How many taste cells are replaced while you sleep (assume 7 hours of sleep)?
2. How would the otolith organs of an astronaut in zero gravity behave? Would the astronaut still have a subjective impression of motion? Would the semicircular canals detect angular acceleration equally well at zero gravity?

47

The Endocrine System

Concept Outline

47.1 Regulation is often accomplished by chemical messengers.

Types of Regulatory Molecules. Regulatory molecules may function as neurotransmitters, hormones, or organ-specific regulators.

Endocrine Glands and Hormones. Endocrine glands secrete molecules called hormones into the blood.

Paracrine Regulation. Paracrine regulators act within organs that produce them.

47.2 Lipophilic and polar hormones regulate their target cells by different means.

Hormones That Enter Cells. Steroid and thyroid hormones are lipophilic hormones that act by entering target cells and stimulating specific genes.

Hormones That Do Not Enter Cells. Other polar hormones act by binding to receptors on the cell surface. Many activate second-messenger molecules within the target cells.

47.3 The hypothalamus controls the secretions of the pituitary gland.

The Posterior Pituitary Gland. The posterior pituitary receives and releases hormones from the hypothalamus.

The Anterior Pituitary Gland. The anterior pituitary produces a variety of hormones under stimulation from hypothalamic releasing hormones.

47.4 Endocrine glands secrete hormones that regulate many body functions.

The Thyroid and Parathyroid Glands. Thyroid hormones regulate metabolism; parathyroid hormones regulate calcium balance.

The Adrenal Glands. The adrenal medulla secretes epinephrine during the fight or flight reaction, while the adrenal cortex secretes steroid hormones that regulate glucose and mineral balance.

The Pancreas. The islets of Langerhans in the pancreas secrete insulin, which acts to lower blood glucose, and glucagon, which acts to raise blood glucose.

Other Endocrine Glands. The gonads, pineal gland, thymus, kidneys, and other organs secrete important hormones that have a variety of functions.

FIGURE 47.1
The endocrine system controls when animals breed. These Japanese macaques live in a close-knit community whose members cooperate to ensure successful breeding and raising of offspring. Not everybody breeds at the same time because hormone levels vary among individuals.

The tissues and organs of the vertebrate body cooperate to maintain homeostasis of the body's internal environment and to control other body functions, such as reproduction. Homeostasis is achieved through the actions of many regulatory mechanisms that involve all the organs of the body. Two systems, however, are devoted exclusively to the regulation of the body organs: the nervous system and the endocrine system (figure 47.1). Both release regulatory molecules that control the body organs by first binding to receptor proteins on or in the cells of those organs. In this chapter, we examine these regulatory molecules, the cells and glands that produce them, and how they function to regulate the body's activities.

Regulation is often accomplished by chemical messengers.

Types of Regulatory Molecules

As we discussed in chapter 45, the axons of neurons secrete chemical messengers called *neurotransmitters* into the synaptic cleft. These chemicals diffuse only a short distance to the postsynaptic membrane, where they bind to their receptor proteins and stimulate the postsynaptic cell (another neuron, or a muscle or gland cell). Synaptic transmission generally affects only the one postsynaptic cell that receives the neurotransmitter.

A **hormone** is a regulatory chemical that is secreted into the blood by an **endocrine gland** or an organ of the body exhibiting an endocrine function. The blood carries the hormone to every cell in the body, but only the **target cells** for a given hormone can respond to it. Thus, the difference between a neurotransmitter and a hormone is not in the chemical nature of the regulatory molecule, but rather in the way it is transported to its target cells, and its distance from these target cells. A chemical regulator called norepinephrine, for example, is released as a neurotransmitter by sympathetic nerve endings and is also secreted by the adrenal gland as a hormone.

Some specialized neurons secrete chemical messengers into the blood rather than into a narrow synaptic cleft. In these cases, the chemical the neurons secrete is sometimes called a **neurohormone.** The distinction between the nervous system and the endocrine system blurs when it comes to such molecules. Indeed, some neurons in the brain secrete hormones that perform endocrine functions.

In addition to the chemical messengers released as neurotransmitters and as hormones, other chemical molecules are released and act *within* an organ as local regulators. In this way, the cells of an organ regulate one another. This type of regulation is not endocrine, because the regulatory molecules work without being transported by the blood, but is otherwise similar to the way that hormones regulate their target cells. Such regulation is called **paracrine.** Another type of chemical messenger that is released into the environment is called a *pheromone*. These messengers aid in the communication between animals, not in the regulation within an animal. A comparison of the different types of chemical messengers used for regulation is given in figure 47.2.

Regulatory molecules released by axons at a synapse are neurotransmitters; those released by endocrine glands into the blood are hormones; and those that act within the organ in which they are produced are paracrine regulators.

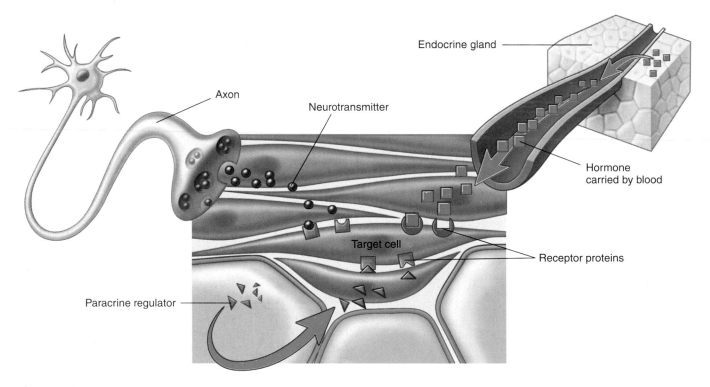

FIGURE 47.2
Different types of chemical messengers. The functions of organs are influenced by neural, paracrine, and endocrine regulators. Each type of chemical regulator binds in a specific fashion to receptor proteins on the surface of or within the cells of target organs.

Endocrine Glands and Hormones

The endocrine system (figure 47.3) includes all of the organs that function exclusively as endocrine glands—the thyroid gland, pituitary gland, adrenal glands, and so on (table 47.1)—as well as organs that secrete hormones in addition to other functions. Endocrine gland cells secrete their chemical signals into surrounding blood capillaries, whereas exocrine glands secrete their products into a duct for transport.

Hormones secreted by endocrine glands belong to four different chemical categories:

1. **Polypeptides** are composed of chains of amino acids that are shorter than about 100 amino acids. Some important examples include insulin and antidiuretic hormone (ADH).
2. **Glycoproteins** are composed of a polypeptide significantly longer than 100 amino acids to which is attached a carbohydrate. Examples include follicle-stimulating hormone (FSH) and luteinizing hormone (LH).
3. **Amines,** derived from the amino acids tyrosine and tryptophan, include hormones secreted by the adrenal medulla, thyroid, and pineal glands.
4. **Steroids** are lipids derived from cholesterol, and include the hormones testosterone, estradiol, progesterone, aldosterone, and cortisol.

Steroid hormones can be subdivided into **sex steroids,** secreted by the testes, ovaries, placenta, and adrenal cortex, and **corticosteroids,** secreted only by the adrenal cortex (the outer portion of the adrenal gland). The corticosteroids include cortisol, which regulates glucose balance, and aldosterone, which regulates salt balance.

The amine hormones secreted by the adrenal medulla (the inner portion of the adrenal gland), known as **catecholamines,** include epinephrine (adrenaline) and norepinephrine (noradrenaline). These are derived from the amino acid tyrosine. Another hormone derived from tyrosine is thyroxine, secreted by the thyroid gland. The pineal gland secretes a different amine hormone, melatonin, derived from tryptophan.

All hormones may be categorized as lipophilic, which are fat-soluble, or lipophobic (polar), which are water-soluble. The **lipophilic hormones** include the steroid hormones and thyroxine; all other hormones are water-soluble. This distinction is important in understanding how these hormones regulate their target cells.

Neural and Endocrine Interactions

The endocrine system is an extremely important regulatory system in its own right, but it also interacts and cooperates with the nervous system to regulate the activities of the other organ systems of the body. The secretory activity of many endocrine glands is controlled by the nervous system.

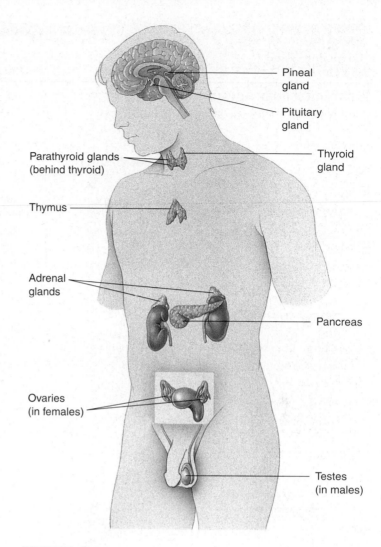

FIGURE 47.3
The human endocrine system.   The major endocrine glands are shown, but many other organs secrete hormones in addition to their primary functions.

Among such glands are the adrenal medulla, posterior pituitary, and pineal gland. These three glands are derived from the neural ectoderm (to be discussed in chapter 51), the same embryonic tissue layer that forms the nervous system. However, the major site that carries out neural regulation of the endocrine system is the brain. As we'll see, the hypothalamus controls the hormonal secretions of the anterior pituitary, which in turn regulates other endocrine glands. On the other hand, the secretion of a number of hormones is largely independent of neural control. For example, the release of insulin by the pancreas and aldosterone by the adrenal cortex is stimulated primarily by increases in the blood concentrations of glucose and potassium (K^+), respectively.

Any organ that secretes a hormone into the blood is part of the endocrine system. Hormones may be any of a variety of different chemicals.

Table 47.1 Principal Endocrine Glands and Their Hormones*

Endocrine Gland and Hormone	Target Tissue	Principal Actions	Chemical Nature
POSTERIOR LOBE OF PITUITARY			
Antidiuretic hormone (ADH)	Kidneys	Stimulates reabsorption of water; conserves water	Peptide (9 amino acids)
Oxytocin	Uterus	Stimulates contraction	Peptide (9 amino acids)
	Mammary glands	Stimulates milk ejection	
ANTERIOR LOBE OF PITUITARY			
Growth hormone (GH)	Many organs	Stimulates growth by promoting protein synthesis and fat breakdown for energy metabolism	Protein
Adrenocorticotropic hormone (ACTH)	Adrenal cortex	Stimulates secretion of adrenal cortical hormones such as cortisol	Peptide (39 amino acids)
Thyroid-stimulating hormone (TSH)	Thyroid gland	Stimulates thyroxine secretion	Glycoprotein
Luteinizing hormone (LH)	Gonads	Stimulates ovulation and corpus luteum formation in females; stimulates secretion of testosterone in males	Glycoprotein
Follicle-stimulating hormone (FSH)	Gonads	Stimulates spermatogenesis in males; stimulates development of ovarian follicles in females	Glycoprotein
Prolactin (PRL)	Mammary glands	Stimulates milk production	Protein
Melanocyte-stimulating hormone (MSH)	Skin	Stimulates color change in reptiles and amphibians; unknown function in mammals	Peptide (two forms; 13 and 22 amino acids)
THYROID GLAND			
Thyroxine (thyroid hormone)	Most cells	Stimulates metabolic rate; essential to normal growth and development	Iodinated amino acid
Calcitonin	Bone	Lowers blood calcium level by inhibiting loss of calcium from bone	Peptide (32 amino acids)
PARATHYROID GLANDS			
Parathyroid hormone	Bone, kidneys, digestive tract	Raises blood calcium level by stimulating bone breakdown; stimulates calcium reabsorption in kidneys; activates vitamin D	Peptide (34 amino acids)

*These are hormones released from endocrine glands. As discussed previously, many hormones are released from other body organs.

Paracrine Regulation

Paracrine regulation occurs in many organs and among the cells of the immune system. Some of these regulatory molecules are known as **cytokines,** particularly if they regulate different cells of the immune system. Other paracrine regulators are called **growth factors,** because they promote growth and cell division in specific organs. Examples include *platelet-derived growth factor, epidermal growth factor,* and the *insulin-like growth factors* that stimulate cell division and proliferation of their target cells. *Nerve growth factor* is a regulatory molecule that belongs to a family of paracrine regulators of the nervous system called **neurotrophins.**

Nitric oxide, which can function as a neurotransmitter (see chapter 45), is also produced by the endothelium of blood vessels. In this context, it is a paracrine regulator because it diffuses to the smooth muscle layer of the blood vessel and promotes vasodilation. The endothelium of blood vessels also produces other paracrine regulators, including *endothelin,* which stimulates vasoconstriction, and *bradykinin,* which pro-

motes vasodilation. This paracrine regulation supplements the regulation of blood vessels by autonomic nerves.

The most diverse group of paracrine regulators are the **prostaglandins.** A prostaglandin is a 20-carbon-long fatty acid that contains a five-member carbon ring. This molecule is derived from the precursor molecule *arachidonic acid,* released from phospholipids in the cell membrane under hormonal or other stimulation. Prostaglandins are produced in almost every organ and participate in a variety of regulatory functions, including:

1. **Immune system.** Prostaglandins promote many aspects of inflammation, including pain and fever. Drugs that inhibit prostaglandin synthesis help alleviate these symptoms.
2. **Reproductive system.** Prostaglandins regulate reproductive functions such as gamete transport, labor, and possibly ovulation. Excessive prostaglandin production may be involved in premature labor, endometriosis, or dysmenorrhea (painful menstrual cramps).

Table 47.1—*continued.* Principal Endocrine Glands and Their Hormones

Endocrine Gland and Hormone	Target Tissue	Principal Actions	Chemical Nature
ADRENAL MEDULLA			
Epinephrine (adrenaline) and norepinephrine (noradrenaline)	Smooth muscle, cardiac muscle, blood vessels	Initiate stress responses; raise heart rate, blood pressure, metabolic rate; dilate blood vessels; mobilize fat; raise blood glucose level	Amino acid derivatives
ADRENAL CORTEX			
Aldosterone	Kidney tubules	Maintains proper balance of Na^+ and K^+ ions	Steroid
Cortisol	Many organs	Adaptation to long-term stress; raises blood glucose level; mobilizes fat	Steroid
PANCREAS			
Insulin	Liver, skeletal muscles, adipose tissue	Lowers blood glucose level; stimulates storage of glycogen in liver	Peptide (51 amino acids)
Glucagon	Liver, adipose tissue	Raises blood glucose level; stimulates breakdown of glycogen in liver	Peptide (29 amino acids)
OVARY			
Estradiol	General	Stimulates development of secondary sex characteristics in females	Steroid
	Female reproductive structures	Stimulates growth of sex organs at puberty and monthly preparation of uterus for pregnancy	
Progesterone	Uterus	Completes preparation for pregnancy	Steroid
	Mammary glands	Stimulates development	
TESTIS			
Testosterone	Many organs	Stimulates development of secondary sex characteristics in males and growth spurt at puberty	Steroid
	Male reproductive structures	Stimulates development of sex organs; stimulates spermatogenesis	
PINEAL GLAND			
Melatonin	Gonads, pigment cells	Function not well understood; influences pigmentation in some vertebrates; may control biorhythms in some animals; may influence onset of puberty in humans	Amino acid derivative

3. **Digestive system.** Prostaglandins produced by the stomach and intestines may inhibit gastric secretions and influence intestinal motility and fluid absorption.

4. **Respiratory system.** Some prostaglandins cause constriction, whereas others cause dilation of blood vessels in the lungs and of bronchiolar smooth muscle.

5. **Circulatory system.** Prostaglandins are needed for proper function of blood platelets in the process of blood clotting.

6. **Urinary system.** Prostaglandins produced in the renal medulla cause vasodilation, resulting in increased renal blood flow and increased excretion of urine.

The synthesis of prostaglandins is inhibited by aspirin. Aspirin is the most widely used of the **nonsteroidal anti-inflammatory drugs (NSAIDs),** a class of drugs that also includes indomethacin and ibuprofen. These drugs produce their effects because they specifically inhibit the enzyme cyclooxygenase-2 (cox-2), needed to produce prostaglandins from arachidonic acid. Through this action, the NSAIDs inhibit inflammation and associated pain. Unfortunately, NSAIDs also inhibit another similar enzyme, cox-1, which helps maintain the wall of the digestive tract, and in so doing can produce severe unwanted side effects, including gastric bleeding and prolonged clotting time. A new kind of pain reliever, celecoxib (Celebrex), inhibits cox-2 but not cox-1, a potentially great benefit to arthritis sufferers and others who must use pain relievers regularly.

The neural and endocrine control systems are supplemented by paracrine regulators, including the prostaglandins, which perform many diverse functions.

47.2 Lipophilic and polar hormones regulate their target cells by different means.

Hormones That Enter Cells

As we mentioned previously, hormones can be divided into those that are lipophilic (lipid-soluble) and those that are lipophobic (water-soluble or hydrophilic). The lipophilic hormones—all of the steroid hormones (figure 47.4) and thyroxine—as well as other lipophilic regulatory molecules (including the retinoids, or vitamin A) can easily enter cells. This is because the lipid portion of the plasma membrane does not present a barrier to the entry of lipophilic regulators, which are of a similar chemical nature. Therefore, all lipophilic regulatory molecules have a similar mechanism of action. Lipophobic (water-soluble) hormones, in contrast, cannot pass through plasma membranes. They must regulate their target cells through membrane receptors.

Steroid hormones are lipids themselves and thus lipophilic; thyroxine is lipophilic because it is made nonpolar during its production from the polar amino acid tyrosine. Because these hormones are not water-soluble, they don't dissolve in plasma but rather travel in the blood

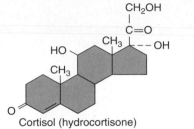

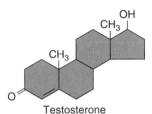

FIGURE 47.4
Chemical structures of some steroid hormones. Steroid hormones are derived from the blood lipid cholesterol. The hormones shown, cortisol, estradiol, and testosterone, differ only slightly in chemical structure yet have widely different effects on the body. Steroid hormones are secreted by the adrenal cortex, testes, ovaries, and placenta.

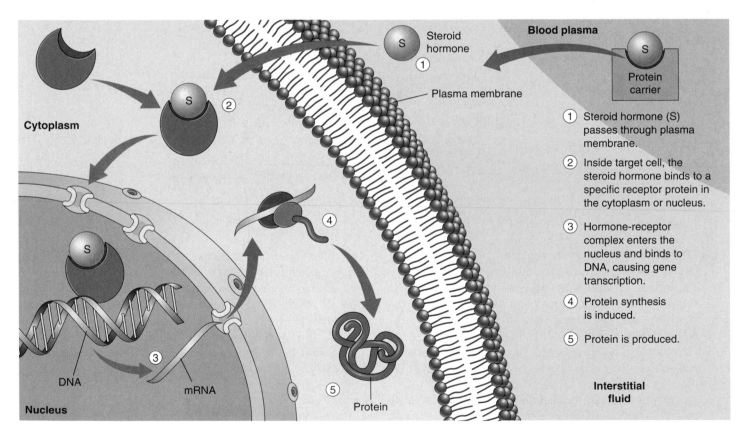

FIGURE 47.5
The mechanism of steroid hormone action. Steroid hormones are lipophilic (lipid-soluble) and thus readily diffuse through the plasma membrane of cells. They bind to receptor proteins in either the cytoplasm or nucleus (not shown). If the steroid binds to a receptor in the cytoplasm, the hormone-receptor complex moves into the nucleus. The hormone-receptor complex then binds to specific regions of the DNA, stimulating the production of messenger RNA (mRNA).

attached to protein carriers. When the hormones arrive at their target cells, they dissociate from their carriers and pass through the plasma membrane of the cell (figure 47.5). Some steroid hormones then bind to very specific receptor proteins in the cytoplasm, and then move as a hormone-receptor complex into the nucleus. Other steroids travel directly into the nucleus before encountering their receptor proteins. Whether the steroid finds its receptor in the nucleus or translocates with its receptor to the nucleus from the cytoplasm, the rest of the story is the same.

The hormone receptor, activated by binding to the lipophilic hormone, is now also able to bind to specific regions of the DNA. These DNA regions are known as the hormone response elements. The binding of the hormone-receptor complex has a direct effect on the level of transcription at that site by activating, or in some cases deactivating, genetic transcription. The activation of transcription produces messenger RNA (mRNA), which then codes for the production of specific proteins. These proteins often have activity that changes the metabolism of the target cell in a specific fashion.

The thyroid hormone's mechanism of action resembles that of the steroid hormones. Thyroxine contains four iodines and so is often abbreviated T_4 (for *tetraiodothyronine*). The thyroid gland also secretes smaller amounts of a similar molecule that has only three iodines, called *triiodothyronine* (and abbreviated T_3). Both hormones enter target cells, but all of the T_4 that enters is changed into T_3 (figure 47.6). Thus, only the T_3 form of the hormone enters the nucleus and binds to nuclear receptor proteins. The hormone-receptor complex, in turn, binds to the appropriate hormone response elements on DNA.

The lipophilic hormones pass through the target cell's plasma membrane and bind to intracellular receptor proteins. The hormone-receptor complex then binds to specific regions of DNA, thereby regulating genes in the target cell.

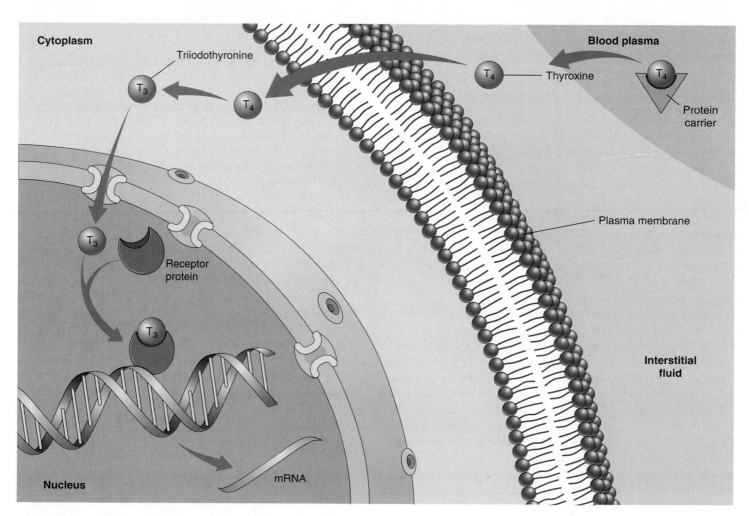

FIGURE 47.6

The mechanism of thyroxine action.   Thyroxine is lipophilic and contains four iodines. When it enters the target cell, thyroxine is changed into triiodothyronine, with three iodines. This hormone moves into the nucleus and binds to nuclear receptors. The hormone-receptor complex then binds to regions of the DNA and stimulates gene transcription.

Hormones That Do Not Enter Cells

Hormones that are too large or too polar to cross the plasma membranes of their target cells include all of the peptide and glycoprotein hormones, as well as the catecholamine hormones epinephrine and norepinephrine. These hormones bind to receptor proteins located on the outer surface of the plasma membrane—the hormones do *not* enter the cell. If you think of the hormone as a messenger sent from an endocrine gland to the target cell, it is evident that a second messenger is needed within the target cell to produce the effects of the hormone. A number of different molecules in the cell can serve as second messengers, as we saw in chapter 7. The interaction between the hormone and its receptor activates mechanisms in the plasma membrane that increase the concentration of the second messengers within the target cell cytoplasm.

The binding of a lipophobic (water-soluble) hormone to its receptor is reversible and usually very brief. After the hormone binds to its receptor and activates a second-messenger system, it dissociates from the receptor and may travel in the blood to another target cell somewhere else in the body. Eventually, enzymes (primarily in the liver) degrade the hormone by converting it into inactive derivatives.

The Cyclic AMP Second-Messenger System

The action of the hormone epinephrine can serve as an example of a second-messenger system. Epinephrine can bind to two categories of receptors, called *alpha* (α)- and *beta* (β)-*adrenergic receptors*. The interaction of epinephrine with each type of receptor activates a different second-messenger system in the target cell.

In the early 1960s, Earl Sutherland showed that cyclic adenosine monophosphate, or **cyclic AMP (cAMP),** serves as a second messenger when epinephrine binds to β-adrenergic receptors on the plasma membranes of liver cells (figure 47.7). The cAMP second-messenger system was the first such system to be described.

The β-adrenergic receptors are associated with membrane proteins called *G proteins* (see chapters 7 and 45). Each G protein is composed of three subunits, and the binding of epinephrine to its receptor causes one of the G protein subunits to dissociate from the other two. This subunit then diffuses within the plasma membrane until it encounters **adenylyl cyclase,** a membrane enzyme that is inactive until it binds to the G protein subunit. When activated by the G protein subunit, adenylyl cyclase catalyzes the formation of cAMP from ATP. The cAMP formed at the inner surface of the plasma membrane diffuses within the cytoplasm, where it binds to and activates **protein kinase-A,** an enzyme that adds phosphate groups to specific cellular proteins.

The identities of the proteins that are phosphorylated by protein kinase-A vary from one cell type to the next, and this variation is one of the reasons epinephrine has such diverse

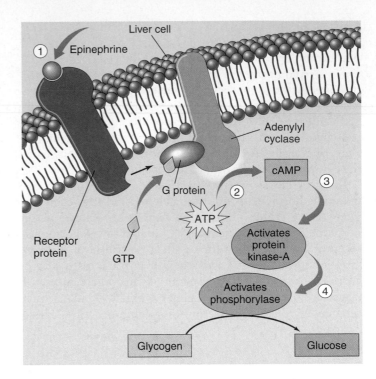

FIGURE 47.7
The action of epinephrine on a liver cell. (*1*) Epinephrine is lipophobic and needs to bind to specific receptor proteins on the cell surface. (*2*) Acting through intermediary G proteins, the hormone-bound receptor activates the enzyme adenylyl cyclase, which converts ATP into cyclic AMP (cAMP). (*3*) Cyclic AMP performs as a second messenger and activates protein kinase-A, an enzyme that was previously present in an inactive form. (*4*) Protein kinase-A phosphorylates and thereby activates the enzyme phosphorylase, which catalyzes the hydrolysis of glycogen into glucose.

effects on different tissues. In liver cells, protein kinase-A phosphorylates and thereby activates another enzyme, phosphorylase, which converts glycogen into glucose. Through this multistep mechanism, epinephrine causes the liver to secrete glucose into the blood during the fight-or-flight reaction, when the adrenal medulla is stimulated by the sympathetic division of the autonomic nervous system (see chapter 45). In cardiac muscle cells, protein kinase-A phosphorylates a different set of cellular proteins, which cause the heart to beat faster and more forcefully.

The IP₃/Ca⁺⁺ Second-Messenger System

When epinephrine binds to α-adrenergic receptors, it doesn't activate adenylyl cyclase and cause the production of cAMP. Instead, through a different type of G protein, it activates another membrane-bound enzyme, phospholipase C (figure 47.8). This enzyme cleaves certain membrane phospholipids to produce the second messenger inositol triphosphate (IP_3). IP_3 diffuses into the cytoplasm

FIGURE 47.8

The IP₃/Ca⁺⁺ second-messenger system. *(1)* The hormone epinephrine binds to specific receptor proteins on the cell surface. *(2)* Acting through G proteins, the hormone-bound receptor activates the enzyme phospholipase C, which converts membrane phospholipids into inositol triphosphate (IP₃). *(3)* IP₃ diffuses through the cytoplasm and binds to receptors on the endoplasmic reticulum. *(4)* The binding of IP₃ to its receptors stimulates the endoplasmic reticulum to release Ca⁺⁺ into the cytoplasm. *(5)* Some of the released Ca⁺⁺ binds to a regulatory protein called calmodulin. *(6)* The Ca⁺⁺/calmodulin complex activates other intracellular proteins, ultimately producing the effects of the hormone.

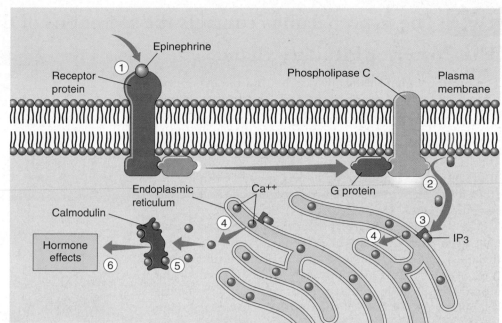

from the plasma membrane and binds to receptors located on the surface of the endoplasmic reticulum.

Recall from chapter 5 that the endoplasmic reticulum is a system of membranous sacs and tubes that serves a variety of functions in different cells. One of its functions is to accumulate Ca⁺⁺ by active transport from the cytoplasm. Other pumps transport Ca⁺⁺ from the cytoplasm through the plasma membrane to the extracellular fluid. These two mechanisms keep the concentration of Ca⁺⁺ in the cytoplasm very low. Consequently, there is an extremely steep concentration gradient for Ca⁺⁺ between the cytoplasm and the inside of the endoplasmic reticulum, and between the cytoplasm and the extracellular fluid.

When IP₃ binds to its receptors on the endoplasmic reticulum, it stimulates the endoplasmic reticulum to release its stored Ca⁺⁺. Calcium channels in the plasma membrane may also open, allowing Ca⁺⁺ to diffuse into the cell from the extracellular fluid. Some of the Ca⁺⁺ that has suddenly entered the cytoplasm then binds to a protein called calmodulin, which has regulatory functions analogous to those of cyclic AMP. One of the actions of calmodulin is to activate another type of protein kinase, resulting in the phosphorylation of a different set of cellular proteins.

What is the advantage of having multiple second-messenger systems? Consider the antagonistic actions of epinephrine and insulin on liver cells. Epinephrine uses cAMP as a second messenger to promote the hydrolysis of glycogen to glucose, while insulin stimulates the synthesis of glycogen from glucose. Clearly, insulin cannot use cAMP as a second messenger. Although the exact mechanism of insulin's action is still not well understood, insulin may act in part through the IP₃/Ca⁺⁺ second-messenger system.

Not all large polar hormones act by increasing the concentration of a second messenger in the cytoplasm of the target cell. Others cause a change in the shape of a membrane protein called an ion channel (see chapters 6 and 45). If these channels are normally "closed," then a change in shape will open them, allowing a particular ion to enter or leave the cell depending on its concentration gradient. If an ion channel is normally open, a chemical messenger can cause it to close. For example, some hormones open Ca⁺⁺ channels on smooth muscle plasma membranes; other hormones close them. This will increase or decrease, respectively, the amount of muscle contraction.

The molecular mechanism for changing the shape of an ion channel is similar to that for activating a second messenger. The hormone first binds to a receptor protein on the outer surface of the target cell. This receptor protein uses a G protein to signal the ion channel to change shape.

Although G proteins play a major role in many hormone functions, they don't seem to be necessary for all identified actions of hormones on target cells. In the cases where G proteins are not involved, the receptor protein may be the ion channel or connected directly to the enzyme or ion channel.

The lipophobic, water-soluble hormones cannot pass through the plasma membrane; they primarily rely on second messengers within the target cells to mediate their actions. Such second messengers include cyclic AMP (cAMP), inositol triphosphate (IP₃), and Ca⁺⁺. In many cases, the second messengers activate previously inactive enzymes.

The Posterior Pituitary Gland

The **pituitary gland** hangs by a stalk from the hypothalamus of the brain posterior to the optic chiasm (see chapter 45). A microscopic view reveals that the gland consists of two parts. The anterior portion appears glandular and is called the **anterior pituitary;** the posterior portion appears fibrous and is the **posterior pituitary.** These two portions of the pituitary gland have different embryonic origins, secrete different hormones, and are regulated by different control systems.

The posterior pituitary appears fibrous because it contains axons that originate in cell bodies within the hypothalamus and extend along the stalk of the pituitary as a tract of fibers. This anatomical relationship results from the way the posterior pituitary is formed in embryonic development. As the floor of the third ventricle of the brain forms the hypothalamus, part of this neural tissue grows downward to produce the posterior pituitary. The hypothalamus and posterior pituitary thus remain interconnected by a tract of axons.

The endocrine role of the posterior pituitary gland first became evident in 1912, when a remarkable medical case was reported: A man who had been shot in the head developed the need to urinate every 30 minutes or so, 24 hours a day. The bullet had lodged in his pituitary gland. Subsequent research demonstrated that removal of this gland produces the same symptoms. Pituitary extracts were found to contain a substance that makes the kidneys conserve water, and in the early 1950s investigators isolated a peptide from the posterior pituitary, **antidiuretic hormone (ADH,** also known as **vasopressin),** that stimulates water retention by the kidneys (figure 47.9). When ADH is missing, the kidneys do not retain water, and excessive quantities of urine are produced. This is why the consumption of alcohol, which inhibits ADH secretion, leads to frequent urination.

The posterior pituitary also secretes **oxytocin,** a second peptide hormone that, like ADH, is composed of nine amino acids. Oxytocin stimulates the milk-ejection reflex, so that contraction of the smooth muscles around the mammary glands and ducts causes milk to be ejected from the ducts through the nipple. During suckling, sensory receptors in the nipples send impulses to the hypothalamus, which triggers the release of oxytocin. Oxytocin is also needed to stimulate uterine contractions in women during childbirth. Oxytocin secretion continues after childbirth in a woman who is breast-feeding, which is why the uterus of a nursing mother returns to its normal size after pregnancy more quickly than does the uterus of a mother who does not breast-feed. In both men and women, oxytocin is thought to be involved in regulating sexual responses, including arousal and orgasm.

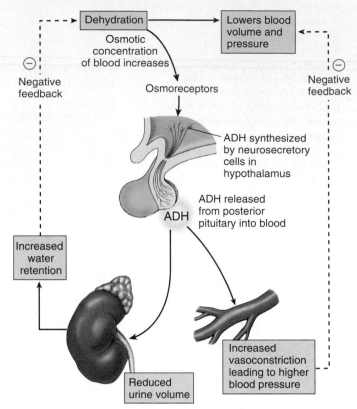

FIGURE 47.9
The effects of antidiuretic hormone (ADH). An increase in the osmotic concentration of the blood stimulates the posterior pituitary gland to secrete ADH, which promotes water retention by the kidneys. This works as a negative feedback loop to correct the initial disturbance of homeostasis.

ADH and oxytocin are actually *produced* by neuron cell bodies located in the hypothalamus. These two hormones are transported along the axon tract that runs from the hypothalamus to the posterior pituitary and are stored in the posterior pituitary. In response to the appropriate stimulation—increased blood plasma osmolality in the case of ADH, the suckling of a baby in the case of oxytocin—the hormones are released by the posterior pituitary into the blood. Because this reflex control involves both the nervous and endocrine systems, the ADH and oxytocin are said to be secreted due to **neuroendocrine reflexes.**

The posterior pituitary gland contains axons originating from neurons in the hypothalamus. These neurons produce ADH and oxytocin, which are stored in and released from the posterior pituitary gland in response to neural stimulation from the hypothalamus.

The Anterior Pituitary Gland

The anterior pituitary, unlike the posterior pituitary, does not develop from a downgrowth of the brain; instead, it develops from a pouch of epithelial tissue that pinches off from the roof of the embryo's mouth. Because it is epithelial tissue, the anterior pituitary is a complete gland—that is, it produces the hormones it secretes. Many, but not all, of these hormones stimulate growth in their target organs, including other hormone production and secretion from endocrine glands. Therefore, the hormones of the anterior pituitary gland are collectively termed *tropic hormones* (Greek *trophe*, "nourishment"), or *tropins*. When the target organ of a tropic hormone is another endocrine gland, that gland is stimulated by the tropic hormone to secrete its own hormones.

The hormones produced and secreted by different cell types in the anterior pituitary gland (figure 47.10, *left*) include the following:

1. **Growth hormone (GH,** or *somatotropin)* stimulates the growth of muscle, bone (indirectly), and other tissues, and is also essential for proper metabolic regulation.
2. **Adrenocorticotropic hormone (ACTH,** or *corticotropin)* stimulates the adrenal cortex to produce corticosteroid hormones, including cortisol (in humans) and corticosterone (in many other vertebrates), which regulate glucose homeostasis.
3. **Thyroid-stimulating hormone (TSH,** or *thyrotropin)* stimulates the thyroid gland to produce thyroxine, which in turn stimulates oxidative respiration.
4. **Luteinizing hormone (LH)** is needed for ovulation and the formation of a corpus luteum in the female menstrual cycle (see chapter 50). It also stimulates the testes to produce testosterone, which is needed for sperm production and for the development of male secondary sexual characteristics.

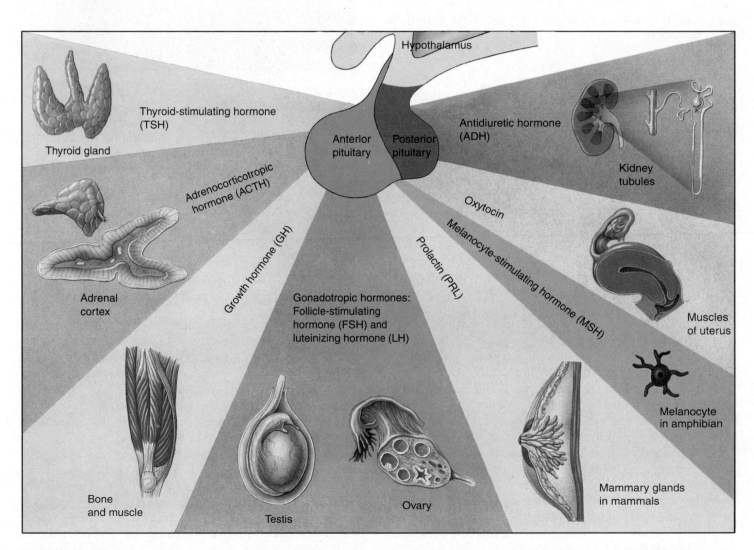

FIGURE 47.10
The major hormones of the anterior and posterior pituitary glands. Only a few of the actions of these hormones are shown.

5. **Follicle-stimulating hormone (FSH)** is required for the development of ovarian follicles in females. In males, it is required for the development of sperm. FSH and LH are both referred to as *gonadotropins* because they stimulate metabolism in gonads.

6. **Prolactin (PRL)** stimulates the mammary glands to produce milk in mammals. It also helps regulate kidney function in vertebrates and the production of "crop milk" (nutritional fluid fed to chicks by regurgitation) in some birds, and it acts on the gills of fish that travel from salt to fresh water to promote sodium retention.

7. **Melanocyte-stimulating hormone (MSH)** stimulates the synthesis and dispersion of melanin pigment, which darkens the epidermis of some fish, amphibians, and reptiles. MSH has no known specific function in mammals, but abnormally high amounts of ACTH can cause skin darkening because it contains the amino acid sequence of MSH within its structure.

Growth Hormone

The importance of the anterior pituitary gland is illustrated by a condition known as *gigantism*, characterized by excessive growth of the entire body or any of its parts. The tallest human being ever recorded, Robert Wadlow, had gigantism (figure 47.11). Born in 1928, he stood 8 feet 11 inches tall, weighed 485 pounds, and was still growing before he died from an infection at the age of 22.

We now know that gigantism is caused by the excessive secretion of growth hormone (GH) by the anterior pituitary gland in a growing child. By contrast, a deficiency in GH secretion during childhood results in *pituitary dwarfism*, failure to achieve normal growth. GH stimulates protein synthesis and growth of muscles and connective tissues; it also indirectly promotes the elongation of bones by stimulating cell division in the cartilaginous epiphyseal growth plates of bones. Researchers found that this stimulation does not occur in the absence of blood plasma, suggesting that bone cells lack receptors for GH and that the stimulation of GH is indirect. We now know that GH stimulates the production of **insulin-like growth factors,** which are produced by the liver and secreted into the blood in response to stimulation by GH. The insulin-like growth factors then stimulate growth of the epiphyseal growth plates and thus elongation of the bones.

When a person's skeletal growth plates have converted from cartilage into bone—that is, the person has become an adult—GH can no longer cause an increase in height. Therefore, excessive GH secretion in an adult results in a form of gigantism called *acromegaly*, characterized by bone and soft tissue deformities such as a protruding jaw and thickening of other facial features. The connection between gigantism and the anterior pituitary gland was first

FIGURE 47.11
The Alton giant. This photograph of Robert Wadlow of Alton, Illinois, taken on his 21st birthday, shows him at home with his father and mother and four siblings. Born normal size, he developed a growth hormone-secreting pituitary tumor as a young child and never stopped growing during his 22 years of life.

noticed in 1909 when a 38-year-old South Dakota farmer was cured of acromegaly by the surgical removal of a pituitary tumor.

Other Anterior Pituitary Hormones

Prolactin is like growth hormone in that it acts on organs that are not endocrine glands. In addition to stimulating milk production in mammals and "crop milk" production in birds, prolactin has varied effects on electrolyte balance by acting on the kidneys, the gills of fish, and the salt glands of marine birds (discussed in chapter 49).

Unlike growth hormone and prolactin, the other anterior pituitary hormones act on specific glands. Some of these hormones have common names, such as thyroid-stimulating hormone (TSH), and alternative names that emphasize the tropic nature of the hormone, such as thyrotropin. TSH stimulates only the thyroid gland, and adrenocorticotropic hormone (ACTH) stimulates only the adrenal cortex (outer portion of the adrenal glands). Follicle-stimulating hormone (FSH) and luteinizing hormone (LH) act only on the gonads (testes and ovaries); hence, they are collectively called *gonadotropic hormones*. Although both FSH and LH act on the gonads, they each target different cells in the gonads of both females and males.

Hypothalamic Control of the Anterior Pituitary

The anterior pituitary gland, unlike the posterior pituitary, is not derived from the brain and does not receive an axon tract from the hypothalamus. Nevertheless, the hypothalamus controls production and secretion of the anterior pituitary hormones. This control is exerted hormonally rather than by means of nerve axons. Neurons in the hypothalamus secrete releasing hormones and inhibiting hormones into blood capillaries at the base of the hypothalamus (figure 47.12). These capillaries drain into small veins that run within the stalk of the pituitary to a second bed of capillaries in the anterior pituitary. This unusual system of vessels is known as the hypothalamohypophysial portal system (another name for the pituitary is the hypophysis). It is called a portal system because it has a second capillary bed downstream from the first; the only other body locations with a similar system are the liver, where capillaries receive blood drained from the gastrointestinal tract (see chapter 43), and the kidneys, where the glomerulus and peritubular capillaries are in series (see chapter 49). Because the second bed of capillaries receives little oxygen from such vessels, the vessels must be delivering something else of importance.

Each hormone released by the hypothalamohypophysial system regulates the secretion of a specific anterior pituitary hormone. For example, *thyrotropin-releasing hormone* (TRH) stimulates the release of TSH; *corticotropin-releasing hormone* (CRH) stimulates the release of ACTH; and *gonadotropin-releasing hormone* (GnRH) stimulates the release of FSH and LH. A releasing hormone for growth hormone, called *growth hormone–releasing hormone* (GHRH), has also been discovered, and a releasing hormone for prolactin has been postulated but has thus far not been identified.

The hypothalamus also secretes hormones that *inhibit* the release of certain anterior pituitary hormones. To date, three such hormones have been discovered: *Somatostatin*, or *growth hormone–inhibiting hormone* (GHIH), inhibits the secretion of GH; *prolactin-inhibiting factor* (PIF), found to be the neurotransmitter dopamine, inhibits the secretion of prolactin; and *melanotropin-inhibiting hormone* (MIH) inhibits the secretion of MSH.

Negative Feedback Control of the Anterior Pituitary

Because hypothalamic hormones control the secretions of the anterior pituitary gland, and the anterior pituitary hormones control the secretions of some other endocrine glands, it may seem that the hypothalamus is in charge of hormonal secretion for the whole body. This idea is not valid, however, for two reasons. First, a number of endocrine organs, such as the adrenal medulla and the pancreas, are not directly regulated by this control system. Second, the hypothalamus and the anterior pituitary gland are themselves partially controlled by the very hormones whose secretion they stimulate! In most cases, this is an inhibitory control, whereby the target gland hormones inhibit the secretions of the hypothalamus and anterior pituitary (figure 47.13). This

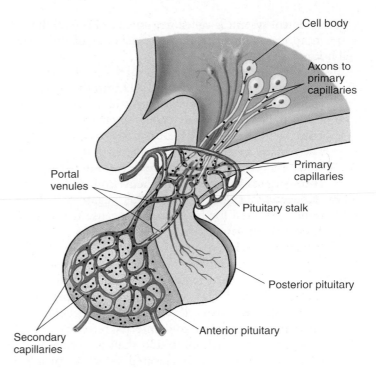

FIGURE 47.12
Hormonal control of the anterior pituitary gland by the hypothalamus. Neurons in the hypothalamus secrete hormones that are carried by short blood vessels directly to the anterior pituitary gland, where they either stimulate or inhibit the secretion of anterior pituitary hormones.

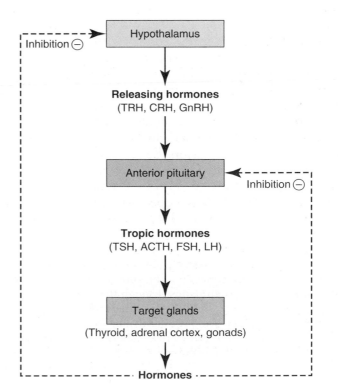

FIGURE 47.13
Negative feedback inhibition. The hormones secreted by some endocrine glands feed back to inhibit the secretion of hypothalamic releasing hormones and anterior pituitary tropic hormones.

type of control system is called negative feedback inhibition and acts to maintain relatively constant levels of the target cell hormone.

Let's consider the hormonal control of the thyroid gland. The hypothalamus secretes TRH into the hypothalamohypophysial portal system, which stimulates the anterior pituitary gland to secrete TSH, which in turn stimulates the thyroid gland to release thyroxine. Among thyroxine's many target organs are the hypothalamus and the anterior pituitary gland themselves. Thyroxine acts upon these organs to inhibit their secretion of TRH and TSH, respectively (figure 47.14). This negative feedback inhibition is essential for homeostasis because it keeps the thyroxine levels fairly constant.

To illustrate the importance of negative feedback inhibition, we will examine a person who lacks sufficient iodine in the diet. Without iodine, the thyroid gland cannot produce thyroxine (which contains four iodines per molecule). As a result, thyroxine levels in the blood fall drastically, and the hypothalamus and anterior pituitary receive far less negative feedback inhibition than is normal. This reduced inhibition causes an elevated secretion of TRH and TSH. The high levels of TSH stimulate the thyroid gland, which swells with intermediate products of the thyroxine metabolic pathway, but it still cannot produce thyroxine without iodine. The consequence of this interruption of the normal inhibition by thyroxine is an enlarged thyroid gland, a condition known as a goiter (figure 47.15). Goiter size can be reduced by providing iodine in the diet.

Positive feedback in the control of the hypothalamus and anterior pituitary by the target glands is not common because positive feedback cannot maintain constancy of the internal environment (homeostasis). Positive feedback accentuates change, driving the change in the same direction. One example of positive control involves the control of labor during birth, an event that culminates in the expulsion of the baby from the uterus. In this case, stretching of the vagina, cervix, and uterus stimulates the release of oxytocin. Oxytocin increases the intensity of uterine contractions, which in turn stimulate further oxytocin release. This is discussed in detail in chapter 50.

The hypothalamus controls the anterior pituitary gland by means of hormones, and the anterior pituitary gland controls some other glands through the hormones it secretes. However, both the hypothalamus and the anterior pituitary gland are controlled by hormones from the endocrine glands that they stimulate through negative feedback inhibition.

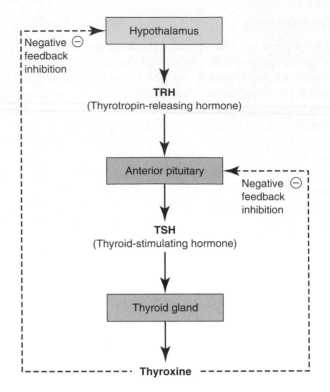

FIGURE 47.14
Regulation of thyroxine secretion. The hypothalamus secretes TRH, which stimulates the anterior pituitary to secrete TSH. The TSH then stimulates the thyroid to secrete thyroxine, which exerts negative feedback control of the hypothalamus and anterior pituitary.

FIGURE 47.15
A person with a goiter. This condition is caused by a lack of iodine in the diet. As a result, thyroxine secretion is low, so there is less negative feedback inhibition of TSH. The elevated TSH secretion, in turn, stimulates the thyroid to swell with intermediate products of the thyroxine pathway, producing the goiter.

47.4 Endocrine glands secrete hormones that regulate many body functions.

The Thyroid and Parathyroid Glands

The endocrine glands that are regulated by the anterior pituitary, and the endocrine glands that are regulated by other means, help control metabolism, electrolyte balance, and reproductive functions. Some of the major endocrine glands are considered in this section.

The Thyroid Gland

The thyroid gland (Greek *thyros*, "shield") is shaped like a shield and lies just below the Adam's apple in the front of the neck. We have already mentioned that the thyroid gland secretes thyroxine and smaller amounts of triiodothyronine (T_3), which stimulate oxidative respiration in most cells in the body and, in so doing, help set the body's basal metabolic rate (see chapter 43). In children, these thyroid hormones also promote growth and stimulate maturation of the central nervous system. Children with underactive thyroid glands are therefore stunted in their growth and suffer severe mental retardation, a condition called cretinism. This condition differs from pituitary dwarfism, which results from inadequate GH and is not associated with abnormal intellectual development.

People who are hypothyroid (whose secretion of thyroxine is too low) can take thyroxine orally, as pills. Only thyroxine and the steroid hormones (as in contraceptive pills) can be taken orally because they are nonpolar (lipophilic) and can pass through the plasma membranes of intestinal epithelial cells without being digested.

There is an additional function of the thyroid gland that is unique to amphibians: Thyroid hormones are needed for the metamorphosis of the larvae into adults (figure 47.16). If the thyroid gland is removed from a tadpole, it will not change into a frog. Conversely, if an immature tadpole is fed pieces of a thyroid gland, it will undergo premature metamorphosis and become a miniature frog!

The thyroid gland also secretes *calcitonin*, a peptide hormone that plays a role in maintaining proper levels of calcium (Ca^{++}) in the blood. When the blood Ca^{++} concentration rises too high, calcitonin stimulates the uptake of Ca^{++} into bones, thus lowering its level in the blood. Although calcitonin may be important in the physiology of

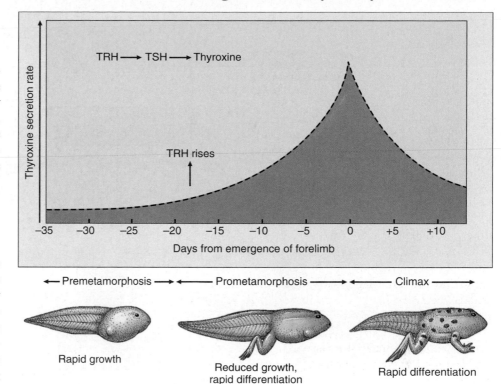

FIGURE 47.16
Thyroxine triggers metamorphosis in amphibians.   In tadpoles at the premetamorphic stage, the hypothalamus releases TRH (thyrotropin-releasing hormone), which causes the anterior pituitary to secrete TSH (thyroid-stimulating hormone). TSH then acts on the thyroid gland, which secretes thyroxine. The hindlimbs then begin to form. As metamorphosis proceeds, thyroxine reaches its maximal level, after which the forelimbs begin to form.

some vertebrates, its significance in normal human physiology is controversial, and it appears less important in the day-to-day regulation of Ca^{++} levels. A hormone that plays a more important role in Ca^{++} homeostasis is secreted by the parathyroid glands, described next.

The Parathyroid Glands and Calcium Homeostasis

The parathyroid glands are four small glands attached to the thyroid. Because of their size, researchers ignored them until well into the twentieth century. The first suggestion that these organs have an endocrine function came from experiments on dogs: If their parathyroid glands were removed, the Ca^{++} concentration in the dogs' blood plummeted to less than half the normal value. The Ca^{++} concentration returned to normal when an extract of parathyroid gland was administered. However, if too much of the extract was administered, the dogs' Ca^{++} levels rose far above normal as the calcium phosphate crystals in their bones were dissolved. It was clear that the parathyroid glands produce a hormone that stimulates the release of Ca^{++} from bone.

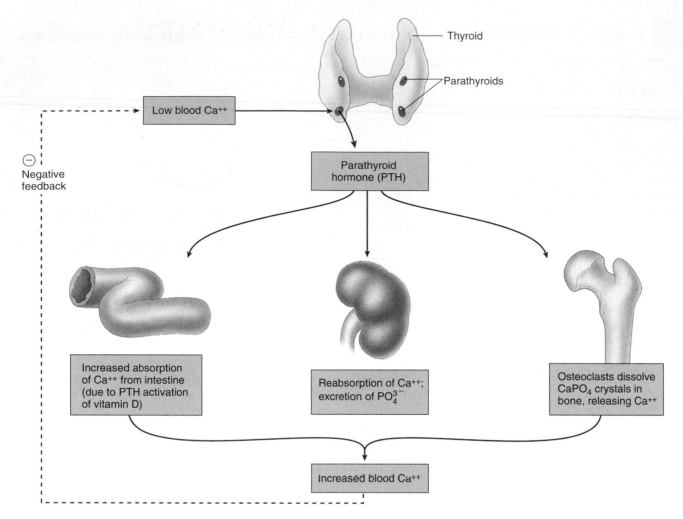

FIGURE 47.17
Regulation of blood Ca++ levels by parathyroid hormone (PTH).   When blood Ca++ levels are low, parathyroid hormone (PTH) is released by the parathyroid glands. PTH directly stimulates the dissolution of bone and the reabsorption of Ca++ by the kidneys. PTH indirectly promotes the intestinal absorption of Ca++ by stimulating the production of the active form of vitamin D.

The hormone produced by the parathyroid glands is a peptide called *parathyroid hormone* (PTH). It is one of only two hormones in humans that are absolutely essential for survival (the other is aldosterone, which will be discussed next). PTH is synthesized and released in response to falling levels of Ca++ in the blood. This cannot be allowed to continue uncorrected, because a significant fall in the blood Ca++ level can cause severe muscle spasms. A normal blood Ca++ level is important for the functioning of muscles, including the heart, and for proper functioning of the nervous and endocrine systems.

PTH stimulates the osteoclasts (bone cells) in bone to dissolve the calcium phosphate crystals of the bone matrix and release Ca++ into the blood (figure 47.17). PTH also stimulates the kidneys to reabsorb Ca++ from the urine and leads to the activation of vitamin D, needed for the absorption of Ca++ from food in the intestine.

Vitamin D is produced in the skin from a cholesterol derivative in response to ultraviolet light. It is called an essential vitamin because in most regions of the world a dietary source is needed to supplement the amount that the skin produces. Secreted into the blood from the skin, vitamin D is actually an inactive form of a hormone. In order to become activated, the molecule must gain two hydroxyl groups (—OH); one of these is added by an enzyme in the liver, the other by an enzyme in the kidneys. The enzyme needed for this final step is stimulated by PTH, thereby producing the active form of vitamin D known as 1, 25-dihydroxyvitamin D. This hormone stimulates the intestinal absorption of Ca++ and thereby helps raise blood Ca++ levels so that bone can become properly mineralized. A diet deficient in vitamin D thus leads to poor bone formation, a condition called rickets.

Thyroxine helps set the basal metabolic rate by stimulating the rate of cellular respiration throughout the body; this hormone is also needed for amphibian metamorphosis. Parathyroid hormone acts to raise the blood Ca++ levels, in part by stimulating osteoclasts to dissolve bone.

The Adrenal Glands

The adrenal glands are located just above each kidney (figure 47.18). Each gland is composed of an inner portion, the adrenal medulla, and an outer layer, the adrenal cortex.

The Adrenal Medulla

The adrenal medulla receives neural input from axons of the sympathetic division of the autonomic nervous system, and it secretes *epinephrine* and *norepinephrine* in response to stimulation by these axons. The actions of these hormones trigger "alarm" responses similar to those elicited by the sympathetic division, helping to prepare the body for "fight or flight." Among the effects of these hormones are an increased heart rate, increased blood pressure, dilation of the bronchioles, elevation in blood glucose, and reduced blood flow to the skin and digestive organs. The actions of epinephrine, released as a hormone, supplement those of norepinephrine, released as a sympathetic nerve neurotransmitter.

The Adrenal Cortex

The hormones from the adrenal cortex are all steroids and are referred to collectively as *corticosteroids*. Cortisol (also called hydrocortisone) and related steroids secreted by the adrenal cortex act on various cells in the body to maintain glucose homeostasis. In mammals, these hormones are referred to as glucocorticoids, and their secretion is primarily regulated by the anterior pituitary hormone ACTH. The glucocorticoids stimulate the breakdown of muscle protein into amino acids, which are carried by the blood to the liver. They also stimulate the liver to produce the enzymes needed for gluconeogenesis, the conversion of amino acids into glucose. This creation of glucose from protein is particularly important during very long periods of fasting or exercise, when blood glucose levels might otherwise become dangerously low.

In addition to regulating glucose metabolism, the glucocorticoids modulate some aspects of the immune response. Glucocorticoids are given medically to suppress the immune system in persons with immune disorders, such as rheumatoid arthritis. Derivatives of cortisol, such as prednisone, have widespread medical use as anti-inflammatory agents.

Aldosterone, the other major corticosteroid, is classified as a mineralocorticoid because it helps regulate mineral balance through two functions. One of its functions is to stimulate the kidneys to reabsorb Na^+ from the urine. (Urine is formed by filtration of blood plasma, so the blood levels of Na^+ decrease if Na^+ is not reabsorbed from the urine; see chapter 49.) Sodium is the major extracellular solute and is needed for the maintenance of normal blood volume and pressure. Without aldosterone, the kidneys would lose excessive amounts of blood Na^+ in the urine, followed by Cl^- and water; this would cause the blood volume and pressure to fall. By stimulating the kidneys to reabsorb salt and water, aldosterone thus maintains the normal blood volume and pressure essential to life.

The other function of aldosterone is to stimulate the kidneys to secrete K^+ into the urine. Thus, when aldosterone levels are too low, the concentration of K^+ in the blood may rise to dangerous levels. Because of these essential functions performed by aldosterone, removal of the adrenal glands, or diseases that prevent aldosterone secretion, are invariably fatal without hormone therapy. The secretion of aldosterone from the adrenal cortex is primarily regulated by angiotensin II, a product of the renin-angiotensin mechanism described in chapter 49.

The adrenal medulla is stimulated by sympathetic neurons to secrete epinephrine and norepinephrine during the fight or flight reaction. The adrenal cortex is stimulated to secrete its glucocorticoids by ACTH from the anterior pituitary, and it is stimulated to secrete its mineralocorticoid by angiotensin II. Cortisol helps regulate blood glucose, and aldosterone acts to regulate blood Na^+ and K^+ levels.

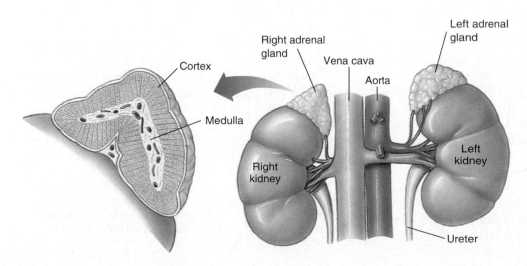

FIGURE 47.18
The adrenal glands. The inner portion of the gland, the adrenal medulla, produces epinephrine and norepinephrine, which initiate a response to stress. The outer portion of the gland, the adrenal cortex, produces steroid hormones that influence many metabolic functions.

42

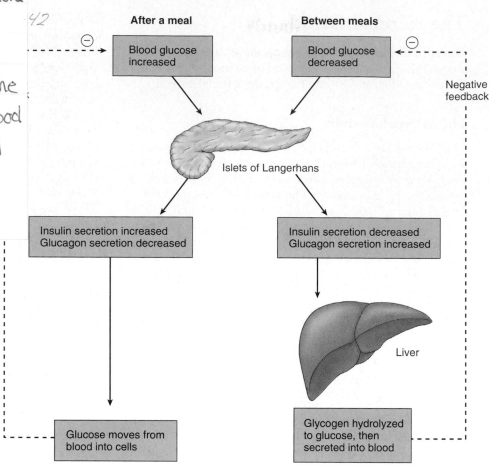

After a meal

Blood glucose increased

Between meals

Blood glucose decreased

Negative feedback

Islets of Langerhans

Insulin secretion increased
Glucagon secretion decreased

Insulin secretion decreased
Glucagon secretion increased

Liver

Glucose moves from blood into cells

Glycogen hydrolyzed to glucose, then secreted into blood

FIGURE 47.19

The antagonistic actions of insulin and glucagon on blood glucose. Insulin stimulates the cellular uptake of blood glucose into skeletal muscles, adipose cells, and the liver after a meal. Glucagon stimulates the hydrolysis of liver glycogen between meals, so that the liver can secrete glucose into the blood. These antagonistic effects help to maintain homeostasis of the blood glucose concentration.

clusters of cells scattered throughout the pancreas; these clusters came to be called *islets of Langerhans*. Laboratory workers later observed that the surgical removal of the pancreas caused glucose to appear in the urine, the hallmark of the disease diabetes mellitus. This suggested that the pancreas, specifically the islets of Langerhans, might be producing a hormone that prevents this disease.

That hormone is *insulin*, secreted by the beta (β) cells of the islets. Insulin was not isolated until 1922, when two young doctors working in a Toronto hospital succeeded where many others had not. On January 11, 1922, they injected an extract purified from beef pancreas into a 13-year-old diabetic boy, whose weight had fallen to 65 pounds and who was not expected to survive. With that single injection, the glucose level in the boy's blood fell 25%. A more potent extract soon brought the level down to near normal. The doctors had achieved the first instance of successful insulin therapy.

Two forms of diabetes mellitus are now recognized. People with *type I*, or insulin-dependent diabetes mellitus, lack the insulin-secreting β cells. Treatment for these patients therefore consists of insulin injections. (Because insulin is a peptide hormone, it would be digested if taken orally and must instead be injected into the blood.) In the past, only insulin extracted from the pancreas of pigs or cattle was available, but today people with insulin-dependent diabetes can inject themselves with human insulin produced by genetically engineered bacteria. Active research on the possibility of transplanting islets of Langerhans holds much promise of a lasting treatment for these patients. Most diabetic patients, however, have *type II*, or non-insulin-dependent diabetes mellitus. They generally have normal or even above-normal levels of insulin in their blood, but their cells have a reduced sensitivity to insulin.

These people do not require insulin injections and can usually control their diabetes through diet and exercise.

The islets of Langerhans produce another hormone; the alpha (α) cells of the islets secrete *glucagon*, which acts antagonistically to insulin (figure 47.19). When a person eats carbohydrates, the blood glucose concentration rises. This stimulates the secretion of insulin by the β cells and inhibits the secretion of glucagon by the α cells. Insulin promotes the cellular uptake of glucose into liver and muscle cells, where it is stored as glycogen, and into adipose cells, where it is stored as fat. Between meals, when the concentration of blood glucose falls, insulin secretion decreases, and glucagon secretion increases. Glucagon promotes the hydrolysis of stored glycogen in the liver and fat in adipose tissue. As a result, glucose and fatty acids are released into the blood and can be taken up by cells and used for energy.

The β cells of the islets of Langerhans secrete insulin, and the α cells secrete glucagon. These two hormones have antagonistic actions on the blood glucose concentration; insulin lowers and glucagon raises blood glucose levels.

Other Endocrine Glands

Molting and Metamorphosis in Insects

As insects grow during postembryonic development, their hardened exoskeletons do not expand. To overcome this problem, insects undergo a series of molts wherein they shed their old exoskeleton (figure 47.20) and secrete a new, larger one. In one molt, a juvenile insect, or larva, often undergoes a radical transformation to the adult. This process is called metamorphosis. Hormonal secretions influence both molting and metamorphosis in insects. Prior to molting, neurosecretory cells on the surface of the brain secrete a small peptide, **brain hormone,** which in turn stimulates a gland in the thorax called the prothoracic gland to produce *molting hormone*, or *ecdysone* (figure 47.21). High levels of ecdysone bring about the biochemical and behavioral changes that cause molting to occur. Another pair of endocrine glands near the brain called the corpora allata produce a hormone called **juvenile hormone.** High levels of juvenile hormone prevent the transformation to the adult and result in a larval-to-larval molt. If the level of juvenile hormone is low, however, the molt will result in metamorphosis.

Sexual Development, Biological Clocks, and Immune Regulation in Vertebrates

The ovaries and testes in vertebrates are important endocrine glands, producing the sex steroid hormones, including estrogens, progesterone, and testosterone, to be described in detail in chapter 50. Estrogen and progesterone are the primary "female" sex steroids, while testosterone and its immediate derivatives are the primary "male" sex steroids, or androgens. During embryonic development, testosterone production in the embryo is critical for the development of male sex organs. In mammals, sex

FIGURE 47.20
A molting cicada. This adult insect is emerging from its old cuticle.

FIGURE 47.21
The hormonal control of metamorphosis in the silkworm moth, *Bombyx mori.* While molting hormone (ecdysone), produced by the prothoracic gland, triggers when molting will occur, juvenile hormone, produced by bodies near the brain called the corpora allata, determines the result of a particular molt. High levels of juvenile hormone inhibit the formation of the pupa and adult forms. At the late stages of metamorphosis, therefore, it is important that the corpora allata not produce large amounts of juvenile hormone.

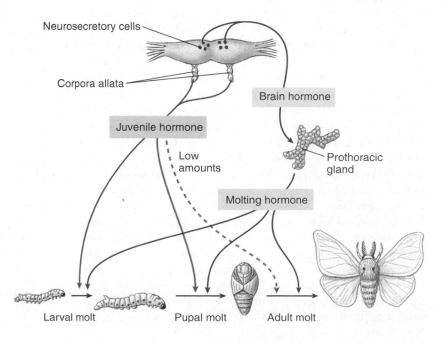

steroids are responsible for the development of secondary sexual characteristics at puberty. These characteristics include breasts in females, body hair, and increased muscle mass in males. Because of this, some bodybuilders illegally take androgens to increase muscle mass. In addition to being illegal, this practice can cause liver disorders as well as a number of other serious side effects. In females, sex steroids are especially important in maintaining the sexual cycle. Estrogen and progesterone produced in the ovaries are critical regulators of the menstrual and ovarian cycles. During pregnancy, estrogen production in the placenta maintains the uterine lining, which protects and nourishes the developing embryo.

Another major endocrine gland is the pineal gland, located in the roof of the third ventricle of the brain in most vertebrates (see figure 45.24). It is about the size of a pea and is shaped like a pinecone (hence its name). The pineal gland evolved from a median light-sensitive eye (sometimes called a "third eye," although it could not form images) at the top of the skull in primitive vertebrates. This pineal eye is still present in primitive fish (cyclostomes) and some reptiles. In other vertebrates, however, the pineal gland is buried deep in the brain and functions as an endocrine gland by secreting the hormone melatonin. One of the actions of melatonin is to cause blanching of the skin of lower vertebrates by reducing the dispersal of melanin granules.

The secretion of melatonin is stimulated by activity of the *suprachiasmatic nucleus* (SCN) of the hypothalamus. The SCN is known to function as the major biological clock in vertebrates, entraining (synchronizing) various body processes to a circadian rhythm (one that repeats every 24 hours). Through regulation by the SCN, the secretion of melatonin by the pineal gland is entrained to cycles of light and dark, decreasing during the day and increasing at night. This daily cycling of melatonin release regulates body cycles such as sleep/wake cycles and temperature cycles. Melatonin helps regulate reproductive physiology in some vertebrate species that have distinct breeding seasons, but the role of melatonin in human reproduction is not clear.

A variety of hormones are secreted by nonendocrine organs. The thymus is the site of production of particular lymphocytes called T cells in many vertebrates and the site of T cell maturation in mammals. It also secretes a number of hormones that function in the regulation of the immune system. The right atrium of the heart secretes atrial natriuretic hormone, which stimulates the kidneys to excrete salt and water in the urine. This hormone, therefore, acts antagonistically to aldosterone, which promotes salt and water retention. The kidneys secrete erythropoietin, a hormone that stimulates the bone marrow to produce red blood cells. Other organs, such as the liver, stomach, and small intestine, also secrete hormones; even the skin has an endocrine function—it secretes vitamin D. The gas nitric oxide, made by many dif-ferent cells, can function as a hormone or a paracrine regulator. One of its major roles involves the control of blood pressure by dilating arteries. The drug sildenafil (Viagra) counters impotence by causing nitric oxide to dilate the blood vessels of the penis.

Endocrine-Disrupting Chemicals

Because target cells are very sensitive to hormones, they occur in very low concentrations in the blood. Therefore, small changes in concentrations can make a big difference in how the target organ functions. Unfortunately, scientists are now finding that the endocrine system is not sufficiently protected from the outside world. Some man-made chemicals (and even a few plant-produced chemicals) can enter the body and interrupt normal endocrine function. These may be chemicals manufactured for some other purpose that accidentally leak into the environment, industrial waste products "thrown away" into the environment, or substances purposefully released into the environment, such as pesticides. These environmental contaminants get into the food we eat and the air we breathe. They are everywhere on earth and cannot be avoided. Some can last for years in the environment and for just as long in an animal's body. Those chemicals that interfere with hormone function are called *endocrine-disrupting chemicals*.

Any chemical that can bind to receptor proteins and mimic the effects of the hormone is called a hormone *agonist*. Any chemical that binds to receptor proteins and has no effect, but blocks the hormone from binding, is called a hormone *antagonist*. Endocrine-disrupting chemicals can also interfere by binding to the hormone's protein carriers in the blood. So far, endocrine-disrupting chemicals have been shown to interfere with reproductive hormones, thyroid hormones, and the immune system chemical messengers. Although these effects may not be lethal, they can cause deformities, interrupt metabolic processes, and cause sterility. If a species is having problems reproducing, maintaining the proper metabolic rate, or fighting off infections, then its numbers will decrease (perhaps even leading to extinction). These environmental contaminants may be harming humans in addition to other species. Laws have been passed requiring the testing of thousands of chemicals to see if they have endocrine-disrupting potential, and the Environmental Protection Agency (EPA) must include this testing in standard protocols before approving any new compounds.

Molting hormone, or ecdysone, and juvenile hormone regulate molting and metamorphosis in insects. Sex steroid hormones from the gonads regulate reproduction, melatonin secreted by the pineal gland helps regulate circadian rhythms, and thymus hormones help regulate the vertebrate immune system.

For interactive testing, visit the Online Learning Center with PowerWeb at www.mhhe.com/Raven7

47.1 Regulation is often accomplished by chemical messengers.

Types of Regulatory Molecules

- A hormone is a regulatory chemical secreted into the blood by an endocrine gland and carried in the bloodstream. (p. 992)
- Paracrine regulators act in the organ in which they were produced and released. (p. 992)

Endocrine Glands and Hormones

- Hormones secreted by the endocrine glands are categorized as polypeptides, glycoproteins, amines, and steroids. (p. 993)
- The secretory activity of many endocrine glands is controlled by the nervous system. (p. 993)

Paracrine Regulation

- Paracrine regulators are numerous, and include cytokines, growth factors, neurotrophins, and prostaglandins. (p. 994)
- Prostaglandins help regulate many systems, including the immune, reproductive, digestive, respiratory, circulatory, and urinary systems. (pp. 994–995)

47.2 Lipophilic and polar hormones regulate their target cells by different means.

Hormones That Enter Cells

- Hormones can be divided into lipophilic (lipid-soluble) and lipophobic (hydrophilic) categories. (p. 996)
- Lipophilic hormones pass through the target cell's plasma membrane and bind to intracellular receptor proteins, thereby regulating target cell genes. (pp. 996–997)

Hormones That Do Not Enter Cells

- Lipophobic hormones cannot pass through plasma membranes, and thus use messengers within the target cells, such as cyclic AMP, IP_3, and Ca^{++}, for mediation. (pp. 998–999)

47.3 The hypothalamus controls the secretions of the pituitary gland.

The Posterior Pituitary Gland

- The posterior pituitary contains axons that originate within the hypothalamus and extend along the pituitary. (p. 1000)
- Antidiuretic hormone (ADH) is a posterior pituitary peptide that stimulates water retention by the kidneys. Oxytocin is a posterior pituitary peptide that stimulates the milk-ejection reflex in mammary glands. Both are produced by neuron cell bodies located in the hypothalamus. (p. 1000)

The Anterior Pituitary Gland

- The hormones produced and secreted by different cell types in the anterior pituitary gland include growth hormone, adrenocorticotropic hormone, thyroid-stimulating hormone, luteinizing hormone, follicle-stimulating hormone, prolactin, and melanocyte-stimulating hormone. (pp. 1001–1002)
- The hypothalamus and the anterior pituitary gland are influenced by negative feedback inhibition from the endocrine glands they stimulate. (pp. 1003–1004)

47.4 Endocrine glands secrete hormones that regulate many body functions.

The Thyroid and Parathyroid Glands

- The thyroid gland secretes thyroxine and triiodothyronine, which stimulate oxidative respiration, and in children, they promote growth and stimulate maturation of the central nervous system. (p. 1005)
- Parathyroid hormone raises blood Ca^{++} levels by stimulating bone dissolution. (pp. 1005–1006)

The Adrenal Glands

- Sympathetic neurons stimulate the adrenal medulla to secrete epinephrine and norepinephrine to trigger alarm responses. (p. 1007)
- The adrenal cortex is stimulated to secrete glucocorticoids by ACTH and mineralocorticoid by angiotensin II. (p. 1007)
- Cortisol helps regulate blood glucose levels, while aldosterone regulates blood sodium and potassium levels. (p. 1007)

The Pancreas

- Beta cells of the islets of Langerhans secrete insulin, which lowers blood glucose levels, while alpha cells secrete glucagon, which raises blood glucose levels. (p. 1008)

Other Endocrine Glands

- In molting insects, metamorphosis and molting are regulated by ecdysone and juvenile hormone. (pp. 1009–1010)
- Gonadal sex steroids regulate reproduction, while melatonin from the pineal gland regulates circadian rhythms, and thymus hormones help regulate the vertebrate immune system. (pp. 1009–1010)

Self Test

1. Which of the following best describes hormones?
   a. Hormones are relatively unstable and work only in the area adjacent to the gland that produced them.
   b. Hormones are stable, long-lasting chemicals released from glands.
   c. All hormones are lipid-soluble.
   d. Hormones are chemical messengers that are released into the environment.

2. The receptor for steroid hormones lies
   a. in the cytoplasm.
   b. within the plasma membrane.
   c. within the nuclear envelope.
   d. in the blood plasma.

3. Second messengers are activated in response to
   a. steroid hormones.
   b. thyroxine.
   c. peptide hormones.
   d. all of these.

4. _____ regulates the kidney's retention of water. Its secretion is suppressed by alcohol.
   a. Prolactin
   b. Oxytocin
   c. Thyroxine
   d. Vasopressin (ADH)

5. Which of the following hormones is not released by the anterior pituitary?
   a. melanocyte-stimulating hormone
   b. gonadotropin-releasing hormone
   c. thyroid-stimulating hormone
   d. growth hormone

6. _____ hormone stimulates the adrenal cortex to produce several of its hormones.
   a. Follicle-stimulating
   b. Luteinizing
   c. Adrenocorticotropic
   d. Growth

7. Parathyroid hormone acts to ensure that
   a. calcium levels in the blood never drop too low.
   b. sodium levels in urine are constant.
   c. potassium levels in the blood don't escalate.
   d. the concentration of water in the blood is sufficient.

8. The adrenal cortex releases _____, which stimulates Na^+ reabsorption by the kidneys.
   a. epinephrine
   b. aldosterone
   c. glucose
   d. cortisol

9. Individuals with type I diabetes
   a. lack β cells in the islets of Langerhans.
   b. produce enough insulin but lack functional receptors on their cells.
   c. can control their diabetes with diet and exercise.
   d. All of these are correct.

10. The hormone melatonin, which is involved in skin blanching in lower vertebrates, is released from the
    a. anterior pituitary gland.
    b. melanocytes.
    c. pineal gland.
    d. suprachiasmatic nucleus of the hypothalamus.

Test Your Visual Understanding

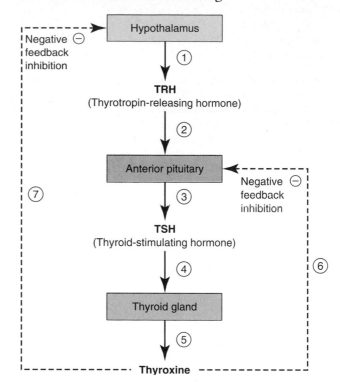

1. A goiter results when the thyroid gland becomes enlarged. The figure above shows the regulation of thyroxine secretion, with the different steps in the process numbered. Indicate which numbered step or steps are disrupted in the process that leads to the formation of a goiter, and explain why this happens and what results.

Applying Your Knowledge

1. The adrenal medulla secretes epinephrine in response to stimulation from the sympathetic division of the autonomic nervous system. Epinephrine triggers the fight or flight response in the body, which causes an increase in the heart rate. In a person at rest, the heart beats about 72 beats per minute and moves about 70 ml of blood per beat. Epinephrine causes the heart rate to double and the volume per beat to triple.
   a. How much blood is moved per minute in a resting heart and in a hormone-stimulated heart?
   b. As a percentage, how much more blood is moved by a hormone-stimulated heart compared to a resting heart?

2. Suppose that two different organs, such as the liver and heart, are sensitive to a particular hormone (such as epinephrine). The cells in both organs have identical receptors for the hormone, and hormone-receptor binding produces the same intracellular second messenger in both organs. However, the hormone produces different effects in the two organs. Explain how this can happen.

3. Many physiological parameters, such as blood Ca^{++} concentration and blood glucose levels, are controlled by two hormones that have opposite effects. What is the advantage of achieving regulation in this manner instead of by using a single hormone that changes the parameters in one direction only?

48

The Immune System

Concept Outline

48.1 Many of the body's most effective defenses are nonspecific.

Skin: The First Line of Defense. The skin provides a barrier and chemical defenses against foreign bodies.

Cellular Counterattack: The Second Line of Defense. Neutrophils and macrophages kill by phagocytosis; natural killer cells kill by making pores in cells.

The Inflammatory Response. Histamines, phagocytotic cells, and fever may all play a role in inflammations.

48.2 Specific immune defenses require the recognition of antigens.

The Immune Response: The Third Line of Defense. Antigens trigger the immune response.

Cells of the Specific Immune System. B cells and T cells serve different functions in the immune response.

Initiating the Immune Response. T cells must be activated by an antigen-presenting cell.

48.3 T cells organize attacks against invading microbes.

T Cells: The Cell-Mediated Immune Response. T cells respond to antigens presented by MHC proteins.

48.4 B cells label specific cells for destruction.

B Cells: The Humoral Immune Response. Antibodies secreted by B cells label pathogens for destruction.

Antibodies. Genetic recombination generates millions of B cells, each specialized to produce one type of antibody.

Antibodies in Medical Diagnosis. Antibodies react against certain blood types and pregnancy hormones.

Evolution of the Immune System. Invertebrates possess immune elements analogous to those of vertebrates.

48.5 The immune system can be defeated.

T Cell Destruction: AIDS. HIV suppresses the immune system by selectively destroying helper T cells.

Antigen Shifting. Some microbes change their surface antigens and thus evade the immune system.

Autoimmunity and Allergy. The immune system sometimes causes disease by attacking its own antigens.

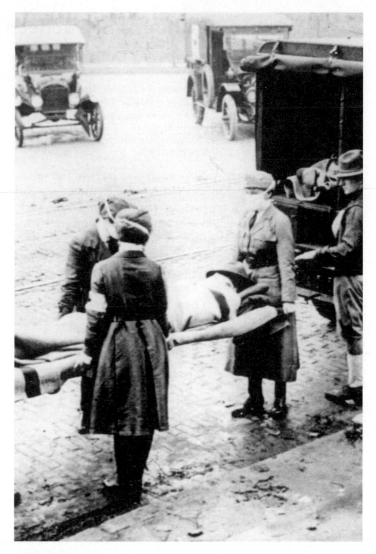

FIGURE 48.1
The influenza epidemic of 1918–1919. The epidemic killed 22 million people worldwide in 18 months. With 25 million Americans infected, the Red Cross often worked around the clock.

When you consider how animals defend themselves, it is natural to think of turtles, armadillos, and other animals covered like tanks with heavy plates of armor. However, armor offers little protection against the greatest dangers vertebrates face—microorganisms and viruses. We live in a world awash with attackers too tiny to see with the naked eye, and no vertebrate could long withstand their onslaught unprotected. We survive because we have evolved a variety of very effective defenses against this constant attack. As we review these defenses, it is important to keep in mind that they are far from perfect. Some 22 million Americans and Europeans died from influenza over an 18-month period in 1918–1919 (figure 48.1), and more than 3 million people will die of malaria this year. Attempts to improve our defenses against infection are among the most active areas of scientific research today.

Many of the body's most effective defenses are nonspecific.

Skin: The First Line of Defense

The vertebrate is defended from infection the same way knights defended medieval cities. "Walls and moats" make entry difficult; "roaming patrols" attack strangers; and "sentries" challenge anyone wandering about and call patrols if a proper "ID" is not presented.

1. **Walls and moats.** The outermost layer of the vertebrate body, the **skin,** is the first barrier to penetration by microbes. Mucous membranes in the respiratory, digestive, and urogenital tracts are also important barriers that protect the body from invasion.
2. **Roaming patrols.** If the first line of defense is penetrated, the response of the body is to mount a **cellular counterattack,** using a battery of cells and chemicals that kill microbes. These defenses act very rapidly after the onset of infection.
3. **Sentries.** Finally, the body is also guarded by mobile cells that patrol the bloodstream, scanning the surfaces of every cell they encounter. They are part of the **immune system.** One kind of immune cell aggressively attacks and kills any cell identified as foreign, whereas the other type marks the foreign cell or virus for elimination by the roaming patrols.

The Skin as a Barrier to Infection

The skin is the largest organ of the vertebrate body, accounting for 15% of an adult human's total weight. The skin not only defends the body by providing a nearly impenetrable barrier, but also reinforces this defense with chemical weapons on the surface. Oil and sweat glands give the skin's surface a pH of 3 to 5, acidic enough to inhibit the growth of many microorganisms. Sweat also contains the enzyme **lysozyme,** which digests bacterial cell walls. In addition to defending the body against invasion by viruses and microorganisms, the skin prevents excessive loss of water to the air through evaporation.

The epidermis of skin is approximately 10 to 30 cells thick, about as thick as this page. The outer layer, called the *stratum corneum,* contains cells that are continuously abraded, injured, and worn by friction and stress during the body's many activities. The body deals with this damage not by repairing the cells, but by replacing them. Cells are shed continuously from the stratum corneum and are replaced by new cells produced in the innermost layer of the epidermis, the *stratum basale,* which contains some of the most actively dividing cells in the vertebrate body. The cells formed in this layer migrate upward and enter a broad, intermediate *stratum spinosum* layer. As they move upward, they form the protein keratin, which makes skin tough and water-resistant. These new cells eventually arrive at the stratum corneum, where they normally remain for about a month before they are shed and replaced by newer cells from below. Psoriasis, which afflicts some 4 million Americans, is a chronic skin disorder in which epidermal cells are replaced every 3 to 4 days, about eight times faster than normal.

The dermis of skin contains connective tissue and is 15 to 40 times thicker than the epidermis. It provides structural support for the epidermis and a matrix for the many blood vessels, nerve endings, muscles, and other structures situated within skin. The wrinkling that occurs as we grow older takes place in the dermis, and the leather used to manufacture belts and shoes is derived from very thick animal dermis.

The layer of subcutaneous tissue below the dermis contains primarily adipose cells. These cells act as shock absorbers and provide insulation, conserving body heat. Subcutaneous tissue varies greatly in thickness in different parts of the body. It is nonexistent in the eyelids, is a half-centimeter thick or more on the soles of the feet, and may be much thicker in other areas of the body, such as the buttocks and thighs.

Other External Surfaces

In addition to the skin, three other potential routes of entry by viruses and microorganisms must be guarded: the *digestive tract,* the *respiratory tract,* and the *urogenital tract.* Recall that each of these tracts opens to the outside, and their surfaces must also protect the body from foreign invaders. Microbes are present in food, but many are killed by saliva (which also contains lysozyme), by the very acidic environment of the stomach, and by digestive enzymes in the intestine. Microorganisms are also present in inhaled air. The cells lining the smaller bronchi and bronchioles secrete a layer of sticky mucus that traps most microorganisms before they can reach the warm, moist lungs, which would provide ideal breeding grounds for them. Other cells lining these passages have cilia that continually sweep the mucus toward the glottis. There it can be swallowed, carrying potential invaders out of the lungs and into the digestive tract. Vaginal secretions are sticky and acidic, and they also promote the growth of bacterial symbionts; all of these characteristics help prevent foreign invasion. In both males and females, acidic urine continually washes potential pathogens from the urinary tract. Occasionally, an infectious agent, called a pathogen, enters one of these systems, and the body then uses defense mechanisms such as vomiting, diarrhea, coughing, sneezing, and mucous secretions to expel the pathogen.

The surface defenses of the body consist of the skin and the mucous membranes lining the digestive, respiratory, and urogenital tracts, which eliminate many microorganisms before they can invade the body tissues.

Cellular Counterattack: The Second Line of Defense

The surface defenses of the vertebrate body are very effective but are occasionally breached, allowing invaders to enter the body. At this point, the body uses a host of nonspecific cellular and chemical devices to defend itself. We refer to this as the second line of defense. These devices all have one property in common: They respond to *any* microbial infection without needing to determine the invader's identity.

Although these cells and chemicals of the nonspecific immune response roam through the body, there is a central location for the collection and distribution of the cells of the immune system; it is called the lymphatic system (see chapter 44). The lymphatic system consists of a network of lymphatic capillaries, ducts, nodes, and lymphatic organs (figure 48.2), and although it has other functions involved with circulation, it also stores cells and other agents used in the immune response. These cells are distributed throughout the body to fight infections, and are also stored in the lymph nodes, where foreign invaders can be identified and eliminated as body fluids pass through.

Cells That Kill Invading Microbes

Perhaps the most important of the vertebrate body's nonspecific defenses are some of the white blood cells called leukocytes that circulate through the body and attack invading microbes within tissues. There are three basic kinds of defending leukocytes, and each kills invading microorganisms differently.

Macrophages ("big eaters") are large, irregularly shaped cells that kill microbes by ingesting them through *phagocytosis*, much as an amoeba ingests a food particle (figure 48.3). Within the macrophage, the membrane-bound vacuole containing the bacterium fuses with a lysosome. Fusion activates lysosomal enzymes that kill the microbe by liberating large quantities of oxygen free radicals. Macrophages also engulf viruses, cellular debris, and dust particles in the lungs. Macrophages roam continuously in the extracellular fluid, and their phagocytic actions supplement those of the specialized phagocytic cells that are part of the structure of the liver, spleen, and bone marrow. In response to an infection, monocytes (an undifferentiated leukocyte) found in the blood squeeze through capillaries to enter the connective tissues. There, at the site of the infection, the monocytes are transformed into additional macrophages.

Neutrophils are the most abundant circulating leukocyte. Like macrophages, they enter infected tissues, where they ingest and kill bacteria by phagocytosis. In addition, neutrophils release chemicals (some of which are identical to household bleach) that kill other bacteria in the neighborhood as well as neutrophils themselves.

Natural killer cells do not attack invading microbes directly. Instead, they kill cells of the body that have been infected with viruses. They kill not by phagocytosis, but

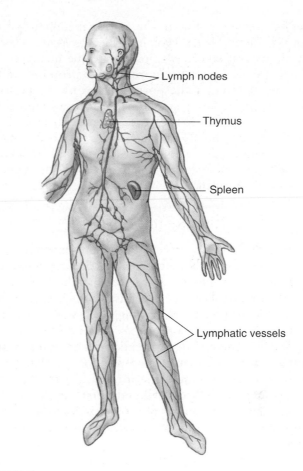

FIGURE 48.2
The lymphatic system. The lymphatic system consists of lymphatic vessels, lymph nodes, and lymphatic organs, including the spleen and thymus gland.

FIGURE 48.3
A macrophage in action. In this scanning electron micrograph (1800×), a macrophage is "fishing" with long, sticky cytoplasmic extensions. Bacterial cells that come in contact with the extensions are drawn toward the macrophage and engulfed.

rather by creating a hole in the plasma membrane of the target cell (figure 48.4). Proteins called *perforins*, released from the natural killer cells, insert into the membrane of the target cell, forming a pore. Other proteins, called *granzymes*, then enter the pores created by the perforin molecules and initiate programmed cell death (apoptosis, described in chapter 19). Natural killer cells also attack cancer cells, often before the cancer cells have had a chance to develop into a detectable tumor. The vigilant surveillance by natural killer cells is one of the body's most potent defenses against cancer.

Proteins That Kill Invading Microbes

The cellular defenses of vertebrates are enhanced by a very effective chemical defense called the *complement system*. This system consists of approximately 20 different proteins that circulate freely in the blood plasma. When they encounter a bacterial or fungal cell wall, these proteins aggregate to form a *membrane attack complex* that inserts itself into the foreign cell's plasma membrane, forming a pore like that produced by natural killer cells (figure 48.5). Fluid enters the foreign cell through this pore, causing the cell to swell and burst. Aggregation of the complement proteins is also triggered by the binding of antibodies to invading microbes, as we will see in a later section.

The proteins of the complement system can augment the effects of other body defenses. Some amplify the inflammatory response (discussed next) by stimulating histamine release; others attract phagocytes (monocytes and neutrophils) to the area of infection; and still others coat invading microbes, roughening the microbes' surfaces so that phagocytes may attach to them more readily.

Another class of proteins that play a key role in body defense are interferons. There are three major categories of interferons: *alpha*, *beta*, and *gamma*. Almost all cells in the body make alpha and beta interferons. These paracrine polypeptides act as messengers that protect normal cells in the vicinity of infected cells from becoming infected. Although viruses are still able to penetrate the neighboring cells, the alpha and beta interferons prevent viral replication and protein assembly in these cells. Gamma interferon is produced only by particular lymphocytes and natural killer cells. The secretion of gamma interferon by these cells is part of the immunological defense against infection and cancer.

A patrolling army of macrophages, neutrophils, and natural killer cells attacks and destroys invading viruses and bacteria and eliminates infected cells. In addition, a system of proteins called complement may be activated to destroy foreign cells, and body cells infected with a virus secrete proteins called interferons that protect neighboring cells.

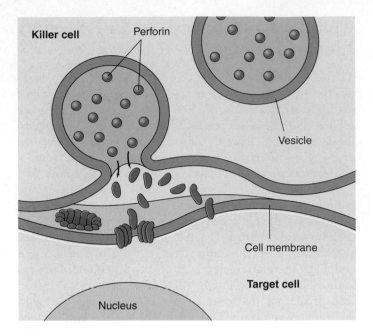

FIGURE 48.4
How natural killer cells kill target cells.   The initial event, the tight binding of the killer cell to the target cell, causes vesicles loaded with *perforin* molecules within the killer cell to move to the plasma membrane and disgorge their contents into the intercellular space over the target cell. The perforin molecules insert into the plasma membrane of the target cell like staves of a barrel, forming a pore that admits fluid and ruptures the cell.

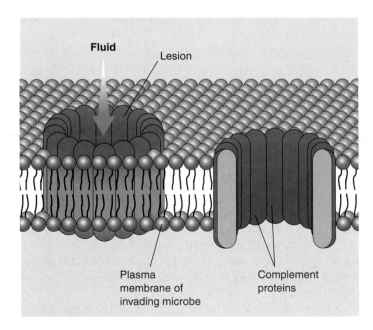

FIGURE 48.5
How complement creates a hole in a cell membrane.   The complement proteins form a complex transmembrane pore resembling the perforin-lined pores formed by natural killer cells. However, the complement proteins are free-floating in the blood plasma and attack the invading microbe directly.

The Inflammatory Response

The inflammatory response is a localized, nonspecific response to infection. Infected or injured cells release chemical alarm signals, most notably histamine and prostaglandins. These chemicals promote the dilation of local blood vessels, which increases the flow of blood to the site of infection or injury and causes the area to become red and warm. They also increase the permeability of capillaries in the area, producing the edema (tissue swelling) so often associated with infection. The more permeable capillaries allow phagocytes (monocytes and neutrophils) to migrate from the blood to the extracellular fluid, where they can attack bacteria. Neutrophils arrive first, spilling out chemicals that kill the bacteria in the vicinity (as well as tissue cells and themselves); the *pus* associated with some infections is a mixture of dead or dying pathogens, tissue cells, and neutrophils. Monocytes follow, become macrophages, and engulf pathogens and the remains of the dead cells (figure 48.6).

The Temperature Response

Macrophages that encounter invading microbes release a regulatory molecule called interleukin-1, which is carried by the blood to the brain. Interleukin-1 and other pyrogens (Greek *pyr*, "fire"), such as bacterial endotoxins, direct neurons in the hypothalamus to raise the body's temperature several degrees above the normal value of 37°C (98.6°F). The elevated temperature that results is called a fever.

Experiments with lizards, which regulate their body temperature by moving to warmer or colder locations, demonstrate that infected lizards choose a warmer environment—they give themselves a fever! Further, if lizards are prevented from elevating their body temperature, they have a slower recovery from their infection. Fever contributes to the body's defense by stimulating phagocytosis and causing the liver and spleen to store iron, reducing blood levels of iron, which bacteria need in large amounts to grow. However, very high fevers are hazardous because excessive heat may denature critical enzymes. In general, temperatures greater than 39.4°C (103°F) are considered dangerous for humans, and those greater than 40.6°C (105°F) are often fatal.

Inflammation aids the fight against infection by increasing blood flow to the site and raising temperature to retard bacterial growth.

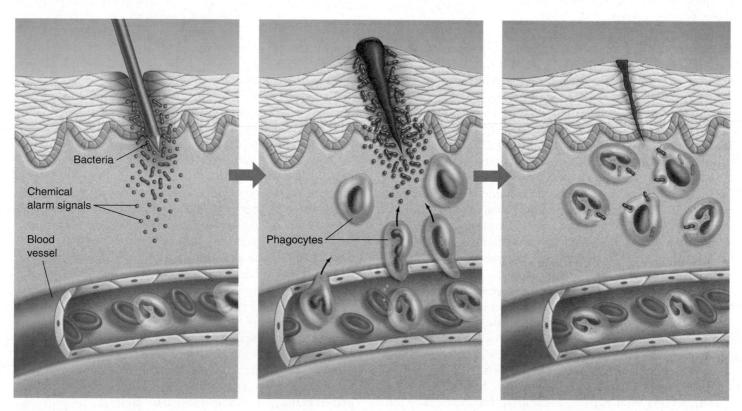

FIGURE 48.6
The events in a local inflammation.   When an invading microbe has penetrated the skin, chemical alarm signals, such as histamine and prostaglandins, cause nearby blood vessels to dilate. Increased blood flow causes swelling and brings a wave of phagocytic cells, which attack and engulf invading bacteria.

48.2 Specific immune defenses require the recognition of antigens.

The Immune Response: The Third Line of Defense

Few of us pass through childhood without contracting some sort of infection. Chicken pox, for example, is an illness that many of us experience before we reach our teens. It is a disease of childhood, because most of us contract it as children and *never catch it again*. Once you have had the disease, you are usually immune to it. Specific immune defense mechanisms provide this immunity.

Discovery of the Immune Response

In 1796, an English country doctor named Edward Jenner carried out an experiment that marks the beginning of the study of immunology. Smallpox was a common and deadly disease in those days. Jenner observed, however, that milkmaids who had caught a much milder form of "the pox" called cowpox (presumably from cows) rarely caught smallpox. Jenner set out to test the idea that cowpox could confer protection against smallpox. He infected people with cowpox (figure 48.7), and as he predicted, many of them became immune to smallpox.

We now know that smallpox and cowpox are caused by two different viruses with similar surfaces. Jenner's patients who were injected with the cowpox virus mounted a defense that was also effective against a later infection of the smallpox virus. Jenner's procedure of injecting a harmless microbe in order to confer resistance to a dangerous one is called **vaccination.** Modern attempts to develop resistance to malaria, herpes, and other diseases often involve delivering antigens via a harmless vaccinia virus related to the cowpox virus.

Many years passed before anyone learned how exposure to an infectious agent can confer resistance to a disease. A key step toward answering this question was taken more than a half-century later by the famous French scientist Louis Pasteur. Pasteur was studying fowl cholera, and he isolated a culture of bacteria from diseased chickens that would produce the disease if injected into healthy birds. Before departing on a two-week vacation, he accidentally left his bacterial culture out on a shelf. When he returned, he injected this old culture into healthy birds and found that it had been weakened; the injected birds became only slightly ill and then recovered. Surprisingly, however, those birds did not get sick when subsequently infected with fresh fowl

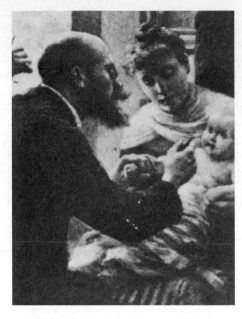

FIGURE 48.7
The birth of immunology. This famous painting shows Edward Jenner inoculating patients with cowpox in the 1790s and thus protecting them from smallpox.

cholera. They remained healthy even if given massive doses of active fowl cholera bacteria that did produce the disease in control chickens. Clearly, something about the bacteria could elicit immunity as long as the bacteria did not kill the animals first. We now know that molecules protruding from the surfaces of the bacterial cells evoked active immunity in the chickens.

Key Concepts of Specific Immunity

An **antigen** is a molecule that provokes a specific immune response. Antigens are large, complex molecules such as proteins; they are generally foreign to the body, and usually present on the surface of pathogens. A large antigen may have several parts, and each part stimulates a different specific immune response. In this case, the different parts are known as **antigenic determinant sites,** and each serves as a different antigen. Particular lymphocytes have receptor proteins on their surfaces that recognize an antigen and direct a specific immune response against either the antigen or the cell that carries the antigen.

Lymphocytes called B cells respond to antigens by producing proteins called **antibodies.** Antibody proteins are secreted into the blood and other body fluids, and thus provide **humoral immunity.** (The term *humor* here is used in its ancient sense, referring to a body fluid.) Other lymphocytes, called T cells, do not secrete antibodies but instead regulate the immune responses of other cells and directly attack the cells that carry the specific antigens. These cells are thus described as producing **cell-mediated immunity.**

The specific immune responses protect the body in two ways. First, an individual can gain immunity by being exposed to a *pathogen* (disease-causing agent) and perhaps getting the disease. This is *acquired immunity*, such as the resistance to the chicken pox that you acquire after having the disease in childhood. Another term for this process is **active immunity.** Second, an individual can gain immunity by obtaining the antibodies from another individual. This happened to you before you were born, as antibodies made by your mother were transferred to your body across the placenta. Immunity gained in this way is called **passive immunity.**

> Antigens are molecules, usually foreign, that provide a specific immune response. This immune response may involve secreted proteins called antibodies, or it may invoke a cell-mediated attack.

Cells of the Specific Immune System

The immune defense mechanisms of the body involve the actions of white blood cells, or leukocytes. Leukocytes include neutrophils, eosinophils, basophils, and monocytes, all of which are phagocytic and are involved in the second line of defense, as well as two types of lymphocytes (*T cells* and *B cells*), which are not phagocytic but are critical to the specific immune response (table 48.1), the third line of defense. T cells direct the cell-mediated response, while B cells direct the humoral response.

After their origin in the bone marrow, T cells migrate to the thymus (hence the designation "T"), a gland just above the heart. There they develop the ability to identify microorganisms and viruses by the antigens exposed on their surfaces. Tens of millions of different T cells are made, each specializing in the recognition of one particular antigen. No invader can escape being recognized by at least a few T cells. There are four principal kinds of T cells: Inducer T cells oversee the development of T cells in the thymus; helper T cells (often symbolized T_H) initiate the immune response; cytotoxic ("cell-poisoning") T cells (often symbolized T_C) lyse cells that have been infected by viruses; and suppressor T cells terminate the immune response.

Unlike T cells, B cells do not travel to the thymus; they complete their maturation in the bone marrow. (B cells are so named because they were originally characterized in a region of chickens called the bursa.) From the bone marrow, B cells are released to circulate in the blood and lymph. Like T cells, individual B cells are specialized to recognize particular foreign antigens. When a B cell encounters the antigen to which it is targeted, it begins to divide rapidly, and its progeny differentiate into plasma cells and memory cells. Each plasma cell is a miniature factory producing antibodies that stick like flags to that antigen wherever it occurs in the body, marking any cell bearing the antigen for destruction. The immunity that Pasteur observed resulted from such antibodies and from the continued presence of the B cells that produced them.

The lymphocytes, T cells and B cells, are involved in the specific immune response. Both are produced in the bone marrow, where the B cells also develop and mature, while T cells complete their development in the thymus.

Table 48.1 Cells of the Immune System

Cell Type	Function
Helper T cell	Commander of the immune response; detects infection and sounds the alarm, initiating both T cell and B cell responses
Inducer T cell	Not involved in the immediate response to infection; mediates the maturation of other T cells in the thymus
Cytotoxic T cell	Detects and kills infected body cells; recruited by helper T cells
Suppressor T cell	Dampens the activity of T and B cells, scaling back the defense after the infection has been checked
B cell	Precursor of plasma cell; specialized to recognize specific foreign antigens
Neutrophil	Engulfs invading bacteria and releases chemicals that kill neighboring bacteria
Plasma cell	Activated B cell that is a biochemical factory devoted to the production of antibodies directed against specific foreign antigens
Mast cell	Initiator of the inflammatory response, which aids the arrival of leukocytes at a site of infection; secretes histamine, which is important in allergic responses
Monocyte	Precursor of macrophage
Macrophage	The body's first cellular line of defense; also serves as antigen-presenting cell to B and T cells and engulfs antibody-covered cells
Natural killer cell	Recognizes and kills infected body cells; natural killer (NK) cell detects and kills cells infected by a broad range of invaders; killer (K) cell attacks only antibody-coated cells

Initiating the Immune Response

To understand how the third line of defense works, imagine you have just come down with the flu. Influenza viruses enter your body in small water droplets inhaled into your respiratory system. If they avoid becoming ensnared in the mucus lining the respiratory membranes (first line of defense) and avoid consumption by macrophages (second line of defense), the viruses infect and kill mucous membrane cells.

At this point, macrophages initiate the immune defense. Macrophages inspect the surfaces of all cells they encounter. The surfaces of most vertebrate cells possess glycoproteins produced by a group of genes called the **major histocompatibility complex (MHC)**. These glycoproteins are called **MHC proteins** or, specifically in humans, **human leukocyte antigens (HLA)**. The genes encoding the MHC proteins are highly polymorphic (have many forms); for example, the human MHC proteins are specified by genes that are the most polymorphic known, with nearly 170 alleles each. Only rarely will two individuals have the same combination of alleles, and the MHC proteins are thus different for each individual, much as fingerprints are. As a result, the MHC proteins on the tissue cells serve as self markers that enable the individual's immune system to distinguish its cells from foreign cells, an ability called **self-versus-nonself recognition**. T cells of the immune system can recognize a cell as self or nonself by the MHC proteins present on the cell surface.

When a foreign particle, such as a virus, infects the body, it is taken in by cells and partially digested. Within the cells, the viral antigens are processed and moved to the surface of the plasma membrane. The cells that perform this function are known as **antigen-presenting cells**. At the membrane, the processed antigens are complexed with the MHC proteins. This enables T cells to recognize antigens presented to them associated with the MHC proteins.

There are two classes of MHC proteins. MHC-I is present on every nucleated cell of the body. MHC-II, however, is found only on macrophages, B cells, and a subtype of T cells called $CD4^+$ T cells (table 48.2). These three cell types work together in one form of the immune response, and their MHC-II markers permit them to recognize one another. Cytotoxic T lymphocytes, which act to destroy infected cells as previously described, can only interact with antigens presented to them with MHC-I proteins. Helper T lymphocytes, whose functions will soon be described, can interact only with antigens presented with MHC-II proteins. These restrictions result from the presence of coreceptors, which are proteins associated with the T cell receptors. The coreceptor known as CD8 is associated with the cytotoxic T cell receptor (these cells can therefore be indicated as $CD8^+$). The CD8 coreceptor can interact only with the MHC-I proteins of an infected cell. The coreceptor known as CD4 is associated with the helper T cell receptor (these cells can thus be indicated as $CD4^+$) and interacts only with the MHC-II proteins of another lymphocyte.

Macrophages encounter foreign particles in the body, partially digest the virus particles, and present the foreign antigens in a complex with the MHC-II proteins on its membrane. This combination of MHC-II proteins and foreign antigens is required for interaction with the receptors on the surface of helper T cells. At the same time, macrophages that encounter antigens or antigen-presenting cells release a protein called *interleukin-1* that acts as a chemical alarm signal (discussed in section 48.3). Helper T cells respond to interleukin-1 by simultaneously initiating two parallel lines of immune system defense: the cell-mediated response carried out by T cells and the humoral response carried out by B cells.

Antigen-presenting cells must present foreign antigens together with MHC-II proteins in order to activate helper T cells, which have the CD4 coreceptor. Cytotoxic T cells use the CD8 coreceptor and must interact with foreign antigens presented on MHC-I proteins.

Table 48.2   Key Cell Surface Proteins of the Immune System

| Cell Type | Immune Receptors | | MHC Proteins | |
	T Receptor	B Receptor	MHC-I	MHC-II
B cell	−	+	+	+
$CD4^+$ T cell	+	−	+	+
$CD8^+$ T cell	+	−	+	−
Macrophage	−	−	+	+

Note: $CD4^+$ T cells include inducer T cells and helper T cells; $CD8^+$ T cells include cytotoxic T cells and suppressor T cells. (+ means present; − means absent.)

T cells organize attacks against invading microbes.

T Cells: The Cell-Mediated Immune Response

The cell-mediated immune response, carried out by T cells, protects the body from virus infection and cancer, killing abnormal or virus-infected body cells.

Once a helper T cell that initiates this response is presented with foreign antigen together with MHC proteins by a macrophage or other antigen-presenting cell, a complex series of steps is initiated. An integral part of this process is the secretion of regulatory molecules known generally as **cytokines,** or more specifically as **lymphokines** if they are secreted by lymphocytes.

When a cytokine is first discovered, it is named according to its biological activity (such as *B cell–stimulating factor*). However, because each cytokine has many different actions, such names can be misleading. Scientists have thus agreed to

use the name *interleukin*, followed by a number, to indicate a cytokine whose amino acid sequence has been determined. Interleukin-1, for example, is secreted by macrophages and can activate the T cell system. B cell–stimulating factor, now called interleukin-4, is secreted by T cells and is required for the proliferation and clone development of B cells. Interleukin-2 is released by helper T cells and, among its effects, is required for the activation of cytotoxic T lymphocytes. We will consider the actions of the cytokines as we describe the development of the T cell immune response.

Cell Interactions in the T Cell Response

When macrophages process the foreign antigens, they secrete **interleukin-1 (IL-1),** which stimulates cell division and proliferation of T cells (figure 48.8). Once the helper T cells have been activated by the antigens presented to

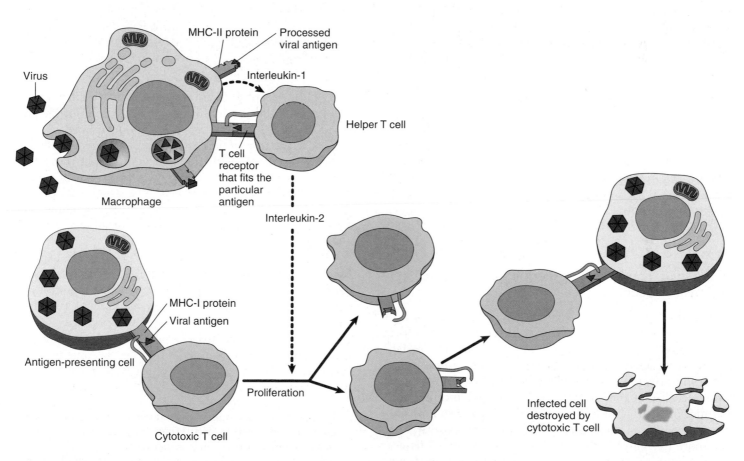

FIGURE 48.8

The T cell immune defense. After a macrophage has processed an antigen, it releases interleukin-1, signaling helper T cells to bind to the antigen-MHC protein complex. This triggers the helper T cell to release interleukin-2, which stimulates the multiplication of cytotoxic T cells. In addition, proliferation of cytotoxic T cells is stimulated when these cells have a receptor that fits the antigen displayed by an antigen-presenting cell, allowing them to bind with the antigen-MHC protein complex. Body cells that have been infected by the antigen are destroyed by the cytotoxic T cells. As the infection subsides, suppressor T cells "turn off" the immune response.

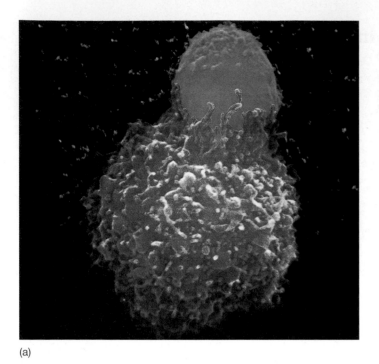 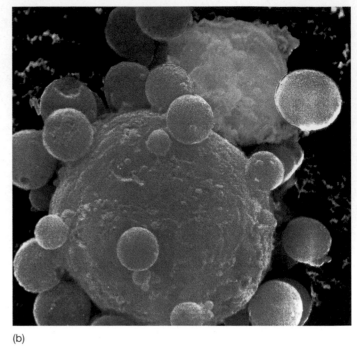

(a)   (b)

FIGURE 48.9
Cytotoxic T cells destroy cancer cells. (*a*) The cytotoxic T cell (*orange*) comes into contact with a cancer cell (*pink*). (*b*) The T cell recognizes that the cancer cell is "nonself" and causes the destruction of the cancer.

them by the macrophages, they secrete the cytokines known as macrophage colony-stimulating factor and gamma interferon, which promote the activity of macrophages. In addition, the helper T cells secrete **interleukin-2 (IL-2),** which stimulates the proliferation of cytotoxic T cells that are specific for the antigen. (Interleukin-2 also stimulates B cells, as we will see in section 48.4.) Cytotoxic T cells can destroy infected cells only if those cells display the foreign antigen together with their MHC-I proteins (see figure 48.8).

T Cells in Transplant Rejection and Surveillance Against Cancer

Cytotoxic T cells will also attack any foreign version of MHC-I as if it signaled a virus-infected cell. Therefore, even though vertebrates did not evolve the immune system as a defense against tissue transplants, their immune systems will attack transplanted tissue and cause graft rejection. Recall that the MHC proteins are polymorphic, but because of their genetic basis, the closer that two individuals are related, the less their MHC proteins will vary and the more likely they will be to tolerate each other's tissues. This is why relatives are often sought for kidney transplants. The drug cyclosporin inhibits graft rejection by inactivating cytotoxic T cells.

As tumors develop, they reveal surface antigens that can stimulate the immune destruction of the tumor cells. Tumor antigens activate the immune system, initiating an attack primarily by cytotoxic T cells (figure 48.9) and natural killer

cells. The concept of **immunological surveillance** against cancer was introduced in the early 1970s to describe the proposed role of the immune system in fighting cancer.

The production of human interferons by genetically engineered bacteria has made large amounts of these substances available for the experimental treatment of cancer. Thus far, interferons have proven to be a useful addition to the treatment of particular forms of cancer, including some types of lymphomas, renal carcinoma, melanoma, Kaposi's sarcoma, and breast cancer.

Interleukin-2, which activates both cytotoxic T cells and B cells, is now also available for therapeutic use through genetic engineering techniques. Particular lymphocytes from cancer patients have been removed, treated with IL-2, and given back to the patients together with IL-2 and gamma interferon. Scientists are also attempting to identify specific antigens and their genes that may become uniquely expressed in cancer cells, in an effort to help the immune system to better target cancer cells for destruction.

Helper T cells are only activated when a foreign antigen is presented together with MHC antigens by a macrophage or other antigen-presenting cells. The helper T cells are also stimulated by interleukin-1 secreted by the macrophages, and when activated, secrete a number of lymphokines. Interleukin-2, secreted by helper T cells, activates both cytotoxic T cells and B cells. Cytotoxic T cells destroy infected cells, transplanted cells, and cancer cells by cell-mediated attack.

B cells label specific cells for destruction.

B Cells: The Humoral Immune Response

B cells also respond to helper T cells activated by interleukin-1. Like cytotoxic T cells, B cells have receptor proteins on their surface, one type of receptor for each type of B cell. B cells recognize invading microbes much as cytotoxic T cells recognize infected cells, but unlike cytotoxic T cells, they do not go on the attack themselves. Rather, they mark the pathogen for destruction by mechanisms that have no "ID check" system of their own. Early in the immune response, the markers placed by B cells alert complement proteins to attack the cells carrying them. Later in the immune response, the markers placed by B cells activate macrophages and natural killer cells.

The way B cells do their marking is simple and foolproof. Unlike the receptors on T cells, which bind only to antigen-MHC protein complexes on antigen-presenting cells, B cell receptors can bind to free, unprocessed antigens. When a B cell encounters an antigen, antigen particles enter the B cell by endocytosis and get processed. Helper T cells that are able to recognize the specific antigen bind to the antigen-MHC protein complex on the B cell and release interleukin-2, which stimulates the B cell to divide. In addition, free, unprocessed antigens stick to antibodies on the B cell surface. This antigen exposure triggers even more B cell proliferation. B cells divide to produce long-lived memory B cells and plasma cells that serve as short-lived antibody factories (figure 48.10). The antibodies are released into the blood plasma, lymph, and other extracellular fluids. Helper T cells are essential in both the cell-mediated and humoral immune responses.

Antibodies are proteins in a class called **immunoglobulins** (abbreviated Ig), which is divided into subclasses based on the structures and functions of the antibodies. The different

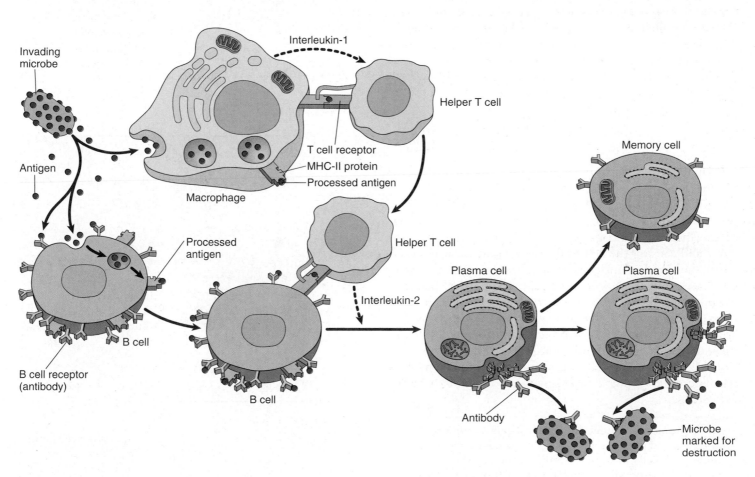

FIGURE 48.10

The B cell immune defense. Invading particles are bound by B cells, which interact with helper T cells and are activated to divide. The multiplying B cells produce memory B cells for future secondary responses and plasma cells that secrete antibodies, which bind to invading microbes and tag them for destruction by macrophages.

immunoglobulin subclasses are summarized in table 48.3 and described as follows:

1. **IgM** is the first type of antibody to be secreted during the primary response, and these antibodies serve as receptors on the lymphocyte surface. They also promote agglutination reactions (causing antigen-containing particles to stick together, or agglutinate).
2. **IgG** is the major form of antibody in the blood plasma and is secreted in a secondary response.
3. **IgD** antibodies serve as receptors for antigens on the B cell surface. Their other functions are unknown.
4. **IgA** is the major form of antibody in external secretions, such as saliva and mother's milk.
5. **IgE** antibodies promote the release of histamine and other agents that aid in attacking a pathogen. Unfortunatcly, they sometimes trigger a full-blown response when a harmless antigen enters thc body, producing allergic symptoms, such as those of hay fever.

Each B cell has on its surface about 100,000 IgM or IgD receptors. Unlike the receptors on T cells, which bind only to antigens presented by certain cells, B cell receptors can bind to *free* antigens. This provokes a primary response in which antibodies of the IgM class are secreted, and also stimulates cell division and clonal expansion. Upon subsequent exposure, the plasma cells secrete large amounts of antibodies that are generally of the IgG class. Although plasma cells live only a few days, they produce a vast number of antibodies. In fact, antibodies constitute about 20% by weight of the total protein in blood plasma. Production of IgG antibodies peaks after about three weeks (figure 48.11).

When IgM (and, to a lesser extent, IgG) antibodies bind to antigens on a cell, they cause the aggregation of complement proteins. As we mentioned earlier, these proteins form a pore that pierces the plasma membrane of the infected cell (see figure 48.5), allowing fluid to enter and causing the cell to burst. In contrast, when IgG antibodies bind to antigens on a cell, they serve as markers that stimulate phagocytosis by macrophages. Because certain complement proteins attract phagocytic cells, activation of complement is generally accompanied by increased phagocytosis. Notice that antibodies don't kill invading pathogens directly; rather, they cause destruction of the pathogens by activating the complement system and by targeting the pathogen for attack by phagocytic cells.

In the humoral immune response, B cells recognize antigens and divide to produce plasma cells and memory B cells, resulting in large numbers of circulating antibodies directed against those antigens. IgM antibodies are produced first, and they activate the complement system. Thereafter, IgG antibodies are produced and promote phagocytosis.

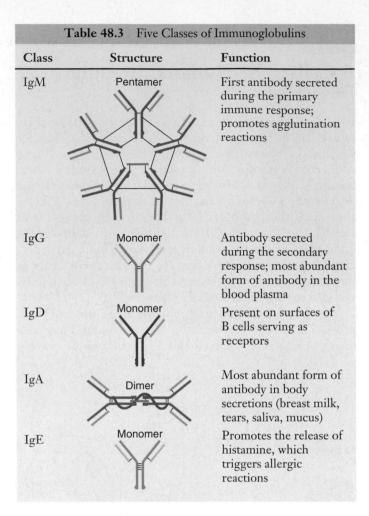

Table 48.3   Five Classes of Immunoglobulins

Class	Structure	Function
IgM	Pentamer	First antibody secreted during the primary immune response; promotes agglutination reactions
IgG	Monomer	Antibody secreted during the secondary response; most abundant form of antibody in the blood plasma
IgD	Monomer	Present on surfaces of B cells serving as receptors
IgA	Dimer	Most abundant form of antibody in body secretions (breast milk, tears, saliva, mucus)
IgE	Monomer	Promotes the release of histamine, which triggers allergic reactions

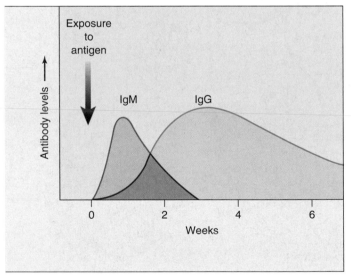

FIGURE 48.11
IgM and IgG antibodies. The first antibodies produced in the humoral immune response are IgM antibodies, which are very effective at activating the complement system. This initial wave of antibody production peaks after about one week and is followed by a far more extended production of IgG antibodies.
Why does the production of IgM antibodies cease after one week?

Antibodies

Structure of Antibodies

Each antibody molecule consists of two identical short polypeptides, called **light chains,** and two identical long polypeptides, called **heavy chains** (figure 48.12). The four chains in an antibody molecule are held together by disulfide (—S—S—) bonds, forming a Y-shaped molecule (figure 48.13).

Comparing the amino acid sequences of different antibody molecules shows that the specificity of antibodies for antigens resides in the two arms of the Y, which have a variable amino acid sequence. The amino acid sequence of the polypeptides in the stem of the Y is constant within a given class of immunoglobulins. Most of the sequence variation between antibodies of different specificity is found in the variable region of each arm. Here, a cleft forms that acts as the binding site for the antigen. Both arms always have exactly the same cleft and so bind to the same antigen.

Antibodies with the same variable segments have identical clefts and therefore recognize the same antigen, but they may differ in the stem portions of the antibody molecule. The stem is formed by the so-called "constant" regions of the heavy chains. In mammals, five different classes of heavy chains form the five classes of immunoglobulins mentioned previously: IgM, IgG, IgA, IgD, and IgE. We have already discussed the roles of IgM and IgG antibodies in the humoral immune response.

IgE antibodies bind to **mast cells (see table 48.1).** The heavy-chain stems of the IgE antibody molecules insert into receptors on the mast cell plasma membrane, in effect creating B cell receptors on the mast cell surface. When these cells encounter the specific antigen recognized by the arms of the antibody, they initiate the inflammatory response by releasing histamine. The resulting vasodilation and increased capillary permeability enable lymphocytes, macrophages, and complement proteins to more easily reach the site where the mast cell encountered the antigen. The IgE antibodies are involved in allergic reactions and will be discussed in more detail in section 48.5.

IgA antibodies are present in secretions such as milk, mucus, and saliva. In milk, these antibodies are thought to provide immune protection to nursing infants, whose own immune systems are not yet fully developed.

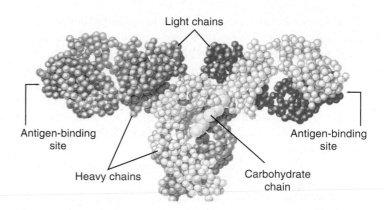

FIGURE 48.12
The structure of an antibody molecule. In this molecular model of an antibody molecule, each amino acid is represented by a small sphere. The heavy chains are colored *blue;* the light chains are *red.* The four chains wind about one another to form a Y shape, with two identical antigen-binding sites at the arms of the Y and a stem region that directs the antibody to a particular portion of the immune response.

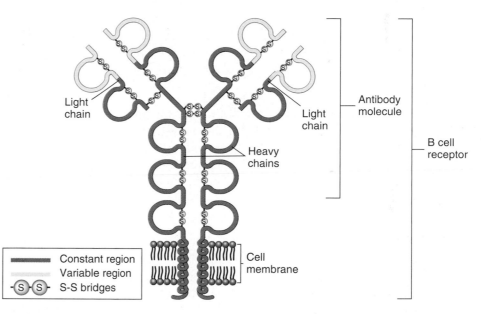

FIGURE 48.13
Structure of an antibody as a B cell receptor. The receptor molecules are characterized by domains of about 100 amino acids (represented as loops) joined by —S—S— covalent bonds. Each receptor has a constant region (*purple*) and a variable region (*yellow*). The receptor binds to antigens at the ends of its two variable regions.

Antibody Diversity

The vertebrate immune system is capable of recognizing as foreign millions of nonself molecules presented to it. Although vertebrate chromosomes contain only a few hundred receptor-encoding genes, it is estimated that human B cells can make between 10^6 and 10^9 different antibody molecules.

How do vertebrates generate millions of different antigen receptors when their chromosomes contain only a few hundred copies of the genes encoding those receptors?

The answer to this question is that in the B cell the millions of immune receptor genes do not have to be inherited at conception because they do not exist as single sequences of nucleotides. Rather, they are assembled by stitching together three or four DNA segments that code for different parts of the receptor molecule. When an antibody is assembled, the different sequences of DNA are brought together to form a composite gene (figure 48.14). This process is called **somatic DNA rearrangement.** For example, combining DNA in different ways can produce 16,000 different heavy chains and about 1200 different light chains (in mouse antibodies).

Two other processes generate even more sequences. First, the DNA segments are often joined together with one or two nucleotides off-register, shifting the reading frame during gene transcription, and so generating a totally different sequence of amino acids in the protein. Second, random "mistakes" occur during successive DNA replications as the lymphocytes divide. Both mutational processes produce changes in amino acid sequences, a phenomenon known as **somatic mutation** because it takes place in a somatic cell, a B cell, rather than in a gamete.

Because a B cell may end up with any heavy-chain gene and any light-chain gene during its maturation, the total number of different antibodies possible is staggering: 16,000 heavy-chain combinations × 1200 light-chain combinations = 19 million different possible antibodies. If we also take into account the changes induced by somatic mutation, the total can exceed 200 million! It should be understood that, although this discussion has centered on B cells and their receptors, the receptors on T cells are as diverse as those on B cells because they are subject to similar somatic rearrangements and mutations.

Immunological Tolerance

A mature animal's immune system normally does not respond to that animal's own tissue. This acceptance of self cells is known as **immunological tolerance.** The immune system of an embryo, on the other hand, is able to respond to both foreign and self molecules, but it loses the ability to respond to self molecules as its development proceeds. Indeed, if foreign tissue is introduced into an embryo before its immune system has developed, the mature animal that results will not recognize that tissue as foreign and will accept grafts of similar tissue without rejection.

There are two general mechanisms for immunological tolerance: clonal deletion and clonal suppression. During the normal maturation of hemopoietic stem cells in an embryo, fetus, or newborn, most lymphocyte clones that have receptors for self antigens are either eliminated (clonal deletion) or suppressed (clonal suppression). The cells "learn" to identify self antigens because the antigens are encountered very frequently. If a receptor is activated frequently, it is as-

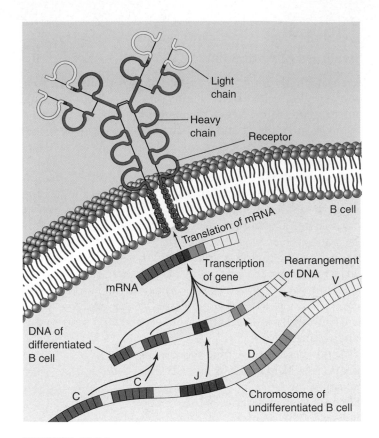

FIGURE 48.14
The lymphocyte receptor molecule is produced by a composite gene. Different regions of the DNA code for different regions of the receptor structure *C*, constant regions; *J*, joining regions; *D*, diversity regions; and *V*, variable regions. These are brought together to make a composite gene that codes for the receptor. Through different somatic rearrangements of these DNA segments, an enormous number of different receptor molecules can be produced.

sumed that the cell is recognizing a self antigen, and the lymphocytes are eliminated or suppressed. Thus, the only clones that survive this phase of development are those that are directed against foreign rather than self molecules.

Immunological tolerance sometimes breaks down, causing either B cells or T cells (or both) to recognize their own tissue antigens. This loss of immune tolerance results in autoimmune disease. Myasthenia gravis, for example, is an autoimmune disease in which individuals produce antibodies directed against acetylcholine receptors on their own skeletal muscle cells, causing paralysis. Autoimmunity will be discussed in more detail in section 48.5.

Active Immunity Through Clonal Selection

As we discussed earlier, B and T cells have receptors on their cell surfaces that recognize and bind to specific antigens. When a particular antigen enters the body, it must, by chance, encounter the specific lymphocyte with the appropriate receptor in order to provoke an immune response.

The first time a pathogen invades the body, only a few B or T cells may have the receptors that can recognize the invader's antigens. Binding of the antigen to its receptor on the lymphocyte surface, however, stimulates cell division and produces a *clone* (a population of genetically identical cells). This process is known as **clonal selection.** In this first encounter, a person gets sick because there are only a few cells that can mount an immune response and at first the response is relatively weak. This is called a **primary immune response** (figure 48.15).

If the primary immune response involves B cells, some become plasma cells that secrete antibodies, and others become memory cells. Because a clone of memory cells specific for that antigen develops after the primary response, the immune response to a second infection by the same pathogen is swifter and stronger. The next time the body is invaded by the same pathogen, the immune system is ready. As a result of the first infection, a large clone of lymphocytes now exists that can recognize that pathogen (figure 48.16). This more effective response, elicited by subsequent exposures to an antigen, is called a **secondary immune response** (figure 48.15).

Memory cells can survive for several decades, which is why people rarely contract chicken pox a second time after they have had it once. Memory cells are also the reason that vaccinations are effective. The vaccine triggers the primary response so that if the actual pathogen is encountered later, the large and rapid secondary response occurs and stops the infection before it can start. The viruses causing childhood diseases have surface antigens that change little from year to year, so the same antibody is effective for decades.

Figure 48.17 summarizes how the cellular and humoral lines of defense work together to produce the body's specific immune response.

An antibody molecule is composed of constant and variable regions. The variable regions recognize a specific antigen because they possess clefts into which the antigen can fit. Lymphocyte receptors are encoded by genes that are assembled by somatic rearrangement and mutation of the DNA. Active immunity is produced by clonal selection and expansion. This occurs because interaction of an antigen with its receptor on the lymphocyte surface stimulates cell division, so that more lymphocytes are available to combat subsequent exposures to the same antigen.

FIGURE 48.16
The clonal selection theory of active immunity. In response to interaction with an antigen that binds specifically to its surface receptors, a B cell divides many times to produce a clone of B cells. Some of these become plasma cells that secrete antibodies for the primary response, while others become memory cells that await subsequent exposures to the antigen for the mounting of a secondary immune response.

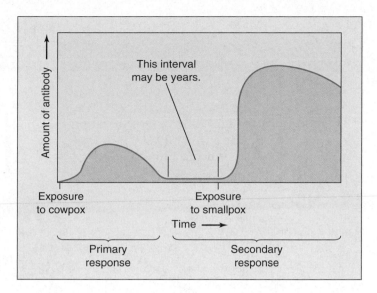

FIGURE 48.15
The development of active immunity. Immunity to smallpox in Jenner's patients occurred because their inoculation with cowpox stimulated the development of lymphocyte clones with receptors that could bind not only to cowpox but also to smallpox antigens. A second exposure stimulates the immune system to produce large amounts of the antibody more rapidly than before. **Would a secondary exposure to cowpox rather than smallpox have produced as strong an antibody response?**

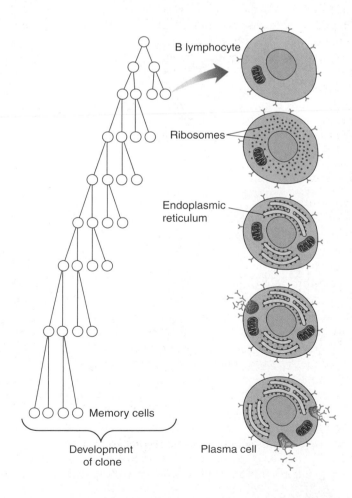

THE IMMUNE RESPONSE

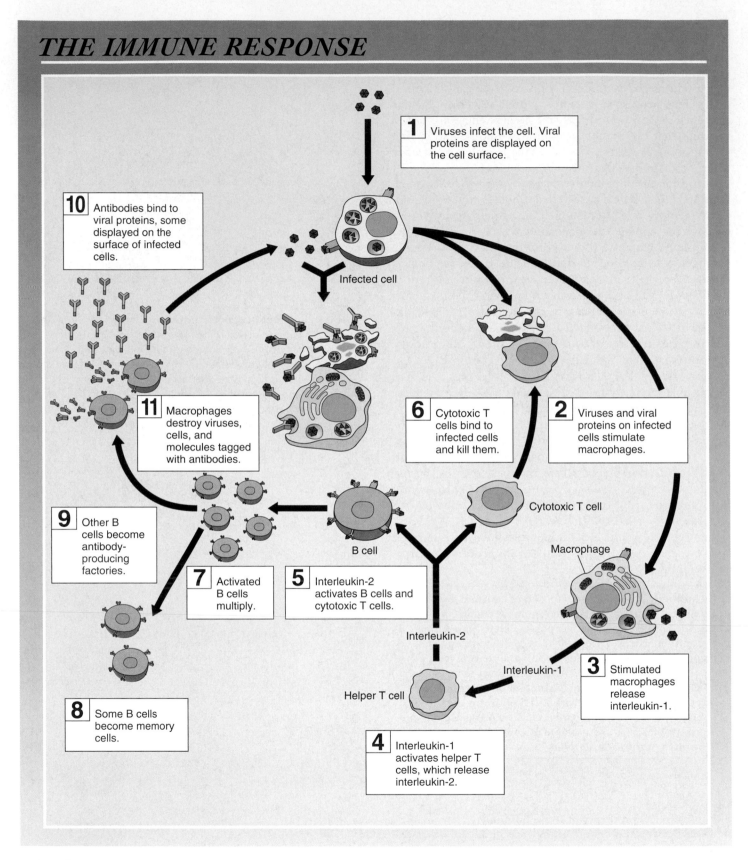

FIGURE 48.17
Overview of the specific immune response.

Antibodies in Medical Diagnosis

Blood Typing

A person's *blood type* denotes the class of antigens found on the red blood cell surface. Red blood cell antigens are clinically important because their types must be matched between donors and recipients for blood transfusions. There are several groups of red blood cell antigens, but the major group is known as the **ABO system.** In terms of the antigens present on the red blood cell surface, a person may be *type A* (with only A antigens), *type B* (with only B antigens), *type AB* (with both A and B antigens), or *type O* (with neither A nor B antigens).

The immune system is tolerant to its own red blood cell antigens. A person who is type A, for example, does not produce anti-A antibodies. Surprisingly, however, people with type A blood do make antibodies against the B antigen, and conversely, people with blood type B make antibodies against the A antigen. This is believed to result from the fact that antibodies made in response to some common bacteria cross-react with the A or B antigens. A person who is type A, therefore, acquires antibodies that can react with B antigens by exposure to these bacteria but does not develop antibodies that can react with A antigens. People who are type AB develop tolerance to both antigens and thus do not produce either anti-A or anti-B antibodies. Those who are type O do not develop tolerance to either antigen and, therefore, have both anti-A and anti-B antibodies in their plasma.

If type A blood is mixed on a glass slide with serum from a person with type B blood, the anti-A antibodies in the serum cause the type A red blood cells to clump together, or **agglutinate** (figure 48.18). These tests allow the blood types to be matched prior to transfusions, so that agglutination will not occur in the blood vessels, where it could lead to inflammation and organ damage.

Rh Factor. Another group of antigens found in most red blood cells is the *Rh factor* (Rh stands for rhesus monkey, in which these antigens were first discovered). People who have these antigens are said to be **Rh-positive,** whereas those who do not are **Rh-negative.** There are fewer Rh-negative people because the Rh-positive allele is more common in humans and is clinically dominant to the Rh-negative allele. The Rh factor is of particular significance when Rh-negative mothers give birth to Rh-positive babies.

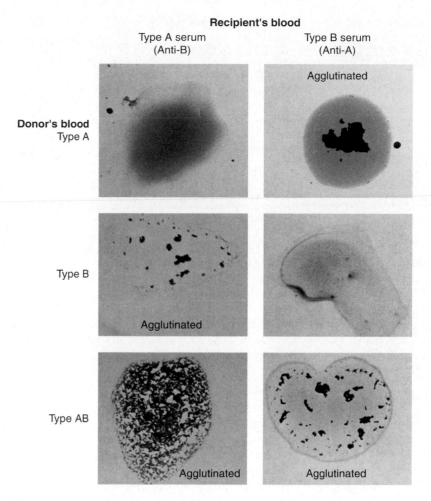

Recipient's blood

Type A serum (Anti-B) · Type B serum (Anti-A)

Donor's blood Type A · Type B · Type AB

FIGURE 48.18

Blood typing. Agglutination of the red blood cells is seen when blood types are mixed with sera containing antibodies against the ABO antigens. Note that no agglutination would be seen if type O blood (not shown) were used.

Because the fetal and maternal blood are normally kept separate across the placenta (see chapter 51), the Rh-negative mother is not usually exposed to the Rh antigen of the fetus during the pregnancy. At the time of birth, however, a variable degree of exposure may occur, and the mother's immune system may become sensitized and produce antibodies against the Rh antigen. If the woman does produce antibodies against the Rh factor, these antibodies can cross the placenta in subsequent pregnancies and cause hemolysis of the Rh-positive red blood cells of the fetus. The baby is therefore born anemic, with a condition called *erythroblastosis fetalis,* or *hemolytic disease of the newborn.*

Erythroblastosis fetalis can be prevented by injecting the Rh-negative mother with an antibody preparation against the Rh factor within 72 hours after the birth of each Rh-positive baby. This is a type of passive immunization in which the injected antibodies inactivate the Rh antigens and thus prevent the mother from becoming actively immunized to them.

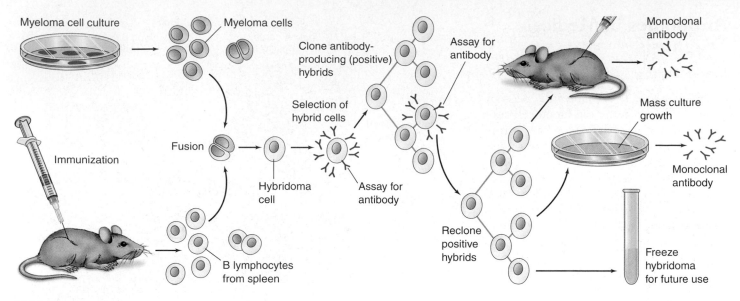

FIGURE 48.19

The production of monoclonal antibodies. These antibodies are produced by cells that arise from successive divisions of a single B cell; hence, all the antibodies target a single antigenic determinant site. Such antibodies are used for a variety of research and clinical applications, including pregnancy testing.

Monoclonal Antibodies

Antibodies are commercially prepared for use in medical diagnosis and research. In the past, antibodies were obtained by chemically purifying a specific antigen and then injecting this antigen into animals. However, because an antigen typically has many different antigenic determinant sites, the antibodies obtained by this method were *polyclonal*; they stimulated the development of different B cell clones with different specificities. This decreased their sensitivity to a particular antigenic site and resulted in some degree of cross-reaction with closely related antigen molecules.

Monoclonal antibodies, by contrast, exhibit specificity for one antigenic determinant only. In the preparation of monoclonal antibodies, an animal (frequently, a mouse) is injected with an antigen and subsequently killed. B lymphocytes are then obtained from the animal's spleen and placed in thousands of different in vitro incubation vessels. These cells would soon die unless hybridized with cancerous multiple myeloma cells. The fusion of a B lymphocyte with a cancerous cell produces a hybrid that undergoes cell division and produces a clone called a *hybridoma*. Each hybridoma secretes large amounts of identical, monoclonal antibodies. From among the thousands of hybridomas produced in this way, the one that produces the desired antibody is cultured for large-scale production, and the rest are discarded (figure 48.19).

The availability of large quantities of pure monoclonal antibodies has resulted in the development of much more sensitive clinical laboratory tests. Modern pregnancy tests, for example, use particles (latex rubber or red blood cells) that are covered with monoclonal antibodies produced against a pregnancy hormone (abbreviated hCG—see

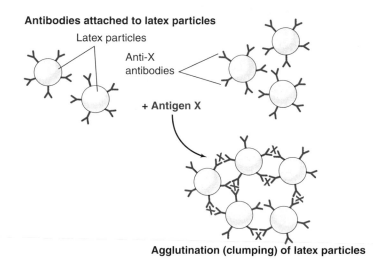

Antibodies attached to latex particles

FIGURE 48.20

Using monoclonal antibodies to detect an antigen. In many clinical tests (such as pregnancy testing), the monoclonal antibodies are bound to particles of latex, which agglutinate in the presence of the antigen.

chapter 50) as the antigen. When these particles are mixed with a sample that contains this hormone antigen from a pregnant woman, the antigen-antibody reaction causes a visible agglutination of the particles (figure 48.20).

Agglutination occurs because different antibodies exist for the ABO and Rh factor antigens on the surface of red blood cells. Monoclonal antibodies are commercially produced antibodies that react against one specific antigen.

Evolution of the Immune System

All organisms possess mechanisms to protect themselves from the onslaught of smaller organisms and viruses. Bacteria defend against viral invasion by means of *restriction endonucleases*, enzymes that degrade any foreign DNA lacking the specific pattern of DNA methylation characteristic of that bacterium. Multicellular organisms face a more difficult problem in defense because their bodies often take up whole viruses, bacteria, or fungi instead of naked DNA.

Invertebrates

Invertebrate animals solve this problem by marking the surfaces of their cells with proteins that serve as "self" labels. Special amoeboid cells in the invertebrate attack and engulf any invading cells that lack such labels. By looking for the absence of specific markers, invertebrates employ a *negative* test to recognize foreign cells and viruses. This method provides invertebrates with a very effective surveillance system, although it has one great weakness: Any microorganism or virus with a surface protein resembling the invertebrate self marker will not be recognized as foreign. An invertebrate has no defense against such a "copycat" invader.

In 1882, the Russian zoologist Elie Metchnikoff became the first to recognize that invertebrate animals possess immune defenses. On a beach in Sicily, he collected the tiny transparent larva of a common starfish. Carefully he pierced it with a rose thorn. When he looked at the larva the next morning, he saw a host of tiny cells covering the surface of the thorn as if trying to engulf it (figure 48.21). The cells were attempting to defend the larva by ingesting the invader by phagocytosis (described in chapter 6). For his discovery of what came to be known as the **cellular immune response**, Metchnikoff was awarded the 1908 Nobel Prize in Physiology or Medicine, along with Paul Ehrlich for his work on the other major part of the immune defense, the antibody or **humoral immune response** (described earlier in this section). The invertebrate immune response shares several elements with the vertebrate one, as listed next.

Phagocytes. All animals possess phagocytic cells that attack invading microbes. These phagocytic cells travel through the animal's circulatory system or circulate within the fluid-filled body cavity. In simple animals such as sponges that lack either a circulatory system or a body cavity, the phagocytic cells roam within the mesoglea.

Distinguishing Self from Nonself. The ability to recognize the difference between cells of one's own body and those of another individual appears to have evolved early in the history of life. Sponges, thought to be the most ancient

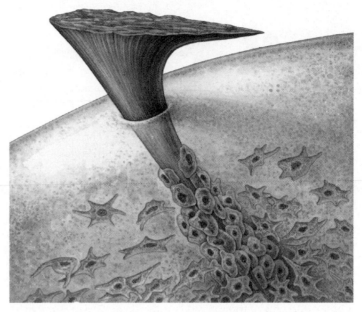

FIGURE 48.21
Discovering the cellular immune response in invertebrates. In a Nobel Prize–winning experiment, the Russian zoologist Metchnikoff pierced the larva of a starfish with a rose thorn and the next day found tiny phagocytic cells covering the thorn.

animals, attack grafts from other sponges, as do insects and starfish. None of these invertebrates, however, exhibit any evidence of immunological memory; apparently, the antibody-based humoral immune defense did not evolve until the vertebrates.

Complement. While invertebrates lack complement proteins, many arthropods (including crabs and a variety of insects) possess an analogous nonspecific defense called the prophenyloxidase (proPO) system. Like the vertebrate complement defense, the proPO defense is activated as a cascade of enzyme reactions, the last of which converts the inactive protein prophenyloxidase into the active enzyme phenyloxidase. Phenyloxidase both kills microbes and aids in encapsulating foreign objects.

Lymphocytes. Invertebrates also lack lymphocytes, but annelid earthworms and other invertebrates do possess lymphocyte-like cells that may be evolutionary precursors of lymphocytes.

Antibodies. All invertebrates possess proteins called lectins that may be the evolutionary forerunners of antibodies. Lectins bind to sugar molecules on cells, making the cells stick to one another. Lectins isolated from sea urchins, mollusks, annelids, and insects appear to tag invading microorganisms, enhancing phagocytosis. The genes encoding vertebrate antibodies are part of a very ancient gene family,

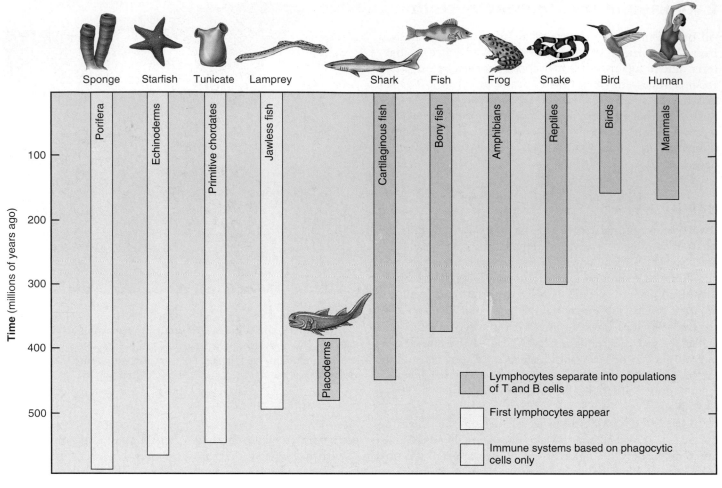

FIGURE 48.22

How immune systems evolved. Lampreys were the first vertebrates to possess an immune system based on lymphocytes, although distinct B and T cells did not appear until the jawed fishes evolved. By the time sharks and other cartilaginous fish appeared, the vertebrate immune response was fully formed.

the immunoglobulin superfamily. Proteins in this group all have a characteristic recognition structure called the Ig fold. The fold probably evolved as a self-recognition molecule in early animals. In moths, grasshoppers, and flies, insect immunoglobulins have been described that bind to microbial surfaces and promote their destruction by phagocytes. The antibody immune response appears to have evolved from these earlier, less complex systems.

Vertebrates

The earliest vertebrates about which we have any clear information, the jawless lampreys that first evolved some 500 million years ago, possess an immune system based on lymphocytes. At this early stage of vertebrate evolution, however, lampreys lack distinct populations of B and T cells such as are found in all higher vertebrates (figure 48.22).

With the evolution of fish with jaws, the modern vertebrate immune system first arose. The oldest surviving group of jawed fishes are the sharks, which evolved some 450 mil-

lion years ago. By then, the vertebrate immune defense had fully evolved. Sharks have an immune response much like that seen in mammals, with a cellular response carried out by T cell lymphocytes and an antibody-mediated humoral response carried out by B cells. The similarities of the cellular and humoral immune defenses are far more striking than the differences. Both sharks and mammals possess a thymus that is important in T cell development and a spleen that is a rich source of B cells. Apparently, 450 million years of evolution did little to change the antibody molecule, because the amino acid sequences of shark and human antibody molecules are very similar. The most notable difference between sharks and mammals is that their antibody-encoding genes are arranged in the genome somewhat differently.

The sophisticated two-part immune defense of mammals evolved about the time jawed fishes appeared. Before then, animals utilized a simpler immune defense based on mobile phagocytic cells.

The immune system can be defeated.

T Cell Destruction: AIDS

One mechanism for defeating the vertebrate immune system is to attack the immune mechanism itself. Many helper T cells and inducer T cells are CD4$^+$ T cells. Therefore, any pathogen that inactivates CD4+ T cells leaves the immune system unable to mount a response to *any* foreign antigen. Acquired immune deficiency syndrome (AIDS) is a deadly disease for just this reason. The AIDS retrovirus, called human immunodeficiency virus (HIV), mounts a direct attack on CD4$^+$ T cells because it recognizes the CD4 coreceptors associated with these cells. HIV is discussed in detail in chapter 26.

HIV's attack on CD4$^+$ T cells cripples the immune system in at least three ways. First, HIV-infected cells die only after releasing replicated viruses that infect other CD4$^+$ T cells, until the entire population of CD4$^+$ T cells is destroyed (figure 48.23). In a normal individual, CD4$^+$ T cells make up 60 to 80% of circulating T cells; in AIDS patients, CD4$^+$ T cells often become too rare to detect (figure 48.24). Second, HIV causes infected CD4$^+$ T cells to secrete a soluble suppressing factor that blocks other T cells from responding to the HIV antigen. Finally, HIV may block transcription of MHC genes, hindering the recognition and destruction of infected CD4$^+$ T cells and thus protecting those cells from any remaining vestiges of the immune system.

The combined effect of these responses to HIV infection is to wipe out the human immune defense. With no defense against infection, any of a variety of otherwise commonplace infections proves fatal. With no ability to recognize and destroy cancer cells when they arise, death by cancer becomes far more likely. Indeed, AIDS was first recognized as a disease because of a cluster of cases of an unusually rare form of cancer. More AIDS victims die of cancer than from any other cause.

Although HIV became a human disease vector only recently, possibly through transmission to humans from chimpanzees in central Africa, AIDS is already clearly one of the most serious diseases in human history (figure 48.25). The fatality rate of AIDS is 100%; no patient exhibiting the symptoms of AIDS has ever been known to survive more than a few years without treatment. Aggressive treatments can prolong life, but how much longer has not been determined. However, the disease is *not* highly contagious, because it is transmitted from one individual to another through the transfer of internal body fluids, typically semen and blood. Not all individuals exposed to HIV (as judged by the presence of anti-HIV antibodies in their blood) have yet acquired the disease.

HIV-positive patients who have not fully exhibited AIDS symptoms can be more effectively treated with drugs. Until recently, the only effective way to slow the progression of the disease involved treatment with drugs

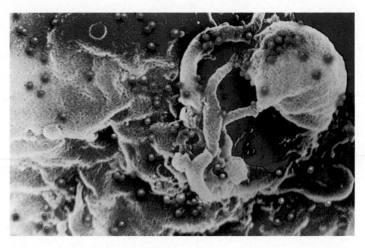

FIGURE 48.23
HIV, the virus that causes AIDS. Viruses released from infected CD4$^+$ T cells soon spread to neighboring CD4$^+$ T cells, infecting them in turn. The individual viruses, colored *red* in this scanning electron micrograph, are extremely small; over 200 million would fit on the period at the end of this sentence.

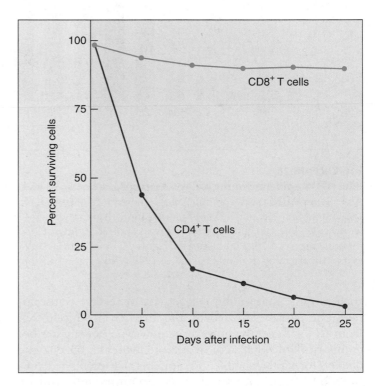

FIGURE 48.24
Survival of T cells in culture after exposure to HIV. The virus has little effect on the number of CD8$^+$ T cells, but it causes the number of CD4$^+$ T cells to decline dramatically.
Are the surviving CD4$^+$ T cells any different from those destroyed? How could you be sure?

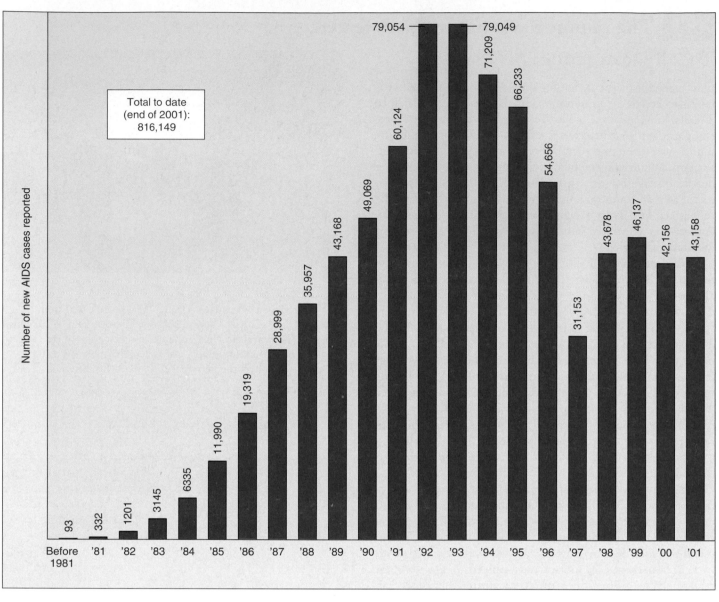

FIGURE 48.25
The AIDS epidemic in the United States: New cases.   The U.S. Centers for Disease Control and Prevention (CDC) reports that
42,156 new AIDS cases were reported in the year 2000 and 43,158 new cases in 2001, with a total of 816,149 cases and 467,910 deaths in
the United States. Over 1.5 million other individuals are thought to be infected with the HIV virus in the United States, and 14 million
worldwide. The 100,000th AIDS case was reported in August 1988, eight years into the epidemic; the next 100,000 cases took just 26
months; the third 100,000 cases took barely 19 months (May 1993), and the fourth 100,000 took only 13 months (June 1994). The
extraordinary jump in numbers seen in 1992 reflects an expansion of the definition of what constitutes an AIDS case.

such as AZT that inhibit the activity of reverse transcriptase, the enzyme needed by the virus to produce DNA
from RNA. Recently, however, a new type of drug has become available that acts to inhibit protease, an enzyme
needed for viral assembly. Treatments that include a combination of reverse transcriptase inhibitors and protease
inhibitors (see chapter 26) appear to lower levels of HIV,
but they are very costly. Efforts to develop a vaccine
against AIDS continue, both by splicing portions of the
HIV surface protein gene into vaccinia virus and by attempting to develop a harmless strain of HIV. These ap-

proaches, while promising, have not yet proved successful
and are limited by the fact that different strains of HIV
seem to possess different surface antigens. Like the influenza virus, HIV has high mutation rates that result in
antigen shifting discussed next, making it difficult to develop an effective vaccine.

**AIDS destroys the ability of the immune system to
mount a defense against any infection. HIV, the virus
that causes AIDS, induces a state of immune deficiency
by attacking and destroying CD4+ T cells.**

Antigen Shifting

A second way that a pathogen may defeat the immune system is to mutate frequently so that it varies the nature of its surface antigens. The virus that causes influenza uses this mechanism, so people have to be immunized against a different strain of this virus periodically. This way of escaping immune attack is known as **antigen shifting,** and is practiced very effectively by trypanosomes, the protists responsible for African sleeping sickness (see chapter 28). Trypanosomes possess several thousand different versions of the genes encoding their surface protein, but the cluster containing these genes has no promoter and so is not transcribed as a unit. The necessary promoter is located within a transposable element that jumps at random from one position to another within the cluster, transcribing a different surface protein gene with every move. Because such moves occur in at least one cell of an infective trypanosome population every few weeks, the vertebrate immune system is unable to mount an effective defense against trypanosome infection. By the time a significant number of antibodies have been generated against one form of trypanosome surface protein, another form is already present in the trypanosome population; this form survives immunological attack, and the infection cycle is renewed. People with sleeping sickness rarely rid themselves of the infection.

Although this mechanism of mutation to alter surface proteins seems very "directed" or intentional on the part of the pathogen, it is actually the process of evolution by natural selection at work. We usually think of evolution as requiring thousands of years to occur, rather than a space of weeks. However, evolution can occur whenever mutations are passed on that provide the offspring organism with a competitive advantage. In the case of viruses, bacteria, and other pathogenic agents, generation times are on the order of hours. Thus, in the time frame of a week, the population has gone through millions of cell divisions. Looking at it from this perspective, it is easy to see how random mutations in the genes for the surface antigens could occur and change the surface of the pathogen in as little as a week's time.

How Malaria Hides from the Immune System

Every year, about a half-million people become infected with the protozoan parasite *Plasmodium falciparum*, which multiplies in their bodies to cause the disease malaria. The plasmodium parasites enter the red blood cells and consume the hemoglobin of their hosts. Normally, this sort of damage to a red blood cell would cause the damaged cell to be transported to the spleen for disassembly, destroying the plasmodium as well. However, the plasmodium avoids this fate by secreting knoblike proteins that extend through the surface of the red blood cell and anchor the cell to the inner surface of the blood vessel.

Over the course of several days, the immune system of the infected person slowly brings the infection under control. During this time, however, a small proportion of the plasmodium parasites change their knob proteins to a form different from those that sensitized the immune system. Cells infected with these individuals survive the immune response, only to start a new wave of infection.

Scientists have recently discovered how the malarial parasite carries out this antigen-shifting defense. About 6% of the total DNA of the plasmodium is devoted to encoding a block of some 150 *var* genes, which are shifted on and off in multiple combinations. Each time a plasmodium divides, it alters the pattern of *var* gene expression about 2%, an incredibly rapid rate of antigen shifting. The exact means by which this occurs is not yet completely understood.

DNA Vaccines May Get Around Antigen Shifting

Vaccination against diseases such as smallpox, measles, and polio involves introducing into your body a dead or disabled pathogen or a harmless microbe with pathogen proteins displayed on its surface. The vaccination triggers an immune response against the pathogen, and the bloodstream and lymphatic system of the vaccinated person contain B cells that will remember and quickly destroy the pathogen in future infections. However, for some diseases, vaccination is nearly impossible because of antigen shifting; the pathogens change over time, and the B cells no longer recognize them. Influenza, as we have discussed, presents different surface proteins yearly. The trypanosomes responsible for sleeping sickness change their surface proteins every few weeks.

A new type of vaccine, based on DNA, may prove effective against almost any disease. The vaccine uses the killer T cells instead of the B cells of the immune system. DNA vaccines consist of a plasmid, a harmless circle of bacterial DNA. This plasmid contains a gene from the pathogen that encodes an internal protein, one that is critical to the function of the pathogen and does not change. When the plasmid is injected into cells, the gene the plasmid carries is transcribed into protein but is not incorporated into the DNA of the cell's nucleus. Fragments of the pathogen protein are then stuck on the cell's membrane, marking it for destruction by T cells. In actual infections later, the immune system will be able to respond immediately. Studies are now under way to isolate the critical, unchanging proteins of pathogens and to investigate fully their use in vaccines for humans.

Antigen shifting refers to the way a pathogen may defeat the immune system by changing its surface antigens and thereby escaping immune recognition. Pathogens that employ this mechanism include influenza viruses, trypanosomes, and the protozoans that cause malaria.

Autoimmunity and Allergy

So far, this section has described ways that pathogens can elude the immune system to cause diseases. There is another way the immune system can fail; it can itself be the agent of disease. Such is the case with autoimmune diseases and allergies, in which the immune system is the cause of the problem, not the cure.

Autoimmune Diseases

Autoimmune diseases are produced by failure of the immune system to recognize and tolerate self antigens. This failure results in the activation of autoreactive T cells and the production of autoantibodies by B cells, causing inflammation and organ damage. There are over 40 known or suspected autoimmune diseases, and they affect 5 to 7% of the population. For reasons that are not understood, two-thirds of the people with autoimmune diseases are women.

Autoimmune diseases can result from a variety of mechanisms. For example, the self antigen may normally be hidden from the immune system so that the immune system treats it as foreign if exposure later occurs. This happens when a protein normally trapped in the thyroid follicles triggers autoimmune destruction of the thyroid (Hashimoto thyroiditis). It also occurs in systemic lupus erythematosus, in which antibodies to nucleoproteins are made. Because the immune attack triggers inflammation, and inflammation causes organ damage, the immune system must be suppressed to alleviate the symptoms of autoimmune diseases. Immune suppression is generally accomplished by administering corticosteroids (including hydrocortisone) and nonsteroidal anti-inflammatory drugs, including aspirin.

Allergy

The term *allergy*, often used interchangeably with *hypersensitivity*, refers to particular types of abnormal immune responses to antigens, which are called *allergens* in these cases. The two major forms of allergy are (1) **immediate hypersensitivity,** which is due to an abnormal B cell response to an allergen that produces symptoms within seconds or minutes, and (2) **delayed hypersensitivity,** which is an abnormal T cell response that produces symptoms within about 48 hours after exposure to an allergen.

Immediate hypersensitivity results from the production of antibodies of the IgE subclass instead of the normal IgG

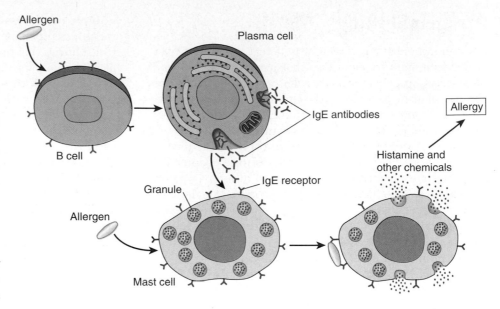

FIGURE 48.26
An allergic reaction. This is an immediate hypersensitivity response, in which B cells secrete antibodies of the IgE class. These antibodies attach to the plasma membranes of mast cells, which secrete histamine in response to antigen-antibody binding.

antibodies. Unlike IgG antibodies, IgE antibodies do not circulate in the blood. Instead, they attach to tissue mast cells and basophils, which have membrane receptors for these antibodies. When the person is again exposed to the same allergen, the allergen binds to the antibodies attached to the mast cells and basophils. This stimulates these cells to secrete various chemicals, including histamine, which produce the symptom of the allergy (figure 48.26).

Allergens that provoke immediate hypersensitivity include various foods, bee stings, and pollen grains. The most common allergy of this type is seasonal hay fever, which may be provoked by ragweed (*Ambrosia*) pollen grains. These allergic reactions are generally mild, but in some allergies (as to penicillin or peanuts in susceptible people), the widespread and excessive release of histamine may cause **anaphylactic shock,** an uncontrolled fall in blood pressure.

In delayed hypersensitivity, symptoms take a longer time (hours to days) to develop than in immediate hypersensitivity. This may be due to the fact that immediate hypersensitivity is mediated by antibodies, whereas delayed hypersensitivity is a T cell response. One of the best-known examples of delayed hypersensitivity is **contact dermatitis,** caused by poison ivy, poison oak, and poison sumac. Because the symptoms are caused by the secretion of lymphokines rather than by the secretion of histamine, treatment with antihistamines provides little benefit. At present, corticosteroids are the only drugs that can effectively treat delayed hypersensitivity.

Autoimmune diseases are produced when the immune system fails to tolerate self antigens.

48.1 Many of the body's most effective defenses are nonspecific.

Skin: The First Line of Defense

- Skin, the largest organ of the vertebrate body, provides a nearly impenetrable barrier and reinforces defense with surface chemical weapons. (p. 1014)
- Mucous membranes of the digestive, respiratory, and urogenital tracts help trap and eliminate microorganisms before they invade body tissues. (p. 1014)

Cellular Counterattack: The Second Line of Defense

- Nonspecific cellular and chemical defenses respond to any microbial infection without the necessity of determining the specific identity of the invader. (p. 1015)
- Macrophages ingest microbes via phagocytosis; neutrophils ingest and kill bacteria by phagocytosis; and natural killer cells kill cells of the body infected with viruses. (p. 1015)
- The complement system consists of approximately 20 different blood proteins that aggregate to form a membrane attack complex, which then forms a pore in the foreign cell's plasma membrane. (p. 1016)
- Body cells infected by a virus secrete interferon to protect neighboring cells. (p. 1016)

The Inflammatory Response

- The inflammatory response is a localized, nonspecific response to infection. (p. 1017)
- Histamine and prostaglandins promote blood vessel dilation and capillary permeability, while a fever stimulates phagocytosis and causes the liver and spleen to store iron. (p. 1017)

48.2 Specific immune defenses require the recognition of antigens.

The Immune Response: The Third Line of Defense

- B cells (B lymphocytes) respond to antigens by producing antibodies that provide humoral immunity. (p. 1018)
- T cells (T lymphocytes) regulate the immune response of other cells, directly attack cells carrying specific antigens, and produce cell-mediated immunity. (p. 1018)
- Active immunity can be gained due to exposure to a pathogen, while passive immunity occurs when one individual obtains antibodies from another individual. (p. 1018)

Cells of the Specific Immune System

- T cells and B cells are both produced in bone marrow, but B cells develop and mature in bone marrow, while T cells complete development in the thymus. (p. 1019)

Initiating the Immune Response

- Helper T cells are activated when antigen-presenting cells present foreign antigens with MHC-II proteins. (p. 1020)
- Cytotoxic T cells interact with foreign antigens presented with MHC-I proteins. (p. 1020)

48.3 T cells organize attacks against invading microbes.

T Cells: The Cell-Mediated Immune Response

- In the cell-mediated immune response, cytotoxic T cells kill abnormal or virus-infected body cells. (p. 1021)
- Interleukin-1 is secreted by macrophages and stimulates helper T cells, while interleukin-2 is secreted by helper T cells and activates cytotoxic T cells and B cells. (pp. 1021–1022)

48.4 B cells label specific cells for destruction.

B Cells: The Humoral Immune Response

- B cells place markers on pathogens to alert complement proteins, macrophages, and natural killer cells to attack those cells. (p. 1023)

- When B cells recognize antigens, they produce plasma cells and large numbers of circulating antibodies to fight the antigens. (p. 1023)
- Antibodies are proteins in a class known as immunoglobulins, which is further divided into subclasses. (p. 1023)
- IgM antibodies are produced first and activate the complement system, while IgG antibodies are produced later to promote phagocytosis. (pp. 1023–1024)

Antibodies

- Each antibody molecule consists of light chains and heavy chains containing variable and constant regions. Most of the sequence variation between antibodies is found in a cleft that only allows a specific antigen to bind. (p. 1025)
- Somatic rearrangement and DNA mutations assemble the genes that encode for the huge numbers of antigen receptors. (pp. 1025–1026)
- Immunological tolerance occurs when an animal's immune system accepts the animal's own tissue; it can occur through clonal deletion or clonal suppression. (p. 1026)
- A primary immune response occurs the first time the body encounters a pathogen and the binding of an antigen stimulates cell division (clonal selection). As a result of the first infection, the body is prepared to recognize and fight off the same pathogen the next time it is encountered (secondary immune response). (pp. 1026–1027)

Antibodies in Medical Diagnosis

- Blood type denotes the class of antibodies found on the surface of red blood cells. The major group is known as the ABO system, while another system is the Rh factor. Mixing different blood types can lead to agglutination. (p. 1029)
- Monoclonal antibodies exhibit specificity for one antigen and are commercially prepared for medical diagnosis and research. (p. 1030)

Evolution of the Immune System

- Early organisms utilized a cellular immune response. (p. 1031)
- Vertebrates appear to be the first group to evolve an immune system based on lymphocytes. (p. 1032)

48.5 The immune system can be defeated.

T Cell Destruction: AIDS

- HIV inactivates CD4$^+$ T cells and thus leaves the immune system incapable of mounting a response to a foreign antigen. (p. 1033)
- HIV exhibits a high mutation rate resulting in antigen shifting, thus making it difficult to develop an effective vaccine. (p. 1034)

Antigen Shifting

- A pathogen may escape recognition by the immune system if it changes its surface antigens. Such antigen shifting is an example of evolution by natural selection. (p. 1035)

Autoimmunity and Allergy

- Autoimmune diseases are produced by a failure of the immune system to recognize and tolerate self-antigens, and can result from a variety of mechanisms. (p. 1036)
- Allergies, or hypersensitivity, can be divided into immediate hypersensitivity and delayed hypersensitivity, both of which can cause the release of histamine. (p. 1036)
- In extreme cases, the widespread release of histamine can lead to anaphylactic shock. (p. 1036)

Self Test

1. The epidermis fights microbial infections by
    a. making the surface of the skin acidic.
    b. excreting lysozyme to attack bacteria.
    c. producing mucus to trap microorganisms.
    d. all of these.
2. Cells that target and kill body cells infected by viruses are:
    a. macrophages.   c. monocytes.
    b. natural killer cells.   d. neutrophils.
3. A molecule present on the plasma membrane that induces an immune response is called
    a. an antigen.   c. an antibody.
    b. an interleukin.   d. a lymphocyte.
4. Which one of the following acts as the "alarm signal" to activate the body's immune system by stimulating helper T cells?
    a. B cells   c. interleukin-2
    b. interleukin-1   d. histamines
5. Cytotoxic T cells are called into action by the
    a. presence of histamine.
    b. presence of interleukin-1.
    c. presence of interleukin-2.
    d. B cell stimulating factor.
6. The immunoglobin _____ is involved in the primary response to an invading microorganism, and _____ is involved in the secondary response.
    a. IgG / IgM
    b. IgM / IgE
    c. IgE / histamines
    d. IgM / IgG
7. How does your body detect millions of different antigens?
    a. The few hundred immunoglobulin genes can be rearranged or can undergo mutations to form millions of antibody molecules.
    b. There are millions of different antibody genes.
    c. The few hundred immunoglobulin genes undergo antigen shifting.
    d. Each B cell has a different set of immunoglobulin genes, and so the activation of different B cells produces different antibodies.
8. The blood from a person with an AB blood type
    a. would agglutinate with anti-A antibodies only.
    b. would agglutinate with anti-B antibodies only.
    c. would agglutinate with both anti-A and anti-B antibodies.
    d. would not agglutinate with either anti-A or anti-B antibodies.
9. The HIV virus is particularly dangerous because it attacks
    a. cells with the CD4$^+$ coreceptor.
    b. helper T cells.
    c. 60 to 80% of circulating T cells in the body.
    d. all of these.
10. Diseases in which the person's immune system no longer recognizes its own MHC proteins are called
    a. allergies.
    b. autoimmune diseases.
    c. immediate hypersensitivity.
    d. delayed hypersensitivity.

Test Your Visual Understanding

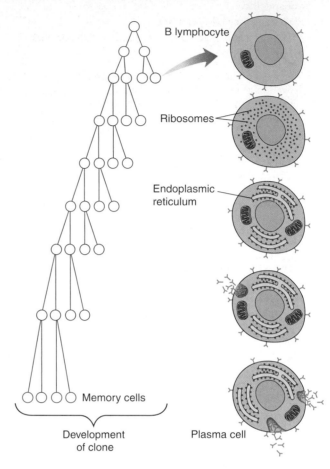

B lymphocyte

Ribosomes

Endoplasmic reticulum

Memory cells

Development of clone

Plasma cell

1. When children have chicken pox, they run a fever and develop a rash that leaves them contagious for 5 to 7 days. In that time, the immune system mounts an attack resulting in the production of antibodies that help the cells of the immune system fight the infection. If these children are exposed to the chicken pox virus again, they do not develop the disease. Using the figure above, explain why a person doesn't develop chicken pox a second time.

Apply Your Knowledge

1. Sometimes a doctor may order a differential blood count to determine whether a person has an infection. The differential blood count detects the levels of white blood cells (WBCs) in the body. Elevated white blood counts indicate an infection. The normal ranges of WBCs in the body are approximately:
    neutrophils—60%
    lymphocytes—25%
    monocytes—8%
    eosinophils—4%
    basophils—3%
A man has a total WBC of 6000/ml, and the differential blood count values are: neutrophils, 3720/ml; lymphocytes, 1500/ml; monocytes, 480/ml; eosinophils, 180/ml; basophils, 120/ml. Is his WBC within normal range?

49

Maintaining the Internal Environment

FIGURE 49.1
Regulating body temperature with water. One of the ways an elephant can regulate its temperature is to spray water on its body. Water also cycles through the elephant's body in enormous quantities each day and helps regulate its internal environment.

Concept Outline

49.1 The regulatory systems of the body maintain homeostasis.

The Need to Maintain Homeostasis. Regulatory mechanisms maintain homeostasis through negative feedback loops.

Antagonistic Effectors and Positive Feedback. Antagonistic effectors cause opposite changes, while positive feedback pushes changes further in the same direction.

49.2 The extracellular fluid concentration is constant in most vertebrates.

Osmolality and Osmotic Balance. Vertebrates have to cope with the osmotic gain or loss of body water.

Osmoregulatory Organs. Invertebrates have a variety of organs to regulate water balance; kidneys are the osmoregulatory organs of most vertebrates.

Evolution of the Vertebrate Kidney. Freshwater bony fish produce a dilute urine, and marine bony fish produce an isotonic urine. Only birds and mammals can retain so much water that they produce a concentrated urine.

Ammonia, Urea, and Uric Acid. The breakdown of protein and nucleic acids yields nitrogen, which is excreted as ammonia in bony fish, as urea in mammals, and as uric acid in reptiles and birds.

49.3 The functions of the vertebrate kidney are performed by nephrons.

The Mammalian Kidney. Each kidney contains nephrons, which produce a filtrate that is modified by reabsorption and secretion to produce urine.

Transport Processes in the Mammalian Nephron. The nephron tubules of birds and mammals have loops of Henle, which function to draw water out of the tubule and back into the blood.

Hormones Control Osmoregulatory Functions. Antidiuretic hormone promotes water retention and the excretion of a highly concentrated urine. Aldosterone stimulates the retention of salt and water, whereas atrial natriuretic hormone promotes the excretion of salt and water.

The first vertebrates evolved in seawater, and the physiology of all vertebrates reflects this origin. Approximately two-thirds of every vertebrate's body is water. If the amount of water in the body of a vertebrate falls much lower than this, the animal will die. In this chapter, we discuss the various mechanisms by which animals avoid gaining or losing too much water. As we shall see, these mechanisms are closely tied to the way animals exploit the varied environments in which they live and to the regulatory systems of the body (figure 49.1).

49.1 The regulatory systems of the body maintain homeostasis.

The Need to Maintain Homeostasis

As the animal body has evolved, specialization has increased. Each cell is a sophisticated machine, finely tuned to carry out a precise role within the body. Such specialization of cell function is possible only when extracellular conditions are kept within narrow limits. Temperature, pH, the concentrations of glucose and oxygen, and many other factors must be held fairly constant for cells to function efficiently and interact properly with one another.

Homeostasis may be defined as the dynamic constancy of the internal environment. The term *dynamic* is used because conditions are never absolutely constant, but fluctuate continuously within narrow limits. Homeostasis is essential for life, and most of the regulatory mechanisms of the vertebrate body that are not devoted to reproduction are concerned with maintaining homeostasis.

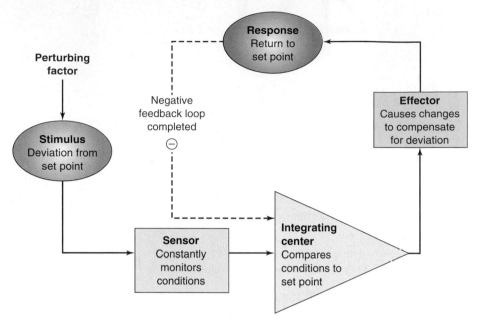

FIGURE 49.2
A generalized diagram of a negative feedback loop. Negative feedback loops maintain a state of homeostasis, or dynamic constancy of the internal environment, by correcting deviations from a set point.

Negative Feedback Loops

To maintain internal constancy, the vertebrate body uses a type of control system known as a *negative feedback loop* (figure 49.2). **Sensors** that are able to measure each condition of the internal environment constantly monitor the extracellular conditions and relay this information (usually via nerve signals) to an **integrating center,** which contains the *set point* (the proper value for that condition). This set point is analogous to the temperature setting on a house thermostat. In a similar manner, there are set points for body temperature, blood glucose concentration, the tension on a tendon, and so on. The integrating center is often a particular region of the brain or spinal cord, but in some cases it can also be cells of endocrine glands. It receives messages from several sensors (the "stimuli"), weighing the relative strengths of each sensor input, and then determines whether the value of the condition is deviating from the set point. When a deviation in a condition occurs, the integrating center sends a message to increase or decrease the activity of particular **effectors.** Effectors are generally muscles or glands, and can change the value of the condition in question back toward the set point value (the "response").

To return to the idea of a home thermostat, suppose you set the thermostat at a set point of 70°F. If the temperature of the house rises sufficiently above the set point, the thermostat (equivalent to an integrating center) receives this input from a temperature sensor, such as a thermometer (a sensor) within the wall unit. It compares the actual temperature to its set point. When these are different, it sends a signal to an effector. The effector in this case may be an air conditioner, which acts to reverse the deviation from the set point.

In a human, if the body temperature exceeds the set point of 37°C, sensors in a part of the brain detect this deviation. Acting via an integrating center (also located in the brain), these sensors stimulate effectors (including sweat glands) that lower the temperature (figure 49.3). We can think of the effectors as "defending" the set points of the body against deviations. It is because the activity of the effectors is influenced by the effects they produce, and because this regulation is in a negative, or reverse, direction that we call this type of control system a negative feedback loop.

The nature of the negative feedback loop becomes clear when we again refer to the analogy of the thermostat and air conditioner. After the air conditioner has been on for some time, the room temperature may fall significantly below the set point of the thermostat. When this occurs, the air conditioner will be turned off. The effector (air conditioner) is turned on by a high temperature; when activated, it produces a negative change (lowering of the temperature) that ultimately causes the effector to be turned off. In this way, constancy is maintained.

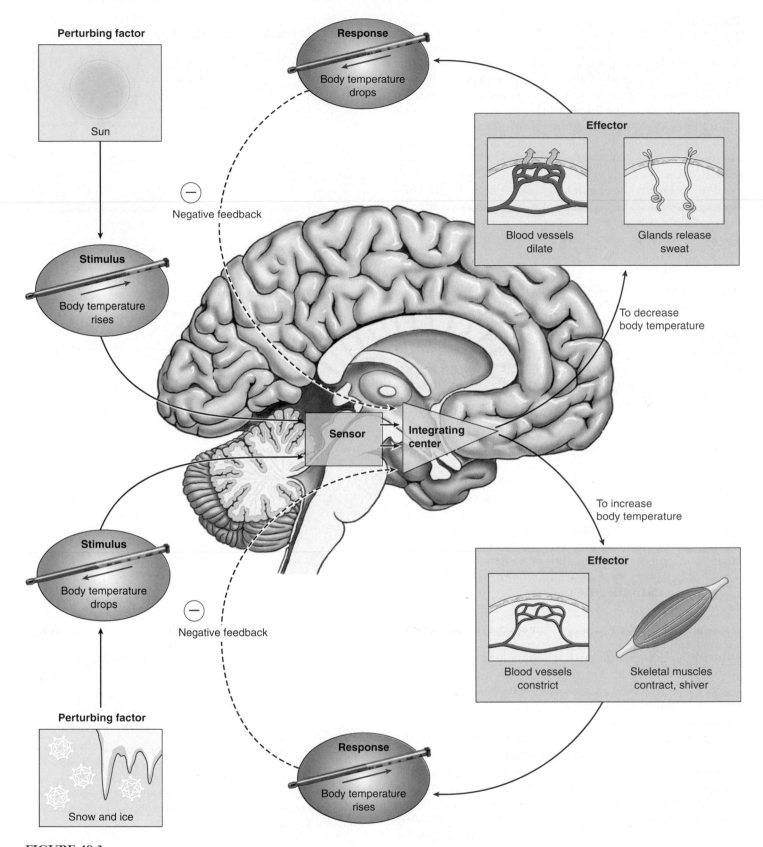

FIGURE 49.3

Negative feedback loops keep the body temperature within a normal range. An increase (*top*) or decrease (*bottom*) in body temperature is sensed by the brain. The integrating center in the brain then processes the information and activates effectors, such as surface blood vessels, sweat glands, and skeletal muscles. When the body temperature returns to normal, negative feedback prevents further stimulation of the effectors by the integrating center.

Regulating Body Temperature

Humans, together with other mammals and with birds, are *endothermic*; they can maintain relatively constant body temperatures independent of the environmental temperature. When the temperature of your blood exceeds 37°C (98.6°F), neurons in a part of the brain called the hypothalamus detect the temperature change. Acting through the control of motor neurons, the hypothalamus responds by promoting the dissipation of heat through sweating, dilation of blood vessels in the skin, and other mechanisms. These responses tend to counteract the rise in body temperature. When body temperature falls, the hypothalamus coordinates a different set of responses, such as shivering and constriction of blood vessels in the skin, which help raise body temperature and correct the initial challenge to homeostasis.

Vertebrates other than mammals and birds are *ectothermic*; their body temperatures are more or less dependent on the environmental temperature. However, to the extent that it is possible, many ectothermic vertebrates attempt to maintain some degree of temperature homeostasis. For example, certain large fish, including tuna, swordfish, and some sharks, can maintain parts of their body at a significantly higher temperature than that of the water. Reptiles attempt to maintain a constant body temperature through behavioral means—by placing themselves in varying locations of sun and shade. That's why you frequently see lizards basking in the sun. Sick lizards even give themselves a "fever" by seeking warmer locations!

Most invertebrates do not employ feedback regulation to physiologically control their body temperature. Instead, they use behavior to adjust their temperature. Many butterflies, for example, must reach a certain body temperature before they can fly. In the cool of the morning, they orient so as to maximize their absorption of sunlight. Moths and many other insects use a shivering reflex to warm their thoracic flight muscles (figure 49.4).

Regulating Blood Glucose   *see also p. 1008*

When you digest a carbohydrate-containing meal, you absorb glucose into your blood. This causes a temporary rise in the blood glucose concentration, which is brought back down in a few hours. What counteracts the rise in blood glucose following a meal?

Glucose levels within the blood are constantly monitored by a sensor, the islets of Langerhans in the pancreas. When levels increase, the islets secrete the hormone *insulin*, which stimulates the uptake of blood glucose into the muscles, liver, and adipose tissue. The islets are, in this case, the sensor and the integrating center. The muscles, liver, and adipose cells are the effectors, taking up glucose to control the levels. The muscles and liver can convert the glucose into the polysaccharide glycogen; adipose cells can convert glucose into fat. These actions lower the blood glucose (figure 49.5) and help to store energy in forms that the body can use later.

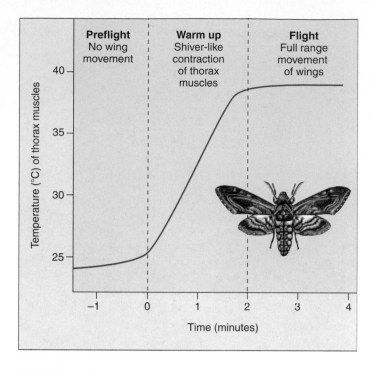

FIGURE 49.4
Thermoregulation in insects. Some insects, such as the sphinx moth, contract their thoracic muscles to warm up for flight. **Why does muscle temperature stop warming after two minutes?**

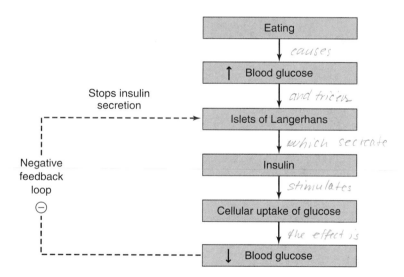

FIGURE 49.5
The negative feedback control of blood glucose. The rise in blood glucose concentration following a meal stimulates the secretion of insulin from the islets of Langerhans in the pancreas. Insulin is a hormone that promotes the entry of glucose in skeletal muscle and other tissue, thereby lowering the blood glucose and compensating for the initial rise.

Negative feedback mechanisms correct deviations from a set point for different internal variables. In this way, body temperature and blood glucose, for example, are kept within a normal range.

Antagonistic Effectors and Positive Feedback

The negative feedback mechanisms that maintain homeostasis often oppose each other to produce a finer degree of control. In a few cases, the body uses positive feedback mechanisms, which push a change further in the same direction.

Antagonistic Effectors

Most factors in the internal environment are controlled by several effectors, which often have antagonistic actions. Control by antagonistic effectors is sometimes described as "push-pull," in which the increasing activity of one effector is accompanied by decreasing activity of an antagonistic effector. This affords a finer degree of control than could be achieved by simply switching one effector on and off.

Room temperature can be maintained, for example, by simply turning an air conditioner on and off, or by just turning a heater on and off. A much more stable temperature, however, can be achieved if the air conditioner and heater are both controlled by a thermostat (figure 49.6). Then the heater is turned on when the air conditioner shuts off, and vice versa. Antagonistic effectors are similarly involved in the control of body temperature and blood glucose. Whereas insulin, for example, lowers blood glucose following a meal, other hormones act to raise the blood glucose concentration between meals, especially when a person is exercising. The heart rate is similarly controlled by antagonistic effectors. Stimulation by sympathetic nerve fibers increases the heart rate, while stimulation by parasympathetic nerve fibers slows the heart rate.

Positive Feedback Loops

Feedback loops that <u>accentuate</u> a disturbance are called positive feedback loops. In a positive feedback loop, <u>perturbations</u> cause the effector to drive the value of the controlled variable even *farther* from the set point. Hence, systems in which there is positive feedback are highly unstable, analogous to a spark that ignites an explosion. <u>They do not help to maintain homeostasis</u>. Nevertheless, such systems are important components of some physiological mechanisms. For example, positive feedback occurs in blood clotting, whereby one clotting factor activates another in a cascade that leads quickly to the formation of a clot. Positive feedback also plays a role in the contractions of the uterus during childbirth (figure 49.7). In this case, stretching of the uterus by the fetus stimulates contraction, and contraction causes further stretching; the cycle continues until the fetus is expelled from the uterus. In the body, most positive feedback systems act as part of some larger mechanism that maintains homeostasis. In the examples we've described, formation of a blood clot stops bleeding and hence tends to keep blood volume constant, and expulsion of the fetus reduces the contractions of the uterus.

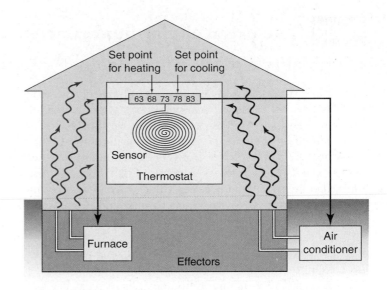

FIGURE 49.6
Room temperature is maintained by antagonistic effectors. If a thermostat senses a low temperature, the heater is turned on, and the air conditioner is turned off. If the temperature is too high, the air conditioner is activated, and the heater is turned off.

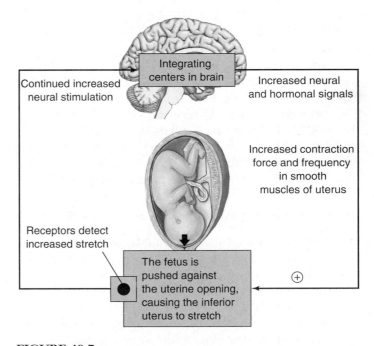

FIGURE 49.7
Positive feedback during childbirth. This is one of the few examples of positive feedback that operate in the vertebrate body.

Effectors that act antagonistically to each other are more effective than effectors that act alone. Positive feedback mechanisms that accentuate changes are less common and have specialized functions in the body.

49.2 The extracellular fluid concentration is constant in most vertebrates.

Osmolality and Osmotic Balance

Water in an animal's body is distributed between the intracellular and extracellular compartments (figure 49.8). In order to maintain osmotic balance, the extracellular compartment of an animal's body (including its blood plasma) must be able to take water from its environment or to excrete excess water into its environment. Inorganic ions must also be exchanged between the extracellular body fluids and the external environment to maintain homeostasis. Such exchanges of water and electrolytes between the body and the external environment occur across specialized epithelial cells and, in most vertebrates, through a filtration process in the kidneys.

Most vertebrates maintain homeostasis in regard to both the total solute concentration of their extracellular fluids and the concentration of specific inorganic ions. Sodium (Na^+) is the major cation in extracellular fluids, and chloride (Cl^-) is the major anion. The divalent cations, calcium (Ca^{++}) and magnesium (Mg^{++}), as well as other ions, also have important functions and must be maintained at their proper concentrations.

Osmolality and Osmotic Pressure

Osmosis is the diffusion of water across a membrane, and it always occurs from a more dilute solution (with a lower solute concentration) to a less dilute solution (with a higher solute concentration). Because the total solute concentration of a solution determines its osmotic behavior, the total moles of solute per kilogram of water are expressed as the **osmolality** of the solution. Solutions that have the same osmolality are *isosmotic*. A solution with a lower or higher osmolality than another is called *hypoosmotic* or *hyperosmotic*, respectively.

If one solution is hyperosmotic compared with another, and if the two solutions are separated by a semipermeable membrane, water may move by osmosis from the more dilute solution to the hyperosmotic one. In this case, the hyperosmotic solution is also **hypertonic** ("higher strength") compared with the other solution, and it has a higher osmotic pressure. The **osmotic pressure** of a solution is a measure of its tendency to take in water by osmosis. An animal cell placed in a hypertonic solution, which has a higher osmotic pressure than the cell cytoplasm, will lose water to the surrounding solution and shrink. In contrast, an animal cell placed in a **hypotonic** solution will gain water and expand.

If a cell is placed in an isosmotic solution, no net water movement may occur. In this case, the isosmotic solution can also be said to be **isotonic.** In medical care, isotonic solutions such as normal saline and 5% dextrose are used to bathe exposed tissues and are given as intravenous fluids.

Osmoconformers and Osmoregulators

Most marine invertebrates are **osmoconformers;** the osmolality of their body fluids is the same as that of seawater (although the concentrations of particular solutes, such as magnesium ion, are not equal). Because the extracellular fluids are isotonic to seawater, there is no osmotic gradient and no tendency for water to leave or enter the body. Therefore, osmoconformers are in osmotic equilibrium with their environment. Among the vertebrates, only the primitive hagfish are strict osmoconformers. The sharks and their relatives in the class Chondrichthyes (cartilaginous fish) are also isotonic to seawater, even though their blood level of NaCl is lower than that of seawater; the difference in total osmolality is made up by retaining urea at a high concentration in their blood plasma.

All other vertebrates are osmoregulators—that is, animals that maintain a relatively constant blood osmolality despite the different concentration in the surrounding environment. The maintenance of a relatively constant body fluid osmolality has permitted vertebrates to exploit a wide variety of ecological niches. Achieving this constancy, however, requires continuous regulation.

Freshwater vertebrates have a much higher solute concentration in their body fluids than that of the surrounding water. In other words, they are hypertonic to their environment. Because of their higher osmotic pressure, water tends to enter their bodies. Consequently, they must prevent water from entering their bodies as much as possible and eliminate the excess water that does enter. In addition, they tend to lose inorganic ions to their environment and so must actively transport these ions back into their bodies.

In contrast, most marine vertebrates are hypotonic to their environment; their body fluids have only about one-third the osmolality of the surrounding seawater. These animals are therefore in danger of losing water by osmosis and must retain water to prevent dehydration. They do this by drinking seawater and eliminating the excess ions through their kidneys and gills.

The body fluids of terrestrial vertebrates have a higher concentration of water than does the air surrounding them. Therefore, they tend to lose water to the air by evaporation from the skin and lungs. All reptiles, birds, and mammals, as well as amphibians during the time when they live on land, face this problem. These vertebrates have evolved urinary/osmoregulatory systems that help them retain water.

Marine invertebrates are isotonic with their environment, but most vertebrates are either hypertonic or hypotonic to their environment and thus tend to gain or lose water. Physiological mechanisms help most vertebrates maintain a constant blood osmolality and constant concentrations of individual ions.

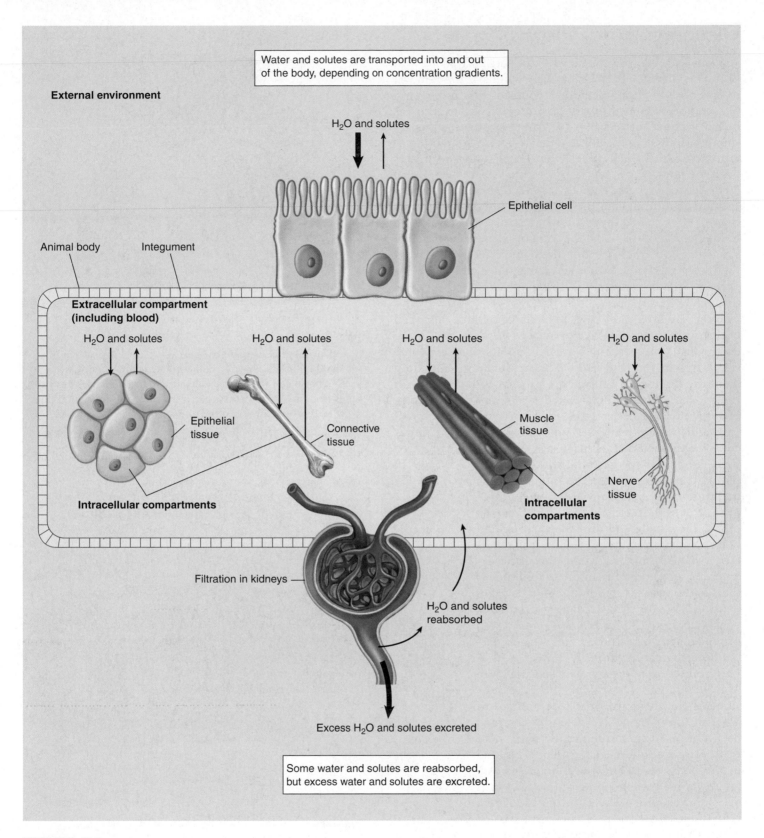

FIGURE 49.8
The interaction between intracellular and extracellular compartments of the body and the external environment. Water can be taken in from the environment or lost to the environment. Exchanges of water and solutes between the extracellular fluids of the body and the environment occur across transport epithelia, and water and solutes can be filtered out of the blood by the kidneys. Overall, the amount of water and solutes that enters and leaves the body must be balanced in order to maintain homeostasis.

Osmoregulatory Organs

Animals have evolved a variety of mechanisms to cope with problems of water balance. In many animals, the removal of water or salts from the body is coupled with the removal of metabolic wastes through the excretory system. Protists employ contractile vacuoles for this purpose, as do sponges. Other multicellular animals have a system of excretory tubules (little tubes) that expel fluid and wastes from the body.

In flatworms, these tubules are called *protonephridia*, and they branch throughout the body into bulblike **flame cells** (figure 49.9). While these simple excretory structures open to the outside of the body, they do not open to the inside of the body. Rather, cilia within the flame cells must draw in fluid from the body. Water and metabolites are then reabsorbed, and the substances to be excreted are expelled through excretory pores.

Other invertebrates have a system of tubules that open both to the inside and to the outside of the body. In the earthworm, these tubules are known as *nephridia* (blue structures in figure 49.10). The nephridia obtain fluid from the body cavity through a process of filtration into funnel-shaped structures called *nephrostomes*. The term *filtration* is used because the fluid is formed under pressure and passes through small openings, so that molecules larger than a certain size are excluded. This filtered fluid is isotonic to the fluid in the coelom, but as it passes through the tubules of the nephridia, NaCl is removed by active transport processes. A general term for transport out of the tubule and into the surrounding body fluids is *reabsorption*. Because salt is reabsorbed from the filtrate, the urine excreted is more dilute than the body fluids—that is, the urine is hypotonic. The kidneys of mollusks and the excretory organs of crustaceans (called *antennal glands*) also produce urine by filtration and reclaim certain ions by reabsorption.

The excretory organs in insects are the Malpighian tubules (figure 49.11), extensions of the digestive tract that branch off anterior to the hindgut. Urine is not formed by filtration in these tubules, because there is no pressure difference between the blood in the body cavity and the tubule. Instead, waste molecules and potassium (K^+) ions are secreted into the tubules by active transport. Secretion is the opposite of reabsorption—ions or molecules are transported from the body fluid into the tubule. The secretion of K^+ creates an osmotic gradient that causes water to enter the tubules by osmosis from the body's open circulatory system. Most of the water and K^+ is then

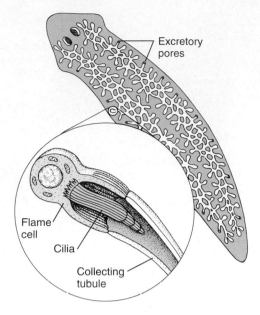

FIGURE 49.9
The protonephridia of flatworms. A branching system of tubules, bulblike flame cells, and excretory pores make up the protonephridia of flatworms. Cilia inside the flame cells draw in fluids from the body by their beating action. Substances are then expelled through pores that open to the outside of the body.

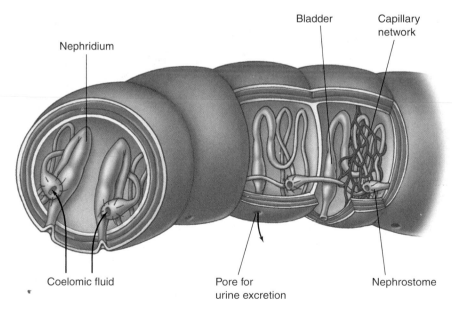

FIGURE 49.10
The nephridia of annelids. Most invertebrates, such as the annelid shown here, have nephridia (*blue*). These consist of tubules that receive a filtrate of coelomic fluid, which enters the funnel-like nephrostomes. Salt can be reabsorbed from these tubules, and the fluid that remains, urine, is released from pores into the external environment.

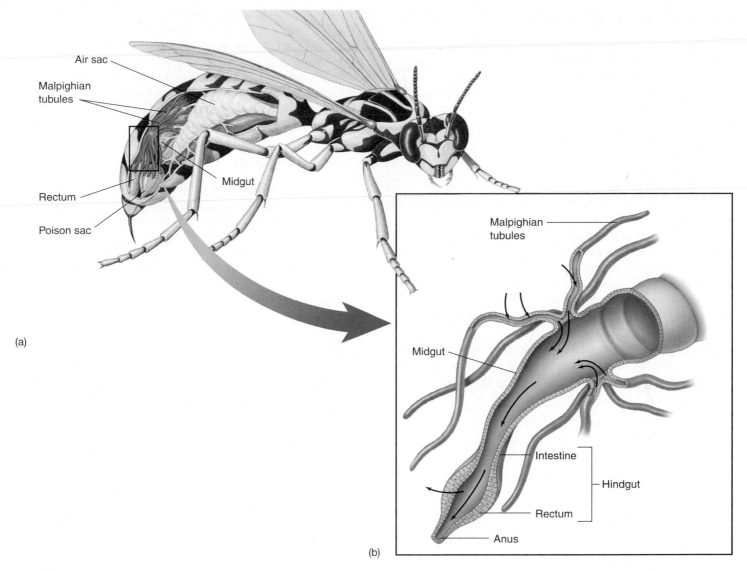

FIGURE 49.11
The Malpighian tubules of insects. (*a*) The Malpighian tubules of insects are extensions of the digestive tract that collect water and wastes from the body's circulatory system. (*b*) K$^+$ is secreted into these tubules, drawing water with it osmotically. Much of this water (see arrows) is reabsorbed across the wall of the hindgut.

reabsorbed into the circulatory system through the epithelium of the hindgut, leaving only small molecules and waste products to be excreted from the rectum along with feces. Malpighian tubules thus provide a very efficient means of water conservation.

The kidneys of vertebrates, unlike the Malpighian tubules of insects, create a tubular fluid by filtering the blood under pressure. In addition to waste products and water, the filtrate contains many small molecules that are of value to the animal, including glucose, amino acids, and vitamins. These molecules and most of the water are reabsorbed from the tubules into the blood, while wastes remain in the filtrate. Additional wastes may be secreted by the tubules and added to the filtrate, and the final waste product, urine, is eliminated from the body.

It may seem odd that the vertebrate kidney should filter out almost everything from blood plasma (except proteins, which are too large to be filtered) and then spend energy to take back or reabsorb what the body needs. But selective reabsorption provides great flexibility, because various vertebrate groups have evolved the ability to reabsorb different molecules that are especially valuable in particular habitats. This flexibility is a key factor underlying the successful colonization of many diverse environments by the vertebrates.

Many invertebrates filter fluid into a system of tubules and then reabsorb ions and water, leaving waste products for excretion. Insects create an excretory fluid by secreting K$^+$ into tubules, which draws water osmotically. The vertebrate kidney produces a filtrate that enters tubules and is modified to become urine.

Evolution of the Vertebrate Kidney

The kidney is a complex organ made up of thousands of repeating units called **nephrons,** each with the structure of a bent tube (figure 49.12). Blood pressure forces the fluid in blood past a filter, called the *glomerulus*, at the top of each nephron. The glomerulus retains blood cells, proteins, and other useful large molecules in the blood but allows the water, and the small molecules and wastes dissolved in it, to pass through and into the bent tube part of the nephron. As the filtered fluid passes through the nephron tube, useful nutrients and ions are recovered from it by both active and passive transport mechanisms, leaving the water and metabolic wastes behind in a fluid urine.

Although the same basic design has been retained in all vertebrate kidneys, a few modifications have occurred. Because the original glomerular filtrate is isotonic to blood, all vertebrates can produce a urine that is isotonic to blood by reabsorbing ions and water in equal proportions. Or they can produce a urine that is hypotonic to blood—more dilute than the blood—by reabsorbing relatively less water. Only birds and mammals can reabsorb enough water from their glomerular filtrate to produce a urine that is hypertonic to blood—more concentrated than the blood—by reabsorbing relatively more water.

Freshwater Fish

Kidneys are thought to have evolved first among the freshwater teleosts, or bony fish. Because the body fluids of a freshwater fish have a greater osmotic concentration than the surrounding water, these animals face two serious problems: (1) Water tends to enter the body from the environment; and (2) solutes tend to leave the body and enter the environment. Freshwater fish address the first problem by *not* drinking water and by excreting a large volume of dilute urine, which is hypotonic to their body fluids. They address the second problem by reabsorbing ions across the nephron tubules, from the glomerular filtrate back into the blood. In addition, they actively transport ions across their gill surfaces from the surrounding water into the blood (figure 49.13, *top*).

Marine Bony Fish

Although most groups of animals seem to have evolved first in the sea, marine bony fish (teleosts) probably evolved from freshwater ancestors, as mentioned in chapter 34. They faced significant new problems in making the transition to the sea because their body fluids are hypotonic to the surrounding seawater. Consequently, water tends to leave their bodies by osmosis across their gills, and they also lose water in their urine. To compensate for this continuous water loss, marine fish drink large amounts of seawater (figure 49.13, *bottom*).

Many of the divalent cations (principally, Ca^{++} and Mg^{++}) in the seawater that a marine fish drinks remain in the digestive tract and are eliminated through the anus. Some, however, are absorbed into the blood, as are the monovalent ions K^+, Na^+, and Cl^-. Most of the monovalent ions are actively transported out of the blood across the gill surfaces, while the divalent ions that enter the blood are secreted into the nephron tubules and excreted in the urine. In these two ways, marine bony fish eliminate the ions they get from the seawater they drink. The urine they excrete is isotonic to their body fluids. It is more concentrated than the urine of freshwater fish, but not as concentrated as that of birds and mammals.

FIGURE 49.12
Organization of the vertebrate nephron. The nephron tubule is a basic design that has been retained in the kidneys of vertebrates. Sugars, amino acids, water, important monovalent ions, and divalent ions are reabsorbed in the proximal arm; water and monovalent ions such as Na^+ and Cl^- are reabsorbed in the loop of Henle; varying amounts of water and monovalent ions (Na^+ and Cl^-) can be reabsorbed in the distal arm and the collecting duct, depending on hormonal influences.

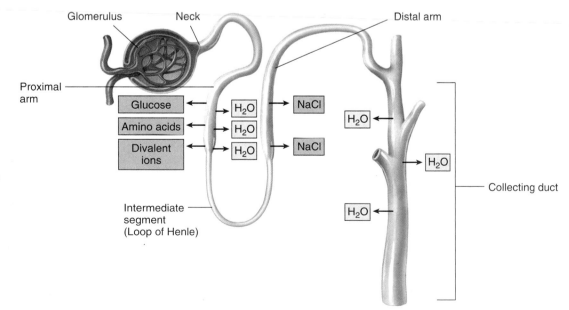

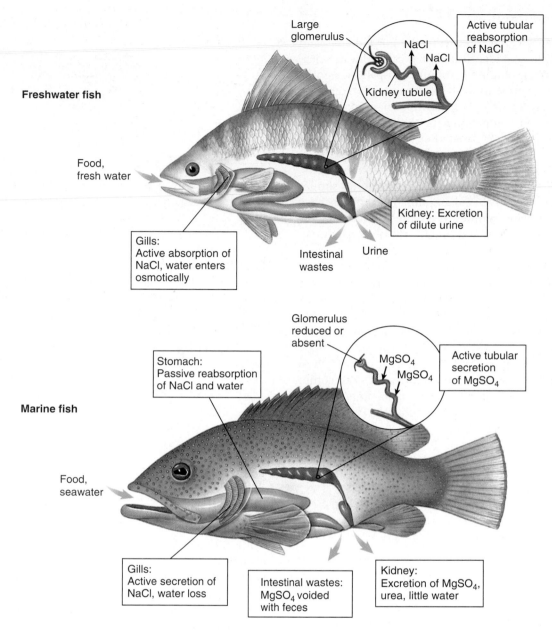

FIGURE 49.13

Freshwater and marine teleosts face different osmotic problems. Whereas the freshwater teleost is hypertonic to its environment, the marine teleost is hypotonic to seawater. To compensate for its tendency to take in water and lose ions, a freshwater fish excretes dilute urine, avoids drinking water, and reabsorbs ions across the nephron tubules. To compensate for its osmotic loss of water, the marine teleost drinks seawater and eliminates the excess ions through active transport across epithelia in the gills and kidneys.

Cartilaginous Fish

The elasmobranchs, including sharks and rays, are by far the most common subclass in the class Chondrichthyes (cartilaginous fish). Elasmobranchs have solved the osmotic problem posed by their seawater environment in a different way than have the bony fish. Instead of having body fluids that are hypotonic to seawater, so that they have to continuously drink seawater and actively pump out ions, the elasmobranchs reabsorb urea from the nephron tubules and maintain a blood urea concentration that is 100 times higher than that of mammals. This added urea makes their blood approximately isotonic to the surrounding sea. Because no net water movement occurs between isotonic solutions, water loss is prevented. Hence, these fishes do not need to drink seawater for osmotic balance, and their kidneys and gills do not have to remove large amounts of ions from their bodies. The enzymes and tissues of the cartilaginous fish have evolved to tolerate the high urea concentrations.

Amphibians and Reptiles

The first terrestrial vertebrates were the amphibians, and the amphibian kidney is identical to that of freshwater fish. This is not surprising, because amphibians spend a significant portion of their time in fresh water, and when on land, they generally stay in wet places. Amphibians produce a very dilute urine and compensate for their loss of Na^+ by actively transporting Na^+ across their skin from the surrounding water.

Reptiles, on the other hand, live in diverse habitats. Those living mainly in fresh water occupy a habitat similar to that of the freshwater fish and amphibians, and thus have similar kidneys. Marine reptiles, including some crocodilians, sea turtles, sea snakes, and one lizard, possess kidneys similar to those of their freshwater relatives but face opposite problems; they tend to lose water and take in salts. Like marine teleosts (bony fish), they drink the seawater and excrete an isotonic urine. Marine teleosts eliminate the excess salt by transport across their gills, while marine reptiles eliminate excess salt through salt glands located near the nose or the eye.

The kidneys of terrestrial reptiles also reabsorb much of the salt and water in their nephron tubules, helping somewhat to conserve blood volume in dry environments. Like fish and amphibians, they cannot produce urine that is more concentrated than the blood plasma. However, they don't really excrete urine, but instead empty it into a cloaca (the common exit of the digestive and urinary tracts) where additional water can be reabsorbed and the wastes excreted with the feces.

Mammals and Birds

Mammals and birds are the only vertebrates able to produce urine with a higher osmotic concentration than their body fluids. This allows these vertebrates to excrete their waste products in a small volume of water, so that more water can be retained in the body. Human kidneys can produce urine that is as much as 4.2 times as concentrated as blood plasma, but the kidneys of some other mammals are even more efficient at conserving water. For example, camels, gerbils, and pocket mice of the genus *Perognathus* can excrete urine 8, 14, and 22 times as concentrated as their blood plasma, respectively. The kidneys of the kangaroo rat are so efficient that it never has to drink water; it can obtain all the water it needs from its food and from water produced in aerobic cellular respiration!

The production of hypertonic urine is accomplished by the *loop of Henle* portion of the nephron (see figure 49.12), found only in mammals and birds. A nephron with a long loop of Henle extends deeper into the renal medulla, where the hypertonic osmotic environment draws out

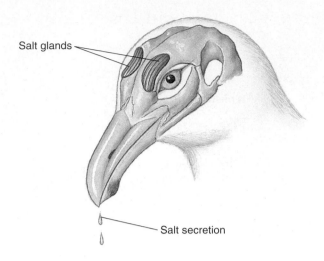

Salt glands

Salt secretion

FIGURE 49.14
How marine birds cope with excess salt. Marine birds drink seawater and then excrete the salt through salt glands. The extremely salty fluid excreted by these glands can then dribble down the beak.

more water, and so can produce more concentrated urine. Most mammals have some nephrons with short loops and other nephrons with much longer loops (see figure 49.16). Birds, however, have relatively few or no nephrons with long loops, so they cannot produce urine that is as concentrated as that of mammals. At most, they can only reabsorb enough water to produce a urine that is about twice the concentration of their blood. Marine birds solve the problem of water loss by drinking salt water and then excreting the excess salt from salt glands near the eyes (figure 49.14).

The moderately hypertonic urine of a bird is delivered to its cloaca, along with the fecal material from its digestive tract. If needed, additional water can be absorbed across the wall of the cloaca to produce a semisolid white paste or pellet, which is excreted.

The kidneys of freshwater fish must excrete copious amounts of very dilute urine, while marine teleosts drink seawater and excrete an isotonic urine. The basic design and function of the nephron of freshwater fishes have been retained in the terrestrial vertebrates. Modifications, particularly the presence of a loop of Henle, allow mammals and birds to reabsorb more water and produce a urine hypertonic to their body fluids.

Ammonia, Urea, and Uric Acid

Amino acids and nucleic acids are nitrogen-containing molecules. When animals catabolize these molecules for energy or convert them into carbohydrates or lipids, they produce nitrogen-containing by-products called **nitrogenous wastes** (figure 49.15) that must be eliminated from the body.

The first step in the metabolism of amino acids and nucleic acids is the removal of the amino ($-NH_2$) group and its combination with H^+ to form **ammonia** (NH_3) in the liver. Ammonia is quite toxic to cells and therefore is safe only in very dilute concentrations. The excretion of ammonia is not a problem for the bony fish and tadpoles, which eliminate most of it by diffusion through the gills and less by excretion in very dilute urine. In elasmobranchs, adult amphibians, and mammals, the nitrogenous wastes are eliminated in the far less toxic form of **urea.** Urea is water-soluble and so can be excreted in large amounts in the urine. It is carried in the bloodstream from its place of synthesis in the liver to the kidneys where it is excreted in the urine.

Reptiles, birds, and insects excrete nitrogenous wastes in the form of **uric acid,** which is only slightly soluble in water. As a result of its low solubility, uric acid precipitates and thus can be excreted using very little water. Uric acid forms the pasty white material in bird droppings called *guano.* The ability to synthesize uric acid in these groups of animals is also important because their eggs are encased within shells, and nitrogenous wastes build up as the embryo grows within the egg. The formation of uric acid, while a lengthy process that requires considerable energy, produces a compound that crystallizes and precipitates. As a precipitate, it is unable to affect the embryo's development even though it is still inside the egg.

Mammals also produce some uric acid, but it is a waste product of the degradation of purine nucleotides (see chapter 3), not of amino acids. Most mammals have an enzyme called *uricase,* which converts uric acid into a more soluble derivative, **allantoin.** Only humans, apes, and the dalmatian dog lack this enzyme and so must excrete the uric acid. In humans, excessive accumulation of uric acid in the joints produces a condition known as *gout.*

The metabolic breakdown of amino acids and nucleic acids produces ammonia as a by-product. Ammonia is excreted by bony fish, but other vertebrates convert nitrogenous wastes into urea and uric acid, which are less toxic nitrogenous wastes.

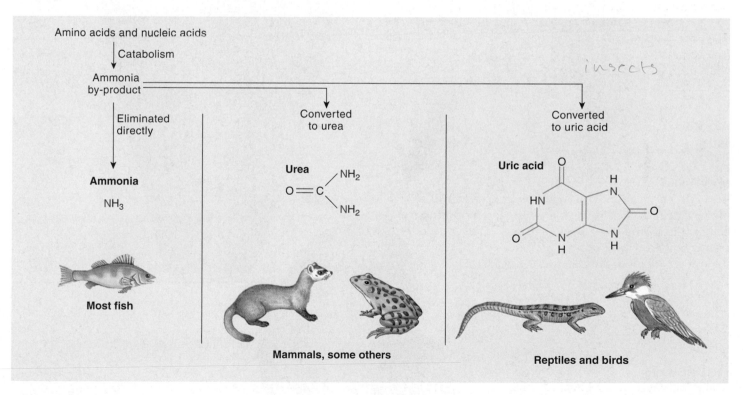

FIGURE 49.15
Nitrogenous wastes. When amino acids and nucleic acids are metabolized, the immediate by-product is ammonia, which is quite toxic but can be eliminated through the gills of teleost fish. Mammals convert ammonia into urea, which is less toxic. Birds and terrestrial reptiles convert it instead into uric acid, which is insoluble in water.

49.3 The functions of the vertebrate kidney are performed by nephrons.

The Mammalian Kidney

In humans, the kidneys are fist-sized organs located in the region of the lower back. Each kidney receives blood from a renal artery, and from this blood, urine is produced. Urine drains from each kidney through a **ureter**, which carries the urine to a **urinary bladder.** Within the kidney, the mouth of the ureter flares open to form a funnel-like structure, the *renal pelvis.* The renal pelvis, in turn, has cup-shaped extensions that receive urine from the renal tissue. This tissue is divided into an outer **renal cortex** and an inner **renal medulla** (figure 49.16). Together, these structures perform filtration, reabsorption, secretion, and excretion.

Nephron Structure and Filtration

On a microscopic level, each kidney contains about one million functioning *nephrons.* Mammalian kidneys contain a mixture of juxtamedullary nephrons, which have long loops that dip deeply into the medulla, and cortical nephrons with shorter loops. The significance of the length of the loops will be explained a little later.

Each nephron consists of a long tubule and associated small blood vessels. First, blood is carried by an *afferent arteriole* to a tuft of capillaries in the renal cortex, the **glomerulus** (figure 49.17). Here the blood is filtered as the blood pressure forces fluid through the porous capillary

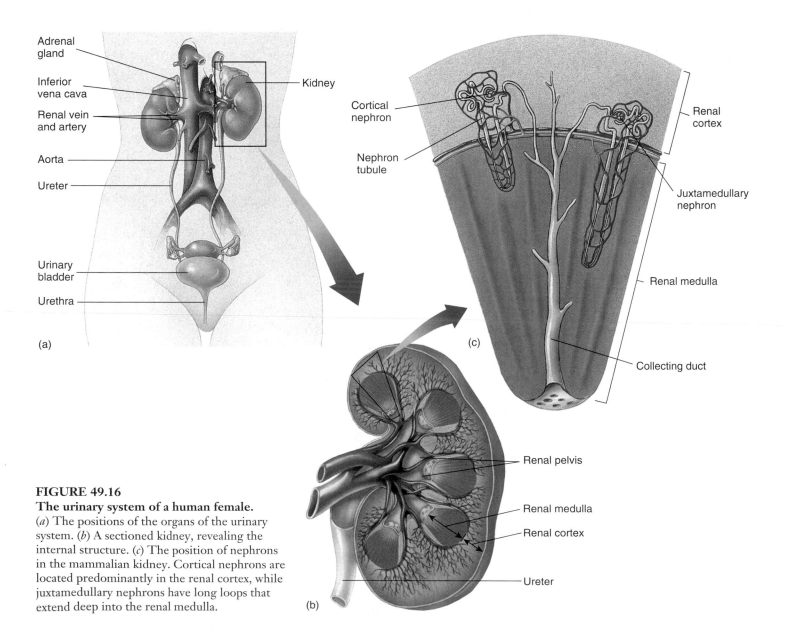

FIGURE 49.16
The urinary system of a human female.
(*a*) The positions of the organs of the urinary system. (*b*) A sectioned kidney, revealing the internal structure. (*c*) The position of nephrons in the mammalian kidney. Cortical nephrons are located predominantly in the renal cortex, while juxtamedullary nephrons have long loops that extend deep into the renal medulla.

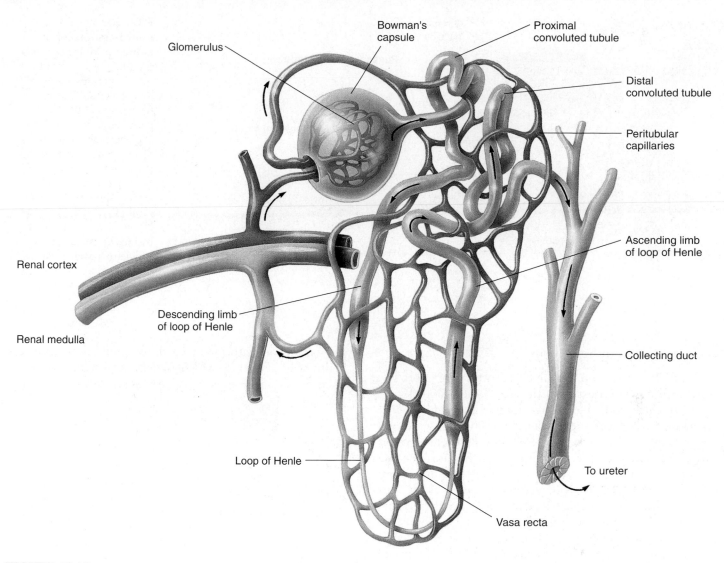

FIGURE 49.17

A nephron in a mammalian kidney. The nephron tubule is surrounded by peritubular capillaries in the cortex, and their vasa recta extensions surround the loop of Henle in the medulla. This capillary bed carries away molecules and ions that are reabsorbed from the filtrate.

walls. Blood cells and plasma proteins are too large to enter this **glomerular filtrate,** but large amounts of the plasma, consisting of water and dissolved molecules, leave the vascular system at this step. The filtrate immediately enters the first region of the nephron tubules. This region, *Bowman's capsule,* envelops the glomerulus much as a large, soft balloon surrounds your hand if you press your fist into it. The capsule has slit openings so that the glomerular filtrate can enter the system of nephron tubules.

After the filtrate enters Bowman's capsule, it goes into a portion of the nephron called the **proximal convoluted tubule,** located in the cortex. In a cortical nephron, the fluid then flows through a loop of Henle that dips only minimally into the medulla before ascending back into the cortex. If it is a juxtamedullary nephron, the loop of Henle extends much deeper into the medulla before ascending back up into the cortex. More water can be reabsorbed from juxtamedullary

nephrons than from cortical nephrons. The fluid then moves deeper into the medulla and back up again into the cortex in a **loop of Henle.** As mentioned in section 49.2, only the kidneys of mammals and birds have loops of Henle, and this is why only birds and mammals have the ability to concentrate their urine. After leaving the loop, the fluid is delivered to a **distal convoluted tubule** in the cortex that next drains into a **collecting duct.** The collecting duct again descends into the medulla, where it merges with other collecting ducts to empty its contents, now called urine, into the renal pelvis.

Blood components that were not filtered out of the glomerulus drain into an *efferent arteriole,* which then empties into a second bed of capillaries called *peritubular capillaries* that surround the tubules. This is only one of several locations in the body where two capillary beds occur in series. As described later, the peritubular capillaries are needed for the processes of reabsorption and secretion.

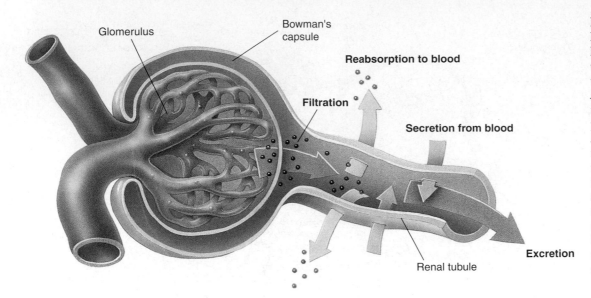

Glomerulus

Bowman's
capsule

Reabsorption to blood

Filtration

Secretion from blood

Excretion

Renal tubule

FIGURE 49.18
Four functions of the kidney. Molecules enter the urine by *filtration* out of the glomerulus and by *secretion* into the tubules from surrounding peritubular capillaries. Molecules that entered the filtrate can be returned to the blood by *reabsorption* from the tubules into surrounding peritubular capillaries, or they may be eliminated from the body by *excretion* through the tubule to a ureter and then to the bladder.

Reabsorption and Secretion

Most of the water and dissolved solutes that enter the glomerular filtrate must be returned to the blood (figure 49.18), or the animal would literally urinate to death. In a human, for example, approximately 2000 liters of blood pass through the kidneys each day, and 180 liters of water leave the blood and enter the glomerular filtrate. Because we only have a total blood volume of about 5 liters and only produce 1 to 2 liters of urine per day, it is obvious that each liter of blood is filtered many times per day, and most of the filtered water is reabsorbed. The reabsorption of water occurs as a consequence of salt (NaCl) reabsorption through mechanisms that will be described shortly.

The reabsorption of glucose, amino acids, and many other molecules needed by the body is driven by active transport and secondary active transport (cotransport) carriers. As in all carrier-mediated transport, a maximum rate of transport is reached whenever the carriers are saturated (see chapter 6). For the renal glucose carriers, saturation occurs when the concentration of glucose in the blood (and thus in the glomerular filtrate) is about 180 milligrams per 100 milliliters of blood. If a person has a blood glucose concentration in excess of this amount, as happens in untreated diabetes mellitus, the glucose left untransported in the filtrate is expelled in the urine. Indeed, the presence of glucose in the urine is diagnostic of diabetes mellitus.

The secretion of foreign molecules and particular waste products of the body involves the transport of these molecules across the membranes of the blood capillaries and kidney tubules into the filtrate. This process is similar to reabsorption, but it proceeds in the opposite direction. Some secreted molecules are eliminated in the urine so rapidly that they may be cleared from the blood in a sin-

gle pass through the kidneys. This rapid elimination explains why penicillin, which is secreted by the nephrons, must be administered in very high doses and several times per day.

Excretion

A major function of the kidney is the elimination of a variety of potentially harmful substances that animals eat and drink. In addition, urine contains nitrogenous wastes, such as urea and uric acid, that are products of the catabolism of amino acids and nucleic acids. Urine may also contain excess K^+, H^+, and other ions that are removed from the blood. Urine's generally high H^+ concentration (pH 5 to 7) helps maintain the acid-base balance of the blood within a narrow range (pH 7.35 to 7.45). Moreover, the excretion of water in urine contributes to the maintenance of blood volume and pressure; the larger the volume of urine excreted, the lower the blood volume.

The purpose of kidney function is therefore homeostasis; the kidneys are critically involved in maintaining the constancy of the internal environment. When disease interferes with kidney function, it causes a rise in the blood concentration of nitrogenous waste products, disturbances in electrolyte and acid-base balance, and a failure in blood pressure regulation. Such potentially fatal changes highlight the central importance of the kidneys in normal body physiology.

The mammalian kidney is divided into a cortex and medulla and contains about a million functional units called nephrons. The nephron tubules receive a blood filtrate from the glomeruli and modify this filtrate to produce urine, which empties into the renal pelvis and is expelled from the kidney through the ureter.

Transport Processes in the Mammalian Nephron

As previously described, in a human approximately 180 liters of isotonic glomerular filtrate enter the Bowman's capsules each day. After passing through the remainder of the nephron tubules, this volume of fluid would be lost as urine if it were not reabsorbed back into the blood. It is clearly impossible to produce this much urine, yet water is only able to pass through a cell membrane by osmosis, and osmosis is not possible between two isotonic solutions. Therefore, some mechanism is needed to create an osmotic gradient between the glomerular filtrate and the blood, allowing reabsorption.

Proximal Convoluted Tubule

Virtually all the nutrient molecules in the filtrate are reabsorbed back into the systemic blood by the proximal convoluted tubule. In addition, approximately two-thirds of the NaCl and water filtered into Bowman's capsule is immediately reabsorbed across the walls of the proximal convoluted tubule. This reabsorption is driven by the active transport of Na^+ out of the filtrate and into surrounding peritubular capillaries. Cl^- follows Na^+ passively because of electrical attraction, and water follows them both because of osmosis. Because NaCl and water are removed from the filtrate in proportionate amounts, the filtrate that remains in the tubule is still isotonic to the blood plasma.

Although only one-third of the initial volume of filtrate remains in the nephron tubule after the initial reabsorption of NaCl and water, it still represents a large volume (60 L out of the original 180 L of filtrate produced per day by both human kidneys). Obviously, no animal can afford to excrete that much urine, so most of this water must also be reabsorbed. It is reabsorbed primarily across the wall of the collecting duct because the interstitial fluid of the renal medulla surrounding the collecting ducts is hypertonic. The hypertonic renal medulla draws water out of the collecting duct by osmosis, leaving behind a hypertonic urine for excretion.

Loop of Henle

The reabsorption of much of the water in the tubular filtrate thus depends on the creation of a hypertonic renal medulla; the more hypertonic the medulla is, the steeper the osmotic gradient will be and the more water will leave the collecting ducts. It is the loops of Henle that create the hypertonic renal medulla in the following manner (figure 49.19):

1. The descending limb of the loop is permeable to water, so water leaves by osmosis as the fluid descends into the hypertonic renal medulla. This water enters the blood vessels of the vasa recta, which are looplike extensions of the peritubular capillaries descending into the medulla, and is carried away in the general circulation.

2. The loss of water from the descending limb multiplies the concentration that can be achieved at each level of the loop through the active extrusion of NaCl by the ascending limb. The longer the loop of Henle, the longer the

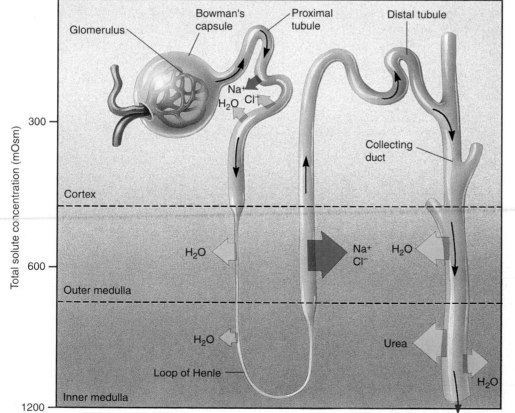

FIGURE 49.19
The reabsorption of salt and water in the mammalian kidney. Active transport of Na^+ out of the proximal tubules is followed by the passive movement of Cl^- and water. Active extrusion of NaCl from the ascending limb of the loop of Henle creates the osmotic gradient required for the reabsorption of water from the descending limb of the loop of Henle and the collecting duct. The changes in osmolality from the cortex to the medulla are indicated to the left of the figure.

region of interaction between the descending and ascending limbs, and the greater the total concentration that can be achieved. In a human kidney, the concentration of filtrate entering the loop is 300 milliosmolal (mOsm), and this concentration is multiplied to more than 1200 milliosmolal at the bottom of the longest loops of Henle in the renal medulla.

3. The ascending limb of the loop actively extrudes Na^+, and Cl^- follows. The mechanism that extrudes NaCl from the ascending limb of the loop differs from that which extrudes NaCl from the proximal tubule, but the most important difference is that the ascending limb is *not permeable to water*. As Na^+ exits, the fluid within the ascending limb becomes increasingly dilute (hypotonic) as it enters the cortex, while the surrounding tissue becomes increasingly concentrated (hypertonic).

4. The NaCl pumped out of the ascending limb of the loop is reabsorbed from the surrounding interstitial fluid into the loops of the *vasa recta*, so that NaCl can diffuse from the blood leaving the medulla to the blood entering the medulla. Thus, the vasa recta functions in a countercurrent exchange, similar to that described for oxygen in the countercurrent flow of water and blood in the gills of fish (see chapter 44). In the case of the renal medulla, the diffusion of NaCl between the blood vessels keeps much of the NaCl within the interstitial fluid, making it hypertonic.

Because fluid flows in opposite directions in the two limbs of the loop, the action of the loop of Henle in creating a hypertonic renal medulla is known as the *countercurrent multiplier system*. The high solute concentration of the renal medulla is primarily the result of NaCl accumulation by the countercurrent multiplier system, but urea also contributes to the total osmolality of the medulla. This is because the descending limb of the loop of Henle and the collecting duct are permeable to urea, which leaves these regions of the nephron by diffusion.

Distal Convoluted Tubule and Collecting Duct

Because NaCl was pumped out of the ascending limb, the filtrate that arrives at the distal convoluted tubule and enters the collecting duct in the renal cortex is hypotonic (with a concentration of only 100 mOsm). The reabsorption of NaCl in these two regions depends on the needs of the body and is under the control of aldosterone. Both ADH and aldosterone influence the distal convoluted tubule and collecting duct, although the latter is more significant. The collecting duct carrying this dilute fluid now plunges into the medulla. As a result of the hypertonic interstitial fluid of the renal medulla, there is a strong osmotic gradient that pulls water out of the collecting duct and into the surrounding blood vessels.

The osmotic gradient is normally constant, but the permeability of the distal convoluted tubule and the col-

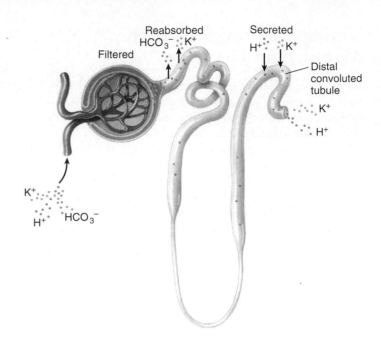

FIGURE 49.20
Controlling salt balance. The nephron controls the amounts of K^+, H^+, and HCO_3^- excreted in the urine. K^+ is completely reabsorbed in the proximal tubule and then secreted in hormonally regulated amounts into the distal tubule. HCO_3^- is filtered but normally completely reabsorbed. H^+ is filtered and also secreted into the distal tubule, so that the final urine has an acidic pH.

lecting duct to water is adjusted by a hormone, **antidiuretic hormone** (**ADH,** also called **vasopressin**), discussed in chapters 44 and 47. When an animal needs to conserve water, the posterior pituitary gland secretes more ADH, and this hormone increases the number of water channels in the plasma membranes of the collecting duct cells. This increases the permeability of the collecting ducts to water so that more water is reabsorbed and less is excreted in the urine. The animal thus excretes a hypertonic urine.

In addition to regulating water balance, the kidneys regulate the balance of electrolytes in the blood by reabsorption and secretion. For example, the kidneys reabsorb K^+ in the proximal tubule and then secrete an amount of K^+ needed to maintain homeostasis into the distal convoluted tubule (figure 49.20). The kidneys also maintain acid-base balance by excreting H^+ into the urine and reabsorbing bicarbonate (HCO_3^-).

The loop of Henle creates a hypertonic renal medulla as a result of the active extrusion of NaCl from the ascending limb and the interaction with the descending limb. The hypertonic medulla then draws water osmotically from the distal convoluted tubule and collecting duct, which are permeable to water under the influence of antidiuretic hormone.

Hormones Control Osmoregulatory Functions

In mammals and birds, the amount of water excreted in the urine, and thus the concentration of the urine, varies according to the changing needs of the body. Acting through the mechanisms described next, the kidneys excrete a hypertonic urine when the body needs to conserve water. If an animal drinks excess water, the kidneys excrete a hypotonic urine. As a result, the volume of blood, the blood pressure, and the osmolality of blood plasma are maintained relatively constant by the kidneys, no matter how much water you drink. The kidneys also regulate the plasma K^+ and Na^+ concentrations and blood pH within very narrow limits. These homeostatic functions of the kidneys are coordinated primarily by hormones (see chapter 47).

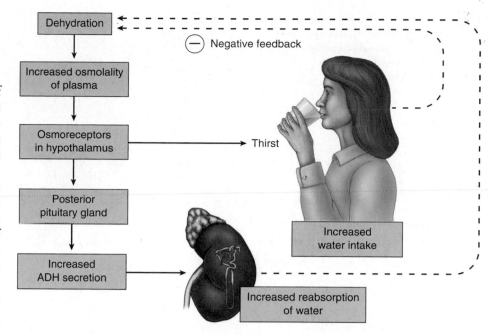

FIGURE 49.21
Antidiuretic hormone stimulates the reabsorption of water by the kidneys. This action completes a negative feedback loop and helps to maintain homeostasis of blood volume and osmolality.

Antidiuretic Hormone

Antidiuretic hormone (ADH) is produced by the hypothalamus and secreted by the posterior pituitary gland. The primary stimulus for ADH secretion is an increase in the osmolality of the blood plasma. The osmolality of plasma increases when a person is dehydrated or when a person eats salty food. Osmoreceptors in the hypothalamus respond to the elevated blood osmolality by sending more nerve signals to the integration center (also in the hypothalamus). This, in turn, triggers a sensation of thirst and an increase in the secretion of ADH (figure 49.21).

ADH causes the walls of the distal convoluted tubules and collecting ducts in the kidney to become more permeable to water. This occurs because water channels are contained within the membranes of intracellular vesicles in the epithelium of the distal convoluted tubules and collecting ducts, and ADH stimulates the fusion of the vesicle membrane with the plasma membrane, similar to the process of exocytosis. When the secretion of ADH is reduced, the plasma membrane pinches in to form new vesicles that contain the water channels, so that the plasma membrane becomes less permeable to water.

Because the extracellular fluid in the renal medulla is hypertonic to the filtrate in the collecting ducts, water leaves the filtrate by osmosis and is reabsorbed into the blood. Under conditions of maximal ADH secretion, a person excretes only 600 milliliters of highly concentrated urine per day. A person who lacks ADH due to pituitary damage has the disorder known as *diabetes insipidus* and constantly excretes a large volume of dilute urine. Such a person is in danger of becoming severely dehydrated and succumbing to dangerously low blood pressure.

Aldosterone and Atrial Natriuretic Hormone

Sodium ion is the major solute in the blood plasma. When the blood concentration of Na^+ falls, therefore, the blood osmolality also falls. This drop in osmolality inhibits ADH secretion, causing more water to remain in the collecting duct for excretion in the urine. As a result, the blood volume and blood pressure decrease. A decrease in extracellular Na^+ also causes more water to be drawn into cells by osmosis, partially offsetting the drop in plasma osmolality but further decreasing blood volume and blood pressure. If Na^+ deprivation is severe, the blood volume may fall so low that blood pressure is insufficient to sustain life. For this reason, salt is necessary for life. Many animals have a "salt hunger" and actively seek salt, such as the deer at "salt licks."

A drop in blood Na^+ concentration is normally compensated by the kidneys under the influence of the hormone aldosterone, which is secreted by the adrenal cortex. Aldosterone stimulates the distal convoluted tubules and collecting ducts to reabsorb Na^+, decreasing the excretion of Na^+ in the urine. Indeed, under conditions of maximal aldosterone secretion, Na^+ may be completely absent from the urine. The reabsorption of Na^+ is followed by Cl^- and

FIGURE 49.22

A lowering of blood volume activates the renin-angiotensin-aldosterone system. (*1*) Low blood volume and a decrease in blood Na⁺ levels reduce blood pressure. (*2*) Reduced blood flow past the juxtaglomerular apparatus triggers (*3*) the release of renin into the blood, which catalyzes the production of angiotensin I from angiotensinogen. (*4*) Angiotensin I converts into a more active form, angiotensin II. (*5*) Angiotensin II stimulates blood vessel constriction and (*6*) the release of aldosterone from the adrenal cortex. (*7*) Aldosterone stimulates the reabsorption of Na⁺ in the distal convoluted tubules. (*8*) Increased Na⁺ reabsorption is followed by the reabsorption of Cl⁻ and water. (*9*) This increases blood volume. An increase in blood volume may also trigger the release of atrial natriuretic hormone, which inhibits the release of aldosterone. These two systems work together to maintain homeostasis.

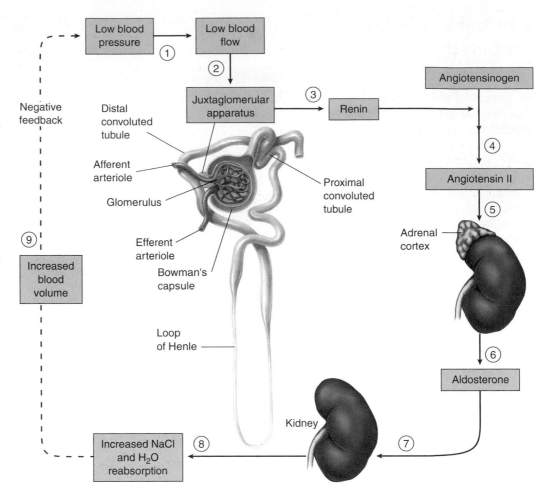

by water, so aldosterone has the net effect of promoting the retention of both salt and water. It thereby helps to maintain blood volume, osmolality, and pressure.

The secretion of aldosterone in response to a decreased blood level of Na⁺ is indirect. Because a fall in blood Na⁺ is accompanied by decreased blood volume, there is a reduced flow of blood past a group of cells called the juxtaglomerular apparatus, located in the region of the kidney between the distal convoluted tubule and the afferent arteriole (figure 49.22). The juxtaglomerular apparatus responds by secreting the enzyme *renin* into the blood, which catalyzes the production of the polypeptide angiotensin I from the protein angiotensinogen. Angiotensin I is then converted by another enzyme into angiotensin II, which stimulates blood vessels to constrict and the adrenal cortex to secrete aldosterone. Thus, homeostasis of blood volume and pressure can be maintained by the activation of this renin-angiotensin-aldosterone system.

In addition to stimulating Na⁺ reabsorption, aldosterone also promotes the secretion of K⁺ into the distal convoluted tubules and collecting ducts. Consequently, aldosterone lowers the blood K⁺ concentration, helping to maintain constant blood K⁺ levels in the face of changing

amounts of K⁺ in the diet. People who lack the ability to produce aldosterone will die if untreated because of the excessive loss of salt and water in the urine and the buildup of K⁺ in the blood.

The action of aldosterone in promoting salt and water retention is opposed by another hormone, atrial natriuretic hormone (ANH, see chapter 44). This hormone is secreted by the right atrium of the heart in response to an increased blood volume, which stretches the atrium. Under these conditions, aldosterone secretion from the adrenal cortex will decrease, and atrial natriuretic hormone secretion will increase, thus promoting the excretion of salt and water in the urine and lowering the blood volume.

ADH stimulates the insertion of water channels into the cells of the distal convoluted tubule and collecting duct, making them more permeable to water. Thus, ADH stimulates the reabsorption of water and the excretion of a hypertonic urine. Aldosterone promotes the reabsorption of NaCl and water across the distal convoluted tubule and collecting duct, as well as the secretion of K⁺ into the tubules. ANH decreases NaCl reabsorption.

49.1 The regulatory systems of the body maintain homeostasis.

The Need to Maintain Homeostasis

- Homeostasis refers to dynamic constancy of the internal environment around set points. (p. 1040)
- A negative feedback loop occurs as the activity of effectors is negatively influenced by the effects they produce. (p. 1040)
- Endothermic organisms can maintain a relatively constant body temperature independent of the environmental temperature. The body temperature of ectothermic organisms is basically dependent on the environmental temperature. (p. 1042)

Antagonistic Effectors and Positive Feedback

- Control by antagonistic effectors occurs when the increasing activity of one effector is accompanied by the decreasing activity of an antagonistic effector, affording a fine degree of control. (p. 1043)
- In a positive feedback loop, perturbations cause the effector to drive the value of the controlled variable even farther from the set point. (p. 1043)

49.2 The extracellular fluid concentration is constant in most vertebrates.

Osmolality and Osmotic Balance

- An animal's body must be able to take water from its environment or excrete excess water into its environment in order to maintain osmolality (total moles of solute per kilogram of water). (p. 1044)
- An animal cell placed in a hypertonic solution tends to lose water, while an animal cell placed in a hypotonic solution tends to gain water. (p. 1044)
- Osmoconformers have body fluids with osmolality equal to that of the surrounding environment (seawater). But osmoregulators maintain a relatively constant blood osmolality despite a different concentration in the surrounding environment. (p. 1044)

Osmoregulatory Organs

- Many invertebrates filter fluids into tubules where water and ions are reabsorbed, leaving waste products to be eliminated. (p. 1046)
- Insects create an osmotic gradient by secreting K^+ into Malpighian tubules, causing water to enter the tubules by osmosis. Water and K^+ are reabsorbed through the epithelium of the hind gut. (pp. 1046–1047)
- Vertebrates create a tubular fluid, urine, by pressurized blood filtration in the kidney. (p. 1047)

Evolution of the Vertebrate Kidney

- The kidney is made of thousands of repeating nephrons. (p. 1048)
- Freshwater fish excrete large amounts of dilute urine, while marine bony fish (teleosts) drink seawater and excrete isotonic urine. (p. 1048)
- Multiple modifications of the nephron, including the loop of Henle, allow birds and mammals to reabsorb water and produce hypertonic urine. (p. 1048)

Ammonia, Urea, and Uric Acid

- Metabolic breakdown of amino acids and nucleic acids produces ammonia in the liver. (p. 1051)
- Reptiles, birds, and insects excrete nitrogenous wastes as low-toxicity uric acid, which is only slightly soluble in water and can be excreted using small amounts of water. (p. 1051)
- Most mammals eliminate nitrogenous wastes in the form of urea, which is water soluble. (p. 1051)

49.3 The functions of the vertebrate kidney are performed by nephrons.

The Mammalian Kidney

- Each kidney receives blood from a renal artery and produces urine, which then drains from each kidney through a ureter and is carried to the urinary bladder. (p. 1052)
- Each kidney contains about one million nephrons, and is divided into the cortex and medulla. (p. 1052)
- Excretion of water in urine eliminates a variety of potentially harmful substances and helps maintain blood volume and pressure. (p. 1054)

Transport Processes in the Mammalian Nephron

- The loop of Henle produces a hypertonic renal medulla due to NaCl extrusion from the ascending limb and interaction with the descending limb. (p. 1055)
- The hypertonic medulla draws water osmotically from the distal convoluted tubule and collecting duct, which are permeable to water under the influence of antidiuretic hormone. (p. 1056)

Hormones Control Osmoregulatory Functions

- The kidneys excrete a hypertonic urine when the body needs to conserve water and a hypotonic urine if the body contains too much water. (p. 1057)
- ADH stimulates the reabsorption of water and the excretion of hypertonic urine. (p. 1058)
- Aldosterone promotes NaCl and water reabsorption across the distal convoluted tubule and collecting ducts, as well as K^+ secretion into the tubules. (p. 1058)

Self Test

1. Which of the following is *not* a method used in maintaining homeostasis in the body?
   a. behavioral changes
   b. negative feedback loops
   c. hormonal actions
   d. positive feedback loops
2. Which of the following is an osmoconformer?
   a. mammals
   b. hagfish
   c. an animal that maintains a relatively constant blood osmolality independent of its surroundings
   d. all of these
3. Which of the following animals use Malpighian tubules for excretion?
   a. ants
   b. birds
   c. mammals
   d. earthworms
4. A shark's blood is isotonic to the surrounding seawater because of the reabsorption of _____ in its blood.
   a. ammonia
   b. uric acid
   c. urea
   d. NaCl
5. Which of the following animals has the least concentrated urine relative to its blood plasma?
   a. bird
   b. freshwater fish
   c. human
   d. camel
6. Which of the following is a function of the kidneys?
   a. The kidneys remove harmful substances from the body.
   b. The kidneys recapture water for use by the body.
   c. The kidneys regulate the levels of salt in the blood.
   d. All of these are functions of the kidneys.
7. The longer loops of Henle and the collecting ducts are located in the
   a. renal cortex.
   b. renal medulla.
   c. renal pelvis.
   d. ureter.
8. Selective reabsorption of components of the glomerular filtrate occurs where?
   a. Bowman's capsule
   b. glomerulus
   c. loop of Henle
   d. collecting duct
9. Humans excrete their excess nitrogenous wastes as
   a. uric acid crystals.
   b. compounds containing protein.
   c. very toxic ammonia.
   d. relatively nontoxic urea.
10. Which of the following statements is *not* true?
    a. ADH makes the collecting duct more permeable to water.
    b. Guano contains high concentrations of uric acid.
    c. Aldosterone is produced by the hypothalamus in response to high levels of sodium ions in the blood.
    d. Uric acid is the least soluble of the nitrogenous waste products.

Test Your Visual Understanding

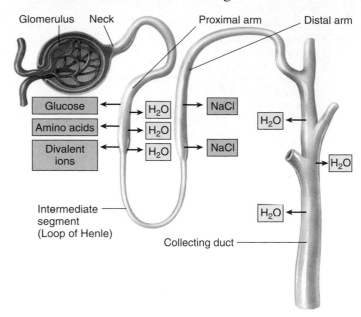

1. Indicate the areas of the nephron that the following hormones target, and describe when and how the hormones elicit their actions.
   a. antidiuretic hormone
   b. aldosterone
   c. atrial natriuretic hormone

Apply Your Knowledge

1. John's doctor is concerned that John's kidneys may not be functioning properly due to a circulatory condition. The doctor wants to determine if the blood volume that is flowing through the kidneys (called renal blood flow rate) is within normal range. Calculate what would be a "normal" renal blood flow rate based on the following information:

   John weighs 90 kg. Assume that a normal total blood volume is 80 ml/kg body weight, and a normal heart pumps the total blood volume through the heart once per minute (cardiac output). Also assume that the normal renal blood flow rate is 21% of cardiac output.
2. The glomeruli filter out a tremendous amount of water and molecules needed by the body, which must later be reclaimed by the energy-requiring process of reabsorption. The Malpighian tubules of insects might seem to function more logically, secreting molecules and ions that need to be excreted. What advantages might a filtration-reabsorption process provide over a strictly secretion process of elimination?

50

Sex and Reproduction

Concept Outline

50.1 Animals employ both sexual and asexual reproductive strategies.

Sexual and Asexual Reproduction. Some animals reproduce asexually, but most reproduce sexually; male and female are usually different individuals, but not always.

50.2 The evolution of reproduction among the vertebrates has led to internalization of fertilization and development.

Fertilization and Development. Among vertebrates that have internal fertilization, the young are nourished by egg yolk or from their mother's blood.

Fish and Amphibians. Most bony fish and amphibians have external fertilization, while most cartilaginous fish have internal fertilization.

Reptiles and Birds. Most reptiles and all birds lay eggs externally, and the young develop inside the egg.

Mammals. Monotremes lay eggs, marsupials have pouches where their young develop, and placental mammals have placentas that nourish the young within the uterus.

50.3 Male and female reproductive systems are specialized for different functions.

Structure and Function of the Male Reproductive System. The testes produce sperm and secrete the male sex hormone testosterone.

Structure and Function of the Female Reproductive System. An egg cell within an ovarian follicle develops and is released from the ovary; the egg cell travels into the female reproductive tract, which undergoes cyclic changes due to hormone secretion.

Birth Control. Various methods of birth control are employed, including barriers to fertilization, prevention of ovulation, and prevention of implantation.

FIGURE 50.1
The bright color of male golden toads serves to attract mates. The rare golden toads of the Monteverde Cloud Forest Reserve of Costa Rica are nearly voiceless and so use bright colors to attract mates. Always rare, they may now be extinct.

The cry of a cat in heat, insects chirping outside the window, frogs croaking in swamps, and wolves howling in a frozen northern forest are all sounds of evolution's essential act, reproduction. These distinct vocalizations, as well as the bright coloration characteristic of some animals, such as the tropical golden toads in figure 50.1, function to attract mates. Few subjects pervade our everyday thinking more than sex, and few urges are more insistent. This chapter deals with sex and reproduction among the vertebrates, including humans.

Sexual and Asexual Reproduction

Most animals, including humans, reproduce sexually. As described in chapter 12, sexual reproduction occurs when a new individual is formed by the union of two sex cells, or **gametes,** a term that includes **sperm** and **eggs** (or ova). The union of sperm and egg cells produces a fertilized egg, or **zygote,** that develops by mitotic division into a new multicellular organism. The zygote and the cells it forms by mitosis are diploid; they contain both members of each homologous pair of chromosomes. The gametes, formed by meiosis in the sex organs, or gonads—the testes and ovaries—are haploid (see section 50.3 for more detail on sperm and egg formation).

Protists and animals including cnidarians, and tunicates as well as many of the more complex animals—reproduce asexually. In asexual reproduction, genetically identical cells are produced from a single parent cell through mitosis. In protists, a single organism divides, a process called **fission,** and then each part becomes a separate but identical organism. Cnidarians commonly reproduce by **budding,** whereby a part of the parent's body becomes separated from the rest and differentiates into a new individual. The new individual may become an independent animal or may remain attached to the parent, forming a colony.

Different Approaches to Sex

Another form of asexual reproduction, **parthenogenesis,** is common in many species of arthropods. In parthenogenesis, females produce offspring from unfertilized eggs. Some species are exclusively parthenogenic (and all female), while others switch between sexual reproduction and parthenogenesis in different generations. In honeybees, for example, a queen bee mates only once and stores the sperm. She then can control the release of the sperm. If no sperm are released, the eggs develop parthenogenetically into drones, which are males; if sperm are allowed to fertilize the eggs, the fertilized eggs develop into other queens or worker bees, which are female.

The Russian biologist Ilya Darevsky reported in 1958 one of the first cases of unusual modes of reproduction

(a)

(b)

FIGURE 50.2
Hermaphroditism and protogyny. (*a*) The hamlet bass (genus *Hypoplectrus*) is a deep-sea fish that is a hermaphrodite—both male and female at the same time. In the course of a single pair-mating, one fish may switch sexual roles as many as four times, alternately offering eggs to be fertilized and fertilizing its partner's eggs. Here the fish acting as a male curves around its motionless partner, fertilizing the upward-floating eggs. (*b*) The bluehead wrasse, *Thalassoma bifasciatium*, is protogynous—females sometimes turn into males. Here, a large male, or sex-changed female, is seen among females, which are typically much smaller.

among vertebrates. He observed that some populations of small lizards of the genus *Lacerta* were exclusively female, and suggested that these lizards could lay eggs that were viable even if they were not fertilized. In other words, they were capable of asexual reproduction in the absence of sperm, a type of parthenogenesis. Further work has shown that parthenogenesis also occurs among populations of other lizard genera.

Another variation in reproductive strategies is **hermaphroditism,** in which one individual has both testes and ovaries, and so can produce both sperm and eggs (figure 50.2*a*). A tapeworm is hermaphroditic and can fertilize itself, a useful strategy because it is unlikely to encounter another tapeworm. Most hermaphroditic animals, however, require another individual in order to reproduce. Two earthworms, for example, are required for reproduction—each functions as both male and female, and each leaves the encounter with fertilized eggs.

Some deep-sea fish are hermaphrodites. Numerous fish genera include species whose individuals can change their sex, a process called *sequential hermaphroditism.* Among

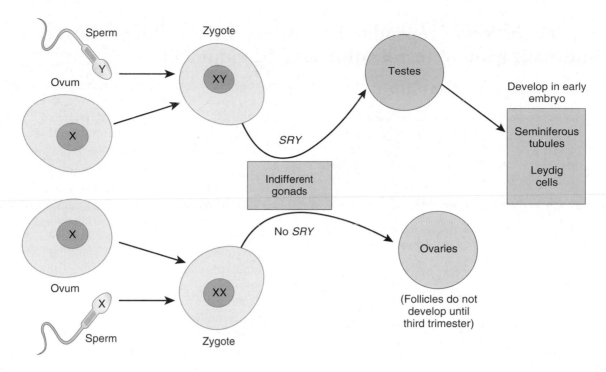

FIGURE 50.3
Sex determination in mammals. The sex-determining region of the mammalian Y chromosome is designated *SRY*. Testes are formed when the Y chromosome and *SRY* are present; ovaries are formed when they are absent.

coral reef fish, for example, both **protogyny** ("first female," a change from female to male) and **protandry** ("first male," a change from male to female) occur. In fish that practice protogyny (figure 50.2*b*), the sex change appears to be under social control. These fish commonly live in large groups, or schools, where successful reproduction is typically limited to one or a few large, dominant males. If those males are removed, the largest female rapidly changes sex and becomes a dominant male.

Sex Determination

Among the fish just described and in some species of reptiles, environmental changes can cause changes in the sex of the animal. In mammals, the sex is determined early in embryonic development. The reproductive systems of human males and females appear similar for the first 40 days after conception. During this time, the cells that will give rise to ova or sperm migrate from the yolk sac to the embryonic gonads, which have the potential to become either ovaries in females or testes in males. For this reason, the embryonic gonads are said to be "indifferent." If the embryo is a male, it will have a Y chromosome with a gene whose prod-

uct converts the indifferent gonads into testes. In females, which lack a Y chromosome, this gene and the protein it encodes are absent, and the gonads become ovaries. An important gene involved in sex determination is known as **SRY** (for "sex determining region of the Y chromosome") (figure 50.3). The *SRY* gene appears to have been highly conserved during the evolution of different vertebrate groups.

Once testes form in the embryo, the testes secrete testosterone and other hormones that promote the development of the male external genitalia and accessory reproductive organs.

If the embryo lacks the *SRY* gene, the embryo develops female external genitalia and sex accessory organs. In other words, all mammalian embryos will develop into females unless a functional *SRY* gene is present.

Sexual reproduction is most common among animals, but many reproduce asexually by fission, budding, or parthenogenesis. Sexual reproduction involves the fusion of gametes derived from different individuals of a species, although some hermaphroditic species can self-fertilize.

50.2 The evolution of reproduction among the vertebrates has led to internalization of fertilization and development.

Fertilization and Development

Vertebrate sexual reproduction evolved in the ocean before vertebrates colonized the land. The females of most species of marine bony fish produce eggs, or ova, in batches and release them into the water. The males generally release their sperm into the water containing the eggs, where the union of the free gametes occurs. This process is known as **external fertilization.**

Although seawater is not a hostile environment for gametes, it does cause the gametes to disperse rapidly, so their release by females and males must be almost simultaneous. Thus, most marine fish restrict the release of their eggs and sperm to a few brief and well-defined periods. Some reproduce just once a year, while others do so more frequently. There are few seasonal cues in the ocean that organisms can use as signals for synchronizing reproduction, but one all-pervasive signal is the cycle of the moon. Once each month, the moon approaches closer to the earth than usual, and when it does, its increased gravitational attraction causes somewhat higher tides. Many marine organisms sense the tidal changes and entrain the production and release of their gametes to the lunar cycle.

Once vertebrates began living on land, they encountered a new danger—desiccation, a problem that was especially severe for the small and vulnerable gametes. On land, the gametes could not simply be released near each other, because they would soon dry up and perish. Consequently, terrestrial vertebrates (as well as some groups of fish) were under intense selective pressure to evolve **internal fertilization**—that is, the introduction of male gametes into the female reproductive tract. By this means, fertilization still occurs in a nondesiccating environment, even when the adult animals are fully terrestrial. The vertebrates that practice internal fertilization have three strategies for embryonic and fetal development:

1. **Oviparity** is found in some bony fish, most reptiles, some cartilaginous fish, some amphibians, a few mammals, and all birds. The eggs, after being fertilized internally, are deposited outside the mother's body to complete their development.

2. **Ovoviviparity** is found in some bony fish (including mollies, guppies, and mosquito fish), some cartilaginous fish, and many reptiles. The fertilized eggs are retained within the mother to complete their development, but the embryos still obtain all of their nourishment from the egg yolk. The young are fully developed when they are hatched and released from the mother.

3. **Viviparity** is found in most cartilaginous fish, some amphibians, a few reptiles, and almost all mammals. The young develop within the mother and obtain nourishment directly from their mother's blood, rather than from the egg yolk (figure 50.4).

Fertilization is external in most fish and amphibians but internal in most other vertebrates. Depending upon the relationship of the developing embryo to the mother and egg, those vertebrates with internal fertilization may be classified as oviparous, ovoviviparous, or viviparous.

**FIGURE 50.4
Viviparous fish carry live, mobile young within their bodies.** The young complete their development within the body of the mother and are then released as small but competent adults. Here, a lemon shark has just given birth to a young shark, which is still attached by the umbilical cord.

Fish and Amphibians

Most fish and amphibians, unlike other vertebrates, reproduce by means of external fertilization.

Fish

Fertilization in most species of bony fish (teleosts) is external, and the eggs contain only enough yolk to sustain the developing embryo for a short time. After the initial supply of yolk has been exhausted, the young fish must seek its food from the waters around it. Development is speedy, and the young that survive mature rapidly. Although thousands of eggs are fertilized in a single mating, many of the resulting individuals succumb to microbial infection or predation, and few grow to maturity.

In marked contrast to the bony fish, fertilization in most cartilaginous fish is internal. The male introduces sperm into the female through a modified pelvic fin. Development of the young in these vertebrates is generally viviparous.

Amphibians

The amphibians invaded the land without fully adapting to the terrestrial environment, and their life cycle is still tied to the water. Fertilization is external in most amphibians, just as it is in most species of bony fish. Gametes from both males and females are released through the cloaca. Among the frogs and toads, the male grasps the female and discharges fluid containing the sperm onto the eggs as they are released into the water (figure 50.5). Although the eggs of most amphibians develop in the water, there are some interesting exceptions (figure 50.6). In two species of frogs, for example, the eggs develop in the vocal sacs and stomach, and the young frogs leave through their father's and mother's mouths, respectively.

The time required for development of amphibians is much longer than that for fish, but amphibian eggs do not include a significantly greater amount of yolk. Instead, the process of development in most amphibians is divided into embryonic, larval, and adult stages, in a way reminiscent of the life cycles found in some insects. The embryo develops within the egg, obtaining nutrients from the yolk. After hatching from the egg, the aquatic larva then functions as a free-swimming, food-gathering machine, often for a considerable period of time. The larvae may increase in size rapidly; some tadpoles, which are the larvae of frogs and toads, grow in a matter of weeks from creatures no bigger than the tip of a pencil into individuals as big as a goldfish. When the larva has grown to a sufficient size, it undergoes a developmental transition, or metamorphosis, into the terrestrial adult form.

The eggs of most bony fish and amphibians are fertilized externally. In amphibians, the eggs develop into a larval stage that undergoes metamorphosis.

FIGURE 50.5
The eggs of frogs are fertilized externally. When frogs mate, as these two are doing, the clasp of the male induces the female to release a large mass of mature eggs, over which the male discharges his sperm.

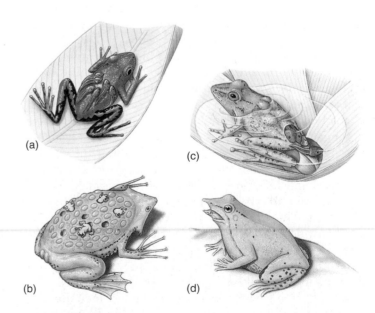

FIGURE 50.6
Different ways young develop in frogs. (*a*) In the poison arrow frog, the male carries the tadpoles on his back. (*b*) In the female Surinam frog, froglets develop from eggs in special brooding pouches on the back. (*c*) In the South American pygmy marsupial frog, the female carries the developing larvae in a pouch on her back. (*d*) Tadpoles of the Darwin's frog develop into froglets in the vocal pouch of the male and emerge from the mouth.

Reptiles and Birds

Most reptiles and all birds are oviparous. After the eggs are fertilized internally, they are deposited outside the mother's body to complete their development. Like most vertebrates that fertilize internally, male reptiles utilize a tubular organ, the penis, to inject sperm into the female (figure 50.7). The penis, containing erectile tissue, can become quite rigid and penetrate far into the female reproductive tract. Most oviparous reptiles lay eggs and then abandon them. These eggs are surrounded by a leathery shell that is deposited as the egg passes through the oviduct, the part of the female reproductive tract leading from the ovary. Other species of reptiles are ovoviviparous, forming eggs that develop into embryos within the body of the mother, and a few species are viviparous.

All birds practice internal fertilization, though most male birds lack a penis. However, in most birds (including swans, geese, and ostriches), the male cloaca extends to function as a penis. As the egg passes along the oviduct, glands secrete albumin proteins (the egg white) and the hard, calcareous shell that distinguishes bird eggs from reptilian eggs. While modern reptiles are *poikilotherms* (animals whose body temperature varies with the temperature of their environment), birds are *homeotherms* (animals that maintain a relatively constant body temperature independent of environmental temperatures). Hence, most birds incubate their eggs after laying them to keep them warm (figure 50.8). The young that hatch from the eggs of most bird species are unable to survive unaided, because their development is still incomplete. These young birds are fed and nurtured by their parents, and they grow to maturity gradually.

The shelled eggs of reptiles and birds constitute one of the most important adaptations of these vertebrates to life on land. Such eggs are known as amniotic eggs because the embryo develops within a fluid-filled cavity surrounded by a membrane called the amnion. The amnion is an extraembryonic membrane—that is, a membrane formed from embryonic cells but located outside the body of the embryo. Other extraembryonic membranes in amniotic eggs include the chorion, which lines the inside of the eggshell, the yolk sac, and the allantois. Together, these extraembryonic membranes in combination with the external calcareous shell provide reptiles and birds with a desiccation-resistant egg that can be laid in dry places. In contrast, the eggs of fish and amphibians contain only one extraembryonic membrane, the yolk sac, and need to be deposited in an aquatic habitat. The viviparous mammals, including humans, also have extraembryonic membranes, as will be described in chapter 51.

Most reptiles and all birds are oviparous, laying amniotic eggs that are protected by watertight membranes from desiccation. Birds, being homeotherms, must keep the eggs warm by incubation.

FIGURE 50.7
The introduction of sperm by the male into the female's body. Reptiles such as these tortoises were the first terrestrial vertebrates to develop this form of reproduction, called copulation, which is particularly suited to a terrestrial environment.

FIGURE 50.8
Crested penguins incubating their egg. This nesting pair is changing the parental guard in a stylized ritual.

Mammals

Some mammals are seasonal breeders, reproducing only once a year, while others have more frequent reproductive cycles. Among the latter, the females generally undergo the reproductive cycles, while the males are more constant in their reproductive activity. Cycling in females involves the periodic release of a mature ovum from the ovary in a process known as ovulation. Most female mammals are "in heat," or sexually receptive to males, only around the time of ovulation. This period of sexual receptivity is called **estrus,** and the reproductive cycle is therefore called an **estrous cycle.** The females continue to cycle until they become pregnant.

In the estrous cycle of most mammals, changes in the secretion of follicle-stimulating hormone (FSH) and luteinizing hormone (LH) by the anterior pituitary gland cause changes in egg cell development and hormone secretion in the ovaries. Humans and apes have menstrual cycles that are similar to the estrous cycles of other mammals in their cyclic pattern of hormone secretion and ovulation. Unlike mammals with estrous cycles, however, human and ape females bleed when they shed the inner lining of their uterus, a process called menstruation, and may engage in copulation at any time during the cycle.

Rabbits and cats differ from most other mammals in that they are induced ovulators. Instead of ovulating in a cyclic fashion regardless of sexual activity, the females ovulate only after copulation as a result of a reflex stimulation of LH secretion (described in section 50.3). This makes these animals extremely fertile.

The most primitive mammals, the **monotremes** (consisting solely of the duck-billed platypus and the echidna), are oviparous, like the reptiles from which they evolved. They incubate their eggs in a nest (figure 50.9*a*) or specialized pouch, and the young hatchlings obtain milk from their mother's mammary glands by licking her skin (because monotremes lack nipples). All other mammals are viviparous, and are divided into two subcategories based on how they nourish their young. The **marsupials,** a group that includes opossums and kangaroos, give birth to fetuses that are incompletely developed. The fetuses complete their development in a pouch of their mother's skin, where they can obtain nourishment from nipples of the mammary glands (figure 50.9*b*). The **placental mammals** (figure 50.9*c*) retain their young for a much longer period of development within the mother's uterus. The fetuses are nourished by a structure known as the placenta, which is derived from both an extraembryonic membrane (the chorion) and the mother's uterine lining. Because the fetal and maternal blood vessels are in very close proximity in the placenta, the fetus can obtain nutrients by diffusion from the mother's blood. The functioning of the placenta is discussed in more detail in chapter 51.

Among mammals that are not seasonal breeders, the females undergo shorter cyclic variations in ovarian function. These are estrous cycles in most mammals and menstrual cycles in humans and apes. Some mammals are induced ovulators, ovulating in response to copulation.

(a)

(b)

(c)

FIGURE 50.9

Reproduction in mammals. (*a*) Monotremes, such as the duck-billed platypus shown here, lay eggs in a nest. (*b*) Marsupials, such as this kangaroo, give birth to small fetuses that complete their development in a pouch. (*c*) In placental mammals, such as this doe nursing her fawn, the young remain inside the mother's uterus for a longer period of time and are born relatively more developed.

50.3 Male and female reproductive systems are specialized for different functions.

Structure and Function of the Male Reproductive System

The structures of the human male reproductive system, typical of mammals, are illustrated in figure 50.10. If testes form in the human embryo, they develop *seminiferous tubules* beginning around 43 to 50 days after conception. The seminiferous tubules are the sites of sperm production. At about 9 to 10 weeks, the Leydig cells, located in the interstitial tissue between the seminiferous tubules, begin to secrete testosterone (the major male sex hormone, or androgen). Testosterone secretion during embryonic development converts indifferent structures into the male external genitalia, the *penis* and the *scrotum*, the latter a sac that contains the testes. In the absence of testosterone, these structures develop into the female external genitalia.

In an adult, each testis is composed primarily of the highly convoluted seminiferous tubules (figure 50.11, *left*). Although the testes are actually formed within the abdominal cavity, shortly before birth they descend through an opening called the inguinal canal into the scrotum, which suspends them outside the abdominal cavity. The scrotum maintains the testes at around 34°C, slightly lower than the core body temperature (37°C). This lower temperature is required for normal sperm development in humans.

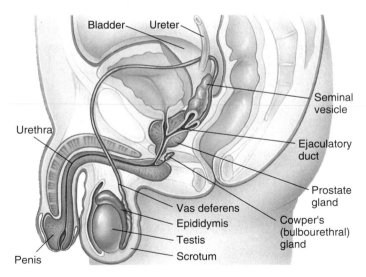

FIGURE 50.10
Organization of the human male reproductive system. The penis and scrotum are the external genitalia, the testes are the gonads, and the other organs are accessory sex organs, aiding the production and ejaculation of semen.

Production of Sperm

The wall of the seminiferous tubule consists of *spermatogonia*, or *germ cells*, which become sperm by meiosis, and supporting *Sertoli cells*. The germ cells near the outer surface of the seminiferous tubule are diploid (with 46 chromosomes in humans), while developing gamete cells located closer to the lumen of the tubule are haploid (with 23 chromosomes each). Each parent spermatogonia cell duplicates by mitosis, and one of the two daughter cells then undergoes meiosis to form sperm; the other remains as a spermatogonia cell. In that way, the male never runs out of spermatogonia cells to produce sperm. Adult males produce an average of 100 to 200 million sperm each day and can continue to do so throughout most of the rest of their lives.

The diploid daughter cell that begins meiosis is called a *primary spermatocyte*. It has 23 pairs of homologous chromosomes (in humans), and each chromosome is duplicated, with two chromatids. The first meiotic division separates the homologous chromosome pairs, producing two haploid *secondary spermatocytes*. However, each chromosome still consists of two duplicate chromatids. Each of these cells then undergoes the second meiotic division to separate the chromatids and produce two haploid cells, the *spermatids*. Therefore, a total of four haploid spermatids are produced by each primary spermatocyte (figure 50.11, *right*). All of these cells constitute the germinal epithelium of the seminiferous tubules because they "germinate" the gametes.

In addition to the germinal epithelium, the walls of the seminiferous tubules contain nongerminal cells known as Sertoli cells. The Sertoli cells nurse the developing sperm and secrete products required for spermatogenesis (sperm production). They also help convert the spermatids into *spermatozoa* by engulfing their extra cytoplasm.

Spermatozoa, or sperm, are relatively simple cells, consisting of a head, body, and tail (figure 50.12). The head encloses a compact nucleus and is capped by a vesicle called an **acrosome,** which is derived from the Golgi complex. The acrosome contains enzymes that aid in the penetration of the protective layers surrounding the egg. The body and tail provide a propulsive mechanism: Within the tail is a flagellum, while inside the body are a centriole, which acts as a basal body for the flagellum, and mitochondria, which generate the energy needed for flagellar movement.

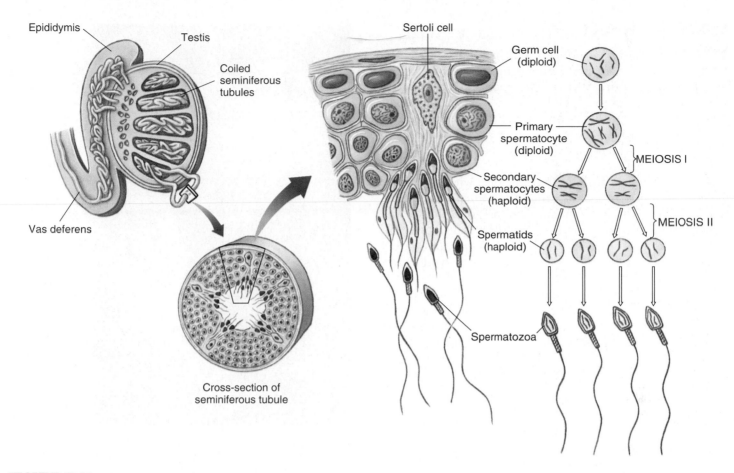

FIGURE 50.11

The testis and spermatogenesis. *Left:* Inside the testis, the seminiferous tubules are the sites of spermatogenesis. Germ cells in the seminiferous tubules give rise to spermatozoa by meiosis. *Right:* Sertoli cells are nongerminal cells within the walls of the seminiferous tubules. They assist spermatogenesis in several ways, such as by helping to convert spermatids into spermatozoa. A primary spermatocyte is diploid. At the end of the first meiotic division, homologous chromosomes have separated, and two haploid secondary spermatocytes form. The second meiotic division separates the sister chromatids and results in the formation of four haploid spermatids.

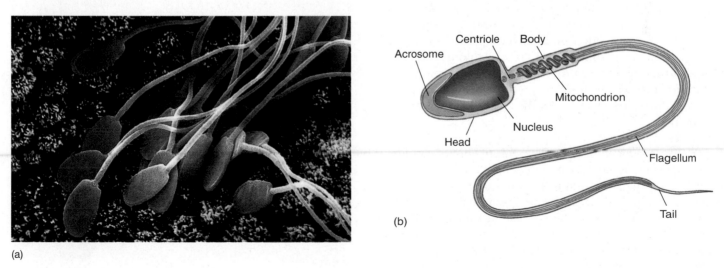

FIGURE 50.12

Human sperm. (*a*) A scanning electron micrograph. (*b*) A diagram of the main components of a sperm cell.

Male Accessory Sex Organs

After the sperm are produced within the seminiferous tubules, they are delivered into a long, coiled tube called the *epididymis* (figure 50.13). The sperm are not motile when they arrive in the epididymis, and they must remain there for at least 18 hours before their motility develops. From the epididymis, the sperm enter another long tube, the *vas deferens*, which passes into the abdominal cavity via the inguinal canal.

The vas deferens from each testis joins with one of the ducts from a pair of glands called the *seminal vesicles* (see figure 50.10), which produce a fructose-rich fluid comprising about 60% of semen volume. From this point, the vas deferens continues as the *ejaculatory duct* and enters the prostate gland at the base of the urinary bladder. In humans, the prostate gland is about the size of a golf ball and is spongy in texture. It contributes up to 30% of the bulk of the semen, the fluid that contains the products of the testes, fluid from the seminal vesicles, and the products of the prostate gland. Within the prostate gland, the ejaculatory duct merges with the urethra from the urinary bladder. The urethra carries the semen out of the body through the tip of the penis. A pair of pea-sized *bulbourethral glands* secrete a fluid that lines the urethra and lubricates the tip of the penis prior to coitus (sexual intercourse).

In addition to the urethra, the penis has two columns of erectile tissue, the *corpora cavernosa*, along its dorsal side and one column, the *corpus spongiosum*, along the ventral side (figure 50.14). Penile erection is produced by neurons in the parasympathetic division of the autonomic nervous system. As a result of the release of nitric oxide by these neurons, arterioles in the penis dilate, causing the erectile tissue to become turgid as it is engorged with blood. This increased pressure in the erectile tissue compresses the veins, so blood flows into the penis but cannot flow out. The drug sildenafil (Viagra) prolongs erection by promoting the action of nitric oxide in the penis. Some mammals, such as the walrus, have a bone in the penis that contributes to its stiffness during erection, but humans do not.

The result of erection and continued sexual stimulation is *ejaculation*, the ejection from the penis of about 2–5 milliliters of semen containing an average of 300 million sperm. Successful fertilization requires such a high sperm count because the odds against any one sperm cell successfully completing the journey to the egg and fertilizing it are extraordinarily high, and the acrosomes of many sperm need to interact with the egg before a single sperm can penetrate the egg. Males with fewer than 20 million sperm per milliliter are generally considered sterile. Despite their large numbers, sperm constitute only about 1% of the volume of the semen ejaculated.

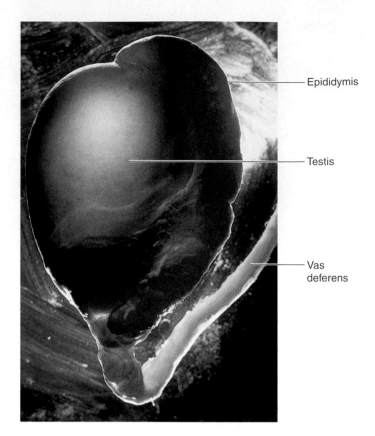

FIGURE 50.13
Photograph of the human testis.   The round object in the center of the photograph is a testis, within which sperm are formed. Cupped around it is the epididymis, a highly coiled passageway in which sperm complete their maturation. Mature sperm are stored in the vas deferens, a long tube that extends from the epididymis.

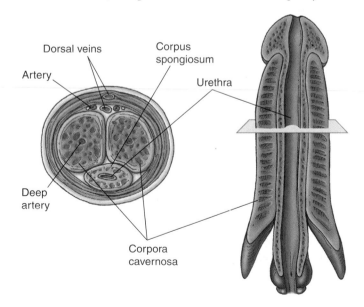

FIGURE 50.14
A penis in cross-section (*left*) and longitudinal section (*right*). Note that the urethra runs through the corpus spongiosum.

Table 50.1	Mammalian Reproductive Hormones
MALE	
Follicle-stimulating hormone (FSH)	Stimulates spermatogenesis
Luteinizing hormone (LH)	Stimulates secretion of testosterone by Leydig cells
Testosterone	Stimulates development and maintenance of male secondary sexual characteristics, accessory sex organs, and spermatogenesis
FEMALE	
Follicle-stimulating hormone (FSH)	Stimulates growth of ovarian follicles and secretion of estradiol
Luteinizing hormone (LH)	Stimulates ovulation, conversion of ovarian follicles into corpus luteum, and secretion of estradiol and progesterone by corpus luteum
Estradiol	Stimulates development and maintenance of female secondary sexual characteristics; prompts monthly preparation of uterus for pregnancy
Progesterone	Completes preparation of uterus for pregnancy; helps maintain female secondary sexual characteristics
Oxytocin	Stimulates contraction of uterus and milk-ejection reflex
Prolactin	Stimulates milk production

Hormonal Control of Male Reproduction

As we saw in chapter 47, the anterior pituitary gland secretes two gonadotropic hormones: FSH and LH. Although these hormones are named for their actions in the female, they are also involved in regulating male reproductive function (table 50.1). In males, FSH stimulates the Sertoli cells to facilitate sperm development, and LH stimulates the Leydig cells to secrete testosterone.

The principle of negative feedback inhibition discussed in chapters 47 and 49 applies to the control of FSH and LH secretion (figure 50.15). The hypothalamic hormone, gonadotropin-releasing hormone (GnRH), stimulates the anterior pituitary gland to secrete both FSH and LH. FSH causes the Sertoli cells to release a peptide hormone called inhibin that specifically inhibits FSH secretion. Similarly, LH stimulates testosterone secretion, and testosterone feeds back to inhibit the release of LH, both directly at the anterior pituitary gland and indirectly by reducing GnRH release. The importance of negative feedback inhibition can be demonstrated by removing the testes; in the absence of testosterone and inhibin, the secretion of FSH and LH from the anterior pituitary is greatly increased.

An adult male produces sperm continuously by meiotic division of germ cells lining the seminiferous tubules. Semen consists of sperm from the testes and fluid contributed by the seminal vesicles and prostate gland. Production of sperm and secretion of testosterone from the testes are controlled by FSH and LH from the anterior pituitary.

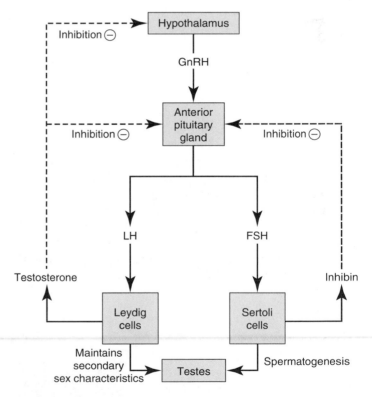

FIGURE 50.15
Hormonal interactions between the testes and anterior pituitary. LH stimulates the Leydig cells to secrete testosterone, and FSH stimulates the Sertoli cells of the seminiferous tubules to secrete inhibin. Testosterone and inhibin, in turn, exert negative feedback inhibition on the secretion of LH and FSH, respectively.

Structure and Function of the Female Reproductive System

The structures of the reproductive system in a human female are shown in figure 50.16. In contrast to the testes, the ovaries develop much more slowly. In the absence of testosterone, the female embryo develops a *clitoris* and *labia majora* from the same embryonic structures that produce a penis and a scrotum in males. Thus, the clitoris and penis, and the labia majora and scrotum, are said to be *homologous structures*. The clitoris, like the penis, contains corpora cavernosa and is therefore erectile. The ovaries contain microscopic structures called *ovarian follicles*, which each contain an egg cell and smaller *granulosa cells*. The ovarian follicles are the functional units of the ovary.

At puberty, the granulosa cells begin to secrete the major female sex hormone, estradiol (also called estrogen), triggering *menarche*, the onset of menstrual cycling. Estradiol also stimulates the formation of the *female secondary sexual characteristics*, including breast development and the production of pubic hair. In addition, estradiol and another steroid hormone, progesterone, help maintain the female accessory sex organs: the fallopian tubes, uterus, and vagina.

Female Accessory Sex Organs

The fallopian tubes (also called uterine tubes or oviducts) transport ova from the ovaries to the uterus. In humans, the uterus is a muscular, pear-shaped organ that narrows to form a neck, the cervix, which leads to the vagina (figure 50.17*a*). The uterus is lined with a simple columnar epithelial membrane called the endometrium. The surface of the endometrium is shed during menstruation, while the underlying portion remains to generate a new surface during the next cycle.

Mammals other than primates have more complex female reproductive tracts, in which part of the uterus divides to form uterine "horns," each of which leads to an oviduct (figure 50.17*b,c*). Cats, dogs, and cows, for example, have one cervix but two uterine horns separated by a septum, or wall. Marsupials, such as opossums, carry the split even further, with two unconnected uterine horns, two cervices, and two vaginas. A male marsupial has a forked penis that can enter both vaginas simultaneously.

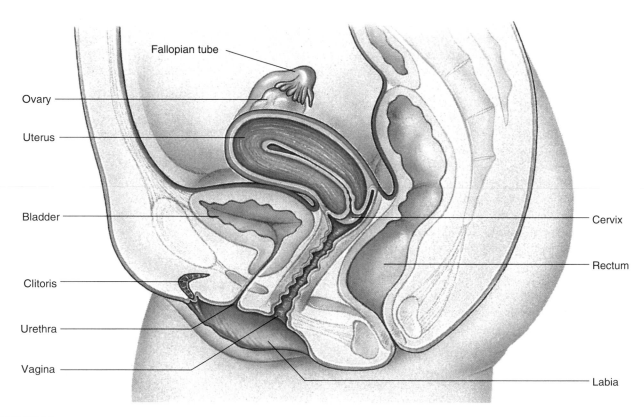

FIGURE 50.16
Organization of the human female reproductive system. The ovaries are the gonads, the fallopian tubes receive the ovulated ova, and the uterus is the womb, the site of development of an embryo if the egg cell becomes fertilized.

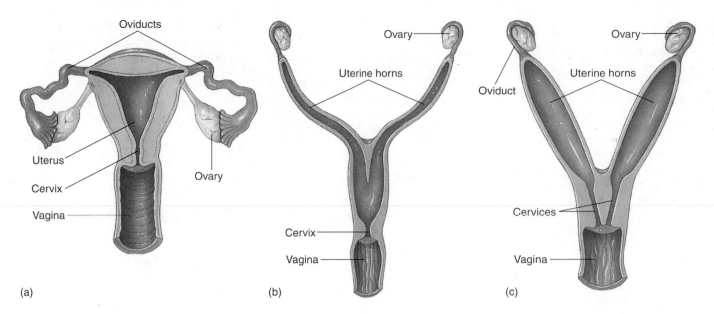

FIGURE 50.17
A comparison of mammalian uteruses. (*a*) Humans and other primates; (*b*) cats, dogs, and cows; and (*c*) rats, mice, and rabbits.

Menstrual and Estrous Cycles

At birth, a female's ovaries contain about 1 million follicles, each with an ovum that has begun meiosis but is arrested in prophase of the first meiotic division. At this stage, the ova are called *primary oocytes*. Some of these primary oocyte–containing follicles are stimulated to develop during each cycle. The human menstrual cycle (Latin *mens*, "month") lasts approximately one month (28 days on the average) and can be divided in terms of ovarian activity into a follicular phase and luteal phase, with the two phases separated by the event of ovulation (see figure 50.21).

Follicular Phase

During each follicular phase, several follicles are stimulated to grow under FSH stimulation, but only one achieves full maturity as a tertiary, or Graafian, follicle by ovulation. This follicle forms a thin-walled blister on the surface of the ovary. The primary oocyte within the Graafian follicle completes the first meiotic division during the follicular phase. Instead of forming two equally large daughter cells, however, it produces one large daughter cell, the *secondary oocyte* (figure 50.18), and one tiny daughter cell, called a *polar body*. Thus, the secondary oocyte acquires almost all of the cytoplasm from the primary oocyte (unequal cytokinesis), increasing its chances of sustaining the early embryo should the oocyte be fertilized. The polar body, on the other hand, disintegrates. The secondary oocyte then begins the second meiotic division, but its progress is arrested at metaphase II. It is in this form that the egg cell is discharged from the ovary at ovulation, and it does not complete the second meiotic division unless it becomes fertilized in the fallopian tube.

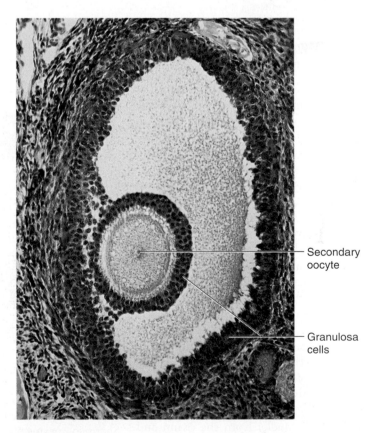

FIGURE 50.18
A mature Graafian follicle in a cat ovary (50×). Note the ring of granulosa cells that surrounds the secondary oocyte. This ring will remain around the egg cell when it is ovulated, and sperm must tunnel through the ring in order to reach the plasma membrane of the egg cell.

FIGURE 50.19

The meiotic events of oogenesis in humans. A primary oocyte is diploid. At the completion of the first meiotic division, one division product is eliminated as a polar body, while the other, the secondary oocyte, is released during ovulation. The secondary oocyte does not complete the second meiotic division until after fertilization; that division yields a second polar body and a single haploid egg, or ovum. Fusion of the haploid egg with a haploid sperm during fertilization produces a diploid zygote.

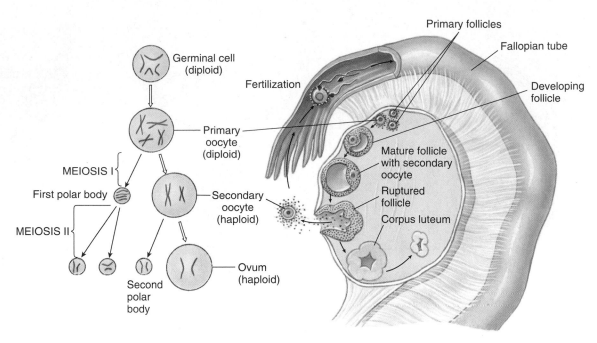

Ovulation

The increasing level of estradiol in the blood during the follicular phase stimulates the anterior pituitary gland to secrete LH about mid-cycle. This sudden secretion of LH causes the fully developed Graafian follicle to burst in the process of ovulation, releasing its secondary oocyte. The released oocyte enters the abdominal cavity near the *fimbriae*, the feathery projections surrounding the opening to the fallopian tube. The ciliated epithelial cells lining the fallopian tube draw in the oocyte and propel it through the fallopian tube toward the uterus. If it is not fertilized, the oocyte disintegrates within a day following ovulation. If it is fertilized, the stimulus of fertilization allows it to complete the second meiotic division, forming a fully mature ovum and a second polar body (figure 50.19). Fusion of the nuclei from the ovum and the sperm produces a diploid zygote. Fertilization normally occurs in the upper one-third of the fallopian tube, and in a human the zygote takes approximately three days to reach the uterus and then another two to three days to implant in the endometrium (figure 50.20).

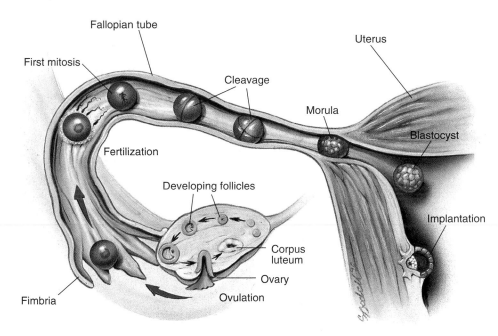

FIGURE 50.20

The journey of an egg. Produced within a follicle and released at ovulation, an egg is swept into a fallopian tube and carried along by waves of ciliary motion in the tube walls. Sperm journeying upward from the vagina fertilize the egg within the fallopian tube. The resulting zygote undergoes several mitotic divisions while still in the tube, so that by the time it enters the uterus, it is a hollow sphere of cells called a blastocyst. The blastocyst implants within the wall of the uterus, where it continues its development. (The egg and its subsequent stages have been enlarged for clarification.)

Luteal Phase

After ovulation, LH stimulation completes the development of the Graafian follicle into a structure called the *corpus luteum* (Latin, "yellow body"). For this reason, the second half of the menstrual cycle is referred to as the **luteal phase.** The corpus luteum secretes both estradiol and another steroid hormone, progesterone. The high blood levels of estradiol and progesterone during the luteal phase now exert negative feedback inhibition of FSH and LH secretion by the anterior pituitary gland. This inhibition during the luteal phase is in contrast to the stimulation exerted by estradiol on LH secretion at mid-cycle, which caused ovulation. The inhibitory effect of estradiol and progesterone on FSH and LH secretion after ovulation acts as a natural contraceptive mechanism, preventing both the development of additional follicles and continued ovulation.

During the follicular phase of the ovarian cycle, the granulosa cells secrete increasing amounts of estradiol, which stimulates the growth of the endometrium. Hence, this portion of the cycle is also referred to as the **proliferative phase** of the endometrium (figure 50.21). During the luteal phase of the cycle, the combination of estradiol and progesterone cause the endometrium to become more vascular, glandular, and enriched with glycogen deposits. Because of the endometrium's glandular appearance and function, this portion of the cycle is known as the **secretory phase** of the endometrium.

In the absence of fertilization, the corpus luteum degenerates due to the decreasing levels of LH and FSH, toward the end of the luteal phase. It does this by secreting hormones (estradiol and progesterone) that inhibit the secretion of LH, the hormone needed for its survival. In many mammals, atrophy of the corpus luteum is assisted by luteolysin, a paracrine regulator believed to be a prostaglandin. The disappearance of the corpus luteum results in an abrupt decline in the blood concentration of estradiol and progesterone at the end of the luteal phase, causing the built-up endometrium to be sloughed off with accompanying bleeding. This process is called *menstruation* and the portion of the cycle in which it occurs is known as the **menstrual phase** of the endometrium.

If the ovulated oocyte is fertilized, however, regression of the corpus luteum and subsequent menstruation are averted by the tiny embryo! It does this by secreting human chorionic gonadotropin (hCG), an LH-like hormone produced by the chorionic membrane of the embryo. By maintaining the corpus luteum, hCG keeps the levels of estradiol and progesterone high and thereby prevents menstruation, which would terminate the pregnancy. Because hCG comes from the embryonic chorion and not from the mother, it is the hormone that is tested for in all pregnancy tests.

Menstruation is absent in mammals with an estrous cycle. Although such mammals do cyclically shed cells from the endometrium, they don't bleed in the process. The estrous cycle is divided into four phases: proestrus, estrus, metestrus, and diestrus, which correspond to the proliferative, mid-cycle, secretory, and menstrual phases of the endometrium in the menstrual cycle.

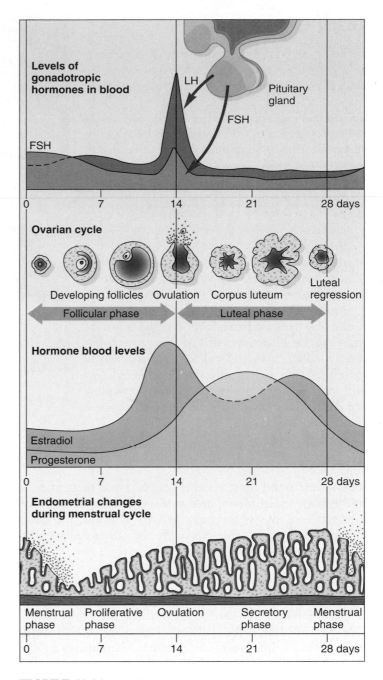

FIGURE 50.21

The human menstrual cycle.   The growth and thickening of the endometrial (uterine) lining is stimulated by estradiol during the proliferative phase. Estradiol and progesterone maintain and regulate the endometrium during the secretory phase. The decline in the levels of these two hormones triggers menstruation, the sloughing off of built-up endometrial tissue.

During each menstrual cycle, several ovarian follicles develop under FSH stimulation, culminating in the ovulation of one follicle under LH stimulation. During the follicular and luteal phases, the hormones secreted by the ovaries stimulate the development of the endometrium, so an embryo can implant there if fertilization occurs. A secondary oocyte is released from an ovary at ovulation, and it only completes meiosis if it is fertilized.

Birth Control

In most vertebrates, copulation is associated solely with reproduction. Reflexive behavior that is deeply ingrained in the female limits sexual receptivity to those periods of the sexual cycle when she is fertile. In humans and a few species of apes, the female can be sexually receptive throughout her reproductive cycle, and this extended receptivity to sexual intercourse serves a second important function—it reinforces pair-bonding, the emotional relationship between two individuals living together.

Not all human couples want to initiate a pregnancy every time they have sexual intercourse, yet sexual intercourse may be a necessary and important part of their emotional lives together. The solution to this dilemma is to find a way to avoid reproduction without avoiding sexual intercourse; this approach is commonly called birth control, or *contraception*. A variety of approaches, differing in effectiveness and in their acceptability to different couples, are commonly taken to achieve birth control (figure 50.22 and table 50.2).

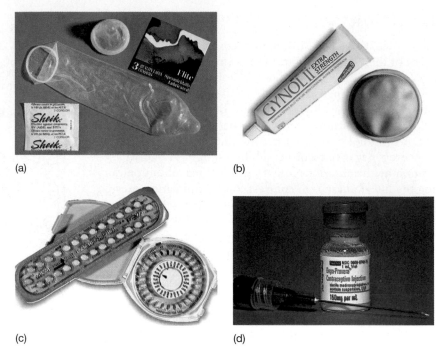

FIGURE 50.22
Four common methods of birth control. (*a*) Condom; (*b*) diaphragm and spermicidal jelly; (*c*) oral contraceptives; (*d*) Depo-Provera.

Abstinence

The simplest and most reliable way to avoid pregnancy is not to have sexual intercourse at all. Of all the methods of birth control, this is the most certain. It is also the most limiting, because it denies a couple the emotional support of a sexual relationship.

Sperm Blockage

If sperm cannot reach the uterus, fertilization cannot occur. One way to prevent the delivery of sperm is to encase the penis within a thin sheath, or condom. Many males do not favor the use of condoms, which tend to decrease their sensory pleasure during intercourse. In principle, this method is easy to apply and foolproof, but in practice it has a failure rate of 3 to 15% because of incorrect or inconsistent use. Nevertheless, it is the most commonly employed form of birth control in the United States. Condoms are also widely used to prevent the transmission of AIDS and other sexually transmitted diseases (STDs). Over a billion condoms are sold in the United States each year.

A second way to prevent the entry of sperm into the uterus is to place a cover over the cervix. The cover may be a relatively tight-fitting cervical cap, which is worn for days at a time, or a rubber dome called a diaphragm, which is inserted immediately before intercourse. Because the dimensions of individual cervices vary, a cervical cap or diaphragm must be fitted by a physician. Failure rates average 4 to 25% for diaphragms, perhaps because of the propensity to insert them carelessly. Failure rates for cervical caps are somewhat lower.

Sperm Destruction

A third general approach to birth control is to eliminate the sperm after ejaculation. This can be achieved in principle by washing out the vagina immediately after intercourse, before the sperm have a chance to enter the uterus. Such a procedure is called a douche (French, "wash"). The douche method is difficult to apply well, because it involves a rapid dash to the bathroom immediately after ejaculation and a very thorough washing. Its failure rate is as high as 40%. Alternatively, sperm delivered to the vagina can be destroyed there with spermicidal jellies or foams. These treatments generally require application immediately before intercourse. Their failure rates vary from 10 to 25%. The use of a spermicide with a condom increases the effectiveness over each method used independently.

Prevention of Ovulation

Since about 1960, a widespread form of birth control in the United States has been the daily ingestion of birth control pills, or oral contraceptives, by women. These pills contain analogues of progesterone, sometimes in combination with

Table 50.2   Methods of Birth Control

Device	Action	Failure Rate[*]	Advantages	Disadvantages
Oral contraceptive	Hormones (progesterone analogue alone or in combination with other hormones) primarily prevent ovulation	1–5, depending on type	Convenient; highly effective; provides significant noncontraceptive health benefits, such as protection against ovarian and endometrial cancers	Must be taken regularly; possible minor side effects, which new formulations have reduced; not for women with cardiovascular risks (mostly smokers over age 35)
Condom	Thin sheath for penis collects semen; "female condoms" sheath vaginal walls	3–15	Easy to use; effective; inexpensive; protects against some sexually transmitted diseases	Requires male cooperation; may diminish spontaneity; may deteriorate on the shelf
Diaphragm	Soft rubber cup covers entrance to uterus, prevents sperm from reaching egg, holds spermicide	4–25	No dangerous side effects; reliable if used properly; provides some protection against sexually transmitted diseases and cervical cancer	Requires careful fitting; some inconvenience associated with insertion and removal; may be dislodged during intercourse
Intrauterine device (IUD)	Small plastic or metal device placed in the uterus; prevents implantation; some contain copper, others release hormones	1–5	Convenient; highly effective; infrequent replacement	Can cause excess menstrual bleeding and pain; risk of perforation, infection, expulsion, pelvic inflammatory disease, and infertility; not recommended for those who eventually intend to conceive or are not monogamous; dangerous in pregnancy
Cervical cap	Miniature diaphragm covers cervix closely, prevents sperm from reaching egg, holds spermicide	Probably similar to that of diaphragm	No dangerous side effects; fairly effective; can remain in place longer than diaphragm	Problems with fitting and insertion; comes in limited number of sizes
Foams, creams, jellies, vaginal suppositories	Chemical spermicides inserted in vagina before intercourse prevent sperm from entering uterus	10–25	Can be used by anyone who is not allergic; protect against some sexually transmitted diseases; no known side effects	Relatively unreliable; sometimes messy; must be used 5–10 minutes before each act of intercourse
Implant (levonorgestrel; Norplant)	Capsules surgically implanted under skin slowly release hormone that blocks ovulation	.03	Very safe, convenient, and effective; very long-lasting (5 years); may have nonreproductive health benefits like those of oral contraceptives	Irregular or absent periods; minor surgical procedure needed for insertion and removal; some scarring may occur
Injectable contraceptive (medroxy-progesterone; Depo-Provera)	Injection every 3 months of a hormone that is slowly released and prevents ovulation	1	Convenient and highly effective; no serious side effects other than occasional heavy menstrual bleeding	Animal studies suggest it may cause cancer, though new studies in humans are mostly encouraging; occasional heavy menstrual bleeding

[*]Failure rate is expressed as pregnancies per 100 actual users per year.

Source: Data from American College of Obstetricians and Gynecologists: Contraception, Patient Education Pamphlet No. AP005. ACOG, Washington, D.C., 1990.

estrogens. As described earlier, progesterone and estradiol act by negative feedback to inhibit the secretion of FSH and LH during the luteal phase of the ovarian cycle, thereby preventing follicle development and ovulation. They also cause a buildup of the endometrium. The hormones in birth control pills have the same effects. Because the pills block ovulation, no ovum is available to be fertilized. A woman generally takes the hormone-containing pills for three weeks; during the fourth week, she takes pills without hormones (placebos), allowing the levels of those hormones in her blood to fall, which causes menstruation. Oral contraceptives provide a very effective means of birth control, with a failure rate of only 1 to 5%. In a variation of the oral contraceptive, hormone-containing capsules are implanted beneath the skin. These implanted capsules have failure rates below 1%.

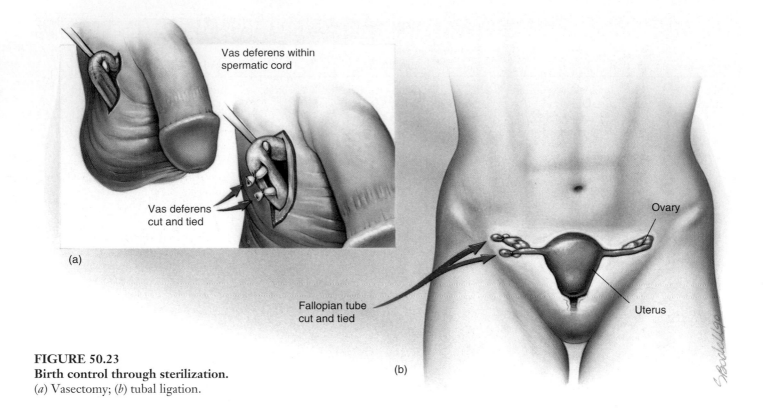

FIGURE 50.23
Birth control through sterilization.
(*a*) Vasectomy; (*b*) tubal ligation.

A small number of women using birth control pills or implants experience undesirable side effects, such as blood clotting and nausea. These side effects have been reduced in newer generations of birth control pills, which contain less estrogen and different analogues of progesterone. Moreover, these new oral contraceptives provide a number of benefits, including reduced risks of endometrial and ovarian cancer, cardiovascular disease, and osteoporosis (for older women). However, they may increase the risk of developing breast cancer and cervical cancer. The risks involved with birth control pills increase somewhat in women who smoke and increase greatly in women over 35 who smoke. The current consensus is that, for many women, the health benefits of oral contraceptives outweigh their risks, although a physician must help each woman determine the relative risks and benefits.

Prevention of Embryo Implantation

The insertion of a coil or other irregularly shaped object into the uterus is an effective means of birth control, because the irritation it produces in the uterus prevents the implantation of an embryo within the uterine wall. Such intrauterine devices (IUDs) have a failure rate of only 1 to 5%. Their high degree of effectiveness probably reflects their convenience; once they are inserted, they can be forgotten. The great disadvantage of this method is that almost a third of the women who attempt to use IUDs experience cramps, pain, and sometimes bleeding and therefore must discontinue using them.

Another method of preventing embryo implantation is the "morning-after pill," which contains 50 times the dose of estrogen present in birth control pills. The pill works by temporarily stopping ovum development, by preventing fertilization, or by stopping the implantation of a fertilized ovum. Its failure rate is 1 to 10%, but many women are uneasy about taking such high hormone doses, because side effects can be severe. This pill is not recommended as a regular method of birth control but rather as a method of emergency contraception.

Sterilization

A completely effective means of birth control is sterilization, the surgical removal of portions of the tubes that transport the gametes from the gonads (figure 50.23). Sterilization may be performed on either males or females, preventing sperm from entering the semen in males and preventing an ovulated oocyte from reaching the uterus in females. In males, sterilization involves a vasectomy, the removal of a portion of the vas deferens from each testis. In females, the comparable operation involves the removal of a section of each fallopian tube.

Fertilization can be prevented by a variety of birth control methods, including barrier contraceptives, hormonal inhibition, surgery, and abstinence. Efficacy rates vary from method to method.

Concept Review

For interactive testing, visit the Online Learning Center with PowerWeb at www.mhhe.com/Raven7

50.1 Animals employ both sexual and asexual reproductive strategies.

Sexual and Asexual Reproduction

- Sexual reproduction occurs when the gametes from different sexes unite. (p. 1062)
- Asexual reproduction can occur through fission, budding, or parthenogenesis. (p. 1062)
- Hermaphroditism occurs when one individual has both testes and ovaries, although most require another individual in order to reproduce. (pp. 1062–1063)
- In mammals, the sex of an individual is determined early in embryonic development. (p. 1063)

50.2 The evolution of reproduction among the vertebrates has led to internalization of fertilization and development.

Fertilization and Development

- Most fish and amphibians engage in external fertilization, while most other vertebrates, including mammals, engage in internal fertilization. (pp. 1064–1065)
- Vertebrates utilizing internal fertilization follow one of three strategies—oviparity, ovoviviparity, or viviparity—based on the degree of dependence of the developing embryo on the mother. (p. 1064)

Fish and Amphibians

- While most bony fish utilize external fertilization, most cartilaginous fish utilize internal fertilization. (p. 1065)
- Most amphibians utilize external fertilization and develop in water into a larval stage before metamorphosing into the terrestrial adult form. (p. 1065)

Reptiles and Birds

- Most reptiles and birds exhibit oviparity and lay amniotic eggs, which contain a fluid-filled cavity surrounded by a membrane called the amnion. (p. 1066)
- Birds are homeothermic and thus must incubate eggs as well as watch after and protect hatchlings until they are developed enough to survive on their own. (p. 1066)

Mammals

- Some mammals are seasonal breeders, while others undergo more frequent reproductive (estrous) cycles; still others are induced ovulators. (p. 1067)

- In mammalian estrous cycles, changes in follicle-stimulating hormone (FSH) and luteinizing hormone (LH) cause changes in egg cell development and ovarian hormone secretions. (p. 1067)
- Mammals fall into one of three categories of reproduction: monotremes, marsupials, or placentals. (p. 1067)

50.3 Male and female reproductive systems are specialized for different functions.

Structure and Function of the Male Reproductive System

- In human males, the scrotum contains two testes; each testis contains seminiferous tubules, which are the sites of sperm production. (p. 1068)
- Semen, which is composed of sperm and fluid from the seminal vesicles and the prostate gland, is ejaculated from the penis following erection and sexual stimulation. (p. 1070)
- FSH stimulates Sertoli cells to facilitate sperm development, and LH stimulates Leydig cells to secrete testosterone. (p. 1071)

Structure and Function of the Female Reproductive System

- In human females, the ovaries contain ovarian follicles, which each contain an egg cell. (p. 1072)
- The human menstrual cycle lasts approximately one month and is divided into two phases (follicular and luteal) which are separated by ovulation. (p. 1073)
- During each menstrual cycle, ovarian follicles begin development due to FSH stimulation and eventually release one secondary oocyte (ovulation) under LH stimulation. (pp. 1073–1075)
- During the follicular and luteal phases of the menstrual cycle, ovaries release hormones to stimulate endometrium thickening, preparing it for the potential implantation of a fertilized egg. (pp. 1073–1075)

Birth Control

- Major forms of birth control include abstinence, sperm blockage, sperm destruction, prevention of ovulation, prevention of embryo implantation, and sterilization. (pp. 1076–1078)

Self Test

1. Sexual reproduction
   a. requires internal fertilization.
   b. does not require meiosis.
   c. increases the genetic diversity within a population.
   d. often occurs between individuals of different species.

2. An animal that is oviparous reproduces by
   a. giving birth to free-living young.
   b. producing internally-fertilized eggs that develop externally.
   c. producing eggs that are fertilized externally.
   d. incubating eggs internally while the fetus develops.

3. The testicles of male mammals are suspended in the scrotum because
   a. the optimum temperature for sperm production is less than the normal core body temperature of the organism.
   b. the optimum temperature for sperm production is higher than the normal core body temperature of the organism.
   c. there is not enough room in the pelvic area for the testicles to be housed internally.
   d. it is easier for the body to expel sperm during ejaculation.

4. Spermatogenesis is *not* directly affected by which hormones?
   a. GnRH
   b. inhibin
   c. FSH
   d. LH

5. Eggs and sperm are genetically very similar, but structurally very different. Why is this so?
   a. Both contain a haploid chromosome number, but eggs must provide nutrients for early development, while sperm must be able to move efficiently.
   b. Both contain a diploid chromosome number, but eggs must provide nutrients for early development, while sperm must be able to move efficiently.
   c. Both contain maternal chromosomes, but only sperm can control which chromosomes are passed on.
   d. Both contain a haploid chromosome number, but only eggs can control which chromosomes are passed on.

6. How would mammalian reproduction be affected if the meiotic strategy of spermatogenesis and oogenesis were reversed?
   a. Not enough eggs would be made each month to ensure reproductive success.
   b. Sperm production would decrease to one-fourth.
   c. Eggs would be diploid while sperm would be haploid.
   d. Sperm would be diploid while eggs would be haploid.

7. Early in the ovarian cycle, estrogen, produced in the follicle, _____ gonadotropin release, while later in the cycle, estrogen _____ gonadotropin release because
   a. inhibits; stimulates; feedback mechanisms are not involved early in the ovarian cycle.
   b. stimulates; inhibits; feedback mechanisms are not involved early in the ovarian cycle.
   c. inhibits; stimulates; the feedback mechanisms are dependent on the concentration of estrogen.
   d. stimulates; inhibits; the feedback mechanisms are dependent on the concentration of estrogen.

8. At what stage of the ovarian cycle are mammalian eggs most likely to become fertilized?
   a. at the beginning of the proliferative phase
   b. immediately after ovulation
   c. during the middle of the secretory phase
   d. during the menstrual phase

9. If we could monitor the amount of total gonadotropin activity in pregnant women, we would expect
   a. high levels of FSH and LH in the uterus to stimulate endometrial thickening.
   b. high levels of circulating FSH and LH to stimulate implantation of the embryo.
   c. high levels of hCG in the uterus to stimulate endometrial thickening.
   d. high levels of circulating hCG to stimulate estrogen and progesterone synthesis.

10. Which contraceptive method is effective at preventing fertilization *and* protecting against transmission of sexually transmitted disease?
    a. oral contraceptives
    b. diaphragm
    c. condom
    d. intrauterine device (IUD)

Test Your Visual Understanding

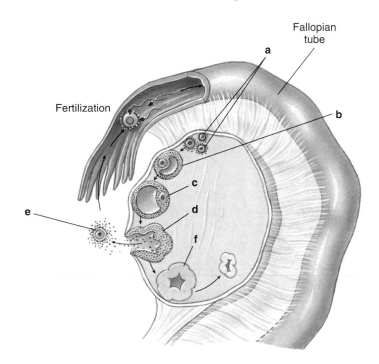

1. The process of oogenesis involves a number of developmental stages and structures. Label this diagram by writing the name of the correct structure next to the appropriate letter:

   a.   d.
   b.   e.
   c.   f.

Apply Your Knowledge

1. Sex determination in mammals is influenced by the *SRY* region of the Y chromosome. If during meiosis, this region of the Y chromosome was accidentally translocated to an X chromosome and this gamete went on to fertilize an egg, what would be the gender of this offspring?

51

Vertebrate Development

Concept Outline

51.1 Fertilization is the initial event in development.

> **Fertilization.** Fertilization of an egg cell by a sperm occurs in three stages: penetration, activation of the egg cell, and fusion of the two haploid nuclei.

51.2 Cell cleavage and the formation of a blastula set the stage for later development.

> **Cell Cleavage Patterns.** The cytoplasm of the zygote is divided into smaller cells in a process called cleavage.

51.3 Gastrulation forms the three germ layers of the embryo.

> **The Process of Gastrulation.** Cells of the blastula invaginate and involute to produce the three primary germ layers: ectoderm, endoderm, and mesoderm.

51.4 Body architecture is determined during the next stages of embryonic development.

> **Developmental Processes During Neurulation.** Mesoderm forms a notochord, ectoderm forms a neural tube, and neural crest forms in vertebrates.
> **How Cells Communicate During Development.** Cell-to-cell contact plays a major role in selecting the paths along which cells develop.
> **Embryonic Development and Vertebrate Evolution.** The embryonic development of a mammal includes stages that are characteristic of many vertebrates.
> **Extraembryonic Membranes.** Embryonic cells form several membranes outside the embryo that provide protection and nourishment for the embryo.

51.5 Human development is divided into trimesters.

> **First Trimester.** A blastocyst implants into the mother's endometrium, and the body organs begin to form during the third and fourth weeks.
> **Second and Third Trimesters.** Further growth and development of the body organs take place during the second and third trimesters.
> **Birth and Postnatal Development.** Birth occurs as a result of uterine contractions. The human brain continues to grow significantly after birth.

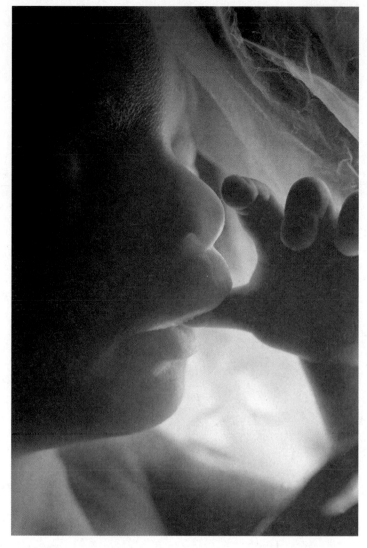

FIGURE 51.1
Development is the process that determines an organism's form and function. A human fetus at 18 weeks is not yet halfway through the 38 weeks—about 9 months—it will spend within its mother, but it has already developed many distinct behaviors, such as the sucking reflex that is so important to survival after birth.

Sexual reproduction, in all but a few vertebrates, unites two haploid gametes to form a single diploid cell called a zygote. The zygote develops by a process of cell division and differentiation into a complex multicellular organism, composed of many different tissues and organs (figure 51.1). Although the process of development is a continuous series of events, with some of the details varying among different vertebrate groups, this chapter examines vertebrate development in stages, concluding with a description of the events that occur during human development.

51.1 Fertilization is the initial event in development.

Fertilization

In vertebrates, as in all sexual animals, the first step in development is the union of male and female gametes, a process called **fertilization.** Fertilization is typically external in fish and amphibians, which reproduce in water, and internal in all other vertebrates. In internal fertilization, small, motile sperm are introduced into the female reproductive tract during mating. The sperm swim up the reproductive tract until they encounter a mature egg, or oocyte, in an oviduct, where fertilization occurs. Fertilization consists of three stages: penetration, activation, and nuclei fusion.

Penetration

As described in chapter 50, the secondary oocyte is released from a fully developed Graafian follicle at ovulation. It is surrounded by the same layer of small granulosa cells that surrounded it within the follicle (figure 51.2). Between the granulosa cells and the egg's plasma membrane is a glycoprotein layer called the zona pellucida. The head of each sperm is capped by an organelle called the acrosome, which contains glycoprotein-digesting enzymes. These enzymes are released as the sperm begin to work their way into the layer of granulosa cells, and the activity of the enzymes enables the sperm to tunnel their way through the zona pellucida to the egg's plasma membrane. In sea urchins, egg cytoplasm bulges out at this point, engulfing the head of the sperm and permitting the sperm nucleus to enter the cytoplasm of the egg (figure 51.3).

Activation

The series of events initiated by sperm penetration are collectively called egg activation. In some frogs, reptiles, and birds, more than one sperm may penetrate the egg, but only one is successful in fertilizing it. In mammals, by contrast, the penetration of the first sperm initiates changes in the egg membrane that prevent the entry of other sperm. As the sperm makes contact with the oocyte membrane, there is a change in the membrane potential (see chapter 45 for discussion of membrane potential) that prevents other

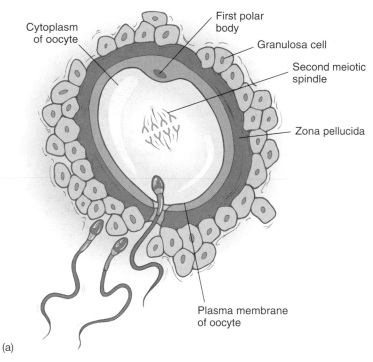

(a)

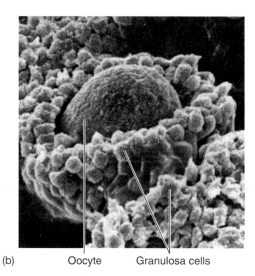

(b)   Oocyte   Granulosa cells

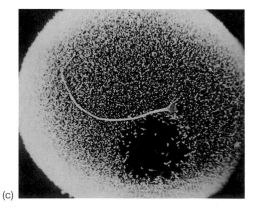

(c)

FIGURE 51.2
Mammalian reproductive cells. (*a*) A sperm must penetrate a layer of granulosa cells and then a layer of glycoprotein called the zona pellucida, before it reaches the oocyte membrane. This penetration is aided by digestive enzymes in the acrosome of the sperm. These scanning electron micrographs show (*b*) a human oocyte (90×) surrounded by numerous granulosa cells, and (*c*) a human sperm on an egg (3000×).

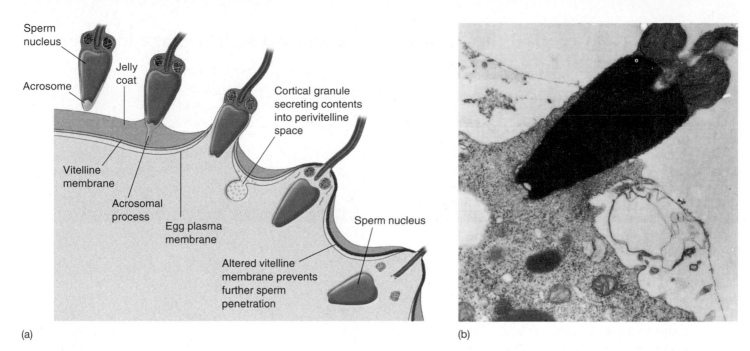

(a)

(b)

FIGURE 51.3

Sperm penetration of a sea urchin egg. (*a*) The stages of penetration. The entire process takes about 20–30 seconds. (*b*) An electron micrograph (50,000×) of penetration. Penetration in both invertebrate and vertebrate eggs is similar.

sperm from fusing with the oocyte membrane. In addition to these changes, sperm penetration can have three other effects on the egg. First, in mammals it stimulates the chromosomes in the egg nucleus to complete the second meiotic division, producing two egg nuclei. One of these nuclei is extruded from the egg as a second polar body (see chapter 50), leaving a single haploid egg nucleus within the egg.

Second, sperm penetration in some animals triggers movements of the egg cytoplasm around the point of sperm entry. These movements ultimately establish the bilateral symmetry of the developing animal. In frogs, for example, sperm penetration causes an outer pigmented cap of egg cytoplasm to rotate toward the point of entry, uncovering a gray crescent of interior cytoplasm opposite the point of penetration (figure 51.4). The position of the gray crescent determines the orientation of the first cell division. A line drawn between the point of sperm entry and the gray crescent would bisect the right and left halves of the future adult. Third, activation is characterized by a sharp increase in protein synthesis and an increase in metabolic activity in general. Experiments demonstrate that the protein synthesis in the activated oocyte is coded by mRNA that was previously produced and already present in the cytoplasm of the unfertilized egg cell.

In some vertebrates, it is possible to activate an egg without the entry of a sperm, simply by pricking the egg membrane. An egg that is activated in this way may go on to develop parthenogenetically. A few kinds of amphibians, fish, and reptiles rely entirely on parthenogenetic reproduction in nature, as we mentioned in chapter 50.

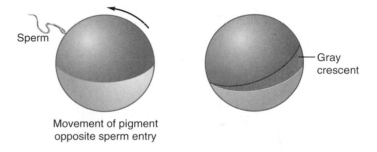

FIGURE 51.4

Gray crescent formation in frog eggs. The gray crescent forms on the side of the egg opposite the point of penetration by the sperm.

Nuclei Fusion

In the third stage of fertilization, the entering sperm nucleus fuses with the haploid egg nucleus to form the diploid nucleus of the zygote. This fusion is triggered by the activation of the egg. If a sperm nucleus is microinjected into an egg without activating the egg, the two nuclei will not fuse. The nature of the signals that are exchanged between the two nuclei, or sent from one to the other, is not known.

The three stages of fertilization are penetration, activation, and nuclei fusion. Penetration initiates a complex series of developmental events, including major movements of cytoplasm, which eventually lead to the fusion of the egg and sperm nuclei.

51.2 Cell cleavage and the formation of a blastula set the stage for later development.

Cell Cleavage Patterns

Following fertilization, the second major event in vertebrate development is the rapid division of the zygote into a larger and larger number of smaller and smaller cells (table 51.1). This period of division, called *cleavage*, is not accompanied by an increase in the overall volume of the zygote cytoplasm. The resulting tightly packed mass of about 32 cells is called a **morula,** and each individual cell in the morula is referred to as a **blastomere.** As the blastomeres continue to divide, they secrete a fluid into the center of the morula. Eventually, a hollow ball of 500 to 2000 cells, the **blastula,** is formed. The fluid-filled cavity within the blastula is known as the *blastocoel.*

The pattern of cleavage division is influenced by the presence and location of yolk, which is abundant in the eggs of many vertebrates (figure 51.5). As we discussed in chapter 50, vertebrates have embraced a variety of reproductive strategies involving different patterns of yolk utilization.

Primitive Chordates

When eggs contain little or no yolk, cleavage occurs throughout the whole egg, a pattern called **holoblastic cleavage** (figure 51.6). This pattern of cleavage was characteristic of the ancestors of the vertebrates and is still seen in groups such as the echinoderms, tunicates, lancelets, and mammals. In these animals, holoblastic cleavage results in the formation of a symmetrical blastula composed of cells of approximately equal size.

Amphibians and Advanced Fish

The eggs of bony fish and frogs contain much more cytoplasmic yolk in one hemisphere than the other. Because yolk-rich cells divide much more slowly than those that have little yolk, holoblastic cleavage in these eggs results in a very asymmetrical blastula with large cells containing a lot of yolk at one pole and a concentrated mass of small cells containing very little yolk at the other. In these blastulas, the pole that is rich in yolk is called the vegetal pole, while the pole that is relatively poor in yolk is called the animal pole (figure 51.7).

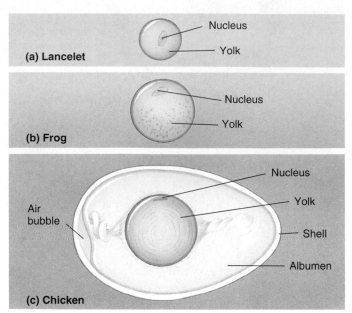

FIGURE 51.5
Yolk distribution in three kinds of eggs. (*a*) In the lancelet, a primitive chordate, the egg consists of a central nucleus surrounded by a small amount of yolk. (*b*) In a frog egg, there is much more yolk, and the nucleus is displaced toward one pole. (*c*) Bird eggs are complexly organized, with the nucleus just under the surface of a large, central yolk.

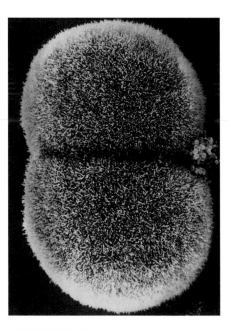

FIGURE 51.6
Holoblastic cleavage (3000×). In this type of cleavage, which is characteristic of eggs with relatively small amounts of yolk, cell division occurs throughout the entire egg.

FIGURE 51.7
Dividing frog eggs. The closest cells in this photo (those near the animal pole) divide faster and are smaller than those near the vegetal pole (below cells of the animal pole).

Table 51.1 Stages of Vertebrate Development (Mammal)

	Fertilization	The haploid male and female gametes fuse to form a diploid-zygote.
	Cleavage	The zygote rapidly divides into many cells, with no overall increase in size. These divisions affect future development because different cells receive different portions of the egg cytoplasm and, hence, different regulatory signals.
	Gastrulation	The cells of the embryo move, forming three primary cell layers: ectoderm, mesoderm, and endoderm.
	Neurulation	In all chordates, the first organ to form is the notochord; second is the dorsal nerve cord.
	Neural crest	During neurulation, the neural crest is produced as the neural goove closes to form the neural tube. The neural crest gives rise to several characteristic structures.
	Organogenesis	Cells from the three primary germ layers combine in various ways to produce the organs of the body.

Reptiles and Birds

The eggs produced by reptiles, birds, and some fish are composed almost entirely of yolk, with a small amount of cytoplasm concentrated at one pole. Cleavage in these eggs occurs only in the tiny disc of polar cytoplasm, called the *blastodisc*, that lies astride the large ball of yolk material. This type of cleavage pattern is called **meroblastic cleavage** (figure 51.8). The resulting embryo is not spherical, but rather has the form of a thin cap perched on the yolk.

Mammals

Mammalian eggs are in many ways similar to the reptilian eggs from which they evolved, except that they contain very little yolk. Because cleavage is not impeded by yolk in mammalian eggs, it is holoblastic, forming a ball of cells surrounding a blastocoel. However, an inner cell mass is concentrated at one pole (figure 51.9). This inner cell mass is analogous to the blastodisc of reptiles and birds, and it goes on to form the developing embryo. The outer sphere of cells, called a **trophoblast,** is analogous to the cells that form the membranes underlying the tough outer shell of the reptilian egg. These cells have changed during the course of mammalian evolution to carry out a very different function: Part of the trophoblast enters the maternal endometrium (the epithelial lining of the uterus) and contributes to the placenta, the organ that permits exchanges between the fetal and maternal bloods. While part of the placenta is composed of fetal tissue (the trophoblast), another part is composed of the modified endometrial tissue (called the *decidua basalis*) of the mother's uterus. The placenta will be discussed in more detail in section 51.5.

The Blastula

Viewed from the outside, the blastula looks like a simple ball of cells all resembling each other. In many animals, this appearance is misleading; unequal distribution of developmental signals from the egg produces some degree of mosaic development, so that the cells are already committed to different developmental paths (see chapter 19). In mammals, however, it appears that all of the blastomeres receive equivalent sets of signals, and body form is determined by cell–cell interactions. In a mammalian blastula, called a *blastocyst*, each cell is in contact with a different set of neighboring cells, and these interactions with neighboring cells are a major factor influencing the developmental fate of each cell. This positional information is particularly important in the orientation of mammalian embryos, setting up different patterns of development along three embryonic axes: anterior-posterior, dorsal-ventral, and proximal-distal.

For a short period of time, just before they implant in the uterus, the cells of the mammalian blastocyst have the

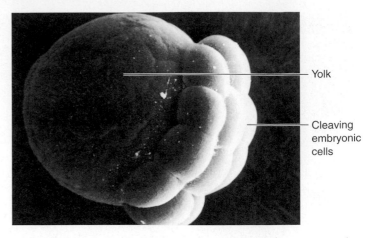

FIGURE 51.8
Meroblastic cleavage (400×). Only a portion of the egg actively divides to form a mass of cells in this type of cleavage, which occurs in eggs with relatively large amounts of yolk.

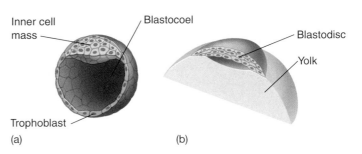

FIGURE 51.9
The embryos of mammals and birds are more similar than they seem. (*a*) A mammalian blastula, called a blastocyst, is composed of a sphere of cells, the trophoblast, surrounding a cavity, the blastocoel, and an inner cell mass. (*b*) An avian (bird) blastula consists of a cap of cells, the blastodisc, resting atop a large yolk mass. The blastodisc will form an upper and a lower layer with a compressed blastocoel in between.

power to develop into many of the 210 different types of cells in the body—and probably all of them. Biologists have long sought to grow these cells, called *embryonic stem cells*, in tissue culture, because such stem cells might in principle be used to produce tissues for human transplant operations. Injected into a patient, for example, they might be able to respond to local signals and produce new tissue (see chapter 20). The first success in growing stem cells in culture was reported in 1998, when researchers isolated cells from the inner cell mass of human blastocysts and successfully grew them in tissue culture. These stem cells continue to grow and divide in culture indefinitely, unlike ordinary body cells, which divide only 50 or so times and then die.

A series of rapid cell divisions called cleavage transforms the zygote into a hollow ball of cells, the blastula. The cleavage pattern is influenced by the amount of yolk and its distribution in the egg.

51.3 Gastrulation forms the three germ layers of the embryo.

The Process of Gastrulation

The first visible results of cytoplasmic distribution and cell position within the blastula can be seen immediately after the completion of cleavage. Certain groups of cells **invaginate** (dent inward) and **involute** (roll inward) from the surface of the blastula in a carefully orchestrated activity called **gastrulation.** The events of gastrulation determine the basic developmental pattern of the vertebrate embryo. By the end of gastrulation, the cells of the embryo have rearranged into three primary **germ layers: ectoderm, mesoderm,** and **endoderm.** The cells in each layer have very different developmental fates. In general, the ectoderm is destined to form the epidermis and neural tissue; the mesoderm gives rise to connective tissue, the skeleton, muscle, and vascular elements; and the endoderm forms the lining of the gut and its derivatives (table 51.2).

How is cell movement during gastrulation brought about? Apparently, migrating cells creep over stationary cells by means of actin filament contractions that change the shapes of the migrating cells, effecting an invagination of blastula tissue. Each cell that moves possesses particular cell surface polysaccharides, which adhere to similar polysaccharides on the surfaces of the other moving cells. This interaction between cell surface molecules enables the migrating cells to adhere to one another and move as a single mass (see chapter 7).

Just as the pattern of cleavage divisions in different groups of vertebrates depends heavily on the amount and distribution of yolk in the egg, so the pattern of gastrulation among vertebrate groups depends on the shape of the blastulas produced during cleavage.

Table 51.2	Developmental Fates of the Primary Germ Layers
Ectoderm	Epidermis, central nervous system, sense organs, neural crest
Mesoderm	Skeleton, muscles, blood vessels, heart, gonads
Endoderm	Lining of digestive and respiratory tracts; liver, pancreas

Gastrulation in Primitive Chordates

In primitive chordates such as lancelets, which develop from symmetrical blastulas, gastrulation begins as the surface of the blastula invaginates into the blastocoel. About half of the blastula's cells move into the interior of the blastula, forming a structure that looks something like an indented tennis ball (figure 51.10a,b). Eventually, the inward-moving wall of cells pushes up against the opposite side of the blastula and then stops moving. The resulting two-layered, cup-shaped embryo is the gastrula (figure 51.10c). The hollow structure resulting from the invagination is called the archenteron, and it is the progenitor of the digestive tube. The opening of the archenteron, the future anus, is known as the blastopore.

This process produces an embryo with two cell layers: an outer ectoderm and an inner endoderm. Soon afterward, a third cell layer, the mesoderm, forms between the ectoderm and endoderm. In lancelets, the mesoderm forms from pouches that pinch off the endoderm. The appearance of these three primary germ layers sets the stage for all subsequent tissue and organ differentiation.

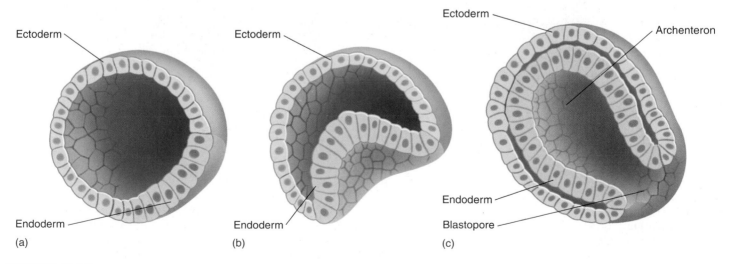

(a) (b) (c)

FIGURE 51.10
Gastrulation in a lancelet. In these chordates, the endoderm is formed by invagination of surface cells (*a*, *b*). This produces the primitive gut, or archenteron (*c*). Mesoderm will later be formed from pouches off the endoderm.

Gastrulation in Most Aquatic Vertebrates

In the blastulas of amphibians and those aquatic vertebrates with asymmetrical yolk distribution, the yolk-laden cells of the vegetal pole are fewer and much larger than the yolk-free cells of the animal pole. Consequently, gastrulation is more complex than it is in the lancelets. First, a layer of surface cells invaginates to form a small, crescent-shaped slit where the blastopore will soon be located. Next, cells from the animal pole involute over the dorsal lip of the blastopore (figure 51.11), at the same location as the gray crescent of the fertilized egg (see figure 51.4). As in the lancelets, the involuting cell layer eventually presses against the inner surface of the opposite side of the embryo, eliminating the blastocoel and producing an archenteron with a blastopore. In this case, however, the blastopore is filled with yolk-rich cells, forming the yolk plug. The outer layer of cells resulting from these movements is the ectoderm, and the inner layer is the endoderm. Other cells that involute over the dorsal lip and ventral lip (the two lips of the blastopore that are separated by the yolk plug) migrate between the ectoderm and endoderm to form the third germ layer, the mesoderm (figure 51.11c).

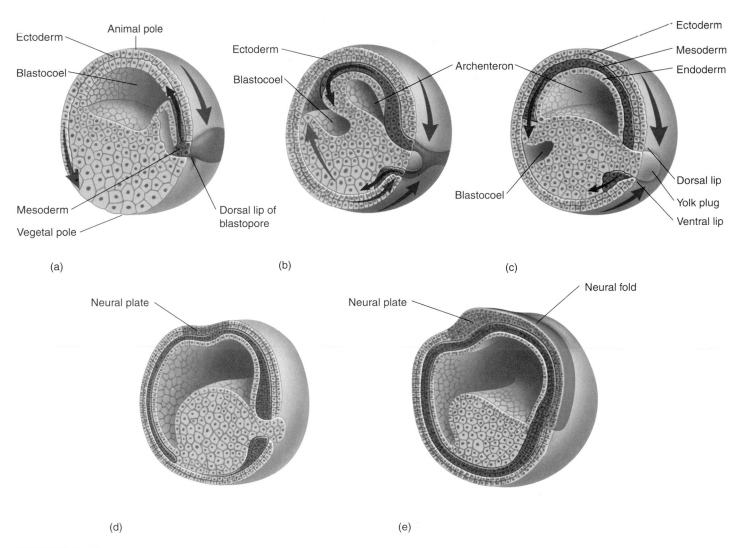

(a) (b) (c)

(d) (e)

FIGURE 51.11

Frog gastrulation.   (*a*) A layer of cells from the animal pole moves toward the vegetal pole, ultimately involuting through the dorsal lip of the blastopore. (*b*) Cells in the dorsal lip zone then involute into the hollow interior, or blastocoel, eventually pressing against the far wall. The three primary germ tissues (ectoderm, mesoderm, and endoderm) become distinguished. Ectoderm is shown in *blue*, mesoderm in *red*, and endoderm in *yellow*. (*c*) The movement of cells in the dorsal lip creates a new internal cavity, the archenteron, and displaces the blastocoel. (*d*) The neural plate later forms from ectoderm. (*e*) This will next form a neural groove and then a neural tube as the embryo begins the process of neurulation. The cells of the neural ectoderm are shown in *green*.

Gastrulation in Reptiles, Birds, and Mammals

In the blastodisc of a bird or reptile and the inner cell mass of a mammal, the developing embryo is a small cap of cells rather than a sphere. No yolk separates the two sides of the embryo, and as a result, the lower cell layer is able to differentiate into endoderm and the upper layer into ectoderm without cell movement. Just after these two primary germ layers form, the mesoderm arises by invagination and involution of cells from the upper layer. The surface cells begin moving to the midline where they involute and migrate laterally to form a mesodermal layer between the ectoderm and endoderm. A furrow along the longitudinal midline marks the site of this involution (figures 51.12 and 51.13). This furrow, analogous to an elongated blastopore, is called the **primitive streak.**

Gastrulation in lancelets involves the formation of ectoderm and endoderm by the invagination of the blastula, and the mesoderm layer forms from pouches pinched from the endoderm. In vertebrates with extensive amounts of yolk, gastrulation requires the involution of surface cells into a blastopore or primitive streak, and the mesoderm is derived from some of these involuted cells.

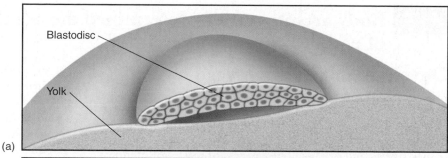

(a)

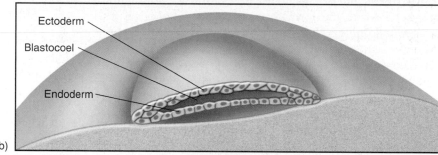

(b)

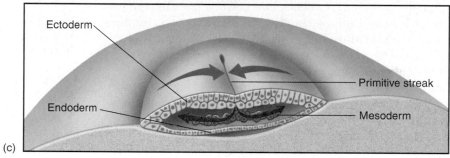

(c)

FIGURE 51.12
Gastrulation in birds. The upper layer of the blastodisc (*a*) differentiates into ectoderm, and the lower layer differentiates into endoderm (*b*). Among the cells that migrate into the interior through the dorsal primitive streak are future mesodermal cells (*c*).

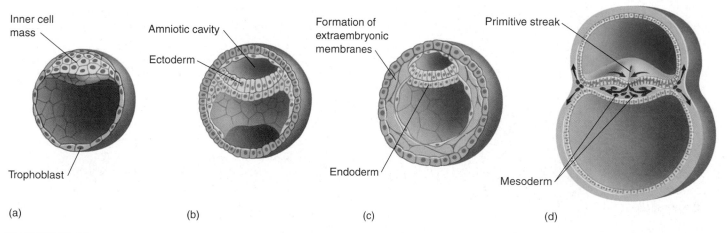

(a)   (b)   (c)   (d)

FIGURE 51.13
Mammalian gastrulation. (*a*) The amniotic cavity forms within the inner cell mass and its base. Layers of ectoderm and endoderm differentiate (*b* and *c*) as in the avian blastodisc. (*d*) A primitive streak develops, through which cells destined to become mesoderm migrate into the interior, again reminiscent of gastrulation in birds. The trophoblast (outer layer of *tan* cells) has now moved further away from the embryo and begins to play a role in forming the placenta.

51.4 Body architecture is determined during the next stages of embryonic development.

Developmental Processes During Neurulation

During the next step in vertebrate development, the three primary germ layers begin their transformation into the body's tissues and organs. The process of tissue differentiation begins with the formation of two morphological features found only in chordates, the **notochord** and the hollow dorsal nerve cord. The development of the dorsal nerve cord is called **neurulation.**

The notochord is first visible soon after gastrulation is complete, forming from mesoderm. It is a flexible rod located along the dorsal midline in the embryos of all chordates, although its function is replaced by the vertebral column when it develops from mesoderm in the vertebrates. After the notochord has been laid down, a layer of ectodermal cells situated above the notochord invaginates, forming a long crease, the **neural groove,** down the long axis of the embryo. The edges of the neural groove then move toward each other and fuse, creating a long, hollow cylinder, the **neural tube** (figure 51.14), which runs beneath the surface of the embryo's back. The neural tube later differentiates into the spinal cord and brain.

The dorsal lip of the blastopore induces the formation of a notochord, and the presence of the notochord induces the overlying ectoderm to differentiate into the neural tube. The process of *induction*, during which one embryonic region of cells influences the development of an adjacent region by changing its developmental pathway, was discussed in chapter 19 and is further examined later in this section.

While the neural tube is forming from ectoderm, the rest of the basic architecture of the body is being determined rapidly by changes in the mesoderm. On either side of the developing notochord, the mesoderm differentiates into a series of segmented blocks called **somitomeres.** As the somitomeres develop, they separate into individual units called **somites.** More somites are added as development continues. Ultimately, the somites give rise to the muscles, vertebrae, and connective tissues. The mesoderm in the head region does not separate into discrete somites but remains connected as somitomeres, which form the striated muscles of the face, jaws, and throat. Some body organs, including the kidneys, adrenal glands, and gonads, develop within another strip of mesoderm that runs alongside the somites. The remainder of the mesoderm moves out and around the endoderm and eventually surrounds it completely. As a result of this movement, the mesoderm becomes separated into two layers. The outer layer is associated with the inner body wall, and the inner layer is associated with the outer lining of the gut. Between these two layers of mesoderm is the coelom (see chapter 31), which becomes the body cavity of the adult.

The Neural Crest

Neurulation occurs in all chordates, and the process in a lancelet is much the same as it is in a human. However, in vertebrates, just before the neural groove closes to form the neural tube, its edges pinch off, forming a small strip of cells, the **neural crest,** which becomes incorporated into the roof of the neural tube (figure 51.14*e*). The cells of the neural crest later move to the sides of the developing embryo. The appearance of the neural crest was a key event in the evolution of the vertebrates because neural crest cells, after migrating to different parts of the embryo, ultimately develop into the structures characteristic of (though not necessarily unique to) the vertebrate body.

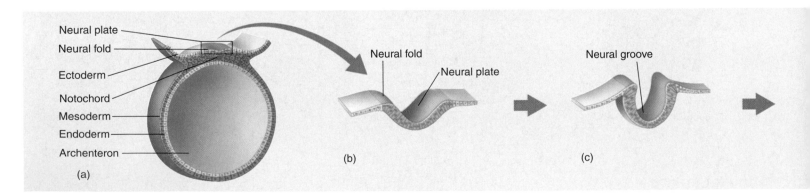

FIGURE 51.14

Mammalian neural tube formation. (*a*) The neural plate forms above the notochord, and the (*b*) cells of the neural plate fold together to form (*c*) the neural groove.

The differentiation of neural crest cells depends on their location. At the anterior end of the embryo, they merge with the anterior portion of the brain, the forebrain, where they contribute to the development of skeletal and connective tissues of the head. Nearby clusters of ectodermal cells associated with the neural crest cells thicken into **placodes,** which are distinct from neural crest cells although they arise from similar cellular interactions. Placodes subsequently develop into parts of the sense organs in the head. The neural crest and associated placodes exist in two lateral strips, which is why the vertebrate sense organs that develop from them are paired.

Neural crest cells located in more posterior positions have very different developmental fates. These cells migrate away from the neural tube to other locations in the head and trunk, where they form connections between the neural tube and the surrounding tissues. At these new locations, they contribute to the development of a variety of structures that are particularly characteristic of the vertebrates, several of which are discussed next. The migration of neural crest cells is unique in that it is not simply a change in the relative positions of cells, such as that seen in gastrulation. Instead, neural crest cells actually pass through other tissues.

The Gill Chamber

Primitive chordates such as lancelets are filter feeders, using the rapid beating of cilia to draw water into their bodies through slits in their pharynx. These pharyngeal slits evolved into the vertebrate gill chamber, a structure that provides a greatly improved means of respiration. The evolution of the gill chamber was certainly a key event in the transition from filter feeding to active predation.

In the development of the gill chamber, some of the neural crest cells form cartilaginous bars between the embryonic pharyngeal slits. Other neural crest cells induce portions of the mesoderm to form muscles along the cartilage, while still others form neurons that carry impulses between the nerve cord and these muscles. A major blood vessel called the aortic arch passes through each of the em-

bryonic bars. Lined by still more neural crest cells, these bars, with their internal blood supply, become highly branched and form the gills of the adult.

Because the stiff bars of the gill chamber can be bent inward by powerful muscles controlled by nerves, the whole structure is a very efficient pump that drives water past the gills. The gills themselves act as highly efficient oxygen exchangers, greatly increasing the respiratory capacity of the animals that possess them.

Elaboration of the Nervous System

Some neural crest cells migrate ventrally toward the notochord and form sensory neurons in the dorsal root ganglia (see chapter 45). Others become specialized as Schwann cells, which insulate nerve fibers and permit the rapid conduction of nerve impulses. Still others form the autonomic ganglia and the adrenal medulla. Cells in the adrenal medulla secrete epinephrine when stimulated by the sympathetic division of the autonomic nervous system during the fight or flight reaction. The similarity in the chemical nature of the hormone epinephrine and the neurotransmitter norepinephrine, released by sympathetic neurons, is understandable—both adrenal medullary cells and sympathetic neurons derive from the neural crest.

Sensory Organs and the Skull

A variety of sense organs develop from the placodes. Included among them are the olfactory (smell) and lateral line (primitive hearing) organs discussed in chapter 46. Neural crest cells contribute to tooth development and to some of the facial and cranial bones of the skull.

The appearance of the neural crest in the developing embryo marks the beginning of the first truly vertebrate phase of development, because many of the structures characteristic of vertebrates derive directly or indirectly from neural crest cells.

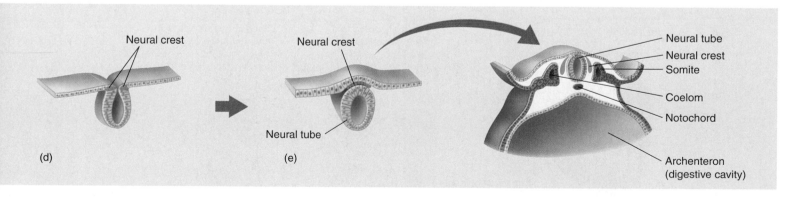

FIGURE 51.14—continued.
(d) The neural groove eventually closes to form a hollow tube. (e) As the tube closes, some of the cells from the dorsal margin of the neural tube differentiate into the neural crest, which is characteristic of vertebrates.

How Cells Communicate During Development

In the process of vertebrate development, the relative position of particular germ layers determines, to a large extent, the organs that develop from them. By now, you may have wondered how these germ layers know where they are. For example, when cells of the ectoderm situated above the developing notochord give rise to the neural groove, how do these cells "know" they are above the notochord?

The solution to this puzzle is one of the outstanding accomplishments of experimental embryology, the study of how embryos form. The great German biologist Hans Spemann and his student Hilde Mangold solved it early in the twentieth century. They removed cells from the dorsal lip of an amphibian blastula and transplanted them to a different location on another blastula (figure 51.15). (The dorsal lip region of amphibian blastulas develops from the gray crescent zone and is the site of origin of those mesoderm cells that later produce the notochord.) The new location corresponded to that of the animal's belly. What happened? Some of the embryos developed *two* notochords, a normal dorsal one and a second one along its belly!

By using genetically different donor and host blastulas, Spemann and Mangold were able to show that the notochord produced by transplanting dorsal lip cells contained host cells as well as transplanted ones. The transplanted dorsal lip cells had acted as **organizers** (see also chapter 19) of notochord development. As such, these cells stimulated a developmental program in the belly cells of the embryos in which they were transplanted: the development of the notochord. The belly cells clearly contained the genetic information (genes) for this developmental program but would not have expressed it in the normal course of their develop-

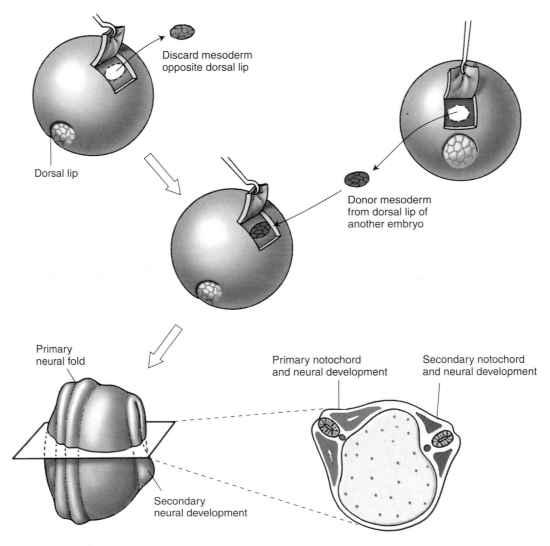

Discard mesoderm opposite dorsal lip

Dorsal lip

Donor mesoderm from dorsal lip of another embryo

Primary neural fold

Secondary neural development

Primary notochord and neural development

Secondary notochord and neural development

FIGURE 51.15
Spemann and Mangold's dorsal lip transplant experiment.

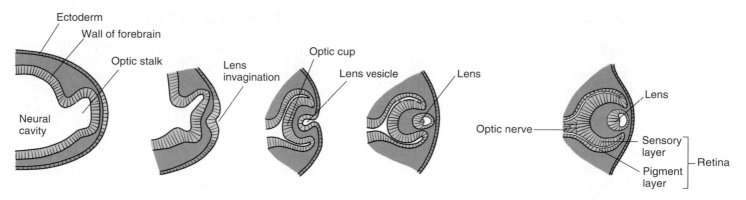

FIGURE 51.16
Development of the vertebrate eye by induction. An extension of the optic stalk grows until it contacts ectoderm, which induces a section of the ectoderm to pinch off and form the lens. Other structures of the eye develop from the optic stalk.

ment. Signals from the transplantation of the dorsal lip cells caused them to do so. These cells had indeed induced the ectoderm cells of the belly to form a notochord. This phenomenon as a whole is known as **induction.**

The process of induction that Spemann discovered appears to be the basic mode of development in vertebrates. Inductions between the three primary tissue types—ectoderm, mesoderm, and endoderm—are referred to as **primary inductions.** Inductions between tissues that have already been differentiated are called **secondary inductions.** The differentiation of the central nervous system during neurulation by the interaction of dorsal ectoderm and dorsal mesoderm to form the neural tube is an example of primary induction. In contrast, the differentiation of the lens of the vertebrate eye from ectoderm by interaction with tissue from the central nervous system is an example of secondary induction.

The vertebrate eye develops as an extension of the forebrain, a stalk that grows outward until it comes into contact with the epidermis (figure 51.16). At a point directly above the growing stalk, a layer of the epidermis pinches off, forming a transparent lens. When the optic stalks of the two eyes have just started to project from the brain and the lenses have not yet formed, one of the budding stalks can be removed and transplanted to a region underneath a different epidermis, such as that of the belly. When Spemann performed this critical experiment, a lens still formed, this time from belly epidermis cells in the region above where the budding stalk had been transplanted.

What is the nature of the inducing signal that passes from one tissue to the other? If a nonporous barrier, such as a layer of cellophane, is imposed between the inducer and the target tissue, no induction takes place. In contrast, a porous filter, through which proteins can pass, does permit induction to occur. The induction process was discussed in detail in chapter 19. In brief, the inducer cells produce a protein factor (paracrine signal molecule) that binds to the cells of the target tissue, initiating changes in gene expression.

The Nature of Developmental Decisions

All of the cells of the body, with the exception of a few specialized ones that have lost their nuclei, have an entire complement of genetic information. Despite the fact that all its cells are genetically identical, an adult vertebrate contains hundreds of cell types, each expressing some unique aspect of the total genetic information for that individual. What factors determine which genes are to be expressed in a particular cell and which are not? In a liver cell, what mechanism keeps the genetic information that specifies nerve cell characteristics turned off? Does the differentiation of that particular cell into a liver cell entail the physical loss of the information specifying other cell types? No, it does not—but cells progressively lose the capacity to *express* ever-larger portions of their genomes. *Development is a process of progressive restriction of gene expression.*

Some cells become **determined** quite early in development. For example, all of the egg cells of the human female are set aside very early in the life of the embryo, although some of these cells will not complete development into functional oocytes for more than 40 years. To a large degree, a cell's location in the developing embryo determines its fate. By changing a cell's location, an experimenter can alter its developmental destiny. However, this is only true up to a certain point in the cell's development. At some stage, every cell's ultimate fate becomes fixed, a process referred to as **commitment.** Commitment is considered irreversible when it occurs during the course of normal development, but researchers have shown that it can be reversed in the laboratory. In fact, entire individuals have been cloned from individual specialized cells, as recounted in chapter 20.

When a cell is "determined," its developmental fate can be predicted; when a cell is "committed," that developmental fate cannot be altered. Determination often occurs very early in development, commitment somewhat later.

Embryonic Development and Vertebrate Evolution

The primitive chordates that gave rise to vertebrates were initially slow-moving, filter-feeding animals with relatively low metabolic rates. Many of the unique vertebrate adaptations that contribute to their varied ecological roles involve structures that arise from neural crest cells. The vertebrates became fast-swimming predators with much higher metabolic rates. This accelerated metabolism permitted a greater level of activity than was possible among the more primitive chordates. Other evolutionary changes associated with the derivatives of the neural crest provided better detection of prey, a greatly improved ability to orient spatially during prey capture, and the means to respond quickly to sensory information. The evolution of the neural crest and of the structures derived from it were thus crucial steps in the evolution of the vertebrates (figure 51.17).

Ontogeny Recapitulates Phylogeny

The patterns of development in the vertebrate groups that evolved most recently reflect in many ways the simpler patterns occurring among earlier forms. Thus, mammalian development and bird development are elaborations of reptile development, which is an elaboration of amphibian development, and so forth (figure 51.18). During the development of a mammalian embryo, traces can be seen of appendages and organs that are apparently relics of more primitive chordates. For example, at certain stages a human embryo possesses pharyngeal slits, which occur in all chordates and are homologous to the gill slits of fish. At later stages, a human embryo also has a tail.

In a sense, the patterns of development in chordate groups have built up in incremental steps over the evolutionary history of those groups. The developmental instructions for each new form seem to have been layered on top of the previous instructions, contributing additional steps in the developmental journey. This hypothesis, pro-

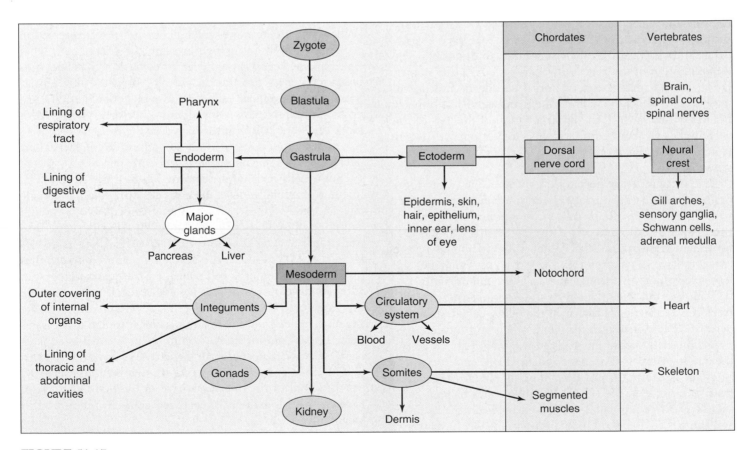

FIGURE 51.17
Derivation of the major tissue types. The three germ layers that form during gastrulation give rise to all the organs and tissues in the body, but the neural crest cells that form from ectodermal tissue give rise to structures that are prevalent in the vertebrate animal, such as gill arches and Schwann cells.

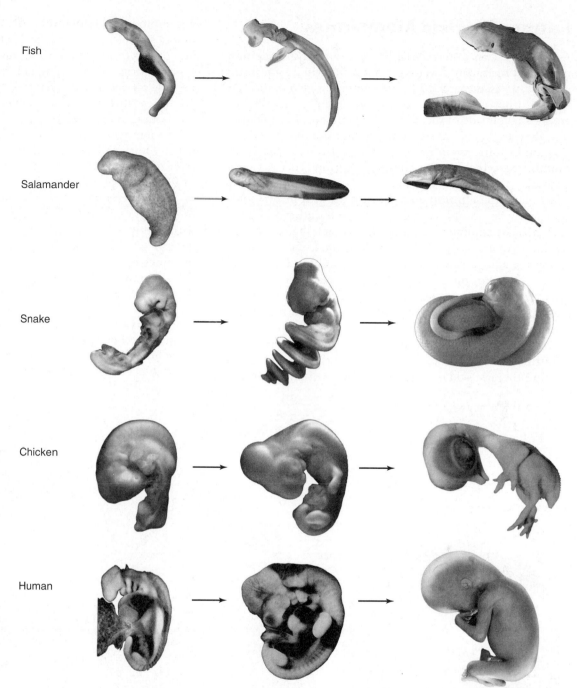

Fish

Salamander

Snake

Chicken

Human

FIGURE 51.18
Embryonic development of vertebrates. Traces of the evolutionary history of the vertebrates can be clearly seen in vertebrate embryos. In these photographs, the early embryonic stages of these vertebrates bear a striking resemblance to each other, even though the individuals are from different classes (fish, amphibians, reptiles, birds, and mammals). All vertebrates start out with an enlarged head region, gill slits, and a tail, regardless of whether these characteristics are retained in the adults.

posed in the nineteenth century by Ernst Haeckel, is referred to as the "biogenic law." It is usually stated as an aphorism: *Ontogeny recapitulates phylogeny;* that is, embryological development (ontogeny) exhibits a progression of changes indicative of evolutionary origin (phylogeny). When stated in this way, the biogenic law is not literally true because embryonic stages are not equivalent to adult ancestral forms. Instead, the embryonic stages of a particular vertebrate often reflect the *embryonic* stages of that vertebrate's ancestors. In other words, the pharyngeal slits of a

mammalian embryo are not like the gill slits its ancestors had when they were *adults.* Rather, they are like the pharyngeal slits its ancestors had when they were *embryos.*

Vertebrates seem to have evolved largely by the addition of new instructions to the developmental program. Development of a mammal thus proceeds through a series of stages, and the earlier stages are unchanged from those that occur in the development of more primitive vertebrates.

Extraembryonic Membranes

As an adaptation to terrestrial life, the embryos of reptiles, birds, and mammals develop within a fluid-filled **amniotic membrane** (see chapter 34). The amniotic membrane and several other membranes form from embryonic cells but are located outside of the body of the embryo. For this reason, they are known as **extraembryonic membranes.** The extraembryonic membranes, later to become the **fetal membranes,** include the amnion, chorion, yolk sac, and allantois.

In birds, the **amnion** and **chorion** arise from two folds that grow to completely surround the embryo (figure 51.19a). The amnion is the inner membrane that surrounds the embryo and suspends it in *amniotic fluid*, thereby mimicking the aquatic environments of fish and amphibian embryos. The chorion is located next to the eggshell and is separated from the other membranes by a cavity, the *extraembryonic coelom*. The **yolk sac** plays a critical role in the nutrition of bird and reptile embryos; it is also present in mammals, though it does not nourish the embryo. The **allantois** is derived as an outpouching of the gut and serves to store the uric acid excreted in the urine of birds. During development, the allantois of a bird embryo expands to form a sac that eventually fuses with the overlying chorion, just under the eggshell. The fusion of the allantois and chorion form a functioning unit in which embryonic blood vessels, carried in the allantois, are brought close to the porous eggshell for gas exchange. The allantois is thus the functioning "lung" of a bird embryo.

In mammals, the embryonic cells form an inner cell mass, which will become the body of the embryo, and a layer of surrounding cells called the trophoblast (see figure 51.9). The trophoblast implants into the endometrial lining of its mother's uterus and becomes the chorionic membrane (figure 51.19b). The part of the chorion in contact with endometrial tissue contributes to the placenta, as is described in more detail in section 51.5. The allantois in mammals contributes blood vessels to the structure that will become the umbilical cord, so that fetal blood can be delivered to the placenta for gas exchange.

The extraembryonic membranes include the yolk sac, amnion, chorion, and allantois. These are derived from embryonic cells and function in a variety of ways to support embryonic development.

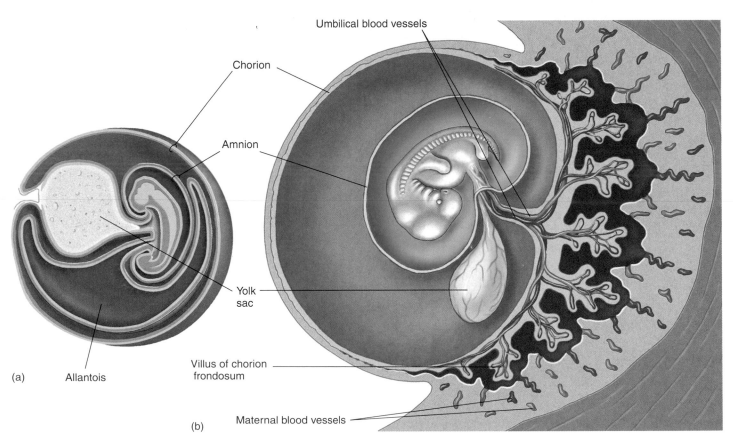

FIGURE 51.19

The extraembryonic membranes. The extraembryonic membranes in (*a*) a chick embyro and (*b*) a mammalian embryo share some of the same characteristics. However, in the chick, the allantois continues to grow until it eventually unites with the chorion just under the eggshell, where it is involved in gas exchange. In the mammalian embryo, the allantois contributes blood vessels to the developing umbilical cord.

51.5 Human development is divided into trimesters.

First Trimester

The development of the human embryo shows its evolutionary origins. Without an evolutionary perspective, we would be unable to account for the fact that human development proceeds in much the same way as development in a bird. In both animals, the embryo develops from a flattened collection of cells—the blastodisc in birds or the inner cell mass in humans. While the blastodisc of a bird is flattened because it is pressed against a mass of yolk, the inner cell mass of a human is flat despite the absence of a yolk mass. In humans as well as in birds, a primitive streak forms and gives rise to the three primary germ layers. Human development, from fertilization to birth, takes an average of 266 days. This time is commonly divided into three periods, called trimesters.

The First Month

About 30 hours after fertilization, the zygote undergoes its first cleavage; the second cleavage occurs about 30 hours after that. By the time the embryo reaches the uterus (6–7 days after fertilization), it is a blastula, which in mammals is referred to as a blastocyst. As we mentioned earlier, the blastocyst consists of an inner cell mass,

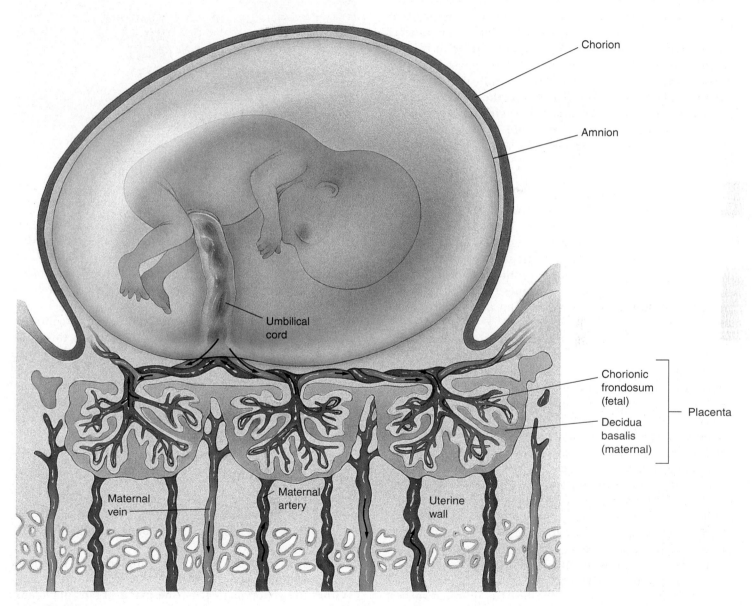

FIGURE 51.20
Structure of the placenta.   The placenta contains a fetal component, the chorionic frondosum, and a maternal component, the decidua basalis. Deoxygenated fetal blood from the umbilical arteries (shown in *blue*) enters the placenta, where it picks up oxygen and nutrients from the mother's blood. Oxygenated fetal blood returns in the umbilical vein (shown in *red*) to the fetus.

which will become the body of the embryo, and a surrounding layer of trophoblast cells (see figure 51.9). The blastocyst begins to grow rapidly and initiates the formation of the amnion and the chorion. The blastocyst digests its way into the endometrial lining of the uterus in the process known as *implantation*.

During the second week after fertilization, the developing chorion forms branched extensions, the *chorionic frondosum* (fetal placenta) that protrude into the endometrium (figure 51.20). These extensions induce the surrounding endometrial tissue to undergo changes and become the *decidua basalis* (maternal placenta). Together, the chorionic frondosum and decidua basalis form a single functioning unit, the placenta (figure 51.21). Within the placenta, the mother's blood and the blood of the embryo come into close proximity but do not mix (see figure 51.19b). Oxygen can thus diffuse from the mother to the embryo, and carbon dioxide can diffuse in the opposite direction. In addition to exchanging gases, the placenta provides nourishment for the embryo, detoxifies certain molecules that may pass into the embryonic circulation, and secretes hormones. Certain substances, such as alcohol, drugs, and antibiotics, are not stopped by the placenta and pass from the mother's bloodstream to the fetus.

One of the hormones released by the placenta is human chorionic gonadotropin (hCG), which was discussed in chapter 50. This hormone is secreted by the trophoblast cells even before they become the chorion, and is the hormone assayed in pregnancy tests. Because its action is almost identical to that of luteinizing hormone (LH), hCG maintains the mother's corpus luteum. The corpus luteum, in turn, continues to secrete estradiol and progesterone, thereby preventing menstruation and further ovulations.

Gastrulation also takes place in the second week after fertilization. The primitive streak can be seen on the surface of the embryo, and the three germ layers (ectoderm, mesoderm, and endoderm) are differentiated.

In the third week, neurulation occurs. This stage is marked by the formation of the neural tube along the dorsal axis of the embryo, as well as by the appearance of the first somites, which give rise to the muscles, vertebrae, and connective tissues. By the end of the third week, over a dozen somites are evident, and the blood vessels and gut have begun to develop. At this point, the embryo is about 2 millimeters long.

Organogenesis (the formation of body organs) begins during the fourth week (figure 51.22a). The eyes form. The tubular heart develops its four chambers and starts to pulsate rhythmically, as it will for the rest of the individual's life. At 70 beats per minute, the heart is destined to beat more than 2.5 billion times during a lifetime of 70 years. Over 30 pairs of somites are visible by the end of the fourth week, and the arm and leg buds have begun to form. The embryo has increased in length to about 5 millimeters. Al-

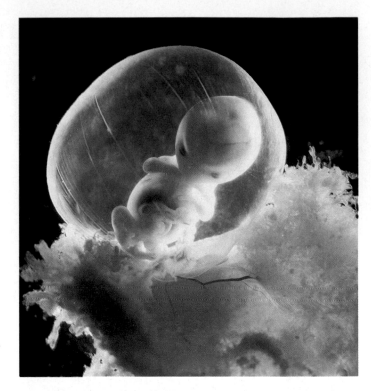

FIGURE 51.21
Placenta and embryo at seven weeks.

though the developmental scenario is now far advanced, many women are still unaware they are pregnant at this stage.

Early pregnancy is a very critical time in development because the proper course of events can be interrupted easily. In the 1960s, for example, many pregnant women took the tranquilizer thalidomide to minimize the discomforts associated with early pregnancy. Unfortunately, this drug had not been adequately tested. It interferes with limb bud development, and its widespread use resulted in many deformed babies. Organogenesis may also be disrupted during the first and second months of pregnancy if the mother contracts rubella (German measles). Most spontaneous abortions occur during this period.

The Second Month

Morphogenesis (the formation of shape) takes place during the second month (figure 51.22b). The miniature limbs of the embryo assume their adult shapes. The arms, legs, knees, elbows, fingers, and toes can all be seen—as well as a short bony tail! The bones of the embryonic tail, an evolutionary reminder of our past, later fuse to form the coccyx. Within the abdominal cavity, the major organs, including the liver, pancreas, and gallbladder, become evident. By the end of the second month, the embryo has grown to about 25 millimeters in length, weighs about one gram, and begins to look distinctly human.

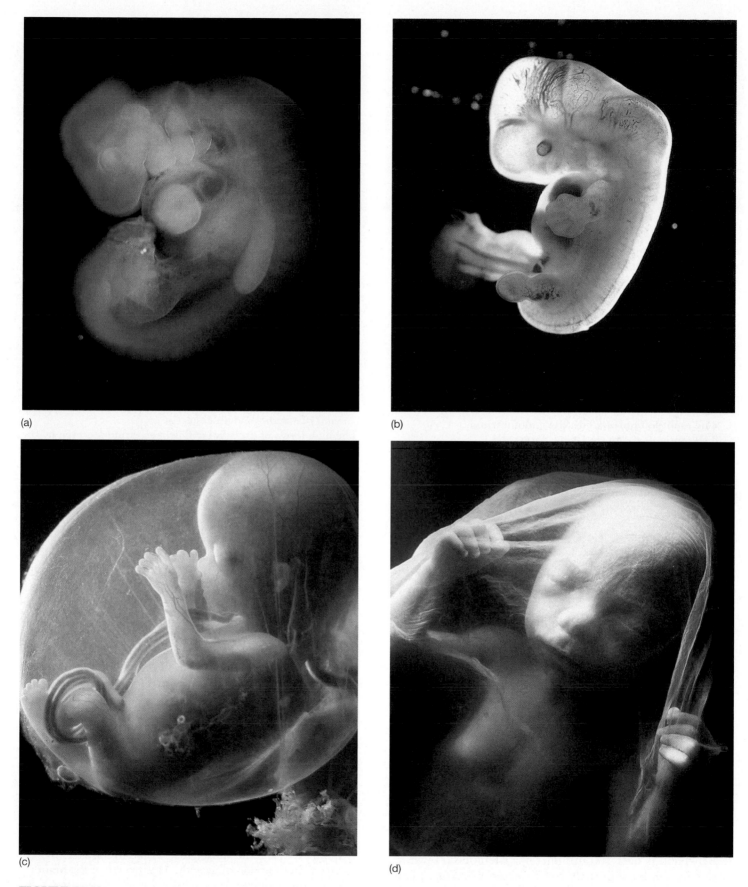

FIGURE 51.22
The developing human. (*a*) Four weeks, (*b*) seven weeks, (*c*) three months, and (*d*) four months.

The eighth week marks the transition from embryo to fetus. At this time, all of the major organs of the body have been established.

The Third Month

The nervous system develops during the third month, and the arms and legs start to move (see figure 51.22c). The embryo begins to show facial expressions and carries out primitive reflexes such as the startle reflex and sucking.

At around 10 weeks, the secretion of human chorionic gonadotropin (hCG) by the placenta declines, and the corpus luteum regresses as a result. However, menstruation does not occur because the placenta itself secretes estradiol and progesterone (figure 51.23). In fact, the amounts of these two hormones secreted by the placenta far exceed the amounts that are ever secreted by the ovaries. The high levels of estradiol and progesterone in the blood during pregnancy continue to inhibit the release of FSH and LH, thereby preventing ovulation. They also help maintain the uterus and eventually prepare it for labor and delivery, and they stimulate the development of the mammary glands in preparation for lactation after delivery.

> **The embryo implants into the endometrium, differentiates the germ layers, forms the extraembryonic membranes, and undergoes organogenesis during the first month and morphogenesis during the second month.**

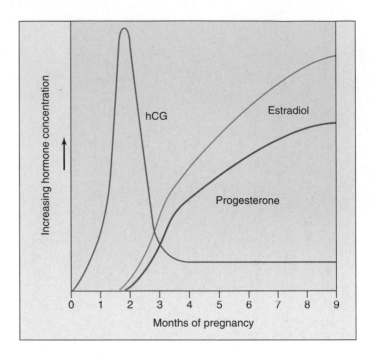

FIGURE 51.23
Hormonal secretion by the placenta. The placenta secretes human chorionic gonadotropin (hCG), which peaks in the second month and then declines. After five weeks, it secretes increasing amounts of estradiol and progesterone.
The high levels of estradiol and progesterone secreted by the placenta prevent ovulation and thus formation of any additional embryos during pregnancy. What would be the expected effect of these high hormone levels in the absence of pregnancy?

Second and Third Trimesters

The second and third trimesters are characterized by the tremendous growth and development required for the viability of the baby after its birth.

Second Trimester

Bones actively enlarge during the fourth month (see figure 51.22d), and by the end of the month, the mother can feel the baby kicking. During the fifth month, the head and body grow a covering of fine hair. This hair, called **lanugo,** is another evolutionary relict but is lost later in development. By the end of the fifth month, the rapid heartbeat of the fetus can be heard with a stethoscope, although it can also be detected as early as 10 weeks with a fetal monitor. The fetus has grown to about 175 millimeters in length and has attained a weight of about 225 grams. Growth begins in earnest in the sixth month; by the end of that month, the baby weighs 600 grams (1.3 lbs) and is over 300 millimeters (1 ft) long. However, most of its prebirth growth is still to come. The baby cannot yet survive outside the uterus without special medical intervention.

Third Trimester

The third trimester is predominantly a period of growth rather than development. The weight of the fetus doubles several times, but this increase in bulk is not the only kind of growth that occurs. Most of the major nerve tracts in the brain, as well as many new neurons (nerve cells), are formed during this period. The developing brain produces neurons at an average rate estimated at more than 250,000 per minute! Neurological growth is far from complete at the end of the third trimester, when birth takes place. If the fetus remained in the uterus until its neurological development was complete, it would grow too large for safe delivery through the pelvis. Instead, the infant is born as soon as the probability of its survival is high, and its brain continues to develop and produce new neurons for months after birth.

> **The critical stages of human development take place quite early, and the subsequent six months are essentially a period of growth. The growth of the brain is not yet complete, however, by the end of the third trimester, and must be completed postnatally.**

Birth and Postnatal Development

In some mammals, changing hormone levels in the developing fetus initiate the process of birth. The fetuses of these mammals have an extra layer of cells in their adrenal cortex, called a fetal zone. Before birth, the fetal pituitary gland secretes corticotropin, which stimulates the fetal zone to secrete steroid hormones. These corticosteroids then induce the uterus of the mother to manufacture prostaglandins, which trigger powerful contractions of the uterine smooth muscles.

The adrenal glands of human fetuses lack a fetal zone, and human birth does not seem to be initiated by this mechanism. In a human, the uterus releases prostaglandins, possibly as a result of the high levels of estradiol secreted by the placenta. Estradiol also stimulates the uterus to produce more oxytocin receptors, and as a result, the uterus becomes increasingly sensitive to oxytocin. Prostaglandins begin the uterine contractions, but then sensory feedback from the uterus stimulates the release of oxytocin from the mother's posterior pituitary gland. Working together, oxytocin and prostaglandins further stimulate uterine contractions, forcing the fetus downward (figure 51.24). This positive feedback mechanism accelerates during labor. Initially, only a few contractions occur each hour, but the rate eventually increases to one contraction every two to three minutes. Finally, strong contractions, aided by the mother's pushing, expel the fetus, which is now a newborn baby, or neonate.

After birth, continuing uterine contractions expel the placenta and associated membranes, collectively called the afterbirth. The umbilical cord is still attached to the baby, and to free the newborn, a doctor or midwife clamps and cuts the cord. Blood clotting and contraction of muscles in the cord prevent excessive bleeding.

Nursing

Milk production, or lactation, occurs in the alveoli of mammary glands when they are stimulated by the anterior pituitary hormone prolactin. Milk from the alveoli is secreted into a series of alveolar ducts, which are surrounded by smooth muscle and lead to the nipple (figure 51.25). During pregnancy, high levels of progesterone stimulate the development of the mammary alveoli, and high levels of

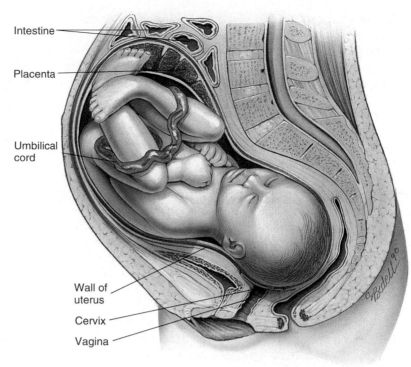

FIGURE 51.24
Position of the fetus just before birth. A developing fetus causes major changes in a woman's anatomy. The stomach and intestines are pushed far up, and considerable discomfort often results from pressure on the lower back. In a natural delivery, the fetus exits through the cervix, which must dilate (expand) considerably to permit passage.

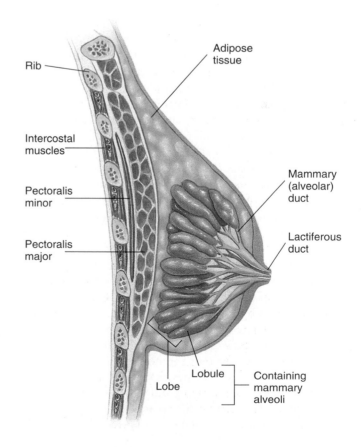

FIGURE 51.25
A sagittal section of a mammary gland. The mammary alveoli produce milk in response to stimulation by prolactin, and milk is ejected through the lactiferous duct in response to oxytocin.

estradiol stimulate the development of the alveolar ducts. However, estradiol blocks the actions of prolactin on the mammary glands, and it inhibits prolactin secretion by promoting the release of prolactin-inhibiting hormone from the hypothalamus. During pregnancy, therefore, the mammary glands are prepared for, but prevented from, lactating.

When the placenta is discharged after birth, the concentrations of estradiol and progesterone in the mother's blood decline rapidly. This decline allows the anterior pituitary gland to secrete prolactin, which stimulates the mammary alveoli to produce milk. Sensory impulses associated with the baby's suckling trigger the posterior pituitary gland to release oxytocin. Oxytocin stimulates contraction of the smooth muscle surrounding the alveolar ducts, thus causing milk to be ejected by the breast. This pathway is known as the milk-ejection reflex. The secretion of oxytocin during lactation also causes some uterine contractions, as it did during labor. These contractions help restore the tone of uterine muscles in mothers who are breast-feeding.

The first milk produced after birth is a yellowish fluid called colostrum, which is both nutritious and rich in maternal antibodies. Milk synthesis begins about three days following the birth and is referred to as the milk "coming in." Many mothers nurse for a year or longer. During this period, important pair-bonding occurs between the mother and child. When nursing stops, the accumulation of milk in the breasts signals the brain to stop secreting prolactin, and milk production ceases.

Postnatal Development

Growth of the infant continues rapidly after birth. Babies typically double their birth weight within two months. Because different organs grow at different rates and cease growing at different times, the body proportions of infants are different from those of adults. The head, for example, is disproportionately large in newborns, but after birth it grows more slowly than the rest of the body. Such a pattern of growth, in which different components grow at different rates, is referred to as *allometric growth*.

In most mammals, brain growth is mainly a fetal phenomenon. In chimpanzees, for instance, the brain and the cerebral portion of the skull grow very little after birth, while the bones of the jaw continue to grow. As a result, the head of an adult chimpanzee looks very different from that of a fetal chimpanzee (figure 51.26). In human infants, on the other hand, the brain and cerebral skull grow at the same rate as the jaw. Therefore, the jaw-skull proportions do not change after birth, and the head of a human adult looks very similar to that of a human fetus. It is primarily for this reason that an early human fetus seems so remarkably adultlike. The fact that the human brain continues to grow significantly for the first few years of postnatal life means that adequate nutrition and a safe environment are particularly crucial during this period for the full development of a person's intellectual potential.

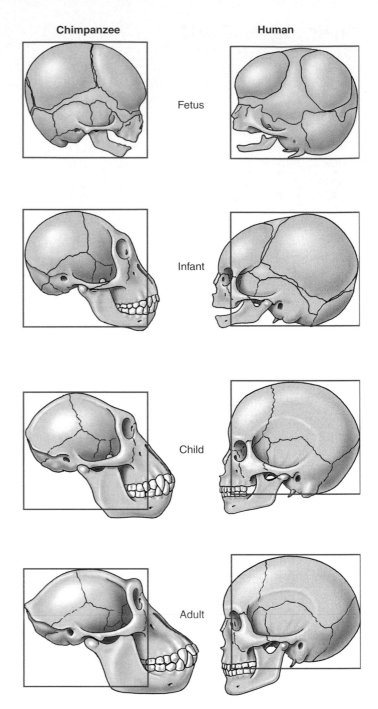

FIGURE 51.26
Allometric growth. In young chimpanzees, the jaw grows at a faster rate than the rest of the head. As a result, the shape of the adult head differs greatly from its shape as a newborn. In humans, the difference in growth between the jaw and the rest of the head is much smaller, and the adult head shape is similar to that of the newborn.

Birth occurs in response to uterine contractions stimulated by oxytocin and prostaglandins. Lactation is stimulated by prolactin and prostaglandins, but the milk-ejection reflex requires the action of oxytocin.

Concept Review

For interactive testing, visit the Online Learning Center with PowerWeb at www.mhhe.com/Raven7

51.1 Fertilization is the initial event in development.

Fertilization

- Fertilization consists of three stages: penetration, activation, and nuclei fusion. (p. 1082)
- The series of events initiated by sperm penetration are collectively called egg activation. (p. 1082)
- Sperm penetration stimulates the second meiotic division in the egg nucleus, may trigger movement of the egg cytoplasm around the point of sperm entry, and leads to an increase in metabolic activity. (pp. 1082–1083)

51.2 Cell cleavage and the formation of a blastula set the stage for later development.

Cell Cleavage Patterns

- Cleavage is a period of cell division that is not accompanied by an increase in the overall volume of zygote cytoplasm and eventually leads to the formation of a hollow blastula. (p. 1084)
- The pattern of cleavage is influenced by the presence and location of yolk. (p. 1084)

51.3 Gastrulation forms the three germ layers of the embryo.

The Process of Gastrulation

- Gastrulation occurs when certain groups of cells invaginate and involute from the surface of the blastula, and determine the basic developmental pattern of the vertebrate embryo. (p. 1088)
- By the end of gastrulation, embryo cells have rearranged themselves into three primary germ layers: ectoderm, mesoderm, and endoderm. (p. 1087)

51.4 Body architecture is determined during the next stages of embryonic development.

Developmental Processes During Neurulation

- Development of the dorsal nerve cord is termed neurulation. (p. 1090)
- The appearance of the neural crest was a key event in vertebrate evolution, ultimately developing into structures characteristic of the vertebrate body. (p. 1090)
- In primitive chordates, pharyngeal slits evolved into the vertebrate gill chamber, improving respiration. (p. 1091)

How Cells Communicate During Development

- The relative position of particular cell layers helps determine the organs that develop. (pp. 1092–1093)
- Inducer cells produce a protein factor that binds to the cells of the target tissue, initiating changes in gene expression. (p. 1093)
- Development is a process of progressive restriction of gene expression. (p. 1093)
- At the stage called commitment, every cell's fate becomes fixed. (p. 1093)

Embryonic Development and Vertebrate Evolution

- Patterns of chordate development have built up in incremental steps over the evolutionary history of those groups. (pp. 1094–1095)
- Biogenic law states that embryological development (ontogeny) exhibits a progression of changes indicative of evolutionary origin (phylogeny). (p. 1095)
- The embryonic stages of a particular vertebrate often reflect the embryonic stages of its ancestors. (p. 1095)

Extraembryonic Membranes

- Extraembryonic membranes become the fetal membranes—amnion, chorion, yolk sac, and allantois—and serve to support embryonic development. (p. 1096)

51.5 Human development is divided into trimesters.

First Trimester

- The embryo reaches the uterus six to seven days after fertilization. Gastrulation takes place in the second week, while neurulation occurs in the third week, and organogenesis begins during the fourth week. (pp. 1097–1098)
- Morphogenesis takes place in the second month as the embryo assumes its adult shape. (p. 1098)
- The eighth week marks the transition from an embryo to a fetus. (p. 1100)
- In the third month, the nervous system develops, and the arms and legs start to move. (p. 1100)

Second and Third Trimesters

- During the fourth month, bones actively enlarge, and the mother can feel the baby kicking. (p. 1100)
- Growth begins in earnest in the sixth month. (p. 1100)
- The third trimester is predominantly a period of growth rather than development. (p. 1100)
- During this period, most of the major nerve tracts in the brain are formed. (p. 1100)

Birth and Postnatal Development

- Oxytocin and prostaglandins help stimulate uterine contractions, assisting the movement of the fetus downward. (p. 1101)
- During pregnancy, the mammary glands are prepared for, but prevented from, lactating. (p. 1101)
- Oxytocin stimulates contraction of the smooth muscle around the alveolar ducts, causing milk to be ejected. (pp. 1101–1102)
- Babies typically double their birth weight within two months. (p. 1102)
- In most mammals, brain growth is predominantly a fetal phenomenon, while in human infants, the brain continues to grow for the first few years of postnatal life. (p. 1102)

For interactive testing, visit the Online Learning Center with PowerWeb at www.mhhe.com/Raven7

Self Test

1. An egg surrounded by a granulosa layer is most likely from a
   a. sea urchin.
   b. human.
   c. frog.
   d. fruit fly.
2. The first thing a sea urchin sperm encounters when it makes contact with the egg is the
   a. plasma membrane.
   b. vitelline membrane.
   c. jelly coat.
   d. zona pellucida.
3. Holoblastic cleavage results in
   a. formation of a symmetrical blastula composed of cells of approximately equal size.
   b. formation of an asymmetrical blastula composed of cells of approximately unequal size.
   c. cell division of only the cells near the animal pole.
   d. cell division of only the cells near the vegetal pole.
4. If the trophoblast layer failed to form in a mammalian embryo, which of the following structures would not develop?
   a. the blastopore
   b. the inner cell mass
   c. the archenteron
   d. the fetal placenta
5. Which forms first during gastrulation of a frog embryo?
   a. mouth
   b. anus
   c. dorsal lip
   d. gut
6. Gastrulation in mammals and birds is similar in that
   a. cells migrate over the dorsal lip to generate the archenteron in the anterior hemisphere of the embryo.
   b. cells migrate inward from the upper layer of the blastodisc to form the mesoderm.
   c. cells migrate outward from the upper layer of the blastodisc to form the mesoderm.
   d. cells migrate inward from the lower layer of the blastodisc to form the mesoderm.
7. Neural crest cells break off from the _____ and later move to the sides of the developing embryo.
   a. placodes
   b. ectoderm
   c. notochord
   d. neural tube
8. The cells of the Spemann organizer are responsible for
   a. notochord development.
   b. cell transplantation.
   c. dorsal lip formation.
   d. development of the eye through an induction mechanism.
9. All of the following are derived from mesoderm *except*
   a. muscles.
   b. the liver.
   c. gonads.
   d. blood vessels.
10. Most fetal growth occurs in
    a. the first trimester.
    b. the second trimester.
    c. the third trimester.
    d. the postnatal period.

Test Your Visual Understanding

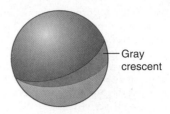

1. In this picture of a newly fertilized frog embryo, where would you predict that the sperm entered the egg? Draw a sperm at the approximate location of sperm entry.

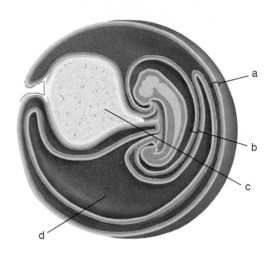

2. Match the following labels with their appropriate letters on this chick embryo and describe their function.
   Allantois
   Amnion
   Chorion
   Yolk sac

Apply Your Knowledge

1. You are preparing for a debate on the use of embryonic stem cells in research. What are the pros and cons of this practice?
2. Comparative embryology of vertebrates has led to the phrase *ontogeny recapitulates phylogeny*. This biogenic law, although not strictly true, supports evolution. Why?
3. What would happen if a mammalian embryo failed to produce sufficient amounts of human chorionic gonadotropin (hCG) in early pregnancy?

52

Behavioral Biology

Chapter Outline

52.1 Many behavioral patterns are innate.

    Approaches to the Study of Behavior. Field biologists focus on evolutionary aspects of behavior.

    Behavioral Genetics. At least some behaviors are genetically determined.

52.2 Learning influences behavior.

    Nonassociative Learning and Conditioning. Conditioning requires association between two stimuli or between a stimulus and a response.

    The Development of Behavior. Parent-offspring interactions influence behavioral development.

    Animal Cognition. It is not clear to what degree animals "think."

    Migratory Behavior. Animals use many cues from the environment to navigate during migrations.

52.3 Communication is a key element of many animal behaviors.

    Courtship. Animals use signals to court one another.

    Communication in Social Groups. Bees and other social animals communicate in complex ways.

52.4 Evolutionary forces shape behavior.

    Behavioral Ecology. Behavior is shaped by natural selection.

    Foraging Behavior. Natural selection favors the most efficient foraging behavior.

    Territorial Behavior. Animals defend territories to increase reproductive advantage and foraging efficiency.

    Reproductive Strategies. The degree of parental investment influences other reproductive behaviors.

    Reproductive Competition and Sexual Selection. Mate choice affects reproductive success, and so is a target of natural selection.

    Mating Systems. Mating systems are reproductive solutions to particular ecological challenges.

52.5 There is considerable controversy about the evolution of social behavior.

    Altruism and Group Living. Many ideas have been put forward to explain the evolution of altruism.

    Group Living and the Evolution of Social Systems. Social organisms exhibit cooperation and altruism.

FIGURE 52.1
Rearing offspring involves complex behavior. Living in groups called prides makes lions better mothers. Females share the responsibilities of nursing and protecting the pride's young, increasing the probability that the youngsters will survive into adulthood.

Organisms interact with their environment in many ways. To understand these interactions, we need to appreciate the internal factors that shape the way an animal behaves, as well as aspects of the external environment that affect individual organisms. In this chapter, we explore the mechanisms that determine an animal's behavior (figure 52.1) and examine the field of behavioral ecology, which investigates how natural selection has molded behavior through evolutionary time.

Part VIII Ecology and Behavior

1105

52.1 Many behavioral patterns are innate.

Approaches to the Study of Behavior

During the past two decades, the study of animal behavior has emerged as an important and diverse science that bridges several disciplines within biology. Evolution, ecology, physiology, genetics, and psychology all have natural and logical linkages with the study of behavior, each discipline adding a different perspective and addressing different questions.

Research in animal behavior has made major contributions to our understanding of nervous system organization, child development, and human communication, as well as the speciation process, community organization, and the mechanism of natural selection itself. The study of the behavior of nonhuman animals has led to the identification of general principles of behavior, which have been applied, often controversially, to humans. This has changed the way we think about the origins of human behavior and the way we perceive ourselves.

Behavior can be defined as the way an animal responds to stimuli in its environment. A stimulus might be as simple as the odor of food. In this sense, a bacterial cell "behaves" by moving toward higher concentrations of the sugar. This behavior is very simple and well suited to the life of bacteria, allowing these organisms to live and reproduce. As animals evolved, they occupied different environments and faced diverse problems that affected their survival and reproduction. Their nervous systems and behavior concomitantly became more complex. Nervous systems perceive and process information concerning environmental stimuli and trigger adaptive motor responses, which we see as patterns of behavior.

When we observe animal behavior, we can explain it in two different ways. First, we might ask *how* it all works—that is, how the animal's senses, nerve networks, or internal state provide a physiological basis for the behavior. In this way, we would be asking a question about *proximate causation*. To analyze the proximate cause of behavior, we might measure hormone levels or record the impulse activity of neurons in the animal. We could also ask *why* the behavior evolved—that is, what is its adaptive value? This is a question concerning *ultimate causation*. To study the ultimate cause of a behavior, we would attempt to determine how it influenced the animal's survival or reproductive success. Thus, a male songbird may sing during the breeding season because it has a level of the steroid sex hormone testosterone, which binds to hormone receptors in the brain and triggers the production of song; this would be the proximate cause of the male bird's song. But the male sings to defend a territory from other males and to attract a female with which to reproduce; this is the ultimate, or evolutionary, explanation for the male's vocalization.

The study of behavior has had a long history of controversy. One source of controversy has been the question of whether behavior is determined more by an individual's genes or by its learning and experience. In other words, is behavior the result of nature (instinct) or nurture (experience)? In the past, this question has been considered an either/or proposition, but we now know that instinct and experience both play significant roles, often interacting in complex ways to produce the final behavior. The scientific study of instinct and learning, as well as their interrelationship, has led to the growth of several scientific disciplines, including ethology (the study of behavior), behavioral genetics, behavioral neuroscience, and comparative psychology.

Innate Behavior

Early research in the field of animal behavior—most notably by Nobel Prize winners Karl von Frisch, Konrad Lorenz, and Niko Tinbergen (figure 52.2)—focused on behavioral patterns that appeared to be instinctive or innate. Because behavior is often *stereotyped* (appearing in the same way in different individuals of a species), these early researchers argued that it must be based on preset paths in the nervous system. In their view, these paths are structured from genetic blueprints and cause animals to show essentially the same behavior from the first time it is produced throughout their lives.

These researchers based their opinions on behaviors such as egg retrieval by geese. Geese incubate their eggs in a nest. If a goose notices that an egg has been knocked out of the nest, it will extend its neck toward the egg, get up, and roll the egg back into the nest with a side-to-side motion of its neck while the egg is tucked beneath its bill (figure 52.3). Even if the egg is removed during retrieval, the goose completes the behavior, as if driven by a program released by the initial sight of the egg outside the nest. According to ethologists, egg retrieval behavior is triggered by a **sign stimulus** (also called a **key stimulus**), the appearance of an egg out of the nest; a component of the goose's nervous system, the **innate releasing mechanism,** provides the neural instructions for the motor program, or **fixed action pattern.** More generally, the sign stimulus is a "signal" in the environment that triggers a behavior. The innate releasing mechanism is the sensory mechanism that detects the signal, and the fixed action pattern is the stereotyped act.

One interesting aspect of sign stimuli is that they are often not very specific; in some situations, a wide variety of objects will trigger a fixed action pattern. For example, geese will attempt to roll baseballs and even beer cans back into their nests. Moreover, once the objects are in the nest, the goose recognizes that they are not eggs and removes them! A simi-

FIGURE 52.2
The founding fathers of ethology. Karl von Frisch, Konrad Lorenz, and Niko Tinbergen pioneered the study of behavioral science. In 1973, they received the Nobel Prize in physiology or medicine for their path-making contributions. Von Frisch led the study of honeybee communication and sensory biology. Lorenz focused on social development (imprinting) and the natural history of aggression. Tinbergen examined the functional significance of behavior and was the first behavioral ecologist.

lar example is provided by male stickleback fish. During the breeding season, males develop bright red coloration on their undersides. Territorial males react aggressively to the approach of other males, performing an aggressive display and even attacking. When Niko Tinbergen observed a male stickleback in a laboratory aquarium displaying aggressively when a red fire truck passed by the window, he realized that the red coloration was the sign stimulus. Subsequent experiments revealed that males would respond to many unfishlike models as long as the models had a red stripe.

This phenomenon is taken one step further by what are termed **supernormal stimuli.** Given a choice between two sign stimuli, one of normal size and the other much larger, many animals will respond to the larger of the two. Thus, geese given a choice of a normal goose egg and one the size of a volleyball will choose to roll the bigger one back to the nest. Why supernormal stimuli exist is not always clear. One aspect to keep in mind, however, is that in many cases, supernormal stimuli do not occur in nature. Thus, geese may prefer eggs the size of volleyballs, but they never encounter eggs of that size. It may be that geese have evolved to respond to the larger object so that they will attend to eggs, rather than smaller, circular rocks. As a result, natural selection may have favored the evolution of a preference for larger objects; this general response may lead to unexpected outcomes in experiments, but probably doesn't often lead to maladaptive behavior.

FIGURE 52.3
Innate egg-rolling response in geese. The series of movements used by a goose to retrieve an egg is a fixed action pattern. Once it detects the sign stimulus (in this case, an egg outside the nest), the goose goes through the entire set of movements: It will extend its neck toward the egg, get up, and roll the egg back into the nest with a side-to-side motion of its neck while the egg is tucked beneath its bill.

Early research in animal behavior emphasized innate behaviors that are the result of preset pathways in the nervous system and thus are likely to be genetically controlled.

Behavioral Genetics

In a famous experiment carried out in the 1940s, Robert Tryon studied the ability of rats to find their way through a maze with many blind alleys and only one exit, where a reward of food awaited. Some rats quickly learned to zip through the maze to the food, making few incorrect turns, while other rats took much longer to learn the correct path. Tryon bred the fast learners with one another to establish a "maze-bright" colony, and he bred the slow learners with one another to establish a "maze-dull" colony. He then tested the offspring in each colony to see how quickly they learned the maze. The offspring of maze-bright rats learned even more quickly than their parents had, while the offspring of maze-dull parents were even poorer at maze learning. After repeating this procedure over several generations, Tryon was able to produce two behaviorally distinct types of rat with very different maze-learning abilities (figure 52.4). Clearly, the ability to learn the maze was to some degree hereditary, governed by genes passed from parent to offspring. Furthermore, those genes appeared to be specific to this behavior, because the two groups of rats did not differ in their ability to perform other behavioral tasks, such as running a completely different kind of maze. Tryon's research demonstrates how a study can reveal that behavior has a heritable component.

Further support for the genetic basis of behavior has come from studies of hybrids. William Dilger of Cornell University examined two species of lovebird (genus *Agapornis*), which differ in the way they carry twigs, paper, and other materials used to build a nest. *Agapornis fischeri* holds nest material in its beak, while *A. roseicollis* carries material tucked under its flank feathers (figure 52.5). When Dilger crossed the two species to produce hybrids, he found that the hybrids carry nest material in a way that seems intermediate between that of the parents: They repeatedly shift material between the beak and the flank feathers. Other studies conducted on courtship songs in crickets and tree frogs also demonstrate the intermediate nature of hybrid behavior.

The role of genetics can also be seen in humans by comparing the behavior of identical twins. Identical twins are, as their name implies, genetically identical. However, most sets of identical twins are raised in the same environment, so it is not possible to determine whether similarities in behavior result from their genetic similarity or from experiences shared as they grew up (the classic nature-versus-nurture debate). However, in some cases, twins have been separated at birth. A recent study of 50 such sets of twins revealed many similarities in personality, temperament, and even leisure-time activities, even though the twins had often been raised in very different environments. These similarities indicate that genetics plays a role in determining behavior even in humans, although the relative importance of genetics versus environment is still hotly debated.

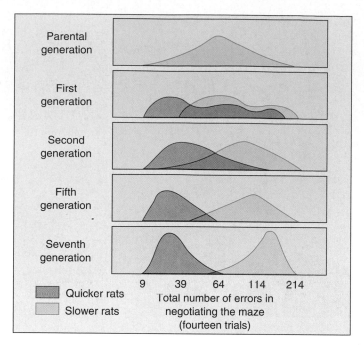

FIGURE 52.4

The genetics of learning. Selection experiments in the laboratory established a genetic basis for differences in the ability to learn to run through a maze.

What would happen if, after the seventh generation, rats were randomly assigned mates regardless of their ability to learn the maze?

FIGURE 52.5

Genetics of lovebird behavior. Some species of lovebirds carry nest material, such as these paper strips, under their flank feathers. When mated with species that carry material in their beaks, the hybrids show intermediate behavior, attempting to carry the material in both ways.

FIGURE 52.6
Genetically caused defect in maternal care. (*a*) In mice, normal mothers take very good care of their offspring, retrieving them if they move away and crouching over them. (*b*) Mothers with the mutant *fosB* allele perform neither of these behaviors, leaving their pups exposed. (*c*) Amount of time female mice were observed crouching in a nursing posture over offspring. (*d*) Proportion of pups retrieved when they were experimentally moved. **Why does the lack of *fosB* alleles lead to maternal inattentiveness?**

(a)
(b)

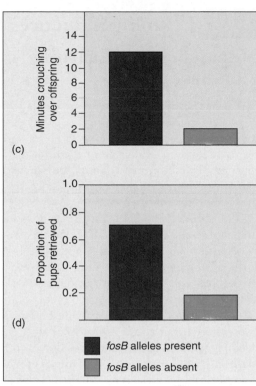
(c)
(d)

fosB alleles present
fosB alleles absent

Single Gene Effects on Behavior

The maze-learning, hybrid, and identical twins studies just described suggest that genes play a role in behavior, but recent research has provided much greater detail on the genetic basis of behavior. In both *Drosophila* and mice, many mutations have been associated with particular behavioral abnormalities.

In fruit flies, for example, individuals that possess alternative alleles for a single gene differ greatly in their feeding behavior as larvae; larvae with one particular allele move around a great deal as they eat, whereas individuals with the alternative allele move hardly at all. A wide variety of mutations at other genes are now known in *Drosophila* that affect almost every aspect of courtship behavior.

The ways in which genetic differences affect behavior have been worked out for several mouse genes. For example, some mice with one mutation have trouble remembering information learned two days earlier about where objects are located. This difference appears to result because the mutant mice do not produce the enzyme α-calcium-calmodulin-dependent kinase II, which plays an important role in the functioning of the hippocampus, a part of the brain important for spatial learning.

Modern molecular biology techniques allow the role of genetics in behavior to be investigated with ever greater precision. For example, male mice genetically engineered to lack the ability to synthesize nitric oxide, a brain neurotransmitter, show increased aggressive behavior.

A particularly fascinating breakthrough occurred in 1996, when scientists discovered a new gene, *fosB*, that seems to determine whether or not female mice will nurture their young. Females with both *fosB* alleles disabled will initially investigate their newborn babies, but then ignore them, in stark contrast to the caring and protective maternal behavior displayed by normal females (figure 52.6).

The cause of this inattentiveness appears to result from a chain reaction. When mothers of new babies initially inspect them, information from their auditory, olfactory, and tactile senses is transmitted to the hypothalamus, where *fosB* alleles are activated, producing a particular protein, which in turn activates other enzymes and genes that affect the neural circuitry within the hypothalamus. These modifications within the brain cause the female to react maternally toward her offspring. In contrast, in mothers lacking the *fosB* alleles, this reaction is stopped midway. No protein is activated, the brain's neural circuitry is not rewired, and maternal behavior does not result.

As these genetic techniques are becoming used more widely, the next few years should see similar dramatic advances in our knowledge of how genes affect various human behaviors.

The genetic basis of behavior is supported by artificial selection experiments, hybridization studies, and studies on the behavior of mutants. Research has also identified specific genes that control behavior.

Nonassociative Learning and Conditioning

Many of the behavioral patterns displayed by animals are not solely the result of instinct. In many cases, animals alter their behavior as a result of previous experiences, a process termed **learning.** The role of learning was first studied intensively in laboratory rodents, but now researchers have learned much about the learning processes and capabilities of a wide range of organisms.

The simplest type of learning, **nonassociative learning,** does not require an animal to form an association between two stimuli or between a stimulus and a response. One form of nonassociative learning is **habituation,** which can be defined as a decrease in response to a repeated stimulus that has no positive or negative consequences. In many cases, the stimulus evokes a strong response when it is first encountered, but the magnitude of the response gradually declines with repeated exposure. For example, young birds see many types of objects moving overhead. At first, they may respond by crouching down and remaining still. Some of the objects, such as falling leaves or members of their own species flying by, are seen very frequently and have no positive or negative consequence to the nestlings. Over time, the young birds may habituate to such stimuli and stop responding. Thus, habituation can be thought of as learning not to respond to a stimulus. Being able to ignore unimportant stimuli is critical for an animal confronting a barrage of stimuli in a complex environment; animals that could not do so would fail to focus their attention on important activities, such as finding food and avoiding predators, and probably would leave few offspring in the next generation.

A change in behavior that involves an association between two stimuli or between a stimulus and a response is termed **associative learning** (figure 52.7). The behavior is modified, or **conditioned,** through the association. This form of learning is more complex than habituation. The two major types of associative learning are **classical conditioning** and **operant conditioning;** they differ in the way the associations are established.

Classical Conditioning

In classical conditioning, the paired presentation of two different kinds of stimuli causes the animal to form an association between the stimuli. Classical conditioning is also called **Pavlovian conditioning,** after Russian psychologist Ivan Pavlov, who first described it. Pavlov presented meat powder, an *unconditioned stimulus,* to a dog and noted that the dog responded by salivating, an *unconditioned response.* If an unrelated stimulus, such as the ring-

(a)

(b)

(c)

FIGURE 52.7
Learning what is edible. Associative learning is involved in predator-prey interactions. (*a*) A naive toad is offered a bumblebee as food. (*b*) The toad is stung, and (*c*) subsequently avoids feeding on bumblebees or any other insects having black-and-yellow coloration. The toad has associated the appearance of the insect with pain, and modifies its behavior.

ing of a bell, was presented at the same time as the meat powder, over repeated trials the dog would salivate in response to the sound of the bell alone. The dog had learned to associate the unrelated sound stimulus with the meat powder stimulus. Its response to the sound stimulus was, therefore, conditioned, and the sound of the bell is referred to as a *conditioned stimulus*.

Operant Conditioning

In operant conditioning, an animal learns to associate its behavioral response with a reward or punishment. American psychologist B. F. Skinner studied operant conditioning in rats by placing them in an apparatus that came to be called a "Skinner box." As the rat explored the box, it would occasionally press a lever by accident, causing a pellet of food to appear. At first, the rat would ignore the lever, eat the food pellet, and continue to move about. Soon, however, it learned to associate pressing the lever (the behavioral response) with obtaining food (the reward). When it was hungry, it would spend all its time pressing the lever. This sort of trial-and-error learning is of major importance to most vertebrates.

Comparative psychologists used to believe that any two stimuli could be linked in classical conditioning and that animals could be conditioned to perform any learnable behavior in response to any stimulus by operant conditioning. As you will see in the following discussion, this view has changed. Today, it is thought that instinct guides learning by determining what type of information can be learned through conditioning.

Instinct and Learning

It is now clear that some animals have innate predispositions toward forming certain associations. For example, if a rat is offered a food pellet at the same time it is exposed to X rays (which later produce nausea), the rat will remember the taste of the food pellet but not its size. Similarly, pigeons can learn to associate *food* with colors but not with sounds, while they can associate *danger* with sounds but not with colors.

These examples of *learning preparedness* demonstrate that what an animal can learn is biologically influenced—that is, learning is possible only within the boundaries set by instinct. Innate programs have evolved because they underscore adaptive responses. In nature, food that is toxic to a rat is likely to have a particular taste; thus, it is adaptive to be able to associate a taste with a feeling of sickness that may develop hours later. The seed a pigeon eats may have a distinctive color that the pigeon can see, but it makes no sound the pigeon can hear. The study of learning has expanded to include its ecological significance, so that we are now able to consider the evolution of learning.

FIGURE 52.8
The Clark's nutcracker has an extraordinary memory. A Clark's nutcracker can remember the locations of up to 2000 seed caches months after hiding them. After conducting experiments, scientists have concluded that the birds use features of the landscape and other surrounding objects as spatial references to memorize the locations of the caches.

An animal's ecology, of course, is key to understanding its mental capabilities. Some species of birds, such as Clark's nutcracker, feed on seeds. Birds store seeds in caches they bury when seeds are abundant so they will have food during the winter. Thousands of seed caches may be buried and then later recovered, sometimes as much as nine months later. One would expect the birds to have an extraordinary spatial memory, and this is indeed what has been found (figure 52.8). Clark's nutcracker, and other seed-hoarding birds, have an unusually large hippocampus, the center for memory storage in the brain (see chapter 45).

Habituation is a simple form of learning in which there is no association between stimuli and responses. In contrast, associative learning (classical and operant conditioning) involves the formation of an association between two stimuli or between a stimulus and a response.

The Development of Behavior

Behavioral biologists now recognize that behavior has both genetic and learned components. Thus far in this chapter, we have discussed the influence of genes and learning separately. But as we will see, these factors interact during development to shape behavior.

Parent-Offspring Interactions

As an animal matures, it may form social attachments to other individuals or develop preferences that will influence behavior later in life. This process, called **imprinting,** is sometimes considered a type of learning. In *filial imprinting,* social attachments form between parents and offspring. For example, young birds of some species begin to follow their mother within a few hours after hatching, and their following response results in a bond between mother and young. However, the young birds' initial experience determines how this imprint is established. The German ethologist Konrad Lorenz showed that birds will follow the first object they see after hatching and direct their social behavior toward that object. Lorenz raised geese from eggs, and when he offered himself as a model for imprinting, the goslings treated him as if he were their parent, following him dutifully (figure 52.9). Black boxes, flashing lights, and watering cans can also be effective imprinting objects (figure 52.10). The success of imprinting is highest during a **sensitive phase,** or a **critical period** (roughly 13 to 16 hours after hatching in geese).

Several studies demonstrate that the social interactions that occur between parents and offspring during the critical period are key to the normal development of behavior. The psychologist Harry Harlow gave orphaned rhesus monkey infants the opportunity to form social attachments with two surrogate "mothers," one made of soft cloth covering a wire frame and the other made only of wire. The infants chose to spend time with the cloth mother, even if only the wire mother provided food, indicating that texture and tactile contact, rather than provision of food, may be among the key qualities in a mother that promote infant social attachment. If infants are deprived of normal social contact, their development is abnormal. Greater degrees of deprivation lead to greater abnormalities in social behavior during childhood and adulthood. Studies of orphaned human infants suggest

FIGURE 52.9
An unlikely parent. The eager goslings follow Konrad Lorenz as if he were their mother. He is the first object they saw when they hatched, and they have used him as a model for imprinting.

(a)

(b)

FIGURE 52.10
How imprinting is studied.
Ducklings will imprint on the first object they see, even (*a*) a black box or (*b*) a white sphere.

that a constant "mother figure" is required for normal growth and psychological development.

Recent research has revealed a biological need for the stimulation that occurs during parent-offspring interactions early in life. Female rats lick their pups after birth, and this stimulation inhibits the release of a hormonelike chemical that can block normal growth. Pups that receive normal tactile stimulation also have more brain receptors for glucocorticoid hormones, longer-lived brain neurons, and a greater tolerance for stress. Premature human infants who are massaged gain weight rapidly. These studies indicate that the need for normal social interaction is based in the brain and that touch and other aspects of contact between parents and offspring are important for physical as well as behavioral development.

Sexual imprinting, is a process in which an individual learns to direct its sexual behavior at members of its own species. **Cross-fostering** studies, in which individuals of one species are raised by parents of another species, reveal that this form of imprinting also occurs early in life. In most species of birds, these studies have shown that the fostered bird will attempt to mate with members of its foster species when it is sexually mature.

Interaction Between Instinct and Learning

The work of Peter Marler and his colleagues on the acquisition of courtship song by white-crowned sparrows provides an excellent example of the interaction between instinct and learning in the development of behavior. Courtship songs are sung by mature males and are species-specific. By rearing male birds in soundproof incubators equipped with speakers and microphones, Marler could control what a bird heard as it matured and then record the song it produced as an adult. He found that white-crowned sparrows that heard no song at all during development or that heard only the song of a different species, the song sparrow, sang a poorly developed song as adults (figure 52.11). But birds that heard the song of their own species or that heard the songs of *both* the white-crowned sparrow and the song sparrow, sang a fully developed, white-crowned sparrow song as adults. These results suggest that these birds have a **genetic template,** or instinctive program, that guides them to learn the appropriate song. During a critical period in development, the template will accept the correct song as a model. Thus, song acquisition depends on learning, but only the song of the correct species can be learned. The genetic template for learning is *selective.* However, learning plays a prominent role as well. If a young white-crowned sparrow becomes deaf after it hears its species' song during the critical period, it will also sing a poorly developed song as an adult. Therefore, the bird must "practice" listening to himself sing, matching what he hears to the model his template has accepted.

Although this explanation of song development stood unchallenged for many years, recent research has shown that white-crowned sparrow males *can* learn another species' song under certain conditions. If a live male strawberry finch is placed in a cage next to a young male sparrow, the young sparrow will learn to sing the strawberry finch's song! This finding indicates that social stimuli may be more effective than a tape-recorded song in overriding the innate program that guides song development. Furthermore, the males of some bird species have no opportunity to hear the song of their own species. In such cases, it appears that the males, instinctively "know" their own species' song. For example, cuckoos are brood parasites; females lay their eggs in the nest of another species of bird, and the young that hatch are reared by the foster parents (figure 52.12). When the cuckoos become adults, they sing the song of their own species rather than that of their foster parents. Because male brood parasites would most likely hear the song of their host species during development, it is adaptive for them to ignore such "incorrect" stimuli. They hear no adult males of their own species singing, so no correct song models are available. In these species, natural selection has programmed the male with a genetically guided song.

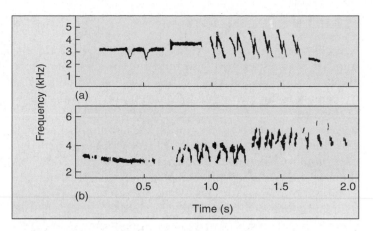

FIGURE 52.11
Song development in birds. (*a*) The sonograms of songs produced by male white-crowned sparrows that had been exposed to their own species' song during development are different from (*b*) those of male sparrows that heard no song during rearing. This difference indicates that the genetic program itself is insufficient to produce a normal song.

FIGURE 52.12
Brood parasite. Cuckoos lay their eggs in the nests of other species of birds. Because the young cuckoos (large bird to the right) are raised by a different species (such as this meadow pipit, smaller bird to the left), they have no opportunity to learn the cuckoo song; the cuckoo song they later sing is innate.

Interactions that occur during sensitive phases of imprinting are critical to normal behavioral development. Physical contact plays an important role in growth and in the development of psychological well-being.

Animal Cognition

To what degree animals "think" is a subject of lively dispute. It is likely each of us could tell an anecdotal story about the behavior of a pet cat or dog that would suggest the animal had a degree of reasoning ability or was capable of thinking. For many decades, however, students of animal behavior flatly rejected the notion that nonhuman animals can think. In fact, behaviorist Lloyd Morgan stated that one should never assume a behavior represents conscious thought if there is any other explanation that precludes the assumption of consciousness. The prevailing approach was to treat animals as though they responded to the environment through instinctive behaviors and simple, innately programmed learning.

In recent years, serious attention has been given to the topic of animal awareness. The central question is whether animals show **cognitive behavior**—that is, do they process information and respond in a manner that suggests thinking (figure 52.13)? What kinds of behavior would demonstrate cognition? Some birds in urban areas remove the foil caps from nonhomogenized milk bottles to get at the cream beneath, and this behavior is known to have spread within a population to other birds. Japanese macaques learn to wash potatoes and float grain to separate it from sand. A chimpanzee pulls the leaves off a tree branch and uses the branch to probe the entrance to a termite nest and gather termites. Vervet monkeys have a vocabulary that identifies specific predators (see figure 52.24).

Only a few experiments have tested the thinking ability of nonhuman animals. Some of these studies suggest that animals may deliberately give false information (that is, they "lie"). Currently, researchers are trying to determine if some primates deceive others to manipulate the behavior of the other members of their troop. Many anecdotal accounts appear to support the idea that deception occurs in some nonhuman primate species, such as baboons and chimpanzees, but it has been difficult to devise field-based experiments to test this idea. Much of this type of research on animal cognition is in its infancy, but it is sure to grow and to raise controversy. In any case, there is nothing to be gained by dogmatically denying the possibility of animal consciousness.

Some examples, particularly those involving problem solving by animals, are hard to explain in any way other

(a)

(b)

FIGURE 52.13
Animal thinking? (*a*) This chimpanzee is stripping the leaves from a twig, which it will then use to probe a termite nest. This behavior strongly suggests that the chimpanzee is consciously planning ahead, with full knowledge of what it intends to do. (*b*) This sea otter is using a rock as an "anvil," against which it bashes a clam to break it open. A sea otter will often keep a favorite rock for a long time, as though it has a clear idea of its future use of the rock. Behaviors such as these suggest that animals have cognitive abilities.

FIGURE 52.14
Problem solving by a chimpanzee. Unable to get the bananas by jumping, the chimpanzee devises a solution.

than as a result of some sort of mental process. For example, in a series of classic experiments conducted in the 1920s, a chimpanzee was left in a room with a banana hanging from the ceiling out of reach. Also in the room were several boxes, each lying on the floor. After some unsuccessful attempts to jump up and grab the bananas, the chimpanzee suddenly looked at the boxes and then immediately proceeded to move them underneath the banana, stack one on top of another, and climb up to claim its prize (figure 52.14).

Perhaps it is not surprising to find obvious intelligence in animals as closely related to us as chimpanzees. But recent studies have found that other animals also show evidence of cognition. Ravens have always been considered among the most intelligent of birds. Bernd Heinrich of the University of Vermont recently conducted an experiment using a group of hand-reared ravens that lived in an outdoor aviary. Heinrich placed a piece of meat on the end of a string and hung it from a branch in the aviary. The birds liked to eat meat, but had never seen string before and were unable to get at the meat. After several hours, during which time the birds periodically looked at the meat but did nothing else, one bird flew to the branch, reached down, grabbed the string, pulled it up, and placed it under his foot. He then reached down and grabbed another piece of the string, repeating this action over and over, each time bringing the meat closer (figure 52.15). Eventually, he brought the meat within reach and grasped it. The raven, presented with a completely novel problem, had devised a solution. Eventually, three of the other five ravens also figured out how to get the meat. Heinrich has conducted other similarly creative experiments that leave little doubt that ravens have advanced cognitive abilities.

Research on the cognitive behavior of animals is in its infancy, but some examples are compelling.

FIGURE 52.15
Problem solving by a raven. Confronted with a problem it has never previously faced, the raven figures out how to get the meat at the end of the string by repeatedly pulling up a bit of string and stepping on it.

Migratory Behavior

Orientation and Migration

Animals may travel to and from a nest to feed or move regularly from one place to another. To do so, they must orient themselves by tracking stimuli in the environment.

Movement toward or away from a stimulus is called **taxis.** The attraction of flying insects to outdoor lights is an example of *positive phototaxis.* Insects that avoid light, such as the common cockroach, exhibit *negative phototaxis.* Other stimuli may be used as orienting cues. For example, trout orient themselves in a stream so as to face against the current. However, not all responses involve a specific orientation. Some animals just become more or less active when stimulus intensity increases; such responses are called **kineses.**

Long-range, two-way movements are known as **migrations.** Ducks and geese migrate along flyways from Canada across the United States each fall and return each spring. Monarch butterflies migrate each fall from central and eastern North America to several small, geographically isolated areas of coniferous forest in the mountains of central Mexico (figure 52.16). Each August, the butterflies begin a flight southward to their overwintering sites. At the end of winter, the monarchs begin the return flight to their summer breeding ranges. What is amazing about the migration of the monarch, however, is that two to five generations may be produced as the butterflies fly north. The butterflies that migrate in the autumn to the precisely located overwintering grounds in Mexico have never been there before.

Recent geographic range expansions by some migrating birds have revealed how migratory patterns change. When colonies of bobolinks became established in the western United States, far from their normal range in the Midwest and East, they did not migrate directly to their winter range in South America. Instead, they migrated east to their ancestral range and then south along the original flyway (figure 52.17). Rather than changing the original migration pattern, they simply added a new pattern. Scientists continue to study the western bobolinks to learn whether, in time, a more efficient migration path will evolve or whether the birds will always follow their ancestral course.

How Migrating Animals Navigate

Biologists have studied migration with great interest, and we now have a good understanding of how these feats of navigation are achieved. It is important to understand the distinction between **orientation** (the ability to follow a bearing) and **navigation** (the ability to set or adjust a bearing, and then follow it). The former is analogous to using a compass, while the latter is like using a compass in conjunction with a map. Experiments on starlings indicate that inexperienced birds migrate by orientation, while older birds that have migrated previously use true navigation (figure 52.18).

Birds and other animals navigate by looking at the sun and the stars. The indigo bunting, which flies during the day and uses the sun as a guide, compensates for the movement of the sun in the sky as the day progresses by reference to the North Star, which does not move in the sky. Buntings also use the positions of the constellations

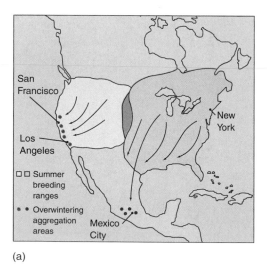

(a)   (b)   (c)

FIGURE 52.16
Migration of monarch butterflies. (*a*) Monarchs from western North America overwinter in areas of mild climate along the Pacific coast. Those from the eastern United States and southeastern Canada migrate to Mexico, a journey of over 3000 kilometers that takes from two to five generations to complete. (*b*) Monarch butterflies arrive at the remote fir forests of the overwintering grounds in Mexico, where they (*c*) form aggregations on the tree trunks.

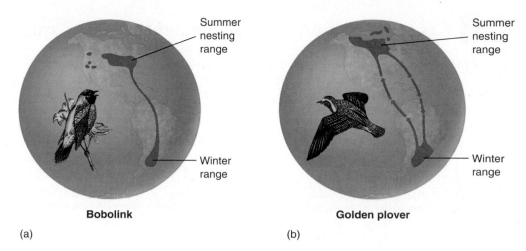

FIGURE 52.17
Birds on the move. (*a*) The summer range of bobolinks recently extended to the far western United States from their more established range in the Midwest. When they migrate to South America in the winter, bobolinks that nested in the West do not fly directly to the winter range; instead, they fly to the Midwest first and then use the ancestral flyway. (*b*) The golden plover has an even longer migration route that is circular. These birds fly from arctic breeding grounds to wintering areas in southeastern South America, a distance of some 13,000 kilometers.

(a)   Bobolink

(b)   Golden plover

and the position of the pole star in the night sky, cues they learn as young birds. Starlings and certain other birds compensate for the sun's apparent movement in the sky by using an internal clock. If such birds are shown an experimental sun in a fixed position while in captivity, they will change their orientation to it at a constant rate of about 15° per hour.

Many migrating birds also have the ability to detect the earth's magnetic field and to orient themselves with respect to it. In a closed indoor cage, they will attempt to move in the correct geographic direction, even though there are no visible external cues. However, the placement of a powerful magnet near the cage can alter the direction in which the birds attempt to move. Researchers have found magnetite, a magnetized iron ore, in the heads of some birds, but have not been able to identify the sensory receptors birds employ to detect magnetic fields.

It appears that the first migration of a bird is innately guided by both celestial cues (the birds fly mainly at night) and the earth's magnetic field. These cues give the same information about the general direction of the migration, but when the two cues are experimentally manipulated to give conflicting directions, the information provided by the stars seems to override the magnetic information. Recent studies, however, indicate that celestial cues tell northern hemisphere birds to move south when they begin their migration, while magnetic cues give them the direction for the specific migratory path (perhaps a southeast turn the bird must make midroute). In short, these new data suggest that celestial and magnetic cues interact during development to fine-tune the bird's navigation.

We know relatively little about how other migrating animals navigate. For instance, green sea turtles migrate from Brazil halfway across the Atlantic Ocean to Ascension Island, where the females lay their eggs. How do they find this tiny island in the middle of the ocean, which they haven't seen for perhaps 30 years? How do

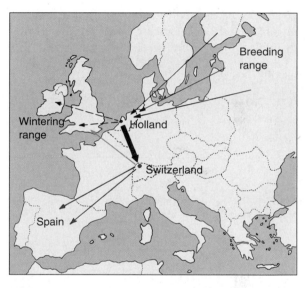

FIGURE 52.18
Migratory behavior of starlings. The navigational abilities of inexperienced birds differ from those of adults that have made the migratory journey before. Starlings were captured in Holland, halfway along their full migratory route from Baltic breeding grounds to wintering grounds in the British Isles; these birds were transported to Switzerland and released. Experienced older birds compensated for the displacement and flew toward the normal wintering grounds (*blue arrow*). Inexperienced young birds kept flying in the same direction, on a course that took them toward Spain (*red arrows*). These observations imply that inexperienced birds fly by orientation, while experienced birds learn true navigation.

the young that hatch on the island know how to find their way to Brazil? Recent studies suggest that wave action is an important cue.

Many animals migrate in predictable ways, navigating by looking at the sun and stars, and in some cases by detecting magnetic fields.

Much of the research in animal behavior is devoted to analyzing the nature of communication signals, determining how they are perceived, and identifying their ecological roles and their evolutionary origins.

Courtship

During courtship, animals produce signals to communicate with potential mates and with other members of their own sex. A *stimulus-response chain* sometimes occurs, in which the behavior of one individual in turn releases a behavior by another individual (figure 52.19).

Courtship Signaling

A male stickleback fish will defend the nest it builds on the bottom of a pond or stream by attacking *conspecific* males (that is, males of the same species) that approach the nest. Niko Tinbergen studied the social releasers responsible for this behavior by making simple clay models. He found that a model's shape and degree of resemblance to a fish were unimportant; any model with a red underside (like the underside of a male stickleback) could release the attack behavior. Tinbergen also used a series of clay models to demonstrate that a male stickleback recognizes a female by her abdomen when it is swollen with eggs.

Courtship signals are often *species-specific*, limiting communication to members of the same species and thus playing a key role in reproductive isolation (see chapter 23). The flashes of fireflies (which are actually beetles) are such species-specific signals. Females recognize conspecific males by their flash pattern (figure 52.20), and males recognize conspecific females by their flash response. This series of reciprocal responses provides a continuous "check" on the species identity of potential mates.

Long-Distance Communication

Chemical signals also mediate interactions between males and females. **Pheromones,** chemical messengers used for communication between individuals of the same species, serve as sex attractants, among other functions, in many animals. Even the human egg produces a chemical attractant to communicate with sperm! Female silk moths (*Bombyx mori*) produce a sex pheromone called *bombykol* in a gland

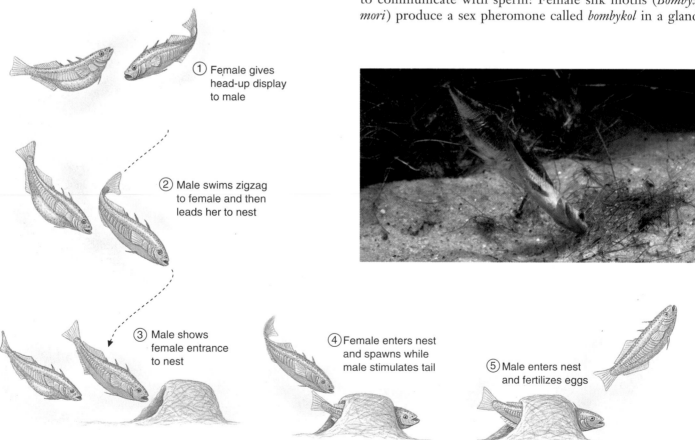

① Female gives head-up display to male

② Male swims zigzag to female and then leads her to nest

③ Male shows female entrance to nest

④ Female enters nest and spawns while male stimulates tail

⑤ Male enters nest and fertilizes eggs

FIGURE 52.19
A stimulus-response chain. Stickleback courtship involves a sequence of behaviors leading to the fertilization of eggs.

associated with the reproductive system. Neurophysiological studies show that the male's antennae contain numerous sensory receptors specific for bombykol. These receptors are extraordinarily sensitive, enabling the male to respond behaviorally to concentrations of bombykol as low as one molecule in 1017 molecules of oxygen in the air.

Many insects, amphibians, and birds produce species-specific acoustic signals to attract mates. Bullfrog males call by inflating and discharging air from their vocal sacs, located beneath the lower jaw. Females can distinguish a conspecific male's call from the call of other frogs that may be in the same habitat and calling at the same time. As mentioned in section 52.2, male birds produce songs, complex sounds composed of notes and phrases, to advertise their presence and to attract females. In many species, variations in the males' songs identify *particular* males in a population. In these species, the song is individually specific as well as species-specific.

Level of Specificity

Different signals provide different levels of information about the sender. The **level of specificity** relates to the function of the signal. Many courtship signals are species-specific to help animals avoid making errors in mating that would produce inviable hybrids or otherwise waste reproductive effort. A male bird's song is individually specific because it allows his presence (as opposed to simply the presence of an unidentifiable member of the species) to be recognized by neighboring birds. When territories are being established, males may sing and aggressively confront neighboring conspecifics to defend their space. Aggression carries the risk of injury, and singing is energetically costly. After territorial borders have been established, intrusions by neighbors are few because the outcomes of the contests have already been determined. Each male then "knows" his neighbor by the song he sings, and also "knows" that male does not constitute a threat because they have already settled their territorial contests. So, all birds in the population can lower their energy costs by identifying their neighbors through their individual songs. Similarly, mammals mark their territories with pheromones that signal individual identity, encoded as a blend of a number of chemicals. Other signals, such as the mobbing and alarm calls of birds, are anonymous, conveying no information about the identity of the sender. These signals may permit communication about the presence of a predator common to several bird species.

Although in most cases signals are used for communication between members of the same species, sometimes two different species are involved. For example, fish with parasites adopt a specific posture in the presence of cleaner fish that signals to the cleaner fish that they are ready to be cleaned (figure 52.21). In a similar vein, some animals send signals to predators. White-tailed deer, for example, raise their tails to display their prominent white undersides while running away from a predator. These *pursuit deterrent* sig-

FIGURE 52.20
Firefly fireworks. The bioluminescent displays of these lampyrid beetles are species-specific and serve as behavioral mechanisms of reproductive isolation. Each number represents the flash pattern of a male of a different species.

FIGURE 52.21
Cleaner fish. This grouper has entered the cleaner fish's "station" and adopted a posture that allows the cleaner fish to enter the mouth and gills and feed on attached parasites.

nals are presumed to indicate to the predator that it has already been seen and thus should not waste its time trying to catch the deer.

Animal communications serve many purposes and are transmitted in many ways.

Communication in Social Groups

Many insects, fish, birds, and mammals live in social groups in which information is communicated between group members. For example, some individuals in mammalian societies serve as "guards." When a predator appears, the guards give an *alarm call*, and group members respond by seeking shelter. Social insects, such as ants and honeybees, produce *alarm pheromones* that trigger attack behavior. Ants also deposit *trail pheromones* between the nest and a food source to lead other colony members to food (figure 52.22). Honeybees have an extremely complex *dance language* that directs hivemates to rich nectar sources.

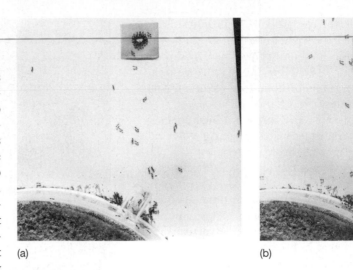

(a)   (b)

FIGURE 52.22
The chemical control of fire ant foraging. Trail pheromones, produced in an accessory gland near the fire ant's sting, organize cooperative foraging. The trails taken by the first ants to travel to a food source (*a*) are soon followed by most of the other ants (*b*).

The Dance Language of the Honeybee

The European honeybee, *Apis mellifera*, lives in hives consisting of 30,000 to 40,000 individuals whose behaviors are integrated into a complex colony. Worker bees may forage for miles from the hive, collecting nectar and pollen from a variety of plants and switching between plant species and populations on the basis of how energetically rewarding their food is. The food sources used by bees tend to occur in patches, and each patch offers much more food than a single bee can transport to the hive. A colony is able to exploit the resources of a patch because of the behavior of scout bees, which locate patches and communicate their location to hivemates through a dance language. Over many years, Nobel laureate Karl von Frisch was able to unravel the details of this communication system.

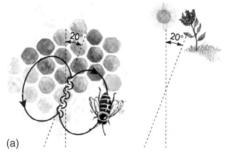

(a)

FIGURE 52.23
The waggle dance of honeybees. (*a*) The angle between the food source, the nest, and the sun is represented by a dancing bee as the angle between the straight part of the dance and vertical. The food is 20° to the right of the sun, and the straight part of the bee's dance on the hive is 20° to the right of vertical. (*b*) A scout bee dances on a comb in the hive.

(b)

After a successful scout bee returns to the hive, she performs a remarkable behavior pattern called a *waggle dance* on a vertical comb (figure 52.23). The path of the bee during the dance resembles a figure-eight. on the straight part of the path, the bee vibrates or waggles her abdomen while producing bursts of sound. She may stop periodically to give her hivemates a sample of the nectar she has carried back to the hive in her crop. As she dances, she is followed closely by other bees, which soon appear as foragers at the new food source.

Von Frisch and his colleagues claimed that the other bees use information in the waggle dance to locate the food source. According to their explanation, the scout bee indicates the *direction* of the food source by representing the angle between the food source, the hive, and the sun as the deviation from vertical of the straight part of the dance performed on the hive wall (that is, if the bee moved straight, then the food source would be in the direction of the sun, but if the food were at a 30° angle relative to the sun's position, then the bee would move upward at a 30° angle from vertical). The *distance* to the food source is indicated by the tempo, or degree of vigor, of the dance.

Adrian Wenner, a scientist at the University of California, did not believe that the dance language communicated anything about the location of food, and he challenged von Frisch's explanation. Wenner maintained that flower odor was the most important cue leading bees to arrive at a new food source. A heated controversy ensued as the two groups of researchers published articles supporting their positions.

(a)

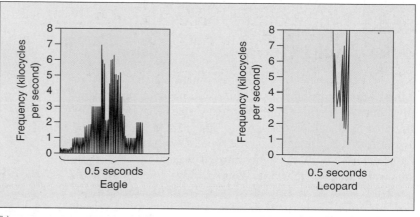

(b)

FIGURE 52.24
Primate semantics. (*a*) Predators, such as this leopard, attack and feed on vervet monkeys. (*b*) The monkeys give different alarm calls when troupe members sight an eagle, leopard, or snake. Each distinctive call elicits a different and adaptive escape behavior.

Such controversies can be very beneficial, because they often generate innovative experiments. In this case, the "dance language controversy" was resolved (in the minds of most scientists) in the mid-1970s by the creative research of James L. Gould. Gould devised an experiment in which hive members were tricked into misinterpreting the directions given by the scout bee's dance. As a result, Gould was able to manipulate where the hive members would go if they were using visual signals. If odor were the cue they were using, hive members would have appeared at the food source, but instead they appeared exactly where Gould had predicted. This confirmed von Frisch's ideas.

Recently, researchers have extended the study of the honeybee dance language by building robot bees whose dances can be completely controlled. Their dances are programmed by a computer and perfectly reproduce the natural honeybee dance; the robots even stop to give food samples! The use of robot bees has allowed scientists to determine precisely which cues direct hivemates to food sources.

Primate Language

Some primates have a "vocabulary" that allows individuals to communicate the identity of specific predators. The vocalizations of African vervet monkeys, for example, distinguish among eagles, leopards, and snakes (figure 52.24). Chimpanzees and gorillas can learn to recognize a large number of symbols and use them to communicate abstract concepts. The complexity of human language would at first appear to defy biological explanation, but closer examination suggests that the differences are in fact superficial—all languages share many basic structural similarities. All of the roughly 3000 languages draw from the same set of 40 consonant sounds (English uses two dozen of them), and any human can learn them. Researchers believe these similarities reflect the way our brains handle abstract information, a genetically determined characteristic of all humans.

Language develops at an early age in humans. Human infants are capable of recognizing the 40 consonant sounds characteristic of speech, including those not present in the particular language they will learn, while ignoring other sounds. In contrast, individuals who have not heard certain consonant sounds as infants can only rarely distinguish or produce them as adults. That is why English speakers have difficulty mastering the throaty French "r," French speakers typically replace the English "th" with "z," and native Japanese often substitute "r" for the unfamiliar English "l." Children go through a "babbling" phase, in which they learn by trial and error how to make the sounds of language. Even deaf children go through a babbling phase using sign language. Next, children quickly and easily learn a vocabulary of thousands of words. Like babbling, this phase of rapid learning seems to be genetically programmed. It is followed by a stage in which children form simple sentences that, though they may be grammatically incorrect, can convey information. Learning the rules of grammar constitutes the final step in language acquisition.

While language is the primary channel of human communication, odor and other nonverbal signals (such as "body language") may also convey information. However, it is difficult to determine the relative importance of these other communication channels in humans.

The study of animal communication involves analysis of the specificity of signals, their information content, and the methods used to produce and receive them.

52.4 Evolutionary forces shape behavior.

Behavioral Ecology

Nobel laureate Niko Tinbergen outlined the different types of questions biologists can ask about animal behavior. In essence, he divided the investigation of behavior into the study of its development, physiological basis, and function (evolutionary significance). One type of evolutionary analysis pioneered by Tinbergen himself was the study of the **survival value** of behavior. That is, how does an animal's behavior allow it to stay alive or keep its offspring alive? For example, Tinbergen observed that after gull nestlings hatch, the parents remove the eggshells from the nest. To understand why this behavior occurs, he camouflaged chicken eggs by painting them to resemble the natural background where they would lie and distributed them throughout the area in which the gulls were nesting (figure 52.25). He placed broken eggshells next to some of the eggs, and as a control, he left other camouflaged eggs alone without eggshells. He then noted which eggs were found more easily by crows. Because the crows could use the white interior of a broken eggshell as a cue, they ate more of the camouflaged eggs that were near eggshells. Thus, Tinbergen concluded that eggshell removal behavior is *adaptive:* It reduces predation and thus increases the offspring's chances of survival.

Tinbergen is credited with being one of the founders of **behavioral ecology,** the study of how natural selection shapes behavior. This branch of ecology examines the **adaptive significance** of behavior, or how behavior may increase survival and reproduction. Current research in behavioral ecology focuses on how behavior contributes to an animal's reproductive success, or **fitness.** As we saw in section 52.1, differences in behavior among individuals often result from genetic differences. Thus, natural selection operating on behavior has the potential to produce evolutionary change. To study the relation between behavior and fitness, then, is to study the process of adaptation itself.

Consequently, the field of behavioral ecology is concerned with two questions. First, is behavior adaptive? Although it is tempting to assume that the behavior produced by individuals must in some way represent an adaptive response to the environment, this need not be the case. As discussed in chapter 21, traits can evolve for many reasons other than natural selection, such as genetic drift, gene flow, or the correlated consequences of selection on other traits. Moreover, traits may be present in a population because they evolved as adaptations in the past, but are no longer useful. These possibilities hold true for behavioral traits as much as for any other kind of trait.

If a trait is adaptive, the question then becomes: How is it adaptive? Although the ultimate criterion is reproductive

FIGURE 52.25
The adaptive value of egg coloration. Niko Tinbergen painted chicken eggs to resemble the mottled brown camouflage of gull eggs. The eggs were used to test the hypothesis that camouflaged eggs are more difficult for predators to find and thus increase the young's chances of survival.

success, behavioral ecologists are interested in *how* a trait can lead to greater reproductive success. By enhancing energy intake, thus increasing the number of offspring produced? By improving success in getting more matings? By decreasing the chance of predation? The job of a behavioral ecologist is to determine the effect of a behavioral trait on each of these activities and then to discover whether increases in, for example, foraging efficiency, translate into increased fitness.

Behavioral ecology is the study of how natural selection shapes behavior.

Foraging Behavior

The best way to introduce behavioral ecology is by examining one well-defined behavior in detail. While many behaviors might be chosen, we will focus on foraging behavior. For many animals, food comes in a variety of sizes. Larger foods may contain more energy but may be harder to capture and less abundant. In addition, some types of food may be farther away than other types. Hence, for these animals foraging involves a trade-off between a food's energy content and the cost of obtaining it. The *net energy* (in calories or Joules) gained by feeding on prey of each size is simply the energy content of the prey minus the energy costs of pursuing and handling it. According to **optimal foraging theory,** natural selection favors individuals whose foraging behavior is as energetically efficient as possible. In other words, animals tend to feed on prey that maximize their net energy intake per unit of foraging time.

A number of studies have demonstrated that foragers do preferentially utilize prey that maximize the energy return. Shore crabs, for example, tend to feed primarily on intermediate-sized mussels, which provide the greatest energetic return; larger mussels yield more energy, but also take considerably more energy to crack open (figure 52.26).

This optimal foraging approach makes two assumptions. First, natural selection will only favor behavior that maximizes energy acquisition if increased energy reserves lead to increases in reproductive success. In some cases, this is true. For example, in both Columbian ground squirrels and captive zebra finches, a direct relationship exists between net energy intake and the number of offspring raised; similarly, the reproductive success of orb-weaving spiders is related to how much food they can capture.

However, animals have other needs besides energy, and sometimes these needs conflict. One obvious alternative is avoiding predators: Often, the behavior that maximizes energy intake is not the one that minimizes predation risk. Thus, the behavior that maximizes fitness often may reflect a trade-off between obtaining the most energy at the least risk of being eaten. Not surprisingly, many studies have shown that a wide variety of animal species alter their foraging behavior—becoming less active, spending more time watching for predators, or staying nearer to cover—when predators are present. Still another alternative is finding mates: Males of many species, for example, will greatly reduce their feeding rate to enhance their ability to attract and defend females. Even within foraging, trade-offs must be made, because maximizing energy is not the sole goal of foraging; particular nutrients are needed as well. Moose, for example, will feed on energy-poor aquatic vegetation to ensure that they get an adequate supply of calcium.

The second assumption of optimal foraging theory is that it has resulted from natural selection. As we have seen, natural selection can lead to evolutionary change

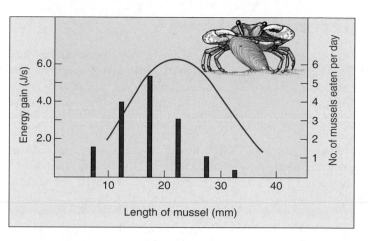

FIGURE 52.26
Optimal diet. The shore crab selects a diet of energetically profitable prey. The curve describes the net energy gain (equal to energy gained minus energy expended) derived from feeding on different sizes of mussels. The bar graph shows the numbers of mussels of each size in the diet. Shore crabs tend to feed on those mussels that provide the most energy.
What factors might be responsible for the slight difference in peak prey length relative to the length optimal for maximum energy gain?

only when differences among individuals have a genetic basis. Few studies have investigated whether differences among individuals in their ability to maximize energy intake are the result of genetic differences, but there are some exceptions. For example, one study found that female zebra finches that were particularly successful in maximizing net energy intake tended to have similarly successful offspring. Because birds were removed from their mothers before they left the nest, this similarity indicated that foraging behavior probably has a genetic basis rather than being a result of young birds learning to forage from their mothers.

Differences in foraging behavior among individuals may also be a function of age. Inexperienced yellow-eyed juncos (a small North American bird), for example, have not learned how to handle large prey items efficiently. As a result, the energetic costs of eating such prey are higher than the benefits, and these birds tend to focus on smaller prey. Only when the birds are older and more experienced do they learn to easily dispatch these prey, which are then included in the diet.

Natural selection may favor the evolution of foraging behaviors that maximize the amount of energy gained per unit time spent foraging. Animals that acquire energy efficiently during foraging may increase their fitness by having more energy available for reproduction, but other considerations, such as avoiding predators, are also important in determining reproductive success.

Territorial Behavior

Animals often move over a large area, or **home range,** during their daily course of activity. In many animal species, the home range of several individuals overlap in time or in space, but each individual defends a portion of its home range and uses it exclusively. This behavior, in which individual members of a species maintain exclusive use of an area that contains some limiting resource, such as foraging ground, food, or potential mates, is called **territoriality** (figure 52.27). The critical aspect of territorial behavior is *defense* against intrusion by other individuals. Territories are defended by displays advertising that the territories are occupied and by overt aggression. A bird sings from its perch within a territory to prevent takeover by a neighboring bird. If an intruder is not deterred by the song, the territory owner may attack and try to drive it away. However, territorial defense has its costs. Singing is energetically expensive, and attacks can lead to injury. In addition, advertisement through song or visual display can reveal one's position to a predator.

Why does an animal bear the costs of territorial defense? Over the past two decades, it has become increasingly clear that an economic approach can be useful in answering this question. Although there are costs to defending a territory, there are also benefits; these benefits may take the form of increased food intake, exclusive access to mates, or access to refuges from predators. Studies of nectar-feeding birds such as hummingbirds and sunbirds provide an example (figure 52.28). A bird benefits from having the exclusive use of a patch of flowers because it can efficiently harvest the nectar they produce. To maintain exclusive use, however, the bird must actively defend the flowers. The benefits of exclusive use outweigh the costs of defense only under certain conditions. Sunbirds, for example, expend 3000 calories per hour chasing intruders from a territory. Whether or not the benefit of defending a territory will exceed this cost depends upon the amount of nectar in the flowers and how efficiently the bird can collect it. For example, if flowers are very scarce or nectar levels are very low, a nectar-feeding bird may not gain enough energy to balance the energy used in defense. Under this circumstance, it is not advantageous to be territorial. Similarly, if flowers are very abundant, a bird can efficiently meet its daily energy requirements without behaving territorially and adding the costs of defense. From an energetic standpoint, defending abundant resources isn't worth the cost. Territoriality thus only occurs at intermediate levels of flower availability and nectar production, where the benefits of defense outweigh the costs.

In many species, exclusive access to females is a more important determinant of territory size for males than is food availability. In some lizards, for example, males maintain enormous territories during the breeding season. These territories, which encompass the territories of several females,

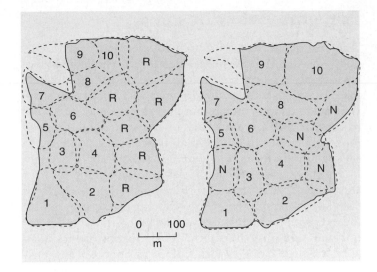

FIGURE 52.27
Competition for space.   Territory size in birds is adjusted according to the number of competitors. When six pairs of great tits (*Parus major*) were removed from their territories (indicated by R in the left figure), their territories were taken over by other birds in the area and by four new pairs (indicated by N in the right figure). Numbers correspond to the birds present before and after.

FIGURE 52.28
The benefit of territoriality.   Sunbirds, found in Africa and ecologically similar to New World hummingbirds, increase nectar availability by defending flowers.

are much larger than what would be required to supply enough food, and they are defended vigorously. In the nonbreeding season, by contrast, male territory size decreases dramatically, as does aggressive territorial behavior.

An economic approach can be used to explain the evolution and ecology of behaviors such as territoriality. This approach assumes that animals that gain more energy from a behavior than they expend will have an advantage in survival and reproduction over animals that behave in less efficient ways.

Reproductive Strategies

During the breeding season, animals make several important "decisions" concerning their choice of mates, how many mates to have, and how much time and energy to devote to rearing offspring. These decisions are all aspects of an animal's **reproductive strategy,** a set of behaviors that presumably have evolved to maximize reproductive success. Reproductive strategies have evolved partly in response to the energetic costs of reproduction and the way food resources, nest sites, and members of the opposite sex are spatially distributed in the environment.

Parental Investment and Mate Choice

Males and females usually differ in their reproductive strategies. Darwin was the first to observe that females often do not simply mate with the first male they encounter, but instead seem to evaluate a male's quality and then decide whether to mate. This behavior, called **mate choice,** has since been described in many invertebrate and vertebrate species.

By contrast, mate choice by males is much less common. Why should this be? Many of the differences in reproductive strategies between the sexes can be understood by comparing the parental investment made by males and females. **Parental investment** refers to the contributions each sex makes in producing and rearing offspring; it is, in effect, an estimate of the energy expended by males and females in each reproductive event.

Many studies have shown that parental investment is high in females. One reason is that eggs are much larger than sperm—195,000 times larger in humans! Eggs contain proteins and lipids in the yolk and other nutrients for the developing embryo, but sperm are little more than mobile DNA. Furthermore, in some groups of animals, females are responsible for gestation and lactation, costly reproductive functions only they can carry out.

The consequence of such great disparities in reproductive investment is that the sexes should face very different selective pressures. Because any single reproductive event is relatively cheap for males, they can best increase their fitness by mating with as many females as possible—male fitness is rarely limited by the amount of sperm they can produce. By contrast, each reproductive event for females is much more costly, and the number of eggs that can be produced often does limit reproductive success. For this reason, a female has an incentive to be choosy, trying to pick the male that can provide the greatest benefit to her offspring. As we shall see, this benefit can take a number of different forms.

These conclusions only hold when female reproductive investment is much greater than that of males. In species with biparental care, males may contribute equally to the cost of raising young; in this case, the degree of mate choice should be equal between the sexes.

Furthermore, in some cases, male investment exceeds that of females. For example, male mormon crickets transfer a protein-containing spermatophore to females during mating. Almost 30% of a male's body weight is made up by the spermatophore, which provides nutrition for the female and helps her develop her eggs. As we might expect, in this case it is the females that compete with each other for access to males that are the choosy sex. Indeed, males are quite selective, favoring heavier females. The selective advantage of this strategy results because heavier females have more eggs; thus, males that choose larger females leave more offspring (figure 52.29).

Another example comes from species in which the males take care of the eggs and developing young. This reversal of the normal sex roles occurs in a wide variety of species, including seahorses and a number of bird and insect species. In such species, as with mormon crickets, males are often choosy, and females must compete for mates.

The relative reproductive investment of the sexes determines reproductive strategies.

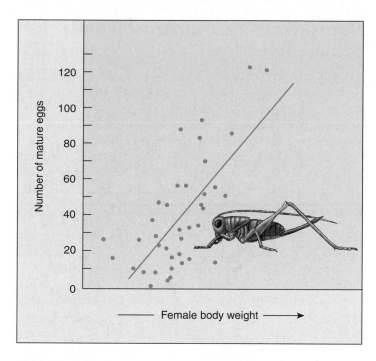

FIGURE 52.29
The advantage of male mate choice. Male mormon crickets choose heavier females as mates, and larger females have more eggs. Thus, male mate selection increases fitness.
Is there a benefit to females for mating with large males?

Reproductive Competition and Sexual Selection

As discussed in chapter 21, the reproductive success of an individual is determined by a number of factors: how long the individual lives, how frequently it mates, and how many offspring it produces per mating. The second of these factors, competition for mating opportunities, has been termed **sexual selection.** Some people consider sexual selection to be distinctive from natural selection, but others see it as a subset of natural selection, just one of the many ways organisms can increase their fitness.

Sexual selection involves both *intrasexual selection*, or interactions between members of one sex ("the power to conquer other males in battle," as Darwin put it), and *intersexual selection*, essentially mate choice ("the power to charm"). Sexual selection leads to the evolution of structures used in combat with other males, such as a deer's antlers and a ram's horns, as well as ornaments used to "persuade" members of the opposite sex to mate, such as long tail feathers and bright plumage (figure 52.30*a,b*). These traits are called **secondary sexual characteristics.**

Intrasexual Selection

In many species, individuals of one sex—usually males—compete with each other for the opportunity to mate with individuals of the other sex. These competitions may take place over ownership of a territory in which females reside or direct control of the females themselves. The latter case is exemplified by many species, such as the impala, in which females travel in large groups with a single male that gets exclusive access to mate with the females and thus strives vigorously to defend this access against other males that would like to supplant him.

In mating systems such as these, a few males may get an inordinate number of matings, and most males do not mate at all. In elephant seals, males control territories on the breeding beaches, and a few dominant males do most of the breeding (see figure 52.32). On one beach, for example, eight males impregnated 348 females, while the remaining males mated rarely, if at all.

For this reason, selection strongly favors any trait that confers greater ability to outcompete other males. In many cases, larger males are able to dominate smaller ones. As a result, in many territorial species, males have evolved to be considerably larger than females, for the simple reason that the largest males are the ones that get to mate. Such differences between the sexes are referred to as **sexual dimorphism.** In other species, males have evolved structures used for fighting, such as horns, antlers, and large canine teeth. These traits are also often sexually dimorphic and may have evolved because of the advantage they give in intrasexual conflicts.

Sometimes competition is not between the males themselves, but between their sperm, a phenomenon called **sperm competition.** In species whose females mate with multiple males, many features have evolved to maximize success at sperm competition. For example, in such species, the testes are often large to produce many sperm per mating, and the sperm themselves are also often larger so that they swim more rapidly and thus enhance the likelihood of fertilizing an egg.

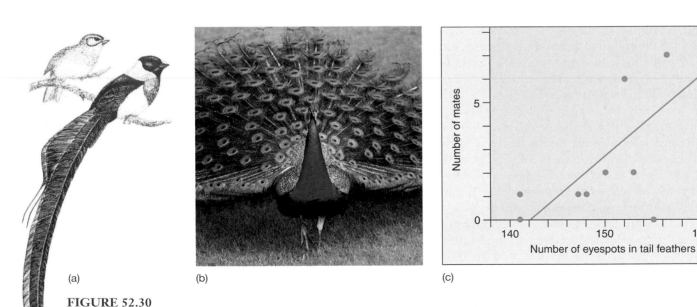

(a)   (b)   (c)

FIGURE 52.30
Products of sexual selection.   Attracting mates with long feathers is common in bird species such as (*a*) the African paradise whydah and (*b*) the peacock, which show pronounced sexual dimorphism. (*c*) Female peahens prefer to mate with males having greater numbers of eyespots in their tail feathers.
Why do females prefer males with more spots?

Intersexual Selection

Peahens prefer to mate with peacocks that have more spots in their long tail feathers (figure 52.30b,c). Similarly, female frogs prefer to mate with males having more complex calls. Why did such mating preferences evolve?

Direct Benefits of Mate Choice.   In some cases, the benefits of mate choice are obvious. In many species of birds and mammals and in some species of other types of animals, males help raise the offspring. In these cases, females would benefit by choosing the male that can provide the best care—the better the parent, the more offspring she is likely to rear.

In other species, males provide no care, but maintain territories that provide food, nesting sites, and predator refuges. In such species, females that choose males with the best territories will maximize their reproductive success.

Indirect Benefits.   In other species, however, males provide no direct benefits of any kind to females. In such cases, it is not intuitively obvious what females have to gain by being choosy. Moreover, what could be the possible benefit of choosing a male with an extremely long tail or a complex song?

A number of theories have been proposed to explain the evolution of such preferences. One idea is that females choose the male that is the healthiest or oldest. Large males, for example, have probably been successful at living long, acquiring a lot of food, and resisting parasites and disease. Similarly, in guppies and some birds, the brightness of a male's color reflects the quality of his diet and overall health. Females may gain two benefits from mating with large or colorful males. First, to the extent that the males' success in living long and prospering is the result of genetic makeup, the female will be ensuring that her offspring receive good genes from their father. Indeed, according to several studies, males that are preferred by females produce offspring that are more vigorous and survive better than offspring of males that are not preferred. Second, healthy males are less likely to be carrying diseases, which might be transmitted to the female during mating.

A variant of this theory goes one step further. In some cases, females prefer mates with traits that are detrimental to survival (figure 52.30). The long tail of the peacock is a hindrance in flying and makes males more vulnerable to predators. Why should females prefer males with such traits? The **handicap hypothesis** states that only genetically superior mates can survive with such a handicap. By choosing a male with the largest handicap, the female is ensuring that her offspring will receive these quality genes. Of course, the male offspring will also inherit the genes for the handicap. For this reason, evolutionary biologists are still debating this hypothesis.

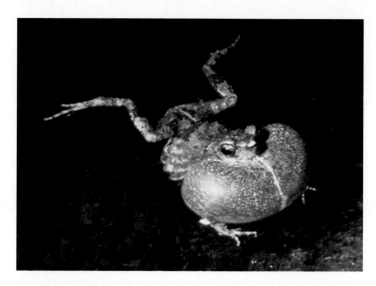

FIGURE 52.31
Male Túngara frog calling.   Female frogs of the genus *Physalaemus* prefer males that include a "chuck" in their call. However, only males of the species *Physalaemus pustulosus* produce such calls.

Other courtship displays appear to have evolved from a predisposition in the female's sensory system toward a certain type of stimulus. For example, females may be better able to detect particular colors or sounds at a certain frequency. **Sensory exploitation** involves the evolution in males of an attractive signal that "exploits" these preexisting biases—for example, if females are particularly adept at detecting red objects, then males will often evolve red coloration. Consider the vocalizations of the Túngara frog (*Physalaemus pustulosus*) (figure 52.31). Unlike related species, males include a "chuck" in their calls. Recent research suggests that not only are females of this species particularly attracted to calls of this sort, but so are females of related species, even though males of these species do not produce "chucks." Why this preference evolved is unknown, but males of the Túngara frog have evolved to take advantage of it.

A great variety of other hypotheses have been proposed to explain the evolution of mating preferences. Many of these hypotheses may be correct in some circumstances, but none seems capable of explaining all of the variation in mating behavior in the animal world. This is an area of vibrant research, with new discoveries appearing regularly.

> Natural selection has favored the evolution of behaviors that maximize the reproductive success of males and females. By evaluating and selecting mates with superior qualities, an animal can increase its reproductive success.

Mating Systems

The number of individuals with which an animal mates during the breeding season varies throughout the animal kingdom. Mating systems include monogamy (one male mates with one female), polygyny (one male mates with more than one female; figure 52.32), and polyandry (one female mates with more than one male). Like mate choice, mating systems have evolved to maximize reproductive fitness. Much research has shown that mating systems are strongly influenced by ecology. For instance, a male may defend a territory that holds nest sites or food sources necessary for a female to reproduce, and the territory may have resources sufficient for more than one female. If males differ in the quality of the territories they hold, a female's fitness is maximized if she mates with a male holding a high-quality territory. Such a male may already have a mate, but it is still more advantageous for the female to breed with that male than with an unmated male that defends a low-quality territory. In this case, natural selection would favor the evolution of polygyny.

Mating systems are also constrained by the needs of offspring. If the presence of both parents is necessary for young to be reared successfully, then monogamy may be favored. This is generally the case in birds, in which over 90% of all species are monogamous. A male may either remain with his mate and provide care for the offspring or desert that mate to search for others; both strategies may increase his fitness. The strategy that natural selection will favor depends upon the requirement for male assistance in feeding or defending the offspring. In some species, offspring are **altricial**—they require prolonged and extensive care. In these species, the need for care by two parents reduces the tendency for the male to desert his mate and seek other matings. In species where the young are **precocial** (requiring little parental care), males may be more likely to be polygynous.

Although polygyny is much more common, polyandrous systems—in which one female mates with several males—are known in a variety of animals. For example, in spotted sandpipers, males take care of all incubation and parenting, and females mate and leave eggs with two or more males.

In recent years, researchers have uncovered many unexpected aspects of animal mating. Some of these discoveries have resulted from the application of new technologies, whereas others have come from detailed and intensive field studies.

Extra-Pair Copulations

Chapter 16 describes how DNA fingerprinting can be used to identify blood samples. Another common use of this technology is to establish paternity. DNA fingerprints are so variable that each individual's pattern is unique. Thus, by comparing the DNA of a man and a child, experts can establish with a relatively high degree of confidence whether the man is the child's father.

This approach is now commonly used in paternity lawsuits, but it has also become a standard weapon in the arsenal of behavioral ecologists. By establishing paternity, researchers can precisely quantify the reproductive success of individual males and thus assess how successful their particular reproductive strategies have been (figure 52.33*a*). In one classic study of red-winged blackbirds (figure 52.33*b*), researchers established that half of all nests contained at least one bird fertilized by a male other than the territory owner; overall, 20% of the offspring were the result of such **extra-pair copulations (EPCs).**

Studies such as this have established that EPCs—"cheating"—are much more pervasive in the bird world than originally suspected. Even in some species that were believed to be monogamous on the basis of behavioral observations, the incidence of offspring being fathered by a male other than the female's mate is sometimes surprisingly high.

Why do individuals have extra-pair copulations? For males, the answer is obvious: increased reproductive success. For females, it is less clear, because in most cases, it does not result in an increased number of offspring. One possibility is that females mate with genetically superior individuals, thus enhancing the genes passed on to their off-

FIGURE 52.32
Female defense polygyny in elephant seals. Male elephant seals fight with each other for possession of territories. Only the largest males can hold territories, which contain many females.

spring. Another possibility is that females can increase the amount of help they get in raising their offspring. If a female mates with more than one male, each male may help raise the offspring. This is exactly what happens in a common English bird, the dunnock. Females mate not only with the territory owner, but also with subordinate males that hang around the edge of the territory. If these subordinates mate enough with a female, they will help raise her young, presumably because they may have fathered some of these young.

Alternative Mating Tactics

Natural selection has led to the evolution of a variety of other means of increasing reproductive success. For example, in many species of fish, there are two genetic classes of males. One group is large and defends territories to obtain matings. The other type of male is small and adopts a completely different strategy. They do not maintain territories, but loiter at the edge of the territories of large males. Just at the end of a male's courtship, when the female is laying her eggs and the territorial male is depositing sperm, the smaller male will dart in and release its own sperm into the water, thus fertilizing some of the eggs. If this strategy is successful, natural selection will favor the evolution of these two different male reproductive strategies.

Similar patterns are seen in other organisms. In some dung beetles, territorial males have large horns that they use to guard the chambers in which females reside, whereas genetically small males don't have horns; instead, they dig side tunnels and attempt to intercept the female inside her chamber. In isopods, there are three genetic size classes. The medium-sized males pass for females and enter a large male's territory in this way; the smallest class are so tiny, they are able to sneak in completely undetected.

This is just a glimpse of the rich diversity in mating systems that have evolved. The bottom line is: If there is a way of increasing reproductive success, natural selection will favor its evolution.

Mating systems represent reproductive adaptations to ecological conditions. The need for parental care and the ability of both sexes to provide it are important influences on the evolution of monogamy, polygyny, and polyandry. Detailed studies of animal mating systems, along with the use of modern molecular techniques, are revealing many surprises. This diversity is a testament to the power of natural selection to favor any trait that increases an animal's fitness.

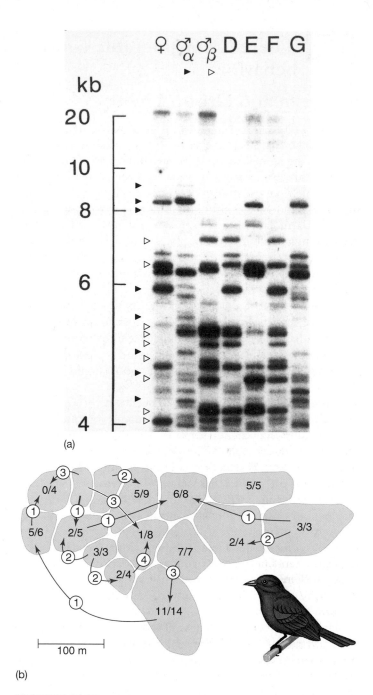

(a)

(b)

FIGURE 52.33

The study of paternity. (*a*) A DNA fingerprinting gel from the dunnock. The bands represent fragments of DNA of different lengths. The four nestlings (D–G) were in the nest of the female. By comparing the bands present in the two males, we can determine which male fathered which offspring. The triangles point to the bands that are diagnostic for one male and not the other. In this case, the β male fathered three (D, E, F, but not G) of the four offspring. (*b*) Results of a DNA fingerprinting study in red-winged blackbirds. Fractions indicate the proportion of offspring fathered by the male in whose territory the nest occurred. Arrows indicate how many offspring were fathered by particular males outside of each territory. Nests on some territories were not sampled.

52.5 There is considerable controversy about the evolution of social behavior.

Altruism and Group Living

Altruism—the performance of an action that benefits another individual at a cost to the actor—occurs in many guises in the animal world. In many bird species, for example, parents are assisted in raising their young by other birds, which are called *helpers at the nest.* In species of both mammals and birds, individuals that spy a predator will give an alarm call, alerting other members of their group, even though such an act would seem to call the predator's attention to the caller. Finally, lionesses with cubs will allow all cubs in the pride to nurse, including cubs of other females.

The existence of altruism has long perplexed evolutionary biologists. If altruism imposes a cost to an individual, how could an allele for altruism be favored by natural selection? One would expect such alleles to be at a disadvantage, and thus their frequency in the gene pool should decrease through time.

A number of explanations have been put forward to explain the evolution of altruism. One suggestion often heard on television documentaries is that such traits evolve for the good of the species. The problem with such explanations is that natural selection operates on individuals within species, not on species themselves. Thus, natural selection will not favor alleles that lead an individual to act in ways that benefit others at a cost to itself; it is even possible for traits to evolve that are detrimental to the species as a whole, as long as they benefit the individual. In some cases, selection can operate on groups of individuals, but such **group selection** is rare. For example, if an allele for cannibalism evolved within a population, individuals with that allele would be favored because they would have more to eat; however, the group might eventually eat itself to extinction, and the allele would be removed from the species. Although group selection can occur, the necessary conditions are rarely met in nature. In most cases, consequently, the "good of the species" cannot explain the evolution of altruistic traits.

Another possibility is that seemingly altruistic acts aren't altruistic after all. For example, helpers at the nest are often young and gain valuable parenting experience by assisting established breeders. Moreover, by hanging around an area, such individuals may inherit the territory when the established breeders die. Similarly, alarm callers may actually benefit by causing other animals to panic. In the ensuing confusion, the caller may be able to slip off undetected. Detailed field studies in recent years have demonstrated that some acts truly are altruistic, but others are not.

Reciprocity

Robert Trivers, now of Rutgers University, proposed that individuals may form "partnerships" in which mutual exchanges of altruistic acts occur because they benefit both participants. In the evolution of such **reciprocal altruism,** "cheaters" (nonreciprocators) are discriminated against and are cut off from receiving future aid. According to Trivers, if the altruistic act is relatively inexpensive, the small benefit a cheater receives by not reciprocating is far outweighed by the potential cost of not receiving future aid. Under these conditions, cheating should not occur.

Vampire bats roost in hollow trees in groups of 8 to 12 individuals. Because these bats have a high metabolic rate, individuals that have not fed recently may die. Bats that have found a host imbibe a great deal of blood, so giving up a small amount to keep a roostmate from starvation presents no great energy cost to the donor. Vampire bats tend to share blood with past reciprocators. If an individual fails to give blood to a bat from which it received blood in the past, it will be excluded from future bloodsharing.

Kin Selection

The most influential explanation for the origin of altruism was presented by William D. Hamilton in 1964. It is perhaps best introduced by quoting a passing remark made in a pub in 1932 by the great population geneticist J. B. S. Haldane. Haldane said that he would willingly lay down his life for *two brothers* or *eight first cousins.* Evolutionarily speaking, Haldane's statement makes sense, because for each allele Haldane received from his parents, his brothers each had a 50% chance of receiving the same allele (figure 52.34). Consequently, it is statistically expected that two of his brothers would pass on as many of Haldane's particular combination of alleles to the next generation as Haldane himself would. Similarly, Haldane and a first cousin would share an eighth of their alleles (see figure 52.34). Their parents, who are siblings, would each share half their alleles, and each of their children would receive half of these, of which half on the average would be in common: $\frac{1}{2} \times \frac{1}{2} \times \frac{1}{2} = \frac{1}{8}$. Eight first cousins would therefore pass on as many of those alleles to the next generation as Haldane himself would. Hamilton saw Haldane's point clearly: Natural selection will favor any strategy that increases the net flow of an individual's alleles to the next generation.

Hamilton showed that by directing aid toward kin, or close genetic relatives, an altruist may increase the reproductive success of its relatives enough to compensate for the reduction in its own fitness. Because the altruist's

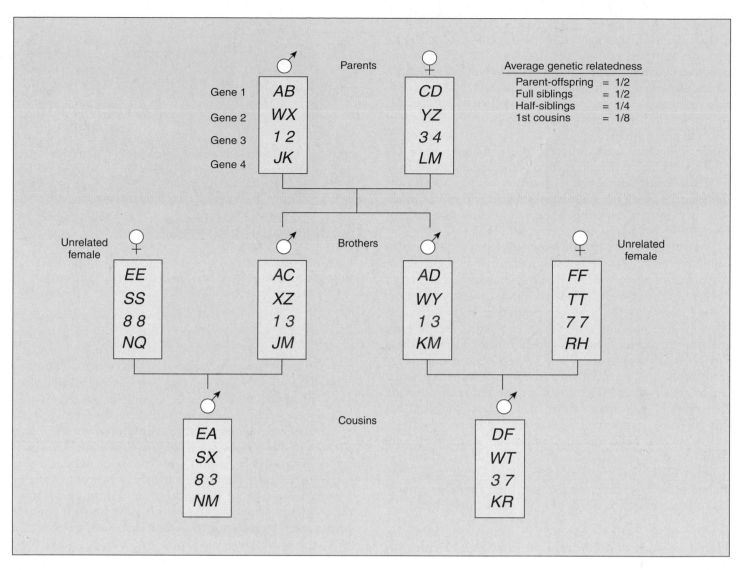

FIGURE 52.34
Hypothetical example of genetic relationships. On average, full siblings share half of their alleles. By contrast, cousins only share one-eighth of their alleles on average.

behavior increases the propagation of alleles in relatives, it will be favored by natural selection. Selection that favors altruism directed toward relatives is called **kin selection.** Although the behaviors being favored are cooperative, the genes are actually "behaving selfishly," because they encourage the organism to support copies of themselves in other individuals. In other words, if an individual has a dominant allele that causes altruism, any action that increases the frequency of this allele in future generations will be favored, even if that action is detrimental to the particular individual taking the action.

Hamilton's kin selection model predicts that altruism is likely to be directed toward close relatives. The more closely related two individuals are, the greater the potential genetic payoff. This relationship is described by

Hamilton's rule, which states that altruistic acts are favored when $rb > c$. In this expression, b and c are the benefits and costs of the altruistic act, respectively, and r is the coefficient of relatedness, the proportion of alleles shared by two individuals through common descent. For example, an individual should be willing to have one less child ($c = 1$) if such actions allow a half-sibling, which shares one-quarter of its genes ($r = 0.25$), to have five or more additional offspring ($b = 5$).

Examples of Kin Selection

Many examples of kin selection are known from the animal world. For example, Belding's ground squirrels give alarm calls when they spot a predator such as a coyote or a badger. Such predators may attack a calling squirrel, so giving a signal places the caller at risk. The social unit of a ground squirrel colony consists of a female and her daughters, sisters, aunts, and nieces. When they mature, males disperse long distances from where they are born; thus, adult males in the colony are not genetically related to the females. By marking all squirrels in a colony with an individual dye pattern on their fur and by recording which individuals gave calls and the social circumstances of their calling, researchers found that females who have relatives living nearby are more likely to give alarm calls than females with no kin nearby. Males tend to call much less frequently, as would be expected because they are not related to most colony members.

Another example of kin selection comes from a bird called the white-fronted bee-eater that lives along rivers in Africa in colonies of 100 to 200 birds. In contrast to ground squirrels, it is the male bee-eaters that usually remain in the colony in which they were born, and the females that disperse to join new colonies. Many bee-eaters do not raise their own offspring, but rather help others. Many of these birds are relatively young, but helpers also include older birds whose nesting attempts have failed. The presence of a single helper, on average, doubles the number of offspring that survive. Two lines of evidence support the idea that kin selection is important in determining helping behavior in this species. First, helpers are normally males, which are usually related to other birds in the colony, and not females, which are not related. Second, when birds have the choice of helping different parents, they almost invariably choose the parents to which they are most closely related.

Haplodiploidy and Hymenopteran Social Evolution

Probably the most famous application of kin selection theory has been to social insects. A hive of honeybees consists of a single queen, who is the sole egg-layer, and up to 50,000 of her offspring, nearly all of whom are female workers with nonfunctional ovaries (figure 52.35). In addition to this reproductive division of labor, honeybees exhibit cooperative care of the brood and overlap of generations, such that queens live alongside their offspring. These are the hallmarks of a **eusocial** system.

The evolutionary origin of eusociality was long a mystery. How could natural selection favor the evolution of sterile workers that provided no offspring? Hamilton explained the origin of eusociality in hymenopterans (that is, bees, wasps, and ants) with his kin selection model. In these insects, males are haploid and females are diploid. This unusual system of sex determination, called *haplodiploidy*, leads to an unusual situation. If the queen is fertilized by a single male, then all female offspring will inherit exactly the same alleles from their father (because he

FIGURE 52.35
Reproductive division of labor in honeybees. The queen (shown here with a red spot painted on her thorax) is the sole egg-layer. Her daughters are sterile workers.

is haploid and has only one copy of each allele). These female offspring will also share among themselves, on average, half of the alleles they get from the queen. Consequently, each female offspring will share, on average, 75% of her alleles with each sister (to verify this, rework figure 52.34, but allow the father to only have one allele for each gene). By contrast, should a female offspring have offspring of her own, she would only share half of her alleles with these offspring (the other half would come from their father). Thus, because of this close genetic relatedness, *workers propagate more of their own alleles by giving up their own reproduction to assist their mother in rearing their sisters, some of whom will be new queens and start new colonies and reproduce.* Thus, this unusual haplodiploid system may have set the stage for the evolution of eusociality in hymenopterans, and indeed, such systems have evolved as many as 12 or more times in the Hymenoptera.

One wrinkle in this theory, however, is that eusocial systems have evolved in several other groups, including thrips, termites, and naked mole rats. Although thrips are also haplodiploid, both termites and naked mole rats are not. Thus, although haplodiploidy may have facilitated the evolution of eusociality, it is not a necessary prerequisite.

Many factors could be responsible for the evolution of altruistic behaviors. Individuals may benefit directly if altruistic acts are reciprocated; kin selection explains how alleles for altruism can increase in frequency if altruistic acts are directed toward relatives. Kin selection is a potent force favoring, in some situations, the evolution of altruism and even complex social systems.

Group Living and the Evolution of Social Systems

Organisms as diverse as prokaryotes, cnidarians, insects, fish, birds, prairie dogs, lions, whales, and chimpanzees exist in social groups. To encompass the wide variety of social phenomena, we can broadly define a **society** as a group of organisms of the same species that are organized in a cooperative manner.

Why have individuals in some species given up a solitary existence to become members of a group? We have just seen that one explanation is kin selection: Groups may be composed of close relatives. In other cases, individuals may benefit directly from social living. For example, a bird that joins a flock may receive greater protection from predators. As flock size increases, the risk of predation decreases because there are more individuals to scan the environment for predators (figure 52.36). A member of a flock may also increase its feeding success if it can acquire information from other flock members about the location of new, rich food sources. In some predators, hunting in groups can increase success and allow the group to tackle prey too large for any one individual.

Insect Societies

In insects, sociality has chiefly evolved in two orders, the Hymenoptera (ants, bees, and wasps) and the Isoptera (termites), although a few other insect groups include social species. As we have just discussed, a number of types of insects have evolved eusocial systems. These social insect colonies are composed of different **castes**, groups of individuals that differ in size and morphology and perform different tasks, such as workers and soldiers (figure 52.37).

In honeybees, the queen maintains her dominance in the hive by secreting a pheromone, called

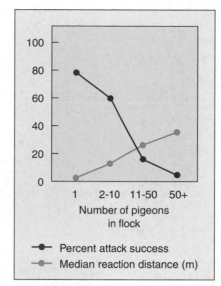

FIGURE 52.36
Flocking behavior decreases predation. As the size of a pigeon flock increases, hawks are less successful at capturing pigeons. Also, when more pigeons are present in the flock, they can detect hawks at greater distances, thus allowing more time for the pigeons to escape.
Would living in a flock affect the time available for foraging in pigeons?

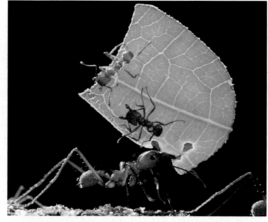

FIGURE 52.37
Castes of ants. These two leaf-cutter ants are members of different castes. The large ant is a worker carrying leaves to the nest, whereas the smaller ants are protecting the worker from attack.

"queen substance," that suppresses development of the ovaries in other females, turning them into sterile workers. Drones (male bees) are produced only for purposes of mating. When the colony grows larger in the spring, some members do not receive a sufficient quantity of queen substance, and the colony begins preparations for swarming. Workers make several new queen chambers, in which new queens begin to develop. Scout workers look for a new nest site and communicate its location to the colony. The old queen and a swarm of female workers then move to the new site. Left behind, a new queen emerges, kills the other potential queens, flies out to mate, and returns to assume "rule" of the hive.

Leafcutter ants provide another fascinating example of the remarkable lifestyles of social insects. Leafcutters live in colonies of up to several million individuals, growing crops of fungi beneath the ground. Their moundlike nests are underground "cities" covering more than 100 square meters, with hundreds of entrances and chambers as deep as 5 meters beneath the ground. The division of labor among the worker ants is related to their size. Every day, workers travel along trails from the nest to a tree or a bush, cut its leaves into small pieces, and carry the pieces back to the nest (see figure 52.37). Smaller workers chew the leaf fragments into a mulch, which they spread like a carpet in the underground fungus chambers. Even smaller workers implant fungal hyphae in the mulch; recent molecular studies suggest that ants have been cultivating these fungi for more than 50 million years! Soon a luxuriant garden of fungi is growing. While other workers weed out undesirable kinds of fungi, nurse ants carry the larvae of the nest to choice spots in the garden, where the larvae graze. Some of these larvae grow into reproductive queens that will disperse from the parent nest and start new colonies, repeating the cycle.

Vertebrate Societies

In contrast to the highly structured and integrated insect societies and their remarkable forms of altruism, vertebrate social groups are usually less rigidly organized and cohesive. It seems paradoxical that vertebrates, which have larger brains and are capable of more complex behaviors, are generally less altruistic than insects. Nevertheless, in some complex vertebrate social systems, individuals may be exhibiting both reciprocity and kin-selected altruism. But vertebrate societies also generally display more conflict and aggression among group members than do insect societies. Conflict in vertebrate societies generally centers on access to food and mates.

Like insect societies, vertebrate societies have particular types of organization. Each social group of vertebrates has a certain size, stability of members, number of breeding males and females, and type of mating system. Behavioral ecologists have learned that the way a group is organized is influenced most often by ecological factors such as food type and predation. For example, meerkats take turns watching for predators while other group members forage for food (figure 52.38).

African weaver birds, which construct nests from vegetation, provide an excellent example of the relationship between ecology and social organization. Their roughly 90 species can be divided according to the type of social group they form. One set of species lives in the forest and builds camouflaged, solitary nests. Males and females are monogamous; they forage for insects to feed their young. The second group of species nests in colonies in trees on the savanna. They are polygynous and feed in flocks on seeds. The feeding and nesting habits of these two sets of species are correlated with their mating systems. In the forest, insects are hard to find, and both parents must cooperate in feeding the young. The camouflaged nests do not call the attention of predators to their brood. On the open savanna, building a hidden nest is not an option. Rather, savanna-dwelling weaver birds protect their young from predators by nesting in trees, which are not very abundant. This shortage of safe nest sites means that birds must nest together in colonies. Because seeds occur abundantly, a female can acquire all the food needed to rear young without a male's help. The male, free from the duties of parenting, spends his time courting many females—a polygynous mating system.

One exception to the general rule that vertebrate societies are not organized like those of insects is the naked mole rat, a small, hairless rodent that lives in and near East Africa. Unlike other kinds of mole rats, which live alone or in small family groups, naked mole rats form large underground colonies with a far-ranging system of tunnels and a central nesting area. It is not unusual for a colony to contain 80 individuals.

Naked mole rats feed on bulbs, roots, and tubers, which they locate by constant tunneling. As in insect soci-

FIGURE 52.38
Foraging and predator avoidance. A meerkat sentinel on duty. Meerkats, *Suricata suricata*, are a species of highly social mongoose living in the semiarid sands of the Kalahari Desert in southern Africa. This meerkat is taking its turn to act as a lookout for predators. Under the security of its vigilance, the other group members can focus their attention on foraging.

eties, there is a division of labor among the colony members, with some mole rats working as tunnelers while others perform different tasks, depending upon the size of their bodies. Large mole rats defend the colony and dig tunnels.

Naked mole rat colonies have a reproductive division of labor similar to the one normally associated with the eusocial insects. All of the breeding is done by a single female, or "queen," who has one or two male consorts. The workers, consisting of both sexes, keep the tunnels clear and forage for food.

Eusocial insects exhibit an advanced social structure that includes reproductive division of labor and workers with different tasks. Social behavior in vertebrates is often characterized by kin-selected altruism. Altruistic behavior is involved in cooperative breeding in birds and alarm-calling in mammals.

52.1 Many behavioral patterns are innate.

Approaches to the Study of Behavior

- Behavior can be defined as the way an animal responds to stimuli in its environment. (p. 1106)
- Proximate causation defines how something works, while ultimate causation addresses why something works. (p. 1106)
- Innate (instinctive) behaviors were argued to be the result of preprogrammed nervous system pathways under genetic control. (p. 1106)

Behavioral Genetics

- Experimental evidence suggests that some behaviors are hereditary, governed by genes passed from one generation to the next. (p. 1108)
- Additional experimentation has identified specific genes that may be associated with behavioral control. (p. 1109)

52.2 Learning influences behavior.

Nonassociative Learning and Conditioning

- Learning occurs when animals alter their behavior as a result of previous experiments. (p. 1110)
- Nonassociative learning, such as habituation, does not require the animal to form an association, while associative learning (classical or operant) requires development of an association between stimuli or between a stimulus and a response. (p. 1110)
- Many experiments have suggested that learning is possible only within the boundaries set by instinct. (p. 1111)

The Development of Behavior

- Behavioral biologists now recognize that behavior has both genetic and learned components. (p. 1112)
- Imprinting occurs when a maturing animal forms social attachments to other individuals or forms preferences that will influence behavior later in life; it is most often successful during a sensitive phase. (p. 1112)
- Experiments with song-learning in birds have suggested that in some cases the genetic template is selective, and learning must also play a prominent role. (p. 1113)

Animal Cognition

- Recent studies have shown compelling evidence of cognition in several species of animals including chimpanzees and ravens. (pp. 1114–1115)

Migratory Behavior

- Birds and other animals navigate (setting a bearing and following it) by looking at the sun and the stars, and potentially by detecting magnetic fields as well. (pp. 1116–1117)

52.3 Communication is a key element of many animal behaviors.

Courtship

- Courtship signals are often species-specific, limiting communication to members of the same species. (p. 1118)
- Different signals provide different levels of information about the sender, and the level of specificity relates to the function of the signal. (p. 1119)

Communication in Social Groups

- Social organisms utilize many modes of communication, including alarm calls, alarm pheromones, trail pheromones, and a dance language. (p. 1120)
- Upon close examination, the complexity of human languages is in fact a superficial difference, and all languages share many basic structural similarities. (p. 1121)

52.4 Evolutionary forces shape behavior.

Behavioral Ecology

- Behavioral ecology is the study of how natural selection shapes behavior. (p. 1122)

Foraging Behavior

- Optimal foraging theory states that natural selection produces animals that tend to feed on prey that maximize their net energy intake per unit foraging time. (p. 1123)

Territorial Behavior

- Territoriality occurs when individuals of a species maintain exclusive use of an area containing some limiting resource. (p. 1124)

Reproductive Strategies

- Because a reproductive event for females is often more costly than it is for males, females tend to be more choosy in mate selection. (p. 1125)

Reproductive Competition and Sexual Selection

- Sexual selection involves both intrasexual and intersexual selection. (pp. 1126–1127)

Mating Systems

- In mating systems such as monogamy, polygyny, and polyandry the number of mates an individual will have during the breeding season varies. (p. 1128)
- Several studies have established that extra-pair copulations are more pervasive in the animal kingdom, especially in birds, than originally suspected. (p. 1128)

52.5 There is considerable controversy about the evolution of social behavior.

Altruism and Group Living

- Altruistic behaviors may have evolved because they impart some benefit to the individual or the individual's relatives. (pp. 1130–1131)
- Kin selection suggests that aiding kin may increase the reproductive success of relatives enough to compensate for the reduction in a helper's own fitness. (pp. 1131–1132)

Group Living and the Evolution of Social Systems

- A society can be broadly defined as a group of organisms of the same species that are organized in a cooperative manner. (p. 1133)
- Social insect colonies are composed of different castes of workers that differ in size and morphology and have different tasks to perform. (p. 1133)
- Vertebrate social behavior is often characterized by kin-selected altruism such as cooperative feeding and rearing of young. (p. 1134)

Self Test

1. A male stickleback's display of aggression at a red object is an example of
   a. innate behavior.
   b. operant conditioning.
   c. associative learning.
   d. habituation.
2. What type of behavior is associated with a "critical period" in which a stimulus must be detected to trigger the response?
   a. cognitive behavior
   b. taxis
   c. imprinting
   d. associative learning
3. The study of song development in sparrows showed that
   a. the acquisition of a species-specific song is innate.
   b. there are two components to this behavior—a genetic template and learning.
   c. song acquisition is an example of associative learning.
   d. All of these are correct.
4. The level of specificity in courtship signaling is often
   a. individual-specific.
   b. anonymous.
   c. species-specific.
   d. any of these.
5. Which of the following is not a function of pheromones?
   a. sex attractants
   b. trigger alarm behaviors
   c. trail markers
   d. All of these are functions of pheromones.
6. Behavioral ecology is the study of
   a. how natural selection shapes behaviors.
   b. how environmental conditions affect animal behaviors.
   c. how feeding opportunities affect animal behaviors.
   d. how interactions with other individuals of the same species affects behaviors.
7. Which of the following would be most likely to occur according to the optimal foraging theory?
   a. Shore crabs eat the largest mussels even though they're harder to crack open.
   b. Columbian ground squirrels eat more because larger squirrels produce more offspring.
   c. Smaller yellow-eyed juncos eat larger prey, although they are harder to manage.
   d. An antelope will venture out on the open plain to feed even though lions are around.
8. Which of the following statements about territories is true?
   a. Territories and home ranges describe the same area of land.
   b. The defense of a territory is always beneficial to the animal.
   c. A territory contains resources exclusively for the animal that defends it.
   d. Territories overlap in time or space with other territories.
9. In the haplodiploidy system of sex determination, males are
   a. haploid.
   b. diploid.
   c. sterile.
   d. not present because bees exist as single-sex populations.
10. According to kin selection, saving the life of your _____ would do the least for increasing your inclusive fitness.
    a. mother
    b. brother
    c. sister-in-law
    d. niece

Test Your Visual Understanding

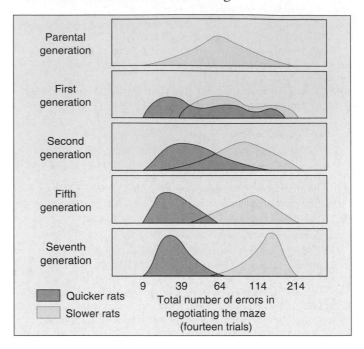

1. Can you predict what the graph in the seventh generation would look like if instead of mating the maze-bright rats with each other and the maze-dull rats with each other, you always mated maze-bright rats with maze-dull rats?

Apply Your Knowledge

1. Imagine an animal society in which the individuals share an average of one-third of their alleles with one another. If the benefits to an individual of performing an altruistic act are twice the costs, does the kin selection model predict that these individuals are likely to be altruistic? Explain your reasoning.
2. If a young animal exhibits a behavior immediately after being born or hatched, is it reasonable to conclude that the behavior is instinctive rather than learned? If not, what sorts of experiments might be performed to distinguish between these two possibilities?
3. Swallows often hunt in groups, while hawks and other predatory birds are usually solitary hunters. What do you think is the basis for this difference?

53

Population Ecology

Concept Outline

53.1 Organisms must cope with a varied environment.

    The Environmental Challenge. Organisms cope with environmental variation with physiological, morphological, and behavioral adaptations.

53.2 Populations are groups of individuals of the same species that live in the same place.

    Population Ranges. Population borders are determined by areas in which individuals cannot survive and reproduce.
    Pattern of Spacing of Individuals in a Population. The distribution of individuals in a population can be random, uniform, or clumped.
    Metapopulations. Sometimes populations are arranged in networks connected by the exchange of individuals.

53.3 Population dynamics depend critically upon age distribution.

    Demography. A population's growth rate is a function of its age structure and age-specific mortality rates.

53.4 Life histories often reflect trade-offs between reproduction and survival.

    The Cost of Reproduction. Evolutionary success is a trade-off between investing in current reproduction and investing in growth that promotes future reproduction.

53.5 Population growth is limited by the environment.

    Population Growth. Populations grow if births exceed deaths until they reach the carrying capacity of their environment.
    Factors that Regulate Populations. Some of the factors that regulate a population's growth depend upon the size of the population; others do not.

53.6 The human population has grown explosively in the last three centuries.

    The Advent of Exponential Growth. Human populations will continue to grow in developing countries because of the number of young people entering their reproductive years.

FIGURE 53.1
Life takes place in populations. This population of gannets is subject to the rigorous effects of reproductive strategy, competition, predation, and other limiting factors.

Ecology, the study of how organisms relate to one another and to their environments, is a complex and fascinating area of biology that has important implications for each of us. In our exploration of ecological principles, we first consider how organisms respond to the abiotic environment in which they occur and how these responses affect the properties of populations, emphasizing population dynamics (figure 53.1). In chapter 54, we discuss communities of coexisting species and the interactions that occur among them. In subsequent chapters, we discuss the functioning of entire ecosystems and of the biosphere, concluding with a consideration of the problems facing our planet and our fellow species.

53.1 Organisms must cope with a varied environment.

The Environmental Challenge

The nature of the physical environment in large measure determines what organisms live there. Key elements include:

Temperature. Most organisms are adapted to live within a relatively narrow range of temperatures and will not thrive if temperatures are colder or warmer. The growing season of plants, for example, is importantly influenced by temperature.

Water. Plants and all other organisms require water. On land, water is often scarce, so patterns of rainfall have a major influence on life.

Sunlight. Almost all ecosystems rely on energy captured by photosynthesis; the availability of sunlight influences the amount of life an ecosystem can support, particularly below the surface in marine communities.

Soil. The physical consistency, pH, and mineral composition of soil often severely limit plant growth, particularly the availability of nitrogen and phosphorus.

Approaches to Coping with Environmental Variation

An individual encountering environmental variation may maintain a "steady-state" internal environment, a condition known as **homeostasis.** Many animals and plants actively employ physiological, morphological, or behavioral mechanisms to maintain homeostasis. The beetle in figure 53.2 is using a behavioral mechanism to cope with drastic changes in water availability. Other animals and plants are known as **conformers** because they conform to the environment in which they find themselves, their bodies adopting the temperature, salinity, and other physical aspects of their surroundings.

Responses to environmental variation can be seen over both the short and the long term. In the short term, spanning periods of a few minutes to an individual's lifetime, organisms have a variety of ways of coping with environmental change. Over longer periods, natural selection can operate to make a population better adapted to the environment.

Individual Responses to Environmental Change.

Physiology. Many organisms are able to adapt to environmental change by making physiological adjustments. Thus, your body constricts the blood vessels on the surface of your face on a cold day, reducing heat loss (and also giving your face a "flush"). Similarly, humans who visit high altitudes initially experience altitude sickness—the symptoms of which include heart palpitations, nausea, fatigue, headache, mental impairment, and in serious

FIGURE 53.2
Meeting the challenge of obtaining moisture. On the dry sand dunes of the Namib Desert in southwestern Africa, the beetle *Onymacris unguicularis* collects moisture from the fog by holding its abdomen up at the crest of a dune to gather condensed water.

cases, pulmonary edema—because of the lower atmospheric pressure and consequent lower oxygen availability in the air. After several days, however, the same people feel fine, because a number of physiological changes have increased the delivery of oxygen to their body tissues (table 53.1).

Some insects avoid freezing in the winter by adding glycerol "antifreeze" to their blood; others tolerate freezing by converting much of their glycogen reserves into alcohols that protect their cell membranes from freeze damage.

Morphology. Animals that maintain a constant internal temperature (endotherms) in a cold environment have adaptations that tend to minimize energy expenditure. Many mammals grow thicker coats during the winter, utilizing their fur as insulation to retain body heat. In general, the thicker the fur, the greater the insulation (figure 53.3). Thus, a wolf's fur is about three times as thick in winter as in summer and insulates more than twice as well. Other mammals escape some of the costs of maintaining a constant body temperature during winter by hibernating during the coldest season, behaving, in effect, like conformers.

Behavior. Many animals deal with variation in the environment by moving from one patch of habitat to another, avoiding areas that are unsuitable. The tropical lizard in figure 53.4 manages to maintain a fairly uniform body temperature in an open habitat by basking in patches of sun and then retreating to the shade when it becomes too hot.

Table 53.1 Physiological Changes at High Altitude

Increased rate of breathing

Increased erythrocyte production, raising the amount of hemoglobin in the blood

Decreased binding capacity of hemoglobin, increasing the rate at which oxygen is unloaded in body tissues

Increased density of mitochondria, capillaries, and muscle myoglobin

By contrast, in shaded forests, the same lizard does not have the opportunity to regulate its body temperature through behavioral means. Thus, it becomes a conformer and adopts the temperature of its surroundings.

Behavioral adaptations can be extreme. Spadefoot toads (genus *Scaphiophus*), which live in the deserts of North America, can burrow nearly a meter below the surface and remain there for as long as nine months of each year, their metabolic rates greatly reduced as they live on fat reserves. When moist, cool conditions return, the toads emerge and breed. The young toads mature rapidly and burrow back underground.

Evolutionary Responses to Environmental Variation. These examples represent different ways in which organisms may adjust to changing environmental conditions. The ability of an individual to alter its physiology, morphology, or behavior is itself an evolutionary adaptation, the result of natural selection. The results of natural selection can also be detected by comparing closely related species that live in different environments. In such cases, species often have evolved striking adaptations to the particular environment in which they live.

For example, animals that live in different climates show many differences. Mammals from colder climates tend to have shorter ears and limbs—a phenomenon termed "Allen's Rule"—which reduces the surface area across which animals lose heat. Lizards that live in different climates exhibit physiological adaptations for coping with life at different temperatures. Desert lizards are unaffected by high temperatures that would kill a lizard from northern Europe, but the northern lizards are capable of running, capturing prey, and digesting food at cooler temperatures at which desert lizards would be completely immobilized.

Many species also exhibit adaptations to living in areas where water is scarce. Everyone knows of the camel and other desert animals that can go extended periods without drinking water. Another example of desert adaptation is seen in frogs. Most frogs have moist skins through which water permeates readily. Such animals could not survive in arid climates because they would rapidly dehydrate and die. However, some frogs have solved this problem by evolving a greatly reduced rate of water loss through the skin. One species, for example, secretes a waxy substance from specialized glands that waterproofs its skin and reduces rates of water loss by 95%.

Adaptation to different environments can also be studied experimentally. For example, when strains of *E. coli* were grown at high temperatures (42°C), the speed at which resources are utilized improved through time. After 2000 generations, this ability increased 30% over what it had been when the experiment started. The mechanism by which efficiency of resource use was increased is still unknown and is the focus of current research.

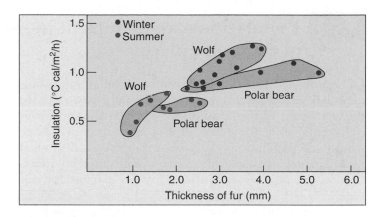

FIGURE 53.3
Morphological adaptation. Fur thickness in North American mammals has a major impact on the degree of insulation the fur provides.
Polar bears are able to live in zoos in warm climates. How thick would you expect the hair of a polar bear to be in a zoo in Miami, Florida?

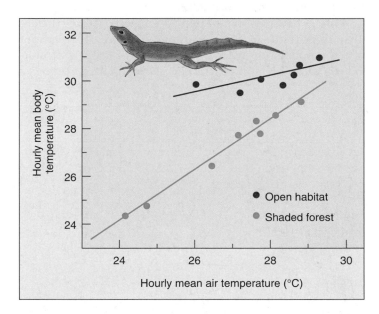

FIGURE 53.4
Behavioral adaptation. The Puerto Rican lizard *Anolis cristatellus* maintains a relatively constant temperature by seeking out and basking in patches of sunlight; in shaded forests, this behavior is not possible, and body temperature conforms to the surroundings.
When given the opportunity, lizards regulate their body temperature to maintain a temperature optimal for physiological functioning. Would lizards in open habitats exhibit different escape behaviors than lizards in shaded forest?

Organisms use a variety of physiological, morphological, and behavioral mechanisms to adjust to environmental variation. Over time, species evolve adaptations to living in different environments.

53.2 Populations are groups of individuals of the same species that live in the same place.

Organisms live as members of **populations**, groups of individuals that occur together at one place and time. In the remainder of this chapter, we consider the properties of populations, focusing on elements that influence whether a population will grow or shrink, and at what rate. The explosive growth of the world's human population in the last few centuries provides a focus for our inquiry.

The term "population" can be defined narrowly or broadly. This flexibility allows us to speak in similar terms of the world's human population, the population of protists in the gut of a termite, or the population of deer that inhabit a forest. Sometimes the boundaries defining a population are sharp, such as the edge of an isolated mountain lake for trout, and sometimes they are fuzzier, as when individual deer readily move back and forth between two forests separated by a cornfield.

Three characteristics of population ecology are particularly important: population range, the area throughout which a population occurs; the pattern of spacing of individuals within that range; and the size a population attains.

Population Ranges

No population, not even one composed of humans, occurs in all habitats throughout the world. Most species, in fact, have relatively limited geographic ranges, and the range of some species is miniscule. The Devil's Hole pupfish, for example, lives in a single hot-water spring in southern Nevada (figure 53.5), and the Socorro isopod is known from a single spring system in New Mexico. At the other extreme, some species are widely distributed. The common dolphin (*Delphinus delphis*), for example, is found throughout all the world's oceans.

As we just saw in section 53.1, organisms must be adapted for the environment in which they occur. Polar bears are exquisitely adapted to survive the cold of the Arctic, but you won't find them in the tropical rain forest. Certain prokaryotes can live in the near-boiling waters of Yellowstone's geysers, but they do not occur in cooler streams nearby. Each population has its own requirements—temperature, humidity, certain types of food, and a host of other factors—that determine where it can live and reproduce and where it can't. In addition, in places that are otherwise suitable, the presence of predators, competitors, or parasites may prevent a population from occupying an area, a topic we will take up in chapter 54.

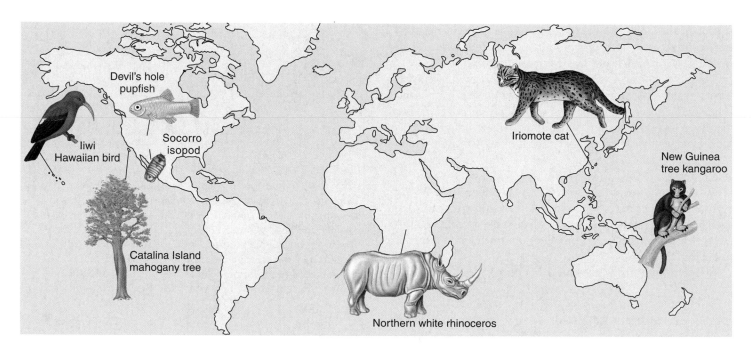

FIGURE 53.5
Species that occur in only one place.   These species, and many others, are only found in a single population. All are endangered species, and should anything happen to their single habitat, the population—and the species—would become extinct.

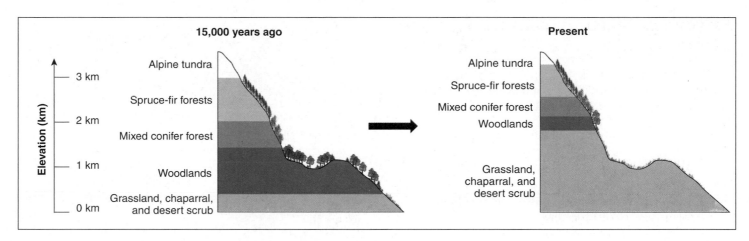

FIGURE 53.6
Altitudinal shifts in population ranges in the mountains of southwestern North America. During the glacial period 15,000 years ago, conditions were cooler than they are now. As the climate has warmed, tree species that require colder temperatures have shifted their range upward in altitude so that they live in the climatic conditions to which they are adapted.

Range Expansions and Contractions

Population ranges are not static; rather, they change through time. These changes occur for two reasons. In some cases, the environment changes. For example, as the glaciers retreated at the end of the last ice age, approximately 10,000 years ago, many North American plant and animal populations expanded northward. At the same time, as climates warmed, species experienced shifts in the elevation at which they could live (figure 53.6).

In addition, populations can expand their ranges when they are able to circumvent inhospitable habitat to colonize suitable, previously unoccupied areas. For example, the cattle egret is native to Africa. Some time in the late 1800s, these birds appeared in northern South America, having made the nearly 2000-mile transatlantic crossing, perhaps aided by strong winds. Since then, they have steadily expanded their range and now can be found throughout most of the United States (figure 53.7).

> A population is a group of individuals of the same species existing together in an area. Its range, the area a population occupies, changes over time.

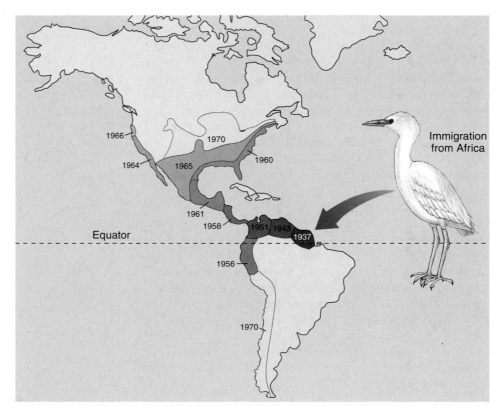

FIGURE 53.7
Range expansion of the cattle egret. The cattle egret—so named because it follows cattle and other hoofed animals, catching any insects or small vertebrates that they disturb—first arrived in South America in the late 1800s. Since the 1930s, the range expansion of this species has been well documented, as it has moved westward and up into much of North America, as well as down the western side of the Andes to near the southern tip of South America.

Pattern of Spacing of Individuals in a Population

Another key characteristic of population structure is the way in which individuals of a population are distributed. They may be randomly spaced, uniformly spaced, or clumped (figure 53.8).

Randomly Spaced

Individuals are randomly spaced within populations when they do not interact strongly with one another or with nonuniform aspects of their environment. Random distributions are not common in nature. Some species of trees, however, appear to exhibit random distributions in Panamanian rain forests (figure 53.8b–d).

Uniformly Spaced

Uniform spacing within a population may often, but not always, result from competition for resources. The means by which it is accomplished, however, varies.

In animals, uniform spacing often results from behavioral interactions, as discussed in chapter 52. In many species, individuals of one or both sexes defend a territory from which other individuals are excluded. These territories provide the owner with exclusive access to resources such as food, water, hiding refuges, or mates and tend to space individuals evenly across the habitat. Even in nonterritorial species, individuals often maintain a defended space into which other animals are not allowed to intrude.

Among plants, uniform spacing is also a common result of competition for resources. In this case, however, the spacing results from direct competition for the resources. Closely spaced individual plants will compete for available sunlight, nutrients, or water. These contests can be direct, as when one plant casts a shadow over another, or indirect, as when two plants compete by extracting nutrients or water from a shared area. In addition, some plants, such as creosote, produce chemicals in the surrounding soil that are toxic to other members of their species. In all of these cases, only plants that are spaced an adequate distance from each other will be able to coexist, leading to uniform spacing.

FIGURE 53.8
Population dispersion. The different patterns of dispersion are exhibited by (*a*) different arrangements of bacterial colonies and (*b-d*) three different species of trees from the same locality in Panama.

Source: Data from Elizabeth Losos, Center for Tropical Forest Science, Smithsonian Tropical Research Institute.

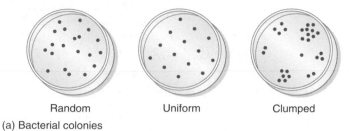

Random   Uniform   Clumped

(a) Bacterial colonies

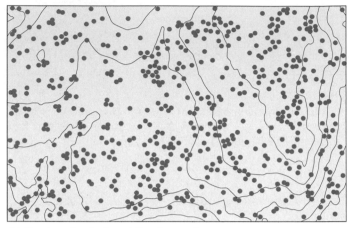

(b) Random distribution of *Brosimum alicastrum*

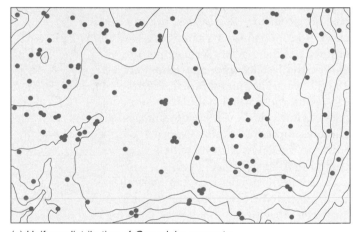

(c) Uniform distribution of *Coccoloba coronata*

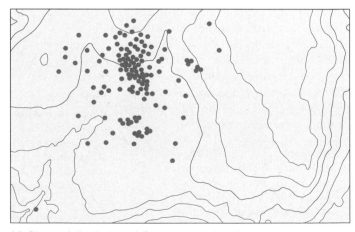

(d) Clumped distribution of *Chamguava schippii*

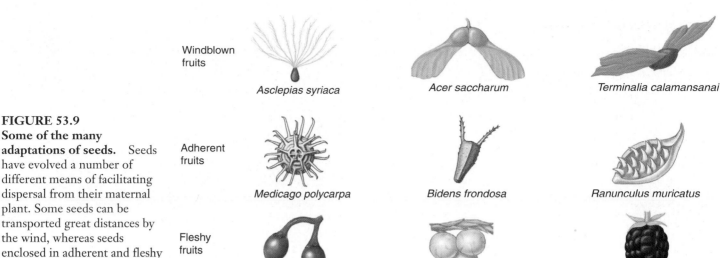

Windblown fruits

Asclepias syriaca

Acer saccharum

Terminalia calamansanai

Adherent fruits

Medicago polycarpa

Bidens frondosa

Ranunculus muricatus

Fleshy fruits

Solanum dulcamara

Juniperus chinensis

Rubus sp.

FIGURE 53.9
Some of the many adaptations of seeds. Seeds have evolved a number of different means of facilitating dispersal from their maternal plant. Some seeds can be transported great distances by the wind, whereas seeds enclosed in adherent and fleshy fruits can be transported by animals.

Clumped Spacing

Individuals clump into groups or clusters in response to uneven distribution of resources in their immediate environments. Clumped distributions are common in nature because individual animals, plants, and microorganisms tend to prefer habitats defined by soil type, moisture, or other aspects of the environment to which they are best adapted.

Social interactions also can lead to clumped distributions. Many species live and move around in large groups, which go by a variety of names (for example, flock, herd, pride). Such groupings can provide many advantages, including increased awareness of and defense against predators, decreased energetic cost of moving through air and water, and access to the knowledge of all group members.

At a broader scale, populations are often most densely populated in the interior of their range and less densely distributed toward the edges. Such patterns usually result from the manner in which the environment changes in different areas. Populations are often best adapted to the conditions in the interior of their distribution. As environmental conditions change, individuals are less well adapted, and thus densities decrease. Ultimately, the point is reached at which individuals cannot persist at all; this marks the edge of a population's range.

The Human Effect

By altering the environment, humans have allowed some species, such as coyotes, to expand their ranges and move into areas they previously did not occupy. Moreover, humans have served as an agent of dispersal for many species. Some of these transplants have been widely successful as is discussed in greater detail in chapter 57. For example, 100 starlings were introduced into New York City in 1896 in a misguided attempt to establish every species of bird mentioned by Shakespeare. Their population steadily spread so that by 1980, they occurred throughout the United States. Similar stories could be told for countless plants and animals, and the list increases every year. Unfortunately, the success of these invaders often comes at the expense of native species.

Dispersal Mechanisms

Dispersal to new areas can occur in many ways. Lizards, for example, have colonized many distant islands, probably due to individuals or their eggs floating or drifting on vegetation. Bats are often the only mammals on distant islands because they can fly to them. Seeds of plants are designed to disperse in many ways (figure 53.9). Some seeds are aerodynamically designed to be blown long distances by the wind. Others have structures that stick to the fur or feathers of animals, so that they are carried long distances before falling to the ground. Still others are enclosed in fleshy fruits. These seeds can pass through the digestive systems of mammals or birds and then germinate at the spot upon which they are defecated. Finally, seeds of *Arceuthobium* are violently propelled from the base of the fruit in an explosive discharge. Although the probability of long-distance dispersal events leading to successful establishment of new populations is slim, over millions of years, many such dispersals have occurred.

The distribution of individuals within a population can be random, uniform, or clumped and is determined in part by the availability of resources.

Metapopulations

Species often exist as a network of distinct populations that interact with each other by exchanging individuals. Such networks, termed **metapopulations**, usually occur in areas in which suitable habitat is patchily distributed and separated by intervening stretches of unsuitable habitat.

To what degree populations within a metapopulation interact depends on the amount of dispersal and is often not symmetrical: Populations increasing in size tend to send out many dispersers, whereas populations at low levels tend to receive more immigrants than they send off. In addition, relatively isolated populations tend to receive relatively few arrivals.

Not all suitable habitats within a metapopulation's area may be occupied at any one time. For various reasons, some individual populations may become extinct, perhaps as a result of an epidemic disease, a catastrophic fire, or inbreeding depression. However, because of dispersal from other populations, such areas may eventually be recolonized. In some cases, the number of habitats occupied in a metapopulation may represent an equilibrium in which the rate of extinction of existing populations is balanced by the rate of colonization of empty habitats.

A species may also exhibit a metapopulation structure in areas in which some habitats are suitable for long-term population maintenance, whereas others are not. In these situations, termed **source-sink metapopulations,** the populations in the better areas (the sources) continually send out dispersers that bolster the populations in the poorer habitats (the sinks). In the absence of such continual replenishment, sink populations would have a negative growth rate and would eventually become extinct.

Metapopulations of butterflies have been studied particularly intensively (figure 53.10). In one study, researchers sampled populations of the glanville fritillary butterfly at 1600 meadows in southwestern Finland. On average, every year, 200 populations became extinct, but 114 empty meadows were colonized. A variety of factors seemed to increase the likelihood of a population's extinction, including small population size, isolation from sources of immigrants, low resource availability (as indicated by the number of flowers on a meadow), and lack of genetic variation within the population. The researchers attribute the greater number of extinctions than colonizations to a string of very dry summers. Because none of the populations is large enough to survive on its own, continued survival of the species in southwestern Finland would appear to require the continued existence of a metapopulation network in which new populations are continually created and existing populations are supplemented by emigrants. Continued bad weather thus may doom the species, at least in this part of its range.

Metapopulations, where they occur, can have two important implications for the range of a species. First, by continually colonizing empty patches, they prevent long-

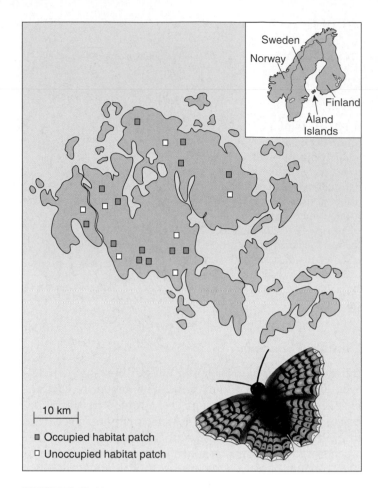

FIGURE 53.10
Metapopulations of butterflies. The glanville fritillary butterfly occurs in metapopulations in southwestern Finland on the Åland Islands. None of the populations is large enough to survive for long on its own, but continual emigration of individuals from other populations allows some populations to survive. In addition, continual establishment of new populations tends to offset extinction of established populations, although in recent years, extinctions have outnumbered colonizations.

term extinction. If no such dispersal existed, then each population might eventually perish, leading to disappearance of the species from the entire area. Moreover, in source-sink metapopulations, the species occupies a larger area than it otherwise might occupy, including marginal areas that could not support a population without a continual influx of immigrants. For these reasons, the study of metapopulations has become very important in conservation biology as natural habitats become increasingly fragmented.

Across broader areas, individuals may occur in populations that are loosely interconnected, termed metapopulations.

53.3 Population dynamics depend critically upon age distribution.

The dynamics of a population—how it changes through time—are affected by many factors. One important factor is the age distribution of individuals—that is, what proportion of individuals are adults, juveniles, and babies.

Demography

Demography (from the Greek *demos*, "the people," + *graphos*, "measurement") is the statistical study of populations. How the size of a population changes through time can be studied at two levels: as a whole or broken down into parts. At the most inclusive level, we can study the whole population to determine whether it is increasing, decreasing, or remaining constant. Populations grow if births outnumber deaths and shrink if deaths outnumber births. Understanding these trends is often easier, however, if we break the population into smaller units composed of individuals of the same age (for example, 1-year-olds) and study the factors affecting birth and death rates for each unit separately.

Factors Affecting Population Growth Rates

Growth rates can be influenced by the population's **sex ratio.** The number of births in a population is usually directly related to the number of females, but may not be as closely related to the number of males in species in which a single male can mate with several females. In many species, males compete for the opportunity to mate with females (a situation discussed in chapter 52); consequently, a few males get many matings, whereas many males do not mate at all. In such species, a female-biased sex ratio would not affect population growth rates; reduction in the number of males simply changes the identities of the reproductive males without reducing the number of births. By contrast, among monogamous species, such as many birds, in which pairs form long-lasting reproductive relationships, a reduction in the number of males can directly reduce the number of births.

Generation time, the average interval between the birth of an individual and the birth of its offspring, can also affect population growth rates. Species differ greatly in generation time. Differences in body size can explain much of this variation—mice go through approximately 100 generations during the course of one elephant generation—but not all of it (figure 53.11). Newts, for example, are smaller than mice, but have considerably longer generation times. Everything else equal, populations with short generations can increase in size more quickly than populations with long generations. Conversely, because generation time and life span are usually closely correlated, populations with short generation times may also diminish in size more rapidly if birthrates suddenly decrease.

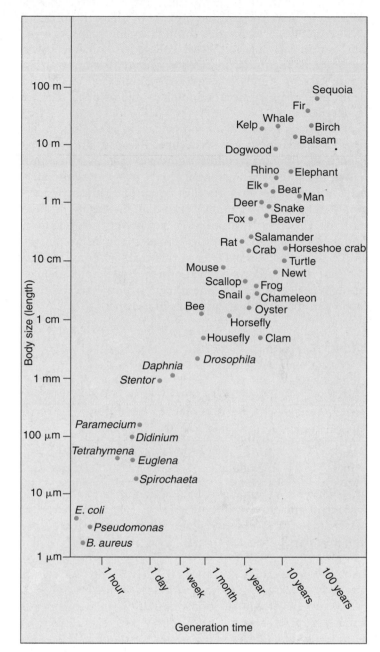

FIGURE 53.11
The relationship between body size and generation time. In general, larger animals have longer generation times, although there are exceptions.
If resources became more abundant, would you expect smaller or larger species to increase in population size more quickly?

Age Structure

A group of individuals of the same age is referred to as a **cohort.** In most species, the probability that an individual will reproduce or die varies through its life span. As a result, within a population, every cohort has a characteristic birthrate, or **fecundity,** defined as the number of offspring produced in a standard time (for example, per year), and death rate, or **mortality,** the number of individuals that die in that period.

The relative number of individuals in each cohort defines a population's **age structure.** Because different cohorts have different fecundity and death rates, age structure has a critical impact on a population's growth rate. Populations with a large proportion of young individuals, for example, tend to grow rapidly because an increasing proportion of their individuals are reproductive. Human populations in many underdeveloped countries are an example, as we will discuss later in this chapter. Conversely, if a large proportion of a population is relatively old, populations may decline. This phenomenon now characterizes Japan and some wealthy countries in Europe.

Life Tables and Population Change Through Time

To assess how populations in nature are changing, ecologists use a **life table,** which tabulates the fate of a cohort from birth until death, showing the number of offspring produced and the number of individuals that die each year. A very nice life table analysis is exhibited in a study of the meadow grass *Poa annua* (table 53.2). This study follows the fate of 843 individuals through time, charting how many survive in each interval and how many offspring each survivor produces.

In table 53.2, the first column indicates the age of the cohort (that is, the number of 3-month intervals from the start of the study). The second and third columns indicate the number of survivors and the proportion of the original cohort still alive at the beginning of that interval. The fifth column presents the **mortality rate,** the proportion of individuals that started that interval alive but died by the end of it. The seventh column indicates the average number of seeds produced by each surviving individual in that interval, and the last column presents the number of seeds produced relative to the size of the original cohort.

Table 53.2   Life Table for a Cohort of the Grass *Poa annua*, Containing 843 Seedlings

Age (in 3-month intervals)	Number Alive at Beginning of Time Interval	Proportion of Cohort Surviving to Beginning of Time Interval (survivorship)	Deaths During Time Interval	Mortality Rate During Time Interval	Seeds Produced During Time Interval	Seeds Produced per Surviving Individual (fecundity)	Seeds Produced per Member of Cohort (fecundity × survivorship)
0	843	1.000	121	0.143	0	0.00	0.00
1	722	0.857	195	0.271	303	0.42	0.36
2	527	0.625	211	0.400	622	1.18	0.74
3	316	0.375	172	0.544	430	1.36	0.51
4	144	0.171	90	0.626	210	1.46	0.25
5	54	0.064	39	0.722	60	1.11	0.07
6	15	0.018	12	0.800	30	2.00	0.04
7	3	0.004	3	1.000	10	3.33	0.01
8	0	0.000	—		Total = 1665		Total = 1.98

Much can be learned by examining life tables. In the case of *P. annua*, we see that both the probability of dying and the number of offspring per surviving individual produced steadily increases with age. By adding up the numbers in the last column, we get the total number of offspring produced per individual in the initial cohort. This number is almost 2, which means that for every original member of the cohort, on average two individuals have been produced. A figure of 1.0 would be the break-even number, the point at which the population was neither growing nor shrinking. In this case, the population appears to be growing rapidly.

In most cases, life table analysis is more complicated than this. First, except for organisms with short life spans, it is difficult to track the fate of a cohort until the death of the last individual. An alternative approach is to construct a cross-sectional study, examining the fate of all cohorts over a single year. In addition, many factors—such as offspring reproducing before all members of their parental generation's cohort have died—complicate the interpretation of whether populations are growing or shrinking.

Survivorship Curves

The percentage of an original population that survives to a given age is called its **survivorship.** One way to express some aspects of the age distribution of populations is through a *survivorship curve*. Examples of different survivorship curves are shown in figure 53.12. In hydra, animals related to jellyfish, individuals are equally likely to die at any age, as indicated by the straight survivorship curve (type II). Oysters produce vast numbers of offspring, only a few of which live to reproduce. However, once they become established and grow into reproductive individuals, their mortality rate is extremely low (type III survivorship curve). Finally, even though human babies are susceptible to death at relatively high rates, mortality rates in humans, as in many other animals and in protists, rise steeply in the postreproductive years (type I survivorship curve). Of course, these are just generalizations, and many organisms show more complicated patterns. Examination of the data for *P. annua*, for example, reveals that it is most similar to a type II survivorship curve (figure 53.13).

The growth rate of a population is a sensitive function of its age structure. The age structure of a population and the manner in which mortality and birthrates vary among different age cohorts determine whether a population will increase or decrease in size.

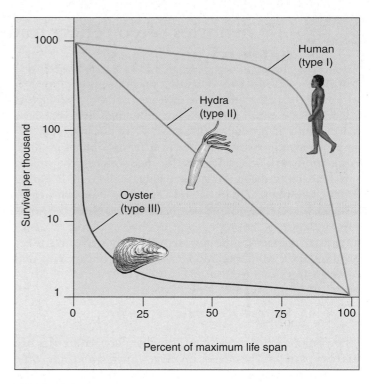

FIGURE 53.12
Survivorship curves. By convention, survival (the vertical axis) is plotted on a log scale. Humans have a type I life cycle, the hydra (an animal related to jellyfish) type II, and oysters type III.

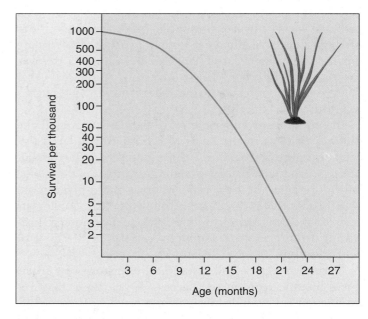

FIGURE 53.13
Survivorship curve for a cohort of the meadow grass *Poa annua*. After several months of age, mortality increases at a constant rate through time.
Suppose you wanted to keep meadow grass in your room as a houseplant. Suppose, too, that you wanted to buy a plant that was likely to live as long as possible. What age plant would you buy?

53.4 Life histories often reflect trade-offs between reproduction and survival.

Natural selection favors traits that maximize the number of surviving offspring left in the next generation. Two factors affect this quantity: how long an individual lives and how many young it produces each year. Why doesn't every organism reproduce immediately after its own birth, produce large families of offspring, care for them intensively, and do this repeatedly throughout a long life, while outcompeting others, escaping predators, and capturing food with ease? The answer is that no one organism can do all of this, simply because not enough resources are available. Consequently, organisms allocate resources either to current reproduction or to increasing their prospects of surviving and reproducing at later life stages.

The Cost of Reproduction

The complete life cycle of an organism constitutes its **life history.** All life histories involve significant trade-offs. Because resources are limited, a change that increases reproduction may decrease survival and reduce future reproduction. Thus, a Douglas fir tree that produces more cones increases its current reproductive success, but it also grows more slowly; because the number of cones produced is a function of how large a tree is, this diminished growth will decrease the number of cones it can produce in the future. Similarly, birds that have more offspring each year have a higher probability of dying during that year or producing smaller clutches the following year (figure 53.14). Conversely, individuals that delay reproduction may grow faster and larger, enhancing future reproduction.

In one elegant experiment, researchers changed the number of eggs in the nests of a bird, the collared flycatcher (figure 53.15). Birds whose clutch size (the number of eggs produced in one breeding event) was decreased laid more eggs the next year, whereas those given more eggs produced fewer eggs the following year. Ecologists refer to the reduction in future reproductive potential resulting from current reproductive efforts as the **cost of reproduction.**

Natural selection will favor the life history that maximizes lifetime reproductive success. When the cost of reproduction is low, individuals should produce as many offspring as possible because there is little cost. Low costs of reproduction may occur when resources are abundant, such that producing offspring does not impair survival or the ability to produce many offspring in subsequent years. Costs of reproduction are also low when overall mortality rates are high. In such cases, individuals may be unlikely to survive to the next breeding season anyway, so the incremental effect of increased reproductive efforts may not make a difference in future survival.

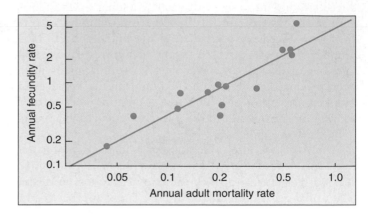

FIGURE 53.14
Reproduction has a price. Increased fecundity in birds correlates with higher mortality in several populations of birds, ranging from the albatross (lowest) to the sparrow (highest). Birds that raise more offspring per year have a higher probability of dying during that year.
Do you think that species in this graph are ordered by body size?

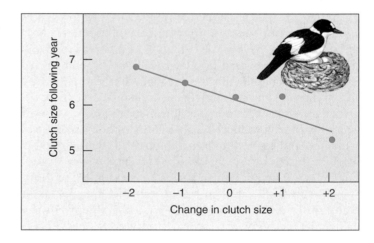

FIGURE 53.15
Reproductive events per lifetime. Adding eggs to nests of collared flycatchers (which increases the reproductive efforts of the female rearing the young) decreases clutch size the following year; removing eggs from the nest increases the next year's clutch size. This experiment demonstrates the trade-off between current reproductive effort and future reproductive success.
Why does this relationship exist?

Alternatively, when costs of reproduction are high, lifetime reproductive success may be maximized by deferring or minimizing current reproduction to enhance growth and survival rates. This may occur when costs of reproduction significantly affect the ability of an individual to survive or decrease the number of offspring that can be produced in the future.

Investment per Offspring

In terms of natural selection, the number of offspring produced is not as important as how many of those offspring themselves survive to reproduce.

A key reproductive trade-off concerns how many resources to invest in producing any single offspring. Assuming that the amount of energy to be invested in offspring is limited, a trade-off must occur between the number of offspring produced and the size of each offspring (figure 53.16). This trade-off has been experimentally demonstrated in the side-blotched lizard, *Uta stansburiana*, which normally lays on average four and a half eggs at a time. When some of the eggs are removed surgically early in the reproductive cycle, the female lizard produces only 1 to 3 eggs, but supplies each of these eggs with greater amounts of yolk, producing eggs and, subsequently, hatchlings that are much larger than normal (figure 53.17).

In the side-blotched lizard and many other species, the size of offspring critically affects their survival prospects—larger offspring have a greater chance of survival. Producing many offspring with little chance of survival might not be the best strategy, but producing only a single, extraordinarily robust offspring also would not maximize the number of surviving offspring. Rather, an intermediate situation, in which several fairly large offspring are produced, should maximize the number of surviving offspring.

Reproductive Events per Lifetime

The trade-off between age and fecundity plays a key role in many life histories. Annual plants and most insects focus all their reproductive resources on a single large event and then die. This life history adaptation is called **semelparity** (from the Latin *semel*, "once," + *parito*, "to beget"). Organisms that produce offspring several times over many seasons exhibit a life history adaptation called **iteroparity** (from the Latin *itero*, "to repeat"). Species that reproduce yearly must avoid overtaxing themselves in any one reproductive episode so that they will be able to survive and reproduce in the future. Semelparity, or "big bang" reproduction, is usually found in short-lived species that have a low probability of staying alive between broods, such as plants growing in harsh climates. Semelparity is also favored when fecundity entails large reproductive cost, as when Pacific salmon migrate upriver to their spawning grounds. In these species, rather than investing some resources in an unlikely bid to survive until the next breeding season, individuals place all their resources into reproduction.

Age at First Reproduction

Among mammals and many other animals, longer-lived species put off reproduction longer than short-lived species

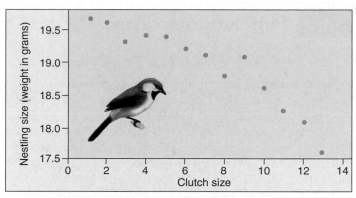

FIGURE 53.16
The relationship between clutch size and offspring size. In great tits, the size of the nestlings is inversely related to the number of eggs laid. The more mouths they have to feed, the less the parents can provide to any one nestling.
Would natural selection favor producing many small young or a few large ones?

FIGURE 53.17
Variation in baby lizard size produced by experimental manipulations. In clutches in which some developing eggs were surgically removed, the remaining offspring were larger (*center*) than lizards produced in control clutches in which all the eggs were allowed to develop (*right*). In experiments in which some of the yolk was removed from the eggs, smaller lizards hatched (*left*).

(relative to expected life span). The advantage of delayed reproduction is that juveniles gain experience before expending the high costs of reproduction. In long-lived animals, this advantage outweighs the energy that is invested in survival and growth rather than reproduction. In shorter-lived animals, on the other hand, time is of the essence; thus, quick reproduction is more critical than juvenile training, and reproduction tends to occur earlier.

Life history adaptations involve many trade-offs between reproductive cost and investment in survival. Different kinds of animals and plants employ quite different approaches.

53.5 Population growth is limited by the environment.

Population Growth

Populations often remain at a relatively constant size, regardless of how many offspring are born. As you saw in chapter 1, Darwin based his theory of natural selection partly on this seeming contradiction. Natural selection occurs because of checks on reproduction, with some individuals producing fewer surviving offspring than others. To understand populations, we must consider how they grow and what factors in nature limit population growth.

The Exponential Growth Model

The rate of population increase, r, is defined as the difference between the birthrate (b) and the death rate (d) corrected for any movement of individuals in or out of the population, whether net emigration (e, movement out of the area) or net immigration (i, movement into the area). Thus,

$$r = (b - d) + (i - e)$$

Movements of individuals can have a major impact on population growth rates. For example, the increase in human population in the United States during the closing decades of the twentieth century was mostly due to immigration. Less than half of the increase came from the reproduction of the people already living there.

The simplest model of population growth assumes that a population grows without limits at its maximal rate and also that rates of immigration and emigration are equal. This rate, called the **biotic potential,** is the rate at which a population of a given species will increase when no limits are placed on its rate of growth. In mathematical terms, this is defined by the following formula:

$$\frac{dN}{dt} = r_iN$$

where N is the number of individuals in the population, dN/dt is the rate of change in its numbers over time, and r_i is the intrinsic rate of natural increase for that population—its innate capacity for growth.

The biotic potential of any population is exponential (red line in figure 53.18). Even when the *rate* of increase remains constant, the actual *number* of individuals accelerates rapidly as the size of the population grows. The result of unchecked exponential growth is a population explosion. A single pair of houseflies, laying 120 eggs per generation, could produce more than 5 trillion descendants in a year. In 10 years, their descendants would form a swarm more than 2 meters thick over the entire surface of the earth! In practice, such patterns of unrestrained growth prevail only for short periods, usually when an organism reaches a new habitat with abundant resources.

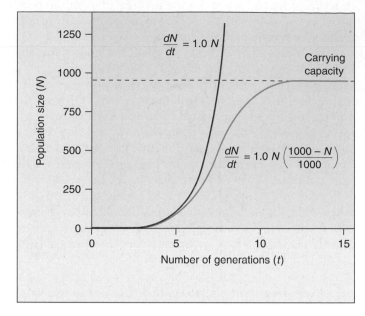

FIGURE 53.18
Two models of population growth. The red line illustrates the exponential growth model for a population with an r of 1.0. The blue line illustrates the logistic growth model in a population with $r = 1.0$ and $K = 1000$ individuals. At first, logistic growth accelerates exponentially; then, as resources become limiting, the death rate increases and growth slows. Growth ceases when the death rate equals the birthrate. The carrying capacity (K) ultimately depends on the resources available in the environment.

Natural examples include dandelions arriving in the fields, lawns, and meadows of North America from Europe for the first time; algae colonizing a newly formed pond; or the first terrestrial immigrants landing on an island recently thrust up from the sea.

Carrying Capacity

No matter how rapidly populations grow, they eventually reach a limit imposed by shortages of important environmental factors, such as space, light, water, or nutrients. A population ultimately may stabilize at a certain size, called the **carrying capacity** of the particular place where it lives. The carrying capacity, symbolized by K, is the maximum number of individuals that the environment can support.

The Logistic Growth Model

As a population approaches its carrying capacity, its rate of growth slows greatly, because fewer resources remain for each new individual to use. The growth curve of such a

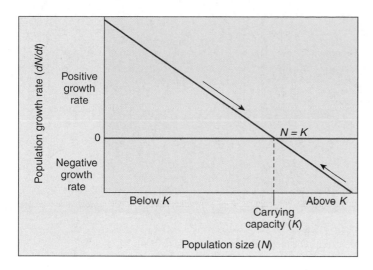

FIGURE 53.19

Relationship between population growth rate and population size. Populations far from the carrying capacity (*K*) will have high growth rates—positive if the population is below *K*, and negative if it is above *K*. As the population approaches *K*, growth rates approach zero.

Why does the growth rate converge on zero?

population, which is always limited by one or more factors in the environment, can be approximated by the following logistic growth equation:

$$\frac{dN}{dt} = rN\left(\frac{K - N}{K}\right)$$

In this model of population growth, the growth rate of the population (*dN/dt*) equals its intrinsic rate of natural increase (*r* multiplied by *N*, the number of individuals present at any one time), adjusted for the amount of resources available. The adjustment is made by multiplying *rN* by the fraction of *K* still unused (*K* minus *N*, divided by *K*). As *N* increases (the population grows in size), the fraction by which *r* is multiplied (the remaining resources) becomes smaller and smaller, and the rate of increase of the population declines.

Graphically, if you plot *N* versus *t* (time), you obtain a **sigmoidal growth curve** characteristic of many biological populations. The curve is called "sigmoidal" because its shape has a double curve like the letter S. As the size of a population stabilizes at the carrying capacity, its rate of growth slows down, eventually coming to a halt (blue line in figure 53.18).

In mathematical terms, as *N* approaches *K*, the *rate* of population growth (*dN/dt*) begins to slow, reaching 0 when *N = K* (figure 53.19). Conversely, if the population size exceeds the carrying capacity, then *K − N* will be negative, and the population will experience a negative growth rate. As the population size declines toward the carrying capacity, the magnitude of this negative growth rate will decrease until it reaches 0 when *N = K*. Notice that the pop-

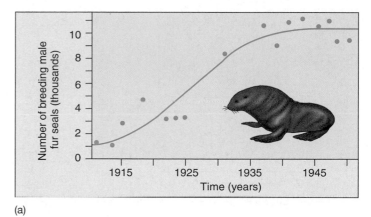

(a)

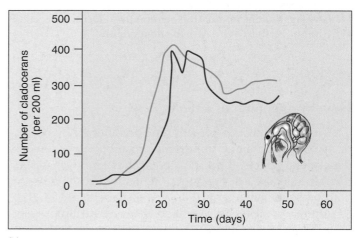

(b)

FIGURE 53.20

Many populations exhibit logistic growth. (*a*) A fur seal (*Callorhinus ursinus*) population on St. Paul Island, Alaska. (*b*) Two laboratory populations of the cladoceran *Bosmina longirostris*. Note that the populations first exceeded the carrying capacity, before decreasing to a size that was then maintained.

Why is there a hump in the population growth curve in (*b*), followed by a decline in the population?

ulation will tend to move toward the carrying capacity regardless of whether it is initially above or below it. For this reason, logistic growth tends to return a population to the same size. In this sense, such populations are considered to be in equilibrium because they would be expected to be at or near the carrying capacity at most times.

In many cases, real populations display trends corresponding to a logistic growth curve. This is true not only in the laboratory, but also in natural populations (figure 53.20*a*). In some cases, however, the fit is not perfect (figure 53.20*b*), and as we shall see shortly, many populations exhibit other patterns.

The size at which a population stabilizes in a particular place is defined as the carrying capacity of that place for that species. Populations often grow to the carrying capacity of their environment.

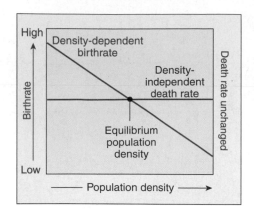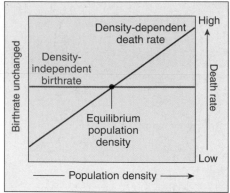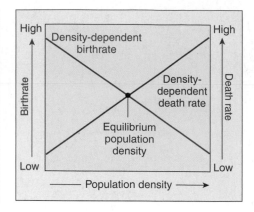

FIGURE 53.21

Density-dependent population regulation. Density-dependent factors can affect birthrates, death rates, or both. **Why might birthrates be density-dependent?**

Factors That Regulate Populations

Density-Dependent Effects

The reason population growth rates are affected by population size is that many important processes have **density-dependent effects.** That is, as population size increases, either reproductive rates decline or mortality rates increase, or both, a phenomenon termed *negative feedback* (figure 53.21).

Populations can be regulated in many different ways. When populations approach their carrying capacity, competition for resources can be severe, leading both to a decreased birthrate and an increased risk of death (figure 53.22). In addition, predators often focus their attention on particularly common prey, which also results in increasing rates of mortality as populations increase. High population densities can also lead to an accumulation of toxic wastes in the environment.

Behavioral changes may also affect population growth rates. Some species of rodents, for example, become antisocial, fighting more, breeding less, and generally acting stressed-out. These behavioral changes result from hormonal actions, but their ultimate cause is not yet clear; most likely, they have evolved as adaptive responses to situations in which resources are scarce. In addition, in crowded populations, the population growth rate may decrease because of an increased rate of emigration of individuals attempting to find better conditions elsewhere (figure 53.23).

FIGURE 53.23

Density-dependent effects. Migratory locusts, *Locusta migratoria*, are a legendary plague of large areas of Africa and Eurasia. At high population densities, the locusts have different hormonal and physical characteristics and take off as a swarm. The most serious infestation of locusts in 30 years occurred in North Africa in 1988.

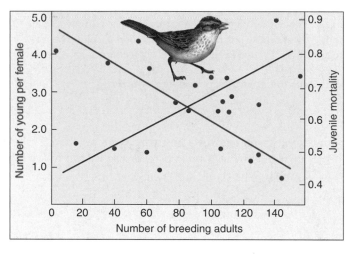

FIGURE 53.22

Density dependence in the song sparrow (*Melospiza melodia*) on Mandarte Island. Reproductive success decreases and mortality rates increase as population size increases.
What would happen if researchers supplemented the food available to the birds?

However, not all density-dependent factors are negatively related to population size. In some cases, growth rates increase with population size. This phenomenon is referred to as the **Allee effect** (after Warder Allee, who first described it), and is an example of *positive feedback*. The Allee effect can take several forms. Most obviously, in populations that are too sparsely distributed, individuals may have difficulty finding mates. Moreover, some species may rely on large groups to deter predators or to provide the necessary stimulation for breeding activities.

Density-Independent Effects

Growth rates in populations sometimes do not correspond to the logistic growth equation. In many cases, such patterns result because growth is under the control of **density-independent effects**. In other words, the rate of growth of a population at any instant is limited by something unrelated to the size of the population.

A variety of factors may affect populations in a density-independent manner. Most of these are aspects of the external environment, such as extremely cold winters, droughts, storms, or volcanic eruptions. Individuals often will be affected by these activities regardless of the size of the population. Populations in areas where such events occur relatively frequently will display erratic growth patterns in which the populations increase rapidly when conditions are benign, but exhibit large reductions whenever the environment turns hostile (figure 53.24). Needless to say, such populations do not produce the sigmoidal growth curves characteristic of the logistic equation.

Population Cycles

In some populations, density-dependent effects lead not to an equilibrium population size but to cyclic patterns of increase and decrease. Ecologists have studied cycles in hare populations since the 1820s. They have found that the North American snowshoe hare (*Lepus americanus*) follows a "10-year cycle" (in reality, it varies from 8 to 11 years). Its numbers fall 10-fold to 30-fold in a typical cycle, and 100-fold changes can occur (figure 53.25). Two factors appear to be generating the cycle: food plants and predators.

Food plants. The preferred foods of snowshoe hares are willow and birch twigs. As hare density increases, the quantity of these twigs decreases, forcing the hares to feed on high-fiber (low-quality) food. Lower birthrates, low juvenile survivorship, and low growth rates follow.

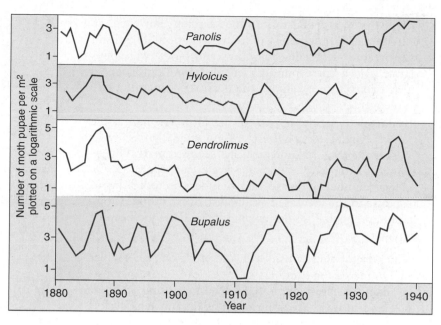

FIGURE 53.24
Fluctuations in the number of pupae of four moth species in Germany. The population fluctuations suggest that density-independent factors are regulating population size. The species concordance in trends through time suggests that the same factors are regulating population size in all species.
What might those factors be?

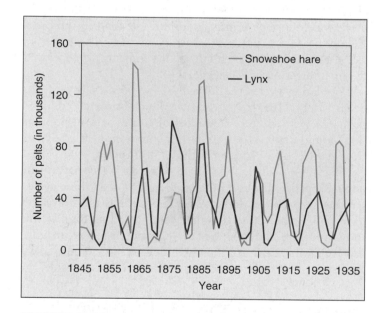

FIGURE 53.25
Linked population cycles of the snowshoe hare and the northern lynx. These data are based on records of fur returns from trappers in the Hudson Bay region of Canada. The lynx population carefully tracks that of the snowshoe hare, but lags behind it slightly.
Suppose experimenters artificially kept the hare population at a high and constant level; what would happen to the lynx population? Conversely, if experimenters artificially kept the lynx population at a high and constant level, what would happen to the hare population?

The hares also spend more time searching for food, an activity that exposes them more to predation. The result is a precipitous decline in willow and birch twig abundance, and a corresponding fall in hare abundance. It takes two to three years for the quantity of mature twigs to recover.

Predators.   A key predator of the snowshoe hare is the Canada lynx (*Lynx canadensis*). The Canada lynx shows a "10-year" cycle of abundance that seems remarkably entrained to the hare abundance cycle (see figure 53.25). As hare numbers increase, lynx numbers do too, rising in response to the increased availability of lynx food. When hare numbers fall, so do lynx numbers, their food supply depleted.

Which factor is responsible for the predator-prey oscillations? Do increasing numbers of hares lead to overharvesting of plants (a hare-plant cycle), or do increasing numbers of lynx lead to overharvesting of hares (a hare-lynx cycle)? Field experiments carried out by C. Krebs and coworkers in 1992 provide an answer. In Canada's Yukon, Krebs set up experimental plots that contained hare populations. If food is added (no food shortage effect) and predators are excluded (no predator effect) in an experimental area, hare numbers increase tenfold and stay there—the cycle is lost. However, the cycle is retained if either of the factors is allowed to operate alone: exclude predators but don't add food (food shortage effect alone), or add food in the presence of predators (predator effect alone). Thus, both factors can affect the cycle, which in practice seems to be generated by the interaction between the two.

Population cycles traditionally have been considered to occur rarely. However, a recent review of nearly 700 long-term (25 years or more) studies of trends within populations found that cycles were not uncommon; nearly 30% of the studies—including birds, mammals, fish, and crustaceans—provided evidence of some cyclic pattern in population size through time, although most of these cycles are nowhere near as dramatic in amplitude as the snowshoe hare and lynx cycles. In some cases, such as that of the snowshoe hare and lynx, density-dependent factors may be involved, whereas in other cases, density-independent factors, such as cyclic climatic patterns, may be responsible.

Population Growth Rates and Life History Models

As we have seen, some species usually maintain stable population sizes near the carrying capacity, whereas the population sizes of other species fluctuate markedly and are often far below carrying capacity. As we saw in our discussion of life histories, the selective factors affecting such species differ markedly. Populations near their carrying capacity may face stiff competition for limited resources. By contrast, resources are abundant in populations far below carrying capacity.

Adaptation	*r*-Selected Populations	*K*-Selected Populations
Age at first reproduction	Early	Late
Life span	Short	Long
Maturation time	Short	Long
Mortality rate	Often high	Usually low
Number of offspring produced per reproductive episode	Many	Few
Number of reproductions per lifetime	Usually one	Often several
Parental care	None	Often extensive
Size of offspring or eggs	Small	Large

Table 53.3   *r*-Selected and *K*-Selected Life History Adaptations

We have already seen the consequences of such differences. When resources are limited, the cost of reproduction often will be very high. Consequently, selection will favor individuals that can compete effectively and utilize resources efficiently. Such adaptations often come at the cost of lowered reproductive rates. Such populations are termed **K-selected** because they are adapted to thrive when the population is near its carrying capacity (*K*). Table 53.3 lists some of the typical features of *K*-selected populations. Examples of *K*-selected species include coconut palms, whooping cranes, whales, and humans.

By contrast, in populations far below the carrying capacity, resources may be abundant. Costs of reproduction will be low, and selection will favor those individuals that can produce the maximum number of offspring. Selection here favors individuals with the highest reproductive rates; such populations are termed **r-selected**. Examples of organisms displaying *r*-selected life history adaptations include dandelions, aphids, mice, and cockroaches.

Most natural populations show life history adaptations that exist along a continuum ranging from completely *r*-selected traits to completely *K*-selected traits. Although these tendencies hold true as generalities, few populations are purely *r*- or *K*-selected and show all of the traits listed in table 53.3. These attributes should be treated as generalities, with the recognition that many exceptions exist.

Density-dependent effects are caused by factors that come into play particularly when the population size is larger; density-independent effects result from factors that operate regardless of population size. Some life history adaptations favor near-exponential growth; others favor the more competitive logistic growth. Most natural populations exhibit a combination of the two.

53.6 The human population has grown explosively in the last three centuries.

The Advent of Exponential Growth

Humans exhibit many *K*-selected life history traits, including small brood size, late reproduction, and a high degree of parental care. These life history traits evolved during the early history of hominids, when the limited resources available from the environment controlled population size. Throughout most of human history, our populations have been regulated by food availability, disease, and predators. Although unusual disturbances, including floods, plagues, and droughts, no doubt affected the pattern of human population growth, the overall size of the human population grew slowly during our early history. Two thousand years ago, perhaps 130 million people populated the earth. It took a thousand years for that number to double, and it was 1650 before it had doubled again, to about 500 million. In other words, for over 16 centuries, the human population was characterized by very slow growth. In this respect, human populations resembled many other species with predominantly *K*-selected life history adaptations.

Starting in the early 1700s, changes in technology gave humans more control over their food supply, enabled them to develop superior weapons to ward off predators, and led to the development of cures for many diseases. At the same time, improvements in shelter and storage capabilities made humans less vulnerable to climatic uncertainties. These changes allowed humans to expand the carrying capacity of the habitats in which they lived, and thus to escape the confines of logistic growth and reenter the exponential phase of the sigmoidal growth curve.

Responding to the lack of environmental constraints, the human population has grown explosively over the last 300 years. While the birthrate has remained unchanged at about 30 per 1000 per year over this period, the death rate has fallen dramatically, from 20 per 1000 per year to its present level of 13 per 1000 per year. The difference between birth and death rates meant that the population grew as much as 2% per year, although the rate has now declined to 1.3% per year.

A 1.3% annual growth rate may not seem large, but it has produced a current human population of 6.3 billion people (figure 53.26)! At this growth rate, 82 million people are added to the world population annually, and the human population will double in 53 years. As we shall see, both the current human population level and the projected growth rate have potentially grave consequences for our future.

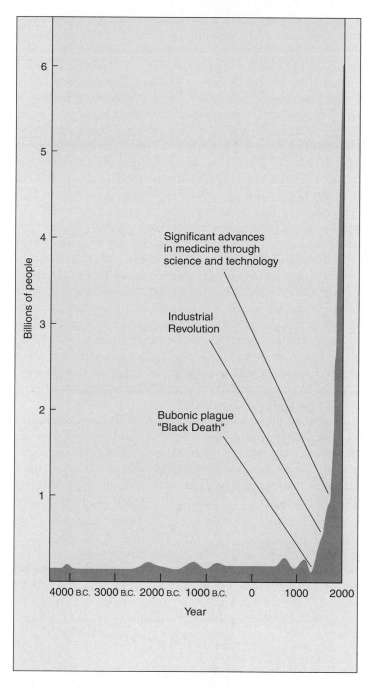

FIGURE 53.26

History of human population size. Temporary increases in death rate, even a severe one such as that occurring during the Black Death of the 1300s, have little lasting impact. Explosive growth began with the industrial revolution in the 1800s, which produced a significant, long-term lowering of the death rate. The current world population is 6.3 billion, and at the present rate, it will double in 53 years.

Based on what we have learned about population growth, what do you predict will happen to human population size?

Population Pyramids

While the human population as a whole continues to grow rapidly at the beginning of the twenty-first century, this growth is not occurring uniformly over the planet. In some countries, such as Mexico, the birthrate is greatly exceeding the death rate (figure 53.27). As a result, if Mexico's population keeps growing at its current rate, it is projected to increase from 102 million in 2002 to 151 million in 2050. Other countries are growing much more slowly. The rate at which a population can be expected to grow in the future can be assessed graphically by means of a *population pyramid*, a bar graph displaying the numbers of people in each age category (figure 53.28). Males are conventionally shown to the left of the vertical age axis, females to the right. A human population pyramid thus displays the age composition of a population by sex. In most human population pyramids, the number of older females is disproportionately large compared to the number of older males, because females in most regions have a longer life expectancy than males.

Viewing such a pyramid, we can predict demographic trends in births and deaths. In general, a rectangular pyramid is characteristic of countries whose populations are stable, their numbers neither growing nor shrinking. A triangular pyramid is characteristic of a country that will exhibit rapid future growth because most of its population has not yet entered the child-bearing years. Inverted triangles are characteristic of populations that are shrinking, usually as a result of sharply declining birthrates.

Examples of population pyramids for Sweden and Kenya in 2000 are shown in figure 53.28. The two countries exhibit very different age distributions. The nearly rectangular population pyramid for Sweden indicates that its population is not expanding. The very triangular pyramid of

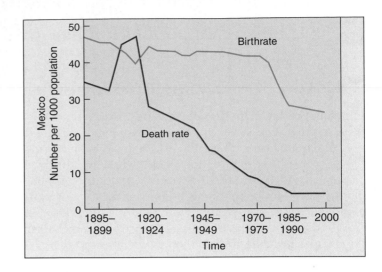

FIGURE 53.27
Why the population of Mexico is growing. The death rate (*red line*) in Mexico fell steadily throughout the last century, while the birthrate (*blue line*) remained fairly steady until 1970. The difference between birth and death rates has fueled a high growth rate. Efforts begun in 1970 to reduce the birthrate have been quite successful, but the growth rate remains high.
Is population growth rate increasing?

Kenya, by contrast, predicts explosive future growth. The difference is most apparent when we consider that only 20% of Sweden's population is less than 15 years old, compared to nearly half of all Kenyans. Moreover, the fertility rate (offspring per woman) in Sweden is 1.6; in Kenya, it is 4.4. As a result, Kenya's population could double in less than 35 years, whereas Sweden's will remain stable.

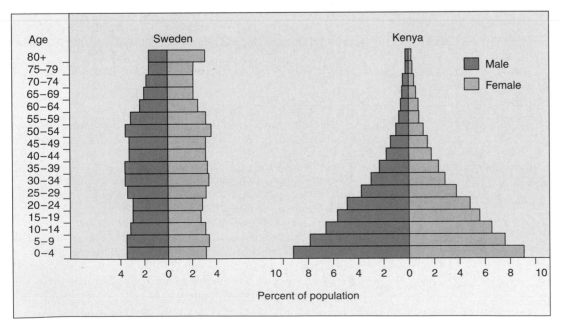

FIGURE 53.28
Population pyramids from 2000. Population pyramids are graphed according to a population's age distribution. Kenya's pyramid has a broad base because of the great number of individuals below childbearing age. When the young people begin to bear children, the population will experience rapid growth. The Swedish pyramid exhibits a slight bulge among middle-aged Swedes, the result of the "baby boom" that occurred in the middle of the twentieth century.
What will the population distributions look like in 20 years?

Table 53.4 A Comparison of 2002 Population Data in Developed and Developing Countries			
	United States (highly developed)	**Brazil (moderately developed)**	**Ethiopia (poorly developed)**
Fertility rate	2.1	2.2	5.9
Doubling time at current rate (yr)	115	53	28
Infant mortality rate (per 1000 births)	6.6	33	97
Life expectancy at birth (yrs)	77	69	52
Per capita GNP (U.S. $)	$34,100	$7300	$660
Population < 15 years old (%)	21	30	44

An Uncertain Future

The earth's rapidly growing human population constitutes perhaps the greatest challenge to the future of the **biosphere,** the world's interacting community of living things. Humanity is adding 82 million people a year to the earth's population—over a million every five days, 150 every minute! In more rapidly growing countries, the resulting population increase is staggering (table 53.4). India, for example, had a population of 1.05 billion in 2002; by 2050, its population will exceed 1.6 billion.

A key element in the world's population growth is its uneven distribution among countries. Of the billion people added to the world's population in the 1990s, 90% live in developing countries (figure 53.29). This is leading to a major reduction in the fraction of the world's population that lives in industrialized countries. In 1950, fully one-third of the world's population lived in industrialized countries; by 1996, that proportion had fallen to one-quarter; and in 2020, the proportion will have fallen to one-sixth. Thus, the world's population growth will be centered in the parts of the world least equipped to deal with the pressures of rapid growth.

Rapid population growth in developing countries has the harsh consequence of increasing the gap between rich and poor. Today, the 19% of the world's population that lives in the industrialized world have a per capita income of $22,060, while 81% of the world's population lives in developing countries and has a per capita income of only $3,580. Further, of the people in the developing world, about one-quarter of the population gets by on $1 per day. Eighty percent of all the energy used today is consumed by the industrialized world, while only 20% is used by developing countries. Perhaps most worrisome for the future, fully 94% of all scientists and engineers reside in the industrialized world, and only 6% live in developing countries. Thus, the problems created by the future's explosive population growth will be faced by countries with little of the world's scientific or technological expertise.

No one knows whether the world can sustain today's population of 6 billion people, much less the far greater numbers expected in the future. As chapter 56 outlines, the world ecosystem is already under considerable stress. We cannot reasonably expect to expand its carrying capacity indefinitely, and indeed we already seem to be stretching the limits. De-

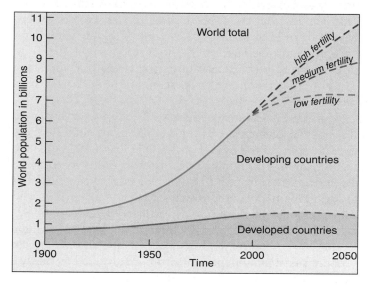

FIGURE 53.29
Distribution of population growth. Most of the worldwide increase in population since 1950 has occurred in developing countries. The age structures of developing countries indicate that this trend will increase in the near future. World population in 2050 likely will be between 7.3 and 10.7 billion, according to a recent United Nations study. Depending on fertility rates, the population at that time will either be increasing rapidly or slightly, or in the best case, declining slightly.
Is this an example of density-dependent population regulation? If so, what factors are regulating population size?

spite using an estimated 45% of the total biological productivity of the earth's landmasses and more than one-half of all renewable sources of fresh water, between one-fourth and one-eighth of all people in the world are malnourished. Moreover, as anticipated by Thomas Malthus in his famous 1798 *Essay on the Principle of Population*, death rates are beginning to rise in some areas. In sub-Saharan Africa, for example, population projections for the year 2025 have been scaled back from 1.33 billion to 1.05 billion (21%) because of the impact of AIDS. Similar decreases are projected for Russia as a result of higher death rates due to disease. If we are to avoid catastrophic increases in the death rate, birthrates must fall dramatically. Faced with this grim dichotomy, significant efforts are underway worldwide to lower birthrates.

Population Growth Rate on the Decline

The world population growth rate is declining, from a high of 2.0% in the period 1965–1970 to 1.3% in 2002. Nonetheless, because of the larger population, this amounts to an increase of 82 million people per year to the world population, compared to 53 million per year in the 1960s.

The United Nations attributes the growth rate decline to increased family planning efforts and the increased economic power and social status of women. The United States has led the world in funding family planning programs abroad, but some groups oppose spending money on international family planning. The opposition states that money is better spent on improving education and the economy in other countries, leading to an increased awareness and lowered fertility rates. The U.N. certainly supports the improvement of education programs in developing countries, but interestingly, it has reported increased education levels *following* a decrease in family size as a result of family planning.

Most countries are devoting considerable attention to slowing the growth rate of their populations, and there are genuine signs of progress. For example, from 1984 to 2000, family planning programs in Kenya succeeded in reducing the fertility rate from 8.0 to 4.4 children per couple, thus lowering the population growth rate from 4.0% per year to 2.1% per year. Because of these efforts, the global population may stabilize at about 8.9 billion people by the middle of the current century. How many people the planet can support sustainably depends on the quality of life that we want to achieve; there are already more people than can be sustainably supported with current technologies.

Consumption in the Developed World Is Also a Problem

Population size is not the only factor that determines resource use; per capita consumption is also important. In this respect, we in the industrialized world need to pay more attention to lessening the impact each of us makes because, even though the vast majority of the world's population is in developing countries, the vast majority of resource consumption occurs in the industrialized countries. Indeed, the wealthiest 20% of the world's population accounts for 86% of the world's consumption of resources and produces 53% of the world's carbon dioxide emissions, whereas the poorest 20% of the world is responsible for only 1.3% of consumption and 3% of carbon dioxide emissions. Looked at another way, in terms of resource use, a child born today in the industrialized world will consume many more resources over the course of his or her life than a child born in the developing world. One way of quantifying this disparity is by calculating what has been termed the **ecological footprint**, which is the amount of productive land required to support an individual at the standard of living of a particular population through the course of his

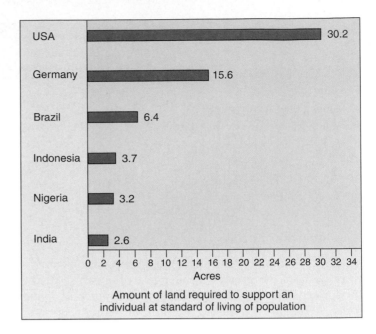

FIGURE 53.30
Ecological footprints of individuals in different countries.
An ecological footprint calculates how much land is required to support a person through his or her life, including the acreage used for production of food, forest products, and housing, in addition to the forest required to absorb the carbon dioxide produced by the combustion of fossil fuels.
Which is a more important cause of resource depletion, overpopulation or overconsumption?

or her life. This figure estimates the acreage used for the production of food (both plant and animal), forest products, and housing, as well as the area of forest required to absorb carbon dioxide produced by the combustion of fossil fuels. As figure 53.30 illustrates, the ecological footprint of an individual in the United States is more than 10 times greater than that of someone in India. Based on these measurements, researchers have calculated that resource use by humans is now one-third greater than the amount that nature can sustainably replace. Moreover, consumption is increasing rapidly in parts of the developing world; if all humans lived at the standard of living in the industrialized world, two additional planet earths, would be needed!

Building a sustainable world is the most important task facing humanity's future. The quality of life available to our children will depend to a large extent on our success in limiting both population growth and the amount of per capita resource consumption.

In 2002, the global human population of 6.3 billion people was growing at a rate of approximately 1.3% annually. At that rate, the population would double in 53 years. Growth rates, however, are declining, but consumption per capita in the developed world is also a significant drain on resources.

53.1 Organisms must cope with a varied environment.

The Environmental Challenge

- Many plants and animals actively employ physiological, morphological, or behavioral mechanisms to maintain homeostasis. Such mechanisms are the result of natural selection. (pp. 1138–1139)

53.2 Populations are groups of individuals of the same species that live in the same space.

Population Ranges

- Boundary edges between populations can range from sharp and impermeable to indistinct and easily permeable. (p. 1140)
- Three aspects of populations are especially important: the range in which a population occurs, the pattern of spacing of individuals throughout their range, and the size a population eventually attains. (p. 1140)
- Population ranges are not static and can change through time, usually as a consequence of environmental change or physiological adaptation. (p. 1141)

Pattern of Spacing of Individuals in a Population

- Individuals are randomly spaced when interactions between individuals are weak or when microenvironmental aspects of the habitat are nonuniform. (p. 1142)
- Uniform spacing often results from resource competition. (p. 1142)
- Clumping or clustering usually occurs as a response to uneven distribution of resources in the immediate environment. (p. 1143)
- Natural dispersal can occur through many avenues, including individuals or their gametes traveling via wind, water, or other organisms. (p. 1143)

Metapopulations

- Metapopulations are a network of distinct populations interacting and exchanging gametes and/or individuals. Such networks usually occur in habitat patches separated by inhospitable habitat. (p. 1144)
- At any given time, not all suitable habitat patches are inhabited, and periodically an inhabited patch will go extinct, only to be colonized later by individuals from a nearby patch. (p. 1144)
- Source-sink metapopulations occur when only some habitats are suitable for long-term population maintenance. In these cases, more permanent populations (sources) continually send individuals to less permanent, poorer habitat patches (sinks), which are periodically purged, often by environmental conditions, and then recolonized from the source population. (p. 1144)

53.3 Population dynamics depend critically upon age distribution.

Demography

- Demography is the statistical study of populations and their changes through time. Populations can be studied as a whole or broken down into constituent parts. (p. 1145)
- Many factors affect population growth, including sex ratio, generation time, and age structure of the populations. (pp. 1145–1146)
- Life tables are used to assess how populations change through time, and are often constructed by following a cohort of individuals from birth until death. (p. 1146)

- Survivorship curves can be constructed by graphing the percentage of an original population, or cohort, that survives to a given age. These curves can be divided into three basic categories based on the period of life in which individuals are most likely to die. (p. 1147)

53.4 Life histories often reflect trade-offs between reproduction and survival.

The Cost of Reproduction

- The complete life cycle of an organism constitutes its life history, and all life histories involve significant trade-offs. (p. 1148)
- Natural selection will favor life histories that maximize lifetime reproductive success. (p. 1148)
- A key reproductive trade-off concerns how many resources to invest in producing any single offspring; thus, a trade-off must exist between the number of offspring produced and the size of each offspring. (p. 1149)
- The trade-off between age and fecundity also plays a key role in many life histories. (p. 1149)
- Organisms that invest all their energy into a single large event and die afterwards are termed semelparous, while iteroparous organisms produce offspring several times over many seasons. (p. 1149)

53.5 Population growth is limited by the environment.

Population Growth

- The actual rate of population increase is defined as the difference between the birthrate and the death rate corrected for movement of animals into and out of a population. (p. 1150)
- The innate capacity of growth for any population is exponential. (p. 1150)
- In logistic growth, as a population approaches carrying capacity, its rate of growth slows as fewer resources remain available for use, resulting in a sigmoidal growth curve. (pp. 1150–1151)

Factors That Regulate Populations

- Density-dependent effects are factors that exert increasing pressure on a population as the population increases in size. (p. 1152)
- Density-independent effects are factors that affect the rate of population growth independent of the size of the population. (p. 1153)

53.6 The human population has grown explosively in the last three centuries.

The Advent of Exponential Growth

- Starting in the early 1700s, changes in technology have given humans more control over their food supply and death rate and made them less vulnerable to climatic uncertainties. The human population has thus been able to escape the confines of logistic growth and has grown explosively over the past 300 years. (p. 1155)
- Population pyramids graphically depict the number of people in each age category. (p. 1156)
- Population growth has been more rapid in developing countries, and the age structures of developing countries indicate that this trend will increase in the near future. (p. 1157)
- Per capita resource consumption is a significant factor determining resource consumption, which is highest in the industrialized countries. (p. 1158)

Self Test

1. The term *homeostasis* refers to
   a. the maintenance of a consistent internal environment.
   b. the ability to conform internal temperature to environmental temperature.
   c. an organism's biotic potential.
   d. the carrying capacity of a population.

2. Which of the following is *not* considered a population?
   a. the ginkgo trees (*Ginkgo biloba*) in New York City
   b. the birds in your hometown
   c. the human inhabitants of Pennsylvania
   d. the grizzly bears (*Ursus arctos*) of Alaska.

3. A clumped population may be due to
   a. weak interactions between the members of a population.
   b. intense competition for uniformly distributed resources.
   c. uneven distribution of resources in the environment.
   d. intense territoriality.

4. The tidewater goby (*Eucyclogobius newberryi*) is an endangered species of fish that occurs as metapopulations in isolated coastal wetlands of California. Large wetlands serve as sources of individuals for populations in small wetlands, which function as sinks due to inferior habitat quality. What effect would most likely be seen on the population of gobies if a barrier to migration between wetlands developed?
   a. The populations in the small and large wetlands would evolve independently of each other.
   b. The populations in the large wetlands would most likely go extinct.
   c. The populations in the small wetlands would most likely go extinct.
   d. The populations in the small and large wetlands would most likely go extinct.

5. Which of the following factors does *not* determine the growth rate of a population?
   a. the population's sex ratio
   b. the species' generation time
   c. the age structure of the population
   d. the optimal temperature at which an organism can reproduce.

6. A population with a larger proportion of older individuals than younger individuals will likely
   a. grow larger and then decline rapidly.
   b. continue to grow larger indefinitely.
   c. grow smaller and may stabilize at a smaller population size.
   d. not experience a change in population size.

7. Humans are an example of an organism with a type I survivorship curve. This means
   a. mortality rates are highest for younger individuals.
   b. mortality rates are highest for older individuals.
   c. mortality rates are constant over the life span of individuals.
   d. the population growth rate is high.

8. According to the Population Reference Bureau (2002), the worldwide intrinsic rate of human population growth (*r*) is currently 1.3%. In the United States, *r* = 0.6%. How will the U.S. population change relative to the world population?
   a. The world population will grow, while the population of the United States will decline.
   b. The world population will grow, while the population of the United States will remain the same.
   c. Both the world and the U.S. populations will grow, but the world population will grow more rapidly.
   d. The world population will decline, while the U.S. population will increase.

9. The logistic population growth model, $dN/dt = rN[(K - N)/K]$, describes a population's growth when an upper limit to growth is assumed. This upper limit to growth is known as the population's _____, and as N gets larger, dN/dt _____.
   a. biotic potential/increases
   b. biotic potential/decreases
   c. carrying capacity/increases
   d. carrying capacity/decreases

10. Which of the following is *not* an example of a density-dependent effect on population growth?
   a. an extremely cold winter
   b. competition for food resources
   c. stress-related illness associated with overcrowding
   d competition for nesting sites

Test Your Visual Understanding

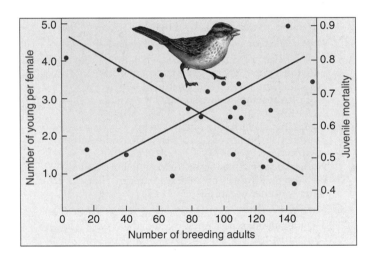

1. The song sparrow *Melospiza melodia* exhibits density-dependent population growth on Mandarte Island. List three reasons that the number of young per female may be sensitive to population size.

Apply Your Knowledge

1. Throughout most of North America, large carnivores have either been extirpated (driven to local extinction) or their populations are at extremely low levels. Explain how the loss of carnivore populations may have changed the vegetation of North America.

2. Do humans show more *r*-selected life-history traits or *K*-selected traits? How does this correlate with current global human population growth?

3. Give your opinion: What constitutes the greatest threat to the future of the planet, the rapidly growing population in developing parts of the world or high resource consumption in the developed world?

54

Community Ecology

Concept Outline

54.1 Biological communities are composed of species that occur together.

    Concepts of Communities. Whether a community is more than the sum of its parts is debated.

54.2 Interactions among competing species shape ecological niches.

    Fundamental and Realized Niches. Interspecific interactions can restrict niche use.

    Gause and the Principle of Competitive Exclusion. If resources are limited, no two species can occupy the same niche indefinitely.

    Resource Partitioning. Species that live together often partition the available resources, reducing competition.

    Detecting Interspecific Competition. Experiments often can detect competition, but they have limitations.

54.3 Predation has ecological and evolutionary effects.

    Predation and Prey Populations. Predators can limit the size of populations.

    Plant Defenses Against Herbivores. Plants use chemicals to defend themselves against animals trying to eat them.

    Animal Defenses Against Predators. Animals defend themselves with camouflage, chemicals, and stings.

    Mimicry. Species can copy the appearance of others.

54.4 Species within a community interact in many ways.

    Coevolution and Symbiosis. Organisms have evolved many adjustments to living together.

    Commensalism. Some organisms use others, neither hurting nor helping their benefactors.

    Mutualism. Often species interactions benefit both.

    Parasitism. Sometimes one organism serves as the food supply of another, much smaller one.

    Interactions Among Ecological Processes. Multiple processes may occur simultaneously.

54.5 Ecological succession may increase the species richness of communities.

    Succession. Species replacements often occur in a regular pattern.

    The Role of Disturbance. Disturbances can disrupt successional change. In some cases, moderate amounts of disturbance increase species diversity.

FIGURE 54.1

Communities involve interactions between disparate groups. This clownfish is one of the few species that can nestle safely among the stinging tentacles of the sea anemone. In addition to safe harbor, the clownfish further benefits by eating scraps of food left over from the anemone's meals. In turn, the sea anemone benefits by ingesting scraps of food dropped by the fish. The sea anemone–clownfish relationship is thus a classic case of symbiosis.

All the organisms that live together in a place are called a community. The myriad of species that inhabit a tropical rain forest are a community. Indeed, every inhabited place on earth supports its own particular array of organisms. Over time, the different species have made many complex adjustments to community living (figure 54.1), evolving together and forging relationships that give the community its character and stability. Both competition and cooperation have played key roles; in this chapter, we look at these and other factors in community ecology.

54.1 Biological communities are composed of species that occur together.

Almost any place on earth is occupied by species, sometimes by many of them, as in the rain forests of the Amazon, and sometimes by only a few, as in the near-boiling waters of Yellowstone's geysers (where a number of microbial species live). The term **community** refers to the species that occur at any particular locality (figure 54.2). Communities can be characterized either by their constituent species or by their properties, such as species richness or primary productivity.

Interactions among community members govern many ecological and evolutionary processes. These interactions, such as predation and mutualism, affect the population biology of particular species—whether a population increases or decreases in abundance, for example—as well as the ways in which energy and nutrients cycle through the ecosystem. Moreover, the community context in many ways affects the patterns of natural selection faced by a species, and thus the evolutionary course it takes.

Scientists study biological communities in many ways, ranging from detailed observations to elaborate, large-scale experiments. In some cases, such studies focus on the entire community, whereas in other cases only a subset of species that are likely to interact with each other are studied. Although scientists sometimes refer to such subsets as communities (for example, the "spider community"), the term **assemblage** is probably more appropriate to connote that the species included are only a portion of those present within the entire community.

FIGURE 54.2
A Tanzanian savanna community. A community consists of all the species—plants, animals, fungi, protists, and prokaryotes—that occur at a locality, in this case Lake Manyara National Park in Tanzania.

Concepts of Communities

Two views exist on the makeup and functioning of communities. The **individualistic concept** of communities, first championed by H. A. Gleason of the University of Chicago early in the twentieth century, holds that a community is nothing more than an aggregation of species that happen to co-occur at one place. By contrast, the **holistic concept** of communities, which can be traced to the work of F. E. Clements, also about a century ago, views communities as an integrated unit. In this sense, the community could be viewed as a superorganism whose constituent species have coevolved to the extent that they function as part of a greater whole, just as the kidneys, heart, and lungs all function together within an animal's body. In this view, then, a community would amount to more than the sum of its parts.

These two views make differing predictions about the integrity of communities across space and time. If, as the individualistic view implies, communities are nothing more than a combination of species that occur together, then moving geographically across the landscape or back through time, we would not expect to see the same community. That is, species should appear and disappear independently, as a function of each species' own unique ecological requirements. By contrast, if a community is an integrated whole, then we would make the opposite prediction: Communities should stay the same through space or time, until being replaced by completely different communities when environmental differences are sufficiently great.

Communities Across Space and Time

Most ecologists today favor the individualistic concept. For the most part, species seem to respond independently to changing environmental conditions. As a result, community composition changes gradually across landscapes as some species appear and become more abundant, while others decrease in abundance and eventually disappear. A famous example of this pattern is the abundance of tree species in the Santa Catalina Mountains of Arizona along a geographic gradient running from very dry to very moist. Figure 54.3 shows that species can change abundance in patterns that are for the most part independent of each other. As a result, tree communities at different localities in these mountains are a continuum, one merging into the next, rather than representing discretely different sets of species.

Similar patterns through time are seen in paleontological studies. For example, a very good fossil record exists for the trees and small mammals that occurred in North America over the past 20,000 years. Examination of prehistoric communities shows little similarity to those that occur today. Many species that occur together today were never found together in the past. Conversely, species that used to occur in the same communities often do not overlap in their geographic ranges today. These findings suggest that as climate has changed during the waxing and waning of the Ice Ages, species have responded independently, rather than shifting their distributions together, as would be expected if the community were an integrated unit.

Nonetheless, in some cases the abundance of species in a community does change geographically in a synchronous pattern. Often, this occurs at **ecotones,** places where the environment changes abruptly. For example, in the western United States, certain patches of habitat have serpentine soils. Such soil differs from normal soil in many ways (for example, high concentrations of nickel, chromium, and iron; low concentrations of copper and calcium). Comparison of the plant species that occur on different soils shows that distinct communities exist on each type, with an abrupt transition from one to the other over a short distance (fig-

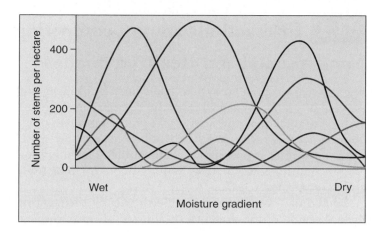

FIGURE 54.3
Abundance of tree species along a moisture gradient in the Santa Catalina Mountains of southeastern Arizona. The species' patterns of abundance are independent of each other. Thus, community composition changes continuously along the gradient.
Why do species exhibit different patterns of response to change in moisture?

ure 54.4). Similar transitions are seen wherever greatly different habitats come into contact, such as at the interface between terrestrial and aquatic habitats or where grassland and forest meet.

A community comprises all species that occur at one site. In most cases, community members vary independently of each other in abundance across space and through time.

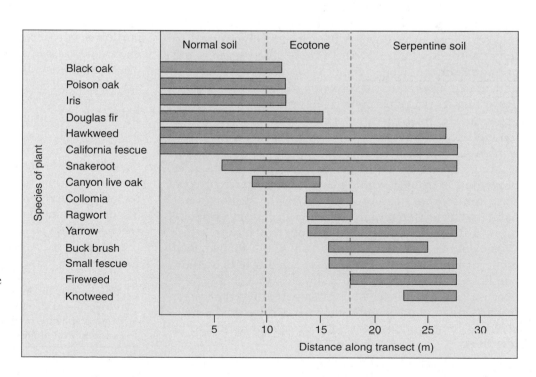

FIGURE 54.4
Change in community composition across an ecotone. The plant communities on normal and serpentine soils are greatly different, and the transition from one community to another occurs over a short distance.
Why is there a sharp transition between the two community types?

54.2 Interactions among competing species shape ecological niches.

Fundamental and Realized Niches

Each organism in an ecosystem confronts the challenge of survival in a different way. The **niche** an organism occupies is the sum total of all the ways it utilizes the resources of its environment. A niche may be described in terms of space utilization, food consumption, temperature range, appropriate conditions for mating, requirements for moisture, and other factors.

Sometimes species are not able to occupy their entire niche because of the presence or absence of other species. Species can interact with each other in a number of ways, and these interactions can either have positive or negative effects. One type of interaction is **interspecific competition,** which occurs when two species attempt to utilize the same resource and there is not enough of the resource to satisfy both. Fighting over resources is referred to as **interference competition;** consuming shared resources is called **exploitative competition.**

The entire niche that a species is capable of using, based on its physiological tolerance limits and resource needs, is called the **fundamental niche.** The actual niche the species occupies is its **realized niche.** Because of interspecific interactions, the realized niche of a species may be considerably smaller than its fundamental niche.

In a classic study, J. H. Connell of the University of California, Santa Barbara, investigated competitive interactions between two species of barnacles that grow together on rocks along the coast of Scotland. Of the two species Connell studied, *Chthamalus stellatus* lives in shallower water, where tidal action often exposes it to air, and *Semibalanus balanoides* (called *Balanus balanoides* prior to 1995) lives lower down, where it is rarely exposed to the atmosphere (figure 54.5). In these areas, space is at a premium. In the deeper zone, *Semibalanus* could always outcompete *Chthamalus* by crowding it off the rocks, undercutting it, and replacing it even where it had begun to grow, an example of interference competition. When Connell removed *Semibalanus* from the area, however, *Chthamalus* was easily able to occupy the deeper zone, indicating that no physiological or other general obstacles prevented it from becoming established there. In contrast, *Semibalanus* could not survive in the shallow-water habitats where *Chthamalus* normally occurs; it evidently does not have the special adaptations that allow *Chthamalus* to occupy this zone. Thus, the fundamental niche of the barnacle *Chthamalus* includes both shallow and deeper zones, but its realized niche is much narrower because *Chthamalus* can be outcompeted by *Semibalanus* in parts of its fundamental niche. By contrast, the realized and fundamental niches of *Semibalanus* appear to be identical.

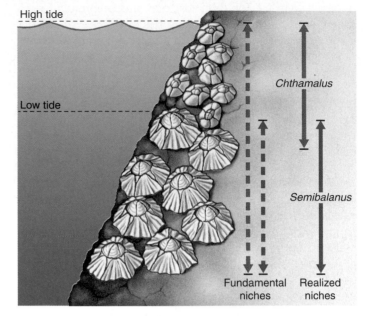

FIGURE 54.5
Competition among two species of barnacles. *Chthamalus* can live in both deep and shallow zones (its fundamental niche), but *Semibalanus* forces *Chthamalus* out of the part of its fundamental niche that overlaps the realized niche of *Semibalanus.*

Processes other than competition can also restrict the realized niche of a species. For example, the plant St. John's-wort was introduced and became widespread in open rangeland habitats in California until a specialized beetle was introduced to control it. Populations of the plant quickly decreased, and it is now only found in shady sites where the beetle cannot thrive. In this case, the presence of a predator limits the realized niche of a plant.

In some cases, the absence of another species leads to a smaller realized niche. For example, many North American plants depend on insects for pollination; indeed, the value of insect pollination for American agriculture has been estimated at greater than $2 billion per year. However, pollinator populations are currently declining for a variety of reasons. Conservationists are concerned that if these insects disappear from some habitats, the niche of many plant species will decrease or even disappear entirely. In this case, then, the absence—rather than the presence—of another species will be the cause of a relatively small realized niche.

A niche may be defined as the way in which an organism utilizes its environment. Interspecific interactions may cause a species' realized niche to be smaller than its fundamental niche.

Gause and the Principle of Competitive Exclusion

In classic experiments carried out in 1934 and 1935, Russian ecologist G. F. Gause studied competition among three species of *Paramecium*, a tiny protist. All three species grew well alone in culture tubes, preying on bacteria and yeasts that fed on oatmeal suspended in the culture fluid (figure 54.6*a*). However, when Gause grew *P. aurelia* together with *P. caudatum* in the same culture tube, the numbers of *P. caudatum* always declined to extinction, leaving *P. aurelia* the only survivor (figure 54.6*b*). Why? Gause found that *P. aurelia* could grow six times faster than its competitor *P. caudatum* because it was able to better utilize the limited available resources, an example of exploitative competition.

From experiments such as this, Gause formulated what is now called the principle of **competitive exclusion**. This principle states that if two species are competing for a limited resource, the species that uses the resource more efficiently will eventually eliminate the other locally. In other words, no two species with the same niche can coexist when resources are limiting.

Niche Overlap

In a revealing experiment, Gause challenged *Paramecium caudatum*—the defeated species in his earlier experiments—with a third species, *P. bursaria*. Because he expected these two species to also compete for the limited bacterial food supply, Gause thought one would win out, as had happened in his previous experiments. But that's not what happened.

Instead, both species survived in the culture tubes, dividing the food resources. How did they do it? In the upper part of the culture tubes, where the oxygen concentration and bacterial density were high, *P. caudatum* dominated because it was better able to feed on bacteria. However, in the lower part of the tubes, the lower oxygen concentration favored the growth of a different potential food, yeast, and *P. bursaria* was better able to eat this food. The fundamental niche of each species was the whole culture tube, but the realized niche of each species was only a portion of the tube. Because the niches of the two species did not overlap too much, both species were able to survive. However, competition did have a negative effect on the participants (figure 54.6*c*). When grown without a competitor, both species reached densities three times greater than when they were grown with a competitor.

Competitive Exclusion

Gause's principle of competitive exclusion can be restated to say that *no two species can occupy the same niche indefinitely when resources are limiting.* Certainly species can and do coexist while competing for some of the same resources. Nevertheless, Gause's hypothesis predicts that when two species coexist on a long-term basis, either resources must not be limited or their niches will always differ in one or more features; otherwise, one species will outcompete the other, and the extinction of the second species will inevitably result.

If resources are limiting, no two species can occupy the same niche indefinitely without competition driving one to local extinction.

FIGURE 54.6
Competitive exclusion among three species of *Paramecium*. In the microscopic world, *Paramecium* is a ferocious predator. (*a*) In his experiments, Gause found that three species of *Paramecium* grew well alone in culture tubes. (*b*) However, *P. caudatum* declined to extinction when grown with *P. aurelia* because they shared the same realized niche, and *P. aurelia* outcompeted *P. caudatum* for food resources. (*c*) *P. caudatum* and *P. bursaria* were able to coexist because the two have different realized niches and thus avoided competition.

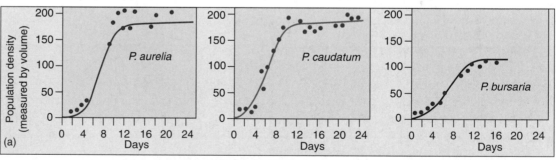

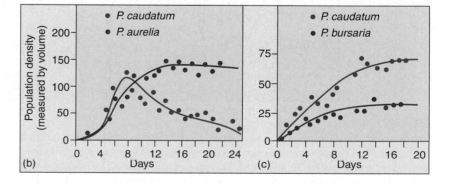

Resource Partitioning

Gause's exclusion principle has a very important consequence: If competition for a limited resource is intense, then either one species will drive the other to extinction, or natural selection will reduce the competition between them. When the late Princeton ecologist Robert MacArthur studied five species of warblers, small insect-eating forest songbirds, he found that they all appeared to be competing for the same resources. However, when he studied them more carefully, he found that each species actually fed in a different part of spruce trees and so ate different subsets of insects. One species fed on insects near the tips of branches, a second within the dense foliage, a third on the lower branches, a fourth high on the trees, and a fifth at the very apex of the trees. Thus, each species of warbler had evolved so as to utilize a different portion of the spruce tree resource. They had *subdivided the niche*, partitioning the available resource to avoid direct competition with one another.

Resource partitioning is often seen in similar species that occupy the same geographic area. Such **sympatric species** often avoid competition by living in different portions of the habitat or by utilizing different food or other resources (figure 54.7). This pattern of resource partitioning is thought to result from the process of natural selection causing initially similar species to diverge in resource use in order to reduce competitive pressures.

Evidence for the role of evolution comes forth by comparing species whose ranges are only partially overlapping. Where the two species co-occur, they tend to exhibit greater differences in morphology (the form and structure of an organism) and resource use than do allopatric populations of the same species. Called **character displacement,**

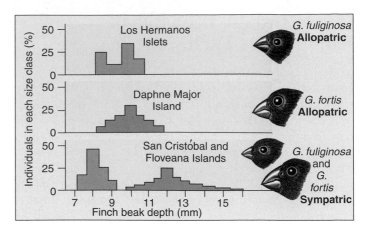

FIGURE 54.8

Character displacement in Darwin's finches. These two species of finches (genus *Geospiza*) have bills of similar size when allopatric, but different size when sympatric.
Why do populations have different bill sizes when they occur with other species?

the differences evident between sympatric species are thought to have been favored by natural selection as a mechanism to facilitate habitat partitioning and thus reduce competition. Thus, the two Darwin's finches in figure 54.8 have bills of similar size where the finches are allopatric, each living on an island where the other does not occur. On islands where they are sympatric, the two species have evolved beaks of different sizes, one adapted to larger seeds and the other to smaller ones.

Sympatric species often partition available resources, reducing competition between them.

FIGURE 54.7

Resource partitioning among sympatric lizard species. Species of *Anolis* lizards on Caribbean islands partition their tree habitats in a variety of ways. Some species of anoles occupy the canopy of trees (*a*), others use twigs on the periphery (*b*), and still others are found at the base of the trunk (*c*). In addition, some use grassy areas in the open (*d*). When two species occupy the same part of the tree, they either utilize different-sized insects as food or partition the thermal microhabitat; for example, one might only be found in the shade, whereas the other would only bask in the sun.

Detecting Interspecific Competition

It is not simple to determine when two species are competing. The fact that two species use the same resources need not imply competition if that resource is not in limited supply. If the population sizes of two species are negatively correlated, such that where one species has a large population, the other species has a small population and vice versa, the two species need not be competing for the same limiting resource. Instead, the two species might be independently responding to the same feature of the environment—perhaps one species thrives best in warm conditions and the other in cool conditions.

Experimental Studies of Competition

Some of the best evidence for the existence of competition comes from experimental field studies. By setting up experiments in which two species occur either alone or together, scientists can determine whether the presence of one species has a negative effect on a population of the second species. For example, a variety of seed-eating rodents occur in North American deserts. In 1988, researchers set up a series of 50 meter × 50 meter enclosures to investigate the effect of kangaroo rats on smaller, seed-eating rodents. Kangaroo rats were removed from half of the enclosures, but not from the other enclosures. The walls of all of the enclosures had holes that allowed rodents to come and go, but in the kangaroo rat removal plots, the holes were too small to allow the kangaroo rats to enter. Over the course of the next three years, the researchers monitored the number of the smaller rodents present in the plots. As figure 54.9 illustrates, the number of other rodents was substantially higher in the absence of kangaroo rats, indicating that kangaroo rats compete with the other rodents and limit their population sizes.

A great number of similar experiments have indicated that interspecific competition occurs between many species of plants and animals. The effects of competition can be seen in aspects of population biology other than population size, such as behavior and individual growth rates. For example, two species of *Anolis* lizards occur on the Caribbean island of St. Maarten. When one of the species, *A. gingivinus*, is placed in 12 meter × 12 meter enclosures without the other species, individual lizards grow faster and perch lower than do lizards of the same species when placed in enclosures in which *A. pogus*, a species normally found near the ground, is also present.

Caution is Necessary

Although experimental studies can be a powerful means of understanding interactions between coexisting species, they have their limitations.

First, care is necessary in interpreting the results of field experiments. Negative effects of one species on another do

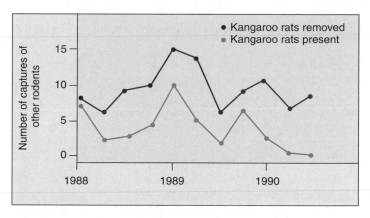

FIGURE 54.9
Detecting interspecific competition. This experiment tests how removal of kangaroo rats affects the population size of other rodents. Immediately after kangaroo rats were removed, the number of other rodents increased relative to the enclosures that still contained kangaroo rats. Notice that population sizes (as estimated by number of captures) changed in synchrony in the two treatments, probably reflecting changes in the weather. **Why are there more individuals of other rodent species when kangaroo rats are excluded?**

not automatically indicate the existence of competition. For example, many similar-sized fish have a negative effect on each other, but it results not from competition, but from the fact that adults of each species prey on juveniles of the other species. In addition, the presence of one species may attract predators, which then also prey on the second species. In this case, the second species may have a lower population size in the presence of the first species due to the predators, even if they are not competing at all. Thus, experimental studies are most effective when combined with detailed examination of the ecological mechanism causing the negative effect of one species on another species.

Second, experimental studies are not always feasible. For example, the coyote population has increased in the United States in recent years simultaneously with the decline of the grey wolf. Is this trend an indication that the species compete? Because of the size of the animals and the large geographic areas occupied by each individual, manipulative experiments involving fenced areas with only one or both species—with each experimental treatment replicated several times for statistical analysis—are not practical. Similarly, studies of slow-growing trees might require many centuries to detect competition between adult trees. In such cases, detailed studies of the ecological requirements of the species are our best bet for understanding interspecific interactions.

Experimental studies can provide strong tests of the hypothesis that interspecific competition occurs, but such studies have limitations. Detailed ecological studies are important regardless of whether experiments are conducted.

54.3 Predation has ecological and evolutionary effects.

Predation is the consuming of one organism by another. In this sense, predation includes everything from a leopard capturing and eating an antelope, to a deer grazing on spring grass. When experimental populations are set up under simple laboratory conditions, the predator often exterminates its prey and then becomes extinct itself, having nothing left to eat (figure 54.10). However, if refuges are provided for the prey, its population drops to low levels but not to extinction. Low prey population levels then provide inadequate food for the predators, causing the predator population to decrease. When this occurs, the prey population can recover.

Predation and Prey Populations

In nature, predators often have large effects on prey populations. Some of the most dramatic examples involve situations in which humans have either added or eliminated predators from an area. For example, the elimination of large carnivores from much of the eastern United States has led to population explosions of white-tailed deer, which strip the habitat of all edible plant life within their reach. Similarly, when sea otters were hunted to near extinction on the western coast of the United States, populations of sea urchins, a principal prey item of the otters, exploded.

Conversely, the introduction of rats, dogs, and cats to many islands around the world has led to the decimation of native faunas. For example, populations of Galápagos tortoises on several islands are endangered by introduced rats, dogs, and cats, which eat the eggs and the young tortoises. Similarly, in New Zealand, several species of birds and reptiles have been eradicated by rat predation and now only occur on a few offshore islands that the rats have not reached. In addition, on Stephens Island, near New Zealand, every individual of the now-extinct Stephens Island wren was killed by a single lighthouse keeper's cat!

A classic example of the role predation can play in a community involves the introduction of prickly pear cactus to Australia in the nineteenth century. In the absence of predators, the cactus spread rapidly, so that by 1925 it occupied 12 million hectares of rangeland in an impenetrable morass of spines that made cattle ranching difficult. To control the cactus, a predator from its natural habitat in Argentina, the moth *Cactoblastis cactorum*, was introduced, beginning in 1926. By 1940, cactus populations had been decimated, and it now generally occurs in small populations.

Predation and Evolution

Predation provides strong selective pressures on prey populations. Any feature that would decrease the probability

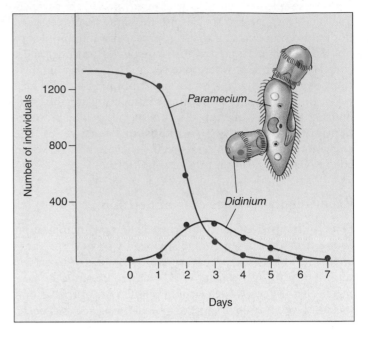

FIGURE 54.10
Predator-prey in the microscopic world. When the predatory *Didinium* is added to a *Paramecium* population, the numbers of *Didinium* initially rise, while the numbers of *Paramecium* steadily fall. When the *Paramecium* population is depleted, however, the *Didinium* individuals also die.
Can you think of any ways this experiment could be changed so that *Paramecium* might not go extinct?

of capture should be strongly favored. In the next three pages, we discuss a number of defense mechanisms in plants and animals. In turn, the evolution of such features causes natural selection to favor counteradaptations in predator populations. In this way, a *coevolutionary arms race* may ensue in which predators and prey are constantly evolving better defenses and better means of circumventing these defenses.

One example comes from the fossil record of mollusks and their predators. During the Mesozoic period (approximately 65 to 225 million years ago), new forms of predatory fish and crustaceans evolved that were able to crush or tear open shells. As a result, a variety of defensive measures evolved in mollusks, including spines, thicker shells, and shells too smooth for predators to grasp. In turn, these adaptations may have pressured predators to evolve more effective predatory adaptations and tactics, such as bigger and stronger claws and the ability to drill into shells.

Predation can have substantial effects on prey populations. As a result, prey species often evolve defensive adaptations.

FIGURE 54.11
Insect herbivores well suited to their hosts. (*a*) The green caterpillars of the cabbage butterfly, *Pieris rapae*, are camouflaged on the leaves of cabbage and other plants on which they feed. Although mustard oils protect these plants against most herbivores, the cabbage butterfly caterpillars are able to break down the mustard oil compounds. (*b*) An adult cabbage butterfly.

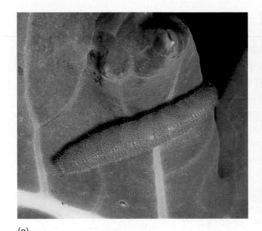
(a)

(b)

Plant Defenses Against Herbivores

Plants have evolved many mechanisms to defend themselves from herbivores. The most obvious are morphological defenses: Thorns, spines, and prickles play an important role in discouraging browsers, and plant hairs, especially those that have a glandular, sticky tip, deter insect herbivores. Some plants, such as grasses, deposit silica in their leaves, both strengthening and protecting themselves. If enough silica is present, these plants are simply too tough to eat.

Chemical Defenses

As significant as morphological adaptations are, the chemical defenses that occur so widely in plants are even more crucial. Best known and perhaps most important in the defenses of plants against herbivores are **secondary chemical compounds.** These are distinguished from primary compounds, which are the components of a major metabolic pathway, such as respiration. Many plants, and apparently many algae as well, contain structurally diverse secondary compounds that are either toxic to most herbivores or disturb their metabolism greatly, preventing, for example, the normal development of larval insects. Consequently, most herbivores tend to avoid the plants that possess these compounds.

The mustard family (Brassicaceae) produces a group of chemicals known as mustard oils. These are the substances that give the pungent aromas and tastes to such plants as mustard, cabbage, watercress, radish, and horseradish. The same tastes we enjoy signal the presence of chemicals that are toxic to many groups of insects. Similarly, plants of the milkweed family (Asclepiadaceae) and the related dogbane family (Apocynaceae) produce a milky sap that deters herbivores from eating them. In addition, these plants usually contain cardiac glycosides, molecules that can produce drastic deleterious effects on the heart function of vertebrates.

The Evolutionary Response of Herbivores

Certain groups of herbivores are associated with each family or group of plants protected by a particular kind of secondary compound. These herbivores are able to feed on these plants without harm, often as their exclusive food source. For example, cabbage butterfly caterpillars (subfamily Pierinae) feed almost exclusively on plants of the mustard and caper families, as well as on a few other small families of plants that also contain mustard oils (figure 54.11). Similarly, caterpillars of monarch butterflies and their relatives (subfamily Danainae) feed on plants of the milkweed and dogbane families. How do these animals manage to avoid the chemical defenses of the plants, and what are the evolutionary precursors and ecological consequences of such patterns of specialization?

We can offer a potential explanation for the evolution of these particular patterns. Once the ability to manufacture mustard oils evolved in the ancestors of the caper and mustard families, the plants were protected for a time against most or all herbivores that were feeding on other plants in their area. At some point, certain groups of insects—for example, the cabbage butterflies—evolved the ability to break down mustard oils and thus feed on these plants without harming themselves. Having developed this ability, the butterflies were able to use a new resource without competing with other herbivores for it. Often, in groups of insects such as cabbage butterflies, sense organs have evolved that are able to detect the secondary compounds their food plants produce. Clearly, the relationship that has formed between cabbage butterflies and the plants of the mustard and caper families is an example of long-term, reciprocal evolutionary adjustment of the characteristics of the members of a biological community, a process termed **coevolution.**

The members of many groups of plants are protected from most herbivores by their secondary compounds. Once the members of a particular herbivore group evolve the ability to feed on one of these types of plants, these herbivores gain access to a new resource, which they can exploit without competition from other herbivores.

Animal Defenses Against Predators

Some animals that feed on plants rich in secondary compounds receive an extra benefit. For example, when the caterpillars of monarch butterflies feed on plants of the milkweed family, they do not break down the cardiac glycosides that protect these plants from herbivores. Instead, the caterpillars concentrate and store the cardiac glycosides in fat bodies; they then pass them through the chrysalis stage to the adult and even to the eggs of the next generation. The incorporation of cardiac glycosides thus protects all stages of the monarch life cycle from predators. A bird that eats a monarch butterfly quickly regurgitates it (figure 54.12) and in the future avoids the conspicuous orange-and-black pattern that characterizes the adult monarch. Some bird species have evolved the ability to tolerate the protective chemicals. These birds eat the monarchs.

Chemical Defenses

Animals also manufacture and use a startling array of substances to perform a variety of defensive functions. Bees, wasps, predatory bugs, scorpions, spiders, and many other arthropods use chemicals to defend themselves and to kill their prey. In addition, various chemical defenses have evolved among marine animals and the vertebrates, including venomous snakes, lizards, fishes, and some birds. The poison-dart frogs of the family Dendrobatidae produce toxic alkaloids in the mucus that covers their brightly colored skin; these alkaloids are deadly to animals that try to eat the frogs (figure 54.13). Some of these toxins are so powerful that a few micrograms will kill a person if injected into the bloodstream. More than 200 different alkaloids have been isolated from these frogs, and some are playing important roles in neuromuscular research. An intensive investigation of marine animals, algae, and flowering plants is under way in search of new drugs to fight cancer and other diseases, or to use as sources of antibiotics.

Defensive Coloration

Many insects that feed on milkweed plants are brightly colored; they advertise their poisonous nature using an ecological strategy known as *warning coloration*, or aposematic coloration. Showy coloration is characteristic of animals that use poisons and stings to repel predators, while organisms that lack specific chemical defenses are seldom brightly colored. In fact, many have *cryptic coloration*—color that blends with the surroundings and thus hides the individual from predators (figure 54.14). Camouflaged animals usually do not live together in groups because a predator that discovers one individual gains a valuable clue to the presence of others.

Animals defend themselves against predators with chemical defenses, warning coloration, and camouflage.

(a)   (b)

FIGURE 54.12
A blue jay learns that monarch butterflies taste bad. (*a*) This cage-reared jay had never seen a monarch butterfly before it tried eating one. (*b*) The same jay regurgitated the butterfly a few minutes later. This bird will probably avoid trying to capture all orange-and-black insects in the future.

FIGURE 54.13
Vertebrate chemical defenses. Frogs of the family Dendrobatidae, abundant in the forests of Latin America, are extremely poisonous to vertebrates. Dendrobatids advertise their toxicity with bright coloration. As a result of either instinct or learning, predators avoid such brightly colored species that might otherwise be suitable prey.

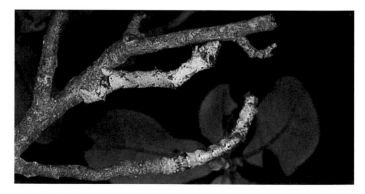

FIGURE 54.14
Cryptic coloration. An inchworm caterpillar (*Necophora quernaria*) (hanging from the upper twig) closely resembles a twig.

Mimicry

During the course of their evolution, many species have come to resemble distasteful ones that exhibit warning coloration. The mimic gains an advantage by looking like the distasteful model. Two types of mimicry have been identified: Batesian and Müllerian mimicry.

Batesian Mimicry

Batesian mimicry is named for Henry Bates, the British naturalist who first brought this type of mimicry to general attention in 1857. In his journeys to the Amazon region of South America, Bates discovered many instances of palatable insects that resembled brightly colored, distasteful species. He reasoned that the mimics would be avoided by predators, who would be fooled by the disguise into thinking the mimic was the distasteful species.

Many of the best-known examples of Batesian mimicry occur among butterflies and moths. Obviously, predators in systems of this kind must use visual cues to hunt for their prey; otherwise, similar color patterns would not matter to potential predators. Increasing evidence indicates that Batesian mimicry can involve nonvisual cues, such as olfaction, although such examples are less obvious to humans.

The kinds of butterflies that provide the models in Batesian mimicry are, not surprisingly, members of groups whose caterpillars feed on only one or a few closely related plant families. The plant families on which they feed are strongly protected by toxic chemicals. The model butterflies incorporate the poisonous molecules from these plants into their bodies. The mimic butterflies, in contrast, belong to groups in which the feeding habits of the caterpillars are not so restricted. As caterpillars, these butterflies feed on a number of different plant families unprotected by toxic chemicals.

One often-studied mimic among North American butterflies is the viceroy, *Limenitis archippus* (figure 54.15*a*). This butterfly, which resembles the poisonous monarch, ranges from central Canada through much of the United States and into Mexico. The caterpillars feed on willows and cottonwoods, and neither caterpillars nor adults were originally thought to be distasteful to birds, although recent findings may dispute this. Interestingly, the Batesian mimicry seen in the adult viceroy butterfly does not extend to the caterpillars: Viceroy caterpillars are camouflaged on leaves, resembling bird droppings, while the monarch's distasteful caterpillars are very conspicuous.

Müllerian Mimicry

Another kind of mimicry, **Müllerian mimicry,** was named for the German biologist Fritz Müller, who first described it in 1878. In Müllerian mimicry, several unrelated but protected animal species come to resemble one another

Danaus plexippus   *Limenitis archippus*

(a) Batesian mimicry: Monarch (*Danaus*) is poisonous; viceroy (*Limenitis*) is palatable mimic

Heliconius erato   *Heliconius melpomene*

Heliconius sapho   *Heliconius cydno*

(b) Müllerian mimicry: two pairs of mimics; all are distasteful

FIGURE 54.15
Mimicry. (*a*) Batesian mimicry. Monarch butterflies (*Danaus plexippus*) are protected from birds and other predators by the cardiac glycosides they incorporate from the milkweeds and dogbanes they feed on as larvae. Adult monarch butterflies advertise their poisonous nature with warning coloration. Viceroy butterflies (*Limenitis archippus*) are Batesian mimics of the poisonous monarch. (*b*) Pairs of Müllerian mimics. *Heliconius erato* and *H. melpomene* are sympatric, and *H. sapho* and *H. cydno* are sympatric. All of these butterflies are distasteful. They have evolved similar coloration patterns in sympatry to minimize predation; predators need only learn one pattern to avoid.

(figure 54.15*b*). If animals that resemble one another are all poisonous or dangerous, they gain an advantage because a predator will learn more quickly to avoid them. In some cases, predator populations even evolve an innate avoidance of species; such evolution may occur more quickly when multiple dangerous prey look alike.

In both Batesian and Müllerian mimicry, mimic and model must not only look alike but also act alike if predators are to be deceived. For example, the members of several families of insects that closely resemble wasps behave surprisingly like the wasps they mimic, flying often and actively from place to place.

In Batesian mimicry, unprotected species resemble others that are distasteful. Both species exhibit warning coloration. In Müllerian mimicry, two or more unrelated but protected species resemble one another, thus achieving a kind of group defense.

Coevolution and Symbiosis

The plants, animals, protists, fungi, and prokaryotes that live together in communities have changed and adjusted to one another continually over a period of millions of years. For example, many features of flowering plants have evolved in relation to the dispersal of the plant's gametes by animals (figure 54.16). These animals, in turn, have evolved a number of special traits that enable them to obtain food or other resources efficiently from the plants they visit, often from their flowers. While doing so, the animals pick up pollen, which they may deposit on the next plant they visit, or seeds, which may be left elsewhere in the environment, sometimes a great distance from the parent plant.

Such interactions are examples of coevolution, a phenomenon we have already seen in predator-prey interactions.

Examples of Symbiosis

Another type of coevolution involves **symbiosis,** in which two or more kinds of organisms interact in often elaborate and more-or-less permanent relationships. All symbiotic relationships carry the potential for coevolution between the organisms involved, and in many instances the results of this coevolution are fascinating. Examples of symbiosis include *lichens*, which are associations of certain fungi with green algae or cyanobacteria. Lichens are discussed in more detail in chapter 30. Another important example are *mycorrhizae*, associations between fungi and the roots of most kinds of plants. The fungi expedite the plant's absorption of certain nutrients, and the plants in turn provide the fungi with carbohydrates. Similarly, root nodules that occur in legumes and certain other kinds of plants contain bacteria that fix atmospheric nitrogen and make it available to their host plants.

In the tropics, leafcutter ants are often so abundant that they can remove a quarter or more of the total leaf surface of the plants in a given area. They do not eat these leaves directly; rather, they take them to underground nests, where they chew them up and inoculate them with the spores of particular fungi. These fungi are cultivated by the ants and brought from one specially prepared bed to another, where they grow and reproduce. In turn, the fungi constitute the primary food of the ants and their larvae. The relationship between leafcutter ants and these fungi is an excellent example of symbiosis. Recent phylogenetic studies using DNA and assuming a molecular clock (see chapter 22) suggest that these symbioses are ancient, perhaps evolving more than 50 million years ago.

FIGURE 54.16
Pollination by a bat. Many flowers have coevolved with other species to facilitate pollen transfer. Insects are widely known as pollinators, but they're not the only ones: birds, bats, and even opossums and lizards serve as pollinators for some species. Notice the cargo of pollen on the bat's snout.

Kinds of Symbiosis

The major kinds of symbiotic relationships include (1) **commensalism,** in which one species benefits while the other neither benefits nor is harmed; (2) **mutualism,** in which both participating species benefit; and (3) **parasitism,** in which one species benefits but the other is harmed. Parasitism can also be viewed as a form of predation, although the organism that is preyed upon does not necessarily die.

> Coevolution is a term that describes the long-term evolutionary adjustments of species to one another. In symbiosis, two or more species interact closely, with at least one species benefitting.

Commensalism

Commensalism is a symbiotic relationship that benefits one species and neither hurts nor helps the other. In nature, individuals of one species are often physically attached to members of another. For example, epiphytes are plants that grow on the branches of other plants. In general, the host plant is unharmed, while the epiphyte that grows on it benefits. Similarly, various marine animals, such as barnacles, grow on other, often actively moving sea animals, such as whales, and thus are carried passively from place to place. These "passengers" presumably gain more protection from predation than they would if they were fixed in one place, and they also reach new sources of food. The increased water circulation that such animals receive as their host moves around may be of great importance, particularly if the passengers are filter feeders. The gametes of the passenger are also more widely dispersed than would be the case otherwise.

When Is Commensalism Not Commensalism?

One of the best-known examples of symbiosis involves the relationships between certain small tropical fishes and sea anemones (see chapter 32). The fish have evolved the ability to live among the stinging tentacles of sea anemones, even though these tentacles would quickly paralyze other fishes that touched them (see figure 54.1). The anemone fishes feed on the detritus left from the meals of the host anemone, remaining uninjured under remarkable circumstances.

On land, an analogous relationship exists between birds called oxpeckers and grazing animals such as cattle or rhinoceroses (figure 54.17). The birds spend most of their time clinging to the animals, picking off parasites and other insects, carrying out their entire life cycles in close association with the host animals.

In each of these instances, it is difficult to be certain whether the second partner receives a benefit or not; there is no clear-cut boundary between commensalism and mutualism. For instance, it may be advantageous to the sea anemone to have particles of food removed from its tentacles because it may then be better able to catch other prey. Similarly, while often thought of as commensalism, the association of grazing mammals and gleaning birds is actually an example of mutualism. The mammal benefits by having parasites and other insects removed from its body, but the birds also benefit by gaining a dependable source of food.

On the other hand, commensalism can easily transform itself into parasitism. For example, oxpeckers are also known to pick not only parasites, but also scabs off their

FIGURE 54.17
Commensalism, mutualism, or parasitism? In this symbiotic relationship, oxpeckers definitely receive a benefit in the form of nutrition from the ticks and other parasites they pick off their host (in this case, an impala) and eat. But the effect on the host is not always clear. If the ticks are harmful, their removal benefits the host, and the relationship is mutually beneficial. If the oxpeckers also pick at scabs, causing blood loss and possible infection, the relationship may be parasitic. If the hosts are unharmed by either the ticks or the oxpeckers, the relationship may be an example of commensalism.

grazing hosts. Once the scab is picked, the birds drink the blood that flows from the wound. Occasionally, the cumulative effect of persistent attacks can greatly weaken the herbivore, particularly when conditions are not favorable, such as during droughts.

Commensalism is the benign use of one organism by another.

Mutualism

Mutualism is a symbiotic relationship among organisms in which both species benefit. Mutualistic relationships are of fundamental importance in determining the structure of biological communities. Some of the most spectacular examples of mutualism occur among flowering plants and their animal visitors, including insects, birds, and bats. As discussed in chapter 29, during the course of their evolution, the characteristics of flowers have evolved in large part in relation to the characteristics of the animals that visit them for food and, in the process, spread their pollen from individual to individual. At the same time, characteristics of the animals have changed, increasing their specialization for obtaining food or other substances from particular kinds of flowers.

Another example of mutualism involves ants and aphids. Aphids, also called greenflies, are small insects that suck fluids from the phloem of living plants with their piercing mouthparts. They extract a certain amount of the sucrose and other nutrients from this fluid, but they excrete much of it in an altered form through their anus. Certain ants have taken advantage of this—in effect, domesticating the aphids. The ants carry the aphids to new plants, where they come into contact with new sources of food, and then consume as food the "honeydew" that the aphids excrete.

Ants and Acacias

A particularly striking example of mutualism involves ants and certain Latin American species of the plant genus *Acacia*. In these species, certain leaf parts, called stipules, are modified as paired, hollow thorns. The thorns are inhabited by stinging ants of the genus *Pseudomyrmex*, which do not nest anywhere else (figure 54.18). Like all thorns that occur on plants, the acacia thorns serve to deter herbivores.

At the tip of the leaflets of these acacias are unique, protein-rich bodies called Beltian bodies, named after the nineteenth-century British naturalist Thomas Belt. Beltian bodies do not occur in species of *Acacia* that are not inhabited by ants, and their role is clear: They serve as a primary food for the ants. In addition, the plants secrete nectar from glands near the bases of their leaves. The ants consume this nectar as well, feeding it and the Beltian bodies to their larvae.

Obviously, this association is beneficial to the ants, and one can readily see why they inhabit acacias of this group. The ants and their larvae are protected within the swollen thorns, and the trees provide a balanced diet, including the sugar-rich nectar and the protein-rich Beltian bodies. What, if anything, do the ants do for the plants?

Whenever any herbivore lands on the branches or leaves of an acacia inhabited by ants, the ants, which continually patrol the acacia's branches, immediately attack and devour the herbivore. The ants that live in the acacias also help their hosts compete with other plants. The ants cut away

FIGURE 54.18
Mutualism: ants and acacias. Ants of the genus *Pseudomyrmex* live within the hollow thorns of certain species of acacia trees in Latin America. The nectaries at the bases of the leaves and the Beltian bodies at the ends of the leaflets provide food for the ants. The ants, in turn, supply the acacias with organic nutrients and protect the acacia from herbivores and shading from other plants.

any branches of other plants that touch the acacia in which they are living. They create, in effect, a tunnel of light through which the acacia can grow, even in the lush tropical rain forests of lowland Central America. In fact, when an ant colony is experimentally removed from a tree, the acacia is unable to compete successfully in this habitat. Finally, the ants bring organic material into their nests. The parts they do not consume, together with their excretions, provide the acacias with an abundant source of nitrogen.

As with commensalism, however, things are not always as they seem. Ant-acacia mutualisms also occur in Africa. In Kenya, several species of acacia ants occur, but only a single species occurs on any tree. One species, *Cremato-gaster nigriceps*, is competitively inferior to two of the other species. To prevent invasion by other ant species, *C. nigriceps* prunes the branches of the acacia, preventing it from coming into contact with branches of other trees, which would serve as a bridge for invaders. Although this behavior is beneficial to the ant, it is detrimental to the tree because it destroys the tissue from which flowers are produced, essentially sterilizing the tree. In this case, what initially evolved as a mutualistic interaction has instead become a parasitic one.

Mutualism involves interactions between species that are mutually beneficial.

Parasitism

Parasitism may be regarded as a special form of symbiosis in which the parasite usually is much smaller than the prey and remains closely associated with it. Parasitism is harmful to the prey organism and beneficial to the parasite. In many cases, the parasite kills its host, and thus the ecological effects of parasitism can be similar to those of predation.

External Parasites

Parasites that feed on the exterior surface of an organism are external parasites, or **ectoparasites** (figure 54.19). Many instances of external parasitism are known in both plants and animals. **Parasitoids** are insects that lay eggs on living hosts. This behavior is common among wasps, whose larvae feed on the body of the unfortunate host, often killing it.

Internal Parasites

Vertebrates are parasitized internally by **endoparasites,** members of many different phyla of animals and protists. Invertebrates also have many kinds of parasites that live within their bodies. Internal parasitism is generally marked by much more extreme specialization than external parasitism, as shown by the many protist and invertebrate parasites that infect humans. The more closely the life of the parasite is linked with that of its host, the more its morphology and behavior are likely to have been modified during the course of its evolution. The same is true of symbiotic relationships of all sorts. Conditions within the body of an organism are different from those encountered outside and are apt to be much more constant. Consequently, the structure of an internal parasite is often simplified, and unnecessary armaments and structures are lost as it evolves (for example, see descriptions of tapeworms in chapter 32).

Parasites Can Manipulate Host Behavior

Many parasites have complex life cycles that require several different hosts for growth to adulthood and reproduction. Recent research has revealed the remarkable adaptations of certain parasites that alter the behavior of the host and thus facilitate transmission from one host to the next. For example, many parasites cause their hosts to behave in ways that make them more vulnerable to their predators; when the host is ingested, the parasite is able to infect the predator. One of the most famous examples involves a parasitic flatworm, *Dicrocoelium dendriticum*, which lives in ants as an intermediate host, but reaches adulthood in large herbivorous mammals such as cattle and deer. Transmission from an ant to a cow might seem difficult, because cows do not normally eat insects. The flatworm, however, has evolved a remarkable adaptation. When an ant is infected, one of the

FIGURE 54.19
An external parasite. The flowering plant dodder (*Cuscuta*) is a parasite that has lost its chlorophyll and its leaves in the course of its evolution. Because it is heterotrophic (unable to manufacture its own food), dodder obtains its food from the host plants it grows on.

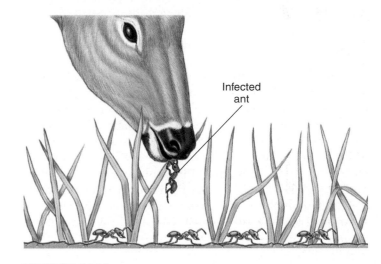

FIGURE 54.20
Parasitic manipulation of host behavior. Due to a parasite in its brain, an ant climbs to the top of a grass blade, where it may be eaten by a grazing herbivore, thus passing the parasite from insect to mammal.

flatworms migrates into the brain and causes the ant to climb to the top of vegetation and lock its mandibles onto a grass blade at the end of the day, just when herbivores are grazing (figure 54.20). The result is that the ant is eaten along with the grass, leading to infection of the grazer.

> **Parasitism is a form of symbiosis that is beneficial to the parasite, but harmful to the host.**

Interactions Among Ecological Processes

We have seen the different ways in which species can interact with each other. In nature, however, more than one type of interaction usually occurs at the same time. In many cases, the outcome of one type of interaction is modified or even reversed when another type of interaction is also occurring.

Predation Reduces Competition

When resources are limiting, a superior competitor can eliminate other species from a community. However, predators can prevent or greatly reduce competitive exclusion by reducing the numbers of individuals of competing species. A given predator may often feed on two, three, or more kinds of plants or animals in a given community. The predator's choice depends partly on the relative abundance of the prey options. In other words, a predator may feed on species *A* when it is abundant and then switch to species *B* when *A* is rare. Similarly, a given prey species may be a primary source of food for increasing numbers of species as it becomes more abundant. In this way, superior competitors may be prevented from outcompeting other species.

Such patterns are often characteristic of biological communities in marine intertidal habitats. For example, in preying selectively on bivalves, starfish prevent bivalves from monopolizing such habitats, opening up space for many other organisms (figure 54.21). When starfish are removed from a habitat, species diversity falls precipitously, and the seafloor community comes to be dominated by a few species of bivalves. Predation tends to reduce competition in natural communities, so it is usually a mistake to attempt to eliminate a major predator, such as wolves or mountain lions, from a community because the result may be a decrease in the biological diversity of the community.

Parasitism May Counter Competition

Parasites may affect sympatric species differently and thus influence the outcome of interspecific interactions. In a classic experiment, Thomas Park of the University of Chicago investigated interactions between two flour beetles, *Tribolium castaneum* and *T. confusum*, with a parasite, *Adelina*. In the absence of the parasite, *T. castaneum* is dominant and *T. confusum* normally goes extinct. When the parasite is present, however, the outcome is reversed and *T. castaneum* perishes. Similar effects of parasites in natural systems have been observed in many species. For example, in the *Anolis* lizards of St. Maarten mentioned previously, the competitively inferior species is resistant to lizard malaria (a form of the disease related to human malaria), whereas the other species is highly susceptible.

(a)

(b)

FIGURE 54.21

Predation reduces competition. (*a*) In a controlled experiment in a coastal ecosystem, Robert Payne of the University of Washington removed a key predator, starfishes (*Pisaster*). (*b*) In response, fiercely competitive mussels exploded in growth, effectively crowding out seven other indigenous species.

Only in areas in which the malaria parasite occurs are the two species capable of coexisting.

Indirect Effects

In some cases, species may not directly interact, yet the presence of one species may affect a second species by way of interactions with a third species. Such effects are termed **indirect effects.** For example, the desert rodents described in section 54.2 (see page 1167) eat seeds, and so do the ants in their community; thus, we might expect them to compete with each other. However, when all rodents were completely

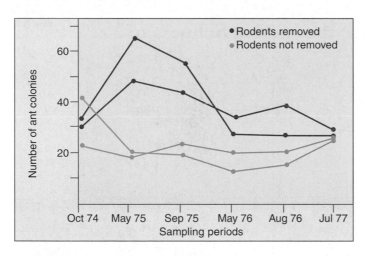

FIGURE 54.22
Change in ant population size after the removal of rodents.
Ants initially increased in population size relative to ants in the
two enclosures from which rodents weren't removed, but then
these ant populations declined.
**Why do ant populations increase and then decrease in the
absence of rodents?**

removed from large experimental enclosures and not allowed
back in (unlike the previous experiment, no holes were
placed in the walls), ant populations first increased but then
declined (figure 54.22). The initial increase was the expected
result of removing a competitor; why did it reverse? The answer reveals the intricacies of natural ecosystems (figure
54.23). Rodents prefer large seeds, whereas ants prefer
smaller seeds. Further, in this system, plants with large seeds
are competitively superior to plants with small seeds. Thus,
the removal of rodents leads to an increase in the number of
plants with large seeds, which reduces the number of small
seeds available to ants, which leads to a decline in ant populations. Thus, the effect of rodents on ants is complicated: a
direct, negative effect of resource competition and an indirect, positive effect mediated by plant competition.

Keystone Species

Species that have particularly strong effects on the composition of communities are termed **keystone species.** Predators, such as the starfish, can often serve as keystone species
by preventing one species from outcompeting others, thus
maintaining high levels of species richness in a community.

There are, however, a wide variety of other types of keystone species. Some species manipulate the environment in
ways that create new habitats for other species. Beavers, for
example, change running streams into small impound-
ments, altering the flow of water and flooding areas (figure
54.24). Similarly, alligators excavate deep holes at the bot-
toms of lakes. In times of drought, these holes are the only
areas where water remains, thus allowing aquatic species
that otherwise would perish to persist until the drought
ends and the lake refills.

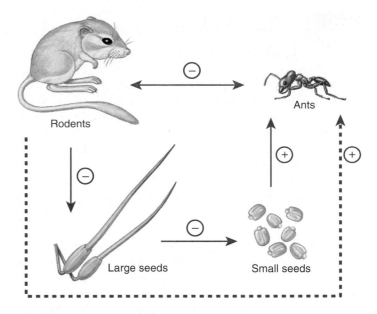

FIGURE 54.23
Rodent-ant interactions. Rodents and ants both eat seeds, so
the presence of rodents has a negative effect on ants, and vice
versa. However, the presence of rodents has a negative effect on
large seeds. In turn, the number of plants with large seeds has a
negative effect on plants that produce small seeds. Hence, the
presence of rodents should increase the number of small seeds. In
turn, the number of small seeds has a positive effect on ant
populations. Thus, indirectly, the presence of rodents has a
positive effect on ant population size.

FIGURE 54.24
Example of a keystone species. Beavers, by constructing dams
and transforming flowing streams into ponds, create new habitats
for many plant and animal species.

**Many different processes are likely to be occurring
simultaneously within communities. Only by
understanding how these processes interact will we be
able to understand how communities function.**

54.5 Ecological succession may increase the species richness of communities.

Even when the climate of an area remains stable year after year, communities have a tendency to change from simple to complex in a process known as **succession.** This process is familiar to anyone who has seen a vacant lot or cleared woods slowly become occupied by an increasing number of plants, or a pond become dry land as vegetation encroaches from the sides.

Succession

If a wooded area is cleared and left alone, plants will slowly reclaim the area. Eventually, all traces of the clearing will disappear, and the area will again be woods. This kind of succession, which occurs in areas where an existing community has been disturbed but soil still remains, is called **secondary succession.**

In contrast, **primary succession** occurs on bare, lifeless substrate, such as rocks, or in open water, where organisms gradually move into an area and change its nature. Primary succession occurs in lakes left behind after the retreat of glaciers, on volcanic islands that rise above the sea, and on land exposed by retreating glaciers (figure 54.25). Primary succession on glacial moraines provides an example (figure 54.26). On the bare, mineral-poor ground exposed when glaciers recede, soil pH is basic as a result of carbonates in the rocks, and nitrogen levels are low. Lichens are the first vegetation able to grow under such conditions. Acidic secretions from the lichens help break down the substrate and reduce the pH, as well as adding to the accumulation of soil. Mosses then colonize these pockets of soil, eventually building up enough nutrients in the soil for alder shrubs to take hold. Over a hundred years, the alders, which have symbiotic bacteria that fix atmospheric nitrogen (see chapter 27), increase soil nitrogen levels, and their acidic leaves further lower soil pH. At this point, spruce are able to become established, and they eventually crowd out the alder and form a dense spruce forest.

In a similar example, an **oligotrophic** lake—one poor in nutrients—may gradually, by the accumulation of organic matter, become **eutrophic**—rich in nutrients. As this occurs, the composition of communities will change, first increasing in species richness and then declining.

Primary succession in different habitats often eventually arrives at the same kinds of vegetation—vegetation characteristic of the region as a whole. This relationship led American ecologist F. E. Clements, as part of his holistic concept of communities, to propose the concept of a final *climax community*. With an increasing realization that (1) the climate keeps changing, (2) the process of succession is often very slow, and (3) the nature of a region's vegetation is being determined to an increasing extent by human

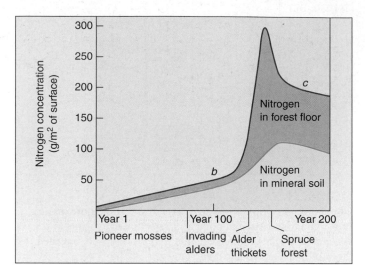

FIGURE 54.25
Plant succession produces progressive changes in the soil. Initially, the glacial moraine at Glacier Bay, Alaska (portrayed in figure 54.26), had little soil nitrogen, but nitrogen-fixing alders led to a buildup of nitrogen in the soil, encouraging the subsequent growth of the coniferous forest. Letters in the graph correspond to photographs in parts *b* and *c* of figure 54.26.

activities, ecologists today do not consider the concept of "climax community" as useful as they once did.

Why Succession Happens

Succession happens because species alter the habitat and the resources available in it in ways that favor other species. Three dynamic concepts are of critical importance in the process: tolerance, facilitation, and inhibition.

1. **Tolerance.** Early successional stages are characterized by weedy, r-selected species that are tolerant of the harsh, abiotic conditions in barren areas.
2. **Facilitation.** The weedy early successional stages introduce local changes in the habitat that favor other, less weedy species. Thus, the mosses in the Glacier Bay succession convert nitrogen to a form that allows alders to invade. The alders in turn lower soil pH as their fallen leaves decompose, and spruce and hemlock, which require acidic soil, are able to invade.
3. **Inhibition.** Sometimes the changes in the habitat caused by one species, while favoring other species, inhibit the growth of the species that caused them. Alders, for example, do not grow as well in acidic soil as the spruce and hemlock that replace them.

Over the course of succession, the number of species typically increases as the environment becomes more hospitable. In some cases, however, as ecosystems mature,

(a)

(b)

(c)

FIGURE 54.26
Primary succession at Alaska's Glacier Bay. (*a*) The sides of the glacier have been retreating at a rate of about 8 meters per year, leaving behind exposed soil from which nitrogen and other minerals have been leached. The first invaders of these exposed sites are pioneer moss species with nitrogen-fixing, mutualistic microbes. (*b*) Within 20 years, young alder shrubs take hold. Rapidly fixing nitrogen, they soon form dense thickets. (*c*) As soil nitrogen levels rise, spruce crowd out the mature alders, forming a forest.

more *K*-selected species replace *r*-selected ones, and superior competitors force out other species, leading ultimately to a decline in species richness.

Succession in Animal Communities

The species of animals present in a community also change through time in a successional pattern. Usually this results because species have different habitat requirements, and as the vegetation changes during succession, habitat disappears for some species and appears for others. A particularly striking example occurred on the Krakatau islands, which were devastated by an enormous volcanic eruption in 1883. Initially composed of nothing but barren ash-fields, the vegetation of the three islands of the group experienced rapid successional change: A few blades of grass appeared the next year, and within 15 years the coastal vegetation was well established and the interior was covered with dense grasslands. By 1930, the islands were almost entirely forested (figure 54.27).

The fauna of Krakatau changed in synchrony with the vegetation. Nine months after the eruption, the only animal found was a single spider, but by 1908, 200 animal species were found in a three-day exploration. For the most part, the first animals were grassland inhabitants, but as trees came to dominate, some of these early colonists, such as the zebra dove and the long-tailed shrike, disappeared and were replaced by forest-inhabiting species, such as fruit bats and frugivorous birds.

Although patterns of succession of animal species have been caused by vegetational succession, changes in the composition of the animal community in turn have affected plant occurrences. In particular, many plant species that are animal-dispersed or pollinated could not colonize Krakatau until their dispersers or pollinators had become established. For example, fruit bats were slow to colonize Krakatau, and until they appeared, few bat-dispersed plant species were present.

FIGURE 54.27
Succession after a volcanic eruption. A major volcanic explosion in 1883 on the island of Krakatau destroyed all life on the island. (*a*) This photo shows a later, much less destructive eruption of the volcano. (*b*) Krakatau, forested and populated by animals.

(a)

(b)

Communities change through time by a process termed succession. Species richness tends to increase through time, though ultimately it may decline.

The Role of Disturbance

Traditionally, many ecologists considered biological communities to be in a state of **equilibrium,** a stable condition that resisted change and fairly quickly returned to its original state if disturbed by humans or natural events. Such stability was usually attributed to the process of interspecific competition.

In recent years, this viewpoint has been reevaluated. Increasingly, scientists are recognizing that communities are constantly changing as a result of climatic changes, species invasions, and disturbance events. As a result, many ecologists now invoke **nonequilibrium** models that emphasize change, rather than stability. A particular focus of ecological research concerns the role that disturbances play in determining the structure of communities.

Disturbances can be widespread or local. Severe disturbances, such as forest fires, drought, and floods, may affect large areas. Animals may also cause severe disruptions. Gypsy moths can devastate a forest by consuming its trees. Unregulated deer populations may grow explosively, the deer overgrazing and so destroying the forest in which they live. On the other hand, local disturbances may affect only a small area, as when a tree falls in a forest or an animal digs a hole and uproots vegetation.

Intermediate Disturbance Hypothesis

In some cases, disturbance may act to increase the species richness of an area. According to the *intermediate disturbance hypothesis*, communities experiencing moderate amounts of disturbance will have higher levels of species richness than communities experiencing either little or great amounts of disturbance. Two factors could account for this pattern. First, in communities where moderate amounts of disturbance occur, patches of habitat will exist at different successional stages. Thus, within the area as a whole, species diversity will be greatest because the full range of species—those characteristic of all stages of succession—will be present. For example, a pattern of intermittent episodic disturbance that produces gaps in the rain forest (as when a tree falls) allows invasion of the gap by other species (figure 54.28). Eventually, the species inhabiting the gap will go through a successional sequence, one tree replacing another, until a canopy tree species comes again to occupy the gap. But if there are many gaps of different ages in the forest, many different species will coexist, some in young gaps and others in older ones.

Second, moderate levels of disturbance may prevent communities from reaching the final stages of succession, in which a few dominant competitors eliminate most of the other species. On the other hand, too much disturbance might leave the community continually in the earliest stages of succession, when species richness is relatively low.

FIGURE 54.28
Intermediate disturbance.   A single fallen tree created a small light gap in the tropical rain forest of Malawi, Africa. Such gaps play a key role in maintaining the high species diversity of the rain forest. In this case, a banana plant is able to sprout up among the dense foliage of trees in the forest.

Ecologists are increasingly realizing that disturbance is common, rather than exceptional, in many communities. As a result, the idea that communities inexorably move along a successional trajectory culminating in the development of a climax community is no longer widely accepted. Rather, predicting the state of a community in the future may be difficult because the unpredictable occurrence of disturbances will often counter successional changes. Understanding the role that disturbances play in structuring communities is currently an important area of investigation in ecology.

> **Change is often common in ecological communities. In some cases, intermediate levels of disturbance may maximize species richness.**

For interactive testing, visit the Online Learning Center with PowerWeb at www.mhhe.com/Raven7

54.1 Biological communities are composed of species that occur together.

Concepts of Communities

- The term community refers to the collection of species interacting in a particular area. (p. 1162)
- The individualistic concept of communities holds that a community is nothing more than an aggregation of species co-occurring in one area. (p. 1162)
- The holistic concept of communities views communities as an integrated unit. (p. 1162)
- Recent studies have found that species changed independently, not synchronously, as climate changed, suggesting the individualistic community view over the holistic view. (pp. 1162–1163)

54.2 Interactions among competing species shape ecological niches.

Fundamental and Realized Niches

- An organism's niche is the sum total of all the ways it utilizes environmental resources. (p. 1164)
- Interspecific competition occurs when two species attempt to utilize the same limited resource. (p. 1164)
- A fundamental niche is the entire niche a species is capable of using, while the realized niche is the actual portion a species occupies. (p. 1164)

Gause and the Principle of Competitive Exclusion

- The principle of competitive exclusion states that if two species are competing for a limited resource, the species that uses the resource more efficiently will eventually limit the other, at least locally. (p. 1165)
- Niche overlap can occur between two or more species, as long as the degree of overlap does not lead to significant negative effects on one or more of the species. (p. 1165)
- Gause's theory predicts when two species coexist on a long-term basis, resources must not be limiting, or else the niches differ by one or more resources. (p. 1165)

Resource Partitioning

- Sympatric species often reduce, or avoid, competition by living in different portions of the habitat or by utilizing different resources. (p. 1166)

Detecting Interspecific Competition

- Although experimental studies can reveal the existence of interspecific competition, and field experiments can often overcome some of the limitations inherent in laboratory conditions, both methods have their limitations that must be overcome. (p. 1167)

54.3 Predation has ecological and evolutionary effects.

Predation and Prey Populations

- Predation provides strong selective pressures on prey populations. (p. 1168)
- Prey species often evolve better defenses against predators, and predators often then evolve better means of circumventing such defenses. (p. 1168)

Plant Defenses Against Herbivores

- Plants have developed many mechanisms for protection against herbivory, including morphological defenses, such as structures to deter herbivory, and chemical defenses, such as secondary chemical compounds. (p. 1169)

Animal Defenses Against Predators

- Defensive coloration can be used either as aposematic (warning) coloring or cryptic coloring (camouflage). (p. 1170)

Mimicry

- Batesian mimicry occurs when palatable insects resemble distasteful or toxic species and thus gain some protection from predation as predators learn to avoid them. (p. 1171)
- Müllerian mimicry occurs when unrelated but distasteful species come to resemble one another. (p. 1171)

54.4 Species within a community interact in many ways.

Coevolution and Symbiosis

- Symbiosis is a form of coevolution in which two or more kinds of organisms live together in a complex, long-term relationship. (p. 1172)

Commensalism

- Commensalism occurs when the relationship between two species benefits one species and neither harms nor benefits the other species. (p. 1173)

Mutualism

- Mutualism occurs when both species involved in the relationship benefit. (p. 1174)

Parasitism

- Parasitism is a special form of symbiosis in which the parasite is usually much smaller than the prey, and the two remain closely associated. Parasitism is harmful to the prey and beneficial to the predator. (p. 1175)
- Parasites can either live outside the host (ectoparasites, such as mosquitoes) or inside the host (endoparasites, such as tapeworms). (p. 1175)
- Recent studies have shown that some parasites can alter the behavior of their host in ways that facilitate transmission between hosts. (p. 1175)

Interactions Among Ecological Processes

- Predators can prevent or greatly reduce competitive exclusion by reducing the numbers of individuals of competing species. (p. 1176)
- Parasites may affect sympatric species differently and thus influence the outcome of interspecific interactions. (p. 1176)
- Keystone species are those that have particularly strong effects on community composition. (p. 1177)

54.5 Ecological succession may increase the species richness of communities.

Succession

- Primary succession occurs when a life-less substrate is gradually inhabited by organisms, which changes the environment. (p. 1178)
- Secondary succession occurs in an area where an existing community has been disturbed. (p. 1178)

The Role of Disturbance

- Increasingly, scientists are recognizing communities as dynamic entities that change due to climatic shifts, species invasions, and disturbance events. (p. 1180)
- The intermediate disturbance hypothesis predicts that communities experiencing moderate levels of disturbance will have higher species richness than communities experiencing either smaller or greater amounts of disturbance. (p. 1180)

Self Test

1. The entire range of factors an organism is able to exploit in its environment is its
   a. community.
   b. realized niche.
   c. fundamental niche.
   d. habitat.
2. The phenomenon known as character displacement is associated with
   a. sympatric species.
   b. allopatric species.
   c. competitive exclusion.
   d. primary succession.
3. As the number of individuals of the predatory species increases, the prey population
   a. increases.
   b. decreases.
   c. stabilizes.
   d. first increases and then begins to decrease.
4. In order for mimicry to be effective in protecting a species from predation, it must
   a. occur in a palatable species that looks like a distasteful species.
   b. have cryptic coloration.
   c. occur such that mimics look and act like models.
   d. occur in only poisonous or dangerous species.
5. Brightly colored poison-dart frogs are an example of
   a. Batesian mimicry.
   b. Müllerian mimicry.
   c. cryptic coloration.
   d. aposematic coloration.
6. Which of the following is an example of commensalism?
   a. a tapeworm living in the gut of its host
   b. a clownfish living among the tentacles of a sea anemone
   c. a lichen
   d. bees feeding on nectar from a flower
7. A parasite that feeds on its host from inside is a(n)
   a. endoparasite.
   b. ectoparasite.
   c. parasitoid.
   d. predator.
8. A species that interacts with many other organisms in its community in critical ways can be called a
   a. predator.
   b. primary succession species.
   c. secondary succession species.
   d. keystone species.
9. Succession that occurs on abandoned agricultural fields is best described as
   a. coevolution.
   b. primary succession.
   c. secondary succession.
   d. prairie succession.
10. Lichen growing on the surface of rocks provides an example of
    a. facilitation.
    b. tolerance.
    c. inhibition.
    d. secondary succession.

Test Your Visual Understanding

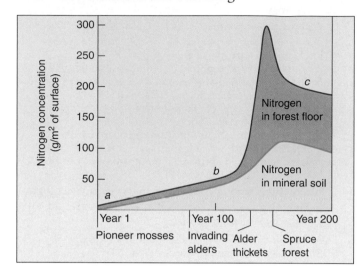

1. The graph above shows three general stages of succession indicated by the labels *a*, *b*, and *c* that occurred after the receding of a glacier at Glacier Bay, Alaska. Using the concepts of tolerance, facilitation, and inhibition, explain what is happening at each stage of succession.

Apply Your Knowledge

1. Through genetic engineering, bacterial toxin genes from a soil bacterium, *Bacillus thuringienis* (Bt), have been inserted into corn plants, making them resistant to certain insects. The corn plants produce a toxin that kills insects but doesn't harm the plant or people. However, because the corn pollen, which also contains the toxins, could spread to neighboring milkweed plants, there was a concern that monarch butterflies, which are killed by the toxin, could be at risk. The lethal dose of toxin consists of 2500 pollen grains/in² of milkweed leaf. Milkweed plants within a field of corn contain on average 500 pollen grains/in² of milkweed leaf. Assume that the level of pollen decreases by 60% with every 10 feet. How much pollen will reach milkweed plants that are 10 feet from the cornfield? 20 feet from the cornfield?

55

Dynamics of Ecosystems

Concept Outline

FIGURE 55.1
Pond in a rain forest. Interactions among species determine the rate at which chemicals and energy move between living elements of the ecosystem and how they reenter the abiotic realm, such as the atmosphere, the ground, or water.

The earth is a closed system with respect to chemicals, but, thanks to input from the sun, an open system in terms of energy. Collectively, the organisms in ecosystems regulate the capture and expenditure of energy and the cycling of chemicals (figure 55.1). As we will see in this chapter, all organisms, including humans, depend on the ability of other organisms—plants, algae, fungi, and some prokaryotes—to recycle the basic components of life. In chapters 56 and 57, we consider the many different types of ecosystems that constitute the biosphere and discuss the many threats to the biosphere and the species it contains.

55.1 Chemicals cycle within ecosystems.

An **ecosystem** includes all the organisms that live in a particular place, plus the abiotic environment in which they live and with which they interact. Two processes occur within ecosystems: (1) Energy enters ecosystems—usually from the sun, but sometimes from other sources—and is converted to chemical energy by photosynthesis or other mechanisms. Subsequently, this energy is passed on in organic compounds to other organisms or is dissipated as heat. (2) Chemical elements move through ecosystems in **biogeochemical cycles.** On a global scale, only a very small portion of these substances is contained within the bodies of organisms; almost all exist in nonliving reservoirs: the atmosphere, water, or rocks. Biogeochemical cycles occur at all levels, from molecular (within an organism) to planetary (within the atmosphere or lithosphere), and at all time scales, from seconds (cellular biochemical reactions) to millennia (weathering of rocks, sedimentation). Most cycles occur simultaneously; we consider them separately only for convenience.

Physical processes such as evaporation and precipitation move these elements through the abiotic portion of the environment. Organisms capture these elements through photosynthesis and transform them into organic compounds, which then may be passed through the biotic portion of the environment. Eventually, they are returned to the abiotic realm through respiration, decomposition, or other pathways.

Both energy and chemical elements move through the biotic environment along the same pathways, primarily photosynthesis, feeding, and decomposition. Nonetheless, important differences exist. In contrast to matter, energy cannot be recycled and is eventually converted to heat. Consequently, continual input from the sun is needed, whereas matter continuously recirculates through the environment and is used and reused.

The Water Cycle

The water cycle (figure 55.2) is the most familiar of all biogeochemical cycles. All life depends directly on the presence of water; the bodies of most organisms consist mainly of this substance. Water is the source of hydrogen ions, whose movements generate ATP in organisms. For that reason alone, water is indispensable to their functioning.

The Path of Free Water

The oceans cover three-fourths of the earth's surface. From the oceans, water evaporates into the atmosphere, a process powered by energy from the sun. Over land, approximately 90% of the water that reaches the atmosphere is moisture that evaporates from the surface of plants through a process called transpiration (see chapter 37). Most precipitation falls directly into the oceans, but some falls on land, where it passes into surface and subsurface bodies of fresh water. Only about 2% of all the water on earth is captured in any form—frozen, held in the soil, or incorporated into the bodies of organisms. All of the rest is free water, circulating between the atmosphere, the land, and the oceans.

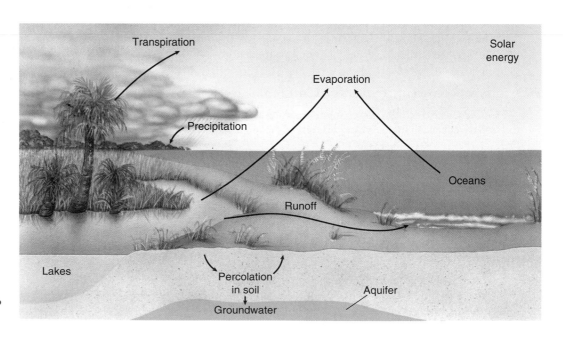

FIGURE 55.2
The water cycle. Water circulates from the atmosphere to the earth and back again.

The Importance of Water to Organisms

Organisms live or die on the basis of their ability to capture water and incorporate it into their bodies. Plants take up water from the earth in a continuous stream. Crop plants require about 1000 kilograms of water to produce one kilogram of food, and the ratio in natural communities is similar. Animals obtain water directly or from the plants or other animals they eat. The amount of free water available at a particular place often determines the nature and abundance of the living organisms present there.

Groundwater

Much less obvious than surface water, which we see in streams, lakes, and ponds, is groundwater, which occurs in aquifers—permeable, saturated, underground layers of rock, sand, and gravel. In many areas, groundwater is the most important reservoir of water. It amounts to more than 96% of all fresh water in the United States. The upper, unconfined portion of the groundwater constitutes the *water table*, which flows into streams and is partly accessible to plants; the lower, confined layers are generally out of reach, although they can be "mined" by humans. The water table is recharged by water that percolates through the soil from precipitation as well as by water that seeps downward from ponds, lakes, and streams. The deep aquifers (underground material reservoirs of water) are recharged very slowly from the water table.

Groundwater flows much more slowly than surface water, anywhere from a few millimeters to a meter or so per day. In the United States, groundwater provides about 25% of the water used for all purposes and supplies about 50% of the population with drinking water. Rural areas tend to depend almost exclusively on wells to access groundwater, and its use is growing at about twice the rate of surface water use. In the Great Plains of the central United States, the extensive use of the Ogallala Aquifer as a source of water for agricultural needs as well as for drinking water is depleting it faster than it can be naturally recharged. This seriously threatens the agricultural production of the area; similar problems are appearing throughout the drier portions of the globe.

Because of the greater rate of groundwater use and because it flows so slowly, the increasing chemical pollution of groundwater is also a very serious problem. It is estimated that about 2% of the groundwater in the United States is already polluted, and the situation is worsening. Pesticides, herbicides, and fertilizers have become a serious threat to water purity. Other key sources of groundwater pollution are the roughly 200,000 surface pits, ponds, and lagoons that are actively used for chemical waste disposal in the United States alone. Because of the large volume of water, its slow rate of turnover, and its inaccessibility, removing pollutants from aquifers is virtually impossible.

FIGURE 55.3
Deforestation breaks the water cycle. As time goes by, the consequences of tropical deforestation may become even more severe, as the extensive erosion in this deforested area on the southeast Asian island of Borneo shows.

Breaking the Water Cycle

In dense forest ecosystems, such as tropical rain forests, more than 90% of the moisture is taken up by plants and then transpired back into the air. Because so many plants in a rain forest are doing this, the vegetation is the primary source of local rainfall. In a very real sense, these plants create their own rain: The moisture that travels up from the plants into the atmosphere falls back to earth as rain.

Where forests are cut down, the water cycle is broken, and moisture is not returned to the atmosphere. Water drains away from the area to the sea instead of rising to the clouds and falling again on the forest. As early as the late 1700s, the great German explorer Alexander von Humbolt reported that stripping the trees from a tropical rain forest in Colombia had prevented water from returning to the atmosphere and created a semiarid desert. It is a tragedy of our time that just such a transformation is occurring in many tropical areas, as tropical and temperate rain forests are being clear-cut or burned (figure 55.3). In the twentieth century, much of Madagascar, a California-sized island off the east coast of Africa that contains many species found nowhere else in the world, was transformed from lush tropical forest into semiarid desert by deforestation. Because the rain no longer falls, there is no practical way to reforest this land. The water cycle, once broken, cannot be reestablished easily.

Water cycles between oceans and the atmosphere. Some 96% of the fresh water in the United States consists of groundwater, which provides 25% of all the water used in this country.

The Carbon Cycle

Carbon dioxide makes up only about 0.03% of the atmosphere. Worldwide, the synthesis of organic compounds from carbon dioxide and water through photosynthesis (see chapter 10) utilizes about 10% of the carbon dioxide in the atmosphere each year. This enormous amount of biological activity takes place as a result of the combined activities of photosynthetic bacteria, protists, and plants. All terrestrial heterotrophic organisms obtain their carbon indirectly from photosynthetic organisms. When the bodies of dead organisms decompose, microorganisms release carbon dioxide, which goes back into the atmosphere or, if in water, settles into the sediment (in fact, marine sediments are by far the largest store of carbon on earth). From there, it can be reincorporated into the bodies of other organisms. Figure 55.4 illustrates the carbon cycle.

Most of the organic compounds formed as a result of carbon dioxide fixation in the bodies of photosynthetic organisms are ultimately broken down and released back into the atmosphere or water. Certain carbon-containing compounds, such as cellulose, are more resistant to breakdown than others, but certain prokaryotes and fungi, as well as a few kinds of insects, are able to accomplish this feat. Some cellulose, however, accumulates as undecomposed organic matter such as peat. The carbon in this cellulose may eventually be incorporated into fossil fuels such as oil or coal.

In addition to the roughly 700 billion metric tons of carbon dioxide in the atmosphere, approximately 1 trillion metric tons are dissolved in the ocean. More than half of this quantity is in the upper layers, where photosynthesis takes place. The fossil fuels, primarily oil and coal, contain more than 5 trillion additional metric tons of carbon, and between 600 million and 1 trillion metric tons are locked up in living organisms at any one time. In global terms, respiration and photosynthesis (see chapters 9 and 10) are approximately balanced, but the balance has been shifted recently because of the consumption of fossil fuels. The combustion of coal, oil, and natural gas has released large stores of carbon into the atmosphere as carbon dioxide. The increased amount of carbon dioxide in the atmosphere appears to be changing global climates, and may do so even more rapidly in the future, as is discussed in chapter 56.

About 10% of the estimated 700 billion metric tons of carbon dioxide in the atmosphere is fixed annually by the process of photosynthesis.

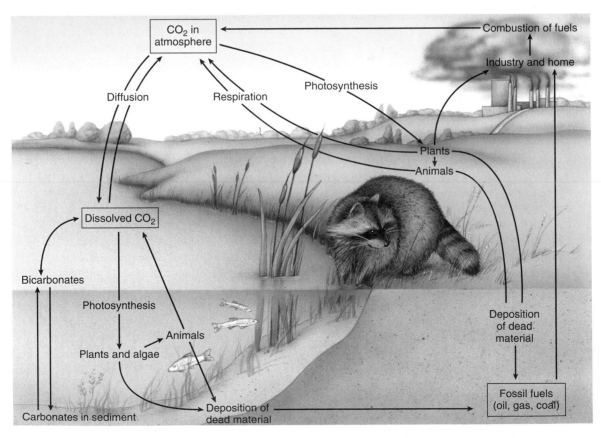

FIGURE 55.4
The carbon cycle.   Photosynthesis captures carbon; respiration returns it to the atmosphere. Dissolved carbon dioxide in aquatic ecosystems mostly reacts with water to form bicarbonates.

The Nitrogen Cycle

Nitrogen is an essential component of organic compounds such as proteins and amino acids. Via the nitrogen cycle, a few kinds of organisms—all of them prokaryotes—can convert, or "fix" (see chapter 27), atmospheric nitrogen (78% of the earth's atmosphere) into forms that can be used for biological processes (figure 55.5). The triple bond that links together the two atoms that make up diatomic atmospheric nitrogen (N_2) makes it a very stable molecule. In living systems, the cleavage of atmospheric nitrogen is catalyzed by a complex of three proteins—ferredoxin, nitrogen reductase, and nitrogenase. This process uses ATP as a source of energy, electrons derived from photosynthesis or respiration, and a powerful reducing agent. The overall reaction of nitrogen fixation is written:

$$N_2 + 3H_2 \rightarrow 2NH_3$$

Some genera of prokaryotes have the ability to fix atmospheric nitrogen. Most are free-living, but some form symbiotic relationships with the roots of legumes (plants of the pea family, Fabaceae) and other plants. Only the symbiotic bacteria fix enough nitrogen to be of major significance in nitrogen production. Because of the activities of such organisms in the past, a large reservoir of ammonia and nitrates now exists in most ecosystems. This reservoir is the immediate source of much of the nitrogen used by organisms. In the oceans, however, no such reservoir exists, and nitrogen is the major limiting nutrient.

Nitrogen-containing compounds, such as proteins in plant and animal bodies, are decomposed rapidly by certain prokaryotes and fungi. These prokaryotes and fungi use the amino acids they obtain through decomposition to synthesize their own proteins and to release excess nitrogen in the form of ammonium ions (NH_4^+), a process known as **ammonification.** The ammonium ions, which also are excreted by many animals, can be converted to soil nitrites and nitrates by certain kinds of organisms and can then be absorbed by plants.

A certain proportion of the fixed nitrogen in the soil is steadily lost. Under anaerobic conditions, nitrate is often converted to nitrogen gas (N_2) and nitrous oxide (N_2O), both of which return to the atmosphere. This process, which several genera of bacteria carry out, is called **denitrification.**

Large amounts of nitrogen are added to the environment in the form of fertilizers applied to agricultural lands. As a result, humans now play an important role in the global nitrogen cycle.

Nitrogen becomes available to organisms almost entirely through the metabolic activities of prokaryotes, some free-living and others that live symbiotically in the roots of legumes and other plants.

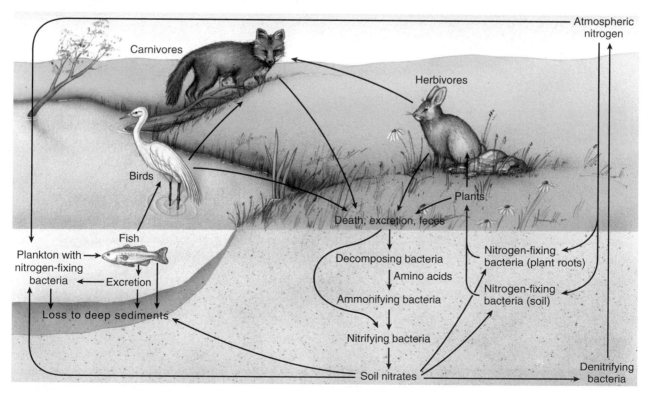

FIGURE 55.5
The nitrogen cycle.   Certain bacteria fix atmospheric nitrogen, converting it to a form living organisms can use. Other bacteria decompose nitrogen-containing compounds from plant and animal materials, returning them to the atmosphere.

The Phosphorus Cycle

In all biogeochemical cycles other than those involving water, carbon, oxygen, and nitrogen, the reservoir of the nutrient exists in mineral form (that is, in rocks), rather than in the atmosphere. The phosphorus cycle (figure 55.6) is presented as a representative example of all other mineral cycles. Phosphorus, a component of ATP, phospholipids, and nucleic acids, plays a critical role in plant nutrition.

Of all the required nutrients other than nitrogen, phosphorus is the most likely to be scarce enough to limit plant growth. Phosphates, in the form of phosphorus anions, exist in soil only in small amounts. This is because they are relatively insoluble and are present only in certain kinds of rocks. As phosphates weather out of soils, they are transported by rivers and streams to the oceans, where they accumulate in sediments. They are naturally brought back up again only by upwelling ocean currents, as occur along the Pacific coast of North and South America. Phosphates brought to the surface are assimilated by algae, which are eaten by fish, which are in turn eaten by birds. Seabirds deposit enormous amounts of guano (feces) rich in phosphorus along certain coasts. Guano deposits have traditionally been used for fertilizer. Crushed phosphate-rich rocks, found in certain regions, are also used for fertilizer. The seas are the only inexhaustible source of phosphorus, making deep-seabed mining increasingly commercially attractive.

Every year, millions of tons of phosphate are added to agricultural lands in the belief that it becomes fixed to and enriches the soil. In general, three times more phosphate than a crop requires is added each year. This is usually in the form of **superphosphate,** which is soluble calcium dihydrogen phosphate, $Ca(H_2PO_4)_2$, derived by treating bones or apatite, the mineral form of calcium phosphate, with sulfuric acid. But the enormous quantities of phosphates that are being added annually to the world's agricultural lands are not leading to proportionate gains in crops. Plants can apparently use only so much of the phosphorus that is added to the soil.

Phosphates are relatively insoluble and are present in most soils only in small amounts. They often are so scarce that their absence limits plant growth.

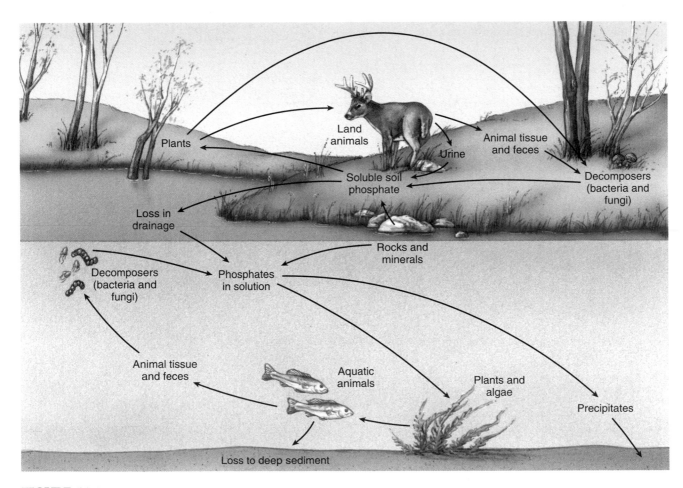

FIGURE 55.6
The phosphorus cycle. Phosphates weather from soils into water, enter plants and animals, and are redeposited in the soil when plants and animals decompose.

Biogeochemical Cycles Illustrated: Recycling in a Forest Ecosystem

An ongoing series of studies conducted at the Hubbard Brook Experimental Forest in New Hampshire has revealed in impressive detail the overall recycling pattern of nutrients in an ecosystem. The way this particular ecosystem functions, and especially the way nutrients cycle within it, has been studied since 1963 by Herbert Bormann of the Yale School of Forestry and Environmental Studies, Gene Likens of the Institute of Ecosystem Studies, and their colleagues. These studies have yielded much of the available information about the cycling of nutrients in forest ecosystems. They have also provided the basis for the development of much of the experimental methodology that is being applied successfully to the study of other ecosystems.

Hubbard Brook is the central stream of a large watershed that drains a region of temperate deciduous forest (figure 55.7a). To measure the flow of water and nutrients within the Hubbard Brook ecosystem, concrete dams with V-shaped notches were built across six tributary streams. All of the water that flowed out of the valleys had to pass through the notches. The researchers measured the precipitation that fell in the six valleys, and determined the amounts of nutrients that were present in the water flowing in the six streams. By these methods, they demonstrated that the undisturbed forests in this area were very efficient at retaining nutrients; the small amounts of nutrients that precipitated from the atmosphere with rain and snow were approximately equal to the amounts of nutrients that ran out of the valleys. These quantities were very low in relation to the total amount of nutrients in the system. There was a small net loss of calcium—about 0.3% of the total calcium in the system per year—and small net gains of nitrogen and potassium.

In 1965 and 1966, the investigators felled all the trees and shrubs in one of the six watersheds and then prevented regrowth by spraying the area with herbicides. The effects were dramatic. The amount of water running out of that valley increased by 40%. This indicated that water that previously would have been taken up by vegetation and ultimately evaporated into the atmosphere was now running off. For the four-month period from June to September 1966, the runoff was four times higher than it had been during comparable periods in the preceding years. The amounts of nutrients running out of the system also greatly increased; for example, the loss of calcium was 10 times higher than it had been previously. Phosphorus, on the other hand, did not increase in the stream water; it apparently was locked up in the soil.

The change in the status of nitrogen in the disturbed valley was especially striking (figure 55.7b). The undisturbed ecosystem in this valley had been accumulating nitrogen at a rate of about 2 kilograms per hectare per year, but the deforested ecosystem *lost* nitrogen at a rate of

(a)

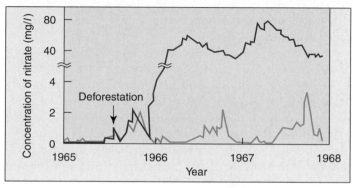

(b)

FIGURE 55.7
The Hubbard Brook experiment. (*a*) A 38-acre watershed was completely deforested, and the runoff monitored for several years. (*b*) Deforestation greatly increased the loss of minerals in runoff water from the ecosystem. The red curve represents nitrate in the runoff water from the deforested watershed; the blue curve, nitrate in runoff water from an undisturbed neighboring watershed.

about 120 kilograms per hectare per year. The nitrate level of the water rapidly increased to a level exceeding that judged safe for human consumption, and the stream that drained the area generated massive blooms of cyanobacteria and algae. In other words, the fertility of this logged-over valley decreased rapidly, while at the same time the danger of flooding greatly increased. This experiment is particularly instructive at the start of the twenty-first century, as large areas of tropical rain forest are being destroyed to make way for cropland, a topic that is discussed further in chapter 56.

When the trees and shrubs in one of the valleys in the Hubbard Brook watershed were cut down and the area was sprayed with herbicide, water runoff and the loss of nutrients from that valley increased. Nitrogen, which had been accumulating at a rate of about 2 kilograms per hectare per year, was lost at a rate of 120 kilograms per hectare per year.

55.2 Energy flows through ecosystems.

Trophic Levels

An ecosystem includes autotrophs and heterotrophs. **Autotrophs** manufacture their own food. Plants, algae, and some prokaryotes use light energy, but other organisms, such as chemoautotrophic prokaryotes that live in deep-water hydrothermal vents and derive energy by oxidizing sulfur compounds, transform inorganic molecules to manufacture their food. To support themselves, **heterotrophs,** which include animals, fungi, most protists and prokaryotes, and nongreen plants, must obtain organic molecules that have been synthesized by autotrophs. Autotrophs are also called **primary producers,** and heterotrophs are also called **consumers.**

Once energy enters an ecosystem, usually as the result of photosynthesis, it is passed, by ingestion or decomposition, from one organism to another in the form of chemical-bond energy. Much of this energy is converted to heat, which cannot be used to sustain life processes. The autotrophs that first acquire this energy provide all the energy heterotrophs use. The organisms that make up an ecosystem delay the release of the energy obtained from the sun back into space.

Green plants, the primary producers of a terrestrial ecosystem, generally capture about 1% of the energy that falls on their leaves, converting it to food energy. In especially productive systems, this percentage may be a little higher. When these plants are consumed by other organisms, only a portion of the plants' accumulated energy is actually converted into the bodies of the organisms that consume them, because most of it is dissipated as heat. The proportion of the ingested energy that is stored as chemical-bond energy varies among organisms. In an invertebrate, for example, about 10% of the food it eats is turned into its own body and thus into potential food for its predators. Although the comparable figure varies from approximately 5% in carnivores to nearly 20% in herbivores, 10% is a good average value for the amount of organic material that reaches the next trophic level.

Several different levels of consumers exist. **Primary consumers,** or herbivores, feed directly on green plants. **Secondary consumers,** carnivores and the parasites of animals, feed in turn on the herbivores. **Decomposers** break down the organic matter accumulated in the bodies of other organisms. Another more general term that includes decomposers is **detritivore.** Detritivores live on the refuse of an ecosystem. They include large scavengers, such as crabs, vultures, and jackals, as well as decomposers.

All of these categories occur in any ecosystem. They represent different **trophic levels,** from the Greek word *trophos,* which means "feeder." Organisms from each

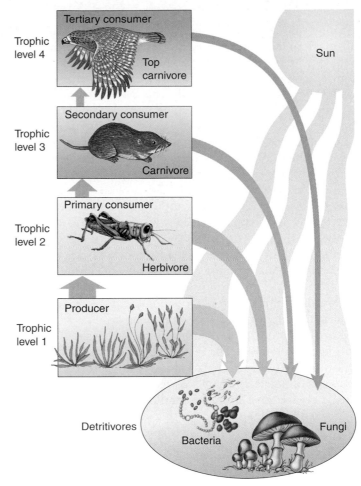

FIGURE 55.8
Trophic levels within a food chain. Plants obtain their energy directly from the sun, placing them at trophic level 1. Animals that eat plants, such as grasshoppers, are primary consumers, or herbivores, and are at trophic level 2. Animals that eat plant-eating animals, such as shrews, are carnivores and are at trophic level 3 (secondary consumers). Animals that eat carnivorous animals, such as hawks, are tertiary consumers at trophic level 4. Detritivores use all trophic levels for food.

trophic level, feeding on one another, make up a series called a **food chain** (figure 55.8). The length and complexity of food chains vary greatly. In real life, it is rather rare for a given organism to feed only on one other type of organism. Usually, each organism feeds on two or more kinds and in turn is eaten by several other kinds of organisms. When diagrammed, the relationship appears as a series of branching lines, rather than a straight line, and is called a **food web.**

Primary Productivity

Approximately 1% of the solar energy that falls on a plant is converted to the chemical bonds of organic material. **Primary production** and **primary productivity** are terms used to describe the amount of energy produced by photosynthetic organisms in a community. *Gross primary productivity* is the total organic matter produced, including that used by the photosynthetic organisms for respiration. *Net primary productivity* (NPP), therefore, is a measure of the amount of organic matter produced in a community in a given time that is available for heterotrophs. It equals the gross primary productivity minus the amount of energy expended by the metabolic activities of the photosynthetic organisms. The total mass of all of the organisms living in an ecosystem, called its **biomass,** increases as a result of its net productivity.

Productive Biological Communities. Some ecosystems have a high net primary productivity. For example, tropical forests and wetlands normally produce between 1500 and 3000 grams of organic material per square meter per year. By contrast, some corresponding figures for other communities are 1200 to 1300 grams for temperate forests, 900 grams for savanna, and 90 grams for deserts (table 55.1).

Secondary Productivity

The rate of biomass production by heterotrophs is called **secondary productivity.** Because herbivores and carnivores cannot carry out photosynthesis, they do not manufacture biomolecules directly from CO_2. Instead, they obtain them by eating plants or other heterotrophs. Secondary productivity by herbivores is approximately an order of magnitude less than the primary productivity upon which it is based. Where does all the energy in plants go? First, much of the biomass is not consumed by herbivores and instead supports the decomposer community (prokaryotes, fungi, and detritivorous animals). Second, some energy is not assimilated by the herbivore's body but is passed on as feces to the decomposers (figure 55.9). Third, not all the chemical-bond energy that herbivores assimilate is retained as chemical-bond energy in the organic molecules of their tissues. Much of it is lost as heat produced by work.

Energy passes through ecosystems, a good deal being converted to heat at each step. Primary productivity occurs as a result of photosynthesis, which is carried out by green plants, algae, and some prokaryotes. Secondary productivity is the production of new biomass by heterotrophs.

Ecosystem Type	Net Primary Productivity (NPP)	
	NPP per Unit Area	World NPP
	(g/m^2)	$(10^{12}$ kg$)$
Algal beds and reefs	2500	1.6
Tropical rain forest	2200	37.4
Wetlands	2000	4.0
Tropical seasonal forest	1600	12.0
Estuaries	1500	2.1
Temperate evergreen forest	1300	6.5
Temperate deciduous forest	1200	8.4
Savanna	900	13.5
Boreal forest	800	9.6
Woodland and shrubland	700	6.0
Cultivated land	650	9.1
Temperate grassland	600	5.4
Continental shelf	360	9.6
Lake and stream	250	0.5
Tundra and alpine	140	1.1
Open ocean	125	41.5
Desert and semidesert shrub	90	1.6
Extreme desert, rock, sand, and ice	3	0.07

Table 55.1 Ecosystem Productivity Per Year

Source: Data in: Begon, Harper, and Townsend, *Ecology* 3/e, Blackwell Science 1996, page 715. Original source: Whittaker, R. H. *Communities and Ecosystems*, 2/e Macmillan, London, 1975.

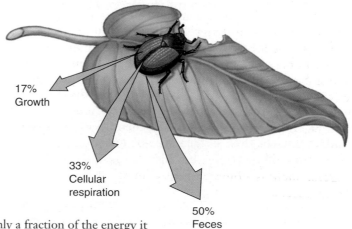

FIGURE 55.9
How heterotrophs utilize food energy. A heterotroph assimilates only a fraction of the energy it consumes. For example, if the bite of an herbivorous insect comprises 500 Joules of energy (1 Joule = 0.239 calories), about 50%, 250 J, is lost in feces, about 33%, 165 J, is used to fuel cellular respiration, and about 17%, 85 J, is converted into insect biomass. Only this 85 J is available to the next trophic level.

17%
Growth

33%
Cellular respiration

50%
Feces

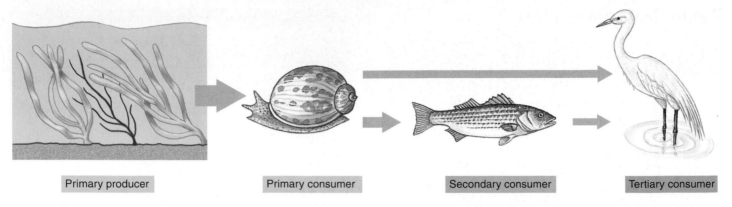

| Primary producer | Primary consumer | Secondary consumer | Tertiary consumer |

FIGURE 55.10
A food chain. Because so much energy is converted to heat at each step, food chains usually consist of just three or four steps.

Energy in Food Chains

Food chains generally consist of only three or four steps (figure 55.10). So much energy is converted to heat at each step that very little usable energy remains in the system after it has been incorporated into the bodies of organisms at four successive trophic levels.

Community Energy Budgets

Lamont Cole of Cornell University studied the flow of energy in a freshwater ecosystem in Cayuga Lake in upstate New York. He calculated that about 150 of each 1000 calories of potential energy fixed by algae and cyanobacteria are transferred into the bodies of small heterotrophs (figure 55.11). Of these, about 30 calories are incorporated into the bodies of smelt, small fish that are the principal secondary consumers of the system. If humans eat the smelt, they gain about 6 of the 1000 calories that originally entered the system. If trout eat the smelt and humans eat the trout, humans gain only about 1.2 calories.

Factors Limiting Community Productivity

Communities with higher productivity can in theory support longer food chains. The limit on a community's productivity is determined ultimately by the amount of sunlight it receives, for this determines how much photosynthesis can occur. This is why in the deciduous forests of North America the net primary productivity increases as the growing season lengthens. NPP is higher in warm climates than cold ones not only because of the longer growing seasons, but also because more nitrogen tends to be available in warm climates, where nitrogen-fixing bacteria are more active.

> Considerable energy is converted to heat at each stage in food chains, which limits their length. In general, more productive communities can support longer food chains.

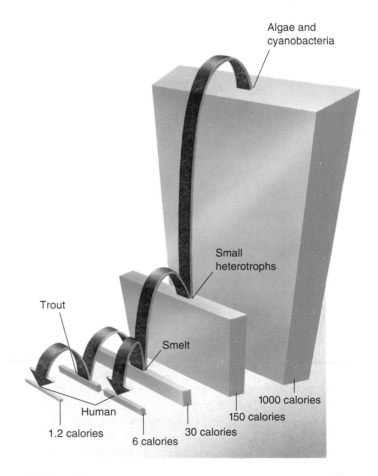

FIGURE 55.11
The food chain in Cayuga Lake. Autotrophic plankton (algae and cyanobacteria) fix the energy of the sun, heterotrophic plankton feed on them, and both are consumed by smelt. The smelt are eaten by trout, with about a fivefold loss in fixed energy. For humans, the amount of smelt biomass is at least five times greater than that available in trout, but humans prefer to eat trout.
Why does it take so many calories of algae to support so few calories of humans?

Ecological Pyramids

As previously mentioned, a plant fixes about 1% of the sun's energy that falls on its green parts. The successive members of a food chain, in turn, process into their own bodies about 10% of the energy available in the organisms on which they feed. For this reason, there are generally far more individuals at the lower trophic levels of any ecosystem than at the higher levels. Similarly, the biomass of the primary producers present in a given ecosystem is greater than the biomass of the primary consumers, with successive trophic levels having less biomass and correspondingly less potential energy.

These relationships, if shown diagrammatically, appear as pyramids (figure 55.12). We can speak of pyramids of biomass, pyramids of energy, pyramids of number, and so forth, as characteristic of ecosystems.

Inverted Pyramids

Some aquatic ecosystems have inverted biomass pyramids. For example, in a planktonic ecosystem—dominated by small organisms floating in water—the turnover of photosynthetic phytoplankton at the lowest level is very rapid, with zooplankton consuming phytoplankton so quickly that the phytoplankton (the producers at the base of the food chain) can never develop a large population size. Because the phytoplankton reproduce very rapidly, the community can support a population of heterotrophs that is larger in biomass and more numerous than the phytoplankton (see figure 55.12b). Although in such cases the pyramid of biomass is inverted, it is important to remember that the pyramid of energy cannot be inverted. Because so little energy in one trophic level can be converted into body mass at the next, higher trophic level, the distribution of energy production in trophic levels will always be pyramidal.

Top Carnivores

The loss of energy that occurs at each trophic level places a limit on how many top-level carnivores a community can support. As we have seen, only about one thousandth of the energy captured by photosynthesis passes all the way through a three-stage food chain to a tertiary consumer. This explains why no predators subsist on lions or eagles— the biomass of these animals is simply insufficient to support another trophic level.

In the pyramid of numbers, top-level predators tend to be fairly large animals. Thus, the small residual biomass available at the top of the pyramid is concentrated in a relatively small number of individuals.

Because energy is converted to heat at every step of a food chain, the biomass of primary producers (photosynthesizers) tends to be greater than that of the herbivores that consume them, and herbivore biomass tends to be greater than the biomass of the predators that consume them.

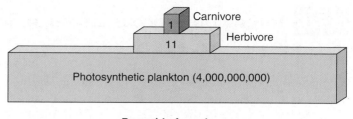

Pyramid of numbers

(a)

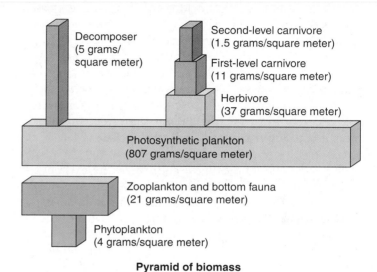

Pyramid of biomass

(b)

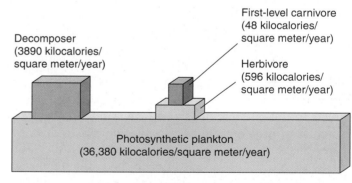

Pyramid of energy

(c)

FIGURE 55.12
Ecological pyramids. Ecological pyramids measure different characteristics of each trophic level. (*a*) Pyramid of numbers. (*b*) Pyramids of biomass, both normal (*top*) and inverted (*bottom*). (*c*) Pyramid of energy.
How can the existence of inverted pyramids of biomass be explained?

Trophic Cascades

The existence of food webs creates the possibility of interactions among species at different trophic levels. Predators not only have effects on the species upon which they prey, but also, indirectly, on the plants eaten by these prey. Conversely, increases in primary productivity not only provide more food for herbivores but, indirectly, lead also to more food for carnivores.

When we look at the world around us, we see a profusion of plant life. Why is this? Why don't herbivore populations increase to the extent that all available vegetation is consumed? The answer, of course, is that predators keep the herbivore populations in check, thus allowing plant populations to thrive. This phenomenon, in which the effect of one trophic level flows up or down to other levels, is called a **trophic cascade**. When the effect flows from predators to lower trophic levels, the cascade is referred to as *top-down*.

Experimental studies have confirmed the existence of trophic top-down cascades. For example, in one study in New Zealand, sections of a stream were isolated with a mesh that prevented fish from entering. In some of the enclosures, brown trout were added, whereas other enclosures were left without large fish. After 10 days, the number of invertebrates in the trout enclosures was one-half the number in the controls (figure 55.13). In turn, the biomass of algae, which invertebrates feed upon, was five times greater in the trout enclosures than in the controls.

The logic of trophic cascades leads to the prediction that a fourth trophic level, carnivores that prey on other carnivores, would also lead to cascading effects. In this case, the top predators would keep lower-level predator populations in check, which should lead to a profusion of herbivores and a paucity of vegetation. In an experiment similar to the one just described, enclosures were created in free-flowing streams in northern California. In this case, large predatory fish were added to some enclosures and not to others. In the large fish enclosures, the number of smaller predators, such as damselfly nymphs, was greatly reduced, leading to an increase in their prey, including algae-eating insects, which led, in turn, to decreases in the biomass of algae (figure 55.14).

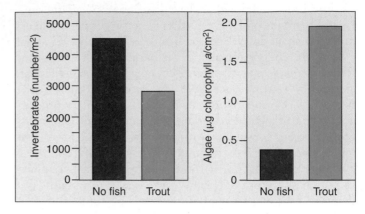

FIGURE 55.13
Trophic cascades. Streams with trout have fewer herbivorous invertebrates and more algae than streams without trout.
Why do streams with trout have more algae?

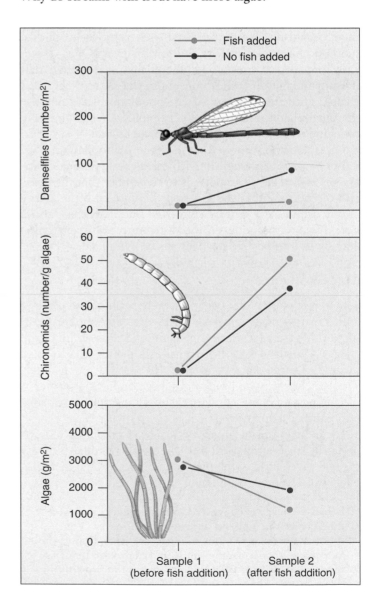

FIGURE 55.14
Four-level trophic cascades. Streams with large, carnivorous fish have fewer lower-level predators, such as damselflies, more herbivorous insects (exemplified by the number of chironomids, a type of aquatic insect), and lower levels of algae.
What might be the effect if snakes that prey on fish were added to the enclosures?

Human Effects on Trophic Cascades

Humans have inadvertently created a test of the trophic cascade hypothesis by removing top predators from ecosystems. The great naturalist Aldo Leopold captured the results long before the trophic cascade hypothesis had ever been scientifically articulated when he wrote in the *Sand County Almanac:*

"I have lived to see state after state extirpate its wolves. I have watched the face of many a new wolfless mountain, and seen the south-facing slopes wrinkle with a maze of new deer trails. I have seen every edible bush and seedling browsed, first to anemic desuetude, and then to death. I have seen every edible tree defoliated to the height of a saddle horn."

Many similar examples exist in which the removal of predators has led to cascading effects on lower trophic levels. On Barro Colorado Island, a hilltop turned into an island by the construction of the Panama Canal at the beginning of the last century, large predators such as jaguars and mountain lions are absent. As a result, smaller predators whose populations are normally held in check—including monkeys, peccaries (a relative of the pig), coatimundis and armadillos—have become extraordinarily abundant. These animals eat almost anything they find. Ground-nesting birds are particularly vulnerable, and many species have declined; at least 15 bird species have vanished from the island entirely. Similarly, in woodlots in the midwestern United States, raccoons, opossums, and foxes have become abundant due to the elimination of larger predators, and populations of ground-nesting birds have declined greatly.

Bottom-Up Effects

In a manner opposite to top-down effects, factors acting at the bottom of food webs may have consequences that ramify to higher trophic levels, leading to what are termed *bottom-up* effects. The basic idea is that when the productivity of an ecosystem is low, herbivore populations will be too small to support any predators. Increases in productivity will be entirely devoured by the herbivores, whose populations will increase in size. At some point, herbivore populations will become large enough that predators can be supported. Thus, further increases in productivity will not lead to increases in herbivore populations, but rather, to increases in predator populations. Again, at some level, top predators will become established that can prey on lower-level predators. With the lower-level predator populations in check, herbivore populations will again increase as productivity increases (figure 55.15).

Experimental evidence for the role of bottom-up effects was provided in an elegant study conducted on the Eel River in northern California. Enclosures were constructed that excluded large fish. A roof was placed above each enclosure. Some roofs were clear and let light pass

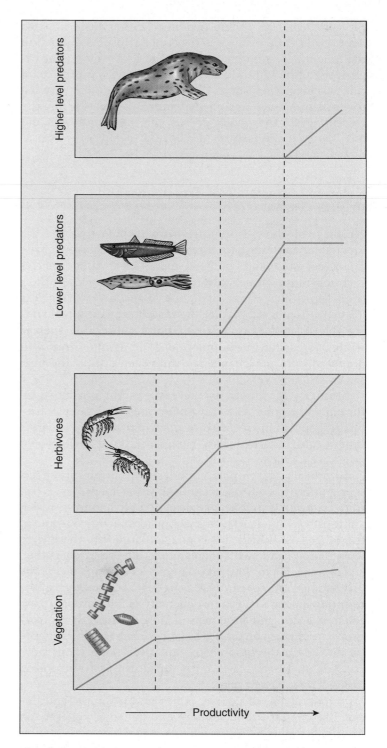

FIGURE 55.15
Bottom-up effects. At low levels of productivity, herbivore populations cannot be maintained. Above some threshold, increases in productivity lead to increases in herbivore biomass; vegetation biomass no longer increases with productivity because it is converted into herbivore biomass. Similarly, above another threshold, herbivore biomass gets converted to carnivore biomass. At this point, vegetation biomass is no longer constrained by herbivores, and so again increases with increasing productivity.
Why are there portions of the curves where vegetation biomass does not increase as productivity increases?

through, whereas others produced either little light or deep shade. The result was that the enclosures differed in the amount of sunlight reaching them. As one might expect, the primary productivity differed and was greatest in the unshaded enclosures. This increased productivity led to both more vegetation and more predators, but the trophic level sandwiched in between, the herbivores, did not increase, precisely as the bottom-up hypothesis predicted (figure 55.16).

Relative Importance of Top-Down and Bottom-Up Effects

Neither top-down nor bottom-up trophic cascades are inevitable. For example, if two species of herbivores exist in an ecosystem and compete strongly, and if one species is much more vulnerable to predation than the other, then top-down effects will not propagate to the next lower trophic level. Rather, increased predation will simply decrease the population of the vulnerable species while increasing the population of its competitor, with potentially no net change on the vegetation in the next lower trophic level.

Similarly, productivity increases might not move up through all trophic levels. In some cases, for example, herbivore populations increase so quickly that their predators cannot control them. Thus, increases in productivity would not move up the food chain.

In other cases, trophic cascades and bottom-up effects may reinforce each other. In one experiment, large fish were removed from one lake, leaving only minnows, which ate most of the algae-eating zooplankton. By contrast, in the other lake, which still contained large fish, there were few minnows and much zooplankton. The researchers then added nutrients to both lakes. In the lake from which large fish had been removed, there were few zooplankton, so the resulting increase in algal productivity did not propagate up the food chain, and large mats of algae formed. By comparison, in the lake containing large fish, increased productivity moved up the food chain, and algae populations were controlled. In this case, both top-down and bottom-up processes were operating.

A food chain, of course, is not always simple and linear. In some cases, species may simultaneously operate on multiple trophic levels, as when a jaguar eats both smaller carnivores and herbivores, or a bear eats both fish and berries. Ecologists are currently working to apply theories of food chain interactions to these more complicated situations.

Because of the linked nature of food webs, species at different trophic levels affect each other, and these effects can promulgate both up and down the food web.

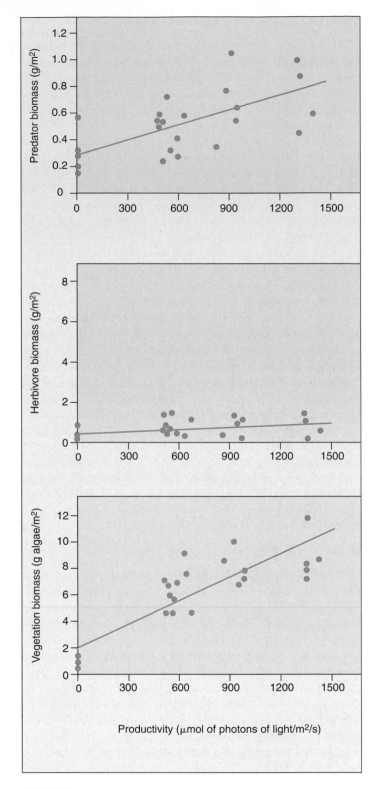

FIGURE 55.16
Bottom-up effects on a stream ecosystem. Increases in the amount of light hitting the stream lead to increases in the amount of vegetation. However, herbivore biomass does not increase with increased productivity because it is converted into predator biomass. **Why is the amount of light an important determinant of predator biomass?**

55.4 Biodiversity promotes ecosystem stability.

Effects of Species Richness

Ecologists have long debated the consequences of differences in species richness among communities. One theory is that species-rich communities are more stable—that is, more constant in composition and better able to resist disturbance. This hypothesis has been elegantly studied by David Tilman and colleagues at the University of Minnesota's Cedar Creek Natural History Area. These workers monitored 207 small rectangular plots of land (8 to 16 m²) for 11 years (figure 55.17a). In each plot, they counted the number of prairie plant species and measured the total amount of plant biomass (that is, the mass of all plants on the plot). Over the course of the study, plant species richness was related to community stability—plots with more species showed less year-to-year variation in biomass. Moreover, in two drought years, the decline in biomass was negatively related to species richness; in other words, plots with more species were less affected. In a related experiment, when seeds of other plant species were added to different plots, the ability of these species to become established was negatively related to species richness (figure 55.17b). More-diverse communities, in other words, are more resistant to invasion by new species, another measure of community stability.

Species richness may also affect other ecosystem processes. Tilman and colleagues monitored 147 experimental plots that varied in number of species to estimate how much growth was occurring and how much nitrogen the growing plants were taking up from the soil. Tilman found that the more species a plot had, the greater the nitrogen uptake and total amount of biomass produced. In his study, increased biodiversity clearly leads to greater productivity.

Laboratory studies on artificial ecosystems have provided similar results. In one elaborate study, ecosystems covering 1 square meter were constructed in growth chambers that controlled temperature, light levels, air currents, and atmospheric gas concentrations. A variety of plants, insects, and other animals were introduced to construct ecosystems composed of 9, 15, or 31 species, with the lower diversity treatments containing a subset of the species in the higher diversity enclosures. As with Tilman's experiments, the amount of biomass produced was related to species richness, as was the amount of carbon dioxide consumed, another measure of the productivity of the ecosystem.

Tilman's conclusion that healthy ecosystems depend on diversity is not accepted by all ecologists. Critics question the validity and relevance of these biodiversity studies, arguing that the more species are added to a plot, the greater the probability that one will be highly productive. To show that high productivity results from high species richness

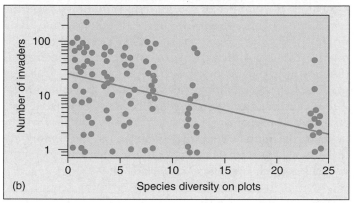

(b) Species diversity on plots

FIGURE 55.17
Effect of species richness on ecosystem stability and productivity. (a) One of the Cedar Creek experimental plots. (b) Community stability can be assessed by looking at the effect of species richness on community invasibility. Each dot represents data from one experimental plot in the Cedar Creek experimental fields. Plots with more species are harder to invade by non-native species, which occur in surrounding agricultural fields.
How could you devise an experiment on invasibility that didn't rely on species from surrounding areas?

per se, rather than from the presence of particular highly productive species, experimental plots would have to exhibit "overyielding"; in other words, plot productivity would have to be greater than that of the single most productive species grown in isolation. Although this point is still debated, recent work at Cedar Creek and elsewhere has provided evidence of overyielding, supporting the claim that species richness of communities enhances community productivity and stability.

Experimental field studies support the hypothesis that species-rich communities are more stable and productive.

Causes of Species Richness

While ecologists still argue about why some ecosystems are more stable than others—that is, better able to avoid permanent change and return to normal after disturbances such as land clearing, fire, invasion by insects, or severe storm damage—most ecologists now accept as a working hypothesis that biologically diverse ecosystems are generally more stable than simple ones. Ecosystems with many different kinds of organisms support a more complex web of interactions, and an alternative species is thus more likely to exist to compensate for the effect of a disruption.

Factors Promoting Species Richness

How does the number of species in a community affect the functioning of the ecosystem? How does ecosystem functioning affect the number of species in a community? It is often extremely difficult to sort out the relative contributions of different factors. Of the many variables that may play a role in determining species richness in a community, we will discuss three: ecosystem productivity, spatial heterogeneity, and climate. Other factors that may play an important role include the evolutionary age of the community (examined later in this chapter) and the degree to which the community has been disturbed (see chapter 54).

Ecosystem Productivity. Ecosystems differ in productivity, which is a measure of how much new growth they can produce. Surprisingly, the relationship between productivity and species richness is not linear. Rather, ecosystems with intermediate levels of productivity tend to have the most species (figure 55.18a). Why this is so is a topic of considerable current debate. One possibility is that levels of produc-

tivity are linked to numbers of predators. At low productivity, there are few predators, and superior competitors eliminate most species, whereas at high productivity, there are so many predators that only the most predation-resistant species survive. At intermediate levels, however, predators may act as keystone species, maintaining species richness (see chapter 54).

Spatial Heterogeneity. Environments that contain more soil types, topographies, and other habitat variations can be expected to accommodate more species because they provide a greater variety of habitats, climates, places to hide from predators, and so on. In general, the species richness of animals tends to reflect the species richness of the plants in their community—that is, the more plant species there are, the greater the variety of microhabitats. In turn, plant species richness reflects the spatial heterogeneity of the ecosystem. Thus, the number of lizard species in the American Southwest mirrors the structural diversity of the plants there (figure 55.18b).

Climate. The role of climate is more difficult to assess. On one hand, more species might be expected to coexist in a seasonal environment than in a constant one, because a changing climate may favor different species at different times of the year. On the other hand, stable environments are able to support specialized species that would be unable to survive where conditions fluctuate. Although other explanations are possible, data for mammal species along the west coast of North America supports the latter prediction (figure 55.18c).

Species richness promotes ecosystem productivity and is fostered by spatial heterogeneity and stable climate.

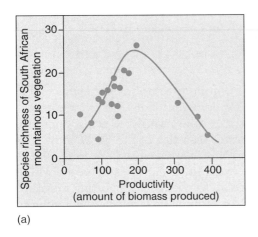

(a)

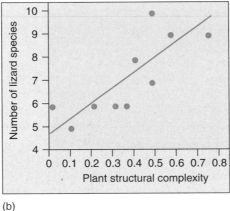

(b)

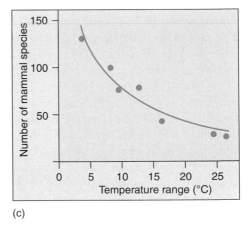

(c)

FIGURE 55.18
Factors that affect species richness. (*a*) *Productivity:* In fynbos plant communities of mountainous areas of South Africa, species richness of plants peaks at intermediate levels of productivity (biomass). (*b*) *Spatial heterogeneity:* The species richness of desert lizards is positively correlated with the structural complexity of the plant cover in desert sites in the American Southwest. (*c*) *Climate:* The species richness of mammals is inversely correlated with monthly mean temperature range along the west coast of North America.
(*a*) **Why is species richness greatest at intermediate levels of productivity?** (*b*) **Why do more structurally complex areas have more species?** (*c*) **Why do areas with less variation in temperature have more species?**

Biogeographic Patterns of Species Diversity

Since before Darwin, biologists have recognized that more different kinds of animals and plants inhabit the tropics than the temperate regions. For many species, there is a steady increase in species richness from the arctic to the tropics. Called a **species diversity cline,** such a biogeographic gradient in numbers of species correlated with latitude has been reported for plants and animals, including birds (figure 55.19), mammals, and reptiles.

Why Are There More Species in the Tropics?

For the better part of a century, ecologists have puzzled over the cline in species diversity from the arctic to the tropics. The difficulty has not been in forming a reasonable hypothesis of why there are more species in the tropics, but rather in sorting through the many reasonable hypotheses that suggest themselves. Here we will consider five of the most commonly discussed suggestions:

Evolutionary Age. It has often been proposed that the tropics have more species than temperate regions because the tropics have existed over long, uninterrupted periods of evolutionary time, while temperate regions have been subject to repeated glaciations. The greater age of tropical communities would have allowed complex population interactions to coevolve within them, fostering a greater variety of plants and animals in the tropics.

However, recent work suggests that the long-term stability of tropical communities has been greatly exaggerated. An examination of pollen within undisturbed soil cores reveals that during glaciations the tropical forests contracted to a few small refuges surrounded by grassland. This suggests that the tropics have not had a continuous record of species richness over long periods of evolutionary time.

Higher Productivity. A second often-advanced hypothesis is that the tropics contain more species because this part of the earth receives more solar radiation than temperate regions do. The argument is that more solar energy, coupled to a year-round growing season, greatly increases the overall photosynthetic activity of plants in the tropics. If we visualize the tropical forest as a pie (total resources) being cut into slices (species niches), we can see that a larger pie accommodates more slices. However, as noted previously, many field studies have indicated that species richness is highest at intermediate levels of productivity. Accordingly, increasing productivity would be expected to lead to lower, not higher, species richness. Perhaps the long column of vegetation down through which light passes in a tropical forest produces a wide range of light frequencies and intensities, creating a greater variety of light environments and so promoting species diversity.

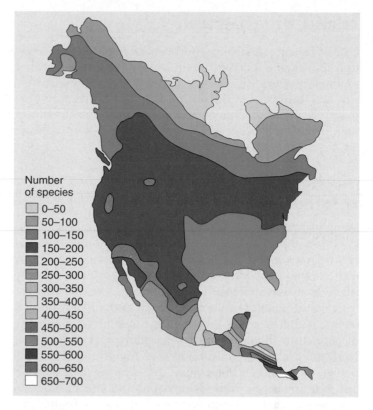

Number of species
- 0–50
- 50–100
- 100–150
- 150–200
- 200–250
- 250–300
- 300–350
- 350–400
- 400–450
- 450–500
- 500–550
- 550–600
- 600–650
- 650–700

FIGURE 55.19
A latitudinal cline in species richness. Among North and Central American birds, a marked increase in the number of species occurs moving toward the tropics. Fewer than 100 species are found at arctic latitudes, while more than 600 species live in southern Central America.

Predictability. Seasonal variation, though it does exist in the tropics, is generally substantially less than in temperate areas. This reduced seasonality might encourage specialization, with niches subdivided to partition resources and so avoid competition. The expected result would be a larger number of more specialized species in the tropics, which is what we see. Many field tests of this hypothesis have been carried out, and almost all support it, reporting larger numbers of narrower niches in tropical communities than in temperate areas.

Predation. Many reports indicate that predation may be more intense in the tropics. In theory, more intense predation could reduce the importance of competition, permitting greater niche overlap and thus promoting greater species richness.

Spatial Heterogeneity. As noted earlier, spatial heterogeneity promotes species richness. Tropical forests, by virtue of their complexity, create a variety of microhabitats and so may foster larger numbers of species.

No one really knows why more species are present in the tropics, but there are plenty of hypotheses.

Island Biogeography

One of the most reliable patterns in ecology is the observation that larger islands contain more species than smaller islands. In 1967, Robert MacArthur of Princeton University and Edward O. Wilson of Harvard University proposed that this **species-area relationship** was a result of the effect of area and isolation on the likelihood of species extinction and colonization.

The Equilibrium Model

MacArthur and Wilson reasoned that species are constantly being dispersed to islands, so islands have a tendency to accumulate more and more species. At the same time that new species are added, however, other species are lost by extinction. As the number of species on an initially empty island increases, the rate of colonization must decrease as the pool of potential colonizing species not already present on the island becomes depleted. At the same time, the rate of extinction should increase—the more species on an island, the greater the likelihood that any given species will perish. As a result, at some point, the number of extinctions and colonizations should be equal and the number of species should then remain constant. Every island of a given size, then, has a characteristic equilibrium number of species that tends to persist through time (the intersection point in figure 55.20*a*), although the species composition will change as some species become extinct and new species colonize.

MacArthur and Wilson's equilibrium model proposes that island species richness is a dynamic equilibrium between colonization and extinction. Both island size and distance from the mainland would play important roles.

We would expect smaller islands to have higher rates of extinction because their population sizes would, on average, be smaller. Also, we would expect fewer colonizers to reach islands that lie farther from the mainland. Thus, small islands far from the mainland would have the fewest species; large islands near the mainland would have the most (figure 55.20*b*).

The predictions of this simple model bear out well in field data. Asian Pacific bird species (figure 55.20*c*) exhibit a positive correlation of species richness with island size, but a negative correlation of species richness with distance from the source of colonists.

Testing the Equilibrium Model

Field studies in which small islands have been censused, cleared, and allowed to recolonize tend to support the equilibrium model. However, long-term experimental field studies are suggesting that the situation is more complicated than MacArthur and Wilson envisioned. Their model predicts a high level of **species turnover** as some species perish and others arrive. However, studies of island birds and spiders indicate that very little turnover occurs from year to year. Moreover, those species that do come and go comprise a subset of species that never attain high populations. A substantial proportion of the species appears to maintain high populations and rarely go extinct. These studies, of course, have only been going on for a relatively short period of time. It is possible that over periods of centuries, the equilibrium model is a good description of what determines island species richness.

Species richness on islands is a dynamic equilibrium between colonization and extinction.

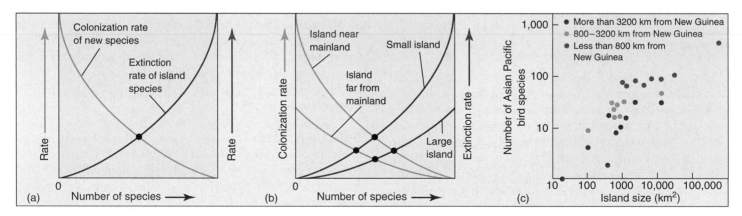

FIGURE 55.20

The equilibrium model of island biogeography. (*a*) Island species richness reaches an equilibrium (*black dot*) when the colonization rate of new species equals the extinction rate of species on the island. (*b*) The equilibrium shifts depending on the rate of colonization, the size of an island, and its distance to sources of colonists. Species richness is positively correlated with island size and inversely correlated with distance from the mainland. Smaller islands have higher extinction rates, shifting the equilibrium point to the left. Similarly, more distant islands have lower colonization rates, again shifting the equilibrium point leftward. (*c*) The effect of distance from a larger island, which can be the source of colonizing species, is readily apparent. More distant islands have fewer species compared to nearer islands of the same size. **Do islands with the same number of species necessarily have the same extinction rate?**

55.1 Chemicals cycle within ecosystems.

- An ecosystem includes all of the organisms living in a particular place, plus the abiotic environment in which they live. (p. 1184)

The Water Cycle

- Water continuously cycles between the bodies of water, land, organisms, and atmosphere. (p. 1184)
- More than 96% of all fresh water in the United States is groundwater. (p. 1185)
- An estimated 2% of the groundwater in the Untied States is already polluted. (p. 1185)
- Cutting down forests disrupts the water cycle as lower amounts of moisture are returned to the atmosphere. (p. 1185)

The Carbon Cycle

- About 10% of the carbon dioxide in the atmosphere is fixed annually by photosynthesis. (p. 1186)
- When the bodies of dead organisms decompose, microorganisms release carbon dioxide into the atmosphere, freeing it for incorporation into other organisms. (p. 1186)
- Large amounts of carbon dioxide are also dissolved in the world's oceans. (p. 1186)

The Nitrogen Cycle

- Certain bacteria and fungi can fix atmospheric nitrogen and decompose nitrogen-containing compounds into forms useful in biological processes. (p. 1187)
- Some nitrogen-fixing bacteria form symbiotic relationships with the roots of legumes and other plants. (p. 1187)

The Phosphorus Cycle

- Phosphates are relatively insoluble and are present only in certain types of rocks. But, other than nitrogen, phosphorus is the required nutrient most likely to be scarce enough to limit plant growth. (p. 1188)

Biogeochemical Cycles Illustrated: Recycling in a Forest Ecosystem

- Deforestation in the Hubbard Brook Experimental Forest led to increased loss of minerals in runoff water from the ecosystem. (p. 1189)

55.2 Energy flows through ecosystems.

Trophic Levels

- Autotrophs (producers) manufacture their own food, while heterotrophs (consumers) must obtain organic molecules that have been synthesized by autotrophs. (p. 1190)
- Once energy enters an ecosystem, it is passed from one organism to another in the form of chemical bonds, and much of the energy is converted to heat. (p. 1190)
- Overall, an average of 10% of organic matter is transferred from one trophic level to the next. (p. 1190)
- Organisms from different trophic levels (feeding levels) make up a food chain, and the length and complexity of food chains vary greatly. Most often, trophic levels exist and are discussed in terms of food webs, or interactions of multiple food chains. (p. 1190)
- Primary productivity describes the amount of organic matter produced from solar energy in a given area during a given period of time, and is usually divided into gross primary productivity and net primary productivity. (p. 1191)
- Secondary productivity refers to the rate of biomass production by heterotrophs. (p. 1191)

Energy in Food Chains

- Food chains are limited in the number of trophic levels that can be supported because so much energy is lost at each step that very little energy usually remains after three or four steps. (p. 1192)
- In theory, communities with higher productivity can support longer food chains. (p. 1192)

Ecological Pyramids

- Due to the loss of useful energy at each trophic level, higher trophic levels generally have fewer individuals than lower levels. (p. 1193)
- Some aquatic ecosystems have inverted biomass pyramids due to extremely rapid turnover of biomass. (p. 1193)
- The loss of energy also places a limit on the number of top-level carnivores that can be supported in a community. (p. 1193)

55.3 Interactions occur among different trophic levels.

Trophic Cascades

- Top-down effects are said to occur when the effect of one trophic level flows down to the next level. (p. 1194)
- Humans have inadvertently tested the trophic cascade hypothesis by removing top predators from the ecosystem. (p. 1195)
- Bottom-up effects are said to occur when factors at the bottom of food webs cause ramifications at higher trophic levels. (p. 1195)
- Neither top-down nor bottom-up effects are inevitable, for many reasons; for example, in some cases, species may simultaneously operate on multiple trophic levels. (p. 1196)

55.4 Biodiversity promotes ecosystem stability.

Effects of Species Richness

- One theory is that species-rich communities are more stable, more constant in composition, and thus better able to resist disturbance. (p. 1197)

Causes of Species Richness

- Three of the main variables determining species richness are ecosystem productivity, spatial heterogeneity, and climate. (p. 1198)

Biogeographic Patterns of Species Diversity

- Five of the most common hypotheses for the cline of low species diversity in the arctic to high species diversity in the tropics are evolutionary age, higher productivity, predictability, predation, and spatial heterogeneity. (p. 1199)

Island Biogeography

- The equilibrium theory of island biogeography proposes that island species richness is a dynamic equilibrium between colonization and extinction. (p. 1200)

Self Test

1. Over land, most of the water in the atmosphere results from _____, whereas over the oceans, most of the water in the atmosphere results from _____.
   a. evaporation/transpiration
   b. evaporation/evaporation
   c. transpiration/evaporation
   d. transpiration/transpiration
2. Which of the statements about groundwater is *not* accurate?
   a. In the U.S., groundwater provides 50% of the population with drinking water.
   b. Groundwaters are being depleted faster than they can be recharged.
   c. Groundwaters are becoming increasingly polluted.
   d. Removal of pollutants from groundwaters is easily achieved.
3. The largest store of carbon molecules on earth is in
   a. the atmosphere.
   b. fossil fuels.
   c. marine sediments.
   d. living organisms.
4. Some bacteria have the ability to "fix" nitrogen. This means
   a. they convert ammonia into nitrites and nitrates.
   b. they convert atmospheric nitrogen gas into biologically useful forms of nitrogen.
   c. they break down nitrogen-rich compounds and release ammonium ions.
   d. they convert nitrate into nitrogen gas.
5. The phosphorus cycle differs from the water, carbon, and nitrogen cycles in that
   a. the reservoir for phosphorus exists in mineral form in rocks rather than in the atmosphere.
   b. phosphorus is far more abundant than water, carbon, or nitrogen.
   c. phosphorus is less important to biological systems than water, carbon, or nitrogen.
   d. phosphorus, once used by an organism, does not cycle back to the environment.
6. In the Hubbard Brook experiments, which of the following situations did *not* occur?
   a. The undisturbed forests were effective at retaining nutrients.
   b. The deforested areas gained nitrogen.
   c. The deforested areas lost more water to runoff than the undisturbed forests.
   d. The deforested areas lost more minerals to runoff than the undisturbed forests.
7. What percentage of the energy that hits a plant's leaves is converted into chemical (food) energy?
   a. 1%
   b. 5%
   c. 10%
   d. 20%
8. According to the trophic cascade hypothesis, the removal of carnivores from an ecosystem may result in
   a. a decline in the number of herbivores and a decline in the amount of vegetation.
   b. a decline in the number of herbivores and an increase in the amount of vegetation.
   c. an increase in the number of herbivores and an increase in the amount of vegetation.
   d. an increase in the number of herbivores and a decrease in the amount of vegetation.
9. Experimental evidence from the Cedar Creek Natural History Area supports the hypothesis that
   a. species richness is related to community stability.
   b. human activities are upsetting the trophic cascade.
   c. bottom-up effects can regulate the number of top carnivores in a system.
   d. some aquatic ecosystems have inverted biomass pyramids.
10. The equilibrium model of island biogeography suggests all of the following *except*
    a. larger islands have more species than smaller islands.
    b. the species richness of an island is determined by colonization and extinction.
    c. smaller islands have lower rates of extinction.
    d. islands closer to the mainland will have higher colonization rates.

Test Your Visual Understanding

Table 55.1 Ecosystem Productivity Per Year		
	Net Primary Productivity (NPP)	
Ecosystem Type	NPP per Unit Area (g/m^2)	World NPP $(10^{12} kg)$
Algal beds and reefs	2500	1.6
Tropical rain forest	2200	37.4
Wetlands	2000	4.0
Tropical seasonal forest	1600	12.0
Estuaries	1500	2.1
Temperate evergreen forest	1300	6.5
Temperate deciduous forest	1200	8.4
Savanna	900	13.5
Boreal forest	800	9.6
Woodland and shrubland	700	6.0
Cultivated land	650	9.1
Temperate grassland	600	5.4
Continental shelf	360	9.6
Lake and stream	250	0.5
Tundra and alpine	140	1.1
Open ocean	125	41.5
Desert and semidesert shrub	90	1.6
Extreme desert, rock, sand, and ice	3	0.07

1. Using this table relate productivity to species richness.

Apply Your Knowledge

1. Nitrogen is a limiting nutrient to plant growth in most ecosystems. To increase plant production, fertilizers that contain nitrogen (and phosphorus) are used. What effects might this influx of nitrogen have on ecosystems?
2. Many U.S. communities struggle with issues of deer overpopulation. Explain how human activities have created this situation.

56

The Biosphere

Concept Outline

56.1 Climate shapes the character of ecosystems.

Effects of the Sun and Atmospheric Circulation. The sun powers major movements in atmospheric circulation.

Atmospheric Circulation, Precipitation, and Climate. Latitude and elevation have important effects on climate, although other factors affect regional climate.

56.2 Biomes are widespread terrestrial ecosystems.

The Major Biomes. Characteristic communities called biomes occur in different climatic regions. Variations in temperature and precipitation are good predictors of which biomes will occur where.

56.3 Aquatic ecosystems cover much of the earth.

Patterns of Circulation in the Oceans. The world's oceans circulate in huge circles deflected by the continents.

Marine Ecosystems. The communities of the ocean are delineated primarily by depth.

Freshwater Habitats. Like miniature oceans, ponds and lakes support different communities at different depths. Freshwater ecosystems are often highly productive.

56.4 Human activity is placing the biosphere under increasing stress.

Pollution. Human industrial and agricultural activity introduces significant levels of many harmful chemicals into ecosystems.

Acid Precipitation. Burning of cheap, high-sulfur coal has introduced sulfur to the upper atmosphere, where it combines with water to form sulfuric acid that falls back to earth, harming ecosystems.

Destruction of the Tropical Forests. Many of the world's tropical forests are being destroyed by human activity.

The Ozone Hole. Industrial chemicals called CFCs are destroying the atmosphere's ozone layer, removing an essential shield from the sun's UV radiation.

Carbon Dioxide and Global Warming. The world's industrialization has led to a marked increase in the atmosphere's level of carbon dioxide, with resulting warming of climates.

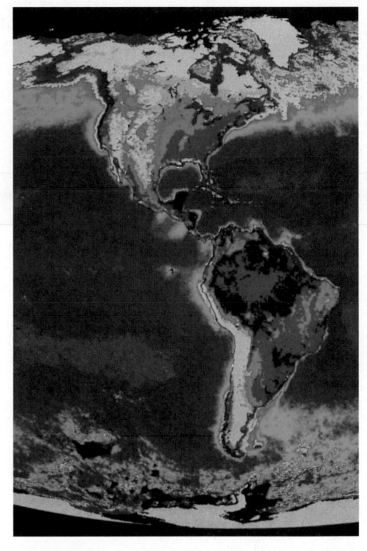

FIGURE 56.1
Life in the biosphere. In this satellite image, yellow zones are largely arid and have relatively little green plant cover, whereas the opposite is true of green zones. Almost every environment on earth can be described in terms of temperature and moisture. These physical parameters have great bearing on the forms of life that are able to inhabit a particular region.

The biosphere includes all living communities on earth, from the profusion of life in the tropical rain forests to the photosynthetic phytoplankton in the world's oceans. In a very general sense, the distribution of life on earth reflects variations in the world's environments, principally in temperature and the availability of water. Figure 56.1 is a satellite image of North and South America, collected over eight years. The colors are keyed to the relative abundance of chlorophyll, a good indicator of rich biological communities. Phytoplankton and algae produce the dark red zones in the oceans and along the seacoasts. Green and dark green areas on land are dense forests, while yellow areas include the deserts of western South America and the tundra of the far north, which have much lower productivity.

56.1 Climate shapes the character of ecosystems.

The distribution of biomes (see section 56.2) results from the interaction of the features of the earth itself, such as different soil types or the occurrence of mountains and valleys, with two key physical factors: (1) the amount of solar heat that reaches different parts of the earth and seasonal variations in that heat; and (2) global atmospheric circulation and the resulting patterns of oceanic circulation. Together, these factors dictate local climate, and so determine the amounts and distribution of precipitation.

Effects of the Sun and Atmospheric Circulation

The earth receives an enormous quantity of heat from the sun in the form of shortwave radiation, and it radiates an equal amount of heat back to space in the form of longwave radiation. About 10^{24} calories arrive at the upper surface of the earth's atmosphere each year, or about 1.94 calories per square centimeter per minute. About half of this energy reaches the earth's surface. The wavelengths that reach the earth's surface are not identical to those that reach the outer atmosphere. Most of the ultraviolet radiation is absorbed by the oxygen (O_2) and ozone (O_3) in the atmosphere. As we will see in section 56.4, the depletion of the ozone layer, apparently as a result of human activities, poses serious ecological problems.

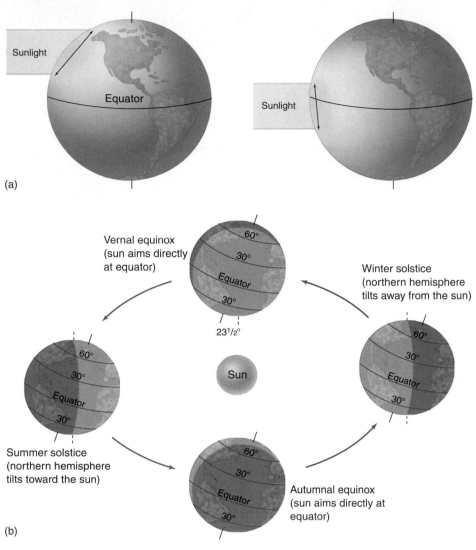

FIGURE 56.2
Relationships between the earth and the sun are critical in determining the nature and distribution of life on earth. (*a*) A beam of solar energy striking the earth in the middle latitudes spreads over a wider area of the earth's surface than a similar beam striking the earth near the equator. (*b*) The rotation of the earth around the sun has a profound effect on climate. In the northern and southern hemispheres, temperatures change in an annual cycle because the earth tilts slightly on its axis in relation to the path around the sun.

How the Sun Affects Climate

The world contains a great diversity of biomes because its climate varies so much from place to place. On a given day, Miami, Florida, and Bangor, Maine, often have very different weather. There is no mystery about this. Because the earth is a sphere, some parts of it receive more energy from the sun than others. This variation is responsible for many of the major climatic differences that occur over the earth's surface, and indirectly, for much of the diversity of biomes. The tropics are warmer than temperate regions because the sun's rays arrive almost perpendicular to regions near the equator. Near the poles, the angle of incidence of the sun's rays spreads them out over a much greater area, providing less energy per unit area (figure 56.2*a*).

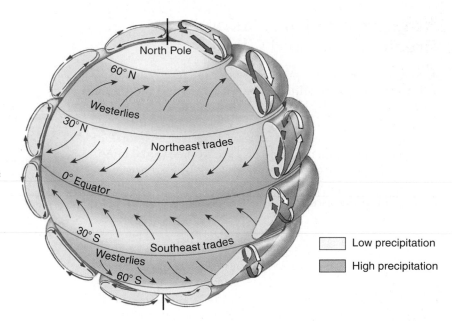

FIGURE 56.3
General patterns of atmospheric circulation. The pattern of air movement toward and away from the earth's surface is shown. Rising air that is cooled creates a band of high precipitation at the equator and at 60°N and 60°S. Reheated, falling air can hold more moisture and creates a band of low precipitation at 30°N and 30°S and near the poles. As air masses change latitudes, they are also deflected by the earth's rotational velocity, producing the trade winds and the westerlies.

The earth's annual orbit around the sun and its daily rotation on its own axis are also important in determining world climates (figure 56.2*b*). Because of the annual cycle, and the inclination of the earth's axis at approximately 23.5° from its plane of revolution around the sun, a progression of seasons occurs in all parts of the earth away from the equator. One pole or the other is tilted closer to the sun at all times, except during the spring and autumn equinoxes.

Major Atmospheric Circulation Patterns

The moisture-holding capacity of air increases when it warms and decreases when it cools. High temperatures near the equator encourage evaporation and create warm, moist air. As this air rises, it is replaced by air that is drawn toward the equator from the north and south. In turn, the movement of this air pulls the heated tropical air away from the equator. As it moves away from the equator, it cools and loses most of its moisture (figure 56.3). Consequently, the greatest amounts of precipitation on earth fall near the equator, which is where earth's tropical rain forests are found. When the air masses that have risen reach about 30° north and south latitude, the dry air, now cooler, sinks and becomes reheated. As the air reheats, its evaporative capacity increases, creating a zone of decreased precipitation and hot deserts. Some of this air mass, still warmer than in the polar regions, continues to flow toward the poles. It rises again at about 60° north and south latitude, producing another zone of high precipitation occupied by relatively wet and cool coniferous forests. At this latitude, there is another low-pressure area, the polar front. Some of this rising air flows back to

the equator, and some continues north and south, descending near the poles and producing another zone of low precipitation before it returns to the equator.

Air Currents Generated by the Earth's Rotation

The earth's rotational velocity is greatest at the equator and decreases toward the poles. Air masses move at the same velocity as the earth does at the same latitude. As an air mass moves toward the equator, its velocity is less than that of the underlying earth. As a result, the air mass is deflected to the northwest south of the equator and to the southwest north of the equator. Conversely, air moving away from the equator meets a slower rotational spin on the earth, and thus is deflected in the opposite direction, to the northeast north of the equator and to the southeast south of the equator.

These deflected air masses between 30° and the equator form the trade winds, which blow all year long and are the steadiest winds found anywhere on earth. They are stronger in winter and weaker in summer. Between 30° and 60° north and south latitude, strong prevailing westerlies blow from west to east and dominate climatic patterns in these latitudes, particularly along the western edges of the continents. Weaker winds, blowing from east to west, occur farther north and south in their respective hemispheres.

Warm air rises near the equator, descends and produces arid zones at about 30° north and south latitude, flows toward the poles, then rises again at about 60° north and south latitude, and moves back toward the equator. Part of this air, however, moves toward the poles, where it produces zones of low precipitation.

Atmospheric Circulation, Precipitation, and Climate

As we have discussed, precipitation is generally low near 30° north and south latitude, where air is falling and warming, and relatively high near 60° north and south latitude, where it is rising and cooling. Partly as a result of these factors, all the great deserts of the world lie near 30° north or south latitude. Some major deserts are formed in the interiors of large continents. These areas have limited precipitation because of their distance from the sea, the ultimate source of most moisture.

Rain Shadows

Other deserts occur because mountain ranges intercept moisture-laden winds from the sea. When this occurs, the air rises and the moisture-holding capacity of the air decreases, resulting in increased precipitation on the windward side of the mountains—the side from which the wind is blowing. As the air descends the other side of the mountains, the leeward side, it is warmed, and its moisture-holding capacity increases, tending to block precipitation. In California, for example, the eastern sides of the Sierra Nevada Mountains are much drier than the western sides, and the vegetation is often very different. This phenomenon is called the **rain shadow effect** (figure 56.4).

Regional Climates

Four relatively small areas, each located on a different continent, share a climate that resembles that of the Mediterranean region. So-called Mediterranean climates are found in portions of southern California and Oregon; in central Chile; in southwestern Australia; and in the Cape region of South Africa. In all of these areas, the prevailing westerlies blow during the summer from a cool ocean onto warm land. As a result, the air's moisture-holding capacity increases, the air absorbing moisture and creating hot, rainless summers. Such climates are unusual on a world scale. In the regions where they occur, many unique kinds of plants and animals, often local in distribution, have evolved. Because of the prevailing westerlies, the great deserts of the world (other than those in the interiors of continents) and the areas of Mediterranean climate lie on the western sides of the continents.

Another kind of regional climate occurs in southern Asia. The monsoon climatic conditions characteristic of India and southern Asia appear during the summer months. During the winter, the trade winds blow from the east-northeast off the cool land onto the warm sea. From June to October, though, when the land is heated, the direction of the air flow reverses, and the winds veer around to blow

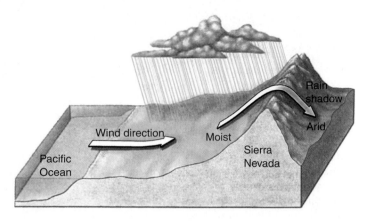

FIGURE 56.4
The rain shadow effect. Moisture-laden winds from the Pacific Ocean rise and are cooled when they encounter the Sierra Nevada Mountains. As their moisture-holding capacity decreases, precipitation occurs, making the middle elevation of the range one of the snowiest regions on earth; it supports tall forests, including those that contain the famous giant sequoias (*Sequoiadendron giganteum*). As the air descends on the east side of the range, its moisture-holding capacity increases again, and the air picks up moisture from its surroundings. As a result, desert conditions prevail on the east side of the mountains.

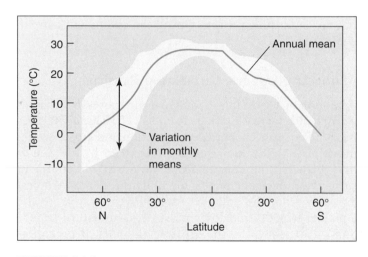

FIGURE 56.5
Temperature varies with latitude. The blue line represents the annual mean temperature at latitudes from the North Pole to Antarctica.
Why is it hotter at low latitudes?

onto the Indian subcontinent and adjacent areas from the southwest, bringing rain. The duration and strength of the monsoon winds spell the difference between food sufficiency and starvation for hundreds of millions of people in this region each year.

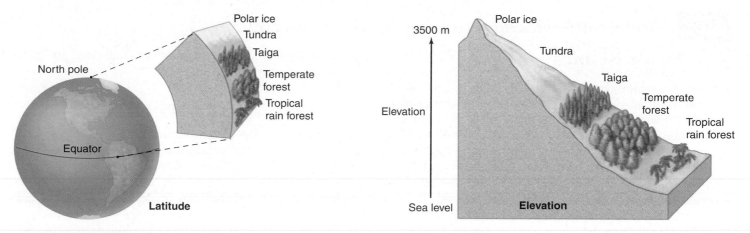

FIGURE 56.6

Elevation affects the distribution of biomes much as latitude does. Biomes that normally occur far north and far south of the equator at sea level also occur in the tropics at high mountain elevations. Thus, on a tall mountain in the tropics, one might see a sequence of biomes like the one illustrated at right.

Seasonal changes in wind circulation also produce changes in ocean currents, sometimes causing nutrient-rich cold water to well up from ocean depths. This produces "blooms" among the plankton and other organisms living near the surface. Similar turnover occurs seasonally in freshwater lakes and ponds, bringing nutrients from the bottom to the surface in the fall and again in the spring.

Latitude

As noted earlier, temperatures are higher in tropical ecosystems for a simple reason: More sunlight per unit area falls on tropical latitudes. Solar radiation is most intense when the sun is directly overhead, and this only occurs in the tropics, where sunlight strikes the equator perpendicularly (see figure 56.2a). As figure 56.5 shows, the highest mean global temperatures occur near the equator (that is, 0 latitude). Because the tropics lie near the equator, there is little variation in mean monthly temperature in tropical ecosystems. As you move from the equator into temperate latitudes, sunlight strikes the earth at a more oblique angle, so that less falls on a given area. As a result, mean temperatures are lower. At temperate latitudes, temperature variation increases because of the increasingly marked seasons.

Elevation

Temperature also varies with elevation, with higher altitudes becoming progressively colder. At any given latitude, air temperature falls about 6°C for every 1000-meter increase in elevation. The ecological consequences of temperature varying with elevation are the same as those for temperature varying with latitude (figure 56.6). Thus, in North America, a 1000-meter increase in elevation results in a temperature drop equal to that of an 880-kilometer increase in latitude. This is one reason "timberline" (the elevation above which trees do not grow) occurs at progressively lower elevations the further the area is from the equator.

Microclimate

Climate also varies on a very fine scale within ecosystems. Within the litter on a forest floor, there is considerable variation in shading, local temperatures, and rates of evaporation from the soil. Called **microclimate,** these very localized climatic conditions can be very different from those of the overhead atmosphere. Gardeners spread straw over newly seeded lawns to create such a moisture-retaining microclimate.

Animals are able to take advantage of these microclimates to meet their needs. For example, many ectothermic animals, such as butterflies and lizards, shuttle between warm and cool microclimates to precisely regulate their body temperature. In some cool, mountainous areas, lizards are able to maintain their body temperature 20°C above air temperature by basking in spots that are bathed by the sun but sheltered from the wind.

The great deserts and associated arid areas of the world mostly lie along the western sides of continents at about 30° north and south latitude. Mountain ranges tend to intercept rain, creating deserts in their shadow. In general, temperatures are warmer in the tropics and at lower elevations.

56.2 Biomes are widespread terrestrial ecosystems.

The Major Biomes

Biomes are major communities of organisms that have a characteristic appearance and are distributed over a wide land area defined largely by regional variations in climate. As you might imagine from such a broad definition, there are many ways to classify biomes, and different ecologists may assign the same community to different biomes. Few disagree, however, about the reality of biomes as major biological communities.

Terrestrial biomes are defined by their characteristic vegetational structure, which in turn, is affected by the climate of a region. Because areas within similar climates impose similar selective conditions, the plants that occur in these regions tend to evolve similar adaptive responses, even if they are not closely related. This is the phenomenon of convergent evolution discussed in chapter 22 and is the reason that deserts around the world are dominated by cactuslike plants, whereas tropical rain forests are composed of tall, lushly vegetated trees.

Distribution of the Major Biomes

Eight major biome categories are presented in this text: tropical rain forest, savanna, desert, temperate grassland, temperate deciduous forest, temperate evergreen forest, taiga, and tundra. These biomes occur worldwide, occupying large regions that can be defined by rainfall and temperature.

Six additional biomes are considered by some ecologists to be subsets of the eight major ones: polar ice, mountain zone, chaparral, warm moist evergreen forest, tropical monsoon forest, and semidesert. They vary remarkably from one another because they have evolved in regions with very different climates. Distributions of these 14 biomes are mapped in figure 56.7

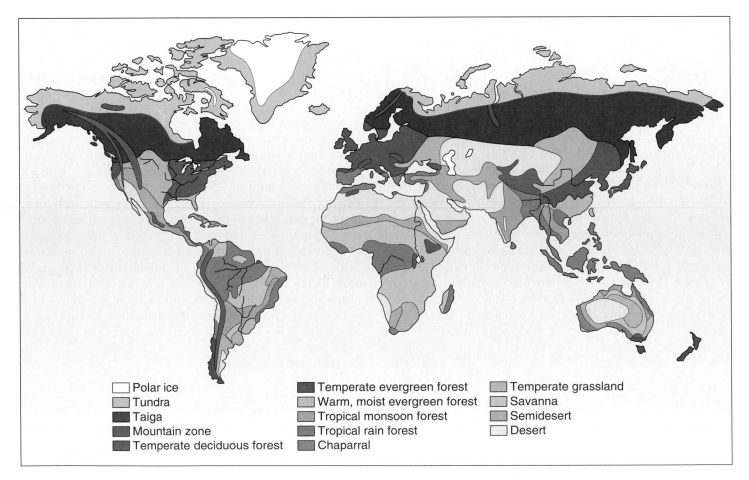

FIGURE 56.7
The distribution of biomes.   Each biome is similar in vegetational structure and appearance wherever it occurs on earth.

Biomes and Climate

Many different environmental factors play a role in determining which biomes are found where. Two key parameters are moisture and temperature, both strong influences on ecosystem productivity. Figure 56.8 presents data on primary productivity as a function of annual precipitation and annual mean temperature. However, different places with the same annual precipitation and temperature sometimes support different biomes, so other factors must also be important. These factors include soil structure and its mineral composition (discussed in detail in chapter 38) and seasonal versus constant climate. Nevertheless, precipitation and temperature do a fine job of predicting which biomes will occur in most places, as figure 56.9 illustrates.

If there were no mountains and no climatic effects caused by the irregular outlines of continents and by different sea temperatures, each biome would form an even belt around the globe, defined largely by latitude. But in truth, these other factors also greatly affect the distribution of biomes. Distance from the ocean has a major impact on rainfall, and elevation affects temperature—for example, the summits of the Rocky Mountains are covered with a vegetation type that resembles the tundra, which normally occurs at a much higher latitude.

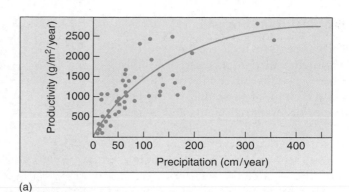

(a)

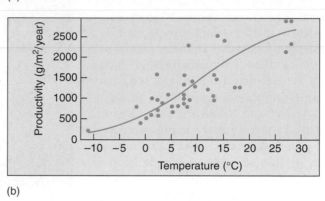

(b)

FIGURE 56.8
The effects of precipitation and temperature on primary productivity. The net primary productivity of ecosystems at 52 locations around the globe depends significantly upon (*a*) mean annual precipitation and (*b*) mean annual temperature.
Why does productivity increase with increasing precipitation and temperature?

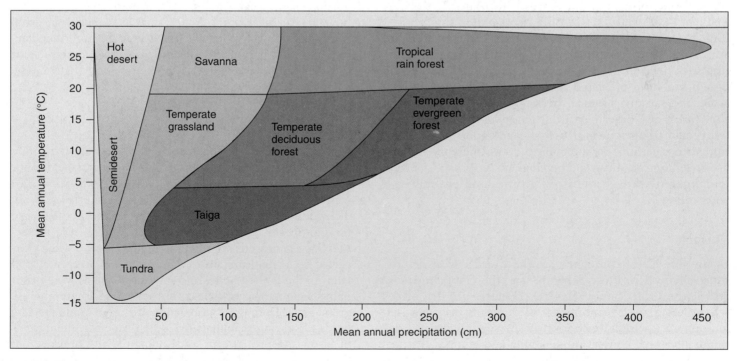

FIGURE 56.9
Predictors of biome distribution. Temperature and precipitation are excellent predictors of biome distribution. At mean annual precipitations between 50 and 150 cm, other factors—such as seasonal drought, fire, and grazing—also have a major influence.

Tropical Rain Forests

Rain forests, which receive 140 to 450 centimeters of rain per year, are the richest ecosystems on earth (figure 56.10). They contain at least half of the earth's species of terrestrial plants and animals— more than 2 million species! In a single square mile of tropical forest in Rondonia, Brazil, there are 1200 species of butterflies—twice the total number found in the United States and Canada combined. The communities that make up tropical rain forests are diverse, and as a result, most kinds of animals, plants, and microorganisms are represented in a given area by very few individuals. Extensive tropical rain forests occur in South America, Africa, and Southeast Asia. But the world's rain forests are being destroyed, and countless species, many of them never seen by humans, are disappearing with them. A quarter of the world's species will disappear with the rain forests during the lifetime of many of us.

FIGURE 56.10
Tropical rain forest.

Savannas

In the dry climates that border the tropics are the world's great grasslands, called **savannas.** Savanna landscapes are open, often with widely spaced trees, and rainfall (75 to 125 centimeters annually) is seasonal. Many of the animals and plants are active only during the rainy season. The huge herds of grazing animals that inhabit the African savanna are familiar to all of us. Such animal communities lived in North America during the Pleistocene epoch but have persisted mainly in Africa. On a global scale, the savanna biome is transitional between tropical rain forest and desert. As these savannas are increasingly converted to agricultural use to feed rapidly expanding human populations in subtropical areas, their inhabitants are struggling to survive. The elephant and rhino are now endangered species; the lion, giraffe, and cheetah will soon follow.

Deserts

In the interior of continents are the world's great deserts, especially in Africa (the Sahara), Asia (the Gobi), and Australia (the Great Sandy Desert). **Deserts** are dry places where less than 25 centimeters of rain falls in a year—an amount so low that vegetation is sparse, and survival depends on water conservation. Plants and animals may restrict their activity to favorable times of the year, when water is present. To avoid high temperatures, most desert vertebrates live in deep, cool, and sometimes even somewhat moist burrows. Those that are active over a greater portion of the year emerge only at night, when temperatures are relatively cool. Some, such as camels, can drink large quantities of water when it is available and then survive long, dry periods. Many animals simply migrate to or through the desert, where they exploit food that may be abundant seasonally.

Temperate Grasslands

Halfway between the equator and the poles are temperate regions where rich **grasslands** grow. These grasslands once covered much of the interior of North America, and they were widespread in Eurasia and South America as well. The roots of perennial grasses characteristically penetrate far into the soil, and grassland soils tend to be deep and fertile. Such grasslands are often highly productive when converted to agricultural use. Many of the rich agricultural lands in the United States and southern Canada were originally occupied by **prairies,** another name for temperate grasslands. Temperate grasslands are often populated by herds of grazing mammals. In North America, huge herds of bison and pronghorns once inhabited the prairies, but these herds are now almost all gone.

Temperate Deciduous Forests

Mild climates (warm summers and cool winters) and plentiful rains promote the growth of **deciduous** (hardwood) **forests** in Eurasia, the northeastern United States, and eastern Canada. A deciduous tree is one that drops its leaves in the winter (figure 56.11). Deer, bears, beavers, and raccoons are familiar animals of the temperate regions. Because the temperate deciduous forests represent the remnants of more extensive forests that stretched across North America and Eurasia several million years ago, the remaining areas in eastern Asia and eastern North America share animals and plants that were once more widespread. The deciduous forest in eastern Asia is rich in species because climatic conditions have historically remained constant. Many perennial herbs live in areas of temperate deciduous forest.

FIGURE 56.11
Temperate deciduous forest.

Temperate Evergreen Forests

Temperate evergreen forests occur in regions where winters are cold and there is a strong, seasonal dry period. The pine forests and oak woodlands of the western United States and California are typical temperate evergreen forests. Temperate evergreen forests are characteristic of regions with nutrient-poor soils. Temperate-mixed evergreen forests represent a broad transitional zone between temperate deciduous forests to the south and taiga to the north. Many of these forests are endangered by overlogging, particularly in the western United States.

Taiga

A great ring of northern forests of coniferous trees (spruce, hemlock, and fir) extends across vast areas of Asia and North America. Coniferous trees have leaves like needles that are kept all year long. This ecosystem, called **taiga,** is one of the largest on earth. The winters are long and cold, and most of the limited precipitation falls in the summer. Because the taiga has too short a growing season for farming, few people live there. Many large herbivores, including elk, moose, deer, and such carnivores as wolves, bears, lynx, and wolverines, live in the taiga. Traditionally, fur trapping has been extensive in this region, as has lumber production. Marshes, lakes, and ponds are common, and are often fringed by willows or birches. Most of the trees occur in dense stands of one or a few species.

Tundra

In the far north, above the great coniferous forests and south of the polar ice, few trees grow. The grassland, called **tundra,** is open, windswept, and often boggy. Enormous in extent, this ecosystem covers one-fifth of the earth's land surface. Very little rain or snow falls. When rain does fall during the brief arctic summer, it sits on frozen ground, creating a sea of boggy ground. **Permafrost,** or permanent ice, usually exists within a meter of the surface. Trees are small and are mostly confined to the margins of streams and lakes. As in taiga, herbs of the tundra are perennials that grow rapidly during the brief summers. Large grazing mammals, including musk-oxen, caribou and reindeer, and carnivores, such as wolves, foxes, and lynx, live in the tundra. Lemming populations rise and fall on a long-term cycle, with important effects on the animals that prey on them.

Major biological communities called biomes can be distinguished in different climatic regions. These communities, which occur in regions of similar climate, are much the same wherever they are found. Variation in annual mean temperature and precipitation are good predictors of which biome will occur where.

56.3 Aquatic ecosystems cover much of the earth.

More than 75% of the world's surface is covered by water. Most of this area is ocean, and the remainder is freshwater area on land. This section considers each in turn.

Patterns of Circulation in the Oceans

Patterns of ocean circulation are determined by the patterns of atmospheric circulation, but they are modified by the locations of landmasses. Ocean circulation is dominated by huge surface gyres (figure 56.12), which move around the subtropical zones of high pressure between approximately 30° north and 30° south latitude. The direction of these gyres is affected by the earth's rotational velocity, in the same way as are the trade winds. Thus, these gyres move clockwise in the northern hemisphere and counterclockwise in the southern hemisphere. The ways they redistribute heat profoundly affect life not only in the oceans but also on coastal lands. For example, the Gulf Stream, in the North Atlantic, swings away from North America near Cape Hatteras, North Carolina, and brings warm water to Europe near the southern British Isles. Because of the Gulf Stream, western Europe is much warmer and more temperate than eastern North America at similar latitudes. As a general principle, western sides of continents in temperate zones of the northern hemisphere are warmer than their eastern sides; the opposite is true of the southern hemisphere.

In South America, the Humboldt Current carries phosphorus-rich cold water northward up the west coast. Phosphorus is brought up from the ocean depths by the upwelling of cool water that occurs as offshore winds blow from the mountainous slopes that border the Pacific Ocean. This nutrient-rich current helps make possible the abundance of marine life that supports the fisheries of Peru and northern Chile. Marine birds, which feed on these organisms, are responsible for the commercially important, phosphorus-rich, guano deposits on the seacoasts of these countries.

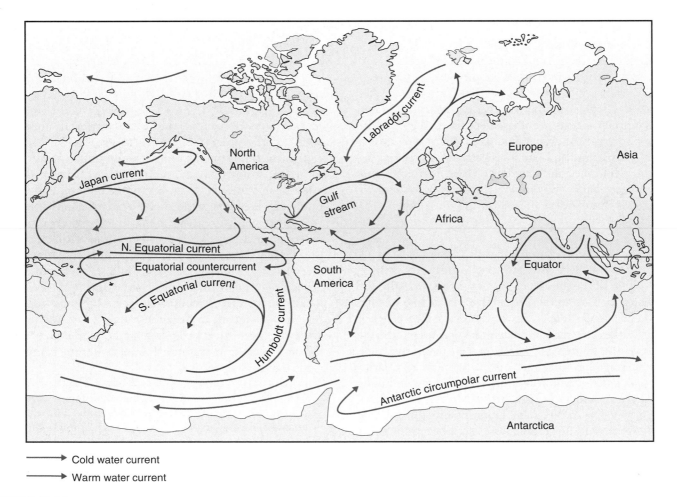

———▶ Cold water current

———▶ Warm water current

FIGURE 56.12
Ocean circulation. Water moving in great surface spiral patterns called gyres profoundly affects the climate on adjacent land masses.

El Niño and Ocean Ecology

Every Christmas, a tepid current sweeps down the coast of Peru and Ecuador from the tropics, reducing the fish population slightly and giving local fishermen some time off. The fishermen named this Christmas current *El Niño* (literally, "the child," after the Christ Child). Now, though, the term is reserved for a catastrophic version of the same phenomenon, one that occurs every two to seven years and is felt not only locally but globally.

Scientists now have a pretty good idea of what goes on in an El Niño. Normally, the Pacific Ocean is fanned by constantly blowing, east-to-west trade winds that push warm surface water away from the ocean's eastern side (Peru, Ecuador, and Chile) and allow cold water to well up from the depths in its place, carrying nutrients that feed plankton and hence fish. This surface water piles up in the west, around Australia and the Philippines, making that area several degrees warmer and a meter or so higher than the eastern side of the ocean. But if the winds slacken briefly, warm water begins to slosh back across the ocean.

Once this happens, ocean and atmosphere conspire to ensure that it keeps happening. The warmer the eastern ocean gets, the warmer and lighter the air above it becomes, and hence the more similar to the air on the western side. This reduces the difference in pressure across the ocean. Because a pressure difference is what makes winds blow, the easterly trade winds weaken further, letting the warm water continue its eastward advance.

The result is to shift the weather systems of the western Pacific Ocean 6000 km eastward. The tropical rainstorms that usually drench Indonesia and the Philippines occur when warm seawater abutting these islands causes the air above it to rise, cool, and condense its moisture into clouds. When the warm water moves east, so do the clouds, leaving the previously rainy areas in drought. Conversely, the western edge of South America, where coastal waters are usually too cold to trigger much rain, gets a soaking, while the upwelling slows down. During an El Niño, commercial fish stocks virtually disappear from the waters of Peru and northern Chile as the fish move south to find colder water, and plankton drop to one-twentieth of their normal abundance.

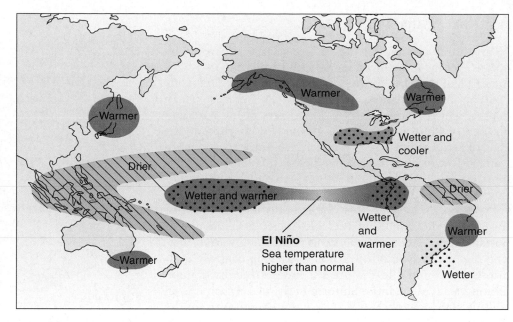

FIGURE 56.13
An El Niño winter. El Niño currents produce unusual weather patterns all over the world as warm waters from the western Pacific move eastward.

That is just the beginning. El Niño's effects are propagated across the world's weather systems (figure 56.13). Violent winter storms lash the coast of California, accompanied by flooding, and El Niño produces colder and wetter winters than normal in Florida and along the Gulf Coast. The American Midwest experiences heavier-than-normal rains, as do Israel and its neighbors.

El Niño can wreak havoc on ecosystems. In the Galápagos Islands, for example, seabird and sea lion populations crash as animals starve due to the lack of fish. By contrast, on land, the heavy rains produce a bumper crop of seeds, and land birds flourish. In Chile, similar effects on seed abundance propagate up the food chain, leading first to increased rodent populations and then to increased predator populations, a nice example of a bottom-up trophic cascade, discussed in chapter 55.

Although the effects of El Niños are now fairly clear, what triggers them remains a mystery. Models of weather patterns suggest that the climatic change that triggers El Niño is "chaotic." Wind and ocean currents return again and again to the same condition, but never in a regular pattern, and small nudges can send them off in many different directions—including an El Niño.

The world's oceans circulate in huge gyres deflected by continental landmasses. Circulation of ocean water redistributes heat, warming the western side of continents. Disturbances in ocean currents, such as El Niño, can have profound influences on world climate.

Marine Ecosystems

Nearly three-quarters of the earth's surface is covered by ocean. Oceans have an average depth of more than 3 kilometers, and they are, for the most part, cold and dark. Heterotrophic organisms inhabit even the greatest ocean depths, which reach nearly 11 kilometers in the Marianas Trench of the western Pacific Ocean. Photosynthetic organisms are confined to the upper few hundred meters of water. Most organisms that live below this level obtain almost all of their food indirectly, as a result of photosynthetic activities that occur above.

As temperature increases, water holds less oxygen. For this reason, the amount of available oxygen becomes an important limiting factor for organisms in some warmer marine regions of the globe. Carbon dioxide, in contrast, is almost never limited in the oceans. Indeed, the distribution of minerals is much more uniform in the ocean than it is on land, where individual soils reflect the composition of the parent rocks from which they have weathered. Nonetheless, this is not true for all minerals, including several—iron, nitrogen, and phosphorus—that affect productivity and are responsible for the differences in plankton density illustrated in figure 56.1 (see also text page 1203).

Some marine habitats, such as coral reefs and estuaries, are remarkably productive (see table 55.1). Overall, approximately 50% of the world's primary production comes from aquatic systems. Because much of this productivity is rapidly grazed and replenished, aquatic systems support about three times more animal production than terrestrial systems. The marine environment consists of three major habitats: (1) the **neritic zone,** the coastal shallow waters; (2) the **pelagic zone,** the area of water above the ocean floor; and (3) the **benthic zone,** the actual ocean floor (figure 56.14). The part of the ocean floor that drops to depths where light does not penetrate is called the **abyssal zone.**

The Neritic Zone

The neritic zone of the ocean is the area less than 300 meters below the surface along the coasts of continents and islands. The zone is small in area, but it is inhabited by large numbers of species (figure 56.15). Intense and sometimes violent interaction between sea and land gives a selective advantage to well-secured organisms that can withstand being washed away by the continual beating of the waves. Part of this zone, the **intertidal (or littoral) region,** is exposed to the air whenever the tides recede.

The world's great fisheries are in shallow waters over continental shelves, either near the continents themselves or in the open ocean, where the seabed is near the surface. Nutrients, derived from land, are much more abundant in coastal and other shallow regions, where upwelling from the depths occurs, than in the open ocean. This accounts for the great productivity of the continental shelf fisheries.

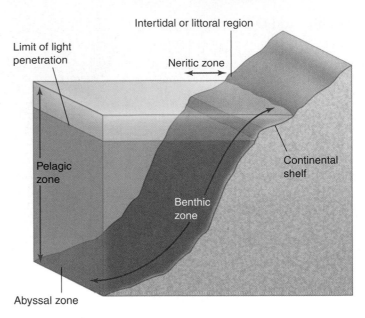

FIGURE 56.14
Marine ecosystems.   Ecologists classify marine communities into neritic, pelagic, and benthic zones, according to depth (which affects how much light penetrates) and distance from shore. The abyssal zone is that portion of the benthic zone that lies beyond the limit of light penetration.

FIGURE 56.15
Diversity in coastal regions.   Fishes and many other kinds of animals find food and shelter among the coral in coastal waters.

The preservation of these fisheries, a source of high-quality protein exploited throughout the world, has become a growing concern. For example, in Chesapeake Bay, environmental stresses have become so severe that they not only threaten the continued existence of formerly highly productive fisheries, but also diminish the quality of human life in these regions.

The Pelagic Zone

Drifting freely in the upper waters of the pelagic zone, a diverse biological community exists, primarily consisting of microscopic organisms called **plankton.** Fish and other larger organisms that swim in these waters constitute the *nekton,* whose members feed on plankton and one another. Some members of the plankton, including protists and some bacteria, are photosynthetic. Collectively, these organisms account for about 40% of all photosynthesis that takes place on earth. Most plankton live in the top 100 meters of the sea, the zone into which light from the surface penetrates freely. Perhaps half of the total photosynthesis in this zone is carried out by organisms less than 10 micrometers in diameter—at the lower limits of size for organisms. These include cyanobacteria and algae, organisms so small that their abundance and ecological importance have been unappreciated until relatively recently.

Many heterotrophic protists and animals live in the plankton and feed directly on photosynthetic organisms and on one another. Gelatinous animals, especially jellyfish and ctenophores, are abundant in the plankton. The largest animals that have ever existed on earth, baleen whales, graze on plankton and nekton, as do a number of other organisms, such as fishes and crustaceans.

Populations of organisms that make up plankton can increase rapidly, and the turnover of nutrients in the sea is great, although the productivity in these systems is quite low. Because nitrogen and phosphorus are often present in only small amounts and organisms may be relatively scarce, the productivity that does occur is more a result of rapid use and recycling rather than an abundance of these nutrients.

The Benthic Zone

The seafloor at depths below 1000 meters has about twice the area of all the land on earth. The seafloor itself is a thick blanket of mud, consisting of fine particles that have settled from the overlying water and accumulated over millions of years. Because of high pressures (an additional atmosphere of pressure for every 10 meters of depth), cold temperatures (2° to 3°C), darkness, and lack of food, biologists originally thought that nothing could live on the seafloor. In fact, recent work has shown that the number of species living at great depth is quite high. Rough estimates of deep-sea diversity have soared to millions of species. Many appear endemic (local). The diversity of species is so high it may rival that of tropical rain forests. Most of these animals are only a few millimeters in size, although larger ones also occur in these regions. Some of the larger ones are bioluminescent (figure 56.16*a*) and thus are able to communicate with one another or attract their prey.

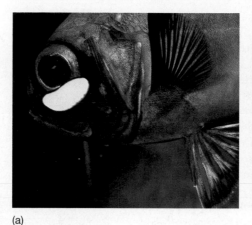

(a)   (b)

FIGURE 56.16
Life in the benthic zone. (*a*) The luminous spot below the eye of this deep-sea fish results from the presence of a symbiotic colony of luminous bacteria. Similar luminous signals are a common feature of mobile deep-sea animals. (*b*) These giant beardworms live along vents where water jets from fissures at 350°C and then cools to the 2°C of the surrounding water.

Animals on the sea bottom depend on the meager leftovers from organisms living kilometers overhead. The low densities and small size of most deep-sea animals is in part a consequence of this limited food supply. In 1977, oceanographers diving in a research submarine were surprised to find dense clusters of large animals living on geothermal energy at a depth of 2500 meters. These deep-sea oases occur where seawater circulates through porous rock at sites where molten material from beneath the earth's crust comes close to the rocky surface. A series of these areas occur on the Mid-Ocean Ridge, where basalt erupts through the ocean floor.

This water is heated to temperatures in excess of 350°C and, in the process, becomes rich in reduced compounds. These compounds, such as hydrogen sulfide, provide energy for bacterial primary production through chemosynthesis instead of photosynthesis. Mussels, clams, and large red-plumed worms in a phylum unrelated to any shallow-water invertebrates cluster around the vents (figure 56.16*b*). Prokaryotes live symbiotically within the tissues of these animals. The animal supplies a place for the prokaryotes to live and transfers CO_2, H_2S, and O_2 to them for their growth; the bacteria supply the animal with organic compounds to use as food. Polychaete worms (see chapter 33), anemones, and limpets live on free-living chemosynthetic bacteria. Crabs act as scavengers and predators, and some of the fish are also predators. This is one of the few ecosystems on earth that does not depend on the sun's energy.

About 40% of the world's photosynthetic productivity is estimated to occur in the oceans. The turnover of nutrients in the plankton is much more rapid than in most other ecosystems, and the total amounts of nutrients are very low.

Freshwater Habitats

Freshwater habitats are distinct from both marine and terrestrial ones, but they are limited in area. Inland lakes cover about 1.8% of the earth's surface, and running water (streams and rivers) covers about 0.3%. All freshwater habitats are strongly connected with terrestrial ones; marshes and swamps constitute intermediate habitats. In addition, a large amount of organic and inorganic material continuously enters bodies of fresh water from communities growing on the land nearby (figure 56.17). Many kinds of organisms are restricted to freshwater habitats. When organisms live in rivers and streams, they must be able to swim against the current or attach themselves in such a way as to resist the effects of the current and avoid being swept away.

Ponds and Lakes

Small bodies of fresh water are called ponds, and larger ones lakes. Because water absorbs light passing through it at wavelengths critical to photosynthesis (every meter absorbs 40% of the red and about 2% of the blue), the distribution of photosynthetic organisms is limited to the upper **photic zone;** only heterotrophic organisms occur in the lower **disphotic** and **aphotic zones,** where very little or no light penetrates.

Like the ocean, ponds and lakes have three zones where organisms occur, distributed according to the depth of the water and its distance from shore (figure 56.18). The *littoral zone* is the shallow area along the shore. The *limnetic zone* is the well-illuminated surface water away from the shore, inhabited by plankton and other organisms that live in open water. The *profundal zone* is the area below the limits where light can effectively penetrate.

Thermal Stratification

Thermal stratification is characteristic of larger lakes in temperate regions (figure 56.19). In summer, warmer water forms a layer at the surface known as the *epilimnion*. Cooler water, called the *hypolimnion* (about 4°C), lies below. An abrupt change in temperature, the thermocline, separates these two layers. Depending on the climate of the particular area, the epilimnion may become as much as 20 meters thick during the summer.

In autumn, the temperature of the epilimnion drops until it is the same as that of the hypolimnion, 4°C. When this occurs, epilimnion and hypolimnion mix—a process called fall overturn. Because water is densest at about 4°C, further cooling of the water as winter progresses creates a layer of cooler, lighter water, which freezes to form a layer of ice at the surface. Below the ice, the water temperature remains between 0° and 4°C, and plants and animals can survive. In spring, the ice melts, and the surface water warms up. When it warms back to 4°C, it again mixes with

FIGURE 56.17
A nutrient-rich stream. Much organic material falls or seeps into streams from communities along the edges. This input increases the stream's biological productivity.

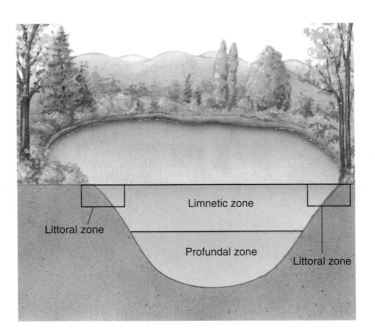

FIGURE 56.18
The three zones in ponds and lakes. A shallow "edge" (littoral) zone lines the periphery of the lake, where attached algae and their insect herbivores live. An open-water surface (limnetic) zone lies across the entire lake and is inhabited by floating algae, plankton, and fish. A dark, deep-water (profundal) zone overlies the sediments at the bottom of the lake and contains numerous prokaryotes and wormlike organisms that consume dead debris settling at the bottom of the lake.

the water below. This process is known as spring overturn. When lake waters mix in the spring and fall, nutrients formerly held in the depths of the lake are returned to the surface, and oxygen from surface waters is carried to the depths. Without such mixing, survival and growth would be impossible for many organisms at all levels of the ecosystem.

Productivity of Freshwater Ecosystems

Some aquatic communities, such as fast-moving streams, are not highly productive. Because the moving water washes away plankton, the photosynthesis that supports the community is limited to algae attached to the surface and to rooted plants.

Productivity of Lakes. Lakes can be divided into two categories based on their production of organic matter. **Eutrophic lakes** contain an abundant supply of minerals and organic matter. As the plentiful organic material drifts below the thermocline from the well-illuminated surface waters of the lake, it provides a source of energy for other organisms. Most of these are oxygen-requiring organisms that can easily deplete the oxygen supply below the thermocline during the summer months. The oxygen supply of the deeper waters cannot be replenished until the layers mix in the fall. This lack of oxygen in the deeper waters of some lakes may have profound effects, as when it allows relatively harmless materials such as sulfates and nitrates to convert into toxic materials such as hydrogen sulfide and ammonia.

In **oligotrophic lakes,** organic matter and nutrients are relatively scarce. Such lakes are often deeper than eutrophic lakes and have very clear, blue water. Their hypolimnetic water is always rich in oxygen.

Human activities can transform oligotrophic lakes into eutrophic ones. In many lakes, phosphorus is in short supply and is the nutrient that limits growth. When excess phosphorus from sources such as fertilizer runoff, sewage, and detergents enters lakes, it can quickly lead to harmful effects. In many cases, this creates perfect conditions for the growth of blue-green algae, which proliferate immensely. Soon, larger plants are outcompeted and disappear, along with the animals that live on them. In addition, as these phytoplankton die and decompose, oxygen in the water is used up, killing the natural fish and invertebrate populations. This situation can be remedied if the continual input of phosphorus is diminished. Given time, lakes can recover and return to prepollution states.

Productivity of Wetlands. Swamps, marshes, bogs, and other **wetlands** covered with water support a wide variety of water-tolerant plants, called hydrophytes ("water plants"), and a rich diversity of invertebrates, birds, and other animals. Wetlands are among the most productive ecosystems on earth (see table 55.1). They also play a key

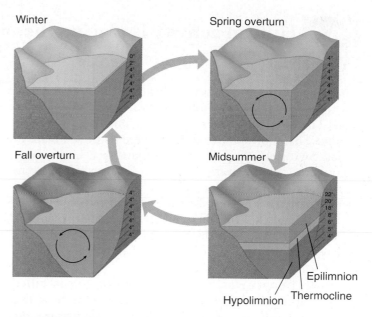

FIGURE 56.19
Stratification in fresh water. The pattern of stratification in a large pond or lake in temperate regions is upset in the spring and fall overturns. Of the three layers of water shown, the hypolimnion consists of the densest water, at 4°C; the epilimnion consists of warmer water that is less dense; and the thermocline is the zone of abrupt change in temperature that lies between them.

ecological role by providing water storage basins that moderate flooding. Many wetlands are being disrupted as humans "develop" what is sometimes perceived as otherwise useless land, but government efforts are now under way to protect the remaining wetlands.

Differences Between Aquatic and Terrestrial Ecosystems

Overall, a number of differences characterize aquatic ecosystems in comparison to terrestrial ecosystems:

1. Almost 100 times more inhabitable space (a conservative limit including only the upper 4000 meters of ocean).
2. A more stable temperature regime.
3. No shortage of water, but light and nutrients can be limiting.
4. Primary producers are generally microscopic, with high rates of turnover, mostly as the result of consumption by grazers.
5. Most aquatic grazers are ectotherms and may have low metabolic costs which allows larger population sizes for a given amount of energy.

The most productive freshwater ecosystems are wetlands. Most lakes are far less productive, limited by lack of nutrients.

Pollution

The Rhine is one of the most beautiful rivers on earth. On the first day of November in 1986, the Rhine almost died. The blow that struck the Rhine did not at first seem deadly. That morning, firefighters were battling a blaze in Basel, Switzerland. The fire was gutting a huge warehouse, into which firefighters shot streams of water to dampen the flames. The warehouse belonged to Sandoz, a giant chemical company. In the rush to contain the fire, no one thought to ask what chemicals were stored in the warehouse. By the time the fire was out, streams of water had washed 30 tons of mercury and pesticides into the Rhine.

Flowing downriver, the deadly wall of poison killed everything it passed. For hundreds of kilometers, dead fish blanketed the surface of the river. Many cities that use the water of the Rhine for drinking had little time to make other arrangements. Even the plants in the river began to die. From Switzerland, all across Germany, to the sea, the river reeked of rotting fish, and not one drop of water was safe to drink.

Six months later, Swiss and German environmental scientists monitoring the effects of the accident were able to report that the blow to the Rhine had not been mortal. Enough small aquatic invertebrates and plants had survived to provide a basis for the eventual return of fish and other water life, and the river was rapidly washing out the remaining residues from the spill. As a lesson difficult to ignore, the spill on the Rhine caused the governments of Germany and Switzerland to intensify efforts to protect the river from future industrial accidents and to regulate the establishment and growth of chemical and industrial plants on its shores.

The Threat of Pollution

Stories similar to the one about the pollution of the Rhine can be told countless times in different places in the industrial world, from the Love Canal in New York to the James River in Virginia to the town of Times Beach in Missouri. But not all pollutants that threaten the sustainability of life are as immediately toxic as those that affected the Rhine. Many forms of pollution arise gradually as by-products of industry. For example, the polymers known as plastics, which we produce in abundance, break down slowly in nature. Much is burned or otherwise degraded to smaller vinyl chloride units. Virtually all the plastic that has ever been produced is still with us, in one form or another. Collectively, it constitutes a new form of pollution.

Widespread agriculture, carried out increasingly by modern methods, introduces large amounts of many new kinds of chemicals into the global ecosystem, including pesticides, herbicides, and fertilizers. Industrialized coun-

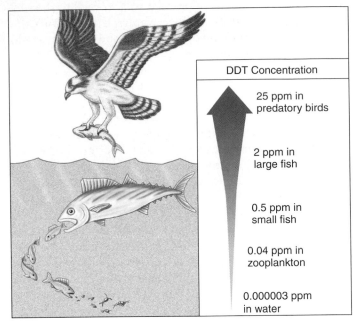

FIGURE 56.20
Biological magnification of DDT. Because DDT accumulates in animal fat, the compound becomes increasingly concentrated in higher levels of the food chain. Before DDT was banned in the United States, predatory bird populations drastically declined because DDT made their eggshells thin and fragile enough to break during incubation.

tries such as the United States now attempt to carefully monitor the side effects of these chemicals. Unfortunately, large quantities of many toxic chemicals no longer manufactured still circulate in the ecosystem.

For example, chlorinated hydrocarbons, a class of compounds that includes DDT, chlordane, lindane, and dieldrin, have been banned from normal use in the United States. They are still manufactured in the United States, however, and exported to other countries, where their use continues. Chlorinated hydrocarbon molecules break down slowly and accumulate in animal fat. Furthermore, as they pass through a food chain, they become increasingly concentrated in a process called **biological magnification** (figure 56.20). DDT caused serious problems by leading to the production of thin, fragile eggshells in many predatory bird species in the United States and elsewhere until the late 1960s, when it was banned in time to save the birds from extinction. Chlorinated compounds have other undesirable side effects and exhibit hormonelike activities in the bodies of animals, disrupting normal hormonal cycles, with sometimes potentially serious consequences.

Pollution is causing ecosystems to accumulate many harmful substances as the result of spills and runoff from agricultural or urban use.

Acid Precipitation

The Four Corners power plant in New Mexico burns coal, sending smoke high into the atmosphere through its smokestacks, each over 65 meters tall. The smoke the stacks belch out contains high concentrations of sulfur dioxide and other sulfates, which produce acid when they combine with water vapor in the air. The intent of those who designed the plant was to release the sulfur-rich smoke high enough that winds would disperse and dilute it, carrying the acids far away. They failed to foresee that the environmental effects of this acidity would be serious.

Sulfur introduced into the upper atmosphere combines with water vapor to produce sulfuric acid, and when the water later falls as rain or snow, the precipitation is acidic. Natural rainwater rarely has a pH lower than 5.6; in the northeastern United States, however, rain and snow now have a pH of 4.2 or below, with occasional storms bringing precipitation with pH as low as 3.0 (figure 56.21).

Acid precipitation destroys life. At pH levels below 5.0, many fish species and other aquatic animals die, unable to reproduce. Thousands of lakes in southern Sweden and Norway no longer support fish; these lakes are now eerily clear. In the northeastern United States and eastern Canada, tens of thousands of lakes are also dying biologically as a result of acid precipitation. In southern Sweden and elsewhere, groundwater now has a pH between 4.0 and 6.0, as acid precipitation slowly filters down into the underground reservoirs.

Enormous forest damage has occurred in the Black Forest in Germany and in the forests of the eastern United States and Canada. It has been estimated that at least 3.5 million hectares of forest in the northern hemisphere are being affected by acid precipitation (figure 56.22), and the problem is clearly growing.

The solution seems obvious: capture and remove the emissions instead of releasing them into the atmosphere. However, there are serious difficulties with this plan. First, it is expensive. The costs of installing and maintaining the necessary industrial equipment in the United States are estimated to be $4 to $5 billion per year. An additional difficulty is that the polluter and the recipient of the pollution are far from each other, and neither wants to pay for what they view as someone else's problem. Nonetheless, the Clean Air Act revisions of 1990 significantly addressed this problem in the United States for the first time, and innovative programs provided incentives to reduce emissions. As a result, the costs have been much less than expected, and the acidity of precipitation is decreasing, at least in some parts of the United States.

Industrial pollutants such as nitric and sulfuric acids, introduced into the upper atmosphere by factory smokestacks, are spread over wide areas by prevailing winds and fall to earth as "acid rain," lowering the pH of water on the ground and killing life.

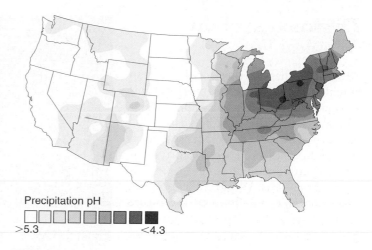

Precipitation pH

>5.3   <4.3

FIGURE 56.21
pH values of rainwater in the United States.   Precipitation in the United States, especially in the Northeast, is more acidic than natural rainwater, which has a pH of about 5.6.

FIGURE 56.22
Damage to trees at Clingman's Dome, Tennessee.   Acid precipitation weakens trees and makes them more susceptible to pests and predators.

Destruction of the Tropical Forests

More than half of the world's human population lives in the tropics, and this percentage is increasing rapidly. For global stability and for the sustainable management of the world ecosystem, it will be necessary to solve the problems of food production and regional stability in these areas. World trade, political and economic stability, and the future of most species of plants, animals, fungi, and microorganisms depend on our addressing these problems.

Loss of Rain Forests

Tropical rain forests are biologically the richest of the world's biomes. Most other kinds of tropical forest, such as seasonally dry forests and savanna forests, have already been largely destroyed, because they tend to grow on more fertile soils and thus were exploited by humans a long time ago. Now the rain forests, which grow on poor soils, are being destroyed. In the mid-1990s, only about 5.5 million square kilometers of tropical rain forest were estimated to still exist in a relatively undisturbed form. This area, about two-thirds the size of the United States (excluding Alaska), represents about half of the original extent of the rain forest. From it, about 160,000 square kilometers are being clear-cut every year, with perhaps an equivalent amount severely disturbed by shifting cultivation, firewood gathering, and the clearing of land for cattle ranching. The total area of tropical rain forest destroyed—and therefore permanently removed from the world total—amounts to an area greater than the size of Indiana each year. At this rate, all of the tropical rain forests in the world would be gone in about 30 years—but in fact, the rate of destruction is even more rapid in many regions. As a result of this overexploitation, experts predict there will be little undisturbed tropical forest left anywhere in the world within a few decades. Many areas now occupied by dense, species-rich forests may still be tree-covered, but the stands will be sparse and species-poor.

Loss of Productivity

Not only does the disappearance of tropical forests lead to a tragic loss of largely unknown biodiversity, but the loss of the forests themselves is ecologically serious.

(a)

(b)

FIGURE 56.23
Destroying the tropical forests. (*a*) When tropical forests are cleared, the ecological consequences can be disastrous. These fires are destroying rain forest in Brazil and clearing it for cattle pasture. (*b*) The consequences of deforestation can be seen on these middle-elevation slopes in Ecuador, which now support only low-grade pastures and permit topsoil to erode into the rivers (note the color of the water, stained brown by high levels of soil erosion). In the 1970s, these areas supported highly productive forest, which protected the watersheds of the area.

Tropical forests are complex, productive ecosystems that function well in the areas where they have evolved. When people cut a forest or open a prairie in the north temperate zone, they provide farmland that we know can be worked for generations. But in most areas of the tropics, people are unable to engage in continuous agriculture. When they clear a tropical forest, they consume natural resources that will never be available again (figure 56.23). The complex ecosystems built up over millions of years are now being dismantled, in almost complete ignorance, by humans.

Biologists must learn more about the construction of sustainable agricultural ecosystems that can meet human needs in tropical and subtropical regions. The ecological concepts we have been reviewing in this chapter and in chapter 55 are universal principles. The undisturbed tropical rain forest has one of the highest rates of net primary productivity of any plant community on earth, and it is therefore imperative to develop ways to harvest it for human purposes in a sustainable, intelligent way.

> **More than half of the tropical rain forests have been destroyed by human activity, and the rate of loss is accelerating.**

The Ozone Hole

The swirling colors of the satellite photo in figure 56.24 represent different concentrations of **ozone** (O_3), a form of oxygen gas that makes up a layer of the stratosphere, where it filters out ultraviolet radiation from the sun. As you can see, over Antarctica there is an "ozone hole" three times the size of the United States, an area where the ozone concentration is much less than elsewhere. The ozone thinning appeared for the first time in 1975, and has reappeared each year since for a few months during the Antarctic winter. Each year, the layer of ozone is thinner and the hole is larger.

The major cause of the ozone depletion had already been suggested in 1974 by Sherwood Roland and Mario Molina, who were awarded the Nobel Prize for their work in 1995. They proposed that chlorofluorocarbons (CFCs), relatively inert chemicals used in cooling systems, fire extinguishers, and Styrofoam containers, were percolating up through the atmosphere and reducing O_3 molecules to O_2. They predicted that one chlorine atom from a CFC molecule could destroy 100,000 ozone molecules in the following mechanism:

UV radiation causes CFCs to release Cl atoms:
$$CCl_3F \xrightarrow{\text{UV}} Cl + CCl_2F$$

UV creates oxygen free radicals:
$$O_2 \longrightarrow 2O$$

Cl atoms and O free radicals interact with ozone:
$$2Cl + 2O_3 \longrightarrow 2ClO + 2O_2$$
$$2ClO + 2O \longrightarrow 2Cl + 2O_2$$

Net reaction: $2O_3 \longrightarrow 3O_2$

Although other factors have also been implicated in ozone depletion, the role of CFCs is so predominant that worldwide agreements have been signed to phase out their production. The United States and most other countries banned the production of CFCs and other ozone-destroying chemicals after 1995, and concentrations of ozone-depleting chemicals in the upper atmosphere are decreasing. Nonetheless, the CFCs that were manufactured earlier are moving slowly upward and will remain in the atmosphere. As a result, the ozone layer will not fully recover until the latter half of this century.

Thinning of the ozone layer in the stratosphere, 25 to 40 kilometers above the surface of the earth, is a matter of serious concern. This layer protects life from the harmful ultraviolet rays of the sun that bombard the earth continuously. Life appeared on land only after the oxygen layer was sufficiently thick to generate enough ozone to shield the surface of the earth from these destructive rays.

Ultraviolet radiation is a serious human health concern. Every 1% drop in atmospheric ozone is estimated to lead to a 6% increase in the incidence of skin cancers. At middle latitudes, the approximately 3% ozone drop that has already occurred worldwide is estimated to have increased skin cancer incidence by as much as 20%. A type of skin cancer called melanoma is one of the more lethal human diseases.

Industrial CFCs released into the atmosphere react with ozone, converting it to oxygen gas. This has the effect of destroying the earth's ozone shield and exposing the earth's surface to increased levels of harmful UV radiation.

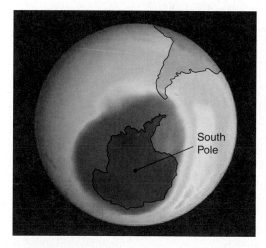

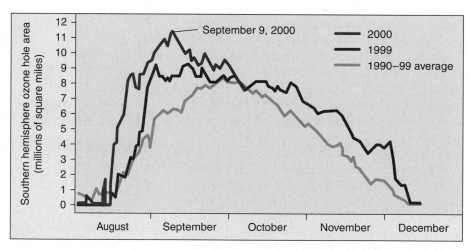

FIGURE 56.24

The growing ozone hole over Antarctica. For decades, NASA satellites have tracked the extent of ozone depletion over Antarctica. Every year since 1975, an ozone "hole" has appeared in August when sunlight triggers chemical reactions in cold air trapped over the South Pole during the Antarctic winter. The hole intensifies during September before tailing off as temperatures rise in November–December. In 2000, the 11.4 million-square-mile hole (*dark purple* in the satellite image) covered an area larger than the United States, Canada, and Mexico combined, the largest hole ever recorded. In September 2000, the hole extended over Punta Arenas, a city of about 120,000 people in southern Chile, exposing residents to very high levels of UV radiation.

Carbon Dioxide and Global Warming

By studying the earth's history and making comparisons with other planets, scientists have determined that concentrations of gases in the atmosphere, particularly carbon dioxide, maintain the average temperature on earth about 25°C higher than it would be if these gases were absent. The particles of carbon dioxide and other gases reflect the longer wavelengths of infrared light, or heat, radiating from the surface of the earth, keeping them in the atmosphere and creating what is known as a **greenhouse effect** (figure 56.25).

Roughly seven times as much carbon dioxide is locked up in fossil fuels as exists in the atmosphere today. Before industrialization, the concentration of carbon dioxide in the atmosphere was approximately 260 to 280 parts per million (ppm). Since the extensive use of fossil fuels began, the amount of carbon dioxide in the atmosphere has been increasing rapidly. During a 25-year period starting in 1958, the concentration of carbon dioxide increased from 315 ppm to more than 340 ppm, and it continues to rise. Climatologists have calculated that the actual mean global temperature has increased about 0.6°C since 1900, a change known as **global warming.** In some areas, however, the increase has been much greater; on Alaska's North Slope, for example, temperatures have increased 2.2° to 3.2°C (figure 56.26). Moreover, the years 1998, 2001, and 2002 were the three warmest on record since instruments started collecting data more than a century ago. Evidence for this warming can be seen in many ways. For example, ice on lakes and rivers forms later and melts sooner than it used to—on average, ice-free seasons are now 2.5 weeks longer than they were a century ago. Also, ice thickness at the North Pole has decreased 40%.

Are these phenomena related? And if the current increase in levels of carbon dioxide is responsible for the increased temperature, what does the future hold? Although the reality of global warming and the effect of greenhouse gases (which include methane, nitrous oxide, ozone, and hydrofluorocarbons as well as carbon dioxide) were hotly debated in the 1990s, the scientific community has, for the most part, reached a consensus that temperatures are increasing and that human-released greenhouse gases are a major cause of this increase. Given that carbon dioxide concentrations in the year 2100 are predicted to be between 490 and 1250 ppm (with somewhere in the middle probably most likely), the question becomes: How much will temperatures rise?

In a recent study, the U.N.-sponsored Intergovernmental Panel on Climate Change, composed of an international team of climate experts, predicted that global temperatures will rise between 1.4° and 5.8°C by the end of the twenty-first century. Such an increase in temperature will have drastic effects—some beneficial, but most detrimental—on both natural ecosystems and human populations.

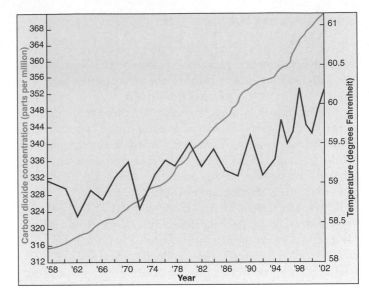

FIGURE 56.25
The greenhouse effect. The concentration of carbon dioxide in the atmosphere has steadily increased since the 1950s (*blue line*). The red line shows the general increase in average global temperature for the same period of time.
Why has temperature change been erratic, even though carbon dioxide levels have risen steadily?

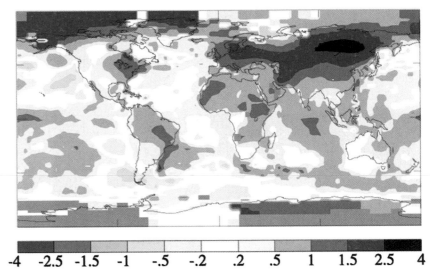

Deviations from mean temperature

FIGURE 56.26
Geographic variation in global warming. The years 2001 and 2002 were the second and third warmest on record, but some areas heated up more than others. Colors indicate how much warming has occurred in different areas relative to the mean temperature from 1951 to 1980.

Effects of Global Warming on Natural Ecosystems

Prehistoric Climate Change. Global warming—and cooling—have occurred in the past, most recently during the ice ages and intervening warm periods that punctuated the last 1.8 million years. During these times, global temperatures changed as much as 10°C from one extreme to the other; sometimes these changes may have occurred rapidly, over periods of only a few decades.

Species responded to climate change by shifting their geographic distributions, tracking their environments. For example, a number of cold-adapted North American tree species that are now found only in the far north or at high elevations lived much further south or at substantially lower elevations 10,000 to 20,000 years ago, when conditions were much cooler.

Range Shifts in Contemporary Species. Present-day global warming is having a similar effect. Examples of range shifts include the expansion of shrubs into areas of the Arctic that were previously shrub-free; northward shifts of at least 39 butterfly species, some as much as 200 kilometers in 27 years (figure 56.27); and northward range changes of 12 British bird species, averaging 400 kilometers over 20 years. Global warming has also led to elevational changes because higher altitudes are now warmer. Such changes include shifts in alpine plants in the European Alps, lowland birds in Costa Rica, and butterflies in California.

Life Cycle Changes. In addition, a changing climate changes the temperature-dependent aspects of the life cycle of many plants and animals. Many migratory birds arrive earlier and depart later from their summer breeding grounds. Many insects and amphibians breed earlier in the year, and many plants flower earlier. Leaves now change color later, and the growing season in Europe has extended 3.6 days later over the past 50 years.

Effects of Global Warming on Species

As many species change their geographic ranges and life cycles, conservation biologists are concerned for several reasons. First, sedentary species, such as some plants and invertebrates, may not be able to disperse rapidly enough to keep up with changing temperatures and environments. Perhaps more importantly, many species may simply not have the option. Although plant and animal species could move freely across the landscape in prehistoric times, that is no longer the case. Now, habitats have been fragmented, and many populations and entire species are isolated in "islands" of habitat surrounded by unsuitable areas that have been altered by humans. If environmental changes render their current habitats uninhabitable, species stuck in these islands may not be able to move and may thus become extinct. Moreover, high-altitude species

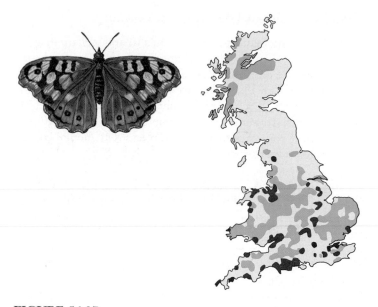

FIGURE 56.27
Butterfly range shift. Distribution of the speckled wood butterfly, *Pararge aegeria*, in Great Britain in 1915–1939 (*black*) and 1970–1997 (*blue*).

may run out of mountain to move up. Indeed, a number of high montane tropical species of frogs have recently disappeared in Costa Rica, and many scientists believe that climate change rendered their habitat unsuitable. Because they already occurred at the highest elevations, they had nowhere to go.

Species also may have trouble adapting to new conditions. One example involves spring breeding in birds. Prior to recent climate changes, the timing of reproduction of many species was finely tuned so that they would produce offspring when food resources were flush. However, the cue that triggers breeding in some species is day length, which of course has not changed. By contrast, as a result of climate change, insects, the food of these birds, appear and reproduce earlier in the year; consequently, the birds are now breeding too late in the season, and not enough food is available to feed their young. Perhaps as a result, populations of many insectivorous migratory birds are declining.

Another problem with global warming concerns species having **temperature-sensitive sex determination**. In other words, the sex of the offspring is determined by the temperature they experience during development. As a result of rising temperatures, sex ratios may be skewed. In turtles, for example, higher temperatures usually produce males, and there is concern that for some species, if temperatures rise enough, no females may be produced.

In many other ways, changing temperatures may affect how species interact with each other and with their environment. Some changes may not be detrimental, or some species may be able to adapt to the new selective pressures they experience. But in other cases, the result may be disruption of ecosystems, population declines, and extinction.

Effects of Global Warming on Humans

Global warming may affect human health and welfare in a variety of ways. Some of these changes may be beneficial, but even if they are detrimental, some countries—particularly the wealthier ones—will be able to adjust. On the other hand, some changes will require extremely costly countermeasures that even wealthy countries will be hardpressed to afford, while poorer countries may find even less detrimental consequences to be very expensive.

Rising Sea Levels. Much of the earth's supply of water is locked up in ice at either high latitudes or high altitudes. As the earth warms, it is likely that some—perhaps much—of this ice will melt. Already, glaciers are retreating in some locations around the world, sometimes at an alarming rate. Similarly, polar ice is melting, and the recent collapse of some Antarctic ice shelves is an ominous indication of what is to come. Most of this ice melt will end up in the oceans, and as a result, sea levels may rise by as much as a meter by the end of this century. Such an increase would cause increased erosion of low-lying land and coastal marshes, and other habitats would also be imperiled. As many as 200 million people would be affected by increased flooding. Should sea levels continue to rise, coastal cities and some entire islands, such as the Maldives in the Indian Ocean, would be endangered.

Other Climatic Effects. Global warming is predicted to have a variety of effects besides increased temperatures. In particular, the frequency of extreme events—such as heat waves, droughts, severe storms, and hurricanes—is expected to increase, and El Niño events, with their attendant climatic effects, may become more common. In addition, rainfall patterns are likely to shift, and those areas that are already water stressed, which are currently home to nearly two billion people, will likely face even graver water shortage problems in the years to come. Some evidence suggests that these effects are already evident in an increase in storms, hurricanes, and the frequency of El Niño events over the past few years.

Effects on Agriculture. Global warming will have both positive and negative effects on agriculture. On the positive side, warmer temperatures and increased atmospheric carbon dioxide tend to increase growth of some crops and thus may increase agricultural yields: other crops, however, will be negatively affected. Further, most crops will be affected by increased frequencies of droughts. Also, on the negative side, changes in rainfall patterns, temperature, pest distributions, and various other factors will require many adjustments. Such changes may come relatively easily for farmers in the developed world, but the associated costs may be devastating for those in the developing countries. Moreover, although crops in north temperate regions may flourish with higher temperatures, many tropical crops are already growing at their maximal temperatures, so increased temperatures may lead to reduced crop yields.

Effects on Human Health. Increasingly frequent storms, flooding, and drought will have adverse consequences on human health. Aside from their direct impact, such events often disrupt the fragile infrastructure of developing countries, leading to the loss of safe drinking water and other problems. As a result, epidemics of cholera and other diseases may be expected to occur more often as a consequence of these events.

In addition, as temperatures rise, areas suitable for tropical organisms will expand northward. Of particular concern are those organisms that cause human diseases. Many diseases currently limited to tropical areas may expand their range and become problematic in non-tropical countries. Diseases transmitted by mosquitoes, such as malaria, dengue fever, and several types of encephalitis, are examples. The distribution of mosquitoes is limited by cold; winter freezes kill many mosquitoes and their eggs. As a result, malaria only occurs in areas where temperatures are usually above 16°C, and yellow fever and dengue fever occur in areas where temperatures are normally above 10°C. (The difference results because the diseases are transmitted by different mosquito species.) Moreover, at higher temperatures, the malaria pathogen matures more rapidly. Malaria already kills one million people every year; some projections suggest that the percentage of the human population at risk for malaria may increase by 33% by the end of the twenty-first century. Moreover, as predicted, malaria already appears to be on the move. By 1980, malaria had been eradicated from all of the United States except California, but in recent years it has appeared in a variety of southern, and even a few northern, states.

Dengue fever (sometimes called "breakbone fever" because of the pain it causes) is also spreading. Previously a disease restricted to the tropics and subtropics, where it infects 50–100 million people, it now occurs in the United States, southern South America, and northern Australia.

One of the most alarming aspects of these diseases is that no vaccines are available. Drug treatment is available (for malaria), but the parasites are rapidly evolving resistance and rendering the drugs ineffective. There is no drug treatment for dengue fever.

Solving the global warming problem will not be easy. It will require large reductions in the amount of carbon dioxide injected into the atmosphere. Some nations are taking steps to reduce their emissions, but others are not. Coordinated international effort is required to slow down the increase in global temperatures. Although the predicted effects of global warming are uncertain, most scientists agree that the impact will be severe.

Global temperatures are rising, at least in part as a result of human activities. Increased temperature will have many detrimental effects on natural ecosystems and human welfare.

56.1 Climate shapes the character of ecosystems.

Effects of the Sun and Atmospheric Circulation

- The tropics are warmer than temperate regions because the sun's rays arrive almost perpendicular to equatorial regions. (p. 1204)
- High temperatures near the equator encourage evaporation and create warm, moist air, causing large amounts of precipitation near the equator. (p. 1205)
- When the air masses reach about 30° north and south latitude, the cool, dry air sinks, forming areas of low precipitation (deserts). (p. 1205)

Atmospheric Circulation, Precipitation, and Climate

- The rain shadow effect helps create deserts at places other than 30° north and south latitude, as mountain ranges intercept moisture-laden winds and cause most of the precipitation to fall on the windward side of the range, blocking moisture from the leeward side. (p. 1206)
- Unique regional climates are also formed due to prevailing winds such as the westerlies and the trade winds. (p. 1206)
- Temperature tends to vary according to latitude—warm in the tropics, and cooler as you move away—and according to elevation, with temperature decreasing as elevation increases. (p. 1207)

56.2 Biomes are widespread terrestrial ecosystems.

The Major Biomes

- Biomes are major communities of organisms with a characteristic appearance that are distributed over areas defined by temperature and precipitation differences. (p. 1208)
- Eight of the major biomes are tropical rain forest, savanna, desert, temperate grassland, temperate deciduous forest, temperate evergreen forest, taiga, and tundra. (p. 1208)

56.3 Aquatic ecosystems cover much of the earth.

Patterns of Circulation in the Oceans

- Ocean circulation patterns are determined by patterns of atmospheric circulation, but are also modified by the location of landmasses. (p. 1212)
- When the east-west trade winds in the Pacific Ocean slacken, warm water begins to move back across the ocean from the coast of South America, causing a phenomenon called El Niño, which widely influences the world's weather systems. (p. 1213)

Marine Ecosystems

- The marine environment consists of three major habitats: the neritic zone (shallow water containing most of the world's major fisheries), the pelagic zone (water above the ocean floor), and the benthic zone (ocean floor). (p. 1214)
- Approximately 40% of all photosynthesis on earth takes place in the oceans. (p. 1215)

Freshwater Habitats

- Freshwater habitats are much more limited in area than marine habitats. (p. 1216)
- Photosynthetic organisms are limited to the upper photic zone of ponds and lakes. (p. 1216)
- Ponds and lakes have three zones where organisms are found, which are distributed according to the distance from shore: the littoral, limnetic, and profundal zones. (p. 1216)

- Thermal stratification is characteristic of larger lakes in temperate regions. (p. 1216)
- Eutrophic lakes contain an abundant supply of organic matter, while organic matter and nutrients are relatively scarce in oligotrophic lakes. (p. 1217)
- Human activities can lead to the eutrophication of oligotrophic waters. (p. 1217)
- Wetlands are often very productive ecosystems that also serve as water storage basins helping to moderate flooding. (p. 1217)

56.4 Human activity is placing the biosphere under increasing stress.

Pollution

- Widespread modern agriculture introduces large amounts of new chemicals into the global ecosystem, including pesticides, herbicides, and fertilizers. (p. 1218)
- Stable, long-lasting chlorinated hydrocarbons such as DDT can become increasingly concentrated in an ecosystem due to biological magnification. (p. 1218)

Acid Precipitation

- Industrial pollution contains many chemicals. For example, sulfur, when introduced into the upper atmosphere, can combine with water vapor to produce sulfuric acid, which can fall to the ground in many forms, including snow and rain. (p. 1219)
- Precipitation with an acidic pH can cause many environmental problems, including lake acidification, groundwater contamination, and forest damage. (p. 1219)

Destruction of the Tropical Forests

- More than half of the world's human population lives in the tropics, and this percentage is increasing. (p. 1220)
- Rain forests grow on poor soil, thus they can become hard to regenerate once they are cleared. (p. 1220)
- Increasingly, larger numbers of people moving into tropical areas are clearing larger areas of rain forest. (p. 1220)

The Ozone Hole

- Industrial chlorofluorocarbons (CFCs) released into the atmosphere for many decades have led to a thinning of the earth's stratospheric ozone layer, which shields the planet from harmful ultraviolet radiation. (p. 1221)

Carbon Dioxide and Global Warming

- The greenhouse effect is caused when carbon dioxide and other gases allow short-wavelength solar radiation into the atmosphere, but trap longer-wavelength heat radiation from escaping, thus warming the earth's atmosphere. (p. 1222)
- In the 1990s, the scientific community reached a consensus that the earth's average temperature is increasing, and human-related greenhouse gases are the major cause of the increase. (p. 1222)
- The effects of global warming on natural ecosystems may include the shifting of the geographic distribution of organisms tracking environmental conditions, as well as life cycle changes to adapt to changing environmental conditions. (p. 1223)
- Continued global warming may also influence human conditions due to rising sea levels, effects on agriculture, and human health issues. (p. 1224)

Self Test

1. A rain shadow results in
   a. extremely wet conditions due to loss of moisture from winds rising over a mountain range.
   b. dry air moving toward the poles that cools and sinks in regions 15° to 30° north/south latitude.
   c. global polar regions that rarely receive moisture from the warmer, tropical regions, and are therefore dryer.
   d. desert conditions on the down-wind side of a mountain due to increased moisture-holding capacity of the winds coming from the seas.

2. What two factors are most important in biome distribution?
   a. temperature and latitude
   b. rainfall and temperature
   c. latitude and rainfall
   d. temperature and soil type

3. Savannas are best described as areas with
   a. extremely dry conditions and sparse vegetation.
   b. cold, dry conditions with herbs and few trees.
   c. warm summers, cool winters, and abundant rainfall that promotes tree growth.
   d. seasonal rainfall, few trees, and abundant grasses.

4. The cacti found in the deserts of North and South America look very much like the euphorbs found in the deserts of Africa. However, these plants are not closely related. The similarities in these plants are due to
   a. convergent evolution as a result of similar environmental pressures.
   b. artificial selection for these similar traits.
   c. differences in rainfall between the two deserts.
   d. differences in pollinator species in the two deserts.

5. Which of the following is not a result of an El Niño event?
   a. The trade winds relax in the central and western Pacific.
   b. The sea surface is about a meter higher at the Philippines than at Ecuador.
   c. A rise in sea surface temperature and a decline in primary productivity adversely affect fisheries in Ecuador and Peru.
   d. Flooding and strong winter storms occur in California.

6. The neritic zone is best described as the
   a. area of water above the ocean floor where a diversity of plankton species are concentrated.
   b. ocean floor that is made up of mud and other fine particles that have settled from the water.
   c. area less than 300 meters below the surface of the oceans along the coasts of continents and islands.
   d. part of the ocean floor that drops to depths where light does not penetrate.

7. The limnetic zone of a lake is best described as the
   a. shallow area along the shore.
   b. area below the limits where light can penetrate.
   c. zone where photosynthesis cannot occur.
   d. well-illuminated surface waters away from the shore.

8. In temperate regions, lakes are thermally stratified, with warm waters at the top and cooler waters at the bottom during the summer. The region of abrupt change between these layers is known as the
   a. thermocline.
   b. hypolimnion.
   c. epilimnion.
   d. fall overturn.

9. Oligotrophic lakes can be turned into eutrophic lakes as a result of human activities such as
   a. overfishing of sensitive species, which disrupts fish communities.
   b. introducing nutrients into the water, which stimulates plant and algal growth.
   c. disrupting terrestrial vegetation near the shore, which causes soil to run into the lake.
   d. spraying pesticides into the water to control aquatic insect populations.

10. The loss of the ozone layer has serious implications for the quality of the environment because
    a. ozone (O_3) protects organisms from ultraviolet radiation that can cause cancer.
    b. a depleted ozone layer causes rainwater to have a lower pH that kills plant life.
    c. loss of the ozone layer causes the sun's rays to get trapped in the atmosphere and increase global temperatures.
    d. a depleted ozone layer can interact with toxic chemicals to increase their effect on organismal health.

Test Your Visual Understanding

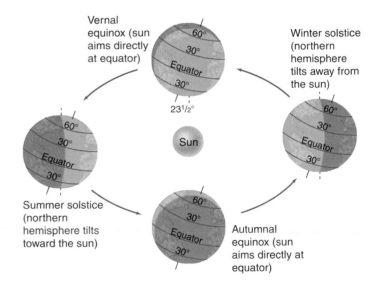

1. Predict what changes to global climate would occur if the earth's axis were not tilted.

Apply Your Knowledge

1. Some vegetarians argue that it is more ethical to eat "lower on the food chain" (eat more grains and vegetables) than to consume meat. Explain this argument in terms of energy conversions in ecosystems.

2. The Clean Water Act protects wetlands. Under the act, there is to be no net loss of wetlands and any wetland losses should be offset through restoration projects or creation of new wetlands. Why are wetlands given this status?

3. The concentration of CO_2 in the atmosphere has steadily increased since the 1950s. A general increase in average global temperature has also occurred. How is global climate change predicted to affect the area where you live?

57

Conservation Biology

Concept Outline

57.1 The new science of conservation biology is focused on conserving biodiversity.

Overview of the Biodiversity Crisis. In prehistoric times, humans decimated the faunas of many areas. Today, worldwide extinction rates are accelerating.

Species Endemism and Hotspots. Some geographic areas are particularly rich in species that occur nowhere else.

What's So Bad About Losing Biodiversity? Biodiversity has considerable direct economic value and provides key support to the biosphere.

57.2 The extinction crisis is a result of many factors.

Factors Responsible for Extinction. Most recorded extinctions can be attributed to a few causes. Sometimes more than one factor can affect a species at the same time.

Habitat Loss. Without a place to live, species cannot survive.

Overexploitation. Species cannot persist if too many individuals are removed by humans.

Detrimental Effects of Introduced Species. Introduced species can wreak havoc on native species and ecosystems.

Disruption of Ecosystems. Extinction of one species can have a cascading effect throughout the food web, making other species vulnerable as well.

The Perils of Small Population Size. Small populations are vulnerable to genetic and demographic problems.

57.3 Successful recovery efforts need to be multidimensional.

Approaches for Preserving Endangered Species. Species preservation efforts take many forms and must be tailored toward the particular threats that species face.

Conservation of Ecosystems. Maintaining large preserves and focusing on the health of the entire ecosystem may be the best means of preserving biodiversity.

FIGURE 57.1
Endangered. The Siberian tiger is in grave danger of extinction, being hunted for its pelt and having its natural habitat greatly reduced. A concerted effort to save it is using many of the approaches discussed in this chapter.

Among the greatest challenges facing the biosphere is the accelerating pace of species extinctions. Not since the Cretaceous period have so many species become extinct in so short a time span (figure 57.1). This challenge has led to the emergence in the last decade of the discipline of conservation biology. Conservation biology is an applied science that seeks to learn how to preserve species, communities, and ecosystems. It studies the causes of declines in species richness and attempts to develop methods for preventing such declines. In this chapter, we first examine the biodiversity crisis and its importance. Then, using case histories, we identify and study factors that have played key roles in many extinctions. We finish with a review of recovery efforts at the species and community levels.

1227

57.1 The new science of conservation biology is focused on conserving biodiversity.

Overview of the Biodiversity Crisis

Extinction is a fact of life. Most species—probably all—become extinct eventually. More than 99% of species known to science (most from the fossil record) are now extinct. However, current rates are alarmingly high. Taking into account the rapid and accelerating loss of habitat that is occurring, especially in the tropics, it has been calculated that as much as 20% of the world's biodiversity may be lost by the middle of this century. In addition, many of these species may be lost before we are even aware of their existence. Scientists estimate that no more than 15% of the world's eukaryotic organisms have been discovered and given scientific names, and this proportion is probably much lower for tropical species.

These losses will not only affect poorly known groups. As many as 50,000 species of the world's total of 250,000 species of plants, 4000 of the world's 20,000 species of butterflies, and nearly 2000 of the world's 9000 species of birds could be lost during this time period. Considering that the human species has been in existence for only 600,000 years of the world's 4.5-billion-year history, and that our ancestors developed agriculture only about 10,000 years ago, this is an astonishing—and dubious—accomplishment.

Extinctions Due to Prehistoric Humans

A great deal can be learned about current rates of extinction by studying the past, and in particular the impact of human-caused extinctions. In prehistoric times, members of *Homo sapiens* wreaked havoc whenever they entered a new area. For example, at the end of the last Ice Age, approximately 12,000 years ago, the fauna of North America was composed of a diversity of large mammals similar to those living in Africa today: mammoths and mastodons, horses, camels, giant ground sloths, saber-toothed cats, and lions, among others (figure 57.2). Shortly after humans arrived, 74 to 86% of the *megafauna* (that is, animals weighing more than 100 pounds) became extinct. These extinctions are thought to have been caused by hunting and, indirectly, by burning and clearing of forests. (Some scientists attribute these extinctions to climate change, but that hypothesis doesn't explain why the end of earlier ice ages was not associated with

FIGURE 57.2
North America before human inhabitants. Animals found in North America prior to the migration of humans included birds and large mammals, such as the ancient North American camel, saber-toothed cat, giant ground sloth, and teratorn vulture.

mass extinctions, nor does it explain why extinctions occurred primarily among larger animals, with smaller species relatively unaffected.)

Around the globe, similar results have followed the arrival of humans. Forty thousand years ago, Australia was occupied by a wide variety of large animals, including marsupials similar in size and ecology to hippos and leopards, a kangaroo nine feet tall, and a 20-foot-long monitor lizard. These all disappeared, at approximately the same time as humans arrived. Smaller islands have also been devastated. Madagascar has seen the extinction of at least 15 species of lemurs, including one the size of a gorilla; a pygmy hippopotamus; and the flightless elephant bird, *Aepyornis*, the largest bird to ever live (more than 3 meters tall and weighing 450 kilograms). On New Zealand, 30 species of birds went extinct, including all 13 species of moas, another group of large, flightless birds. Interestingly, one continent that seems to have been spared these megafaunal extinctions is Africa. Scientists speculate that this lack of extinction in prehistoric Africa may have resulted because much of human evolution occurred in Africa. Consequently, other African species had been coevolving with humans for several million years and thus had evolved counteradaptations to human predation.

Extinctions in Historical Time

Historical extinction rates are best known for birds and mammals because these species are conspicuous—that is, relatively large and well studied. Estimates of extinction rates for other species are much rougher. The data presented in table 57.1, based on the best available evidence, show recorded extinctions from 1600 to the present. These estimates indicate that about 85 species of mammals and 113 species of birds have become extinct since the year 1600. That is about 2.1% of known mammal species and 1.3% of known birds. The majority of extinctions have occurred in the last 150 years. The extinction rate for birds and mammals was about one species every decade from 1600 to 1700, but it rose to one species every year during the period from 1850 to 1950, and to four species per year between 1986 and 1990 (figure 57.3). This increase in the rate of extinction is the heart of the biodiversity crisis.

Unfortunately, the biodiversity crisis seems to be worsening. For example, the number of bird species recognized as "critically endangered" increased 8% from 1996 to 2000, and a 2002 report suggested that as many as half of the earth's plant species may be threatened with extinction. Some researchers predict that two-thirds of all vertebrate species could perish by the end of this century.

The majority of historic extinctions—though by no means all of them—have occurred on islands. For example, of the 90 species of mammals that have gone extinct in the last 500 years, 73% lived on islands (and another 19% in Australia). The particular vulnerability of island species probably results from a number of factors: Such species have often evolved in the absence of predators, and so have lost their ability to escape both humans and introduced predators such as rats and cats. In addition, humans have introduced competitors and diseases; avian malaria, for example has devastated the bird fauna of the Hawaiian Islands. Finally, island populations are often relatively small, and thus particularly vulnerable to extinction, as we shall see later in this chapter.

In recent years, the extinction crisis has moved from islands to continents. Most species now threatened with extinction occur on continents, and these areas will bear the brunt of the extinction crisis in this century.

Some people have argued that we should not be concerned, because extinctions are a natural event and mass extinctions (the extinction of large numbers of species within a geologically short period of time) have occurred in the past. Indeed, mass extinctions have taken place several times over the past half-billion years. However, the current mass extinction event is notable in several respects. First, it

is the only such event triggered by a single species. Moreover, although species diversity usually recovers after a few million years, this is a long time to deny our descendants the benefits and joys of biodiversity. In addition, it is not clear that biodiversity will rebound this time. After previous mass extinctions, new species have evolved to utilize resources available due to extinctions of the species that previously used them. Today, however, such resources are unlikely to be available, because humans are destroying the habitats and taking the resources for their own use.

> Since prehistoric times, humans have had a devastating effect on biodiversity almost everywhere in the world. Most historical extinctions have occurred on islands, but most future extinctions will occur on continents.

Taxon	Recorded Extinctions				Approximate Number of Species	Percent of Taxon Extinct
	Mainland	Island	Ocean	Total		
Mammals	30	51	4	85	4,000	2.1
Birds	21	92	0	113	9,000	1.3
Reptiles	1	20	0	21	6,300	0.3
Amphibians*	2	0	0	2	4,200	0.05
Fish	22	1	0	23	19,100	0.1
Invertebrates	49	48	1	98	1,000,000+	0.01
Flowering plants	245	139	0	384	250,000	0.2

Table 57.1 Recorded Extinctions Since 1600 A.D.

*An alarming decline in amphibian populations has occurred recently, and many species may be on the verge of extinction.

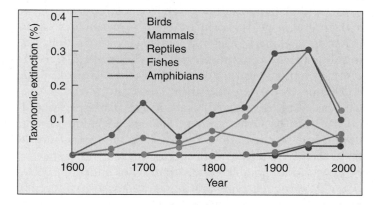

FIGURE 57.3
Trends in species loss. These graphs present data on recorded animal extinctions since 1600. The majority of extinctions have occurred on islands, with birds and mammals particularly affected (although this may reflect to some degree our more limited knowledge of other groups).
Why are extinction rates highest for birds and mammals?

Species Endemism and Hotspots

A species found naturally in only one geographic area and no place else is said to be **endemic** to that area. The area over which an endemic species is found may be very large. For example, the black cherry tree (*Prunus serotina*) is endemic to all of temperate North America. More typically, however, endemic species occupy restricted ranges. The Komodo dragon (*Varanus komodoensis*) lives only on a few small islands in the Indonesian archipelago, while the Mauna Kea silversword (*Argyroxiphium sandwicense*) lives in a single volcano crater on the island of Hawaii.

Isolated geographic areas, such as oceanic islands, lakes, and mountain peaks, often have high percentages of endemic species, many in significant danger of extinction. The number of endemic plant species varies greatly in the United States from one state to another. For example, 379 plant species are found in Texas and nowhere else, whereas New York has only one endemic plant species. California, with its varied array of habitats, including deserts, mountains, seacoast, old-growth forests, and grasslands, is home to more endemic plant species than any other state.

Worldwide, notable concentrations of endemic species occur in particular regions. Conservationists have recently identified areas, termed **hotspots**, that have high endemism and are disappearing at a rapid rate. Such hotspots include Madagascar, a variety of tropical rain forests, the eastern Hi-

malayas, areas with Mediterranean climates such as California, South Africa, and Australia, and in several other climatic areas (figure 57.4 and table 57.2). Overall, 25 such hotspots have been identified, which in total contain nearly half of all the terrestrial species in the world.

Why these areas contain so many endemic species is a topic of active scientific research. Some of these hotspots occur in areas of high species diversity, and the explanations for high species diversity in general (see chapter 55), such as high productivity, probably apply to these hotspots as well. In addition, some hotspots occur on isolated islands, such as New Zealand, New Caledonia, and Polynesia (including the Hawaiian Islands), where evolutionary diversification over long periods of time has resulted in rich biotas composed of plant and animal species found nowhere else in the world.

Table 57.2 Numbers of Endemic Species in Some Hotspot Areas

Region	Mammals	Reptiles	Amphibians	Plants
Atlantic coastal Brazil	160	60	253	6,000
South American Chocó	60	63	210	2,250
Philippines	115	159	65	5,832
Tropical Andes	68	218	604	20,000
Southwestern Australia	7	50	24	4,331
Madagascar	84	301	187	9,704
Cape region (South Africa)	9	19	19	5,682
California Floristic Province	30	16	17	2,125
New Caledonia	6	56	0	2,551
South-central China	75	16	51	3,500

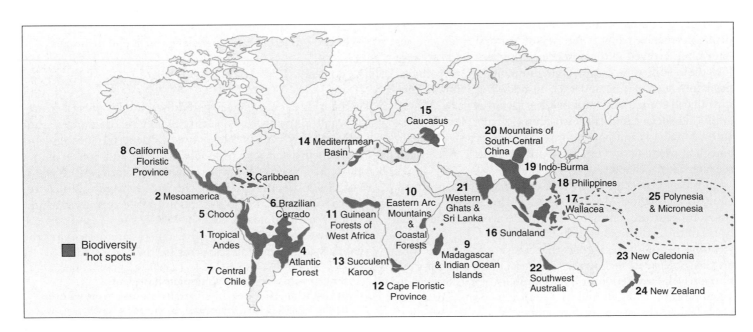

FIGURE 57.4
Hotspots of high endemism. These areas are rich in endemic species under threat of imminent extinction.

Population Growth in Hotspots

Because of the great number of endemic species that hotspots contain, conserving their biological diversity must be an important component of efforts to safeguard the world's biological heritage. Or, to look at it another way, by protecting just 1.4% of the world's land surface, 44% of the world's vascular plants and 35% of its terrestrial vertebrates can be preserved.

Unfortunately, hotspots contain not only many endemic species, but also growing human populations. In 1995, these areas contained 1.1 billion people—20% of the world's population—sometimes at high densities (figure 57.5*a*). More importantly, human populations were growing in all but one of these hotspots as the result of both high birth and immigration rates; overall, the rate of growth exceeded the global average in 19 hotspots (figure 57.5*b*). Indeed, in some hotspots, the rate of growth is nearly twice that of the rest of the world.

Not surprisingly, many of these areas are experiencing high rates of habitat destruction as land is cleared for agriculture, housing, and economic development. More than 70% of the original area of each hotspot has already disappeared, and in 14 hotspots, 15% or less of the original habitat remains. In Madagascar, it is estimated that 90% of the original forest has already been lost—this on an island where 85% of the species are found nowhere else in the world. In the forests of the Atlantic coast of Brazil, the extent of deforestation is even higher: 95% of the original forest is gone.

Of course, population pressure is not the only cause of habitat destruction in hotspots. Commercial exploitation to meet the demands of more affluent people in the developed world also plays an important role. For example, large-scale logging of tropical rain forests occurs in countries around the world to provide lumber, most of which ends up in the United States, western Europe, and Japan. Similarly, many forests in Central and South America are cleared to make way for cattle ranches that produce cheap meat for fast-food restaurants. Hotspots in more affluent countries are often at risk because they occur in areas where land has great value for real estate and commercial purposes.

Regardless of its cause, decimation of hotspots will take a great toll on the world's biological diversity. As we shall see shortly, such massive rates of habitat destruction result in extremely high rates of species extinction.

Some areas of the earth have particularly high levels of species endemism. Unfortunately, many of these areas are currently in great jeopardy due to human population growth and habitat destruction, with correspondingly high rates of species extinction.

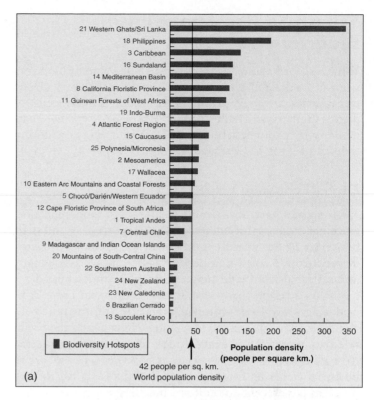

(a)

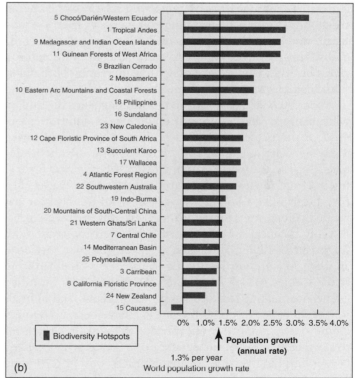

(b)

FIGURE 57.5

Human populations in hotspots. The rich biodiversity in many hotspots is under pressure from (*a*) dense and (*b*) rapidly growing human populations.

Why do population density and growth rates differ among hotspots?

What's So Bad About Losing Biodiversity?

What's so bad about losing species? The value of biodiversity can be divided into three principal components: (1) *direct economic value* of products we obtain from species of plants, animals, and other groups; (2) *indirect economic value* of benefits produced by species without our consuming them; and (3) *ethical and aesthetic values*.

Direct Economic Value

Many species have direct value as sources of food, medicine, clothing, biomass (for energy and other purposes), and shelter. Most of the world's food crops, for example, are each derived from a small number of plants that were originally domesticated from wild plants in tropical and semiarid regions. As a result, many of our most important crops contain relatively little genetic variation (equivalent to a "founder effect" discussed in chapter 21), whereas their wild relatives have great diversity. In the future, genetic variation from wild strains of these species may be needed if we are to improve yields or find a way to breed resistance to new pests. In fact, recent agricultural breeding experiments have illustrated the value of conserving wild relatives of common crops. For example, by breeding commercial varieties with a small, oddly colored wild species of the tomato from the mountains of Peru, scientists were able to increase crop yields by 50%, while at the same time increasing both nutritional content and color.

About 70% of the world's population depends directly on wild plants as their source of medicine. In addition about 40% of the prescription and nonprescription drugs used today have active ingredients extracted from plants or animals. Aspirin, the world's most widely used drug, was first extracted from the leaves of the tropical willow, *Salix alba*. The rosy periwinkle, *Catharanthus roseus*, from Madagascar has yielded potent drugs for combating leukemia (figure 57.6).

Only recently have biologists perfected the techniques that make possible the transfer of genes from one species to another. We are just beginning to use genes obtained from other species to our advantage (see chapter 16). So-called "gene prospecting" of the genomes of plants and animals for useful genes has only begun. We have been able to examine only a minute proportion of the world's organisms to see whether any of their genes have useful properties. By conserving biodiversity, we maintain the option of finding useful benefits in the future; unfortunately, many of the most promising species occur in habitats that are being destroyed at an alarming rate, such as tropical rain forests.

Indirect Economic Value

Diverse biological communities are of vital importance to healthy ecosystems. They help maintain the chemical

FIGURE 57.6
The rosy periwinkle. Two drugs extracted from the Madagascar periwinkle *Catharanthus roseus*, vinblastine and vincristine, effectively treat common forms of childhood leukemia, increasing chances of survival from 20% to over 95%.

quality of natural water, buffer ecosystems against floods and drought, preserve soils and prevent loss of minerals and nutrients, moderate local and regional climate, absorb pollution, and promote the breakdown of organic wastes and the cycling of minerals. By destroying biodiversity, we are creating conditions of instability and lessened productivity and promoting desertification, waterlogging, mineralization, and many other undesirable outcomes throughout the world.

Economists have recently been able to compare the societal value, in monetary terms, of intact habitats compared with the value of destroying those habitats. Surprisingly, in most studies conducted so far, intact ecosystems are more valuable than the products derived by destroying them. For example, in Thailand, coastal mangrove habitats are commonly cleared so that shrimp farms can be established. Although the shrimp produced are valuable, their value is vastly outweighed by the benefits in timber, charcoal production, offshore fisheries, and storm protection provided by the mangroves (figure 57.7*a*). Similarly, intact tropical rain forest in Cameroon, West Africa, provides fruit and other forest materials. Clearing the forest for agriculture or palm plantations leads to stream-polluting erosion as well as increased flooding. Combining all the costs and benefits of the three options, maintaining intact forests has the highest economic value (figure 57.7*b*).

Probably the most famous example of the value of intact ecosystems is provided by the watersheds of New York City. Ninety percent of the water for the New York area's nine million residents comes from the Catskill Mountains and the nearby headwaters of the Delaware River (figure 57.8). Water that runs off from over 1600 square miles of rural, mountainous areas is collected into reservoirs and then transported by aqueduct more than 85 miles to New York City at a rate of 1.3 billion gallons per day.

In the 1990s, New York City faced a dilemma. New federal water regulations were requiring ever cleaner water, even as development and pollution in the source areas of the water were threatening to compromise water quality. The city had two choices: either work to protect the functioning ecosystem so that it could produce clean water, or construct filtration plants to clean it upon arrival. Economic analysis made the choice clear: Building the plants would cost $6 billion, with annual operating costs of $300 million, whereas spending a billion dollars over ten years could preserve the ecosystem and maintain water purity. The decision was easy.

These examples provide some idea of the value of the services that ecosystems provide. But maintaining ecosystems is not always more valuable than converting them to other uses. Certainly, when the United States was being settled and land was plentiful, ecosystem conversion was beneficial. Even today, habitat destruction may sometimes be economically desirable. Nonetheless, we still have only a rudimentary knowledge of the many ways intact ecosystems provide services. In many cases, it is not until they are lost that the value becomes clear, as unexpected negative effects, such as increased flooding and pollution or decreased rainfall, become apparent.

The same argument can be made for preserving particular species within ecosystems. Given how little we know about the biology of most species, particularly in the tropics, it is impossible to predict all the consequences of removing a species. Imagine taking a parts list for an airplane and randomly changing a digit in one of the part numbers: You might change a cushion to a roll of toilet paper—but you might just as easily change a key bolt holding up the wing to a pencil. By removing biodiversity, we are gambling with the future of the ecosystems upon which we depend and whose functioning we understand very little.

Ethical and Aesthetic Values

Many people believe that preserving biodiversity is an ethical issue because every species is of value in its own right, even if humans are not able to exploit or benefit from it. These people feel that along with the power to exploit and destroy other species comes responsibility: As the only organisms capable of eliminating large numbers of species and entire ecosystems, and as the only organisms capable of reflecting upon what we are doing, humans should act as guardians or stewards for the diversity of life around us.

Almost no one would deny the aesthetic value of biodiversity—a beautiful flower or a noble elephant—but

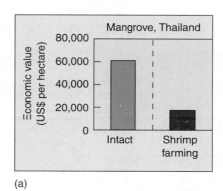

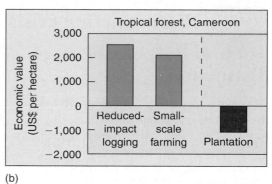

(a)   (b)

FIGURE 57.7

The economic value of maintaining habitats. (*a*) Mangroves in Thailand and (*b*) rain forests in Cameroon provide more economic benefits if they are left standing than if they are destroyed and the land used for other purposes.

If shrimp farms established on cleared mangrove habitats make money, how can clearing mangroves not be an economic plus?

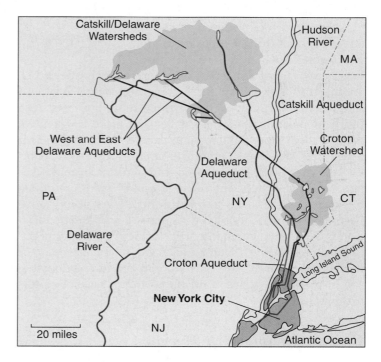

FIGURE 57.8

New York City's water source. New York gets its water from distant rain catchments. Preserving the ecological integrity of these areas is cheaper than building new water treatment plants.

how do we place a value on beauty? Perhaps the best we can do is to appreciate the deep sense of loss we would feel if it no longer existed.

Biodiversity is of great value in its own right, as well as for the products it provides, its contributions to the health of the ecosystems we depend on, and the beauty it offers.

Factors Responsible for Extinction

A variety of causes, independently or in concert, are responsible for extinctions (table 57.3). Historically, overexploitation was the major cause of extinction; although overexploitation is still a factor, habitat loss is the major problem for most groups today, while introduced species rank second. Many other factors can contribute to species extinctions as well, including disruption of ecosystem interactions, pollution, loss of genetic variation, and catastrophic disturbances, either natural or man-made.

More than one of these factors may affect a species. In fact, a chain reaction is possible in which the action of one factor predisposes a species to be more severely affected by another factor. For example, habitat destruction may lead to decreased birthrates and increased mortality rates. As a result, populations become smaller and more fragmented, making them more vulnerable to disasters such as floods or forest fires, which may eliminate populations. As the habitat becomes more fragmented, populations become isolated, so that genetic interchange ceases and areas devastated by disasters are not recolonized. As populations become very small, inbreeding increases, and genetic variation is lost through genetic drift, further decreasing population fitness. Which factor causes the final coup de grace may be irrelevant; many factors, and the interactions between them, may have contributed to a species' eventual extinction.

Case Study: Amphibian Declines

In 1963, herpetologist Jay Savage was hiking through pristine cloud forest in Costa Rica. Reaching a wind-swept ridge, he couldn't believe his eyes. Before him was a huge aggregation of breeding toads. What was so amazing was the color of the toads: bright, eye-dazzling orange, unlike anything he had ever seen before (figure 57.9). The color of the toads was so amazing and unexpected that Savage briefly considered the possibility that colleagues had played a practical joke, getting to the clearing before him and somehow coloring normal toads orange. Realizing that this could not be, he went on to study the toads, eventually describing a species new to science, the golden toad, *Bufo periglenes*.

For the next 24 years, large numbers of toads were seen during the breeding season each spring. Their home

Table 57.3 Causes of Extinctions

	Percentage of Species Influenced by the Given Factor*				
Group	Habitat Loss	Overexploitation	Species Introduction	Other	Unknown
EXTINCTIONS					
Mammals	19	23	20	2	36
Birds	20	11	22	2	45
Reptiles	5	32	42	0	21
Fish	35	4	30	4	48
THREATENED EXTINCTIONS					
Mammals	68	54	6	20	—
Birds	58	30	28	2	—
Reptiles	53	63	17	9	—
Fish	78	12	28	2	—

*Some species may be influenced by more than one factor; thus, some rows may exceed 100%.

FIGURE 57.9
An extinct species. The golden toad was last seen in the wild in 1989.

was legally recognized as the Monteverde Cloud Forest Reserve, a well-protected, intact, and functioning ecosystem, seemingly a successful model of conservation. Then, in 1988, few toads were seen, and in 1989, only a single male was observed. Since then, despite exhaustive efforts, no more golden toads have been found. Despite living in a well-protected ecosystem, with no obvious threats from pollution, introduced species, overexploitation, or any other factor, the species appears to have gone extinct,

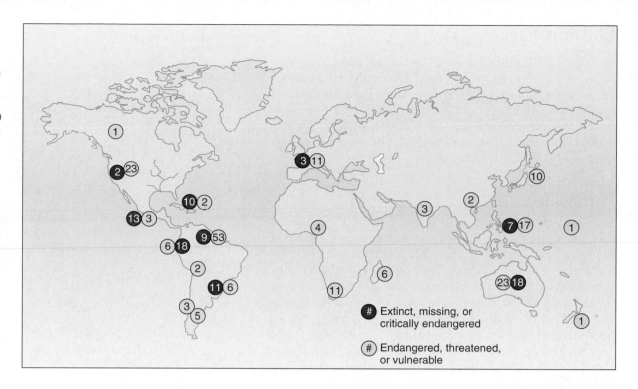

FIGURE 57.10
Amphibian extinction crisis. Circles indicate the number of extinct (*red*) and endangered (*yellow*) species around the world. These numbers are rapidly being revised upward as scientists focus their attention on little-known species, many of which turn out to be in grave danger.

right under the eyes of watchful scientists and conservationists. How could this happen?

Frogs in Trouble. At the first World Herpetological Congress in 1989 in Canterbury, England, frog experts from around the world met to discuss conservation issues relating to frogs and toads. At this meeting, it became clear that the golden toad story was not unique. Experts reported case after case of similar stories: Frog populations that had once been abundant were now decreasing or entirely gone.

Since then, scientists have devoted a great deal of time and effort to determining whether frogs and other amphibian species truly are in trouble and, if so, why. Unfortunately, the situation appears to be even worse than originally suspected. In 2002, experts associated with the University of California at Berkeley reported that at least 32 amphibian species have gone extinct in recent years; another 26 are "missing in action," not having been seen for many years and possibly extinct; and 91 species, in countries as different as Ecuador, Venezuela, Australia, and the United States, are critically endangered (figure 57.10). Moreover, these numbers are probably underestimates; little information exists from many areas of the world, such as Southeast Asia and central Africa. Indeed, in that same year, researchers suggested that as many as 100 species from the island nation of Sri Lanka have recently gone extinct, news that is perhaps not surprising given that 95% of the nation's rain forests have also disappeared in recent times.

Cause for Concern. Amphibian declines are worrisome for several reasons. First, many of the species—

including the golden toad—have declined in pristine, well-protected habitats. If species are going extinct in such areas, it bodes ill for our ability to preserve global biological diversity. Second, many amphibian species are particularly sensitive to the state of the environment because of their moist skin, which allows chemicals from the environment to pass into the body, and their use of aquatic habitats for larval stages (such as the tadpoles of frogs), which requires unpolluted water. In other words, amphibians may be analogous to the canaries formerly used in coal mines to detect problems with air quality: If the canaries keeled over, the miners knew they had to get out.

Third, no single cause for amphibian declines is apparent. Although a single cause would be of concern, it would also suggest that a coordinated global effort could reverse the trend, as happened with chlorofluorocarbons and decreasing ozone levels (see chapter 56). However, different species are afflicted by different problems, including habitat destruction, global warming–induced environmental changes, pollution, decreased ozone levels, parasite epidemics, and introduced species. This is an area of active scientific research, and the implication is that the global environment is deteriorating in many different ways. Could amphibians be global "canaries," serving as indicators that the world's environment is in serious trouble?

Many factors are responsible for extinction. Many amphibian species are in trouble around the world. No single cause is responsible, which raises worry about the overall state of the global environment.

Habitat Loss

As table 57.3 indicates, habitat loss is the most important cause of modern-day extinction. Given the tremendous amounts of ongoing destruction of all types of habitat, from rain forest to ocean floor, this should come as no surprise. Natural habitats may be adversely affected by humans in four ways: (1) destruction, (2) pollution, (3) disruption, and (4) habitat fragmentation.

Destruction

A proportion of the habitat available to a particular species may simply be destroyed. This is a common occurrence in the "clear-cut" harvesting of timber, in the burning of tropical forest to produce grazing land, and in urban and industrial development. Forest clearance has been, and continues to be, by far the most pervasive form of habitat disruption (figure 57.11). Many tropical forests are being cut or burned at a rate of 1% or more per year.

To estimate the effect of reductions in habitat available to a species, biologists often use the well-established observation that larger areas support more species (see figure 55.20). Although this relationship varies according to geographic area, type of organism, and type of area (for example, oceanic islands or patches of habitat on the mainland), in general a tenfold increase in area leads to approximately a doubling in the number of species. This relationship suggests, conversely, that if the area of a habitat is reduced by 90%, so that only 10% remains, then half of all species will be lost. Evidence for this theory comes from a study of extinction rates of birds on habitat islands (that is, islands of a particular type of habitat surrounded by unsuitable habitat) in Finland where the extinction rate was found to be inversely proportional to island size (figure 57.12).

Pollution

Habitat may be degraded by pollution to the extent that some species can no longer survive there. Degradation occurs as a result of many forms of pollution, from acid rain to pesticides. Aquatic environments are particularly vulnerable; for example, many northern lakes in both Europe and North America have been essentially sterilized by acid rain.

Disruption

Human activities may disrupt a habitat enough to make it untenable for some species. For example, visitors to caves in Alabama and Tennessee caused significant population declines in bats over an eight-year period, some as great as 100%. When visits were fewer than one per month, less

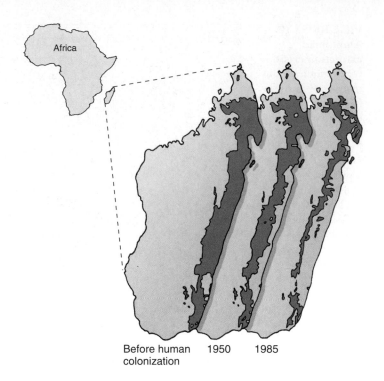

FIGURE 57.11
Extinction and habitat destruction. The rain forest covering the eastern coast of Madagascar, an island off the coast of East Africa, has been progressively destroyed as the island's human population has grown. Ninety percent of the original forest cover is now gone. Many species have become extinct, and many others are threatened, including 16 of Madagascar's 31 primate species.

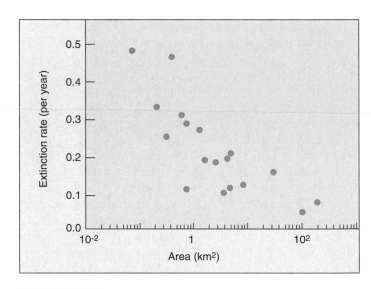

FIGURE 57.12
Extinction and the species-area relationship. The data present percent extinction rates as a function of habitat area for birds on a series of Finnish islands. Smaller islands experience far greater local extinction rates.
Why does extinction rate increase with decreasing island size?

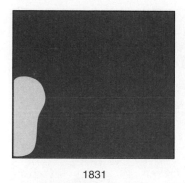
1831

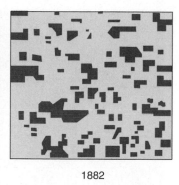
1882

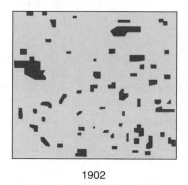
1902

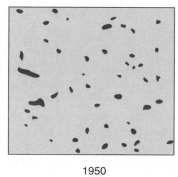
1950

FIGURE 57.13
Fragmentation of woodland habitat. From the time of settlement of Cadiz Township, Wisconsin, the forest has been progressively reduced from a nearly continuous cover to isolated woodlots covering less than 1% of the original area.

than 20% of bats were lost, but caves having more than four visits per month suffered population declines of 86–95%.

Habitat Fragmentation

Loss of habitat by a species frequently results not only in lowered population numbers, but also in fragmentation of the population into unconnected patches (figure 57.13).

A habitat may become fragmented in nonobvious ways, as when roads and habitation intrude into forest. The effect is to carve the populations living in the habitat into a series of smaller populations, often with disastrous consequences because of the relationship between range size and extinction rate. Although detailed data are not available, fragmentation of wildlife habitat in developed temperate areas is thought to be very substantial.

As habitats become fragmented and shrink in size, the relative proportion of the habitat that occurs on the boundary, or edge, increases. **Edge effects** can significantly degrade a population's chances of survival. Changes in microclimate (temperature, wind, humidity, etc.) near the edge may reduce appropriate habitat for many species more than the physical fragmentation suggests. In isolated fragments of rain forest, for example, trees on the edge are exposed to direct sunlight and, consequently, hotter and drier conditions than they are accustomed to in the cool, moist forest interior. As a result, in one study, the biomass of trees within 100 meters of the forest edge decreased by 36% in the first 17 years after fragment isolation.

Also, increasing habitat edges opens up opportunities for parasites and predators, which are both more effective at edges. As fragments decrease in size, the proportion of habitat that is distant from any edge decreases, and consequently, more and more of the habitat is within the range of these predators. Habitat fragmentation is blamed for local extinctions in a wide range of species.

FIGURE 57.14
A study of habitat fragmentation. Biodiversity was monitored in the isolated patches of rain forest in Manaus, Brazil, before and after logging. Fragmentation led to significant species loss within patches.

The impact of habitat fragmentation can be seen clearly in a major study done in Manaus, Brazil, where the rain forest was commercially logged. Landowners agreed to preserve patches of rain forest of various sizes, and censuses of these patches were taken before the logging started, while they were still part of a continuous forest. After logging, species began to disappear from the now-isolated patches (figure 57.14). First to go were the monkeys, which have large home ranges. Birds that prey on ant colonies followed, disappearing from patches too small to maintain enough ant colonies to support them.

Because some species, such as monkeys, require large patches, large fragments are indispensable if we wish to preserve high levels of biodiversity. The take-home lesson is that preservation programs will need to provide suitably large habitat fragments to avoid this impact.

Case Study: Songbirds

Every year since 1966, the U.S. Fish and Wildlife Service has organized thousands of amateur ornithologists and bird-watchers in an annual bird count called the Breeding Bird Survey. In recent years, a shocking trend has emerged. While year-round residents that prosper around humans, such as robins, starlings, and blackbirds, have increased their numbers and distribution over the last 30 years, forest songbirds have declined severely. The decline has been greatest among long-distance migrants such as thrushes, orioles, tanagers, catbirds, vireos, buntings, and warblers. These birds nest in northern forests in the summer, but spend their winters in South or Central America or the Caribbean Islands.

In many areas of the eastern United States, more than three-quarters of the tropical migrant bird species have declined significantly. Rock Creek Park in Washington, D.C., for example, has lost 90% of its long-distance migrants in the past 20 years. Nationwide, American redstarts declined about 50% in the single decade of the 1970s. Studies of radar images from National Weather Service stations in Texas and Louisiana indicate that only about half as many birds fly over the Gulf of Mexico each spring compared to the numbers in the 1960s. This suggests a total loss of about half a billion birds.

The culprit responsible for this widespread decline appears to be habitat fragmentation and loss. Fragmentation of breeding habitat and nesting failures in the summer nesting grounds of the United States and Canada have had a major negative impact on the breeding of woodland songbirds. Many of the most threatened species are adapted to deep woods and need an area of 25 acres or more per pair to breed and raise their young. As woodlands are broken up by roads and developments, it is becoming increasingly difficult for them to find enough contiguous woods to nest successfully.

A second and perhaps even more important factor is the availability of critical winter habitat in Central and South America. Studies of the American redstart clearly indicate that birds with better winter habitat have a superior chance of successfully migrating back to their breeding grounds in the spring. Peter Marra and Richard Holmes of Dartmouth College and Keith Hobson of the Canadian Wildlife Service captured birds, took blood samples, and measured the levels of the stable carbon isotope ^{13}C. Plants growing in the best overwintering habitats in Jamaica and Honduras (mangroves and wetland forests) have low levels of ^{13}C, and so do the redstarts that feed on them. Of these wet-forest birds, 65% maintained or gained weight over the winter. By contrast, plants growing in substandard dry scrub have high levels of ^{13}C, and so do the redstarts that feed on them. Scrub-dwelling birds lost up to 11% of their body mass over the winter. Now here's the key: Birds that winter in the substandard scrub leave later in the spring on the

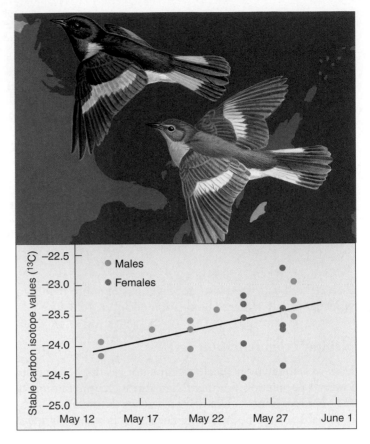

FIGURE 57.15

The American redstart, a migratory songbird. The numbers of this species are in serious decline. The graph presents data on the level of ^{13}C in redstarts arriving at summer breeding grounds. Early arrivals, with the best shot at reproductive success, have lower levels of ^{13}C, indicating they wintered in more favorable mangrove-wetland forest habitats.

long flight to northern breeding grounds, arrive later at their summer homes, and have fewer young (figure 57.15). The proportion of ^{13}C in birds arriving in New Hampshire breeding grounds increases as spring wears on and scrub-overwintering stragglers belatedly arrive. Thus, loss of mangrove habitat in the neotropics is having a real negative impact. As the best habitat disappears, overwintering birds fare poorly, and this leads to population declines. Unfortunately, the Caribbean lost about 10% of its mangroves in the 1980s, and continues to lose about 1% per year. This loss of key habitat appears to be a driving force in the looming extinction of songbirds.

As habitats are destroyed, remaining habitat becomes fragmented, increasing the threat to many species. Fragmentation of summer breeding grounds and loss of high-quality overwintering habitat seem to be contributing to a marked decline in migratory songbird species.

Overexploitation

Species that are hunted or harvested by humans have historically been at grave risk of extinction, even when the species is initially very abundant. A century ago, the skies of North America were darkened by huge flocks of passenger pigeons, but after being hunted as free and tasty food, they were driven to extinction. The bison that used to migrate in enormous herds across the central plains of North America only narrowly escaped the same fate.

The existence of a commercial market often leads to overexploitation of a species. The international trade in furs, for example, has severely reduced the numbers of chinchilla, vicuña, otter, and many cat species. The harvesting of commercially valuable trees provides another example: Almost all West Indies mahogany trees (*Swietenia mahogani*) have been logged, and the extensive cedar forests of Lebanon, once widespread at high elevations, now survive in only a few isolated groves.

A particularly telling example of overexploitation is the commercial harvesting of fish in the North Atlantic. During the 1980s, fishing fleets continued to harvest large amounts of cod off Newfoundland, even as the population numbers declined precipitously. By 1992, the cod population had dropped to less than 1% of its original numbers. The American and Canadian governments have closed the fishery, but no one can predict whether the fish populations will recover. The Atlantic bluefin tuna has experienced a 90% population decline in the past 10 years. The swordfish has declined even further. In both cases, the drop has led to even more intense fishing of the remaining populations.

Case Study: Whales

Whales, the largest living animals that ever evolved, are rare in the world's oceans today, their numbers driven down by commercial whaling. Before the advent of cheap, high-grade oils manufactured from petroleum in the early twentieth century, oil made from whale blubber was an important commercial product in the worldwide marketplace. In addition, the fine, latticelike structure termed "baleen" used by baleen whales to filter-feed plankton from seawater was used in undergarments. Because a whale is such a large animal, each individual captured is of significant commercial value.

Right whales were the first to bear the brunt of commercial whaling. They were called "right" whales because they were slow and easy to capture, and they provided up to 150 barrels of blubber oil and abundant baleen, making them the right whale for a commercial whaler to hunt.

As the right whale declined in the eighteenth century, whalers turned to the gray, humpback, and bowhead whales. As their numbers declined, whalers turned to the blue, the largest of all whales, and when those were decimated, to the fin, then the Sei, and then the sperm whales.

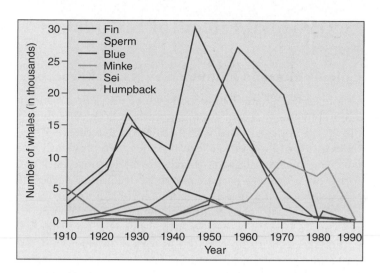

FIGURE 57.16
World catch of whales in the twentieth century. Each species is hunted in turn until its numbers fall so low that hunting it becomes commercially unprofitable.
Why might whale populations fail to recover once hunting is stopped?

As each species of whale became the focus of commercial whaling, its numbers began a steep decline (figure 57.16).

Hunting of right whales was made illegal in 1935. By then, they had been driven to the brink of extinction, their numbers less than 5% of what they had been. Although protected ever since, their numbers have not recovered in either the North Atlantic or the North Pacific. By 1946, several other whale species faced imminent extinction, and whaling nations formed the International Whaling Commission (IWC) to regulate commercial whale hunting. Like a fox guarding the henhouse, the IWC for decades did little to limit whale harvests, and whale numbers continued to decline steeply. Finally, in 1974, when the numbers of all but the small minke whales had been driven down, the IWC banned hunting of blue, gray, and humpback whales, and instituted partial bans on other species. The rule was violated so often, however, that the IWC in 1986 instituted a worldwide moratorium on all commercial killing of whales. While some commercial whaling continues, often under the guise of harvesting for scientific studies, annual whale harvests have dropped dramatically in the last 20 years.

Some species appear to be recovering, while others do not. Humpback numbers have more than doubled since the early 1960s, increasing nearly 10% annually, and Pacific gray whales have fully recovered to their previous numbers of about 20,000 animals after being hunted to less than 1000. Right, sperm, fin, and blue whales have not recovered, and no one knows whether they will.

Overharvesting has driven most large whale species to the brink of extinction. Stopping the harvest has allowed recovery of some, but not all, species.

Detrimental Effects of Introduced Species

Colonization and Extinction as Natural Processes

Colonization, a natural process by which a species expands it geographic range, occurs in many ways: A flock of birds gets blown off course, a bird eats a fruit and defecates its seed miles away, or lowered sea levels connect two previously isolated landmasses, allowing species to freely move back and forth. Such events—particularly those leading to successful establishment of a new population—probably occur rarely, but when they do, the resulting change to natural communities can be large. The reason is that colonization brings together species with no previous history of interaction. Consequently, ecological interactions may be particularly strong because the species have not evolved ways of adjusting to the presence of each other, such as adaptations to avoid predation or minimize competitive effects.

The paleontological record documents many cases in which geologic changes brought previously isolated species together, such as when the Isthmus of Panama emerged above the sea approximately three million years ago, connecting the previously isolated faunas and floras of North and South America. In some cases, the result has been an increase in species diversity, but in other cases, invading species have led to the extinction of natives.

Human Influence on the Process

Unfortunately, what was naturally a rare process has become all too common in recent years, thanks to the actions of humans. Species introductions due to human activities occur in many ways, sometimes intentionally, but usually not. Plants and animals can be transported in the ballast of large ocean vessels, in nursery plants, as stowaways in boats, cars, and planes, as beetle larvae within wood products—even as seeds in the mud stuck to the bottom of a shoe! Overall, some researchers estimate that as many as 50,000 species have been introduced into the United States. The results of such introductions are occasionally catastrophic.

The effects of introductions on humans have been enormous. In the United States alone, nonnative species cost the economy an estimated $140 billion per year. For example, dozens of foreign weeds in Colorado have covered more than a million acres. Just three of these species cost wheat farmers tens of millions of dollars. At the same time, leafy spurge, a plant from Europe, outcompetes native grasses, ruining rangeland for cattle at a price tag of $144 million per year. The zebra mussel, a mollusk native to the Black Sea, is a huge problem throughout much of the eastern and central United States, where it can attain densities as high as $700,000/m^2$, clogging pipes, including those for water and power plants, and causing an estimated $3–5 billion damage a year.

Introduced species can have impacts on human health as well. For example, West Nile fever, the death rate of which

FIGURE 57.17

The akiapolaau, an endangered Hawaiian bird. More than two-thirds of Hawaii's native bird species are now extinct or have been greatly reduced in population size. Bird faunas on islands around the world have experienced similar declines after human arrival.

is steadily mounting, was probably introduced from Africa or the Middle East in the late 1990s.

The effect of species introductions on native ecosystems is equally dramatic. Islands have been particularly affected. For example, a lighthouse keeper's cat singlehandedly (singlepawedly?) wiped out an entire species, the Stephens Island wren. Rats had a devastating effect throughout the South Pacific where bird species nested on the ground and had no defense against the voracious predators to which they were evolutionarily naive. More recently, the brown tree snake, introduced to the island of Guam, essentially eliminated all species of forest birds. In Hawaii, the problem has been slightly different: Introduced mosquitoes brought with them avian malaria, to which the native species had evolved no resistance. The result is that more than 100 species (more than 70% of the native fauna) either went extinct or are now restricted to higher and cooler elevations where the mosquitoes don't occur (figure 57.17).

The effects of introduced species are not always direct, but instead may reverberate throughout an ecosystem. For example, the Argentine ant has spread through much of the southern United States, eliminating most native ant species with which it comes in contact. The extinction of these ant species has had a dramatic negative effect on the coastal horned lizard, which specializes on the larger native species. In their absence, the lizards have shifted to less-preferred prey species. In addition, the native species consume seeds, and in the process, play an important role in seed dispersal. Argentine ants, by contrast, do not eat seeds. In South Africa, where the Argentine ant has also appeared, at least one plant species has experienced decreased reproductive success due to the loss of its dispersal agent.

The most dramatic effects of introduced species, however, occur when entire ecosystems are transformed. Some plant species can completely overrun a habitat, displacing all native species and turning the area into a monoculture (that is, an area occupied by a single species). In California, the yellow star thistle now covers 4 million hectares of what was once highly productive grassland. In Hawaii, a small tree native to the Canary Islands, *Myrica faya*, has spread widely. Because it is able to fix nitrogen at high rates, it has caused a 90-fold increase in the nitrogen content of the soil, thus allowing other, nitrogen-requiring species to invade. In South Africa, several introduced tree species have such high water uptake rates that nearby rivers are now dry.

Efforts to Combat Introduced Species

Once an introduced species becomes established, eradicating it is often extremely difficult, expensive, and time-consuming. Some efforts—such as the removal of goats and rabbits from certain small islands—have been successful, but many other efforts have failed. The best hope for stopping the ravages of introduced species is to prevent species from being introduced in the first place. Although easier said than done, government agencies are now working strenuously to put into place procedures that can intercept species in transit, before they have the opportunity to become established.

Case Study: Lake Victoria Cichlids

Lake Victoria, an immense, shallow, freshwater sea about the size of Switzerland in the heart of equatorial East Africa, had until 1954 been home to an incredibly diverse collection of over 300 species of cichlid fishes (see figure 23.14). These small, perchlike fish range from 2 to 10 inches in length, with males having endless varieties of color. Today, most of these cichlid species are threatened, endangered, or extinct.

What happened to bring about the abrupt loss of so many endemic cichlid species? In 1954, the Nile perch, *Lates niloticus*, a commercial fish with a voracious appetite, was introduced on the Ugandan shore of Lake Victoria. Nile perch, which grow to over 4 feet in length, were to form the basis of a new fishing industry (figure 57.18). For decades, these perch did not seem to have a significant impact; over 30 years later, in 1978, Nile perch still made up less than 2% of the fish harvested from the lake.

Then something happened to cause the Nile perch population to explode and to spread rapidly through the lake, eating their way through the cichlids. By 1986, Nile perch constituted nearly 80% of the total catch of fish from the lake and the endemic cichlid species were virtually gone. Over 70% of cichlid species disappeared, including all open-water species.

So what happened to kick-start the mass extinction of the cichlids? The trigger seems to have been eutrophication. Before 1978, Lake Victoria had high oxygen levels at

FIGURE 57.18
Victor and vanquished. The Nile perch (larger fishes in foreground), a commercial fish introduced into Lake Victoria as a potential food source, is responsible for the virtual extinction of hundreds of species of cichlid fishes (smaller fishes in tub).

all depths, down to the bottom layers more than 60 meters deep. However, by 1989 high inputs of nutrients from agricultural runoff and sewage from towns and villages had led to algal blooms that severely depleted oxygen levels in deeper parts of the lake. Cichlids feed on algae, and initially their population numbers are thought to have risen in response to this increase in their food supply, but unlike the conditions during similar algal blooms of the past, the Nile perch was present to take advantage of the situation. With a sudden increase in its food supply (cichlids), the numbers of Nile perch exploded, and they simply ate all available cichlids.

Since 1990, the situation has been compounded by the introduction into Lake Victoria of a floating water weed from South America, the water hyacinth *Eichornia crassipes*. Extremely fecund under eutrophic conditions, thick mats of water hyacinth soon covered entire bays and inlets, choking off the coastal habitats of non-open-water cichlids.

Human activities are increasingly introducing many species all over the world. Such introductions often have catastrophic effects on the native fauna and flora, and can cause enormous economic damage.

Disruption of Ecosystems

Species often become vulnerable to extinction when their web of ecological interactions becomes seriously disrupted. Because of the many relationships linking species in an ecosystem (see chapter 55), human activities that affect one species can ramify throughout an ecosystem, ultimately affecting many other species.

A recent case in point are the sea otters that live in the cold waters off Alaska and the Aleutian Islands. Otter populations have declined sharply in recent years. In a 500-mile stretch of coastline, otter numbers have dropped from 53,000 in the 1970s to an estimated 6000, a plunge of nearly 90%. Investigating this catastrophic decline, marine ecologists uncovered a chain of interactions among the species of the ocean and kelp forest ecosystems, a falling-domino series of lethal effects that illustrates the concepts of both top-down and bottom-up trophic cascades discussed in chapter 55.

Case Study: Alaskan Near-Shore Habitat

The first in a series of events leading to the sea otter's decline seems to have been the heavy commercial harvesting of whales (see the case study earlier in this section, page 1239). Without whales to keep their numbers in check, ocean zooplankton thrived, leading in turn to proliferation of a species of fish called pollock that feeds on the abundant zooplankton. Given this ample food supply, the pollock proved to be very successful competitors of other northern Pacific fish, such as herring and ocean perch, so that levels of these other fish fell steeply in the 1970s.

Then the falling chain of dominoes began to accelerate. The decline in the nutritious forage fish led to an ensuing crash in Alaskan populations of Steller's sea lions and harbor seals, for which pollock did not provide sufficient nourishment. Numbers of these pinniped species have fallen precipitously since the 1970s.

Pinnipeds are the major food of orcas, also called killer whales. Faced with a food shortage, some orcas turned to the next best thing: sea otters. In one bay where the entrance from the sea was too narrow and shallow for orcas to enter, only 12% of the sea otters have disappeared, while in a similar bay that orcas could enter easily, two-thirds of the otters disappeared in a year's time.

Without otters to eat them, the population of sea urchins exploded, eating the kelp and thus "deforesting" the kelp forests and denuding the ecosystem (figure 57.19). As a result, fish species that live in the kelp forest, such as sculpins and greenlings (a cod relative), are declining.

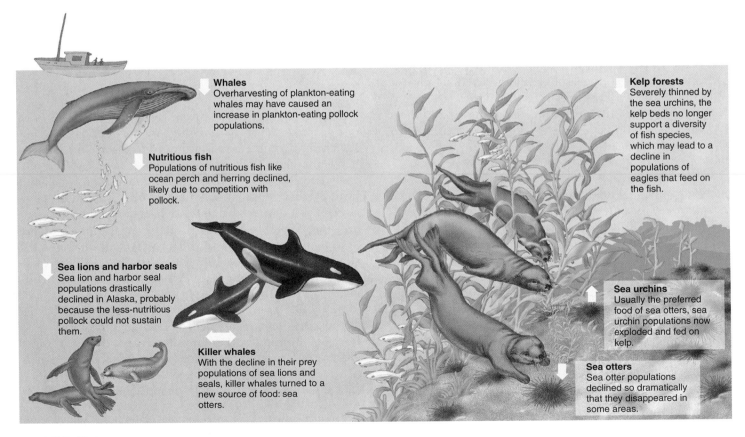

FIGURE 57.19
Disruption of the kelp forest ecosystem.   Overharvesting by commercial whalers altered the balance of fish in the ocean ecosystem, inducing killer whales to feed on sea otters, a keystone species of the kelp forest ecosystem.

Preserving Keystone Species

As discussed in chapter 54, a keystone species is a species that exerts a particularly strong influence on the structure and functioning of a certain ecosystem. The sea otters of figure 57.19 are a keystone species of the kelp forest ecosystem, and their removal can have disastrous consequences. There is no hard-and-fast line that allows us to clearly identify keystone species. Rather, it is a qualitative concept, a statement that a species plays a particularly important role in its community. Keystone species are usually characterized by the strength of their impact on their community. *Community importance* measures the change in some quantitative aspect of the ecosystem (for example, species richness, productivity, or nutrient cycling) per unit of change in the abundance of a species.

Case Study: Flying Foxes. The severe decline of many species of pteropodid bats, or "flying foxes," in the Old World tropics is an example of how the loss of a keystone species can dramatically affect the other species living within an ecosystem, sometimes even leading to a cascade of further extinctions (figure 57.20). These bats have very close relationships with important plant species on the islands of the Pacific and Indian Oceans. The family Pteropodidae contains nearly 200 species, approximately one-quarter of them in the genus *Pteropus*, and is widespread on the islands of the South Pacific, where they are the most important—and often the only—pollinators and seed dispersers. A study in Samoa found that 80 to 100% of the seeds landing on the ground during the dry season were deposited by flying foxes. Many species are entirely dependent on these bats for pollination. Some have evolved features such as night-blooming flowers that prevent any other potential pollinators from taking over the role of the fruit bats.

In Guam, where the two local species of flying fox have recently been driven extinct or nearly so, the impact on the ecosystem appears to be substantial. Botanists have found that some plant species are not fruiting or are doing so only marginally, producing fewer fruits than normal. Fruits are not being dispersed away from parent plants, so offspring shoots are being crowded out by the adults.

Flying foxes are being driven to extinction by human hunters who kill them for food and for sport, and by or-

FIGURE 57.20

Preserving keystone species. The flying fox is a keystone species on many Old World tropical islands. It pollinates many of the plants, and is a key disperser of seeds. Its elimination due to hunting and habitat loss is having a devastating effect on the ecosystems of many South Pacific Islands.

chard farmers who consider them pests. Flying foxes are particularly vulnerable because they live in large, easily seen groups of up to a million individuals. Because they move in regular and predictable patterns and can be easily tracked to their home roost, hunters can easily bag thousands at a time.

Programs aimed at preserving particular species of flying foxes are only just beginning. One particularly successful example is the program to save the Rodrigues fruit bat, *Pteropus rodricensis*, which occurs only on Rodrigues Island in the Indian Ocean near Madagascar. The population dropped from about 1000 individuals in 1955 to fewer than 100 by 1974, largely due to the loss of the fruit bat's forest habitat to farming. Since 1974, the species has been legally protected, and the forest area of the island is being increased through a tree-planting program. Eleven captive-breeding colonies have been established, and the bat population is now increasing rapidly. The combination of legal protection, habitat restoration, and captive breeding has in this instance produced a very effective preservation program.

Because of the interrelationships among species within an ecosystem, activities that directly harm one species may indirectly have large impacts on many other species as well.

The Perils of Small Population Size

Because of the factors just discussed, populations of many species are fragmented and reduced in size. Such populations are particularly prone to extinction.

Demographic Factors

Small populations are vulnerable to a variety of demographic factors. By nature of their small size, they are ill-equipped to withstand a catastrophic event, such as a flood, forest fire, or disease epidemic. One example is the heath hen. Although the species was once common throughout the eastern United States, hunting pressure in the eighteenth and nineteenth centuries eventually eliminated all but one population, on the island of Martha's Vineyard near Cape Cod, Massachusetts. Protected in a nature preserve, the population was increasing in number until a fire destroyed most of the preserve's habitat. The small surviving population was then ravaged the next year by an unusual congregation of predatory birds, followed shortly thereafter by a disease epidemic.

When populations become extremely small, bad luck can spell the end. For example, the dusky seaside sparrow (figure 57.21), a now-extinct subspecies that was found on the east coast of Florida, dwindled to a population of five individuals, all of which happened to be males. In a large population, the probability that all individuals will be of one sex is infinitesimal. But in small populations, just by the luck of the draw, it is possible that five or ten or even 20 consecutive births will all be individuals of one sex, and that can be enough to send a species to extinction. In addition, when populations are small, individuals may have trouble finding each other (the Allee effect discussed in chapter 53), thus leading the population into a downward spiral toward extinction.

Lack of Genetic Variability

Small populations face a second dilemma. Because of their low numbers, such populations are prone to the loss of genetic variation as a result of genetic drift (figure 57.22). Indeed, many small populations contain little or no genetic variability. The result of such genetic homogeneity can be catastrophic. Genetic variation is beneficial to a population both because of heterozygote advantage (see chapter 21) and because genetically variable individuals tend not to have two copies of deleteriously recessive alleles. As a result, populations lacking variation are often composed of sickly, unfit, or sterile individuals. Indeed, in the laboratory, groups of rodents and fruit flies that are maintained at small population sizes often perish after a few generations as each generation becomes less robust and fertile than the preceding one. Although it is difficult to demonstrate that a species has gone

FIGURE 57.21
Alive no more. This male was one of the last dusky seaside sparrows, a subspecies that no longer exists.

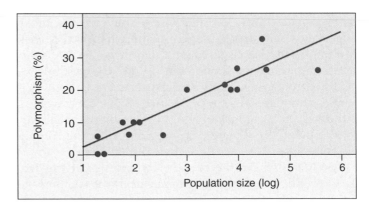

FIGURE 57.22
Loss of genetic variability in small populations. The percentage of polymorphic genes in isolated populations of the tree *Halocarpus bidwillii* in the mountains of New Zealand is a sensitive function of population size.
Why do small populations lose genetic variation?

extinct because of lack of genetic variation, studies of both zoo and natural populations clearly reveal that more genetically variable individuals have greater fitness. Furthermore, in the longer term, populations with limited genetic variation have diminished ability to adapt to changing environments.

Demographic and Genetic Factors Interact

As populations decrease in size, demographic and genetic factors feed on each other, causing what has been termed an "extinction vortex." That is, as a population gets smaller, it becomes more vulnerable to demographic catastrophes. In turn, genetic variation starts to be lost, causing reproductive rates to decline and population numbers to decline even further, and so on. Eventually, the population disappears entirely, but attributing its demise to one particular factor would be misleading.

Case Study: Prairie Chickens

The greater prairie chicken (*Tympanuchus cupido pinnatus*) is a showy, two-pound wild bird renowned for its flamboyant mating rituals (figure 57.23). Abundant in many midwestern states, the prairie chickens in Illinois have in the past six decades undergone a population collapse. Once, enormous numbers of birds occurred throughout the state, but with the 1837 introduction of the steel plow, the first that could slice through the deep, dense root systems of prairie grasses, the Illinois prairie began to be replaced by farmland, and by the turn of the century, the prairie had vanished. By 1931, a related subspecies known as the heath hen (*Tympanuchus cupido cupido*) had become extinct in Illinois. The greater prairie chicken fared little better, its numbers falling to 25,000 statewide in 1933 and then to 2000 in 1962. In surrounding states with less intensive agriculture, it continued to prosper.

In 1962 and 1967, sanctuaries were established in Illinois to attempt to preserve the prairie chicken. But privately owned grasslands kept disappearing, along with their prairie chickens, and by the 1980s the birds were extinct in Illinois except for the two preserves. And even there, their numbers kept falling. By 1990, the egg hatching rate, which at one time had averaged between 91 and 100%, had dropped to an extremely low 38%. By the mid-1990s, the count of males had dropped to as low as six in each sanctuary.

What was wrong with the sanctuary populations? One suggestion was that because of very small population sizes and a mating ritual whereby one male may dominate a flock, the Illinois prairie chickens had lost so much genetic variability as to create serious inbreeding problems. To test this idea, biologists at the University of Illinois compared DNA from frozen tissue samples of birds that had died in Illinois between 1974 and 1993, and found that in recent

FIGURE 57.23
A mating ritual. The male prairie chicken inflates bright orange air sacs, part of his esophagus, into balloons on each side of his head. As air is drawn into the sacs, it creates a three-syllable "boom-boom-boom" that can be heard for miles.

years Illinois birds had indeed become genetically less diverse. After extracting DNA from tissue in the roots of feathers from stuffed birds collected in the 1930s from the same population, the researchers found that Illinois birds had lost fully one-third of the genetic diversity of birds living in the same place before the population collapse of the 1970s. By contrast, prairie chicken populations in other states still contained much of the genetic variation that had disappeared from Illinois populations.

Now the stage was set to halt the Illinois prairie chicken's race toward extinction. Wildlife managers began to transplant birds from genetically diverse populations of Minnesota, Kansas, and Nebraska to Illinois. Between 1992 and 1996, a total of 518 out-of-state prairie chickens were brought in to interbreed with the Illinois birds, and hatching rates were back up to 94% by 1998. It looks as though the Illinois prairie chickens have been saved from extinction.

The key lesson to be learned is the importance of not allowing things to go too far—not to drop down to a single isolated population. Without the outlying genetically different populations, the prairie chickens in Illinois could not have been saved. For example, when the last population of the dusky seaside sparrow lost its last female, there was no other source of females and the subspecies went extinct.

Small populations face a variety of perils: They have less ability to rebound from catastrophes and are vulnerable to loss of genetic variation. Worst of all, these problems build on each other, magnifying their impact.

57.3 Successful recovery efforts need to be multidimensional.

Approaches for Preserving Endangered Species

Once the cause of a species' endangerment is known, it becomes possible to design a recovery plan. If the cause is commercial overharvesting, regulations can be issued to lessen the impact and protect the threatened species. If the cause is habitat loss, plans can be instituted to restore the habitat. Loss of genetic variability in isolated subpopulations can be countered by transplanting individuals from genetically different populations. Populations in immediate danger of extinction can be captured, introduced into a captive-breeding program, and later reintroduced to other suitable habitat.

Of course, all of these solutions are extremely expensive. As Bruce Babbitt, Secretary of the Interior in the Clinton administration, noted, it is much more economical to prevent "environmental trainwrecks" from occurring than to clean them up afterwards. Preserving ecosystems and monitoring species before they are threatened is the most effective means of protecting the environment and preventing extinctions.

Habitat Restoration

Conservation biology typically concerns itself with preserving populations and species in danger of decline or extinction. Conservation, however, requires that there be something left to preserve, while in many situations, conservation is no longer an option. Species, and in some cases whole communities, have disappeared or been irretrievably modified. The clear-cutting of the temperate forests of Washington State leaves little behind to conserve, as does converting a piece of land into a wheat field or an asphalt parking lot. Redeeming these situations requires restoration rather than conservation.

Three quite different sorts of habitat restoration programs might be undertaken, depending on the cause of the habitat loss.

Pristine Restoration.
In ecosystems where all species have been effectively removed, conservationists might attempt to restore the plants and animals that are the natural inhabitants of the area, if these species can be identified. When abandoned farmland is to be restored to prairie, as in figure 57.24, how would you know what to plant? Although it is in principle possible to reestablish each of the original species in their original proportions, rebuilding a community requires knowing the identities of all the original inhabitants and the ecologies of each of the species. We rarely have this much information, so no restoration is ever truly pristine.

(a)

(b)

FIGURE 57.24
Habitat restoration. The University of Wisconsin–Madison Arboretum has pioneered restoration ecology. (*a*) The restoration of the prairie was at an early stage in November 1935. (*b*) The prairie as it looks today. This picture was taken at approximately the same location as the 1935 photograph.

Removing Introduced Species.
Sometimes the habitat of a species has been destroyed by a single introduced species. In such a case, habitat restoration involves removing the introduced species. Restoration of the once-diverse cichlid fishes to Lake Victoria will require more than breeding and restocking the endangered species. The introduced water hyacinth and Nile perch populations will have to be brought under control or removed, and eutrophication will have to be reversed.

It is important to act quickly if an introduced species is to be removed. When aggressive African bees (the so-called "killer bees") were inadvertently released in Brazil, they remained confined to the local area only one season. Now they occupy much of the western hemisphere.

Cleanup and Rehabilitation.
Habitats seriously degraded by chemical pollution cannot be restored until the pollution is cleaned up. The successful restoration of the Nashua River in New England is one example of how a concerted effort can succeed in restoring a heavily polluted habitat to a relatively pristine condition.

Captive Breeding

Recovery programs, particularly those focused on one or a few species, must sometimes involve direct intervention in natural populations to avoid an immediate threat of extinction.

The Peregrine Falcon.   American populations of birds of prey, such as the peregrine falcon (*Falco peregrinus*), began an abrupt decline shortly after World War II. Of the approximately 350 breeding pairs east of the Mississippi River in 1942, all had disappeared by 1960. The culprit proved to be the chemical pesticide DDT (dichlorodiphenyl-trichloroethane) and related organochlorine pesticides. Birds of prey are particularly vulnerable to DDT because they feed at the top of the food chain, where DDT becomes concentrated. DDT interferes with the deposition of calcium in the bird's eggshells, causing most of the eggs to break before they are ready to hatch.

The use of DDT was banned by federal law in 1972, causing levels in the eastern United States to fall quickly. However, there were no peregrine falcons left in the eastern United States to reestablish a natural population. Falcons from other parts of the country were used to establish a captive-breeding program at Cornell University in 1970, with the intent of reestablishing the peregrine falcon in the eastern United States by releasing offspring of these birds. By the end of 1986, over 850 birds had been released in 13 eastern states, producing an astonishingly strong recovery (figure 57.25).

The California Condor.   Numbers of the California condor (*Gymnogyps californianus*), a large, vulturelike bird with a wingspan of nearly 3 meters, have been declining gradually for the past 200 years. By 1985, condor numbers had dropped so low that the bird was on the verge of extinction. Six of the remaining 15 wild birds disappeared that year alone. The entire breeding population of the species consisted of the birds remaining in the wild and an additional 21 birds in captivity. In a last-ditch attempt to save the condor from extinction, the remaining birds were captured and placed in a captive-breeding population. The breeding program was set up in zoos, with the goal of releasing offspring on a large, 5300-ha ranch in prime condor habitat. Birds were isolated from human contact as much as possible, and closely related individuals were prevented from breeding. By early 2002, the captive population of California condors had reached over 120 individuals. Thirty-two captive-reared condors have been released successfully in California at two sites in the mountains north of Los Angeles, after extensive prerelease training to avoid power poles and people. All of the released birds seem to be doing well, and another 21 birds released into the Grand Canyon have adapted successfully. Biologists are particularly excited by breeding activities that resulted in the first-ever offspring produced in the wild by captive-reared parents.

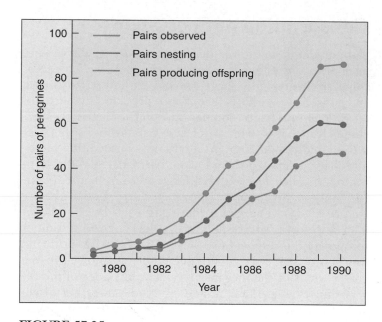

FIGURE 57.25
Captive breeding.   The peregrine falcon has been reestablished in the eastern United States by releasing captive-bred birds over a period of 10 years.

Yellowstone Wolves.   The ultimate goal of captive-breeding programs is not simply to preserve interesting species, but rather to restore ecosystems to a balanced, functional state. Yellowstone Park has been an ecosystem out of balance, due in large part to the systematic extermination of the gray wolf (*Canis lupus*) in the park early in this century. Without these predators to keep their numbers in check, herds of elk and deer expanded rapidly, damaging vegetation so that the elk themselves starve in time of scarcity. In an attempt to restore the park's natural balance, two complete wolf packs from Canada were released into the park in 1995 and 1996. The wolves adapted well, breeding so successfully that by 2002 the park contained 16 free-ranging packs and more than 200 wolves.

While ranchers near the park have been unhappy about the return of the wolves, little damage to livestock has been noted, and the ecological equilibrium of Yellowstone Park seems well on the way to being regained. Elk are congregating in larger herds, and their populations are not growing as rapidly as in years past. Importantly, wolves are killing coyotes and their pups, driving them out of some areas. Coyotes, the top predators in the absence of wolves, are known to attack cattle on surrounding ranches, so reintroduction of wolves to the park may actually benefit the cattle ranchers who are opposed to it.

Efforts to preserve endangered species are as diverse as the causes of endangerment. Although captive breeding is not a solution in all, or even most, cases, it has helped restore several vertebrate species.

Conservation of Ecosystems

Habitat fragmentation is one of the most pervasive enemies of biodiversity conservation efforts. As we have seen, some species require large patches of habitat to thrive, and conservation efforts that cannot provide suitable habitat of such a size are doomed to failure. Since it has become clear that isolated patches of habitat lose species far more rapidly than large preserves do, conservation biologists have promoted the creation, particularly in the tropics, of so-called *megareserves*, large areas of land containing a core of one or more undisturbed habitats (figure 57.26). The key to devoting such large tracts of land to reserves successfully over a long period of time is to operate the reserve in a way compatible with local land use. Thus, while no economic activity is allowed in the core regions of the megareserve, the remainder of the reserve may be used for nondestructive harvesting of resources. Linking preserved areas to carefully managed land zones creates a much larger total "patch" of habitat than would otherwise be economically practical, and thus addresses the key problem created by habitat fragmentation. Pioneering these efforts, eight such megareserves have been created in Costa Rica (figure 57.27) to jointly manage biodiversity and economic activity.

In addition to this focus on maintaining large enough reserves, in recent years, conservation biologists also have recognized that the best way to preserve biodiversity is to focus on preserving intact ecosystems, rather than particular species. For this reason, attention in many cases is turning to identifying those ecosystems most in need of preservation and devising the means to protect not only the species within the ecosystem, but the functioning of the ecosystem itself.

Efforts are being undertaken worldwide to preserve biodiversity in megareserves designed to counter the influences of habitat fragmentation. Focusing on the health of entire ecosystems, rather than particular species, can often be a more effective means of preserving biodiversity.

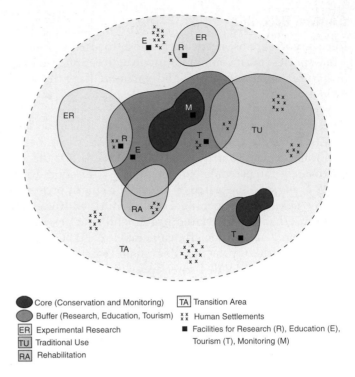

● Core (Conservation and Monitoring)
◐ Buffer (Research, Education, Tourism)
[ER] Experimental Research
[TU] Traditional Use
[RA] Rehabilitation

[TA] Transition Area
×× Human Settlements
■ Facilities for Research (R), Education (E), Tourism (T), Monitoring (M)

FIGURE 57.26

Design of a megareserve. A megareserve, or biosphere reserve, recognizes the need for people to have access to resources. Critical ecosystems are preserved in the core zone. Research and tourism are allowed in the buffer zone. Sustainable resource harvesting and permanent habitation are allowed in the multiple-use areas surrounding the buffer.

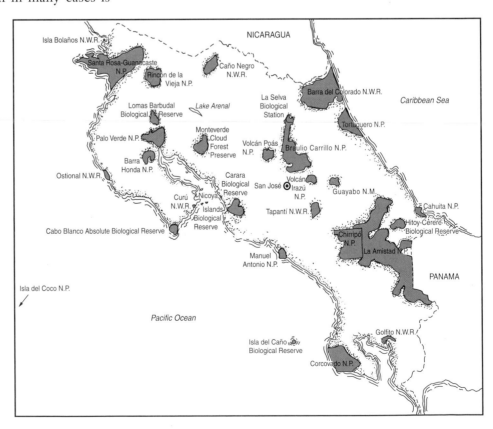

FIGURE 57.27

Biosphere reserves in Costa Rica. Costa Rica, a country the size of West Virginia, has placed about 12% of its land into national parks and eight megareserves.

For interactive testing, visit the Online Learning Center with PowerWeb at www.mhhe.com/Raven7

57.1 The new science of conservation biology is focused on conserving biodiversity.

Overview of the Biodiversity Crisis

- More than 99% of species known to science are now extinct. (p. 1228)
- It is estimated that as much as 20% of the world's biodiversity may be lost during the next 50 years. (p. 1228)
- Shortly after humans arrived, 74–86% of the earth's megafauna became extinct. (p. 1228)
- Estimates indicate that about 85 species of mammals and 113 species of birds have become extinct in the past 400 years. (p. 1229)
- The majority of historic extinctions have occurred on islands, although most species now threatened by extinction live on continents. (p. 1229)

Species Endemism and Hotspots

- An endemic species is found naturally in only one geographic area. (p. 1230)
- Isolated, geographic areas often have high percentages of endemic species. (p. 1230)
- Worldwide, concentrations of endemic species occur in several hotspots that must be the focus of conservation efforts. Unfortunately, these areas currently also contain 20% of the world's human population. (pp. 1230–1231)
- Population pressure as well as habitat loss due to commercial exploitation play an important role in biodiversity loss. (p. 1231)

What's So Bad About Losing Biodiversity?

- The value of biodiversity can be divided into direct economic value (food, clothing), indirect economic value (ecosystem maintenance), and ethical and aesthetic value (beauty). (pp. 1232–1233)

57.2 The extinction crisis is a result of many factors.

Factors Responsible for Extinction

- Historically, overexploitation was the major factor responsible for extinction, but habitat loss is the most significant problem for most groups today. (p. 1234)
- Other factors, such as ecosystem disruption, pollution, loss of genetic diversity, catastrophic disturbance, and habitat fragmentation, can also contribute to species extinction. (p. 1234)
- Many amphibian species around the world are in decline, and because no single cause can be determined, fears are rising as to the overall health of the global environment. (p. 1235)

Habitat Loss

- Natural habitats may be adversely affected by humans in four ways: destruction, pollution, human disruption, and fragmentation. (pp. 1236–1237)
- Fragmentation of summer breeding grounds and loss of overwintering habitats appear to be contributing to an overall decline in the numbers of migratory songbirds. (p. 1238)

Overexploitation

- Species that are harvested by people have historically been at grave risk of extinction. (p. 1239)
- The existence of a commercial market often leads to overexploitation of a species. (p. 1239)
- Overharvesting has brought most species of large whales to the brink of extinction. (p. 1239)

Detrimental Effects of Introduced Species

- Although colonization and range expansion is a natural process, humans are increasingly causing the introduction of species to new, unintended areas. (p. 1240)
- Some species introductions can lead to the transformation of an ecosystem, the displacement of native species, and the production of a monoculture. (pp. 1240–1241)
- The introduction of Nile perch to Lake Victoria, along with eutrophication, caused a large reduction in the number of endemic cichlid species in Lake Victoria. (p. 1241)

Disruption of Ecosystems

- Species often become vulnerable to extinction when their web of ecological interactions becomes seriously disrupted. (p. 1242)
- Activities that detrimentally affect one species may indirectly affect other species as well. (pp. 1242–1243)

The Perils of Small Population Size

- Small populations are ill-equipped to withstand a catastrophic event, and single, natural events can lead to the extinction of a population or a species. (p. 1244)
- Small populations are also prone to the loss of genetic variation due to genetic drift. (p. 1244)
- As a population gets smaller, it becomes more vulnerable to demographic catastrophes, in turn causing a decline in genetic variation and a subsequent decline in reproductive rates, which set in motion a cascading effect. (p. 1245)

57.3 Successful recovery efforts need to be multidimensional.

Approaches for Preserving Endangered Species

- Many approaches exist for preserving and fostering endangered species, and these include habitat restoration and captive propagation. (p. 1246)
- Habitat restoration can involve many actions, including pristine restoration, removal of introduced species, and cleanup and rehabilitation. (p. 1246)
- Captive propagation has produced many success stories, including the peregrine falcon, the california condor, and wolves in Yellowstone National Park. (p. 1247)

Conservation of Ecosystems

- Conservation efforts are increasingly being turned toward preserving large tracts of land over long periods of time in order to preserve intact ecosystems rather than individual or particular species. (p. 1248)

Self Test

1. Historically, island species have tended to become extinct faster than species living on a mainland. Which of the following reasons can not be used to explain this phenomenon?
   a. Island species have often evolved in the absence of predators and have no natural avoidance strategies.
   b. Humans have introduced diseases and competitors to islands, which negatively impacts island populations.
   c. Island populations are usually smaller than mainland populations.
   d. Island populations are usually less fit than mainland populations.

2. An endemic species is
   a. one found in many different geographic areas.
   b. one found naturally in just one geographic area.
   c. one found only on islands.
   d. one that has been introduced to a new geographic area.

3. Conservation hotspots are best described as
   a. areas with large numbers of endemic species that are disappearing rapidly.
   b. areas where people are particularly active supporters of biological diversity.
   c. islands that are experiencing high rates of extinction.
   d. areas where native species are being replaced with introduced species.

4. Which of the following does *not* distinguish the current mass extinction event from past events?
   a. This is the only extinction event to be triggered by a single species.
   b. This is the only extinction event from which biodiversity may not recover.
   c. This is the only extinction event in which the majority of losses are occurring in mainland areas.
   d. This is the only extinction event in which the majority of losses are occurring on islands.

5. The ability of an intact ecosystem, such as a wetland, to buffer against flooding and filter pollutants from water is a(n) _____ value of biodiversity.
   a. direct economic
   b. indirect economic
   c. ethical
   d. aesthetic

6. Which of the following is currently considered the leading cause of extinction?
   a. overexploitation of species
   b. competition from introduced species
   c. habitat loss
   d. pollution

7. What percentage of Madagascar's original forests have been lost due to human activity?
   a. 10%
   b. 30%
   c. 60%
   d. 90%

8. Numbers of migratory songbirds are declining in North America. Which of the following factors is important in this decline?
   a. pollution
   b. human disruption of breeding behavior
   c. habitat fragmentation in the United States
   d. global climate change

9. Most large whale species have been driven to the brink of extinction. Which of the following is the most accepted explanation for this situation?
   a. overexploitation
   b. habitat loss
   c. pollution
   d. competition from introduced species

10. A keystone species is one that
    a. has a higher likelihood of extinction than a nonkeystone species.
    b. exerts a strong influence on an ecosystem.
    c. causes other species to become extinct.
    d. has a weak influence on an ecosystem.

Test Your Visual Understanding

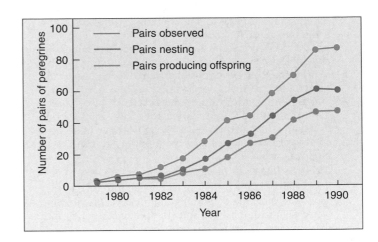

1. List the possible factors that explain the trend in this graph.

Apply Your Knowledge

1. Do you know of any endangered or threatened species that live near you? What factors have led to their decline? What efforts are being made to preserve any endangered species in your area?

2. Exotic species can have detrimental impacts on native species. Are all exotics bad? Can you name any instances in which exotics have not had detrimental effects?

3. Recent advances in genetic technologies have allowed scientists to develop microbes that can metabolize toxins in the environment, such as pesticides and oil slicks. What are the dangers of releasing these genetically modified organisms into the wild?

Glossary

A

ABO blood group A set of four phenotypes produced by different combinations of three alleles at a single locus; blood types are A, B, AB, and O, depending on which alleles are expressed as antigens on the red blood cell surface.

abscission (L. *ab*, away, off, + *scissio*, dividing) In vascular plants, the dropping of leaves, flowers, fruits, or stems at the end of the growing season, as the result of the formation of a layer of specialized cells (the abscission zone) and the action of a hormone (ethylene).

acquired immune deficiency syndrome (AIDS) An infectious and usually fatal human disease caused by a retrovirus, HIV (human immunodeficiency virus), which attacks T cells. The affected individual is helpless in the face of microbial infections.

actin (Gr. *actis*, a ray) One of the two major proteins that make up vertebrate muscle; the other is myosin.

action potential A transient, all-or-none reversal of the electric potential across a membrane; in neurons, an action potential initiates transmission of a nerve impulse.

activation energy The energy that must be processed by a molecule in order for it to undergo a specific chemical reaction.

active site The region of an enzyme surface to which a specific set of substrates binds, lowering the activation energy required for a particular chemical reaction and so facilitating it.

active transport The pumping of individual ions or other molecules across a cellular membrane from a region of lower concentration to one of higher concentration (that is, against a concentration gradient); this transport process requires energy, which is typically supplied by the expenditure of ATP.

adaptation (L. *adaptare*, to fit) A peculiarity of structure, physiology, or behavior that promotes the likelihood of an organism's survival and reproduction in a particular environment.

adaptive radiation The evolution of several divergent forms from a primitive and unspecialized ancestor.

adenosine diphosphate (ADP) A nucleotide consisting of adenine, ribose sugar, and two phosphate groups; formed by the removal of one phosphate from an ATP molecule.

adenosine triphosphate (ATP) A nucleotide consisting of adenine, ribose sugar, and three phosphate groups; ATP is the energy currency of cellular metabolism in all organisms.

adenylyl cyclase An enzyme that produces large amounts of cAMP from ATP; the cAMP acts as a second messenger in a target cell.

adipose cells Fat cells, found in loose connective tissue, usually in large groups that form adipose tissue. Each adipose cell can store a droplet of fat (triacylglyceride).

adventitious (L. *adventicius*, not properly belonging to) Referring to a structure arising from an unusual place, such as stems from roots or roots from stems.

aerobic (Gr. *aer*, air, + *bios*, life) Requiring free oxygen; any biological process that can occur in the presence of gaseous oxygen (O_2).

alga, pl. **algae** A unicellular or simple multicellular photosynthetic organism lacking multicellular sex organs.

allantois (Gr. *allas*, sausage, + *eidos*, form) A membrane of the amniotic egg that functions in respiration and excretion in birds and reptiles and plays an important role in the development of the placenta in most mammals.

allele (Gr. *allelon*, of one another) One of two or more alternative states of a gene.

allometric growth (Gr. *allos*, other, + *ergon*, action) A pattern of growth in which different components grow at different rates.

allopatric speciation The differentiation of geographically isolated populations into distinct species.

allosteric site A part of an enzyme, away from its active site, that serves as an on/off switch for the function of the enzyme.

alternation of generations A reproductive cycle in which a haploid (*n*) phase, the gametophyte, gives rise to gametes, which, after fusion to form a zygote, germinate to produce a diploid (*2n*) phase, the sporophyte. Spores produced by meiotic division from the sporophyte give rise to new gametophytes, completing the cycle.

altruism Self-sacrifice for the benefit of others; in formal terms, the behavior that increases the fitness of the recipient while reducing the fitness of the altruistic individual.

alveolus, pl. **alveoli** (L., a small cavity) One of many small, thin-walled air sacs within the lungs in which the bronchioles terminate.

amino acid The subunit structure from which proteins are produced, consisting of a central carbon atom with a carboxyl group (—COOH), an amino group (—NH$_2$), a hydrogen, and a side group (*R* group); only the side group differs from one amino acid to another.

amniocentesis (Gr. *amnion*, membrane around the fetus, + *centes*, puncture) Examination of a fetus indirectly, by tests on cell cultures grown from fetal cells obtained from a sample of the amniotic fluid surrounding the developing embryo or tests on the fluid itself.

amnion (Gr., membrane around the fetus) The innermost of the extraembryonic membranes; the amnion forms a fluid-filled sac around the embryo in amniotic eggs.

amniotic egg An egg that is isolated and protected from the environment by a more or less impervious shell during the period of its development and that is completely self-sufficient, requiring only oxygen.

amyloplast (Gr. *amylon*, starch, + *plastos*, formed) A plant organelle called a plastid that specializes in storing starch.

anabolism (Gr. *ana*, up, + *bolein*, to throw) The biosynthetic or constructive part of metabolism; those chemical reactions involved in biosynthesis.

anaerobic (Gr. *an*, without, + *aer*, air, + *bios*, life) Any process that can occur without oxygen, such as anaerobic fermentation or H$_2$S photosynthesis.

analogous (Gr. *analogos*, proportionate) Structures that are similar in function but different in evolutionary origin, such as the wing of a bat and the wing of a butterfly.

anaphase In mitosis and meiosis II, the stage initiated by the separation of sister chromatids, during which the daughter chromosomes move to opposite poles of the cell; in meiosis I, marked by separation of replicated homologous chromosomes.

androecium (Gr. *andros*, man, + *oilos*, house) The floral whorl that comprises the stamens.

aneuploidy (Gr. *an*, without, + *eu*, good, + *ploid*, multiple of) The condition in an organism whose cells have lost or gained a chromosome; Down syndrome, which results from an extra copy of human chromosome 21, is an example of aneuploidy in humans.

angiosperms The flowering plants, one of five phyla of seed plants. In angiosperms, the ovules at the time of pollination are completely enclosed by tissues.

animal pole In fish and other aquatic vertebrates with asymmetrical yolk distribution in their eggs, the hemisphere of the blastula comprising cells relatively poor in yolk.

anion (Gr. *anienae*, to go up) A negatively charged ion.

anther (Gr. *anthos*, flower) In angiosperm flowers, the pollen-bearing portion of a stamen.

antheridium, pl. **antheridia** A sperm-producing organ.

antibody (Gr. *anti*, against) A protein called immunoglobulin that is produced by lymphocytes in response to a foreign substance (antigen) and released into the bloodstream.

anticodon The three-nucleotide sequence at the end of a transfer RNA molecule that is complementary to, and base-pairs with, an amino-acid–specifying codon in messenger RNA.

antigen (Gr. *anti*, against, + *genos*, origin) A foreign substance, usually a protein or polysaccharide, that stimulates an immune response.

anus The terminal opening of the gut; the solid residues of digestion are eliminated through the anus.

aorta (Gr. *aeirein*, to lift) The major artery of vertebrate systemic blood circulation; in mammals, carries oxygenated blood away from the heart to all regions of the body except the lungs.

apical meristem (L. *apex*, top, + Gr. *meristos*, divided) In vascular plants, the growing point at the tip of the root or stem.

apoptosis A process of programmed cell death, in which dying cells shrivel and shrink; used in all animal cell development to produce planned and orderly death of cells not destined to be present in the final tissue.

aposematic coloration An ecological strategy of some organisms that "advertise" their poisonous nature by the use of bright colors.

aquifers Permeable, saturated, underground layers of rock, sand, and gravel, which serve as reservoirs for groundwater.

archaebacteria A group of bacteria that are among the most primitive still in existence; characterized by the absence of peptidoglycan in their cell walls, a feature that distinguishes them from all other bacteria.

archegonium, pl. archegonia (Gr. *archegonos*, first of a race) The multicellular egg-producing organ in bryophytes and some vascular plants.

archenteron (Gr. *arche*, beginning, + *enteron*, gut) The principal cavity of a vertebrate embryo in the gastrula stage; lined with endoderm, it opens up to the outside and represents the future digestive cavity.

arteriole A smaller artery, leading from the arteries to the capillaries.

arteriosclerosis Hardening and thickening of the wall of an artery.

ascomycetes A large group comprising part of the "true fungi." They are characterized by separate hyphae, asexually produced conidiospores, and sexually produced ascospores within asci.

ascospore A fungal spore produced within an ascus.

ascus, pl. asci (Gr. *askos*, wineskin, bladder) A specialized cell, characteristic of the ascomycetes, in which two haploid nuclei fuse to produce a zygote that divides immediately by meiosis; at maturity, an ascus contains ascospores.

asexual reproduction The process by which an individual inherits all of its chromosomes from a single parent, thus being genetically identical to that parent; cell division is by mitosis only.

aster In animal cell mitosis, a radial array of microtubules extending from the centrioles toward the plasma membrane,

possibly serving to brace the centrioles for retraction of the spindle.

atrioventricular (AV) node A slender connection of cardiac muscle cells that receives the heartbeat impulses from the sinoatrial node and conducts them by way of the bundle of His.

atrium (L., vestibule or courtyard) An antechamber; in the heart, a thin-walled chamber that receives venous blood and passes it on to the thick-walled ventricle; in the ear, the tympanic cavity.

autonomic nervous system (Gr. *autos*, self, + *nomos*, law) The involuntary neurons and ganglia of the peripheral nervous system of vertebrates; regulates the heart, glands, visceral organs, and smooth muscle.

autosome (Gr. *autos*, self, + *soma*, body) Any eukaryotic chromosome that is not a sex chromosome; autosomes are present in the same number and kind in both males and females of the species.

autotroph (Gr. *autos*, self, + *trophos*, feeder) An organism able to build all the complex organic molecules that it requires as its own food source, using only simple inorganic compounds.

auxin (Gr. *auxein*, to increase) A plant hormone that controls cell elongation, among other effects.

axon (Gr., axle) A process extending out from a neuron that conducts impulses away from the cell body.

B

bacteriophage (Gr. *bakterion*, little rod, + *phagein*, to eat) A virus that infects bacterial cells; also called a *phage*.

Barr body A deeply staining structure, seen in the interphase nucleus of a cell of an individual with more than one X chromosome, that is a condensed and inactivated X. Only one X remains active in each cell after early embryogenesis.

basal body A self-reproducing, cylindrical, cytoplasmic organelle composed of nine triplets of microtubules from which the flagella or cilia arise.

base-pair A complementary pair of nucleotide bases, consisting of a purine and a pyrimidine.

basidiospore A spore of the basidiomycetes, produced within and borne on a basidium after nuclear fusion and meiosis.

basidium, pl. basidia (L., a little pedestal) A specialized reproductive cell of the basidiomycetes, often club shaped,

in which nuclear fusion and meiosis occur.

basophil A leukocyte containing granules that rupture and release chemicals that enhance the inflammatory response. Important in causing allergic responses.

Batesian mimicry A situation in which a palatable or nontoxic organism resembles another kind of organism that is distasteful or toxic. Both species exhibit warning coloration.

B cell A type of lymphocyte that, when confronted with a suitable antigen, is capable of secreting a specific antibody protein.

behavioral ecology The study of how natural selection shapes behavior.

biennial A plant that normally requires two growing seasons to complete its life cycle. Biennials flower in the second year of their lives.

bile salts A solution of organic salts that is secreted by the vertebrate liver and temporarily stored in the gallbladder; emulsifies fats in the small intestine.

binary fission (L. *binarius*, consisting of two things or parts, + *fissus*, split) Asexual reproduction by division of one cell or body into two equal or nearly equal parts.

binomial distribution The distribution of phenotypes seen among the progeny of a cross in which there are only two alternative alleles.

biodiversity The number of species and their range of behavioral, ecological, physiological, and other adaptations, in an area.

biomass (Gr. *bios*, life, + *maza*, lump or mass) The total mass of all the living organisms in a given population, area, or other unit being measured.

biome One of the major terrestrial ecosystems, characterized by climatic and soil conditions; the largest ecological unit.

bipedal Able to walk upright on two feet.

bipolar cell A specialized type of neuron connecting cone cells to ganglion cells in the visual system. Bipolar cells receive a hyperpolarized stimulus from the cone cell and then transmit a depolarization stimulus to the ganglion cell.

blade The broad, expanded part of a leaf; also called the lamina.

blastocoel (Gr. *blastos*, sprout, + *koilos*, hollow) The central cavity of the blastula stage of vertebrate embryos.

blastodisc (Gr. *blastos*, sprout, + *discos*, a round plate) In the development of birds, a disclike

area on the surface of a large, yolky egg that undergoes cleavage and gives rise to the embryo.

blastomere (Gr. *blastos*, sprout, + *meros*, part) One of the cells of a blastula.

blastopore (Gr. *blastos*, sprout, + *poros*, a path or passage) In vertebrate development, the opening that connects the archenteron cavity of a gastrula stage embryo with the outside.

blastula (Gr., a little sprout) In vertebrates, an early embryonic stage consisting of a hollow, fluid-filled ball of cells one layer thick; a vertebrate embryo after cleavage and before gastrulation.

Bohr effect The release of oxygen by hemoglobin molecules in response to elevated ambient levels of CO_2.

Bowman's capsule In the vertebrate kidney, the bulbous unit of the nephron, which surrounds the glomerulus.

β-oxidation In the cellular respiration of fats, the process by which two-carbon acetyl groups are removed from a fatty acid and combined with coenzyme A to form acetyl-CoA, until the entire fatty acid has been broken down.

bronchus, pl. bronchi (Gr. *bronchos*, windpipe) One of a pair of respiratory tubes branching from the lower end of the trachea (windpipe) into either lung.

bryophytes Nonvascular plants, including mosses, hornworts, and liverworts.

bud An asexually produced outgrowth that develops into a new individual. In plants, an embryonic shoot, often protected by young leaves; buds may give rise to branch shoots.

C

C₃ photosynthesis The main cycle of the dark reactions of photosynthesis, in which CO_2 binds to ribulose 1,5-bisphosphate (RuBP) to form two three-carbon phosphoglycerate (PGA) molecules.

C₄ photosynthesis A process of CO_2 fixation in photosynthesis by which the first product is the four-carbon oxaloacetate molecule.

callus (L. *callos*, hard skin) Undifferentiated tissue; a term used in tissue culture, grafting, and wound healing.

Calvin cycle The dark reactions of C₃ photosynthesis; also called the Calvin-Benson cycle.

calyx (Gr. *kalyx*, a husk, cup) The sepals collectively; the outermost flower whorl.

capillary (L. *capillaris*, hairlike) The smallest of the blood vessels;

the very thin walls of capillaries are permeable to many molecules, and exchanges between blood and the tissues occur across them; the vessels that connect arteries with veins.

capsid The outermost protein covering of a virus.

carapace (Fr. from Sp. *carapacho*, shell) Shieldlike plate covering the cephalothorax of decapod crustaceans; the dorsal part of the shell of a turtle.

carbohydrate (L. *carbo*, charcoal, + *hydro*, water) An organic compound consisting of a chain or ring of carbon atoms to which hydrogen and oxygen atoms are attached in a ratio of approximately 2:1; having the generalized formula $(CH_2O)_n$; carbohydrates include sugars, starch, glycogen, and cellulose.

carbon fixation The conversion of CO_2 into organic compounds during photosynthesis; the first stage of the dark reactions of photosynthesis, in which carbon dioxide from the air is combined with ribulose 1,5-bisphosphate.

carpel (Gr. *karpos*, fruit) A leaflike organ in angiosperms that encloses one or more ovules.

carrying capacity The maximum population size that a habitat can support.

cartilage (L. *cartilago*, gristle) A connective tissue in skeletons of vertebrates. Cartilage forms much of the skeleton of embryos, very young vertebrates, and some adult vertebrates, such as sharks and their relatives.

catabolism (Gr. *ketabole*, throwing down) In a cell, those metabolic reactions that result in the breakdown of complex molecules into simpler compounds, often with the release of energy.

catalysis The process by which chemical subunits of larger organic molecules are held and positioned by enzymes that stress their chemical bonds, leading to the disassembly of the larger molecule into its subunits, often with the release of energy.

cation (Gr. *katienai*, to go down) A positively charged ion.

cecum (L. *caecus*, blind) In vertebrates, a blind pouch at the beginning of the large intestine.

cell cycle The repeating sequence of growth and division through which cells pass each generation.

cell plate The structure that forms at the equator of the spindle during early telophase in the dividing cells of plants and a few green algae.

cell surface receptor A cell surface protein that binds a signal molecule and converts the extracellular signal into an intracellular one.

cellular respiration The metabolic harvesting of energy by oxidation, ultimately dependent on molecular oxygen; carried out by the Krebs cycle and oxidative phosphorylation.

cellulose (L. *cellula*, a little cell) The chief constituent of the cell wall in all green plants, some algae, and a few other organisms; an insoluble complex carbohydrate formed of microfibrils of glucose molecules.

cell wall The rigid, outermost layer of the cells of plants, some protists, and most bacteria; the cell wall surrounds the plasma membrane.

central nervous system (CNS) That portion of the nervous system where most association occurs; in vertebrates, it is composed of the brain and spinal cord; in invertebrates, it usually consists of one or more cords of nervous tissue, together with their associated ganglia.

centriole (Gr. *kentron*, center, + L. *olus*, little one) A cytoplasmic organelle located outside the nuclear membrane, identical in structure to a basal body; found in animal cells and in the flagellated cells of other groups; divides and organizes spindle fibers during mitosis and meiosis.

centromere (Gr. *kentron*, center, + *meros*, a part) Condensed region on a eukaryotic chromosome where sister chromatids are attached to each other after replication. *See* kinetochore.

cerebellum (L., little brain) The hindbrain region of the vertebrate brain that lies above the medulla (brain stem) and behind the forebrain; it integrates information about body position and motion, coordinates muscular activities, and maintains equilibrium.

cerebral cortex The thin surface layer of neurons and glial cells covering the cerebrum; well developed only in mammals, and particularly prominent in humans. The cerebral cortex is the seat of conscious sensations and voluntary muscular activity.

cerebrum (L., brain) The portion of the vertebrate brain (the forebrain) that occupies the upper part of the skull, consisting of two cerebral hemispheres united by the corpus callosum. It is the primary association center of the brain. It coordinates and processes sensory input and coordinates motor responses.

chelicera, pl. chelicerae (Gr. *chele*, claw, + *keras*, horn) The first pair of appendages in horseshoe crabs, sea spiders, and arachnids—the chelicerates, a group of arthropods. Chelicerae usually take the form of pincers or fangs.

chemiosmosis The mechanism by which ATP is generated in mitochondria and chloroplasts; energetic electrons excited by light (in chloroplasts) or extracted by oxidation in the Krebs cycle (in mitochondria) are used to drive proton pumps, creating a proton concentration gradient; when protons subsequently flow back across the membrane, they pass through channels that couple their movement to the synthesis of ATP.

chiasma An X-shaped figure that can be seen in the light microscope during meiosis; evidence of crossing over, where two chromatids have exchanged parts; chiasmata move to the ends of the chromosome arms as the homologues separate.

chitin (Gr. *chiton*, tunic) A tough, resistant, nitrogen-containing polysaccharide that forms the cell walls of certain fungi, the exoskeleton of arthropods, and the epidermal cuticle of other surface structures of certain other invertebrates.

chloroplast (Gr. *chloros*, green, + *plastos*, molded) A cell-like organelle present in algae and plants that contains chlorophyll (and usually other pigments) and carries out photosynthesis.

chorion (Gr., skin) The outer member of the double membrane that surrounds the embryo of reptiles, birds, and mammals; in placental mammals, it contributes to the structure of the placenta.

chromatid (Gr. *chroma*, color, + L. *-id*, daughters of) One of the two daughter strands of a duplicated chromosome that is joined by a single centromere.

chromatin (Gr. *chroma*, color) The complex of DNA and proteins of which eukaryotic chromosomes are composed; chromatin is highly uncoiled and diffuse in interphase nuclei, condensing to form the visible chromosomes in prophase.

chromosome (Gr. *chroma*, color, + *soma*, body) The vehicle by which hereditary information is physically transmitted from one generation to the next; in a bacterium, the chromosome consists of a single naked circle of DNA; in eukaryotes, each chromosome consists of a single linear DNA molecule and associated proteins.

cilium A short cellular projection from the surface of a eukaryotic cell, having the same internal structure of microtubules in a 9 + 2 arrangement as seen in a flagellum.

circadian rhythm An endogenous cyclical rhythm that oscillates on a daily (24-hour) basis.

cisterna A small collecting vessel that pinches off from the end of a Golgi body to form a transport vesicle that moves materials through the cytoplasm.

cladistics A taxonomic technique used for creating hierarchies of organisms that represent true phylogenetic relationship and descent.

class A taxonomic category between phyla and orders. A class contains one or more orders, and belongs to a particular phylum.

classical conditioning The repeated presentation of a stimulus in association with a response that causes the brain to form an association between the stimulus and the response, even if they have never been associated before.

clathrin A protein located just inside the plasma membrane in eukaryotic cells, in indentations called clathrin-coated pits.

cleavage In vertebrates, a rapid series of successive cell divisions of a fertilized egg, forming a hollow sphere of cells, the blastula.

climax vegetation Vegetation encountered in a self-perpetuating community of plants that has proceeded through all the stages of succession and stabilized.

cloaca (L., sewer) In some animals, the common exit chamber from the digestive, reproductive, and urinary system; in others, the cloaca may also serve as a respiratory duct.

cloning Producing a cell line or culture all of whose members contain identical copies of a particular nucleotide sequence; an essential element in genetic engineering.

coacervate (L. *coacervatus*, heaped up) A spherical aggregation of lipid molecules in water, held together by hydrophobic forces.

cochlea (Gr. *kochlios*, a snail) In terrestrial vertebrates, a tubular cavity of the inner ear containing the essential organs for hearing.

codon (L., code) The basic unit of the genetic code; a sequence of three adjacent nucleotides in DNA or mRNA that codes for one amino acid.

coenzyme (L. *co-*, together, + Gr. *en*, in, + *zyme*, leaven) A nonprotein organic molecule such as NAD^+ that plays an accessory role in enzyme-catalyzed processes, often by acting as a donor or acceptor of electrons.

coevolution (L. *co-*, together, + *e-*, out, + *volvere*, to fill) The simultaneous development of adaptations in two or more populations, species, or other categories that interact so closely that each is a strong selective force on the other.

cofactor One or more nonprotein components required by enzymes in order to function; many cofactors are metal ions, others are organic coenzymes.

collenchyma cell (Gr. *kolla*, glue, + *en*, in, + *chymein*, to pour) In plants, the cells that form a supporting tissue called collenchyma; often found in regions of primary growth in stems and in some leaves.

commensalism (L. *cum*, together with, + *mensa*, table) A relationship in which one individual lives close to or on another and benefits, and the host is unaffected; a kind of symbiosis.

community (L. *communitas*, community, fellowship) All of the species inhabiting a common environment and interacting with one another.

companion cell A specialized parenchyma cell that is associated with each sieve-tube member in the phloem of a plant.

competitive exclusion The hypothesis that two species with identical ecological requirements cannot exist in the same locality indefinitely, and that the more efficient of the two in utilizing the available scarce resources will exclude the other; also known as Gause's principle.

complementary Describes genetic information in which each nucleotide base has a complementary partner with which it forms a base-pair.

complement system The chemical defense of a vertebrate body that consists of a battery of proteins that become activated by the walls of bacteria and fungi.

concentration gradient The concentration difference of a substance across a distance; in a cell, a greater concentration of its molecules in one region than in another.

cone (1) In plants, the reproductive structure of a conifer. (2) In vertebrates, a type of light-sensitive neuron in the retina, concerned with the perception of color and with the most acute discrimination of detail.

conidia An asexually produced fungal spore.

conjugation (L. *conjugare*, to yolk together) Temporary union of two unicellular organisms, during which genetic material is transferred from one cell to the other; occurs in bacteria, protists, and certain algae and fungi.

contractile vacuole In protists and some animals, a clear fluid-filled vacuole that takes up water from within the cell and then contracts, releasing it to the outside through a pore in a cyclical manner; functions primarily in osmoregulation and excretion.

conus arteriosus The anteriormost chamber of the embryonic heart in vertebrate animals.

convergent evolution The independent development of similar structures in organisms that are not directly related; often found in organisms living in similar environments.

cork cambium The lateral meristem that forms the periderm, producing cork (phellem) toward the surface (outside) of the plant and phelloderm toward the inside.

cornea (L. *corneus*, horny) The transparent outer layer of the vertebrate eye.

corolla (L., a little crown) The petals, collectively; usually the conspicuously colored flower whorl.

corpus callosum The band of nerve fibers that connects the two hemispheres of the cerebrum in humans and other primates.

corpus luteum (L., yellowish body) A structure that develops from a ruptured follicle in the ovary after ovulation.

cortex (L., bark) The outer layer of a structure; in animals, the outer, as opposed to the inner, part of an organ; in vascular plants, the primary ground tissue of a stem or root.

cotyledon (Gr. *kotyledon*, a cup-shaped hollow) A seed leaf that generally stores food in dicots or absorbs it in monocots, providing nourishment used during seed germination.

crassulacean acid metabolism (CAM) A mode of carbon dioxide fixation by which CO_2 enters open leaf stomata at night and is used in photosynthesis during the day, when stomata are closed to prevent water loss.

cross-current flow In bird lungs, the latticework of capillaries arranged across the air flow, at a 90° angle.

crossing over In meiosis, the exchange of corresponding chromatid segments between homologous chromosomes; responsible for genetic recombination between homologous chromosomes.

cuticle (L. *cuticula*, little skin) A waxy or fatty, noncellular layer (formed of a substance called cutin) on the outer wall of epidermal cells.

cyanobacteria A group of photosynthetic bacteria, sometimes called the "blue-green algae," that contain the chlorophyll pigments most abundant in plants and algae, as well as other pigments.

cyclic AMP (cAMP) A form of adenosine monophosphate (AMP) in which the atoms of the phosphate group form a ring; found in almost all organisms, cAMP functions as an intracellular second messenger that regulates a diverse array of metabolic activities.

cystic fibrosis An autosomal disorder that produces the most common fatal genetic disease in Caucasians, characterized by secretion of thick mucus that clogs passageways in the lungs, liver, and pancreas.

cytochrome (Gr. *kytos*, hollow vessel, + *chroma*, color) Any of several iron-containing protein pigments that serve as electron carriers in transport chains of photosynthesis and cellular respiration.

cytokinesis (Gr. *kytos*, hollow vessel, + *kinesis*, movement) Division of the cytoplasm of a cell after nuclear division.

cytoplasm (Gr. *kytos*, hollow vessel, + *plasma*, anything molded) The material within a cell, excluding the nucleus; the protoplasm.

cytoskeleton A network of protein microfilaments and microtubules within the cytoplasm of a eukaryotic cell that maintains the shape of the cell, anchors its organelles, and is involved in animal cell motility.

cytotoxic T cell (Gr. *kytos*, hollow vessel, + *toxin*) A special T cell activated during cell-mediated immune response that recognizes and destroys infected body cells.

D

dehydration synthesis (L. *co-*, together, + *densare*, to make dense) A type of chemical reaction in which two molecules join to form one larger molecule, simultaneously splitting out a molecule of water; one molecule is stripped of a hydrogen atom, and another is stripped of a hydroxyl group (—OH), resulting in the joining of the two molecules, while the H^+ and —OH released may combine to form a water molecule.

demography (Gr. *demos*, people, + *graphein*, to draw) The properties of the rate of growth and the age structure of populations.

denaturation The loss of the native configuration of a protein or nucleic acid as a result of excessive heat, extremes of pH, chemical modification, or changes in solvent ionic strength or polarity that disrupt hydrophobic interactions; usually accompanied by loss of biological activity.

dendrite (Gr. *dendron*, tree) A process extending from the cell body of a neuron, typically branched, that conducts impulses toward the cell body.

deoxyribonucleic acid (DNA) The genetic material of all organisms; composed of two complementary chains of nucleotides wound in a double helix.

depolarization The movement of ions across a plasma membrane that locally wipes out an electrical potential difference.

derived character A characteristic used in taxonomic analysis representing a departure from the primitive form.

desmosome A type of anchoring junction that links adjacent cells by connecting their cytoskeletons with cadherin proteins.

diaphragm (Gr. *diaphrassein*, to barricade) (1) In mammals, a sheet of muscle tissue that separates the abdominal and thoracic cavities and functions in breathing. (2) A contraceptive device used to block the entrance to the uterus temporarily and thus prevent sperm from entering during sexual intercourse.

diastolic pressure In the measurement of human blood pressure, the minimum pressure between heartbeats (repolarization of the ventricles). *Compare with* systolic pressure.

dicot Short for dicotyledon; a class of flowering plants generally characterized as having two cotyledons, net-veined leaves, and flower parts usually in fours or fives.

differentiation A developmental process by which a relatively unspecialized cell undergoes a progressive change to a more specialized form or function.

diffusion (L. *diffundere*, to pour out) The net movement of dissolved molecules or other particles from a region where they are more concentrated to a region where they are less concentrated.

digestion (L. *digestio*, separating out, dividing) The breakdown of complex, usually insoluble foods into molecules that can be absorbed into cells, where they are degraded to yield energy and the raw materials for synthetic processes.

dihybrid (Gr. *dis*, twice, + *hibridia*, mixed offspring) An individual heterozygous at two different loci; for example *A/a B/b*.

dikaryotic (Gr. *di*, two, + *karyon*, kernel) In fungi, having pairs of nuclei within each cell.

dioecious (Gr. *di*, two, + *oikos*, house) Having the male and female elements on different individuals.

diploid (Gr. *diploos*, double, + *eidos*, form) Having two sets of chromosomes (*2n*); in animals, twice the number characteristic of gametes; in plants, the chromosome number characteristic of the sporophyte generation; in contrast to haploid (*n*).

directional selection A form of selection in which selection acts to eliminate one extreme from an array of phenotypes.

disaccharide A carbohydrate formed of two simple sugar molecules bonded covalently.

disruptive selection A form of selection in which selection acts to eliminate rather than favor the intermediate type.

diurnal (L. *diurnalis*, day) Active during the day.

DNA ligase The enzyme that links together Okazaki fragments in DNA replication of the lagging strand; it also links other broken areas of the DNA backbone.

domain (1) A distinct modular region of a protein that serves a particular function in the action of the protein, such as a regulatory domain or a DNA-binding domain. (2) In taxonomy, the level higher than kingdom. The three domains currently recognized are Bacteria, Archaea, and Eukarya.

dominant An allele that is expressed when present in either the heterozygous or the homozygous condition.

double fertilization The fusion of the egg and sperm (resulting in a *2n* fertilized egg, the zygote) and the simultaneous fusion of the second male gamete with the polar nuclei (resulting in a primary endosperm nucleus, which is often triploid, *3n*); a unique characteristic of all angiosperms.

Down syndrome A congenital syndrome caused by the presence of an extra copy of chromosome 21.

duodenum (L. *duodeni*, 12 each—from its length, about 12 fingers' breadth) In vertebrates, the upper portion of the small intestine.

E

ecdysis (Gr. *ekdysis*, stripping off) Shedding of outer, cuticular layer; molting, as in insects or crustaceans.

ecdysone (Gr. *ekdysis*, stripping off) Molting hormone of arthropods, which triggers when ecdysis occurs.

ecology (Gr. *oikos*, house, + *logos*, word) The study of interactions of organisms with one another and with their physical environment.

ecosystem (Gr. *oikos*, house, + *systems*, that which is put together) A major interacting system that includes organisms and their nonliving environment.

ecotype (Gr. *oikos*, house, + L. *typus*, image) A locally adapted variant of an organism; differing genetically from other ecotypes.

ectoderm (Gr. *ecto*, outside, + *derma*, skin) One of the three embryonic germ layers of early vertebrate embryos; ectoderm gives rise to the outer epithelium of the body (skin, hair, nails) and to the nerve tissue, including the sense organs, brain, and spinal cord.

ectomycorrhizae Externally developing mycorrhizae that do not penetrate the cells they surround.

ectotherms Animals such as reptiles, fish, or amphibians, whose body temperature is regulated by their behavior or by their surroundings.

electronegativity A property of atomic nuclei that refers to the affinity of the nuclei for valence electrons; a nucleus that is more electronegative has a greater pull on electrons than one that is less electronegative.

electron transport chain The passage of energetic electrons through a series of membrane-associated electron-carrier molecules to proton pumps embedded within mitochondrial or chloroplast membranes. *See* chemiosmosis.

endergonic Describes a chemical reaction in which the products contain more energy than the reactants, so that free energy must be put into the reaction from an outside source to allow it to proceed.

endocrine gland (Gr. *endon*, within, + *krinein*, to separate) Ductless gland that secretes hormones into the extracellular spaces, from which they diffuse into the circulatory system.

endocytosis (Gr. *endon*, within, + *kytos*, hollow vessel) The uptake of material into cells by inclusion within an invagination of the plasma membrane; the uptake of solid material is phagocytosis, while that of dissolved material is pinocytosis.

endoderm (Gr. *endon*, within, + *derma*, skin) One of the three embryonic germ layers of early vertebrate embryos, destined to give rise to the epithelium that lines internal structures and most of the digestive and respiratory tracts.

endodermis (Gr. *endon*, within, + *derma*, skin) In vascular plants, a layer of cells forming the innermost layer of the cortex in roots and some stems.

endometrium (Gr. *endon*, within, + *metrios*, of the womb) The lining of the uterus in mammals; thickens in response to secretion of estrogens and progesterone and is sloughed off in menstruation.

endomycorrhizae Mycorrhizae that develop within cells.

endoplasmic reticulum (ER) An internal membrane system that forms a netlike array of channels and interconnections of organelles within the cytoplasm of eukaryotic cells.

endorphin One of a group of small neuropeptides produced by the vertebrate brain; like morphine, endorphins modulate pain perception.

endosperm (Gr. *endon*, within, + *sperma*, seed) A storage tissue characteristic of the seeds of angiosperms, which develops from the union of a male nucleus and the polar nuclei of the embryo sac. The endosperm is digested by the growing sporophyte either before maturation of the seed or during its germination.

endosymbiosis Theory that proposes that eukaryotic cells evolved from a symbiosis between different species of prokaryotes.

endotherm An animal capable of maintaining a constant body temperature. *See* homeotherm.

energy The capacity to do work.

enhancer A site of regulatory protein binding on the DNA molecule distant from the promoter and start site for a gene's transcription.

enthalpy In a chemical reaction, the energy contained in the chemical bonds of the molecule, symbolized as H; in a cellular reaction, the free energy is equal to the enthalpy of the reactant molecules in the reaction.

entropy (Gr. *en*, in, + *tropos*, change in manner) A measure of the randomness or disorder of a system; a measure of how much energy in a system has become so dispersed (usually as evenly distributed heat) that it is no longer available to do work.

enzyme (Gr. *enzymes*, leavened, from *en*, in, + *zyme*, leaven) A protein that is capable of speeding up specific chemical reactions by lowering the required activation energy.

epicotyl The region just above where the cotyledons are attached.

epidermis (Gr. *epi*, on or over, + *derma*, skin) The outermost layers of cells; in plants, the exterior primary tissue of leaves, young stems, and roots; in vertebrates, the nonvascular external layer of skin, of ectodermal origin; in invertebrates, a single layer of ectodermal epithelium.

epididymis (Gr. *epi*, on, + *didymos*, testicle) A sperm storage vessel; a coiled part of the sperm duct that lies near the testis.

epistasis (Gr. *epi*, on, + *stasis*, standing still) Interaction between two nonallelic genes in which one of them modifies the phenotypic expression of the other.

epithelium (Gr. *epi*, on, + *thele*, nipple) In animals, a type of tissue that covers an exposed surface or lines a tube or cavity.

equilibrium (L. *aequus*, equal, + *libra*, balance) A stable condition; the point at which a chemical reaction proceeds as rapidly in the reverse direction as it does in the forward direction, so that there is no further net change in the concentrations of products or reactants. In ecology, a stable condition that resists change and fairly quickly returns to its original state if disturbed by humans or natural events.

erythrocyte (Gr. *erythros*, red, + *kytos*, hollow vessel) Red blood cell, the carrier of hemoglobin.

erythropoiesis The manufacture of blood cells in the bone marrow.

estrous cycle The periodic cycle in which periods of estrus correspond to ovulation events.

estrus (L. *oestrus*, frenzy) The period of maximum female sexual receptivity, associated with ovulation of the egg.

ethology (Gr. *ethos*, habit or custom, + *logos*, discourse) The study of patterns of animal behavior in nature.

euchromatin (Gr. *eu*, good, + *chroma*, color) That portion of a eukaryotic chromosome that is transcribed into mRNA; contains active genes that are not tightly condensed during interphase.

eukaryote (Gr. *eu*, good, + *karyon*, kernel) A cell characterized by membrane-bounded organelles, most notably the nucleus, and one that possesses chromosomes whose DNA is associated with proteins; an organism composed of such cells.

eutrophic (Gr. *eutrophos*, thriving) Refers to a lake in which an abundant supply of minerals and organic matter exists.

evolution (L. *evolvere*, to unfold) Genetic change in a population of organisms; in general, evolution leads to progressive change from simple to complex.

exergonic Describes a chemical reaction in which the products contain less free energy than the reactants, so that free energy is released in the reaction.

exocrine gland (Gr. *ex*, out of, + *krinein*, to separate) A type of gland that releases its secretion through a duct, such as a digestive gland or a sweat gland.

exocytosis (Gr. *ex*, out of, + *kytos*, hollow vessel) A type of bulk transport out of cells in which a vacuole fuses with the plasma membrane, discharging the vacuole's contents to the outside.

exon (Gr. *exo*, outside) A segment of DNA that is both transcribed into RNA and translated into protein.

exoskeleton (Gr. *exo*, outside, + *skeletos*, hand) An external skeleton, as in arthropods.

exteroceptor (L. *exter*, outward, + *capere*, to take) A receptor that is excited by stimuli from the external world.

F

facilitated diffusion Carrier-assisted diffusion of molecules across a cellular membrane through specific channels from a region of higher concentration to one of lower concentration; the process is driven by the concentration gradient and does not require cellular energy from ATP.

fat A molecule composed of glycerol and three fatty acid molecules.

feedback inhibition Control mechanism whereby an increase in the concentration of some molecules inhibits the synthesis of that molecule.

fermentation (L. *fermentum*, ferment) The enzyme-catalyzed extraction of energy from organic compounds without the involvement of oxygen.

fertilization The fusion of two haploid gamete nuclei to form a diploid zygote nucleus.

fibroblast (L. *fibra*, fiber, + Gr. *blastos*, sprout) A flat, irregularly branching cell of connective tissue that secretes structurally strong proteins into the matrix between the cells.

First Law of Thermodynamics Energy cannot be created or destroyed, but can only undergo conversion from one form to another; thus, the amount of energy in the universe is unchangeable.

fitness The genetic contribution of an individual to succeeding generations. relative fitness refers to the fitness of an individual relative to other individuals in a population.

fixed action pattern A stereotyped animal behavior response, thought by ethologists to be based on programmed neural circuits.

foraging behavior A collective term for the many complex, evolved behaviors that influence what an animal eats and how the food is obtained.

founder effect The effect by which rare alleles and combinations of alleles may be enhanced in new populations.

fovea (L., a small pit) A small depression in the center of the retina with a high concentration of cones; the area of sharpest vision.

free energy Energy available to do work.

free radical An ionized atom with one or more unpaired electrons, resulting from electrons that have been energized by ionizing radiation being ejected from the atom; free radicals react violently with other molecules, such as DNA, causing damage by mutation.

fruit In angiosperms, a mature, ripened ovary (or group of ovaries), containing the seeds.

functional genomics Research into the function of specific genes in an organism.

fundamental niche Also referred to as the hypothetical niche, this is the entire niche an organism could fill if there were no other interacting factors (such as competition or predation).

G

gametangium, pl. **gametangia** (Gr. *gamein*, to marry, + L. *tangere*, to touch) A cell or organ in which gametes are formed.

gamete (Gr., wife) A haploid reproductive cell.

gametocytes Cells in the malarial sporozoite life cycle capable of giving rise to gametes when in the correct host.

gametophyte In plants, the haploid (*n*), gamete-producing generation, which alternates with the diploid (*2n*) sporophyte.

ganglion, pl. **ganglia** (Gr., a swelling) An aggregation of nerve cell bodies; in invertebrates, ganglia are the integrative centers; in vertebrates, the term is restricted to aggregations of nerve cell bodies located outside the central nervous system.

gap junction A junction between adjacent animal cells that allows the passage of materials between the cells.

gastrula (Gr., little stomach) In vertebrates, the embryonic stage in which the blastula with its single layer of cells turns into a three-layered embryo made up of ectoderm, mesoderm, and endoderm.

gene (Gr. *genos*, birth, race) The basic unit of heredity; a sequence of DNA nucleotides on a chromosome that encodes a protein, tRNA, or rRNA molecule, or regulates the transcription of such a sequence.

gene conversion Alteration of one homologous chromosome by the cell's error-detection and repair system to make it resemble the sequence on the other homologue.

genetic drift Random fluctuation in allele frequencies over time by chance.

genome The entire DNA sequence of an organism.

genomics The study of genomes as opposed to individual genes.

genotype (Gr. *genos*, birth, race, + *typos*, form) The genetic constitution underlying a single trait or set of traits.

genus, pl. **genera** (L., race) A taxonomic group that ranks below a family and above a species.

germination (L. *germinare*, to sprout) The resumption of growth and development by a spore or seed.

germ layers The three cell layers formed at gastrulation of the embryo that foreshadow the future organization of tissues; the layers, from the outside inward, are the ectoderm, the mesoderm, and the endoderm.

gill (1) In aquatic animals, a respiratory organ, usually a thin-walled projection from some part of the external body surface, endowed with a rich capillary bed and having a large surface area. (2) In basidiomycete fungi, the plates on the underside of the cap.

glomerular filtrate The fluid that passes out of the capillaries of each glomerulus.

glomerulus (L., a little ball) A cluster of capillaries enclosed by Bowman's capsule.

glucagon (Gr. *glykys*, sweet, + *ago*, to lead forward) A vertebrate hormone produced in the pancreas that acts to initiate the breakdown of glycogen to glucose subunits.

gluconeogenesis The synthesis of glucose from noncarbohydrates (such as proteins or fats).

glucose A common six-carbon sugar ($C_6H_{12}O_6$); the most common monosaccharide in most organisms.

glycocalyx A "sugar coating" on the surface of a cell resulting from the presence of polysaccharides on glycolipids and glycoproteins embedded in the outer layer of the plasma membrane.

glycogen (Gr. *glykys*, sweet, + *gen*, of a kind) Animal starch; a complex branched polysaccharide that serves as a food reserve in animals, bacteria, and fungi.

glycolipid Lipid molecule modified within the Golgi complex by having a short sugar chain (polysaccharide) attached.

glycolysis (Gr. *glykys*, sweet, + *lyein*, to loosen) The anaerobic breakdown of glucose; this enzyme-catalyzed process yields two molecules of pyruvate with a net of two molecules of ATP.

glycoprotein Protein molecule modified within the Golgi complex by having a short sugar chain (polysaccharide) attached.

glyoxysome A small cellular organelle or microbody containing enzymes necessary for conversion of fats into carbohydrates.

Golgi apparatus A collection of flattened stacks of membranes (each called a Golgi body) in the cytoplasm of eukaryotic cells; functions in collection, packaging, and distribution of molecules synthesized in the cell.

G protein A protein that binds guanosine triphosphate (GTP) and assists in the function of cell surface receptors. When the receptor binds its signal molecule, the G protein binds GTP and is activated to start a chain of events within the cell.

gravitropism (L. *gravis*, heavy, + *tropes*, turning) Growth response to gravity in plants; formerly called geotropism.

ground meristem (Gr. *meristos*, divisible) The primary meristem, or meristematic tissue, that gives rise to the plant body (except for the epidermis and vascular tissues).

guttation (L. *gutta*, a drop) The exudation of liquid water from leaves due to root pressure.

gymnosperm (Gr. *gymnos*, naked, + *sperma*, seed) A seed plant with seeds not enclosed in an ovary; conifers are gymnosperms.

gynoecium (Gr. *gyne*, woman, + *oikos*, house) The aggregate of carpels in the flower of a seed plant.

H

habitat (L. *habitare*, to inhabit) The environment of an organism; the place where it is usually found.

habituation (L. *habitus*, condition) A form of learning; a diminishing response to a repeated stimulus.

haplodiploidy A phenomenon occurring in certain organisms such as wasps, wherein both haploid (male) and diploid (female) individuals are encountered.

haploid (Gr. *haploos*, single, + *ploion*, vessel) Having only one set of chromosomes (n), in contrast to diploid ($2n$).

Hardy-Weinberg equilibrium A mathematical description of the fact that allele and genotype frequencies remain constant in a random-mating population in the absence of inbreeding, selection, or other evolutionary forces; usually stated: if the frequency of allele a is p and the frequency of allele b is q, then the genotype frequencies after one generation of random mating will always be $p^2 + 2pq + q^2 = 1$.

Haversian canal After Clopton Havers, English anatomist. Narrow channels that run parallel to the length of a bone and contain blood vessels and nerve cells.

helper T cell A class of white blood cells that initiates both the cell-mediated immune response and the humoral immune response; helper T cells are the targets of the AIDS virus (HIV).

hemoglobin (Gr. *haima*, blood, + L. *globus*, a ball) A globular protein in vertebrate red blood cells and in the plasma of many invertebrates that carries oxygen and carbon dioxide.

hemophilia (Gr. *haima*, blood, + *philios*, friendly) A group of hereditary diseases characterized by failure of the blood to clot.

hemopoietic stem cell (Gr. *haima*, blood, + *poiesis*, a making) The cells in bone marrow where blood cells are formed.

hermaphroditism (Gr. *hermaphroditos*, containing both sexes) Condition in which an organism has both male and female functional reproductive organs.

heterochromatin (Gr. *heteros*, other, + *chroma*, color) The portion of a eukaryotic chromosome that is not transcribed into RNA; remains condensed in interphase and stains intensely in histological preparations.

heterokaryotic (Gr. *heteros*, other, + *karyon*, kernel) In fungi, having two or more genetically distinct types of nuclei within the same mycelium.

heterosporous (Gr. *heteros*, other, + *sporos*, seed) In vascular plants, having spores of two kinds, namely, microspores and megaspores.

heterotroph (Gr. *heteros*, other, + *trophos*, feeder) An organism that cannot derive energy from photosynthesis or inorganic chemicals, and so must feed on other plants and animals, obtaining chemical energy by degrading their organic molecules.

heterozygous Having two different alleles of the same gene; the term is usually applied to one or more specific loci, as in "heterozygous with respect to the *W* locus" (that is, the genotype is *W/w*).

histone (Gr. *histos*, tissue) One of a group of relatively small, very basic polypeptides, rich in arginine and lysine, forming the core of nucleosomes around which DNA is wrapped in the first stage of chromosome condensation.

holoblastic cleavage (Gr. *holos*, whole, + (*blastos*, germ) Process in vertebrate embryos in which the cleavage divisions all occur at the same rate, yielding a uniform cell size in the blastula.

homeobox A sequence of 180 nucleotides located in homeotic genes that produces a 60-amino-acid peptide sequence (the homeodomain) active in transcription factors.

homeostasis (Gr. *homos*, similar, + *stasis*, standing) The maintenance of a relatively stable internal physiological environment in an organism; usually involves some form of feedback self-regulation.

homeotherm (Gr. *homos*, same or similar, + *therme*, heat) An organism, such as a bird or mammal, capable of maintaining a stable body temperature independent of the environmental temperature. *See* endotherm.

homeotic gene One of a series of "master switch" genes that determine the form of segments developing in the embryo.

hominid (L. *homo*, man) Any primate in the human family, Hominidae. *Homo sapiens* is the only living representative.

hominoid (L. *homo*, man) Collectively, hominids and apes; the monkeys and hominoids constitute the anthropoid primates.

homokaryotic (Gr. *homos*, same, + *karyon*, kernel) In fungi, having nuclei with the same genetic makeup within a mycelium.

homologue One of a pair of chromosomes of the same kind located in a diploid cell; one copy of each pair of homologues comes from each gamete that formed the zygote.

homologous structure (Gr. *homologia*, agreement) A condition in which the similarity between two structures or functions is indicative of a common evolutionary origin.

homosporous (Gr. *homos*, same or similar, + *sporos*, seed) In some plants, production of only one type of spore rather than differentiated types. *Compare with* heterosporous.

homozygous Being a homozygote, having two identical alleles of the same gene; the term is usually applied to one or more specific loci, as in "homozygous with respect to the *W* locus" (that is, the genotype is *W/W* or *w/w*).

hormone (Gr. *hormaein*, to excite) A molecule, usually a peptide or steroid, that is produced in one part of an organism and triggers a specific cellular reaction in target tissues and organs some distance away.

human immunodeficiency virus (HIV) The virus responsible for AIDS, a deadly disease that destroys the human immune system. HIV is a retrovirus (its genetic material is RNA) that is thought to have been introduced to humans from chimpanzees.

hybridization The mating of unlike parents.

hydrostatic skeleton (Gr. *hydro*, water, + *skeletos*, hard) The skeleton of most soft-bodied invertebrates that have neither an internal nor an external skeleton. They use the relative incompressibility of the water within their bodies as a kind of skeleton.

hyperosmotic The condition in which a (hyperosmotic) solution has a higher osmotic concentration than that of a second solution. *Compare with* hypoosmotic.

hyperpolarization Above-normal negativity of a cell membrane during its resting potential.

hypersensitive response Plants respond to pathogens by selectively killing plant cells to block the spread of the pathogen.

hypha, pl. hyphae (Gr. *hyphe*, web) A filament of a fungus or oomycete; collectively, the hyphae comprise the mycelium.

hypocotyl The region immediately below where the cotyledons are attached.

hypoosmotic The condition in which a (hypoosmotic) solution has a lower osmotic concentration than that of a second solution. *Compare with* hyperosmotic.

hypothalamus (Gr. *hypo*, under, + *thalamos*, inner room) A region of the vertebrate brain just below the cerebral hemispheres, under the thalamus; a center of the autonomic nervous system, responsible for the integration and correlation of many neural and endocrine functions.

I

imaginal disk One of about a dozen groups of cells set aside in the abdomen of a larval insect and committed to forming key parts of the adult insect's body.

immune response In vertebrates, a defensive reaction of the body to invasion by a foreign substance or organism. *See* antibody and B cell.

immunoglobulin (L. *immunis*, free, + *globus*, globe) An antibody.

inbreeding The breeding of genetically related plants or animals; inbreeding tends to increase homozygosity.

inclusive fitness Describes the sum of the number of genes directly passed on in an individual's offspring and those genes passed on indirectly by kin (other than offspring) whose existence results from the benefit of the individual's altruism.

induction Process by which one group of embryonic cells affects an adjacent group, thereby inducing those cells to differentiate in a manner they otherwise would not have.

industrial melanism (Gr. *melas*, black) Phrase used to describe the evolutionary process in which initially light-colored organisms become dark as a result of natural selection.

inflammatory response (L. *inflammare*, to flame) A generalized nonspecific response to infection that acts to clear an infected area of infecting microbes and dead tissue cells so that tissue repair can begin.

initiation factor One of several proteins involved in the formation of an initiation complex in prokaryote polypeptide synthesis.

insertional inactivation Destruction of a gene's function by the insertion of a transposon.

instar A larval developmental stage in insects.

interferon In vertebrates, a protein produced in virus-infected cells that inhibits viral multiplication.

interneuron (association neuron) A nerve cell found only in the middle of the spinal cord that acts as a functional link between sensory neurons and motor neurons.

internode In plants, the region of a stem between two successive nodes.

interoceptor A receptor that senses information related to the body itself, its internal condition, and its position.

interphase The period between two mitotic or meiotic divisions in which a cell grows and its DNA replicates; includes G_1, S, and G_2 phases.

intron (L. *intra*, within) Portion of mRNA as transcribed from eukaryotic DNA that is removed by enzymes before the mature mRNA is translated into protein. *See* exon.

inversion (L. *invertere*, to turn upside down) A reversal in order of a segment of a chromosome; also, to turn inside out, as in embryogenesis of sponges or discharge of a nematocyst.

ion Any atom or molecule containing an unequal number of electrons and protons and therefore carrying a net positive or net negative charge.

ionizing radiation High-energy radiation that is highly mutagenic, producing free radicals that react with DNA; includes X rays and gamma rays.

isomer (Gr. *isos*, equal, + *meros*, part) One of a group of molecules identical in atomic composition but differing in structural arrangement; for example, glucose and fructose.

isosmotic The condition in which the osmotic concentrations of two solutions are equal, so that no net water movement occurs between them by osmosis.

K

karyotype (Gr. *karyon*, kernel, + *typos*, stamp or print) The morphology of the chromosomes of an organism as viewed with a light microscope.

keratin (Gr. *kera*, horn, + *in*, suffix used for proteins) A tough, fibrous protein formed in epidermal tissues and modified into skin, feathers, hair, and hard structures such as horns and nails.

kidney In vertebrates, the organ that filters the blood to remove nitrogenous wastes and regulates the balance of water and solutes in blood plasma.

kilocalorie Unit describing the amount of heat required to raise the temperature of a kilogram of water by one degree Celsius (°C); sometimes called a Calorie, equivalent to 1000 calories.

kinesis (Gr. *kinesis*, motion) Changes in activity level in an animal that are dependent on stimulus intensity. *See* kinetic energy.

kinetic energy The energy of motion.

kinetochore (Gr. *kinetikos*, putting in motion, + *choros*, chorus) Disc-shaped protein structure within the centromere to which the spindle fibers attach during mitosis or meiosis. *See* centromere.

kingdom The second highest commonly used taxonomic category.

kin selection Selection favoring relatives; an increase in the frequency of related individuals (kin) in a population, leading to an increase in the relative frequency in the population of those alleles shared by members of the kin group.

Krebs cycle Another name for the citric acid cycle; also called the tricarboxylic acid (TCA) cycle.

L

labrum (L., a lip) The upper lip of insects and crustaceans situated above or in front of the mandibles.

larynx The voice box; a cartilaginous organ that lies between the pharynx and trachea and is responsible for sound production in vertebrates.

lateral line system A sensory system encountered in fish, through which mechanoreceptors in a line down the side of the fish are sensitive to motion.

lateral meristems (L. *latus*, side, + Gr. *meristos*, divided) In vascular plants, the meristems that give rise to secondary tissue; the vascular cambium and cork cambium.

Law of Independent Assortment Mendel's second law of heredity, stating that genes located on nonhomologous chromosomes assort independently of one another.

Law of Segregation Mendel's first law of heredity, stating that alternative alleles for the same gene segregate from each other in production of gametes.

leaf primordium, pl. primordia (L. *primordium*, beginning) A lateral outgrowth from the apical meristem that will eventually become a leaf.

lenticels (L. *lenticella*, a small window) Spongy areas in the cork surfaces of stem, roots, and other plant parts that allow interchange of gases between internal tissues and the atmosphere through the periderm.

leucoplast (Gr. *leukos*, white, + *plasein*, to form) In plant cells, a colorless plastid in which starch grains are stored; usually found in cells not exposed to light.

leukocyte (Gr. *leukos*, white, + *kytos*, hollow vessel) A white blood cell; a diverse array of nonhemoglobin-containing blood cells, including phagocytic macrophages and antibody-producing lymphocytes.

lichen Symbiotic association between a fungus and a photosynthetic organism such as a green alga or cyanobacterium.

limbic system The hypothalamus, together with the network of neurons that link the hypothalamus to some areas of the cerebral cortex. Responsible for many of the most deep-seated drives and emotions of vertebrates, including pain, anger, sex, hunger, thirst, and pleasure.

lipase (Gr. *lipos*, fat, + *-ase*, enzyme suffix) An enzyme that catalyzes the hydrolysis of fats.

lipid (Gr. *lipos*, fat) A nonpolar hydrophobic organic molecule that is insoluble in water (which is polar) but dissolves readily in nonpolar organic solvents; includes fats, oils, waxes, steroids, phospholipids, and carotenoids.

lipid bilayer The structure of a cellular membrane, in which two layers of phospholipids spontaneously align so that the hydrophilic head groups are exposed to water, while the hydrophobic fatty acid tails are pointed toward the center of the membrane.

locus The position on a chromosome where a gene is located.

loop of Henle In the kidney of birds and mammals, a hairpin-shaped portion of the renal tubule in which water and salt are reabsorbed from the glomerular filtrate by diffusion.

lophophore A ring or U-shaped arrangement of tentacles, often ciliated, into which the coelom of the animal extends. The organ functions in feeding and gas exchange and is encountered in only a few phyla of small marine invertebrates.

luteal phase The second phase of the female reproductive cycle, during which the mature eggs are released into the fallopian tubes, a process called ovulation.

lymph (L. *lympha*, clear water) In animals, a colorless fluid derived from blood by filtration through capillary walls in the tissues.

lymphatic system In animals, an open vascular system that reclaims water that has entered interstitial regions from the bloodstream (lymph); includes the lymph nodes, spleen, thymus, and tonsils.

lymphocyte (L. *lympha*, water, + Gr. *kytos*, hollow vessel) A type of white blood cell. Lymphocytes are responsible for the immune response; there are two principal classes—B cells and T cells.

lymphokine A regulatory molecule that is secreted by lymphocytes. In the immune response, lymphokines secreted by helper T cells unleash the cell-mediated immune response.

lysis (Gr., a loosening) Disintegration of a cell by rupture of its plasma membrane.

M

macroevolution (Gr. *makros*, large, + L. *evolvere*, to unfold) The creation of new species and the extinction of old ones.

macromolecule (Gr. *makros*, large, + L. *moleculus*, a little mass) An extremely large biological molecule; refers specifically to proteins, nucleic acids, polysaccharides, lipids, and complexes of these.

macronutrients (Gr. *makros*, large, + L. *nutrire*, to nourish) Inorganic chemical elements required in large amounts for plant growth, such as nitrogen, potassium, calcium, phosphorus, magnesium, and sulfur.

macrophage A large phagocytic cell that is able to engulf and digest cellular debris and invading bacteria.

major histocompatibility complex (MHC) A set of protein cell-surface markers anchored in the plasma membrane, which the immune system uses to identify "self." All the cells of a given individual have the same "self" marker, called an MHC protein.

Malpighian tubules Blind tubules opening into the hindgut of terrestrial arthropods; they function as excretory organs.

mandibles (L. *mandibula*, jaw) In crustaceans, insects, and myriapods, the appendages immediately posterior to the antennae; used to seize, hold, bite, or chew food.

mantle The soft, outermost layer of the body wall in mollusks; the mantle secretes the shell.

marsupial (L. *marsupium*, pouch) A mammal in which the young are born early in their development, sometimes as soon as eight days after fertilization, and are retained in a pouch.

mass flow hypothesis The overall process by which materials move in the phloem of plants.

matrix (L. *mater*, mother) In mitochondria, the solution in the interior space surrounded by the cristae that contains the enzymes and other molecules involved in oxidative respiration; more generally, that part of a tissue within which an organ or process is embedded.

menstrual cycle (L. *mens*, month) In humans and higher primates, the cyclic changes in the ovaries and uterine endometrium; lasts about a month in humans.

menstruation Periodic sloughing off of the blood-enriched lining of the uterus when pregnancy does not occur.

meristem (Gr. *merizein*, to divide) Undifferentiated plant tissue from which new cells arise.

meroblastic cleavage (Gr. *meros*, part, + *blastos*, sprout) A type of cleavage in the eggs of reptiles, birds, and some fish. Occurs only on the blastodisc.

mesoderm (Gr. *mesos*, middle, + *derma*, skin) One of the three embryonic germ layers that form in the gastrula; gives rise to muscle, bone and other connective tissue, the peritoneum, the circulatory system, and most of the excretory and reproductive systems.

mesophyll (Gr. *mesos*, middle, + *phyllon*, leaf) The photosynthetic parenchyma of a leaf, located within the epidermis.

messenger RNA (mRNA) The RNA transcribed from structural genes; RNA molecules complementary to a portion of one strand of DNA, which are translated by the ribosomes to form protein.

metabolism (Gr. *metabole*, change) The sum of all chemical processes occurring within a living cell or organism.

metamorphosis (Gr. *meta*, after, + *morphe*, form, + *osis*, state of) Process in which a marked change in form takes place during postembryonic development as, for example, from tadpole to frog.

metaphase (Gr. *meta*, after, + *phasis*, form) The stage of mitosis or meiosis during which microtubules become organized into a spindle and the chromosomes come to lie in the spindle's equatorial plane.

metastasis The process by which cancer cells move from their point of origin to other locations in the body; also, a population of cancer cells in a secondary location, the result of movement from the primary tumor.

methanogens Obligate, anaerobic archaebacteria that produce methane.

microarray DNA sequences are placed on a microscope slide or chip with a robot. The microarray can then be probed with RNA from specific tissues to identify expressed DNA.

microbody A cellular organelle bounded by a single membrane and containing a variety of enzymes; generally derived from endoplasmic reticulum; includes peroxisomes and glyoxysomes.

microevolution (Gr. *mikros*, small, + L. *evolvere*, to unfold) Refers to the evolutionary process itself.

Evolution within a species. Also called adaptation.

micronutrient (Gr. *mikros*, small, + L. *nutrire*, to nourish) A mineral required in only minute amounts for plant growth, such as iron, chlorine, copper, manganese, zinc, molybdenum, and boron.

micropyle In the ovules of seed plants, an opening in the integuments through which the pollen tube usually enters.

microtubule (Gr. *mikros*, small, + L. *tubulus*, little pipe) In eukaryotic cells, a long, hollow protein cylinder, composed of the protein tubulin; these influence cell shape, move the chromosomes in cell division, and provide the functional internal structure of cilia and flagella.

microvillus (Gr. *mikros*, small, + L. *villus*, shaggy hair) Cytoplasmic projection from epithelial cells; microvilli greatly increase the surface area of the small intestine.

middle lamella The layer of intercellular material, rich in pectic compounds, that cements together the primary walls of adjacent plant cells.

mimicry (Gr. *mimos*, mime) The resemblance in form, color, or behavior of certain organisms (mimics) to other more powerful or more protected ones (models).

mitosis (Gr. *mitos*, thread) Somatic cell division; nuclear division in which the duplicated chromosomes separate to form two genetically identical daughter nuclei.

monocot Short for monocotyledon; flowering plant in which the embryos have only one cotyledon, the floral parts are generally in threes, and the leaves typically are parallel-veined.

monocyte (Gr. *monos*, single, + *kytos*, hollow vessel) A type of leukocyte that becomes a phagocytic cell (macrophage) after moving into tissues.

monoecious (Gr. *monos*, single, + *oecos*, house) A plant in which the staminate and pistillate flowers are separate, but borne on the same individual.

monophyletic In phylogenetic classification, a group that includes the most recent common ancestor of the group and all its descendants. A clade is a monophyletic group.

monosaccharide (Gr. *monos*, single, + L. *saccharum*, sugar) A simple sugar that cannot be decomposed into smaller sugar molecules.

monotreme (Gr. *monos*, single, + *treme*, hole) An egg-laying mammal.

morphogen A signal molecule produced by an embryonic

organizer region that informs surrounding cells of their distance from the organizer, thus determining relative positions of cells during development.

morphology The form and structure of an organism.

morula (L., a little mulberry) Solid ball of cells in the early stage of embryonic development.

mosaic development A pattern of embryonic development in which initial cells produced by cleavage divisions contain different developmental signals (determinants) from the egg, setting the individual cells on different developmental paths.

motor (efferent) neuron Neuron that transmits nerve impulses from the central nervous system to an effector, which is typically a muscle or gland.

Müllerian mimicry After Fritz Müller, German biologist. A phenomenon in which two or more unrelated but protected species resemble one another, thus achieving a kind of group defense.

multigene family A collection of related genes on a single chromosome or on different chromosomes.

muscle fiber A long, cylindrical, multinucleated cell containing numerous myofibrils, which is capable of contraction when stimulated.

mutagen (L. *mutare*, to change, + Gr. *genaio*, to produce) An agent that induces changes in DNA (mutations); includes physical agents that damage DNA and chemicals that alter DNA bases.

mutant (L. *mutare*, to change) A mutated gene; alternatively, an organism carrying a gene that has undergone a mutation.

mutation A permanent change in a cell's DNA; includes changes in nucleotide sequence, alteration of gene position, gene loss or duplication, and insertion of foreign sequences.

mutualism (L. *mutuus*, lent, borrowed) A symbiotic association in which two (or more) organisms live together, and both members benefit.

mycelium, pl. **mycelia** (Gr. *mykes*, fungus) In fungi, a mass of hyphae.

mycorrhiza, pl. **mycorrhizae** (Gr. *mykes*, fungus, + *rhiza*, root) A symbiotic association between fungi and the roots of a plant.

myelin sheath (Gr. *myelinos*, full of marrow) A fatty layer surrounding the long axons of motor neurons in the peripheral nervous system of vertebrates.

myofilament (Gr. *myos*, muscle, + L. *filare*, to spin) A contractile

microfilament, composed largely of actin and myosin, within muscle.

myosin (Gr. *myos*, muscle, + *in*, belonging to) One of the two protein components of microfilaments (the other is actin); a principal component of vertebrate muscle.

N

natural killer cell A cell that does not kill invading microbes, but rather, the cells infected by them.

natural selection The differential reproduction of genotypes; caused by factors in the environment; leads to evolutionary change.

negative feedback A homeostatic control mechanism whereby an increase in some substance or activity inhibits the process leading to the increase; also known as feedback inhibition.

nephridium, pl. **nephridia** (Gr. *nephros*, kidney) In invertebrates, a tubular excretory structure.

nephrid organ A filtration system of many freshwater invertebrates in which water and waste pass from the body across the membrane into a collecting organ, from which they are expelled to the outside through a pore.

nephron (Gr. *nephros*, kidney) Functional unit of the vertebrate kidney; one of numerous tubules involved in filtration and selective reabsorption of blood; each nephron consists of a Bowman's capsule, an enclosed glomerulus, and a long attached tubule; in humans, called a renal tubule.

nerve A group or bundle of nerve fibers (axons) with accompanying neurological cells, held together by connective tissue; located in the peripheral nervous system.

neural crest A special strip of cells that develops just before the neural groove closes over to form the neural tube in embryonic development.

neural groove The long groove formed along the long axis of the embryo by a layer of ectodermal cells.

neural tube The dorsal tube, formed from the neural plate, that differentiates into the brain and spinal cord.

neuroglia (Gr. *neuron*, nerve, + *glia*, glue) Nonconducting nerve cells that are intimately associated with neurons and appear to provide nutritional support.

neuromuscular junction The structure formed when the tips of axons contact (innervate) a muscle fiber.

neuron (Gr., nerve) A nerve cell specialized for signal transmission; includes cell body, dendrites, and axon.

neurotransmitter (Gr. *neuron*, nerve, + L. *trans*, across, + *mittere*, to send) A chemical released at the axon terminal of a neuron that travels across the synaptic cleft, binds a specific receptor on the far side, and depending on the nature of the receptor, depolarizes or hyperpolarizes a second neuron or a muscle or gland cell.

neurulation A process in early embryonic development by which a dorsal band of ectoderm thickens and rolls into the neural tube.

neutrophil An abundant type of granulocyte capable of engulfing microorganisms and other foreign particles; neutrophils comprise about 50 to 70% of the total number of white blood cells.

niche The role played by a particular species in its environment.

nicotinamide adenine dinucleotide (NAD⁺) A molecule that becomes reduced (to NADH) as it carries high-energy electrons from oxidized molecules and delivers them to ATP-producing pathways in the cell.

nociceptor A naked dendrite that acts as a receptor in response to a pain stimulus.

nocturnal (L. *nocturnus*, night) Active primarily at night.

node (L. *nodus*, knot) The part of a plant stem where one or more leaves are attached. *See* internode.

node of Ranvier After L.A. Ranvier, French histologist. A gap formed at the point where two Schwann cells meet and where the axon is in direct contact with the surrounding intercellular fluid.

nonassociative learning A learned behavior that does not require an animal to form an association between two stimuli, or between a stimulus and a response.

nonsense codon One of three codons (UAA, UAG, and UGA) that are not recognized by tRNAs, thus serving as "stop" signals in the mRNA message and terminating translation.

notochord (Gr. *noto*, back, + L. *chorda*, cord) In chordates, a dorsal rod of cartilage that runs the length of the body and forms the primitive axial skeleton in the embryos of all chordates.

nucellus (L. *nucella*, a small nut) Tissue composing the chief pair of young ovules, in which the embryo sac develops; equivalent to a megasporangium.

nucleic acid A nucleotide polymer; chief types are deoxyribonucleic acid (DNA), which is double-stranded, and ribonucleic acid (RNA), which is typically single-stranded.

nucleolus (L., a small nucleus) In eukaryotes, the site of rRNA synthesis; a spherical body composed chiefly of rRNA in the process of being transcribed from multiple copies of rRNA genes.

nucleosome (L. *nucleus*, kernel, + *soma*, body) The fundamental packaging unit of eukaryotic chromosomes; a complex of DNA and histone proteins in which the double-helical DNA winds around eight molecules of histone; chromatin is composed of long sequences of nucleosomes.

nucleotide A single unit of nucleic acid, composed of a phosphate, a five-carbon sugar (either ribose or deoxyribose), and a purine or a pyrimidine.

nucleus In atoms, the central core, containing positively charged protons and (in all but hydrogen) electrically neutral neutrons; in eukaryotic cells, the membranous organelle that houses the chromosomal DNA; in the central nervous system, a cluster of nerve cell bodies.

O

ocellus, pl. **ocelli** (L., little eye) A simple light receptor common among invertebrates.

Okazaki fragment A short segment of DNA produced by discontinuous replication elongating in the 5'→ 3' direction away from the replication.

olfaction (L. *olfactum*, smell) The process or function of smelling.

ommatidium, pl. **ommatidia** (Gr., little eye) The visual unit in the compound eye of arthropods; contains light-sensitive cells and a lens able to form an image.

oncogene (Gr. *oncos*, cancer, + *genos*, birth) A mutant form of a growth-regulating gene that is inappropriately "on," causing unrestrained cell growth and division.

oocyst The zygote in a sporozoan life cycle. It is surrounded by a tough cyst to prevent dehydration or other damage.

operant conditioning A learning mechanism in which the reward follows only after the correct behavioral response.

operator A site of gene regulation; a sequence of nucleotides overlapping the promoter site and recognized by a repressor protein; binding of the repressor prevents

binding of the polymerase to the promoter site and so blocks transcription of the structural gene.

operculum A flat, bony, external protective covering over the gill chamber in fish.

operon (L. *operis*, work) A cluster of adjacent structural genes transcribed as a unit into a single mRNA molecule.

order A category of classification above the level of family and below that of class.

organ (Gr. *organon*, tool) A body structure composed of several different tissues grouped in a structural and functional unit.

organelle (Gr. *organella*, little tool) Specialized part of a cell; literally, a small cytoplasmic organ.

orthologs Genes that reflect the conservation of a single gene found in an ancestor.

osmoconformer An animal that maintains the osmotic concentration of its body fluids at about the same level as that of the medium in which it is living.

osmosis (Gr. *osmos*, act of pushing, thrust) The diffusion of water across a selectively permeable membrane (a membrane that permits the free passage of water but prevents or retards the passage of a solute); in the absence of differences in pressure or volume, the net movement of water is from the side containing a lower concentration of solute to the side containing a higher concentration.

osmotic pressure The potential pressure developed by a solution separated from pure water by a differentially permeable membrane. The higher the solute concentration, the greater the osmotic potential of the solution; also called *osmotic potential*.

osteoblast (Gr. *osteon*, bone, + *blastos*, bud) A bone-forming cell.

osteocyte (Gr. *osteon*, bone, + *kytos*, hollow vessel) A mature osteoblast.

outcrossing Breeding with individuals other than oneself or one's close relatives.

ovary (L. *ovum*, egg) (1) In animals, the organ in which eggs are produced. (2) In flowering plants, the enlarged basal portion of a carpel, which contains the ovule(s); the ovary matures to become the fruit.

oviduct (L. *ovum*, egg, + *ductus*, duct) In vertebrates, the passageway through which ova (eggs) travel from the ovary to the uterus.

oviparity (L. *ovum*, egg, + *parere*, to bring forth) Refers to a type of reproduction in which the eggs are developed after leaving the body of the mother, as in reptiles.

ovoviviparity Refers to a type of reproduction in which young hatch from eggs that are retained in the mother's uterus.

ovulation In animals, the release of an egg or eggs from the ovary.

ovum, pl. **ova** (L., egg) The egg cell; female gamete.

oxidation (Fr. *oxider*, to oxidize) Loss of an electron by an atom or molecule; in metabolism, often associated with a gain of oxygen or a loss of hydrogen.

oxidative respiration Process of cellular activity in which glucose or other molecules are broken down to water and carbon dioxide with the release of energy.

oxygen debt The amount of oxygen required to convert the lactic acid generated in the muscles during exercise back into glucose.

oxytocin (Gr. *oxys*, sharp, + *tokos*, birth) A hormone of the posterior pituitary gland that affects uterine contractions during childbirth and stimulates lactation.

ozone O₃, a stratospheric layer of the earth's atmosphere responsible for filtering out ultraviolet radiation supplied by the sun.

P

pacemaker A patch of excitatory tissue in the vertebrate heart that initiates the heartbeat.

palisade parenchyma (L. *palus*, stake) In plant leaves, the columnar, chloroplast-containing parenchyma cells of the mesophyll. Also called *palisade cells*.

papilla A small projection of tissue.

paracrine A type of chemical signaling between cells in which the effects are local and short-lived.

paralogs Two genes within an organism that arose from the duplication of one gene in an ancestor.

paraphyletic In phylogenetic classification, a group that includes the most recent common ancestor of the group, but not all its descendants.

parapodia (Gr. *para*, beside, + *pous*, foot) One of the paired lateral processes on each side of most segments in polychaete annelids.

parasexuality In certain fungi, the fusion and segregation of heterokaryotic haploid nuclei to produce recombinant nuclei.

parasitism (Gr. *para*, beside, + *sitos*, food) A living arrangement in which an organism lives on or in an organism of a different species and derives nutrients from it.

parenchyma cell The most common type of plant cell; characterized by large vacuoles, thin walls, and functional nuclei.

parthenogenesis (Gr. *parthenos*, virgin, + *genesis*, birth) The development of an egg without fertilization, as in aphids, bees, ants, and some lizards.

partial pressure The components of each individual gas—such as nitrogen, oxygen, and carbon dioxide—that together comprise the total air pressure.

pelagic Free-swimming, usually in open water.

pellicle A tough, flexible covering in ciliates and euglenoids.

peptide bond (Gr. *peptein*, to soften, digest) The type of bond that links amino acids together in proteins through a dehydration reaction.

perianth (Gr. *peri*, around, + *anthos*, flower) In flowering plants, the petals and sepals taken together.

pericycle (Gr. *peri*, around, + *kykos*, circle) In vascular plants, one or more cell layers surrounding the vascular tissues of the root, bounded externally by the endodermis and internally by the phloem.

periderm (Gr. *peri*, around, + *derma*, skin) Outer protective tissue in vascular plants that is produced by the cork cambium and functionally replaces epidermis when it is destroyed during secondary growth; the periderm includes the cork, cork cambium, and phelloderm.

peristalsis (Gr. *peristaltikos*, compressing around) In animals, a series of alternating contracting and relaxing muscle movements along the length of a tube such as the oviduct or alimentary canal that tend to force material such as an egg cell or food through the tube.

peroxisome A microbody that plays an important role in the breakdown of highly oxidative hydrogen peroxide by catalase.

petal A flower part, usually conspicuously colored; one of the units of the corolla.

petiole (L. *petiolus*, a little foot) The stalk of a leaf.

phagocyte (Gr. *phagein*, to eat, + *kytos*, hollow vessel) Any cell that engulfs and devours microorganisms or other particles.

phagocytosis (Gr., cell-eating) Endocytosis of a solid particle; the plasma membrane folds inward around the particle (which may be another cell) and engulfs it to form a vacuole.

pharynx (Gr., gullet) In vertebrates, a muscular tube that connects the mouth cavity and the esophagus; it serves as the gateway to the digestive tract and to the trachea.

phenotype (Gr. *phainein*, to show, + *typos*, stamp or print) The realized expression of the genotype; the physical appearance or functional expression of a trait.

pheromone (Gr. *pherein*, to carry, + *hormonos*, exciting, stirring up) Chemical substance released by one organism that influences the behavior or physiological processes of another organism of the same species. Some pheromones serve as sex attractants, as trail markers, and as alarm signals.

phloem (Gr. *phloos*, bark) In vascular plants, a food-conducting tissue basically composed of sieve elements, various kinds of parenchyma cells, fibers, and sclereids.

phosphodiester bond The type of bond that links nucleotides in a nucleic acid; formed when the phosphate group of one nucleotide binds to the 3′ hydroxyl group of the sugar of another.

phospholipid Similar in structure to a fat, but having only two fatty acids attached to the glycerol backbone, with the third space linked to a phosphorylated molecule; contains a polar hydrophilic "head" end (phosphate group) and a nonpolar hydrophobic "tail" end (fatty acids).

photoperiodism (Gr. *photos*, light, + *periodos*, a period) The tendency of biological reactions to respond to the duration and timing of day and night; a mechanism for measuring seasonal time.

photoreceptor (Gr. *photos*, light) A light-sensitive sensory cell.

photosystem An organized collection of chlorophyll and other pigment molecules embedded in the thylakoid of chloroplasts; traps photon energy and channels it as energetic electrons to the thylakoid membrane.

phototropism (Gr. *photos*, light, + *trope*, turning) In plants, a growth response to a light stimulus.

phycologist (Gr. *phykos*, seaweed) The scientist who studies algae.

phylogeny The evolutionary history of an organism, including which species are closely related and in what order related species evolved; often represented in the form of an evolutionary tree.

phylum, pl. phyla (Gr. *phylon*, race, tribe) A major category, between kingdom and class, of taxonomic classifications.

phytochrome (Gr. *phyton*, plant, + *chroma*, color) A plant pigment that is associated with the absorption of light; photoreceptor for red to far-red light.

phytoremediation The process of removing contamination from soil or water using plants.

pigment (L. *pigmentum*, paint) A molecule that absorbs light.

pilus, pl. pili Extensions of a bacterial cell enabling it to transfer genetic materials from one individual to another or to adhere to substrates.

pinocytosis (Gr. *pinein*, to drink, + *kytos*, hollow vessel, + *osis*, condition) The process of fluid uptake by endocytosis in a cell.

pistil (L. *pistillum*, pestle) Central organ of flowers, typically consisting of ovary, style, and stigma; a pistil may consist of one or more fused carpels and is more technically and better known as the gynoecium.

pith The ground tissue occupying the center of the stem or root within the vascular cylinder.

placenta, pl. placentae (L., a flat cake) (1) In flowering plants, the part of the ovary wall to which the ovules or seeds are attached. (2) In mammals, a tissue formed in part from the inner lining of the uterus and in part from other membranes, through which the embryo (later the fetus) is nourished while in the uterus and through which wastes are carried away.

plankton (Gr. *planktos*, wandering) Free-floating, mostly microscopic, aquatic organisms.

plasma (Gr., form) The fluid of vertebrate blood; contains dissolved salts, metabolic wastes, hormones, and a variety of proteins, including antibodies and albumin; blood minus the blood cells.

plasma cell An antibody-producing cell resulting from the multiplication and differentiation of a B lymphocyte that has interacted with an antigen.

plasma membrane The membrane surrounding the cytoplasm of a cell; consists of a single phospholipid bilayer with embedded proteins.

plasmid (Gr. *plasma*, a form or mold) A small fragment of extrachromosomal DNA, usually circular, that replicates independently of the main chromosome, although it may have been derived from it.

plasmodesmata In plants, cytoplasmic connections between adjacent cells.

plasmodium (Gr. *plasma*, a form or mold, + *eidos*, form) Stage in the life cycle of myxomycetes (plasmodial slime molds); a multinucleate mass of protoplasm surrounded by a membrane.

plastid (Gr. *plastos*, formed or molded) An organelle in the cells of photosynthetic eukaryotes that is the site of photosynthesis and, in plants and green algae, of starch storage.

platelet (Gr. dim. of *plattus*, flat) In mammals, a fragment of a white blood cell that circulates in the blood and functions in the formation of blood clots at sites of injury.

pleiotropy Condition in which an individual allele has more than one effect on production of the phenotype.

plumule The epicotyl of a plant with its two young leaves.

point mutation An alteration of one nucleotide in a chromosomal DNA molecule.

polar body Minute, nonfunctioning cell produced during the meiotic divisions leading to gamete formation in vertebrates.

pollen tube A tube formed after germination of the pollen grain; carries the male gametes into the ovule.

pollination The transfer of pollen from an anther to a stigma.

polyandry The condition in which a female mates with more than one male.

polyclonal antibody An antibody response in which an antigen elicits many different antibodies, each fitting a different portion of the antigen surface.

polygyny (Gr. *poly*, many, + *gyne*, woman, wife) A mating choice in which a male mates with more than one female.

polymer (Gr. *poly*, many, + *meris*, part) A molecule composed of many similar or identical molecular subunits; starch is a polymer of glucose.

polymerase chain reaction (PCR) A process by which DNA polymerase is used to copy a sequence of interest repeatedly, making millions of copies of the same DNA.

polymorphism (Gr. *poly*, many, + *morphe*, form) The presence in a population of more than one allele of a gene at a frequency greater than that of newly arising mutations.

polypeptide (Gr. *poly*, many, + *peptein*, to digest) A molecule consisting of many joined amino acids; not usually as complex as a protein.

polyphyletic In phylogenetic classification, a group that does not include the most recent common ancestor of all members of the group.

polyploidy Condition in which one or more entire sets of chromosomes is added to the diploid genome.

polysaccharide (Gr. *poly*, many, + *sakcharon*, sugar, from Latin *sakara*, gravel, sugar) A carbohydrate composed of many monosaccharide sugar subunits linked together in a long chain; examples are glycogen, starch, and cellulose.

polyunsaturated fat A fat molecule having at least two double bonds between adjacent carbons in one or more of the fatty acid chains.

population (L. *populus*, the people) Any group of individuals, usually of a single species, occupying a given area at the same time.

posttranscriptional control A mechanism of control over gene expression that operates after the transcription of mRNA is complete.

potential energy Energy that is not being used, but could be; energy in a potentially usable form; often called "energy of position."

precapillary sphincter A ring of muscle that guards each capillary loop and that, when closed, blocks flow through the capillary.

primary endosperm nucleus In flowering plants, the result of the fusion of a sperm nucleus and the (usually) two polar nuclei.

primary growth In vascular plants, growth originating in the apical meristems of shoots and roots; results in an increase in length.

primary immune response The first response of an immune system to a foreign antigen. If the system is challenged again with the same antigen, the memory cells created during the primary response will respond more quickly.

primary induction Inductions between the three primary tissue types—ectoderm, mesoderm, and endoderm.

primary nondisjunction Failure of chromosomes to separate properly at meiosis I.

primary phloem In plant phloem, the cells involved in food conduction.

primary productivity The amount of energy produced by photosynthetic organisms in a community.

primary structure The specific amino acid sequence of a protein.

primary tissues Tissues that comprise the primary plant body.

primary transcript The initial mRNA molecule copied from a gene by RNA polymerase, containing a faithful copy of the entire gene, including introns as well as exons.

primary wall In plants, the wall layer deposited during the period of cell expansion.

primate Monkeys and apes (including humans).

primitive streak (L. *primus*, first) In the early embryos of birds, reptiles, and mammals, a dorsal, longitudinal strip of ectoderm and mesoderm that is equivalent to the blastopore in other forms.

principle of parsimony Principle stating that scientists should favor the hypothesis that requires the fewest assumptions.

prions Infectious proteinaceous particles.

procambium (Gr. *pro*, before, + *cambiare*, to exchange) In vascular plants, a primary meristematic tissue that gives rise to primary vascular tissues.

prokaryote (Gr. *pro*, before, + *karyon*, kernel) A bacterium; a cell lacking a membrane-bounded nucleus or membrane-bounded organelles.

promoter A specific nucleotide sequence to which RNA polymerase attaches to initiate transcription of mRNA from a gene.

prophase (Gr. *pro*, before, + *phasis*, form) An early stage in nuclear division, characterized by the formation of a microtubule spindle along the future axis of division, the shortening and thickening of the chromosomes, and their movement toward the equator of the spindle (the "metaphase plate").

proprioceptor (L. *proprius*, one's own) In vertebrates, a sensory receptor that senses the body's position and movements.

prostaglandins (Gr. *prostas*, a porch or vestibule, + *glans*, acorn) A group of modified fatty acids that function as chemical messengers.

prostate gland (Gr. *prostas*, a porch or vestibule) In male mammals, a mass of glandular tissue at the base of the urethra that secretes an alkaline fluid that has a stimulating effect on the sperm as they are released.

protein (Gr. *proteios*, primary) A chain of amino acids joined by peptide bonds.

protein kinase An enzyme that adds phosphate groups to proteins, changing their activity.

proteomics The study of all proteins in an organism.

proton pump A protein channel in a membrane of the cell that expends energy to transport protons against a concentration gradient; involved in the chemiosmotic generation of ATP.

proto-oncogene A normal gene that promotes cell division, so called because mutations that cause these genes to become overexpressed convert them into oncogenes that produce excessive cellular proliferation (i.e., cancer).

protozoa (Gr. *protos*, first, + *zoon*, animal) The traditional name given to heterotrophic protists.

pseudogene (Gr. *pseudos*, false, + *genos*, birth) A copy of a gene that is not transcribed.

pseudopod (Gr. *pseudes*, false, + *pous*, foot) A nonpermanent cytoplasmic extension of the cell body. Also called a *pseudopodium*.

punctuated equilibrium A hypothesis about the mechanism of evolutionary change proposing that long periods of little or no change are punctuated by periods of rapid evolution.

pupa (L., girl, doll) A developmental stage of some insects in which the organism is nonfeeding, immotile, and sometimes encapsulated or in a cocoon; the pupal stage occurs between the larval and adult phases.

purine (Gr. *purinos*, fiery, sparkling) The larger of the two general kinds of nucleotide base found in DNA and RNA; a nitrogenous base with a double-ring structure, such as adenine or guanine.

pyrimidine (alt. of pyridine, from G. *pyr*, fire) The smaller of two general kinds of nucleotide base found in DNA and RNA; a nitrogenous base with a single-ring structure, such as cytosine, thymine, or uracil.

Q

quaternary structure The structural level of a protein composed of more than one polypeptide chain, each of which has its own tertiary structure; the individual chains are called subunits.

R

radicle (L. *radicula*, root) The part of the plant embryo that develops into the root.

radioactivity The emission of nuclear particles and rays by unstable atoms as they decay into more stable forms.

radula (L., scraper) Rasping tongue found in most mollusks.

realized niche The actual niche occupied by an organism when all biotic and abiotic interactions are taken into account.

receptor-mediated endocytosis Process by which specific macromolecules are transported into eukaryotic cells at clathrin-coated pits, after binding to specific cell-surface receptors.

receptor protein A highly specific cell-surface receptor embedded in a cell membrane that responds only to a specific messenger molecule.

recessive An allele that is only expressed when present in the homozygous condition, while being "hidden" by the expression of a dominant allele in the heterozygous condition.

reciprocal altruism Performance of an altruistic act with the expectation that the favor will be returned. A key and very controversial assumption of many theories dealing with the evolution of social behavior. *See* altruism.

reciprocal recombination A mechanism of genetic recombination that occurs only in eukaryotic organisms, in which two chromosomes trade segments; can occur between nonhomologous chromosomes as well as the more usual exchange between homologous chromosomes in meiosis.

recombinant DNA Fragments of DNA from two different species, such as a bacterium and a mammal, spliced together in the laboratory into a single molecule.

reduction (L. *reductio*, a bringing back; originally "bringing back" a metal from its oxide) The gain of an electron by an atom, often with an associated proton.

reflex (L. *reflectere*, to bend back) In the nervous system, a motor response subject to little associative modification; a reflex is among the simplest neural pathways, involving only a sensory neuron, sometimes (but not always) an interneuron, and one or more motor neurons.

reflex arc The nerve path in the body that leads from stimulus to reflex action.

refractory period The recovery period after membrane depolarization during which the membrane is unable to respond to additional stimulation.

replication fork The Y-shaped end of a growing replication bubble in a DNA molecule undergoing replication.

repolarization Return of the ions in a nerve to their resting potential distribution following depolarization.

repressor (L. *reprimere*, to press back, keep back) A protein that regulates DNA transcription by preventing RNA polymerase from attaching to the promoter and transcribing the structural gene. *See* operator.

residual volume The amount of air remaining in the lungs after the maximum amount of air has been exhaled.

resting membrane potential The charge difference (difference in electric potential) that exists across a neuron at rest (about 70 millivolts).

restriction endonuclease An enzyme that cleaves a DNA duplex molecule at a particular base sequence, usually within or near a palindromic sequence; also called a restriction enzyme.

restriction fragment length polymorphism (RFLP) Restriction enzymes recognize very specific DNA sequences. Alleles of the same gene or surrounding sequences may have base-pair differences, so that DNA near one allele is cut into a different length fragment than DNA near the other allele. These different fragments separate based on size on gels.

retina (L., a small net) The photosensitive layer of the vertebrate eye; contains several layers of neurons and light receptors (rods and cones); receives the image formed by the lens and transmits it to the brain via the optic nerve.

retrovirus (L. *retro*, turning back) An RNA virus. When a retrovirus enters a cell, a viral enzyme (reverse transcriptase) transcribes viral RNA into duplex DNA, which the cell's machinery then replicates and transcribes as if it were its own.

Rh blood group A set of cell surface markers (antigens) on the surface of red blood cells in humans and rhesus monkeys (for which it is named); although there are several alleles, they are grouped into two main types, called Rh-positive and Rh-negative.

rhizome (Gr. *rhizoma*, mass of roots) In vascular plants, a usually more or less horizontal underground stem; may be enlarged for storage or may function in vegetative reproduction.

ribonucleic acid (RNA) A class of nucleic acids characterized by the presence of the sugar ribose and the pyrimidine uracil; includes mRNA, tRNA, and rRNA.

ribosomal RNA (rRNA) A class of RNA molecules found, together with characteristic proteins, in ribosomes; transcribed from the DNA of the nucleolus.

ribosome The molecular machine that carries out protein synthesis; the most complicated aggregation of proteins in a cell, also containing three different rRNA molecules.

ribozyme An RNA molecule that can behave as an enzyme, sometimes catalyzing its own assembly; rRNA also acts as a ribozyme in the polymerization of amino acids to form protein.

RNA polymerase An enzyme that catalyzes the assembly of an mRNA molecule, the sequence of which is complementary to a DNA molecule used as a template. *See* transcription.

RNA primer In DNA replication, a sequence of about 10 RNA nucleotides complementary to unwound DNA that attaches at a replication fork; the DNA polymerase uses the RNA primer as a starting point for addition of DNA nucleotides to form the new DNA strand; the RNA primer is later removed and replaced by DNA nucleotides.

RNA splicing A nuclear process by which intron sequences of a primary mRNA transcript are cut out and the exon sequences spliced together to give the correct linkages of genetic information that will be used in protein construction.

rod Light-sensitive nerve cell found in the vertebrate retina; sensitive to very dim light; responsible for "night vision."

root The usually descending axis of a plant, normally below ground, which anchors the plant and serves as the major point of entry for water and minerals.

rumen An "extra stomach" in cows and related mammals wherein digestion of cellulose occurs and from which partially digested material can be ejected back into the mouth.

S

saltatory conduction A very fast form of nerve impulse conduction in which the impulses leap from node to node over insulated portions.

saprobes Heterotrophic organisms that digest their food externally (such as most fungi).

sarcolemma The specialized cell membrane in a muscle cell.

sarcoma A cancerous tumor arising from cells of connective tissue, bone, or muscle.

sarcomere (Gr. *sarx*, flesh, + *meris*, part of) Fundamental unit of contraction in skeletal muscle; repeating bands of actin and myosin that appear between two Z lines.

sarcoplasmic reticulum (Gr. *sarx*, flesh, + *plassein*, to form, mold; L. *reticulum*, network) The endoplasmic reticulum of a muscle cell. A sleeve of membrane that wraps around each myofilament.

satellite DNA A nontranscribed region of the chromosome with a distinctive base composition; a short nucleotide sequence repeated tandemly many thousands of times.

saturated fat A fat composed of fatty acids in which all the internal carbon atoms contain the maximum possible number of hydrogen atoms.

Schwann cells The supporting cells associated with projecting axons, along with all the other nerve cells that make up the peripheral nervous system.

sclereid (Gr. *skleros*, hard) In vascular plants, a sclerenchyma cell with a thick, lignified, secondary wall having many pits; not elongate like a fiber.

sclerenchyma cell (Gr. *skleros*, hard, + *en*, in, + *chymein*, to pour) A type of tissue made up of sclerenchyma cells.

scrotum (L., bag) The pouch that contains the testes in most mammals.

scuttellum The modified cotyledon in cereal grains.

secondary cell wall In plants, the innermost layer of the cell wall. Secondary walls have a highly organized microfibrillar structure and are often impregnated with lignin.

secondary growth In vascular plants, an increase in stem and root diameter made possible by cell division of the lateral meristems.

secondary immune response The swifter response of the body the second time it is invaded by the same pathogen because of the presence of memory cells, which quickly become antibody-producing plasma cells.

secondary induction An induction between tissues that have already differentiated.

secondary structure In a protein, hydrogen-bonding interactions between CO and NH groups of the primary structure.

Second Law of Thermodynamics A statement concerning the transformation of potential energy into heat; it says that disorder (entropy) is continually increasing in the universe as energy changes occur, so disorder is more likely than order.

second messenger A small molecule or ion that carries the message from a receptor on the target cell surface into the cytoplasm.

segregation The process by which alternative forms of traits are expressed in offspring rather than blending each trait of the parents in the offspring.

selection The process by which some organisms leave more offspring than competing ones, and their genetic traits tend to appear in greater proportions among members of succeeding generations than the traits of those individuals that leave fewer offspring.

selectively permeable Condition in which a membrane is permeable to some substances but not to others.

self-fertilization The union of egg and sperm produced by a single hermaphroditic organism.

semen (L., seed) In reptiles and mammals, sperm-bearing fluid expelled from the penis during male orgasm.

semicircular canal Any of three fluid-filled canals in the inner ear that help to maintain balance.

semiconservative replication DNA replication in which each strand of the original duplex serves as the template for construction of a totally new complementary strand, so the original duplex is partially conserved in each of the two new DNA molecules.

senescent Aged, or in the process of aging.

sensory (afferent) neuron A neuron that transmits nerve impulses from a sensory receptor to the central nervous system or central ganglion.

sepal (L. *sepalum*, a covering) A member of the outermost floral whorl of a flowering plant.

septum, pl. **septa** (L., fence) A wall between two cavities.

seta, pl. **setae** (L., bristle) In an annelid, bristles of chitin that help anchor the worm during locomotion or when it is in its burrow.

sex-linked A trait determined by a gene carried on the X chromosome and absent on the Y chromosome.

sexual reproduction The process of producing offspring through an alternation of fertilization (producing diploid cells) and meiotic reduction in chromosome number (producing haploid cells).

sexual selection A type of differential reproduction that results from variable success in obtaining mates.

shoot In vascular plants, the aboveground portions, such as the stem and leaves.

sieve cell In the phloem of vascular plants, a long, slender sieve element with relatively unspecialized sieve areas and with tapering end walls that lack sieve plates.

sinoatrial (SA) node *See* pacemaker.

sinus (L., curve) A cavity or space in tissues or in bone.

sister chromatid One of two identical copies of each chromosome, still linked at the centromere, produced as the chromosomes duplicate for mitotic division; similarly, one of two identical copies of each homologous chromosome present in a tetrad at meiosis.

sodium-potassium pump Transmembrane channels engaged in the active (ATP-driven) transport of sodium ions, exchanging them for potassium ions, where both ions are being moved against their respective concentration gradients; maintains the resting membrane potential of neurons and other cells.

solute A molecule dissolved in some solution; as a general rule, solutes dissolve only in solutions of similar polarity; for example, glucose (polar) dissolves in (forms hydrogen bonds with) water (also polar), but not in vegetable oil (nonpolar).

solvent The medium in which one or more solutes is dissolved.

somatic cell Any of the cells of a multicellular organism except those that are destined to form gametes (germ-line cells).

somatic mutation A change in genetic information (mutation) occurring in one of the somatic cells of a multicellular organism, not passed from one generation to the next.

somatic nervous system (Gr. *soma*, body) In vertebrates, the neurons of the peripheral nervous system that control skeletal muscle.

somite (Gr. *soma*, body) One of the blocks, or segments, of tissue into which the mesoderm is divided during differentiation of the vertebrate embryo.

Southern blot A procedure used for identifying a specific gene, in which DNA from the source being tested is cut into fragments with restriction enzymes and separated by gel electrophoresis, then blotted onto a sheet of nitrocellulose and probed with purified, labeled, single-stranded DNA corresponding to a specific gene; if the DNA matching the specific probe is present in the source DNA, it is visible as a band of radioactive label on the sheet.

species, pl. species (L., kind, sort) A kind of organism; species are designated by binomial names written in italics.

spectrin A scaffold of proteins that links plasma membrane proteins to actin filaments in the cytoplasm

of red blood cells, producing their characteristic biconcave shape.

sperm (Gr. *sperma*, seed) A mature male gamete, usually motile and smaller than the female gamete.

spermatid (Gr. *sperma*, seed) In animals, each of four haploid (*n*) cells that result from the meiotic divisions of a spermatocyte; each spermatid differentiates into a sperm cell.

spermatozoa The male gamete, usually smaller than the female gamete, and usually motile.

sphincter (Gr. *sphinkter*, band, from *sphingein*, to bind tightly) In vertebrate animals, a ring-shaped muscle capable of closing a tubular opening by constriction (such as between stomach and small intestine or between anus and exterior).

spindle apparatus The assembly that carries out the separation of chromosomes during cell division; composed of microtubules (spindle fibers) and assembled during prophase at the equator of the dividing cell.

spiracle (L. *spiraculum*, from *spirare*, to breathe) External opening of a trachea in arthropods.

spongy parenchyma A leaf tissue composed of loosely arranged, chloroplast-bearing cells. *See* palisade parenchyma.

sporangium, pl. sporangia (Gr. *spora*, seed, + *angeion*, a vessel) A structure in which spores are produced.

spore A haploid reproductive cell, usually unicellular, capable of developing into an adult without fusion with another cell.

sporophyte (Gr. *spora*, seed, + *phyton*, plant) The spore-producing, diploid (*2n*) phase in the life cycle of a plant having alternation of generations.

stabilizing selection A form of selection in which selection acts to eliminate both extremes from a range of phenotypes.

stamen (L., thread) The organ of a flower that produces the pollen; usually consists of anther and filament; collectively, the stamens make up the androecium.

starch (Mid. Eng. *sterchen*, to stiffen) An insoluble polymer of glucose; the chief food storage substance of plants.

statocyst (Gr. *statos*, standing, + *kystis*, sac) A sensory receptor sensitive to gravity and motion.

stele The central vascular cylinder of stems and roots.

stem cell A relatively undifferentiated cell in animal tissue that can divide to produce more differentiated tissue cells.

stereoscopic vision (Gr. *stereos*, solid, + *opitkos*, pertaining to the eye) Ability to perceive a single, three-dimensional image from the simultaneous but slightly divergent two-dimensional images delivered to the brain by each eye.

stigma (Gr., mark, tattoo mark) (1) In angiosperm flowers, the region of a carpel that serves as a receptive surface for pollen grains. (2) Light-sensitive eyespot of some algae.

stipules Leaflike appendages that occur at the base of some flowering plant leaves or stems.

stolon (L. *stolo*, shoot) A stem that grows horizontally along the ground surface and may form adventitious roots, such as runners of the strawberry plant.

stoma, pl. stomata (Gr., mouth) In plants, a minute opening bordered by guard cells in the epidermis of leaves and stems; water passes out of a plant mainly through the stomata.

stratum corneum The outer layer of the epidermis of the skin of the vertebrate body.

striated muscle (L. *striare*, to groove) Skeletal voluntary muscle and cardiac muscle.

stromatolite A fossilized mat of ancient bacteria formed as long as 2 billion years ago, in which the bacterial remains individually resemble some modern-day bacteria.

style (Gr. *stylos*, column) In flowers, the slender column of tissue that arises from the top of the ovary and through which the pollen tube grows.

substrate (L. *substratus*, strewn under) (1) The foundation to which an organism is attached. (2) A molecule upon which an enzyme acts.

succession In ecology, the slow, orderly progression of changes in community composition that takes place through time.

summation Repetitive activation of the motor neuron resulting in maximum sustained contraction of a muscle.

surface tension A tautness of the surface of a liquid, caused by the cohesion of the molecules of liquid. Water has an extremely high surface tension.

surface-area-to-volume ratio Relationship of the surface area of a structure, such as a cell, to the volume it contains.

swim bladder An organ encountered only in the bony fish that helps the fish regulate its buoyancy by increasing or decreasing the amount of gas in

the bladder via the esophagus or a specialized network of capillaries.

symbiosis (Gr. *syn*, together with, + *bios*, life) The condition in which two or more dissimilar organisms live together in close association; includes parasitism (harmful to one of the organisms), commensalism (beneficial to one, of no significance to the other), and mutualism (advantageous to both).

sympatric speciation The differentiation of populations within a common geographic area into species.

synapomorphy In systematics, a derived character that is shared by clade members.

synapse (Gr. *synapsis*, a union) A junction between a neuron and another neuron or muscle cell; the two cells do not touch, the gap being bridged by neurotransmitter molecules.

synapsis (Gr., contact, union) The point-by-point alignment (pairing) of homologous chromosomes that occurs before the first meiotic division; crossing over takes place during synapsis.

synaptic cleft The space between two adjacent neurons.

synaptic vesicle A vesicle of a neurotransmitter produced by the axon terminal of a nerve. The filled vesicle migrates to the presynaptic membrane, fuses with it, and releases the neurotransmitter into the synaptic cleft.

synaptonemal complex A protein lattice that forms between two homologous chromosomes in prophase I of meiosis, holding the replicated chromosomes in precise register with each other so that base-pairs can form between nonsister chromatids for crossing over that is usually exact within a gene sequence.

syncytial blastoderm A structure composed of a single large cytoplasm containing about 4000 nuclei in embryonic development of insects such as *Drosophila*.

syngamy (Gr. *syn*, together with, + *gamos*, marriage) The process by which two haploid cells (gametes) fuse to form a diploid zygote; fertilization.

synteny Extensive conserved regions of DNA among different species.

systematics The reconstruction and study of evolutionary relationships.

systolic pressure A measurement of how hard the heart is contracting. When measured during a blood pressure reading, ventricular systole (contraction) is what is being monitored.

T

tagma, pl. **tagmata** (Gr., arrangement, order, row) A compound body section of an arthropod resulting from embryonic fusion of two or more segments; for example, head, thorax, abdomen.

taxis, pl. **taxes** (Gr., arrangement) An orientation movement by a (usually) simple organism in response to an environmental stimulus.

taxonomy The science of classifying living things. By agreement among taxonomists, no two organisms can have the same name, and all names are expressed in Latin.

T cell A type of lymphocyte involved in cell-mediated immunity and interactions with B cells; the "T" refers to the fact that T cells are produced in the thymus.

telencephalon (Gr. *telos*, end, + *encephalon*, brain) The most anterior portion of the brain, including the cerebrum and associated structures.

telomere A specialized nontranscribed structure that caps each end of a chromosome.

tendon (Gr. *tendon*, stretch) A strap of cartilage that attaches muscle to bone.

tertiary structure The folded shape of a protein, produced by hydrophobic interactions with water, ionic and covalent bonding between side chains of different amino acids, and van der Waal's forces; may be changed by denaturation so that the protein becomes inactive.

testcross A mating between a phenotypically dominant individual of unknown genotype and a homozygous "tester," done to determine whether the phenotypically dominant individual is homozygous or heterozygous for the relevant gene.

testis, pl. **testes** (L., witness) In mammals, the sperm-producing organ.

tetanus Sustained forceful muscle contraction with no relaxation.

thalamus (Gr. *thalamos*, chamber) That part of the vertebrate forebrain just posterior to the cerebrum; governs the flow of information from all other parts of the nervous system to the cerebrum.

thermodynamics (Gr. *therme*, heat, + *dynamis*, power) The study of transformations of energy, using heat as the most convenient form of measurement of energy.

thigmotropism In plants, unequal growth in some structure that comes about as a result of physical contact with an object.

threshold The minimum amount of stimulus required for a nerve to fire (depolarize).

thylakoid (Gr. *thylakos*, sac, + *-oides*, like) A saclike membranous structure containing chlorophyll in cyanobacteria and the chloroplasts of eukaryotic organisms.

tight junction Region of actual fusion of plasma membranes between two adjacent animal cells that prevents materials from leaking through the tissue.

tissue (L. *texere*, to weave) A group of similar cells organized into a structural and functional unit.

totipotent A cell that possesses the full genetic potential of the organism.

trachea, pl. **tracheae** (L., windpipe) A tube for breathing; in terrestrial vertebrates, the windpipe that carries air between the larynx and bronchi (which leads to the lungs); in insects and some other terrestrial arthropods, a system of chitin-lined air ducts.

tracheids In plant xylem, dead cells that taper at the ends and overlap one another.

transcription (L. *trans*, across, + *scribere*, to write) The enzyme-catalyzed assembly of an RNA molecule complementary to a strand of DNA.

transcription factor One of a set of proteins required for RNA polymerase to bind to a eukaryotic promoter region, become stabilized, and begin the transcription process.

transfection The transformation of eukaryotic cells in culture.

transfer RNA (tRNA) (L. *trans*, across, + *ferre*, to bear or carry) A class of small RNAs (about 80 nucleotides) with two functional sites; at one site, an "activating enzyme" adds a specific amino acid, while the other site carries the nucleotide triplet (anticodon) specific for that amino acid.

translation (L. *trans*, across, + *latus*, that which is carried) The assembly of a protein on the ribosomes, using mRNA to specify the order of amino acids.

translocation (L. *trans*, across, + *locare*, to put or place) (1) In plants, the long-distance transport of soluble food molecules (mostly sucrose), which occurs primarily in the sieve tubes of phloem tissue. (2) In genetics, the interchange of chromosome segments between nonhomologous chromosomes.

transpiration (L. *trans*, across, + *spirare*, to breathe) The loss of water vapor by plant parts; most transpiration occurs through the stomata.

transposition Type of genetic recombination in which transposable elements (transposons) move from one site in the DNA sequence to another, apparently randomly.

transposon (L. *transponere*, to change the position of) A DNA sequence capable of transposition.

triglyceride (triacylglycerol) An individual fat molecule, composed of a glycerol and three fatty acids.

triploid Possessing three sets of chromosomes.

trochophore A free-swimming larval stage unique to the mollusks and annelids.

trophic level (Gr. *trophos*, feeder) A step in the movement of energy through an ecosystem.

trophoblast (Gr. *trephein*, to nourish, + *blastos*, germ) In vertebrate embryos, the outer ectodermal layer of the blastodermic vesicle; in mammals, it is part of the chorion and attaches to the uterine wall.

tropism (Gr. *trope*, a turning) A response to an external stimulus.

tropomyosin (Gr. *tropos*, turn, + *myos*, muscle) Low-molecular-weight protein surrounding the actin filaments of striated muscle.

troponin Complex of globular proteins positioned at intervals along the actin filament of skeletal muscle; thought to serve as a calcium-dependent "switch" in muscle contraction.

tubulin (L. *tubulus*, small tube, + *in*, belonging to) Globular protein subunit forming the hollow cylinder of microtubules.

tumor-suppressor gene A gene that normally functions to inhibit cell division; mutated forms can lead to the unrestrained cell division of cancer, but only when both copies of the gene are mutant.

turgor (L. *turgor*, a swelling) The pressure within a cell resulting from the movement of water into the cell; a cell with high turgor pressure is said to be turgid. *See* osmotic pressure.

U

unequal crossing over A process by which a crossover in a small region of misalignment at synapsis causes two homologous chromosomes to exchange segments of unequal length.

unsaturated fat A fat molecule in which one or more of the fatty acids contain fewer than the maximum number of hydrogens attached to their carbons.

urea (Gr. *ouron*, urine) An organic molecule formed in the vertebrate liver; the principal form of disposal of nitrogenous wastes by mammals.

urethra (Gr. from *ourein*, to urinate) The tube carrying urine from the bladder to the exterior of mammals.

uric acid Insoluble nitrogenous waste products produced largely by reptiles, birds, and insects.

urine (Gr. *ouron*, urine) The liquid waste filtered from the blood by the kidney and stored in the bladder pending elimination through the urethra.

uterus (L., womb) In mammals, a chamber in which the developing embryo is contained and nurtured during pregnancy.

V

vacuole A membrane-bounded sac in the cytoplasm of some cells, used for storage or digestion purposes in different kinds of cells; plant cells often contain a large central vacuole that stores water, proteins, and waste materials.

vascular cambium In vascular plants, a cylindrical sheath of meristematic cells, the division of which produces secondary phloem outwardly and secondary xylem inwardly; the activity of the vascular cambium increases stem or root diameter.

vascular tissue (L. *vasculum*, a small vessel) Containing or concerning vessels that conduct fluid.

vas deferens (L. *vas*, a vessel, + *ferre*, to carry down) In mammals, the tube carrying sperm from the testes to the urethra.

vasopressin A posterior pituitary hormone that regulates the kidney's retention of water.

vegetal pole The hemisphere of the zygote comprising cells rich in yolk.

vein (L. *vena*, a blood vessel) (1) In plants, a vascular bundle forming a part of the framework of the conducting and supporting tissue of a stem or leaf. (2) In animals, a blood vessel carrying blood from the tissues to the heart.

veliger The second larval stage of mollusks following the trochophore stage, during which the beginning of a foot, shell, and mantle can be seen.

ventricle   A muscular chamber of the heart that receives blood from an atrium and pumps blood out to either the lungs or the body tissues.

vesicle   (L. *vesicula*, a little bladder) A small intracellular, membrane-bounded sac in which various substances are transported or stored.

vessel element   In vascular plants, a typically elongated cell, dead at maturity, which conducts water and solutes in the xylem.

vestibular apparatus   The complicated sensory apparatus of the inner ear that provides for balance and orientation of the head in vertebrates.

villus, pl. villi   (L., a tuft of hair) In vertebrates, one of the minute, fingerlike projections lining the small intestine that serve to increase the absorptive surface area of the intestine.

visceral mass   (L., internal organs) Internal organs in the body cavity of an animal.

vitamin   (L. *vita*, life, + *amine*, of chemical origin) An organic substance that cannot be synthesized by a particular organism but is required in small amounts for normal metabolic function.

viviparity   (L. *vivus*, alive, + *parere*, to bring forth) Refers to reproduction in which eggs develop within the mother's body and young are born free-living.

voltage-gated ion channel   A transmembrane pathway for an ion that is opened or closed by a change in the voltage, or charge difference, across the plasma membrane.

W

water potential   The potential energy of water molecules. Regardless of the reason (e.g., gravity, pressure, concentration of solute particles) for the water potential, water moves from a region where water potential is greater to a region where water potential is lower.

wild type   In genetics, the phenotype or genotype that is characteristic of the majority of individuals of a species in a natural environment.

X

xylem   (Gr. *xylon*, wood) In vascular plants, a specialized tissue, composed primarily of elongate, thick-walled conducting cells, which transports water and solutes through the plant body.

Y

yolk plug   A plug occurring in the blastopore of amphibians during formation of the archenteron in embryological development.

Z

zona pellucida   An outer membrane that encases a mammalian egg.

zoospore   A motile spore.

zooxanthellae—Symbiotic photosynthetic protists in the tissues of corals.

zygomycetes   (Gr. *zygon*, yoke, + *mykes*, fungus) A type of fungus whose chief characteristic is the production of sexual structures called zygosporangia, which result from the fusion of two of its simple reproductive organs.

zygote   (Gr. *zygotos*, paired together) The diploid ($2n$) cell resulting from the fusion of male and female gametes (fertilization).

Credits

Bettmann; **13.24:** © Science Photo Library/Photo Researchers; **13.26:** © Alfred Paseika/Science Photo Library/Photo Researchers; **13.28 (both):** © Cabisco/Phototake; **13.34:** © Leonard Lessin/Peter Arnold; **13.36:** © Hans Reinhard/Okapia/Photo Researchers; **13.37 (left):** Courtesy of Loris McGavaran, Denver Children's Hospital; **13.37 (right):** © Richard Hutchings/Photo Researchers Inc.

Chapter 14

Figure 14.1: © PhotoDisc/Volume 29; **14.9 (both):** From "The Double Helix," by J.D. Watson, Atheneum Press, N.Y. 1968; **14.10a:** © Barrington Brown/Photo Researchers Inc.; **14.11:** From M. Meselson and F.W. Stahl/ *Proceedings of the Nat. Acad. of Sci.* 44 (1958):671; **14.17 (both):** From *Biochemistry* 4/e by Stryer © 1995 by Lubert Stryer. Used with permission of W.H. Freeman and Company; **14.19:** © Prof. Ulrich Laemmli/Photo Researchers Inc.; **14.20a:** Courtesy of Dr. David Wolstenholme

Chapter 15

Figure 15.1: © K.G. Murti/Visuals Unlimited; **15.3:** N. Ban, P. Nissen, J. Hansen, P.B. Moore & T.A. Steitz, "The Complete Atomic Structure of the Large Ribosomal subunit at 2.4A Resolution," Reprinted with permission from *Science*, v. 289 #5481, p917, © 2000 American Association for the Advancement of Science; **15.7:** From R.C. Williams; *Proc. Nat. Acad. of Sci.* 74 (1977):2313; **15.13:** Courtesy of Dr. Oscar L. Miller; **15.18b:** Courtesy of Dr. Bert O'Malley, Baylor College of Medicine

Chapter 16

Figure 16.1: © Stanley Cohen/Science Photo Library/Photo Researchers Inc.; **16.7b:** Courtesy of Bio-Rad Laboratories; **16.15:** Courtesy of Lifecodes Corp., Stamford, CT; **16.16:** AP/Wide World Photos; **16.17:** R.L. Brinster, U. of Pennsylvania Sch. of Vet. Med.; **16.20:** Courtesy of Monsanto

Chapter 17

Figure 17.1: Courtesy of Robert D. Fleischmann, The Institute for Genomic Research; **17.4:** Courtesy of Celera Genomics; **17.11 (all):** Reproduced with permission from Altpeter et al, *Plant Cell Reports* 16:12-17, 1996, photos

provided by Indra Vasil; **17.12:** Courtesy of Research Collaboratory for Structural Bioinformatics; **17.13:** © Corbis/R-F Website; **17.14:** © Grant Heilman/Grant Heilman Photography

Chapter 18

Figure 18.1: Courtesy of Dr. Claus Pelling; **18.15 (both):** Courtesy of Dr. Harrison Echols; **18.18a:** Courtesy of Dr. Victoria Foe

Chapter 19

Figure 19.1: © Cabisco/Visuals Unlimited; **19.3:** Photo Lennart Nilsson/Albert Bonniers Forlag AB, *A Child is Born*, Dell Publishing Company; **19.4 (all):** © Cabisco/Phototake; **19.6:** © Ed Lewis; **19.15 (top left):** Dr. Christiane Nusslein-Volhard/Max Planck Institute as published in From Egg to Adult, © 1992 HHMI; **19.15b-d:** James Langeland, Stephen Paddock & Sean Carroll as published in *From Egg to Adult*, © 1992 HHMI; **19.16 (all):** Courtesy of Manfred Frasch; **19.17:** Courtesy of E.B. Lewis

Chapter 20

Figure 20.1: © University of Wisconsin-Madison News & Public Affairs, Photo by Jeff Miller; **20.3:** Courtesy of Dr. Charles Brinton; **20.10:** © Custom Medical Stock Photo; **20.18:** Courtesy of American Cancer Society; **20.23:** AP/Wide World Photos; **20.25:** © University of Wisconsin-Madison News & Public Affairs

Chapter 21

Figure 21.1: © Corbis/R-F Website; **21.3:** Biological Photo Service; **21.19:** Courtesy of H. Rodd

Chapter 22

Figure 22.1: © PhotoDisc Website; **22.4 (both):** © Breck P. Kent/Animals Animals/Earth Scenes; **22.9 (both):** Courtesy of Lyudmilla N. Trut, Institute of Cytology & Genetics, Siberian Dept. of the Russian Academy of Sciences

Chapter 23

Figure 23.1: Jonathan Losos; **23.3:** © Porterfield/Chickering/Photo Researchers Inc.; **23.4:** © Barbara Gerlach/Visuals Unlimited; **23.6 (top left):** © John Shaw/Tom Stack & Associates; **23.6 (bottom left):** © Rob & Ann

Simpson/Visuals Unlimited; **23.6 (top right):** © Suzanne L. Collins & Joseph T. Collins/National Audubon Society Collection/Photo Researchers; **23.6 (bottom right):** © Phil A. Dotson/National Audubon Society Collection/Photo Researchers; **23.8 (left):** © Chas. McRae/Visuals Unlimited; **23.8 (right):** Jonathan Losos; **23.12 (left):** © Jeffrey Taylor; **23.12 (right):** © William P. Mull; **23.15:** © G.R. Roberts

Chapter 24

Figure 24.1: Courtesy of Richard P. Elinson; **p. 492a:** © Dr. R. Clark & M. Goff/Science Photo Library/Photo Researchers; **p. 492b:** © PhotoDisc Green/Getty Images; **p. 492c:** © BIOS(C. Ruoso)/Peter Arnold, Inc.; **p. 492d:** © Darwin Dale/Photo Researchers; **p. 492e:** Centers for Disease Control; **p. 492f:** © Paul G. Young/The Institute for Genomic Research; **p. 492g:** © Dr. Jeremy Burgess/Science Photo Library/Photo Researchers; **p. 492h:** © Science Photo Library/Photo Researchers; **p. 492I:** © PhotoDisc/Getty Images; **p. 492j:** © CNRI/Science Photo Library/Photo Researchers; **24.4:** © PhotoDisc/Getty Images; **24.10:** Courtesy of Anna Di Gregorio; **24.13 (top left):** © Image Bank/Getty Images; **24.13 (top right):** © Darwin Dale/Photo Researchers; **24.13 (bottom left):** © Aldo Brando/Peter Arnold, Inc.; **24.13 (bottom right):** © Tom E. Adams/Peter Arnold, Inc.; **24.14 (both):** Courtesy of Walter Gehring, reprinted with permission from Induction of Ectopic Eyes by Targeted Expression of the Eyeless Gene in Drosophila, G. Halder, P. Callaerts, Walter J. Gehring, *Science* Vol. 267, © 24 March 1995 American Association for the Advancement of Science; **24.15 (both):** Courtesy of Dr. William Jeffery

Chapter 25

Figure 25.1: © Corbis/Volume 8; **25.2 (top left):** © Corbis/Volume 102; **25.2 (top right):** © PhotoDisc/Volume 56; **25.2 (bottom left):** © PhotoDisc/Volume 1; **25.2 (bottom right):** © Corbis/Volume 8; **25.6:** Image # 5789, photo by D. Finnin/American Museum of Natural History

Chapter 26

Figure 26.1: © K.G. Murti/Visuals Unlimited; **26.3a:** © Dept. of Microbiology, Biozentrum/Science

Photo Library/Photo Researchers; **26.7:** Courtesy of Katherine Sutliff, from *Science* 300: 1377, May 30, 2003, p. 1377, © American Association for the Advancement of Science, Reprinted with permission; **26.8:** © Corbis/Volume 40

Chapter 27

Figure 27.1: © David M. Phillips/Visuals Unlimited; **p. 547 (above):** © Abraham & Beachey/BPS/Tom Stack & Associates; **p. 547 (below):** © F. Widell/Visuals Unlimited; **27.3a:** © Science Photo Library/Photo Researchers; **27.3b:** © University of Regensburg, Courtesy of Reinhard Rachel; **27.3c:** © Andrew Syred/Science Photo Library/Photo Researchers; **27.3d:** © Microfield Scientific Ltd./Science Photo Library/ Photo Researchers; **27.3e:** © Alfred Paseika/Science Photo Library/Photo Researchers; **27.3g:** © S.W. Watson/ Visuals Unlimited; **27.3h:** © Dennis Kunkel Microscopy, Inc.; **27.3I:** Courtesy of Dr. Hans Reichenbach; **27.4:** © G.W. Willis; **27.5:** © Julius Adler/Visuals Unlimited; **27.6 (left):** © W. Watson/ Visuals Unlimited; **27.6 (right):** © Norma J. Lang/Biological Photo Service; **27.8:** © CNRI/SPL/Photo Researchers; **27.10:** © R. Calentine/ Visuals Unlimited; **27.11:** © Science/Visuals Unlimited

Chapter 28

Figure 28.1: © John D. Cunningham/Visuals Unlimited; **28.2:** Courtesy of Dr. Edward W. Daniels, Argonne National Lab, the University of Illinois College of Medicine at Chicago; **28.5:** © John D. Cunningham/Visuals Unlimited; **28.6a:** © Michael Abby/ Visuals Unlimited; **28.7 (left):** © Manfred Kage/Peter Arnold Inc.; **28.7 (right):** © Edward S. Ross; **28.11a:** © Brian Parker/ Tom Stack & Associates; **28.12:** © Gregory Ochocki/Photo Researchers; **28.13:** © D.P. Wilson/Photo Researchers; **28.15:** © John D. Cunningham/ Visuals Unlimited; **28.16:** © Phil A. Harrington/Peter Arnold, Inc.; **28.18:** © Richard Rowan/Photo Researchers; **28.17:** © Manfred Kage/ Peter Arnold; **28.19:** © Edward S. Ross; **28.20a:** © John Shaw/Tom Stack & Associates; **28.20b:** © John Shaw/Tom Stack & Associates; **28.21:** © Genichiro Higuchi,

Higuchi Science Laboratory; **p.578:** © D.P. Wilson/Photo Researchers

Chapter 29

Figure 29.1: © Stephen J. Krasemann/DRK Photo; **29.4:** © Edward S. Ross; **29.6:** © Kirtley Perkins/Visuals Unlimited; **29.7:** © Kingsley R. Stern; **29.8:** Courtesy of Hans Steur, The Netherlands; **29.9:** Darrell Vodopich; **29.10:** © Kingsley R. Stern; **29.11, 29.12, 29.14:** © Edward S. Ross; **29.16 (left):** © Walter H. Hodge/Peter Arnold Inc.; **29.16 (center):** © Kjell Sandved/ Butterfly Alphabet; **29.16c:** © Runk/Schoenberger/Grant Heilman Photography; **29.17:** Courtesy of David Dilcher & Ge Sun; **29.18:** Courtesy of Sandra Floyd

Chapter 30

Figure 30.1: © Bill Keogh/Visuals Unlimited; **30.2a:** Courtesy of Dr. Peter Daszak; **30.2b:** © Cabisco/ Visuals Unlimited; **30.2c:** © Robert Simpson/Tom Stack & Associates; **30.2d:** © Michael & Patricia Fogden; **30.3:** Courtesy of E.C. Setliff & W. L. MacDonald; **30.4:** © Kjell Sandved/ Butterfly Alphabet; **30.5:** © Microfield Scientific Ltd/Photo Researchers; **30.6:** © L. West/Photo Researchers; **30.7:** Ralph Williams/USDA Forest Service; **30.9a:** © 1997 Regents of the University of Michigan; **30.10:** © Cabisco/Phototake; **30.11a:** © Alexandra Lowry/The National Audubon Society Collection/ Photo Researchers; **30.12a:** © Ed Pembleton; **30.12b:** © Kjell Sandved/ Butterfly Alphabet; **30.13:** © David M. Phillips/ Visuals Unlimited; **30.14a:** © James Castner; **30.14b:** © Edward S. Ross; **30.15:** © Ed Reschke; **30.16b:** © D.H. Marx/ Visuals Unlimited; **30.17:** © Scott Camazine/Photo Researchers; **30.18:** Courtesy of Zoology Department/University of Canterbury, New Zealand; **30.19a:** Courtesy of Laura E. Sweets, University of Missouri; **30.19b:** © Manfred Kage/Peter Arnold, Inc.

Chapter 31

Figure 31.1: © Corbis/Volume 53; **p. 619a:** © Corbis/Volume 86; **p. 619b:** © Corbis/Volume 65; **p. 619c:** © David M. Phillips/Visuals Unlimited; **p. 619d:** © Royalty-Free/Corbis **p. 619e:** © Edward S.

Ross; **p. 619f:** © Corbis; **p. 619g:** © Cleveland P. Hickman Jr.; **p. 619h:** © Cabisco/Phototake; p. 619I: © Ed Reschke; **31.2 (top left):** © Corbis/Volume 53; **31.2 (top right):** © Corbis; **31.2 (bottom left):** © Corbis; **31.2d:** © Corbis; **31.8:** Courtesy of Dr. Igor Eeckhaut

Chapter 32

Figure 32.1: © Denise Tackett/Tom Stack & Associates **32.4:** © Andrew J. Martinez/Photo Researchers; **32.9:** © Gwen Fidler/Tom Stack & Associates; **32.10:** © Kelvin Aitken/Peter Arnold Inc.; **32.11:** © Daniel Gotshall; **32.12:** © David Wrobel/Visuals Unlimited; **32.16:** © Kjell Sandved/Butterfly Alphabet; **32.17:** © Biology Media/R. Knauf/Photo Researchers; **32.19:** © T.E. Adams/Visuals Unlimited; **32.20:** Courtesy of Matthias Obst

Chapter 33

Figure 33.1: © James H. Robinson/Animals Animals/Earth Scenes; **33.2:** © Alex Kerstich/Visuals Unlimited; **33.4:** © A. Flowers & L. Newman/Photo Researchers; **33.7b:** © Kjell Sandved/Butterfly Alphabet; **33.8:** © Milton Rand/Tom Stack & Associates; **33.10:** © Fred Bavendam/Peter Arnold Inc.; **33.11:** © Visuals Unlimited; **33.13:** © Kjell Sandved/Butterfly Alphabet; **33.14:** © David Dennis/Tom Stack & Associates; **33.15:** © Cleveland P. Hickman; **33.17b:** © Robert Brons/Biological PhotoService; **33.18b:** © Fred Bavendam/Peter Arnold Inc.; **33.26:** © T.E. Adams/Visuals Unlimited; **33.27:** © Kjell Sandved/Butterfly Alphabet; **33.28a:** © Rod Planck/Tom Stack & Associates; **33.28b:** © Ann Moreton/Tom Stack & Associates; **33.29:** © Alex Kerstich/Visuals Unlimited; **33.30:** © Edward S. Ross; **33.31a:** © Cleveland P. Hickman; **33.31b:** © Kjell Sandved/Butterfly Alphabet; **33.31c:** © Norm Thomas/Photo Researchers; **33.31d:** © Valorie Hodgson/Visuals Unlimited; **33.31e:** © Corbis/Volume 6; **33.31f:** © Kjell Sandved/Butterfly Alphabet; **33.33:** © John Shaw/Tom Stack & Associates; **33.34:** © Kjell Sandved/Butterfly Alphabet; **33.35a:** © Alex Kerstitch/Visuals Unlimited; **33.35b:** © Randy Morse/Tom Stack & Associates; **33.35c:** © Daniel W. Gotshall/Visuals Unlimited;

33.38: © Kjell Sandved/Butterfly Alphabet; **33.40:** © Daniel W. Gotshall/Visuals Unlimited; **33.41:** © Kjell Sandved/Visuals Unlimited

Chapter 34

Figure 34.1: © Corbis; **34.3:** © Eric N. Olson, PhD/The University of Texas MD Anderson Cancer Center; **34.5a:** © Rick Harbo Marine Images; **34.6:** © Heather Angel; **34.10:** © Corbis; **34.15:** © Corbis/Volume 53; **34.16:** © John D. Cunningham/Visuals Unlimited; **34.18:** © 1979 Peter Scoones/Contact Press Images/Woodfin Camp & Associates Inc.; **34.21:** © Natural History Museum/J. Sibbick; **34.22a:** © John Shaw/Tom Stack & Associates; **34.22b:** © Suzanne L. Collins & Joseph T. Collins/Photo Researchers; **34.22c:** © Jany Sauvanet/Photo Researchers; **34.24:** © Natural History Museum/J. Sibbick; **34.27:** Photo by Paul Sareno/University of Chicago; **34.31:** © William J. Weber/Visuals Unlimited; **34.32a:** © John Cancalosi/Tom Stack & Associates; **34.32b:** © Rod Planck/Tom Stack & Associates; **34.33:** © Corbis/ Volume 6; **34.37:** © Corbis; **34.38:** © PhotoDisc/Volume 44; **34.41:** © Stephen Dalton/National Audubon Society Collection/Photo Researchers; **34.42:** © B.J. Alcock/Visuals Unlimited; **34.42b:** © Corbis/ Volume 6; **34.42c:** © Stephen J. Krasemann/DRK Photo; **34.43:** © Alan Nelson/Animals Animals/Earth Scenes; **34.45 (all):** © David L. Brill; **34.47:** © 1985 David L. Brill/Fossil credit: National Museums of Kenya, Nairobi; **34.49:** AP/Wide World Photos

Chapter 35

Figure 35.1: Photo by Susan Singer; **35.2:** © Terry Ashley/Tom Stack & Associates; **35.4a:** © Hart-Davis/Science Photo Library/Photo Researchers; **35.4b:** © R. Calentine/Visuals Unlimited; **35.8 (both):** © Heidi Mullen; **35.9:** Courtesy of Liming Zhao & Fred Sack, Ohio State University; **35.10:** © Andrew Syred/Science Photo Library/Photo Researchers; **35.11 (both):** Courtesy of Alan Lloyd; **35.12a:** © Biophoto Associates/Photo Researchers Inc.; **35.12b:** © John D. Cunningham/Visuals Unlimited; **35.12c:** © Lawrence Mellinchamp/Visuals Unlimited;

35.13c: Courtesy of Wilfred Cote, Suny College of Environmental Forestry; **35.14:** © Randy Moore/Visuals Unlimited; **35.15:** © E.J. Cable/Tom Stack & Associates; **35.16a:** Courtesy John Schiefelbein, from Myeong Min Lee & John Schiefelbein, "WEREWolf, MYB-related Protein in Arabidopsis," *Cell* V. 99:473-483, Nov. 24, 1999; **35.16b:** Courtesy of Dr. Philip Benfey, from Wysocka-Diller, J.W., Helariutta, Y., Fukaki, H., Malamy, J.E. and P.N. Benfey (2000) Molecular analysis of SCARECROW function reveals a radial patterning mechanism common to root and shoot Development, *Cell* 127, 595-603; **35.17:** Photomicrograph by G.S. Ellmore; **35.19a:** © Kingsley R. Stern; **35.19b:** Photomicrograph by G.S. Ellmore; **35.21a:** © Walter H. Hodge/Peter Arnold, Inc.; **35.21b:** Courtesy of Robert A. Schlising; **35.21c:** © Kingsley R. Stern; **35.22:** Courtesy of J.H. Troughton and L. Donaldson and Industrial Research Ltd.; **35.24:** © John D. cunningham/ Visuals Unlimited; **35.25 (both):** © Ed Reschke; **35.26:** © Ed Reschke/Peter Arnold Inc.; **35.27a:** © Ed Reschke; **35.27b:** © Jack M. Bostrack/ Visuals Unlimited; **35.29:** Richard Waites & Andrew Hudson, from Phantastica: a gene required for dorsoventrality of leaves in Antirrhinum majus, *Development* 121, 2143-2154 (1995) © The Company of Biologists Limited 1995; **35.30a:** © Kjell Sandved/Butterfly Alphabet; **35.30b:** © Pat Anderson/Visuals Unlimited; **35.31a:** © Edward S. Ross; **35.31b:** © Glenn M. Oliver/ Visuals Unlimited; **35.31c:** © Joel Arrington/Visuals Unlimited; **35.34:** © Ed Reschke; **35.35 (above):** © Michael P. Godomski/Photo Researchers; **35.35 (below):** © Ed Reschke/Peter Arnold, Inc.; **p.753:** © Kingsley R. Stern

Chapter 36

Figure 36.1: © Norm Thomas/The National Audubon Society Collection/Photo Researchers; **36.4:** Courtesy of E.C. Yeung & D.W. Meinke; **36.5:** Courtesy of Dr. Chun-Ming Liu; **36.6:** Courtesy of Kathy Barton & Jeff Lang; **36.7a(above):** © Runk/ Schoenberger/Grant Heilman Photography; **36.7b:** © Kevin & Betty Collins/Visuals Unlimited; **36.8:** © Jack M. Bostrack/Visuals Unlimited; **36.10a:** © Ed Reschke/Peter Arnold, Inc.;

36.10b: © David Sieren/Visuals Unlimited; **36.11 (top left):** © Kingsley R. Stern; **36.11 (top center):** © James Richardson/Visuals Unlimited; **36.11 (top right):** © Kingsley R. Stern; **36.11 (middle left):** © Kingsley R. Stern; **36.11 (center):** © Kingsley R. Stern; **36.11 (middle right):** © Barry L. Runk/Grant Heilman Photography; **36.11 (bottom left):** Courtesy of Robert A. Schlising; **36.11 (bottom right):** © Charles D.Winters/Photo Researchers; **36.12a:** © Edward S. Ross; **36.12b, 36.13, 36.14:** © James Castner; **36.16:** Courtesy of Prof. Tuan-hua David Ho

Chapter 37

Figure 37.1: © Richard Rowan's Collection, Inc./Photo Researchers; **37.7:** © John D. Cunningham/Visuals Unlimited; **37.8:** © Terry Ashley/Tom Stack & Associates; **37.9:** © Grant Heilman/Grant Heilman Photography; **37.10:** © Bruce Iverson Photomicrography; **37.12a:** © Andrew Syred/Science Photo Library/Photo Researchers; **37.12b:** © Bruce Iverson/Science Photo Library/Photo Researchers

Chapter 38

Figure 38.1: © Scott T. Smith/Corbis **38.2 (all):** Courtesy of Dr. Emmanuel Epstein; **38.4:** © Michael P. Gadomski/Photo Researchers; **38.6 (both):** Courtesy of Nicholas School of the Environment and Earth Sciences, Duke University; **38.7:** Courtesy of Sharon Long; **38.9:** © Kjell Sandved/Butterfly Alphabet; **38.10:** © Runk/Schoenberger/Grant Heilman Photography; **38.11:** © Barry Rice; **38.12:** © Don Albert; **38.14:** U.S. EPA National Research Lab, Cincinnati; **38.15 (all):** © AP/Wide World Photos; **p.793:** U.S. EPA National Research Lab, Cincinnati

Chapter 39

Figure 39.1: © Holt Studios International (Nigel Cattlin)/Photo Researchers; **39.2:** © Jane Grushow/Grant Heilman Photography; **39.3:** USDA/Agricultural Research Service; **39.4 (both):** USDA/Agricultural Research Service; **39.6:** © C. Allan Morgan/Peter Arnold, Inc.; **39.7:** © Runk/Schoenberger/Grant Heilman Photography

Chapter 40

Figure 40.1: © John D. Cunningham/Visuals Unlimited; **40.5:** © Runk/Schoenberger/Grant Heilman Photography; **40.6:** © John D. Cunningham/Visuals Unlimited; **40.7:** © S.J. Krasemann/Peter Arnold Inc.; **40.8:** Courtesy of Frank B. Salisbury; **40.9a:** © T. Walker; **40.9b:** © R.J. Delorit, Agronomy Publications; **40.10a:** © Jim Zipp/The National Audubon Society Collection/Photo Researchers; **40.10b:** © Runk/Schoenberger/Grant Heilman Photography; **40.11:** © Prof. Malcolm B. Wilkins, Botany Dept., Glasgow University; **40.18 (all):** © Prof. Malcolm B. Wilkins, Botany Dept., Glasgow University; **40.20:** © Robert Calentine/Visuals Unlimited; **40.21:** ©Runk/Schoenberger/Grant Heilman Photography; **40.22:** © Sylvan H. Wittwer/Visuals Unlimited; **40.25a:** © John Solden/Visuals Unlimited; **40.25b:** Courtesy of Donald R. McCarty, from "Molecular Analysis of viviparous-1: An Abscisic Acid-Insensitive Mutant of Maize," *The Plant Cell*, v.1, 523-532, © 1989 American Society of Plant Physiologists; **40.25c:** © David M. Phillips/Visuals Unlimited;

Chapter 41

Figure 41.1: © Richard La Val/Animals Animals; **41.3a:** © Stephen G. Maka/DRK Photo; **41.3b:** © Heidi Mullen **41.4:** Courtesy of Lingjing Chen & Renee Sung; **41.5 (both):** Detlef Weigel & Ove Nilsson, The Salk Institute for Biological Studies; **41.7:** © Jim Strawser/Grant Heilman Photography; **41.12 (all):** Courtesy of John L. Bowman; **41.14:** © John Bishop/Visuals Unlimited; **41.15:** © Paul Gier/Visuals Unlimited; **41.16:** Courtesy of Enrico Coen; **41.18a:** Courtesy of William F. Chissoe, Noble Microscopy Lab, U. of Oklahoma; **41.18b:** Courtesy of Dr. Joan Nowicke, Smithsonian Institution; **41.19:** © Kingsley R. Stern; **41.20:** © Edward S. Ross; **41.21:** © Michael & Patricia Fogden; **41.22 (both):** © Thomas Eisner; **41.23** © John D. Cunningham/Visuals Unlimited; **41.24:** © Edward S. Ross; **41.25a:** © David Sieren/Visuals Unlimited; **41.25b:** © Barbara Gerlach/Visuals Unlimited; **41.28:** © Jerome Wexler/Photo Researchers; **41.29 (all):** Courtesy of Dr. Hans Ulrich Koop, from

Plant Cell Reports, 17:601-604; **41.30 (both):** © Edward S. Ross

Chapter 42

Figure 42.1: Photo Lennart Nilsson/Albert Bonniers Forlag AB, *Behold Man*, Little Brown & Co.; **p. 861 (top 3):** © Ed Reschke; **p. 861 (4th from top):** © Fred Hossler/visuals Unlimited; **p. 861 (bottom):** © Ed Reschke; **42.6:** © J. Gross/Science Photo Library/Photo Researchers; **42.7:** © Biophoto Associates/Photo Researchers; **p. 863 (top):** © Biophoto Associates/Photo Researchers; **p. 863 (2nd from top):** © Cleveland P. Hickman; **p. 863 (3rd from top):** © Chuck Brown/Photo Researchers; **p. 863 (4th from top):** © Ed Reschke; **p. 863 (bottom):** © Ken Edward/Science Source/Photo Researchers; **42.8b:** © Ed Reschke; **42.9:** © Ed Reschke; **42.10:** © David M. Phillips/Visuals Unlimited; **p. 867 (all):** © Ed Reschke; **42.13:** © Anthony Bannister/Animals Animals/Earth Scenes; **42.19, 42.20:** © Dr. H.E. Huxley; **42.31:** © Treat Davidson/Photo Researchers; **p. 886:** © Dr. H.E. Huxley

Chapter 43

Figure 43.1: © John Gerlach/Animals Animals/Earth Scenes; **43.20:** From O.T. Avery, C.M. Macleod & M. McCarty, "Studies on the chemical nature of the substance inducing transformation of pneumococcal types," reproduced from the *Journal of Experimental Medicine* 79 (1944): 137-158, fig. 1 by copyright permission of the Rockefeller University Press, reproduced by permission. Photograph made by Mr. Joseph B. Haulenbeek

Chapter 44

Figure 44.1: © Professors P.M. Motta & S. Correr/Science Photo Library/Photo Researchers; **44.11:** © Ed Reschke; **44.17 (all):** Courtesy of Frank P. Sloop, Jr.; **44.18:** © Frans Lanting/Minden Pictures

Chapter 45

Figure 45.1: Courtesy of David I. Vaney, University of Queensland, Australia; **45.4:** © C.S. Raines/Visuals Unlimited; **45.14:** © John Heuser, Washington University School of Medicine, St. Louis, MO; **45.16:** © Ed Reschke; **45.18b:** © E.R. Lewis, YY Zeevi, T.E. Everhart, U. of

California/Biological Photo Service; **45.27:** Dr. Marcus E. Rachle, Washington University, McDonnell Center for High Brain Function; **45.28:** Photo Lennart Nilsson/Albert Bonniers Forlag AB, *Behold Man*, Little Brown & Co.; **45.31:** © E.R. Lewis/Biological Photo Service

Chapter 46

Figure 46.1: © Omikron/Photo Researchers; **46.6:** © Ed Reschke; **46.23:** © Leonard L. Rue, III

Chapter 47

Figure 47.1: © Francois Gohier/Science Source/Photo Researchers; **47.11:** © Corbis/Bettmann; **47.15:** © John Paul Kay/Peter Arnold, Inc.; **47.20:** © Robert & Linda Mitchell

Chapter 48

Figure 48.1: National Library of Medicine; **48.3:** © Manfred Kage/Peter Arnold, Inc.; **48.7:** © Visuals Unlimited; **48.9 (both):** © Dr.Andrejs LiepinsScience Photo Library/Photo Researchers; **48.18:** © Stuart Fox; **48.23:** © CDC/Science Source/Photo Researchers

Chapter 49

Figure 49.1: © Belinda Wright/DRK Photo

Chapter 50

Figure 50.1: © Michael Fogden/DRK Photo; **50.2a:** © Chuck Wise/Animals Animals/Earth Scenes; **50.2b:** © Fred McConnaughey/The National Audubon Society Collection/Photo Researchers; **50.4:** © David Doubilet; **50.5:** © Hans Pfletschinger/Peter Arnold Inc.; **50.7:** © Cleveland P. Hickman Jr.; **50.8:** © Frans Lanting/Minden Pictures; **50.9a:** © Jean Phillippe Varin/Jacana/Photo Researchers; **50.9b:** © Tom McHugh/The National Audubon Society Collection/Photo Researchers; **50.9c:** © Corbis/Volume 86; **50.12a:** © David M. Phillips/Photo Researchers; **50.13:** Photo Lennart Nilsson/Bonniers Forlag AB, *A Child is Born*, Dell Publishing Co.; **50.18:** © Ed Reschke; **50.22 (all):** © McGraw-Hill Higher Education/Bob Coyle, photographer

Chapter 51

Figure 51.1: Photo Lennart Nilsson/Bonniers Forlag AB, *A*

Child is Born, Dell Publishing Co.; **51.2b:** © P. Bagavandoss/Science Source/Photo Researchers; **51.2c:** © David M. Phillips/ Visuals Unlimited; **51.3b:** Courtesy of Dr. Everett Anderson; **51.6:** © David M. Phillips/Visuals Unlimited; **51.7:** © Cabisco/ Phototake; **51.8:** © David M. Phillips/Visuals Unlimited; **51.21:** Photo Lennart Nilsson/ Albert Bonniers Forlag AB, *A Child is Born*, Dell Publishing Company; **51.22 (all):** Photo Lennart Nilsson/Albert Bonniers Forlag AB, *A Child is Born*, Dell Publishing Company

Chapter 52

Figure 52.1: © K. Ammann/Bruce Coleman Inc.; **52.2 (left):** © UPI/ Corbis Bettmann; **52.2 (center):** © Corbis/ Bettmann; **52.2 (right):** © UPI/ Corbis-Bettmann; **52.3:** From J.L. Gould, Ehology, Norton 1982; **52.5:** © William C. Dilger, Cornell University; **52.6:** From J.R. Brown et al, "A defect in nurturing mice lacking . . . gene for fosB" *Cell* v. 86, 1996 pp 297-308, © Cell Press; **52.7 (all):** © Lee Boltin Picture Library; **52.8:** © William Grenfell/Visuals Unlimited; **52.9a:** Thomas McAvoy, Life Magazine/© Time, Inc.; **52.10 (both):** Grzimek's *Encyclopedia of Ethology* Van Nostrand-Reinhold Co.; **52.12:** © Roger Wilmshurst/ The National Audubon Society Collection/ Photo Researchers; **52.13a:** © Linda Koebner/Bruce Coleman; **52.13b:** © Jeff Foott/Tom Stack & Associates; **52.14 (all):** Superstock; **52.15:** Courtesy of Bernd Heinrich; **52.16b:** © Fred Bruenner/Peter Arnold Inc.; **52.16c:** © James L. Amos/Peter Arnold Inc.; **52.19:** © Dwight R. Kuhn/DRK Photo; **52.21** © Corbis/ Volume 53; **52.22:** © Sol Mednick; **52.23b:** © Dr. Mark Moffett/Minden Pictures; **52.24a:** © S. Osolinski /OSF/Animals Animals/Earth Scenes; **52.25:** Nina Leen, Life Magazine, © Time Inc.; **52.28:** © Bios(C.Thouvenin)/Peter Arnold, Inc.; **52.30b:** © B. Chudleigh/Vireo; **52.32:** © George D. Lepp/Corbis; **52.33a:** Courtesy of T.A. Burke, Reprinted by permission from *Nature*, "Parental care and mating behavior of polyandrous dunnocks," 338: 247-251, 1989; **52.35:** © Edward S. Ross; **52.37:** © Mark Moffett/ Minden Pictures; **52.38:** © Nigel Dennis/National Audubon Society Collection/Photo Researchers

Chapter 53

Figure 53.1: © PhotoDisc/Volume 44; **53.2:** Courtesy of William J.

Hamilton III; **53.17:** Courtesy of Barry Sinervo; **53.23 (left):** © Jean Vie/Gamma; **53.23 (right):** © Gianni Tortole/National Audubon Society Collection/Photo Researchers

Chapter 54

Figure 54.1: © Corbis; **54.2:** © Tim Davis/Photo Researchers; **54.7 (all):** J.B. Losos; **54.11 (both):** © Edward S. Ross; **54.12 (both):** © Lincoln P. Brower; **54.13:** © Michael & Patricia Fogden/Corbis; **54.14:** © James L. Castner; **54.16:** © Merlin D. Tuttle/Bat Conservation International; **54.17:** © PhotoDisc/Volume 44; **54.18:** © Michael Fogden/DRK Photo; **54.19:** © Edward S. Ross; **54.21a:** © F. Stuart Westmorland/ Photo Researchers; **54.21b:** © Anne Wertheim/ Animals Animals/Earth Scenes; **54.24:** © David Hosking/National Audubon Society Collection/ Photo Researchers **(all):** © Tom Bean; **54.27 (both):** Courtesy of Robert Whittaker; **54.28:** © Edward S. Ross

Chapter 55

Figure 55.1: © Corbis/Volume 46; **55.3:** © Martin Harvey, Gallo Images/Corbis; **55.7a:** U.S. Forest Service; **55.17a:** © Layne Kennedy/Corbis

Chapter 56

Figure 56.1: NASA; **56.10:** © Michael Graybill & Jan Hodder/Biological Photo Service; **56.11:** © IFA/Peter Arnold, Inc.; **56.15:** © Digital Vision/Picture Quest; **56.16a:** © Jim Church; **56.16b:** Courtesy of J. Frederick Grassel, Woods Hole Oceanographic Institution; **56.17:** © Edward S. Ross; **56.22:** © Gilbert S. Grant/ National Audubon Society Collection/Photo Researchers; **56.23a:** © Peter May/Peter Arnold Inc.; **56.23b:** © Frans Lanting/Minden Pictures; **56.24a:** NASA

Chapter 57

Figure 57.1: © Tom & Pat Leeson/Photo Researchers; **57.2:** Photo by Tom McHugh, © Natural History Museum of Los Angeles County/Photo Researchers; **57.6:** © Edward S. Ross; **57.9:** © Michael Fogden/ DRK Photo; **57.14:** © 1990 R.O. Bierregaard; **57.15:** Reprinted with permission from *Science*, Vol 282 Dec. © 1998 American Association for the Advancement of Science; **57.17:** © Jack Jeffrey; **57.18:** Mark Chandler; **57.20:** Merlin D.

Tuttle/Bat Conservation International; **57.21:** U.S. Fish & Wildlife Service; **57.23:** © Wm. J. Weber/Visuals Unlimited; **57.24 (both):** University of Wisconsin-Madison Arboretum

Text

Chapter 1

Box 1.1: From Howard Neverov, "The Consent" in *The Collected Works of Howard Nemerov*, 7th Edition, 1981. Reprinted by permission.

Chapter 5

Figure 5.6: Copyright © 2002 from *Molecular Biology of the Cell* by Bruce Alberts, et al. Reproduced by permission of Routledge/Taylor & Francis Books, Inc.

Chapter 6

Figure 6.10: Modified from Alberts, et al., *Molecular Biology of the Cell*, 3rd Edition, 1994 Garland Publishing, New York, NY.

Chapter 10

Figure 10.7: From Raven, et al., *Biology of Plants*, 5th edition. Reprinted by permission of Worth Publishers. **Figure 10.13:** From Lincoln Taiz and Eduardo Zeiger, *Plant Physiology*, 1991 Benjamin-Cummings Publishing. Reprinted with permission of the authors.

Chapter 11

Figure 11.4: Copyright © 2002 from *Molecular Biology of the Cell* by Bruce Alberts, et al. Reproduced by permission of Routledge/Taylor & Francis Books, Inc.

Chapter 14

Table 14.1: Data from E. Chargaff and J. Davidson (editors), *The Nucleic Acids*, 1955, Academic Press, New York, NY.

Chapter 16

TA 16.1: CALVIN AND HOBBES © 1995 Watterson. Reprinted with permission of Universal Press Syndicate. All rights reserved.

Chapter 17

Table 17.1: Reprinted with permission from *Nature*. Copyright Macmillan Magazines Limited. **Figure 17.9:** G. More, K.M. Devos, Z. Wang, and M.D. Gale: "Grasses, line up and form a circle," *Current Biology*, 1995, vol. 5, pp. 737-739. **Figure 17.11:** Modified from Keho Villiard and Sommerville, "DNA Microarrays for studies of

Photosynthetic Organisms," *Trends in Plant Science*, 1999.

Chapter 18

Figure 18.22: From an *Introduction of Genetic Analysis* 5/e by Anthony J.F. Griffiths, et al., Copyright © 1976, 1981, 1986, 1989, 1993 by W.H. Freeman and Company. Used with permission.

Chapter 19

Figure 19.5: Copyright © John Kochik for Howard Hughes Medical Institute. **Figure 19.9 (text):** From H. Robert Horvitz, *From Egg to Adult*, published by Howard Hughes Medical Institute. Copyright © 1992. Reprinted by permission. **Figure 19.9 (top):** M.E. Challinor illustration. From Howard Hughes Medical Institute © as published in *From Egg to Adult*, 1992. Reprinted by permission. **Figure 19.9 (bottom):** Illustration by: The studio of Wood Ronsaville Harlin, Inc. **Figure 19.20:** Modified from John Kochik for Howard Hughes Medical Institute.

Chapter 20

Table 20.2: Data from the American Cancer Society, Inc., 2002.

Chapter 21

Figure 21.8: Data from P.A. Powers, et al., "A Multidisciplinary Approach to the Selectionist/Neutralist Controversy." *Oxford Surveys in Evolutionary Biology*. Oxford University Press, 1993. **Figure 21.10:** From R.F. Preziosi and D.J. Fairbairn, "Sexual Size Dimorphism and Selection in the Wild in the Waterstrider *Aquarius remigis*: Lifetime Fecundity Selection on Female Total Length and Its Components," *Evolution, International Journal of Organic Evolution* 51:467-474, 1997. **Figure 21.11:** Data from M.R. MacNair in J.M. Bishops & L.M. Cook, *Genetic Consequences of Man-Made Change*, Academic Press, 1981, p. 177-207. **Figure 21.12:** Adapted from Clark, B. "Balanced Polymorphism and the Diversity of Sympatric Species." Syst. Assoc. Publ., Vol. 4, 1962.

Chapter 22

Figure 22.3A: Data from Grant, "Natural Selection and Darwin's Finches" in *Scientific American*, October 1991. **Figure 22.3b:** Data from Grant, "Natural Selection and Darwin's Finches" in *Scientific American*, October 1991. **Figure 22.5:** Data from Grant, et al., "Parallel Rise and Fall of Melanic Peppered Moths" in *Journal of*

Heredity, vol. 87, 1996, Oxford University Press. **Figure 22.6:** Data from G. Dayton and A. Roberson, *Journal of Genetics*, Vol. 55, p. 154, 1957.

Chapter 23

Figure 23.2: Data from R. Conant & J.T. Collins, *Reptiles & Amphibians of Eastern/Central North America*, 3rd edition, 1991. Houghton Mifflin Company. **Figure 23.10:** Data from B.M. Bechler, et al., *Birds of New Guinea*, 1986, Princeton University Press. **Figure 23.18:** Data from D. Futuyma, *Evolutionary Biology*, 1998, Sinauer.

Chapter 25

Figure 25.8: From Richard O. Prum and Alan H. Brush, "The Evolutionary Origin and Diversification of Feathers," *Quarterly Review of Biology* 77 (3): 261-295, September 2002. Reprinted with permission of The University of Chicago Press.

Chapter 27

Figure 27.9: Data from U.S. Centers for Disease Control and Prevention, Atlanta, GA.

Chapter 31

Figure 31.09: Courtesy of Joel Cracraft.

Chapter 36

Figure 36.3: From Ralph Quantrano, Washington University. **Figure 36.16:** From G.B. Fincher, "Molecular and Cellular Biology Associated with Endosperm Mobilization in Germinating Cereal Grains." Reprinted with permission from the *Annual Review of Plant Physiology and Plant Molecular Biology*, Volume 40. Copyright © 1989 by Annual Reviews www.annualreviews.org.

Chapter 40

Figure 40.10: Data from Hong et al., Arabidopsis *not* Mutants Define Multiple Functions Required for Acclimation to High Temperatures, *Plant Physiology*, Vol. 132, pp. 757–767, 2003.

Chapter 41

Figure 41.8: After McDaniel, 1996. **Figure 41.9:** After McDaniel, 1996. **Figure 41.10:** After McDaniel, 1996.

Chapter 42

Figure 42.32: From *New York Times*, December 15, 1998. Copyright © 1998 *New York Times*. Reprinted with permission.

Chapter 48

Figure 48.21: From Beck & Habicht, "Immunity and the Invertebrates" in *Scientific American*, November 1996. Reprinted by permission of Roberto Osti Illustrations. **Figure 48.25:** Data from U.S. Centers for Disease Control and Prevention, Atlanta, GA.

Chapter 49

Figure 49.4: Data from B. Heinrich, *Science*, American Association for the Advancement of Science.

Chapter 50

Table 50.2: Data from American College of Obstetricians and Gynecologists: Contraception, Patient Education Pamphlet No. AP005.ACOG, Washington, D.C., 1990.

Chapter 52

Figure 52.6: Data from J.R. Brown et al, "A Defect in Nurturing in Mice Lacking the Immediate Early Gene for fosB", Cell, 1996. **Figure 52.11:** Reprinted from *Animal Behaviour*, Vol. 51, M.D. Beecher, P.K. Stoddard, S.E. Campbell, and C.L. Horning, "Repertoire Matching Between Neighbouring Song Sparrows," pp. 917-923. Copyright © 1996, with permission from Elsevier. **Figure 52.20:** From John Alcock, *Animal Behavior*, 1989. **Figure 52.24b:** From John Alcock, *Journal of Animal Behavior*, 1988. Reprinted by permission of Academic Press, Ltd., London. **Figure 52.30C:** Data from M. Petrie, et al. "Peahens Prefer Peacocks with Elaborate Trains, *Animal Behavior*, 1991. **Figure 52.33B:** Data from H.L. Gibbs et al., Realized Reproductive Excess of Polygynous Red-Winged Blackbirds Revealed by NDA Markers," *Science*, 1990.

Chapter 53

Figure 53.24: From G.C. Varley, "Population Changes in German Forest Pests," *Journal of Animal Ecology*, Vol. 18, May 1949. Reprinted with permission of British Ecology Society. **Table 53.1:** Based on Table 14-11 in A.J. Vander, J.H. Sherman, and D.S. Luciano, *Human Physiology*, 5th ed. Copyright © 1997 McGraw-Hill Companies, Inc., Dubuque, Iowa. All Rights Reserved. **Figure 53.5:** After E.R. Pianka, *Evolutionary Ecology.*, 4th edition, New York, Harper & Row, 1987. **Figure 53.6:** Data from Brown & Lomolino, *Biogeography*, 3rd edition, 1998, Sinauer Associates, Inc. **Figure 53.7:** Data from Brown & Lomolino, *Biogeography*, 3rd edition, 1998, Sinauer Associates, Inc. After A.T. Smith *Ecology*, 1974. **Figure 53.8:** Data from Elizabeth Losos, Center for Tropical Forest Science, Smithsonian Tropical Research Institute. **Figure 53.10:** Data from *Patch Occupation and Population Size of the Glanville Fritillary in the Anland Islands*, Metapopulation Research Group, Helsinki, Finland. **Figure 53.11:** Data from Bonner, 1965. **Figure 53.2:** Modified from Ricklefs, 1997. **Figure 53.13:** Modified from Ricklefs, 1997. **Figure 53.16:** Data from C.M. Perrins, *Animal Ecology*, 1995. **Figure 53.20B:** Data from C.E. Goulden, L.L. Henry, and A.J. Tessier, *Ecology*, 1982. **Figure 53.22:** Data from Arcese & Smith J. *Animal Biology*, vol. 57, pp. 119-136, 1989 and Smith et al. 1991. **Table 53.3:** After E.R. Pianka, *Evolutionary Ecology*, 4th edition, 1987, New York, Harper & Row, 1987. **Figure 53.30:** Data from National Geographic, July 2001.

Chapter 54

Figure 54.3: From *The Economy of Nature* 4/e, by Robert E. Ricklefs. Copyright © 1973, 1979 by Chiron Press Inc.; Copyright © 1990, 2000 by W.H. Freeman and Company. Used with permission. **Figure 54.4:** From *The Economy of Nature 4/e*, by Robert E. Ricklefs. Copyright © 1973, 1979 by Chiron Press Inc.; Copyright © 1990, 2000 by W.H. Freeman and Company. Used with permission. **Figure 54.6:** Data from Begon et al., *Ecology*, 1996. After: W.B. Clapham, *Natural Ecosystems*, Clover, Macmillan. **Figure 54.8:** Data from E.J. Heske, et al., *Ecology*, 1994. **Figure 54.9:** Data from E.J. Heske, et al., *Ecology*, 1994. **Figure 54.22:** Data from D.W. Davidson et al., "Granivory in a Desert Ecosystem." *Ecology*, 1984.

Chapter 55

Figure 55.18: Data from F. Morrin, *Community Ecology*, Blackwell, 1999. **Figure 55.16:** Data from J.T. Wooten & M.E. Power, "Productivity, Consumers & the Structure of a River Food Chain," *Proceedings National Academic Sciences*, 1993. **Figure 55.13:** Data from Flecker, A.S. and Townsend, C.R., "Community-Wide Consequences of Trout Introduction in New Zealand Streams." In *Ecosystem Management: Selected Readings*, F.B. Samson and F.L. Knopf eds., Springer-Verlag, New York, 1996. **Figure 55.14:** Data from M. Power "Habitat Heterogenieity and the Functional Significance of Fish," *Ecology*, 1997. **Table 55.1:** After Whittaker, 1975.

Chapter 56

Figure 56.13: Map from "An Act of God," July 19, 1997, *The Economist*. Copyright © 1997 *The Economist* Newspaper Ltd. All rights reserved. Reprinted with permission. Further reproduction prohibited. www.economist.com. **Figure 56.25:** Data from Geophysical Monograph, American Geophysical Union, National Academy of Sciences, and National Center for Atmospheric Research.

Chapter 57

Figure 57.3: After Smith et al., 1993. **Figure 57.10:** IUCN 2000, AmphibiaWeb, Hero J.M. & L. Shoo, 2003. Chapter 7 in *Amphibian Conservation*, Smithsonian Press. Background biodiversity hotspots map from Myers et al, 2000. *Nature* 403:853-858 c/o Conservation International. Prepared by J.M. Hero, April 2002. **Figure 57.11:** After Green and Sussman, 1990. **Figure 57.12:** Data assembled by Pima, 1991. **Figure 57.15:** Data from Marra, Hobson, Holmes, "Linking Winter & Summer Events," in *Science*, Dec. 1998. **Figure 57.16:** Data from UNEP, *Environmental Data Report*, 1993, 1994. **Figure 57.22:** Data from H.L. Billington, "Effects of Population Size on a Genetic Variation in a Dioecious Conifer" in *Conservation Biology*, Blackwell Scientific Publication, Inc., 1991. **Figure 57.25:** Data from The Peregrine Fund. **Figure 57.27:** From "Quetzal and The Macaw," The Story of Costa Rica's National Parks, by David Rains Wallace, 1992 Sierra Club. Reprinted courtesy of David Rains Wallce. **Figure 57.4:** From *Nature's Place: Human Population and the Future of Biological Diversity*, by Richard P. Cincotta and Robert Engelman, 2000. Reprinted with permission of Population Action International. **Figure 57.5:** From *Nature's Place: Human Population and the Future of Biological Diversity*, by Richard P. Cincotta and Robert Engelman, 2000. Reprinted with permission of Population Action International. **Figure 57.26:** From *Nature and Resources, Vol. XXII*, 1986. Copyright © 1986 UNESCO Press. Reproduced by permission of UNESCO.

Index

B

O